Goro Shimura

Collected Papers II

1967 – 1977

Reprint of the 2002 Edition

 Springer

Goro Shimura
Mathematics Department
Princeton University
Princeton, NJ 08544-1000
USA

ISSN 2194-9875
ISBN 978-1-4939-1832-4   (Softcover)
         978-0-387-95416-5   (Hardcover)
DOI 10.1007/978-1-4612-2076-3
Springer New York Heidelberg Dordrecht London

Library of Congress Control Number: 2012954381

Contents

Notes II   **825**

List of Articles

The numbers in parentheses are the articles not included in this collection.

Volume I

[(52)] On a certain ideal of the center of a Frobeniusean algebra, Scientific Papers of the College of General Education, University of Tokyo, 2 (1952), 117–124.

[54] A note on the normalization-theorem of an integral domain, Scientific Papers of the College of General Education, University of Tokyo, 4 (1954), 1–8.

[55] Reduction of algebraic varieties with respect to a discrete valuation of the basic field, American Journal of Mathematics, 77 (1955), 134–176.

[56] On complex multiplications, Proceedings of the International Symposium on Algebraic Number Theory, Tokyo-Nikko, 1955, Science Council of Japan, Tokyo (1956), 23–30.

[57a] La fonction ζ du corps des fonctions modulaires elliptiques, Comptes rendus des séances de l'Académie des Sciences, 244 (1957), 2127–2130.

[(57b)] Kindai-teki Seisu-ron (Modern Number Theory, in Japanese, with Yutaka Taniyama), Kyoritsu Shuppan, 1957.

[58a] Correspondances modulaires et les fonctions zeta de courbes algébriques, Journal of the Mathematical Society of Japan, 10 (1958), 1–28.

[58b] Modules des variétés abéliennes polarisées et fonctions modulaires, Séminaire Henri Cartan, École Normale Supérieure, 1957/58, *Fonctions Automorphes*, Exposé 18–20 (1958).

[(58c)] Fonctions automorphes et variétés abéliennes, Séminaire Bourbaki, 1957/58, Exposé 167 (1958).

[59a] Fonctions automorphes et correspondances modulaires, Proceedings of the International Congress of Mathematicians, Edinburgh, 1958 (1959), 330–338.

[59b] On the theory of automorphic functions, Annals of Mathematics, 70 (1959), 101–144.

[59c] Sur les intégrales attachées aux formes automorphes, Journal of the Mathematical Society of Japan, 11 (1959), 291–311.

[59d] On specializations of abelian varieties (with Shoji Koizumi), Scientific Papers of the College of General Education, University of Tokyo, 9 (1959), 187–211.

[60a] On vector differential forms attached to automorphic forms (with Michio Kuga), Journal of the Mathematical Society of Japan, 12 (1960), 258–270.

[(60b)] Automorphic functions and number theory: I (in Japanese), Sugaku, 11 (1960), 193–205.

[(61a)] Complex multiplication of abelian varieties and its applications to number theory (with Yutaka Taniyama), Publications of the Mathematical Society of Japan, No. 6 (1961).

[61b] On the zeta functions of the algebraic curves uniformized by certain automorphic functions, Journal of the Mathematical Society of Japan, 13 (1961), 275–331.

[(61c)] Automorphic functions and number theory: II (in Japanese), Sugaku, 13 (1961), 65–80.

[62a] On Dirichlet series and abelian varieties attached to automorphic forms, Annals of Mathematics, 76 (1962), 237–294.

[62b] On the class-fields obtained by complex multiplication of abelian varieties, Osaka Mathematical Journal, 14 (1962), 33-44.

[63a] Arithmetic of alternating forms and quaternion hermitian forms, Journal of the Mathematical Society of Japan, 15 (1963), 33–65.

[63b] On analytic families of polarized abelian varieties and automorphic functions, Annals of Mathematics, 78 (1963), 149–192.

[63c] On the cohomology groups attached to certain vector valued differential forms on the product of the upper half planes (with Yozo Matsushima), Annals of Mathematics, 78 (1963), 417–449.

[63d] On modular correspondences for $Sp(n, \mathbf{Z})$ and their congruence relations, Proceedings of the National Academy of Sciences, 49 (1963), 824–828.

[63e] On the fields of definition for fields of automorphic functions, Proceedings of the 1963 Number Theory Conference, Boulder, Colorado, August 5–24, 1963, 26–32.

[64a] Arithmetic of unitary groups, Annals of Mathematics, 79 (1964), 369–409.

[64b] On the field of definition for a field of automorphic functions, Annals of Mathematics, 80 (1964), 160–189.

Volume II

[69] Local representations of Galois groups, Annals of Mathematics, 89 (1969), 99–124.

[70a] On canonical models of arithmetic quotients of bounded symmetric domains, Annals of Mathematics, 91 (1970), 144–222.

[70b] On canonical models of arithmetic quotients of bounded symmetric domains: II, Annals of Mathematics, 92 (1970), 528–549.

[71a] On arithmetic automorphic functions, Proceedings of the International Congress of Mathematicians, Nice, 1970, vol. 2, 343–348 (1971).

[71b] On the zeta-function of an abelian variety with complex multiplication, Annals of Mathematics, 94 (1971), 504–533.

[71c] Class fields over real quadratic fields in the theory of modular functions, in *Several Complex Variables* II, Maryland 1970, Lecture notes in mathematics 185 (1971), 169–188.

[(71d)] Introduction to the arithmetic theory of automorphic functions, Publications of the Mathematical Society of Japan, No. 11, Iwanami Shoten and Princeton University Press, 1971.

[71e] On elliptic curves with complex multiplication as factors of the Jacobians of modular function fields, Nagoya Mathematical Journal, 43 (1971), 199–208.

[72a] On the field of rationality for an abelian variety, Nagoya Mathematical Journal, 45 (1972), 167–178.

[72b] Class fields over real quadratic fields and Hecke operators, Annals of Mathematics, 95 (1972), 130–190.

[73a] On modular forms of half integral weight, Annals of Mathematics, 97 (1973), 440–481.

[(73b)] Complex multiplication, Proceedings of the International Summer School of Modular Functions of One Variable, Antwerp, 1972, Lecure notes in mathematics, 320 (1973), 37–56.

[(73c)] Modular forms of half integral weight, Proceedings of the International Summer School of Modular Functions of One Variable, Antwerp, 1972, Lecure notes in mathematics, 320 (1973), 57–74.

[73d] On the factors of the jacobian variety of a modular function field, Journal of the Mathematical Society of Japan, 25 (1973), 523–544.

[74] On the trace formula for Hecke operators, Acta mathematica, 132 (1974), 245–281.

[75a] On the holomorphy of certain Dirichlet series, Proceedings of the London Mathematical Society, 3rd ser. 31 (1975), 79–98.

[75b] On the real points of an arithmetic quotient of a bounded symmetric domain, Mathematische Annalen, 215 (1975), 135–164.

[81d] Arithmetic of differential operators on symmetric domains, Duke Mathematical Journal 48 (1981), 813–843.

[(81e)] Corrections to "The special values of the zeta functions associated with Hilbert modular forms," vol 45 (1978), 637–679, Duke Mathematical Journal 48 (1981), 697.

[82a] Models of an abelian variety with complex multiplication over small fields, Journal of Number Theory, 15 (1982), 25–35.

[82b] The periods of certain automorphic forms of arithmetic type, Journal of the Faculty of Science, University of Tokyo, Sec. IA, 28 (1982), 605–632.

[82c] Confluent hypergeometric functions on tube domains, Mathematische Annalen, 260 (1982), 269–302.

[83a] Algebraic relations between critical values of zeta functions and inner products, American Journal of Mathematics, 105 (1983), 253–285.

[83b] On Eisenstein series, Duke Mathematical Journal, 50 (1983), 417–476.

[84a] Differential operators and the singular values of Eisenstein series, Duke Mathematical Journal, 51 (1984), 261–329.

[84b] On differential operators attached to certain representations of classical groups, Inventiones mathematicae, 77 (1984), 463–488.

[85a] On Eisenstein series of half-integral weight, Duke Mathematical Journal, 52 (1985), 281–314.

[85b] On the Eisenstein series of Hilbert modular groups, Revista Matemática Iberoamericana, 1 (1985), 1–42.

[86] On a class of nearly holomorphic automorphic forms, Annals of Mathematics, 123 (1986), 347–406.

[87a] Nearly holomorphic functions on hermitian symmetric spaces, Mathematische Annalen, 278 (1987), 1–28.

[87b] On Hilbert modular forms of half-integral weight, Duke Mathematical Journal, 55 (1987), 765–838.

[88] On the critical values of certain Dirichlet series and the periods of automorphic forms, Inventiones mathematicae, 94 (1988), 245–305.

Volume IV

[89a] Yutaka Taniyama and his time, Bulletin of the London Mathematical Society, 21 (1989), 186–196.

[89b] *L*-functions and eigenvalue problems, *Algebraic analysis, geometry, and number theory*, Proceedings of the JAMI Inaugural Conference 1988, Supplement to the American Journal of Mathematics, 1989, 341–396.

[90a] Invariant differential operators on hermitian symmetric spaces, Annals of Mathematics, 132 (1990), 237–272.

[90b] On the fundamental periods of automorphic forms of arithmetic type, Inventiones mathematicae, 102 (1990), 399–428.

[(90c)] Some old and recent arithmetical results concerning modular forms and related zeta functions, University of Istanbul, Faculty of Science, Journal of Mathematics, 49 (1990), 45–56.

[91] The critical values of certain Dirichlet series attached to Hilbert modular forms, Duke Mathematical Journal, 63 (1991), 557–613.

[(93a)] Arithmeticity of the special values of various zeta functions and the periods of abelian integrals (in Japanese), Sugaku, 45 (1993), 111–127; English translation by Toshitsune Miyake, Sugaku expositions, 8 (1995), 17–38.

[93b] On the transformation formulas of theta series, American Journal of Mathematics, 115 (1993), 1011–1052.

[93c] On the Fourier coefficients of Hilbert modular forms of half-integral weight, Duke Mathematical Journal, 71 (1993), 501–557.

[94a] Fractional and trigonometric expressions for matrices, American Mathematical Monthly, 101 (1994), 744–758.

[94b] Euler products and Fourier coefficients of automorphic forms on symplectic groups, Inventiones mathematicae, 116 (1994), 531–576.

[94c] Differential operators, holomorphic projection, and singular forms, Duke Mathematical Journal, 76 (1994), 141–173.

[95a] Eisenstein series and zeta functions on symplectic groups, Inventiones mathematicae, 119 (1995), 539–584.

[95b] Zeta functions and Eisenstein series on metaplectic groups, Inventiones mathematicae, 121 (1995), 21–60.

[96a] Convergence of zeta functions on symplectic and metaplectic groups, Duke Mathematical Journal, 82 (1996), 327–347.

[96b] Response, Notices of the American Mathematical Society, vol. 43, No. 11 (November 1996), 1344–1347.

[(97a)] Euler Products and Eisenstein series, CBMS Regional Conference Series in Mathematics, No. 93, American Mathematical Society, 1997.

[97b] Zeta functions and Eisenstein series on classical groups, Proceedings of the National Academy of Sciences, 94, 11133–11137 (1997).

[(98)] Abelian varieties with complex multiplication and modular functions, Princeton University Press, 1998.

[99a] An exact mass formula for orthogonal groups, Duke Mathematical Journal, 97 (1999), 1–66.

[99b] The number of representations of an integer by a quadratic form, Duke Mathematical Journal, 100 (1999), 59–92.

[99c] Generalized Bessel functions on symmetric spaces, Journal für die reine und angewandte Mathematik, 509 (1999), 35–66.

[99d] Some exact formulas on quaternion unitary groups, Journal für die reine und angewandte Mathematik, 509 (1999), 67–102.

[99e] André Weil as I knew him, Notices of the American Mathematical Society, vol. 46, No. 4 (April 1999), 428–433

[(00)] Arithmeticity in the theory of automorphic forms, Mathematical Surveys and Monographs, vol. 82, American Mathematical Society, 2000.

[(01a)] Letter to the editor, Notices of the American Mathematical Society, vol. 48, No. 7 (August 2001), 678.

[01b] Arithmeticity of Dirichlet series and automorphic forms on unitary groups, unpublished.

[01c] The relative regulator of an algebraic number field, unpublished.

67a

Discontinuous groups and abelian varieties

Mathematische Annalen, 168 (1967), 171-199

In a paper on symplectic geometry, CARL LUDWIG SIEGEL, to whom this memoir is most respectfully dedicated, introduced a remarkable class of discontinuous groups Γ operating on the unit ball $\mathscr{S}_+$ of complex symmetric matrices. This class is not included, except in some special cases, in the class of discontinuous groups related to the families of abelian varieties characterized by the type of polarization and endomorphisms, which I have investigated in a series of previous papers. The present research will show that there is still a family of abelian varieties, of which the moduli-space is the quotient $\mathscr{S}_+/\Gamma$ of Siegel's type, and the members are characterized by a certain non-holomorphic endomorphism. Such a family is not unique for a given $\mathscr{S}_+/\Gamma$. Each $\mathscr{S}_+/\Gamma$ represents the moduli-spaces for infinitely many distinct families of abelian varieties. We shall consider in this paper also discontinuous groups operating on the unit ball $\mathscr{S}_-$ of complex skew-symmetric matrices.

To describe the discontinuous groups and the families of abelian varieties more closely, let F be a totally real algebraic number field of degree g, and K a totally imaginary quadratic extension of F. Let L be a central division algebra over K with a positive involution ϱ which induces the non-trivial automorphism of K over F. We put $[L:K]=q^2$, and assume that $q=1$ or 2. Let ι denote the identity mapping of K or the main involution of L according as $q=1$ or 2. Let T and S be invertible matrices of degree $m \geq 1$ with coefficients in L such that

$$S^{\iota\varrho}=\kappa S^{-1}, \quad {}^tT^\varrho=\eta T, \quad T\cdot{}^tS^\varrho=-S^{\iota\varrho}T^{\iota\varrho}$$

with an element κ of F and $\eta=1$ or -1. Define algebraic groups $\mathfrak{G}$, $\mathfrak{G}_0$ over $\mathbf{Q}$ and Lie groups $\mathfrak{G}_\mathbf{R}$, $\mathfrak{G}_{0\mathbf{R}}$ by

$$\mathfrak{G}_\mathbf{R} = \{X \in GL_m(L_\mathbf{R}) | X T \cdot {}^tX^\varrho = T\},$$
$$\mathfrak{G} = \mathfrak{G}_\mathbf{R} \cap GL_m(L),$$
$$\mathfrak{G}_{0\mathbf{R}} = \{X \in \mathfrak{G}_\mathbf{R} | X^{\iota\varrho}S = SX\},$$
$$\mathfrak{G}_0 = \mathfrak{G}_{0\mathbf{R}} \cap GL_m(L),$$

where $L_\mathbf{R}=L\otimes_\mathbf{Q}\mathbf{R}$. Let P_∞^+ (resp. P_∞^-) denote the set of archimedean prime spots of F where $\eta\kappa$ is positive (resp. negative). We assume that T is "definite" at every prime spot in P_∞^-. Let h be the number of prime spots belonging to P_∞^+,

1

and let $\varepsilon = (-1)^{q-1}\eta$, $2r = mq$. Then the quotient of $\mathfrak{G}_{\mathbf{R}}$ (resp. $\mathfrak{G}_{0\mathbf{R}}$) by a maximal compact subgroup with natural complex structure may be identified with a bounded symmetric domain $\mathscr{H}$ (resp. $\mathscr{H}_\varepsilon$) defined as follows:

$\mathscr{H} = \mathscr{S} \times \cdots \times \mathscr{S}$, $\mathscr{H}_\varepsilon = \mathscr{S}_\varepsilon \times \cdots \times \mathscr{S}_\varepsilon$ (h copies in each product),

$\mathscr{S} =$ the set of all complex square matrices z of size r such that $1 - {}^t\bar{z}z$ is a positive definite hermitian matrix,

$\mathscr{S}_\varepsilon = \{z \in \mathscr{S} \,|\, \varepsilon^t z = z\}$.

Here one should note that the natural holomorphic injection of $\mathscr{H}_\varepsilon$ into $\mathscr{H}$ is compatible with the injection of $\mathfrak{G}_{0\mathbf{R}}$ into $\mathfrak{G}_{\mathbf{R}}$. Let $\mathfrak{G}$ act on the vector space L^m of all m-dimensional row vectors by the right multiplication. For a free $\mathbf{Z}$-module $\mathfrak{M}$ of maximal rank in L^m, we can define a discrete subgroup Γ of $\mathfrak{G}_{\mathbf{R}}$ (resp. Γ_0 of $\mathfrak{G}_{0\mathbf{R}}$) by

$$\Gamma = \{X \in \mathfrak{G} \,|\, \mathfrak{M}X = \mathfrak{M}\} \,,$$

$$\Gamma_0 = \Gamma \cap \mathfrak{G}_0 \,.$$

We assume $0 < h < g$. Then $\mathscr{H}/\Gamma$ and $\mathscr{H}_\varepsilon/\Gamma_0$ are compact. Now $\mathscr{H}/\Gamma$ is the variety of moduli for a family

$$\Sigma = \{A(z) \,|\, z \in \mathscr{H}\}$$

of abelian varieties $A(z)$ of the type discussed in [5] and characterized by polarization and endomorphisms. From the injection of $\mathscr{H}_\varepsilon$ into $\mathscr{H}$, we obtain naturally a morphism μ of $\mathscr{H}_\varepsilon/\Gamma_0$ to $\mathscr{H}/\Gamma$. Therefore $\mathscr{H}_\varepsilon/\Gamma_0$ is a moduli-variety of a subfamily $\Sigma_0 = \{A(z) \,|\, z \in \mathscr{H}_\varepsilon\}$ of Σ.

Our main purpose is: (i) to obtain a characterization of the members of Σ_0; (ii) to show that the space $\mathscr{H}_\varepsilon/\Gamma_0$, as a projective variety, can be defined over an algebraic number field.

The characterization can be done in the following way. First we note that every member $A(z)$ of Σ has L as a subalgebra of the endomorphism algebra. Let $\mathfrak{U}/D$ be a complex torus isomorphic to A_z, with a complex vector space $\mathfrak{U}$ and a lattice D in $\mathfrak{U}$. Then from the action of the endomorphisms on $\mathfrak{U}$, one obtains a representation Φ of L by complex linear transformations on $\mathfrak{U}$. Restrict Φ to F, and decompose $\mathfrak{U}$ into the direct sum $\mathfrak{U} = \mathfrak{U}^+ + \mathfrak{U}^-$ of invariant subspaces $\mathfrak{U}^+$ and $\mathfrak{U}^-$ corresponding respectively to P_∞^+ and P_∞^- in an obvious sense. Change the complex structure of $\mathfrak{U}^-$ for its complex conjugate and keep the complex structure of $\mathfrak{U}^+$ as it is. Then we obtain a complex torus $\mathfrak{U}'/D$ with the same underlying real torus as $\mathfrak{U}/D$, and with a modified complex structure. Now the parametrization of Σ by the points of $\mathscr{H}$ is so chosen that there is a certain isogeny of $A(\varepsilon^t z)$ to $\mathfrak{U}'/D$. (A more precise formulation of this fact is (2.15.6) in the text.) Therefore each member of the subfamily Σ_0 has a real analytic endomorphism $\mathfrak{d}$ such that $\mathfrak{d}^2 = \kappa$, $\mathfrak{d}$ is complex linear on $\mathfrak{U}^+$ and complex anti-linear on $\mathfrak{U}^-$. This gives a characterization of Σ_0. If $\eta = 1$, we can substitute for $\mathfrak{d}$ a certain rational cohomology class, which is of $(1, 1)$-type in $\mathfrak{U}^+$, and of $(2, 0)$- or $(0, 2)$-type in $\mathfrak{U}^-$ (2.26).

Let $V = \mathscr{H}/\Gamma$ and $V_0 = \mathscr{H}_\varepsilon/\Gamma_0$. In view of the results of [5, 8], we may assume that V is defined over an algebraic number field. We shall prove that V_0

and the morphism μ of V_0 to V can be defined over an algebraic number field k (Th. 4.3, 4.4). If $q=1$ and $\varepsilon=1$ (the case including Siegel's class), k may be chosen so as to be abelian over the field K' generated over the rational number field $\mathbf{Q}$ by $\mathrm{tr}\,\Phi(\alpha)$ for all $\alpha \in K$ (Th. 4.10). The proofs rely on the following two facts: (i) as is shown in [8], the model V can be chosen so that the coordinates of each point of V generate the field of moduli of the corresponding polarized abelian variety; (ii) there is a dense subset Z of $\mathscr{H}_\varepsilon$ such that $A(z)$ for $z \in Z$ has sufficiently many complex multiplications. We shall prove the above result also for the congruence subgroups of Γ_0. As for further discussion and the possibility of generalizing this result, we refer the reader to the *concluding remarks* at the end of the paper.

Advice to the reader. There are four types of families Σ_0, depending on q and ε. If the reader wishes to avoid troublesome calculation, it is suggested that he assumes throughout $q=1$ and $\varepsilon=1$, hence $\eta=1$, $L=K$; accordingly ι is the identity mapping of K, and ϱ is the nontrivial automorphism of K over F. Not much of substance will be lost by making this easement. In fact, the strongest result (Th. 4.10) will be proved only in this case.

Notation. As usual, $\mathbf{Z}$, $\mathbf{Q}$, $\mathbf{R}$ and $\mathbf{C}$ denote respectively the ring of rational integers, the rational number field, the real number field, and the complex number field. Further $\mathbf{K}$ denotes the division ring of real quaternions. If S is an associative ring with identity element and V is an S-module, the ring of all S-linear endomorphisms of V is denoted by $\mathrm{End}(V, S)$, and the group of all S-linear automorphisms of V by $GL(V, S)$. For a positive integer r, $M_r(S)$ denotes the ring of all matrices of size r with entries in S, and $GL_r(S)$ the group of all invertible elements in $M_r(S)$. The identity element of $M_r(S)$ is denoted by 1_r. If S is a semi-simple (separable) algebra over a field F, $\mathrm{Tr}_{S/F}$ means the reduced trace from S to F. If x is a complex number or a complex matrix, $\bar{x}$ means the complex conjugate of x. We denote by $\mathscr{H}/\Gamma$ the quotient of a symmetric domain $\mathscr{H}$ by a discontinuous group Γ, though we define the action of elements of Γ on the left of the points of $\mathscr{H}$. Since we do not need the right multiplication by elements of the Lie group (such as the elements of a maximal compact subgroup), no confusion will be produced by this convention.

1. Involutions and hermitian forms

1.1. Let A be a semi-simple algebra over a field. By an *involution* of A, we understand an anti-automorphism σ of A such that $(x^\sigma)^\sigma = x$ for every $x \in A$. When A is simple, we say that σ is of *the first* or *second kind* according as σ is the identity mapping on the center of A or not.

1.2. Proposition. *Let L be a central simple algebra over a field K, and ι (resp. ϱ) an involution of the first (resp. second) kind of L. Suppose that $\iota\varrho = \varrho\iota$. Put*

$$F = \{x \in K \mid x^\varrho = x\}, \quad B = \{x \in L \mid x^\iota = x^\varrho\}.$$

Then B is a central simple algebra over F, and $L = B \otimes_F K$. Moreover, the restriction of ι to B is an involution of B of the first kind, and

$$\{x \in B \mid xx^\iota = 1\} = \{x \in L \mid xx^\iota = xx^\varrho = 1\}.$$

Proof. Since K is a separable quadratic extension of F, we can find an element λ of K such that $K = F(\lambda)$ and $\lambda^\varrho = 1 - \lambda$ (regardless of the characteristic of K). For $x \in L$, put $y = (2\lambda - 1)^{-1}(x - x^{\iota\varrho})$, $z = x - \lambda y$. Then we see easily that $y \in B$, $z \in B$, $x = z + \lambda y \in B + B\lambda$, and $B \cap B\lambda = \{0\}$. Hence $L = B \otimes_F K$. This implies that B is a central simple algebra over F. If $u \in B$, then $(u^\iota)^\varrho = (u^\varrho)^\iota = (u^\iota)^\iota$. Hence $B^\iota = B$. The last equality of our proposition is obvious.

1.3. Proposition. *Let L, K, F, ι, ϱ be as in 1.2. Let r and t be invertible elements of L such that $r^\iota = \varepsilon r$, $t^\varrho = \eta t$, with $\varepsilon = \pm 1$, $\eta = \pm 1$. Define involutions α and β of L by $x^\alpha = rx^\iota r^{-1}$, $x^\beta = tx^\varrho t^{-1}$ $(x \in L)$. Then $\alpha\beta = \beta\alpha$ if and only if there exists an element c of F such that $t(r^\varrho)^{-1}t^\iota = cr$.*

Proof. We have $x^{\alpha\beta} = t(r^\varrho)^{-1}x^{\iota\varrho}r^\varrho t^{-1}$ and $x^{\beta\alpha} = r(t^\iota)^{-1}x^{\varrho\iota}t^\iota r^{-1}$. Hence $\alpha\beta = \beta\alpha$ if and only if there exists an element c of K such that $t(r^\varrho)^{-1} = cr(t^\iota)^{-1}$. Applying ϱ to this equality, and taking the inverse, we obtain $\eta t^{-1}r = (c^\varrho)^{-1} \times (r^\varrho)^{-1}\eta t^\iota$, so that $c^\varrho = r^{-1}t(r^\varrho)^{-1}t^\iota = c$, which shows $c \in F$.

1.4. Proposition. *Let B be a central simple algebra over a field F, with an involution γ of the first kind. Let K be a separable quadratic extension of F, and let $L = B \otimes_F K$. Then there exist involutions ι and ϱ of L of the first and second kind, respectively, such that $\iota\varrho = \varrho\iota$, $B = \{x \in L \mid x^\iota = x^\varrho\}$, $F = \{x \in K \mid x^\varrho = x\}$, and $x^\gamma = x^\iota$ for every $x \in B$.*

Proof. Let α be the non-trivial automorphism of K over F. We can define involutions ι and ϱ of L so that $x^\iota = x^\varrho = x^\gamma$ for every $x \in B$, and $y^\iota = y$, $y^\varrho = y^\alpha$ for every $y \in K$. Then our assertions can be easily verified.

1.5. Let A be a semi-simple algebra over a field F, and V a left A-module. Let σ be an F-linear involution of A. By an *A-valued σ-hermitian (resp. σ-anti-hermitian) form* on V, we understand an F-bilinear mapping $h : V \times V \to A$ such that $h(ax, by) = ah(x, y)b^\sigma$, $h(x, y)^\sigma = h(y, x)$ (resp. $h(x, y)^\sigma = -h(y, x)$) for $a \in A$, $b \in A$, $(x, y) \in V \times V$. We call h *non-degenerate* if $h(x, V) = \{0\}$ implies $x = 0$.

1.6. Proposition. *Let A, F, σ, V be as above. Suppose that all the simple components of the center of A are separable over F. Denote by $\mathrm{Tr}_{A/F}$ the reduced trace of A to F. Let f be an F-bilinear mapping of $V \times V$ to F such that $f(ax, y) = f(x, a^\sigma y)$, $f(x, y) = f(y, x)$ (resp. $f(x, y) = -f(y, x)$) for $a \in A$, $(x, y) \in V \times V$. Then there exists one and only one A-valued σ-hermitian (resp. σ-anti-hermitian) form h on V such that $f(x, y) = \mathrm{Tr}_{A/F}(h(x, y))$. Conversely, if f is defined by $f(x, y) = \mathrm{Tr}_{A/F}(h(x, y))$ with an A-valued σ-hermitian-(resp. σ-anti-hermitian) form h on V, then f has the above property.*

This was proved in [6, I, 1.2] for σ-anti-hermitian forms. The same argument applies to the σ-hermitian case.

1.7. *The algebra (L, ω, κ).* Let L, K, ι, ϱ, F and B be as in 1.2. Let κ be a non-zero element of F. We can construct an associative algebra E by $E = L + L\omega$ under the rules $\omega a = a^{\iota\varrho}\omega$ for $a \in L$ and $\omega^2 = \kappa$. We denote by (L, ω, κ) this algebra E with a particular embedding of L into E and a particular element ω. We can extend ϱ to an involution of E (which is denoted again by ϱ) by $(a + b\omega)^\varrho = a^\varrho - b^\iota\omega$ for $a, b \in L$. Put $B' = K + K\omega$. Then B' is a quaternion algebra over F. Moreover we see easily that E can be identified with $B \otimes_F B'$, and hence E is central simple over F. Further we note that ϱ induces the main involution of B'.

Let V be a left L-module. Suppose that an F-linear map $u: V \to V$ is given and satisfies $u(ax) = a^{\iota\varrho}(ux)$ and $u(ux) = \kappa x$ for $x \in V$, $a \in L$. Then putting $\omega x = ux$, we obtain a structure of left E-module on V. Conversely, every structure of E-module on V, compatible with the operation of L, can be obtained by giving such a "semi-linear" endomorphism of V. We have obviously

$$\mathrm{End}(V, E) = \{\beta \in \mathrm{End}(V, L) \mid (\omega x)\beta = \omega(x\beta)\}.$$

1.8. Proposition. *Let L, K, ι, ϱ, F, ω and E be as in 1.7. Let V be a left E-module. Let $\mathfrak{H}$ (resp. $\mathfrak{H}'$) denote the set of all E-valued ϱ-hermitian (resp. ϱ-anti-hermitian) forms on V. Let $\mathfrak{T}$ (resp. $\mathfrak{T}'$) denote the set of all L-valued ϱ-hermitian (ϱ-anti-hermitian) forms T on V such that $T(x, \omega y) = -T(\omega x, y)^{\iota\varrho}$. Then there exists an F-linear isomorphism $\mathfrak{T} \to \mathfrak{H}$ (resp. $\mathfrak{T}' \to \mathfrak{H}'$) which maps an element T of $\mathfrak{T}$ (resp. $\mathfrak{T}'$) to an element H of $\mathfrak{H}$ (resp. $\mathfrak{H}'$) under the relation*

$$(1.8.1) \qquad \mathrm{Tr}_{L/F}(T(x, y)) = \mathrm{Tr}_{E/F}(H(x, y)) \qquad ((x, y) \in V \times V).$$

Moreover T is non-degenerate if and only if H is non-degenerate.

This follows easily from 1.6.

1.9. The notation being as in 1.7, let V and V' be left E-modules of the same dimension over F. Then they are isomorphic as E-modules, so that there exists an L-linear isomorphism φ of V to V' such that $(\omega x)\varphi = \omega(x\varphi)$ $(x \in V)$. Further let H and H' be E-valued ϱ-hermitian (or ϱ-anti-hermitian) forms on V and V' respectively. Let T and T' be L-valued ϱ-hermitian (or ϱ-anti-hermitian) forms on V and V' respectively such that

$$\mathrm{Tr}_{L/F}(T(x, y)) = \mathrm{Tr}_{E/F}(H(x, y)) \qquad ((x, y) \in V \times V),$$
$$\mathrm{Tr}_{L/F}(T'(x, y)) = \mathrm{Tr}_{E/F}(H'(x, y)) \qquad ((x, y) \in V' \times V').$$

Let α be an L-linear isomorphism of V to V'. Then α *is an E-linear isomorphism of V to V' such that $H'(x\alpha, y\alpha) = H(x, y)$, if and only if $(\omega x)\alpha = \omega(x\alpha)$ and $T'(x\alpha, y\alpha) = T(x, y)$.* In particular, assuming T to be non-degenerate, we have

$$(1.9.1) \qquad \{\alpha \in GL(V, E) \mid H(x\alpha, y\alpha) = H(x, y)\}$$
$$= \{\alpha \in GL(V, L) \mid T(x\alpha, y\alpha) = T(x, y), (\omega x)\alpha = \omega(x\alpha)\}.$$

1.10. Let us take V to be the vector space L^m of all m-dimensional row vectors with components in L, where m is a positive integer. Then $\mathrm{End}(V, L)$ may be identified with $M_m(L)$, the action of $M_m(L)$ on L^m being defined by the right multiplication. The action of ω on L^m can be expressed by $\omega x = x^{\iota\varrho}S$ $(x \in L^m)$ with an element S of $M_m(L)$ such that $S^{\iota\varrho}S = \kappa 1_m$. Then $\mathrm{End}(V, E)$ can be identified with $\{X \in M_m(L) \mid SX = X^{\iota\varrho}S\}$. We can paraphrase the first three lines of 1.9 as follows. *For any two elements S and S_1 of $M_m(L)$ satisfying $S^{\iota\varrho}S = S_1^{\iota\varrho}S_1 = \kappa 1_m$, there exists an element Z of $GL_m(L)$ such that $S = Z^{\iota\varrho}S_1 Z^{-1}$.*

Let T_0 be a K-valued ϱ-hermitian form (resp. ϱ-anti-hermitian) form on L^m. Then there exists an element T of $M_m(L)$ such that ${}^t T^\varrho = T$ (resp. ${}^t T^\varrho = -T$) and $T_0(x, y) = xT \cdot {}^t y^\varrho$ $((x, y) \in L^m \times L^m)$. We see easily that, S being as above,

$$T \cdot {}^t S^\varrho = -S^{\iota\varrho}T^{\iota\varrho} \Leftrightarrow T_0(x, \omega y) = -T_0(\omega x, y)^{\iota\varrho}.$$

If T satisfies this, then we obtain an E-valued ϱ-hermitian (or ϱ-anti-hermitian) form H on L^m such that $\mathrm{Tr}_{E/F}(H(x, y)) = \mathrm{Tr}_{L/F}(x\,T \cdot {}^ty^\varrho)$, in view of 1.8. Therefore, *one can introduce into L^m a structure of E-module together with an E-valued ϱ-hermitian (or ϱ-anti-hermitian) form, if and only if one has two elements S and T of $M_m(L)$ such that*

$$(1.10.1) \qquad S^{\iota\varrho}S = \kappa 1_m, \quad {}^tT^\varrho = \eta T, \quad T \cdot {}^tS^\varrho = -S^{\iota\varrho}T^{\iota\varrho},$$

according as $\eta = 1$ or -1. By virtue of 1.9, the structure of E-module with a ϱ-hermitian (or ϱ-anti-hermitian) form obtained from S and T is isomorphic to another structure obtained from S' and T' if and only if there exists an element Y of $GL_m(L)$ such that

$$Y^{\iota\varrho}S\,Y^{-1} = S', \quad YT \cdot {}^tY^\varrho = T'.$$

1.11. Proposition. *Let κ be a real number, and let T and S be elements of $GL_{2r}(\mathbf{C})$ satisfying (1.10.1) with complex conjugation as ϱ, the identity mapping as $\cdot\iota$, and $m = 2r$ for a positive integer r.*

(1.11.1) If $\kappa > 0$ and $\eta = 1$, there exists an element Y of $GL_{2r}(\mathbf{C})$ such that

$$Y^\varrho S\,Y^{-1} = \sqrt{\kappa}\begin{pmatrix} 0 & 1_r \\ 1_r & 0 \end{pmatrix}, \quad YT \cdot {}^tY^\varrho = \begin{pmatrix} 1_r & 0 \\ 0 & -1_r \end{pmatrix}.$$

(1.11.2) If $\kappa < 0$, $\eta = 1$ and T is positive definite, then there exists an element Y of $GL_{2r}(\mathbf{C})$ such that $Y^\varrho S\,Y^{-1} = \sqrt{-\kappa}\begin{pmatrix} 0 & -1_r \\ 1_r & 0 \end{pmatrix}$, $YT \cdot {}^tY^\varrho = 1_{2r}$.

(1.11.3) If $\kappa < 0$ and $\eta = -1$, there exists an element Y of $GL_{2r}(\mathbf{C})$ such that

$$Y^\varrho S\,Y^{-1} = \sqrt{-\kappa}\begin{pmatrix} 0 & -1_r \\ 1_r & 0 \end{pmatrix}, \quad YT \cdot {}^tY^\varrho = \sqrt{-1}\begin{pmatrix} 1_r & 0 \\ 0 & -1_r \end{pmatrix}.$$

(1.11.4) If $\kappa > 0$, $\eta = -1$ and $-\sqrt{-1}\,T$ is positive definite, then there exists an element Y of $GL_{2r}(\mathbf{C})$ such that $Y^\varrho S\,Y^{-1} = \sqrt{\kappa}\begin{pmatrix} 0 & 1_r \\ 1_r & 0 \end{pmatrix}$, $YT\,{}^tY^\varrho = \sqrt{-1}\cdot 1_{2r}$.

Proof. In the present case we have $L = K = \mathbf{C}$ and $F = \mathbf{R}$. Let $E = \mathbf{C} + \mathbf{C}\omega = (\mathbf{C}, \omega, \kappa)$. We regard $\mathbf{C}^{2r}$ as a left E-module by means of the action $(a + b\omega)x = ax + bx^\varrho S$ for $a, b \in \mathbf{C}$, $x \in \mathbf{C}^{2r}$. By 1.8, there exists an E-valued $\mathbf{R}$-bilinear form H on $\mathbf{C}^{2r}$ such that $\mathrm{Tr}_{E/\mathbf{R}}(H(x, y)) = \mathrm{Tr}_{\mathbf{C}/\mathbf{R}}(x\,T \cdot {}^ty^\varrho)$. We note that ϱ is the main involution of E.

(i) If $\kappa > 0$ and $\eta = 1$, then E is isomorphic to $M_2(\mathbf{R})$ and H is ϱ-hermitian. It is well-known that any two non-degenerate $M_2(\mathbf{R})$-valued ϱ-hermitian forms, with the same number of variables, are equivalent. This fact combined with the last remark of 1.10 proves (1.11.1).

(ii) If $\kappa < 0$ and $\eta = 1$, then E is isomorphic to the division ring $\mathbf{K}$ of real quaternions, and H is ϱ-hermitian. We have $H(x, x) = xT \cdot {}^tx^\varrho$, hence H is positive definite if T is positive definite. Any two $\mathbf{K}$-valued positive definite ϱ-hermitian forms, with the same number of variables, are equivalent. Hence, by the same reasoning as in (i) we get (1.11.2).

(iii) If $\kappa < 0$ and $\eta = -1$, then E is isomorphic to $\mathbf{K}$, and H is ϱ-anti-hermitian. There is only one, up to equivalence, non-degenerate $\mathbf{K}$-valued

ϱ-anti-hermitian form with a given number of variables. Hence we get (1.11.3).

(iv) If $\kappa > 0$ and $\eta = -1$, then E is isomorphic to $M_2(\mathbf{R})$, and H is ϱ-anti-hermitian. Put

$$P^+ = \{x \in \mathbf{C}^{2r} | \omega x = \sqrt{\kappa}\, x\}, \quad P^- = \{x \in \mathbf{C}^{2r} | \omega x = -\sqrt{\kappa}\, x\}.$$

We see easily that P^+ and P^- are real vector subspaces of V, $V = P^+ + P^-$, $\{0\} = P^+ \cap P^-$. Let $\mathfrak{T}'$ be as in 1.4, and let $U \in \mathfrak{T}'$. Then $\sqrt{-1}\, U(x, y)$ defines an $\mathbf{R}$-valued symmetric form on P^+, and on P^-. In fact, if x and y are elements of P^+, then we have $\sqrt{-1}\, U(x, y) = \sqrt{-1}\, \sqrt{\kappa}^{-1} U(x, \omega y)$, so that $\sqrt{-1}\, U(x, y)$ is symmetric on P^+. Furthermore we have $\sqrt{-1}\, U(x, y) = -\sqrt{-1}\, U(y, x)^\varrho = -\sqrt{-1}\, U(x, y)^\varrho$ for $(x, y) \in P^+ \times P^+$, which proves that $\sqrt{-1}\, U(x, y) \in \mathbf{R}$ for $(x, y) \in P^+ \times P^+$. The same proof applies to the behavior of $\sqrt{-1}\, U$ on P^-. Moreover we see that $\sqrt{-1}\, P^+ = P^-$ and $\sqrt{-1}\, U(x, y) = \sqrt{-1}\, U(\sqrt{-1}\, x, \sqrt{-1}\, y)$. Now let U and U' be two elements of $\mathfrak{T}'$ such that $\sqrt{-1}\, U$ and $\sqrt{-1}\, U'$ are positive definite. Then there exists an element β of $GL(P^+, \mathbf{R})$ such that $\sqrt{-1}\, U'(x\beta, y\beta) = \sqrt{-1}\, U(x, y)$ for $(x, y) \in P^+ \times P^+$. Define an element α of $GL(V, \mathbf{R})$ by $(x + \sqrt{-1}\, y)\alpha = x\beta + \sqrt{-1}\,(y\beta)$ for $x \in P^+$, $y \in P^+$. Then α is $\mathbf{C}$-linear, $\omega(x\alpha) = (\omega x)\alpha$ and $\sqrt{-1}\, U'(x\alpha, y\alpha) = \sqrt{-1}\, U(x, y)$ for $(x, y) \in V \times V$. This proves (1.11.4).

1.12. *Remark.* Let $\mathfrak{T}'$ be as in 1.8, with $F = \mathbf{R}$, $L = K = \mathbf{C}$, $\kappa > 0$. Define P^+ and P^- as in (iv) of Proof of 1.11. Then *the correspondence* $T \to \sqrt{-1}\, T$ *gives an $\mathbf{R}$-linear isomorphism of $\mathfrak{T}'$ to the vector space of all $\mathbf{R}$-valued symmetric forms on P^+*. In fact, this is an injection, since T is uniquely determined by its behavior on P^+. Furthermore the space of all $\mathbf{R}$-valued symmetric forms on P^+ is of dimension $2r^2 + r$. On the other hand, by 1.4, $\mathfrak{T}'$ is isomorphic to $\mathfrak{H}'$; and $\mathfrak{H}'$ is of dimension $2r^2 + r$ over $\mathbf{R}$. This proves our assertion. We see also that $\sqrt{-1}\, T$ is positive definite on V if and only if $\sqrt{-1}\, T$ is positive definite on P^+.

1.13. Let L, K, ι, ϱ, F and B be as in 1.2. Suppose that K is an algebraic number field and L is a division algebra. Put $[L : K] = q^2$. Since L has an involution of the first kind, we have $q = 1$ or 2. Let κ, ω, E and B' be as in 1.7. If $q = 1$, then $L = K$, $B = F$, $E = B'$, and ι is the identity mapping. If $q = 2$, L (resp. B) is a quaternion algebra over K (resp. F); and E is isomorphic to $M_2(B_0)$ with a quaternion algebra B_0 over F. If we denote by $\{X\}$ the class of a central simple algebra X over F in the Brauer group over F, then, in either case $q = 1, 2$, we have

(1.13.1) $E = B \otimes_F B' \cong M_q(B_0)$ $(B = F, B_0 = B'$ if $q = 1)$,

(1.13.2) $\{B\} \cdot \{B'\} \cdot \{B_0\} = 1$.

Here we admit the case $B_0 \cong M_2(F)$.

Assuming $q = 2$, put $W = \{x \in E | x^\varrho = x\}$. Then W is of dimension 10 or 6 over F according as ι is the main involution of L or not. Identify E with $M_2(B_0)$, and denote by σ the main involution of B_0. Then there exists an element u of $M_2(B_0)$ such that ${}^t u^\sigma = \pm u$ and $x^\varrho = u \cdot {}^t x^\sigma u^{-1}$ for $x \in E = M_2(B_0)$. Computing the dimension of W over F by means of this expression of ϱ, we find that ${}^t u^\sigma = -u$ or u according as ι is the main involution of L or not.

1.14. The notation being as in 1.13, let T, $S \in GL_m(L)$, and let $R = T \cdot {}^t S^\varrho$. Then one can easily verify that (1.10.1) is equivalent to

$$(1.14.1) \qquad {}^t R' = -\eta R, \quad {}^t T^\varrho = \eta T, \quad T \cdot {}^t (R^{-1})^\varrho \cdot {}^t T' = -\kappa^{-1} R.$$

Assuming this, define involutions α and β of $M_m(L)$ by

$$(1.14.2) \qquad X^\alpha = R \cdot {}^t X' R^{-1}, \quad X^\beta = T \cdot {}^t X^\varrho T^{-1}.$$

By (1.14.1) and 1.3, we have $\alpha\beta = \beta\alpha$. Let us apply 1.2 with $M_m(L)$, α and β in place of L, ι and ϱ. Assume that the algebra B_0 in (1.13.1) is a division algebra. We have obviously

$$\{X \in M_m(L) | X^\alpha = X^\beta\} = \{X \in M_m(L) | X^{\iota\varrho} S = SX\} \cong \mathrm{End}(L^m, E) \cong M_r(B_0),$$

where $r = qm/2$. Identify the algebra $\{X \in M_m(L) | X^\alpha = X^\beta\}$ with $M_r(B_0)$. By 1.2, we have

$$(1.14.3) \qquad\qquad M_m(L) = M_r(B_0) \otimes_F K,$$

and the restriction of α to $M_r(B_0)$ is an involution of the first kind. Hence there exists an element Z of $M_r(B_0)$ such that ${}^t Z^\sigma = \varepsilon Z$ with $\varepsilon = \pm 1$ and $X^\alpha = Z \cdot {}^t X^\sigma Z^{-1}$ for every $X \in M_r(B_0)$, where σ is the main involution of B_0. Computing the dimension of the vector space $\{X \in M_m(L) | X^\alpha = X\}$, we find, in view of (1.14.2, 3), that $\varepsilon = \eta$ if $q = 1$; if $q = 2$, one has $\varepsilon = -\eta$ or $\varepsilon = \eta$ according as ι is the main involution of L or not. In any case we have, by 1.2,

$$\{X \in M_m(L) | X X^\alpha = X X^\beta = 1\} = \{X \in M_m(L) | XT \cdot {}^t X^\varrho = T, X^{\iota\varrho} S = SX\}$$

$$= \{X \in M_r(B_0) | XZ \cdot {}^t X^\sigma = Z\}.$$

In this way the group (1.9.1) may be identified with the unitary group of a hermitian or anti-hermitian form over B_0 with respect to the main involution. (If $q = 1$ and hence $L = K$, this is obvious from (1.9.1), since $E = B_0$, and $\varrho = \sigma$.) One can verify this fact in the case $q = 2$ more directly from (1.9.1) without using 1.2 or 1.3, but by means of the relation $x^\varrho = u \cdot {}^t x^\sigma u^{-1}$ discussed at the end of 1.13.

1.15. Proposition. *Let L, F, K, ι, ϱ and κ be as above. Suppose that $L = K$, ι is the identity mapping, and K is an algebraic number field of finite degree. Let S and T be elements of $GL_{2r}(K)$ satisfying (1.10.1) with $\eta = 1$ and $m = 2r$, where r is a positive integer. Suppose further that T^τ is positive definite for every isomorphism τ of K into $\mathbf{C}$ such that $F^\tau \subset \mathbf{R}$, $K^\tau \not\subset \mathbf{R}$ and $\kappa^\tau < 0$. Then there exists an element Y of $GL_{2r}(K)$ such that $Y^\varrho S Y^{-1} = \begin{pmatrix} 0 & 1_r \\ \kappa 1_r & 0 \end{pmatrix}$, $YT \cdot {}^t Y^\varrho$*

$$= \begin{pmatrix} 1_r & 0 \\ 0 & -\kappa 1_r \end{pmatrix}.$$

Proof. Let $E = (K, \omega, \kappa)$ be as in 1.7. We regard K^{2r} as a left E-module by the rule $(a + b\omega)x = ax + bx^\varrho S$ for $a, b \in K$, $x \in K^{2r}$. By 1.8, there exists an E-valued ϱ-hermitian form H on K^{2r} such that $\mathrm{Tr}_{E/F}(H(x, y)) = \mathrm{Tr}_{K/F}(xT \cdot {}^t y^\varrho)$. Let v be an archimedean prime spot of F defined by an isomorphism τ of K into $\mathbf{C}$. Let F_v denote the completion of F with respect to v, and let $K_v = K \otimes_F F_v$, $E_v = E \otimes_F F_v$. If $F^\tau \subset \mathbf{R}$, $K^\tau \not\subset \mathbf{R}$ and $\kappa^\tau < 0$, we may identify F_v, K_v, E_v with $\mathbf{R}$,

C, K, respectively. Since $H(x, x) = xT \cdot {}^t x^\varrho$, H is positive definite at v, in view of our assumption on T. By the Hasse principle for quaternion hermitian forms, any two such H's are equivalent. From this fact and the discussion of 1.10, we obtain our proposition.

1.16. *Remark.* Let the notation be as in 1.14. If $L = K$, $\eta = 1$ and the arithmetical conditions on the fields and numbers are disregarded, then the condition (1.14.1) is equivalent to SIEGEL [10, (67)]. Therefore, if $\mathfrak{r}$ denotes the ring of integers in K, the group

$$\{X \in M_{2r}(\mathfrak{r}) \mid X T \cdot {}^t X^\varrho = T, \; X^\varrho S = S X\}$$

is the same as $\Lambda(\mathfrak{G}, \mathfrak{H})$ introduced in [10, No. 25]. The present F, K, κ, R and T correspond respectively to K, K_0, s^{-1}, $\mathfrak{G}$ and $\mathfrak{H}$ of [10, No. 25] (cf. also [10, No. 27]). One may also notice that (1.11.1) is essentially equivalent to [10, Lemma 3], and 1.15 is a generalization of [10, Lemma 5]. One can formulate and prove a result similar to 1.15 also in the case $L \neq K$, for which precise statement will be left to the reader.

2. Analytic families of abelian varieties

2.1. Hereafter we fix the notation as follows. (See *Advice to the reader*.)

F: a totally real algebraic number field of finite degree.

K: a totally imaginary quadratic extension of F.

L: a central division algebra over K.

$g = [F : \mathbf{Q}]$.

$q^2 = [L : K]$; $q = 1$ or 2, cf. 1.13.

ϱ: a positive involution of L which induces the non-trivial automorphism of K over F.

ι: the identity mapping of K or the main involution of L according as $q = 1$ or 2. (It should be noted that, when $q = 2$, the main involution of L commutes with every involution of L over F.)

$B = \{x \in L \mid x^\iota = x^\varrho\}$; $B = F$ if $q = 1$.

κ: a non-zero element of F.

$E = (L, \omega, \kappa) = L + L\omega$, $\omega^2 = \kappa$, $\omega a = a^{\iota\varrho}\omega$ for $a \in L$; ϱ is extended to E by $(a + b\omega)^\varrho = a^\varrho - b^\iota\omega$ (cf. 1.7).

$L_\mathbf{R} = L \otimes_\mathbf{Q} \mathbf{R}$, $E_\mathbf{R} = E \otimes_\mathbf{Q} \mathbf{R}$; the $\mathbf{R}$-linear extensions of ι and ϱ are denoted again by ι and ϱ.

We note that, if $q = 2$, B should be a totally definite quaternion algebra over F, since the main involution of B ($= \iota = \varrho$) is a positive involution. Therefore, in view of 1.2 and 1.4, the above L, ϱ, B, with $q = 2$, can be obtained as follows. Choose a totally definite quaternion algebra B over F, which does not split over K, and put $L = B \otimes_F K$. Let γ be the main involution of B. Define ϱ so that $x^\varrho = x^\gamma$ for $x \in B$ and ϱ induces the non-trivial automorphism of K over F. It can be easily seen that ϱ is positive.

2.2. In view of [5, Lemma 1], we can find g inequivalent absolutely irreducible $\mathbf{R}$-linear representations $\varphi_v : L_\mathbf{R} \to M_q(\mathbf{C})$ ($v = 1, \ldots, g$) so that

12*

(2.2.1) $\varphi_v(a^\varrho) = {}^t\overline{\varphi_v(a)}$ *for every* $a \in L_\mathbf{R}$ *and every* v.

(2.2.2) $\{\varphi_1, \ldots, \varphi_g, \bar{\varphi}_1, \ldots, \bar{\varphi}_g\}$ *is a complete set of inequivalent absolutely irreducible representations of* $L_\mathbf{R}$.

We can define an isomorphism τ_v of K into $\mathbf{C}$ by

$$(2.2.3) \qquad\qquad \varphi_v(a) = a^{\tau_v} \cdot 1_q \qquad\qquad (a \in K).$$

The restrictions of $\tau_1, \ldots, \tau_g$ to F are distinct, so that $(K; \{\tau_v\})$ is a CM-type in the sense of [9].

Suppose that $q = 2$. Since ι is the main involution of L, we have

$$(2.2.4) \qquad \varphi_v(a^\iota) = j \cdot {}^t\varphi_v(a)j^{-1}, \quad j = \begin{pmatrix} 0 & 1 \\ -1 & 0 \end{pmatrix} \qquad (a \in L_\mathbf{R}),$$

$$(2.2.5) \qquad \varphi_v(a^{\iota\varrho}) = j\overline{\varphi_v(a)}j^{-1} \qquad\qquad (a \in L_\mathbf{R}).$$

For every positive integer m, we define a representation $\varphi_v^m : M_m(L_\mathbf{R}) \to M_{qm}(\mathbf{C})$ as follows. Let $a = (a_{ij}) \in M_m(L_\mathbf{R})$ with $a_{ij} \in L_\mathbf{R}$. If $q = 1$, we put $\varphi_v^m(a) = (\varphi_v(a_{ij}))$. If $q = 2$, put $\varphi_v(a_{ij}) = \begin{pmatrix} b_{ij} & c_{ij} \\ d_{ij} & e_{ij} \end{pmatrix}$ with $b_{ij}, c_{ij}, d_{ij}, e_{ij} \in \mathbf{C}$, and

$$\varphi_v^m(a) = \begin{pmatrix} B & C \\ D & E \end{pmatrix}, \quad B = (b_{ij}), \quad C = (c_{ij}), \quad D = (d_{ij}), \quad E = (e_{ij}).$$

In either case ($q = 1, 2$), we have, on account of (2.2.1, 5),

$$(2.2.6) \qquad \varphi_v^m({}^t a^\varrho) = {}^t\overline{\varphi_v^m(a)} \qquad\qquad (a \in M_m(L_\mathbf{R})),$$

$$(2.2.7) \qquad \varphi_v^m(a^{\iota\varrho}) = J_m\overline{\varphi_v^m(a)}J_m^{-1} \qquad\qquad (a \in M_m(L_\mathbf{R})),$$

$$(2.2.8) \qquad J_m = \begin{cases} 1_m & \text{if } q = 1, \\ \begin{pmatrix} 0 & 1_m \\ -1_m & 0 \end{pmatrix} & \text{if } q = 2. \end{cases}$$

2.3. We denote by ε a number which can be $+1$ or -1. We shall construct four types, depending on ε and q, of analytic families of abelian varieties. We put $\varepsilon_q = (-1)^{q-1}$ and $\eta = \varepsilon\varepsilon_q$. Let h be an integer such that $0 < h < g$. We assume hereafter that the number κ satisfies

$$(2.3.1) \qquad \begin{aligned} \eta\kappa^{\tau_v} &> 0 & (v = 1, \ldots, h), \\ \eta\kappa^{\tau_v} &< 0 & (v = h+1, \ldots, g). \end{aligned}$$

Let $B' = K + K\omega$. Then the assumption (2.3.1) implies

$$(2.3.2) \quad B' \otimes_\mathbf{Q} \mathbf{R} \cong \begin{cases} \underbrace{M_2(\mathbf{R}) \times \cdots \times M_2(\mathbf{R})}_{h} \times \underbrace{\mathbf{K} \times \cdots \times \mathbf{K}}_{g-h} & \text{if } \eta = 1, \\ \underbrace{\mathbf{K} \times \cdots \times \mathbf{K}}_{h} \times \underbrace{M_2(\mathbf{R}) \times \cdots \times M_2(\mathbf{R})}_{g-h} & \text{if } \eta = -1, \end{cases}$$

where $\mathbf{K}$ denotes the division ring of real quaternions, and the v-th factor corresponds to τ_v. Let B_0 be as in (1.13.1). If $q = 1$, we have $B_0 = B'$. Suppose that $q=2$. Since B is totally definite, we obtain, in view of (1.13.2),

$$(2.3.3) \qquad B_0 \otimes_{\mathbf{Q}} \mathbf{R} \cong \begin{cases} \underbrace{M_2(\mathbf{R}) \times \cdots \times M_2(\mathbf{R})}_{h} \times \underbrace{\mathbf{K} \times \cdots \times \mathbf{K}}_{g-h} & \text{if } \varepsilon = 1, \\[2ex] \underbrace{\mathbf{K} \times \cdots \times \mathbf{K}}_{h} \times \underbrace{M_2(\mathbf{R}) \times \cdots \times M_2(\mathbf{R})}_{g-h} & \text{if } \varepsilon = -1, \end{cases}$$

where the v-th factor corresponds to τ_v.

2.4. Let m be a positive integer, and L^m the vector space of all m-dimensional row vectors with components in L. Let S and T be two elements of $GL_m(L)$ such that

$$(2.4.1) \qquad S^{\prime\varrho}S = \kappa 1_m, \quad {}^t T^{\varrho} = \eta T, \quad T \cdot {}^t S^{\varrho} = -S^{\prime\varrho} T^{\prime\varrho},$$

$$(2.4.2) \qquad \sqrt{\eta}^{-1} \varphi_v^m(T) \quad \text{is positive definite for} \quad v > h.$$

Here $\sqrt{\eta}$ means 1 or the imaginary unit according as $\eta = 1$ or -1. We regard L^m as a left E-module by $(a+b\omega)x = ax + bx^{\prime\varrho}S$ for $a, b \in L$, $x \in L^m$ (cf. 1.7, 1.10). We take an element ζ of K such that $\zeta^\varrho = -\zeta$ and $\mathrm{Im}(\zeta^{\tau v}) > 0$ for every v. We put then

$$(2.4.3) \qquad T^* = \begin{cases} \zeta T, \\ T, \end{cases} \quad T^*(x, y) = \begin{cases} \zeta x T \cdot {}^t y^\varrho & \text{if } \eta = 1, \\ x T \cdot {}^t y^\varrho & \text{if } \eta = -1, \end{cases} \quad ((x, y) \in L^m \times L^m).$$

2.5. Put $qm = 2r$. By (2.3.2), we see that B' is a division algebra. If $q = 1$, we have $L = K$, so that K^m is a B'-module. It follows that m should be an even number. Hence r is always a positive integer.

We define a representation Φ of $L_{\mathbf{R}}$ by complex matrices of size $2qrg$ as follows:

$$\Phi(a) = \begin{pmatrix} \Phi_1(a) & & \\ & \ddots & \\ & & \Phi_g(a) \end{pmatrix}, \quad \Phi_v(a) = \begin{cases} \begin{pmatrix} \varphi_v^r(a \cdot 1_r) & 0 \\ 0 & \overline{\varphi_v^r(a \cdot 1_r)} \end{pmatrix} & (1 \leqq v \leqq h), \\[2ex] \varphi_v^{2r}(a \cdot 1_{2r}) & (h < v \leqq g). \end{cases}$$

We denote by $\mathfrak{U}$ (resp. $\mathfrak{U}_v$) the representation space of Φ (resp. Φ_v) and put

$$\mathfrak{U}^+ = \mathfrak{U}_1 + \cdots + \mathfrak{U}_h, \quad \mathfrak{U}^- = \mathfrak{U}_{h+1} + \cdots + \mathfrak{U}_g.$$

We shall often regard $\mathfrak{U}$ (resp. $\mathfrak{U}_v$) as the vector space of all complex column vectors of dimension $2qrg$ (resp. $2qr$).

2.6. Proposition. *There exists an element* ψ *of* $\mathrm{End}(\mathfrak{U}, \mathbf{R})$ *such that* $\psi^2 = \Phi(\kappa)$, $\psi\Phi(a) = \Phi(a^{\prime\varrho})\psi$ *for every* $a \in L$, *and*

$$\psi(\sqrt{-1}\,x) = \begin{cases} \sqrt{-1}\,\psi(x) & \text{for } x \in \mathfrak{U}^+, \\ -\sqrt{-1}\,\psi(x) & \text{for } x \in \mathfrak{U}^-. \end{cases}$$

Moreover, if ψ' *is another element of* $\mathrm{End}(\mathfrak{U}, \mathbf{R})$ *satisfying the same conditions, then there exists an element* Y *of* $GL(\mathfrak{U}, \mathbf{C})$ *such that* $Y\psi' = \psi Y$ *and* $Y\Phi(a) = \Phi(a)Y$ *for every* $a \in L$.

Proof. Define the action of ψ on each $\mathfrak{U}_v$ by

$$(2.6.1) \qquad \psi(y) = \begin{cases} \sqrt{|\kappa^{\tau_v}|}\begin{pmatrix} 0 & J_r \\ \varepsilon J_r & 0 \end{pmatrix} y & (y \in \mathfrak{U}_v;\, 1 \leq v \leq h), \\[2ex] \sqrt{|\kappa^{\tau_v}|}\begin{pmatrix} 0 & 1_r \\ -\varepsilon 1_r & 0 \end{pmatrix} \bar{y} & (q=1;\, y \in \mathfrak{U}_v;\, h < v \leq g), \\[2ex] \sqrt{|\kappa^{\tau_v}|}\begin{pmatrix} 0 & j_\varepsilon \\ -j_\varepsilon & 0 \end{pmatrix} \bar{y} & (q=2;\, y \in \mathfrak{U}_v;\, h < v \leq g), \end{cases}$$

where J_r is as in (2.2.8) with r in place of m, and $j_\varepsilon = \begin{pmatrix} 0 & 1_r \\ -\varepsilon 1_r & 0 \end{pmatrix}$. In view of

(2.2.5, 7), it can be easily verified that ψ has the required properties. To prove the last assertion, we consider $\mathfrak{U}$ as an $E_{\mathbf{R}}$-module by means of the action $(a + b\omega)x = \Phi(a)x + \Phi(b)\,\psi(x)$ for $a, b \in L_{\mathbf{R}}$, $x \in \mathfrak{U}$. Then $\mathfrak{U}^+$ and $\mathfrak{U}^-$ are $E_{\mathbf{R}}$-submodules of $\mathfrak{U}$. Since ψ is C-linear on $\mathfrak{U}^+$, we can consider $\mathfrak{U}^+$ as a module over $E_{\mathbf{C}} = E \otimes_{\mathbf{Q}} \mathbf{C}$. Now a representation of $E_{\mathbf{C}}$ (resp. $E_{\mathbf{R}}$) can be determined, up to equivalence, by its restriction to the center $F_{\mathbf{C}} = F \otimes_{\mathbf{Q}} \mathbf{C}$ (resp. $F_{\mathbf{R}} = F \otimes_{\mathbf{Q}} \mathbf{R}$). Therefore, if there are two elements ψ and ψ' of $GL(\mathfrak{U}, \mathbf{R})$ with the above properties, then we find an element Y^+ (resp. Y^-) of $GL(\mathfrak{U}^+, \mathbf{C})$ (resp. $GL(\mathfrak{U}^-, \mathbf{R})$) such that $Y^\pm \psi = \psi' Y^\pm$ and $Y^\pm \Phi(a) = \Phi(a) Y^\pm$ for every $a \in L_{\mathbf{R}}$. In view of our definition of Φ, we can find an element b of $K_{\mathbf{R}}$ such that $\Phi_v(b) = \sqrt{-1} \cdot 1_{2qr}$ for $h < v \leq g$. Since Y^- is commutative with this operation, we see that Y^- is C-linear. This completes the proof.

2.7. Let ψ be as in 2.6. We call an **R**-linear isomorphism $\mathfrak{y}$ of $L_{\mathbf{R}}^m$ onto $\mathfrak{U}$ a *conductive map* of $(L_{\mathbf{R}}^m, \omega)$ to $(\mathfrak{U}, \psi)$ if $\mathfrak{y}(\omega x) = \psi(\mathfrak{y}(x))$ and $\mathfrak{y}(ax) = \Phi(a)\,\mathfrak{y}(x)$ for every $a \in L$ and $x \in L_{\mathbf{R}}^m$.

2.8. Let A be an abelian variety (with **C** as universal domain). Regarding A as a real Lie group, we denote by $\mathrm{End}^0(A)$ the ring of all real analytic endomorphisms of A, and put $\mathrm{End}_{\mathbf{Q}}^0(A) = \mathrm{End}^0(A) \otimes_{\mathbf{Z}} \mathbf{Q}$. If A is of complex dimension n, it is clear that $\mathrm{End}^0(A)$ and $\mathrm{End}_{\mathbf{Q}}^0(A)$ are isomorphic to $M_{2n}(\mathbf{Z})$ and $M_{2n}(\mathbf{Q})$, respectively.

2.9. Let $\mathfrak{M}$ be a free **Z**-submodule of L^m of maximal rank. We assume that T^* satisfies

$$(2.9.1) \qquad \mathrm{Tr}_{L/\mathbf{Q}}(T^*(\mathfrak{M}, \mathfrak{M})) = \mathbf{Z}.$$

For a given T, we can always replace it by a suitable positive rational multiple of it which satisfies (2.9.1), retaining the condition (2.4.1, 2). Define a PEL-type Ω in the sense of [8] by

$$(2.9.2) \qquad \Omega = (L, \Phi, \varrho; T^*, \mathfrak{M}; v_1, \ldots, v_s)$$

with a set of elements $v_1, \ldots, v_s$ of L^m. Now we put

$$(2.9.3) \qquad \Omega_0 = (\Omega, \omega) = (L, \Phi, \varrho; T^*, \mathfrak{M}; v_1, \ldots, v_s; \omega),$$

and call it a *hyper-PEL-type.*

2.10. Let $\mathscr{Q} = (A, \mathscr{C}, \theta; t_1, \ldots, t_s)$ be a PEL-structure, having L as endomorphism algebra, in the sense of [8, 1.5], and let $\mathfrak{b}$ be an element of $\mathrm{End}_{\mathbf{Q}}^0(A)$ (cf. 2.8). We put

$$\mathscr{R} = (\mathscr{Q}, \mathfrak{b}) = (A, \mathscr{C}, \theta, \mathfrak{b}; t_1, \ldots, t_s),$$

and call it a *hyper-PEL-structure*. Let us fix an element ψ of $GL(\mathfrak{U}, \mathbf{R})$ with the properties of 2.6. Now $\mathscr{R}$ is said to be of type Ω_0, if there exist a lattice D in $\mathfrak{U}$ and a commutative diagram

$$(2.10.1) \qquad \begin{array}{ccccccccc} 0 & \longrightarrow & \mathfrak{M} & \longrightarrow & L_{\mathbf{R}}^m & \longrightarrow & L_{\mathbf{R}}^m/\mathfrak{M} & \longrightarrow & 0 \quad \text{(exact)} \\ & & \downarrow{\scriptstyle \eta} & & \downarrow{\scriptstyle \eta} & & \downarrow & & \\ 0 & \longrightarrow & D & \longrightarrow & \mathfrak{U} & \overset{\xi}{\longrightarrow} & A & \longrightarrow & 0 \quad \text{(exact)} \end{array}$$

satisfying the following conditions (2.10.2—6).

(2.10.2) ξ gives a biregular isomorphism of $\mathfrak{U}/D$ to A, and $\xi(\Phi(a)u) = \theta(a)\,\xi(u)$ for $a \in L$, $u \in \mathfrak{U}$, provided that $\theta(a) \in \mathrm{End}(A)$.

(2.10.3) η is a conductive map of $(L_{\mathbf{R}}^m, \omega)$ to $(\mathfrak{U}, \psi)$, and $\eta(\mathfrak{M}) = D$.

(2.10.4) $\xi \circ \psi = \mathfrak{d} \circ \xi$.

(2.10.5) The polarization $\mathscr{C}$ contains a divisor which determines a Riemann form $\mathfrak{E}$ on $\mathfrak{U}/D$ such that $\mathfrak{E}(\eta(x), \eta(x')) = \mathrm{Tr}_{L/\mathbf{Q}}(T^*(x, x'))$ for $(x, x') \in L^m \times L^m$.

(2.10.6) $t_i = \xi(\eta(v_i))$.

If these conditions are satisfied, $\mathscr{Q}$ is of type Ω in the sense of [8, 3.2], and $\mathfrak{d}^2 = \theta(\kappa)$, $\theta(a)\mathfrak{d} = \mathfrak{d}\theta(a'^\varrho)$ for $a \in L$. Moreover, from (2.10.5) we obtain

$$(2.10.7) \qquad \mathfrak{E}(\Phi(a)u, v) = \mathfrak{E}(u, \Phi(a^\varrho)v) \qquad (a \in L; (u, v) \in \mathfrak{U} \times \mathfrak{U}),$$

and $\mathfrak{E}(\eta(x), \psi(\eta(x'))) = \mathrm{Tr}_{L/\mathbf{Q}}(T^*(x, \omega x'))$, hence by 1.10,

$$(2.10.8) \qquad \mathfrak{E}(u, \psi v) = \eta\mathfrak{E}(\psi u, v) = -\eta\mathfrak{E}(v, \psi u) \qquad ((u, v) \in \mathfrak{U} \times \mathfrak{U}).$$

2.11. Let $\mathscr{R} = (\mathscr{Q}, \mathfrak{d})$ and $\mathscr{R}' = (\mathscr{Q}', \mathfrak{d}')$ be hyper-PEL-structures. An isomorphism (resp. isogeny) λ of $\mathscr{Q}$ to $\mathscr{Q}'$ is called an *isomorphism* (resp. *isogeny*) of $\mathscr{R}$ to $\mathscr{R}'$ if $\lambda\mathfrak{d} = \mathfrak{d}'\lambda$.

Let $\Omega_0' = (L, \Phi, \varrho; T^{*\prime}, \mathfrak{M}'; v_1', \ldots, v_s'; \omega)$ be another hyper-PEL-type, with the same $L, \Phi, \varrho, s, \omega$ as before. We say that Ω_0 and Ω_0' are equivalent if there exists an E-linear automorphism α of L^m such that $T^{*\prime}(x\alpha, y\alpha) = T^*(x, y)$, $\mathfrak{M}\alpha = \mathfrak{M}'$ and $v_i\alpha \equiv v_i' \bmod \mathfrak{M}'$ for every i. Let $\mathscr{R}$ be of type Ω_0. Then it is clear that $\mathscr{R}$ is of type Ω_0' if and only if Ω_0 and Ω_0' are equivalent.

2.12. Let J_m be as in (2.2.8). Put

$$(2.12.1) \qquad T_v = \varphi_v^m(T), \quad T_v^* = \varphi_v^m(T^*), \quad S_v = J_m^{-1}\varphi_v^m(S) \qquad (1 \le v \le g).$$

By (2.4.1) and (2.2.6, 7), we have

$$(2.12.2) \qquad \bar{S}_v S_v = \varepsilon_q \kappa^{\tau v} 1_{2r}, \quad {}^t\bar{T}_v = \eta T_v, \quad T_v \cdot {}^t\bar{S}_v = -\varepsilon_q \bar{S}_v \bar{T}_v.$$

Then we can find an element W_v of $M_{2r}(\mathbf{C})$ for each v so that

$$(2.12.3) \qquad \sqrt{-1}\, W_v(T_v^*)^{-1} \cdot {}^t\bar{W}_v = \begin{cases} \begin{pmatrix} 1_r & 0 \\ 0 & -1_r \end{pmatrix} & (1 \le v \le h), \\[1em] 1_{2r} & (h < v \le g). \end{cases}$$

$$(2.12.4) \qquad {}^tW_v^{-1} S_v \cdot {}^t\bar{W}_v = \begin{cases} \sqrt{|\kappa^{\tau v}|} \begin{pmatrix} 0 & \varepsilon 1_r \\ 1_r & 0 \end{pmatrix} & (1 \le v \le h), \\[1em] \sqrt{|\kappa^{\tau v}|} \begin{pmatrix} 0 & -\varepsilon 1_r \\ 1_r & 0 \end{pmatrix} & (h < v \le g). \end{cases}$$

The matrices W_ν can be obtained as follows:

Case $q = 1, \varepsilon = 1$. Apply (1.11.1, 2) to S_ν and T_ν, and adopt $|\zeta^{\tau_\nu}|^{1/2} \cdot {}^t\bar{Y}^{-1}$ as W_ν.

Case $q = 2, \varepsilon = 1$. Apply (1.11.1, 2) to S_ν and $(\zeta^{\tau_\nu})^{-1} T_\nu$, and adopt $|\zeta^{\tau_\nu}|^{1/2} \cdot {}^t\bar{Y}^{-1}$ as W_ν.

Case $\varepsilon = -1$. Apply (1.11.3, 4) to S_ν and T_ν^*, and adopt ${}^t\bar{Y}^{-1}$ as W_ν.

2.13. Let $\mathscr{S}$ be the space of all complex matrices z of size r such that $1 - {}^t\bar{z}z$ is positive definite, and $\mathscr{H}$ the product of h copies of $\mathscr{S}$. In [5] (cf. also [8]) we have obtained a "universal" family of PEL-structures of type Ω, $\Sigma(\Omega) = \{\mathscr{Q}_z | z \in \mathscr{H}\}$. We are going to construct a certain subfamily of $\Sigma(\Omega)$ whose members have hyper-PEL-structure of type Ω_0. For this purpose we first recall some results of [5].

Let $\mathscr{Q} = (A, \mathscr{C}, \theta; t_1, \ldots, t_s)$ be a PEL-structure of type Ω. Then there exists a commutative diagram (2.10.1) satisfying (2.10.2—6) except the conditions concerning ω, ψ and $\mathfrak{d}$. Let d_j, for $j = 1, \ldots, m$, be the element of L^m whose j-th component is 1 and all the other components are 0. Put

$$(2.13.1) \qquad {}^t\eta(d_i) = ({}^tx_i^1 \ldots {}^tx_i^\nu \ldots {}^tx_i^g) \qquad (x_i^\nu \in \mathfrak{U}_\nu; 1 \leq i \leq m),$$

$$(2.13.2) \qquad {}^tx_i^\nu = \begin{cases} ({}^tu_i^\nu \, {}^tv_i^\nu) & (q = 1; 1 \leq \nu \leq h), \\ ({}^tu_i^\nu \, {}^tu_i^{*\nu} \, {}^tv_i^\nu \, {}^tv_i^{*\nu}) & (q = 2; 1 \leq \nu \leq h), \\ ({}^tu_i^\nu \, {}^tu_i^{*\nu}) & (q = 2; h < \nu \leq g). \end{cases}$$

Here $u_i^\nu, v_i^\nu \in \mathbf{C}^r$ and $u_i^\nu, u_i^{*\nu}, v_i^\nu, v_i^{*\nu} \in \mathbf{C}^r$ for $1 \leq \nu \leq h$; $u_i^\nu, u_i^{*\nu} \in \mathbf{C}^{2r}$ for $q = 2$, $h < \nu \leq g$. Then we define a square matrix X_ν of size $2r$ for each ν by

$$(2.13.3) \qquad X_\nu = \begin{cases} \begin{pmatrix} u_1^\nu \ldots u_{2r}^\nu \\ \bar{v}_1^\nu \ldots \bar{v}_{2r}^\nu \end{pmatrix} & (q = 1; 1 \leq \nu \leq h), \\ (x_1^\nu \ldots x_{2r}^\nu) & (q = 1; h < \nu \leq g), \\ \begin{pmatrix} u_1^\nu \ldots u_r^\nu \, u_1^{*\nu} \ldots u_r^{*\nu} \\ \bar{v}_1^\nu \ldots \bar{v}_r^\nu \, \bar{v}_1^{*\nu} \ldots \bar{v}_r^{*\nu} \end{pmatrix} & (q = 2; 1 \leq \nu \leq h), \\ (u_1^\nu \ldots u_r^\nu \, u_1^{*\nu} \ldots u_r^{*\nu}) & (q = 2; h < \nu \leq g). \end{cases}$$

As is seen [5, pp. 158—162], there exist h elements $z_1, \ldots, z_h$ of $\mathscr{S}$ such that

$$(2.13.4) \qquad X_\nu \bar{W}_\nu^{-1} = \begin{pmatrix} p_\nu & 0 \\ 0 & p_\nu' \end{pmatrix} \begin{pmatrix} 1_r & z_\nu \\ {}^t\bar{z}_\nu & 1_r \end{pmatrix} \qquad (1 \leq \nu \leq h),$$

with elements p_ν and p_ν' of $GL_r(\mathbf{C})$. From a given $\mathscr{Q}$, we get in this way a point $z = (z_1, \ldots, z_h)$ of $\mathscr{H}$.

2.14. Conversely, for a given point $z = (z_1, \ldots, z_h)$ of $\mathscr{H}$, define $X_1, \ldots, X_g$ by

$$(2.14.1) \qquad X_\nu \bar{W}_\nu^{-1} = \begin{cases} \begin{pmatrix} 1_r & z_\nu \\ {}^t\bar{z}_\nu & 1_r \end{pmatrix} & (1 \leq \nu \leq h), \\ 1_{2r} & (h < \nu \leq g), \end{cases}$$

and $\eta(d_1), \ldots, \eta(d_m)$ by going through (2.13.1—3) backward. Further, for $\sum_{i=1}^m a_i d_i \in L_{\mathbf{R}}^m$ with $a_i \in L_{\mathbf{R}}$, put

$$(2.14.2) \qquad \eta(\textstyle\sum_{i=1}^m a_i d_i; z) = \eta(\sum_{i=1}^m a_i d_i) = \sum_{i=1}^m \Phi(a_i) \eta(d_i),$$

and put $D = \eta(\mathfrak{M})$. Then $\mathfrak{U}/D$ has a structure of abelian variety, from which we obtain naturally a PEL-structure $\mathcal{Q}$ of type Ω satisfying (2.10.1—6) except the conditions on ω, ψ and $\mathfrak{d}$. We write this $\mathcal{Q}$ as $\mathcal{Q}_z$, and put

$$\mathcal{Q}_z = (A_z, \mathscr{C}_z, \theta_z; t_{1z}, \ldots, t_{sz}),$$
$$\mathcal{P}_z = (A_z, \mathscr{C}_z, \theta_z).$$

(Here $\mathcal{P}_z$ and $\mathcal{Q}_z$ are defined only up to isomorphisms.)

2.15. Let $\mathcal{Q}$ and η be as in 2.13. Define X_v as in (2.13.3). Now we define an **R**-linear mapping η' of $L_{\mathbf{R}}^m$ to $\mathfrak{U}$ by

$$(2.15.1) \qquad\qquad \eta'(x) = \psi^{-1}\eta(\omega x) \qquad\qquad (x \in L_{\mathbf{R}}^m).$$

We see easily that $\eta'(ax) = \Phi(a)\,\eta'(x)$ for every $a \in L$. Let $x_i'^{\nu}$ and X_v' be the vectors and matrices defined by (2.13.1—3) with η' in place of η. Let $S = (\sigma_{ij})$ with $\sigma_{ij} \in L$. From (2.15.1) we obtain $\psi\eta'(d_i) = \eta(d_i S) = \sum_{j=1}^m \Phi(\sigma_{ij})\,\eta(d_j)$, so that

$$(2.15.2) \qquad\qquad \psi x_i'^{\nu} = \sum_{j=1}^m \Phi_{\nu}(\sigma_{ij}) x_j^{\nu}.$$

Let us take ψ to be defined by (2.6.1). Then from (2.13.3) and (2.15.2), we obtain

$$(2.15.3) \qquad |\kappa^{\tau_\nu}|^{-1/2} \cdot X_v \cdot {}^t\varphi_{\nu}^m(S) J_m = \begin{cases} \begin{pmatrix} 0 & 1_r \\ \varepsilon 1_r & 0 \end{pmatrix} \cdot \overline{X}_v' & (1 \leqq \nu \leqq h), \\[2ex] \begin{pmatrix} 0 & 1_r \\ -\varepsilon 1_r & 0 \end{pmatrix} \cdot \overline{X}_v' & (h < \nu \leqq g). \end{cases}$$

By (2.12.4) and (2.13.4) we have

$$(2.15.4) \qquad\qquad X_v' \overline{W}_v^{-1} = \begin{pmatrix} \overline{p}_v' & 0 \\ 0 & \overline{p}_v \end{pmatrix}\begin{pmatrix} 1_r & \varepsilon {}^t z_v \\ \varepsilon \overline{z}_v & 1_r \end{pmatrix} \qquad (1 \leqq \nu \leqq h),$$

$$(2.15.5) \qquad X_v \overline{W}_v^{-1} \cdot \begin{pmatrix} 0 & 1_r \\ -\varepsilon 1_r & 0 \end{pmatrix} = \begin{pmatrix} 0 & 1_r \\ -\varepsilon 1_r & 0 \end{pmatrix} \cdot \overline{X}_v' W_v^{-1} \qquad (h < \nu \leqq g).$$

Since $\varepsilon {}^t z_v \in \mathscr{S}$, the relation (2.15.4) shows that $\mathfrak{U}/\eta'(\mathfrak{M})$ has a structure of abelian variety, from which we obtain naturally a PEL-structure $\mathcal{Q}'$ of type Ω, corresponding to the point $(\varepsilon {}^t z_1, \ldots, \varepsilon {}^t z_h)$ of $\mathscr{H}$ (cf. 2.14 and [5, 2.5]). More explicitly, take X_v to be defined by (2.14.1). Then (2.15.4, 5) implies that X_v' is defined by (2.14.1) with $\varepsilon {}^t z_v$ in place of z_v. Therefore, by means of the function $\eta(x, z)$ defined by (2.14.2), the relation (2.15.1) can be expressed in the form

$$(2.15.6) \qquad\qquad \psi\eta(x, \varepsilon {}^t z) = \eta(\omega x, z) \qquad\qquad (x \in L_{\mathbf{R}}^m, z \in \mathscr{H}),$$

where $\varepsilon {}^t z = (\varepsilon {}^t z_1, \ldots, \varepsilon {}^t z_h)$ for $z = (z_1, \ldots, z_h)$.

2.16. Let $\mathscr{R} = (\mathcal{Q}, \mathfrak{d})$ be a hyper-PEL-structure of type Ω_0, and η a conductive map of $(L_{\mathbf{R}}^m, \omega)$ to $(\mathfrak{U}, \psi)$ satisfying the conditions of 2.10 for the present $\mathscr{R}$. Define x_i^{ν}, X_v and z_v through (2.13.1—4) with this η. Since $\psi\eta(x) = \eta(\omega x)$, we obtain the relations (2.15.4, 5) with $X_v = X_v'$, so that ${}^t z_v = \varepsilon z_v$, $p_v' = \overline{p}_v$ for

$1 \leqq v \leqq h$. Define an element f of $GL(\mathfrak{U}, \mathbf{C})$ by $f(y) = P_v^{-1} y$ for $y \in \mathfrak{U}_v$, $1 \leqq v \leqq g$, where

$$(2.16.1) \qquad P_v = \begin{cases} \begin{pmatrix} p_v & 0 \\ 0 & p_v \end{pmatrix} & (q=1, 1 \leqq v \leqq h), \\[1em] X_v \bar{W}_v^{-1} & (q=1, h < v \leqq g), \\[1em] \begin{pmatrix} p_v & & & 0 \\ & p_v & & \\ & & p_v & \\ 0 & & & p_v \end{pmatrix} & (q=2, 1 \leqq v \leqq h), \\[2em] \begin{pmatrix} X_v \bar{W}_v^{-1} & 0 \\ 0 & X_v \bar{W}_v^{-1} \end{pmatrix} & (q=2, h < v \leqq g). \end{cases}$$

Then $f\psi = \psi f$ and $f\Phi(a) = \Phi(a)f$ for every $a \in L$, so that $f\eta$ is again a conductive map of $(L_{\mathbf{R}}^m, \omega)$ to $(\mathfrak{U}, \psi)$. Replace η, ξ, D and $\mathfrak{E}$ by $f\eta$, ξf^{-1}, $f(D)$ and $\mathfrak{E}(f^{-1}(u), f^{-1}(v))$, and denote them again by η, ξ, D and $\mathfrak{E}$. This amounts to the exchange of the X_v by the ones satisfying (2.14.1).

2.17. Let, for $\varepsilon = \pm 1$,

$$(2.17.1) \qquad \mathscr{S}_\varepsilon = \{ z \in \mathscr{S} \mid {}^t z = \varepsilon z \},$$

$$(2.17.2) \qquad \mathscr{H}_\varepsilon = \mathscr{S}_\varepsilon \times \cdots \times \mathscr{S}_\varepsilon \quad (h \text{ copies}).$$

Consider $\mathcal{Q}_z$ for a point z of $\mathscr{H}_\varepsilon$, and the mapping $\eta(x) = \eta(x, z)$ defined by (2.14.2). By (2.15.6), η is a conductive map of $(L_{\mathbf{R}}^m, \omega)$ to $(\mathfrak{U}, \psi)$. Define an element $\mathfrak{d}_z$ of $\mathrm{End}_{\mathbf{Q}}^0(A_z)$ by $\mathfrak{d}_z \circ \xi = \xi \circ \psi$. Then $(\mathcal{Q}_z, \mathfrak{d}_z)$ is a hyper-PEL-structure of type Ω_0. Put

$$(2.17.3) \qquad \mathscr{R}_z = (\mathcal{Q}_z, \mathfrak{d}_z).$$

Thus we have obtained an analytic family

$$(2.17.4) \qquad \Sigma_0 = \Sigma(\Omega_0) = \{ \mathscr{R}_z \mid z \in \mathscr{H}_\varepsilon \}$$

of hyper-PEL-structures of type Ω_0. The consideration of 2.16 shows:

(2.17.5) *Every hyper-PEL-structure of type Ω_0 is isomorphic to a member of $\Sigma(\Omega_0)$.*

2.18. Our next task is to determine all the isomorphism-classes of hyper-PEL-structures. Let

$$(2.18.1) \quad \mathfrak{G}_{\mathbf{R}}(T) = \{ U \in GL_m(L_{\mathbf{R}}) \mid U T \cdot {}^t U^\varrho = T \}, \quad \mathfrak{G}(T) = \mathfrak{G}_{\mathbf{R}}(T) \cap M_m(L).$$

Let H be an E-valued ϱ-hermitian or ϱ-anti-hermitian form on L^m such that $\mathrm{Tr}_{L/F}(T(x, y)) = \mathrm{Tr}_{E/F}(H(x, y))$ for $(x, y) \in L^m \times L^m$, according as $\eta = 1$ or -1 (cf. 1.8). Let

$$(2.18.2) \qquad \mathfrak{G}_{\mathbf{R}}(H) = \{ \alpha \in GL(L_{\mathbf{R}}^m, E_{\mathbf{R}}) \mid H(x\alpha, y\alpha) = H(x, y) \}.$$

For the reason explained in 1.9 and 1.10, the group $\mathfrak{G}_{\mathbf{R}}(H)$ can be identified with a subgroup

$$(2.18.3) \qquad \{U \in \mathfrak{G}_{\mathbf{R}}(T) \,|\, U'^{\varrho} S = S U\}$$

of $\mathfrak{G}_{\mathbf{R}}(T)$. Hereafter, we always regard $\mathfrak{G}_{\mathbf{R}}(H)$ as a subgroup of $\mathfrak{G}_{\mathbf{R}}(T)$ by means of this identification, and put $\mathfrak{G}(H) = \mathfrak{G}_{\mathbf{R}}(H) \cap M_m(L)$.

For every $U \in \mathfrak{G}_{\mathbf{R}}(T)$, put

$$(2.18.4) \qquad {}^t W_v^{-1} \overline{\varphi_v^m(U)} \cdot {}^t W_v = \begin{pmatrix} a_v & b_v \\ c_v & d_v \end{pmatrix} \qquad (1 \leqq v \leqq h).$$

We can let U act on $\mathscr{H}$ by the rule (cf. [5, 2.7])

$$(2.18.5) \qquad U(z_1, \ldots, z_h) = (w_1, \ldots, w_h) \qquad (z_v, w_v \in \mathscr{S}),$$
$$w_v = (a_v z_v + b_v)(c_v z_v + d_v)^{-1} \qquad (1 \leqq v \leqq h).$$

One can easily verify that

$$(2.18.6) \qquad {}^t w_v = (\bar{d}_v \cdot {}^t z_v + \bar{c}_v)(\bar{b}_v \cdot {}^t z_v + \bar{a}_v)^{-1} \qquad (1 \leqq v \leqq h),$$
$$(2.18.7) \qquad U(\mathscr{H}_\varepsilon) = \mathscr{H}_\varepsilon \quad \text{if} \quad U \in \mathfrak{G}_{\mathbf{R}}(H).$$

It can be also shown that $\mathscr{H}_\varepsilon$ is isomorphic to the quotient of $\mathfrak{G}_{\mathbf{R}}(H)$ by a maximal compact subgroup.

For every $U \in \mathfrak{G}_{\mathbf{R}}(T)$, put

$$(2.18.8) \qquad U^* = S^{-1} U'^{\varrho} S.$$

In view of (2.4.1) we see that $U^* \in \mathfrak{G}_{\mathbf{R}}(T)$. With the symbols of (2.18.4), we have

$$(2.18.9) \quad {}^t W_v^{-1} \overline{\varphi_v^m(U^*)} \cdot {}^t W_v = \begin{pmatrix} 0 & 1_r \\ \varepsilon 1_r & 0 \end{pmatrix} \begin{pmatrix} \bar{a}_v & \bar{b}_v \\ \bar{c}_v & \bar{d}_v \end{pmatrix} \begin{pmatrix} 0 & \varepsilon 1_r \\ 1_r & 0 \end{pmatrix} = \begin{pmatrix} \bar{d}_v & \varepsilon \bar{c}_v \\ \varepsilon \bar{b}_v & \bar{a}_v \end{pmatrix}.$$

Let J_ε be the holomorphic automorphism of $\mathscr{H}$ defined by

$$(2.18.10) \qquad J_\varepsilon(z_1, \ldots, z_h) = (\varepsilon^t z_1, \ldots, \varepsilon^t z_h) \qquad (z_v \in \mathscr{S}).$$

By (2.18.6, 9) we have

$$(2.18.11) \qquad U^*(J_\varepsilon(z)) = J_\varepsilon(U(z)) \qquad (z \in \mathscr{H}).$$

For two lattices $\mathfrak{M}$ and $\mathfrak{N}$ in L^m such that $\mathfrak{M} \subset \mathfrak{N}$, we define discrete subgroups of $\mathfrak{G}_{\mathbf{R}}(T)$ and $\mathfrak{G}_{\mathbf{R}}(H)$ by

$$(2.18.12) \quad \begin{cases} \Gamma(T, \mathfrak{M}) = \{U \in GL_m(L) \,|\, U T \cdot {}^t U^{\varrho} = T, \ \mathfrak{M} U = \mathfrak{M}\}, \\ \Gamma(T, \mathfrak{N}/\mathfrak{M}) = \{U \in \Gamma(T, \mathfrak{M}) \,|\, \mathfrak{N}(1 - U) \subset \mathfrak{M}\}, \\ \Gamma(H, \mathfrak{M}) = \{U \in \Gamma(T, \mathfrak{M}) \,|\, U'^{\varrho} S = S U\}, \\ \Gamma(H, \mathfrak{N}/\mathfrak{M}) = \{U \in \Gamma(H, \mathfrak{M}) \,|\, \mathfrak{N}(1 - U) \subset \mathfrak{M}\}. \end{cases}$$

In view of (2.4.2), we see that $\mathscr{H}/\Gamma(T, \mathfrak{N}/\mathfrak{M})$ and $\mathscr{H}_\varepsilon/\Gamma(H, \mathfrak{N}/\mathfrak{M})$ are compact.

2.19. Let U be an element of $GL_m(L)$ such that $U T \cdot {}^t U^{\varrho} = aT$ with a totally positive element a of F. We can find an element R of $\mathfrak{G}_{\mathbf{R}}(T)$ such that $(a^{\tau_v})^{1/2} \varphi_v^m(R) = \varphi_v^m(U)$ for every v. Then we define the action of U on $\mathscr{H}$ by

$U(z) = R(z)$. Now there exists an element $\Lambda(U, z)$ of $GL(\mathfrak{U}, \mathbf{C})$ such that

$$(2.19.1) \qquad \Lambda(U, z)\, \mathfrak{y}(x, z) = \mathfrak{y}(xU, U^{-1}(z)) \qquad (x \in L_{\mathbf{R}}^m,\ z \in \mathscr{H}).$$

This has been shown in [5, 2.8] (cf. also [6, II, 4.2]). For reader's convenience, here we reproduce the construction of $\Lambda(U, z)$. Call $X_\nu(z)$ the matrix X_ν defined by (2.14.1), and put $w = U^{-1}(z)$. By our definition of $U(z)$, we have

$$\begin{pmatrix} 1_r & z_\nu \\ {}^t\bar{z}_\nu & 1_r \end{pmatrix} \begin{pmatrix} p_\nu & 0 \\ 0 & p'_\nu \end{pmatrix} = {}^t W_\nu^{-1}\, \overline{\varphi_\nu^m(U)} \cdot {}^t W_\nu \begin{pmatrix} 1_r & w_\nu \\ {}^t\bar{w}_\nu & 1_r \end{pmatrix} \qquad (1 \leqq \nu \leqq h)$$

with elements p_ν and p'_ν of $GL_r(\mathbf{C})$. Then

$$(2.19.2) \qquad X_\nu(w) \cdot {}^t\varphi_\nu^m(U) = \begin{cases} \begin{pmatrix} {}^t\bar{p}_\nu & 0 \\ 0 & {}^t\bar{p}'_\nu \end{pmatrix} \cdot X_\nu(z) & (1 \leqq \nu \leqq h), \\[2ex] \bar{W}_\nu \cdot {}^t\varphi_\nu^m(U)\, \bar{W}_\nu^{-1} X_\nu(z) & (h < \nu \leqq g). \end{cases}$$

If $q = 1$, define an element Λ of $GL(\mathfrak{U}, \mathbf{C})$ by $\Lambda y = \Lambda_\nu y$ for $y \in \mathfrak{U}_\nu$, with $\Lambda_\nu = \begin{pmatrix} {}^t\bar{p}_\nu & 0 \\ 0 & {}^t p'_\nu \end{pmatrix}$ or $\bar{W}_\nu \cdot {}^t\varphi_\nu^m(U)\, \bar{W}_\nu^{-1}$ according as $\nu \leqq h$ or $\nu > h$. Then (2.19.2) shows that $\Lambda\mathfrak{y}(d_i, z) = \mathfrak{y}(d_i U, w)$ for $1 \leqq i \leqq m$. Therefore we get (2.19.1) with this Λ as $\Lambda(U, z)$. The case $q = 2$ can be treated similarly by "doubling" the matrices as in (2.16.1).

Let $\mathfrak{M}$ and the v_i be as in (2.9.2), and let $\mathfrak{N} = \mathfrak{M} + \sum_{i=1}^s \mathbf{Z} v_i$. In [5, 2.8], we have seen that $\Lambda(U, z)$ has the following properties.

(2.19.3) *If a is an integer and $\mathfrak{M}U \subset \mathfrak{M}$, then $\Lambda(U, z)$ defines naturally an isogeny of $\mathscr{P}_z$ to $\mathscr{P}_w$. Every isogeny of $\mathscr{P}_z$ to another member $\mathscr{P}_y$ can be obtained from $\Lambda(U, z)$ with an element U of $GL_m(L)$ such that $UT \cdot {}^tU^\varrho = aT$ with a positive integer a, $\mathfrak{M}U \subset \mathfrak{M}$, and $y = U^{-1}(z)$.*

(2.19.4) *$\Lambda(U, z)$ defines an isomorphism of $\mathscr{P}_z$ to $\mathscr{P}_w$ (resp. $\mathscr{Q}_z$ to $\mathscr{Q}_w$), with $w = U^{-1}(z)$, if and only if $U \in \Gamma(T, \mathfrak{M})$ (resp. $U \in \Gamma(T, \mathfrak{N}/\mathfrak{M})$).*

It follows especially that $\mathscr{Q}_z$ and $\mathscr{Q}_y$ are isomorphic if and only if $z = U(y)$ for some $U \in \Gamma(T, \mathfrak{N}/\mathfrak{M})$.

2.20. Proposition. *Let $U \in \Gamma(H, \mathfrak{N}/\mathfrak{M})$, $z \in \mathscr{H}_\varepsilon$ and $U^{-1}(z) = w$. Then $\Lambda(U, z)$ gives an isomorphism of $\mathscr{R}_z$ to $\mathscr{R}_w$. Conversely, every isomorphism of a member $\mathscr{R}_z$ of $\Sigma(\Omega_0)$ to another member $\mathscr{R}_w$ can be obtained from $\Lambda(U, z)$ with an element U of $\Gamma(H, \mathfrak{N}/\mathfrak{M})$ such that $U(w) = z$.*

Proof. If $U \in \Gamma(H, \mathfrak{N}/\mathfrak{M})$, we have $(\omega x)U = \omega(xU)$, so that

$$\Lambda(U, z)\psi\mathfrak{y}(x, z) = \Lambda(U, z)\, \mathfrak{y}(\omega x, z) = \mathfrak{y}((\omega x)U, U^{-1}(z))$$

$$= \mathfrak{y}(\omega(xU), U^{-1}(z)) = \psi \cdot \Lambda(U, z)\, \mathfrak{y}(x, z),$$

hence $\Lambda(U, z)\psi = \psi \cdot \Lambda(U, z)$. Therefore $\Lambda(U, z)$ gives an isomorphism of $\mathscr{R}_z$ to $\mathscr{R}_w$. The converse part can be proved by reversing the argument, on account of (2.19.3, 4).

2.21. Proposition. *Let U be an element of $\mathfrak{G}_{\mathbf{R}}(T)$ and let $Y = U^{-1}S^{-1}U^\varrho S$. Suppose that $U(z) = w$ for some points z and w of $\mathscr{H}_\varepsilon$. Then $Y \in \mathfrak{G}_{\mathbf{R}}(T)$ and $Y(z) = z$. Moreover if $U \in \Gamma(T, \mathfrak{M})$ and the commutor of $\theta_z(L)$ in $\mathrm{End}_{\mathbf{Q}}(A_z)$ is $\theta_z(K)$, then $Y = \alpha 1_m$ with a root of unity α in K.*

Proof. The first assertion follows directly from (2.18.11). As for the second assertion, we notice that $\Lambda(Y, z)$ defines an element of $\mathrm{End}_{\mathbf{Q}}(A_z)$ commuting with every element of $\theta_z(L)$. Hence $Y = \alpha 1_m$ with an element α of K. We have then $\alpha^{mq} = N(Y) = N(U)^{-1} N(U)^{\varrho}$, where N is the reduced norm of $M_m(L)$ to K. Since $U \in \Gamma(T, \mathfrak{M})$, $N(U)$ is a root of unity. Hence α is a root of unity.

2.22. Proposition. *For any two lattices $\mathfrak{M}$ and $\mathfrak{N}'$ in L^m, there exists a lattice $\mathfrak{N}$ in L^m with the following properties.*

(2.22.1) $\mathfrak{N} \supset \mathfrak{M} + \mathfrak{N}'$.

(2.22.2) *If* $U \in \Gamma(T, \mathfrak{N}/\mathfrak{M})$ *and* $U(z) = w$ *for some points z and w of $\mathscr{H}_\varepsilon$, then* $U \in \Gamma(H, \mathfrak{N}/\mathfrak{M})$.

Proof. We can find a lattice $\mathfrak{N}_1$ so that $\mathfrak{N}_1 \supset \mathfrak{M} + \mathfrak{N}'$ and $\Gamma(T, \mathfrak{N}_1/\mathfrak{M})$ has no element of finite order other than the identity element. Then there exists a lattice $\mathfrak{N}$ such that $\mathfrak{N} \supset \mathfrak{N}_1$ and $U^{-1} S^{-1} U^{\prime \varrho} S \in \Gamma(T, \mathfrak{N}_1/\mathfrak{M})$ if $U \in \Gamma(T, \mathfrak{N}/\mathfrak{M})$. Let U be as in (2.22.2), and let $Y = U^{-1} S^{-1} U^{\prime \varrho} S$. Then by 2.21, $Y(z) = z$, and $Y \in \Gamma(T, \mathfrak{N}_1/\mathfrak{M})$, hence Y is of finite order. By our choice of $\mathfrak{N}_1$, we have $Y = 1_m$, so that $SU = U^{\prime \varrho} S$, which shows that $U \in \mathfrak{G}(H)$. Hence $U \in \Gamma(H, \mathfrak{N}/\mathfrak{M})$.

2.23. Proposition. *Let $\mathscr{R} = (\mathscr{Q}, \mathfrak{d})$ and $\mathscr{R}' = (\mathscr{Q}, \mathfrak{d}')$ be hyper-PEL-structures of type Ω_0, with the same underlying PEL-structure $\mathscr{Q} = (A, \mathscr{C}, \theta; t_1, \ldots, t_s)$. Suppose that the commutor of $\theta(L)$ in $\mathrm{End}_{\mathbf{Q}}(A)$ is $\theta(K)$. Then there exists a root of unity α in K such that $\mathfrak{d}' = \theta(\alpha) \cdot \mathfrak{d}$.*

Proof. By (2.17.5), there exists an isomorphism λ of $\mathscr{R}$ to a member $\mathscr{R}_z$ of $\Sigma(\Omega_0)$, and an isomorphism λ' of $\mathscr{R}'$ to a member $\mathscr{R}_w$ with $z, w \in \mathscr{H}_\varepsilon$. Since $\lambda' \lambda^{-1}$ is an isomorphism of $\mathscr{Q}_z$ to $\mathscr{Q}_w$, there exists an element U of $\Gamma(T, \mathfrak{N}/\mathfrak{M})$ such that $U(w) = z$ and $\lambda' \lambda^{-1}$ is obtained from $\Lambda(U, z)$. By 2.21, $U^{-1} S^{-1} U^{\prime \varrho} S = \alpha 1_m$ with a root of unity α in K. We have then $\Phi(\alpha) \eta(x, w) = \eta(x U^{-1} S^{-1} U^{\prime \varrho} S, w) = \psi \Lambda(U, z) \psi^{-1} \Lambda(U, z)^{-1} \eta(x, w)$, so that $\Phi(\alpha) \Lambda(U, z) \psi = \psi \Lambda(U, z)$. Shifting this relation to $\mathscr{R}$ and $\mathscr{R}'$ by λ and λ', we get $\theta(\alpha)\mathfrak{d} = \mathfrak{d}'$.

2.24. Proposition. *Let $\mathfrak{M}$ and the v_i be as in (2.9.2). Let $(\mathscr{Q}, \mathfrak{d})$ and $(\mathscr{Q}, \mathfrak{d}')$ be as in 2.23 without any assumption on $\mathrm{End}_{\mathbf{Q}}(A)$. Suppose that (2.22.2) is satisfied with $\mathfrak{N} = \mathfrak{M} + \sum_{i=1}^{s} \mathbf{Z} v_i$. Then $\mathfrak{d} = \mathfrak{d}'$.*

Proof. Let U be as in the proof of 2.23. By (2.22.2), $U \in \Gamma(H, \mathfrak{N}/\mathfrak{M})$, so that $\Lambda(U, z)$ provides an isomorphism of $\mathscr{R}_z$ to $\mathscr{R}_w$. Hence $\mathfrak{d} = \mathfrak{d}'$.

2.25. Proposition. *Suppose that either $\varepsilon = 1$ or $r \geq 3$. Then there exists a point z of $\mathscr{H}_\varepsilon$ such that the commutor of $\theta_z(L)$ in $\mathrm{End}_{\mathbf{Q}}(A_z)$ is $\theta_z(K)$.*

Proof. Let μ be an element of $\mathrm{End}_{\mathbf{Q}}(A_z)$, and λ the element of $\mathrm{End}(\mathfrak{U}, \mathbf{C})$ corresponding to μ. If μ commutes with every element of $\theta_z(L)$, then there exists an element U of $M_m(L)$ such that $\lambda \eta(x, z) = \eta(x U, z)$ for every $x \in L_{\mathbf{R}}^m$. Then repeating the argument of [5, Prop. 16, Proof], we find a join $\mathfrak{W}$ of countably many analytic subsets of $\mathscr{H}_\varepsilon$, other than $\mathscr{H}_\varepsilon$ itself, such that the commutor of $\theta_z(L)$ in $\mathrm{End}_{\mathbf{Q}}(A_z)$ is $\theta_z(K)$ if $z \in \mathscr{H}_\varepsilon - \mathfrak{W}$.

2.26. Let $\mathscr{R} = (\mathscr{Q}, \mathfrak{d}) = (A, \mathscr{C}, \ldots)$ be a hyper-PEL-structure of type Ω_0. Let η, D and $\mathfrak{E}$ be as in 2.10. Put

$$(2.26.1) \qquad\qquad P(u, v) = \mathfrak{E}(u, \psi v) \qquad\qquad ((u, v) \in \mathfrak{U} \times \mathfrak{U}).$$

It can be easily verified that P satisfies the following conditions.

$$(2.26.2) \qquad P(u, v) = -\eta P(v, u) \qquad ((u, v) \in \mathfrak{U} \times \mathfrak{U}),$$

$$(2.26.3) \qquad P(D, D) \subset \mathbf{Q},$$

$$P(u, \sqrt{-1}\,v) = \eta P(v, \sqrt{-1}\,u) \qquad ((u, v) \in \mathfrak{U}^+ \times \mathfrak{U}^+),$$

$$(2.26.4) \qquad P(u, \sqrt{-1}\,v) = -\eta P(v, \sqrt{-1}\,u) \qquad ((u, v) \in \mathfrak{U}^- \times \mathfrak{U}^-),$$

$$P(\mathfrak{U}^+, \mathfrak{U}^-) = 0,$$

$$(2.26.5) \qquad P(\Phi(a)u, v) = P(u, \Phi(a)v) \qquad (a \in L;(u, v) \in \mathfrak{U} \times \mathfrak{U}).$$

Let $g_1, \ldots, g_{2n}$ be a basis of D over $\mathbf{Z}$, and let $x_1(u), \ldots, x_{2n}(u)$ be the real coordinate functions on $\mathfrak{U}$ defined by $u = \sum_{i=1}^{n} x_i(u)g_i$ for $u \in \mathfrak{U}$, where $n = \dim(A)$. Suppose that $\eta = 1$. Define a 2-form $\mathfrak{p}$ on $\mathfrak{U}/D$ by

$$(2.26.6) \qquad \mathfrak{p} = \sum_{i<j} P(g_i, g_j)\, dx_i \wedge dx_j.$$

Then we see that this definition of $\mathfrak{p}$ is independent of the choice of the g_i, and the cohomology class of $\mathfrak{p}$ is *rational* on account of (2.26.3). The property (2.26.4) shows that $\mathfrak{p}$ can be written in the form

$$(2.26.7) \qquad \mathfrak{p} = \mathfrak{p}^+ + \mathfrak{p}^-,$$

where $\mathfrak{p}^+$ is a 2-form of (1,1)-type involving only the coordinates in $\mathfrak{U}^+$, and $\mathfrak{p}^-$ is a 2-form, without terms of (1, 1)-type, involving only the coordinates in $\mathfrak{U}^-$. In this way $\mathfrak{d}$ (together with $\mathscr{C}$) determines a 2-cohomology class $\mathfrak{p}$ on A. Conversely, $\mathfrak{d}$ is determined by $\mathfrak{p}$ through (2.26.1) and (2.26.6). Thus the structure $(\mathscr{Q}, \mathfrak{d})$ is equivalent to a structure $(\mathscr{Q}, \mathfrak{p})$ formed by $\mathscr{Q}$ and a 2-cohomology class $\mathfrak{p}$ on A. Therefore if $\eta = 1$, we can formulate the hyper-PEL-structure in terms of rational 2-cohomology class of a special type, which is not of (1, 1)-type.

2.27. Proposition. *Let $(\mathscr{Q}, \mathfrak{d})$ be a hyper-PEL-structure of type $(T^*, \mathfrak{M};$ $v_1, \ldots, v_s; \omega)$, and $\mathscr{Q}'$ a PEL-structure of type $(T^*, \mathfrak{M}'; v'_1, \ldots, v'_s)$. Suppose that there exists an isogeny λ of $\mathscr{Q}$ to $\mathscr{Q}'$. Then one can define a hyper-PEL-structure $(\mathscr{Q}', \mathfrak{d}')$ and a hyper-PEL-type $(cT^*, \mathfrak{N}; w_1, \ldots, w_s; \omega) = \Omega'_0$ with a positive integer c, so that $(\mathscr{Q}', \mathfrak{d}')$ is of type Ω'_0 and λ is an isogeny of $(\mathscr{Q}, \mathfrak{d})$ to $(\mathscr{Q}', \mathfrak{d}')$. Moreover, if $\varepsilon = 1$, one can take c to be 1.*

Here and in the following proof, we omit the symbols L, Φ, ϱ, which are common to all PEL- or hyper-PEL-types in question.

Proof. Let η, D and ξ be defined for $(\mathscr{Q}, \mathfrak{d})$ as in 2.10. Let η', D' and ξ' be defined for $\mathscr{Q}'$ similarly, with $\mathfrak{M}'$ and v'_i in place of $\mathfrak{M}$ and v_i, the conditions on ω, ψ, $\mathfrak{d}$ being disregarded. Then λ corresponds to an element μ of $GL(\mathfrak{U}, \mathbf{C})$ by $\lambda \circ \xi = \xi' \circ \mu$. Since λ is an isogeny of $\mathscr{Q}$ to $\mathscr{Q}'$, there exists an element U of $GL_m(L)$ such that $\mu\eta(x) = \eta'(xU)$ for $x \in L_{\mathbf{R}}^m$, and $T^*(xU, yU) = c \cdot T^*(x, y)$ with a positive integer c. Put $\mathfrak{N} = \mathfrak{M}'U^{-1}$, $w_i = v'_i U^{-1}$, $D'' = \eta(\mathfrak{N})$. Define $\mathfrak{d}'$ on $\mathscr{Q}'$ by $\mathfrak{d}' \circ \lambda = \lambda \circ \mathfrak{d}$. Then η, $\xi' \circ \mu$, D'', $\mathfrak{N}$, w_i, cT^* satisfy all the properties (2.10.1—6) for $(\mathscr{Q}', \mathfrak{d}')$. This proves the first assertion. Suppose that $\varepsilon = 1$. Let $H(x, y)$ be an E-valued bilinear form on L^m satisfying (1.8.1) for the present T. Then there exists an element Z of $GL(L^m, E)$ such that $H(xZ, yZ) = cH(x, y)$. This fact can be verified by applying Hasse's principle (for quaternion hermitian forms) to H and cH. Therefore $(cT^*, \mathfrak{N}; w_1, \ldots, w_s; \omega)$ is equivalent to $(T^*, \mathfrak{N}Z; w_1 Z, \ldots, w_s Z; \omega)$. This proves the last assertion.

3. Existence of singular members of Σ_0

The purpose of this section is to prove

3.1. Proposition. *There exist a member $\mathcal{R}_z = (A_z, \ldots)$ of Σ_0, an extension M of F, and an isomorphism θ of $L \otimes_F M$ to $\mathrm{End}_Q(A_z)$ satisfying the following conditions.*

(3.1.1) *$[M : F] = 2r$, and M is a totally imaginary quadratic extension of a totally real field.*

(3.1.2) *$\theta_z(a \otimes 1) = \theta(a)$ for $a \in L$.*

(3.1.3) *$K \not\subseteq M$ in case $\varepsilon = 1$.*

The point z will be obtained as a fixed point of a sufficiently general "elliptic element" U of $\mathfrak{G}(H)$. By the classical Cayley transformation $U = (1 - Y) \times \times (1 + Y)^{-1}$, our problem is reduced to the construction of an element Y whose characteristic roots are all purely imaginary. To get such a Y, we begin with two lemmas.

3.2. Lemma. *Let S and A be elements of $GL_{2r}(\mathbf{C})$ such that ${}^t\bar{S} = S$, ${}^t\bar{A} = -A$. Suppose that S is positive or negative definite. Then the characteristic roots of $(SA)^2$ are all negative real numbers.*

Proof. There is an element V of $GL_{2r}(\mathbf{C})$ such that $\pm S = V \cdot {}^t\bar{V}$. Put $Y = {}^t\bar{V} A V$. Then $V^{-1}(SA)^2 V = -Y \cdot {}^t\bar{Y}$, q. e. d.

3.3. Lemma. *Let P be a field of characteristic 0. Let S and A be elements of $GL_{2r}(P)$ such that ${}^tS = S$ and ${}^tA = -A$. Let $X = (x_{ij})$ be a variable matrix of degree $2r$ with $4r^2$ independent variables x_{ij} over P. Put*

$$f(y) = f(y, x_{ij}) = \det [y \cdot 1_{2r} - X S {}^t X A]$$

with a variable y over $P(x_{ij})$. Then there exists a polynomial $\varphi(y) = \varphi(y, x_{ij})$ in $P[y, x_{ij}]$ such that $f(y) = \varphi(y^2)$. Further φ is absolutely irreducible, and the Galois group of $\varphi(y)$ (as polynomial in y) over $P(x_{ij})$ is isomorphic to the symmetric group of r letters.

Proof. Putting $U = XS \cdot {}^tX$, we have $A^{-1} \cdot {}^t(UA)A = -UA$, so that $f(y) = f(-y)$, which proves the first assertion. To prove the second assertion, we may assume that P is algebraically closed. Then we can find elements V and W of $GL_{2r}(P)$ such that $S = V \cdot {}^tV$, $A = {}^tW J W$ with

$$J = \begin{pmatrix} j & & & 0 \\ & \ddots & & \\ & & \ddots & \\ 0 & & & j \end{pmatrix}, \quad j = \begin{pmatrix} 0 & 1 \\ -1 & 0 \end{pmatrix}.$$

Put $Y = W X V$. Then $f(y) = \det [y \cdot 1_{2r} - Y \cdot {}^tYJ]$. Thus the problem is reduced to the case $S = 1_{2r}$, $A = J$. Then we can prove the irreducibility of φ and the assertion concerning the Galois group of φ, by induction on r, using the same argument as in the proof of [6, I, 4.11].

3.4. Let B_0 be as in (1.13.1) and 2.3. Let σ denote the main involution of B_0. Let Ψ_v, for each v, be an irreducible representation: $M_r(B_0) \otimes_{\mathbf{Q}} \mathbf{C} \to M_{2r}(\mathbf{C})$ such that $\Psi_v(a \cdot 1_r) = a^{\tau_v} 1_{2r}$, for every $a \in F$. Let $B_0^{(v)}$ denote the v-th factor of the right hand side of (2.3.3). We can assume

(3.4.1) $\Psi_v(M_r(B_0)) \subset M_{2r}(\mathbf{R})$ if $B_0^{(v)} = M_2(\mathbf{R})$,

(3.4.2) $\Psi_v({}^\iota X^\sigma) = \overline{{}^t \Psi_v(X)}$ for every $X \in M_r(B_0)$ if $B_0^{(v)} = \mathbf{K}$.

Since σ is the main involution of B_0, there exists, for each v, an element A_v of $GL_{2r}(\mathbf{C})$ such that ${}^t A_v = -A_v$, $\Psi_v({}^\iota X^\sigma) = A_v \cdot {}^t \Psi_v(X) A_v^{-1}$ for every $X \in M_r(B_0)$. We may assume that A_v is a real matrix if $B_0^{(v)} = M_2(\mathbf{R})$. These follow from the fact that the main involution of $M_2(P)$ over a field P is given by $x \to j \cdot {}^t x j^{-1}$ with $j = \begin{pmatrix} 0 & 1 \\ -1 & 0 \end{pmatrix}$.

3.5. As remarked in 1.14, there exists a K-linear isomorphism $\mathfrak{J}$ of $M_r(B_0) \otimes \otimes_F K$ to $M_m(L)$, which maps the group $\{U \in M_r(B_0) | UZ \cdot {}^t U^\sigma = Z\}$ onto $\{U \in M_m(L) | U T \cdot {}^t U^\varrho = T, \; U^{\iota \varrho} S = S U\}$, where Z is an element of $GL_r(B_0)$ such that ${}^t Z^\sigma = \varepsilon Z$. (Recall that ι is the main involution of L, if $q = 2$.) We see that $\Psi_v(Z)$ is hermitian if $\varepsilon = 1$ and $B_0^{(v)} = \mathbf{K}$, and $\Psi_v(Z)A_v$ is real symmetric if $\varepsilon = -1$ and $B_0^{(v)} = M_2(\mathbf{R})$. In view of the assumption: (2.4.2), we observe that

(3.5.1) *$\Psi_v(Z)$ is positive or negative definite, if $\varepsilon = 1$ and $B_0^{(v)} = \mathbf{K}$; $\Psi_v(Z)A_v$ is positive or negative definite, if $\varepsilon = -1$ and $B_0^{(v)} = M_2(\mathbf{R})$.*

We choose and fix any element W of $M_r(B_0)$ such that ${}^t W^\sigma = -\varepsilon W$ and the following conditions are satisfied.

(3.5.2) *$\Psi_v(W)A_v$ is positive or negative definite, if $\varepsilon = 1$ and $B_0^{(v)} = M_2(\mathbf{R})$; $\Psi_v(W)$ is positive or negative definite, if $\varepsilon = -1$ and $B_0^{(v)} = \mathbf{K}$.*

(3.5.3) *If $\varepsilon = 1$ and N denotes the reduced norm of $M_r(B_0)$ to F, then $(-1)^r N(WZ^{-1})\zeta^2$ is not a square of any element of F, where ζ is as in 2.4.*

In fact, if $\varepsilon = -1$, we can take the identity matrix to be W. If $\varepsilon = 1$, the existence of W follows easily from the fact that there are infinitely many non-isomorphic totally imaginary quadratic extensions of F contained in B_0.

3.6. Let $\{e_i | 1 \leq i \leq 4r^2\}$ be a basis of $M_r(B_0)$ over F, and $\{a_1, \ldots, a_g\}$ a basis of F over $\mathbf{Q}$. Let x_{pi}, for $1 \leq p \leq g$, $1 \leq i \leq 4r^2$, be $4r^2 g$ independent variables over $\mathbf{C}$. We denote simply by (x) the set of all these variables. Extend Ψ_v to a $\mathbf{C}(x)$-linear representation of $M_r(B_0) \otimes_{\mathbf{Q}} \mathbf{C}(x)$ to $M_{2r}(\mathbf{C}(x))$, and extend σ to a $\mathbf{C}(x)$-linear involution of $B_0 \otimes_{\mathbf{Q}} \mathbf{C}(x)$. Put $X = \sum_{p=1}^{g} \sum_{i=1}^{4r^2} a_p e_i x_{pi}$, and

$$(3.6.1) \qquad f_v(y) = f_v(y, x) = \det[y \cdot 1_{2r} - \Psi_v(XW \cdot {}^t X^\sigma Z^{-1})] \qquad (1 \leq v \leq g).$$

Then f_v is a polynomial with coefficients in F^{τ_v}, and $\{f_1, \ldots, f_g\}$ is a complete set of conjugates of f_1 over $\mathbf{Q}$. Put

$$Z_v = (\Psi_v(Z)A_v)^{-1}, \; W_v = \Psi_v(W)A_v, \; U_v = \Psi_v(X) = \sum_{i=1}^{4r^2} (\sum_{p=1}^{g} a_p^{\tau_v} x_{pi}) \Psi_v(e_i).$$

Then ${}^t Z_v = -\varepsilon Z_v$, ${}^t W_v = \varepsilon W_v$, and

$$(3.6.2) \qquad f_v(y) = \det[y \cdot 1_{2r} - U_v W_v \cdot {}^t U_v Z_v].$$

3.7. Now suppose that $\varepsilon = 1$. Apply 3.3 to (3.6.1) with U_1, W_1 and Z_1 in place of X, S and A. Then there exists a polynomial $\varphi(y, x)$, with coefficients in F, irreducible over $\mathbf{C}(x)$, such that

$$\varphi(y^2, x)^{\mathfrak{r}_v} = \det [y \cdot 1_{2r} - \Psi_v(XW \cdot {}^tX^\sigma Z^{-1})].$$

By Hilbert's irreducibility theorem, there exists a set of rational numbers (b_{pi}) such that $\varphi(y, b)$ is irreducible over K. Put $X_b = \sum_{p=1}^{g} \sum_{i=1}^{4r^2} a_p b_{pi} e_i$, $Y = X_b W \times \times {}^tX_b^\sigma Z^{-1}$. Then $Y \in M_r(B_0)$, and

$$(3.7.1) \qquad\qquad YZ = -Z \cdot {}^tY^\sigma,$$

$$(3.7.2) \qquad\qquad \varphi(y^2, b)^{\mathfrak{r}_v} = \det [y \cdot 1_{2r} - \Psi_v(Y)] \qquad\qquad (1 \le v \le g).$$

If $h < v \le g$, we have $B_0^{(v)} = \mathbf{K}$, and $\Psi_v(X_b^{-1} Y X_b) = \Psi_v(W) \cdot \overline{{}^t\Psi_v(X_b)} \Psi_v(Z)^{-1} \cdot \Psi_v(X_b)$. In view of (3.5.1) and 3.2, we see that the roots of $\varphi(y, b)$ for $h < v \le g$ are all negative. If $1 \le v \le h$, we have $B_0^{(v)} = M_2(\mathbf{R})$, and

$$\Psi_v(Y) = [\Psi_v(X_b) \Psi_v(W) A_v \cdot {}^t\Psi_v(X_b)] \cdot (\Psi_v(Z) A_v)^{-1}.$$

Therefore, from (3.5.2) and 3.2, we see that the roots of $\varphi(y, b)^{\mathfrak{r}_v}$ for $1 \le v \le h$ are all negative.

Let α be a root of $\varphi(y^2, b) = 0$. Put $M = F(\alpha)$. Then α^2 is totally negative, hence M is a totally imaginary quadratic extension of $F(\alpha^2)$. Since $\varphi(y, b)$ is irreducible, $[M : F] = 2r$. Put $R = (1 - Y)(1 + Y)^{-1}$. Notice that $1 + Y$ is invertible, since $\varphi(y, b)$ is irreducible over F. By (3.7.1), we have $RZ^tR^\sigma = Z$, so that $\mathfrak{I}(R) \in \mathfrak{G}(H)$, where $\mathfrak{I}$ is as in 3.5. Observe that $\varphi_v^m \circ \mathfrak{I}$ and Ψ_v are equivalent as representations of $M_r(B_0)$, since they coincide on the center. Put $U = \mathfrak{I}(R)$. Then every characteristic root of $\varphi_v^m(U)$ can be written in the form $(1 - \beta)(1 + \beta)^{-1}$ with a characteristic root β of $\Psi_v(Y)$. Since β is purely imaginary, we have $|(1 - \beta)(1 + \beta)^{-1}| = 1$.

Let z_0 be a point of $\mathscr{H}_\varepsilon$, and let $\mathfrak{K} = \{V \in \mathfrak{G}_{\mathbf{R}}(H) | V(z_0) = z_0\}$. Then $\mathfrak{K}$ is a maximal compact subgroup of $\mathfrak{G}_{\mathbf{R}}(H)$. On account of the above property of roots of $\varphi_v^m(U)$, there exists an element E of $\mathfrak{G}_{\mathbf{R}}(H)$ such that $E^{-1}UE \in \mathfrak{K}$. Put $z = E(z_0)$. Then $U(z) = z$, and $z \in \mathscr{H}_\varepsilon$. By (2.19.1) we have $\Lambda(U, z) \eta(x, z) = \eta(xU, z)$, so that $\Lambda(U, z)$ gives an element λ of $\mathrm{End}_{\mathbf{Q}}(A_z)$. Now $\mathfrak{I}$ maps $K \otimes_F F[R]$ isomorphically onto the subring $K[U]$ of $M_m(L)$. For $\sum_{p=0}^{2r-1} a_p U^p \in K[U]$ with $a_p \in K$, we have

$$\eta(x \cdot \sum_{p=0}^{2r-1} a_p U^p, z) = \sum_{p=0}^{2r-1} \Phi(a_p) \Lambda(U, z)^p \eta(x, z),$$

so that $\sum_{p=0}^{2r-1} a_p U^p$ defines an element of $\mathrm{End}_{\mathbf{Q}}(A_z)$. Since $F[R]$ is isomorphic to M, we get in this way an isomorphism θ of $K \otimes_F M$ into $\mathrm{End}_{\mathbf{Q}}(A_z)$ so that $\theta(a \otimes 1) = \theta_z(a)$ for $a \in K$ and $\theta(1 \otimes \alpha) = \lambda$. Since $\Lambda(U, z)$ commutes with every element of $\Phi(L)$, we see that every element of $\theta(K \otimes_F M)$ commutes with every element of $\theta_z(L)$. Let $L' = \theta_z(L)$ and $M' = \theta(1 \otimes M)$. Then $L'M'$ is a homomorphic image of $L \otimes_F M$. The algebra $L \otimes_F M$ is simple or a direct sum of two simple algebras; in either case, $K \otimes_F M$ is the center of $L \otimes_F M$. Now the center of $L'M'$ contains $\theta(K \otimes_F M)$, an isomorphic image of $K \otimes_F M$. Hence $L'M'$ should be isomorphic to $L \otimes_F M$. We can therefore extend θ to an isomorphism of $L \otimes_F M$ into $\mathrm{End}_{\mathbf{Q}}(A_z)$ as in (3.1.2).

To see (3.1.3), assume that $K \subset M$. Then $\varphi(y^2, b)$ is reducible over K, hence there exists a polynomial $g(y)$ in $K[y]$ such that $\varphi(y^2, b) = (-1)^r g(y) g(-y)$.

Let c be the constant term of g. Then we have $N(X_b)^2 \cdot N(WZ^{-1}) = (-1)^r c^2$, where N is as in (3.5.3). Since $N(X_b)^2 \cdot N(WZ^{-1}) \in F$, we have $c = d\zeta$ with an element $d \in F$ and the element ζ of K which we have fixed in 2.4. Hence $N(X_b)^2 \cdot N(WZ^{-1}) = (-1)^r d^2 \zeta^2$. This contradicts (3.5.3) and completes the proof of 3.1 in the case $\varepsilon = 1$.

3.8. Next suppose that $\varepsilon = -1$. The notation being as in 3.6, we have

$$(3.8.1) \qquad f_v(y, x) = \det [y \cdot 1_{2r} - {}^tU_v Z_v U_v W_v].$$

Apply 3.3 to (3.8.1) with ${}^tU_v$, Z_v and W_v in place of X, S and A. Then repeating almost the same argument as in 3.7, we obtain 3.1 in the case $\varepsilon = -1$. It should be noted that, in the course of proof, one has to combine 3.2 with the property (3.5.1, 2) in the case $\varepsilon = -1$.

3.9. Let $\mathscr{R}_z$, M and θ be as in 3.1. Take any subfield P of L containing K such that $[P : K] = q$, and put $R = \theta(P \otimes_F M)$. Then R is a commutative semi-simple subalgebra of $\mathrm{End}_\mathbf{Q}(A_z)$, whose rank over $\mathbf{Q}$ is $2 \cdot \dim(A_z)$ ($= 4qrg$). Let $R_1, \ldots, R_v$ be the simple components of R, and let e_i be the identity element of R_i. Then $A_z = \sum_i e_i A_z$, and $\mathrm{End}_\mathbf{Q}(e_i A_z)$ contains an isomorphic image of R_i. In view of [9, 5.1, Prop. 1], we have $2 \cdot \dim(e_i A_z) = [R_i : \mathbf{Q}]$, hence $e_i A_z$ belongs to a CM-type. It follows from this that *the field of moduli of $\mathscr{R}_z$ is algebraic over* $\mathbf{Q}$. If $\varepsilon = 1$, we can choose P so that $P \otimes_F M$ is a field, on account of (3.1.3). If $q = 2$ ($\varepsilon = \pm 1$), A_z is not simple, since $\mathrm{End}_\mathbf{Q}(A_z)$ is not commutative.

3.10. If $r = 1$ and $\varepsilon = -1$, $\mathscr{H}_\varepsilon$ consists of only one point, hence Ω_0 has the only member, which should coincide with the one obtained in 3.1. Therefore, the discussion of 3.9 shows that *the only member of Ω_0 in the case $r = 1$ and $\varepsilon = -1$ is obtained from an abelian variety with sufficiently many complex multiplications.*

4. Some properties of $\mathscr{H}_\varepsilon / \Gamma(H, \mathfrak{N}/\mathfrak{M})$

4.1. Let $\mathfrak{M}$ and $\mathfrak{N}$ be two lattices in L^m such that $\mathfrak{M} \subset \mathfrak{N}$. We can embed $\mathscr{H}/\Gamma(T, \mathfrak{N}/\mathfrak{M})$ (resp. $\mathscr{H}_\varepsilon/\Gamma(H, \mathfrak{N}/\mathfrak{M})$) into a projective space. Here we remind the reader that these quotients are compact. Let V (resp. V_0) be the image variety, and φ (resp. φ_0) the holomorphic mapping of $\mathscr{H}$ to V (resp. $\mathscr{H}_\varepsilon$ to V_0) which induces a biregular isomorphism of $\mathscr{H}/\Gamma(T, \mathfrak{N}/\mathfrak{M})$ to V (resp. $\mathscr{H}_\varepsilon/\Gamma(H, \mathfrak{N}/\mathfrak{M})$ to V_0). We call such a mapping φ (resp. φ_0) the *projection* of $\mathscr{H}$ to V (resp. $\mathscr{H}_\varepsilon$ to V_0). From the injection of $\mathscr{H}_\varepsilon$ into $\mathscr{H}$, we obtain a morphism μ of V_0 to V such that the following diagram is commutative:

$$(4.1.1) \qquad \begin{array}{ccc} \mathscr{H}_\varepsilon & \longrightarrow & \mathscr{H} \\ {\scriptstyle \varphi_0}\downarrow & & \downarrow{\scriptstyle \varphi} \\ V_0 & \xrightarrow{\mu} & V \end{array}$$

Let Ω be the PEL-type (2.9.2) with the v_i such that $\mathfrak{N} = \mathfrak{M} + \sum_{i=1}^s \mathbf{Z}v_i$. We shall put

$$V = V(T, \mathfrak{N}/\mathfrak{M}) = V(\mathfrak{N}/\mathfrak{M}) = V(\Omega),$$
$$V_0 = V_0(H, \mathfrak{N}/\mathfrak{M}) = V_0(\mathfrak{N}/\mathfrak{M}) = V_0(\Omega_0),$$

when it is necessary to specify T, $\mathfrak{N}$, $\mathfrak{M}$ or Ω. Let k_Ω, or $k(\Omega)$, denote the algebraic number field defined in [8, 5.1]. We may assume that $V(\Omega)$ is actually a "moduli-variety" with the properties of [8, 6.2]; here we recall the following two of those properties.

(4.1.2) *$V(\Omega)$ is defined over k_Ω.*

(4.1.3) *For every (isomorphism-class of) PEL-structure $\mathcal{Q}$ of type Ω, we can assign a point $\mathfrak{v}(\mathcal{Q})$ of V so that $k_\Omega(\mathfrak{v}(\mathcal{Q}))$ is the field of moduli of $\mathcal{Q}$; for a member $\mathcal{Q}_z$ of $\Sigma(\Omega)$, one has $\varphi(z) = \mathfrak{v}(\mathcal{Q}_z)$.*

4.2. Theorem. *Let V and φ be as above. Let J_ε be as in (2.18.10), and let*

$$Z_\varepsilon = \{\varphi(z) \times \varphi(J_\varepsilon(z)) \mid z \in \mathcal{H}\}\,(\subset V \times V)\,.$$

Then Z_ε is an algebraic subvariety of $V \times V$ defined over an algebraic number field.

Proof. Let $\Gamma = \Gamma(T, \mathfrak{N}/\mathfrak{M})$, $\Gamma_1 = \Gamma(T, \mathfrak{M})$. By (2.18.11), $J_\varepsilon \Gamma J_\varepsilon$ and $S^{-1}\Gamma_1^{\prime\varrho}S$ give the same group of transformations on $\mathcal{H}$. Since

$$S^{-1}\Gamma_1^{\prime\varrho}S = \{U \in \mathfrak{G}(T) \mid \mathfrak{M}^{\prime\varrho}SU = \mathfrak{M}^{\prime\varrho}S\}\,,$$

the transformation groups given by $J_\varepsilon \Gamma J_\varepsilon$ and Γ are commensurable. It follows that Z_ε is an analytic subvariety of $V \times V$, hence an algebraic subvariety. Now we observe that A_z and $A_{U(z)}$ are isogenous if $U \in \mathfrak{G}(T)$. Let w be a point of $\mathcal{H}$ such that $\varphi(w)$ is algebraic. The field of moduli of $\mathcal{Q}_{U(w)}$ is algebraic over the field of moduli of $\mathcal{Q}_w$. In view of (4.1.3), $\varphi(U(w))$ is algebraic. Now by 3.1, we can find a point u of $\mathcal{H}$, an algebraic extension M of F of degree $2r$, and an isomorphism θ of $L \otimes_F M$ into $\mathrm{End}_\mathbf{Q}(A_u)$ so that $\theta = \theta_u$ on L. (If $K = L$, this is included, as a special case, in a more general result [6, I, 5.7] and [6, III, § 2].) Let $L' = L \otimes_F M$. Then Φ can be extended to an isomorphism Φ' of L' into $\mathrm{End}(\mathfrak{U}, \mathbf{C})$ so that, for every $\alpha \in L'$, the linear transformation $\Phi'(\alpha)$ represents $\theta(\alpha)$. If $\alpha \in M$, $\Phi'(\alpha)$ commutes with the elements of $\Phi(L)$, so that $\Phi'(\alpha)\mathfrak{U}_v \subset \mathfrak{U}_v$ for every v. Hence $\Phi'(L')\mathfrak{U}_v \subset \mathfrak{U}_v$ for every v. Then we see easily that $\psi^{-1}\Phi'(L')\psi \subset \mathrm{End}(\mathfrak{U}, \mathbf{C})$. Put $v = J_\varepsilon(u)$. By (2.15.6) we have, for every $\alpha \in L'$,

$$\psi^{-1}\Phi'(\alpha)\psi\eta(\mathbf{Q}\mathfrak{M}, v) \subset \eta(\mathbf{Q}\mathfrak{M}, v)\,.$$

This shows that $\psi^{-1}\Phi'(\alpha)\psi$ defines an element of $\mathrm{End}_\mathbf{Q}(A_v)$. Hence there exists an isomorphism of L' into $\mathrm{End}_\mathbf{Q}(A_v)$. For the same reason as explained in 3.9, the field of moduli of $\mathcal{Q}_v$ is algebraic, so that $\varphi(J_\varepsilon(u))$ is algebraic. Let $Y \in \mathfrak{G}(T)$. Since $J_\varepsilon Y(u) = Y^* J_\varepsilon(u)$, we see that $\varphi(J_\varepsilon Y(u))$ is algebraic, taking the above point w to be $J_\varepsilon(u)$. Put

$$Z' = \{\varphi(Y(u)) \times \varphi(J_\varepsilon Y(u)) \mid Y \in \mathfrak{G}(T)\}\,.$$

Since $\mathfrak{G}(T)$ is dense in $\mathfrak{G}_\mathbf{R}(T)$, Z' is a dense subset of Z_ε. The above consideration shows that all the points of Z' are algebraic. Now let σ be an automorphism of $\mathbf{C}$ which leaves all the algebraic numbers invariant. Then $Z_\varepsilon^\sigma \supset Z'^\sigma = Z'$. Since Z' is dense in Z_ε, we have $Z_\varepsilon^\sigma \supset Z_\varepsilon$, so that $Z_\varepsilon^\sigma = Z_\varepsilon$, which completes the proof.

4.3. Theorem. *Let the notation be as in (4.1.1). Then the subvariety $\mu(V_0)$ of V is defined over an algebraic number field.*

13*

Proof. The point u in the proof of 4.2 can be taken in $\mathcal{H}_\varepsilon$ (see 3.1). We see that $\mu(V_0) = \varphi(\mathcal{H}_\varepsilon)$, hence the set $\{\varphi(Y(u)) \mid Y \in \mathfrak{G}(H)\}$ is dense in $\mu(V_0)$. As observed in the proof of 3.2, the points $\varphi(Y(u))$ are all algebraic. Therefore, by means of the argument at the end of the proof of 3.2, we get the present assertion.

Second proof. If $\mathfrak{N}$ is large enough, (in other words, if the "level" is high enough), one can prove, without difficulty, that $\mu(V_0)$ is the projection, to V, of a component of $\Delta \cap Z_\varepsilon$, where Δ is the diagonal on $V \times V$. This proves 4.3 in such a case. The same proof will probably apply to the general case. At any rate, one can reduce the problem to the case of "high level".

4.4. Theorem. *The quotient $\mathcal{H}_\varepsilon / \Gamma(H, \mathfrak{N}/\mathfrak{M})$ has a projective embedding whose image variety is defined over an algebraic number field.*

The difference of 4.4 from 4.3 lies in that the morphism μ in (4.1.1) is not necessarily a biregular embedding of $\mathcal{H}_\varepsilon / \Gamma(H, \mathfrak{N}/\mathfrak{M})$.

Proof. We observe that if $\mathfrak{N}$ satisfies (2.22.1, 2), then every lattice containing $\mathfrak{N}$ satisfies (2.22.1, 2). Therefore we can find a positive integer b such that the lattice $\mathfrak{N}' = b^{-1}\mathfrak{M}$ satisfies: (i) $\mathfrak{M} + \mathfrak{N} \subset \mathfrak{N}'$; (ii) the natural map $\mathcal{H}_\varepsilon / \Gamma(H, \mathfrak{N}'/\mathfrak{M}) \to \mathcal{H} / \Gamma(T, \mathfrak{N}'/\mathfrak{M})$ is injective; (iii) $\Gamma(T, \mathfrak{N}'/\mathfrak{M})$ has no element of finite order other than the identity. Then the injection of (ii) maps $\mathcal{H}_\varepsilon / \Gamma(H, \mathfrak{N}'/\mathfrak{M})$ biregularly onto a subvariety of $V(\mathfrak{N}'/\mathfrak{M})$. We identify $\mathcal{H}_\varepsilon / \Gamma(H, \mathfrak{N}'/\mathfrak{M})$ with this subvariety, which is defined over an algebraic number field, as is shown in 4.3. Since $\mathfrak{N}' = b^{-1}\mathfrak{M}$, $\Gamma(T, \mathfrak{N}'/\mathfrak{M})$ (resp. $\Gamma(H, \mathfrak{N}'/\mathfrak{M})$) is a normal subgroup of $\Gamma(T, \mathfrak{M})$ (resp. $\Gamma(H, \mathfrak{M})$). Let $U \in \Gamma(T, \mathfrak{M})$. Then $z \to U(z)$ defines a biregular automorphism α of $V(N'/\mathfrak{M})$. It has been shown in [8, 6.3] that α is defined over $k(\Omega')$ with a PEL-type $\Omega' = (L, \Phi, \varrho; T, \mathfrak{M}; v'_1, \ldots, v'_n)$ such that $\mathfrak{N}' = \mathfrak{M} + \sum_{i=1}^n \mathbf{Z}v'_i$. If $U \in \Gamma(H, \mathfrak{M})$, α induces a biregular automorphism of $\mathcal{H}_\varepsilon / \Gamma(H, \mathfrak{N}'/\mathfrak{M})$. Now $\mathcal{H}_\varepsilon / \Gamma(H, \mathfrak{N}/\mathfrak{M})$ can be obtained as a quotient of $\mathcal{H}_\varepsilon / \Gamma(H, \mathfrak{N}'/\mathfrak{M})$ by a finite group of biregular automorphisms composed by the elements α obtained from all $U \in \Gamma(H, \mathfrak{N}/\mathfrak{M})$. This proves our assertion.

4.5. *Remark.* If $r = 1$ and $\varepsilon = 1$, we have $\mathcal{H} = \mathcal{H}_\varepsilon$, so that μ is surjective. Therefore, in this case, V_0 is a covering of V. If $\mathfrak{N}$ is sufficiently large, V_0 can be identified with V by μ.

4.6. Proposition. *If $\varepsilon = 1$, one has $\theta_z(L) = \mathrm{End}_\mathbf{Q}(A_z)$ for every point z of $\mathcal{H}_\varepsilon$ such that $\dim_\mathbf{Q}(\varphi(z)) = \dim(\mathcal{H}_\varepsilon)$.*

Proof. Let y be a generic point of $\mu(V_0)$ over the algebraic closure of $\mathbf{Q}$. Take a point z of $\mathcal{H}_\varepsilon$ so that $y = \varphi(z)$. By 2.25, the commutor of $\theta_z(L)$ in $\mathrm{End}_\mathbf{Q}(A_z)$ is $\theta_z(K)$. Now we can repeat the discussion of [5, pp. 181—2] for A_z. Then we find that A_z is isogenous to a product of copies of a simple abelian variety A'. Define any PEL-structure $\mathcal{Q}' = (A', \ldots)$ with this A'. Then the reasoning of [5, p. 182] shows that the field of moduli of $\mathcal{Q}'$ is of dimension $\leq r^2 h/2$, unless $\theta_z(L) = \mathrm{End}_\mathbf{Q}(A_z)$. If $\varepsilon = 1$, one has $\dim_\mathbf{Q}(y) = \dim(\mathcal{H}_\varepsilon) = hr(r+1)/2 > hr^2/2$. This proves our proposition.

4.7. Suppose that $\Gamma(T, \mathfrak{N}/\mathfrak{M})$ has no element of finite order other than the identity. By [8, 5.3], we can construct, over k_Ω, a fibre system of abelian varieties

$\mathfrak{F} = \{V, W, \mathfrak{h}, \mathfrak{f}\}$ such that $\mathfrak{h}^{-1}(u)$ is isomorphic to A_z if $u = \varphi(z)$. (For precise definition and results, see [8, 1.1 and 5.3].) Now define W_0, $\mathfrak{h}_0$ and $\mathfrak{f}_0$ by

$$W_0 = \{(x, y) \in V_0 \times W \mid \mu(x) = \mathfrak{h}(y)\},$$

$$\mathfrak{h}_0(x, y) = x \quad \text{for} \quad (x, y) \in W_0,$$

$$\mathfrak{f}_0(x) = (x, \mathfrak{f}(\mu(x))) \quad \text{for} \quad x \in V_0.$$

By [8, 1.3], we see that $\mathfrak{F}_0 = \{V_0, W_0, \mathfrak{h}_0, \mathfrak{f}_0\}$ is a fibre system of abelian varieties. By 4.3, 4.4 and their proofs, V_0 and μ can be chosen so as defined over an algebraic number field. Therefore $\mathfrak{F}_0$ can be defined over an algebraic number field.

4.8. In 1.14, we have seen that $\mathfrak{G}(H)$ can be identified with the group

$$(4.8.1) \qquad \{X \in M_r(B_0) \mid XZ \cdot {}^tX^\sigma = Z\}.$$

This group can be defined independently of the field K. Hence there are infinitely many hyper-PEL-types Ω_0 with distinct K, ω, etc., to which the quotient $\mathscr{H}_\varepsilon / \Gamma(H, \mathfrak{N}/\mathfrak{M})$ is common. To see this more closely, let us discuss the case $q = \varepsilon = 1$. Let B_0 be a quaternion algebra over F satisfying (2.3.3). Let K be a totally imaginary quadratic extension of F such that $B_0 \otimes_F K \cong M_2(K)$. For a given B_0, there are infinitely many such K. With any choice of K, B_0 can be written as $B_0 = (K, \omega, \kappa)$ with an element κ of F satisfying (2.3.1). Let V be a left B_0-module whose dimension over K is $2r$. Take a B_0-valued σ-hermitian form H on V, where σ is the main involution of B_0. One has to assume that H is definite at every archimedean prime spot of F corresponding to τ_v for $v > h$. Define T by (1.8.1). (Notice that $E = B_0$ in view of (1.13.1).) Then we have an identification (1.9.1). This means that, there are infinitely many embeddings μ of the *same* $\mathscr{H}_\varepsilon / \Gamma(H, \mathfrak{N}/\mathfrak{M})$ into distinct $\mathscr{H} / \Gamma(T, \mathfrak{N}/\mathfrak{M})$. By 4.6, we have $\operatorname{End}_\mathbf{Q}(A_z) = \theta_z(K)$ for a generic member A_z. We thus get infinitely many distinct families of abelian varieties, depending on various K, parametrized by the same quotient space $\mathscr{H}_\varepsilon / \Gamma(H, \mathfrak{N}/\mathfrak{M})$.

4.9. Let K' be the field generated over $\mathbf{Q}$ by the elements $\sum_{v=1}^h (x^{\tau_v} + x^{\tau_v \varrho}) + 2 \cdot \sum_{v=h+1}^g x^{\tau_v}$ for all $x \in K$. Then k_Ω is abelian over K'. This has been shown in [6, II, 9.4] when $\mathfrak{r}\mathfrak{M} \subset \mathfrak{M}$ holds, where $\mathfrak{r}$ denotes the ring of integers in K. The assertion for the general case follows from this result by virtue of [6, II, 1.12] and [8, 5.9].

4.10. Theorem. *Let the notation be as in 4.1 and 4.9. Suppose that $q = \varepsilon = 1$. Then one can choose V_0 and μ so as to be defined over an abelian extension of K'.*

4.11. Before proving this theorem, we make a simple remark on complex multiplication. Let F_0 be a totally real algebraic number field, and K_0 a totally imaginary quadratic extension of F_0. We denote by ϱ_0 the non-trivial automorphism of K_0 over F_0. Let $(K_0; \{\varphi_i\})$ be a *CM*-type, and $(K_0^*; \{\psi_j\})$ the dual of $(K_0; \{\varphi_i\})$ (cf. [9]). Let Φ_0 be a representation of K_0 by complex matrices, which is equivalent to the direct sum of φ_i, and let ζ_0 be an element of K_0 such that $\zeta_0^{\varrho_0} = -\zeta_0$ and $\operatorname{Im}(\zeta_0^{\varphi_i}) > 0$ for every i. Further take a $\mathbf{Z}$-lattice $\mathfrak{m}$ in K_0, and elements $w_1, \ldots, w_s$ of K_0. Let $\mathscr{Q} = (A, \mathscr{C}, \theta; t_1, \ldots, t_s)$ be a PEL-structure of type $(K_0, \Phi_0, \varrho_0; \zeta_0, \mathfrak{m}; w_1, \ldots, w_s)$. Then the result of [9, §§ 15—17] shows that *the field of moduli of* $\mathbf{Q}$ *is abelian over* K_0^*. In [9, §§ 15—17], it is assumed that the abelian variety in question is simple. But the discussion is valid so far as the polarization $\mathscr{C}$ determines the involution ϱ_0 in K_0 (so that it is obtained from an element ζ_0 of K_0 as in [9, 6.2, Th. 4]). It should also be noted that the field $k_0^*(F(t))$ in [9, 16.3, Main th. 3] is exactly the field of moduli of $(A, \mathscr{C}, \iota; t)$ (ι is employed in [9] instead of θ). If $\mathfrak{r}_0$ denotes the ring of integers of K_0 and $\iota(\mathfrak{r}_0) = \iota(K_0) \cap \operatorname{End}(A)$, then, for every positive integer N, we can always find a point t on A so that $\iota(\mathfrak{r}_0)t = \{u \in A | Nu = 0\}$ [9, 7.5, Prop. 21]. In view of these facts, we see that the field of moduli of $\mathscr{2}$ is abelian over K_0^* in the general case.

4.12. *Proof of 4.10.* First we note that $L = K$, $E = B_0$, $B = F$ if $q = 1$. Let H be an E-valued ϱ-hermitian form on K^{2r} such that

$$\operatorname{Tr}_{E/F}(H(x, y)) = \operatorname{Tr}_{K/F}(T(x, y)) \quad \text{(cf. 1.8)}.$$

Let us write T^1, H^1, S^1, $\mathfrak{U}^1$, Φ^1, ψ^1 for the corresponding symbols defined in the case $r = 1$. In view of 1.15, we can regard the structure (K^{2r}, T, S, H) as the direct sum of r copies of (K^2, T^1, S^1, H^1) in an obvious sense, and the representation space $(\mathfrak{U}, \Phi, \psi)$ of $E_{\mathbf{R}}$ as the direct sum of r copies of $(\mathfrak{U}^1, \Phi^1, \psi^1)$. Let $\mathfrak{r}$ be the ring of integers in K, and let $\mathfrak{M}^1$ be an $\mathfrak{r}$-lattice in K^2. Put

$$\Omega^1 = (K, \Phi^1, \varrho; T^1, \mathfrak{M}^1; 0), \quad \Omega_0^1 = (\Omega^1, \omega).$$

Let k be the smallest field of rationality for $\mu(V_0)$ containing k_Ω, and k' a finite Galois extension of K', containing k. In [6, I, 5.7; III, § 2] we obtained a PEL-structure $(A^1, \mathscr{C}^1, \bar{\theta}; 0)$ of a type $(K_0, \Phi_0, \varrho_0; \zeta_0, \mathfrak{m}; 0)$ discussed in 4.11, such that: (i) K_0 is an extension of K; (ii) if θ^1 denotes the restriction of $\bar{\theta}$ to K, then $(A^1, \mathscr{C}^1, \theta^1; 0)$ is of type Ω^1; (iii) K_0^* and k' are linearly disjoint over K', where K_0^* is as in 4.11. Since $\mathscr{H}_\varepsilon = \mathscr{H}$ if $r = \varepsilon = 1$, we can define a hyper-PEL-structure $\mathscr{R}^1 = (A^1, \mathscr{C}^1, \theta^1, \mathfrak{d}^1; 0)$ of type Ω_0^1. Let $\mathfrak{y}^1$, D^1, ξ^1 be the symbols satisfying (2.10.1—6) for $\mathscr{R}^1$. Put

$$\mathfrak{y}(x) = \mathfrak{y}^1(x_1) \times \cdots \times \mathfrak{y}^1(x_r) \quad \text{for} \quad x = (x_1, \ldots, x_r) \in K_{\mathbf{R}}^2 \times \cdots \times K_{\mathbf{R}}^2 = K_{\mathbf{R}}^{2r},$$

and $\mathfrak{y}(\mathfrak{M}) = D$. Then, from the complex torus $\mathfrak{U}/D$, we obtain naturally a hyper-PEL-structure $\mathscr{R} = (A, \mathscr{C}, \theta, \mathfrak{d}; t_1, \ldots, t_s)$ of type Ω_0, for which $\mathfrak{y}$ and D satisfy (2.10.1—6). Let $\mathscr{R}_z$ be a member of $\Sigma(\Omega_0)$ isomorphic to $\mathscr{R}$. Then for every $Y \in \mathfrak{G}(H)$, we see that $\mathscr{R}_{Y(z)}$ is isogenous to the product of r copies of $\mathscr{R}^1$. Therefore the field of moduli of $\mathscr{2}_{Y(z)}$ is contained in the field of moduli of $(A^1, \mathscr{C}^1, \theta^1; u_1, \ldots, u_p)$ for suitable points $u_1, \ldots, u_p$ of A^1. Hence by the remark in 4.11, the field of moduli of $\mathscr{2}_{Y(z)}$ is contained in an abelian extension of K_0^*. Let k_1 be the maximal abelian extension of K_0^*. Then the points of the set $\{\varphi(Y(u)) | Y \in \mathfrak{G}(H)\}$ in the proof of 4.3 are all rational over k_1. Therefore that proof shows that $\mu(V_0)$ is defined over k_1, so that $k \cdot K_0^* \subset k_1$. Let σ be an automorphism of k' over K'. Since K_0^* and k' are linearly disjoint over K', we can extend σ to an automorphism τ of $k' \cdot K_0^*$ over K_0^*. Then we have $k^\sigma \cdot K_0^* = (k \cdot K_0^*)^\tau = k \cdot K_0^*$, so that $k^\sigma = k$, which proves that k is normal over K'. Since $k \cdot K_0^*$ is abelian over K_0^*, it follows that k is abelian over K'. Thus we have shown that $\mu(V_0)$ is defined over an abelian extension of K'. Now let us repeat the proof of 4.4 with this result in mind. Since $k(\Omega')$ and $k(\Omega)$ are abelian over K' as remarked in 4.9, we can form the quotient of $\mathcal{H}_\varepsilon / \Gamma(H, \mathfrak{N}'/\mathfrak{M})$ by a finite group *rationally* over an abelian extension of K'. This completes our proof.

4.13. *Concluding remarks.* One can conjecture that the same conclusion as 4.10 is true in the remaining three cases (i) $q = 2$, $\varepsilon = 1$; (ii) $q = 1$, $\varepsilon = -1$; (iii) $q = 2$, $\varepsilon = -1$, except when $\varepsilon = -1$ and $r = 1$. In fact, we obtained in 3.1 and 3.9 a member $\mathcal{R}_z$ of $\Sigma(\Omega_0)$ with sufficiently many complex multiplications. Therefore, by means of the same reasoning as in 4.12, we can prove that V_0 and μ can be defined over an abelian extension of a certain number field M^*, where M^* is determined only by R_z. If we can take M^* in a sufficiently general position, then the argument at the end of 4.12 will prove the desired result. Moreover, a careful analysis of these singular members will lead to a more precise (class-field-theoretical) characterization of the field of rationality, similar to the result [6]. One can also expect a theorem concerning the bottom field for $\mathcal{H}_\varepsilon / \Gamma(H, \mathfrak{N}/\mathfrak{M})$, which will include [7] and [6, II, § 10] as a special case. For the problems of this kind, it is not absolutely necessary to have a characterization of the abelian varieties in the family, be it algebro-geometric or not, though the characterization itself is very interesting and useful. We could actually avoid the use of hyper-PEL-type or -structure in the discussion of § 4. Therefore one may note that the method of proof of our theorems and the above comments apply to other types of family of abelian varieties which have been obtained recently by KUGA and SATAKE [1, 3].

Regarding the characterization, MUMFORD and TATE have investigated the families of abelian varieties A with a prescribed type of rational cohomology classes of (p, p)-type on the product $A \times \cdots \times A$ [2]. It does not seem that our family Σ_0 can be characterized by such a means (see also 2.26).

References

[1] KUGA, M.: Fibre varieties over a symmetric space whose fibres are abelian varieties. Lecture note, University of Chicago 1963—64 (1965).

[2] MUMFORD, D.: Families of abelian varieties, Proceedings of Symposia in Pure Mathematics, vol. **9**, Algebraic Groups, Amer. Math. Soc. (1966).

[3] SATAKE, I.: Symplectic representations of algebraic groups satisfying a certain analylicity condition. To appear.

[4] SHIMURA, G.: On the theory of automorphic functions. Ann. Math. **70**, 101—144 (1959).

[5] -- On analytic families of polarized abelian varieties and automorphic functions. Ann. Math. **78**, 149—192 (1963).

[6] — On the field of definition for a field of automorphic functions, I, II, III. Ann. Math. **80**, 160—189 (1964), **81**, 124—165 (1965), **83**, 377—385 (1966).

[7] -- Class-fields and automorphic functions. Ann. Math. **80**, 444—463 (1964).

[8] — Moduli and fibre systems of abelian varieties. Ann. Math. **83**, 294—338 (1966).

[9] — and Y. TANIYAMA: Complex multiplication of abelian varieties and its applications to number theory. Publ. Math. Soc. Japan, No. 6 (1961).

[10] SIEGEL, C. L.: Symplectic geometry. Am. J. Math. **65**, 1—86 (1943).

Professor G. SHIMURA
Department of Mathematics, Princeton University
Princeton, N.J.

(Received October 15, 1965)

67b

Construction of class fields and zeta functions of algebraic curves

Annals of Mathematics, 85 (1967), 58-159

As the title indicates, the present investigation deals with two problems which are closely related. The first one is the construction of class fields by means of special values of automorphic functions with respect to a discontinuous group obtained from a quaternion algebra. To be precise, let F be a totally real algebraic number field of degree g, and B a quaternion algebra over F, which may or may not be a division algebra. The tensor product B_R of B with the real number field R over the rational number field Q may be identified with $M_2(R)^r \times K^{g-r}$, the product of r copies of the total matric algebra $M_2(R)$ of degree 2 over R and $g - r$ copies of the division ring K of real quaternions, where r is an integer such that $0 \leq r \leq g$. We assume that $r > 0$. Let B^+ denote the group of elements of B whose projections to the first r factors $M_2(R)$ of B_R have positive determinants. The elements of B^+ act naturally on the product $\mathfrak{H}^r$ of r copies of the upper half complex plane $\mathfrak{H}$. Take a maximal order $\mathfrak{o}$ in B and denote by $\Gamma(\mathfrak{o})$ the group of units in $\mathfrak{o}$ which belong to B^+. Then $\Gamma(\mathfrak{o})$ can be considered as a properly discontinuous group acting on $\mathfrak{H}^r$, and $\mathfrak{H}^r/\Gamma(\mathfrak{o})$ is of finite measure. We note that $\mathfrak{H}^r/\Gamma(\mathfrak{o})$ is compact unless $B = M_2(F)$. Now our first result, in a specialized form, can be stated as follows.

Suppose that $r = 1$. Then there exist a complete non-singular algebraic curve V and a holomorphic mapping φ of $\mathfrak{H}$ into V with the following properties.

(1) φ gives a biregular morphism of $\mathfrak{H}/\Gamma(\mathfrak{o})$ into V.

(2) V is defined over the field C_F, which is the maximal abelian extension of F in which no primes of F, except for archimedean primes, are ramified.

(3) Let M be a totally imaginary quadratic extension of F, isomorphic to a quadratic subfield of B over F. If z is a regular fixed point of M on $\mathfrak{H}$, then the composite $C_F(\varphi(z)) \cdot M$ is the maximal unramified abelian extension C_M of M.

Here a *regular fixed point* of M on $\mathfrak{H}^r$ $(r \geq 1)$ is defined as follows. If f is an F-linear isomorphism of M into B, then $f(M - \{0\}) \subset B^+$, and there is a unique fixed point z on $\mathfrak{H}^r$ of the elements of $f(M - \{0\})$. Moreover, for a given M, there always exists an f such that the image of the ring of integers of M by f is contained in $\mathfrak{o}$. We call z a regular fixed point of M if f has this last property.

Thus φ affords a generalization of the classical modular function $j(\tau)$, with F in place of $\mathbf{Q}$. Furthermore V and φ are uniquely characterized by the above properties, up to biregular equivalence over C_F. More generally, we shall show that every ray class field over M can be generated, over a ray class field over F, by the special values of automorphic functions with respect to a congruence subgroup of $\Gamma(\mathfrak{o})$ at a regular fixed point of M (Main Theorem I (3.2)).

An explicit reciprocity law for these class fields will be given as Main Theorem II (3.5). To avoid excessive details, we state the result at this point only in the simplest case where $C_F = F$ and the congruence subgroup in question is $\Gamma(\mathfrak{o})$ itself. *If φ, M, f, z and C_M are as above, and $\sigma = \left(\dfrac{C_M/M}{\mathfrak{b}} \right)$ with an ideal $\mathfrak{b}$ in M, then we have*

$$(4) \qquad\qquad \varphi(z)^\sigma = \varphi\big(\alpha^{-1}(z)\big)$$

with an element α of B^+ such that $f(\mathfrak{b})\mathfrak{o} = \alpha\mathfrak{o}$.

A completely different type of results will be obtained when $r = g$. In this case, instead of $\Gamma(\mathfrak{o})$, it is more natural to take a subgroup $\Gamma_1(\mathfrak{o})$ of $\Gamma(\mathfrak{o})$ consisting of the elements whose projections to the factors $M_2(\mathbf{R})$ of B_R have determinant 1. We shall obtain an algebraic variety V of dimension g, defined over $\mathbf{Q}$, and a holomorphic mapping φ of $\mathfrak{H}^g$ onto V which gives a biregular morphism of $\mathfrak{H}^g/\Gamma_1(\mathfrak{o})$ onto V. Then, for every totally imaginary quadratic extension M of F, isomorphic to a subfield of B, and for every regular fixed point z of M on $\mathfrak{H}^g$, we shall show that $\varphi(z)$ generates an unramified class field over a certain algebraic number field M'. Here M' is determined by the injection of M into B, and may or may not coincide with M. We also obtain results involving ramified extensions and congruence subgroups of $\Gamma_1(\mathfrak{o})$. Complete descriptions of the corresponding ideal groups and an explicit reciprocity law analogous to (4) will be stated as Main Theorems III and IV (9.3, 9.4), of which we do not try to give any details here, since the precise statement is rather lengthy.

The case where $r = g = 2$ and $B = M_2(F)$ was studied by Hecke in his dissertation. Our theorems, specialized to the case of Hilbert modular functions, give a sharpening and a generalization of this result of Hecke. It may be observed that, for the construction of class fields, the automorphic functions of one variable provide a more transparent and complete result than the functions of Hilbert-Hecke type. In fact, the result for $r = 1$ is far deeper than that for $r = g$, for reasons which we shall explain later. However, we can actually regard both results as two extreme cases of a more general theory which is expected to hold without restrictions on r. The conjectures about this will be discussed in 9.12.

Our second subject is the determination of the zeta function, in the sense of Hasse and Weil, of the curve V with the above properties (1–3). In a previous paper [8], we have developed, with this aim, a theory of Hecke operators on the space of automorphic forms with respect to $\Gamma(\mathfrak{o})$, and obtained zeta functions of the form

$$(5) \qquad\qquad D_k(s) = \sum_{\mathfrak{a}} T_k(\mathfrak{a})N(\mathfrak{a})^{-s} ,$$

where $\mathfrak{a}$ runs over all integral ideals in F, and the $T_k(\mathfrak{a})$ are Hecke operators defined for the cusp forms of weight k. It has been shown that $D_k(s)$ can be continued holomorphically to the whole s-plane and satisfies a functional equation; moreover, $D_k(s)$ has an Euler product of Hecke type. Now take a prime ideal $\mathfrak{p}$ in F, and a prime factor $\mathfrak{P}$ of $\mathfrak{p}$ in C_r. For almost all $\mathfrak{P}$, the reduction $\mathfrak{P}(V)$ of V modulo $\mathfrak{P}$ is a complete non-singular curve. Let $Z(u, \mathfrak{P}(V))$ denote the numerator part of the zeta function of $\mathfrak{P}(V)$ over the residue field of C_r modulo $\mathfrak{P}$. Then our Main Theorem V (12.2) asserts that, *for almost all $\mathfrak{p}$, the product* $\prod_{\mathfrak{P}|\mathfrak{p}} Z(N(\mathfrak{P})^{-s}, \mathfrak{P}(V))$ *coincides with the $\mathfrak{p}$-Euler factor of* $\det (D_2(s))^{-1}$. Therefore we can verify the Hasse conjecture for V, thus carrying out the program proposed in a previous paper [14, 6.3, II].

In the proofs of these results, the following facts are crucial.

(A) There exists a family Σ of abelian varieties A_z parametrized by the points z on $\mathfrak{H}^r$ such that two members A_z and A_w (endowed with structures of polarization and endomorphisms) are isomorphic if and only if $z = \gamma(w)$ for some $\gamma \in \Gamma_1(\mathfrak{o})$.

(B) Given such a family Σ, there exist an algebraic variety W, an algebraic number field k_Σ, and a holomorphic mapping ψ of $\mathfrak{H}^r$ onto W such that

(i) ψ gives a biregular morphism of $\mathfrak{H}^r/\Gamma_1(\mathfrak{o})$ onto W;

(ii) W is defined over k_Σ;

(iii) $k_\Sigma(\psi(z))$ is the field of moduli of A_z for every $z \in \mathfrak{H}^r$.

(C) If M is a totally imaginary quadratic extension of F, isomorphic to a quadratic subfield of B, and z is a regular fixed point of M on $\mathfrak{H}^r$, then A_z is isogenous to a product of copies of an abelian variety A such that $\mathrm{End}_Q(A)$ contains an extension S of M, and $[S : Q] = 2 \cdot \dim(A)$.

(D) If A is as in (C), then the field of moduli of A is a class field over a certain algebraic number field S' which is determined by the representation of S in the tangent space of A at the origin.

The last point (D) has been investigated in [15] in detail; (A) resp. (B) is a special case of the general theory [9] resp. [13]; the nature of the field k_Σ has been studied in [11]. The special member A_z mentioned in (C), and the corresponding class field, will be discussed in §§ 5, 6 of the present paper.

The results in the case $r = g$ follow almost immediately from the combination of these facts. The family Σ consists of abelian varieties A_z such that $\dim (A_z) = 2g$ and $\mathrm{End}_Q (A_z)$ contains B. Then it is sufficient to take simply W and ψ as V and φ. The field k_Σ in this case (with level one) turns out to be Q; and one has $M = S$ in (C).

When $r = 1$, the circumstances are far more complicated. In general if $r < g$, one can find, for any totally imaginary quadratic extension K of F, a family Σ_K of abelian varieties A_z such that $\mathrm{End}_Q (A_z)$ contains $B \otimes_F K$. Therefore, for a given $\Gamma(\mathfrak{o})$, there are infinitely many distinct families Σ_K, depending on the choice of K, and hence infinitely many distinct (W, ψ) of (B). Furthermore, the field k_Σ is not necessarily abelian over F, but a class field over a certain number field K'. However, the class field C_F is the intersection of the maximal unramified abelian extensions $C_{K'}$ for all these K'. Thus the main point of the proof in the case $r = 1$ lies in the procedure of constructing from these (W, ψ) a model (V, φ) defined over C_F. In each family Σ_K, the class field over S' mentioned in (D) is shown to be contained in the composite $C_M C_{K'}$. The property (3) of φ can be established by means of the fact that C_M is the intersection of the $C_M \cdot C_{K'}$ for all K'. Therefore it is noteworthy that *the field $C_F(\varphi(z))$ in (3) has no meaning as the field of moduli of an abelian variety.* We should also remark that the notion of weak polarization (§ 4) is necessary in order to dispose of the difference between $\Gamma(\mathfrak{o})$ and $\Gamma_1(\mathfrak{o})$.

As for the determination of the zeta function of the curve V, we consider, for every prime ideal $\mathfrak{p}$ in F, an algebraic correspondence X on $V \times V$ defined by

$$X = \{\varphi(z) \times \varphi(\alpha(z)) \mid z \in \mathfrak{H}\} \,,$$

with an element α of $\mathfrak{o} \cap B^+$ such that $\mathfrak{p}$ is generated by the reduced norm of α to F. Here we assume, for the sake of simplicity, that $C_F = F$. We can show that X is rational over F and independent of the choice of α. Let M and z be as in the above assertion (3). Suppose that $\mathfrak{p}$ has a prime factor $\mathfrak{q}$ in M such that $N(\mathfrak{p}) = N(\mathfrak{q})$. Then the reciprocity law (4) shows that $\varphi(z)^\sigma \times \varphi(z)$ is contained in X with $\sigma = \left(\dfrac{C_M/M}{\mathfrak{q}} \right)$. Taking reduction modulo $\mathfrak{p}$, we find a point u on $\mathfrak{p}(V)$ such that $u^{N(\mathfrak{p})} \times u$ is contained in $\mathfrak{p}(X)$. This process can produce infinitely many distinct such points. Therefore $\mathfrak{p}(X)$ must contain the transpose of the $N(\mathfrak{p})^{\text{th}}$ power Frobenius correspondence Π. Since X is symmetric, we obtain thus a congruence ralation

$$(6) \qquad\qquad \mathfrak{p}(X) = \Pi + {}^{t}\Pi \;,$$

from which we can easily derive the connection of the zeta function of $\mathfrak{p}(V)$ with the Hecke operators. The congruence relation of the form (6) has been first found by Eichler for certain modular function fields, and generalized by the author to some other cases. Our Theorem 11.17 includes all these as special cases.

One final remark may be worth adding. The groups $\Gamma(\mathfrak{o})$ of our type exhaust all the arithmetic discontinuous groups acting on $\mathfrak{H}$, at least up to commensurability. In this sense, no direct generalization of our results in the one-dimensional case may be expected. Some further general remarks, together with some examples, will be discussed in 3.18, 3.19, 9.12, and 12.4 of the text.

This work is most cordially dedicated to André Weil, who initiated the modern theory of abelian varieties, and has had a major influence in the theory of zeta functions of algebraic varieties.

Notation. We denote by Z, Q, R, C and K, respectively, the ring of rational integers, the rational number field, the real number field, the complex number field, and the division ring of real quaternions. For an associative ring S with an identity element, $M_n(S)$ denotes the ring of all matrices of size n with entries in S, and $\mathrm{GL}_n(S)$ the group of invertible elements in $M_n(S)$. The identity element of $M_n(S)$ is denoted by 1_n. We denote by $\mathfrak{H}/\Gamma$ the quotient of a symmetric domain $\mathfrak{H}$ by a discontinuous group Γ, though the action of the elements of Γ is defined on the left of the points of $\mathfrak{H}$. If K is a Galois extension of a field F, $G(K/F)$ means the Galois group of K over F. The terminology and the symbols employed in the above introduction will not necessarily be retained in the text.

1. Preliminaries on number fields

1.1. By an algebraic number field, we always understand an algebraic extension of Q contained in C. Let F be an algebraic number field of finite degree. We denote by $\mathfrak{r}_F$ the ring of all algebraic integers in F. For every prime ideal $\mathfrak{p}$ in F, we denote by $F_{\mathfrak{p}}$ the $\mathfrak{p}$-completion of F. If $\mathfrak{a}$ is an ideal in F, then $\mathfrak{a}_{\mathfrak{p}}$ denotes the $\mathfrak{p}$-closure of $\mathfrak{a}$ in $F_{\mathfrak{p}}$.

Let $\mathfrak{c}$ be an integral ideal in F, and $\mathfrak{u}$ a formal product of real archimedean prime spots of F. For non-zero elements x and y of F, we write $x \equiv y \bmod^* \mathfrak{c}\mathfrak{u}$

if $x^{-1}y$ is positive at every real archimedean prime spot involved in $\mathfrak{u}$, and $x^{-1}y - 1 \in \mathfrak{c}_\mathfrak{p}$ for every prime ideal $\mathfrak{p}$ dividing $\mathfrak{c}$. We denote by $I(F, \mathfrak{c})$ the group of all ideals in F prime to $\mathfrak{c}$, and by $P(F, \mathfrak{cu})$ the group of all principal ideals $x \cdot \mathfrak{r}_F$ such that $x \in F$ and $x \equiv 1 \bmod^* \mathfrak{cu}$. We put $I(F) = I(F, \mathfrak{c})$ if $\mathfrak{c} = \mathfrak{r}_F$.

Let K be a finite algebraic extension of F. The different of K relative to F is denoted by $\mathfrak{d}(K/F)$. We put $D(K/F) = N_{K/F}(\mathfrak{d}(K/F))$. Suppose that K is normal over F. Let $\mathfrak{p}$ be a prime ideal in F, and $\mathfrak{P}$ a prime ideal in K dividing $\mathfrak{p}$. If $\mathfrak{p}$ is unramified in K, we denote by $[K/F, \mathfrak{P}/\mathfrak{p}]$ the Frobenius automorphism of $\mathfrak{P}$ over F, i.e., an element σ of $G(K/F)$ such that $a^\sigma \equiv a^{N(\mathfrak{p})} \bmod \mathfrak{P}$ for every $a \in \mathfrak{r}_K$. When K is abelian over F, we write $[K/F, \mathfrak{P}/\mathfrak{p}]$ simply as $[K/F, \mathfrak{p}]$, and define $[K/F, \mathfrak{a}]$ for every $\mathfrak{a} \in I(F, D(K/F))$ so that $\mathfrak{a} \mapsto [K/F, \mathfrak{a}]$ is a homomorphism of $I(F, D(K/F))$ into $G(K/F)$.

For every integral ideal $\mathfrak{c}$ in F, we shall denote by $C(F, \mathfrak{c})$ the class field over F corresponding to the ideal group $P(F, \mathfrak{cu}_0)$, where $\mathfrak{u}_0$ is the product of *all* real archimedean prime spots of F. The field $C(F, \mathfrak{c})$ has the following property: writing K for $C(F, \mathfrak{c})$, for every integral ideal $\mathfrak{m}$ in F, one has

$$P(F, \mathfrak{cu}_0) \cap I(F, \mathfrak{m}) = \{\mathfrak{a} \in I(F, \mathfrak{cm}) \mid [K/F, \mathfrak{a}] = 1\} .$$

Conversely, if a finite abelian extension K of F satisfies this equality for at least one integral ideal $\mathfrak{m}$ in F, then $K = C(F, \mathfrak{c})$.

1.2. Lemma. *Let F, L, M be algebraic number fields of finite degree such that $F \subset L \cap M$. Suppose that L and M are linearly disjoint over F, and $\mathfrak{d}(LM/L) = \mathfrak{d}(M/F)\mathfrak{r}_{LM}$. Then $\mathfrak{r}_{LM} = \mathfrak{r}_L \cdot \mathfrak{r}_M$.*

Proof. For every element z of $\mathfrak{r}_M$ such that $M = F(z)$, let $f_z(X) = 0$ be the minimum equation for z over F, and let

$$\mathfrak{f}_z = f_z'(z)\mathfrak{d}(M/F)^{-1} , \qquad \mathfrak{g}_z = f_z'(z)\mathfrak{d}(LM/L)^{-1} .$$

Then it is well known that $\mathfrak{f}_z \subset \mathfrak{r}_F[z] \subset \mathfrak{r}_M$, $\mathfrak{g}_z \subset \mathfrak{r}_L[z] \subset \mathfrak{r}_{LM}$. By our assumption, we have $\mathfrak{g}_z = \mathfrak{f}_z \cdot \mathfrak{r}_{LM}$. Now the greatest common divisor of the $\mathfrak{f}_z$ for all z is $\mathfrak{r}_M$. Therefore $\mathfrak{r}_{LM}$ is generated by the $\mathfrak{r}_L[z]$ for all $z \in \mathfrak{r}_M$, hence $\mathfrak{r}_{LM} = \mathfrak{r}_L \cdot \mathfrak{r}_M$.

1.3. Lemma. *Let F, L, M be algebraic number fields of finite degree such that $F \subset L \cap M$. Suppose that*
   (i) *L is a Galois extension of F;*
   (ii) *$\mathfrak{d}(L/F)$ is prime to $\mathfrak{d}(M/F)$;*
   (iii) *L and M are linearly disjoint over F.*
Then $C(F, \mathfrak{c}) \supset C(M, \mathfrak{c}) \cap L$ for every integral ideal $\mathfrak{c}$ in F.

Proof. Put $S = C(M, \mathfrak{c}) \cap L$. Then SM is an abelian extension of M. Let σ be an automorphism of L over F. We can extend σ to an automorphism τ of LM over M. Then $S^\sigma M = (SM)^\tau = SM$. Since L and M are linearly dis-

35

joint over F, we have $S^\sigma = S$. Hence S is a Galois extension of F. Moreover S is abelian over F, since SM is abelian over M. Furthermore we have $\mathfrak{d}(SM/M) = \mathfrak{d}(S/F)$. Now identify $G(SM/M)$ with $G(S/F)$, and take a character χ of $G(S/F)$. Let T_χ be the subfield of S corresponding to the kernel of χ. Let $\mathfrak{f}(T_\chi/F)$ (resp. $\mathfrak{f}(T_\chi M/M)$) be the finite part of the conductor of T_χ over F (resp. $T_\chi M$ over M). By the conductor-discriminant theorem, we have

$$D(S/F) = \prod_\chi \mathfrak{f}(T_\chi/F) , \qquad D(SM/M) = \prod_\chi \mathfrak{f}(T_\chi M/M) .$$

Now $\mathfrak{f}(T_\chi/F) \subset \mathfrak{f}(T_\chi M/M)$ for every χ. Hence we must have $\mathfrak{f}(T_\chi/F) = \mathfrak{f}(T_\chi M/M)$. Since $T_\chi M \subset C(M, \mathfrak{c})$, we have $\mathfrak{f}(T_\chi M/M) \supset \mathfrak{c}$, so that $\mathfrak{f}(T_\chi/F) \supset \mathfrak{c}$. This shows that the conductor of S is a divisor of $\mathfrak{c}$, hence $S \subset C(F, \mathfrak{c})$.

1.4. Lemma. *Let F, L, M be algebraic number fields of finite degree such that $F \subset L \cap M$, and $\mathfrak{c}$ an integral ideal in F. Let L_0 be the smallest Galois extension of F containing L. Suppose that*

(i) *$\mathfrak{d}(M/F)$ is prime to $\mathfrak{c} \cdot \mathfrak{d}(L_0/F)$;*

(ii) *M and $C(L_0, \mathfrak{c})$ are linearly disjoint over F.*

Then $C(F, \mathfrak{c}) = C(M, \mathfrak{c}) \cap C(L, \mathfrak{c})$.

Proof. We see that $C(L_0, \mathfrak{c})$ is a Galois extension of F. Therefore, applying 1.3 to $C(L_0, \mathfrak{c})$ and M, we get $C(F, \mathfrak{c}) \supset C(M, \mathfrak{c}) \cap C(L_0, \mathfrak{c}) \supset C(M, \mathfrak{c}) \cap C(L, \mathfrak{c})$. Since $C(F, \mathfrak{c}) \subset C(M, \mathfrak{c}) \cap C(L, \mathfrak{c})$ is obvious, we get the desired equality.

1.5. Lemma. *Let F be an algebraic number field of finite degree. Let S_1, S_2, S_3 be mutually disjoint finite sets of prime ideals in F. Let R_1, R_2 be disjoint sets of real archimedean prime spots of F. Then there are infinitely many quadratic extensions M of F satisfying the following conditions.*

(1.5.1) Every prime in $R_1 \cup S_1$ decomposes in M.

(1.5.2) Every prime in $R_2 \cup S_2$ is ramified in M.

(1.5.3) Every prime in S_3 remains prime in M.

(1.5.4) There exists only one prime ideal in F which is ramified in M and not contained in S_2.

Proof. Let $\mathfrak{u}_1$ resp. $\mathfrak{u}_2$ be the product of all primes in R_1 resp. R_2. Let S_3' be the set of all prime ideals which divide 2 and are not contained in $S_1 \cup S_2$. We can find a positive integer e such that if $\mathfrak{p} \in S_1 \cup S_3 \cup S_3'$, $x \in F$ and $x \equiv 1 \bmod^* \mathfrak{p}^e$, then $F_\mathfrak{p}(\sqrt{x}) = F_\mathfrak{p}$. For each $\mathfrak{p} \in S_3 \cup S_3'$, there exists an element $a_\mathfrak{p}$ of F such that $F_\mathfrak{p}(\sqrt{a_\mathfrak{p}})$ is a quadratic unramified extension of $F_\mathfrak{p}$. We can then find an element a of F such that $a \equiv 1 \bmod^* \mathfrak{u}_1$, $a \equiv -1 \bmod^* \mathfrak{u}_2$, $a \equiv a_\mathfrak{p} \bmod^* \mathfrak{p}^e$ for every $\mathfrak{p} \in S_3 \cup S_3'$, and $a \equiv 1 \bmod^* \mathfrak{p}^e$ for every $\mathfrak{p} \in S_1$. Let $\mathfrak{x}$ resp. $\mathfrak{y}$ be the product of all the prime ideals in $S_1 \cup S_3 \cup S_3'$ resp. S_2. Take a prime ideal $\mathfrak{q}$

belonging to the same class as $a\mathfrak{y}^{-1}$ modulo $P(F, \mathfrak{x}^e u_1 u_2)$. Then $a^{-1}\mathfrak{y}\mathfrak{q} = (b)$ with an element b of F such that $b \equiv 1 \bmod^* \mathfrak{x}^e u_1 u_2$. We may assume that $\mathfrak{q} \notin S_1 \cup S_2 \cup S_3 \cup S_3'$. Then it can be easily verified that $F(\sqrt{ab})$ has all the required properties. Since there are infinitely many choices of $\mathfrak{q}$, there are infinitely many such M.

1.6. Let W be a vector space of finite dimension over an algebraic number field F of finite degree. By a *lattice* in W, we understand a finitely generated Z-submodule of W which generates W over Q. A lattice $\mathfrak{M}$ in W is called an $\mathfrak{r}_F$-*lattice* if $\mathfrak{r}_F\mathfrak{M} \subset \mathfrak{M}$. For every prime ideal $\mathfrak{p}$ in F, we put $W_\mathfrak{p} = W \otimes_F F_\mathfrak{p}$, and denote by $\mathfrak{M}_\mathfrak{p}$ (for an $\mathfrak{r}_F$-lattice $\mathfrak{M}$) the $\mathfrak{p}$-closure of $\mathfrak{M}$ in $W_\mathfrak{p}$, i.e., $\mathfrak{M}_\mathfrak{p} = \mathfrak{r}_\mathfrak{p}\mathfrak{M}$ if $\mathfrak{r}_\mathfrak{p}$ is the ring of $\mathfrak{p}$-integers in $F_\mathfrak{p}$.

2. Quaternion algebras and quadratic subfields

2.1. In this section F denotes a totally real algebraic number field of finite degree, and B a quaternion algebra over F, i.e., a central simple algebra over F such that $[B : F] = 4$. We put $g = [F : Q]$. For every archimedean or non-archimedean prime spot v of F, let F_v denote the completion of F with respect to v, and let $B_v = B \otimes_F F_v$. We say that v is *ramified* in B, or B is *ramified* at v, if B_v is a division algebra. The following lemmas are special cases of Hasse's theorems on central simple algebras over algebraic number fields.

2.2. LEMMA. *The number of archimedean or non-archimedean prime spots of F ramified in B is finite and even. Conversely, for every finite set P of an even number of prime spots of F, there exists a quaternion algebra over F, unique up to F-linear isomorphism, which is ramified exactly at the primes of P.*

2.3. LEMMA. *Let K be a quadratic extension of F. The following three conditions are equivalent to each other.*

(2.3.1) *$B \otimes_F K$ is isomorphic to $M_2(K)$.*

(2.3.2) *There exists an F-linear isomorphism of K into B.*

(2.3.3) *Every archimedean or non-archimedean prime spot of F ramified in B is ramified in K or remains prime in K.*

We shall denote by $J(B)$ the set of all totally imaginary quadratic extensions K of F (regarded as subfields of C) such that $B \otimes_F K$ is isomorphic to $M_2(K)$.

2.4. Let $\tau_{01}, \cdots, \tau_{0g}$ denote all the isomorphisms of F into R. We denote F_v by $F^{(\nu)}$ or $F_\mathfrak{p}$ according as v corresponds to $\tau_{0\nu}$ or a prime ideal $\mathfrak{p}$ in F. We write B_v accordingly $B^{(\nu)}$ or $B_\mathfrak{p}$. A mapping

$$(2.4.1) \qquad F \ni a \longmapsto (a^{\tau_{01}}, \cdots, a^{\tau_{0g}}) \in \boldsymbol{R}^g \text{ (the product of } g \text{ copies of } \boldsymbol{R})$$

can be extended to an isomorphism of $F \otimes_{\boldsymbol{Q}} \boldsymbol{R}$ to $\boldsymbol{R}^g$. Then the νth factor of $\boldsymbol{R}^g$ can be identified with $F^{(\nu)}$. We observe that $B^{(\nu)}$, being a quaternion algebra over $\boldsymbol{R}$, is isomorphic to $M_2(\boldsymbol{R})$ or the division ring of real quaternions, $\boldsymbol{K}$. We reorder $\tau_{01}, \cdots, \tau_{0g}$ so that $B^{(\nu)} \cong M_2(\boldsymbol{R})$ for $1 \leq \nu \leq r$ and $B^{(\nu)} \cong \boldsymbol{K}$ for $r < \nu \leq g$. We fix, once and for all, these isomorphisms, and identify the algebras each one with another as follows:

$$(2.4.2) \qquad B \otimes_{\boldsymbol{Q}} \boldsymbol{R} = B^{(1)} \times \cdots \times B^{(g)} = M_2(\boldsymbol{R}) \times \cdots \times M_2(\boldsymbol{R}) \times \boldsymbol{K} \times \cdots \times \boldsymbol{K}.$$

We assume throughout $r > 0$. We denote by $D(B/F)$ the product of all prime ideals of F which are ramified in B. If s is the number of such prime ideals, $r + s$ should be even in view of 2.2.

Define a subset $\mathfrak{F}_r$ of $\boldsymbol{C}^r$ by

$$\mathfrak{F}_r = \{(z_1, \cdots, z_r) \in \boldsymbol{C}^r \mid \mathrm{Im}(z_1) \neq 0, \cdots, \mathrm{Im}(z_r) \neq 0\} \ .$$

Obviously $\mathfrak{F}_r$ has exactly 2^r connected components. We define the action of an invertible element α of $B \otimes_{\boldsymbol{Q}} \boldsymbol{R}$ on $\mathfrak{F}_r$ by

$$\alpha = (\alpha^{(1)}, \cdots, \alpha^{(g)}) \ , \qquad \alpha^{(\nu)} = \begin{bmatrix} a_\nu & b_\nu \\ c_\nu & d_\nu \end{bmatrix} \in M_2(\boldsymbol{R}) \qquad (\nu = 1, \cdots, r) \ ,$$

$$\alpha(z_1, \cdots, z_r) = (w_1, \cdots, w_r) \ , \qquad w_\nu = (a_\nu z_\nu + b_\nu)/(c_\nu z_\nu + d_\nu) \ .$$

2.5. The algebra B has an involution $x \longmapsto x'$ which is uniquely determined by the property that $x + x'$ is the reduced trace of x to the center F. We put

$$N_{B/F}(x) = xx' \ , \qquad\qquad \mathrm{Tr}_{B/F}(x) = x + x'$$
$$N_{B/Q}(x) = N_{F/Q}(N_{B/F}(x)) \ , \qquad \mathrm{Tr}_{B/Q}(x) = \mathrm{Tr}_{F/Q}(\mathrm{Tr}_{B/F}(x)) \qquad (x \in B) \ .$$

Let $\mathfrak{o}$ be a maximal order in B and $\mathfrak{y}$ a right $\mathfrak{o}$-ideal. (As for the terminology, see for example [8, §1].) We denote by $N_{B/F}(\mathfrak{y})$ the ideal in F generated over $\mathfrak{r}_F$ by the elements $N_{B/F}(x)$ for all $x \in \mathfrak{y}$. We say that $\mathfrak{y}$ is *prime to* an integral ideal $\mathfrak{c}$ in F if $\mathfrak{y}_\mathfrak{p} = \mathfrak{o}_\mathfrak{p}$ for every prime factor $\mathfrak{p}$ of $\mathfrak{c}$ (cf. 1.6).

2.6. PROPOSITION. *Let $M \in J(B)$, and let f be an F-linear isomorphism of M into B. Let a be an element of M not contained in F. Then $f(a)$ has exactly one fixed point on each connected component of $\mathfrak{F}_r$. The fixed point depends only on $f(M)$ and is independent of the choice of a. Conversely, suppose that an element α of B, not contained in F, has a fixed point on $\mathfrak{F}_r$. Then $F(\alpha)$ is isomorphic to a member of $J(B)$.*

Each fixed point on a connected component $\mathfrak{H}'$ of $\mathfrak{F}_r$ will be called the *fixed point of $f(M)$ on $\mathfrak{H}'$*.

PROOF. Put $f(a) = \alpha = (\alpha^{(1)}, \cdots, \alpha^{(g)})$ with $\alpha^{(\nu)} \in B^{(\nu)}$. Since M is totally imaginary, we see that each $\alpha^{(\nu)}$ for $\nu = 1, \cdots, r$ gives an elliptic transformation on the upper and lower half planes on C. Therefore $f(a)$ has exactly one fixed point z on each connected component $\mathfrak{H}'$ of $\mathfrak{F}_r$. Let b be another non-zero element of M, and let $\beta = f(b)$. Then $\beta(z) = \beta\alpha(z) = \alpha(\beta(z))$. Since M is totally imaginary, $N_{B/F}(\beta)$ is totally positive, hence $\beta(z) \in \mathfrak{H}'$. Since α has the only fixed point on $\mathfrak{H}'$, we have $\beta(z) = z$. This proves the first assertion. Conversely, let $\alpha \in B$, $\alpha \notin F$, $\alpha(z) = z$ with $z \in \mathfrak{F}_r$. If $\alpha = (\alpha^{(1)}, \cdots, \alpha^{(g)})$ with $\alpha^{(\nu)}$ in $B^{(\nu)}$, we see that the characteristic roots of $\alpha^{(\nu)}$, for $\nu = 1, \cdots, r$, are not real, hence the first r archimedean prime spots of F corresponding to $\tau_{01}, \cdots, \tau_{0r}$ are ramified in $F(\alpha)$. On the other hand, from (2.4.2) we see that the remaining $g - r$ archimedean prime spots of F are ramified in every quadratic subfield of B. Hence $F(\alpha)$ is totally imaginary.

2.7. Let $M \in J(B)$. Let f_1 and f_2 be two F-linear isomorphisms of M into B. Then there exists an element λ of B such that $f_1(x) = \lambda f_2(x)\lambda^{-1}$. The element λ is uniquely determined up to the left multiplication by elements of $f_1(M)$. Especially the signature of $N_{B/F}(\lambda)^{\tau_{0\nu}}$ for each ν is uniquely determined by f_1 and f_2. For example, if f_2 is obtained from f_1 by $f_2(x) = f_1(x)'$, then $N_{B/F}(\lambda)^{\tau_{0\nu}} < 0$ for $\nu = 1, \cdots, r$, since M is totally imaginary and $B^{(\nu)} = M_2(\mathbf{R})$ for $\nu = 1, \cdots, r$. On a given component of $\mathfrak{F}_r$, this pair f_1 and f_2 correspond to the same fixed point. Now, if we classify all the F-linear isomorphisms of M into B by the inner automorphisms $u \mapsto \lambda u \lambda^{-1}$ of B with totally positive $N_{B/F}(\lambda)$, then there are exactly 2^r classes. We shall show that each class determines r isomorphisms of M into C.

Fix an F-linear isomorphism f of M into B. Let $\mathfrak{H}^r$ denote the product of r copies of the upper half plane $\mathfrak{H} = \{z \in C \mid \mathrm{Im}(z) > 0\}$, and $z_0 = (z_1^0, \cdots, z_r^0)$ the fixed point of $f(M)$ on $\mathfrak{H}^r$. Let ψ_ν, for $\nu = 1, \cdots, r$, be a one-to-one holomorphic mapping of $\mathfrak{H}$ onto the unit disk $\{w \in C \mid |w| < 1\}$ such that $\psi_\nu(z_\nu^0) = 0$. Then, for every $a \in M$, $a \neq 0$, one has

$$(2.7.1) \qquad \psi_\nu \circ f(a)^{(\nu)} \circ \psi_\nu^{-1}(w) = (a^{\tau_\nu \rho}/a^{\tau_\nu}) \cdot w \qquad (|w| < 1) .$$

Here $f(a)^{(\nu)}$ is the projection of $f(a)$ to the νth factor of (2.4.2); ρ is the complex conjugation; and τ_ν is an isomorphism of M into C which does not depend on a. Thus f determines r isomorphisms $\tau_1, \cdots, \tau_r$ of M into C. We note that τ_ν coincides with $\tau_{0\nu}$ on F for each ν. The ordered set $\{\tau_1, \cdots, \tau_r\}$ does not depend on the choice of the ψ_ν; but it depends on the choice of the identification (2.4.2), i.e., the choice of r isomorphisms of B into $M_2(\mathbf{R})$. Define $f_1 \colon M \to B$ by $f_1(x) = \alpha f(x)\alpha^{-1}$ for $x \in M$ with an invertible element α of B. Then f_1 determines the same $\{\tau_\nu\}$ as f, if and only if $N_{B/F}(\alpha)$ is totally positive.

Especially suppose that $r = 1$, and τ_{01} is the identity mapping of F. We shall call f *normalized* if τ_1 is the identity mapping of M. (Recall that, by our agreement, F resp. M is a subfield of $\boldsymbol{R}$ resp. $\boldsymbol{C}$.) If f is not normalized, then one can get a normalized map $f_1 \colon M \to B$ by $f_1(x) = f(x)'$, with the same fixed point on $\mathfrak{H}$.

If $r = g$, we see that $(M; \{\tau_1, \cdots, \tau_g\})$ is a CM-type in the sense of [15] (see also §5 below). We shall call $(M; \{\tau_1, \cdots, \tau_g\})$ *the* CM-*type determined by* f.

2.8. PROPOSITION. *Let $M \in J(B)$. Then, for every maximal order $\mathfrak{o}$ in B, there exists an F-linear isomorphism f of M into B such that $f(\mathfrak{r}_M) \subset \mathfrak{o}$.*

PROOF. Let $\mathfrak{u}$ be the product of $g - r$ archimedean prime spots of F corresponding to $\tau_{0,r+1}, \cdots, \tau_{0g}$, and let S be the class field over F corresponding to $P(F, \mathfrak{u})$. Since M is totally imaginary, M is not contained in S, so that S and M are linearly disjoint over F. Let h be any F-linear isomorphism of M into B. We can find a maximal order $\mathfrak{o}_1$ in B containing $h(\mathfrak{r}_M)$. Let $\mathfrak{o}$ be an arbitrary maximal order in B. There exists a right $\mathfrak{o}_1$-ideal $\mathfrak{x}$ in B which is a left $\mathfrak{o}$-ideal at the same time. Let $\mathfrak{a} = N_{B/F}(\mathfrak{x})$ and $\sigma = [S/F, \mathfrak{a}]$. Since M and S are linearly disjoint over F, we can extend σ to an automorphism τ of SM over M, and find a prime ideal $\mathfrak{q}$ in M such that $\tau = [SM/M, \mathfrak{q}]$. Then $N_{M/F}(\mathfrak{q})$ belongs to the same ideal class as $\mathfrak{a}$ modulo $\mathfrak{u}$. Let $\mathfrak{o}_2$ be the left order of $h(\mathfrak{q})\mathfrak{o}_1$. Then $\mathfrak{o}_2$ is maximal, $h(\mathfrak{r}_M) \subset \mathfrak{o}_2$, and $N_{B/F}(h(\mathfrak{q})\mathfrak{o}_1) = N_{M/F}(\mathfrak{q})$. By Eichler's theorem, the right $\mathfrak{o}_1$-ideals $h(\mathfrak{q})\mathfrak{o}_1$ and $\mathfrak{x}$ are equivalent. Hence there exists an element μ of B such that $h(\mathfrak{q})\mathfrak{o}_1 = \mu\mathfrak{x}$. Comparing the left orders of $h(\mathfrak{q})\mathfrak{o}_1$ and $\mu\mathfrak{x}$, we have $\mathfrak{o}_2 = \mu\mathfrak{o}\mu^{-1}$. Define $f \colon M \to B$ by $f(x) = \mu^{-1}h(x)\mu$. Since $h(\mathfrak{r}_M) \subset \mathfrak{o}_2$, we have $f(\mathfrak{r}_M) \subset \mathfrak{o}$. This completes the proof.

2.9. PROPOSITION. *Let $\mathfrak{o}$ be a maximal order in B, and $\mathfrak{x}$ a right $\mathfrak{o}$-ideal. Let $M \in J(B)$. Let T be the set of all prime ideals $\mathfrak{p}$ in F such that $\mathfrak{x}_\mathfrak{p} \neq \mathfrak{o}_\mathfrak{p}$. Let T_1 resp. T_2 be the set of all $\mathfrak{p}$ in T which are unramified resp. ramified in B. Suppose that every $\mathfrak{p}$ in T_1 decomposes in M, and every $\mathfrak{p}$ in T_2 is ramified in M. Then there exist an F-linear isomorphism f of M into B, and an ideal $\mathfrak{b}$ in M such that $f(\mathfrak{r}_M) \subset \mathfrak{o}$ and $\mathfrak{x} = f(\mathfrak{b})\mathfrak{o}$.*

PROOF. Let $\mathfrak{p} \in T_1$. We can identify $B_\mathfrak{p}$ resp. $\mathfrak{o}_\mathfrak{p}$ with $M_2(F_\mathfrak{p})$ resp. $M_2(\mathfrak{r}_\mathfrak{p})$, where $\mathfrak{r}_\mathfrak{p}$ is the ring of $\mathfrak{p}$-adic integers in $F_\mathfrak{p}$. We can take an element y of $M_2(F_\mathfrak{p})$ so that $\mathfrak{x}_\mathfrak{p} = y\mathfrak{o}_\mathfrak{p}$ and then take units u, v of $\mathfrak{o}_\mathfrak{p}$ so that $uyv = \begin{bmatrix} \pi^a & 0 \\ 0 & \pi^b \end{bmatrix}$ with a prime element π of $F_\mathfrak{p}$ and integers a, b. Let $\mathfrak{p}\mathfrak{r}_M = \mathfrak{q}_1\mathfrak{q}_2$ be the prime decomposition of $\mathfrak{p}$ in M. Define an ideal $\mathfrak{b}_1$ in M by $\mathfrak{b}_1 = \prod_{\mathfrak{p} \in T_1} \mathfrak{q}_1^a \mathfrak{q}_2^b$. Next let $\mathfrak{p} \in T_2$. If $\mathfrak{m}_\mathfrak{p}$ is the maximal ideal of $\mathfrak{o}_\mathfrak{p}$, we have $\mathfrak{x}_\mathfrak{p} = \mathfrak{m}_\mathfrak{p}^c$ with an integer c. Let $\mathfrak{p}\mathfrak{r}_M = \mathfrak{q}^2$ with a prime ideal $\mathfrak{q}$ in M. Define an ideal $\mathfrak{b}_2$ in M by $\mathfrak{b}_2 = \prod_{\mathfrak{x} \in T_2} \mathfrak{q}^c$. By 2.8, there exists an F-linear isomorphism h of M into B such that $h(\mathfrak{r}_M) \subset \mathfrak{o}$.

Then we see that $h(\mathfrak{b}_1\mathfrak{b}_2)\mathfrak{o}$ has the same set of invariants as $\mathfrak{x}$ in the sense of [8, 1.3] (see also 10.1 below). By [8, Prop. 1.4], there exists a unit γ of $\mathfrak{o}$ such that $N_{B/F}(\gamma) = 1$, and $\gamma h(\mathfrak{b}_1\mathfrak{b}_2)\mathfrak{o} = \mathfrak{x}$. Define f by $f(x) = \gamma h(x)\gamma^{-1}$. Then f and $\mathfrak{b}_1\mathfrak{b}_2$ have all the required properties.

2.10. PROPOSITION. *Let $\mathfrak{o}_1, \cdots, \mathfrak{o}_s$ be maximal orders in B, and let X_μ, for $\mu = 1, \cdots, s$, be a finite set of right $\mathfrak{o}_\mu$-ideals, and R a finite algebraic extension of F. Then there exist infinitely many M in $J(B)$ satisfying the following two conditions.*

(2.10.1) M and R are linearly disjoint over F.

(2.10.2) For every μ and $\mathfrak{x} \in X_\mu$, there exist an F-linear isomorphism f of M into B and an ideal $\mathfrak{b}$ in M such that $f(\mathfrak{r}_M) \subset \mathfrak{o}_\mu$ and $\mathfrak{x} = f(\mathfrak{b})\mathfrak{o}_\mu$.

Moreover, if every ideal in $\bigcup_\mu X_\mu$ is prime to $D(B/F)$, then, for a given integral ideal $\mathfrak{h}$ in F, there are infinitely many M in $J(B)$ satisfying (2.10.1–2) and

(2.10.3) $D(M/F)$ is prime to $\mathfrak{h}$.

PROOF. Let T be the set of all prime ideals $\mathfrak{p}$ in F such that $\mathfrak{x}_\mathfrak{p} \neq \mathfrak{o}_{\mu\mathfrak{p}}$ for some μ and some $\mathfrak{x} \in X_\mu$. Let T_1 resp. T_2 be the set of all $\mathfrak{p}$ in T which are unramified resp. ramified in B. With a given $\mathfrak{h}$, let S_3 be the set of all prime ideals in F dividing $D(B/F)\mathfrak{h}$ and not contained in T. Let $\mathfrak{q}$ be a prime ideal in F which decomposes completely in R and is not contained in $T \cup S_3$. Put $S_1 = T_1$, $S_2 = T_2 \cup \{\mathfrak{q}\}$. By 1.5, there are infinitely many quadratic extensions M of F satisfying (1.5.1–4) with these S_i, $R_1 = \varnothing$, and $R_2 = $ the set of all archimedean prime spots of F. In view of 2.3, $M \in J(B)$. By our choice of S_i and 2.9, it can be easily seen that M has the required properties.

2.11. LEMMA (Chevalley-Hasse-Noether). *Let $\mathfrak{o}_1$ and $\mathfrak{o}_2$ be maximal orders in B, and let $M \in J(B)$. Let f be an F-linear isomorphism of M into B. Suppose that $f(\mathfrak{r}_M) \subset \mathfrak{o}_1$. Then one has $f(\mathfrak{r}_M) \subset \mathfrak{o}_2$ if and only if there exists an ideal $\mathfrak{a}$ in M such that $\mathfrak{o}_1 f(\mathfrak{a}) = f(\mathfrak{a})\mathfrak{o}_2$.*

A proof in a more general case (with non-maximal orders), together with references to the papers of Chevalley, Hasse, and Noether, is given in Eichler [2, Satz 7].

2.12. PROPOSITION. *Let $\mathfrak{o}_1, \mathfrak{o}_2$, M and f be as in 2.11. Suppose that $f(\mathfrak{r}_M) \subset \mathfrak{o}_1 \cap \mathfrak{o}_2$. Let T be the set of two-sided integral $\mathfrak{o}_1$-ideals $\mathfrak{Q}$ such that $N_{B/F}(\mathfrak{Q})$ is square-free and divisible only by prime ideals in F, which are ramified in B and not ramified in M. (If s is the number of such prime ideals, then T contains 2^s elements.) Let $\mathfrak{x}$ be a right $\mathfrak{o}_1$-ideal which is a left*

$\mathfrak{o}_2$-*ideal. Then there exist an ideal* $\mathfrak{b}$ *in* M *and a member* $\mathfrak{O}$ *of* T *such that* $\mathfrak{x} = f(\mathfrak{b})\mathfrak{O}$. *Such* $\mathfrak{b}$ *and* $\mathfrak{O}$ *are uniquely determined by* $\mathfrak{x}$.

PROOF. By 2.11, there exists an ideal $\mathfrak{a}$ in M such that $\mathfrak{o}_1 f(\mathfrak{a}) = f(\mathfrak{a})\mathfrak{o}_2$. Then $f(\mathfrak{a})\mathfrak{x}$ is a two-sided $\mathfrak{o}_1$-ideal. Decompose $f(\mathfrak{a})\mathfrak{x}$ into the product of prime $\mathfrak{o}_1$-ideals. Then we obtain an expression $f(\mathfrak{a})\mathfrak{x} = \mathfrak{c}\mathfrak{O}_0\mathfrak{O}$ with an ideal $\mathfrak{c}$ in F, $\mathfrak{O} \in T$, and a two-sided integral $\mathfrak{o}_1$-ideal $\mathfrak{O}_0$ such that $N_{B/F}(\mathfrak{O}_0)$ is divisible only by the prime ideals in F which are ramified in M. We can find an integral ideal $\mathfrak{e}$ in M such that $N_{M/F}(\mathfrak{e}) = N_{B/F}(\mathfrak{O}_0)$. Put $\mathfrak{b} = \mathfrak{a}^{-1}\mathfrak{c}\mathfrak{e}$. Then $\mathfrak{x} = f(\mathfrak{b})\mathfrak{O}$. To prove the uniqueness, assume that $f(\mathfrak{b})\mathfrak{O} = f(\mathfrak{b}_1)\mathfrak{O}_1$ with an ideal $\mathfrak{b}_1$ in M and a member $\mathfrak{O}_1$ of T. If $\mathfrak{O} \neq \mathfrak{O}_1$, there exists a prime ideal $\mathfrak{p}$ in F whose exponent in $N_{B/F}(\mathfrak{O}^{-1}\mathfrak{O}_1)$ is ± 1. Since $M \in J(B)$, $\mathfrak{p}$ should remain prime in M, in view of our definition of T. Hence $\mathfrak{p}$ is involved in $N_{M/F}(\mathfrak{b}_1^{-1}\mathfrak{b})$ only with an even exponent. This is a contradiction, and hence the uniqueness is proved.

2.13. *Representatives for orders and ideals in* B. Let $\mathfrak{u}_0$ be the product of all archimedean prime spots of F, and let $\mathfrak{u}_1 = \mathfrak{u}_0 \cdot (\mathfrak{p}_{\infty 1} \cdots \mathfrak{p}_{\infty r})^{-1}$, where $\mathfrak{p}_{\infty \nu}$ denotes the archimedean prime spot of F corresponding to $\tau_{0\nu}$ (see 2.4). Let h_0 resp. h_1 denote the class number of F modulo $\mathfrak{u}_0$ resp. $\mathfrak{u}_1$, i.e.,

$$h_0 = [I(F) : P(F, \mathfrak{u}_0)] , \qquad h_1 = [I(F) : P(F, \mathfrak{u}_1)] ,$$

with the notation of 1.1. If E_0 resp. E_1 denotes the group of all units $\equiv 1 \bmod^* \mathfrak{u}_0$ resp. $\mathfrak{u}_1$ in F, then $[E_1 : E_0]$ is a divisor of 2^r, and $h_0 = 2^r h_1/[E_1 : E_0]$.

For every maximal order $\mathfrak{o}$ in B and an integral two-sided $\mathfrak{o}$-ideal $\mathfrak{e}$, we put

$$\Gamma(\mathfrak{o}) = \{\gamma \in \mathfrak{o} \mid N_{B/F}(\gamma) \in E_1\} ,$$
$$\Gamma(\mathfrak{o}, 1) = \{\gamma \in \mathfrak{o} \mid N_{B/F}(\gamma) \in E_0\} ,$$
$$\Gamma(\mathfrak{o}, \mathfrak{e}) = \{\gamma \in \Gamma(\mathfrak{o}, 1) \mid \gamma - 1 \in \mathfrak{e}\} .$$

If $\mathfrak{e} = \mathfrak{c}\mathfrak{o}$ or $\mathfrak{e} = d\mathfrak{o}$ with an ideal $\mathfrak{c}$ in F or an integer d, we write $\Gamma(\mathfrak{o}, \mathfrak{e})$ also as $\Gamma(\mathfrak{o}, \mathfrak{c})$, or $\Gamma(\mathfrak{o}, d)$ accordingly.

Let us fix a maximal order $\mathfrak{o}$ in B, and classify all the right $\mathfrak{o}$-ideals modulo the left multiplication by all the invertible elements of B. By Eichler's theorem, we get exactly h_1 classes.

Let B^+ denote the set of all elements α of B such that $N_{B/F}(\alpha)$ is totally positive. Instead of taking all invertible elements, we can consider the equivalence modulo the left multiplication by elements in B^+. We call such an equivalence class a *narrow class of right* $\mathfrak{o}$-*ideals*. Then there are exactly h_0 narrow classes. Let the $\mathfrak{x}_\lambda$, for $\lambda = 1, \cdots, h_0$, be representatives for narrow classes of right $\mathfrak{o}$-ideals. Put

$$\mathfrak{x}_{\lambda\mu} = \mathfrak{x}_\lambda\mathfrak{x}_\mu^{-1} , \qquad \mathfrak{o}_\lambda = \mathfrak{x}_{\lambda\lambda} \qquad\qquad (\lambda, \mu = 1, \cdots, h_0) .$$

Then we see easily that

(2.13.1) *The $\mathfrak{o}_\lambda$ are maximal orders in B.*

(2.13.2) $\mathfrak{x}_{\lambda\mu}$ *is a right $\mathfrak{o}_\mu$-ideal and a left $\mathfrak{o}_\lambda$-ideal.*

(2.13.3) $\mathfrak{x}_{\lambda\mu}\mathfrak{x}_{\mu\nu} = \mathfrak{x}_{\lambda\nu}.$

(2.13.4) *For every μ, the $\mathfrak{x}_{\lambda\mu}$ for $\lambda = 1, \cdots, h_0$ form a set of representatives for narrow classes of right $\mathfrak{o}_\mu$-ideals.*

We call a system $\Delta = \{\mathfrak{o}_\lambda, \mathfrak{x}_{\lambda\mu}\,(\lambda, \mu = 1, \cdots, h_0)\}$ a *representative set of orders and ideals in B*, if (2.13.1–4) are satisfied. From these conditions we can easily derive

(2.13.5) $\mathfrak{x}_{\lambda\lambda} = \mathfrak{o}_\lambda,\ \mathfrak{x}_{\lambda\mu}^{-1} = \mathfrak{x}_{\mu\lambda}.$

(2.13.6) *For each μ, the $N_{B/F}(\mathfrak{x}_{\lambda\mu})$ for $\lambda = 1, \cdots, h_0$ form a set of representatives for $I(F)/P(F, \mathfrak{u}_0)$.*

For every integral ideal $\mathfrak{c}$ in F, we can choose Δ so that

(2.13.7) *Every $\mathfrak{x}_{\lambda\mu}$ is prime to $\mathfrak{c}$.*

We say that Δ is $\mathfrak{c}$-*regular* if this is satisfied. Further we have

(2.13.8) *If $\{\mathfrak{o}_\lambda, \mathfrak{x}_{\lambda\mu}\}$ and $\{\bar{\mathfrak{o}}_\lambda, \bar{\mathfrak{x}}_{\lambda\mu}\}$ are two representative sets of orders and ideals in B, then there exist elements $\alpha_\lambda\ (\lambda = 1, \cdots, h_0)$ in B^+ and a permutation $\lambda \mapsto \bar\lambda$ of $\{1, \cdots, h_0\}$ such that $\bar{\mathfrak{x}}_{\bar\lambda\bar\mu} = \alpha_\lambda \mathfrak{x}_{\lambda\mu} \alpha_\mu^{-1}.$*

To show this, take a right $\bar{\mathfrak{o}}_1$-ideal $\mathfrak{y}$ which is a left $\mathfrak{o}_1$-ideal. Then, for each λ, $\mathfrak{x}_{\lambda 1}\mathfrak{y}$ is a right $\bar{\mathfrak{o}}_1$-ideal. Hence there is a unique $\bar\lambda$ such that $\alpha_\lambda \mathfrak{x}_{\lambda 1}\mathfrak{y} = \bar{\mathfrak{x}}_{\bar\lambda 1}$ with an element α_λ of B^+. We obtain then our assertion.

2.14. *Multiplicative congruence in B.* Let $\mathfrak{o}$ be a maximal order in B, $\mathfrak{e}$ an integral two-sided $\mathfrak{o}$-ideal, and let $\mathfrak{c} = \mathfrak{r}_F \cap \mathfrak{e}$. We say that an invertible element α of B is *prime to* $\mathfrak{e}$ if α is a unit of $\mathfrak{o}_\mathfrak{p}$ for every prime ideal $\mathfrak{p}$ in F dividing $\mathfrak{c}$, i.e., if $\alpha\mathfrak{o}$ is prime to $\mathfrak{c}$ in the sense of 2.5. (It is meaningless to say that α is prime to an integral ideal $\mathfrak{a}$ in F. However, as we have defined, one can meaningfully say that α is prime to $\mathfrak{a}\mathfrak{o}$, or $\alpha\mathfrak{o}$ is prime to $\mathfrak{a}$.) We write

$$\alpha \equiv 1 \qquad \mathrm{mod}^* \mathfrak{e}$$

if α is prime to $\mathfrak{e}$, and $\alpha - 1 \in \mathfrak{e}_\mathfrak{p}$ for every prime ideal $\mathfrak{p}$ in F dividing $\mathfrak{c}$, and $\alpha \equiv \beta \bmod^* \mathfrak{e}$ if $\alpha^{-1}\beta \equiv 1 \bmod^* \mathfrak{e}$. If $\alpha \equiv \beta \bmod^* \mathfrak{e}$, then $\alpha' \equiv \beta' \bmod^* \mathfrak{e}$, and $N_{B/F}(\alpha) \equiv N_{B/F}(\beta) \bmod^* \mathfrak{c}$. Moreover, we have

(2.14.1) *For every element α of B prime to $\mathfrak{e}$, there exists an element β of $\mathfrak{o} \cap B^+$ such that $\alpha \equiv \beta \bmod^* \mathfrak{e}$.*

This follows easily from the following 2.15, which is a special case of Eichler's approximation theorem [1, Satz 5].

2.15. LEMMA. *Let $\mathfrak{o}$ be a maximal order in B, and $\mathfrak{e}$ an integral two-sided $\mathfrak{o}$-ideal. Let b be an element of $\mathfrak{r}_F$ and α an element of $\mathfrak{o}$, such that $b \equiv 1 \bmod^* \mathfrak{u}_1$ and $N_{B/F}(\alpha) \equiv b \bmod^* \mathfrak{e} \cap \mathfrak{r}_F$. Then there exists an element β of $\mathfrak{o}$ such that $\beta \equiv \alpha \bmod \mathfrak{e}$ and $N_{B/F}(\beta) = b$.*

2.16.[1] Let $\Delta = \{\mathfrak{o}_\lambda, \mathfrak{x}_{\lambda\mu} (\lambda, \mu = 1, \cdots, h_0)\}$ be a representative set of orders and ideals in B which is $\mathfrak{c}$-regular for an integral ideal $\mathfrak{c}$ in F. Let $M \in J(B)$. By an *optimal embedding of M into B with respect to Δ*, we mean a couple $\mathfrak{f} = (f, \mathfrak{o}_\mu)$ formed by an F-linear isomorphism f of M into B and a maximal order $\mathfrak{o}_\mu$ among $\mathfrak{o}_1, \cdots, \mathfrak{o}_{h_0}$, such that $f(\mathfrak{r}_M) \subset \mathfrak{o}_\mu$. Let $\mathfrak{f} = (f, \mathfrak{o}_\mu)$ and $\mathfrak{f}_1 = (f_1, \mathfrak{o}_\lambda)$ be two optimal embeddings of M into B with respect to Δ. Then we can find an element α of B such that $f_1(x) = \alpha^{-1} f(x) \alpha$ for all $x \in M$. Then we see that $f(\mathfrak{r}_M) \subset \mathfrak{o}_\mu \cap \alpha \mathfrak{o}_\lambda \alpha^{-1}$. Let T_μ denote the set T defined in 2.12 with $\mathfrak{o}_\mu$ in place of $\mathfrak{o}_1$ of 2.12. Since $\alpha \mathfrak{o}_\lambda \alpha^{-1}$ is the left order of $\alpha \mathfrak{x}_{\lambda\mu}$, there exist, by 2.12, an ideal $\mathfrak{a}$ in M and a member $\mathfrak{Q}$ of T_μ such that $\alpha \mathfrak{x}_{\lambda\mu} = f(\mathfrak{a})\mathfrak{Q}$. We see that $\mathfrak{Q}$ and the ideal-class of $\mathfrak{a}$ in M are uniquely determined by $\mathfrak{f}$ and $\mathfrak{f}_1$. We say that $\mathfrak{f}$ and $\mathfrak{f}_1$ belong to the same *branch* if $N_{B/F}(\alpha)$ is totally positive (cf. 2.7) and $\mathfrak{Q} = \mathfrak{o}_\mu$. If that is so, we define a symbol $\{\mathfrak{f}_1 : \mathfrak{f}\}$ by

$$\{\mathfrak{f}_1 : \mathfrak{f}\} = \text{the ideal class of } \mathfrak{a} \text{ in } M.$$

We say also that $\mathfrak{f}$ and $\mathfrak{f}_1$ are *equivalent* if $\lambda = \mu$ and $f_1(x) = \gamma^{-1} f(x) \gamma$ with an element γ of $\Gamma(\mathfrak{o}_\lambda, 1)$. In this case $\mathfrak{f}$ and $\mathfrak{f}_1$ belong to the same branch, and $\{\mathfrak{f}_1 : \mathfrak{f}\} = 1$. By 2.7, 2.12, and a straightforward argument, one can prove easily

2.17. PROPOSITION. *Let $M \in J(B)$. Let s be the number of prime ideals in F which are ramified in B but not ramified in M, and h_M the class number of M. Then there are exactly 2^{s+r} branches, each of which contains exactly h_M optimal embeddings of M into B, up to equivalence, with respect to a fixed representative set of orders and ideals in B. Moreover, if $\mathfrak{f}, \mathfrak{f}_1, \mathfrak{f}_2$ are optimal embeddings of M into B of the same branch, then*

(2.17.1) $\{\mathfrak{f}_2 : \mathfrak{f}_1\}\{\mathfrak{f}_1 : \mathfrak{f}\} = \{\mathfrak{f}_2 : \mathfrak{f}\}.$

(2.17.2) $\{\mathfrak{f}_1 : \mathfrak{f}\} = 1$ *if and only if $\mathfrak{f}$ and $\mathfrak{f}_1$ are equivalent.*

(2.17.3) $\mathfrak{f}_1 \mapsto \{\mathfrak{f}_1 : \mathfrak{f}\}$ *gives a one-to-one correspondence between the branch of $\mathfrak{f}$ and the ideal class group in M.*

3. First two main theorems and their consequences

3.1. Throughout this section, F denotes a totally real algebraic number field of degree g, and B a quaternion algebra over F. We assume that $r = 1$, and τ_{01} is the identity mapping of F with the notation of 2.4. We let $\mathfrak{H}$ denote the upper half plane:

---

[1] Discussions of the same type have been made in Eichler [2] and Shimizu [6, 3.2].

$$\mathfrak{H} = \{z \in C \mid \mathrm{Im}(z) > 0\} \, .$$

For every maximal order $\mathfrak{o}$ in B and a two-sided integral $\mathfrak{o}$-ideal $\mathfrak{e}$, the group $\Gamma(\mathfrak{o}, \mathfrak{e})$ defined in 2.13 gives a properly discontinuous group of transformations on $\mathfrak{H}$, and $\mathfrak{H}/\Gamma(\mathfrak{o}, \mathfrak{e})$ is of finite measure. It is well known that $\mathfrak{H}/\Gamma(\mathfrak{o}, \mathfrak{e})$ is not compact if and only if B is isomorphic to $M_2(Q)$. (This fact is essentially proved in Fricke-Klein [3].) Now we state the first two main theorems of this paper, of which the proofs will be given in later sections.

3.2. MAIN THEOREM I. *Let $\mathfrak{o}$ be a maximal order in B, $\mathfrak{e}$ a two-sided integral $\mathfrak{o}$-ideal, and $\mathfrak{c} = \mathfrak{e} \cap \mathfrak{r}_F$. Then there exist a complete non-singular algebraic curve V, and a holomorphic mapping φ of $\mathfrak{H}$ into V satisfying the following conditions.*

(3.2.1) *V is defined over $C(F, \mathfrak{c})$.*

(3.2.2) *φ gives a biregular isomorphism of $\mathfrak{H}/\Gamma(\mathfrak{o}, \mathfrak{e})$ into V.*

(3.2.3) *Let $M \in J(B)$, and let f be an F-linear isomorphism of M into B such that $f(\mathfrak{r}_M) \subset \mathfrak{o}$. Let z be the fixed point of $f(M)$ on $\mathfrak{H}$, and $\mathfrak{e}_M = f^{-1}(\mathfrak{e} \cap f(\mathfrak{r}_M))$. Then $M(\varphi(z)) \cdot C(F, \mathfrak{c}) = C(M, \mathfrak{e}_M)$.*

As for the notation, see 1.1, 2.3, and 2.4. As for (3.2.3), see also 2.6 and 2.8. We note that $\mathfrak{e}_M = (\mathfrak{e}_M)^\rho$ for the non-trivial automorphism ρ of M over F, and $\mathfrak{c} = \mathfrak{r}_F \cap \mathfrak{e}_M$, $C(F, \mathfrak{c}) \subset C(M, \mathfrak{e}_M) \subset C(M, \mathfrak{c})$. If $\mathfrak{e} = \mathfrak{a}\mathfrak{o}$ with an integral ideal $\mathfrak{a}$ in F, then $\mathfrak{c} = \mathfrak{a}$ and $\mathfrak{e}_M = \mathfrak{a}\mathfrak{r}_M$. It should also be observed that φ is surjective unless $B = M_2(Q)$. If $B = M_2(Q)$, then one can show that the image $\varphi(\mathfrak{H})$ is a Zariski open subset of V rational over $C(F, \mathfrak{c})$ (see 9.11). We call (V, φ) a *canonical model for* $\mathfrak{H}/\Gamma(\mathfrak{o}, \mathfrak{e})$ if the above conditions (3.2.1–3) are satisfied. The uniqueness can be shown as follows.

3.3. THEOREM. *Let (V, φ) and (V', φ') be two canonical models for $\mathfrak{H}/\Gamma(\mathfrak{o}, \mathfrak{e})$. Then there exists a biregular morphism j of V onto V', defined over $C(F, \mathfrak{c})$, such that $\varphi' = j \circ \varphi$.*

Proof will be given in 7.12.

3.4. Let $\Delta = \{\mathfrak{o}_\lambda, \mathfrak{x}_{\lambda\mu} \, (\lambda, \mu = 1, \cdots, h_0)\}$ be a representative set of orders and ideals in B (2.13), and $\mathfrak{c}$ an integral ideal in F. Suppose that Δ is $\mathfrak{c}$-regular. Let $\mathfrak{e}_1$ be an integral two-sided $\mathfrak{o}_1$-ideal such that $\mathfrak{c} = \mathfrak{e}_1 \cap \mathfrak{r}_F$. Put $\mathfrak{e}_\lambda = \mathfrak{x}_{\lambda 1} \mathfrak{e}_1 \mathfrak{x}_{1\lambda}$ for each λ. Then $\mathfrak{o}_{\lambda\mathfrak{p}} = \mathfrak{o}_{1\mathfrak{p}}$ and $\mathfrak{e}_{\lambda\mathfrak{p}} = \mathfrak{e}_{1\mathfrak{p}}$ for all λ and all $\mathfrak{p}$ dividing $\mathfrak{c}$. Therefore an element α of B is prime to $\mathfrak{e}_\lambda$ (in the sense of 2.14) if and only if α is prime to $\mathfrak{e}_1$. Further $\alpha \equiv \beta \bmod^* \mathfrak{e}_\lambda$ if and only if $\alpha \equiv \beta \bmod^* \mathfrak{e}_1$. We shall denote by $U(\mathfrak{e})$ the set of all elements α of B^+ (see 2.13) prime to $\mathfrak{e}_1$, and put $U_0(\mathfrak{e}) = \{\alpha \in B^+ \mid \alpha \equiv 1 \bmod^* \mathfrak{e}_1\}$. Here we can take $\mathfrak{e}_\lambda$ in place of $\mathfrak{e}_1$ for any λ, for the reason mentioned above. Obviously $U_0(e)$ is a normal subgroup of $U(e)$. By (2.14.1), we see that $U(e)/U_0(e)$ is canonically isomorphic to $GL_1(o_\lambda/e_\lambda)$ for every λ.

3.5. MAIN THEOREM II. *Let the notation and assumption be as in 3.1 and 3.4. Then there exists a system $\{V_\lambda, \varphi_\lambda, R_\sigma^{\mu\lambda}(\alpha)\ (\lambda, \mu = 1, \cdots, h_0; \alpha \in U(e))\}$ formed by the objects satisfying the following conditions.*

(3.5.1) $(V_\lambda, \varphi_\lambda)$ is a canonical model for $\mathfrak{H}/\Gamma(o_\lambda, e_\lambda)$.

(3.5.2) $R_\sigma^{\mu\lambda}(\alpha)$ is defined for every $\alpha \in U(e)$ and for $\sigma = [C(F, c)/F, N_{B/F}(\alpha \mathfrak{x}_{\lambda\mu})]$, and is a biregular morphism of V_λ to V_μ^σ, rational over $C(F, c)$.

(3.5.3) $R_\sigma^{\mu\lambda}(\alpha) = R_\sigma^{\mu\lambda}(\beta)$ if $\alpha \equiv \beta \bmod^* e$, and $R_\tau^{\nu\mu}(\gamma)^\sigma \circ R_\sigma^{\mu\lambda}(\alpha) = R_{\tau\sigma}^{\nu\lambda}(\gamma\alpha)$.

(3.5.4) Let $M \in J(B)$ and let f be a normalized (see 2.7) F-linear isomorphism of M into B such that $f(\mathfrak{r}_M) \subset o_\mu$. Let z be the fixed point of $f(M)$ on $\mathfrak{H}$, and $e_M = f^{-1}(e_\mu \cap f(\mathfrak{r}_M))$. Let $\mathfrak{b}$ be an ideal in M, prime to c, and $\tau = [C(M, e_M)/M, \mathfrak{b}]$. Then one has $\alpha \mathfrak{x}_{\lambda\mu} = f(\mathfrak{b})o_\mu$ with a unique λ, and an element α of $U(e)$. With such an element α, one has

$$\varphi_\mu(z)^\tau = R_\sigma^{\mu\lambda}(\alpha)[\varphi_\lambda(\alpha^{-1}(z))]\ ,$$

where $\sigma = [C(F, c)/F, N_{B/F}(\alpha \mathfrak{x}_{\lambda\mu})]$.

In (3.5.4), we see easily that σ is the restriction of τ to $C(F, c)$. It should also be observed that *the ideal e_M does not depend on the choice of f and μ.* In fact, let f_1 be a normalized F-linear isomorphism of M into B such that $f_1(\mathfrak{r}_M) \subset o_\nu$. Let $\mathfrak{b} = f_1^{-1}(e_\nu \cap f_1(\mathfrak{r}_M))$. By 2.7, there exists an element β of B^+ such that $f_1(x) = \beta^{-1} f(x) \beta$. By 2.12 and 2.16, there exist an ideal $\mathfrak{y}$ in M and a two-sided o_μ-ideal $\mathfrak{Q}$ such that $\beta \mathfrak{x}_{\nu\mu} = f(\mathfrak{y})\mathfrak{Q}$. Then

$$e_\nu = \mathfrak{x}_{\nu\mu} e_\mu \mathfrak{x}_{\mu\nu} = \beta^{-1} f(\mathfrak{y}) \mathfrak{Q} e_\mu \mathfrak{Q}^{-1} f(\mathfrak{y}^{-1}) \beta$$
$$= \beta^{-1} f(\mathfrak{y}) e_\mu f(\mathfrak{y}^{-1}) \beta \supset \beta^{-1} f(\mathfrak{y} e_M \mathfrak{y}^{-1}) \beta = f_1(e_M)\ ,$$

hence $e_M \subset \mathfrak{b}$. Exchanging f for f_1, we get $\mathfrak{b} \subset e_M$, so that $\mathfrak{b} = e_M$.

The notation being as in (3.5.4), define f_α by $f_\alpha(x) = \alpha^{-1} f(x) \alpha$ for $x \in M$. Then we see easily that $f_\alpha(\mathfrak{r}_M) \subset o_\lambda$, and $\alpha^{-1}(z)$ is the fixed point of $f_\alpha(M)$.

3.6. THEOREM. *Let $\{\bar{V}_\lambda, \bar{\varphi}_\lambda, \bar{R}_\sigma^{\mu\lambda}(\alpha)\}$ be another system satisfying the conditions of 3.5. Then there exists, for each λ, a biregular morphism P_λ of V_λ to $\bar{V}_\lambda$, defined over $C(F, c)$, such that*

$$\bar{\varphi}_\lambda = P_\lambda \circ \varphi_\lambda\ , \qquad \bar{R}_\sigma^{\mu\lambda}(\alpha) = P_\mu^\sigma \circ R_\sigma^{\mu\lambda}(\alpha) \circ P_\lambda^{-1}\ .$$

PROOF. By 3.3, there exists a biregular morphism P_λ of V_λ to $\bar{V}_\lambda$, defined over $C(F, c)$, such that $\bar{\varphi}_\lambda = P_\lambda \circ \varphi_\lambda$. Let $\alpha \in U(e)$. Fix λ and μ. By 2.10, there exist infinitely many $M \in J(B)$, for each of which there exists an

F-linear isomorphism f of M into B and an ideal $\mathfrak{b}$ in M such that $f(\mathfrak{r}_M) \subset \mathfrak{o}_\mu$, and $\alpha \mathfrak{x}_{\lambda\mu} = f(\mathfrak{b})\mathfrak{o}_\mu$. Define σ, τ and z as in (3.5.4). Then we have

$$\varphi_\mu(z)^\tau = R_\sigma^{\mu\lambda}(\alpha)[\varphi_\lambda(\alpha^{-1}(z))] \, ,$$
$$P_\mu^\sigma(\varphi_\mu(z)^\tau) = \bar{R}_\sigma^{\mu\lambda}(\alpha)[P_\lambda \circ \varphi_\lambda(\alpha^{-1}(z))] \, .$$

Put $S = (P_\mu^\sigma)^{-1} \circ \bar{R}_\sigma^{\mu\lambda}(\alpha) \circ P_\lambda$. Then S is a biregular morphism of V_λ to V_μ^σ, and $S(\varphi_\lambda(\alpha^{-1}(z))) = R_\sigma^{\mu\lambda}(\alpha)[\varphi_\lambda(\alpha^{-1}(z))]$. If we take distinct M's, then the corresponding points $\varphi_\lambda(\alpha^{-1}(z))$ are obviously distinct. Hence we have $S(u) = R_\sigma^{\mu\lambda}(\alpha)(u)$ for infinitely many distinct points u, so that $S = R_\sigma^{\mu\lambda}(\alpha)$, which completes the proof.

3.7. PROPOSITION. *Let the notation be as in 3.5. If $\gamma \in \Gamma(\mathfrak{o}_\lambda, 1)$, then $R_1^{\lambda\lambda}(\gamma)$ is a biregular automorphism of V_λ obtained from the mapping $w \mapsto \gamma(w)$ on $\mathfrak{H}$, i.e., $R_1^{\lambda\lambda}(\gamma)[\varphi_\lambda(w)] = \varphi_\lambda(\gamma(w))$ for all $w \in \mathfrak{H}$.*

PROOF. In (3.5.4), let $\mathfrak{b} = \mathfrak{r}_M$, $\alpha = \gamma$, $\lambda = \mu$, and $\tau = $ identity map. Then we obtain $\varphi_\lambda(z) = R_1^{\lambda\lambda}(\gamma)[\varphi_\lambda(\gamma^{-1}(z))]$. Note that if we take distinct M's, then the corresponding points $\varphi_\lambda(z)$ on V_λ are distinct. Since there are infinitely many distinct M's, we obtain the equality of our proposition.

3.8. Let $\{\bar{\mathfrak{o}}_\lambda, \bar{\mathfrak{x}}_{\lambda\mu}\}$ be another representative set of orders and ideals in B, which is $\mathfrak{c}$-regular. In view of (2.13.8), by re-ordering the $\mathfrak{o}_\lambda$ if necessary, we can find elements η_λ of B^+ such that $\bar{\mathfrak{x}}_{\lambda\mu} = \eta_\lambda^{-1}\mathfrak{x}_{\lambda\mu}\eta_\mu$. Put $\bar{e}_\lambda = \eta_\lambda^{-1}e_\lambda\eta_\lambda$. Define $U(\bar{e})$ and $U_0(\bar{e})$ as in 3.4 with respect to the $\bar{e}_\lambda$. We see that

$$\Gamma(\bar{\mathfrak{o}}_\lambda, \bar{e}_\lambda) = \eta_\lambda^{-1}\Gamma(\mathfrak{o}_\lambda, e_\lambda)\eta_\lambda \, .$$

Let $\mathfrak{o}_\mathfrak{p} = \mathfrak{o}_{1\mathfrak{p}}$, and $\bar{\mathfrak{o}}_\mathfrak{p} = \bar{\mathfrak{o}}_{1\mathfrak{p}}$. Since $\bar{\mathfrak{x}}_{\lambda\mu} = \eta_\lambda^{-1}\mathfrak{x}_{\lambda\mu}\eta_\mu$, we have $\bar{\mathfrak{o}}_\mathfrak{p} = \eta_\lambda^{-1}\mathfrak{o}_\mathfrak{p}\eta_\mu$ for every λ, μ and every prime ideal $\mathfrak{p}$ dividing $\mathfrak{c}$. It follows that

$$\eta_\lambda^{-1}\eta_\mu \in U(\bar{e}) \, , \qquad \eta_\lambda\eta_\mu^{-1} \in U(e) \, ,$$
$$U(\bar{e}) = \eta_\lambda^{-1}U(e)\eta_\mu \, , \qquad U_0(\bar{e}) = \eta_\lambda^{-1}U_0(e)\eta_\lambda$$

for every λ and μ. Now put $\bar{\varphi}_\lambda(w) = \varphi_\lambda(\eta_\lambda(w))$ for $w \in \mathfrak{H}$, $\bar{R}_\sigma^{\mu\lambda}(\alpha) = R_\sigma^{\mu\lambda}(\eta_\mu\alpha\eta_\lambda^{-1})$ for $\alpha \in U(\bar{e})$ and $\sigma = [C(F, \mathfrak{c})/F, N_{B/F}(\alpha\bar{\mathfrak{x}}_{\lambda\mu})]$. Then it can be easily verified that $\{V_\lambda, \bar{\varphi}_\lambda, \bar{R}_\sigma^{\mu\lambda}(\alpha)\}$ satisfies the conditions (3.5.1–4) for $\{\bar{\mathfrak{o}}_\lambda, \bar{\mathfrak{x}}_{\lambda\mu}\}$. In this sense, the assertion of 3.5 is essentially independent of the choice of $\{\mathfrak{o}_\lambda, \mathfrak{x}_{\lambda\mu}\}$. In other words, to prove 3.5, it is sufficient to show the existence of $\{V_\lambda, \varphi_\lambda, R_\sigma^{\mu\lambda}(\alpha)\}$ only for a specially chosen $\{\mathfrak{o}_\lambda, \mathfrak{x}_{\lambda\mu}\}$.

3.9. THEOREM. *Let the notation and assumption be as in 3.5. The system $\{V_\lambda, \varphi_\lambda, R_\sigma^{\mu\lambda}(\alpha)\}$ can be chosen so that the following conditions are satisfied.*

(3.9.1) $V_\lambda = V_\mu^\sigma$ *if* $\sigma = [C(F, \mathfrak{c})/F, N_{B/F}(\mathfrak{x}_{\lambda\mu})]$.

(3.9.2) $R_\sigma^{\mu\lambda}(1)$ *is the identity mapping of* V_λ.

PROOF. Put $T_\lambda = R_\sigma^{1\lambda}(1)$, $W_\lambda = V_1^\sigma$ for $\sigma = [C(F, \mathfrak{c})/F, N_{B/F}(\mathfrak{x}_{\lambda 1})]$. Then T_λ is a biregular morphism of V_λ to W_λ, rational over $C(F, \mathfrak{c})$. Put $S_\tau^{\mu\lambda}(\alpha) = T_\mu^\tau \circ R_\tau^{\mu\lambda}(\alpha) \circ T_\lambda^{-1}$ with $\tau = [C(F, \mathfrak{c})/F, N_{B/F}(\alpha \mathfrak{x}_{\lambda \mu})]$, and $\psi_\lambda = T_\lambda \circ \varphi_\lambda$. Then it can be easily verified that $\{W_\lambda, \psi_\lambda, S_\tau^{\mu\lambda}(\alpha)\}$ satisfies the conditions (3.5.1–4). If $\rho = [C(F, \mathfrak{c})/F, N_{B/F}(\mathfrak{x}_{\lambda\mu})]$ and $\sigma = [C(F, \mathfrak{c})/F, N_{B/F}(\mathfrak{x}_{\lambda 1})]$, then

$$\sigma \rho^{-1} = [C(F, \mathfrak{c})/F, N_{B/F}(\mathfrak{x}_{\mu 1})] \,,$$

hence $W_\mu^\rho = V_1^{\sigma\rho^{-1}\rho} = V_1^\sigma = W_\lambda$, and

$$S_\rho^{\mu\lambda}(1) = T_\mu^\rho \circ R_\rho^{\mu\lambda}(1) \circ T_\lambda^{-1} = R_{\sigma\rho^{-1}}^{1\mu}(1)^\rho \circ R_\rho^{\mu\lambda}(1) \circ T_\lambda^{-1} = R_\sigma^{1\lambda}(1) \circ T_\lambda^{-1}$$
$$= \text{the identity mapping of } W_\lambda \,.$$

This proves our theorem.

3.10. REMARK. The notation being as in 3.5, suppose that $\mathfrak{c} = \mathfrak{r}_F$ and the conditions (3.9.1, 2) are satisfied. Then the relation (3.5.4) can be written in the form

(3.10.1) $\quad \varphi_\mu(z)^\tau = \varphi_\lambda(\alpha^{-1}(z))$, *where* $\tau = [C(M, 1)/M, \mathfrak{b}]$, $\alpha \mathfrak{x}_{\lambda\mu} = f(\mathfrak{b})\mathfrak{o}_\mu$.

Let us fix an $M \in J(B)$. Let $\mathfrak{K}$ be a branch of optimal embeddings of M into B in the sense of 2.16. Suppose that for at least one of $\mathfrak{f} = (f, \mathfrak{o}_\mu) \in \mathfrak{K}$, f is normalized. Then this is so for *every* $\mathfrak{f} \in \mathfrak{K}$. Now take an element ξ of $C(F, 1)$ which generates $C(F, 1)$ over F. For each $\mathfrak{f} = (f, \mathfrak{o}_\mu)$, put $u(\mathfrak{f}) = (\varphi_\mu(z), \xi^\rho)$ with the fixed point z of $f(M)$ on $\mathfrak{H}$ and $\rho = [C(F, 1)/F, N(\mathfrak{x}_{\mu 1})]$. Then the following three assertions hold.

(3.10.2) *For every* $\mathfrak{f} \in \mathfrak{K}$, *$u(\mathfrak{f})$ generates $C(M, 1)$ over M.*

(3.10.3) *Let* $\mathfrak{f}, \mathfrak{f}_1 \in \mathfrak{K}$. *If $\mathfrak{b}$ is an ideal in M and $\{\mathfrak{f}_1 : \mathfrak{f}\}$ is the ideal class of $\mathfrak{b}$, then $u(\mathfrak{f})^\tau = u(\mathfrak{f}_1)$ with $\tau = [C(M, 1)/M, \mathfrak{b}]$.*

(3.10.4) *The $u(\mathfrak{f})$ for $\mathfrak{f} \in \mathfrak{K}$ form a complete set of conjugates of each $u(\mathfrak{f})$ over F.*

The first assertion is simply a reformulation of (3.2.3) with $\mathfrak{e} = \mathfrak{o}$. From (3.10.1), 2.16, and 2.17, we obtain easily (3.10.3). The last (3.10.4) follows from (3.10.2) and (3.10.3).

3.11. The field $C(F, \mathfrak{c})$ is not necessarily the field of moduli of the curve V_λ. In [12, 4.3 and 4.9], it has been shown that the field of moduli of V_λ (or the bottom field of V_λ, cf. [12, 2.1]) is contained in a certain unramified class field $C^*(F, B)$ over F, whose description will be given in 3.12 below. Actually one can derive [12, 4.3 and 4.9] in the one-dimensional case from the above 3.5. In a certain case, one can show that V_λ can actually be chosen so as to be defined over $C^*(F, B)$. However, for the question of this type, it is more natural to consider a somewhat bigger group

$$\Gamma^*(\mathfrak{o}_\lambda, \mathfrak{e}_\lambda) = \{\alpha \in U(\mathfrak{e}) \mid \alpha\mathfrak{o}_\lambda = \mathfrak{o}_\lambda\alpha\}$$

in place of $\Gamma(\mathfrak{o}_\lambda, \mathfrak{e}_\lambda)$. We shall now discuss the problem of constructing a canonical model for $\mathfrak{H}/\Gamma^*(\mathfrak{o}_\lambda, \mathfrak{e}_\lambda)$ defined over $C^*(F, B)$ in the special case $\mathfrak{e}_\lambda = \mathfrak{o}_\lambda$.

3.12. Let $\mathfrak{o}$ be a maximal order in B, and L_0 the subgroup of $I(F, 1)$ consisting of the $N_{B/F}(\mathfrak{y})$ for all two-sided $\mathfrak{o}$-ideals $\mathfrak{y}$. If $\mathfrak{p}_1, \cdots, \mathfrak{p}_s$ are the prime ideals in F which are ramified in B, then L_0 is generated by $\mathfrak{p}_1, \cdots, \mathfrak{p}_s$ and the squares of all the ideals in F. Therefore L_0 does not depend on the choice of $\mathfrak{o}$. Let $L_1 = L_0 \cap P(F, \mathfrak{u}_0)$, and let L_2 be the group of all principal ideals of the form (a^2) with $a \in F$. Then we see easily that $[L_1 : L_2]$ is finite. Define a subgroup $\Gamma^*(\mathfrak{o})$ of B^+ by

$$(3.12.1) \qquad \Gamma^*(\mathfrak{o}) = \{\alpha \in B^+ \mid \alpha\mathfrak{o} = \mathfrak{o}\alpha\} \ .$$

If we denote by F^* the group of non-zere elements in F, then

$$(3.12.2) \qquad [\Gamma^*(\mathfrak{o}) : F^* \cdot \Gamma(\mathfrak{o}, 1)] = [L_1 : L_2] \ .$$

In fact, consider a map $\Gamma^*(\mathfrak{o}) \ni \alpha \mapsto N_{B/F}(\alpha)\mathfrak{r}_F \in L_1$, which is surjective in view of Eichler's theorem. If $N_{B/F}(\alpha)\mathfrak{r}_F \in L_2$, then there exists an element a of F such that $N_{B/F}(\alpha)\mathfrak{r}_F = a^2\mathfrak{r}_F$, hence $N_{B/F}(a^{-1}\alpha\mathfrak{o}) = \mathfrak{r}_F$. Since $a^{-1}\alpha\mathfrak{o}$ is a two-sided $\mathfrak{o}$-ideal, we have $a^{-1}\alpha\mathfrak{o} = \mathfrak{o}$, so that $a^{-1}\alpha = \gamma$ with a unit γ of $\mathfrak{o}$. Since $\alpha \in B^+$, we have $\gamma \in \Gamma(\mathfrak{o}, 1)$, hence $\alpha = a\gamma \in \Gamma(\mathfrak{o}, 1) \cdot F^*$. This proves (3.12.2).

We observe that $\Gamma^*(\mathfrak{o})$, as a group of transformations on $\mathfrak{H}$, defines a fuchsian group.

Now define a subgroup $I^*(F, B)$ of $I(F, 1)$ by $I^*(F, B) = L_0 \cdot P(F, \mathfrak{u}_0)$. Let $C^*(F, B)$ be the class field over F corresponding to $I^*(F, B)$. (In [12], we have written $I(B/F)$ and $k(B)$ for $I^*(F, B)$ and $C^*(F, B)$ with D in place of B.) For every $M \in J(B)$, let $I^*(M, B)$ denote the subgroup of $I(M, 1)$ generated by $P(M, 1)$, $I(F, 1)$, and the prime factors of $D(B/F)$ in M. Further let $C^*(M, B)$ denote the class field over M corresponding to $I^*(M, B)$. If $\mathfrak{a} \in I^*(M, B)$, then $N_{M/F}(\mathfrak{a}) \in I^*(F, B)$, hence $C^*(F, B) \subset C^*(M, B)$.

3.13. THEOREM. *Let the notation be as in* 3.12. *Then, for every maximal order $\mathfrak{o}$ in B, there exists a complete non-singular curve W, and a holomorphic mapping ψ of $\mathfrak{H}$ into W satisfying the following conditions.*

(3.13.1) W is defined over $C^*(F, B)$.

(3.13.2) ψ gives a biregular isomorphism of $\mathfrak{H}/\Gamma^*(\mathfrak{o})$ into W.

(3.13.3) Let $M \in J(B)$ and let f be an F-linear isomorphism of M into B such that $f(\mathfrak{r}_M) \subset \mathfrak{o}$. Let z be the fixed point of $f(M)$ on $\mathfrak{H}$. Then

$$M\big(\psi(z)\big) \cdot C^*(F, B) = C^*(M, B) \ .$$

The proof will be given in the following 3.14–16.

3.14. Let $\{V_\lambda, \varphi_\lambda, R_\sigma^{\mu\lambda}(\alpha)\}$ be as in 3.5 with $e_\lambda = o_\lambda$. We normalize it as in 3.9. For each μ and λ, let

$$(3.14.1) \qquad\qquad \Theta_{\mu\lambda} = \{\beta \in B^+ \mid \beta o_\lambda = o_\mu \beta\} \ .$$

Obviously $\Theta_{\lambda\lambda} = \Gamma^*(o_\lambda)$. For the sake of simplicity, we put $\Gamma_\lambda = \Gamma(o_\lambda, 1)$ and $\Gamma_\lambda^* = \Gamma^*(o_\lambda)$. Now we have

$(3.14.2)$ $\Theta_{\mu\lambda} \neq \varnothing$ *if and only if* $N_{B/F}(\mathfrak{x}_{\lambda\mu}) \in I^*(F, B)$.

In fact if $\beta \in \Theta_{\mu\lambda}$, then one has $\beta o_\lambda = o_\mu \beta$, so that $\beta \mathfrak{x}_{\lambda\mu}$ is a two-sided o_μ-ideal. Hence

$$N_{B/F}(\mathfrak{x}_{\lambda\mu}) = N_{B/F}(\beta^{-1})N_{B/F}(\beta \mathfrak{x}_{\lambda\mu}) \in P(F, \mathfrak{u}_0)\cdot L_0 \ .$$

Conversely, if $N_{B/F}(\mathfrak{x}_{\lambda\mu}) \in I^*(F, B)$, then there exists a two-sided o_μ-ideal q such that $N_{B/F}(\mathfrak{x}_{\lambda\mu})N_{B/F}(\mathfrak{q})^{-1} \in P(F, \mathfrak{u}_0)$. Therefore, by Eichler's theorem, $\mathfrak{x}_{\lambda\mu}\mathfrak{q}^{-1} = \beta o_\mu$ with an element β of B^+. Then $o_\lambda = \beta o_\mu \beta^{-1}$, which proves (3.14.2).

In the next place we shall show

$(3.14.3)$ *For every* $\beta \in \Theta_{\mu\lambda}$*, there exists a biregular morphism* $S^{\mu\lambda}(\beta)$ *of* V_μ *to* V_λ*, defined over* $C(F, 1)$ *such that* $S^{\mu\lambda}(\beta)(\varphi_\lambda(z)) = \varphi_\mu(\beta(z))$.

To show this, define $\varphi'\colon \mathfrak{H} \to V_\mu$ by $\varphi'(z) = \varphi_\mu(\beta(z))$. We see easily that (V_μ, φ') is a canonical model for $\mathfrak{H}/\Gamma_\lambda$, hence 3.3 shows the existence of $S^{\mu\lambda}(\beta)$ as described in (3.14.3).

$(3.14.4)$ *If* $\beta \in \Theta_{\mu\lambda}$ *and* $\gamma \in \Theta_{\nu\mu}$*, then* $\gamma\beta \in \Theta_{\nu\lambda}$ *and* $S^{\nu\mu}(\gamma) \circ S^{\mu\lambda}(\beta) = S^{\nu\lambda}(\gamma\beta)$.

This can be verified in a straightforward way.

$(3.14.5)$ *Let* $\sigma = [C(F, 1)/F, N_{B/F}(\mathfrak{x}_{\lambda\nu})] = [C(F, 1)/F, N_{B/F}(\mathfrak{x}_{\mu\kappa})]$ *and* $\beta \in \Theta_{\mu\lambda}$*. Then there exists an element* ε *of* $\Theta_{\kappa\nu}$ *such that* $\beta \mathfrak{x}_{\lambda\nu} = \mathfrak{x}_{\mu\kappa}\varepsilon$*. With such an element* ε*, one has* $S^{\kappa\nu}(\varepsilon)^\sigma = S^{\mu\lambda}(\beta)$.

To prove this, we first observe that $\mathfrak{x}_{\kappa\mu}\beta \mathfrak{x}_{\lambda\nu}$ is a right o_ν-ideal, and $N_{B/F}(\mathfrak{x}_{\kappa\mu}\beta \mathfrak{x}_{\lambda\nu}) \in P(F, \mathfrak{u}_0)$. Hence there exists an element ε of B^+ such that $\mathfrak{x}_{\kappa\mu}\beta \mathfrak{x}_{\lambda\nu} = \varepsilon o_\nu$. Then $\varepsilon \in \Theta_{\kappa\nu}$. Let M, f, z, τ and $\mathfrak{b}$ be as in (3.5.4) with β, λ, and ν in place of α, λ, and μ. By 2.10, such M and $\mathfrak{b}$ exist. Since we have normalized $\{V_\lambda, \varphi_\lambda, R_\sigma^{\mu\lambda}(\alpha)\}$ as in 3.9, we have $V_\nu^\sigma = V_\lambda$ and $\varphi_\nu(z)^\tau = \varphi_\lambda(\beta^{-1}(z))$. Define $f_1\colon M \to B$ by $f_1(x) = \varepsilon f(x)\varepsilon^{-1}$ for $x \in M$. Then $f_1(\mathfrak{r}_M) \subset o_\kappa$ and $f_1(\mathfrak{b})o_\kappa = \varepsilon f(\mathfrak{b})o_\nu\varepsilon^{-1} = \varepsilon\beta \mathfrak{x}_{\lambda\nu}\varepsilon^{-1} = \varepsilon \mathfrak{x}_{\mu\kappa}$. Since $\varepsilon(z)$ is the fixed point of $f_1(M)$, we have $\varphi_\kappa(\varepsilon(z))^\tau = \varphi_\mu(z)$ by (3.5.4). Hence

$$\begin{aligned}
S^{\mu\lambda}(\beta)\big(\varphi_\lambda(\beta^{-1}(z))\big) &= \varphi_\mu(z) = \varphi_\kappa(\varepsilon(z))^\tau \\
&= S^{\kappa\nu}(\varepsilon)^\sigma\big(\varphi_\nu(z)^\tau\big) = S^{\kappa\nu}(\varepsilon)^\sigma\big(\varphi_\lambda(\beta^{-1}(z))\big) \ .
\end{aligned}$$

Since there are infinitely many distinct M's (2.10), there are infinitely many distinct points of the form $\varphi_\lambda(\beta^{-1}(z))$. Hence we have $S^{\mu\lambda}(\beta) = S^{\kappa\nu}(\varepsilon)^\sigma$.

Specialize (3.14.5) to the case $\lambda = \mu$ and $\nu = \kappa$. Then

$$(3.14.6) \quad S^{\nu\nu}(\varepsilon)^\sigma = S^{\lambda\lambda}(\beta) \quad if \quad \sigma = [C(F, 1)/F, N(\mathfrak{x}_{\lambda\nu})], \ \beta \in \Gamma_\lambda^*, \ \varepsilon \in \Gamma_\nu^* \quad and$$

$\beta\mathfrak{x}_{\lambda\nu} = \mathfrak{x}_{\lambda\nu}\varepsilon.$

3.15. We see that $\Gamma_\mu^* \ni \varepsilon \mapsto S^{\mu\mu}(\varepsilon)$ gives an isomorphism of $\Gamma_\mu^*/(F^* \cdot \Gamma_\mu)$ onto a group A_μ of biregular automorphisms of V_μ. Since these automorphisms are rational over $C(F, 1)$ (3.14.3), we can construct the quotient V_μ^* of V_μ by A_μ so that V_μ^* and the natural projection P_μ of V_μ to V_μ^* are rational over $C(F,1)$. If $\sigma = [C(F, 1)/F, N_{B/F}(\mathfrak{x}_{\lambda\mu})]$, we have $V_\mu^\sigma = V_\lambda$ (3.9.1), and $A_\mu^\sigma = A_\lambda$ in view of (3.14.6). Therefore, we can define the V_μ^* and P_μ for $\mu = 1, \cdots, h_0$ in such a way that $V_\mu^{*\sigma} = V_\lambda^*$ and $P_\mu^\sigma = P_\lambda$ for $\sigma = [C(F, 1)/F, N_{B/F}(\mathfrak{x}_{\lambda\mu})]$.

Now suppose that $N_{B/F}(\mathfrak{x}_{\lambda\mu}) \in I^*(F, B)$. Let $\sigma = [C(F, 1)/F, N_{B/F}(\mathfrak{x}_{\lambda\mu})]$ and $\beta \in \Theta_{\lambda\mu}$. Since $V_\mu^\sigma = V_\lambda$, $S^{\lambda\mu}(\beta)$ gives a biregular morphism of V_μ to V_μ^σ. By (3.14.4) and (3.14.6), we have $S^{\lambda\mu}(\beta)A_\mu S^{\lambda\mu}(\beta)^{-1} = A_\lambda = A_\mu^\sigma$. Hence there exists a biregular morphism T_σ^μ of V_μ^* to $V_\mu^{*\sigma}$ such that

$$(3.15.1) \qquad\qquad T_\sigma^\mu \circ P_\mu = P_\mu^\sigma \circ S^{\lambda\mu}(\beta) \ .$$

(Note also that $P_\mu^\sigma = P_\lambda$.) Let α be another element of $\Theta_{\lambda\mu}$ and let $\delta = \beta^{-1}\alpha$. Then $\delta \in \Gamma_\mu^*$, hence

$$P_\mu^\sigma \circ S^{\lambda\mu}(\alpha) = P_\mu^\sigma \circ S^{\lambda\mu}(\beta\delta)$$
$$= P_\mu^\sigma \circ S^{\lambda\mu}(\beta) \circ S^{\mu\mu}(\delta) = T_\sigma^\mu \circ P_\mu \circ S^{\mu\mu}(\delta) = T_\sigma^\mu \circ P_\mu \ .$$

This shows that T_σ^μ does not depend on the choice of β.

Suppose that $N_{B/F}(\mathfrak{x}_{\mu 1}) \in I^*(F, B)$. Let $\tau = [C(F,1)/F, N_{B/F}(\mathfrak{x}_{\mu 1})]$ and $\varepsilon \in \Theta_{\mu 1}$. Then with the above σ and β, we have $\beta\varepsilon \in \Theta_{\lambda 1}$ and $\sigma\tau = [C(F,1)/F, N_{B/F}(\mathfrak{x}_{\lambda 1})]$. Let ν be an index such that $\sigma = [C(F, 1)/F, N_{B/F}(\mathfrak{x}_{\nu 1})]$. Then

$$\tau = \sigma\tau\sigma^{-1} = [C(F, 1)/F, N_{B/F}(\mathfrak{x}_{\lambda\nu})] \ .$$

Hence, by (3.14.5), we have $S^{\lambda\mu}(\beta) = S^{\nu 1}(\delta)^\tau$ with an element δ of $\Theta_{\nu 1}$. Therefore

$$T_{\sigma\tau}^1 \circ P_1 = P_1^{\sigma\tau} \circ S^{\lambda 1}(\beta\varepsilon)$$
$$= P_1^{\sigma\tau} \circ S^{\lambda\mu}(\beta) \circ S^{\mu 1}(\varepsilon) = P_1^{\sigma\tau} \circ S^{\nu 1}(\delta)^\tau \circ S^{\mu 1}(\varepsilon)$$
$$= (T_\sigma^1 \circ P_1)^\tau \circ S^{\mu 1}(\varepsilon) = (T_\sigma^1)^\tau \circ T_\tau^1 \circ P_1$$

by successive application of (3.15.1). Hence we obtain

$$(3.15.2) \qquad\qquad T_{\sigma\tau}^1 = (T_\sigma^1)^\tau \circ T_\tau^1 \ .$$

Let G denote the Galois group of $C(F, 1)$ over F, and N the subgroup of G corresponding to $C^*(F, B)$. Obviously N consists of the $[C(F, 1)/F, N_{B/F}(\mathfrak{x}_{\lambda\mu})]$ with $N_{B/F}(\mathfrak{x}_{\lambda\mu}) \in I^*(F, B)$. The formula (3.15.2) has been proved for every σ and τ in N.

51

Now apply Weil's criterion [18] to V_1^* and T_σ^1. Then we obtain a curve W_1, defined over $C^*(F, B)$, and a biregular morphism T of W_1 to V_1^*, defined over $C(F, 1)$, such that

$$(3.15.3) \qquad\qquad T_\sigma^1 = T^\sigma \circ T^{-1} \qquad\qquad \textit{for every } \sigma \in N .$$

Put $W_\mu = W_1^\tau$ and $Q_\mu = (T^{-1} \circ P_1)^\tau$ for $\tau = [C(F, 1)/F, N_{B/F}(\mathfrak{x}_{\mu 1})]$. Then $W_\mu^\rho = W_\nu$ and $Q_\mu^\rho = Q_\nu$ if $\rho = [C(F, 1)/F, N(\mathfrak{x}_{\nu\mu})]$. Now we have

$$(3.15.4)\quad W_\mu^\sigma = W_\mu \ and \ Q_\mu^\sigma \circ S^{\lambda\mu}(\beta) = Q_\mu \ if \ \sigma = [C(F, 1)/F, N_{B/F}(\mathfrak{x}_{\lambda\mu})] \in N$$
$and \ \beta \in \Theta_{\lambda\mu}.$

To see this, let κ be an index such that $\sigma = [C(F, 1)/F, N_{B/F}(\mathfrak{x}_{\kappa 1})]$. By (3.14.5), $S^{\lambda\mu}(\beta) = S^{\kappa 1}(\delta)^\tau$ with an element δ of $\Theta_{\kappa 1}$ and

$$\tau = [C(F, 1)/F, N_{B/F}(\mathfrak{x}_{\mu 1})] .$$

Then

$$Q_1^\sigma \circ S^{\kappa 1}(\delta) = (T^{-1})^\sigma \circ P_1^\sigma \circ S^{\kappa 1}(\delta) = (T^{-1})^\sigma \circ T_\sigma^1 \circ P_1 = T^{-1} \circ P_1 = Q_1 .$$

Hence

$$Q_\mu^\sigma \circ S^{\lambda\mu}(\beta) = Q_1^{\tau\sigma} \circ S^{\kappa 1}(\delta)^\tau = [Q_1^\sigma \circ S^{\kappa 1}(\delta)]^\tau = Q_1^\tau = Q_\mu , \qquad\qquad \text{q.e.d.}$$

3.16. Define $\psi_\mu : \mathfrak{H} \to W_\mu$ by $\psi_\mu(z) = Q_\mu(\varphi_\mu(z))$. Then ψ_μ gives a biregular mapping of $\mathfrak{H}/\Gamma_\mu^*$ into V_μ^*. We shall now prove that (W_μ, ψ_μ) has the property (3.13.3) with $\mathfrak{o} = \mathfrak{o}_\mu$. Let $M, f, z, \mathfrak{b}, \alpha, \sigma,$ and τ be as in (3.5.4). Then we have

$$(3.16.1) \qquad\qquad \psi_\mu(z)^\tau = \psi_\lambda(\alpha^{-1}(z)) .$$

In fact,

$$\psi_\mu(z)^\tau = Q_\mu^\sigma(\varphi_\mu(z)^\tau) = Q_\lambda(\varphi_\lambda(\alpha^{-1}(z))) = \psi_\lambda(\alpha^{-1}(z)) .$$

Now suppose that τ is the identity mapping on $M(\psi_\mu(z)) \cdot C^*(F, B)$. Then $\sigma \in N$ and $\psi_\mu(z)^\tau = \psi_\mu(z)$. Therefore $W_\lambda = W_\mu^\sigma = W_\mu$. Let β be an arbitrary element of $\Theta_{\lambda\mu}$. By (3.15.4), we have

$$\psi_\mu(z) = Q_\mu(\varphi_\mu(z))$$
$$= Q_\mu^\sigma \circ S^{\lambda\mu}(\beta)(\varphi_\mu(z)) = Q_\lambda \circ \varphi_\lambda(\beta(z)) = \psi_\lambda(\beta(z)) ,$$

so that $\psi_\lambda(\beta(z)) = \psi_\lambda(\alpha^{-1}(z))$ by (3.16.1). Hence there exists an element γ of Γ_λ^* such that $\gamma\beta(z) = \alpha^{-1}(z)$. By 2.6, there exists an element y of M such that $\alpha\gamma\beta = f(y)$. It follows that

$$f(y^{-1}\mathfrak{b})\mathfrak{o}_\mu = f(y^{-1})\alpha\mathfrak{x}_{\lambda\mu} = (\gamma\beta)^{-1}\mathfrak{x}_{\lambda\mu} .$$

Since $(\gamma\beta)^{-1} \in \Theta_{\mu\lambda}$, we see that $f(y^{-1}\mathfrak{b})\mathfrak{o}_\mu$ is a two-sided $\mathfrak{o}_\mu$-ideal. Now it can be easily verified that

$(3.16.2)$ *An ideal $\mathfrak{a}$ in M belongs to $I^*(M, B)$, if and only if there exists an element x in M such that $f(x\mathfrak{a})\mathfrak{o}_\mu$ is a two-sided $\mathfrak{o}_\mu$-ideal.*

Therefore $\mathfrak{b} \in I^*(M, B)$. This shows that τ is the identity mapping on $C^*(M, B)$.

Conversely suppose that τ is the identity mapping on $C^*(M, B)$. Then $\sigma \in N$ and $\mathfrak{b} \in I^*(M, B)$. By (3.16.2), there exists an element u in M such that $f(u^{-1}\mathfrak{b})\mathfrak{o}_\mu$ is a two-sided $\mathfrak{o}_\mu$-ideal. Let ε be an element of B^+ such that $f(u^{-1}\mathfrak{b})\mathfrak{o}_\mu = \varepsilon\mathfrak{x}_{\lambda\mu}$. Then $\varepsilon \in \Theta_{\mu\lambda}$. We have (3.16.1) with $f(u)\varepsilon$ as α. Therefore we have

$$\psi_\mu(z)^\tau = \psi_\lambda\big(\alpha^{-1}(z)\big)$$
$$= \psi_\lambda\big(\varepsilon^{-1}(z)\big) = Q_\lambda \circ S^{\lambda\mu}(\varepsilon^{-1})\big(\varphi_\mu(z)\big) = Q_\mu\big(\varphi_\mu(z)\big) = \psi_\mu(z) ,$$

so that τ is the identity mapping on $C^*(F, B) \cdot M(\psi_\mu(z))$. This completes the proof of 3.13.

3.17. Theorem. *The notation being as in 3.5, let $M \in J(B)$. Let f be an F-linear isomorphism of M into B, and w the fixed point of $f(M)$. Let τ be an automorphism of C over F such that $\tau = [C(F, \mathfrak{c})/F, N_{B/F}(\alpha\mathfrak{x}_{\lambda\mu})]$ on $C(F, \mathfrak{c})$ with $\alpha \in U(\mathfrak{e})$. Then there exists an F-linear isomorphism f_1 of M into B such that $\varphi_\mu(w)^\tau = R_\sigma^{\mu\lambda}(\alpha)[\varphi_\lambda(w_1)]$ with the fixed point w_1 of $f_1(M)$ and $f^{-1}[f(M) \cap \mathfrak{o}_\mu] = f_1^{-1}[f_1(M) \cap \mathfrak{o}_\lambda].$*

Here we do not assume $f(\mathfrak{r}_M) \subset \mathfrak{o}_\mu$. One should also notice that τ may not be the identity mapping on M. Proof will be given in 8.9 and 9.11.

3.18. *Examples.* Let E_0, E_1, $\Gamma(\mathfrak{o})$ and $\Gamma(\mathfrak{o}, 1)$ be as in 2.13. By 2.15, we have $[\Gamma(\mathfrak{o}) : \Gamma(\mathfrak{o}, 1)] = [E_1 : E_0]$. Therefore, from the formula of Shimizu [6, p. 192, (5)], we obtain, with $\Gamma = \Gamma(\mathfrak{o}, 1)$,

$$(3.18.1) \quad \int_{\mathfrak{H}/\Gamma} y^{-2}dxdy$$
$$= 2^{3-3g}\pi^{1-2g}[E : E_0]D(F/\mathbf{Q})^{3/2}\zeta_F(2) \prod_{\mathfrak{p}|D(B/F)} \big(N(\mathfrak{p}) - 1\big) .$$

Here $z = x + iy$ is the variable on $\mathfrak{H}$; E denotes the group of *all* units in F, and ζ_F the Dedekind zeta function of F. (The formula of Shimizu concerns the measure of $\mathfrak{F}_1/\Gamma(\mathfrak{o})$. See also the remark at [6, p. 192, 5].) On the other hand, if $\mathfrak{g}$ is the genus of $\mathfrak{H}/\Gamma$ and $e_1, \cdots, e_t$ are the orders of inequivalent elliptic elements in Γ (modulo the center), then one has a well known formula

$$(3.18.2) \qquad (2\pi)^{-1}\int_{\mathfrak{H}/\Gamma} y^{-2}dxdy = 2\mathfrak{g} - 2 + \sum_{i=1}^{t} (e_i - 1)/e_i .$$

Here we assume that $\mathfrak{H}/\Gamma$ is compact. The numbers e_i can be determined by investigating the embedding of orders of quadratic extensions of F into $\mathfrak{o}$, cf. Eichler [2]. (In a certain case, the value of (3.18.2) determines $\mathfrak{g}$ and the e_i uniquely.) The following table describes a few examples of $\Gamma^*(\mathfrak{o})$ (defined by (3.12.1)) which are triangle groups, i.e., groups of genus 0 with exactly three inequivalent elliptic elements.

F	$N_{F/Q}(D(B/F))$	$[\Gamma^*(\mathfrak{o}) : F^* \cdot \Gamma(\mathfrak{o}, 1)]$	$\{e_i\}$ for $\Gamma^*(\mathfrak{o})$
$Q(\sqrt{2})$	2	2	$\{2, 3, \ 8\}$
$Q(\sqrt{3})$	2	1	$\{2, 3, 12\}$
$Q(\sqrt{3})$	3	1	$\{2, 4, 12\}$
$Q(\sqrt{5})$	4	2	$\{2, 4, \ 5\}$
$Q(\sqrt{5})$	5	2	$\{2, 3, 10\}$
$Q(\sqrt{5})$	9	2	$\{2, 5, \ 6\}$
F_7	1	1	$\{2, 3, \ 7\}$
F_9	1	1	$\{2, 3, \ 9\}$
F_{11}	1	1	$\{2, 3, 11\}$
F_{16}	2	2	$\{2, 3, 16\}$
F_{24}	2	1	$\{2, 3, 24\}$
F_{30}	5	1	$\{2, 3, 30\}$

Here $F_m = Q(\zeta + \zeta^{-1})$ with $\zeta = e^{2\pi i/m}$ (cf. Fricke-Klein [3, pp. 606–621]). A few more examples with higher genera will be given in 12.4.

Now let us confine ourselves to the cases $F = F_m$ with $m = 7, 9, 11$. Then $h_0 = 1$, and $\Gamma(\mathfrak{o}, 1)$ modulo the center is a triangle group with distinct indices e_1, e_2, e_3. If u_1, u_2, u_3 are the representatives of the elliptic points, there exists a holomorphic mapping ω of $\mathfrak{H}$ onto the Riemann sphere S such that the $\omega(u_i)$ are all rational over Q, and gives a biregular morphism of $\mathfrak{H}/\Gamma(\mathfrak{o}, 1)$ onto S. (For example, one may put $\omega(u_i) = 0, 1, \infty$ for $i = 1, 2, 3$.) Regarding S as a curve of genus 0, let us now prove

(3.18.3) (S, ω) *is a canonical model for* $\mathfrak{H}/\Gamma(\mathfrak{o}, 1)$.

Let (V, φ) be as in 3.2. Obviously, there exists a biregular morphism q of V to S such that $q \circ \varphi = \omega$. The assertion (3.18.3) will be proved if we show that q is defined over F. (Since $h_0 = 1$, we have $C(F, 1) = F$.) Let τ be an automorphism of C over F. By 3.17, we see that $\varphi(u_i)^\tau = \varphi(w)$ with an elliptic point w whose order is the same as u_i. Since the e_j are distinct, w should be equivalent to u_i with respect to $\Gamma(\mathfrak{o}, 1)$, hence $\varphi(u_i)^\tau = \varphi(u_i)$. This shows that the $\varphi(u_i)$ are all rational over F. Therefore we have

$$q^\tau\big(\varphi(u_i)\big) = \omega(u_i)^\tau = \omega(u_i) = q\big(\varphi(u_i)\big)$$

for every i, so that $q^\tau = q$. This completes the proof of (3.18.3).

Since $D(B/F) = (1)$ in these three cases, *every* totally imaginary quadratic extension M of F can be embedded in B. For such an M, by 2.8, there exists an F-linear isomorphism f of M into B such that $f(\mathfrak{r}_M) \subset \mathfrak{o}$, where $\mathfrak{o}$ is a fixed maximal order in B. Let z be the fixed point of $f(M)$ on $\mathfrak{H}$. If h_M denotes the class-number of M, then there are exactly h_M such points z_i $(i = 1, \cdots, h_M)$

which are inequivalent modulo $\Gamma(\mathfrak{o}, 1)$. This follows easily from 2.17. By 2.17 and (3.5.4), we obtain the following assertion (cf. also 3.10).

(3.18.4) *The $\omega(z_i)$ for $i = 1, \cdots, h_M$ form a complete set of conjugates of $\omega(z_1)$ over F. For each i, $M(\omega(z_i))$ is the absolute class field $C(M, 1)$ over M.*

Thus the triangle function ω gives an analogue of the classical modular function $J(z)$ for the totally real algebraic number field F_m ($m = 7, 9, 11$).

3.19. REMARK. The case $F = F_7$ of the above table yields a famous group with the smallest measure of fundamental domain. Moreover, in this case, if Γ' is a normal subgroup of $\Gamma = \Gamma(\mathfrak{o}, 1)$ without elliptic elements, then $\mathfrak{H}/\Gamma'$ is a compact Riemann surface with $84(\mathfrak{g}' - 1)$ holomorphic automorphisms, $\mathfrak{g}'$ being the genus of Γ'. In fact $\Gamma/\Gamma'E$ is the group of automorphisms and $[\Gamma : \Gamma'E] = 84(\mathfrak{g}' - 1)$ by (3.18.2). (The converse is true: if Γ' is a fuchsian group such that $\mathfrak{H}/\Gamma'$ has $84(\mathfrak{g}' - 1)$ holomorphic automorphisms, then Γ' is conjugate to a normal subgroup of Γ modulo the center.) For example, one can take $\Gamma(\mathfrak{o}, \mathfrak{c})$ as Γ' for an integral ideal $\mathfrak{c}$ in F. The index $[\Gamma : E\Gamma']$ can be computed as follows. First we note that $E_0 = \{\varepsilon^2 \mid \varepsilon \in E\}$ in the present case. Therefore, if we put $\Gamma_1 = \{\gamma \in \mathfrak{o} \mid N_{B/F}(\gamma) = 1\}$ and $\Gamma_{\mathfrak{c}} = \Gamma_1 \cap \Gamma(\mathfrak{o}, \mathfrak{c})$ then $\Gamma = \Gamma_1 E$, hence

$$[\Gamma : \Gamma'E] = [\Gamma_1 : \Gamma_1 \cap \Gamma'E] = [\Gamma_1 : \Gamma_{\mathfrak{c}}]/[\Gamma_1 \cap \Gamma'E : \Gamma_{\mathfrak{c}}] .$$

Let $\mathrm{SL}(\mathfrak{o}/\mathfrak{co})$ be the subgroup of $\mathrm{GL}_1(\mathfrak{o}/\mathfrak{co})$ consisting of the residue classes of $\alpha \bmod \mathfrak{co}$ such that $N_{B/F}(\alpha) \equiv 1 \bmod \mathfrak{c}$. By 2.15, $\mathrm{SL}(\mathfrak{o}/\mathfrak{co})$ is isomorphic to $\Gamma_1/\Gamma_{\mathfrak{c}}$. Hence $[\Gamma_1 : \Gamma_{\mathfrak{c}}] = N(\mathfrak{c})^3 \prod_{\mathfrak{p}|\mathfrak{c}} (1 - N(\mathfrak{p})^{-2})$. On the other hand, if

$$E_{\mathfrak{c}} = \{\varepsilon \in E \mid \varepsilon \equiv 1 \bmod \mathfrak{c}\} \quad \text{and} \quad E'_{\mathfrak{c}} = \{\varepsilon \in E \mid \varepsilon^2 \equiv 1 \bmod \mathfrak{c}\} ,$$

then $[\Gamma_1 \cap \Gamma'E : \Gamma_{\mathfrak{c}}] = [E'_{\mathfrak{c}} : E_{\mathfrak{c}}]$. Hence

$$(3.19.1) \qquad [\Gamma(\mathfrak{o}, 1) : E \cdot \Gamma(\mathfrak{o}, \mathfrak{c})] = [E'_{\mathfrak{c}} : E_{\mathfrak{c}}]^{-1} N(\mathfrak{c})^3 \prod_{\mathfrak{p}|\mathfrak{c}} (1 - N(\mathfrak{p})^{-2}) .$$

If $\mathfrak{c}$ is a prime ideal, $[E'_{\mathfrak{c}} : E_{\mathfrak{c}}] = 1$ or 2 according as $\mathfrak{c} = (2)$ or not. In this way we obtain the following table for smaller values of $N(\mathfrak{c})$:

$N(\mathfrak{c})$	genus of $\mathfrak{H}/\Gamma(\mathfrak{o}, \mathfrak{c})$	number of automorphisms
7	3	168
8	7	504
13	14	1092
27	118	9828
29	146	12180

It should be observed that $\Gamma(\mathfrak{o}, \mathfrak{c})$ has no elliptic elements unless $\mathfrak{c} = (1)$. The first two cases with genera 3 and 7 are previously known. The remaining cases seem to be new. (If R is a compact Riemann surface of genus $\mathfrak{g}$ with

$84(\mathfrak{g} - 1)$ automorphisms, every Galois covering of R has the same property, provided that the Galois group is the product of $2\mathfrak{g}$ copies of a cyclic group. The above examples can not be obtained in this manner, since the group of automorphisms is simple.)

It is well-known that the Riemann surface of genus 3 with 168 automorphisms can be obtained as the quotient of $\mathfrak{H}$, together with cusps, modulo $\{\alpha \in \mathrm{SL}_2(\boldsymbol{Z}) \mid \alpha \equiv 1 \bmod (7)\}$. The above result tells us that the *fundamental group* of this Riemann surface can be obtained as an arithmetically defined discontinuous group $\Gamma(\mathfrak{o}, \mathfrak{c})$. We shall now prove

(3.19.2) *If* $\mathfrak{c}$ *and* $\mathfrak{b}$ *are distinct integral ideals in* F, *then* $\mathfrak{H}/\Gamma_\mathfrak{c}$ *and* $\mathfrak{H}/\Gamma_\mathfrak{b}$ *are not biregularly isomorphic.*

Let $p(\alpha)$ denote the projection of an element α of B to the first factor $M_2(\boldsymbol{R})$ of (2.4.2). If $\mathfrak{H}/\Gamma_\mathfrak{c}$ and $\mathfrak{H}/\Gamma_\mathfrak{b}$ are biregularly isomorphic, there exists an element β of $\mathrm{SL}_2(\boldsymbol{R})$ such that

$$\beta \cdot p(\Gamma_\mathfrak{c} \cdot \{\pm 1\}) \beta^{-1} = p(\Gamma_\mathfrak{b} \cdot \{\pm 1\}) .$$

Since B is spanned by $\Gamma_\mathfrak{c}$ over F, we obtain an automorphism τ of B over F such that $p(\alpha^\tau) = \beta p(\alpha)\beta^{-1}$ $(\alpha \in B)$. We can find an element δ of B such that $\alpha^\tau = \delta\alpha\delta^{-1}$ for every $\alpha \in B$. Then $\delta\Gamma_\mathfrak{c}\delta^{-1} \cdot \{\pm 1\} = \Gamma_\mathfrak{b} \cdot \{\pm 1\}$. Since $\mathfrak{o}$ is spanned by $\Gamma_\mathfrak{c}$ over $\mathfrak{r}_F$, we have $\delta\mathfrak{o}\delta^{-1} = \mathfrak{o}$, hence

$$\Gamma_\mathfrak{c} \cdot \{\pm 1\} = \delta\Gamma_\mathfrak{c}\delta^{-1} \cdot \{\pm 1\} = \Gamma_\mathfrak{b} \cdot \{\pm 1\} .$$

If $\mathfrak{c} \neq \mathfrak{b}$, one can easily find an element γ of $\Gamma_\mathfrak{c}$ such that $\gamma \not\equiv \pm 1 \bmod \mathfrak{b}\mathfrak{o}$. This proves (3.19.2).

There are exactly three prime ideals $\mathfrak{p}$ in F such that $N(\mathfrak{p}) = 13$. Note that $\Gamma_\mathfrak{p}$ and $\Gamma(\mathfrak{o}, \mathfrak{p})$ give the same transformation group on $\mathfrak{H}$. Therefore, in view of (3.19.2), we know that there are three biregularly inequivalent compact Riemann surfaces of genus 14 each of which has $84 \cdot 13$ holomorphic automorphisms.

3.20. *Reduction of proofs of Main Theorems* I *and* II *to a special case.* Let the notation be as in 3.4. Let $\mathfrak{h}_1$ be an integral two-sided $\mathfrak{o}_1$-ideal such that $\mathfrak{h}_1 \supset \mathfrak{e}_1$, and let $\mathfrak{h}_\lambda = \mathfrak{x}_{\lambda 1}\mathfrak{h}_1\mathfrak{x}_{1\lambda}$. Let us call $\{V_\lambda, \varphi_\lambda, R_\sigma^{\mu\lambda}(\alpha)\}$ of 3.5 a *canonical system of level* $\mathfrak{e}$ *with respect to* Δ. We shall now prove the following assertion.

(3.20.1) *If a canonical system of level* $\mathfrak{e}$ *with respect to* Δ *exists, then a canonical system of level* $\mathfrak{h}$ *with respect to* Δ *exists.*

In view of 3.8, this implies

(3.20.2) *To prove 3.2 and 3.5, it is sufficient to show the existence of a canonical system of level* (c) *with respect to* Δ *for every positive integer* c, *where Δ is a (c)-regular representative set of orders and ideals in B.*

PROOF OF (3.20.1). Let $c = e_1 \cap r_F$ and $b = \mathfrak{h}_1 \cap r_F$. Let $\{V_\lambda, \varphi_\lambda, R_\sigma^{\mu\lambda}(\alpha)\}$ be as in 3.5. We note that $\Gamma(o_\lambda, e_\lambda)$ is a normal subgroup of $\Gamma(o_\lambda, \mathfrak{h}_\lambda)$, and $\gamma \mapsto R_1^{\lambda\lambda}(\gamma)$ is a homomorphism of $\Gamma(o_\lambda, \mathfrak{h}_\lambda)/\Gamma(o_\lambda, e_\lambda)$ onto a finite group of bi-regular automorphisms of V_λ, rational over $C(F, c)$. Construct a quotient V_λ' of V_λ by this group. Let P_λ denote the natural projection of V_λ to V_λ'. We can assume that V_λ' and P_λ are rational over $C(F, c)$. Now let $\alpha \in U(e)$, $\sigma = [C(F, c)/F, N_{B/F}(\alpha \mathfrak{x}_{\lambda\mu})]$, and $\gamma \in \Gamma(o_\lambda, \mathfrak{h}_\lambda)$. By 2.15, we can find an element δ of o_μ such that $\delta \equiv \alpha\gamma\alpha^{-1} \bmod^* e$, and $N_{B/F}(\delta) = N_{B/F}(\gamma)$. Then $\delta \equiv 1 \bmod^* \mathfrak{h}$, hence $\delta \in \Gamma(o_\mu, \mathfrak{h}_\mu)$. By (3.5.3), we have

$$R_\sigma^{\mu\lambda}(\alpha) \circ R_1^{\lambda\lambda}(\gamma) = R_\sigma^{\mu\lambda}(\alpha\gamma) = R_\sigma^{\mu\lambda}(\delta\alpha) = R_1^{\mu\mu}(\delta)^\sigma \circ R_\sigma^{\mu\lambda}(\alpha) \; .$$

It follows that there exists a biregular morphism $S_\sigma^{\mu\lambda}(\alpha)$ of V_λ' to $V_\mu'^\sigma$, defined over $C(F, c)$, such that $P_\mu^\sigma \circ R_\sigma^{\mu\lambda}(\alpha) = S_\sigma^{\mu\lambda}(\alpha) \circ P_\lambda$. By our definition of V_λ' and P_λ, we see that

(3.20.3) $S_1^{\lambda\lambda}(\gamma)$ *is the identity mapping if* $\gamma \in \Gamma(o_\lambda, \mathfrak{h}_\lambda)$.

Further, if $\beta \in U(e)$ and $\tau = [C(F, c)/F, N_{B/F}(\beta \mathfrak{x}_{\mu\nu})]$, we can easily verify

(3.20.4) $$S_{\tau\sigma}^{\nu\lambda}(\beta\alpha) = S_\tau^{\nu\mu}(\beta)^\sigma \circ S_\sigma^{\mu\lambda}(\alpha) \; .$$

Moreover

(3.20.5) *If* $\alpha, \beta \in U(e)$, $\alpha \equiv \beta \bmod^* \mathfrak{h}$ *and* $(N_{B/F}(\beta^{-1}\alpha)) \in P(F, \mathfrak{cu}_0)$, *then* $S_\sigma^{\mu\lambda}(\alpha) = S_\sigma^{\mu\lambda}(\beta)$.

In fact, let ε be a totally positive unit of F such that $N_{B/F}(\beta^{-1}\alpha) \equiv \varepsilon \bmod^* \mathfrak{cu}_0$. By 2.15, there exists an element γ of o_λ such that $\gamma \equiv \beta^{-1}\alpha \bmod^* e$ and $N_{B/F}(\gamma) = \varepsilon$. Then $\gamma \in \Gamma(o_\lambda, \mathfrak{h}_\lambda)$. Therefore

$$S_\sigma^{\mu\lambda}(\alpha) = S_\sigma^{\mu\lambda}(\beta\gamma) = S_\sigma^{\mu\lambda}(\beta) \circ S_1^{\lambda\lambda}(\gamma) = S_\sigma^{\mu\lambda}(\beta) \qquad \text{by (3.20.3)} \; .$$

Let $X = U_0(\mathfrak{h}) \cap U(e)$ and $Y = \{\alpha \in X \mid (N_{B/F}(\alpha)) \in P(F, \mathfrak{cu}_0)\}$. Then Y is a normal subgroup of X, and

(3.20.6) $\alpha \mapsto [C(F, c)/F, (N_{B/F}(\alpha))]$ *gives an isomorphism of X/Y onto the Galois group $G(C(F, c)/C(F, b))$ of $C(F, c)$ over $C(F, b)$.*

The only non-trivial point is the surjectivity which can be shown as follows. Let $\sigma \in G(C(F, c)/C(F, b))$. Then we can find an element a of F, prime to c, such that $\sigma = [C(F, c)/F, (a)]$, $a \equiv 1 \bmod^* \mathfrak{bu}_0$. We may assume that a is an integer. Since $a \equiv 1 \equiv N_{B/F}(1) \bmod^* \mathfrak{bu}_0$, there exists, by 2.15, an element α of o_λ such that $\alpha \equiv 1 \bmod \mathfrak{h}_\lambda$ and $N_{B/F}(\alpha) = a$. Since a is prime to c, α is prime to e_λ, hence $\alpha \in X$. This proves the desired surjectivity.

Now we consider the $S_\sigma^{\lambda\lambda}(\alpha)$ for $\alpha \in X$. By (3.20.5), $S_\sigma^{\lambda\lambda}(\alpha)$ depends only on the class of α modulo Y, and hence it depends only on

$$\sigma = [C(F, \mathfrak{c})/F, (N_{B/F}(\alpha))] \qquad \text{by (3.20.6)} .$$

Let us put $S_\sigma^\lambda = S_\sigma^{\lambda\lambda}(\alpha)$. Then from (3.20.4) we obtain $S_{\tau\sigma}^\lambda = (S_\tau^\lambda)^\sigma \circ S_\sigma^\lambda$ for $\sigma, \tau \in G(C(F, \mathfrak{c})/C(F, \mathfrak{b}))$. By Weil's criterion [18], we can find a curve V_λ'', defined over $C(F, \mathfrak{b})$, and a biregular morphism T_λ of V_λ'' to V_λ', defined over $C(F, \mathfrak{c})$, such that $S_\sigma^\lambda = T_\lambda^\sigma \circ T_\lambda^{-1}$.

Let $\xi \in U(\mathfrak{e})$, $\beta \in X$, $\sigma = [C(F, \mathfrak{c})/F, N_{B/F}(\xi \mathfrak{x}_{\lambda\mu})]$, $\tau = [C(F, \mathfrak{c})/F, (N_{B/F}(\beta))]$, and $\alpha = \xi\beta\xi^{-1}$. Then $\alpha \in X$ and $\alpha\beta^{-1} \in Y$. Since $\sigma\tau = \tau\sigma$, we have

$$\begin{aligned} S_\sigma^{\mu\lambda}(\xi)^\tau \circ S_\tau^\lambda &= S_\sigma^{\mu\lambda}(\xi)^\tau \circ S_\tau^{\lambda\lambda}(\beta) \\ &= S_{\sigma\tau}^{\mu\lambda}(\xi\beta) = S_{\tau\sigma}^{\mu\lambda}(\alpha\xi) = S_\tau^{\mu\mu}(\alpha)^\sigma \circ S_\sigma^{\mu\lambda}(\xi) = (S_\tau^\mu)^\sigma \circ S_\sigma^{\mu\lambda}(\xi) . \end{aligned}$$

Hence

$$(3.20.7) \qquad S_\sigma^{\mu\lambda}(\xi) = ((S_\tau^\mu)^\sigma)^{-1} \circ S_\sigma^{\mu\lambda}(\xi)^\tau \circ S_\tau^\lambda .$$

Applying σ to $S_\tau^\mu = T_\mu^\tau \circ T_\mu^{-1}$, we find $(T_\mu^{\sigma\tau})^{-1} = (T_\mu^\sigma)^{-1} \circ ((S_\tau^\mu)^{-1})^\sigma$, so that by (3.20.7)

$$\begin{aligned} [(T_\mu^\sigma)^{-1} \circ S_\sigma^{\mu\lambda}(\xi) \circ T_\lambda]^\tau &= (T_\mu^\sigma)^{-1} \circ ((S_\tau^\mu)^{-1})^\sigma \circ S_\sigma^{\mu\lambda}(\xi)^\tau \circ S_\tau^\lambda \circ T_\lambda \\ &= (T_\mu^\sigma)^{-1} \circ S_\sigma^{\mu\lambda}(\xi) \circ T_\lambda . \end{aligned}$$

This shows that $(T_\mu^\sigma)^{-1} \circ S_\sigma^{\mu\lambda}(\xi) \circ T_\lambda$ is defined over $C(F, \mathfrak{b})$.

Now replace V_λ', P_λ, $S_\sigma^{\mu\lambda}(\xi)$ by V_λ'', $T_\lambda^{-1} \circ P_\lambda$, $(T_\mu^\sigma)^{-1} \circ S_\sigma^{\mu\lambda}(\xi) \circ T_\lambda$ respectively, and for the sake of simplicity, denote them again by the same symbols as before. Then by this replacement, we may now assume, besides the property (3.20.4), that V_λ' and $S_\sigma^{\mu\lambda}(\xi)$ are defined over $C(F, \mathfrak{b})$, and $S_\sigma^{\lambda\lambda}(\alpha)$ is the identity mapping for every $\alpha \in X$.

Let $\alpha, \beta \in U(\mathfrak{e})$, $\alpha \equiv \beta \bmod^* \mathfrak{h}$,

$$\begin{aligned} \sigma &= [C(F, \mathfrak{c})/F, N_{B/F}(\alpha \mathfrak{x}_{\lambda\mu})] , \\ \tau &= [C(F, \mathfrak{c})/F, N_{B/F}(\beta \mathfrak{x}_{\lambda\mu})] . \end{aligned}$$

Then $\sigma = \tau$ on $C(F, \mathfrak{b})$. Let $\delta = \alpha^{-1}\beta$, $\rho = \sigma^{-1}\tau$. Since $\delta \in X$ and ρ is the identity mapping on $C(F, \mathfrak{b})$, we have

$$S_\tau^{\mu\lambda}(\beta) = S_{\sigma\rho}^{\mu\lambda}(\alpha\delta) = S_\sigma^{\mu\lambda}(\alpha)^\rho \circ S_\rho^{\lambda\lambda}(\delta) = S_\sigma^{\mu\lambda}(\alpha)^\rho = S_\sigma^{\mu\lambda}(\alpha) .$$

This means that $S_\sigma^{\mu\lambda}(\alpha)$ depends only on λ, μ, and the class of $\alpha \bmod^* \mathfrak{h}$. Since $U(\mathfrak{h})/U_0(\mathfrak{h})$ is isomorphic to $U(\mathfrak{e})/X$, we can now understand that $S_\sigma^{\mu\lambda}(\alpha)$ is defined for every $\alpha \in U(\mathfrak{h})$ and $\sigma = [C(F, \mathfrak{b})/F, N_{B/F}(\alpha \mathfrak{x}_{\lambda\mu})]$.

Define $\varphi_\lambda' : \mathfrak{H} \to V_\lambda'$ by $\varphi_\lambda' = P_\lambda \circ \varphi_\lambda$. From our construction of V_λ' and P_λ, we see that φ_λ' gives a biregular morphism of $\mathfrak{H}/\Gamma(\mathfrak{o}_\lambda, \mathfrak{h}_\lambda)$ to V_λ'. We shall now prove that $\{V_\lambda', \varphi_\lambda', S_\sigma^{\mu\lambda}(\alpha)\}$ satisfies the conditions (3.2.3) and (3.5.4). Let M, f and z be as in (3.5.4). Put $\mathfrak{e}_M = f^{-1}(f(\mathfrak{r}_M) \cap \mathfrak{e}_\mu)$, and $\mathfrak{h}_M = f^{-1}(f(\mathfrak{r}_M) \cap \mathfrak{h}_\mu)$. Let

$\rho = [C(M, \mathfrak{h}_M)/M, \mathfrak{a}]$ with an ideal $\mathfrak{a}$ in M prime to $\mathfrak{b}$. By (2.13.4), there exist an index λ and an element ξ of B^+ such that $\xi \mathfrak{x}_{\lambda\mu} = f(\mathfrak{a})\mathfrak{o}_\mu$. Extend ρ to an automorphism τ of $C(M, e_M)$, and let $\tau = [C(M, e_M)/M, \mathfrak{b}]$ with an ideal $\mathfrak{b}$ in M, prime to $\mathfrak{c}$. Then $\mathfrak{b} = t\mathfrak{a}$ with an element t of M such that $t \equiv 1 \bmod^* \mathfrak{h}_M$. Put $\alpha = f(t)\cdot\xi$. Then $\alpha \mathfrak{x}_{\lambda\mu} = f(\mathfrak{b})\mathfrak{o}_\mu$. By (3.5.4) we have

$$\varphi_\mu(z)^\tau = R_\sigma^{\mu\lambda}(\alpha)[\varphi_\lambda(\alpha^{-1}(z))] \, ,$$

where $\sigma = [C(F, \mathfrak{c})/F, N_{B/F}(\alpha \mathfrak{x}_{\lambda\mu})]$. It follows that

$$\begin{aligned}
\varphi_\mu'(z)^\tau = P_\mu^\tau(\varphi_\mu(z)^\tau) &= P_\mu^\sigma \circ R_\sigma^{\mu\lambda}(\alpha)[\varphi_\lambda(\alpha^{-1}(z))] \\
&= S_\sigma^{\mu\lambda}(\alpha) \circ P_\lambda \circ \varphi_\lambda(\alpha^{-1}(z)) = S_\sigma^{\mu\lambda}(\alpha)[\varphi_\lambda'(\alpha^{-1}(z))] \, .
\end{aligned}$$

Since $\alpha \equiv \xi \bmod^* \mathfrak{h}$, we have $S_\sigma^{\mu\lambda}(\alpha) = S_\sigma^{\mu\lambda}(\xi)$. Since z is the fixed point of $f(M)$, we have $\alpha^{-1}(z) = \xi^{-1}(z)$, so that $\varphi_\mu'(z)^\tau = S_\sigma^{\mu\lambda}(\xi)[\varphi_\lambda'(\xi^{-1}(z))]$. If τ is the identity mapping on $C(M, \mathfrak{h}_M)$, then $\lambda = \mu$ and $\mathfrak{a} = u\mathfrak{r}_M$ with an element u of M such that $u \equiv 1 \bmod^* \mathfrak{h}_M$. Taking ξ to be $f(u)$, we find $\varphi_\mu'(z)^\tau = \varphi_\mu'(z)$. This shows that $\varphi_\mu'(z)$ is rational over $C(M, \mathfrak{h}_M)$. Hence we can now write

(3.20.8)
$$\varphi_\mu'(z)^\rho = S_\sigma^{\mu\lambda}(\xi)[\varphi_\lambda'(\xi^{-1}(z))] \, .$$

Suppose conversely that ρ is the identity mapping on $M(\varphi_\mu'(z))\cdot C(F, \mathfrak{b})$. Then $\lambda = \mu$ and $\varphi_\mu'(z) = S_1^{\mu\mu}(\xi)[\varphi_\mu'(\xi^{-1}(z))]$. Since $\sigma = [C(F, \mathfrak{b})/F, (N_{B/F}(\xi))]$, there exists a unit y in F such that $N_{B/F}(\xi) \equiv y \bmod^* \mathfrak{b}\mathfrak{u}_0$. By 2.15, there exists an element γ of $\mathfrak{o}_\mu$ such that $\gamma \equiv \xi \bmod^* \mathfrak{h}$ and $N_{B/F}(\gamma) = y$. Then γ is a unit of $\mathfrak{o}_\mu$, hence $\xi\gamma^{-1}\mathfrak{o}_\mu = \xi\mathfrak{o}_\mu = f(\mathfrak{a})\mathfrak{o}_\mu$. Therefore we can take $\xi\gamma^{-1}$ in place of ξ. Putting $\eta = \xi\gamma^{-1}$, we have $\varphi_\mu'(z) = S_1^{\mu\mu}(\eta)[\varphi_\mu'(\eta^{-1}(z))]$. Since $\eta \equiv 1 \bmod^* \mathfrak{h}$, $S_1^{\mu\mu}(\eta)$ is the identity mapping, hence $\varphi_\mu'(z) = \varphi_\mu'(\eta^{-1}(z))$. Therefore we can find an element δ of $\Gamma(\mathfrak{o}_\mu, \mathfrak{h}_\mu)$ such that $\delta(z) = \eta^{-1}(z)$. By 2.6, $\eta\delta = f(v)$ with an element v of M. We have then $f(v)\mathfrak{o}_\mu = \eta\delta\mathfrak{o}_\mu = f(\mathfrak{a})\mathfrak{o}_\mu$, so that $\mathfrak{a} = v\mathfrak{r}_M$ (see below). Since $f(v) = \eta\delta \equiv 1 \bmod^* \mathfrak{h}$, we have $v \equiv 1 \bmod^* \mathfrak{h}_M$, so that ρ is the identity mapping on $C(M, \mathfrak{h}_M)$. This proves $C(M, \mathfrak{h}_M) = C(F, \mathfrak{b})\cdot M(\varphi_\mu'(z))$ and completes the proof. Here we have made use of the following fact.

(3.20.9) *Let $\mathfrak{o}$ be a maximal order in B, and $M \in J(B)$. Let f be an F-linear isomorphism of M into B such that $f(\mathfrak{r}_M) \subset \mathfrak{o}$. If $f(\mathfrak{a})\mathfrak{o} = f(\mathfrak{b})\mathfrak{o}$ for ideals $\mathfrak{a}$ and $\mathfrak{b}$ in M, then $\mathfrak{a} = \mathfrak{b}$.*

In fact we have $f(\mathfrak{a}^{-1}\mathfrak{b})\mathfrak{o} = \mathfrak{o}$, so that $\mathfrak{a}^{-1}\mathfrak{b} \subset f^{-1}(\mathfrak{o} \cap f(M)) = \mathfrak{r}_M$, hence $\mathfrak{b} \subset \mathfrak{a}$. Interchanging $\mathfrak{a}$ and $\mathfrak{b}$, we get $\mathfrak{a} = \mathfrak{b}$. The assertion (3.20.9) is obviously true even if $r > 1$.

4. The moduli variety for weak PEL-structures

4.1. Let L be an algebra over $\mathbf{Q}$ with identity, and Φ a $\mathbf{Q}$-linear isomorphism of L into the ring of $\mathbf{C}$-linear transformations in $\mathbf{C}^n$, which maps identity to identity. Let A be an abelian variety of dimension n, defined over a subfield of C. Suppose that there exists an isomorphism θ of L into $\mathrm{End}_Q(A)$. We say that (A, θ) *is of type* (L, Φ) if there exists a complex torus C^n/D, with a lattice D in C^n, isomorphic to A, such that $\Phi(a)$ represents $\theta(a)$ for every $a \in L$. This being so, we can regard $Q \cdot D$ as a module over L by the action of $\Phi(L)$. Let us take a left L-module W isomorphic to $Q \cdot D$, and fix an L-isomorphism $\mathfrak{y}$ of W to $Q \cdot D$. Put $L_R = L \otimes_Q R$, $W_R = W \otimes_Q R$. Then $\mathfrak{y}$ can be extended to an L_R-linear isomorphism of W_R to $R \cdot D = C^n$. Put $\mathfrak{M} = \mathfrak{y}^{-1}(D)$. Thus we obtain a commutative diagram

$$(4.1.1) \qquad \begin{array}{ccccccccc} 0 & \longrightarrow & \mathfrak{M} & \longrightarrow & W_R & \longrightarrow & W_R/\mathfrak{M} & \longrightarrow & 0 \\ & & \downarrow & & \downarrow{\scriptstyle \mathfrak{y}} & & \downarrow & & \\ 0 & \longrightarrow & D & \longrightarrow & C^n & \overset{\xi}{\longrightarrow} & A & \longrightarrow & 0 \end{array}$$

where ξ is a holomorphic map of C^n to A which gives a biregular isomorphism of C^n/D to A. We extend Φ to L_R so that

$$(4.1.2) \qquad \Phi(a)\mathfrak{y}(u) = \mathfrak{y}(au) \qquad\qquad (a \in L_R,\, u \in W_R)\,.$$

Since $\Phi(L)$ maps $Q \cdot D$ to $Q \cdot D$, we see easily that

$\quad$ (4.1.3) *The direct sum of Φ and its complex conjugate $\bar{\Phi}$ is equivalent to a rational representation.*

Now let $\mathcal{C}$ be a polarization of A. Suppose that L is a semi-simple algebra with a positive involution ρ, and the involution of $\mathrm{End}_Q(A)$ determined by $\mathcal{C}$ coincides with the map $\theta(a) \to \theta(a^\rho)$ on $\theta(L)$. Let $\mathfrak{E}(x, y)$ be the Riemann form on C^n determined by a divisor X in $\mathcal{C}$. By [11, I, 1.2], there exists a non-degenerate L-valued ρ-anti-hermitian form $T(x, y)$ on W in the sense of [11, I, 1.1] such that

$$(4.1.4) \qquad \mathfrak{E}\big(\mathfrak{y}(u),\, \mathfrak{y}(v)\big) = \mathrm{tr}\big(T(u, v)\big) \qquad\qquad (u, v \in W_R)\,.$$

Here tr means the reduced trace of L_R to R, and T is extended to an R-bilinear L_R-valued form on W_R. We note that

$$(4.1.5) \qquad \mathrm{tr}\big(T(\mathfrak{M}, \mathfrak{M})\big) = Z$$

holds if and only if the divisor X is a *basic polar divisor* in $\mathcal{C}$ to the effect that every divisor in $\mathcal{C}$ is algebraically equivalent to a multiple μX with a positive integer μ.

Now suppose that L is simple. Then W is isomorphic to a direct sum of several copies of an irreducible L-module. Therefore W is determined by $\dim(A)$. Especially if $2n$ is divisible by $[L : Q]$, W is isomorphic to the (left) L-module L^m of all m-dimensional row vectors with components in L, where

$m = 2n/[L : Q]$. This is always so if L is a division algebra. Hereafter we always assume

(4.1.6) *L is a simple algebra belonging to the four types of* [9, Prop. 1] *and $2n$ is divisible by* $[L : Q]$;

and W will be often identified with L^m. We make this assumption simply because the theory developed in [9] is restricted to this case. Actually one can discuss a more general case without difficulty, though we shall not do it in the present paper. At least it should be observed that the results of [13] are valid so far as (4.1.6) is satisfied, even if L is not a division algebra. All the proofs of [13] remain true without any change.

Let $v_1, \cdots, v_s$ be elements of W, and let $t_1, \cdots, t_s$ be elements of A. We call $\mathfrak{Q} = (A, \mathcal{C}, \theta; t_1, \cdots, t_s)$ a PEL-*structure* of type

$$\Omega = (L, \Phi, \rho; T, \mathfrak{M}; v_1, \cdots, v_s) ,$$

if one can find a commutative diagram (4.1.1) so that (4.1.2), (4.1.4) are satisfied and $t_i = \xi(\mathfrak{y}(v_i))$ for $i = 1, \cdots, s$. The collection of data, Ω, is called a PEL-*type*. We assume, besides (4.1.3), (4.1.5), and (4.1.6), the condition [9, p. 160, (25)] for T so that there exists at least one PEL-structure of type Ω.

4.2. Given a PEL-type Ω as above, let $\mathfrak{o} = \{a \in L \mid a\mathfrak{M} \subset \mathfrak{M}\}$. Then $\mathfrak{o}$ is an order in L. Define an algebraic group $\mathrm{U}(T)$, a Lie group $\mathrm{U}_R(T)$, discrete subgroups $\Gamma(T, \mathfrak{M})$ and $\Gamma(T, \mathfrak{N}/\mathfrak{M})$ of $\mathrm{U}_R(T)$ by

$$\mathrm{U}(T) = \{\alpha \in \mathrm{GL}_m(L) \mid T(x\alpha, y\alpha) = T(x, y)\} ,$$
$$\mathrm{U}_R(T) = \{\alpha \in \mathrm{GL}_m(L_R) \mid T(x\alpha, y\alpha) = T(x, y)\} ,$$
$$\Gamma(T, \mathfrak{M}) = \{\alpha \in \mathrm{U}(T) \mid \mathfrak{M}\alpha = \mathfrak{M}\} ,$$
$$\mathfrak{N} = \mathfrak{M} + \textstyle\sum_{i=1}^{s} \mathfrak{o}v_i ,$$
$$\Gamma(T, \mathfrak{N}/\mathfrak{M}) = \{\alpha \in \Gamma(T, \mathfrak{M}) \mid \mathfrak{N}(1 - \alpha) \subset \mathfrak{M}\} .$$

Let $\mathcal{K}$ be a bounded symmetric domain isomorphic to the quotient of $\mathrm{U}_R(T)$ by a maximal compact subgroup. Then $\mathrm{U}_R(T)$ acts on $\mathcal{K}$. (For an explicit form of the action in terms of matrices, see [9, 2.7].) In [9, §2] and [13, 3.3], we obtained a real analytic mapping $\mathfrak{y}: W_R \times \mathcal{K} \to C^n$ with the following properties:

(4.2.1) $\mathfrak{y}(x, z)$, *with $x \in W_R$ and $z \in \mathcal{K}$, is holomorphic in z.*

(4.2.2) *For every $z \in \mathcal{K}$, if we put $\mathfrak{y}_z(x) = \mathfrak{y}(x, z)$, then $\mathfrak{y}_z$ is an R-linear isomorphism of W_R to C^n, and $\mathfrak{y}_z(ax) = \Phi(a)\mathfrak{y}_z(x)$ for every $a \in L$.*

(4.2.3) *For each $z \in \mathcal{K}$, if we put $\mathfrak{E}_z(u, v) = \mathrm{tr}\big(T(\mathfrak{y}_z^{-1}(u), \mathfrak{y}_z^{-1}(v))\big)$, then $\mathfrak{E}_z$ is a Riemann form on the complex torus $C^n/\mathfrak{y}_z(\mathfrak{M})$.*

Let A_z be any projective embedding of $C^n/\mathfrak{y}_z(\mathfrak{M})$, and $\mathcal{C}_z$ the polarization of A_z determined by $\mathfrak{E}_z$. Let $\theta_z(a)$ be an element of $\mathrm{End}_Q(A_z)$ corresponding to $\Phi(a)$ for every $a \in L$. Further let t_{iz} be the point on A_z corresponding to $\mathfrak{y}_z(v_i) \bmod \mathfrak{y}_z(\mathfrak{M})$. Put $\mathfrak{Q}_z = (A_z, \mathcal{C}_z, \theta_z; t_{1z}, \cdots, t_{sz})$. Then:

(4.2.4) $\mathfrak{Q}_z$ *is a* PEL-*structure of type* Ω.

(4.2.5) *Every* PEL-*structure of type* Ω *is isomorphic to* $\mathfrak{Q}_z$ *for some* $z \in \mathfrak{K}$.

We put $\mathcal{P}_z = (A_z, \mathcal{C}_z, \theta_z)$ and often identify $\mathcal{P}_z$ with a PEL-structure $(A_z, \mathcal{C}_z, \theta_z; 0)$. Further we put

(4.2.6) $$\Sigma(\Omega) = \Sigma_\Omega = \{ \mathfrak{Q}_z \mid z \in \mathfrak{K} \} \ .$$

In the following treatment, we always *exclude* the bad case where the following three conditions in (4.2.7) are simultaneously satisfied.

(4.2.7) $L \neq Q$, $\dim(\mathfrak{K}) = 1$, *and* $\mathfrak{K}/\Gamma(T, \mathfrak{M})$ *is not compact.*

4.3. Let F be the set of all elements x in the center of L such that $x^\rho = x$. Then F is a totally real algebraic number field. We call an L-linear automorphism α of W a *similitude* of T if $T(x\alpha, y\alpha) = \mu T(x, y)$ with an element μ of F. We put $\mu = \mu(\alpha)$, and denote by $G(T)$ the group of all similitudes of T. Let $G^+(T)$ be the subgroup of $G(T)$ consisting of all α such that $\mu(\alpha)$ is totally positive. We can define the action of $G^+(T)$ on $\mathfrak{K}$ as follows: for every $\alpha \in G^+(T)$, we can find an element q of $F_R = F \otimes_Q R$ so that $q^2 = \mu(\alpha)$; then $q^{-1}\alpha \in \mathrm{U}_R(T)$; we put $\alpha(z) = (q^{-1}\alpha)(z)$ for $z \in \mathfrak{K}$ (cf. [9, 5.3]).

4.4. PROPOSITION. *For every* α *of* $G^+(T)$ *and* $z \in \mathfrak{K}$, *there exists a* C-*linear automorphism* $\Lambda(\alpha, z)$ *of* C^n *such that*

$$\Lambda(\alpha, z)\mathfrak{y}(x, z) = \mathfrak{y}\big(x\alpha, \alpha^{-1}(z)\big) \qquad\qquad (x \in W_R) \ .$$

PROOF. This fact, in substance, has been shown in [9, 2.8]. In fact, put $w = \alpha^{-1}(z)$. From our definition of the action of elements of $G^+(T)$, we can conclude the existence of matrices Λ_ν such that $\Lambda_\nu X_\nu(z, T) = X_\nu(w, T)^t\omega_\nu(\alpha)$, the notation being as in [9, 2.6 and 2.4]. Then it can be verified that Λ_ν is of the form [9, (34)]. Define Λ by [9, (32), (33)]. Then

$$\Lambda \mathfrak{x}_i(z, T) = \sum_{j=1}^{m} \Phi(a_{ij})\mathfrak{x}_j(w, T)$$

with $\alpha = (a_{ij})$ and the $\mathfrak{x}_i(z, T)$ of [9, 2.6]. It follows that $\Lambda\mathfrak{y}(x, z) = \mathfrak{y}(x\alpha, w)$ for every $x \in W_R$.

4.5. Let $\mathrm{Pic}(A)$ denote the Picard variety of an abelian variety A. Take non-degenerate divisors X and Y on A. Then X and Y determine isogenies φ_X and φ_Y of A to $\mathrm{Pic}(A)$ in the usual manner, and we have $\varphi_Y = \varphi_X \cdot \varepsilon$ with an element ε of $\mathrm{End}_Q(A)$. We write then $Y \equiv X \cdot \varepsilon$.

4.6. PROPOSITION. *Let $(A, \mathcal{C}, \theta; t_1, \cdots, t_s)$ be of type Ω, and β a totally positive element of F such that $\mathrm{tr}(\beta T(\mathfrak{M}, \mathfrak{M})) \subset \mathbf{Z}$. Then, for a basic polar divisor X in $\mathcal{C}$, there exists a positive non-degenerate divisor Y on A such that $Y \equiv X \cdot \theta(\beta)$. Moreover, if a divisor Z on A corresponds to a Riemann form and $Z \equiv X \cdot \theta(\gamma)$ with an element γ of F, then γ is totally positive and $\mathrm{tr}(\gamma T(\mathfrak{M}, \mathfrak{M})) \subset \mathbf{Z}$.*

PROOF. We may assume that $(A, \mathcal{C}, \theta; t_1, \cdots, t_s) = \mathfrak{Q}_z$ for some $z \in \mathcal{H}$. Then X corresponds to a Riemann form $\mathfrak{E}_z$. Now we can find an element η of F_R so that $\beta = \eta^2$. Put $\mathfrak{E}(x, y) = \mathfrak{E}_z(x, \Phi(\beta)y)$. Then

$$\mathfrak{E}(x, y) = \mathfrak{E}_z(\Phi(\eta)x, \Phi(\eta)y) .$$

It is easy to verify that $\mathfrak{E}$ is a Riemann form, and hence $\mathfrak{E}$ corresponds to a positive non-degenerate divisor Y on A. Then we have obviously $\varphi_Y = \varphi_X \cdot \theta(\beta)$, so that $Y \equiv X \cdot \theta(\beta)$ (cf. [15, 3.3]). Next let Z and γ be as above. Let ψ be an L_R-linear automorphism of W_R determined by $\mathfrak{h}_z(x\psi) = \sqrt{-1}\mathfrak{h}_z(x)$ for all $x \in W_R$. Then

$$\mathfrak{E}_z\big(\mathfrak{h}_z(u), \sqrt{-1}\mathfrak{h}_z(v)\big) = \mathrm{tr}\big(T(u, v\psi)\big) ,$$

and

$$\mathfrak{E}_z\big(\Phi(\gamma)\mathfrak{h}_z(u), \sqrt{-1}\mathfrak{h}_z(v)\big) = \mathrm{tr}\big(\gamma T(u, v\psi)\big) .$$

Now $\mathfrak{E}_z(\Phi(\gamma)x, y)$ is the Riemann form determined by Z. Therefore

$$\mathrm{tr}\big(\gamma T(\mathfrak{M}, \mathfrak{M})\big) \subset \mathbf{Z} ,$$

and $\mathrm{tr}(\gamma T(u, u\psi)) > 0$, $\mathrm{tr}(T(u, u\psi)) > 0$ for $0 \neq u \in W_R$. It follows from this that γ is totally positive.

4.7. Let $\mathfrak{Q} = (A, \mathcal{C}, \theta; t_1, \cdots, t_s)$ be a PEL-structure of type Ω, and X a basic polar divisor in $\mathcal{C}$. By 4.6, for every totally positive element β of F, there exists a positive integer q and a positive non-degenerate divisor Y on A such that $Y \equiv X \cdot \theta(q\beta)$. We denote by $\mathcal{C} \cdot \theta(\beta)$ the polarization of A containing Y, and by $\mathfrak{Q} \cdot \beta$ the PEL-structure $(A, \mathcal{C} \cdot \theta(\beta), \theta; t_1, \cdots, t_s)$. If r is a positive rational number such that $\mathrm{tr}(r\beta T(\mathfrak{M}, \mathfrak{M})) = \mathbf{Z}$, then $\mathfrak{Q} \cdot \beta$ is of type $(L, \Phi, \rho; r\beta T, \mathfrak{M}; v_1, \cdots, v_s)$.

4.8. Let F^+ denote the group of all totally positive elements in F. Let us fix a subgroup E of F^+ containing all positive rational numbers. The notation being as in 4.7, let $\mathfrak{D}$ denote the join of $\mathcal{C} \cdot \theta(\beta)$ for all $\beta \in E$. We call $\mathfrak{D}$ a *weak polarization* of A (with respect to E), and

$$\mathfrak{R} = (A, \mathfrak{D}, \theta; t_1, \cdots, t_s)$$

a weak PEL-structure. We write $\mathfrak{D} = \mathcal{C} \cdot \theta(E)$ and $\mathfrak{R} = \mathfrak{Q} \cdot E$. A collection of data

$$\Omega_0 = (L, \Phi, \rho; E \cdot T, \mathfrak{M}; v_1, \cdots, v_s)$$

is called a *weak* PEL-*type*. We understand that $E \cdot T = E \cdot T'$ if $T = \varepsilon T'$ for an element ε of E. We say that $\mathcal{R}$ is of type Ω_0, if $\mathcal{R}$ is obtained by the above procedure, i.e., if $\mathfrak{D}$ contains a polarization $\mathcal{C}$ such that $(A, \mathcal{C}, \theta; \{t_i\})$ is of type $(L, \Phi, \rho; \beta T, \mathfrak{M}; \{v_i\})$ for some $\beta \in E$. If σ is an automorphism of C, we can define $\mathcal{R}^\sigma = (A^\sigma, \mathfrak{D}^\sigma, \theta^\sigma; \{t_i^\sigma\}) = \mathfrak{Q}^\sigma \cdot E$ with $\mathfrak{D}^\sigma = \mathcal{C}^\sigma \cdot \theta^\sigma(E)$. Let

$$\Omega_0' = (L, \Phi, \rho; E \cdot T', \mathfrak{M}'; v_1', \cdots, v_s')$$

be another weak PEL-type with the same L, Φ, ρ, W and s. We say that Ω_0 is equivalent to Ω_0' if there exist an L-linear automorphism α of W and an element γ of E such that $T'(x\alpha, y\alpha) = \gamma T(x, y)$, $\mathfrak{M}\alpha = \mathfrak{M}'$ and $v_i\alpha \equiv v_i' \bmod \mathfrak{M}'$ for every i.

4.9. Let $\mathcal{R}$ be as above, and let $\mathcal{R}' = (A', \mathfrak{D}', \theta'; t_1', \cdots, t_s')$ be a weak PEL-structure. Let X (resp. X') be a divisor in $\mathfrak{D}$ (resp. $\mathfrak{D}'$). We call an isomorphism (resp. isogeny) λ of A to A' an *isomorphism* (resp. *isogeny*) of $\mathcal{R}$ to $\mathcal{R}'$ if $\lambda^{-1}(X') \equiv X \cdot \theta(\gamma)$ for some $\gamma \in E$, $\lambda\theta(a) = \theta'(a)\lambda$ for all $a \in L$, and $\lambda t_i = t_i'$ for every i.

4.10. For every $z \in \mathcal{H}$, we can define naturally a weak PEL-structure

$$\mathcal{R}_z = (A_z, \mathfrak{D}_z, \theta_z; t_{1z}, \cdots, t_{sz}) = \mathfrak{Q}_z \cdot E$$

with $\mathfrak{D}_z = \mathcal{C}_z \cdot \theta_z(E)$. Then the $\mathcal{R}_z$ are all of type Ω_0. We see easily that every weak PEL-structure of type Ω_0 is isomorphic to $\mathcal{R}_z$ for some z. We put

$$\Sigma(\Omega_0) = \{\mathcal{R}_z \mid z \in \mathcal{H}\}.$$

Define subgroups $\Gamma_0(T, \mathfrak{M})$ and $\Gamma_0(T, \mathfrak{N}/\mathfrak{M})$ of $G^+(T)$ by

$$\Gamma_0(T, \mathfrak{M}) = \{\alpha \in G^+(T) \mid \mathfrak{M}\alpha = \mathfrak{M}, \mu(\alpha) \in E\},$$
$$\Gamma_0(T, \mathfrak{N}/\mathfrak{M}) = \{\alpha \in \Gamma_0(T, \mathfrak{N}/\mathfrak{M}) \mid \mathfrak{N}(1 - \alpha) \subset \mathfrak{M}\}.$$

If $\alpha \in \Gamma_0(T, \mathfrak{M})$, $\mu(\alpha)$ is a totally positive unit in E.

4.11. PROPOSITION. *If $\alpha \in \Gamma_0(T, \mathfrak{N}/\mathfrak{M})$, the linear transformation $\Lambda(\alpha, z)$ of C^n obtained in 4.4 gives an isomorphism of $\mathcal{R}_z$ to $\mathcal{R}_w$ with $w = \alpha^{-1}(z)$. Conversely, every isomorphism of $\mathcal{R}_z$ to $\mathcal{R}_w$, with z and w in $\mathcal{H}$, can be obtained from $\Lambda(\alpha, z)$ with an element α of $\Gamma_0(T, \mathfrak{N}/\mathfrak{M})$ such that $\alpha(w) = z$.*

This follows easily from our definition of weak PEL-structures and an obvious modification of the discussion of [9, p. 165].

4.12. PROPOSITION. *Let $\mathfrak{Q}$ be a PEL-structure of type Ω, and let β be a totally positive element of F. Then $\mathfrak{Q}$ and $\mathfrak{Q} \cdot \beta$ have the same field of moduli.*

PROOF. For every automorphism σ of C, one can verify in a straightforward way that $\mathfrak{A}$ is isomorphic to $\mathfrak{A}^\sigma$ if and only if $\mathfrak{A} \cdot \beta$ is isomorphic to $\mathfrak{A}^\sigma \cdot \beta$. Then we obtain our assertion.

4.13. PROPOSITION. *Let $\mathfrak{A}$ and $\mathfrak{A}'$ be PEL-structures with the same L, Φ, ρ. If $\mathfrak{A} \cdot E$ and $\mathfrak{A}' \cdot E$ are isomorphic, $\mathfrak{A}$ and $\mathfrak{A}'$ have the same field of moduli.*

PROOF. Our assumption implies that $\mathfrak{A}'$ is isomorphic to $\mathfrak{A} \cdot \varepsilon$ for an element ε of E. Then we get our assertion immediately from 4.12.

4.14. PROPOSITION. *Let Ω and Ω_0 be as above, and let k_Ω be the algebraic number field determined in [13, 5.1]. Let K' be the field generated over $\mathbf{Q}$ by* $\mathrm{tr}\,\Phi(a)$ *for all a in the center of L. Then there exists a field $k(\Omega_0)$ with the following properties.*

(4.14.1) $K' \subset k(\Omega_0) \subset k_\Omega$.

(4.14.2) *Let $\mathfrak{R}$ be a weak PEL-structure of type Ω_0, and σ an automorphism of C. Then $\mathfrak{R}^\sigma$ is of type Ω_0 if and only if σ is the identity mapping on $k(\Omega_0)$.*

Moreover the field $k(\Omega_0)$ can be uniquely characterized by the property (4.14.2).

PROOF. Let k_1 be the smallest Galois extension of $\mathbf{Q}$ containing k_Ω. For every automorphism σ of C, Ω^σ is determined, up to equivalence, only by the effect of σ on k_Ω (see [13, 5.1, Proof] and [11, III, 3.1]). Let N be the subgroup of $G(k_1/\mathbf{Q})$ consisting of the elements σ such that Ω^σ is equivalent to $(L, \Phi, \rho; \beta T, \mathfrak{M}; v_1, \cdots, v_s)$ for some $\beta \in E$. Let $k(\Omega_0)$ be the subfield of k_1 corresponding to N. Then one can easily verify that $k(\Omega_0)$ satisfies (4.14.2). The property (4.14.1) and the last assertion are obvious.

4.15. Let β be a totally positive element of F, and $\mathfrak{N}$ a lattice in W such that $\mathrm{tr}(\beta T(\mathfrak{N}, \mathfrak{N})) = \mathbf{Z}$. Let

$$\Omega' = (L, \Phi, \rho; \beta T, \mathfrak{N}; u_1, \cdots, u_s)$$

with elements $u_1, \cdots, u_s$ in W. Let $\mathfrak{h}(x, z)$ be as in 4.2. Then we can construct a family $\Sigma(\Omega') = \{\mathfrak{A}'_z \mid z \in \mathcal{H}\}$ of PEL-structures of type Ω' by means of the same $\mathfrak{h}$. In fact, the mapping $\mathfrak{h}: W_R \times \mathcal{H} \to C^n$ has been defined in [13, 3.3] by $\mathfrak{h}(x, z) = \sum_{i=1}^m \Phi(x_i)\mathfrak{x}_i(z, T)$ for $x = (x_1, \cdots, x_m) \in L_R^m = W_R$ with the vectors $\mathfrak{x}_i(z, T)$ of [9, p. 163]. The $\mathfrak{x}_i(z, T)$ depend on the choice of the matrices W_ν in [9, pp. 160–161]. If we replace T by βT, then it is sufficient to replace W_ν by $c_\nu W_\nu$ with a suitable positive real number c_ν for each ν. Then we may assume that $\mathfrak{x}_i(z, \beta T) = P\mathfrak{x}_i(z, T)$ with a real diagonal matrix P commuting with the elements of $\Phi(L)$; P depends only on β, and not on $\mathfrak{M}, \mathfrak{N}, z$. Therefore, a map $\mathfrak{h}': W_R \times \mathfrak{H} \to C^n$ defined by $\mathfrak{h}'(x, z) = P\mathfrak{h}(x, z)$

plays the role of $\mathfrak{y}$ for the PEL-type Ω'. Then, in order to construct $\Sigma(\Omega')$, we can employ $\mathfrak{y}$ instead of $\mathfrak{y}'$, since the only difference is the transformation by P.

Especially, suppose that $\mathfrak{M} = \mathfrak{N}$ and $u_i = v_i$. Then it can be easily verified that $\mathfrak{Q}'_z$ is isomorphic to $\mathfrak{Q}_z \cdot \beta$. Therefore we may understand that $\Sigma(\Omega') = \{\mathfrak{Q}_z \cdot \beta \mid z \in \mathfrak{K}\}$.

4.16. PROPOSITION. *The notation being as in 4.15, suppose that $\mathfrak{M} = \mathfrak{N}$ and $u_i = v_i$. Then $k(\Omega) = k(\Omega')$.*

PROOF. Let σ be an automorphism of C over $k(\Omega)$. Let $\mathfrak{Q}$ be of type Ω. Then $\mathfrak{Q}^\sigma$ is of type Ω, so that $\mathfrak{Q}^\sigma \cdot \beta$ is of type Ω'. Since $(\mathfrak{Q} \cdot \beta)^\sigma = \mathfrak{Q}^\sigma \cdot \beta$, σ must be the identity on $k(\Omega')$. Exchanging Ω for Ω' and repeating the same argument, we have $k(\Omega) = k(\Omega')$.

4.17. The PEL-types Ω and Ω' being as in 4.15, suppose that $\mathfrak{M} = \mathfrak{N}$ and $u_i = v_i$. Let $(V, \mathfrak{v})$ satisfy the property of [13, 6.2] for Ω, and let φ be a holomorphic mapping of $\mathfrak{K}$ to V satisfying [13, (6.2.5)]. We call $(V, \mathfrak{v})$ a *moduli-variety for* PEL-*structures of type Ω*, and $(V, \mathfrak{v}, \varphi)$ a *moduli-system for the family Σ_Ω*. For every PEL-structure $\mathfrak{Q}'$ of type Ω', there exists a PEL-structure $\mathfrak{Q}$ of type Ω such that $\mathfrak{Q}'$ is isomorphic to $\mathfrak{Q} \cdot \beta$. This follows from the consideration of 4.15. Put $\mathfrak{v}'(\mathfrak{Q}') = \mathfrak{v}(\mathfrak{Q})$. By 4.12 and 4.16, we see that V and $\mathfrak{v}'$ satisfy the conditions (6.2.1) and (6.2.4) of [13, 6.2] with Ω' in place of Ω. Now the consideration of 4.15 shows that V, $\mathfrak{v}'$ and φ satisfy [13, (6.2.5)] with $\Sigma(\Omega')$ in place of Σ_Ω. Here we note that $G(\beta T) = G(T)$, $\mathrm{U}(\beta T) = \mathrm{U}(T)$, $\Gamma(\beta T, \mathfrak{N}/\mathfrak{M}) = \Gamma(T, \mathfrak{N}/\mathfrak{M})$, and the action of elements of $G^+(T)$ on $\mathfrak{K}$ is invariant by the exchange of T for βT. (It is not suffered by the exchange of W_ν for $c_\nu W_\nu$.) In view of the uniqueness [13, 6.8], we see that $(V, \mathfrak{v}', \varphi)$ has all the properties of [13, 6.2] for Ω'. In other words, $(V, \mathfrak{v}', \varphi)$ is a moduli-system for the family $\Sigma(\Omega')$.

4.18. Coming back to the weak PEL-type Ω_0, let σ be an automorphism of C over the field $k(\Omega_0)$ determined by 4.14. By (4.14.2), Ω^σ is equivalent to $(L, \Phi, \rho; \varepsilon T, \mathfrak{M}; \{v_i\})$ with an element ε of E. By 4.16, we have $k(\Omega) = k(\Omega^\sigma) = k(\Omega)^\sigma$ (cf. [11, III, 3.1]). It follows that $k(\Omega)$ is a Galois extension of $k(\Omega_0)$.

Let $V, \mathfrak{v}$ and φ be as in 4.17. For every $\mathfrak{Q}_1$ of type Ω^σ, define $\mathfrak{v}^\sigma$ by $\mathfrak{v}^\sigma(\mathfrak{Q}_1) = \mathfrak{v}(\mathfrak{Q}_1^{\sigma^{-1}})^\sigma$. If τ is another automorphism of C such that $\tau = \sigma$ on k_Ω, $\mathfrak{v}(\mathfrak{Q}^{\sigma\tau^{-1}}) = \mathfrak{v}(\mathfrak{Q})^{\sigma\tau^{-1}}$, hence

$$\mathfrak{v}(\mathfrak{Q}_1^{\tau^{-1}})^\tau = \mathfrak{v}\big((\mathfrak{Q}_1^{\sigma^{-1}})^{\sigma\tau^{-1}}\big)^\tau = \mathfrak{v}(\mathfrak{Q}_1^{\sigma^{-1}})^\sigma .$$

Therefore $\mathfrak{v}^\sigma$ depends only on the effect of σ on k_Ω. We see easily that $(V^\sigma, \mathfrak{v}^\sigma)$ is a moduli-variety for PEL-structures of type Ω^σ. Now take ε to be the element β in the consideration of 4.17, and define $\mathfrak{v}'$. By the uniqueness of moduli-variety [13, 6.7], there exists a biregular morphism f_σ of V onto V^σ, defined over k_Ω, such that $\mathfrak{v}^\sigma = f_\sigma \circ \mathfrak{v}'$. Let $\varphi_\sigma = f_\sigma \circ \varphi$. Then $(V^\sigma, \mathfrak{v}^\sigma, \varphi_\sigma)$ is a moduli-system for $\Sigma(\Omega^\sigma)$.

For brevity, let us put $k = k_\Omega$, $k_0 = k(\Omega_0)$. Let Δ be the group of all $\alpha \in \Gamma_0(T, \mathfrak{R}/\mathfrak{M})$ which give the identity mapping on $\mathfrak{H}$. We see that $\Delta \cdot \Gamma(T, \mathfrak{R}/\mathfrak{M})$ is a normal subgroup of $\Gamma_0(T, \mathfrak{R}/\mathfrak{M})$. Put

$$\mathfrak{A} = \Gamma_0(T, \mathfrak{R}/\mathfrak{M})/\Delta \cdot \Gamma(T, \mathfrak{R}/\mathfrak{M}) \; ;$$

then $\mathfrak{A}$ is a finite group. For every $g \in \mathfrak{A}$, take an element α of $\Gamma_0(T, \mathfrak{R}/\mathfrak{M})$ which represents g. Then we obtain a biregular morphism S_g of V to V such that $S_g(\varphi(z)) = \varphi(\alpha(z))$. Then S_g is determined only by g, and independent of the choice of α. Since $\mathfrak{R}_{\alpha(z)}$ is isomorphic to $\mathfrak{R}_z$, $\mathfrak{Q}_z$ and $\mathfrak{Q}_{\alpha(z)}$ have the same field of moduli, hence $k(\varphi(z)) = k(\varphi(\alpha(z)))$. It follows that S_g is defined over k. Therefore we can construct a quotient variety V_1 of V by $\{S_g \mid g \in \mathfrak{A}\}$ over k. We may assume that V_1 is a Zariski open subset of a projective variety, and V_1 is normal. Let π be the projection of V to V_1 and $\varphi_1 = \pi \circ \varphi$. Then φ_1 induces a biregular morphism of $\mathfrak{H}/\Gamma_0(T, \mathfrak{R}/\mathfrak{M})$ onto V_1.

Put $\delta = \mu(\alpha)^{-1}$. Since $\mathfrak{Q}_{\alpha(z)}$ is isomorphic to $\mathfrak{Q}_z \cdot \delta$, we have $S_g(\mathfrak{v}(\mathfrak{Q}_z)) = \mathfrak{v}(\mathfrak{Q}_z \cdot \delta)$, hence $S_g(\mathfrak{v}(\mathfrak{Q})) = \mathfrak{v}(\mathfrak{Q} \cdot \delta)$ for every $\mathfrak{Q}$ of type Ω. Applying σ, we get $S_g^\sigma(\mathfrak{v}^\sigma(\mathfrak{Q}')) = \mathfrak{v}^\sigma(\mathfrak{Q}' \cdot \delta)$ for every $\mathfrak{Q}'$ of type Ω^σ. If $\mathfrak{Q}' = \mathfrak{Q}'_z = \mathfrak{Q}_z \cdot \varepsilon$, then $\mathfrak{Q}' \cdot \delta$ is isomorphic to $\mathfrak{Q}'_{\alpha(z)}$. Therefore

$$S_g^\sigma\big(\varphi_\sigma(z)\big) = S_g^\sigma\big(\mathfrak{v}^\sigma(\mathfrak{Q}')\big) = \mathfrak{v}^\sigma(\mathfrak{Q}' \cdot \delta) = \mathfrak{v}^\sigma(\mathfrak{Q}'_{\alpha(z)}) = \varphi_\sigma\big(\alpha(z)\big) \; .$$

It follows that $S_g^\sigma \circ f_\sigma = f_\sigma \circ S_g$. Hence there exists a biregular morphism h_σ of V_1 to V_1^σ, defined over k, such that $h_\sigma \circ \pi = \pi^\sigma \circ f_\sigma$.

Now for every weak PEL-structure $\mathfrak{R}'$ of type Ω_0, we can find a member $\mathfrak{R}_z$ of $\Sigma(\Omega_0)$ isomorphic to $\mathfrak{R}'$. Define $\mathfrak{v}_1(\mathfrak{R}')$ by $\mathfrak{v}_1(\mathfrak{R}') = \varphi_1(z)$. From our construction of V_1, we see that $\mathfrak{v}_1(\mathfrak{R}')$ is uniquely determined by $\mathfrak{R}'$, and independent of the choice of z. Furthermore, $\mathfrak{v}_1(\mathfrak{R}') = \mathfrak{v}_1(\mathfrak{R}'')$ if and only if $\mathfrak{R}'$ and $\mathfrak{R}''$ are isomorphic. Since σ is the identity on k_0, $\mathfrak{R}^\sigma$ is of type Ω_0 if $\mathfrak{R}$ is of type Ω_0. Define $\mathfrak{v}_1^\sigma$ by $\mathfrak{v}_1^\sigma(\mathfrak{R}) = \mathfrak{v}_1(\mathfrak{R}^{\sigma^{-1}})^\sigma$ for every $\mathfrak{R}$ of type Ω_0. If $\mathfrak{Q}$ is of type Ω^σ and $\mathfrak{R} = \mathfrak{Q} \cdot E$, we have $\mathfrak{v}_1(\mathfrak{R}^{\sigma^{-1}})^\sigma = \pi(\mathfrak{v}(\mathfrak{Q}^{\sigma^{-1}}))^\sigma = \pi^\sigma(\mathfrak{v}^\sigma(\mathfrak{Q}))$, so that $\mathfrak{v}_1^\sigma$ is determined only by the effect of σ on k. Now, for every $\mathfrak{Q}$ of type Ω, we have

$$
\begin{aligned}
h_\sigma\big(\mathfrak{v}_1(\mathfrak{Q} \cdot E)\big) &= h_\sigma\big(\pi(\mathfrak{v}(\mathfrak{Q}))\big) \\
&= \pi^\sigma\big(f_\sigma(\mathfrak{v}(\mathfrak{Q}))\big) = \pi^\sigma\big(f_\sigma(\mathfrak{v}'(\mathfrak{Q} \cdot \varepsilon))\big) = \pi^\sigma\big(\mathfrak{v}^\sigma(\mathfrak{Q} \cdot \varepsilon)\big) = \mathfrak{v}_1^\sigma(\mathfrak{Q} \cdot E) \; .
\end{aligned}
$$

We obtain therefore

$$(4.18.1) \qquad\qquad h_\sigma \circ \mathfrak{v}_1 = \mathfrak{v}_1^\sigma \; .$$

As is seen above, we can understand that h_σ and $\mathfrak{v}_1^\sigma$ are determined for the element σ of the Galois group $G(k/k_0)$. Then, from (4.18.1), we get $h_{\sigma\tau} = (h_\sigma)^\tau \circ h_\tau$ for every $\sigma, \tau \in G(k/k_0)$. By Weil [17], we obtain a variety V_0 defined over k_0 and a biregular morphism h of V_0 to V_1, defined over k, such that

$$(4.18.2) \qquad h_\sigma = h^\sigma \circ h^{-1} \qquad\qquad \textit{for every } \sigma \in G(k/k_0) \,.$$

We may assume that V_0 is a Zariski open subset of a projective variety. Now we are ready to state and prove the main results of this section.

4.19. PROPOSITION. *Let $\mathfrak{R}$ be a weak* PEL*-structure of type Ω_0. Then there exists a subfield l of C which is characterized by the following property: an automorphism σ of C is the identity mapping on l if and only if $\mathfrak{R}^\sigma$ is isomorphic to $\mathfrak{R}$.*

We call l the field of moduli of $\mathfrak{R}$.

4.20. THEOREM. *Let Ω be as in 4.1, and let E, Ω_0 be as in 4.8. Further let $\Gamma_0(T, \mathfrak{R}/\mathfrak{M})$ be as in 4.10, and $k(\Omega_0)$ be as in 4.14. Put $k_0 = k(\Omega_0)$. Then one can define a variety V_0 and assign, to every weak* PEL*-structure $\mathfrak{R}$ of type Ω_0, exactly one point $\mathfrak{v}_0(\mathfrak{R})$ of V_0 so that the following conditions are satisfied.*

(4.20.1) V_0 is defined over k_0, and everywhere normal.

(4.20.2) $\mathfrak{v}_0(\mathfrak{R}) = \mathfrak{v}_0(\mathfrak{R}')$ if and only if $\mathfrak{R}$ and $\mathfrak{R}'$ are isomorphic.

(4.20.3) Let $\mathfrak{Q} = (A, \mathcal{C}, \theta; t_1, \cdots, t_s)$ be a PEL*-structure of type Ω, defined over a field K, and $\mathfrak{p}$ a C-valued discrete place of K. Suppose that A is without defect for $\mathfrak{p}$, and $\mathfrak{p}(c) = c$ for every $c \in k_0$. Let $\mathfrak{R} = \mathfrak{Q} \cdot E$, $\mathfrak{p}(\mathfrak{R}) = \mathfrak{p}(\mathfrak{Q}) \cdot E$. Then $\mathfrak{p}(\mathfrak{R})$ is of type Ω_0, $\mathfrak{v}_0(\mathfrak{R})$ is rational over K, and $\mathfrak{p}(\mathfrak{v}_0(\mathfrak{R})) = \mathfrak{v}_0(\mathfrak{p}(\mathfrak{R}))$.*

(4.20.4) The field of moduli of $\mathfrak{R}$ is $k_0(\mathfrak{v}_0(\mathfrak{R}))$.

(4.20.5) There exists a holomorphic mapping φ_0 of $\mathcal{K}$ onto V_0, which induces a biregular isomorphism of $\mathcal{K}/\Gamma_0(T, \mathfrak{R}/\mathfrak{M})$ onto V_0, and such that $\mathfrak{v}_0(\mathfrak{R}_z) = \varphi_0(z)$ for every member $\mathfrak{R}_z$ of $\Sigma(\Omega_0)$ defined in 4.10.

(4.20.6) V_0 is a Zariski open subset of a projective variety.

PROOFS OF 4.19 AND 4.20. Let the notation be as in 4.18. Define $\mathfrak{v}_0$ by $\mathfrak{v}_0(\mathfrak{R}) = h^{-1}(\mathfrak{v}_1(\mathfrak{R}))$. Put $\varphi_0 = h^{-1} \circ \varphi_1$. Then $(V_0, \mathfrak{v}_0, \varphi_0)$ satisfies the conditions (4.20. 1,2,5,6). Now let σ be an automorphism of C over k_0. Then

(4.20.7) $\mathfrak{v}_0(\mathfrak{R})^\sigma = \mathfrak{v}_0(\mathfrak{R}^\sigma)$ for every $\mathfrak{R}$ of type Ω_0.

In fact, by (4.18.1) and (4.18.2), we have

$$\mathfrak{d}_0(\mathcal{R})^\sigma = (h^\sigma)^{-1}\big(\mathfrak{d}_1(\mathcal{R})^\sigma\big)$$
$$= (h^\sigma)^{-1}\big(\mathfrak{d}_1^\sigma(\mathcal{R}^\sigma)\big) = h^{-1} \circ h_\sigma^{-1}\big(\mathfrak{d}_1^\sigma(\mathcal{R}^\sigma)\big) = h^{-1}\big(\mathfrak{d}_1(\mathcal{R}^\sigma)\big) = \mathfrak{d}_0(\mathcal{R}^\sigma) \; .$$

Let $l = k_0(\mathfrak{d}_0(\mathcal{R}))$. If σ is the identity on l, we have $\mathfrak{d}_0(\mathcal{R}) = \mathfrak{d}_0(\mathcal{R})^\sigma = \mathfrak{d}_0(\mathcal{R}^\sigma)$, so that $\mathcal{R}$ is isomorphic to $\mathcal{R}^\sigma$. Conversely, given an automorphism σ of C, suppose that $\mathcal{R}$ is isomorphic to $\mathcal{R}^\sigma$. Then $\mathfrak{d}_0(\mathcal{R}) = \mathfrak{d}_0(\mathcal{R}^\sigma)$. By 4.14, σ is the identity on k_0, hence $\mathfrak{d}_0(\mathcal{R}^\sigma) = \mathfrak{d}_0(\mathcal{R})^\sigma$ by (4.20.7). Therefore we have $\mathfrak{d}_0(\mathcal{R})^\sigma = \mathfrak{d}_0(\mathcal{R})$, so that σ is the identity mapping on l. This proves 4.19, and (4.20.4) at the same time.

Let the notation be as in (4.20.3). As in [13, 6.4], we can find a finite set x of elements of K with the property

(4.20.8) $\mathfrak{Q}$ *is defined over* $k_0(x)$, *and for every* C-*valued discrete places* $\mathfrak{p}'$ *and* $\mathfrak{p}''$ *of* K, *one has* $\mathfrak{p}'(\mathfrak{Q}) = \mathfrak{p}''(\mathfrak{Q})$ *if* $\mathfrak{p}'(x) = \mathfrak{p}''(x)$.

Let $y = (x, \mathfrak{d}_0(\mathcal{R}), \mathfrak{d}(\mathfrak{Q}))$ and $w = \mathfrak{p}(y)$. Since w is a specialization of y over k_0, we can find a generic specialization y' of y over k_0 so that w is a specialization of y' over k. Let τ be an isomorphism of $k_0(y)$ to $k_0(y')$ over k_0 such that $y' = y^\tau$. Since w is a specialization of y' over k, we can find a discrete place $\mathfrak{q}$ of $k(y')$ so that $\mathfrak{q}(y') = w$ and $\mathfrak{q}(c) = c$ for every $c \in k$. Since $\mathfrak{q}(y') = \mathfrak{p}(y)$, we have

(4.20.9) $\qquad \mathfrak{q}\big(\mathfrak{d}_0(\mathcal{R})^\tau\big) = \mathfrak{p}\big(\mathfrak{d}_0(\mathcal{R})\big) \;, \qquad \mathfrak{q}(\mathfrak{Q}^\tau) = \mathfrak{p}(\mathfrak{Q}) \;,$

in view of (4.20.8). In fact this can be seen by putting $\mathfrak{p}'(a) = \mathfrak{q}(a^\tau)$ for every $a \in k_0(x)$. Since $\mathfrak{Q}^\tau$ is of type Ω^τ, $\mathfrak{q}(\mathfrak{Q}^\tau)$ is of type Ω^τ, and hence $\mathfrak{p}(\mathcal{R}) = \mathfrak{p}(\mathfrak{Q}) \cdot E$ is of type Ω_0. Now we have $\mathfrak{d}_0(\mathcal{R}) = h^{-1} \circ \pi(\mathfrak{d}(\mathfrak{Q}))$ and so $\mathfrak{d}_0(\mathcal{R})^\tau = (h^{-1} \circ \pi)^\tau(\mathfrak{d}(\mathfrak{Q})^\tau)$. Since $(h^{-1} \circ \pi)^\tau$ is defined over k and $\mathfrak{q}(c) = c$ for every $c \in K$, we obtain

(4.20.10) $\qquad \mathfrak{q}\big(\mathfrak{d}_0(\mathcal{R})^\tau\big) = (h^{-1} \circ \pi)^\tau\big(\mathfrak{q}(\mathfrak{d}(\mathfrak{Q})^\tau)\big) \;.$

Since $(V^\tau, \mathfrak{d}^\tau)$ is a moduli-variety for Ω^τ, we have

$$\mathfrak{q}\big(\mathfrak{d}(\mathfrak{Q})^\tau\big) = \mathfrak{q}\big(\mathfrak{d}^\tau(\mathfrak{Q}^\tau)\big) = \mathfrak{d}^\tau\big(\mathfrak{q}(\mathfrak{Q}^\tau)\big) = \mathfrak{d}^\tau\big(\mathfrak{p}(\mathfrak{Q})\big) \;.$$

Hence, by (4.20.9) and (4.20.10), we have

$$\mathfrak{p}\big(\mathfrak{d}_0(\mathcal{R})\big) = \mathfrak{q}\big(\mathfrak{d}_0(\mathcal{R})^\tau\big) = (h^{-1} \circ \pi)^\tau\big(\mathfrak{d}^\tau(\mathfrak{p}(\mathfrak{Q}))\big) \;.$$

Extend τ to an automorphism of C. Then

$$(h^{-1} \circ \pi)^\tau\big(\mathfrak{d}^\tau(\mathfrak{p}(\mathfrak{Q}))\big) = \big(h^{-1} \circ \pi \circ \mathfrak{d}(\mathfrak{p}(\mathfrak{Q})^{\tau^{-1}})\big)^\tau = \mathfrak{d}_0\big(\mathfrak{p}(\mathcal{R})^{\tau^{-1}}\big)^\tau = \mathfrak{d}_0\big(\mathfrak{p}(\mathcal{R})\big)$$

in view of (4.20.7). Thus we get $\mathfrak{p}(\mathfrak{d}_0(\mathcal{R})) = \mathfrak{d}_0(\mathfrak{p}(\mathcal{R}))$, which completes the proof.

4.21. PROPOSITION. *The notation and assumption being as in 4.20, let* V' *be a variety and* $\mathfrak{d}'$ *an "assignment" which assigns a point* $\mathfrak{d}'(\mathcal{R})$ *of* V' *to every weak* PEL-*structure* $\mathcal{R}$ *of type* Ω_0, *satisfying the conditions* (4.20.1–4) *(with V', $\mathfrak{v}'$ in place of V_0, $\mathfrak{v}_0$). Suppose that every point of V' is of the form $\mathfrak{v}'(\mathcal{R})$ with a weak PEL-structure $\mathcal{R}$ of type Ω_0. Then there exists a biregular morphism j of V onto V', defined over k_0, such that $\mathfrak{v}'(\mathcal{R}) = j(\mathfrak{v}(\mathcal{R}))$.*

PROOF. By our assumption, we can define a one-to-one mapping j of V onto V' such that $j(\mathfrak{v}_0(\mathcal{R})) = \mathfrak{v}'(\mathcal{R})$ for every weak PEL-structure $\mathcal{R}$ of type Ω_0. For every PEL-structure $\mathfrak{Q}$ of type Ω, put $\mathfrak{v}''(\mathfrak{Q}) = \mathfrak{v}'(\mathfrak{Q} \cdot E)$. Then, with V and $\mathfrak{v}$ as before, there exists, by [13, 6.7], a morphism j_1 of V to V', defined over k, such that $\mathfrak{v}''(\mathfrak{Q}) = j_1(\mathfrak{v}(\mathfrak{Q}))$. We see easily that $j \circ h^{-1} \circ \pi = j_1$. Let x be a generic point of V over k, and $y = h^{-1} \circ \pi(x)$. Let $b \in V_0$. Let (b, c) be a specialization of $(y, j(y))$ over k. Extend this to a specialization $(x, y, j(y)) \to (a, b, c)$ over k. Since V_0 is the quotient variety by a finite group, we have $a \in V$, and hence $b = h^{-1} \circ \pi(a)$. Since $j(y) = j_1(x)$, we have

$$c = j_1(a) = j \circ h^{-1} \circ \pi(a) = j(b) \ .$$

It follows that $j(y)$ has the only specialization $j(b)$ over the specialization $y \to b$ ref. k. This implies that j is a morphism of V_0 to V' defined over k. By (4.20.4), we have $k_0(j(y)) = k_0(y)$. Hence j is defined over k_0. Since V' is normal and j is one-to-one, j is biregular. This completes the proof.

4.22. We call $(V_0, \mathfrak{v}_0)$ obtained in 4.20 a *moduli-variety for weak PEL-structures of type Ω_0*. Further, with φ_0 satisfying (4.20.5), we call $(V_0, \mathfrak{v}_0, \varphi_0)$ a *moduli-system* for the family $\Sigma(\Omega_0)$ of weak PEL-structures.

4.23. PROPOSITION. *Let Ω_0 be a weak PEL-type as in 4.20, and $k(\Omega_0)$ the field defined in 4.14. For every automorphism σ of C, there exists a weak PEL-type Ω_0^σ which is characterized, up to equivalence, by the following property.*

(4.23.1) *If $\mathcal{R}$ is of type Ω_0, then $\mathcal{R}^\sigma$ is of type Ω_0^σ.*

Moreover, the following assertions hold.

(4.23.2) *Ω_0^σ is equivalent to Ω_0^τ if and only if $\sigma = \tau$ on $k(\Omega_0)$.*

(4.23.3) *$k(\Omega_0^\sigma) = k(\Omega_0)^\sigma$.*

(4.23.4) *Let $(V_0, \mathfrak{v}_0)$ be a moduli-variety for weak PEL-structures of type Ω_0. Define $\mathfrak{v}_0^\sigma$ by $\mathfrak{v}_0^\sigma(\mathcal{R}) = \mathfrak{v}_0(\mathcal{R}^{\sigma^{-1}})^\sigma$. Then $\mathfrak{v}_0^\sigma$ depends only on the effect of σ on $k(\Omega_0)$, and $(V_0^\sigma, \mathfrak{v}_0^\sigma)$ is a moduli-variety for weak PEL-structures of type Ω_0^σ.*

PROOF. Let Ω be as in 4.1. We can define Ω^σ as in [13, 5.1] and [11, III, 3.1]. Then Ω^σ is of the form $\Omega^\sigma = (L, \Phi^\sigma, \rho; T', \mathfrak{M}'; v_1', \cdots, v_s')$ with the same L, ρ and s. Define Ω_0^σ by $\Omega_0^\sigma = (L, \Phi^\sigma, \rho; E \cdot T, \mathfrak{M}'; v_1', \cdots, v_s')$. Then Ω_0^σ has the property (4.23.1). It is obvious that Ω_0^σ can be characterized by (4.23.1). The property (4.23.2) follows immediately from (4.23.1) and (4.14.2). Then

(4.23.3) is obvious. Define $\mathfrak{v}_0^\sigma$ as in (4.23.4). If $\sigma = \tau$ on $k(\Omega_0)$, then

$$\mathfrak{v}_0(\mathcal{R}^{\tau^{-1}})^\tau = \left(\mathfrak{v}_0(\mathcal{R}^{\tau^{-1}})^{\tau\sigma^{-1}}\right)^\sigma = \mathfrak{v}_0(\mathcal{R}^{\tau^{-1}\tau\sigma^{-1}})^\sigma = \mathfrak{v}_0(\mathcal{R}^{\sigma^{-1}})^\sigma \ .$$

Therefore $\mathfrak{v}_0^\sigma$ depends only on the effect of σ on $k(\Omega_0)$. It can be verified in a straightforward way that $(V_0^\sigma, \mathfrak{v}_0^\sigma)$ satisfies (4.20.1–4) for Ω_0^σ. This completes the proof.

4.24. Let Z denote the center of L. Then either $Z = F$, or Z is a quadratic extension of L. Let $\mathfrak{r}$ denote the ring of integers in Z. Suppose that our lattice $\mathfrak{M}$ and points $v_1, \cdots, v_s$ of W satisfy

$$\mathfrak{r}\mathfrak{M} \subset \mathfrak{M} \ , \qquad \mathfrak{c}^{-1}\mathfrak{M} = \mathfrak{M} + \sum_{i=1}^s \mathfrak{r}v_i$$

with an integral ideal $\mathfrak{c}$ in Z. Let $\mathfrak{a}$ be an ideal in Z, prime to $\mathfrak{c}$, and $y_1, \cdots, y_q$ be points in W such that $\mathfrak{c}^{-1}\mathfrak{a}\mathfrak{M} = \mathfrak{a}\mathfrak{M} + \sum_{i=1}^q \mathfrak{r}y_i$. Let Ω_0 be as before, and let $\Omega_0' = (L, \Phi, \rho; E \cdot T, \mathfrak{a}\mathfrak{M}; y_1, \cdots, y_q)$. We can construct the family $\Sigma(\Omega_0') = \{\mathcal{R}_z' \mid z \in \mathcal{H}\}$ by using the same parametrizing mapping $\mathfrak{y}(x, z)$ for $\Sigma(\Omega_0)$ (cf. 4.2 and 4.15). Namely we have

$$\mathcal{R}_z' = (A_z', \mathfrak{D}_z', \theta_z'; u_{1z}, \cdots, u_{qz})$$

with an abelian variety A_z' isomorphic to $C^n/\mathfrak{y}(\mathfrak{a}\mathfrak{M}, z)$. We see easily

$$(4.24.1) \quad \begin{aligned} \sum_{i=1}^s \theta_z(\mathfrak{r})t_{iz} &= \{t \in A_z \mid \theta_z(\mathfrak{c})t = 0\} \ , \\ \sum_{i=1}^q \theta_z'(\mathfrak{r})u_{iz} &= \{u \in A_z' \mid \theta_z'(\mathfrak{c})u = 0\} \ . \end{aligned}$$

Let $(V, \mathfrak{b}, \varphi)$ be a moduli-system for the family $\Sigma(\Omega_0)$. For every weak PEL-structure $\mathcal{R}'$ of type Ω_0', we can find a point z on $\mathcal{H}$ so that $\mathcal{R}'$ is isomorphic to $\mathcal{R}_z'$. Define $\mathfrak{v}'(\mathcal{R}')$ by $\mathfrak{v}'(\mathcal{R}') = \varphi(z)$.

4.25. PROPOSITION. *The notation and assumption being as in 4.24, $(V, \mathfrak{v}', \varphi)$ is a moduli-system for the family $\Sigma(\Omega_0')$, and $k(\Omega_0) = k(\Omega_0')$.*

PROOF. First we note that $\Gamma_0(T, \mathfrak{c}^{-1}\mathfrak{M}/\mathfrak{M}) = \Gamma_0(T, \mathfrak{c}^{-1}\mathfrak{a}\mathfrak{M}/\mathfrak{a}\mathfrak{M})$. Let d be an element of Z such that $d\mathfrak{a}^{-1}$ is integral and prime to $\mathfrak{c}$. We observe that $\Lambda(d, z)$ of 4.4 gives an isogeny λ_z of A_z to A_z', whose kernel is isomorphic to $d^{-1}\mathfrak{a}\mathfrak{M}/\mathfrak{M}$. We see that

$$\mathrm{Ker}(\lambda_z) = \{t \in A_z \mid \theta_z(d\mathfrak{a}^{-1})t = 0\} \ ,$$
$$\lambda_z\left(\sum_{i=1}^s \theta_z(\mathfrak{r})t_{iz}\right) = \sum_{i=1}^q \theta_z'(\mathfrak{r})u_{iz} \ .$$

The last equality follows easily from (4.24.1). Therefore we have

$$u_{iz} = \sum_{j=1}^s \lambda_z \cdot \theta_z(b_{ij})t_{jz}$$

with $b_{ij} \in \mathfrak{r}$. Now let us show that

(4.25.1) $\mathcal{R}_z$ *and* $\mathcal{R}_z'$ *have the same field of moduli.*

Let l_z resp. l_z' denote the field of moduli of $\mathcal{R}_z$ resp. $\mathcal{R}_z'$. Let σ be an automorphism of C over l_z. Then there exists an isomorphism ε of $\mathcal{R}_z$ to $(\mathcal{R}_z)^\sigma$. We see that λ_z^σ is an isogeny of A_z^σ to $A_z'^\sigma$, and

$$\mathrm{Ker}(\lambda_z^\sigma) = \{t \in A_z^\sigma \mid \theta_z^\sigma(d\mathfrak{a}^{-1})t = 0\}\,, \qquad u_{iz}^\sigma = \sum_{j=1}^s \lambda_z^\sigma\theta_z^\sigma(b_{ij})t_{jz}^\sigma\,.$$

We observe that $\mathrm{Ker}(\lambda_z^\sigma \circ \varepsilon) = \mathrm{Ker}(\lambda_z)$, hence there exists an isomorphism η of A_z' to $A_z'^\sigma$ such that $\eta \circ \lambda_z = \lambda_z^\sigma \circ \varepsilon$. Then we have $\eta \cdot \theta_z'(x) = \theta_z'^\sigma(x)\eta$ for every $x \in \mathfrak{r}$, and $\eta^{-1}(X'^\sigma) \equiv X' \cdot \theta_z'(\gamma)$ with $\gamma \in E$ for every divisor X' in $\mathfrak{D}_z'$, since the isogenies λ_z, λ_z^σ and ε have the corresponding properties. Furthermore, we have

$$\begin{aligned}
\eta u_{iz} &= \sum_{j=1}^s \eta\lambda_z\theta_z(b_{ij})t_{jz} \\
&= \sum_j \lambda_z^\sigma\varepsilon\theta_z(b_{ij})t_{jz} = \sum_j \lambda_z^\sigma\theta_z^\sigma(b_{ij})\varepsilon t_{jz} = \sum_j \lambda_z^\sigma\theta_z^\sigma(b_{ij})t_{jz}^\sigma = u_{iz}^\sigma\,.
\end{aligned}$$

It follows that η is an isomorphism of $\mathcal{R}_z'$ to $\mathcal{R}_z'^\sigma$. Therefore σ is the identity on l_z'. This shows $l_z' \subset l_z$. Now we have $\mathfrak{M} = \mathfrak{a}^{-1}(\mathfrak{a}\mathfrak{M})$. Therefore, exchanging $\mathfrak{M}$ and $\mathfrak{a}\mathfrak{M}$, and considering a suitable isogeny of $\mathcal{R}_z'$ to $\mathcal{R}_z$ with the same nature as λ_z, we can show that $l_z \subset l_z'$. Thus we have proved (4.25.1). Let $(V_1, \mathfrak{b}_1, \varphi_1)$ be a moduli-system for the family $\Sigma(\Omega_0')$. Since both V and V_1 are isomorphic to $\mathcal{H}/\Gamma(T, \mathfrak{c}^{-1}\mathfrak{M}/\mathfrak{M})$, there exists a biregular morphism f of V to V_1 such that $\varphi_1 = f \circ \varphi$. Let $k = k(\Omega_0)$ and $k' = k(\Omega_0')$. Let x be a generic point of V over k. Let z be a point on $\mathcal{H}$ such that $\varphi(z) = x$. Then $k' \subset l_z' = l_z = k(x)$. Since k is the algebraic closure of Q in $k(x)$, we have $k' \subset k$. Similarly $k \subset k'$, hence $k = k'$. Now we have

$$k(x) = k\big(\varphi(z)\big) = l_z = l_z' = k\big(\varphi_1(z)\big) = k\big(f(x)\big)\,.$$

This shows that f is defined over k. Since $\mathfrak{b}_1(\mathcal{R}_z') = \varphi_1(z) = f(\varphi(z)) = f(\mathfrak{b}'(\mathcal{R}_z'))$ for every z on $\mathcal{H}$, we have $\mathfrak{b}_1 = f \circ \mathfrak{b}'$, so that f gives an "isomorphism" of $(V, \mathfrak{b}', \varphi)$ to $(V_1, \mathfrak{b}_1, \varphi_1)$, defined over $k(\Omega_0')$. Therefore $(V, \mathfrak{b}', \varphi)$ is a moduli-system for $\Sigma(\Omega_0')$.

5. Generalized CM-types and class fields

5.1. Let K be an algebraic number field of finite degree, and $p = [K:Q]$. Let $\tau_1, \cdots, \tau_p$ be all the isomorphisms of K into C, and Φ a Q-linear isomorphism of K into $M_n(C)$, such that $\Phi(1)$ is the identity matrix, with any positive integer n. Then we have $\mathrm{tr}(\Phi(x)) = \sum_{\nu=1}^p r_\nu x^{\tau_\nu}$ for every $x \in K$, with non-negative integers r_ν. We express this by writing

$$(5.1.1) \qquad\qquad \Phi \sim \sum_{\nu=1}^p r_\nu \tau_\nu\,.$$

We shall now construct another couple (K', Φ') out of a given (K, Φ).

Let R be a Galois extension of Q containing K. Denote by G the Galois group of R over Q, and G_Q the group ring of G over Q. For an element $\xi = \sum_{\gamma \in G} a_\gamma\gamma$ of G_Q, we put $\xi^* = \sum_\gamma a_\gamma\gamma^{-1}$. For each ν, take an element of G

whose restriction to K is τ_ν, and denote it again by τ_ν. Then $G = \bigcup_{\nu=1}^{p} H\tau_\nu$. Define elements η and ω of G_Q by

$$(5.1.2) \qquad \eta = \sum_{\gamma \in H} \gamma , \qquad \omega = \eta \cdot \sum_{\nu=1}^{p} r_\nu \tau_\nu .$$

Let K' be the field generated over Q by the elements $\mathrm{tr}(\Phi(x))$ for all $x \in K$, and H' the subgroup of G corresponding to K'. One can easily verify the following two assertions.

(5.1.3) $H' = \{\sigma \in G \mid \omega\sigma = \omega\}$.

(5.1.4) *For every $x \in K$, the characteristic polynomial of $\Phi(x)$ has coefficients in K'.*

Let $G = \bigcup_{\mu=1}^{q} H'\sigma_\mu$ be a disjoint union, with $q = [K' : Q]$. Define an element η' of G_Q by $\eta' = \sum_{\gamma \in H'} \gamma$. In view of (5.1.3), one can find q non-negative integers $s_1, \cdots, s_q$ such that $\omega^* = \eta' \sum_{\mu=1}^{q} s_\mu \sigma_\mu$. Define a representation Φ' of K' by complex matrices so that $\Phi' \sim \sum_{\mu=1}^{q} s_\mu \sigma_\mu$. It can be easily verified that Φ' is uniquely determined, up to equivalence, by (K, Φ), and independent of the choice of R. We call (K', Φ') *the dual of* (K, Φ).

5.2. *Examples.* First suppose that the r_ν are all equal to 1. Then $K' = Q$ and Φ' is the identity mapping.

Next, suppose that τ_1 is the identity mapping, $r_1 = 1, r_2 = \cdots = r_p = 0$. Then $K' = K$ and $\Phi' = \Phi$.

These are extreme (but still meaningful) cases. More typical examples will be seen shortly.

5.3. The notation being as in 5.1, for every ideal $\mathfrak{x}$ in K, we define $\det \Phi(\mathfrak{x})$ by

$$(5.3.1) \qquad \det \Phi(\mathfrak{x}) = \prod_{\nu=1}^{p} (\mathfrak{x}^{\tau_\nu})^{r_\nu} .$$

The right hand side should be considered as an ideal (or a divisor) in an extension of K. In general, let M be a finite extension of an algebraic number field F and $\mathfrak{a}$ an ideal in M. For convenience, we say that $\mathfrak{a}$ *is an ideal in* F if $\mathfrak{a} = \mathfrak{b}\mathfrak{r}_M$ with an ideal $\mathfrak{b}$ in F, and often identify $\mathfrak{a}$ with $\mathfrak{b}$ when there is no fear of confusion.

5.4. PROPOSITION. *Let (K, Φ) and (K', Φ') be as in 5.1. Then, for every element u of K and every ideal $\mathfrak{x}$ in K, $\det \Phi(u) \in K'$ and $\det \Phi(\mathfrak{x})$ is an ideal in K'. Furthermore, for every element v of K' and every ideal $\mathfrak{x}$ in K', $\det \Phi'(v) \in K$, and $\det \Phi'(\mathfrak{y})$ is an ideal in K.*

This is almost obvious. The proof in the special case [15, 8.3, Prop. 29] applies to the general case.

5.5. Proposition. *Let (K, Φ) and (K', Φ') be as in 5.1, and (K'', Φ'') the dual of (K', Φ'). Then $K'' \subset K'$, and $\operatorname{tr} \Phi(x) = \operatorname{tr} \Phi''(\operatorname{Tr}_{K/K''}(x))$ for every $x \in K$. In particular, if $K'' = K$, then $(K'', \Phi'') = (K, \Phi)$.*

This can be verified in a straightforward way.

5.6. Proposition. *Let (K', Φ') resp. (L', Ψ') be the dual of (K, Φ) resp. (L, Ψ). Suppose that $K \subset L$ and $\operatorname{tr} \Phi(x) = \operatorname{tr} \Psi(x)$ for every $x \in K$. Then $K' \subset L'$. Moreover, for every element w of L' and every ideal $\mathfrak{y}$ in L', one has*

$$N_{L/K}(\det \Psi'(w)) = \det \Phi'(N_{L'/K'}(w)) ,$$
$$N_{L/K}(\det \Psi'(\mathfrak{y})) = \det \Phi'(N_{L'/K'}(\mathfrak{y})) .$$

Proof. Let $R, G, H, H', \tau_\nu, r_\nu, \eta, \eta', \omega$ be as in 5.1. We can take R so that $L \subset R$. Let E resp. E' be the subgroup of G corresponding to L resp. L', and let $G = \bigcup_{i=1}^{k} E\alpha_i$ and $G = \bigcup_{j=1}^{m} E'\beta_j$ be disjoint unions, where $k = [L:Q]$ and $m = [L':Q]$. Let $\Psi \sim \sum_{i=1}^{k} t_i \alpha_i$ with non-negative integers t_i. Define elements ξ, ξ', χ of G_Q by

$$\xi = \sum_{\gamma \in E} \gamma , \qquad \xi' = \sum_{\gamma \in E'} \gamma , \qquad \chi = \xi \cdot \sum_{i=1}^{k} t_i \alpha_i .$$

Then $\chi^* = \xi' \cdot \sum_{j=1}^{m} u_j \beta_j$, $\Psi' \sim \sum_{j=1}^{m} u_j \beta_j$ with non-negative integers u_j. Now the first assertion $K' \subset L'$ is obvious. Since $\operatorname{tr} \Phi(x) = \operatorname{tr} \Psi(x)$ for $x \in K$, we have $\eta \cdot \sum_{i=1}^{k} t_i \alpha_i = \omega$. Let $H = \bigcup_\lambda E\varphi_\lambda$ be a disjoint union. Then we have

$$\begin{aligned}
\omega^* &= (\textstyle\sum_i t_i \alpha_i)^* \cdot \eta \\
&= (\textstyle\sum_i t_i \alpha_i)^* \cdot \xi \cdot (\textstyle\sum_\lambda \varphi_\lambda) = \chi^* \cdot (\textstyle\sum_\lambda \varphi_\lambda) = \xi' \cdot (\textstyle\sum_j u_j \beta_j) \cdot (\textstyle\sum_\lambda \varphi_\lambda) .
\end{aligned}$$

Our formulas follow from this easily.

5.7. Now we shall confine ourselves to a special type of (K, Φ) called a generalized CM-type. We denote by x^ρ or $\bar{x}$ the complex conjugate of a complex number x. By a CM-*field*, we understand a totally imaginary quadratic extension of a totally real algebraic number field of finite degree. The following three propositions can be proved in a straightforward way.

5.8. Proposition. *An algebraic number field K of finite degree is a* CM-*field if and only if the following two conditions are satisfied.*

(5.8.1) *ρ induces a non-trivial automorphism of K.*

(5.8.2) *$\rho\tau = \tau\rho$ for every isomorphism τ of K into C.*

5.9. Proposition. *Let K be a* CM-*field and L the smallest Galois extension of Q containing K. Then L is a* CM-*field.*

5.10. Proposition. *The composite of a finite number of* CM-*fields is a* CM-*field.*

5.11. We call a Q-linear isomorphism Φ of a CM-field K into $M_n(C)$, with any positive integer n, a CM-*representation* of K if the following two conditions are satisfied.

(5.11.1) Φ *maps the identity of K to the identity matrix.*

(5.11.2) *The direct sum of Φ and Φ^ρ is equivalent to a rational representation of K.*

Let $2g = [K:Q]$. Let $\tau_1, \cdots, \tau_{2g}$ be all the isomorphisms of K into C, and $\Phi \sim \sum_{\nu=1}^{2g} r_\nu \tau_\nu$ with non-negative integers r_ν. Arrange the τ_ν so that $\tau_{g+\nu} = \tau_\nu \rho$ for $\nu = 1, \cdots, g$. Then (5.11.2) is equivalent to

$$(5.11.3) \qquad\qquad r_\nu + r_{g+\nu} = n/g \qquad\qquad (\nu = 1, \cdots, g) .$$

Therefore n/g should be an integer. We call (K, Φ) a *generalized* CM-*type* if Φ is a CM-representation of a CM-field K. The integer n/g is called *the index* of (K, Φ), or of Φ. A generalized CM-type of index one is exactly a CM-type in the sense of [15, 5.2]. We observe that

$$(5.11.4) \qquad \det\left(\Phi(x)\right)\det\left(\Phi(x)\right)^\rho = N_{K/Q}(x)^{n/g} \qquad \textit{for every } x \in K .$$

5.12. Let (K, Φ) be a generalized CM-type, and (K', Φ') the dual of (K, Φ). Let τ_ν and r_ν be as above, and let σ be an automorphism of C. Then

$$\left(\operatorname{tr}\Phi(x)\right)^{\sigma\rho} = \sum_\nu r_\nu x^{\tau_\nu\sigma\rho} = \sum_\nu r_\nu x^{\rho\tau_\nu\sigma} = \sum_\nu r_\nu x^{\tau_\nu\rho\sigma} = \left(\operatorname{tr}\Phi(x)\right)^{\rho\sigma}$$

in view of 5.8. Hence $\sigma\rho = \rho\sigma$ on K'. Applying 5.8 to K', we obtain

(5.12.1) K' *is Q or a* CM-*field according as Φ^ρ is equivalent to Φ or not.*

We call (K, Φ) *effective* if Φ^ρ is not equivalent to Φ. Every (K, Φ) of an odd index is effective.

5.13. PROPOSITION. *Let (K, Φ) be an effective generalized* CM-*type, and (K', Φ') the dual of (K, Φ). Then (K', Φ') is an effective generalized* CM-*type of the same index as (K, Φ).*

PROOF. Let R, G and ω be as in 5.1. We can take R so as to be a CM-field, in view of 5.9. Then, ρ, as an element of G, belongs to the center of G. If m is the index of (K, Φ), we have $\omega + \omega\rho = m \cdot \sum_{\gamma \in G} \gamma$, so that $\omega^* + \omega^*\rho = m \cdot \sum_{\gamma \in G} \gamma$. This shows that Φ' is a CM-representation of index m. Furthermore, since (K, Φ) is effective, we have $\omega \neq \omega\rho$, hence $\omega^* \neq \omega^*\rho$. Therefore (K', Φ') is effective.

5.14. Let (K, Φ) be a generalized CM-type of index 2, and let $2g = [K:Q]$. Then we can find g isomorphisms $\tau_1, \cdots, \tau_g$ of K into C and an integer r such that

$$(5.14.1) \qquad\qquad \Phi \sim \sum_{\nu=1}^r (\tau_\nu + \tau_\nu\rho) + 2 \cdot \sum_{\nu=r+1}^g \tau_\nu .$$

(5.14.2) $$0 \leq r \leq g .$$

(5.14.3) $\{\tau_1, \cdots, \tau_g, \tau_1\rho, \cdots, \tau_g\rho\}$ *is the set of all isomorphisms of* K *into* C. (*In other words,* $(K; \{\tau_1, \cdots, \tau_g\})$ *is a* CM-*type in the sense of* [15].)

Hereafter until the end of 5.22, we fix such a (K, Φ), $\{\tau_\nu\}$, and assume

(5.14.4) $$0 < r < g .$$

This implies especially that (K, Φ) is effective, and $g \geq 2$. We shall denote by F the maximal totally real subfield of K. Obviously $[K : F] = 2$ and $F = \{x \in K \mid x^\rho = x\}$.

Let (K', Φ') be the dual of (K, Φ), and $[K' : Q] = 2h$. Since (K', Φ') is of index 2, we can find an expression for Φ' similar to (5.14.1). Thus there exist h isomorphisms $\{\sigma_1, \cdots, \sigma_h\}$ of K' into C and an integer s such that

(5.14.5) $$\Phi' \sim \sum_{\lambda=1}^{s} (\sigma_\lambda + \sigma_\lambda\rho) + 2 \cdot \sum_{\lambda=s+1}^{h} \sigma_\lambda , \qquad 0 \leq s \leq h ,$$

and $\{\sigma_1, \cdots, \sigma_h, \sigma_1\rho, \cdots, \sigma_h\rho\}$ is the set of all isomorphisms of K' into C. Let R, G, H, H', η, η' be as in 5.1. We take R to be a CM-field. We extend τ_ν and σ_λ to elements of G and denote them again by τ_ν and σ_λ. By the definition of Φ' (5.1), we see easily that

(5.14.6) $$H' = \{\gamma \in G \mid (\textstyle\bigcup_{\nu=r+1}^{g} H\tau_\nu)\gamma = \bigcup_{\nu=r+1}^{g} H\tau_\nu\} ,$$

(5.14.7) K' *is generated over* Q *by* $\sum_{\nu=r+1}^{g} x^{\tau_\nu}$ *for all* $x \in K$.

(5.14.8) $$\textstyle\bigcup_{\nu=r+1}^{g} \tau_\nu^{-1} H = \bigcup_{\lambda=s+1}^{h} H'\sigma_\lambda ,$$

(5.14.9) $$\textstyle\bigcup_{\nu=1}^{r} (H\tau_\nu \cup H\tau_\nu\rho)^{-1} = \bigcup_{\lambda=1}^{s} (H'\sigma_\lambda \cup H'\sigma_\lambda\rho) .$$

Hence we have $0 < s < h$, so that (K', Φ') is effective. From these relations, we obtain the following assertions.

(5.14.10) *For every element* x *in* K', $\prod_{\lambda=1}^{s} x^{\sigma_\lambda} x^{\sigma_\lambda\rho}$ *is an element of* F, *and* $\prod_{\lambda=s+1}^{h} x^{\sigma_\lambda}$ *is an element of* K.

(5.14.11) *For every ideal* $\mathfrak{y}$ *in* K', $\prod_{\lambda=1}^{s} \mathfrak{y}^{\sigma_\lambda} \mathfrak{y}^{\sigma_\lambda\rho}$ *is an ideal in* F, *and* $\prod_{\lambda=s+1}^{h} \mathfrak{y}^{\sigma_\lambda}$ *is an ideal in* K.

5.15. Let M be a totally imaginary quadratic extension of F, different from K, and let $\chi_1, \cdots, \chi_g$ be g isomorphisms of M into C such that

(5.15.1) $\chi_\nu = \tau_\nu$ on F $(\nu = 1, \cdots, g)$.

Define a CM-representation Ψ of M by

(5.15.2) $$\Psi \sim 2 \cdot \sum_{\nu=1}^{r} \chi_\nu + \sum_{\nu=r+1}^{g} (\chi_\nu + \chi_\nu\rho) .$$

Let (M', Ψ') be the dual of (M, Ψ), and $2k = [M' : Q]$. Then we can find k isomorphisms $\varepsilon_1, \cdots, \varepsilon_k$ of M' into C and an integer t such that $0 < t < k$ and

$$(5.15.3) \qquad \Psi' \sim 2 \cdot \sum_{\lambda=1}^{t} \varepsilon_\lambda + \sum_{\lambda=t+1}^{k} (\varepsilon_\lambda + \varepsilon_\lambda \rho) \,.$$

Let $S = KM$. Define $2g$ isomorphisms $\alpha_1, \cdots, \alpha_g, \beta_1, \cdots, \beta_g$ of S into C by

$(5.15.4a)$ $\alpha_\nu = \tau_\nu$ *on* K, $\alpha_\nu = \chi_\nu$ *on* M *for* $\nu = 1, \cdots, g,$

$(5.15.4b)$ $\beta_\nu = \tau_\nu \rho$ *on* K, $\beta_\nu = \chi_\nu$ *on* M *for* $\nu = 1, \cdots, r,$

$(5.15.4c)$ $\beta_\nu = \tau_\nu$ *on* K, $\beta_\nu = \chi_\nu \rho$ *on* M *for* $\nu = r + 1, \cdots, g.$

Let Ξ be a CM-representation of S such that

$$(5.15.5) \qquad \Xi \sim \sum_{\nu=1}^{g} (\alpha_\nu + \beta_\nu) \,.$$

The equivalence class of Ξ is characterized by the property that the restriction of Ξ to K resp. M is equivalent to Φ resp. Ψ. We investigated the CM-type (S, Ξ) of this kind in [7]. The following 5.16–22 will provide a sharpening of the results of [7, § 4].

5.16. PROPOSITION. *Let the notation be as in 5.14 and 5.15. Let* (S', Ξ') *be the dual of* (S, Ξ). *Then* $S' = K'M'$. *Further, for every element* y *of* S' *and every ideal* $\mathfrak{z}$ *in* S', *one has*

$$\det \Xi'(y) = \prod_{\lambda=1}^{t} N_{S'/M'}(y)^{\varepsilon_\lambda} \cdot \prod_{\mu=s+1}^{h} N_{S'/K'}(y)^{\sigma_\mu} \,,$$
$$\det \Xi'(\mathfrak{z}) = \prod_{\lambda=1}^{t} N_{S'/M'}(\mathfrak{z})^{\varepsilon_\lambda} \cdot \prod_{\mu=s+1}^{h} N_{S'/K'}(\mathfrak{z})^{\sigma_\mu} \,.$$

PROOF. Applying (5.14.7) to (M, Ψ), we see that

$(5.16.1)$ M' *is generated over* $\mathbf{Q}$ *by* $\sum_{\nu=1}^{r} x^{\chi_\nu}$ *for all* $x \in M$.

For every $x \in S$, we have

$$\operatorname{tr} \Xi(x) = \sum_{\nu=1}^{r} (x^{\alpha_\nu} + x^{\beta_\nu}) + \sum_{\nu=r+1}^{g} (x^{\alpha_\nu} + x^{\beta_\nu})$$
$$= \sum_{\nu=1}^{r} \operatorname{Tr}_{S/M}(x)^{\chi_\nu} + \sum_{\nu=r+1}^{g} \operatorname{Tr}_{S/K}(x)^{\tau_\nu} \,.$$

In view of (5.14.7) and (5.16.1) we have $S' \subset M'K'$. Since the restriction of Ξ to K resp. M is equivalent to Φ resp. Ψ, we have $M'K' \subset S'$, so that $S' = M'K'$. Let R, G, H, H' be as in 5.1. We take R so that $S \subset R$, and R is a CM-field. Let J, J', E, E' be the subgroups of G corresponding to M, M', S, S', respectively. Extend $\tau_\nu, \chi_\nu, \alpha_\nu, \beta_\nu, \sigma_\lambda, \varepsilon_\mu$ to elements of G, and denote them again by the same letters. Applying (5.14.8) to (M, Ψ), we obtain

$$(5.16.2) \qquad \bigcup_{\nu=1}^{r} \chi_\nu^{-1} J = \bigcup_{\lambda=1}^{t} J' \varepsilon_\lambda \,.$$

By our definition of α_ν and β_ν, we have

$$E\alpha_\nu \cup E\beta_\nu = J\chi_\nu \qquad\qquad \text{for } \nu = 1, \cdots, r \,,$$
$$= H\tau_\nu \qquad\qquad \text{for } \nu = r + 1, \cdots, g \,.$$

Therefore, by (5.14.8) and (5.16.2), we have

$$(5.16.3) \qquad \bigcup_{\nu=1}^{g} (E\alpha_\nu \cup E\beta_\nu)^{-1} = \left(\bigcup_{\nu=1}^{r} \chi_\nu^{-1} J \right) \cup \left(\bigcup_{\nu=r+1}^{g} \tau_\nu^{-1} H \right)$$
$$= \left(\bigcup_{\lambda=1}^{t} J' \varepsilon_\lambda \right) \cup \left(\bigcup_{\lambda=s+1}^{h} H' \sigma_\lambda \right) \,.$$

Our formulas for $\det \Xi'(y)$ and $\det \Xi'(\mathfrak{z})$ follow easily from this relation.

5.17. PROPOSITION. *Let the notation be as in* 5.14., 5.15, *and* 5.16. *Suppose that* $r = 1$, *and* τ_1 *is the identity mapping on* F. *Then* $F \subset K'$, $M = M'$, $S' = MK'$, $t = 1$, *and* $\varepsilon_1 = \chi_1$ *on* M. *Moreover, for every element* x *of* K' *and every ideal* $\mathfrak{y}$ *in* K', *one has*

$$N_{K'/F}(x) = \prod_{\lambda=1}^{s} x^{\sigma_\lambda} x^{\sigma_\lambda \rho} , \qquad N_{K'/F}(\mathfrak{y}) = \prod_{\lambda=1}^{s} \mathfrak{y}^{\sigma_\lambda} \mathfrak{y}^{\sigma_\lambda \rho} .$$

Further for every element w *of* S' *and every ideal* $\mathfrak{z}$ *in* S', *one has*

$$\det \Xi'(w) = N_{S'/M}(w)^{\chi_1} \cdot \prod_{\lambda=s+1}^{h} N_{S'/K'}(w)^{\sigma_\lambda} ,$$
$$\det \Xi'(\mathfrak{z}) = N_{S'/M}(\mathfrak{z})^{\chi_1} \cdot \prod_{\lambda=s+1}^{h} N_{S'/K'}(\mathfrak{z})^{\sigma_\lambda} .$$

PROOF. If $x \in F$, we have $x = \mathrm{Tr}_{F/Q}(x) - \sum_{\nu=2}^{g} x^{\tau_\nu} \in K'$. In view of (5.15.1), χ_1 is the identity mapping or ρ on M. By (5.16.1), $M' = M^{\chi_1} = M$. Therefore, in the proof of 5.16, one has $J = J'$. Hence (5.16.2) (with $r = 1$) shows that $t = 1$ and $\chi_1^{-1} J = J' \varepsilon_1$. Since χ_1 is the identity mapping or ρ on M, ε_1 should coincide with χ_1 on M. The remaining part of our proposition follows from (5.14.9) and 5.16.

5.18. PROPOSITION. *Let the notation and the assumption be as in* 5.17. *Then* (K, Φ) *resp.* (M, Ψ) *is the dual of* (K', Φ') *resp.* (M', Ψ'). *Moreover, if* M *is not contained in* K', *then* (S, Ξ) *is the dual of* (S', Ξ'), *and the restriction of* Ξ' *to* K' *resp.* M *is equivalent to* Φ' *resp.* $([K' : F]/2) \cdot \Psi'$.

PROOF. Let (K'', Φ''), (M'', Ψ''), (S'', Ξ'') be the duals of (K', Φ'), (M', Ψ'), (S', Ξ') respectively. Since $t = 1$ and ε_1 is the identity mapping or ρ on $M = M'$, we get $M'' = M'$, so that $\Psi'' = \Psi$ by 5.5. Let H'' be the subgroup of G corresponding to K''. By 5.5, we have $K'' \subset K$, so that $H \subset H''$. Applying (5.14.9) to (K', Φ') and (K'', Φ''), we obtain

$$\bigcup_{\lambda=1}^{s} (H'\sigma_\lambda \cup H'\sigma_\lambda \rho)^{-1} = \bigcup_{\mu=1}^{p} (H''\delta_\mu \cup H''\delta_\mu \rho)$$

with elements δ_μ of G. By (5.14.9), we have

$$H\tau_1 \cup H\tau_1\rho = \bigcup_{\mu=1}^{p} (H''\delta_\mu \cup H''\delta_\mu \rho) .$$

Since $H \subset H''$, we must have $p = 1$, $H = H''$, hence $K = K''$. By 5.5, $\Phi = \Phi''$. By (5.16.3), we have, for every $z \in S'$.

$$\mathrm{tr}\,\Xi'(z) = \mathrm{Tr}_{S'/M}(z)^{\chi_1} + \sum_{\lambda=s+1}^{h} \mathrm{Tr}_{S'/K'}(z)^{\sigma_\lambda} .$$

Suppose that M is not contained in K'. If $x \in K'$, we have

$$\mathrm{tr}\,\Xi'(x) = \mathrm{Tr}_{K'/F}(x) + 2\sum_{\lambda=s+1}^{h} x^{\sigma_\lambda} = \mathrm{tr}\,\Phi'(x) .$$

Hence $K = K'' \subset S''$. Now note that $\sum_{\lambda=s+1}^{h} y^{\sigma_\lambda} \in F$ for every $y \in F$. If $z \in M$, we have $\mathrm{tr}\,\Xi'(z) = [K' : F]z^{\chi_1} + \sum_{\lambda=s+1}^{h} \mathrm{Tr}_{M/F}(z)^{\sigma_\lambda} \in M$. Applying 5.6 to (M, Ψ) and (S, Ξ), we obtain $\mathrm{Tr}_{S/M}(\mathrm{tr}\,\Xi'(z)) = \mathrm{tr}\,\Psi'(\mathrm{Tr}_{S'/M}(z))$, hence

$$2\cdot \mathrm{tr}\, \Xi'(z) = [K':F]\cdot \mathrm{tr}\, \Psi'(z) \qquad\qquad \text{if } z \in M .$$

Therefore we have $M = M'' \subset S''$, so that $S = KM \subset S''$. By 5.5, we obtain $S = S''$, and $\Xi = \Xi''$. This completes the proof.

5.19. Let $(K,\Phi), (K',\Phi'), (M,\Psi), (M',\Psi'), (S,\Xi), (S',\Xi')$ be as in 5.14–16. Suppose that $r = 1$ and τ_1 is the identity mapping on F. Let $\mathfrak{c}$ be an integral ideal in F, and e the smallest positive integer divisible by $\mathfrak{c}$. Let $I(S', \Xi', \mathfrak{c})$ be the group of ideals $\mathfrak{x}$ in S', prime to e, satisfying the following condition.

(5.19.1) *There exists an element y of S such that $\det \Xi'(\mathfrak{x}) = y\mathfrak{r}_S$, $y \equiv 1 \bmod^* \mathfrak{c}$, and $N(\mathfrak{x})^{-1}yy^\rho$ is a unit of F.* (For the notation, see 1.1.)

By (5.11.4), we have $I(S', e) \subset I(S', \Xi', \mathfrak{c})$. Hence $I(S', \Xi', \mathfrak{c})$ corresponds to a class field over S', which we denote by $C(S', \Xi', \mathfrak{c})$.

5.20. PROPOSITION. $C(S', \Xi', \mathfrak{c}) \subset C(K', e)\cdot C(M, \mathfrak{c})$.

PROOF. The composite $S'\cdot C(K', e)$ resp. $S'\cdot C(M, \mathfrak{c})$ is a class field over S' corresponding to the ideal group of ideals $\mathfrak{x}$ in S', prime to e, such that $N_{S'/K'}(\mathfrak{x}) \in I(K', e)$ resp. $N_{S'/M}(\mathfrak{x}) \in I(M, \mathfrak{c})$. Let $\mathfrak{x}$ satisfy the last two relations. Then $N_{S'/K'}(\mathfrak{x}) = (a)$, $a \equiv 1 \bmod^* (e)$ with an element a of K', and $N_{S'/M}(\mathfrak{x}) = (b)$, $b \equiv 1 \bmod^* \mathfrak{c}$ with an element b of M. By 5.17, $\prod_{\lambda=1}^{s} a^{\sigma_\lambda} a^{\sigma_\lambda \rho} = N_{K'/F}(a) \in F$, $bb^\rho \in F$, and $(bb^\rho) = N_{S'/F}(\mathfrak{x}) = N_{K'/F}(N_{S'/K'}(\mathfrak{x})) = (\prod_{\lambda=1}^{s} a^{\sigma_\lambda} a^{\sigma_\lambda \rho})$, hence $bb^\rho = u \prod_{\lambda=1}^{s} a^{\sigma_\lambda} a^{\sigma_\lambda \rho}$ with a unit u in F. Put $y = b^{\chi_1} \prod_{\lambda=s+1}^{h} a^{\sigma_\lambda}$. By (5.14.10) and 5.17, we see that $y \in S$, $\det \Xi'(\mathfrak{x}) = (y)$, and $y \equiv 1 \bmod^* \mathfrak{c}$. Furthermore, $yy^\rho = u \prod_{\lambda=1}^{h} a^{\sigma_\lambda} a^{\sigma_\lambda \rho} = u\cdot N_{K'/\mathbf{Q}}(a) = u\cdot N(\mathfrak{x})$. Therefore $\mathfrak{x} \in I(S', \Xi', \mathfrak{c})$. This proves our assertion.

5.21. Let d be an element of K such that $d^\rho = -d \neq 0$. The notation and assumption being as in 5.19, we consider the following two conditions on K and M.

(5.21.1) *K contains no roots of unity other than ± 1.*

(5.21.2) *There is no ideal $\mathfrak{a}$ in M such that $\mathfrak{a}\mathfrak{r}_S = d\mathfrak{r}_S$.*

The latter condition does not depend on the choice of d.

5.22. PROPOSITION. *The notation and assumption being as in 5.19, suppose that both (5.21.1) and (5.21.2) are satisfied. Then*

$$C(K', e)\cdot C(S', \Xi', \mathfrak{c}) = C(K', e)\cdot C(M, \mathfrak{c}) .$$

PROOF. In view of 5.20, it is sufficient to prove

$$C(M, \mathfrak{c}) \subset C(K', e)\cdot C(S', \Xi', \mathfrak{c}) .$$

Let $\mathfrak{x}$ be an ideal in $I(S', \Xi', \mathfrak{c})$, and let $N_{S'/K'}(\mathfrak{x}) \in I(K', e)$. Then there exists an element y of S and an element a of K' such that

$$N_{S'/M}(\mathfrak{x})^{\chi_1}\cdot \prod_{\lambda=s+1}^{h} N_{S'/K'}(\mathfrak{x})^{\sigma_\lambda} = (y) ,$$

$y \equiv 1 \bmod^* \mathfrak{c}$, $yy^\rho = u \cdot N(\mathfrak{x})$ with a unit u in F, $N_{S'/K'}(\mathfrak{x}) = (a)$, $a \equiv 1 \bmod^* (e)$. Then $\prod_{\lambda=1}^{s} a^{\sigma_\lambda} a^{\sigma_\lambda \rho} = N_{K'/F}(a)$, and $\prod_{\lambda=1}^{h} a^{\sigma_\lambda} a^{\sigma_\lambda \rho} = N_{K'/Q}(a) = N(\mathfrak{x})$. Let ψ denote the identity mapping or ρ on S according as χ_1 is the identity mapping or ρ on M. Define b by $b^\psi = y \cdot (\prod_{\lambda=s+1}^{h} a^{\sigma_\lambda})^{-1}$. Then $b \in S$, $b \equiv 1 \bmod^*_1 \mathfrak{c}$, and

$$bb^\rho = yy^\rho (\prod_{\lambda=s+1}^{h} a^{\sigma_\lambda} a^{\sigma_\lambda \rho})^{-1}$$
$$= u \cdot N(\mathfrak{x}) \cdot (\prod_{\lambda=s+1}^{h} a^{\sigma_\lambda} a^{\sigma_\lambda \rho})^{-1} = u \cdot \prod_{\lambda=1}^{s} a^{\sigma_\lambda} a^{\sigma_\lambda \rho} = u \cdot N_{K'/F}(a) \in F .$$

We have further $(b) = N_{S'/M}(\mathfrak{x})$. Therefore, our proposition will be proved if we can show $b \in M$. To see this, let σ be the automorphism of S over M other than the identity mapping. Then $(b^\sigma) = N_{S'/M}(\mathfrak{x}) = (b)$, hence $b^\sigma = bf$ with a unit f in S. We have $ff^\sigma = 1$, and moreover, since $bb^\rho \in F$, we have $bb^\rho = (bb^\rho)^\sigma = bfb^\rho f^\rho$, so that $ff^\rho = 1$. It follows that $f^\rho = f^\sigma$, hence $f \in K$. Therefore f is a root of unity in K, and hence $f = \pm 1$ in view of (5.21.1). Assume that $f = -1$. Since $d \in K$, we have $d^\sigma = d^\rho = -d$, so that $(d/b)^\sigma = d/b$, hence $d/b \in M$. It follows that $(d) = (d/b) \cdot N_{S'/M}(\mathfrak{x})$, which contradicts (5.21.2). Therefore we have $f = 1$, and $b^\sigma = b$, so that $b \in M$. This completes our proof.

5.23. Let K be a CM-field and Ψ a representation of $M_r(K)$ by complex matrices. Let A be an abelian variety defined over a subfield of C, and θ an isomorphism of $M_r(K)$ into $\mathrm{End}_Q(A)$ such that (A, θ) is of type $(M_r(K), \Psi)$ in the sense of 4.1. Let e_{ij} be the matrix units of $M_r(K)$. Take a positive integer p so that $\theta(pe_{ij}) \in \mathrm{End}(A)$ for every i and j. Put $A_i = \theta(pe_{ii})A$. Then we have isogenies

$$\lambda: A \longrightarrow A_1 \times \cdots \times A_r, \ \lambda(x) = (\theta(pe_{11})x, \cdots, \theta(pe_{rr})x) \qquad (x \in A),$$
$$\mu_{ij}: A_j \longrightarrow A_i, \qquad\qquad \mu_{ij} = \text{the restriction of } \theta(pe_{ij}) \text{ to } A_j.$$

Define an isomorphism θ_i of K into $\mathrm{End}_Q(A_i)$ so that $\theta_i(b)$ is the restriction of $\theta(be_{ii})$ to A for $b \in K$. Then $\mu_{ij}\theta_j(b) = \theta_i(b)\mu_{ij}$ for every i, j and $b \in K$. Therefore, we see that the (A_i, θ_i) are of type (K, Φ) with the same representation Φ of K, and hence *the restriction of Ψ to K is equivalent to the direct sum of r copies of Φ.*

Let us now assume that the following two conditions are satisfied.

(5.23.1) $$2 \cdot \dim(A) = r \cdot [K : Q] .$$

(5.23.2) $$\theta(\mathfrak{r}_K \cdot 1_r) \subset \mathrm{End}(A) .$$

Then we can take the above abelian varieties A_i and θ_i so that

(5.23.3) $$A = A_1 \times \cdots \times A_r , \qquad \theta_i(\mathfrak{r}_K) \subset \mathrm{End}(A_i) .$$

In fact, let us consider the diagram (4.1.1) in the present case. In view of (5.23.1), W is a vector space of dimension r over K, and, in view of (5.23.2), $\mathfrak{M}$ is an $\mathfrak{r}_K$-lattice. By a well-known theorem, we can find r ideals $\mathfrak{a}_i$ in K and

r linearly independent elements u_i in W so that $\mathfrak{M} = \mathfrak{a}_1 u_1 + \cdots + \mathfrak{a}_r u_r$. (In reality one can choose $\mathfrak{a}_1 = \cdots = \mathfrak{a}_{r-1} = \mathfrak{r}_K$, and $\mathfrak{a}_r$ to be any ideal belonging to an ideal class which is determined by $\mathfrak{M}$.) Since θ is an isomorphism of $M_r(K)$ into $\mathrm{End}_Q(A)$, we note that, for every K-linear endomorphism α of W, there exists a C-linear endomorphism β of C^n such that $\beta \cdot \mathfrak{h} = \mathfrak{h} \cdot \alpha$, the notation being as in (4.1.1). In particular, let ε_i be a K-linear endomorphism of W such that $\varepsilon_i u_j = \delta_{ij} u_i$ with Kronecker's δ_{ij}, and let β_i be the corresponding C-linear endomorphism of C^n. Then one has

$$\mathfrak{h}(K_R u_i) = \mathfrak{h}(\varepsilon_i W_R) = \beta_i \mathfrak{h}(W_R) = \beta_i(C^n) \ ,$$

hence $\mathfrak{h}(K_R u_i)$ is a complex vector subspace of C^n. Put $U_i = \mathfrak{h}(K_R u_i)$, $D_i = \mathfrak{h}(\mathfrak{a}_i u_i)$ for $i = 1, \cdots, r$. Then C^n (resp. D) is the direct sum of the U_i (resp. the D_i) for $i = 1, \cdots, r$, and β_i gives a homomorphism of C^n/D onto U_i/D_i, so that U_i/D_i has a structure of abelian variety. With the previous notation e_{ij}, we may consider $\varepsilon_i = e_{ii}$ with respect to the new basis $\{u_i\}$. Hence writing A_i for U_i/D_i, we have naturally a commutative diagram

$$
\begin{array}{ccccccccc}
0 & \longrightarrow & \mathfrak{a}_i & \longrightarrow & K_R & \longrightarrow & K_R/\mathfrak{a}_i & \longrightarrow & 0 \\
 & & \downarrow & & \downarrow{\scriptstyle \mathfrak{h}_i} & & \downarrow & & \\
0 & \longrightarrow & D_i & \longrightarrow & U_i & \overset{\xi_i}{\longrightarrow} & A_i & \longrightarrow & 0
\end{array}
$$

where $\mathfrak{h}_i$ is defined by $\mathfrak{h}_i(x) = \mathfrak{h}(x u_i)$ for $x \in K_R$. Since $\mathfrak{a}_i$ is an ideal in K, for every $a \in \mathfrak{r}_K$, we obtain an element $\theta_i(a)$ of $\mathrm{End}(A_i)$ such that $\theta_i(a)\xi_i(\mathfrak{h}_i(x)) = \xi_i(\mathfrak{h}_i(ax))$. This proves our assertion.

5.24. Still under the assumption (5.23.1, 2), let $\mathcal{C}$ be a polarization of A, and X a basic polar divisor in $\mathcal{C}$. For every integral ideal $\mathfrak{c}$ in K, put $\mathfrak{g}(\mathfrak{c}, A) = \{t \in A \mid \theta(\mathfrak{c})t = 0\}$. Let (K', Φ') be the dual of (K, Φ), and e an arbitrary positive integer. We shall now prove

(5.24.1) *Let $\mathfrak{a}$ be an integral ideal in K', prime to e, and σ an automorphism of C such that $\sigma = [C(K', e)/K', \mathfrak{a}]$ on $C(K', e)$ (see 1.1). Put $\mathfrak{b} = \det \Phi'(\mathfrak{a})$. Then there exists an isogeny λ of A to A^σ such that:*
(i) *$\alpha^\sigma \lambda = \lambda\alpha$ for every $\alpha \in \mathrm{End}_Q(A)$;*
(ii) *$\mathrm{Ker}(\lambda) = \mathfrak{g}(\mathfrak{b}, A)$;*
(iii) *$\lambda^{-1}(X^\sigma) \equiv N(\mathfrak{a})X$;*
(iv) *$\lambda(u) = u^\sigma$ for every $u \in \mathfrak{g}(e, A)$.*

Let (K'', Φ'') be the dual of (K', Φ'). By [15, 8.3], (K', Φ') is the dual of (K'', Φ''). By 5.5, $K'' \subset K$, and the restriction of Φ to K'' is equivalent to s times Φ'', where $s = [K : K'']$. (This can be also seen from [15, 8.3].) Moreover, A_i is isogenous to a product of s copies of a simple abelian variety B

with an isomorphism $\omega: K'' \to \mathrm{End}_Q(B)$ such that (B, ω) is of type (K'', Φ''). This fact is implicit in [15, Ch. II]. One easy way to prove this is as follows. Construct any (B, ω) of type (K'', Φ'') and take the product B^s of s copies of B. Then we have an isomorphism $\omega': M_s(K'') \to \mathrm{End}_Q(B^s)$. Take any K''-linear injection $\iota: K \to M_s(K'')$. Then we see that $(B^s, \omega' \circ \iota)$ is of type (K, Φ). By [15, p. 45, Cor.], $(B^s, \omega' \circ \iota)$ is isogenous to (A_i, θ_i). Hence θ_i can be extended to an isomorphism of $M_s(K'')$ onto $\mathrm{End}_Q(A_i)$. Since $\theta_i(\mathfrak{r}_{K''}) \subset \mathrm{End}(A_i)$, we can apply the argument of 5.23 to A_i, and find that A_i is the product of s simple abelian varieties. Therefore, to prove (5.24.1), we may assume, after replacing the A_i and K by these simple abelian varieties and K'', that the A_i are simple, $\theta_i(K) = \mathrm{End}(A_i)$, and (K, Φ) is the dual of (K', Φ'). Note that (K', Φ') remains the same. Now in [15, 16.3], the following fact has been proved for a certain finite extension k of $C(K', e)$, normal over K', and a certain integral ideal $\mathfrak{n}$ in K'.

(5.24.2) *If $\mathfrak{p}$ is a prime ideal in K', prime to $\mathfrak{n}$, and $\sigma = [k/K', \mathfrak{P}/\mathfrak{p}]$ with a prime ideal $\mathfrak{P}$ in k dividing $\mathfrak{p}$, then there exists an isogeny μ_i of A_i to A_i^σ, for each i, such that:*
    (i) *$\theta_i^\sigma(a)\mu_i = \mu_i\theta_i(a)$ for every $a \in K$;*
    (ii) *$\mathrm{Ker}(\mu_i) = \mathfrak{g}(\mathfrak{q}, A_i)$ where $\mathfrak{q} = \det \Phi'(\mathfrak{p})$;*
    (iii) *$\mu_i^{-1}(X_i) \equiv N(\mathfrak{p})X_i$;*
    (iv) *$\mu_i(u) = u^\sigma$ for every $u \in \mathfrak{g}(e, A_i)$.*

Here X_i is a divisor on A_i which defines a polarization on A_i. Of course k and $\mathfrak{n}$ are independent of $\mathfrak{p}$. Let $\mathfrak{a}, \mathfrak{b}$ and σ be as in (5.24.1). We can find a prime ideal $\mathfrak{p}$ in K' and a prime ideal $\mathfrak{P}$ in k dividing $\mathfrak{p}$ so that $\sigma = [k/K', \mathfrak{P}/\mathfrak{p}]$ on k and $\mathfrak{p}$ does not divide $\mathfrak{n}$. Then we have $\sigma = [C(K', e)/K', \mathfrak{p}]$, so that $\mathfrak{p}$ and $\mathfrak{a}$ belong to the same ideal class modulo (e) in K'. We find therefore an element γ of K' such that $\mathfrak{a} = \gamma\mathfrak{p}$ and $\gamma \equiv 1 \bmod^* (e)$. Applying (5.24.2) to this $\mathfrak{p}$, we obtain isogenies μ_i as described in (5.24.2). Put $\delta = \det \Phi'(\gamma)$. Then $\delta \in K$, $\delta \equiv 1 \bmod^* (e)$, and $\mathfrak{b} = \delta\mathfrak{q}$. By [15, p. 58, Prop. 13], there exists an isogeny λ_i of A_i to A_i^σ such that $\lambda_i = \mu_i\theta_i(\delta)$. (Note that $\mathfrak{b}$ is an integral ideal, but δ may not be integral.) Define an isogeny λ of A to A^σ so that $\lambda = \lambda_i$ on A_i. Obviously λ satisfies (i), (ii), and (iv) of (5.24.1). To see (iii), let

$$Y = \sum_{i=1}^{r} (A_1 \times \cdots \times A_{i-1} \times X_i \times A_{i+1} \times \cdots \times A_r).$$

Then $\lambda^{-1}(Y^\sigma) \equiv N(\mathfrak{a})Y$, since $\delta\delta^\rho = N(\gamma)$ (5.11.4). Define isogenies φ_X and φ_Y of A to $\mathrm{Pic}(A)$ attached canonically to X and Y. Since $\mathrm{End}_Q(A) = \theta(M_r(K))$, we have $\varphi_X = \varphi_Y \cdot \theta(\varepsilon)$ with an element ε of $M_r(K)$. Then

$${}^t\lambda\varphi_X^\sigma\lambda = {}^t\lambda\varphi_Y^\sigma \cdot \theta^\sigma(\varepsilon)\lambda = {}^t\lambda\varphi_Y^\sigma\lambda\theta(\varepsilon) = N(\mathfrak{a})\varphi_Y\theta(\varepsilon) = N(\mathfrak{a})\varphi_X\,,$$

hence $\lambda^{-1}(X^\sigma) \equiv N(\mathfrak{a})X$. Thus λ has all the properties (i–iv) of (5.24.1), q.e.d.

6. Singular members of a family of abelian varieties

6.1. Now we consider PEL-structures in the following special case. Let F, g and $\tau_{01}, \cdots, \tau_{0g}$ be as in 2.1 and 2.4, and K a totally imaginary quadratic extension of F. Let $\tau_1, \cdots, \tau_g$ be extensions of $\tau_{01}, \cdots, \tau_{0g}$ to K. Let L be a central simple algebra over K with a positive involution ρ. We put $[L : K] = q^2$ and assume that $q = 1$ or 2. In either case, ρ induces the non-trivial automorphism of K over F. (The case $L = M_2(K)$ is admitted.) By [9, Lemma 1], we can identify L_R with a product $M_q(C)^g$ of g copies of $M_q(C)$ so that, if $\omega_\nu: L_R \to M_q(C)$ denotes the projection to the νth factor, then, for each ν,

$$(6.1.1) \qquad \omega_\nu(x^\rho) = {}^t\overline{\omega_\nu(x)} \qquad (x \in L) ,$$

$$(6.1.2) \qquad \omega_\nu(a) = a^{\tau_\nu} \cdot 1_q \qquad (a \in K) .$$

We are going to apply the general theory of §4 to this L and ρ with $m = 2/q$, so that the dimension n of abelian varieties will be $2gq$. We take a representation Φ of L_R by complex matrices of size n such that

$$(6.1.3) \quad \textit{the restriction of } \Phi \textit{ to } K \sim q \cdot \left(\sum_{\nu=1}^{r} (\tau_\nu + \tau_\nu\rho) + 2 \sum_{\nu=r+1}^{g} \tau_\nu \right),$$

with the notation of 5.1.1. Let L^m (with $m = 2/q$) be as in 4.1, and T an element of $M_m(L)$ such that ${}^tT^\rho = -T$. (In this section, we use L^m instead of W.) We assume

$(6.1.4)$ *The complex hermitian matrix* $-\sqrt{-1} \cdot \omega_\nu(T)$ *has the same signature as* $\begin{bmatrix} 1 & 0 \\ 0 & -1 \end{bmatrix}$ *or* $\begin{bmatrix} 1 & 0 \\ 0 & 1 \end{bmatrix}$ *according as* $\nu \leqq r$ *or* $\nu > r$.

If $q = 2$, then $T \in L$, so that $\omega_\nu(T)$ is meaningful. If $q = 1$, we put, here and in the following, $\omega_\nu(X) = X^{\tau_\nu}$ for every $X \in M_2(K)$. We define an L_R-valued ρ-anti-hermitian form $T(x, y)$ on L^m by $T(x, y) = xT \cdot {}^t y^\rho$ for $x, y \in L^m$. Let $\mathfrak{M}$ be a Z-lattice in L^m and $v_1, \cdots, v_s$ be elements of L^m. For every totally positive element β of F such that $\operatorname{tr}(\beta T(\mathfrak{M}, \mathfrak{M})) = Z$, we can consider a PEL-type

$$(6.1.5) \qquad \Omega = (L, \Phi, \rho; \beta T, \mathfrak{M}; v_1, \cdots, v_s) ,$$

and a family $\Sigma_\Omega = \{ \mathfrak{Q}_z \mid z \in \mathcal{H} \}$ of PEL-structures

$$\mathfrak{Q}_z = (A_z, C_z, \theta_z; t_{1z}, \cdots, t_{sz}) .$$

In the present case, the domain $\mathcal{H}$ is nothing but the product $\mathfrak{D}^r$ of r copies of the unit disk $\mathfrak{D} = \{ z \in C \mid |z| < 1 \}$. (With the terminology and notation of [9], L is of (Type IV), and $\sum_{\nu=1}^{g} r_\nu s_\nu$ is r.) We denote by F^+ the set of all totally positive elements in F, and define a weak PEL-type

$$(6.1.6) \qquad \Omega_0 = (L, \Phi, \rho; F^+ \cdot T, \mathfrak{M}; v_1, \cdots, v_s)$$

and a family $\Sigma(\Omega_0)$ of weak PEL-structures $\mathcal{R}_z = \mathcal{Q}_z \cdot F^+$ as in 4.10. (In the present case, $F^+ = E$.)

6.2. Define the groups $\mathrm{U}(T)$, $\mathrm{U}_R(T)$ as in 4.2, and $G(T)$, $G^+(T)$ as in 4.3. The action of elements of $\mathrm{U}_R(T)$ on $\mathcal{H}$ can be defined as follows [9, 2.7]. For each $\nu \leq r$, take a matrix W_ν of $\mathrm{GL}_2(C)$ so that

$$W_\nu(\sqrt{-1}\omega_\nu(T)^{-1}) \cdot {}^t\bar{W}_\nu = \begin{bmatrix} 1 & 0 \\ 0 & -1 \end{bmatrix}.$$

For $\gamma \in \mathrm{U}_R(T)$, put

$$ {}^tW_\nu^{-1}\overline{\omega_\nu(\gamma)} \cdot {}^tW_\nu = \begin{bmatrix} a_\nu & b_\nu \\ c_\nu & d_\nu \end{bmatrix} \qquad (\nu = 1, \cdots, r),$$

and define the action of γ on $\mathcal{H} = \mathfrak{D}^r$ by

$$\gamma(z_1, \cdots, z_r) = (u_1, \cdots, u_r) \qquad u_\nu = (a_\nu z_\nu + b_\nu)/(c_\nu z_\nu + d_\nu).$$

Since $M_m(L) = M_2(K)$ or L according as $q = 1$ or 2, $M_m(L)$ is a quaternion algebra over K. Let $\alpha \mapsto \alpha'$ denote the main involution of $M_m(L)$, and let

$$(6.2.1) \qquad \mathfrak{B} = \{\alpha \in M_m(L) \mid T(x\alpha', y) = T(x, y\alpha)\}.$$

By [10, 1.5 and proof of 2.6], we know that $\mathfrak{B}$ is a quaternion algebra over F, and

$$\mathfrak{B} \otimes_F K = M_m(L), \qquad G(T) = \mathfrak{B}^* \cdot K^*,$$
$$\mathfrak{B}^* = \{\alpha \in G(T) \mid \mu(\alpha) = \alpha\alpha'\},$$

where K^* resp. $\mathfrak{B}^*$ denotes the group of invertible elements of K resp. $\mathfrak{B}$, and $\mu(\alpha)$ is as in 4.3. Here is an essential connection of the present families of abelian varieties with a quaternion algebra over F. In this section, however, we do not identify $\mathfrak{B}$ with the quaternion algebra B of §2.

6.3. PROPOSITION. *Let $\mathfrak{B}$ be as in (6.2.1), $\mathfrak{c}$ an integral ideal in F, and $\mathfrak{o} = \{a \in \mathfrak{B} \mid \mathfrak{M}a \subset \mathfrak{M}\}$. Let $\Gamma(\mathfrak{o}, \mathfrak{c})$ be the set of all units γ in $\mathfrak{o}$ such that $\mu(\gamma)$ is totally positive and $\gamma \equiv 1 \bmod \mathfrak{co}$. Let $\Gamma_0(T, \mathfrak{c}^{-1}\mathfrak{M}/\mathfrak{M})$ be as in 4.10 with $E = F^+$. Suppose that $\mathfrak{o}$ is a maximal order and the following two conditions are satisfied.*

(6.3.1) K has no roots of unity other than ± 1.

(6.3.2) There exists a prime ideal in F which is ramified in K and does not divide $2 \cdot D(B/F)$.

Then $\Gamma(\mathfrak{o}, \mathfrak{c}) = \Gamma_0(T, \mathfrak{c}^{-1}\mathfrak{M}/\mathfrak{M})$.

This is, in substance, a generalization of [12, 1.16].

PROOF. We have obviously $\Gamma(\mathfrak{o}, \mathfrak{c}) \subset \Gamma_0(T, \mathfrak{c}^{-1}\mathfrak{M}/\mathfrak{M})$ in view of (6.2.1). Put $\lambda(\xi) = \xi\xi'$ for every $\xi \in M_m(L)$. Then $\lambda(\xi)\lambda(\xi)^\rho = \mu(\xi)^2$. Let $\alpha \in \Gamma_0(T, \mathfrak{c}^{-1}\mathfrak{M}/\mathfrak{M})$. Then $\mu(\alpha)$ is a totally positive unit of F, and $\lambda(\alpha)$ is a unit of K. Therefore $\mu(\alpha)^{-1}\lambda(\alpha)$ is a unit of K whose absolute value is one. In view of (5.8.2), every conjugate of $\mu(\alpha)^{-1}\lambda(\alpha)$ has the same property. Hence $\mu(\alpha)^{-1}\lambda(\alpha)$ is a root of unity. By (6.3.1), $\mu(\alpha)^{-1}\lambda(\alpha) = \pm 1$. If $\mu(\alpha) = \lambda(\alpha)$, then $\alpha \in \mathfrak{B}$. Since $\mathfrak{c}^{-1}\mathfrak{M}(1 - \alpha) \subset \mathfrak{M}$, we have $\mathfrak{c}^{-1}(1 - \alpha) \subset \mathfrak{o}$, so that $\alpha \in \Gamma(\mathfrak{o}, \mathfrak{c})$. Assume that $\mu(\alpha) = -\lambda(\alpha)$. Since $G(T) = \mathfrak{B}^* \cdot K^*$, we have $\alpha = b \cdot c$ with $b \in \mathfrak{B}$ and $c \in K$. Then $c^2\lambda(b) = \lambda(\alpha) = -\mu(\alpha) = -\mu(b)cc^\rho$. Since $\lambda(b) = \mu(b)$, we obtain $c^\rho = -c$, so that $K = F(c)$. On the other hand, $\mathfrak{M}bc = \mathfrak{M}\alpha = \mathfrak{M}$, therefore $\mathfrak{M}b^{-1}\mathfrak{o}b = c\mathfrak{M}bb^{-1}\mathfrak{o}b = c\mathfrak{M}b = \mathfrak{M}$. It follows that $b^{-1}\mathfrak{o}b = \mathfrak{o}$, hence $b\mathfrak{o}$ is a two-sided $\mathfrak{o}$-ideal. We can find an ideal $\mathfrak{a}$ in F and prime ideals $\mathfrak{p}_1, \cdots, \mathfrak{p}_s$ in F dividing $D(B/F)$ so that $\lambda(b)\mathfrak{r}_F = \mathfrak{a}^2\mathfrak{p}_1 \cdots \mathfrak{p}_s$. Since $\lambda(\alpha)$ is a unit, we have $\mathfrak{r}_K = \lambda(\alpha)\mathfrak{r}_K = c^2\lambda(b)\mathfrak{r}_K = (c\mathfrak{a})^2\mathfrak{p}_1 \cdots \mathfrak{p}_s$. This implies that a prime ideal $\mathfrak{p}$ in F can be ramified in $K = F(c)$ only if $\mathfrak{p}$ divides $2 \cdot D(B/F)$. This contradicts (6.3.2), and hence completes the proof.

6.4. Our aim is to consider the special members of $\Sigma(\Omega_0)$. Let $M \in J(\mathfrak{B})$ (see 2.3), and let f be an F-linear isomorphism of M into $\mathfrak{B}$. By 2.6, there exists a point z on $\mathcal{H}$ which is fixed by the elements of $f(M) - F$. (The present $\mathcal{H}$ is $\mathfrak{D}^r$ and not a connected component of $\mathfrak{F}_r$. But the assertion is still true for an obvious reason.) Let $\Lambda(\alpha, z)$ (with $\alpha \in G^+(T)$) be as in 4.4. Suppose that $M \neq K$. Then MK is a field, and we can identify $L \otimes_K KM$ with $M_q(KM)$. Now, if $0 \neq b \in M$, we see that $\Lambda(f(b), z)$ defines an element of $\mathrm{End}_Q(A_z)$. Therefore we obtain an isomorphism θ of $M_q(KM)$ into $\mathrm{End}_Q(A_z)$ such that $\theta(a) = \theta_z(a)$ for $a \in L$, and $\theta(b)$ corresponds to $\Lambda(f(b), z)$ for $b \in M$.

Since $G(T) = K^* \cdot \mathfrak{B}^*$, every non-trivial fixed point of an element of $G^+(T)$ on $\mathcal{H}$ is a fixed point of an element of $\mathfrak{B}$. By 2.6, such a fixed point z defines a totally imaginary quadratic extension M of F, and thus an isomorphism θ of $M_q(KM)$ into $\mathrm{End}_Q(A_z)$.

Put $S = KM$. Extend ρ to $L \otimes_K KM = M_q(S)$ so that ρ is the complex conjugation on KM. Then the polarization $\mathcal{C}_z$ of A_z induces an involution of $\mathrm{End}_Q(A_z)$ which coincides with $\theta(x) \mapsto \theta(x^\rho)$ on $\theta(M_q(S))$. In fact, this is obvious if $x \in L$. If $b \in M$, then $f(b) \in \mathfrak{B}$, hence

$$T(uf(b), v) = T(u, vf(b)') = T(u, vf(b^\rho)) .$$

Therefore $\theta(b^\rho)$ is the transform of $\theta(b)$ by the involution determined by $\mathcal{C}_z$.

6.5. To discuss the representation of $\theta(M_q(S))$ with respect to an analytic coordinate system of A_z, let us first consider the case $q = 2$ (and hence $m = 1$). By 5.23, A_z is isogenous to a product of two copies of an abelian variety A_1

with an isomorphism θ_1 of S into $\mathrm{End}_Q(A_1)$. Let Ξ be a representation of S such that (A_1, θ_1) is of type (S, Ξ) in the sense of 4.1. Let Ψ resp. Φ_1 be the restriction of Ξ to M resp. K. Then (K, Φ_1), (M, Ψ), (S, Ξ) are generalized CM-types of index respectively 2, 2, 1 in the sense of 5.11. By the consideration of 5.23, the restriction of Φ to K is equivalent to the direct sum of two copies of Φ_1. Hence, in view of (6.1.3), one has

$$\Phi_1 \sim \sum_{\nu=1}^{r} (\tau_\nu + \tau_\nu \rho) + 2 \sum_{\nu=r+1}^{g} \tau_\nu \ .$$

Let

(6.5.1) $$\Xi \sim \sum_{\nu=1}^{g} (\alpha_\nu + \beta_\nu)$$

with isomorphisms α_ν and β_ν of S into C which coincide with $\tau_{\nu 0}$ on F. Since $\Xi = \Phi_1$ on K, we may assume $\alpha_\nu = \tau_\nu$, $\beta_\nu = \tau_\nu \rho$ on K for $1 \le \nu \le r$, and $\alpha_\nu = \beta_\nu = \tau_\nu$ on K for $\nu > r$. Since $\alpha_\nu = \beta_\nu$ on F, α_ν coincides with β_ν or $\beta_\nu \rho$ on M. Now we note that, among the $2g$ isomorphisms α_ν, β_ν, there are no two which coincide or which are complex conjugate to each other. Therefore we have $\alpha_\nu = \beta_\nu$ on M for $\nu \le r$, and $\alpha_\nu = \beta_\nu \rho$ on M for $\nu > r$. Let χ_ν be the restriction of α_ν to M for each ν. Then

(6.5.2) $$\Psi \sim \sum_{\nu=1}^{r} 2\chi_\nu + \sum_{\nu=r+1}^{g} (\chi_\nu + \chi_\nu \rho) \ .$$

Therefore (K, Φ_1), (M, Ψ), (S, Ξ) are exactly in the same situation as in 5.15. (The structure of Ψ can be also seen more directly from an investigation of $\Lambda(\beta, z)$.)

Next suppose that $q = 1$. Then the above discussion is true with $A_1 = A_z$, $\theta_1 = \theta_z$. In both cases $q = 1, 2$, we denote by (K', Φ'), (M', Ψ'), and (S', Ξ') the duals of (K, Φ_1), (M, Ψ), and (S, Ξ) respectively (see §5).

6.6. Now suppose that $\mathfrak{r}_K \mathfrak{M} \subset \mathfrak{M}$. Then $\theta_w(\mathfrak{r}_K) \subset \mathrm{End}(A_w)$ for every $w \in \mathcal{H}$. Let M, f and z be as in 6.4. Let $\mathfrak{r}' = \{a \in M \mid \mathfrak{M}f(a) \subset \mathfrak{M}\}$. Then $\mathfrak{r}'$ is an order in M, and obviously $\theta(\mathfrak{r}_K \mathfrak{r}') \subset \mathrm{End}(A_z)$. Therefore, if $\mathfrak{M}f(\mathfrak{r}_M) \subset \mathfrak{M}$ and $\mathfrak{r}_S = \mathfrak{r}_K \mathfrak{r}_M$, then $\theta(\mathfrak{r}_S) \subset \mathrm{End}(A_z)$. In view of 1.2, we obtain the following assertion.

(6.6.1) *The notation being as in 6.4 and 6.5, one has* $\theta(\mathfrak{r}_S) \subset \mathrm{End}(A_z)$, *if* $\mathfrak{r}_K \mathfrak{M}f(\mathfrak{r}_M) \subset \mathfrak{M}$ *and* $D(M/F)$ *is prime to* $D(K/F)$.

Further, if $\theta(\mathfrak{r}_S) \subset \mathrm{End}(A_z)$ and $q = 2$, one can write $A_z = A_1 \times A_2$ and obtain isomorphisms $\theta_i \colon S \longrightarrow \mathrm{End}_Q(A_i)$ such that $\theta_i(\mathfrak{r}_S) \subset \mathrm{End}(A_i)$ (5.23).

6.7. PROPOSITION. *The notation being as in 6.4 and 6.5, suppose that* $g > 1$, $r = 1$, $M \not\subset K'$, $\theta(\mathfrak{r}_S) \subset \mathrm{End}(A_z)$, $\mathfrak{r}_K \mathfrak{M} \subset \mathfrak{M}$, *and* $\mathfrak{c}^{-1}\mathfrak{M} = \mathfrak{M} + \sum_{i=1}^{s} \mathfrak{r}_K v_i$ *with an integral ideal* $\mathfrak{c}$ *in* F. *Then the following assertions hold.*

(6.7.1) A_1 *is simple,* $\theta_1(S) = \mathrm{End}_Q(A_1)$, *and* $\theta_1(\mathfrak{r}_S) = \mathrm{End}(A_1)$.

(6.7.2) *The field* $C(S', \Xi', \mathfrak{c})$ *defined in 5.19 is exactly the composite of*

M and the field of moduli of $\mathcal{R}_z$.

(6.7.3) *Let e be the smallest positive integer divisible by $\mathfrak{c}$, and let $k = C(S', e)$. Let $\mathfrak{a}$ be an integral ideal in S', prime to e, and σ an automorphism of C such that $\sigma = [k/S', \mathfrak{a}]$ on k (see 1.1). Put $\mathfrak{b} = \det \Xi'(\mathfrak{a})$. Let X be a basic polar divisor in $\mathcal{C}_z$. Then there exists an isogeny λ of A_z to A_z^σ such that:*

(i) $\theta^\sigma(a)\lambda = \lambda\theta(a)$ *for every $a \in M_q(S)$;*

(ii) $\mathrm{Ker}(\lambda) = \{t \in A_z \mid \theta(\mathfrak{b})t = 0\}$;

(iii) $\lambda^{-1}(X^\sigma) \equiv N(\mathfrak{a})X$;

(iv) $\lambda(u) = u^\sigma$ *for every u on A_z satisfying $\theta(\mathfrak{c})u = 0$.*

PROOF. By 5.18, (S, Ξ) is a primitive CM-type in the sense of [15, 8.2], hence (6.7.1) holds. The assertion (6.7.3) follows immediately from (5.24.1). To prove (6.7.2), let the notation be as in (6.7.3). Suppose that σ is the identity mapping on $C(S', \Xi', \mathfrak{c})$. Then there exists an element y in S such that $\mathfrak{b} = (y)$, $y \equiv 1 \bmod^* \mathfrak{c}$, $N(\mathfrak{a}) = vyy^\rho$ with a unit v in F. By [15, §7, Prop. 7 and Prop. 11], there exists an isomorphism η of A_z to A_z^σ such that $\lambda = \eta \cdot \theta(y)$. Then

$$\eta^{-1}(X^\sigma) \equiv X \cdot \theta(v) , \qquad \eta(u) = u^\sigma$$

for every $u \in A_z$ such that $\theta(\mathfrak{c})u = 0$, and $\eta \cdot \theta(a) = \theta^\sigma(a)\eta$ for all $a \in L$. Therefore η is an isomorphism of $\mathcal{R}_z$ to $\mathcal{R}_z^\sigma$, hence σ is the identity mapping on the field of moduli of $\mathcal{R}_z$.

Conversely, take any automorphism σ of C over the composite of M and the field of moduli of $\mathcal{R}_z$. Then σ is the identity on $S' = MK'$, since K' is contained in the field of moduli of $\mathcal{R}_z$ (4.14). Take an integral ideal $\mathfrak{a}$ in S', prime to (e), so that $\sigma = [k/S', \mathfrak{a}]$, and apply (6.7.3) to this $\mathfrak{a}$ and σ. Then we obtain λ with the properties (i–iv). By the definition of the field of moduli of $\mathcal{R}_z$, there exists an isomorphism η of $\mathcal{R}_z$ to $\mathcal{R}_z^\sigma$. We have $\eta^{-1}(X^\sigma) \equiv X \cdot \theta(v)$ with a totally positive element v in F. By (6.7.1), $\lambda = \eta \cdot \theta(y)$ with an element y of $M_q(S)$. Since λ and η commute with the action of elements of L, y is contained in S. By (ii) of (6.7.3), we have $\mathfrak{b} = (y)$, and

$$N(\mathfrak{a})X \equiv \lambda^{-1}(X^\sigma) \equiv X \cdot \theta(vyy^\rho) ,$$

so that $N(\mathfrak{a}) = vyy^\rho$. Since $(yy^\rho) = N(\mathfrak{a})\mathfrak{r}_s$ by (5.11.4), v is a unit. Moreover, if $u \in A_z$ and $\theta(\mathfrak{c})u = 0$, then $\theta(y)u = \eta^{-1}\lambda u = \eta^{-1}u^\sigma = u$. Hence we have $y - 1 \in \mathfrak{c}\mathfrak{r}_s$, so that $\mathfrak{a} \in I(S', \Xi', \mathfrak{c})$ (see 5.19). Therefore σ is the identity mapping on $C(S', \Xi', \mathfrak{c})$. This proves (6.7.2) and completes the whole proof.

6.8. **PROPOSITION.** *Let Ω_0 be a weak PEL-type defined by (6.1.6). Suppose that $g > 1$, $r = 1$, $\mathfrak{r}_K\mathfrak{M}\mathfrak{o} = \mathfrak{M}$ with a maximal order $\mathfrak{o}$ in $\mathfrak{B}$, and $\mathfrak{c}^{-1}\mathfrak{M} = \mathfrak{M} + \sum_{i=1}^s \mathfrak{r}_K v_i$ with an integral ideal $\mathfrak{c}$ in F. Let e be the smallest positive* *integer divisible by* $\mathfrak{c}$. *Let* $k(\Omega_0)$ *be the field characterized by* 4.14. *Then* $k(\Omega_0) \subset C(K', e)$.

PROOF. Let R be a finite Galois extension of F containing $k(\Omega_0)$, K, and $C(K', e)$. Applying 2.10 to $\mathfrak{B}$, we can find an $M \in J(B)$ such that M is not contained in R, and $D(M/F)$ is prime to $D(R/F)$. By 2.8, there exists an F-linear isomorphism f of M into B such that $f(\mathfrak{r}_M) \subset \mathfrak{o}$. Let z be the fixed point on $\mathcal{H}$ of the elements of $f(M) - F$. Consider $\mathfrak{R}_z$ and θ of 6.4 and 6.5. By (6.6.1) we have $\theta(\mathfrak{r}_s) \subset \mathrm{End}(A_z)$. Therefore, the notation being as in 6.7, we have $k(\Omega_0) \subset C(S', \Xi', \mathfrak{c}) \subset C(K'M, e)$. Taking K', R and $K'M$ to be the fields F, L and M of 1.3, we find $k(\Omega_0) \subset C(K'M, e) \cap R \subset C(K', e)$, which proves our proposition.

6.9. PROPOSITION. *Let* $M \in J(B)$ *and* $S = K \otimes_F M$. *Suppose that* $K \neq M$. *Let* w *be a point of* $\mathcal{H}$. *Suppose that there exists an isomorphism* θ *of* $L \otimes_K S$ *into* $\mathrm{End}_Q(A_w)$ *satisfying the following conditions.*

(6.9.1) $\theta(a) = \theta_w(a)$ *for* $a \in L$,

(6.9.2)   *The involution of* $\mathrm{End}_Q(A_w)$ *determined by* $\mathcal{C}_w$ *coincides with* $\theta(x) \mapsto \theta(x^\rho)$ *on* $\theta(S)$, *where* ρ *is the complex conjugation in* S.
Then there exists an F-*linear isomorphism* f *of* M *into* $\mathfrak{B}$ *such that* w *is the fixed point of* $f(M)$, *and* θ *is obtained by the procedure of* 6.4.

PROOF. Let β be an element of M such that $M = F(\beta)$ and $\theta(\beta) \in \mathrm{End}(A_w)$. Put $\lambda = \theta(\beta)$. If X is a basic polar divisor in $\mathcal{C}_w$, we have $\lambda^{-1}(X) \equiv X \cdot \theta(\beta\beta^\rho)$ in view of (6.9.2). From the consideration of [9, p. 165], we see that λ can be obtained from $\Lambda(\xi, w)$ of 4.4 with an element ξ of $G^+(T)$ such that $\xi^{-1}(w) = w$. Since $G(T) = K^* \cdot \mathfrak{B}^*$, we have $\xi = \eta\zeta$ with $\eta \in K^*$, and $\zeta \in \mathfrak{B}^*$. Since $\lambda \notin \theta_w(K)$, we have $\xi \notin K$, so that $\zeta \notin F$. Since $\zeta(w) = \xi(w) = w$, $F(\zeta)$ should be isomorphic to a member of $J(B)$ by 2.6. Now $\Lambda(\zeta, w)$ defines an element of $\mathrm{End}_Q(A_w)$, which is obviously $\theta(\eta^{-1}\beta)$. Hence $F(\eta^{-1}\beta)$ is a totally imaginary quadratic extension of F, different from K. Since M is the only such subfield of S, we have $M = F(\eta^{-1}\beta)$, so that $\eta \in F$. Therefore we can define an F-linear isomorphism f of M into $\mathfrak{B}$ so that $f(\beta) = \eta\zeta$. Then we obtain the desired result.

7. Proof of Main Theorem I: Case $g > 1$

7.1. Throughout this section, we use the symbols F, g, $\tau_{01}, \cdots, \tau_{0g}$, B, r with the same meaning as in 2.1 and 2.4. After 7.4, we shall be assuming $g > 1$ and $r = 1$.

7.2. PROPOSITION. *Let* K *be a totally imaginary quadratic extension of* F, *and let* $L = B \otimes_F K$. *Let* $x \mapsto x'$ *denote the main involution of* L *over*

K. Then there exist a positive involution ρ in L and an invertible element v of L such that $v^\rho = -v$ and $B = \{x \in L \mid x' = vx^\rho v^{-1}\}$.

PROOF. Define an involution σ of L so that $x^\sigma = x'$ for $x \in B$ and σ induces the non-trivial automorphism of K over F. We can identify $L_R = L \otimes_Q R$ with the product $M_2(C)^g$ of g copies of $M_2(C)$ so that, if $\psi_\nu : L_R \to M_2(C)$ denotes the projection to the νth factor, then $\psi_\nu(a) = a^{\tau_\nu} 1_2$ for $a \in F$. Since $x \mapsto x'$ is the main involution of B, we may assume, in view of (2.4.2),

$$(7.2.1) \qquad \psi_\nu(a^\sigma) = j^{-1} \cdot {}^t\overline{\psi_\nu(a)} j , \qquad j = \begin{bmatrix} 0 & 1 \\ -1 & 0 \end{bmatrix} \qquad \text{for } 1 \leq \nu \leq r ,$$

$$= {}^t\overline{\psi_\nu(a)} \qquad \text{for } r < \nu \leq g ,$$

for every $a \in B$. Then, by our definition of σ on K, we see that (7.2.1) holds for all $a \in L$. Put

$$W = \{w \in L \mid w^\sigma = -w\} , \qquad W_R = W \otimes_Q R ,$$
$$W_\nu = \{w \in M_2(C) \mid {}^t\overline{(jw)} = jw\} \qquad \text{for } 1 \leq \nu \leq r ,$$
$$= \{w \in M_2(C) \mid {}^t\overline{w} = -w\} \qquad \text{for } r < \nu \leq g .$$

Then W_R can be identified with $W_1 \times \cdots \times W_g$, and W is dense in W_R. Therefore we can find an element v of W so that $\det[j \cdot \psi_\nu(v)]$ is positive for $1 \leq \nu \leq r$, and negative for $r < \nu \leq g$. Define ρ by $x^\rho = v^{-1}x^\sigma v$ for $x \in L$. Since $v^\sigma = -v$, ρ is an involution of L, and $\psi_\nu(x^\rho) = (j \cdot \psi_\nu(v))^{-1} \cdot {}^t\overline{\psi_\nu(x)}(j \cdot \psi_\nu(v))$ for $\nu \leq r$, and $\psi_\nu(x^\rho) = \psi_\nu(v)^{-1} \cdot {}^t\overline{\psi_\nu(x)}\psi_\nu(v)$ for $\nu > r$. This shows that ρ is a positive involution. If $x \in B$, then $x' = x^\sigma = vx^\rho v^{-1}$, hence $B \subset \{x \in L \mid x' = vx^\rho v^{-1}\}$. By [10, 1.5] the last set is a quaternion algebra over F, hence we obtain our proposition.

7.3. Let K, $\tau_1, \cdots, \tau_g$, L, ρ and T be as in 6.1, with $q = 2$. Define $\mathfrak{B}$ by (6.2.1). Now Prop. 7.2 shows that the quaternion algebra B of 2.4 can be obtained as such an algebra $\mathfrak{B}$ with $L = B \otimes_F K$ for an arbitrarily given totally imaginary quadratic extension K of F. In fact, we choose ρ and v as in 7.2 and adopt cv as T with a suitable c in F such that (6.1.4) is satisfied.

Let $\alpha^{(\nu)}$ be defined as in 2.4 for $\alpha \in B$. Let $\mathfrak{H}$ (resp. $\mathfrak{K}$) denote the product of r copies of the upper half plane (resp. the unit disk). Then every element of B^+ (resp. $\mathfrak{B}^+$) acts on $\mathfrak{H}$ (resp. $\mathfrak{K}$) as is defined in 2.4 (resp. 6.2). Let us now show

(7.3.1) *There exist an F-linear isomorphism ι of B onto $\mathfrak{B}$ and a holomorphic isomorphism j of $\mathfrak{H}$ to $\mathfrak{K}$ such that $j(\alpha(z)) = \iota(\alpha)(j(z))$ for every $\alpha \in B^+$ and every $z \in \mathfrak{H}$.*

The only non-trivial point of this assertion is concerned with the distinction between the upper and lower half planes. Let ω_ν and W_ν be as in 6.1 and 6.2. Let $U = \begin{bmatrix} 1 & i \\ 1 & -i \end{bmatrix}$, $J = \begin{bmatrix} 0 & 1 \\ -1 & 0 \end{bmatrix}$ $(i = \sqrt{-1})$ and

$$\varphi_\nu(x) = U^{-1} \cdot {}^t\bar{W}_\nu^{-1} \omega_\nu(x) \cdot {}^t\bar{W}_\nu U \qquad\qquad \text{for } x \in L .$$

Then we see easily that if $\nu \leq r$, $x \in G(T)$ and $\mu(x)^{\tau_\nu} = \mu_\nu$, then

$$\varphi_\nu(x) J \cdot {}^t\overline{\varphi_\nu(x)} = \mu_\nu J .$$

If $x \in \mathfrak{B}$, then $xx' = \mu(x)$ and $\varphi_\nu(x') = J \cdot {}^t\varphi_\nu(x) J^{-1}$, hence

$$\varphi_\nu(x) J \cdot {}^t\overline{\varphi_\nu(x)} = \varphi_\nu(x) \cdot \varphi_\nu(x') J = \varphi_\nu(x) J \cdot {}^t\varphi_\nu(x) .$$

It follows that $\varphi_\nu(x) \in M_2(\boldsymbol{R})$ for every $x \in B$ and $\nu \leq r$. Let ι be an arbitrary F-linear isomorphism of B to $\mathfrak{B}$. Then we can find an element Y_ν of $\mathrm{GL}_2(\boldsymbol{R})$ so that $x^{(\nu)} = Y_\nu^{-1} \varphi_\nu(\iota(x)) Y_\nu$ for all $x \in B$ and $\nu \leq r$. We can take ι so that $\det(Y_\nu) > 0$ for all $\nu \leq r$. In fact, take an element z of B so that $\det(Y_\nu \cdot \varphi_\nu(z)) > 0$ and replace $\iota(x)$ by $z\iota(x)z^{-1}$, and Y_ν by $\varphi_\nu(z) Y$. Now after this modification of ι and Y_ν, define a mapping j of $\mathfrak{H}$ to $\mathfrak{K}$ by

$$j(z_1, \cdots, z_r) = (w_1, \cdots, w_r) ,$$

$$w_\nu = (a_\nu z_\nu + b_\nu)/(c_\nu z_\nu + d_\nu) , \qquad \bar{U} Y_\nu = \begin{bmatrix} a_\nu & b_\nu \\ c_\nu & d_\nu \end{bmatrix} \qquad (\nu = 1, \cdots, r) .$$

If $x \in B$, then $\varphi_\nu(x)$ is real, hence

$$x^{(\nu)} = Y_\nu^{-1} \varphi_\nu(\iota(x)) Y_\nu = Y_\nu^{-1} \bar{U}^{-1} \cdot {}^t W_\nu^{-1} \cdot \overline{\omega_\nu(\iota(x))} \cdot {}^t W_\nu \bar{U} Y_\nu .$$

Recalling the definition of the action of elements of $G(T)$ on $\mathfrak{H}$ (6.2), we obtain (7.3.1).

Hereafter, for each choice of K and $\tau_1, \cdots, \tau_g$, we choose and fix L, ρ and T so that (6.1.4) is satisfied, and B is isomorphic to the algebra $\mathfrak{B}$ defined by (6.2.1). Further we fix ι and j as in (7.3.1).

7.4. Let $M \in J(B)$ and f be an F-linear injection of M into B. Let z_0 be the fixed point of $f(M)$ on $\mathfrak{H}$, and $z = j(z_0)$. Define $S, \theta, \Xi, \Psi, \chi_\nu$ as in 6.5 for this z, identifying B with $\mathfrak{B}$ through ι. Let us now prove

(7.4.1) *If $r = 1$ and f is normalized in the sense of 2.7, then χ_1 is the identity mapping on M.*

In fact, let $b \in M - F$, $\beta = \iota(f(b))$. Then $\Lambda(\beta, z)$ of 4.4 can be obtained as a matrix Λ of [9, p. 161, (32)], which is determined by [9, p. 162, (33), (34)] with matrices Λ_ν such that

$$\Lambda_\nu \cdot \begin{bmatrix} 1 & z_\nu \\ \bar{z}_\nu & 1 \end{bmatrix} = \begin{bmatrix} 1 & z_\nu \\ \bar{z}_\nu & 1 \end{bmatrix} \cdot \bar{W}_\nu \cdot {}^t\omega_\nu(\beta) \bar{W}_\nu^{-1} \qquad [9, \text{p. } 165, (39)] .$$

By the consideration of 6.5 and the formulas [9, p. 162, (33), (34)], we see that $\Lambda_1 = \begin{bmatrix} b^{\chi_1} & 0 \\ 0 & b^{\chi_1 \rho} \end{bmatrix}$. Therefore, if $P = \begin{bmatrix} 1 & z_1 \\ \bar{z}_1 & 1 \end{bmatrix}$, then

$$(7.4.2) \qquad {}^t W_\nu^{-1} \overline{\omega_\nu(\beta)} \cdot {}^t W_\nu = P \begin{bmatrix} b^{\chi_1 \rho} & 0 \\ 0 & b^{\chi_1} \end{bmatrix} P^{-1} .$$

In view of the definition of the action of β on $\mathcal{H}$, and by (7.3.1), the left hand side of (7.4.2) represents the transformation $j^{-1} \circ f(b) \circ j$ on $\mathcal{H}$ (= the unit disk). Now the matrix P defines a holomorphic automorphism of the unit disk. Therefore if f is normalized, then χ_1 should be the identity mapping, in view of the definition of 2.7.

7.5. PROPOSITION. *Let the notation be as in 7.3, and let $\mathfrak{o}$ be a maximal order in B. Then there exists an $\mathfrak{r}_K$-lattice $\mathfrak{M}$ in L such that*

$$\mathfrak{o} = \{a \in B \mid \mathfrak{M} a(a) \subset \mathfrak{M}\} .$$

PROOF. Let $\mathfrak{N}$ be an arbitrary $\mathfrak{r}_K$-lattice in L. Put $\mathfrak{M} = \mathfrak{N} a(\mathfrak{o})$. Then $\mathfrak{M}$ has the required property.

7.6. PROPOSITION. *Suppose that τ_{01} is the identity mapping on F and $g > 1$. Let R be a finite algebraic extension of F, and $\mathfrak{h}$ an integral ideal in F. Then there exist a totally imaginary quadratic extension K of F and extensions $\tau_1, \cdots, \tau_g$ of $\tau_{01}, \cdots, \tau_{0g}$ to K satisfying the following conditions.*

(7.6.1) *K is not contained in R.*

(7.6.2) *$D(K/F)$ is prime to $\mathfrak{h}$.*

(7.6.3) *K is generated over $\mathbf{Q}$ by $\sum_{\nu=2}^{g} x^{\tau_\nu}$ for all $x \in K$.*

(7.6.4) *K has no roots of unity other than ± 1.*

(7.6.5) *There exists a prime ideal in F which is ramified in K and does not divide $2 \cdot D(B/F)$.*

PROOF. Let p be a rational prime which does not divide

$$N(\mathfrak{h} \cdot D(B/F)) \cdot D(R/\mathbf{Q})$$

and such that $p \equiv -1 \bmod (4)$. Let $\lambda = \sqrt{-p}$ and $K = F(\lambda)$. Define τ_ν so that $\lambda^{\tau_\nu} = \lambda$ and $\tau_\nu = \tau_{0\nu}$ on F. For every a and b of F, we have

$$\sum_{\nu=2}^{g} (a + b\lambda)^{\tau_\nu} = [\mathrm{Tr}_{F/\mathbf{Q}}(a) - a] + \lambda \cdot [\mathrm{Tr}_{F/\mathbf{Q}}(b) - b] .$$

From this we obtain (7.6.3). We see that $\mathfrak{d}(\mathbf{Q}(\lambda)/\mathbf{Q}) = (\lambda)$. By our choice of p, we have $\mathfrak{d}(K/F) = (\lambda)$, so that (7.6.1, 2 and 5) are satisfied. Now there are only a finite number of quadratic extensions of F containing roots of unity other than ± 1. Therefore (7.6.4) is satisfied under a suitable choice of p.

7.7. LEMMA. *Let V be a variety defined over a field k, and X a one-dimensional subvariety of V, defined over a separably generated extension* *of k. Suppose that, for every finite algebraic extension K of k, X contains infinitely many points u algebraic over k, such that K and $k(u)$ are linearly disjoint over k. Then X is rational over k.*

PROOF. We can find a finite separably algebraic extension k_1 of k such that X is defined over a regular extension k_2 of k_1. Let τ be an isomorphism of k_2 into the universal domain, over k. By assumption, X contains an infinite set Y of points u, algebraic over k, such that k_1 and $k(u)$ are linearly disjoint over k for every $u \in Y$. Since k_2 is regular over k_1, we see that k_2 and $k(u)$ are linearly disjoint over k. Therefore, for each $u \in Y$, we can extend τ to an isomorphism σ of $k_2(u)$ into the universal domain so that $u^\sigma = u$. (σ may depend on u.) Then we have $u = u^\sigma \in X \cap X^\tau$, so that $Y \subset X \cap X^\tau$. Since X is one-dimensional and Y is an infinite set, we have $X = X^\tau$, which proves the lemma.

7.8. LEMMA. *Let V and W be algebraic curves defined over a field k, and f a morphism of V to W, rational over a separably generated extension of k. Suppose that, for every finite algebraic extension K of k, there exist infinitely many points u on V such that K and $k(u, f(u))$ are linearly disjoint over k, and $u, f(u)$ are algebraic over k. Then f is rational over k.*

This can be seen easily by taking the graph of f to be the X of 7.7.

7.9. Let us now start the proof of Main Theorem I in the case $g > 1$. Let $\mathfrak{o}, \mathfrak{e}$ and $\mathfrak{c}$ be as in 3.2. In view of 3.20, it is sufficient to consider the case where $\mathfrak{e} = c\mathfrak{o}$ with a positive integer c. We assume this throughout the remaining part of this section except in 7.12. We have $\mathfrak{c} = c\mathfrak{r}_F$ obviously. First we state a weaker form of 3.2 and a variation of 3.3.

7.10. PROPOSITION. *Let $k_0 = C(F, \mathfrak{c})$. Then there exist a non-singular curve V_0, a holomorphic mapping φ_0 of $\mathfrak{H}$ to V_0 and a finite algebraic extension R_0 of F satisfying the following conditions.*

(7.10.1) *V_0 is defined over k_0.*

(7.10.2) *φ_0 gives a biregular isomorphism of $\mathfrak{H}/\Gamma(\mathfrak{o}, \mathfrak{c})$ onto V_0.*

(7.10.3) *Let M, f and z be as in (3.2.3). Suppose that M is not contained in R_0, and $D(M/F)$ is prime to $D(R_0/F)$. Then $M \cdot k_0(\varphi_0(z)) = C(M, \mathfrak{c})$.*

7.11. PROPOSITION. *Let (V_0, φ_0, R_0) and (V_1, φ_1, R_1) be two systems satisfying the conditions (7.10.1–3) with the same $\mathfrak{o}$ and $\mathfrak{c}$. Then there exists a biregular isomorphism q of V_0 to V_1, defined over k_0, such that $\varphi_1 = q \circ \varphi_0$.*

PROOF. Since both V_0 and V_1 are isomorphic to $\mathfrak{H}/\Gamma(\mathfrak{o}, \mathfrak{c})$ (through φ_0 and φ_1), we find a biregular morphism q of V_0 to V_1 such that $\varphi_1 = q \circ \varphi_0$. Let R

be any finite Galois extension of F containing k_0, R_0 and R_1. Let M, f and z be as in (3.2.3). Suppose that M is not contained in R, and $D(M/F)$ is prime to $D(R/F)$. By 2.10, there are infinitely many such M. Put $u = \varphi_0(z)$. By our assumption, we have $M \cdot k_0(u) = M \cdot k_0(q(u)) = C(M, \mathfrak{c})$. By 1.3, we have

$$C(M, \mathfrak{c}) \cap R = C(F, \mathfrak{c}) = k_0 \ .$$

Since R is normal over k_0, we see that $C(M, \mathfrak{c})$ and R are linearly disjoint over k_0. If we take distinct fields M, then we get distinct points u. By 7.8, q is rational over k_0. This completes the proof.

7.12. PROOF OF 3.3. In 3.3, $\mathfrak{e}$ may be different from ∞. If $\mathfrak{e} = \infty$, 3.3 follows from 7.11. Even if $\mathfrak{e} \neq \infty$, we can repeat the reasoning of the proof of 7.11. The only difference is that we have $M \cdot k_0(u) = M \cdot k_0(q(u)) = C(M, \mathfrak{e}_M)$ with the ideal $\mathfrak{e}_M$ in M (see (3.2.3)). Since $\mathfrak{cr}_M \subset \mathfrak{e}_M$, we have $C(M, \mathfrak{e}_M) \subset C(M, \mathfrak{c})$. By 1.3, we have again $k_0 = k_0(u, q(u)) \cap R$. Therefore, by the same reasoning as in the proof of 7.11, we obtain 3.3.

7.13. To prove 7.10, let us take K, $\tau_1, \cdots, \tau_g$, L, ρ and T as in 7.3. Let $\mathfrak{o}$ and $\mathfrak{c}$ be as in 7.9. By 7.5, we can find a lattice $\mathfrak{M}$ in L such that $\mathfrak{r}_K \mathfrak{M} \mathfrak{c}(\mathfrak{o}) \subset \mathfrak{M}$. Take a representation Φ of L as in (6.1.3) with $q = 2$ and $r = 1$, and elements $v_1, \cdots, v_s$ of L so that $\mathfrak{c}^{-1}\mathfrak{M} = \mathfrak{M} + \sum_{i=1}^{s} \mathfrak{r}_K v_i$. Let Ω_0 be a weak PEL-type defined by (6.1.6) with these choices of symbols. Let $(V, \mathfrak{b}, \varphi)$ be a moduli-system for the family $\Sigma(\Omega_0)$ of weak PEL-structures in the sense of 4.22. Now we assume that K and $\tau_1, \cdots, \tau_g$ satisfy (7.6.3–5). Then, by 6.3, we have $\iota(\Gamma(\mathfrak{o}, \mathfrak{c})) = \Gamma_0(T, \mathfrak{c}^{-1}\mathfrak{M}/\mathfrak{M})$. Now in this situation we obtain

(7.13.1) *Let M, f and z be as in (3.2.3). Suppose that $M \neq K$ and $D(M/F)$ is prime to $D(K/F)$. Then $M(\varphi(j(z))) \cdot C(K, \mathfrak{c}) = C(M, \mathfrak{c}) \cdot C(K, \mathfrak{c})$.*

In fact, by 6.8, $k(\Omega_0) \subset C(K, \mathfrak{c})$. (Note that $K' = K$ in view of (5.14.7) and (7.6.3).) Therefore, by (4.20.4, 5), (6.6.1), (6.7.2) and 5.22, we have

$$M(\varphi(j(z))) \cdot C(K, \mathfrak{c}) = C(S', \Xi', \mathfrak{c}) \cdot C(K, \mathfrak{c}) = C(M, \mathfrak{c}) \cdot C(K, \mathfrak{c}) \ .$$

Here it should be observed that (5.21.2) is satisfied since $D(M/F)$ is prime to $D(K/F)$ and K satisfies (7.6.5).

7.14. After choosing K and the τ_ν as in 7.13, we can find, by 7.6, another totally imaginary quadratic extension K^- of F, and extensions $\tau_1^-, \cdots, \tau_g^-$ of $\tau_{01}, \cdots, \tau_{0g}$ to K^- satisfying (7.6.3–5), and

(7.14.1) K^- *is not contained in* $C(K, \mathfrak{c})$,

(7.14.2) $D(K^-/F)$ *is prime to* $\mathfrak{c} \cdot D(K/F)$.

Then, by 1.4, we have

$$(7.14.3) \qquad C(K, \mathfrak{c}) \cap C(K^-, \mathfrak{c}) = C(F, \mathfrak{c}) \ .$$

Now we consider the objects for this K^- and τ_v^- corresponding to L, T, j, etc., as in 7.12, and denote them by L^-, T^-, j^-, etc., with $-$ on the upper right to specify that they are defined with respect to K^-. We define a weak PEL-type Ω_0^- and take a moduli-system $(V^-, \mathfrak{v}^-, \varphi^-)$ for $\Sigma(\Omega_0^-)$. We have again $\iota^-(\Gamma(\mathfrak{o}, \mathfrak{c})) = \Gamma_0(T^-, \mathfrak{c}^{-1}\mathfrak{M}^-/\mathfrak{M}^-)$. Therefore we can define a biregular morphism h of V to V^- so that $h \circ \varphi \circ j = \varphi^- \circ j^-$. Let $R = C(K, \mathfrak{c}) \cdot C(K^-, \mathfrak{c})$. We are going to prove

$(7.14.4)$ *h is defined over R.*

To show this, let Y be an arbitrary finite Galois extension of F containing R. By 2.10, there are infinitely many M in $J(B)$ such that M is not contained in Y, and $D(M/F)$ is prime to $D(Y/F)$. For such an M, let f and z be as in (3.2.3). Put $u = \varphi(j(z))$. Then $h(u) = \varphi^-(j^-(z))$. Applying (7.13.1) to K and K^-, we obtain

$$(7.14.5) \qquad C(K, \mathfrak{c}) \cdot M(u) = C(K, \mathfrak{c}) \cdot C(M, \mathfrak{c}) \ ,$$

$$(7.14.6) \qquad C(K^-, \mathfrak{c}) \cdot M(h(u)) = C(K^-, \mathfrak{c}) \cdot C(M, \mathfrak{c}) \ ,$$

so that $R \cdot M(u) = R \cdot C(M, \mathfrak{c}) = R \cdot M(h(u))$. Hence $R(u, h(u)) \subset R \cdot C(M, \mathfrak{c})$. By 1.3 and our choice of M, we have $C(F, \mathfrak{c}) = C(M, \mathfrak{c}) \cap Y$. Since Y is normal over $C(F, \mathfrak{c})$, Y and $C(M, \mathfrak{c})$ are linearly disjoint over $C(F, \mathfrak{c})$. It follows that $R \cdot C(M, \mathfrak{c})$ and Y are linearly disjoint over R, hence $R(u, h(u))$ and Y are linearly disjoint over R. Since there are infinitely many such points u, we obtain (7.14.4) by means of 7.8.

7.15. Note that $C(K, \mathfrak{c})$ is normal over F. Therefore (7.14.3) shows that $C(K, \mathfrak{c})$ and $C(K^-, \mathfrak{c})$ are linearly disjoint over $C(F, \mathfrak{c})$. Let G be the Galois group of R over $C(K^-, \mathfrak{c})$. We can identify G with the Galois group of $C(K, \mathfrak{c})$ over $C(F, \mathfrak{c})$. For every $\sigma \in G$, put $h_\sigma = (h^\sigma)^{-1} \circ h$. Then h_σ is a biregular morphism of V to V^σ, defined over R. Let us prove

$(7.15.1)$ *h_σ is defined over $C(K, \mathfrak{c})$.*

To see this, let Y, M and u be as in 7.14. Since $C(M, \mathfrak{c})$ and Y are linearly disjoint over $C(F, \mathfrak{c})$, we can extend σ to an automorphism λ of C over $C(K^-, \mathfrak{c}) \cdot C(M, \mathfrak{c})$. Then, by (7.14.6), we have $h(u) = h(u)^\lambda = h^\sigma(u^\lambda)$, so that $u^\lambda = h_\sigma(u)$. From (7.14.5) we see that u is rational over $C(K, \mathfrak{c}) \cdot C(M, \mathfrak{c})$. Since $C(K, \mathfrak{c}) \cdot C(M, \mathfrak{c})$ is normal over $C(F, \mathfrak{c})$, applying λ to (7.14.5), we see that u^λ is rational over $C(K, \mathfrak{c}) \cdot C(M, \mathfrak{c})$. Now Y and $C(K, \mathfrak{c}) \cdot C(M, \mathfrak{c})$ are linearly disjoint over $C(K, \mathfrak{c})$. Moreover, for a given Y, there are infinitely many M. Hence there are infinitely many points u on V such that $(u, h_\sigma(u))$

is rational over an extension of $C(K, \mathfrak{c})$ which is linearly disjoint with Y over $C(K, \mathfrak{c})$. By 7.8 we obtain (7.15.1).

7.16. From our definition of h_σ, we obtain $h_{\sigma\tau} = (h_\sigma)^\tau \circ h_\tau$ for

$$\sigma, \tau \in G\big(C(K, \mathfrak{c})/C(F, \mathfrak{c})\big) \, .$$

By Weil [18], there exist a non-singular curve V_1, defined over $C(F, \mathfrak{c})$, and a biregular morphism π of V_1 to V, defined over $C(K, \mathfrak{c})$, such that $h_\sigma = \pi^\sigma \circ \pi^{-1}$ for every $\sigma \in G(C(K, \mathfrak{c})/C(F, \mathfrak{c}))$.

Now for every automorphism σ of R over $C(K^-, \mathfrak{c})$, we have $(h \circ \pi)^\sigma = h^\sigma \circ \pi^\sigma = h \circ h_\sigma^{-1} \circ \pi^\sigma = h \circ \pi$, hence $h \circ \pi$ is defined over $C(K^-, \mathfrak{c})$. Put $\varphi_1 = \pi^{-1} \circ \varphi \circ j$. Let us again consider M and u of 7.14. Put $w = \varphi_1(z)$. Then $u = \pi(w)$ and $h(u) = h(\pi(w))$. Since $h \circ \pi$ is defined over $C(K^-, \mathfrak{c})$, we have, by (7.14.6),

(7.16.1) $C(K^-, \mathfrak{c}) \cdot M(w) = C(K^-, \mathfrak{c}) \cdot M(h(u)) = C(K^-, \mathfrak{c}) \cdot C(M, \mathfrak{c}) \, .$

Since π is defined over $C(K, \mathfrak{c})$, we have, by (7.14.5),

(7.16.2) $C(K, \mathfrak{c}) \cdot M(w) = C(K, \mathfrak{c}) \cdot M(u) = C(K, \mathfrak{c}) \cdot C(M, \mathfrak{c}) \, .$

From (7.16.1, 2) we obtain

(7.16.3) $C(F, \mathfrak{c}) \cdot M(w) \subset [C(K, \mathfrak{c}) \cdot C(M, \mathfrak{c})] \cap [C(K^-, \mathfrak{c}) \cdot C(M, \mathfrak{c})] \, .$

Now $C(M, \mathfrak{c})$ is linearly disjoint with $C(K, \mathfrak{c}) \cdot C(K^-, \mathfrak{c})$ over $C(F, \mathfrak{c})$. Therefore, from (7.14.3) and (7.16.3) we obtain $C(F, \mathfrak{c}) \cdot M(w) \subset C(M, \mathfrak{c})$. Since $C(K, \mathfrak{c})$ and $C(M, \mathfrak{c})$ are linearly disjoint over $C(F, \mathfrak{c})$, from the last inclusion and (7.16.2), we get $C(F, \mathfrak{c}) \cdot M(w) = C(M, \mathfrak{c})$. This shows that V_1 and φ_1 have the properties (7.10.1–3) with $C(K, \mathfrak{c}) \cdot C(K^-, \mathfrak{c})$ as R_0. The proof of 7.10 is thus completed.

7.17. Let us now show that (V_0, φ_0) of 7.10 has the property (3.2.3). Let M, f and z be as in (3.2.3). Let $x = \varphi_0(z)$. Let P be a finite Galois extension of F, which contains the algebraic closure of F in $C(M, \mathfrak{c}) \cdot k_0(x)$, where $k_0 = C(F, \mathfrak{c})$. (At the moment, we do not know whether x is algebraic over F or not.) By 7.6, we can take, in the discussion of 7.13–16, the field K so that K is not contained in P, and $D(K/F)$ is prime to $D(P/F)$. Then, as is shown in 7.16, we obtain (V_1, φ_1) satisfying the conditions of 7.10 with a certain R_0. By 1.3, we have $C(F, \mathfrak{c}) = C(K, \mathfrak{c}) \cap P$, hence $C(K, \mathfrak{c})$ and P are linearly disjoint over $C(F, \mathfrak{c})$. Let $(V, \mathfrak{v}, \varphi)$ be as in 7.13, and π be as in 7.16. By 7.11, there exists a biregular morphism q of V_1 to V_0, defined over k_0, such that $q \circ \varphi_1 = \varphi_0$. Then $\varphi_1(z) = q^{-1}(x)$, and $\pi(q^{-1}(x)) = \varphi(j(z))$. Since π is defined over $C(K, \mathfrak{c})$, we have, by (7.13.1),

(7.17.1) $M \cdot k_0(x) \cdot C(K, \mathfrak{c}) = M(\varphi(j(z))) \cdot C(K, \mathfrak{c}) = C(M, \mathfrak{c}) \cdot C(K, \mathfrak{c}) \, .$

This shows especially that x is algebraic over F, hence both $M \cdot k_0(x)$ and $C(M, \mathfrak{c})$ are contained in P. Since $C(K, \mathfrak{c})$ is linearly disjoint with P over $C(F, \mathfrak{c})$, from (7.17.1) we obtain $M \cdot k_0(x) = C(M, \mathfrak{c})$. Therefore (V_0, φ_0) satisfies (3.2.3). This completes the proof of 3.2 in the case $g > 1$ and $\mathfrak{e} = c\mathfrak{o}$. The result of 7.16 shows also

(7.17.2) *Let the notation and assumption be as in 7.13, and (V_0, φ_0) a canonical model for $\mathfrak{H}/\Gamma(\mathfrak{o}, \mathfrak{c})$. Then there exists a biregular morphism π of V_0 to V, defined over $C(K, \mathfrak{c})$, such that $\pi \circ \varphi_0 = \varphi \circ j$.*

8. Proof of Main Theorem II: Case $g > 1$

8.1. In this section, we assume $g > 1$ and $r = 1$. Let $\mathfrak{o}_\lambda$, $\mathfrak{x}_{\lambda\mu}$, $\mathfrak{e}_\lambda$, $U(\mathfrak{e})$ and $U_0(\mathfrak{e})$ be as in 3.4. In view of 3.20, in order to prove 3.5, it is sufficient to treat the case where $\mathfrak{e}_\lambda = c\mathfrak{o}_\lambda$ and $\mathfrak{c} = c\mathfrak{r}_F$ with a positive integer c. We assume this throughout §8. Then, in (3.5.4), we always have $\mathfrak{e}_M = c\mathfrak{r}_M = c\mathfrak{r}_M$. Further we assume that the ideals $\mathfrak{x}_{\lambda\mu}$ are prime to $D(B/F)$. In view of 3.8, this has no effect on generality.

8.2. PROPOSITION. *Let $\{V_\lambda, \varphi_\lambda, R_\sigma^{\mu\lambda}(\alpha)\}$ (resp. $\{\bar{V}_\lambda, \bar{\varphi}_\lambda, \bar{R}_\sigma^{\mu\lambda}(\alpha)\}$) be a system satisfying the conditions (3.5.1–3) and Y (resp. $\bar{Y}$) a finite algebraic extension of F. Suppose that (3.5.4) is satisfied under the assumption that M is not contained in Y (resp. $\bar{Y}$) and $D(M/F)$ is prime to $D(Y/F)$ (resp. $D(\bar{Y}/F)$). Then there exists, for each λ, a biregular morphism P_λ of V_λ to $\bar{V}_\lambda$, defined over $C(F, \mathfrak{c})$, such that $\bar{\varphi}_\lambda = P_\lambda \circ \varphi_\lambda$ and $\bar{R}_\sigma^{\mu\lambda}(\alpha) = P_\mu^\sigma \circ R_\sigma^{\mu\lambda}(\alpha) \circ P_\lambda^{-1}$.*

PROOF. We can repeat the reasoning of the proof of 3.6 only with a slight modification. In fact first exchanging α for a suitable element $\equiv \alpha \bmod^* \mathfrak{e}$, we may assume that $\alpha\mathfrak{o}_\lambda$ is prime to $D(B/F)$. Then, by 2.10, we can take the field M in the proof of 3.6 so that M is not contained in $Y\bar{Y}$, and $D(M/F)$ is prime to $D(Y\bar{Y}/F)$. Now the argument of the proof of 3.6 applies to the present case with this choice of M.

8.3. Let $K, \tau_1, \cdots, \tau_g, L, T, \rho, \iota$, and j be as in 7.3. For brevity, we identify B with $\iota(B)$ by ι. We assume the conditions (7.6.3–5) and

(8.3.1) *K is not contained in $C(F, \mathfrak{c})$.*

Take a representation Φ of L as in (6.1.3) with $q = 2$ and $r = 1$. Let us fix a lattice $\mathfrak{M}$ in L such that $\mathfrak{r}_K \mathfrak{M}\mathfrak{o}_1 \subset \mathfrak{M}$. Put $\mathfrak{M}_\lambda = \mathfrak{M}\mathfrak{x}_{1\lambda}$. Then $\mathfrak{r}_K \mathfrak{M}_\lambda \mathfrak{o}_\lambda \subset \mathfrak{M}_\lambda$. Further we fix a set $\mathfrak{U}$ of representatives for the ideal-classes modulo $\mathfrak{c}$ in K. Now we consider all weak PEL-types of the form

(8.3.2) $$\Omega = (L, \Phi, \rho;\ F^+ \cdot T,\ \mathfrak{a}\mathfrak{M}_\lambda;\ x_1, x_2)$$

with $\mathfrak{a} \in \mathfrak{U}$ and elements x_1, x_2 of L such that

$$(8.3.3) \qquad \mathfrak{c}^{-1}\mathfrak{a}\mathfrak{M}_\lambda = \mathfrak{a}\mathfrak{M}_\lambda + \mathfrak{r}_K x_1 + \mathfrak{r}_K x_2 \,.$$

We see that $(\mathfrak{a}\mathfrak{M}_\lambda)_\mathfrak{p} = \mathfrak{M}_\mathfrak{p}$ for every λ, every $\mathfrak{a} \in \mathfrak{U}$, and every prime ideal $\mathfrak{p}$ in F dividing $\mathfrak{c}$. (For the notation, see 1.6.) Hence if y_1 and y_2 are elements of L such that $(\mathfrak{c}^{-1}\mathfrak{M})_\mathfrak{p} = (\mathfrak{M} + \mathfrak{r}_K y_1 + \mathfrak{r}_K y_2)_\mathfrak{p}$ for all $\mathfrak{p}$ dividing $\mathfrak{c}$, then we can find[1] elements x_1 and x_2 of L so that $x_i \equiv y_i \bmod \mathfrak{M}_\mathfrak{p}$ $(i = 1, 2)$ for all such $\mathfrak{p}$, and (8.3.3) is satisfied. Therefore, we express Ω also as

$$\Omega = (L, \Phi, \rho; F^+ \cdot T, \mathfrak{a}\mathfrak{M}_\lambda; y_1, y_2)$$

for such y_1 and y_2 (even if y_1, y_2 may not be contained in $\mathfrak{c}^{-1}\mathfrak{a}\mathfrak{M}_\lambda$), since Ω is certainly determined by these data, up to equivalence. The symbols L, Φ, ρ, $F^+ \cdot T$ will be common to all the weak PEL-types considered in this section. Therefore we write also $\Omega = (\mathfrak{a}\mathfrak{M}_\lambda; y_1, y_2)$. Taking a suitable set of indices I_λ for each λ, we write $\Omega = \Omega_{\lambda i}$ with $i \in I_\lambda$ so that $\{\Omega_{\lambda i} \mid i \in I_\lambda\}$ is the set of all weak PEL-types of the form (8.3.2).

Now we define a family $\Sigma_{\lambda i} = \{\mathcal{R}_{\lambda i}(z) \mid z \in \mathcal{K}\}$ of weak PEL-structures of type $\Omega_{\lambda i}$, by means of one and the same parametrizing function $\mathfrak{y}(x, z)$ (see 4.1). We put

$$(8.3.4) \qquad \mathcal{R}_{\lambda i}(z) = \mathcal{R}_{\lambda i z} = (A_{\lambda i z}, C_{\lambda i z}, \theta_{\lambda i z}; t_{\lambda i z 1}, t_{\lambda i z 2}) \,.$$

Let $(V_{\lambda i}, \mathfrak{b}_{\lambda i}, \varphi_{\lambda i})$ be a moduli-system for the family $\Sigma_{\lambda i}$. By 4.25, we may assume that $V_{\lambda i} = V_{\lambda 1}$ and $\varphi_{\lambda i} = \varphi_{\lambda 1}$ for all $i \in I_\lambda$. Therefore we write simply V_λ and φ_λ for $V_{\lambda i}$ and $\varphi_{\lambda i}$.

8.4. Define $\sigma_1, \cdots, \sigma_g$ as in (5.14.5) for the present K and $\tau_1, \cdots, \tau_g$ with $r = 1$. In view of (7.6.3) we have $K = K'$, so that by (5.14.9), we observe that the integers h and s of (5.14.5) are equal to g and 1, respectively. We shall now prove

(8.4.1) *Let* $\mathfrak{t}$ *be an ideal in* K, *prime to* $\mathfrak{c}$, *and let* $\sigma = [C(K, \mathfrak{c})/K, \mathfrak{t}]$, *and* $\alpha \in U(\mathfrak{e})$. *Suppose that* $\sigma = [C(F, \mathfrak{c})/F, N_{B/F}(\alpha \chi_{\lambda \mu})]$ *on* $C(F, \mathfrak{c})$. *Let* $\Omega_{\mu i} = (\mathfrak{a}\mathfrak{M}_\mu; x_1, x_2)$ *with* $\mathfrak{a} \in \mathfrak{U}$, *and let* $\mathfrak{b}$ *be an element of* $\mathfrak{U}$ *which is equivalent to* $\mathfrak{a}\mathfrak{z}^{-1}$, *where* $\mathfrak{z} = \prod_{\nu=2}^g \mathfrak{t}^{\sigma_\nu}$. *Then* $(\Omega_{\mu i})^\sigma$ *is equivalent to a weak* PEL-*type* $\Omega_{\lambda j}$ *of the form* $\Omega_{\lambda j} = (\mathfrak{b}\mathfrak{M}_\lambda; x_1 \alpha, x_2 \alpha)$.

Here we observe that $(\Omega_{\mu i})^\sigma$ is meaningful in view of 6.8 and (4.23.2).

8.5. Given $\alpha \in U(\mathfrak{e})$, we can find an element β of B^+ so that $\beta \equiv \alpha \bmod^* \mathfrak{e}$, and $\beta \mathfrak{o}_\lambda$ is prime to $\mathfrak{c}D(B/F)$. By 2.10, given λ and μ as in (8.4.1), there are infinitely many M in $J(B)$ satisfying the following conditions:

(8.5.1) *M is not contained in* $C(K, \mathfrak{c})$.

(8.5.2) *$D(M/F)$ is prime to* $\mathfrak{c} \cdot D(K/F)$.

(8.5.3) *There exist an F-linear isomorphism f of M into B and an ideal $\mathfrak{m}$ in M such that $f(\mathfrak{r}_M) \subset \mathfrak{o}_\mu$ and $\beta \mathfrak{X}_{\lambda\mu} = f(\mathfrak{m})\mathfrak{o}_\mu$.*

Here we may assume that f is normalized in the sense of 2.7, by exchanging $f(x)$ for $f(x^\rho)$ if necessary. Take such M, f and $\mathfrak{m}$. By 1.4, we have

$$C(M, \mathfrak{c}) \cap C(K, \mathfrak{c}) = C(F, \mathfrak{c}) ,$$

hence $C(M, \mathfrak{c})$ and $C(K, \mathfrak{c})$ are linearly disjoint over $C(F, \mathfrak{c})$. Therefore we can extend σ of (8.4.1) to an automorphism τ of C so that $\tau = [C(M, \mathfrak{c})/M, \mathfrak{m}]$ on $C(M, \mathfrak{c})$. Put $S = KM$ and $k = C(S, \mathfrak{c})$ with the notation of 6.4–6.7. (By our assumption (7.6.3), we have $K = K'$ and hence $S = S'$.) Take an *integral* ideal $\mathfrak{q}$ in S, prime to $\mathfrak{c}$ such that $\tau = [k/S, \mathfrak{q}]$ on k. Put

$$\mathfrak{g} = N_{S/M}(\mathfrak{q}) , \qquad \mathfrak{h} = \prod_{\nu=2}^{g} N_{S/K}(\mathfrak{q})^{\sigma_\nu} .$$

We have then $\tau = [C(M, \mathfrak{c})/M, \mathfrak{g}]$, so that $\mathfrak{g}$ and $\mathfrak{m}$ belong to the same ideal class modulo $\mathfrak{c}$ in M. Hence there exists an element a of M such that $a \equiv 1 \bmod^* \mathfrak{c}$ and $\mathfrak{g} = a\mathfrak{m}$. Similarly there exists an element b of K such that $b \equiv 1 \bmod^* \mathfrak{c}$, $N_{S/K}(\mathfrak{q}) = b\mathfrak{t}$. Put $s = \prod_{\nu=2}^{g} b^{\sigma_\nu}$. Then $s \in K$, $s \equiv 1 \bmod^* \mathfrak{c}$, and $\mathfrak{h} = s\mathfrak{z}$. Let $\mathfrak{b}$ be as in (8.4.1). There exists an element d of K such that $d \equiv 1 \bmod^* \mathfrak{c}$ and $\mathfrak{b} = da\mathfrak{z}^{-1}$. Put $\varepsilon = sd \cdot f(a)\beta$. Then $\varepsilon \in G^+(T)$ and $\mathfrak{b}\mathfrak{M}_\lambda \varepsilon^{-1} = (sd)^{-1}\mathfrak{b}\mathfrak{M}_\mu(f(a)\beta \mathfrak{X}_{\lambda\mu})^{-1} = \mathfrak{h}^{-1}a\mathfrak{M}_\mu f(\mathfrak{g})^{-1}$. Since $\mathfrak{h}$ and $\mathfrak{g}$ are integral ideals, we see that $a\mathfrak{M}_\mu \varepsilon \subset \mathfrak{b}\mathfrak{M}_\lambda$.

Let z_0 be the fixed point of $f(M)$ on $\mathfrak{H}$, and let $z = j(z_0)$, $w = \beta^{-1}(z)$. Then $w = \varepsilon^{-1}(z)$. Now put $\Omega_{\lambda j} = (\mathfrak{b}\mathfrak{M}_\lambda; x_1\alpha, x_2\alpha)$. Consider the member $\mathcal{R}_{\lambda j w}$ of $\Sigma_{\lambda j}$ and $\mathcal{R}_{\mu i z}$ of $\Sigma_{\mu i}$ corresponding to these points w and z. Then $\Lambda(\varepsilon, z)$ of 4.4 gives an isogeny ξ of $\mathcal{R}_{\mu i z}$ to $\mathcal{R}_{\lambda j w}$. In fact, for $\nu = 1, 2$ and for every prime ideal $\mathfrak{p}$ in F dividing $\mathfrak{c}$, we have $x_\nu \varepsilon \equiv sdx_\nu f(a)\beta \equiv x_\nu\alpha \bmod \mathfrak{M}_\mathfrak{p}$. Moreover $\mathrm{Ker}(\xi)$ is isomorphic to $\mathfrak{b}\mathfrak{M}_\lambda \varepsilon^{-1}/a\mathfrak{M}_\mu = \mathfrak{h}^{-1} \cdot (a\mathfrak{M}_\mu) \cdot f(\mathfrak{g})^{-1}/a\mathfrak{M}_\mu$. Hence

$$\mathrm{Ker}(\xi) = \{t \in A_{\mu i z} \mid \theta(\mathfrak{g}\mathfrak{h})t = 0\} ,$$

where θ is an extension of $\theta_{\mu i z}$ to $M_2(S)$ (cf. 6.4–5). On the other hand, by 6.7, there exists an isogeny η of $\mathcal{R}_{\mu i z}$ to $(\mathcal{R}_{\mu i z})^\tau$ with the properties (i–iv) of (6.7.3). (The present $\mathfrak{q}$ and τ play the role of $\mathfrak{a}$ and σ of (6.7.3).) By 5.17, we have $\det \Xi'(\mathfrak{q}) = \mathfrak{g}\mathfrak{h}$, so that $\mathrm{Ker}(\eta) = \mathrm{Ker}(\xi)$. (Note that χ_1 is the identity mapping of M in view of (7.4.1), since f is normalized.) Hence there exists an isomorphism δ of $(\mathcal{R}_{\mu i z})^\tau$ to $\mathcal{R}_{\lambda j w}$ such that $\xi = \delta \circ \eta$. (Here observe that $\delta(t_{\mu i z}^\tau) = \delta\eta(t_{\mu i z}) = \xi(t_{\mu i z}) = t_{\lambda j w}$, so far as $\xi = \delta \circ \eta$.) This implies that $(\Omega_{\mu i})^\sigma$ is equivalent to $\Omega_{\lambda j}$, and hence the proof of (8.4.1) is completed.

8.6. Let the notation be as in 8.4. By 4.23 and 4.21, there exists a biregular morphism ψ, defined over $k(\Omega_{\lambda j})$, of V_μ^σ to V_λ such that $\mathfrak{v}_{\lambda j} = \psi \circ \mathfrak{v}_{\mu i}^\sigma$. Let us now show that

(8.6.1) *ψ depends only on λ, μ, σ and the class of α mod* $\mathfrak{e}$, and not on the choice of i.*

Let $\Omega_{\mu p} = (\mathfrak{a}^*\mathfrak{M}_\mu; x_1^*, x_2^*)$ be another weak PEL-type with the same μ, and let α^* be an element of B such that $\alpha \equiv \alpha^*$ mod* $\mathfrak{e}$. Define

$$\Omega_{\lambda q} = (\mathfrak{b}^*\mathfrak{M}_\lambda; x_1^*\alpha, x_2^*\alpha)$$

so that $\mathfrak{b}^* \in \mathfrak{U}$ and $(\Omega_{\mu p})^\sigma$ is equivalent to $\Omega_{\lambda q}$. Let ψ^* be a biregular morphism of V_μ^σ to V_λ such that $\mathfrak{b}_{\lambda q} = \psi^* \circ \mathfrak{b}_{\mu p}^\sigma$. Our aim is to prove $\psi = \psi^*$. By our choice of $(V_\lambda, \mathfrak{b}_{\lambda i}, \varphi_\lambda)$, we have $\mathfrak{b}_{\mu i}(\mathcal{R}_{\mu i z}) = \varphi_\mu(z) = \mathfrak{b}_{\mu p}(\mathcal{R}_{\mu p z})$ for every $z \in \mathcal{H}$. (This z is not necessarily the one considered in 8.5.) Extend σ to an automorphism of C over K. Then $\mathfrak{b}_{\mu i}^\sigma(\mathcal{R}_{\mu i z}^\sigma) = \varphi_\mu(z)^\sigma = \mathfrak{b}_{\mu p}^\sigma(\mathcal{R}_{\mu p z}^\sigma)$. Let d and d^* be elements of K such that $\mathfrak{b} = d\mathfrak{a}\mathfrak{z}^{-1}$, $\mathfrak{b}^* = d^*\mathfrak{a}^*\mathfrak{z}^{-1}$, $d \equiv d^* \equiv 1$ mod* $\mathfrak{c}$. Let r be an integer such that $r\mathfrak{a} \subset \mathfrak{a}^*$. Since $\mathfrak{a}$ and $\mathfrak{a}^*$ are prime to $\mathfrak{c}$, we may assume r to be prime to $\mathfrak{c}$. Let ζ_z be an isogeny of $A_{\mu i z}$ to $A_{\mu p z}$ obtained from $\Lambda(r, z)$. Then we see that

$$(8.6.2) \qquad \operatorname{Ker}(\zeta_z) = \{u \in A_{\mu i z} \mid \theta_{\mu i z}(r\mathfrak{a}^{*-1}\mathfrak{a})u = 0\} ,$$

$$(8.6.3) \qquad t_{\mu p z \nu} = \textstyle\sum_{\kappa=1}^{2} \theta_{\mu p z}(a_{\nu\kappa})\zeta_z t_{\mu i z \kappa} \qquad\qquad (\nu = 1, 2)$$

with elements $a_{\nu\kappa}$ of $\mathfrak{r}_K$. The $a_{\nu\kappa}$ are determined modulo $\mathfrak{c}$ by the relation

$$(8.6.4) \qquad x_\nu^* \equiv \textstyle\sum_{\kappa=1}^{2} a_{\nu\kappa} r x_\kappa \qquad \operatorname{mod} \mathfrak{M}_p \qquad\qquad (\nu = 1, 2)$$

for all prime factors $\mathfrak{p}$ of $\mathfrak{c}$ in F. Hence the $a_{\nu\kappa}$ are independent of z.

Fix z, and let w be a point on $\mathcal{H}$ such that $\mathcal{R}_{\mu i z}^\sigma$ is isomorphic to $\mathcal{R}_{\lambda j w}$. Such a point w exists, since $\Omega_{\mu i}^\sigma = \Omega_{\lambda j}$. Since $(d^{-1}d^*r)\mathfrak{b} \subset \mathfrak{b}^*$, $\Lambda(d^{-1}d^*r, w)$ defines an isogeny ω of $A_{\lambda j w}$ to $A_{\lambda q w}$. We see that

$$(8.6.5) \qquad \operatorname{Ker}(\omega) = \{u \in A_{\lambda j w} \mid \theta_{\lambda j w}(r\mathfrak{a}^{*-1}\mathfrak{a})u = 0\} .$$

Let ε be an isomorphism of $\mathcal{R}_{\lambda j w}$ to $\mathcal{R}_{\mu i z}^\sigma$. Then by (8.6.2) and (8.6.5), we have $\operatorname{Ker}(\omega) = \operatorname{Ker}(\zeta_z^\sigma \circ \varepsilon)$. Hence there exists an isomorphism η of $A_{\lambda q w}$ to $A_{\mu p z}^\sigma$ such that $\eta \circ \omega = \zeta_z^\sigma \circ \varepsilon$. We see that $\eta \circ \theta_{\lambda q w}(a) = \theta_{\mu p z}^\sigma(a) \circ \eta$ for every $a \in L$, since ω, ζ_z^σ and ε have the corresponding property. Since $d \equiv d^* \equiv 1$ mod $\mathfrak{c}$ and $\alpha \equiv \alpha^*$ mod* $\mathfrak{e}$, we have $d^{-1}d^*r \cdot x_\nu \equiv r \cdot x_\nu$ mod $\mathfrak{M}_p$, so that by (8.6.4)

$$x_\nu^*\alpha^* \equiv x_\nu^*\alpha \equiv \textstyle\sum_\kappa a_{\nu\kappa} r x_\kappa \alpha \equiv \textstyle\sum_\kappa a_{\nu\kappa} d^{-1}d^* r x_\kappa \alpha \qquad \operatorname{mod} \mathfrak{M}_p$$

for all prime factors $\mathfrak{p}$ of $\mathfrak{c}$. Hence $t_{\lambda q w \nu} = \sum_\kappa \theta_{\lambda q w}(a_{\nu\kappa})\omega(t_{\lambda j w \kappa})$. Therefore

$$\begin{aligned}
\eta t_{\lambda q w \nu} &= \textstyle\sum_\kappa \theta_{\mu p z}^\sigma(a_{\nu\kappa})\zeta_z^\sigma \circ \varepsilon(t_{\lambda j w \kappa}) \\
&= [\textstyle\sum_\kappa \theta_{\mu p z}(a_{\nu\kappa})\zeta_z(t_{\mu i z \kappa})]^\sigma = t_{\mu p z \nu}^\sigma \qquad\qquad \text{by (8.6.3) .}
\end{aligned}$$

It follows that η is an isomorphism of $\mathcal{R}_{\lambda q w}$ to $\mathcal{R}_{\mu p z}^\sigma$. Therefore

$$\begin{aligned}
\psi^*(\varphi_\mu(z)^\sigma) &= \psi^*(\mathfrak{b}_{\mu p}^\sigma(\mathcal{R}_{\mu p z}^\sigma)) \\
&= \mathfrak{b}_{\lambda q}(\mathcal{R}_{\lambda q w}) = \varphi_\lambda(w) = \mathfrak{b}_{\lambda j}(\mathcal{R}_{\lambda j w}) = \psi(\mathfrak{b}_{\mu i}^\sigma(\mathcal{R}_{\mu i z}^\sigma)) = \psi(\varphi_\mu(z)^\sigma) .
\end{aligned}$$

Since z is an arbitrary point on $\mathfrak{H}$, this proves $\psi^* = \psi$, and hence (8.6.1).

8.7. Let us now write $\psi_\sigma^{\mu\lambda}(\alpha)$ for ψ^{-1}. In view of 6.8, we note that $\psi_\sigma^{\mu\lambda}(\alpha)$ is defined over $C(K, \mathfrak{c})$. If $\gamma \in U(\mathfrak{e})$, $\sigma' \in G(C(K, \mathfrak{c})/K)$, and

$$\sigma' = [C(F, \mathfrak{c})/F, N(\gamma \mathfrak{x}_{\mu\nu})]$$

on $C(F, \mathfrak{c})$, then we have

(8.7.1)
$$\psi_{\sigma'}^{\nu\mu}(\gamma)^\sigma \circ \psi_\sigma^{\mu\lambda}(\alpha) = \psi_{\sigma'\sigma}^{\nu\lambda}(\gamma\alpha) .$$

This follows easily from the relation $\mathfrak{b}_{\lambda j} = \psi \circ \mathfrak{b}_{\mu i}^\sigma$ in a straightforward way. Further if $\alpha \equiv \alpha_1 \bmod^* \mathfrak{e}$, we have

(8.7.2)
$$\psi_\sigma^{\mu\lambda}(\alpha) = \psi_\sigma^{\mu\lambda}(\alpha_1) .$$

Let M, β, τ, z and w be as in 8.5. Then with $\psi^{-1} = \psi_\sigma^{\mu\lambda}(\alpha)$, we have

$$\psi\big(\varphi_\mu(z)^\tau\big) = \psi\big(\mathfrak{b}_{\mu i}^\sigma(\mathcal{R}_{\mu iz}^\tau)\big) = \mathfrak{b}_{\lambda j}(\mathcal{R}_{\lambda jw}) = \varphi_\lambda(w) = \varphi_\lambda\big(\beta^{-1}(z)\big) .$$

Namely we have

(8.7.3)
$$\psi_\sigma^{\mu\lambda}(\alpha)\big[\varphi_\lambda\big(\beta^{-1}(z)\big)\big] = \varphi_\mu(z)^\tau .$$

Now let $(V_\lambda^0, \varphi_\lambda^0)$ be a canonical model for $\mathfrak{H}/\Gamma(\mathfrak{o}_\lambda, \mathfrak{c})$ for each λ. By (7.17.2), there exists a biregular morphism ω_λ of V_λ to V_λ^0, defined over $C(K, \mathfrak{c})$, such that $\varphi_\lambda^0 = \omega_\lambda \circ \varphi_\lambda \circ j$. Now put

(8.7.4)
$$R_\sigma^{\mu\lambda}(\alpha) = \omega_\mu^\sigma \circ \psi_\sigma^{\mu\lambda}(\alpha) \circ \omega_\lambda^{-1}$$

for every $\lambda, \mu, \sigma \in G(C(K, \mathfrak{c})/K)$, $\alpha \in U(\mathfrak{e})$ such that $\sigma = [C(F, \mathfrak{c})/F, N_{B/F}(\alpha \mathfrak{x}_{\lambda\mu})]$ on $C(F, \mathfrak{c})$. If $\sigma' \in G(C(K, \mathfrak{c})/K)$ and $\sigma' = [C(F, \mathfrak{c})/F, N_{B/F}(\gamma \mathfrak{x}_{\mu\nu})]$, then, from (8.7.1), we obtain

(8.7.5)
$$R_{\sigma'}^{\nu\mu}(\gamma)^\sigma \circ R_\sigma^{\mu\lambda}(\alpha) = R_{\sigma'\sigma}^{\nu\lambda}(\gamma\alpha) .$$

Further, by (8.7.2), we have $R_\sigma^{\mu\lambda}(\alpha) = R_\sigma^{\mu\lambda}(\alpha_1)$ if $\alpha \equiv \alpha_1 \bmod^* \mathfrak{e}$. Now, from (8.7.3), we obtain

(8.7.6)
$$R_\sigma^{\mu\lambda}(\alpha)\big[\varphi_\lambda^0\big(\beta^{-1}(z_0)\big)\big] = \varphi_\mu^0(z_0)^\tau ,$$

where $z_0 = j^{-1}(z)$. Since (V_μ^0, φ_μ^0) is a canonical model for $\mathfrak{H}/\Gamma(\mathfrak{o}_\mu, \mathfrak{c})$, $\varphi_\mu^0(z_0)^\tau$ is rational over $C(M, \mathfrak{c})$. Since $\beta \mathfrak{x}_{\lambda\mu} = f(\mathfrak{m})\mathfrak{o}_\mu$ with an ideal $\mathfrak{m}$ in M (see 8.5), if we define an F-linear injection f_1 of M into B by $f_1(x) = \beta^{-1} f(x)\beta$, then $f_1(\mathfrak{r}_M) \subset \mathfrak{o}_\lambda$, and $\beta^{-1}(z_0)$ is the fixed point of $f_1(M)$ on $\mathfrak{H}$. Therefore $\varphi_\lambda^0(\beta^{-1}(z_0))$ is rational over $C(M, \mathfrak{c})$. Note that the definition of $R_\sigma^{\mu\lambda}(\alpha)$ does not depend on the choice of M. Let P be any finite Galois extension of F containing $C(F, \mathfrak{c})$. Then there are infinitely many M satisfying (8.5.1–3) and such that M is not contained in P and $D(M/F)$ is prime to $D(P/F)$. By 1.3 we have $C(F, \mathfrak{c}) = C(M, \mathfrak{c}) \cap P$, so that P and $C(M, \mathfrak{c})$ are linearly disjoint over $C(F, \mathfrak{c})$. Applying 7.8 to $R_\sigma^{\mu\lambda}(\alpha)$, we find that $R_\sigma^{\mu\lambda}(\alpha)$ is defined over $C(F, \mathfrak{c})$.

Now suppose that σ is the identity mapping on $C(F, \mathfrak{c})$, $\alpha \equiv 1 \bmod^* \mathfrak{e}$. Then we have $\lambda = \mu$. Moreover, at the beginning of 8.5, we could have taken $\beta = 1$. With this choice of β, we have $w = z$, $\mathfrak{m} = \mathfrak{r}_M$, so that τ is the identity on $C(M, \mathfrak{c})$. Then, from (8.7.6), we obtain

$$(8.7.7) \qquad\qquad R_\sigma^{\lambda\lambda}(\alpha)[\varphi_\lambda^0(z_0)] = \varphi_\lambda^0(z_0) \ .$$

It follows that $R_\sigma^{\lambda\lambda}(\alpha)$ has infinitely many fixed points. Therefore $R_\sigma^{\lambda\lambda}(\alpha)$ is the identity mapping. Applying this fact to (8.7.5), we have $R_{\sigma'}^{\nu\mu}(\gamma) = R_{\sigma'\sigma}^{\nu\mu}(\gamma\alpha) = R_{\sigma'\sigma}^{\nu\mu}(\gamma)$. This means that $R_{\sigma'}^{\nu\mu}(\gamma)$ is determined by μ, ν, γ and the action of σ' on $C(F, \mathfrak{c})$.

In view of (8.3.1), we observe that $R_\sigma^{\mu\lambda}(\alpha)$ can be defined for all possible choices of $\lambda, \mu, \sigma, \alpha$ under the conditions $\alpha \in U(\mathfrak{e})$ and

$$\sigma = [C(F, \mathfrak{c})/F, N_{B/F}(\alpha \mathfrak{x}_{\lambda\mu})] \ .$$

8.8. To complete the final step of the proof of 3.5, let $M \in J(B)$, and let f be a normalized F-linear isomorphism of M into B such that $f(\mathfrak{r}_M) \subset \mathfrak{o}_\mu$. Let z_0 be the fixed point of $f(M)$ on $\mathfrak{H}$, $\mathfrak{l}$ an ideal in M, prime to $\mathfrak{c}$, and

$$\tau = [C(M, \mathfrak{c})/M, \mathfrak{l}] \ .$$

Since $\{\mathfrak{x}_{1\mu}, \cdots, \mathfrak{x}_{h_0\mu}\}$ is a set of representatives for the narrow classes of right $\mathfrak{o}_\mu$-ideals, there exist an index λ and an element α of B^+ such that $\alpha \mathfrak{x}_{\lambda\mu} = f(\mathfrak{l})\mathfrak{o}_\mu$. Since $\mathfrak{l}$ is prime to $\mathfrak{c}$ and $\mathfrak{x}_{\lambda\mu}$ is prime to $\mathfrak{c}$, we have $\alpha \in U(\mathfrak{e})$. Suppose that M satisfies

(8.8.1) *M is not contained in $C(K, \mathfrak{c})$, $\mathfrak{r}_{KM} = \mathfrak{r}_K \mathfrak{r}_M$, and $C(M, \mathfrak{c})$ is linearly disjoint with $C(K, \mathfrak{c})$ over $C(F, \mathfrak{c})$.*

This is satisfied, for example, if (8.5.1) and (8.5.2) are satisfied. We notice that the reasoning of 8.5 is valid with $\alpha = \beta$ and $\mathfrak{m} = \mathfrak{l}$. (Observe that we do not need the fact $\beta \mathfrak{o}_\lambda$ is prime to $D(B/F)$ any more.) After repeating it, we find $R_\sigma^{\mu\lambda}(\alpha)[\varphi_\lambda^0(\alpha^{-1}(z_0))] = \varphi_\mu^0(z_0)^\tau$ as in (8.7.6). This proves that our system $\{V_\lambda^0, \varphi_\lambda^0, R_\sigma^{\mu\lambda}(\alpha)\}$ satisfies (3.5.1–3) and (3.5.4) under the assumption (8.8.1).

Proposition 8.2 shows the uniqueness of such a system. Fix such a system and take an arbitrary $M \in J(B)$. We can find a totally imaginary quadratic extension K^- of F and $\{\tau_1^-, \cdots, \tau_g^-\}$ with the same properties as K and $\{\tau_1, \cdots, \tau_g\}$, such that K^- is not contained in $C(M, \mathfrak{c})$, and $D(K^-/M)$ is prime to $\mathfrak{c} \cdot D(M/F)$ (7.6). By 1.2 and 1.4, the condition (8.8.1) is satisfied with K^- in place of K. Using this K^-, we can construct another system $\{V_\lambda^-, \varphi_\lambda^-, R_\sigma^{\mu\lambda}(\alpha)\}$. Since (8.8.1) is satisfied (with K^- in place of K), we have $R_{\sigma^-}^{\mu\lambda}(\alpha)[\varphi_\lambda^-(\alpha^{-1}(z_0))] = \varphi_\mu^-(z_0)^\tau$, where α, z_0 and τ are as above. Now $\{V_\lambda^-, \varphi_\lambda^-, R_{\sigma^-}^{\mu\lambda}(\alpha)\}$ is biregularly equivalent, over $C(F, \mathfrak{c})$, with $\{V_\lambda^0, \varphi_\lambda^0, R_\sigma^{\mu\lambda}(\alpha)\}$ as stated in 8.2. Then we see easily that $R_\sigma^{\mu\lambda}(\alpha)[\varphi_\lambda^0(\alpha^{-1}(z_0))] = \varphi_\mu^0(z_0)^\tau$ holds. This completes the proof of Main Theorem II in the case $g > 1$.

8.9. PROOF OF 3.17. Let the notation be as in 3.17. Let $z = j(w)$. Consider the member $\mathcal{R}_{\mu i z}$ of $\Sigma_{\mu i}$ (with any choice of K). Then $\mathcal{R}_{\mu i z}$ has sufficiently many complex multiplications. It follows that $\varphi_\mu(w)$ is algebraic over F. Take any finite Galois extension P of F containing $M(\varphi_\mu(w))$ and $C(F, \mathfrak{c})$. In the discussion of 8.3–8.7, we can take K so that $K \not\subset P$. Then we get $\{V_\lambda^0, \varphi_\lambda^0, R_\sigma^{\mu\lambda}(\alpha)\}$ as above using this K. By the uniqueness, we may consider this system as the system in 3.17. (Note that $C(F, \mathfrak{c}) \cdot M(\varphi_\mu(w))$ is the same in both systems.) We can find an automorphism ξ of KP such that $\xi = \tau$ on P and ξ is the identity mapping on K. Consider again $\mathcal{R}_{\mu i z}$ of the family $\Sigma_{\mu i}$ with this choice of K and with any i. By (8.4.1), we have $(\Omega_{\mu i})^\xi = \Omega_{\lambda j}$ with certain λ and j, hence $(\mathcal{R}_{\mu i z})^\xi$ is isomorphic to a member $\mathcal{R}_{\lambda j y}$ with a point y on $\mathcal{K}$. Applying 6.9 to $\mathcal{R}_{\lambda j y}$, we find an F-linear isomorphism f_1 of M into B such that $y = j(w_1)$ with the fixed point w_1 of $f_1(M)$ on $\mathfrak{H}$. Now, using the notation of 8.7, we have

$$\varphi_\lambda^0(w_1) = \omega_\lambda \circ \varphi_\lambda \circ j(w_1)$$
$$= \omega_\lambda \circ \varphi_\lambda(y) = \omega_\lambda(\mathfrak{v}_{\lambda j}(\mathcal{R}_{\lambda j y})) = \omega_\lambda(\mathfrak{v}_{\lambda j}((\mathcal{R}_{\mu i z})^\xi)) .$$

By (8.7.4),

$$R_\sigma^{\mu\lambda}(\alpha) \circ \varphi_\lambda^0(w_1) = \omega_\mu^\xi \circ \psi_\xi^{\mu\lambda}(\alpha)(\mathfrak{v}_{\lambda j}(\mathcal{R}_{\mu i z}^\xi))$$
$$= \omega_\mu^\xi(\mathfrak{v}_{\mu i}^\xi((\mathcal{R}_{\mu i z})^\xi)) = \omega_\mu^\xi(\varphi_\mu(z)^\xi) = \varphi_\mu^0(w)^\xi = \varphi_\mu^0(w)^\tau .$$

This proves the first assertion of 3.17. Now define $\theta: L \otimes_K KM \to \mathrm{End}_Q(A_{\mu i z})$ as in 6.4. If $\mathfrak{r}' = f^{-1}(f(M) \cap \mathfrak{o}_\mu)$, then $\mathfrak{r}' = \theta^{-1}(\theta(M) \cap \mathrm{End}(A_{\mu i z}))$. Therefore $\mathfrak{r}' = (\theta^\sigma)^{-1}(\theta^\sigma(M) \cap \mathrm{End}(A_{\mu i z}^\sigma))$. From this we obtain, in view of the last assertion of 6.9, $\mathfrak{r}' = f_1^{-1}(f_1(M) \cap \mathfrak{o}_\lambda)$. This completes the proof of 3.17, in the case $g > 1$.

9. Main Theorems III, IV, and conjectures

9.1. We call the quaternion algebra B of 3.4 *totally indefinite* if $r = g$, i.e., if every archimedean prime of F is unramified in B. Especially $M_2(F)$ belongs to this category. In this section, except in the last two paragraphs, we always assume that B is totally indefinite.

Let us define subgroups D and D^+ of $\mathrm{GL}_1(B)$ by

$$D = \{\alpha \in B \mid 0 \neq N_{B/F}(\alpha) \in Q\} ,$$
$$D^+ = \{\alpha \in B \mid 0 < N_{B/F}(\alpha) \in Q\} .$$

For a maximal order $\mathfrak{o}$ in B and an integral two-sided $\mathfrak{o}$-ideal $\mathfrak{e}$, we put

$$\Gamma_1(\mathfrak{o}) = \{\gamma \in \mathfrak{o} \mid N_{B/F}(\gamma) = 1\} .$$
$$\Gamma_1(\mathfrak{o}, \mathfrak{e}) = \{\gamma \in \Gamma_1(\mathfrak{o}) \mid \gamma \equiv 1 \bmod \mathfrak{e}\} .$$

Further we denote by $D(\mathfrak{e})$ the group of all elements α of D^+ which are prime to $\mathfrak{e}$, and put $D_0(\mathfrak{e}) = \{\alpha \in D(\mathfrak{e}) \mid \alpha \equiv 1 \bmod^* \mathfrak{e}\}$ (see 2.14). As before, we denote by $\mathfrak{H}$ the upper half plane, and $\mathfrak{H}^g$ the product of g copies of $\mathfrak{H}$.

9.2. To state our theorems, it is necessary to introduce a special type of class fields. Let $K \in J(B)$. Let f be an F-linear isomorphism of K into B. By 2.7, f determines a CM-type $(K; \{\tau_1, \cdots, \tau_g\})$. Let Ψ be a representation of K such that $\Psi \sim \sum_{\nu=1}^g \tau_\nu$ in the sense of (5.1.1), and let (K', Ψ') be the dual of (K, Ψ) (see 5.1). For every integral ideal $\mathfrak{a}$ in K, we define a group of ideals $I_0(K', \Psi', \mathfrak{a})$ as follows. Let c be the smallest positive integer divisible by $\mathfrak{a}$. Then $I_0(K', \Psi', \mathfrak{a})$ is the set of all ideals $\mathfrak{x}$ in K', prime to c, satisfying

(9.2.1) *There exists an element y of K such that* $\det \Psi'(\mathfrak{x}) = y\mathfrak{r}_K, y \equiv 1$ $\bmod^* \mathfrak{a}$, *and* $N(\mathfrak{x}) = yy^\rho$.

Here ρ denotes the complex conjugation. By (5.11.4), we have $I(K', c) \subset I_0(K', \Psi', \mathfrak{a})$. Therefore $I_0(K', \Psi', \mathfrak{a})$ corresponds to a class field over K', which we denote by $C(K', \Psi', \mathfrak{a})$. If $\mathfrak{x} \in I_0(K', \Psi', \mathfrak{a})$, then $N(\mathfrak{x}) \equiv 1 \bmod^* (c)$, hence $C_0(K', \Psi', \mathfrak{a})$ contains a primitive c^{th} root of unity. The ideal group $I_0(K', \Psi', \mathfrak{a})$ and the class field $C_0(K', \Psi', \mathfrak{a})$ are exactly the ones considered in [15, §§ 15, 16].

9.3. MAIN THEOREM III. *Let the notation and assumption be as in 9.1. Let c be the smallest positive integer divisible by $\mathfrak{e}$, and let $\zeta = e^{2\pi i/c}$. Then there exist a Zariski open subset V of a projective variety, and a holomorphic mapping φ of $\mathfrak{H}^g$ onto V satisfying the following conditions.*

(9.3.1) V *is normal and defined over* $\mathbf{Q}(\zeta)$.

(9.3.2) φ *gives a biregular isomorphism of* $\mathfrak{H}^g/\Gamma_1(\mathfrak{o}, \mathfrak{e})$ *onto* V.

(9.3.3) *Let $K \in J(B)$ and let f be an F-linear isomorphism of K into B such that $f(\mathfrak{r}_K) \subset \mathfrak{o}$. Let z be the fixed point of $f(K)$ on $\mathfrak{H}^g$, and $\mathfrak{e}_K = f^{-1}(\mathfrak{e} \cap f(\mathfrak{r}_M))$. Further let (K', Ψ') be the dual of the CM-type determined by f (see 2.7 and 9.2). Then*

$$K'\big(\zeta, \varphi(z)\big) = C_0(K', \Psi', \mathfrak{e}_K) \ .$$

9.4. MAIN THEOREM IV. *The couple (V, φ) of 9.3 can be chosen in such a way that there exist biregular isomorphisms $R_\sigma(\alpha)$ of V to V^σ which are defined for every $\alpha \in D(\mathfrak{e})$ and $\sigma = [\mathbf{Q}(\zeta)/\mathbf{Q}, (N_{B/F}(\alpha))]$, and satisfy the following conditions.*

(9.4.1) $R_\sigma(\alpha)$ *is rational over* $\mathbf{Q}(\zeta)$.

(9.4.2) $R_\sigma(\alpha) = R_\sigma(\beta)$ *if $\alpha \equiv \beta \bmod^* \mathfrak{e}$, and $R_\sigma(\alpha)^\tau \circ R_\tau(\beta) = R_{\sigma\tau}(\alpha\beta)$ for every α and β of $D(\mathfrak{e})$.*

(9.4.3) *If $\gamma \in \Gamma_1(\mathfrak{o})$, the $R_1(\gamma)[\varphi(w)] = \varphi(\gamma(w))$ for every $w \in \mathfrak{H}^g$.*

(9.4.4) *Let K, f, z, K', Ψ' and e_K be as in (9.3.3). Let $\mathfrak{x}$ be an ideal in K', prime to c, and $\tau = [C_0(K', \Psi', e_K)/K', \mathfrak{x}]$. Put $\mathfrak{y} = \det \Psi'(\mathfrak{x})$ (see (5.10.1)). Then there exists an element α of $D(e)$ such that $f(\mathfrak{y})\mathfrak{o} = \alpha\mathfrak{o}$. With such an element α, one has $\varphi(z)^\tau = R_\sigma(\alpha)[\varphi(\alpha^{-1}(z))]$.*

Proofs of these theorems will be completed at 9.10. It should be observed that, if $g = 1$, the statements of 9.3 and 9.4 coincide with those of 3.2 and 3.5 (modulo some minor points, cf. 9.11). Therefore, by proving 9.3 and 9.4 in general, we will be able to complete the proof of 3.2 and 3.5 in the case $g = 1$, which has been excluded in the discussion of §§ 7, 8.

9.5. LEMMA. *Let $\mathfrak{z}$ be a right $\mathfrak{o}$-ideal such that $N_{B/F}(\mathfrak{z}) = a\mathfrak{x}_F$ with a positive rational number a. Then there exists an element α of D^+ such that $\mathfrak{z} = \alpha\mathfrak{o}$ and $N_{B/F}(\alpha) = a$.*

PROOF. By Eichler's theorem, there exists an element β of B such that $\mathfrak{z} = \beta\mathfrak{o}$. Then $aN_{B/F}(\beta)^{-1}$ is a unit of F. Hence by 2.15, there exists an element γ of $\mathfrak{o}$ such that $N_{B/F}(\gamma) = aN_{B/F}(\beta)^{-1}$. Putting $\alpha = \beta\gamma$ we get the desired result.

9.6. By [9, Prop. 2], B has a positive involution ρ; and every positive involution ρ of B can be obtained by $x^\rho = ax'a^{-1}$ with an element a of B such that a^2 is a totally negative number of F, where $x \mapsto x'$ is the main involution of B. Fix such ρ and a. Then

(9.6.1) *If u is an invertible element of B such that $u^\rho = -u$, then $u' = -u$ and $x^\rho = u^{-1}x'u$ for every $x \in B$.*

In fact $-u = u^\rho = au'a^{-1}$. Since $a' = -a$, we have $(ua)' = -au' = ua$, so that $ua = b$ with an element b of F. Then we get our assertion immediately. We can also verify easily

(9.6.2) *If u is as in (9.6.1), $xux^\rho = N_{B/F}(x)u$ for every $x \in B$.*

9.7. Fix a positive involution ρ of B. By [9, Lem. 1], we can find g representations $\omega_\nu \colon B \to M_2(\mathbf{R})$ $(\nu = 1, \cdots, g)$ such that $\omega_\nu(x^\rho) = {}^t\omega_\nu(x)$ for $x \in B$, and $\omega_\nu(a) = a^{\sigma_\nu}1_2$ for $a \in F$. We fix such $\omega_1, \cdots, \omega_g$. Let Φ be a reduced representation of B over $\mathbf{Q}$, i.e., a representation equivalent to the direct sum of $\omega_1, \cdots, \omega_g$. Fix an invertible element u of B such that $u^\rho = -u$, and define $T(x, y)$ by $T(x, y) = xuy^\rho$ for $x, y \in B$. Since $\omega_\nu(u)$ is an alternating matrix, we have

$$\omega_\nu(u) = b_\nu \cdot \begin{bmatrix} 0 & -1 \\ 1 & 0 \end{bmatrix} \qquad (\nu = 1, \cdots, g)$$

with a real number b_ν. We may assume that $b_\nu > 0$ for all ν, by exchanging u for du with an element d of F, if necessary. Now we can consider a PEL-type

$$\Omega = (B, \Phi, \rho;\ T, \mathfrak{M};\ v_1, \cdots, v_s)$$

with a Z-lattice $\mathfrak{M}$ in B and elements v_i in B. Here we have to assume $\mathrm{Tr}_{B/Q}(T(\mathfrak{M}, \mathfrak{M})) = Z$. This is satisfied again by an exchange of u for su with a positive rational number s.

The general theory [9] tells us that the domain $\mathcal{H}$ in this case is $\mathfrak{H}^g$. Moreover, we have $G^+(T) = B^+$ in view of (9.6.2). Let $W_\nu = \begin{bmatrix} 1 & 0 \\ 0 & b_\nu \end{bmatrix}$. Then $W_\nu \cdot \omega_\nu(u)^{-1} \cdot {}^t W_\nu = \begin{bmatrix} 0 & 1 \\ -1 & 0 \end{bmatrix}$. Therefore we can take these W_ν as the matrices W_ν considered in [9, p. 160]. Now we take the identification (2.4.2) to be defined by $x^{(\nu)} = W_\nu^{-1} \cdot \omega_\nu(x) W_\nu$, where $x^{(\nu)}$ denotes the projection of an element x of B to the νth factor of (2.4.2). Then the action of elements of B^+ on $\mathfrak{H}^g$ defined in 2.4 coincides with the action of elements of $G^+(T)$ defined in [9, p. 164] and 4.3.

9.8. For our purpose, it is convenient to take $\mathfrak{M}$ to be a maximal order in B. Thus, the notation being as in 9.1, we consider a PEL-type

$$\Omega = (B, \Phi, \rho;\ T, \mathfrak{o};\ y)$$

with an element y of B such that $\mathfrak{e}^{-1} = \mathfrak{o}y + \mathfrak{o}$. Then we get a family $\Sigma = \{\mathfrak{Q}_z \mid z \in \mathfrak{H}^g\}$ of PEL-structures

$$\mathfrak{Q}_z = (A_z, \mathcal{C}_z, \theta_z;\ t_z)$$

of type Ω. We construct this family as in [13, 3.3]. Then A_z is a projective variety in the same projective space P^N, and there is a map $\xi: B_R \times \mathfrak{H}^g \to P^N$ such that, for each $z \in \mathfrak{H}^g$, $\xi(x, z)$ is the point on A_z corresponding to $\mathfrak{y}(x, z)$, where $\mathfrak{y}$ is the parametrizing function as in 4.2. The point t_z can be obtained as $t_z = \xi(y, z)$. Let $(V, \mathfrak{v}, \varphi)$ be the moduli system for Σ. We shall prove 9.3 and 9.4 with these V and φ. In view of (9.6.2), we see that the group $\Gamma(T, \mathfrak{N}/\mathfrak{M})$ introduced in 4.2 coincides with $\Gamma_1(\mathfrak{o}, \mathfrak{e})$, since $\mathfrak{M} = \mathfrak{o}$ and $\mathfrak{N} = \mathfrak{e}^{-1}$ in the present case. Therefore (9.3.2) is certainly satisfied.

For every α in B prime to $\mathfrak{e}$, we can find an element β of $\mathfrak{o}$ such that $\beta \equiv \alpha \bmod^* \mathfrak{e}$. Put $y' = y\beta$. Then $\mathfrak{e}^{-1} = \mathfrak{o}y' + \mathfrak{o}$ and $y\alpha \equiv y' \bmod \mathfrak{o}_\mathfrak{p}$ for every prime ideal $\mathfrak{p}$ in F dividing $N_{B/F}(\mathfrak{e})$. With such an element y', define a PEL-type Ω_α by $\Omega_\alpha = (B, \Phi, \rho;\ T, \mathfrak{o};\ y')$. Ω_α is uniquely determined, up to equivalence, by Ω and $\alpha \bmod^* \mathfrak{e}$. Define $\mathfrak{Q}_{z\alpha}$ by $\mathfrak{Q}_{z\alpha} = (A_z, \mathcal{C}_z, \theta_z;\ t_{z\alpha})$ with $t_{z\alpha} = \xi(y', z)$. Then $\Sigma_\alpha = \{\mathfrak{Q}_{z\alpha} \mid z \in \mathfrak{H}^g\}$ is a family of PEL-structures of type Ω_α. By 4.25, if we define $\mathfrak{v}_\alpha$ by $\mathfrak{v}_\alpha(\mathfrak{Q}_{z\alpha}) = \varphi(z)$, then $(V, \mathfrak{v}_\alpha, \varphi)$ is the moduli-system for Σ_α.

(In this case we take the group of positive rational numbers to be E considered in 4.8.) Let $k(\Omega)$ be the field determined by [13, 5.1]. By [11, II, 6.8], we have $k(\Omega) = k(\Omega_\alpha) = Q(\zeta)$ (see also [13, (7.2.2)]). Therefore V is defined over $Q(\zeta)$. Now

(9.8.1) *Let* $\alpha \in D(e)$ *and* $\sigma = [Q(\zeta)/Q, N_{B/F}(\alpha)]$. *Then* Ω^σ *is equivalent to* Ω_α.

This is only a restatement of a special case of [11, II, 6.8]. By [13, 6.7], there exists a biregular morphism $R_\sigma(\alpha)$ of V onto V^σ, rational over $Q(\zeta)$, such that $\mathfrak{v}^\sigma = R_\sigma(\alpha) \circ \mathfrak{v}_\alpha$ (cf. also (4.23.4) and 4.21). If τ is an automorphism of C which coincides with σ on $Q(\zeta)$, we have

$$\varphi(z)^\tau = \mathfrak{v}(\mathfrak{A}_z)^\tau = \mathfrak{v}^\sigma(\mathfrak{A}_z^\tau) = R_\sigma(\alpha) \circ \mathfrak{v}_\alpha(\mathfrak{A}_z^\tau) \ .$$

Therefore we obtain

(9.8.2) $\varphi(z)^\tau = R_\sigma(\alpha)[\varphi(w)]$ *if* $\mathfrak{A}_z^\tau$ *is isomorphic to* $\mathfrak{A}_{w\alpha}$ *and* $\tau = \sigma = [Q(\zeta)/Q, N_{B/F}(\alpha)]$ *on* $Q(\zeta)$.

By a straightforward argument, we can verify that the $R_\sigma(\alpha)$ satisfy (9.4.2). This fact is also included in [11, II, 6.8].

Let $w \in \mathfrak{H}^g$ and $z = \gamma(w)$ with an element γ of $\Gamma_1(\mathfrak{o})$. Then $\Lambda(\gamma, z)$ of 4.4 gives an isomorphism of (A_z, C_z, θ_z) to (A_w, C_w, θ_w), which sends t_z to $t_{w\gamma}$. It follows that $\mathfrak{A}_z$ is isomorphic to $\mathfrak{A}_{w\gamma}$. Taking τ of (9.8.2) to be the identity mapping, we obtain (9.4.3).

9.9. Let us now investigate the special members of Σ. Let $K \in J(B)$. Let f be an F-linear isomorphism of K into B, and z the fixed point of $f(K)$ on $\mathfrak{H}^g$. For every $b \in K$, $b \neq 0$, we see that $\Lambda(f(b), z)$ of 4.4 defines an element of $\mathrm{End}_Q(A_z)$. Since $B \otimes_F K$ is isomorphic to $M_2(K)$, we obtain an isomorphism θ of $M_2(K)$ into $\mathrm{End}_Q(A_z)$ such that $\theta(x) = \theta_z(x)$ for $x \in B$, and $\theta(b)$ corresponds to $\Lambda(f(b), z)$ for $b \in K$. Extend ρ to $M_2(K)$ so that b^ρ is the conjugate of b over F for every $b \in K$. Then we see easily that the polarization C_z induces an involution of $\mathrm{End}_Q(A_z)$ which coincides with $\theta(x) \mapsto \theta(x^\rho)$ on $\theta(M_2(K))$.

By 5.23, A_z is isogenous to a product $A_1 \times A_2$ of abelian varieties A_i with an isomorphism $\theta_i \colon K \to \mathrm{End}_Q(A_i)$ $(i = 1, 2)$; and the restriction of Φ to K is equivalent to the sum of 2 copies of a representation Ψ of K such that (A_i, θ_i) is of type (K, Ψ) in the sense of 4.1. Now we have

(9.9.1) *If* $(K, \{\tau_1, \cdots, \tau_g\})$ *is the* CM-*type determined by* f *in the sense of* 2.7, *then* $\Psi \sim \sum_{\nu=1}^g \tau_\nu$ *in the sense of* (5.1.1).

In fact, let $z = (z_1, \cdots, z_g)$ and $Y_\nu = \begin{bmatrix} z_\nu & 1 \\ \bar{z}_\nu & 1 \end{bmatrix}$. Apply the result of [9, p. 165] to $\Lambda = \Lambda(f(b), z)$ with $b \in K$. If $x = f(b)$, the formula [9, p. 165, (39)] can be written as $\Lambda_\nu Y_\nu W_\nu = Y_\nu W_\nu \cdot {}^t\omega_\nu(x)$, hence ${}^t\bar{\Lambda}_\nu = {}^t\bar{Y}_\nu^{-1} x^{(\nu)} \cdot {}^t\bar{Y}_\nu$. We see that the fractional linear transformation obtained from the matrix ${}^t\bar{Y}_\nu$ represents a holomorphic mapping of the unit disk onto $\mathfrak{H}$. Therefore, in view of (2.7.1) and [9, p. 162, (33), (34)], we see that the matrix Λ of [9, p. 161, (32)] (which coincides with the present $\Lambda(f(b), z)$) has the characteristic roots $b^{\tau_1}, \cdots, b^{\tau_g}$, each with multiplicity 2. This proves (9.9.1).

9.10. Now assume that $f(\mathfrak{r}_M) \subset \mathfrak{o}$. Then $\theta(\mathfrak{r}_K) \subset \mathrm{End}(A_z)$, and hence (5.24.1) is applicable. Before applying (5.24.1) to A_z, we need a few more considerations. Let e_K be as in (9.3.3) and let $\mathfrak{x}, \mathfrak{y}$ and τ be as in (9.4.4). Consider a right $\mathfrak{o}$-ideal $f(\mathfrak{y})\mathfrak{o}$. By (5.11.4), we have $N_{B/F}(f(\mathfrak{y})\mathfrak{o}) = \mathfrak{y}\mathfrak{y}^\rho = N(\mathfrak{x})\mathfrak{r}_F$, hence, by 9.5, there exists an element α of D such that $f(\mathfrak{y})\mathfrak{o} = \alpha\mathfrak{o}$ and $N_{B/F}(\alpha) = N(\mathfrak{x})$. Sinch $\mathfrak{y}$ is prime to c, $\alpha \in D(\mathfrak{e})$. We note that

$$(9.10.1) \qquad \tau = [\mathbf{Q}(\zeta)/\mathbf{Q}, (N_{B/F}(\alpha))] \qquad on \ \mathbf{Q}(\zeta) .$$

Extend τ to an automorphism of C, and denote it again by τ. Take an integral ideal $\mathfrak{a}$ in K', prime to c, such that $\tau = [C(K', c)/K', \mathfrak{a}]$ on $C(K', c)$. Put $\mathfrak{b} = \det \Psi'(\mathfrak{a})$. Since $\tau = [C_0(K', \Psi', e_K)/K', \mathfrak{x}]$, there exists an element a of K such that $\mathfrak{b}\mathfrak{y}^{-1} = a\mathfrak{r}_K$, $a \equiv 1 \bmod^* e_K$, and $N(\mathfrak{a}\mathfrak{x}^{-1}) = aa^\rho$. Put $\beta = f(a)\alpha$. Then $\beta \in D(\mathfrak{e}) \cap \mathfrak{o}$, $\mathfrak{o}\beta^{-1} = \mathfrak{o}f(\mathfrak{b})^{-1}$ and $\beta \equiv \alpha \bmod^* \mathfrak{e}$. Put $w = \alpha^{-1}(z)$. Then $\beta^{-1}(z) = w$, and hence $\Lambda(\beta, z)$ of 4.4 defines an isogeny μ of A_z to A_w. From the formula $\Lambda(\beta, z)\mathfrak{y}(x, z) = \mathfrak{y}(x\beta, w)$ of 4.4, we see that

$$\mathrm{Ker}(\mu) = \{t \in A_z \mid \theta(\mathfrak{b})t = 0\}$$

(since $\mathfrak{o}\beta^{-1} = \mathfrak{o}f(\mathfrak{b})^{-1}$), and $\mu(t_z) = t_{w\beta}$. Therefore μ is an isogeny of $\mathfrak{Q}_z$ to $\mathfrak{Q}_{w\alpha}$. Now apply (5.24.1) to $\mathfrak{Q}_z$. Then we obtain an isogeny λ of $\mathfrak{Q}_z$ to $\mathfrak{Q}_z^\tau$ such that $\mathrm{Ker}(\lambda) = \mathrm{Ker}(\mu)$. Hence $\mathfrak{Q}_z^\tau$ is isomorphic to $\mathfrak{Q}_{w\alpha}$. (The discussion of this part is similar to and simpler than 8.5.) It follows from this and (9.8.2) that one has

$$(9.10.2) \qquad \varphi(z)^\tau = R_\sigma(\alpha)[\varphi(\alpha^{-1}(z))] .$$

Suppose that τ is the identity mapping on $C(K', \Psi', e_K)$. Then $\mathfrak{y} = d\mathfrak{r}_K$, $N(\mathfrak{x}) = dd^\rho$, $d \equiv 1 \bmod^* e_K$ with an element d of K. Since $f(\mathfrak{y})\mathfrak{o} = f(d)\mathfrak{o}$ and $f(d) \in D(\mathfrak{e})$, we can take $f(d)$ as α in (9.10.2). Then we obtain $\varphi(z)^\tau = \varphi(z)$, so that τ is the identity mapping on $K'(\varphi(z))$. Conversely suppose that τ is the identity mapping on $K'(\varphi(z), \zeta)$. By (9.10.1), we have $N_{B/F}(\alpha) \equiv 1 \bmod^* (c)$, and by (9.10.2), $\varphi(z) = R_1(\alpha)[\varphi(\alpha^{-1}(z))]$. Since $N_{B/F}(\alpha) \equiv 1 \bmod^* (c)$, there exists, by 2.15, an element γ of $\mathfrak{o}$ such that $\gamma \equiv \alpha \bmod^* \mathfrak{e}$ and $N_{B/F}(\gamma) = 1$. Then $R_1(\alpha) = R_1(\gamma)$, so that $\varphi(z) = R_1(\gamma)[\varphi(\alpha^{-1}(z))] = \varphi(\gamma\alpha^{-1}(z))$ by (9.4.3) which has been proved. Hence there exists an element δ of $\Gamma_1(\mathfrak{o}, \mathfrak{e})$ such that $\gamma\alpha^{-1}(z) = \delta(z)$. By 2.6, there exists an element v of K such that $\alpha\gamma^{-1}\delta = f(v)$.

Then $f(\mathfrak{h})\mathfrak{o} = \alpha\mathfrak{o} = \alpha\gamma^{-1}\delta\mathfrak{o} = f(v)\mathfrak{o}$, hence $\mathfrak{h} = v\mathfrak{r}_K$ by (3.20.9). Further $vv^\rho = N_{B/F}(\alpha\gamma^{-1}\delta) = N_{B/F}(\alpha) = N(\mathfrak{x})$, $\alpha\gamma^{-1}\delta \equiv 1 \bmod^* \mathfrak{e}$, and hence $v \equiv 1 \bmod^* \mathfrak{e}_K$. Therefore $\mathfrak{h} \in I_0(K', \Psi', \mathfrak{e}_K)$, hence τ is the identity mapping on $C_0(K', \Psi', \mathfrak{e}_K)$. This proves (9.3.3). Then the formula (9.10.2) implies (9.4.4). We have thus proved 9.3 and 9.4.

9.11. As mentioned earlier, this completes also the proofs of 3.2 and 3.5 in the case $g = 1$. As for the uniqueness 3.3, the proof at 7.12 applies also to the case $g = 1$.

The variety V of 9.3 is complete unless $B = M_2(F)$. If $B = M_2(F)$, V is not complete, since we have formulated the theorem so that φ is a surjective mapping. Contrary to this, in 3.2, we have taken V to be always a complete curve. It should be observed, however, that there is no essential difference between 9.3 and 3.2 (or, 9.4 and 3.5) in the case $g = 1$. At any rate, if $g = 1$, 3.2 resp. 3.5 can be derived from 9.3 resp. 9.4. Therefore, when $B = M_2(Q)$, this proves, in view of the uniqueness, that $\varphi(\mathfrak{H})$ in 3.2 is a Zariski open subset of V, rational over $Q(\zeta)$.

If $B = M_2(F)$, we could use, instead of the above Ω, a family of abelian varieties of dimension g whose endomorphism algebras contain F (i.e. of (Type I) of [9]). This makes the study of special members somewhat simpler.

As for 3.17, one can formulate a similar result for totally indefinite B, with a proof which is similar to (and simpler than) that in 6.9 and 8.9. A precise statement may be left to the reader.

9.12. *Problems and conjectures.* (A) In our main theorems, we have considered only the fixed points corresponding to the embeddings of $M \in J(B)$ which embed $\mathfrak{r}_M$ into the maximal order in question. It is certainly possible to discuss a more general case of embeddings without this restriction. One can prove without difficulty that $\varphi(z)$, with such a fixed point z, generates again a class field over M. The theory of [15, § 17] tells us more or less what the nature of this class field would be. It will be worth while working it out, since such a question will become more natural when one seeks, for example, a development of the idea of Ihara [4] for the curves of our type. In this connection, it will be interesting to investigate the special members of each family of abelian varieties in more detail, especially over fields of positive characteristic, by taking, for example, reduction modulo $\mathfrak{p}$.

(B) In the case $r = g$, as our proof shows, V and φ can be obtained as a natural variety of moduli of a family of abelian varieties. This makes the treatment much simpler than the case $r = 1$, where one has to construct the canonical model out of infinitely many varieties of moduli, by lowering down the field of rationality. There is no *standard* family of abelian varieties which is associated with (V, φ) in the case $1 = r < g$. Therefore one may ask whether there exists any family of geometric structures of which our canonical model (V, φ) is a natural variety of moduli, though it is unlikely that such structures can be found at hand.

One can ask also more plainly: is it possible to obtain a proof without abelian varieties? In general, the connection of abelian varieties with arithmetic discontinuous groups does not seem accidental. But it may be worthwhile to find a proof without, or with less use of, abelian varieties, because such a proof may bring a new outlook and the possibility of generalization.

(C) Our main theorems I–IV are concerned with two extreme cases $r = 1$ and $r = g$. As to the general case, one can actually make conjectures which include these theorems as special cases.

Take B and $\tau_{01}, \cdots, \tau_{0g}$ as in 2.4. We fix our attention to the first r isomorphisms $\tau_{01}, \cdots, \tau_{0r}$ corresponding to archimedean primes of F which are unramified in B. Let Θ be a representation of F by complex matrices such that $\Theta \sim \sum_{\nu=1}^{r} \tau_{0\nu}$, and let (F', Θ') be the dual of (F, Θ) in the sense of 5.1. Then F'' is a totally real algebraic number field. Let A denote the set of all elements of the form $\det \Theta'(a)$ with $0 \neq a \in F'$, and A^+ the set of all totally positive elements of A. Put

$$
\begin{aligned}
D &= \{\alpha \in B \mid N_{B/F}(\alpha) \in A \}, \\
D^+ &= \{\alpha \in B \mid N_{B/F}(\alpha) \in A^+\}.
\end{aligned}
$$

Let $\mathfrak{o}$ be a maximal order in B and $\mathfrak{e}$ an integral two-sided $\mathfrak{o}$-ideal. Put

$$
\Gamma[\mathfrak{o}, \mathfrak{e}] = \{\gamma \in \mathfrak{o} \cap D^+ \mid \gamma\mathfrak{o} = \mathfrak{o}, \gamma \equiv 1 \bmod \mathfrak{e}\}.
$$

The difference of this group from the one defined in 2.13 is not great, but there is a subtle distinction between them.

Let $M \in J(B)$. Every F-linear isomorphism f of M into B determines r isomorphisms $\chi_1, \cdots, \chi_r$ of M into C (2.7). Let Ψ be a representation of M such that $\Psi \sim \sum_{\nu=1}^{r} \chi_\nu$, and (M', Ψ') the dual of (M, Ψ). By 5.6, we have $F' \subset M'$, and $N_{M/F}(\det \Psi'(x)) = \det \Theta'(N_{M'/F'}(x))$ for every $x \in M'$. ($[M' : F']$ is not necessarily equal to 2.) For an integral ideal $\mathfrak{c}$ in F, let d be the smallest positive integer divisible by $\mathfrak{c}$, and let $I(F', \Theta', \mathfrak{c})$ denote the group of all ideals $\mathfrak{a}$ in F', prime to d, satisfying

(9.12.1) *There exists an element u of A^+ such that $\det \Theta'(\mathfrak{a}) = (u)$ and* $u \equiv 1 \bmod^* \mathfrak{c}$. (For the notation, see 5.3.)

Since $I(F', d\mathfrak{u}) \subset I(F', \Theta', \mathfrak{c})$, the group $I(F', \Theta', \mathfrak{c})$ corresponds to a class field over F', which we denote by $C(F', \Theta', \mathfrak{c})$. Similarly, if $\mathfrak{h}$ is an integral ideal in M and k is the smallest positive integer divisible by $\mathfrak{h}$, we denote by $I(M', \Psi', \mathfrak{h})$ the group of all ideals $\mathfrak{b}$ in M' prime to k, satisfying

(9.12.2) *There exists an element w of M such that* $\det \Psi'(\mathfrak{b}) = (w)$, $w \equiv 1 \bmod^* \mathfrak{h}$, *and* $ww^\rho \in A^+$.

We denote by $C(M', \Psi', \mathfrak{h})$ the class field over M' corresponding to $I(M', \Psi', \mathfrak{h})$. After these preparations, we can now state

CONJECTURE I. *Let $\mathfrak{o}$ be a maximal order in B, $\mathfrak{e}$ an integral two-sided $\mathfrak{o}$-ideal, and $\mathfrak{c} = \mathfrak{e} \cap F$. Then there exist an algebraic variety V and a holomorphic mapping φ of $\mathfrak{H}^r$ onto V satisfying the following conditions.*

(I.1) *V is defined over $C(F', \Theta', \mathfrak{c})$.*

(I.2) *φ gives a biregular isomorphism of $\mathfrak{H}^r/\Gamma[\mathfrak{o}, \mathfrak{e}]$ onto V.*

(I.3) *Let $M \in J(B)$ and let f be an F-linear isomorphism of M into B such that $f(\mathfrak{r}_M) \subset \mathfrak{o}$. Let z be the fixed point of $f(M)$ on $\mathfrak{H}^r$, and $\mathfrak{e}_M = f^{-1}(\mathfrak{e} \cap f(M))$. Define M' and Ψ' as above. Then*

$$M'\bigl(\varphi(z)\bigr) \cdot C(F', \Theta', \mathfrak{c}) = C(M', \Psi', \mathfrak{e}_M) \ .$$

CONJECTURE II. *The couple (V, φ) satisfying (I.1–3) is unique up to biregular isomorphisms rational over $C(F', \Theta', \mathfrak{c})$.*

Let us consider the set of all right $\mathfrak{o}$-ideals $\mathfrak{x}$ such that $N_{B/F}(\mathfrak{x}) = \det \Theta'(\mathfrak{a})$ with an ideal $\mathfrak{a}$ in F''. Classify them by left multiplication of elements of D^+. Let $\{\mathfrak{x}_1, \cdots, \mathfrak{x}_q\}$ be a set of representatives for the classes of such right $\mathfrak{o}$-ideals in this sense. We can take them so that $\mathfrak{o} = \mathfrak{x}_1$, the $\mathfrak{x}_\nu$ are prime to $\mathfrak{e}$, and $N_{B/F}(\mathfrak{x}_\nu) = \det \Theta'(\mathfrak{a}_\nu)$ with an ideal $\mathfrak{a}_\nu$ in F'' prime to $N(\mathfrak{c})$, for each ν. Put $\mathfrak{x}_{\lambda\mu} = \mathfrak{x}_\lambda \mathfrak{x}_\mu^{-1}$, $\mathfrak{o}_\lambda = \mathfrak{x}_{\lambda\lambda}$, $\mathfrak{e}_\lambda = \mathfrak{x}_{\lambda 1}\mathfrak{e}\mathfrak{x}_{1\lambda}$ for $\lambda, \mu = 1, \cdots, q$. Let $D(\mathfrak{e})$ denote the set of all elements α of D, prime to $\mathfrak{e}$, and $D_0(\mathfrak{e}) = \{\alpha \in D(\mathfrak{e}) \mid \alpha \equiv 1 \bmod^* \mathfrak{e}\}$.

CONJECTURE III. *There exists a system*

$$\{V_\lambda, \varphi_\lambda, R_\sigma^{\mu\lambda}(\alpha)\ (\lambda, \mu = 1, \cdots, q; \alpha \in D(\mathfrak{e}))\}$$

with the following properties.

(III.1) *For each λ, $(V_\lambda, \varphi_\lambda)$ satisfies (I.1–3) with $\mathfrak{o}_\lambda$ and $\mathfrak{e}_\lambda$ in place of $\mathfrak{o}$ and $\mathfrak{e}$.*

(III.2) *Let $\alpha \in D(\mathfrak{e})$, $N_{B/F}(\alpha \mathfrak{x}_{\lambda\mu}) = \det \Theta'(\mathfrak{a})$ with an ideal $\mathfrak{a}$ in F' prime to $N(\mathfrak{c})$, and $\sigma = [C(F', \Theta', \mathfrak{c})/F', \mathfrak{a}]$. For such α, λ, μ and σ, $R_\sigma^{\mu\lambda}(\alpha)$ is defined and a biregular morphism of V_λ to V_μ^σ, rational over $C(F', \Theta', \mathfrak{c})$.*

(III.3) *$R_\sigma^{\mu\lambda}(\alpha) = R_\sigma^{\mu\lambda}(\beta)$ if $\alpha \equiv \beta \bmod^* \mathfrak{e}$, and $R_\tau^{\nu\mu}(\beta)^\sigma \circ R_\sigma^{\mu\lambda}(\alpha) = R_{\tau\sigma}^{\nu\lambda}(\beta\alpha)$ for every α and β in $D(\mathfrak{e})$.*

(III.4) *If $\gamma \in \Gamma[\mathfrak{o}_\lambda, \mathfrak{o}_\lambda]$, then $R_1^{\lambda\lambda}(\gamma)[\varphi(w)] = \varphi(\gamma(w))$ for every $w \in \mathfrak{H}^r$.*

(III.5) *Let M, f, z and $\mathfrak{e}_M$ be as in (I.3) with $\mathfrak{o}_\mu$ in place of $\mathfrak{o}$. Let $\mathfrak{a}$ be an* *ideal in M' prime to $N(\mathfrak{c})$, and $\tau = [C(M', \Psi', e_M)/M', \mathfrak{a}]$. Put $\mathfrak{b} = \det \Psi'(\mathfrak{a})$. Then one has $f(\mathfrak{b})\mathfrak{o}_\mu = \alpha \mathfrak{x}_{\lambda\mu}$ with a unique λ and an element α of $D(\mathfrak{e})$. With such an element α, one has*

$$\varphi_\mu(z)^\tau = R_\sigma^{\mu\lambda}(\alpha)[\varphi_\lambda(\alpha^{-1}(z))] .$$

Our main theorems show that these conjectures are true if $r = 1$ or $r = g$ (except for Conjecture II in the case $r = g$). One can justify the above conjectures heuristically by investigating the special members of the family of abelian varieties considered in §6, and generalizing Propositions 5.20 and 5.22 accordingly. The results of [12] may be regarded as another support of the conjectures.

The above consideration suggests also a new type of zeta-functions attached to the algebraic group D and a new type of class numbers. The subjects are, however, beyond the scope of the present paper.

10. Automorphic forms and Hecke operators

10.1. Let F, B, $F_\mathfrak{p}$ and $B_\mathfrak{p}$ be as in 2.4. Let $\mathfrak{o}$ be a maximal order in B, and $\mathfrak{a}$ a right $\mathfrak{o}$-ideal. We shall now define the set of invariants of $\mathfrak{a}$. Take a prime ideal $\mathfrak{p}$ in F, and suppose first that $\mathfrak{p}$ does not divide $D(B/F)$. Let $\mathfrak{r}_\mathfrak{p}$ denote the ring of integers in $F_\mathfrak{p}$. We can identify $B_\mathfrak{p}$ resp. $\mathfrak{o}_\mathfrak{p}$ with $M_2(F_\mathfrak{p})$ resp. $M_2(\mathfrak{r}_\mathfrak{p})$, and find an element β of $\mathfrak{o}_\mathfrak{p}$ so that $\mathfrak{a} = \beta\mathfrak{o}_\mathfrak{p}$. Then we find two units δ and ε of $\mathfrak{o}_\mathfrak{p}$ so that $\delta\beta\varepsilon = \begin{bmatrix} \pi^{c_1} & 0 \\ 0 & \pi^{c_2} \end{bmatrix}$, $c_1 \leqq c_2$, where π is a prime element of $\mathfrak{r}_\mathfrak{p}$, and c_1, c_2 are integers. Next, suppose that $\mathfrak{p}$ divides $D(B/F)$. Put $N_{B/F}(\mathfrak{a})_\mathfrak{p} = \mathfrak{p}^c$ with an integer c. Then we put $e_\mathfrak{p}(\mathfrak{a}) = \{\mathfrak{p}^c\}$ or $\{\mathfrak{p}^{c_1}, \mathfrak{p}^{c_2}\}$ according as $\mathfrak{p}$ divides $D(B/F)$ or not, and $e(\mathfrak{a}) = \bigcup_\mathfrak{p} e_\mathfrak{p}(\mathfrak{a})$ (the join for all $\mathfrak{p}$). We call $e(\mathfrak{a})$ *the set of invariants of* $\mathfrak{a}$, and $e_\mathfrak{p}(\mathfrak{a})$ *the $\mathfrak{p}$-part of* $e(\mathfrak{a})$. We denote by $\mathfrak{E}$ the set of all possible types of sets e such that $e = e(\mathfrak{a})$ for a right $\mathfrak{o}$-ideal $\mathfrak{a}$. For $e \in \mathfrak{E}$, we denote by $e_\mathfrak{p}$ the $\mathfrak{p}$-part of e. $\mathfrak{E}$ does not depend on the choice of $\mathfrak{o}$. For an element e of $\mathfrak{E}$, we write $e = \{\mathfrak{p}^a, \mathfrak{p}^b\}$ resp. $\{\mathfrak{p}^c\}$ with a prime ideal $\mathfrak{p}$, if $e_\mathfrak{p} = \{\mathfrak{p}^a, \mathfrak{p}^b\}$ resp. $\{\mathfrak{p}^c\}$ and $e_\mathfrak{q} = e_\mathfrak{q}(\mathfrak{o})$ for all $\mathfrak{q} \neq \mathfrak{p}$. We say that e is *prime to* an integral ideal $\mathfrak{c}$ in F, if $e_\mathfrak{p} = e_\mathfrak{p}(\mathfrak{o})$ for all $\mathfrak{p}$ dividing $\mathfrak{c}$. We call e *integral*, if $e = e(\mathfrak{a})$ with an integral right $\mathfrak{o}$-ideal $\mathfrak{a}$. We say that a right $\mathfrak{o}$-ideal $\mathfrak{a}$ and a right $\mathfrak{o}_1$-ideal $\mathfrak{a}_1$ (with maximal orders $\mathfrak{o}$ and $\mathfrak{o}_1$) are *similar* if $e(\mathfrak{a}) = e(\mathfrak{a}_1)$. (Here $\mathfrak{o}$ and $\mathfrak{o}_1$ may or may not be the same.) Then we may regard $\mathfrak{E}$ as the set of all such similarity classes (cf. [8, 1.3]). We can easily verify

(10.1.1) $e(\xi\mathfrak{a}) = e(\mathfrak{a}\xi)$ *and* $e(\xi\mathfrak{a}\xi^{-1}) = e(\mathfrak{a})$ *for every invertible ξ of B.*

For every $e \in \mathfrak{E}$ we define $N_{B/F}(e)$ by $N_{B/F}(e) = N_{B/F}(\mathfrak{a})$ with any $\mathfrak{a}$ such that $e = e(\mathfrak{a})$. If $e = \{\mathfrak{p}^a, \mathfrak{p}^b\}$, $N_{B/F}(e) = \mathfrak{p}^{a+b}$.

10.2. PROPOSITION. *Let $\mathfrak{o}_1$ and $\mathfrak{o}_2$ be maximal orders in B, and let $M \in J(B)$. Let f_1 and f_2 be F-linear isomorphisms of M into B such that $f_i(\mathfrak{r}_M) \subset \mathfrak{o}_i$ $(i = 1, 2)$. Then $e(f_1(\mathfrak{b})\mathfrak{o}_1) = e(f_2(\mathfrak{b})\mathfrak{o}_2)$ for every ideal $\mathfrak{b}$ in M.*

This can be proved easily by the standard local argument.

10.3. Hereafter until the end of the paper, we use the symbols $\Delta = \{\mathfrak{o}_\lambda, \mathfrak{x}_{\lambda\mu}\}$, e_λ, $\mathfrak{c}$, B^+, $U(\mathfrak{e})$, $U_0(\mathfrak{e})$, $\Gamma(\mathfrak{o}_\lambda, 1)$, $\Gamma(\mathfrak{o}_\lambda, e_\lambda)$ as introduced in 2.13, 3.1 and 3.4. For brevity, we put

$$\Gamma_\lambda^1 = \Gamma(\mathfrak{o}_\lambda, 1) , \qquad \Gamma_\lambda^{\mathfrak{e}} = \Gamma(\mathfrak{o}_\lambda, e_\lambda) .$$

It can be easily seen that $\Gamma_\lambda^1 \subset U(\mathfrak{e})$ and $\Gamma_\lambda^{\mathfrak{e}} = U_0(\mathfrak{e}) \cap \Gamma_\lambda^1$ for every λ. We assume, without losing generality, that the $\mathfrak{x}_{\lambda\mu}$ are prime to $\mathfrak{c} \cdot D(B/F)$ (see 3.8). We assume $r = 1$ throughout, though the assertions of some propositions are true even if $r > 1$.

(10.3.1) *For every index μ and every $e \in \mathfrak{E}$, there exist an index λ and an element $\alpha \in B^+$ such that $e(\alpha\mathfrak{x}_{\lambda\mu}) = e$.*

In fact, take a right $\mathfrak{o}_\mu$-ideal $\mathfrak{a}$ so that $e = e(\mathfrak{a})$. By (2.13.4), there exists an element α of B^+ such that $\mathfrak{a} = \alpha\mathfrak{x}_{\lambda\mu}$ with a suitable λ.

10.4. PROPOSITION. *Let $\alpha \in B^+$. If $\alpha\mathfrak{x}_{\lambda\mu}$ is prime to $\mathfrak{c}$, $\Gamma_\mu^1\alpha\Gamma_\lambda^1 = \Gamma_\mu^{\mathfrak{e}}\alpha\Gamma_\lambda^1$.*

PROOF. Let $\mathfrak{f}$ be an integral ideal in F such that $\mathfrak{f}\mathfrak{o}_\mu \subset \alpha\mathfrak{o}_\lambda\alpha^{-1}$. Since $\alpha\mathfrak{x}_{\lambda\mu}$ is prime to $\mathfrak{c}$ and $\alpha\mathfrak{o}_\lambda\alpha^{-1} = \alpha\mathfrak{x}_{\lambda\mu}\mathfrak{o}_\mu(\alpha\mathfrak{x}_{\lambda\mu})^{-1}$, we can assume $\mathfrak{f}$ to be prime to $\mathfrak{c}$. Let $\beta \in \Gamma_\mu^{\mathfrak{e}}\alpha\Gamma_\lambda^1$. By [8, Prop. 1.5], $e(\beta\mathfrak{x}_{\lambda\mu}) = e(\alpha\mathfrak{x}_{\lambda\mu})$. By [8, Prop. 1.4], there exists an element γ of $\mathfrak{o}_\mu$ such that $N_{B/F}(\gamma) = 1$ and $\beta\mathfrak{x}_{\lambda\mu} = \gamma\alpha\mathfrak{x}_{\lambda\mu}$. Then $\beta = \gamma\alpha\delta$ with a unit δ of $\mathfrak{o}_\lambda$. Since α, β, γ belong to B^+, we have $\delta \in B^+$, hence $\delta \in \Gamma_\lambda^1$. Now we can find an element ε of $\mathfrak{o}_\mu$ such that $\varepsilon \equiv \gamma \bmod \mathfrak{c}\mathfrak{o}_\mu$, $\varepsilon \equiv 1 \bmod \mathfrak{f}\mathfrak{o}_\mu$. Then $N_{B/F}(\varepsilon) \equiv 1 \bmod \mathfrak{c}\mathfrak{f}\mathfrak{o}_\mu$. By 2.15, there exists an element ζ of $\mathfrak{o}_\mu$ such that $N_{B/F}(\zeta) = 1$ and $\zeta \equiv \varepsilon \bmod \mathfrak{c}\mathfrak{f}\mathfrak{o}_\mu$. Then $\zeta \in \Gamma_\mu^{\mathfrak{f}}$ and $\gamma\zeta^{-1} \in \Gamma_\mu^{\mathfrak{e}}$. Now $\alpha^{-1}\zeta\alpha - 1 = \alpha^{-1}(\zeta - 1)\alpha \in \mathfrak{o}_\lambda$, hence $\alpha^{-1}\zeta\alpha \in \Gamma_\lambda^1$, $\zeta \in \alpha\Gamma_\lambda^1\alpha^{-1}$. Therefore $\beta = \gamma\alpha\delta = (\gamma\zeta^{-1}) \cdot \zeta \cdot (\alpha\delta) \in \Gamma_\mu^{\mathfrak{e}} \cdot \alpha\Gamma_\lambda^1\alpha^{-1} \cdot \alpha\delta = \Gamma_\mu^{\mathfrak{e}}\alpha\Gamma_\lambda^1$. Since β is an arbitrary element of $\Gamma_\mu^1\alpha\Gamma_\lambda^1$, this proves our assertion.

10.5. PROPOSITION. *Let α and β be elements of $U(\mathfrak{e})$. Then*

(10.5.1) $\Gamma_\mu^{\mathfrak{e}}\alpha\Gamma_\lambda^{\mathfrak{e}} = \{\xi \in \Gamma_\mu^1\alpha\Gamma_\lambda^1 \mid \xi \equiv \alpha \bmod^* \mathfrak{e}\}.$

(10.5.2) *One has $\Gamma_\mu^{\mathfrak{e}}\alpha\Gamma_\lambda^{\mathfrak{e}} = \Gamma_\mu^{\mathfrak{e}}\beta\Gamma_\lambda^{\mathfrak{e}}$ if and only if $\Gamma_\mu^1\alpha\Gamma_\lambda^1 = \Gamma_\mu^1\beta\Gamma_\lambda^1$ and $\alpha \equiv \beta \bmod^* \mathfrak{e}$.*

(10.5.3) *One has $\alpha\Gamma_\mu^{\mathfrak{e}} = \beta\Gamma_\mu^{\mathfrak{e}}$ if and only if $\alpha\Gamma_\mu^1 = \beta\Gamma_\mu^1$ and $\alpha \equiv \beta \bmod^* \mathfrak{e}$.*

(10.5.4) *If $\Gamma_\mu^{\mathfrak{e}}\alpha\Gamma_\lambda^{\mathfrak{e}} = \bigcup_i \alpha_i\Gamma_\lambda^{\mathfrak{e}}$ is a disjoint union, then $\Gamma_\mu^1\alpha\Gamma_\lambda^1 = \bigcup_i \alpha_i\Gamma_\lambda^1$ is a disjoint union.*

PROOF. The assertion (10.5.3) is obvious. Then (10.5.4) follows immediately from 10.4 and (10.5.3). To prove (10.5.1), let $\xi \in \Gamma_\mu^1 \alpha \Gamma_\lambda^1$ and $\xi \equiv \alpha \bmod^* \mathfrak{e}$. By 10.4, there exists an element γ of $\Gamma_\mu^\mathfrak{e}$ and δ of Γ_λ^1 such that $\xi = \gamma \alpha \delta$. Then we have $\alpha \equiv \xi \equiv \alpha \delta \bmod^* \mathfrak{e}$, so that $\delta \equiv 1 \bmod^* \mathfrak{e}$, hence $\delta \in \Gamma_\lambda^\mathfrak{e}$. Therefore $\xi \in \Gamma_\mu^\mathfrak{e} \alpha \Gamma_\lambda^\mathfrak{e}$, and we get (10.5.1). The assertion (10.5.2) is an immediate consequence of (10.5.1).

10.6. PROPOSITION. *Let $\mathfrak{x}$ and $\mathfrak{y}$ be right $\mathfrak{o}_\mu$-ideals prime to $\mathfrak{e}$ such that $\mathfrak{e}(\mathfrak{x}) = \mathfrak{e}(\mathfrak{y})$. Then there exists an element γ of $\Gamma_\mu^\mathfrak{e}$ such that $\gamma \mathfrak{x} = \mathfrak{y}$.*

PROOF. By [8, Prop. 1.4], there exists an element δ of Γ_μ^1 such that $\delta \mathfrak{x} = \mathfrak{y}$. Take an element α of B^+ and λ so that $\mathfrak{x} = \alpha \mathfrak{x}_{\lambda\mu}$. By 10.4, $\delta \alpha \in \Gamma_\mu^\mathfrak{e} \alpha \Gamma_\lambda^1$. Hence there exist elements γ of $\Gamma_\mu^\mathfrak{e}$ and ε of Γ_λ^1 such that $\delta \alpha = \gamma \alpha \varepsilon$. Then $\mathfrak{y} = \delta \alpha \mathfrak{x}_{\lambda\mu} = \gamma \alpha \mathfrak{x}_{\lambda\mu} = \gamma \mathfrak{x}$,   q.e.d.

10.7. PROPOSITION. *Let $U_1(\mathfrak{e})$ be the set of all α in $U(\mathfrak{e})$ such that $N_{B/F}(\alpha) \equiv u \bmod^* \mathfrak{c}$ with a totally positive unit u in F. Then $U_1(\mathfrak{e}) = \Gamma_\lambda^1 U_0(\mathfrak{e})$, and $U_1(\mathfrak{e})/U_0(\mathfrak{e})$ is isomorphic to $\Gamma_\lambda^1/\Gamma_\lambda^\mathfrak{e}$ for every λ.*

PROOF. Let $\alpha \in U_1(\mathfrak{e})$, and let u be a totally positive unit in F such that $N_{B/F}(\alpha) \equiv u \bmod^* \mathfrak{c}$. We can find an element β of $\mathfrak{o}_\lambda$ such that $\beta \equiv \alpha \bmod^* \mathfrak{e}$. By 2.15, there exists an element γ of $\mathfrak{o}_\lambda$ such that $\gamma \equiv \beta \bmod \mathfrak{e}_\lambda$ and $N_{B/F}(\gamma) = u$. Then $\gamma \in \Gamma_\lambda^1$, $\alpha = \gamma \cdot \gamma^{-1} \alpha$ and $\gamma^{-1} \alpha \equiv 1 \bmod^* \mathfrak{e}$, hence $\gamma^{-1} \alpha \in U_0(\mathfrak{e})$. This proves our assertions.

10.8. Now let us consider the free $\mathbf{Z}$-module $\mathfrak{R}_{\mu\lambda}$ generated by all double cosets $\Gamma_\mu^\mathfrak{e} \alpha \Gamma_\lambda^\mathfrak{e}$ with $\alpha \in B^+$, and introduce a law of multiplication $\mathfrak{R}_{\nu\mu} \times \mathfrak{R}_{\mu\lambda} \to \mathfrak{R}_{\nu\lambda}$ as in [8, 2.1]. Further, for every $e \in \mathfrak{E}$, we define $T(e)$ as in [8, 2.3]. This symbol has the following property.

(10.8.1) *If* $(\Gamma_\nu^1 \beta \Gamma_\mu^1) \cdot (\Gamma_\mu^1 \alpha \Gamma_\lambda^1) = \sum_k c_k \cdot (\Gamma_\nu^1 \xi_k \Gamma_\lambda^1)$ *and* $e = e(\alpha \mathfrak{x}_{\lambda\mu})$, $f = e(\beta \mathfrak{x}_{\mu\nu})$, $g_k = e(\xi_k \mathfrak{x}_{\lambda\nu})$, *then* $T(f) T(e) = \sum_k c_k T(g_k)$.

There is a slight difference of (10.8.1) from [8, Prop. 2.4]. In [8], we considered the group of *all* units of a maximal order. Here Γ_λ^1 is the group of units of $\mathfrak{o}_\lambda$ with *totally positive* norm. However, the assertion is still true in view of [8, Prop. 1.4].

10.9. PROPOSITION. *Let $\alpha, \beta \in U(\mathfrak{e})$. Suppose that*

$$(\Gamma_\nu^1 \beta \Gamma_\mu^1) \cdot (\Gamma_\mu^1 \alpha \Gamma_\lambda^1) = \sum c_\xi \cdot (\Gamma_\nu^1 \xi \Gamma_\lambda^1)$$

with elements ξ satisfying $\xi \equiv \beta \alpha \bmod^* \mathfrak{e}$. Then

$$(\Gamma_\nu^\mathfrak{e} \beta \Gamma_\mu^\mathfrak{e}) \cdot (\Gamma_\mu^\mathfrak{e} \alpha \Gamma_\lambda^\mathfrak{e}) = \sum c_\xi \cdot (\Gamma_\nu^\mathfrak{e} \xi \Gamma_\lambda^\mathfrak{e}) .$$

This can be proved by means of 10.4, (10.5.2–4), and the argument of the proof of [14, Prop. 1.15].

10.10. For every $\alpha \in B^+$, define $j(\alpha, z)$ by

$$(10.10.1) \qquad j(\alpha, z) = \det(\alpha^{(1)})^{1/2}(cz + d)^{-1}, \qquad \alpha^{(1)} = \begin{bmatrix} * & * \\ c & d \end{bmatrix},$$

where $\alpha^{(1)}$ is as in 2.4. For every integer $k \geq 0$, let $S_k(\Gamma_\lambda^c)$ denote the vector space of all cusp forms of weight k on $\mathfrak{H}$ with respect to Γ_λ^c, i.e., the vector space of all holomorphic functions $f(z)$ on $\mathfrak{H}$ satisfying

$$f(\gamma(z)) = f(z)j(\gamma, z)^{-k} \qquad \text{for all } \gamma \in \Gamma_\lambda^c .$$

When $B = M_2(\mathbf{Q})$, we impose on f the usual condition of vanishing at cusps. Let $\alpha \in B^+$ and let $\Gamma_\mu^c \alpha \Gamma_\lambda^c = \bigcup_{i=1}^d \alpha_i \Gamma_\lambda^c$ be a disjoint union. We can define a C-linear mapping $(\Gamma_\mu^c \alpha \Gamma_\lambda^c)_k$ of $S_k(\Gamma_\lambda^c)$ to $S_k(\Gamma_\mu^c)$ by

$$(\Gamma_\mu^c \alpha \Gamma_\lambda^c)_k f = \sum_{i=1}^d f(\alpha_i^{-1}(z))j(\alpha_i^{-1}, z)^k$$

for $f \in S_k(\Gamma_\lambda^c)$.

10.11. Let ρ be a representation of $U(e)/U_0(e)$ in a complex vector space W. We regard ρ as a representation of $U(e)$, and denote by $S_{k,\rho}(\Gamma_\lambda^1)$ the vector space of all holomorphic mappings $\mathfrak{f}$ of $\mathfrak{H}$ into W satisfying

$$(10.11.1) \qquad \mathfrak{f}(\gamma(z)) = \rho(\gamma)\mathfrak{f}(z)j(\gamma, z)^{-k} \qquad \textit{for all } \gamma \in \Gamma_\lambda^1 .$$

When $B = M_2(\mathbf{Q})$, we assume further that $\mathfrak{f}$ vanishes at cusps. (In any case, the components of $\mathfrak{f}$ belong to $S_k(\Gamma_\lambda^c)$.) Let $\alpha \in U(e)$ and let $\Gamma_\mu^1 \alpha \Gamma_\lambda^1 = \bigcup_{i=1}^d \alpha_i \Gamma_\lambda^1$ be a disjoint union. We define a C-linear mapping $(\Gamma_\mu^1 \alpha \Gamma_\lambda^1)_{k,\rho}$ of $S_{k,\rho}(\Gamma_\lambda^1)$ to $S_{k,\rho}(\Gamma_\mu^1)$ by

$$(10.11.2) \qquad (\Gamma_\mu^1 \alpha \Gamma_\lambda^1)_{k,\rho}\mathfrak{f} = \sum_{i=1}^d \rho(\alpha_i)\mathfrak{f}(\alpha_i^{-1}(z))j(\alpha_i^{-1}, z)^k$$

for $\mathfrak{f} \in S_{k,\rho}(\Gamma_\lambda^1)$. Now put

$$(10.11.3) \qquad S_{k,\rho} = S_{k,\rho}(\Gamma_1^1) \times \cdots \times S_{k,\rho}(\Gamma_{h_0}^1) .$$

Let E_1 and E_0 be as in 2.13, and let $\Gamma_\lambda^* = \{\gamma \in \mathfrak{o}_\lambda \mid N_{B/F}(\gamma) \in E_1\}$. As is observed in [8, p. 255, Remark 2], if $[E_1 : E_0] = 2$, then $S_{k,\rho}(\Gamma_\lambda^1)$ is isomorphic to $S(\Gamma_\lambda^*; \rho; \{k, 0, \cdots, 0\})$ which is defined in [8, p. 253]. If $E_1 = E_0$, $S(\Gamma_\lambda^*; \rho; \{k, 0, \cdots, 0\})$ is isomorphic to the direct sum of $S_{k,\rho}(\Gamma_\lambda^1)$ and $S_{k,\rho}(\Gamma_\mu^1)$ with an index μ such that $N_{B/F}(\mathfrak{x}_{\lambda\mu})$ is a principal ideal in F. Therefore, in either case, $S_{k,\rho}$ can be identified with $S(\rho; \{k, 0, \cdots, 0\})$ of [8, p. 254]. We also note that $S_{k,\rho}(\Gamma_\lambda^1) = \{0\}$ unless the restriction of ρ to E_0 contains the identity representation.

Let $\mathfrak{E}_c$ denote the set of all e in $\mathfrak{E}$ prime to c. For every $e \in \mathfrak{E}_c$, we define a linear endomorphism $T(e)_{k,\rho}$ on $S_{k,\rho}$ as follows. For each λ, we choose an element α of $U(e)$ and an index μ so that $e(\alpha \mathfrak{x}_{\lambda\mu}) = e$, and define $T(e)_{k,\rho} = (\Gamma_\mu^1 \alpha \Gamma_\lambda^1)_{k,\rho}$ on $S_{k,\rho}(\Gamma_\lambda^1)$. Identifying $S_{k,\rho}$ with $S(\rho; \{k, 0, \cdots, 0\})$, we see easily that $T(e)_{k,\rho}$ can be identified with $\mathfrak{T}(e)$ of [8, 3.4].

For an integral ideal $\mathfrak{a}$ in F, let $T(\mathfrak{a})$ (resp. $T(\mathfrak{a})_{k,\rho}$) denote the sum of the $T(e)$ (resp. $T(e)_{k,\rho}$) for all integral e such that $N_{B/F}(e) = \mathfrak{a}$. Now we define a Dirichlet series $D(s; k, \rho)$ by

$$(10.11.4) \qquad D(s; k, \rho) = \sum_e T(e)_{k,\rho} N\big(N_{B/F}(e)\big)^{-s} = \sum_{\mathfrak{a}} T(\mathfrak{a})_{k,\rho} N(\mathfrak{a})^{-s} ,$$

where $\sum_e$ is the sum extended over all integral e in $\mathfrak{E}$ prime to $\mathfrak{c}$, and $\sum_{\mathfrak{a}}$ over all integral ideals $\mathfrak{a}$ in F prime to $\mathfrak{c}$. Our $D(s; k, \rho)$ can be identified with the Dirichlet series $D(s; \xi)$ of [8, p. 273, Th. 1] with the identity character as ξ. Therefore, $D(s; k, \rho)$ can be continued holomorphically to the whole s-plane, and satisfies a functional equation. (For an explicit form of the functional equation in a special case, see 12.4 below.) Moreover, $D(s; k, \rho)$ has an Euler product:

$$(10.11.5) \qquad \begin{aligned} D(s; k, \rho) = {\prod}' &[1 - T(\mathfrak{p})_{k,\rho} N(\mathfrak{p})^{-s}]^{-1} \\ \times {\prod}'' &[1 - T(\mathfrak{p})_{k,\rho} N(\mathfrak{p})^{-s} + T(\mathfrak{p}, \mathfrak{p})_{k,\rho} N(\mathfrak{p})^{1-2s}]^{-1} . \end{aligned}$$

Here $\prod'$ is the product extended over all prime ideals $\mathfrak{p}$ in F dividing $D(B/F)$ but not $\mathfrak{c}$, and $\prod''$ over all $\mathfrak{p}$ prime to $\mathfrak{c} \cdot D(B/F)$; $T(\mathfrak{p}, \mathfrak{p})$ stands for $T(e)$ with $e = \{\mathfrak{p}, \mathfrak{p}\}$.

10.12. PROPOSITION. *Let $U_1(e)$ be as in 10.7, E the group of all units in F, and ψ a representation of $U_1(e)/U_0(e)E$ by complex matrices, and ρ a representation of $U(e)/U_0(e)E$ induced from ψ. Put $\chi(\alpha) = \mathrm{tr}(\rho(\alpha))$. Let $N = [U_1(e) : U_0(e)E]$, and let $\{\gamma_1, \cdots, \gamma_N\}$ be a set of representatives for $\Gamma_\lambda^1/\Gamma_\lambda^e E$. Then, for every λ and for every $\xi \in U(e)$,*

$$\mathrm{tr}[(\Gamma_\lambda^1 \xi \Gamma_\lambda^1)_{k,\rho}] = N^{-1} \sum_{i=1}^N \chi(\xi\gamma_i) \cdot \mathrm{tr}[(\Gamma_\lambda^e \gamma_i \Gamma_\lambda^e)_k \cdot (\Gamma_\lambda^e \xi \Gamma_\lambda^e)_k] .$$

Moreover, $\mathrm{tr}[(\Gamma_\lambda^1 \xi \Gamma_\lambda^1)_{k,\rho}]$ is a real number.

Here observe that $\Gamma_\lambda^1/\Gamma_\lambda^e E$ is (canonically) isomorphic to $U_1(e)/U_0(e)E$ in view of 10.7. The assertions can be proved in the same way as in [5, 1.15, 1.19].

Correction to [5, 1.15]. In the proof, the congruences $\xi \equiv 1 \bmod \mathfrak{b}$ or $\xi \not\equiv 1 \bmod \mathfrak{b}$ should read $\det(\xi) \equiv 1 \bmod (b_0)$ or $\det(\xi) \not\equiv 1 \bmod (b_0)$.

10.13. With indeterminates x, y, u over C, define polynomials $F_n(x, y)$ in $Z[x, y]$ for $n = 1, 2, \cdots$ by

$$-(d/du) \log (1 - xu + yu^2) = \sum_{i=1}^\infty F_n(x, y)u^{n-1} .$$

If $x = r + s$ and $y = rs$ with indeterminates r and s, then $F_n(x, y) = r^n + s^n$. Further if X and Y are commuting matrices, then

$$(d/du) \log [\det(1 - Xu + Yu^2)^{-1}] = \sum_{n=1}^\infty \mathrm{tr}(F_n(X, Y))u^{n-1} .$$

Let $T(\mathfrak{p}^a, \mathfrak{p}^b)$ denote $T(e)$ with $e = \{\mathfrak{p}^a, \mathfrak{p}^b\}$ as in [8, p. 249]. We can express $F_n(T(1, \mathfrak{p}), N(\mathfrak{p}) T(\mathfrak{p}, \mathfrak{p}))$ in the form

$$(10.13.1) \qquad F_n(T(1, \mathfrak{p}), N(\mathfrak{p}) T(\mathfrak{p}, \mathfrak{p})) = \sum_{a+b=n} c_{ab} T(\mathfrak{p}^a, \mathfrak{p}^b) \qquad (0 \leqq a \leqq b)$$

with integers c_{ab}.

11. Congruence relations of algebraic correspondences

11.1. Let U and V be complete non-singular algebraic curves defined over a field k of characteristic $p \geqq 0$, and X a one-dimensional algebraic cycle on $U \times V$. We denote by ^tX the cycle on $V \times U$ which is the transform of X by the transformation $(u, v) \to (v, u)$ of $U \times V$ to $V \times U$. For every 0-cycle c on U, we define a 0-cycle $X[c]$ by $X[c] = \mathrm{pr}_V[X \cdot (c \times V)]$ (see [16, Ch. VIII, § 4] and [17]). If W is a complete non-singular algebraic curve, and Y is a one-dimensional algebraic cycle on $V \times W$, we define a cycle $Y \circ X$ on $U \times W$ by

$$Y \circ X = \mathrm{pr}_{U \times W}[(X \times W) \cdot (U \times Y)]$$

(cf. [16, p. 235], [17], [5, 2.1]).

11.2. LEMMA. *Let U and V be as in 11.1. Let X and Y be positive one-cycles on $U \times V$. Suppose that Y has no component of the form $w \times V$ with $w \in U$, and there are infinitely many points u on U such that $X[u] - Y[u]$ is a positive cycle. Then $X - Y$ is a positive cycle. (For the definition of positive cycles, see [16, Ch. VIII, § 1].)*

PROOF. Write $X - Y$ in the form $X - Y = \sum_i a_i X_i$ with integers a_i and absolutely irreducible curves X_i. Assume that $X_i \neq X_j$ for $i \neq j$, and $a_1 < 0$. Observe that $\bigcup_{i \neq 1} (X_i \cap X_1)$ is a finite set. Hence, by our assumption, there exists a point u on U such that $X[u] - Y[u]$ is positive and u does not belong to the set-theoretical projection of $\bigcup_{i \neq 1} (X_i \cap X_1)$ to U. Then $X_1[u]$ and $X_i[u]$ have no common components if $i \neq 1$. Since $a_1 < 0$, $\sum_i a_i X_i[u]$ can not be positive. This is a contradiction,   q.e.d.

11.3. Let us fix a system $\{V_\lambda, \varphi_\lambda, R_\sigma^{\mu\lambda}(\alpha)\}$ which is established by 3.5. The mapping φ_λ is surjective unless $B = M_2(Q)$. If $B = M_2(Q)$ and if we denote by $\mathfrak{H}^*$ the union of $\mathfrak{H}$, Q (regarded as a subset of the complex plane), and the point at infinity, φ_λ can be naturally extended to a mapping $\varphi_\lambda^*: \mathfrak{H}^* \to V_\lambda$. Now we define a subset $X(\Gamma_\mu^c \alpha \Gamma_\lambda^c) = X_{\mu\lambda}(\alpha)$ of $V_\lambda \times V_\mu$, for every $\alpha \in B^+$, by

$$\begin{aligned}
X_{\mu\lambda}(\alpha) = X(\Gamma_\mu^c \alpha \Gamma_\lambda^c) &= \{\varphi_\mu^*(z) \times \varphi_\lambda^*(\alpha(z)) \mid z \in \mathfrak{H}^*\} \qquad \text{if } B = M_2(Q), \\
&= \{\varphi_\lambda(z) \times \varphi_\mu(\alpha(z)) \mid z \in \mathfrak{H}\} \qquad \text{otherwise.}
\end{aligned}$$

It can be easily seen that $X_{\mu\lambda}(\alpha)$ is an algebraic correspondence. Prop. 3.7 shows

$$(11.3.1) \qquad R_1^{\lambda\lambda}(\gamma) = X_{\lambda\lambda}(\gamma) \qquad\qquad \textit{for every } \gamma \in \Gamma_\lambda^1 .$$

Furthermore, we see easily

$$(11.3.2) \qquad {}^t X_{\mu\lambda}(\alpha) = X_{\lambda\mu}(\alpha') = X_{\lambda\mu}(\alpha^{-1}) ;$$

(11.3.3) *If* $(\Gamma_\nu^\epsilon \beta \Gamma_\mu^\epsilon) \cdot (\Gamma_\mu^\epsilon \alpha \Gamma_\lambda^\epsilon) = \sum_\xi c_\xi \cdot (\Gamma_\nu^\epsilon \xi \Gamma_\lambda^\epsilon)$ *with positive integers* c_ξ, *then* $X_{\nu\mu}(\beta) \circ X_{\mu\lambda}(\alpha) = \sum_\xi c_\xi X_{\nu\lambda}(\xi)$, *and* $X_{\lambda\mu}(\alpha') \circ X_{\mu\nu}(\beta') = \sum_\xi c_\xi X_{\lambda\nu}(\xi')$.

11.4. PROPOSITION. *If an element* α *of* B^+ *is prime to* $c \cdot D(B/F)\mathfrak{o}_1$, *then* $X_{\mu\lambda}(\alpha)$ *is rational over* $C(F, c)$.

PROOF. Let S be a finite Galois extension of F containing $C(F, c)$. By 2.10, there are infinitely many $M \in J(B)$ such that

(i) M is not contained in S;

(ii) $D(M/F)$ is prime to $D(S/F)$;

(iii) there exist an F-linear isomorphism f of M into B and an ideal $\mathfrak{b}$ in M such that $f(\mathfrak{r}_M) \subset \mathfrak{o}_\mu$ and $\alpha\mathfrak{x}_{\lambda\mu} = f(\mathfrak{b})\mathfrak{o}_\mu$. Let z_0 be the fixed point of $f(M)$ on $\mathfrak{H}$. By (3.2.3) and (3.5.4), both $\varphi_\mu(z_0)$ and $\varphi_\lambda(\alpha^{-1}(z_0))$ are rational over $C(M, e_M)$, hence over $C(M, c)$. By 1.3, we have $C(M, c) \cap S = C(F, c)$, so that $C(M, c)$ and S are linearly disjoint over $C(F, c)$. We see that $\varphi_\lambda(\alpha^{-1}(z_0)) \times \varphi_\mu(z_0)$ lies on $X_{\mu\lambda}(\alpha)$. Applying 7.7 to $X_{\mu\lambda}(\alpha)$, we get our proposition.

11.5. For $\alpha \in U(e)$ and $\sigma = [C(F, c)/F, N_{B/F}(\alpha\mathfrak{x}_{\lambda\mu})]$, define $Z_\sigma^\mu(\alpha)$ by

$$(11.5.1) \qquad Z_\sigma^\mu(\alpha) = R_\sigma^{\mu\lambda}(\alpha) \circ X_{\lambda\mu}(\alpha') .$$

Here observe that λ is determined by μ, σ and α. Now we have

$$(11.5.2) \qquad Z_\sigma^\mu(\alpha) = Z_\sigma^\mu(\beta) \ \textit{if} \ \Gamma_\mu^1 \alpha \Gamma_\lambda^1 = \Gamma_\mu^1 \beta \Gamma_\lambda^1.$$

In fact, by 10.4, we have $\beta = \gamma\alpha\delta$ with $\gamma \in \Gamma_\mu^\epsilon$ and $\delta \in \Gamma_\lambda^1$. Hence

$$\begin{aligned}
Z_\sigma^\mu(\beta) &= R_\sigma^{\mu\lambda}(\gamma\alpha\delta) \circ X_{\lambda\mu}(\delta'\alpha'\gamma') \\
&= R_\sigma^{\mu\lambda}(\alpha\delta) \circ X_{\lambda\mu}(\delta'\alpha') \\
&= R_\sigma^{\mu\lambda}(\alpha) \circ R_1^{\lambda\lambda}(\delta) \circ X_{\lambda\lambda}(\delta') \circ X_{\lambda\mu}(\alpha') = R_\sigma^{\mu\lambda}(\alpha) \circ X_{\lambda\mu}(\alpha')
\end{aligned}$$

in view of (3.5.3) and (11.3.1–3). This proves (11.5.2). Therefore, with an element e of $\mathfrak{E}$ such that $e = e(\alpha\mathfrak{x}_{\lambda\mu})$ we put $Z_\sigma^\mu(e) = Z_\sigma^\mu(\alpha)$, since this does not depend on the choice of α.

11.6. PROPOSITION. *Let* $M \in J(B)$. *Let* f *be a normalized* F-*linear isomorphism of* M *into* B *such that* $f(\mathfrak{r}_M) \subset \mathfrak{o}_\mu$, *and* z *the fixed point of* $f(M)$ *on* $\mathfrak{H}$. *Let* $\mathfrak{b}$ *be an ideal in* M, *prime to* c, *and* $\tau = [C(M, c)/M, \mathfrak{b}]$. *If* $e = e(f(\mathfrak{b})\mathfrak{o}_\mu)$ *and* $\sigma = [C(F, c)/F, N_{B/F}(e)]$, *then* $\varphi_\mu(z) \times \varphi_\mu(z)^\tau \in Z_\sigma^\mu(e)$.

This follows immediately from (3.5.4) and the definition of $Z_\sigma^\mu(e)$.

11.7. PROPOSITION. $R_\tau^{\mu\nu}(\beta)^\sigma \circ Z_\sigma^\nu(e) = Z_\sigma^\mu(e)^\tau \circ R_\tau^{\mu\nu}(\beta)$.

PROOF. Let α be an element of B^+ such that $e(\alpha \mathfrak{x}_{\lambda\mu}) = e$ with an index λ (see (10.3.1)). By 2.10, there exist an $M \in J(B)$ and a normalized F-linear isomorphism f of M into B such that $f(\mathfrak{r}_M) \subset \mathfrak{o}_\mu$, $\alpha \mathfrak{x}_{\lambda\mu} = f(\mathfrak{a})\mathfrak{o}_\mu$, $\beta \mathfrak{x}_{\nu\mu} = f(\mathfrak{b})\mathfrak{o}_\mu$ with ideals $\mathfrak{a}$ and $\mathfrak{b}$ in M. Let w be the fixed point of $f(M)$ on $\mathfrak{H}$. Define f_1 by $f_1(x) = \beta^{-1}f(x)\beta$, and put $z = \beta^{-1}(w)$. Then $f_1(\mathfrak{r}_M) \subset \mathfrak{o}_\nu$, and z is the fixed point of $f_1(M)$ on $\mathfrak{H}$. We can find an element γ of B^+ and an index κ such that $\gamma \mathfrak{x}_{\kappa\nu} = f_1(\mathfrak{a})\mathfrak{o}_\nu$. By 10.2, $e(\gamma \mathfrak{x}_{\kappa\nu}) = e(\alpha \mathfrak{x}_{\lambda\mu}) = e$. Let $\xi = [C(M, \mathfrak{c})/M, \mathfrak{a}]$ and $\eta = [C(M, \mathfrak{c})/M, \mathfrak{b}]$. By 11.6, we have $\varphi_\nu(z) \times \varphi_\nu(z)^\xi \in Z_\sigma^\nu(e)$, $\varphi_\mu(w) \times \varphi_\mu(w)^\xi \in Z_\sigma^\mu(e)$, and by (3.5.4), $\varphi_\nu(z) \times \varphi_\mu(w)^\eta \in R_\tau^{\mu\nu}(\beta)$. Hence $\varphi_\nu(z) \times \varphi_\mu(w)^{\xi\eta}$ is common to $R_\tau^{\mu\nu}(\beta)^\sigma \circ Z_\sigma^\nu(e)$ and $Z_\sigma^\mu(e)^\tau \circ R_\tau^{\mu\nu}(\beta)$. Since there are infinitely many such points and these algebraic correspondences are irreducible, we obtain the desired equality.

11.8. PROPOSITION. *Let α, α_1, β, γ be elements of $U(e)$ such that $e(\alpha \mathfrak{x}_{\lambda\mu}) = e(\alpha_1 \mathfrak{x}_{\kappa\nu})$ and $\alpha\gamma \equiv \beta\alpha_1 \bmod^* e$. Then $X_{\mu\lambda}(\alpha)^\tau \circ R_\tau^{\lambda\kappa}(\gamma) = R_\tau^{\mu\nu}(\beta) \circ X_{\nu\kappa}(\alpha_1)$.*

PROOF. Applying (11.5.1) to both sides of 11.7, we get

$$R_\tau^{\mu\nu}(\beta)^\sigma \circ R_\sigma^{\nu\kappa}(\alpha_1) \circ {}^t X_{\nu\kappa}(\alpha_1) = R_\sigma^{\mu\lambda}(\alpha)^\tau \circ {}^t X_{\mu\lambda}(\alpha)^\tau \circ R_\tau^{\mu\nu}(\beta) .$$

Now, by (3.5.3),

$$R_\tau^{\mu\nu}(\beta)^\sigma \circ R_\sigma^{\nu\kappa}(\alpha_1) = R_{\tau\sigma}^{\mu\kappa}(\beta\alpha_1) = R_{\sigma\tau}^{\mu\kappa}(\alpha\gamma) = R_\sigma^{\mu\lambda}(\alpha)^\tau \circ R_\tau^{\lambda\kappa}(\gamma) .$$

From these relations, we obtain the equality of our proposition.

11.9. For every $\mathfrak{x} \subset B$, let $\mathfrak{x}'$ denote the transform of $\mathfrak{x}$ by the main involution of B. We observe that $(\mathfrak{x}_{\lambda 1}')^{-1}$ is a right $\mathfrak{o}_1$-ideal. In view of (2.13.4), we find a permutation $\lambda \mapsto \bar\lambda$ of $\{1, \cdots, h_0\}$ and an element δ_λ of B^+, for each λ, such that $\delta_\lambda(\mathfrak{x}_{\lambda 1}')^{-1} = \mathfrak{x}_{\bar\lambda 1}$. Then we see easily

$$(11.9.1) \qquad \delta_\lambda \mathfrak{o}_\lambda \delta_\lambda^{-1} = \mathfrak{o}_{\bar\lambda} , \qquad \Gamma_{\bar\lambda}^e \delta_\lambda \Gamma_\lambda^e = \delta_\lambda \Gamma_\lambda^e = \Gamma_{\bar\lambda}^e \delta_\lambda .$$

$$(11.9.2) \qquad \mathfrak{x}_{\bar\mu\bar\lambda} = \delta_\mu \mathfrak{x}_{\lambda\mu}' \delta_\lambda^{-1} .$$

$$(11.9.3) \qquad \delta_\lambda \alpha' \delta_\mu^{-1} \mathfrak{x}_{\bar\mu\bar\lambda} = \delta_\lambda (\mathfrak{x}_{\lambda\mu}\alpha)' \delta_\lambda^{-1} \qquad\qquad \textit{for every } \alpha \in B^+ .$$

$$(11.9.4) \qquad e(\delta_\lambda \alpha' \delta_\mu^{-1} \mathfrak{x}_{\bar\mu\bar\lambda}) = e(\alpha \mathfrak{x}_{\lambda\mu}) \qquad\qquad \textit{for every } \alpha \in B^+ .$$

From (11.9.1), we see that $X_{\bar\lambda\lambda}(\delta_\lambda)$ is a biregular morphism of V_λ to $V_{\bar\lambda}$, which is rational over $C(F, \mathfrak{c})$ in view of 11.4. Now we shall prove

$$(11.9.5) \qquad X_{\mu\bar\mu}(\delta_\mu^{-1})^\sigma \circ Z_\sigma^{\bar\mu}(e) \circ X_{\bar\mu\mu}(\delta_\mu) = Z_\sigma^\mu(e) .$$

Let $M, f, z, \mathfrak{b}, \tau$ be as in 11.6. Then $\varphi_\mu(z) \times \varphi_\mu(z)^\tau \in Z_\sigma^\mu(e)$. Define $f_1 \colon M \to B$ by $f_1(x) = \delta_\mu f(x)\delta_\mu^{-1}$ for $x \in M$ and put $w = \delta_\mu(z)$. Then $f_1(\mathfrak{r}_M) \subset \mathfrak{o}_{\bar\mu}$ and w is the fixed point of $f_1(M)$. Since $e(f_1(\mathfrak{b})\mathfrak{o}_\mu) = e(f(\mathfrak{b})\mathfrak{o}_\mu) = e$ by 10.2, we have $\varphi_{\bar\mu}(w) \times \varphi_{\bar\mu}(w)^\tau \in Z_\sigma^{\bar\mu}(e)$, so that $\varphi_\mu(z) \times \varphi_\mu(z)^\tau$ belongs to the left hand side of (11.9.5). Therefore both sides of (11.9.5) has infinitely many common points. Since they are irreducible cycles, we obtain the equality.

For every $\beta \in U(e)$ and $\sigma = [C(F, \mathfrak{c})/F, N_{B/F}(\beta' \mathfrak{x}_{\lambda\mu})]$, define a biregular morphism $S_\sigma^{\lambda\mu}(\beta)$ of V_μ to V_λ^σ by

$$(11.9.6) \qquad S_\sigma^{\lambda\mu}(\beta) = X_{\lambda\bar{\lambda}}(\delta_\lambda^{-1})^\sigma \circ R_\sigma^{\bar{\lambda}\bar{\mu}}(\delta_\lambda \beta \delta_\mu^{-1}) \circ X_{\bar{\mu}\mu}(\delta_\mu) \ .$$

One can verify in a straightforward way

$$(11.9.7) \qquad S_\sigma^{\lambda\mu}(\beta)^\tau \circ S_\tau^{\mu\nu}(\gamma) = S_{\sigma\tau}^{\lambda\nu}(\beta\gamma) \ .$$

Further, if $e = e(\alpha \mathfrak{x}_{\lambda\mu})$, then by (11.9.4) and (11.5.1) we have

$$Z_\sigma^{\bar{\lambda}}(e) = R_\sigma^{\bar{\lambda}\bar{\mu}}(\delta_\lambda \alpha' \delta_\mu^{-1}) \circ X_{\bar{\mu}\bar{\lambda}}(\delta_\mu'^{-1} \alpha \delta_\lambda') \ .$$

Since $X_{\lambda\lambda}(\delta_\lambda') = X_{\lambda\bar{\lambda}}(\delta_\lambda^{-1})$, we have, by (11.9.5),

$$(11.9.8) \qquad Z_\sigma^{\lambda}(e) = S_\sigma^{\lambda\mu}(\alpha') \circ X_{\mu\lambda}(\alpha) \ if \ e = e(\alpha \mathfrak{x}_{\lambda\mu}).$$

11.10. PROPOSITION. *If* $\alpha \in U(e)$, $\mathfrak{a} = N_{B/F}(\alpha \mathfrak{x}_{\lambda\mu})$, $e_1 = e(\mathfrak{a} \mathfrak{o}_\mu)$, *and* $\sigma = [C(F, \mathfrak{c})/F, \mathfrak{a}]$, *then* $R_\sigma^{\mu\lambda}(\alpha)^\sigma \circ S_\sigma^{\lambda\mu}(\alpha') = Z_{\sigma^2}^\mu(e_1)$.

PROOF. Take $\beta \in B^+$ and ν so that $\beta \mathfrak{x}_{\nu\mu} = \mathfrak{a}\mathfrak{o}_\mu$. Then $\beta \mathfrak{o}_\nu = \mathfrak{o}_\mu \beta$, and

$$\begin{aligned} \alpha^{-1}\beta \mathfrak{x}_{\nu\lambda} &= \alpha^{-1}\beta \mathfrak{x}_{\nu\mu}\mathfrak{x}_{\mu\lambda} \\ &= \alpha^{-1}\mathfrak{a}\mathfrak{x}_{\mu\lambda} = \alpha^{-1}\alpha\alpha' \mathfrak{x}_{\lambda\mu}' \mathfrak{x}_{\lambda\mu}\mathfrak{x}_{\mu\lambda} = \alpha' \mathfrak{x}_{\lambda\mu}' = (\mathfrak{x}_{\lambda\mu}\alpha)' \ , \end{aligned}$$

hence $e(\alpha^{-1}\beta \mathfrak{x}_{\nu\lambda}) = e(\alpha \mathfrak{x}_{\lambda\mu})$. By 2.10, there exist an $M \in J(B)$ and a normalized F-linear isomorphism f of M into B such that $f(\mathfrak{r}_M) \subset \mathfrak{o}_{\bar{\lambda}}$ and $f(\mathfrak{b})\mathfrak{o}_{\bar{\lambda}} = \delta_\lambda \alpha' \delta_\mu^{-1} \mathfrak{x}_{\bar{\mu}\bar{\lambda}}$ with an ideal $\mathfrak{b}$ in M. Define $f_1: M \to B$ by $f_1(x) = \delta_\lambda^{-1} f(x)\delta_\lambda$. Then

$$f_1(\mathfrak{r}_M) \subset \delta_\lambda^{-1}\mathfrak{o}_{\bar{\lambda}}\delta_\lambda = \mathfrak{o}_\lambda \ ,$$

and

$$f_1(\mathfrak{b})\mathfrak{o}_\lambda = \delta_\lambda^{-1} f(\mathfrak{b})\mathfrak{o}_{\bar{\lambda}}\delta_\lambda = \alpha' \delta_\mu^{-1} \mathfrak{x}_{\bar{\mu}\bar{\lambda}}\delta_\lambda = \alpha' \mathfrak{x}_{\lambda\mu}' = \alpha^{-1}\beta \mathfrak{x}_{\nu\lambda} \ .$$

Let z be the fixed point of $f(M)$ on $\mathfrak{H}$. Then with $\tau = [C(M, \mathfrak{c})/M, \mathfrak{b}]$, we have

$$\begin{aligned} \varphi_\mu\big(\alpha\delta_\lambda^{-1}(z)\big) \times \varphi_{\bar{\mu}}\big(\delta_\mu\alpha\delta_\lambda^{-1}(z)\big) &\in X_{\bar{\mu}\mu}(\delta_\mu) \ , \\ \varphi_{\bar{\mu}}\big(\delta_\mu\alpha\delta_\lambda^{-1}(z)\big) \times \varphi_{\bar{\lambda}}(z)^\tau &\in R_\sigma^{\bar{\lambda}\bar{\mu}}(\delta_\lambda \alpha' \delta_\mu^{-1}) \qquad (3.5.4) \ , \\ \varphi_{\bar{\lambda}}(z)^\tau \times \varphi_\lambda\big(\delta_\lambda^{-1}(z)\big)^\tau &\in X_{\lambda\bar{\lambda}}(\delta_\lambda^{-1})^\sigma \ , \\ \varphi_\lambda\big(\delta_\lambda^{-1}(z)\big)^\tau \times \varphi_\nu\big(\beta^{-1}\alpha\delta_\lambda^{-1}(z)\big) &\in R_\sigma^{\lambda\nu}(\alpha^{-1}\beta)^{-1} \qquad (3.5.4) \ . \end{aligned}$$

As for the last inclusion, note that $\delta_\lambda^{-1}(z)$ is the fixed point of $f_1(M)$. Therefore, putting $w = \alpha\delta_\lambda^{-1}(z)$, we see that

$$\varphi_\mu(w) \times \varphi_\nu\big(\beta'(w)\big) \in R_\sigma^{\lambda\nu}(\alpha^{-1}\beta)^{-1} \circ S_\sigma^{\lambda\mu}(\alpha') \ .$$

Obviously $\varphi_\mu(w) \times \varphi_\nu(\beta'(w)) \in X_{\nu\mu}(\beta')$. Since there are infinitely many choices of M, this shows $X_{\nu\mu}(\beta') = R_\sigma^{\lambda\nu}(\alpha^{-1}\beta)^{-1} \circ S_\sigma^{\lambda\mu}(\alpha')$. Now we have $R_\sigma^{\lambda\nu}(\alpha^{-1}\beta) = R_{\sigma^{-1}}^{\lambda\mu}(\alpha^{-1})^{\sigma^2} \circ R_{\sigma^2}^{\mu\nu}(\beta)$ and $R_{\sigma^{-1}}^{\lambda\mu}(\alpha^{-1})^\sigma = {}^t R_\sigma^{\mu\lambda}(\alpha)$, hence

$$Z_{\sigma^2}^\mu(e_1) = R_{\sigma^2}^{\mu\nu}(\beta) \circ X_{\nu\mu}(\beta') = R_\sigma^{\mu\lambda}(\alpha)^\sigma \circ S_\sigma^{\lambda\mu}(\alpha') \ , \qquad\qquad \text{q.e.d.}$$

11.11. Specialize 11.10 to the case $\lambda = \mu$ and $\sigma = [C(F, \mathfrak{c})/F, (\alpha\alpha')]$. Then
$$R_\sigma^{\mu\mu}(\alpha)^\sigma \circ S_\sigma^{\mu\mu}(\alpha') = Z_{\sigma^2}^\mu(e_1)$$
$$= R_{\sigma^2}^{\mu\mu}(\alpha\alpha') \circ X_{\mu\mu}(\alpha\alpha') = R_{\sigma^2}^{\mu\mu}(\alpha\alpha') = R_\sigma^{\mu\mu}(\alpha)^\sigma \circ R_\sigma^{\mu\mu}(\alpha') \ ,$$
hence, exchanging α for α', we obtain

$$(11.11.1) \qquad\qquad S_\sigma^{\mu\mu}(\alpha) = R_\sigma^{\mu\mu}(\alpha) \ .$$

11.12. Fix a prime ideal $\mathfrak{p}$ in F prime to $\mathfrak{c} \cdot D(B/F)$. For every positive integer n, let $\Theta_\mathfrak{p}^n$ denote the set of all ordered pairs (λ, μ) with λ, μ in $\{1, \cdots, h_0\}$ such that $N_{B/F}(\mathfrak{x}_{\lambda\mu})\mathfrak{p}^{-n} \in P(F, \mathfrak{u}_0)$, where $\mathfrak{u}_0$ is as in 2.13 (see also 1.1). If n is given, μ is uniquely determined by λ, and vice versa. For each $(\lambda, \mu) \in \Theta_\mathfrak{p}^1$, we fix once for all, an element β_λ of $U(e)$ such that $e(\beta_\lambda\mathfrak{x}_{\lambda\mu}) = \{1, \mathfrak{p}\}$. Such a β_λ always exists in view of (10.3.1). For every positive integer n, we define $\beta_{n\lambda}$ by $\beta_{n\lambda} = \beta_\kappa \cdots \beta_\nu\beta_\mu\beta_\lambda$ with a chain of n pairs (λ, μ), (μ, ν), $\cdots$, (κ, ω) in $\Theta_\mathfrak{p}^1$. We put $\beta_{0\lambda} = 1$ for every λ. We shall now prove

(11.12.1) *If $e = \{\mathfrak{p}^a, \mathfrak{p}^b\} \in \mathfrak{E}$, $a \geq 0$, $b \geq 0$ and $a + b = n$, then there exist an element ε of B^+ and an index ω such that $e(\varepsilon\mathfrak{x}_{\lambda\omega}) = e$, $(\lambda, \omega) \in \Theta_\mathfrak{p}^n$, and $\varepsilon \equiv \beta_{n\lambda} \bmod^* e$.*

By (10.3.1), there exist an element δ of B^+ and an index ω such that $e(\delta\mathfrak{x}_{\lambda\omega}) = e$, and $(\lambda, \omega) \in \Theta_\mathfrak{p}^n$. Using the notation of [8, Prop. 2.6], $T(\mathfrak{p}^a, \mathfrak{p}^b)$ occurs as a component of $T(1, \mathfrak{p})^n$. It follows that
$$\Gamma_\omega^1\delta\Gamma_\lambda^1 \subset \Gamma_\omega^1\beta_\kappa\Gamma_\kappa^1 \cdots \Gamma_\nu^1\beta_\mu\Gamma_\mu^1\beta_\lambda\Gamma_\lambda^1$$
with a chain (λ, μ), $\cdots$, (κ, ω) as above. By 10.4 we have
$$\Gamma_\omega^1\beta_\kappa\Gamma_\kappa^1 \cdots \Gamma_\nu^1\beta_\mu\Gamma_\mu^1\beta_\lambda\Gamma_\lambda^1 = \Gamma_\kappa^e\beta_\kappa\Gamma_\kappa^e \cdots \Gamma_\nu^e\beta_\mu\Gamma_\mu^e\beta_\lambda\Gamma_\lambda^1 \ .$$
Hence we can find elements γ_ω of Γ_ω^e, γ_κ of Γ_κ^e, $\cdots$, γ_ν of Γ_ν^e, γ_μ of Γ_μ^e and γ_λ of Γ_λ^1 such that $\delta = \gamma_\omega\beta_\kappa\gamma_\kappa \cdots \gamma_\nu\beta_\mu\gamma_\mu\beta_\lambda\gamma_\lambda$. Put $\delta\gamma_\lambda^{-1} = \varepsilon$. Then $\varepsilon\mathfrak{x}_{\lambda\omega} = \delta\mathfrak{x}_{\lambda\omega}$ and $\varepsilon \equiv \beta_\kappa \cdots \beta_\mu\beta_\lambda = \beta_{n\lambda} \bmod^* e$, q.e.d.

11.13. Let $\mathfrak{E}_\mathfrak{p}$ denote the set of all $e \in \mathfrak{E}$ of the form $e = \{\mathfrak{p}^a, \mathfrak{p}^b\}$, $a \geq 0$, $b \geq 0$. For every $e \in \mathfrak{E}_\mathfrak{p}$, take ε as in (11.12.1) and put

$$(11.13.1) \qquad\qquad X_{\lambda\omega}(e) = X_{\lambda\omega}(\mathfrak{p}^a, \mathfrak{p}^b) = X_{\lambda\omega}(\varepsilon') \ .$$

By virtue of (10.5.2), $X_{\lambda\omega}(e)$ depends only on λ, e and the choice of the β_λ, and not on the choice of ε. By (11.5.1), we obtain

$$(11.13.2) \qquad X_{\lambda\omega}(e) = {}^tR_\tau^{\omega\lambda}(\beta_{n\lambda}) \circ Z_\tau^\omega(e) \qquad (\tau = [C(F, \mathfrak{c})/F, \mathfrak{p}^n]) \ .$$

Further

(11.13.3) *If $e, f, g \in \mathfrak{E}_\mathfrak{p}$ and $T(e)T(f) = \sum_g c_g T(g)$, then $X_{\nu\mu}(f) \circ X_{\mu\lambda}(e) = \sum_g c_g X_{\nu\lambda}(g)$.*

In fact, let $X_{\mu\lambda}(e) = X_{\mu\lambda}(\varepsilon')$, $X_{\nu\mu}(f) = X_{\nu\mu}(\eta')$, $X_{\nu\lambda}(g) = X_{\nu\lambda}(\gamma')$ with $\varepsilon \equiv \beta_{n\mu}$, $\eta \equiv \beta_{m\nu}$, $\gamma \equiv \beta_{k\nu}$ mod* e, where $k = m + n$. From $T(e) \cdot T(f) = \sum_g c_g T(g)$, we obtain $(\Gamma_\lambda^1 \varepsilon \Gamma_\mu^1) \cdot (\Gamma_\mu^1 \eta \Gamma_\nu^1) = \sum_g c_g \cdot (\Gamma_\lambda^1 \gamma \Gamma_\nu^1)$ (10.8.1). Since $\gamma \equiv \varepsilon\eta$ mod* e, we obtain (11.13.3) by means of 10.9 and (11.3.3). (Since $T(e) T(f) = T(f) T(e)$ [8, Prop. 2.2], the order of the products is not important.)

11.14. We shall now consider reduction modulo $\mathfrak{P}$ of V_λ, $X_{\mu\lambda}(\alpha)$, etc. with respect to a prime ideal $\mathfrak{P}$ in $C(F, \mathfrak{c})$. We shall denote by $\mathfrak{P}(X)$ the reduction of X modulo $\mathfrak{P}$. If $\mathfrak{P}$ is fixed, $\mathfrak{P}(X)$ will be also denoted by $\tilde{X}$. To simplify the matter, we may assume that the curves V_λ are projective.

Let M, f, z be as in (3.5.4), and let M_1, f_1, z_1 be symbols with the same property. Our next problem is to find a condition under which the points $\mathfrak{q}(\varphi_\mu(z))$ and $\mathfrak{q}(\varphi_\mu(z_1))$ are distinct, where $\mathfrak{q}$ is a prime factor of $\mathfrak{P}$ in an extension of $C(F, \mathfrak{c})$. This will be done by investigating the reduction modulo $\mathfrak{q}$ of corresponding abelian varieties. Here we shall prove only a coarse result which suffices our present need, though the refinement is possible and actually an important question.

Take a positive integer d so that $d \in \mathfrak{c}$ and the group

$$(11.14.1) \qquad \{\gamma \in \Gamma(\mathfrak{o}_\lambda, 1) \mid N_{B/F}(\gamma) = 1, \gamma - 1 \in d\mathfrak{o}_\lambda\}$$

has no elements of finite order, other than the identity element, for all λ. Let $K, \tau_1, \cdots, \tau_g, L, \rho, T, \iota$ and j be as in 6.1 and 7.3. For simplicity, we identify B with $\iota(B)$. Take a representation Φ of L as in (6.1.3) with $q = 2$ and $r = 1$. Let $\mathfrak{M}$ be a lattice in L such that $\mathfrak{r}_K \mathfrak{M} \mathfrak{o}_1 \subset \mathfrak{M}$. Put $\mathfrak{M}_\lambda = \mathfrak{M} \mathfrak{r}_{1\lambda}$. Take elements $y_{\lambda 1}$ and $y_{\lambda 2}$ of L so that $d^{-1}\mathfrak{M}_\lambda = \mathfrak{M}_\lambda + \mathfrak{r}_K y_{\lambda 1} + \mathfrak{r}_K y_{\lambda 2}$. Let

$$\Omega_\lambda = \{L, \Phi, \rho; T, \mathfrak{M}_\lambda; y_{\lambda 1}, y_{\lambda 2}\} \, .$$

We take K and $\tau_1, \cdots, \tau_g$ so that (7.6.3–5) are satisfied. By 6.3, we have $\Gamma(\mathfrak{o}_\lambda, d) = \Gamma_0(T, d^{-1}\mathfrak{M}_\lambda/\mathfrak{M}_\lambda)$, so that $\Gamma(T, d^{-1}\mathfrak{M}_\lambda/\mathfrak{M}_\lambda)$ coincides with the group (11.14.1). Put $\mathfrak{o}_L = \{a \in L \mid a\mathfrak{M} \subset \mathfrak{M}\}$. By [13, 5.3], there exists a fibre system of PEL-structures

$$\mathfrak{F}_\lambda = \{U_\lambda, W_\lambda, h_\lambda, f_\lambda, Y_\lambda, S_\lambda(a) \, (a \in \mathfrak{o}_L), f_{\lambda 1}, f_{\lambda 2}\} \, ,$$

whose fibres $\mathfrak{Q}_{\lambda u} = (A_{\lambda u}, C_{\lambda u}, \theta_{\lambda u}; \{f_{\lambda i}(u)\})$ $(u \in U_\lambda, i = 1, 2)$ are PEL-structures of type Ω_λ, such that the conditions [13, (5.3.0–5)] are satisfied. Let $\psi_\lambda \colon \mathcal{H} \to U_\lambda$ be the mapping with the property [13, (5.3.5)].

Since $\Gamma(\mathfrak{o}_\lambda, d) \subset \Gamma(\mathfrak{o}_\lambda, e_\lambda)$, there exists a morphism ρ_λ of U_λ to V_λ such that $\rho_\lambda \circ \psi_\lambda \circ j = \varphi_\lambda$. Then we see easily that ρ_λ is defined over an algebraic number field k of finite degree. In fact, let M, f, z be as in (3.5.4). Then the point $\varphi_\lambda(z)$ is algebraic. On the other hand, the coordinates of $\psi_\lambda \circ j(z)$ generate, over $k(\Omega_\lambda)$, the field of moduli of $\mathfrak{Q}_{\lambda u}$, where $u = \psi_\lambda \circ j(z)$. As observed in

6.4, $A_{\lambda u}$ has sufficiently many complex multiplications, so that $\psi_\lambda \circ j(z)$ is algebraic. By 7.8, this proves that ρ_λ is defined over the algebraic closure of Q.

Fix such an algebraic number field k of finite degree, containing $C(F, \mathfrak{c})$ and $k(\Omega_\lambda)$ for all λ. Let P' be the set of all prime ideals $\mathfrak{q}$ in k such that the following conditions are satisfied.

(11.14.2) *For every λ and every automorphism σ of $C(F, \mathfrak{c})$ over F, V_λ^σ is $\mathfrak{q}$-simple in the sense of [15, 9.4], and $\mathfrak{q}(V_\lambda^\sigma)$ is a complete non-singular curve of the same genus as V_λ.*

(11.14.3) *For every λ, μ and $\alpha \in U(\mathfrak{e})$, $R_\sigma^{\mu\lambda}(\alpha)$ resp. $S_\sigma^{\lambda\mu}(\alpha)$ is defined everywhere on $\mathfrak{q}(V_\lambda)$ resp. $\mathfrak{q}(V_\mu)$ in the sense of [15, 10.1]. (Accordingly, $\mathfrak{q}(R_\sigma^{\mu\lambda}(\alpha))$ is a biregular morphism of $\mathfrak{q}(V_\lambda)$ to $\mathfrak{q}(V_\mu^\sigma)$.*

(11.14.4) *For every λ, the fibre system $\{U_\lambda, W_\lambda, h_\lambda, f_\lambda\}$ behaves well for $\mathfrak{q}$ in the sense of [5, 5.3].*

(11.14.5) *ρ_λ is everywhere defined on $\mathfrak{q}(U_\lambda)$ for every λ in the sense of [15, 10.1].*

By [15, §12] and [5, 5.3], P' contains all except a finite number of prime ideals in k. We denote by P the set of all prime ideals $\mathfrak{p}$ in F, prime to $\mathfrak{c} \cdot D(B/F) \cdot D(K/Q)$, such that all prime factors of $\mathfrak{p}$ in k are contained in P'. Of course P contains all except a finite number of prime ideals in F.

11.15. First we treat the case $g > 1$. Let $\mathfrak{p} \in P$. By 1.5, there exist infinitely many $M \in J(B)$ satisfying the following conditions.

(11.15.1) $M \neq K$.

(11.15.2) *$\mathfrak{p}$ decomposes into two distinct primes in M.*

(11.15.3) *$D(M/F)$ is prime to $D(K/F)N(\mathfrak{p})$.*

Fix an index μ. By 2.8, there exists a normalized F-linear isomorphism f of M into B such that $f(\mathfrak{r}_M) \subset \mathfrak{o}_\mu$. Let z_0 be the fixed point of $f(M)$ on $\mathfrak{H}$. Put $z = j(z_0)$, $u = \psi_\mu(z)$, $v = \varphi_\mu(z_0)$. Then $\rho_\mu(u) = v$. As discussed in 6.4, we can extend θ_u to an isomorphism θ of $M_2(S)$ to $\mathrm{End}_Q(A_u)$, where $S = KM$. Moreover, by (6.6.1), we have $\theta(\mathfrak{r}_S) \subset \mathrm{End}(A_u)$, hence, by the results of 5.23, 6.6 and 6.7, we can write $A_u = A_1 \times A_2$ and obtain an isomorphism $\theta_i : S \to \mathrm{End}_Q(A_i)$ such that $\theta_i(\mathfrak{r}_S) = \mathrm{End}(A_i)$. Let R be a finite extension of $k \cdot S$ over which u, v and all elements of $\theta_i(\mathfrak{r}_S)$ and $\mathrm{End}(A_u)$ are rational. Let $\mathfrak{q}$ be a prime ideal in R which divides $\mathfrak{p}$. Let us denote the action of reduction modulo $\mathfrak{q}$ by tilde. By (11.14.4), $\tilde{A}_u$ is an abelian variety. If ε_i denotes the projection of A_u to A_i for $i = 1, 2$, then $\tilde{\varepsilon}_i$ is a meaningful element of $\mathrm{End}(\tilde{A}_u)$

by [15, 11.1, Prop. 12]. Then $\tilde{A}_u = \tilde{A}_1 \times \tilde{A}_2$. Define $\tilde{\theta}_i$ by $\tilde{\theta}_i(a) = q(\theta_i(a))$ for every $a \in \mathfrak{r}_S$. Let π (resp. π_i) denote the $N(q)^{th}$ power endomorphism of $\tilde{A}_u$ (resp. $\tilde{A}_i$). By [15, §13, Th. 1], $\pi_i = \tilde{\theta}_i(\mu_i)$ with an element μ_i of S, and

$$(11.15.4) \qquad\qquad (\mu_i) = \det \Xi'(N_{R/S}(q)) ,$$

where Ξ' is as in 6.5. Now π commutes with every element of $\tilde{\theta}(M_2(S))$. Hence $\pi \in \tilde{\theta}(S) \cap \mathrm{End}(\tilde{A}_u) = \tilde{\theta}(\mathfrak{r}_S)$. (In general, $\mathrm{End}_Q(\tilde{A}_u)$ may be larger than $\tilde{\theta}(M_2(S))$. However, if we take any maximal subfield S' of $M_2(S)$, we have $\pi \in \tilde{\theta}(S')$ by [15, 5.1, Prop. 1].) This shows that $\mu_1 = \mu_2$, hence $\pi = \tilde{\theta}(\mu_1)$. Put $q_M^\nu = N_{R/M}(q)$ with a prime ideal q_M in M and a positive integer ν. By 5.17 and (5.14.11), we can express (11.15.4) in the form $(\mu_1) = q_M^\nu \cdot t$ with an ideal t in K. Note that $K = K'$, $S = S'$, $\chi_1 =$ the identity mapping on M in the present case. Since $\mathfrak{p}$ decomposes in M (11.15.2), we see that $\mu_1^m \notin K$ for any positive integer m. Hence $\tilde{\theta}(KM)$ can be generated by $\tilde{\theta}_u(K)$ and π^m, for every positive integer m. Therefore, if we take distinct M's under the conditions (11.15.1–3), the corresponding abelian varieties $q(A_u)$ are distinct, and hence *the points* $q(\varphi_\mu(z))$ *are distinct.*

This conclusion is true also in the case $g = 1$. One way of seeing this is to consider the family of abelian varieties as in 9.7 instead of the family of §6. Then, without any auxiliary field K, one gets the same conclusion.

11.16. Let $\mathfrak{p} \in P$. Let us fix a prime ideal $\mathfrak{P}$ in $C(F, \mathfrak{c})$ which divides $\mathfrak{p}$, and denote by tilde the action of reduction modulo $\mathfrak{P}$. For every λ and an integer $n \geq 0$, let Π_λ^n denote the Frobenius correspondence of degree $N(\mathfrak{p})^n$ on $\tilde{V}_\lambda \times \tilde{V}_\lambda^q$, where $q = N(\mathfrak{p})^n$, i.e., the locus of $x \times x^q$ on $\tilde{V}_\lambda \times \tilde{V}_\lambda^q$ with $x \in \tilde{V}_\lambda$. For brevity, let us put $[\mathfrak{a}] = [C(F, \mathfrak{c})/F, \mathfrak{a}]$ for every ideal $\mathfrak{a}$ in F prime to $\mathfrak{c}$.

11.17. THEOREM. *The notation being as in 11.16, let $(\lambda, \mu) \in \Theta_\mathfrak{p}^1$, and let α be an element of $U(\mathfrak{e})$ such that $e(\alpha \chi_{\lambda\mu}) = \{1, \mathfrak{p}\}$. Then one has*

$$\tilde{X}_{\lambda\mu}(\alpha') = {}^t\tilde{R}_\sigma^{\mu\lambda}(\alpha) \circ \Pi_\mu^1 + {}^t\Pi_\lambda^1 \circ \tilde{S}_\sigma^{\lambda\mu}(\alpha') ,$$

where $\sigma = [\mathfrak{p}]$.

PROOF. Let M, f, z, u, v be as in 11.15. Let $\mathfrak{Q}$ be a prime divisor of the algebraic closure of Q whose restriction to $C(F, \mathfrak{c})$ is $\mathfrak{P}$. Denote by tilde the action of reduction modulo $\mathfrak{Q}$. Let q be the restriction of $\mathfrak{Q}$ to $C(M, \mathfrak{c})$, and $\tau = [C(M, \mathfrak{c})/M, q]$. By 11.6, we have $v \times v^\tau \in Z_\sigma^\mu(\mathfrak{e}) = R_\sigma^{\mu\lambda}(\alpha) \circ X_{\lambda\mu}(\alpha')$, where $e = \{1, \mathfrak{p}\}$. In view of (11.15.2), we have $N(\mathfrak{p}) = N(q)$, so that

$$\tilde{v} \times \tilde{v}^{N(\mathfrak{p})} \in \tilde{R}_\sigma^{\mu\lambda}(\alpha) \circ \tilde{X}_{\lambda\mu}(\alpha') .$$

As observed in 11.15, there are infinitely many points $\tilde{v}$ with this property.

By 11.2, this shows $\Pi_\mu^1 \subset \tilde{R}_\sigma^{\mu\lambda}(\alpha) \circ \tilde{X}_{\lambda\mu}(\alpha')$, hence ${}^t\tilde{R}_\sigma^{\mu\lambda}(\alpha) \circ \Pi_\mu^1 \subset \tilde{X}_{\lambda\mu}(\alpha')$. Similarly, taking λ in place of μ, we find points w on V_λ such that

$$w \times w^\tau \in Z_\sigma^\lambda(e) = S_\sigma^{\lambda\mu}(\alpha') \circ X_{\mu\lambda}(\alpha) \qquad \text{(see (11.9.8))}.$$

Considering modulo $\mathfrak{O}$, and applying 11.2, we find $\Pi_\lambda^1 \subset \tilde{S}_\sigma^{\lambda\mu}(\alpha') \circ \tilde{X}_{\mu\lambda}(\alpha)$, hence ${}^t\Pi_\lambda^1 \circ \tilde{S}_\sigma^{\lambda\mu}(\alpha') \subset {}^t\tilde{X}_{\mu\lambda}(\alpha)$. Put $X = \tilde{X}_{\lambda\mu}(\alpha')$, $U = {}^t\tilde{R}_\sigma^{\mu\lambda}(\alpha) \circ \Pi_\mu^1$, $U' = {}^t\Pi_\lambda^1 \circ \tilde{S}_\sigma^{\lambda\mu}(\alpha')$. Using the symbols d, d' of Weil [17, p. 31], we have

$$d(U) = d'(U') = 1 , \qquad d'(U) = d(U') = N(\mathfrak{p}) ,$$

hence $U \neq U'$. It follows that $X - (U + U')$ is positive. But $d(X) = d'(X) = 1 + N(\mathfrak{p})$. (This follows easily from [8, Prop. 1.5] and (10.5.4). Note also that d and d' do not change by reduction modulo $\mathfrak{P}$.) Therefore we obtain $X = U + U'$,   q.e.d.

11.18. The elements $\beta_{n\lambda}$ being as in 11.12, define algebraic correspondences U_μ^n and $U_\mu'^n$ on $\tilde{V}_\mu \times \tilde{V}_\lambda$ by

$$U_\mu^n = {}^t\tilde{R}_\tau^{\mu\lambda}(\beta_{n\lambda}) \circ \Pi_\mu^n , \qquad U_\mu'^n = {}^t\Pi_\lambda^n \circ \tilde{S}_\tau^{\lambda\mu}(\beta_{n\lambda}') ,$$

where $(\lambda, \mu) \in \Theta_{\mathfrak{p}}^n$ and $\tau = [\mathfrak{p}^n]$. Then

(11.18.1) *If $(\lambda, \mu) \in \Theta_{\mathfrak{p}}^n$ and $(\mu, \nu) \in \Theta_{\mathfrak{p}}^m$, then $U_\mu^n \circ U_\nu^m = U_\nu^{n+m}$, $U_\mu'^m \circ U_\nu'^n = U_\nu'^{m+n}$.*

This can be easily obtained from (3.5.3), (11.9.7) and the definition of $\beta_{n\lambda}$.

11.19. PROPOSITION. *If (λ, μ), $(\mu, \nu) \in \Theta_{\mathfrak{p}}^1$, then*

$$U_\mu^1 + U_\mu'^1 = \tilde{X}_{\lambda\mu}(1, \mathfrak{p}) ,$$
$$U_\mu^1 \circ U_\nu'^1 = U_\mu'^1 \circ U_\nu^1 = N(\mathfrak{p}) \cdot \tilde{X}_{\lambda\nu}(\mathfrak{p}, \mathfrak{p}) .$$

PROOF. The first equality is only a restatement of 11.17 with $\alpha = \beta_\lambda$. As to the second, by 11.10, we have $S_\sigma^{\mu\nu}(\beta_\mu') = {}^tR_\sigma^{\nu\mu}(\beta_\mu)^\sigma \circ Z_{\sigma^2}^\nu(e_1)$, where $e_1 = \{\mathfrak{p}, \mathfrak{p}\}$, $\sigma = [\mathfrak{p}]$. Hence

$$U_\mu^1 \circ U_\nu^{\mu 1} = {}^t\tilde{R}_\sigma^{\mu\lambda}(\beta_\lambda) \circ \Pi_\mu^1 \circ {}^t\Pi_\mu^1 \circ \tilde{S}_\sigma^{\mu\nu}(\beta_\mu')$$
$$= N(\mathfrak{p}) \cdot {}^t\mathfrak{P}\big(R_\sigma^{\nu\mu}(\beta_\mu)^\sigma \circ R_\sigma^{\mu\lambda}(\beta_\lambda)\big) \circ \tilde{Z}_{\sigma^2}^\nu(e_1)$$
$$= N(\mathfrak{p}) \cdot {}^t\tilde{R}_{\sigma^2}^{\nu\lambda}(\beta_\mu\beta_\lambda) \circ \tilde{Z}_{\sigma^2}^\nu(e_1) .$$

Since $\beta_{2\lambda} = \beta_\mu\beta_\lambda$, we obtain $U_\mu^1 \circ U_\nu'^1 = N(\mathfrak{p}) \cdot \tilde{X}_{\lambda\nu}(e_1)$ in view of (11.13.2). Similarly we have

$$U_\mu'^1 \circ U_\nu^1 = {}^t\Pi_\lambda^1 \circ \tilde{S}_\sigma^{\lambda\mu}(\beta_\lambda') \circ {}^t\tilde{R}_\sigma^{\nu\mu}(\beta_\mu) \circ \Pi_\nu^1$$
$$= {}^t\Pi_\lambda^1 \circ \mathfrak{P}\big({}^tR_\sigma^{\mu\lambda}(\beta_\lambda)^\sigma \circ Z_{\sigma^2}^\mu(e_1) \circ {}^tR_\sigma^{\nu\mu}(\beta_\mu)\big) \circ \Pi_\nu^1 \qquad (11.10)$$
$$= {}^t\Pi_\lambda^1 \circ \mathfrak{P}\big({}^tR_\sigma^{\mu\lambda}(\beta_\lambda)^\sigma \circ {}^tR_\sigma^{\nu\mu}(\beta_\mu)^{\sigma^2} \circ Z_{\sigma^2}^\nu(e_1)^\sigma\big) \circ \Pi_\nu^1 \qquad (11.7)$$
$$= {}^t\Pi_\lambda^1 \circ \mathfrak{P}\big([{}^tR_{\sigma^2}^{\nu\lambda}(\beta_\mu\beta_\lambda) \circ Z_{\sigma^2}^\nu(e_1)]^\sigma\big) \circ \Pi_\nu^1 \qquad (3.5.3)$$
$$= {}^t\Pi_\lambda^1 \circ \Pi_\lambda^1 \circ \tilde{X}_{\lambda\nu}(e_1) \qquad (11.13.2)$$
$$= N(\mathfrak{p}) \cdot \tilde{X}_{\lambda\nu}(\mathfrak{p}, \mathfrak{p}).$$

We employed also the fact that the action of σ modulo $\mathfrak{P}$ amounts to the $N(\mathfrak{p})^{\text{th}}$ power.

11.20. PROPOSITION. *Let c_{ab} be as in* (10.13.1). *Then*

$$U_\mu^n + U_\mu'^n = \sum_{a+b=n} c_{ab} \widetilde{X}_{\lambda\mu}(\mathfrak{p}^a, \mathfrak{p}^b)\,,$$

where $(\lambda, \mu) \in \Theta_\mathfrak{p}^n$.

This follows easily from 11.19, (11.18.1) and (11.13.3).

12. The zeta function and L-functions of V_λ

12.1. Let V be a complete non-singular curve defined over a field of an arbitrary characteristic. For a 1-cycle X on $V \times V$, we shall denote by $I(X)$ the intersection number of X with the diagonal on $V \times V$.

Let $\{V_\lambda, \varphi_\lambda, R_\sigma^{\mu\lambda}(\alpha)\}$ be as in 3.5. We see easily that $\gamma \mapsto R_1^{\lambda\lambda}(\gamma)$ gives a homomorphism of Γ_λ^1 onto a group of biregular automorphisms of V_λ. Let $U(\mathfrak{e})$ and $U_0(\mathfrak{e})$ be as in 3.4, and $U_1(\mathfrak{e})$ as in 10.7. Let E denote the group of units in F. Since $R_1^{\lambda\lambda}(\gamma)[\varphi_\lambda(z)] = \varphi_\lambda(\gamma(z))$ for every $z \in \mathfrak{H}$, we see that $R_1^{\lambda\lambda}(\gamma)$ is the identity mapping if and only if $\gamma \in \Gamma_\lambda^\mathfrak{e} E$. By 10.7, $\Gamma_\lambda^1/(\Gamma_\lambda^\mathfrak{e} E)$ is canonically isomorphic to $U_1(\mathfrak{e})/(U_0(\mathfrak{e})E)$. Let ψ be a complex-valued character of $U_1(\mathfrak{e})/(U_0(\mathfrak{e})E)$, and χ the character of $U(\mathfrak{e})/(U_0(\mathfrak{e})E)$ induced from ψ. We regard χ resp. ψ also as a character of $U(\mathfrak{e})$ resp. $U_1(\mathfrak{e})$ in a natural way. For brevity, let us put $G = U(\mathfrak{e})/(U_0(\mathfrak{e})E)$, $G_1 = U_1(\mathfrak{e})/(U_0(\mathfrak{e})E)$ and $R_\sigma^\lambda(a) = R_\sigma^{\lambda\lambda}(\alpha)$ for an element a of G represented by an element α of $U(\mathfrak{e})$. (Note that σ is determined by the class of $\alpha \bmod^* \mathfrak{e}$.)

Define P as in 11.15. Let $\mathfrak{p} \in P$, and let $\mathfrak{P}$ be a prime ideal in $C(F, \mathfrak{c})$ which divides $\mathfrak{p}$. Denote by $\kappa_\mathfrak{P}$ the residue field of $C(F, \mathfrak{c})$ modulo $\mathfrak{P}$. Now we define an L-function of $\mathfrak{P}(V_\lambda)$, $L(u) = L(u; \mathfrak{P}(V_\lambda), \kappa_\mathfrak{P}, \psi)$, as a formal power-series in u, with the leading coefficient 1, satisfying

$$(d/du) \log L(u) = N^{-1} \sum_{n=1}^\infty u^{n-1} \sum_{a \in G_1} \psi(a) \cdot I[{}^t\mathfrak{P}(R_1^\lambda(a)) \circ \Pi_{\mathfrak{P}\lambda}^n]\,.$$

Here $N = [G_1 : 1]$, and $\Pi_{\mathfrak{P}\lambda}^n$ is the Frobenius correspondence of degree $N(\mathfrak{P})^n$ on $\mathfrak{P}(V_\lambda) \times \mathfrak{P}(V_\lambda)$.

12.2. MAIN THEOREM V. *The notation being as above, let ρ be a representation of $U(\mathfrak{e})/(U_0(\mathfrak{e})E)$ whose character is χ. Suppose that $\chi(\alpha) = \chi(\alpha')$ for every α in $U(\mathfrak{e})$. Then, for every prime ideal $\mathfrak{p}$ in P, and for each λ, one has*

$$\prod_{\mathfrak{P}|\mathfrak{p}} L(u^f; \mathfrak{P}(V_\lambda), \kappa_\mathfrak{P}, \psi)$$
$$= \prod_{\mathfrak{P}|\mathfrak{p}} [(1 - u^f)(1 - N(\mathfrak{P})u^f)]^{-r} \cdot \det [1 - T(\mathfrak{p})_{2,\rho} u + T(\mathfrak{p}, \mathfrak{p})_{2,\rho} N(\mathfrak{p}) u^2]\,.$$

Here r is the multiplicity of the identity character in ψ; f is the integer *determined by* $N(\mathfrak{P}) = N(\mathfrak{p})^f$; $T(\mathfrak{p})_{2,\rho}$ *and* $T(\mathfrak{p}, \mathfrak{p})_{2,\rho}$ *are the operators defined in* 10.10–11 *(with* $k = 2$*).*

It should be observed that the right hand side of our equality is independent of λ, and P contains all except a finite number of prime ideals in F.

PROOF. First we see that $\alpha\alpha' \in U_0(\mathfrak{e})E$ for every $\alpha \in U_1(\mathfrak{e})$, hence $\psi(\alpha') = \psi(\alpha^{-1})$. It follows that $\chi(\alpha') = \chi(\alpha^{-1})$ for every $\alpha \in U(\mathfrak{e})$. Therefore the assumption $\chi(\alpha) = \chi(\alpha')$ is equivalent to

(12.2.1) $\chi(\alpha)$ *is a real number for every* $\alpha \in U(\mathfrak{e})$.

Now, for brevity, put $L(x, \mathfrak{P}(V_\lambda)) = L(x; \mathfrak{P}(V_\lambda), \kappa_{\mathfrak{P}}, \psi)$. We are going to prove

$$(12.2.2) \quad \begin{aligned} (d/du) &\log \left[\textstyle\prod_{\mathfrak{P}|\mathfrak{p}} L\big(u^f, \mathfrak{P}(V_\lambda)\big) \right] \\ &= N^{-1} \textstyle\sum_{n=1}^\infty u^{n-1} \sum_{\mu=1}^{h_0} \sum_a \chi(a) \cdot I\big[{}^t\mathfrak{P}_0\big(R_\sigma^\mu(a)\big) \circ \Pi_\mu^n\big] , \end{aligned}$$

where $\sum_a$ is the sum over all the elements a of G such that $\sigma = [C(F, \mathfrak{c})/F, \mathfrak{p}^n]$, and $\mathfrak{P}_0$ is a fixed one among the $\mathfrak{P}$. To prove this, let W be a set of representatives for G/G_1. Then $\chi(a) = 0$ or $= \sum_{b \in W} \psi(b^{-1}ab)$ according as $a \notin G_1$ or $a \in G_1$. Therefore we have

$$(12.2.3) \quad \begin{aligned} N^{-1} &\textstyle\sum_{n=1}^\infty u^{n-1} \sum_a \chi(a) \cdot I\big[{}^t\mathfrak{P}_0\big(R_\sigma^\mu(a)\big) \circ \Pi_\mu^n\big] \\ &= N^{-1} \textstyle\sum_{m=1}^\infty u^{mf-1} \sum_{b \in W} \sum_{a \in G_1} \psi(b^{-1}ab) \cdot I\big[{}^t\mathfrak{P}_0\big(R_1^\mu(a)\big) \circ \Pi_{\mathfrak{P}_0\mu}^m\big] . \end{aligned}$$

If β is an element of $U(\mathfrak{e})$ representing b, and $\tau = [C(F, \mathfrak{c})/F, (N_{B/F}(\beta))]$, then one can easily verify

$$I\big[{}^t\mathfrak{P}_0\big(R_1^\mu(a)\big) \circ \Pi_{\mathfrak{P}_0\mu}^m\big] = I\big[{}^t\mathfrak{Q}\big(R_1^\mu(b^{-1}ab)\big) \circ \Pi_{\mathfrak{Q}\mu}^m\big] \text{ with } \mathfrak{Q} = \mathfrak{P}_0^\tau ,$$

by the same reasoning as in [5, (6.8.5)]. Observe that G/G_1 is isomorphic to the Galois group of $C(F, \mathfrak{c})$ over $C(F, 1)$. Hence (12.2.3) is equal to

$$\begin{aligned} N^{-1}u^{f-1} &\textstyle\sum_{m=1}^\infty u^{f(m-1)} \sum'_{\mathfrak{Q}} \sum_{a \in G_1} \psi(a) \cdot I\big[{}^t\mathfrak{Q}\big(R_1^\mu(a)\big) \circ \Pi_{\mathfrak{Q}\mu}^m\big] \\ &= \textstyle\sum'_{\mathfrak{Q}} f^{-1}(d/du) \log L\big(u^f, \mathfrak{Q}(V_\mu)\big) , \end{aligned}$$

where $\sum'_{\mathfrak{Q}}$ is the sum extended over $\mathfrak{Q} = \mathfrak{P}_0^\tau$ with all automorphisms τ of $C(F, \mathfrak{c})$ over $C(F, 1)$. Now observe that $R_\sigma^{\mu\lambda}(1)$ is a biregular morphism of V_λ to V_μ^σ, where $\sigma = [C(F, \mathfrak{c})/F, N_{B/F}(\mathfrak{x}_{\lambda\mu})]$. Moreover,

$$R_\sigma^{\mu\lambda}(1) \circ R_1^{\lambda\lambda}(\alpha) = R_\sigma^{\mu\lambda}(\alpha) = R_1^{\mu\mu}(\alpha)^\sigma \circ R_\sigma^{\mu\lambda}(1)$$

for every $\alpha \in U_1(\mathfrak{e})$. From these facts, one can easily derive

$$L\big(x, \mathfrak{Q}(V_\mu)\big) = L\big(x, \mathfrak{Q}^\sigma(V_\lambda)\big)$$

for every prime ideal $\mathfrak{Q}$ dividing $\mathfrak{p}$. Let Z_μ denote the sum of (12.2.3). Then we have

$$\sum_{\mu=1}^{h_0} Z_\mu = \sum_{\mu=1}^{h_0} f^{-1} \sum_{\mathfrak{Q}}' (d/du) \log L(u^s, \mathfrak{Q}(V_\mu))$$
$$= \sum_{\mathfrak{P}\mid\mathfrak{p}} (d/du) \log L(u^s, \mathfrak{P}(V_\lambda)) ,$$

which proves (12.2.2).

Let $\{\gamma_{\mu 1}, \cdots, \gamma_{\mu N}\}$ be a set of representatives for $\Gamma_\mu^1/(\Gamma_\mu^c E)$. Denote by tilde the action of reduction modulo $\mathfrak{P}_0$. Then the right hand side of (12.2.2) is equal to each of the following sums:

$$(12.2.4) \qquad N^{-1} \sum_{\mu=1}^{h_0} \sum_n' u^{n-1} \sum_{j=1}^N \chi(\beta_{n\mu}\gamma_{\mu j}) \cdot I[{}^t\widetilde{R}_1^{\mu\mu}(\beta_{n\mu}\gamma_{\mu j}) \circ \Pi_\mu^n] ,$$

$$(12.2.5) \qquad N^{-1} \sum_{\mu=1}^{h_0} \sum_n' u^{n-1} \sum_{j=1}^N \chi(\gamma_{\mu j}'\beta_{n\mu}') \cdot I[{}^t\Pi_\mu^n \circ \widetilde{S}_1^{\mu\mu}(\gamma_{\mu j}'\beta_{n\mu}')] .$$

Here $\sum_n'$ is the sum extended over all positive n divisible by f. Also note that $\beta_{n\mu} \in U_1(\mathfrak{e})$ if and only if n is divisible by f; if $\beta_{n\mu} \in U_1(\mathfrak{e})$, then $(N_{B/F}(\beta_{n\mu})) = \mathfrak{p}^n$. To get (12.2.5), we have employed the fact $I({}^tX) = I(X)$ and (11.11.1). Since $\chi(\alpha) = \chi(\alpha')$, we have

$$(d/du) \log [\textstyle\prod_{\mathfrak{P}\mid\mathfrak{p}} L(u^s, \mathfrak{P}(V_\lambda))]$$
$$= (1/2) \cdot [(12.2.4) + (12.2.5)]$$
$$= (2N)^{-1} \sum_{\mu=1}^{h_0} \sum_n' u^{n-1} \sum_{j=1}^N \chi(\beta_{n\mu}\gamma_{\mu j}) \cdot I[{}^t\widetilde{R}_1^{\mu\mu}(\gamma_{\mu j}) \circ (U_\mu^n + U_\mu'^n)] .$$

By (11.20), this is equal to

$$(2N)^{-1} \sum_{\mu=1}^{h_0} \sum_n' u^{n-1} \sum_{j=1}^N \chi(\beta_{n\mu}\gamma_{\mu j}) \cdot I[\widetilde{X}_{\mu\mu}(\gamma_{\mu j}^{-1}) \circ \sum_{a+b=n} c_{ab}\widetilde{X}_{\mu\mu}(\mathfrak{p}^a, \mathfrak{p}^b)] .$$

In the last sum, we can remove tildes, since the intersection number does not change by reduction modulo $\mathfrak{P}_0$ [5, 6.3]. By the Lefschetz fixed point formula, we have

$$2^{-1} I(X(\Gamma_\mu^c \alpha \Gamma_\mu^c)) = d(\Gamma_\mu^c \alpha \Gamma_\mu^c) - \mathrm{Re}[\mathrm{tr}(\Gamma_\mu^c \alpha' \Gamma_\mu^c)_2] ,$$

where $d(\Gamma_\mu^c \alpha \Gamma_\mu^c)$ is the number of right cosets in $\Gamma_\mu^c \alpha \Gamma_\mu^c$. For each μ and integers a, b such that $a + b = n$, we can find, by (11.12.1), an element ε_{ab}^λ of $U(\mathfrak{e})$ such that $e(\varepsilon_{ab}^\lambda \zeta_{\lambda\mu}) = \{\mathfrak{p}^a, \mathfrak{p}^b\}$ and $\varepsilon_{ab}^\lambda \equiv \beta_{n\lambda} \bmod^* \mathfrak{e}$. Then $X_{\lambda\mu}(\mathfrak{p}^a, \mathfrak{p}^b) = X(\Gamma_\lambda^c \varepsilon_{ab}^\lambda{}' \Gamma_\mu^c)$, and

$$\sum_{a+b=n} c_{ab} d(\Gamma_\mu^c \varepsilon_{ab}^\mu \Gamma_\mu^c) = 1 + N(\mathfrak{p})^n \qquad\qquad \text{if } n \equiv 0 \bmod (f) .$$

Therefore we have

$$(12.2.6) \qquad (d/du) \log [\textstyle\prod_{\mathfrak{P}\mid\mathfrak{p}} L(u^s, \mathfrak{P}(V_\lambda))] = A_1 - A_2 ,$$

where

$$A_1 = N^{-1} \sum_{\mu=1}^{h_0} \sum_n' u^{n-1} \sum_{j=1}^N \chi(\beta_{n\mu}\gamma_{\mu j})(1 + N(\mathfrak{p})^n) ,$$
$$A_2 = N^{-1} \sum_{\mu=1}^{h_0} \sum_n' u^{n-1} \sum_{j=1}^N \chi(\varepsilon_{ab}^\mu \gamma_{\mu j})$$
$$\times \mathrm{Re}\{\mathrm{tr}[(\Gamma_\mu^c \gamma_{\mu j} \Gamma_\mu^c)_2 \cdot \sum_{a+b=n} c_{ab}(\Gamma_\mu^c \varepsilon_{ab}^\mu \Gamma_\mu^c)_2]\} .$$

By (12.2.1) and 10.12, we have

$$A_2 = \sum_{\mu=1}^{h_0} \sum_n' u^{n-1} \sum_{a+b=n} c_{ab} \, \mathrm{tr}[(\Gamma_\mu^1 \varepsilon_{ab}^\mu \Gamma_\mu^1)_{2,\rho}]$$
$$= \sum_n' u^{n-1} \sum_{a+b=n} c_{ab} \, \mathrm{tr}[\,T(\mathfrak{p}^a, \mathfrak{p}^b)_{2,\rho}]\,.$$

Here observe that, if $a + b \not\equiv 0 \bmod (f)$, $T(\mathfrak{p}^a, \mathfrak{p}^b)_{k,\rho}$ maps $S_{k,\rho}(\Gamma_\mu^1)$ to $S_{k,\rho}(\Gamma_\lambda^1)$ with $\lambda \neq \mu$, so that $\mathrm{tr}[\,T(\mathfrak{p}^a, \mathfrak{p}^b)_{k,\rho}] = 0$. Therefore

$$(12.2.7) \qquad \begin{aligned} A_2 &= \sum_{n=1}^\infty u^{n-1} \, \mathrm{tr}[F_n(T(1, \mathfrak{p})_{2,\rho},\, N(\mathfrak{p}) T(\mathfrak{p}, \mathfrak{p})_{2,\rho})] \\ &= (d/du) \log \{\det[1 - T(1, \mathfrak{p})_{2,\rho} u + N(\mathfrak{p}) T(\mathfrak{p}, \mathfrak{p})_{2,\rho} u^2]^{-1}\}\,, \end{aligned}$$

in view of the consideration of 10.13. To compute A_1, note that $N^{-1} \sum_{j=1}^N \chi(\xi\gamma_j)$ equals $r\cdot[U(\mathfrak{e}) : U_1(\mathfrak{e})]$ or 0 according as $\xi \in U_1(\mathfrak{e})$ or $\xi \notin U_1(\mathfrak{e})$. Since

$$h_0 \cdot [U(\mathfrak{e}) : U_1(\mathfrak{e})] = [C(F, \mathfrak{c}) : F]\,,$$

we have

$$(12.2.8) \qquad \begin{aligned} A_1 &= rf^{-1}[C(F, \mathfrak{c}) : F](du'/du) \sum_{m=1}^\infty u^{f(m-1)}(1 + N(\mathfrak{P}_0)^m) \\ &= (d/du) \log \big[\textstyle\prod_{\mathfrak{P}|\mathfrak{p}} (1 - u')(1 - N(\mathfrak{P})u')\big]^{-r}\,. \end{aligned}$$

From (12.2.6, 7, 8), we obtain the equality of our theorem.

12.3. *The global L-function of V_λ (in the sense of Hasse-Weil) over* $C(F, \mathfrak{c})$ is an infinite product

$$L\big(s; V_\lambda, C(F, \mathfrak{c}), \psi\big) = \textstyle\prod_{\mathfrak{P}} Z(N(\mathfrak{P})^{-s}; \mathfrak{P}(V_\lambda), \kappa_\mathfrak{p}, \psi)\,,$$

where the product is extended over all prime ideals $\mathfrak{P}$ in $C(F, \mathfrak{c})$ dividing a prime ideal $\mathfrak{p}$ in P. From 12.2 and (10.11.5), we obtain immediately

$$(12.3.1) \qquad \begin{aligned} &L\big(s; V_\lambda, C(F, \mathfrak{c}), \psi\big) \\ &= \textstyle\prod_{\mathfrak{p} \notin P} f_\mathfrak{p}(N(\mathfrak{p})^{-s})\cdot[Z(s; C(F, \mathfrak{c}))Z(s-1; C(F, \mathfrak{c}))]^r \cdot \det[D(s; 2, \rho)^{-1}]\,. \end{aligned}$$

Here $f_\mathfrak{p}(u)$ is a rational function in u, $D(s; 2, \rho)$ is the Dirichlet series defined in (10.11.4) (with $k = 2$), and $Z(s; C(F, \mathfrak{c}))$ is the Dedekind zeta function of $C(F, \mathfrak{c})$. Therefore the global L-function can be continued analytically to the whole s-plane and satisfies a functional equation. In particular, if ψ is the character of a regular representation of $U_1(\mathfrak{e})/U_0(\mathfrak{e})E$, then ρ is a regular representation of $U(\mathfrak{e})/U_0(\mathfrak{e})E$. In this case, $L\big(s; V_\lambda, C(F, \mathfrak{c}), \psi\big)$ is the zeta function of V_λ over $C(F, \mathfrak{c})$.

12.4. *Remarks and examples.* The quotient $\mathfrak{H}/\Gamma_\lambda^\mathfrak{c}$ can be regarded as a moduli-space of abelian varieties, though infinitely many distinct families of abelian varieties can be associated with a single $\mathfrak{H}/\Gamma_\lambda^\mathfrak{c}$. For each choice of such a family, one can certainly develop a theory which generalizes [5]. If $g > 1$, one has to take a class field over K' as a field of definition of the fibre space in question, K' being the field defined in §5 and 6.5. Therefore a little complexity may be expected in such a generalization.

One can also make some considerations similar to [5, 6.12–15] in the present case. But here we do not go into detail, since it is an almost straightforward translation. As for previously known results on the zeta function of a curve over a number field, we refer the reader to [5, 7.11] and [14].

Now let η be a Grössen-character of F defined modulo $\mathfrak{c}$, considered as a character of the ideal group $I(F, \mathfrak{c})$ (and not as a character of the adele group). We can define a zeta-function $D(s; k, \rho, \eta)$ by

$$D(s; k, \rho, \eta) = \sum_{\mathfrak{a}} \eta(\mathfrak{a}) \cdot T(\mathfrak{a})_{k,\rho} N(\mathfrak{a})^{-s},$$

where $\mathfrak{a}$ runs over all integral ideals in F prime to $\mathfrak{c}$. By [8, Th. 1], $D(s; k, \rho, \eta)$ can be continued to the whole s-plane and satisfies a functional equation. In view of a recent result of Weil [19], it will be worth while giving an explicit form of the functional equation. For the sake of simplicity, we assume

(12.4.1) $h_0 = 1$; $\mathfrak{c}$ *is prime to* $D(B/F)\mathfrak{d}(F/\mathbf{Q})$; $\mathfrak{c}$ *is the conductor of* η; ρ *is the identity representation of* $U(\mathfrak{e})/U_0(\mathfrak{e})E$.

The $\tau_{0\nu}$ being as in 2.4, put $x_\nu = x^{\tau_{0\nu}}$ for $x \in F$ and $\nu = 1, \cdots, g$. By the property of Grössen-character, we find integers b_ν and real numbers a_ν such that

$$\eta((x))^{-1} = \prod_{\nu=1}^{g} |x_\nu|^{ia_\nu} \cdot (x_\nu/|x_\nu|)^{b_\nu} \qquad \text{if } x \in F, \ x \equiv 1 \bmod^* \mathfrak{c}, \ (i = \sqrt{-1}).$$

Put

$$\xi_\infty(x) = \prod_{\nu=1}^{g} |x_\nu|^{ia_\nu} \cdot (x_\nu/|x_\nu|)^{b_\nu}$$

for *all* non-zero $x \in F$, and $\xi_\mathfrak{c}(x) = \eta((x))^{-1}\xi_\infty(x)^{-1}$ for every x in F, prime to $\mathfrak{c}$. Take an element c of F so that $\mathfrak{c} = (c)$ and define a gaussian sum G_η by

$$G_\eta = \xi_\infty(c) \cdot \sum_{w} \exp\left[2\pi i \operatorname{Tr}_{F/\mathbf{Q}}(c^{-1}w)\right] \cdot \xi_\mathfrak{c}(w),$$

where w runs over all residue-classes of $\mathfrak{r}_F$ modulo $\mathfrak{c}$. Now define $\Gamma(s; k, \eta)$ and $\Lambda(s; k, \eta)$ by

$$\Gamma(s; k, \eta) = A^s (2\pi)^{-gs - ia} \Gamma(s + (k/2) - 1 + ia_1) \prod_{\nu=2}^{g} \Gamma(s + ia_\nu),$$

$$A = N_{F/\mathbf{Q}}(\mathfrak{c}) D(F/\mathbf{Q}) N_{F/\mathbf{Q}}(D(B/F))^{1/2},$$

$$a = \sum_{\nu=1}^{g} a_\nu,$$

$$\Lambda(s; k, \eta) = \Gamma(s; k, \eta) D(s; k, \rho_0, \eta),$$

where ρ_0 is the identity representation. Then we have

(12.4.2)
$$\Lambda(s; k, \eta) = C \cdot \Lambda(2 - s; k, \bar{\eta}),$$
$$C = i^k N_{F/\mathbf{Q}}(\mathfrak{c}) G_\eta^{-2} \eta(D(B/F)\mathfrak{d}(F/\mathbf{Q})^2) \cdot T_0,$$

where T_0 is a linear operator on $S_{k,\rho}$ such that $T_0^2 = 1$. These results can be obtained as a special case of [8, p. 273, Th. 1, p. 276, Case (II)]. T_0 is the operator $\mathfrak{T}(\mathfrak{d}_\nu)$ of [8, p. 273, (40)].

As a numerical example, let us take $F = Q(5^{1/2})$, and $D(B/F)$ to be a prime ideal q in F. By 2.2, B is uniquely determined by the choice of q. Let o be a maximal order in B and $\Gamma = \Gamma(o, 1)$. Then the canonical model V for $\mathfrak{H}/\Gamma$ can be defined over F. The following table shows the genus of V for the first few values of $N(q)$.

$N(q)$	2^2	3^2	7^2	13^2	17^2	5	11	19	29	31	41	59	61	71	79	89	101
genus of V	0	0	1	3	5	0	0	0	0	1	1	0	2	1	1	1	2

If the genus of V is one, then we can easily verify $T_0 = -1$, so that $C = N(\mathfrak{c})G_\eta^{-2}\eta(5\mathfrak{q})$ in the case $k = 2$.

Finally let us consider the curve of genus 3 with 168 automorphisms. As observed in 3.19, this curve is obtained as a canonical model V of $\mathfrak{H}/\Gamma(o, \mathfrak{p})$ for a certain quaternion algebra over a cubic field $F_7 = Q(\zeta + \zeta^{-1})$, where $\zeta = e^{2\pi i/7}$ and $N(\mathfrak{p}) = 7$. On the other hand, a canonical model V' of $\mathfrak{H}/\Gamma$ with $\Gamma = \{\alpha \in SL_2(Z) \mid \alpha \equiv 1(7)\}$ gives rise to a curve biregularly equivalent to V over C. It can be easily verified that $C(F_7, \mathfrak{p}) = Q(\zeta) = C(Q, 7)$, hence both V and V' have $Q(\zeta)$ as their natural field of definition. It is an open question whether V and V' are biregularly equivalent over $Q(\zeta)$ or not. However, we can at least make the following observation which seems rather remarkable. Take an integral ideal $\mathfrak{c}$ in F_7 divisible by $\mathfrak{p}$ and take ψ to be the character of $\Gamma(o, 1)/\Gamma(o, \mathfrak{c})E$ induced from the identity character of $\Gamma(o, \mathfrak{p})E/\Gamma(o, \mathfrak{c})E$. Then the L-function with this $\mathfrak{c}$ and ψ is exactly the global zeta function of V over $C(F_7, \mathfrak{c})$ (not over $C(F_7, \mathfrak{p})$). In this way one can verify the Hasse-Weil conjecture for V over $C(F_7, \mathfrak{c})$ for every multiple $\mathfrak{c}$ of $\mathfrak{p}$. This result can not be covered by the result obtained from the congruence subgroups of $SL_2(Z)$, since the latter gives only the zeta functions of V' over the abelian extensions of Q. There are many more examples of the same nature.

PRINCETON UNIVERSITY

REFERENCES

1. M. EICHLER, *Allgemeine Kongruenzklasseneinteilungen der Ideale einfacher Algebren über algebraischen Zahlkörpern und ihre L-Reihen*, J. Reine Angew. Math., 179 (1938), 227-251.
2. ———, *Zur Zahlentheorie der Quaternionen-Algebren*, J. Reine Angew. Math., 195 (1956), 127-151.
3. R. FRICKE und F. KLEIN, Vorlesungen über die Theorie der automorphen Funktionen, I, Leipzig, Teubner, 1897.
4. Y. IHARA, *Algebraic curves* mod p *and algebraic groups*, to appear in Proceedings of Symposia in Pure Mathematics series.
5. M. KUGA and G. SHIMURA, *On the zeta function of a fibre variety whose fibres are abelian varieties*, Ann. of Math., 82 (1965), 478-539.
6. H. SHIMIZU, *On zeta functions of quaternion algebras*, Ann. of Math., 81 (1965), 166-193.

7. G. SHIMURA, *On the class-fields obtained by complex multiplication of abelian varieties*, Osaka Math. J., 14 (1962), 33–44.

8. ———, *On Dirichlet series and abelian varieties attached to automorphic forms*, Ann. of Math., 76 (1962), 237–294.

9. ———, *On analytic families of polarized abelian varieties and automorphic functions*, Ann. of Math., 78 (1963), 149–192.

10. ———, *Arithmetic of unitary groups*, Ann. of Math., 79 (1964), 369–409.

11. ———, *On the field of definition for a field of automorphic functions* I, II, III, Ann. of Math., 80 (1964), 160–189, 81 (1965), 124–165, 83 (1966) 377–385.

12. ———, *Class-fields and automorphic functions*, Ann. of Math., 80 (1964), 444–463.

13. ———, *Moduli and fibre systems of abelian varieties*, Ann. of Math., 83 (1966), 294–338.

14. ———, *On the zeta-functions of the algebraic curves uniformized by certain automorphic functions*, J. Math. Soc. Japan, 13 (1961), 275–331.

15. ——— and Y. TANIYAMA, Complex multiplication of abelian varieties and its applications to number theory, Publ. Math. Soc. Japan, No. 6, 1961.

16. A. WEIL, Foundations of Algebraic Geometry, 2nd edition, Providence, 1962.

17. ———, Sur les courbes algébriques et les variétés qui s'en déduisent, Hermann, Paris, 1948.

18. ———, *The field of definition of a variety*, Amer. J. Math., 78 (1956), 509–524.

19 ———, *Über die Bestimmung Dirichletscher Reihen durch Funktionalgleichungen*, to appear in Math. Ann.

(Received June 6, 1966)

67c

Number fields and zeta functions associated with discontinuous groups and algebraic varieties

Proceedings of the International Congress of Mathematicians,
Moscow, 1966, 100-107 (1967)*

1. Introduction. Let Γ be an arithmetically defined discontinuous group operating on a bounded symmetric domain H, and φ a holomorphic mapping of H into a projective space which induces a biregular morphism of H/Γ onto a Zariski open subset V of a projective variety. Then one can set the following problems.

(A) *Find a model V defined over an algebraic number field.*

(B) *With a suitably chosen V and φ, characterize number-theoretically the field generated by the coordinates of $\varphi(z_0)$ for an isolated fixed point z_0 of an analytic automorphism α of H such that $\alpha\Gamma\alpha^{-1}$ is commensurable with Γ.*

(C) *Find a connection between the zeta function of V and the Hecke operators defined for the automorphic forms with respect to Γ.*

A typical example is the case where Γ is a principal congruence subgroup of the modular group $SL_2(\mathbf{Z})$ and H is the upper half plane. Then one finds a curve V so as to be defined over the rational number field $\mathbf{Q}$. For such a Γ, the classical theory of complex multiplication solves Problem (B); the last problem (C) has been investigated by Eichler [1] and myself [8], [9].

The main purpose of this lecture is to answer the above questions for all possible arithmetic Γ (at least up to commensurability) acting on the upper half plane. This provides a complete analogue to the results obtained for $SL_2(\mathbf{Z})$. It turns out especially that the maximal abelian extension of a totally imaginary quadratic extension of a totally real algebraic number field F can be generated, over the maximal abelian extension of F, by the special values of automorphic functions of one variable. Although the principal part of this lecture is thus concerned with the one-dimensional H, I shall first discuss Problems (A) and (B) in the higher-dimensional case, because it will give a clear perspective to the whole theory.

2. The moduli-variety for a family of abelian varieties. In many cases, we can attach to H and Γ a family $\Sigma = \left\{ Q_z \,\middle|\, z \in H \right\}$, parametrized by the points on H, of structures Q_z, each of which is formed by an abelian variety A_z, a polarization of A_z, endomorphisms of A_z, and points of finite order on A_z. Two members Q_z and Q_w of Σ are isomorphic if and only if $\gamma(z) = w$ for some $\gamma \in \Gamma$. We say that H/Γ is of type (P), if the members of Σ can be characterized only by the type of polarization, endomorphisms, and points of finite order. In such a case, one can approach Problems (A) and (B) as follows. First the algebro-geometric characterization of the members enables us to show the existence of an algebraic number field k with the property that: *for every*

132

$Q_z \in \Sigma$ *and an automorphism* σ *of* **C**, σ *is the identity map on* k *if and only if* $(Q_z)^\sigma$ *is isomorphic to a member of* Σ. Then one can choose V and φ so that the following conditions are satisfied [13]:

(2.1) V *is defined over* k.

(2.2) *If* Q *and* Q' *are structures isomorphic to members* Q_z *and* Q_w *of* Σ *respectively, and if* Q' *is a specialization of* Q *over* k, *then* $(\varphi(w), Q')$ *is a specialization of* $(\varphi(z), Q)$ *over* k.

For a given Σ, such V and φ are unique up to biregular morphisms over k. The field $k(\varphi(z))$, for each point $z \in H$, has an invariant meaning for the structure Q_z, since it can be characterized by the following property:

(2.3) *An automorphism* σ *of* **C** *is the identity mapping on* $k(\varphi(z))$ *if and only if* $(Q_z)^\sigma$ *is isomorphic to* Q_z.

We call $k(\varphi(z))$ *the field of moduli of* Q_z.

Now, if z is an isolated fixed point of an analytic automorphism of H described in (B), then A_z has sufficiently many complex multiplications. Therefore the determinantion of the number field $k(\varphi(z))$ can be essentially done by applying the theory of complex multiplication of abelian varieties [16] to A_z. However, it is not always a simple task but actually an interesting problem to determine the structure of A_z. The study of some special ones among these A_z is very effectively employed to characterize the field k number-theoretically [12].

It should be observed that this recognition of H/Γ as a moduli-variety does not necessarily give the ultimate answer to our problems, for some reasons which will be exlained in the following section. Nevertheless, by this means, one can obtain at least the first approximation to the solution, which is really the best possilbe in a number of cases, as is seen in the classical case of $SL_2(\mathbf{Z})$ and elliptic curves.

3. Reflections on the idea of Section 2. There are some spaces H'/Γ' which are not of type (P), but can be embedded holomorphically into H/Γ of type (P) through injections $H' \to H$ and $\Gamma' \to \Gamma$. I found nontrivial examples for this in the case of quaternion unitary groups defined by Siegel [17], and communicated the idea of generalization to Kuga, who, together with Satake, has investigated such an embedding in a general framework (cf. [4], [6], [7]). In such a case, we can still show that H'/Γ' has a model defined over an algebraic number field. To show this, first we take V and φ for H/Γ so as to satisfy (2.1) and (2.2), and then choose a point z on H' so that A_z has sufficiently many complex multiplications. The points w on H' such that A_w is isogenous to A_z form a dense subset X of H'. Since $k(\varphi(w))$ is the field of moduli of Q_w, we see that the coordinates of the points in $\varphi(X)$ are all algebraic numbers. Therefore, if σ is an automorphism of **C** over the algebraic closure of **Q**, then one has $\varphi(X) = \varphi(X)^\sigma \subset \varphi(H')^\sigma$, so that $\varphi(H') = \varphi(H')^\sigma$. It follows that $\varphi(H')$, a model for H'/Γ', is defined over an algebraic number field. The case of unitary groups of quaternion hermitian or anti-hermitian forms will be discussed in detail in [14]. In this way, we know that H'/Γ' has a model defined over an algebraic number field, if there is a family of abelian varieties attached to H'/Γ' in the sense of Kuga and Satake. A careful analysis of the special members Q_z and the fields $k(\varphi(z))$ may yield a more precise result.

The field k determined for the family Σ of §2 is not necessarily the smallest field of definition for the field of automorphic functions with respect to Γ. In reality, k is a number field which has an essential meaning for the family Σ of abelian varieties, but not always so for H/Γ. For example, to a certain H/Γ, one can attach infinitely many distinct families of abelian varieties; the field k may depend on the choice of a family. These phenomena can happen both in the case of type (P) and in the case not of type (P). Therefore if we ask for "absolute" statements concerning H/Γ without reference to abelian varieties, the answer of §2 to our questions may be considered incomplete in such a case. However, by studying these families more closely, it is possible to obtain an "absolute result" at least for the one-dimensional H, which is our object in the following sections.

4. Construction of class fields by automorhic functions with respect to an arithmetic Fuchsian group. Let B be a quaternion algebra over a totally real algebraic number field F of degree g. If r is the number of archimedean primes of F ramified in B, one can identify $B_{\mathbf{R}} = B \otimes_{\mathbf{Q}} \mathbf{R}$ with the product $M_2(\mathbf{R})^r \times \mathbf{K}^{g-r}$ of r copies of the total matrix ring $M_2(\mathbf{R})$ of degree 2 over the real number field $\mathbf{R}$, and $g - r$ copies of the division ring $\mathbf{K}$ of real quaternions. For every $\alpha \in B_{\mathbf{R}}$, let $\alpha = (\alpha_1, \ldots, \alpha_g)$ with $\alpha_\nu \in M_2(\mathbf{R})$ or $\mathbf{K}$ according as $\nu \leq r$ or $\nu > r$. Denote by B^+ the set of all α in B such that $\det(\alpha_\nu) > 0$ for $\nu \leq r$. If $\alpha \in B^+$, the action of α on the product $\mathfrak{H}^r$ of r copies of the upper half plane $\mathfrak{H} = \{ z \in \mathbf{C} \,|\, \mathrm{Im}(z) > 0 \}$ is defined by

$$\alpha(z_1, \ldots, z_r) = (w_1, \ldots, w_r), \quad w_\nu = (a_\nu z_\nu + b_\nu)/(c_\nu z_\nu + d_\nu), \quad \alpha_\nu = \begin{bmatrix} a_\nu & b_\nu \\ c_\nu & d_\nu \end{bmatrix}.$$

Let $\mathfrak{o}$ be a maximal order in B, and $\mathfrak{c}$ an integral ideal in F. Let $\Gamma(\mathfrak{o}, \mathfrak{c})$ denote the set of all units γ in $\mathfrak{o}$, belonging to B^+, such that $1 - \gamma \in \mathfrak{c}\mathfrak{o}$. Then $\Gamma(\mathfrak{o}, \mathfrak{c})$ can be considered as a discontinuous group acting on $\mathfrak{H}^r$. It is well-known that $\mathfrak{H}^r/\Gamma(\mathfrak{o}, \mathfrak{c})$ is compact if B is a division ring, especially if $r < g$. Hereafter let us assume that $r = 1$, and the archimedean prime of F corresponding to the unique factor $M_2(\mathbf{R})$ of $B_{\mathbf{R}}$ is obtained from the identity mapping of F.

Lemma 1. *Let M be a totally imaginary quadratic extension of F which is isomorphic to a quadratic subfield of B over F. Then the following assertions hold.*

(1) If f is an F-linear isomorphism of M into B, then $f(M) - \{0\} \subset B^+$, and every element in $f(M) - F$ has exactly one fixed point on $\mathfrak{H}$, which is common to all elements of $f(M) - F$.

(2) If $\mathfrak{r}_M$ denotes the ring of algebraic integers in M, then there exists an F-linear isomorphism f of M into B such that $f(\mathfrak{r}_M) \subset \mathfrak{o}$.

For an algebraic number field K of finite degree and an integral ideal $\mathfrak{a}$ in K, we denote by $C(K, \mathfrak{a})$ the abelian extension of K in which a prime ideal $\mathfrak{p}$ of K, prime to $\mathfrak{a}$, decomposes completely if and only if $\mathfrak{p} = (b)$ with an element

b of K which is congruent multiplicatively to 1 modulo $\mathfrak{a}$ and positive at every real archimedean prime of K.

Theorem I. *The notation and assumptions being as above, there exists a complete non-singular algebraic curve V and a holomorphic mapping φ of $\mathfrak{H}$ into V satisfying the following three conditions:*

(I.1) V *is defined over* $C(F, \mathfrak{c})$.

(I.2) φ *gives a biregular isomorphism of* $\mathfrak{H}/\Gamma(\mathfrak{o}, \mathfrak{c})$ *into* V.

(I.3) *Let M and f be as in (2) of Lemma 1, and z the fixed point of the elements of $f(M) - F$ on $\mathfrak{H}$. Then $C(M, \mathfrak{c})$ is the composite of $M(\varphi(z))$ and $C(F, \mathfrak{c})$.*

We call such (V, φ) a *canonical model* for $\mathfrak{H}/\Gamma(\mathfrak{o}, \mathfrak{c})$. If (V, φ) and (V', φ') are two canonical models for $\mathfrak{H}/\Gamma(\mathfrak{o}, \mathfrak{c})$, then one can show the existence of a biregular morphism j of V to V', defined over $C(F, \mathfrak{c})$, such that $j \circ \varphi = \varphi'$. In this sense, (V, φ) is uniquely determined for $\mathfrak{H}/\Gamma(\mathfrak{o}, \mathfrak{c})$.

The reciprocity-law for the extension $C(M, \mathfrak{c})/M$ can be described explicitly in terms of the special points $\varphi(z)$. For simplicity, let us consider the case $\mathfrak{c} = (1)$. Classify all the right $\mathfrak{o}$-ideals with respect to left multiplication by the elements of B^+. Let $\{\mathfrak{r}_1, \ldots, \mathfrak{r}_h\}$ be a set of representatives for the classes of right $\mathfrak{o}$-ideals in this sense. Let $\mathfrak{o}_\lambda$ be the left order of $\mathfrak{r}_\lambda$. Set $\mathfrak{r}_{\lambda\mu} = \mathfrak{r}_\lambda \mathfrak{r}_\mu^{-1}$. For every right $\mathfrak{o}_\mu$-ideal $\mathfrak{r}$, let $N(\mathfrak{r})$ denote the ideal in F generated by the elements $N(\xi)$ for all $\xi \in \mathfrak{r}$, where $N(\xi)$ denotes the reduced norm of ξ to F. We notice that, for each μ, $\{N(\mathfrak{r}_{1\mu}), \ldots, N(\mathfrak{r}_{h\mu})\}$ is a set of representatives for the ideal-classes modulo the product $\mathfrak{u}$ of all archimedean primes of F, and $h = [C(F, 1) : F]$.

Theorem II. *The notation being as above, there exists a system*

$$\left\{ V_\lambda, \ \varphi_\lambda \ (\lambda = 1, \ldots, h) \right\}$$

satisfying the following conditions:

(II.1) *For each λ, $(V_\lambda, \varphi_\lambda)$ is a canonical model for $\mathfrak{H}/\Gamma(\mathfrak{o}_\lambda, 1)$.*

(II.2) $V_\lambda = V_\mu^\sigma$ *if* $\sigma = \left(\dfrac{C(F, 1)/F}{N(\mathfrak{r}_{\lambda\mu})} \right)$.

(II.3) *Let M and f be as in Lemma 1, under the condition $f(\mathfrak{r}_M) \subset \mathfrak{o}_\mu$. Let z be the fixed point on $\mathfrak{H}$ of the elements of $f(M) - F$, and $\mathfrak{b}$ a fractional ideal in M. Suppose that f is normalized in the sense defined below. Then there exist an element α of B^+ and a unique λ such that $\alpha \mathfrak{r}_{\lambda\mu} = f(\mathfrak{b})\mathfrak{o}_\mu$. With such an element α and $\tau = \left(\dfrac{C(M, 1)/M}{\mathfrak{b}} \right)$, we have $\varphi_\mu(z)^\tau = \varphi_\lambda(\alpha^{-1}(z))$.*

Here we say that f is normalized if $(d/dw)\big[f(a)(w)\big]_{w=z} = \bar{a}/a$ for every $a \in M, \neq 0$. If we define $\bar{f}$ by $\bar{f}(a) = f(\bar{a})$ for $a \in M$, then we see that either f or $\bar{f}$ is normalized.

Example. Let $d = 7$, 9, or 11, and let $F = F_d = \mathbf{Q}(\zeta + \zeta^{-1})$ with $\zeta = e^{2\pi i/d}$. Then there exists a quaternion algebra B over F which is unramified at exactly one archimedean prime of F corresponding to the identity mapping of F, and unramified at every finite prime spot of F. Such a B is unique up to isomorphisms over F. In this case we have $h = 1$ and $C(F, 1) = F$. Let $\mathfrak{o}$ be a maximal order in B. For every maximal order $\mathfrak{o}'$ in B, there exists an element β of B^+ such that $\mathfrak{o}' = \beta\mathfrak{o}\beta^{-1}$. One can show that $\Gamma(\mathfrak{o}, 1)$ modulo the center is a triangle group with three classes of elliptic elements of order 2, 3, d, and hence $\mathfrak{H}/\Gamma(\mathfrak{o}, 1)$ is of genus 0. Let w_2, w_3, w_d be corresponding elliptic points on $\mathfrak{H}$, which are unique modulo $\Gamma(\mathfrak{o}, 1)$. Then there is an automorphic function φ on $\mathfrak{H}$ with respect to $\Gamma(\mathfrak{o}, 1)$ which can be characterized by the property that $\varphi(w_2) = 1$, $\varphi(w_3) = 0$, $\varphi(w_d) = \infty$, and φ gives a biregular morphism of $\mathfrak{H}/\Gamma(\mathfrak{o}, 1)$ onto the complex projective line V. Then it can be shown that this (V, φ) is a canonical model for $\mathfrak{H}/\Gamma(\mathfrak{o}, 1)$. In this case, for every totally imaginary quadratic extension M of F, there exists an F-linear isomorphism f of M into B such that $f(\mathfrak{r}_M) \subset \mathfrak{o}$. Let $\{z_1, \ldots, z_q\}$ be a set of representatives for the $\Gamma(\mathfrak{o}, 1)$-equivalence classes of the fixed points of $f(M) - F$ for all such f. Then q is exactly the class number of M; and from the statements (I.3) and (II.3) we can derive the following assertion:

The values $\varphi(z_1), \ldots, \varphi(z_q)$ form a complete set of conjugates of $\varphi(z_1)$ over F, and $M\big(\varphi(z_i)\big)$ is the maximal unramified abelian extension of M for each i.

Thus φ affords a complete analogue of the classical modular function $j(\tau)$, with the field F_d in place of $\mathbf{Q}$.

Remark 1. We can actually prove certain existence-theorems of a canonical model (V, φ) for $\mathfrak{H}^r/\Gamma(\mathfrak{o}, \mathfrak{c})$ with any $r \geq 1$, which include the above theorems as particular cases. The cases where $r = 1$ and $r = g$ will be treated in [15]. We can also release the condition $f(\mathfrak{r}_M) \subset \mathfrak{o}$ in (I.3) by replacing $C(M, \mathfrak{c})$ by a suitably defined class field over M. I shall discuss the most general theorems including all these in a paper subsequent to [15].

Remark 2. It may be worth inserting here a short history of the Fuchsian group of our type. In the case $F = \mathbf{Q}$, Poicaré found that the integral transformations of an indefinite ternary quadratic form generate a Fuchsian group [5]. He realized this when *he was walking on a cliff*, as he described it as an example of mathematical discoveries in his "Science et Méthode". Then Fricke treated the case of F of higher degree. A considerable part of two volumes of Fricke-Klein [2] is devoted to the investigation of this group and its tranformation-equations. The possibility of transformation theory was already noticed in [5]. If $B = M_2(F)$, the group $\Gamma(\mathfrak{o}, 1)$ is essentially the so-called Hilbert modular group. The case $B = M_2(F)$ with $g = 2$ was studied by Hecke in his dissertation [3]. In this work, discussing the analytic functions meaningful for the construction of abelian extensions, Hecke, curiously enough, referred to the results of Fricke mentioned above as being *without specific meaning in number theory* (Werke, p.23). Actually, as I have shown, Fricke's automorphic functions provide a more transparent result for the construction of class fields than the functions of Hilbert-Hecke type, though we can regard both as special cases of a more general theory as noted in Remark 1. The discontinuous groups defined by Siegel [17] may be considered as a generalization of the Poincaré-Fricke group to the higher-dimensional case. These are of course special cases of what recent authors call arithmetically defined discontinuous groups.

5. The zeta functions of B and V. Let V_λ, φ_λ, and $\mathfrak{o}_\lambda$ be as in Theorem II. Set $\Gamma_\lambda = \Gamma(\mathfrak{o}_\lambda, 1)$. We regard B as a subring of $M_2(\mathbf{R})$ through the projection of $B_\mathbf{R}$ to $M_2(\mathbf{R})$. Then $N(\alpha) = \det(\alpha)$ for every $\alpha \in B$. For $\xi = \begin{bmatrix} a & b \\ c & d \end{bmatrix} \in GL_2(\mathbf{R})$ with $\det(\xi) > 0$ and for $z \in \mathfrak{H}$, set $j(\xi, z) = \det(\xi)^{-1/2}(cz + d)$. Let $S_m(\Gamma_\lambda)$, for a positive integer m, denote the vector space of all cusp forms of weight m with respect to Γ_λ. Every element of $S_m(\Gamma_\lambda)$ is a holomorphic function $f(z)$ on $\mathfrak{H}$ such that $f\big(\gamma(z)\big) = f(z)j(\gamma, z)^m$ for all $\gamma \in \Gamma_\lambda$. If $\alpha \in B^+$, we can define a linear map $(\Gamma_\mu \alpha \Gamma_\lambda)_m : S_m(\Gamma_\lambda) \to S_m(\Gamma_\mu)$ as follows: Let $\Gamma_\mu \alpha \Gamma_\lambda = \bigcup_{i=1}^e \alpha_i \Gamma_\lambda$ be a disjoint union. Then, for $f \in S_m(\Gamma_\lambda)$, we put

$$(\Gamma_\mu \alpha \Gamma_\lambda)_m f = \sum_{i=1}^e f\big(\alpha_i^{-1}(z)\big) j(\alpha_i^{-1}, z)^{-m}.$$

For an integral ideal $\mathfrak{a}$ in F, let $T_{\mu\lambda}(\mathfrak{a})_m$ denote the sum of all the $(\Gamma_\mu \alpha \Gamma_\lambda)_m$ such that $\alpha \mathfrak{r}_{\lambda\mu}$ is integral and $N(\alpha \mathfrak{r}_{\lambda\mu}) = \mathfrak{a}$. It should be observed that λ and $\mathfrak{a}$ determine μ uniquely. Further let $R_{\nu\lambda}(\mathfrak{a})_m = (\Gamma_\nu \beta \Gamma_\lambda)_m$ with an element $\beta \in B^+$ such that $\beta \mathfrak{r}_{\lambda\nu} = \mathfrak{a}\mathfrak{o}_\nu$. Then we let $T(\mathfrak{a})_m$ $\big($resp. $R(\mathfrak{a})_m\big)$ denote the linear trahsformation on

$$S_m = S_m(\Gamma_1) \oplus \cdots \oplus S_m(\Gamma_h)$$

whose restriction to $S_m(\Gamma_\lambda)$ coincides with $T_{\mu\lambda}(\mathfrak{a})_m$ (resp. $R_{\nu\lambda}(\mathfrak{a})_m$).

Now, define a Dirichlet series $D_m(s)$ by

$$D_m(s) = \sum_\mathfrak{a} T(\mathfrak{a})_m N(\mathfrak{a})^{-s},$$

where $\mathfrak{a}$ runs over all integral ideals in F. It can be shown that $D_m(s)$ converges for sufficiently large $\mathrm{Re}(s)$ and has an Euler product:

$$D_m(s) = \prod \big[1 - T(\mathfrak{p})_m N(\mathfrak{p})^{-s}\big]^{-1} \prod \big[1 - T(\mathfrak{p})_m N(\mathfrak{p})^{-s} + R(\mathfrak{p})_m N(\mathfrak{p})^{1-2s}\big]^{-1},$$

where the first product is taken over all the prime ideals $\mathfrak{p}$ ramified in B, and the second over all $\mathfrak{p}$ unramified in B. Moreover, $D_m(s)$ is holomorphically continued to the whole s-plane and satisfies a functional equation:

$$R_m(s) = \Gamma_m(s)D_m(s) = ZYR_m(2 - s)Y^{-1},$$
$$\Gamma_m(s) = (2\pi)^{-gs}d^{s/2}\Gamma(s)^{g-1}\Gamma(s - 1 + m/2),$$

where Z and Y are certain invertible linear transformations on S_m, and d is a certain integer; for details, see [11].

Let (V, φ) be as in Theorem I. Let $\mathfrak{p}$ be a prime ideal in F, and $\mathfrak{P}$ a prime ideal in $C(F, 1)$ dividing $\mathfrak{p}$. Let us denote by $\mathfrak{P}(V)$ the reduction of V modulo $\mathfrak{P}$, and assume that $\mathfrak{P}(V)$ is a non-singular curve. The zeta function $Z_\mathfrak{P}(u)$ of the curve $\mathfrak{P}(V)$, over the residue field mod $\mathfrak{P}$, has the form

$$Z_{\mathfrak{P}}(u) = Z^1_{\mathfrak{P}}(u)/\big[(1-u)(1-N(\mathfrak{P})u)\big],$$

where $Z^1_{\mathfrak{P}}(u)$ is a polynomial in u of degree $2t$, if t denotes the genus of V.

Theorem III. *The notation being as above, we have, for almost all* $\mathfrak{p}$ *unramified in* B,

$$\prod_{\mathfrak{P}|\mathfrak{p}} Z^1_{\mathfrak{P}}\big(N(\mathfrak{P})^{-s}\big) = \det\big[1 - T(\mathfrak{p})_2 N(\mathfrak{p})^{-s} + R(\mathfrak{p})_2 N(\mathfrak{p})^{1-2s}\big].$$

It follows that if we define the global zeta function $\zeta(s, V)$ of V, in the sense of Hasse and Weil, by

$$\zeta(s, V) = \prod'_{\mathfrak{P}} Z^1_{\mathfrak{P}}\big(N(\mathfrak{P})^{-s}\big)^{-1},$$

the product being taken over all $\mathfrak{P}$ with good reduction, then $\zeta(s, V)$ differs from $\det\big[D_2(s)\big]$ only by a finite number of $\mathfrak{p}$-factors. Therefore $\zeta(s, V)$ can be continued holomorphically to the whole s-plane and satisfies a functional equation. We can actually obtain a more general result on certain L-functions for the coverings of V defined by the congruence subgroups of Γ_λ [15].

6. Ideas of proofs. One can attach two types of families of abelian varieties to the quotient $\mathfrak{H}^r/\Gamma(\mathfrak{o}, \mathfrak{c})$. Let K be a totally imaginary quadratic extension of F and let $\tau_1, \ldots, \tau_g$ be isomorphic embeddings of K into $\mathbf{C}$ such that $\{\tau_1, \ldots, \tau_g, \tau_1\rho, \ldots, \tau_g\rho\}$ is the set of all isomorphisms of K into $\mathbf{C}$, where ρ means complex conjugation. Let $L = B \otimes_F K$. We consider an abelian variety A belonging to one of the following two types:

(i) $\dim(A) = 2g$; there exists an isomorphism θ of K into $\mathrm{End}_{\mathbf{Q}}(A)$ such that the analytic representation (i.e. the representation in the tangent space of A at the origin) of $\theta(a)$ is equivalent to $\sum_{\nu=1}^r (\tau_\nu + \tau_\nu\rho) + 2\sum_{\nu=r+1}^g \tau_\nu$.

(ii) $\dim(A) = 4g$; there exists an isomorphism θ of L into $\mathrm{End}_{\mathbf{Q}}(A)$ such that the analytic representation of $\theta(a)$ for $a \in K$ is equivalent to $2\sum_{\nu=1}^r (\tau_\nu + \tau_\nu\rho) + 4\sum_{\nu=r+1}^g \tau_\nu$.

With a suitable polarization and points of finite order, one obtains a family $\Sigma = \{Q_z \mid z \in \mathfrak{H}^r\}$ of the type described in §1, with $\Gamma = \Gamma(\mathfrak{o}, \mathfrak{c})$. For the families of type (1), one has to assume that B splits over K. Let us assume $r = 1$, though one can investigate the general case $r \geq 1$ by the same method. Let K' denote the field generated over $\mathbf{Q}$ by $\sum_{\nu=2}^g x^{\tau_\nu}$ for all $x \in K$. Define the field k as in §1 for this Σ. It can be shown that $k \subset C(K', \mathfrak{c})$. Here and in the following we assume that $\mathfrak{c}$ is generated by a rational integer. (The general case can be easily reduced to this case.) Take V and φ satisfying (2.1–2) for the present Σ. Let M and z be as in (I.3). Then either KM or $B \otimes_F KM$ is contained in the endomorphism algebra of the special member Q_z according as Q_z is of the above type (i) or (ii). In any case Q_z has sufficiently many complex multiplications. The general theory [16] tells us that the field of moduli of Q_z, which coincides with $k\big(\varphi(z)\big)$ by virtue of (2.3), is an abelian extension E of $K'M$. The nature of the class field E has been investigated in [10]. Up to this point, we have chosen one K, and constructed (V, φ) with respect to this fixed K. Now there are infinitely many choices of K. The class field $C(M, \mathfrak{c})$ can be obtained as the intersection of the composite $E \cdot C(K', \mathfrak{c})$ for all K. This fact enables us to apply Weil's criterion [18] to lower the field of definition to $C(F, \mathfrak{c})$, and then to find a new model (V, φ) satsifying (I.1–3). Theorem II can be proved by investigating more closely the isogenies beween Q_z and its transforms by Frobenius automorphisms of E.

To explain the proof of Theorem III, let us first define some algebraic correspondences. Let $\{V_\lambda, \varphi_\lambda\}$ be as in Theorem II, and let

$$X_{\mu\lambda}(\mathfrak{p}) = \left\{ \big(\varphi_\lambda(z), \varphi_\mu(\alpha(z))\big) \in V_\lambda \times V_\mu \,\big|\, z \in \mathfrak{H} \right\},$$
$$Y_{\mu\nu}(\mathfrak{p}) = \left\{ \big(\varphi_\nu(z), \varphi_\mu(\beta(z))\big) \in V_\nu \times V_\mu \,\big|\, z \in \mathfrak{H} \right\},$$

with elements α and β of B^+ such that $\alpha\mathfrak{x}_{\lambda\mu}$ is integral and $N(\alpha\mathfrak{x}_{\lambda\mu}) = \mathfrak{p}$, $\beta\mathfrak{x}_{\nu\mu} = \mathfrak{po}_\mu$. Further let $\lambda \mapsto \lambda'$ be a permutation of $\{1, \ldots, h\}$ such that $N(\mathfrak{x}_{\lambda'})$ and $N(\mathfrak{x}_\lambda^{-1})$ belong to the same ideal class modulo $\mathfrak{u}$ in F, and let δ_λ be an element of B^+ such that $\delta_\lambda\mathfrak{x}_\lambda = N(\mathfrak{x}_\lambda)\mathfrak{x}_{\lambda'}$ (cf. [11, p.257]). Let

$$U_\lambda = \left\{ \big(\varphi_\lambda(z), \varphi_{\lambda'}(\delta_\lambda(z))\big) \in V_\lambda \times V_{\lambda'} \,\big|\, z \in \mathfrak{H} \right\}.$$

Then $X_{\mu\lambda}(\mathfrak{p})$, $Y_{\mu\nu}(\mathfrak{p})$, and U_λ are algebraic correspondences defined over $C(F, 1)$, and U_λ is birational. The operators $T_{\mu\lambda}(\mathfrak{p})_2$ and $R_{\mu\nu}(\mathfrak{p})_2$ are nothing but the representations of $X_{\mu\lambda}(\mathfrak{p})$ and $Y_{\mu\nu}(\mathfrak{p})$ by differential forms of the first kind on the curves. Let $\mathfrak{P}$ be a prime ideal in $C(F, 1)$ dividing $\mathfrak{p}$. Denote by tilde the reduction modulo $\mathfrak{P}$. Then we have

$$(5.1) \qquad \widetilde{X_{\mu\lambda}(\mathfrak{p})} = {}^t\Pi_\mu + \widetilde{U}_\mu^{-1} \circ \Pi_{\lambda'} \circ \widetilde{U}_\lambda,$$

$$(5.2) \qquad \widetilde{Y_{\mu\nu}(\mathfrak{p})} = \widetilde{U}_\mu^{-1} \circ \widetilde{U}_\lambda^{N(\mathfrak{p})}.$$

Here Π_λ denotes the locus of $(x, x^{N(\mathfrak{p})})$ on $\widetilde{V}_\lambda \times \widetilde{V}_\lambda^{N(\mathfrak{p})}$, and ${}^t\Pi_\lambda$ its transpose. These congruence relations are obtained from the reciprocity law (II.3). By a simple computation, we can easily derive Theorem III from (5.1-2).

REFERENCES

1. M. Eichler, Quaternäre quadratische Formen und die Riemannsche Vermutung für die Kongruenzzetafunktion, Arch. Math. **5** (1954), 355-366.

2. R. Fricke und F. Klein, Vorlesungen über die Theorie der automorphen Funktionen I, II, Leipzig, Teubner, 1897-1912.

3. E. Hecke, Höhere Modulfunktionen und ihre Anwendung auf die Zahlentheorie, Math. Ann. **71** (1912), 1-37, (Werke, 21-57).

4. M. Kuga, Fibre varieties over a symmetric space whose fibres are abelian varieties, lecture notes, Univ. of Chicago, 1963-64.

5. H. Poincaré, Les fonctions fuchsiennes et l'arithmétique, J. de math. (4) 3 (1887), 405-464 (œuvre II, 463-511).

6. I. Satake, Holomorphic imbeddings of symmetric domains into a Siegel space, Amer. J. Math. **87** (1965), 425-461.

7. ______ , Symplectic representations of algebraic groups satisfying a certain analyticicity condition, to appear.

8. G. Shimura, Correspondances modulaires et les fonctions zeta de courbes algébriques, J. Math. Soc. Japan, **10** (1958), 1-28.

9. _____ , On the zeta functions of the algebraic curves uniformized by certain automorphic functions, J. Math. Soc. Japan, **13** (1961), 275-331.

10. _____ , On the class-fields obtained by complex multiplication of abelian varieties, Osaka Math. J. **14** (1962), 33-44.

11. _____ , On Dirichlet series and abelian varieties attached to automorphic forms, Ann. of Math, **76** (1962), 237-294.

12. _____ , On the field of definition for a field of automorphic functions I, II, III, Ann. of Math, **80** (1964), 160-189; **81** (1965), 124-165; **83** (1966), 377-385.

13. _____ , Moduli and fibre system of abelian varieties, Ann. of Math, **83** (1966), 294-338.

14. _____ , Discontinuous groups and abelian varieties, Math. Ann. **168** (1967), 171-199.

15. _____ , Construction of class fields and zeta functions of algebraic curves, Ann. of Math. **85** (1967), 58-159.

16. G. Shimura and Y. Taniyama, Complex multiplication of abelian varieties and its applications to number theory, Pub. Math. Soc. Japan, No. 6 (1961).

17. C. L. Siegel, Symplectic geometry, Amer. J. Math. **65** (1943), 1-86.

18. A. Weil, The field of definition of a variety, Amer. J. Math. **78** (1956), 509-524.

Algebraic number fields and symplectic discontinuous groups

Annals of Mathematics, 86 (1967), 503-592

In a previous paper [9], quoted hereafter as [C], we have constructed a certain canonical model for the quotient of the product of copies of the upper half complex plane by an arithmetic discontinuous group. The purpose of the present work is to extend this result to the case of higher dimensional Siegel spaces. To be more specific, let B be a quaternion algebra over a totally real algebraic number field F of degree g, and ι the main involution of B. Let $\mathfrak{G}^+$ denote the group of all elements α of $GL_n(B)$ such that $\alpha \cdot {}^t\alpha^\iota = \nu(\alpha)1_n$ with a totally positive element $\nu(\alpha)$ of F. Take a maximal order $\mathfrak{o}$ in B and an integral ideal $\mathfrak{c}$ in F, and put

$$\Gamma_\mathfrak{c} = \{\gamma \in GL_n(\mathfrak{o}) \cap \mathfrak{G}^+ \mid 1 - \gamma \in \mathfrak{c} \cdot M_n(\mathfrak{o})\} \ .$$

Then $\Gamma_\mathfrak{c}$ can be considered as a discontinuous group acting on the product $\mathfrak{H}_n^r$ of r copies of the space $\mathfrak{H}_n$ of complex symmetric matrices of degree n with positive definite imaginary parts. Here r is the number of archimedean primes of F unramified in B. The groups of this type exhaust all the arithmetic discontinuous groups acting on $\mathfrak{H}_n^r$, up to commensurability. In particular, if $B = M_2(\mathbf{Q})$, $\Gamma_{(1)}$ is the Siegel modular group. Now our first Main Theorem (5.3) shows the existence of a *canonical model* $(V_\mathfrak{c}, \varphi_\mathfrak{c})$ formed by an algebraic variety $V_\mathfrak{c}$ and a holomorphic map $\varphi_\mathfrak{c}$ of $\mathfrak{H}_n^r$ onto $V_\mathfrak{c}$ which gives a biregular isomorphism of $\mathfrak{H}_n^r/\Gamma_\mathfrak{c}$ onto $V_\mathfrak{c}$, satisfying the following two conditions (1), (2). Since the precise formulation is lengthy, we give here only a somewhat vague and weaker statement.

(1) *$V_\mathfrak{c}$ is defined over a class field C over a certain algebraic number field F'.*

(2) *Every isolated fixed point z of $\mathfrak{G}^+$ on $\mathfrak{H}_n^r$ determines a certain algebraic number field P', and the composite $P' \cdot C(\varphi_\mathfrak{c}(z))$ is a class field C' over P'.*

The field F' is generated by the elements $\sum_{\nu=1}^r x^{\tau_{0\nu}}$ over $\mathbf{Q}$ for all $x \in F$, where $\tau_{01}, \cdots, \tau_{0r}$ are the isomorphisms of F into $\mathbf{R}$ corresponding to the archimedean primes of F unramified in B. Further, a point z of $\mathfrak{H}_n^r$ is called an *isolated fixed point of* $\mathfrak{G}^+$, if z is the only common fixed point of $S_z = \{\alpha \in \mathfrak{G}^+ \mid \alpha(z) = z\}$. For such a point z, the subalgebra Y of $M_n(B)$ generated by S_z over F is of the form $Y = Y_1 \oplus \cdots \oplus Y_t$, where Y_i is a central simple algebra over a totally imaginary quadratic extension P_i of a totally real algebraic number field containing F. If $[Y_i : P_i] = q_i^2$ and $[P_i : F] = 2m_i$, then $n = \sum_{i=1}^t m_i q_i$. Let $z = (z_1, \cdots, z_r)$ with $z_\nu \in \mathfrak{H}_n$. For every $\alpha \in \mathfrak{G}^+$, we can speak of its νth component α_ν acting on $\mathfrak{H}_n$. Then we can find, for each $\nu = 1, \cdots, r$, a holomorphic isomorphism π_ν of $\mathfrak{H}_n$ onto the unit ball $\mathfrak{D}_n$ of complex symmetric matrices of degree n and a representation Ψ_ν of Y by complex matrices of degree n such that $\pi_\nu(z_\nu) = 0$ and

$$\pi_\nu \alpha_\nu \pi_\nu^{-1}(w) = \overline{\Psi_\nu(\alpha)} \cdot w \cdot \Psi_\nu(\alpha)^{-1} \qquad (\alpha \in S_z,\ w \in \mathfrak{D}_n) .$$

Now P' in (2) is the field generated by the numbers $\sum_{\nu=1}^r \operatorname{tr} \Psi_\nu(\alpha)$ over $\mathbf{Q}$ for all $\alpha \in P_1 \oplus \cdots \oplus P_t$. The ideal groups corresponding to the class fields C and C' will be determined completely.

The second Main Theorem (5.7) is concerned with a system of several canonical models. It describes the Galois group of V_c over $V_{(1)}$ as well as an explicit reciprocity law in C and C'. The first four main theorems of [C] have been stated with regard to the cases $n = r = 1$ and $n = r/g = 1$. Our present results include all these as special cases, and verify Conjectures I and III of [C, 9.12]. Moreover, in [C], we have discussed only the isolated fixed points for which $Y \cap M_n(\mathfrak{o})$ is the maximal order in Y. (Note that $Y = P_1$ and $[Y : F] = 2$ if $n = 1$.) *Arbitrary* isolated fixed points will be considered in the present paper. In §10, we shall give another system of canonical models for $\mathfrak{H}_n^r/\Gamma^*$ with a group Γ_c^* somewhat larger than Γ. This is a generalization of [C, 3.13].

Our method of proof is essentially the same as in [C]. Principles similar to (A), (B), (C), (D) in the Introduction of [C] apply to the higher dimensional case. However, there is a feature which is evident only when $n > 1$. As in [C], we construct a family Σ of abelian varieties whose space of moduli is $\mathfrak{H}_n^r/\Gamma_c$. We can embed Σ into a larger family Σ' of abelian varieties whose space of moduli is $\mathfrak{S}^r/\Gamma'$, where $\mathfrak{S}$ is the unit ball of complex square matrices of degree n, and Γ' is an arithmetic discontinuous group acting on $\mathfrak{S}^r$ and containing Γ_c. If $n = 1$, disregarding some minor differences, we can identify $\mathfrak{H}_n^r/\Gamma_c$ with $\mathfrak{S}^r/\Gamma'$. This of course breaks down for $n > 1$. If $n > 1$ and $r < g$, there is no known algebro-geometric way of characterizing the members of Σ, while the family Σ' can be characterized by endomorphisms and polarization. One can overcome this difficulty by taking a dense subset of $\mathfrak{H}_n^r$ which corresponds to abelian varieties with sufficiently many complex multiplications. Further, it should be noted that, if $r < g$, we have infinitely many families Σ for a given $\mathfrak{H}_n^r/\Gamma_c$, and use *all* of them in the proof. On the other hand, if $r = g$, it suffices to take a single family which is characterized by endomorphisms and polarization (cf. Introduction of [C]).

In the final § 11, we shall consider an infinite sequence of Galois coverings $V_c \leftarrow V_{cl} \leftarrow V_{cl^2} \leftarrow \cdots$ for a prime ideal $\mathfrak{l}$ in F. Given an algebraic point y on V_c, take all the points on V_{cl^m} lying on y, and adjoin their coordinates to C. Then we obtain an infinite normal extension $\mathfrak{K}$ of $C(y)$ whose Galois group has a certain $\mathfrak{l}$-adic representation. We shall prove a few elementary properties of this representation for the Frobenius automorphisms of $\mathfrak{K}$, which are analogous to those known for the points of finite order on an abelian variety. Actually if $r = g$, the former is essentially a paraphrase of the latter. But in the case $r < g$, our extension $\mathfrak{K}$ and its $\mathfrak{l}$-adic representation are new, though they are loosely connected with abelian varieties. When $y = \varphi_c(z)$ with an isolated fixed point z of $\mathfrak{G}^+$ on $\mathfrak{H}_n^r$, we know that $P'\mathfrak{K}$ is abelian over P', and the $\mathfrak{l}$-adic representation of the Frobenius automorphisms can be described by means of a Grössen-character of the field C' of (2). However, if y is an arbitrary algebraic point on V_c, the nature of $\mathfrak{K}$ is quite mysterious. We have no way of even obtaining an example in the case $r < g$, for which the Frobenius automorphisms are explicitly given. It might be too early to predict how far one would be able to proceed in this direction. But it is at least certain that our canonical models should not be regarded only as a tool of constructing *abelian extensions*, since it is hoped that they will play some role in the study of *non-abelian* extensions. In this sense, the title of the previous paper may not have been the most appropriate one.

It will of course be possible to generalize our whole theory to the case of an arbitrary arithmetic discontinuous group acting on a bounded symmetric domain, by the same method, whenever there exists a family of abelian varieties. Although the present investigation deals with only one type of symmetric domain, the author believes that this case is general enough to reveal almost all possible aspects of the theory.

CONTENTS

Notation and terminology

By a ring or an algebra, we always mean an associative one. If S is a ring and X is an S-module, we denote by $\mathrm{End}(X, S)$ the ring of all S-linear endomorphisms of X, and by $\mathrm{GL}(X, S)$ the group of all S-linear automorphisms of X. For a left S-module X, the elements of $\mathrm{End}(X, S)$ act on the right of X. If S is a separable algebra over a field F, $N_{S/F}$ and $\mathrm{Tr}_{S/F}$ denote respectively the reduced norm and trace from S to F. The trace of a matrix A is denoted by $\mathrm{tr}\,(A)$. The notation $[Y : Z]$ means the index of a subgroup Z of a group Y, and also the dimension of a vector space Y over a field Z. The distinction will be clear from the context.

By an algebraic variety, we understand a Zariski open subset of a projective variety which is absolutely irreducible. We take the complex number field C as the universal domain. Therefore, any variety V is defined over a subfield of C, and a point x of V has coordinates in C. If k is a subfield of C, $k(x)$ denotes the field generated over k by the (quotients of the projective) coordinates of x. The distinction of this from other symbols, such as the value of a function at a point, will be clear from the context. For an isomorphism σ of a field of rationality for V into C, V^{σ} denotes the variety whose defining equations are the transforms by σ of the defining equations for V. If $K_1, \cdots, K_t$ are subfields of C, $K_1 \cdots K_t$ means the composite of the K_i, i.e., the smallest subfield of C containing the K_i.

We shall also use the symbols introduced in [C], some of which are listed as follows for reader's convenience.

K: the division ring of Hamilton quaternions.

$M_n(S)$: the ring of all matrices of size n with entries in a ring S.

$\mathrm{GL}_n(S)$: the group of all invertible elements of $M_n(S)$.

$I(F), I(F, (1))$: the group of all fractional ideals in an algebraic number field F.

$I(F, \mathfrak{c})$: the group of all ideals in $I(F)$ prime to an integral ideal $\mathfrak{c}$ in F.

$P(F, \mathfrak{cu})$: the group of all principal ideals (a) in F such that $a \equiv 1$ mod* $\mathfrak{cu}$, where $\mathfrak{u}$ is a product of archimedean primes of F, and mod* means the multiplicative congruence, see [C, 1.1].

$C(F, \mathfrak{c})$: the class field over F corresponding to $P(F, \mathfrak{cu}_0)$, where $\mathfrak{u}_0$ is the product of *all* archimedean primes of F, see [C, 1.1].

$D(K/F)$: the relative discriminant of a finite algebraic extension K of F.

$\mathfrak{d}(K/F)$: the relative different of K over F.

$G(K/F)$: the Galois group of a normal extension K over F.

$[K/F, \mathfrak{a}]$: the Artin symbol for an abelian extension K of F and an ideal $\mathfrak{a}$ in F prime to $D(K/F)$, see [C, 1.1].

$D(B/F)$: the product of all prime ideals in F which are ramified in a quaternion algebra B over F, see [C, 2.1-4].

End (A): the ring of endomorphisms of an abelian variety A.

$\text{End}_{Q}(A)$: the tensor product of End (A) with Q over Z.

1. Involutions and hermitian forms

1.1. Let S be an algebra over a field F. By an *involution* of S, we understand an anti-automorphism σ of S (not necessarily F-linear) such that $(a^{\sigma})^{\sigma} = a$ for all $a \in S$. For $\alpha = (a_{ij}) \in M_n(S)$ with $a_{ij} \in S$, we shall write $'\alpha^{\sigma} = {}'(a_{ij}^{\sigma})$. Let X be a left S-module, and σ an F-linear involution of S. By an *S-valued σ-hermitian* (resp. *σ-anti-hermitian*) *form* on X, we understand an F-bilinear map $h: X \times X \longrightarrow S$ such that $h(ax, by) = ah(x, y)b^{\sigma}$, $h(x, y)^{\sigma} = h(y, x)$ (resp. $h(x, y)^{\sigma} = -h(y, x)$) for $a \in S$, $b \in S$, $(x, y) \in X \times X$. We call h *non-degenerate* if $h(y, X) = 0$ implies $y = 0$.

1.2. PROPOSITION. *Let Z be a field of characteristic 0, S a central simple algebra over Z with an involution σ, X a left S-module, and h a non-degenerate S-valued σ-hermitian or σ-anti-hermitian form on X. Suppose that every element α of $\mathrm{GL}(X, S)$ satisfying $h(x\alpha, y\alpha) = h(x, y)$ can be obtained from the multiplication by an element of Z. Then X is an irreducible S-module, and S can be identified with End (X, Z).*

PROOF. Let $W = \{a \in Z \mid a^{\sigma} = a\}$, and $A = \text{End}\,(X, S)$. We see that the map $(x, y) \mapsto \text{Tr}_{S/W}\,(h(x, y))$ defines a non-degenerate W-bilinear form on X. Therefore, for every $\gamma \in A$, we can find a unique $\gamma^{\tau} \in \text{End}\,(X, W)$ so that

$$\text{Tr}_{S/W}\,\big(h(x\gamma^{\tau}, y)\big) = \text{Tr}_{S/W}\,\big(h(x, y\gamma)\big) \qquad \big((x, y) \in X \times X\big).$$

We can easily verify that $\gamma^{\tau} \in A$, and τ is an involution of A over W. Let $G = \{\alpha \in A \mid \alpha\alpha^{\tau} = 1\}$. Note that A is a central simple algebra over Z, and every α in G satisfies $h(x\alpha, y\alpha) = h(x, y)$. Let $U = \{\lambda \in A \mid \lambda^{\tau} = -\lambda\}$. If $U \not\subset Z$, there are infinitely many λ in U, not contained in Z, such that $N_{A/W}(1 + \lambda) \neq 0$. For such λ, we have $(1 - \lambda)(1 + \lambda)^{-1} \in G \subset Z$, hence $\lambda \in Z$, which is a contradiction. Therefore $U \subset Z$. If $W = Z$, this implies $U = \{0\}$, hence $\lambda^{\tau} = \lambda$ for all $\lambda \in A$. It follows that A is commutative, i.e., $A = Z$. If $W \neq Z$, we have $[U : W] = [A : Z] = n^2$ with a positive integer n. Since $U \subset Z$, we must have $n = 1$, hence $A = Z$ in any case. Therefore X must be an irreducible S-module, and $S = \text{End}\,(X, Z)$,   q.e.d.

1.3. Let S be an algebra over $\boldsymbol{Q}$ (resp. $\boldsymbol{R}$), σ an involution of S, and φ a regular representation of S over $\boldsymbol{Q}$ (resp. $\boldsymbol{R}$). We say that σ is *positive* if $\operatorname{tr}(\varphi(aa^\sigma)) > 0$ for all non-zero $a \in S$. It makes no difference to take a reduced representation of S over $\boldsymbol{Q}$ (resp. $\boldsymbol{R}$) in place of φ.

We call an algebraic number field P a CM-*field*, if P is a totally imaginary quadratic extension of a totally real algebraic number field of finite degree (cf. [C, 5.7–10]).

1.4. PROPOSITION. *Let Y be a central simple algebra over a* CM-*field P. Suppose that Y has an involution whose restriction to P coincides with the complex conjugation. Then Y has a positive involution.*

PROOF. Put $Y_{\boldsymbol{R}} = Y \otimes_{\boldsymbol{Q}} \boldsymbol{R}$, $[Y : P] = n^2$, $[P : \boldsymbol{Q}] = 2m$. Then $Y_{\boldsymbol{R}}$, as an algebra over $\boldsymbol{R}$, can be identified with the direct sum of m copies of $M_n(\boldsymbol{C})$. Let p_ν, for $\nu = 1, \cdots, m$, denote the projection of $Y_{\boldsymbol{R}}$ to the νth factor $M_n(\boldsymbol{C})$. Let σ be an involution of Y which coincides with the complex conjugation on P. Extend σ $\boldsymbol{R}$-linearly to $Y_{\boldsymbol{R}}$. Then we can find an element u_ν of $\mathrm{GL}_n(\boldsymbol{C})$ such that $^t\bar{u}_\nu = u_\nu$ and $p_\nu(x^\sigma) = u_\nu^{-1} \cdot {}^t\overline{p_\nu(x)} u_\nu$ for all $x \in Y_{\boldsymbol{R}}$. Put

$$T = \{z \in Y \mid z^\sigma = z\}, \qquad T_\nu = \{w \in M_n(\boldsymbol{C}) \mid {}^t\overline{(u_\nu w)} = u_\nu w\}.$$

Then T is a vector space over $\boldsymbol{Q}$, hence T is dense in $T \otimes_{\boldsymbol{Q}} \boldsymbol{R}$, and $T \otimes_{\boldsymbol{Q}} \boldsymbol{R}$ can be identified with $T_1 \times \cdots \times T_m$. Therefore we can find an element z in T such that $u_\nu p_\nu(z)$ is a positive definite hermitian matrix for every ν. Define an involution ρ of Y by $x^\rho = z^{-1} x^\sigma z$ for $x \in Y$. From our choice of z, it is easy to see that ρ is positive.

1.5. PROPOSITION. *Every* CM-*field P contains infinitely many elements x such that $xx^\rho = 1$ and $P = \boldsymbol{Q}(x)$, where ρ denotes the complex conjugation.*

PROOF. By the approximation theorem in a number field, we can find an element y of P such that y^σ / y is not real for every isomorphism σ of P into $\boldsymbol{C}$ other than the identity map. Put $x = y^\rho / y$. Then $xx^\rho = 1$. If $x^\sigma = x$, we have $y^\sigma / y = (y^\rho / y)^\sigma$, hence σ must be the identity map. Therefore $P = \boldsymbol{Q}(x)$. Moreover y can be chosen so that y^ρ / y is different from any given finite number of elements. This proves our assertion.

1.6. PROPOSITION. *Let $P_1, \cdots, P_t$ be* CM-*fields, and $Y_1, \cdots, Y_t$ be central simple algebras over $P_1, \cdots, P_t$, respectively. Put $Y = Y_1 \oplus \cdots \oplus Y_t$. Let δ be an involution of Y whose restriction to each P_i is the complex conjugation. Then Y can be spanned linearly over $\boldsymbol{Q}$ by the set*

$$(Y, \delta)_1 = \{x \in Y \mid xx^\delta = 1\}.$$

PROOF. Let U be the subalgebra of Y spanned by $(Y, \delta)_1$ over $\boldsymbol{Q}$. Let e_i be the identity element of P_i. Since $e_i^\delta = e_i$, we see that

$$e_1 + \cdots + e_{i-1} - e_i + e_{i+1} + \cdots + e_t \in (Y, \delta)_1 ,$$

hence $e_i \in U$. Put $(Y_i, \delta)_1 = \{x \in Y_i \mid xx^\delta = e_i\}$. Then $(Y_i, \delta)_1 = (Y, \delta)_1 e_i$. Therefore, to prove $U = Y$, it suffices to show that Y_i is spanned by $(Y_i, \delta)_1$ over $\boldsymbol{Q}$. In other words, we may assume that Y is simple. Now, by 1.5, $P \subset U$. Put $D = \{a \in Y \mid a^\delta = -a, 1 + a \text{ is invertible}\}$. For $a \in D$, put $b = (1 - a)(1 + a)^{-1}$. Then $b \in (Y, \delta)_1$, hence $a = (1 - b)(1 + b)^{-1} \in U$. Now P contains an element ζ such that $\zeta^\delta = -\zeta$. Therefore U contains all elements x of Y such that $x^\delta = \pm x$. This proves $U = Y$.

2. Orders and ideals in a semi-simple algebra

2.1. Let X be a vector space of finite dimension over $\boldsymbol{Q}$. By a *lattice* in X, we understand a finitely generated $\boldsymbol{Z}$-submodule of X which generates X over $\boldsymbol{Q}$. Let S be a semi-simple algebra of finite rank over $\boldsymbol{Q}$. By an *order* in S, we understand a subring of S containing $\boldsymbol{Z}$ which is a lattice in S. If S is commutative, we denote by $\mathfrak{r}_S$ the unique maximal order in S, and define the *conductor* $\mathfrak{h}$ of an order $\mathfrak{g}$ in S by $\mathfrak{h} = \{x \in \mathfrak{g} \mid \mathfrak{r}_S x \subset \mathfrak{g}\}$. When S is an algebraic number field of finite degree, by an *ideal in S* or an *$\mathfrak{r}_S$-ideal*, we always mean a fractional ideal in S with respect to $\mathfrak{r}_S$ in the usual sense.

In this paper, we fix a basic algebraic number field F of finite degree, and consider algebras of finite rank over F. An order $\mathfrak{g}$ in a semi-simple algebra over F is called an *$\mathfrak{r}_F$-order* if $\mathfrak{r}_F \subset \mathfrak{g}$. (From §3 forward, F will be assumed totally real; this assumption is unnessary in this section.) The letter $\mathfrak{p}$ will always mean a prime ideal in F, and $F_\mathfrak{p}$ (resp. $\mathfrak{r}_\mathfrak{p}$) the $\mathfrak{p}$-completion of F (resp. $\mathfrak{r}_F$).

Let X be a vector space of finite dimension over F. A lattice $\mathfrak{x}$ in X is called an *$\mathfrak{r}_F$-lattice* if $\mathfrak{r}_F \mathfrak{x} \subset \mathfrak{x}$. For such X and $\mathfrak{x}$, we put $X_\mathfrak{p} = X \otimes_F F_\mathfrak{p}$ and $\mathfrak{x}_\mathfrak{p} = \mathfrak{r}_\mathfrak{p} \mathfrak{x}\ (\subset X_\mathfrak{p})$.

2.2. Let S be a semi-simple algebra over F, and $\mathfrak{g}$ an $\mathfrak{r}_F$-order in S. We call an $\mathfrak{r}_F$-lattice $\mathfrak{a}$ in S a *right* (resp. *left*) $\mathfrak{g}$-*ideal* if $\mathfrak{a}_\mathfrak{p} = x_\mathfrak{p} \mathfrak{g}_\mathfrak{p}$ (resp. $\mathfrak{a}_\mathfrak{p} = \mathfrak{g}_\mathfrak{p} x_\mathfrak{p}$) with an element $x_\mathfrak{p}$ of $S_\mathfrak{p}$ for every $\mathfrak{p}$. For such an $\mathfrak{a}$, we define $\mathfrak{a}^{-1}$ to be a left (resp. right) $\mathfrak{g}$-ideal such that $(\mathfrak{a}^{-1})_\mathfrak{p} = \mathfrak{g}_\mathfrak{p} x_\mathfrak{p}^{-1}$ (resp. $= x_\mathfrak{p}^{-1} \mathfrak{g}_\mathfrak{p}$). Note that $\mathfrak{g}$ is uniquely determined by $\mathfrak{a}$; we call $\mathfrak{g}$ the *right* (resp. *left*) *order* of $\mathfrak{a}$. Further, $\mathfrak{a}$ is called a *two-sided $\mathfrak{g}$-ideal* if $\mathfrak{a}$ is both a right $\mathfrak{g}$-ideal and a left $\mathfrak{g}$-ideal; $\mathfrak{a}$ is called *integral* if $\mathfrak{a} \subset \mathfrak{g}$. We say that $\mathfrak{a}$ is *prime to* an integral ideal $\mathfrak{c}$ in F if $\mathfrak{a}_\mathfrak{p} = \mathfrak{g}_\mathfrak{p}$ for every $\mathfrak{p}$ dividing $\mathfrak{c}$. It can easily be seen that for every ideal $\mathfrak{b}$ in F, $\mathfrak{b}\mathfrak{g}$ is a two-sided $\mathfrak{g}$-ideal.

Two $\mathfrak{r}_F$-orders $\mathfrak{g}_1$ and $\mathfrak{g}_2$ in S are called *locally conjugate*, if, for every prime ideal $\mathfrak{p}$ in F, there exists an invertible element $x_\mathfrak{p}$ of $S_\mathfrak{p}$ such that

$x_\mathfrak{p} g_{1\mathfrak{p}} x_\mathfrak{p}^{-1} = g_{2\mathfrak{p}}$. Two maximal orders in S are always locally conjugate. We see easily that g_1 and g_2 are locally conjugate if and only if there exists a right g_1-ideal which is a left g_2-ideal.

Let $\mathfrak{X}_0$ be a maximal set of locally conjugate orders in S. Let $\mathfrak{X}$ be the set of all $\mathfrak{r}_F$-lattices in S which are right g-ideals for some $g \in \mathfrak{X}_0$. For two members $\mathfrak{a}$ and $\mathfrak{b}$ of $\mathfrak{X}$, call $\mathfrak{a}\mathfrak{b}$ a *proper product* if the right order of $\mathfrak{a}$ coincides with the left order of $\mathfrak{b}$. Then it is easy to verify that $\mathfrak{X}$ forms a groupoid with respect to the proper product, and $\mathfrak{X}_0$ is the set of *units* of $\mathfrak{X}$.

2.3. Let S and g be as in 2.2, and c an integral ideal in F. For two invertible elements x and y of S, we write

$$(2.3.1) \qquad\qquad x \equiv y \bmod^* cg$$

if x and y are units of $g_\mathfrak{p}$ and $x - y \in (cg)_\mathfrak{p}$ for every $\mathfrak{p}$ dividing c.

2.4. Suppose that S is commutative. Let g be an $\mathfrak{r}_F$-order in S, $\mathfrak{h}$ the conductor of g, and $\mathfrak{h}_0 = \mathfrak{h} \cap F$. Obviously $\mathfrak{h}_0$ is an integral ideal in F. Let c be an integral ideal in F. For two invertible elements x and y of S, we write

$$(2.4.1) \qquad\qquad x \equiv y \bmod^* (g; c)$$

if x and y are units of $g_\mathfrak{p}$ and $x - y \in (cg)_\mathfrak{p}$ for every $\mathfrak{p}$ dividing $c\mathfrak{h}_0$. We see easily that

(2.4.2) *If* $x \equiv 1 \bmod^* \mathfrak{b}\mathfrak{r}_S$ *for an* $\mathfrak{r}_F$-*ideal* $\mathfrak{b}$ *contained in* $c\mathfrak{h}_0$, *then* $x \equiv 1 \bmod^* (g; c)$.

If S is a field and $g = \mathfrak{r}_S$, the congruence (2.3.1) and (2.4.1) coincide with the usual multiplicative congruence $x \equiv y \bmod^* c\mathfrak{r}_S$.

2.5. Let S, g, $\mathfrak{h}$ and $\mathfrak{h}_0$ be as in 2.4. Note that $(\mathfrak{r}_S)_\mathfrak{p} \neq g_\mathfrak{p}$ if and only if $\mathfrak{p}$ divides $\mathfrak{h}_0$. For every $\mathfrak{r}_S$-ideal $\mathfrak{a}$ prime to $\mathfrak{h}_0$, we define a g-ideal $\mathfrak{a}'$ so that $\mathfrak{a}'_\mathfrak{p} = g_\mathfrak{p}$ or $\mathfrak{a}_\mathfrak{p}$ according as $\mathfrak{p}$ divides $\mathfrak{h}_0$ or not. We write $\mathfrak{a}'$ as $[\mathfrak{a}]_g$. Then $\mathfrak{a} \mapsto [\mathfrak{a}]_g$ is a multiplicative map, and the following assertions can easily be verified.

(2.5.1) $[\mathfrak{a}]_g = \mathfrak{a} \cap g$ *if* $\mathfrak{a}$ *is integral.*

(2.5.2) $\mathfrak{a} = [\mathfrak{a}]_g \mathfrak{r}_S$.

(2.5.3) *If* $x \in S$ *and* $x \equiv 1 \bmod^* (g; \mathfrak{r}_F)$, *then* $[x\mathfrak{r}_S]_g = xg$.

2.6. PROPOSITION. *Let* S *be a quadratic extension of* F, g *an* $\mathfrak{r}_F$-*order in* S, $\mathfrak{h}$ *the conductor of* g, *and* $\mathfrak{h}_0 = \mathfrak{h} \cap F$. *Then* $g = \mathfrak{r}_F + \mathfrak{h}$, *and* $\mathfrak{h} = \mathfrak{r}_S \mathfrak{h}_0$. *Conversely, for every integral ideal* $\mathfrak{h}_0$ *in* F, $\mathfrak{r}_F + \mathfrak{h}_0 \mathfrak{r}_S$ *is an* $\mathfrak{r}_F$-*order in* S *with conductor* $\mathfrak{h}_0 \mathfrak{r}_S$.

This is well-known (at least when $F = \mathbf{Q}$), and can be proved by the standard local argument.

2.7. PROPOSITION. *Let* S, g, $\mathfrak{h}$, *and* $\mathfrak{h}_0$ *be as in 2.6. Let* c *be an integral*

ideal in F. Then, for $x \in S$, one has $x \equiv 1 \bmod^* (\mathfrak{g}; \mathfrak{c})$ if and only if there exists an element w of F prime to $\mathfrak{ch}_0$ such that $x \equiv w \bmod^* \mathfrak{ch}$ and $w \equiv 1 \bmod^* \mathfrak{c}$. In particular, $x \equiv 1 \bmod^* (\mathfrak{g}; \mathfrak{r}_F)$ if and only if there exists an element w of F prime to $\mathfrak{h}_0$ such that $x \equiv w \bmod^* \mathfrak{h}$.

This can also be proved in a straightforward way, using the fact that $\mathfrak{cg} = \mathfrak{c} + \mathfrak{ch}$. The following proposition (well-known at least when $F = \boldsymbol{Q}$) may be worth inserting here, though we shall not make any explicit use of it. The proof can be given by the standard local argument.

2.8. PROPOSITION. *Let S and $\mathfrak{g}$ be as in 2.6, and $\mathfrak{a}$ an $\mathfrak{r}_F$-lattice in S. Then $\mathfrak{a}$ is a $\mathfrak{g}$-ideal if and only if $\mathfrak{g} = \{x \in S \mid x\mathfrak{a} \subset \mathfrak{a}\}$.*

2.9. PROPOSITION. *Let X be a vector space over F of finite dimension, K a finite algebraic extension of F, and $X_K = K \otimes_F X$. Let $\mathfrak{x}$ and $\mathfrak{h}$ be $\mathfrak{r}_F$-lattices in X, and α an element of $\mathrm{End}\,(X_K, K)$ such that $X\alpha \subset X$. Then $\mathfrak{r}_K\mathfrak{x}\alpha \subset \mathfrak{r}_K\mathfrak{h}$ if and only if $\mathfrak{x}\alpha \subset \mathfrak{h}$.*

PROOF. If m is the dimension of X over F, there exist m ideals $\mathfrak{a}_i$ in F and a basis $\{u_i\}$ of X over F such that $\mathfrak{h} = \sum_{i=1}^m \mathfrak{a}_i u_i$. By the linear independence of $\{u_i\}$ over K, we have

$$\mathfrak{r}_K\mathfrak{h} \cap X = \sum_i (\mathfrak{r}_K\mathfrak{a}_i \cap F)u_i = \sum_i \mathfrak{a}_i u_i = \mathfrak{h} \, .$$

Therefore, if $\mathfrak{r}_K\mathfrak{x}\alpha \subset \mathfrak{r}_K\mathfrak{h}$, then $\mathfrak{x}\alpha \subset \mathfrak{r}_K\mathfrak{h} \cap X = \mathfrak{h}$. This proves our assertion.

3. The group of similitudes of a quaternion hermitian form

3.1. Throughout the rest of the paper, F will denote a totally real algebraic number field of finite degree, and B a quaternion algebra over F. We put $g = [F:\boldsymbol{Q}]$, and denote by F^+ the set of all totally positive elements of F. Let $\tau_{01}, \cdots, \tau_{0g}$ denote all the isomorphisms of F into $\boldsymbol{R}$. We denote by $\mathfrak{p}_{\infty\nu}$ the archimedean prime of F obtained from $\tau_{0\nu}$. After rearranging the $\tau_{0\nu}$, we assume that B is unramified at $\mathfrak{p}_{\infty 1}, \cdots, \mathfrak{p}_{\infty r}$, and ramified at $\mathfrak{p}_{\infty r+1}, \cdots, \mathfrak{p}_{\infty g}$. We assume throughout $r > 0$, and put $\mathfrak{u} = \prod_{\nu > r} \mathfrak{p}_{\infty\nu}$. The main involution of B will be denoted by ι.

Let X be a left B-module isomorphic to the direct sum of n copies of B, where n is a positive integer. Let $h(x, y)$ be a non-degenerate B-valued ι-hermitian form on X. We assume that the following condition is satisfied:

(3.1.1) $h(x, x)^{\tau_{0\nu}} > 0$ *for every* $\nu > r$ *and* $0 \neq x \in X$.

Here note that $h(x, x) \in F$ for every $x \in X$. By the Hasse principle, for a given n, there is one and only one such h, up to isomorphisms. We can also express this fact in the following way:

(3.1.2) *If h satisfies (3.1.1), there exists a basis $\{x_1, \cdots, x_n\}$ of X over B such that $h(x_i, x_j) = 1$ or 0 according as $i = j$ or $i \neq j$.*

Let $\mathfrak{G}$ denote the group of all similitudes of h, i.e., the group of all elements α of $\mathrm{GL}(X, B)$ such that $h(x\alpha, y\alpha) = ch(x, y)$ with an element c of F. We put $c = \nu(\alpha)$, and define a subgroup $\mathfrak{U}$ of $\mathfrak{G}$ by $\mathfrak{U} = \{\alpha \in \mathfrak{G} \mid \nu(\alpha) = 1\}$.

For every prime ideal $\mathfrak{p}$ in F, we extend h to $X_\mathfrak{p}$ $F_\mathfrak{p}$-bilinearly, and denote by $\mathfrak{G}_\mathfrak{p}$ the group of all elements α of $\mathrm{GL}(X_\mathfrak{p}, B_\mathfrak{p})$ such that $h(x\alpha, y\alpha) = \nu(\alpha)h(x, y)$ with $\nu(\alpha) \in F_\mathfrak{p}$. We put $\mathfrak{U}_\mathfrak{p} = \{\alpha \in \mathfrak{G}_\mathfrak{p} \mid \nu(\alpha) = 1\}$. Further, put $F_R = F \otimes_Q R$, $B_R = B \otimes_Q R$, $X_R = X \otimes_Q R$, and consider X_R as a B_R-module. We extend h to X_R R-bilinearly, and denote by $\mathfrak{G}_R$ the group of all elements α of $\mathrm{GL}(X_R, B_R)$ such that $h(x\alpha, y\alpha) = \nu(\alpha)h(x, y)$ with $\nu(\alpha) \in F_R$. We put $\mathfrak{U}_R = \{\alpha \in \mathfrak{G}_R \mid \nu(\alpha) = 1\}$.

3.2. *Remark.* When B is isomorphic to $M_2(F)$, an arbitrarily given left B-module X (of finite dimension over F) is not necessarily isomorphic to the direct sum of copies of B. However, if X has a non-degenerate ι-hermitian form, X is always isomorphic to such a direct sum. In fact, if h is such a form, we can find an element x_0 of X such that $h(x_0, x_0) \neq 0$. Put $X_1 = \{y \in X \mid h(y, x_0) = 0\}$. By the standard argument, together with the fact $h(x_0, x_0) \in F$, we see that $X = Bx_0 \oplus X_1$, and Bx_0 is isomorphic to B. Applying induction to X_1, we get our assertion.

3.3. PROPOSITION. *For an element c of F, there exists an element α of $\mathfrak{G}$ such that $\nu(\alpha) = c$ if and only if $c \equiv 1 \bmod^* \mathfrak{u}$.*

This can be obtained by applying the Hasse principle to the form h and ch.

3.4. Let $\mathfrak{x}$ and $\mathfrak{y}$ be $\mathfrak{r}_F$-lattices in X. We denote by $\nu(\mathfrak{x}/\mathfrak{y})$ the ideal in F generated by the $\nu(\xi)$ for all $\xi \in \mathfrak{G}$ such that $\mathfrak{y}\xi \subset \mathfrak{x}$. We say that $\mathfrak{x}$ is *similar* to $\mathfrak{y}$ if there exists, for every $\mathfrak{p}$, an element $u_\mathfrak{p}$ of $\mathfrak{G}_\mathfrak{p}$ such that $\mathfrak{x}_\mathfrak{p} = \mathfrak{y}_\mathfrak{p} u_\mathfrak{p}$. If so, $\nu(\mathfrak{x}/\mathfrak{y})_\mathfrak{p} = (\nu(u_\mathfrak{p}))$. A maximal set of similar $\mathfrak{r}_F$-lattices is called a *genus* of $\mathfrak{r}_F$-lattices in X.

3.5. Let $\mathfrak{x}$ be an $\mathfrak{r}_F$-lattice in X, and $\mathfrak{c}$ an integral ideal in F. For $\alpha, \beta \in \mathfrak{G}$, we write

$$\alpha \equiv \beta \qquad \bmod^* (\mathfrak{x}; \mathfrak{c})$$

if $\mathfrak{x}_\mathfrak{p}\alpha = \mathfrak{x}_\mathfrak{p}\beta = \mathfrak{x}_\mathfrak{p}$ and $\mathfrak{x}_\mathfrak{p}(\alpha - \beta) \subset (\mathfrak{c}\mathfrak{x})_\mathfrak{p}$ for every $\mathfrak{p}$ dividing $\mathfrak{c}$. We shall now show:

(3.5.1) *If $\alpha, \beta \in \mathfrak{G}$ and $\alpha \equiv \beta \bmod^* (\mathfrak{x}; \mathfrak{c})$, then $\nu(\alpha) \equiv \nu(\beta) \bmod^* \mathfrak{c}$.*

In fact, put $f(x, y) = \mathrm{Tr}_{B/F}(h(x, y))$. Let $\mathfrak{a}$ be the ideal in F generated over $\mathfrak{r}_F$ by the numbers $f(x, y)$ for all x and y in $\mathfrak{x}$. If $x, y \in \mathfrak{x}$, we have

$$(\nu(\alpha) - \nu(\beta))f(x, y) = f(x\alpha, y\alpha) - f(x\beta, y\beta)$$
$$= f(x\alpha - x\beta, y\alpha) + f(x\beta, y\alpha - y\beta) \, .$$

This shows that $(\nu(\alpha) - \nu(\beta)) \cdot \mathfrak{a}_\mathfrak{p} \subset (\mathfrak{c}\mathfrak{a})_\mathfrak{p}$ for every $\mathfrak{p}$ dividing $\mathfrak{c}$, hence $\nu(\alpha) - \nu(\beta) \in \mathfrak{c}_\mathfrak{p}$. For such $\mathfrak{p}$, both $\nu(\alpha)$ and $\nu(\beta)$ are $\mathfrak{p}$-units. Therefore we obtain (3.5.1).

3.6. Let S be a semi-simple algebra over F, and f an F-linear isomorphism of S into End (X, B) which maps identity to identity. Fix an $\mathfrak{r}_F$-lattice $\mathfrak{x}$ in X, and put $\mathfrak{g} = \{a \in S \mid \mathfrak{x} f(a) \subset \mathfrak{x}\}$. Then $\mathfrak{g}$ is an $\mathfrak{r}_F$-order in S, and we have

(3.6.1) *Let $\mathfrak{a}$ be a two-sided $\mathfrak{g}$-ideal, and v an element of S. If $\mathfrak{x} f(\mathfrak{a}) = \mathfrak{x} f(v)$, one has $\mathfrak{a} = v\mathfrak{g} = \mathfrak{g}v$.*
In fact we have $\mathfrak{x} f(v\mathfrak{a}^{-1}) = \mathfrak{x} = \mathfrak{x} f(\mathfrak{a}v^{-1})$. Hence we have $v\mathfrak{a}^{-1} \subset \mathfrak{g}$ and $\mathfrak{a}v^{-1} \subset \mathfrak{g}$, so that $v\mathfrak{g} \subset \mathfrak{a} \subset \mathfrak{g}v$. But $v^{-1}\mathfrak{g}v$ can not contain $\mathfrak{g}$ properly, hence $v\mathfrak{g} = \mathfrak{a} = \mathfrak{g}v$.

(3.6.2) *For $v \in S$, one has $f(v) \equiv 1 \bmod^* (\mathfrak{x}; \mathfrak{c})$ if and only if $v \equiv 1 \bmod^* \mathfrak{c}\mathfrak{g}$.*
This follows immediately from our definition.

3.7. We say that an $\mathfrak{r}_F$-lattice $\mathfrak{x}$ in X is *ample* if the following two conditions are satisfied for every $\mathfrak{p}$.

(3.7.1) *For every non-zero element y of $\mathfrak{r}_\mathfrak{p}$, there exists an element β of $\mathfrak{G}_\mathfrak{p}$ such that $\mathfrak{x}_\mathfrak{p}\beta \subset \mathfrak{x}_\mathfrak{p}$ and $\nu(\beta) = y$.*

(3.7.2) *Let α be an element of $\mathfrak{G}_\mathfrak{p}$ and u a unit of $\mathfrak{r}_\mathfrak{p}$ such that $\mathfrak{x}_\mathfrak{p}\alpha = \mathfrak{x}_\mathfrak{p}$ and $\nu(\alpha) \equiv u \bmod \mathfrak{p}^\lambda$ with an integer $\lambda \geqq 0$. Then there exists an element β of $\mathfrak{G}_\mathfrak{p}$ such that $\mathfrak{x}_\mathfrak{p}(\alpha - \beta) \subset \mathfrak{p}^\lambda \mathfrak{x}_\mathfrak{p}$ and $\nu(\beta) = u$.*

We observe that an $\mathfrak{r}_F$-lattice similar to an ample $\mathfrak{r}_F$-lattice is also ample. Further, from (3.7.1), we can easily derive that

(3.7.3) *Let $\mathfrak{x}$ be an ample $\mathfrak{r}_F$-lattice in X, and $\mathfrak{c}$ an integral ideal in F. Then there exists an $\mathfrak{r}_F$-lattice $\mathfrak{y}$ similar to $\mathfrak{x}$ such that $\mathfrak{y} \subset \mathfrak{x}$ and $\nu(\mathfrak{y}/\mathfrak{x}) = \mathfrak{c}$.*

3.8. *Example of ample $\mathfrak{r}_F$-lattices.* Let $\mathfrak{x}$ be an $\mathfrak{r}_F$-lattice in X, and $\mathfrak{o} = \{a \in B \mid a\mathfrak{x} \subset \mathfrak{x}\}$. Then $\mathfrak{o}$ is an $\mathfrak{x}_F$-order in B. Note that $\mathfrak{o}^\iota = \mathfrak{o}$, since $\mathfrak{r}_F \subset \mathfrak{o}$. We shall now prove that $\mathfrak{x}$ is ample if the following three conditions are satisfied for every $\mathfrak{p}$.

(3.8.1) *For every $y \in \mathfrak{r}_\mathfrak{p}$, there exists an element a of $\mathfrak{o}_\mathfrak{p}$ such that $aa^\iota = y$.*

(3.8.2) *For every $z \in \mathfrak{r}_\mathfrak{p}$, there exists an element b of $\mathfrak{o}_\mathfrak{p}$ such that $b + b^\iota = z$.*

(3.8.3) *There exists a basis $\{x_1, \cdots, x_n\}$ of $X_\mathfrak{p}$ over $B_\mathfrak{p}$ such that $\mathfrak{x}_\mathfrak{p} = \mathfrak{o}_\mathfrak{p}x_1 + \cdots + \mathfrak{o}_\mathfrak{p}x_n$ and $h(x_i, x_j) \in F_\mathfrak{p}$ for every i and j.*

Let $0 \neq y \in \mathfrak{r}_\mathfrak{p}$. By (3.8.1), we can take an element a of $\mathfrak{o}_\mathfrak{p}$ so that $aa^\iota = y$. Define an element β of $GL(X_\mathfrak{p}, B_\mathfrak{p})$ by $x_i\beta = ax_i$. Then $\beta \in \mathfrak{G}_\mathfrak{p}$, $\mathfrak{x}_\mathfrak{p}\beta \subset \mathfrak{x}_\mathfrak{p}$ and $\nu(\beta) = y$. Thus $\mathfrak{x}$ satisfies (3.7.1). Next let α, u and λ be as in (3.7.2). We may assume $\lambda > 0$, since (3.7.1) implies (3.7.2) if $\lambda = 0$. Put $v = u \cdot \nu(\alpha)^{-1}$. Then v is a unit of $\mathfrak{r}_\mathfrak{p}$, and $v \equiv 1 \bmod \mathfrak{p}^\lambda$. We shall now construct inductively a sequence $\{b_0, b_1, \cdots, b_n, \cdots\}$ of units of $\mathfrak{o}_\mathfrak{p}$ such that $b_0 = 1$, $b_n b_n^\iota \equiv v \bmod \mathfrak{p}^{n+\lambda}$,

$b_{n+1} \equiv b_n \bmod \mathfrak{p}^{n+\lambda}\mathfrak{o}_\mathfrak{p}$. Assume that b_n has been defined. Put $v - b_n b_n' = \pi^{n+\lambda}w$ with a prime element π of $F_\mathfrak{p}$ and an element w of $\mathfrak{r}_\mathfrak{p}$. By (3.8.2), we have $\mathrm{Tr}_{B/F}(b_n\mathfrak{o}_\mathfrak{p}) = \mathfrak{r}_\mathfrak{p}$, so that $\mathrm{Tr}_{B/F}(b_n c') = w$ with an element c of $\mathfrak{o}_\mathfrak{p}$. Put $b_{n+1} = b_n + \pi^{n+\lambda}c$. Thus we get a sequence $\{b_n\}$ with the required properties. Obviously this sequence converges to an element b of $\mathfrak{o}_\mathfrak{p}$ such that $bb' = v$ and $b \equiv 1 \bmod \mathfrak{p}^\lambda\mathfrak{o}_\mathfrak{p}$. Now define an element β of $\mathrm{End}\,(X_\mathfrak{p}, B_\mathfrak{p})$ by $x_i\beta = bx_i\alpha$ for every i. Then $\beta \in \mathfrak{G}_\mathfrak{p}$ and $\nu(\beta) = u$. Further $x_i(\alpha - \beta) = (1 - b)x_i\alpha \in \mathfrak{p}^\lambda\mathfrak{x}_\mathfrak{p}$, hence β has the required properties of (3.7.2).

Now every maximal order $\mathfrak{o}$ satisfies (3.8.1, 2). A simple example of non-maximal order satisfying (3.8.1, 2) is given as follows: If $\mathfrak{p}$ is unramified in B, $B_\mathfrak{p}$ can be identified with $M_2(F_\mathfrak{p})$. Fix non-negative integers m and n, and put

$$\mathfrak{o}_\mathfrak{p} = \left\{ \begin{bmatrix} a & b \\ c & d \end{bmatrix} \in M_2(\mathfrak{r}_\mathfrak{p}) \,\middle|\, b \in \mathfrak{p}^m,\ c \in \mathfrak{p}^n \right\}.$$

If $\mathfrak{o}$ is an order in B such that $\mathfrak{o}_\mathfrak{p}$ is given in this way for every $\mathfrak{p}$ unramified in B (with m and n depending on $\mathfrak{p}$), and $\mathfrak{o}_\mathfrak{p}$ is maximal for every $\mathfrak{p}$ ramified in B, then $\mathfrak{o}$ satisfies (3.8.1, 2).

3.9. PROPOSITION. *Let P be a finite set of prime ideals in F. For each $\mathfrak{p}$ in P, let $\alpha_\mathfrak{p}$ be an element of $\mathfrak{U}_\mathfrak{p}$. Let $\mathfrak{x}$ and $\mathfrak{y}$ be $\mathfrak{r}_F$-lattices in X. Then there exists an element β of $\mathfrak{U}$ such that $\mathfrak{x}_\mathfrak{p}(\beta - \alpha_\mathfrak{p}) \subset \mathfrak{y}_\mathfrak{p}$ for all $\mathfrak{p} \in P$ and $\mathfrak{x}_\mathfrak{p}\beta = \mathfrak{x}_\mathfrak{p}$ for all $\mathfrak{p} \notin P$. Moreover, let $\mathfrak{R}$ be a non-empty open subset of $\mathfrak{U}_R$, and $\mathfrak{q}$ a prime ideal in F which is unramified in B. Then there exists an element γ of $\mathfrak{U}$ such that $\gamma \in \mathfrak{R}$ and $\mathfrak{x}_\mathfrak{p}\gamma = \mathfrak{x}_\mathfrak{p}$ for all $\mathfrak{p} \neq \mathfrak{q}$.*

This is a special case of Kneser's approximation theorem [3]. The first part is included in [6, 6.7].

3.10. PROPOSITION. *Let $\mathfrak{x}$ and $\mathfrak{y}$ be ample $\mathfrak{r}_F$-lattices in X belonging to the same genus, and c an element of F such that $\nu(\mathfrak{x}/\mathfrak{y}) = (c)$ and $c \equiv 1 \bmod^* \mathfrak{u}$, where $\mathfrak{u}$ is as in 3.1. Then there exists an element ζ of $\mathfrak{G}$ such that $\mathfrak{x} = \mathfrak{y}\zeta$ and $\nu(\zeta) = c$.*

PROOF. By 3.3, we can find an element γ of $\mathfrak{G}$ so that $\nu(\gamma) = c$. For every $\mathfrak{p}$, there exists an element $\alpha_\mathfrak{p}$ of $\mathfrak{G}_\mathfrak{p}$ such that $\mathfrak{x}_\mathfrak{p} = \mathfrak{y}_\mathfrak{p}\gamma\alpha_\mathfrak{p}$. We see that $\nu(\alpha_\mathfrak{p})$ is a unit of $\mathfrak{r}_\mathfrak{p}$. By (3.7.1), there exists an element $\beta_\mathfrak{p}$ of $\mathfrak{G}_\mathfrak{p}$ such that $\mathfrak{x}_\mathfrak{p}\beta_\mathfrak{p} = \mathfrak{x}_\mathfrak{p}$ and $\nu(\beta_\mathfrak{p}) = \nu(\alpha_\mathfrak{p})^{-1}$. Put $\delta_\mathfrak{p} = \alpha_\mathfrak{p}\beta_\mathfrak{p}$. Then $\mathfrak{x}_\mathfrak{p} = \mathfrak{y}_\mathfrak{p}\gamma\delta_\mathfrak{p}$ and $\nu(\delta_\mathfrak{p}) = 1$. Let P be the set of all $\mathfrak{p}$ such that $\mathfrak{x}_\mathfrak{p} \neq \mathfrak{y}_\mathfrak{p}\gamma$. By 3.9, there exists an element ε of $\mathfrak{U}$ such that $\mathfrak{x}_\mathfrak{p}\varepsilon = \mathfrak{x}_\mathfrak{p}$ for $\mathfrak{p} \notin P$, and $\mathfrak{y}_\mathfrak{p}\gamma(\varepsilon - \delta_\mathfrak{p}) \subset \mathfrak{p}\mathfrak{y}_\mathfrak{p}\gamma\delta_\mathfrak{p}$ for $\mathfrak{p} \in P$. Then $\mathfrak{y}_\mathfrak{p}\gamma\varepsilon = \mathfrak{y}_\mathfrak{p}\gamma\delta_\mathfrak{p}$ for all $\mathfrak{p}$, hence $\mathfrak{y}\gamma\varepsilon = \mathfrak{x}$. Putting $\zeta = \gamma\varepsilon$, we obtain our proposition.

3.11. PROPOSITION. *Let $\mathfrak{x}$ be an ample $\mathfrak{r}_F$-lattice in X, $\mathfrak{c}$ an integral ideal in F, b an element of $\mathfrak{r}_F$ prime to $\mathfrak{c}$, and α an element of $\mathfrak{G}$. Suppose that $\mathfrak{x}_\mathfrak{p}\alpha = \mathfrak{x}_\mathfrak{p}$ for every $\mathfrak{p}$ dividing $\mathfrak{c}$, $b \equiv 1 \bmod^* \mathfrak{u}$, and $\nu(\alpha) \equiv b \bmod^* \mathfrak{c}$. Then there exists an element β of $\mathfrak{G}$ such that $\mathfrak{x}\beta \subset \mathfrak{x}$, $\nu(\beta) = b$, and $\beta \equiv \alpha \bmod^* (\mathfrak{x}; \mathfrak{c})$; especially $\mathfrak{x}\beta = \mathfrak{x}$ if b is a unit of $\mathfrak{r}_F$.*

PROOF. By 3.3, we can find an element γ of $\mathfrak{G}$ so that $\nu(\gamma) = b$. By (3.7.2), there exists, for each $\mathfrak{p}$ dividing $\mathfrak{c}$, an element $\zeta_\mathfrak{p}$ of $\mathfrak{G}_\mathfrak{p}$ such that $\mathfrak{x}_\mathfrak{p}(\zeta_\mathfrak{p} - \alpha) \subset \mathfrak{c}_\mathfrak{p}\mathfrak{x}_\mathfrak{p}$ and $\nu(\zeta_\mathfrak{p}) = b$. Let $P = \{\mathfrak{p} \mid \mathfrak{x}_\mathfrak{p}\gamma \neq \mathfrak{x}_\mathfrak{p}\}$. By (3.7.1), for each $\mathfrak{p} \in P$ not dividing $\mathfrak{c}$, we can find an element $\xi_\mathfrak{p}$ of $\mathfrak{G}_\mathfrak{p}$ so that $\mathfrak{x}_\mathfrak{p}\xi_\mathfrak{p} \subset \mathfrak{x}_\mathfrak{p}$ and $\nu(\xi_\mathfrak{p}) = b$. By 3.9, there exists an element δ of $\mathfrak{U}$ such that

$$
\begin{aligned}
\mathfrak{x}_\mathfrak{p}(\delta - \zeta_\mathfrak{p}\gamma^{-1}) &\subset (\mathfrak{c}\mathfrak{x}\gamma^{-1})_\mathfrak{p} && \text{if } \mathfrak{p} \mid \mathfrak{c}\,, \\
\mathfrak{x}_\mathfrak{p}(\delta - \xi_\mathfrak{p}\gamma^{-1}) &\subset \mathfrak{x}_\mathfrak{p}\gamma^{-1} && \text{if } \mathfrak{p} \nmid \mathfrak{c} \text{ and } \mathfrak{p} \in P\,, \\
\mathfrak{x}_\mathfrak{p}\delta &= \mathfrak{x}_\mathfrak{p} && \text{if } \mathfrak{p} \nmid \mathfrak{c} \text{ and } \mathfrak{p} \notin P\,.
\end{aligned}
$$

Put $\beta = \delta\gamma$. Then β has all the required properties.

4. The isolated fixed points of $\mathfrak{G}^+$

4.1. For every positive integer n, we define spaces $\mathfrak{H}_n$ and $\mathfrak{D}_n$ by

$$
\mathfrak{H}_n = \{z \in M_n(C) \mid {}^t z = z,\ \mathrm{Im}\,(z) > 0\}\,,
$$
$$
\mathfrak{D}_n = \{z \in M_n(C) \mid {}^t z = z,\ 1 - {}^t\bar{z}z > 0\}\,,
$$

where $x > 0$ means that x is a positive definite hermitian matrix, and groups $\mathrm{Gp}\,(n, \boldsymbol{R})$, $\mathrm{Sp}\,(n, \boldsymbol{R})$, $\mathrm{Gp}^+\,(n, \boldsymbol{R})$ by

$$
\mathrm{Gp}\,(n, \boldsymbol{R}) = \{U \in \mathrm{GL}_{2n}\,(\boldsymbol{R}) \mid {}^t U J_n U = \nu(U)J_n \text{ with } \nu(U) \in \boldsymbol{R}\}\,,
$$
$$
\mathrm{Sp}\,(n, \boldsymbol{R}) = \{U \in \mathrm{Gp}\,(n, \boldsymbol{R}) \mid \nu(U) = 1\}\,,
$$
$$
\mathrm{Gp}^+\,(n, \boldsymbol{R}) = \{U \in \mathrm{Gp}\,(n, \boldsymbol{R}) \mid \nu(U) > 0\}\,,
$$

where

$$
J_n = \begin{bmatrix} 0 & 1_n \\ -1_n & 0 \end{bmatrix}.
$$

We define the action of an element $U = \begin{bmatrix} a & b \\ c & d \end{bmatrix}$ of $\mathrm{Gp}^+\,(n, \boldsymbol{R})$ (with a, b, c, d in $M_n(\boldsymbol{R})$) on $\mathfrak{H}_n$ by $U(z) = (az + b)(cz + d)^{-1}$ for $z \in \mathfrak{H}_n$.

4.2. PROPOSITION. *For every $z \in \mathfrak{H}_n$, put*

$$
\mathfrak{G}_z = \{U \in \mathrm{Sp}\,(n, \boldsymbol{R}) \mid U(z) = z\}\,.
$$

Then for each fixed z, there exist a matrix $T = \begin{bmatrix} a & b \\ c & d \end{bmatrix}$, with a, b, c, d in $M_n(C)$, and a representation ψ of $\mathfrak{G}_z$ by complex unitary matrices of size n with the following properties:

(4.2.1) Define $T(w) = (aw + b)(cw + d)^{-1}$ for $w \in \mathfrak{H}_n$. Then the map

$w \mapsto T(w)$ *gives a biregular map of* $\mathfrak{H}_n$ *onto* $\mathfrak{D}_n$ *such that* $T(z) = 0$.

$$(4.2.2) \qquad TUT^{-1} = \begin{bmatrix} \overline{\psi(U)} & 0 \\ 0 & \psi(U) \end{bmatrix} \qquad \textit{for every } U \in \mathfrak{S}_z .$$

The representation ψ *of* $\mathfrak{S}_z$ *is uniquely determined up to equivalence.* *Moreover,* T *can be chosen so that* $TJ_n \cdot {}^t\overline{T} = 2i \cdot \begin{bmatrix} 1_n & 0 \\ 0 & -1_n \end{bmatrix}$.

PROOF. It suffices to take $T = \begin{bmatrix} 1_n & -i1_n \\ 1_n & i1_n \end{bmatrix} \cdot S$ with an element S of Sp (n, R) such that $S(z) = i1_n$. Cf. Siegel [11].

4.3. PROPOSITION. *Let* z, $\mathfrak{S}_z$ *and* ψ *be as in 4.2, and let* $U \in \mathfrak{S}_z$. *Let* $\zeta_1, \cdots, \zeta_n$ *be all the characteristic roots of* $\psi(U)$. *Then* z *is the only fixed point of* U *if and only if* $\zeta_j \zeta_k \neq 1$ *for every* j *and* k ($j = k$ *or* $j \neq k$).

PROOF. Through the map $w \mapsto T(w)$ of (4.2.1), the question can be reduced to the investigation of the fixed points of the map $v \mapsto \overline{\psi(U)} \cdot v \cdot \psi(U)^{-1}(v \in \mathfrak{D}_n)$. Then our assertion is obvious.

4.4. Let F, B, X, h, $\mathfrak{G}$, $\mathfrak{U}$, $\mathfrak{G}_R$, $\mathfrak{U}_R$ be as in 3.1. Take a basis $\{x_1, \cdots, x_n\}$ of X over B as in (3.1.2). With respect to this basis, X can be identified with the space B^n of all n-dimensional row vectors with components in B, and End (X, B) with $M_n(B)$, the action of $M_n(B)$ on B^n being defined by the right multiplication. Then we have $h(x, y) = x \cdot {}^t y^\iota$, and

$$(4.4.1) \qquad h(x \cdot {}^t \alpha^\iota, y) = h(x, y\alpha) \qquad ((x, y) \in B^n \times B^n; \alpha \in M_n(B)) ;$$

further $\mathfrak{G}_R$ can be identified with the group

$$\{\alpha \in \mathrm{GL}_n(B_R) \mid \alpha \cdot {}^t \alpha^\iota = \nu(\alpha)1_n \text{ with } \nu(\alpha) \in F_R\} .$$

Let $\tau_{01}, \cdots, \tau_{0g}$ be as in 3.1. We see that B_R is isomorphic to $M_2(R)^r \times K^{g-r}$, the νth factor corresponding to $\tau_{0\nu}$ (see [C, 2.4]), hence $M_n(B_R)$ (or End (X_R, B_R)) is isomorphic to $M_{2n}(R)^r \times M_n(K)^{g-r}$. Now we shall obtain an R-linear isomorphism ω of End (X_R, B_R) onto $M_{2n}(R)^r \times M_n(K)^{g-r}$ with the following properties:

(4.4.2) *If* $\omega(\xi) = (\xi^{(1)}, \cdots, \xi^{(g)})$, *one has* $({}^t\xi^\iota)^{(\nu)} = J_n \cdot {}^t\xi^{(\nu)} J_n^{-1}$ *for* $\nu = 1, \cdots, r$ *and for every* $\xi \in M_n(B_R)$.

(4.4.3) $\omega(a1_n) = (a^{\tau_{01}}1_{2n}, \cdots, a^{\tau_{0r}}1_{2n}, a^{\tau_{0r+1}}1_n, \cdots, a^{\tau_{0g}}1_n)$ $(a \in F)$.

(4.4.4) $\omega(\mathfrak{G}_R) = \mathrm{Gp}\,(n, R)^r \times \mathrm{Gp}^*\,(n, K)^{g-r}$,

where

$$\mathrm{Gp}^*\,(n, K) = \{\alpha \in \mathrm{GL}_n\,(K) \mid \alpha \cdot {}^t\bar{\alpha} = \nu(\alpha)1_n \text{ with } \nu(\alpha) \in R\} ,$$

and the bar means the quaternion conjugation. To see this, take any R-linear isomorphism $\omega_0: B_R \to M_2(R)^r \times K^{g-r}$ such that $\omega_0(a) = (a^{\tau_{01}}, \cdots, a^{\tau_{0g}})$ for $a \in F$. If $x \in B_R$ and $\omega_0(x) = (x^{(1)}, \cdots, x^{(g)})$, then $(x^\iota)^{(\nu)} = J_1 \cdot {}^t x^{(\nu)} J_1^{-1}$ for $\nu \leq r$

and $(x^\iota)^{(\nu)} = \overline{x^{(\nu)}}$ for $\nu > r$. For $\xi = (x_{ij}) \in M_n(B_R)$ with $x_{ij} \in B_R$, define $\xi^{(\nu)}$ by

$$(4.4.5) \qquad \xi^{(\nu)} = \begin{cases} \begin{bmatrix} (a_{ij}^\nu) & (b_{ij}^\nu) \\ (c_{ij}^\nu) & (d_{ij}^\nu) \end{bmatrix} & \\ (x_{ij}^{(\nu)}) & \end{cases} \qquad \text{where } x_{ij}^{(\nu)} = \begin{bmatrix} a_{ij}^\nu & b_{ij}^\nu \\ c_{ij}^\nu & d_{ij}^\nu \end{bmatrix} \quad (\nu \leq r) , \qquad (\nu > r) ,$$

and ω by $\omega(\xi) = (\xi^{(1)}, \cdots, \xi^{(g)})$. Then we see easily that ω has all the required properties.

Now put

$$\mathfrak{G}_R^+ = \{\alpha \in \mathfrak{G}_R \mid \alpha^{(\nu)} \in \mathrm{Gp}^+ (n, R) \text{ for all } \nu \leqq r\} ,$$
$$\mathfrak{G}^+ = \mathfrak{G} \cap \mathfrak{G}_R^+ = \{\alpha \in \mathfrak{G} \mid \nu(\alpha) \in F^+\} ,$$

where F^+ is as in 3.1. Then every α in $\mathfrak{G}_R^+$ operates on the product $\mathfrak{H}_n^r$ of r copies of $\mathfrak{H}_n$ in a natural way,

$$\alpha(z_1, \cdots, z_r) = \big(\alpha^{(1)}(z_1), \cdots, \alpha^{(r)}(z_r)\big)$$

for every $(z_1, \cdots, z_r) \in \mathfrak{H}_n^r$ with $z_\nu \in \mathfrak{H}_n$. Of course this action depends on the choice of ω.

4.5. PROPOSITION. *Fix a point z of $\mathfrak{H}_n^r$, and put $S = \{\alpha \in \mathfrak{G}^+ \mid \alpha(z) = z\}$. Let Y denote the subalgebra of $M_n(B)$ generated by S over F, and δ the restriction of the involution $\xi \mapsto {}^t\xi^\iota$ to Y. Then δ is a positive involution of Y, and $S = \{\alpha \in Y \mid \alpha\alpha^\delta \in F^+\}$.*

PROOF. If $\alpha \in S$, then $\alpha \cdot {}^t\alpha^\iota \in F^+$, hence ${}^t\alpha^\iota \in S$. Therefore Y is stable under δ. Let $z = (z_1, \cdots, z_r)$ with $z_\nu \in \mathfrak{H}_n$. For each z_ν, take a matrix T_ν and a representation ψ_ν which play the roles of T and ψ of 4.2 and such that $T_\nu J_n \cdot {}^t\bar{T}_\nu = 2i \begin{bmatrix} 1_n & 0 \\ 0 & -1_n \end{bmatrix}$. We can extend ψ_ν to a representation of Y, which we write again ψ_ν, by means of the relation $T_\nu \alpha^{(\nu)} T_\nu^{-1} = \begin{bmatrix} \overline{\psi_\nu(\alpha)} & 0 \\ 0 & \psi_\nu(\alpha) \end{bmatrix}$ for $\alpha \in Y$, $\psi_\nu(a) = a^{\tau_0\nu} 1_n$ for $a \in F$. Since $(\alpha^\delta)^{(\nu)} = J_n {}^t\alpha^{(\nu)} J_n^{-1}$ for $\alpha \in Y$ and $\nu \leqq r$, we have

$$\begin{aligned} T_\nu(\alpha\alpha^\delta)^{(\nu)} T_\nu^{-1} &= T_\nu \alpha^{(\nu)} J_n \cdot {}^t\alpha^{(\nu)} J_n^{-1} T_\nu^{-1} \\ &= (T_\nu \alpha^{(\nu)} T_\nu^{-1})(T_\nu J_n \cdot {}^t\bar{T}_\nu) \cdot {}^t(\bar{T}_\nu \alpha^{(\nu)} \bar{T}_\nu^{-1})(T_\nu J_n \cdot {}^t\bar{T}_\nu)^{-1} \\ &= \begin{bmatrix} \overline{\psi_\nu(\alpha)} & 0 \\ 0 & \psi_\nu(\alpha) \end{bmatrix}\begin{bmatrix} 1_n & 0 \\ 0 & -1_n \end{bmatrix}\begin{bmatrix} {}^t\psi_\nu(\alpha) & 0 \\ 0 & {}^t\overline{\psi_\nu(\alpha)} \end{bmatrix}\begin{bmatrix} 1_n & 0 \\ 0 & -1_n \end{bmatrix} \\ &= \begin{bmatrix} \overline{\psi_\nu(\alpha)} \cdot {}^t\psi_\nu(\alpha) & 0 \\ 0 & \psi_\nu(\alpha) \cdot {}^t\overline{\psi_\nu(\alpha)} \end{bmatrix} . \end{aligned}$$

Therefore, if $\alpha \in Y$ and $\alpha\alpha^\delta = b \in F^+$, then $(b^{(\nu)})^{-1/2}\psi_\nu(\alpha)$ is a unitary matrix for $\nu \leqq r$, hence $\alpha(z) = z$. Further we see that $\mathrm{tr}\,((\alpha\alpha^\delta)^{(\nu)}) > 0$ if $0 \neq \alpha \in Y$ for $\nu \leqq r$. (Note that $Y \ni \alpha \neq 0$ implies $\alpha^{(\nu)} \neq 0$, hence $\psi_\nu(\alpha) \neq 0$.) But $\mathrm{tr}\,((\alpha \cdot {}^t\alpha^\iota)^{(\nu)}) > 0$ for every non-zero $\alpha \in M_n(B)$ and for $\nu > r$. Hence if Tr denotes the reduced trace of $M_n(B)$ to $\boldsymbol{Q}$, we have $\mathrm{Tr}\,(\alpha\alpha^\delta) > 0$ for $0 \neq \alpha \in Y$. Hence δ is a positive involution of Y.

4.6. Theorem. *Let z, S, Y, δ be as in 4.5. Suppose that z is the only common fixed point of the elements of S. Then Y is a direct sum of simple algebras $Y_1, \cdots, Y_t$ whose centers are CM-fields. Moreover, if P_k denotes the center of Y_k and $[Y_k : P_k] = q_k^2$, $[P_k : F] = 2m_k$, then $n = \sum_{k=1}^{t} m_k q_k$.*

Proof. First note that Y is semi-simple, since it has a positive involution. Let $Y_1, \cdots, Y_t$ be the simple components of Y, and ε_k the identity element of Y_k. As in 4.4, identify X with B^n and $\mathrm{End}\,(X, B)$ with $M_n(B)$. Put $X_k = X\varepsilon_k$, $S_k = S\varepsilon_k$. Let $h_k(x, y)$ denote the restriction of h to X_k. Since ${}^t\varepsilon_k^\iota = \varepsilon_k^\delta = \varepsilon_k$, we see, by (4.4.1), that $h(x\varepsilon_j, y\varepsilon_k) = 0$ if $j \neq k$. It follows that h_k is non-degenerate on X_k. By 3.2, X_k is isomorphic to B^{n_k} with a positive integer n_k. Since $\varepsilon_1 + \cdots + \varepsilon_t$ is the identity element of $M_n(B)$, we have $n = n_1 + \cdots + n_t$. Obviously h_k satisfies (3.1.1) on X_k. Let $\mathfrak{G}_k$ be the group of all similitudes of h_k. Put

$$\mathfrak{U}_k = \{\alpha \in \mathfrak{G}_k \mid \nu(\alpha) = 1\},$$
$$S^* = \mathfrak{U}_R \cap (F_R \cdot S), \qquad S_k^* = \mathfrak{U}_{kR} \cap (F_R \cdot S_k).$$

Then, changing the coordinate-sytem of X suitably, we obtain diagonal injections

$$\begin{aligned}
&Y = Y_1 \oplus \cdots \oplus Y_t \to M_{n_1}(B) \oplus \cdots \oplus M_{n_t}(B) \to M_n(B), \\
(4.6.1) \qquad &S^* \to S_1^* \times \cdots \times S_t^* \to \mathfrak{U}_{1R} \times \cdots \times \mathfrak{U}_{tR} \to \mathfrak{U}_R,
\end{aligned}$$

which are compatible with the involutions in question. Without losing generality, we may assume $z = (i1_n, \cdots, i1_n)$. Then we can easily find a biregular (diagonal) injection of $\mathfrak{H}_{n_1}^r \times \cdots \times \mathfrak{H}_{n_t}^r$ into $\mathfrak{H}_n^r$ which is compatible with (4.6.1). Let $(z^1, \cdots, z^t)$, with $z^k \in \mathfrak{H}_{n_k}^r$, be the point on $\mathfrak{H}_{n_1}^r \times \cdots \times \mathfrak{H}_{n_t}^r$ corresponding to z. We see easily that

(i) z^k is the only common fixed point of S_k;

(ii) Y_k is spanned by S_k over F;

(iii) $\{\alpha \in \mathfrak{U}_k \mid \alpha(z^k) = z^k\} \subset S_k$.

We do not know, however, whether or not $\{\alpha \in \mathfrak{G}_k \mid \alpha(z^k) = z^k\} \subset Y_k$.

Fix our attention to one k. Let P_k denote the center of Y_k. Since Y_k has a positive involution, P_k is either totally real, or a CM-field. We can find an element d of B such that d^2 is a totally negative element of F. Define an involution π of B by $v^\pi = d^{-1}v'd$ for $v \in B$. Then we can easily verify that

$$(4.6.2) \qquad\qquad \mathrm{Tr}_{B/F}\,(vv^\pi)^{\iota_0\iota} > 0 \qquad\qquad \text{for } 0 \neq v \in B$$

Put $f(x, y) = \mathrm{Tr}_{B/F}\,(h_k(x, y)d)$ for $(x, y) \in X_k \times X_k$. Then we have $f(x, y) =$

$-f(y, x)$, and

$$(4.6.3) \qquad f(vxw, y) = f(x, v^{\pi}yw^{\delta}) \qquad (v \in B,\ w \in Y_k,\ (x, y) \in X_k \times X_k) .$$

Let $A = B \otimes_F Y_k$. Then A is a central simple algebra over P_k. For every $a = \sum_i v_i \otimes w_i \in A$ with $v_i \in B$, $w_i \in Y_k$, define the action of a on X_k by $ax = \sum_i v_i x w_i^{\delta}$ $(x \in X_k)$. Then X_k becomes a left A-module. Further we can define an involution σ of A by $a^{\sigma} = \sum_i v_i^{\tau} \otimes w_i^{\delta}$. From (4.6.3) we obtain $f(ax, y) = f(x, a^{\sigma}y)$ for $a \in A$. By [8, 1.6], there exists a non-degenerate A-valued σ-anti-hermitian form $\varphi(x, y)$ on X_k such that $f(x, y) = \mathrm{Tr}_{A/F}(\varphi(x, y))$. Let α be an element of $\mathrm{GL}(X_k, A)$ such that $\varphi(x\alpha, y\alpha) = \varphi(x, y)$. By our definition of the action of A on X_k, we see that $\alpha \in \mathrm{GL}(X_k, B)$. Moreover, for every $v \in B$, we have

$$\mathrm{Tr}_{B/F}\big(vh_k(x\alpha, y\alpha)d\big) = \mathrm{Tr}_{A/F}\big(\varphi(vx\alpha, y\alpha)\big) = \mathrm{Tr}_{A/F}\big(\varphi(vx, y)\big)$$
$$= \mathrm{Tr}_{B/F}\big(vh_k(x, y)d\big) ,$$

hence $h_k(x\alpha, y\alpha) = h_k(x, y)$. Thus $\alpha \in \mathfrak{U}_k$. Since α commutes with the action of the elements of Y_k, we have, for every $\xi \in S_k$, $\xi\alpha(z^k) = \alpha\xi(z^k) = \alpha(z^k)$. But z^k is the only common fixed point of S_k, hence we have $\alpha(z^k) = z^k$, so that $\alpha \in Y_k$ by the above (iii). Therefore α must be contained in the center of Y_k, i.e., $\alpha \in P_k$. By 1.2, A must be isomorphic to $M_s(P_k)$ with a positive integer s, and X_k is of dimension s over P_k.

Now assume that P_k is totally real. Let τ be an isomorphism of P_k into R, which coincides with τ_{01} on F. Identify P_k with P_k^{τ} through τ, and put $X^* = X_k \otimes_{P_k} R$, $Y^* = Y_k \otimes_{P_k} R$, $A^* = A \otimes_{P_k} R$. Then A^* is isomorphic to $M_s(R)$, and X^* is of dimension s over R. Extend φ and σ R-linearly. Then σ is a positive involution of A^*. This follows from (4.6.2) and the positivity of δ on Y_k. By [5, Lem. 1], we can find an R-linear isomorphism χ of A^* onto $M_s(R)$ so that $\chi(x^{\sigma}) = {}^t\chi(x)$. Let e_{ij}, for $i, j = 1, \cdots, s$, be the matrix units of $M_s(R)$. Put $\mu_{ij} = \chi^{-1}(e_{ij})$, $W_1 = \mu_{11}X^*$. Then $\mu_{ij}^{\sigma} = \mu_{ji}$. We can define an R-valued R-bilinear form $\psi(x, y)$ on W_1 so that $\mu_{11}\psi(x, y) = \varphi(x, y)$. Since φ is σ-anti-hermitian, ψ is alternating. Moreover, we have $\varphi(\mu_{ii}x, \mu_{jj}y) = \mu_{i1}\varphi(\mu_{1i}x, \mu_{1j}y)\mu_{1j} = \mu_{ij}\psi(\mu_{1i}x, \mu_{1j}y)$, hence ψ is different from 0. But this is a contradiction, since W_1 is of dimension one over R.

Therefore P_k must be a CM-field. Put $2m_k = [P_k : F]$, $q_k^2 = [Y_k : P_k]$. Then $2m_k s^2 = [A : F] = [B : F][Y_k : F] = 8m_k q_k^2$, hence $s = 2q_k$. It follows that $4n_k = [X_k : F] = s \cdot [P_k : F] = 2m_k s = 4m_k q_k$, hence $n_k = m_k q_k$. This completes the proof.

4.7. **Theorem.** *Let $P_1, \cdots, P_t$ be* CM-*fields containing F, and $Y_1, \cdots, Y_t$ be central simple algebras over $P_1, \cdots, P_t$, respectively. Let $Y = Y_1 \oplus \cdots \oplus Y_t$, $P = P_1 \oplus \cdots \oplus P_t$, $[Y_i : P_i] = q_i^2$, $[P_i : F] = 2m_i$. Suppose that $n = \sum_{i=1}^t m_i q_i$.*

Then the following assertions hold.

(4.7.1) *If there exists an F-linear isomorphism f of Y into $M_n(B)$, each Y_j belongs to the same Brauer class as $B \otimes_F P_i$. Further f maps the identity element of Y to the identity element of $M_n(B)$, and $f(Y)$ and $f(P)$ are the commutors of each other in $M_n(B)$.*

(4.7.2) *If Y_i belongs to the same Brauer class as $B \otimes_F P_i$, Y_i has a positive involution.*

(4.7.3) *Suppose that Y_i belongs to the same Brauer class as $B \otimes_F P_i$ for each i; let δ be a positive involution of Y. Then there exists an F-linear isomorphism f of Y into $M_n(B)$, such that $f(a^\delta) = {}^tf(a)^\iota$ for every $a \in Y$.*

(4.7.4) *Let f and f' be F-linear isomorphisms of Y into $M_n(B)$. Then there exists an element α of $\mathrm{GL}_n(B)$ such that $f'(a) = \alpha^{-1}f(a)\alpha$ for all $a \in Y$. Moreover, if $f(a^\delta) = {}^tf(a)^\iota$ and $f'(a^\delta) = {}^tf'(a)^\iota$ for all $a \in Y$, the element α can be taken so that $\alpha \cdot {}^t\alpha^\iota \in \{f(a) \mid a \in P, a^\delta = a\}$.*

(4.7.5) *Let δ and f be as in (4.7.3). Put*

$$(Y, \delta)_0 = \{a \in Y \mid aa^\delta \in F^+\}, \qquad (Y, \delta)_1 = \{a \in Y \mid aa^\delta = 1\},$$
$$(P, \delta)_0 = P \cap (Y, \delta)_0, \qquad (P, \delta)_1 = P \cap (Y, \delta)_1.$$

Then $f((P, \delta)_1)$ contains an element which has one and only one fixed point on $\mathfrak{H}_n^r$; and this point is the only common fixed point of the elements of $f((Y, \delta)_0)$. Moreover, Y (resp. P) is spanned by $(Y, \delta)_1$ (resp. $(P, \delta)_1$) over $\mathbf{Q}$.

(4.7.6) *The notation being as in (4.7.5), let $z = (z_1, \cdots, z_r)$, with $z_\nu \in \mathfrak{H}_n$, be the fixed point of $f((Y, \delta)_0)$, and ψ_ν the representation of $\mathfrak{S}_{z_\nu}$ determined as in 4.2. Then $\psi_\nu \circ f$ can be extended uniquely to a representation $\Psi^{(\nu)}: Y \to M_n(C)$ such that $\Psi^{(\nu)}(a) = a^{\tau_0 \nu}1_n$ for $a \in F$. Further let $\{\varphi_{ik}^{(\nu)} \mid k = 1, \cdots, 2m_i\}$, for each i, be a set of inequivalent absolutely irreducible representations of Y_i such that $\varphi_{ik}^{(\nu)}(a) = a^{\tau_0 \nu}1_{q_i}$ for $a \in F$. Then, for every $\nu \leq r$, the direct sum of $\Psi^{(\nu)}$ and its complex conjugate is equivalent to the direct sum of $\varphi_{ik}^{(\nu)}$ for $i = 1, \cdots, t$ and $k = 1, \cdots, 2m_i$.*

PROOF. To prove (4.7.1), let f be an F-linear isomorphism of Y into $M_n(B)$, and ε_i be the identity element of Y_i for each i. Put $X_i = Xf(\varepsilon_i)$. Since the X_i form a direct sum, we have $\sum_{i=1}^t [X_i : F] \leq 4n$. Put $A_i = B \otimes_F Y_i$. Then A_i is a central simple algebra over P_i, and $[A_i : P_i] = 4q_i^2$. For every $a = \sum_k b_k \otimes c_k \in A_i$ with $b_k \in B$, $c_k \in Y_i$, we define the action of a on X_i by $xa = \sum_k b_k' xf(c_k)$ for $x \in X_i$. Then X_i becomes a *right* A_i-module. The dimension of such a (non-zero) module over P_i is at least $[A_i : P_i]^{1/2} = 2q_i$. Hence we have $[X_i : F] \geq 4m_iq_i$. By our assumption $n = \sum_{i=1}^t m_iq_i$, we must have $[X_i : F] = 4m_iq_i$, and $X = X_1 + \cdots + X_t$. This implies especially that f maps the identity element of Y to the identity element of $M_n(B)$. Moreover, since $[X_i : P_i] = 2q_i$, A_i must be isomorphic to $M_{2q_i}(P_i)$. Now A_i may be identified with $(B \otimes_F P_i) \otimes_{P_i} Y_i$. Since $B \otimes_F P_i$ is a quaternion algebra over P_i, we see that Y_i belongs to the same Brauer class as $B \otimes_F P_i$. Now let Z be the commutor of $f(P)$ in $M_n(B)$. Let $\alpha \in Z$. Then α maps X_i into X_i for each i, and α commutes with the action of $B \otimes_F P_i$ on X_i. Since $(B \otimes_F P_i) \otimes_{P_i} Y_i$ is isomorphic to $M_{2q_i}(P_i)$ and $[X_i : P_i] = 2q_i$, we see that the action of α on X_i can be represented by an element of $f(Y_i)$. This proves $Z = f(Y)$. Let U be the commutor of $f(Y)$ in $M_n(B)$. Since $P \subset Y$, we have $U \subset Z = f(Y)$, hence U is contained in the center of $f(Y)$. It follows easily that $U = f(P)$. This finishes the proof of (4.7.1).

If Y_i belongs to the same Brauer class as $B \otimes_F P_i$, Y_i is isomorphic to a matrix algebra over either P_i or $B \otimes_F P_i$, In both cases, we can find easily an involution of Y_i whose restriction to P_i is the complex conjugation. Therefore (4.7.2) follows from 1.4.

Let the notation and the assumption be as in (4.7.3). Put $A_i = B \otimes_F Y_i$. Then A_i is isomorphic to $M_{2q_i}(P_i)$. Define an involution σ of A_i by $(b \otimes c)^\sigma = b^\iota \otimes c^\delta$ for $b \in B$, $c \in Y_i$. Identify A_i with $M_{2q_i}(P_i)$. Denote by ρ the complex conjugation in P_i. Then there exists an element u of $M_{2q_i}(P_i)$ such that ${}^t u^\rho = u$ and $a^\sigma = u \cdot {}^t a^\rho u^{-1}$ for every $a \in A_i$. Let W_i be the space of all $(2q_i)$-dimensional row vectors with components in P_i. Put $\varphi_i(x, y) = \mathrm{Tr}_{P_i/F}(xu \cdot {}^t y^\rho)$ for $x, y \in W_i$. Regard W_i as a left B-module and a right Y_i-module by defining $bxc = x(b^\iota \otimes c)$ for $b \in B$, $c \in Y_i$, $x \in W_i$. Then we see that $\varphi_i(b^\iota x c^\delta, y) = \varphi_i(x, byc)$. By [8, 1.6], there exists a non-degenerate B-valued ι-hermitian form $h_i(x, y)$ on W_i such that $\varphi_i(x, y) = \mathrm{Tr}_{B/F}(h_i(x, y))$. We shall now prove $h_i(x, x)^{\tau_{0\nu}} > 0$ for $0 \neq x \in W_i$ and $\nu > r$. Fix our attention to some $\tau_{0\nu}$ with $\nu > r$, and take an isomorphism τ of P_i into C which coincides with $\tau_{0\nu}$ on F; further extend τ to an isomorphism of $M_{2q_i}(P_i)$ into $M_{2q_i}(C)$ in a natural way. From our definition of σ, we see easily that $\mathrm{tr}(ww^\sigma)^\tau > 0$ for $0 \neq w \in A_i = M_{2q_i}(P_i)$. (Note that B is ramified at $\tau_{0\nu}$, and δ is a positive involution of Y_i.) Since $w^\sigma = u \cdot {}^t x^\rho u^{-1}$, we see that the complex hermitian matrix u^τ is definite, hence $\varphi_i(x, x)^{\tau_{0\nu}} > 0$ for $0 \neq x \in W_i$. It follows that $h_i(x, x)^{\tau_{0\nu}} > 0$ for $0 \neq x \in W_i$. Now observe that W_i is of dimension $4m_i q_i$ over F, $h_i(xc, y) = h_i(x, yc^\delta)$ for $c \in Y_i$, and the action of every c of Y_i on W_i commutes with the action of B. Therefore, in view of (3.1.2), putting $n_i = m_i q_i$, we can find an F-linear isomorphism f_i of Y_i into $M_{n_i}(B)$ such that $f_i(x^\delta) = {}^t f_i(x)^\iota$. Taking the diagonal embedding

$$M_{n_1}(B) \oplus \cdots \oplus M_{n_t}(B) \to M_n(B) ,$$

we obtain an isomorphism f of Y into $M_n(B)$ with the properties of (4.7.3).

Let f and f' be as in (4.7.4). Let ε_i and A_i be as above. Put $X_i = Xf(\varepsilon_i)$, $X_i' = Xf'(\varepsilon_i)$, and regard X_i, X_i' as A_i-modules. We have seen that they are irreducible A_i-modules. Hence there exists, for each i, an A_i-isomorphism α_i of X_i onto X_i'. Define $\alpha : X \to X$ so that $\alpha = \alpha_i$ on X_i. Then $\alpha \in \mathrm{End}\,(X, B)$, and $x\alpha f'(c) = xf(c)\alpha$ for $x \in X_i$ and $c \in Y_i$. It follows that $f'(a) = \alpha^{-1} f(a)\alpha$ for every $a \in Y$. Suppose now that $f(a^\delta) = {}^{\iota}f(a)^{\iota}$, $f'(a^\delta) = {}^{\iota}f'(a)^{\iota}$ for all $a \in Y$. Then $\alpha^{-1}f(a)\alpha = {}^{\iota}f'(a^\delta)^{\iota} = {}^{\iota}(\alpha^{-1}f(a^\delta)\alpha)^{\iota} = {}^{\iota}\alpha^{\iota}f(a)({}^{\iota}\alpha^{\iota})^{-1}$, hence $\alpha \cdot {}^{\iota}\alpha^{\iota}$ commutes with the elements of $f(Y)$. By (4.7.1), we have $\alpha \cdot {}^{\iota}\alpha^{\iota} = f(b)$ with an element b of P. Then $f(b^\delta) = {}^{\iota}f(b)^{\iota} = f(b)$, hence $b^\delta = b$. This completes the proof of (4.7.4).

Let the notation be as in (4.7.5). The last assertion is included in 1.6. Since δ is a positive involution, we see that $f((Y, \delta)_1)$ is contained in a compact subgroup of $\mathfrak{U}_R$. Therefore, there exists at least one fixed point z on $\mathfrak{H}_n^r$ common to the elements of $f((Y, \delta)_1)$. Let z_ν, ψ_ν and $\varphi_{ik}^{(\nu)}$ be as in (4.7.6). Since Y is generated by $(Y, \delta)_1$ over F, $\psi_\nu \circ f$ can be extended uniquely to a representation $\Psi^{(\nu)}$ of Y as described in (4.7.6). From (4.2.2) we see that $\Psi^{(\nu)} + \bar{\Psi}^{(\nu)}$ is equivalent to the representation: $Y \ni a \mapsto f(a)^{(\nu)}$ (in the notation of 4.4; see also the proof of 4.5). Hence $\Psi^{(\nu)}$ is faithful, and $\mathrm{tr}(\Psi^{(\nu)}(a) + \bar{\Psi}^{(\nu)}(a)) \in F'^{\tau_{0\nu}}$ for every $a \in Y$. Therefore the $\varphi_{ik}^{(\nu)}$ for all i and k occur in $\Psi^{(\nu)} + \bar{\Psi}^{(\nu)}$ with positive multiplicities. Since $n = \sum_{i=1}^{t} m_i q_i$, the multiplicities should all be equal to one. This proves (4.7.6), and implies also that, $\Psi^{(\nu)}$ can not contain two absolutely irreducible representations of Y which are complex conjugate to each other. The restriction of $\Psi^{(\nu)}$ to P has the same property, since every representation of Y can be determined, up to equivalence, by its restriction to P. By 1.5, we can find, for each i, an element b_i of P_i such that $b_i b_i^\delta = 1$, $P_i = Q(b_i)$, and b_i is different from any conjugate of b_j over Q if $i \neq j$. Let $c = b_1 + \cdots + b_t$. Then, every characteristic root of $\Psi^{(\nu)}(c)$ is imaginary, and $\Psi^{(\nu)}(c)$ can not have two characteristic roots which are complex conjugate to each other. By 4.3, z is the only fixed point of $f(c)$. If $\alpha \in f((Y, \delta)_0)$, α commutes with $f(c)$, hence $f(c)(\alpha(z)) = \alpha(f(c)(z)) = \alpha(z)$. But z is the only fixed point of $f(c)$, hence $\alpha(z) = z$. This proves (4.7.5) and ends the whole proof of 4.7.

4.8. PROPOSITION. *Let the notation be as in* (4.7.3, 5, 6). *Let* $S = \{\alpha \in \mathfrak{G}^+ \mid \alpha(z) = z\}$. *Then* $S = f((Y, \delta)_0)$ *if and only if the following condition is satisfied.*

(4.8.1) *For every proper subalgebra R of P containing F, which is a direct sum of* CM-*fields, the restriction of $\Psi^{(\nu)}$ to R, for some ν, contains two isomorphisms of a simple component of R into C which are complex* *conjugate.*

PROOF. Obviously $f((Y, \delta)_0) \subset S$. Let $\mathfrak{Y}$ be the subalgebra of $M_n(B)$ generated by S over F. By (4.7.5), we have $f(Y) \subset \mathfrak{Y}$. Let $\mathfrak{P}$ be the center of $\mathfrak{Y}$. By (4.7.1), $\mathfrak{P} \subset f(P)$. By 4.6, $\mathfrak{Y}$ and $\mathfrak{P}$ are algebras of the same type as Y and P. Hence by (4.7.1), $\mathfrak{Y}$ is the commutor of $\mathfrak{P}$ in $M_n(B)$. Therefore, if $S \neq f((Y, \delta)_0)$, we have $\mathfrak{Y} \neq f(Y)$ by 4.5, so $\mathfrak{P} \neq f(P)$. Applying (4.7.6) to $\mathfrak{Y}$ and $\mathfrak{P}$, we see that (4.8.1) can not hold for $R = f^{-1}(\mathfrak{P})$. This proves the *if* part. Conversely, suppose that P has a proper subalgebra R containing F which is a direct sum of CM-fields, and such that the restriction of $\Psi^{(\nu)}$ to R, for any ν, can not contain two complex conjugate isomorphisms of any simple component of R into C. By the same reasoning as in the end of the proof of 4.7, we can find an element d of R such that $dd^\delta = 1$ and z is the only fixed point of $f(d)$. Let $\mathfrak{Z}$ be the commutor of $f(R)$ in $M_n(B)$. Obviously $\mathfrak{Z}$ is stable under the involution $\alpha \mapsto {}^t\alpha^\iota$, and contains $f(Y)$. Let $\mathfrak{Z}_0 = \{\alpha \in \mathfrak{Z} \mid \alpha \cdot {}^t\alpha^\iota \in F^+\}$. If $\alpha \in \mathfrak{Z}_0$, we have $\alpha(z) = \alpha f(d)(z) = f(d)(\alpha(z))$, so that $\alpha(z) = z$. Therefore $f((Y, \delta)_0) \subset \mathfrak{Z}_0 \subset S$. Since $\mathfrak{Z}$ is the commutor of $f(R)$ in $M_n(B)$, we see easily that $\mathfrak{Z}$ is a direct sum of simple algebras whose centers are simple components of $f(R)$. By 1.6, $\mathfrak{Z}$ is generated by $\mathfrak{Z}_0$ over F. Therefore, if $f((Y, \delta)_0) = S$, we have $f(Y) = \mathfrak{Z}$, hence $P = R$. This proves the *only if* part.

4.9. Let $P, Y, P_1, \cdots, P_t, Y_1, \cdots, Y_t$ be as in 4.7, and δ a positive involution of Y. Further let f be an F-linear isomorphism of Y into $M_n(B)$ such that $f(a^\delta) = {}^t f(a)^\iota$ for every $a \in Y$. For such (Y, P, δ, f), we can always define $(Y, \delta)_0$ as in (4.7.5). Let z be the only common fixed point of $f((Y, \delta)_0)$ whose existence is ensured by (4.7.5). We call (Y, P, δ, f) *primitive* if

$$f((Y, \delta)_0) = \{\alpha \in \mathfrak{G}^+ \mid \alpha(z) = z\} \, ,$$

and denote by $J_n(B)$ the *set* (by abuse of language) of all primitive (Y, P, δ, f). From 4.8, we see, for example, that (Y, P, δ, f) is primitive if $[P_i : F] = 2$ for $i = 1, \cdots, t$, and no two of the P_i are isomorphic over F. In particular, if $n = 1$, (Y, P, δ, f) is always primitive.

Now releasing the primitivity, let $\Psi^{(\nu)}$ be as in (4.7.6). Fix our attention to a simple component P_i of P. By (4.7.6), we can write, for each $\nu \leq r$,

$$(4.9.1) \qquad \operatorname{tr} \Psi^{(\nu)}(a) = q_i \sum_{k=1}^{m_i} a^{\chi_{\nu k}} \qquad (a \in P_i)$$

with m_i isomorphisms $\chi_{\nu k}$ of P_i into C which, together with their complex conjugates, form the set of all extensions of $\tau_{0\nu}$ to P_i. Let Ψ_i denote a representation: $P_i \to M_{rm_i}(C)$ such that

$$(4.9.2) \qquad \Psi_i \sim \sum_{\nu=1}^{r} \sum_{k=1}^{m_i} \chi_{\nu k}$$

with the notation of [C, 5.1]. We denote symbolically by (P, Ψ) the collection of t couples (P_i, Ψ_i). Let (P'_i, Ψ'_i) be the dual of (P_i, Ψ_i) in the sense of [C, 5.1], and P' the composite of $P'_1, \cdots, P'_t$. We denote by (P', Ψ') the collection of these t couples (P'_i, Ψ'_i), and call it *the dual of* (P, Ψ), or *the dual of* (Y, P, δ, f). Note that P' itself is a CM-field, while P is a direct sum of CM-fields; (P'_i, Ψ'_i) is a CM-type in the sense of [C, 5.11] or [10] if and only if $r = g$. Here and henceforth, by a couple like (P_i, Ψ_i), we understand a couple formed by a field and an *equivalence-class* of representations.

For every element x of P', we define an element $\Psi'(x)$ of P by

$$(4.9.3) \qquad \Psi'(x) = \sum_{i=1}^{t} \det \Psi'_i(N_{P'/P'_i}(x)) .$$

Here note that $\det \Psi'_i(u) \in P_i$ for every $u \in P'_i$, by [C, 5.4]. Further, for every ideal $\mathfrak{y}$ in P', we denote by $\Psi'(\mathfrak{y})$ the $\mathfrak{r}_P$-ideal generated by the elements $\Psi'(x)$ for all $x \in \mathfrak{y}$ over $\mathfrak{r}_P$. Using the notation introduced in [C, 5.3], we can write

$$(4.9.4) \qquad \Psi'(\mathfrak{y}) = \sum_{i=1}^{t} \det \Psi'_i(N_{P'/P'_i}(\mathfrak{y})) ,$$

where the ith term is an ideal in P_i.

4.10. Let $\tau_{01}, \cdots, \tau_{0g}$ be as in 3.1. We fix our attention to the first r of them which are unramified in B, and denote by Θ a representation of F such that $\Theta \sim \sum_{\nu=1}^{r} \tau_{0\nu}$, and by (F', Θ') the dual of (F, Θ) in the sense of [C, 5.1]. Since $\operatorname{tr} \Psi_i(x) = m_i \cdot \operatorname{tr} \Theta(x)$ for $x \in F$, we have $F' \subset P'_i$. Now we have

$$(4.10.1) \qquad \det \Psi'_i(u) \cdot \det \Psi'_i(u)^\rho = \det \Theta'(N_{P'_i/F'}(u)) \qquad (u \in P'_i),$$

where ρ is the complex conjugation. To show this, take the smallest Galois extension R of $\mathbf{Q}$ containing P_i, and denote by G the Galois group of R over $\mathbf{Q}$. Extend $\tau_{0\nu}$ and $\chi_{\nu k}$ to elements of G, and denote them by the same letters. Let H, H', J, J' be the subgroups of G corresponding to F, F', P_i, P'_i respectively. By the definition of *the dual* in [C, 5.1], we have

$$(4.10.2) \qquad \Theta' \sim \sum_{\lambda=1}^{s} \sigma_\lambda , \qquad \Psi'_i \sim \sum_{\mu=1}^{p} \omega_\mu$$

with elements σ_λ and ω_μ of G such that

$$(4.10.3) \qquad \bigcup_{\nu=1}^{r} (H\tau_{0\nu})^{-1} = \bigcup_{\lambda=1}^{s} H'\sigma_\lambda , \qquad \bigcup_{\nu=1}^{r} \bigcup_{k=1}^{m_i} (J\chi_{\nu k})^{-1} = \bigcup_{\mu=1}^{p} J'\omega_\mu .$$

Since the $\chi_{\nu k}$ and $\chi_{\nu k}\rho$, for $k = 1, \cdots, m_i$, form the set of all extensions of $\tau_{0\nu}$ to P_i, we have

$$\bigcup_{\nu=1}^{r} \bigcup_{k=1}^{m_i} (J\chi_{\nu k} \cup J\chi_{\nu k}\rho) = \bigcup_{\nu=1}^{r} H\tau_{0\nu} ,$$

hence

$$\bigcup_{\mu=1}^{p} (J'\omega_\mu \cup J'\omega_\mu\rho) = \bigcup_{\lambda=1}^{s} H'\sigma_\lambda ,$$

from which follows (4.10.1). Substituting $N_{P'/P'_i}(x)$ for u, we obtain

$$(4.10.4) \qquad \Psi'(x)\Psi'(x)^\delta = \det \Theta'(N_{P'/F'}(x)) \qquad (x \in P').$$

This formula holds obviously under the substitution of an $\mathfrak{r}_P$-ideal $\mathfrak{y}$ for x.

4.11. The notation being as in 4.9 and 4.10, let $\mathfrak{x}$ be an $\mathfrak{r}_F$-lattice in X, and $\mathfrak{g} = \{a \in P \mid \mathfrak{x}f(a) \subset \mathfrak{x}\}$. Let $\mathfrak{h}$ be the conductor of $\mathfrak{g}$, and $\mathfrak{h}_0 = \mathfrak{h} \cap F$. We shall now prove

(4.11.1) *If $\mathfrak{a}$ is an ideal in P' prime to $N(\mathfrak{h}_0)$, then $\mathfrak{x}f([\Psi'(\mathfrak{a})]_\mathfrak{g})$ is similar to $\mathfrak{x}$, and $\nu(\mathfrak{x}f([\Psi'(\mathfrak{a})]_\mathfrak{g})/\mathfrak{x}) = \det \Theta'(N_{P'/F'}(\mathfrak{a}))$.*

Put $\mathfrak{y} = \mathfrak{x}f([\Psi'(\mathfrak{a})]_\mathfrak{g})$. If $\mathfrak{p}$ divides $N(\mathfrak{h}_0)$, we have $\mathfrak{x}_\mathfrak{p} = \mathfrak{y}_\mathfrak{p}$. Suppose $\mathfrak{p}$ does not divide $N(\mathfrak{h}_0)$. We can find an element b of P' so that $b^{-1}\mathfrak{a}$ is prime to $N(\mathfrak{p})$. Put $d = \Psi'(b)$. Then $d \in P$, and by (4.10.4), $dd^\delta = \det \Theta'(N_{P'/F'}(b)) \in F$, hence $f(d) \in \mathfrak{G}$ and $\nu(f(d)) = dd^\delta$. Obviously $\mathfrak{y}_\mathfrak{p} = \mathfrak{x}_\mathfrak{p}f(d)$. This proves (4.11.1).

5. Main theorems

5.1. Let F, B, g, r, X, h, $\mathfrak{G}$, $\mathfrak{U}$, $\mathfrak{G}_R$, $\mathfrak{U}_R$ be as in 3.1. Throughout the rest of the paper, we fix an identification of $\mathrm{End}(X_R, B_R)$ with $M_{2n}(R)^r \times M_n(K)^{g-r}$ as in 4.4; hence the action of $\mathfrak{G}_R^+$ on $\mathfrak{H}_n^r$ is fixed once and for all. Define (F', Θ') as in 4.10. Let $\mathfrak{F}$ denote the set of all ideals $\mathfrak{a}$ in F which can be written in the form $\mathfrak{a} = \det \Theta'(\mathfrak{b})$ (see [C, 5.3]) with an ideal $\mathfrak{b}$ in F'. Similarly, let W_0 denote the set of all elements x of F which can be written in the form $x = \det \Theta'(y)$ with a totally positive element y of F'. Further let

$$W_1 = \{u \in F^+ \mid u\mathfrak{r}_F \in \mathfrak{F}\} \ .$$

Take any subgroup W of W_1 containing W_0, and define D_W by

$$D_W = \{\alpha \in \mathfrak{G} \mid \nu(\alpha) \in W\} \ .$$

Then, D_W, as a subgroup of $\mathfrak{G}^+$, operates on $\mathfrak{H}_n^r$. For every $\mathfrak{r}_F$-lattice $\mathfrak{x}$ in X and an integral ideal $\mathfrak{c}$ in F, we put

$$\Gamma_W(\mathfrak{x}) = \{\gamma \in D_W \mid \mathfrak{x}\gamma = \mathfrak{x}\} \ ,$$
$$\Gamma_W(\mathfrak{x}, \mathfrak{c}) = \{\gamma \in \Gamma_W(\mathfrak{x}) \mid \mathfrak{x}(1 - \gamma) \subset \mathfrak{c}\mathfrak{x}\} \ .$$

These (modulo their centers) give discontinuous groups acting on $\mathfrak{H}_n^r$. The quotient $\mathfrak{H}_n^r/\Gamma_W(\mathfrak{x}, \mathfrak{c})$ is of finite measure; it is compact except for the following two cases:

(i) B is isomorphic to $M_2(F)$;

(ii) B is a division algebra, $r = g$, and $n > 1$.

This result follows easily from a theorem of Borel and Harish-Chandra [1]. If $r = 1$, $\Gamma_W(\mathfrak{x})$ is commensurable with the group introduced in Siegel [11].

Our main theorems will be concerned with a *canonical model* for $\mathfrak{H}_n^r/\Gamma_W(\mathfrak{x}, \mathfrak{c})$ in the case where $\mathfrak{x}$ is an ample $\mathfrak{r}_F$-lattice in the sense of 3.7. W will be arbitrary so far as $W_0 \subset W \subset W_1$. The change of W has only a small (but not negligible) effect on our theorems. Actually, once we obtain our results for a special choice of W, then it is easy to derive them for an arbitrary W. The case $W = W_0$ may be the most natural one, though there is no strong objective reason for this.

5.2. For every integral ideal c in F, denote by $I_W(F', \Theta', c)$ the group of all ideals $\mathfrak{a}$ in F' prime to $N(c)$, satisfying the following condition:

(5.2.1) *There exists an element v of W such that* $\det \Theta'(\mathfrak{a}) = (v)$ *and* $v \equiv 1 \bmod^* c$.

If u is a totally positive element of F' such that $u \equiv 1 \bmod^* (N(c))$, then $(u) \in I_W(F', \Theta', c)$. Hence $I_W(F', \Theta', c)$ corresponds to a class field over F', which we denote by $C_W(F', \Theta', c)$. We note that

(5.2.2) $I_W(F', \Theta', c) = \{\mathfrak{a} \in I(F', N(c)) \mid \det \Theta'(\mathfrak{a}) \in P(F, \mathfrak{cu}_0)\}$ *if* $W = W_1$, where $\mathfrak{u}_0$ denotes the product of all archimedean primes of F.

Let (Y, P, δ, f) be a member of $J_\pi(B)$, and let (P', Ψ') be its dual (see 4.9). Put, for a given $\mathfrak{r}_F$-lattice $\mathfrak{x}$ in X,

$$\mathfrak{g}_r = \{a \in Y \mid \mathfrak{x}f(a) \subset \mathfrak{x}\},$$
$$\mathfrak{g} = \mathfrak{g}_P = \mathfrak{g}_r \cap P = \{a \in P \mid \mathfrak{x}f(a) \subset \mathfrak{x}\}.$$

Then $\mathfrak{g}_r$ (resp. $\mathfrak{g}$) is an $\mathfrak{r}_F$-order in Y (resp. P). Let $\mathfrak{h}$ be the conductor of $\mathfrak{g}$ (see 2.1), and $\mathfrak{h}_0 = \mathfrak{h} \cap F$. We denote by $I_W(P', \Psi', \mathfrak{g}_r, c)$ the group of all ideals $\mathfrak{b}$ in P', prime to $N(\mathfrak{ch}_0)$, satisfying the following condition:

(5.2.3) *There exists an element v of Y such that* $[\Psi'(\mathfrak{b})]_\mathfrak{g}\mathfrak{g}_r = v\mathfrak{g}_r$, $vv^\delta \in W$ *and* $v \equiv 1 \bmod^* \mathfrak{cg}_r$ (for the notation, see 2.3, 2.5 and 4.9).

If $Y = P$, in view of (2.5.2) and (2.5.3), we see that (5.2.3) is equivalent to the following condition:

(5.2.4) *There exists an element v of P such that* $\Psi'(\mathfrak{b}) = \mathfrak{r}_P v$, $vv^\delta \in W$ *and* $v \equiv 1 \bmod^* (\mathfrak{g}; c)$.

Note that, both in (5.2.3) and (5.2.4), v is a unit of $\mathfrak{g}_\mathfrak{p}$ for every $\mathfrak{p}$ dividing $\mathfrak{ch}_0$.

Coming back to the general case, we see that, if $y \in P'$ and $y \equiv 1 \bmod^* (N(\mathfrak{ch}_0))$, then $\Psi'(y) \equiv 1 \bmod^* \mathfrak{cg}_r$ by (2.4.2), and $\Psi'(y)\Psi'(y)^\delta = \det \Theta'(N_{P'/F'}(y)) \in W$ by virtue of (4.10.4), hence $(y) \in I_W(P', \Psi', \mathfrak{g}_r, c)$ by (2.5.3). Therefore $I_W(P', \Psi', \mathfrak{g}_r, c)$ corresponds to a class field over P', which we denote by $C_W(P', \Psi', \mathfrak{g}_r, c)$. If $\mathfrak{b} \in I_W(P', \Psi', \mathfrak{g}_r, c)$, we have $\mathfrak{x}f([\Psi'(\mathfrak{b})]_\mathfrak{g}) = \mathfrak{x}f(v)$ and $vv^\delta \in W$, $v \equiv 1 \bmod^* \mathfrak{cg}_r$ with $v \in Y$, so that by (4.11.1) and (4.10.4),

$$vv^\delta \mathfrak{r}_P = \Psi'(\mathfrak{b})\Psi'(\mathfrak{b})^\delta = \det \Theta'(N_{P'/F'}(\mathfrak{b})) \cdot \mathfrak{r}_P,$$

hence $N_{P'/F'}(\mathfrak{b}) \in I_W(F', \Theta', c)$ by (3.6.2) and (3.5.1). It follows that

$$C_W(F', \Theta', c) \subset C_W(P', \Psi', \mathfrak{g}_r, c) \subset C(P', N(\mathfrak{ch}_0)).$$

5.3. MAIN THEOREM I. *Let the notation be as in 5.1 and 5.2. Let $\mathfrak{x}$ be an ample $\mathfrak{r}_F$-lattice in X, and c an integral ideal in F. Fix any subgroup W*

of W_1 containing W_0. Then there exist an algebraic variety V and a holomorphic mapping φ of $\mathfrak{H}_n^r$ onto V, invariant under $\Gamma_W(\mathfrak{x}, c)$, satisfying the following conditions:

(5.3.1) *φ gives a biregular isomorphism of $\mathfrak{H}_n^r/\Gamma_W(\mathfrak{x}, c)$ onto V.*

(5.3.2) *V is defined over $C_W(F', \Theta', c)$.*

(5.3.3) *Let $(Y, P, \delta, f) \in J_n(B)$, and let (P', Ψ') be the dual of (Y, P, δ, f). Put $\mathfrak{g}_r = \{a \in Y \mid \mathfrak{x} f(a) \subset \mathfrak{x}\}$. Let z be the fixed point of $f((Y, \delta)_0)$ on $\mathfrak{H}_n^r$ (see 4.9 and (4.7.5)). Then $C_W(P', \Psi', \mathfrak{g}_r, c)$ is exactly the composite of $C_W(F', \Theta', c)$ and $P'(\varphi(z))$.*

The proof will be completed in §8 or §9 according as $r < g$ or $r = g$. It should be noted that $\mathfrak{g}_r$ depends on the choice of $\mathfrak{x}$ and f. In general, given a discontinuous group Γ acting on $\mathfrak{H}_n^r$, we call a couple (V, φ) a *model for* $\mathfrak{H}_n^r/\Gamma$, if V is an algebraic variety defined over a subfield of C and φ is a holomorphic mapping of $\mathfrak{H}_n^r$ into V, invariant under Γ, which induces a biregular isomorphism of $\mathfrak{H}_n^r/\Gamma$ onto V. When $\Gamma = \Gamma_W(\mathfrak{x}, c)$, (V, φ) is called a *canonical model for* $\mathfrak{H}_n^r/\Gamma_W(\mathfrak{x}, c)$ if (5.3.1–3) are satisfied. This notion depends on the choice of the identification of $M_n(B)$ with $M_{2n}(R)^r \times K^{g-r}$, which amounts to the choice of the action of $\mathfrak{G}^+$ on $\mathfrak{H}_n^r$. (We fixed it at the beginning of this section.) Up to automorphisms of $\mathfrak{H}_n^r$, there are exactly 2^r different ways of choosing the action of $\mathfrak{G}^+$ on $\mathfrak{H}_n^r$. However, it can easily be shown that the change of the action of $\mathfrak{G}^+$ can be compensated by the change of the lattice $\mathfrak{x}$ with the original choice of the action. Therefore, we lose nothing by fixing one particular action of $\mathfrak{G}^+$ on $\mathfrak{H}_n^r$.

5.4. Let $\mathfrak{J}$ be as in 5.1. By an *$\mathfrak{J}$-genus of $\mathfrak{x}_F$-lattices* in X, we understand a maximal subset Λ of a genus (see 3.4) such that $\nu(\mathfrak{x}/\mathfrak{y}) \in \mathfrak{J}$ whenever $\mathfrak{x}$ and $\mathfrak{y}$ belong to Λ. By a *D_W-class of $\mathfrak{x}_F$-lattices* in X, we understand a maximal set M of $\mathfrak{x}_F$-lattices such that if $\mathfrak{x}$ and $\mathfrak{y}$ belong to M, $\mathfrak{x}\alpha = \mathfrak{y}$ for some $\alpha \in D_W$. An $\mathfrak{J}$-genus is obviously a disjoint union of several D_W-classes. For an integral ideal c in F, we denote by $\mathfrak{J}_c$ the set of all ideals $\mathfrak{a}$ in F of the form $\mathfrak{a} = \det \Theta'(\mathfrak{b})$ with an ideal $\mathfrak{b}$ in F' prime to $N(c)$.

5.5. PROPOSITION. *Let Λ be an $\mathfrak{J}$-genus consisting of ample $\mathfrak{x}_F$-lattices, and let $q = [I(F') : I_W(F', \Theta', (1))]$. Then the following assertions hold.*

(5.5.1) $q = [C_W(F', \Theta', (1)) : F'] = [\mathfrak{J} : \{u\mathfrak{x}_F \mid u \in W\}]$.

(5.5.2) *Let $\mathfrak{x} \in \Lambda$. Then the map $\Lambda \ni \mathfrak{y} \mapsto \nu(\mathfrak{y}/\mathfrak{x})$ gives a one-to-one correspondence between all D_W-classes of $\mathfrak{x}_F$-lattices contained in Λ and $\mathfrak{J}$ modulo $\{u\mathfrak{x}_F \mid u \in W\}$.*

(5.5.3) *For every integral ideal c in F, there exists a set of representatives $\{\mathfrak{x}_1, \cdots, \mathfrak{x}_q\}$ for the D_W-classes in Λ such that $\nu(\mathfrak{x}_\lambda/\mathfrak{x}_\mu) \in \mathfrak{J}_c$ and $\mathfrak{x}_{\lambda\mathfrak{p}} = \mathfrak{x}_{\mu\mathfrak{p}}$ for* *all $\mathfrak{p}$ dividing $\mathfrak{c}$, for $\lambda, \mu = 1, \cdots, q$.*

PROOF. The assertion (5.5.1) follows from the definition of $C_W(F', \Theta', (1))$ and the fact that $\mathfrak{a} \mapsto \det \Theta'(\mathfrak{a})$ gives an isomorphism of $I(F')/I_W(F', \Theta', (1))$ onto $\mathfrak{J}/\{u\mathfrak{r}_F \mid u \in W\}$. Now take a set of representatives $\{\mathfrak{b}_1, \cdots, \mathfrak{b}_q\}$ for $I(F')$ modulo $I_W(F', \Theta', (1))$. We can assume the $\mathfrak{b}_\lambda$ to be integral and prime to $N(\mathfrak{c})$. By (3.7.3), we can find, for each λ, an $\mathfrak{r}_F$-lattice $\mathfrak{x}_\lambda$ similar to $\mathfrak{x}$ so that $\mathfrak{x}_\lambda \subset \mathfrak{x}$ and $\nu(\mathfrak{x}_\lambda/\mathfrak{x}) = \det \Theta'(\mathfrak{b}_\lambda)$. Since $\mathfrak{b}_\lambda$ is integral and prime to $N(\mathfrak{c})$, we have $\nu(\mathfrak{x}_\lambda/\mathfrak{x}_\mu) \in \mathfrak{J}_{\mathfrak{c}}$ and $\mathfrak{x}_{\lambda\mathfrak{p}} = \mathfrak{x}_{\mu\mathfrak{p}}$ for every $\mathfrak{p}$ dividing $\mathfrak{c}$. This proves (5.5.2) and (5.5.3) in view of 3.10.

5.6. From now on, we fix an $\mathfrak{J}$-genus Λ of ample $\mathfrak{r}_F$-lattices. Let $\Delta = \{\mathfrak{x}_1, \cdots, \mathfrak{x}_q\}$ be a set of representatives for D_W-classes of $\mathfrak{r}_F$-lattices in Λ. From (5.5.2), we see easily that

(5.6.1) *For each μ, $\{\nu(\mathfrak{x}_1/\mathfrak{x}_\mu), \cdots, \nu(\mathfrak{x}_q/\mathfrak{x}_\mu)\}$ is a set of representatives for* $\mathfrak{J}/\{u\mathfrak{r}_F \mid u \in W\}$.

Given an integral ideal $\mathfrak{c}$ in F, we say that Δ is $\mathfrak{c}$-*regular* if $\nu(\mathfrak{x}_\lambda/\mathfrak{x}_\mu) \in \mathfrak{J}_{\mathfrak{c}}$ and $\mathfrak{x}_{\lambda\mathfrak{p}} = \mathfrak{x}_{\mu\mathfrak{p}}$ for all $\mathfrak{p}$ dividing $\mathfrak{c}$, for every λ and μ.

Suppose that Δ is $\mathfrak{c}$-regular. We denote by $D_W(\Delta, \mathfrak{c})$ the set of all α in D_W such that $\nu(\alpha)\mathfrak{r}_F \in \mathfrak{J}_{\mathfrak{c}}$ and $\mathfrak{x}_{\lambda\mathfrak{p}}\alpha = \mathfrak{x}_{\lambda\mathfrak{p}}$ for all $\mathfrak{p}$ dividing $\mathfrak{c}$. Observe that this group is independent of λ. For $\alpha, \beta \in D_W(\Delta, \mathfrak{c})$, we write

$$(5.6.2) \qquad\qquad \alpha \equiv \beta \qquad \mathrm{mod}^* (\Delta; \mathfrak{c})$$

if $\alpha \equiv \beta \,\mathrm{mod}^* (\mathfrak{x}_\lambda, \mathfrak{c})$ for some λ. If so, it holds for *all* λ. By (3.5.1), we have $\nu(\alpha) \equiv \nu(\beta) \,\mathrm{mod}^* \mathfrak{c}$ if $\alpha \equiv \beta \,\mathrm{mod}^* (\Delta; \mathfrak{c})$. Further, $\nu(\mathfrak{x}_\mu/\mathfrak{x}_\lambda\alpha) \in \mathfrak{J}_{\mathfrak{c}}$ for every $\alpha \in D_W(\Delta, \mathfrak{c})$ and every λ and μ. Let us now prove

(5.6.3) *For every λ and every automorphism σ of $C_W(F', \Theta', \mathfrak{c})$ over F', there exist an element α of $D_W(\Delta, \mathfrak{c})$, an index μ, and an ideal $\mathfrak{a}$ in F' prime to $N(\mathfrak{c})$ such that $\sigma = [C_W(F', \Theta', \mathfrak{c})/F', \mathfrak{a}]$ and $\det \Theta'(\mathfrak{a}) = \nu(\mathfrak{x}_\mu\alpha/\mathfrak{x}_\lambda)$.*

Take an integral ideal $\mathfrak{a}$ in F' prime to $N(\mathfrak{c})$ so that $\sigma = [C_W(F', \Theta', \mathfrak{c})/F', \mathfrak{a}]$. By (3.7.3), there exists an $\mathfrak{r}_F$-lattice $\mathfrak{y}$ in X similar to $\mathfrak{x}_\lambda$ such that $\mathfrak{y} \subset \mathfrak{x}_\lambda$ and $\nu(\mathfrak{y}/\mathfrak{x}_\lambda) = \det \Theta'(\mathfrak{a})$. Since $\nu(\mathfrak{y}/\mathfrak{x}_\lambda) \in \mathfrak{J}$, there exist an index μ and an element α of D_W such that $\mathfrak{y} = \mathfrak{x}_\mu\alpha$. Since $\mathfrak{a}$ is prime to $N(\mathfrak{c})$, we have $(\nu(\alpha)) \in \mathfrak{J}_{\mathfrak{c}}$ and $\mathfrak{y}_{\mathfrak{p}} = \mathfrak{x}_{\lambda\mathfrak{p}}$ for all $\mathfrak{p}$ dividing $\mathfrak{c}$. This shows $\alpha \in D_W(\Delta, \mathfrak{c})$.

5.7. MAIN THEOREM II. *Let $\mathfrak{c}$ be an integral ideal in F, and $\Delta = \{\mathfrak{x}_1, \cdots, \mathfrak{x}_q\}$ a $\mathfrak{c}$-regular set of representatives of D_W-classes in an $\mathfrak{J}$-genus of ample $\mathfrak{r}_F$-lattices in X (see 5.4 and 5.6). Then there exists a system*

$$\{V_\lambda, \varphi_\lambda, R_\sigma^{\mu\lambda}(\alpha) \,(\lambda, \mu = 1, \cdots, q; \alpha \in D_W(\Delta, \mathfrak{c}))\}$$

formed by the objects satisfying the following conditions:

(5.7.1) *For each λ, $(V_\lambda, \varphi_\lambda)$ is a canonical model for $\mathfrak{H}_n/\Gamma_W(\mathfrak{x}_\lambda; \mathfrak{c})$.*

(5.7.2) *Let* $\alpha \in D_W(\Delta, \mathfrak{c})$, $\nu(\mathfrak{x}_\mu \alpha/\mathfrak{x}_\lambda) = \det \Theta'(\mathfrak{a})$ *with an ideal* $\mathfrak{a}$ *in* F' *prime to* $N(\mathfrak{c})$, *and let* $\sigma = [C_W(F', \Theta', \mathfrak{c})/F', \mathfrak{a}]$. $R_\sigma^{\mu\lambda}(\alpha)$ *is a biregular isomorphism of* V_λ *onto* V_μ^σ, *rational over* $C_W(F', \Theta', \mathfrak{c})$, *defined for such* α, λ, μ *and* σ.

(5.7.3) $R_\sigma^{\mu\lambda}(\alpha) = R_\sigma^{\mu\lambda}(\beta)$ *if* $\alpha \equiv \beta \bmod^* (\Delta; \mathfrak{c})$.

(5.7.4) $R_\tau^{\nu\mu}(\beta)^\sigma \circ R_\sigma^{\mu\lambda}(\alpha) = R_{\tau\sigma}^{\nu\lambda}(\beta\alpha)$.

(5.7.5) *If* $\gamma \in \Gamma_W(\mathfrak{x}_\lambda)$, $R_1^{\lambda\lambda}(\gamma)[\varphi_\lambda(w)] = \varphi_\lambda(\gamma(w))$ *for every* $w \in \mathfrak{H}_n^r$.

(5.7.6) *Let* (Y, P, δ, f), (P', Ψ') *and* z *be as in* (5.3.3). *Put*

$$\mathfrak{g}_r = \{a \in Y \mid \mathfrak{x}_\mu f(a) \subset \mathfrak{x}_\mu\}, \, \mathfrak{g} = \mathfrak{g}_r \cap P.$$

Let $\mathfrak{h}$ *be the conductor of* $\mathfrak{g}$, *and* $\mathfrak{h}_0 = \mathfrak{h} \cap F$. *Further let* $\mathfrak{b}$ *be an ideal in* P' *prime to* $N(\mathfrak{c}\mathfrak{h}_0)$ *and let*

$$\tau = [C_W(P', \Psi', \mathfrak{g}_r, \mathfrak{c})/P', \mathfrak{b}].$$

Then one has $\mathfrak{x}_\mu f([\Psi'(\mathfrak{b})]_\mathfrak{g})^{-1} = \mathfrak{x}_\lambda \alpha^{-1}$ *with an element* α *of* $D_W(\Delta, \mathfrak{c})$ *and a unique* λ. *With such an element* α, *one has*

(5.7.6a) $$\varphi_\mu(z)^\tau = R_\sigma^{\mu\lambda}(\alpha)[\varphi_\lambda(\alpha^{-1}(z))].$$

We shall complete the proof in §§ 8, 9.

5.8. A few remarks may help the understanding of the above theorem. First, in (5.7.2), it should be observed that σ is uniquely determined by λ, μ, and the class of $\alpha \bmod^* (\Delta; \mathfrak{c})$; further, λ (resp. μ) and σ determine μ (resp. λ) uniquely by virtue of (5.6.1).

Next, in (5.7.6), the existence of α and λ can be shown as follows. By (4.11.1), $\mathfrak{x}_\mu f([\Psi'(\mathfrak{b})]_\mathfrak{g})^{-1}$ is similar to $\mathfrak{x}_\mu$, and

$$\nu(\mathfrak{x}_\mu f([\Psi'(\mathfrak{b})]_\mathfrak{g})^{-1}/\mathfrak{x}_\mu) = \det \Theta'(N_{P'/F'}(\mathfrak{b}))^{-1} \in \mathfrak{R}_\mathfrak{c}.$$

Therefore, we find a unique λ and an element α of D_W such that $\mathfrak{x}_\mu f([\Psi'(\mathfrak{b})]_\mathfrak{g})^{-1} = \mathfrak{x}_\lambda \alpha^{-1}$. Then we see easily that $(\nu(\alpha)) \in \mathfrak{R}_\mathfrak{c}$ and $\alpha \in D_W(\Delta, \mathfrak{c})$. By (4.10.4), we obtain the following formula

(5.8.1) $$\nu(\mathfrak{x}_\mu \alpha/\mathfrak{x}_\lambda) \cdot \mathfrak{r}_P = \Psi'(\mathfrak{b})\Psi'(\mathfrak{b})^\delta = \det \Theta'(N_{P'/F'}(\mathfrak{b})) \cdot \mathfrak{r}_P.$$

This shows especially that σ of (5.7.6a) is the restriction of τ to $C_W(F', \Theta', \mathfrak{c})$. Note also that $f(\Psi'(x)) \in D_W$ for $0 \neq x \in P'$ by (4.10.4).

Still in (5.7.6), if we define $f_\alpha: Y \to M_n(B)$ by $f_\alpha(x) = \alpha^{-1} f(x)\alpha$ for $x \in Y$, then (Y, P, δ, f_α) determines the same (P', Ψ') as before; and $\alpha^{-1}(z)$ is the fixed point of $f_\alpha((Y, \delta)_0)$. Moreover, $\mathfrak{g}_r = \{a \in Y \mid \mathfrak{x}_\lambda f_\alpha(a) \subset \mathfrak{x}_\lambda\}$. In fact, put $\mathfrak{z} = [\Psi'(\mathfrak{b})]_\mathfrak{g}$. For an element $a \in Y$, we have

$$\mathfrak{x}_\lambda f_\alpha(a) \subset \mathfrak{x}_\lambda \rightleftharpoons \mathfrak{x}_\lambda \alpha^{-1} f(a) \subset \mathfrak{x}_\lambda \alpha^{-1} \rightleftharpoons \mathfrak{x}_\mu f(\mathfrak{z})^{-1} f(a) \subset \mathfrak{x}_\mu f(\mathfrak{z})^{-1} \rightleftharpoons \mathfrak{x}_\mu f(a) \subset \mathfrak{x}_\mu, \quad \text{q.e.d.}$$

We call $\{V_\lambda, \varphi_\lambda, R_\sigma^{\mu\lambda}(\alpha)\}$ satisfying (5.7.1-6) a *canonical system of level* $\mathfrak{c}$ *with respect to* Δ *and* W.

The above theorem may be viewed in (at least) two rather different, but intimately related, aspects. On the one hand, it gives an explicit law of reciprocity for the abelian extension $C_W(P', \Psi', \mathfrak{g}_r, \mathfrak{c})$ of P'. On the other hand, one sees that $\mathfrak{H}_n^r/\Gamma_W(\mathfrak{x}_\lambda, \mathfrak{c})$ is a Galois covering of $\mathfrak{H}_n^r/\Gamma_W(\mathfrak{x}_\lambda)$, for which the behavior of the Galois group can be described in terms of the $R_\sigma^{\mu\lambda}(\alpha)$ (see also 5.9 below). We shall discuss the latter point in more detail in the last section of this paper.

5.9. THEOREM. *The notation and the assumption being as in 5.7, let $\{V_\lambda, \varphi_\lambda, R_\sigma^{\mu\lambda}(\alpha)\}$ be a canonical system of level $\mathfrak{c}$ with respect to Δ and W. Let $\mathfrak{m}$ be an integral ideal in F which divides $\mathfrak{c}$, and $\{U_\lambda, \psi_\lambda, S_\sigma^{\mu\lambda}(\alpha)\}$ a canonical system of level $\mathfrak{m}$ with respect to the same Δ and W. Then there exists, for each λ, a morphism T_λ of V_λ onto U_λ rational over $C_W(F', \Theta', \mathfrak{c})$ such that*

$$\psi_\lambda = T_\lambda \circ \varphi_\lambda, \qquad S_{\bar\sigma}^{\mu\lambda}(\alpha) \circ T_\lambda = T_\mu^\sigma \circ R_\sigma^{\mu\lambda}(\alpha)$$

for every μ, λ, and every $\alpha \in D_W(\Delta, \mathfrak{c})$, where $\bar\sigma$ is the restriction of σ to $C_W(F', \Theta', \mathfrak{m})$.

In particular, if $\mathfrak{c} = \mathfrak{m}$, this implies the uniqueness of a canonical system of level $\mathfrak{c}$, up to biregular isomorphisms over $C_W(F', \Theta', \mathfrak{c})$. The proof of 5.9 will be given in 8.2.

5.10. *Cases $r = 1$ and $r = g$.* First let us assume that $r = g$, so that B is totally indefinite. Then $F' = \boldsymbol{Q}$ and $\Theta'(v) = v$ for $v \in \boldsymbol{Q}$. (see [C, 5.2]). Therefore, if E^+ denotes the group of all totally positive units of F, then $W_1 = \boldsymbol{Q}^+ \cdot E^+$ and $W_0 = \boldsymbol{Q}^+$. Moreover, we have $q = 1$, so that we can use the symbol $R_\sigma(\alpha)$ instead of $R_\sigma^{\mu\lambda}(\alpha)$. In particular, if $W = W_0$, we have $C_W(F', \Theta', \mathfrak{c}) = \boldsymbol{Q}(\zeta)$ for a root of unity $\zeta = e^{2\pi i/c}$ with a positive integer c such that $c\boldsymbol{Z} = \mathfrak{c} \cap \boldsymbol{Q}$. In this case, $R_\sigma(\alpha)$ is defined for $\sigma = [\boldsymbol{Q}(\zeta)/\boldsymbol{Q}, (\nu(\alpha))]$, and (5.8.1) can be written as

$$(5.10.1) \qquad \nu(\alpha)\mathfrak{r}_P = \Psi'(\mathfrak{b})\Psi'(\mathfrak{b})^\delta = N(\mathfrak{b})\mathfrak{r}_P .$$

For further discussions in the case $r = g$, see §9, especially 9.5 and 9.8.

Next assume that $r = 1$, and τ_{01} is the identity map of F. We have then $F' = F$ and $\Theta'(v)$ for $v \in F$, so that $W_0 = W_1 = F^+$ and $D_W = \mathfrak{G}^+$. Therefore $C_W(F', \Theta', \mathfrak{c}) = C(F, \mathfrak{c})$. The set $\mathfrak{I}_\mathfrak{c}$ consists of all ideals in F prime to $N(\mathfrak{c})$. Now let $D_W^0(\Delta, \mathfrak{c})$ denote the group of all α in $\mathfrak{G}^+$ such that $\mathfrak{x}_{\lambda\mathfrak{p}}\alpha = \mathfrak{x}_\mathfrak{p}$ for all $\mathfrak{p}$ dividing $\mathfrak{c}$. Then $D_W(\Delta, \mathfrak{c})$ consists of all α in $D_W^0(\Delta, \mathfrak{c})$ such that $\nu(\alpha)$ is prime to $N(\mathfrak{c})$. The symbol $R_\sigma^{\mu\lambda}(\alpha)$ in 5.7 is defined for $\alpha \in D_W(\Delta, \mathfrak{c})$. We shall now extend it to the elements of $D_W^0(\Delta, \mathfrak{c})$. For every $\alpha \in D_W^0(\Delta, \mathfrak{c})$, we can find an element β of $D_W(\Delta, \mathfrak{c})$ such that $\beta \equiv \alpha \bmod^* (\Delta, \mathfrak{c})$. In fact it suffices to apply 3.11 to the present case with $\mathfrak{x} = \mathfrak{x}_\lambda$ and a totally positive element b of $\mathfrak{r}_F$

prime to $N(\mathfrak{c})$ such that $\nu(\alpha) \equiv b \bmod^* \mathfrak{c}$. Define $R_\sigma^{\mu\lambda}(\alpha)$ to be the same as $R_\sigma^{\mu\lambda}(\beta)$, where $\sigma = [C(F, \mathfrak{c})/F, (\nu(\alpha))]$. Then (5.7.3, 4) are still true. Since $F = F'$, (4.10.4) can now be written as $\Psi'(x)\Psi'(x)^\delta = N_{P'/F}(x)$ for every $x \in P'$. Therefore, if an ideal $\mathfrak{b}$ in P' is prime to an ideal $\mathfrak{a}$ in F, $\Psi'(\mathfrak{b})$ is prime to $\mathfrak{a}$. Let $I_W^0(P', \Psi', \mathfrak{g}_Y, \mathfrak{c})$ be the group of all ideals $\mathfrak{b}$ in P', prime to $\mathfrak{c}\mathfrak{h}_0$, satisfying (5.2.3). Then both $I_W^0(P', \Psi', \mathfrak{g}_Y, \mathfrak{c})$ and $I_W(P', \Psi', \mathfrak{g}_Y, \mathfrak{c})$ correspond to the same class field $C_W(P', \Psi', \mathfrak{g}_Y, \mathfrak{c})$, and

$$C_W(P', \Psi', \mathfrak{g}_Y, \mathfrak{c}) \subset C(P', \mathfrak{c}\mathfrak{h}_0) .$$

Now the assertion (5.7.6) holds for every ideal $\mathfrak{b}$ in P' prime to $\mathfrak{c}\mathfrak{h}_0$, which is not necessarily prime to $N(\mathfrak{c}\mathfrak{h}_0)$, with α in $D_W^0(\Delta, \mathfrak{c})$. In fact, let $\mathfrak{b}$ be an ideal in P' prime to $\mathfrak{c}\mathfrak{h}_0$. The discussion of 5.8 shows the existence of λ and an element α of $\mathfrak{G}^+$ such that $\mathfrak{x}_\mu f([\Psi'(\mathfrak{b})]_\mathfrak{g})^{-1} = \mathfrak{x}_\lambda \alpha^{-1}$. Then $\alpha \in D_W^0(\Delta, \mathfrak{c})$. Take an element u of P' so that $u \equiv 1 \bmod^* \mathfrak{c}\mathfrak{h}_0$ and $u\mathfrak{b}$ is prime to $N(\mathfrak{c}\mathfrak{h}_0)$. Put $\beta = f(\Psi'(u))\alpha$. Then $\beta \in \mathfrak{G}^+, \mathfrak{x}_\mu f([\Psi'(u\mathfrak{b})]_\mathfrak{g})^{-1} = \mathfrak{x}_\lambda \beta^{-1}, \beta^{-1}(z) = \alpha^{-1}(z)$, and $\nu(\mathfrak{x}_\mu \beta/\mathfrak{x}_\lambda) = N_{P'/F}(u\mathfrak{b})$, hence $\beta \in D_W(\Delta, \mathfrak{c})$. Further $\beta \equiv \alpha \bmod^* (\Delta, \mathfrak{c})$ and $R_\sigma^{\mu\lambda}(\beta) = R_\sigma^{\mu\lambda}(\alpha)$. Therefore we obtain the desired result from the assertion (5.7.6) with $u\mathfrak{b}$ and β in place of $\mathfrak{b}$ and α.

5.11. *Case $n = 1$.* If $n = 1$, X can be identified with B, and $\mathfrak{G}$ with the group of invertible elements of B. We have $\nu(\alpha) = N_{B/F}(\alpha)$ for $\alpha \in \mathfrak{G}$, and

$$\mathfrak{G}^+ = B^+ = \{\alpha \in B \mid \nu(\alpha) \text{ is totally positive}\} .$$

The notation being as in 4.7, we see that $t = 1$ and $Y = P = P_1$, P is isomorphic to a quadratic subfield of B, and δ is the complex conjugation of P. Therefore $J_1(B)$ may be regarded as the set of all embeddings of such fields. Thus the discussions of 4.6, 4.7, and 4.9 are essentially equivalent to those of [C, 2.6 and 2.7].

Now Main Theorems I, II, III, IV of [C] are special cases of the above 5.3 and 5.7, up to a certain minor point which will be explained below. Moreover, the latter show that Conjectures I and III of [C, 9.12] are true (again up to a minor point). In fact, the symbols A^+ and D^+ of [C, p. 137] are certainly one choice of W and D_W. Further, the $\mathfrak{o}_\lambda$ and $\mathfrak{x}_{\lambda\mu}$ being as in [C, p. 138], take $\mathfrak{x}_{11}$ as $\mathfrak{x}_\lambda$ in the present theorem 5.7. Then $\nu(\mathfrak{x}_\mu \alpha/\mathfrak{x}_\lambda) = N_{B/F}(\alpha\mathfrak{x}_{\lambda\mu})$, and the group $\Gamma[\mathfrak{o}_\lambda, \mathfrak{co}_\lambda]$ of [C, p. 137] coincides with the present $\Gamma_W(\mathfrak{x}_\lambda, \mathfrak{c})$ with $W = A^+$. The only point is that the main theorems and the conjectures of [C] have been formulated with an integral two-sided ideal $\mathfrak{e}$ as level. But the case of arbitrary $\mathfrak{e}$ can easily be derived from the special case $\mathfrak{e} = \mathfrak{co}$. Finally it may be noted that our present main theorems solve the question raised in [C, 9.12, (A)], and even include the case where the right or left orders of $\mathfrak{x}_\lambda$ are not necessarily maximal. Since P is quadratic over F, the definition of the ideal group $I_W(P', \Psi', \mathfrak{g}_r, \mathfrak{c})$ can be simplified by combining 2.7 with (5.2.4).

5.12. As for the uniqueness of canonical model, we shall prove it only in the form of 5.9, which, in the case $n = 1$, is not as strong as Conjecture II of [C, 9.12]. In general, the notation and the assumption being as in 5.3, let (V, φ) and (V', φ') be two canonical models for $\mathfrak{H}_n^r / \Gamma_W(\mathfrak{x}, \mathfrak{c})$. Then there exists obviously a biregular morphism S of V onto V' such that $\varphi' = S \circ \varphi$. It is conjecturable that S is defined over $C_W(F', \Theta', \mathfrak{c})$. We shall at least prove in 8.9 the following assertion.

(5.12.1) S *is defined over* $C_W(F', \Theta', \mathfrak{c})$, *if* $r = 1$ *and* $\mathfrak{x}$ *is similar to an* $\mathfrak{r}_F$-*lattice* $\mathfrak{o}x_1 + \cdots + \mathfrak{o}x_n$ *with a maximal order* $\mathfrak{o}$ *in* B *and a basis* $\{x_i\}$ *of* X *over* B *as in* (3.1.2).

5.13. PROPOSITION. *Let* $\mathfrak{c}$ *be an integral ideal in* F. *Let* $\Delta = \{\mathfrak{x}_1, \cdots, \mathfrak{x}_q\}$ *and* $\bar{\Delta} = \{\bar{\mathfrak{x}}_1, \cdots, \bar{\mathfrak{x}}_q\}$ *be* $\mathfrak{c}$-*regular sets of representatives for* D_W-*classes of* $\mathfrak{r}_F$-*lattices in the same* $\mathfrak{J}$-*genus. If there exists a canonical system of level* $\mathfrak{c}$ *with respect to* Δ *and* W, *then there exists a canonical system of level* $\mathfrak{c}$ *with respect to* $\bar{\Delta}$ *and* W.

PROOF. For each λ, there exists an element η_λ of D_W and a unique $\bar{\lambda}$ such that $\bar{\mathfrak{x}}_{\bar{\lambda}} = \mathfrak{x}_\lambda \eta_\lambda$. We see that $\lambda \mapsto \bar{\lambda}$ is a permutation of $\{1, \cdots, q\}$. By re-ordering the $\mathfrak{x}_\lambda$, we may assume $\lambda = \bar{\lambda}$. Then $\Gamma_W(\bar{\mathfrak{x}}_\lambda, \mathfrak{c}) = \eta_\lambda^{-1} \Gamma_W(\mathfrak{x}_\lambda, \mathfrak{c}) \eta_\lambda$ and $\nu(\eta_\lambda^{-1} \eta_\mu) \in \mathfrak{J}_\mathfrak{c}$. For every $\mathfrak{p}$ dividing $\mathfrak{c}$, we have $\mathfrak{x}_{\lambda \mathfrak{p}} \eta_\lambda = \mathfrak{x}_{\mu \mathfrak{p}} \eta_\mu$, hence $\eta_\mu \eta_\lambda^{-1} \in D_W(\Delta, \mathfrak{c})$. Similarly $\eta_\mu^{-1} \eta_\lambda \in D_W(\bar{\Delta}, \mathfrak{c})$, and $D_W(\bar{\Delta}, \mathfrak{c}) = \eta_\lambda^{-1} D_W(\Delta, \mathfrak{c}) \eta_\lambda$ for every λ and μ. Now let $\{V_\lambda, \varphi_\lambda, R_\sigma^{\mu\lambda}(\alpha)\}$ be as in 5.7. Put $\bar{\varphi}_\lambda(w) = \varphi_\lambda(\eta_\lambda(w))$ for $w \in \mathfrak{H}_n^r$, $\bar{R}_\sigma^{\mu\lambda}(\alpha) = R_\sigma^{\mu\lambda}(\eta_\mu \alpha \eta_\lambda^{-1})$ for $\alpha \in D_W(\bar{\Delta}, \mathfrak{c})$ and $\sigma = [C_W(F', \Theta', \mathfrak{c})/F', \mathfrak{a}]$ with an ideal $\mathfrak{a}$ in F' prime to $N(\mathfrak{c})$, such that $\nu(\bar{\mathfrak{x}}_\mu \alpha / \bar{\mathfrak{x}}_\lambda) = \det \Theta'(\mathfrak{a})$. Then it can be verified in a straightforward way that $\{V_\lambda, \bar{\varphi}_\lambda, \bar{R}_\sigma^{\mu\lambda}(\alpha)\}$ is a canonical system of level $\mathfrak{c}$ with respect to $\bar{\Delta}$ and W.

5.14. PROPOSITION. *Let* $\mathfrak{c}$ *and* $\Delta = \{\mathfrak{x}_\lambda\}$ *be as in* 5.13, *and* $\mathfrak{m}$ *an integral ideal in* F *dividing* $\mathfrak{c}$. *Let* β *be an element of* $D_W(\Delta, \mathfrak{m})$. *Then there exists an element* α *of* $D_W(\Delta, \mathfrak{c})$ *such that* $\alpha \equiv \beta \bmod^* (\Delta; \mathfrak{m})$.

PROOF. Since $\nu(\beta) \in \mathfrak{J}_\mathfrak{m}$, there exists an ideal $\mathfrak{b}$ in F', prime to $N(\mathfrak{m})$ such that $(\nu(\beta)) = \det \Theta'(\mathfrak{b})$. Take a totally positive element x of F' so that $x \equiv 1 \bmod^* N(\mathfrak{m})$ and $x\mathfrak{b}$ is an integral ideal prime to $N(\mathfrak{c})$. Put $y = \det \Theta'(x)$. Then $y \in F^+$, $y\nu(\beta) \equiv \nu(\beta) \bmod^* N(\mathfrak{m})$, and $(y\nu(\beta)) = \det \Theta'(x\mathfrak{b})$. Since $x\mathfrak{b}$ is integral, $y\nu(\beta) \in \mathfrak{r}_F$. By 3.11, there exists an element α of $\mathfrak{G}$ such that $\mathfrak{x}_\lambda \alpha \subset \mathfrak{x}_\lambda$, $\nu(\alpha) = y\nu(\beta)$, and $\alpha \equiv \beta \bmod^* (\mathfrak{x}_\lambda; \mathfrak{m})$. Since $x\mathfrak{b}$ is prime to $N(\mathfrak{c})$, we see that $\alpha \in D_W(\Delta, \mathfrak{c})$,   q.e.d.

5.15. PROPOSITION. *Let* c *and* $\Delta = \{\mathfrak{x}_\lambda\}$ *be as in* 5.13. *Suppose that a system*

$$\{V_\lambda,\ \varphi_\lambda,\ R_\sigma^{\mu\lambda}(\alpha)\ (\lambda,\ \mu = 1,\ \cdots,\ q;\ \alpha \in D_W(\Delta,\ c))\}$$

satisfies the following three conditions and (5.7.2–5).

(5.15.1) $(V_\lambda,\ \varphi_\lambda)$ *is a model for* $\mathfrak{H}_n^r/\Gamma_W(\mathfrak{x}_\lambda,\ c)$.

(5.15.2) V_λ *is defined over* $C_W(F',\ \Theta',\ c)$.

(5.15.3) *The notation being as in* (5.7.6), $\varphi_\mu(z)$ *is rational over* $C_W(P',$ $\Psi',\ \mathfrak{g}_r,\ c)$, *and* (5.7.6$_a$) *holds.*

Then, for each λ, $(V_\lambda,\ \varphi_\lambda)$ *is a canonical model for* $\mathfrak{H}_n^r/\Gamma_W(\mathfrak{x}_\lambda,\ c)$.

PROOF. The condition (5.15.3) implies the inclusion

(5.15.4) $$P'\big(\varphi_\mu(z)\big)\cdot C_W(F',\ \Theta',\ c) \subset C_W(P',\ \Psi',\ \mathfrak{g}_r,\ c)\ .$$

Our only task is to show that this is actually an equality. For this purpose, suppose that, in (5.7.6$_a$), τ is the identity map on the left hand side of (5.15.4). Then we have $\lambda = \mu$ and

(5.15.5) $$\varphi_\lambda(z) = R_1^{\lambda\lambda}(\alpha)[\varphi_\lambda(\alpha^{-1}(z))]\ .$$

Since

$$\tau = [C_W(F',\ \Theta',\ c)/F',\ N_{P'/F'}(\mathfrak{b})]$$

on $C_W(F',\ \Theta',\ c)$, there exists an element u of W such that

$$\det \Theta'\big(N_{P'/F'}(\mathfrak{b})\big) = (u),\ u \equiv 1 \bmod^* c\ .$$

By (5.8.1), we have $(\nu(\alpha)) = (u)$, hence, if we put $y = u^{-1}\nu(\alpha)$, then $y \in W$, y is a unit of F, and $\nu(\alpha) \equiv y \bmod^* c$. By 3.11, there exists an element γ of $\mathfrak{G}$ such that $\mathfrak{x}_\lambda\gamma = \mathfrak{x}_\lambda,\ \nu(\gamma) = y$, and $\gamma \equiv \alpha \bmod^* (\mathfrak{x}_\lambda;\ c)$. Then $\gamma \in \Gamma_W(\mathfrak{x}_\lambda)$. Put $\beta = \alpha\gamma^{-1}$. Since $\mathfrak{x}_\lambda\beta^{-1} = \mathfrak{x}_\lambda\alpha^{-1}$, we can take β in place of α. Substituting β for α in (5.15.5), we obtain $\varphi_\lambda(z) = R_1^{\lambda\lambda}(\beta)[\varphi_\lambda(\beta^{-1}(z))]$. Since $\beta \equiv 1 \bmod^* (\Delta;\ c)$, $R_1^{\lambda\lambda}(\beta)$ is the identity map, hence $\varphi_\lambda(z) = \varphi_\lambda(\beta^{-1}(z))$. Therefore we can find an element ε of $\Gamma_W(\mathfrak{x}_\lambda,\ c)$ so that $\varepsilon(z) = \beta^{-1}(z)$. Since

$$f\big((Y,\ \delta)_0\big) = \{\xi \in \mathfrak{G}^+\ |\ \xi(z) = z\}\ ,$$

$\beta\varepsilon = f(t)$ with an element t of Y. We have then

$$\mathfrak{x}_\lambda f(t^{-1}) = \mathfrak{x}_\lambda\beta^{-1} = \mathfrak{x}_\lambda\alpha^{-1} = \mathfrak{x}_\lambda f\big([\Psi'(\mathfrak{b})]_\mathfrak{g}\big)^{-1}\ .$$

By (3.6.1), we have $[\Psi'(\mathfrak{b})]_\mathfrak{g}\mathfrak{g}_r = t\mathfrak{g}_r$, and $t \equiv 1 \bmod^* c\mathfrak{g}_r$ by (3.6.2). Furthermore, since $f(t) = \beta\varepsilon \in D_W$, we have $tt^\delta = \nu(\beta\varepsilon) \in W$. It follows that $\mathfrak{b} \in I_W(P',\ \Psi',\ \mathfrak{g}_r,\ c)$ (see (5.2.3)), hence τ is the identity map. This completes the proof.

5.16. We shall now show that the proofs of 5.3 and 5.7 can be reduced to the special case $W = W_1$, by choosing a sufficiently *high* level c. In the course of our proof, we shall have to consider the objects $D_W,\ \Gamma_W(\mathfrak{x},\ c)$, $C_W(F',\ \Theta',\ c)$ etc., defined with respect to a fixed W and also with respect to

W_1. For simplicity, to indicate that they are defined with respect to W_1, we drop the subscript W_1 from the symbols and use D, $\Gamma(\mathfrak{x}, \mathfrak{c})$, etc., instead of D_{W_1}, $\Gamma_{W_1}(\mathfrak{x}, \mathfrak{c})$, etc. Further we shall write $C_W(\mathfrak{c})$ (resp. $C(\mathfrak{c})$) for $C_W(F', \Theta', \mathfrak{c})$ (resp. $C_{W_1}(F', \Theta', \mathfrak{c})$). Since W_1 contains all the totally positive units of F, we have obviously

$$(5.16.1) \qquad \begin{aligned} \Gamma(\mathfrak{x}) &= \{\gamma \in \mathfrak{G}^+ \mid \mathfrak{x}\gamma = \mathfrak{x}\} \,, \\ \Gamma(\mathfrak{x}, \mathfrak{c}) &= \{\gamma \in \Gamma(\mathfrak{x}) \mid \gamma \equiv 1 \bmod^* (\mathfrak{x}; \mathfrak{c})\} \,. \end{aligned}$$

Let $\mathfrak{c}$ and Δ be as in 5.7. By a theorem of Chevalley [2], there exists an integral ideal $\mathfrak{b}$ in F, divisible by $\mathfrak{c}$, with the following property:

(5.16.2) *If ε is a unit of F and $\varepsilon \equiv 1 \bmod \mathfrak{b}$, there exists a unit η of F such that $\eta \equiv 1 \bmod \mathfrak{c}$ and $\eta^2 = \varepsilon$.*

Fix such a $\mathfrak{b}$. We obtain then, for every $\mathfrak{r}_F$-lattice $\mathfrak{x}$ in X,

$$(5.16.3) \qquad \Gamma(\mathfrak{x}, \mathfrak{b}) \subset E \cdot \Gamma_W(\mathfrak{x}_\lambda, \mathfrak{c}) \,,$$

where E denotes the group of all units of F. We are going to construct a canonical system of level $\mathfrak{c}$ with respect to Δ and W, assuming that 5.3 and 5.7 are true with $\mathfrak{b}$ and W_1 in place of $\mathfrak{c}$ and W. In view of 5.13, we may assume that Δ is $\mathfrak{b}$-regular. Consider now D-classes of $\mathfrak{r}_F$-lattices in the $\mathfrak{J}$-genus in question. Since $W \subset W_1$, we have $D_W \subset D$, hence we can choose the representatives for such D-classes among $\{\mathfrak{x}_1, \cdots, \mathfrak{x}_q\}$. After re-arranging the $\mathfrak{x}_\lambda$ suitably, we may assume, with a positive integer $p \leq q$, that $\{\mathfrak{x}_1, \cdots, \mathfrak{x}_p\}$ is a set of representatives for them. Put $\Delta_1 = \{\mathfrak{x}_1, \cdots, \mathfrak{x}_p\}$.

Let $\{U_\kappa, \psi_\kappa, S_\sigma^{\nu\kappa}(\alpha)\ (\nu, \kappa = 1, \cdots, p;\ \alpha \in D(\Delta_1, \mathfrak{b}))\}$ be a canonical system of level $\mathfrak{b}$ with respect to Δ_1 and W_1. For each $\lambda \in \{1, \cdots, q\}$, there exists a unique $\kappa \in \{1, \cdots, p\}$ such that $\mathfrak{x}_\kappa$ belongs to the same D-class as $\mathfrak{x}_\lambda$. Choose and fix an element ζ_λ of D such that $\mathfrak{x}_\kappa \zeta_\lambda = \mathfrak{x}_\lambda$. Then $\zeta_\lambda \in D(\Delta_1, \mathfrak{b})$, $\Gamma(\mathfrak{x}_\kappa, \mathfrak{b}) = \zeta_\lambda \Gamma(\mathfrak{x}_\lambda, \mathfrak{b}) \zeta_\lambda^{-1}$. If $\lambda \leq p$, then $\kappa = \lambda$, and we put $\zeta_\lambda = 1$. Now define $(U_\lambda, \psi_\lambda)$ by $U_\lambda = U_\kappa$ and $\psi_\lambda(z) = \psi_\kappa(\zeta_\lambda(z))$ for $z \in \mathfrak{H}_n^r$. Further, for $\lambda, \mu \in \{1, \cdots, q\}$ and $\alpha \in D(\Delta_1, \mathfrak{b})$, we define $S_\sigma^{\mu\lambda}(\alpha)$ by $S_\sigma^{\mu\lambda}(\alpha) = S_\sigma^{\nu\kappa}(\zeta_\mu \alpha \zeta_\lambda^{-1})$ where $\mathfrak{x}_\nu \zeta_\mu = \mathfrak{x}_\mu$, $\mathfrak{x}_\kappa \zeta_\lambda = \mathfrak{x}_\lambda$, $\nu \leq p$, $\kappa \leq p$. Note that $\nu(\mathfrak{x}_\mu \alpha / \mathfrak{x}_\lambda) = \nu(\mathfrak{x}_\nu \zeta_\mu \alpha \zeta_\lambda^{-1} / \mathfrak{x}_\kappa)$. Thus we have a system

$$\{U_\lambda, \psi_\lambda, S_\sigma^{\mu\lambda}(\alpha)\ (\lambda, \mu = 1, \cdots, q;\ \alpha \in D(\Delta_1, \mathfrak{b}))\}$$

which contains the original system as a part of it. It can easily be verified that this new system satisfies (5.7.1–6) with $\mathfrak{b}$ and W_1 in place of $\mathfrak{c}$ and W, except for the following point: with the new system, in (5.7.6), λ is not necessarily unique.

In view of (5.7.5), we see that the $S_1^{\lambda\lambda}(\gamma)$, for $\gamma \in \Gamma_W(\mathfrak{x}_\lambda, \mathfrak{c})$ form a group of biregular automorphisms of U_λ isomorphic to $\Gamma_W(\mathfrak{x}_\lambda, \mathfrak{c})E/\Gamma(\mathfrak{x}_\lambda, \mathfrak{d})E$. Since they are rational over $C(\mathfrak{d})$, we can construct a quotient $\bar{V}_\lambda$ of U_λ by this group and the projection $Q_\lambda \colon U_\lambda \to \bar{V}_\lambda$ so as to be both rational over $C(\mathfrak{d})$. Obviously $(\bar{V}_\lambda, Q_\lambda \circ \psi_\lambda)$ is a model for $\mathfrak{H}_n^r/\Gamma_W(\mathfrak{x}_\lambda, \mathfrak{c})$. Let $\alpha \in D(\Delta, \mathfrak{d})$ and $\gamma \in \Gamma_W(\mathfrak{x}_\lambda, \mathfrak{c})$. By 3.11, we can find an element ε of $\mathfrak{G}$ such that $\mathfrak{x}_\mu \varepsilon = \mathfrak{x}_\mu$, $\nu(\varepsilon) = \nu(\gamma)$ and $\varepsilon \equiv \alpha\gamma\alpha^{-1} \bmod^* (\mathfrak{x}_\mu; \mathfrak{c})$. Then $\varepsilon \in \Gamma_W(\mathfrak{x}_\mu, \mathfrak{c})$. By (5.7.3,4), we have

$$S_\sigma^{\mu\lambda}(\alpha) \circ S_1^{\lambda\lambda}(\gamma) = S_\sigma^{\mu\lambda}(\alpha\gamma) = S_\sigma^{\mu\lambda}(\varepsilon\alpha) = S_1^{\mu\mu}(\varepsilon)^\sigma \circ S_\sigma^{\mu\lambda}(\alpha) .$$

Hence there exists a biregular morphism $\bar{R}_\sigma^{\mu\lambda}(\alpha)$ of $\bar{V}_\lambda$ to $\bar{V}_\mu^\sigma$, rational over $C(\mathfrak{d})$, such that $\bar{R}_\sigma^{\mu\lambda}(\alpha) \circ Q_\lambda = Q_\mu^\sigma \circ S_\sigma^{\mu\lambda}(\alpha)$. We see easily

$$(5.16.4) \qquad\qquad \bar{R}_{\tau\sigma}^{\nu\lambda}(\beta\alpha) = \bar{R}_\tau^{\nu\mu}(\beta)^\sigma \circ \bar{R}_\sigma^{\mu\lambda}(\alpha) ,$$

(5.16.5) $\bar{R}_1^{\lambda\lambda}(\gamma)$ *is the identity map of* $\bar{V}_\lambda$ *if* $\gamma \in \Gamma_W(\mathfrak{x}_\lambda, \mathfrak{c})$.

Now put

$$\mathfrak{X} = \{\alpha \in D_W(\Delta, \mathfrak{d}) \mid \alpha \equiv 1 \bmod^* (\Delta; \mathfrak{c})\} ,$$
$$\mathfrak{Y} = \{\alpha \in \mathfrak{X} \mid \nu(\alpha) \equiv \varepsilon \bmod^* \mathfrak{d} \text{ for some } \varepsilon \in E \cap W\} .$$

Then we have

(5.16.6) $\bar{R}_\sigma^{\mu\lambda}(\alpha) = \bar{R}_\sigma^{\mu\lambda}(\beta)$ *if* $\alpha, \beta \in D_W(\Delta, \mathfrak{d})$ *and* $\beta^{-1}\alpha \in \mathfrak{Y}$.
To see this, let $\nu(\beta^{-1}\alpha) \equiv \varepsilon \bmod^* \mathfrak{d}$ with $\varepsilon \in E \cap W$. By 3.11, there exists an element γ of $\mathfrak{G}$ such that $\mathfrak{x}_\lambda\gamma = \mathfrak{x}_\lambda$, $\nu(\gamma) = \varepsilon$ and $\gamma \equiv \beta^{-1}\alpha \bmod^* (\mathfrak{x}_\lambda; \mathfrak{d})$. Since $\beta^{-1}\alpha \equiv 1 \bmod^* (\mathfrak{x}_\lambda; \mathfrak{c})$, we have $\gamma \in \Gamma_W(\mathfrak{x}_\lambda, \mathfrak{c})$, hence

$$\bar{R}_\sigma^{\mu\lambda}(\alpha) = \bar{R}_\sigma^{\mu\lambda}(\beta\gamma) = \bar{R}_\sigma^{\mu\lambda}(\beta) \circ \bar{R}_1^{\lambda\lambda}(\gamma) = \bar{R}_\sigma^{\mu\lambda}(\beta) \qquad \text{by (5.16.5)} .$$

For every $\alpha \in \mathfrak{X}$, take an ideal $\mathfrak{a}$ in F' prime to $N(\mathfrak{d})$, such that $(\nu(\alpha)) = \det \Theta'(\mathfrak{a})$, and let $\bar{\sigma} = [C_W(\mathfrak{d})/F', \mathfrak{a}]$. Then

(5.16.7) *The map* $\alpha \mapsto \bar{\sigma}$ *gives an isomorphism of* $\mathfrak{X}/\mathfrak{Y}$ *onto* $G(C_W(\mathfrak{d})/C_W(\mathfrak{c}))$. The assertion is trivial except for the surjectivity, which can be proved as follows. Let $\bar{\sigma} \in G(C_W(\mathfrak{d})/C_W(\mathfrak{c}))$. Take an integral ideal $\mathfrak{b}$ in F', prime to $N(\mathfrak{d})$, such that $\bar{\sigma} = [C_W(\mathfrak{d})/F', \mathfrak{b}]$. Since $\bar{\sigma}$ is the identity map on $C_W(\mathfrak{c})$, there exists an element u of W such that $u \equiv 1 \bmod^* \mathfrak{c}$ and $\det \Theta'(\mathfrak{b}) = (u)$. Note that u is integral. By 3.11, we can find an element β of $\mathfrak{G}$ such that $\mathfrak{x}_1\beta \subset \mathfrak{x}_1$, $\nu(\beta) = u$, $\beta \equiv 1 \bmod^* (\mathfrak{x}_1; \mathfrak{c})$. Since (u) is integral and prime to $N(\mathfrak{d})$, we see that $\beta \in D_W(\Delta, \mathfrak{d})$, hence $\beta \in \mathfrak{X}$. This proves the surjectivity.

Now, for every $\bar{\sigma} \in G(C_W(\mathfrak{d})/C_W(\mathfrak{c}))$, take an element α of $\mathfrak{X}$ corresponding to $\bar{\sigma}$ by the map of (5.16.7), and put $Z_\sigma^\lambda = \bar{R}_\sigma^{\lambda\lambda}(\alpha)$. Note that σ is the restriction of $\bar{\sigma}$ to $C(\mathfrak{d})$, and Z_σ^λ does not depend on the choice of α, in view of (5.16.6). From (5.16.4) we obtain $Z_{\tau\sigma}^\lambda = (Z_\tau^\lambda)^{\bar{\sigma}} \circ Z_\sigma^\lambda$ for $\bar{\sigma}, \bar{\tau} \in G(C_W(\mathfrak{d})/C_W(\mathfrak{c}))$. By Weil's criterion [12], we can find a variety V_λ, defined over $C_W(\mathfrak{c})$, and a biregular morphism T_λ of V_λ onto $\bar{V}_\lambda$, defined over $C_W(\mathfrak{d})$, such that $Z_\sigma^\lambda = (T_\lambda)^{\bar{\sigma}} \circ T_\lambda^{-1}$.

536

GORO SHIMURA

For every $\xi \in D_W(\Delta, \mathfrak{b})$, put $R^{\mu\lambda}_{\bar\sigma}(\xi) = (T^{\bar\sigma}_\mu)^{-1} \circ \bar R^{\mu\lambda}_\sigma(\xi) \circ T_\lambda$, where

$$\bar\sigma = [C_W(\mathfrak{b})/F', \mathfrak{a}], \quad \det \Theta'(\mathfrak{a}) = \nu(\mathfrak{x}_\mu \xi/\mathfrak{x}_\lambda)$$

with an ideal $\mathfrak{a}$ in F' prime to $N(\mathfrak{b})$, and σ is the restriction of $\bar\sigma$ to $C(\mathfrak{b})$. Then $R^{\mu\lambda}_{\bar\sigma}(\xi)$ is a biregular morphism of V_λ onto $V^{\bar\sigma}_\mu$ defined over $C_W(\mathfrak{b})$. Further we see that $R^{\mu\lambda}_{\bar\sigma}(\xi)$ depends only on λ, μ, and the class of $\alpha \bmod^* (\Delta; \mathfrak{b})$, and

$$R^{\nu\lambda}_{\bar\tau\bar\sigma}(\eta\xi) = R^{\nu\mu}_{\bar\tau}(\eta)^{\bar\sigma} \circ R^{\mu\lambda}_{\bar\sigma}(\xi) \qquad \text{for } \xi, \eta \in D_W(\Delta; \mathfrak{b}) .$$

If $\xi \in \mathfrak{X}$, we have $\bar R^{\lambda\lambda}_{\bar\sigma}(\xi) = Z^\lambda_{\bar\sigma} = T^{\bar\sigma}_\lambda \circ T^{-1}_\lambda$, hence we obtain

(5.16.8) $R^{\lambda\lambda}_{\bar\sigma}(\xi)$ *is the identity map of* V_λ *if* $\xi \in \mathfrak{X}$, *i.e., if* $\xi \equiv 1 \bmod^* (\Delta; \mathfrak{c})$.

Let $\alpha \in \mathfrak{X}$, $\beta \in D_W(\Delta, \mathfrak{b})$, and $\gamma = \beta\alpha\beta^{-1}$. Then $\gamma \in \mathfrak{X}$, and by (5.16.8) we have

$$R^{\mu\lambda}_{\bar\tau}(\beta) = R^{\mu\mu}_{\bar\sigma}(\gamma)^{\bar\tau} \circ R^{\mu\lambda}_{\bar\tau}(\beta) = R^{\mu\lambda}_{\bar\sigma\bar\tau}(\gamma\beta) = R^{\mu\lambda}_{\bar\sigma\bar\tau}(\beta\alpha)$$
$$= R^{\mu\lambda}_{\bar\tau}(\beta)^{\bar\sigma} \circ R^{\lambda\lambda}_{\bar\sigma}(\alpha) = R^{\mu\lambda}_{\bar\tau}(\beta)^{\bar\sigma} .$$

By virtue of (5.16.7), this shows that $R^{\mu\lambda}_{\bar\tau}(\beta)$ is rational over $C_W(\mathfrak{c})$.

In the next place, let $\alpha, \beta \in D_W(\Delta, \mathfrak{b})$. Take ideals $\mathfrak{a}, \mathfrak{b}$ in F' prime to $N(\mathfrak{b})$ so that $\det \Theta'(\mathfrak{a}) = \nu(\mathfrak{x}_\mu \alpha/\mathfrak{x}_\lambda)$, $\det \Theta'(\mathfrak{b}) = \nu(\mathfrak{x}_\mu \beta/\mathfrak{x}_\lambda)$, and put $\bar\sigma = [C_W(\mathfrak{b})/F', \mathfrak{a}]$, $\bar\tau = [C_W(\mathfrak{b})/F', \mathfrak{b}]$. Suppose $\alpha \equiv \beta \bmod^* (\Delta; \mathfrak{c})$. Put $\delta = \beta\alpha^{-1}$, $\bar\rho = \bar\tau\bar\sigma^{-1}$. Then $\bar\rho$ is the identity map on $C_W(\mathfrak{c})$, and

$$R^{\mu\lambda}_{\bar\tau}(\beta) = R^{\mu\lambda}_{\bar\rho\bar\sigma}(\delta\alpha) = R^{\mu\mu}_{\bar\rho}(\delta)^{\bar\sigma} \circ R^{\mu\lambda}_{\bar\sigma}(\alpha) = R^{\mu\lambda}_{\bar\sigma}(\alpha) \qquad \text{by (5.16.8)} .$$

This implies that $R^{\mu\lambda}_{\bar\sigma}(\alpha)$ depends only on λ, μ and the class of $\alpha \bmod^* (\Delta; \mathfrak{c})$; and only the effect of $\bar\sigma$ on $C_W(\mathfrak{c})$ is essential. By 5.14, for every $\varepsilon \in D_W(\Delta, \mathfrak{c})$, we can find $\alpha \in D_W(\Delta, \mathfrak{b})$ so that $\alpha \equiv \varepsilon \bmod^* (\Delta; \mathfrak{c})$. Define $R^{\mu\lambda}_{\bar\sigma}(\varepsilon)$ to be the same as $R^{\mu\lambda}_{\bar\sigma}(\alpha)$. In the symbol $R^{\mu\lambda}_{\bar\sigma}(\varepsilon)$, $\bar\sigma$ should now be considered as the automorphism $[C_W(\mathfrak{c})/F', \mathfrak{a}]$ with an ideal $\mathfrak{a}$ in F', prime to $N(\mathfrak{c})$, such that $\det \Theta'(\mathfrak{a}) = \nu(\mathfrak{x}_\mu \varepsilon/\mathfrak{x}_\lambda)$. Put $\varphi_\lambda = T^{-1}_\lambda \circ Q_\lambda \circ \psi_\lambda$. Then $(V_\lambda, \varphi_\lambda)$ is a model for $\mathfrak{H}^r_n/\Gamma_W(\mathfrak{x}_\lambda, \mathfrak{c})$. We are going to show that $\{V_\lambda, \varphi_\lambda, R^{\mu\lambda}_{\bar\sigma}(\varepsilon)\}$ is a canonical system of level $\mathfrak{c}$ with respect to Δ and W.

First we observe that (5.7.2–4) are satisfied. Let $\xi \in D_W(\Delta, \mathfrak{b})$, and let $\mathfrak{z}$ be an ideal in F' prime to $N(\mathfrak{b})$ such that $\det \Theta'(\mathfrak{z}) = \nu(\mathfrak{x}_\mu \xi/\mathfrak{x}_\lambda)$. Put $\omega = [C_W(\mathfrak{b})/F', \mathfrak{z}]$. Denote the restriction of ω to any subfield of $C_W(\mathfrak{b})$ by the same latter ω. Then we have the following commutative diagram:

$$
\begin{array}{ccccc}
\mathfrak{H}^r_n & \xrightarrow{\ \psi_\lambda\ } & U_\lambda & \xrightarrow{\ S^{\mu\lambda}_\omega(\xi)\ } & U^\omega_\mu \\
& & \downarrow{\scriptstyle Q_\lambda} & & \downarrow{\scriptstyle Q^\omega_\mu} \\
\text{id}\ \downarrow & & \bar V_\lambda & \xrightarrow{\ \bar R^{\mu\lambda}_\omega(\xi)\ } & \bar V^\omega_\mu \\
& & \downarrow{\scriptstyle T^{-1}_\lambda} & & \downarrow{\scriptstyle (T^{-1}_\mu)^\omega} \\
\mathfrak{H}^r_n & \xrightarrow{\ \varphi_\lambda\ } & V_\lambda & \xrightarrow{\ R^{\mu\lambda}_\omega(\xi)\ } & V^\omega_\mu
\end{array}
$$

174

If $\gamma \in \Gamma_W(\mathfrak{x}_\lambda), S_1^{\lambda\lambda}(\gamma)[\psi_\lambda(w)] = \psi_\lambda(\gamma(w))$ for all $w \in \mathfrak{H}_n^r$. From the above diagram (with $\lambda = \mu$ and $\omega = $ identity), we obtain $R_1^{\lambda\lambda}(\gamma)[\varphi_\lambda(w)] = \varphi_\lambda(\gamma(w))$, hence (5.7.5) is satisfied.

Now let (Y, P, δ, f), (P', Ψ'), z, $\mathfrak{g}_r$, $\mathfrak{g}$, $\mathfrak{h}$, $\mathfrak{h}_0$, $\mathfrak{b}$, τ, μ, λ and α be as in (5.7.6). Let η be an *arbitrary* extension of τ to $C_W(P', \Psi', \mathfrak{g}_r, \mathfrak{b})$, and $\mathfrak{a}$ an ideal in P', prime to $N(\mathfrak{b}\mathfrak{h}_0)$, such that

$$\eta = [C_W(P', \Psi', \mathfrak{g}_r, \mathfrak{b})/P', \mathfrak{a}] .$$

Since $\eta = \tau$ on $C_W(P', \Psi', \mathfrak{g}_r, \mathfrak{c})$, there exists an element v of Y such that

$$[\Psi'(\mathfrak{a}\mathfrak{b}^{-1})]_\mathfrak{g}\mathfrak{g}_r = v\mathfrak{g}_r, vv^\delta \in W, v \equiv 1 \bmod^* \mathfrak{c}\mathfrak{g}_r .$$

Put $\xi = f(v)\alpha$. Then $\mathfrak{x}_\mu f([\Psi'(\mathfrak{a})]_\mathfrak{g})^{-1} = \mathfrak{x}_\lambda \xi^{-1}$. Since $\mathfrak{a}$ is prime to $N(\mathfrak{b})$, we see that $\xi \in D_W(\Delta, \mathfrak{b})$. Let ω be the restriction of η to $C_W(\mathfrak{b})$. Then $S_\omega^{\mu\lambda}(\xi)$ is defined, and $\psi_\mu(z)^\eta = S_\omega^{\mu\lambda}(\xi)[\psi_\lambda(\xi^{-1}(z))]$. By the above diagram, we obtain $\varphi_\mu(z)^\eta = R_\omega^{\mu\lambda}(\xi)[\varphi_\lambda(\xi^{-1}(z))]$. We have $\xi \equiv \alpha \bmod^* (\Delta; \mathfrak{c})$ by (3.6.2), hence $R_\omega^{\mu\lambda}(\xi) = R_\omega^{\mu\lambda}(\alpha)$. Further $\xi^{-1}(z) = \alpha^{-1}(z)$. Therefore

$$(5.16.9) \qquad \varphi_\mu(z)^\eta = R_\omega^{\mu\lambda}(\alpha)[\varphi_\lambda(\alpha^{-1}(z))] .$$

If τ is the identity map, we can take $\mathfrak{b} = (1)$, $\lambda = \mu$ and $\alpha = 1$. Then (5.16.9) becomes $\varphi_\mu(z)^\eta = \varphi_\mu(z)$. This shows that $\varphi_\mu(z)$ is rational over $C_W(P', \Psi', \mathfrak{g}_r, \mathfrak{c})$. Therefore we can put τ in place of η in (5.16.9). Then (5.16.9) implies (5.7.6). By 5.15, this proves that $\{V_\lambda, \varphi_\lambda, R_\sigma^{\mu\lambda}(\alpha)\}$ is a canonical system of level $\mathfrak{c}$ with respect to Δ and W.

6. Families of abelian varieties and their special members

6.1. Let F and B be as in 3.1, and K a totally imaginary quadratic extension of F. Put $L = B \otimes_F K$. Then L is a quaternion algebra over K. We denote also by ι the main involution of L, and by L^n the vector space of all n-dimensional row vectors with components in L. We consider L^n as a left L-module and as a right $M_n(L)$-module as usual.

6.2. PROPOSITION. *The notation being as above, there exists a positive involution ρ of L, two invertible elements S and T of $M_n(L)$, and an element κ of F satisfying the following conditions:*

(6.2.1) ${}^t T^\rho = -T, S^{\iota\rho}S = \kappa 1_n, T \cdot {}^t S^\rho = -S^{\iota\rho}T^{\iota\rho}.$

(6.2.2) *If i_0 is the natural injection of $M_n(B)$ into $M_n(L)$, and $\mathfrak{G} = \{U \in M_n(B) \mid U \cdot {}^t U^\iota = \nu(U)1_n$ with $\nu(U) \in F\}$, then one has*

(6.2.2$_a$) $i_0(M_n(B)) = \{U \in M_n(L) \mid U^{\iota\rho}S = SU\},$

(6.2.2$_b$) $i_0(\mathfrak{G}) = \{U \in GL_n(L) \mid U^{\iota\rho}S = SU, UT \cdot {}^t U^\rho = \nu(U)T$ with $\nu(U) \in F\},$

where $\nu(i_0(U)) = \nu(U)$.

PROOF. By [C, 7.2], L has a positive involution ρ and an invertible element v such that

$$v^\rho = -v \quad \text{and} \quad B = \{x \in L \mid x^\iota = vx^\rho v^{-1}\} \,.$$

Put $T = v1_n$, $S = v^{-1}1_n$, $\kappa = -(vv^\iota)^{-1}$. Then $\kappa \in K$ and

$$\kappa^\rho = -(v^{\iota\rho}v^\rho)^{-1} = -(v^\iota v)^{-1} = \kappa \,,$$

hence $\kappa \in F$. The verification of the above formulas can be made in a straight-forward way.

6.3. Let $\tau_{01}, \cdots, \tau_{0g}$ be as in 3.1, and $\tau_1, \cdots, \tau_g$ be extensions of $\tau_{01}, \cdots, \tau_{0g}$ to K. There are 2^g choices for the set $\{\tau_\nu\}$. We take any one of them. Put $K_R = K \otimes_Q R$, $L_R = L \otimes_Q R$. Take ρ, T and S as in 6.2. By [5, Lem. 1 and 2.1], we can find R-linear representations $\varphi_\nu \colon L_R \to M_2(C)$ for $\nu = 1, \cdots, g$ such that

$$(6.3.1) \qquad\qquad \varphi_\nu(a) = a^{\tau_\nu}1_2 \qquad\qquad (a \in K) \,,$$

$$(6.3.2) \qquad\qquad \varphi_\nu(a^\rho) = \overline{{}^t\varphi_\nu(a)} \qquad\qquad (a \in L_R) \,,$$

(6.3.3) $\{\varphi_1, \cdots, \varphi_g, \bar\varphi_1, \cdots, \bar\varphi_g\}$ *is a complete set of inequivalent absolutely irreducible R-linear representations of L_R.*

Since ι is the main involution of L, we have

$$(6.3.4) \qquad \varphi_\nu(a^\iota) = J_1 \cdot {}^t\varphi_\nu(a)J_1^{-1}, \; J_1 = \begin{bmatrix} 0 & 1 \\ -1 & 0 \end{bmatrix} \qquad (a \in L_R) \,.$$

In (6.3.2, 4) we extend ι and ρ to L_R R-linearly.

For every positive integer m, we define $\varphi_\nu^m \colon M_m(L_R) \to M_{2m}(C)$ as follows: let $\xi = (x_{ij}) \in M_m(L_R)$ with $x_{ij} \in L_R$; put

$$\varphi_\nu(x_{ij}) = \begin{bmatrix} a_{ij} & b_{ij} \\ c_{ij} & d_{ij} \end{bmatrix}$$

and

$$\varphi_\nu^m(\xi) = \begin{bmatrix} A & B \\ C & D \end{bmatrix} \qquad \text{with } A = (a_{ij}),\; B = (b_{ij}),\; C = (c_{ij}),\; D = (d_{ij}) \,.$$

Then from (6.3.2, 4) we obtain

$$(6.3.5) \qquad\qquad \varphi_\nu^m({}^t\xi^\rho) = {}^t\overline{\varphi_\nu^m(\xi)} \qquad\qquad (\xi \in M_m(L_R)) \,,$$

$$(6.3.6) \qquad\qquad \varphi_\nu^m({}^t\xi^\iota) = J_m \cdot {}^t\varphi_\nu^m(\xi)J_m^{-1} \qquad\qquad (\xi \in M_m(L_R)) \,,$$

where J_m is as in 4.1. Now, in 6.2, we can take T and κ so that

(6.3.7) $\kappa^{\tau_{0\nu}} < 0$ *or* > 0 *according as* $\nu \leqq r$ *or* $\nu > r$.

(6.3.8) *The complex hermitian matrix* $\sqrt{-1}\varphi_\nu^n(T)^{-1}$ *has exactly n or $2n$ positive characteristic roots according as* $\nu \leqq r$ *or* $\nu > r$.

In fact, let v be as in the proof of 6.2. We see that $\varphi_\nu(v)$ is skew-hermitian, and indefinite or definite according as $\nu \leqq r$ or $\nu > r$, since B is unramified or ramified at the corresponding archimedean prime of F. Therefore, replacing T by cT with a suitable $c \in F$, we obtain (6.3.8). Since $\kappa = -(vv')^{-1}$, we have $\kappa^{\tau_0\nu} = -\det \varphi_\nu(v)^{-1}$, hence (6.3.7).

6.4. Define an R-linear representation $\Phi: L_R \longrightarrow M_{4ng}(C)$ as follows:

$$\Phi(a) = \begin{bmatrix} \Phi_1(a) & & \\ & \ddots & \\ & & \Phi_g(a) \end{bmatrix}, \quad \Phi_\nu(a) = \begin{cases} \begin{bmatrix} \varphi_\nu^n(a1_n) & 0 \\ 0 & \overline{\varphi_\nu^n(a1_n)} \end{bmatrix} & (1 \leqq \nu \leqq r)\,, \\[6pt] \varphi_\nu^{2n}(a1_{2n}) & (r < \nu \leqq g)\,. \end{cases}$$

Let $\mathfrak{M}$ be a lattice in L^n (in the sense of 2.1), and $v_1, \cdots, v_s$ be elements of L^n. We consider a weak PEL-type in the sense of [C, 4.7] defined by

$$(6.4.1) \qquad \Omega = (L, \Phi, \rho; F^+ \cdot T, \mathfrak{M}; v_1, \cdots, v_s)\,.$$

Let $\mathbb{S}$ be the space of all $z \in M_n(C)$ such that $1 - {}^t\bar{z}z$ is positive definite, and $\mathbb{S}^r$ the product of r copies of $\mathbb{S}$. Now we can construct a family

$$(6.4.2) \qquad \textstyle\sum_\Omega = \{\mathcal{R}_z \mid z \in \mathbb{S}^r\}$$

of weak PEL-structures of type Ω in the sense of [C, 4.8] in the following way. First we can find an element W_ν of $M_{2n}(C)$ for each ν so that

$$(6.4.3) \qquad \sqrt{-1}\, W_\nu \varphi_\nu^n(T)^{-1} \cdot {}^t\bar{W}_\nu = \begin{cases} \begin{bmatrix} 1_n & 0 \\ 0 & -1_n \end{bmatrix} & (1 \leqq \nu \leqq r)\,, \\[6pt] 1_{2n} & (r < \nu \leqq g)\,, \end{cases}$$

$$(6.4.4) \qquad {}^t W_\nu^{-1} J_n^{-1} \varphi_\nu^n(S) \cdot {}^t\bar{W}_\nu = \begin{cases} \sqrt{|\kappa^{\tau_\nu}|}\, \begin{bmatrix} 0 & 1_n \\ 1_n & 0 \end{bmatrix} & (1 \leqq \nu \leqq r)\,, \\[6pt] \sqrt{|\kappa^{\tau_\nu}|}\, \begin{bmatrix} 0 & -1_n \\ 1_n & 0 \end{bmatrix} & (r < \nu \leqq g)\,. \end{cases}$$

(See [8, 1.11, 2.12]. The present discussion is included in [8] as a special case $\varepsilon = 1$, $\eta = -1$, $q = 2$. In the proof of 6.2, we have obtained T and S as $T = v1_n$, $S = v^{-1}1_n$ with an element v of L such that $v^\rho = -v$. Transforming $\varphi_\nu(v)$ into a diagonal form, the existence of W_ν can be shown more directly.)

For every $z = (z_1, \cdots, z_r) \in \mathbb{S}^r$ with $z_\nu \in \mathbb{S}$, define $X_1, \cdots, X_g$ by

$$X_\nu \bar{W}_\nu^{-1} = \begin{cases} \begin{bmatrix} 1_n & z_\nu \\ {}^t\bar{z}_\nu & 1_n \end{bmatrix} & (1 \leqq \nu \leqq r)\,, \\[6pt] 1_{2n} & (r < \nu \leqq g)\,, \end{cases}$$

and define complex column vectors u_i^ν, $u_i^{*\nu}$, v_i^ν, $v_i^{*\nu}$ by

$$X_\nu = \begin{cases} \begin{bmatrix} u_1^\nu & \cdots & u_n^\nu & u_1^{*\nu} & \cdots & u_n^{*\nu} \\ \bar{v}_1^\nu & \cdots & \bar{v}_n^\nu & \bar{v}_1^{*\nu} & \cdots & \bar{v}_n^{*\nu} \end{bmatrix} & (1 \leqq \nu \leqq r)\,, \\[6pt] \begin{bmatrix} u_1^\nu & \cdots & u_n^\nu & u_1^{*\nu} & \cdots & u_n^{*\nu} \end{bmatrix} & (r < \nu \leqq g)\,. \end{cases}$$

Here u_i^ν and $u_i^{*\nu}$ are n-dimensional or $2n$-dimensional according as $\nu \leqq r$ or $\nu > r$. Further define a column vector $\mathfrak{y}_i(z)$ in $C^{4n g}$ by

$$
{}^t x_i^\nu =
\begin{cases}
({}^t u_i^\nu \quad {}^t u_i^{*\nu} \quad {}^t v_i^\nu \quad {}^t v_i^{*\nu}) & (1 \leqq \nu \leqq r) , \\
({}^t u_i^\nu \quad {}^t u_i^{*\nu}) & (r < \nu \leqq g) ,
\end{cases}
$$

$$
{}^t\mathfrak{y}_i(z) = ({}^t x_i^1 \quad \cdots \quad {}^t x_i^\nu \quad \cdots \quad {}^t x_i^g) \qquad (i = 1, \cdots, n) .
$$

Let d_i, for $i = 1, \cdots, n$, be the element of L^n, whose i-component is 1 and other components are 0. Define a map $\mathfrak{y}: L_R^n \times \mathfrak{S}^r \to C^{4ng}$ by

$$
\mathfrak{y}\left(\textstyle\sum_{i=1}^n a_i d_i, z\right) = \sum_{i=1}^n \Phi(a_i)\mathfrak{y}_i(z) \qquad (a_i \in L_R, z \in \mathfrak{S}^r) .
$$

Put $\mathfrak{y}_z(x) = \mathfrak{y}(x, z)$ for $x \in L_R^n$ and $z \in \mathfrak{S}^r$. Then $\mathfrak{y}_z$ is an R-linear isomorphism of L_R^n onto C^{4ng} satisfying

$$
\mathfrak{y}_z(ax) = \Phi(a)\mathfrak{y}_z(x) \qquad (a \in L_R, x \in L_R^n) .
$$

The complex torus $C^{4ng}/\mathfrak{y}_z(\mathfrak{M})$ has a structure of abelian variety for every $z \in \mathfrak{S}^r$. In fact, define an R-bilinear form $\mathfrak{E}_z(u, v)$ on $C^{4ng} \times C^{4ng}$ by

$$
(6.4.5) \qquad
\begin{aligned}
\mathfrak{E}_z(u, v) &= \mathrm{Tr}_{L_R/R}\big(T(\mathfrak{y}_z^{-1}(u), \mathfrak{y}_z^{-1}(v)) \big) , & \cdot(u, v \in C^{4ng}) , \\
T(x, y) &= x T \cdot {}^t y^\rho & (x, y \in L_R^n) .
\end{aligned}
$$

For a suitable positive integer c, $c\mathfrak{E}_z$ defines a Riemann form on the complex torus. (For details, see [5, §2] and [8, §2].) Thus we obtain, for each $z \in \mathfrak{S}^r$, an abelian variety A_z with a polarization $\mathcal{C}_z$. Moreover, for every $a \in L$, $\Phi(a)$ defines an element of $\mathrm{End}_Q(A_z)$, which we write $\theta_z(a)$. Putting $\mathfrak{D}_z = \mathcal{C}_z \cdot \theta_z(F^+)$ with the notation of [C, 4.8], and denoting by t_{iz} the point on A_z represented by $\mathfrak{y}_z(v_i)$, we obtain a weak PEL-structure

$$
\mathcal{R}_z = (A_z, \mathfrak{D}_z, \theta_z; t_{1z}, \cdots, t_{nz})
$$

of type Ω. This finishes the construction of the family (6.4.2). It should be noted that every weak PEL-structure of type Ω is isomorphic to $\mathcal{R}_z$ for some $z \in \mathfrak{S}^r$. We can actually embed the A_z for all $z \in \mathfrak{S}^r$ into the same projective space so that everything depends on z holomorphically (see [7, §3].) In this paper, however, we shall not need this result. We only have to understand that A_z is a projective variety, which is fixed once for all for each $z \in \mathfrak{S}^r$.

6.5. Define groups $G(T), G_R(T), G^0(T), G_R^0(T)$ as follows:

$$
\begin{aligned}
G_R(T) &= \{\alpha \in GL_n(L_R) \mid T(x\alpha, y\alpha) = \nu(\alpha) T(x, y) \text{ with } \nu(\alpha) \in F_R\} , \\
G(T) &= G_R(T) \cap M_n(L) , \\
G_R^0(T) &= \{\alpha \in G_R(T) \mid \alpha^{t\rho} S = S\alpha\} , \\
G^0(T) &= G_R^0(T) \cap M_n(L) .
\end{aligned}
$$

Let F_R^+ denote the identity component of the group of invertible elements of F_R. Let $G_R^+(T)$ denote the set of all α in $G_R(T)$ such that $\nu(\alpha) \in F_R^+$. Then

$G_R^+(T)$ acts on $\mathfrak{S}^r$ as follows: for every $\alpha \in G_R(T)$ and $z = (z_1, \cdots, z_r) \in \mathfrak{S}^r$, put

$$(6.5.1) \qquad {}^t W_\nu^{-1} \overline{\varphi_\nu^n(\alpha)} \cdot {}^t W_\nu = \begin{bmatrix} a_\nu & b_\nu \\ c_\nu & d_\nu \end{bmatrix} \qquad (1 \leqq \nu \leqq r) ;$$

Then

$$\alpha(z) = (w_1, \cdots, w_r) , \qquad w_\nu = (a_\nu z_\nu + b_\nu)(c_\nu z_\nu + d_\nu)^{-1} \qquad \text{(see [5, 2.7])} .$$

Now, with $\mathfrak{M}$ and the v_i of (6.4.1), assuming $\mathfrak{r}_K \mathfrak{M} \subset \mathfrak{M}$, put $\mathfrak{N} = \mathfrak{M} + \sum_{i=1}^{s} \mathfrak{r}_K v_i$, and

$$\Gamma(T, \mathfrak{M}) = \{\alpha \in G(T) \mid \nu(\alpha) \in F^+, \ \mathfrak{M}\alpha = \mathfrak{M}\} ,$$
$$\Gamma(T, \mathfrak{N}/\mathfrak{M}) = \{\alpha \in \Gamma(T, \mathfrak{M}) \mid \mathfrak{N}(1 - \alpha) \subset \mathfrak{M}\} ,$$
$$\Gamma^0(T, \mathfrak{M}) = \Gamma(T, \mathfrak{M}) \cap G^0(T) ,$$
$$\Gamma^0(T, \mathfrak{N}/\mathfrak{M}) = \Gamma(T, \mathfrak{N}/\mathfrak{M}) \cap G^0(T) .$$

Then $\mathfrak{R}_z$ and $\mathfrak{R}_w$, with z and w in $\mathfrak{S}^r$, are isomorphic if and only if $z = \gamma(w)$ for some $\gamma \in \Gamma(T, \mathfrak{N}/\mathfrak{M})$. For details see [5, 2.8] and [C, 4.11].

Up to this point, we have not used the formula (6.4.4), whose meaning will be explained in the following proposition. First note that the space $\mathfrak{D}_n$ defined in 4.1 is a subset of $\mathfrak{S}$. We shall show

$(6.5.2)$ $\xi(\mathfrak{D}_n^r) = \mathfrak{D}_n^r$ if $\xi \in G_R^0(T)$ and $\nu(\xi) \in F_R^+$.

6.6. PROPOSITION. *Let $X(=B^n)$ and $\mathfrak{G}(\subset M_n(B))$ be as in 4.4. Then there exist an F-linear isomorphism i of $M_n(B)$ into $M_n(L)$, a B-linear isomorphism i' of X into L^n, and a holomorphic isomorphism j of $\mathfrak{H}_n^r$ onto $\mathfrak{D}_n^r$ such that $i(M_n(B)) = M_n(B)$, $i(\mathfrak{G}) = G^0(T)$, $\nu(i(\alpha)) = \nu(\alpha)$, $j(\alpha(z)) = i(\alpha)(j(z))$ for $\alpha \in \mathfrak{G}^+$ and $z \in \mathfrak{H}_n^r$, and $i'(bx\alpha) = b \cdot i'(x)i(\alpha)$, for $b \in B, x \in X, \alpha \in M_n(B)$.*

PROOF. Put

$$U = \begin{bmatrix} 1_n & -i1_n \\ 1_n & i1_n \end{bmatrix} \qquad\qquad \text{with } i = \sqrt{-1} ,$$

and

$$\psi_\nu(\xi) = U^{-1} \cdot {}^t W_\nu^{-1} \overline{\varphi_\nu^n(\xi)} \cdot {}^t W_\nu U \qquad\qquad \text{for } \xi \in M_n(L_R) .$$

By (6.4.3) and (6.3.5), we can easily verify that, if $\xi \in G_R(T)$,

$$(6.6.1) \qquad \psi_\nu(\xi) J_n \cdot \overline{{}^t \psi_\nu(\xi)} = \nu(\xi)^{(\nu)} J_n \qquad (1 \leqq \nu \leqq r) ,$$

where $\nu(\xi)^{(\nu)}$ is the νth component of $\nu(\xi)$. If $\xi \in M_n(B_R)$, we have $\xi = S^{-1} \xi^{t\rho} S$, hence by (6.3.5, 6), $\varphi_\nu^n(\xi) = \varphi_\nu^n(S)^{-1} J_n \overline{\varphi_\nu^n(\xi)} J_n^{-1} \varphi_\nu^n(S)$. By (6.4.4), we have

$$\begin{aligned} {}^t W_\nu^{-1} \overline{\varphi_\nu^n(\xi)} \cdot {}^t W_\nu &= {}^t W_\nu^{-1} \overline{\varphi_\nu^n(S^{-1})} J_n \varphi_\nu^n(\xi) J_n^{-1} \overline{\varphi_\nu^n(S)} \cdot {}^t W_\nu \\ &= \begin{bmatrix} 0 & 1_n \\ 1_n & 0 \end{bmatrix} {}^t \overline{W}_\nu^{-1} \varphi_\nu^n(\xi) \cdot {}^t \overline{W}_\nu \begin{bmatrix} 0 & 1_n \\ 1_n & 0 \end{bmatrix} . \end{aligned}$$

Therefore

$$\psi_{|\nu}(\xi) = U^{-1}\begin{bmatrix} 0 & 1_n \\ 1_n & 0 \end{bmatrix}\bar{U}\overline{\psi_\nu(\xi)}\,\bar{U}^{-1}\begin{bmatrix} 0 & 1_n \\ 1_n & 0 \end{bmatrix}U = \overline{\psi_\nu(\xi)}\ .$$

This shows that $\psi_\nu(\xi)$ is a real matrix, and hence, from (6.6.1), we obtain $\psi_\nu(\xi)J_n \cdot {}^t\psi_\nu(\xi) = \nu(\xi)^{(\nu)}J_n$ for $\xi \in G_R^0(T)$ and $\nu = 1, \cdots, r$. Therefore, if $\nu(\xi)^{(\nu)} > 0$, $\psi_\nu(\xi)$ acts on $\mathfrak{H}_n$, hence $U\psi_\nu(\xi)U^{-1} = {}^tW_\nu^{-1}\overline{\varphi_\nu^n(\xi)} \cdot {}^tW_\nu$ maps $\mathfrak{D}_n$ onto itself. This proves (6.5.2). Let i_0 be the natural injection of $M_n(B)$ into $M_n(L)$ as in 6.2. Then $i_0(\mathfrak{G}) = G^0(T)$, and $\psi_\nu \circ i_0$ is an R-linear representation: $M_n(B) \rightarrow M_{2n}(R)$ such that $\psi_\nu \circ i_0(a) = a^{\tau_0\nu}1_{2n}$ for $a \in F$. Therefore, if $\alpha \in M_n(B)$ and $\alpha^{(\nu)}$ denotes the projection of α defined in (4.4.2), there exists, for each $\nu \leqq r$, an element Y_ν of $\mathrm{GL}_{2n}(R)$ such that $\alpha^{(\nu)} = Y_\nu^{-1}\psi_\nu(i_0(\alpha))Y_\nu$ for all $\alpha \in M_n(B)$. If $\alpha \in \mathfrak{U}_R$, both $\psi_\nu(i_0(\alpha))$ and $\alpha^{(\nu)}$ belong to $\mathrm{Sp}(n, R)$. Therefore Y_ν can be taken from the group $\mathrm{Gp}(n, R)$ defined in 4.1. Now choose an element β of $\mathfrak{G}$ so that $\nu(\beta)^{\tau_0\nu} \cdot \nu(Y_\nu) > 0$ for $\nu = 1, \cdots, r$, and define i by $i(\alpha) = i_0(\beta\alpha\beta^{-1})$. Put $Z_\nu = \psi_\nu(i_0(\beta))Y_\nu$. Then $\nu(Z_\nu) > 0$, and

$$\xi^{(\nu)} = Z_\nu^{-1}\psi_\nu\big(i(\xi)\big)Z_\nu = Z_\nu^{-1}U^{-1} \cdot {}^tW_\nu^{-1}\overline{\varphi_\nu^n(i(\xi))} \cdot {}^tW_\nu UZ_\nu\ .$$

Now define a map $j\colon \mathfrak{H}_n^r \longrightarrow \mathfrak{D}_n^r$ by

$$j(z_1, \cdots, z_r) = (w_1, \cdots, w_r) \qquad\qquad (z_\nu \in \mathfrak{H}_n,\ w_\nu \in \mathfrak{D}_n)\ ,$$

$$w_\nu = (a_\nu z_\nu + b_\nu)(c_\nu z_\nu + d_\nu)^{-1}, \quad UZ_\nu = \begin{bmatrix} a_\nu & b_\nu \\ c_\nu & d_\nu \end{bmatrix} \qquad (\nu = 1, \cdots, r)\ .$$

Further define $i'\colon B^n \longrightarrow B^n(\subset L^n)$ by $i'(x) = x\beta^{-1}$ for $x \in B^n$. Then i, i' and j have all the required properties.

In [8], we have shown that the members $\mathcal{R}_z$ of Σ_Ω with $z \in \mathfrak{D}_n^r$, can be characterized by the possession of a certain real analytic endomorphism. We shall not use this fact in the present paper.

6.7. Now for every $\alpha \in G_R^+(T)$ and $w \in \mathfrak{S}^r$, there exists a C-linear automorphism $\Lambda(\alpha, w)$ of $C^{4n\rho}$ satisfying

$$(6.7.1) \qquad\qquad \Lambda(\alpha, w)\mathfrak{y}(x, w) = \mathfrak{y}\big(x\alpha, \alpha^{-1}(w)\big) \qquad\qquad (x \in L_R^n)\ .$$

For the proof, see [5, 2.5–8], [C, 4.4], [8, 2.19]. Using this $\Lambda(\alpha, w)$, let us now investigate the special members of Σ_Ω corresponding to the fixed points dissussed in §4.

Take i, i', and j as in 6.6. Let Y, P, δ, f be as in (4.7.3, 5), and z the fixed point of $f((Y, \delta)_0)$ on $\mathfrak{H}_n^r$, and let $w = j(z)$. Let $a \in (Y, \delta)_0$ and $\alpha = i(f(a))$. With the present w and α, (6.7.1) becomes $\Lambda(\alpha, w)\mathfrak{y}_w(x) = \mathfrak{y}_w(x\alpha)$. Since Y is spanned by $(Y, \delta)_0$ over Q, we can define an anti-isomorphism $\lambda\colon Y \longrightarrow$ End $(C^{4n\rho}, C)$ so that $\lambda(c)\mathfrak{y}_w(x) = \mathfrak{y}_w(x \cdot i(f(c)))$ for every $c \in Y$ and $\lambda(a) =$

$\Lambda(\alpha, w)$. If $\mathfrak{E}_w$ denotes the Riemann form defined by (6.4.5), we see easily that

$$(6.7.2) \qquad \mathfrak{E}_w(\lambda(a)u, v) = \mathfrak{E}_w(u, \lambda(a^\delta)v) \qquad (u, v \in C^{4ng}) .$$

Now let $Z = L \otimes_F Y$. Then λ can be extended to Z so that $\lambda(b \otimes c)\mathfrak{y}_w(x) = \mathfrak{y}_w(b'x \cdot i(f(c)))$ for $b \in L$, $c \in Y$, $x \in L_R^n$. Still λ is an anti-isomorphism, since $L \otimes_F i(f(Y))$ can be regarded as a subalgebra of $L \otimes_K M_n(L)$, and L^n can be regarded as a faithful $L \otimes_K M_n(L)$-module in an obvious way. Let $\theta_0(a)$ denote, for $a \in Z$, the element of $\mathrm{End}_Q(A_w)$ corresponding to $\lambda(a)$. Then θ_0 is an anti-isomorphism of Z into $\mathrm{End}_Q(A_w)$, and the involution of $\mathrm{End}_Q(A_w)$ determined by $\mathfrak{D}_w$ sends $\theta_0(b \otimes c)$ to $\theta_0(b^\rho \otimes c^\delta)$ for $b \in L$ and $c \in Y$, by virtue of (6.7.2).

Now let $Y_i, P_i, m_i, q_i (i = 1, \cdots, t)$ be as in 4.7. Let S denote the center of Z. By (4.7.1), $B \otimes_F Y_i$ is isomorphic to $M_{2q_i}(P_i)$. Sence we can write $Z = K \otimes_F (B \otimes_F Y)$, Z is isomorphic to

$$M_{2q_1}(K \otimes_F P_1) \oplus \cdots \oplus M_{2q_t}(K \otimes_F P_t) .$$

Assume hereafter that

(6.7.3) *K is not contained in the composite* $P_1 \cdots P_t$.

Then $K \otimes_F P_i$ can be identified with the composite KP_i, and S with

$$KP_1 \oplus \cdots \oplus KP_t = K \otimes_F P .$$

6.8. Put $S_i = KP_i$. Denete by θ the restriction of θ_0 to S. Since θ_0 is an anti-isomorphism of Z $(\cong M_{2q_1}(S_1) \oplus \cdots \oplus M_{2q_t}(S_t))$ into $\mathrm{End}_Q(A_w)$, A_w should be isogenous to a product

$$A_1^{2q_1} \times \cdots \times A_t^{2q_t}$$

where $A_i^{2q_i}$ denotes the product of $2q_i$ copies of an abelian variety A_i with an isomorphism $\theta_i \colon S_i \to \mathrm{End}_Q(A_i)$ (not necessarily surjective), which corresponds to the restriction of θ to S_i. By [10, §5, Prop. 1], we have

$$\dim(A_i) \geq [S_i : Q]/2 = 2m_i g .$$

Since $n = \sum_{i=1}^t m_i q_i$ (as assumed in 4.7), we must have

$$(6.8.1) \qquad \dim(A_i) = 2m_i g = [S_i : Q]/2 .$$

Let (A_i, θ_i) be of type (S_i, Ξ_i) in the sense of [C, 4.1]; by definition, Ξ_i is a representation of S_i (through θ_i) in the tangent space of A_i at the origin. The relation (6.8.1) shows that (S_i, Ξ_i) is a generalized CM-type of index one in the sense of [C, 5.11], i.e., a CM-type in the sense of [10].

Let us fix our attention to one particular i. Let Ψ_i and $\chi_{\nu k}$, for $\nu \leq r$ and $k = 1, \cdots, m_i$, be as in (4.9.2). Let ρ denote the complex conjugation. For each $\nu > r$, choose m_i isomorphisms $\chi_{\nu k}(k = 1, \cdots, m_i)$ of P_i into C so that $\chi_{\nu k}, \chi_{\nu k}\rho$ for $k = 1, \cdots, m_i$ form the set of all the extensions of $\tau_{0\nu}$ to P_i. We

544　　　　　　　　　　　　　GORO SHIMURA

shall now prove

(6.8.2) *The restriction of Ξ_i to P_i is equivalent to*

$$\sum_{\nu=1}^{r}\sum_{k=1}^{m_i} 2\chi_{\nu k} + \sum_{\nu=r+1}^{g}\sum_{k=1}^{m_i}(\chi_{\nu k} + \chi_{\nu k}\rho) \,.$$

For this purpose, we recall that $\Lambda(\alpha, w)$ can be obtained as a matrix Λ of [5, p. 161, (32)] which is determined as follows.

Let $a \in (Y, \delta)_0$ and $\alpha = i(f(a))$. For this α, let $a_\nu, b_\nu, c_\nu, d_\nu$ be as in (6.5.1). In the proof of 6.6, we have shown that

$$\begin{bmatrix} 1_n & -\sqrt{-1}\cdot 1_n \\ 1_n & \sqrt{-1}\cdot 1_n \end{bmatrix}^{-1} \begin{bmatrix} a_\nu & b_\nu \\ c_\nu & d_\nu \end{bmatrix} \begin{bmatrix} 1_n & -\sqrt{-1}\cdot 1_n \\ 1_n & \sqrt{-1}\cdot 1_n \end{bmatrix}$$

is a real matrix. It follows that $\bar{a}_\nu = d_\nu$ and $c_\nu = \bar{b}_\nu$. Now put $\Sigma_\nu = {}^t(\bar{b}_\nu w_\nu + \bar{a}_\nu)$. Since $\alpha(w) = w$, we have $w_\nu = (a_\nu w_\nu + b_\nu)(\bar{b}_\nu w_\nu + \bar{a}_\nu)^{-1}$, so that

$$(6.8.3) \qquad \begin{bmatrix} \Sigma_\nu & 0 \\ 0 & \bar{\Sigma}_\nu \end{bmatrix} \begin{bmatrix} 1_n & w_\nu \\ \bar{w}_\nu & 1_n \end{bmatrix} = \begin{bmatrix} 1_n & w_\nu \\ \bar{w}_\nu & 1_n \end{bmatrix} \bar{W}_\nu \cdot {}^t\varphi_\nu^n(\alpha) \bar{W}_\nu^{-1} \qquad (\nu \leqq r) \,.$$

Now put

$$(6.8.4) \qquad \Lambda'_\nu = \begin{cases} \begin{bmatrix} \Sigma_\nu & & & \\ & \Sigma_\nu & & \\ & & \Sigma_\nu & \\ & & & \Sigma_\nu \end{bmatrix} & (\nu \leqq r) \,, \\[2em] \begin{bmatrix} \Lambda_\nu & 0 \\ 0 & \Lambda_\nu \end{bmatrix}, \quad \Lambda_\nu = \bar{W}_\nu \cdot {}^t\varphi_\nu^n(\alpha) \bar{W}_\nu^{-1} & (\nu > r) \,. \end{cases}$$

Then Λ is obtained as

$$(6.8.5) \qquad \Lambda = \begin{bmatrix} \Lambda'_1 & & \\ & \ddots & \\ & & \Lambda'_g \end{bmatrix} \,.$$

This is a special case of the discussion of [5, pp. 161–165]. Our present L is of (Type IV) of [5], and the above formulas correspond to [5, (32), (33), (34), (39)]. In the present situation, since $\alpha \in G^0(T)$, the matrix Δ_ν of [5, (33), (34)] must coincide with Σ_ν, as shown in (6.8.3).

Now let $x \mapsto (p_\nu x + q_\nu)(\bar{q}_\nu x + \bar{p}_\nu)^{-1}$ be an analytic automorphism of $\mathfrak{D}_n$ which sends 0 to w_ν for each $\nu \leqq r$. Then $q_\nu = w_\nu \bar{p}_\nu$. Put

$$U_\nu = \begin{bmatrix} p_\nu & w_\nu \bar{p}_\nu \\ \bar{w}_\nu p_\nu & \bar{p}_\nu \end{bmatrix} \,.$$

From (6.8.3) we obtain

$$(6.8.6) \qquad {}^tW_\nu^{-1}\overline{\varphi_\nu^n(\alpha)} \cdot {}^tW_\nu = U_\nu \begin{bmatrix} p_\nu^{-1} \cdot {}^t\bar{\Sigma}_\nu p_\nu & 0 \\ 0 & \bar{p}_\nu^{-1} \cdot {}^t\Sigma_\nu \bar{p}_\nu \end{bmatrix} U_\nu^{-1} \qquad (\nu \leqq r) \,.$$

The left hand side represents the transformation of $\alpha = i(f(a))$ on $\mathfrak{D}_n$. Therefore, if $j_\nu : \mathfrak{H}_n \to \mathfrak{D}_n$ denotes the νth part of the map $j : \mathfrak{H}_n^r \to \mathfrak{D}_n^r$, the formula (6.8.6) shows that $j_\nu \circ f(a)^{(\nu)} \circ j_\nu^{-1}$ is represented by the right hand side of (6.8.6), where $f(a)^{(\nu)}$ is as in (4.4.2). This implies that the representation $a \mapsto {}^t\Sigma_\nu$ is equivalent to the representation $\Psi^{(\nu)}$ of (4.7.6). Therefore, if we restrict a to P, the representation $a \mapsto \Lambda = \lambda(a)$ contains $\Psi^{(\nu)}$ exactly with multiplicity 4 for each $\nu \leq r$. By (4.9.1), $\Psi^{(\nu)}$ contains $\chi_{\nu k}$ exactly q_i times. Since $A_i^{2q_i}$ is the factor of A_w in question, each $\chi_{\nu k}$ with $\nu \leq r$ should occur in Ξ_i exactly with multiplicity 2. If $\nu > r$, (6.8.4, 5) show that Ξ_i contains the $\chi_{\nu k}$ and $\chi_{\nu k}\rho$ for $k = 1, \cdots, m_i$ with positive multiplicities. Comparing the dimensions, we see that the multiplicities are all equal to one. Thus we obtain (6.8.2).

6.9. In view of (6.8.2), we can find $2m_i g$ isomorphisms $\alpha_{\nu k}, \beta_{\nu k}$ ($\nu = 1, \cdots, g; k = 1, \cdots, m_i$) of S_i into C so that

$$
\begin{aligned}
\Xi_i &\sim \sum_{\nu=1}^{g} \sum_{k=1}^{m_i} (\alpha_{\nu k} + \beta_{\nu k}) \,, \\
\alpha_{\nu k} &= \beta_{\nu k} = \chi_{\nu k} && \text{on } P_i \text{ for } \nu \leq r \,, \\
\alpha_{\nu k} &= \chi_{\nu k}, \ \beta_{\nu k} = \chi_{\nu k}\rho && \text{on } P_i \text{ for } \nu > r \,.
\end{aligned}
$$

(6.9.1)

Let F_i denote the maximal real subfield of P_i, and let $K_i = KF_i$. Then $\alpha_{\nu k} = \beta_{\nu k}$ on F_i for every ν and every k. Therefore $\alpha_{\nu k}$ coincides with $\beta_{\nu k}$ or $\beta_{\nu k}\rho$ on K_i. Now, among $2m_i g$ isomorphisms $\alpha_{\nu k}, \beta_{\nu k}$, there are no two which coincide or which are complex conjugate to each other, since (S_i, Ξ_i) is a CM-type. Therefore, in view of (6.9.1), we have $\alpha_{\nu k} = \beta_{\nu k}\rho$ on K_i if $\nu \leq r$, and $\alpha_{\nu k} = \beta_{\nu k}$ on K_i, if $\nu > r$. Denoting by $\tau_{\nu k}$ the restriction of $\alpha_{\nu k}$ to K_i, we have

(6.9.2) *The restriction of Ξ_i to K_i is equivalent to*

$$
\sum_{\nu=1}^{r} \sum_{k=1}^{m_i} (\tau_{\nu k} + \tau_{\nu k}\rho) + \sum_{\nu=r+1}^{g} \sum_{k=1}^{m_i} 2\tau_{\nu k} \,.
$$

The restriction of $\tau_{\nu k}$ to K coincides with τ_ν or $\tau_\nu \rho$. By (6.3.1) and our definition of Φ at the beginning of 6.4, $\tau_\nu \rho$ can not occur in Φ if $\nu > r$. Therefore $\tau_{\nu k} = \tau_\nu$ on K if $\nu > r$.

6.10. PROPOSITION. *Let (P_i, Ψ_i) and (S_i, Ξ_i) be as in (4.9.2) and 6.8. Assume that $r < g$. Let Φ_0 denote a representation of K such that $\Phi_0 \sim \sum_{\nu=r+1}^{g} \tau_\nu$. Let (K', Φ_0'), (P_i', Ψ_i') and (S_i', Ξ_i') be the duals of (K, Φ_0), (P_i, Ψ_i) and (S_i, Ξ_i) respectively. Then $S_i' = K'P_i'$, and*

$$
\det \Xi_i'(y) = \det \Psi_i'(N_{S_i'/P_i'}(y)) \cdot \det \Phi_0'(N_{S_i'/K'}(y))
$$

for every element or ideal y in S_i'. When $r = g$, the formula is still valid, if we understand that $K' = Q$, $S_i' = P_i'$, and $\det \Phi_0'(x)$ is the identity element or ideal in K for every element or ideal x in Q.

PROOF. Let Φ_i be a representation of K_i such that $\Phi_i \sim \sum_{\nu=r+1}^{g} \sum_{k=1}^{m_i} \tau_{\nu k}$, and (K_i', Φ_i') the dual of (K_i, Φ_i). Since $\tau_{\nu k} = \tau_\nu$ on K for $\nu > r$, we see easily that the sums ω of [C, (5.1.2)] defined for (K, Φ_0) and (K_i, Φ_i) (with respect to the same Galois group) should coincide. Therefore $(K', \Phi_0') = (K_i', \Phi_i')$. Now, in view of (6.8.2) and (6.9.2), we can apply [C, 5.16] to the present case, taking K_i, P_i and S_i to be K, M and S of that proposition. Note that Φ_i, Φ_i', Ψ_i, Ψ_i', Ξ_i and Ξ_i' correspond respectively to the representations $\sum_{\nu=r+1}^{g} \tau_\nu$, $\sum_{\lambda=s+1}^{h} \sigma_\lambda$, $\sum_{\nu=1}^{r} \chi_\nu$, $\sum_{\lambda=1}^{t} \varepsilon_\lambda$, Ξ and Ξ' of [C, 5.14–15]; we need also the relations (5.14.7), (5.14.8), (5.16.1), (5.16.2) of [C]. Then we obtain the desired results for $r < g$. If $r = g$, we have $\operatorname{tr} \Xi_i(x) = \operatorname{tr} \Psi_i(\operatorname{Tr}_{S_i/P_i}(x))$, hence $S_i' = P_i'$ and $\Xi_i' = \Psi_i'$.

6.11. Let (S_i, Ξ_i) for $i = 1, \cdots, t$ be arbitrary CM-types (i.e., generalized CM-types of index one in the sense of [C, 5.11]) which are not necessarily those considered above. For each i, let (A_i, θ_i) be a couple formed by an abelian variety A_i and an isomorphism θ_i of S_i into $\operatorname{End}_Q(A_i)$, such that (A_i, θ_i) is of type (S_i, Ξ_i) in the sense of [C, 4.1]. Let $n_1, \cdots, n_t$ be positive integers, and let A be an abelian variety isogenous to the product

$$(6.11.1) \qquad\qquad A' = A_1^{n_1} \times \cdots \times A_t^{n_t} .$$

Put $S = S_1 \oplus \cdots \oplus S_t$. Fix an isogeny of A to A', and let θ denote the isomorphism of S into $\operatorname{End}_Q(A)$ such that $\theta(a_i)$ corresponds to $\theta_i(a_i)$ for every i and every $a_i \in S_i$ in an obvious sense. The couple (A_w, θ) considered in 6.7–10 is an example of such an (A, θ). We consider the following condition on a positive non-degenerate divisor X on A.

(6.11.2) *If* * *denotes the involution of* $\operatorname{End}_Q(A)$ *determined by* X, *then* $\theta(a)^* = \theta(a^\rho)$ *for every* $a \in S_i (i = 1, \cdots, t)$, *where* ρ *is the complex conjugation.*

Put now

$$(6.11.3) \qquad\qquad \mathfrak{t} = \theta^{-1}[\theta(S) \cap \operatorname{End}(A)] .$$

Then $\mathfrak{t}$ is an order in S. Let (S_i', Ξ_i') be the dual of (S_i, Ξ_i), and S' the composite of $S_1', \cdots, S_t'$. Further let $\mathfrak{f}$ be the conductor of $\mathfrak{t}$, and f_0 the smallest positive integer such that $f_0 1_S \in \mathfrak{f}$, where 1_S denotes the identity element of S. Take an arbitrary positive integer e.

6.12. PROPOSITION. *The notation being as in 6.11, let* $\mathfrak{a}$ *be an integral ideal in* S', *prime to* ef_0, *and* σ *an automorphism of* C *such that*

$$\sigma = [C(S', ef_0)/S', \mathfrak{a}]$$

on $C(S', ef_0)$. *Put* $\mathfrak{b}_i = \det \Xi_i'(N_{S'/S_i'}(\mathfrak{a}))$, $\mathfrak{b} = \mathfrak{b}_1 \oplus \cdots \oplus \mathfrak{b}_t (\subset S)$. *Let* ε_i *be the identity element of* S_i. *Then there exists an isogeny* λ *of* A *to* A^σ *such that*
(6.12.1) $\alpha^\sigma \lambda = \lambda \alpha$ *for every* $\alpha \in \operatorname{End}_Q(A)$ *commuting with the* $\theta(\varepsilon_i)$;

(6.12.2) $\mathrm{Ker}(\lambda) = \{u \in A \mid \theta(\mathfrak{b} \cap \mathfrak{k})u = 0\}$;

(6.12.3) $\lambda^{-1}(X^\sigma) \equiv N(\mathfrak{a})X$ *for every positive non-degenerate divisor X on A satisfying* (6.11.2);

(6.12.4) $\lambda v = v^\sigma$ *for every $v \in A$ such that $ev = 0$.*

PROOF. Let $\mathfrak{r}_i$ denote the ring of all algebraic integers in S_i, and $\mathfrak{r}_s = \mathfrak{r}_1 \oplus \cdots \oplus \mathfrak{r}_t$. Let (A_0, θ_0) be an $\mathfrak{f}$-transform of (A, θ), and κ an $\mathfrak{f}$-multiplication of (A, θ) to (A_0, θ_0) in the sense of [10, 7.1]. Then $\theta_0(\mathfrak{r}_S) = \theta_0(S) \cap \mathrm{End}(A)$. We can find an isogeny μ of A_0 to A such that $\mu\kappa = f_0\theta(1_S)$. We have obviously $f_0 u = 0$ for every $u \in \mathrm{Ker}(\mu)$. Put $A_{0i} = \theta_0(\varepsilon_i)A_0$, and define $\theta_{0i}: S_i \to \mathrm{End}_Q(A_{0i})$ so that $\theta_{0i}(a) = \theta_0(a)$ on A_{0i} for every $a \in S_i$. Then $\theta_{0i}(\mathfrak{r}_i) = \theta_{0i}(S_i) \cap \mathrm{End}_Q(A_{0i})$. Put $\mathfrak{a}_i = N_{S'_i/S'_i}(\mathfrak{a})$. Then $\sigma = [C(S'_i, ef_0)/S'_i, \mathfrak{a}_i]$ on $C(S'_i, ef_0)$. Now we can apply [C, (5.24.1)] to (A_{0i}, θ_{0i}) with ef_0 in place of the integer e in that assertion. Take any polar divisor X_i on A_{0i}. Then we obtain an isogeny λ_{0i} of A_{0i} to A_{0i}^σ satisfying the following conditions:

(i) $\beta^\sigma \lambda_{0i} = \lambda_{0i}\beta$ for every $\beta \in \mathrm{End}_Q(A_{0i})$;

(ii) $\mathrm{Ker}(\lambda_{0i}) = \{s \in A_{0i} \mid \theta(\mathfrak{b}_i)s = 0\}$;

(iii) $\lambda_{0i}^{-1}(X_i^\sigma) \equiv N(\mathfrak{a})X_i$ (note that $N(\mathfrak{a}) = N(\mathfrak{a}_i)$);

(iv) $\lambda_{0i}v = v^\sigma$ for every $v \in A_{0i}$ such that $ef_0 v = 0$.

Define $\lambda_0: A_0 \to A_0^\sigma$ so that $\lambda_0 = \lambda_{0i}$ on A_i. If $x \in \mathrm{Ker}(\mu)$, we have $f_0 x = 0$, hence $\lambda_0 x = x^\sigma$ so that $\mu^\sigma \lambda_0 x = \mu^\sigma x^\sigma = (\mu x)^\sigma = 0$. It follows that

$$\mathrm{Ker}(\mu) \subset \mathrm{Ker}(\mu^\sigma \lambda_0) \, .$$

Therefore we can find an isogeny λ of A to A^σ so that $\lambda\mu = \mu^\sigma \lambda_0$. It is easy to verify that λ satisfies (6.12.1), and $\lambda^{-1}(Y^\sigma) \equiv N(\mathfrak{a})Y$, where

$$Y = \kappa^{-1}\left[\sum_{i=1}^{t}(A_{01} \times \cdots \times A_{0,i-1} \times X_i \times A_{0,i+1} \times \cdots \times A_{0t})\right] \, .$$

Now, by [10, 6.2, Th. 4], we can take X_i so that the involution of $\mathrm{End}_Q(A_{0i})$ determined by X_i sends $\theta_{0i}(a)$ to $\theta_{0i}(a^\rho)$ for every $a \in S_i$. Then Y satisfies (6.11.2). Let X be a divisor on A satisfying (6.11.2). Let φ_X and φ_Y be the isogenies of A to $\mathrm{Pic}(A)$ associated to X and Y. Put $\zeta = (\varphi_Y)^{-1}\varphi_X$. Since X and Y determine the same involution on $\theta(S)$, we see easily that ζ commutes with the elements of $\theta(S)$. By (6.12.1), we have $\zeta^\sigma \lambda = \lambda\zeta$, hence

$${}^t\lambda\varphi_X^\sigma\lambda = {}^t\lambda\varphi_Y^\sigma\zeta^\sigma\lambda = {}^t\lambda\varphi_Y^\sigma\lambda\zeta = N(\mathfrak{a})\varphi_Y\zeta = N(\mathfrak{a})\varphi_X \, ,$$

i.e., $\lambda^{-1}(X^\sigma) \equiv N(\mathfrak{a})X$, which proves (6.12.3).

Let $x \in A$ be such that $ex = 0$. Take an element y of A_0 so that $x = \mu y$. Then $ef_0 y = e\kappa x = 0$. Therefore, by (iv), we have $\lambda_0 y = y^\sigma$, hence

$$\lambda x = \lambda\mu y = \mu^\sigma \lambda_0 y = \mu^\sigma y^\sigma = x^\sigma \, ,$$

which proves (6.12.4). Since $\lambda\mu = \mu^\sigma \lambda_0$, $\mathrm{Ker}(\lambda)$ has the same order as $\mathrm{Ker}(\lambda_0)$, which is $\prod_{i=1}^{t} N(\mathfrak{b}_i)^{n_i}$ by (ii). Call this number b, and write simply b for $b \cdot 1_S$.

Then we see that both $\mathrm{Ker}\,(\lambda)$ and $\{v \in A \mid \theta(\mathfrak{b} \cap \mathfrak{l})v = 0\}$ are subgroups of $\mathrm{Ker}\,(\theta(\mathfrak{b}))$. Since $\mathfrak{a}$ is prime to f_0, $\mathfrak{b}$ is prime to f_0. It follows that μ gives an isomorphism of $\mathrm{Ker}\,(\theta_0(\mathfrak{b}))$ onto $\mathrm{Ker}\,(\theta(\mathfrak{b}))$. Hence, if $v \in A$ and $\theta(\mathfrak{b} \cap \mathfrak{l})v = 0$, then there exists an element z of $\mathrm{Ker}\,(\theta_0(\mathfrak{b}))$ such that $\mu z = v$. Let $a \in_{\mathfrak{l}} \mathfrak{b} \cap \mathfrak{l}$. Then $\mu \cdot \theta_0(a)z = \theta(a)\mu z = 0$. Since $\theta_0(a)z \in \mathrm{Ker}\,(\theta_0(\mathfrak{b}))$, we have $\theta_0(a)z = 0$. Now observe that $\mathfrak{b} = (\mathfrak{b} \cap \mathfrak{l})\mathfrak{r}_S$ by (2.5.2), since $\mathfrak{b}$ is prime to f_0. Therefore we have $\theta_0(\mathfrak{b})z = 0$, so that $\lambda_0 z = 0$, hence $\lambda v = \lambda \mu z = \mu^\sigma \lambda_0 z = 0$, i.e., $v \in \mathrm{Ker}\,(\lambda)$. Conversely, let $v \in \mathrm{Ker}\,(\lambda)$. Take again an element z of $\mathrm{Ker}\,(\theta_0(\mathfrak{b}))$ so that $\mu z = v$. Then $\mu^\sigma \lambda_0 z = \lambda \mu z = \lambda v = 0$. Since μ^σ is an isomorphism of $\mathrm{Ker}\,(\theta_0^\sigma(\mathfrak{b}))$ onto $\mathrm{Ker}\,(\theta^\sigma(\mathfrak{b}))$, and since $\lambda_0 z \in \mathrm{Ker}\,(\theta_0^\sigma(\mathfrak{b}))$, we have $\lambda_0 z = 0$, so that $\theta_0(\mathfrak{b})z = 0$ by (ii). Therefore if $a \in \mathfrak{b} \cap \mathfrak{l}$, we have $\theta(a)v = \theta(a)\mu z = \mu \cdot \theta_0(a)z = 0$, so that $\theta(\mathfrak{b} \cap \mathfrak{l})v = 0$. This proves (6.12.2) and completes the proof.

6.13. Let us now consider the case where (A, θ) and S are the ones obtained from $\mathcal{R}_w$ as in 6.7–8. The lattice $\mathfrak{M}$ being as in 6.4, put

$$\mathfrak{g} = \{a \in P \mid \mathfrak{M}i(f(a)) \subset \mathfrak{M}\} \ .$$

Let us assume that $\mathfrak{r}_K \mathfrak{M} \subset \mathfrak{M}$, and define $\mathfrak{l}$ as in (6.11.3). Obviously $\mathfrak{r}_K \mathfrak{g} \subset \mathfrak{l}$. Further assume

$$(6.13.1) \qquad \mathfrak{r}_S = \mathfrak{r}_K \mathfrak{r}_P \ .$$

This is the case, for example, if $D(K/F)$ is prime to $D(P_i/F)$ for every i, see [C, 1.2]. Let $\mathfrak{h}$ and $\mathfrak{f}$ be the conductors of $\mathfrak{g}$ and $\mathfrak{l}$ respectively, and $\mathfrak{h}_0 = \mathfrak{h} \cap F$. Under the assumption (6.13.1), we have $\mathfrak{h} \subset \mathfrak{f}$. Let (K', Φ_0'), (P_i', Ψ_i'), (S_i', Ξ_i') be as in 6.10. Further let $\mathfrak{a}$, $\mathfrak{b}$, $\mathfrak{b}_i$, σ and λ be as in 6.12. Put

$$\mathfrak{a}_{1i} = \det \Psi_i'\big({}_{S'/P_i'}(\mathfrak{a})\big), \quad \mathfrak{a}_2 = \det \Phi_0'\big(N_{S'/K'}(\mathfrak{a})\big) \ ,$$
$$\mathfrak{a}_1 = \mathfrak{a}_{11} \oplus \cdots \oplus \mathfrak{a}_{1t} \ (\subset P_1 \oplus \cdots \oplus P_t = P) \ .$$

(As mentioned in 6.10. $\mathfrak{a}_2 = \mathfrak{r}_K$ if $r = g$.) Assume that $\mathfrak{a}$ is prime to $N(\mathfrak{h}_0)$. We shall now show that (6.12.2) can be written as

$$(6.13.2) \qquad \mathrm{Ker}\,(\lambda) = \{u \in A_w \mid \theta\big((\mathfrak{a}_1 \cap \mathfrak{g}) \cdot \mathfrak{a}_2\big)u = 0\} \ .$$

Here we consider $(\mathfrak{a}_1 \cap \mathfrak{g}) \cdot \mathfrak{a}_2$ as a submodule of $S = K \otimes_F P$.

First we observe, in view of 6.10, that $\mathfrak{b}_i = \mathfrak{a}_{1i}\mathfrak{a}_2$. This not only is the divisor-theoretical equality, but also means that $\mathfrak{b}_i$ consists of the sums of the elements cd with $c \in \mathfrak{a}_{1i}$ and $d \in \mathfrak{a}_2$, since $\mathfrak{r}_{S_i} = \mathfrak{r}_K \mathfrak{r}_{P_i}$. Therefore $\mathfrak{b} = \mathfrak{a}_1 \mathfrak{a}_2 = \mathfrak{a}_1 \otimes_{\mathfrak{r}_F} \mathfrak{a}_2$. Note that $\mathfrak{a}_1$ and $\mathfrak{a}_2$ are integral. Now we have

$$(6.13.3) \qquad (\mathfrak{a}_1 \mathfrak{a}_2) \cap \mathfrak{l} = \mathfrak{l} \cdot (\mathfrak{a}_1 \cap \mathfrak{g}) \cdot \mathfrak{a}_2 \ .$$

In fact, if $\mathfrak{p}$ is a prime ideal in F which does not divide $N(\mathfrak{h}_0)$, we have

$$\big((\mathfrak{a}_1 \mathfrak{a}_2) \cap \mathfrak{l}\big)_\mathfrak{p} = \mathfrak{a}_{1\mathfrak{p}}\mathfrak{a}_{2\mathfrak{p}} = (\mathfrak{a}_1 \cap \mathfrak{g})_\mathfrak{p}\mathfrak{a}_{2\mathfrak{p}} \ .$$

If $\mathfrak{p}$ divides $N(\mathfrak{h}_0)$, then we have $(\mathfrak{a}_1\mathfrak{a}_2)_{\mathfrak{p}} = (\mathfrak{r}_S)_{\mathfrak{p}}$ since $\mathfrak{a}$ is prime to $N(\mathfrak{h}_0)$, hence $(\mathfrak{a}_1\mathfrak{a}_2 \cap \mathfrak{t})_{\mathfrak{p}} = \mathfrak{t}_{\mathfrak{p}}$ and $(\mathfrak{a}_1 \cap \mathfrak{g})_{\mathfrak{p}} = \mathfrak{g}_{\mathfrak{p}}$. Therefore the localization of (6.13.3) at every $\mathfrak{p}$ is true, hence (6.13.3) itself is true. Since $\theta(\mathfrak{t}) \subset \mathrm{End}\,(A_w)$, we see from (6.13.3) that, for $u \in A_w$, $\theta(\mathfrak{a}_1\mathfrak{a}_2 \cap \mathfrak{t})u = 0 \Leftrightarrow \theta((\mathfrak{a}_1 \cap \mathfrak{g})\cdot\mathfrak{a}_2)u = 0$. Therefore (6.13.2) follows from (6.12.2).

6.14. Let i and i' be as in 6.6, and $\mathfrak{x}$ an $\mathfrak{r}_F$-lattice in X. Put $\mathfrak{M} = \mathfrak{r}_K\cdot i'(\mathfrak{x})$. Since $L^* = K \otimes_F i'(X)$, $\mathfrak{M}$ is an $\mathfrak{r}_K$-lattice in L^*. In the following sections, we shall deal only with $\mathfrak{r}_K$-lattices in L^* of this type. If Y, P, δ, f are as in 6.7, we have

$$
(6.14.1) \qquad
\begin{aligned}
\{a \in Y \mid \mathfrak{M}i(f(a)) \subset \mathfrak{M}\} &= \{a \in Y \mid \mathfrak{x}f(a) \subset \mathfrak{x}\}\,, \\
\{a \in P \mid \mathfrak{M}i(f(a)) \subset \mathfrak{M}\} &= \{a \in P \mid \mathfrak{x}f(a) \subset \mathfrak{x}\}\,.
\end{aligned}
$$

This follows immediately from 2.9. For an integral ideal $\mathfrak{c}$ in F, put

$$
\Gamma(\mathfrak{x}, \mathfrak{c}) = \{\alpha \in \mathfrak{G}^+ \mid \mathfrak{x}\alpha = \mathfrak{x},\ \mathfrak{x}(1 - \alpha) \subset \mathfrak{c}\mathfrak{x}\}\,.
$$

Again by 2.9, we have

$$
(6.14.2) \qquad\qquad i\big(\Gamma(\mathfrak{x}, \mathfrak{c})\big) = \Gamma^0(T,\, \mathfrak{c}^{-1}\mathfrak{M}/\mathfrak{M})\,.
$$

We shall now show that for every integral ideal $\mathfrak{a}$ in F, there always exists an integral ideal $\mathfrak{c}$ in F such that $\mathfrak{c} \subset \mathfrak{a}$ and the following condition is satisfied:

(6.14.3) *If* $\alpha \in \Gamma(T, \mathfrak{c}^{-1}\mathfrak{M}/\mathfrak{M})$ *and* $\alpha(\mathfrak{D}_n^r) \cap \mathfrak{D}_n^r \neq \varnothing$, *then* $\alpha \in i(\Gamma(\mathfrak{x}, \mathfrak{c}))$.

To see this, put

$$
\begin{aligned}
\Gamma_1(T, \mathfrak{a}^{-1}\mathfrak{M}/\mathfrak{M}) &= \{\alpha \in \Gamma(T, \mathfrak{a}^{-1}\mathfrak{M}/\mathfrak{M}) \mid \nu(\alpha) = 1\}\,, \\
\Gamma_1^0(T, \mathfrak{a}^{-1}\mathfrak{M}/\mathfrak{M}) &= G^0(T) \cap \Gamma_1(T, \mathfrak{a}^{-1}\mathfrak{M}/\mathfrak{M})\,.
\end{aligned}
$$

Then [8, 2.22] shows the existence of an integral ideal $\mathfrak{b}$ in F such that if $\alpha \in \Gamma_1(T, \mathfrak{b}^{-1}\mathfrak{M}/\mathfrak{M})$ and $\alpha(\mathfrak{D}_n^r) \cap \mathfrak{D}_n^r \neq \varnothing$, then $\alpha \in \Gamma_1^0(T, \mathfrak{b}^{-1}\mathfrak{M}/\mathfrak{M})$. In [8], we have assumed $r < g$. But the reasoning of [8, 2.22] is obviously true also in the case $r = g$. Now, by Chevalley [2], we can find an integral ideal $\mathfrak{c}$ in F so that, if ε is a unit of K such that $\varepsilon \equiv 1 \bmod \mathfrak{c}$, then there exists a unit η of F such that $\varepsilon = \eta^2$ and $\eta \equiv 1 \bmod \mathfrak{b}$. We can of course take $\mathfrak{c}$ so that $\mathfrak{c} \subset \mathfrak{a} \cap \mathfrak{b}$. Let $\alpha \in \Gamma(T, \mathfrak{c}^{-1}\mathfrak{M}/\mathfrak{M})$. Then $\nu(\alpha)$ is a unit of K, and $\nu(\alpha) \equiv 1 \bmod \mathfrak{c}$ (for a proof see below). Therefore $\nu(\alpha) = \eta^2$, $\eta \equiv 1 \bmod \mathfrak{b}$ with a unit η of F. Then $\eta^{-1}\alpha \in \Gamma_1(T, \mathfrak{b}^{-1}\mathfrak{M}/\mathfrak{M})$. Suppose $\alpha(\mathfrak{D}_n^r) \cap \mathfrak{D}_n^r \neq \varnothing$. Then $\eta^{-1}\alpha \in G^0(T)$, hence $\alpha \in G^0(T) \cap \Gamma(T, \mathfrak{c}^{-1}\mathfrak{M}/\mathfrak{M}) = \Gamma^0(T, \mathfrak{c}^{-1}\mathfrak{M}/\mathfrak{M})$. In view of (6.14.2), this proves (6.14.3).

To show $\nu(\alpha) \equiv 1 \bmod \mathfrak{c}$, let $\mathfrak{b}$ denote the $\mathfrak{r}_K$-ideal generated by the numbers $T(x, y)$ for all x, y in $\mathfrak{M}$. Since

$$
\big(\nu(\alpha) - 1\big)T(x, y) = T(x\alpha - x, y\alpha) + T(x, y\alpha - y)\,,
$$

we have $(\nu(\alpha) - 1)\mathfrak{b} \subset \mathfrak{c}\mathfrak{b}$, hence $\nu(\alpha) - 1 \in \mathfrak{c}$, q.e.d.

7. Construction of models over $K' \cdot C(F', \Theta', \mathfrak{c})$
(The first descent of the field of rationality)

7.1. Let the notation be as in 5.1. In view of the result of 5.16, to prove our main theorems 5.3 and 5.7, it is sufficient to prove them in the special case $W = W_1$. Therefore, until the end of 8.7, we assume $W = W_1$, and denote $D_W(\Delta, \mathfrak{c})$, $\Gamma_W(\mathfrak{x}, \mathfrak{c})$, etc., simply by $D(\Delta, \mathfrak{c})$, $\Gamma(\mathfrak{x}, \mathfrak{c})$ etc., without the subscript W. Further we denote $C_W(F', \Theta', \mathfrak{c})$ simply by C_0. The purpose of this section is to prove the following proposition which is a weaker form of 5.3 and 5.7. We shall be assuming $r < g$ except in 7.4 and 7.5. For $P = P_1 \oplus \cdots \oplus P_t$ as in 4.7, we shall put $D(P/F) = \prod_{i=1}^{t} D(P_i/F)$.

7.2. PROPOSITION. *Let the notation and the assumption be as in 5.1–7 and 7.1. Let Z be an arbitrary finite algebraic extension of C_0, and l a positive integer. Then there exist an integral ideal $\mathfrak{k}$ in F, a quadratic extension M of F', and a system*

$$\{V_\lambda,\ \varphi_\lambda,\ R_\sigma^{\mu\lambda}(\alpha)\ (\lambda,\ \mu = 1,\ \cdots,\ q;\ \alpha \in D(\Delta, \mathfrak{c}))\}$$

satisfying the following conditions:

(7.2.1) *M and Z are linearly disjoint over F', $D(M/F') \neq (1)$, and $D(M/F')\mathfrak{k}$ is prime to $l \cdot D(Z/F')$.*

(7.2.2) *$(V_\lambda, \varphi_\lambda)$ is a model for $\mathfrak{H}_n^r/\Gamma(\mathfrak{x}_\lambda, \mathfrak{c})$.*

(7.2.3) *V_λ is defined over MC_0.*

(7.2.4) *Let $\alpha \in D(\Delta, \mathfrak{c})$, $\nu(\mathfrak{x}_\mu\alpha/\mathfrak{x}_\lambda) = \det \Theta'(\mathfrak{a})$ with an ideal $\mathfrak{a}$ in F' prime to $N(\mathfrak{c})$, and $\sigma = [C_0/F', \mathfrak{a}]$. $R_\sigma^{\mu\lambda}(\alpha)$ is a biregular isomorphism of V_λ onto $V_\mu^{\bar\sigma}$ rational over MC_0, defined for such α, λ, μ and σ, where $\bar\sigma$ is the unique automorphism of MC_0 over M, which coincides with σ on C_0. (Note that $\bar\sigma$ exists by virtue of (7.2.1).)*

(7.2.5) *$R_\sigma^{\mu\lambda}(\alpha) = R_\sigma^{\mu\lambda}(\beta)$ if $\alpha \equiv \beta \bmod^* (\Delta, \mathfrak{c})$.*

(7.2.6) *$R_\tau^{\nu\mu}(\beta)^{\bar\sigma} \circ R_\sigma^{\mu\lambda}(\alpha) = R_{\tau\sigma}^{\nu\lambda}(\beta\alpha)$, where $\bar\sigma$ is as in (7.2.4).*

(7.2.7) *If $\gamma \in \Gamma(\mathfrak{x}_\lambda)$, $R_1^{\lambda\lambda}(\gamma)[\varphi_\lambda(w)] = \varphi_\lambda(\gamma(w))$ for every $w \in \mathfrak{H}_n^r$.*

(7.2.8) *Let (Y, P, δ, f), (P', Ψ'), z, $\mathfrak{g}_r$, $\mathfrak{g}$, $\mathfrak{h}$, $\mathfrak{h}_0$, $\mathfrak{b}$, τ, and α be as in (5.7.6) with $W = W_1$. Suppose that $D(P/F)$ is prime to $\mathfrak{k}$. Then $\varphi_\mu(z)$ is rational over $M \cdot C(P', \Psi', \mathfrak{g}_r, \mathfrak{c})$. Moreover, suppose that τ can be extended to an automorphism ω of $M \cdot C(P', \Psi', \mathfrak{g}_r, \mathfrak{c})$ over M. Then $\varphi_\mu(z)^\omega = R_\sigma^{\mu\lambda}(\alpha)[\varphi_\lambda(\alpha^{-1}(z))]$.*

First we note that, if 7.2 is true for an ideal $\mathfrak{b}$ divisible by $\mathfrak{c}$, it is also true for $\mathfrak{c}$. This can be shown by the same argument as in 5.16. The present situation is simpler than 5.16. The first part of 5.16 concerning the extended system for Δ_1 is unnecessary, since we can simply put $\Delta = \Delta_1$ in the present case. Then always adjoining M to the basic field, we obtain the desired result.

7.3 PROPOSITION. *Let R be a finite algebraic extension of F', and m a positive integer. Then there exist a totally imaginary quadratic extension K of F and extensions $\tau_1, \cdots, \tau_g$ of $\tau_{01}, \cdots, \tau_{0g}$ to K satisfying the following conditions:*

(7.3.1) *If K' is the field generated over Q by $\sum_{\nu=r+1}^{g} x^{\tau_\nu}$ for all $x \in K$, then $[K' : F'] = 2$, K' is not contained in R, $D(K'/F')$ is prime to m, and $D(K'/F') \neq (1)$.*

(7.3.2) *$D(K/F)$ is prime to m, and $D(K/F) \neq (1)$.*

(7.3.3) *K has no roots of unity other than ± 1.*

PROOF. Let p be a rational prime which does not divide $m \cdot D(R/Q) \cdot D(F/Q)$, and such that $p \equiv 3 \bmod (4)$. Let $\lambda = \sqrt{-p}$ and $K = F(\lambda)$. Define τ_ν, for $\nu = 1, \cdots, g$ so that $\lambda^{\tau_\nu} = \lambda$ and $\tau_\nu = \tau_{0\nu}$ on F. For every a and b of F, we have

$$\sum_{\nu=r+1}^{g} (a + b\lambda)^{\tau_\nu} = \sum_{\nu=r+1}^{g} a^{\tau_{0\nu}} + \lambda \cdot \sum_{\nu=r+1}^{g} b^{\tau_{0\nu}} \,.$$

This shows $K' = F'(\lambda)$. By our choice of p, we have $\mathfrak{d}(K'/F') = \mathfrak{d}(Q(\lambda)/Q) = (\lambda)$, so that (7.3.1) is satisfied. Further we have $\mathfrak{d}(K/F) = (\lambda)$, hence (7.3.2). Since there are infinitely many choices of p, (7.3.3) is satisfied by a suitably large p.

7.4 Let P be a totally imaginary quadratic extension of F such that $B \otimes_F P$ is isomorphic to $M_2(P)$, and let $Y = M_n(P)$. Define a positive involution δ of Y by $x^\delta = {}^t x^\rho$ for $x \in Y$ with the complex conjugation ρ on P. By 4.7, we can find an F-linear isomorphism f of $M_n(P)$ into $M_n(B)$ so that $f(x^\delta) = {}^t f(x)^\iota$. Then $(Y, P, \delta, f) \in J_n(B)$ with the notation of 4.9. Let (P, Ψ) be as in 4.9, and (P', Ψ') the dual of (P, Ψ). In the present case, Ψ is considered as a representation of P such that $\Psi \sim \sum_{\nu=1}^{r} \chi_\nu$ with isomorphisms χ_ν of P into C which coincide with $\tau_{0\nu}$ on F, respectively. Therefore P' is the field generated over Q by $\sum_{\nu=1}^{r} u^{\chi_\nu}$ for all $u \in P$.

7.5 PROPOSITION. *Let F_0 be the smallest Galois extension of Q containing F, R a finite algebraic extension of F', $\mathfrak{b}$ an integral ideal in F' prime to $D(F_0/F')$, and $\mathfrak{a}$ an integral ideal in F. Then there exists a totally imaginary quadratic extension P of F such that $B \otimes_F P$ is isomorphic to $M_2(P)$ and the following condition is satisfied.*

(7.5.1) *Let (P', Ψ') be as in 7.4 with any f. Then P' and R are linearly disjoint over F', $D(P'/F')$ is prime to $\mathfrak{b}$, and $D(P/F)$ is prime to $\mathfrak{a}$.*

PROOF. We may assume that R is a normal extension of Q containing F_0. Let p be a rational prime which is prime to $\mathfrak{ab} \cdot D(B/F)$ and decomposes completely in R. Take a prime factor $\mathfrak{p}$ of p in F. By [C, 1.5], there exists a totally imaginary quadratic extension P of F such that every prime factor of

$D(R/Q)D(B/F)\mathfrak{a}N(\mathfrak{b})$ in F remains prime in P, and $\mathfrak{p}$ is the only prime ideal in F which is ramified in P. Then B splits over P, and $D(P/F)$ is prime to $\mathfrak{a}N(\mathfrak{b})D(R/Q)$. Let P' and $\chi_\nu(\nu \leq r)$ be as in 7.4. For each $\nu = r+1, \cdots, g$, let χ_ν be an extension of $\tau_{0\nu}$ to P. Let S be the smallest Galois extension of Q containing P. For every ν, we have $\mathfrak{d}(P^{\chi_\nu}F_0/F_0) = \mathfrak{d}(P/F)^{\chi_\nu}$, since $D(P/F)$ is prime to $D(F_0/Q)$. Now S is the composite of the $P^{\chi_\nu}F_0$ for $\nu = 1, \cdots, g$. Therefore $\mathfrak{d}(S/F_0)$ is a divisor of $\prod_{\nu=1}^{g}\mathfrak{d}(P/F)^{\chi_\nu}$, hence $\mathfrak{d}(S/F_0)$ is prime to $N(\mathfrak{b})$. Since $\mathfrak{b}$ is prime to $D(F_0/F')$, we see that $D(S/F')$ is prime to $\mathfrak{b}$, hence $D(P'/F')$ is prime to $\mathfrak{b}$, for $P' \subset S$. Now we have $\mathfrak{d}(P^{\chi_\nu}R/R) = \mathfrak{d}(P/F)^{\chi_\nu}$, since $D(P/F)$ is prime to $D(R/Q)$. By our choice of $\mathfrak{p}$, the $\mathfrak{d}(P/F)^{\chi_\nu}$ for $\nu = 1, \cdots, g$ are pairwise relatively prime. It follows that $[SR : R] = 2^g = [S : F_0]$. This implies that S and R are linearly disjoint over F_0. Therefore, to complete our proof, it suffices to show that P' and F_0 are linearly disjoint over F'. For this purpose, take an element d of P so that $d^2 \in F$ and $P = F(d)$. Put $c = d^2$, $c_\nu = c^{\chi_\nu}$, $d_\nu = d^{\chi_\nu}$ for $\nu = 1, \cdots, g$. Let G (resp. G_S) denote the Galois group of F_0 (resp. S) over Q. Let $\alpha \in G_S$, and let β be the restriction of α to F_0. Obviously β determines a permutation of $\{c_1, \cdots, c_g\}$. If $c_\nu^\beta = c_\mu$, then $d_\nu^\alpha = \varepsilon_\nu d_\mu$ with $\varepsilon_\nu = \pm 1$. Observe that α is completely determined by β and $\varepsilon_1, \cdots, \varepsilon_g$. Since $[S : F_0] = 2^g$, we can conclude that, for every $\beta \in G$, and for every arrangement $(\varepsilon_1, \cdots, \varepsilon_g)$ of ± 1, there exists an element α of G_S such that $\alpha = \beta$ on F_0 and $d_\nu^\alpha = \varepsilon_\nu d_\mu$ for $c_\nu^\beta = c_\mu$. We apply this to the case where β is the identity map on F'. Then, by our definition of F', $\{\chi_1\beta, \cdots, \chi_r\beta\}$ coincides with $\{\chi_1, \cdots, \chi_r\}$ on F, as a whole. For this β, we can find an element α of G_S so that $\alpha = \beta$ on F_0 and $d_\nu^\alpha = d_\mu$ for $c_\nu^\beta = c_\mu$. Then, for every $x, y \in F$, we have

$$(7.5.2) \qquad \left(\textstyle\sum_{\nu=1}^{r}(x + yd)^{\chi_\nu}\right)^\alpha = \sum_{\nu=1}^{r} x^{\chi_\nu\beta} + \sum_{\nu=1}^{r} y^{\chi_\nu\beta} d_\nu^\alpha .$$

If $\chi_\nu\beta = \chi_\mu$ on F, then $c_\nu^\beta = c_\mu$, hence $d_\nu^\alpha = d_\mu = d^{\chi_\mu}$. Therefore, (7.5.2) is equal to

$$\sum_{\mu=1}^{r} x^{\chi_\mu} + \sum_{\mu=1}^{r} (yd)^{\chi_\mu} = \sum_{\mu=1}^{r} (x + yd)^{\chi_\mu} .$$

This shows that α is the identity map on P'. Thus every automorphism of F_0 over F' can be extended to an automorphism of S over P'. It follows that P' and F_0 are linearly disjoint over F'. This completes the proof.

7.6   Let us now start the proof of 7.2. Take $K, L, \tau_1, \cdots, \tau_g, \rho, T$ as in 6.1–3. Further let i, i', and j be as in 6.6. For the sake of simplicity, we identify α with $i(\alpha)$ for every $\alpha \in \mathfrak{G}$, and x with $i'(x)$ for every $x \in X$. Then $\mathfrak{G} \subset G(T)$, $L_n = K \otimes_F X$, and $\mathfrak{G}^+$ acts both on $\mathfrak{H}_n^r$ and $\mathfrak{S}^r$, and $j(\alpha(z)) = \alpha(j(z))$ for $\alpha \in \mathfrak{G}^+$ and $z \in \mathfrak{H}_n^r$. Let Φ_0 and (K', Φ_0') be as in 6.10. Let Z be a field and l a positive integer given in 7.2. We may assume that Z is normal over Q and contains F. In view of 7.3, we can take K and the τ_ν so that the following condition is satisfied.

(7.6.1) $[K' : F'] = 2$, K' *is not contained in* Z, $D(K'/F') \neq (1)$, $D(K/F) \neq (1)$, *and* $D(K/F)D(K'/F')$ *is prime to* $l \cdot D(Z/F')$.

Now define $\mathfrak{r}_K$-lattices $\mathfrak{M}_\lambda$ and $\mathfrak{N}_\lambda$ in L^* by

$$\mathfrak{M}_\lambda = \mathfrak{r}_K \mathfrak{x}_\lambda, \quad \mathfrak{N}_\lambda = \mathfrak{c}^{-1}\mathfrak{r}_K \mathfrak{x}_\lambda \qquad (\lambda = 1, \cdots, q) \; .$$

By (6.14.2), we have

$$\Gamma(\mathfrak{x}_\lambda, \mathfrak{c}) = \Gamma(T, \mathfrak{N}_\lambda/\mathfrak{M}_\lambda) \cap \mathfrak{G} \qquad (\lambda = 1, \cdots, q) \; .$$

As remarked above, we can replace $\mathfrak{c}$ by its multiple (for the proof of 7.2). Therefore we may assume, in view of the result of 6.14, that, for every λ, the following two conditions are satisfied.

(7.6.2) *If* $\alpha \in \Gamma(T, \mathfrak{N}_\lambda/\mathfrak{M}_\lambda)$ *and* $\alpha(\mathfrak{D}_n^r) \cap \mathfrak{D}_n^r \neq \varnothing$, *then* $\alpha \in \Gamma(\mathfrak{x}_\lambda, \mathfrak{c})$.

(7.6.3) $\Gamma(T, \mathfrak{N}_\lambda/\mathfrak{M}_\lambda)$ *modulo its center has no elements of finite order other than the identity element.*

This means that, from the natural injection $\mathfrak{D}_n^r \to \mathfrak{S}^r$, we obtain a natural biregular embedding of $\mathfrak{D}_n^r/\Gamma(\mathfrak{x}_\lambda, \mathfrak{c})$ into $\mathfrak{S}^r/\Gamma(T, \mathfrak{N}_\lambda/\mathfrak{M}_\lambda)$.

7.7. Now fix a set $\mathfrak{C}$ of representatives for the ideal classes modulo $N(\mathfrak{c})$ in K. Define a representation Φ of L as in 6.4, and consider a weak PEL-type of the form

(7.7.1) $$\Omega = (L, \Phi, \rho; F^+ \cdot T, \mathfrak{aM}_\lambda; x_1, \cdots, x_{4n}) \; ,$$

with $\mathfrak{a} \in \mathfrak{C}$ and elements x_ν of L^* such that

(7.7.2) $$\mathfrak{aN}_\lambda = \mathfrak{aM}_\lambda + \sum_{\nu=1}^{4n} \mathfrak{r}_K x_\nu \; .$$

Since L^* is of dimension $4n$ over K, we can find such x_ν.

We understand that the symbols x_ν in (7.7.1) indicate only their classes modulo $\mathfrak{aM}_\lambda$. In other words, we identify Ω with the one obtained by the change of x_ν for $x_\nu + u_\nu$ with $u_\nu \in \mathfrak{aM}_\lambda$. We see that $(\mathfrak{aM}_\lambda)_\mathfrak{p} = \mathfrak{M}_{\lambda\mathfrak{p}}$ and $(\mathfrak{aN}_\lambda)_\mathfrak{p} = \mathfrak{N}_{\lambda\mathfrak{p}}$ for every $\mathfrak{p}$ dividing $\mathfrak{c}$ (as for the subscript $\mathfrak{p}$, see 2.1). Hence if $y_1, \cdots, y_{4n}$ are elements of L^* such that $\mathfrak{N}_{\lambda\mathfrak{p}} = (\mathfrak{M}_\lambda + \sum_{\nu=1}^{4n} \mathfrak{r}_K y_\nu)_\mathfrak{p}$ for all $\mathfrak{p}$ dividing $\mathfrak{c}$, then we can find elements x_ν of L^* so that $x_\nu \equiv y_\nu \bmod \mathfrak{M}_{\lambda\mathfrak{p}}$ $(\nu = 1, \cdots, 4n)$ for all such $\mathfrak{p}$, and (7.7.2) is satisfied. Therefore we express Ω also as

$$\Omega = (L, \Phi, \rho; F^+ \cdot T, \mathfrak{aM}_\lambda; y_1, \cdots, y_{4n})$$

for such y_ν (even if y_ν may not be contained in $\mathfrak{aN}_\lambda$), since Ω is certainly determined by these data. The symbols $L, \Phi, \rho, F^+ \cdot T$ will be common to all the weak PEL-types considered in this section. Thus we write also

(7.7.3) $$\Omega = (\mathfrak{aM}_\lambda; \{y_\nu\}) \; .$$

Taking a suitable set of indices I_λ for each λ, we write $\Omega = \Omega_{\lambda i}$ with $i \in I_\lambda$ so that $\{\Omega_{\lambda i} \mid j \in I_\lambda\}$ is the set of all weak PEL-types of the form (7.7.1) or (7.7.3) with $\mathfrak{a} \in \mathfrak{A}$ under the condition (7.7.2).

Now we define a family $\sum_{\lambda i} = \{\mathcal{R}_{\lambda i z} \mid z \in \mathfrak{S}^r\}$ of weak PEL-structures of type $\Omega_{\lambda i}$, by means of one and the same parametrizing function $\mathfrak{y}(x, z)$ as in 6.4. We put

$$\mathcal{R}_{\lambda i}(z) = \mathcal{R}_{\lambda i z} = \left(A_{\lambda i z}, \mathfrak{D}_{\lambda i z}, \theta_{\lambda i z}; \{t_{\lambda i z \nu}\} \right) \qquad (z \in \mathfrak{S}^r) .$$

We remind the reader that the map $x \mapsto \mathfrak{y}(x, z)$ is, for each $z \in \mathfrak{S}^r$, an R-linear isomorphism of L_R^n onto $C^{4n g}$; $A_{\lambda i z}$ is isomorphic to $C^{4n g}/\mathfrak{y}(\mathfrak{a}\mathfrak{M}_\lambda, z)$; and the point $t_{\lambda i z \nu}$ corresponds to $\mathfrak{y}(x_\nu, z)$ if $\Omega_{\lambda i} = (\mathfrak{a}\mathfrak{M}_\lambda; \{x_\nu\})$. Here note that $\Omega_{\lambda i}$ and $\Omega_{\lambda j}$ should be distinguished, except the above noted identification, even if they are equivalent in the sense of [C, 4.8].

Let $(U_{\lambda i}, \mathfrak{b}_{\lambda i}, \psi_{\lambda i})$ be a moduli-system for the family $\sum_{\lambda i}$ in the sense of [C, 4.22]. By [C, 4.24–25], we may assume that $U_{\lambda i} = U_{\lambda 1}$ and $\psi_{\lambda i} = \psi_{\lambda 1}$ for all $i \in I_\lambda$. Therefore we write $U_{\lambda i}$ and $\psi_{\lambda i}$ simply as U_λ and ψ_λ.

7.8. PROPOSITION. *Let $\Omega_{\mu i} = (\mathfrak{a}\mathfrak{M}_\mu; \{x_\nu\})$, and let $k(\Omega_{\mu i})$ be the field determined by [C, 4.14]. Then $k(\Omega_{\mu i}) \subset C(K', N(\mathfrak{c}))$. Moreover, let $\mathfrak{t}$ be an ideal in K' prime to $N(\mathfrak{c})$ and let $\sigma = [C(K', N(\mathfrak{c}))/K', \mathfrak{t}]$, $\mathfrak{z} = \det \Phi_0'(\mathfrak{t})$. Let $\mathfrak{w}$ be an ideal in F' prime to $N(\mathfrak{c})$, α an element of $D(\Delta, \mathfrak{c})$ and λ an index such that*

(7.8.1) $\sigma = [C_0/F', \mathfrak{w}]$ on C_0, $\nu(\mathfrak{x}_\mu \alpha / \mathfrak{x}_\lambda) = \det \Theta'(\mathfrak{w})$.

Further let $\mathfrak{b}$ be an element of $\mathfrak{A}$ equivalent to $\mathfrak{a}\mathfrak{z}^{-1} \bmod N(\mathfrak{c})$. Then $(\Omega_{\mu i})^\sigma$ is equivalent to a weak PEL-type $\Omega_{\lambda j}$ of the form $\Omega_{\lambda j} = (\mathfrak{b}\mathfrak{M}_\lambda; \{x, \alpha\})$.

Here note that $\mathfrak{b}$ is determined only by $\mathfrak{a}$ and σ, and independent of the choice of $\mathfrak{t}$. The proof will be completed in 7.9. First we remark the following fact:

(7.8.2) *If $\varepsilon \in D(\Delta, \mathfrak{c})$ and $\nu(\varepsilon) \equiv u \bmod^* \mathfrak{c}$ with a totally positive unit u of F, then $\Omega_{\mu i}$ is equivalent to $(\mathfrak{a}\mathfrak{M}_\mu; \{x_\nu \varepsilon\})$.*

In fact, by 3.11, we can find an element γ of $\mathfrak{G}$ such that $\nu(\gamma) = u$, $\mathfrak{x}_\mu \gamma = \mathfrak{x}_\mu$, $\gamma \equiv \varepsilon \bmod^* (\mathfrak{x}_\mu; \mathfrak{c})$. Then $\gamma \in \Gamma(\mathfrak{x}_\mu)$. Let $\Omega_{\mu j} = (\mathfrak{a}\mathfrak{M}_\mu; \{x_\nu \varepsilon\})$. Since $\gamma \equiv \varepsilon \bmod^* (\mathfrak{x}_\mu; \mathfrak{c})$, we have $\Omega_{\mu j} = (\mathfrak{a}\mathfrak{M}_\mu; \{x_\nu \gamma\})$. Take any point z of $\mathfrak{S}^r$, and put $w = \gamma^{-1}(z)$. Then $\Lambda(\gamma, z)$ of 6.7 defines an isomorphism of $\mathcal{R}_{\mu i z}$ to $\mathcal{R}_{\mu j w}$, since $\Lambda(\gamma, z)\mathfrak{y}(x_\nu, z) = \mathfrak{y}(x_\nu \gamma, w)$. This proves that $\Omega_{\mu i}$ is equivalent to $\Omega_{\mu j}$, hence (7.8.2).

7.9. Let (Y, P, δ, f) and (P', Ψ') be as in (5.3.3), and let

$$\mathfrak{g}_r = \{a \in Y \mid \mathfrak{x}_\mu f(a) \subset \mathfrak{x}_\mu\}, \quad \mathfrak{g} = \mathfrak{g}_r \cap P .$$

Let $\mathfrak{h}$ be the conductor of $\mathfrak{g}$, and $\mathfrak{h}_0 = \mathfrak{h} \cap F$. Assume that the following conditions are satisfied.

(7.9.1) $D(P/F)$ *is prime to* $D(K/F)$. (*This implies that K is not isomor-*

phic over F to any subfield of any simple component of P, since $D(K/F) \neq$ (1).)

(7.9.2) *P' and $k(\Omega_{\mu i}) \cdot C(K', N(\mathfrak{c}))$ are linearly disjoint over F' for every μ.* (Note that $k(\Omega_{\mu i})$ depends only on μ by [C, 4.25].)

Such a (Y, P, δ, f) exists, by virtue of 7.5.

Let the notation be as in 7.8. By (7.9.2), we can extend σ to an automorphism τ of C over $K'P'$. Take an ideal $\mathfrak{m}$ in P' prime to $N(\mathfrak{c}\mathfrak{h}_0)$, such that

$$\tau = [C(P', \Psi', \mathfrak{g}_r, \mathfrak{c})/P', \mathfrak{m}] \qquad \text{on } C(P', \Psi, \mathfrak{g}_r, \mathfrak{c}) \ .$$

Further let $S' = K'P'$, and let $\mathfrak{q}$ be an *integral* ideal in S' prime to $N(\mathfrak{c}\mathfrak{h}_0)$ such that

$$\tau = [C(S', N(\mathfrak{c}\mathfrak{h}_0))/S', \mathfrak{q}] \qquad \text{on } C(S', N(\mathfrak{c}\mathfrak{h}_0)) \ .$$

As is shown in 5.8, we can find an element β of $D(\Delta, \mathfrak{c})$ and an index κ such that $\mathfrak{x}_\nu f([\Psi'(\mathfrak{m})]_{\mathfrak{g}})^{-1} = \mathfrak{x}_\kappa \beta^{-1}$. By (5.8.1), we have $\nu(\mathfrak{x}_\mu \beta/\mathfrak{x}_\kappa) = \det \Theta'(N_{P'/F'}(\mathfrak{m}))$, so that

$$\nu(\mathfrak{x}_\mu \alpha/\mathfrak{x}_\lambda)^{-1}\, \nu(\mathfrak{x}_\mu \beta/\mathfrak{x}_\kappa) = \det \Theta'(\mathfrak{w}^{-1} N_{P'/F'}(\mathfrak{m})) \ .$$

Since $\tau = \sigma$ on C_0, we have $\mathfrak{w}^{-1} N_{P'/F'}(\mathfrak{m}) \in I(P', \Theta', \mathfrak{c})$, hence $\nu(\alpha^{-1}\beta)\, \nu(\mathfrak{x}_\lambda/\mathfrak{x}_\kappa) = (y)$ with an element y of W_1 such that $y \equiv 1 \bmod^* \mathfrak{c}$. In view of (5.6.1), we have $\kappa = \lambda$ and $(\nu(\alpha^{-1}\beta)) = (y)$. Since both $\nu(\alpha^{-1}\beta)$ and y are totally positive,

(7.9.3) $\nu(\beta^{-1}\alpha) \equiv u \bmod^* \mathfrak{c}$ *with a totally positive unit u of F.*

Now $\tau = [C(P', \Psi', \mathfrak{g}_r, \mathfrak{c})/P', N_{S'/P'}(\mathfrak{q})]$ on $C(P', \Psi', \mathfrak{g}_r, \mathfrak{c})$. Therefore $\mathfrak{m}^{-1} N_{S'/P'}(\mathfrak{q})$ belongs to the ideal group $I(P', \Psi', \mathfrak{g}_r, \mathfrak{c})$ defined in 5.2; so we have $[\Psi'(\mathfrak{m}^{-1} N_{S'/P'}(\mathfrak{q}))]_{\mathfrak{g}} \cdot \mathfrak{g}_r = a\mathfrak{g}_r$ with an element a of Y such that $aa^\delta \in W_1$ and $a \equiv 1 \bmod^* \mathfrak{c}\mathfrak{g}_r$. Since $\tau = \sigma$ on $C(K', N(\mathfrak{c}))$, there exists an element b of K' such that $\mathfrak{t}^{-1} N_{S'/K'}(\mathfrak{q}) = (b)$, and $b \equiv 1 \bmod^* (N(\mathfrak{c}))$. Put $s = \det \Phi_0'(b)$, $\mathfrak{q}_1 = \Psi'(N_{S'/P'}(\mathfrak{q}))$, $\mathfrak{q}_2 = \det \Phi_0'(N_{S'/K'}(\mathfrak{q}))$. Then $s \in K$, $s \equiv 1 \bmod^* (N(\mathfrak{c}))$, and $\mathfrak{q}_2 = s\mathfrak{z}$. Since b is equivalent to $a\mathfrak{z}^{-1}$ modulo $N(\mathfrak{c})$, there exists an element d of K such that $da = \mathfrak{z}b$ and $d \equiv 1 \bmod^* (N(\mathfrak{c}))$. Put $\varepsilon = sd \cdot f(a)\beta$. Since $sd \in K$ and $f(a)\beta \in \mathfrak{G}$, ε belongs to the group $G(T)$. We have then

$$\begin{aligned} b\mathfrak{M}_\lambda \varepsilon^{-1} &= s^{-1}d^{-1}\, b\mathfrak{x}_\lambda \beta^{-1} f(a)^{-1} \\ &= \mathfrak{q}_2^{-1} a\mathfrak{x}_\mu f([\Psi'(\mathfrak{m})]_{\mathfrak{g}})^{-1} f(a)^{-1} = \mathfrak{q}_2^{-1} a\mathfrak{M}_\mu f([\mathfrak{q}_1]_{\mathfrak{g}})^{-1}. \end{aligned}$$

Since $\mathfrak{q}$ is an integral ideal, $\mathfrak{q}_1$ and $\mathfrak{q}_2$ are integral, hence $a\mathfrak{M}_\mu \varepsilon \subset b\mathfrak{M}_\lambda$.

Let z_0 be the fixed point of $f((Y, \delta)_0)$ on $\mathfrak{H}_n^r$, and let $z = j(z_0)$, $w = \beta^{-1}(z)$. Then $w = \varepsilon^{-1}(z)$. We have $(b\mathfrak{M}_\lambda; \{x_\nu\beta\}) = \Omega_{\lambda k}$ with an index k. Consider the member $\mathcal{R}_{\lambda k w}$ of $\sum_{\lambda k}$ and the member $\mathcal{R}_{\mu i z}$ of $\sum_{\mu i}$ corresponding to these points w and z. Then $\Lambda(\varepsilon, z)$ of 6.7 defines an isogeny ξ of $\mathcal{R}_{\mu i z}$ to $\mathcal{R}_{\lambda k w}$. Here note that, for every $\mathfrak{p}$ dividing $\mathfrak{c}$, we have $x_\nu \varepsilon \equiv x_\nu \beta \bmod \mathfrak{M}_{\lambda \mathfrak{p}}$. Moreover, $\mathrm{Ker}(\xi)$ is isomorphic to

$$\mathfrak{b}\mathfrak{M}_\lambda \varepsilon^{-1}/\mathfrak{a}\mathfrak{M}_\mu = \left(\mathfrak{q}_2^{-1}\,\mathfrak{a}\mathfrak{M}_\mu f([\mathfrak{q}_1]_\mathfrak{g})^{-1}\right)/\mathfrak{a}\mathfrak{M}_\mu\ .$$

Therefore, if we denote by θ the isomorphism of $S = K \otimes_F P$ into $\mathrm{End}_Q(A_{\mu i z})$ as in 6.8, we have

$$\mathrm{Ker}\,(\xi) = \{t \in A_{\mu i z} \mid \theta\big((\mathfrak{q}_1 \cap \mathfrak{g})\cdot \mathfrak{q}_2\big)t = 0\}\ .$$

We are going to apply (6.12) and (6.13.2) to the present case. By (6.14.1), the present $\mathfrak{g}$ coincides with $\mathfrak{g}$ of 6.13–14. In view of (7.9.1), the conditions (6.7.3) and (6.13.1) are satisfied. Thus we can apply (6.12) and (6.13.2) with $A_{\mu i z}$, $\mathfrak{q}$ and τ as A, $\mathfrak{a}$ and σ. Then we obtain an isogeny η of $\mathcal{R}_{\mu i z}$ to $(\mathcal{R}_{\mu i z})^\tau$ such that $\mathrm{Ker}(\eta)$ coincides with $\mathrm{Ker}(\xi)$. Hence there exists an isomorphism ζ of $(A_{\mu i z})^\tau$ to $A_{\lambda k w}$ such that $\zeta\eta = \xi$. Then $\zeta(t_{\mu i z v}^\tau) = \zeta\eta(t_{\mu i z v}) = t_{\lambda k w v}$. It follows that ζ is an isomorphism of $(\mathcal{R}_{\mu i z})^\tau$ to $\mathcal{R}_{\lambda k w}$. This implies especially that $(\Omega_{\mu i})^\tau$ is equivalent to $\Omega_{\lambda k} = (\mathfrak{b}\mathfrak{M}_\lambda; \{x_\nu \beta\})$. By (7.9.3) and (7.8.2), we see that $\Omega_{\lambda k}$ is equivalent to $(\mathfrak{b}\mathfrak{M}_\lambda; \{x_\nu \beta\beta^{-1}\alpha\})$. Hence $(\Omega_{\mu i})^\tau$ is equivalent to $(\mathfrak{b}\mathfrak{M}_\lambda; \{x_\nu \alpha\})$.

By virtue of the assumption (7.9.2), for any isomorphism σ_0 of

$$k(\Omega_{\mu i})\cdot C(K',\,N(\mathfrak{c}))$$

into C over K', we can extend σ_0 to an automorphism τ of C, to which the above result is applicable. If σ_0 is the identity on $C(K',\,N(\mathfrak{c}))$, then we can take $\mathfrak{b} = \mathfrak{a}$ and $\alpha = 1$. Then $(\Omega_{\mu i})^\tau$ is equivalent to $\Omega_{\mu i}$. Hence τ is the identity on $k(\Omega_{\mu i})$ (see [C, 4.14]). This shows $k(\Omega_{\mu i}) \subset C(K',\,N(\mathfrak{c}))$, and completes the proof of 7.8.

7.10. Let $\lambda, \mu, \alpha, \sigma, \Omega_{\mu i}$ and $\Omega_{\lambda j}$ be as in 7.8. By [C, 4.21, 4.23], there exists a biregular isomorphism T, defined over $k(\Omega_{\lambda j})$ of U_λ to U_μ^σ such that $T\circ\mathfrak{v}_{\lambda j} = \mathfrak{v}_{\mu i}^\sigma$. (This T will easily be distinguished from the anti-hermitian form T in the definition of Ω in (7.7.1), since the latter will never appear in the rest of this section.) Let us now prove

(7.10.1) *T depends only on $\lambda,\ \mu,\ \sigma$ and the class of $\alpha \bmod^*(\Delta;\ \mathfrak{c})$, and not on the choice of i.*

Let $\Omega_{\mu h} = (\mathfrak{a}^*\mathfrak{M}_\mu; \{x_\nu^*\})$ be another weak PEL-type with any $\mathfrak{a}^* \in \mathcal{C}$ and the same μ, and let α^* be an element of $D(\Delta,\ \mathfrak{c})$ such that $\alpha^* \equiv \alpha \bmod^*(\Delta;\ \mathfrak{c})$. Take an ideal $\mathfrak{w}^*$ in F' prime to $N(\mathfrak{c})$ so that $\nu(\chi_\mu \alpha^*/\chi_\lambda) = \det \Theta'(\mathfrak{w}^*)$. By our definition of C_0, we have $\sigma = [C_0/F',\ \mathfrak{w}^*]$ on C_0. By 7.8, $(\Omega_{\mu h})^\sigma$ is equivalent to $\Omega_{\lambda k} = (\mathfrak{b}^*\mathfrak{M}_\lambda; \{x_\nu^*\alpha^*\})$ with an element $\mathfrak{b}^*$ of $\mathcal{C}$ equivalent to $\mathfrak{a}^*\mathfrak{z}^{-1} \bmod N(\mathfrak{c})$. Let T^* be a biregular isomorphism of U_λ to U_μ^σ such that $T^*\circ\mathfrak{v}_{\lambda k} = \mathfrak{v}_{\mu h}^\sigma$. Our aim is to prove $T = T^*$. Let d and d^* be elements of K such that $d\mathfrak{a} = \mathfrak{z}\mathfrak{b}$, $d^*\mathfrak{a}^* = \mathfrak{z}\mathfrak{b}^*$, $d \equiv d^* \equiv 1 \bmod N(\mathfrak{c})$. Let e be a positive integer such that $e\mathfrak{a} \subset \mathfrak{a}^*$. Since $\mathfrak{a}$ and $\mathfrak{a}^*$ are prime to $N(\mathfrak{c})$, we may assume e to be prime to $N(\mathfrak{c})$. Take an arbitrary $z \in \mathfrak{S}^r$ which is not necessarily the one considered in 7.9. Let ζ

be an isogeny of $A_{\mu i z}$ to $A_{\mu h z}$ obtained from $\Lambda(e, z)$. From (6.7.1), we see that

$$(7.10.2) \qquad \mathrm{Ker}(\zeta) = \{u \in A_{\mu i z} \mid \theta_{\mu i z}(ea^{*-1}\mathfrak{a})u = 0\} \ .$$

We can find elements $a_{\nu \kappa}$ of $\mathfrak{r}_K$ so that

$$(7.10.3) \qquad x_\nu^* \equiv \sum_{\kappa=1}^{4n} a_{\nu \kappa} e x_\kappa \quad \mathrm{mod} \ \mathfrak{M}_\mathfrak{p} \qquad (\nu = 1, \cdots, 4n)$$

for every $\mathfrak{p}$ dividing $\mathfrak{c}$. Again from (6.7.1) we obtain

$$(7.10.4) \qquad t_{\mu h z \nu} = \sum_{\kappa=1}^{4n} \theta_{\mu h z}(a_{\nu \kappa}) \zeta t_{\mu i z \kappa} \qquad (\nu = 1, \cdots, 4n) \ .$$

Extend σ to an automorphism of C over K'. Since $\Omega_{\mu i}^\sigma$ is equivalent to $\Omega_{\lambda j}$, $(\mathcal{R}_{\mu i z})^\sigma$ is isomorphic to $\mathcal{R}_{\lambda j w}$ for some $w \in \mathfrak{S}^r$. Since $d^{-1}d^*eb \subset \mathfrak{b}^*$, $\Lambda(d^{-1}d^*e, w)$ defines an isogeny π of $A_{\lambda j w}$ to $A_{\lambda k w}$. We see that

$$(7.10.5) \qquad \mathrm{Ker}(\pi) = \{u \in A_{\lambda j w} \mid \theta_{\lambda j w}(ea^{*-1}\mathfrak{a})u = 0\} \ .$$

Let ε be an isomorphism of $\mathcal{R}_{\lambda j w}$ to $(\mathcal{R}_{\mu i z})^\sigma$. By (7.10.2) and (7.10.5), we have $\varepsilon(\mathrm{Ker}(\pi)) = \mathrm{Ker}(\zeta^\sigma)$. Hence there exists an isomorphism η of $A_{\lambda k w}$ to $A_{\mu h z}^\sigma$ such that $\eta\pi = \zeta^\sigma\varepsilon$. We see easily that $\eta \cdot \theta_{\lambda k w}(a) = \theta_{\mu h z}^\sigma(a)\eta$ for every $a \in L$, since π, ζ^σ and ε have the corresponding property. Since $d \equiv d^* \equiv 1 \ \mathrm{mod}^* N(\mathfrak{c})$ and $\alpha \equiv \alpha^* \ \mathrm{mod}^*(\Delta; \mathfrak{c})$, we have, by (7.10.3), $x_\nu^*\alpha^* \equiv \sum_\kappa a_{\nu \kappa}d^{-1}d^*ex_\kappa\alpha \ \mathrm{mod} \ \mathfrak{M}_\mathfrak{p}$ for every $\mathfrak{p}$ dividing $\mathfrak{c}$, so that $t_{\lambda k w \nu} = \sum_\kappa \theta_{\lambda k w}(a_{\nu \kappa})\pi t_{\lambda j w \kappa}$. Therefore

$$\begin{aligned} \eta t_{\lambda k w \nu} &= \sum_\kappa \theta_{\mu h z}^\sigma(a_{\nu \kappa})\eta\pi t_{\lambda j w \kappa} = \sum_\kappa \theta_{\mu h z}^\sigma(a_{\nu \kappa})\zeta^\sigma\varepsilon t_{\lambda j w \kappa} \\ &= [\sum_\kappa \theta_{\mu h z}(a_{\nu \kappa})\zeta t_{\mu i z \kappa}]^\sigma = (t_{\mu h z \nu})^\sigma \qquad \text{by (7.10.4).} \end{aligned}$$

It can easily be verified that η^{-1} sends $(\mathfrak{D}_{\mu h z})^\sigma$ to $\mathfrak{D}_{\lambda k w}$. Thus η is an isomorphism of $\mathcal{R}_{\lambda k w}$ to $(\mathcal{R}_{\mu h z})^\sigma$. By the property of $(U_\mu, \mathfrak{v}_{\mu i}, \psi_\mu)$, we have $\psi_\mu(z) = \mathfrak{v}_{\mu h}(\mathcal{R}_{\mu h z})$, hence

$$\begin{aligned} T^{*-1}(\psi_\mu(z)^\sigma) &= T^{*-1}(\mathfrak{v}_{\mu h}^\sigma(\mathcal{R}_{\mu h z}^\sigma)) = \mathfrak{v}_{\lambda k}(\mathcal{R}_{\lambda k w}) = \psi_\lambda(w) \\ &= \mathfrak{v}_{\lambda j}(\mathcal{R}_{\lambda j w}) = T^{-1}(\mathfrak{v}_{\mu i}^\sigma(\mathcal{R}_{\mu i z}^\sigma)) = T^{-1}(\psi_\mu(z)^\sigma) \ . \end{aligned}$$

Since z is an arbitrary point of $\mathfrak{S}^r$, this shows $T = T^*$, hence (7.10.1)

    7.11 Now let us write $T_\sigma^{\mu \lambda}(\alpha)$ for T. We have then

$$(7.11.1) \qquad T_\sigma^{\mu \lambda}(\alpha) \circ \mathfrak{v}_{\lambda j} = \mathfrak{v}_{\mu i}^\sigma \ .$$

The symbol $T_\sigma^{\mu \lambda}(\alpha)$ is defined for indices λ, μ, an element α of $D(\Delta, \mathfrak{c})$ and $\sigma \in G(C(K', N(\mathfrak{c}))/K')$ such that

$$\sigma = [C_0/F', \mathfrak{w}], \qquad \nu(\mathfrak{x}_\mu\alpha/\mathfrak{x}_\lambda) = \det \Theta'(\mathfrak{w})$$

for an ideal $\mathfrak{w}$ in F' prime to $N(\mathfrak{c})$.

    If μ, ν, γ, τ are such that $T_\tau^{\nu \mu}(\gamma)$ is defined, then

$$(7.11.2) \qquad T_\tau^{\nu \mu}(\gamma) \circ T_\sigma^{\mu \lambda}(\alpha) = T_{\tau \sigma}^{\nu \lambda}(\gamma\alpha) \ .$$

This follows from (7.11.1) in a straightforward way. From (7.10.1), we obtain

$$(7.11.3) \qquad T_\sigma^{\mu\lambda}(\alpha) = T_\sigma^{\mu\lambda}(\alpha_1) \quad if \; \alpha \equiv \alpha_1 \bmod^* (\Delta; c) \; .$$

Let us consider the case where $\lambda = \mu$, $\alpha \in \Gamma(\mathfrak{x}_\mu)$ and $\sigma = $ identity. Then $\Omega_{\mu i} = (\mathfrak{a}\mathfrak{M}_\mu; \{\mathfrak{x}_\nu\})$ is equivalent to $(\mathfrak{a}\mathfrak{M}_\mu; \{x_\nu\alpha\})$ by (7.8.2). Moreover we have seen in the proof of (7.8.2) that if $z \in \mathfrak{S}^r$, $w = \alpha^{-1}(z)$ and $\Omega_{\mu j} = (\mathfrak{a}\mathfrak{M}_\mu; \{x_\nu\alpha\})$, then $\mathcal{R}_{\mu i z}$ is isomorphic to $\mathcal{R}_{\mu j w}$. Therefore

$$\psi_\mu(z) = \mathfrak{v}_{\mu i}(\mathcal{R}_{\mu i z}) = \mathfrak{v}_{\mu i}(\mathcal{R}_{\mu j w}) = T_1^{\mu\mu}(\alpha)[\mathfrak{v}_{\mu j}(\mathcal{R}_{\mu j w})] = T_1^{\mu\mu}(\alpha)[\psi_\mu(w)] \; .$$

This proves

(7.11.4)　*If* $\alpha \in \Gamma(\mathfrak{x}_\mu)$, $T_1^{\mu\mu}(\alpha)[\psi_\mu(w)] = \psi_\mu(\alpha(w))$ *for every* $w \in \mathfrak{S}^r$.

7.12. Let (Y, P, δ, f), (P', Ψ'), $\mathfrak{g}_Y$, $\mathfrak{g}$, $\mathfrak{h}_0$ be as in 7.9. We assume the condition (7.9.1), but release (7.9.2). Let τ be an automorphism of C over $K'P'$. Take an ideal $\mathfrak{m}$ in P' prime to $N(c\mathfrak{h}_0)$ such that

$$(7.12.1) \qquad \tau = [C(P', \Psi', \mathfrak{g}_Y, c)/P', \mathfrak{m}] \qquad on \; C(P', \Psi', \mathfrak{g}_Y, c) \; .$$

Further let σ be the restriction of τ to $C(K', N(c))$. By 5.8, we can find an element β of $D(\Delta, c)$ and an index λ such that

$$(7.12.2) \qquad \mathfrak{x}_\mu f\big([\Psi'(\mathfrak{m})]_\mathfrak{g}\big)^{-1} = \mathfrak{x}_\lambda \beta^{-1} \; .$$

Let z be the fixed point of $f((Y, \delta)_0)$ on $\mathfrak{D}_n^r$. Then

$$(7.12.3) \qquad \psi_\mu(z)^\tau = T_\sigma^{\mu\lambda}(\beta)[\psi_\lambda(\beta^{-1}(z))] \; .$$

In fact, we can repeat the argument of 7.9 by taking $\mathfrak{w} = N_{P'/F'}(\mathfrak{m})$ and $\alpha = \beta$. Note that we do not need the linear disjointness (7.9.2), since we start with an automorphism τ of C over $K'P'$. In 7.9, we have shown that $\mathcal{R}_{\mu i z}^\tau$ is isomorphic to $R_{\lambda k w}$ with $w = \beta^{-1}(z)$. Therefore

$$\psi_\mu(z)^\tau = \mathfrak{v}_{\mu i}^\tau(\mathcal{R}_{\mu i z}^\tau) = T_\sigma^{\mu\lambda}(\beta)[\mathfrak{v}_{\lambda k}(\mathcal{R}_{\lambda k w})] = T_\sigma^{\mu\lambda}(\beta)[\psi_\lambda(w)] \; ,$$

which proves (7.12.3).

7.13. Put $V_\lambda = \psi_\lambda(\mathfrak{D}_n^r)$, and denote by φ_λ the restriction of ψ_λ to $\mathfrak{D}_n^r$. By virtue of (7.6.2, 3), $\mathfrak{D}_n^r/\Gamma(\mathfrak{x}_\lambda, c)$ can be biregularly embedded in $\mathfrak{S}^r/\Gamma(T, \mathfrak{N}_\lambda/\mathfrak{M}_\lambda)$. Therefore $(V_\lambda, \varphi_\lambda)$ is certainly a model for $\mathfrak{D}_n^r/\Gamma(\mathfrak{x}_\lambda, c)$. We shall now show that

(7.13.1)　V_λ *is defined over* $C(K', N(c))$ *for every* λ, *and* $T_\sigma^{\mu\lambda}(\alpha)$ *maps* V_λ *onto* V_μ^σ.

Let the notation be as in 7.8 and 7.9. By 3.11 and (7.9.3), we can find an element γ of $\mathfrak{G}$ such that $\mathfrak{x}_\lambda\gamma = \mathfrak{x}_\lambda$, $\nu(\gamma) = u$, $\gamma \equiv \beta^{-1}\alpha \bmod^*(\Delta; c)$. Then $\gamma \in \Gamma(\mathfrak{x}_\lambda)$. We can replace β of (7.12.2) by $\beta\gamma$. Since $\alpha \equiv \beta\gamma \bmod^* (\Delta; c)$, we have $T_\sigma^{\mu\lambda}(\alpha) = T_\sigma^{\mu\lambda}(\beta\gamma)$. Substituting $\beta\gamma$ for β in (7.12.3), we obtain

$$T_\sigma^{\mu\lambda}(\alpha)[\varphi_\lambda(\gamma^{-1}\beta^{-1}(z))] = \varphi_\mu(z)^\tau \; ,$$

so that $\varphi_\mu(z)^\tau \in T_\sigma^{\mu\lambda}(\alpha)(V_\lambda)$. Now keeping (Y, P, δ) the same, change f for the map $f_\xi : Y \to M_n(B)$ defined by $f_\xi(v) = \xi f(v)\xi^{-1}$ with an element ξ of $\mathfrak{G}^+$. Then the objects $\mathfrak{g}_Y$, $\mathfrak{g}$, $\mathfrak{h}_0$ will change accordingly. But still we can make the same consideration as before with the same P', Ψ', λ, μ, α, σ, and τ. Since $f_\xi((Y, \delta)_0)$ leaves $\xi^{-1}(z)$ fixed, we have $\varphi_\mu(\xi^{-1}(z))^\tau \in T_\sigma^{\mu\lambda}(\alpha)(V_\lambda)$ for every $\xi \in \mathfrak{G}^+$. Now the points $\varphi_\mu(\xi^{-1}(z))^\tau$ for all $\xi \in \mathfrak{G}^+$ form a Zariski dense subset of V_μ^τ. It follows that $V_\mu^\tau \subset T_\sigma^{\mu\lambda}(\alpha)(V_\lambda)$. Since V_λ has the same dimension as V_μ, we obtain $V_\mu^\tau = T_\sigma^{\mu\lambda}(\alpha)(V_\lambda)$. This is true for any automorphism τ of C over $K'P'$ which coincides with σ on $C(K', N(\mathfrak{c}))$. This implies that V_μ is defined over $P' \cdot C(K', N(\mathfrak{c}))$. Now this conclusion is valid so far as (7.9.1, 2) are satisfied. Therefore we take new (Y_0, P_0, δ_0) and (P_0', Ψ_0'), etc., so that P_0' is linearly disjoint with $P' \cdot C(K', N(\mathfrak{c}))$ over F'. Then V_μ should be defined over $P_0' \cdot C(K', N(\mathfrak{c}))$, hence over

$$P' \cdot C(K', N(\mathfrak{c})) \cap P_0' \cdot C(K', N(\mathfrak{c})) = C(K', N(\mathfrak{c})) \ .$$

This completes the proof of (7.13.1).

Let $R_\sigma^{\mu\lambda}(\alpha)$ denote the restriction of $T_\sigma^{\mu\lambda}(\alpha)$ to V_λ. Then $R_\sigma^{\mu\lambda}(\alpha)$ is a biregular isomorphism of V_λ to V_μ^σ, rational over $C(K', N(\mathfrak{c}))$. Further, (7.11.2–4) and (7.12.3) can now be written as follows:

(7.13.2) $R_\tau^{\nu\mu}(\gamma)^\sigma \circ R_\sigma^{\mu\lambda}(\alpha) = R_{\tau\sigma}^{\nu\lambda}(\gamma\alpha)$, and $R_\sigma^{\mu\lambda}(\alpha) = R_\sigma^{\mu\lambda}(\alpha_1)$ if $\alpha \equiv \alpha_1 \bmod^* (\Delta; \mathfrak{c})$.

(7.13.3) If $\gamma \in \Gamma(\chi_\mu)$, $R_1^{\mu\mu}(\gamma)[\varphi_\mu(w)] = \varphi_\mu(\gamma(w))$ for every $w \in \mathfrak{D}_n^\tau$.

(7.13.4) $\varphi_\mu(z)^\tau = R_\sigma^{\mu\lambda}(\beta)[\varphi_\lambda(\beta^{-1}(z))]$ in the situation of 7.12.

7.14. Let $G_0 = G(C(K', N(\mathfrak{c}))/C_0K')$. We observe that $R_\tau^{\mu\mu}(1)$ is defined for every $\tau \in G_0$, and is a biregular morphism of V_μ to V_μ^τ. By (7.13.2), we have $R_\eta^{\mu\mu}(1)^\tau \circ R_\tau^{\mu\mu}(1) = R_{\eta\tau}^{\mu\mu}(1)$ for every η and τ of G_0. By Weil's criterion [12], there exist a variety V_μ', rational over C_0K', and a biregular isomorphism S_μ of V_μ to V_μ', rational over $C(K', N(\mathfrak{c}))$, such that $R_\tau^{\mu\mu}(1) = (S_\mu^\tau)^{-1} \circ S_\mu$ for every $\tau \in G_0$. Now replace V_μ, φ_μ, $R_\sigma^{\mu\lambda}(\alpha)$ by V_μ', $S_\mu \circ \varphi_\mu$, $S_\mu^\sigma \circ R_\sigma^{\mu\lambda}(\alpha) \circ S_\lambda^{-1}$ respectively, and, for the sake of simplicity, denote them again by the same symbols as before. (The objects U_λ, ψ_λ, $T_\sigma^{\mu\lambda}(\alpha)$ will not be needed any longer.) By this replacement, we can now assume, besides (7.13. 2–4),

(7.14.1) V_λ is rational over $K'C_0$,

(7.14.2) $R_\tau^{\mu\mu}(1)$ is the identity map of V_μ if $\tau \in G_0$.

For every $\sigma \in G(C(K', N(\mathfrak{c}))/K')$ and $\tau \in G_0$, we have

$$R_{\tau\sigma}^{\mu\lambda}(\alpha) = R_\tau^{\mu\mu}(1)^\sigma \circ R_\sigma^{\mu\lambda}(\alpha) = R_\sigma^{\mu\lambda}(\alpha) \ .$$

Therefore $R_\sigma^{\mu\lambda}(\alpha)$ depends only on λ, μ, α and the action of σ on $K'C_0$. Moreover, still with τ in G_0, we have

$$R_\sigma^{\mu\lambda}(\alpha)^\tau = R_\sigma^{\mu\lambda}(\alpha)^\tau \circ R_\tau^{\lambda\lambda}(1) = R_{\sigma\tau}^{\mu\lambda}(\alpha) = R_\sigma^{\mu\lambda}(\alpha) \ ,$$

hence

(7.14.3) $R_\sigma^{\mu\lambda}(\alpha)$ *is rational over* $K'C_0$.

By (7.6.1), $G(K'C_0/K')$ can be identified with $G(C_0/F'')$, hence we can now understand that the subscript σ of $R_\sigma^{\mu\lambda}(\alpha)$ is an element of $G(C_0/F'')$. Instead of the map $\varphi_\lambda : \mathfrak{D}_n^r \to V_\lambda$, consider the map $\varphi_\lambda \circ j : \mathfrak{H}_n^r \to V_\lambda$, and denote $\varphi_\lambda \circ j$ again by φ_λ, since we shall not consider $\mathfrak{D}_n^r$ any longer. We are going to show that the system $\{V_\lambda, \varphi_\lambda, R_\sigma^{\mu\lambda}(\alpha)\}$ satisfies (7.2. 1–8) with $M = K'$ and $\mathfrak{t} = D(K/F)$.

First (7.2.1) follows from (7.6.1). The conditions (7.2. 2–7) are certainly satisfied. Now we have the formula (7.13.4) for an automorphism τ of C over MP' ($= K'P'$). Suppose that τ is the identity map on $M \cdot C(P', \Psi', \mathfrak{g}_Y, \mathfrak{c})$, the notation being as in 7.12 (or in (7.2.8)). Then we can take the ideal $\mathfrak{m}$ of (7.12.1) to be (1), hence $\lambda = \mu$ and $\beta = 1$ in (7.12.2). Therefore, from (7.13.4) we obtain $\varphi_\mu(z)^\tau = R_1^{\mu\mu}(1)[\varphi_\mu(z)] = \varphi_\mu(z)$, so that $\varphi_\mu(z)$ is rational over $M \cdot C(P', \Psi', \mathfrak{g}_Y, \mathfrak{c})$. This fact together with (7.13.4) shows that (7.2.8) is satisfied. The proof of 7.2 is thus completed.

8. The second descent of the field of rationality

8.1. PROPOSITION. *Let* $\mathfrak{x}$ *be an* $\mathfrak{r}_F$-*lattice in* X, *and* $\mathfrak{q}$ *a prime ideal in* F *which is unramified in* B. *Let* P, Y *and* δ *be as in* 7.4. *For every* F-*linear isomorphism* f *of* Y *into* $M_n(B)$ *such that* $f(a^\delta) = {}^t f(a)^\iota$, *let*

$$\mathfrak{g}_f = \{a \in P \mid \mathfrak{x} f(a) \subset \mathfrak{x}\},$$

$\mathfrak{h}_f$ *the conductor of* $\mathfrak{g}_f$, $\mathfrak{h}_{0f} = F \cap \mathfrak{h}_f$, *and* z_f *the fixed point of* $f((Y, \delta)_0)$ *on* $\mathfrak{H}_n^r$. *Then there exists an integral ideal* $\mathfrak{e}$ *in* F, *determined only by* $\mathfrak{x}$, *and independent of* P *and* $\mathfrak{q}$, *with the following property: denote by* $Z(P, \mathfrak{q})$ *the set of* z_f *for all* f *such that* $\mathfrak{h}_{0f}$ *is divisible only by the prime factors of* $\mathfrak{e}\mathfrak{q}$; *then* $Z(P, \mathfrak{q})$ *is dense on* $\mathfrak{H}_n^r$.

PROOF. Let $\mathfrak{o}$ be a maximal order in B. Let $\{x_i\}$ be the basis of X over B as in (3.1.2) which we considered in 4.4. Put $\mathfrak{y} = \mathfrak{o}x_1 + \cdots + \mathfrak{o}x_n$. Let $\mathfrak{e}$ be the product of all $\mathfrak{p}$ such that $\mathfrak{y}_\mathfrak{p} \neq \mathfrak{x}_\mathfrak{p}$. By [C, 2.8], there exists an F-linear isomorphism f_0 of P into B such that $f_0(\mathfrak{r}_P) \subset \mathfrak{o}$. Define $f_1 : Y \to M_n(B)$ by $f_1((a_{ij})) = (f_0(a_{ij}))$ for $(a_{ij}) \in M_n(P) = Y$ with $a_{ij} \in P$. We see easily that $f_1(x^\delta) = {}^t f_1(x)^\iota$. Let w be the fixed point of $f_1((Y, \delta)_0)$. By 3.9, for every non-empty open subset N of $\mathfrak{H}_n^r$, there exists an element α of $\mathfrak{U}$ such that $\alpha(w) \in N$ and $\mathfrak{x}_\mathfrak{p} \alpha = \mathfrak{x}_\mathfrak{p}$ for every $\mathfrak{p} \neq \mathfrak{q}$. Define $f : Y \to M_n(B)$ by $f(a) = \alpha f_1(a) \alpha^{-1}$. Then $z_f = \alpha(w)$. If $\mathfrak{p}$ does not divide $\mathfrak{e}\mathfrak{q}$, we have $\mathfrak{x}_\mathfrak{p} \alpha = \mathfrak{x}_\mathfrak{p} = \mathfrak{y}_\mathfrak{p}$, so that

$$\mathfrak{x}_\mathfrak{p} f(\mathfrak{r}_P)_\mathfrak{p} \subset \mathfrak{x}_\mathfrak{p},$$

hence $(\mathfrak{r}_P)_\mathfrak{p} = (\mathfrak{g}_f)_\mathfrak{p}$. Therefore $\mathfrak{h}_{0f}$ is not divisible by such a $\mathfrak{p}$, so $z_f \in Z(P, \mathfrak{q})$,

$N \cap Z(P, \mathfrak{q}) \neq \varnothing$. This proves our proposition.

8.2. Proof of 5.9. Let $\{V_\lambda, \varphi_\lambda, R_\sigma^{\mu\lambda}(\alpha)\}$ and $\{U_\lambda, \psi_\lambda, S_\sigma^{\mu\lambda}(\alpha)\}$ be as in 5.9. Since $\Gamma_W(\mathfrak{x}_\lambda, \mathfrak{c}) \subset \Gamma_W'(\mathfrak{x}_\lambda, \mathfrak{m})$, there exists, for each λ, a morphism T_λ of V_λ onto U_λ such that $\psi_\lambda = T_\lambda \circ \varphi_\lambda$. We can find a finite algebraic extension Z of $C_W(F', \Theta', \mathfrak{c})$ so that the T_λ are defined over a regular extension Z_1 of Z. Let α, λ, μ, σ be such that $R_\sigma^{\mu\lambda}(\alpha)$ is defined. Namely we have

$$(8.2.1) \qquad \sigma = [C_W(F', \Theta', \mathfrak{c})/F', \mathfrak{a}] , \qquad \nu(\mathfrak{x}_\mu\alpha/\mathfrak{x}_\lambda) = \det \Theta'(\mathfrak{a}) ,$$

with an ideal $\mathfrak{a}$ in F' prime to $N(\mathfrak{c})$. We shall denote the restriction of σ to $C_W(F', \Theta', \mathfrak{m})$ also by σ. Extend σ to an isomorphism τ of Z_1 into C. Let (Y, P, δ, f), (P', Ψ'), $\mathfrak{g}_Y$, $\mathfrak{g}$, $\mathfrak{h}_0$, and z be as in (5.7.6). By 7.5, we can take P so that

$(8.2.2)$ P' and Z are linearly disjoint over F'.

Then we can extend τ to an automorphism ω of C over P'. Take an ideal $\mathfrak{b}$ in P' prime to $N(\mathfrak{c}\mathfrak{h}_0)$ so that

$$\omega = [C_W(P', \Psi', \mathfrak{g}_Y, \mathfrak{c})/P', \mathfrak{b}] \qquad \qquad \text{on } C_W(P', \Psi', \mathfrak{g}_Y, \mathfrak{c}) ,$$

and take an element β of $D_W(\Delta, \mathfrak{c})$ so that $\mathfrak{x}_\mu f([\Psi'(\mathfrak{b})]_\mathfrak{g})^{-1} = \mathfrak{x}_\kappa\beta$ with an index κ (see 5.8). By (5.8.1), we have

$$\nu(\mathfrak{x}_\mu\alpha/\mathfrak{x}_\lambda)^{-1}\, \nu(\mathfrak{x}_\mu\beta/\mathfrak{x}_\kappa) = \det \Theta'(\mathfrak{a}^{-1}N_{P'/F'}(\mathfrak{b})) .$$

Since $\omega = \sigma$ on $C_W(F', \Theta', \mathfrak{c})$, we have $\mathfrak{a}^{-1}N_{P'/F'}(\mathfrak{b}) \in I_W(F', \Theta', \mathfrak{c})$, hence

$$\nu(\alpha^{-1}\beta)\, \nu(\mathfrak{x}_\lambda/\mathfrak{x}_\kappa) = (y)$$

with an element y of W such that $y \equiv 1 \bmod{}^* \mathfrak{c}$. In view of (5.6.1), we have $\kappa = \lambda$, and $(\nu(\alpha^{-1}\beta)) = (y)$. Put $u = y^{-1}\nu(\alpha^{-1}\beta)$. Then u is a unit of F contained in W, and $u \equiv \nu(\alpha^{-1}\beta) \bmod{}^* \mathfrak{c}$. By 3.11, there exists an element γ of $\mathfrak{G}$ such that $\nu(\gamma) = u$, $\mathfrak{x}_\lambda\gamma = \mathfrak{x}_\lambda$ and $\gamma \equiv \alpha^{-1}\beta \bmod{}^* (\Delta; \mathfrak{c})$. Then $\gamma \in \Gamma_W(\mathfrak{x}_\lambda)$. Now, by $(5.7.6_a)$, we have $\varphi_\mu(z)^\omega = R_\sigma^{\mu\lambda}(\beta)[\varphi_\lambda(\beta^{-1}(z))]$ and $\psi_\mu(z)^\omega = S_\sigma^{\mu\lambda}(\beta)[\psi_\lambda(\beta^{-1}(z))]$. Note that

$$R_\sigma^{\mu\lambda}(\alpha)^{-1} = R_\sigma^{\mu\lambda}(\beta\gamma^{-1})^{-1} = R_1^{\lambda\lambda}(\gamma) \circ R_\sigma^{\mu\lambda}(\beta)^{-1} .$$

Similarly $S_\sigma^{\mu\lambda}(\alpha)^{-1} = S_1^{\lambda\lambda}(\gamma) \circ S_\sigma^{\mu\lambda}(\beta)^{-1}$. Therefore

$$(8.2.3) \qquad \begin{aligned} T_\lambda \circ R_\sigma^{\mu\lambda}(\alpha)^{-1}(\varphi_\mu(z)^\omega) &= T_\lambda \circ R_1^{\lambda\lambda}(\gamma)[\varphi_\lambda(\beta^{-1}(z))] = \psi_\lambda(\gamma\beta^{-1}(z)) \\ &= S_1^{\lambda\lambda}(\gamma) \circ S_\sigma^{\mu\lambda}(\beta)^{-1}[\psi_\mu(z)^\omega] = S_\sigma^{\mu\lambda}(\alpha)^{-1} \circ T_\mu^\omega[\varphi_\mu(z)^\omega] . \end{aligned}$$

Now observe that, after fixing (Y, P, δ) and ω, we can change f for $f_\xi : Y \to M_n(B)$ defined by $f_\xi(a) = \xi f(a)\xi^{-1}$ with an element ξ of $\mathfrak{U}$. This does not change (P', Ψ'); and $f_\xi((Y, \delta)_0)$ has $\xi(z)$ as its fixed point. The points $\xi(z)$ for all $\xi \in \mathfrak{U}$ form a dense subset of $\mathfrak{H}_n^r$, hence the $\varphi_\mu(\xi(z))^\omega$ for all $\xi \in \mathfrak{U}$ form at least a

Zariski dense subset of V_μ^σ. Substituting $\xi(z)$ for z in (8.2.3), we see that $T_\lambda \circ R_\sigma^{\mu\lambda}(\alpha)^{-1}$ and $S_\sigma^{\mu\lambda}(\alpha)^{-1} \circ T_\mu^\omega$ have the same action on this subset. (Note that β and γ may depend on ξ, but this has no effect on our conclusion.) It follows that

$$(8.2.4) \qquad\qquad T_\lambda \circ R_\sigma^{\mu\lambda}(\alpha)^{-1} = S_\sigma^{\mu\lambda}(\alpha)^{-1} \circ T_\mu^\omega .$$

Specialize this to the case $\lambda = \mu$, $\alpha = 1$ and $\sigma = $ the identity map. Then our result shows that $T_\mu^\tau = T_\mu$ for every isomorphism τ of $Z_1 P'$ into C over $P'C_0$, where $C_0 = C_W(F', \Theta', \mathfrak{c})$. This shows that T_μ is defined over $P'C_0$. By (8.2.2) we have $C_0 = P'C_0 \cap Z_1$, so that T_μ is defined over C_0. Therefore (8.2.4) can now be written as $S_\sigma^{\mu\lambda}(\alpha) \circ T_\lambda = T_\mu^\sigma \circ R_\sigma^{\mu\lambda}(\alpha)$. This completes the proof.

8.3. Now we shall consider the case $W = W_1$ and employ the conventional notation introduced in 7.1, especially $C_0 = C(F', \Theta', \mathfrak{c})$. Our purpose is to construct a canonical system satisfying the conditions of 5.7 from a number of systems which satisfy the weaker conditions of 7.2.

8.4. PROPOSITION. *Let M (resp. $\bar{M}$) be a finite algebraic extension of F', $\mathfrak{t}$ (resp. $\bar{\mathfrak{t}}$) an integral ideal in F, and $\{V_\lambda, \varphi_\lambda, R_\sigma^{\mu\lambda}(\alpha)\}$ (resp. $\{\bar{V}_\lambda, \bar{\varphi}_\lambda, \bar{R}_\sigma^{\mu\lambda}(\alpha)\}$) a system such that the conditions (7.2. 2–8) are satisfied. (Note that (7.2.1) is omitted; M and $\bar{M}$ may not be quadratic over F'.) Suppose that $M\bar{M}$ and C_0 are linearly disjoint over F'. Then there exists, for each λ, a biregular isomorphism T_λ of V_λ onto $\bar{V}_\lambda$, rational over $M\bar{M}C_0$, such that*

$$T_\lambda \circ \varphi_\lambda = \bar{\varphi}_\lambda, \quad T_\mu^\tau \circ R_\sigma^{\mu\lambda}(\alpha) \circ T_\lambda^{-1} = \bar{R}_\sigma^{\mu\lambda}(\alpha) ,$$

where τ is the unique automorphism of $M\bar{M}C_0$ over $M\bar{M}$ which coincides with σ on C_0.

The proof can be given exactly in the same manner as in 8.2. We have only to consider all isomorphisms or automorphisms over $M\bar{M}C_0$ and choose P so that $D(P/F)$ is prime to $\mathfrak{t}\bar{\mathfrak{t}}$, and (8.2.2) is satisfied. This is possible by virtue of 7.5.

8.5. Let Z be a finite Galois extension of $\mathbf{Q}$ containing F and C_0. Take an integral ideal $\mathfrak{e}$ in F with the property of 8.1 for the $\mathfrak{r}_F$-lattices $\mathfrak{x}_1, \cdots, \mathfrak{x}_q$. Let $M, \mathfrak{t}, \{V_\lambda, \varphi_\lambda, R_\sigma^{\mu\lambda}(\alpha)\}$ and $\bar{M}, \bar{\mathfrak{t}}, \{\bar{V}_\lambda, \bar{\varphi}_\lambda, \bar{R}_\sigma^{\mu\lambda}(\alpha)\}$ be as in 8.4. By virtue of 7.2, we can choose them so that the following four conditions are satisfied:

(8.5.1) $D(\bar{M}/F')\bar{\mathfrak{t}}$ *is prime to* $N(\mathfrak{c}\mathfrak{e})D(Z/F')$, *and* $D(\bar{M}/F') \neq (1)$.

(8.5.2) $D(M/F')\mathfrak{t}$ *is prime to* $N(\mathfrak{c}\mathfrak{e})D(\bar{M}Z/F')$, *and* $D(M/F') \neq (1)$.

(8.5.3) $\bar{M}$ *is linearly disjoint with* C_0 *over* F', *and* $[\bar{M} : F'] = 2$.

(8.5.4) M *is linearly disjoint with* $\bar{M}C_0$ *over* F', *and* $[M : F'] = 2$.

From the last two conditions, it follows that $M\bar{M}$ and C_0 are linearly disjoint over F'. Therefore we can find a biregular isomorphism T_λ of V_λ onto $\bar{V}_\lambda$ as in 8.4.

Let G_0 be the Galois group of $M\bar{M}C_0$ over $\bar{M}C_0$. We can identify G_0 with the Galois group of MC_0 over C_0. Put $T^\mu_\tau = (T^{-1}_\mu)^\tau \circ T_\mu$ for every $\tau \in G_0$ and every μ. Then T^μ_τ is a biregular isomorphism of V_μ to V^τ_μ rational over $M\bar{M}C_0$.

Let (Y, P, δ, f), (P', Ψ'), μ, λ, $\mathfrak{g}_Y$, $\mathfrak{g}$, $\mathfrak{h}_0$ and z be as in (5.7.6.). Let $u = \varphi_\mu(z)$, $v = \bar{\varphi}_\mu(z)$. Let us assume

(8.5.5)　$D(P'/F')$ *is prime to* $D(M\bar{M}/F')$, *and* $D(P/F)$ *is prime to* $\bar{\mathfrak{t}}\mathfrak{t}$.

By 7.5 and (8.5.1,2), such a choice of (Y, P, δ, f) is always possible. By (8.5.1, 2), we have

$$\mathfrak{d}(M\bar{M}C_0/\bar{M}C_0) = \mathfrak{d}(M/F') \neq (1), \qquad \mathfrak{d}(M\bar{M}C_0/MC_0) = \mathfrak{d}(\bar{M}/F') \neq (1) .$$

By our choice of $\mathfrak{e}$, and by (8.5.1, 2), we can take f so that $N(\mathfrak{h}_0)$ is prime to $D(M\bar{M}/F')$. Put $C_1 = C(P', \Psi', \mathfrak{g}_Y, \mathfrak{c})$. Then every prime factor of $D(C_1/P')$ is a factor of $N(\mathfrak{ch}_0)$, and hence $\mathfrak{d}(C_1/F')$ is prime to $\mathfrak{d}(M\bar{M}/F')$ in view of (8.5.2) and (8.5.5). Therefore $\mathfrak{d}(\bar{M}C_1/\bar{M}C_0)$ is prime to $\mathfrak{d}(M\bar{M}C_0/\bar{M}C_0)$, and $\mathfrak{d}(MC_1/MC_0)$ is prime to $\mathfrak{d}(M\bar{M}C_0/MC_0)$. Since $\mathfrak{d}(M\bar{M}C_0/\bar{M}C_0) \neq (1)$ and $[M\bar{M}C_0 : \bar{M}C_0] = 2$, we see that $\bar{M}C_1$ and $M\bar{M}C_0$ are linearly disjoint over $\bar{M}C_0$. Similarly MC_1 and $M\bar{M}C_0$ are linearly disjoint over MC_0. Therefore, we can extend every given τ of G_0 to an automorphism ω of C over $\bar{M}C_1$. By (7.2.8), $v(= \bar{\varphi}_\mu(z))$ is rational over $\bar{M}C_1$, hence we have $T_\mu(u) = v = v^\omega = T^\tau_\mu(u^\omega)$, so that $T^\mu_\tau(u) = u^\omega$. Since u is rational over MC_1 (by (7.2.8)), and MC_1 is stable under ω, we see that u^ω is rational over MC_1. Let σ be an automorphism of $M\bar{M}C_0$ over MC_0. Since MC_1 and $M\bar{M}C_0$ are linearly disjoint over MC_0, we can extend σ to an automorphism π of C over MC_1. Then $(T^\mu_\tau)^\sigma(u) = T^\mu_\tau(u)^\pi = u^{\omega\pi} = u^\omega = T^\mu_\tau(u)$. This implies that $(T^\mu_\tau)^{-1} \circ (T^\mu_\tau)^\sigma$ has a fixed point u. By 8.1, the points u, for which this conclusion is true, form a dense subset of V_μ. (Note that ω and π may depend on u; but σ and τ are independent of u.) Therefore we have $T^\mu_\tau = (T^\mu_\tau)^\sigma$ for every automorphism σ of $M\bar{M}C_0$ over MC_0, hence

(8.5.6)　T^μ_τ *is defined over* MC_0.

From our definition of T^μ_τ, we see easily that $T^\mu_{\eta\tau} = (T^\mu_\eta)^\tau \circ T^\mu_\tau$ for $\tau, \eta \in G_0$. Therefore, by Weil's criterion [12], there exist, for each μ, an algebraic variety $V_{0\mu}$ rational over C_0, and a biregular isomorphism S_μ of $V_{0\mu}$ to V_μ rational over MC_0, such that $T^\mu_\tau = S^\tau_\mu \circ S^{-1}_\mu$ for every $\tau \in G_0$. We see that

$$(T_\mu \circ S_\mu)^\tau = T^\tau_\mu \circ S^\tau_\mu = T_\mu \circ (T^\mu_\tau)^{-1} \circ S^\tau_\mu = T_\mu \circ S_\mu$$

for every $\tau \in G_0$. This implies

(8.5.7)　$T_\mu \circ S_\mu$ *is rational over* $\bar{M}C_0$.

8.6. For a given $R^{\mu\lambda}_\sigma(\alpha)$, let ω be the unique automorphism of $M\bar{M}C_0$ over

$M\bar{M}$ which coincides with σ on C_0. Then $T_\mu^\omega \circ R_\sigma^{\mu\lambda}(\alpha) \circ T_\lambda^{-1} = \bar{R}_\sigma^{\mu\lambda}(\alpha)$ as stated in 8.4. Now define a biregular isomorphism $Q_\sigma^{\mu\lambda}(\alpha)$ of $V_{0\lambda}$ to $V_{0\mu}^\sigma$ by

$$Q_\sigma^{\mu\lambda}(\alpha) = (S_\mu^\omega)^{-1} \circ R_\sigma^{\mu\lambda}(\alpha) \circ S_\lambda .$$

Then $Q_\sigma^{\mu\lambda}(\alpha)$ is rational over MC_0. Further we have

$$Q_\sigma^{\mu\lambda}(\alpha) = ((T_\mu \circ S_\mu)^\omega)^{-1} \circ \bar{R}_\sigma^{\mu\lambda}(\alpha) \circ (T_\lambda \circ S_\lambda) .$$

Therefore, in view of (8.5.7), we see that $Q_\sigma^{\mu\lambda}(\alpha)$ is rational over $\bar{M}C_0$. Since $C_0 = MC_0 \cap \bar{M}C_0$, $Q_\sigma^{\mu\lambda}(\alpha)$ must be rational over C_0.

8.7. Define $\varphi_{0\lambda} : \mathfrak{H}_n^r \to V_{0\lambda}$ by $\varphi_{0\lambda} = S_\lambda^{-1} \circ \varphi_\lambda$. Then $(V_{0\lambda}, \varphi_{0\lambda})$ is a model for $\mathfrak{H}_n^r / \Gamma(\mathfrak{r}_\lambda, \mathfrak{c})$. Thus we have obtained a system $\{V_{0\lambda}, \varphi_{0\lambda}, Q_\sigma^{\mu\lambda}(\alpha)\}$ satisfying (7.2.2–7) with F' as M. We shall now show that this system satisfies (7.2.8) with F'' and (1) in place of M and $\mathfrak{r}$.

Let us fix (Y, P, δ, f), (P', Ψ') and z as in (5.7.6). The field $C(P', \Psi', \mathfrak{g}_Y, \mathfrak{c})$ in (5.7.6) may depend on the choice of μ. Let C_2 be the composite of the $C(P', \Psi', \mathfrak{g}_Y, \mathfrak{c})$ for $\mu = 1, \cdots, q$. By 7.2, we can find a quadratic extension M^* of F', an integral ideal $\mathfrak{r}^*$ in F, and a system $\{V_\lambda^*, \varphi_\lambda^*, R_\sigma^{\mu\lambda*}(\alpha)\}$ satisfying (7.2.2–8) and the following condition:

(8.7.1) M^* *and* $C_2 M\bar{M}(\varphi_{01}(z), \cdots, \varphi_{cq}(z))$ *are linearly disjoint over* F', *and* $\mathfrak{r}^*$ *is prime to* $D(P/F)$.

By 8.4, there exists, for each λ, a biregular isomorphism U_λ (resp. $\bar{U}_\lambda$) of V_λ^* onto V_λ (resp. $\bar{V}_\lambda$) rational over M^*MC_0 (resp. $M^*\bar{M}C_0$) such that

$$U_\lambda \circ \varphi_\lambda^* = \varphi_\lambda, \qquad U_\mu^\omega \circ R_\sigma^{\mu\lambda*}(\alpha) \circ U_\lambda^{-1} = R_\sigma^{\mu\lambda}(\alpha) ,$$

(resp. $\bar{U}_\lambda \circ \varphi_\lambda^* = \bar{\varphi}_\lambda$, $\bar{U}_\mu^\omega \circ R_\sigma^{\mu\lambda*}(\alpha) \circ \bar{U}_\lambda^{-1} = \bar{R}_\sigma^{\mu\lambda}(\alpha)$), where ω is the unique automorphism of M^*MC_0 over M^*M (resp. $M^*\bar{M}C_0$ over $M^*\bar{M}$) which coincides with σ on C_0. Put $Z_\lambda = U_\lambda^{-1} \circ S_\lambda$. Now we have a commutative diagram:

(8.7.2)

$$
\begin{array}{ccc}
V_{0\lambda} & \xrightarrow{\;Z_\lambda\;} & V_\lambda^* \\
{\scriptstyle S_\lambda}\downarrow & \swarrow{\scriptstyle U_\lambda} & \downarrow{\scriptstyle \bar{U}_\lambda} \\
V_\lambda & \xrightarrow[\;T_\lambda\;]{} & \bar{V}_\lambda
\end{array}
$$

Since S_λ and U_λ are rational over M^*MC_0, Z_λ is rational over M^*MC_0. On the other hand, $T_\lambda \circ S_\lambda$ and $\bar{U}_\lambda$ are rational over $M^*\bar{M}C_0$, hence Z_λ is rational over M^*MC_0. By (8.5.3,4) and (8.7.1), we have

$$(M^*MC_0) \cap (M^*\bar{M}C_0) = M^*C_0 ,$$

hence Z_λ is rational over M^*C_0. Further we have

(8.7.3) $$Z_\lambda \circ \varphi_{0\lambda} = \varphi_\lambda^*, \qquad Z_\mu^\omega \circ Q_\sigma^{\mu\lambda}(\alpha) \circ Z_\lambda^{-1} = R_\sigma^{\mu\lambda*}(\alpha) ,$$

where ω is the unique automorphism of M^*C_0 over M^* which coincides with

σ on C_0. (The last relation is obvious if ω is the unique automorphism of $M^*M\bar{M}C_0$ over $M^*M\bar{M}$ which coincides with σ on C_0. Since Z_μ is rational over M^*C_0, one can restrict ω to M^*C_0.)

Now coming back to (P', Ψ') and z, let μ, λ, σ, τ, $\mathfrak{b}$, and α be as in (5.7.6). Put $C_1 = C(P', \Psi', \mathfrak{g}_\Upsilon, \mathfrak{c})$. By our assumption (8.7.1), we can extend τ to an automorphism π of M^*C_1 over M^*. Since $D(P/F)$ is prime to $\mathfrak{l}^*$, by (7.2.8), we see that $\varphi_\mu^*(z)$ is rational over M^*C_1, and

$$\varphi_\mu^*(z)^\pi = R_\sigma^{\mu\lambda*}(\alpha)[\varphi_\lambda^*(\alpha^{-1}(z))] .$$

Hence, by (8.7.3), $\varphi_{0\mu}(z)$ is rational over M^*C_1, and

$$(8.7.4) \qquad \varphi_{0\mu}(z)^\pi = Q_\sigma^{\mu\lambda}(\alpha)[\varphi_{0\lambda}(\alpha^{-1}(z))] .$$

By (8.7.1), M^* and $C_1(\varphi_{0\mu}(z))$ are linearly disjoint over F'. Therefore, from the equality $M^*C_1 = M^*C_1(\varphi_{0\mu}(z))$, we obtain $C_1 = C_1(\varphi_{0\mu}(z))$, hence $\varphi_{0\mu}(z)$ is rational over C_1. Then, in (8.7.4), we can put τ instead of π. Thus we have shown that $\{V_{0\lambda}, \varphi_{0\lambda}, Q_\sigma^{\mu\lambda}(\alpha)\}$ satisfies (7.2. 2–8) with F' and (1) as M and $\mathfrak{l}$. Applying 5.15 to this system, we know that $(V_{0\lambda}, \varphi_{0\lambda})$ is a canonical model for $\mathfrak{H}_n^r/\Gamma(\chi_\lambda, \mathfrak{c})$. This, together with the result of 5.16, completes the proof of our main theorems 5.3 and 5.7 in the case $r < g$.

8.8. PROPOSITION. *Let the notation be as in 8.1. Let $\mathfrak{o}$, $\{x_i\}$, and $\mathfrak{y}$ be as in the proof of 8.1. If χ is similar to $\mathfrak{y}$, one can take $\mathfrak{r}_F$ as the ideal $\mathfrak{e}$ of 8.1.*

PROOF. Since $\mathfrak{o}$ is maximal, there exists a left $\mathfrak{o}$-ideal $\mathfrak{a}$ such that

$$N_{B/F}(\mathfrak{a}) = \nu(\chi/\mathfrak{y}) .$$

Put $t = \sum_{i=1}^n \mathfrak{a}x_i$. We see easily that t is similar to $\mathfrak{y}$, and $\nu(t/\mathfrak{y}) = \nu(\chi/\mathfrak{y})$. By 3.10, there exists an element β of $\mathfrak{U}$ such that $t = \chi\beta$. Let $\mathfrak{o}_1$ be the right order of $\mathfrak{a}$. By [C, 2.8], there exists an F-linear isomorphism f_0 of P into B such that $f_0(\mathfrak{r}_P) \subset \mathfrak{o}_1$. Define $f_1 : Y \to M_n(B)$ so that $f_1((a_{ij})) = \beta(f_0(a_{ij}))\beta^{-1}$ for $(a_{ij}) \in M_n(P) = Y$ with $a_{ij} \in P$. Then $\chi f_1(\mathfrak{r}_P) = \chi$. Now repeating, with this f_1, the last part of the proof of 8.1, we obtain our assertion.

8.9. PROOF OF (5.12.1). Assume that $r = 1$. We have then $F' = F^{\tau_{01}}$, $W = F^+$, $D_W = \mathfrak{G}^+$. For every ideal $\mathfrak{a}$ in F, let us put $\mathfrak{a}' = \mathfrak{a}^{\tau_{01}}$. Then

$$C_W(F', \Theta', \mathfrak{c}) = C(F', \mathfrak{c}') .$$

We can find a finite normal extension k of Q containing F and $C(F', \mathfrak{c}')$, and a regular extension k_1 of k over which S is defined. Let P, Y, δ be as in 7.4 and 8.1. Let $f, \mathfrak{g}_f, \mathfrak{y}_{0f}$, and z_f be as in 8.1. Define (P', Ψ') as in 4.9. Since $r = 1$, we have $P' = P^\pi$ with an extension π of τ_{01} to P. By [C, 1.5], we can take P so that $P \not\subset k$, and $D(P/F)$ is prime to $D(k/Q)$. Then $P' \not\subset k$, and $D(P'/F')$ is prime to $D(k/Q)$. Let us take a prime ideal q in F prime to

$D(k/Q)D(B/F)$. By 8.1 and 8.8, the points z_f for all f such that $\mathfrak{h}_{0f}$ is a power of q form a dense subset of $\mathfrak{H}_n^r$. For such an f, we have

$$P'(\varphi(z_f), \varphi'(z_f)) \subset C_{W'}(P', \Psi', \mathfrak{g}_{fr}, \mathfrak{c}) \subset C(P', \mathfrak{c}'q'^m)$$

for some positive integer m, where $\mathfrak{g}_{fr} = \{a \in Y \mid \mathfrak{x}f(a) \subset \mathfrak{x}\}$. Put

$$C(L, \mathfrak{c}'q'^\infty) = \bigcup_{m=1}^{\infty} C(L, \mathfrak{c}'q'^m)$$

for every finite extension L of F'. By [C, 1.3], we have

$$k \cap C(P', \mathfrak{c}'q'^\infty) \subset C(F', \mathfrak{c}'q'^\infty) \ .$$

Let σ be an automorphism of k over $k \cap C(P', \mathfrak{c}'q'^\infty)$. We can extend σ to an automorphism of C over $C(P', \mathfrak{c}'q'^\infty)$, which we write again by σ. Since $\varphi(z_f)$ and $\varphi'(z_f)$ are rational over $C(P', \mathfrak{c}'q'^\infty)$, we have $S(\varphi(z_f)) = \varphi'(z_f) = \varphi'(z_f)^\sigma = S^\sigma(\varphi(z_f))$. This shows that the fixed points of $S^{-1} \circ S^\sigma$ form a dense subset of V, hence $S^\sigma = S$. It follows that S is defined over $k \cap C(P', \mathfrak{c}'q'^\infty)$, hence over $C(F', \mathfrak{c}'q'^\infty)$. Now we have

$$C(F', \mathfrak{c}'q'^\infty) \cap C(F', \mathfrak{c}'q_1'^\infty) = C(F', \mathfrak{c}')$$

if q and q_1 are distinct prime ideals in F prime to c. This completes the proof.

9. The case of totally indefinite quaternion algebras
(Case $r = g$)

9.1. In this section we assume throughout $r = g$. Then $F' = Q$. Now we can repeat the discussions of §7 by putting $K' = Q$, and understanding that $\det \Phi'_0(x)$ is the identity element or ideal in K for every element or ideal x in Q. It suffices to disregard all the conditions on K'. We obtain 7.2 with $M = Q$ and $\mathfrak{t} = D(K/F)$. Of course the condition $D(M/F') \neq (1)$ should be ignored. Then we repeat the reasoning of 8.5–7 with $M = \bar{M} = Q$, and complete the proof of 5.3 and 5.7 in the case $r = g$. Since $M = \bar{M} = Q$, the discussion can be considerably simplified.

9.2. One can actually prove our main theorems in the case $r = g$ more directly, by showing that the moduli-variety for a certain family of abelian varieties affords, without any descent, a canonical model. This is the main purpose of this section.

By [5, p. 153, Prop. 2] and [C, 9.6], B has a positive involution ρ defined by $x^\rho = e^{-1}x'e$ for $x \in B$ with an element e of B such that $-e^2$ is a totally positive element of F. Let $\omega_0(x) = (x^{(1)}, \cdots, x^{(g)})$ for $x \in B_R$ and $\omega(\xi) = (\xi^{(1)}, \cdots, \xi^{(g)})$ for $\xi \in M_n(B_R)$ be defined as in 4.4. By [5, Lem. 1], we may assume that $(x^\rho)^{(\nu)} = {}^t(x^{(\nu)})$ for $x \in B_R$ and $\nu \leq r$. Then we see that $e^{(\nu)}$ is an alternating matrix of size 2, hence

$$e^{(\nu)} = b_\nu \cdot \begin{bmatrix} 0 & -1 \\ 1 & 0 \end{bmatrix} \qquad\qquad \text{for } \nu = 1, \cdots, g,$$

with real numbers b_ν. We may assume that $b_\nu > 0$ for all ν, by changing e for $d \cdot e$ with an element d of F. In view of (4.4.2,5), we see that $({}^t\xi^\rho)^{(\nu)} = {}^t\xi^{(\nu)}$ for every $\xi \in M_n(B_R)$. Put $T_0 = e \cdot 1_n$. Then $T_0^{(\nu)} = b_\nu \cdot J_n^{-1}$ for $\nu = 1, \cdots, g$.

Now identify X with B^n, and put

$$T(u, v) = u T_0 \cdot {}^t v^\rho = h(u, v)e \qquad\qquad (u, v \in X).$$

Then T is a B-valued ρ-anti-hermitian form on X. Define a representation Φ of B_R by

$$\Phi(x) = \begin{bmatrix} (x \cdot 1_n)^{(1)} & & \\ & \ddots & \\ & & (x \cdot 1_n)^{(g)} \end{bmatrix} \qquad\qquad (x \in B_R).$$

Let $\mathfrak{x}$ be an $\mathfrak{r}_F$-lattice in X. Let us now consider a PEL-type

(9.2.1) $$\Omega = (B, \Phi, \rho;\, T, \mathfrak{x};\, v_1, \cdots, v_s)$$

with elements v_i of X such that

(9.2.2) $$c^{-1}\mathfrak{x} = \mathfrak{x} + \sum_{i=1}^{s} \mathfrak{r}_F v_i.$$

Changing e for $p \cdot e$ with a suitable positive rational number p, we may assume that $\mathrm{Tr}_{B/Q}(T(\mathfrak{x}, \mathfrak{x})) = Z$. Now we can construct a family $\sum_\Omega = \{\mathfrak{Q}_z \mid z \in \mathfrak{H}_n^g\}$ of PEL-structures of type Ω as follows.

Put

$$W_\nu = \begin{bmatrix} 1_n & 0 \\ 0 & b_\nu 1_n \end{bmatrix}.$$

Then

$$W_\nu T_0^{(\nu)-1} \cdot {}^t W_\nu = \begin{bmatrix} 0 & 1_n \\ -1_n & 0 \end{bmatrix} \qquad\qquad \text{for } \nu = 1, \cdots, g.$$

Therefore, with these W_ν and $\xi^{(\nu)}$ as the symbols W_ν and $\omega_\nu(\xi)$ in [5, § 2], we can apply the general theory of [5] to the present case. Namely, for $z = (z_1, \cdots, z_g) \in \mathfrak{H}_n^g$ with $z_\nu \in \mathfrak{H}_n$, define complex column vectors u_i^ν, v_i^ν of dimension n by

$$\begin{bmatrix} u_1^\nu \cdots u_n^\nu & v_1^\nu \cdots v_n^\nu \\ \bar{u}_1^\nu \cdots \bar{u}_n^\nu & \bar{v}_1^\nu \cdots \bar{v}_n^\nu \end{bmatrix} = \begin{bmatrix} z_\nu & 1_n \\ \bar{z}_\nu & 1_n \end{bmatrix} \cdot W_\nu \qquad\qquad (\nu = 1, \cdots, g).$$

Further define complex column vectors $\mathfrak{y}_i(z)$ by

$${}^t\mathfrak{y}_i(z) = ({}^t u_i^1\, {}^t v_i^1 \cdots {}^t u_i^g\, {}^t v_i^g) \qquad\qquad (i = 1, \cdots, n).$$

Let d_i be the element of B^n whose i-component is 1 and other components are 0. Define a map $\mathfrak{y}: B_R^n \times \mathfrak{H}_n^g \to C^{2ng}$ by

$$\mathfrak{y}\left(\sum_{i=1}^n b_i d_i, z\right) = \sum_{i=1}^n \Phi(b_i)\mathfrak{y}_i(z) \qquad (b_i \in B_R, z \in \mathfrak{H}_n^g).$$

Put $\mathfrak{y}_z(x) = \mathfrak{y}(x, z)$ for $x \in B_R^n$, $z \in \mathfrak{H}_n^g$. Then $\mathfrak{y}_z$ is an R-linear isomorphism of B_R^n onto C^{2ng} satisfying $\mathfrak{y}_z(ax) = \Phi(a)\mathfrak{y}_z(x)$ $(a \in B_R, x \in B_R^n)$. Define an R-bilinear form $\mathfrak{E}_z(u, v)$ on $C^{2ng} \times C^{2ng}$ by

$$\mathfrak{E}_z(u, v) = \mathrm{Tr}_{B_R/R}\big(T\big(\mathfrak{y}_z^{-1}(u), \mathfrak{y}_z^{-1}(v)\big)\big),$$

where T should be extended R-bilinearly to B_R^n. Then $\mathfrak{E}_z$ is a Riemann form on $C^{2ng}/\mathfrak{y}_z(\mathfrak{x})$. Thus we obtain, for each $z \in \mathfrak{H}_n^g$, an abelian variety A_z with a a polarization $\mathcal{C}_z$. For every $a \in B$, $\Phi(a)$ defines an element of $\mathrm{End}_Q(A_z)$, which we write $\theta_z(a)$. Let t_{iz} denote the point of A_z represented by $\mathfrak{y}_z(v_i) \bmod \mathfrak{y}_z(\mathfrak{x})$. In this way we obtain a PEL-structure

$$\mathfrak{Q}_z = (A_z, \mathcal{C}_z, \theta_z; t_{1z}, \cdots, t_{sz})$$

of type Ω, and hence the family $\sum_\Omega = \{\mathfrak{Q}_z \mid z \in \mathfrak{H}_n^g\}$.

Now observe that a similitude of T is a similitude of h, and *vice versa*. Therefore $\mathfrak{U}_R$ coincides with the group $G(T)$ of [5, p. 164, (38)]. The action of $\mathfrak{U}_R$ on $\mathfrak{H}_n^g$ defined in [5, p.164] differs from the one in 4.4 by the effect of $W_1, \cdots, W_g$. Hereafter, we shall consider only the former, and construct our canonical model with respect to this choice of the action of $\mathfrak{U}_R$ on $\mathfrak{H}_n^g$. (Of course the action of $\mathfrak{G}_R^+$ should be modified accordingly.)

9.3. To investigate the special members of $\sum_\Omega$, let us recall that, for every α of $\mathfrak{G}_R^+$ and $z \in \mathfrak{H}_n^g$, there exists a C-linear automorphism $\Lambda(\alpha, z)$ of C^{2ng} satisfying

$$(9.3.1) \qquad \Lambda(\alpha, z)\mathfrak{y}(x, z) = \mathfrak{y}\big(x\alpha, \alpha^{-1}(z)\big) \qquad (x \in B_R^n, z \in \mathfrak{H}_n^g)$$

(see [5, 2.5–8], [C, 4.4]). Let Y, P, δ be as in 4.7, and f an F-linear isomorphism of Y into $M_n(B)$ such that $f(a^\delta) = {}'f(a)^\iota$ for $a \in Y$. Let z be the fixed point of $f((Y, \delta)_0)$ on $\mathfrak{H}_n^g$. Let $\alpha = f(c)$ with $c \in (Y, \delta)_0$. From (9.3.1) with these α and z, we obtain $\Lambda(\alpha, z)\mathfrak{y}_z(x) = \mathfrak{y}_z(x\alpha)$. Since Y is spanned by $(Y, \delta)_0$ over F, we can define an anti-isomorphism λ of Y into $\mathrm{End}\,(C^{2ng}, C)$ so that $\lambda(b)\mathfrak{y}_z(x) = \mathfrak{y}_z(xf(b))$ for every $b \in Y$ and $\lambda(c) = \Lambda(f(c), z)$. Let $Z = B \otimes_F Y$. By the same method as in 6.7, we can find an anti-isomorphism θ_0 of Z into $\mathrm{End}_Q(A_z)$ such that $\theta_0(a \otimes 1) = \theta_z(a^\iota)$ for $a \in B$, and $\theta_0(1 \otimes b)$ is represented by $\lambda(b)$ for $b \in Y$. Further we see that the involution of $\mathrm{End}_Q(A_z)$ determined by $\mathcal{C}_z$ sends $\theta_0(a \otimes b)$ to $\theta_0(a^\rho \otimes b^\delta)$ for $a \in B$, $b \in Y$.

Now let Y_i, P_i, m_i, q_i $(i = 1, \cdots, t)$ be as in 4.7. By (4.7.1), Z is isomorphic to $M_{2q_1}(P_1) \oplus \cdots \oplus M_{2q_t}(P_t)$, and P is the center of Z. Let θ denote the restriction of θ_0 to P. Then A_z must be isogenous to a product $A_1^{2q_1} \times \cdots \times A_t^{2q_t}$, where $A_i^{2q_i}$ is the product of $2q_i$ copies of an abelian variety A_i with an injection $\theta_i: P_i \rightarrow \mathrm{End}_Q(A_i)$, which corresponds to the restriction of θ to P_i. By [10, §5, Prop. 1], we have $\dim(A_i) \geq [P_i:Q]/2 = m_i g$. Since $n = \sum_{i=1}^{t} m_i q_i$, we have $\dim(A_i) = m_i g = [P_i:Q]/2$. Let Ψ_i be the representation of P_i defined in (4.9.2). Now we can show

$\quad$ (9.3.2) (A_i, θ_i) *is of type* (P_i, Ψ_i) *in the sense of* [C, 4.1.].

$\quad$ To see this, let $z = (z_1, \cdots, z_g)$. The above $\Lambda(\alpha, z)$ can be obtained as the matrix Λ of [5, p. 161, (32)]. Define Λ_ν and Σ_ν by [5, p. 162, (33), (34)]. (The present B is of (Type II) of [5].) Let $Y_\nu = \begin{bmatrix} z_\nu & 1_n \\ \bar{z}_\nu & 1_n \end{bmatrix}$. Then the formula [5, p. 165, (39)] can now be written as $\Lambda_\nu Y_\nu W_\nu = Y_\nu W_\nu \cdot {}^t\alpha^{(\nu)}$, hence ${}^t\bar{\Lambda}_\nu = {}^t\bar{Y}_\nu^{-1} W_\nu^{-1} \alpha^{(\nu)} W_\nu \cdot {}^t\bar{Y}_\nu$. Take a positive definite real symmetric matrix p_ν so that p_ν^2 is the imaginary part of z_ν. Put

$$Q_\nu = \begin{bmatrix} \bar{z}_\nu p_\nu^{-1} & z_\nu p_\nu^{-1} \\ p_\nu^{-1} & p_\nu^{-1} \end{bmatrix} = {}^t\bar{Y}_\nu \cdot \begin{bmatrix} p_\nu^{-1} & 0 \\ 0 & p_\nu^{-1} \end{bmatrix}.$$

Then the fractional linear transformation $w \mapsto (\bar{z}_\nu p_\nu^{-1} w + z_\nu p_\nu^{-1})(p_\nu^{-1} w + p_\nu^{-1})^{-1}$ represented by the matrix Q_ν maps $\mathfrak{D}_n$ onto $\mathfrak{H}_n$ and the origin to z_ν; and

$$\begin{bmatrix} p_\nu \cdot {}^t\bar{\Sigma}_\nu p_\nu^{-1} & 0 \\ 0 & p_\nu \cdot {}^t\Sigma_\nu p_\nu^{-1} \end{bmatrix} = Q_\nu^{-1} W_\nu^{-1} \alpha^{(\nu)} W_\nu Q_\nu .$$

This implies that the representation $(Y, \delta)_0 \ni c \mapsto f(c) \mapsto {}^t\Sigma_\nu$ is equivalent with the representation $\Psi^{(\nu)}$ of (4.7.6). Restricting c to P, we see that the representation $c \mapsto \Lambda(f(c), z)$ contains $\Psi^{(\nu)}$ with multiplicity 2 for $\nu = 1, \cdots, g$. Since the restriction of $\sum_{\nu=1}^{g} \Psi^{(\nu)}$ to P_i contains Ψ_i with multiplicity q_i, we obtain (9.3.2).

$\quad$ **9.4.** Let W_0, W_1, and D_W be as in 5.1. Since $r = g$, we have $W_0 = Q^+$, and $W_1 = Q^+ E^+$, where E^+ is the group of all totally positive units of F. For the sake of simplicity, we shall prove 5.3 and 5.7 only in the case $W = W_0 = Q^+$, and denote D_W, $\Gamma_W(\mathfrak{x}, \mathfrak{c})$ etc., simply by D, $\Gamma(\mathfrak{x}, \mathfrak{c})$, etc. without the subscript W. (This is different from the convention of 7.1.) The general case can be derived by an argument similar to 5.16. We note that

$$D = \{\alpha \in \mathfrak{G}^+ \mid \nu(\alpha) \in Q\} ,$$

$$\Gamma(\mathfrak{x}) = \{\gamma \in \mathfrak{U} \mid \mathfrak{x}\gamma = \mathfrak{x}\} , \qquad \Gamma(\mathfrak{x}, \mathfrak{c}) = \{\gamma \in \Gamma(\mathfrak{x}) \mid \mathfrak{x}(1 - \gamma) \subset \mathfrak{c}\mathfrak{x}\} .$$

Further, any ample $\mathfrak{x}$, only by itself, constitutes a set Δ of representatives for D-classes of $\mathfrak{r}_F$-lattices. We fix an ample $\mathfrak{r}_F$-lattice $\mathfrak{x}$, and denote $D(\Delta, \mathfrak{c})$ simply by $D_\mathfrak{c}$. Observe that $D_\mathfrak{c}$ is the set of all α in D such that $\nu(\alpha)$ is prime to $N(\mathfrak{c})$ and $\mathfrak{x}_\mathfrak{p}\alpha = \mathfrak{x}_\mathfrak{p}$ for all $\mathfrak{p}$ dividing $\mathfrak{c}$.

$\quad$ Now let us write the PEL-type (9.2.1) as $\Omega(v_1, \cdots, v_s)$, other symbols

$B, \Phi, \rho, T, \mathfrak{x}$ being fixed. We identify $\Omega(v_1, \cdots, v_s)$ with $\Omega(w_1, \cdots, w_s)$ if $v_i \equiv w_i \bmod \mathfrak{x}$ for every i, and consider only those PEL-types satifying (9.2.2). If $u_1, \cdots, u_s$ are elements of B^n such that $(\mathfrak{c}^{-1}\mathfrak{x})_\mathfrak{p} = (\mathfrak{x} + \sum_{i=1}^s \mathfrak{r}_F u_i)_\mathfrak{p}$ for all $\mathfrak{p}$ dividing $\mathfrak{c}$, we can find elements w_i so that $\mathfrak{c}^{-1}\mathfrak{x} = \mathfrak{x} + \sum_{i=1}^s \mathfrak{r}_F w_i$ and $w_i \equiv u_i \bmod \mathfrak{x}_\mathfrak{p}$ for all such $\mathfrak{p}$. Then we understand that $\Omega(u_1, \cdots, u_s)$ denotes $\Omega(w_1, \cdots, w_s)$. In the following discussion, we fix $v_1, \cdots, v_s$, and put $\Omega = \Omega(v_1, \cdots, v_s)$, $\Omega_\alpha = \Omega(v_1\alpha, \cdots, v_s\alpha)$ for every $\alpha \in D_\mathfrak{c}$. By means of the above $\mathfrak{y}(x, z)$, we can construct a family

$$\Sigma_\alpha = \Sigma(\Omega_\alpha) = \{ \mathfrak{Q}_{z\alpha} \mid z \in \mathfrak{H}_n^g \}$$

as described in 9.2. We obtain then

$$\mathfrak{Q}_{z\alpha} = (A_z, C_z, \theta_z; \{ t_{iz\alpha} \})$$

with A_z, C_z, θ_z independent of α, and the points $t_{iz\alpha}$ on A_z represented by $\mathfrak{y}_z(w_i)$, where the w_i are such that $\mathfrak{c}^{-1}\mathfrak{x} = \mathfrak{x} + \sum_i \mathfrak{r}_F w_i$ and $w_i \equiv v_i\alpha \bmod \mathfrak{x}_\mathfrak{p}$ for every $\mathfrak{p}$ dividing $\mathfrak{c}$.

Let $(V, \mathfrak{v}, \varphi)$ be a moduli-system for the family $\Sigma(\Omega)$ in the sense of [C, 4.17] and [7, 6.2]. By [C, 4.25], if we define $\mathfrak{v}_\alpha$ by $\mathfrak{v}_\alpha(\mathfrak{Q}_{z\alpha}) = \varphi(z)$, then $(V, \mathfrak{v}_\alpha, \varphi)$ becomes a moduli-system for Σ_α. We can take this couple (V, φ) as a canonical model for $\mathfrak{H}_n^g / \Gamma(\mathfrak{x}, \mathfrak{c})$. More precisely, we shall prove

9.5. Theorem. *Let Ω_α and $(V, \mathfrak{v}_\alpha, \varphi)$ be as above, k_Ω the field determined by* [7, 5.1] *or* [C, 4.14]*, and $\zeta = e^{2\pi i/c}$ with the smallest positive integer c divisible by $\mathfrak{c}$. Then*

(9.5.1) $k_\Omega = \boldsymbol{Q}(\zeta)$.

(9.5.2) *Let $\alpha \in D_\mathfrak{c}$ and $\sigma = [\boldsymbol{Q}(\zeta)/\boldsymbol{Q}, (\nu(\alpha))]$. Then Ω^σ is equivalent to Ω_α.*

(9.5.3) *There exists a biregular isomorphism $R_\sigma(\alpha)$ of V onto V^σ rational over $\boldsymbol{Q}(\zeta)$ such that $\mathfrak{v}^\sigma = R_\sigma(\alpha) \circ \mathfrak{v}_\alpha$.*

(9.5.4) $\{V, \varphi, R_\sigma(\alpha)(\alpha \in D_\mathfrak{c})\}$ *satisfies* (5.7.1-6) *(with $W = \boldsymbol{Q}^+$).*

As observed in 5.10, we have $\boldsymbol{Q}(\zeta) = C(F', \Theta', \mathfrak{c})$. The proof of 9.5 will be completed in 9.7.

9.6. Let $(Y, P, \delta, f), (P', \Psi')$ and $\mathfrak{g}_\Gamma$ be as in (5.3.3), and $\mathfrak{g} = \mathfrak{g}_\Gamma \cap P$. Let $\mathfrak{h}$ be the conductor of $\mathfrak{g}$, and $\mathfrak{h}_0 = \mathfrak{h} \cap F$. Let us assume

(9.6.1) *P' and $k_\Omega \cdot \boldsymbol{Q}(\zeta)$ are linearly disjoint over $\boldsymbol{Q}$.*

By 7.5, such a (Y, P, δ, f) exists. Let $\alpha \in D_\mathfrak{c}$ and $\sigma = [\boldsymbol{Q}(\zeta)/\boldsymbol{Q}, (\nu(\alpha))]$. Then σ can be extended to an automorphism τ of $\boldsymbol{Q}$ over P'. Let $\mathfrak{b}$ be an ideal in P' prime to $N(\mathfrak{c}\mathfrak{h}_0)$ such that

$$\tau = [C(P', \Psi', \mathfrak{g}_\Gamma, \mathfrak{c})/P', \mathfrak{b}] \qquad\qquad \text{on } C(P', \Psi', \mathfrak{g}_\Gamma, \mathfrak{c}).$$

By 5.8, we find an element β of $D_\mathfrak{c}$ such that $\mathfrak{x}f([\Psi'(\mathfrak{b})]_\mathfrak{g})^{-1} = \mathfrak{x}\beta^{-1}$; and by (5.8.1), $\nu(\beta) = N(\mathfrak{b})$. Since $\sigma = \tau$ on $\boldsymbol{Q}(\zeta)$, we have $\nu(\beta^{-1}\alpha) \equiv 1 \bmod^*(\mathfrak{c})$. Take an integral ideal q in P' so that

$$\tau = [C(P', N(\mathfrak{c}\mathfrak{h}_0))/P', \mathfrak{q}] \qquad \text{on } C(P', N(\mathfrak{c}\mathfrak{h}_0)) \,.$$

Then $\mathfrak{b}^{-1}\mathfrak{q} \in I(P', \Psi', \mathfrak{g}_Y, \mathfrak{c})$, hence there exists an element a of Y such that $aa^\delta \in Q^+$, $a \equiv 1 \bmod^* \mathfrak{c}\mathfrak{g}_Y$, and $[\Psi'(\mathfrak{b}^{-1}\mathfrak{q})]_\mathfrak{g}\mathfrak{g}_Y = a\mathfrak{g}_Y$. Put $\varepsilon = f(a)\beta$. Then $\varepsilon \in D$ and $\mathfrak{x}\varepsilon^{-1} = \mathfrak{x}f([\Psi'(\mathfrak{q})]_\mathfrak{g})^{-1}$.

Let z be the fixed point of $f((Y, \delta)_0)$ on $\mathfrak{H}_n^g$, and $w = \beta^{-1}(z)$. Then $w = \varepsilon^{-1}(z)$. Consider the members $\mathfrak{A}_{z1}$ of Σ_1 and $\mathfrak{A}_{w\beta}$ of Σ_β. Then $\Lambda(\varepsilon, z)$ of (9.3.1) defines an isogeny μ of $\mathfrak{A}_{z1}$ to $\mathfrak{A}_{w\beta}$. Here note that $\nu(\varepsilon) \in Q^+$, and $v_i\varepsilon \equiv v_i\beta$ $\bmod \mathfrak{x}_\mathfrak{p}$ for every $\mathfrak{p}$ dividing $\mathfrak{c}$. Moreover, $\mathrm{Ker}\,(\mu)$ is isomorphic to $\mathfrak{x}\varepsilon^{-1}/\mathfrak{x}$. Therefore if we denote by θ the isomorphism of P into $\mathrm{End}_Q(A_z)$ as in 9.3, we have $\theta(\mathfrak{g}) = \theta(P) \cap \mathrm{End}(A_z)$, and

$$\mathrm{Ker}\,(\mu) = \{t \in A_z \mid \theta(\Psi'(\mathfrak{q}) \cap \mathfrak{g})t = 0\} \,.$$

Now we can apply 6.12 to the present case with $(S'_i, \Xi'_i) = (P'_i, \Psi'_i)$, $S' = P'$, $\mathfrak{l} = \mathfrak{g}$, $\mathfrak{b} = \Psi'(\mathfrak{q})$. Then we find an isogeny λ of $\mathfrak{A}_{z1}$ to $\mathfrak{A}_{z1}^\mathfrak{r}$ such that $\mathrm{Ker}\,(\lambda) = \mathrm{Ker}\,(\mu)$. It follows that $\mathfrak{A}_{z1}^\mathfrak{r}$ is isomorphic to $\mathfrak{A}_{w\beta}$. Since $\nu(\beta^{-1}\alpha) \equiv 1 \bmod^* (c)$, we can find, by 3.11, an element γ of $\mathfrak{G}$ such that $\nu(\gamma) = 1$, $\mathfrak{x}\gamma = \mathfrak{x}$, $\gamma \equiv \beta^{-1}\alpha$ $\bmod^* (\mathfrak{x}; \mathfrak{c})$. Put $w' = \gamma^{-1}(w)$. We see that $\Lambda(\gamma, w)$ gives an isomorphism of $\mathfrak{A}_{w\beta}$ to $\mathfrak{A}_{w'\beta\gamma}$. Since $\beta\gamma \equiv \alpha \bmod^* (\mathfrak{x}; \mathfrak{c})$, we see that $\Sigma_{\beta\gamma} = \Sigma_\alpha$. Thus $\mathfrak{A}_{z1}^\mathfrak{r}$ is isomorphic to a member of Σ_α, hence $\Omega^\mathfrak{r}$ is equivalent to Ω_α.

By (9.6.1), any isomorphism σ of $k_\Omega \cdot Q(\zeta)$ into C can be extended to an automorphism τ of C to which the above result is applicable. If σ is the identity on $Q(\zeta)$, we can take α to be 1, so that Ω^σ is equivalent to Ω, hence σ is the identity on k_Ω. This shows $k_\Omega \subset Q(\zeta)$, and completes the proof of (9.5.2).

To prove (9.5.1), σ and α being as in (9.5.2), assume that Ω_α is equivalent to Ω. By the definition of equivalence of PEL-types [7, 3.1], there exists an element γ of $\Gamma(\mathfrak{x})$ such that $v_i\alpha \equiv v_i\gamma \bmod \mathfrak{x}_\mathfrak{p}$ for all $\mathfrak{p}$ dividing $\mathfrak{c}$. Then $\alpha \equiv \gamma$ $\bmod^* (\mathfrak{x}, \mathfrak{c})$, hence $\nu(\alpha) \equiv \nu(\gamma) = 1 \bmod^* \mathfrak{c}$. It follows that $[Q(\zeta)/Q, (\nu(\alpha))] = 1$, hence $Q(\zeta) \subset k_\Omega$. This proves (9.5.1).

9.7. The existence of $R_\sigma(\alpha)$ as in (9.5.3) follows immediately from [C, 4.21, 4.23]. We see easily that the $R_\sigma(\alpha)$ satisfy (5.7.3,4). Let $\gamma \in \Gamma(\mathfrak{x})$, $w \in \mathfrak{H}_n^g$, $z = \gamma(w)$. Then $\Lambda(\gamma, z)$ gives an isomorphism of $\mathfrak{A}_{z1}$ to $\mathfrak{A}_{w\gamma}$. Hence

$$\varphi(\gamma(w)) = \varphi(z) = \mathfrak{v}(\mathfrak{A}_{z1}) = R_1(\gamma)[\mathfrak{v}_\gamma(\mathfrak{A}_{w\gamma})] = R_1(\gamma)[\varphi(w)] \,.$$

This proves (5.7.5).

Now let (Y, P, δ, f), (P', Ψ'), $\mathfrak{g}_Y$, $\mathfrak{g}$, $\mathfrak{h}$, $\mathfrak{h}_0$ be as in 9.6. We release the condition (9.6.1), and start with *any* automorphism τ of C over P'. Let σ be the restriction of τ to $Q(\zeta)$. Let z, β and w be as in 9.6. Repeating the reasoning of 9.6, we see that $\mathfrak{A}_{z1}^{\mathfrak{r}}$ is isomorphic to $\mathfrak{A}_{w\beta}$. Therefore

$$\varphi(z)^{\mathfrak{r}} = \mathfrak{v}^{\sigma}(\mathfrak{A}_{z1}^{\mathfrak{r}}) = R_{\sigma}(\beta)[\mathfrak{v}_{\beta}(\mathfrak{A}_{w\beta})] = R_{\sigma}(\beta)[\varphi(\beta^{-1}(z))] .$$

In particular, if τ is the identity on $C(P', \Psi', \mathfrak{g}_Y, \mathfrak{c})$, we could have taken $\mathfrak{b} = (1)$ and $\beta = 1$. Then we have $\varphi(z)^{\mathfrak{r}} = \varphi(z)$, so that $\varphi(z)$ is rational over $C(P', \Psi', \mathfrak{g}_Y, \mathfrak{c})$. Thus our system satisfies (5.15.3). Therefore, by 5.15, we obtain 5.3 and 5.7 in the present case, and hence (9.5.4).

9.8. Suppose that $B = M_2(F)$. Then $\Gamma(\mathfrak{x})$ is (commensurable with) the so-called Hilbert-Siegel modular group. In this case, we can define $\mathfrak{G}$ in the following way. Let e_{ij} $(i = 1, 2; j = 1, 2)$ be the matrix units of $M_2(F)$. Put $X_1 = e_{11}X$. Then X_1 is a vector space over F of dimension $2n$. We can easily find a non-degenerate F-bilinear alternating form $h_1 : X_1 \times X_1 \to F$ so that $h(u, v) = h_1(u, v)e_{12}$ for $(u, v) \in X_1 \times X_1$. Then the restriction of every element of $GL(X, B)$ to X_1 defines an isomorphism of $\mathfrak{G}$ onto the group

$$\mathfrak{G}_1 = \{\alpha \in GL(X_1, F) \mid h_1(u\alpha, v\alpha) = \nu(\alpha)h_1(u, v) \text{ with } \nu(\alpha) \in F\} .$$

(See for example [4, 2.4].) Thus we can define $\mathfrak{G}$ as the group of similitudes of an alternating form with $2n$ variables.

Let $\mathfrak{x}_1$ be an $\mathfrak{r}_F$-lattice in X_1. Put $\mathfrak{x} = \mathfrak{x}_1 + e_{21}\mathfrak{x}_1$, $\mathfrak{o} = M_2(\mathfrak{r}_F)$. Then $\mathfrak{o}\mathfrak{x} = \mathfrak{x}$ and $\mathfrak{x}_1 = e_{11}\mathfrak{x}$. We shall now show that $\mathfrak{x}$ *is ample*. First observe that for every prime ideal $\mathfrak{p}$ in F, an element α of $\mathfrak{G}_\mathfrak{p}$ satisfies $\mathfrak{x}_\mathfrak{p}\alpha \subset \mathfrak{x}_\mathfrak{p}$ if and only if $\mathfrak{x}_{1\mathfrak{p}}\alpha \subset \mathfrak{x}_{1\mathfrak{p}}$. By [4, Prop. 1.3], there exist $\mathfrak{r}_F$-ideals $\mathfrak{a}_1, \cdots, \mathfrak{a}_n$ and a basis

$$\{u_1, \cdots, u_n, v_1, \cdots, v_n\}$$

of X_1 over F such that $h(u_i, u_j) = h(v_i, v_j) = 0$, $h(u_i, v_j) = 1$ or 0 according as $i = j$ or $i \neq j$, $\mathfrak{x} = \sum_{i=1}^{n} (\mathfrak{r}_F u_i + \mathfrak{a}_i v_i)$, $\mathfrak{a}_1 \supset \mathfrak{a}_2 \supset \cdots \supset \mathfrak{a}_n$. To show (3.7.1), let y be a non-zero element of $\mathfrak{r}_\mathfrak{p}$. Define an element α_1 of $GL(X_1, F)$ by $u_i\alpha_1 = yu_i$, $v_i\alpha_1 = v_i$. Then $\alpha_1 \in \mathfrak{G}_1$, $\mathfrak{x}_\mathfrak{p}\alpha_1 \subset \mathfrak{x}_\mathfrak{p}$, and $\nu(\alpha_1) = y$, and hence (3.7.1). Next let α, u, and λ be as in (3.7.2). Put $t = \nu(\alpha)^{-1}u$. Then t is a unit of $\mathfrak{r}_\mathfrak{p}$, and $t \equiv 1 \bmod \mathfrak{p}^{\lambda}$. Define an element β of $\mathfrak{G}$ so that $u_i\beta = tu_i\alpha$, $v_i\beta = v_i\alpha$. Then $\nu(\beta) = u$ and $\mathfrak{x}_\mathfrak{p}(\alpha - \beta) \subset \mathfrak{p}^{\lambda}\mathfrak{x}_\mathfrak{p}$, q.e.d.

Identifying $\mathfrak{G}$ with $\mathfrak{G}_1$, we have

$$\Gamma(\mathfrak{x}) = \{\alpha \in \mathfrak{G}_1 \mid \mathfrak{x}_1\alpha = \mathfrak{x}_1, \nu(\alpha) = 1\} ,$$
$$\Gamma(\mathfrak{x}, \mathfrak{c}) = \{\alpha \in \Gamma(\mathfrak{x}) \mid \mathfrak{x}_1(1 - \alpha) \subset \mathfrak{c}\mathfrak{x}_1\} .$$

Therefore, if $B = M_2(F)$, we can formulate our main theorems with respect to an alternating form h_1 in X_1 and an (arbitrary) $\mathfrak{r}_F$-lattice $\mathfrak{x}_1$ in X_1. Now take a positive rational number d so that $\mathrm{Tr}_{F/\mathbf{Q}}(d \cdot h_1(\mathfrak{x}_1, \mathfrak{x}_1)) = \mathbf{Z}$, and elements $w_1, \cdots, w_s$ of X_1 so that $\mathfrak{c}^{-1}\mathfrak{x}_1 = \mathfrak{x}_1 + \sum_{i=1}^{s} \mathfrak{r}_F w_i$. Then define a PEL-type

$$\Omega' = (F, \Phi', \rho; d \cdot h_1, \mathfrak{x}_1; w_1, \cdots, w_s) ,$$

where ρ is the identity map of F, and Φ' is a representation of F such that $\Phi' \sim n \cdot \sum_{\nu=1}^{g} \tau_{0\nu}$. By the same method as in 9.2-7, we obtain a family $\Sigma' = \{\mathfrak{Q}'_z \mid z \in \mathfrak{H}^g_n\}$ of PEL-structures of type Ω', whose moduli-variety affords a canonical model for $\mathfrak{H}^g_n/\Gamma(\mathfrak{x}, \mathfrak{c})$. One can easily show that each member of the previous family Σ_0 (with $B = M_2(F)$) is isogenous to the product of two copies of a member of Σ'.

10. Another canonical system
(Further descent of the field of rationality)

10.1. In our Main Theorems I and II, one may notice the following three points:

(A) *It may happen that* $\Gamma_W(\mathfrak{x}_\lambda, \mathfrak{c})$ *and* $\Gamma_W(\mathfrak{x}_\mu, \mathfrak{c})$ *are transformed to each other by an inner automorphism of* $\mathfrak{G}^+$ *even if* $\lambda \neq \mu$.

(B) $\mathfrak{H}^r_n/\Gamma_W(\mathfrak{x}_\lambda, \mathfrak{c})$ *may have a model defined over a proper subfield of* $C_W(F', \Theta', \mathfrak{c})$.

(C) $\Gamma_W(\mathfrak{x}_\lambda)$ *may not be a maximal discontinuous group even if* $W = W_1$.

In the case $n = r = 1$, we have shown in [C, 3.13] the existence of a meaningful model for $\mathfrak{H}_1/\Gamma^*$ defined over a certain subfield of $C(F, 1)$, where

$$\Gamma^* = \{\alpha \in B^+ \mid \alpha \mathfrak{o} \alpha^{-1} = \mathfrak{o}\}$$

with a maximal order $\mathfrak{o}$ in B. It is our purpose of this section to generalize this result to the case of arbitrary n and r. We shall obtain a new *canonical system* for a set of representatives of conjugacy classes of certain discontinuous groups $\Gamma^*_W(\mathfrak{x}_\lambda, \mathfrak{c})$ which are somewhat larger than $\Gamma_W(\mathfrak{x}_\lambda, \mathfrak{c})$ and generalize the above Γ^*. Although this result can be derived directly from Main Theorems I and II, it will give a further understanding of our theory in connection with the above three points.

10.2. We shall make no assumption on n and r, but restrict our discussion to a special type of lattices in X.

Let us fix an $\mathfrak{F}$-genus Λ of ample $\mathfrak{r}_F$-lattices in X (see 3.7 and 5.4). Put $\mathfrak{o} = \{a \in B \mid a\mathfrak{x} \subset \mathfrak{x}\}$ with any $\mathfrak{x}$ in Λ. Then $\mathfrak{o}$ is an $\mathfrak{r}_F$-order in B which depends only on Λ. For every right, left or two-sided $\mathfrak{o}$-ideal $\mathfrak{t}$, we put $N_{B/F}(\mathfrak{t}) = \mathfrak{t}\mathfrak{t}' \cap F$. *In this section we assume that* Λ *contains a lattice satisfying* (3.8.1-3) *for every* $\mathfrak{p}$. Then *every* lattice in Λ satisfies (3.8.1-3). Moreover, for every $\mathfrak{x} \in \Lambda$ and every two-sided $\mathfrak{o}$-ideal $\mathfrak{t}$, $\mathfrak{t}\mathfrak{x}$ is similar to $\mathfrak{x}$, and $N_{B/F}(\mathfrak{t}) = \nu(\mathfrak{t}\mathfrak{x}/\mathfrak{x})$.

Let $\mathfrak{c}$ be an integral ideal in F, and $\Delta = \{\mathfrak{x}_1, \cdots, \mathfrak{x}_q\}$ a $\mathfrak{c}$-regular set of representatives of D_W-classes in Λ. Here we choose and fix any W as in 5.1. Let $\mathfrak{X}_\mathfrak{c}$ denote the group of all two-sided $\mathfrak{o}$-ideals $\mathfrak{t}$ prime to $\mathfrak{c}$ such that $N_{B/F}(\mathfrak{t}) \in \mathfrak{F}_\mathfrak{c}$. Put

$$\Theta_{\mu\lambda}(\mathfrak{c}) = \{\alpha \in D_W(\Delta, \mathfrak{c}) \,|\, \mathfrak{x}_\mu\alpha = \mathfrak{t}\mathfrak{x}_\lambda \text{ for some } \mathfrak{t} \in \mathfrak{T}_\mathfrak{c}\}\,,$$
$$\Gamma_W^*(\mathfrak{x}_\lambda) = \Theta_{\lambda\lambda}((1))\,,$$
$$\Gamma_W^*(\mathfrak{x}_\lambda, \mathfrak{c}) = \{\alpha \in \Theta_{\lambda\lambda}(\mathfrak{c}) \,|\, \alpha \equiv 1 \bmod^* (\Delta; \mathfrak{c})\}\,.$$

Then we have

(10.2.1)
$$[F^* \cdot \Gamma_W^*(\mathfrak{x}_\lambda) : F^* \cdot \Gamma_W(\mathfrak{x}_\lambda)] < \infty\,,$$

where F^* is the group of non-zero elements in F (for a proof, see 10.12 below).
Therefore $\Gamma_W^*(\mathfrak{x}_\lambda, \mathfrak{c})$ gives a discontinuous group of transformations on $\mathfrak{H}_n^r$.

10.3. Let $I_W^*(F', B, \mathfrak{c})$ denote the group of all ideals $\mathfrak{a}$ in F' prime to $N(\mathfrak{c})$
such that $\det \Theta'(\mathfrak{a}) = u \cdot N_{B/F}(\mathfrak{t})$ with an element u of W satisfying $u \equiv 1 \bmod^* \mathfrak{c}$
and an element $\mathfrak{t}$ of $\mathfrak{T}_\mathfrak{c}$. Observe that $I_W^*(F', B, \mathfrak{c})$ contains the following two
types of ideals.

(i) *All the principal ideals (a) with totally positive elements a of F'
such that $a \equiv 1 \bmod^* (c)$, where c is the smallest positive integer divisible by* $\mathfrak{c}$.

(ii) *The squares of ideals in F' prime to* (c).

Therefore $I_W^*(F', B, \mathfrak{c})$ corresponds to a class field over F', which we shall
denote by $C_W^*(F', B, \mathfrak{c})$. Obviously $C_W^*(F', B, \mathfrak{c}) \subset C_W(F', \Theta', \mathfrak{c})$, and the Galois
group of $C_W^*(F', B, \mathfrak{c})$ over F' is a product of cyclic groups of order 2.

Let $(Y, P, \delta, f) \in J_n(B)$, and let (P', Ψ') be its dual. For a given $\mathfrak{x} \in \Lambda$,
put

$$\mathfrak{g}_Y = \{a \in Y \,|\, \mathfrak{x}f(a) \subset \mathfrak{x}\}\,, \qquad\qquad \mathfrak{g} = \mathfrak{g}_Y \cap P\,.$$

Let $\mathfrak{h}$ be the conductor of $\mathfrak{g}$, and $\mathfrak{h}_0 = \mathfrak{h} \cap F$. We denote by $I_W^*(P', \Psi', \mathfrak{x}, \mathfrak{c})$ the
group of all ideals $\mathfrak{b}$ in P' prime to $N(\mathfrak{c}\mathfrak{h}_0)$ satisfying the following condition

(10.3.1) *There exist an element $\mathfrak{t}$ of $\mathfrak{T}_\mathfrak{c}$ and an element v of Y such that*
$vv^\delta \in W$, $v \equiv 1 \bmod^* \mathfrak{c}\mathfrak{g}_Y$, and $\mathfrak{x}f([\Psi'(\mathfrak{b})]_\mathfrak{g}) = \mathfrak{t}\mathfrak{x}f(v)$.
Since $I_W(P', \Psi', \mathfrak{g}_Y, \mathfrak{c}) \subset I_W^*(P', \Psi', \mathfrak{x}, \mathfrak{c})$, the ideal group $I_W^*(P', \Psi', \mathfrak{x}, \mathfrak{c})$ corresponds to a class field over P', which we denote by $C_W^*(P', \Psi', \mathfrak{x}, \mathfrak{c})$. We see
easily that

$$C_W^*(F', B, \mathfrak{c}) \subset C_W^*(P', \Psi', \mathfrak{x}, \mathfrak{c}) \subset C_W(P', \Psi', \mathfrak{g}_Y, \mathfrak{c})\,.$$

10.4. Let us consider a minimal subset $\{\mathfrak{x}_1', \cdots, \mathfrak{x}_s'\}$ of $\{\mathfrak{x}_1, \cdots, \mathfrak{x}_q\}$ with
the following property.

(10.4.1) *Every $\mathfrak{x}$ in Λ can be written in the form $\mathfrak{x} = \mathfrak{t}\mathfrak{x}_\kappa'\alpha$ with an index*
$\kappa \leq s$, some $\mathfrak{t} \in \mathfrak{T}_{(1)}$ and some $\alpha \in D_W$.
For simplicity, after re-arranging the $\mathfrak{x}_\lambda$ suitably, we take $\{\mathfrak{x}_1, \cdots, \mathfrak{x}_s\}$ as such
a set $\{\mathfrak{x}_1', \cdots, \mathfrak{x}_s'\}$.

Let $\mathfrak{R}_\mathfrak{c}$ denote the subgroup of $\mathfrak{T}_{(1)} \times D_W$ consisting of all the $(\mathfrak{t}, \alpha)$ with
$\mathfrak{t} \in \mathfrak{T}_{(1)}$ and $\alpha \in D_W$ such that $(\mathfrak{x}_i\alpha)_\mathfrak{p} = (\mathfrak{t}\mathfrak{x}_i)_\mathfrak{p}$ for every $\mathfrak{p}$ dividing $\mathfrak{c}$ and
$N_{B/F}(\mathfrak{t})^{-1}\nu(\alpha) \in \mathfrak{J}_\mathfrak{c}$. For such a $(\mathfrak{t}, \alpha)$, we see that $\mathfrak{t} \in \mathfrak{T}_\mathfrak{c}$ if and only if $\alpha \in D_W(\Delta, \mathfrak{c})$

10.5. Main Theorem III. *The notation and the assumption being as above, there exists a system*

$$\{U_\lambda,\ \psi_\lambda,\ S_\sigma^{\mu\lambda}(\mathfrak{t},\ \alpha)\ (\lambda,\ \mu = 1,\ \cdots,\ s;\ (\mathfrak{t},\ \alpha) \in \mathfrak{R}_\mathrm{c})\}$$

with the following properties:

(10.5.1) *$(U_\lambda,\ \psi_\lambda)$ is a model for $\mathfrak{H}_n^r/\Gamma_W^*(\mathfrak{x}_\lambda,\ \mathfrak{c})$.*

(10.5.2) *U_λ is defined over $C_W^*(F',\ B,\ \mathfrak{c})$.*

(10.5.3) *Let $(\mathfrak{t},\ \alpha) \in \mathfrak{R}_\mathrm{c}$, $\nu(\mathfrak{x}_\mu\alpha/\mathfrak{t}\mathfrak{x}_\lambda) = \det \Theta'(\mathfrak{a})$ with an ideal $\mathfrak{a}$ in F' prime to $N(\mathfrak{c})$, and let $\sigma = [C_W^*(F',\ B,\ \mathfrak{c})/F',\ \mathfrak{a}]$. $S_\sigma^{\mu\lambda}(\mathfrak{t},\ \alpha)$ is defined for such $(\mathfrak{t},\ \alpha)$, $\lambda,\ \mu$ and σ, and is a biregular isomorphism of U_λ onto U_μ^σ rational over $C_W^*(F',\ B,\ \mathfrak{c})$.*

(10.5.4) *$S_\sigma^{\mu\lambda}(\mathfrak{t},\ \alpha) = S_\sigma^{\mu\lambda}(\mathfrak{m},\ \beta)$ if $\mathfrak{t}\mathfrak{m}^{-1} \in \mathfrak{T}_\mathrm{c}$ and $\alpha\beta^{-1} \equiv 1 \bmod^* (\Delta;\ \mathfrak{c})$.*

(10.5.5) *$S_\tau^{\nu\mu}(\mathfrak{m},\ \beta)^\sigma \circ S_\sigma^{\mu\lambda}(\mathfrak{t},\ \alpha) = S_{\tau\sigma}^{\nu\lambda}(\mathfrak{m}\mathfrak{t},\ \beta\alpha)$.*

(10.5.6) *$S_1^{\lambda\lambda}(\mathfrak{t},\ \alpha)(\psi_\lambda(w)) = \psi_\lambda(\alpha(w))$ for every $w \in \mathfrak{H}_n^r$, if $\alpha \in \Gamma_W^*(\mathfrak{x}_\lambda)$, $\mathfrak{t} \in \mathfrak{T}_{(1)}$ and $\mathfrak{x}_\lambda\alpha = \mathfrak{t}\mathfrak{x}_\lambda$.*

(10.5.7) *Let $(Y,\ P,\ \delta,\ f)$, $(P',\ \Psi')$ and z be as in (5.3.3). Then*

(10.5.7$_\mathrm{a}$)
$$C_W^*(F',\ B,\ \mathfrak{c}) \cdot P'\big(\psi_\mu(z)\big) = C_W^*(P',\ \Psi',\ \mathfrak{x}_\mu,\ \mathfrak{c})\ .$$

Further let $\mathfrak{g} = \{a \in P\ |\ \mathfrak{x}_\mu f(a) \subset \mathfrak{x}_\mu\}$, $\mathfrak{h}$ be the conductor of $\mathfrak{g}$, and $\mathfrak{h}_0 = \mathfrak{h} \cap F$. Let $\mathfrak{b}$ be an ideal in P' prime to $N(\mathfrak{c}\mathfrak{h}_0)$, and let $\tau = [C_W^*(P',\ \Psi',\ \mathfrak{x}_\mu,\ \mathfrak{c})/P',\ \mathfrak{b}]$. Then one has

(10.5.7$_\mathrm{b}$)
$$\mathfrak{x}_\mu f\big([\Psi'(\mathfrak{b})]_\mathfrak{g}\big)^{-1} = \mathfrak{t}\mathfrak{x}_\nu\xi^{-1}$$

with a unique $\nu \leqq s$ and some $(\mathfrak{t},\ \xi) \in \mathfrak{R}_\mathrm{c}$. With such a $(\mathfrak{t},\ \xi)$, one has

(10.5.7$_\mathrm{c}$)
$$\psi_\mu(z)^\tau = S_\sigma^{\mu\nu}(\mathfrak{t},\ \xi)[\psi_\nu(\xi^{-1}(z))]\ .$$

10.6. We can make a few remarks of the same nature as those in 5.8. First, the existence of $(\mathfrak{t},\ \xi)$ in (10.5.7) can be shown as follows. Take λ and α as in (5.7.6). By our choice of $\{\mathfrak{x}_1,\ \cdots,\ \mathfrak{x}_s\}$, we have $\mathfrak{x}_\lambda = \mathfrak{t}\mathfrak{x}_\nu\beta$ with a unique $\nu \leqq s$, some $\mathfrak{t} \in \mathfrak{T}_{(1)}$ and $\beta \in D_{W'}$. Put $\xi = \alpha\beta^{-1}$. Then we find (10.5.7$_\mathrm{b}$), and $(\mathfrak{t},\ \xi) \in \mathfrak{R}_\mathrm{c}$. Further, from (5.8.1) we obtain a relation

(10.6.1)
$$\begin{aligned}
\nu(\mathfrak{x}_\mu\xi/\mathfrak{x}_\nu) &= N_{B/F}(\mathfrak{t})\big(F \cap \Psi'(\mathfrak{b})\Psi'(\mathfrak{b})^\delta\big) \\
&= N_{B/F}(\mathfrak{t}) \det \Theta'(N_{P'/F'}(\mathfrak{b}))\ ,
\end{aligned}$$

hence σ is the restriction of τ to $C_W^*(F',\ B,\ \mathfrak{c})$.

Suppose that $n = 1$. Then we can identify X with B, $\mathfrak{G}$ with the group of invertible elements in B, and $\nu(\alpha) = N_{B/F}(\alpha)$ for $\alpha \in \mathfrak{G}$. The order $\mathfrak{o}$ being as in 10.2, we see, by (3.8.3), that the $\mathfrak{x}_\lambda$ are left $\mathfrak{o}$-ideals. Let $\mathfrak{o}_\lambda$ be the right order of $\mathfrak{x}_\lambda$. Then

$$\Theta_{\mu\lambda}(\mathfrak{c}) = \{\alpha \in D_{W'}(\Delta,\ \mathfrak{c})\ |\ \mathfrak{o}_\mu\alpha = \alpha\mathfrak{o}_\lambda\}\ .$$

Let $(Y, P, \delta, f) \in J_1(B)$. Then $Y = P$, $[P:F] = 2$, and δ is the complex conjugation in P. The order $\mathfrak{g}$ defined in (10.5.7) can be given by $\mathfrak{g} = f^{-1}(\mathfrak{o}_\mu \cap f(P))$. Suppose that $\mathfrak{g} = \mathfrak{r}_P$. Denote by $I_W(P, \mathfrak{o}_\mu, \mathfrak{c})$ the ideal group in P generated by the following two types of ideals.

(i) $v\mathfrak{r}_P$ *with an element* v *of* P *prime to* $N(\mathfrak{c})$ *such that* $vv^\delta \in W$ *and* $v \equiv 1 \bmod^* \mathfrak{c}$;

(ii) *the ideals* $\mathfrak{a}$ *in* P *prime to* $N(\mathfrak{c})$ *such that* $f(\mathfrak{a})\mathfrak{o}_\mu$ *is a two-sided* $\mathfrak{o}_\mu$-*ideal*. We see easily that an ideal $\mathfrak{b}$ in P' belongs to $I_W^*(P', \Psi', \mathfrak{x}_\mu, \mathfrak{c})$ if and only if $\Psi'(\mathfrak{b}) \in I_W(P, \mathfrak{o}_\mu, \mathfrak{c})$. If $\mathfrak{o}$ is a maximal order in B, we can further simplify the definition of $I_W(P, \mathfrak{o}_\mu, \mathfrak{c})$.

When $r = 1$ and τ_{01} is the identity map of F, we can make a discussion similar to that of 5.10, of which no detailed description will be necessary, because it is almost a direct translation to the present setting. It may be remarked only that one can take $\mathfrak{X}_\mathfrak{c}$ to be the group of all two-sided $\mathfrak{o}$-ideals prime to $\mathfrak{c}$. See also [C, 3.11–13, (3.16.2)].

10.7. Let us now start the proof of 10.5. For simplicity, put

$$C = C_W(F', \Theta', \mathfrak{c}), \qquad C^* = C_W^*(F', B, \mathfrak{c}),$$
$$\mathfrak{X} = \mathfrak{X}_{(1)}, \qquad \Theta_{\mu\lambda} = \Theta_{\mu\lambda}(\mathfrak{r}_F).$$

Take $\{V_\lambda, \varphi_\lambda, R_\sigma^{\mu\lambda}(\alpha)\}$ as in 5.7. Let $\beta \in \Theta_{\mu\nu}$. Since $\beta^{-1}\Gamma_W(\mathfrak{x}_\mu, \mathfrak{c})\beta = \Gamma_W(\mathfrak{x}_\nu, \mathfrak{c})$, we can define a biregular isomorphism $T^{\mu\nu}(\beta)$ of V_ν onto V_μ by $T^{\mu\nu}(\beta)(\varphi_\nu(z)) = \varphi_\mu(\beta(z))$. We shall now prove

(10.7.1) $T^{\mu\nu}(\beta)$ *is rational over* C.

(10.7.2) *Let* $\alpha \in D_W(\Delta, \mathfrak{c})$ *and* $t \in \mathfrak{X}$, $t\mathfrak{x}_\nu = \mathfrak{x}_\mu\beta$, $t\mathfrak{x}_\kappa = \mathfrak{x}_\lambda\varepsilon$ *with* $\beta \in \Theta_{\mu\nu}$, $\varepsilon \in \Theta_{\lambda\kappa}$. *Then* $R_\sigma^{\mu\lambda}(\beta\alpha\varepsilon^{-1}) \circ T^{\lambda\kappa}(\varepsilon) = T^{\mu\nu}(\beta)^\sigma \circ R_\sigma^{\nu\kappa}(\alpha)$.
(Note that $\nu(\mathfrak{x}_\mu\beta\alpha\varepsilon^{-1}/\mathfrak{x}_\lambda) = \nu(\mathfrak{x}_\nu\alpha/\mathfrak{x}_\kappa)$, and $\beta\alpha\varepsilon^{-1} \in D_W(\Delta, \mathfrak{c})$.)

For every $t \in \mathfrak{X}$, put $\Delta_t = \{t\mathfrak{x}_1, \cdots, t\mathfrak{x}_q\}$. We see easily that Δ_t is a $\mathfrak{c}$-regular set of representatives for D_W-classes in Λ (even if t may not be prime to $\mathfrak{c}$). Therefore we find a permutation $\lambda \mapsto \bar\lambda$ of $\{1, \cdots, q\}$ and an element η_λ of D_W, for each λ, such that $t\mathfrak{x}_{\bar\lambda} = \mathfrak{x}_\lambda\eta_\lambda$. Put $\bar\varphi_{\bar\lambda}(z) = \varphi_\lambda(\eta_\lambda(z))$ for $z \in \mathfrak{H}_n^r$, and $\bar{R}_\sigma^{\mu\lambda}(\alpha) = R_\sigma^{\mu\lambda}(\eta_\mu\alpha\eta_\lambda^{-1})$ for every $\alpha \in D_W(\Delta, \mathfrak{c})$, and $\sigma = [C/F', \mathfrak{a}]$ with an ideal $\mathfrak{a}$ in F' such that

$$\det \Theta'(\mathfrak{a}) = \nu(\mathfrak{x}_{\bar\mu}\alpha/\mathfrak{x}_{\bar\lambda}) = \nu(\mathfrak{x}_\mu\eta_\mu\alpha\eta_\lambda^{-1}/\mathfrak{x}_\lambda).$$

Put $\bar{V}_{\bar\lambda} = V_\lambda$. As is seen in the proof of 5.13, $\{\bar{V}_{\bar\lambda}, \bar\varphi_{\bar\lambda}, \bar{R}_\sigma^{\mu\lambda}(\alpha)\}$ is a canonical system of level $\mathfrak{c}$ with respect to Δ_t and W. On the other hand, we see easily that $\{V_\lambda, \varphi_\lambda, R_\sigma^{\mu\lambda}(\alpha)\}$ is a canonical system of level $\mathfrak{c}$ with respect to Δ_t and W, since $\Gamma_W(\mathfrak{x}_\lambda, \mathfrak{c}) = \Gamma_W(t\mathfrak{x}_\lambda, \mathfrak{c})$ and $D_W(\Delta_t, \mathfrak{c}) = D_W(\Delta, \mathfrak{c})$. Therefore, by 5.9, there exists, for each λ, a biregular isomorphism T_λ of $V_{\bar\lambda}$ to $\bar{V}_{\bar\lambda}(= V_\lambda)$ rational over

C such that $T_\lambda \circ \varphi_{\bar\lambda} = \bar\varphi_{\bar\lambda}$ and $\bar{R}_\sigma^{\bar\mu\bar\lambda}(\alpha) = T_\mu^\sigma \circ R_\sigma^{\bar\mu\bar\lambda}(\alpha) \circ T_\lambda^{-1}$. Hence $T_\lambda \circ \varphi_{\bar\lambda}(z) = \varphi_\lambda(\eta_\lambda(z))$ and $R_\sigma^{\mu\lambda}(\eta_\mu \alpha \eta_\lambda^{-1}) = T_\mu^\sigma \circ R_\sigma^{\bar\mu\bar\lambda}(\alpha) \circ T_\lambda^{-1}$. We use this result to prove (10.7.1,2). Let the notation be as in (10.7.2). In the above discussion, we may take β and ε as η_μ and η_λ with $\nu = \bar\mu$ and $\kappa = \bar\lambda$. Then $T_\mu = T^{\mu\nu}(\beta)$ and $T_\lambda = T^{\lambda\kappa}(\varepsilon)$, hence we obtain (10.7.1,2).

(10.7.3) *If $\xi \in \Theta_{\nu\mu}$ and $\eta \in \Theta_{\mu\lambda}$, then $\xi\eta \in \Theta_{\nu\lambda}$ and $T^{\nu\mu}(\xi) \circ T^{\mu\lambda}(\eta) = T^{\nu\lambda}(\xi\eta)$.* This is an immediate consequence of the definition of $T^{\nu\mu}(\xi)$.

(10.7.4) *Let $\beta \in \Gamma_W^*(\mathfrak{x}_\mu, \mathfrak{c})$, $\sigma = [C/F', \mathfrak{a}]$, $\det \Theta'(\mathfrak{a}) = \nu(\mathfrak{x}_\mu \alpha / \mathfrak{x}_\lambda)$ with an ideal $\mathfrak{a}$ in F' prime to $N(\mathfrak{c})$, and an element α of $D_W(\Delta, \mathfrak{c})$. Then there exists an element ε of $\Gamma_W^*(\mathfrak{x}_\lambda, \mathfrak{c})$ such that $T^{\mu\mu}(\beta)^\sigma = R_\sigma^{\mu\lambda}(\alpha) \circ T^{\lambda\lambda}(\varepsilon) \circ R_\sigma^{\mu\lambda}(\alpha)^{-1}$.*

In fact, we have $\mathfrak{x}_\mu \beta = t\mathfrak{x}_\mu$ for some $t \in \mathfrak{X}_\mathfrak{c}$. Then $N_{B/F}(t) = (\nu(\beta))$. Since $\nu(t\mathfrak{x}_\lambda/\mathfrak{x}_\lambda) = (\nu(\beta))$, there exists, by 3.10, an element ζ of $\mathfrak{G}$ such that $t\mathfrak{x}_\lambda = \mathfrak{x}_\lambda \zeta$ and $\nu(\zeta) = \nu(\beta)$. Then $\zeta \in D_W(\Delta, \mathfrak{c})$ and $\nu(\beta\zeta^{-1}) = 1$. By 3.11, there exists an element γ of $\mathfrak{G}$ such that $\mathfrak{x}_\lambda \gamma = \mathfrak{x}_\lambda$, $\nu(\gamma) = 1$, $\gamma \equiv \beta\zeta^{-1} \bmod^* (\Delta; \mathfrak{c})$. Put $\varepsilon = \gamma\zeta$. Then $\varepsilon \in \Gamma_W^*(\mathfrak{x}_\lambda, \mathfrak{c})$. Since $\beta\alpha\varepsilon^{-1} \equiv \alpha \bmod^* (\Delta; \mathfrak{c})$, we obtain (10.7.4) from (10.7.2).

10.8. We observe that the $T^{\mu\mu}(\beta)$ for all $\beta \in \Gamma_W^*(\mathfrak{x}_\mu, \mathfrak{c})$ form a finite group A_μ of biregular automorphisms of V_μ. In view of (10.7.1), we can construct a quotient U_μ of V_μ by A_μ so that U_μ and the natural projection Z_μ of V_μ to U_μ are both rational over C. In view of (10.7.4), we can define a biregular isomorphism $S_\sigma^{\mu\lambda}(\alpha)$ of U_λ to U_μ^σ, rational over C so that

$$(10.8.1) \qquad Z_\mu^\sigma \circ R_\sigma^{\mu\lambda}(\alpha) = S_\sigma^{\mu\lambda}(\alpha) \circ Z_\lambda \; .$$

Note that $S_\sigma^{\mu\lambda}(\alpha)$ is defined if and only if $R_\sigma^{\mu\lambda}(\alpha)$ is defined. Further, by virtue of (10.7.3), we can define, for every $\zeta \in \Theta_{\mu\lambda}$, a biregular isomorphism $Y^{\mu\lambda}(\zeta)$ of U_λ to U_μ rational over C so that

$$(10.8.2) \qquad Y^{\mu\lambda}(\zeta) \circ Z_\lambda = Z_\mu \circ T^{\mu\lambda}(\zeta) \; .$$

Obviously we have

$$(10.8.3) \qquad S_\tau^{\nu\mu}(\beta)^\sigma \circ S_\sigma^{\mu\lambda}(\alpha) = S_{\tau\sigma}^{\nu\lambda}(\beta\alpha) \; ,$$

$$(10.8.4) \qquad S_\sigma^{\mu\lambda}(\alpha) = S_\sigma^{\mu\lambda}(\beta) \qquad\qquad \textit{if } \alpha \equiv \beta \bmod^* (\Delta; \mathfrak{c}) \; ,$$

(10.8.5) *$Y^{\mu\mu}(\xi)$ is the identity map if $\xi \in \Gamma_W^*(\mathfrak{x}_\mu, \mathfrak{c})$.*
From our definition of $T^{\mu\lambda}(\beta)$, we see that $T^{\lambda\lambda}(\gamma) = R_1^{\lambda\lambda}(\gamma)$ if $\gamma \in \Gamma_W(\mathfrak{x}_\lambda)$, hence

$$(10.8.6) \qquad Y^{\lambda\lambda}(\gamma) = S_1^{\lambda\lambda}(\gamma) \qquad\qquad \textit{if } \gamma \in \Gamma_W(\mathfrak{x}_\lambda) \; .$$

Further, from (10.8.1, 2) we obtain easily

(10.8.7) *The notation being as in (10.7.2), one has*

$$S_\sigma^{\mu\lambda}(\beta\alpha\varepsilon^{-1}) \circ Y^{\lambda\kappa}(\varepsilon) = Y^{\mu\nu}(\beta)^\sigma \circ S_\sigma^{\nu\kappa}(\alpha) \; .$$

10.9. Fix an index μ. Let $\sigma \in G(C/C^*)$. Take an ideal $\mathfrak{a}$ in F' prime to $N(\mathfrak{c})$ so that $\sigma = [C/F', \mathfrak{a}]$. By our definition of C^*, we have $\det \Theta'(\mathfrak{a}) =$

$u \cdot N_{B/F}(\mathfrak{t})$ with $u \in W$, $u \equiv 1 \bmod^* \mathfrak{c}$, and $\mathfrak{t} \in \mathfrak{X}_\mathfrak{c}$. Then $\mathfrak{t}^{-1}\mathfrak{x}_\mu = \mathfrak{x}_\lambda \zeta^{-1}$ for some λ and some $\zeta \in \Theta_{\mu\lambda}(\mathfrak{c})$. We see that λ is uniquely determined by σ and μ. Since $\nu(\mathfrak{x}_\mu \zeta / \mathfrak{x}_\lambda) = u^{-1} \det \Theta'(\mathfrak{a})$, $R^{\mu\lambda}_\sigma(\zeta)$ can be defined. Put $Q^\mu_\sigma = S^{\mu\lambda}_\sigma(\zeta) \circ Y^{\lambda\mu}(\zeta^{-1})$. Then Q^μ_σ is a biregular isomorphism of U_μ to U^σ_μ rational over C; it is independent of the choice of $\mathfrak{a}$, $\mathfrak{t}$, and ζ. In fact, let $\sigma = [C/F', \mathfrak{b}]$, $\det \Theta'(\mathfrak{b}) = v \cdot N_{B/F}(\mathfrak{q})$, $v \in W$, $v \equiv 1 \bmod^* \mathfrak{c}$, $\mathfrak{q} \in \mathfrak{X}_\mathfrak{c}$, $\mathfrak{q}^{-1}\mathfrak{x}_\mu = \mathfrak{x}_\lambda \xi^{-1}$, $\xi \in \Theta_{\mu\lambda}(\mathfrak{c})$. Then we have $\det \Theta'(\mathfrak{a}^{-1}\mathfrak{b}) = u^{-1}v \cdot N_{B/F}(\mathfrak{t}^{-1}\mathfrak{q}) = u^{-1}v \cdot \nu(\mathfrak{x}_\mu \zeta / \mathfrak{x}_\lambda)^{-1}\nu(\mathfrak{x}_\mu \xi / \mathfrak{x}_\lambda) = (u^{-1}v \cdot \nu(\zeta^{-1}\xi))$. Since $[C/F', \mathfrak{a}^{-1}\mathfrak{b}] = 1$, we have $\det \Theta'(\mathfrak{a}^{-1}\mathfrak{b}) = (w)$ with an element $w \in W$ such that $w \equiv 1 \bmod^* \mathfrak{c}$. It follows that $\nu(\zeta^{-1}\xi) \equiv y \bmod^* \mathfrak{c}$ with a unit y of F contained in W. By 3.11, there exists an element γ of $\mathfrak{G}$ such that $\gamma \equiv \zeta^{-1}\xi \bmod^* (\Delta; \mathfrak{c})$, $\mathfrak{x}_\lambda \gamma = \mathfrak{x}_\lambda$, and $\nu(\gamma) = y$. By (10.7.3), (10.8.3), and (10.8.6), we have

$$S^{\mu\lambda}_\sigma(\xi) \circ Y^{\lambda\mu}(\xi^{-1}) = S^{\mu\lambda}_\sigma(\zeta\gamma) \circ Y^{\lambda\mu}(\xi^{-1}) = S^{\mu\lambda}_\sigma(\zeta) \circ S^{\lambda\lambda}_1(\gamma) \circ Y^{\lambda\mu}(\xi^{-1})$$
$$= S^{\mu\lambda}_\sigma(\zeta) \circ Y^{\lambda\lambda}_1(\gamma) \circ Y^{\lambda\mu}(\xi^{-1}) = S^{\mu\lambda}_\sigma(\zeta) \circ Y^{\lambda\mu}(\zeta^{-1}) \circ Y^{\mu\mu}(\zeta\gamma\xi^{-1}) \ .$$

Now $\zeta\gamma\xi^{-1} \in \Gamma^*_W(\mathfrak{x}_\mu, \mathfrak{c})$, hence $Y^{\mu\mu}(\zeta\gamma\xi^{-1})$ is the identity map. This shows that Q^μ_σ is determined only by μ and σ.

Let σ, μ, λ, $\mathfrak{t}$, and ζ be as above. Let $\tau = [C/F', \mathfrak{e}] \in G(C/C^*)$, $\det \Theta'(\mathfrak{e}) = s \cdot N_{B/F}(\mathfrak{m})$, $s \in W$, $s \equiv 1 \bmod^* \mathfrak{c}$, $\mathfrak{m} \in \mathfrak{X}_\mathfrak{c}$, $\mathfrak{m}^{-1}\mathfrak{x}_\mu = \mathfrak{x}_\kappa \eta^{-1}$, $\eta \in \Theta_{\mu\kappa}(\mathfrak{c})$. Then $Q^\mu_\tau = S^{\mu\kappa}_\tau(\eta) \circ Y^{\kappa\mu}(\eta^{-1})$. Now let $\mathfrak{t}^{-1}\mathfrak{x}_\kappa = \mathfrak{x}_\nu \xi^{-1}$ with some ν and $\xi \in \Theta_{\kappa\nu}(\mathfrak{c})$. By (10.8.7), we have $Y^{\lambda\mu}(\zeta^{-1})^\tau \circ S^{\mu\kappa}_\tau(\eta) = S^{\lambda\nu}(\zeta^{-1}\eta\xi) \circ Y^{\nu\kappa}(\xi^{-1})$, and $\mathfrak{t}^{-1}\mathfrak{m}^{-1}\mathfrak{x}_\mu = \mathfrak{x}_\nu \xi^{-1}\eta^{-1}$. Therefore

$$(Q^\mu_\sigma)^\tau \circ Q^\mu_\tau = S^{\mu\lambda}_\sigma(\zeta)^\tau \circ Y^{\lambda\mu}(\zeta^{-1})^\tau \circ S^{\mu\kappa}_\tau(\eta) \circ Y^{\kappa\mu}(\eta^{-1})$$
$$= S^{\mu\lambda}_\sigma(\zeta)^\tau \circ S^{\lambda\nu}_\tau(\zeta^{-1}\eta\xi) \circ Y^{\nu\kappa}(\xi^{-1}) \circ Y^{\kappa\mu}(\eta^{-1}) = S^{\mu\nu}_{\sigma\tau}(\eta\xi) \circ Y^{\nu\mu}(\xi^{-1}\eta^{-1}) = Q^\mu_{\sigma\tau} \ .$$

Applying Weil's criterion [12] to U_μ and Q^μ_σ, we obtain a variety U^*_μ, rational over C^*, and a biregular isomorphism Q_μ of U^*_μ to U_μ, rational over C, such that $Q^\mu_\sigma = (Q_\mu)^\sigma \circ Q^{-1}_\mu$ for every $\sigma \in G(C/C^*)$.

Now replace U_μ, Z_μ, $S^{\mu\lambda}_\sigma(\alpha)$, $Y^{\mu\lambda}(\beta)$ by U^*_μ, $Q^{-1}_\mu \circ Z_\mu$, $(Q^{-1}_\mu)^\sigma \circ S^{\mu\lambda}_\sigma(\alpha) \circ Q_\lambda$, $Q^{-1}_\mu \circ Y^{\mu\lambda}(\beta) \circ Q_\lambda$ respectively, and denote them again by the original letters. By this replacement, we can now assume, retaining the properties (10.8.1–7),

    (10.9.1) U_μ *is rational over* C^*,

    (10.9.2) $S^{\mu\lambda}_\sigma(\xi) = Y^{\mu\lambda}(\xi)$ *if* $\xi \in \Theta_{\mu\lambda}(\mathfrak{c})$.

Suppose that $\alpha \in \Theta_{\nu\kappa}(\mathfrak{c})$ in the relation (10.8.7). Then $\beta\alpha\varepsilon^{-1} \in \Theta_{\mu\lambda}(\mathfrak{c})$, hence by (10.9.2), we have $Y^{\mu\nu}(\beta) = Y^{\mu\nu}(\beta)^\sigma$. Since every automorphism of C over C^* can be obtained as such a σ (see the beginning of 10.9), this shows that $Y^{\mu\nu}(\beta)$ is rational over C^* for every $\beta \in \Theta_{\mu\nu}$.

Combining (10.9.2) with (10.8.7) and assuming $\beta \in \Theta_{\mu\nu}(\mathfrak{c})$, we obtain

$$S^{\mu\lambda}_\sigma(\beta\alpha\varepsilon^{-1}) \circ S^{\lambda\kappa}_\tau(\varepsilon) = S^{\mu\nu}_\tau(\beta)^\sigma \circ S^{\nu\kappa}_\sigma(\alpha) = S^{\mu\kappa}_{\tau\sigma}(\beta\alpha) \ .$$

Here $\tau = [C/F', \mathfrak{a}]$ with an ideal $\mathfrak{a}$ in F' such that $\det \Theta'(\mathfrak{a}) = N_{B/F}(\mathfrak{t}) =$

$\nu(\mathfrak{x}_\lambda\varepsilon/\mathfrak{x}_\kappa) = \nu(\mathfrak{x}_\mu\beta/\mathfrak{x}_\nu)$, t being as in (10.7.2). Now $S^{\mu\kappa}_{\tau\sigma}(\beta\alpha) = S^{\mu\lambda}_\sigma(\beta\alpha\varepsilon^{-1})^\tau \circ S^{\lambda\kappa}_\tau(\varepsilon)$. It follows that $S^{\mu\lambda}_\sigma(\beta\alpha\varepsilon^{-1})^\tau = S^{\mu\lambda}_\sigma(\beta\alpha\varepsilon^{-1})$. Observe that every automorphism of C over C^* can be obtained as such a τ. Therefore $S^{\mu\lambda}_\sigma(\xi)$ is rational over C^* for every $\xi \in D_W(\Delta, \mathfrak{c})$.

Since $S^{\mu\lambda}_\sigma(\xi)$ is a biregular isomorphism of U_λ to U^σ_μ, and U_μ is rational over C^*, we can now understand that the subscript σ of $S^{\mu\lambda}_\sigma(\xi)$ is an element of $G(C^*/F')$ determined by $\sigma = [C^*/F', \mathfrak{a}]$ and $\det \Theta'(\mathfrak{a}) = \nu(\mathfrak{x}_\mu\xi/\mathfrak{x}_\lambda)$.

10.10. Let $(t, \alpha) \in \mathfrak{R}_\mathfrak{c}$ and $\sigma = [C^*/F', \mathfrak{a}]$ with an ideal $\mathfrak{a}$ in F' prime to $N(\mathfrak{c})$ such that $\det \Theta'(\mathfrak{a}) = \nu(\mathfrak{x}_\mu\alpha/t\mathfrak{x}_\lambda)$. We have $t\mathfrak{x}_\lambda\alpha^{-1} = \mathfrak{x}_\kappa\xi^{-1}$ for a unique κ and some $\xi \in D_W$. By our definition of $\mathfrak{R}_\mathfrak{c}$, we see that $\xi \in D_W(\Delta, \mathfrak{c})$, and $S^{\mu\kappa}_\sigma(\xi)$ is meaningful. Define $S^{\mu\lambda}_\sigma(t, \alpha)$ by

$$(10.10.1) \qquad\qquad S^{\mu\lambda}_\sigma(t, \alpha) = S^{\mu\kappa}_\sigma(\xi) \circ Y^{\kappa\lambda}(\xi^{-1}\alpha) .$$

This does not depend on the choice of ξ, in view of (10.8.6).

To show (10.5.5), let $m\mathfrak{x}_\mu\beta^{-1} = \mathfrak{x}_\omega\eta^{-1}$ with an index ω and some $\eta \in D_W$. Further let $mt\mathfrak{x}_\lambda\alpha^{-1}\beta^{-1} = \mathfrak{x}_\rho\zeta^{-1}$ with an index ρ and some $\zeta \in D_W$. Then $m\mathfrak{x}_\kappa = \mathfrak{x}_\rho\zeta^{-1}\beta\xi$, hence by (10.8.7), we have

$$S^{\omega\rho}_\sigma(\eta^{-1}\zeta) \circ Y^{\rho\kappa}(\zeta^{-1}\beta\xi) = Y^{\omega\mu}(\eta^{-1}\beta)^\sigma \circ S^{\mu\kappa}_\sigma(\xi) .$$

Therefore

$$S^{\nu\mu}_\tau(m, \beta)^\sigma \circ S^{\mu\lambda}_\sigma(t, \alpha) = S^{\nu\omega}_\tau(\eta)^\sigma \circ Y^{\omega\mu}(\eta^{-1}\beta)^\sigma \circ S^{\mu\kappa}_\sigma(\xi) \circ Y^{\kappa\lambda}(\xi^{-1}\alpha)$$
$$= S^{\nu\omega}_\tau(\eta)^\sigma \circ S^{\omega\rho}_\sigma(\eta^{-1}\zeta) \circ Y^{\rho\kappa}(\zeta^{-1}\beta\xi) \circ Y^{\kappa\lambda}(\xi^{-1}\alpha) = S^{\nu\rho}_{\tau\sigma}(\zeta) \circ Y^{\rho\lambda}(\zeta^{-1}\beta\alpha) .$$

This proves (10.5.5).

For each $\lambda \le q$, define $\psi_\lambda: \mathfrak{H}^r_n \to U_\lambda$ by $\psi_\lambda = Z_\lambda \circ \varphi_\lambda$. Then $(U_\lambda, \psi_\lambda)$ is a model for $\mathfrak{H}^r_n/\Gamma^*_W(\mathfrak{x}_\lambda, \mathfrak{c})$. Now consider the part of $\{U_\lambda, \psi_\lambda, S^{\mu\lambda}_\sigma(t, \alpha)\}$ consisting of the objects with $\lambda, \mu \le s$. We shall show that this system satisfies all the conditions of 10.5. We have seen that $(10.5.1, 2, 3, 5)$ are satisfied. Let (t, α), μ, λ, κ and ξ be as above. Suppose $\mu = \lambda$, $t \in \mathfrak{T}_\mathfrak{c}$ and $\alpha \equiv 1 \bmod^* (\Delta; \mathfrak{c})$. Then σ is the identity map on C^*, and $Y^{\kappa\mu}(\xi^{-1}\alpha) = S^{\kappa\mu}_1(\xi^{-1}\alpha) = S^{\kappa\mu}_1(\xi^{-1})$ by (10.8.4) and (10.9.2), hence $S^{\mu\mu}_1(t, \alpha)$ is the identity map. This, together with (10.5.5), proves (10.5.4).

Let $\alpha \in \Gamma^*_W(\mathfrak{x}_\lambda)$ and $\mathfrak{x}_\lambda\alpha = t\mathfrak{x}_\lambda$ with $t \in \mathfrak{T}$. Then $(t, \alpha) \in \mathfrak{R}_\mathfrak{c}$ and $S^{\lambda\lambda}_1(t, \alpha) = Y^{\lambda\lambda}(\alpha)$. Therefore, from (10.8.2), we obtain (10.5.6).

To prove (10.5.7), let (Y, P, δ, f), (P', Ψ'), z, $\mathfrak{g}_Y$, $\mathfrak{g}$, $\mathfrak{h}$, $\mathfrak{h}_0$, $\mathfrak{b}$, and τ be as in (5.3.3) and (5.7.6). By (5.7.6), we have $\mathfrak{x}_\mu f([\Psi'(\mathfrak{b})]_\mathfrak{g})^{-1} = \mathfrak{x}_\lambda\alpha^{-1}$ for a unique $\lambda \le q$ and some $\alpha \in D_W(\Delta, \mathfrak{c})$. As shown in 10.6, we have $(t, \xi) \in \mathfrak{R}_\mathfrak{c}$ satisfying $(10.5.7_b)$. For such a (t, ξ), put $\beta = \xi^{-1}\alpha$. Then $\mathfrak{x}_\lambda = t\mathfrak{x}_\nu\beta$. By $(5.7.6_a)$, we have

$$\psi_\mu(z)^\tau = Z_\mu^\tau\big(\varphi_\mu(z)^\tau\big) = Z_\sigma^\sigma \circ R_\sigma^{\mu\lambda}(\alpha)\big[\varphi_\lambda(\alpha^{-1}(z))\big] = S_\sigma^{\mu\lambda}(\alpha) \circ Z_\lambda\big[\varphi_\lambda(\alpha^{-1}(z))\big]$$
$$= S_\sigma^{\mu\lambda}(\alpha)\big[\psi_\lambda(\alpha^{-1}(z))\big] .$$

Now we have $\mathfrak{t}\mathfrak{x}_\nu\xi^{-1} = \mathfrak{x}_\lambda\alpha^{-1}$, hence $S_\sigma^{\mu\nu}(\mathfrak{t}, \xi) = S_\sigma^{\mu\lambda}(\alpha) \circ Y^{\lambda\nu}(\beta^{-1})$ by (10.10.1). Therefore

$$(10.10.2) \qquad S_\sigma^{\mu\nu}(\mathfrak{t}, \xi)\big[\psi_\nu(\xi^{-1}(z))\big] = S_\sigma^{\mu\lambda}(\alpha)\big[\psi_\lambda(\alpha^{-1}(z))\big] = \psi_\mu(z)^\tau .$$

Suppose that τ is the identity map on $C_W^*(P', \Psi', \mathfrak{x}_\mu, \mathfrak{c})$. Then $\mu = \nu$, $\mathfrak{x}_\mu f([\Psi'(\mathfrak{b})]_\mathfrak{g})^{-1} = \mathfrak{t}\mathfrak{x}_\mu f(v)^{-1}$ with some $\mathfrak{t} \in \mathfrak{T}_\mathfrak{c}$ and $v \in Y$ such that $vv^\delta \in W$, $v \equiv 1 \bmod^* \mathfrak{cg}_Y$. Put $\xi = f(v)$. Then $\xi \equiv 1 \bmod^* (\Delta; \mathfrak{c})$, hence $S_\sigma^{\mu\nu}(\mathfrak{t}, \xi)$ is the identity map by (10.5.4). By (10.10.2) with this $(\mathfrak{t}, \xi)$, we have $\psi_\mu(z) = \psi_\mu(z)^\tau$. This shows that $\psi_\mu(z)$ is rational over $C_W^*(P', \Psi', \mathfrak{x}_\mu, \mathfrak{c})$.

Conversely, in the setting of (10.10.2), suppose that τ is the identity map on $C^* \cdot P'(\psi_\mu(z))$. Then $\nu = \mu$, and $\psi_\mu(z) = S_1^{\mu\mu}(\mathfrak{t}, \xi)\big[\psi_\mu(\xi^{-1}(z))\big]$. Since $\tau = [C^*/F', N_{P'/F'}(\mathfrak{b})]$, there exist an element u of W and an element $\mathfrak{t}'$ of $\mathfrak{T}_\mathfrak{c}$ such that $\det \Theta'(N_{P'/F'}(\mathfrak{b})) = u \cdot N_{B/F}(\mathfrak{t}')$, $u \equiv 1 \bmod^* \mathfrak{c}$. By (10.6.1) we have $\nu(\mathfrak{t}'\mathfrak{t}\mathfrak{x}_\mu\xi^{-1}/\mathfrak{x}_\mu) = (u^{-1})$, hence there exists, by 3.10, an element η of $\mathfrak{G}$ such that $\nu(\eta) = u^{-1}$ and $\mathfrak{t}'\mathfrak{t}\mathfrak{x}_\mu\xi^{-1} = \mathfrak{x}_\mu\eta$. Since $\nu(\eta) \equiv 1 \bmod^* \mathfrak{c}$, there exists, by 3.11, an element γ of $\mathfrak{G}$ such that $\mathfrak{x}_\mu\gamma = \mathfrak{x}_\mu$, $\nu(\gamma) = 1$ and $\gamma \equiv \eta \bmod^* (\Delta; \mathfrak{c})$. Put $\zeta = \gamma^{-1}\eta\xi$. Then $\mathfrak{t}'\mathfrak{t}\mathfrak{x}_\mu = \mathfrak{x}_\mu\zeta$, $\zeta \in \Gamma_W^*(\mathfrak{x}_\mu)$, $\zeta\xi^{-1} \equiv 1 \bmod^* (\Delta; \mathfrak{c})$. By (10.5.4) and (10.5.6), we have

$$\psi_\mu(z) = S_1^{\mu\mu}(\mathfrak{t}, \xi)\big[\psi_\mu(\xi^{-1}(z))\big] = S_1^{\mu\mu}(\mathfrak{t}'\mathfrak{t}, \zeta)\big[\psi_\mu(\xi^{-1}(z))\big] = \psi_\mu(\zeta\xi^{-1}(z)) ,$$

so that $\zeta\xi^{-1}(z) = \varepsilon(z)$ for some $\varepsilon \in \Gamma_W^*(\mathfrak{x}_\mu, \mathfrak{c})$. We have therefore $\varepsilon^{-1}\zeta\xi^{-1} = f(v)$ with an element v of Y. Then $vv^\delta = \nu(\varepsilon^{-1}\zeta\xi^{-1}) \in W$, $f(v) \equiv 1 \bmod^* (\Delta; \mathfrak{c})$, hence $v \equiv 1 \bmod^* \mathfrak{cg}_Y$ by (3.6.2), and $\mathfrak{x}_\mu\varepsilon = \mathfrak{t}''\mathfrak{x}_\mu$ with some $\mathfrak{t}'' \in \mathfrak{T}_\mathfrak{c}$. Hence

$$\mathfrak{x}_\mu f([\Psi'(\mathfrak{b})]_\mathfrak{g})^{-1} = \mathfrak{t}\mathfrak{x}_\mu\xi^{-1} = \mathfrak{t}\mathfrak{x}_\mu\zeta^{-1}\varepsilon f(v) = \mathfrak{t}'^{-1}\mathfrak{t}''\mathfrak{x}_\mu f(v) .$$

It follows that $\mathfrak{b} \in I_W^*(P', \Psi', \mathfrak{x}_\mu, \mathfrak{c})$, hence τ is the identity map on $C_W^*(P', \Psi', \mathfrak{x}_\mu, \mathfrak{c})$. This proves (10.5.7$_\mathrm{a}$), and completes the proof of 10.5.

10.11. THEOREM. *Let $\mathfrak{b}$ be an integral ideal in F which divides $\mathfrak{c}$. Let $\{U_\lambda, \psi_\lambda, S_\sigma^{\mu\lambda}(\mathfrak{t}, \alpha)\}$ be a system satisfying (10.5.1–7), and $\{\bar{U}_\lambda, \bar{\psi}_\lambda, \bar{S}_\sigma^{\mu\lambda}(\mathfrak{t}, \alpha)\}$ be a system satisfying (10.5.1–7) with $\mathfrak{b}$ in place of $\mathfrak{c}$ (and with the same W and $\{\mathfrak{x}_1, \cdots, \mathfrak{x}_s\}$). Then there exists, for each $\lambda \leq s$, a morphism T_λ of U_λ onto $\bar{U}_\lambda$, rational over $C_W^*(F', B, \mathfrak{c})$, such that*

$$\bar{\psi}_\lambda = T_\lambda \circ \psi_\lambda , \qquad \bar{S}_{\bar\sigma}^{\mu\lambda}(\mathfrak{t}, \alpha) \circ T_\lambda = T_\mu^\sigma \circ S_\sigma^{\mu\lambda}(\mathfrak{t}, \alpha)$$

for every μ, λ and every $(\mathfrak{t}, \alpha) \in \mathfrak{R}_\mathfrak{c}$, where $\bar\sigma$ is the restriction of σ to $C_W^*(F', B, \mathfrak{b})$.

The proof is quite similar to that in 8.2, and therefore may be omitted.

10.12. PROOF OF (10.2.1). Let $\mathfrak{S}$ be the group of all two-sided $\mathfrak{o}$-ideals, and $\mathfrak{S}_0$ (resp. $\mathfrak{S}_F$) the subgroup of $\mathfrak{S}$ consisting of $a\mathfrak{o}$ (resp. $\mathfrak{ao}$) for all invertible elements a of F (resp. all ideals $\mathfrak{a}$ in F). We shall now prove (without assuming (3.8.1, 2))

(10.12.1) $[\mathfrak{S} : \mathfrak{S}_0]$ *is finite.*

Since $[\mathfrak{S}_F : \mathfrak{S}_0] < \infty$, it suffices to show $[\mathfrak{S} : \mathfrak{S}_F] < \infty$. For every prime ideal $\mathfrak{p}$ in F, let $C_\mathfrak{p}$ denote the group of all invertible elements x of $B_\mathfrak{p}$ such that $x\mathfrak{o}_\mathfrak{p} = \mathfrak{o}_\mathfrak{p}x$, and $U_\mathfrak{p}$ the group of all units of $\mathfrak{o}_\mathfrak{p}$. Let us now prove

(10.12.2) $[C_\mathfrak{p} : U_\mathfrak{p}F_\mathfrak{p}^*] < \infty$ *for every* $\mathfrak{p}$; $[C_\mathfrak{p} : U_\mathfrak{p}F_\mathfrak{p}^*] = 1$ *for all except a finite number of* $\mathfrak{p}$.

Here $F_\mathfrak{p}^*$ is the group of non-zero elements of $F_\mathfrak{p}$. Take a maximal order $\mathfrak{o}_\mathfrak{p}'$ in $B_\mathfrak{p}$ containing $\mathfrak{o}_\mathfrak{p}$, and denote by $U_\mathfrak{p}'$ the group of all units of $\mathfrak{o}_\mathfrak{p}'$. We have $\mathfrak{p}^e\mathfrak{o}_\mathfrak{p}' \subset \mathfrak{o}_\mathfrak{p}$ for a suitably large e, so that $U_\mathfrak{p}$ contains every x of $U_\mathfrak{p}'$ such that $x \equiv 1 \bmod \mathfrak{p}^e\mathfrak{o}_\mathfrak{p}'$, hence $[U_\mathfrak{p}' : U_\mathfrak{p}] < \infty$. First suppose that $B_\mathfrak{p}$ is a division algebra. Then $[B_\mathfrak{p}^* : U_\mathfrak{p}'F_\mathfrak{p}^*] = 2$, hence $[C_\mathfrak{p} : U_\mathfrak{p}F_\mathfrak{p}^*] \leq 2[U_\mathfrak{p}' : U_\mathfrak{p}] < \infty$. Next suppose that $B_\mathfrak{p} = M_2(F_\mathfrak{p})$. We may assume $\mathfrak{o}_\mathfrak{p}' = M_2(\mathfrak{r}_\mathfrak{p})$. Let $x \in C_\mathfrak{p}$. Fix a prime element π of $F_\mathfrak{p}$. We can find two elements u and v of $U_\mathfrak{p}'$ and an element w of $F_\mathfrak{p}$ so that

$$uxvw = \begin{bmatrix} 1 & 0 \\ 0 & \pi^c \end{bmatrix}$$

with a non-negative integer c. Then $\mathfrak{o}_\mathfrak{p} = x\mathfrak{o}_\mathfrak{p}x^{-1} \subset \mathfrak{o}_\mathfrak{p}' \cap x\mathfrak{o}_\mathfrak{p}'x^{-1}$. Since $\mathfrak{p}^e\mathfrak{o}_\mathfrak{p}' \subset \mathfrak{o}_\mathfrak{p}$, we have

$$\begin{bmatrix} 0 & 0 \\ \pi^e & 0 \end{bmatrix} \in u^{-1}\mathfrak{o}_\mathfrak{p}u \;,$$

hence $c \leq e$. Let $U_\mathfrak{p}' = \bigcup_{i=1}^{m} U_\mathfrak{p}z_i$ be a coset decomposition. Then we have

$$C_\mathfrak{p} \subset \bigcup_{i,j=1}^{m} \bigcup_{c=0}^{e} F_\mathfrak{p}^* U_\mathfrak{p}z_i \begin{bmatrix} 1 & 0 \\ 0 & \pi^c \end{bmatrix} z_j^{-1} U_\mathfrak{p} \;,$$

hence $[C_\mathfrak{p} : F_\mathfrak{p}^*U_\mathfrak{p}] < \infty$, since $F_\mathfrak{p}^*U_\mathfrak{p}$ is a normal subgroup of $C_\mathfrak{p}$. Moreover, $C_\mathfrak{p} = F_\mathfrak{p}^*U_\mathfrak{p}$ if $\mathfrak{o}_\mathfrak{p} = \mathfrak{o}_\mathfrak{p}'$. Thus we obtain (10.12.2).

Let $\mathfrak{t} \in \mathfrak{S}$. For every $\mathfrak{p}$, take an element $x_\mathfrak{p}$ of $B_\mathfrak{p}$ so that $\mathfrak{t}_\mathfrak{p} = \mathfrak{o}_\mathfrak{p}x_\mathfrak{p}$. Then $x_\mathfrak{p} \in C_\mathfrak{p}$. We see easily that the map $\mathfrak{t} \mapsto x_\mathfrak{p}$ defines an isomorphism of $\mathfrak{S}/\mathfrak{S}_F$ onto $\prod_\mathfrak{p} (C_\mathfrak{p}/U_\mathfrak{p}F_\mathfrak{p}^*)$. Therefore (10.12.1) follows from (10.12.2).

To prove (10.2.1), let E denote the group of all units of F, and E^2 the set of ε^2 for all $\varepsilon \in E$. Further put

$$\Gamma = \{\alpha \in \mathfrak{G} \mid \mathfrak{x}_\lambda \alpha = \mathfrak{t}\mathfrak{x}_\lambda \text{ for some } \mathfrak{t} \in \mathfrak{S}\} \;,$$
$$\Gamma^0 = \{\alpha \in \mathfrak{G} \mid \mathfrak{x}_\lambda \alpha = \mathfrak{x}_\lambda\} \;,$$
$$\Gamma^1 = \{\alpha \in \mathfrak{U} \mid \mathfrak{x}_\lambda \alpha = \mathfrak{x}_\lambda\} \;.$$

Let $\alpha \in \Gamma$ and $\mathfrak{x}_\lambda \alpha = t\mathfrak{x}_\lambda$ with $t \in \mathfrak{S}$. From the correspondence $\alpha \mapsto t$, we obtain an isomorphism of $F^*\Gamma/F^*\Gamma^0$ into $\mathfrak{S}/\mathfrak{S}_0$. Further the map $\alpha \mapsto \nu(\alpha)$ defines an isomorphism of $\Gamma^0/E\Gamma^1$ into E/E^2. Thus $[F^*\Gamma : F^*\Gamma^1] \leqq [\mathfrak{S} : \mathfrak{S}_0][E : E^2] < \infty$. Since $\Gamma^1 \subset \Gamma_W(\mathfrak{x}_\lambda)$ and $\Gamma_W^*(\mathfrak{x}_\lambda) \subset \Gamma$, this completes the proof of (10.2.1).

11. Galois groups and Frobenius automorphisms

11.1. As was already noticed in §5, a canonical model $V_\mathfrak{c}$ for $\mathfrak{H}_n^r/\Gamma_W(\mathfrak{x}, \mathfrak{c})$ is a Galois covering of a canonical model $V_{(1)}$ for $\mathfrak{H}_n^r/\Gamma_W(\mathfrak{x})$. The Galois group can be determined by means of 5.7 and 5.9 (see 11.3 below). If we take a point on $V_{(1)}$ whose coordinates are algebraic numbers, and consider points on $V_\mathfrak{c}$ lying on it, we obtain a Galois extension of an algebraic number field. By means of an infinite sequence of coverings, we can define an infinite Galois extension whose Galois group has a certain l-adic representation. It is the purpose of this section to investigate such a representation and the behavior of Frobenius automorphisms.

Let $\mathfrak{c}$ and $\mathfrak{e}$ be integral ideals in F such that $\mathfrak{e} \subset \mathfrak{c}$, and $\Delta = \{\mathfrak{x}_1, \cdots, \mathfrak{x}_q\}$ be an $\mathfrak{e}$-regular set of representatives of D_W-classes in an $\mathfrak{F}$-genus of ample $\mathfrak{r}_F$-lattices in X (see 5.4 and 5.6). Let

$$\{V_\lambda, \varphi_\lambda, R_\sigma^{\mu\lambda}(\alpha)\,(\lambda, \mu = 1, \cdots, q; \alpha \in D_W(\Delta, \mathfrak{c}))\}$$

(resp. $\{U_\lambda, \psi_\lambda, S_\sigma^{\mu\lambda}(\alpha)\}$) be a canonical system of level $\mathfrak{c}$ (resp. $\mathfrak{e}$) with respect to Δ and W. For simplicity, put $k_\mathfrak{a} = C_W(F', \Theta', \mathfrak{a})$ for every integral ideal $\mathfrak{a}$ in F. By 5.9, there exists, for each λ, a morphism T_λ of U_λ onto V_λ, rational over $k_\mathfrak{e}$, such that

$$(11.1.1) \qquad \varphi_\lambda = T_\lambda \circ \psi_\lambda, \quad R_\sigma^{\mu\lambda}(\alpha) \circ T_\lambda = T_\mu^\sigma \circ S_\sigma^{\mu\lambda}(\alpha) ,$$

where we denote the restriction of an automorphism σ of $k_\mathfrak{e}$ to $k_\mathfrak{c}$ by the same letter σ. Let us fix our attention to $\mathfrak{x}_1, V_1, \varphi_1, R_\sigma^{11}(\alpha), U_1, \psi_1, S_\sigma^{11}(\alpha), T_1$ and denote them by $\mathfrak{x}, V, \varphi, R_\sigma(\alpha), U, \psi, S_\sigma(\alpha), T$. Observe that, over the universal domain, $T\colon U \to V$ defines a Galois covering, whose Galois group is isomorphic to $E \cdot \Gamma_W(\mathfrak{x}, \mathfrak{c})/E \cdot \Gamma_W(\mathfrak{x}, \mathfrak{e})$, where E is the group of all units in F. Therefore, one has $T(y_1) = T(y_2)$ with y_1 and y_2 on U if and only if $y_1 = S_1(\gamma)(y_2)$ for some $\gamma \in \Gamma_W(\mathfrak{x}, \mathfrak{c})$, by virtue of (5.7.5).

11.2. PROPOSITION. *The notation being as above, let $y' \in U$ and $y = T(y')$. Then $k_\mathfrak{e}(y')$ depends only on y and is a finite Galois extension of $k_\mathfrak{c}(y)$. For every $\tau \in G(k_\mathfrak{e}(y')/k_\mathfrak{c}(y))$, there exists an element β of $D_W(\Delta, \mathfrak{e})$ such that $y'^\tau = S_\sigma(\beta)(y')$ and $\beta \equiv 1 \bmod^* (\mathfrak{x}; \mathfrak{c})$, where σ is the restriction of τ to $k_\mathfrak{e}$.*

Here it should be noted that y may or may not be algebraic over $k_\mathfrak{c}$.

PROOF. If $T(y'') = y$ with some $y'' \in U$, then $y'' = S_1(\gamma)(y')$ for some

$\gamma \in \Gamma_W(\mathfrak{x}, \mathfrak{c})$. Since $S_1(\gamma)$ is rational over $k_\mathfrak{e}$, we have $k_\mathfrak{e}(y'') = k_\mathfrak{e}(y')$, which proves the first assertion. Let τ be an automorphism of C over $k_\mathfrak{c}(y)$, and σ the restriction of τ to $k_\mathfrak{e}$. Take an *integral* ideal $\mathfrak{a}$ in F' prime to $N(\mathfrak{e})$ so that $\sigma = [k_\mathfrak{e}/F', \mathfrak{a}]$. Since σ is the identity on $k_\mathfrak{c}$, we have $\det \Theta'(\mathfrak{a}) = (v)$, $v \equiv 1 \bmod^* \mathfrak{c}$ with some $v \in W$. Since $\mathfrak{a}$ is integral, $v \in \mathfrak{r}_F$. By 3.11, there exists an element α of $\mathfrak{G}$ such that $\mathfrak{x}\alpha \subset \mathfrak{x}$, $\nu(\alpha) = v$, and $\alpha \equiv 1 \bmod^* (\mathfrak{x}; \mathfrak{c})$. Then $\alpha \in D_W(\Delta, \mathfrak{e})$, and $S_\sigma(\alpha)$ is meaningful. We have $T = T^\sigma \circ S_\sigma(\alpha)$ by (11.1.1). Now $y'^\tau \in U^\sigma$. Put $y'' = S_\sigma(\alpha)^{-1}(y'^\tau)$. Then $T(y'') = T^\sigma(y'^\tau) = T(y')^\tau = y^\tau = y$, hence $y'' = S_1(\gamma)(y')$ with some $\gamma \in \Gamma_W(\mathfrak{x}, \mathfrak{c})$. Therefore $y'^\tau = S_\sigma(\alpha\gamma)(y')$. Since $S_\sigma(\alpha\gamma)$ is rational over $k_\mathfrak{e}$, y'^τ is rational over $k_\mathfrak{e}(y)$. This proves the second assertion. Putting $\beta = \alpha\gamma$, we obtain the last assertion.

11.3. PROPOSITION. *Let y and y' be as in 11.2, and z a point on $\mathfrak{H}_n^r$ such that $\psi(z) = y'$. Put*

$$\Gamma^z = \{\gamma \in \Gamma_W(\mathfrak{x}, \mathfrak{c}) \mid \gamma(z) = z\},$$
$$\mathfrak{S} = \{\gamma \in \Gamma_W(\mathfrak{x}, \mathfrak{c}) \mid S_1(\gamma)(y') = y'\},$$
$$\mathfrak{R} = \{\alpha \in D_W(\Delta, \mathfrak{e}) \mid \alpha \equiv 1 \bmod^* (\mathfrak{x}; \mathfrak{c})\},$$
$$\mathfrak{R}_1 = \{\alpha \in D_W(\Delta, \mathfrak{e}) \mid \alpha \equiv 1 \bmod^* (\mathfrak{x}; \mathfrak{e})\},$$

$$E = the\ group\ of\ all\ units\ of\ F,$$

$$E_W(\mathfrak{c}) = \{e \in E \mid e^2 \in W,\ e \equiv 1 \bmod^* \mathfrak{c}\}.$$

Then $E_W(\mathfrak{c}) = \Gamma^z \cap E$ and $\mathfrak{S} = \Gamma^z \cdot \Gamma_W(\mathfrak{x}, \mathfrak{e})$. The correspondence

$$G\big(k_\mathfrak{e}(y')/k_\mathfrak{c}(y)\big) \ni \tau \mapsto \beta \in \mathfrak{R}$$

determined by 11.2 gives an isomorphism of $G(k_\mathfrak{e}(y')/k_\mathfrak{c}(y))$ into $\mathfrak{R}/(\mathfrak{R}_1 \cdot \Gamma^z)$. Moreover, if y is generic on V over $k_\mathfrak{c}$, then this isomorphism is surjective, and $\Gamma^z = E_W(\mathfrak{c})$.

PROOF. The first equality is obvious. The second one follows from (5.7.5). Let τ, σ, and β be as in 11.2. Then $\sigma = [k_\mathfrak{e}/F', \mathfrak{a}]$ and $\det \Theta'(\mathfrak{a}) = (\nu(\beta))$ with an ideal $\mathfrak{a}$ in F' prime to $N(\mathfrak{e})$. Suppose that τ is the identity on $k_\mathfrak{e}(y')$. Then $\det \Theta'(\mathfrak{a}) = (t)$ and $t \equiv 1 \bmod^* \mathfrak{e}$ with an element t of W. Put $u = t^{-1}\nu(\beta)$. Then u is a unit of F, $u \in W$, and $u \equiv \nu(\beta) \bmod^* \mathfrak{e}$. By 3.11, there exists an element γ of $\mathfrak{G}$ such that $\mathfrak{x}\gamma = \mathfrak{x}$, $\nu(\gamma) = u$, and $\gamma \equiv \beta \bmod^* (\mathfrak{x}; \mathfrak{e})$. Then $\gamma \in \Gamma_W(\mathfrak{x}, \mathfrak{c})$ and $S_1(\gamma) = S_1(\beta)$, so that $y' = y'^\tau = S_1(\gamma)(y')$. Therefore we have $\gamma \in \mathfrak{S}$, hence $\beta \in \mathfrak{R}_1\mathfrak{S} = \mathfrak{R}_1\Gamma^z$. Suppose that y is generic on V over $k_\mathfrak{c}$. Then obviously $\Gamma^z = E_W(\mathfrak{c})$. Take any $\alpha \in \mathfrak{R}$, and put $y'' = S_\sigma(\alpha)(y')$. Observe that $\sigma \in G(k_\mathfrak{e}/k_\mathfrak{c})$. Since y'' is generic on U^σ over $k_\mathfrak{e}$, we can find an isomorphism τ of $k_\mathfrak{e}(y')$ onto $k_\mathfrak{e}(y'')$ such that $y'^\tau = y''$, and $\tau = \sigma$ on $k_\mathfrak{e}$. Since $S_\sigma(\alpha)$ is birational, we have $k_\mathfrak{e}(y') = k_\mathfrak{e}(y'^\tau)$, and

$$y^\tau = T^\sigma(y'^\tau) = T^\sigma \circ S_\sigma(\alpha)(y') = R_\sigma(\alpha) \circ T(y') = y .$$

Thus $\tau \in G(k_\mathfrak{c}(y')/k_\mathfrak{c}(y))$. This proves the desired surjectivity.

11.4. Fix an integral ideal $\mathfrak{c}$ in F, and take a prime ideal $\mathfrak{l}$ in F prime to $\mathfrak{c}$. Put $k_m = C_W(F', \Theta', \mathfrak{cl}^m)$ for $m = 0, 1, 2, \cdots$. Taking $\mathfrak{cl}^m$ to be $\mathfrak{e}$ in the above discussion, denote $U, \psi, S_\sigma(\alpha)$ by $V^m, \varphi^m, R_\sigma^m(\alpha)$ respectively. We understand $V^0 = V$, $\varphi^0 = \varphi$, $R_\sigma^0(\alpha) = R_\sigma(\alpha)$. We obtain a morphism T^m of V^m onto V^{m-1} rational over k_m such that

$$(11.4.1) \qquad (T^m)^\sigma \circ R_\sigma^m(\alpha) = R_\sigma^{m-1}(\alpha) \circ T^m \qquad (\alpha \in D_W(\Delta; \mathfrak{cl})) .$$

Let $y = y_0 \in V = V^0$. Take a sequence of points $y_0, y_1, y_2, \cdots, y_m, \cdots$ such that $y_m \in V^m, y_{m-1} = T^m(y_m)$. (For example, choose a point z of $\mathfrak{H}_n^r$ so that $\varphi(z) = y$, and put $y_m = \varphi^m(z)$.) Further put

$$(11.4.2) \qquad \begin{aligned} \mathfrak{K} &= \mathfrak{K}(\mathfrak{l}, y) = \textstyle\bigcup_{m=0}^\infty k_m(y_m) , \\ k_\mathfrak{l}(\mathfrak{c}) &= \textstyle\bigcup_{m=0}^\infty k_m , \qquad k = k_0 . \end{aligned}$$

In view of 11.2, $\mathfrak{K}$ is a normal extension of $k(y)$; it depends only on y and is independent of the choice of $y_1, y_2, \cdots$.

We are going to investigate the Galois group of $\mathfrak{K}$ over $k(y)$, and the behavior of the Frobenius automorphisms of $\mathfrak{K}$. First let $\mathfrak{G}_\mathfrak{l}$ be as in 3.1, and $\mathfrak{G}_\mathfrak{l}(\mathfrak{x})$ the set of all $\alpha \in \mathfrak{G}_\mathfrak{l}$ such that $\mathfrak{x}_\mathfrak{l}\alpha = \mathfrak{x}_\mathfrak{l}$. For simplicity, we assume

$$(11.4.3) \qquad \{\gamma \in \Gamma_W(\mathfrak{x}, \mathfrak{c}) \mid \gamma(z) = z\} = E_W(\mathfrak{c}) ,$$

where $E_W(\mathfrak{c})$ is as in 11.3, and z is a point of $\mathfrak{H}_n^r$ such that $\varphi(z) = y$. This condition does not depend on the choice of z. We shall denote by $\mathfrak{E}_\mathfrak{l}'(\mathfrak{c})$ the closure of $E_W(\mathfrak{c})$ in $F_\mathfrak{l}$.

Now, for every $\tau \in G(\mathfrak{K}/k(y))$, we can find a sequence of elements α_m of $D_W(\Delta; \mathfrak{cl})$ so that

$$(11.4.4) \qquad \begin{aligned} (y_m)^\tau &= R_\tau^m(\alpha_m)(y_m) , \\ \alpha_{m+1} &\equiv \alpha_m \bmod{}^* (\mathfrak{x}; \mathfrak{l}^m), \ \alpha_m \equiv 1 \bmod{}^* (\mathfrak{x}; \mathfrak{c}) . \end{aligned}$$

In fact, suppose that $\alpha_0, \alpha_1, \cdots, \alpha_m$ have been obtained. Take an element β of $D_W(\Delta; \mathfrak{cl})$ so that $y_{m+1}^\tau = R_\tau^{m+1}(\beta)(y_{m+1})$. By (11.4.1), we have $R_\tau^m(\beta)(y_m) = y_m^\tau = R_\tau^m(\alpha_m)(y_m)$, hence $\beta^{-1}\alpha_m \equiv \varepsilon \bmod{}^* (\mathfrak{x}; \mathfrak{cl}^m)$ with an element ε of $E_W(\mathfrak{c})$ by 11.3 and (11.4.3). Taking $\beta\varepsilon$ as α_{m+1}, we obtain (11.4.4).

Such a sequence $\{\alpha_m\}$ satisfying (11.4.4) converges to an element α of $\mathfrak{G}_\mathfrak{l}(\mathfrak{x})$. We shall now show that α is determined modulo $\mathfrak{E}_\mathfrak{l}'(\mathfrak{c})$ uniquely by τ. In fact, if $\{\alpha_m'\}$ is another sequence satisfying (11.4.4), we see that $\alpha_m' \equiv \alpha_m\varepsilon_m \bmod{}^* (\mathfrak{x}; \mathfrak{cl}^m)$ with an element ε_m of $E_W(\mathfrak{c})$ by 11.3 and (11.4.3). If $\alpha' = \lim \alpha_m'$, we have $\lim \varepsilon_m = \alpha^{-1}\alpha' \in \mathfrak{E}_\mathfrak{l}'(\mathfrak{c})$. Thus from the correspondence $\tau \mapsto \alpha$, we obtain a homomorphism $J_\mathfrak{l}$ of $G(\mathfrak{K}/k(y))$ into $\mathfrak{G}_\mathfrak{l}(\mathfrak{x})/\mathfrak{E}'_\mathfrak{l}(\mathfrak{c})$.

Similarly, for every $\sigma \in G(k_\mathfrak{l}(\mathfrak{c})/k)$, we can find a sequence of elements b_m of W and ideals $\mathfrak{a}_m$ in F', prime to $N(\mathfrak{cl})$ such that

$$\sigma = [k_m/F', \mathfrak{a}_m], \qquad \det \Theta'(\mathfrak{a}_m) = (b_m), \qquad\qquad b_m \in W,$$
$$b_{m+1} \equiv b_m \mod^* \mathfrak{l}^m, \qquad b_m \equiv 1 \mod^* \mathfrak{c}.$$

Let $\mathfrak{r}_\mathfrak{l}^*$ denote the group of units of $\mathfrak{r}_\mathfrak{l}$, and $\mathfrak{E}_\mathfrak{l}(\mathfrak{c})$ the closure of $\{e \in E \cap W \mid e \equiv 1 \bmod \mathfrak{c}\}$ in $F_\mathfrak{l}$. Then $\{b_m\}$ converges to an element b of $\mathfrak{r}_\mathfrak{l}^*$. From the correspondence $\sigma \mapsto b$, we obtain an *isomorphism* of $G(k_\mathfrak{l}(\mathfrak{c})/k)$ into $\mathfrak{r}_\mathfrak{l}^*/\mathfrak{E}_\mathfrak{l}(\mathfrak{c})$. Moreover, if σ is the restriction of τ to $k_\mathfrak{l}(\mathfrak{c})$, we can take $\nu(\alpha_m)$ as b_m, hence $\nu(\alpha)$ is an element of $\mathfrak{r}_\mathfrak{l}^*$ corresponding to σ.

Now we shall show that $J_\mathfrak{l}$ *is injective*. Suppose that $J_\mathfrak{l}(\tau) = \alpha = \lim \alpha_m \in \mathfrak{E}'_\mathfrak{l}(\mathfrak{c})$. Then $\nu(\alpha) \in \mathfrak{E}_\mathfrak{l}(\mathfrak{c})$, hence the restriction of τ to $k_\mathfrak{l}(\mathfrak{c})$ is the identity map. We can find a sequence $\{\xi_m\}$ of elements of $E_W(\mathfrak{c})$ so that $\xi_{m+1} \equiv \xi_m \bmod \mathfrak{cl}^m$ and $\alpha^{-1} = \lim \xi_m$. For $\mu \geq m$, we have

$$y_m^\tau = R_\mathfrak{l}^m(\alpha_\mu)(y_m) = R_\mathfrak{l}^m(\alpha_\mu \xi_\mu)(y_m).$$

Since $\lim_{\mu \to \infty} \alpha_\mu \xi_\mu = 1$, we have $\alpha_\mu \xi_\mu \equiv 1 \mod^* (\mathfrak{x}; \mathfrak{cl}^m)$ for sufficiently large μ, hence $y_m^\tau = y_m$. This proves the injectivity of $J_\mathfrak{l}$.

The map $J_\mathfrak{l}$ depends on the choice of $\{y_m\}$. Take another sequence $\{u_m\}$ such that $u_m \in V^m$, $T^m(u_m) = u_{m-1}$, $u_0 = y$. By 11.2, we have $\mathfrak{K}(\mathfrak{l}, y) = \bigcup_{m=0}^\infty k_m(u_m)$. As remarked in 11.1, we find elements γ_m of $\Gamma_W(\mathfrak{x}, \mathfrak{c})$ so that $u_m = R_\mathfrak{l}^m(\gamma_m)(y_m)$. We can take the γ_m so that $\gamma_{m+1} \equiv \gamma_m \mod^* (\mathfrak{x}; \mathfrak{l}^m)$. Let $\gamma = \lim \gamma_m \in \mathfrak{G}_\mathfrak{l}(\mathfrak{x})$. If $\{\alpha_m\}$ is as in (11.4.4), we have

$$u_m^\tau = R_\mathfrak{l}^m(\gamma_m)^\tau \circ R_\tau^m(\alpha_m) \circ R_\mathfrak{l}^m(\gamma_m^{-1})(u_m) = R_\mathfrak{l}^m(\gamma_m \alpha_m \gamma_m^{-1})(u_m).$$

This shows that, if we define an injection $J'_\mathfrak{l}$ of $\mathfrak{G}(\mathfrak{K}/k(y))$ into $\mathfrak{G}_\mathfrak{l}(\mathfrak{x})/\mathfrak{E}'_\mathfrak{l}(\mathfrak{c})$ with respect to $\{u_m\}$, then $J'_\mathfrak{l}(\tau) = \gamma \cdot J_\mathfrak{l}(\tau) \cdot \gamma^{-1}$. Thus $J_\mathfrak{l}$ is *unique up to an inner automorphism*.

11.5. PROPOSITION. *Suppose that y is algebraic over k. For a given $\mathfrak{l}$, there are only a finite number of prime ideals in $k(y)$ which are ramified in $\mathfrak{K}(\mathfrak{l}, y)$.*

Proof will be given in 11.11.

11.6. From now on we assume that y is algebraic over k. Let $\mathfrak{p}$ be a prime ideal in $k(y)$ unramified in $\mathfrak{K} = \mathfrak{K}(\mathfrak{l}, y)$, $\mathfrak{P}$ a prime divisor of $\mathfrak{K}$ which divides $\mathfrak{p}$, and $\sigma_\mathfrak{P}$ the Frobenius automorphism of $\mathfrak{K}$ over $k(y)$ for $\mathfrak{P}$. Then $J_\mathfrak{l}(\sigma_\mathfrak{P})$ is an element of $\mathfrak{G}_\mathfrak{l}(\mathfrak{x})/\mathfrak{E}'_\mathfrak{l}(\mathfrak{c})$, whose conjugacy class is uniquely determined only by $\mathfrak{p}$. It is now important to investigate the $J_\mathfrak{l}(\sigma_\mathfrak{P})$ or their conjugacy classes for all possible choices of $\mathfrak{l}$ and $\mathfrak{p}$. Let us assume that $\mathfrak{l}$ does not divide $D(B/F)$. Then $B_{\mathfrak{l}}$ can be identified with $M_2(F_{\mathfrak{l}})$, hence $\mathfrak{G}_{\mathfrak{l}}$ may be considered as a subgroup of $GL_{2n}(F_{\mathfrak{l}})$. Therefore, it is meaningful to speak of the quotients of the characteristic roots of $J_{\mathfrak{l}}(\sigma_{\mathfrak{P}})$, which are certainly determined only by $k(y)$, $\mathfrak{l}$ and $\mathfrak{p}$.

11.7. PROPOSITION. *The notation being as above, let ζ be the quotient of any two characteristic roots of (a matrix representing) $J_{\mathfrak{l}}(\sigma_{\mathfrak{P}})$. Then, for almost all $\mathfrak{p}$, ζ is algebraic over F, and every conjugate of ζ over $\mathbf{Q}$ has absolute value one.*

11.8. To prove 11.7, we consider an $\mathfrak{l}$-*adic coordinate system* on an abelian variety. Let A be an abelian variety, F an algebraic number field of degree g, and θ an isomorphism of F into $\mathrm{End}_{\mathbf{Q}}(A)$. Suppose that $\theta(\mathfrak{r}_F) \subset \mathrm{End}(A)$, and $\theta(1)$ is the identity map of A. For every prime ideal $\mathfrak{l}$ in F, put

$$\mathfrak{g}_{\mathfrak{l}} = \cup_{m=1}^{\infty} \{t \in A \mid \theta(\mathfrak{l}^m)t = 0\} .$$

Suppose that $\mathfrak{l}$ is prime to the characteristic of a field of definition for A. Then one can easily show that $\mathfrak{g}_{\mathfrak{l}}$ is isomorphic to the product of s copies of $F_{\mathfrak{l}}/\mathfrak{r}_{\mathfrak{l}}$, where $s = 2 \cdot \dim(A)/g$. Let Y be the commutor of $\theta(F)$ in $\mathrm{End}_{\mathbf{Q}}(A)$. Then we obtain naturally an F-linear representation

$$R_{\mathfrak{l}} : Y \longrightarrow M_s(F_{\mathfrak{l}}) .$$

11.9. PROPOSITION. *For every $x \in Y$, the characteristic polynomial f_x of $R_{\mathfrak{l}}(x)$ has coefficients in F, and is independent of $\mathfrak{l}$. Moreover, $N_{F/\mathbf{Q}}(f_x)$ (understood in an obvious way) is exactly the characteristic polynomial of x in the usual sense.*

PROOF. Identify F with $\theta(F)$. To prove our proposition, we may assume that $F[x]$ is semi-simple. Let l be the rational prime divisible by $\mathfrak{l}$. Put $F_l = F \otimes_{\mathbf{Q}} \mathbf{Q}_l$, $T = F[x]$, $T_l = T \otimes_{\mathbf{Q}} \mathbf{Q}_l$, $T_{\mathfrak{l}} = T \otimes_F F_{\mathfrak{l}}$. Then $F_{\mathfrak{l}}$ may be identified with a simple component of F_l. Let ε be the identity element of this simple component, considered as an element of F_l. Then $T_{\mathfrak{l}}$ may be identified with εT_l, and $F_{\mathfrak{l}}^s$ with εF_l^s. Let

$$\mathfrak{g}_l = \cup_{m=1}^{\infty} \{t \in A \mid l^m t = 0\} .$$

Then we can find a commutative diagram

$$\begin{array}{ccc}
F_{\mathfrak{l}}^s & \longrightarrow & F_l^s \cong \mathbf{Q}_l^{sg} \\
\downarrow & & \downarrow \\
\mathfrak{g}_{\mathfrak{l}} & \longrightarrow & \mathfrak{g}_l
\end{array}$$

where the horizontal arrows are the natural injections. Let R_l be the representation $Y \to M_{sg}(\mathbf{Q}_l)$ obtained from the action of Y on $\mathfrak{g}_l$. It is a classical fact that, for every $b \in Y$, the characteristic polynomial of $R_l(b)$ has rational coefficients, and is independent of l. Since T is a commutative semi-simple algebra, we can find a T-module U, independent of l, so that F_l^* is T_l-isomorphic to $U \otimes_{\mathbf{Q}} \mathbf{Q}_l$. Then F_l^* is T_l-isomorphic to $\varepsilon(U \otimes_{\mathbf{Q}} \mathbf{Q}_l)$. Obviously $\varepsilon(U \otimes_{\mathbf{Q}} \mathbf{Q}_l)$ is T_l-isomorphic to $U \otimes_F F_l$. Therefore the restriction of R_l (resp. R_l^*) to T is equivalent with the representation of T on U over $\mathbf{Q}$ (resp. F). Our assertions follow easily from this fact.

11.10. Suppose that A and the elements of $\theta(\mathfrak{r}_F)$ are defined over an algebraic number field k. Let $k_{(l)}$ be the field generated over k by the coordinates of the points t in $\mathfrak{g}_l$. For every prime ideal $\mathfrak{p}$ in k, take a prime divisor $\mathfrak{P}$ of $k_{(l)}$ which divides $\mathfrak{p}$ and a Frobenius automorphism σ of $k_{(l)}$ over k with respect to $\mathfrak{P}$ and $\mathfrak{p}$. Then σ gives an automorphism of $\mathfrak{g}_l$ commuting with the elements of $\theta(\mathfrak{r}_F)$, hence σ can be represented by an element $R_l(\sigma)$ of $M_s(\mathfrak{r}_l)$. Now we have

(11.10.1) *Suppose that A has a non-degenerate reduction modulo $\mathfrak{p}$, and $\mathfrak{p}$ is prime to $N(l)$. Then $\mathfrak{p}$ is unramified in $k_{(l)}$; the characteristic polynomial of $R_l(\sigma)$ has coefficients in $\mathfrak{r}_F$, and is independent of l; moreover, every conjugate of a characteristic root of $R_l(\sigma)$ over $\mathbf{Q}$ has absolute value $N(\mathfrak{p})^{1/2}$.*

This can be obtained by applying 11.9 to the $N(\mathfrak{p})^{\mathrm{th}}$ power endomorphism on the reduction of A modulo $\mathfrak{p}$.

11.11. To prove 11.5 and 11.7, we consider only the case $W = W_1$. The general case can easily be reduced to this special case. Let K, L, Φ, ρ, T be as in §7. Let i, i', and j be as in 6.6. For simplicity, we identify α with $i(\alpha)$ for every $\alpha \in \mathfrak{G}$, and u with $i'(u)$ for every $u \in X$. For every integer $m \geqq 0$, define a weak PEL-type

$$\Omega^m = (L, \Phi, \rho; F^+ \cdot T, \mathfrak{M}; x_1^m, \cdots, x_{4n}^m) ,$$

with $\mathfrak{M} = \mathfrak{r}_K \cdot \mathfrak{x} (\subset L^n)$ and elements $x_1^m, \cdots, x_{4n}^m$ of L^n such that

$$(\mathfrak{cl}^m)^{-1}\mathfrak{M} = \mathfrak{M} + \sum_{\nu=1}^{4n} \mathfrak{r}_K x_\nu^m .$$

(There are many choices for the x_ν^m. We fix any choice.) We construct a family $\Sigma^m = \{\mathfrak{R}_w^m \mid w \in \mathfrak{S}^r\}$ of weak PEL-structures

$$\mathfrak{R}_w^m = \left(A_w, \mathfrak{D}_w, \theta_w; \{t_{w\nu}^m\}\right)$$

by means of the same parametrizing function $\mathfrak{y}(x, z)$. Observe that $(A_w, \mathfrak{D}_w, \theta_w)$ does not depend on m, and $t_{w\nu}^m$ corresponds to $\mathfrak{y}(x_\nu^m, w)$.

Let $(U^m, \mathfrak{v}^m, \psi^m)$ be a moduli-system for the family Σ^m. The discussion of §§7, 8 shows that there exists a morphism Z^m of V^m into U^m, defined over an algebraic number field, such that $Z^m(\varphi^m(z)) = \psi^m(j(z))$ for every $z \in \mathfrak{H}_n^r$. In fact, if m is sufficiently large, the condition (7.6.2) is satisfied with $\mathfrak{cl}^m$ in place of c. Hence Z^m is a biregular embedding of V^m into U^m defined over $C(K', N(\mathrm{cl}^m))$. For smaller m, Z^m may not be biregular, but at least its existence can be derived from the result for larger m.

Let y be a point of V algebraic over k. Take $z \in \mathfrak{H}_n^r$ so that $\varphi(z) = y$, and put $w = j(z)$. Since $\psi^0(w) = Z^0(y)$ is algebraic over k, we may assume that $\mathfrak{R}_w^0$ is defined over a finite algebraic extension k' of $k(y)$. We assume $C(K', N(\mathfrak{c})) \subset k'$ and take $\varphi^m(z)$ as y_m. Let $\mathfrak{f}_m$ be the field of moduli of $\mathfrak{R}_w^m$. We see that $k_m(y_m)$ is contained in the composite of $\mathfrak{f}_m$ and $k' \cdot C(K', N(\mathrm{cl}^m))$. Further, $\mathfrak{f}_m$ is contained in the field generated over $\mathfrak{f}_0$ by the coordinates of the points t on A_w such that $\theta_w(\mathrm{cl}^m)t = 0$. From this fact we obtain easily 11.5.

Let $\tau \in G(\mathfrak{K}/k(y))$. Take $\{\alpha_m\}$ as in 11.4. Suppose that τ can be extended to an automorphism σ of C over k'. Let $\mathfrak{g}_\mathfrak{l}$ be defined as in 11.8 with A_w as A. Then $\mathfrak{g}_\mathfrak{l}$ is isomorphic to $L_\mathfrak{l}^n/\mathfrak{M}_\mathfrak{l}$. In fact, if we put $L' = \bigcup_{m=1}^\infty \{x \in L^n \mid \mathfrak{l}^m x \in \mathfrak{M}\}$, we have $L_\mathfrak{l}^n = L' + \mathfrak{M}_\mathfrak{l}$, and $\mathfrak{M} = L' \cap \mathfrak{M}_\mathfrak{l}$, hence $L' \ni x \mapsto \mathfrak{y}(x, w)$ gives a homomorphism of $L_\mathfrak{l}^n$ onto $\mathfrak{g}_\mathfrak{l}$ with kernel $\mathfrak{M}_\mathfrak{l}$. Call this homomorphism h. Then we can find an element β of $\mathrm{GL}_n(L_\mathfrak{l})$ such that

$$(11.11.1) \qquad \mathfrak{M}_\mathfrak{l}\beta = \mathfrak{M}_\mathfrak{l}, \quad h(x)^\sigma = h(x\beta) \qquad (x \in L_\mathfrak{l}^n) \,.$$

We shall now prove

(11.11.2) *There exists an element ε of $K_\mathfrak{l}$ such that $\varepsilon^{-1}\beta \in \mathfrak{G}_\mathfrak{l}(\mathfrak{x})$, and $J_\mathfrak{l}(\tau)$ is represented by $\varepsilon^{-1}\beta$.*

Take an ideal $\mathfrak{t}_m$ in K' prime to $N(\mathrm{cl})$ so that

$$\sigma = [C(K', N(\mathrm{cl}^m))/K', \mathfrak{t}_m] \qquad \text{on } C(K', N(\mathrm{cl}^m)) \,,$$

and put $\mathfrak{z}_m = \det \Phi_0'(\mathfrak{t}_m)$, where Φ_0' is as in 6.10 and 7.8. Taking α_m as α in 7.8, we see that $(\Omega^m)^\sigma$ is equivalent to

$$\Omega' = \left(L, \Phi, \rho; F^+ \cdot T, u\mathfrak{z}_m^{-1}\mathfrak{M}; \{v_\nu\}\right)$$

with an element u of K and elements v_ν of L^n such that $u \equiv 1 \bmod^* N(\mathrm{cl}^m)$ and $v_\nu - x_\nu^m \alpha_m \in M_\mathfrak{p}$ for all $\mathfrak{p}$ dividing cl. Since σ is the identity map on $C(K', N(\mathfrak{c}))$, we have $\mathfrak{t}_m = (t_m)$ with an element t_m of K' such that $t_m \equiv 1 \bmod^* (N(\mathfrak{c}))$. Put $b_m = u^{-1} \det \Phi_0'(\mathfrak{t}_m)$. Then $b_m \equiv 1 \bmod^* (N(\mathfrak{c}))$, and hence Ω' is equivalent to

$$\Omega'' = \left(L, \Phi, \rho; F^+ \cdot T, \mathfrak{M}; \{v_\nu'\}\right) \,,$$

with elements v_ν' of $(\mathrm{cl}^m)^{-1}\mathfrak{M}$ such that $v_\nu' \equiv b_m v_\nu \bmod \mathfrak{M}_\mathfrak{p}$ for all $\mathfrak{p}$ dividing cl. Construct families $\Sigma(\Omega')$ and $\Sigma(\Omega'')$ of weak PEL-structures of type Ω' and Ω'' respectively. By [C, 4.24–25], we can write moduli-systems for $\Sigma(\Omega')$ and $\Sigma(\Omega'')$ in the form $(U^m, \mathfrak{v}', \psi^m)$ and $(U^m, \mathfrak{v}'', \psi^m)$ with the same U^m and ψ^m as before. As is seen in 7.10, there exists a biregular isomorphism $T(\alpha_m)$ of U^m

to $(U^m)^\sigma$, rational over $C(K', N(\mathrm{cl}^m))$, such that $T(\alpha_m) \circ \mathfrak{v}' = \mathfrak{v}^{m\sigma}$. This $T(\alpha_m)$ satisfies $T(\alpha_m) \circ Z^m = (Z^m)^\sigma \circ R^m_\sigma(\alpha_m)$ for sufficiently large m.

Let $s_m = \psi^m(w)$. Then $Z^m(y_m) = s_m = \mathfrak{v}^m(\mathcal{R}^m_w) = \mathfrak{v}'(\mathcal{R}'_w)$, where $\mathcal{R}'_w$ is *the* member of $\Sigma(\Omega')$ corresponding to w. We have now

$$\mathfrak{v}^{m\sigma}((\mathcal{R}^m_w)^\sigma) = \mathfrak{v}^m(\mathcal{R}^m_w)^\sigma = s^\sigma_m = (Z^m)^\sigma(y^\sigma_m) = (Z^m)^\sigma \circ R^m_\sigma(\alpha_m)(y_m)$$
$$= T(\alpha_m) \circ Z^m(y_m) = T(\alpha_m)(s_m) = T(\alpha_m)(\mathfrak{v}'(\mathcal{R}'_w)) = \mathfrak{v}^{m\sigma}(\mathcal{R}'_w).$$

It follows that $(\mathcal{R}^m_w)^\sigma$ is isomorphic to $\mathcal{R}'_w$. Since Ω' is equivalent to Ω'', $\mathcal{R}'_w$ is isomorphic to

$$\left(A_w, \mathfrak{D}_w, \theta_w; \{t''_{wv}\}\right)$$

with the elements $t''_{wv} = \mathfrak{y}(v'_v, w)$. Let λ_m be an isomorphism of $\mathcal{R}'_w$ to $(\mathcal{R}^m_w)^\sigma$. Since σ is the identity on k', λ_m is an automorphism of $\mathcal{R}^0_w$; and $t^\sigma_{wv} = \lambda_m t''_{wv}$. By our assumption (11.4.3), $\lambda_m = \theta_w(e_m)$ with an element e_m of $E_W(\mathfrak{c})$. If β is as in (11.11.1), we have $h(x^m_v\beta) = t^\sigma_{wv} = \theta_w(e_m)t''_{wv}$ hence $x^m_v\beta \equiv e_m v'_v \equiv e_m b_m x^m_v \alpha_m \bmod \mathfrak{M}_\mathrm{I}$. Since $\{\alpha_m\}$ converges to an element α of $\mathfrak{G}_\mathrm{I}(\mathfrak{x})$, $\{e_m b_m\}$ converges to an element ε of K_I such that $\beta = \varepsilon\alpha$. This proves (11.11.2).

11.12. Coming back to $\mathfrak{p}$, $\mathfrak{P}$ and $\sigma_\mathfrak{P}$ of 11.6, let $\mathfrak{K}'$ be the field generated over $k'\mathfrak{K}$ by the coordinates of the points in $\mathfrak{g}_\mathrm{I}$. Take a prime divisor $\mathfrak{Q}$ in $\mathfrak{K}'$ which divides $\mathfrak{P}$. Let $\sigma_\mathfrak{Q}$ be the Frobenius automorphism of $\mathfrak{K}'$ over k' for $\mathfrak{Q}$. If $\mathfrak{q}$ is the prime ideal in k' divisible by $\mathfrak{Q}$, and $N_{k'/k(\mathfrak{y})}(\mathfrak{q}) = \mathfrak{p}'$, then $\sigma_\mathfrak{P}^f = \sigma_\mathfrak{Q}$ on $\mathfrak{K}$. Now apply (11.11.2) to $\sigma_\mathfrak{Q}$. Observe that β is essentially the same as $R_\mathrm{I}(\sigma_\mathfrak{Q})$ of (11.10.1). Therefore we obtain 11.7.

11.13. If $y = \varphi(z)$ with an isolated fixed point z on $\mathfrak{H}^r_n$ of $\mathfrak{G}^+$, one has further information on $J_\mathrm{I}(\sigma_\mathfrak{P})$. Let (Y, P, δ, f), (P', Ψ'), z, $\mathfrak{g}_r$ be as in (5.3.3). We assume that the following two conditions are satisfied.

(11.13.1) $Y = P$, *and* P *is a field.*

(11.13.2) *Every unit* u *of* P *such that* $uu^\delta \in W$ *and* $u \equiv 1 \bmod \mathfrak{cr}_P$ *is real.* The first condition implies $[P : F] = 2n$. Now write $\mathfrak{g}$ for $\mathfrak{g}_r$; let $\mathfrak{h}$ be the conductor of $\mathfrak{g}$, and $\mathfrak{h}_0 = \mathfrak{h} \cap F$. Let $C = C_W(P', \Psi', \mathfrak{g}, \mathfrak{c})$. We can define a Grössen-character χ of C as follows. Let $\mathfrak{a}$ be an ideal in C prime to $N(\mathfrak{ch}_0)$. By class field theory, we have $N_{C/P'}(\mathfrak{a}) \in I_W(P', \Psi', \mathfrak{g}, \mathfrak{c})$, so that $\Psi'(N_{C/P'}(\mathfrak{a})) = (v)$ with an element v of P such that $vv^\delta \in W$ and $v \equiv 1 \bmod^* (\mathfrak{g}; \mathfrak{c})$ (see (5.2.4)). Put $\chi(\mathfrak{a}) = v^\delta/v$. By (11.13.2), $\chi(\mathfrak{a})$ does not depend on the choice of v. It is easily seen that χ is actually a Grössen-character. If every unit u of P such that $uu^\delta \in W$ and $u \equiv 1 \bmod \mathfrak{cr}_P$ is totally positive, we can define another Grössen-character χ_0 by putting $\chi_0(\mathfrak{a}) = |v|/v$. Then $\chi = \chi^2_0$.

Let us take $\varphi(z)$ to be the point y, and $\varphi^m(z) = y_m$. By (5.3.3), $\mathfrak{K}P'$ is an abelian extension of P', and the behavior of $G(\mathfrak{K}P'/P')$ can be described by means of (5.7.6). But this does not settle our question completely, since P' is not necessarily contained in $k(y)$. For example, consider the case where $B = M_2(\mathbf{Q})$, $n = 1$, $\mathfrak{c} = (1)$, φ is the classical modular function j, and P is an imaginary quadratic field. Then $P' = P$, and P is not contained in $\mathbf{Q}(j)$. However, one can still investigate $J_{\mathfrak{l}}(\sigma_{\mathfrak{P}})$ by means of (5.7.6). Let $\mathfrak{p}$, $\mathfrak{P}$, and $\sigma_{\mathfrak{P}}$ be as in 11.6. Take a prime divisor $\mathfrak{Q}$ of $P'\mathfrak{R}$ which divides $\mathfrak{P}$, and let $\mathfrak{q}$ be the prime ideal of C divisible by $\mathfrak{Q}$, and τ the Frobenius automorphism of $P'\mathfrak{R}$ over C for $\mathfrak{Q}$. Note that $C = k(y)P'$. If $N_{C/k(y)}(\mathfrak{q}) = \mathfrak{p}^\mu$, we have $\sigma_{\mathfrak{P}}^\mu = \tau$ on $\mathfrak{R}$. Observe that

$$\tau = [C(P', \Psi', \mathfrak{g}, \mathfrak{cl}^m)/P', N_{C/P'}(\mathfrak{q})] \qquad \text{on } C(P', \Psi', \mathfrak{g}, \mathfrak{cl}^m) .$$

We have $\Psi'(N_{C/P'}(\mathfrak{q})) = (v)$ with an element v of P such that $vv^\delta \in W$ and $v \equiv 1 \bmod^* (\mathfrak{g}; \mathfrak{c})$. By (5.7.6), we have $y_m^\tau = R_\sigma^m(f(v))(y_m)$. Hence $J_{\mathfrak{l}}(\sigma_{\mathfrak{P}}^\mu)$ *can be represented by $f(v)$.*

In particular, suppose that $n = 1$. Then $[P : F] = 2$, and $\mathfrak{G}_{\mathfrak{l}}(\mathfrak{x})$ may be embedded in $\mathrm{GL}_2(F_{\mathfrak{l}})$. The quotients of the characteristic roots of $J_{\mathfrak{l}}(\sigma_{\mathfrak{P}}^\mu)$ are v^δ/v and v/v^δ, i.e., $\chi(\mathfrak{q})$ and $\chi(\mathfrak{q})^{-1}$. Further if $r = 1$ and τ_{01} is the identity map, we have $P = P'$, hence $\mu = 1$ or 2. In the special case $B = M_2(\mathbf{Q})$, one can show that if $\mu = 2$, the quotients of the characteristic roots of $J_{\mathfrak{l}}(\sigma_{\mathfrak{P}})$ are -1. It is an open question whether or not this is true in the general case whenever $\mu = 2$.

11.14. With the characters χ and χ_0, we can associate L-functions of Hecke's type. Therefore we are tempted to ask whether there exist zeta functions analogous to such L-functions, associated to $\mathfrak{R}(\mathfrak{l}, y)$ and $J_{\mathfrak{l}}(\sigma_{\mathfrak{P}})$, in the general case where y is not necessarily obtained from an isolated fixed point. Before asking this, one has to prove that $J_{\mathfrak{l}}(\sigma_{\mathfrak{P}})$ has a certain invariant property independent of $\mathfrak{l}$, as suggested by 11.8 and (11.11.2). If the quotients $\zeta_{\mathfrak{p}}$ and $\zeta_{\mathfrak{p}}^{-1}$ of two characteristic roots of $J_{\mathfrak{l}}(\sigma_{\mathfrak{P}})$ (in the case $n = 1$) depend only on $\mathfrak{p}$, one can define an Euler product

$$\prod_{\mathfrak{p}} \left[(1 - \zeta_{\mathfrak{p}} N(\mathfrak{p})^{-s})(1 - \zeta_{\mathfrak{p}}^{-1} N(\mathfrak{p})^{-s}) \right]^{-1}$$

and ask whether this satisfies a functional equation, though there is a slight doubt that this may not be the most natural one. At any rate, it is quite certain that the investigation of the extension $\mathfrak{R}$ and the $J_{\mathfrak{l}}(\sigma_{\mathfrak{P}})$ will enrich our knowledge of non-abelian extensions.

If $r = g$, there is a standard family of abelian varieties associated with $\mathfrak{H}_n^r/\Gamma_W(\mathfrak{x}, \mathfrak{c})$, as discussed in §9. In this case, our consideration is essentially the same, up to some minor points, as the investigation of the points of finite order on an abelian variety and its zeta function. However, if $r < g$, our theory has no such interpretation. The extension $\mathfrak{R}$ seems to have a nature rather different from the field obtained from the points of finite order on an abelian variety. At the present moment, we have no way of knowing it, if y does not correspond to an isolated fixed point, except that a few weaker results can be derived from the elementary properties of abelian varieties, as we have actually done in 11.8-12 for the proof of 11.7.

Appendix

Corrections to [C]

p. 66: In Prop. 2.6, before "Conversely" insert "Moreover, every element of B^+ leaving this point fixed belongs to $f(M)$." The proof should be supplied accordingly. See 4.5, 4.6 and 4.7 of the present paper which include this as a special case.

p. 77, line 2: "$U(\mathfrak{e})$" shoul be "$U_0(\mathfrak{e})$".

p. 78, line 16: "V_μ to V_λ" should be "V_λ to V_μ".

p. 80, line 6: "$N(\mathfrak{x}_{\nu\mu})$" should be "$N_{B/F}(\mathfrak{x}_{\nu\mu})$".

p. 92, line 11 from bottom: "$\alpha \in \Gamma_0(T, \mathfrak{N}/\mathfrak{M})$" should be "$\alpha \in \Gamma_0(T, \mathfrak{M})$".

p. 98, line 3 from bottom: "$E \cdot T$" should be "$E \cdot T'$".

p. 99, line 7: "L" should be "F".

p. 100, line 6: "$\mathfrak{r}$" should be "L".

p. 102, line 2: "$K'' \subset K'$" should be "$K'' \subset K$".

p. 108, line 25: "A" should be "A_i".

p. 110, line 19: "$\mu_i^{-1}(X_i)$" should be "$\mu_i^{-1}(X_i^\sigma)$".

p. 110, lines 26, 27: "Applying $\cdots$ described in (5.24.2)" should be "Take μ_1 so as to satisfy (5.24.2) with this $\mathfrak{p}$ and $i = 1$. Take any isogeny ζ_i of (A_1, θ_1) to (A_i, θ_i). By [15, Ch. II, Prop. 11, Prop. 23], there exists an isogeny μ_i of A_i to A_i^σ such that $\mu_i\zeta_i = \zeta_i^\sigma\mu_1$. Since $\mathrm{End}_Q(A_i) = \theta_i(K)$, we see easily that each μ_i satisfies (5.24.2), and moreover $\mu_i\alpha = \alpha^\sigma\mu_j$ for every isogeny α of A_j to A_i".

p. 112, line 3 from bottom; p. 113, lines 13, 15; p. 116, lines 4, 6, 12, 26; p. 118, lines 6, 9: "B" should be "$\mathfrak{B}$".

p. 118, line 14: "$\varphi_\nu(x)$" should be "$\varphi_\nu(\iota(x))$".

p. 118, line 16: "$\mathfrak{H}$" should be "$\mathcal{K}$".

pp. 124-127: In the definition of Ω of (8.3.2), one should take *four* elements, instead of two elements x_1, x_2. This change has no effect on the validity of the discussion in 8.3-8.9. See 7.7 of the present paper.

p. 125: One should distinguish two weak PEL-types even if they are equivalent. Therefore the phrase "up to equivalence" in line 9 should be modified suitably. See 7.7 of the present paper.

p. 127, line 21: "$\Omega_{\mu i}^{\sigma} = \Omega_{\lambda j}$" should be "$\Omega_{\mu i}^{\sigma}$ is equivalent to $\Omega_{\lambda j}$".

p. 130, line 12: "we have $(\Omega_{\mu i})^{\xi} = \Omega_{\lambda j}$" should be "$(\Omega_{\mu i})^{\xi}$ is equivalent to $\Omega_{\lambda j}$".

p. 132, line 1: "the" should be "then".

p. 138, line 2 from bottom: Both "φ" should be "φ_{λ}".

p. 143, line 4 from bottom: "$\sum_{i=1}^{\infty}$" should be "$\sum_{n=1}^{\infty}$".

p. 149, line 24: For the definition of Ω_{λ}, one should take *four* elements in place of $y_{\lambda 1}$, $y_{\lambda 2}$; see the above correction to pp. 124–127.

p. 158, Reference 4: "algebraic groups" should be "arithmetic groups".

PRINCETON UNIVERSITY

REFERENCES

1. A. BOREL and HARISH-CHANDRA, *Arithmetic subgroups of algebraic groups*, Ann. of Math., 75 (1962), 485–535.
2. C. CHEVALLEY, *Deux théorèmes d'arithmétique*, J. Math. Soc. Japan, 3 (1951), 36–44.
3. M. KNESER, *Starke Approximation in algebraischen Gruppen*: I, J. Reine Angew. Math., 218 (1965), 190–203.
4. G. SHIMURA, *Arithmetic of alternating forms and quaternion hermitian forms*, J. Math. Soc. Japan, 15 (1963), 33–65.
5. ———, *On analytic families of polarized abelian varieties and automorphic functions*, Ann. of Math., 78 (1963), 149–192.
6. ———, *Arithmetic of unitary groups*, Ann. of Math., 79 (1964), 369–409.
7. ———, *Moduli and fibre systems of abelian varieties*, Ann. of Math., 83 (1966), 294–338.
8. ———, *Discontinuous groups and abelian varieties*, Math. Ann., 168 (1967), 171–199.
9. ———, *Construction of class fields and zeta functions of algebraic curves*, Ann. of Math., 85 (1967), 58–159, referred to as [C].
10. ——— and Y. TANIYAMA, Complex multiplication of abelian varieties and its applications to number theory, Publ. Math. Soc. Japan, No. 6, 1961.
11. C. L. SIEGEL, *Symplectic geometry*, Amer. J. Math., 65 (1943), 1–86.
12. A. WEIL, *The field of definition of a variety*, Amer. J. Math., 78 (1956), 509–524.

(Received June 13, 1967)

Algebraic varieties without deformation and the Chow variety

Journal of the Mathematical Society of Japan, 20 (1968), 336-341

(Received May 25, 1967)

The purpose of this note is to indicate two simple facts, of which the first one is almost obvious, once it is formulated:

THEOREM 1. *If a complex projective non-singular variety V has no deformation, then V is biregularly equivalent to a projective variety defined over an algebraic number field.*

Here we say that V has *no deformation* if $H^1(V, \Theta) = 0$, where Θ denotes the sheaf of germs of holomorphic sections of the tangent bundle of V. Calabi and Vesentini [1] have proved that $H^1(V, \Theta) = 0$ if V is the quotient S/Γ of an irreducible bounded symmetric domain S of dimension > 1 by a discontinuous group Γ operating freely on S such that S/Γ is compact. (See also [4], [5], [11].) Therefore Th. 1 shows that such a quotient has a model defined over an algebraic number field.

To state the second fact, let $C_p(N, m, d)$ denote the set of all the Chow points of positive cycles of dimension m and degree d in the projective space of dimension N, defined with respect to a universal domain of characteristic $p \geqq 0$. It is well known that $C_p(N, m, d)$ is a Zariski closed set in a certain projective space, which is defined by equations with coefficients in the prime field. Then one can ask the following question:

(Q) *Is every absolutely irreducible component of $C_p(N, m, d)$ defined over the prime field?*[1]

The answer is negative if the characteristic is 0:

THEOREM 2. *There exist positive integers N, m, d such that $C_0(N, m, d)$ has a component which is not defined over the rational number field.*

Such a component will be obtained so as to contain the Chow point of a certain variety without deformation. This is why we present these two theorems together. We shall also show in the last section that the answer to the question (Q) is still negative even if the characteristic is positive.

---

1) I thank S. Lichtenbaum for reminding me of this question.

1. We shall denote by $c(X)$ the Chow point of a positive cycle X in a projective space, and by P^N the projective space of dimension N. A variety or a curve will always mean an absolutely irreducible one. In this section, the universal domain is the complex number field C.

PROPOSITION 1. *Let V be a non-singular variety of dimension m and degree d in P^N, and B a component of $C_0(N, m, d)$ containing $c(V)$. Suppose that V has no deformation. Then there exists a Zariski open subset B' of B such that if $c(X) \in B'$, X is a non-singular variety biregularly isomorphic to V.*

PROOF. Let k_0 be a finitely generated extension of the rational number field Q over which $c(V)$ and B are rational. Let x be a generic point of B over k_0, and W the cycle such that $c(W) = x$. We shall now prove

(1.1) *W is a non-singular variety biregularly isomorphic to V.*

The non-singularity is obvious, since V is a specialization of W over k_0. Put $u = c(V)$. We can find a generic specialization y of x over k_0 and a finitely generated extension k of k_0, such that u is a specialization of y over k and $k(y)$ is a regular extension of k of dimension one. Let E be a complete non-singular curve with a generic point z over k such that $k(z) = k(y)$, and f a morphism of E into B, rational over k, such that $f(z) = y$. Take a point v on E such that $f(v) = u$. For each $t \in E$, let W_t denote the cycle such that $c(W_t) = f(t)$. Then W_t is the unique specialization of W_z compatible with the specialization $z \to t$ over k. Let E' be the set of all t on E such that W_t is a non-singular variety. Then E' is a Zariski open subset of E rational over k, containing both z and v. Let w be a generic point of W_z over $k(z)$. (Note that W_z is defined over $k(z)$, since $c(W_z) = f(z)$.) Let H be the locus of $z \times w$ on $E' \times P^N$ over k. Then we see easily that the intersection product of H with $t \times P^N$ on $E' \times P^N$ is $t \times W_t$ for every $t \in E'$. By [**10**, Ch. VI, § 2, Th. 6], every point of $t \times W_t$ is simple on H. Thus we have a non-singular variety H which may be regarded as a fibre bundle with E' as base and the W_t as fibres. By the theorem of Frölicher and Nijenhuis [**2**] (cf. also [**3**]), our assumption $H^1(V, \Theta) = 0$ implies that there exists a neighborhood M of v on E' such that $V (= W_v)$ is biregularly isomorphic to W_t for every $t \in M$. We can find a generic point s of E over k lying in M. Let q be a biregular morphism of W_s to V. Since $f(s)$ is a generic specialization of x over k_0, there exists an automorphism σ of C over k_0 such that $f(s)^\sigma = x$. Then q^σ is a biregular morphism of $W (= (W_s)^\sigma)$ to V. This proves (1.1).

Now for every $c(X) \in B$, $(c(X), c(V))$ is a specialization of $(c(W), c(V))$ over k_0. Over this specialization, the graph of q^σ can be specialized to the graph of a biregular morphism of X onto V, at least for all $c(X)$ in some Zariski open subset of B (see for example [**9**, § 12.3, Prop. 24]). This proves our proposition.

PROOF OF THEOREM 1. Let V, B and B' be as in Prop. 1, and k the smallest field of definition for B. Then k is an algebraic number field of finite degree. We can find a point v of B' algebraic over k. Let X be such that $c(X) = v$. Then X is biregularly isomorphic to V and defined over $k(v)$. This proves Th. 1.

One could actually prove Th. 1 more directly. For example, take the locus U of $c(V)$ over the algebraic closure of Q, and construct a fibre variety of which the base is a Zariski open subset of U and the generic fibre at $c(V)$ is V. Specializing $c(V)$ to an algebraic point on U close to $c(V)$, one obtains the desired conclusion.

2. The universal domain being still C, let V be a projective variety. A subfield h of C is called *the bottom field* (resp. *the strong bottom field*) for V if the following condition is satisfied.

   (2.1) *An automorphism σ of C is the identity mapping on h if and only if V^σ is birationally (resp. biregularly) equivalent to V.*

Such a field h is unique for V if it exists. The following three assertions are immediate consequences of this definition.

   (2.2) *If W is birationally (resp. biregularly) equivalent to V, and h is the bottom field (resp. the strong bottom field) for V, then h is the bottom field (resp. the strong bottom field) for W.*

   (2.3) *Every field of definition for V contains the bottom field (resp. the strong bottom field) for V, if the latter exists.*

   (2.4) *If V is defined over an algebraic number field, then the bottom field and the strong bottom field for V exist and are algebraic number fields of finite degree.*

We say that V is a *minimal model* if every rational mapping of a variety X into V is defined at all the simple points of X. Then we can easily verify

   (2.5) *The bottom field and the strong bottom field for V exist and coincide, if either one of them exists and V is a non-singular minimal model.*

PROPOSITION 2. *Let V and B be as in Prop. 1. Suppose that V has no deformation. Then the bottom field and the strong bottom field for V exist and are contained in the smallest field of definition for B.*

PROOF. By (2.2), (2.4) and Th. 1, we know that the strong bottom field for V exists and is an algebraic number field of finite degree. Denote it by h. Let k_0 be a finitely generated extension of h over which B and $c(V)$ are rational. Let x be a generic point of B over k_0, and W the variety such that $c(W) = x$. By (1.1) and (2.2), h is the strong bottom field for W. Let k be the smallest field of definition for B. Since W is defined over $k(x)$, we have $h \subset k(x)$ by (2.3). Since both h and k are algebraic over Q and k is algebraically closed in $k(x)$, we have $h \subset k$, which completes the proof for the strong bottom field. The same reasoning applies also to the bottom field.

3. In order to prove Th. 2, it is sufficient, in view of Prop. 2, to show the existence of a non-singular projective variety without deformation whose strong bottom field is not Q. We shall obtain such a variety as a quotient $\mathfrak{H}^r/\Gamma$ of the product of r copies of the upper half plane $\mathfrak{H} = \{z \in C \,|\, \mathrm{Im}\,(z) > 0\}$ by a discontinuous group Γ. To define Γ, we consider a totally real algebraic number field F of degree g, and a division quaternion algebra D over F. Let $p_{\infty 1}, \cdots, p_{\infty g}$ be the archimedean primes of F. Suppose that D is unramified at $p_{\infty 1}, \cdots, p_{\infty r}$ and ramified at $p_{\infty r+1}, \cdots, p_{\infty g}$. Put $D_R = D \underset{Q}{\otimes} R$. Then D_R can be identified with $M_2(R)^r \times K^{g-r}$, the product of r copies of the total matrix algebra of degree 2 over the real number field R and $g-r$ copies of the division ring K of real quaternions. Let $\mathfrak{o}$ be a maximal order in D, and $\Gamma(\mathfrak{o})$ the group of all units in $\mathfrak{o}$ whose reduced norm to F is 1. For every $\alpha \in \Gamma(\mathfrak{o})$, let $\begin{bmatrix} a_\nu & b_\nu \\ c_\nu & d_\nu \end{bmatrix}$ $(\nu = 1, \cdots, r)$ be the projections of α to the first r factors $M_2(R)$ of D_R. Define the action of α on $\mathfrak{H}^r$ by

$$\alpha(z_1, \cdots, z_r) = (w_1, \cdots, w_r), \quad w_\nu = (a_\nu z_\nu + b_\nu)(c_\nu z_\nu + d_\nu)^{-1} \quad (z_\nu \in \mathfrak{H}).$$

Then $\Gamma(\mathfrak{o})$ gives a discontinuous group operating on $\mathfrak{H}^r$. Now we assume

(3.1) *The multiplicative group of invertible elements of D has no element of finite order other than ± 1.*

Then $\mathfrak{H}^r/\Gamma(\mathfrak{o})$ is biregularly isomorphic to a non-singular projective variety V. By [5, Th. 7.4], V has no deformation if $r > 1$. Assume further

(3.2) $r = g-1 > 1$, *and* $p_{\infty g}$ *corresponds to the identity mapping of* F.

The results of [7, 4.9 and 4.10] tell us that, under the assumptions (3.1) and (3.2), the composite of F and the bottom field for V is an unramified class field over F corresponding to an ideal group $I(D/F)$ in F generated by the following three types of ideals:

(I) the principal ideals (a) such that a is totally positive;

(II) the squares of all ideals in F;

(III) the prime ideals in F which are ramified in D.

In view of these facts, we construct our counterexample in the following way. Take any totally real algebraic number field F of degree $g > 2$ whose class number is even. (For example, $F = Q(\sqrt{10}, \sqrt{13})$.) We can find a prime ideal $\mathfrak{p}$ in F which is generated by a totally positive element and decomposes in every quadratic extension of F containing non-trivial roots of unity. There exists a quaternion algebra D over F which is ramified exactly at $\mathfrak{p}$ and $p_{\infty g}$, where $p_{\infty g}$ is as in (3.2). By our choice of $\mathfrak{p}$, we see that (3.1) is satisfied, and $I(D/F)$ is different from the whole ideal group of F. Hence a model for $\mathfrak{H}^r/\Gamma(\mathfrak{o})$,

234

with any maximal order $\mathfrak{o}$ in D, has the bottom field different from Q. This completes the proof of Th. 2.

It is possible to obtain many more examples of varieties without deformation and with bottom fields $\neq Q$, by considering the quotient of a bounded symmetric domain, other than $\mathfrak{H}^r$, by a discontinuous group. This can be done, for example, by making the results of [7, § 2] more precise, or extending them to the case of congruence subgroup of the discontinuous group in question (cf. also [8]).

4.[2] Another type of counterexample can be obtained by considering the topological structure of varieties. We first prove an elementary

PROPOSITION 3. *Let V and V' be non-singular projective varieties whose Chow points belong to the same component of $C_0(N, m, d)$. Then there exists a diffeomorphism of V onto V'.*

PROOF. Let B, k_0, z, E', f, w and H be as in the proof of Prop. 1. We see that V, as a fibre of H, is diffeomorphic to a generic fibre W_z with a generic point z of E' over k. (Note that we do not need the condition $H^1(V, \Theta) = 0$.) Let B_0 be the set of all simple points r of B such that $r = c(X)$ with a non-singular variety X. Then B_0 is a Zariski open subset of B rational over k_0. Let K be the locus of $f(z) \times w$ over k_0 on $B_0 \times P^N$. By the same type of argument as in the proof of Prop. 1, we see that K is non-singular, and $K \to B_0$ defines a fibre bundle whose fibres are the varieties X such that $c(X) \in B_0$. If $c(X)$ and $c(Y)$ belong to B_0, then X and Y are diffeomorphic. Since $c(W_z)$ $(=f(z))$ belongs to B_0, we see that V is diffeomorphic to the fibres of K. Applying the same reasoning to V', we obtain our assertion.

Now Serre [6] gave an example of a non-singular projective variety V defined over an algebraic number field k such that V is not homeomorphic to V^σ for some isomorphism σ of k into C. Then the component of $C_0(N, m, d)$ containing $c(V)$ can not be defined over Q, since it can not contain $c(V^\sigma)=c(V)^\sigma$ in view of Prop. 3. It may be interesting to investigate the deformation of this V.

5. Let us now consider $C_p(N, m, d)$ for positive p. Since $C_0(N, m, d)$ is defined by equations with rational coefficients, we can naturally speak of its reduction modulo p for a prime number p.

PROPOSITION 4. *There exists a finite set $S(N, m, d)$ of rational primes such that $C_p(N, m, d)$ is exactly the reduction of $C_0(N, m, d)$ modulo p if p is not contained in $S(N, m, d)$.*

This can easily be verified by applying the following two simple principles

---

2) The possibility of using Serre's example was kindly suggested by G. Washnitzer.

to the defining equations of $C_0(N, m, d)$.

(5.1) *Reduction modulo p of the intersection of algebraic sets is the intersection of the reduced sets for all except a finite number of p.*

(5.2) *Reduction modulo p of the projection of an algebraic set is the projection of the reduced set for all except a finite number of p.*

For more precise statements of these facts, see, for example, [9, Ch. III, Prop. 19, Prop. 20].

Now let B be a component of $C_0(N, m, d)$ which is not defined over $\mathbf{Q}$. Let $B_1, \cdots, B_r$ be all the components of $C_0(N, m, d)$ with $B_1 = B$. Let $y_i = c(B_i)$ for $i = 1, \cdots, r$. Then both $\mathbf{Q}(y_1)$ and $\mathbf{Q}(y_1, \cdots, y_r)$ are algebraic number fields of degree > 1. We can find a prime ideal $\mathfrak{p}$ in $\mathbf{Q}(y_1, \cdots, y_r)$ such that: (i) y_1 modulo $\mathfrak{p}$ is not rational over the prime field; (ii) $\mathfrak{p}(B_1)^{3)}$ is absolutely irreducible and not contained in $\mathfrak{p}(B_i)$ for $i > 1$; and (iii) the rational prime p divisible by $\mathfrak{p}$ does not belong to $S(N, m, d)$. Then $\mathfrak{p}(B_1)$ is a component of $C_p(N, m, d)$ which is not defined over the prime field. Thus the answer to the question (Q) is negative at least for infinitely many prime characteristics.

Princeton University

References

[1] E. Calabi and E. Vesentini, On compact, locally symmetric Kähler manifolds, Ann. of Math., 71 (1960), 472–507.

[2] A. Frölicher and A. Nijenhuis, A theorem on stability of complex structures, Proc. Nat. Acad. Sci. U. S. A., 43 (1957), 239–241.

[3] K. Kodaira and D. C. Spencer, On deformations of complex analytic structures, Ann. of Math., 67 (1958), 328–466.

[4] Y. Matsushima and S. Murakami, On certain cohomology groups attached to hermitian symmetric spaces, Osaka J. Math., 2 (1965), 1–35.

[5] Y. Matsushima and G. Shimura, On the cohomology groups attached to certain vector valued differential forms on the product of the upper half planes, Ann. of Math., 78 (1963), 417–449.

[6] J.-P. Serre, Exemples de variétés projectives conjuguées non homéomorphes, C. R. Acad. Sci. Paris, 258 (1964), 4194–4196.

[7] G. Shimura, Class-fields and automorphic functions, Ann. of Math., 80 (1964), 444–463.

[8] G. Shimura, On the field of definition for a field of automorphic functions, I, II, III, Ann. of Math., 80 (1964), 160–189; 81 (1965), 124–165; 83 (1966), 377–385.

[9] G. Shimura and Y. Taniyama, Complex multiplication of abelian varieties and its applications to number theory, Publ. Math. Soc. Japan, No. 6, 1961.

[10] A. Weil, Foundations of algebraic geometry, 2nd edition, Providence, 1962.

[11] A. Weil, On discrete subgroups of Lie groups, I, II, Ann. of Math., 72 (1960), 369–384; 75 (1962), 578–602.

---

3) $\mathfrak{p}(X)$ means the cycle obtained from X by reduction modulo $\mathfrak{p}$.

68c

An ℓ-adic method in the theory of automorphic forms

The text of a lecture at the conference Automorphic functions for arithmetically defined groups,
Oberwolfach, Germany, July 28-August 3, 1968

Introduction

Let Γ be a Fuchsian group of the first kind acting on the upper half complex plane $\mathfrak{H}$, and $\mathfrak{H}^*$ the union of $\mathfrak{H}$ and the cusps of Γ. Further let $S_m(\Gamma)$, for a positive integer m, denote the vector space of all cusp forms with respect to Γ, with the factor of automorphy of the form $(cz+d)^m$. Denote by $H_P^1(\Gamma, \mathbf{Z})$ the module of all homomorphisms of Γ into $\mathbf{Z}$ vanishing at all parabolic elements of Γ. Then $H_P^1(\Gamma, \mathbf{Z}) \otimes \mathbf{R}$ can be identified with the first cohomology group of $\Gamma \backslash \mathfrak{H}^*$ with real coefficients. On the other hand, $S_2(\Gamma)$ is essentially the same as the vector space of all holomorphic differential forms on $\Gamma \backslash \mathfrak{H}^*$, and hence we have a canonical $\mathbf{R}$-linear isomorphism

$$(1) \qquad S_2(\Gamma) \cong H_P^1(\Gamma, \mathbf{Z}) \otimes \mathbf{R}.$$

If J denotes the jacobian variety of $\Gamma \backslash \mathfrak{H}^*$, then there exists, for every rational prime ℓ, an isomorphism

$$(2) \qquad H_P^1(\Gamma, \mathbf{Z}) \otimes (\mathbf{Q}_\ell/\mathbf{Z}_\ell) \cong \bigcup_{\nu=1}^{\infty} \left\{ t \in J \,\middle|\, \ell^\nu t = 0 \right\},$$

where $\mathbf{Q}_\ell$ and $\mathbf{Z}_\ell$ denote the ℓ-adic completions of $\mathbf{Q}$ and $\mathbf{Z}$ respectively.

The isomorphism (2) is of great significance for arithmetically defined Γ such as the congruence subgroups of $SL_2(\mathbf{Z})$. In fact, for such a Γ, one can define Hecke operators on $S_m(\Gamma)$ with which some important zeta functions are defined. If $m = 2$, we can find, by means of the above two isomorphisms, a connection of a Hecke operator of a prime degree p with the p-th power endomorphism on J modulo p (cf. Eichler [E1] and the author [S1], [S3], S6]). This enables us to apply the results of Weil on such Frobenius endomorphisms to the zeta functions in question.

Our main purpose is to present a generalization of these facts for the case $m \geq 3$. For simplicity, let us explain our ideas in the special case $\Gamma = SL_2(\mathbf{Z})$, although a more general case, including not only congruence subgroups of higher level but also other types of arithmetic Γ with compact $\Gamma \backslash \mathfrak{H}$, can be treated. We first consider a symmetric tensor representation of $SL_2(\mathbf{R})$ of degree n. Restricting this representation to Γ, we can let Γ act on $X = \mathbf{Z}^{n+1}$. Then a *parabolic cocycle* of Γ with values in X is, by definition, a map $u : \Gamma \to X$ such that

$$(3) \qquad u(\alpha\beta) = u(\alpha) + \alpha u(\beta) \quad \text{for every} \quad \alpha, \beta \in \Gamma,$$

$$(4) \qquad u(\gamma) \in (\gamma - 1)X \quad \text{for every parabolic} \quad \gamma \in \Gamma.$$

237

We call u a *coboundary* if $u(\alpha) = (\alpha - 1)y$ with an element y of X independent of α. Then we define the parabolic cohomology group $H_P^1(\Gamma, X)$ to be the parabolic cocycles modulo the coboundaries. This type of cohomology, in connection with automorphic forms, was first introduced by Eichler in [E2] with $X_{\mathbf{C}} = \mathbf{C}^{n+1}$ in place of X. He showed especially that $H_P^1(\Gamma, X_{\mathbf{C}})$ has twice as much dimension as $S_{n+2}(\Gamma)$ over $\mathbf{C}$. In his theory, an important role was played by automorphic forms of the second kind, which will not be discussed in the present paper. I have shown in [S2] that there exists a canonical $\mathbf{R}$-linear isomorphism

$$(5) \qquad S_{n+2}(\Gamma) \cong H_P^1(\Gamma, X) \otimes \mathbf{R}$$

which is analogous to (1), and which commutes with the natural action of Hecke operators on $S_{n+2}(\Gamma)$ and $H_P^1(\Gamma, X)$.

In the present paper we shall define a certain module $\mathfrak{W}_\ell$ for each rational prime ℓ, and establish a surjective homomorphism with finite kernel

$$(6) \qquad H_P^1(\Gamma, X) \otimes (\mathbf{Q}_\ell / \mathbf{Z}_\ell) \longrightarrow \mathfrak{W}_\ell.$$

Furthermore we shall obtain an anti-representation μ^ℓ and a representation μ_*^ℓ of $\mathrm{Gal}(\overline{\mathbf{Q}}/\mathbf{Q})$ into the group of automorphisms of $\mathfrak{W}_\ell$, where $\overline{\mathbf{Q}}$ denotes the algebraic closure of $\mathbf{Q}$, with the following property:

Let p be a rational prime, σ_p a Frobenius automorphism of $\overline{\mathbf{Q}}$ for any prime divisor of p, and $\Lambda_\ell(p)$ the action of the Hecke operator of degree p on $\mathfrak{W}_\ell$ transferred by the isomorphism of (5) and the map of (6). Then

$$(7) \qquad \Lambda_\ell(p) = \mu^\ell(\sigma_p) + \mu_*^\ell(\sigma_p),$$
$$(8) \qquad p^{n+1} = \mu^\ell(\sigma_p) \circ \mu_*^\ell(\sigma_p).$$

Since the module $\mathfrak{W}_\ell$ is isomorphic to $(\mathbf{Q}_\ell / \mathbf{Z}_\ell)^e$ with $0 \le e \in \mathbf{Z}$, we can view ${}^t\mu^\ell$ and μ_*^ℓ as $GL_e(\mathbf{Z}_\ell)$-valued representations of $\mathrm{Gal}(\overline{\mathbf{Q}}/\mathbf{Q})$.

The module $\mathfrak{W}_\ell$ can be obtained as follows. Let $\Gamma(a)$ denote the principal congruence subgroup of Γ of level a, and $J(a)$ the jacobian variety of $\Gamma(a)\backslash \mathfrak{H}^*$. Let $u_1, \dots, u_{n+1}$ be the components of a parabolic cocycle u defined as above. Since $(\gamma - 1)X \subset aX$ for $\gamma \in \Gamma(a)$, we have

$$u_\nu(\alpha\beta) \equiv u_\nu(\alpha) + u_\nu(\beta) \pmod{a\mathbf{Z}} \quad \text{for every} \quad \alpha, \beta \in \Gamma(a),$$
$$u_\nu(\gamma) \equiv 0 \pmod{a\mathbf{Z}} \quad \text{for every parabolic} \quad \gamma \in \Gamma(a).$$

Therefore we can define characters u_ν^* of $\Gamma(a)$ by

$$u_\nu^*(\alpha) = \exp\left(2\pi i \cdot u_\nu(\alpha)/a\right) \qquad (\alpha \in \Gamma(a); \ \nu = 1, \dots, n+1).$$

Each u_ν^* determines a point t_ν of $J(a)$ such that $at_\nu = 0$. We thus obtain, from u, a point $t = (t_1, \dots, t_{n+1})$ of a finite abelian group $Y(a)$ defined by

$$Y(a) = \left\{ t \in J(a)^{n+1} \,\middle|\, at = 0 \right\}.$$

If a divides b, then we have a canonical homomorphism $J(a) \to J(b)$ associated with the projection $\Gamma(b)\backslash\mathfrak{H}^* \to \Gamma(a)\backslash\mathfrak{H}^*$. Restrict this homomorphism to $Y(a)$, and consider the case $a = \ell^m$ and $b = \ell^{m+1}$. Then we obtain a homomorphism $Y(\ell^m) \to Y(\ell^{m+1})$, by which we can form a direct limit $\varinjlim Y(\ell^m)$. The above procedure of constructing t from u defines a map

$$H^1_P(\Gamma, X) \otimes (\mathbf{Q}_\ell/\mathbf{Z}_\ell) \longrightarrow \varinjlim Y(\ell^m).$$

Denoting by $\mathfrak{W}_\ell$ the image of this map, we obtain (6).

It was shown in [S2] that if L denotes the image of $H^1_P(\Gamma, X) \otimes 1$ in $S_{n+2}(\Gamma)$ by the isomorphism of (5), then the complex torus $S_{n+2}(\Gamma)/L$ has a structure of an abelian variety for every even $n \geq 0$. The above construction of $\mathfrak{W}_\ell$ presents an algebro-geometric interpretation of the division points of this abelian variety, which may be regarded as a partial answer to the question raised in [S2]. We end this introduction by stating a principle emerging from our theory:

In order to understand various problems concerning Γ, we need all congruence subgroups of Γ.

1. Parabolic cohomology of groups

1.1. Let G be a group and X a left G-module. Then one can define the cohomology groups $H^i(G, X)$ for nonnegative integers i. In this paper we shall discuss the cohomology groups of this type for $i \leq 2$ by taking G to be a Fuchsian group Γ or its factor group by a normal subgroup of finite index. If Γ has parabolic elements, it is necessary to modify the definition of 1-cocyle and 2-coboundary. Let us first recall the definition of $H^i(G, X)$, and then define the modified cohomology groups in a somewhat more general setting than necessary. We fix a commutative ring R with identity element, and assume that X is an $R[G]$-module, where $R[G]$ is the group-ring of G over R. In later applications R will be $\mathbf{Z}$ or a field.

We denote by $C^i(G, X)$ the R-module of all maps of G^i into X, where G^i denotes the product of i copies of G; we understand that $C^0(G, X) = X$. For $u \in C^i(G, X)$ define an element ∂u of $C^{i+1}(G, X)$ by

$$\partial u(\alpha) = (\alpha - 1)u \quad \text{if} \quad i = 0,$$
$$\partial u(\alpha_1, \alpha_2, \ldots, \alpha_{i+1}) = \alpha_1 \cdot u(\alpha_2, \ldots, \alpha_{i+1})$$
$$+ \sum_{j=1}^{i}(-1)^j u(\alpha_1, \ldots, \alpha_{j-1}, \alpha_j\alpha_{j+1}, \ldots, \alpha_{i+1})$$
$$+ (-1)^{i+1}u(\alpha_1, \ldots, \alpha_i) \quad \text{if} \quad i > 0.$$

Put then

$$Z^i(G, X) = \{\, u \in C^i(G, X) \mid \partial u = 0 \,\},$$
$$B^i(G, X) = \begin{cases} 0 & \text{if} \quad i = 0, \\ \partial(C^{i-1}(G, X)) & \text{if} \quad i > 0, \end{cases}$$
$$H^i(G, X) = Z^i(G, X)/B^i(G, X),$$
$$X^G = \{\, x \in X \mid \alpha x = x \ \text{for all} \ \alpha \in G \,\}.$$

Then $H^0(G, X)$ and $Z^0(G, X)$ can be identified with X^G; also, $Z^1(G, X)$ (resp. $B^1(G, X)$) consists of the maps $u : G \to X$ such that $u(\alpha\beta) = u(\alpha) + \alpha u(\beta)$, (resp. $u(\alpha) = (\alpha - 1)x$ with an element x independent of α).

Now we fix a system $\mathfrak{P} = (P;\ X_\pi\ (\pi \in P))$ consisting of a subset P of G and a set of R-submodules X_π of X, defined for each $\pi \in P$, such that

$$(1.1.1) \qquad\qquad (\pi - 1)X \subset X_\pi.$$

Denote by $C^1(G, X, \mathfrak{P})$ the R-submodule of $C^1(G, X)$ consisting of the elements u with the following property:

$$(1.1.2) \qquad\qquad u(\pi) \in X_\pi \quad \text{for every} \quad \pi \in P.$$

Then we put

$$Z^1(G, X, \mathfrak{P}) = Z^1(G, X) \cap C^1(G, X, \mathfrak{P}),$$
$$B^2(G, X, \mathfrak{P}) = \partial(C^1(G, X, \mathfrak{P})),$$
$$H^0(G, X, \mathfrak{P}) = H^0(G, X),$$
$$H^1(G, X, \mathfrak{P}) = Z^1(G, X, \mathfrak{P})/B^1(G, X),$$
$$H^2(G, X, \mathfrak{P}) = Z^2(G, X)/B^2(G, X, \mathfrak{P}).$$

Notice that $B^1(G, X) \subset Z^1(G, X, \mathfrak{P})$ in view of (1.1.1). We say that $\mathfrak{P}$ is *elementary* if $X_\pi = (\pi-1)X$ for all $\pi \in P$. In this case we denote $C^1(G, X, \mathfrak{P})$, $Z^1(G, X, \mathfrak{P})$, $B^2(G, X, \mathfrak{P})$, and $H^i(G, X, \mathfrak{P})$ by $C_P^1(G, X)$, $Z_P^1(G, X)$, $B_P^2(G, X)$, and $H_P^i(G, X)$, respectively. The cohomology groups $H_P^i(G, X)$ are quite adequate to the main purposes of this paper. But some simple properties of the usual $H^i(G, X)$ can be extended to $H^i(G, X, \mathfrak{P})$, but not to $H_P^i(G, X)$, especially when R is not a field. For example, we have:

1.2. Proposition. *Let* $0 \longrightarrow W \xrightarrow{g} X \xrightarrow{f} Y \longrightarrow 0$ *be an exact sequence of* $R[G]$-*modules, and let* $\mathfrak{P} = (P; \{X_\pi\})$ *be as above. Define* $\mathfrak{Q}$ *and* $\mathfrak{S}$ *by*

$$\mathfrak{Q} = (P; \{Y_\pi\}), \quad Y_\pi = f(X_\pi), \quad \mathfrak{S} = (P; \{W_\pi\}), \quad W_\pi = g^{-1}(X_\pi \cap g(W)).$$

Then we have an exact sequence

$$0 \longrightarrow W^G \longrightarrow X^G \longrightarrow Y^G \longrightarrow H^1(G, W, \mathfrak{S}) \longrightarrow H^1(G, X, \mathfrak{P})$$
$$\longrightarrow H^1(G, Y, \mathfrak{Q}) \longrightarrow H^2(G, W, \mathfrak{S}) \longrightarrow H^2(G, X, \mathfrak{P}) \longrightarrow H^2(G, Y, \mathfrak{Q}).$$

This can easily be verified by the standard argument. Such an exact sequence does not necessarily hold with elementary $\mathfrak{P}$, $\mathfrak{Q}$, $\mathfrak{S}$, since $(\pi - 1)g(W)$ does not necessarily coincide with $g(W) \cap (\pi - 1)X$. Thus it seems rather natural and also convenient to consider $H^i(G, X, \mathfrak{P})$ with general $\mathfrak{P}$. However, it is often cumbersome to define $\mathfrak{P}$ for each G-module X. In the following treatment, we shall therefore minimize the use of $H^i(G, X, \mathfrak{P})$ with non-elementary $\mathfrak{P}$, and formulate the results in a less general form.

1.3. Let G, R, X, and $\mathfrak{P}$ be as in §1.1. Take R to be $\mathbf{Z}$, and suppose that X is a free $\mathbf{Z}$-module of finite rank. Put $X_\mathbf{Q} = X \otimes_\mathbf{Z} \mathbf{Q}$ and regard $X_\mathbf{Q}$ as a $\mathbf{Q}[G]$-module. For every positive integer a we can consider $a^{-1}X/X$ a G-module, and hence $H_P^1(G, a^{-1}X/X)$ is meaningful. If a divides b, the natural injection $a^{-1}X/X \to b^{-1}X/X$ defines a map

$$(1.3.1) \qquad i_{a,b} : H_P^1(G, a^{-1}X/X) \longrightarrow H_P^1(G, b^{-1}X/X).$$

Form the direct limit $\varinjlim H_P^1(G, a^{-1}X/X)$ with respect to the maps $i_{a,b}$. We can now define a map

$$(1.3.2) \qquad j : H_P^1(G, X) \otimes (\mathbf{Q}/\mathbf{Z}) \longrightarrow \varinjlim H_P^1(G, a^{-1}X/X).$$

as follows. Let $c \in \mathbf{Q}$, $u \in Z_P^1(G, X)$, and let ξ be the cohomology class of u. Take an integer $a > 0$ so that $ac \in \mathbf{Z}$, and define an element v of $Z_P^1(G, a^{-1}X/X)$ by $v(\alpha) = c \cdot u(\alpha) \pmod{X}$. Then we define $j(\xi \otimes (c \pmod{\mathbf{Z}}))$ to be the element of $\varinjlim H_P^1(G, a^{-1}X/X)$ represented by the cohomology class of v.

Put $\mathfrak{Q}_a = (P; \{Y_\pi^a\})$ with $Y_\pi^a = X \cap (\pi - 1)a^{-1}X$. Then we can form $\varinjlim H^2(G, X, \mathfrak{Q}_a)$ with respect to the natural surjective map of $H^2(G, X, \mathfrak{Q}_a)$ to $H^2(G, X, \mathfrak{Q}_b)$, defined whenever a divides b. Let $\eta \in \varinjlim H_P^1(G, a^{-1}X/X)$, and let v be an element of $Z_P^1(G, a^{-1}X/X)$, for some a, that represents η. We can find an element u of $C_P^1(G, X)$ such that $a^{-1}u(\alpha) \pmod{X} = v(\alpha)$. Then $a^{-1}\partial u \in Z^2(G, X)$. Denote by $k(\eta)$ the element of $\varinjlim H^2(G, X, \mathfrak{Q}_a)$ represented by $a^{-1}\partial u$. It can easily be verified that $k(\eta)$ depends only on η, and we thus obtain a $\mathbf{Z}$-linear map

$$k : \varinjlim H_P^1(G, a^{-1}X/X) \longrightarrow \varinjlim H^2(G, X, \mathfrak{Q}_a).$$

Then it is easy to see that

$$(1.3.3) \quad H_P^1(G, X) \otimes (\mathbf{Q}/\mathbf{Z}) \xrightarrow{\ j\ } \varinjlim H_P^1(G, a^{-1}X/X) \xrightarrow{\ k\ } \varinjlim H^2(G, X, \mathfrak{Q}_a)$$

$$\text{is exact.}$$

Observe that $\varinjlim H^2(G, X, \mathfrak{Q}_a)$ is a homomorphic image of $H_P^2(G, X)$.

1.4. Proposition. *If $H_P^2(G, X)$ is annihilated by a positive integer N, then the image of j coincides with $N \cdot \varinjlim H_P^1(G, a^{-1}X/X)$.*

PROOF. Since $N \cdot (\mathbf{Q}/\mathbf{Z}) = \mathbf{Q}/\mathbf{Z}$, the image of j is contained in the module in question. Let $\xi \in \varinjlim H_P^1(G, a^{-1}X/X)$. By our assumption we have $k(N\xi) = 0$. Therefore, by the exactness of (1.3.3), $N\xi$ is contained in the image of j.

1.4a. Proposition. *Suppose that there exists a positive integer M such that $X \cap (\pi - 1)X_{\mathbf{Q}} \subset M^{-1}(\pi - 1)X$ for every $\pi \in P$. Then $M \cdot \mathrm{Ker}(j) = \{0\}$. (Notice that such an M exists if there exists a finite set P_0 such that every element of P is of the form $\alpha \pi^\mu \alpha^{-1}$ with $\alpha \in G$, $\pi \in P_0$, and $\mu \in \mathbf{Z}$.)*

PROOF. Let ξ be the cohomology class of $u \in Z_P^1(G, X)$, and s a positive integer. Suppose that $j(\xi \otimes (s^{-1} \pmod{\mathbf{Z}})) = 0$. Then there exists a positive multiple a of s such that $s^{-1}u \pmod{X}$ belongs to $B^1(G, a^{-1}X/X)$. Put $v = as^{-1}u$. Then there exists an element y of $a^{-1}X$ such that $s^{-1}u(\alpha) = a^{-1}v(\alpha) \equiv (\alpha - 1)y \pmod{X}$ for all $\alpha \in G$. If $\pi \in P$, then we have $a^{-1}v(\pi) - (\pi - 1)y \in X \cap (\pi - 1)X_{\mathbf{Q}} \subset M^{-1}(\pi - 1)X$. Put $w(\alpha) = a^{-1}Mv(\alpha) - (\alpha - 1)My$ for

$\alpha \in G$. Then we see that $w \in Z_P^1(G, X)$, and $\cdot aw - Mv \in B^1(G, X)$. Let η be the cohomology class of w. Then $Mas^{-1}\xi = a\eta$, so that $M\xi \otimes \left(s^{-1} \pmod{Z}\right) = Mas^{-1}\xi \otimes \left(a^{-1} \pmod{Z}\right) = a\eta \otimes \left(a^{-1} \pmod{Z}\right) = \eta \otimes \left(1 \pmod{Z}\right) = 0$, Q.E.D.

1.5. Let G, R, X, and P be as in §1.1, and G' a normal subgroup of G. Put $P' = P \cap G'$. Denote by $Z_P^1(G', X)^G$ the R-submodule of $Z_{P'}^1(G', X)$ consisting of all the u in $Z_{P'}^1(G', X)$ with the following property:

(1.5.1) *There exists an element* y *of* $C_P^1(G, X)$ *such that* $\alpha u(\alpha^{-1}\gamma\alpha) - u(\gamma) = (\gamma - 1)y(\alpha)$ *for all* $\gamma \in G'$ *and all* $\alpha \in G$.

We easily see that $B^1(G', X) \subset Z_P^1(G', X)^G$. We then put

$$H_P^1(G', X)^G = Z_P^1(G', X)^G / B^1(G', X).$$

If P is empty, $H_P^1(G', X)^G$ coincides with the module of all G-invariant elements of $H^1(G', X)$, the action of G on $C^1(G, X)$ being defined by $u \mapsto {}^\alpha u$ with $({}^\alpha u)(\gamma) = \alpha u(\alpha^{-1}\gamma\alpha)$ for $u \in C^1(G, X)$, $\alpha \in G$, and $\gamma \in G'$.

Every element v of $Z_P^1(G, X)$ satisfies

(1.5.2) $\qquad \alpha v(\alpha^{-1}\gamma\alpha) - v(\gamma) = (\gamma - 1)v(\alpha)$ *for every* $\alpha \in G$ *and every* $\gamma \in G'$;

hence the restriction to G' defines a map

$$\mathrm{Res} : H_P^1(G, X) \longrightarrow H_P^1(G', X)^G.$$

Now viewing $X^{G'}$ as a (G/G')-module, we can define $H^i(G/G', X^{G'})$. If P is empty, there is a well-known exact sequence

(1.5.3) $\qquad 0 \longrightarrow H^1(G/G', X^{G'}) \overset{\mathrm{Inf}}{\longrightarrow} H^1(G, X) \overset{\mathrm{Res}}{\longrightarrow}$

$$H^1(G', X)^G \overset{\lambda}{\longrightarrow} H^2(G/G', X^{G'}) \overset{\mathrm{Inf}}{\longrightarrow} H^2(G, X),$$

with the usual Inf(lation) and a certain map λ (see below). This is not necessarily true with nontrivial P, since $(\pi - 1)X^{G'}$ does not necessarily coincide with $(\pi - 1)X \cap X^{G'}$, and the latter module may not be determined by the coset $\pi G'$.

1.6. To discuss the case of nontrivial P, we assume that there exists a subset Q of P satisfying the following two conditions:

(1.6.1) *Every element of* P *is of the form* $\alpha\pi^\mu\alpha^{-1}$ *with* $\pi \in Q$, $\alpha \in G$, *and* $\mu \in \mathbf{Z}$;

(1.6.2) *No two distinct elements of* $Q \cup \{1\}$ *belong to the same coset modulo* G'.

Let us denote by ν the natural map of G to G/G'. Put $\overline{Q} = \nu(Q)$, $\overline{\pi} = \nu(\pi)$, $Y_{\overline{\pi}} = (\pi - 1)X \cap X^{G'}$ for every $\pi \in Q$, and

(1.6.3) $\qquad\qquad\qquad \mathfrak{Q} = (\overline{Q}, \{Y_{\overline{\pi}}\}).$

1.7. Proposition. *If* Q *satisfies (1.6.1), then* $Z_P^1(G, X) = Z_Q^1(G, X)$ *for every* G-module X, *and consequently* $H_P^1(G, X) = H_Q^1(G, X)$.

PROOF. Obviously $Z_P^1(G, X) \subset Z_Q^1(G, X)$. Let $u \in Z_Q^1(G, X)$, and for a fixed $\pi \in Q$ let $u(\pi) = (\pi - 1)x$ with $x \in X$. Then $u(\pi^m) = \sum_{i=0}^{m-1} \pi^i u(\pi) = (\pi^m - 1)x$ for $0 < m \in \mathbf{Z}$, and $\mu(\pi^{-m}) = -\pi^{-m}u(\pi^m) = (\pi^{-m} - 1)x$. Further, for every $\alpha \in G$ and $\mu \in \mathbf{Z}$ we have $u(\alpha\pi^\mu\alpha^{-1}) = (\alpha\pi^\mu\alpha^{-1} - 1)(\alpha x - u(\alpha))$. Thus $u \in Z_P^1(G, X)$. This proves our proposition.

1.8. Proposition. *Let the notation and the assumption be as in §§1.5 and 1.6. Then there exists a map λ such that the following sequence is exact:*

$$0 \longrightarrow H^1(G/G', X^{G'}, \mathfrak{Q}) \xrightarrow{\text{Inf}} H^1_P(G, X)$$

$$\xrightarrow{\text{Res}} H^1_P(G', X)^G \xrightarrow{\lambda} H^2(G/G', X^{G'}, \mathfrak{Q}).$$

PROOF. Let $u \in Z^1(G/G', X^{G'}, \mathfrak{Q})$. Put $v = u \circ \nu$. Then $v \in Z^1_Q(G, X) = Z^1_P(G, X)$. Assigning the cohomology class of v to the class of u, we obtain the map Inf : $H^1(G/G', X^{G'}, \mathfrak{Q}) \to H^1_P(G, X)$. If $v(\alpha) = (\alpha - 1)x$ for all $\alpha \in G$ with a fixed x, then $0 = u(\nu(1)) = v(\delta) = (\delta - 1)x$ for all $\delta \in G'$, and hence $x \in X^{G'}$. Therefore this map is injective. Next let $t \in Z^1_P(G, X)$. Suppose that the restriction of t to G' belongs to $B^1(G', X)$, i.e., $t(\delta) = (\delta-1)y$ for all $\delta \in G'$ with an element y of X independent of δ. Put $v(\alpha) = t(\alpha)-(\alpha-1)y$ for $\alpha \in G$. Then $v \in Z^1_P(G, X)$, and $v(\delta) = 0$ for all $\delta \in G'$. If $\alpha \in G$ and $\delta \in G'$, we have $\delta v(\alpha) = v(\delta) + \delta v(\alpha) = v(\delta\alpha) = v(\alpha\alpha^{-1}\delta\alpha) = v(\alpha) + \alpha v(\alpha^{-1}\delta\alpha) = v(\alpha)$; hence $v(\alpha) \in X^{G'}$, and $v(\alpha)$ depends only on $G'\alpha$. Therefore $v = u \circ \nu$ with an element u of $C^1(G/G', X^{G'})$. Clearly $u \in Z^1(G/G', X^{G'}, \mathfrak{Q})$. This shows that Ker(Res) = Im(Inf). To define the map λ, let $u \in Z^1_P(G', X)^G$. By our definition, there exists an element y of $C^1_P(G, X)$ such that

$$(1.8.1) \quad \alpha u(\alpha^{-1}\delta\alpha) - u(\delta) = (\delta - 1)y(\alpha) \text{ for all } \delta \in G' \text{ and all } \alpha \in G.$$

For $\delta, \varepsilon \in G'$ we have $\varepsilon u(\varepsilon^{-1}\delta\varepsilon) - u(\delta) = (\delta - 1)u(\varepsilon)$. Therefore we can take y so that $y = u$ on G'. (This does not interfere with the condition $y \in C^1_P(G, X)$.) For $\alpha, \beta \in G$ and $\delta \in G'$ we have

$$(\delta - 1)y(\alpha\beta) = \alpha\beta u(\beta^{-1}\alpha^{-1}\delta\alpha\beta) - u(\delta)$$
$$= \alpha[\beta u(\beta^{-1}\alpha^{-1}\delta\alpha\beta) - u(\alpha^{-1}\delta\alpha)] + \alpha u(\alpha^{-1}\delta\alpha) - u(\delta)$$
$$= \alpha(\alpha^{-1}\delta\alpha - 1)y(\beta) + (\delta - 1)y(\alpha) = (\delta - 1)(\alpha y(\beta) + y(\alpha)).$$

This shows that $\partial y(\alpha, \beta) \in X^{G'}$.

Now we can find a map $v : G \to X$ satisfying

$(1.8.2) \quad \alpha u(\alpha^{-1}\delta\alpha) - u(\delta) = (\delta - 1)v(\alpha)$ *for every* $\alpha \in G$ *and every* $\delta \in G'$,

$(1.8.3) \quad v(\delta) = u(\delta)$ *for every* $\delta \in G'$,

$(1.8.4) \quad v(\alpha\delta) = v(\alpha) + \alpha u(\delta)$ *for every* $\alpha \in G$ *and every* $\delta \in G'$,

$(1.8.5) \quad v(\pi) \in (\pi - 1)X$ *for every* $\pi \in Q$.

Indeed, take an element y of $C^1_P(G, X)$ as in (1.8.1) such that $y = u$ on G'; then $y(1) = 0$. In view of (1.6.2), we can find a complete set S of representatives for G/G' containing $Q \cup \{1\}$. Put $v(\alpha\gamma) = y(\alpha) + \alpha u(\gamma)$ for $\alpha \in S$ and $\gamma \in G'$. Then it can easily be seen that the above (1.8.2-5) are satisfied. We see that $\partial v(\alpha, \beta) \in X^{G'}$ as above. Moreover, from (1.8.2) and (1.8.5) we can easily derive

$(1.8.6) \quad v(\delta\alpha) = u(\delta) + \delta v(\alpha)$ *for every* $\alpha \in G$ *and every* $\delta \in G'$.

Further, for $\alpha, \beta \in G$ and $\gamma, \delta \in G'$ we have

$$\partial v(\alpha\gamma, \beta\delta) = \alpha\gamma v(\beta\delta) - v(\alpha\gamma\beta\delta) + v(\alpha\gamma)$$
$$= \alpha\gamma v(\beta) + \alpha\gamma\beta u(\delta) - v(\alpha\beta) - \alpha\beta u(\beta^{-1}\gamma\beta\delta) + v(\alpha) + \alpha u(\gamma)$$
$$= \alpha v(\beta) - v(\alpha\beta) + v(\alpha) = \partial v(\alpha, \beta).$$

Therefore we can define an element z of $Z^2(G/G', X^{G'})$ by $z\big(\nu(\alpha), \nu(\beta)\big) = \partial v(\alpha, \beta)$ for $\alpha, \beta \in G$. We define a map $\lambda : H^1_P(G', X)^G \to H^2(G/G', X^{G'}, \mathfrak{Q})$ by

$$(1.8.7) \qquad \lambda(\text{the cohomology class of } u) = \text{the cohomology class of } z.$$

To see that this is well-defined, let u' be an element of $Z^1_P(G', X)^G$ such that $u - u' \in B^1(G', X)$, and let $v' : G \to X$ be a map satisfying (1.8.2-5) for u'. Then $u'(\gamma) - u(\gamma) = (\gamma - 1)x$ with an element x of X independent of γ. Put $w(\alpha) = v'(\alpha) - v(\alpha) - (\alpha - 1)x$ for $\alpha \in G$. From (1.8.4) and (1.8.6) we see that $w(\alpha\gamma) = w(\alpha)$ and $w(\delta\alpha) = \delta w(\alpha)$ for $\alpha \in G$ and $\gamma, \delta \in G'$. Hence $w(\alpha) \in X^{G'}$, and $w(\alpha)$ depends only on $\alpha G'$. Further $w(\pi) \in (\pi - 1)X \cap X^{G'}$ for every $\pi \in Q$. Therefore we find an element s of $C^1(G/G', X^{G'}, \mathfrak{Q})$ such that $s\big(\nu(\alpha)\big) = w(\alpha)$. Since $\partial w = \partial v' - \partial v$, this shows that the class of z is independent of the choice of u and v.

Suppose that u is the restriction of an element y of $Z^1_P(G, X)$. Then we can take y as v, so that $\partial v = 0$, and hence $\lambda \circ \mathrm{Res} = 0$. Suppose next that $z \in B^2(G/G', X^{G'}, \mathfrak{Q})$. Then there exists an element t of $C^1(G/G', X^{G'}, \mathfrak{Q})$ such that $z = \partial t$. Define a map $w : G \to X$ by $w(\alpha) = t\big(\nu(\alpha)\big)$. Put $y = v - w$. Then $\partial y = 0$, $y(\pi) \in (\pi - 1)X$ for every $\pi \in Q$, and hence $y \in Z^1_Q(G, X) = Z^1_P(G, X)$. We see that $t(1) = \partial t(1, 1) = \partial v(1, 1) = v(1) = 0$, and hence $w(\gamma) = t(1) = 0$ if $\gamma \in G'$. Therefore $y(\gamma) = v(\gamma) = u(\gamma)$ for $\gamma \in G'$, so that u is the restriction of y to G'. This completes the proof of our proposition.

1.9. Let us now assume that $X^{G'} = X$ in the setting of §§1.5 and 1.6. Then $\mathfrak{Q}$ is elementary (see §1.1), and hence $H^2(G/G', X^{G'}, \mathfrak{Q}) = H^2_Q(G/G', X)$. Moreover, $B^1(G', X) = \{0\}$, and $H^1_P(G', X)^G$ can be identified with the module of all homomorphisms $u : G' \to X$ such that $u(\pi) = 0$ for all $\pi \in P'$, and $\alpha u(\alpha^{-1}\delta\alpha) = u(\delta)$ for all $\alpha \in G$ and all $\delta \in G'$. Therefore the map $v : G \to X$ defined by

$$(1.9.1) \qquad\qquad v(\alpha\delta) = \alpha u(\delta) \quad \text{for } \alpha \in S \text{ and } \delta \in G'$$

satisfies the above (1.8.2-5). Notice that $v(\alpha) = 0$ for every $\alpha \in S$. Now, given $\alpha, \beta \in S$, take $\gamma \in S$ so that $\nu(\alpha\beta) = \nu(\gamma)$. Then $\lambda(u)$ is represented by a cocycle z defined by

$$(1.9.2) \qquad z\big(\nu(\alpha), \nu(\beta)\big) = \partial v(\alpha, \beta) = -v(\alpha\beta) = -\gamma u(\gamma^{-1}\alpha\beta).$$

2. Cohomology of Fuchsian groups

2.1. In this section G will denote a Fuchsian group of the first kind acting on the upper half complex plane $\mathfrak{H} = \big\{ z \in \mathbf{C} \,\big|\, \mathrm{Im}(z) > 0 \big\}$, by which we mean that G is a discrete subgroup of $SL_2(\mathbf{R})/\{\pm 1_2\}$ such that $G \backslash \mathfrak{H}$ has finite measure.

We shall consider $H_Q^i(G, X)$ with a certain subset Q of the set P of all parabolic elements of G; X will be an arbitrary $R[G]$-module, where R is a commutative ring with identity element. The main purpose of this section is to establish an isogeny of $H_Q^i(G, X)$ onto a certain cohomology group defined with respect to a simplicial complex on $\mathfrak{H}$, and thereby compute $H_Q^i(G, X)$. If $G\backslash\mathfrak{H}$ is compact and G acts freely on $\mathfrak{H}$, such an isogeny is actually an isomorphism and a special case of a well-known isomorphism due to Hopf, Eilenbeerg, MacLane, and Eckmann. It is therefore our task to modify the standard argument so that the difficulty arising from parabolic and elliptic elements of G can be avoided.

2.2. To construct a simplicial complex on $\mathfrak{H}$, we first choose a set $\{\varepsilon_1, \ldots, \varepsilon_\lambda\}$ of representatives of elliptic elements of G, i.e., a minimal set such that every elliptic elements of G is conjugate (in G) to a power of some ε_j. Let m_j be the order of ε_j, and M the least common multiple of $m_1, \ldots, m_\lambda$. We put $M = 1$ if $\{\varepsilon_j\}$ is empty. Let $\mathfrak{H}^*$ be the union of $\mathfrak{H}$ and the cusps of G. Then $G\backslash\mathfrak{H}^*$ has a structure of a compact Riemann surface. Let $r_1, \ldots, r_\mu$ be the points of $G\backslash\mathfrak{H}^*$ corresponding to the cusps of $\mathfrak{H}^*$. Take a small open disk D_k on $G\backslash\mathfrak{H}^*$ containing r_k so that the closures of $D_1, \ldots, D_\mu$ are mutually disjoint. For example, if the point at infinity is a cusp of G corresponding to r_k, we can take D_k to be the image of $\{\infty\} \cup \{z \in \mathfrak{H} \mid \mathrm{Im}(z) > y_0\}$ under the natural map $\mathfrak{H}^* \to G\backslash\mathfrak{H}^*$ for a suitably large y_0.

Let $\mathfrak{H}_0$ be the inverse image of $(G\backslash\mathfrak{H}^*) - \left(\bigcup_{k=1}^\mu D_k\right)$ under the map $\mathfrak{H}^* \to G\backslash\mathfrak{H}^*$. We make a simplicial complex K with the underlying space $\mathfrak{H}_0$ so that the following conditions (2.2.1-4) are satisfied:

(2.2.1) *Every element of G induces a simplicial map of K onto itself.*
(2.2.2) *The fixed point of ε_j on $\mathfrak{H}$ is a 0-simplex of K; we denote it by t_j.*
(2.2.3) *There exists a 1-chain d_k of K that is mapped onto the boundary of D_k.*
(2.2.4) *There exists a fundamental domain for $G\backslash\mathfrak{H}_0$ whose closure consists of a finite number of simplexes of K.*

We can construct such a K, for example, by taking a fundamental domain for $G\backslash\mathfrak{H}^*$ and removing the parts corresponding to the D_k.

Let $(A_i, \partial, \mathbf{a})$ be the chain complex with coefficients in R obtained from K with the usual boundary operator ∂ and the (unit) augmentation $\mathbf{a}$. Since $\mathfrak{H}_0$ is homeomorphic to $\mathbf{R}^2$, we have an exact sequence

$$(2.2.5) \qquad 0 \longrightarrow A_2 \xrightarrow{\partial} A_1 \xrightarrow{\partial} A_0 \xrightarrow{\mathbf{a}} R \longrightarrow 0.$$

By (2.2.1), A_i becomes an $R[G]$-module, and ∂ commutes with the action of $R[G]$. By (2.2.3) we have

$$(2.2.6) \qquad \partial d_k = \pi_k(q_k) - q_k \qquad (1 \leq k \leq \mu)$$

with a 0-simplex q_k and an element π_k of P. Then every parabolic element of G is conjugate to a power of some π_k. Put $Q = \{\pi_1, \ldots, \pi_\mu\}$.

Let $A^i(G, X)$ denote the module of all $R[G]$-linear maps of A_i into X, and let $\partial : A^i(G, X) \to A^{i+1}(G, X)$ be defined by $\partial u = u\partial$ for $u \in A^i(G, X)$. Further let $A_Q^1 G, X)$ be the submodule consisting of all $u \in A^1(G, X)$ such that $u(d_k) \in (\pi_k - 1)X$ for all k. Then we put

10

$$Z^i(K, G, X) = \{\, u \in A^i(G, X) \mid \partial u = 0 \,\}, \quad B^i(K, G, X) = \partial A^{i-1}(G, X),$$
$$Z_Q^1(K, G, X) = Z^1(K, G, X) \cap A_Q^1(G, X), \quad B_Q^2(K, G, X) = \partial A_Q^1(G, X),$$
$$H_Q^0(K, G, X) = Z^0(K, G, X),$$
$$H_Q^1(K, G, X) = Z_Q^1(K, G, X)/B^1(K, G, X),$$
$$H_Q^2(K, G, X) = Z^2(K, G, X)/B_Q^2(K, G, X).$$

2.3. Proposition. *There exist, for $i = 0, 1, 2$, an R-homomorphism g^i of $H_Q^i(K, G, X)$ into $H_Q^i(G, X)$ and an R-homomorphism f^i of $H_Q^i(G, X)$ into $H_Q^i(K, G, X)$ such that*

$$g^i \circ f^i = M \cdot \text{the identity map of } H_Q^i(G, X),$$
$$f^i \circ g^i = M \cdot \text{the identity map of } H_Q^i(K, G, X).$$

In particular, if R is a field whose characteristic is 0 or prime to M, then $H_Q^i(K, G, X)$ is isomorphic to $H_Q^i(G, X)$.

2.4. To prove this proposition, we employ a well-known chain complex $(E_i, \partial, \mathbf{a})$ consisting of the following:

(2.4.1) E_i for $0 \leq i \in \mathbf{Z}$ is the free R-module generated by all ordered sets $[\alpha_0, \alpha_1, \dots, \alpha_i]$ of $i + 1$ elements of G.

(2.4.2) $\partial : E_i \to E_{i-1}$ is defined by
$$\partial[\alpha_0, \alpha_1, \dots, \alpha_i] = \sum_{\nu=0}^{i} (-1)^\nu [\alpha_0, \dots, \alpha_{\nu-1}, \alpha_{\nu+1}, \dots, \alpha_i].$$

(2.4.3) $\mathbf{a}\big(\sum_\nu b_\nu [\alpha_\nu]\big) = \sum_\nu b_\nu$ for $\sum_\nu b_\nu [\alpha_\nu] \in E_0$ with $b_\nu \in R$.

(2.4.4) G acts on E_i by the rule $\beta[\alpha_0, \alpha_1, \dots, \alpha_i] = [\beta\alpha_0, \beta\alpha_1, \dots, \beta\alpha_i]$.

It is well-known that

(2.4.5) $\qquad \cdots \xrightarrow{\partial} E_2 \xrightarrow{\partial} E_1 \xrightarrow{\partial} E_0 \xrightarrow{\mathbf{a}} R \longrightarrow 0 \quad$ is exact.

Denote by $E^i(X)$ the module of all $R[G]$-linear maps of E_i into X, and define $\partial : E^i(X) \to E^{i+1}(X)$ by $\partial u = u\partial$. Let $C^i(G, X)$ be as in §1.1. For every $u \in C^i(G, X)$ put $\bar{u}([\alpha_0, \dots, \alpha_i]) = \alpha_0 u(\alpha_0^{-1}\alpha_1, \alpha_1^{-1}\alpha_2, \dots, \alpha_{i-1}^{-1}\alpha_i)$. Then we see that $u \mapsto \bar{u}$ gives an R-isomorphism of $C^i(G, X)$ onto $E^i(X)$, and $\partial\bar{u} = \overline{\partial u}$.

We are going to define an R-linear map $f : A_i \to E_i$ such that:

(2.4.6) $\mathbf{a}f = M\mathbf{a}$, $f\partial = \partial f$, and $f\alpha = \alpha f$ for every $\alpha \in G$,

(2.4.7) $f(t_j) = (M/m_j) \sum_{\nu=0}^{m_j - 1} [\varepsilon_j^\nu]$ for $j = 1, \dots, \lambda$,

(2.4.8) $f(d_k) = M[1, \pi_k]$ for $k = 1, \dots, \mu$.

Such an f can be obtained by the standard argument using induction on i, with some care about t_j and d_k. We first define $f(t_j)$ by (2.4.7), and put $f(\alpha(t_j)) = \alpha f(t_j)$ for all $\alpha \in G$. Then take a finite set S_0 of 0-simplexes so that every 0-simplex, other than the elliptic points of G, can be written $\alpha(p)$ with a unique $p \in S_0$ and a unique $\alpha \in G$. We include the points q_k of (2.2.6) in S_0. Then we put $f(\alpha(p)) = M[\alpha]$ for every $\alpha \in G$ and every $p \in S_0$. Similarly we fix a finite set S_i of i-simplexes for $i = 1, 2$ so that the $\alpha(s)$ for all $\alpha \in G$

and all $s \in S_i$ form a free R-basis of A_i. We include the d_k in S_1. Obviously $\mathbf{a}f = M\mathbf{a}$. Therefore $\mathbf{a}f(\partial s) = 0$ for every $s \in S_1$. By virtue of (2.4.5) we can define $f(s)$ so that $\partial f(s) = f(\partial s)$. In particular, we can put $f(d_k) = M[1, \pi_k]$ without contradiction. Then we put $f(\alpha(s)) = \alpha f(s)$ for every $\alpha \in G$. Next, for $s \in S_2$ we have $\partial f(\partial s) = 0$; hence we can define $f(s)$ so that $\partial f(s) = f(\partial s)$, in view of (2.4.5). Then we put $f(\alpha(s)) = \alpha f(s)$. Thus we obtain the desired f.

To an element $u \in C^i(G, X)$, we assign an element w of $A^i(G, X)$ by $w = \overline{u} \circ f$. If $u \in C^1_Q(G, X)$, then $w(d_k) = Mu(\pi_k) \in (\pi_k - 1)X$, and hence $u \in A^1_Q(G, X)$. Moreover it can easily be seen that the correspondence $u \mapsto w$ commutes with ∂, and hence defines a homomorphism f^i of $H^i_Q(G, X)$ into $H^i_Q(K, G, X)$.

By a similar argument we can define an R-linear map $g : E_i \to A_i$ satisfying the following two conditions:

(2.4.9) $\mathbf{a}g = \mathbf{a}$, $g\partial = \partial g$, and $g\alpha = \alpha g$ for every $\alpha \in G$;

(2.4.10) $g([1, \pi_k]) = d_k + (\pi_k - 1)e_k$ with a 1-chain e_k such that $\partial e_k = p_0 - q_k$.

Here p_0 is a fixed 0-simplex in S_0. To define such a g, first put $g([\alpha]) = \alpha(p_0)$ for every $\alpha \in G$. Then define $g[1, \alpha])$ so that $\partial g([1, \alpha]) = \alpha(p_0) - p_0$, and put $g([\alpha, \beta]) = \alpha g([1, \alpha^{-1}\beta])$. In particular, we can define $g[1, \pi_k])$ as in (2.4.10). Since $\partial g(\partial[1, \alpha, \beta]) = 0$, we can define $g([1, \alpha, \beta])$ so that $\partial g([1, \alpha, \beta]) = g(\partial[1, \alpha, \beta])$ as (2.2.5) is exact. Then we put $g([\alpha, \beta, \gamma]) = \alpha g([1, \alpha^{-1}\beta, \alpha^{-1}\gamma])$.

Now, to an element $x \in A^i(G, X)$ we assign an element y of $C^i(G, X)$ so that $\overline{y} = x \circ g$. If $x \in A^1_Q(G, X)$, then $y(\pi_k) = \overline{y}([1, \pi_k]) = x(d_k) + (\pi_k - 1)x(e_k) \in (\pi_k - 1)X$, so that $y \in C^1_Q(G, X)$. Moreover, we can easily verify that the correspondence $x \mapsto y$ commutes with ∂, and hence defines a homomorphism g^i of $H^i_Q(K, G, X)$ into $H^i_Q(G, X)$.

2.5. Continuing the proof of Proposition 2.3, let us now construct an R-linear map $U : E_i \to E_{i+1}$ with the following properties:

(2.5.1) $U\alpha = \alpha U$ for every $\alpha \in G$;

(2.5.2) $f \circ g - M \cdot$ (the identity map) $= \partial U + U\partial$;

(2.5.3) $U([1, \pi_k]) \in (\pi_k - 1)E_2$.

We first observe that $f(g(x)) = Mx$ for $x \in E_0$. Defining $U = 0$ on E_0, we see that (2.5.2) is satisfied on E_0. Let α be an element of G other than the π_k. Since

$$\partial\{f(g([1, \alpha])) - M[1, \alpha]\} = f(g(\partial[1, \alpha])) - M\partial[1, \alpha] = 0,$$

we can define, in view of (2.4.5), $U([1, \alpha])$ so that $\partial U([1, \alpha]) = f(g([1, \alpha])) - M[1, \alpha]$. If $\alpha = \pi_k$, we have to choose $U([1, \alpha])$ more specifically. Since $\partial f(e_k) = 0$, we can find an element c_k of E_2 so that $\partial c_k = f(e_k)$. Put $U([1, \pi_k]) = (\pi_k - 1)c_k$. In view of (2.4.8) and (2.4.10), we have $f(g([1, \pi_k])) - M[1, \pi_k] = (\pi_k - 1)f(e_k) = \partial U([1, \pi_k])$. Now we put $U([\alpha, \beta]) = \alpha U([1, \alpha^{-1}\beta])$. Then (2.5.2) is true on E_1. Further we have to define $U([1, \alpha, \beta])$ so that

$$\partial U\big([1,\,\alpha,\,\beta]\big) = f\big(g([1,\,\alpha,\,\beta])\big) - M[1,\,\alpha,\,\beta] - U\big(\partial[1,\,\alpha,\,\beta]\big).$$

This is feasible, since the boundary of the right-hand side is 0. Putting

$$U\big([\alpha,\,\beta,\,\gamma]\big) = \alpha U\big([1,\,\alpha^{-1}\beta,\,\alpha^{-1}\gamma]\big),$$

we obtain the desired U.

Let $x \in Z^i(G,\,X)$. Then there exists an element y of $C^{i-1}(G,\,X)$ such that $\overline{y} = \overline{x} \circ U$. By (2.5.2) we obtain $\overline{x} \circ f \circ g - M\overline{x} = \partial \overline{y}$. (If $i \leq 1$, then $y = 0$.) If $i = 2$, then $y(\pi_k) = \overline{x}\big(U([1,\,\pi_k])\big) = (\pi_k - 1)\overline{x}(c_k) \in (\pi_k - 1)X$, and hence $y \in C_Q^1(G,\,X)$. This shows that $g^i \circ f^i = M \cdot$ (the identity map) for $i = 0,\,1,\,2$.

Similarly we obtain an R-linear map $V : A_i \to A_{i+1}$ with the following properties:

(2.5.4) $V\alpha = \alpha V$ for every $\alpha \in G$;

(2.5.5) $g \circ f - M \cdot$ (the identity map) $= \partial V + V\partial$;

(2.5.6) $V(d_k) = 0$.

Since $\mathbf{a} \circ g \circ f = M \cdot \mathbf{a}$ on A_0, we can define $V(s)$ for $s \in S_0$ so that $\partial V(s) = g\big(f(s)\big) - Ms$. In particular, we can put $V(q_k) = Me_k$. As for t_j, we take a 1-chain h_j in A_1 so that $\partial h_j = p_0 - t_j$, and put $V(t_j) = (M/m_j)\sum_{\nu=0}^{m_j-1}\varepsilon_j^\nu(h_j)$. Then we can put $V\big(\alpha(p)\big) = \alpha V(p)$ for $\alpha \in G$ and an arbitrary 0-simplex p without contradiction. By a procedure similar to the construction of U, we define V on S_1 and S_2 so that (2.5.5) is satified, and put $V\big(\alpha(s)\big) = \alpha V(s)$ for $\alpha \in G$ and $s \in S_i$. The choice of $V(d_k)$ as in (2.5.6) is possible in view of (2.4.8) and (2.4.10). Notice that $V = 0$ on A_2. Now let $u \in Z^i(K,\,G,\,X)$. Then $u \circ g \circ f - Mu = \partial(u \circ V)$. If $i = 2$, we have $u\big(V(d_k)\big) = 0$, and hence $u \circ V \in A_Q^1(G,\,X)$. This proves that $f^i \circ g^i = M \cdot$ (the identity map), and completes the proof of Proposition 2.3.

2.6. Actually the isomorphism between $H_Q^0(G,\,X)$ and $H_Q^0(K,\,G,\,X)$ can be seen easily. Indeed, if $w \in Z^0(K,\,G,\,X)$, then $w(p)$ is independent of p. Therefore $\gamma w(p) = w\big(\gamma(p)\big) = w(p)$ for all $\gamma \in G$, and hence $w(p) \in X^G = H_Q^0(G,\,X)$. Conversely, any element of X^G corresponds to an element of $H_Q^0(K,\,G,\,X)$. Thus $H_Q^0(K,\,G,\,X)$ is *always* isomorphic to $X^G = H_Q^0(G,\,X)$.

2.7. Proposition. *Let Y be the R-submodule of X generated by $(\alpha - 1)X$ for all $\alpha \in G$. Then $H_Q^2(K,\,G,\,X)$ is isomorphic to X/Y.*

PROOF. Take a fundamental domain F for $G\backslash\mathfrak{H}_0$ as described in (2.2.4). We may assume:

(2.7.1) F *is simply connected;*

(2.7.2) *If $a_1,\,\dots,\,a_r$ are the 2-simplexes contained in F, then*

$$\partial\textstyle\sum_{i=1}^{r} a_i = \sum_{k=1}^{\mu}\alpha_k(d_k) + \sum_{l=1}^{\kappa}(\beta_l - 1)b_l$$

with some $\alpha_k,\,\beta_l \in G$ and some $b_l \in A_1$.

Then G is generated by the γ_l and $\alpha_k \pi_k \alpha_k^{-1}$, and A_2 is generated by the $\gamma(a_i)$ for all i and all $\gamma \in G$. Therefore an element u of $Z^2(K, G, X)$ is determined by the values $u(a_i)$. Let us put $F = \sum_{i=1}^{r} a_i$. Suppose $u(F) \in Y$. Then there exist elements y_k and z_l of X such that

$$(2.7.3) \qquad u(F) = \sum_{k=1}^{\mu} (\alpha_k \pi_k \alpha_k^{-1} - 1) y_k + \sum_{l=1}^{\kappa} (\beta_l - 1) z_l.$$

We can find an element w of $A_Q^1(G, X)$ such that $u = \partial w$, $w(d_k) = (\pi_k - 1)\alpha_k^{-1} y_k$, and $w(b_l) = z_l$. Indeed, we first define the values of w at d_k and b_l as specified. Then we set the values of w at the 1-simplexes lying inside F one by one so that $u(a_j) = w(\partial a_j)$. This is feasible in view of (2.7.3). Then extend w to the whole A_1 by the property $w\gamma = \gamma w$ for all $\gamma \in G$. Thus $u \in B_Q^2(K, G, X)$ if $u(F) \in Y$. Conversely, if $u = \partial w$ with $w \in A_Q^1(G, X)$, then $u(F) = w(\partial F) = \sum_{k=1}^{\mu} w(\alpha_k(d_k)) + \sum_{l=1}^{\kappa} (\beta_l - 1) w(b_l) \in Y$. This completes the proof.

2.8. Proposition. *Suppose that R is a field and X is a finite-dimensional vector space over R. Put*

$$h = \text{the genus of } G\backslash\mathfrak{H}^*, \quad \zeta = \dim(X^G), \quad \zeta' = \dim(X/Y),$$
$$\xi_j = \dim\left(\{\, x \in X \,|\, \varepsilon_j x = x \,\}\right) \qquad (1 \le j \le \lambda),$$
$$\eta_k = \dim\left((\pi_k - 1)X\right) \qquad (1 \le k \le \mu),$$

where $\dim(\ )$ *denotes the dimension over R, the ε_j and π_k are as in §2.2, and Y is as in Proposition 2.7. Then*

$$\dim\left(H_Q^1(K, G, X)\right) = (2h - 2)\dim(X) + \sum_{k=1}^{\mu} \eta_k + \sum_{j=1}^{\lambda} \left(\dim(X) - \xi_j\right) + \zeta + \zeta'.$$

PROOF. Let N_i be the number of G-inequivalent i-simplexes in K. Then we easily see that $N_0 - N_1 + N_2 + \mu = 2 - 2h$,

$$\dim\left(A^0(G, X)\right) = N_0 \cdot \dim(X) - \sum_{j=1}^{\lambda} \left(\dim(X) - \xi_j\right),$$
$$\dim\left(A_Q^1(G, X)\right) = N_1 \cdot \dim(X) - \sum_{k=1}^{\mu} \left(\dim(X) - \eta_k\right),$$
$$\dim\left(A^2(G, X)\right) = N_2 \cdot \dim(X),$$
$$\sum_{i=0}^{2}(-1)^i \dim\left(H_Q^i(K, G, X)\right)$$
$$= \dim\left(A^0(G, X)\right) - \dim\left(A_Q^1(G, X)\right) + \dim\left(A^2(G, X)\right).$$

The desired formula follows immediately from these relations, the result of §2.6, and Proposition 2.7.

2.9. Let P be the set of all parabolic elements of G and let $Q = \{\pi_k\}$ as above. Then P and Q satisfy (1.6.1), and hence $H_P^1(G, X) = H_Q^1(G, X)$ by Proposition 1.7. Thus, in view of Proposition 2.3, $\dim\left(H_P^1(G, X)\right)$ can be computed by means of the formula of Proposition 2.8, provided that R is a field whose characteristic is 0 or prime to M.

We cannot have $H_P^2(G, X) = H_Q^2(G, X)$ in general. It seems natural to consider the submodule of $Z^2(G, X)$ consisting of the cocycles vanishing at all parabolic subgroups of G. With such a submodule, we would be able to obtain a result which does not depend on the choice of Q, although it would make the discussion of §§2.4-2.5 longer and more complicated.

3. Automorphic forms

3.1. For $\begin{bmatrix} u \\ v \end{bmatrix} \in \mathbf{C}^2$ and $0 \leq n \in \mathbf{Z}$ we define a column vector $\begin{bmatrix} u \\ v \end{bmatrix}^n \in \mathbf{C}^{n+1}$ and a representation $\rho_n : GL_2(\mathbf{C}) \to GL_{n+1}(\mathbf{C})$ by

$$\begin{bmatrix} u \\ v \end{bmatrix}^n = {}^t(u^n, u^{n-1}v, \ldots, u^{n-k}v^k, \ldots, uv^{n-1}, v^n),$$

$$\rho_n(\alpha) \begin{bmatrix} u \\ v \end{bmatrix}^n = \left(\begin{bmatrix} u \\ v \end{bmatrix} \right)^n \qquad (\alpha \in GL_2(\mathbf{C})).$$

There exists a unique nondegenerate bilinear form on $\mathbf{C}^{n+1}$ represented by a real matrix Θ_n such that

$$ {}^t\begin{bmatrix} u \\ v \end{bmatrix}^n \cdot \Theta_n \cdot \begin{bmatrix} z \\ w \end{bmatrix}^n = \det \begin{bmatrix} u & z \\ v & w \end{bmatrix}^n.$$

We easily see that

$$ {}^t\Theta_n = (-1)^n \Theta_n, \qquad {}^t\rho_n(\alpha)\Theta_n\rho_n(\alpha) = \det(\alpha)^n \Theta_n,$$

$$ GL_2(\mathbf{R}) \cap \mathrm{Ker}(\rho_n) = \begin{cases} \{1_2\} & \text{if } n \text{ is odd}, \\ \{\pm 1_2\} & \text{if } n \text{ is even}. \end{cases}$$

If $n = 0$, we understand that $\rho_0(\alpha) = \Theta_0 = 1$ for every $\alpha \in GL_2(\mathbf{C})$.

3.2. It is necessary for our later discussion to define a Fuchsian group from a subgroup of $GL_2^+(\mathbf{R})$ which is not necessarily contained in $SL_2(\mathbf{R})$, where $GL_2^+(\mathbf{R}) = \{\alpha \in GL_2(\mathbf{R}) \mid \det(\alpha) > 0\}$. Let Γ be a subgroup of $GL_2^+(\mathbf{R})$ and let $\Delta = \Gamma \cap \mathbf{R}1_2$. We put $\overline{\Gamma} = \Gamma/\Delta$ and assume that $\overline{\Gamma}$ is a Fuchisian group of the first kind. Let X be a Γ-module on which every element of Δ acts as the identity map. Then X can be viewed as a $\overline{\Gamma}$-module. Let P (resp. $\overline{P}$) be the set of parabolic elements of Γ (resp. $\overline{\Gamma}$). Put

$$Z_P^1(\Gamma, \Delta, X) = \{u \in Z_P^1(\Gamma, X) \mid u(e) = 0 \text{ for all } e \in \Delta\}.$$

If ν is the natural map of Γ to $\overline{\Gamma}$, then we obtain an isomorphism of $Z_{\overline{P}}^1(\overline{\Gamma}, X)$ onto $Z_P^1(\Gamma, \Delta, X)$ by $w \mapsto w \circ \nu$. This sends $B^1(\overline{\Gamma}, X)$ onto $B^1(\Gamma, X)$, and so $H_{\overline{P}}^1(\overline{\Gamma}, X)$ is canonically isomorphic to $Z_P^1(\Gamma, \Delta, X)/B^1(\Gamma, X)$. If $X^\Gamma = \{0\}$, we easily see that $Z_P^1(\Gamma, \Delta, X) = Z_P^1(\Gamma, X)$. In such a case, $H_{\overline{P}}^1(\overline{\Gamma}, X)$ can be identified with $H_P^1(\Gamma, X)$.

3.3. For $\alpha = \begin{bmatrix} a & b \\ c & d \end{bmatrix} \in GL_2(\mathbf{R})$ and z in the upper half plane $\mathfrak{H}$ we put

(3.3.1) $\qquad \alpha(z) = (az+b)/(cz+d), \qquad j_\alpha(z) = j(\alpha, z) = cz + d.$

Then we have

(3.3.2) $$\begin{bmatrix} \alpha(z) \\ 1 \end{bmatrix}^n = j_\alpha(z)^{-n} \rho_n(\alpha) \begin{bmatrix} z \\ 1 \end{bmatrix}^n.$$

Let Γ, Δ, and $\overline{\Gamma}$ be as in §3.2. We consider a representation $\Psi : \Gamma \to GL_r(\mathbf{R})$ with any $r \geq 0$ satisfying the following two conditions:

(3.3.3) *Ψ maps Γ into a compact subgroup of $GL_2(\mathbf{R})$;*

(3.3.4) *$\mathrm{Ker}(\Psi)$ is of finite index in Γ if Γ has cusps.*

Then for $0 < m \in \mathbf{Z}$ we denote by $S_m(\Gamma, \Psi)$ the vector space of all holomorphic maps $f : \mathfrak{H} \to \mathbf{C}^r$ satisfying the following two conditions:

(3.3.5) $\det(\alpha)^{m/2} j_\alpha(z)^{-m} f\big(\alpha(z)\big) = \Psi(\alpha) f(z)$ *for all* $\alpha \in \Gamma$;

(3.3.6) *The components of f are cusp forms with respect to $\mathrm{Ker}(\Psi)$ if Γ has cusps. (This is meaningful in view of (3.3.4).)*

If Ψ is absolutely irreducible, then $S_m(\Gamma, \Psi) \neq \{0\}$ only when $\Psi(c) = (c/|c|)^m$ for every $c \in \Delta$. Further, if Ψ is the direct sum of of two representations Ψ_1 and Ψ_2, then $S_m(\Gamma, \Psi)$ can be identified with the direct sum of of $S_m(\Gamma, \Psi_1)$ and $S_m(\Gamma, \Psi_2)$. Therefore, without losing generality, we shall hereafter assume

(3.3.7) $\Psi(c) = (c/|c|)^m$ *for every* $c \in \Delta$.

We can find a positive definite real symmetric matrix W of size r such that $^t\Psi(\alpha) W \Psi(\alpha) = W$ for all $\alpha \in \Gamma$. Then we define a positive definite hermitian metric $\langle f, g \rangle$ on $S_m(\Gamma, \Psi)$ (depending on W) by

$$\langle f, g \rangle = \int_{\Gamma \backslash \mathfrak{H}} {}^t f(z) W \overline{g(z)} y^{m-2} dx dy \qquad \big(f, g \in S_m(\Gamma, \Psi);\ z = x + iy\big).$$

We now consider $S_{n+2}(\Gamma, \Psi)$ with a nonnegative integer n. Our principal aim of this section is to find an isomorphism of $S_{n+2}(\Gamma, \Psi)$ onto $H^1_{\overline{P}}(\overline{\Gamma}, X)$ with a suitable $\overline{\Gamma}$-module X. We first define, for $f \in S_{n+2}(\Gamma, \Psi)$, a holomorphic differential form $\iota(f)$ with values in $\mathbf{C}^r \otimes \mathbf{C}^{n+1}$ by

$$(3.3.8) \qquad \iota(f) = f(z) \otimes \begin{bmatrix} z \\ 1 \end{bmatrix}^n dz.$$

If $n = 0$, we understand that $\iota(f) = f(z) dz$. Put

$$(3.3.9) \qquad Z = W \otimes \Theta_n, \qquad \chi(\alpha) = \det(\alpha)^{-n/2} \Psi(\alpha) \otimes \rho_n(\alpha) \qquad (\alpha \in \Gamma).$$

Then $^t\chi(\alpha) Z \chi(\alpha) = Z$ for every $\alpha \in \Gamma$. In view of (3.3.2) and (3.3.5), we obtain

$$(3.3.10) \qquad \iota(f) \circ \alpha = \chi(\alpha) \iota(f) \qquad \text{for every}\ \ \alpha \in \Gamma,$$

$$(3.3.11) \qquad \mathrm{Re}\big(\iota(f)\big) \circ \alpha = \chi(\alpha) \mathrm{Re}\big(\iota(f)\big) \qquad \text{for every}\ \ \alpha \in \Gamma,$$

where $\circ \alpha$ means the transform of a form by α, and Re means the real part. Therefore we can define an $\mathbf{R}$-valued $\mathbf{R}$-bilinear form $A(f, g)$ on $S_{n+2}(\Gamma, \Psi)$ by

$$(3.3.12) \qquad A(f, g) = \int_{\Gamma \backslash \mathfrak{H}} {}^t\mathrm{Re}\big(\iota(f)\big) \wedge Z \cdot \mathrm{Re}\big(\iota(g)\big).$$

Since $^t\iota(f) \wedge Z \cdot \overline{\iota(g)} = -(2i)^{n+1} \cdot {}^t f(z) W \overline{g(z)} y^n dx \wedge dy$, we have

$$A(f, g) = (2i)^{n-1} \big[\langle f, g \rangle + (-1)^{n+1} \langle g, f \rangle \big],$$

$$A(f, g) = (-1)^{n+1} A(g, f), \qquad A(f, i^{n-1} g) = 2^n \cdot \mathrm{Re}\{\langle f, g \rangle\}.$$

Therefore $A(f, g)$ is nondegenerate.

3.4. We view $\mathbf{R}^r \otimes \mathbf{R}^{n+1}$ as an $\mathbf{R}[\Gamma]$-module through χ, and also as an $\mathbf{R}[\overline{\Gamma}]$-module, in view of (3.3.7). Hereafter we denote this module by X_n^Ψ. Let us now establish an $\mathbf{R}$-linear isomorphism of $S_{n+2}(\Gamma, \Psi)$ onto $H^1_{\overline{P}}(\overline{\Gamma}, X_n^\Psi)$. For every $f \in S_{n+2}(\Gamma, \Psi)$ define a map $u : \Gamma \to X_n^\Psi$ by

$$(3.4.1) \qquad u(\alpha) = \int_z^{\alpha(z)} \mathrm{Re}\big(\iota(f)\big)$$

with an arbitrarily fixed point z of $\mathfrak{H}$. Clearly $u(\alpha)$ does not depend on the path of integration, $u(c) = 0$ for $c \in \Delta$, and $u \in Z^1_P(\Gamma, X_n^\Psi)$ in view of (3.3.11). Moreover, the cohomology class of u depends only on f, and is independent of the choice of z; furthermore $u(\pi) \in (\pi - 1)X_n^\Psi$ for every $\pi \in P$ because of (3.3.6). (For the detailed proof of these, see [S2, p.298].) Consequently $u \in Z^1_P(\Gamma, \Delta, X_n^\Psi)$. Thus we obtain an $\mathbf{R}$-linear map

$$(3.4.2) \quad \varphi : S_{n+2}(\Gamma, \Psi) \longrightarrow H^1_{\overline{P}}(\overline{\Gamma}, X_n^\Psi), \quad \varphi(f) = \text{the cohomology class of } u.$$

In [S2, pp.300-301, formula (19) in particular] we have shown that if u resp. v are cocycles representing $\varphi(f)$ resp. $\varphi(g)$, then

$$
\begin{aligned}
(3.4.3) \qquad A(f, g) = & \sum_{i=1}^h {}^t u(\alpha_i^{-1}) Z\big[v(\alpha_i^{-1}\beta_i\alpha_i\tau_{i-1}) - v(\tau_{i-1})\big] \\
& + \sum_{i=1}^h {}^t u(\beta_i) Z\big[v(\beta_i\alpha_i\tau_{i-1}) - v(\alpha_i^{-1}\beta_i\alpha_i\tau_{i-1})\big] \\
& + \sum_{j=1}^\mu {}^t u(\varepsilon_j^{-1}) Z\big[m_j^{-1} \textstyle\sum_{\nu=0}^{m_j-1} v(\varepsilon_j^\nu) - v(\varepsilon_{j-1}\cdots\varepsilon_1\tau_g)\big] \\
& - \sum_{k=1}^\lambda {}^t u(\pi_k^{-1}) Z v(\pi_{k-1}\cdots\pi_1\varepsilon_\mu\cdots\varepsilon_1\tau_g)\big] \\
& + \sum_{k=1}^\lambda {}^t u(\pi_k^{-1}) Z w_k.
\end{aligned}
$$

Here $\big\{ \alpha_1, \beta_1, \ldots, \alpha_h, \beta_h, \varepsilon_1, \ldots, \varepsilon_\mu, \pi_1, \ldots, \pi_\lambda \big\}$ is a set of generators of $\overline{\Gamma}$ and w_k is an element of X_n such that:

$(3.4.4) \quad \pi_\lambda \cdots \pi_1 \varepsilon_\mu \cdots \varepsilon_1 \beta_h^{-1}\alpha_h^{-1}\beta_h\alpha_h \cdots \beta_1^{-1}\alpha_1^{-1}\beta_1\alpha_1 = 1;$

$(3.4.5)$ *The ε_j and π_k are suitably chosen elliptic and parabolic elements of $\overline{\Gamma}$ as in §2.2; m_j is the order of ε_j;*

$(3.4.6) \quad \tau_i = \beta_i^{-1}\alpha_i^{-1}\beta_i\alpha_i \cdots \beta_1^{-1}\alpha_1^{-1}\beta_1\alpha_1 \qquad (i = 1, \ldots, h);$

$(3.4.7) \quad (1 - \pi_k)w_k = v(\pi_k) \qquad (k = 1, \ldots, \lambda).$

Observe that ${}^t u(\pi_k^{-1}) Z w_k$ does not depend on the choice of w_k.

3.5. Theorem. *For every $n \geq 0$ and Ψ such that $\Psi(c) = (c/|c|)^n$ for every $c \in \Delta$, the map φ of (3.4.2) is an isomorphism.*

PROOF. The proof in some special cases was given in the previous papers: (i) [S2, Theorem 1] when n is even and Ψ is trivial. (ii) [S4, Theorem 2] when n is even and Γ has no cusps; this method applies also to the case of odd n. (iii) [MS, Proposition 4.4] when Γ has no cusps; this includes also the case of the product of several copies of $\mathfrak{H}$.

Here we prove the case where the kernel of Ψ is of finite index in Γ. This together with the previously known results proves our theorem. First, the injectivity of φ can be shown without any assumption. Indeed, suppose $\varphi(f) = 0$.

Then (3.4.3) shows that $A(f, g) = 0$ for all g. Since A is nondegenerate, f must be 0.

Let us now compute the dimension of $H^1_{\overline{P}}(\overline{\Gamma}, X_n)$ by means of Proposition 2.8, where X_n means X_n^{Ψ} with trivial Ψ. Let ε_j and π_k be defined for $\overline{\Gamma}$ as in §§2.2 and 3.4, and let ξ_j, η_k, ζ, and ζ' be as in Proposition 2.8. Suppose first that $n = 0$. We have $\eta_k = 0$ and $\xi_j = \zeta = \zeta' = \dim(X_0) = 1$, and hence $\dim\left(H^1_{\overline{P}}(\overline{\Gamma}, X_0)\right) = 2h$.

Next suppose that $n > 0$. We have $\zeta = \zeta' = 0$ in view of the irreducibility of the restriction of ρ_n to Γ (see [B1]). Let m_j be the order of ε_j. Then the action of ε_j on X_n has $n + 1$ characteristic roots $\{\omega^{n-2i}\}_{i=0}^n$ with a root of unity ω whose order is m_j or $2m_j$ according as n is odd or even. Therefore $\dim(X_n) - \xi_j$ is the number of these roots of unity different from 1. Now we can show that

$$\dim(X_n) - \xi_j = 2\left[(n + 2)(1 - m_j^{-1})/2\right],$$

where $[x]$ means the largest integer $\leq x$. The verification is somewhat intricate but quite elementary. To determine η_k, we take an element π'_k of Γ that represents π_k. We see that the Jordan form of π'_k is $\begin{bmatrix} \pm 1 & 1 \\ 0 & \pm 1 \end{bmatrix}$, and hence

$$\eta_k = \begin{cases} n + 1 & \text{if } n \text{ is odd and } \operatorname{tr}(\pi'_k) = -2, \\ n & \text{otherwise.} \end{cases}$$

By Proposition 2.8 we thus obtain

$$\dim\left(H^1_{\overline{P}}(\overline{\Gamma}, X_n)\right)$$

$$= \begin{cases} 2h & \text{if } n = 0, \\ 2(h-1)(n+1) + \lambda n + \lambda' + 2\sum_{j=1}^{\mu}\left[(n+2)(1-m_j^{-1})/2\right] & \text{if } n > 0, \end{cases}$$

where λ denotes the number of inequivalent cusps of Γ, and

$$\lambda' = \begin{cases} \text{the number of } \pi'_k \text{ such that } \operatorname{tr}(\pi'_k) = -2 & \text{if } n \text{ is odd,} \\ 0 & \text{if } n \text{ is even.} \end{cases}$$

Therefore we find that

$$(3.5.1) \qquad \dim\left(H^1_{\overline{P}}(\overline{\Gamma}, X_n)\right) = 2 \cdot (\text{the complex dimension of } S_{n+2}(\Gamma, \Psi_0)),$$

where Ψ_0 means the trivial representation. Indeed, the dimension of $S_{n+2}(\Gamma, \Psi_0)$ can easily be computed by the Riemann-Roch theorem, and agrees with the half of $\dim\left(H^1_{\overline{P}}(\overline{\Gamma}, X_n)\right)$ given as above. Equality (3.5.1) was first proved by Eichler [E2] for even n. Now the injectivity of φ together with (3.5.1) proves our theorem for trivial Ψ. Before discussing the case of nontrivial Ψ, let us consider Hecke operators.

3.6. Let Γ_1 and Γ_2 be subgroups of $GL_2^+(\mathbf{R})$ of the above type, Δ_i the group of all scalar matrices in Γ_i, $\overline{\Gamma}_i = \Gamma_i/\Delta_i$, and P_i resp. $\overline{P}_i$ the set of all parabolic elements of Γ_i resp. $\overline{\Gamma}_i$. Further let η be an element of $GL_2^+(\mathbf{R})$ such that Γ_2 and $\eta\Gamma_1\eta^{-1}$ are commensurable. Then $\Gamma_2\eta\Gamma_1$ can be expressed as a disjoint union

$$(3.6.1) \qquad \Gamma_2 \eta \Gamma_1 = \bigcup_{i=1}^{d} \eta_i \Gamma_1$$

with finitely many elements η_i. Let Γ' be the smallest semigroup in $GL_2(\mathbf{R})$ containing Γ_1, Γ_2, and η and let $R[\Gamma']$ be the semi-group-ring of Γ' over a ring R as in §1.1. Further let X be an $R[\Gamma']$-module such that every element of $\Delta_1 \cup \Delta_2$ acts on X as the identity map. Then X may be viewed as an $R[\overline{\Gamma}_i]$-module. We are going to define an R-linear map

$$(3.6.2) \qquad (\Gamma_2 \eta \Gamma_1)_X : H^1_{\overline{P}_1}(\overline{\Gamma}_1, X) \longrightarrow H^1_{\overline{P}_2}(\overline{\Gamma}_2, X).$$

First, for every $u \in Z^1(\Gamma_1, X)$, define a map $v : \Gamma_2 \to X$ as follows: Given $\alpha \in \Gamma_2$ and η_i, let $\alpha^{-1}\eta_i = \eta_j \alpha_i^{-1}$ with some j and $\alpha_i \in \Gamma_1$; obviously $\eta_i \mapsto \eta_j$ is a permutation of $\{\eta_i\}_{i=1}^{d}$; put then

$$(3.6.3) \qquad v(\alpha) = \sum_{i=1}^{d} \eta_i u(\alpha_i).$$

It can easily be seen that $v \in Z^1(\Gamma_2, X)$; moreover, $v \in B^1(\Gamma_2, X)$ if $u \in B^1(\Gamma_1, X)$. Now the cohomology class of v does not depend on the choice of the η_i. Indeed, let $\zeta_i = \eta_i \gamma_i$ with $\gamma_i \in \Gamma_1$. Then $\alpha^{-1}\zeta_i = \zeta_j(\gamma_i^{-1}\alpha_i\gamma_j)^{-1}$ and

$$(3.6.4) \qquad \sum_i \zeta_i u(\gamma_i^{-1}\alpha_i\gamma_j) = v(\alpha) + (\alpha - 1)\sum_j \eta_j u(\gamma_j).$$

Thus we obtain the same cohomology class as before. We shall now show that $v \in Z^1_{P_2}(\Gamma_2, X)$ if $u \in Z^1_{P_1}(\Gamma_1, X)$. Let $\pi \in P_2$ and for a given i let $\pi^{-1}\eta_i = \eta_j \beta_i^{-1}$ with some j and $\beta_i \in \Gamma_1$. In view of (3.6.4) it is sufficient to show that $\sum_i \eta_i u(\beta_i) \in (\pi - 1)X$ with a suitable choice of the η_i. (Notice that we may choose the η_i depending on π.) Take the subgroup Λ generated by π, and consider disjoint decompositions

$$\Gamma_2 \eta \Gamma_1 = \bigcup_\zeta \Lambda\zeta\Gamma_1, \qquad \Lambda\zeta\Gamma_1 = \bigcup_{\nu=0}^{k-1} \pi^\nu \zeta\Gamma_1,$$

where k is the smallest positive integer such that $\pi^k \in \zeta\Gamma_1\zeta^{-1}$, which may depend on ζ. Then we take $\{\pi^\nu \zeta\}$ to be $\{\eta_i\}$. Since $\pi^{-1}\pi^\nu\zeta = \pi^{\nu-1}\zeta$ for $\nu > 0$ and $\pi^{-1}\zeta = \pi^{k-1}\zeta(\zeta^{-1}\pi^k\zeta)^{-1}$, we have $\sum_i \eta_i u(\beta_j) = \sum_\zeta \zeta u(\zeta^{-1}\pi^k\zeta)$. Since $\zeta^{-1}\pi^k\zeta \in P_1$, we have $u(\zeta^{-1}\pi^k\zeta) = (\zeta^{-1}\pi^k\zeta - 1)y_\zeta$ with an element y_ζ of X, and hence $\sum_i \eta_i u(\beta_i) = \sum_\zeta(\pi^k - 1)\zeta y_\zeta \in (\pi - 1)X$, Q.E.D.

A similar and simpler argument shows that $v \in Z^1_{P_2}(\Gamma_2, \Delta_2, X)$ if $u \in Z^1_{P_1}(\Gamma_1, \Delta_1, X)$, the notation being as in §3.2. Assigning the cohomology class of v to the cohomology class of u, we thus obtain an R-linear map $(\Gamma_2 \eta \Gamma_1)_X$.

3.7. The notation being the same as in §3.6, let Ψ be a multiplicative map of Γ' into $GL_r(\mathbf{R})$ which maps Γ_1 and Γ_2 into compact subgroups of $GL_r(\mathbf{R})$. Define $\chi(\alpha)$ for $\alpha \in \Gamma'$ by (3.3.9). We can now define a $\mathbf{C}$-linear map

$$(3.7.1) \qquad (\Gamma_2 \eta \Gamma_1)_{n+2, \Psi} : S_{n+2}(\Gamma_1, \Psi) \longrightarrow S_{n+2}(\Gamma_2, \Psi) \quad \text{by}$$

$$(\Gamma_2 \eta \Gamma_1)_{n+2, \Psi} f = \sum_{i=1}^{d} \det(\eta_i)^{-(n+2)/2}\Psi(\eta_i)f(\eta_i^{-1}(z))j(\eta_i^{-1}, z)^{-n-2}$$

for $f \in S_{n+2}(\Gamma_1, \Psi)$. It can easily be seen that this is well-defined independently of the choice of the η_i; moreover,

$$(3.7.2) \qquad \iota\big((\Gamma_2 \eta \Gamma_1)_{n+2, \Psi} f\big) = \sum_{i=1}^{d} \chi(\eta_i)\iota(f) \circ \eta_i^{-1}.$$

We see by a straightforward calculation that if a cocycle u represents $\varphi(f)$, then the cocycle v defined by (3.6.3) represents $\varphi\big((\Gamma_2\eta\Gamma_1)_{n+2,\Psi}f\big)$ (cf. [S2, pp.307-308]). We can express this fact by an equality

$$(3.7.3) \qquad (\Gamma_2\eta\Gamma_1)_X \circ \varphi = \varphi \circ (\Gamma_2\eta\Gamma_1)_{n+2,\Psi}, \quad \text{where} \quad X = X_n^{\Psi}.$$

3.8. Coming back to the original Γ and Ψ such that $[\Gamma : \mathrm{Ker}(\Psi)] < \infty$, let Γ_0 be the kernel of Ψ, P_0 the set of all parabolic elements of Γ_0, and Ψ_0 the trivial representation of Γ_0. Consider $\Gamma\eta\Gamma_0$ by taking η to be the identity element. We see that $S_{n+2}(\Gamma_0, \Psi)$ is the direct sum of r copies of $S_{n+2}(\Gamma_0, \Psi_0)$; hence the map $\varphi : S_{n+2}(\Gamma_0, \Psi) \to H^1_{\overline{P}_0}(\overline{\Gamma}_0, X_n^{\Psi})$ is surjective by what we have already proved. In view of this fact and (3.7.3), in order to prove Theorem 3.5 for Γ and Ψ, it is sufficient to show that $(\Gamma\Gamma_0)_X$ is surjective. To show this, let $\Gamma = \bigcup_{i=1}^d \eta_i\Gamma_0$, and $t \in Z^1_P(\Gamma, \Delta, X)$. Let u be the restriction of t to Γ_0. Define v by (3.6.3) with $\alpha \in \Gamma$ and $\alpha_i \in \Gamma_0$. Then

$$v(\dot\alpha) = \sum_{i=1}^d \eta_i t(\eta_i^{-1}\alpha\eta_j) = d \cdot t(\alpha) + (\alpha - 1)\sum_{i=1}^d t(\eta_i).$$

This means that v belongs to the same cohomology class as $d \cdot t$, and hence $(\Gamma\Gamma_0)_X$ is surjective. This completes the proof of Theorem 3.5.

4. The algebro-geometric realization of $H^1_{\overline{P}}(\overline{\Gamma}, X) \otimes (\mathbf{Q}/\mathbf{Z})$

4.1. Let Γ, Δ, $\overline{\Gamma}$, P, and $\overline{P}$ be as in §§3.2 and 3.4. We consider a free $\mathbf{Z}$-module X of a finite rank m on which Γ acts, and a family of normal subgroups $\{\Gamma_a\}$ of Γ whose member Γ_a is defined for each positive integer a. Put $P_a = P\cap\Gamma_a$, $\Delta_a = \Delta\cap\Gamma_a$, $\overline{\Gamma}_a = \Gamma_a/\Delta_a$. We assume the following conditions on these:

(4.1.1) $\Gamma_a \subset \Gamma_b$ if b divides a; $\Gamma_1 = \Gamma$.

(4.1.2) $[\Gamma : \Gamma_a] < \infty$.

(4.1.3) *There exists a positive integer a_0 such that $\overline{\Gamma}_a$ has no element of finite order other than the identity element if $a \geq a_0$.*

(4.1.4) *Every element of Δ acts as the identity map on X; hence X can be viewed as a $\overline{\Gamma}$-module.*

(4.1.5) *$(\gamma - 1)X \subset aX$ for every $\gamma \in \Gamma_a$.*

In the next section we shall discuss $\{\Gamma_a\}$ and X given more explicitly. In this section, however, we shall treat them axiomatically only under the above conditions and (4.3.1) below.

We denote by V_a the compact Riemann surface isomorphic to $\Gamma_a\backslash\mathfrak{H}^*$, and fix a projection map φ_a of $\mathfrak{H}^*$ onto V_a. We take the Jacobian variety J_a of V_a, and for a divisor $\mathfrak{d}$ on V_a of degree 0 we denote by $\psi_a(\mathfrak{d})$ the point of J_a corresponding to $\mathfrak{d}$. For every $\alpha \in \Gamma$ we define an automorphism $R_a(\alpha)$ of V_a and an automorphism $r_a(\alpha)$ of J_a so that $R_a(\alpha) \circ \varphi_a = \varphi_a \circ \alpha$ and $r_a(\alpha) \circ \psi_a = \psi_a \circ R_a(\alpha)$. We can naturally define $r_a(\alpha)$ and $R_a(\alpha)$ also for $\alpha \in \overline{\Gamma}$, since $R_a(c)$ is the identity map for every $c \in \Delta$.

4.2. Observe that $X \otimes J_a$ can be viewed as an abelian variety, which is isomorphic to the product of m copies of J_a. For α, $\beta \in \Gamma$ we see that $\alpha \otimes r_a(\beta)$ is an automorphism of this abelian variety. Put

$$(4.2.1) \qquad W_a = W(a)$$
$$= \{\, t \in X \otimes J_a \mid at = 0 \text{ and } (\alpha \otimes r_a(\alpha))t = t \text{ for every } \alpha \in \Gamma \,\}.$$

We assume hereafter $a \geq a_0$ with a_0 of (4.1.3). Let us now define an isomorphism

$$(4.2.2) \qquad \kappa_a : H^1_{\overline{P}}\big(\overline{\Gamma}_a,\, a^{-1}X/X\big)^{\overline{\Gamma}} \longrightarrow W_a,$$

where we consider $a^{-1}X$ in $X_{\mathbf{Q}} = X \otimes_{\mathbf{Z}} \mathbf{Q}$. By (4.1.5), $\overline{\Gamma}_a$ acts trivially on $a^{-1}X/X$, and hence $B^1\big(\overline{\Gamma}_a,\, a^{-1}X/X\big) = \{0\}$, and $H^1_{\overline{P}}\big(\overline{\Gamma}_a,\, a^{-1}X/X\big)^{\overline{\Gamma}}$ can be identified with the module of all homomorphisms u of $\overline{\Gamma}_a$ into $a^{-1}X/X$ such that

$$(4.2.3) \qquad \alpha u(\alpha^{-1}\gamma\alpha) = u(\gamma) \quad \text{for every } \gamma \in \overline{\Gamma}_a \text{ and } \alpha \in \overline{\Gamma},$$

$$(4.2.4) \qquad u(\pi) = 0 \quad \text{for every } \pi \in \overline{P} \cap \overline{\Gamma}_a.$$

Next we fix coordinate functions $\theta_i : X_{\mathbf{Q}} \to \mathbf{Q}$ so that

$$(4.2.5) \qquad x \mapsto \big(\theta_1(x), \,\ldots,\, \theta_m(x)\big)$$

gives an isomorphism of $X_{\mathbf{Q}}$ onto $\mathbf{Q}^m$ which maps X onto $\mathbf{Z}^m$. Given $u \in \mathrm{Hom}\big(\overline{\Gamma}_a,\, a^{-1}X/X\big)$ satisfying (4.2.4), we can define, for $1 \leq \nu \leq m$, a character u_ν^* of $\overline{\Gamma}_a$ by

$$(4.2.6) \qquad u_\nu^*(\gamma) = \exp\big[2\pi i \theta_\nu\big(u(\gamma)\big)\big] \qquad (\gamma \in \overline{\Gamma}_a).$$

Then there exists a meromorphic function f_ν on $\mathfrak{H}$, meromorphic even at the cusps, such that

$$(4.2.7) \qquad f_\nu\big(\gamma(z)\big) = u_\nu^*(\gamma) f_\nu(z) \quad \text{for every } \gamma \in \overline{\Gamma}_a.$$

We can define the divisor of f_ν on V_a, written $\mathrm{div}(f_\nu)$, whose degree is 0. By virtue of (4.1.3) this is a divisor with integral coefficients in the usual sense; no fractional coefficient is necessary. Put $t_\nu = \psi_a\big(\mathrm{div}(f_\nu)\big)$. Then $at_\nu = 0$, since f_ν^a is a meromorphic function on V_a; t_ν is determined only by u, independently of the choice of f_ν. Conversely, every a-division point of J_a is obtained in this manner from a character of $\overline{\Gamma}_a$, trivial on $\overline{P} \cap \overline{\Gamma}_a$, whose values are a-th roots of unity. Thus u determines a point $t = (t_1, \ldots, t_m)$ of $J_a^m = X \otimes J_a$, which is obviously independent of the choice of θ_ν. We write $t = \kappa_a(u)$, which is the definition of the map of (4.2.2). (Note: $\kappa_a(u)$ can be defined without (4.2.3).)

Let us now show that u satisfies (4.2.3) if and only if $t \in W_a$. First define a representation $\rho : \mathbf{Z}[\overline{\Gamma}] \to M_m(\mathbf{Z})$ by

$$(4.2.8) \quad \rho(\alpha) = \big(\rho_{\mu\nu}(\alpha)\big), \quad \theta_\mu(\alpha x) = \textstyle\sum_{\nu=1}^m \rho_{\mu\nu}(\alpha)\theta_\nu(x) \qquad (\alpha \in \overline{\Gamma},\, x \in X).$$

For m-tuples $f = (f_\nu)_{\nu=1}^m$ and $g = (g_\nu)_{\nu=1}^m$ of meromorphic functions, we write $f \cdot g = (f_\nu g_\nu)_{\nu=1}^m$ and define f^α for $\alpha \in \mathbf{Z}[\overline{\Gamma}]$ by $f^\alpha = (h_\nu)_{\nu=1}^m$ with $h_\mu = \prod_{\nu=1}^m f_\nu^{\rho_{\mu\nu}(\alpha)}$. Then we have

$$(4.2.9) \qquad (f^\alpha)^\beta = f^{\beta\alpha} \qquad (\alpha,\, \beta \in \mathbf{Z}[\overline{\Gamma}]).$$

If $f = (f_\nu)_{\nu=1}^m$ is obtained from u through (4.2.6) and (4.2.7), then it can easily be seen that $(f \circ \alpha^{-1})^\alpha$ corresponds to u' defined by $u'(\gamma) = \alpha u(\alpha^{-1}\gamma\alpha)$. Thus $\kappa_a(u) \in W_a$ if and only if (4.2.3) is satified. In this way we see that the map κ_a of (4.2.2) is indeed an isomorphism.

4.3. We fix a finite set Q of parabolic elements of $\overline{\Gamma}$ so that every element of $\overline{P}$ is of the form $\alpha\pi^i\alpha^{-1}$ with $\pi \in Q$, $\alpha \in \overline{\Gamma}$, and $i \in \mathbf{Z}$. Further we assume, in addition to (4.1.1-5), the following condition:

(4.3.1) *There exists a positive integer a_1 such that no two distinct elements of $Q \cup \{1\}$ belong to the same coset modulo $\overline{\Gamma}_a$ if $a \geq a_1$.*

We assume hereafter $a \geq a_1$. Let $\nu : \overline{\Gamma} \to \overline{\Gamma}/\overline{\Gamma}_a$ be the natural map, and let $T_a = \nu(Q)$. In the following treatment T_a will appear only as a subscript. Therefore we shall write it simply T, since no confusion is expected. By Proposition 1.8 we have an exact sequence

$$(4.3.2) \quad H^1_{\overline{P}}(\overline{\Gamma},\, a^{-1}X/X) \xrightarrow{\mathrm{Res}} H^1_{\overline{P}}(\overline{\Gamma}_a,\, a^{-1}X/X)^{\overline{\Gamma}} \xrightarrow{\lambda_a} H^2_T(\overline{\Gamma}/\overline{\Gamma}_a,\, a^{-1}X/X),$$

where λ_a is the map defined by (1.8.7) with respect to Q. Let E_a^m denote the product of m copies of a group

$$E_a = \big\{\, \zeta \in \mathbf{C} \,\big|\, \zeta^a = 1 \,\big\}.$$

Then the map $x \mapsto \big(\exp\big[2\pi i\theta_\nu(x)\big]\big)_{\nu=1}^m$ gives an isomorphism of $a^{-1}X/X$ onto E_a^m, which induces an isomorphism

$$\iota : H^2_T(\overline{\Gamma}/\overline{\Gamma}_a,\, a^{-1}X/X) \longrightarrow H^2_T(\overline{\Gamma}/\overline{\Gamma}_a,\, E_a^m).$$

We are going to establish a commutative diagram

$$(4.3.3) \qquad \begin{array}{ccc} H^1_{\overline{P}}(\overline{\Gamma}_a,\, a^{-1}X/X)^{\overline{\Gamma}} & \xrightarrow{\ -\lambda_a\ } & H^2_T(\overline{\Gamma}/\overline{\Gamma}_a,\, a^{-1}X/X) \\[2mm] \kappa_a \downarrow & & \downarrow \iota \\[4mm] W_a & \xrightarrow{\ \lambda'_a\ } & H^2_T(\overline{\Gamma}/\overline{\Gamma}_a,\, E_a^m) \end{array}$$

with a map λ'_a defined in an algebro-geometric way. Let K_a denote the field of all meromorphic functions on V_a. Then $\{\, f \circ \varphi_a \,|\, f \in K_a \,\}$ coincides with the field of all automorphic functions with respect to Γ_a. To define λ'_a, we fix a set of representatives S_a for $\overline{\Gamma}$ modulo $\overline{\Gamma}_a$, containing $Q \cup \{1\}$. Now, for $t = (t_\nu)_{\nu=1}^m \in W_a$ we define an element y of $Z^2(\overline{\Gamma}/\overline{\Gamma}_a,\, E_a^m)$ through the following procedure:

(i) Take m divisors $\mathfrak{d}_\nu$ of degree 0 on V_a involving no cusps of Γ so that $t_\nu = \psi_a(\mathfrak{d}_\nu)$; this is of course possible.

(ii) Since $at_\nu = 0$, we can find m elements g_ν of K_a such that $a\mathfrak{d}_\nu = \mathrm{div}(g_\nu)$. Put $g = (g_\nu)_{\nu=1}^m$.

(iii) Since $\big(\alpha \otimes r_a(\alpha)\big)t = t$ for $\alpha \in \overline{\Gamma}$, the sum $\sum_\nu \rho_{\mu\nu}(\alpha)R_a(\alpha)(\mathfrak{d}_\nu)$ is linearly equivalent to $\mathfrak{d}_\mu$, and hence $g = h_\alpha^a \cdot \big(g \circ R_a(\alpha^{-1})\big)^\alpha$, where $h_\alpha = (h_{\alpha\nu})_{\nu=1}^m$ with $h_{\alpha\nu} \in K_a$.

(iv) For each $\alpha \in S_a$, fix h_α as above. In particular, if $\alpha = \pi \in Q$, we consider the cusp s fixed by π. In view of (i), we can find an m-tuple $c_\pi = (c_{\pi\nu})_{\nu=1}^m$ with $c_{\pi\nu} \in \mathbf{C}^\times$ so that $c_\pi^a = g(\varphi_a(s))$. Then we choose h_π so that $h_\pi\big(\varphi_a(s)\big) = (c_\pi)^{1-\pi}$.

(v) Given $\alpha, \beta \in S_a$, put $\alpha\beta = \gamma\delta$ with $\delta \in \overline{\Gamma}_a$ and $\gamma \in S_a$. Define an element y of $C^2(\overline{\Gamma}/\overline{\Gamma}_a, E_a^m)$ by

$$(4.3.4) \qquad y\big(\nu(\alpha), \nu(\beta)\big) = h_\alpha \cdot \big(h_\beta \circ R_a(\alpha^{-1})\big)^\alpha \cdot h_\gamma^{-1} \cdot \big(g \circ R_a(\gamma^{-1})\big)^{(\alpha\beta-\gamma)/a}.$$

Observe here that $[\rho(\alpha\beta) - \rho(\gamma)]/a \in M_m(\mathbf{Z})$.

To see that y is a cocyle corresponding to $-\iota\big(\lambda_a(\kappa_a^{-1}(t))\big)$ with values in E_a^m, we take an element u of $H_{\overline{P}}^1(\overline{\Gamma}_a, a^{-1}X/X)^{\overline{\Gamma}}$ corresponding to t, and $f = (f_\nu)_{\nu=1}^m$ as in (4.2.7). Then $\mathrm{div}(f_\nu) = \mathfrak{d}_\nu + \mathrm{div}(b_\nu)$ with an element b_ν of K_a. Therefore $f^a = (c \cdot b^a \cdot g) \circ \varphi_a$, where $b = (b_\nu)_{\nu=1}^m$ and $c = (c_\nu)_{\nu=1}^m$ with $c_\nu \in \mathbf{C}^\times$. Taking $c_\nu^{-1/a} \cdot (b_\nu^{-1} \circ \varphi_a) \cdot f_\nu$ in place of f_ν, we may assume that $f^a = g \circ \varphi_a$. Since $\big(\alpha \otimes r_a(\alpha)\big)t = t$, we have $f = (d_\alpha \circ \varphi_a) \cdot (f \circ \alpha^{-1})^\alpha$ with $d_\alpha = (d_{\alpha\nu})_{\nu=1}^m$, $d_{\alpha\nu} \in K_a$. Then $h_\alpha = e_\alpha d_\alpha$ with $e_\alpha \in E_a^m$. If $\alpha = \pi \in Q$ and $\pi(s) = s$ with a cusp s, we have $d_\pi\big(\varphi_a(s)\big) = f(s)^{1-\pi}$, and hence $e_\pi = \big(c_\pi \cdot f(s)^{-1}\big)^{1-\pi}$. Since $c_\pi^a = g\big(\varphi_a(s)\big) = f(s)^a$, we see that $c_\pi \cdot f(s)^{-1} \in E_a^*$. Let $u^* = (u_\nu^*)_{\nu=1}^m$ with the u_ν^* defined by (4.2.6). Then

$$\begin{aligned}
f &= d_\gamma \cdot (f \circ \gamma^{-1})^\gamma = d_\gamma \cdot (f \circ \delta\beta^{-1}\alpha^{-1})^{\alpha\beta} \cdot (f \circ \gamma^{-1})^{\gamma-\alpha\beta} \\
&= d_\gamma \cdot u^*(\delta)^{\alpha\beta} \cdot (f \circ \beta^{-1}\alpha^{-1})^{\alpha\beta} \cdot (f \circ \gamma^{-1})^{\gamma-\alpha\beta} \\
&= u^*(\delta)^\gamma \cdot d_\alpha^{-1} \cdot (d_\beta \circ \alpha^{-1})^{-\alpha} \cdot d_\gamma \cdot f \cdot (f \circ \gamma^{-1})^{\gamma-\alpha\beta},
\end{aligned}$$

and hence

$$u^*(\gamma^{-1}\alpha\beta)^\gamma = d_\alpha \cdot (d_\beta \circ \alpha^{-1})^\alpha \cdot d_\gamma^{-1} \cdot (f \circ \gamma^{-1})^{\alpha\beta-\gamma},$$

where we write d_α for $d_\alpha \circ \varphi_a$ for simplicity. Thus

$$(4.3.5) \qquad y\big(\nu(\alpha), \nu(\beta)\big) = e_\alpha \cdot e_\beta^\alpha \cdot e_\gamma^{-1} \cdot u^*(\gamma^{-1}\alpha\beta)^\gamma.$$

Now, by (1.9.2), $\lambda_a(u)$ is represented by a cocycle η such that $\eta\big(\nu(\alpha), \nu(\beta)\big) = -\gamma u(\gamma^{-1}\alpha\beta)$. Observe that $\nu(\alpha) \mapsto e_\alpha$ gives an element of $C_T^1(\overline{\Gamma}/\overline{\Gamma}_a, E_a^m)$. Therefore y is an element of $Z^2(\overline{\Gamma}/\overline{\Gamma}_a, E_a^m)$, whose class modulo $B_T^2(\overline{\Gamma}/\overline{\Gamma}_a, E_a^m)$ is the same as $-\iota\big(\lambda_a(u)\big)$. Taking the class of y to be $\lambda_a'(t)$, we obtain (4.3.3).

4.4. If a divides b, there exists, by (4.1.1), a rational map $R_{a,b} : V_b \to V_a$ such that $R_{a,b} \circ \varphi_b = \varphi_a$. We can then define a homomorphism

$$(4.4.1) \qquad\qquad r_{b,a}^* : J_a \longrightarrow J_b$$

such that $r_{b,a}^* \circ \psi_a = \psi_b \circ (R_{a,b})^{-1}$ and $r_{b,a}^* \circ r_a(\alpha) = r_b(\alpha) \circ r_{b,a}^*$ for every $\alpha \in \Gamma$. Obviously the map $1 \otimes r_{b,a}^* : X \otimes J_a \to X \otimes J_b$ sends W_a into W_b.

Let us now form the direct limit $\varinjlim W_a$ with respect to the maps $1 \otimes r_{b,a}^*$. From the injection $a^{-1}X/X \to b^{-1}X/X$ and the restriction of a cocycle of $\overline{\Gamma}_a$ to $\overline{\Gamma}_b$ we obtain a map

$$H_{\overline{P}}^1(\overline{\Gamma}_a, a^{-1}X/X)^{\overline{\Gamma}} \longrightarrow H_{\overline{P}}^1(\overline{\Gamma}_b, b^{-1}X/X)^{\overline{\Gamma}}.$$

This map makes the following diagram commutative:

$$\begin{array}{ccccc}
H_{\overline{P}}^1(\overline{\Gamma}, a^{-1}X/X) & \xrightarrow{\ \mathrm{Res}\ } & H_{\overline{P}}^1(\overline{\Gamma}_a, a^{-1}X/X)^{\overline{\Gamma}} & \xrightarrow{\ \kappa_a\ } & W_a \\
\Big\downarrow & & \Big\downarrow & & \Big\downarrow \\
H_{\overline{P}}^1(\overline{\Gamma}, b^{-1}X/X) & \xrightarrow{\ \mathrm{Res}\ } & H_{\overline{P}}^1(\overline{\Gamma}_b, b^{-1}X/X)^{\overline{\Gamma}} & \xrightarrow{\ \kappa_b\ } & W_b
\end{array}$$

Therefore we obtain a sequence of direct limits:

$$(4.4.2) \qquad \varinjlim H^1_{\overline{P}}(\overline{\Gamma},\, a^{-1}X/X) \xrightarrow{\mathfrak{R}} \varinjlim H^1_{\overline{P}}(\overline{\Gamma}_a,\, a^{-1}X/X)^{\overline{\Gamma}} \xrightarrow{\kappa} \varinjlim W_a.$$

In these direct limits we assume $a \geq \operatorname{Max}(a_0, a_1)$. Combining (4.4.2) with the map $H^1_{\overline{P}}(\overline{\Gamma},\, X) \otimes (\mathbf{Q}/\mathbf{Z}) \to \varinjlim H^1_{\overline{P}}(\overline{\Gamma},\, a^{-1}X/X)$ defined in (1.3.2), we obtain a map

$$(4.4.3) \qquad \omega : H^1_{\overline{P}}(\overline{\Gamma},\, X) \otimes (\mathbf{Q}/\mathbf{Z}) \longrightarrow \varinjlim W_a.$$

5. Connection of Hecke operators with Frobenius automorphisms

5.1. We shall now proceed to the main part of our theory by restricting our discussion to the case of arithmetically defined Fuchsian groups. Let F be a totally real algebraic number field of finite degree, and B a quaternion algebra over F. Let $g = [F : \mathbf{Q}]$, and let $\tau_1, \ldots, \tau_g$ be the isomorphic embeddings of F into $\mathbf{R}$. We identify F with F^{τ_1} by τ_1; in other words, we understand that $F \subset \mathbf{R}$, and τ_1 is the identity map. We assume that B is unramified at τ_1 and ramified at the remaining τ_i. Put $B_{\mathbf{R}} = B \otimes_{\mathbf{Q}} \mathbf{R}$. Then $B_{\mathbf{R}}$ can be identified with $M_2(\mathbf{R}) \times \mathbf{K}^{g-1}$, where $\mathbf{K}$ denotes the division ring of Hamilton quaternions. Notice that B under this assumption is either a division algebra or isomorphic to $M_2(\mathbf{Q})$. For $\alpha \in B \otimes_F L$, with any extension L of F, we denote by α^ι the image of α under the main involution of $B \otimes_F L$; we put then $\nu(\alpha) = \alpha\alpha^\iota$, $\operatorname{tr}(\alpha) = \alpha + \alpha^\iota$, and

$$(5.1.1) \qquad B_0^\times = \{\, \alpha \in B \mid N_{F/\mathbf{Q}}(\nu(\alpha)) > 0 \,\}.$$

Notice that $N_{F/\mathbf{Q}}(\nu(\alpha)) > 0$ if and only if $\nu(\alpha) > 0$, in which case $\nu(\alpha)$ is totally positive.

5.2. We fix a maximal order $\mathfrak{o}$ in B, and put

$$\Gamma = \{\, \gamma \in B_0^\times \mid \gamma\mathfrak{o} = \mathfrak{o} \,\}, \qquad \Gamma_a = \{\, \gamma \in \Gamma \mid \gamma - 1 \in a\mathfrak{o} \,\} \qquad (0 < a \in \mathbf{Z}).$$

Further, for every rational prime p we put $B_p = B \otimes_{\mathbf{Q}} \mathbf{Q}_p$, $\mathfrak{o}_p = \mathfrak{o} \otimes_{\mathbf{Z}} \mathbf{Z}_p$, and

$$D_a = \{\, \alpha \in B_0^\times \mid \alpha\mathfrak{o}_p = \mathfrak{o}_p \text{ for every } p|a \,\},$$
$$D_a^0 = \{\, \alpha \in D_a \mid (\alpha - 1)\mathfrak{o}_p \subset a\mathfrak{o}_p \text{ for every } p|a \,\}.$$

Denote by k_a the maximal class field over F whose conductor divides $a \cdot \mathfrak{u}$, where $\mathfrak{u}$ denotes the product of all archimedean primes of F. For $\alpha \in D_a$ we put

$$\sigma_a(\alpha) = \big[k_a/F, \nu(\alpha)\mathfrak{r}\big] \qquad \text{(Artin symbol)},$$

where $\mathfrak{r}$ denotes the maximal order of F. Let Δ denote the group of units of $\mathfrak{r}$. Put $\Delta_a = \Gamma_a \cap \Delta$, $\overline{\Gamma} = \Gamma/\Delta$, and $\overline{\Gamma}_a = \Gamma_a/\Delta_a$. We let $B_0^\times$ act on $\mathfrak{H}$ through the projection of $B_{\mathbf{R}}$ to $M_2(\mathbf{R})$. Then $\overline{\Gamma}_a$ is a Fuchsian group of the first kind, and $\mathfrak{H}^* = \mathfrak{H}$ unless $B = M_2(\mathbf{Q})$.

Now, by [S6, Theorems 3.2 and 3.5] and [S7, Theorems 5.3, 5.7, and 5.9], there exists a system $\{V_a,\, \varphi_a,\, R_a(\alpha),\, R_{a,b}\}$ formed by the objects satisfying the following conditions:

(5.2.1) V_a is a complete nonsingular algebraic curve defined over k_a.

(5.2.2) φ_a is a Γ_a-invariant holomorphic map of $\mathfrak{H}^*$ onto V_a that gives a biregular map of $\Gamma_a \backslash \mathfrak{H}^*$ onto V_a.

(5.2.3) $R_a(\alpha)$, defined for $\alpha \in D_a$, is a k_a-rational biregular isomorphism of V_a onto $V_a^{\sigma_a(\alpha)}$.

(5.2.4) $R_a(\alpha)^{\sigma_a(\beta)} \circ R_a(\beta) = R_a(\alpha\beta)$.

(5.2.5) $R_a(\alpha) = R_a(\beta)$ if $\alpha^{-1}\beta \in D_a^0$.

(5.2.6) $R_a(\gamma)[\varphi_a(z)] = \varphi_a(\gamma(z))$ for every $\gamma \in \Gamma$ and $z \in \mathfrak{H}^*$.

(5.2.7) A certain property at isolated fixed points of $B_0^\times$ on $\mathfrak{H}$ as described in [S6, (3.2.3), (3.5.4)] and [S7, (5.3.3), (5.7.6)], which will not be needed in this paper.

(5.2.8) $R_{a,b}$, defined if $a|b$, is a k_b-rational morphism of V_b onto V_a such that $R_{a,b} \circ \varphi_b = \varphi_a$ and $R_{a,b}^{\sigma_a(\alpha)} \circ R_b(\alpha) = R_a(\alpha) \circ R_{a,b}$ for every $\alpha \in D_b$.

We can find a jacobian variety J_a of V_a rational over k_a. Then, as in §4.1, we denote by $\psi_a(\mathfrak{d})$ the point of J_a corresponding to a divisor $\mathfrak{d}$ on V_a of degree 0. Further, for every $\alpha \in D_a$ we can define a k_a-rational isomorphism $r_a(\alpha)$ of J_a onto $J_a^{\sigma_a(\alpha)}$ so that $r_a(\alpha) \circ \psi_a = \psi_a^{\sigma_a(\alpha)} \circ R_a(\alpha)$. In view of (5.2.6), we can define $R_a(\alpha)$ and $r_a(\alpha)$ also for $\alpha \in \overline{\Gamma}$. (In §4.1 we defined $R_a(\alpha)$ and $r_a(\alpha)$ only for $\alpha \in \Gamma$, while here these symbols are defined on a larger set D_a.) The $r_a(\alpha)$ satisfy the formulas similar to (5.2.4) and (5.2.5).

5.3. We now consider a representation $\rho : B^\times \to GL_m(\mathbf{Q})$ with any positive integer m satisfying the following conditions:

(5.3.1) ρ is rational over $\mathbf{Q}$, if we view $B^\times$ as (the group of $\mathbf{Q}$-rational points of) a linear algebraic group defined over $\mathbf{Q}$ by means of a regular representation of B over $\mathbf{Q}$.

(5.3.2) The restriction of ρ to $\{\, \alpha \in B \,|\, \nu(\alpha) = 1 \,\}$ is $\mathbf{Q}$-irreducible and nontrivial.

(5.3.3) $\rho(c) = N_{F/\mathbf{Q}}(c)^h$ for every $c \in F^\times$ with an even positive integer $h = h_\rho$ independent of c.

(5.3.4) There is a $\mathbf{Z}$-lattice X in $\mathbf{Q}^m$ such that $\rho(\alpha)X \subset X$ for every $\alpha \in \mathfrak{o} \cap B^\times$ and $[\rho(\alpha) - \rho(\beta)]X \subset aX$ if $\alpha, \beta \in \mathfrak{o} \cap D_a$ and $\alpha - \beta \in a\mathfrak{o}$.

For example, we can take the representation ρ_n of §3.1 and $\mathbf{Z}^{n+1}$ as ρ and X if $B = M_2(\mathbf{Q})$ and $\mathfrak{o} = M_2(\mathbf{Z})$. We shall later construct ρ and X in the general case.

If ρ and X are as above, we have, by [B2, Theorem 1],

(5.3.5) The restriction of ρ to Γ_a is irreducible and nontrivial for every a.

We exclude the case of trivial representation, since that case concerns the space of cusp forms of weight 2, and the ℓ-adic representations attached to such forms are established in [S6].

5.4. Let ρ, X, and m be as in (5.3.1-4); put $X_{\mathbf{Q}} = X \otimes_{\mathbf{Z}} \mathbf{Q}$. By (5.3.3) we can view X as a $\overline{\Gamma}$-module, and observe that the present Γ_a and X satisfy (4.1.1-5) and (4.3.1), and hence the whole results of Section 4 are applicable to the present case. Define W_a as in (4.2.1) for every positive integer a, and denote it also by $W(a)$; namely

$$(5.4.1) \qquad W_a = W(a)$$
$$= \big\{ t \in X \otimes J_a \,\big|\, at = 0,\ \big(\rho(\gamma) \otimes r_a(\gamma)\big)t = t \ \text{for every}\ \gamma \in \Gamma \big\}.$$

Let $\overline{\mathbf{Q}}$ denote the algebraic closure of $\mathbf{Q}$. We are going to define two types of action of $\mathrm{Gal}(\overline{\mathbf{Q}}/k_1)$ on W_a, where k_1 is the field k_a in §5.2 with $a = 1$.

Let $\sigma \in \mathrm{Gal}(\overline{\mathbf{Q}}/k_1)$. Take an element α of $\mathfrak{o} \cap D_a$ so that $\sigma = \sigma_a(\alpha)$ on k_a. This is possible since σ is the identity map on k_1. Define two automorphisms $\mu_a(\sigma)$ and $\mu_a^*(\sigma)$ of W_a by

$$(5.4.2\mathrm{a}) \qquad \mu_a(\sigma)t = \big(\rho(\alpha^\iota) \otimes r_a(\alpha)^{-1}\big)t^\sigma,$$

$$(5.4.2\mathrm{b}) \qquad \mu_a^*(\sigma)t = N_{F/\mathbf{Q}}\big(\nu(\alpha)\big)\big[\big(\rho(\alpha) \otimes r_a(\alpha)\big)t\big]^{\sigma^{-1}},$$

for $t \in W_a$. Here observe that $\rho(\beta) \otimes r_a(\alpha)$ is an isogeny of $X \otimes J_a$ onto $X \otimes J_a^{\sigma_a(\alpha)}$ for any $\alpha,\ \beta \in \mathfrak{o} \cap D_a$. To see that $\mu_a(\sigma)$ and $\mu_a^*(\sigma)$ actually send W_a onto itself, take $\gamma \in \Gamma$. Then there exists an element δ of Γ such that $\alpha\gamma - \delta\alpha \in a\mathfrak{o}$. (In fact, apply [S7, 3.11] to $\alpha\gamma\alpha^{-1}$.) Then $r_a(\alpha) \circ r_a(\gamma) = r_a(\delta)^\sigma \circ r_a(\alpha)$, and hence $r_a(\gamma) \circ r_a(\alpha)^{-1} = r_a(\alpha)^{-1} \circ r_a(\delta)^\sigma$, and $\gamma\alpha^\iota - \alpha^\iota\delta \in a\mathfrak{o}$. Therefore, if $t \in W_a$, we have, in view of (5.3.4),

$$\big(\rho(\gamma) \otimes r_a(\gamma)\big)\mu_a(\sigma)t = \mu_a(\sigma)\big[\big(\rho(\delta) \otimes r_a(\delta)\big)t\big] = \mu_a(\sigma)t,$$

so that $\mu_a(\sigma)t \in W_a$. A similar and simpler argument applies to $\mu_a^*(\sigma)$.

Now $\mu_a(\sigma)$ and $\mu_a^*(\sigma)$ depend only on σ, and are independent of the choice of α. In fact, let β be another element of $\mathfrak{o} \cap D_a$ such that $\sigma = \sigma_a(\beta)$. Then $\nu(\alpha)e - \nu(\beta) \in a\mathfrak{r}$ with a totally positive unit e in F. Therefore, by [S7, 3.11], there exists an element γ of $\mathfrak{o}$ such that $\nu(\gamma) = e$ and $\alpha\gamma - \beta \in a\mathfrak{o}$. Then $\gamma \in \Gamma$, $\gamma^\iota\alpha^\iota - \beta^\iota \in a\mathfrak{o}$, and

$$\big(\rho(\beta^\iota) \otimes r_a(\beta)^{-1}\big)t^\sigma = \big(\rho(\gamma^\iota\alpha^\iota) \otimes r_a(\gamma)^{-1}r_a(\alpha)^{-1}\big)t^\sigma$$
$$= \big(\rho(\gamma) \otimes r_a(\gamma)\big)^{-1}\big(\rho(\alpha^\iota) \otimes r_a(\alpha)^{-1}\big)t^\sigma = \big(\rho(\alpha^\iota) \otimes r_a(\alpha)^{-1}\big)t^\sigma.$$

A similar argument applies to $\mu_a^*(\sigma)$.

We easily see that

$$(5.4.3) \quad \mu_a(\sigma\tau) = \mu_a(\tau)\mu_a(\sigma), \quad \mu_a^*(\sigma\tau) = \mu_a^*(\sigma)\mu_a^*(\tau) \qquad \big(\sigma,\ \tau \in \mathrm{Gal}(\overline{\mathbf{Q}}/k_1)\big).$$

Moreover, if q is an element of $\mathfrak{r}$ such that $\sigma = [k_a/F, q\mathfrak{r}]$, then $\nu(\alpha) - cq \in a\mathfrak{r}$ with a totally positive unit c in F; hence, by (5.3.3), the action of $\rho(\alpha\alpha^\iota)$ on X coincides with that of $N_{F/\mathbf{Q}}(q)^h$ modulo aX. Therefore we have

$$(5.4.4) \quad \mu_a(\sigma)\mu_a^*(\sigma) = \mu_a^*(\sigma)\mu_a(\sigma) = N_{F/\mathbf{Q}}(q)^{h+1} \ \text{if}\ q \in \mathfrak{r} \ \text{and}\ \sigma = [k_a/F, q\mathfrak{r}]$$
$$\text{on}\ k_a.$$

For a divisor a of b, define $r_{b,a}^* : J_a \longrightarrow J_b$ as in §4.4. Then

$$(5.4.5) \qquad (r_{b,a}^*)^{\sigma_a(\alpha)} \circ r_a(\alpha) = r_b(\alpha) \circ r_{b,a}^* \qquad (\alpha \in D_b).$$

Therefore we obtain

$$(5.4.6\mathrm{a}) \qquad (1 \otimes r_{b,a}^*) \circ \mu_a(\sigma) = \mu_b(\sigma) \circ (1 \otimes r_{b,a}^*),$$

$$(5.4.6\mathrm{b}) \qquad (1 \otimes r_{b,a}^*) \circ \mu_a^*(\sigma) = \mu_b^*(\sigma) \circ (1 \otimes r_{b,a}^*).$$

Let us now consider $\varinjlim W_a$ and the map ω as in (4.4.3) for the present system. In view of (5.4.6a, b), we can define two automorphisms $\mu(\sigma)$ and $\mu^*(\sigma)$ of $\varinjlim W_a$ as the direct limits of $\mu_a(\sigma)$ and $\mu_a^*(\sigma)$. From (5.4.3) we obtain

$$(5.4.7) \quad \mu(\sigma\tau) = \mu(\tau)\mu(\sigma), \quad \mu^*(\sigma\tau) = \mu^*(\sigma)\mu^*(\tau) \quad (\sigma, \tau \in \mathrm{Gal}(\overline{\mathbf{Q}}/k_1)).$$

Furthermore, we easily see that

$(5.4.8)$ $\mu(1)$ and $\mu^*(1)$ are the identity map.

Let $\mathfrak{W}$ denote the image of $H_{\overline{P}}^1(\overline{\Gamma}, X) \otimes (\mathbf{Q}/\mathbf{Z})$ by ω. Then we obtain the following theorem which is one of the crucial points of our theory.

5.5. Theorem. *The maps $\mu(\sigma)$ and $\mu^*(\sigma)$ induce automorphisms of $\mathfrak{W}$.*

PROOF. Let $\mathfrak{W}'$ be the image of $\varinjlim H_{\overline{P}}^1(\overline{\Gamma}, a^{-1}X/X)$ under the map $\kappa \circ \mathfrak{R}$ of (4.4.2). Then

$(5.5.1)$ *There exists a positive integer N such that $N \cdot \mathfrak{W}' = \mathfrak{W}$.*

To show this, let Q be a set of representatives for the generators of all arabolic subgroups of $\overline{\Gamma}$, as considered in §§2.2 and 4.3. Let Y be the submodule of X generated by $(\rho(\alpha) - 1)X$ for all $\alpha \in \Gamma$. By (5.3.5) we have $Y \otimes_{\mathbf{Z}} \mathbf{Q} = X_{\mathbf{Q}}$, and hence X/Y is finite. Therefore, by Propositions 2.3 and 2.7 there exists a positive integer N which annihilates $H_Q^2(\overline{\Gamma}, X)$. Consider the map

$$j : H_Q^1(\overline{\Gamma}, X) \otimes (\mathbf{Q}/\mathbf{Z}) \longrightarrow \varinjlim H_Q^1(\overline{\Gamma}, a^{-1}X/X)$$

obtained in §1.3. By Proposition 1.4 the image of j coincides with

$$N \cdot \varinjlim H_Q^1(\overline{\Gamma}, a^{-1}X/X).$$

Now $H_{\overline{P}}^1(\overline{\Gamma}, L) = H_Q^1(\overline{\Gamma}, L)$ for any $\overline{\Gamma}$-module L by Proposition 1.7. Thus we obtain (5.5.1).

In view of (5.5.1), (5.4.7), and (5.4.8), it is sufficient to show that $\mu(\sigma)$ and $\mu^*(\sigma)$ map $\mathfrak{W}'$ into itself. This can be done by means of the map λ_a' defined in §4.3. We take a positive integer a_1 so that (4.3.1) is satisfied for the present Q, and consider the integers $a \geq \mathrm{Max}(a_0, a_1)$. Let $u \in Z_{\overline{P}}^1(\overline{\Gamma}, a^{-1}X/X)$ for such an a, and let t be the element of W_a corresponding to u. Let u_ν^*, f_ν, and t_ν be defined as in §4.2. Define $g = (g_\nu)_{\nu=1}^m$ nd $h_\alpha = (h_{\alpha\nu})_{\nu=1}^m$ for this t as in §4.3. Since t is rational over $\overline{\mathbf{Q}}$, we may assume that g and h are rational over $\overline{\mathbf{Q}}$. Further we may assume that $f^a = g \circ \varphi_a$ as in §4.3. Fix a set S_a of representatives for $\overline{\Gamma}$ modulo $\overline{\Gamma}_a$, containing $Q \cup \{1\}$. We have $g = h_\alpha^a \cdot \left[g \circ R(\alpha^{-1})\right]^\alpha$ for $\alpha \in S_a$, and hence

$$(5.5.2) \qquad g^\sigma = (h_\alpha^\sigma)^a \cdot \left[g^\sigma \circ R(\alpha^{-1})^\sigma\right]^\alpha \qquad (\alpha \in S_a).$$

Here and in the following discussion, we write $R(\alpha)$ for $R_a(\alpha)$ for simplicity.

Now take an element ξ of $\mathfrak{o} \cap D_a$ so that $\sigma_a(\xi) = \sigma$ on k_a. Put

$$\overline{g} = (\overline{g}_\nu)_{\nu=1}^m = \left[g^\sigma \circ R(\xi^\iota)\right]^\xi.$$

Then we easily see that $\psi_a(a^{-1} \cdot \mathrm{div}(\overline{g}_\nu))$ determines the ν-th component of $\left[\xi \otimes r(\xi^\iota)^{-1}\right]t^\sigma = \mu_a(\sigma)t$. We can define a permutation $\alpha \mapsto \overline{\alpha}$ of S_a by $\xi\alpha \equiv \pm\overline{\alpha}\xi \pmod{a\mathfrak{o}}$. For $\varepsilon = \overline{\alpha}$, put

$$\overline{h}_\varepsilon = \left[h_\alpha^\sigma \circ R(\xi^\iota) \right]^\xi \cdot \left[g^\sigma \circ R(\xi^\iota \varepsilon^{-1}) \right]^{(\xi\alpha - \varepsilon\xi)/a}.$$

Then $\overline{g} = \overline{h}_\varepsilon^a \cdot \left[\overline{g} \circ R(\varepsilon^{-1}) \right]^\varepsilon$. Define an element y of $Z^2(\overline{\Gamma}/\overline{\Gamma}_a, E_a^m)$ by (4.3.4) with the present g and h_α. Further define similarly a cocycle $\overline{y}$, with respect to the same S_a, by using $\overline{g}$ and $\overline{h}_\varepsilon$.

Let us disregard temporarily the condition at the cusps described in (iv) of §4.3. Let $\alpha\beta = \gamma\delta$ with $\alpha, \beta, \gamma \in S_a$ and $\delta \in \overline{\Gamma}_a$. Let $\overline{\alpha} = \varepsilon$, $\overline{\beta} = \zeta$, and $\overline{\gamma} = \eta$. Put

$$h_\alpha^* = h_\alpha^\sigma \circ R(\xi^\iota) \quad \text{and} \quad g^* = g^\sigma \circ R(\xi^\iota).$$

Then denoting by ν_a the natural map of $\overline{\Gamma}$ to $\overline{\Gamma}/\overline{\Gamma}_a$, we have

$$\overline{y}\big(\nu_a(\varepsilon), \nu_a(\zeta)\big) = \overline{h}_\varepsilon \cdot \left[\overline{h}_\zeta \circ R(\varepsilon^{-1}) \right]^\varepsilon \cdot \overline{h}_\eta^{-1} \cdot \left[\overline{g} \circ R(\eta^{-1}) \right]^{(\varepsilon\zeta - \eta)/a}$$

$$= (h_\alpha^*)^\xi \cdot \left[h_\beta^* \circ R(\varepsilon^{-1}) \right]^{\varepsilon\xi} \cdot (h_\gamma^*)^{-\xi} \cdot \left[g^* \circ R(\varepsilon^{-1}) \right]^{(\xi\alpha - \varepsilon\xi)/a} \cdot \left[g^* \circ R(\zeta^{-1}\varepsilon^{-1}) \right]^{\varepsilon(\xi\beta - \zeta\xi)/a}$$

$$\cdot \left[g^* \circ R(\eta^{-1}) \right]^{(\eta\xi - \xi\gamma)/a} \cdot \left[g^* \circ R(\eta^{-1}) \right]^{(\varepsilon\zeta - \eta)\xi/a}.$$

Now, from (4.3.4) we obtain

$$\overline{y}\big(\nu_a(\alpha), \nu_a(\beta)\big)^{\sigma\xi} = (h_\alpha^*)^\xi \cdot \left[h_\beta^* \circ R(\varepsilon^{-1}) \right]^{\xi\alpha} \cdot (h_\gamma^*)^{-\xi} \cdot \left[g^* \circ R(\eta^{-1}) \right]^{\xi(\alpha\beta - \gamma)/a}.$$

Therefore

$$y\big(\nu_a(\alpha), \nu_a(\beta)\big)^{-\sigma\xi} \cdot \overline{y}\big(\nu_a(\varepsilon), \nu_a(\zeta)\big)$$

$$= \left[h_\beta^* \circ R(\varepsilon^{-1}) \right]^{\varepsilon\xi - \xi\alpha} \cdot \left[g^* \circ R(\varepsilon^{-1}) \right]^{(\xi\alpha - \varepsilon\xi)/a} \cdot \left[g^* \circ R(\eta^{-1}) \right]^\chi,$$

where $\chi = \left[\xi(\gamma - \alpha\beta) + \varepsilon(\xi\beta - \zeta\xi) + (\eta\xi - \xi\gamma) + (\varepsilon\zeta - \eta)\xi \right]/a = (\varepsilon\xi - \xi\alpha)\beta/a$. From (5.5.2) we obtain $g^* = (h_\beta^*)^a \cdot \left[g^* \circ R(\zeta^{-1}) \right]^\beta$, and hence

$$\left[h_\beta^* \circ R(\varepsilon^{-1}) \right]^a = \left[g^* \circ R(\varepsilon^{-1}) \right] \cdot \left[g^* \circ R(\eta^{-1}) \right]^{-\beta}.$$

It follows that

$$(5.5.3) \qquad\qquad \overline{y}\big(\nu_a(\varepsilon), \nu_a(\zeta)\big) = y\big(\nu_a(\alpha), \nu_a(\beta)\big)^{-\sigma\xi}.$$

Suppose now that Γ has no cusps. We have started from an element u of $Z^1(\overline{\Gamma}, a^{-1}X/X)$. Therefore, by Proposition 1.8 and the commutativity of (4.3.3), y belongs to $B_T^2(\overline{\Gamma}/\overline{\Gamma}_a, E_a^m)$, and hence so does $\overline{y}$ by virtue of (5.5.3). It follows that $\mu_a(\sigma)t$ belongs to the image of $H^1(\overline{\Gamma}, a^{-1}X/X)$, again by Proposition 1.8 and the commutativity of (4.3.3). Then, from (5.4.3) we see that $\mu_a^*(\sigma)t$ belongs to the image of $H^1(\overline{\Gamma}, a^{-1}X/X)$. This shows that $\mathfrak{W}'$ is stable under both $\mu(\sigma)$ and $\mu^*(\sigma)$ if Γ has no cusps.

Next suppose that Γ has cusps. Then we may assume that $B = M_2(\mathbf{Q})$, $\mathfrak{o} = M_2(\mathbf{Z})$, and $\Gamma = SL_2(\mathbf{Z})$. In this case we have to choose h_π for $\pi \in Q$ as in (iv) of §4.3. Now our task is to verify that $\overline{h}_\pi$ has this property, that is, $\overline{g}\big(\varphi_a(s)\big) = \overline{c}_\pi^a$ and $\overline{h}_\pi\big(\varphi_a(s)\big) = (\overline{c}_\pi)^{1-\pi}$ with an m-tuple $\overline{c}_\pi = (\overline{c}_{\pi\nu})_{\nu=1}^m$ of nonzero constants, where s is the cusp fixed by π. Obviously $\mathrm{div}(\overline{g}_\nu)$ involves no cusps. We can assume that Q consists of a single element $\pi = \begin{bmatrix} 1 & 1 \\ 0 & 1 \end{bmatrix}$ (modulo ±1), and S_a contains π^v for $0 \le v < a$. Further we can take our system $\big(V_a, \varphi_a, R_a(\alpha)\big)$ so that

(5.5.4) V_a and $\varphi_a(\infty)$ are rational over $\mathbf{Q}$;

(5.5.5) $R_a(\alpha)$ is the identity map if $\alpha = \begin{bmatrix} 1 & 0 \\ 0 & d \end{bmatrix}$, $(d, a) = 1$.

Indeed, it is sufficient to take the function field of V_a to be $\mathbf{Q}\big(\mathbf{j}(z), \mathbf{j}(az), \mathbf{f}_{10}(z)\big)$ of [S1, §14] (see also [S7, Theorem 9.5]). Now we can take ξ in the form $\xi = \begin{bmatrix} q & 0 \\ 0 & 1 \end{bmatrix}$ with an integer q prime to a. Then we have $\xi\pi^r \equiv \pi\xi \pmod{a}$ with an integer r such that $rq \equiv 1 \pmod{a}$ and $0 < r < a$, and hence $\overline{\pi^r} = \pi$ with the above notation. Let f and d_α be as in §4.3. We can take d_α as h_α, and hence

$$(5.5.6) \qquad f = (h_\alpha \circ \varphi_a) \cdot (f \circ \alpha^{-1})^\alpha \qquad (\alpha \in \Gamma).$$

Now $\xi^\iota = \begin{bmatrix} 1 & 0 \\ 0 & q \end{bmatrix}$, so that $R(\xi^\iota)$ is the identity map. Writing β for π^r, we have thus $\overline{h}_\pi = (h_\beta^\sigma)^\xi \cdot \big[g^\sigma \circ R(\pi^{-1})\big]^{(\xi\beta - \pi\xi)/a}$, and hence

$$\overline{h}_\pi\big(\varphi_a(\infty)\big) = \big[h_\beta\big(\varphi_a(\infty)\big)^\sigma\big]^\xi \cdot \big[g\big(\varphi_a(\infty)\big)^\sigma\big]^{(\xi\beta - \pi\xi)/a}.$$

By (5.5.6), $f(\infty) = h_\beta\big(\varphi_a(\infty)\big) \cdot f(\infty)^\beta$. Since $f^a = g \circ \varphi_a$, we have

$$\overline{h}_\pi\big(\varphi_a(\infty)\big) = \big[f(\infty)^\sigma\big]^{\xi(1-\beta)+(\xi\beta - \pi\xi)} = \big[(f(\infty)^\sigma)^\xi\big]^{1-\pi},$$

and $\overline{g}\big(\varphi_a(\infty)\big) = \big[g\big(\varphi_a(\infty)\big)^\sigma\big]^\xi = \big[f(\infty)^\sigma\big]^{a\xi}$. Thus $\overline{h}_\pi$ has the required property. Repeating the same argument as in the case without cusps, we can complete the proof of our theorem.

5.6. So far we have been considering all positive integers a. We now fix a prime number ℓ, and take a to be powers of ℓ. Namely we consider the map $H^1_{\overline{P}}(\overline{\Gamma}, \ell^{-n}X/X) \to H^1_{\overline{P}}(\overline{\Gamma}, \ell^{-m}X/X)$ and also the map $W(\ell^n) \to W(\ell^m)$ for $m > n$; we then take the direct limits $\varinjlim H^1_{\overline{P}}(\overline{\Gamma}, \ell^{-n}X/X)$ and $\varinjlim W(\ell^n)$ with respect to these maps. Since $\mathbf{Q}_\ell/\mathbf{Z}_\ell$ can be identified with $\bigcup_{n=0}^{\infty} \ell^{-n}\mathbf{Z}/\mathbf{Z}$, we have a map

$$(5.6.1) \qquad j_\ell : H^1_{\overline{P}}(\overline{\Gamma}, X) \otimes (\mathbf{Q}_\ell/\mathbf{Z}_\ell) \longrightarrow \varinjlim H^1_{\overline{P}}(\overline{\Gamma}, \ell^{-n}X/X)$$

instead of (1.3.2). In this way we obtain a map

$$(5.6.2) \qquad \omega_\ell : H^1_{\overline{P}}(\overline{\Gamma}, X) \otimes (\mathbf{Q}_\ell/\mathbf{Z}_\ell) \longrightarrow \varinjlim W(\ell^n),$$

which is "the ℓ-part" of the map ω of (4.4.3) in an obvious sense. Clearly the image of ω_ℓ can be identified with

$$(5.6.3) \qquad \mathfrak{W}_\ell = \bigcup_{n=1}^{\infty} \big\{ t \in \mathfrak{W} \,\big|\, \ell^n t = 0 \big\}.$$

5.7. Theorem. $\mathrm{Ker}(\omega_\ell)$ is finite.

Proof. Denote Γ_a and $\overline{\Gamma}_a$ also by $\Gamma(a)$ and $\overline{\Gamma}(a)$. By Proposition 1.4a, $\mathrm{Ker}(j_\ell)$ is finite; also κ_a of (4.2.2) is an isomorphism, and so κ of (4.4.2) is an isomorphism. Therefore, to prove our theorem, it is sufficient to show that $\mathrm{Ker}(\mathfrak{R}_\ell)$ is finite, where $\mathfrak{R}_\ell$ is the direct limit of the restriction map $H^1_{\overline{P}}(\overline{\Gamma}, \ell^{-n}X/X) \to$

$H^1_{\overline{P}}(\overline{\Gamma}(\ell^n), \ell^{-n}X/X)^{\overline{\Gamma}}$. By Proposition 1.8 the kernel of this map is the isomorphic image of $H^1(\overline{\Gamma}/\overline{\Gamma}(\ell^n), \ell^{-n}X/X, \mathfrak{Q})$. Therefore our assertion follows from the following fact

(5.7.1) $H^1(\overline{\Gamma}/\overline{\Gamma}(\ell^n), \ell^{-n}X/X)$ has order less than an integer depending only on Γ, X, and ℓ.

This will be proven in §5.16.

5.8. Let us now consider the action of Hecke operators on W_a and $\mathfrak{W}_\ell$. As shown in §3.6, $\Gamma\eta\Gamma$, for every $\eta \in \mathfrak{o} \cap B^\times$, acts on $H^1_{\overline{P}}(\overline{\Gamma}, X)$, and hence on $H^1_{\overline{P}}(\overline{\Gamma}, X) \otimes (\mathbf{Q}_\ell/\mathbf{Z}_\ell)$. Denote this action by $(\Gamma\eta\Gamma)_{X,\ell}$. On the other hand, for every $\eta \in B_0^\times$ we can define an algebraic correspondence $\mathfrak{T}(\Gamma_a\eta\Gamma_a)$ on $V_a \times V_a$ by

$$\mathfrak{T}(\Gamma_a\eta\Gamma_a) = \left\{ \varphi_a(z) \times \varphi_a\big(\eta(z)\big) \,\big|\, z \in \mathfrak{H}^* \right\}.$$

From this we obtain an element $\mathfrak{t}(\Gamma_a\eta\Gamma_a)$ of $\mathrm{End}(J_a)$ which sends $\psi_a\big(\mathrm{div}(\mathfrak{d})\big)$ to $\psi_a\big(\mathrm{div}(\mathfrak{e})\big)$, if a divisor $\mathfrak{d}$ of degree 0 on V_a is sent to $\mathfrak{e}$ by $\mathfrak{T}(\Gamma_a\eta\Gamma_a)$, that is, if $\mathfrak{e}$ is the right projection of $\mathfrak{T}(\Gamma_a\eta\Gamma_a) \cdot (\mathfrak{d} \times V_a)$. Let us now assume

(5.8.1) $\eta \in \mathfrak{o}$ and $\nu(\eta)$ is prime to $a \cdot D(B/F)$, where $D(B/F)$ is the discriminant of B over F.

Then by [S6, Proposition 11.4], the curve $\mathfrak{T}(\Gamma_a\eta\Gamma_a)$ is k_a-rational, and hence so is $\mathfrak{t}(\Gamma_a\eta\Gamma_a)$. Furthermore, if $\eta\alpha \equiv \beta\eta \pmod{a\mathfrak{o}}$ with $\alpha, \beta \in \mathfrak{o} \cap D_a$, then

(5.8.2) $$\mathfrak{t}(\Gamma_a\eta\Gamma_a)^{\sigma_a(\alpha)} \circ r_a(\alpha) = r_a(\beta) \circ \mathfrak{t}(\Gamma_a\eta\Gamma_a).$$

This is an immediate consequence of [S6, Proposition 11.8]. By means of (5.8.2) we can easily verify that $\rho(\eta) \otimes \mathfrak{t}(\Gamma_a\eta\Gamma_a)$ maps W_a into itself. Moreover, we have

(5.8.3) *The action of $\rho(\eta) \otimes \mathfrak{t}(\Gamma_a\eta\Gamma_a)$ on W_a depends only on $\Gamma\eta\Gamma$.*

Indeed, if $\Gamma\xi\Gamma = \Gamma\eta\Gamma$, then $\xi \equiv \eta\gamma \pmod{a\mathfrak{o}}$ with an element γ of Γ. By [S6, Proposition 10.5], we have $\Gamma_a\xi\Gamma_a = \Gamma_a\eta\gamma\Gamma_a = (\Gamma_a\eta\Gamma_a)(\Gamma_a\gamma\Gamma_a)$. Since $\big(\rho(\eta\gamma) - \rho(\xi)\big)X \subset aX$ and $\mathfrak{t}(\Gamma_a\gamma\Gamma_a) = r_a(\gamma)$, we have, for $t \in W_a$,

$$\big[\rho(\xi) \otimes \mathfrak{t}(\Gamma_a\xi\Gamma_a)\big]t = \big[\rho(\eta\gamma) \otimes \mathfrak{t}(\Gamma_a\eta\Gamma_a) \cdot r_a(\gamma)\big]t = \big[\rho(\eta) \otimes \mathfrak{t}(\Gamma_a\eta\Gamma_a)\big]t,$$

which proves (5.8.3).

If c divides a, then $\nu(\eta)$ is prime to c, and $\mathfrak{t}(\Gamma_a\xi\Gamma_a) \circ r_{a,c}^* = r_{a,c}^* \circ \mathfrak{t}(\Gamma_c\eta\Gamma_c)$, and hence we can define the direct limit of $\rho(\eta) \otimes \mathfrak{t}(\Gamma_a\eta\Gamma_a)$ with $a = \ell^n$ on $\varinjlim W(\ell^n)$. We denote this limit by $\Lambda_\ell(\Gamma\eta\Gamma)$, since it depends only on $\Gamma\eta\Gamma$, as shown in (5.8.3).

5.9. Theorem. *If $\eta \in \mathfrak{o} \cap B_0^\times$ and $\nu(\eta)$ is prime to $\ell \cdot D(B/F)$, then*

(5.9.1) $$\omega_\ell \circ (\Gamma\eta\Gamma)_{X,\ell} = \Lambda_\ell(\Gamma\eta\Gamma) \circ \omega_\ell,$$

and consequently $\mathfrak{W}_\ell$ is stable under $\Lambda_\ell(\Gamma\eta\Gamma)$.

PROOF. Let $w \in Z^1_{\overline{P}}(\overline{\Gamma}, X)$, and let u be the restriction of $a^{-1}w$ to $\overline{\Gamma}_a$. Define u_ν^*, $f = (f_\nu)_{\nu=1}^m$, $t = (t_\nu)_{\nu=1}^m$, and $\rho_{\mu\nu}(\alpha)$ as in §4.2. Let $\Gamma_a\eta\Gamma_a =$

$\bigcup_{\iota=1}^{d} \eta_\iota \Gamma_a$ be a disjoint union. Then $\Gamma \eta \Gamma = \bigcup_{\iota=1}^{d} \eta_\iota \Gamma$ is also a disjoint union. Put

$$\overline{f} = (\overline{f}_\nu)_{\nu=1}^{m} = \prod_{\iota=1}^{d} (f \circ \eta_\iota^{-1})^\eta, \quad \overline{t} = (\overline{t}_\nu)_{\nu=1}^{m}, \quad \overline{t}_\mu = \psi_a(\operatorname{div}(\overline{f}_\mu)).$$

Consider w as an element of $Z_P^1(\Gamma, \Delta, X)$ as in §3.2, and define an element $\overline{w}$ of $Z_P^1(\Gamma, \Delta, X)$ by $\overline{w}(\alpha) = \sum_{\iota=1}^{d} \eta_\iota w(\alpha_\iota)$, where α_ι is an element of Γ such that $\alpha^{-1}\eta_\iota = \eta_\jmath \alpha_\iota^{-1}$ (see §3.6). Define characters $\overline{u}_\nu^*$ of Γ_a by

$$\overline{u}_\nu^*(\alpha) = \exp\left[2\pi i \theta_\nu\big(\overline{w}(\alpha)\big)/a\right] \qquad (\alpha \in \Gamma_a).$$

Since $\rho(\eta) \equiv \rho(\eta_\iota) \pmod{a}$, we have

$$\overline{f} \circ \alpha = \prod_{\iota=1}^{d} (f \circ \alpha_\iota \eta_\iota^{-1})^\eta = \prod_{\iota=1}^{d} u^*(\alpha_\iota)^{\eta_\jmath} \cdot \overline{f} = u^*(\alpha)\overline{f} \qquad (\alpha \in \Gamma_a).$$

Therefore $\overline{t}$ is the point of W_a corresponding to $a^{-1}\overline{w}$. Now we have

$$\operatorname{div}(\overline{f}_\mu) = \sum_{\nu=1}^{n+1} \rho_{\mu\nu}(\eta)\operatorname{div}\big(\prod_{\iota=1}^{d}(f_\nu \circ \eta_\iota^{-1})\big) = \sum_{\nu=1}^{n+1} \rho_{\mu\nu}(\eta)\mathfrak{T}(\Gamma_a \eta \Gamma_a)\big(\operatorname{div}(f_\nu)\big),$$

and hence $\overline{t} = \left[\rho(\eta) \otimes \mathfrak{t}(\Gamma_a \eta \Gamma_a)\right]t$. Since t and $\overline{t}$ correspond to $a^{-1}w$ and $a^{-1}\overline{w}$ respectively, we obtain our theorem by taking a to be ℓ^n.

5.10. Since $\overline{\Gamma}$ is finitely generated, $H_{\frac{1}{P}}^1(\overline{\Gamma}, X)$ is isomorphic to the product of finite groups and $\mathbf{Z}^e$ with $0 \leq e \in \mathbf{Z}$. Then $H_{\frac{1}{P}}^1(\overline{\Gamma}, X) \otimes (\mathbf{Q}_\ell/\mathbf{Z}_\ell)$ is isomorphic to $(\mathbf{Q}_\ell/\mathbf{Z}_\ell)^e$. Since $\operatorname{Ker}(\omega_\ell)$ is finite, $\mathfrak{W}_\ell$ is also isomorphic to $(\mathbf{Q}_\ell/\mathbf{Z}_\ell)^e$.

Let $\mu^\ell(\sigma)$ and $\mu_*^\ell(\sigma)$ denote the restriction of $\mu(\sigma)$ and $\mu_*(\sigma)$ to $\mathfrak{W}_\ell$. Then, with respect to a fixed isomorphism of $\mathfrak{W}_\ell$ to $(\mathbf{Q}_\ell/\mathbf{Z}_\ell)^e$, the endomorphisms $\Lambda_\ell(\Gamma \eta \Gamma)$, $\mu^\ell(\sigma)$, and $\mu_*^\ell(\sigma)$ of $\mathfrak{W}_\ell$ can be represented by matrices in $M_e(\mathbf{Z}_\ell)$. Obviously the matrices corresponding to $\mu^\ell(\sigma)$ and $\mu_*^\ell(\sigma)$ belong to $GL_e(\mathbf{Z}_\ell)$. Thus we have two representations

$$(5.10.1) \qquad {}^t\mu^\ell, \ \mu_*^\ell : \operatorname{Gal}(\overline{\mathbf{Q}}/k_1) \longrightarrow GL_e(\mathbf{Z}_\ell).$$

Notice that we have to take the transpose of μ^ℓ in view of (5.4.7). Since $\operatorname{Gal}(\overline{\mathbf{Q}}/k_1)$ and $GL_e(\mathbf{Z}_\ell)$ are compact topological groups, it is meaningful to speak of the continuity of the maps μ^ℓ and μ_*^ℓ. In fact, we have:

5.11. Proposition. *The maps μ^ℓ and μ_*^ℓ are continuous.*

PROOF. Fixing a positive integer n, put $M = \{t \in \mathfrak{W}_\ell \mid \ell^n t = 0\}$. We can find a positive integer $s > n$ such that every point of M is represented by a point of $W(\ell^s)$. Put $a = \ell^s$. Let k' be the smallest subfield of $\overline{\mathbf{Q}}$ containing k_a over which the points of W_a are rational. Then $[K' : \mathbf{Q}] < \infty$. Let $\sigma \in \operatorname{Gal}(\overline{\mathbf{Q}}/k_1 k')$. Then we can take $\alpha = 1$ in (5.4.2) to define $\mu_a(\sigma)$ and $\mu_a^*(\sigma)$, so that $\mu_a(\sigma)$ and $\mu_a^*(\sigma)$ are the identity map. Therefore $\mu^\ell(\sigma)$ and $\mu_*^\ell(\sigma)$ leave the points of M invariant, which proves our theorem.

5.12. Main Theorem. *Suppose that $k_1 = F$, that is, every ideal in F is a principal ideal generated by a totally positive element. Then there exists a finite set S of prime ideals in F, depending only on the quaternion algebra B, with the following properties:*

(5.12.1) *Let K_ℓ (resp. K_ℓ^*) be the subfield of $\overline{\mathbf{Q}}$ corresponding to the kernel of μ^ℓ (resp. μ_*^ℓ) by Galois theory. Then a prime ideal $\mathfrak{p}$ in F is unramified in K_ℓ and K_ℓ^*, provided $\mathfrak{p} \notin S$ and $\mathfrak{p} \nmid \ell$.*

(5.12.2) *Let $\mathfrak{p}$ be a prime ideal in F, $\mathfrak{P}$ a prime divisor of $\overline{\mathbf{Q}}$ dividing $\mathfrak{p}$, $\sigma_\mathfrak{P}$ a Frobenius automorphism of $\overline{\mathbf{Q}}$ over F with respect to $\mathfrak{P}$ and $\mathfrak{p}$, and α an element of $\mathfrak{o} \cap B_0^\times$ such that $\nu(\alpha)\mathfrak{r} = \mathfrak{p}$. (Such an α always exists.) If $\mathfrak{p} \notin S$ and $\mathfrak{p} \nmid \ell$, then*

(5.12.2a)
$$\Lambda_\ell(\Gamma\alpha\Gamma) = \mu^\ell(\sigma_\mathfrak{P}) + \mu_*^\ell(\sigma_\mathfrak{P}),$$

(5.12.2b)
$$N(\mathfrak{p})^{h+1} = \mu^\ell(\sigma_\mathfrak{P}) \circ \mu_*^\ell(\sigma_\mathfrak{P}) = \mu_*^\ell(\sigma_\mathfrak{P}) \circ \mu^\ell(\sigma_\mathfrak{P}).$$

(5.12.3) $S = \varnothing$ *if* $B = M_2(\mathbf{Q})$.

PROOF. Let $\mathfrak{p}$, $\mathfrak{P}$, $\sigma_\mathfrak{P}$, and α be as in (5.12.2). Denote by $\mathfrak{P}(Y)$ the reduction modulo $\mathfrak{P}$ of an algebro-geometric object Y rational over $\overline{\mathbf{Q}}$. Relation (5.12.2b) follows immediately from (5.4.4). Now, by [S6, Theorem 11.17] (see also [S1] and [S3] for earlier results of the same type), we have, for almost all $\mathfrak{p}$,

$$\mathfrak{P}\big(\mathfrak{T}(\Gamma_a\alpha^\iota\Gamma_a)\big) = \mathfrak{P}\big({}^tR_a(\alpha)\big) \circ \Pi_a + {}^t\Pi_a \circ \mathfrak{P}\big(R_a(\alpha^\iota)\big)$$

on $\mathfrak{P}(V_a \times V_a)$, where Π_a and ${}^t\Pi_a$ are respectively the locus of $x \times x^{N(\mathfrak{p})}$ on $\mathfrak{P}(V_a) \times \mathfrak{P}(V_a)^{N(\mathfrak{p})}$ and its transpose. Then we have

(5.12.4)
$$\mathfrak{P}\big(\mathfrak{t}(\Gamma_a\alpha^\iota\Gamma_a)\big) = \mathfrak{P}\big(r_a(\alpha)^{-1}\big) \circ \pi_a + \pi_a^* \circ \mathfrak{P}\big(r_a(\alpha^\iota)\big)$$

on $\mathfrak{P}(J_a)$, where π_a is the Frobenius isogeny of $\mathfrak{P}(J_a)$ to $\mathfrak{P}(J_a)^{N(\mathfrak{p})}$, and π_a^* its transpose. We have not specified for what a and $\mathfrak{p}$ these relations hold. In fact, there exists a finite set S of prime ideals in F such that:

(5.12.5) *If $\mathfrak{p} \notin S$ and $\mathfrak{p} \nmid a$, then J_a is without defect for $\mathfrak{P}$ in the sense of [ST, §11.1], and (5.12.4) holds.*

Igusa [I] proved that one can take $S = \varnothing$ if $B = M_2(\mathbf{Q})$. If B is a division algebra, we can construct a fibre variety Y, of which the base is V_a for some a, and the fibres are certain abelian varieties (see [S5]). Then V_b, for a multiple b of a, can be obtained as a section of Y obtained from the points of the fibres annihilated by b. Therefore, if Y has "good" reduction modulo $\mathfrak{P}$, so does V_b provided $\mathfrak{P}$ does not divide b. The existence of S can be shown in this way. We do not discuss this point in detail here, since we shall treat the same type of problem in a more general case on a future occasion.

From (5.12.4) we obtain

$$\mathfrak{P}\big(\rho(\alpha^\iota) \otimes \mathfrak{t}(\Gamma_a\alpha^\iota\Gamma_a)\big) = \big[\rho(\alpha^\iota) \otimes \mathfrak{P}(r_a(\alpha^{-1}))\big] \circ (1 \otimes \pi_a)$$
$$+ (1 \otimes \pi_a^*) \circ \big[\rho(\alpha^\iota) \otimes \mathfrak{P}(r_a(\alpha^\iota))\big] \quad \text{on} \quad \mathfrak{P}(X \otimes J_a).$$

Since $\mathfrak{p}$ does not divide ℓ, the map $t \mapsto \mathfrak{P}(t)$ for $t \in W_a$ is injective. Moreover, since $\sigma_\mathfrak{P}$ is a Frobenius automorphism, we have

$$\mathfrak{P}(\mu_a(\sigma_\mathfrak{P})t) = \big[\rho(\alpha^\iota) \otimes \mathfrak{P}(r_a(\alpha^{-1}))\big]\mathfrak{P}(t)^{N(\mathfrak{p})} \quad \text{and}$$

$$\mathfrak{P}(\mu_a^*(\sigma_\mathfrak{P})t) = N(\mathfrak{p})\big\{\big[\rho(\alpha^\iota) \otimes \mathfrak{P}(r_a(\alpha^\iota))\big]\mathfrak{P}(t)\big\}^{1/N(\mathfrak{p})}$$

for $t \in W_a$. This together with the injectivity of tha map $t \mapsto \mathfrak{P}(t)$ proves that

$$\rho(\alpha^\iota) \otimes \mathfrak{t}(\Gamma_a\alpha^\iota\Gamma_a) = \mu_a(\sigma_\mathfrak{P}) + \mu_a^*(\sigma_\mathfrak{P})$$

on W_a. Taking the direct limit, we obtain (5.12.2a), since $\Lambda_\ell(\Gamma\alpha\Gamma) = \Lambda_\ell(\Gamma\alpha^\iota\Gamma)$.

To prove (5.12.1), let $K(a)$ denote the smallest field over which the points u of J_a such that $au = 0$ are all rational. A prime ideal $\mathfrak{p}$ is unramified in $K(a)$ if $\mathfrak{p} \notin S$ and $\mathfrak{p} \nmid a$. The proof of Proposition 5.11 shows that both K_ℓ and K_ℓ^* are contained in $\bigcup_{n=1}^\infty K(\ell^n)$, and hence we obtain (5.12.1).

The point of the above theorem is that $\Lambda_\ell(\Gamma\alpha\Gamma)$ on the left-hand side of (5.12.2a) is essentially $(\Gamma\alpha\Gamma)_{X,\ell}$ defined on $H^1_{\overline{P}}(\overline{\Gamma}, X) \otimes (\mathbf{Q}_\ell/\mathbf{Z}_\ell)$. Indeed, fix an isomorphism I of $H^1_{\overline{P}}(\overline{\Gamma}, X) \otimes (\mathbf{Q}_\ell/\mathbf{Z}_\ell)$ onto $(\mathbf{Q}_\ell/\mathbf{Z}_\ell)^e$ and also an isomorphism I' of $\mathfrak{W}_\ell$ to $(\mathbf{Q}_\ell/\mathbf{Z}_\ell)^e$. Since $\mathrm{Ker}(\omega_\ell)$ is finite, ω_ℓ is represented by an element U of $GL_n(\mathbf{Q}_\ell)$ with entries in $\mathbf{Z}_\ell$ with respect to I and I'. Then (5.9.1) shows that the matrix representing $\Lambda_\ell(\Gamma\alpha\Gamma)$ is the transform of the matrix representing $(\Gamma\alpha\Gamma)_{X,\ell}$ by U.

5.13. To define $\mathfrak{W}$ and $\mu(\sigma)$ etc., we have fixed a lattice X in the representation space of ρ. Let us now examine to what degree these objects depend on the choice of X. Let X' be another $\mathbf{Z}$-lattice in $X_{\mathbf{Q}}$ satisfying (5.3.4). Then we find two $B^\times$-automorphisms f and g of $X_{\mathbf{Q}}$ such that $f(X) \subset X'$, $g(X') \subset X$, and $g\big(f(x)\big) = f\big(g(x)\big) = cx$ for all $x \in X_{\mathbf{Q}}$ with a positive integer c. Define $\mathfrak{W}$ and ω with respect to X' in place of X, and denote them by $\mathfrak{W}'$ and ω'. Then from f and g we naturally obtain a commutative diagram

$$H^1_{\overline{P}}(\overline{\Gamma}, X) \otimes (\mathbf{Q}/\mathbf{Z}) \longrightarrow H^1_{\overline{P}}(\overline{\Gamma}, X') \otimes (\mathbf{Q}/\mathbf{Z}) \longrightarrow H^1_{\overline{P}}(\overline{\Gamma}, X) \otimes (\mathbf{Q}/\mathbf{Z})$$

$$\downarrow \qquad\qquad\qquad\qquad \downarrow \qquad\qquad\qquad\qquad \downarrow$$

$$\mathfrak{W} \quad \xrightarrow{\ f^*\ } \quad \mathfrak{W}' \quad \xrightarrow{\ g^*\ } \quad \mathfrak{W}$$

such that $g^* \circ f^*$ (resp. $f^* \circ g^*$) is c times the identity map of $\mathfrak{W}$ (resp. $\mathfrak{W}'$). The maps f^* and g^* commute with the action of $\mathrm{Gal}(\overline{\mathbf{Q}}/k_1)$ and $\Gamma\eta\Gamma$. Therefore the ℓ-adic representations μ^ℓ and μ_*^ℓ of $\mathrm{Gal}(\overline{\mathbf{Q}}/k_1)$ defined with respect to X' and $\mathfrak{W}'$ are equivalent to the original ones defined with respect to X and $\mathfrak{W}$, in the group $GL_e(\mathbf{Q}_\ell)$.

5.14. Let us now construct some representations ρ satisfying the conditions of §5.3. For simplicity, we consider here only two types of representations, though the method explained below can produce many more representations. These two types will be referred to as Type (I) and Type (II) in later discussions.

(I) Take a finite Galois extension S of $\mathbf{Q}$ so that $B_S = B \otimes_{\mathbf{Q}} S$ is isomorphic to $M_2(S)^g$. Let h_ν be the projection map of B_S to the νth factor $M_2(S)$. Then

$$(5.14.1) \qquad h_\nu(a^\iota) = v \cdot {}^t h_\nu(a) v^{-1} \quad \text{with} \quad v = \begin{bmatrix} 0 & -1 \\ 1 & 0 \end{bmatrix} \qquad (a \in B_S).$$

Fix a positive integer n, and define a representation

$$\chi : B_S^\times \longrightarrow GL_q(S), \quad q = (n+1)^g,$$

by $\chi = \bigotimes_{\nu=1}^g (\rho_n \circ h_\nu)$. Obviously χ is absolutely irreducible, and for every $\sigma \in \mathrm{Gal}(S/\mathbf{Q})$, χ^σ is equivalent to χ over S. Consequently there exists a unique automorphism a_σ of $M_q(S)$ over S such that $\chi^\sigma = a_\sigma \circ \chi$. Then $a_{\sigma\tau} = (a_\sigma)^\tau \circ a_\tau$. Therefore we can find an algebra Y over $\mathbf{Q}$ and an isomorphism

$$(5.14.2) \qquad\qquad f : M_q(S) \longrightarrow Y \otimes_{\mathbf{Q}} S$$

over S such that $f = f^\sigma \circ a_\sigma$. (The simplest, if not the best, way to obtain Y is to put $Y = \{ x \in M_q(S) \mid x^\sigma = a_\sigma(x) \text{ for all } \sigma \in \mathrm{Gal}(S/\mathbf{Q}) \}$. Then it can easily be seen that $Y \otimes_{\mathbf{Q}} S = M_q(S)$.) Putting $\lambda = f \circ \chi$, we obtain a $\mathbf{Q}$-rational representation $\lambda : B^\times \to Y^\times$. In view of (5.14.2), we see that Y is central simple over $\mathbf{Q}$. Moreover, by (5.14.1), we have an involution φ of $M_q(S)$ over S such that $\chi(a^\iota) = \varphi(\chi(a))$. Then $a_\sigma \circ \varphi = \varphi^\sigma \circ a_\sigma$ for every $\sigma \in \mathrm{Gal}(S/\mathbf{Q})$. Therefore we obtain an involution ψ of Y over $\mathbf{Q}$ such that $\lambda(a^\iota) = \lambda(a)^\psi$. It follows that Y is either isomorphic to $M_q(\mathbf{Q})$, or to $M_{q/2}(D)$ with a division quaternion algebra D over $\mathbf{Q}$; hence there is a $\mathbf{Q}$-linear isomorphism of Y into $M_{\mu q}(\mathbf{Q})$ with $\mu = 1$ or 2. Combining this with λ, we finally obtain a $\mathbf{Q}$-rational representation

$$\rho : B^\times \longrightarrow GL_{\mu q}(\mathbf{Q}), \quad q = (n+1)^g, \quad \mu = 1 \text{ or } 2.$$

We call this. ρ a representation of Type (I). From our construction we see that $\rho(c) = N_{F/\mathbf{Q}}(c)^n$ for $c \in F^\times$.

To obtain a lattice X as in (5.3.4), take a maximal order R in $M_q(S)$ so that $\chi(\mathfrak{o} \cap B^\times) \subset R$. This is of course possible. Put $T = f(R) \cap Y$. Then T is an order in Y, and $\lambda(\mathfrak{o} \cap B^\times) \subset T$. Take a T-stable lattice X in $\mathbf{Q}^{\mu q}$. With a suitable choice of R, we can assume that $\chi(\alpha) - \chi(\beta) \in \mathfrak{a}R$ if $\alpha,\ \beta \in \mathfrak{o} \cap B^\times$ and $\alpha - \beta \in \mathfrak{a}\mathfrak{o}$. Then $\lambda(\alpha) - \lambda(\beta) \in \mathfrak{a}T$, and hence X satisfies (5.3.4).

(II) Take an F-linear isomorphism $\lambda : B \longrightarrow M_2(K)$ with a quadratic extension K of F unrafmified at τ_1 (see §5.1). If σ denotes the nontrivial automorphism of K over F, we have $\lambda(x)^\sigma = \zeta \lambda(x) \zeta^{-1}$ with an element ζ of $GL_2(K)$. Then $\zeta^\sigma \zeta = c$ with an element c of F. Fix a positive integer n, and put

$$Y = \{ u \in M_{n+1}(K) \mid u^\sigma = \rho_n(\zeta) u \rho_n(\zeta)^{-1} \}.$$

It can easily be seen that $M_{n+1}(K) = Y \otimes_F K$; hence Y is a central simple algebra over F. (To define Y, we can also use the same trick as in (I).) Since Y splits over a quadratic extension of F, Y is isomorphic either to $M_{n+1}(F)$ or to $M_{(n+1)/2}(D)$ with a quaternion algebra D over F. In particular, if n is even, Y is isomorphic to $M_{n+1}(F)$. Now $\rho_n \circ \lambda$ gives an F-rational map of $B^\times$ into $Y^\times$. Combining this with an F-linear isomorphism of Y into $M_q(F)$, where $q = \mu(n+1)$ with $\mu = 1$ or 2, we obtain an F-rational representation χ of $B^\times$ into $GL_q(F)$. Assuming n to be even, define $\rho' : B^\times \to GL_{n+1}(F)$ by

$$\rho'(\alpha) = \big(N_{F/\mathbf{Q}}(\nu(\alpha))/\nu(\alpha) \big)^{n/2} \chi(\alpha).$$

Combining further ρ' with a $\mathbf{Q}$-linear injection of $M_{n+1}(F)$ into $M_{g(n+1)}(\mathbf{Q})$, we obtain a $\mathbf{Q}$-rational representation $\rho : B^\times \longrightarrow GL_{g(n+1)}(\mathbf{Q})$ such that $\rho(c) = N_{F/\mathbf{Q}}(c)^n$ for every $c \in F^\times$. This is a representation of Type (II). The existence of a lattice X satisfying (5.3.4) can be shown in the same manner as for Type (I). It should be noted that if $F = \mathbf{Q}$, a representation of Type (I) is $\mathbf{Q}$-equivalent to one of Type (II), and vice versa.

5.15. Let p_ν denote the projection map of $B_{\mathbf{R}}$ to its νth factor; then $p_\nu(B_{\mathbf{R}})$ is $M_2(\mathbf{R})$ or K according as $\nu = 1$ or $\nu > 1$. Further we may assume that

34

$p_\nu(a) = a^{\tau_\nu}$ for $a \in F$ (see §5.1). Let J be an $\mathbf{R}$-linear isomorphism of $\mathbf{K}$ into $M_2(\mathbf{C})$. Taking $\mathbf{R}$, $\mathbf{C}$, $\mathbf{K}$ in place of F, K, B in (II) of §5.14, we obtain, for even n, a representation ρ_n^* of $\mathbf{K}^\times$ into $GL_{n+1}(\mathbf{R})$, which is equivalent to $\rho_n \circ J$. Put $\chi_n(y) = \nu(y)^{-n/2} \rho_n^*(y)$ for $y \in \mathbf{K}^\times$, and $N(\alpha) = \prod_{\mu=1}^g \nu(p_\mu(\alpha))$ for $\alpha \in B_\mathbf{R}^\times$. Fix a positive even integer n, and take a representation ρ of Type (I) or (II) for n as given in §5.14. Define a representation Ψ_I and Ψ_{II} of $B_\mathbf{R}^\times$ by

$$\Psi_I = \text{the direct sum of } \mu \text{ copies of } N^{n/2} \cdot \bigotimes_{\nu=2}^g (\chi_n \circ p_\nu), \quad \mu = 1 \text{ or } 2,$$

$$\Psi_{II} = N^{n/2} \cdot \left[\bigoplus_{\nu=2}^g (\chi_n \circ p_\nu) \right],$$

where μ is as in §5.14. Then both Ψ_I and Ψ_{II} are trivial on Δ, and map Γ onto compact groups. Moreover, ρ is equivalent to

$$(\nu \circ p_1)^{-n/2} \cdot (\rho_n \circ p_1) \otimes \Psi_I \quad \text{if } \rho \text{ is of Type (I)},$$

$$(\nu \circ p_1)^{-n/2} \cdot (\rho_n \circ p_1) \otimes N^{n/2} \oplus (\rho_0 \circ p_1) \otimes \Psi_{II} \quad \text{if } \rho \text{ is of Type (II)}.$$

Take X as in (5.3.4). Since $\overline{\Gamma}$ is finitely generated, $H_{\overline{P}}^1(\overline{\Gamma}, X)$ is a finitely generated $\mathbf{Z}$-module, and $H_{\overline{P}}^1(\overline{\Gamma}, X) \otimes \mathbf{R}$ can be identified with $H_{\overline{P}}^1(\overline{\Gamma}, X \otimes \mathbf{R})$. Now, by Theorem 3.5 we have an $\mathbf{R}$-linear isomorphism

$$(5.15.1) \quad H_{\overline{P}}^1(\overline{\Gamma}, X) \otimes \mathbf{R} \cong \begin{cases} S_{n+2}(\Gamma, \Psi_I) & \text{if } \rho \text{ is of Type (I)}, \\ S_{n+2}(\Gamma, N^{n/2}) \oplus S_2(\Gamma, \Psi_{II}) & \text{if } \rho \text{ is of Type (II)}. \end{cases}$$

Indeed, if ρ is of Type (I), for example, then ρ and Ψ_I correspond to χ and Ψ of §3.3. (Notice that $N^{n/2}$ is trivial on Γ.)

On the other hand, we have a surjective map with finite kernel

$$(5.15.2) \qquad\qquad \omega_\ell : H_{\overline{P}}^1(\overline{\Gamma}, X) \otimes (\mathbf{Q}_\ell/\mathbf{Z}_\ell) \longrightarrow \mathfrak{W}_\ell.$$

Let $\mathcal{R}_\ell$ be the Hecke ring generated by the $\Gamma\eta\Gamma$ for all $\eta \in \mathfrak{o} \cap D_c$ with $c = \ell \cdot D(B/F)$. By virtue of (3.7.3) and Theorem 5.9, the isomorphism of (5.15.1) and ω_ℓ commute with the action of $\mathcal{R}_\ell$ on the modules.

5.16. *Sketch of the proof of (5.7.1).* Let E be a division quaternion algebra over a totally real algebraic number field K of degree q such that $E \otimes_\mathbf{Q} \mathbf{R}$ is isomorphic to $M_2(\mathbf{R})^p \times \mathbf{K}^{q-p}$ with $p > 2$. Let $\mathfrak{O}$ be a maximal order of E and let

$$\Gamma' = \left\{ \gamma \in \mathfrak{O}^\times \,\middle|\, N_{E/K}(\gamma) = 1, \ \gamma - 1 \in N\mathfrak{O} \right\}$$

with a positive integer N. Choose N so that Γ' has no elements of finite order other than 1. Take a $\mathbf{Q}$-rational representation $R : E^\times \to GL(V)$ with a finite-dimensional vector space V over $\mathbf{Q}$, and take a $\mathbf{Z}$-lattice Y in V such that $R(\alpha)Y \subset Y$ for every $\alpha \in \mathfrak{o} \cap E^\times$; we assume that $R(c) = N_{K/\mathbf{Q}}(c)^t$ for every $c \in K^\times$ with $0 \leq t \in \mathbf{Z}$. Define $H^i(\Gamma', V)$ and $H^i(\Gamma', Y)$ as in §1.1. Now, by [MS, Theorem 7.1], $H^i(\Gamma', V) = \{0\}$ if $i < p$ for "almost all" R; to be precise, the vanishing holds provided R can be decomposed over $\mathbf{C}$ into the direct sum of representations of the types described in that theorem. For every integral ideal $\mathfrak{a}$ in K, we have a standard exact sequence

$$\longrightarrow H^1(\Gamma', \mathfrak{a}^{-1}Y) \longrightarrow H^1(\Gamma', \mathfrak{a}^{-1}Y/Y) \longrightarrow H^2(\Gamma', Y) \longrightarrow$$

Since $H^1(\Gamma', V) = \{0\}$ and $H^2(\Gamma', V) = \{0\}$, both $H^1(\Gamma', \mathfrak{a}^{-1}Y)$ and $H^2(\Gamma', Y)$ are finite. The structure of the Γ-module $\mathfrak{a}^{-1}Y$ depends only on the ideal class of $\mathfrak{a}$, and hence the above exact sequence shows that there exists an integer M depending only on R and Y such that $H^1(\Gamma', \mathfrak{a}^{-1}Y/Y)$ has order $< M$. Put

$$\Gamma'(\mathfrak{a}) = \left\{ \gamma \in \Gamma' \,\middle|\, \gamma - 1 \in \mathfrak{a}\mathfrak{O} \right\}.$$

By (1.5.3), $H^1(\Gamma'/\Gamma'(\mathfrak{a}), \mathfrak{a}^{-1}Y/Y)$ is injectively mapped into $H^1(\Gamma', \mathfrak{a}^{-1}Y/Y)$, so that $H^1(\Gamma'/\Gamma'(\mathfrak{a}), \mathfrak{a}^{-1}Y/Y)$ has order $< M$.

Now, to prove (5.7.1), we first assume that B is a division algebra, and take a suitable totally real algebraic number field L linearly disjoint with F over $\mathbf{Q}$; then we take $K = F \otimes_{\mathbf{Q}} L$ and $E = B \otimes_{\mathbf{Q}} L$. We choose L so that ℓ splits completely in L. Take a prime factor $\mathfrak{l}_0$ of ℓ in L, and put $\mathfrak{l} = \mathfrak{l}_0 \mathfrak{r}_K$, where $\mathfrak{r}_K$ is the maximal order of K. Given $\{\rho, X\}$, we can find $\{R, Y\}$ defined with such K and E so that $H^1(\overline{\Gamma}/\overline{\Gamma}(\ell^n), \ell^{-n}X/X)$ is isomorphic to $H^1(\Gamma'/\Gamma'(\mathfrak{l}^n), \mathfrak{l}^{-n}Y/Y)$. Without going into the details of how we choose L, R, and Y, let us just say that the main reason for the feasibility is that the localization of L resp. E at $\mathfrak{l}_0$ is isomorphic to $\mathbf{Q}_\ell$ and $B \otimes_{\mathbf{Q}} \mathbf{Q}_\ell$. In this way we obtain (5.7.1) when B is a division algebra. It remains to prove the case $B = M_2(\mathbf{Q})$. Again the structure of $H^1(\overline{\Gamma}/\overline{\Gamma}(\ell^n), \ell^{-n}X/X)$ is a local problem concerning $M_2(\mathbf{Q}_\ell)$. Therefore we take a division quaternion algebra B' over $\mathbf{Q}$ that splits over $\mathbf{Q}_\ell$. Since (5.7.1) holds for B', the same is true for $M_2(\mathbf{Q})$. This completes the proof.

5.17. We have assumed that F has class number 1 in the narrow sense in The Main Theorem 5.12; also our Γ defined in §5.2 is a group of "level 1." These assumptions are not absolutely necessary, and in fact we can treat the case of an arbitrary congruence subgroup with no assumption on the class number. We do not dicuss such genral cases in this paper, because our principal aim is to present some new ideas in a relatively simple case without introducing many symbols. At any rate, the main theorems of [S6] and [S7] invoked in §5.2 apply to B as in §5.1 over any totally real F, and we believe that the reader can easily find formulations in the general case.

References

[B1] A. Borel, Density properties for certain subgroups of semi-simple groups without compact components, Ann. of Math.,**72** (1960),

[B2] A. Borel, Density and maximality of arithmetic subgroups, J. reine angew. math. **224** (1966), 78–89.

[E1] M. Eichler, Quaternäre quadratische Formen und die Riemannsche Vermutung für die Kongruenzzetafunktion, Arch. Math., **5** (1954), 355–366.

[E2] M. Eichler, Eine Verallgemeinerung der Abelschen Integrale, Math. Z., **67** (1957), 267–298.

[I] J. Igusa, Kroneckerian model of fields of elliptic modular functions, Amer. J. Math., **81** (1959), 561–577.

[MS] Y. Matsushima and G. Shimura, On the cohomology groups attached to certain vector valued differential forms on the product of the upper half planes, Ann. of Math., **78** (1963), 417–449.

[S1] G. Shimura, Correspondances modulaires et les fonctions zeta de courbes algébriques, J. Math. Soc. Japan, **10** (1958), 1–28.

[S2] ______ , Sur les intégrales attachées aux formes automorphes, J. Math. Soc. Japan, **11** (1959), 291–311.

[S3] ______ , On the zeta functions of the algebraic curves uniformized by certain automorphic functions, J. Math. Soc. Japan, **13** (1961), 275–331.

[S4] ______ , On Dirichlet series and abelian varieties attached to automorphic forms, Ann. of Math., **76** (1962), 237–294.

[S5] ______ , Moduli and fibre systems of abelian varities, Ann. of Math. **83** (1966), 294–338.

[S6] ______ , Construction of class fields and zeta functions of algebraic curves, Ann. of Math. **85,** (1967), 58–159.

[S7] ______ , Algebraic number fields and symplectic discontinuous groups, Ann. of Math. 86 (1967), 503–592.

[ST] G. Shimura and Y. Taniyama, Complex multiplication of abelian varieties and its applications to number theory, Publ. Math. Soc. Japan, No.6, 1961.

Local representations of Galois groups

Annals of Mathematics, 89 (1969), 99-124

Introduction

The purpose of this paper is to present some infinite Galois extensions of algebraic number fields , whose Galois groups possess some matrix representations in local fields, through which the Frobenius automorphisms have algebraic numbers of absolute value $N(\mathfrak{p})^{\kappa/2}$ as characteristic roots. Here $\mathfrak{p}$ is the prime ideal of the basic field, for which the Frobenius automorphism is considered, and κ is an integer determined by the representation. These Galois extensions will be generated by the coordinates of the points lying on some Galois coverings of an algebraic variety, which is isomorphic to the quotient of the Siegel upper half space $\mathfrak{H}_n$ by an arithmetic discontinuous group. In the last section of the previous paper [7], we have discussed certain local representations of the Galois groups of such extensions. The present results will give a re-formulation and a refinement of those obtained there.[1]

To be more specific, we consider a certain reductive algebraic group $\mathfrak{G}$ defined over Q which has the following properties:

(1) The center of $\mathfrak{G}_Q$ is isomorphic to the multiplicative group $F^\times$ of a totally real algebraic number field F.

(2) $\mathfrak{G}_Q$ acts faithfully on a vector space X over F, where the action of $F^\times$ is the same as the center of $\mathfrak{G}_Q$ through the isomorphism of (1).

(3) The semi-simple part of $\mathfrak{G}_R$ is isomorphic to the product of $\mathrm{Sp}(n, R)$ and a compact group.

By (3), we can find a subgroup $\mathfrak{G}^+$ of $\mathfrak{G}_Q$ of index 2, which acts on $\mathfrak{H}_n$. Let $\mathfrak{r}_F$ denote the ring of algebraic integers in F, and $\mathfrak{x}$ a finitely generated $\mathfrak{r}_F$-submodule of X which spans X over F. Define subgroups Γ and $\Gamma_\mathfrak{a}$ of $\mathfrak{G}^+$, for every integral ideal $\mathfrak{a}$ in F, by

$$\Gamma = \{\gamma \in \mathfrak{G}^+ \mid \mathfrak{x}\gamma = \mathfrak{x}\} \,,$$
$$\Gamma_\mathfrak{a} = \{\gamma \in \Gamma \mid \mathfrak{x}(1 - \gamma) \subset \mathfrak{a}\mathfrak{x}\} \,.$$

Then Γ and $\Gamma_\mathfrak{a}$ can be considered as discontinuous groups acting on $\mathfrak{H}_n$. These

---

* During the preparation of this paper the author received partial support from the National Science Foundation.

[1] The one-dimensional case has been discussed, without proofs, in an expository article [8].

include $\mathrm{Sp}(n, Z)$ and its congruence subgroups as special cases. By virtue of the results of [7], we can obtain, for a suitable choice of $\mathfrak{x}$, an algebraic variety $V_{\mathfrak{a}}$ and a $\Gamma_{\mathfrak{a}}$-invariant holomorphic map $\varphi_{\mathfrak{a}}$ of $\mathfrak{H}_n$ onto $V_{\mathfrak{a}}$ so that the following conditions are satisfied:

(4) $V_{\mathfrak{a}}$ is defined over the maximal class field $k_{\mathfrak{a}}$ over F whose conductor divides $\mathfrak{a}\mathfrak{u}_0$, where $\mathfrak{u}_0$ is the product of all archimedean primes of F.

(5) $\varphi_{\mathfrak{a}}$ gives a biregular isomorphism of $\mathfrak{H}_n/\Gamma_{\mathfrak{a}}$ onto $V_{\mathfrak{a}}$.

(6) If $\mathfrak{a} \subset \mathfrak{b}$, there exists a morphism $T_{\mathfrak{b},\mathfrak{a}}$ of $V_{\mathfrak{a}}$ onto $V_{\mathfrak{b}}$, rational over $k_{\mathfrak{a}}$, such that $T_{\mathfrak{b},\mathfrak{a}} \circ \varphi_{\mathfrak{a}} = \varphi_{\mathfrak{b}}$.

The couple $(V_{\mathfrak{a}}, \varphi_{\mathfrak{a}})$ can be characterized by (4), (5), and a certain property at isolated fixed points of $\mathfrak{G}^+$ on $\mathfrak{H}_n$ [7]. Let us write $V_{\mathfrak{a}}$, $\varphi_{\mathfrak{a}}$, and $k_{\mathfrak{a}}$ simply as V, φ, and k, when $\mathfrak{a} = \mathfrak{r}_F$.

Now let y be an arbitrary point of V whose coordinates may or may not be algebraic. Take a point z of $\mathfrak{H}_n$ so that $\varphi(z) = y$. Let $\mathfrak{R}_y$ be the composite of the fields $k_{\mathfrak{a}}(\varphi_{\mathfrak{a}}(z))$ for all $\mathfrak{a}$. Then $\mathfrak{R}_y$ is an infinite Galois extension of $k(y)$ independent of the choice of z. It is this Galois extension which interests us here. For simplicity, let us assume that $\{\gamma \in \Gamma \mid \gamma(z) = z\}$ coincides with the group E of units of F, i.e., z is not a non-trivial fixed point of Γ. For every rational prime l, let $\mathfrak{G}_l$ denote the group of the points of $\mathfrak{G}$ rational over Q_l, and let $\mathfrak{x}_l = \mathfrak{x} \otimes_Z Z_l$, where Q_l (resp. Z_l) denotes the l-adic completion of Q (resp. Z). Put

$$\mathfrak{u}_l = \{\gamma \in \mathfrak{G}_l \mid \mathfrak{x}_l\gamma = \mathfrak{x}_l\} \, ,$$
$$\mathfrak{u} = \prod_l \mathfrak{u}_l \qquad\qquad \text{(the product for all } l) \, .$$

Then we can establish a continuous isomorphism J of the Galois group $G(\mathfrak{R}_y/k(y))$ into $\mathfrak{u}/\bar{E}$, where $\bar{E}$ is the closure of E in $\mathfrak{u}$. This isomorphism is surjective if y is generic on V.

Let us now assume that the coordinates of y are algebraic numbers. To study the Frobenius automorphisms of $\mathfrak{R}_y$, we consider a representation

$$\omega : \mathfrak{G} \longrightarrow \mathrm{GL}_r \, ,$$

rational over Q, satisfying the following two conditions:

(7) $\omega(a) = N_{F/Q}(a)^{\kappa}$ for every $a \in F^{\times}$ with an integer κ independent of a.

(8) $\omega(e) = 1$ for every $e \in E$.

For each l, we obtain from ω a representation

$$\omega_l : \mathfrak{G}_l \longrightarrow \mathrm{GL}_r(Q_l) \, .$$

Combining J with the projection of $\mathfrak{u}$ to $\mathfrak{u}_l$, we obtain a homomorphism

$$J_l : G(\mathfrak{R}_y/k(y)) \longrightarrow \mathfrak{u}_l/E_l \, ,$$

where E_l is the closure of E in $\mathfrak{U}_l$. In view of (8), we now have an l-adic representation

$$\omega_l \circ J_l \colon G\big(\mathfrak{R}_y/k(y)\big) \longrightarrow \mathrm{GL}_r(\boldsymbol{Q}_l) \ .$$

Let $\mathfrak{p}$ be a prime ideal in $k(y)$, σ a Frobenius automorphism of $\mathfrak{R}_y$ over $k(y)$ for any prime divisor of $\mathfrak{R}_y$ extending $\mathfrak{p}$. We can then prove:

There exists an integral ideal $\mathfrak{t}$ *in* $k(y)$ *such that the conjugacy class of* $\omega_l(J_l(\sigma))$ *is determined by* l *and* $\mathfrak{p}$, *and independent of the choice of* σ, *if* $\mathfrak{p}$ *is prime to* $l\mathfrak{t}$.

In this setting, our main result is stated as follows.[2]

The integral ideal $\mathfrak{t}$ *can be chosen so that the characteristic roots of* $\omega_l(J_l(\sigma))$ *are algebraic numbers of absolute value* $N(\mathfrak{p})^{\kappa/2}$ *if* $\mathfrak{p}$ *is prime to* $l\mathfrak{t}$. *Moreover, there exists a positive integer* μ, *independent of* $\mathfrak{p}$ *and* l, *such that the characteristic polynomial of* $\omega_l(J_l(\sigma^\mu))$ *has rational coefficients, and is independent of* l, *so long as* $\mathfrak{p}$ *is prime to* $l\mathfrak{t}$.

This may be regarded as a generalization of the previously known results concerning the Galois extension generated by the points of finite order on an abelian variety. In fact, if $\mathfrak{G}$ is the group of similitudes of an alternating form of $2n$ variables with rational coefficients, then Γ is commensurable with $\mathrm{Sp}(n, \boldsymbol{Z})$. It is a well-known fact that $\mathfrak{H}_n/\mathrm{Sp}(n, \boldsymbol{Z})$ is the variety of moduli for certain polarized abelian varieties of dimension n. Therefore every point z on $\mathfrak{H}_n$ determines an isomorphism class of polarized abelian varieties. Then the above field $\mathfrak{R}_y$ is very close to the field $\mathfrak{R}'$ generated by the points of finite order on any abelian variety A in the isomorphism class. The representation $\omega_l \circ J_l$ is almost equivalent to the representation ψ_l of $G(\mathfrak{R}'/k')$ in the module $\bigcup_{\nu=0}^{\infty} \{t \in A \mid l^\nu t = 0\}$, where k' is a field of definition of A, although there is a subtle difference between $\omega_l \circ J_l$ and ψ_l, or between $\mathfrak{R}_y$ and $\mathfrak{R}'$, which will be discussed in the text. In this special case, the fact that the characteristic roots have magnitude $N(\mathfrak{p})^{\kappa/2}$ is essentially equivalent to the Riemann hypothesis for abelian varieties over finite fields, proved by A. Weil. Thus our assertion concerning the characteristic roots of $\omega_l(J_l(\sigma))$ may be viewed as a generalization of this result. It should be noted here that, if the degree of F over $\boldsymbol{Q}$ is greater than one, the characteristic roots of $\omega_l(J_l(\sigma))$ can not be interpreted, so far as our theory tells, as those of the Frobenius endomorphism of an abelian variety over a finite field, although one may hope to obtain some geometric interpretation of $\omega_l(J_l(\sigma))$ in the future. For further comments and conjectures, the reader is referred to § 2 of the text.

---

[2] In the text we shall prove somewhat more general theorems than this statement, see 1.11 and 1.13.

To prove the above result, we consider the families of abelian varieties investigated in [7]. Let K be an arbitrary totally imaginary quadratic extension of F. Then there exists a family $\Sigma_K = \{A_w^K \mid w \in \mathfrak{H}_n\}$ of abelian varieties, whose variety of moduli is isomorphic to V over the complex number field. Each A_w^K has K as a subalgebra of its endomorphism algebra. With the point $y = \varphi(z)$ on V as above, we can thus associate infinitely many abelian varieties A_z^K depending on K. Let $\mathfrak{P}$ be a prime divisor of the algebraic closure of $\mathbf{Q}$ which divides $\mathfrak{p}$, and let β_l be the l-adic representation of the Frobenius endomorphism of degree $N(\mathfrak{p})^f$ on A_z^K modulo $\mathfrak{P}$, for a sufficiently large positive integer f. By means of the technique developed in [6, 7], we can show that $\beta_l = s \cdot \alpha_l$, with a number s of K independent of l and z, and an element α_l of $\mathfrak{U}_l$ which represents $J_l(\sigma^f)$. The decomposition $\beta_l = s \cdot \alpha_l$ is essentially of the same nature as the decomposition of the Frobenius isogeny, which was crucial in [6, 7] (cf. 5.17, 6.7, (8.4.1) of [6], and 6.10, 6.12, 7.8 of [7]). Since the number s satisfies $N_{K/\mathbf{Q}}(s) = N(\mathfrak{p})^{\mu(g-1)}$, where $g = [F : \mathbf{Q}]$, we can derive the desired conclusion from the known properties of β_l.

Finally it may be worth noting that every arithmetically defined discontinuous group acting on $\mathfrak{H}_n$ is commensurable with a group Γ of our type.

Notation

We shall mainly use the same notation as in [6, 7]. For an associative ring with an identity element, we denote by $S^\times$ the group of invertible elements of S. If $\mathfrak{G}$ is an algebraic group defined over a field k, and Y is an algebra over k, then $\mathfrak{G}_Y$ or $\mathfrak{G}(Y)$ denotes the group of the points of $\mathfrak{G}$ rational over Y. For a finite or an infinite Galois extension K of a field F, we denote by $G(K/F)$ the Galois group of K over F. If F is an algebraic number field of finite degree, we denote by $\mathfrak{r}_F$ the ring of algebraic integers in F. Further, for an integral ideal $\mathfrak{c}$ in F, $C(F, \mathfrak{c})$ denotes the class field over F characterized by the following property: a prime ideal $\mathfrak{p}$ in F decomposes completely in $C(F, \mathfrak{c})$ if and only if $\mathfrak{p} = b\mathfrak{r}_F$ with an element b of F satisfying $b \equiv 1 \bmod^* \mathfrak{c}\mathfrak{u}_0$, where $\bmod^*$ means the multiplicative congruence, and $\mathfrak{u}_0$ is the product of all archimedean primes of F. If K is a finite abelian extension of F, and $\mathfrak{a}$ is an integral ideal in F unramified in K, then $[K/F, \mathfrak{a}]$ means the Artin symbol.

1. Statement of results

1.1. *The group of similitudes of a quaternion hermitian form.* We construct an algebraic group $\mathfrak{G}$ mentioned in the Introduction as follows. Let F be a totally real algebraic number field of finite degree, and B a quaternion algebra over F which may or may not be a division algebra. We put $g =$

$[F:Q]$, and denote by F^+ the group of all totally positive elements of F, and by ι the main involution of B. Let B^n denote the module of all n-dimensional row vectors with components in B. We let B and $\mathrm{GL}_n(B)$ act on the left and right of B^n, respectively. Define a B-valued ι-hermitian form $h(x, y)$ on B^n by $h(x, y) = x \cdot {}^t y^\iota$ for $(x, y) \in B^n \times B^n$. Then we define $\mathfrak{G}$ (resp. $\mathfrak{G}^+$) as the group of similitudes (resp. positive similitudes) of h:

$$\mathfrak{G}_Q = \{\alpha \in \mathrm{GL}_n(B) \mid h(x\alpha, y\alpha) = \nu(\alpha)h(x, y) \text{ with } \nu(\alpha) \in F\},$$
$$\mathfrak{G}^+ = \{\alpha \in \mathfrak{G}_Q \mid \nu(\alpha) \in F^+\}.$$

We consider $\mathfrak{G}_Q$ as a subgroup of $\mathrm{GL}_{4ng}(Q)$ by means of a matrix representation with respect to a basis of B^n over Q.

Let $\tau_{01}, \cdots, \tau_{0g}$ be the isomorphisms of F into R. We assume that B is unramified at τ_{01}, and ramified at $\tau_{02}, \cdots, \tau_{0g}$; further we take τ_{01} to be the identity mapping of F. Then $B_R = B \otimes_Q R$ can be identified with $M_2(R) \times K^{g-1}$, where K denotes the division ring of Hamilton quaternions, and this identification is compatible with the map

$$F \ni a \longmapsto (a^{\tau_{01}}, \cdots, a^{\tau_{0g}}) \in R^g.$$

For every $\mathfrak{r}_F$-lattice $\mathfrak{x}$ in B^n and every integral ideal $\mathfrak{a}$ in F, we put

$$\Gamma(\mathfrak{x}) = \{\gamma \in \mathfrak{G}^+ \mid \mathfrak{x}\gamma = \mathfrak{x}\},$$
$$\Gamma(\mathfrak{x}, \mathfrak{a}) = \{\gamma \in \Gamma(\mathfrak{x}) \mid \mathfrak{x}(1 - \gamma) \subset \mathfrak{a}\mathfrak{x}\}.$$

As is shown in [7, 4.4], $\mathfrak{G}^+$ acts on the Siegel upper half space $\mathfrak{H}_n$, and $\Gamma(\mathfrak{x}, \mathfrak{a})$ (modulo its center) gives a discontinuous group acting on $\mathfrak{H}_n$. In [7], we discussed a more general case where B_R is isomorphic to $M_2(R)^r \times K^{g-r}$, hence $\mathfrak{G}^+$ acts on $\mathfrak{H}_n^r$. In the present paper, however, we assume throughout $r = 1$, in order to avoid complexity, although a large part of our discussion can be extended to such a general case (cf. [7, § 11]).

Let us fix $\mathfrak{x}$ and assume that $\mathfrak{x}$ is ample in the sense of [7, 3.7], and denote $\Gamma(\mathfrak{x})$ and $\Gamma(\mathfrak{x}, \mathfrak{a})$ simply by Γ and $\Gamma_{\mathfrak{a}}$, respectively. We note that $\mathfrak{x}$ is ample, for example, if $\mathfrak{x} = \mathfrak{o}^n = \{(a_1, \cdots, a_n) \in B^n \mid a_i \in \mathfrak{o} \text{ for all } i\}$ with a maximal order $\mathfrak{o}$ in B. We denote by $D^{\mathfrak{a}}$ the group of all α in $\mathfrak{G}^+$ such that $\mathfrak{x}_{\mathfrak{p}}\alpha = \mathfrak{x}_{\mathfrak{p}}$ for all prime ideals $\mathfrak{p}$ in F dividing $\mathfrak{a}$, where $\mathfrak{x}_{\mathfrak{p}}$ is the $\mathfrak{p}$-closure of $\mathfrak{x}$ as defined in [7, 2.1.]. Further, we denote by $D_0^{\mathfrak{a}}$ the normal subgroup of $D^{\mathfrak{a}}$ consisting of all α such that $\mathfrak{x}_{\mathfrak{p}}(1 - \alpha) \subset \mathfrak{a}\mathfrak{x}_{\mathfrak{p}}$ for all $\mathfrak{p}$ dividing $\mathfrak{a}$. We put, for simplicity,

$$k_{\mathfrak{a}} = C(F, \mathfrak{a}) \qquad \text{(see Notation)},$$
$$\sigma_{\mathfrak{a}}(\alpha) = [k_{\mathfrak{a}}/F, (\nu(\alpha))] \qquad (\alpha \in D^{\mathfrak{a}}).$$

Remark. Since B is unramified only at τ_{01}, B is not a division algebra if and only if $F = Q$ and B is isomorphic to $M_2(Q)$. In this case $\mathfrak{G}$ can be identified with a group of similitudes of an alternating form in the following way. Let e_{ij} be the matrix units of $M_2(Q)$. Put $X_1 = e_{11}B^n$. Then $[X_1 : F] = 2n$, and there exists a non-degenerate alternating form $h_1 : X_1 \times X_1 \to Q$ such that $h(u, v) = h_1(u, v)e_{12}$ for $(u, v) \in X_1 \times X_1$. The restriction of every element of $GL_n(B)$ to X_1 defines an isomorphism of $\mathfrak{G}_Q$ onto

$$\mathfrak{G}_Q^0 = \{\alpha \in GL(X_1, F) \mid h_1(x\alpha, y\alpha) = \nu(\alpha)h_1(x, y) \text{ with } \nu(\alpha) \in Q\} .$$

Take any $\mathfrak{r}_F$-lattice $\mathfrak{x}_1$ in X_1, and put $\mathfrak{x} = \mathfrak{x}_1 + e_{21}\mathfrak{x}_1$. Then $\mathfrak{x}$ is ample, and the groups $\Gamma(\mathfrak{x})$ and $\Gamma(\mathfrak{x}, \mathfrak{a})$ can be identified with

$$\Gamma^0(\mathfrak{x}) = \{\gamma \in \mathfrak{G}_Q^0 \mid \nu(\gamma) = 1, \mathfrak{x}_1\gamma = \mathfrak{x}_1\} ,$$
$$\Gamma^0(\mathfrak{x}, \mathfrak{a}) = \{\gamma \in \Gamma^0(\mathfrak{x}) \mid \mathfrak{x}_1(1 - \gamma) \subset \mathfrak{a}\mathfrak{x}_1\} ,$$

(see [7, 9.8]). Further we can define $D^\mathfrak{a}$ and $D_0^\mathfrak{a}$ in terms of the properties with respect to $\mathfrak{x}_{1\mathfrak{p}}$. Thus the discussion in the case $B = M_2(Q)$ can be somewhat simplified. However, we shall not repeat such a simplification in the following treatment except in 2.3.

1.2. *Canonical system of models for $\mathfrak{H}_n/\Gamma_\mathfrak{a}$.* By [7, 5.3, 5.7], for every integral ideal $\mathfrak{a}$ in F, there exists a system

$$(1.2.0) \qquad\qquad \{V_\mathfrak{a}, \varphi_\mathfrak{a}, R_\mathfrak{a}(\alpha) \qquad (\alpha \in D^\mathfrak{a})\}$$

formed by the objects satisfying the following conditions:

(1.2.1) *$(V_\mathfrak{a}, \varphi_\mathfrak{a})$ is a model for $\mathfrak{H}_n/\Gamma_\mathfrak{a}$ in the sense that $V_\mathfrak{a}$ is a Zariski open subset of a projective variety, and $\varphi_\mathfrak{a}$ is a $\Gamma_\mathfrak{a}$-invariant holomorphic map of $\mathfrak{H}_n$ onto $V_\mathfrak{a}$ which gives a biregular isomorphism of $\mathfrak{H}_n/\Gamma_\mathfrak{a}$ onto $V_\mathfrak{a}$.*

(1.2.2) *$V_\mathfrak{a}$ is defined over $k_\mathfrak{a}$.*

(1.2.3) *For every $\alpha \in D^\mathfrak{a}$, $R_\mathfrak{a}(\alpha)$ is a biregular isomorphism of $V_\mathfrak{a}$ onto $V_\mathfrak{a}^{\sigma_\mathfrak{a}(\alpha)}$, rational over $k_\mathfrak{a}$.*

(1.2.4) *$R_\mathfrak{a}(\beta)^{\sigma_\mathfrak{a}(\alpha)} \circ R_\mathfrak{a}(\alpha) = R_\mathfrak{a}(\beta\alpha)$.*

(1.2.5) *$R_\mathfrak{a}(\alpha) = R_\mathfrak{a}(\beta)$ if $\alpha^{-1}\beta \in D_0^\mathfrak{a}$.*

(1.2.6) *If $\gamma \in \Gamma$, then $R_\mathfrak{a}(\gamma)[\varphi_\mathfrak{a}(w)] = \varphi_\mathfrak{a}(\gamma(w))$ for every $w \in \mathfrak{H}_n$.*

(1.2.7) *A certain property at isolated fixed points of $\mathfrak{G}^+$ on $\mathfrak{H}_n$ as described in [7, (5.3.3), (5.7.6)], which we do not need in this section.*

The system (1.2.0) can be characterized by these properties [7, 5.9]. Moreover, if $\mathfrak{a} \subset \mathfrak{b}$, we can find, by [7, 5.9], a morphism $T_{\mathfrak{b},\mathfrak{a}}$ of $V_\mathfrak{a}$ onto $V_\mathfrak{b}$, rational over $k_\mathfrak{a}$ such that

$$(1.2.8) \qquad \varphi_\mathfrak{b} = T_{\mathfrak{b},\mathfrak{a}} \circ \varphi_\mathfrak{a} , \qquad R_\mathfrak{b}(\alpha) \circ T_{\mathfrak{b},\mathfrak{a}} = T_{\mathfrak{b},\mathfrak{a}}^{\sigma_\mathfrak{a}(\alpha)} \circ R_\mathfrak{a}(\alpha) \qquad (\alpha \in D^\mathfrak{a}) .$$

(We have to find $V_\mathfrak{a}, \varphi_\mathfrak{a}, R_\mathfrak{a}(\alpha), T_{\mathfrak{b},\mathfrak{a}}$ simultaneously for all $\mathfrak{a}$ and $\mathfrak{b}$. For the discussion about this point, see 3.1 below.)

Let $E_\mathfrak{b}$ denote the group of units e of $\mathfrak{r}_F$ such that $e \equiv 1 \bmod \mathfrak{b}$. We can consider $V_\mathfrak{a}$ as a Galois covering of $V_\mathfrak{b}$ through the map $T_{\mathfrak{b},\mathfrak{a}}$ if $\mathfrak{a} \subset \mathfrak{b}$. If the varieties are considered over C, the Galois group of this covering is obviously $\Gamma_\mathfrak{b}/\Gamma_\mathfrak{a} E_\mathfrak{b}$. Therefore, from (1.2.6), we obtain

(1.2.9) *For two points x and y of $V_\mathfrak{a}$, one has $T_{\mathfrak{b},\mathfrak{a}}(x) = T_{\mathfrak{b},\mathfrak{a}}(y)$ if and only if $R_\mathfrak{a}(\gamma)(x) = y$ for some $\gamma \in \Gamma_\mathfrak{b}$.*

1.3. *Localization of $\mathfrak{G}$.* For every prime ideal $\mathfrak{l}$ in F, let $F_\mathfrak{l}$ (resp. $\mathfrak{r}_\mathfrak{l}$) denote the $\mathfrak{l}$-adic completion of F (resp. $\mathfrak{r}_F$). Put $B_\mathfrak{l} = B \otimes_F F_\mathfrak{l}$, $\mathfrak{x}_\mathfrak{l} = \mathfrak{x} \otimes_{\mathfrak{r}_F} \mathfrak{r}_\mathfrak{l} \, (\subset B_\mathfrak{l}^n)$. Then we can extend h to $B_\mathfrak{l}^n$, and define the localization $\mathfrak{G}_\mathfrak{l}$ of $\mathfrak{G}$ at $\mathfrak{l}$ by

$$\mathfrak{G}_\mathfrak{l} = \{\alpha \in \mathrm{GL}_n(B_\mathfrak{l}) \mid h(x\alpha, y\alpha) = \nu(\alpha)h(x, y) \text{ with } \nu(\alpha) \in F_\mathfrak{l}\}\,.$$

(If $\mathfrak{G}^*$ denotes the algebraic group over F such that $\mathfrak{G}_F^*$ can be identified with $\mathfrak{G}_Q$, then $\mathfrak{G}_\mathfrak{l} = \mathfrak{G}^*(F_\mathfrak{l})$.) Let

$$\mathfrak{U}_\mathfrak{l} = \{\alpha \in \mathfrak{G}_\mathfrak{l} \mid \mathfrak{x}_\mathfrak{l}\alpha = \mathfrak{x}_\mathfrak{l}\}\,,$$

$$\mathfrak{U} = \prod_\mathfrak{l} \mathfrak{U}_\mathfrak{l} \qquad\qquad \text{(the product over all prime ideals } \mathfrak{l} \text{ in } F)\,.$$

With the usual product topology, $\mathfrak{U}$ becomes a compact group. For every integral ideal $\mathfrak{a}$ in F, we denote by $\mathfrak{U}(\mathfrak{a})$ the subgroup of $\mathfrak{U}$ consisting of the elements $(\alpha_\mathfrak{l})$ with $\alpha_\mathfrak{l}$ in $\mathfrak{U}_\mathfrak{l}$, such that $\mathfrak{x}_\mathfrak{l}(\alpha_\mathfrak{l} - 1) \subset \mathfrak{a}\mathfrak{x}_\mathfrak{l}$ for all $\mathfrak{l}$. We consider Γ as a subgroup of $\mathfrak{U}$ by the diagonal embedding, and denote by $\bar{E}_\mathfrak{a}$ the closure of $E_\mathfrak{a}$ in $\mathfrak{U}$. Now we have

(1.3.1) *$\mathfrak{U}/\mathfrak{U}(\mathfrak{a})$ is canonically isomorphic to $D^\mathfrak{a}/D_0^\mathfrak{a}$. Moreover every element of $D^\mathfrak{a}/D_0^\mathfrak{a}$ can be represented by an element β of $\mathfrak{G}^+$ such that $\mathfrak{x}\beta \subset \mathfrak{x}$.*

To show this, define an injective map $j: D^\mathfrak{a} \to \mathfrak{U}$ so that $j(\alpha)$ has α or 1 as its $\mathfrak{l}$-component according as $\mathfrak{l}$ divides $\mathfrak{a}$ or not. Then $j(D_0^\mathfrak{a}) = \mathfrak{U}(\mathfrak{a}) \cap j(D^\mathfrak{a})$. Therefore it suffices to show $\mathfrak{U} = \mathfrak{U}(\mathfrak{a}) \cdot j(D^\mathfrak{a})$. Let $(\alpha_\mathfrak{l}) \in \mathfrak{U}$ with $\alpha_\mathfrak{l} \in \mathfrak{U}_\mathfrak{l}$. By the approximation theorem in F, we can find an element b of F^+ such that $b \equiv \nu(\alpha_\mathfrak{l}) \bmod \mathfrak{a}\mathfrak{r}_\mathfrak{l}$ for all $\mathfrak{l}$ dividing $\mathfrak{a}$. Repeat the proof of [7, 3.11] with the present b, $\mathfrak{a}$, and $\alpha_\mathfrak{l}$ in place of b, c, and α in that proposition. Then we obtain an element β of $\mathfrak{G}$ such that $\mathfrak{x}\beta \subset \mathfrak{x}$, $\nu(\beta) = b$ and $\mathfrak{x}_\mathfrak{l}(\beta - \alpha_\mathfrak{l}) \subset \mathfrak{a}\mathfrak{x}_\mathfrak{l}$ for all $\mathfrak{l}$ dividing $\mathfrak{a}$. Then $\beta \in D^\mathfrak{a}$ and $j(\beta^{-1})(\alpha_\mathfrak{l}) \in \mathfrak{U}(\mathfrak{a})$, q.e.d.

Now, for every u in $\mathfrak{U}$, we take an element α of $D^\mathfrak{a}$ corresponding to u by the isomorphism stated in (1.3.1), and define $R_\mathfrak{a}(u)$ (resp. $\sigma_\mathfrak{a}(u)$) to be the same as $R_\mathfrak{a}(\alpha)$ (resp. $\alpha_\mathfrak{a}(\alpha)$). Then the properties (1.2.3–5) of $R_\mathfrak{a}(\alpha)$ are still true with $\mathfrak{U}$ and $\mathfrak{U}(\mathfrak{a})$ instead of $D^\mathfrak{a}$ and $D_0^\mathfrak{a}$.

1.4. Let us now fix an integral ideal $\mathfrak{c}$ in F and an arbitrary point y on $V_\mathfrak{c}$, whose coordinates may or may not be algebraic. We consider all integral ideals $\mathfrak{a}$ in F such that $\mathfrak{a} \subset \mathfrak{c}$. We call a set of points $\{y_\mathfrak{a}\}_{\mathfrak{a} \subset \mathfrak{c}}$ a *coherent set* (or *sequence*) *of points lying above y* if the following condition is satisfied.

$$(1.4.1) \qquad y_\mathfrak{a} \in V_\mathfrak{a}, \qquad y_\mathfrak{c} = y, \qquad T_{\mathfrak{b},\mathfrak{a}}(y_\mathfrak{a}) = y_\mathfrak{b} \qquad \text{if } \mathfrak{a} \subset \mathfrak{b}.$$

Such a sequence can be obtained, for example, by putting $y_\mathfrak{a} = \varphi_\mathfrak{a}(z)$ for every $\mathfrak{a} \subset \mathfrak{c}$ with a point z on $\mathfrak{H}_n$ such that $y = \varphi_\mathfrak{c}(z)$. We call this type of coherent set *standard*. We are going to state a few theorems, proofs of which will be given in later sections.

1.5. THEOREM. *Let $\{y_\mathfrak{a}\}_{\mathfrak{a} \subset \mathfrak{c}}$ be a coherent set of points lying above a point y on $V_\mathfrak{c}$, and $\mathfrak{K}_y$ the union of the fields $k_\mathfrak{a}(y_\mathfrak{a})$ for all $\mathfrak{a} \subset \mathfrak{c}$. Then $\mathfrak{K}_y$ is an infinite Galois extension of $k_\mathfrak{c}(y)$, depending only on y and not on the choice of $\{y_\mathfrak{a}\}$. Moreover, for every $\tau \in G(\mathfrak{K}_y/k_\mathfrak{c}(y))$, there exists an element u of $\mathfrak{U}(\mathfrak{c})$ such that $y_\mathfrak{a}^\tau = R_\mathfrak{a}(u)(y_\mathfrak{a})$ and $\sigma_\mathfrak{a}(u) = \tau$ on $k_\mathfrak{a}$ for every $\mathfrak{a} \subset \mathfrak{c}$.*

1.6. THEOREM. *The notation being as in 1.5, let z be a point of $\mathfrak{H}_n$ such that $\varphi_\mathfrak{c}(z) = y$, and $S = \{\gamma \in \Gamma_\mathfrak{c} \mid \gamma(z) = z\}$. Further let $\bar{S}$ denote the closure of S in $\mathfrak{U}$, and Z the normalizer of $\bar{S}$ in $\mathfrak{U}(\mathfrak{c})$. Then there exists an element d of $\mathfrak{U}(\mathfrak{c})$, dependent on $\{y_\mathfrak{a}\}$, such that the element u of 1.5 corresponding to $\tau \in G(\mathfrak{K}_y/k_\mathfrak{c}(y))$ belongs to dZd^{-1}, and is uniquely determined by τ modulo $d\bar{S}d^{-1}$. One thus obtains a continuous isomorphism of $G(\mathfrak{K}_y/k_\mathfrak{c}(y))$ into $dZd^{-1}/d\bar{S}d^{-1}$. Moreover, if $y_\mathfrak{a} = \varphi_\mathfrak{a}(z)$ for all $\mathfrak{a} \subset \mathfrak{c}$, one can put $d = 1$.*

We denote by J the isomorphism of $G(\mathfrak{K}_y/k_\mathfrak{c}(y))$ into $dZd^{-1}/d\bar{S}d^{-1}$. This is determined up to an inner automorphism of $\mathfrak{U}(\mathfrak{c})$. We observe that S contains $E_\mathfrak{c}$ as a subgroup of finite index, and $S/E_\mathfrak{c}$ is the stability group at z of the transformation group $\Gamma_\mathfrak{c}/E_\mathfrak{c}$. Therefore $S = E_\mathfrak{c}$ and $Z = \mathfrak{U}(\mathfrak{c})$ unless z is a non-trivial fixed point of $\Gamma_\mathfrak{c}$.

1.7. THEOREM. *If y is generic on $V_\mathfrak{c}$ over $k_\mathfrak{c}$, J is a surjective isomorphism of $G(\mathfrak{K}_y/k_\mathfrak{c}(y))$ onto $\mathfrak{U}(\mathfrak{c})/\bar{E}_\mathfrak{c}$.*

1.8. Hereafter, to the end of this section, we assume that the coordinates of the point y on $V_\mathfrak{c}$ are algebraic numbers. For every set M of prime ideals in F, we put

$$\mathfrak{U}_M = \prod_{\mathfrak{l} \in M} \mathfrak{U}_\mathfrak{l}, \qquad \mathfrak{U}_M(\mathfrak{c}) = \mathfrak{U}(\mathfrak{c}) \cap \mathfrak{U}_M,$$

where $\mathfrak{U}_M$ is considered as a subgroup of $\mathfrak{U}$ by the natural injection. Further, we denote by S_M (resp. $E_{\mathfrak{c}M}, Z_M$) the image of $\bar{S}$ (resp. $\bar{E}_\mathfrak{c}, Z$) by the projection map of $\mathfrak{U}$ to $\mathfrak{U}_M$. Then we obtain a natural homomorphism

$$(1.8.1) \qquad\qquad Z/\bar{S} \longrightarrow Z_M/S_M.$$

Let $\{y_\mathfrak{a}\}, \mathfrak{K}_y, d$, and J be as in 1.5-6. For simplicity, let us assume $d = 1$. Combining J with (1.8.1), we obtain a continuous homomorphism

$$(1.8.2) \qquad\qquad J_M \colon G(\mathfrak{K}_y/k_\mathfrak{c}(y)) \longrightarrow Z_M/S_M.$$

1.9. THEOREM. *Let $\mathfrak{R}_y^M$ be the subfield of $\mathfrak{R}_y$ corresponding to the kernel of J_M. Suppose that M consists of all prime factors in F of a rational integer. Then there exists an integral ideal $\mathfrak{t}$ in $k_c(y)$, independent of M, such that every prime ideal in $k_c(y)$, prime to $\mathfrak{t}$ and the ideals in M, is unramified in $\mathfrak{R}_y^M$.*

1.10. *Local representations obtained from a rational representation of* $\mathfrak{G}$. Our main interest is in the behavior of the Frobenius substitutions of $\mathfrak{R}_y$ over $k_c(y)$. For this purpose, let us take a representation

$$\omega\colon \mathfrak{G} \longrightarrow \mathrm{GL}_r\,,$$

with any degree r, satisfying the following three conditions (1.10.1–3):

(1.10.1) *ω is rational over Q.*

(1.10.2) *$\omega(a) = N_{F/Q}(a)^\kappa$ for every $a \in F^\times$, with an integer κ independent of a.*

(1.10.3) *$\omega(s) = 1$ for all $s \in S$.*

The last two conditions are dependent on each other in the following way. Since S contains E_c, we observe on one hand

(1.10.4) *If ω is irreducible over Q, (1.10.2) follows from (1.10.3).* On the other hand, $S = E_c$ for all y, except for the points $y = \varphi_c(z)$ with a nontrivial fixed point z of Γ_c, and

(1.10.5) *Suppose $S = E_c$. If κ is even, or every element of E_c is totally positive, then (1.10.3) follows from (1.10.2).*

It should be noted that the condition (1.10.3) depends only on y, and not on the choice of z.

For every rational prime l, let $F_l = F \otimes_Q Q_l$, $B_l = B \otimes_Q Q_l$, and

$$\mathfrak{G}_l = \mathfrak{G}(Q_l) = \{\alpha \in \mathrm{GL}_n(B_l) \mid \alpha \cdot {}^t\alpha^\iota = \nu(\alpha)1_n \text{ with } \nu(\alpha) \in F_l\}\,.$$

Then we obtain naturally from ω a representation

$$\omega_l\colon \mathfrak{G}_l \longrightarrow \mathrm{GL}(Q_l)$$

rational over Q_l. Let us write $\mathfrak{U}_M$, $\mathfrak{U}_M(\mathfrak{c})$, S_M, E_{cM}, Z_M, J_M as $\mathfrak{U}_l$, $\mathfrak{U}_l(\mathfrak{c})$, S_l, $E_{c,l}$, Z_l, J_l, when M consists of all prime factors of l in F. Observe that $\mathfrak{U}_l$ can be considered as a subgroup of $\mathfrak{G}_l$. In view of (1.10.3), we see that the composed map $\omega_l \circ J_l$ is meaningful and thus gives a representation

$$\omega_l \circ J_l\colon G\big(\mathfrak{R}_y/k_c(y)\big) \longrightarrow \mathrm{GL}_r(Q_l)\,.$$

Let $\mathfrak{t}$ be as in Theorem 1.9, and $\mathfrak{p}$ a prime ideal in $k_c(y)$ prime to $l\mathfrak{t}$. Take any prime divisor $\mathfrak{P}$ of $\mathfrak{R}_y$ which divides $\mathfrak{p}$, and a Frobenius automorphism $\sigma_\mathfrak{P}$ of $\mathfrak{R}_y$ over $k_c(y)$ with respect to $\mathfrak{P}$ and $\mathfrak{p}$. By 1.9, $\mathfrak{p}$ is unramified in the subfield of $\mathfrak{R}_y$ corresponding to the kernel of $\omega_l \circ J_l$. Therefore $\omega_l(J_l(\sigma_\mathfrak{P}))$ is determined by $\mathfrak{P}$, and independent of the choice of $\sigma_{\mathfrak{P}}$. Now we have

(1.10.6) *The conjugacy-class (and hence the characteristic polynomial) of $\omega_l(J_l(\sigma_{\mathfrak{P}}))$ is determined by y, $\mathfrak{p}$, and l, and independent of the choice of $\{y_\alpha\}$ and $\mathfrak{P}$, so long as $\mathfrak{p}$ is prime to lt.*

In fact, the change of $\{y_\alpha\}$ and $\mathfrak{P}$ transforms $\omega_l(J_l(\sigma_{\mathfrak{P}}))$ only by an inner automorphism of $\mathrm{GL}_r(Q_l)$.

1.11. Main theorem (the first form). *There exists an integral ideal $\mathfrak{t}'$ in $k_c(y)$, divisible by $\mathfrak{t}$, and independent of ω, $\mathfrak{p}$, and l, such that the following statements hold.*

(1.11.1) *If $\mathfrak{p}$ is prime to $l\mathfrak{t}'$, every element of $\mathfrak{U}_l$ representing $J_l(\sigma_{\mathfrak{P}})$ is semi-simple, and $\omega_l(J_l(\sigma_{\mathfrak{P}}))$ is semi-simple.*

(1.11.2) *The characteristic roots of $\omega_l(J_l(\sigma_{\mathfrak{P}}))$ are algebraic numbers of absolute value $N(\mathfrak{p})^{\kappa/2}$, if $\mathfrak{p}$ is prime to $l\mathfrak{t}'$.*

(1.11.3) *There exists a positive integer μ, independent of ω, $\mathfrak{p}$ and l, but dependent on y, such that the matrices $\omega_l(J_l(\sigma_{\mathfrak{P}}^\mu))$, for all l prime to $\mathfrak{p}\mathfrak{t}'$, have the same characteristic polynomial (dependent on $\mathfrak{p}$) with rational coefficients.*

1.12. *Local representations obtained from a rational representation of a maximal torus of $\mathfrak{G}$.* Instead of taking a representation of the whole group $\mathfrak{G}$, it seems also meaningful, and probably even more reasonable, to consider a representation of a maximal torus of $\mathfrak{G}$. Let us fix an arbitrary maximal algebraic torus $\mathfrak{T}$ of $\mathfrak{G}$ defined over Q. We consider a representation $\xi: \mathfrak{T} \to \mathrm{GL}_m$ satisfying the following four conditions:

(1.12.1) ξ *is rational over Q.*

(1.12.2) *The equivalence class of ξ is invariant under the Weyl group of $\mathfrak{T}$ (over the universal domain); i.e., if $\alpha \in \mathfrak{G}$, $\alpha\mathfrak{T}\alpha^{-1} = \mathfrak{T}$, and $\xi^\alpha(x) = \xi(\alpha x \alpha^{-1})$, then ξ^α is equivalent to ξ.*

(1.12.3) $\xi(a) = N_{F/Q}(a)^\kappa$ *for every $a \in F^\times$, with an integer κ independent of a.*

(1.12.4) $\xi(e) = 1$ *for all $e \in E_c$.*

Observe that $F^\times \subset \mathfrak{T}_Q$, since $\mathfrak{T}$ is a maximal torus. In connection with (1.12.2), we notice a simple fact:

(1.12.5) *Let k be an algebraically closed field, and let*

$$G = \{\alpha \in \mathrm{GL}_{2n}(k) \mid {}^t\alpha j \alpha = \nu(\alpha)j \text{ with } \nu(\alpha) \in k\}, \qquad j = \begin{bmatrix} 0 & -1_n \\ 1_n & 0 \end{bmatrix}.$$

If two semi-simple elements of G have the same value of ν and the same characteristic polynomial, then they are conjugate in G. Further if they *belong to the same maximal torus T of G, then they are transformed to each other by an inner automorphism of G stable on T.*

We also observe that $\mathfrak{G}(k)$ is isomorphic to the product of g copies of the above group G, for any algebraically closed extension k of $\mathbf{Q}$.

Let $\overline{Q}_l$ denote the algebraic closure of Q_l. Then we obtain from ξ a representation $\xi_l : T(\overline{Q}_l) \to \mathrm{GL}_m(\overline{Q}_l)$. To study the behavior of the Frobenius automorphisms of $G(\mathfrak{K}_y/k_c(y))$ with respect to ξ, we assume, for simplicity,

(1.12.6) $S = E_c$, i.e., z *is not a non-trivial fixed point of* Γ_c.

Let $\mathfrak{p}$, $\mathfrak{P}$, $\sigma_\mathfrak{P}$ be as in 1.10, and let $\mathfrak{t}'$ be as in 1.11. Suppose $\mathfrak{p}$ does not divide $l\mathfrak{t}'$. By (1.11.1), $J_l(\sigma_\mathfrak{P})$ is represented by a semi-simple element u_l of $\mathfrak{G}_l$. Therefore, we can find an element v_l of $\mathfrak{T}(\overline{Q}_l)$ with the same conjugacy-class as u_l. The element v_l is uniquely determined by $\mathfrak{p}$ modulo $E_{c,l}$ and modulo the action of the Weyl group of $\mathfrak{T}$, on account of (1.12.5). Therefore we obtain

(1.12.7) *The characteristic polynomial of* $\xi_l(v_l)$ *is uniquely determined by* $\mathfrak{p}$, *and independent of the choice of* $\{y_a\}$, $\mathfrak{P}$, $\sigma_\mathfrak{P}$, *and* v_l.

1.13. MAIN THEOREM (the second form). *There exists an integral ideal* $\mathfrak{t}''$ *in* $k_c(y)$, *divisible by* $\mathfrak{t}$, *and independent of* ω, $\mathfrak{p}$, *and* l, *such that the following statements hold.*

(1.13.1) *The characteristic roots of* $\xi_l(v_l)$ *are algebraic numbers of absolute value* $N(\mathfrak{p})^{\kappa/2}$, *if* $\mathfrak{p}$ *is prime to* $l\mathfrak{t}''$.

(1.13.2) *There exists a positive integer* μ, *independent of* ξ, $\mathfrak{p}$ *and* l, *but dependent on* y, *such that the* $\xi_l(v_l)^\mu$, *for all* l *prime to* $\mathfrak{p}\mathfrak{t}''$, *have the same characteristic polynomial (dependent on* $\mathfrak{p}$) *with rational coefficients.*

1.14. THEOREM. *If* $F = \mathbf{Q}$, *one can set* $\mu = 1$ *in* (1.11.3) *and* (1.13.2).

2. Some remarks and conjectures

2.1. *Conjectures.* The above theorems tempt us to make a few conjectures. We first state the conjectures in a somewhat less exact form, and discuss their refinement in the next paragraph.

Conjecture I. One can always set $\mu = 1$ *in* (1.11.3) *and* (1.13.2).

As is shown in 1.14, this is true if $F = \mathbf{Q}$. Suppose this conjecture is true. Let $\mathfrak{t}'$ and $\omega_l(J_l(\sigma_\mathfrak{P}))$ be as in 1.11, and let $\mathfrak{t}''$ and $\xi_l(v_l)$ be as in 1.13. Let $f_\mathfrak{p}^\omega$ (resp. $f_\mathfrak{p}^\xi$) denote the characteristic polynomial of $\omega_l(J_l(\sigma_\mathfrak{P}))$ (resp. $\xi_l(v_l)$). These are independent of l, under *Conjecture I.* Then we can define Euler products:

$$(2.1.1) \qquad \zeta\big(s, \mathfrak{K}_y/k_c(y), \omega\big) = \prod\nolimits_{\mathfrak{p} \nmid \mathfrak{t}'} N(\mathfrak{p})^{r \cdot} \cdot f_\mathfrak{p}^\omega(N(\mathfrak{p})^s)^{-1} ,$$

$$(2.1.2) \qquad \zeta\big(s, \mathfrak{K}_y/k_c(y), \xi\big) = \prod\nolimits_{\mathfrak{p} \nmid \mathfrak{t}''} N(\mathfrak{p})^{m \cdot} \cdot f_\mathfrak{p}^\xi(N(\mathfrak{p})^s)^{-1} ,$$

where the products are taken over all prime ideals $\mathfrak{p}$ in $k_c(y)$, prime to $\mathfrak{t}'$ and $\mathfrak{t}''$, respectively. Note that r (resp. m) is the degree of the representation ω (resp. ξ), hence, of $f_{\mathfrak{p}}^{\omega}$ (resp. $f_{\mathfrak{p}}^{\xi}$).

Conjecture II. *Each of these Euler products can be analytically continued to the whole complex s-plane, and satisfies a functional equation.*

The latter Euler product seems more natural than the former, which might be factored into many zeta functions of various kinds.

In general, one can not expect, without any assumption on ω or ξ, that the characteristic roots of $\omega_l(J_l(\sigma_{\mathfrak{P}}))$ or $\xi_l(v_l)$ are algebraic integers. In fact, if this is true for some ω with positive κ, then it is not true for ${}^t\omega^{-1}$. Therefore we impose the following condition on ω and ξ.

(2.1.3) $\omega(\alpha)$ *is integral over Z for every α of $\mathfrak{G}_Q$ such that $\mathfrak{x}\alpha \subset \mathfrak{x}$.*

(2.1.4) $\xi(\alpha)$ *is integral over Z for every α of $\mathfrak{X}$ integral over Z.*

Under these conditions, the integer κ of (1.10.2) or (1.12.3) is non-negative.

Conjecture III. *If ω (resp. ξ) satisfies (2.1.3) (resp. (2.1.4)), then the characteristic roots of $\omega_l(J_l(\sigma_{\mathfrak{P}}))$ (resp. $\xi_l(v_l)$) are algebraic integers.*

By (1.3.1), every element of $\mathfrak{U}_l$ can be approximated by elements β of $\mathfrak{G}_Q$ satisfying $\mathfrak{x}\beta \subset \mathfrak{x}$. Therefore, we can at least assert that the characteristic polynomials of $\omega_l(J_l(\sigma_{\mathfrak{P}}))$ and $\xi_l(v_l)$ have coefficients in Z_l (under (2.1.3) and (2.1.4)).

It is plausible that the Euler factors are obtained from Hecke operators defined with respect to some algebraic groups. This is certainly true in some special cases. Whenever such a relation exists, it yields a certain reciprocity law (for abelian and non-abelian extensions), as the author remarked in [3, 6.3] and [5].

2.2. *Refinement of conjectures.* Let $\mathfrak{t}_0$ be the maximal one among the integral ideals $\mathfrak{t}$ of $k_c(y)$ with the property that every prime ideal $\mathfrak{p}$ in $k_c(y)$ prime to $l\mathfrak{t}$ is unramified in $\mathfrak{R}_y$. Then one can make a stronger form of *Conjecture* I:

Conjecture I'. *The assertions (1.11.1-3), (1.13.1-2) hold with $\mathfrak{t}_0$ as $\mathfrak{t}'$ and $\mathfrak{t}''$, and with $\mu = 1$.*

If this is true, we can take all prime ideals in $k_c(y)$ prime to $\mathfrak{t}_0$ in the Euler products (2.1.1,2), and also make a stronger form of *Conjecture* III. One can also ask a question

Is the assertion of 1.9 true, with $\mathfrak{t}_0$ as $\mathfrak{t}$, and without the assumption that M consists of all prime factors in F of a rational integer?

Furthermore, it seems meaningful to consider an arbitrary finite extension

k' of $k_c(y)$, and discuss the Frobenius automorphisms of $k'\mathfrak{R}_v$ over k'; one can then define $\zeta(s, k'\mathfrak{R}_v, \omega)$ in a similar way.

2.3. *Relation with the conjecture of Hasse and Weil.* If $F = \mathbf{Q}$, there exists a family of polarized abelian varieties, for which V_c is the variety of moduli (see 3.11 below). Therefore a point y on V_c (or a point z on $\mathfrak{H}_n$) determines an isomorphism class of structures $\mathscr{P} = (A, \mathcal{C}, \theta; t_1, \cdots, t_s)$. Here A is an abelian variety; $\mathcal{C}$ is a polarization of A; θ is an isomorphism of B or $\mathbf{Q}$ into $\mathrm{End}_\mathbf{Q}(A)$ according as B is a division algebra or $B = M_2(\mathbf{Q})$; $\dim(A) = 2n$ or n accordingly; the t_i are points of A which generate the subgroup $\{t \in A \mid ct = 0\}$ of A, where c is the positive integer such that $\mathfrak{c} = c\mathbf{Z}$. Let us write V_c and k_c as V_c and k_c. Any field of definition for (any isomorphic model of) $\mathscr{P}$ must contain $k_c(y)$ since $k_c(y)$ is the field of moduli of $\mathscr{P}$. If y is algebraic, we can take $\mathscr{P}$ so as to be defined over a finite algebraic extension k' of $k_c(y)$. Fix any such $\mathscr{P}$ and k'. Let $\mathfrak{R}'$ be the field generated over k' by the coordinates of the points of finite order on A. Then we can easily show $\mathfrak{R}_v \subset \mathfrak{R}'$. Further, for each rational prime l, let $\mathfrak{o} = \{a \in B \mid a\mathfrak{x} \subset \mathfrak{x}\}$, and $A(l) = \bigcup_{\nu=1}^\infty \{t \in A \mid l^\nu t = 0\}$. Then we see that $A(l)$ is isomorphic to $B_l^n/\mathfrak{x}_l$ (over $\mathfrak{o}$) or to $X_{1l}/\mathfrak{x}_{1l}$ according as B is a division algebra or $B = M_2(\mathbf{Q})$, where X_1 and $\mathfrak{x}_1$ are as in *Remark* of 1.1. Let $\sigma \in G(\mathfrak{R}'/k_c(y))$. Since σ is the identity on $k_c(y)$, there exists an isomorphism λ of $\mathscr{P}^\sigma$ to $\mathscr{P}$. Then the map

$$A(l) \ni t \longmapsto \lambda t^\sigma \in A(l)$$

is an automorphism of $A(l)$, which is represented by an element β_l of $\mathrm{GL}_n(B_l)$ or $\mathrm{GL}(X_{1l}, \mathbf{Q}_l)$. We note that λ is unique up to automorphisms of $\mathscr{P}$. Further, every automorphism of $\mathscr{P}$ corresponds to an element of $S = \{\gamma \in \Gamma_c \mid \gamma(z) = z\}$ (see 3.11 below). Thus β_l is unique up to S_l. We shall show in 3.11 that if σ' is the restriction of σ to $\mathfrak{R}_v$, then β_l represents $J_l(\sigma')$ with respect to a suitable coordinate system. If S consists only of the identity element, we can take $\mathscr{P}$ so as to be defined over $k_c(y)$; such a $\mathscr{P}$ is unique up to isomorphism over $k_c(y)$ [4, 1.5]. In this sense our representation J_l is equivalent to the l-adic representation of $G(\mathfrak{R}'/k_c(y))$ on A. If $k' = k_c(y)$, $B = M_2(\mathbf{Q})$, and ω is the identity map of $\mathfrak{G}^0$ (see 1.1), then $\zeta(s, \mathfrak{R}_v/k_c(y), \omega)$ is nothing other than (the one-dimensional part of the zeta function of A over $k_c(y)$ in the sense of Hasse and Weil. There is a notable difference, however. For example, in the case $n = g = 1$, V is a variety of moduli for elliptic curves. For every y on V, there are infinitely many elliptic curves defined over $\mathbf{Q}(y)$, with the same invariant y, which are isomorphic over C, but not over $\mathbf{Q}(y)$. They have distinct $\mathfrak{R}'$ and distinct zeta functions. Moreover, with $k' = \mathbf{Q}(y)$, the l-adic representation of $G(\Re'/k')$ is faithful, while the kernel of J contains -1. Therefore our representation J describes only the property of the isomorphism classes of the structures $\mathscr{P}$, and not of each individual one, although the difference diminishes when one takes V_c with $c \geqq 3$ as the basic variety.

2.4. *Further remarks and questions.* We notice that $\mathfrak{U}(c)/E_c$ is similar to the idèle group of a number field modulo the connected component. It may also be remarked that the representation ω and ξ are analogous to the Grössen-character of type (A_0) in the sense of Taniyama [10] and Weil [11]. If $y = \varphi(z)$ with an isolated fixed point z of $\mathfrak{G}^+$ in the sense of [7, § 4], the main theorems of [7] tell us the nature of $\Re_y$ and J. There exists a certain number field P' such that $P'\Re_y$ is abelian over P'. In particular, if $n = 1$, P' is a totally imaginary quadratic extension of F embeddable in B, and $\Re_y$ is the maximal abelian extension of P'. Under a certain condition, we can show that $\zeta(s, \Re_y/k_c(y), \xi)$ is a product of several zeta functions of $k_c(y)$ with Grössen-characters (cf. [7, 11.13, 11.14]), of which we do not want to give any details here, since we shall find a more suitable place to discuss the topic on a future occasion.

As is shown in 2.3, if $F = Q$, our local representations are essentially the same as the l-adic representations on certain abelian varieties. In § 3, we shall see that certain abelian varieties can be associated with $\Re_y$ and J, even if $[F:Q] > 1$. But the relationship is not as close as in the case $F = Q$. One can therefore ask whether there exists, in the case $[F:Q] > 1$, any geometric structure in which $J_i(\sigma_\mathfrak{P})$ or $\omega_i(J_i(\sigma_\mathfrak{P}))$ has some natural interpretation.

In this paper we have confined ourselves to the case of quotients of the Siegel half spaces. (Even in this case we abandoned the generality by assuming that B is unramified at *only one* archimedean prime of F.) It is of course possible and also worth while to investigate the same kind of questions for other types of algebraic groups. Another interesting question is to consider the Galois coverings of varieties over finite fields, which are obtained from the above V_a by reduction modulo a prime divisor, and to study the Frobenius automorphisms, as Ihara [1,2] did in the one-dimensional case. In order to proceed in this direction, it is indispensable to make a careful analysis at the isolated fixed points of $\mathfrak{G}^+$. Some of these questions will be discussed in subsequent papers.

3. Proof of theorems

3.1. *Simultaneous canonical systems for all levels.* Before proving our theorems, let us re-formulate the main theorems of [7]. The notion of a canonical system of level $\mathfrak{a}$ has been defined in [7, 5.7, 5.8] with respect to an $\mathfrak{a}$-*regular* set of representatives of classes of lattices in B^n. We shall now show how one can avoid a somewhat artificial notion of $\mathfrak{a}$-regularity. Since we need canonical systems of level $\mathfrak{a}$ simultaneously for *all* integral ideals $\mathfrak{a}$ in F, the elimination of $\mathfrak{a}$-regularity is rather indispensable.

Let $\Delta = \{\mathfrak{x}_1, \cdots, \mathfrak{x}_q\}$ be a set of representatives for D_W-classes of $\mathfrak{r}_F$-lattices in Λ, the notation being as in [7, 5.6], without the assumption $r = 1$. For every λ and μ, put

$$(3.1.1) \quad \begin{aligned} D_W^{\mu\lambda}(\mathfrak{a}) &= D_W^{\mu\lambda}(\Delta, \mathfrak{a}) \\ &= \{\alpha \in D_W \mid \nu(\mathfrak{x}_\mu\alpha/\mathfrak{x}_\lambda) \in \mathfrak{I}_\mathfrak{a}, \; \mathfrak{x}_{\mu\mathfrak{p}}\alpha = \mathfrak{x}_{\lambda\mathfrak{p}} \text{ for all } \mathfrak{p} \text{ dividing } \mathfrak{a}\}, \end{aligned}$$

where $\mathfrak{I}_\mathfrak{a}$ is as in [7, 5.4]. Then one can easily verify the following facts.

$$(3.1.2) \qquad \alpha \in D_W^{\mu\lambda}(\mathfrak{a}), \qquad \beta \in D_W^{\lambda\kappa}(\mathfrak{a}) \Longrightarrow \alpha\beta \in D_W^{\mu\kappa}(\mathfrak{a}).$$

$$(3.1.3) \qquad \alpha \in D_W^{\mu\lambda}(\mathfrak{a}) \Longrightarrow \alpha^{-1} \in D_W^{\lambda\mu}(\mathfrak{a}).$$

For two elements α and β of $D_W^{\mu\lambda}(\mathfrak{a})$, we write

$$(3.1.4) \qquad \alpha \equiv \beta \; \mathrm{mod}^*(\Delta; \mathfrak{a})_{\mu\lambda}$$

if $\mathfrak{x}_{\mu\mathfrak{p}}(\alpha - \beta) \subset \mathfrak{a}\mathfrak{x}_{\lambda\mathfrak{p}}$ for all $\mathfrak{p}$ dividing $\mathfrak{a}$. We see easily that

$$(3.1.5) \qquad \Gamma_W(\mathfrak{x}_\lambda, \mathfrak{a}) = \{\gamma \in \Gamma_W(\mathfrak{x}_\lambda) \mid \gamma \equiv 1 \; \mathrm{mod}^* \, (\Delta; \mathfrak{a})_{\lambda\lambda}\},$$

$$(3.1.6) \qquad \alpha \equiv \beta \; \mathrm{mod}^* \, (\Delta; \mathfrak{a})_{\mu\lambda} \Longrightarrow \alpha\beta^{-1} \equiv 1 \; \mathrm{mod}^* \, (\Delta; \mathfrak{a})_{\mu\mu}.$$

$$(3.1.7) \qquad \alpha \equiv \beta \; \mathrm{mod}^* \, (\Delta; \mathfrak{a})_{\mu\lambda} \Longrightarrow \nu(\mathfrak{x}_\mu\alpha/\mathfrak{x}_\lambda) \cdot \nu(\mathfrak{x}_\mu\beta/\mathfrak{x}_\lambda)^{-1} \in P(F, \mathfrak{a}\mathfrak{u}_0).$$

Here $\mathfrak{u}_0$ is the product of all archimedean primes of F.

If the set Δ is $\mathfrak{a}$-regular, the present $D_W^{\mu\lambda}(\Delta, \mathfrak{a})$ coincides with $D_W(\Delta, \mathfrak{a})$ of [7, 5.6] for all μ and λ, and the congruence (3.1.5) agrees with that of [7, (5.6.2)] (and with that of [6, 3.4] if $n = r = 1$).

Now Main Theorem II of [7, 5.7] is true without the assumption of $\mathfrak{c}$-regularity, through the following modification:

(i) $D_W(\Delta, \mathfrak{c})$ should be replaced by $D_W^{\mu\lambda}(\Delta, \mathfrak{c})$ (in three instances).

(ii) In [7, (5.7.3)], $\mathrm{mod}^*(\Delta; \mathfrak{c})$ should be replaced by $\mathrm{mod}^*(\Delta; \mathfrak{c})_{\mu\lambda}$.

This can be shown by the same reasoning as in [7, 5.13]. Similarly, [7, 5.9] is true without $\mathfrak{c}$-regularity, and with the above modification (i). Further, if $r = 1$ and τ_{01} is the identity mapping, we can replace $D_W^{\mu\lambda}(\Delta, \mathfrak{c})$ by a simpler group

$$D_{W0}^{\mu\lambda}(\Delta, \mathfrak{c}) = \{\alpha \in \mathfrak{G}^+ \mid \mathfrak{x}_{\mu\mathfrak{p}}\alpha = \mathfrak{x}_{\lambda\mathfrak{p}} \text{ for all } \mathfrak{p} \text{ dividing } \mathfrak{c}\},$$

for the same reason as in [7, 5.10].

We can thus obtain canonical systems $\{V_\lambda, \varphi_\lambda, R_\sigma^{\mu\lambda}(\alpha)(\alpha \in D_{W0}^{\mu\lambda}(\Delta, \mathfrak{a}))\}$ of level $\mathfrak{a}$, with respect to the *same* Δ (and W), for all integral ideals $\mathfrak{a}$ in F. A given ample $\mathfrak{r}_F$-lattice $\mathfrak{x}$ can be taken as the first member of Δ. Writing $V_\mathfrak{a}, \varphi_\mathfrak{a}, R_\mathfrak{a}(\alpha), D^\mathfrak{a}$ for $V_1, \varphi_1, R_\sigma^{11}(\alpha), D_{W0}^{11}(\Delta, \mathfrak{a})$, we obtain a simultaneous system satisfying the conditions of 1.2.

3.2. Proof of 1.5. Let the notation be as in 1.4 and 1.5. In [7, 11.2, 11.3], we have proved the following assertions for two integral ideals $\mathfrak{a}$ and $\mathfrak{b}$ in F such that $\mathfrak{a} \subset \mathfrak{b}$.

(3.2.1) $k_\mathfrak{a}(y_\mathfrak{a})$ *is a Galois extension of* $k_\mathfrak{b}(y_\mathfrak{b})$, *which depends only on* y *and* $\mathfrak{a}$, *and not on the choice of* $y_\mathfrak{a}$.

(3.2.2) *For every* $\tau \in G(k_\mathfrak{a}(y_\mathfrak{a})/k_\mathfrak{b}(y_\mathfrak{b}))$, *there exists an element* β *of* $D^\mathfrak{a} \cap D_0^\mathfrak{b}$ *such that* $y_\mathfrak{a} = R_\mathfrak{a}(\beta)(y_\mathfrak{a})$ *and* $\tau = \sigma_\mathfrak{a}(\beta)$ *on* $k_\mathfrak{a}$.

(3.2.3) *Let* z *be a point of* $\mathfrak{H}_n$ *such that* $\varphi_\mathfrak{a}(z) = y_\mathfrak{b}$, *and let* $S_\mathfrak{b} = \{\gamma \in \Gamma_\mathfrak{b} \mid \gamma(z) = z\}$. *Then*

$$D_0^\mathfrak{a} S_\mathfrak{b} = \{\alpha \in D^\mathfrak{a} \cap D_0^\mathfrak{b} \mid R_\mathfrak{a}(\alpha)(y_\mathfrak{a}) = y_\mathfrak{a}, \sigma_\mathfrak{a}(\alpha) = 1\} \, ,$$

$$\Gamma_\mathfrak{a} S_\mathfrak{b} = \{\gamma \in \Gamma_\mathfrak{b} \mid R_\mathfrak{a}(\gamma)(y_\mathfrak{a}) = y_\mathfrak{a}\} \, ,$$

and the element β *of* (3.2.2) *belongs to the normalizer* N *of* $D_0^\mathfrak{a} S_\mathfrak{b}$ *in* $D^\mathfrak{a} \cap D_0^\mathfrak{b}$.

(3.2.4) *From the correspondence* $\tau \mapsto \beta$ *of* (3.2.2), *one obtains an isomorphism of* $G(k_\mathfrak{a}(y_\mathfrak{a})/k_\mathfrak{b}(y_\mathfrak{b}))$ *into* $N/(D_0^\mathfrak{a} S_\mathfrak{b})$.

(3.2.5) *If* $y_\mathfrak{b}$ *is generic on* $V_\mathfrak{b}$ *over* $k_\mathfrak{b}$, *then* $S_\mathfrak{b} = E_\mathfrak{b}$, $N = D^\mathfrak{a} \cap D_0^\mathfrak{b}$, *and this isomorphism is surjective.*

(In [7, 11.3], $\mathfrak{R}$ should have been replaced by the normalizer of $\mathfrak{R}_1 \cdot \Gamma^z$ in $\mathfrak{R}$, for an obvious reason.)

Now the first assertion of 1.5 follows immediately from (3.2.1). To find a desired element α of $\mathfrak{U}(\mathfrak{c})$, we first prove that there exists, for $\tau \in G(\mathfrak{R}_y/k_\mathfrak{c}(y))$, a set of elements $\{\xi(\mathfrak{a})\}_{\mathfrak{a} \subset \mathfrak{c}}$ satisfying the following condition:

$$
\begin{aligned}
&\xi(\mathfrak{a}) \in D^\mathfrak{a} \cap D_0^\mathfrak{c} \, , \qquad y_\mathfrak{a}^\tau = R_\mathfrak{a}(\xi(\mathfrak{a}))(y_\mathfrak{a}) \, , \\
(3.2.6) \quad &\tau = \sigma_\mathfrak{a}(\xi(\mathfrak{a})) \text{ on } k_\mathfrak{a} \, , \qquad \xi(\mathfrak{a})^{-1}\xi(\mathfrak{b}) \in D_0^\mathfrak{b} \qquad\qquad \text{if } \mathfrak{a} \subset \mathfrak{b} \, .
\end{aligned}
$$

To show this, take an infinite sequence of ideals $\mathfrak{a}_n$ in F so that

$$\mathfrak{c} = \mathfrak{a}_1 \supset \mathfrak{a}_2 \supset \cdots \supset \mathfrak{a}_n \supset \mathfrak{a}_{n+1} \supset \cdots$$

and every integral ideal in F divides some $\mathfrak{a}_n$. We are going to define a sequence of elements ζ_n so that $\zeta_n \in D^{\mathfrak{a}_n} \cap D_0^\mathfrak{c}$, $(y_{\mathfrak{a}_n})^\tau = R_{\mathfrak{a}_n}(\zeta_n)(y_{\mathfrak{a}_n})$, $\tau = \sigma_{\mathfrak{a}_n}(\zeta_n)$ on $k_{\mathfrak{a}_n}$, $\zeta_{n+1}^{-1}\zeta_n \in D_0^{\mathfrak{a}_n}$. This can be done by induction on n. First put $\zeta_1 = 1$. Assume that ζ_{n-1} has been defined. By (3.2.2), there exists an element β of $D^{\mathfrak{a}_n} \cap D_0^\mathfrak{c}$ such that $(y_{\mathfrak{a}_n})^\tau = R_{\mathfrak{a}_n}(\beta)(y_{\mathfrak{a}_n})$ and $\tau = \sigma_{\mathfrak{a}_n}(\beta)$ on $k_{\mathfrak{a}_n}$. Then $\tau = \sigma_{\mathfrak{a}_{n-1}}(\beta)$ on $k_{\mathfrak{a}_{n-1}}$, and $R_{\mathfrak{a}_{n-1}}(\zeta_{n-1})(y_{\mathfrak{a}_{n-1}}) = (y_{\mathfrak{a}_{n-1}})^\tau = R_{\mathfrak{a}_{n-1}}(\beta)(y_{\mathfrak{a}_{n-1}})$. Take a point w of

$\mathfrak{H}_n$ so that $\varphi_{\mathfrak{a}_n}(w) = y_{\mathfrak{a}_n}$. By (3.2.3), there exists an element γ of $\Gamma_{\mathfrak{c}}$ such that $\zeta_{n-1}^{-1}\beta\gamma \in D_0^{\mathfrak{a}_{n-1}}$ and $\gamma(w) = w$. Put $\zeta_n = \beta\gamma$. Thus we have obtained the desired sequence $\{\zeta_n\}$. Now, for every integral ideal $\mathfrak{b} \subset \mathfrak{c}$, take the smallest n such that $\mathfrak{a}_n \subset \mathfrak{b}$, and put $\xi(\mathfrak{b}) = \zeta_n$. Then the set $\{\xi(\mathfrak{b})\}$ satisfies (3.2.6).

For a prime ideal $\mathfrak{l}$ in F, we see that $\{\xi(\mathfrak{l}^n)\}_{n=1,2,\dots}$ converges to an element of $\mathfrak{U}_{\mathfrak{l}}$. Let $u_{\mathfrak{l}}$ be this limit, and u the element of $\mathfrak{U}$ with components $u_{\mathfrak{l}}$. Since $\xi(\mathfrak{l}^n) \in D_0^{\mathfrak{c}}$, we have $u \in \mathfrak{U}(\mathfrak{c})$. We see that $\mathfrak{x}_{\mathfrak{l}}(u_{\mathfrak{l}} - \xi(\mathfrak{a})) \subset \mathfrak{a}\mathfrak{x}_{\mathfrak{l}}$ for every $\mathfrak{l}$ dividing $\mathfrak{a}$. Therefore u corresponds to $\xi(\mathfrak{a})$ by the isomorphism in (1.3.1), hence $R_{\mathfrak{a}}(u) = R_{\mathfrak{a}}(\xi(\mathfrak{a}))$, and $\sigma_{\mathfrak{a}}(u) = \sigma_{\mathfrak{a}}(\xi(\mathfrak{a}))$. This completes the proof of 1.5.

3.3. PROOF OF 1.6. Let τ and u be as in 1.5. Further let S, $\bar{S}$, and Z be as in 1.6. Let us first assume that $y_{\mathfrak{a}} = \varphi_{\mathfrak{a}}(z)$ for all $\mathfrak{a} \subset \mathfrak{c}$. Suppose that τ is the identity mapping. For every $\mathfrak{a}$, take an element ξ of $D^{\mathfrak{a}}$ corresponding to u by the isomorphism (1.3.1). Then $R_{\mathfrak{a}}(\xi)(y_{\mathfrak{a}}) = y_{\mathfrak{a}}$ and $\sigma_{\mathfrak{a}}(\xi) = 1$. By (3.2.4), $\xi \in D_0^{\mathfrak{c}}S$, hence $u \in \mathfrak{U}(\mathfrak{a})S$. Since the $\mathfrak{U}(\mathfrak{a})$, for all $\mathfrak{a}$, form a basis of neighborhoods of the identity element of $\mathfrak{U}$, this shows that $u \in \bar{S}$. Since this reasoning is reversible, we see that τ is the identity if and only if $u \in \bar{S}$. Now let v be an element of $\mathfrak{U}(\mathfrak{c})$ corresponding to an element σ of $G(\mathfrak{K}_y/k_{\mathfrak{c}}(y))$. Then, for every $u \in \bar{S}$, we see that $\sigma_{\mathfrak{a}}(v^{-1}uv) = 1$, and $R_{\mathfrak{a}}(u)(y_{\mathfrak{a}}) = y_{\mathfrak{a}}$, hence

$$R_{\mathfrak{a}}(v^{-1}uv)(y_{\mathfrak{a}}) = R_{\mathfrak{a}}(v^{-1}u)^{\sigma}(y_{\mathfrak{a}}^{\sigma}) = R_{\mathfrak{a}}(v^{-1})^{\sigma}(y_{\mathfrak{a}}^{\sigma}) = y_{\mathfrak{a}} \,,$$

so that $v^{-1}uv \in \bar{S}$. It follows that $v \in Z$. Thus one obtains an isomorphism J of $G(\mathfrak{K}_y/k_{\mathfrak{c}}(y))$ into $Z/\bar{S}$. As is seen from above, if u corresponds to τ, and τ is the identity on $k_{\mathfrak{a}}(y_{\mathfrak{a}})$, then $u \in \mathfrak{U}(\mathfrak{a})S$. This shows the continuity of J.

Let $\{y_{\mathfrak{a}}'\}$ be another coherent set lying above y. By (1.2.9), there exists, for each $\mathfrak{a}$, an element $\gamma_{\mathfrak{a}} = \gamma(\mathfrak{a})$ of $\Gamma_{\mathfrak{c}}$ such that $R_{\mathfrak{a}}(\gamma_{\mathfrak{a}})(y_{\mathfrak{a}}) = y_{\mathfrak{a}}'$. By the same procedure as in 3.2, we may take $\{\gamma_{\mathfrak{a}}\}$ so that $\gamma_{\mathfrak{a}}^{-1}\gamma_{\mathfrak{b}} \in D_0^{\mathfrak{b}}$ if $\mathfrak{a} \subset \mathfrak{b}$. Then $\{\gamma(\mathfrak{l}^n)\}$ converges to an element $d_{\mathfrak{l}}$ of $\mathfrak{U}_{\mathfrak{l}}$. Let d be the element of $\mathfrak{U}$ with components $d_{\mathfrak{l}}$. Then $d \in \mathfrak{U}(\mathfrak{c})$ since $\gamma_{\mathfrak{a}} \in \Gamma_{\mathfrak{c}}$. Let $\tau \in G(\mathfrak{K}_y/k_{\mathfrak{c}}(y))$, and let u be an element of $\mathfrak{U}(\mathfrak{c})$ such that $\sigma_{\mathfrak{a}}(u) = \tau$ on $k_{\mathfrak{a}}$, and $R_{\mathfrak{a}}(u)(y_{\mathfrak{a}}) = y_{\mathfrak{a}}^{\tau}$. Then we have $\sigma_{\mathfrak{a}}(dud^{-1}) = \tau$ on $k_{\mathfrak{a}}$, and $R_{\mathfrak{a}}(dud^{-1})(y_{\mathfrak{a}}') = y_{\mathfrak{a}}'^{\tau}$. Therefore, we obtain, with respect to $\{y_{\mathfrak{a}}'\}$, an isomorphism J' of $G(\mathfrak{K}_y/k_{\mathfrak{c}}(y))$ into $dZd^{-1}/d\bar{S}d^{-1}$, satisfying $J'(\tau) = d \cdot J(\tau)d^{-1}$, which completes the proof of 1.6.

3.4. PROOF OF 1.7. Suppose that y is generic on $V_{\mathfrak{c}}$ over $k_{\mathfrak{c}}$. For every $u \in \mathfrak{U}$ and for every $\mathfrak{a} \subset \mathfrak{c}$, take an element ξ of $D^{\mathfrak{a}}$ corresponding to u by the isomorphism (1.3.1). By (3.2.5), there exists an element σ of $G(k_{\mathfrak{a}}(y_{\mathfrak{a}})/k_{\mathfrak{c}}(y))$ such that $\sigma = \sigma_{\mathfrak{a}}(\xi)$ on $k_{\mathfrak{a}}$, and $y_{\mathfrak{a}}^{\sigma} = R_{\mathfrak{a}}(\xi)(y_{\mathfrak{a}})$. Extend σ to an element τ of $G(\mathfrak{K}_y/k_{\mathfrak{c}}(y))$. If v is an element of $\mathfrak{U}(\mathfrak{c})$ corresponding to τ, then $uv^{-1} \in \mathfrak{U}(\mathfrak{a})E_{\mathfrak{c}}$. (Note that $S = E_{\mathfrak{c}}$ in the present case.) This shows that the image of J is dense in $\mathfrak{U}(\mathfrak{c})/\bar{E}_\mathfrak{c}$, hence 1.7.

3.5. Let us now start the proof of 1.9 and 1.11. Our argument is a refinement of that of [7, 11.11]. We first take an imaginary quadratic extension K_0 of Q and put $K = K_0 F$, $L = B \otimes_F K$. Define extensions $\tau_1, \cdots, \tau_g$ of $\tau_{01}, \cdots, \tau_{0g}$ to K so that the τ_ι are identity on K_0.

Take Φ, ρ, T as in [7, § 6](always with $r = 1$), and i, i', j as in [7, 6.6]. For simplicity, we identify α with $i(\alpha)$ for every $\alpha \in \mathfrak{G}$, and x with $i'(x)$ for every $x \in B^n$. Define an $\mathfrak{r}_\Lambda$-lattice $\mathfrak{M}$ in L^n by $\mathfrak{M} = \mathfrak{r}_K \cdot \mathfrak{x}$. Let $\mathfrak{D}_n$ and $\mathfrak{S}$ be as in [7, 4.1, 6.4]. By [7, 6.14], we can find an integral ideal $\mathfrak{e}$ so that the following conditions are satisfied.

(3.5.1) $\mathfrak{e} \subset \mathfrak{c}$, *and* $\Gamma_\mathfrak{e}$ *modulo its center has no elements of finite order.*

(3.5.2) *For every integral ideal* $\mathfrak{a} \subset \mathfrak{e}$, *the injection* $\mathfrak{D}_n \rightarrow \mathfrak{S}$ *induces a biregular embedding of* $\mathfrak{D}_n/\Gamma_\mathfrak{a}$ *into* $\mathfrak{S}/\Gamma(T, \mathfrak{a}^{-1}\mathfrak{M}/\mathfrak{M})$, *the notation being as in* [7, § 6].

For every integral ideal $\mathfrak{a} \subset \mathfrak{e}$, we consider a weak PEL-type

$$\Omega_\mathfrak{a} = (L, \Phi, \rho; F^+ \cdot T, \mathfrak{M}; x_1^\mathfrak{a}, \cdots, x_{4n}^\mathfrak{a})$$

with elements $x_\iota^\mathfrak{a}$ of L^n such that $\mathfrak{a}^{-1}\mathfrak{M} = \mathfrak{M} + \sum_{\nu=1}^{4n} \mathfrak{r}_K x_\nu^\mathfrak{a}$. We construct a family $\sum_\mathfrak{m}^\mathfrak{a} = \{Q_w^\mathfrak{a} \mid w \in \mathfrak{S}\}$ of weak PEL-structures

$$Q_w^\mathfrak{a} = \left(A_w, \mathfrak{D}_w, \theta_w; \{t_{w\nu}^\mathfrak{a}\}\right)$$

by means of the parametrizing function $\mathfrak{y}(x, z)$ common to all $\mathfrak{a}$ (see [7, 6.4]). Observe that, for a fixed w, $(A_w, \mathfrak{D}_w, \theta_w)$ is common to the $Q_w^\mathfrak{a}$ for all $\mathfrak{a}$, and $t_{w\nu}^\mathfrak{a}$ corresponds to $\mathfrak{y}(x_\nu^\mathfrak{a}, w)$. Let $(U_\mathfrak{a}, \mathfrak{v}_\mathfrak{a}, \psi_\mathfrak{a})$ be a moduli-system for the family $\Sigma_\mathfrak{a}$. The discussion of [7, §§ 7,8] shows that there exists a morphism $Z_\mathfrak{a}$ of $V_\mathfrak{a}$ into $U_\mathfrak{a}$, rational over $C(K, N(\mathfrak{a}))$ such that $Z_\mathfrak{a} \circ \varphi_\mathfrak{a} = \psi_\mathfrak{a} \circ j$.

To prove 1.9 and 1.11, we may assume that $\{y_\mathfrak{a}\}$ is standard, i.e., $y_\mathfrak{a} = \varphi_\mathfrak{a}(z)$ with a point z of $\mathfrak{H}_n$. Put $w = j(z)$. Then $Z_\mathfrak{a}(y_\mathfrak{a}) = \psi_\mathfrak{a}(w)$. We may assume that $Q_w^\mathfrak{c}$ is defined over a finite Galois extension k' of $k_\mathfrak{c}(y)$, containing $C(K, (1))$. Let $\mathfrak{f}_\mathfrak{a}'$ be the field generated over k' by the coordinates of the points t on A_w such that $\theta_w(\mathfrak{a})t = 0$. Let $\mathfrak{f}_\mathfrak{a}$ be the field of moduli of $Q_w^\mathfrak{a}$. Then we see that

(3.5.3) $k_\mathfrak{a}(y_\mathfrak{a}) \cdot C\big(K, N(\mathfrak{a})\big) = \mathfrak{f}_\mathfrak{a} \cdot C\big(K, N(\mathfrak{a})\big) \subset \mathfrak{f}_\mathfrak{a}' \cdot C\big(K, N(\mathfrak{a})\big)$.

It is well known (see for example [9, p. 109, p. 150]) that there exists an integral ideal $\mathfrak{n}$ of k', dependent only on A_w, such that:

(3.5.4.) A_w *has no defect* (i.e., A_w *has a good reduction*) *for every prime ideal of* k' *prime to* $\mathfrak{n}$, *in the sense of* [9, 11.1].

(3.5.5) *Every prime ideal in* k', *prime to* $N(\mathfrak{a})\mathfrak{n}$ *is unramified in* $\mathfrak{f}_\mathfrak{a}'$.

Let c be a rational integer, and M the set of all prime factors of c in F. Define $\mathfrak{K}_y^M$ as in 1.9. Then $\mathfrak{K}_y^M$ is the join of the fields $k_\mathfrak{a}(y_\mathfrak{a})$ for all $\mathfrak{a}$ such that $N(\mathfrak{a})$ is a divisor of a power of c. Therefore 1.9 follows from (3.5.3) and (3.5.5), with $\mathfrak{t} = D(k'/k_\mathfrak{c}(y)) \cdot N_{k'/k_\mathfrak{c}(y)}(\mathfrak{n}) \cdot N(\mathfrak{e})$.

3.6. To prove 1.11, fix a rational prime l. We use the same notation as in 3.5. Write $V_\mathfrak{a}$, $\varphi_\mathfrak{a}$, $R_\mathfrak{a}(\alpha)$, $y_\mathfrak{a}$, $U_\mathfrak{a}$, $\mathfrak{v}_\mathfrak{a}$, $\psi_\mathfrak{a}$, $\Omega_\mathfrak{a}$, $Q_w^\mathfrak{a}$, $x_w^\mathfrak{a}$, $Z_\mathfrak{a}$ as V_m, φ_m, $R_m(\alpha)$, y_m, U_m, $\mathfrak{v}_m$, ψ_m, Ω_m, Q_w^m, x_w^m, Z_m for $\mathfrak{a} = l^m\mathfrak{e}$. Let $\mathfrak{K}'$ be the composite of $\mathfrak{K}_y$ and $\mathfrak{f}'_\mathfrak{a}$ for $\mathfrak{a} = l^m\mathfrak{e}$, $m = 1, 2, \cdots$. Let $\mathfrak{p}$, $\mathfrak{P}$ and $\sigma_\mathfrak{P}$ be as in 1.10. Suppose that $\mathfrak{p}$ is prime to $l \cdot N(\mathfrak{e}) \cdot \mathfrak{t}$. Take a prime divisor $\mathfrak{P}'$ of $\mathfrak{K}'$ dividing $\mathfrak{P}$. Let $\mathfrak{p}'$ be the restriction of $\mathfrak{P}'$ to k', and τ the Frobenius automorphism of $\mathfrak{K}'$ over k' with respect to $\mathfrak{P}'$ and $\mathfrak{p}'$. If the degree of $\mathfrak{p}'$ relative to $k_\mathfrak{c}(y)$ is f, then $\varepsilon\tau = \sigma_\mathfrak{P}^f$ on $\mathfrak{K}_y$ with an element ε of $\mathrm{Ker}\,(J_l)$. Now

$$\tau = [C(K, N(l^m\mathfrak{e}))/K, N_{k'/K}(\mathfrak{p}')] \qquad \text{on } C(K, N(l^m\mathfrak{e})) \,.$$

Since k' contains $C(K, (1))$, we can find an element t of K so that $N_{k'/K}(\mathfrak{p}') = (t)$. Put $s = \det \Phi'_0(t)$, where Φ'_0 is as in [7, 6.10, 7.8]. We observe that s is prime to $N(l\mathfrak{e})$. Let $\{\xi(\mathfrak{a})\}$ be defined for τ as in (3.2.6). Put $\xi_m = \xi(N(l^m\mathfrak{e}))$. Then

$$\tau = [C(F, N(l^m\mathfrak{e}))/F, (\nu(\xi_m))] \qquad \text{on } C(F, N(l^m\mathfrak{e})) \,.$$

By [7, 7.8], Ω_m^τ is equivalent to

$$\Omega' = (L, \Phi, \rho; F^+ \cdot T, bs^{-1}\mathfrak{M}; \{x'_\nu\}) \,,$$

where b is an element of K such that $b \equiv 1 \bmod^*(N(l^m\mathfrak{e}))$, and $\{x'_\nu\}$ are elements of $(l^m\mathfrak{e})^{-1}\mathfrak{M}$ such that $x'_\nu - x_\nu^m\xi_m \in \mathfrak{M}_\mathfrak{q}$ for all prime ideals q in F dividing $l^m\mathfrak{e}$. Let x''_ν be an element of $(l^m\mathfrak{e})^{-1}\mathfrak{M}$ such that $x''_\nu - b^{-1}sx'_\nu \in \mathfrak{M}_\mathfrak{q}$ for all prime ideals q in F dividing $l^m\mathfrak{e}$. Then Ω' is equivalent to

$$\Omega'' = (L, \Phi, \rho; F^+ \cdot T, \mathfrak{M}; \{x''_\nu\}) \,.$$

Now construct families $\Sigma(\Omega')$ and $\Sigma(\Omega'')$ of weak PEL-structures of type Ω' and Ω'', respectively. By [6, 4.24–25], we can write moduli-systems for $\Sigma(\Omega')$ and $\Sigma(\Omega'')$ in the form $(U_m, \mathfrak{v}', \psi_m)$ and $(U_m, \mathfrak{v}'', \psi_m)$ with the above U_m and ψ_m. The discussion of [7, 7.10] shows that there exists a biregular isomorphism of U_m to U_m^τ, rational over $C(K, N(l^m\mathfrak{e}))$, such that $T(\xi_m) \circ \mathfrak{v}' = \mathfrak{v}_m^\tau$ and $Z_m^\tau \circ R_m(\xi_m) = T(\xi_m) \circ Z_m$. Let $v_m = \psi_m(w)$. Then $Z_m(y_m) = v_m = \mathfrak{v}_m(Q_w^m) = \mathfrak{v}'(Q'_w)$, where Q'_w is the member of $\Sigma(\Omega')$ corresponding to w. We have then

$$\mathfrak{v}_m^\tau((Q_w^m)^\tau) = \mathfrak{v}_m(Q_w^m)^\tau = v_m^\tau = Z_m^\tau(y_m^\tau) = Z_m^\tau(R_m(\xi_m)(y_m))$$
$$= T(\xi_m)(Z_m(y_m)) = T(\xi_m)(\mathfrak{v}'(Q'_w)) = \mathfrak{v}_m^\tau(Q'_w) \,.$$

It follows that $(Q_w^m)^\tau$ is isomorphic to Q'_w. Since Ω' is equivalent to Ω'', Q'_w is isomorphic to $(A_w, \mathfrak{D}_w, \theta_w; \{t''_{w\nu}\})$ with the elements $t''_{w\nu}$ represented by $\mathfrak{y}(x''_\nu, w)$. Let λ_m be an isomorphism of Q'_w to $(Q^m_w)^\tau$. Since τ is the identity on k', λ_m is an automorphism of Q^0_w. Therefore $(t^m_{w\nu})^\tau$ and $t''_{w\nu}$ are points of A_w, and $\lambda_m t''_{w\nu} = (t^m_{w\nu})^\tau$.

Now every automorphism of Q^0_w can be obtained from an element γ of $\Gamma(T, e^{-1}\mathfrak{M}/\mathfrak{M})$ such that $\gamma(w) = w$ (see [6, 4.11]). By (3.5.2), γ is an element of $\Gamma_\mathfrak{e}$ such that $\gamma(z) = z$, hence $\gamma \in E_\mathfrak{e}$ by (3.5.1). Let γ_m be the element of $E_\mathfrak{e}$ corresponding to λ_m in this sense; i.e., λ_m is represented by $\Lambda(\gamma_m, w)$ of [7, (6.7.1)]. Let

$$A_w(l) = \bigcup_{\nu=1}^\infty \{t \in A_w \mid l^\nu t = 0\} \,,$$

$$L^n_l = L^n \otimes_Q Q_l \,, \qquad \mathfrak{M}_l = \mathfrak{M} \otimes_Z Z_l \,(\subset L^n_l) \,, \qquad \mathfrak{X} = \bigcup_{\nu=1}^\infty \{x \in L^n \mid l^\nu x \in \mathfrak{M}\} \,.$$

Then we have $L^n_l = \mathfrak{X} + \mathfrak{M}_l$, $\mathfrak{M} = \mathfrak{X} \cap \mathfrak{M}_l$, hence $L^n_l/\mathfrak{M}_l$ is isomorphic to $\mathfrak{X}/\mathfrak{M}$, and the map $\mathfrak{X} \ni x \mapsto \mathfrak{y}(x, w)$ gives a homomorphism of L^n_l onto $A_w(l)$ with kernel $\mathfrak{M}_l$. Call this homomorphism h. We thus have an l-adic coordinate system on $A_w(l)$:

$$0 \longrightarrow \mathfrak{M}_l \longrightarrow L^n_l \overset{h}{\longrightarrow} A_w(l) \longrightarrow 0 \qquad\qquad \text{(exact)} \,.$$

Since τ is an automorphism of $A_w(l)$ commuting with the action of L, we find an element β_l of $GL_n(L_l)$ such that

$$(3.6.1) \qquad\qquad \mathfrak{M}_l \beta_l = \mathfrak{M}_l \,, \qquad h(x)^\tau = h(x\beta_l) \qquad\qquad (x \in L^n_l) \,.$$

Now we have $h(x^m_\nu) = t^m_{w\nu}$ and $h(x^m_\nu \beta_l) = (t^m_{w\nu})^\tau = \lambda_m t''_{w\nu} = h(x''_\nu \gamma_m)$, hence

$$(3.6.2) \qquad\qquad x^m_\nu \beta_l \equiv x''_\nu \gamma_m \equiv x^m_\nu \xi_m s \gamma_m \qquad \mod \mathfrak{M}_l \,.$$

Now $\{\xi_m\}$ converges to an element u_l of $\mathfrak{G}_l$, which represents $J_l(\tau)$. Since $(l^m e)^{-1}\mathfrak{M} = \mathfrak{M} + \sum_{\nu=1}^{4n} \mathfrak{r}_K x^m_\nu$, it follows from (3.6.2) that $\{\gamma_m\}$ converges to an element δ of $E_{\mathfrak{e},l}$ such that $\beta_l = u_l s \delta$. Put $u'_l = u_l \delta$. Then $\beta_l = s u'_l$, and u'_l represents $J_l(\tau)$.

Now Φ'_0 is a representation of K obtained in [7, 6.10]. In the present situation, we have $K' = K$ on account of our choice of the τ_i. Therefore, the notation being as in [6, 5.14], we have $\Phi'_0 \sim \sum_{\lambda=2}^g \sigma_\lambda$. (This fact follows from [6, (5.14.8)]; see the proof of [7, 6.10]. In the present case, the numbers s and h of [6, (5.14.8)] become 1 and g, since $K' = K$ and $r = 1$.) Let ρ denote the complex conjugation. Then $s = \det \Phi'_0(t) = \prod_{\lambda=2}^g t^{\sigma_\lambda}$, hence

$$(3.6.3) \qquad\qquad ss^\rho = \prod_{\lambda=2}^g t^{\sigma_\lambda} t^{\sigma_\lambda \rho} = N_{K/Q}(t)/N_{K/F}(t) \,,$$

since $\{\sigma_1, \cdots, \sigma_g, \sigma_1\rho, \cdots, \sigma_g\rho\}$ is the set of all isomorphisms of K into C, and σ_1 is the identity map. Put $s_0 = N_{K/K_0}(s)$. Then $s_0 \in K_0$, and

$$(3.6.4) \qquad\qquad s_0 s_0^\rho = N_{K/K_0}(ss^\rho) = K_{K/Q}(t)^{g-1} = N(\mathfrak{p}')^{g-1} \,.$$

3.7. Let $\mathfrak{G}'$ be the subgroup of $GL_n(L)$ generated by $\mathfrak{G}$ and $K^\times$. We consider $\mathfrak{G}'$ and $GL_n(L)$ as algebraic subgroups of GL_{8gn} defined over Q by means of a matrix representation with respect to a basis of L^n over Q. Let η be a regular representation of K_0 over Q, and ω be as in 1.10. Define a representation $\omega' \colon \mathfrak{G}' \to GL_{2r}$ by

$$\omega'(ab) = \eta\big(N_{K/K_0}(a)^\kappa\big) \otimes \omega(b) \qquad (a \in K^\times, b \in \mathfrak{G}_Q) \ .$$

Since $F^\times = \mathfrak{G}_Q \cap K^\times$, ω' is well defined in view of (1.10.2). Let $\mathfrak{G}'_l = \mathfrak{G}'(Q_l)$, and define $\omega'_l \colon \mathfrak{G}'_l \to GL_{2r}(Q_l)$ in an obvious way. Since $s \in K^\times$, we have $\beta_l \in \mathfrak{G}'_l$, and

$$(3.7.1) \qquad\qquad \omega'_l(\beta_l) = \eta(s_0^\kappa) \otimes \omega_l(u'_l) \ .$$

Now (3.6.1) shows that β_l is exactly the l-adic representation of τ on $A_w(l)$ in the usual sense (see for example [9, 18.5]). Therefore, in view of (3.5.4) and [9, p. 150,(7)], Weil's theorem implies that the characteristic roots of β_l are all algebraic integers of absolute value $N(\mathfrak{p}')^{1/2}$; moreover β_l is semi-simple, hence u'_l and $\omega'_l(\beta_l)$ are semi-simple.

Let X be an algebraic torus in $\mathfrak{G}'_l$ containing β_l. Restrict ω'_l to X. If x is a generic element of X, we see that every characteristic root of $\omega'_l(x)$ is of the form $x_1^{\mu_1} \cdots x_j^{\mu_j}$ with characteristic roots $x_1, \cdots, x_j$ of x and integers μ_i. Specializing x to an element of $Q^\times$, we see, in view of (1.10.2), that $\mu_1 + \cdots + \mu_j = \kappa g$. Therefore every characteristic root of $\omega'_l(\beta_l)$ is an algebraic number of absolute value $N(\mathfrak{p}')^{\kappa g/2}$. It follows from this, (3.6.4), and (3.7.1) that every characteristic root of $\omega_l(u'_l)$ is an algebraic number whose absolute value is $N(\mathfrak{p}')^{\kappa g/2}N(\mathfrak{p}')^{\kappa(1-g)/2} = N(\mathfrak{p}')^{\kappa/2}$. This proves (1.11.2), since $N(\mathfrak{p}') = N(\mathfrak{p})^f$ and $\omega_l(u'_l) = \omega_l(J_l(\tau)) = \omega_l(J_l(\sigma_{\mathfrak{P}}^f))$.

Let v (resp. v_l) be an arbitrary element of $\mathfrak{U}$ (resp. $\mathfrak{U}_l$) which represents $J(\sigma_{\mathfrak{P}})$ (resp. $J_l(\sigma_{\mathfrak{P}})$). Then $v_l^f = u'_l\zeta$ with an element ζ of S_l. If $S = E_c$, we see that $u'_l\zeta$ is semi-simple, hence v_l is semi-simple.

In case $S \neq E_c$, we take an integral ideal $\mathfrak{b}$ in F, prime to $N(\mathfrak{p})$, so that $\mathfrak{b} \subset \mathfrak{e}$ and $E_{\mathfrak{b}} = \{\gamma \in \Gamma_{\mathfrak{b}} \mid \gamma(z) = z\}$. Then we obtain an injection

$$J' \colon G\big(\mathfrak{R}_v/k_{\mathfrak{b}}(y_{\mathfrak{b}})\big) \longrightarrow \mathfrak{U}(\mathfrak{b})/\bar{E}_{\mathfrak{b}} \ .$$

Let $\mathfrak{q}$ be the restriction of $\mathfrak{P}$ to $k_{\mathfrak{b}}(y_{\mathfrak{b}})$, and b the degree of $\mathfrak{q}$ relative to $k_c(y_c)$. Since $\mathfrak{U}(c)/\mathfrak{U}(\mathfrak{b})$ is a finite group, we can find a positive integer m so that $v^m \in \mathfrak{U}(\mathfrak{b})$, and $\sigma_{\mathfrak{P}}^m$ is the identity map on $k_{\mathfrak{b}}(y_{\mathfrak{b}})$. Then v^{mb} represents $J'(\sigma_{\mathfrak{P}}^{mb})$, and the above result shows that v_l^{mb} is semi-simple, since $\sigma_{\mathfrak{P}}^{mb}$ is the mth power of the Frobenius automorphism of $\mathfrak{R}_v$ over $k_{\mathfrak{b}}(y_{\mathfrak{b}})$ with respect to $\mathfrak{P}$ and $\mathfrak{q}$. Therefore v_l is semi-simple, hence (1.11.1).

To prove (1.11.3), we first state a well-known fact as

3.8. Lemma. *Let U be a finite dimensional vector space over an algebraic number field K of degree h, and let ζ be a K-linear endomorphism of U. Let $\tau_1, \cdots, \tau_h$ be all the isomorphisms of K into the algebraic closure $\overline{\mathbf{Q}}$ of $\mathbf{Q}$, and let $e_1, \cdots, e_h$ be the idempotents of $K \otimes_{\mathbf{Q}} \overline{\mathbf{Q}}$ such that*

$$K \otimes_{\mathbf{Q}} \overline{\mathbf{Q}} = \sum_{i=1}^{h} e_i(1 \otimes \overline{\mathbf{Q}}), \ a \otimes 1 = \sum_{i=1}^{h} e_i(1 \otimes a^{\tau_i})$$

for $a \in K$. Consider $\bar{U} = U \otimes_{\mathbf{Q}} \overline{\mathbf{Q}}$ as a module over $K[\zeta] \otimes_{\mathbf{Q}} \overline{\mathbf{Q}}$ in a natural way. Let q be the characteristic polynomial of ζ on U over K. Then q^{τ_i} is the characteristic polynomial of ζe_i on $e_i \bar{U}$ over $\overline{\mathbf{Q}}$, and $N_{K/\mathbf{Q}}(q) = q^{\tau_1} \cdots q^{\tau_h}$ is the characteristic polynomial of ζ on U over $\mathbf{Q}$.

3.9. We observe that $\mathfrak{G}'(\overline{\mathbf{Q}})$ can be identified with the product of g copies of a group

$$Y' = \{(c_1\alpha, c_2\alpha) \in \mathrm{GL}_{2n}(\overline{\mathbf{Q}})^2 \mid c_i \in \overline{\mathbf{Q}}, \alpha \in \mathrm{Sp}(n, \overline{\mathbf{Q}})\} ,$$

and $\mathfrak{G}(\overline{\mathbf{Q}})$ with the product of g copies of

$$Y = \{(c_1\alpha, c_2\alpha) \in Y' \mid c_1 = c_2\} .$$

It is necessary to distinguish the number field K and the subgroup $K^\times$ of $\mathfrak{G}'_{\mathbf{Q}}$. Let us denote the latter by $\mathfrak{D}_{\mathbf{Q}}$ with an algebraic subgroup $\mathfrak{D}$ of $\mathfrak{G}'$ defined over $\mathbf{Q}$, and denote by λ the isomorphism of $K^\times$ onto $\mathfrak{D}_{\mathbf{Q}}$. Then we can identify $\lambda(a)$, for $a \in K^\times$, with the element of Y'' whose i-component is $(a^{\tau_i}, a^{\tau_i \rho})$. Further, with this notation, we should now write $\beta_l = \lambda(s)u'_l$. Take any isomorphism ε of $\overline{\mathbf{Q}}$ into $\overline{\mathbf{Q}}_l$. Since u'_l is semi-simple and the characteristic roots of u'_l are algebraic numbers, there exists an element γ of $\mathfrak{G}(\overline{\mathbf{Q}}_l)$ such that $\gamma u'_l \gamma^{-1} \in \mathfrak{G}(\overline{\mathbf{Q}}^\varepsilon)$. Let $(\alpha_{li}, \alpha_{li})$ ($\in Y$) be the i-component of $(\gamma u'_l \gamma^{-1})^{\varepsilon^{-1}}$. Then $(\gamma\beta_l\gamma^{-1})^{\varepsilon^{-1}} \in \mathfrak{G}'(\overline{\mathbf{Q}}) = Y'^g$, and $(s^{\tau_i}\alpha_{li}, s^{\tau_i\rho}\alpha_{li})$ ($\in Y'$) is the i-component of $(\gamma\beta_l\gamma^{-1})^{\varepsilon^{-1}}$. (Note that $\lambda(s)^\varepsilon = \lambda(s)$, since $\lambda(s) \in \mathfrak{G}'_{\mathbf{Q}}$.)

Let $\tilde{A}$ be the reduction of A_w modulo $\mathfrak{P}'$, and π the $N(\mathfrak{p}')^{\text{th}}$ power endomorphism of $\tilde{A}$. Then β_l may be considered as the l-adic representation of π (see [9, p. 150, (7)]). Let H be the subalgebra of $\mathrm{End}_{\mathbf{Q}}(A)$ generated by K and π, where $\widetilde{\theta(K)}$ is identified with K. Now there exists an H-module U, independent of l, such that $U \otimes_{\mathbf{Q}} \overline{\mathbf{Q}}$ is isomorphic to $L^* \otimes_{\mathbf{Q}} \overline{\mathbf{Q}}$ over $H \otimes_{\mathbf{Q}} \overline{\mathbf{Q}}$. Here the action of π on $L^* \otimes_{\mathbf{Q}} \overline{\mathbf{Q}}$ is defined by that of $(\gamma\beta_l\gamma^{-1})^{\varepsilon^{-1}}$. This follows from the fact that H is a commutative semi-simple algebra, and the characteristic polynomial of every element of H, considered on L^*_l, is rational over $\mathbf{Q}$ and independent of l. Let q be the characteristic polynomial of $s^{-1}\pi$ on U over F, and q_{li} the characteristic polynomial of α_{li}. By the above lemma, we have $q^{\tau_{0i}} = q^i_{li}$ for every i. (Observe that, with respect to a basis of L^* over $\mathbf{Q}$, every element of Y should be repeated twice.) Therefore we can find a polynomial q_0, with coefficients in F, such that $q_0^s = q$ and $q_0^{\tau_0 i} = q_{1i}$. (If a power of a monic polynomial q_0 has coefficients in F, so does q_0 itself.)

Now consider the anti-hermitian form T on L^n which was used for the definition of a weak PEL-type Ω_a at the beginning of 3.5. Let T_l be the Q_l-linear extension of T to L_l^n. Then we have, by [4, (3.2.4)],

$$T_l(x\beta_l, x'\beta_l) = N(\mathfrak{p}') T_l(x, x') \qquad\qquad ((x, x') \in L_l^n \times L_l^n)\,,$$

since $\zeta^\tau = \zeta^{N(\mathfrak{p}')}$ for every root of unity ζ whose order is a power of l. By [7, 6.6], we have $T_l(xu_l', x'u_l') = \nu(u_l') T_l(x, x')$, hence $\nu(u_l') = (ss^\rho)^{-1} N(\mathfrak{p}') = N_{K/F}(t)$ by (3.6.3). It follows that

$${}^t\alpha_{li} j_n \alpha_{li} = N_{K/F}(t)^{\tau_0 i} j_n \qquad\qquad \text{with } j_n = \begin{bmatrix} 0 & 1_n \\ -1_n & 0 \end{bmatrix}.$$

Therefore, in view of (1.12.5), the conjugacy class of $(\gamma u_l' \gamma^{-1})^{\epsilon^{-1}}$ in $\mathfrak{G}(\overline{Q})$ is completely determined by t and π, hence independent of l. Hence there exists a polynomial d with coefficients in $\overline{Q}$, independent of l and ε, such that d^ϵ is the characteristic polynomial of $\omega_l(u_l')$. Since d^ϵ has coefficients in Q_l for every isomorphism ε of $\overline{Q}$ into $\overline{Q}_l$ and for almost all l, we see that d is rational over Q. Since $\omega_l(u_l') = \omega_l(J_l(\sigma_\mathfrak{P}^f))$ and f is a divisor of $[k': k_\mathfrak{c}(y)]$, this proves (1.11.3) with $\mu = [k': k_\mathfrak{c}(y)]$.

3.10. Let the notation be as in 1.12 and 1.13. Define a representation $\xi': \mathfrak{X}\cdot\mathfrak{D} \to \mathrm{GL}_{2m}$ by $\xi'(ab) = \eta(N_{K/K_0}(a)^\epsilon) \otimes \xi(b)$ for $a \in \mathfrak{D}_Q = K^\times$ and $b \in \mathfrak{X}_Q$, where η is as in 3.7. Extending the basic field to $\overline{Q}_l$, we obtain a representation $\xi_l': \mathfrak{X}(\overline{Q}_l)\cdot\mathfrak{D}(\overline{Q}_l) \to \mathrm{GL}_{2m}(\overline{Q}_l)$. Let β_l, u_l', s, s_0 be as in 3.7. Take an element w of $\mathfrak{G}(\overline{Q}_l)$ so that $wu_l'w^{-1} \in \mathfrak{X}(\overline{Q}_l)$. Put $x_l = wu_l'w^{-1}$. Then $w\beta_l w^{-1} = x_l s$, and $\xi'(w\beta_l w^{-1}) = \eta(s_0^\epsilon) \otimes \xi_l(x_l)$. Applying to $w\beta_l w^{-1}$ the same argument as in 3.7, we obtain (1.13.1).

Since u_l' is rational over Q_l, x_l^σ is conjugate to x_l in $\mathfrak{G}(\overline{Q}_l)$ for every automorphism σ of $\overline{Q}_l$ over Q_l. Hence the characteristic polynomial of $\xi_l(x_l)$ is rational over Q_l. Now we can find an element α of $\mathfrak{X}(\overline{Q})$ belonging to the same conjugacy class as the element $(\gamma u_l' \gamma^{-1})^{\epsilon^{-1}}$ obtained in 3.9. Then $\xi(\alpha)^\epsilon$ has the same characteristic polynomial as $\xi_l(x_l)$. By the same argument as in the end of 3.9, we obtain (1.13.2).

3.11. To discuss the case $F = Q$, we consider a family of abelian varieties of [7, § 9]. Let

$$\Omega = \Omega(v_1, \cdots, v_s) = (B, \Phi, \rho; T, \mathfrak{x}; v_1, \cdots, v_s)$$

be as in [7, (9.2.1)]. We can write $\mathfrak{c} = cZ$ with a positive integer c, and $k_\mathfrak{c} = Q(e^{2\pi i/c})$. Fix a rational prime l. For every integer $m \geqq 0$, we consider a

PEL-type $\Omega_m = \Omega(v_1^m, \cdots, v_s^m)$ with elements v_i^m of B^n such that $(cl^m)^{-1}\mathfrak{x} = \mathfrak{x} + \sum_{i=1}^s \mathfrak{x}_F v_i$. We construct a family $\Sigma_m = \{Q_w^m \mid w \in \mathfrak{H}_n\}$ of PEL-structures

$$Q_w^m = \left(A_w, \mathcal{C}_w, \theta_w; \{t_{wv}^m\}\right)$$

of type Ω_m as in [7, § 9.2.]. Let $(V_m, \mathfrak{v}_m, \varphi_m)$ be a moduli-system for Σ_m. By [7, 9.5], we can take (V_m, φ_m) as $(V_\mathfrak{a}, \varphi_\mathfrak{a})$ for $\mathfrak{a} = cl^m Z$; let us write k_m and $R_m(\alpha)$ for $k_\mathfrak{a}$ and $R_\mathfrak{a}(\alpha)$. Moreover, for every $\alpha \in D^{cl}$, we can define

$$\Omega_{m\alpha} = \Omega(v_1^m\alpha, \cdots, v_s^m\alpha), \quad \Sigma_{m\alpha} = \{Q_{w\alpha}^m \mid w \in \mathfrak{H}_n\},$$

and $\mathfrak{v}_{m\alpha}$ as in [7, 9.4].

Now consider y, $\{y_\mathfrak{a}\}$, and $\mathfrak{R}_y$ as before. Write y_m for $y_\mathfrak{a}$ with $\mathfrak{a} = cl^m Z$. We may assume that $y_m = \varphi_m(z)$ with a point z of $\mathfrak{H}_n$ common to all m. Let us consider the member Q_z^m with this z. (To show the fact stated in 2.3, it suffices to take $\mathscr{P}$ as Q_z^m.) Since y is algebraic, we may assume that A_z is defined over a finite algebraic extension k' of $k_\mathbb{C}(y)$. Then we find an integral ideal $\mathfrak{n}$ of k' such that

(3.11.1) A_z *has no defect for every prime ideal of* k' *prime to* $\mathfrak{n}$. Let $\mathfrak{p}$, $\mathfrak{P}$, $\sigma_\mathfrak{P}$ be as in 1.10. Suppose that $\mathfrak{p}$ is prime to $clN(\mathfrak{n})$, and take a prime divisor $\mathfrak{P}_0$ of $\overline{Q}$ dividing $\mathfrak{P}$. Then $\sigma_\mathfrak{P}$ can be extended to a Frobenius automorphism τ of $\overline{Q}$ over $k_\mathbb{C}(y)$ with respect to $\mathfrak{P}_0$ and $\mathfrak{p}$. (Note that τ is not necessarily the identity on k'.) Let $\{\xi(\mathfrak{a})\}$ be defined for τ as in (3.2.6). Put $\xi_m = \xi(cl^m)$. Then $\tau = [k_m/Q, (\nu(\xi_m))]$ on k_m. By [7, (9.5.2), (9.5.3)], Ω_m^τ is equivalent to $\Omega_{m\xi_m}$, and $\mathfrak{v}_m^\tau = R_m(\xi_m) \circ \mathfrak{v}_{m\xi_m}$. In view of the definition of $\mathfrak{v}_{m\xi_m}$ (see [7, p.570]), we have $y_m = \varphi_m(z) = \mathfrak{v}_{m\xi_m}(Q_{z\xi_m}^m)$, hence

$$R_m(\xi_m)\left(\mathfrak{v}_{m\xi_m}(Q_{z\xi_m}^m)\right) = R_m(\xi_m)(y_m) = y_m^\tau = \varphi_m(z)^\tau = \mathfrak{v}_m(Q_z^m)^\tau = \mathfrak{v}_m^\tau(Q_z^{m\tau})$$
$$= R_m(\xi_m)\left(\mathfrak{v}_{m\xi_m}(Q_z^{m\tau})\right).$$

Hence there exists an isomorphism λ_m of $Q_z^{m\tau}$ to $Q_{z\xi_m}^m$. If we put

$$Q_{z\xi_m}^m = \left(A_z, \mathcal{C}_z, \theta_z; \{t_\nu'\}\right),$$

then $\lambda_m t_{z\nu}^{m\tau} = t_\nu'$. If $\mathfrak{y}(x, z)$ is the parametrizing function considered in [7, 9.2], t_ν' corresponds to $\mathfrak{y}(v_\nu', z)$, where v_ν' is an element of $(cl^m)^{-1}\mathfrak{x}$ such that $v_\nu' - v_\nu^m \xi_m \in \mathfrak{x}_p$ for all rational primes p dividing cl.

Let $A_z(l) = \bigcup_{\nu=1}^\infty \{t \in A_z \mid l^\nu t = 0\}$, $B_l^n = B^n \otimes_Q Q_l$, $\mathfrak{x}_l = \mathfrak{x} \otimes_Z Z_l \ (\subset B_l^n)$. From the map $B^n \ni x \mapsto \mathfrak{y}(x, z)$ we obtain a homomorphism h with an exact sequence

$$0 \longrightarrow \mathfrak{x}_l \longrightarrow B_l^n \stackrel{h}{\longrightarrow} A_z(l) \longrightarrow 0.$$

Now we see that $A_z(l) \ni t \mapsto \lambda_0 t^\tau$ is an automorphism of $A_z(l)$, commuting with the action of B. Therefore we find an element β_l of $GL_n(B_l)$ such that $\mathfrak{x}_l \beta_l = \mathfrak{x}_l$, $\lambda_0 h(x)^\tau = h(x\beta_l)$ $(x \in B_l^n)$. Since $\lambda_0 \lambda_m^{-1}$ is an automorphism of Q_z^0, there exists an element γ_m of Γ_c such that $\gamma_m(z) = z$, and $\lambda_0\lambda_m^{-1}$ is obtained from $\Lambda(\gamma_m, z)$ of [7, 6.7](see [6, 4.4, 4.11]). Then we have $h(v_\nu^m) = t_{z\nu}^m$ and

$$h(v_\nu^m\beta_l) = \lambda_0 h(v_\nu^m)^\tau = \lambda_0 t_{z\nu}^{m\tau} = \lambda_0\lambda_m^{-1}t_\nu' = \lambda_0\lambda_m^{-1}h(v_\nu') = \lambda_0\lambda_m^{-1}h(v_\nu^m\xi_m) = h(v_\nu^m\xi_m\gamma_m) \ .$$

Therefore $v_\nu^m\beta_l \equiv v_\nu^m\xi_m\gamma_m \bmod \mathfrak{x}_l$. Since ξ_m converges to an element u_l of $\mathfrak{G}_l$, we see that γ_m converges to an element δ of S_l, and $\beta_l = u_l\delta$.

Let $\widetilde{A}$ be the reduction of A_z modulo $\mathfrak{P}_0$, and π the $N(\mathfrak{p})^{\text{th}}$ power isogeny of $\widetilde{A}$ to $\widetilde{A}^{N(\mathfrak{p})}$. Denoting reduction modulo $\mathfrak{P}_0$ by tilde, we have $\widetilde{\lambda_0\pi}\widetilde{h(x)} = \widetilde{h(x\beta_l)}$ for every $x \in B_l^n$, and the map $B_l^n \ni x \mapsto \widetilde{h(x)} \in \widetilde{A}$ defines an l-adic coordinate system on $\widetilde{A}$. Therefore β_l is the representation of $\widetilde{\lambda_0}\pi$ with respect to this coordinate system. Since $\widetilde{\lambda_0}\pi$ is independent of l, the characteristic polynomial of β_l has rational coefficients, and is independent of l. Here we identify β_l with its matrix representation with respect to a basis of B^n over Q.

Since $[k_m/Q, (\nu(\xi_m))] = [k_m/Q, (N(\mathfrak{p}))]$, and since $\nu(\xi_m)$ is positive, we have $\nu(\xi_m) \equiv N(\mathfrak{p}) \bmod^*(cl^m)$. Further we have $\nu(\gamma_m) = 1$, hence $\nu(\beta_l) = N(\mathfrak{p})$. Let $\mathfrak{T}$ be a maximal torus of $\mathfrak{G}$ defined over Q. Then there exists an element η of $\mathfrak{G}(\overline{Q}_l)$ such that $\eta\beta_l\eta^{-1} \in \mathfrak{T}(\overline{Q})$. Now $\mathfrak{G}(\overline{Q})$ can be identified, with respect to a suitable basis of $B^n \otimes_Q \overline{Q}$ over $\overline{Q}$, with the group of all matrices of the form $\alpha = \begin{bmatrix} ab & 0 \\ 0 & ab \end{bmatrix}$ with $a \in \overline{Q}^\times$ and $b \in \mathrm{Sp}(n, \overline{Q})$; note that $\nu(\alpha) = a^2$. By (1.12.5), the conjugacy class of $\eta\beta_l\eta^{-1}$ in $\mathfrak{G}(\overline{Q})$ is independent of l. Therefore, if ω is as in 1.10, the characteristic polynomial of $\omega_l(J_l(\sigma\mathfrak{P})) = \omega_l(\beta_l)$ is independent of l. Put $\varepsilon = \eta\beta_l\eta^{-1}$. If σ is an automorphism of $\overline{Q}$ over Q, ε^σ and ε have the same rational characteristic polynomial, and $\nu(\varepsilon^\sigma) = N(\mathfrak{p}) = \nu(\varepsilon)$. We find, by (1.12.5), an element ζ of $\mathfrak{G}(\overline{Q})$ such that $\varepsilon^\sigma = \zeta\varepsilon\zeta^{-1}$. It follows that the characteristic polynomial of $\omega(\varepsilon)$ is invariant under σ. Hence the characteristic polynomial of $\omega_l(\beta_l)$ has coefficients in Q. This proves the assertion of 1.14 concerning (1.11.3). The same reasoning applies to the part for (1.13.2).

PRINCETON UNIVERSITY

REFERENCES

[1] Y. IHARA, "Algebraic curves mod p and arithmetic groups", in Proc. Symp. Pure Math. IX, Algebraic groups and discontinuous subgroups, Amer. Math. Soc., Providence, 1966, 265-271.

[2] ————, On congruence monodromy problems, lecture notes, Univ. of Tokyo, 1968.

[3] G. SHIMURA, On the zeta-functions of the algebraic curves uniformized by certain automorphic functions, J. Math. Soc. Japan **13** (1961), 275-331.

[4] ————, On the field of definition for a field of automorphic functions: II, Ann. of Math. **81** (1965), 124-165.

[5] ————, A reciprocity law in non-solvable extensions, J. reine angew. Math. **221** (1966), 209-220.

[6] ————, Construction of class fields and zeta functions of algebraic curves, Ann. of

Math. **85** (1967), 58–159.

[7] ————, *Algebraic number fields and symplectic discontinuous groups*, Ann. of Math. **86** (1967), 503–592.

[8] ————, Automorphic functions and number theory, Lecture Notes in Mathematics, 54, Springer, 1968.

[9] ———— and Y. TANIYAMA, Complex multiplication of abelian varieties and its applications to number theory, Publ. Math. Soc. Japan, No. 6, 1961.

[10] Y. TANIYAMA, *L-functions of number fields and zeta functions of abelian varieties*, J. Math. Soc. Japan **9** (1957), 330–366.

[11] A. WEIL, "On a certain type of characters of the idèle class group of an algebraic number-field", in Proc. Int. Symp. Alg. N. Th., Tokyo-Nikko, 1955, 1-7.

(Received April 22, 1968)

On canonical models of arithmetic quotients of bounded symmetric domains

Annals of Mathematics, 91 (1970), 144-222

The main theme of this paper is the construction of a simultaneous system of models, over algebraic number fields, for the quotients of a bounded symmetric domain by all arithmetic subgroups, of congruence type, of a semi-simple algebraic group. Our solution of this question, in the case specified below, leads us naturally to an analogue of class field theory, with a certain reductive algebraic group in place of an algebraic number field, although too much stress on this point may narrow down the subject to a somewhat traditional and unimaginative framework. Before presenting the formulation in which this analogy is apparent, let us first discuss a rather naive problem setting.

We start with a semi-simple algebraic group U defined over the rational number field. Suppose that $U_R{}^1$ modulo a maximal compact subgroup defines a bounded symmetric domain $\mathcal{H}$. For an arithmetic subgroup Γ of U, we consider a model (V, φ) of $\mathcal{H}/\Gamma$, by which we mean a Zariski open subset V of a projective variety, and a Γ-invariant holomorphic map φ of $\mathcal{H}$ onto V that induces a biregular isomorphism of $\mathcal{H}/\Gamma$ to V. The existence of such a (V, φ) is assured by the work of Baily and Borel. Then one may ask the following question:

(I) *To every arithmetic subgroup Γ of U defined by some congruence conditions, assign a model $(V_\Gamma, \varphi_\Gamma)$ of $\mathcal{H}/\Gamma$ and an algebraic number field k_Γ so that the following statements (I. 1-4) hold.*

(I. 1) *V_Γ is defined over k_Γ.*

(I. 2) *If $\Delta \subset \Gamma$, and $P_{\Gamma\Delta}$ is the projection map of V_Δ to V_Γ defined by $P_{\Gamma\Delta} \circ \varphi_\Delta = \varphi_\Gamma$, then $k_\Gamma \subset k_\Delta$, and $P_{\Gamma\Delta}$ is rational over k_Δ.*

(I. 3) *If $\alpha \in U_Q$ and $\alpha\Gamma\alpha^{-1} = \Delta$, then $k_\Gamma = k_\Delta$, and the biregular isomorphism $J_{\Delta\Gamma}(\alpha)$ of V_Γ to V_Δ, defined by $J_{\Delta\Gamma}(\alpha)\big(\varphi_\Gamma(z)\big) = \varphi_\Delta\big(\alpha(z)\big)$ for $z \in \mathcal{H}$, is rational over k_Γ.*

(I. 4) *If w is an isolated fixed point of an element of U_Q on $\mathcal{H}$, the field*

---

* This work was partially supported by the National Science Foundation Grant GP 8520.
[1] For the notation, see § 0 below.

$k_\Gamma(\varphi_\Gamma(w))$ *generated over k_Γ by the coordinates of $\varphi_\Gamma(w)$ has a certain class-field-theoretical property like that of singular values of elliptic modular functions.*

One may also require that the system of models $(V_\Gamma, \varphi_\Gamma)$ for all Γ is characterized by these properties.

This question (I) has essentially been answered in the previous works [11, 12] for the groups U of the following type:

(Sp) U *is* **Q**-*simple, and U_C is isomorphic to a product of copies of* Sp(n, C).

However, our theorems in these papers were stated in terms of the relationship between a fixed Γ and its principal congruence subgroups. Since such subgroups form a cofinal subfamily of the whole family of Γ, one can easily obtain a meaningful model $(V_\Gamma, \varphi_\Gamma)$ for an arbitrarily given Γ. But this may be considered incomplete, if one seeks a system $\{V_\Gamma, \varphi_\Gamma, k_\Gamma\}$ *defined simultaneously for the whole family of Γ.*

Assuming that such a system exists, take a point z of $\mathcal{K}$ so that the coordinates of $\varphi_\Gamma(z)$ are algebraic numbers, and let $\mathfrak{K}$ denote the union of the fields $k_\Delta(\varphi_\Delta(z))$ for all $\Delta \subset \Gamma$. For the group of the above type, if $\mathcal{K}$ is irreducible, we have shown in [13] that the Galois group of $\mathfrak{K}$ over $k_\Gamma(\varphi_\Gamma(z))$ has various l-adic representations, through which the Frobenius automorphisms have the characteristic roots whose absolute values are of the Riemann-Ramanujan-Weil type. One can therefore propose:

(II) *Investigate the Galois extensions and representations of the same type for a more general class of group.*

In the study of these problems in the previous papers, we have noticed the following two aspects: (i) it is necessary to consider a certain reductive group, of which the semi-simple part is U, and the torus part is non-trivial; (ii) it is more natural to associate the model (V, φ) and the number field k with the congruence condition by which Γ is defined, rather than with the discontinuous group Γ itself. Such a condition can most conveniently be described in terms of adeles. Therefore, in this paper, we shall answer the above questions for the group of type (Sp) by reformulating them in the adelization of a suitable reductive group.

It is (well) known that every group U of type (Sp) can be obtained as the unitary group of a hermitian form over a quaternion algebra B. Let F be the center of B, and G the group of all similitudes of the hermitian form. Writing hereafter G^u for U, and ν for the multiplier of a similitude, we obtain an exact sequence

(1) $$1 \longrightarrow G^u \longrightarrow G \overset{\nu}{\longrightarrow} F^\times \longrightarrow 1 .$$

We consider G and G^u as algebraic groups defined over $\mathbf{Q}$. The center of $G_\mathbf{Q}$ can be identified with $F^\times$. Since $G_\mathbf{R}^u$ modulo a maximal compact subgroup should be a bounded symmetric domain, we observe that $G_\mathbf{R}^u$ is isomorphic to the product of a compact group and r copies of $\mathrm{Sp}(n, \mathbf{R})$ for a positive integer r. Therefore the identity component $G_{\mathbf{R}+}$ of $G_\mathbf{R}$ acts on the product $\mathfrak{H}_n^r$ of r copies of the Siegel half space $\mathfrak{H}_n$ of degree n. Let G_A denote the adelization of G. We consider a subgroup S of G_A of the type $S = S_0 G_{\mathbf{R}+}$, where $G_\mathbf{R}$ is identified with the archimedean part of G_A, and S_0 is a compact subgroup of the non-archimedean part G_0 of G_A, which describes the congruence condition. Put $\Gamma_S = S \cap G_\mathbf{Q}$. Then Γ_S modulo its center is a discontinuous group acting on $\mathfrak{H}_n^r$. We shall associate with each S, under a certain restriction, a model $\langle V_S, \varphi_S \rangle$ of $\mathfrak{H}_n^r/\Gamma_S$ and an algebraic number field k_S so that the conditions (I. 1–4), suitably adelized, are satisfied.

Since the description of k_S and all these properties in the most general case is lengthy, let us assume, for simplicity, $r = 1$. (This includes, for example, the case $G_\mathbf{Q}^u = \mathrm{Sp}(n, \mathbf{Q})$.) Although the case $r = 1$ is by no means typical, it will be able to give a rough idea of what can be done. Under this assumption, we consider the family $\mathfrak{Z}$ of *all* subgroups S of G_A of the type $S = S_0 G_{\mathbf{R}+}$ with any open compact subgroup S_0 of G_0. For every algebraic number field K of finite degree, let K_{ab} denote the maximal abelian extension of K, and $K_A^\times$ the idele group of K. Further, for $t \in K_A^\times$, denote by $[t, K]$ the element of $\mathrm{Gal}\,(K_{ab}/K)$ canonically corresponding to t. Put $G_{A+} = G_0 G_{\mathbf{R}+}$, and $G_{\mathbf{Q}+} = G_\mathbf{Q} \cap G_{A+}$. The symbol ν in (1) can be extended to a map of G_A into $F_A^\times$. Put $\sigma(x) = [\nu(x)^{-1}, F]$ for $x \in G_A$. Let k_S be the subfield of F_{ab} corresponding to the subgroup $F^\times \cdot \nu(S)$ of $F_A^\times$ by class field theory. Now our main theorem, specialized to the case $r = 1$, can be stated as follows.

There exists a system

(2) $$\{V_S, \varphi_S, J_{TS}(x), (S, T \in \mathfrak{Z}; x \in G_{A+})\}$$

satisfying the following conditions (i–iv).

   (i) *$\langle V_S, \varphi_S \rangle$ is a model of $\mathfrak{H}_n/\Gamma_S$.*

   (ii) *V_S is defined over k_S.*

   (iii) *$J_{TS}(x)$, defined if $xSx^{-1} \subset T$, is a morphism of V_S onto $V_T^{\sigma(x)}$ rational over k_S, and has the following properties:*

$$\begin{cases} (\mathrm{iii_a}) & J_{SS}(x) \text{ is the identity map of } V_S \text{ if } x \in S; \\ (\mathrm{iii_b}) & J_{TS}(x)^{\sigma(y)} \circ J_{SR}(y) = J_{TR}(xy); \\ (\mathrm{iii_c}) & J_{TS}(\alpha)[\varphi_S(z)] = \varphi_T(\alpha(z)) \text{ for } z \in \mathfrak{H}_n \text{ if } \alpha \in G_{\mathbf{Q}+}. \end{cases}$$

(iv) *A certain reciprocity law holds at every isolated fixed point of G_{Q+} on $\mathfrak{H}_n$.*

For simplicity, let us give the precise statement of the most important property (iv) only in the case $n = 1$, though, again, this should not be considered typical. We have $G_Q = B^{\times}$ if $n = 1$. Let K be a totally imaginary quadratic extension of F, and f an F-linear isomorphism of K into B. Then $f(K^{\times})$ has a unique common fixed point w on $\mathfrak{H}_1$. We can define an isomorphism π of K into C by

$$(d/dz)[f(a)(z)]_{z=w} = \overline{a^{\pi}}/a^{\pi} \qquad \text{for all } a \in K^{\times},$$

where the bar means the complex conjugation. Let $c \in K_A^{\times}$, and $\mu = [c^{\pi}, K^{\pi}]$. Then, *for every $S \in \mathfrak{Z}$, the point $\varphi_S(w)$ is rational over $(K^{\pi})_{ab}$, and*

(iv$_a$) $\quad \varphi_S(w)^{\mu} = J_{ST}(f(c)^{-1})(\varphi_T(w)), \; T = f(c)Sf(c)^{-1}.$

This is the adelized version of (a special case of) the formulas [11, (3.5.4)] and [12, (5.7.6$_s$)].

The nature of the symbol $J_{TS}(x)$ may not easily be understood at sight. Leaving the detailed explanation to the text (especially 2.6), we content ourselves here only with the following observation. Let $\mathfrak{L}_T$ be the field of all automorphic functions d on $\mathfrak{H}_n$ with respect to Γ_T of the form $d = e \circ \varphi_T$ with a function e on V_T rational over k_T, and let $\mathfrak{L}$ be the union of the $\mathfrak{L}_T$ for all $T \in \mathfrak{Z}$. Further let Aut $(\mathfrak{L}/F)$ denote the group of all automorphisms of $\mathfrak{L}$ over F. Then every element x of G_{A+} gives an element $\tau(x)$ of Aut $(\mathfrak{L}/F)$ by means of the rule $d^{\tau(x)} = e^{\sigma(x)} \circ J_{TS}(x) \circ \varphi_S$, where $S = x^{-1}Tx$. If $x = \alpha \in G_{Q+}$, the relation (iii$_c$) implies $d^{\tau(\alpha)}(z) = d(\alpha(z))$. Thus the symbol $J_{TS}(x)$ may be regarded as a geometric form of the automorphism $\tau(x)$ of $\mathfrak{L}$. Moreover, if either $\mathfrak{H}_n/\Gamma_S$ is compact or $G_Q = GL_2(Q)$, we can show that

(3) Aut $(\mathfrak{L}/F)$ *is isomorphic to G_{A+}/C, where C is the closure of $F^{\times} \cdot G_{R+}$ in G_A (see 2.8).*

Here is an obvious analogy with class field theory, mentioned at the beginning. By means of $\tau(x)$, the reciprocity-law (iv$_a$) can be written as

(iv$_b$) $\quad d(w)^{\mu} = d^{\tau(f(c)^{-1})}(w)$ *for every d in $\mathfrak{L}$ defined at w.*

The answer to the question (II) is also given in terms of these V_S and $J_{TS}(x)$. To state the result, let us remove the assumption $r = 1$. We have a system of $V_S, \varphi_S, J_{TS}(x)$ and k_S for G like the above one. Fix S and a point u of V_S whose coordinates are algebraic numbers. Take a point z of $\mathfrak{H}_n^r$ so that $\varphi_S(z) = u$, and let $\mathfrak{K}_z$ denote the union of the fields $k_T(\varphi_T(z))$ for all T. Then we shall obtain a continuous injective homomorphism

$$h \colon \mathrm{Gal}\,\big(\mathfrak{K}_z/k_S(u)\big) \longrightarrow S_0/\overline{E}.$$

Here we assume, for simplicity, that the isotropy group

$$E = \{\gamma \in \Gamma_s \mid \gamma(z) = z\}$$

is contained in F, and denote by $\bar{E}$ the closure of the projection of E to G_0. To study the Frobenius automorphisms, we consider an irreducible representation $\omega \colon G \to GL_m$, rational over Q, with any positive integer m, which is trivial on E, and such that $\omega(a) = N_{F/Q}(a)^{\kappa} \cdot 1_m$ for every $a \in F^{\times}$ with an integer κ independent of a. Let l be a rational prime, and G_l (resp. S_l) the l-part of G_A (resp. S), and E_l the closure of the projection of E to S_l. Then, combining h with the projection of $S_0/\bar{E}$ to S_l/E_l, we obtain a homomorphism

$$h_l \colon \mathrm{Gal}\left(\Re_z/k_S(u)\right) \longrightarrow S_l/E_l \ .$$

Further we obtain naturally from ω a representation

$$\omega_l \colon G_l \longrightarrow GL_m(Q_l) \ .$$

where Q_l denotes the l-adic number field. Since ω is trivial on E, we can define an l-adic representation

$$\omega_l \circ h_l \colon \mathrm{Gal}\left(\Re_z/k_S(u)\right) \longrightarrow GL_m(Q_l) \ .$$

Let $\mathfrak{p}$ be a prime ideal in $k_S(u)$, and σ a Frobenius automorphism of $\Re_z$ over $k_S(u)$ for any prime divisor of $\Re_z$ extending $\mathfrak{p}$. We can show that there exists a finite set Y of prime ideals in $k_S(u)$ such that the conjugacy class of $\omega_l(h_l(\sigma))$ is determined by l and $\mathfrak{p}$, and independent of the choice of σ, if $\mathfrak{p} \nmid l$ and $\mathfrak{p} \notin Y$. Moreover, the following assertions hold with a suitable choice of Y.

(i) *There exists a positive integer μ, independent of $\mathfrak{p}$ and l, such that the characteristic polynomial of $\omega_l(h_l(\sigma^{\mu}))$ has rational coefficients, and is independent of l, so long as $\mathfrak{p} \nmid l$ and $\mathfrak{p} \notin Y$.*

(ii) *The characteristic roots of $\omega_l(h_l(\sigma))$ are algebraic numbers of absolute value $N(\mathfrak{p})^{r\kappa/2}$.*

All these results can be proved by the same method as in [11, 12, 13]. Actually we have almost translated the proof in these papers in the adele language. The new proof is still intricate and long; for the moment, we may be satisfied with it, considering that our present theorems are far stronger than the previous ones. In the course of our treatment, we have also given an adelized version of the fundamental theorem on abelian varieties of CM-type (4.3), which by itself may be of an independent interest. Further, the congruence relation for modular correspondences in the one-dimensional case will be reformulated in our present language (2.23).

Our theory is restricted to the groups acting on $\mathfrak{H}_n^r$. We can now propose a generalization of our theory to the whole family of groups which are connected with all four main types of bounded symmetric domains. The completion of this task does not seem so difficult, since the ideas of the present paper are certainly applicable to other types of groups. For further comments on this and other questions, the reader is referred to 2.6–16 of the text.

Finally it should be noted that there is an important aspect of our theory, which we left untouched in this paper, and which is closely related to a recent work of Ihara [2]. It is very probable that a system like (2), consisting of varieties and morphisms defined over *finite fields*, can be associated with a certain subgroup of G_{A+}. If one formulates an isomorphism of the type (3) in this case (see 2.9 of the text), one can lay more emphasis on the analogy with class field theory. Actually Ihara has already made an essential contribution to this subject in the one-dimensional case. The author hopes to discuss this question in detail on some future occasion.

CONTENTS

0. Notation and terminology

0.0. We denote by $Z, Q, R, C,$ and H, respectively, the ring of rational integers, the rational number field, the real number field, the complex number field, and the division ring of Hamilton quaternions.

0.1. For an associative ring P with identity element, we denote by $P^{\times}$ the group of all invertible elements of P. $M_n(P)$ denotes the ring of all square matrices of size n with entries in P, and $GL_n(P) = M_n(P)^{\times}$. The identity element of $M_n(P)$ is denoted by 1_n. If X is a P-module, End (X, P) denotes the ring of all P-linear endomorphisms of X, and $GL(X, P) = \text{End}\,(X, P)^{\times}$. For a left P-module X, we let the elements of $\text{End}(X, P)$ act on the right of X. If P is a separable algebra over a field F, $N_{P/F}$ and $\text{Tr}_{P/F}$ denote respectively the reduced norm and trace from P to F. If P is a Galois extension of F, $\text{Gal}(P/F)$ denotes the Galois group of P over F.

0.2. By an *algebraic variety*, we understand a Zarisky open subset of a projective variety which is absolutely irreducible. We take the complex number field C as the universal domain (except in 2.22–23). If V is an algebraic variety, we often use the same symbol V for the set of all C-rational points of V. (This does not apply to an *algebraic group*, see below. Therefore our notation is not completely consistent. However, since the distinction between *algebraic variety* and *algebraic group* in our usage will be clear from the context, no confusion will be expected.) Thus a point x of V has coordinates in C. If k is a subfield of C over which V is defined, $k(x)$ denotes the field generated by the (quotients of the projective) coordinates of x. If π is an isomorphism of k into C, V^π denotes the variety whose defining equations are the transforms by π of the defining equations for V. If $K_1, \cdots, K_t$ are subfields of C, $K_1 \cdots K_t$ means the *composite* of the K_i, i.e., the smallest subfield of C containing the K_i. For an abelian variety A, $\mathrm{End}(A)$ denotes the ring of all endomorphisms of A, and $\mathrm{End}_Q(A) = \mathrm{End}(A) \otimes_Z Q$. We regard A as an algebraic variety, but not as an algebraic group in the sense below.

0.3. By an *algebraic number field*, we always understand a subfield of C algebraic over Q. The algebraic closure of Q in C will be denoted by $\overline{Q}$. For a rational prime l, Q_l and Z_l denote respectively the l-adic number field and the ring of l-adic integers. As is explained in the Introduction, we shall use the language of adeles, for which the basic references are Weil [16, 18]. If F is an algebraic number field of finite degree, we denote by $\mathfrak{r}_F$ the ring of algebraic integers in F, and by $F_A^\times$ the group of ideles of F. Then $F_\infty^\times$ denotes the archimedean part of $F_A^\times$, $F_{\infty+}^\times$ the identity component of $F_\infty^\times$, and F_{ab} the maximal abelian extension of F in C. For every $u \in F_A^\times$, we denote by $[u, F]$ the element of $\mathrm{Gal}(F_{ab}/F)$ canonically associated with u. We write $u \gg 0$ if the archimedean part of u belongs to $F_{\infty+}^\times$, and put $F_+^\times = \{a \in F^\times \mid a \gg 0\}$. For a positive integer c, we write $u \equiv 1 \bmod_0 (c)$ if, for every non-archimedean prime p of F, the p-component u_p of u is a p-unit, and $(u_p - 1)/c$ is a p-integer. If K is a finite algebraic extension of F, $D(K/F)$ and $\mathfrak{b}(K/F)$ denote respectively the relative discriminant and different of K over F.

0.4. By an *algebraic group*, we always mean a linear one. If G is an algebraic group defined over Q, G_A denotes the adelization of G. For an extension K of Q, the group of K-rational points on G is denoted by G_K or $G(K)$. We denote by G_0 and G_∞ the non-archimedean and archimedean parts of G_A, respectively; we have $G_A = G_0 G_\infty$. We shall often identify G_∞ with G_R. The identity component of G_∞ is denoted by $G_{\infty+}$. We put then $G_{A+} = G_0 G_{\infty+}$, $G_{Q+} = G_Q \cap G_{A+}$. Let G' be another Q-rational algebraic group, and $f : G \to G'$

a Q-rational homomorphism. Then we obtain, from f, a map of G_A to G'_A in a natural way, which we denote by the same letter f. If Y is an algebra of finite rank over Q, (for example, if Y is an algebraic number field of finite degree), we shall often consider $Y^\times$ as the group of Q-rational points of an algebraic group defined over Q. But, without introducing any new notation, we shall use the same letter $Y^\times$ to denote this algebraic group, since no confusion is feared.

0.5. Let V be a vector space over Q, and G a Q-rational algebraic subgroup of $GL(V, Q)$ (by abuse of language). Let $\mathfrak{m}$ be a Z-lattice in V, and $x \in G_A$. Then a Z-lattice $\mathfrak{m}x$ is meaningfully defined. Moreover, the multiplication by x defines an isomorphism of $V/\mathfrak{m}$ to $V/\mathfrak{m}x$. In fact, put $V_p = V \otimes_Q Q_p$, $\mathfrak{m}_p = \mathfrak{m} \otimes_Z Z_p$ for every rational prime p. Let x_p denote the p-component of x. Then $V/\mathfrak{m}$ is canonically isomorphic to the direct sum of $V_p/\mathfrak{m}_p$ for all p, and the multiplication by x_p defines an isomorphism of $V_p/\mathfrak{m}_p$ to $V_p/\mathfrak{m}_p x_p$, hence an isomorphism of $V/\mathfrak{m}$ to $V/\mathfrak{m}x$. Thus, for an element u of $V/\mathfrak{m}$, we denote by ux the corresponding element of $V/\mathfrak{m}x$. If c is a positive integer, we write $x \equiv 1 \bmod_0 (\mathfrak{m}, c)$ if $\mathfrak{m}x = \mathfrak{m}$ and $\mathfrak{m}_p(x_p - 1) \subset c\mathfrak{m}_p$ for all p.

0.6. Let $\mathfrak{H}$ be a bounded symmetric domain, and Γ a properly discontinuous group of transformations on $\mathfrak{H}$. We call (V, φ) a *model* of $\mathfrak{H}/\Gamma$ if V is an algebraic variety in the sense of 0.2, and φ is a Γ-invariant holomorphic map of $\mathfrak{H}$ into V, which induces a biregular isomorphism of $\mathfrak{H}/\Gamma$ to V. Although we let every element of Γ act on the left of $\mathfrak{H}$, we use the notation $\mathfrak{H}/\Gamma$ for the orbit space. The group Γ may often have the non-trivial center which acts trivially on $\mathfrak{H}$.

1. Algebraic preliminaries

1.1. Let F be an algebraic number field of finite degree, and Θ an absolute equivalence-class of Q-linear representations of F (as a Q-algebra) by complex matrices, which send the identity element of F to the identity matrix. If $p = [F : Q]$, and $\tau_1, \cdots, \tau_p$ denote all the isomorphisms of F into C, then Θ is determined by the multiplicity, say r_ν, of each τ_ν in any representation in the class Θ. We write then $\Theta \sim \sum_{\nu=1}^p r_\nu \tau_\nu$, and put

$$\det \Theta(x) = \prod_{\nu=1}^p (x^{\tau_\nu})^{r_\nu}, \quad \operatorname{tr} \Theta(x) = \sum_{\nu=1}^p r_\nu x^{\tau_\nu} \qquad (x \in F) .$$

In [11, § 5], we discussed some elementary properties of a couple (F, Θ) of this type, especially the relation of (F, Θ) with another couple (F', Θ') obtained from (F, Θ). Although the results obtained in [11] are quite adequate for our need in the present paper, we shall present here a simple way to define (F', Θ').

Let K be another algebraic number field of finite degree, and V a module over $F \otimes_Q K$, of finite dimension over Q. Then we may regard V as a left F-module and a right K-module by putting

$$axb = (a \otimes b)x \qquad\qquad (a \in F, b \in K, x \in V) .$$

We call V, with such a structure, an (F, K)-module. Given an (F, K)-module, we find two representations

$(1.1.1_a)$ $F \to \mathrm{End}\,(V, K)$,

$(1.1.1_b)$ $K \to \mathrm{End}\,(V, F)$,

in an obvious way. If V is of dimension n over K, $\mathrm{End}\,(V, K)$ is isomorphic to $M_n(K)$, hence we obtain an equivalence class Θ of representations of F of degree n. Similarly we obtain a couple (K, Ψ) from $(1.1.1_b)$. We say that V is of type (F, Θ) and of type (K, Ψ).

1.2. PROPOSITION. *Let F and K be algebraic number fields of finite degree, and Θ an equivalence class of representations of F in the sense of 1.1. Then there exists an (F, K)-module V of type (F, Θ) if and only if* $\mathrm{tr}\,\Theta(x) \in K$ *for all $x \in F$. Moreover, such a V is unique up to isomorphisms.*

PROOF. The *only-if*-part is obvious. To prove the *if*-part, let $F = Q(y)$ with an element y of F, and $h(T) = \prod_{\nu=1}^{p}(T - y^{\tau_\nu})^{r_\nu}$ with an indeterminate T. If $\mathrm{tr}\,\Theta(x) \in K$ for all $x \in F$, $h(T)$ has coefficients in K. Let $h = g_1 \cdots g_s$ be a decomposition into irreducible factors over K. Put $V = K[T]/(g_1) + \cdots + K[T]/(g_s)$, and consider V as a K-module, and also as an F-module by defining the action of y to be the same as the multiplication by T. Then V is of type (F, Θ). It is easy to see that every (F, K)-module of type (F, Θ) is isomorphic to this V.

1.3. Given a couple (F, Θ), let F' be the field generated over Q by $\mathrm{tr}\,\Theta(x)$ for all $x \in F$. By 1.2, there exists a unique (F, F')-module V of type (F, Θ). Let V be of type (F', Θ'). Thus from a couple (F, Θ), we obtain another couple (F', Θ'), which we call the *reflex* of (F, Θ). (In the previous papers, (F', Θ') was called the *dual* of (F, Θ), which does not seem appropriate.) Let us now show that (F', Θ') coincides with the one obtained by the procedure of [11,5.1].

Let R be a finite Galois extension of Q containing F, $G = \mathrm{Gal}\,(R/Q)$, and $Z[G]$ the group ring of G over Z. For an element $\xi = \sum_{\gamma \in G} a_\gamma \gamma$ of $Z[G]$ with $a_\gamma \in Z$, we put $\xi^* = \sum_\gamma a_\gamma \gamma^{-1}$. Let H and H' be the subgroups of G corresponding to F and F' respectively, and let

$$G = \bigcup_{\nu=1}^{p} H\tau_\nu = \bigcup_{\mu=1}^{q} H'\sigma_\mu \qquad\qquad (p = [F:Q], q = [F':Q])$$

be disjoint coset decompositions. Then we can write $\Theta \sim \sum_{\nu=1}^{p} r_\nu \tau_\nu$, $\Theta' \sim \sum_{\mu=1}^{p} s_\mu \sigma_\mu$ with non-negative integers r_ν and s_μ. Define elements $\eta, \eta', \omega, \omega'$ of $Z[G]$ by

$$\eta = \sum_{\gamma \in H} \gamma \,, \qquad \omega = \sum_{\nu=1}^{p} r_\nu \eta \tau_\nu \,,$$
$$\eta' = \sum_{\gamma \in H'} \gamma \,, \qquad \omega' = \sum_{\mu=1}^{q} s_\mu \eta' \sigma_\mu \,.$$

From our definition of F', we obtain $H' = \{\gamma \in G \mid \omega\gamma = \omega\}$, hence

$$(1.3.1) \qquad\qquad \omega\eta' = [H':1]\omega \,.$$

Now our task is to prove

$$(1.3.2) \qquad\qquad \omega^* = \omega' \,.$$

Let V be an (F, F')-module of type (F, Θ), and let $W = V \otimes_{F'} R$, $n = \sum_{\nu=1}^{p} r_\nu$. Then we can find n elements π_i of G and a basis $\{x_1, \cdots, x_n\}$ of W over R so that $ax_i = x_i a^{\pi_i}$ for every $a \in F$, and $\sum_{i=1}^{n} \eta\pi_i = \omega$. Let $\{e_1, \cdots, e_t\}$ be a basis of R over F. We see that $x_i R = x_i \sum_{h=1}^{t} F^{x_i} e_h^{\pi_i} = \sum_{h=1}^{t} F x_i e_h^{\pi_i}$. Let $b \in R$, and $e_h b^{\pi_i^{-1}} = \sum_{k=1}^{t} c_{hk}^i e_k$ with $c_{hk}^i \in F$. Then we have $x_i e_h^{\pi_i} b = \sum_{k=1}^{t} c_{hk}^i x_i e_k^{\pi_i}$. Since the $x_i e_h^{\pi_i}$, for all i and all h, form a basis of W over F, the right multiplication by b has the polynomial

$$\prod_{i=1}^{n} \det [T - (c_{hk}^i)_{h,k}] = \prod_{i=1}^{n} \prod_{\gamma \in H} [T - b^{\pi_i^{-1}\gamma}]$$

as its characteristic polynomial. Therefore

$$[R : F']\omega' = \sum_{i=1}^{n} \eta' \pi_i^{-1} \eta = \eta'\omega^* = [R : F']\omega^* \,,$$

in view of (1.3.1). This proves (1.3.2).

1.4. PROPOSITION. *Let (F', Θ') be the reflex of (F, Θ), and $\Theta \sim \sum_{\nu=1}^{p} r_\nu \tau_\nu$ as above. Put $r = \sum_{\nu=1}^{p} r_\nu$. Then*

$$N_{F/Q}(\det \Theta'(x)) = N_{F'/Q}(x)^r \qquad\qquad (x \in F') \,.$$

PROOF. Take an (F, F')-module V of type (F, Θ). Then V is of type (F', Θ'). Consider V as a vector space over Q. For $x \in F'$, let S denote the Q-linear endomorphism of V obtained from the multiplication by x. Compute $\det(S)$ in two different ways, by considering V as an F-module, and as an F'-module. Note that r is the dimension of V over F'. Then we obtain the desired equality. It is also easy to verify the formula by means of ω and ω' of 1.3.

1.5. PROPOSITION. *Let (F', Θ') be the reflex of (F, Θ), and (F_1, Θ_1) the reflex of (F', Θ'). Then*

(1.5.1) $F_1 \subset F$, *and* $\operatorname{tr} \Theta(x) = \operatorname{tr} \Theta_1(\operatorname{Tr}_{F/F_1}(x))$ *for every* $x \in F$.

(1.5.2) (F', Θ') *is the reflex of* (F_1, Θ_1).

PROOF. Let V be an (F, F')-module of type (F, Θ). Then V is of type (F', Θ'). By 1.2, we have $F_1 \subset F$. Let W be an (F_1, F')-module of type (F', Θ'). By the uniqueness of (F, F')-module of type (F', Θ'), V and $F \otimes_{F_1} W$ are isomorphic, hence (1.5.1). Let (F_2, Θ_2) be the reflex of (F_1, Θ_1). By (1.5.1), we have $F_2 \subset F'$, and $\operatorname{tr} \Theta'(x) = \operatorname{tr} \Theta_2(\operatorname{Tr}_{F'/F_2}(x))$ for $x \in F'$. From the last assertion of (1.5.1) we obtain $F' \subset F_2$, hence $F' = F_2$ and $\Theta' = \Theta_2$.

1.6. *Remark.* We can consider a couple like (F, Θ) and its reflex for a more general type of field. In fact, fix a basic field k, and take two algebraically closed fields L_1 and L_2 containing k. Let F be a finite separable extension of k contained in L_1, and Θ an equivalence class of k-linear representations of F by matrices with coefficients in L_2. Then the reflex (F', Θ') of (F, Θ) is defined with a subfield F' of L_2 and an equivalence class of representations of F' by matrices over L_1.

1.7. PROPOSITION. *Let (F', Θ') be the reflex of (F, Θ). Put $\lambda = \det \Theta'$, and consider $F^\times$ and $F'^\times$ algebraic groups defined over $\mathbf{Q}$. Then the following assertions hold.*

(1.7.1) λ *is a $\mathbf{Q}$-rational homomorphism of $F'^\times$ into $F^\times$.*

(1.7.2) *If $x \in F_A'^\times$ and $x \equiv 1 \bmod_0 (b)$ for a positive integer b, then $\lambda(x) \equiv 1 \bmod_0 (b)$.* (For the notation, see 0.3 and 0.4.)

PROOF. Take V as in the proof of 1.5. Then we have a representation $F' \to \operatorname{End}(V, F)$ in the class of Θ', which is obviously $\mathbf{Q}$-rational, hence its determinant, λ, is $\mathbf{Q}$-rational. Further let N be a $\mathbf{Z}$-lattice in V, and $M = \mathfrak{r}_F N \mathfrak{r}_{F'}$. Then M is a $\mathbf{Z}$-lattice in V stable under $\mathfrak{r}_F$ and $\mathfrak{r}_{F'}$. For every rational prime p, put $V_p = V \otimes_{\mathbf{Q}} \mathbf{Q}_p, F_p = F \otimes_{\mathbf{Q}} \mathbf{Q}_p, F_p' = F' \otimes_{\mathbf{Q}} \mathbf{Q}_p, M_p = M \otimes_{\mathbf{Z}} \mathbf{Z}_p, \mathfrak{r}_p = \mathfrak{r}_F \otimes_{\mathbf{Z}} \mathbf{Z}_p, \mathfrak{r}_p' = \mathfrak{r}_{F'} \otimes_{\mathbf{Z}} \mathbf{Z}_p$. For $x_p \in F_p'^\times$, represent the action of x_p on V_p by a matrix T_p with coefficients in F_p with respect to a basis of M_p over $\mathfrak{r}_p$. If $x_p \equiv 1 \bmod b\mathfrak{r}_p'$, we have $M_p(x_p - 1) \subset bM_p$, hence $\det(T_p) \equiv 1 \bmod b\mathfrak{r}_p$, which proves (1.7.2).

1.8. By a CM-*field*, we understand a totally imaginary quadratic extension of a totally real algebraic number field of finite degree (cf. [11, 5.7–10]). Let K be a CM-field, and Φ a class of representations of K. We call (K, Φ) a CM-type if the sum of Φ and its complex conjugate is the class of regular representations of K over $\mathbf{Q}$. This coincides with the definition of [14] and [11, 5.11]. The reflex of a CM-type is also a CM-type [11, 5.13].

1.9. PROPOSITION. *Let (K, Φ) be a CM-type, (K', Φ') the reflex of (K, Φ), and $2g = [K : \mathbf{Q}]$. Then the following assertions hold:*

(1.9.1) $[K' : \mathbf{Q}] \leq 2^g$.

(1.9.2) *There exists an element b of K such that $\det \Phi(b)$ generates K' over $\mathbf{Q}$.*

(1.9.3) *If $[K' : \mathbf{Q}] = 2^g$, (K, Φ) is the reflex of (K', Φ').*

(1.9.4) *Let (K, Ψ) be a CM-type, and (K^*, Ψ^*) the reflex of (K, Ψ). If $[K': Q] = 2^g$, there exists an automorphism π of C such that $\Psi = \Phi^\pi$, $K^* = K'^\pi$, and $\Psi^*(x^\pi) = \Phi'(x)$ for $x \in K'$.*

PROOF. Let L be the smallest Galois extension of Q containing K. Let $\Phi \sim \sum_{\nu=1}^g \sigma_\nu$. For every $\tau \in \mathrm{Gal}\,(L/Q)$, we can define Φ^τ by $\Phi^\tau \sim \sum_{\nu=1}^g \sigma_\nu \tau$. Then (K, Φ^τ) is a CM-type. Observe that there are exactly 2^g CM-types with the same K, and K' is the subfield of L corresponding to $\{\tau \in \mathrm{Gal}\,(L/Q)\,|\,\Phi^\tau = \Phi\}$. This proves (1.9.1). If $[K': Q] = 2^g$, then the (K, Φ^τ) for all $\tau \in \mathrm{Gal}\,(L/Q)$ exhaust all the CM-types defined with K. Therefore we obtain (1.9.4). Let (K_1, Φ_1) be the reflex of (K', Φ'). By 1.5, $K_1 \subset K$, and (K', Φ') is the reflex of (K_1, Φ_1). Therefore, if $2h = [K_1: Q]$, we have $[K': Q] \leq 2^h \leq 2^g$. Therefore, if $[K': Q] = 2^g$, we have $K = K_1$, hence (1.9.3), in view of (1.5.1). To prove (1.9.2), take a rational prime p which fully decomposes in L. Let $\mathfrak{q}$ be a prime ideal in K which divides p. We can find an element b of K such that $(b) = \mathfrak{q}^n$ for some positive integer n. Put $c = \det \Phi(b)$. Then it is easy to see that, for $\tau \in \mathrm{Gal}\,(L/Q)$, c^τ and c generate the same ideal in L if and only if $\Phi^\tau = \Phi$, hence (1.9.2).

1.10. Let F be a totally real algebraic number field of degree g. Let us now construct a totally imaginary quadratic extension K of F with the following property:

(1.10.1) *For every CM-type (K, Φ) and its reflex (K', Φ'), one has $[K': Q] = 2^g$* (cf. (1.9.4)).

Let R be the smallest Galois extension of Q containing F. Take a rational prime p which fully decomposes in R. Let $\mathfrak{p}$ be a prime factor of p in F. By [11,1.5], we can find a totally imaginary quadratic extension K of F, such that $D(K/F)$ is divisible only by $\mathfrak{p}$ and a prime ideal $\mathfrak{q}$ in F prime to $p \cdot D(R/Q)$. Let (K, Φ) be a CM-type, with $\Phi \sim \sum_{\nu=1}^g \sigma_\nu$, and (K', Φ') the reflex of (K, Φ). Let $\{u_1, \cdots, u_g\}$ be a basis of F over Q. Then, for every $c \in K$, $\sum_{\nu=1}^g c^{\sigma_\nu} u_\nu^{\sigma_\nu} \in K'$. Since $u_\nu^{\sigma_\nu} \in R$, we see that $c^{\sigma_\nu} \in RK'$, hence RK' is the composite of $K^{\sigma_1}, \cdots, K^{\sigma_g}$. Observe that $\mathfrak{p}^{\sigma_1}, \cdots, \mathfrak{p}^{\sigma_g}$ are pairwise relatively prime. Now we have $\mathfrak{d}(K^{\sigma_\nu}R/R) = \mathfrak{d}(K/F)^{\sigma_\nu}$, hence $D(K^{\sigma_\nu}R/R)$ is divisible by $\mathfrak{p}^{\sigma_\nu}$, but prime to $\mathfrak{p}^{\sigma_\mu}$ for $\mu \neq \nu$. It follows that $[K'R: R] = 2^g$, hence $[K': Q] \geq 2^g$. From this and (1.9.1), we obtain (1.10.1).

2. Main theorems on canonical system

2.1. The semi-simple part of the group to be discussed is a Q-simple group which becomes, over C, a product of copies of a symplectic group. The classification of simple groups tells us that such a group is obtained as a quaternion unitary group (see Weil [17]). Therefore we start with an algebraic number field F of finite degree, and a quaternion algebra B over F, which may or may not be a division algebra. Put $g = [F : Q]$. We denote by ι the main involution of B, and consider a non-degenerate B-valued ι-hermitian form $h(x, y)$ on a non-trivial left B-module X. The form h is, by definition, an F-bilinear map of $X \times X$ to B satisfying $h(ax, by) = ah(x, y)b^{\iota}$, $h(x, y)^{\iota} = h(y, x)$ for $a \in B$, $b \in B$, $x \in X$, $y \in X$. Then X must be isomorphic to the product B^n of n copies of B, with a positive integer n, (even if $B = M_2(F)$, see [12,3.2]). Now we define an algebraic group G over Q so that

$$G_Q = \{\alpha \in GL(X, B) \mid h(x\alpha, y\alpha) = \nu(\alpha)h(x, y) \text{ with } \nu(\alpha) \in F\} \,.$$

We consider G an algebraic subgroup of GL_{4ng} with respect to a basis of X over Q. We shall often consider the multiplier ν as a homomorphism $\nu : G \to F^{\times}$, and put

$$(2.1.1) \qquad\qquad G^u = \{\alpha \in G \mid \nu(\alpha) = 1\} \,.$$

As we said, G_C^u is isomorphic to $\mathrm{Sp}(n, C)^r$.

We are interested only in the case where

(2.1.2) G_R^u *modulo a maximal compact subgroup becomes a bounded symmetric domain.*

Then it is easily verified that F *must be totally real.* Therefore, $B_R = B \otimes_Q R$ can be identified with $M_2(R)^r \times H^{g-r}$, where $0 < r \leq g$, and H denotes the division ring of Hamilton quaternions. Let $\tau_1, \cdots, \tau_g$ be all the isomorphisms of F into R. We understand that the r factors $M_2(R)$ of B_R correspond to τ_1, $\cdots, \tau_r$, and the $g - r$ factors H to $\tau_{r+1}, \cdots, \tau_g$. Then the condition (2.1.2) is satisfied if and only if

(2.1.3) $h(x,x)^{\tau_\nu} > 0$ *for every* $\nu > r$, *and every* $x \in X$, $x \neq 0$.

Note that $h(x, x) \in F$ for every $x \in X$. By the Hasse principle, such an h is unique for n, up to isomorphisms. In other words,

(2.1.4) *If h satisfies (2.1.3), there exists a basis $\{x_1, \cdots, x_n\}$ of X over B such that $h(x_i, x_j) = 1$ or 0 according as $i = j$ or $i \neq j$.*

Therefore G_Q can be identified with

$$\{\alpha \in GL_n(B) \mid \alpha \cdot {}^t\alpha^{\iota} = \nu(\alpha)1_n \text{ with } \nu(\alpha) \in F\} \,.$$

The quotient space of G_R^u by a maximal compact subgroup is isomorphic to the product $\mathfrak{H}_n^r$ of r copies of the Siegel upper half space $\mathfrak{H}_n$ of degree n. Therefore, $G_{\infty+}$, identified with the connected component of G_R, acts on $\mathfrak{H}_n^r$ (for details, see [12, 4.4]). We fix this action of $G_{\infty+}$ on $\mathfrak{H}_n^r$ throughout the paper.

2.2. Let us consider a couple (F, Θ) in the sense of 1.1, with the above F, and $\Theta \sim \sum_{\nu=1}^r \tau_\nu$. Let (F', Θ') be the reflex of (F, Θ) in the sense of 1.3,

311

and $\lambda = \det \Theta'$. Then λ is a homomorphism of $F''^{\times}$ into $F^{\times}$. Define[2] an algebraic subgroup $\tilde{D}$ of $G \times F''^{\times}$, rational over $\mathbf{Q}$, by

$$\tilde{D} = \{(x, c) \in G \times F''^{\times} \mid \nu(x) = \lambda(c)\} \ .$$

Let us now consider the adelization G_A of G, and put

$$D = \{x \in G_A \mid \nu(x) \in \lambda(F''^{\times}_A)\} \ ,$$
$$D_+ = D \cap G_{A+} \ .$$

Obviously D is the projection of $\tilde{D}_A$ to G_A. Observe also that

$$G_{A+} = \{x \in G_A \mid \nu(x) \gg 0\} \ .$$

Define subgroups $\mathcal{G}$ and $\mathcal{G}_+$ of G_A by[3]

$$\mathcal{G} = D \cdot G_{\mathbf{Q}} \cdot G_{\infty+} \ , \qquad \mathcal{G}_+ = \mathcal{G} \cap G_{A+} \ .$$

We see easily that

$$(2.2.1) \quad \mathcal{G}_+ = D_+ G_{\mathbf{Q}+} G_{\infty+} = \{x \in G_{A+} \mid \nu(x) \in \lambda(F''^{\times}_A)F^{\times}F^{\times}_{\infty+}\} \quad \text{(see 3.1 below)} \ .$$

2.3. Let $\mathfrak{Z}$ denote the set of all subgroups S of G_{A+} satisfying the following two conditions:

(2.3.1) $S = S_0 G_{\infty+}$ *with a compact subgroup S_0 of G_0.*

(2.3.2) *S contains the projection to G_A of an open subgroup of $\tilde{D}_A$.*

(For example, this is satisfied if S_0 is open in G_0.)

The following two properties of $\mathfrak{Z}$ are obvious.

(2.3.3) *If $x \in G_A$ and $S \in \mathfrak{Z}$, then $xSx^{-1} \in \mathfrak{Z}$.*

(2.3.4) *If $S \in \mathfrak{Z}$ and $T \in \mathfrak{Z}$, then $S \cap T \in \mathfrak{Z}$.*

The members of $\mathfrak{Z}$ are commensurable with each other if $r = 1$ (see 2.6 below), but not so in the general case.

For every $S \in \mathfrak{Z}$, put $\Gamma_S = S \cap G_{\mathbf{Q}}$. Then Γ_S, modulo its center, can be considered a discontinuous group of transformations on $\mathfrak{H}^r_n$. Our theorem will concern a simultaneous system of models of $\mathfrak{H}^r_n/\Gamma_S$ for all $S \in \mathfrak{Z}$. These models will be defined over the number fields k_S defined as follows. First put

$$\mathfrak{X}(S) = \mathfrak{X}_S = \{c \in F''^{\times}_A \mid \lambda(c) \in F^{\times} \cdot \nu(S)\} \ .$$

It can be shown that $\mathfrak{X}_S$ is an open subgroup of $F''^{\times}_A$ containing $F''^{\times}F''^{\times}_{\infty+}$ (see 2.10 below). Therefore $\mathfrak{X}_S$ corresponds to an abelian extension of F' of finite degree, which we denote by k_S or $k(S)$. We see easily that

---

[2] If the reader is not interested in the most general case, but wishes to grasp the idea quickly, he may be advised to assume throughout that $r = 1$ and τ_1 is the identity map. Under this assumption, all definitions and propositions are considerably simplified, see (5) of 2.6. However, as we said in the Introduction, this case should not be considered typical.

[3] It seems more natural to consider, instead of $\mathcal{G}_+$, the subgroup of G_{A+} consisting of all x such that $\nu(x)$ belongs to the closure of $\lambda(F''^{\times}_A) \cdot F^{\times} \cdot F^{\times}_{\infty+}$ in $F^{\times}_A$.

(2.3.5) $k_S \subset k_T$ *if* $T \subset S$.

(2.3.6) $k(xSx^{-1}) = k(S)$ *for every* $x \in G_A$.

Let $\mathfrak{k}$ denote the composite of the k_S for all $S \in \mathfrak{Z}$. We define a homomorphism $\sigma\colon \mathfrak{G} \to \mathrm{Gal}\,(\mathfrak{k}/F')$ by

(2.3.7) $\sigma(x) = $ *the restriction of* $[c^{-1}, F']$ *to* $\mathfrak{k}$,

with an element c of $F'^{\times}_A$ such that $\lambda(c)/\nu(x) \in F^{\times}F^{\times}_{\infty+}$. This definition is meaningful, since, if $\lambda(c) \in F^{\times}F^{\times}_{\infty+}$, $[c, F']$ is the identity on k_S for all $S \in \mathfrak{Z}$.

2.4. As is explained in the Introduction, it is most important to discuss the property of our models of $\mathfrak{H}^r_n/\Gamma_S$ at the isolated fixed points of G_{Q+} on $\mathfrak{H}^r_n$. Therefore let us now recall the results of [12, § 4] on such points. Let $P_1, \cdots,$ P_t be CM-fields containing F, in the sense of 1.8, and $Y_1, \cdots, Y_t$ be central simple algebras over $P_1, \cdots, P_t$, respectively. Put $Y = Y_1 \times \cdots \times Y_t$, $P = P_1 \times \cdots \times P_t$, $[Y_i : P_i] = q_i^2$, $[P_i : F] = 2m_i$. We assume the following two conditions:

(2.4.1) $n = \sum_{i=1}^t m_i q_i$.

(2.4.2) *For each* i, Y_i *and* $B \otimes_F P_i$ *belong to the same Brauer algebra class over* P_i.

Then Y has a positive involution. Moreover, for every positive involution δ of Y, there exists an F-linear isomorphism f of Y into $M_n(B)$ such that $f(a^\delta) = {}^t f(a)^\iota$. Put

$$\{Y, \delta\} = \{a \in Y \mid aa^\delta \in F^{\times}_+\}\,.$$

Then $f(\{Y, \delta\})$ is contained in G_{Q+}, and has a unique common fixed point z on $\mathfrak{H}^r_n$. Conversely, every isolated fixed point is obtained in this way, with some Y, δ, and f. We call (Y, P, δ, f) *primitive* if

$$f(\{Y, \delta\}) = \{\alpha \in G_{Q+} \mid \alpha(z) = z\}\,.$$

(If this is not satisfied, (Y, P, δ, f) can be extended to a larger (Y', P', δ', f') which is primitive.)

We can define, through the procedure of [12, (4.7.6), 4.9], a certain representation class Ψ_i of P_i, hence a couple (P_i, Ψ_i) as treated in 1.1. Let (P'_i, Ψ'_i) be the reflex of (P_i, Ψ_i), and P' the composite of $P'_1, \cdots, P'_t$. Then we can define a map $\eta\colon P'^{\times} \to G$ by

(2.4.3) $\qquad \eta(v) = f(\det \Psi'_1(N_{P'/P'_1}(v)), \cdots, \det \Psi'_t(N_{P'/P'_t}(v)))$ $\qquad (v \in P'^{\times})\,.$

If Ψ' is as in [12, (4.9.3)], we have $\eta = f \circ \Psi'$. We see that η is a Q-rational homomorphism, and, by [12, (4.10.4)],

(2.4.4) $\qquad\qquad \nu(\eta(v)) = \lambda(N_{P'/F}(v))$ $\qquad\qquad (v \in P'^{\times})\,.$

Therefore, by our definition of the group D_+, we see that

(2.4.5) $$\eta(P'^{\times}_A) \subset D_+ .$$

2.5. Main Theorem. *There exists a system*

$$\{V_S, \varphi_S, J_{TS}(x), (S, T \in \mathfrak{Z}, x \in \mathcal{G}_+)\}$$

formed by the objects satisfying the following conditions:

(2.5.1) *For each* $S \in \mathfrak{Z}$, (V_S, φ_S) *is a model of* $\mathfrak{H}^r_n/\Gamma_S$ *in the sense of* 0.6.

(2.5.2) V_S *is defined over* k_S.

(2.5.3) $J_{TS}(x)$, *defined if and only if* $xSx^{-1} \subset T$, *is a morphism of* V_S *onto* $V_T^{\sigma(x)}$, *rational over* k_S, *and has the following properties:*

$\quad$ (2.5.3$_a$) $J_{SS}(x)$ *is the identity map if* $x \in S \cap \mathcal{G}_+$;

$\quad$ (2.5.3$_b$) $J_{TS}(x)^{\sigma(y)} \circ J_{SR}(y) = J_{TR}(xy)$;

$\quad$ (2.5.3$_c$) $J_{TS}(\alpha)[\varphi_S(z)] = \varphi_T(\alpha(z))$ *if* $\alpha \in G_{Q+}$ *(and* $\alpha S\alpha^{-1} \subset T$).

(2.5.4) *Let* (Y, P, δ, f), P', z, *and* η *be as in* 2.4. *Then, for every* $S \in \mathfrak{Z}$, *the point* $\varphi_S(z)$ *is rational over* P'_{ab}. *Further, for every* $v \in P'^{\times}_A$, *one has* $\varphi_T(z)^{\tau} = J_{TS}(\eta(v)^{-1})[\varphi_S(z)]$, *where* $\tau = [v, P']$, *and* $T = \eta(v)^{-1}S\eta(v)$.

Note that $\sigma(\eta(v)^{-1}) = [v, P']$ on $\mathfrak{k}$. From (2.5.4), we can easily derive (see 3.7 below)

(2.5.5) *Let* $C(S, f)$ *be the abelian extension of* P' *corresponding to the subgroup* $\{v \in P'^{\times}_A \mid \eta(v) \in f(\{Y, \delta\}) \cdot S\}$ *of* $P'^{\times}_A$. *Then* $P' \cdot k_S(\varphi_S(z)) \subset C(S, f)$. *If* (Y, P, δ, f) *is primitive,* $P' \cdot k_S(\varphi_S(z)) = C(S, f)$.

The proof of the above theorem will be completed in §§ 5, 6. We call $\{V_S, \varphi_S, J_{TS}(x)\}$ a *canonical system* for G. *This is unique* to the effect that if $\{V'_S, \varphi'_S, J'_{TS}(x)\}$ is another canonical system for G, then there exists, for each $S \in \mathfrak{Z}$, a biregular isomorphism M_S of V_S onto V'_S such that $\varphi'_S = M_S \circ \varphi_S$, and $M_T^{\sigma(x)} \circ J_{TS}(x) = J'_{TS}(x) \circ M_S$ for every $x \in \mathcal{G}_+$ satisfying $xSx^{-1} \subset T$. For proof, see 3.9.

2.6. The theorem may require some comments, particularly about the nature of the symbol $J_{TS}(x)$.

(1) If $S \subset T$, we obtain $J_{TS}(1) \circ \varphi_S = \varphi_T$ from (2.5.3$_c$). Therefore, $J_{TS}(1)$ is the morphism obtained from the natural projection of $\mathfrak{H}^r_n/\Gamma_S$ to $\mathfrak{H}^r_n/\Gamma_T$. The theorem asserts that this is defined over k_S.

If $xSx^{-1} = T$, we have

$$J_{TS}(x) \circ J_{ST}(x^{-1})^{\sigma(x)} = J_{TT}(1)^{\sigma(x)} = \text{the identity map of } V_T^{\sigma(x)},$$

$$J_{ST}(x^{-1})^{\sigma(x)} \circ J_{TS}(x) = J_{SS}(1) = \text{the identity map of } V_S.$$

Therefore $J_{TS}(x)$ is a biregular isomorphism of V_S onto $V_T^{\sigma(x)}$.

If, more generally, $xSx^{-1} \subset T$, we have, putting $R = x^{-1}Tx$,

(2.6.1) $$J_{TS}(x) = J_{TR}(x) \circ J_{RS}(1) .$$

Thus $J_{TS}(x)$ is the composed map of a biregular isomorphism and a project ion morphism.

(2) If $\alpha S \alpha^{-1} = T$ with some $\alpha \in G_{Q+}$, we have $\alpha \Gamma_S \alpha^{-1} = \Gamma_T$, hence it is obvious that V_S and V_T are biregularly isomorphic under the morphism $J_{TS}(\alpha)$ which sends $\varphi_S(z)$ to $\varphi_T(\alpha(z))$ as described in (2.5.3$_c$). The fact that this is rational over k_S is not trivial. If $x S x^{-1} = T$ with an element x of $\mathcal{G}$ which is not necessarily in G_{Q+}, the existence of the biregular isomorphism $J_{TS}(x)$ of V_S to $V_T^{\sigma(x)}$ is far from triviality. Indeed, there is no obvious relation between Γ_S and Γ_T, even in the case $B = M_2(Q)$. This is especially so if the class number of F is greater than one.

(3) Suppose that T is a normal subgroup of S. Then Γ_T is a normal subgroup of Γ_S, hence V_T is a Galois covering of V_S. The Galois group of this covering, considered over C, is isomorphic to $\Gamma_S/[\Gamma_T \cdot (\Gamma_S \cap F^{\times})]$, and formed by the $J_{TT}(\gamma)$ with $\gamma \in \Gamma_S$. If we consider, more precisely, V_S and V_T over k_S and k_T, we obtain a larger group as the Galois group, which is formed by the $J_{TT}(x)$ with $x \in S \cap \mathcal{G}_+$. A more detailed discussion will be made in § 7.

(4) We may have $\Gamma_S = \Gamma_T$ for two distinct S and T. This is natural, because we are associating with S not only a variety V_S but also a number field k_S. For example, in the case $G = GL_2(Q)$, let N be a positive integer, and let

$$U = G_{\infty+} \cdot \prod_p GL_2(Z_p) \qquad \text{(the product over all rational primes } p\text{),}$$
$$S = \{x \in U \mid x_p \equiv 1_2 \bmod N \cdot M_2(Z_p)\} \,,$$
$$T = \left\{ x \in U \mid x_p \equiv \begin{bmatrix} 1 & 0 \\ 0 & d \end{bmatrix} \bmod N \cdot M_2(Z_p) \text{ with } d \in Z_p \right\} ,$$

where x_p denotes the p-component of x. Then

$$\Gamma_U = SL_2(Z) \,, \qquad \Gamma_S = \Gamma_T = \{\alpha \in SL_2(Z) \mid \alpha \equiv 1_2 \bmod (N)\} \,,$$
$$k_U = k_T = Q \,, \qquad k_S = Q(e^{2\pi i/N}) \,.$$

Coming back to the general case, we have (see 3.6 below)

(2.6.2) If $T \subset S \subset \mathcal{G}$, $\Gamma_S = \Gamma_T$, and $k_S = k_T$, then $S = T$.

(5) Suppose that $r = 1$. Then $F' = F^{\tau_1}$, and $\lambda = \tau_1^{-1}$. Therefore $\tilde{D}$ and G are isomorphic over Q, and $D = G_A$, $\mathcal{G}_+ = G_{A+}$. If we take, for simplicity, τ_1 to be the identity map, then $F' = F$, $\sigma(x) = [\nu(x)^{-1}, F]$ for $x \in G_{A+}$, and $\mathfrak{X}_S = F^{\times} \cdot \nu(S)$. Therefore $\mathfrak{k} = F_{ab}$, $\mathfrak{Z}$ is the family of all subgroups $S = S_0 G_{\infty+}$ with *any* open compact subgroup S_0 of G_0. Consequently any two members of $\mathfrak{Z}$ are commensurable with each other. Thus everything can be simplified to a considerable degree in the case $r = 1$.

(6) Let F_1 be a subfield of F. Suppose that $\tau_1, \cdots, \tau_r$ are exactly all the isomorphisms of F into R which are the identity map on F_1. Then $F' = F_1$, and λ is the natural injection of F_1 into F. Therefore

$$D = \{x \in G_A \mid \nu(x) \in F_{1A}^\times\} ,$$
$$D \cap G_Q = \{x \in G_Q \mid \nu(x) \in F_1^\times\} .$$

Further, $\sigma(x) = [\nu(x)^{-1}, F_1]$ for $x \in D$, $\mathfrak{X}_S = F_1^\times \cdot \nu(S)$ if $S \subset D$, and $\mathfrak{k} = (F_1)_{ab}$. Obviously this always applies to the case $r = g$, in which we have $F_1 = Q$.

(7) In the reciprocity law (2.5.4), T depends on S and τ. This might give an impression that infinitely many models V_T are involved, even for a fixed S and z. Actually it is not so. In fact, take any member S' of $\mathfrak{Z}$ containing S as a normal subgroup. (For example, we can take S itself as S'.) Consider a double coset decomposition

$$\mathfrak{S}_+ = \bigcup_{\nu=1}^p G_{Q+} x_\nu (S' \cap \mathfrak{S}) .$$

The number of cosets is finite, see 3.5 below. Write (V_ν, φ_ν) for (V_R, φ_R) if $R = x_\nu S x_\nu^{-1}$, and put $J_{\nu\mu}(u) = J_{RT}(x_\nu u x_\mu^{-1})$ for $u \in S' \cap \mathfrak{S}$, $R = x_\nu S x_\nu^{-1}$, $T = x_\mu S x_\mu^{-1}$. Then we obtain a system

$$(2.6.3) \qquad \{V_\nu, \varphi_\nu, J_{\nu\mu}(u), (u \in S' \cap \mathfrak{S};\ \mu, \nu = 1, \cdots, p)\} .$$

Let the notation be as in (2.5.4). Take any index ν. Since $\eta(v)x_\mu \in \mathfrak{S}_+$, we have $\eta(v)x_\nu = \alpha x_\mu s$ for some μ, $\alpha \in G_{Q+}$ and $s \in S' \cap \mathfrak{S}$. If R and T are as above, and $U = \eta(v)R\eta(v)^{-1}$, we have $\varphi_R(z)^\tau = J_{RU}\big(\eta(v)^{-1}\big)\big(\varphi_U(z)\big) = J_{RT}(x_\nu s^{-1} x_\mu^{-1})\big(\varphi_T(\alpha^{-1}(z))\big)$, hence $\varphi_\nu(z)^\tau = J_{\nu\mu}(s^{-1})\big(\varphi_\mu(\alpha^{-1}(z))\big)$. Thus (2.6.3) is exactly (a generalization of) the canonical system obtained in [11, 12]. The relation [12, (5.7.3, 4, 5)] can easily be derived from $(2.5.3_{a,b,c})$.

2.7. Let L_S be the field of all functions on V_S defined over k_S. (A *function on V_S* is a rational map of V_S into the projective line, other than the constant ∞, see Weil [15, p. 246].) Define a field $\mathfrak{L}$ of functions on $\mathfrak{H}_n^r$ by

$$\mathfrak{L} = \bigcup_{S \in \mathfrak{Z}} \mathfrak{L}_S , \qquad \mathfrak{L}_S = \{f \circ \varphi_S \mid f \in L_S\} .$$

We are going to define a map τ of $\mathfrak{S}_+$ into the group Aut $(\mathfrak{L}/F')$ of all automorphisms of $\mathfrak{L}$ over F'. Let $x \in \mathfrak{S}_+$. Observe that, for every $f \in L_S$, $f^{\sigma(z)}$ is a function on $V_S^{\sigma(z)}$ rational over k_S. Put

$$(f \circ \varphi_S)^{\tau(z)} = f^{\sigma(z)} \circ J_{ST}(x) \circ \varphi_T \qquad\qquad (T = x^{-1}Sx) .$$

Then we see easily that τ is a homomorphism of $\mathfrak{S}_+$ into Aut $(\mathfrak{L}/F')$, and has the following properties:

(2.7.1) $\tau(x) = \sigma(x)$ on $\mathfrak{k}$.

(2.7.2) $h^{\tau(\alpha)} = h(\alpha(z))$ for $\alpha \in G_{Q+}$, $h \in \mathfrak{L}$, $z \in \mathfrak{H}_n^r$.

Now (2.5.4) can be re-formulated as follows:

(2.7.3) *Let (Y, P, δ, f), P', z, and η be as in 2.4. Then, for every $h \in \mathfrak{L}$*

defined at z, $h(z)$ is rational over P'_{ab}. Moreover, if $v \in P'^{\times}_A$, $\omega = [v, P']$, and $x = \eta(v)^{-1}$, then $h^{\tau(x)}$ is also defined at z, and $h(z)^{\omega} = h^{\tau(x)}(z)$.

We can introduce a Hausdorff topology into Aut $(\mathfrak{L}/F')$ by taking

$$\{\pi \in \text{Aut}\,(\mathfrak{L}/F') \mid q_1^{\pi} = q_1, \cdots, q_m^{\pi} = q_m\}$$

to be an open neighborhood of the identity, for any finite set $\{q_i\}$ of elements of $\mathfrak{L}$.

2.8. **THEOREM.** *The notation being as in 2.7, let $\mathcal{C} = \bigcap_{S \in \mathfrak{Z}} S \cdot F^{\times}$. Then $\mathcal{C}$ is the kernel of τ. Moreover, if $r = 1$, τ is continuous, and $\mathcal{C}$ is the closure of $F^{\times} \cdot G_{\infty+}$ in G_A. Furthermore, if $r = 1$ and either $\mathfrak{H}_n/\Gamma_S$ is compact or $G_Q = GL_2(Q)$, then τ is surjective, and induces a topological isomorphism of $G_{A+}/\mathcal{C}$ onto Aut $(\mathfrak{L}/F')$.*

Note that the compactness of $\mathfrak{H}_n/\Gamma_S$ is independent of S, see 2.16 below. One may conjecture that our assertion is true so long as $r = 1$, without the assumption of compactness. In order to obtain the same type of conclusion in the case $r > 1$, it will probably be necessary to modify the formulation (cf. footnote 3). It should also be mentioned that Ihara [2], and Pjateckii-Shapiro and Shafarevic [4] investigated, in different context from each other and from ours, the group of automorphisms of a field of automorphic functions with respect to infinitely many commensurable discontinuous groups.

PROOF. The first assertion follows from

$$(2.8.1) \qquad F^{\times} \cdot (S \cap \mathfrak{G}) = \{x \in \mathfrak{G}_+ \mid \tau(x) \text{ is the identity map on } \mathfrak{L}_S\}\,.$$

To show this, suppose that $\tau(x)$ is the identity map on $\mathfrak{L}_S$. Then $\sigma(x)$ is the identity map on k_S. By 3.5 below, $x = s\alpha$ for some $s \in S \cap \mathfrak{G}$ and $\alpha \in G_{Q+}$. Then for every $f \in L_S$, we have $f \circ \varphi_S = f \circ J_{ST}(s\alpha) \circ \varphi_T$, so that $\varphi_S(z) = \varphi_S(\alpha(z))$. Therefore $\alpha \in F^{\times} \cdot \Gamma_S$, hence $x \in F^{\times} \cdot (S \cap \mathfrak{G})$. Conversely it is straightforward to verify that $\tau(x)$ is the identity on $\mathfrak{L}_S$ if $x \in F^{\times} \cdot (S \cap \mathfrak{G})$. Now suppose that $r = 1$. Let S_0 denote the projection of S to G_0 (cf. 2.3). Then the sets S_0, for all $S \in \mathfrak{Z}$, form a basis of neighborhoods of the identity of G_0. Therefore $\mathcal{C}$ is the closure of $F^{\times} \cdot G_{\infty+}$. Moreover, (2.8.1) shows that τ is continuous, and $\tau(F^{\times} \cdot S)$ is open in $\tau(\mathfrak{G}_+)$. Since the sets $F^{\times} \cdot S/\mathcal{C}$, for all $S \in \mathfrak{Z}$, form a basis of neighborhoods of the identity of $G_{A+}/\mathcal{C}$, τ is a homeomorphism of $G_{A+}/\mathcal{C}$ onto $\tau(G_{A+})$. Since $G_{A+}/\mathcal{C}$ is locally compact, $\tau(G_{A+})$ is a locally compact subgroup of Aut $(\mathfrak{L}/F')$, hence closed in Aut $(\mathfrak{L}/F')$. (Note that any locally compact subgroup of a Hausdorff topological group is closed.) Therefore, to complete the proof, it suffices to prove that $\tau(G_{A+})$ is dense in Aut $(\mathfrak{L}/F')$.

Let $\zeta \in \text{Aut}\,(\mathfrak{L}/F')$. By 3.3 below, there exists an element y of G_{A+} such that $\sigma(y) = \zeta$ on $\mathfrak{k}$. Put $\pi = \zeta \cdot \tau(y)^{-1}$. Then π is the identity map on $\mathfrak{k}$. Let $\mathfrak{L}_S^*$ (resp. $\mathfrak{L}^*$) be the composite of $\mathfrak{L}_S$ (resp. $\mathfrak{L}$) and C. Since $\mathfrak{L}$ and C are linearly disjoint over $\mathfrak{k}$, we can extend π to an automorphism of $\mathfrak{L}^*$ over C, which we denote again by π. Take an arbitrary $\mathfrak{r}_F$-lattice $\mathfrak{m}$ in B^n. We can find a positive integer b such that

(2.8.2) *If $R = \{x \in G_{A+} \mid x \equiv 1 \bmod_0 (\mathfrak{m}, b)\}$, then Γ_R modulo its center has no element of finite order other than the identity element.*

(2.8.3) *If $\varepsilon \in \mathfrak{r}_F^\times$ and $\varepsilon \equiv 1 \bmod b\mathfrak{r}_F$, then $\varepsilon = \delta^2$ for some $\delta \in \mathfrak{r}_F^\times$* (see Chevalley [1]).

We can find members S and T of $\mathfrak{Z}$ so that $T \subset S \subset R$, $(\mathfrak{L}_R)^{\pi^{-1}} \subset \mathfrak{L}_S$, $\mathfrak{L}_S^\pi \subset \mathfrak{L}_T$, and T is a normal subgroup of R. Then $\mathfrak{L}_R^* \subset \mathfrak{L}_S^{*\pi} \subset \mathfrak{L}_T^*$. Now Gal $(\mathfrak{L}_T^*/\mathfrak{L}_R^*)$ is isomorphic to $\Gamma_R/\Gamma_T \cdot (\Gamma_R \cap F^\times)$. Let Δ be the subgroup of Γ_R such that $\Delta/\Gamma_T \cdot (\Gamma_R \cap F^\times)$ corresponds to $\mathfrak{L}_S^{*\pi}$. Put $U = \Delta \cdot T$. Then we see that $U \in \mathfrak{Z}$, $\Gamma_U = \Delta$, and π gives an isomorphism of $\mathfrak{L}_S^*$ onto $\mathfrak{L}_U^*$ over C. Therefore we have a birational map ξ of V_U onto V_S such that $(f \circ \varphi_S)^\pi = f \circ \xi \circ \varphi_U$ for every $f \in L_S$. We are going to show that there exists an element α of G_{Q+} such that $\xi \circ \varphi_U = \varphi_S \circ \alpha$.

(I) If $\mathfrak{H}_n/\Gamma_S$ is compact, both V_S and V_U are complete non-singular minimal models, hence ξ is everywhere biregular. Therefore we can find an element β of $\mathrm{Sp}(n, R)$ such that $\xi \circ \varphi_U = \varphi_S \circ \beta$, and

$$\beta^{-1} \cdot p(\Gamma_S)\beta \subset R^\times \cdot p(\Gamma_U) \, ,$$

where p denotes the projection of $M_n(B)_A$ to the first factor $M_{2n}(R)$ of the archimedean part $M_n(B) \otimes_Q R$ of $M_n(B)_A$. Let $\gamma \in \Gamma_S$. Then there exists an element $\delta \in \Gamma_U$ and an element c of R such that $\beta^{-1} \cdot p(\gamma)\beta = c \cdot p(\delta)$. Then $\nu(\gamma)^{\tau_1} = c^2\nu(\delta)^{\tau_1}$. By (2.8.3), we have $c = e^{\tau_1}$ with an element e of $\mathfrak{r}_F$. It follows that $\beta^{-1} \cdot p(\Gamma_S)\beta \subset p(M_n(B))$. Since Γ_S spans $M_n(B)$ over F, we have $\beta^{-1} \cdot p(M_n(B))\beta = p(M_n(B))$. Then we obtain an automorphism μ of $M_n(B)$ by $\beta^{-1} \cdot p(w)\beta = p(w^\mu)$ for $w \in M_n(B)$. Since μ is the identity map on F, there exists an element α of $GL_n(B)$ such that $w^\mu = \alpha^{-1}w\alpha$, hence $\beta^{-1} \cdot p(w)\beta = p(\alpha)^{-1}p(w)p(\alpha)$ for all $w \in M_n(B)$. Therefore $\beta = d \cdot p(\alpha)$ with $d \in R^\times$, hence $\alpha \in G_{Q+}$. We have then $\xi \circ \varphi_U = \varphi_S \circ \alpha$.

(II) If $G_Q = GL_2(Q)$, let $\mathfrak{H}'$ be the union of $\mathfrak{H}_1$ and the cusps of $SL_2(Z)$, i.e., $\mathfrak{H}' = \mathfrak{H}_1 \cup Q \cup \{\infty\}$. Let V_S' be the complete non-singular curve isomorphic to $\mathfrak{H}'/\Gamma_S$. We can consider V_S a Zariski open subset of V_S', and extend φ_S to a map φ_S' of $\mathfrak{H}'$ onto V_S'. (V_S' is actually defined over k_S. But, in the present discussion, it is enough to consider V_S and V_S' over C.) Now ξ can be extended to a biregular isomorphism ξ' of V_U' onto V_S'. Let $t \in \mathfrak{H}' - \mathfrak{H}_1$. Suppose that $\xi'(\varphi_U'(t)) \in V_S$. By considering a principal congruence subgroup of a suitably high level, we can find a member P of $\mathfrak{Z}$ so that $P \subset U$ and V'_P is a covering of V'_U ramified at $\varphi'_U(t)$. Therefore, if v is a discrete valuation of $\mathfrak{L}^*_U$ corresponding to the point $\varphi'_U(t)$, then v has a ramified extension in $\mathfrak{L}^*$. Define a valuation u of $\mathfrak{L}^*_S$ by $u(a) = v(a^\xi)$ for $a \in \mathfrak{L}^*_S$. Then u corresponds to the point $\xi'(\varphi'_U(t))$ of V_S. Therefore every extension of u to $\mathfrak{L}^*$ is unramified, which is a contradiction. Thus we must have $\xi'(\varphi'_U(t)) \notin V_S$, so that ξ maps V_U onto V_S. Hence $\xi \circ \varphi_U = \varphi_S \circ \beta$ and $\beta^{-1}(\{\pm 1\} \cdot \Gamma_S)\beta = \{\pm 1\} \cdot \Gamma_U$ with an element β of $SL_2(\mathbf{R})$. By the same argument as in (I), we obtain an element α of G_{Q+} such that $\xi \circ \varphi_U = \varphi_S \circ \alpha$.

Now, in both cases (I, II), we have $(f \circ \varphi_S)^\xi = f \circ \varphi_S \circ \alpha$ for every $f \in L_S$, hence $\pi = \tau(\alpha)$ on $\mathfrak{L}_S$, and $\zeta = \tau(\alpha y)$ on $\mathfrak{L}_R$. Recall that R is defined by (2.8.2) with a positive integer b, and therefore $\mathfrak{L}$ is the union of the $\mathfrak{L}_R$ for all positive integers b. This means that $\tau(G_{A+})$ is dense in $\mathrm{Aut}(\mathfrak{L}/F')$, and therefore our proof is completed.

2.9. As is mentioned in the Introduction, it is conjecturable that one can obtain a system similar to the system of 2.5 over finite fields. For simplicity, let us consider here only the case $n = r = 1$, and τ_1 is the identity map of F. Then G_Q may be identified with $B^\times$. Let $\mathfrak{o}$ be a fixed maximal order in B, and $\mathfrak{p}$ a prime ideal in F where B is unramified. For every integral ideal $\mathfrak{a}$ in F, let $S(\mathfrak{o}, \mathfrak{a})$ denote the set of all x in G_{A+} such that $\mathfrak{o}x = \mathfrak{o}$ and $\mathfrak{o}_q(x_q - 1) \subset \mathfrak{a}\mathfrak{o}_q$ for all prime factors q of $\mathfrak{a}$, where x_q means the q-component of x in an obvious sense. Let $\mathfrak{P}$ be a prime divisor of F_{ab} such that $F \cap \mathfrak{P} = \mathfrak{p}$. One can conjecture that, if $S = S(\mathfrak{o}, \mathfrak{a})$ and $\mathfrak{a}$ is prime to $\mathfrak{p}$, the curve V_S can be chosen so as to have a non-singular reduction modulo $\mathfrak{P}$. Let $\mathfrak{A}_S$ be the function-field of the reduced curve over the residue field of k_S modulo $k_S \cap \mathfrak{P}$. It is meaningful to consider the composite $\mathfrak{A}$ of the fields $\mathfrak{A}_S$ for all such S (with a fixed $\mathfrak{o}$). Let $F_\mathfrak{p}$ denote the completion of F at $\mathfrak{p}$, $\mathfrak{r}_\mathfrak{p}$ the maximal compact subring of $F_\mathfrak{p}$, $\kappa_\mathfrak{p}$ the residue field of F modulo $\mathfrak{p}$, and $\mathfrak{o}_\mathfrak{p}$ the $\mathfrak{p}$-closure of $\mathfrak{o}$ in $B \otimes_F F_\mathfrak{p}$. Further let $\mathrm{Aut}(\mathfrak{A}/\kappa_\mathfrak{p})$ denote the group of all automorphisms of $\mathfrak{A}$ over $\kappa_\mathfrak{p}$, furnished with a topology like that of $\mathrm{Aut}(\mathfrak{L}/F')$ defined in 2.7. Then, as an analogy with 2.8, we can conjecture that $\mathrm{Aut}(\mathfrak{A}/\kappa_\mathfrak{p})$ is isomorphic to $G^{(\mathfrak{p})}/C_\mathfrak{p}$, where

$$G^{(\mathfrak{p})} = \{x \in G_{A+} \mid x_\mathfrak{p} \in F_\mathfrak{p}^\times \cdot \mathfrak{o}_\mathfrak{p}^\times, \nu(x) \in M_\mathfrak{p}\},$$
$$M_\mathfrak{p} = \text{the closure of } F_\mathfrak{p}^\times \cdot F^\times \cdot F_{\infty+}^\times \text{ in } G_A,$$
$$C_\mathfrak{p} = \text{the closure of } F^\times \cdot \mathfrak{o}_\mathfrak{p}^\times \cdot G_{\infty+} \text{ in } G_A.$$

Further let $N_\mathfrak{p}$ denote the closure of $\mathfrak{r}_\mathfrak{p}^\times \cdot F^\times \cdot F_{\infty+}^\times$ in G_A, and $\bar{\kappa}_\mathfrak{p}$ the algebraic closure of $\kappa_\mathfrak{p}$. Then $M_\mathfrak{p}/N_\mathfrak{p}$ is canonically isomorphic to $\mathrm{Gal}(\bar{\kappa}_\mathfrak{p}/\kappa_\mathfrak{p})$. An element of $G^{(\mathfrak{p})}$ should act on $\bar{\kappa}_\mathfrak{p}$ through the map ν composed with this isomorphism. Moreover, one would be able to obtain a system of curves over finite fields, similar to the system of 2.5, with $G^{(\mathfrak{p})}$ in place of $\mathfrak{G}_+$. This subject is closely connected with the work of Ihara [2]. A detailed discussion about the relationship, together with some other ideas, will be made in a subsequent paper.

2.10. To study the nature of $\mathfrak{X}_S$, let us now define a special type of S. Let $\mathfrak{m}$ be a $\mathbf{Z}$-lattice in X, and b a positive integer. Put

$$(2.10.1) \qquad \tilde{S}_0(\mathfrak{m}, b) = \{(x, c) \in \tilde{D}_0 \mid x \equiv 1 \bmod_0 (\mathfrak{m}, b),\ c \equiv 1 \bmod_0 (b)\}$$

(For the notation, see 0.3, 0.5). Let $S_0(\mathfrak{m}, b)$ be the projection of $\tilde{S}_0(\mathfrak{m}, b)$ to G_0, and

$$(2.10.2) \qquad S(\mathfrak{m}, b) = S_0(\mathfrak{m}, b) \cdot G_{\infty+}.$$

Then $S(\mathfrak{m}, b) \in \mathfrak{Z}$. By our definition of $\mathfrak{Z}$, every member of $\mathfrak{Z}$ contains $S(\mathfrak{m}, b)$ for some b. (Note that we can fix $\mathfrak{m}$.) In [12, 3.7, 3.8], we have shown the existence of a $\mathbf{Z}$-lattice $\mathfrak{m}$ with the following property:

$(2.10.3)$ *For every positive integer b, and for every $d \in F_A^\times$ such that $d \equiv 1 \bmod_0 (b)$, there exists an element x of G_0 such that $\nu(x)/d \in F_\infty^\times$ and $x \equiv 1 \bmod_0 (\mathfrak{m}, b)$.*

Now consider an open subgroup

$$\mathfrak{u}(b) = \{c \in F_A'^\times \mid c \equiv 1 \bmod_0 (b),\ c \gg 0\}$$

of $F_A'^\times$. If $c \in \mathfrak{u}(b)$, we have $\lambda(c) \equiv 1 \bmod_0 (b)$ by (1.7.2), and $\lambda(c) \gg 0$, hence $\lambda(c) = \nu(x)$ with an element x of G_{A+} such that $x \equiv 1 \bmod_0 (\mathfrak{m}, b)$. Then $x \in S(\mathfrak{m}, b)$. Therefore $\mathfrak{u}(b) \subset \mathfrak{X}(S(\mathfrak{m}, b))$, hence $\mathfrak{X}(S(\mathfrak{m}, b))$ corresponds to a class field over F' whose conductor divides b. This shows that $\mathfrak{X}_S$, for every $S \in \mathfrak{Z}$, contains an open subgroup of $F_A'^\times$, and therefore corresponds to a class field over F' of finite degree.

Suppose especially that $r = 1$ and τ_1 is the identity map of F. By [12, (3.5.1)], we have $\nu(S(\mathfrak{m}, b)) \subset \mathfrak{u}(b)$ so that $\nu(S(\mathfrak{m}, b)) = \mathfrak{u}(b)$. It follows that k_S, for $S = S(\mathfrak{m}, b)$, is the class field over F corresponding to $F^\times \cdot \mathfrak{u}(b)$, i.e., the maximal class field over F whose conductor divides b.

2.11. In our canonical system, we have considered only G_{A+}, but not the whole G_A. It is more natural and actually practicable to formulate a theorem involving the whole G_A. Let us now briefly indicate how this can be done.

We consider the lower half space $\mathfrak{H}_n^- = \{-z \mid z \in \mathfrak{H}_n\}$, and put $\mathfrak{F}_n = \mathfrak{H}_n \cup \mathfrak{H}_n^-$. Then G_∞ acts on $\mathfrak{F}_n^r$ in an obvious way. Let Σ be the group of order 2, consisting of ± 1. Define a map

$$\mathrm{sgn}\colon G_A \longrightarrow \Sigma^r = \Sigma \times \cdots \times \Sigma \qquad (r \text{ copies})$$

by taking $\mathrm{sgn}(x)$ to be the signatures of the archimedean components of $\nu(x)$. The space $\mathfrak{F}_n^r$ has 2^r connected components, and each element x of G_∞ permutes these components according to sgn (x). Observe that the restriction of the map sgn to G_Q is surjective.

Let $\mathfrak{Z}'$ be the family of all subgroups S of G_A satisfying (2.3.2) and

(2.11.1) $S = S_0 G'$ *with a compact subgroup* S_0 *of* G_0 *and a subgroup* G' *of* G_∞ *containing* $G_{\infty+}$.

(A more appropriate notation would have been $\mathfrak{Z}$ and $\mathfrak{Z}_+$ instead of $\mathfrak{Z}'$ and $\mathfrak{Z}$.)

Define Γ_S and k_S in the same way as before, for every $S \in \mathfrak{Z}'$. Now $\mathfrak{F}_n^r / \Gamma_S$ is a disjoint union of 2^{r-e} irreducible algebraic varieties, if sgn (Γ_S) is of order 2^e. We say that (W, ψ) is a *model* of $\mathfrak{F}_n^r / \Gamma_S$, if W is a disjoint union of 2^{r-e} algebraic varieties which are Zariski open subsets of projective subvarieties of the *same* projective space, and ψ is a Γ_S-invariant map of $\mathfrak{F}_n^r$ onto W, which gives a biregular isomorphism of $\mathfrak{F}_n^r / \Gamma_S$ onto W. We say also that W is rational over a field k, if W, as a whole, is invariant under all automorphisms of C over k. Now we have

2.12. Theorem. *There exists a system*

$$\{ W_S, \psi_S, K_{TS}(x), (S, T \in \mathfrak{Z}'; x \in \mathfrak{G}) \}$$

formed by the objects satifying the conditions exactly like (2.5.1–4) *under the following modification:*

(i) *The symbols* $V_S, \varphi_S, J_{TS}(x), \mathfrak{Z}, \mathfrak{G}_+, G_{Q+}, \mathfrak{H}_n^r$ *should be replaced by* $W_S, \psi_S, K_{TS}(x), \mathfrak{Z}', \mathfrak{G}, G_Q, \mathfrak{F}_n^r$.

(ii) *The fixed point* z *of* (2.5.4) *can be taken not only on* $\mathfrak{H}_n^r$, *but also on any component of* $\mathfrak{F}_n^r$.

Such a system can be obtained from the canonical system of 2.5 by the standard way of taking the quotient, which is similar to the argument of 3.10 below. We do not go into details of the proof, since it is almost straightforward and reveals no new aspect of the theory.

2.13. Now we can propose the generalization of our theorem in the following form.

Problem. Find a family $\mathfrak{F}$, *as large as possible, of reductive groups defined over* Q *so that for each group in* $\mathfrak{F}$, *we have a system exactly like the system obtained in* 2.5.

We call such a system a *canonical system* for the group in question. We have to assume that the semi-simple part G^u of each G in $\mathfrak{F}$ defines a bounded symmetric domain. It is expected that for every (non-exceptional) semi-simple group G^u over Q of this type, $\mathfrak{F}$ contains at least one G whose semi-simple part is isogenous to G^u. Moreover, the canonical systems for two distinct groups are not only defined and characterized individually, but also tied up with each other in many cases. Namely, if G' and G are in $\mathcal{F}$, and f is a Q-rational homomorphism of G' into G, then it is often that f defines a holomorphic embedding ε of the corresponding symmetric domain $\mathcal{K}'$ into $\mathcal{K}$. Let $\{V_S, \varphi_S, k_S, J_{TS}(x)\}$ and $\{V'_L, \varphi'_L, k'_L, J'_{ML}(x)\}$ be canonical systems for G and G' respectively. Here S and T (resp. L and M) are subgroups of G_A (resp. G'_A) of a certain type. If $f(L) \subset S$, we have $f(\Gamma_L) \subset \Gamma_S$, so that we obtain a morphism E_{SL} of V'_L into V_S such that $E_{SL} \circ \varphi'_L = \varphi_S \circ \varepsilon$. In some cases, one can prove

(2.13.1) $k_S \subset k'_L$, and E_{SL} is rational over k'_L.

(2.13.2) *If further* $f(M) \subset T$, $x \in G'_A$, $xLx^{-1} \subset M$, $f(x)Sf(x)^{-1} \subset T$, *then* $E_{TM}^{\sigma'(x)} \circ J'_{ML}(x) = J_{TS}(f(x)) \circ E_{SL}$, *provided that* $J'_{ML}(x)$ *and* $J_{TS}(f(x))$ *are defined, where* σ' *is the symbol defined for* G' *corresponding to* σ.

We may consider (2.5.4) an example of such a relation. Namely, we put $G' = \{Y, \delta\}$, and take $\mathcal{K}'$ to be the space consisting of one point, and $\varepsilon(\mathcal{K}') = z$. In § 8, we shall discuss an interesting example in which both G and G' are groups of the type of our theorem.

If G_1 and G_2 are groups belonging to $\mathcal{F}$, one can expect that $G_1 \times G_2 \in \mathcal{F}$. For example, if $G_1 = G_2 = GL_2(Q)$, we can define the canonical system for $G_1 \times G_2$ as follows. Put $G^* = G_1 \times G_2$,

$$D^* = \{(a, b) \in G_1 \times G_2 \mid \det(a) = \det(b)\},$$
$$\mathcal{G}^* = D_A^* \cdot G_Q^* \cdot D_{R+}^*, \qquad \mathcal{G}_+^* = G^* \cap G_{A+}^*.$$

For $u = rst \in \mathcal{G}^*$ with $r = (a, b) \in D_A^*$, $s \in G_Q^*$, $t \in G_{R+}^*$, put $\sigma(u) = [\det(a)^{-1}, Q]$. Then it is easy to state and prove a theorem about the canonical system for G^*, the details of which may be left to the reader.

2.14. The family $\mathcal{F}$ may contain many groups whose semi-simple parts are isogenous. For example, both GL_2 and the projective linear group PGL_2 can belong to $\mathcal{F}$, and have distinct canonical systems. More generally, let us now discuss an algebraic group G' over Q defined by $G' = G/F^\times$, with the group G of our type. We can consider G'_Q as the subgroup of $GL(M_n(B), F)$(see 0.1) consisting of all inner automorphisms of $M_n(B)$ obtained from the elements of G_Q. Put $F_A^{\times 2} = \{a^2 \mid a \in F_A^\times\}$. Combining ν and λ with the natural map of $F_A^\times$ to $F_A^\times/F_A^{\times 2}$, we obtain two maps

$$\nu': G'_A \longrightarrow F_A^\times/F_A^{\times 2},$$
$$\lambda': F_A'^\times \longrightarrow F_A^\times/F_A^{\times 2}.$$

$G'_{\infty+}$ acts on $\mathfrak{H}_n^r$ in an obvious way. Put

$$D' = \{x \in G'_A \mid \nu'(x) \in \lambda'(F'^{\times}_A)\}$$
$$D'_+ = D' \cap G'_{A+} , \qquad \mathcal{G}' = D'G'_Q G'_{\infty+} , \qquad \mathcal{G}'_+ = D'_+ G'_{Q+} G'_{\infty+} = \mathcal{G}' \cap G'_{A+} .$$

Obviously $\mathcal{G}'_+$ (resp. D'_+) is isomorphic to $F^{\times}_A \mathcal{G}_+/F^{\times}_A$ (resp. $F^{\times}_A D_+/F^{\times}_A$). Denote by $\mathcal{Z}'$ the set of all subgroups S of G'_{A+} satisfying the following two conditions:

(2.14.1) *$S = S_0 G'_{\infty+}$ with a compact subgroup S_0 of G'_0.*

(2.14.2) *S contains the projection to G'_A of a member of $\mathcal{Z}$.*

For each $S \in \mathcal{Z}'$, put $\Gamma'_S = S \cap G'_Q$, and

$$\mathcal{X}'_S = \{c \in F'^{\times}_A \mid \lambda'(c) \in (F^{\times} F^{\times}_A/F^{\times 2}_A) \cdot \nu'(S)\} .$$

Then $\mathcal{X}'_S$ corresponds to a class field k'_S over F', which is obviously a composite of a finite number of quadratic extensions of F'. Let $\mathfrak{k}'$ denote the composite of the k'_S for all $S \in \mathcal{Z}'$. Define a homomorphism

$$\sigma': G' \longrightarrow \mathrm{Gal}\,(\mathfrak{k}'/F')$$

by $\sigma'(x) = [c^{-1}, F']$ on $\mathfrak{k}'$ with an element c of $F'^{\times}_A$ such that $\nu'(x)/\lambda'(c) \in (F^{\times} F^{\times}_{\infty+} F^{\times}_A)/F^{\times 2}_A$. If (Y, P, δ, f), P', η are given as in 2.4, we define $\eta': P'^{\times}_A \to D'_+$ by combining η with the projection of D_+ to D'_+. Now we have

2.15. Theorem. *The notation being as above, there exists a system*

$$\{V'_S, \varphi'_S, J'_{TS}(x), (S, T \in \mathcal{Z}'; x \in \mathcal{G}'_+)\}$$

satisfying the conditions exactly like (2.5.1–4) under the replacement of $\mathcal{Z}$, $\mathcal{G}_+$, G_{Q+}, V_S, φ_S, $J_{TS}(x)$, Γ_S, k_S, $\sigma(x)$, η *by* $\mathcal{Z}'$, $\mathcal{G}'_+$, G'_{Q+}, V'_S, φ'_S, $J'_{TS}(x)$, Γ'_S, k'_S, $\sigma'(x)$, η'.

This may be regarded as an adelization of [12,10.5] and [11,3.13]. We can derive it from 2.5 by a standard argument which is similar to and simpler than that of § 3 below.

There is a notable difference between the family $\{\Gamma_S\}$ of discontinuous groups obtained from G and the family $\{\Gamma'_S\}$ obtained from G', even if we consider Γ_S modulo its center. For example, let us take the simplest case where $G = GL_2$. For a positive integer N, put

$$\mathfrak{o} = \left\{ \begin{bmatrix} a & b \\ c & d \end{bmatrix} \in M_2(\mathbf{Z}) \,\middle|\, c \equiv 0 \bmod (N) \right\} ,$$
$$S = \{x \in G'_{A+} \mid x \mathfrak{o} x^{-1} = \mathfrak{o}\} ,$$
$$\Gamma_0(N) = \mathfrak{o} \cap SL_2(\mathbf{Z}) .$$

Then we see that Γ'_S is generated by the projection of $\Gamma_0(N)$ to G'_Q and an element represented by $\begin{bmatrix} 0 & -1 \\ N & 0 \end{bmatrix}$. But this type of discontinuous group can not be obtained as Γ_S with $S \in \mathfrak{Z}$. On the other hand, the principal congruence subgroup of $SL_2(\mathbf{Z})$ of level N can not be obtained as Γ'_S, if N has two distinct odd prime factors.

2.16. The quotient $\mathfrak{H}^r_n/\Gamma_S$ is not compact if and only if $M_n(B)$ is of one of the following two types:

(i) $r = g$ and $n > 1$;

(ii) $B = M_2(F)$.

In these cases, it is natural to ask the existence of a canonical system with models (V^*_S, φ^*_S) of the compactifications of $\mathfrak{H}^r_n/\Gamma_S$ instead of the models (V_S, φ_S) of $\mathfrak{H}^r_n/\Gamma_S$. If such a system exists, it is also natural and interesting to consider the behavior of the components of $V^*_S - V_S$ under the automorphisms of $\bar{Q}$. If $G = GL_2$, it is easy to settle these questions. In the general case, one may, for example, pose a question: *are all these components of $V^*_S - V_S$ rational over $\mathbf{Q}_{ab}$?*

2.17. We shall now show the usefulness of the symbol $J_{rS}(x)$ in the theory of modular correspondences on V_S. We first consider correspondences on arbitrary algebraic varieties, following the terminology of Weil [15]. Let U and V be algebraic varieties of the same dimension m, and X an algebraic m-cycle on $U \times V$. We denote by ^tX the n-cycle on $V \times U$ which is the transform of X by the map $(u, v) \mapsto (v, u)$ of $U \times V$ to $V \times U$. We call X a *proper correspondence*, or a *proper m-cycle*, on $U \times V$, if the following two conditions are satisfied:

(2.17.1) *X is everywhere complete over U and over V.*

(2.17.2) *For every $a \in U$ and $b \in V$, neither $|X| \cap (a \times V)$ nor $|X| \cap (U \times a)$ has components of dimension >0.*

The latter condition is equivalent to

(2.17.3) *If k is a field of rationality for U, V, and X, then, for every (a, b) in $|X|$, $k(a, b)$ is algebraic over $k(a)$ and over $k(b)$.*

For a 0-cycle c on U, we define a 0-cycle $X[c]$ on V by

$$X[c] = \mathrm{pr}_V[X \cdot (c \times V)] .$$

Let W be another variety of dimension m, X a proper m-cycle on $U \times V$, and Y a proper m-cycle on $V \times W$. We can easily verify that $X \times W$ and $U \times Y$ intersect properly on $U \times V \times W$, and if we put

$$Z = \mathrm{pr}_{U \times W}[(X \times W) \cdot (U \times Y)] ,$$

then Z is a proper m-cycle on $U \times W$. We write $Z = Y \circ X$. This law of composition is associative.

Now let us consider our V_S and $J_{TS}(x)$. We mean by the same symbol $J_{TS}(x)$ the subvariety of $V_S \times V_T^{\sigma(x)}$, which is the graph of $J_{TS}(x)$. In view of (2.6.1), $J_{TS}(x)$ is a proper correspondence in the above sense.

Let $S \in \mathfrak{Z}$, $T \in \mathfrak{Z}$, and $x \in \mathcal{G}_+$. Put $U = S \cap x^{-1}Tx$. Then we can define a proper correspondence $X_{TS}(x)$ on $V_S \times V_T^{\sigma(x)}$ by

(2.17.4) $\quad X_{TS}(x) = J_{TU}(x) \circ {}^t J_{SU}(1)$.

2.18. THEOREM. *The correspondences $X_{TS}(x)$ have the following properties. (We put $S_1 = S \cap \mathcal{G}_+$ for every $S \in \mathfrak{Z}$.)*

(2.18.1) $\quad X_{TS}(x)$ *is absolutely irreducible, and rational over k_U, where* $U = S \cap x^{-1}Tx$.

(2.18.2) $\quad X_{TS}(x) = J_{TS}(x)$ *if* $S \subset x^{-1}Tx$.

(2.18.3) $\quad X_{TS}(x) = {}^t J_{ST}(x^{-1})^{\sigma(x)}$ *if* $x^{-1}Tx \subset S$.

(2.18.4) $\quad X_{TS}(x)$ *depends only on* $T_1 x \Gamma_S$.

(2.18.5) $\quad$ *If* $k_S = k_U$ *for* $U = S \cap x^{-1}Tx$, $X_{TS}(x)$ *depends only on* $T_1 x S_1$.

(2.18.6) $\quad X_{TS}(\alpha) = \{\varphi_S(z) \times \varphi_T(\alpha(z)) \mid z \in \mathfrak{H}_n^r\}$ *if* $\alpha \in G_{\mathbf{Q}+}$.

(2.18.7) $\quad X_{TR}(xy) = X_{TS}(x)^{\sigma(y)} \circ J_{SR}(y)$ *if* $y \in \mathcal{G}_+$ *and* $R = y^{-1}Sy$.

(2.18.8) $\quad X_{RS}(yx) = J_{RT}(y)^{\sigma(x)} \circ X_{TS}(x)$ *if* $y \in \mathcal{G}_+$ *and* $R = yTy^{-1}$.

(2.18.9) $\quad {}^t X_{TS}(x) = X_{ST}(x^{-1})^{\sigma(x)}$.

PROOF. The assertions (2.18.1, 2, 6) are obvious from our definition. The relation (2.18.8) can be verified in a straightforward way. To show (2.18.9), put $U = S \cap x^{-1}Tx$ and $W = xUx^{-1} = T \cap xSx^{-1}$. Then

$$
\begin{aligned}
X_{ST}(x^{-1})^{\sigma(x)} &= J_{SW}(x^{-1})^{\sigma(x)} \circ {}^t J_{TW}(1)^{\sigma(x)} \\
&= J_{SU}(1) \circ J_{UW}(x^{-1})^{\sigma(x)} \circ {}^t J_{TW}(1)^{\sigma(x)} \\
&= J_{SU}(1) \circ {}^t J_{WU}(x) \circ {}^t J_{TW}(1)^{\sigma(x)} \\
&= J_{SU}(1) \circ {}^t J_{TU}(x) = {}^t X_{TS}(x) \ .
\end{aligned}
$$

We obtain (2.18.7) from (2.18.8, 9). Then (2.18.4) and (2.18.5) follow from (2.18.7, 8); (2.18.3) from (2.18.2, 9).

In connection with (2.18.4, 5), we have

2.19. PROPOSITION. *Let $S \in \mathfrak{Z}$, $T \in \mathfrak{Z}$, $x \in \mathcal{G}_+$, $U = S \cap x^{-1}Tx$. Then the following three conditions are equivalent to each other.*

(2.19.1) $\quad k_U = k_S$.

(2.19.2) $\quad S_1 = U_1 \Gamma_S$.

(2.19.3) $\quad T_1 x S_1 = T_1 x \Gamma_S$.

PROOF. The equivalence of the last two conditions are obvious. By 3.5 below, we have $k_U = k_S$ if and only if $S_1 G_{\mathbf{Q}+} = U_1 G_{\mathbf{Q}+}$, hence the equivalence of the first two conditions.

2.20. To develop further our theory of correspondences, let us restrict ourselves, for simplicity, to the case $r = 1$. Observe, then, that any two members of $\mathfrak{Z}$ are commensurable. For S and T in $\mathfrak{Z}$, let $\mathfrak{M}_{TS}$ denote the free Z-module consisting of all formal finite sums $\sum c_x \cdot (TxS)$ of double cosets TxS with $x \in \mathfrak{S}_+$ and $c_x \in Z$. We can define a bilinear map of $\mathfrak{M}_{TS} \times \mathfrak{M}_{SR}$ into $\mathfrak{M}_{TR}$ in the following way. Let

$$TxS = \bigcup_{i=1}^{d} Tx_i \,, \qquad SyR = \bigcup_{j=1}^{e} Sy_j \,,$$

be disjoint unions. For every $w \in TxSyR$, the number of pairs (i, j), such that $Tx_i y_j = Tw$, depends only on three double cosets TxS, SyR, TwR, and is independent of the choice of x_i, y_j and w. Denote this number by $\mu(TxS, SyR; TwR)$, and put

$$(TxS) \cdot (SyR) = \sum \mu(TxS, SyR; TwR) \cdot TwR \,,$$

where the sum is taken over all TwR contained in $TxSyR$. We can extend this law of composition to a bilinear map of $\mathfrak{M}_{TS} \times \mathfrak{M}_{SR}$ into $\mathfrak{M}_{TR}$.

2.21. THEOREM. *Let* $R, S, T \in \mathfrak{Z}$, *and* $x, y \in \mathfrak{S}_+$. *Suppose that* $r = 1$, $R \cap F^\times = S \cap F^\times = T \cap F^\times$, $TxS = Tx\Gamma_S$, $SyR = Sy\Gamma_R$, *and* $TwR = Tw\Gamma_R$ *for every* $w \in TxSyR$. *Let* $(TxS) \cdot (SyR) = \sum c_w \cdot (TwR)$ *with* $c_w \in Z$. *Then*

$$X_{TS}(x)^{\sigma(y)} \circ X_{SR}(y) = \sum c_w \cdot X_{TR}(w) \,.$$

PROOF. Put $Q = y^{-1}Sy, P = y^{-1}x^{-1}Txy, M = x^{-1}Tx, R \cap Q = W, Q \cap P = Z$. Then $P \cap F^\times = Q \cap F^\times = R \cap F^\times$, and $k_R = k_W, k_Q = k_Z$, hence $QR = Q\Gamma_R$, $PQ = P\Gamma_Q$. Therefore we have $(PQ) \cdot (QR) = \sum c_\gamma \cdot (P\gamma R)$ with elements γ of Γ_Q. In a straightforward way we can show

$$(TxS) \cdot (SyR) = \sum c_\gamma \cdot (Txy\gamma R) \,.$$

By (2.18.8), we have $X_{TS}(x) = J_{TM}(x) \circ X_{MS}(1)$ and $X_{SR}(y) = J_{SQ}(y) \circ X_{QR}(1)$, hence

$$\begin{aligned}
X_{TS}(x)^{\sigma(y)} \circ X_{SR}(y) &= J_{TM}(x)^{\sigma(y)} \circ X_{MQ}(y) \circ X_{QR}(1) \\
&= J_{TM}(x)^{\sigma(y)} \circ J_{MP}(y) \circ X_{PQ}(1) \circ X_{QR}(1) \\
&= J_{TP}(xy) \circ X_{PQ}(1) \circ X_{QR}(1) \,.
\end{aligned}$$

Now let $\Gamma_R = \bigcup_j \Gamma_W \beta_j$ and $\Gamma_Q = \bigcup_i \Gamma_Z \alpha_i$ be disjoint unions. Then $QR = \bigcup_j Q\beta_j$ and $PQ = \bigcup_i P\alpha_i$ are disjoint unions. Let $z \in \mathfrak{H}_n$. Since $\Gamma_R \cap F^\times = \Gamma_W \cap F^\times$, we have ${}^t J_{RW}(1)[\varphi_R(z)] = \sum_j \varphi_W(\beta_j(z))$, so that $X_{QR}(1)[\varphi_R(z)] = \sum_j \varphi_Q(\beta_j(z))$. For the same reason, we have

$$X_{PQ}(1)[X_{QR}(1)[\varphi_R(z)]] = \sum_{i,j} \varphi_P(\alpha_i \beta_j(z)) \,.$$

From our definition of $(PQ) \cdot (QR)$ in 2.20, it follows that $X_{PQ}(1) \circ X_{QR}(1) = \sum c_\gamma \cdot X_{PR}(\gamma)$, hence

$$X_{TS}(x)^{\sigma(y)} \circ X_{SR}(y) = \sum c_\gamma \cdot J_{TP}(xy) \circ X_{PR}(\gamma) = \sum c_\gamma \cdot X_{TR}(xy\gamma) \ .$$

By our assumption and (2.18.5), $X_{TR}(w)$, for every $w \in TxSyR$, depends only on TwR. Therefore we obtain our theorem.

2.22. Let us now assume that $n = r = 1$, and τ_1 is the identity map of F. Then G_Q can be identified with $B^\times$. Fix a member S of $\mathfrak{Z}$. Then we can find a maximal order $\mathfrak{o}$ in B and an integral ideal $\mathfrak{a}$ in F so that

$$(2.22.1) \qquad\qquad S(\mathfrak{o}, \mathfrak{a}) \subset S \subset S(\mathfrak{o}, \mathfrak{r}_F) \ ,$$

where the notation $S(\mathfrak{o}, \mathfrak{a})$ is as in 2.9. Among the ideals $\mathfrak{a}$ satisfying (2.22.1), there is a maximal one, which we denote again by $\mathfrak{a}$. Let $\mathfrak{p}$ be a prime ideal in F, which is prime to $\mathfrak{a}$ and unramified in B. Let $F_\mathfrak{p}$, $\mathfrak{r}_\mathfrak{p}$, and $\mathfrak{o}_\mathfrak{p}$ be as in 2.9.

Let us fix a prime divisor $\mathfrak{P}$ of $\bar{Q}$ which divides $\mathfrak{p}$. If X is a variety, or a point, etc., rational over $\bar{Q}$, we denote by $\tilde{X}$ or $\mathfrak{P}(X)$ the reduction of X modulo $\mathfrak{P}$. Let π denote the $N(\mathfrak{p})^{\text{th}}$ power automorphism of the residue field of $\bar{Q}$ modulo $\mathfrak{P}$, and Φ_S the Frobenius correspondence on $\tilde{V}_S \times \tilde{V}_S^\pi$, i.e., the locus of $a \times a^{N(\mathfrak{p})}$ on $\tilde{V}_S \times \tilde{V}_S^\pi$. Now the congruence relation [11, 11.17] can be adelized as follows.

2.23. THEOREM. *The notation and assumption being as above, let $w_\mathfrak{p}$ be an element of $\mathfrak{o}_\mathfrak{p}$ such that $\nu(w_\mathfrak{p})$ is a prime element of $F_\mathfrak{p}$, and w an element of G_A of which the $\mathfrak{p}$-component is $w_\mathfrak{p}$ and other components are all equal to 1. Then, for almost all $\mathfrak{p}$, one has*

$$\tilde{X}_{SS}(w^{-1}) = \Phi_S + {}^t\Phi_S^\pi \circ \tilde{J}_{SS}(\nu(w)^{-1}) \ .$$

Here we have to assume that $\tilde{V}_S$ is a non-singular curve, and, for every $c \in F_A^\times$, $\tilde{J}_{SS}(c)$ is a biregular isomorphism of $\tilde{V}_S$ onto $\mathfrak{P}(V_S^{\sigma(c)})$. Since $F_A^\times/(F^\times \cap S)$ is finite, these are satisfied for almost all $\mathfrak{p}$.

Before proving our theorem, let us give a variation of it in the case where $\mathfrak{p}$ is a principal ideal generated by a totally positive element. Under such a condition, we can find an element α of $\mathfrak{o}$ so that $\nu(\alpha)$ is totally positive and generates $\mathfrak{p}$. Then we take α as $w_\mathfrak{p}$. Let y be an element of G_A, of which the $\mathfrak{p}$-component is 1 and all other components are α. Then $\alpha = wy$. Put $R = y^{-1}Sy$. By (2.18.7) and (2.18.9), we have

$$X_{SR}(\alpha) = X_{SS}(w)^{\sigma(y)} \circ J_{SR}(y) = {}^tX_{SS}(w^{-1}) \circ J_{SR}(y) \ ,$$

so that

$$\tilde{X}_{SR}(\alpha) = {}^t\Phi_S \circ \tilde{J}_{SR}(y) + {}^t\tilde{J}_{SS}(\nu(w)^{-1}) \circ \Phi_S^\pi \circ \tilde{J}_{SR}(y) \ .$$

Now ${}^t\tilde{J}_{SS}(\nu(w)^{-1}) \circ \Phi_S^\pi \circ \tilde{J}_{SR}(y) = {}^t\tilde{J}_{SS}(\nu(w)^{-1}) \circ \tilde{J}_{SR}(y)^\pi \circ \Phi_R = \tilde{J}_{SR}(\nu(w)y)^\pi \circ \Phi_R$, and $\tilde{J}_{SR}(\nu(w)y)^\pi \circ \tilde{J}_{RS}(y^t) = \tilde{J}_{SS}(\nu(\alpha)) = \tilde{J}_{SS}(1)$, hence

$$(2.23.1) \quad \tilde{X}_{SR}(\alpha) = {}^t\Phi_S \circ \tilde{J}_{SR}(y) + {}^t\tilde{J}_{RS}(y^t) \circ \Phi_R \ .$$

PROOF OF 2.23. We can find infinitely many imaginary quadratic extensions K of F such that $\mathfrak{p}$ decomposes in K, and K is F-linearly embeddable in B. Take such a K and an F-linear isomorphism f of K into B. Then $f(K^\times)$ has a fixed point z on $\mathfrak{H}_1$. By [11, 2.8], we can take f so that $f(\mathfrak{r}_K) \subset \mathfrak{o}$, and further f is normalized in the sense of [11, 2.7]. Then K, f, z, K and the identity map correspond to P, f, z, P' and Ψ' of 2.4. Let $\mathfrak{P} \cap K = \mathfrak{q}$. By (2.5.5), $K \cdot k_S(\varphi_S(z))$ is a class field over K corresponding to $K^\times \cdot \{v \in K_A^\times \mid f(v) \in S\}$. Since $f(\mathfrak{r}_K) \subset \mathfrak{o}$, $S(\mathfrak{o}, \mathfrak{a}) \subset S$, and $\mathfrak{p}$ is prime to $\mathfrak{a}$, we see that $\mathfrak{q}$ is unramified in $K \cdot k_S(\varphi_S(z))$. Let $K_\mathfrak{q}$ be the completion of K at $\mathfrak{q}$, $u_\mathfrak{q}$ a prime element of $K_\mathfrak{q}$, and u an element of $K_A^\times$ of which the $\mathfrak{q}$-component is $u_\mathfrak{q}$ and other components are all equal to 1. Let τ be a Frobenius automorphism of $\bar{Q}$ over K with respect to $\mathfrak{P}$ and $\mathfrak{q}$. Since $\tau = [u, K]$ on $K \cdot k_S(\varphi_S(z))$, we have, by (2.5.4), $\varphi_S(z)^\tau = J_{ST}(f(u)^{-1})[\varphi_T(z)]$, where $T = f(u)Sf(u)^{-1}$. Since $f(\mathfrak{r}_K) \subset \mathfrak{o}$, we see that $f(u)_\mathfrak{p}$ is contained in $\mathfrak{o}_\mathfrak{p}$ and has the same elementary divisors as $w_\mathfrak{p}$. Therefore $Sf(u)^{-1}S = Sw^{-1}S$. Put $U = S \cap T$. If S_0 and U_0 denote the projections of S and U on G_0, respectively, we can write $S_0 = \mathfrak{o}_\mathfrak{p}^\times \cdot S'$ with a subgroup S' of $\prod_{\mathfrak{l} \neq \mathfrak{p}} G_\mathfrak{l}$, hence

$$U_0 = \left(\mathfrak{o}_\mathfrak{p}^\times \cap f(u)_\mathfrak{p}\mathfrak{o}_\mathfrak{p}^\times f(u)_\mathfrak{p}^{-1}\right)S' \ .$$

Then we see easily that $k_U = k_S$. By 2.19, (2.18.5), and (2.17.4), we have

$$\begin{aligned}
X_{SS}(w^{-1}) &= X_{SS}(f(u)^{-1}) = J_{SU}(f(u)^{-1}) \circ {}^t J_{SU}(1) \\
&= J_{ST}(f(u)^{-1}) \circ J_{TU}(1) \circ {}^t J_{SU}(1) \ .
\end{aligned}$$

It follows that $\varphi_S(z) \times \varphi_S(z)^\tau$ is contained in $X_{SS}(w^{-1})$. Put $a = \mathfrak{P}(\varphi_S(z))$. Then $a \times a^\tau \in \tilde{X}_{SS}(w^{-1})$. As is explained in [11, 11.5], we obtain infinitely many distinct such points a on $\tilde{V}_S$. (More precisely, this holds for almost all $\mathfrak{p}$.) Therefore $\Phi_S \subset \tilde{X}_{SS}(w^{-1})$. By (2.18.9), we have $X_{SS}(w) = {}^t X_{SS}(w^{-1})^{\sigma(w)}$, hence ${}^t \tilde{X}_{SS}(w)^\tau = \tilde{X}_{SS}(w^{-1})$. Put $e = \nu(w)$. Then we have $Sw^{-1}S = Sw^t e^{-1}S = Swe^{-1}S$, so that $X_{SS}(w^{-1}) = X_{SS}(w)^{\sigma(e^{-1})} \circ J_{SS}(e^{-1})$ by (2.18.7). Therefore we obtain

$$(2.23.2) \qquad \tilde{X}_{SS}(w^{-1}) = {}^t \tilde{X}_{SS}(w^{-1})^\tau \circ \tilde{J}_{SS}(e^{-1}) \supset {}^t \Phi_S^\tau \circ \tilde{J}_{SS}(e^{-1}) \ .$$

For a correspondence X on $U \times V$ as in 2.17, put $d(X) = \deg(X[c])$, $d'(X) = d({}^t X)$ with any generic point c on U (cf. Weil, "Courbes Algébriques", p. 31). Then $d(\Phi_S) = 1$, $d'(\Phi_S) = N(\mathfrak{p})$, $d({}^t \Phi_S^\tau \circ \tilde{J}_{SS}(e^{-1})) = N(\mathfrak{p})$, $d'({}^t \Phi_S^\tau \circ \tilde{J}_{SS}(e^{-1})) = 1$. Since Φ_S and ${}^t \Phi_S^\tau \circ \tilde{J}_{SS}(e^{-1})$ are distinct and irreducible, we see that

$$(2.23.3) \qquad \Phi_S + {}^t \Phi_S^\tau \circ \tilde{J}_{SS}(e^{-1}) \subset \tilde{X}_{SS}(w^{-1}) \ .$$

Put $R = wSw^{-1} \cap S$. Then $R_0 = (\mathfrak{o}_\mathfrak{p}^\times \cap w\mathfrak{o}_\mathfrak{p}^\times w^{-1})S'$, hence we see easily that $S \cap F^\times = R \cap F^\times$ and $k_S = k_R$. Therefore $S = R\Gamma_S$ by 2.19. By (2.17.4), we have

$$d\big(\tilde{X}_{SS}(w^{-1})\big) = d\big(X_{SS}(w^{-1})\big) = [\Gamma_S : \Gamma_R] = [S : R]$$
$$= [\mathfrak{o}_{\mathfrak{p}}^{\times} : \mathfrak{o}_{\mathfrak{p}}^{\times} \cap w\mathfrak{o}_{\mathfrak{p}}^{\times}w^{-1}] = 1 + N(\mathfrak{p}) .$$

From this and (2.23.2), we obtain $d'\big(\tilde{X}_{SS}(w^{-1})\big) = d\big(\tilde{X}_{SS}(w^{-1})\big) = 1 + N(\mathfrak{p})$. Thus both sides of (2.23.3) have the same values of d and d'. Therefore the inclusion (2.23.3) must actually be an equality, q.e.d.

3. Some lemmas and reduction of the proof

3.0. In order to obtain our canonical system, it is sufficient to construct V_S, φ_S and $J_{TS}(x)$ for sufficiently small S and T. The objects for larger S and T are obtained by considering the quotients by a finite group, and by lowering the field of rationality. If B is totally indefinite, the canonical system can be obtained more or less directly from the modulus-variety for a family of abelian varieties, taking account of this reduction process. In the case $r < g$, however, we need another type of reduction. We shall first construct a number of systems with weaker properties than required, and then obtain a canonical system from them by lowering the field of definition. In this section, we shall discuss these two types of reduction, together with some fundamental lemmas about our group G. Actual construction of V_S, φ_S and $J_{TS}(x)$, for smaller S and T, or with weaker properties, will be made in §§ 5, 6.

3.1. Let H be an arbitrary finite algebraic extension of F'. Put

$$(3.1.1) \qquad \begin{aligned} \lambda_H(c) &= \lambda\big(N_{H/F'}(c)\big) & (c \in H_A^{\times}) , \\ D_H &= \{x \in G_A \mid \nu(x) \in \lambda_H(H_A^{\times})\} , \\ D_{H+} &= D_H \cap G_{A+} , \\ \mathcal{G}_H &= D_H G_Q G_{\infty+}, \quad \mathcal{G}_{H+} = \mathcal{G}_H \cap G_{A+} . \end{aligned}$$

If $H = F'$, these symbols are the same as those without the subscript H. Now we have

$$(3.1.2) \qquad \mathcal{G}_{H+} = D_{H+} G_{Q+} G_{\infty+} = \{x \in G_{A+} \mid \nu(x) \in \lambda_H(H_A^{\times}) F^{\times} F_{\infty+}^{\times}\} .$$

To see this, let $x \in G_{A+}$, $\nu(x) = \lambda_H(c)ab$ with $c \in H_A^{\times}$, $a \in F^{\times}$, $b \in F_{\infty+}^{\times}$. Take $d \in H^{\times}$ so that $N_{H/F'}(cd) \gg 0$. Then $\nu(x) = \lambda_H(cd)\lambda_H(d)^{-1}ab$, hence $\lambda_H(d)^{-1}a \gg 0$. We can find elements α of G_{Q+} and u of $G_{\infty+}$ so that $\nu(\alpha) = \lambda_H(d)^{-1}a$ and $\nu(u) = b$. Then $\nu(xu^{-1}\alpha^{-1}) = \lambda_H(cd)$, hence $xu^{-1}\alpha^{-1} \in D_{H+}$, q.e.d. This proves especially (2.2.1).

For every $S \in \mathcal{Z}$, put

$$\mathfrak{X}_H(S) = \mathfrak{X}_{HS} = \{b \in H_A^{\times} \mid \lambda_H(b) \in F^{\times} \cdot \nu(S)\}$$
$$= \{b \in H_A^{\times} \mid N_{H/F'}(b) \in \mathfrak{X}_S\} .$$

Then, $k_S H$, as a class field over H, corresponds to the subgroup $\mathfrak{X}_{HS}$ of $H_A^{\times}$.

We have

(3.1.3)
$$\mathfrak{X}_{\mathit{IIS}} = \{b \in H_A^\times \mid \lambda_{\mathit{II}}(b) \in F^\times \cdot \nu(S \cap \mathcal{G}_{\mathit{II}})\} \ .$$

To show this, let $\lambda_{\mathit{II}}(b) = c \cdot \nu(x)$ with $c \in F^\times$ and $x \in S$. Then $x \in G_{A+}$ and $\nu(x) \in \lambda_{\mathit{II}}(H_A^\times)F^\times$. By (3.1.2), we have $x \in \mathcal{G}_{\mathit{II}}$, hence (3.1.3).

Now we can define a homomorphism

$$\sigma_{\mathit{II}} \colon \mathcal{G}_{\mathit{II}} \longrightarrow \mathrm{Gal}\,(\mathfrak{k}H/H)$$

by $\sigma_{\mathit{II}}(x) = [b^{-1}, H]$ on $\mathfrak{k}H$ with an element b of $H_A^\times$ such that $\nu(x)/\lambda_H(b) \in F^\times F_{\infty+}^\times$. We see that $\sigma_{\mathit{II}}(x) = \sigma(x)$ on $\mathfrak{k}$, and $\sigma_{\mathit{II}}(x)$ is the identity if $x \in G_Q G_{\infty+}$.

3.2. We call a subset $\mathcal{W}$ of $\mathfrak{Z}$ *normal* if $xSx^{-1} \in \mathcal{W}$ for every $S \in \mathcal{W}$ and every $x \in \mathcal{G}$. Let H^* be an arbitrary finite algebraic extension of F. For a normal subset $\mathcal{W}$ of $\mathfrak{Z}$, we call a system

$$\{V_S, \varphi_S, J_{TS}(x), (S,\, T \in \mathcal{W}; \ x \in \mathcal{G}_{\mathit{II}+})\}$$

a *weak canonical system relative to* $\{\mathcal{W}, H, H^*\}$ if these objects satisfy the conditions (2.5.1–4) under the following modification (3.2.1–4).

(3.2.1) (V_S, φ_S) *is defined only for* $S \in \mathcal{W}$.

(3.2.2) $\mathcal{G}$, k_S, *and* $\sigma(x)$ *are replaced by* $\mathcal{G}_{\mathit{II}}$, $k_S H$, *and* $\sigma_H(x)$, *respectively.*

(3.2.3) $J_{TS}(x)$ *is defined if and only if* $xSx^{-1} = T$ *(equality!)*, $x \in \mathcal{G}_{\mathit{H}+}$, $S \in \mathcal{W}$, $T \in \mathcal{W}$.

(3.2.4) The modification of (2.5.4) is as follows. *Let* (Y, P, δ, f), P', z, *and* η *be as in 2.4. Suppose that* $P_i \otimes_F H^*$ *is a field for every simple component* P_i *of* P. *Then, for every* $S \in \mathcal{W}$, $\varphi_S(z)$ *is rational over* $(P'H)_{ab}$. *Further, let* $u \in (P'H)_A^\times$, $v = N_{P'H/P'}(u)$, $\tau = [u, P'H]$. *Then* $\varphi_S(z)^\tau = J_{ST}(\eta(v)^{-1})[\varphi_T(z)]$, *where* $S = \eta(v)^{-1}T\eta(v)$.

With this element u of $(P'H)_A^\times$, put $w = N_{P'H/H}(u)$. By (2.4.4), we have

(3.2.5)
$$\nu\big(\eta(v)\big) = \lambda_H(w) \ ,$$

hence $\eta(v) \in D_{\mathit{H}+}$. Moreover,

(3.2.6)
$$[u, P'H] = \sigma_H\big(\eta(v)^{-1}\big) = [w, H] \qquad\qquad \text{on } \mathfrak{k}H \ .$$

From (3.2.4), we can derive (see 3.7 below)

(3.2.7) *Let* $C(S, f)$ *be as in* (2.5.5). *Then* $P' \cdot H \cdot k_S\big(\varphi_S(z)\big) \subset C(S, f) \cdot H$. *If* (Y, P, δ, f) *is primitive,* $P' \cdot H \cdot k_S\big(\varphi_S(z)\big) = C(S, f) \cdot H$. *It follows especially that* $\varphi_S(z)$ *is rational over* $P'_{ab}H$.

If L (resp. L^*) is a finite algebraic extension of H (resp. H^*), we see that $\mathcal{G}_L \subset \mathcal{G}_{\mathit{H}}$, and a weak canonical system relative to $\{\mathcal{W}, H, H^*\}$ can be viewed as a weak canonical system relative to $\{\mathcal{W}, L, L^*\}$ by means of the restriction of x to $\mathcal{G}_{L+}$.

3.3. Proposition. *The homomorphism* σ_H, *even restricted to* D_{H+}, *is surjective.*

Proof. Let $\tau \in \mathrm{Gal}\,(\mathfrak{k}H/H)$. Take an element b of $H_A^\times$ so that $\tau = [b^{-1}, H]$ on $\mathfrak{k}H$, and an element c of $H^\times$ so that $bc \gg 0$. For any rational prime p, there exists an element x_p of $G_p = G(\mathbf{Q}_p)$ such that $\nu(x_p) = \lambda_H(bc)_p$, where the subscript p means the p-component. Now $\lambda_H(bc)_p$ is a p-unit for almost all p. Therefore we can take x_p so that $\mathfrak{m}_p x_p = \mathfrak{m}_p$ for all such p, with a $\mathbf{Z}$-lattice $\mathfrak{m}$ satisfying (2.10.3). Thus we obtain an element x of G_{A+} such that $\nu(x) = \lambda_H(bc)$. Then $x \in D_{H+}$, and $\sigma_H(x) = [b^{-1}, H] = \tau$ on $\mathfrak{k}H$.

3.4. Proposition. *For every* $S \in \mathfrak{Z}$, *one has* $G_A^u = G_{\mathbf{Q}}^u \cdot (S \cap G_A^u)$, *where* G^u *is as in* (2.1.1).

This follows from the strong approximation theorem for G^u, see Kneser [3], and the author [6].

3.5. Proposition. *Put* $S_H = S \cap \mathfrak{G}_H$ *for each* $S \in \mathfrak{Z}$. *Then the following assertions hold.*

(3.5.1) $G_{\mathbf{Q}+}S_H$ *is a normal subgroup of* $\mathfrak{G}_{H+}$, *and* $x \mapsto \sigma_H(x)$ *gives an isomorphism of* $\mathfrak{G}_{H+}/G_{\mathbf{Q}+}S_H$ *onto* $\mathrm{Gal}\,(k_S H/H)$.

(3.5.2) *For every* $y \in \mathfrak{G}_{H+}$, *one has*

$$G_{\mathbf{Q}+}S_H y = G_{\mathbf{Q}+}y S_H = S_H G_{\mathbf{Q}+} y = S_H y G_{\mathbf{Q}+} = y G_{\mathbf{Q}+}S_H = y S_H G_{\mathbf{Q}+}$$
$$= \{x \in \mathfrak{G}_{H+} \mid \sigma_H(x) = \sigma_H(y) \text{ on } k_S H\}\,.$$

(3.5.3) $[\mathfrak{G}_{H+} : G_{\mathbf{Q}+}S_H] = [H_A^\times : \mathfrak{X}_{HS}] = [k_S H : H]$.

Proof. If $y \in G_{\mathbf{Q}+}S_H$, $\sigma_H(y)$ is obviously the identity map on $k_S H$. Let $x \in D_{H+}$, and $\nu(x) = \lambda_H(b)$ with $b \in H_A^\times$. Suppose that $\sigma_H(x)$ is the identity on $k_S H$. Then $\nu(x) = c \cdot \nu(w)$ with $c \in F^\times$ and $w \in S_H$ by (3.1.3). Since $c \gg 0$, there exists an $\alpha \in G_{\mathbf{Q}+}$ such that $c = \nu(\alpha)$. Then $\nu(\alpha^{-1}xw^{-1}) = 1$. By 3.4, $\alpha^{-1}xw^{-1} = \beta z$ with $\beta \in G_{\mathbf{Q}}^u$ and $z \in S \cap G_A^u$. Therefore $x = \alpha\beta z w \in G_{\mathbf{Q}+}S_H$. This together with 3.3 proves (3.5.1). It follows that $G_{\mathbf{Q}+}S_H = S_H G_{\mathbf{Q}+}$. Since $k(ySy^{-1}) = k(S)$, we have $G_{\mathbf{Q}+}S_H = G_{\mathbf{Q}+}yS_Hy^{-1}$, hence $G_{\mathbf{Q}+}S_H y = G_{\mathbf{Q}+}yS_H$, and similarly $y^{-1}S_H y G_{\mathbf{Q}+} = S_H G_{\mathbf{Q}+}$, hence (3.5.2). The last assertion follows immediately from (3.5.1) and the definition of $\mathfrak{X}_{HS}$.

3.6. Proof of (2.6.2). If $T \subset S \subset \mathfrak{G}$ and $k_S = k_T$, we have $G_{\mathbf{Q}+}S = G_{\mathbf{Q}+}T$ by 3.5, so that $S \subset (G_{\mathbf{Q}+} \cap S)T = \Gamma_S T$. Therefore, if further $\Gamma_S = \Gamma_T$, we obtain $S = T$.

3.7. Proof of (3.5.5) and (3.2.7). The notation being as in (3.2.4), suppose $\eta(v) = \zeta t$ with $\zeta \in f(\{Y,\delta\})$ and $t \in S$. Then $\zeta \in G_{\mathbf{Q}+}$, and $t \in S_H$. Therefore, if $\tau = [u, P'H]$, $\sigma_H(\eta(v))$ is the identity on $k_S H$, and by (2.5.3$_{\mathrm{a,c}}$) and (3.2.4), $\varphi_S(z)^\tau = J_{ST}(\eta(v)^{-1})(\varphi_T(z)) = J_{ST}(t^{-1}\zeta^{-1})(\varphi_T(z)) = \varphi_S(z)$.

Conversely suppose that $\varphi_S(z)^\tau = \varphi_S(z)$, and τ is the identity on k_S. Since $\tau = \sigma_H(\eta(v)^{-1})$ on $k_S H$, we have, by (3.5.1), $\eta(v)^{-1} = s\alpha$ with $\alpha \in G_{Q+}$ and $s \in S_H$. Then we have $\varphi_S(z) = J_{ST}(s\alpha)[\varphi_T(z)] = \varphi_S(\alpha(z))$, hence $z = \gamma\alpha(z)$ with $\gamma \in \Gamma_S$. If (Y, P, δ, f) is primitive, $\gamma\alpha = f(b)$ with $b \in \{Y, \delta\}$. Then $\eta(v)^{-1} = s\gamma^{-1}f(b) \in S \cdot f(\{Y, \delta\})$, hence $\eta(v) \in f(\{Y, \delta\})S$. This proves (3.2.7). Taking F' for H, we obtain (2.5.5).

3.8. PROPOSITION. *Let $\mathcal{W}$ and $\mathcal{W}'$ be normal subsets of $\mathcal{Z}$. Let*

$$\{V_S, \varphi_S, J_{TS}(x), (S, T \in \mathcal{W}; x \in \mathcal{G}_{H+})\},$$
$$\{V'_L, \varphi'_L, J'_{ML}(x), (L, M \in \mathcal{W}'; x \in \mathcal{G}_{H+})\}$$

be weak canonical systems relative to $\{\mathcal{W}, H, H^*\}$ and $\{\mathcal{W}', H, H^*\}$, respectively. Then, for every $S \in \mathcal{W}$ and $L \in \mathcal{W}'$ such that $S \subset L$, there exists a morphism E_{LS} of V_S onto V'_L, rational over $k_S H$, such that $E_{LS} \circ \varphi_S = \varphi'_L$, and

$$(3.8.1) \qquad E_{MT}^{\sigma_H(x)} \circ J_{TS}(x) = J'_{ML}(x) \circ E_{LS} \qquad (x \in \mathcal{G}_{H+}, xSx^{-1} = T, xLx^{-1} = M).$$

(This proposition and the following proof are essentially the adelized version of [12, 5.9 and 8.2].)

PROOF. If $S \subset L$, we have $\Gamma_S \subset \Gamma_L$, so that a morphism E_{LS} of V_S onto V'_L is uniquely determined by $E_{LS} \circ \varphi_S = \varphi'_L$. E_{LS} is rational over the algebraic closure $\bar{Q}$ of Q, since the points $\varphi_S(z)$ described in (3.2.4) form a dense subset of V_S, and the corresponding fact holds on V_L. Fix L and S, and let k be a finite extension of $k_S H$, over which E_{LS} is rational. Let $x \in \mathcal{G}_{H+}$, and let τ be the restriction of $\sigma_H(x)$ to $k_S H$. Consider (Y, P, δ, f), P', z, and η as in (2.5.4). By [12, 7.5], we can take them so that P' and k are linearly disjoint over F', and $P_i \otimes_F H^*$ is a field for every simple component P_i of P. Extend τ to an automorphism π of $\bar{Q}$ over P'. Take an element u of $(P'H)_A^\times$ so that $\pi = [u, P'H]$ on $(P'H)_{ab}$. Put $v = N_{P'H/H}(u)$. Then $\sigma_H(x) = \sigma_H(\eta(v)^{-1})$ on $k_S H$, hence $\eta(v)^{-1} = \alpha x s$ with $\alpha \in G_{Q+}$ and $s \in S_H$ by (3.5.2). Let $T = xSx^{-1}$, $M = xLx^{-1}$, $U = \alpha T\alpha^{-1} = \eta(v)^{-1}S\eta(v)$, $N = \alpha M\alpha^{-1} = \eta(v)^{-1}L\eta(v)$. By (3.2.4), we have $\varphi_U(z)^\pi = J_{US}(\eta(v)^{-1})(\varphi_S(z)) = J_{UT}(\alpha)^\pi(J_{TS}(x)(\varphi_S(z)))$, so that $\varphi_T(\alpha^{-1}(z))^\pi = J_{TS}(x)(\varphi_S(z))$, and similarly $\varphi'_M(\alpha^{-1}(z))^\pi = J'_{ML}(x)(\varphi'_L(z))$. Therefore

$$(3.8.2) \qquad E_{MT}^\pi(J_{TS}(x)(\varphi_S(z))) = \varphi'_M(\alpha^{-1}(z))^\pi = J'_{ML}(x)(E_{LS}(\varphi_S(z))).$$

Now, after fixing x, Y, P, δ, and π, we can change f for $f_\beta: Y \to M_n(B)$ defined by $f_\beta(a) = \beta f(a)\beta^{-1}$ with any element β of G_{Q+}. This does not change P'. The fixed point of $f_\beta(\{Y, \delta\})$ is $\beta(z)$. Then we obtain (3.8.2) with $\beta(z)$ in place of z, and with the same x and π. The points $\varphi_S(\beta(z))$ form a dense subset of V_S. Therefore we obtain

$$(3.8.3) \qquad E_{MT}^\pi \circ J_{TS}(x) = J'_{ML}(x) \circ E_{LS}.$$

178   GORO SHIMURA

Take x to be the identity element. Then we have $E_{LS}^{\tau} = E_{LS}$ for any iso-morphism π of k into $\bar{\boldsymbol{Q}}$ over $k_S H$. This shows that E_{LS} is rational over $k_S H$. Therefore we can substitute $\sigma_H(x)$ for π in (3.8.3). This completes the proof.

3.9. Let $\mathcal{W}$ be a normal subset of $\mathfrak{Z}$, and

$$\{V_S,\ \varphi_S,\ J_{TS}(x),\ (S,\ T \in \mathcal{W};\ x \in \mathcal{G}_{H+})\}$$

a weak canonical system relative to $\{\mathcal{W},\ H,\ H^*\}$. Apply 3.8 to the case $\mathcal{W} = \mathcal{W}'$. Then we find, for R and S of $\mathcal{W}$ satisfying $S \subset R$, a morphism E_{RS} of V_S onto V_R, rational over $k_S H$, such that $E_{RS} \circ \varphi_S = \varphi_R$, and

$$(3.9.1) \qquad\qquad E_{UT}^{\sigma_H(x)} \circ J_{TS}(x) = J_{UR}(x) \circ E_{RS}$$

where $x \in \mathcal{G}_{H+}$, $xSx^{-1} = T$, $xRx^{-1} = U$. Define $J_{US}(x)$ to be the morphism of V_S to $V_U^{\sigma_H(x)}$ given by both sides of (3.9.1). $J_{US}(x)$ is defined whenever $x \in \mathcal{G}_{H+}$ and $xSx^{-1} \subset U$. It is now easy to verify that this extended system $\{J_{US}(x)\}$ satisfies $(2.5.3_{\mathrm{a,b,c}})$ under the modification of 3.2. Therefore, hereafter, if a weak canonical system relative to $\{\mathcal{W},\ H,\ H^*\}$ is given, we shall always use the symbol $J_{TS}(x)$ in this extended sense.

Coming back to 3.8, it is now easy to verify the formula (3.8.1) in the case $xSx^{-1} \subset T$, $xLx^{-1} \subset M$. Especially this means *the uniqueness of canonical system for G.*

3.10. PROPOSITION. *Let $\mathcal{W}$ be a normal subset of $\mathfrak{Z}$, and*

$$\mathfrak{S} = \{V_S,\ \varphi_S,\ J_{TS}(x),\ (S,\ T \in \mathcal{W};\ x \in \mathcal{G}_{H+})\}$$

be a weak canonical system relative to $\{\mathcal{W},\ H,\ H^*\}$. Suppose that every member of $\mathfrak{Z}$ contains at least one member of $\mathcal{W}$. Let $\mathcal{W}'$ be the set of all members of $\mathfrak{Z}$ which contain some member of $\mathcal{W}$ as a normal subgroup. Then $\mathcal{W}'$ is normal, and $\mathfrak{S}$ can be extended to a weak canonical system relative to $\{\mathcal{W}',\ H,\ H^*\}$.

PROOF. The normality of $\mathcal{W}'$ is obvious. Let us put, for simplicity, $\pi(x) = \sigma_H(x)$ for $x \in \mathcal{G}_{H+}$. Let $L \in \mathcal{W}'$. We are going to construct a model $(V_L,\ \varphi_L)$ of $\mathfrak{H}_n^r/\Gamma_L$ satisfying the following two conditions.

(3.10.1) V_L *is defined over* $k_L H$.

(3.10.2) *For every* $S \in \mathcal{W}$ *contained in* L, *let* E_{LS} *be the morphism of* V_S *to* V_L *defined by* $E_{LS} \circ \varphi_S = \varphi_L$. *Then* E_{LS} *is rational over* $k_S H$, *and* $E_{LS}^{\pi(x)} \circ J_{ST}(x) = E_{LT}$ *whenever* $x \in L_H$ *and* $T = x^{-1}Sx$.

If L is already contained in $\mathcal{W}$, the original (V_L, φ_L) satisfies these conditions. Therefore let us assume $L \notin \mathcal{W}$. To obtain (V_L, φ_L), first fix any member R of $\mathcal{W}$, which is a normal subgroup of L. Let $\mathfrak{A}$ be the group formed by all biregular automorphisms $J_{RR}(\gamma)$ with $\gamma \in \Gamma_L$. We can construct a quotient

333

V of V_R by $\mathfrak{A}$ so that both V and the projection map E of V_R to V are rational over $k_R H$. Let $x \in L_H$ and $\gamma \in \Gamma_L$. Then $\pi(x\gamma x^{-1}) = 1$, hence $x\gamma x^{-1} = \delta y$ with $\delta \in G_{Q+}$ and $y \in R_H$, by (3.5.1). We see that $\delta \in \Gamma_L$. Since $J_{RR}(y)$ is the identity map, we have

$$J_{RR}(x) \circ J_{RR}(\gamma) = J_{RR}(x\gamma) = J_{RR}(\delta y x) = J_{RR}(\delta)^{\pi(x)} \circ J_{RR}(x) \ .$$

This implies $J_{RR}(x) \circ \mathfrak{A} \circ J_{RR}(x)^{-1} = \mathfrak{A}^{\pi(x)}$. Therefore we can find a biregular map $J(x)$ of V to $V^{\pi(x)}$, rational over $k_R H$, such that $J(x) \circ E = E^{\pi(x)} \circ J_{RR}(x)$. By 3.5, the map $\pi = \sigma_H$ induces an isomorphism of $G_{Q+}L_H/G_{Q+}R_H$ onto $\mathrm{Gal}\,(k_R H/k_L H)$, hence an isomorphism of $L_H/\Gamma_L R_H$ onto $\mathrm{Gal}\,(k_R H/k_L H)$. Let y be an element of L_H such that $\pi(x) = \pi(y)$ on $k_R H$. Then $y = \varepsilon u x$ with $\varepsilon \in \Gamma_L$ and $u \in R_H$. It follows that $J_{RR}(y) = J_{RR}(\varepsilon)^{\pi(x)} \circ J_{RR}(x)$. Since $E = E \circ J_{RR}(\varepsilon)$, we have $J(y) \circ E = E^{\pi(x)} \circ J_{RR}(y) = E^{\pi(x)} \circ J_{RR}(x) = J(x) \circ E$, hence $J(y) = J(x)$. Therefore $J(x)$ depends only on the effect of $\pi(x)$ on $k_R H$. We can thus put $J_\xi = J(x)$ for every $\xi \in \mathrm{Gal}\,(k_R H/k_L H)$, when ξ is the restriction of $\pi(x)$ to $k_R H$. It is easily verified that $J_{\xi\zeta} = J_\xi^\zeta \circ J_\zeta$. By the Weil criterion, we can find a variety V_L, rational over $k_L H$, and a biregular map Q of V_L to V, rational over $k_R H$, such that $J_\zeta = Q^\zeta \circ Q^{-1}$ for all $\zeta \in \mathrm{Gal}\,(k_R H/k_L H)$. Put $E_{LR} = Q^{-1} \circ E$ and $\varphi_L = E_{LR} \circ \varphi_R$. Then (V_L, φ_L) is a model of $\mathfrak{H}_n^r/\Gamma_L$, and $E_{LR}^{\pi(x)} \circ J_{RR}(x) = E_{LR}$ for every $x \in L_H$. Now let S and E_{LS} be as in (3.10.2). By our assumption, we can find a member Y of $\mathfrak{W}$ contained in $R \cap S$. Fix an element x of L_H, and put $T = x^{-1}Sx$, $W = x^{-1}Yx$. Then $E_{LY} = E_{LS} \circ J_{SY}(1) = E_{LR} \circ J_{RY}(1)$. Therefore E_{LY} is rational over $k_Y H$, and

$$E_{LY}^{\pi(x)} \circ J_{YW}(x) = E_{LR}^{\pi(x)} \circ J_{RY}(1)^{\pi(x)} \circ J_{YW}(x) = E_{LR}^{\pi(x)} \circ J_{RW}(x)$$
$$= E_{LR}^{\pi(x)} \circ J_{RR}(x) \circ J_{RW}(1) = E_{LR} \circ J_{RW}(1) = E_{LW} \ .$$

Substituting $E_{LS} \circ J_{SY}(1)$ for E_{LY}, and $E_{LT} \circ J_{TW}(1)$ for E_{LW}, we obtain

$$E_{LT} \circ J_{TW}(1) = E_{LS}^{\pi(x)} \circ J_{SY}(1)^{\pi(x)} \circ J_{YW}(x) = E_{LS}^{\pi(x)} \circ J_{SW}(x)$$
$$= E_{LS}^{\pi(x)} \circ J_{ST}(x) \circ J_{TW}(1) \ ,$$

hence $E_{LT} = E_{LS}^{\pi(x)} \circ J_{ST}(x)$. In particular, $E_{LS} = E_{LS}^{\pi(x)}$ if $x \in S_H$. This shows, on account of (3.5.1), that E_{LS} is rational over $k_S H$, hence (V_L, φ_L) satisfies (3.10.2).

After constructing (V_L, φ_L) (which satisfies (3.10.1, 2)) for each $L \in \mathfrak{W}'$, let us now construct $J_{ML}(y)$ for $M, L \in \mathfrak{W}'$, $y \in \mathfrak{S}_{H+}$, $M = yLy^{-1}$. Take a member R of $\mathfrak{W}$, which is a normal subgroup of L, and put $S = yRy^{-1}$. Let $\gamma \in \Gamma_L$. Since $\pi(y\gamma y^{-1}) = 1$, we have $y\gamma y^{-1} = \delta x$ with $\delta \in G_{Q+}$ and $x \in S_H$. Then $\delta \in \Gamma_M$, and $J_{SR}(y) \circ J_{RR}(\gamma) = J_{SR}(y\gamma) = J_{SR}(\delta x y) = J_{SS}(\delta)^{\pi(y)} \circ J_{SR}(y)$. Therefore $E_{MS}^{\pi(y)} \circ J_{SR}(y) \circ J_{RR}(\gamma) = E_{MS}^{\pi(y)} \circ J_{SR}(y)$ for every $\gamma \in \Gamma_M$, hence there exists a morphism $J_{ML}(y)$ of V_L to $V_M^{\pi(y)}$, rational over $k_R H$, such that $J_{ML}(y) \circ E_{LR} =$

$E_{MS}^{\pi(y)} \circ J_{SR}(y)$. If $y \in L_H$, we have $J_{LL}(y) \circ E_{LR} = E_{LR}^{\tau(y)} \circ J_{RR}(y) = E_{LR}$, hence $J_{LL}(y)$ is the identity map. If $y = \alpha \in G_{Q+}$,

$$J_{ML}(\alpha)[\varphi_L(z)] = J_{ML}(\alpha)\big(E_{LR}(\varphi_R(z))\big) = E_{MS}\big(J_{SR}(\alpha)(\varphi_R(z))\big)$$
$$= E_{MS}\big(\varphi_S(\alpha(z))\big) = \varphi_M(\alpha(z)) \ .$$

Let T be an arbitrary member of $\mathcal{W}$ contained in L. We are going to show that $J_{ML}(y)$ satisfies

$$(3.10.3) \qquad\qquad J_{ML}(y) \circ E_{LT} = E_{MU}^{\pi(y)} \circ J_{UT}(y) \qquad\qquad (U = yTy^{-1}) \ .$$

By our assumption, $R \cap T$ contains a member X of $\mathcal{W}$. Put $Y = yXy^{-1}$. Then

$$J_{ML}(y) \circ E_{LT} \circ J_{TX}(1) = J_{ML}(y) \circ E_{LX} = J_{ML}(y) \circ E_{LR} \circ J_{RX}(1)$$
$$= E_{MS}^{\pi(y)} \circ J_{SR}(y) \circ J_{RX}(1) = E_{MS}^{\pi(y)} \circ J_{SX}(y)$$
$$= E_{MS}^{\pi(y)} \circ J_{SY}(1)^{\pi(y)} \circ J_{YX}(y) = E_{MY}^{\pi(y)} \circ J_{YX}(y)$$
$$= E_{MU}^{\pi(y)} \circ J_{UY}(1)^{\pi(y)} \circ J_{YX}(y) = E_{MU}^{\pi(y)} \circ J_{UX}(y)$$
$$= E_{MU}^{\pi(y)} \circ J_{UT}(y) \circ J_{TX}(1) \ .$$

This proves (3.10.3).

Now let $x \in \mathcal{G}_{H+}$, $N = xMx^{-1}$, $W = xSx^{-1}$. Then we can define $J_{NM}(x)$ and $J_{NL}(xy)$. By (3.10.3), we have $J_{NM}(x) \circ E_{MS} = E_{NW}^{\pi(x)} \circ J_{WS}(x)$, $J_{NL}(xy) \circ E_{LR} = E_{NW}^{\pi(xy)} \circ J_{WR}(xy)$. Therefore,

$$J_{NM}(x)^{\pi(y)} \circ J_{ML}(y) \circ E_{LR} = J_{NM}(x)^{\pi(y)} \circ E_{MS}^{\pi(y)} \circ J_{SR}(y)$$
$$= E_{NW}^{\pi(xy)} \circ J_{WS}(x)^{\pi(y)} \circ J_{SR}(y) = E_{NW}^{\pi(xy)} \circ J_{WR}(xy)$$
$$= J_{NL}(xy) \circ E_{LR} \ ,$$

hence $J_{NM}(x)^{\pi(y)} \circ J_{ML}(y) = J_{NL}(xy)$. Then we see that $J_{ML}(y) = J_{ML}(uy)$ for every $u \in M_H$. Therefore, if $y \in M$, we have $J_{NM}(x)^{\pi(y)} = J_{NM}(xy) = J_{NM}(xyx^{-1}x) = J_{NM}(x)$, since $xyx^{-1} \in N$. Now the restriction of $\pi(y)$ to $k_R H$ for all $y \in M_H$, exhausts the elements of $\mathrm{Gal}\,(k_R H / k_M H)$. Therefore $J_{NM}(x)$ is rational over $k_M H$.

Let (Y, P, δ, f), P', z, η, u, v, τ be as in 2.4 and (3.2.4). Then $\varphi_L(z) = E_{LR}(\varphi_R(z))$ is rational over $P'_{ab}H$. Let $M = \eta(v)^{-1}L\eta(v)$, $S = \eta(v)^{-1}R\eta(v)$. Since $\tau = \pi(\eta(v)^{-1})$ on $k_R H$, we have

$$\varphi_M(z)^\tau = E_{MS}^\tau\big(\varphi_S(z)^\tau\big) = E_{MS}^\tau\big(J_{SR}(\eta(v)^{-1})(\varphi_R(z))\big)$$
$$= J_{ML}(\eta(v)^{-1})\big(E_{LR}(\varphi_R(z))\big) = J_{ML}(\eta(v)^{-1})(\varphi_L(z)) \ .$$

This completes the proof.

3.11. Proposition. *Let $\mathcal{W}$ be a normal subset of $\mathcal{Z}$, and U a member of $\mathcal{Z}$ such that $U \cap G_0$ is open in G_0. Suppose that there exists a weak canonical system relative to $\{\mathcal{W}, H, H^*\}$. Suppose further that, for every $S \in \mathcal{Z}$, there exists a member of $\mathcal{W}$, which is a normal subgroup of U, and contained in*

S. Then there exists a weak canonical system relative to $\{\mathfrak{Z}, H, H^*\}$.

PROOF. Define $\mathcal{W}'$ as in 3.10 for $\mathcal{W}$. By 3.10, there exists a weak canonical system relative to $\{\mathcal{W}', H, H^*\}$. Let $R \in \mathfrak{Z}$. Put $S = \bigcap_{x \in R} x(R \cap U)x^{-1}$. Then S is a normal subgroup of R. Since $U \cap G_0$ is open in G_0, $R \cap U$ is open in R, hence $R \cap U$ is of finite index in R. Therefore $S \in \mathfrak{Z}$. By our assumption, S contains some $T \in \mathcal{W}$, which is a normal subgroup of U, hence $S \in \mathcal{W}'$. Thus R has a normal subgroup which is a member of $\mathcal{W}'$. Applying 3.10 to $\mathcal{W}'$ and $\mathfrak{Z}$, we obtain a weak canonical system relative to $\{\mathfrak{Z}, H, H^*\}$.

3.12. Let P be a totally imaginary quadratic extension of F such that $B \otimes_F P$ is isomorphic to $M_2(P)$, and let $Y = M_n(P)$. Define a positive involution δ of Y by $x^\delta = {}^t x^\rho$ for $x \in Y$ with the complex conjugation ρ on P. By [12, 4.7], we can find an F-linear isomorphism f of $M_n(P)$ into $M_n(B)$ so that $f(x^\delta) = {}^t f(x)^\iota$. Define P', Ψ' and η for this (Y, P, δ, f) as in 2.4. P' and Ψ' remain the same even if f is replaced by $f'(x) = \alpha f(x)\alpha^{-1}$ with any $\alpha \in G_{Q+}$.

3.13. PROPOSITION. *Let* $S \in \mathfrak{Z}$, *and let* $\mathfrak{q}$ *be a prime ideal in* F, *which is unramified in* B. *Let* $P, Y,$ *and* δ *be as in 3.12. For every* F-*linear isomorphism* $f: Y \to M_n(B)$ *such that* $f(a^\delta) = {}^t f(a)^\iota$, *let* z_f *be the fixed point of* $f(\{Y, \delta\})$ *on* $\mathfrak{H}_n^r$, *and* $C(S, f)$ *the class field defined in* (2.5.5). *Then there exists a positive integer* e, *depending only on* S, *and independent of* P *and* $\mathfrak{q}$, *with the following property: if* $Z(P, \mathfrak{q})$ *denotes the set of* z_f *for all* f *such that* $C(S, f)$ *is ramified over* P' *only at the prime factors of* $eN(\mathfrak{q})$, *then* $Z(P, \mathfrak{q})$ *is dense in* $\mathfrak{H}_n^r$.

PROOF. Although this is essentially the same as [12, 8.1], we reproduce here the proof in the adelized form. Let $\mathfrak{o}$ be a maximal order in B, and $\{x_i\}$ a basis of X over B as in (2.1.4). Put $\mathfrak{m} = \mathfrak{o}x_1 + \cdots + \mathfrak{o}x_n$. By [11, 2.8], there exists an F-linear isomorphism f_0 of P into B such that $f_0(\mathfrak{r}_P) \subset \mathfrak{o}$. We can find a positive integer e so that $S(\mathfrak{m}, e) \subset S$, where $S(\mathfrak{m}, e)$ is as in (2.10.2). Define $f_1: Y \to M_n(B)$ by $f_1((a_{ij})) = (f_0(a_{ij}))$ for $(a_{ij}) \in M_n(P) = Y$ with $a_{ij} \in P$. We see easily that $f_1(x^\delta) = {}^t f_1(x)^\iota$. Let w be the fixed point of $f_1(\{Y, \delta\})$ on $\mathfrak{H}_n^r$. By [12, 3.9], for every non-empty open subset $\mathfrak{N}$ of $\mathfrak{H}_n^r$, there exists an element α of G_Q^u such that $\alpha(w) \in \mathfrak{N}$, and $\mathfrak{m}_\mathfrak{p} \alpha = \mathfrak{m}_\mathfrak{p}$ for every prime ideal $\mathfrak{p}$ in F other than $\mathfrak{q}$. Then $\mathfrak{q}^d \mathfrak{m}\alpha \subset \mathfrak{m} \subset \mathfrak{q}^{-d}\mathfrak{m}\alpha$ for a suitable positive integer d. Define $f: Y \to M_n(B)$ by $f(a) = \alpha f_1(a)\alpha^{-1}$. Then $z_f = \alpha(w)$. Let $u \in P_A'^\times$ be such that $u \equiv 1 \bmod_0 (eN(\mathfrak{q})^{2d+1})$. Define $\eta_1: P'^\times \to G$ with respect to f_1. Then $\eta_1(u) \equiv 1 \bmod_0 (\mathfrak{m}, eN(\mathfrak{q})^{2d+1})$, hence $\alpha\eta_1(u)\alpha^{-1} \equiv 1 \bmod_0 (\mathfrak{m}, e)$, and $\nu(\alpha\eta_1(u)\alpha^{-1}) = \lambda(N_{P'/F'}(u))$, $N_{P'/F'}(u) \equiv 1 \bmod_0 (e)$. Therefore $\alpha\eta_1(u)\alpha^{-1} \in S$. It follows that the class-field $C(S, f)$ is ramified over P' only at the prime factors of $eN(\mathfrak{q})$, hence

$z_f \in Z(P, \mathfrak{q})$. Therefore $\mathfrak{N} \cap Z(P, \mathfrak{q}) \neq \varnothing$. This proves our proposition.

3.14. PROPOSITION. *Suppose that, for every finite algebraic extension k of FF' and every positive integer m, there exist an extension H of F', a finite algebraic extension H^* of F, and a weak canonical system relative to $\{\mathfrak{Z}, H, H^*\}$ satisfying the following two conditions:*

(3.14.1) $[H:F'] = 2$, $D(H/F')$ is prime to m, and $D(H/F') \neq (1)$.

(3.14.2) H^* is linearly disjoint with k over F.

Then there exists a canonical system for G.

PROOF. Take a member W_1 of $\mathfrak{Z}$, and put $\mathfrak{W} = \{xW_1x^{-1} \mid x \in \mathfrak{G}\}$. In view of 3.9 and 3.8, it is sufficient to construct a weak canonical system relative to $\{\mathfrak{W}, F', F\}$. Since $\mathfrak{G}/[G_{\boldsymbol{Q}+} \cdot (\mathfrak{G} \cap W_1)]$ is finite, we can find a finite subset $\mathfrak{W}'$ of $\mathfrak{W}$ so that $\mathfrak{W} = \{\alpha W\alpha^{-1} \mid W \in \mathfrak{W}', \alpha \in G_{\boldsymbol{Q}+}\}$. Take a positive integer e which has the property of 3.13 for $\bigcap_{W \in \mathfrak{W}'} W$. For simplicity, put $k = k(W_1)$. Then $k = k_S$ for every $S \in \mathfrak{W}$. Take a finite Galois extension k' of $\boldsymbol{Q}$ containing k. By our assumption, there exist finite algebraic extensions H of F', H^* of F, and a weak canonical system

$$\{V_S^H, \varphi_S^H, J_{TS}^H(x), (S, T \in \mathfrak{W}; x \in \mathfrak{G}_{H+})\}$$

relative to $\{\mathfrak{W}, H, H^*\}$ such that

(3.14.3) $[H:F'] = 2$, $D(H/F')$ is prime to $e \cdot D(k'/F')$, and $D(H/F') \neq (1)$.

Further there exist finite algebraic extensions K of F', K^* of F, and a weak canonical system

$$\{V_S^K, \varphi_S^K, J_{TS}^K(x), S, T \in \mathfrak{W}; x \in \mathfrak{G}_{K+})\}$$

relative to $\{\mathfrak{W}, K, K^*\}$ such that

(3.14.4) $[K:F'] = 2$, $D(K/F')$ is prime to $e \cdot D(k'H/F')$ and $D(K/F') \neq (1)$.

From (3.14.3) and (3.14.4) we can easily derive

(3.14.5) k and HK are linearly disjoint over F'.

By 3.8, there exists, for each $S \in \mathfrak{W}$, a biregular isomorphism E_S of V_S^H onto V_S^K, rational over kHK, such that $\varphi_S^K = E_S \circ \varphi_S^H$, $E_T^{\tau(x)} \circ J_{TS}^H(x) = J_{TS}^K(x) \circ E_S$ for every $x \in \mathfrak{G}_{HK+}$, where $\tau(x) = \sigma_{HK}(x)$, $T = xSx^{-1}$. For every $\xi \in \mathrm{Gal}(kHK/kK)$, put $E_{S\xi} = (E_S^{-1})^\xi \circ E_S$. Then $E_{S\xi}$ is a biregular isomorphism of V_S^H to $V_S^{H\xi}$, rational over kHK. We are going to show that

(3.14.6) $E_{S\xi}$ is defined over kH.

We first prove this for the members of $\mathfrak{W}'$. Let $W \in \mathfrak{W}'$. Take Y, P, δ, f as in 3.12. By [12, 7.5], these can be chosen so that

(3.14.7) $D(P'/F')$ is prime to $D(HK/F')$.

(3.14.8) P is linearly disjoint with H^*K^* over F.

Let z be the fixed point of $f(\{Y, \delta\})$ on $\mathfrak{H}_n$, and let C denote the class field

$C(W, f)$ defined as in (2.5.5). By 3.13, (3.14.3) and (3.14.4), we can take f so that

(3.14.9) $D(C/P')$ *is prime to* $D(HK/F')$.

By (3.14.3, 4, 7, 9), CK (resp. CH) is linearly disjoint with HK over K (resp. H), and CK (resp. CH) is linearly disjoint with kHK over kK (resp. kH). Therefore ξ can be extended to an element π of Gal $(\overline{Q}/CK)$. Put $a = \varphi_W^H(z)$, $b = \varphi_W^K(z)$. Then $b = E_W(a)$. By (3.2.7) and (3.14.8), b is rational over CK. We have therefore $E_W(a) = b = b^\pi = E_W^\xi(a^\pi)$, so that $E_{W\xi}(a) = a^\pi$. Now, by (3.2.7) and (3.14.8), a is rational over CH. Since $(CH)^\pi = CH$, a^π is rational over CH. Let $\sigma \in$ Gal (kHK/kH). Since CH and kHK are linearly disjoint over kH, σ can be extended to an automorphism τ of CHK over CH. Then $E_{W\xi}^\sigma(a) = E_{W\xi}(a)^\tau = a^{\pi\tau} = a^\pi = E_{W\xi}(a)$. By 3.13, the points $a = \varphi_W^H(z)$, for which this conclusion holds, form a dense subset of V_W^H. Therefore $E_{W\xi}^\sigma = E_{W\xi}$ for every $\sigma \in$ Gal (kHK/kH), hence $E_{W\xi}$ is rational over kH, for every $W \in \mathcal{W}'$.

Let $S \in \mathcal{W}$. Then $S = \alpha W \alpha^{-1}$ with $\alpha \in G_{Q+}$ and $W \in \mathcal{W}'$. Then $E_S \circ J_{SW}^H(\alpha) = J_{SW}^K(\alpha) \circ E_W$, and $E_S^\xi \circ J_{SW}^H(\alpha)^\xi = J_{SW}^K(\alpha) \circ E_W^\xi$. Therefore

$$\left(J_{SW}^H(\alpha)^\xi\right)^{-1} \circ E_{S\xi} \circ J_{SW}^H(\alpha) = E_{W\xi} .$$

Now $J_{SW}^H(\alpha)$ is rational over kH, and $(kH)^\xi = kH$. Therefore $J_{SW}^H(\alpha)^\xi$ is rational over kH, hence (3.14.6).

From our definition of $E_{S\xi}$, we obtain $E_{S\xi\zeta} = E_{S\xi}^\zeta \circ E_\zeta$ for $\xi, \zeta \in$ Gal (kH/k). Therefore, by the Weil criterion, there exists, for each $S \in \mathcal{W}$, an algebraic variety V_S, rational over k, and a biregular isomorphism M_S of V_S onto V_S^H, rational over kH, such that $E_{S\xi} = M_S^\xi \circ M_S^{-1}$ for every $\xi \in$ Gal (kH/k). We have then, for every $\zeta \in$ Gal(kHK/kK), $(E_S \circ M_S)^\zeta = E_S^\zeta \circ M_S^\zeta = E_S^\zeta \circ E_{S\zeta} \circ M_S = E_S \circ M_S$, hence

(3.14.10) $E_S \circ M_S$ *is rational over* kK.

For $x \in \mathcal{G}_{H+}$ and $T = xSx^{-1}$, put $J_{TS}(x) = (M_T^\sigma)^{-1} \circ J_{TS}^H(x) \circ M_S$, where $\sigma = \sigma_H(x)$. Further, put $\varphi_S = M_S^{-1} \circ \varphi_S^H$. Then

$$\{V_S, \varphi_S, J_{TS}(x), (S, T \in \mathcal{W}; x \in \mathcal{G}_{H+})\}$$

is a weak canonical system relative to $\{\mathcal{W}, H, H^*\}$. If $x \in \mathcal{G}_{HK+}$, we have

$$(E_T \circ M_T)^\tau \circ J_{TS}(x) = J_{TS}^K(x) \circ (E_S \circ M_S) ,$$

where $\tau = \sigma_{HK}(x)$. This, together with (3.14.10), shows that $J_{TS}(x)$ is rational over kK, hence, over k, if $x \in \mathcal{G}_{HK+}$.

Let $S^* = S \cap \mathcal{G}$, $S_{HK} = S \cap \mathcal{G}_{HK}$. Then $G_{Q+}S^* \cap \mathcal{G}_{HK} = G_{Q+}S_{HK}$. By (3.5.3) and (3.14.5), we have

$$[k : F'] = [\mathcal{G}_+ : G_{Q+}S^*] \geqq [\mathcal{G}_{HK+}S^* : G_{Q+}S^*] = [\mathcal{G}_{HK+} : G_{Q+}S_{HK}]$$
$$= [kHK : HK] = [k : F'] .$$

It follows that $\mathcal{G}_+ = \mathcal{G}_{HK+}S^*$ for every $S \in \mathcal{W}$.

Let $x \in \mathcal{G}_+$, and $T = xSx^{-1}$. Then $x = x's$ with $x' \in \mathcal{G}_{HK+}$ and $s \in S^*$. Define $J_{TS}(x)$ by $J_{TS}(x) = J_{TS}(x')$. This does not depend on the choice of x'. If $x \in \mathcal{G}_{H+}$, this coincides with the original $J_{TS}(x)$. Let $y \in \mathcal{G}_+$, $U = yTy^{-1}$, $y = y't$ with $y' \in \mathcal{G}_{HK+}, t \in T^*$. Then $yx = y'tx's = y'x'x'^{-1}tx's$. Since $x'^{-1}tx' \in S^*$ and $\sigma(x') = \sigma(x)$ on k, we have

$$J_{US}(yx) = J_{US}(y'x') = J_{UT}(y')^{\sigma(x')} \circ J_{TS}(x') = J_{UT}(y)^{\sigma(x)} \circ J_{TS}(x).$$

By our definition, $J_{SS}(x)$ is the identity if $x \in S \cap \mathcal{G}$.

To prove that $\{V_S, \varphi_S, J_{TS}(x), (S, T \in \mathcal{W}; x \in \mathcal{G}_+)\}$ is a weak canonical system relative to $\{\mathcal{W}, F', F\}$, it remains to show that this satisfies (2.5.4). Let P, P', η, and z be as in (2.5.4). Fix a member R of $\mathcal{W}$, and put $C_1 = C(R, f)$. By our assumption, we can find finite algebraic extensions L of F', L^* of F, and a weak canonical system

$$\{V_S^L, \varphi_S^L, J_{TS}^L(x), (S, T \in \mathcal{W}; x \in \mathcal{G}_{L+})\}$$

relative to $\{\mathcal{W}, L, L^*\}$ such that

(3.14.11) $[L:F'] = 2$, and L is linearly disjoint with C_1HK over F';

(3.14.12) $L^* \otimes_F P_i$ is a field for every simple component P_i of P.

By 3.8, there exists, for each $S \in \mathcal{W}$, a biregular isomorphism Q_S of V_S^L onto V_S, rational over kHL, such that $Q_S \circ \varphi_S^L = \varphi_S$, and

(3.14.13) $\qquad Q_T^\tau \circ J_{TS}^L(x) = J_{TS}(x) \circ Q_S \qquad (x \in \mathcal{G}_{HL+}, \tau = \sigma_{HL}(x), T = xSx^{-1}).$

Now $E_S \circ M_S \circ Q_S \circ \varphi_S^L = \varphi_S^K$, hence $E_S \circ M_S \circ Q_S$ is rational over kKL by 3.8. This combined with (3.14.10) shows that Q_S is rational over kKL, hence over kL.

Since k and HL are linearly disjoint over F', we have $\mathcal{G}_+ = \mathcal{G}_{HL+}S^*$. Let $x \in \mathcal{G}_{L+}$, and $T = xSx^{-1}$. Then $x = x's$ with $x' \in \mathcal{G}_{HL+}$ and $s \in S^*$. We see that $s \in S_L$. Then $J_{TS}(x) = J_{TS}(x')$ and $J_{TS}^L(x) = J_{TS}^L(x')$. This implies that (3.14.13) is also true for $x \in \mathcal{G}_{L+}$ and $\tau = \sigma_L(x)$.

Coming back to P', η, z, and R, we see that $\varphi_R(z) = Q_R(\varphi_R^L(z))$ is rational over both C_1H and C_1L. Since L and C_1H are linearly disjoint over F', we have $C_1 = C_1H \cap C_1L$, so that $\varphi_R(z)$ is rational over C_1. Let $v \in P_A'^\times$. We can find an element π of $\mathrm{Gal}(\overline{Q}/L)$ such that $\pi = [v, P']$ on C_1. Take an element u of $(P'L)_A^\times$ so that $[u, P'L] = \pi$ on $(P'L)_{ab}$. Put $v_1 = N_{P'L/P'}(u)$, $T = \eta(v_1)R\eta(v_1)^{-1}$. Since $\eta(v_1) \in D_{L+}$, we have

$$\varphi_R(z)^\pi = Q_R^\pi\left(\varphi_R^L(z)^\pi\right) = Q_R^\pi\left(J_{RT}^L(\eta(v_1)^{-1})(\varphi_T^L(z))\right) = J_{RT}\left(\eta(v_1)^{-1}\right)(\varphi_T(z)).$$

Since $[v_1, P'] = [v, P']$ on $C(R, f)$, we have $\eta(v^{-1}v_1) = \alpha t$ with $\alpha \in f(\{Y, \delta\})$ and $t \in R$. Then $t \in \mathcal{G} \cap R$, and $\eta(v_1)^{-1} = t^{-1}\eta(v)^{-1}\alpha^{-1}$, since $\eta(v)$ commutes with every element of $f(Y)$. Put $U = \eta(v)R\eta(v)^{-1}$. Since $U = \alpha^{-1}T\alpha$ and $\alpha(z) = z$, we have

$$J_{RT}\big(\eta(v_1)^{-1}\big)\big(\varphi_T(z)\big) = J_{RT}\big(t^{-1}\eta(v)^{-1}\alpha^{-1}\big)\big(\varphi_T(z)\big) = J_{RU}\big(\eta(v)^{-1}\big)\big(\varphi_U(z)\big) \ .$$

This proves (2.5.4) for our system. Thus we have obtained a weak caninical system relative to $\{\mathfrak{W}, F', F\}$, and therefore completed the proof.

4. A fundamental theorem on abelian varieties of CM-type

4.1. Let K_i, for $i = 1, \cdots, q$, be CM-fields in the sense of 1.8, and let

$$Z = Z_1 \times \cdots \times Z_q \ , \qquad Z_i = M_{n_i}(K_i)$$

with positive integers n_i. Let A be an abelian variety defined over a subfield of C, and θ an isomorphism of Z into $\mathrm{End}_Q(A)$. Let (A, θ) be of type (Z, Ψ), in the sense of [11, 4.1], with an equivalence class Ψ of representations. Then A is isogenous to a product

(4.1.1) $$A_1^{n_1} \times \cdots \times A_q^{n_q}$$

with q abelian varieties A_i. For each i, there exists an isomorphism θ_i of K_i into $\mathrm{End}_Q(A_i)$ such that θ can be obtained from the θ_i in an obvious way with respect to a fixed isogeny of A to (4.1.1). Further, if (A_i, θ_i) is of type (K_i, Φ_i), then the restriction of Ψ to K_i is equivalent to $n_i \cdot \Phi_i$ plus a zero representation (see [11, 5.23]).

Let us now assume

(4.1.2) $$2 \cdot \dim (A) = \sum_{i=1}^{q} n_i [K_i : Q] \ .$$

By [14, Prop. 1 of 5.1], we have $\dim (A_i) = [K_i : Q]/2$, hence (K_i, Φ_i) is a CM-type in the sense of 1.8, for every i. Let $\mathcal{C}$ be a polarization of A satisfying the following condition:

(4.1.3) *If* * *denotes the involution of* $\mathrm{End}_Q(A)$ *determined by* $\mathcal{C}$, *then* $\theta(a)^* = \theta(a^\rho)$ *for every* $a \in K_i$, *where* ρ *is the complex conjugation.*

This is automatically satisfied if $\theta(Z)^* = \theta(Z)$, especially if $\theta(Z) = \mathrm{End}_Q(A)$. Let $d = \dim(A)$. We can find a lattice Λ in C^d and an exact sequence

$$0 \longrightarrow \Lambda \longrightarrow C^d \overset{\xi}{\longrightarrow} A \longrightarrow 0$$

so that ξ defines a biregular isomorphism of the complex torus C^d/Λ to A. Take a representation of Z in the class Ψ, and denote it again by Ψ. We may assume

(4.1.4) $$\theta(a) \circ \xi = \xi \circ \Psi(a) \qquad\qquad \textit{for every } a \in Z \ .$$

Let $W_i = K_i^{n_i}$, $W = W_1 \times \cdots \times W_q$, $W_R = W \otimes_Q R$. Consider W as a left Z-module in an obvious way. Now the Q-linear span of Λ can be viewed as a Z-module through Ψ. In view of (4.1.2), it is easy to see that this is isomorphic to W. Since Λ spans C^d over R, we can find an R-linear isomorphism $\mathfrak{y}$ of W_R onto C^d such that

(4.1.5) $\mathfrak{y}(ax) = \Psi(a)\mathfrak{y}(x)$ *for every $a \in Z$ and every $x \in W_R$.*

Put $\mathfrak{m} = \mathfrak{y}^{-1}(\Lambda)$. Then $\mathfrak{m}$ is a Z-lattice in W, and we obtain a commutative diagram:

(4.1.6)

$$
\begin{array}{ccccccccc}
0 & \longrightarrow & \mathfrak{m} & \longrightarrow & W_R & \longrightarrow & W_R/\mathfrak{m} & \longrightarrow & 0 \\
 & & \downarrow & & \downarrow{\scriptstyle\mathfrak{y}} & & \downarrow & & \\
0 & \longrightarrow & \Lambda & \longrightarrow & C^d & \overset{\xi}{\longrightarrow} & A & \longrightarrow & 0
\end{array}
$$

Let $\mathfrak{E}(x, y)$ be a Riemann form on C^d/Λ corresponding to a basic polar divisor in $\mathcal{C}$. Put

(4.1.7) $U(x, y) = \mathfrak{E}\big(\mathfrak{y}(x), \mathfrak{y}(y)\big)$ $(x, y \in W)$.

Then U is a Q-valued alternating form on W. The condition (4.1.3) implies

(4.1.8) $U(ax, y) = U(x, a^\rho y)$ $(a \in K_i,\ i = 1, \cdots, q)$.

Further let $t_1, \cdots, t_s$ be points of finite order on A. Then we find elements $v_1, \cdots, v_s$ of $W/\mathfrak{m}$ so that

(4.1.9) $\xi\big(\mathfrak{y}(v_j)\big) = t_j$ $(j = 1, \cdots, s)$.

Thus, from $\mathfrak{Q} = (A, \mathcal{C}, \theta; t_1, \cdots, t_s)$, we obtain a collection of data

$$\omega = (Z, \Psi, \mathfrak{m}, U; v_1, \cdots, v_s)$$

which is of the same nature as what we called a PEL-type in the previous papers. More precisely, we say that $\mathfrak{Q}$ *is of type* ω, if there exists a commutative diagram (4.1.6) such that (4.1.4, 5, 7, 9) hold. Now ω has the following important property:

(4.1.10) *Two structures $\mathfrak{Q}$ and $\mathfrak{Q}'$ of the same type ω are isomorphic.* To see this, it is sufficient to prove

(4.1.11) *Let $\mathfrak{y}$ and $\mathfrak{y}'$ be R-linear isomorphisms of W_R onto C^d satisfying (4.1.5). Then there exists a C-linear automorphism f of C^d such that $\mathfrak{y}' = f \circ \mathfrak{y}$.*

We can of course define an R-linear automorphism f of C^d such that $\mathfrak{y}' = f \circ \mathfrak{y}$. Our task is to show that f is C-linear. In view of (4.1.2), we observe that $\Psi(Z) \otimes_Q R$ contains a maximal commutative semi-simple subalgebra of $\operatorname{End}(C^d, R)$. Define an R-linear automorphism g of C^d by $g(x) = \sqrt{-1}x$. Then g commutes with every element of $\Psi(Z)$, hence $g \in \Psi(Z) \otimes_Q R$. Now f commutes with every element of $\Psi(Z) \otimes_Q R$, hence with g, q.e.d.

4.2. Let (K_i', Φ_i') be the reflex of (K_i, Φ_i), and K' the composite of

$K'_1, \cdots, K'_q$. Put $\zeta_i = \det \Phi'_i$, and define a map $\eta: K' \to Z$ by

$$\eta(a) = \sum_{i=1}^{q} \zeta_i(N_{K'/K'_i}(a))\varepsilon_i \qquad\qquad (a \in K') \, ,$$

where ε_i denotes the identity element of Z_i. Obviously

(4.2.1) *K' is generated by* tr $\Psi(x)$ *over $\mathbf{Q}$ for all x in the center of Z.*

Let σ be an (algebraic) automorphism of C. Then we can define a structure $\mathfrak{Q}^\sigma = (A^\sigma, C^\sigma, \theta^\sigma; t_1^\sigma, \cdots, t_s^\sigma)$ in an obvious way. In view of (4.2.1), $(A^\sigma, \theta^\sigma)$ is of type (Z, Ψ), if σ is the identity on K'.

4.3. THEOREM. *Let $\mathfrak{Q}$ be of type ω, and $c \in K'^\times_A$. Let σ be an automorphism of C such that $\sigma = [c, K']$ on K'_{ab}. Let $\mathfrak{c}$ be the ideal in K' associated with c. Then $\mathfrak{Q}^\sigma$ is of type*

$$\left(Z, \Psi, \eta(c)^{-1}\mathfrak{m}, N_{K'/\mathbf{Q}}(\mathfrak{c})U; \eta(c)^{-1}v_1, \cdots, \eta(c)^{-1}v_s\right) \, .$$

(For the notation, see 0.5.)

This is an adelized version of [12,6.12]. In the special case $Z = K_1$, our assertion is essentially included in [14, § 16]. Here let us derive the theorem from [12, 6.12], although a more direct proof is desirable.

PROOF. Let $L = \sum_{i=1}^{q} K_i\varepsilon_i$, $\mathfrak{l} = \theta^{-1}[\theta(L) \cap \mathrm{End}\,(A)]$. Let $\mathfrak{r}_L$ denote the maximal order in L. Take a positive integer e so that $e\mathfrak{r}_L \subset \mathfrak{l}$, and $ev_i = 0$ for $i = 1, \cdots, s$. We can find an element a of $K'^\times$ so that $a\mathfrak{c}$ is integral, and the p-component of ac is $\equiv 1 \bmod (e^2)$ for all prime factors p of e. Then $[c, K'] = [ac, K']$. Now we can apply [12, 6.12] to the present case with σ, e^2, and ac as σ, ef_0, and a of that proposition. The present L and (K_i, Φ_i) correspond S and (S_i, Ξ_i) of [12, 6.12]. Therefore we find an isogeny λ of A to A^σ such that the following (4.3.1–4) are satisfied.

(4.3.1) $\theta^\sigma(x)\lambda = \lambda\theta(x)$ for $x \in Z$.

(4.3.2) $\mathrm{Ker}\,(\lambda) = \{t \in A \mid \theta(\mathfrak{l} \cap \mathfrak{b})t = 0\}$, where $\mathfrak{b} = \mathfrak{b}_1 \oplus \cdots \oplus \mathfrak{b}_q$, $\mathfrak{b}_i = \det \Phi'_i(N_{K'/K'_i}(ac))$.

(4.3.3) $\lambda^{-1}(X^\sigma) = N(ac)X$ for a basic polar divisor X in C.

(4.3.4) $\lambda t = t^\sigma$ for every t on A such that $et = 0$.

We see easily that $\mathfrak{b} = \eta(ac)\mathfrak{r}_L$. Since $e\mathfrak{r}_L \subset \mathfrak{l}$, we see easily that, for every rational prime p, $(\eta(ac)\mathfrak{l})_p$ coincides with $\mathfrak{l}_p$ or $\mathfrak{b}_p$ according as p divides e or not. Similarly $(\mathfrak{b} \cap \mathfrak{l})_p$ coincides with $\mathfrak{l}_p$ or $\mathfrak{b}_p$ according as p divides e or not. Therefore $\eta(ac)\mathfrak{l} = \mathfrak{b} \cap \mathfrak{l}$. Put $\mathfrak{m}' = \eta(c)^{-1}\mathfrak{m}$, $\Lambda' = \mathfrak{y}(\mathfrak{m}')$. Changing $\mathfrak{m}$ and Λ for $\mathfrak{m}'$ and Λ' in the diagram (4.1.6), we obtain an abelian variety A' and an isogeny μ of A to A' such that the following diagrams are commutative:

(4.3.5)

$$\begin{array}{ccccccccc}
0 & \longrightarrow & \mathfrak{m}' & \longrightarrow & W_R & \longrightarrow & W_R/\mathfrak{m} & \longrightarrow & 0 \\
 & & \downarrow & & {\scriptstyle \mathfrak{y}}\downarrow & & \downarrow & & \\
0 & \longrightarrow & \Lambda' & \longrightarrow & C^d & \overset{\xi'}{\longrightarrow} & A' & \longrightarrow & 0
\end{array}$$

$$W_R \xrightarrow{\;\mathfrak{h}\;} C^d \xrightarrow{\;\xi\;} A$$
$$\eta(a)\downarrow \qquad \Psi(a)\downarrow \qquad \mu\downarrow$$
$$W_R \xrightarrow[\mathfrak{h}]{} C^d \xrightarrow[\xi']{} A'$$

We see that Ker (μ) corresponds to $\eta(ac)^{-1}\mathfrak{m}/\mathfrak{m}$. Since $\eta(ac)\mathfrak{t} = \mathfrak{b} \cap \mathfrak{t}$, we see that $w \in \eta(ac)^{-1}\mathfrak{m}$ if and only if $(\mathfrak{b} \cap \mathfrak{t})w \subset \mathfrak{m}$. In other words,

$$(4.3.6) \qquad \mathrm{Ker}\,(\mu) = \{t \in A \mid \theta(\mathfrak{b} \cap \mathfrak{t})t = 0\} = \mathrm{Ker}\,(\lambda)\,.$$

Further we have $\mu t_i = \xi'(\mathfrak{h}(\eta(a)v_i))$. Since $\eta(ac)_p - 1 \in e\mathfrak{t}_p$ for every prime p dividing e, we have $(\eta(a)_p - \eta(c^{-1})_p)\mathfrak{m}_p \subset e\mathfrak{m}'_p$ for such a p, so that $\eta(a)v_i \equiv \eta(c)^{-1}v_i \bmod \mathfrak{m}'$. By (4.3.6), we can find an isomorphism π of A' to A^σ so that $\lambda = \pi \circ \mu$. Then we obtain a commutative diagram by replacing A' and ξ' of (4.3.5) by A^σ and $\pi \circ \xi'$. We have $\pi(\xi'(\mathfrak{h}(\eta(c)^{-1}v_i))) = \pi\mu(t_i) = \lambda t_i = t_i^\sigma$, and $\pi \circ \xi' \circ \Psi(\alpha) = \theta^\sigma(\alpha) \circ \pi \circ \xi'$ for $\alpha \in Z$. Therefore, Q^σ is of type

$$(Z, \Psi, \mathfrak{m}', U'; \eta(c)^{-1}v_1, \cdots, \eta(c)^{-1}v_s)$$

with a suitable alternating form U' on W. Since $\lambda \circ \xi \circ \mathfrak{h} = \pi \circ \xi' \circ \mathfrak{h} \circ \eta(a)$ and $\lambda^{-1}(X^\sigma) \equiv N(ac)X$, we have

$$N(ac)U(x, y) = U'(\eta(a)x, \eta(a)y) = U'(\eta(a)\eta(a)^\rho x, y)\,.$$

Since (K'_i, Φ'_i) is a CM-type, we have, by our definition of η, $\eta(a)\eta(a)^\rho = N_{K'/Q}(a)$, hence $U' = N(c)U$. This completes the proof.

4.4. Let us insert here a generalization of the result of [11, 4.24, 4.25]. Let

$$\Omega = (L, \Phi, \rho; E \cdot T, \mathfrak{M}; v_1, \cdots, v_s)$$

be a weak PEL-type in the sense of [11, 4.8], where the symbols are the same as given there. In [11], the v_i are taken to be elements of the L-module W of [11, 4.1]. But in this paper, we rather consider them as elements of $W/\mathfrak{M}$, as in the above consideration. This has no effect on the results of the previous papers.

Let Z be the center of L. Let $a \in L_A^\times$, $c \in Z_A^\times$. Suppose that $c\mathfrak{M} \subset \mathfrak{M}$. Consider another weak PEL-type

$$\Omega' = (L, \Phi, \rho; E \cdot \kappa T, a\mathfrak{M}; y_1, \cdots, y_q)$$

with a totally positive element κ of F, where $F = \{x \in Z \mid x^\rho = x\}$. Suppose that

$$c^{-1}\mathfrak{M}/\mathfrak{M} = \sum_{i=1}^s Zv_i\,, \qquad c^{-1}a\mathfrak{M}/a\mathfrak{M} = \sum_{i=1}^q Zy_i\,.$$

Now construct families $\Sigma(\Omega) = \{\mathcal{R}_z \mid z \in \mathcal{K}\}$, $\Sigma(\Omega') = \{\mathcal{R}'_z \mid z \in \mathcal{K}\}$ of weak PEL-structures of type Ω and Ω' respectively, as in [11, 4.10, 4.15], by means of the same parametrizing function $\mathfrak{y}$. Let $(V, \mathfrak{b}, \varphi)$ be a moduli-system for $\Sigma(\Omega)$. For every weak PEL-structure $\mathcal{R}'$ of type Ω', we can find a point z on $\mathcal{H}$ so that $\mathcal{R}'$ is isomorphic to $\mathcal{R}'_z$. Define $\mathfrak{b}'(\mathcal{R}')$ by $\mathfrak{b}'(\mathcal{R}') = \varphi(z)$. Then

(4.4.1) $(V, \mathfrak{b}', \varphi)$ *is a moduli-system for* $\Sigma(\Omega')$, *and* $k(\Omega) = k(\Omega')$.

This fact was given in [11, 4.24, 4.25] under a somewhat more restrictive condition. The proof of [11, 4.25] is actually valid in the above setting. The only necessary changes are the following three points.

(i) Let $\mathfrak{o} = \{x \in L \mid x\mathfrak{M} \subset \mathfrak{M}\}$, $\mathfrak{a} = \mathfrak{a}\mathfrak{o}, \mathfrak{c} = c(\mathfrak{o} \cap Z)$. Take an element d of L so that $d\mathfrak{M} \subset \mathfrak{a}\mathfrak{M}$, and $(d\mathfrak{M})_p = (\mathfrak{a}\mathfrak{M})_p$ for all the rational primes p such that $(c\mathfrak{M})_p \neq \mathfrak{M}_p$. Define λ_z as in the proof of [11, 4.25]. Then $\mathrm{Ker}\,(\lambda_z) = \{t \in A_z \mid \theta_z(\mathfrak{o}a^{-1}d)t = 0\}$. (One can put $\mathfrak{o}a^{-1}d = \mathfrak{a}^{-1}d$.)

(ii) Let $\mathfrak{D}_z$, $\mathfrak{D}'_z$, σ, ϵ, η be as in the proof of [11, 4.25]. Let $X \in \mathfrak{D}_z$, $X' \in \mathfrak{D}'_z$. Then $\lambda_z^{-1}(X') \equiv X \cdot \theta_z(\kappa d^\rho d)$, and $(\lambda_z^\sigma)^{-1}(X'^\sigma) \equiv X^\sigma \cdot \theta_z^\sigma(\kappa d^\rho d)$. Now $\epsilon^{-1}(X^\sigma) \equiv X \cdot \theta_z(\gamma)$ with $\gamma \in E$. Therefore $\eta^{-1}(X'^\sigma) \equiv X' \cdot \theta'_z(\gamma)$, as stated in [11, p. 100, line 6].

(iii) Replace $\mathfrak{r}$ by Z. (Read "L" for "$\mathfrak{r}$" in [11, p. 100, line 6].)

5. The case of totally indefinite quaternion algebras

5.1. Let us now assume that $r = g$, i.e., B is totally indefinite. Then $F' = Q$, λ is the natural injection of $Q^\times$ into $F^\times$, and

$$D = \{x \in G_A \mid \nu(x) \in Q_A^\times\} \, .$$

If $x \in \mathcal{G}_+$, we have $\nu(x) = abc$ with $a \in Q_A^\times, b \in F_+^\times, c \in F_{\infty+}^\times$. Let a^* be the positive integer which generates the ideal associated with a. We can then define a homomorphism $\mu: \mathcal{G}_+ \to F_+^\times$ by

(5.1.1) $$\mu(x) = a^* b \, .$$

It can easily be verified that this does not depend on the choice of a, b, c. Obviously $\mu = \nu$ on G_{Q+}, and $\nu(x)/\mu(x) \in Q_A^\times \cdot F_{\infty+}^\times$.

Define a positive involution ρ of B, a B-valued ρ-anti-hermitian form T on X, and a representation Φ of B_R as in [12, 9.2]. We consider a PEL-type

$$\Omega = (B, \Phi, \rho; \kappa T, \mathfrak{m}; d_1, \cdots, d_s)$$

with a Z-lattice $\mathfrak{m}$ in X, elements $d_1, \cdots, d_s$ of $X/\mathfrak{m}$, and a totally positive element κ of F such that $\mathrm{Tr}_{B/Q}\big(\kappa T(\mathfrak{m}, \mathfrak{m})\big) = Z$. Since B, Φ, ρ, T are common to all PEL-types considered in this section, we write simply $\Omega = \big(\kappa, \mathfrak{m}, \{d_i\}\big)$. We construct a family $\Sigma(\Omega) = \{\mathfrak{A}_z \mid z \in \mathfrak{H}_n^g\}$ of PEL-structures

$$\mathfrak{A}_z = (A_z, C_z, \theta_z; t_{1z}, \cdots, t_{sz})$$

by means of the parametrizing function $\mathfrak{y}$ as in [12, 9.2], common to all Ω

(see also [11, 4.15]). Put $\mathfrak{y}_z(u) = \mathfrak{y}(u, z)$ for $u \in X_R$, $z \in \mathfrak{H}_n^a$. We denote by k_Ω the field determined by [9, 5.1]. Put

$$(5.1.2) \qquad S = S(\Omega) = G_{\infty+} \cdot \{y \in D_+ \mid \mathfrak{m}y = \mathfrak{m}, \, d_i y = d_i \, (1 \leq i \leq s)\} \, .$$

We are going to prove

(5.1.3) *Let the notation be as above. Then $k_\Omega = k_S$. Moreover, for every* $x \in \mathcal{G}_+$, $\Omega^{\sigma(x)}$ *is equivalent to* $\Omega' = \big(\mu(x)^{-1}\kappa, \, \mathfrak{m}x, \, \{d_i x\}\big)$.

5.2. Let k' be an arbitrary finite algebraic extension of k_S, and let $(Y, P, \delta, f), (P_i, \Psi_i), P, \eta$, and z be as in 2.4. By [12, 7.5], we can take these so that

(5.2.1) P' *and* k' *are linearly disjoint over* $\mathbf{Q}$.

Let τ be an isomorphism of k' into C such that $\tau = \sigma(x)$ on k_S. We can extend τ to an automorphism π of C over P'. Take an element v of $P_A'^\times$ so that $[v, P'] = \pi$ on P'_{ab}. Since $\sigma(x) = \sigma(\eta(v)^{-1})$ on k_S, we have, by (3.5.2), $\eta(v)^{-1} = ux\alpha$ with $\alpha \in G_{\mathbf{Q}+}$ and $u \in S$. Consider the member $\mathfrak{Q}_z$ of Σ_Ω for the fixed point z. By [12, 9.3], θ_z can be extended to an anti-isomorphism θ of $B \otimes_F Y$ into $\mathrm{End}_\mathbf{Q}(A_z)$, and (A_z, θ) belongs to the type discussed in § 4. With the terminology there, $\mathfrak{Q}^* = (A_z, C_z, \theta; t_{1z}, \cdots, t_{sz})$ is of type

$$(Z, \Xi, \mathfrak{m}, U; d_1, \cdots, d_s) \, ,$$

where $Z = B \otimes_F Y = M_{2q_1}(P_1) \oplus \cdots \oplus M_{2q_t}(P_t)$, with the notation of 2.4, $U(a, b) = \mathrm{Tr}_{B/\mathbf{Q}}\big(\kappa T(a, b)\big)$, and Ξ is such that its restriction to P_i is equivalent to $2q_i\Psi_i$ plus a zero representation (see [12, 9.3]). (Here we are considering X as a Z-module. Therefore X plays the role of the module W in 4.1. We obtain a diagram (4.1.6) with $\mathfrak{y}_z$ and A_z as $\mathfrak{y}$ and A.) Observe that the conditions (4.1.2) and (4.1.3) are satisfied. Therefore, by 4.3, $\mathfrak{Q}^{*\pi}$ is of type

$$(5.2.2) \qquad (Z, \Xi, \mathfrak{m}\eta(v)^{-1}, \xi U; d_1\eta(v)^{-1}, \cdots, d_s\eta(v)^{-1}) \, ,$$

with $\xi = N(\mathfrak{a})$, where $\mathfrak{a}$ is the ideal in P' associated with v. Let

$$\Omega'' = \big(\xi\kappa, \, \mathfrak{m}\eta(v)^{-1}, \, \{d_i\eta(v)^{-1}\}\big) \, .$$

Consider the member $\mathfrak{Q}_z'' = (A_z'', \cdots)$ of $\Sigma(\Omega'')$ with the same z. Since A_z'' is obtained as $C^n/\mathfrak{y}_z(\mathfrak{m}\eta(v)^{-1})$ by means of the same $\mathfrak{y}_z$ as for A_z, the structure $\mathfrak{Q}_z''$, furnished with an anti-isomorphism of Z into $\mathrm{End}_\mathbf{Q}(A_z'')$, is of type (5.2.2). By (4.1.10), $\mathfrak{Q}_z^\pi$ and $\mathfrak{Q}_z''$ must be isomorphic. Since $\eta(v)^{-1} = ux\alpha$ and $u \in S$, we have $\Omega'' = \big(\xi\kappa, \, \mathfrak{m}x\alpha, \, \{d_i x\alpha\}\big)$. We see that $\mu(u) = 1$, hence $\mu(x)^{-1} = \nu(\alpha)\xi$ by (2.4.4). Put $w = \alpha(z)$, and consider the member $\mathfrak{Q}_w'$ of $\Sigma(\Omega')$ at w, where Ω' is as in (5.1.3). The map $\Lambda(\alpha, w)$ of [12, (9.3.1)] defines an isomorphism of $\mathfrak{Q}_w'$ to $\mathfrak{Q}''$. Therefore $\mathfrak{Q}_z^\pi$ is isomorphic to $\mathfrak{Q}_w'$, hence Ω^π is equivalent to Ω'.

Now taking $k_\Omega k_S$ as k', we can apply this result to any isomorphism τ of $k_\Omega k_S$ into C such that $\tau = \sigma(x)$ on k_S. Taking x to be the identity, we see that

Ω^τ is equivalent to Ω for any isomorphism τ of $k_\Omega k_S$ into C over k_S. This shows that $k_\Omega \subset k_S$, by [9, 5.1].

To prove $k_S \subset k_\Omega$, let τ be an isomorphism of $k_\Omega k_S$ into C over k_Ω. By (3.5.1), there exists an element x of D_+ such that $\sigma(x) = \tau$ on k_S. Then Ω is equivalent to Ω'. Therefore, there exists an element ζ of G_Q such that $\mathfrak{m}x = \mathfrak{m}\zeta$, $d_i x = d_i \zeta$, $\nu(\zeta) = \mu(x)$. Then $x\zeta^{-1} \in S$, hence $\sigma(x)$ is the identity map on k_S. This proves $k_S \subset k_\Omega$, and completes the proof of (5.1.3).

5.3. Let Ω, x, and Ω' be as in (5.1.3). Let $(V, \mathfrak{b}, \varphi)$ and $(V', \mathfrak{b}', \varphi')$ be moduli-systems for $\Sigma(\Omega)$ and $\Sigma(\Omega')$ respectively, in the sense of [9] and [11, 4. 17]. Observe that

$$\Gamma_S = \{\gamma \in G_Q \mid \nu(\gamma) = 1,\ \mathfrak{m}\gamma = \mathfrak{m},\ d_i\gamma = d_i\ (1 \leq i \leq s)\}\,,$$

hence $\Gamma_S = \Gamma(T, \mathfrak{n}/\mathfrak{m})$ with the notation of [9, 3.3], where $\mathfrak{n}$ is the $\dot{Z}$-lattice in X such that $\mathfrak{n}/\mathfrak{m} = \sum_i Zd_i$. Therefore (V, φ) is a model of $\mathfrak{H}_\mathfrak{n}^q/\Gamma_S$. Since $\Omega^{\sigma(x)}$ is equivalent to Ω', there exists, by [11, 4.21 and 4.23], a biregular isomorphism J of V' to $V^{\sigma(x)}$, rational over k_S, such that $J \circ \mathfrak{b}' = \mathfrak{b}^{\sigma(x)}$. Let π, α, z and w be as in 5.2. Since $\mathfrak{A}_z^\tau$ is isomorphic to $\mathfrak{A}'_w$, we have

$$(5.3.1) \qquad \varphi(z)^\tau = \mathfrak{b}^\tau(\mathfrak{A}_z^\tau) = J(\mathfrak{b}'(\mathfrak{A}'_w)) = J(\varphi'(\alpha(z)))\,.$$

5.4. Let $\Omega_1 = (\kappa_1, \mathfrak{m}_1, \{d_{1i}\})$ be another PEL-type such that $S = S(\Omega_1)$ and $(V_1, \mathfrak{b}_1, \varphi_1)$ a moduli-system for $\Sigma(\Omega_1)$. Then there exists a biregular isomorphism E of V to V_1 such that $\varphi_1 = E \circ \varphi$. Let us now prove

(5.4.1) *E is rational over k_S.*

We first observe that the point $\varphi(z)$, for the above z, is algebraic, since A_z has sufficiently many complex multiplications. The points of this type form a dense subset of V. Since $E(\varphi(z)) = \varphi_1(z)$, E is rational over a finite algebraic extension k_1 of k_S. Repeat the reasoning of 5.2 with k_1 as k' and $x = 1$. Then we have $\Omega' = \Omega$, so that $\varphi(z)^\tau = \varphi(\alpha(z))$. Similarly $\varphi_1(z)^\tau = \varphi_1(\alpha(z))$ with the same α, hence $E^\tau(\varphi(z)^\tau) = \varphi_1(z)^\tau = E(\varphi(z)^\tau)$. Now we can change z for $\beta(z)$ with any $\beta \in G_{Q+}$ without changing π. (α may change.) Since the points $\varphi(\beta(z))^\tau$ form a Zariski dense subset of V, we obtain $E^\tau = E$ for any isomorphism τ of k_1 into C over k_S, hence (5.4.1).

5.5. Let $\mathfrak{W}$ be the set of all members S of $\mathfrak{Z}$ of the form $S = S(\Omega)$ with some Ω as given in 5.1. $\mathfrak{W}$ is obviously normal. For each $S \in \mathfrak{W}$, take Ω so that $S = S(\Omega)$, and take $(V, \mathfrak{b}, \varphi)$ as in 5.3. We put $V = V_S$, $\varphi = \varphi_S$. The argument of 5.4 shows that this does not depend on the choice of Ω.

Let x, Ω', $(V', \mathfrak{b}', \varphi')$, and J be as in (5.1.3) and 5.3. Put $T = x^{-1}Sx$. Then $T = S(\Omega')$, and we may put $(V', \varphi') = (V_T, \varphi_T)$. From (5.3.1), we obtain

(5.5.1) $\quad \varphi_S(z)^\tau = J(\varphi_T(\alpha(z)))$. $\quad (\pi = [v, P']$ on P'_{ab}, $\eta(v)^{-1} = ux\alpha$, $u \in S$.)

This relation characterizes J. In fact, suppose J_1 is a biregular isomorphism of V_T to V_S^z satisfying the same relation. Then we have $J^{-1}(\varphi_S(z)^z) = \varphi_T(\alpha(z)) = J_1^{-1}(\varphi_S(z)^z)$. We can substitute $\beta(z)$ for z with any $\beta \in G_{Q+}$ without changing π. Since the points $\varphi_S(\beta(z)^\pi)$ form a Zariski dense subset of V_S, we obtain $J = J_1$. Therefore we can put $J = J_{ST}(x)$ without worrying about the choice of Ω. We are going to show that

$$\{V_S,\ \varphi_S,\ J_{ST}(x),\ (S,\ T \in \mathcal{W};\ x \in \mathcal{G}_+)\}$$

is a weak canonical system relative to $\{\mathcal{W}, Q, F\}$. The rules $(2.5.3_a)$ and $(2.5.3_b)$ follow from our definition of $J_{ST}(x)$ in a straightforward way. Now take an element γ of G_{Q+} as x. If $w = \gamma^{-1}(z)$ with any $z \in \mathfrak{H}_n^g$, the map $\Lambda(\gamma, z)$ of [12, (9.3.1)] gives an isomorphism of the member $\mathfrak{A}_z$ of $\Sigma(\Omega)$ to the member $\mathfrak{A}_w'$ of $\Sigma(\Omega')$, where $\Omega' = (\nu(\gamma)^{-1}\kappa,\ \mathfrak{m}\gamma,\ \{d_i\gamma\})$. Therefore $\varphi_S(\gamma(w)) = \mathfrak{b}(\mathfrak{A}_z) = J_{ST}(\gamma)(\mathfrak{b}'(\mathfrak{A}_w')) = J_{ST}(\gamma)(\varphi_T(w))$. This proves $(2.5.3_c)$.

5.6. To prove $(2.5.4)$, let us come back again to the situation of 5.2. We do not assume $(5.2.1)$, but start with an automorphism π of C over P'. Repeat the argument of 5.2 with $\eta(v)^{-1}$ as x. Then $\Omega' = \Omega'' = (\xi\kappa,\ \mathfrak{m}\eta(v)^{-1},\ \{d_i\eta(v)^{-1}\})$. Put $T = \eta(v)S\eta(v)^{-1}$. Then $S(\Omega') = T$. From $(5.5.1)$, we obtain $\varphi_S(z)^\pi = J_{ST}(\eta(v)^{-1})(\varphi_T(z))$. This shows that $\varphi_S(z)^\pi$ depends only on the effect of π on P_{ab}'. It follows that $\varphi_S(z)$ is rational over P_{ab}'. Thus we obtain $(2.5.4)$.

5.7. Fix any Z-lattice $\mathfrak{m}$ in X, and put

$$U = \{x \in G_{A+} \mid \mathfrak{m}x = \mathfrak{m}\}\ ,$$

and, for every positive integer e,

$$S(\mathfrak{m}, e) = G_{\infty+} \cdot \{x \in D_+ \mid x \equiv 1\ \mathrm{mod}_0\ (\mathfrak{m}, e)\}\ .$$

Then we see that $S(\mathfrak{m}, e)$ belongs to $\mathcal{W}$, and is a normal subgroup of U. Moreover, for every $S \in \mathcal{Z}$, there exists a positive integer e such that $S(\mathfrak{m}, e) \subset S$. Thus $\mathcal{W}$ and U satisfy the condition of 3.11. Since we have established a weak canonical system relative to $\{\mathcal{W}, Q, F\}$, the assertion of 3.11 assures the existence of a canonical system for G.

6. Proof of main theorem in the case $r < g$ [4]

6.1. Let K be a totally imaginary quadratic extension of F. We choose extensions of $\tau_1, \cdots, \tau_g$ to K, and denote them again by $\tau_1, \cdots, \tau_g$. Put $L =$

---

[4] In this section we assume throughout $r < g$. However, our reasoning is actually valid even in the case $r = g$, with a suitable interpretation of symbols. For example, we should understand that $\pi(a) = 1$ if $r = g$.

$B \otimes_F K$, $L_R = L \otimes_Q R$. Take a representation Φ of L_R, a positive involution ρ of L, and an L-valued anti-hermitian form T on L^n as in [12, 6.2–4]. The restriction of ρ to K coincides with the complex conjugation. In this section, the complex conjugation in any number field will also be denoted by ρ. Define the group $G(T)$ of similitudes of T, and $G_R(T)$ as in [12, 6.5]. We may put

$$G(T) = G_Q^*, \; G_R(T) = G_R^*$$

with an algebraic group G^* defined over Q, (and contained in GL_{8gn} with respect to a basis of L^n over Q). We have a homomorphism

$$\nu: G^* \longrightarrow F^\times$$

which gives the multiplier $\nu(x)$ of each similitude $x \in G^*$.

Let us consider a couple (K, Φ_K) with

$$\Phi_K \sim n \sum_{\nu=1}^r (\tau_\nu + \tau_\nu \rho) + 2n \sum_{\nu=r+1}^g \tau_\nu .$$

Let (K', Φ') be the reflex of (K, Φ_K), and $2h = [K': Q]$. Then, as was seen in [11, 5.14], we have

$$\Phi' \sim n \sum_{\nu=1}^s (\sigma_\nu + \sigma_\nu \rho) + 2n \sum_{\nu=s+1}^h \sigma_\nu$$

with a positive integer s such that $0 < s < h$, and a suitable choice of h isomorphisms $\sigma_1, \cdots, \sigma_h$, which, together with $\sigma_1 \rho, \cdots, \sigma_h \rho$, form the set of all distinct isomorphisms of K' into C. Let $\Phi_0 \sim \sum_{\nu=r+1}^g \tau_\nu$. Then the reflex of (K, Φ_0) is given by (K', Φ_0') with the same K' and $\Phi_0' = \sum_{\nu=s+1}^h \sigma_\nu$ (see [11, (5.14.7, 8)]).

Put $\zeta = \det \Phi'$, $\pi = \det \Phi_0'$. Then

$$(6.1.1) \qquad \zeta(a) = [N_{K'/Q}(a)\pi(a)/\pi(a)^\rho]^n \qquad (a \in K') .$$

We shall denote by $\delta(x)$ the reduced norm of an element x of $M_n(L)$ to K. Then

$$(6.1.2) \qquad N_{K/F}\big(\delta(x)\big) = \nu(x)^{2n} \qquad (x \in G^*) ,$$

$$(6.1.3) \qquad \lambda(N_{K'/F'}(y))\pi(y)\pi(y)^\rho = N_{K'/Q}(y) \qquad (y \in K') .$$

The latter formula follows from [11, 5.6].

6.2. The notation G_{A+}^* and $G_{\infty+}^*$ being as in 0.4, we have

$$G_{A+}^* = \{x \in G_A^* \mid \nu(x) \gg 0\} .$$

We let $G_{\infty+}^*$ act on the domain $\mathfrak{S}^r$ as in [12, 6.5]. Define an algebraic subgroup $\tilde{C}$ of $G^* \times K'^\times$ defined over Q, by

$$\tilde{C} = \{(x, a) \in G^* \times K'^\times \mid \delta(x) = \zeta(a), \nu(x) = N_{K'/Q}(a)\} ,$$

and denote by C the projection of $\tilde{C}_A$ to G_A^*. Then we put

$$C_+ = C \cap G_{A+}^* , \qquad \mathcal{G}_+^* = C_+ G_{Q+}^* G_{\infty+}^* .$$

We can extend the map $\nu\colon G_{Q+}^* \to F_+^\times$ to a homomorphism

$$\mu\colon \mathcal{G}_+^* \longrightarrow F_+^\times$$

as follows. By our definition of $\mathcal{G}_+^*$, we have $\nu(x) = abc$ with $a \in Q_A^\times$, $b \in F_+^\times$, $c \in F_{\infty+}^\times$. Let a_1 be the positive integer which generates the ideal associated with a. We put then

$$(6.2.1) \qquad\qquad \mu(x) = a_1 b .$$

This is independent of the choice of a, b, c.

6.3. Let $\mathfrak{D}_n$ be the domain defined in [12, 4.1]. In [12, 6.6], we have obtained the following maps:

    a Q-rational injection $i\colon G \to G^*$,

    a B-linear injection $i'\colon B^n = X \to L^n$,

    a holomorphic bijection $j\colon \mathfrak{H}_n^r \to \mathfrak{D}_n^r \subset \mathcal{S}^r$,

such that $\nu(i(\alpha)) = \nu(\alpha), j(\alpha(z)) = i(\alpha)(j(z)), i'(x\alpha) = i'(x)i(\alpha)$ for $\alpha \in G_Q, z \in \mathfrak{H}_n^r$, and $x \in B^n$. Hereafter we identify G with $i(G)$, and understand that $G_{\infty+}$ acts both on $\mathfrak{H}_n^r$ and $\mathfrak{D}_n^r$. We note

$$(6.3.1) \qquad\qquad \delta(x) = \nu(x)^n \qquad\qquad (x \in G) .$$

Put $K' = H$, and define λ_H, D_{H+}, and $\mathcal{G}_{H+}$ as in 3.1. If $y \in D_{H+}$, there exists an element h of $H_A^\times$ such that $\nu(y) = \lambda_H(h)$. Put $w = \pi(h)y$. Since $\pi(h) \in K_A^\times$, we see that $w \in G_A^*$. Moreover, we have, by (6.1.1), (6.1.3) and (6.3.1),

$$(6.3.2) \qquad\qquad \nu(w) = \pi(h)\pi(h)^\rho \nu(y) = N_{H/Q}(h) ,$$

$$(6.3.3) \qquad\qquad \delta(w) = \pi(h)^{2n}\nu(y)^n = \zeta(h) ,$$

so that $(w, h) \in \tilde{C}_A$, hence

$$(6.3.4) \qquad\qquad \pi(h)y \in C_+ .$$

In the next place, consider an x of $\mathcal{G}_{H+}$. We can find an element d of $H_A^\times$ such that $\nu(x)/\lambda_H(d) \in F_+^\times F_{\infty+}^\times$. Then

    (6.3.5) *There exists an element γ of G_{Q+}, an element x_1 of D_{H+}, and an element q of $G_{\infty+}$ such that $\nu(x_1) = \lambda_H(d)$, and $x = qx_1\gamma$. It follows especially that $\pi(d)x \in \mathcal{G}_+^*$.*

To show this, put $\nu(x) = \lambda_H(d)cf$ with $c \in F_+^\times, f \in F_{\infty+}^\times$. Then we can find elements γ of G_Q and q of G_∞ so that $\nu(\gamma) = c$ and $\nu(q) = f$. Put $x_1 = q^{-1}x\gamma^{-1}$. Then $\nu(x_1) = \lambda_H(d)$. By (6.3.4), $\pi(d)x_1 \in C_+$, hence $\pi(d)x \in \mathcal{G}_+^*$.

6.4. For every $\mathfrak{r}_K$-lattice $\mathfrak{N}$ in L^n, and for every positive integer a, put

$$\Gamma^*(\mathfrak{N}, a) = \{\gamma \in G_\mathbf{Q}^* \mid \nu(\gamma) = 1, \mathfrak{N}\gamma = \mathfrak{N}, \mathfrak{N}(1 - \gamma) \subset a\mathfrak{N}\} \; .$$

Take an $\mathfrak{r}_F$-lattice $\mathfrak{m}$ in B^* satisfying (2.10.3), and put $\mathfrak{M} = \mathfrak{r}_K \cdot i'(\mathfrak{m})$. For every positive integer a, define $S(\mathfrak{m}, a)$ as in 2.10, and put, with two positive integers b and c,

$$(6.4.1) \qquad S = S(\mathfrak{m}, c) \cdot \{x \in S(\mathfrak{m}, b) \mid \nu(x) = 1\} \; .$$

Let us now show that, for a given positive integer a, we can find b and c so that the following three conditions are satisfied:

(6.4.2) $c\mathbf{Z} \subset b\mathbf{Z} \subset a\mathbf{Z}$, hence $S \subset S(\mathfrak{m}, a)$.

(6.4.3) *Put* $E = \mathfrak{r}_F^\times$. *Then, for every* $u \in G_A$,

$$E \cdot \Gamma(u^{-1}Su) = \{\alpha \in E \cdot \Gamma^*(\mathfrak{M}u, b) \mid \alpha(\mathfrak{D}_n^r) \cap \mathfrak{D}_n^r \neq \varnothing\} \; .$$

(6.4.4) *For every* $u \in G_A$, $\Gamma^*(\mathfrak{M}u, b)$ *has no element of finite order other than the identity element.*

To show this, we first take a finite set $\mathcal{Y}$ of $\mathfrak{r}_F$-lattices in B^* so that

$$\{\mathfrak{m}u \mid u \in G_A\} = \{\mathfrak{n}\beta \mid \mathfrak{n} \in \mathcal{Y}, \beta \in G_\mathbf{Q}\} \; .$$

By [10, 2.22] and [12, 2.9], there exists a positive integer b divisible by a, such that, for every $\mathfrak{n} \in \mathcal{Y}$, if $\alpha \in \Gamma^*(\mathfrak{r}_K\mathfrak{n}, b)$ and $\alpha(\mathfrak{D}_n^r) \cap \mathfrak{D}_n^r \neq \varnothing$, then $\alpha \in G_\mathbf{Q}$. Further we can choose this b so that $\Gamma^*(\mathfrak{r}_K\mathfrak{n}, b)$ has no element of finite order other than the identity, for every $\mathfrak{n} \in \mathcal{Y}$. Now, by Chevalley [1], there exists a positive integer c, divisible by b, such that, if $\varepsilon \in E$ and $\varepsilon \equiv 1 \bmod (c)$, then $\varepsilon = \eta^2$, $\eta \equiv 1 \bmod (b)$ for some $\eta \in E$. Let $u \in G_A$. Then $\mathfrak{m}u = \mathfrak{n}\beta$ for some $\mathfrak{n} \in \mathcal{Y}, \beta \in G_\mathbf{Q}$. If $\alpha \in \Gamma^*(\mathfrak{M}u, b)$ and $\alpha(\mathfrak{D}_n^r) \cap \mathfrak{D}_n^r \neq \varnothing$, then $\beta\alpha\beta^{-1} \in \Gamma^*(\mathfrak{r}_K\mathfrak{n}, b)$. By our choice of b, we have $\beta\alpha\beta^{-1} \in G_\mathbf{Q}$, so that $\alpha \in G_\mathbf{Q}$, hence $\alpha \in \Gamma(u^{-1}Su)$ on account of [12, 2.9]. Conversely, let $\gamma \in \Gamma(u^{-1}Su)$. Then $u\gamma u^{-1} = vw$ with $v \in S(\mathfrak{m}, c)$, $w \in S(\mathfrak{m}, b), \nu(w) = 1$, hence $\nu(\gamma) \equiv 1 \bmod (c)$. Therefore $\nu(\gamma) = \xi^2, \xi \equiv 1 \bmod (b)$ with $\xi \in E$. Then $\gamma = \xi \cdot \xi^{-1}\gamma, \nu(\xi^{-1}\gamma) = 1$. Since $\xi^{-1}\gamma \equiv 1 \bmod_0 (\mathfrak{M}u, b)$, we have $\xi^{-1}\gamma \in \Gamma^*(\mathfrak{M}u, b)$, so that $\gamma \in E \cdot \Gamma^*(\mathfrak{M}u, b)$, q.e.d.

6.5. Let a, b, c, and S be as in 6.4. Let M_c denote the class field over H corresponding to the subgroup $H^\times \cdot \{h \in H_A^\times \mid h \equiv 1 \bmod_0 (c)\}$ of $H_A^\times$. Then M_c *contains* $k_S H$. In fact, let $h \in H_A^\times, h \equiv 1 \bmod_0 (c)$. Then $\lambda_H(h) \equiv 1 \bmod_0 (c)$. Since $\mathfrak{m}$ satisfies (2.10.3), there exists an element s of G_{A+} such that $\nu(s) = \lambda_H(h)$ and $s \equiv 1 \bmod_0 (\mathfrak{m}, c)$. Then $s \in S(\mathfrak{m}, c) \subset S$, hence $h \in \mathfrak{X}_{HS}$. Therefore $k_S H \subset M_c$.

6.6. Let $\mathfrak{m}, \mathfrak{M}, a, b, c$, and S be as in 6.4. Let $\mathfrak{N}$ be a $\mathbf{Z}$-lattice in L^* of the form $\mathfrak{N} = p\mathfrak{M}y$ with $p \in K_A^\times$ and $y \in G_A$. Observe that

$$\Gamma^*(\mathfrak{N}, b) = \Gamma^*(\mathfrak{M}y, b) \; .$$

Let us now consider a PEL-type

$$\Omega = (L, \Phi, \rho; \kappa T, \mathfrak{N}; q_1, \cdots, q_s)$$

with elements q_i of $L^n/\mathfrak{N}$, and a totally positive element κ of F such that $b^{-1}\mathfrak{N}/\mathfrak{N} = \sum_{i=1}^s Zq_i$, $\mathrm{tr}_{L/Q}(\kappa T(\mathfrak{N}, \mathfrak{N})) = Z$. Since L, Φ, ρ, and T are common to all these PEL-types, we write simply $\Omega = (\kappa, \mathfrak{N}, \{q_i\})$. We construct a family $\Sigma(\Omega) = \{\mathfrak{Q}_z \mid z \in \mathfrak{S}^r\}$ of PEL-structures

$$\mathfrak{Q}_z = (A_z, \mathcal{C}_z, \theta_z; t_{1z}, \cdots, t_{sz})$$

by means of the parametrizing function $\mathfrak{y}$ as in [12, 6.4], common to all Ω of this type (see [11, 4.15]).

Let $x \in \mathfrak{G}_{H+}$, and let d be an element of $H_A^\times$ such that $\nu(x)/\lambda_H^\cdot(d) \in F^\times F_{\infty+}^\times$. We are going to prove:

(6.6.1) *The notation being as above, one has $k_\Omega \subset M_c$. Moreover, if $\sigma = [d^{-1}, H]$, Ω^σ is equivalent to*

$$\Omega' = \left(\mu(\pi(d)x)^{-1}\kappa, \pi(d)\mathfrak{N}x, \{\pi(d)q_ix\}\right) .$$

6.7. Let k' be an arbitrarily fixed finite algebraic extension of M_c. Take $Y, P, (P_i, \Psi_i), P', \eta$, and z as in 2.4. Put $y = j(z)$. By [12, 7.4, 7.5], we can take these so that

(6.7.1) *P' and k' are linearly disjoint over F'*;

(6.7.2) *$P_i \otimes_F K$ is a field for every i.*

Let τ be an isomorphism of k' into C such that $\tau = [d^{-1}, H]$ on M_c. Then τ can be extended to an automorphism π of C over $P'H$. Take an element u of $(P'H)_A^\times$ so that $\pi = [u, P'H]$ on $(P'H)_{ab}$. Put $v = N_{P'H/P'}(u)$, $w = N_{P'H/H}(u)$. By (2.4.4), we have

$$(6.7.3) \qquad\qquad\qquad \nu(\eta(v)) = \lambda_H(w) .$$

Consider the member $\mathfrak{Q}_y$ of $\Sigma(\Omega)$ at the point y. By [12, 6.7–9], θ_z can be extended to an anti-isomorphism θ^* of $Z = L \otimes_F Y$ into $\mathrm{End}_Q(A_z)$. Define Ξ_i as in [12, 6.8]. Then $\mathfrak{Q}^* = (A_y, \mathcal{C}_y, \theta^*; t_{1y}, \cdots, t_{sy})$ is of the type discussed in § 4. The present (KP_i, Ξ_i) corresponds to (K_i, Φ_i) of 4.1. With the terminology of 4.1, $\mathfrak{Q}^*$ is of type

$$(Z, \Xi, \mathfrak{N}, U; q_1, \cdots, q_s)$$

where $U(a, a') = \mathrm{tr}_{L/Q}(\kappa T(a, a'))$, and Ξ is such that its restriction to KP_i is equivalent to $2q_i \cdot \Xi_i$ plus a zero representation. (Here we consider L^n as a Z-module.) Observe that the conditions (4.1.2) and (4.1.3) are satisfied. We can therefore apply 4.3 to $\mathfrak{Q}^*$. By [12, 6.10], $P'H$ and $[u, P'H]$ correspond to K' and $[c, K']$ of 4.3. Further the formula of [12, 6.10] shows that $\pi(w)\eta(v)$ corresponds to $\eta(c)$ of 4.3. Therefore $\mathfrak{Q}^{*\tau}$ is of type

(6.7.4)
$$\left(Z,\ \Xi,\ \pi(w)^{-1}\mathfrak{N}\eta(v)^{-1},\ \xi U;\ \{\pi(w)^{-1}q_i\eta(v)^{-1}\}\right),$$

with $\xi = N_{P'H/Q}(\mathfrak{u})$, where $\mathfrak{u}$ is the ideal in $P'H$ associated with u. Now let

$$\Omega'' = \left(\xi\kappa,\ \pi(w)^{-1}\mathfrak{N}\eta(v)^{-1},\ \{\pi(w)^{-1}q_i\eta(v)^{-1}\}\right).$$

Consider the member $\mathfrak{A}_y'' = (A_y'', \cdots)$ of $\Sigma(\Omega'')$ with the same y. Then $\mathfrak{A}_y''$, furnished with an anti-isomorphism of Z into $\mathrm{End}_Q(A_y'')$, is of type (6.7.4). By (4.1.10), $(\mathfrak{A}_v)^\tau$ and $\mathfrak{A}_y''$ must be isomorphic.

6.8. As shown in (6.3.5), we can find an element γ of G_{Q+}, an element q of $G_{\infty+}$, and an element x_1 of D_{H+} so that $\nu(x_1) = \lambda_H(d)$ and $x = qx_1\gamma$. Since $[d^{-1}, H] = [w, H]$ on M_c, we have $dw = eh$ with $e \in H^\times$, $h \in H_A^\times$, $h \equiv 1 \bmod_0 (c)$. Let $\mathfrak{N} = f\mathfrak{M}p$ with $f \in K_A^\times$ and $p \in G_A$. Put $T = p^{-1}Sp$. Then $\mathfrak{N}t = \mathfrak{N}$ for every $t \in T$. Since $\mathfrak{m}$ satisfies (2.10.3), there exists an element s of G_A such that $\nu(s) = \lambda_H(h)$, and $s \equiv 1 \bmod_0 (\mathfrak{m}p, c)$. Then $s \in T$. Since $\lambda_H(e) \in F_+^\times$, there exists an element ε of G_{Q+} such that $\nu(\varepsilon) = \lambda_H(e)$. By (6.7.3), we have

$$\nu\big(x_1^{-1}s\eta(v)^{-1}\varepsilon\big) = \lambda_H(d^{-1}hw^{-1}e) = 1 \ .$$

By 3.4, $x_1^{-1}s\eta(v)^{-1}\varepsilon = m\psi$ with $\psi \in G_Q^{\mathfrak{u}}$ and $m \in G_A^{\mathfrak{u}} \cap (x_1^{-1}Tx_1)$. Put $\alpha = \gamma^{-1}\psi\varepsilon^{-1}$, $t = s^{-1}x_1 m x_1^{-1}$. Then

(6.8.1)
$$\eta(v)^{-1} = tq^{-1}x\alpha,\ t \in T,\ \nu(t) = \lambda_H(h)^{-1},\ \alpha \in G_{Q+}\ ,$$

(6.8.2)
$$\pi(w)^{-1}\eta(v)^{-1} = \big(\pi(h)^{-1}t\big)\cdot\big(\pi(d)q^{-1}x\big)\cdot\big(\pi(e)^{-1}\alpha\big)\ .$$

By (6.3.2), we have $\nu\big(\pi(h)^{-1}t\big) = N_{H/Q}(h)^{-1}$. Since $h \equiv 1 \bmod_0 (c)$, we obtain $\mu\big(\pi(h)^{-1}t\big) = 1$. Similarly $\nu\big(\pi(w)^{-1}\eta(v)^{-1}\big) = N_{H/Q}(w)^{-1} = N_{P'H/Q}(u)^{-1}$, hence $\mu\big(\pi(w)^{-1}\eta(v)^{-1}\big) = \xi^{-1}$. Therefore from (6.8.2), we obtain

(6.8.3)
$$\xi^{-1} = \mu\big(\pi(d)x\big)\cdot\nu\big(\pi(e)^{-1}\alpha\big)\ .$$

Since $\pi(h) \equiv 1 \bmod_0 (c)$, and $tq^{-1} \equiv 1 \bmod_0 (\mathfrak{N}, \mathfrak{b})$, we have

$$\Omega'' = \big(\xi\kappa,\ \pi(e^{-1}d)\mathfrak{N}x\alpha,\ \{\pi(e^{-1}d)q_i x\alpha\}\big)\ .$$

Put $y_1 = \alpha(y)$, and consider the member $\mathfrak{A}_{y_1}'$ of $\Sigma(\Omega')$, where Ω' is as in (6.6.1). The map $\Lambda\big(\pi(e)^{-1}\alpha, y_1\big)$ of [12, (6.7.1)] and [10, 2.19] defines an isomorphism of $\mathfrak{A}_{y_1}'$ to $\mathfrak{A}_y''$ on account of (6.8.3). Therefore $(\mathfrak{A}_v)^\tau$ is isomorphic to $\mathfrak{A}_{y_1}'$, hence Ω^τ is equivalent to Ω'.

Now take $k_\Omega M_c$ as k', and let $x = 1, d = 1$. Then the above result shows that Ω^τ is equivalent to Ω for every isomorphism τ of $k_\Omega M_c$ into C over M_c. This proves that $k_\Omega \subset M_c$. Thus we obtain (6.6.1).

6.9. Let $\mathfrak{W} = \{p^{-1}Sp \mid p \in G_A\}$. Our aim is to construct a weak canonical system relative to $\{\mathfrak{W}, H, K\}$. Fix an element p of G_A, and put $T = p^{-1}Sp$. Consider $\Omega = (\kappa, f\mathfrak{M}p, \{q_i\})$ with any $f \in K_A^\times$ and any elements $q_1, \cdots, q_s$ of

$L^{\kappa}/f\mathfrak{M}p$ such that $b^{-1}\mathfrak{N}/\mathfrak{N} = \sum_{i=1}^{\cdot} \mathbf{Z}q_i$, where $\mathfrak{N} = f\mathfrak{M}p$. Let $(V, \mathfrak{v}, \varphi)$ be a moduli-system for $\Sigma(\Omega)$ in the sense of [11, 4.17] and [9, 6.2]. Then (V, φ) is a model of $\mathfrak{S}^r/\Gamma^*(\mathfrak{N}, b)$. Put $V_T = \varphi(\mathfrak{D}_n^r)$ and $\varphi_T = \varphi \circ j$. By (6.4.3) and (6.4.4), we see that (V_T, φ_T) is a model of $\mathfrak{S}_n^r/\Gamma_T$.

Let x, d, and Ω' be as in (6.6.1), and $(V', \mathfrak{v}', \varphi')$ be the moduli-system for $\Sigma(\Omega')$. Put $U = x^{-1}Tx, V_U = \varphi'(\mathfrak{D}_n^r)$, and $\varphi_U = \varphi' \circ j$. Then (V_U, φ_U) is a model of $\mathfrak{S}_n^r/\Gamma_U$. By (6.6.1), if $\sigma = [d^{-1}, H]$, Ω^σ is equivalent to Ω'. Therefore, by [11, 4.21, 4.23], there exists a biregular isomorphism J of V' to V^σ, rational over M_c, such that $\mathfrak{v}^\sigma = J \circ \mathfrak{v}'$.

Let P', π, z, y, α, and y_1 be as in 6.7 and 6.8. Then we have shown that if π is an automorphism of C over $P'H$ which coincides with $[d^{-1}, H]$ on M_c, then $\mathfrak{A}'_{y_1}$ is isomorphic to $(\mathfrak{A}_y)^\pi$. Therefore $\varphi_T(z)^\pi = \varphi(y)^\pi = \mathfrak{v}^\pi(\mathfrak{A}_y)^\pi = J(\mathfrak{v}'(\mathfrak{A}'_{y_1})) = J(\varphi'(y_1)) = J(\varphi_U(\alpha(z)))$. We have thus

$$(6.9.1) \qquad\qquad \varphi_T(z)^\pi = J(\varphi_U(\alpha(z))) \,.$$

Now the point $\varphi_T(z) = \varphi(j(z))$ is algebraic over $\mathbf{Q}$, since (6.9.1) shows that $\varphi_T(z)^\pi$ depends only on the effect of π on $(P'H)_{ab}$; one can also notice that A_y has sufficiently many complex multiplications, hence $\varphi(y) = \varphi_T(z)$ is algebraic. The points of this type form a dense subset of V_T. Therefore V_T is rational over a finite algebraic extension k_1 of M_c. Take k_1 as k' in 6.7 and 6.8. Let τ be an isomorphism of k_1 into C such that $\tau = [d^{-1}, H]$ on M_c. Extending τ to π, we obtain (6.9.1). Now we can change z for $\beta(z)$ with any $\beta \in G_{Q+}$ without changing P' and π. The points $\beta(z)$ form a dense subset of $\mathfrak{S}_n^r$, hence J^{-1} sends V_T^π into V_U. Take both x and d to be the identity element. Then we see that $V_T = V_T^\tau$ for every isomorphism τ of k_1 into C over M_c. This shows that

(6.9.2) *V_T is rational over M_c.*

Similarly V_U is rational over M_c. Therefore, if $\sigma = [d^{-1}, H]$, J gives a biregular isomorphism of V_U onto V_T^σ rational over M_c.

6.10. To define (V_T, φ_T), we have chosen f, p, and $\{q_i\}$. Let us now show that it is actually independent of the choice of f, p, and $\{q_i\}$. Let p_1 be another element of G_A such that $T = p_1^{-1}Sp_1$, and consider $\Omega_1 = (\kappa_1, f_1\mathfrak{M}p_1, \{q_{1i}\})$ of the same nature as Ω. Define (V_{1T}, φ_{1T}) by the same procedure as before, by using $\Sigma(\Omega_1)$ in place of $\Sigma(\Omega)$. We have obviously a biregular isomorphism Q of V_T to V_{1T} such that $\varphi_{1T} = Q \circ \varphi_T$. Our task is to show that Q is rational over M_c. Since $Q(\varphi_T(z)) = \varphi_{1T}(z)$ for the points z of the above type, we see that Q is defined over a finite algebraic extension k_2 of M_c. Take k_2 as k' in 6.7 and 6.8. Further let $x = 1$ and $d = 1$. Let τ be an isomorphism of k_2 into C over

M_c. Extending τ to π, we obtain $\varphi_T(z)^\pi = \varphi_T(\alpha(z))$ as before. Similarly we have $\varphi_{1T}(z)^\pi = \varphi_{1T}(\alpha(z))$ with *the same* α, π and z. (Note that $f_1 \mathfrak{M} p_1$ and q_{1i} are invariant under the action of any element t of T.) Therefore $Q^\tau(\varphi_T(z)^\pi) = \varphi_{1T}(z)^\pi = Q(\varphi_T(z)^\pi)$. We can again substitute $\beta(z)$ for z with any $\beta \in G_{Q+}$ and with the same π. Since the points $\varphi_T(\beta(z))^\pi$ form a Zariski dense subset of V_T^σ, we obtain $Q^\tau = Q$, so that Q is rational over M_c. Thus (V_T, φ_T) does not depend on the choice of f, p and $\{q_i\}$.

6.11. Denote by $J_{TU}(x, d)$ the restriction of J to V_U. This is defined for every $x \in \mathfrak{S}_{H+}$ and an element d of $H_A^\times$ such that $\nu(x)/\lambda_H(d) \in F^\times F_{\infty+}^\times$. The map $J_{TU}(x, d)$ does not depend on the choice of f, p, and $\{q_i\}$, since it is characterized by the property (6.9.1). In fact, let J_1 be another biregular isomorphism of V_U to V_T^σ satisfying (6.9.1). Then $J_1^{-1}(\varphi_T(z)^\pi) = \varphi_U(\alpha(z)) = J^{-1}(\varphi_T(z)^\pi)$. Substituting $\beta(z)$ for z with any $\beta \in G_{Q+}$, and observing that the points $\varphi_T(z)^\pi$ form a Zariski dense subset of V_T^σ, we can conclude that $J_1^{-1} = J^{-1}$, q.e.d.

6.12. Let us now show that $J_{TU}(x, d)$ depends only on the coset xU and the effect of $[d, H]$ on M_c. Let $x^* = xs$, $s \in U$, $\nu(x^*)/\lambda_H(d^*) \in F^\times F_{\infty+}^\times$, $d^* \in H_A^\times$, $[d, H] = [d^*, H]$ on M_c. Then $d^{-1}d^* = gh$, $g \in H^\times$, $h \in H_A^\times$, $h \equiv 1 \bmod_0 (c)$, $\pi(d^*)x^* = \pi(d)x \cdot \pi(gh)s$, and $\nu(s)/\lambda_H(gh) = ae$ with $a \in F^\times$, $e \in F_{\infty+}^\times$. Since $a \cdot \lambda_H(g) \gg 0$, we have $a\lambda_H(g) = \nu(\varepsilon)$ with $\varepsilon \in G_{Q+}$, and $e = \nu(r)$ with $r \in G_{\infty+}$. Since $\mathfrak{m}$ satisfies (2.10.3), there exists an element s' of G_A such that $\nu(s') = \lambda_H(h)$ and $s' \equiv 1 \bmod_0 (\mathfrak{m}px, c)$. Then $s' \in U$, and $\nu(s'^{-1}r^{-1}s\varepsilon^{-1}) = 1$. By 3.4, $s'^{-1}r^{-1}s\varepsilon^{-1} = s''\gamma$ with $\gamma \in G_Q^u$ and $s'' \in U \cap G_A^u$. We see that $\gamma\varepsilon = s''^{-1}s'^{-1}r^{-1}s \in G_Q \cap U = \Gamma_U$, and $\pi(gh)s = (\pi(h)rs's'')(\pi(g)\gamma\varepsilon)$. By (6.3.2), we have $\nu(\pi(h)s') = N_{H/Q}(h)$, so that $\mu(\pi(h)rs's'') = 1$. Therefore $\mu(\pi(gh)s) = \nu(\pi(g)\gamma\varepsilon)$, hence

(6.12.1) $$\mu(\pi(d^*)x^*) = \mu(\pi(d)x) \cdot \nu(\pi(g)\gamma\varepsilon) \,.$$

Let Ω', $\Sigma(\Omega')$, and $(V', \mathfrak{v}', \varphi')$, J, and (V_U, φ_U) be as in (6.6.1) and 6.9. Define $\Sigma(\Omega^*) = \{\mathfrak{Q}_y^* \mid y \in \mathfrak{S}^r\}$ for

$$\Omega^* = \left(\mu(\pi(d^*)x^*)^{-1}\kappa, \pi(d^*)\mathfrak{N}x^*, \{\pi(d^*)q_i x^*\}\right) \,.$$

Observe that $\mathfrak{N}x = \mathfrak{N}x^*$ and $q_i x = q_i x^*$. By 4.4, we can take the moduli-system for $\Sigma(\Omega^*)$ in the form $(V', \mathfrak{v}^*, \varphi')$. By (6.6.1), Ω^σ is equivalent to Ω^*. Therefore, we obtain a biregular isomorphism J^* of V' to V^σ such that $J^* \circ \mathfrak{v}^* = \mathfrak{v}^\sigma$. Put $\omega = \pi(g)\gamma\varepsilon$. Observe that ω sends $\pi(d)\mathfrak{N}x$ and $\pi(d)q_i x$ to $\pi(d^*)\mathfrak{N}x^*$ and $\pi(d^*)q_i x^*$. Therefore, in view of (6.12.1), the map $\Lambda(\omega, y)$ of [12, (6.7.1)], for any $y \in \mathfrak{S}^r$, gives an isomorphism of the member $\mathfrak{Q}_y'$ of $\Sigma(\Omega')$ to $\mathfrak{Q}_{y'}^*$, where $y' = \omega^{-1}(y)$. Take y to be $j(z)$ with $z \in \mathfrak{S}_x^r$. Then we have

$$J(\varphi_U(z)) = J(\mathfrak{v}'(\mathfrak{A}'_{\mathfrak{v}'})) = J^*(\mathfrak{v}^*(\mathfrak{A}^*_{\mathfrak{v}^*})) = J^*(\varphi'(\omega^{-1}(y))) = J^*(\varphi_U(z)) \,,$$

since $\gamma\varepsilon \in \Gamma_U$ and $\pi(g) \in K^\times$. This proves $J = J^*$, and thus the desired property of $J_{TU}(x, d)$.

6.13. Let T, x, d, U be as above. Let $y \in \mathcal{G}_{H+}$, $e \in H_A^\times$, $\nu(y)/\lambda_H(e) \in F^\times F_{\infty+}^\times$, $R = y^{-1}Uy$. We are going to prove

$$(6.13.1) \qquad J_{TR}(xy, de) = J_{TU}(x, d)^{\tau} \circ J_{UR}(y, e) \,, \qquad \tau = [e^{-1}, H] \,.$$

Consider again $\Omega, \Omega', J, (V, \mathfrak{v}, \varphi)$ and $(V', \mathfrak{v}', \varphi')$ as in 6.9. Let

$$\Omega'' = \left(\mu(\pi(de)xy)^{-1}\kappa, \pi(de)\mathfrak{R}xy, \{\pi(de)q_ixy\}\right) \,,$$

and let $(V'', \mathfrak{v}'', \varphi'')$ be the moduli-system for $\Sigma(\Omega'')$. Then we may put $V_R = \varphi''(\mathfrak{D}_n^r)$, $\varphi_R = \varphi'' \circ j$. Further we have a biregular isomorphism J' of V'' to V'^τ such that $J' \circ \mathfrak{v}'' = \mathfrak{v}'^\tau$. Then $J^\tau \circ J' \circ \mathfrak{v}'' = J^\tau \circ \mathfrak{v}'^\tau = \mathfrak{v}^{\sigma\tau}$. By our definition, the restriction of $J^\tau \circ J'$ (resp. J') to V_R is exactly $J_{TR}(xy, de)$(resp. $J_{UR}(y, e)$), hence (6.13.1).

Let us now consider the special case where $d = 1$, and $x = \alpha \in G_{Q+}$. We have then $\Omega' = \left(\nu(\alpha)^{-1}\kappa, \mathfrak{R}\alpha, \{q_i\alpha\}\right)$. Let $w \in \mathbb{S}^r$ and $w' = \alpha(w)$. Now the map $\Lambda(\alpha, w')$ of [12, (6.7.1)] gives an isomorphism of the member $\mathfrak{A}_w$ of $\Sigma(\Omega)$ to the member $\mathfrak{A}'_w$ of $\Sigma(\Omega')$. Therefore $J(\varphi'(w)) = J(\mathfrak{v}'(\mathfrak{A}'_w)) = \mathfrak{v}(\mathfrak{A}_{w'}) = \varphi(w')$. Substituting $j(z)$ for w, we obtain

$$(6.13.2) \qquad J_{TU}(\alpha, 1)(\varphi_U(z)) = \varphi_T(\alpha(z)) \qquad (z \in \mathfrak{H}_n^r) \,.$$

6.14. Let $(Y, P, \delta, f), P', \eta$, and z be as in 2.4. Assume (6.7.2), but disregard (6.7.1). Let $u \in (P'H)_A^\times$, $v = N_{P'H/P'}(u)$, $w = N_{P'H/H}(u)$, $\pi = [u, P'H]$, $U = \eta(v)T\eta(v)^{-1}$. By (6.7.3), $J_{TU}(\eta(v)^{-1}, w^{-1})$ is meaningful. Taking $\eta(v)^{-1}$ and w^{-1} as x and d, we obtain, from (6.9.1),

$$(6.14.1) \qquad \varphi_T(z)^\pi = J_{TU}(\eta(v)^{-1}, w^{-1})(\varphi_U(z)) \,.$$

6.15. In view of the definition of $\mathfrak{X}_{HS}$ in 3.1, we observe that

$(6.15.1)$ *For every $\tau \in \mathrm{Gal}\,(M_c/k_T H)$, there exist an element d of $H_A^\times$ and an element x of T such that $\tau = [d^{-1}, H]$ on M_c, and $\nu(x)/\lambda_H(d) \in F^\times$.*

If τ, x, d are in this situation, $J_{TT}(x, d)$ is meaningful. Now the result of 6.12 shows that this depends only on τ, and not on the choice of d and x. Put therefore $J_T(\tau) = J_{TT}(x, d)$. By (6.13.1), $J_T(\sigma\tau) = J_T(\sigma)^\tau \circ J_T(\tau)$ for σ, $\tau \in \mathrm{Gal}\,(M_c/k_T H)$. By the Weil criterion, we can find an algebraic variety $\bar{V}_T$, rational over $k_T H$, and a biregular map $Q_T: V_T \to \bar{V}_T$, rational over M_c, such that $Q_T^\tau \circ J_T(\tau) = Q_T$.

Now we consider $\bar{V}_T$ and Q_T for all $T \in \mathfrak{W}$, and put $\bar{\varphi}_T = Q_T \circ \varphi_T$, $\bar{J}_{TU}(x, d) = Q_T^\tau \circ J_{TU}(x, d) \circ Q_U^{-1}$, where $\tau = [d^{-1}, H]$. Then the system $(\bar{V}_T, \bar{\varphi}_T, \bar{J}_{TU}(x, d))$ is biregularly equivalent to $\left(V_T, \varphi_T, J_{TU}(x, d)\right)$ over M_c in an obvious sense. Writing again $\left(\bar{V}_T, \bar{\varphi}_T, \bar{J}_{TU}(x, d)\right)$ as $\left(V_T, \varphi_T, J_{TU}(x, d)\right)$, we now have a system satisfying (6.13.1), (6.13.2), (6.14.1), and besides,

(6.15.2) V_T *is rational over* $k_T H$;

(6.15.3) $J_{TT}(x, d)$ *is the identity map if* $x \in T$.

The relation of V_T with $\Sigma(\Omega)$ and its moduli-system will no longer be needed.

6.16. Take x of the relation (6.13.1) to be an element of T. Then $J_{TR}(xy, de) = J_{TR}(y, e)$. This means that $J_{TR}(y, e)$ depends only on Ty, in particular, only on y. Let us therefore put $J_{TU}(x) = J_{TU}(x, d)$. Then (6.13.1) can be written in the form

$$(6.16.1)\quad J_{TR}(xy) = J_{TU}(x)^\tau \circ J_{UR}(y)\,, \qquad \tau = [e^{-1}, H],\ \nu(y)/\lambda_H(e) \in F^\times F^\times_{\infty+}\,.$$

Take y in U. Then $U = R$, and $xyx^{-1} \in T$, hence $J_{TU}(xy) = J_{TU}(xyx^{-1}x) = J_{TU}(x)$. Therefore, in view of (6.15.1), we see that $J_{TU}(x) = J_{TU}(x)^\tau$ for every $\tau \in \mathrm{Gal}\,(M_c/k_T H)$. It follows that $J_{TU}(x)$ is rational over $k_T H$. We can therefore substitute $\sigma_H(y)$ for τ in (6.16.1).

Our consideration shows that

$$\{V_T, \varphi_T, J_{TU}(x), (T, U \in \mathcal{W};\ x \in \mathcal{G}_{H+})\}$$

is a weak canonical system relative to $\{\mathcal{W}, H, K\}$. In fact, the conditions corresponding to $(2.5.3_{a,b})$ have been verified. We obtain $(2.5.3_c)$ from (6.13.2), and (3.2.4) from (6.14.1).

6.17. For every positive integer a, we have defined a member S of $\mathcal{Z}$ by (6.4.1) with suitable multiples b and c of a, and put $\mathcal{W} = \{p^{-1}Sp \mid p \in G_A\}$. With the same $\mathfrak{m}$, put $U = \{x \in G_{A+} \mid \mathfrak{m}x = \mathfrak{m}\}$. Then we see that S is a normal subgroup of U. If we consider $\mathcal{W}$ for all positive integers a, and denote by $\mathcal{W}'$ the union of all these $\mathcal{W}$, then $\mathcal{W}'$ and U satisfy the condition of 3.11. Therefore, by 3.11. there exists a weak canonical system relative to $\{\mathcal{Z}, H, K\}$.

Now $H = K'$ depends on the choice of K and $\{\tau_\nu\}$ as in 6.1. By [12, 7.3], we can choose K and $\{\tau_\nu\}$ so that the conditions (3.14.1, 2) are satisfied for any given k and $\mathfrak{m}$. Therefore, by 3.14, there exists a canonical system for G.

**7. Galois extension of a number field obtained
from coverings of a canonical model**

7.1. The aim of this section is to reformulate and generalize the results of [12, § 11] and [13]. In the latter paper, only the case $r = 1$ was considered. We shall treat here the general case without any condition on r.

Let S and T be members of $\mathcal{Z}$. Suppose that T is a normal subgroup of S. Then Γ_T is a normal subgroup of Γ_S. Therefore we can consider V_T as a

Galois covering of V_S by means of the projection map $J_{ST}(1)$ of V_T to V_S. Hereafter we shall write simply C_{ST} for $J_{ST}(1)$. We note

(7.1.1) $C_{ST}^{\sigma(x)} \circ J_{TT}(x) = C_{ST}$ *for every* $x \in S \cap \mathfrak{S}$.

(7.1.2)   *If* $C_{ST}(u) = C_{ST}(v)$ *with two points* u *and* v *of* V_T, *then* $u = J_{TT}(\gamma)(v)$ *for some* $\gamma \in \Gamma_S$.

We shall always denote by S_0 the projection of S to G_0. Then

(7.1.3) $S = S_0 G_{\infty+}$, $S \cap \mathfrak{S} = (S_0 \cap \mathfrak{S}) \cdot G_{\infty+}$.

Further, for every $z \in \mathfrak{H}_n^r$, we put

(7.1.4) $\Delta(z, S) = \{\gamma \in \Gamma_S \mid \gamma(z) = z\}$,

and denote by $\Delta(z, S)_0$ the projection of $\Delta(z, S)$ to S_0.

7.2. PROPOSITION. *Let* $S \in \mathfrak{Z}$, $T \in \mathfrak{Z}$, $u_S \in V_S$, $u_T \in V_T$. *Suppose that* T *is a normal subgroup of* S, *and* $C_{ST}(u_T) = u_S$. *Then*

(7.2.1) $k_T(u_T)$ *is a finite Galois extension of* $k_S(u_S)$, *which depends only on* u_S *and* T, *and not on the choice of* u_T.

(7.2.2)   *For every* $\tau \in \mathrm{Gal}\left(k_T(u_T)/k_S(u_S)\right)$, *there exists an element* x *of* $S_0 \cap \mathfrak{S}$ *such that* $\tau = \sigma(x)$ *on* k_T, *and* $u_T^\tau = J_{TT}(x)(u_T)$.

PROOF. Let τ be an isomorphism of $k_T(u_T)$ into C over $k_S(u_S)$. By 3.5, $\tau = \sigma(y)$ on k_T with some $y \in S \cap \mathfrak{S}$. Since $y T y^{-1} = T$, $J_{TT}(y)$ maps V_T onto V_T^τ. Therefore $u_T^\tau = J_{TT}(y)(v)$ for some $v \in V_T$. Then

$$u_S = u_S^\tau = C_{ST}^\tau(u_T^\tau) = C_{ST}^\tau\big(J_{TT}(y)(v)\big) = C_{ST}(v) .$$

By (7.1.2), $v = J_{TT}(\gamma)(u_T)$ for some $\gamma \in \Gamma_S$. Put $x = y\gamma$. Then $x \in S \cap \mathfrak{S}$, $u_T^\tau = J_{TT}(x)(u_T)$, and $\tau = \sigma(x)$ on k_T. Since $J_{TT}(x)$ is rational over k_T, we have $k_T(u_T^\tau) = k_T(u_T)$, so that $k_T(u_T)$ is a finite Galois extension of $k_S(u_S)$. This extension depends only on T and u_S, on account of (7.1.2). Further we can replace x by its projection to S_0. This completes the proof.

7.3. PROPOSITION. *Let the notation be as in 7.2, and let* z *be a point of* $\mathfrak{H}_n^r$ *such that* $\varphi_T(z) = u_T$. *Let* $\mathfrak{Y} = \Delta(z, S)_0 \cdot (T_0 \cap \mathfrak{S})$, *and let* $\mathfrak{X}$ *be the normalizer of* $\mathfrak{Y}$ *in* $S_0 \cap \mathfrak{S}$. *Then the correspondence* $\tau \mapsto x$ *of (7.2.2) gives an isomorphism of* $\mathrm{Gal}\left(k_T(u_T)/k_S(u_S)\right)$ *into* $\mathfrak{X}/\mathfrak{Y}$.

PROOF. Let τ and x be as in (7.2.2). Suppose that τ is the identity. By (3.5.1), $x = s\alpha$ with $\alpha \in G_Q$ and $s \in T \cap \mathfrak{S}$. Then $\alpha \in \Gamma_S$, and

$$\varphi_T(z) = u_T = J_{TT}(s\alpha)(u_T) = J_{TT}(\alpha)\big(\varphi_T(z)\big) = \varphi_T(\alpha(z)) .$$

Therefore $z = \gamma\alpha(z)$ with $\gamma \in \Gamma_T$. Then $\gamma\alpha \in \Delta(z, S)$, and $x = s\gamma^{-1}\gamma\alpha \in \mathfrak{Y}$. Conversely, we see immediately that $J_{TT}(x)(u_T) = u_T$ if $x \in \mathfrak{Y}$. The remaining part of our proposition can be verified in a straightforward way.

7.4. Proposition. *Let the notation be as in 7.2 and 7.3. Suppose that u_T is generic on V_T over k_T. Then $\Delta(z, S) = F^\times \cap S$, $\mathfrak{X} = S_0 \cap \mathfrak{S}$, and the isomorphism of 7.3 is a surjective map to $(S_0 \cap \mathfrak{S})/\mathfrak{Y}$.*

Proof. The first two assertions are obvious. To prove the last one, let $x \in S_0 \cap \mathfrak{S}$, and $v = J_{TT}(x)(u_T)$. Then v is generic on $V_T^{\sigma(x)}$ over k_T. Therefore we can find an isomorphism τ of $k_T(u_T)$ onto $k_T(v)$ so that $u_T^\tau = v$ and $\tau = \sigma(x)$ on k_T. Then

$$u_S^\tau = C_{ST}^\tau(u_T^\tau) = C_{ST}^\tau\big(J_{TT}(x)(u_T)\big) = C_{ST}(u_T) = u_S ,$$

hence $\tau \in \mathrm{Gal}\big(k_T(u_T)/k_S(u_S)\big)$, q.e.d.

7.5. Let us denote by $\mathfrak{Z}'$ the set of all $S \in \mathfrak{Z}$ satisfying

(7.5.1) S_0 *is the projection to G_0 of an open compact subgroup of $\tilde{D}_A$.*

In view of (2.3.2), we see easily that

(7.5.2)   *If $S \in \mathfrak{Z}'$, $T \in \mathfrak{Z}$, and $T \subset S$, then $T \in \mathfrak{Z}'$, and T is open in S.*

(7.5.3)   *Two members of $\mathfrak{Z}'$ are commensurable with each other.*

For example, the groups of type (2.10.2) belong to $\mathfrak{Z}'$. We have $\mathfrak{Z} = \mathfrak{Z}'$ if $r = 1$. Further, if the map λ is the injection of F'' into F as in (6) of 2.6, $\mathfrak{Z}'$ is the set of all S such that $S_0 \subset D$.

Under the above assumption (7.5.1), we have $S_0 = S_0 \cap \mathfrak{S}$, so that the statements of 7.2, 7.3, and 7.4 are simplified accordingly.

Let us now fix a member S of $\mathfrak{Z}'$ and a point u of V_S, and consider the set $\mathfrak{W}$ of all normal subgroups T of S which are members of $\mathfrak{Z}$. We call a set of points $\{u_T\}_{T \in \mathfrak{W}}$ a *coherent set* (or *sequence*) *of points lying above u*, if

$$(7.5.4) \qquad\qquad u_T \in V_T,\ u_S = u,\ C_{TR}(u_R) = u_T \qquad\qquad (R \subset T) .$$

Take a point z of $\mathfrak{H}_n^r$ so that $\varphi_S(z) = u$, and put $u_T = \varphi_T(z)$ for all $T \in \mathfrak{W}$. Then $\{u_T\}$ satisfies (7.5.4). We call this type of coherent set *standard*.

7.6. Proposition. *Let $\{u_T\}_{T \in \mathfrak{W}}$ be a coherent set of points lying above a point u of V_S, and $\mathfrak{R}_u$ the union of the fields $k_T(u_T)$ for all $T \in \mathfrak{W}$. Then $\mathfrak{R}_u$ is an infinite Galois extension of $k_S(u)$, depending only on u, and not on the choice of $\{u_T\}$. Moreover, for every $\tau \in \mathrm{Gal}\,(\mathfrak{R}_u/k_S(u))$, there exists an element x of S_0 such that $\sigma(x) = \tau$ on $\mathfrak{k}$ and $u_T^\tau = J_{TT}(x)(u_T)$ for all $T \in \mathfrak{W}$.*

Proof. The first assertion follows immediately from (7.2.1). Now for a given $\tau \in \mathrm{Gal}\,(\mathfrak{R}_u/k_S(u))$, we find, by (7.2.2), an element x_T of S_0 such that $u_T^\tau = J_{TT}(x_T)(u_T)$ and $\sigma(x_T) = \tau$ on k_T. Let P_T denote the set of all such x_T. Then P_T is a closed subset of S_0. (One has $P_T = x_T \Delta(z, S)_0 T_0$ if $u_T = \varphi_T(z)$.) Obviously $\{P_T\}_{T \in \mathfrak{W}}$ has the finite intersection property. Since S_0 is compact, there is an element x common to the P_T for all $T \in \mathfrak{W}$, q.e.d.

7.7. *Remark.* Let $\{u_T\}$ and u be as in 7.6. Then we can define u_R for *all* $R \in \mathfrak{Z}$ contained in S (not necessarily normal in S), by $u_R = C_{RT}(u_T)$ with any $T \in \mathfrak{W}$ such that $T \subset R$. This does not depend on the choice of T. We see easily that (7.5.4) is still true for this extended sequence. Moreover, if x and τ are as in 7.6, we have

$$(7.7.1) \qquad\qquad u_R^\tau = J_{RQ}(x)(u_Q) \qquad\qquad (Q = x^{-1}Rx) \ .$$

7.8. PROPOSITION. *The notation being as in 7.6, let z be a point of $\mathfrak{H}_n^r$ such that $\varphi_S(z) = u$, $\mathfrak{B}$ the closure of $\Delta(z, S)_0$ in S_0, and $\mathfrak{A}$ the normalizer of $\mathfrak{B}$ in S_0. Then there exists an element d of S_0, depending on $\{u_T\}$, such that the element x of 7.6 corresponding to $\tau \in \mathrm{Gal}\,(\mathfrak{R}_u/k_S(u))$ belongs to $d\mathfrak{A}d^{-1}$, and is uniquely determined by τ modulo $d\mathfrak{B}d^{-1}$. The correspondence $\tau \mapsto x$ defines a continuous isomorphism of $\mathrm{Gal}\,(\mathfrak{R}_u/k_S(u))$ into $d\mathfrak{A}d^{-1}/d\mathfrak{B}d^{-1}$. Moreover, if $u_T = \varphi_T(z)$ for all $T \in \mathfrak{W}$, one can put $d = 1$.*

PROOF. Let us first assume $u_T = \varphi_T(z)$ for all $T \in \mathfrak{W}$. Let τ and x be as in 7.6. By 7.3, τ is the identity if and only if $x \in \Delta(z, S)_0 T_0$ for all $T \in \mathfrak{W}$. Now, by (7.5.2), the T_0 form a basis of neighborhoods of the identity element of S_0. Therefore τ is the identity if and only if $x \in \mathfrak{B}$. Let $\{v_T\}_{T \in \mathfrak{W}}$ be another coherent set lying above u. By (7.1.2), there exists, for each $T \in \mathfrak{W}$, an element y_T of S_0 such that $J_{TT}(y_T)(u_T) = v_T$ and $\sigma(y_T) = 1$. All such y_T form a closed subset Q_T of S_0. Then $\{Q_T\}_{T \in \mathfrak{W}}$ has the finite intersection property. Since S_0 is compact, there exists a point d common to all Q_T. Then $J_{TT}(d)(u_T) = v_T$ for all $T \in \mathfrak{W}$. With this d, our assertions can be verified in a straightforward way.

7.9. We denote by h the isomorphism of $\mathrm{Gal}\,(\mathfrak{R}_u/k_S(u))$ into $d\mathfrak{A}d^{-1}/d\mathfrak{B}d^{-1}$ obtained in 7.8. Put

$$(7.9.1) \qquad\qquad\qquad E_S = F^\times \cap S \ .$$

We observe that $\Delta(z, S)$ contains E_S as a subgroup of finite index, and $\Delta(z, S)/E_S$ is the stability group at z of the transformation group Γ_S/E_S. Therefore $\Delta(z, S) = E_S$ and $\mathfrak{A} = S_0$ unless z is a non-trivial fixed point of Γ_S.

7.10. PROPOSITION. *Let E_{S0} be the projection of E_S to G_0, and $\bar{E}_{S0}$ the closure of E_{S0} in S_0. If u is generic on V_S over k_S, the map h is a surjective isomorphism of $\mathrm{Gal}\,(\mathfrak{R}_u/k_S(u))$ to $S_0/\bar{E}_{S0}$.*

This follows immediately from 7.4.

7.11. We assume hereafter, until the end of this section, that *the coordinates of the point u are algebraic numbers.* Therefore, $\mathfrak{R}_u$ is an infinite Galois extension of an algebraic number field $k_S(u)$. For every prime ideal $\mathfrak{l}$ in F, the $\mathfrak{l}$-component $G_{\mathfrak{l}}$ of G_A is meaningful. (If G' is the algebraic group over F such that $R_{F/Q}(G') = G$, (with the notation of [16]), then $G_{\mathfrak{l}}$ can be identified with $G'(F_{\mathfrak{l}})$, where $F_{\mathfrak{l}}$ denotes the $\mathfrak{l}$-completion of F.) For every finite set M of prime ideals in F, we put $G_M = \prod_{\mathfrak{l} \in M} G_{\mathfrak{l}}$, and denote by $S_M, \mathfrak{A}_M, \mathfrak{B}_M$ the projections of $S, \mathfrak{A}, \mathfrak{B}$, to G_M, respectively. Then we obtain a natural homomorphism of $\mathfrak{A}/\mathfrak{B}$ onto $\mathfrak{A}_M/\mathfrak{B}_M$. Combining h with this, we obtain a continuous homomorphism

$$(7.11.1) \qquad h_M \colon \mathrm{Gal}\left(\mathfrak{R}_u/k_S(u)\right) \longrightarrow \mathfrak{A}_M/\mathfrak{B}_M \ .$$

7.12. THEOREM. *The notation and the assumption being as above, let $\mathfrak{R}_u^M$ be the subfield of $\mathfrak{R}_u$ corresponding to the kernel of h_M. Then there exists a finite set Y of prime ideals of $k_S(u)$, depending only on u, and independent of M, with the following property: If M consists of all prime factors in F of a rational integer, then every prime ideal in $k_S(u)$, prime to the members of M and Y, is unramified in $\mathfrak{R}_u^M$.*

Proof will be given in 7.16–20.

7.13. To study the behavior of the Frobenius automorphisms of $\mathfrak{R}_u$ over $k_S(u)$, we consider a representation

$$(7.13.1) \qquad \omega \colon G \longrightarrow GL_m$$

with any degree m, satisfying the following three conditions:

(7.13.2) ω *is rational over* Q.

(7.13.3) $\omega(a) = N_{F/Q}(a)^{\kappa} 1_m$ *for every* $a \in F^{\times}$, *with an integer* κ *independent of* a.

(7.13.4) $\omega(\delta) = 1$ *for all* $\delta \in \Delta(z, S)$.

A few explicit examples of such representations will be given in §8. As for the interdependence of the conditions (7.13.3) and (7.13.4), see [13, (1.10.4), (1.10.5)].

For every rational prime l, we write $G_l, S_l, \mathfrak{A}_l, \mathfrak{B}_l$ for $G_M, S_M, \mathfrak{A}_M, \mathfrak{B}_M$ if M consists of all prime factors of l in F. Then G_l can be identified with the group of Q_l-rational points of G. Therefore we obtain naturally from ω a representation

$$\omega_l \colon G_l \longrightarrow GL_m(Q_l) \ .$$

In view of (7.13.4), we obtain a representation

$$\omega_l \circ h_l \colon \mathrm{Gal}\left(\mathfrak{R}_u/k_S(u)\right) \longrightarrow GL_m(Q_l) \ .$$

Let Y be as in 7.12, and $\mathfrak{p}$ a prime ideal in $k_S(u)$ such that $\mathfrak{p} \nmid l, \mathfrak{p} \notin Y$. Take any prime divisor $\mathfrak{P}$ of $\mathfrak{R}_u$ which divides $\mathfrak{p}$, and a Frobenius automorphism $\sigma_{\mathfrak{P}}$ of $\mathfrak{R}_u$ over $k_S(u)$ for $\mathfrak{P}$ and $\mathfrak{p}$. By 7.12, we observe

(7.13.5) *The conjugacy-class of $\omega_l(h_l(\sigma_{\mathfrak{P}}))$ in $GL_m(\mathbf{Q}_l)$ is determined by u, $\mathfrak{p}$, and l, and independent of the choice of $\{u_T\}$ and $\mathfrak{P}$, so long as $\mathfrak{p} \nmid l$ and $\mathfrak{p} \notin Y$.*

7.14. THEOREM. *Let the notation and the assumption be the same as in 7.11–13. Then there exists a finite set Y_1 of prime ideals in $k_S(u)$, containing Y, and independent of ω, $\mathfrak{p}$, l, such that the following statements hold.*

(7.14.1) *If $\mathfrak{p} \nmid l$ and $\mathfrak{p} \notin Y_1$, every element of G_l representing $h_l(\sigma_{\mathfrak{P}})$ is semi-simple, and $\omega_l(h_l(\sigma_{\mathfrak{P}}))$ is semi-simple.*

(7.14.2) *The characteristic roots of $\omega_l(h_l(\sigma_{\mathfrak{P}}))$ are algebraic numbers of absolute value $N(\mathfrak{p})^{\kappa r/2}$, if $\mathfrak{p} \nmid l$ and $\mathfrak{p} \notin Y_1$, where r is the number of archimedean primes in F unramified in B (see 2.1).*

(7.14.3) *There exists a positive integer μ, independent of ω, $\mathfrak{p}$, and l, but dependent on u, such that the matrices $\omega_l(h_l(\sigma_{\mathfrak{P}}^\mu))$, for all l prime to $\mathfrak{p}$ and the members of Y_1, have the same characteristic polynomial (dependent on $\mathfrak{p}$) with rational coefficients.*

7.15. THEOREM. *If $r = g$, and S is defined by (5.1.2) with any $\mathbf{Z}$-lattice $\mathfrak{m}$ in $B^\ast$ and any elements d_i of $B^\ast/\mathfrak{m}$, then one can put $\mu = 1$ in (7.14.3).*

The proof of these two theorems will be given in the following 7.16–20. We can re-formulate 7.14 and 7.15 by taking a representation of a maximal torus of G instead of the representation ω, as we did in the case $r = 1$ in [13]. For this and further comments, we refer the reader to [13, 1.12–13, and § 2].

7.16. *Reduction to a special case.* Let R be a member of $\mathscr{Z}'$ satisfying

$$(7.16.1) \qquad\qquad \Delta(w, R) \subset F^\times \qquad\qquad \text{for all } w \in \mathfrak{H}_n^r.$$

We shall now show that if 7.12 and 7.14 hold for R, then they are also true for any member S of $\mathscr{Z}'$ containing R.

Let u and $\{u_T\}$ be as above. To prove 7.12 and 7.14, we may assume that $u_T = \varphi_T(z)$ for all $T \in \mathscr{W}$ with a point z of $\mathfrak{H}_n^r$. Define $\mathfrak{A}$, $\mathfrak{B}$, and h for R with the same $\{u_T\}$, and denote them by $\mathfrak{A}'$, $\mathfrak{B}'$, and h'. Then we see that $\mathfrak{B}' = \mathfrak{B} \cap R_0$. Let $\mathfrak{C}$ be the subgroup of $\mathfrak{A}'$ such that $\mathfrak{C}/\mathfrak{B}'$ is the image of $\mathrm{Gal}\,(\mathfrak{R}_u/k_R(u_R))$ by h'. Then $\mathfrak{C}\mathfrak{B}/\mathfrak{B}$ is the image by h of the subgroup of $\mathrm{Gal}\,(\mathfrak{R}_u/k_S(u)$ corresponding to $k_R(u_R)$.

Let $\tau \in \mathrm{Ker}\,(h'_M)$, and let x be an element of $\mathfrak{C}$ which represents $h'(\tau)$. Then x represents also $h(\tau)$. Since the M-component (in an obvious sense) of x belongs to $\mathfrak{B}'_M$, we see that $\tau \in \mathrm{Ker}\,(h_M)$. Therefore $\mathfrak{R}_u^M$ is contained in the corresponding field defined with respect to R. Thus the assertion of 7.12 for S follows from that for R.

Let ω be as in 7.13. If (7.13.4) is satisfied for S, then it is satisfied with

R in place of S. The notation $\mathfrak{p}$, $\mathfrak{P}$, and $\sigma_{\mathfrak{P}}$ being as in 7.13, let $\mathfrak{q}$ be the restriction of $\mathfrak{P}$ to $k_R(u_R)$. Let Y_1' be the set of prime ideals in $k_R(u_R)$, such that the assertion of 7.14 for R is true with Y_1'. Let Y be as in 7.12, and Y_0 the set of all prime ideals in $k_S(u)$ which are either ramified in $k_R(u_R)$, or have some divisors contained in Y_1'. Put $Y_1 = Y \cup Y_0$. Suppose that $\mathfrak{p} \nmid l$, $\mathfrak{p} \notin Y_1$. Then $\mathfrak{q} \notin Y_1'$. Let f be the degree of $\mathfrak{q}$ over $k_S(u)$, and $\sigma_{\mathfrak{P}}'$ a Frobenius element of $\mathrm{Gal}\left(\mathfrak{R}_u/k_R(u_R)\right)$ for $\mathfrak{P}$ and $\mathfrak{q}$. Then $\sigma_{\mathfrak{P}}'$ coincides with $\sigma_{\mathfrak{P}}^f$ on $\mathfrak{R}_u^1$, hence $\sigma_{\mathfrak{P}}' = \zeta \sigma_{\mathfrak{P}}^f$ with an element ζ of $\mathrm{Ker}\,(h_l)$. Let x be an element of S_l which represents $h_l(\sigma_{\mathfrak{P}})$, and y an element of R_l which represents $h_l'(\sigma_{\mathfrak{P}}')$. Then $y = tx^f$ with an element t of $\mathfrak{B}_l$. Therefore

$$(7.16.2) \qquad \omega_l\big(h_l'(\sigma_{\mathfrak{P}}')\big) = \omega_l\big(h_l(\sigma_{\mathfrak{P}})\big)^f \;.$$

We can find a positive integer a so that $x^a \in R_l$, and $\sigma_{\mathfrak{P}}^a$ is the identity map on $k_R(u_R)$. Then x^{af} is an element of R_l representing $h'(\sigma_{\mathfrak{P}}')^a$. Therefore $y^a = sx^{af}$ with an element s of $\mathfrak{B}_l'$. By (7.16.1), $s \in F_l^\times$. Now the assertion (7.14.1) for R implies that y is semi-simple, hence x is semi-simple. Therefore we obtain (7.14.1) and (7.14.2) for S from this and (7.16.2).

Since we can find a positive integer which is divisible by all possible values of f, the assertion (7.14.3) for S follows from that for R, and the relation (7.16.2).

7.17. *The case $r < g$.* Let us now assume $r < g$, and prove 7.12 and 7.14 for a member S of $\mathcal{Z}'$ of a special type. Let K, L, H, and G^* be as in §6. Take $\mathfrak{m}$ and $\mathfrak{M}$ as in 6.4, and S in the form (6.4.1) with suitable positive integers b and c, so that the conditions (6.4.3, 4) and the following condition are satisfied.

$$(7.17.1) \quad \{\beta \in G_{\mathbf{Q}+}^* \mid \beta(w) = w, \ \beta \equiv 1 \bmod_0 (\mathfrak{M}, b)\} \subset F^\times \qquad \text{for every } w \in \mathcal{S}^r \;.$$

For every positive integer e which is a multiple of b, consider a PEL-type

$$\Omega_e = (L, \Phi, \rho; \kappa T, \mathfrak{M}; q_1^e, \cdots, q_s^e)$$

with elements $q_1^e, \cdots, q_s^e$ of $L^n/\mathfrak{M}$ and a totally positive element κ of F such that $e^{-1}\mathfrak{M}/\mathfrak{M} = \sum_{i=1}^s \mathbf{Z} q_i^e$ and $\mathrm{tr}_{L/\mathbf{Q}}\big(\kappa T(\mathfrak{M}, \mathfrak{M})\big) = \mathbf{Z}$. Let $\Sigma(\Omega_e)$ be as in 6.6, and let $(U_e, \mathfrak{b}_e, \psi_e)$ be the moduli-system for $\Sigma(\Omega_e)$. For each e, we can take, by virtue of Chevalley [1], a positive integer e' so that, if $\varepsilon \in \mathfrak{r}_F^\times$ and $\varepsilon \equiv 1 \bmod (e')$, then $\varepsilon = \eta^2$, $\eta \equiv 1 \bmod (e)$ for some $\eta \in \mathfrak{r}_F^\times$. Choose any such e' and put

$$S_e = S(\mathfrak{m}, e') \cdot \{x \in S(\mathfrak{m}, e) \mid \nu(x) = 1\} \;.$$

We put $e' = c$ and $S_b = S$ if $e = b$. Then (6.4.3,4) hold with S_e and e in place of S and b. We write V_e, φ_e, k_e, and $J_e(x)$ for V_T, φ_T, k_T, and $J_{TT}(x)$ if $T = S_e$

and $x \in S$. The result of § 6, together with 3.8, shows that the biregular embedding Z_e of V_e into U_e defined by $Z_e \circ \varphi_e = \psi_e \circ j$ is rational over $M_{e'}$, where $M_{e'}$ is a subfield of H_{ab} defined as in 6.5 (with e' in place of c). Let $\{u_T\}$ be the coherent set of points lying above u in question. To prove our theorems, we may assume that $u_T = \varphi_T(z)$ with a point z of $\mathfrak{H}_n^r$ common to all $T \in \mathfrak{W}$. Put $u_e = u_T$ for $T = S_e$. Obviously $\mathfrak{K}_u$ is the union of $k_e(u_e)$ for all e. Put $w = j(z)$. Now the member of $\Sigma(\Omega_e)$ at w is of the form

$$\mathfrak{A}_w^e = \left(A_w, \mathcal{C}_w, \theta_w; \{t_i^e\} \right)$$

with $(A_w, \mathcal{C}_w, \theta_w)$ common to all e; the point t_i^e corresponds to $\mathfrak{y}(q_i^e, w)$ with the parametrizing function $\mathfrak{y}$ of [12, 6.4]. We may assume that $\mathfrak{A}_w^b$ is defined over a finite algebraic extension k' of $M_c(u)$.

Let us now consider l-adic coordinate systems on A_w in the following way. Put $A_w^0 = \bigcup_{s=1}^{\infty} \{t \in A_w \mid et = 0\}$. Then the map $L^* \ni a \mapsto \mathfrak{y}(a, w)$ gives a homomorphism $\mathfrak{t}$ of L^* onto A_w^0 with kernel $\mathfrak{M}$ such that $\mathfrak{t}(q_i^e) = t_i^e$. We can define a homomorphism

$$\eta: \mathrm{Gal}\,(\bar{Q}/k') \longrightarrow GL_n(L_A)_0 \qquad \text{(For notation, see 0.4)}$$

by $\mathfrak{t}(a)^\tau = \mathfrak{t}(a\eta(\tau))$ for $a \in L^*/\mathfrak{M}$. The l-component of $\eta(\tau)$ is exactly the l-adic representation of $\mathrm{Gal}\,(\bar{Q}/k')$ on A_w; one has $\mathfrak{M}\eta(\tau) = \mathfrak{M}$. We are going to prove

(7.17.2) *Let $\tau \in \mathrm{Gal}\,(\bar{Q}/k')$. Let d be an element of $H_A^\times$ such that $\tau = [d^{-1}, H]$ on H_{ab} and $d \equiv 1 \bmod_0 (c)$. Then there exists an element x of S_0 which represents $h(\tau)$ such that $\eta(\tau)$ is the non-archimedean part of $\pi(d)x$.* (Since τ is the identity on M_c, such an element d always exists.)

Since $\mathfrak{m}$ satisfies (2.10.3), there exists an element y of S such that $\nu(y) = \lambda_H(d)$. Repeating the proof of 7.2 with S_e in place of T, we find an element γ_e of Γ_S such that $u_e^\tau = J_e(y\gamma_e)(u_e)$, and $\tau = \sigma(y\gamma_e)$ on k_e. Put $x_e = y\gamma_e$. Then $\nu(x_e)/\lambda_H(d) \in F^\times$. Applying (6.6.1) to the present case, we see that Ω_e^τ is equivalent to

$$\Omega_e' = \left(\mu(\pi(d)x_e)^{-1}\kappa, \mathfrak{M}, \{\pi(d)q_i^e x_e\} \right) .$$

(Observe that $\pi(d)\mathfrak{M}x_e = \mathfrak{M}$.) In view of the result of 4.4, we can take the moduli-system for Ω_e' in the form $(U_e, \mathfrak{v}_e', \psi_e)$ with the same U_e and ψ_e as before. The argument of 6.9–16 shows that there exists a biregular isomorphism J_e' of U_e to U_e^τ such that $\mathfrak{v}_e^\tau = J_e' \circ \mathfrak{v}_e'$ and $Z_e^\tau \circ J_e(x_e) = J_e' \circ Z_e$. Let $\mathfrak{A}_w'$ be the member of Ω_e' at w. We have then

$$Z_e(u_e) = Z_e(\varphi_e(z)) = \psi_e(w) = \mathfrak{v}_e(\mathfrak{A}_w^e) = \mathfrak{v}_e'(\mathfrak{A}_w') ,$$

so that

$$\mathfrak{v}_e^\tau(\mathfrak{A}_w^{e\tau}) = Z_e^\tau(u_e^\tau) = Z_e^\tau(J_e(x_e)(u_e)) = J_e'(Z_e(u_e)) = J_e'(\mathfrak{v}_e'(\mathfrak{A}_w')) = \mathfrak{v}_e^\tau(\mathfrak{A}_w') .$$

Therefore $(\mathfrak{A}_w^e)^\tau$ is isomorphic to $\mathfrak{A}_w'$. Since τ is the identity on k', we have

$$(\mathfrak{A}_w^e)^\tau = \big(A_w,\, \mathcal{C}_w,\, \theta_w;\, \{t_i^{e\tau}\}\big)\,.$$

On the other hand, we have $\mathfrak{A}_w' = \big(A_w,\, \mathcal{C}_w',\, \theta_w;\, \{t_i'\}\big)$ with elements $t_i' = \mathfrak{t}\big(\pi(d)q_i^e x_e\big)$. Therefore we have an automorphism α_e of (A_w, θ_w) which sends $\mathcal{C}_w'$ to $\mathcal{C}_w$, and t_i' to $t_i^{e\tau}$. Such an α_e is always obtained as the map $\Lambda(\beta_e, w)$ of $[12, (6.7.1)]$ with an element β_e of G_{Q+}^* such that $\beta_e(w) = w$, and $\mathfrak{M}\beta_e = \mathfrak{M}$. Then

$$\mathfrak{t}(q_i^e)^\tau = \alpha_e\big(\mathfrak{t}\big(\pi(d)q_i^e x_e\big)\big) = \mathfrak{t}\big(\pi(d)q_i^e x_e \beta_e\big)\,,$$

hence $q\eta(\tau) = \pi(d)qx_e\beta_e$ for every $q \in e^{-1}\mathfrak{M}/\mathfrak{M}$. Since τ is the identity map on k', $\eta(\tau)$ is the identity map on $b^{-1}\mathfrak{M}/\mathfrak{M}$. Further, since $d \equiv 1 \bmod_0 (c)$ and $x_e \in S$, $\pi(d)x_e$ gives the identity map on $b^{-1}\mathfrak{M}/\mathfrak{M}$. It follows that $\beta_e \equiv 1 \bmod_0 (\mathfrak{M}, b)$. In view of $(7.17.1)$, $\beta_e \in F^\times$. Put $x_e' = x_e\beta_e$. We have then $J_e(x_e')(u_e) = u_e^\tau$, $\tau = \sigma(x_e')$ on k_e, and $q\eta(\tau) = \pi(d)qx_e'$ for $q \in e^{-1}\mathfrak{M}/\mathfrak{M}$. Therefore the non-archimedean part of x_e' converges to an element x of S_0, which has the required property of $(7.17.2)$.

7.18. Let Y_0 be the set of all prime ideals $\mathfrak{p}$ in k' for which A_w has defect, i.e., A_w has no good reduction modulo $\mathfrak{p}$. Let f be a positive integer, and $\mathfrak{K}'$ the field generated over k' by the coordinates of points t of $\bigcup_{m=1}^{\infty}\{t \in A_w \mid f^m t = 0\}$. Now it is well-known that a prime ideal $\mathfrak{q}$ in k' is unramified in $\mathfrak{K}'$ if $\mathfrak{q} \notin Y_0$ and $\mathfrak{q} \nmid f$ (see $[14, 18.5]$). Let $\mathfrak{K}''$ be the composite of the class fields over H corresponding to

$$H^\times \cdot \{d \in H_A^\times \mid d \equiv 1 \bmod_0 (f^m)\}$$

for $m = 1, 2, \cdots$. Let M be the set of all prime divisors of f in $k_S(u)$. We shall now show that

$$(7.18.1) \qquad\qquad \mathfrak{K}_u^M \subset \mathfrak{K}'\mathfrak{K}''\,.$$

Let $\tau \in \mathrm{Gal}\,(\overline{Q}/\mathfrak{K}'\mathfrak{K}'')$. Then the l-component of $\eta(\tau)$ is the identity for all $l \mid f$. Further we can find an element d of $H_A^\times$ so that $\tau = [d^{-1}, H]$ on H_{ab}, $d \equiv 1 \bmod_0 (c)$, and the l-component of d is the identity for every $l \mid f$. Then $(7.17.2)$ implies that $\tau \in \mathrm{Ker}\,(h_M)$, hence $(7.18.1)$.

Now 7.12 follows from $(7.18.1)$ if we take Y to be the set of prime ideals in $k_S(u)$ which are either ramified in k' or have some extensions belonging to Y_0.

7.19. Let K'' be the field generated over Q by the elements $\prod_{i=1}^{g} a^{\tau_i}$ for all $a \in K$, and χ a regular representation of K'' over Q. Put $\xi(a) = \chi(a^{\tau_1}\cdots a^{\tau_g})^e$ for $a \in K^\times$. Then ξ is a Q-rational representation of $K^\times$, and

$$(7.19.1) \qquad\qquad \xi(a) = N_{F/Q}(a)^e \qquad\qquad\qquad \textit{for every } a \in F^\times\,.$$

Now observe that $K^\times$ and G generate an algebraic subgroup G' of G^*, rational over Q, such that $G'_Q = K^\times \cdot G_Q$. Define a representation ω' of G' by $\omega'(ax) = \xi(a) \otimes \omega(x)$ for $a \in K^\times$, $x \in G$. This is well defined in view of (7.13.3) and (7.19.1). Further we can define the localizations ξ_l and ω'_l in an obvious way.

Let $\mathfrak{p}$, $\mathfrak{P}$, and $\sigma_\mathfrak{P}$ be as in 7.13. Suppose that $\mathfrak{p} \nmid cl$, $\mathfrak{p} \notin Y$. Let $\mathfrak{Q}$ be a prime divisor of $\bar{Q}$ which divides $\mathfrak{P}$, $\mathfrak{q}$ the restriction of $\mathfrak{Q}$ to k', and τ a Frobenius element of $\mathrm{Gal}(\bar{Q}/k')$ for $\mathfrak{Q}$. If f is the degree of $\mathfrak{q}$ relative to $k_s(u)$, we have $\tau = \varepsilon \sigma_\mathfrak{P}^f$ on $\mathfrak{R}_u$ with an element ε of $\mathrm{Ker}\,(h_l)$.

Since $M_c \subset k'$, we have $N_{k'/H}(\mathfrak{q}) = (t)$ with an element t of H, such that $t \equiv 1 \bmod c\mathfrak{r}_H$. By (6.1.3), we have

$$(7.19.2) \qquad \lambda_H(t)\pi(t)\pi(t)^\rho = N_{H/Q}(t) = N(\mathfrak{q}) \,.$$

By 1.4, we have $N_{F/Q}(\lambda(a)) = N_{F'/Q}(a)^r$ for every $a \in F'$. Applying $N_{F/Q}$ to (7.19.2), we obtain $N_{H/Q}(t)^r N_{K/Q}(\pi(t)) = N(\mathfrak{q})^g$. Put $s = \prod_{\nu=1}^g \pi(t)^{\tau_\nu}$. Then $s \in K''$, and $ss^\rho = N_{K/Q}(\pi(t)) = N(\mathfrak{q})^{g-r}$. Therefore every conjugate of s over Q is of absolute value $N(\mathfrak{q})^{(g-r)/2}$, hence every characteristic root of $\xi(\pi(t)) = \chi(s)^\kappa$ is of absolute value $N(\mathfrak{q})^{\kappa(g-r)/2}$.

Let d be an element of $H_A^\times$ such that $\tau = [d^{-1}, H]$ on H_{ab} and $d \equiv 1 \bmod_0 (c)$. Let t^* be the element of $H_A^\times$, of which the component at $\mathfrak{q} \cap H$ is the identity, and all other components are t. Then $t^* \equiv 1 \bmod_0 (c)$, and $\tau = [t^{*-1}, H]$ on any abelian extension of H in which $\mathfrak{q} \cap H$ is unramified. Therefore, for every positive integer ν, we can find an element e_ν of $H^\times$ and an element h_ν of $H_A^\times$ so that $d^{-1}t^* = h_\nu e_\nu$, and $h_\nu \equiv 1 \bmod_0 (cl^\nu)$. Then $e_\nu \in \mathfrak{r}_H^\times$, and $e_\nu \equiv 1 \bmod c\mathfrak{r}_H$. By (7.17.1) and our choice of c, we see that if e is a unit of $\mathfrak{r}_K$ and $e \equiv 1 \bmod c\mathfrak{r}_K$, then e is a totally positive element of F. Therefore we have $\xi(\pi(e_\nu)) = 1$. If d_l denotes the l-component of d, we have $\xi_l(\pi(d_l)) = \xi_l(\pi(t))$.

Let x be an element of S_l which represents $h_l(\sigma_\mathfrak{P})$, and $\eta_l(\tau)$ the l-component of $\eta(\tau)$. By (7.17.2), we have $\eta_l(\tau) = \pi(d_l)yx^f$ with an element y of $\mathfrak{B}_l$. Now $\pi(d_l)y$ is contained in $(K \otimes_Q Q_l)^\times$. Since $\eta_l(\tau)$ is semi-simple, this shows that x is semi-simple. Furthermore, we have $\omega_l(h_l(\sigma_\mathfrak{P}^f)) = \omega_l(x)^f$, and

$$(7.19.3) \qquad \omega'_l(\eta_l(\tau)) = \xi_l(\pi(t)) \otimes \omega_l(h_l(\sigma_\mathfrak{P}^f)) \,.$$

Let $\mathcal{T}$ be an algebraic torus in $G'_l = K_l^\times \cdot G_l$ containing $\eta_l(\tau)$. If a is a generic element of $\mathcal{T}$, every characteristic root of $\omega'_l(a)$ is of the form $a_1^{\mu_1} \cdots a_j^{\mu_j}$ with characteristic roots $a_1, \cdots, a_j$ of a and integers μ_i. Specializing a to an element of $Q^\times$, we see, in view of (7.13.3) and (7.19.1), that $\mu_1 + \cdots + \mu_j = \kappa g$. Now the Weil theorem implies that every characteristic root of $\eta_l(\tau)$ is an algebraic integer of absolute value $N(\mathfrak{q})^{1/2}$, hence every characteristic root of $\omega'(\eta_l(\tau))$ is of absolute value $N(\mathfrak{q})^{\kappa g/2}$. It follows from this and (7.19.3) that every characteristic root of $\omega_i(h_i(\sigma_{\mathfrak{P}}^f))$ is an algebraic number of absolute value $N(\mathfrak{q})^{\epsilon g/2} N(\mathfrak{q})^{\epsilon(r-g)/2} = N(\mathfrak{q})^{\epsilon r/2}$. This proves (7.14.2).

The assertion (7.14.3) can now be proved in the same way as in [13, 3.9]. The present symbols G, G', $\eta_i(\tau)$, $\pi(t)$, $\pi(t^{-1}d_i)yx'$, $\mathfrak{q}$, $k_S(u)$ correspond respectively to $\mathfrak{G}$, $\mathfrak{G}'$, β_i, s (or $\lambda(s)$), u_i', $\mathfrak{p}'$, $k_c(y)$ considered there. (In [13, p. 120, line 10 from bottom], read $\tilde{A}$ for A.) The only necessary modification occurs at the following point. First we can easily show that $\nu(\eta_i(\tau)) = N(\mathfrak{q})$ as in [13, p. 121]. Then, by (7.19.2), we obtain

$$\nu(\pi(t^{-1}d_i)yx') = \nu(\eta_i(\tau))/(\pi(t)\pi(t)^\rho) = \lambda_H(t) \ .$$

This relation corresponds to the equality $\nu(u_i') = N_{K/F}(t)$ of [13, p. 121, line 8]. (Here $\lambda_H(t)$ must be considered as an element of the center of $G(Q)$.)

We thus obtain 7.12 and 7.14 for a member S of $\mathcal{Z}'$ of type (6.4.1). This combined with the result of 7.16 completes the proof of 7.12 and 7.14 in the case $r < g$.

7.20. *The case $r = g$.* Let us assume that $r = g$, and S is of the type (5.1.2) with a PEL-type $\Omega = (\kappa, \mathfrak{m}, \{d_i\})$. Let $\mathfrak{n}$ be a Z-lattice in B^n such that $\mathfrak{n}/\mathfrak{m} = \sum_i Zd_i$. For every positive integer $e > 0$, we fix a PEL-type $\Omega_e = (\kappa, \mathfrak{m}, \{d_i^e\})$ with elements d_i^e of $B^n/\mathfrak{m}$ such that $\sum_i Zd_i^e = (e^{-1}\mathfrak{m} + \mathfrak{n})/\mathfrak{m}$. Construct a family $\Sigma(\Omega_e) = \{\mathfrak{A}_w^e \mid w \in \mathfrak{H}_n^g\}$ of PEL-structures $\mathfrak{A}_w^e = (A_w, \mathcal{C}_w, \theta_w; \{t_{w\nu}^e\})$ of type Ω_e as in 5.1 (see [12, 9.2]). Put $S_e = S(\Omega_e)$. Further, when $T = S_e$, put $k_e = k_T$, $(V_e, \varphi_e) = (V_T, \varphi_T)$, $\Gamma_e = \Gamma_T$, $J_e(x) = J_{TT}(x)$ for $x \in S$. The result of §5 shows that the moduli-system for $\Sigma(\Omega_e)$ can be taken in the form $(V_e, \mathfrak{v}_e, \varphi_e)$.

Let u and $\{u_T\}$ be the points in question. We may assume that $u_T = \varphi_T(z)$ with a point z common to all T. Put $u_e = \varphi_e(z)$. Since u is algebraic, we may assume that $\mathfrak{A}_z^1$ is defined over a finite algebraic extension k' of $k_S(u)$. Let Y_0 be the set of all prime ideals in k' for which A_z has defect, i.e., A_z has no good reduction.

Put $A_z^0 = \bigcup_{e=1}^\infty \{t \in A_z \mid et = 0\}$. If $\mathfrak{y}$ is the parametrizing function defined in [12, 9.2], the map $B^n \ni a \mapsto \mathfrak{y}(a, z)$ gives a homomorphism $\mathfrak{t}$ of B^n onto A_z^0 with kernel $\mathfrak{m}$.

Let $\tau \in \mathrm{Gal}\,(\overline{Q}/k_S(u))$. Since $k_S(u)$ is the field of moduli of $\mathfrak{A}_z^1$, there exists an isomorphism β of $\mathfrak{A}_z^{1\tau}$ to $\mathfrak{A}_z^1$. Then the map $A_z^0 \ni t \mapsto \beta t^\tau$ is an automorphism of A_z^0, commuting with the action of B. Therefore we obtain an element $\eta(\tau)$ of $GL_n(B)_0$ ($=$ the non-archimedean part of $GL_n(B)_A$) such that

$$\beta\mathfrak{t}(a)^\tau = \mathfrak{t}(a\eta(\tau)) \qquad\qquad (a \in B^n/\mathfrak{m}) \ .$$

If τ is the identity on k', we can take β to be the identity map, so that the l-component of $\eta(\tau)$ is exactly the l-adic representation of τ on A_z.

Now let x be an element of S_0 which represents $h(\tau)$. By (5.1.3), Ω_e^r is equivalent to $\Omega_e' = (\kappa, \mathfrak{m}, \{d_e^r x\})$. (Observe that $\mathfrak{m}x = \mathfrak{m}$, and $\mu(x) = 1$.) By (4.4.1), we can take the moduli-system for Ω_e' in the form $(V_e, \mathfrak{v}_e', \varphi_e)$ with the same (V_e, φ_e) as before. Moreover, the reasoning of §5 shows that $\mathfrak{v}_e^r = J_e(x) \circ \mathfrak{v}_e'$. Let $\mathfrak{A}_z'^e$ be the member of Ω_e' at z. Then $\mathfrak{v}_e(\mathfrak{A}_z^e) = \varphi_e(z) = u_e = \mathfrak{v}_e'(\mathfrak{A}_z'^e)$, hence

$$\mathfrak{v}_e^r(\mathfrak{A}_z^{er}) = \mathfrak{v}_e(\mathfrak{A}_z^e)^r = u_e^r = J_e(x)(u_e) = J_e(x)\big(\mathfrak{v}_e'(\mathfrak{A}_z'^e)\big) = \mathfrak{v}_e^r(\mathfrak{A}_z'^e) \, .$$

It follows that there exists an isomorphism α_e of $\mathfrak{A}_z'^e$ to $\mathfrak{A}_z^{er}$. Now $\mathfrak{A}_z'^e$ is of the form

$$\mathfrak{A}_z'^e = \big(A_z, C_z, \theta_z; \{t_i'^e\}\big) \, ,$$

where $t_i'^e = t(d_e^r x)$. We have therefore $\alpha_e t(ax) = t(a)^r$ for $a \in (\mathfrak{n} + e^{-1}\mathfrak{m})/\mathfrak{m}$, hence $\beta \alpha_e t(a) = t(a)$ for $a \in \mathfrak{n}/\mathfrak{m}$. This means that $\beta \alpha_e$ is an automorphism of $\mathfrak{A}_z^1$. By [11, 4.11], there exists an element γ_e of $\Delta(z, S)$ such that $\beta \alpha_e$ is obtained from $\Lambda(\gamma_e, z)$. Then $\beta \alpha_e t(a) = t(a \gamma_e)$ for $a \in (\mathfrak{n} + e^{-1}\mathfrak{m})/\mathfrak{m}$, hence

$$t\big(a\eta(\tau)\big) = \beta \alpha_e t(ax) = t(a x \gamma_e) \qquad\qquad \text{for } a \in (\mathfrak{n} + e^{-1}\mathfrak{m})/\mathfrak{m} \, .$$

Therefore, the non-archimedean part of γ_e converges to an element ε of $\mathfrak{B}$, and we have $\eta(\tau) = x\varepsilon$. Now we can prove 7.12 by the same reasoning as in 7.18. The present situation is much simpler, since it is unnecessary to consider an extension like $\mathfrak{K}''$. .

Let $\sigma_{\mathfrak{P}}$ be as in 7.13. Taking τ to be $\sigma_{\mathfrak{P}}$ in the above consideration, we see, from the relation $\eta(\tau) = x\varepsilon$, that the l-component $\eta_l(\sigma_{\mathfrak{P}})$ of $\eta(\sigma_{\mathfrak{P}})$ represents $h_l(\sigma_{\mathfrak{P}})$. Therefore we obtain (7.14.2) from the Weil theorem. Since $\eta_l(\sigma_{\mathfrak{P}})$ is semi-simple, we obtain (7.14.1) if $\Delta(z, S) \subset F^\times$. The general case can be reduced to this special case by the consideration of 7.16. Furthermore, 7.14 can be verified in the same way as in [13, p. 123 and 3.9]. The symbol β_l of [13, p. 123] corresponds to the present $\eta_l(\sigma_{\mathfrak{P}})$.

8. Some representations of G and an embedding of a canonical system into another canonical system

8.1. We shall now investigate some examples of representations ω of G satisfying the conditions of 7.13. In a special case, such a representation defines an embedding of our canonical system into another canonical system defined with respect to a certain algebraic group, again of our type. We shall discuss the nature of this embedding, from which we can derive some information about the matrix $\omega_l(h_l(\sigma_{\mathfrak{P}}))$ of 7.14.

Define a group $Gp(n, \bar{Q})$ by

$$Gp(n, \overline{Q}) = \{U \in GL_{2n}(\overline{Q}) \mid {}^t U J_n U = \nu(U) J_n \text{ with } \nu(U) \in \overline{Q}\} ,$$

(8.1.1)
$$J_n = \begin{bmatrix} 0 & 1_n \\ -1_n & 0 \end{bmatrix}.$$

Then $G_{\overline{Q}}$ can be identified with $Gp(n, \overline{Q})^g$. Let $p_1, \cdots, p_g$ denote the projections of $G_{\overline{Q}}$ to the factors $Gp(n, \overline{Q})$. Let χ be an arbitrary irreducible $\overline{Q}$-rational representation of $Gp(n, \overline{Q})$ into GL_d with any degree d. Define a representation $\mu: G_{\overline{Q}} \to GL_h(\overline{Q})$, with $h = d^g$, by $\mu(x) = \chi(p_1(x)) \otimes \cdots \otimes \chi(p_g(x))$. We see that μ is irreducible, and μ^τ is equivalent to μ for every $\tau \in \mathrm{Gal}\,(\overline{Q}/Q)$. Therefore we can find an automorphism α_τ of $M_h(\overline{Q})$ over $\overline{Q}$ such that $\mu^\tau = \alpha_\tau \circ \mu$. Since μ is irreducible, α_τ is unique for τ. We see easily that $\alpha_{\sigma\tau} = \alpha_\sigma^\tau \circ \alpha_\tau$ for $\sigma, \tau \in \mathrm{Gal}\,(\overline{Q}/Q)$. As is well-known, we can find a central simple algebra E over Q, and a Q-linear isomorphism f of E into $M_h(\overline{Q})$ so that $M_h(\overline{Q}) = f(E) \otimes_Q \overline{Q}$, and $\alpha_\tau = f^\tau \circ f^{-1}$ (cf. Weil [17]; it is sufficient to put

$$E = \{x \in M_h(\overline{Q}) \mid x^\tau = \alpha_\tau(x) \text{ for all } \tau \in \mathrm{Gal}\,(\overline{Q}/Q\}.)$$

Then we see that $f^{-1} \circ \mu: G \to E^\times$ is a Q-rational representation. Take any Q-rational representation $\xi: E^\times \to GL_m$ such that $\xi(a) = a^\kappa$ for $a \in Q^\times$. (For example, a regular representation of E over Q satisfies this with $\kappa = 1$.) Put $\omega = \xi \circ f^{-1} \circ \mu$. Then ω satisfies (7.13.2, 3).

To study E and ω more closely in a special case, let us first prove

8.2. PROPOSITION. *Let K be an algebraic number field of finite degree, A a central simple algebra over K, and $[A : K] = m^2$. Let $p_1, \cdots, p_s$ be inequivalent absolutely irreducible Q-linear representations of A into $M_m(C)$, and L the algebraic number field generated over Q by $\sum_{i=1}^s \mathrm{tr}(p_i(x))$ for all $x \in K$. Then there exist a central simple algebra E over L, and an L-rational homomorphism β of $(A \otimes_Q L)^\times$ into $E^\times$, satisfying the following conditions:*

(8.2.1) $[E : L] = m^{2s}$.

(8.2.2) *If q is an L-linear representation of E into $M_f(C)$ with $f = m^s$, then $q \circ \beta$ is equivalent to $p_1 \otimes \cdots \otimes p_s$.*

PROOF. We can find a finite Galois extension T of Q containing K, and assume, without losing generality, that $p_1, \cdots, p_s$ are T-linear homomorphisms of $A \otimes_Q T$ into $M_m(T)$. Put $p = p_1 \otimes \cdots \otimes p_s$. For any $\sigma \in \mathrm{Gal}\,(T/L)$, p^σ is equivalent to p over C. Then, by a well-known principle, they must be equivalent over T, hence $p^\sigma = \alpha_\sigma \circ p$ with a unique automorphism α_σ of $M_f(T)$ over T. Then $\alpha_{\sigma\tau} = \alpha_\sigma^\tau \circ \alpha_\tau$ for $\sigma, \tau \in \mathrm{Gal}\,(T/L)$. Put

$$E = \{x \in M_f(T) \mid x^\sigma = \alpha_\sigma(x) \text{ for all } \sigma \in \mathrm{Gal}\,(T/L)\} .$$

We see easily that $M_f(T) = E \otimes_L T$, hence E is central simple over L, and

$[E:L] = f^2$. Let q denote the natural injection of E into $M_f(T)$, and $\beta = q^{-1} \circ p$. Then E and β have the required properties.

The uniqueness of E and β can be stated as follows.

8.3. PROPOSITION. *The notation being as in 8.2, let L' be a subfield of C, E' a central simple algebra over L', and β' an L'-rational homomorphism of $(A \otimes_Q L')^\times$ into $E'^\times$. Suppose that, for an absolutely irreducible L'-linear representation q' of E' by complex matrices, $q' \circ \beta'$ is equivalent to $p_1 \otimes \cdots \otimes p_s$. Then $L \subset L'$, and there exists an isomorphism h of $E \otimes_L L'$ onto E' over L' such that $\beta' = h \circ \beta$.*

PROOF. If $f = m^s$, we see that $[E':L'] = f^2$. Take a finite Galois extension S of L' containing L so that $E' \otimes_{L'} S = M_f(S)$, and $E \otimes_L S = M_f(S)$. Let $p = p_1 \otimes \cdots \otimes p_s$. For every $\sigma \in \mathrm{Gal}\,(S/L')$, p^σ is equivalent to p, hence σ is the identity map on L. Therefore $L \subset L'$. Let q (resp. q') be the injection of $E \otimes_L L'$ (resp. E') into $E \otimes_L S = M_f(S)$ (resp. $E' \otimes_{L'} S = M_f(S)$). Now $q \circ \beta$ and $q' \circ \beta'$ are S-rational representations of $(A \otimes_Q S)^\times$, which are equivalent over C. Then they must be equivalent over S, hence $q' \circ \beta' = \pi \circ q \circ \beta$ with an automorphism π of $M_f(S)$ over S. For the same reason, there exists, for each $\sigma \in \mathrm{Gal}\,(S/L')$, an automorphism λ_σ (resp. μ_σ) of $M_f(S)$ over S, such that $(q \circ \beta)^\sigma = \lambda_\sigma \circ (q \circ \beta)$ (resp. $(q' \circ \beta')^\sigma = \mu_\sigma \circ (q' \circ \beta')$). Then we have $\mu_\sigma \circ \pi = \pi^\sigma \circ \lambda_\sigma$, and

$$q(E \otimes_L L') = \{x \in M_f(S) \mid x^\sigma = \lambda_\sigma(x)\}\,,$$
$$q'(E') = \{x \in M_f(S) \mid x^\sigma = \mu_\sigma(x)\}\,.$$

(In fact, the left hand side is contained in the right hand side which is an algebra of degree f^2 over L'.) It follows that π maps $q(E \otimes_L L')$ onto $q'(E')$. Therefore we can define an isomorphism h of $E \otimes_L L'$ onto E' over L' by $\pi \circ q = q' \circ h$. Then $\beta' = h \circ \beta$.

8.4. *Remark.* Instead of $p_1 \otimes \cdots \otimes p_s$, we can also take $p_1 \oplus \cdots \oplus p_s$, and make a similar consideration. It may be interesting to investigate this in a more general case in connection with the notion of reflex introduced in § 1.

8.5. PROPOSITION. *The notation being as in 8.2, suppose that A has an involution ζ over K. Then E has an involution π over L such that $\beta(x^\zeta) = \beta(x)^\pi$.*

PROOF. The notation being as in the proof of 8.2, we can find an involution θ_i of $M_m(T)$ over T such that $\theta_i \circ p_i = p_i \circ \zeta$. Then we can define an involution θ of $M_f(T)$ over T by $\theta = \theta_1 \otimes \cdots \otimes \theta_s$. We see that $\alpha_\sigma \circ \theta = \theta^\sigma \circ \alpha_\sigma$ for every $\sigma \in \mathrm{Gal}(T/L)$. Put $\pi = q^{-1} \circ \theta \circ q$. Then $\pi^\sigma = \pi$ for every $\sigma \in \mathrm{Gal}(T/L)$, hence π is an involution of E over L, and $\beta \circ \zeta = \pi \circ \beta$.

8.6. Let us apply 8.2 and 8.5 to the case where $A = M_n(B)$ and $x^\zeta = d \cdot {}^t x^\iota d^{-1}$ with an element d of $GL_n(B)$ such that ${}^t d^\iota = \varepsilon d$, $\varepsilon = \pm 1$. We take s to be g, so that $L = Q$, and $\{p_1, \cdots, p_g\}$ is a complete set of inequivalent absolutely irreducible representations of $M_n(B)$. Then we obtain a central simple algebra E over Q, an involution π of E over Q, and a Q-rational representation $\beta: GL_n(B) \to E^\times$, such that $E \otimes_Q C$ is isomorphic to $M_m(C)$, and $\beta(d \cdot {}^t x^\iota d^{-1}) = \beta(x)^\pi$ for $x \in GL_n(B)$, where $m = (2n)^g$. Since π is an involution of the first kind, we have $E = M_q(C)$ with a quaternion algebra C over Q, and $q = 2^{g-1} n^g$. Let ι denote the main involution of C. Then there exists an element v of $E^\times$ such that $y^\pi = v \cdot {}^t y^\iota v^{-1}$ for $y \in E^\times$, and ${}^t v^\iota = \varepsilon' v$, $\varepsilon' = \pm 1$. Put

$$G(B, d) = \{x \in GL_n(B) \mid xd \cdot {}^t x^\iota = \nu(x)d \text{ with } \nu(x) \in F^\times\},$$

$$G(C, v) = \{y \in GL_q(C) \mid yv \cdot {}^t y^\iota = \nu(y)v \text{ with } \nu(y) \in Q^\times\}.$$

We observe that β maps $G(B, d)$ into $G(C, v)$, and

$$(8.6.1) \qquad\qquad \nu\big(\beta(x)\big) = N_{F/Q}\big(\nu(x)\big) \qquad\qquad \text{for } x \in G(B, d).$$

Let us now show

(8.6.2) $\varepsilon' = -(-\varepsilon)^g$;

(8.6.3) C *is definite or indefinite according as $g - r$ is odd or even.*

By virtue of the property of the main involution, we can find an element j_i of $GL_{2n}(\bar{Q})$ so that $p_i({}^t x^\iota) = j_i \cdot {}^t p_i(x) j_i^{-1}$ for $x \in GL_n(B)$, ${}^t j_i = -j_i$. Then ${}^t(p_i(d)j_i) = -\varepsilon p_i(d)j_i$. Put $c = p_1(d)j_1 \otimes \cdots \otimes p_g(d)j_g$. Then $p(d \cdot {}^t x^\iota d^{-1}) = c \cdot {}^t p(x)c^{-1}$ for $x \in GL_n(B)$. By the property (8.2.2) of β, there exists a Q-linear injection f of E into $M_{2q}(\bar{Q})$ such that $p = f \circ \beta$ on $GL_n(B)$. Then $f(y^\pi) = c \cdot {}^t f(y)c^{-1} = f(v \cdot {}^t y^\iota v^{-1}) = f(v)f({}^t y^\iota)f(v^{-1})$. We can find an element h of $GL_{2q}(\bar{Q})$ such that ${}^t h = -h$ and $f({}^t y^\iota) = h \cdot {}^t f(y)h^{-1}$. Then $f(v)h = bc$ with an element b of $\bar{Q}^\times$. It follows that

$$\varepsilon' f(v) = f({}^t v^\iota) = h \cdot {}^t f(v)h^{-1} = -b \cdot {}^t ch^{-1} = -(-\varepsilon)^g f(v),$$

hence (8.6.2).

To show (8.6.3), identify $M_n(B)_R = M_n(B) \otimes_Q R$ with a product $M_{2n}(R)^r \times M_n(H)^{g-r}$ as in 2.1 and [12, 4.4], where H denotes the Hamilton quaternions, and let $p'_1, \cdots, p'_g$ be the projections of $M_n(B)_R$ into these g factors $M_{2n}(R)$ and $M_n(H)$. Let E' be the tensor product of r copies of $M_{2n}(R)$ and $g - r$ copies of $M_n(H)$ over R, and put $\beta' = p'_1 \otimes \cdots \otimes p'_g$. Observe that E', R, and β' have the properties of E', L' and β' of 8.3. Therefore 8.3 implies that $E \otimes_Q R$ is isomorphic to E', hence (8.6.3).

8.7. Let us now consider the following two special cases:

(I) $r = 1$ and $d = 1_n$, hence $\varepsilon' = (-1)^{g-1}$.

(II) $r = g - 1$, $\varepsilon = -1$, and $p_i(d)j_i$ is definite for all $i \leq r$. (We may assume that p_i maps $GL_n(B)$ into $GL_{2n}(R)$ if $i \leq r$. Then $p_i(d)j_i$ is a real symmetric matrix.)

By (8.6.2, 3), we see that, in both cases, C is definite or indefinite according as $\varepsilon' = -1$ or 1. Therefore $G(C, v)$ is a group of the type discussed in [5]; $G(B, d)$ belongs to the groups considered in [10]. The semi-simple part of $G(B, d)_R$ (resp. $G(C, v)_R$) modulo a maximal compact subgroup has a structure of a bounded symmetric domain $\mathcal{K}$ (resp. $\mathcal{K}'$). Moreover, by a suitable choice of such compact subgroups and complex structure, we can show in a straightforward way that the representation β defines a holomorphic embedding e of $\mathcal{K}$ into $\mathcal{K}'$. (An explicit form of e in a special case is given in 8.10 below.) Since we can associate a (standard) family of abelian varieties with the group $G(C, v)$ parametrized by $\mathcal{K}'$ as given in [5], we obtain a holomorphic family[5] of abelian varieties associated with $G(B, d)$ and parametrized by $\mathcal{K}$, by means of the "pull back". In general, except in some special cases, this family does not belong to those considered in [5], [10].

8.8. Let us further specialize to the case where $r = 1$, $d = 1_n$, and g is odd. Then $G(B, d)$ is exactly a special case of the group G considered in the previous sections. Further we have $\varepsilon' = 1$, and C is indefinite, hence $G(C, v)$ is also of the same type as G (with Q in place of F). (The algebra C can be isomorphic to $M_2(Q)$ even if B is a division algebra.) In fact, since the basic field is Q, the condition (2.1.3) is automatically satisfied. Therefore we can put $v = 1_q$, losing no generality. Thus we may put $\mathcal{K} = \mathfrak{H}_n$ and $\mathcal{K}' = \mathfrak{H}_q$, where $q = 2^{g-1}n^g$. Put $G = G(B, 1_n)$, $G^* = G(C, 1_q)$. We have maps

$$\beta : G \longrightarrow G^* , \qquad e : \mathfrak{H}_n \longrightarrow \mathfrak{H}_q$$

such that $\beta(a)e(z) = e(a(z))$ for $a \in G_{R+}$ and $z \in \mathfrak{H}_n$. Let

$$\{ V_S, \varphi_S, J_{TS}(x), (S, T \in \mathcal{Z}; x \in G_{A+}) \} ,$$

$$\{ V_L^*, \varphi_L^*, J_{ML}^*(y), (L, M \in \mathcal{Z}^*; y \in G_{A+}^*) \}$$

be canonical systems for G and G^*, respectively. For every $S \in \mathcal{Z}$ (resp. $L \in \mathcal{Z}^*$), let k_S (resp. k_L^*) denote the field defined in 2.3. Further, put $\Gamma_S = S \cap G_Q$, $\Gamma_L^* = L \cap G_Q^*$. If $\beta(S) \subset L$, we have $\beta(\Gamma_S) \subset \Gamma_L^*$, so that there exists a unique rational map E_{LS} of V_S into V_L^* such that $E_{LS} \circ \varphi_S = \varphi_L^* \circ e$.

8.9. THEOREM. *Let the notation and the assumption be as in 8.8. If*

---

[5] This type of family of abelian varieties was first found by D. Mumford in the case $[F : Q] = 3$, $C = M_2(Q)$, $r = 1$, and $d = 1$. The author wishes to express his thanks to Mumford for his kind communications about this (and other topics).

$L \in \mathfrak{Z}^*$, $S \in \mathfrak{Z}$, and $\beta(S) \subset L$, then $k_L^* \subset k_S$, and E_{LS} is rational over k_S. Further, let $M \in \mathfrak{Z}^*$, $T \in \mathfrak{Z}$, $\beta(T) \subset M$, $x \in G_{A+}$, $xSx^{-1} \subset T$, $\beta(x)L\beta(x)^{-1} \subset M$. Then

$$E_{MT}^{\sigma(x)} \circ J_{TS}(x) = J_{ML}^*(\beta(x)) \circ E_{LS} .$$

The assertion $k_L^* \subset k_S$ follows immediately from (8.6.1). The remaining part will be proved in the following 8.10–12.

8.10. Let us first study more closely the embedding e. Identify $M_n(B)_R$ with $M_{2n}(R) \times M_n(H)^{g-1}$, and let p_1 (resp. $p_2, \cdots, p_g$) be the projection of $M_n(B)_R$ to $M_{2n}(R)$ (resp. $M_n(H)$). We may assume that

$$p_1({}^\iota x^\iota) = J_n \cdot {}^t p_1(x) J_n^{-1} , \qquad J_n = \begin{bmatrix} 0 & 1_n \\ -1_n & 0 \end{bmatrix} ,$$

$$p_i({}^\iota x^\iota) = {}^t p_i(x)^\iota \qquad\qquad (i = 2, \cdots, g) ,$$

where the main involution of H is also denoted by ι (see [12, 4.4]). Since g is odd, we can find an isomorphism δ_0 of $M_n(H) \otimes \cdots \otimes M_n(H)$ ($g-1$ copies) onto $M_k(R)$ over R, where $k = (2n)^{g-1}$. Then

$$\delta_0({}^\iota x_2^\iota \otimes \cdots \otimes {}^\iota x_g^\iota) = S \cdot {}^t \delta_0(x_2 \otimes \cdots \otimes x_g) S^{-1} \qquad (x_i \in M_n(H))$$

with a positive definite real symmetric matrix S. Changing δ_0 for its combination with a suitable automorphism of $M_k(R)$, we may assume that $S = 1_k$, hence

$$\delta_0({}^\iota x_2^\iota \otimes \cdots \otimes {}^\iota x_g^\iota) = {}^t \delta_0(x_2 \otimes \cdots \otimes x_g) \qquad (x_i \in M_n(H)) .$$

Put

$$W = \{x \in GL_n(H) \mid x \cdot {}^\iota x^\iota \in R^\times\} ,$$
$$W_1 = \{x \in GL_n(H) \mid x \cdot {}^\iota x^\iota = 1_n\} ,$$
$$\delta(x_2, \cdots, x_g) = \delta_0(x_2 \otimes \cdots \otimes x_g) \qquad (x_i \in M_n(H)) .$$

Then $\delta(W_1^{g-1}) \subset SO(k, R)$. Further define a map ξ of $M_{2n}(R) \times M_n(H)^{g-1}$ into $M_{2nk}(R)$ by $\xi(u, v) = u \otimes \delta(v)$ for $u \in M_{2n}(R)$, $v \in M_n(H)^{g-1}$. Put $p(x) = (p_1(x), \cdots, p_g(x))$ for $x \in M_n(B)_R$. Then

$$\xi(p({}^\iota x^\iota)) = (J_n \otimes 1_k) \cdot {}^t \xi(p(x))(J_n \otimes 1_k)^{-1} \qquad (x \in M_n(B)_R) .$$

On the other hand, we can find an isomorphism γ of $M_q(C)_R$ into $M_{2q}(R)$ such that $\gamma({}^\iota x^\iota) = J_q \cdot {}^t \gamma(x) J_q^{-1}$. Choosing a suitable coordinate system, we may put $J_q = J_n \otimes 1_k$, since $q = nk$. By 8.3, there exists an automorphism α of $M_{2q}(R)$ over R such that $\gamma \circ \beta = \alpha \circ \xi \circ p$. Changing γ for $\alpha^{-1} \circ \gamma$, we obtain $\gamma \circ \beta = \xi \circ p$, hence

$$(8.10.1) \qquad \gamma(\beta(x)) = p_1(x) \otimes \delta(p_2(x), \cdots, p_g(x)) \qquad (x \in GL_n(B)_R) .$$

Now we have a commutative diagram

$$
\begin{array}{ccc}
G_R & \xrightarrow{\ \beta\ } & G_R^* \\
{\scriptstyle p}\Big\downarrow & & \Big\downarrow{\scriptstyle r} \\
Gp(n,\,R) \times W^{g-1} & \xrightarrow{\ \xi\ } & Gp(q,\,R)
\end{array}
$$

where $Gp(n,\,R)$ is defined by (8.1.1) with R in place of $\bar{Q}$. Let $U_n = SO(2n,\,R) \cap Gp(n,\,R)$. Then $\xi(U_n \times W_1^{g-1}) \subset U_q$. Define an embedding e of $\mathfrak{H}_n$ into $\mathfrak{H}_q$ by $e(z) = z \otimes 1_q$ for $z \in \mathfrak{H}_n$. This is compatible with the map β, i.e., $\xi(a \times 1)(e(z)) = e(a(z))$ for every $a \in Gp(n,\,R)$ with positive $\nu(a)$, hence $\beta(a)(e(z)) = e(a(z))$ for every $a \in G_{Q+}$.

8.11. Let us now consider (Y, P, δ, f), Ψ, z, and η as in 3.12 and 2.4. Since $r = 1$, $\Psi \sim \chi_1$ with an extension χ_1 of τ_1 to P, and $P' = P^{\chi_1}$. Take arbitrary extensions $\chi_2, \cdots, \chi_g$ of $\tau_2, \cdots, \tau_g$ to P, and let (P^*, Φ^*) be the reflex of (P, Φ) with $\Phi \sim \sum_{\nu=1}^{g} \chi_\nu$. By 1.10, we can take P so that $[P^*\!:Q] = 2^g$. Further, by (1.9.2), there exists an element c of the form $c = \det \Phi(b)$, for some $b \in P$, such that $P^* = Q(c)$. By our definition of β, we see that every characteristic root of $\beta(f(b))$ is of the form $\det \Phi_1(b)$ with a suitable Φ_1 such that (P, Φ_1) is a CM-type. By (1.9.4), $\det \Phi_1(b)$ is a conjugate of c over Q. (Moreover, $\det \Phi_1(b) = \det \Phi(b)$ if and only if $\Phi_1 = \Phi$.) Therefore we can define a Q-linear isomorphism f^* of P^* into $M_q(C)$ by $f^*(c) = \beta(f(b))$. Let $\mathfrak{Y}^*$ be the commutor of $f^*(P^*)$ in $M_q(C)$. Then $\mathfrak{Y}^*$ is a central simple algebra over $f^*(P^*)$, and $[\mathfrak{Y}^*\!:f^*(P^*)] = n^{2g}$. Therefore we can consider a central simple algebra Y^* over P^* isomorphic to $\mathfrak{Y}^*$, and extend f^* to an isomorphism of Y^* to $\mathfrak{Y}^*$, which we denote again by f^*. Since every element of $\beta(f(Y^\times))$ commutes with $f^*(c)$, we have $\beta(f(Y^\times)) \subset f^*(Y^*)$, so that we can define a Q-rational map $\alpha\colon Y^\times \to Y^{*\times}$ by $\beta \circ f = f^* \circ \alpha$. Then $\alpha(x) = \det \Phi(x)$ for $x \in P^\times$, especially $\alpha(b) = c$. We have therefore $\alpha(x^\rho) = \alpha(x)^\rho$ for $x \in P^\times$, if ρ denotes the complex conjugation. It follows that

$$
{}^t\!f^*(c){}^\iota = \beta({}^t\!f(b){}^\iota) = \beta(f(b^\rho)) = f^*(\alpha(b^\rho)) = f^*(\alpha(b)^\rho) = f^*(c^\rho)\,,
$$

hence ${}^t\!f^*(u){}^\iota = f^*(u^\rho)$ for all $u \in P^*$. Taking the commutor, we obtain ${}^t\!f^*(Y^*){}^\iota = f^*(Y^*)$. Therefore we obtain an involution ε of Y^* such that ${}^t\!f^*(y){}^\iota = f^*(y^\varepsilon)$ for every $y \in Y^*$. Then

$$
f^*(\alpha(x^\delta)) = \beta(f(x^\delta)) = {}^t\!\beta(f(x)){}^\iota = {}^t\!f^*(\alpha(x)){}^\iota = f^*(\alpha(x)^\varepsilon)\,,
$$

hence $\alpha(x^\delta) = \alpha(x)^\varepsilon$ for $x \in Y^\times$.

Now let $z^* = e(z)$. Then $f^*(c)(z^*) = \beta(f(b))(z^*) = e(f(b)(z)) = e(z) = z^*$, hence z^* is fixed by $f^*(c)$. Let T be the map of $\mathfrak{H}_n$ into $\mathfrak{D}_n$ such that $T(z) = 0$ as in [12, 4.2], and Ψ be the representation of Y defined by

$$T \cdot f(x) T^{-1} = \begin{bmatrix} \overline{\Psi(x)} & 0 \\ 0 & \Psi(x) \end{bmatrix} \qquad (x \in Y)$$

(see [12, 4.2 and 4.7]). We obtain similarly a representation Ξ of P^* at the fixed point $e(z)$. From (8.10.1) and the fact that $\delta(W_1^{g-1}) \subset SO(n, R)$, we see that $\Xi(\alpha(x))$ is equivalent to

$$\Psi(x) \otimes p_2(f(x)) \otimes \cdots \otimes p_g(f(x)) .$$

Especially, $\Xi(c)$ is equivalent to

$$b^{\chi_1} \cdot \begin{bmatrix} b^{\chi_2} & 0 \\ 0 & b^{\chi_2 \rho} \end{bmatrix} \otimes \cdots \otimes \begin{bmatrix} b^{\chi_g} & 0 \\ 0 & b^{\chi_g \rho} \end{bmatrix} \otimes 1_d , \qquad\qquad d = n^g .$$

Therefore the characteristic roots of $\Xi(c)$ are of the form

(8.11.1) $$b^{\chi_1} \cdot \prod_{\chi \in \mathfrak{a}} b^{\chi} \cdot \prod_{\tau \in \mathfrak{b}} b^{\chi\rho} , \qquad\qquad \mathfrak{a} \cup \mathfrak{b} = \{\chi_2, \cdots, \chi_g\} .$$

Define isomorphisms ξ_i, for $1 \leq i \leq 2^{g-1}$, of P^* into C so that the c^{ξ_i} are the numbers of (8.11.1). Then $(P^*, \sum_i \xi_i)$ is a CM-type. Let $(P_1, \sum_\nu \sigma_\nu)$ be its reflex. Take the smallest Galois extension L of Q that contains P. An element γ of Gal (L/Q) is the identity on P_1 if and only if $\sum_i \xi_i \gamma = \sum_i \xi_i$; this occurs if and only if γ permutes the elements of (8.11.1). This is equivalent to the condition $\chi_1 \gamma = \chi_1$. Therefore $P_1 = P^{\chi_1}$. Let H and H^* be the subgroups of Gal (L/Q) corresponding to P and P^* respectively. Take an element of Gal (L/Q) whose restriction to P^* is ξ_i, and denote it again by the same letter. Then we observe that $\bigcup_i H^* \xi_i$ is the set of all elements γ of Gal (L/Q) such that $\chi_1 \in \{\chi_1 \gamma, \cdots, \chi_g \gamma\}$. This is so if and only if $\gamma^{-1} = \chi_1^{-1} \chi_\nu$ on P^{χ_1} for some ν. It follows that $(P^{\chi_1}, \sum \chi_1^{-1} \chi_\nu)$ is the reflex of $(P^*, \sum_i \xi_i)$. Since P has the property (1.9.4), P^{χ_1} has the same property. It follows that $(P^*, \sum_i \xi_i)$ is the reflex of $(P^{\chi_1}, \sum \chi_1^{-1} \chi_\nu)$.

Since $(P^*, \sum_i \xi_i)$ is a CM-type, and Ξ is a multiple of $\sum_i \xi_i$, we see that z^* is the *unique* fixed point of $f^*(c)$. Now every element of $f^*(Y^*)$ commutes with $f^*(c)$, hence z^* is a common fixed point of $f^*(\{Y^*, \varepsilon\})$. Now the map $\eta: P'^\times \to G$ is given by $\eta(x) = f(x^{\chi_1^{-1}})$. (Note that $P' = P^{\chi_1}$.) With respect to $(Y^*, P^*, \varepsilon, f^*)$, we obtain the same P', and the map $\eta^*: P'^\times \to G^*$ is given by

$$\eta^*(y) = f^*(\det \Phi(y^{\chi_1^{-1}})) .$$

Therefore we have

(8.11.2) $$\beta(\eta(x)) = f^*(\alpha(x^{\chi_1^{-1}})) = \eta^*(x) \qquad\qquad (x \in P'^\times) .$$

8.12. Now we have $E_{LS}(\varphi_S(z)) = \varphi_L^*(z^*)$ with the above fixed points z and z^*. Since $\varphi_S(z)$ and $\varphi_L^*(z^*)$ are algebraic, and the points $\varphi_S(z)$ of this type are dense in V_S, we see that E_{LS} is rational over a finite algebraic extension k of

k_S. In view of the construction of P in 1.9, we can take P so that $P' = P^{\chi_1}$ is linearly disjoint with k over $F' = F^{\tau_1}$, still under the condition $[P^*: Q] = 2^g$.

Let x, M, and T be as in Theorem 8.9 to be proved. Let τ be an isomorphism of k into $\bar{Q}$ such that $\tau = \sigma(x)$ on k_S. Then we can extend τ to an element π of $\mathrm{Gal}\,(\bar{Q}/P')$. Take an element v of P'_A so that $\pi = [v, P']$ on P'_{ab}. Then $\sigma(x) = \sigma(\eta(v^{-1}))$ on k_S, hence, by (3.5.2), $\eta(v)^{-1} = \alpha x s$ for some $\alpha \in G_{Q+}$ and $s \in S$. Let $U = \eta(v)^{-1} S \eta(v)$. Then $\alpha^{-1} U \alpha = x S x^{-1} \subset T$, hence

$$\varphi_T\big(\alpha^{-1}(z)\big)^\pi = J_{TU}(\alpha^{-1})^\pi \big(\varphi_U(z)^\pi\big) = J_{TU}(\alpha^{-1})^\pi\big[J_{US}\big(\eta(v)^{-1}\big)\big(\varphi_S(z)\big)\big] = J_{TS}(x)\big(\varphi_S(z)\big) \ .$$

By (8.11.2), we have $\eta^*(v)^{-1} = \beta(\alpha)\beta(x)\beta(s)$. Since $\beta(s) \in L$, we have, by the same type of computation, $\varphi_M\big(\beta(\alpha)^{-1}(z^*)\big)^\pi = J^*_{ML}\big(\beta(x)\big)\big(\varphi_L^*(z^*)\big)$. Therefore

$$E^\pi_{MT}\big[J_{TS}(x)\big(\varphi_S(z)\big)\big] = J^*_{ML}\big(\beta(x)\big)\big[E_{LS}\big(\varphi_S(z)\big)\big] \ .$$

Now we can change z for $\gamma(z)$ with any $\gamma \in G_{Q+}$ without changing P', x, and π. Since the points $\varphi_S\big(\gamma(z)\big)$ are dense in V_S, we obtain $E^\pi_{MT} \circ J_{TS}(x) = J^*_{ML}\big(\beta(x)\big) \circ E_{LS}$. Take x to be the identity element, and $T = S$, $M = L$. Then we have $E^\tau_{LS} = E_{LS}$ for every isomorphism τ of k into $\bar{Q}$ over k_S. This shows that E_{LS} is rational over k_S, and completes the proof of 8.9.

8.13. Let us now apply 8.9 to the study of the representation h and $\omega_l \circ h_l$ considered in § 7. Fix members S of $\mathfrak{Z}$ and L of $\mathfrak{Z}^*$. Let $\mathfrak{W}$ (resp. $\mathfrak{W}^*$) be the set of all normal subgroups of S (resp. L) which are members of $\mathfrak{Z}$ (resp. $\mathfrak{Z}^*$). Let u be a point of V_S, and $\{u_T\}_{T \in \mathfrak{W}}$ a coherent set of points lying above u. Put $v = E_{LS}(u)$, and $v_M = E_{MT}(u_T)$ for every $M \in \mathfrak{W}^*$, where $T \in \mathfrak{W}$, and $\beta(T) \subset L$. This definition of v_M does not depend on the choice of T. We thus obtain a coherent set of points $\{v_M\}_{M \in \mathfrak{W}^*}$ lying above v. Let $\mathfrak{K}_u$ (resp. $\mathfrak{K}_v^*$) be the composite of $k_T(u_T)$ for all $T \in \mathfrak{W}$ (resp. $k_M^*(v_M)$ for all $M \in \mathfrak{W}^*$). Then $\mathfrak{K}_v^* \subset \mathfrak{K}_u$. For simplicity, let us assume that $u_T = \varphi_T(z)$ with a point z of $\mathfrak{H}_n$ common to all $T \in \mathfrak{W}$. Put $w = e(z)$, and

$$\Delta^*(w, L) = \{\gamma \in \Gamma_L^* \mid \gamma(w) = w\} \ .$$

Now we obtain continuous isomorphisms

$$h \ : \mathrm{Gal}\,\big(\mathfrak{K}_u/k_S(u)\big) \longrightarrow \mathfrak{A}/\mathfrak{B} \ ,$$
$$h^* : \mathrm{Gal}\,\big(\mathfrak{K}_v^*/k_L^*(v)\big) \longrightarrow \mathfrak{A}^*/\mathfrak{B}^* \ ,$$

where $\mathfrak{A}$, $\mathfrak{B}$, $\mathfrak{A}^*$, $\mathfrak{B}^*$ are defined as in 7.8. Let ξ be the natural homomorphism of $\mathrm{Gal}\,\big(\mathfrak{K}_u/k_S(u)\big)$ into $\mathrm{Gal}\,\big(\mathfrak{K}_v^*/k_L(v)\big)$; $\xi(\tau)$ is the restriction of τ to $\mathfrak{K}_v^*$. Then we obtain, as an immediate consequence of 8.6,

(8.13.1) *If $\tau \in \mathrm{Gal}\,\big(\mathfrak{K}_u/k_S(u)\big)$ and x is an element of S_0 which represents $h(\tau)$, then $\beta(x)$ is an element of L_0 which represents $h^*(\xi(\tau))$.*

Now let $\omega^*: G^* \to GL_m$ be a Q-rational representation satisfying the following two conditions.

(8.13.2) *$\omega^*(a) = a^\kappa \cdot 1_m$ for every $a \in Q^\times$, with an integer κ independent of a.* (For example, the restriction of a regular representation of $M_q(C)$ over Q to G^* satisfies this with $\kappa = 1$.)

$$(8.13.3) \qquad\qquad \omega^*(\delta) = 1 \qquad\qquad \textit{for all } \delta \in \Delta^*(w, L) \, .$$

Put $\omega = \omega^* \circ \beta$. Then ω is a representation satisfying (7.13. 2–4). Let Y_1 (resp. Y_1^*) be a set of prime ideals in $k_S(u)$ (resp. $k_L^*(v)$) with the properties of 7.14 for the point u (resp. v). Let $\mathfrak{p}$, $\mathfrak{P}$, and $\sigma_\mathfrak{P}$ be as in 7.13. Let $\mathfrak{q} = \mathfrak{p} \cap k_L^*(v)$, $\mathfrak{Q} = \mathfrak{P} \cap \mathfrak{R}_v^*$, and $\sigma_\mathfrak{Q}^*$ be a Frobenius automorphism of $\mathfrak{R}_v^*$ over $k_L^*(v)$ with respect to $\mathfrak{Q}$ and $\mathfrak{q}$. Let l be a rational prime. Suppose that $\mathfrak{p} \nmid l$, $\mathfrak{p} \notin Y_1$, $\mathfrak{q} \notin Y_1^*$. If f is the degree of $\mathfrak{p}$ over $\mathfrak{q}$, we have $\sigma_\mathfrak{P} = \zeta \sigma_\mathfrak{Q}^{*f}$ with an element ζ of $\mathrm{Ker}\,(h_l^*)$. From (8.13.1), we obtain

$$(8.13.4) \qquad\qquad \omega_l\big(h_l(\sigma_\mathfrak{P})\big) = \omega_l^*\big(h_l^*(\sigma_\mathfrak{Q}^*)\big)^f \, .$$

If L is of the form (5.1.2) (defined with respect to C and G^*), we can apply 7.15 to $\omega_l^*\big(h_l^*(\sigma_\mathfrak{Q})\big)^f$, since $r = g = 1$ for C. Therefore we obtain

8.14. THEOREM. *Suppose that $\beta(S) \subset L$ with an $L \in \mathfrak{Z}^*$ of the form (5.1.2). If a representation ω of G is of the form $\omega = \omega^* \circ \beta$ with a representation ω^* of G^* satisfying (8.13.2, 3), then the assertion (7.14.3) holds with $\mu = 1$.*

As was remarked in [13, 2.3], and as is shown in 7.20, $h_l^*(\sigma_\mathfrak{Q}^*)$ is essentially the Frobenius map on a certain abelian variety (up to a subtle difference). Therefore the characteristic roots of $\omega_l\big(h_l(\sigma_\mathfrak{P})\big)$ have a geometric interpretation if ω is of the above type. It is an open question whether or not a similar interpretation can be made in the case $1 < r < g$.

8.15. Let the notation be as in 8.9. Put $W_L = E_{LS}(V_S)$. Then W_L is a subvariety of V_L rational over $(F^{\tau_1})_{ab}$. This is obviously independent of the choice of S. From 8.9, we obtain

$$(8.15.1) \qquad W_M^{\sigma(x)} = J_{ML}^*\big(\beta(x)\big)(W_L) \qquad\qquad \big(x \in G_{A+}, \, M = \beta(x)L\beta(x)^{-1}\big) \, .$$

This relation is analogous to (2.5.4). Here the behavior of a *subvariety* W_M under an automorphism of $\bar{Q}$ is described, while, in (2.5.4), the behavior of a *point* is given. One interesting feature of (8.15.1) is that it concerns the reciprocity-law for the class fields over a *totally real* field. We have treated the question of this type in the previous papers [8] and [7, II, § 10]. It may be interesting to investigate the relation (8.15.1) in more detail, in connection with the results of these papers. Here we content ourselves with only indicating that, if $n = 1$ and τ_1 is the identity map of F, the smallest field of definition of W_L containing k_L^* is the class field over F corresponding to the subgroup

$$\nu\big(B^\times \cdot \{x \in B_A^\times \mid \beta(x) \in Q^\times \cdot L\}\big)$$

of $F_A^\times$.

Corrections to [12]

p. 552, line 1. For "prime ideal" read "prime factor of p".

p. 552, lines 10, 11. For "the $\mathfrak{d}(P/F)^{\chi_\nu} \cdots$ relatively prime" read "$D(P/F)^{\chi_\nu}$ is divisible by $\mathfrak{p}^{\chi_\nu}$, but prime to $\mathfrak{p}^{\chi_\mu}$ if $\mu \neq \nu$".

PRINCETON UNIVERSITY

REFERENCES

[1] C. CHEVALLEY, *Deux théorèmes d'arithmétique*, J. Math. Soc. Japan **3** (1951), 36-44.

[2] Y. IHARA, On Congruence Monodromy Problems, vol. 1, Univ. of Tokyo, 1968.

[3] M. KNESER, *Starke Approximation in algebraischen Gruppen*: I, J. Reine u. Angew. Math. **218** (1965), 190-203.

[4] PJATECKII-SHAPIRO and SHAFAREVIC, *Galois theory of transcendental extensions and uniformization*, Izv. Akad. Nauk SSSR Ser. Mat. **30** (1966), 671-704. (Amer. Math. Soc. Translations **69** (1968), 111-145.)

[5] G. SHIMURA, *On analytic families of polarized abelian varieties and automorphic functions*, Ann. of Math. **78** (1963), 149-192.

[6] ———, *Arithmetic of unitary groups*, Ann. of Math. **79** (1964), 369-409.

[7] ———, *On the field of definition for a field of automorphic functions*: I, II, III, Ann. of Math. **80** (1964), 160-189, **81** (1965), 124-165, **83** (1966), 377-385.

[8] ———, *Class fields and automorphic functions*, Ann. of Math. **80** (1964), 444-463.

[9] ———, *Moduli and fibre systems of abelian varieties*, Ann. of Math. **83** (1966), 294-338.

[10] ———, *Discontinuous groups and abelian varieties*, Math. Ann. **168** (1967), 171-199.

[11] ———, *Construction of class fields and zeta functions of algebraic curves*, Ann. of Math. **85** (1967), 58-159.

[12] ———, *Algebraic number fields and symplectic discontinuous groups*, Ann. of Math. **86** (1967), 503-592.

[13] ———, *Local representations of Galois groups*, Ann. of Math. **89** (1969), 99-124.

[14] ——— and Y. TANIYAMA, Complex multiplication of abelian varieties and its applications to number theory, Publ. Math. Soc. Japan, No. 6, 1961.

[15] A. WEIL, Foundations of Algebraic Geometry, 2nd edition, Providence, 1962,

[16] ———, Adeles and algebraic groups, lecture notes, Institute for Advanced Study, Princeton, 1961.

[17] ———, *Algebras with involutions and the classical groups*, J. Ind. Math. Soc. **24** (1960), 589-623.

[18] ———, Basic Number Theory, Springer, Berlin-Heidelberg-New York, 1967.

(Received April 15, 1969)

On canonical models of arithmetic quotients of bounded symmetric domains: II

Annals of Mathematics, 92 (1970), 528-549

This is a continuation of the previous paper with the same title, which will be referred to as [C]. We assume that the reader is familiar with the results of [C], and use the same notation. In [C, 2.7, 2.8], we have defined a certain function field $\mathfrak{L}$, and discussed its group of automorphisms Aut $(\mathfrak{L}/F')$. We have shown that if $r = 1$, and if either the quotients $\mathfrak{H}_n^r/\Gamma_S$ are compact, or $G_Q = GL_2(Q)$, then Aut $(\mathfrak{L}/F')$ is isomorphic to $G_{A+}/\mathcal{C}$, where $\mathcal{C}$ denotes the closure of $F^\times G_{\infty+}$ in G_A. One of our main purposes is to extend this result to the case $r > 1$. As was mentioned in [C, footnote 3], it is natural to consider the subgroup $\bar{\mathcal{G}}_+$ of G_{A+} consisting of all x such that $\nu(x)$ belongs to the closure of $\lambda(F_A'^\times)F^\times F_{\infty+}^\times$ in $F_A^\times$. It will be shown that $\dot{\mathcal{G}}_+$ is actually the closure of the group $\mathcal{G}_+$ defined in [C, 2.2]. Then we shall define a homomorphism of $\bar{\mathcal{G}}_+$ into Aut $(\mathfrak{K}/F')$ with kernel $\mathcal{C}$, where $\mathfrak{K}$ is a certain subfield of $\mathfrak{L}$ which has $\mathfrak{L}$ as a "scalar extension". To describe the whole Aut $(\mathfrak{K}/F')$, $\dot{\mathcal{G}}_+$ is still insufficient. In fact, if $r > 1$, Aut $(\mathfrak{K}/F')$ may have some elements arising from "permutations" of the factors of $\mathfrak{H}_n^r$. Therefore we consider a certain extension $\mathfrak{A}$ of $\bar{\mathcal{G}}_+/\mathcal{C}$ by a group of automorphisms of F, and formulate the result about canonical models in terms of all open compact subgroups of $\mathfrak{A}$. Although this extension $\mathfrak{A}$ is certainly natural, the group $\bar{\mathcal{G}}_+$ is more essential. Therefore, when one seeks the generalization of this theory for other types of groups, it is most important to find the analogue of $\bar{\mathcal{G}}_+$; the construction of $\mathfrak{A}$ is rather of secondary importance.

As we already mentioned in [C], one can expect such a generalization for all reductive algebraic groups connected with all four types of classical bounded symmetric domains. Recently K. Miyake [2] has shown that this is actually the case for all unitary groups over algebras with involutions of the second kind. He has been able to construct analogues of $\bar{\mathcal{G}}_+$ and $\mathfrak{A}$, thus establishing a theory for such algebraic groups, completely parallel to ours.

* This work was partially supported by the National Science Foundation Grant GP 14590.

Therefore the reader with particular interest in this topic is advised to read also [2] and compare all definitions and results.

1. A generalization of Galois theory for an arbitrary extension

1.1. Let us first make a few elementary observations about the group $\mathrm{Aut}\,(K/k)$ of all automorphisms of an arbitrary field K over a subfield k. Although the following propositions are not absolutely necessary for the proofs of the main theorems, our later discussion will be clearly understood only from the viewpoint of this section, especially that of 1.3.

We make $\mathrm{Aut}\,(K/k)$ a Hausdorff topological group by taking, as a basis of neighborhoods of the identity element, all subgroups of the form

$$\{\sigma \in \mathrm{Aut}\,(K/k) \mid x_1^\sigma = x_1, \cdots, x_n^\sigma = x_n\}$$

for any finite set $\{x_1, \cdots, x_n\}$ of elements of K. In the following propositions, we shall fix K and k, and for a subfield F of K containing k, put

$$\mathfrak{g}(F) = \{\sigma \in \mathrm{Aut}\,(K/k) \mid x^\sigma = x \text{ for all } x \in F\} = \mathrm{Aut}\,(K/F)\,,$$

and for every subgroup H of $\mathrm{Aut}\,(K/k)$,

$$\mathfrak{f}(H) = \{x \in K \mid x^\sigma = x \text{ for all } \sigma \in H\}\,.$$

Observe that the topology of $\mathrm{Aut}\,(K/F)$ is the same as the topology induced from that of $\mathrm{Aut}\,(K/k)$.

1.2. PROPOSITION. *Let $\mathfrak{C}$ denote the set of all compact subgroups of $\mathrm{Aut}\,(K/k)$, and $\mathfrak{S}$ the set of all subfields of K containing k, over which K is a (finite or an infinite) Galois extension. Then $\mathfrak{g}(\mathfrak{f}(H)) = H$ and $\mathfrak{f}(H) \in \mathfrak{S}$ for every $H \in \mathfrak{C}$; $\mathfrak{f}(\mathfrak{g}(F)) = F$ and $\mathfrak{g}(F) \in \mathfrak{C}$ for every $F \in \mathfrak{S}$. Thus there is a one-to-one correspondence between $\mathfrak{C}$ and $\mathfrak{S}$.*

1.3. PROPOSITION. *The notation being as in 1.2, let $\mathfrak{C}'$ be the set of all open compact subgroups of $\mathrm{Aut}\,(K/k)$, and $\mathfrak{S}'$ the subset of $\mathfrak{S}$ consisting of all $F \in \mathfrak{S}$ which are finitely generated over $\mathfrak{f}(\mathrm{Aut}\,(K/k))$. Suppose that $\mathfrak{S}'$ is not empty. Then $\mathrm{Aut}\,(K/k)$ is locally compact, and the one-to-one correspondence between $\mathfrak{C}$ and $\mathfrak{S}$ induces a one-to-one correspondence between $\mathfrak{C}'$ and $\mathfrak{S}'$.*

1.4. PROPOSITION. *Let H be a subgroup of $\mathrm{Aut}\,(K/k)$, $F = \mathfrak{f}(H)$, and F_1 the algebraic closure of F in K. Then F_1 is a Galois extension of F. If moreover $\mathfrak{g}(F) = H$, then $\mathfrak{g}(F_1)$ is a normal subgroup of H, and $H/\mathfrak{g}(F_1)$, as an abstract group, is canonically isomorphic to a dense subgroup of $\mathrm{Gal}\,(F_1/F)$.*

1.5. PROPOSITION. *If $\mathfrak{f}(\mathfrak{g}(F)) = F$ for a subfield F of K containing k, then $\mathfrak{f}(\mathfrak{g}(M)) = M$ for every finite algebraic extension M of F contained in K.*

The proofs of these propositions are given in [5, § 6.3].

2. Lemmas on the idele group of an algebraic number field

2.1. In this section, F will denote an arbitrary algebraic number field of finite degree, and $F_A^\times$ the group of ideles of F. We use the same notation as introduced in [C, 0.3], and denote by F_c the closure of $F^\times F_{\infty+}^\times$ in $F_A^\times$.

2.2. LEMMA. *Let E be the group of all the units of F which are positive at all real archimedean primes of F, E_0 the projection of E to the non-archimedean part of $F_A^\times$, and $\bar E_0$ the closure of E_0 in $F_A^\times$. For a positive integer m, put*

$$\bar E_0^m = \{x^m \mid x \in \bar E_0\}\,, \qquad F_c^m = \{x^m \mid x \in F_c\}\,.$$

Then $F_c = \bar E_0 F^\times F_{\infty+}^\times$, $\bar E_0 = E_0 \bar E_0^m$, and $F_c = F^\times F_c^m$ for every positive integer m.

PROOF. If $x \in F_c$, we can find, for every m, an element a_m of $F^\times$ so that $a_m^{-1}x \equiv 1 \bmod_0 (m!)$, and $a_m^{-1}x \gg 0$. Put $e_m = a_1^{-1}a_m$. Then $e_m \in E$, and $e_m^{-1}a_1^{-1}x \equiv 1 \bmod_0 (m!)$. Therefore the non-archimedean part of $a_1^{-1}x$ belongs to $\bar E_0$. This shows $F_c \subset \bar E_0 F^\times F_{\infty+}^\times$. Since the opposite inclusion is obvious, we obtain the first assertion. Next, since $[E_0 : E_0^m] < \infty$, we see that $[E_0 \bar E_0^m : \bar E_0^m] < \infty$. We also observe that $\bar E_0^m$ is closed, hence $E_0 \bar E_0^m$ is closed. This proves the second assertion. The last assertion follows immediately from the first and second ones.

2.3. LEMMA. *For every positive integer b, there exists a positive integer e such that*

$$\{x \in F_c \mid x \equiv 1 \bmod_0 (e),\ x \gg 0\} \subset \{x^2 \mid x \in F_c,\ x \equiv 1 \bmod_0 (b)\}\,.$$

PROOF. For a positive integer d, put

$$U(d) = \{x \in F_A^\times \mid x \equiv 1 \bmod_0 (d),\ x \gg 0\}\,,$$
$$E(d) = F^\times \cap U(d)\,.$$

By Chevalley [1], given b, there exists a positive integer e such that $E(e) \subset \{x^2 \mid x \in E(b)\}$. Let d be a multiple of e. Since $F^\times U(d)$ is open in $F_A^\times$, we have $F_c \subset F^\times U(d)$. Therefore, if $x \in F_c \cap U(e)$, we have $x = yz$ with $y \in F^\times$ and $z \in U(d)$. Then $y \in U(e) \cap F^\times = E(e)$, and $z \in F_c \cap U(d)$. This shows $F_c \cap U(e) = E(e) \cdot (F_c \cap U(d))$. Therefore

$$F_c \cap U(e) \subset \{x^2 \mid x \in F_c \cap U(b)\} \cdot U(d)$$

for every multiple d of eb. Since $\{x^2 \mid x \in F_c \cap U(b)\}$ is closed, and the sets $U(d)$, for all d, form a basis of neighborhoods of $F_{\infty+}^\times$, we have $F_c \cap U(e) \subset \{x^2 \mid x \in F_c \cap U(b)\}$.   q.e.d.

2.4. Let F' be another algebraic number field of finite degree, and Θ' a $\boldsymbol{Q}$-linear representation of F' by complex matrices such that $\operatorname{tr}(\Theta'(x)) \in F$ for every $x \in F'$. (Then Θ' is equivalent to a $\boldsymbol{Q}$-linear representation of F' by matrices with coefficients in F, see [C, 1.2].) Put $\lambda(x) = \det(\Theta'(x))$ for $x \in F'$. Then λ defines a $\boldsymbol{Q}$-rational homomorphism of $F'^{\times}$ to $F^{\times}$, and hence can be extended to a continuous homomorphism of $F_A'^{\times}$ to $F_A^{\times}$, which we denote also by λ.

2.5. LEMMA. *The notation being as above, $\lambda(F_A'^{\times})F_c$ is closed in $F_A^{\times}$. Moreover, for every open subgroup U of $F_A'^{\times}$, $\lambda(U)F_c$ is open and of finite index in $\lambda(F_A'^{\times})F_c$.*

PROOF. From λ we obtain a continuous homomorphism

$$\lambda^*\colon F_A'^{\times}/F_c' \longrightarrow \lambda(F_A'^{\times})F_c/F_c \subset F_A^{\times}/F_c \ .$$

Our first assertion follows easily from the fact that $F_A'^{\times}/F_c'$ is compact. Further, if U is an open subgroup of $F_A'^{\times}$, $[F_A'^{\times}: UF_c']$ is finite, hence $[\lambda(F_A'^{\times})F_c\colon \lambda(U)F_c]$ is finite. Now $\lambda(U)F_c/F_c$, being the image of UF_c'/F_c' by λ^*, must be compact, hence the last assertion.

3. The closure $\bar{\mathcal{G}}_+$ of the group $\mathcal{G}_+$

3.1. Let us now consider the group G of similitudes of a hermitian form over a quaternion algebra B defined in [C, 2.1]. Recall that there is an exact sequence

$$1 \longrightarrow G^{u} \longrightarrow G \xrightarrow{\ \nu\ } F^{\times} \longrightarrow 1$$

where $\nu(x)$ denotes the multiplier of $x \in G$, and F is the center of B, which is a totally real algebraic number field of degree g. Then we can associate with G a couple (F, Θ) with a class Θ of representations of F and its reflex (F', Θ'). We define $\lambda\colon F_A'^{\times} \longrightarrow F_A^{\times}$ by $\lambda(y) = \det(\Theta'(y))$ for $y \in F_A'^{\times}$, and put

$$
\begin{aligned}
D &= \left\{x \in G_A \mid \nu(x) \in \lambda(F_A'^{\times})\right\}, \\
D_+ &= D \cap G_{A+}, \\
\mathcal{G} &= DG_{Q}G_{\infty+}, \\
\mathcal{G}_+ &= D_+G_{Q+}G_{\infty+} = \mathcal{G} \cap G_{A+} && \text{(see [C, 2.2, 0.4])}.
\end{aligned}
$$

In [C] we have stated the main results with respect to this group $\mathcal{G}_+$. For instance, the morphisms $J_{TS}(x)$ have been defined for the elements x of $\mathcal{G}_+$. As we have already mentioned in [C, Footnote 3], it is more natural to take the closure $\bar{\mathcal{G}}_+$ of $\mathcal{G}_+$ in G_A instead of $\mathcal{G}_+$ itself. Therefore we are going to study the structure of $\bar{\mathcal{G}}_+$.

Let $\tau_1, \cdots, \tau_g$ be the isomorphisms of F into R (see [C, 2.1]). For $y \in F_A^{\times}$, we denote by $y^{(i)}$ the component of y at the archimedean prime of F corresponding to τ_i. We have of course $y \gg 0$ if and only if $y^{(i)} > 0$ for all i. Furthermore, by [4, 3.3], we have

$$(3.1.1) \qquad \nu(G_Q) = \nu(G_A) \cap F^\times \ = \{a \in F^\times \mid a^{(i)} > 0 \text{ for } i = 1, \cdots, r\} \,,$$

$$(3.1.2) \qquad \nu(G_{Q+}) = \nu(G_{A+}) \cap F^\times = \{a \in F^\times \mid a \gg 0\} \,.$$

3.2. PROPOSITION. *Let G_c (resp. G_{c+}) denote the closure of $G_Q G_{\infty+}$ (resp. $G_{Q+}G_{\infty+}$) in G_A. Then*

$$G_c = F_c G_Q G_A^u \ = \{x \in G_A \mid \nu(x) \in F_c\} \,,$$
$$G_{c+} = F_c G_{Q+} G_A^u = \{x \in G_{A+} \mid \nu(x) \in F_c\} = G_c \cap G_{A+} \,.$$

RROOF. Since $F^\times \subset G_{Q+}$, we have $F_c \subset G_{c+}$. By [C, 3.4], we have $G_A^u \subset G_{Q+}G_{\infty+}S_0$ for any open subgroup S_0 of G_0, so that $G_A^u \subset G_{c+}$. Therefore we obtain

$$F_c G_{Q+} G_A^u \subset G_{c+} \subset \{x \in G_{A+} \mid \nu(x) \in F_c\} \,,$$
$$F_c G_Q G_A^u \subset G_c \ \subset \{x \in G_A \mid \nu(x) \in F_c\} \,.$$

Let $x \in G_A$ and $\nu(x) \in F_c$. By 2.2, $\nu(x) = ab^2$ with $a \in F^\times$, $b \in F_c$. Observe that $a = \nu(b^{-1}x) \in \nu(G_A) \cap F^\times$. By (3.1.1), $a = \nu(\alpha)$ for some $\alpha \in G_Q$. Then $\nu(b^{-1}\alpha^{-1}x) = 1$, so that $x = b\alpha \cdot (\alpha^{-1}b^{-1}x) \in F_c G_Q G_A^u$, which proves the first line of equalities. If $x \in G_{A+}$, we have $a \gg 0$, so that $\alpha \in G_{Q+}$, which proves the second line of equalities.

3.3. PROPOSITION. *$F_c G_{\infty+}$ is the closure of $F^\times G_{\infty+}$ in G_A, and $G_{Q+} \cap F_c G_{\infty+} = F^\times$.*

PROOF. The first assertion is obvious. To see the second one, it is sufficient to observe that, if $x \in G_{Q+} \cap F_c G_{\infty+}$, every non-archimedean component of x belongs to the center of G_{Q+}.

3.4. PROPOSITION. *Let $\bar{\mathcal{G}}$ (resp. $\bar{\mathcal{G}}_+$) denote the closure of $\mathcal{G}$ (resp. $\mathcal{G}_+$). Then*

$$\bar{\mathcal{G}} = G_c D \ \ = F_c \mathcal{G} \ = F_c G_Q D \ \ = \{x \in G_A \mid \nu(x) \in \lambda(F_A'^\times) F_c\} \,,$$
$$\bar{\mathcal{G}}_+ = G_{c+} D_+ = F_c \mathcal{G}_+ = F_c G_{Q+} D_+ = \{x \in G_{A+} \mid \nu(x) \in \lambda(F_A'^\times) F_c\} = \bar{\mathcal{G}} \cap G_{A+} \,.$$

PROOF. Since $G_A^u \subset D_+$, 3.2 implies that

$$G_c D = F_c G_Q D \subset F_c \mathcal{G} \subset \bar{\mathcal{G}} \,,$$
$$G_{c+} D_+ = F_c G_{Q+} D_+ \subset F_c \mathcal{G}_+ \subset \bar{\mathcal{G}}_+ \,.$$

Since $\lambda(F_A'^\times) F_c$ is closed by 2.5, we have

$$\bar{\mathfrak{S}} \subset \{x \in G_A \mid \nu(x) \in \lambda(F_A'^\times)F_c\} \,,$$
$$\bar{\mathfrak{S}}_+ \subset \{x \in G_{A+} \mid \nu(x) \in \lambda(F_A'^\times)F_c\} \,.$$

Now let $x \in G_A$ and $\nu(x) \in \lambda(F_A'^\times)F_c$. Then $\nu(x) = \lambda(a)b$ with $a \in F_A'^\times$, $b \in F_c$. Take $e \in F'^\times$ so that $ea \gg 0$. Then $\nu(x) = \lambda(ae)\lambda(e)^{-1}b$, and $(\lambda(e)^{-1}b)^{(i)} > 0$ for all $i \leq r$. Therefore $\lambda(e)^{-1}b = \nu(y)$ for some $y \in G_A$ (cf. [C, proof of 3.3]). By 3.2, $y \in G_c$, since $\lambda(e)^{-1}b \in F_c$. Then we have $\nu(y^{-1}x) = \lambda(ae)$, so that $y^{-1}x \in D$, and $x = y \cdot y^{-1}x \in G_c D$, which completes the proof of the first line of equalities. If $x \in G_{A+}$, we have $\lambda(e)^{-1}b \gg 0$, so that $y \in G_{A+}$ and $y^{-1}x \in D_+$, and hence $x = y \cdot y^{-1}x \in G_{c+}D_+$, which proves the second line of equalities.

3.5. In [C], we associated with a certain subgroup S of G_{A+} a model of $\mathfrak{H}_n^r/(S \cap G_{Q+})$. We modify this by taking subgroups of $\bar{\mathfrak{S}}_+$ as follows. Let $\mathfrak{Z}^*$ denote the set of all the subgroups M of $\bar{\mathfrak{S}}_+$ satisfying the following two conditions.

(3.5.1) $F_c G_{\infty+} \subset M$;

(3.5.2) $M/F_c G_{\infty+}$ *is open and compact in* $\bar{\mathfrak{S}}_+/F_c G_{\infty+}$, *(hence M is open in $\bar{\mathfrak{S}}_+$).*

(We consider *open compact subgroups* for the reason explained by 1.3. This will be justified by 6.3 and 6.5 below.)

We see easily that

(3.5.3) *If $M \in \mathfrak{Z}^*$ and $N \in \mathfrak{Z}^*$, then $M \cap N \in \mathfrak{Z}^*$;*

(3.5.4) *If $x \in G_A$ and $M \in \mathfrak{Z}^*$, then $xMx^{-1} \in \mathfrak{Z}^*$;*

(3.5.5) *Any two members of $\mathfrak{Z}^*$ are commensurable.*

For every $M \in \mathfrak{Z}^*$, put $\Gamma_M = M \cap G_Q$.

3.6. PROPOSITION. *Let $\mathfrak{Z}$ be the set of certain subgroups of G_{A+} defined in [C, 2.3]. For every $S \in \mathfrak{Z}$, put $S' = S \cap \bar{\mathfrak{S}}_+$ and $S'' = S \cap \mathfrak{S}_+$. Then $F_c S'$ and $F_c S''$ are contained in $\mathfrak{Z}^*$.*

PROOF. Let S_0 be the projection of S to G_0, and $S_0' = S_0 \cap \bar{\mathfrak{S}}_+$. By [C, (2.3.1)], S_0' is compact, and $S' = S_0' G_{\infty+}$. Now $F_c S'/F_c G_{\infty+}$ is the image of S_0' by the natural map of $\bar{\mathfrak{S}}_+$ to $\bar{\mathfrak{S}}_+/F_c G_{\infty+}$. Therefore $F_c S'/F_c G_{\infty+}$ is compact. Let $S(\mathfrak{m}, b)$ be as in [C, 2.10], with a Z-lattice $\mathfrak{m}$ in $X = B^n$ and a positive integer b. Then S contains $S(\mathfrak{m}, b)$ for sufficiently large b. Fix such a b, and take a positive integer e as in 2.3. Put

$$U'(e) = \{a \in F_A'^\times \mid a \equiv 1 \bmod_0 (e), \, a \gg 0\} \,,$$
$$T = \{x \in \bar{\mathfrak{S}}_+ \mid x \equiv 1 \bmod_0 (\mathfrak{m}, e)\} \,,$$
$$R = \{x \in \bar{\mathfrak{S}}_+ \mid \nu(x) \in \lambda(U'(e))F_c\} \,.$$

Then T is open in $\bar{\mathfrak{S}}_+$. Observe that R is the inverse image of $\lambda(U'(e))F_c$ by the map ν of $\bar{\mathfrak{S}}_+$ to $\lambda(F_A'^\times)F_c$. By 2.5, $\lambda(U'(e))F_c$ is open in $\lambda(F_A'^\times)F_c$, hence R

is open in $\bar{\mathcal{G}}_+$. Let $x \in T \cap R$. Then $\nu(x) = \lambda(v)w$ with $v \in U'(\mathfrak{e})$ and $w \in F_c$. Since $x \in T$, we have $\nu(x) \equiv 1 \bmod_0 (\mathfrak{e})$, so that $w \equiv 1 \bmod_0 (\mathfrak{e})$. By 2.3, $w = z^2$ with an element z of F_c such that $z \equiv 1 \bmod_0 (\mathfrak{b})$. Then we have $\nu(z^{-1}x) = \lambda(v)$, so that $z^{-1}x \in S(\mathfrak{m}, \mathfrak{b})$, hence $x = z \cdot z^{-1}x \in F_c S(\mathfrak{m}, \mathfrak{b})$. This shows that

$$T \cap R \subset F_c S(\mathfrak{m}, \mathfrak{b}) \subset F_c S'' \subset F_c S' \ .$$

Therefore $F_c S'$ and $F_c S''$ are open in $\bar{\mathcal{G}}_+$. It follows that $F_c S''/F_c G_{\infty+}$, being an open subgroup of a compact group $F_c S'/F_c G_{\infty+}$, must be compact.

3.7. Proposition. *Let S, S', and S'' be as in 3.6. If S' is open in $\bar{\mathcal{G}}_+$ (especially if S is open in G_{A+}), one has $F_c S' = F_c S'' = F^\times S'$, $F_c S' \cap \mathcal{G}_+ = F^\times S''$, $G_Q \cap F_c S' = F^\times \cdot \Gamma_S$, and $F_c \cdot \nu(S') = F^\times \cdot \nu(S')$.*

Proof. Since $F^\times S'$ is an open subgroup of $\bar{\mathcal{G}}_+$ containing $F^\times F^\times_{\infty+}$, we have $F_c \subset F^\times S'$, so that $F_c S' \subset F^\times S'$. Since the opposite inclusion is obvious, we have $F_c S' = F^\times S'$. Then $F_c S' = F_c (S \cap F_c \mathcal{G}_+) = F_c S \cap F_c \mathcal{G}_+ = F_c (F_c S \cap \mathcal{G}_+) = F_c (F^\times S \cap \mathcal{G}_+) = F_c F^\times (S \cap \mathcal{G}_+) = F_c S''$, and $F_c S' \cap \mathcal{G}_+ = F^\times S' \cap \mathcal{G}_+ = F^\times (S' \cap \mathcal{G}_+) = F^\times S''$. Further $G_Q \cap F_c S' = G_Q \cap F^\times S' = F^\times (G_Q \cap S') = F^\times \cdot \Gamma_S$. Now [C, (2.10.3)] implies that ν is an open mapping of G_A to $F^\times_A$, hence $\nu(S')$ is open in $\nu(\bar{\mathcal{G}}_+)$. Observe that $F_c^2 = \{a^2 \mid a \in F_c\}$ is the closure of $(F^\times F^\times_{\infty+})^2 = \{a^2 \mid a \in F^\times F^\times_{\infty+}\}$, and contained in $\nu(\bar{\mathcal{G}}_+)$. Since $F^{\times 2} \cdot \nu(S')$ is a closed subgroup of $\nu(\bar{\mathcal{G}}_+)$ containing $(F^\times F^\times_{\infty+})^2$, we have $F_c^2 \subset F^{\times 2} \cdot \nu(S')$, so that $F_c = F^\times F_c^2 \subset F^\times \cdot \nu(S')$, and $F_c \cdot \nu(S') \subset F^\times \cdot \nu(S')$. Since the opposite inclusion is obvious, we obtain $F_c \cdot \nu(S') = F^\times \cdot \nu(S')$.

3.8. Proposition. *Let M be an open subgroup of $\bar{\mathcal{G}}_+$ containing $F_c G_{\infty+}$. Then there exists an open compact subgroup S_0 of G_0 such that M contains $F_c(S_0 G_{\infty+} \cap \bar{\mathcal{G}}_+)$. (Note that $S_0 G_{\infty+}$ is a member of $\mathcal{Z}$.)*

This is obvious.

3.9. Proposition. *For every $M \in \mathcal{Z}^*$, $\Gamma_M/F^\times$ is a properly discontinuous group of transformations on $\mathfrak{H}_n^r$, and $\mathfrak{H}_n^r/\Gamma_M$ is of finite measure.*

Proof. Take S_0 as in 3.8 for the present M, and put $S = S_0 G_{\infty+}$, $N = F_c(S \cap \bar{\mathcal{G}}_+)$. By (3.5.5), $[M : N]$ is finite, hence $[\Gamma_M : \Gamma_N]$ is finite. Since $\Gamma_N = F^\times \Gamma_S$ by 3.7, we obtain our assertion.

3.10. Consider the map

$$\lambda^* : F_A'^\times \longrightarrow \lambda(F_A'^\times) F_c/F_c$$

which is obtained by composing λ with the natural map of $F_A^\times$ to $F_A^\times/F_c$. Denote by $\mathfrak{k}^*$ the subfield of F_{ab}' corresponding to the kernel of λ^*. Observe that

$$(3.10.1) \qquad \mathrm{Gal}\,(\mathfrak{k}^*/F') \cong F_A'^{\times}/\mathrm{Ker}\,(\lambda^*) \cong \lambda(F_A'^{\times})F_c/F_c \qquad \text{(canonically)} .$$

On the other hand, we have an exact sequence

$$(3.10.2) \qquad 1 \longrightarrow G_{c+} \longrightarrow \bar{\mathcal{G}}_+ \xrightarrow{\;\nu^*\;} \lambda(F_A'^{\times})F_c/F_c \longrightarrow 1 ,$$

where ν^* is the composed map of ν with the natural map of $F_A^{\times}$ to $F_A^{\times}/F_c$.
The exactness follows from 3.2 and the surjectivity of ν^*, which can be seen
as follows: Let $b \in F_A'^{\times}$. Take an element d of $F'^{\times}$ so that $bd \gg 0$. As was
shown in [C, proof of 3.3], there exists an element x of G_{A+} such that $\nu(x) = \lambda(bd)$. Then $x \in \bar{\mathcal{G}}_+$, and $\nu^*(x)$ is the coset $\lambda(b)F_c$, which proves the desired
surjectivity.

By (3.10.2), we see that $\bar{\mathcal{G}}_+/G_{c+}$ is isomorphic to the group of (3.10.1).
Therefore we can define an isomorphism

$$\rho: \bar{\mathcal{G}}_+/G_{c+} \longrightarrow \mathrm{Gal}\,(\mathfrak{k}^*/F')$$

so that $\rho(x \cdot G_{c+}) = [u^{-1}, F']$ for $x \in \bar{\mathcal{G}}_+$ with an element u of $F_A'^{\times}$ such that
$\nu(x) \in \lambda(u)F_c$. For convenience, we shall regard ρ also as a homomorphism of
$\bar{\mathcal{G}}_+$ to $\mathrm{Gal}\,(\mathfrak{k}^*/F')$ with kernel G_{c+}, so that $\rho(x) = \rho(x \cdot G_{c+})$ for $x \in \bar{\mathcal{G}}_+$.

For each $M \in \mathcal{Z}^*$, denote by k_M the subfield of $\mathfrak{k}^*$ corresponding to $\rho(M)$.
Since $\rho(M) = \rho(MG_{c+})$ and $G_{Q+}M = G_{c+}M = MG_{c+} = MG_{Q+}$ (note that $G_A^u \subset MG_{Q+}$
by [C, 3.4]), we have

$$G_{Q+}M = MG_{Q+} = \{x \in \bar{\mathcal{G}}_+ \mid \rho(x) \text{ is the identity map on } k_M\} .$$

More generally, for any $y \in \bar{\mathcal{G}}_+$, we have

$$(3.10.3) \quad \begin{aligned} yMG_{Q+} &= yG_{Q+}M = MyG_{Q+} = MG_{Q+}y = G_{Q+}yM = G_{Q+}My \\ &= \{x \in \bar{\mathcal{G}}_+ \mid \rho(x) = \rho(y) \text{ on } k_M\} . \end{aligned}$$

This follows easily from the fact that $k_M = k_N$ if $N = yMy^{-1}$ (cf. [C, 3.5]).
We also see that

(3.10.4) k_M *is the subfield of* F_{ab}' *corresponding to the subgroup*

$$\mathfrak{X}_M = \{u \in F_A'^{\times} \mid \lambda(u) \in \nu(M)F_c\}$$

of $F_A'^{\times}$;

$$(3.10.5) \qquad [k_M : F'] = [F_A'^{\times} : \mathfrak{X}_M] = [\lambda(F_A'^{\times})F_c : \nu(M)F_c] = [\bar{\mathcal{G}}_+ : MG_{c+}] .$$

In [C, 2.3], we defined certain subfields $\mathfrak{k}$ and k_S $(S \in \mathcal{Z})$ of F_{ab}', and a
homomorphism σ of $\mathcal{G}_+$ onto $\mathrm{Gal}\,(\mathfrak{k}/F')$. The relationship between these fields
and the above $\mathfrak{k}^*$, k_M can be explained as follows.

3.11. PROPOSITION. *The field* $\mathfrak{k}^*$ *is the union of* k_M *for all* $M \in \mathcal{Z}^*$, *and*
$\mathfrak{k}^* \subset \mathfrak{k}$. *For every* $x \in \mathcal{G}_+$, $\rho(x)$ *is the restriction of* $\sigma(x)$ *to* $\mathfrak{k}^*$. *Moreover, if*

$S \in \mathcal{Z}$ *and* $M = F_c S'$ *(see 3.6), one has* $k_M \subset k_S$. *If, further,* S' *is open in* $\bar{\mathcal{G}}_+$, *then* $k_M = k_S$.

PROOF. Let $M = F_c S'$ with $S \in \mathcal{Z}$. By [C, (3.1.3)], k_S corresponds to $\{u \in F_A'^\times \mid \lambda(u) \in F^\times \cdot \nu(S \cap \mathcal{G}_+)\}$. Since $F^\times \cdot \nu(S \cap \mathcal{G}_+) \subset F_c \cdot \nu(M)$, we have $k_M \subset k_S$. If S' is open in $\bar{\mathcal{G}}_+$, we have $F_c \cdot \nu(M) = F_c \cdot \nu(S') = F^\times \cdot \nu(S') \subset F^\times \cdot \nu(S)$ by 3.7. Therefore, in view of the definition of k_S, we have $k_S \subset k_M$, so that $k_M = k_S$. The remaining assertions are now obvious.

It is now easy to state (and prove) the reformulation of [C, Th. 2.5] about the canonical system for G, with $\bar{\mathcal{G}}_+$ and $\mathcal{Z}^*$ in place of $\mathcal{G}_+$ and $\mathcal{Z}$. We shall actually prove a somewhat stronger result than this, by considering a certain extension of $\bar{\mathcal{G}}_+/F_c G_{\infty+}$, for the reason explained in the Introduction.

4. The extension $\mathfrak{A}$ of $\bar{\mathcal{G}}_+/F_c G_{\infty+}$ by certain automorphisms of F

4.1. Let $D(B/F)$ be the product of all prime ideals in F which are ramified in B, and $\mathfrak{g}$ the group of all automorphisms γ of F such that $D(B/F)^\gamma = D(B/F)$, and $\{\gamma\tau_1, \cdots, \gamma\tau_r\}$ coincides with $\{\tau_1, \cdots, \tau_r\}$ as a whole, where the τ_i are as in [C, 2.1] and 3.1. Further let Aut (B) denote the group of all Q-linear automorphisms of B (which are not necessarily F-linear). Then it is easy to see that every element of $\mathfrak{g}$ can be extended to an element of Aut (B). Thus we have an exact sequence

$$1 \longrightarrow B^\times/F^\times \longrightarrow \text{Aut } (B) \longrightarrow \mathfrak{g} \longrightarrow 1 \, ,$$

where the image of $B^\times/F^\times$ in Aut (B) is the group of all inner automorphisms.

Remark. Aut (B) does not necessarily contain a subgroup which is mapped isomorphically onto $\mathfrak{g}$. For example, let $[F: Q] = r = 2$, and $D(B/F) = (pq)$ with two rational primes p and q which remain prime in F. Then $\mathfrak{g}$ is of order 2. Suppose that there is an element π of Aut (B) which is of order 2, and non-trivial on F. Put $U = \{x \in B \mid x^\pi = x\}$. Then we see easily that $B = U \otimes_Q F$, and U is a quaternion algebra over Q. If F_p denotes the completion of F at p, we have $B \otimes_F F_p = U \otimes_Q F_p = (U \otimes_Q Q_p) \otimes_{Q_p} F_p$. Since $[F_p: Q_p] = 2$, $B \otimes_F F_p$ is isomorphic to $M_2(F_p)$, a contradiction.

4.2. Now let us consider the involution $x \mapsto {}^\iota x^\iota$ of $M_n(B)$. Let Aut $(M_n(B), \iota)$ denote the group of all Q-linear automorphisms of $M_n(B)$ which commutes with this involution. Recall that G_Q can be identified with the group

$$\{x \in GL_n(B) \mid x \cdot {}^\iota x^\iota \in F^\times\} \, .$$

As Weil [6] observed, $G_Q/F^\times$ can be identified with the group of all F-linear elements of Aut $(M_n(B), \iota)$. In fact, for every $b \in G_Q$, we can define an F-linear element β of Aut $(M_n(B), \iota)$ by $x^\beta = b^{-1}xb$; conversely every F-linear element of Aut $(M_n(B), \iota)$ can be obtained from an element of G_Q, or rather of $G_Q/F^\times$. Writing $\zeta(b)$ for β, we obtain the first half of the following assertion.

(4.2.1) *There is an exact sequence*

$$1 \longrightarrow G_Q/F^\times \overset{\zeta}{\longrightarrow} \mathrm{Aut}\,(M_n(B), \iota) \overset{\eta}{\longrightarrow} \mathfrak{g} \longrightarrow 1 \, ,$$

where ζ is defined as above, and η is the restriction of the action to F.

To show the surjectivity of η, let $\pi \in \mathrm{Aut}\,(B)$. We see easily that $a^{x\iota} = a^{\iota\pi}$ for every $a \in B$, hence $({}^\iota x')^\pi = {}^\iota(x^\pi)^\iota$ for every $x \in M_n(B)$.   q.e.d.

4.3. For brevity, let us put

$$(4.3.1) \qquad\qquad A = \mathrm{Aut}\,(M_n(B), \iota) \, , \quad A^1 = \zeta(G_Q/F^\times) \, .$$

We are going to define the action of a certain subgroup of A on $\mathfrak{H}_n^r$. First recall the direct sum decomposition

$$M_n(B_R) = M_n(B) \otimes_Q R = M_{2n}(R)^r \times M_n(H)^{g-r}$$

considered in [4, 4.4] (K was used instead of H in [4]). For $x \in M_n(B_R)$, let $x^{(1)}, \cdots, x^{(r)}$ denote the projection of x to the first r factors $M_{2n}(R)$. As in [4, 4.4], we fix the coordinate system so that

$$(4.3.2) \qquad ({}^\iota x')^{(\nu)} = J_n \cdot {}^\iota x^{(\nu)} J_n^{-1}, \quad J_n = \begin{bmatrix} 0 & 1_n \\ -1_n & 0 \end{bmatrix} \quad (x \in M_n(B_R); \nu = 1, \cdots, r) ,$$

$$(4.3.3) \qquad\qquad (a \cdot 1_n)^{(\nu)} = a^{\tau_\nu} \cdot 1_n \qquad\qquad (a \in F; \nu = 1, \cdots, r) .$$

If $\alpha \in A$ and $\alpha\tau_\nu = \tau_\mu$, we can find an element ξ_μ of $GL_{2n}(R)$ so that $(x^\alpha)^{(\nu)} = \xi_\mu^{-1} x^{(\mu)} \xi_\mu$ for all $x \in M_n(B_R)$; the correspondence $\nu \mapsto \mu$ is a permutation of $\{1, \cdots, r\}$; the element ξ_μ is unique up to constant factors. Moreover, in view of (4.3.2), we see that ${}^\iota\xi_\mu J_n \xi_\mu = c_\mu J_n$ with $c_\mu \in R^\times$. Let A_+ denote the set of all α in A such that $c_\mu > 0$ for $j = 1, \cdots, r$, and let $A_+^1 = A_+ \cap A^1$. For every $\alpha \in A_+$, we define the action of α on $\mathfrak{H}_n^r$ by

$$\alpha(z_1, \cdots, z_r) = (w_1, \cdots, w_r) , \quad w_\mu = \xi_\mu(z_\nu) \qquad\qquad (z_\nu \in \mathfrak{H}_n) .$$

We see easily that A_+^1 can be identified with $G_{Q+}/F^\times$, and the action of an element of A_+^1 is the same as that of a corresponding element of G_{Q+}. Further, we see easily that $[A^1 : A_+^1] \leq [A : A_+] \leq 2^r$. By (3.1.1) and (3.1.2), we have $[A^1 : A_+^1] = 2^r$, so that $[A : A_+] = 2^r$. It follows that $A = A^1 A_+$, and the following sequence is exact:

$$1 \longrightarrow A_+^1 \longrightarrow A_+ \longrightarrow \mathfrak{g} \longrightarrow 1 .$$

4.4. PROPOSITION. *Let γ be an element of* $\mathrm{Gal}\,(\bar{Q}/Q)$ *whose restriction to* F *belongs to* $\mathfrak{g}$. *If* (F'', Θ'') *is the reflex of* (F', Θ'), *then* γ *is the identity map on* F''. *Especially,* $\lambda(c)^\gamma = \lambda(c)$ *for every* $c \in F''^\times$.

PROOF. Let K be the smallest Galois extension of Q containing F. Let H and H' be the subgroups of $\mathrm{Gal}\,(K/Q)$ corresponding to F and F', respectively. Extend each τ_i to an element of $\mathrm{Gal}\,(K/Q)$, and denote it again by τ_i. Then $\Theta \sim \sum_{i=1}^{r} \tau_i$, $\Theta' \sim \sum_{j=1}^{s} \sigma_j$ with elements σ_j such that $\bigcup_{i=1}^{r} \tau_i^{-1} H = \bigcup_{j=1}^{s} H' \sigma_j$. If γ is an element of $\mathrm{Gal}\,(K/Q)$ whose restriction belongs to $\mathfrak{g}$, we have $\bigcup_i \tau_i^{-1} H \gamma = \bigcup_i \tau_i^{-1} \gamma H = \bigcup_i \tau_i^{-1} H$, so that $\sum_{j=1}^{s} \sigma_j \gamma \sim \sum_{j=1}^{s} \sigma_j$ on F'. Since F'' is generated over Q by the elements $\sum_{j=1}^{s} a^{\sigma_j}$ for all $a \in F'$, we obtain our assertion.

4.5. Observe that every element α of A induces an automorphism of G such that $\nu(x^\alpha) = \nu(x)^\alpha$. This action can be extended to G_A in a natural way. Then, by virtue of 4.4, we see easily that the groups D_+, $\mathfrak{G}_+$, $\bar{\mathfrak{G}}_+$, and $F_c G_{\infty+}$ are stable under A. Therefore each α of A acts on $\bar{\mathfrak{G}}_+/F_c G_{\infty+}$; we write the action by the same letter α. Moreover $\rho(x^\alpha) = \rho(x)$ for every $x \in \bar{\mathfrak{G}}_+$ and $\alpha \in A$.

By 3.3, $G_{Q+}/F^\times$ is canonically isomorphic to a subgroup of $G_{c+}/F_c G_{\infty+}$. Since the latter group is a subgroup of $\bar{\mathfrak{G}}_+/F_c G_{\infty+}$, identifying A_+^1 with $G_{Q+}/F^\times$ we can thus define an injective homomorphism $\psi^1 \colon A_+^1 \to \bar{\mathfrak{G}}_+/F_c G_{\infty+}$ such that $y \cdot \psi(\alpha) = \psi(\alpha) y^\alpha$ for every $y \in \bar{\mathfrak{G}}_+/F_c G_{\infty+}$. Again for brevity we put

$$(4.5.1) \qquad\qquad \mathfrak{A}^1 = \bar{\mathfrak{G}}_+/F_c G_{\infty+} \, .$$

4.6. PROPOSITION. *There exist a group* $\mathfrak{A}$ *containing* $\mathfrak{A}^1$ *as a subgroup, and an injective homomorphism* ψ *of* A_+ *into* $\mathfrak{A}$ *satisfying the following conditions:*

    (4.6.1) $\psi = \psi^1$ *on* A_+^1.

    (4.6.2) $\mathfrak{A}^1$ *is a normal subgroup of* $\mathfrak{A}$.

    (4.6.3) $\psi(A_+^1) = \psi(A_+) \cap \mathfrak{A}^1$.

    (4.6.4) $\mathfrak{A} = \psi(A_+)\mathfrak{A}^1$.

    (4.6.5) $x \cdot \psi(\alpha) = \psi(\alpha) x^\alpha$ *for every* $x \in \mathfrak{A}^1$ *and* $\alpha \in A_+$.

Moreover, $\mathfrak{A}$ *and* ψ *are uniquely determined by these conditions up to isomorphisms.*

PROOF. Let $A_+ = \bigcup_{\alpha \in U} \alpha A_+^1$ be a coset decomposition with any set of representatives U, which is fixed in the proof. If $\alpha \beta A_+^1 = \gamma A_+^1$ with $\alpha, \beta, \gamma \in U$, we have $\alpha \beta = \gamma \zeta_{\alpha\beta,\gamma}$ with an element $\zeta_{\alpha\beta,\gamma}$ of A_+^1. Put $\mathfrak{A} = U \times \mathfrak{A}^1$, and define a law of composition in $\mathfrak{A}$ by

$$(\alpha, x)(\beta, y) = \left(\gamma,\ \psi^1(\zeta_{\alpha\beta,\gamma})x^\beta y\right) \qquad (\alpha, \beta, \gamma \in U; x, y \in \mathfrak{A}^1)\ ,$$

and define $\psi \colon A_+ \to \mathfrak{A}$ by

$$\psi(\alpha\varepsilon) = \left(\alpha,\ \psi^1(\varepsilon)\right) \qquad (\alpha \in U; \varepsilon \in A_+^1)\ .$$

Then it is straightforward to verify the desired properties. The uniqueness is also obvious.

4.7. The notation being as in 4.6, we see that the diagram

$$
\begin{array}{ccccccccc}
& & \overset{\displaystyle G_{Q+}/F^\times}{\underset{\|}{}} & & & & & & \\
1 & \longrightarrow & A_+^1 & \longrightarrow & A_+ & \longrightarrow & \mathfrak{g} & \longrightarrow & 1 \\
& & \downarrow{\scriptstyle \psi^1} & & \downarrow{\scriptstyle \psi} & & \downarrow & & \\
1 & \longrightarrow & \mathfrak{A}^1 & \longrightarrow & \mathfrak{A} & \longrightarrow & \mathfrak{g} & \longrightarrow & 1 \\
& & \underset{\displaystyle \bar{\mathfrak{G}}_+/F_c G_{\infty+}}{\overset{\|}{}} & & & & & &
\end{array}
$$

is commutative, and the rows are exact. Hereafter we identify A_+ and A_+^1 with $\psi(A_+)$ and $\psi^1(A_+^1)$ respectively, so that $G_{c+}/F_c G_{\infty+}$ is the closure of A_+^1 in $\mathfrak{A}^1$ by 3.2. We can make $\mathfrak{A}$ a topological group by taking $\mathfrak{A}^1$ as an open subgroup of $\mathfrak{A}$. Let $\bar{A}_+$ and $\bar{A}_+^1$ denote the closures of A_+ and A_+^1 respectively. Then $\bar{A}_+ = \bar{A}_+^1 A_+$, and $A_+^1 = \bar{A}_+^1 \cap A_+$.

Observe that we have (canonical) isomorphisms

$$(4.7.1) \qquad \mathfrak{A}/\bar{A}_+ \cong \mathfrak{A}^1/\bar{A}_+^1 \cong \bar{\mathfrak{G}}_+/G_{c+} \overset{\rho}{\cong} \mathrm{Gal}\,(\mathfrak{k}^*/F')\ .$$

Therefore we can define a homomorphism

$$\bar{\rho} \colon \mathfrak{A} \longrightarrow \mathrm{Gal}\,(\mathfrak{k}^*/F')$$

with kernel $\bar{A}_+$. In other words, $\bar{\rho}(x \cdot F_c G_{\infty+}) = \rho(x)$ for $x \in \bar{\mathfrak{G}}_+$, and ρ is trivial on A_+. Since there will be no fear of confusion, we shall write $\bar{\rho}$ again as ρ.

4.8. Let $\mathfrak{Z}$ denote the set of all open compact subgroups of $\mathfrak{A}$. For every $W \in \mathfrak{Z}$, denote by k_W the subfield of $\mathfrak{k}^*$ corresponding to the subgroup $\rho(W)$ of $\mathrm{Gal}\,(\mathfrak{k}^*/F')$. Since $\rho(W) = \rho(WA_+)$ and $WA_+ = W\bar{A}_+$, we observe that

$(4.8.1)$ $\quad WA_+ = \{y \in \mathfrak{A} \mid \rho(y)$ is the identity map on $k_W\}$.

Also we see easily that

$(4.8.2)$ *If $W \in \mathfrak{Z}$ and $X \in \mathfrak{Z}$, then $W \cap X \in \mathfrak{Z}$.*

$(4.8.3)$ *If $u \in \mathfrak{A}$ and $W \in \mathfrak{Z}$, then $uWu^{-1} \in \mathfrak{Z}$.*

$(4.8.4)$ *Any two members of $\mathfrak{Z}$ are commensurable.*

$(4.8.5)$ *If $M \in \mathfrak{Z}^*$, then $M/F_c G_{\infty+} \in \mathfrak{Z}$; moreover, every member of $\mathfrak{Z}$ contained in $\mathfrak{A}^1$ is of the form $M/F_c G_{\infty+}$ for some $M \in \mathfrak{Z}^*$.*

$(4.8.6)$ *If $W \in \mathfrak{Z}$, then $W \cap \mathfrak{A}^1 \in \mathfrak{Z}$.*

If $W = M/F_cG_{\infty+}$ with $M \in \mathfrak{Z}^*$, then k_W coincides with the field k_M defined in 3.10.

Now, for every $W \in \mathfrak{Z}$, put $\Gamma_W = A_+ \cap W$. If $W = M/F_cG_{\infty+}$ with $M \in \mathfrak{Z}^*$, we have $\Gamma_W = \Gamma_M/F^\times$. Therefore, in view of 3.9 and (4.8.6), we see that, for any $W \in \mathfrak{Z}$, Γ_W is a properly discontinuous group of transformations on $\mathfrak{H}_n^r$, and $\mathfrak{H}_n^r/\Gamma_W$ is of finite measure.

5. Canonical system of models for $\mathfrak{H}_n^r/\Gamma_X$

5.1. Now we can state our main theorem as a reformulation and a strengthening of [C, 2.5], by taking $\mathfrak{A}$ and $\mathfrak{Z}$ instead of the previous $\mathfrak{G}_+$ and $\mathfrak{Z}$. Although the new theorem is stronger than the old one, the essential feature is not much changed.

5.2. THEOREM. *There exists a system*

$$\{V_X, \varphi_X, J_{WX}(u), (X, W \in \mathfrak{Z}; u \in \mathfrak{A})\}$$

formed by the objects satisfying the following conditions.

(5.2.1) *For each $X \in \mathfrak{Z}$, (V_X, φ_X) is a model of $\mathfrak{H}_n^r/\Gamma_X$.*

(5.2.2) *V_X is rational over k_X.*

(5.2.3) *$J_{WX}(u)$, defined if and only if $uXu^{-1} \subset W$, is a morphism of V_X onto $V_W^{\rho(u)}$, rational over k_X, and has the following properties:*

(5.2.3a) *$J_{XX}(u)$ is the identity map if $u \in X$;*

(5.2.3b) *$J_{WX}(u)^{\rho(v)} \circ J_{XT}(v) = J_{WT}(uv)$;*

(5.2.3c) *$J_{WX}(\alpha)[\varphi_X(z)] = \varphi_W(\alpha(z))$ for every $\alpha \in A_+$ and $z \in \mathfrak{H}_n^r$ (if $\alpha X\alpha^{-1} \subset W$).*

(5.2.4) *Let (Y, P, δ, f), P', z, and η be as in [C, 2.4]. Then, for every $X \in \mathfrak{Z}$, $\varphi_X(z)$ is rational over P'_{ab}. Further, for every $v \in P'^\times_A$, one has $\varphi_W(z)^\tau = J_{WX}(\eta^*(v)^{-1})[\varphi_X(z)]$, where $\tau = [v, P']$, and $W = \eta^*(v)^{-1}X\eta^*(v)$, and $\eta^*(v)$ means the coset of $\eta(v)$ modulo $F_cG_{\infty+}$ (so that $\eta^*(v)$ is an element of $\mathfrak{A}^1 = \bar{\mathfrak{G}}_+/F_cG_{\infty+}$).*

5.3. We prove the above theorem in several steps. As in 3.5, we put $S' = S \cap \bar{\mathfrak{G}}_+$ and $S'' = S \cap \mathfrak{G}_+$ for every $S \in \mathfrak{Z}$. Let

$$\{V_S, \varphi_S, J_{TS}(x), (S, T \in \mathfrak{Z}; x \in \mathfrak{G}_+)\}$$

be as in [C, 2.5]. Let S and T be members of $\mathfrak{Z}$ such that $F^\times S'' \subset F^\times T''$. Then $F^\times \Gamma_S \subset F^\times \Gamma_T$, so that there exists a morphism P of V_S onto V_T such that $P \circ \varphi_S = \varphi_T$. Moreover, by [C, 3.5]. $k_T \subset k_S$. Let us now prove that P is rational over k_S. To show this, take a point w of $\mathfrak{H}_n^r$ so that $\varphi_S(w)$ is generic on V_S over k_S. Put $R = S \cap T$, $a_R = \varphi_R(w)$, $a_S = \varphi_S(w)$, $a_T = \varphi_T(w)$. Then both $k_S(a_S)$ and $k_T(a_T)$ are contained in $k_R(a_R)$. Let τ be an isomorphism of $k_R(a_R)$ into C over $k_S(a_S)$. By [C, 3.5], there exists an element y of S'' such that $\tau = \sigma(y)$ on k_R. Then a_R^τ belongs to $V_R^{\sigma(y)}$. Since $S'' \subset F^\times T''$, we observe that

$uRu^{-1} = R$ for every $u \in S''$, hence $J_{RR}(u)$ is meaningful. Therefore we find a point b of V_R such that $J_{RR}(y)(b) = a_R^\tau$. Then $a_S = a_S^\tau = J_{SR}(1)^\tau(a_R^\tau) = J_{SR}(y)(b) = J_{SR}(1)(b)$ since $y \in S \cap \mathcal{G}_+$. Thus both b and u_R are mapped to a_S by the projection $J_{SR}(1)$. Therefore $b = \varphi_R(\gamma(w))$ for some $\gamma \in \Gamma_S$. Put $x = y\gamma$. Then $a_R^\tau = J_{RR}(y)(J_{RR}(\gamma)(a_R)) = J_{RR}(x)(a_R)$, and $x \in S''$. Since $x \in F^\times T''$, we have $a_T^\tau = J_{TR}(1)^\tau(a_R^\tau) = J_{TR}(x)(a_R) = J_{TR}(1)(a_R) = a_T$, so that τ is the identity map on $k_T(a_T)$. This implies that $k_T(a_T) \subset k_S(a_S)$, hence P is rational over k_S.

5.4. We are going to construct a system $\{V_X, \varphi_X, J_{WX}(u)\}$ only for $W, X \subset \mathfrak{A}^1$ and $u \in \mathfrak{A}^1$. Let $\mathcal{Y}$ denote the set of all S in $\mathcal{Z}$ which are open in G_A, and put

$$\mathfrak{Y} = \{F_c S'/F_c G_{\infty+} \mid S \in \mathcal{Y}\} .$$

Let $W = F_c S'/F_c G_{\infty+}$ with $S \in \mathcal{Y}$. We define V_W, φ_W by $V_W = V_S$, $\varphi_W = \varphi_S$. If we put $M = F_c S'$, we have $k_W = k_M = k_S$ by 3.11. Further, by 3.7, $\Gamma_W = \Gamma_M/F^\times = F^\times \cdot \Gamma_S/F^\times$, so that (V_W, φ_W) is a model of $\mathfrak{H}_n^\tau/\Gamma_W$. The discussion of 5.3, together with 3.7, shows that this definition of (V_W, φ_W) is independent of the choice of S (up to biregular isomorphisms over k_S).

Let $X = uWu^{-1}$ with $u \in \mathfrak{A}^1$. By 3.4, we can take $y \in \mathcal{G}_+$ so that $u = yF_c G_{\infty+}$. Put $T = ySy^{-1}$. Then $X = F_c T'/F_c G_{\infty+}$. By our definition, (V_X, φ_X) can be identified with (V_T, φ_T). Note that $\rho(u) = \rho(y) = \sigma(y)$ on k_W. Now we define $J_{XW}(u)$ to be the same as $J_{TS}(y)$. We are going to show that $J_{XW}(u)$ does not depend on the choice of S and y. If y' is another element of $\mathcal{G}_+$ such that $u = y'F_c G_{\infty+}$, we see that $y^{-1}y' \in F_c G_{\infty+} \cap \mathcal{G}_+ \subset F^\times S \cap \mathcal{G}_+ = F^\times S''$, hence $J_{TS}(y) = J_{TS}(y')$. Thus our definition does not depend, at least, on the choice of y.

Let (Y, P, δ, f), P', z, and η be as in [C, 2.4]. By [4, 7.5], we can take these so that P' and k_X are linearly disjoint over F'. Then extend $\rho(u)$ to an automorphism π of $\bar{Q}$. Take $v \in P_A'^\times$ so that $\pi = [v, P']$ on P_{ab}'. Put $N = F_c T'$. (Note that N depends only on W and u.) Since $\rho(\eta(v)^{-1}) = \rho(y)$ on $k_X = k_N$, we have, by (3.10.3), $\eta(v)^{-1} = my\alpha$ with $m \in N$, and $\alpha \in G_{Q+}$. Put $R = \alpha^{-1}S\alpha$. Then $\eta(v)T\eta(v)^{-1} = \alpha^{-1}y^{-1}Ty\alpha = R$, hence $\varphi_T(z)^\pi = J_{TR}(\eta(v)^{-1})[\varphi_R(z)] = J_{TR}(my\alpha)[\varphi_R(z)] = J_{TS}(y)[\varphi_S(\alpha(z))]$, namely

$$(5.4.1) \qquad \varphi_X(z)^\pi = J_{XW}(u)[\varphi_W(\alpha(z))] .$$

Now we can change z for $\beta(z)$ with any $\beta \in G_{Q+}$ without changing π and α. Since the points $\varphi_W(\alpha\beta(z))$, with $\beta \in G_{Q+}$, form a dense subset of V_W, the relation (5.4.1) characterizes the map $J_{XW}(u)$. Therefore $J_{XW}(u)$ is independent of the choice of (y and) S.

5.5. We have thus obtained a system

$$\{V_X,\ \varphi_X,\ J_{WX}(u),\ (W,\ X \in \mathfrak{Y};\ u \in \mathfrak{A}^1)\}$$

where $J_{WX}(u)$ is defined when and only when $uXu^{-1} = W$. Now it is easy to verify the conditions (5.2.1–4) for this system.

Furthermore, we can prove propositions exactly like [C, 3.8 and 3.10] in the present setting, and give a discussion of the same type as in [C, 3.9]. Fix a Z-lattice $\mathfrak{m}$ in B^n, and put, for every positive integer b,

$$U(\mathfrak{m},\ b) = \{x \in G_{A+} \mid x \equiv 1\ \mathrm{mod}_0\ (\mathfrak{m},\ b)\}\ ,$$
$$W_b = F_c \cdot U(\mathfrak{m},\ b)'/F_c G_{\infty+}\ .$$

Then W_b is a normal subgroup of W_1, and every open compact subgroup of $\mathfrak{A}^1$ contains W_b for some b, on account of 3.8. Therefore, by means of the same type of argument as in [C, 3.11], we can extend our system so that the symbols are defined for all W, X of $\mathfrak{Z}$ contained in $\mathfrak{A}^1$, and for $u \in \mathfrak{A}^1$, and the desired conditions (5.2.1–4) are satisfied.

5.6. Still with W, X in $\mathfrak{A}^1$, we shall now define $J_{WX}(u)$ for all $u \in \mathfrak{A}$. For this purpose, let us first define $J_{WX}(\alpha)$ for $\alpha \in A_+$. Fix $\alpha \in A_+$. For each $X \in \mathfrak{Z}$ contained in $\mathfrak{A}^1$, put $\bar{X} = \alpha X \alpha^{-1}$, $\bar{V}_X = V_{\bar{X}}$, and $\bar{\varphi}_X = \varphi_{\bar{X}} \circ \alpha$. Since $\Gamma_X = \alpha^{-1} \Gamma_{\bar{X}} \alpha$, we see that $(\bar{V}_X, \bar{\varphi}_X)$ is a model of $\mathfrak{H}_n^r/\Gamma_X$. Further put $\bar{J}_{WX}(u) = J_{\bar{W}\bar{X}}(\alpha u \alpha^{-1})$ for $W = uXu^{-1}$, $u \in \mathfrak{A}^1$.

Let (Y, P, δ, f), P', and z be as in [C, 2.4]. Consider the restriction of α^{-1} to F, and extend it to an isomorphism β of Y to an algebra $\bar{Y}$ over F. Define an involution $\bar{\delta}$ of $\bar{Y}$ by $y^{\bar{\delta}} = y^{\beta^{-1}\delta\beta}$ for $y \in \bar{Y}$, and $\bar{P} = P^\beta$, and $\bar{f}(x^\beta) = f(x)^{\alpha^{-1}} = \alpha f(x)\alpha^{-1}$ for $x \in Y$. Then $\alpha(z)$ is the fixed point of $\bar{f}(\{\bar{Y}, \bar{\delta}\})$. Define ξ_μ as in 4.3 for the present α. Then $\alpha(z) = (w_1, \cdots, w_r)$ with $w_\mu = \xi_\mu(z_\nu)$, and

$$(5.6.1) \qquad f(x)^{(\nu)} = \big(\bar{f}(x^\beta)^\alpha\big)^{(\nu)} = \xi_\mu^{-1}\bar{f}(x^\beta)^{(\mu)}\xi_\mu \qquad \text{if } \alpha\tau_\nu = \tau_\mu\ .$$

Define $\Psi^{(\nu)}$ as in [4, (4.7.6)], and similarly $\bar{\Psi}^{(\nu)}$ for $(\bar{Y}, \bar{P}, \bar{\delta}, \bar{f})$. In view of (5.6.1), we see that $\bar{\Psi}^{(\mu)}(x^\beta)$ is equivalent to $\Psi^{(\nu)}(x)$. Define (P'_i, Ψ'_i), for $i = 1, \cdots, t$, P', and η as in [4, 4.9] and [C, 2.4]. Now β could have been chosen in such a way that $\beta = \gamma$ on P_i for every i, with an element γ of Gal $(\bar{Q}/Q)$ independent of i. Then we obtain (P'_i, Ψ'^γ_i) and P' as the corresponding objects for $(\bar{Y}, \bar{P}, \bar{\delta}, \bar{f})$. Further if $\bar{\eta}$ is the map corresponding to η, defined with respect to $\bar{f}$, we see that

$$(5.6.2) \qquad \bar{\eta}(v) = \eta(v)^{\alpha^{-1}} \text{ for every } v \in P'^\times_A\ .$$

It is now easy to check all the properties (5.2.1–4) for

$$\{\bar{V}_X,\ \bar{\varphi}_X,\ \bar{J}_{WX}(u),\ (W,\ X \in \mathfrak{Z};\ W,\ X \subset \mathfrak{A}^1;\ u \in \mathfrak{A}^1)\}\ .$$

We especially need formula (5.6.2) to verify (5.2.4). For the same reason as in [C, 3.8], there exists a biregular map Q_X of V_X to $\bar{V}_X$, *rational over* k_X, such that $Q_X \circ \varphi_X = \bar{\varphi}_X$, and

$$(5.6.3) \qquad Q_X^{\rho(u)} \circ J_{XW}(u) = \bar{J}_{XW}(u) \circ Q_W \qquad (X = uWu^{-1}, u \in \mathfrak{A}^1) .$$

Put $Q_X = J_{\bar{X}X}(\alpha)$. Then we have

$$(5.6.4) \qquad J_{\bar{X}X}(\alpha)[\varphi_X(z)] = \bar{\varphi}_X(z) = \varphi_{\bar{X}}(\alpha(z)) \qquad (\bar{X} = \alpha X \alpha^{-1}, z \in \mathfrak{H}_n^r) .$$

Therefore, if $\alpha \in A_+^1$, the symbol $J_{\bar{X}X}(\alpha)$ coincides with the original one. Now for an arbitrary element v of $\mathfrak{A}$, we decompose v in the form $v = u\alpha$ with $u \in \mathfrak{A}^1$ and $\alpha \in A_+$, and put

$$J_{WX}(v) = J_{W\bar{X}}(u) \circ J_{\bar{X}X}(\alpha) \qquad (\bar{X} = \alpha X \alpha^{-1}, vXv^{-1} \subset W) .$$

Then we obtain, without any conflict, the symbol $J_{WX}(v)$ for all $v \in \mathfrak{A}$. It is now straightforward to check the properties $(5.2.3_{a,b,c})$ for this extended symbol, by virtue of (5.6.3) and (5.6.4).

Finally, by means of the same principle as in [C, 3.10], we can extend our system to the desired one as stated in 5.2, since every member X of $\mathfrak{Z}$ contains $X \cap \mathfrak{A}^1$ as a normal subgroup of finite index. The proof of 5.2 is thus completed.

5.7. *Remark.* If $r = 1$, we see that $A_+ = A_+^1 = G_{Q+}/F^\times$, $\bar{\mathfrak{G}}_+ = \mathfrak{G}_+ = G_{A+}$, and $\mathfrak{A} = \mathfrak{A}^1 = G_{A+}/F_c G_{\infty+}$. Therefore, for every $S \in \mathfrak{Z}$, if we put $W = F_c S/F_c G_{\infty+}$, then $W \in \mathfrak{Z}$, and $\Gamma_W = F^\times \Gamma_S/F^\times$ as is shown in 3.6. Our construction in 5.4 shows that we can put $(V_W, \varphi_W) = (V_S, \varphi_S)$. However, an arbitrary member W of $\mathfrak{Z}$ is not necessarily of the form $W = F_c S/F_c G_{\infty+}$. For example, consider the case $G = GL_2$. For a positive integer N, put

$$\mathfrak{o} = \left\{ \begin{bmatrix} a & b \\ c & d \end{bmatrix} \in M_2(\mathbf{Z}) \mid c \equiv 0 \bmod (N) \right\} ,$$

$$L = \{ x \in G_{A+} \mid x_p \mathfrak{o}_p = \mathfrak{o}_p x_p \text{ for all primes } p \} ,$$

$$\Gamma_0(N) = \mathfrak{o} \cap SL_2(\mathbf{Z}) , \quad W = L/\mathbf{Q}^\times G_{\infty+} .$$

It can easily be shown that $W \in \mathfrak{Z}$, and Γ_W is generated by $\mathbf{Q}^\times \cdot \Gamma_0(N)/\mathbf{Q}^\times$ and the element represented by $\begin{bmatrix} 0 & -1 \\ N & 0 \end{bmatrix}$. This group Γ_W, as a transformation group, cannot be obtained as Γ_S with any $S \in \mathfrak{Z}$, as was remarked in [C, 2.15].

5.8. The system $\{V_X, \varphi_X, J_{WX}(u)\}$ of Th. 5.2 is *unique* in the same sense as in [C, 2.5]. The proof can be given by the same method as in [C, 3.8, 3.9].

6. The group $\mathfrak{A}$ as a group of automorphisms of a function field

6.1. For each $X \in \mathfrak{Z}$, let K_X denote the field of all functions on V_X (in the sense of algebraic geometry) rational over k_x, and let

$$\Re_x = \{f \circ \varphi_x \mid f \in K_x\},$$
$$\Re = \bigcup_{x \in \mathfrak{Z}} \Re_x.$$

We call $\Re$ *the field of arithmetic automorphic functions on $\mathfrak{H}_n^r$ with respect to G.* Now we can define a homomorphism

$$\tau \colon \mathfrak{A} \longrightarrow \text{Aut}\,(\Re/F')$$

in the following way. Let $u \in \mathfrak{A}$. Observe that, for every $f \in K_x$, $f^{\rho(u)}$ is a function on $V_x^{\rho(u)}$ rational over k_x. Put

$$(f \circ \varphi_x)^{\tau(u)} = f^{\rho(u)} \circ J_{XW}(u) \circ \varphi_W \qquad (W = u^{-1}Xu).$$

Then we see easily that τ is a homomorphism as stated above.

6.2. Theorem. *The homomorphism τ has the following properties.*

(6.2.1)　$\tau(u) = \rho(u)$ *on* $\mathfrak{k}^*$.

(6.2.2)　$h^{\tau(\alpha)} = h(\alpha(z))$ *for* $\alpha \in A_+$, $h \in \Re$, $z \in \mathfrak{H}_n^r$.

(6.2.3)　*Let* (Y, P, δ, f), P', z, *and* η^* *be as in* [C, 2.4] *and* (5.2.4). *Then, for every $h \in \Re$ defined at z, $h(z)$ is rational over P'_{ab}. Moreover, if $v \in P'^{\times}_A$, $\omega = [v, P']$, and $u = \eta^*(v)^{-1}$, then $h^{\tau(u)}$ is also defined at z, and $h(z)^{\omega} = h^{\tau(u)}(z)$.*

Proof. The first property is obvious from the definition of $\tau(u)$; (6.2.2) follows from $(5.2.3_c)$, and (6.2.3) from (5.2.4).

6.3. Theorem. *For every $W \in \mathfrak{Z}$, $\Re$ is an infinite Galois extension of $\Re_W$, and $\tau(W) = \text{Gal}\,(\Re/\Re_W)$.*

Proof. First we prove a somewhat weaker

(6.3.1)　　　　　　　$W = \{u \in \mathfrak{A} \mid \tau(u) \text{ is the identity map on } \Re_W\}.$

Suppose that $\tau(u)$ is the identity map on $\Re_W$. Then, by (6.2.1), $\rho(u)$ must be the identity map on k_W, so that, by (4.8.1), $u = t\alpha$ with $t \in W$ and $\alpha \in A_+$. Then $J_{WX}(u) = J_{WX}(\alpha)$, and for every $f \in \Re_W$, we have $f \circ \varphi_W = (f \circ \varphi_W)^{\tau(u)} = f \circ J_{WX}(\alpha) \circ \varphi_X$ with $X = u^{-1}Wu$, so that $\varphi_W = J_{WX}(\alpha) \circ \varphi_X = \varphi_W \circ \alpha$. Therefore $\alpha \in \Gamma_W$, hence $u \in W$. Conversely, it is straightforward to verify that $\tau(u)$ is the identity map on $\Re_W$ if $u \in W$. Thus we obtain (6.3.1). Before completing the proof of 6.3, we prove

6.4. Proposition. *Let W and X be members of $\mathfrak{Z}$ such that $X \subset W$. Then the following assertions hold.*

(6.4.1)　$[W \colon X] = [\Re_X \colon \Re_W]$;

(6.4.2)　*If X is a normal subgroup of W, $\Re_X$ is a Galois extension of $\Re_W$, and the sequence*

$$1 \longrightarrow X \longrightarrow W \overset{\tau}{\longrightarrow} \mathrm{Gal}\,(\mathfrak{K}_X/\mathfrak{K}_W) \longrightarrow 1$$

is exact.

PROOF. Since $W \cap XA_+ = X \cdot (W \cap A_+) = X \cdot \Gamma_W$, we have $[W:X] = [W:X\Gamma_W][X\Gamma_W:X] = [WA_+:XA_+][\Gamma_W:\Gamma_X]$. By (4.8.1), $[WA_+:XA_+] = [k_X:k_W]$. Observe that $C\mathfrak{K}_W$ (resp. $C\mathfrak{K}_X$) is the field of all automorphic functions on $\mathfrak{H}_n^r$ with respect to Γ_W (resp. Γ_X). Further, $\mathfrak{K}_W$ (resp. $\mathfrak{K}_X$) is linearly disjoint with C over k_W (resp. k_X). Therefore $[\Gamma_W:\Gamma_X] = [C\mathfrak{K}_X:C\mathfrak{K}_W] = [\mathfrak{K}_X:k_X\mathfrak{K}_W]$, and $[k_X:k_W] = [k_X\mathfrak{K}_W:\mathfrak{K}_W]$. Combining these we obtain (6.4.1). If X is a normal subgroup of W, we see easily that, for every $u \in W$, $\tau(u)$ gives an automorphism of $\mathfrak{K}_X$ over $\mathfrak{K}_W$. By (6.3.1), (with X in place of W), $\tau(u)$ is the identity map on $\mathfrak{K}_X$ if and only if $u \in X$. Therefore W/X is isomorphic to a subgroup of $\mathrm{Aut}\,(\mathfrak{K}_X/\mathfrak{K}_W)$. By virtue of (6.4.1), $\mathfrak{K}_X$ must be a Galois extension of $\mathfrak{K}_W$, and hence (6.4.2).

Coming back to the proof of 6.3, we observe, in view of (4.8.4), that, for every $T \in \mathfrak{Z}$, there exists a member X of $\mathfrak{Z}$ which is a normal subgroup of W and contained in T. It follows that $\mathfrak{K}$ is the union of the fields $\mathfrak{K}_X$ for all normal subgroups X of W which are members of $\mathfrak{Z}$. Therefore (6.4.2) implies that $\mathfrak{K}$ is a Galois extension of $\mathfrak{K}_W$, and $\tau(W)$ is dense in $\mathrm{Gal}\,(\mathfrak{K}/\mathfrak{K}_W)$. Now the members of $\mathfrak{Z}$ form a basis of neighborhoods of the identity element of $\mathfrak{A}$. Therefore (6.3.1) implies that τ is injective and continuous. Then $\tau(W)$, being the continuous image of a compact group, must be compact. Hence we have $\tau(W) = \mathrm{Gal}\,(\mathfrak{K}/\mathfrak{K}_W)$. One can also prove (6.4.2) and 6.3 by means of the argument of [C, 7.2–7.10].

6.5. THEOREM. *The map τ is a topological isomorphism of $\mathfrak{A}$ onto an open subgroup of $\mathrm{Aut}\,(\mathfrak{K}/F')$. If $\mathfrak{H}_n^r/\Gamma_W$ for any $W \in \mathfrak{Z}$ is compact, or $G_Q = GL_2(Q)$, then τ is surjective.*

Note that the compactness of $\mathfrak{H}_n^r/\Gamma_W$ does not depend on the choice of W.

PROOF. This is a generalization of [C, 2.8], and can be proved more or less in the same manner, except that one has to be careful about the difference between $\mathfrak{A}$ and $\mathfrak{A}^1$ in the case $r > 1$.

We have seen that τ is injective and continuous. Since the subgroups $\mathrm{Gal}\,(\mathfrak{K}/\mathfrak{K}_W)$ of $\mathrm{Aut}\,(\mathfrak{K}/F')$ for all $W \in \mathfrak{Z}$ form a basis of neighborhoods of the identity element, we obtain the first assertion from 6.3. Therefore, to prove the second assertion, it is sufficient to show that $\tau(\mathfrak{A})$ is dense in $\mathrm{Aut}\,(\mathfrak{K}/F')$, if $\mathfrak{H}_n^r/\Gamma_W$ is compact, or $G_Q = GL_2(Q)$. In the latter case we have $\mathfrak{A} = G_{A+}/F_cG_{\infty+}$ (see 5.7), and our assertion has been proved in [C, 2.8]. Therefore we assume hereafter that $\mathfrak{H}_n^r/\Gamma_W$ is compact.

Let $\zeta \in \mathrm{Aut}\,(\Re/F')$. By (4.7.1), there exists an element y of $\mathfrak{A}$ such that $\rho(y) = \zeta$ on $\mathfrak{l}^*$. Put $\pi = \zeta \cdot \tau(y)^{-1}$. Then π is the identity map on $\mathfrak{l}^*$. Let $\Re_W^*$ (resp. $\Re^*$) be the composite of $\Re_W$ (resp. $\Re$) and C. Since $\Re$ and C are linearly disjoint over $\mathfrak{l}^*$, we can extend π to an automorphism of $\Re^*$ over C, which we denote again by π. Take an arbitrary $\mathfrak{r}_F$-lattice $\mathfrak{m}$ in B^*. We can find a positive integer b so that the following two conditions are satisfied.

(6.5.1) *If* $R = \{x \in \bar{\mathfrak{S}}_+ \mid x \equiv 1 \mod_0 (\mathfrak{m}, b)\}$, *then* Γ_R *modulo its center has no element of finite order other than the identity element.*

(6.5.2) *If* $\varepsilon \in \mathfrak{r}_F^\times$ *and* $\varepsilon \equiv 1 \mod b\mathfrak{r}_F$, *then* $\varepsilon = \eta^2$ *for some* $\eta \in \mathfrak{r}_F^\times$ (see Chevalley [1]).

Put $W = F_c R / F_c G_{\infty+}$. We can find members X and Y of $\mathfrak{Z}$ so that $Y \subset X \subset W$, $(\Re_W)^{\pi^{-1}} \subset \Re_X$, $(\Re_X)^\pi \subset \Re_Y$, and Y is a normal subgroup of W. Then $\Re_W^* \subset (\Re_X^*)^\pi \subset \Re_Y^*$. Now $\mathrm{Gal}\,(\Re_Y^*/\Re_W^*)$ is isomorphic to Γ_W/Γ_Y. Let Δ be the subgroup of Γ_W such that Δ/Γ_Y corresponds to $(\Re_X^*)^\pi$. Put $T = \Delta Y$. Then we see that $T \in \mathfrak{Z}$, $\Gamma_T = \Delta$, and π gives an isomorphism of $\Re_X^*$ onto $\Re_T^*$ over C. Therefore we have a birational map ξ of V_T onto V_X such that $(f \circ \varphi_X)^\pi = f \circ \xi \circ \varphi_T$ for every $f \in \Re_X$. We are going to show that there exists an element α of A_+ such that $\xi \circ \varphi_T = \varphi_X \circ \alpha$.

In view of (6.5.1), we see that both V_X and V_T are complete non-singular minimal models, hence ξ is everywhere biregular. Therefore we can find a holomorphic automorphism β of $\mathfrak{H}_n^r$ such that $\xi \circ \varphi_T = \varphi_X \circ \beta$, and $\beta^{-1}\Gamma_X\beta$, as a transformation group of $\mathfrak{H}_n^r$, coincides with Γ_T. Now we can find r elements β_μ of $\mathrm{Sp}\,(n, \mathbf{R})$ such that

$$(6.5.3) \qquad \beta(z_1, \cdots, z_r) = (w_1, \cdots, w_r)\,, \qquad w_\mu = \beta_\mu(z_\nu)\,,$$

where $\nu \mapsto \mu$ is a permutation of $\{1, \cdots, r\}$. Since X and T are contained in W, and $\Gamma_W = F^\times \Gamma_R$, every element of Γ_X and Γ_T is represented by an element of Γ_R. Let γ be an element of Γ_R which represents an element of Γ_X. Then $\beta^{-1}\gamma\beta$, as a transformation of $\mathfrak{H}_n^r$, coincides with an element δ of Γ_R which represents an element of Γ_T. Thus

$$\beta\delta(z_1, \cdots, z_r) = \gamma\beta(z_1, \cdots, z_r) = \big(\gamma^{(1)}(w_1), \cdots, \gamma^{(r)}(w_r)\big)\,,$$

where $\gamma^{(\nu)}$ denotes the projection of γ to the ν-th factor $M_{2n}(\mathbf{R})$ of $M_n(B)_\mathbf{R}$ as defined in 4.3, and w_ν is as in (6.5.3). Therefore we have $\beta_\mu \delta^{(\nu)}(z_\nu) = \gamma^{(\mu)}(w_\mu) = \gamma^{(\mu)}\beta_\mu(z_\nu)$, so that $\beta_\mu^{-1}\gamma^{(\mu)}\beta_\mu = c_\mu \delta^{(\nu)}$ with $c_\mu \in \mathbf{R}$. We have then $\nu(\gamma)^{\tau_\mu} = c_\mu^2 \cdot \nu(\delta)^{\tau_\nu}$. Let

$$\Lambda = \{\gamma \in \Gamma_R \mid \nu(\gamma) = 1, \gamma \text{ represents an element of } \Gamma_X\}\,.$$

If $\gamma \in \Lambda$, we have $c_\mu^{-2} = \nu(\delta)^{\tau_\nu}$. By (6.5.2), $\nu(\delta) = \eta^2$ with $\eta \in \mathfrak{r}_F^\times$. Then $c_\mu^{-1} = \pm \eta^{\tau_\nu} \in F^{\tau_\nu}$. It follows that $\beta_\mu^{-1}\Lambda^{(\mu)}\beta_\mu \subset M_n(B)^{(\nu)}$. Since Λ spans $M_n(B)$ over $\mathbf{Q}$

(see below), we obtain an automorphism α_μ of $M_n(B)$ such that $\beta_\mu^{-1} x^{(n)} \beta_\mu = (x^{\alpha_\mu})^{(\nu)}$. If we take the above γ as x, we see that $\gamma^{\alpha_\mu} = \pm \eta \delta$, hence $(\gamma^2)^{\alpha_\mu} = (\eta \delta)^2$. Now $\{\gamma^2 \mid \gamma \in \Lambda\}$ generates $M_n(B)$ over Q (see below). Therefore α_μ is independent of μ. Writing α for α_μ, we obtain $\beta_\mu^{-1} x^{(\mu)} \beta_\mu = (x^\alpha)^{(\nu)}$. Substituting the elements of F for x, we see that $\alpha \tau_\nu = \tau_\mu$. Moreover, by virtue of (4.3.2), we see easily that α commutes with the involution $x \mapsto {}^t x'$ of $M_n(B)$. It follows that $\alpha \in A$. Since $\beta_\mu \in \mathrm{Sp}\,(n,\,R)$, we see that $\alpha \in A_+$, and $\alpha(z) = \beta(z)$. Thus we have obtained an element α of A_+ with the property $\xi \circ \varphi_T = \varphi_X \circ \alpha$, so that $(f \circ \varphi_X)^\tau = f \circ \varphi_X \circ \alpha$ for every $f \in K_X$, hence $\pi = \tau(\alpha)$ on $\Re_X$, and $\zeta = \tau(\alpha y)$ on $\Re_W$. Now $W = F_c R / F_c G_{\infty+}$ with R defined by (6.5.1). Since the argument is applicable to R with an arbitrarily large b, this shows that $\tau(\mathfrak{A})$ is dense in $\mathrm{Aut}\,(\Re/F')$, and therefore completes the proof.

6.6. In the above, we needed the fact that if we put, for a positive integer d,

$$\Lambda_d = \{\gamma \in G_Q^\nu \mid \gamma \equiv 1 \bmod_0 (\mathfrak{m},\, d)\}\,,$$

then $\{\gamma^2 \mid \gamma \in \Lambda_d\}$ generates $M_n(B)$ over Q, for any d. (Observe that the above Λ contains Λ_d for a sufficiently large d.) To show this, fix d, and take a rational prime p so that.

(i) p does not divide d;

(ii) p decomposes completely in F; and

(iii) B is unramified at every prime factor of p.

Then $M_n(B) \otimes_Q Q_p$ can be identified with $M_{2n}(Q_p)^g$, and the p-component G_p^ν of G_A with $\mathrm{Sp}\,(n,\,Q_p)^g$. By choosing a suitable p and a suitable coordinate system, we may assume that

$$\mathrm{Sp}\,(n,\,Z_p) = \{x \in G_p^\nu \mid \mathfrak{m}_p x = \mathfrak{m}_p\}\,.$$

We can easily find $4n^2 g$ elements x_i of $\mathrm{Sp}\,(n,\,Z_p)^g$ whose squares span $M_{2n}(Q_p)^g$ over Q_p. Put $\mathfrak{M}_p = \sum_{i=1}^{4n^2 g} Z_p x_i^2$. There exists a positive integer e such that if $y \in M_{2n}(Q_p)^g = M_n(B) \otimes_Q Q_p$ and $\mathfrak{m}_p y \subset p^e \mathfrak{m}_p$, then $y \in p \mathfrak{M}_p$. By the strong approximation theorem, we can find elements γ_i of G_Q^ν such that

$$\mathfrak{m}_p(\gamma_i - x_i) \subset p^e \mathfrak{m}_p\,,$$
$$\mathfrak{m}_q(\gamma_i - 1) \subset d\mathfrak{m}_q \qquad\qquad \text{for all primes } q \neq p\,.$$

Then $\gamma_i \in \Lambda_d$, and $\mathfrak{m}_p(\gamma_i^2 - x_i^2) \subset p^e \mathfrak{m}_p$, so that $\gamma_i^2 - x_i^2 \in p \mathfrak{M}_p$. It follows that $4n^2 g$ elements γ_i^2 span $M_{2n}(Q_p)^g$ over Q_p, and hence they span $M_n(B)$ over Q.

q.e.d.

6.7. PROPOSITION. *If $\mathfrak{L}$ is the field defined in* [C, 2.7] *and* $\mathfrak{k}$ *is as in* [C, 2.3], *then* $\mathfrak{L} = \mathfrak{k} \cdot \Re$.

PROOF. We have seen that $\mathfrak{k}^* \subset \mathfrak{k}$ in 3.11. Define $\mathfrak{L}_S$ as in [C, 2.7]. Then we see that $\mathfrak{R}_X = \mathfrak{L}_S$ for $X = F_c S'/F_c G_{\infty+}$, S open in G_{A+}, since we may put $\langle V_X, \varphi_X \rangle = \langle V_S, \varphi_S \rangle$ in such a case. Therefore $\mathfrak{R} \subset \mathfrak{L}$. Let $S = S(\mathfrak{m}, b)$ be as in [C, 2.10]. By Chevalley [1], we can find a positive multiple c of b so that if $\varepsilon \in \mathfrak{r}_F^\times$ and $\varepsilon \equiv 1 \bmod c\mathfrak{r}_F$, then $\varepsilon = \eta^2$ with an element η of $\mathfrak{r}_F^\times$ such that $\eta \equiv 1 \bmod b\mathfrak{r}_F$. Put

$$T = \{x \in G_{A+} \mid x \equiv 1 \bmod_0 (\mathfrak{m}, c)\} .$$

If $\gamma \in \Gamma_T$, we have $\nu(\gamma) \equiv 1 \bmod c\mathfrak{r}_F$, hence $\nu(\gamma) = \eta^2$ with an element η of $\mathfrak{r}_F^\times$ such that $\eta \equiv 1 \bmod b\mathfrak{r}_F$. Then $\nu(\eta^{-1}\gamma) = 1$ and $\eta^{-1}\gamma \equiv 1 \bmod_0 (\mathfrak{m}, c)$, so that $\eta^{-1}\gamma \in \Gamma_S$. This shows that $\Gamma_T \subset F^\times \Gamma_S$. Put $R = S \cap F^\times T$. Then $F^\times R'' \subset F^\times T''$. As is observed in 5.3, there exists a morphism P of V_R to V_T rational over k_R such that $P \circ \varphi_R = \varphi_T$. We observe that $F^\times \Gamma_R = F^\times \Gamma_T$. Therefore P is biregular. Now $J_{ST}(1) \circ P^{-1}$ is a morphism of V_T onto V_S, rational over k_R. Our construction shows that $V_T = V_X$ if we put $X = F_c T'/F_c G_{\infty+}$. Therefore we see that $\mathfrak{L}_S \subset k_R \mathfrak{R}_X$. Since $\mathfrak{L}$ is the union of the $\mathfrak{L}_S$ for all S of the type $S = S(\mathfrak{m}, b)$, we have $\mathfrak{L} \subset \mathfrak{k} \cdot \mathfrak{R}$, so that $\mathfrak{L} = \mathfrak{k} \cdot \mathfrak{R}$.

7. Specialization of the function field

Let us now assume that $r = 1$, and $g = [F : \mathbf{Q}]$ is odd. In [C, § 8], we have constructed a homomorphism β of G into another algebraic group G^*. Here G^* and β are rational over $\mathbf{Q}$, and $G_{\mathbf{Q}}^*$ can be identified with

$$\{x \in GL_q(C) \mid x \cdot {}^t x^\iota \in \mathbf{Q}\} ,$$

where $q = 2^{g-1} n^g$, and C is an indefinite quaternion algebra over $\mathbf{Q}$. Thus G^* is of the same type as G. We have also considered an embedding e of $\mathfrak{H}_n$ into $\mathfrak{H}_q$ such that $\beta(a)(e(z)) = e(a(z))$ for every $a \in G_{\infty+}$ and $z \in \mathfrak{H}_n$. Let $\mathfrak{R}$ and $\mathfrak{R}^*$ be the fields of arithmetic automorphic functions on $\mathfrak{H}_n$ with respect to G, and on $\mathfrak{H}_q$ with respect to G^*, respectively, in the sense of 6.1. We are going to reformulate the result of [C, 8.9] in terms of $\mathfrak{R}$ and $\mathfrak{R}^*$. As was remarked in 5.7, we have $\mathfrak{A} = G_{A+}/F_c G_{\infty+}$. Therefore, we consider the map τ a homomorphism of G_{A+} into Aut $(\mathfrak{R}/F')$ with kernel $F_c G_{\infty+}$. Similarly, we have a homomorphism τ^* of G_{A+}^* into Aut $(\mathfrak{R}^*/\mathbf{Q})$.

7.1. THEOREM. *Let $\mathfrak{M}^*$ be the ring of all the elements h of $\mathfrak{R}^*$ which are holomorphic at some point of $e(\mathfrak{H}_n)$. Then $h \circ e \in \mathfrak{R}$ for all $h \in \mathfrak{M}^*$. Moreover, for every $x \in G_{A+}$, $\mathfrak{M}^*$ is stable under $\tau^*(\beta(x))$, and $h^{\tau^*(\beta(x))} \circ e = (h \circ e)^{\tau(x)}$ for all $h \in \mathfrak{M}^*$.*

This is obviously analogous to (6.2.3).

PROOF. Consider canonical systems

$$\{V_S, \varphi_S, J_{TS}(x), \quad (S, T \in \mathfrak{Z}; x \in G_{A+})\},$$

$$\{V_L^*, \varphi_L^*, J_{ML}^*(y), (L, M \in \mathfrak{Z}^*; y \in G_{A+}^*)\}$$

for G and G^*, respectively. Here it is sufficient to consider them in the sense of [C, 2.5]. Let k_S and k_L^* denote respectively the subfields of F_{ab}' and Q_{ab} corresponding to S and L as before. Suppose that $\beta(S) \subset L$. Then $\beta(\Gamma_S) \subset \Gamma_L^* = L \cap G_Q^*$, so that there exists a unique morphism E_{LS} of V_S into V_L^* such that $E_{LS} \circ \varphi_S = \varphi_L^* \circ e$. By [C, 8.9], $k_L^* \subset k_S$, and E_{LS} is rational over k_S. Now let h be an arbitrary element of $\mathfrak{M}^*$. Then $h = f \circ \varphi_L^*$ with a function f on V_L^* rational over k_L for some L. Take S so that $\beta(S) \subset L$. Then $h \circ e = f \circ \varphi_L^* \circ e = f \circ E_{LS} \circ \varphi_S$. Observe that $f \circ E_{LS} \in k_S(V_S)$. Therefore, for a given $x \in G_{A+}$, putting $T = x^{-1}Sx$ and $M = \beta(x)^{-1}L\beta(x)$, we obtain, by [C, 8.9],

$$\begin{aligned}
(h \circ e)^{\tau(x)} &= (f \circ E_{LS} \circ \varphi_S)^{\tau(x)} = (f \circ E_{LS})^{\sigma(x)} \circ J_{ST}(x) \circ \varphi_T \\
&= f^{\sigma(x)} \circ E_{LS}^{\sigma(x)} \circ J_{ST}(x) \circ \varphi_T = f^{\sigma(x)} \circ J_{LM}^*\big(\beta(x)\big) \circ E_{MT} \circ \varphi_T \\
&= f^{\sigma^*(\beta(x))} \circ J_{LM}^*\big(\beta(x)\big) \circ \varphi_M^* \circ e = h^{\tau^*(\beta(x))} \circ e,
\end{aligned}$$

where σ (resp. σ^*) is the map of G_{A+} (resp. G_{A+}^*) onto F_{ab}' (resp. Q_{ab}). Our assertions now follow immediately from this result.

PRINCETON UNIVERSITY

REFERENCES

[1] C. CHEVALLEY, *Deux théorèmes d'arithmétique*, J. Math. Soc. Japan **3** (1951), 36–44.

[2] K. MIYAKE, *On models of certain automorphic function fields*, to appear.

[3] =[C] G. SHIMURA, *On canonical models of arithmetic quotients of bounded symmetric domains*, Ann. of Math. **91** (1970), 144–222.

[4] ———, *Algebraic number fields and symplectic discontinuous groups*, Ann. of Math. **86** (1967), 503–592.

[5] ———, *Introduction to the arithmetic theory of automorphic functions*, to appear.

[6] A. WEIL, *Algebras with involutions and the classical groups*, J. Ind. Math. Soc. **24** (1960), 589–623.

(Received January 27, 1970)

On arithmetic automorphic functions

Proceedings of the International Congress of Mathematicians,
Nice, 1970, vol. 2, 343-348 (1971)

1. – To explain our problems, we start with a semi-simple algebraic group G^1 defined over $\mathbf{Q}$ satisfying the following condition :

(∗) *The quotient of* $G_{\mathbf{R}}^1$ *(see Notation below) by a maximal compact subgroup defines a bounded symmetric domain* $\mathfrak{H}$.

If Γ is an arithmetic subgroup of G^1, there is a Γ-invariant holomorphic map φ of $\mathfrak{H}$ into a projective space which induces a biregular isomorphism of $\mathfrak{H}/\Gamma$ onto a Zariski open subset V of a projective variety (see Baily and Borel [1]). We call such a couple (V, φ) a *model of* $\mathfrak{H}/\Gamma$. Then our first problem, in a naive form, is as follows.

(P) *Associate with each* Γ *of congruence type a model* $(V_\Gamma, \varphi_\Gamma)$ *and an algebraic number field* k_Γ *of finite degree so that the following conditions are satisfied :*

(P.1) V_Γ *is defined over* k_Γ.

(P.2) *If* $\alpha \in G_{\mathbf{Q}}^1$ *and* $\alpha \Gamma \alpha^{-1}$ *is contained in another arithmetic group* Δ, *then* $k_\Delta \subset k_\Gamma$, *and the morphism* $J_{\Delta\Gamma}(\alpha)$ *of* V_Γ *onto* V_Δ, *defined by* $J_{\Delta\Gamma}(\alpha) \circ \varphi_\Gamma = \varphi_\Delta \circ \alpha$, *is rational over* k_Γ.

(P.3) *If w is an isolated fixed point on $\mathfrak{H}$ of an element of $G_{\mathbf{Q}}^1$, the field $k_\Gamma(\varphi_\Gamma(w))$ generated over k_Γ by the coordinates of $\varphi_\Gamma(w)$ has a certain class field theoretical property similar to that of singular values of elliptic modular functions.*

My purpose of this lecture is to give a brief survey of the recent development in this and other related problems.

As Weil [9] showed, any $\mathbf{Q}$-simple algebraic group G^1 can be represented, up to isogeny and with few exceptions, as the group of automorphisms of an involutorial algebra. Under the condition (∗), such a group belongs to one of the following four types (modulo the center in each case).

(I) The unitary group of a hermitian form over a quaternion algebra B whose center is a totally real algebraic number field F. (B can be the matrix algebra $M_2(F)$.)

(II) The special unitary group of a hermitian form over a division algebra, with respect to an involution of the second kind, whose center is a totally imaginary quadratic extension of a totally real algebraic number field.

(III) The orthogonal group of a quadratic form over a totally real algebraic number field.

(IV) The unitary group of an anti-hermitian form over a quaternion algebra whose center is a totally real algebraic number field.

(In each case, certain conditions on the signature of the quadratic or hermitian form are necessary).

At present, the above problem has been settled for the groups of type I (by the author [8]), and of type II (by K. Miyake [3]). Actually, instead of assigning

$$(V_\Gamma , \varphi_\Gamma , k_\Gamma)$$

to a discontinuous group Γ, we consider the adelization G_A of a certain reductive group G whose semi-simple part is G^1, and assign a model (V_S , φ_S) and a number field k_S to a subgroup S of G_A of a certain type. This formulation is almost inevitable for practical, aesthetic, and philosophical reasons, and suggests an analogy with class field theory, as will be seen later.

Notation. — For an algebraic group G over $\mathbf{Q}$, we denote by G_A its adelization. The archimedean and non-archimedean parts of G_A are denoted by G_∞ and G_0, respectively, so that $G_A = G_0 G_\infty$. We identify G_∞ with $G_\mathbf{R}$, the group of all $\mathbf{R}$-rational points of G. Then $G_{\infty+}$ denotes the identity component of G_∞. We put $G_{A+} = G_0 G_{\infty+}$, $G_{\mathbf{Q}+} = G_{A+} \cap G_\mathbf{Q}$. A $\mathbf{Q}$-rational homomorphism λ of G to another $\mathbf{Q}$-rational algebraic group G' defines naturally a continuous homomorphism of G_A to G'_A, which we denote again by λ. For any algebraic number field F of finite degree, we view $F^\times = F - \{0\}$ as (the $\mathbf{Q}$-rational points of) an algebraic group over $\mathbf{Q}$. Then $F_A^\times$ denotes the idele group, and $F_{\infty+}^\times$ the identity component of $F_\infty^\times = (F \otimes_\mathbf{Q} \mathbf{R})^\times$. Further F_{ab} denotes the maximal abelian extension of F, and F_c the closure of $F^\times F_{\infty+}^\times$ in $F_A^\times$.

2. — First let us make preliminay considerations about an object (F, Θ) consisting of an algebraic number field F of finite degree, and an absolute equivalence class Θ of $\mathbf{Q}$-linear representations of F by complex matrices. If $g = [F : \mathbf{Q}]$ and $\tau_1 , \ldots , \tau_g$ denote all the isomorphisms of F into $\mathbf{C}$, then Θ is determined by the multiplicity, say r_i, of each τ_i in Θ. We write then $\Theta \sim \sum_{i=1}^{g} r_i \tau_i$, and put

$$\det \Theta(x) = \prod_{i=1}^{g} (x^{\tau_i})^{r_i}, \quad \operatorname{tr} \Theta(x) = \sum_{i=1}^{g} r_i x^{\tau_i} \quad \text{for } x \in F.$$ Given (F, Θ), let F' denote the field generated over $\mathbf{Q}$ by $\operatorname{tr} \Theta(x)$ for all $x \in F$. Then there is a unique $(F \otimes_\mathbf{Q} F')$-module V such that Θ is equivalent to the representation of F into the ring of F'-linear endomorphisms of V. Mapping F' into the ring of F-linear endomorphisms of V, we obtain an equivalence class Θ' of representations of F. The couple (F', Θ') is called *the reflex of* (F, Θ). We see easily that $\det \Theta'(y)$ and $\operatorname{tr} \Theta'(y)$ belong to F for every $y \in F'$.

3. — To state our results for the group of type I, let F be a totally real algebraic number field of degree g, and B a quaternion algebra over F, which may or may not be a division algebra. Define an algebraic group G over $\mathbf{Q}$ so that

$$G_\mathbf{Q} = \{\alpha \in GL_n(B) \mid {}^t\alpha^\iota \cdot \alpha = \nu(\alpha) \cdot 1_n \quad \text{with} \quad \nu(\alpha) \in F^\times\} ,$$

where ι is the main involution of B. The semi-simple part of G is

$$G^1 = \{\alpha \in G \mid \nu(\alpha) = 1\} .$$

Now $B \otimes_Q R = M_2(R)^r \oplus H^{g-r}$ with an integer $r \geqq 0$, where H denotes the Hamilton quaternions. Accordingly we have $G_R^1 = Sp(n, R)^r \times Sp(n)^{g-r}$, so that G_R^1 modulo a maximal compact subgroup can be identified with the product $\mathfrak{H}_n^r$ of r copies of the Siegel upper half space $\mathfrak{H}_n$ of degree n. Let $\tau_1, \ldots, \tau_r$ be the isomorphisms of F into R corresponding to the first r factors $M_2(R)$ of $B \otimes_Q R$. Consider (F, Θ) with $\Theta \sim \sum_{i=1}^{r} \tau_i$, and let (F', Θ') be the reflex of (F, Θ). Put $\lambda = \det \Theta'$, and define two subgroups $\mathscr{G}_+$ and $\overline{\mathscr{G}}_+$ of G_{A+} by

$$\mathscr{G}_+ = G_{Q+} G_{\infty+} \cdot \{x \in G_{A+} \mid \nu(x) \in \lambda(F_A'^\times)\},$$

$$\overline{\mathscr{G}}_+ = \{x \in G_{A+} \mid \nu(x) \in \lambda(F_A'^\times) F_c\}.$$

It can be shown that $\overline{\mathscr{G}}_+$ is the closure of $\mathscr{G}_+$.

Let $\mathfrak{Z}$ denote the set of all the subgroups S of $\overline{\mathscr{G}}_+$ satisfying the following two conditions : (i) $F_c G_{\infty+} \subset S$; (ii) $S/F_c G_{\infty+}$ *is open and compact in* $\overline{\mathscr{G}}_+/F_c G_{\infty+}$. (If especially $r = 1$, we have $\overline{\mathscr{G}}_+ = G_{A+}$, so that $\mathfrak{Z}$ is in one-to-one correspondence with the set of all open compact subgroups of $G_0/(G_0 \cap F_c)$.) For every $S \in \mathfrak{Z}$, put $\Gamma_S = S \cap G_Q$. Then $F^\times \subset \Gamma_S$, and Γ_S, or rather $\Gamma_S/F^\times$, is a discontinuous group of transformations on $\mathfrak{H}_n^r$. We are going to assign, to each $S \in \mathfrak{Z}$, a model (V_S, φ_S) of $\mathfrak{H}_n^r/\Gamma_S$ and a number field k_S. To define k_S, we consider the canonical isomorphism of $F_A'^\times/F_c'$ onto $\mathrm{Gal}(F_{ab}'/F')$. Let $\mathfrak{k}$ denote the subfield of F_{ab}' corresponding to the kernel of the map $\lambda^* : F_A'^\times \to \lambda(F_A'^\times) F_c/F_c$ which is naturally obtained from λ. Then $\mathrm{Gal}(\mathfrak{k}/F')$ is canonically isomorphic to $\lambda(F_A'^\times) F_c/F_c$. Since $\nu(\overline{\mathscr{G}}_+) \subset \lambda(F_A'^\times) F_c$, composing this isomorphism with ν, we obtain a homomorphism $\sigma : \overline{\mathscr{G}}_+ \to \mathrm{Gal}(\mathfrak{k}/F')$. Now k_S is defined to be the subfield of $\mathfrak{k}$ corresponding to $\sigma(S)$. After these preparations, our first main theorem can be stated as follows.

THEOREM 1. — *There exists a system*

$$\{V_S, \varphi_S, J_{TS}(x), (S, T \in \mathfrak{Z} ; x \in \overline{\mathscr{G}}_+)\}$$

formed by the objects satisfying the following conditions.

(1) *For each* $S \in \mathfrak{Z}$, (V_S, φ_S) *is a model of* $\mathfrak{H}_n^r/\Gamma_S$.

(2) V_S *is rational over* k_S.

(3) $J_{TS}(x)$, *defined if and only if* $xSx^{-1} \subset T$, *is a morphism of* V_S *onto* $V_T^{\sigma(x)}$, *rational over* k_S, *and has the following properties :*

(3a) $J_{SS}(x)$ *is the identity map if* $x \in S$;

(3b) $J_{TS}(x)^{\sigma(y)} \circ J_{SR}(y) = J_{TR}(xy)$;

(3c) $J_{TS}(\alpha) \circ \varphi_S = \varphi_T \circ \alpha$ *for every* $\alpha \in G_{Q+}$ *if* $\alpha S \alpha^{-1} \subset T$.

(4) *A certain reciprocity law holds at every isolated fixed point of* G_{Q+}.

Thus the symbol $J_{TS}(x)$ includes $J_{\Delta\Gamma}(\alpha)$ of (P.2) as a special case. For an element x of $\overline{\mathscr{G}}_+$ not contained in G_{Q+}, the existence of the morphism $J_{TS}(x)$ of V_S to $V_T^{\sigma(x)}$ is quite a non-trivial fact. This is especially so when the class number of F is greater than one.

For simplicity, we give the precise statement of (4) in the special case where $r = 1$, τ_1 is the identity map, and the isotropy group at the fixed point spans a number field. (See [6], [8] for the general case.) Let P_0 be a totally real extension of F of degree n, and P a totally imaginary quadratic extension of P_0 that splits B. Then there exists an F-linear isomorphism f of P into $M_n(B)$ such that $f(a^\rho) = {}^t f(a)^\iota$ for all $a \in P$, where ρ denotes the complex conjugation. If $a \in P^\times$ and $aa^\rho \in F$, we see that $f(a) \in G_{\mathbf{Q}+}$, and these elements $f(a)$ have a unique common fixed point z on $\mathfrak{H}_n$. We can then find a Q-linear representation $\Psi : P \to M_n(\mathbf{C})$ and a holomorphic isomorphism π of $\mathfrak{H}_n$ onto the unit ball $\mathfrak{D}_n$ of complex symmetric matrices of size n so that

$$(\pi \cdot f(a) \cdot \pi^{-1})\,(w) = \overline{\Psi(a)} \cdot w \cdot \Psi(a)^{-1} \qquad (a \in P^\times, aa^\rho \in F\,; w \in \mathfrak{D}_n)$$

Let (P', Ψ') be the reflex of (P, Ψ). Put $\eta(x) = f(\det \Psi'(x))$ for $x \in P'$. It can be shown that $F' \subset P'$, η maps $P'^\times$ into G, and $\nu \circ \eta = \lambda \circ N_{P'/F'}$, hence $\eta(P_{\mathbf{A}}'^\times) \subset \mathbf{G}_+$. Now the property (4) in this special case is as follows : *For each $S \in \mathfrak{Z}$, the point $\varphi_S(z)$ is rational over P'_{ab}, and for every $v \in P_{\mathbf{A}}'^\times$ and $T = \eta(v) S \eta(v)^{-1}$ one has*

$$(4a) \qquad\qquad \varphi_S(z)^{[v]} = J_{ST}(\eta(v)^{-1})\,(\varphi_T(z))\,,$$

where $[v]$ denotes the element of $\mathrm{Gal}(P'_{ab}/P')$ corresponding to v.

4. — Let K_S denote the field of all functions (in the sense of algebraic geometry) on V_S rational over k_S. Define a field $\mathfrak{K}$ of meromorphic functions on $\mathfrak{H}_n^r$ by

$$\mathfrak{K} = \bigcup_{S \in \mathfrak{Z}} \mathfrak{K}_S\,, \qquad \mathfrak{K}_S = \{f \circ \varphi_S \mid f \in K_S\}\,.$$

The elements of $\mathfrak{K}$ may be called *arithmetic automorphic functions with respect to G*. Denote by $\mathrm{Aut}(\mathfrak{K}/\mathfrak{H})$ the group of all automorphisms of $\mathfrak{K}$ over any subfield $\mathfrak{H}$, and make it a topological group by taking as a basis of neighborhoods of the identity the subgroups $\mathrm{Aut}(\mathfrak{K}/\mathfrak{L})$ for all subfields $\mathfrak{L}$ finitely generated over $\mathfrak{H}$. For each $x \in \overline{\mathbf{G}}_+$, we can define an element $\tau(x)$ of $\mathrm{Aut}(\mathfrak{K}/F')$ by the rule

$$(f \circ \varphi_S)^{\tau(x)} = f^{\sigma(x)} \circ J_{ST}(x) \circ \varphi_T \qquad (T = x^{-1} S x\,, f \in K_S)\,.$$

Theorem 2. — *The map $x \to \tau(x)$ is a continuous homomorphism of $\overline{\mathbf{G}}_+$ into $\mathrm{Aut}(\mathfrak{K}/F')$, and has the following properties :*

(1) The kernel of τ is $F_c \cdot G_{\infty+}$.

(2) $\tau(x) = \sigma(x)$ on $\mathfrak{k}$.

(3) $h^{\tau(\alpha)} = h \circ \alpha$ for every $\alpha \in G_{\mathbf{Q}+}$ and $h \in \mathfrak{K}$

(4) Let P, f, P', z, and η be as above (assuming $r = 1$). Then, for every $h \in \mathfrak{K}$ defined at z, $h(z)$ is rational over P'_{ab}. Moreover, if $v \in P_{\mathbf{A}}'^\times$ and $u = \eta(v)^{-1}$, then $h^{\tau(u)}$ is also defined at z, and $h(z)^{[v]} = h^{\tau(u)}(z)$.

The last property (or (4a) of Th.1) is a generalization of the wellknown behavior of the singular values of the classical modular function j.

Theorem 3. — *The map τ induces a topological isomorphism of $\overline{\mathbf{G}}_+/F_c\,G_{\infty+}$ onto an open subgroup of $\mathrm{Aut}(\mathfrak{K}/F')$. For every $S \in \mathfrak{Z}$, $\mathfrak{K}$ is an infinite Galois* *extension of* $\Re_S$, *and* $\tau(S) = \mathrm{Gal}(\Re/\Re_S)$. *Moreover, if* $r = 1$ *and* $\mathcal{H}_n/\Gamma_S$ *for any* $S \in \mathfrak{Z}$ *is compact, or* $G_Q = GL_2(Q)$, *then* τ *is surjective.*

One can conjecture that τ is surjective so long as $r = 1$. If $r > 1$, $\mathrm{Aut}(\Re/F')$ may be larger than $\tau(\overline{\mathcal{G}_+})$, since the "permutations" of the factors of $\check{\varrho}_n^r$ may give rise to automorphisms of $\Re$, see [8, II].

We notice that the relation (4) of Th. 2 explains the deep arithmetic meaning of the map τ, exactly similar to the fact that the canonical isomorphism of $F_A^\times/F_c$ onto $\mathrm{Gal}(F_{ab}/F)$, for any number field F, is defined locally by the Frobenius automorphisms. Thus the above two theorems provide an analogue of class field theory for the field $\Re$ which is of Kroneckerian dimension > 1.

The construction of V_S is based on the existence of families of abelian varieties $\Sigma = \{A_z \,|\, z \in \mathcal{S}_n\}$ with the property that A_z and A_w (endowed with structures of polarization and endomorphisms) are isomorphic if and only if $z = \gamma(w)$ for some $\gamma \in \Gamma_S$. We shall not discuss further the ideas of the proofs beyond this remark, since they are explained rather in detail in [4] and the Introduction of [5].

5. — As is already mentioned, Miyake [3] has obtained the results completely parallel to Theorems 1, 2, 3 for the groups of type II by the same method. One can naturally propose the generalization to a wider class of groups. Since there are families of abelian varieties, similar to the above Σ, for each of the groups belonging to the remaining types III and IV, it is quite certain that one will be able to construct, without much difficulties, a system like that of Th. 1 for a suitably chosen reductive group G' containing the given semi-simple group of any type, excluding exceptional ones. We call such a system a *canonical system for* G', and can define *arithmetic automorphic functions with respect to* G' in the same manner as above. Now there is an important aspect of this framework : *canonical systems for two groups are not only defined and characterized individually, but also consistent with each other.* To be more specific, let

$$\{V_S', \varphi_S', J_{TS}'(x), k_S'\} \quad \text{and} \quad \{V_L'', \varphi_L'', J_{ML}''(x), k_L''\}$$

be canonical systems for G' and G'', respectively, and f a Q-rational homomorphism of G' into G''. Suppose that f defines a holomorphic embedding ϵ of the corresponding bounded domain $\mathcal{H}'$ into $\mathcal{H}''$. If $f(S) \subset L$, we have $f(\Gamma_S) \subset \Gamma_L$, so that there is a morphism E_{LS} of V_S' into V_L'' such that $E_{LS} \circ \varphi_S' = \varphi_L'' \circ \epsilon$. In some cases we can prove : (i) $k_L'' \subset k_S'$; (ii) E_{LS} is rational over k_S' ; (iii) *If further* $f(T) \subset M$, $x \in G_A'$, $xSx^{-1} \subset T$, $f(x) Lf(x)^{-1} \subset M$, *then*

$$E_{MT}^{\sigma'(x)} \circ J_{TS}'(x) = J_{ML}''(f(x)) \circ E_{LS} \quad ,$$

provided that $J_{TS}'(x)$ *and* $J_{ML}''(f(x))$ *are defined, where* σ' *is the symbol defined for* G' *corresponding to* σ. If $G' = G''$, and f is the identity map, this means the unicity of the canonical system. We may consider (4, 4a) of Th. 1 as an example of such a relation. ($\mathcal{H}'$ is a single point in this case). More examples, in which G' and G'' are of type I, are given in [8, I, § 8]. The consistency of this kind can also be formulated in terms of $\mathrm{Aut}(\Re/F')$ and τ . Let $\Re'\ \tau'$, $\Re''$, τ'' be the corresponding symbols for G' and G'', respectively. If an element h of $\Re''$ is holomorphic at some point of $\epsilon(\mathcal{H}')$, then $h \circ \epsilon \in \Re'$, and $h^{\tau''(f(x))} \circ \epsilon = (h \circ \epsilon)^{\tau'(x)}$ for every $x \in G'_\mathbf{A}$ such that $\tau'(x)$ and $\tau''(f(x))$ are meaningful. Again (4) of Th. 2 is a special case.

Because of the restriction on the length, I have to give up the discussion of two important topics related to the above theory, namely :

(A) *Construction of a system of varieties over finite fields,* as Ihara [2] has given in the one-dimensional case. The reader is referred to his lecture and [8, I, § 2.8].

(B) *Local representations of the Galois group of an infinite extension of a number field obtained from the points of the coverings lying on an algebraic point of V_S.* The characteristic roots of the Frobenius automorphisms, defined through these representations, have absolute values of the Riemann-Ramanujan-Weil type. Thus they provide generalizations of l-adic representations of abelian varieties. See [7] and [8, I, §§7, 8].

REFERENCES

[1] Baily W.L. and Borel A. — Compactification of arithmetic quotients of bounded symmetric domains, *Ann. of Math.,* 84, 1966, p. 442-528.
[2] Ihara Y. — On Congruence Monodromy Problems, *lecture notes,* University of Tokyo, I, 1968; II, 1969.
[3] Miyake K. — *On models of certain automorphic function fields,* to appear.
[4] Shimura G. — Number fields and zeta functions associated with discontinuous groups and algebraic varieties, *Proc. Int. Congress of Math. Moscow,* 1966, p. 290-298.
[5] Shimura G. — Construction of class fields and zeta functions of algebraic curves, *Ann. of Math.,* 85, 1967, p. 58-159.
[6] Shimura G. — Algebraic number fields and symplectic discontinuous groups, *Ann. of Math.,* 86, 1967, p. 503-592.
[7] Shimura G. — Local representations of Galois groups, *Ann. of Math.,* 89, 1969, p. 99-124.
[8] Shimura G. — On canonical models of arithmetic quotients of bounded symmetric domains, I, *Ann. of Math.,* 91, 1970, p. 144-222; II, 92, 1970, p. 528-549.
[9] Weil A. — Algebras with involutions and the classical groups, *J. Ind. Math. Soc..* 24, 1960, p. 589-623.

Princeton University
Dept. of Mathematics,
Princeton
New Jerwey 08 540 (USA)

71b

On the zeta-function of an abelian variety with complex multiplication

Annals of Mathematics, 94 (1971), 504-533

Our object of study is an abelian variety A of dimension n, defined over an algebraic number field k of finite degree, whose endomorphism algebra $\mathrm{End}_Q(A)$ contains an isomorphic image $\theta(K)$ of an algebraic number field K of degree $2n$. Here θ denotes a fixed isomorphism of K into $\mathrm{End}_Q(A)$. We assume throughout, without losing much generality, that K is a totally imaginary quadratic extension of a totally real algebraic number field. It is well-known that the zeta-function of A over k can be given by several L-functions with Grössen-characters, which was first proved by Deuring [1] in the one-dimensional case, and by Taniyama [7] in the higher-dimensional case. In the present paper, we shall discuss the following two problems:

(I) Determination of the zeta-function of A over k in the case where the elements of $\theta(K) \cap \mathrm{End}(A)$ are not necessarily defined over k.

(II) Construction of an abelian variety A with a given Grössen-character, and the study of the relationship between the isomorphism-class or the isogeny-class of A and the character.

As to the first problem, examples of such abelian varieties are provided by the factors of the jacobians of the curves treated by Weil [9]. If A is an elliptic curve and K is an imaginary quadratic field, we see that the elements of $\mathrm{End}(A)$ are rational over k if and only if $K \subset k$. Then Deuring [1] determined the zeta-function of A in both cases $K \subset k$ and $K \not\subset k$. However, the previous investigations in the higher-dimensional case (Taniyama [7], Shimura-Taniyama [6], Serre-Tate [2]) were all concerned with the case in which the elements of $\theta(K) \cap \mathrm{End}(A)$ are rational over k. Therefore our purpose is to loosen this condition. We fix our attention to the maximal real subfield F of K. Let k_1 be an algebraic number field of finite degree, over which A and the elements of $\theta(F) \cap \mathrm{End}(A)$ are rational. Then one can define (the one-dimensional part of) the zeta-function of A over k_1 relative to F, denoted by $\zeta(s; A/k_1, F)$, by means of the $\mathfrak{l}$-adic representations on A for the prime ideals $\mathfrak{l}$ in F. Now there is an algebraic number field K', determined by (A, θ), with

* During the preparation of this paper, the author was a John Simon Guggenheim Fellow, and partially supported by NSF Grant GP-23855.

406

the property that the elements of $\theta(K) \cap \mathrm{End}(A)$ are rational over the composite $k_1 K'$. (K' is defined in terms of the "reflex" of a CM-type, see text; $K' = K$ if $n = 1$.) We shall show that if $K' \not\subset k_1$, then $\zeta(s; A/k_1, F)$ is exactly the L-function $L(s, \psi)$ with a Grössen-character ψ of $k_1 K'$ (Theorem 7). Actually we have proved a weaker result of the same type in [5, § 7.8]. But there were two unsatisfactory points: (i) A was assumed simple; (ii) a question was left open concerning possible bad factors for the primes of $k_1 K'$ ramified over k_1. Therefore our task here is to remove the assumption that A is simple, and settle the second point by showing that A (over $k_1 K'$) has no good reduction for such primes. The proof (given in § 6) is rather simple, and in fact simpler than Deuring's proof [1, IV, § 4] in the one-dimensional case.

It should be noted that those l-adic representations on A over k_1 are of non-abelian type, and $\zeta(s; A/k_1, F)$ is similar to the Artin L-function with a character induced from an abelian character. Probably this aspect of the theory will become more important when one looks deep into the nature of such a zeta-function. This is one of the reasons why we take up the first problem.

To explain our results about the second problem, let ψ be a Grössen-character of k or $K'k$, according as $K' \subset k$ or $K' \not\subset k$. We shall give a necessary and sufficient condition for ψ that there exists an abelian variety A over k whose zeta-function is given by ψ (Theorems 6, 10).[1] If $K' \not\subset k$, we have to make an assumption that k has at least one real archimedean prime, without which the theorem is shown to be false by an example. There is one shortcoming: We do not know whether a given field k or $K'k$ has a Grössen-character with the required properties. By our theorems, the existence of such a character is equivalent to the existence of an abelian variety rational over k of a given type. If A is an elliptic curve, the last point produces no problem, since it reduces to the assumption that k contains the invariant of A. In the higher-dimensional case, it is still unknown whether a polarized abelian variety has a model defined over its field of moduli. We shall, however, construct a Grössen-character and an abelian variety with that character over its field of moduli (taking account of endomorphisms), under a certain condition which is satisfied, for example, if $\theta(K) \cap \mathrm{End}(A)$ is the maximal order in $\theta(K)$ (Theorem 11, Proposition 7). Again, the question will be treated in both cases $K' \subset k$ and $K' \not\subset k$. Incidentally, the model so obtained has a special property that its points of finite order are rational over the maximal abelian extension of K'. It should also be noted that there were previously

---

[1] Theorem 6 is due to W. Casselman, to whom we would like to express our hearty thanks for his permission to include his result in the present paper.

no known examples of abelian varieties with complex multiplication defined over their fields of moduli, except for elliptic curves and the factors of the jacobians of the curves considered in [9].

In the final section, we generalize these results by taking a more general type of subfield of K instead of F. We could have given an exposition once and for all in the most general case (in the text as well as in this introduction). But to avoid blurring the reader's insight by too much technical detail, we first discuss the relatively simple case with F, and then the general case in § 6. In treating our problems, it becomes necessary to study the relationship between the various fields of moduli of A with or without taking account of endomorphisms in $\theta(K)$ or $\theta(F)$, and also the fields of definition for those objects. This will be done in § 3. For the reader's convenience, we have restated a few known results as Lemmas 1, 2, and Theorems 1, 2, 3, 4.

0. Notation and terminology

0.1. As usual, Z, Q, R, C denote the ring of rational integers, the rational number field, the real number field, the complex number field, respectively. For an associative ring S with an identity element, we denote by $S^\times$ the group of all invertible elements of S.

0.2. If x is a complex number, or a complex matrix, etc., x^ρ or $\bar{x}$ denotes the complex conjugate of x. However, $\bar{Q}$ denotes, as the only exception, the algebraic closure of Q in C. By an *algebraic number field*, we always mean a subfield of $\bar{Q}$. The restriction of the complex conjugation to any algebraic number field is also denoted by the same symbol ρ.

0.3. If F is an algebraic number field of finite degree, we denote by $F_A^\times$, $F_\infty^\times$, and F_{ab} the idele group of F, the archimedean part of $F_A^\times$, and the maximal abelian extension of F, respectively. For $x \in F_A^\times$, we denote by $[x, F]$ the element of $\mathrm{Gal}(F_{ab}/F)$ canonically associated with x. Further $|x|_0$ denotes the absolute norm of the ideal associated with x, which is a positive rational number. The archimedean component of x (i.e., the projection of x to $F_\infty^\times$) is denoted by x_∞. For an automorphism σ of F, we denote by the same symbol σ the natural extension of σ to $F_A^\times$. This applies especially to ρ, if $F^\rho = F$. By a *Grössen-character* of $F_A^\times$, we understand a continuous homomorphism of $F_A^\times$ into $C^\times$ whose kernel contains $F^\times$.

0.4. By a *lattice* in an algebraic number field F of finite degree, we understand a finitely generated Z-submodule of F that spans F over Q. If $\mathfrak{a}$ is a lattice in F, and p is a rational prime, we put $F_p = F \otimes_Q Q_p$, $\mathfrak{a}_p = \mathfrak{a} \otimes_Z Z_p$, where Q_p and Z_p denote the p-adic completions of Q and Z, respectively. For $x \in F_A^\times$, we denote by $x\mathfrak{a}$ a unique lattice in F such that $(x\mathfrak{a})_p =$ $x_p \mathfrak{a}_p$ for all p, where x_p denotes the p-component of x. (Note that $x_p \in F_p^\times$.) Further we can obtain an isomorphism of $F/\mathfrak{a}$ onto $F/x\mathfrak{a}$ in the following way. Observe that $F/\mathfrak{a}$ is the direct sum of the modules $F_p/\mathfrak{a}_p$ for all p. Multiplication by x_p defines an isomorphism of $F_p/\mathfrak{a}_p$ onto $F_p/x_p\mathfrak{a}_p$. Combining these isomorphisms for all p, we obtain the desired isomorphism of $F/\mathfrak{a}$ onto $F/x\mathfrak{a}$. The image of an element u of $F/\mathfrak{a}$ by this isomorphism is denoted by xu. If $x \in F_\infty^\times$, then $x\mathfrak{a} = \mathfrak{a}$, and the action of x on $F/\mathfrak{a}$ is trivial. (See also [4, 0.5], [5, § 5.2].)

0.5. For an abelian variety A, we denote by $\mathrm{End}(A)$ the ring of all endomorphisms of A (over the universal domain), and put $\mathrm{End}_Q(A) = \mathrm{End}(A) \otimes_Z Q$. By *the zeta-function of A* over an algebraic number field k of finite degree, we always mean "the one-dimensional part". More precisely, it is the product

$$(*) \qquad\qquad \prod_\mathfrak{p} \det[1 - N(\mathfrak{p})^{-s} \cdot \varphi_\mathfrak{p}]^{-1}$$

taken over all prime ideals $\mathfrak{p}$ in k for which A has good reduction. Here $\varphi_\mathfrak{p}$ denotes the l-adic representation of the Frobenius endomorphism of degree $N(\mathfrak{p})$ on the variety obtained from A by reduction modulo $\mathfrak{p}$, l being any rational prime not divisible by $\mathfrak{p}$.

1. The Grössen-character determined by an abelian variety of *CM*-type

First let us recall some basic definitions and known results. We call an algebraic number field of finite degree a *CM-field*, if it is a totally imaginary quadratic extension of a totally real subfield. Every isomorphism of such a field into C commutes with the complex conjugation. By a *CM-type*, we understand a couple (K, Φ) formed by a CM-field K and an absolute equivalence class Φ of representations of K (as a Q-algebra) by complex matrices such that the direct sum of Φ and its complex conjugate is equivalent to the class of regular representations of K over Q. We shall often denote by the same symbol Φ any representation belonging to the class Φ. For a given CM-type (K, Φ), let K' be the field generated over Q by $\mathrm{tr}\,\Phi(x)$ for all $x \in K$. Then there is a unique $(K \otimes_Q K')$-module V such that Φ is equivalent to the representation of K into the ring of K'-linear endomorphisms of V. Mapping K' into the ring of K-linear endomorphisms of V, we obtain an equivalence class Φ' of representations of K'. The couple (K', Φ') is also a CM-type, and called *the reflex* (or *dual*) *of* (K, Φ). For details, see [6], [4, § 1].

Let us now fix a CM-type (K, Φ), and denote by n the degree of Φ. Then $[K:Q] = 2n$. We shall be concerned with a structure (A, θ), or $(A, \mathcal{C}, \theta)$, formed by an abelian variety A of dimension n, an isomorphism θ of K into $\mathrm{End}_Q(A)$, and a polarization $\mathcal{C}$ of A. We say that (A, θ) *is of type* (K, Φ) if the representation of K, through θ, on the tangent space of A at the origin is equivalent to Φ.

LEMMA 1. *If* (A, θ) *is of type* (K, Φ), *the following three conditions are equivalent:*

  (i) A *is simple;*

  (ii) $\theta(K) = \mathrm{End}_Q(A)$;

  (iii) (K, Φ) *is the reflex of its reflex.*

Moreover, if these are satisfied, and an automorphism π *of* K *satisfies* $\mathrm{tr}\,\Phi(a^\pi) = \mathrm{tr}\,\Phi(a)$ *for all* $a \in K$, *then* π *is the identity map.*

For the proof, see [6, § 8]. The last assertion implies that θ is uniquely determined, if A is simple and (K, Φ) is fixed. We call (K, Φ) *primitive* if it satisfies (iii) of Lemma 1.

LEMMA 2. *Let* (K', Φ') *be the reflex of* (K, Φ), *and* (S, Ω) *the reflex of* (K', Φ'). *Then* (K', Φ') *is primitive, and coincides with the reflex of* (S, Ω). *Moreover,* $S \subset K$, *and the restriction of* Φ *to* S *is equivalent to* $[K:S]$ *copies of* Ω. *If* (A, θ) *is of type* (K, Φ), *then* $\theta(S)$ *is the center of* $\mathrm{End}_Q(A)$.

For the proof, see [6, § 8], [4, § 1].

To obtain an explicit analytic form of (A, θ), consider Φ as a matrix representation acting on the complex n-dimensional vector space C^n, and extend it R-linearly to $K_R = K \otimes_Q R$. Take any R-linear isomorphism $\mathfrak{y}$ of K_R onte C^n such that $\mathfrak{y}(ab) = \Phi(a)\mathfrak{y}(b)$ for all $a \in K_R$, $b \in K_R$. (Obviously, $\mathfrak{y}$ is obtained by $\mathfrak{y}(a) = \Phi(a)z$ with $z = \mathfrak{y}(1)$.) If (A, θ) is of type (K, Φ), we obtain an exact sequence

$$0 \longrightarrow \mathfrak{y}(\mathfrak{a}) \longrightarrow C^n \overset{\xi}{\longrightarrow} A \longrightarrow 0$$

with a lattice $\mathfrak{a}$ in K, and a holomorphic map ξ such that $\theta(a) \circ \xi = \xi \circ \Phi(a)$ for $a \in K$, $\theta(a) \in \mathrm{End}(A)$. Now we assume, throughout the paper, that A has a polarization $\mathcal{C}$ satisfying

(1.1) $\theta(K)$ *is stable under the involution of* $\mathrm{End}_Q(A)$ *determined by* $\mathcal{C}$.

This is satisfied whenever A is simple, because of Lemma 1.

If $P(x, y)$ is the Riemann form of a basic polar divisor in $\mathcal{C}$, then (1.1) is equivalent to

$$(1.2) \qquad P\big(\mathfrak{y}(u), \mathfrak{y}(v)\big) = \mathrm{Tr}_{K/Q}(\varepsilon u v^\rho) \qquad\qquad ((u, v) \in K \times K)$$

with an element ε *of* K *such that* $\varepsilon^\rho = -\varepsilon$.

Here ρ denotes the complex conjugation, see 0.2. In this situation, we say that (A, C, θ) *is of type* $(K, \Phi; \mathfrak{a}, \varepsilon)$ (with respect to ξ and $\mathfrak{y}$). It can easily be seen that (K, Φ) determines the isogeny-class of (A, θ) over C; $(K, \Phi; \mathfrak{a})$ the isomorphism-class of (A, θ) over C, and $(K, \Phi; \mathfrak{a}, \varepsilon)$ the isomorphism-class of (A, C, θ) over C. For details, see [4, §4], [5, §5.5], [6]. The polarization C can be shown to be rational over any field k of definition for (A, θ) (see Prop. 3 below).

To state the main theorem of complex multiplication, we consider the reflex (K', Φ') of (K, Φ). Observe that $\det \Phi'$ defines a map of $K'^{\times}$ into $K^{\times}$, and extend it to a continuous homomorphism

$$(1.3) \qquad f: K_A'^{\times} \longrightarrow K_A^{\times} \qquad\qquad \left(f(x) = \det \Phi'(x) \text{ if } x \in K'^{\times}\right) .$$

If (S, Ω) is the reflex of (K', Φ'), we see from Lemma 2 that f actually maps $K_A'^{\times}$ into $S_A^{\times}$.

THEOREM 1. *Let* (A, C, θ) *be of type* $(K, \Phi; \mathfrak{a}, \varepsilon)$. *Let* σ *be an element of* $\mathrm{Aut}(C/K')$, *and* z *an element of* $K_A'^{\times}$ *such that* $\sigma = [z, K']$ *on* K_{ab}'. *Then there is an exact sequence*

$$0 \longrightarrow \mathfrak{y}\left(f(z)^{-1}\mathfrak{a}\right) \longrightarrow C^n \overset{\xi'}{\longrightarrow} A^\sigma \longrightarrow 0$$

with the following properties:

    (i) $(A^\sigma, C^\sigma, \theta^\sigma)$ *is of type* $\left(K, \Phi; f(z)^{-1}\mathfrak{a}, |z|_0 \cdot \varepsilon\right)$ *with respect to* ξ' *and* $\mathfrak{y}$.

    (ii) $\xi\left(\mathfrak{y}(v)\right)^\sigma = \xi'\left(\mathfrak{y}\left(f(z)^{-1}v\right)\right)$ *for all* $v \in K/\mathfrak{a}$.

For the proof, see [4, 4.3], [5, §5.5]. (For the notation, see 0.3, 0.4.)

Suppose that (A, θ) is defined over an algebraic number field k of finite degree. By [6, §8.5, Prop. 30], $K' \subset k$ (see also Proposition 2 below). Define a map $g: k_A^{\times} \to K_A^{\times}$ by

$$(1.4) \qquad g = f \circ N_{k/K'} ,$$

and a map $\omega: K_R \to A$ by

$$(1.5) \qquad \omega = \xi \circ \mathfrak{y} .$$

We often consider ω also to be an isomorphism of $K_R/\mathfrak{a}$ onto A. Then $\omega(K/\mathfrak{a})$ is the module of all points of finite order on A. Since the direct sum of Φ' and its complex conjugate is equivalent to a regular representation of K', we obtain

$$(1.6) \qquad g(x)g(x)^\rho = N_{k/Q}(x) \qquad\qquad (x \in k_A^{\times}) .$$

THEOREM 2. *The notation and assumptions being as above, every point of finite order on A is rational over k_{ab}. Moreover, there exists a homomorphism* $\alpha: k_A^{\times} \to K^{\times}$ *with the following properties:*

(i) $\omega(v)^{[x,k]} = \omega(\alpha(x)g(x)^{-1}v)$ *for all* $x \in k_A^\times$, $v \in K/\mathfrak{a}$;

(ii) $\alpha(x)g(x)^{-1}\mathfrak{a} = \mathfrak{a}$;

(iii) $\alpha(x)\alpha(x)^\rho = |x|_0$;

(iv) $\mathrm{Ker}(\alpha)$ *is open in* $k_A^\times$.

This can easily be obtained from Theorem 1, see [5, § 7.8, Prop. 7.40]. For the reason explained above, it is not necessary to assume that $\mathcal{C}$ is rational over k. The map α depends only on the isogeny-class of (A, θ) over k, and is independent of the choice of ξ, $\mathfrak{y}$, and $\mathfrak{a}$. It can easily be shown that α coincides with the map ε of Serre-Tate [2, Th. 10]. We see easily that the element $\alpha(x)$ is uniquely determined for x by (i). Therefore we have

(1.7) $\alpha(x) = g(x)$ *if* $x \in k^\times$,

(1.8) $\alpha(x) = 1$ *if* $x \in k_\infty^\times$.

For $u \in K_A^\times$, denote by $u_{\infty\nu}$ the component of u at the ν-th archimedean prime of K. Fix any identification of C with the completion of K at this prime. Define a map $\psi_\nu \colon k_A^\times \to C^\times$ by

$$(1.9) \qquad\qquad \psi_\nu(x) = (\alpha(x)/g(x))_{\infty\nu} \qquad\qquad (\nu = 1, \cdots, n)\,.$$

From (1.7), (1.8), and (iv) of Theorem 2, we see that ψ_ν is a Grössen-character of $k_A^\times$, and $\psi_\nu(x) = 1/g(x)_{\infty\nu}$ for $x \in k_\infty^\times$. For details, see [5, § 7.8].

As we said in 0.2, we consider K as a subfield of C. Take the natural injection $K \to C$ so as to correspond to the first archimedean prime, and write $\psi_1(x)$ simply $\psi(x)$:

$$(1.11) \qquad\qquad \psi(x) = (\alpha(x)/g(x))_{\infty 1} \qquad\qquad (x \in k_A^\times)\,.$$

We call ψ *the Grössen-character of* $k_A^\times$ *determined by the isomorphism-class of* (A, θ) *over* k, or simply, *determined by* (A, θ). By (1.8) and Theorem 2, ψ satisfies the following two conditions:

(1.12) $\psi(y) = 1/g(y)_{\infty 1}$ *for all* $y \in k_\infty^\times$;

(1.13) *If* $x \in k_A^\times$ *and* $x_\infty = 1$, *then* $\psi(x) \in K^\times$, $\psi(x)\psi(x)^\rho = |x|_0$, *and* $\psi(x)\mathfrak{a} = g(x)\mathfrak{a}$.

The map α can be recovered from ψ by $\alpha(x) = \psi(x)$ for $x_\infty = 1$, and $\alpha(x) = 1$ for $x \in k_\infty^\times$.

Remark 1. If $n = 1$, then A is an elliptic curve, K is an imaginary quadratic field, and $K = K'$. In this case, we can disregard the polarization. Moreover, we can normalize θ so that Φ is the identity map of K to K. Then $g(x) = N_{k/K}(x)$. We may also assume that $\mathfrak{y}$ is the identity injection of K into $C^\times$, so that $\xi(u) = \omega(u) = u$ for every $u \in K$.

PROPOSITION 1. *Let (K, Φ), (A, θ), and k be as above, σ an automorphism of K, and γ an automorphism of $\bar{\mathbf{Q}}$. Put $\Phi^*(a) = \Phi(a^\sigma)^\gamma$ and $\theta^*(a) = \theta(a^\sigma)^\gamma$ for $a \in K$. Let α and g be the maps defined for (A, θ) as above, and let α^* and g^* be the corresponding maps for (A^γ, θ^*) defined with respect to k^γ. Then (A^γ, θ^*) is of type (K, Φ^*), $g^*(x) = g(x^{\gamma^{-1}})^{\sigma^{-1}}$, and $\alpha^*(x) = \alpha(x^{\gamma^{-1}})^{\sigma^{-1}}$.*

Proof. The first assertion is obvious. Extend σ to an automorphism of $\bar{\mathbf{Q}}$, and denote it again by σ. Let (K', Φ') be the reflex of (K, Φ). Put $\Phi^{*\prime}(a) = \Phi'(a^{\gamma^{-1}})^{\sigma^{-1}}$ for $a \in K'^\gamma$. Then it can easily be verified that $(K'^\gamma, \Phi^{*\prime})$ is the reflex of (K, Φ^*), hence $g^*(x) = g(x^{\gamma^{-1}})^{\sigma^{-1}}$. Take $\omega\colon K_R/\mathfrak{a} \to A$ and $\omega^*\colon K_R/\mathfrak{a}^* \to A^\gamma$ as in §1 with lattices $\mathfrak{a}$ and $\mathfrak{a}^*$ in K. Then the map

$$u \mapsto \omega^{*-1}\big(\omega(u^\sigma)^\gamma\big)$$

is an isomorphism of $K_R/\mathfrak{a}^{\sigma^{-1}}$ onto $K_R/\mathfrak{a}^*$ commuting with the action of elements of K. Therefore $\mathfrak{a}^* = t\mathfrak{a}^{\sigma^{-1}}$ and $\omega^{*-1}\big(\omega(u^\sigma)^\gamma\big) = tu$ with an element t of $K_A^\times$. Now, for every $x \in K_A^\times$, we have

$$\omega^*(tu)^{[x^\gamma, k^\gamma]} = \omega^*(u^\sigma)^{\gamma[x^\gamma, k^\gamma]} = \omega^*(u^\sigma)^{[x, k]^\gamma} = \omega\big(\alpha(x)g(x)^{-1}u^\sigma\big)^\gamma$$
$$= \omega^*\big(\alpha(x)^{\sigma^{-1}}\big(g(x)^{\sigma^{-1}}\big)^{-1}tu\big) \,,$$

so that $\alpha^*(x^\gamma) = \alpha(x)^{\sigma^{-1}}$, q.e.d.

2. The zeta-function of A over a field containing K'

We can define as usual an L-function $L(s, \psi_\nu)$ for each ψ_ν of (1.9) (see, e.g., [8]). Now the main theorem concerning the zeta-function of A, first proved by Taniyama [7] in a somewhat weaker form, can be stated as follows:

THEOREM 3. *Let (A, θ) be defined over an algebraic number field k of finite degree. Then the zeta function of A over k coincides exactly with the product $\prod_{\nu=1}^{n} L(s, \psi_\nu)L(s, \bar\psi_\nu)$, where the ψ_ν are defined by (1.9) and $\bar\psi_\nu$ is the complex conjugate of ψ_ν.*

The point that A has good reduction at a prime $\mathfrak{p}$ of k if and only if ψ_ν is unramified at $\mathfrak{p}$ is due to Serre and Tate [2]. For a simple proof of the theorem, see Shimura and Taniyama [6, § 18.3] (in a weaker form), and Shimura [5, § 7.8] (in the above form).

In [5, § 7.8], we have also proved

THEOREM 4. *Let H be a subfield of k containing K'. Then the following two conditions are equivalent.*

(i) There exists a Grössen-character φ of $H_A^\times$ such that $\psi = \varphi \circ N_{k/H}$.

(ii) All the points of A of finite order are rational over $H_{ab}k$. Moreover, if these conditions are satisfied, the number of Grössen-characters φ of $H_A^\times$ as in (i) is exactly $[H_{ab} \cap k : H]$.

We shall now prove that the isomorphism class of (A, θ) over k is completely determined by $(K, \Phi; \mathfrak{a})$ and α:

THEOREM 5. *Let (A, θ) and (A', θ') be two structures defined over a field k, which are isogenous over C, and determine the same Grössen-character of $k_A^\times$. Then every isogeny of (A, θ) to (A', θ') is rational over k.*

Proof. Let (A, θ), $\mathfrak{a}$, ω be as above, and let (A', θ') be of type $(K, \Phi; \mathfrak{a}')$. Define $\omega': K_R/\mathfrak{a}' \to A'$. Let λ be an isogeny of (A, θ) to (A', θ'). Then there exists an element c of K such that $c\mathfrak{a} \subset \mathfrak{a}'$ and $\lambda(\omega(v)) = \omega'(cv)$. If $\sigma \in \mathrm{Aut}(C/k)$ and $\sigma = [x, k]$ on k_{ab} with $x \in k_A^\times$, our assumption implies

$$\omega(v)^\sigma = \omega\big(\alpha(x)g(x)^{-1}v\big) \qquad\qquad (v \in K/\mathfrak{a}) ,$$
$$\omega'(v)^\sigma = \omega'\big(\alpha(x)g(x)^{-1}v'\big) \qquad\qquad (v' \in K/\mathfrak{a}')$$

with the same $\alpha(x)$. It follows that $\lambda^\sigma(\omega(v)^\sigma) = \lambda(\omega(v)^\sigma)$ for all $v \in K/\mathfrak{a}$, hence $\lambda = \lambda^\sigma$. This shows that λ is rational over k, q.e.d.

Let us now discuss whether there exists a structure $(A, \mathcal{C}, \theta)$ over a given field k with a given Grössen-character ψ of $k_A^\times$. We consider exclusively the structures of a fixed type $(K, \Phi; \mathfrak{a}, \varepsilon)$. Denote by Z the group of all roots of unity in K, and put

$$(2.1) \qquad\qquad Z(\mathfrak{a}) = \{\zeta \in Z \mid \zeta\mathfrak{a} \subset \mathfrak{a}\} .$$

If $(A, \mathcal{C}, \theta)$ is of type $(K, \Phi; \mathfrak{a}, \varepsilon)$, the elements $\theta(\zeta)$ with $\zeta \in Z(\mathfrak{a})$ form its group of all automorphisms.

Let k be a finite extension of K', and ψ a Grössen-character of $k_A^\times$. If ψ is determined by some (A, θ), it must satisfy (1.12) and (1.13). Therefore (1.12) and (1.13) are "necessary conditions" for ψ. Casselman has shown that they are actually sufficient:

THEOREM 6 (Casselman). *Let k be a finite extension of K', and ψ a Grössen-character of $k_A^\times$ satisfying (1.12) and (1.13). Then there exists a structure $(A, \mathcal{C}, \theta)$ of type $(K, \Phi; \mathfrak{a}, \varepsilon)$ rational over k that determines ψ.*

Before proving this, it may be worthwhile restating the result in the one-dimensional case. As noticed in Remark 1, we can take Φ to be the identity map. Then (1.12) and (1.13) become

$$(2.2) \qquad\qquad \psi(y) = N_{k/K}(y)^{-1} \qquad\qquad \text{for all } y \in k_\infty^\times ;$$

(2.3) *If $x \in k_A^\times$ and $x_\infty = 1$, then $\psi(x) \in K^\times$, and $\psi(x)\mathfrak{a} = N_{k/K}(x)\mathfrak{a}$.*

Note that the equality $\psi(x)\psi(x)^\rho = |x|_0$ follows from other two properties of $\psi(x)$ of (2.3). Therefore we obtain the following

COROLLARY. *Let K be an imaginary quadratic field, $\mathfrak{a}$ a lattice in K, j the invariant of an elliptic curve isomorphic to $C/\mathfrak{a}$, and k a finite extension* *of $K(j)$. Let ψ be a Grössen-character of $k_A^\times$ satisfying (2.2) and (2.3). Then there exists an elliptic curve defined over k that determines ψ.*

Proof of Theorem 6. First we can take a structure (A, C, θ) of type $(K, \Phi; \mathfrak{a}, \varepsilon)$ rational over $\bar{Q}$. For every $\sigma \in \mathrm{Gal}(\bar{Q}/k)$, take an element x of $k_A^\times$ so that $[x, k] = \sigma$ on k_{ab}. Define $\omega \colon K_R \to A$ as above for this A. By Theorem 1, there exists an exact sequence

$$0 \longrightarrow \mathfrak{y}\big(g(x)^{-1}\mathfrak{a}\big) \longrightarrow C^n \xrightarrow{\ \xi'\ } A^\sigma \longrightarrow 0$$

such that: (i) $(A^\sigma, C^\sigma, \theta^\sigma)$ is of type $\big(K, \Phi, g(x)^{-1}\mathfrak{a}, |x|_0 \cdot \varepsilon\big)$; (ii) $\omega(v)^\sigma = \omega'\big(g(x)^{-1}v\big)$ for all $v \in K/\mathfrak{a}$, where we put $\omega' = \xi' \circ \mathfrak{y}$. Now define $\alpha \colon k_A^\times \to K^\times$ so that $\alpha(x) = \psi(x)$ for $x_\infty = 1$ and $\alpha(x) = 1$ for $x \in k_\infty^\times$. By (1.13), $\alpha(x)g(x)^{-1}\mathfrak{a} = \mathfrak{a}$ and $\alpha(x)\alpha(x)^\rho = |x|_0$. Therefore $\alpha(x)$ gives rise to an isomorphism λ of A onto A^σ such that $\lambda\big(\omega(v)\big) = \omega'\big(\alpha(x)^{-1}v\big)$ for all $v \in K_R/\mathfrak{a}$. Since $\alpha(x)\alpha(x)^\rho = |x|_0$, we see that λ is an isomorphism of (A, C, θ) onto $(A^\sigma, C^\sigma, \theta^\sigma)$. Let c_σ denote the non-archimedean part of $\alpha(x)g(x)^{-1}$, and put $\lambda = \lambda_\sigma$. We see easily that c_σ is uniquely determined for $\sigma \in \mathrm{Gal}(\bar{Q}/k)$, independently of the choice of x. Since $\omega(v)^\sigma = \lambda_\sigma\big(\omega(c_\sigma v)\big)$ for all $v \in K/\mathfrak{a}$, λ_σ is also uniquely determined by σ. Moreover, we have $c_{\sigma\tau} = c_\sigma c_\tau$, so that $\lambda_{\sigma\tau} = \lambda_\sigma^\tau \circ \lambda_\tau$ for $\sigma, \tau \in \mathrm{Gal}(\bar{Q}/k)$. Therefore we find a structure (A_1, C_1, θ_1), rational over k, and an isomorphism μ of (A, C, θ) onto (A_1, C_1, θ_1), rational over $\bar{Q}$, such that $\mu = \mu^\sigma \circ \lambda_\sigma$ for all $\sigma \in \mathrm{Gal}(\bar{Q}/k)$. Put $\omega_1 = \mu \circ \omega$. Then

$$\omega_1(v)^\sigma = \mu^\sigma\big(\lambda_\sigma\big(\omega(c_\sigma v)\big)\big) = \omega_1\big(\alpha(x)g(x)^{-1}v\big)$$

for all $v \in K/\mathfrak{a}$, hence (A_1, θ_1) determines ψ, q.e.d.

Remark 2. If there exists a Grössen-character ψ of $k_A^\times$ satisfying (1.12) and (1.13), then k contains the field of moduli of any structure of type $(K, \Phi; \mathfrak{a}, \varepsilon)$. This is an immediate consequence either of the above theorem, or of [5, Cor. 5.16]. Therefore one can ask the following natural question:

For a finite algebraic extension k of the field of moduli, does there always exist a Grössen-character of $k_A^\times$ satisfying (1.12) and (1.13)?

By Theorems 2 and 6, the existence of such a Grössen-character is equivalent to the existence of a structure (A, C, θ) of a given type rational over k. Especially one can ask this existence problem with the field of moduli as k. This is trivial in the one-dimensional case, since any isomorphism class of elliptic curves has a model defined over its field of moduli. But in the higher dimensional case, the question seems difficult. In § 5, we shall give an affirmative answer under a certain condition on $\mathfrak{a}$. If k is sufficiently large, the existence of ψ or (A, C, θ) is rather trivial (see Remark 6 of § 5).

3. Fields of moduli and fields of definition

Let (S, Ω) be the reflex of (K', Φ'). Let F, F', T denote the maximal real subfields of K, K', S, respectively. Then $T = F \cap S$ and $K = FS$. For a given $(A, \mathcal{C}, \theta)$ as in §1, we denote by θ_X, for any subfield X of K, the restriction of θ to X, by M_X the field of moduli of $(A, \mathcal{C}, \theta_X)$, and by M the field of moduli of $(A, \mathcal{C})$, i.e., $M = M_Q$. The field M_X is the subfield of C which is characterized by the property that *an automorphism σ of C is the identity map on M_X if and only if $(A, \mathcal{C}, \theta_X)$ is isomorphic to $(A^\sigma, \mathcal{C}^\sigma, \theta_X^\sigma)$* (see [3, II, 1.4]). We have obviously $M \subset M_T \subset M_F \subset M_K$, and $M_T \subset M_S \subset M_K$.

PROPOSITION 2. *The notation being as above, one has* $M_T = F'M$, $M_S = K'M$, $M_K = K'M_F$.

Proof. Let $\tau \in \mathrm{Aut}(C/M)$. Then there is an isomorphism h of $(A, \mathcal{C})$ to $(A, \mathcal{C})^\tau$. Since $\theta(S)$ is the center of $\mathrm{End}_Q(A)$ by Lemma 2, we obtain an automorphism π of S by

$$(3.0) \qquad\qquad h^{-1}\theta(a)^\tau h = \theta(a^\pi) \qquad\qquad (a \in S) .$$

Obviously π depends only on τ, and not on the choice of h. Now we have

(3.1) *M contains the smallest subfield of K' over which K' is normal;*

(3.2) *K' is normal over $M \cap K'$;*

(3.3) *M_S is normal over M;*

(3.4) *$M_S = K'M = K'M_T$;*

(3.5) *π depends only on the restriction of τ to M_S;*

(3.6) *The map $\tau \mapsto \pi$ defines an isomorphism of $\mathrm{Gal}(M_S/M)$ into $\mathrm{Aut}(S)$, and also an isomorphism of $\mathrm{Gal}(M_S/M_T)$ into $\mathrm{Gal}(S/T)$.*

(These were proved in [5, Prop. 5.17 and its proof] under the assumption $K = S$.) To show these, observe that (3.0) implies $\mathrm{tr}\,\Phi(a)^\tau = \mathrm{tr}\,\Phi(a^\pi)$ for all $a \in S$, hence $K'^\tau = K'$, since (K', Φ') is the reflex of (S, Ω). Let K'' denote the fixed subfield of K' by $\mathrm{Aut}(K')$. Then every element of $\mathrm{Aut}(C/M)$ induces the identity map on K'', so that $K'' \subset M$. This proves (3.1) and (3.2). Let $\tau \in \mathrm{Aut}(C/M_S)$. Then we can take h so that π is the identity map. Then $\mathrm{tr}\,\Phi(a)^\tau = \mathrm{tr}\,\Phi(a)$ for all $a \in S$, hence τ is the identity map on K'. This proves $K' \subset M_S$. Conversely, if $\tau \in \mathrm{Aut}(C/K'M)$, then $\mathrm{tr}\,\Phi(a^\pi) = \mathrm{tr}\,\Phi(a)$ for all $a \in S$. By Lemmas 1 and 2, π must be the identity map, so that h is an isomorphism of $(A, \mathcal{C}, \theta_S)$ to $(A^\tau, \mathcal{C}^\tau, \theta_S^\tau)$. Therefore τ is the identity map on M_S, so that $M_S \subset K'M$. This proves (3.4), and hence (3.3), (3.5), and the first part of (3.6). The second part of (3.6) is obvious. Now, to show that $M_K = K'M_F$ and $M_T =$

$F'M$, let $\tau \in \mathrm{Aut}(C/K'M_F)$. Then the above h can be taken so that $h^{-1}\theta(a)^\tau h = \theta(a)$ for all $a \in F$. Moreover, since τ is the identity map on M_S, π is the identity map on S by (3.5), hence $\theta(a)^\tau h = h \cdot \theta(a)$ for all $a \in K$. Therefore h is an isomorphism of (A, C, θ) to $(A, C, \theta)^\tau$, so that τ is the identity map on M_K. This proves that $M_K \subset K'M_F$. Since $K' \subset M_S \subset M_K$ and $M_F \subset M_K$, we obtain $M_K = K'M_F$. Now (3.6) implies that $[K' : M_T \cap K'] = [M_S : M_T] \leqq 2$. We are going to prove:

(3.7) $F' \subset M_T$.

If $M_S = M_T$, this is obvious. If $M_S \neq M_T$, we have $\pi = \rho$ for some $\tau \in \mathrm{Gal}(M_S/M_T)$. Then $\mathrm{tr}\,\Phi(a)^\tau = \mathrm{tr}\,\Phi(a^\rho) = \mathrm{tr}\,\Phi(a)^\rho$ for all $a \in S$, so that $\tau = \rho$ on K'. Since τ is the identity map on M_T, this implies that ρ is the identity map on $M_T \cap K'$, so that $M_T \cap K' \subset F'$. Since $[K' : M_T \cap K'] = 2$, we obtain $M_T \cap K' = F'$, hence (3.7). In the next place, let $\tau \in \mathrm{Aut}(C/F'M)$, and consider h and π as above. If τ is the identity map on K', then π is the identity map on S. If $\tau = \rho$ on K', $\mathrm{tr}\,\Phi(a^{\rho\pi}) = \mathrm{tr}\,\Phi(a^\rho)^\tau = \mathrm{tr}\,\Phi(a)^{\rho\tau} = \mathrm{tr}\,\Phi(a)$ for $a \in S$. By Lemma 1, we have $\rho = \pi$ on S, so that π is the identity map on T. This means that π is the identity map for *every* $\tau \in \mathrm{Aut}(C/F'M)$. Then h gives an isomorphism of (A, C, θ_T) to $(A, C, \theta_T)^\tau$, so that τ is the identity map on M_T. It follows that $M_T \subset F'M$. This together with (3.7) completes the proof.

Let $\mathfrak{o} = \theta^{-1}[\mathrm{End}(A) \cap \theta(K)]$. From the above discussion, we see easily that

(3.8) $\mathfrak{o}^\rho = \mathfrak{o}$ *if* $K' \not\subset M_F$.

PROPOSITION 3. *The notation being as above, let k be a field of definition for (A, C). Then $F'k$ and $K'k$ are the smallest fields of definition, containing k, for (A, C, θ_T) and (A, C, θ_S), respectively. Moreover, if (A, C, θ_F) is rational over k, then $K'k$ is the smallest field of definition, containing k, for (A, C, θ).*

Proof. We have $M \subset k$, so that $K'k = M_S k$ and $F'k = M_T k$. If (A, C, θ_T) is defined over an extension k' of k, then $M_T \subset k'$, so that $M_T k \subset k'$. Conversely, let $\sigma \in \mathrm{Aut}(C/M_T k)$. Then we obtain an automorphism ψ of K by $\theta(a)^\sigma = \theta(a^\psi)$. Now the map $\tau \mapsto \pi$ in the proof of Proposition 2 gives an isomorphism of $\mathrm{Gal}(M_S/M_T)$ into $\mathrm{Aut}(S/T)$. This implies that if τ is the identity map on M_T, then π is the identity map on T. Taking σ to be τ, we see that ψ is the identity map on T, hence (A, C, θ_T) is invariant under σ. This shows that (A, C, θ_T) is rational over $M_T k$. This proves our assertion concerning (A, C, θ_T). The remaining part can be proved in a similar way.

PROPOSITION 4. *The fields of moduli of (A, C, θ) and (A, C, θ_F) are inde-*

pendent of the choice of the polarization $\mathcal{C}$ satisfying (1.1). *Moreover, a polarization of A satisfying* (1.1) *is rational over any field of definition for* (A, θ_F).

Proof. Let $\mathcal{C}$ and $\mathcal{C}'$ be polarizations of A satisfying (1.1), and X and X' be divisors in $\mathcal{C}$ and $\mathcal{C}'$, respectively. If φ_X and $\varphi_{X'}$ are homomorphisms of A into $\mathrm{Pic}(A)$ attached to X and X', we have $\varphi_{X'} = \varphi_X \circ \mu$ with an element μ of $\mathrm{End}_0(A)$. From (1.1), we obtain

$$\varphi_X^{-1} \cdot {}^t\theta(a) \cdot \varphi_X = \varphi_{X'}^{-1} \cdot {}^t\theta(a) \cdot \varphi_{X'} = \theta(a^\rho)$$

for every $a \in K$. It follows that μ commutes with every element of $\theta(K)$, hence $\mu = \theta(b)$ with $b \in K$. It can easily be seen that $\theta(b)$ is invariant under the positive involution of $\theta(K)$, so that $b \in F$. Let σ be an automorphism of C such that there is an isomorphism λ of $(A, \mathcal{C}, \theta_F)$ onto $(A, \mathcal{C}, \theta_F)^\sigma$. Then $\lambda^{-1}(X^\sigma)$ is algebraically equivalent to X, so that ${}^t\lambda \circ \varphi_{X^\sigma} \circ \lambda = \varphi_X$. Then

$${}^t\lambda \circ \varphi_{X'^\sigma} \circ \lambda = {}^t\lambda \circ \varphi_{X^\sigma} \circ \theta(b)^\sigma \circ \lambda = {}^t\lambda \circ \varphi_{X^\sigma} \circ \lambda \circ \theta(b) = \varphi_X \circ \theta(b) = \varphi_{X'},$$

so that $\lambda^{-1}(X'^\sigma)$ is algebraically equivalent to X'. Therefore λ is an isomorphism of $(A, \mathcal{C}', \theta_F)$ to $(A, \mathcal{C}', \theta_F)^\sigma$. This means that the field of moduli of $(A, \mathcal{C}, \theta_F)$ contains that of $(A, \mathcal{C}', \theta_F)$. Since the converse is also true, we obtain the first assertion, because of the equality $M_K = M_F K'$. Now suppose that (A, θ_F) is rational over a field k. We may assume that X and X' are rational over the algebraic closure k^* of k. Let $\tau \in \mathrm{Aut}(k^*/k)$. Then $\theta(a)^\tau = \theta(a)$ for $a \in F$, and $\theta(a)^\tau = \theta(a^\pi)$ for $a \in S$ with $\pi \in \mathrm{Gal}(S/T)$. Therefore $\theta(a)^\tau = \theta(a^\psi)$ for $a \in K$ with $\psi \in \mathrm{Gal}(K/F)$. Apply τ to the equality ${}^t\theta(a) \cdot \varphi_X = \varphi_X \cdot \theta(a^\rho)$. Then we find ${}^t\theta(a) \cdot \varphi_{X^\tau} = \varphi_{X^\tau} \cdot \theta(a^\rho)$. Let U be the sum of all distinct X^τ with $\tau \in \mathrm{Gal}(k^*/k)$. Then ${}^t\theta(a) \cdot \varphi_U = \varphi_U \cdot \theta(a^\rho)$, so that the polarization determined by U satisfies (1.1) and is rational over k. To prove that $\mathcal{C}'$ is rational over k, take U as X in the relation $\varphi_{X'} = \varphi_X \circ \theta(b)$ obtained above. Let $\sigma \in \mathrm{Gal}(k^*/k)$, and apply σ to the last equality. Then we find $\varphi_{X'^\sigma} = \varphi_{X'}$, so that X'^σ is algebraically equivalent to X'. Let Y be the sum of all distinct X'^σ with $\sigma \in \mathrm{Gal}(k^*/k)$. Then Y is rational over k, and belongs to $\mathcal{C}'$, q.e.d.

Remark 3. The field of moduli of $(A, \mathcal{C})$ does depend on $\mathcal{C}$ even if A is simple. For example, let A be the jacobian of the curve $y^2 = 1 - x^p$ with an odd prime p, and $K = \mathbf{Q}(\zeta)$, $F = \mathbf{Q}(\zeta + \zeta^{-1})$ with $\zeta = e^{2\pi i/p}$. Define $\theta: K \to \mathrm{End}_0(A)$ so that $\theta(\zeta)$ corresponds to the map $(x, y) \mapsto (\zeta x, y)$ on the curve. Then A is simple, defined over $\mathbf{Q}$, and has a canonical divisor X rational over $\mathbf{Q}$. Moreover, $K' = K$, and $\theta(a)^\sigma = \theta(a^\sigma)$ for every $\sigma \in \mathrm{Gal}(K/\mathbf{Q})$. Now we can find a totally positive element s that generates a prime ideal of F of degree 1, prime to p. Then there is a positive non-degenerate divisor Y such that $\varphi_Y = \varphi_X \circ \theta(s)$. Since $\operatorname{Pic}(A) = A$ and φ_X is the identity map, we have $\varphi_Y = \theta(s)$. If $\mathcal{C}'$ is the polarization containing Y, the field of moduli of $(A, \mathcal{C}')$ is F. In fact, let λ be an isomorphism of $(A, \mathcal{C}')$ to $(A, \mathcal{C}')^\sigma$ for some $\sigma \in \operatorname{Gal}(K/\mathbf{Q})$. Then $\lambda = \theta(e)$ with a unit e in K, and $\lambda^{-1}(Y^\sigma)$ is algebraically equivalent to Y, so that ${}^t\lambda \circ \varphi_{Y^\sigma} \circ \lambda = \varphi_Y$. Therefore $\theta(e^\sigma) \cdot \theta(s^\sigma) \cdot \theta(e) = \theta(s)$, hence $e^\sigma s^\sigma e = s$. This is possible if and only if σ is the identity map on F. Thus the field of moduli of $(A, \mathcal{C}')$ is F.

Remark 4. Suppose that $n = 2$ and K is not normal over $\mathbf{Q}$. Then $[K':\mathbf{Q}] = 4$, and K' is not normal over $\mathbf{Q}$ (see [6, § 8.4]). Therefore, it follows from (3.1) that F' is contained in the field of moduli of $(A, \mathcal{C})$, hence $M_F = M$.

4. The zeta-function of A relative to F over a field not containing K'

Let $\mathfrak{o}_F$ denote the ring of all algebraic integers in F. We now impose the following condition on $(A, \mathcal{C}, \theta)$:

(4.1) $\quad \theta(\mathfrak{o}_F) \subset \operatorname{End}(A)$.

If $(A, \mathcal{C}, \theta)$ is of type $(K, \Phi; \mathfrak{a}, \varepsilon)$, this is equivalent to

(4.2) $\quad \mathfrak{o}_F \mathfrak{a} \subset \mathfrak{a}$.

Under this assumption, we see that, for every prime ideal $\mathfrak{l}$ in F, the $\mathfrak{o}_F$-module

$$(4.3) \qquad \qquad \bigcup_{m=1}^{\infty} \{t \in A \mid \theta(\mathfrak{l}^m)t = 0\}$$

is isomorphic to the $\mathfrak{l}$-part of $(F/\mathfrak{o}_F)^2$. If $(A, \mathcal{C}, \theta_F)$ is rational over an algebraic number field k_1 of finite degree, we can define an "$\mathfrak{l}$-adic representation" of the commutor of $\theta(\mathfrak{o}_F)$ in $\operatorname{End}(A)$ on this module, and a similar representation on the variety obtained from A by reduction modulo a prime $\mathfrak{p}$ of k_1, whenever A has good reduction modulo $\mathfrak{p}$, and $\mathfrak{p}$ is prime to $N(\mathfrak{l})$. Then we define *the zeta-function of A over k_1 relative to F*, denoted by $\zeta(s; A/k_1, F)$, by taking the $\mathfrak{l}$-adic representation of the Frobenius map in place of $\varphi_\mathfrak{p}$ in $(*)$ of 0.5. For details, the reader is referred to [5, § 7.6].

If $K' \subset k_1$, then $(A, \mathcal{C}, \theta)$ is rational over k_1 by Proposition 3. In [5, Th. 7.43], we have shown that

$$(4.4) \qquad \qquad \zeta(s; A/k_1, F) = L(s, \psi)L(s, \bar{\psi})$$

with the Grössen-character ψ determined by (A, θ). The case of a basic field not containing K' is more interesting; we obtain then

THEOREM 7. *Let k_1 be a field of definition for (A, θ_F). Suppose that*

$\theta(\mathfrak{o}_F) \subset \mathrm{End}(A)$, *and k_1 does not contain K'. Let ψ be the Grössen-character of $(k_1K')_A^\times$ determined by (A, θ). Then $\zeta(s; A/k_1, F)$ coincides exactly with $L(s, \psi)$.*

In the case where A is an elliptic curve, this was proved by Deuring [1, IV].

By virtue of Proposition 3, (A, θ) is rational over k_1K', and the assumption $K' \not\subset k_1$ implies that $F' = k_1 \cap K'$. Put $k = k_1K'$. In [5, Th. 7.46, Prop. 7.47], under the assumption that A is simple, we have proved a slightly weaker result to the effect that $\zeta(s; A/k_1, F)$ coincides with $L(s, \psi)$ up to finitely many possible bad factors, which may occur only for the prime ideals in k ramified over k_1. The assumption that A is simple was necessary only to show the following property:

(4.5) $\theta(a)^\gamma = \theta(a^\rho)$ *for all $a \in K$, if γ is the generator of $\mathrm{Gal}(k/k_1)$.*

Actually this is true even if A is not simple. In fact, $\theta(a)^\gamma = \theta(a^\pi)$ for all $a \in S$ with an element π of $\mathrm{Gal}(S/T)$. Since $\gamma = \rho$ on K', we have $\pi = \rho$ on S. On the other hand $\theta(a)^\gamma = \theta(a)$ for all $a \in F$. Since $K = FS$, we obtain (4.5). Then we can repeat the discussion of [5, Th. 7.46, Prop. 7.47] without assuming that A is simple. Here we do not give a detailed proof, since we shall prove in § 6 a stronger result than the above theorem, by taking a more general type of subfield of K in place of F. The study of the behavior of A modulo the primes of k ramified over k_1 will also be done there in such a general case.

Let us now ask how many isomorphism-classes of $(A, \mathcal{C}, \theta_F)$, or isogeny-classes of (A, θ_F), over k_1 can be determined by a given ψ.

THEOREM 8. *Let (A, θ) and (A', θ') be two structures of the same type (K, Φ) which determine the same Grössen-character of $(k_1K')_A^\times$, where k_1 is a field of definition for (A, θ_F) and (A', θ'_F) not containing K'. Then there is an isogeny of (A, θ) onto (A', θ') rational over k_1.*

Proof. Put $k = k_1K'$. Let λ be an isogeny of (A, θ) to (A', θ'). By Theorem 5, λ is rational over k. Let γ be the generator of $\mathrm{Gal}(k/k_1)$. By (4.5), we see that λ^γ is also an isogeny of (A, θ) to (A', θ'), hence $\lambda^\gamma = \lambda \circ \theta(b)$ with $b \in K$. Applying γ to the equality, we find $b^\rho b = 1$, so that $b = c/c^\rho$ for some $c \in K$. Take a positive integer m so that $\theta(mc) \in \mathrm{End}(A)$, and put $\mu = \lambda \circ \theta(mc)$. Then μ is rational over k, and $\mu^\gamma = \mu$, hence μ is rational over k_1.

In the next place, we fix $(K, \Phi; \mathfrak{a}, \varepsilon)$ under the condition (4.2), and a finite extension k_1 of F' such that $k_1 \cap K' = F'$. Put $k = k_1K'$. Let Σ be the set of all $(A, \mathcal{C}, \theta_F)$ rational over k_1 such that $(A, \mathcal{C}, \theta)$ is-of type $(K, \Phi;$

$\mathfrak{a}, \varepsilon$) and determines ψ over k, where ψ is a fixed Grössen-character of $k_A^\times$.

THEOREM 9. *If Σ is not empty, Σ consists of exactly two isomorphism-classes over k_1.*

Proof. Fix a member $(A, \mathcal{C}, \theta_F)$ of Σ. For an arbitrary member $(A', \mathcal{C}', \theta'_F)$ of Σ, take an isomorphism λ of $(A, \mathcal{C}, \theta)$ onto $(A', \mathcal{C}', \theta')$. By Theorem 5, λ is rational over k. Let γ be the generator of $\mathrm{Gal}(k/k_1)$. Then $\lambda^\gamma = \lambda \circ \theta(\zeta)$ with $\zeta \in Z(\mathfrak{a})$. For every $a \in K$, we have

$$\left(\lambda \circ \theta(a)\right)^\gamma = \left(\lambda \circ \theta(a)\right) \circ \theta(a^{-1}a^\rho \zeta) \ ,$$

on account of (4.5). Put

$$Z^2(\mathfrak{a}) = \{\zeta^2 \,|\, \zeta \in Z(\mathfrak{a})\} \ .$$

Then $[Z(\mathfrak{a}) : Z^2(\mathfrak{a})] = 2$. We see easily that ζ modulo $Z^2(\mathfrak{a})$ depends only on $(A', \mathcal{C}', \theta'_F)$, and not on the choice of λ. By virtue of the well-known principle of Galois cohomology, we see that the map

$$(A', \mathcal{C}', \theta'_F) \mapsto \zeta \bmod Z^2(\mathfrak{a})$$

gives a one-to-one correspondence between the isomorphism-classes in Σ over k_1 and $Z(\mathfrak{a})/Z^2(\mathfrak{a})$, hence our assertion.

Let the notation and the assumptions be as in Theorem 7. By Proposition 1, we have

(4.6) $\psi(x^\gamma) = \psi(x)^\rho$ *for all* $x \in (k_1 K')_A^\times$, *if γ is the generator of* $\mathrm{Gal}(k_1 K'/k_1)$,

which forms a necessary condition for the existence of $(A, \mathcal{C}, \theta_F)$ rational over k_1. Actually it is almost sufficient:

THEOREM 10. *Let k_1 be a finite algebraic extension of F' not containing K', and $k = k_1 K'$. Let ψ be a Grössen-character of $k_A^\times$ satisfying (1.12), (1.13), and (4.6). Suppose that*

(i) *$\mathfrak{o}_F \mathfrak{a} \subset \mathfrak{a}$;*

(ii) *k_1 contains the field of moduli of a structure $(A', \mathcal{C}', \theta'_F)$ such that $(A', \mathcal{C}', \theta')$ is of type $(K, \Phi; \mathfrak{a}, \varepsilon)$;*

(iii) *k_1 has at least one real archimedean prime.*
Then there exists a structure $(A, \mathcal{C}, \theta_F)$ rational over k_1 such that $(A, \mathcal{C}, \theta)$ is of type $(K, \Phi; \mathfrak{a}, \varepsilon)$ and determines ψ.

In the one-dimensional case, we obtain

COROLLARY. *Let K be an imaginary quadratic field, $\mathfrak{a}$ a lattice in K, j the invariant of an elliptic curve isomorphic to $C/\mathfrak{a}$, k_1 a finite algebraic extension of $Q(j)$ with at least one real archimedean prime, and let $k = k_1 K$. Let ψ be a Grössen-character of $k_A^\times$, satisfying (2.2), (2.3), and (4.6). Then there exists an elliptic curve rational over k_1 that determines ψ over k.*

It is well-known that $Q(j)$ has at least one real archimedean prime (see, for example, [5, pp. 123-124]). Therefore one can take $Q(j)$ as k_1 in the above corollary.

To prove Theorem 10, we need

LEMMA 3. *Let (A, C, θ) be of type $(K, \Phi; \mathfrak{a}, \varepsilon)$. Define*

$$\theta^*: K \to \mathrm{End}_Q(A^\rho) \ by \ \theta^*(a) = \theta(a^\rho)^\rho \ .$$

Then $(A^\rho, C^\rho, \theta^*)$ is of type $(K, \Phi; \mathfrak{a}^\rho, \varepsilon)$.

Proof. We can take Φ and the map $\mathfrak{y}$ of §1 in the form

$$\Phi(a) = \begin{bmatrix} a^{\varphi_1} & & \\ & \ddots & \\ & & a^{\varphi_n} \end{bmatrix}, \qquad \mathfrak{y}(a) = \begin{bmatrix} a^{\varphi_1} \\ \vdots \\ a^{\varphi_n} \end{bmatrix} \qquad (a \in K)$$

with n isomorphisms φ_i of K into C (see [6, § 6.1]). Then $\Phi(a)^\rho = \Phi(a^\rho)$, $\mathfrak{y}(a)^\rho = \mathfrak{y}(a^\rho)$. Consider an exact sequence

$$0 \longrightarrow \mathfrak{y}(\mathfrak{a}) \longrightarrow C^n \overset{\xi}{\longrightarrow} A \longrightarrow 0 \ .$$

Define a map ξ' of C^n into A^ρ by $\xi'(u) = \xi(\bar{u})^\rho$ for $u \in C^n$, where $\bar{u} = (\bar{u}_1, \cdots, \bar{u}_n)$ if $u = (u_1, \cdots, u_n)$ with $u_i \in C$. Clearly ξ' is holomorphic and gives an isomorphism of $C^n/\mathfrak{y}(\mathfrak{a}^\rho)$ onto A^ρ. Then (A^ρ, θ^*) is of type $(K, \Phi; \mathfrak{a}^\rho)$ with respect to ξ' and $\mathfrak{y}$. Put $\omega = \xi \circ \mathfrak{y}$, $\omega' = \xi' \circ \mathfrak{y}$. Then $\omega'(\bar{u}) = \omega(u)^\rho$. Let Y be a basic polar divisor in C, and P the Riemann form of Y. If we define the symbol $e_{Y,N}(s, t)$ as in Weil [10, § XI], we have, by [6, p. 25, (7)].

$$e_{Y,N}\big(\xi(u), \xi(v)\big) = \exp[-2\pi i \cdot N \cdot P(u, v)]$$

for $u, v \in N^{-1}\mathfrak{y}(\mathfrak{a})$, where N is any positive integer. For a given $\mathfrak{y}$, P is uniquely determined by this relation. Applying ρ to this, we have

$$e_{Y^\rho,N}\big(\xi'(\bar{u}), \xi'(\bar{v})\big) = \exp[2\pi i \cdot N \cdot P(u, v)] \ .$$

Therefore, if we put $P'(u, v) = -P(\bar{u}, \bar{v})$, then P' is the Riemann form of Y^ρ. By (1.2), we have

$$P\big(\mathfrak{y}(a), \mathfrak{y}(b)\big) = -P\big(\mathfrak{y}(a^\rho), \mathfrak{y}(b^\rho)\big) = -\mathrm{Tr}_{K/Q}(\varepsilon a^\rho b) = \mathrm{Tr}_{K/Q}(\varepsilon a b^\rho) \ ,$$

since $\varepsilon^\rho = -\varepsilon$, so that Y^ρ corresponds to ε. This completes the proof.

Proof of Theorem 10. By Theorem 6, there exists a structure $\mathscr{P} = (A, C, \theta)$ rational over k that determines ψ. Now put $\mathscr{P}^* = (A^\gamma, C^\gamma, \theta^*)$, where γ is the generator of $\mathrm{Gal}(k/k_1)$, and $\theta^*(a) = \theta(a^\rho)^\gamma$. By (4.6) and Proposition 1, we see that $\mathscr{P}$ and $\mathscr{P}^*$ determine the same Grössen-character, so that every isogeny of (A, θ) to (A^γ, θ^*) is rational over k. By the assumption (iii), there exists an isomorphism δ of $\bar{Q}$ into C such that $k_1^\delta \subset R$. Then every isogeny of

$\mathscr{P}^{\delta}$ to $\mathscr{P}^{*\delta}$ is rational over k^{δ}. Since k_1 contains the field of moduli of $(A, \mathcal{C}, \theta_F)$, there is an isomorphism μ of $(A, \mathcal{C}, \theta_F)$ onto $(A^\tau, \mathcal{C}^\tau, \theta_F^\tau)$. Then it can easily be seen that μ is an isomorphism of $\mathscr{P}$ onto $\mathscr{P}^*$. Therefore $\mathscr{P}^\delta$ and $\mathscr{P}^{*\delta}$ are isomorphic. Let $\mathscr{P}^\delta$ be of type $(K, \Phi; \mathfrak{b}, \eta)$. Since $\gamma\delta = \delta\rho$ on k, we have $(A, \mathcal{C})^{\tau\delta} = (A, \mathcal{C})^{\delta\rho}$, and $\theta^*(a)^\delta = \theta(a^\rho)^{\delta\rho}$. By Lemma 3, $\mathscr{P}^{*\delta}$ is of type $(K, \Phi; \mathfrak{b}^\rho, \eta)$. Since $\mathscr{P}^\delta$ and $\mathscr{P}^{*\delta}$ are isomorphic, there exists an element α of K such that $\mathfrak{b}^\rho = \alpha\mathfrak{b}$ and $\alpha\alpha^\rho = 1$. Take an element β of K so that $\alpha = \beta/\beta^\rho$. Then $(\beta\mathfrak{b})^\rho = \beta\mathfrak{b}$. Replacing $\mathfrak{b}$ and η by $\beta\mathfrak{b}$ and $(\beta\beta^\rho)^{-1}\eta$, we can assume that $\mathfrak{b}^\rho = \mathfrak{b}$. Define ξ and ξ' as in Lemma 3 with A^δ and $\mathfrak{b}$ in place of A and $\mathfrak{a}$. Then we find an isomorphism λ of $\mathscr{P}^\delta$ onto $\mathscr{P}^{*\delta}$ such that $\xi' = \lambda \circ \xi$. Now we have

$$\xi(u) = \xi'(\bar{u})^\rho = \lambda^\rho\big(\xi(\bar{u})^\rho\big) = \lambda^\rho\big(\lambda\big(\xi(u)\big)\big)\,,$$

so that $\lambda^\rho\lambda = 1$. Put $\lambda_1 = \lambda^{\delta^{-1}}$. Then λ_1 is an isomorphism of $\mathscr{P}$ onto $\mathscr{P}^*$ rational over k such that $\lambda_1^\tau\lambda_1 = 1$. Therefore we can find a structure $(A', \mathcal{C}', \theta_F')$ rational over k_1 and isomorphic to $(A, \mathcal{C}, \theta_F)$ over k, q.e.d.

Remark 5. Theorem 10, without the condition (iii), is false, even in the one-dimensional case. To get a counter-example, let $K = Q(\sqrt{-7})$, $k_1 = Q(\sqrt{-2})$, $k = k_1K$, and let $\mathfrak{a}$ be any fractional ideal in K. Since the class number of K is one, the invariant j of $C/\mathfrak{a}$ belongs to Q, so that there is an elliptic curve E', rational over Q, with the invariant j. By class field theory, we find a unique cyclic extension k^* of k_1 of degree 4, and of conductor (7); k^* contains k. Let γ be the generator of $\mathrm{Gal}(k^*/k_1)$, and $\sigma = \gamma^2$. Then we can find an elliptic curve E rational over k, and an isomorphism λ of E to E' rational over k^*, such that $\lambda^\sigma = -\lambda$. Let ψ and ψ' be the Grössen-characters of $k_A^\times$, determined by E and E', respectively. Then we see easily that $\psi = \psi'\chi$ with a character χ of $k_A^\times$ of order 2, corresponding to the quadratic extension k^* of k. Now ψ' satisfies (4.6). It can easily be seen that $\chi(x^\tau) = \chi(x)^\rho$, hence ψ satisfies (4.6). Assume that there exists an elliptic curve E'' with j as its invariant, rational over k_1, which determines ψ over k. Then, by Theorem 5, there is an isomorphism μ of E'' to E rational over k. We see that $\mu \circ \lambda$ is rational over k^*, and $(\mu \circ \lambda)^\tau = \pm\mu \circ \lambda$, since $\mathrm{Aut}(E) = \{\pm 1\}$. Therefore we have $\mu \circ \lambda = (\mu \circ \lambda)^\sigma = \mu^\sigma \circ \lambda^\sigma = -\mu \circ \lambda$, which is a contradiction.

With regard to condition (iii) of Theorem 10 in the higher-dimensional case, we have

PROPOSITION 5. *If $\mathfrak{a}^\rho = \mathfrak{a}$, there exists a structure $(A, \mathcal{C}, \theta_F)$ of type $(K, \Phi; \mathfrak{a}, \varepsilon)$ such that $(A, \mathcal{C}, \theta_F)$ is rational over an algebraic number field contained in R.*

Proof. Take any $(A, \mathcal{C}, \theta)$ of a given type defined over an algebraic number field k. Taking kk^ρ instead of k if necessary, we may assume that $k^\rho = k$. By Lemma 3, $(A^\rho, \mathcal{C}^\rho, \theta^*)$, with $\theta^*(a) = \theta(a^\rho)^\rho$, is isomorphic to $(A, \mathcal{C}, \theta)$. With the same notation as in the proof of Lemma 3, we can define an isomorphism λ of $(A, \mathcal{C}, \theta)$ to $(A^\rho, \mathcal{C}^\rho, \theta^*)$ by $\lambda \circ \xi = \xi'$. Replacing k by a larger field if necessary, we may assume that λ is rational over k. As in the proof of Theorem 10, we have $\lambda^\rho \lambda = 1$, so that we obtain a desired structure over the maximal real subfield of k.

PROPOSITION 6. *Suppose that* $(f(x)\mathfrak{a})^\rho = f(x)\mathfrak{a}$ *for some* $x \in K_A'^\times$. *Then there exists a structure* $(A, \mathcal{C}, \theta)$ *of type* $(K, \Phi; \mathfrak{a}, \varepsilon)$ *such that* $(A, \mathcal{C}, \theta_F)$ *is rational over an algebraic number field of finite degree with at least one real archimedean prime.*

This follows immediately from Theorem 1 and Proposition 5.

5. Construction of $(A, \mathcal{C}, \theta)$ over the field of moduli

Fix $(K, \Phi; \mathfrak{a}, \varepsilon)$, and let L (instead of M_K) denote the field of moduli of any $(A, \mathcal{C}, \theta)$ of type $(K, \Phi; \mathfrak{a}, \varepsilon)$, and M_F the field of moduli of $(A, \mathcal{C}, \theta_F)$. In this section, we shall, under a certain condition on K and $\mathfrak{a}$, construct $(A, \mathcal{C}, \theta)$ rational over L, and also $(A, \mathcal{C}, \theta_F)$ over M_F, which has the following property:

(5.1) *Every point of A of finite order is rational over K_{ab}'.*

By Theorem 4, if ψ is the Grössen-character of $L_A^\times$ determined by such a structure, we have $\psi = \varphi \circ N_{L/K'}$, with a Grössen-character φ of $K_A'^\times$. Therefore we start our discussion by defining a Grössen-character of $K_A'^\times$ of a certain special type.

Let U denote the maximal compact subgroup of the non-archimedean part of $K_A^\times$, and let

$$U(\mathfrak{a}) = \{x \in U \mid x\mathfrak{a} = \mathfrak{a}\} \, .$$

For every element x of $K_A^\times$, denote by x_0 the non-archimedean part of x. Define $Z(\mathfrak{a})$ by (2.1), and put

$$Z(\mathfrak{a})_0 = \{\zeta_0 \mid \zeta \in Z(\mathfrak{a})\} \, .$$

Now we consider the following conditions on $\mathfrak{a}$:

(5.2) $U(\mathfrak{a})$ *has an open subgroup W such that $U(\mathfrak{a}) = Z(\mathfrak{a})_0 \times W$.*

(5.2') *There exists a continuous homomorphism h of $U(\mathfrak{a})$ onto $Z(\mathfrak{a})$ such that $h(\zeta_0) = \zeta$ for every $\zeta \in Z(\mathfrak{a})$.*

These two conditions are equivalent. In fact, the subgroup W and the map h are in one-to-one correspondence by the relationship

(5.3) $W = \mathrm{Ker}(h)$.

Now put

$$(5.4) \qquad \mathfrak{o} = \{x \in K \mid x\mathfrak{a} \subset \mathfrak{a}\} \ .$$

If $\mathfrak{o}_F\mathfrak{a} \subset \mathfrak{a}$, then $\mathfrak{o}_F \subset \mathfrak{o}$, so that $\mathfrak{o}^\rho = \mathfrak{o}$, and $U(\mathfrak{a})^\rho = U(\mathfrak{a})$. Under the assumption $\mathfrak{o}_F\mathfrak{a} \subset \mathfrak{a}$, if W and h satisfy (5.2′) and (5.3), the following two additional conditions are equivalent:

(5.5) $\ W^\rho = W.$

(5.5′) $\ h(x^\rho) = h(x)^\rho$ *for all* $x \in U(\mathfrak{a})$.

We note that L is a finite abelian extension of K', by virtue of [6, § 15.3, Main Th. 1], or [5, § 5.5]. Let P be the subgroup of $K_A'^\times$ corresponding to L by class field theory. Then $P = K'^\times \cdot N_{L/K'}(L_A^\times)$. Moreover, by [5, Cor. 5.16], P consists of all the elements z of $K_A'^\times$ such that

$$(5.6) \qquad \beta\mathfrak{a} = f(z)\mathfrak{a} \ , \qquad \beta\beta^\rho = |z|_0$$

for some $\beta \in K^\times$, where f is defined by (1.3). Then we see that $(f(z)/\beta)_0 \in U(\mathfrak{a})$. With any h as in (5.2′), define a homomorphism φ_h of P into $C^\times$ by

$$(5.7) \qquad \varphi_h(z) = h\big((f(z)/\beta)_0\big) \cdot (\beta/f(z))_{\infty 1} \qquad\qquad (z \in P) \ .$$

To show that this does not depend on the choice of β, let $\gamma\mathfrak{a} = f(z)\mathfrak{a}$, $\gamma\gamma^\rho = |z|_0$ with $\gamma \in K^\times$. Then we have $\beta\beta^\rho = \gamma\gamma^\rho$ and $\beta^{-1}\gamma\mathfrak{a} = \mathfrak{a}$, so that $\beta^{-1}\gamma \in Z(\mathfrak{a})$. Therefore

$$
\begin{aligned}
h\big((f(z)/\gamma)_0\big) \cdot (\gamma/f(z))_{\infty 1} &= h\big((f(z)/\beta)_0(\beta/\gamma)_0\big) \cdot (\gamma/f(z))_{\infty 1} \\
&= h\big((f(z)/\beta)_0\big) \cdot (\beta/\gamma) \cdot (\gamma/f(z))_{\infty 1} \\
&= h\big((f(z)/\beta)_0\big) \cdot (\beta/f(z))_{\infty 1} \ , \qquad\qquad \text{q.e.d.}
\end{aligned}
$$

If $z \in K'^\times$, we can take $f(z)$ to be β, so that $\varphi_h(z) = 1$. Moreover, if $z \in K_\infty'^\times$, we can take $\beta = 1$, hence

(5.8) $\ \varphi_h(z) = f(z)_{\infty 1}^{-1}$ *for* $z \in K_\infty'^\times.$

Further, if $z_\infty = 1$ and $f(z) \in W$, we can again take $\beta = 1$, hence $\varphi_h(z) = 1$. Thus φ_h is continuous on P, and trivial on $K'^\times$. Now we can extend φ_h to a Grössen-character of $K_A'^\times$. The number of Grössen-characters of $K_A'^\times$ which coincide with φ_h on P is exactly

$$[K_A'^\times : P] = [L : K'] \ .$$

THEOREM 11. *Suppose that condition* (5.2′) *is satisfied. Define* φ_h *as above with any* h *of* (5.2′). *Then there exists a structure* (A, C, θ) *of type* $(K, \Phi; \mathfrak{a}, \varepsilon)$ *such that*

    (i) (A, C, θ) *is rational over* L;

    (ii) $\varphi_h \circ N_{L/K'}$ *is the Grössen-character of* $L_A^\times$ *determined by* (A, C, θ);

    (iii) A *satisfies* (5.1).

Moreover, if $\mathfrak{o}_F\mathfrak{a} \subset \mathfrak{a}$ and (5.5') is satisfied, then the structure can be chosen so that

(iv) $(A, \mathcal{C}, \theta_F)$ *is rational over* M_F.

Actually (iii) follows immediately from (ii), by virtue of Theorem 4.

Proof. Put $\psi = \varphi \circ N_{L/K'}$. Then we see easily that ψ satisfies (1.12) and (1.13). Therefore the existence of $(A, \mathcal{C}, \theta)$ over L follows immediately from Theorem 6. The structure $(A, \mathcal{C}, \theta_F)$ over M_F is obtained by Theorem 10, if M_F has at least one real archimedean prime. Since we do not know whether or not M_F has this property, we make use of a different method. We may assume that $M_F \neq L$, since the question is trivial otherwise. Take $(A, \mathcal{C}, \theta)$ of type $(K, \Phi; \mathfrak{a}, \varepsilon)$ over L that determines ψ, under the conditions (5.5') and $\mathfrak{o}_F\mathfrak{a} \subset \mathfrak{a}$. It is well-known that the latter condition implies that $\mathfrak{a} = s\mathfrak{o}$ with an element s of $K_A^\times$, where $\mathfrak{o}$ is as in (5.4) (cf. for example [5, Prop. 4.11, Ex. 4.13, (5.4.2)]). Put $r = s/s^\rho$. Then $rr^\rho = 1$, and $r\mathfrak{a}^\rho = \mathfrak{a}$. Let τ be an element of $\mathrm{Gal}(\bar{Q}/M_F)$ which is not trivial on L. Then there exists an isomorphism λ of $(A, \mathcal{C}, \theta_F)$ to $(A, \mathcal{C}, \theta_F)^\tau$. The map

$$v \mapsto \omega^{-1}\big(\lambda^{-1}(\omega(rv^\rho)^\tau)\big)$$

is an $\mathfrak{o}$-linear automorphism of $K/\mathfrak{a}$, so that

$$\omega^{-1}\big(\lambda^{-1}(\omega(rv^\rho)^\tau)\big) = ev$$

with $e \in U(\mathfrak{a})$. On account of (5.2), replacing λ by $\lambda \cdot \theta(\zeta)$ with a suitable $\zeta \in Z(\mathfrak{a})$, we may assume that $e \in W$. Thus we have

$$\omega(rv^\rho)^\tau = \lambda\big(\omega(ev)\big) \qquad\qquad (v \in K/\mathfrak{a}) \ .$$

From (5.5'), we obtain $\psi(x^\tau) = \psi(x)^\rho$. Now λ is an isomorphism of $(A, \mathcal{C}, \theta)$ to $(A^\tau, \mathcal{C}^\tau, \theta^*)$. By Proposition 1, $(A^\tau, \mathcal{C}^\tau, \theta^*)$ determines the same character ψ as $(A, \mathcal{C}, \theta)$, so that λ is rational over L by Theorem 5. If $\pi = \tau^2$, we have, as $rr^\rho = 1$,

$$\omega(v)^\pi = \big(\omega(r \cdot (rv^\rho)^\rho)^\tau\big)^\tau = \big(\lambda(\omega(erv^\rho))\big)^\tau = (\lambda^\tau\lambda)(\omega(ee^\rho v)) \ .$$

Now $\lambda^\tau\lambda$ is an automorphism of $(A, \mathcal{C}, \theta)$, hence $\lambda^\tau\lambda = \theta(\zeta)$ with $\zeta \in Z(\mathfrak{a})$, so that $\omega(v)^\pi = \omega(\zeta ee^\rho v)$ for all $v \in K/\mathfrak{a}$. Take $x \in L_A^\times$ so that $x_\infty = 1$ and $\pi = [x, L]$ on L_{ab}. Define $\alpha(x)$ as in Theorem 2, and put $z = N_{L/K'}(x)$. Define β as in (5.7) for this z. Then we see that

$$(\alpha(x)/f(z))_0 = h\big((f(z)/\beta)_0\big)_0 \cdot (\beta/f(z))_0 \in W \ .$$

Therefore $\omega(v)^{[x, L]} = \omega(\alpha(x)f(z)^{-1}v) = \omega(dv)$ for all $v \in K/\mathfrak{a}$ with an element d of W. Then $\zeta_0 ee^\rho = d$, hence $\zeta = 1$, since $ee^\rho \in W$ and $Z(\mathfrak{a})_0 \cap W = \{1\}$. Thus we obtain $\lambda^\tau\lambda = 1$, which ensures the existence of $(A, \mathcal{C}, \theta_F)$ over M_F.

PROPOSITION 7. *Let Z be the group of all roots of unity in K. Then the condition (5.2) is satisfied if there exists a subgroup Z' of Z such that Z is the direct product of $Z(\mathfrak{a})$ and Z'.*

Such a Z' exists, for example, if either the order of Z is squarefree, or $Z = Z(\mathfrak{a})$. To prove Proposition 7, we need

LEMMA 4. *Let m be the order of Z. Then K has infinitely many prime ideals $\mathfrak{q}$ of absolute degree one such that $N(\mathfrak{q}) \equiv 1 \mod (m)$, and $(N(\mathfrak{q}) - 1)/m$ is prime to m.*

Proof. Let $p_1, \cdots, p_r$ be the prime factors of m. Let ζ and ζ' be primitive m-th and $(mp_1 \cdots p_r)$-th roots of unity, respectively. Then $Q(\zeta')$ is a cyclic extension of $Q(\zeta)$ of degree $p_1 \cdots p_r$, so that every intermediate field between $Q(\zeta)$ and $Q(\zeta')$ can be written as $Q(\zeta'')$ with a root of unity ζ''. Apply this fact to $K \cap Q(\zeta')$. Since ζ generates the group of all roots of unity in K, we must have $Q(\zeta) = K \cap Q(\zeta')$, so that K and $Q(\zeta')$ are linearly disjoint over $Q(\zeta)$. Take a finite Galois extension S of $Q(\zeta)$ containing both K and $Q(\zeta')$. Then we can find an element σ of $\mathrm{Gal}(S/K)$ whose restriction to $Q(\zeta')$ generates $\mathrm{Gal}(Q(\zeta')/Q(\zeta))$. By the Tschebotareff Density Theorem, there exist infinitely many prime ideals in S which are unramified over Q and have σ as its Frobenius element relative to $Q(\zeta)$. Take any one of them, say $\mathfrak{P}$, and put $\mathfrak{q} = \mathfrak{P} \cap K$, $\mathfrak{p} = \mathfrak{P} \cap Q(\zeta)$. We can take $\mathfrak{P}$ so that $\mathfrak{p}$ is of absolute degree one. Put $p = N(\mathfrak{p})$. Then $p \equiv 1 \mod (m)$. By our choice of σ, we have $N(\mathfrak{q}) = N(\mathfrak{p}) = p$. On the other hand, $\mathfrak{p}$ remains prime in $Q(\zeta')$, so that $p - 1$ is not divisible by mp_i for $i = 1, \cdots, r$, q.e.d.

Proof of Proposition 7. Take $\mathfrak{q}$ as in Lemma 4 so that $\mathfrak{q}$ is prime to m, and put $p = N(\mathfrak{q})$. Let $\mathfrak{o}_\mathfrak{q}$ be the ring of $\mathfrak{q}$-adic integers, and let

$$\mathfrak{h} = \{a \in \mathfrak{o}_\mathfrak{q} \mid a \equiv 1 \mod \mathfrak{q}\} \ .$$

Taking the projection to the place $\mathfrak{q}$, we obtain a homomorphism j of U onto $\mathfrak{o}_\mathfrak{q}^\times/\mathfrak{h}$. By our choice of $\mathfrak{q}$, $\mathfrak{o}_\mathfrak{q}^\times$ has a subgroup $\mathfrak{g}$ of index m containing $\mathfrak{h}$. Let $W = j^{-1}(\mathfrak{g})$. Since m is prime to $(p - 1)/m$, it is easy to verify that $U = Z_0 \times W$. Put $W' = (Z_0' \times W) \cap U(\mathfrak{a})$. Then $U(\mathfrak{a}) = Z(\mathfrak{a})_0 \times W'$, q.e.d.

Example 1. Let us consider the case $n = 2$, under the condition that (K, Φ) is primitive. Then K is either cyclic over Q, or not normal over Q (see [6, § 8.4]). The order of Z is either 2 or 10. Therefore, by Proposition 7, the condition (5.2) is satisfied for every Z-lattice $\mathfrak{a}$ in K, so that we obtain a structure satisfying (i, ii, iii) of Theorem 11.

Example 2. Suppose that K is normal over Q, $\mathfrak{a}$ is a fractional ideal in K, and (K, Φ) is primitive. Take a prime ideal $\mathfrak{q}$ in K as in Lemma 4, and define h as in the proof of Proposition 7. Now we can find a set of automorphisms $\{\tau_1, \cdots, \tau_n\}$ of K such that $\Phi(a)$ has $a^{\tau_1}, \cdots, a^{\tau_n}$ as its characteristic roots. We have $K' = K$, and $f(z) = \det \Phi'(z) = \prod_{i=1}^{n} z^{\tau_i^{-1}}$. Therefore it can easily be seen that the conductor of φ_h is exactly $\prod_{i=1}^{n} \mathfrak{q}^{\tau_i}$.

Remark 6. Take $v_i \in K/\mathfrak{a}$ for $i = 1, \cdots, s$ so that

(5.10) *There is no element ζ of $Z(\mathfrak{a})$ other than the identity element such that $\zeta v_i = v_i$ for $i = 1, \cdots, s$.*

Let $(A, \mathcal{C}, \theta)$ be of type $(K, \Phi; \mathfrak{a}, \varepsilon)$, and let t_i be the element of A corresponding to v_i. Put $\mathcal{P} = (A, \mathcal{C}, \theta; t_1, \cdots, t_s)$. Then (5.10) is equivalent to

(5.11) *$\mathcal{P}$ has no automorphism other than the identity element.*

Let k be the field of moduli of $\mathcal{P}$. By [3, II, 1.5], we can take $\mathcal{P}$ so as to be rational over k; moreover such a $\mathcal{P}$ is unique up to isomorphisms over k. On the other hand, by [5, Cor. 5.16], k is the subfield of K'_{ab} corresponding to the subgroup R of $K'^{\times}_A$ which consists of the elements z such that

(5.12) $$|z|_0 = \beta\beta^\rho, \qquad \beta^{-1}f(z)\mathfrak{a} = \mathfrak{a}, \qquad \beta^{-1}f(z)v_i = v_i \qquad (i = 1, \cdots, s)$$

for some $\beta \in K^{\times}$. On account of (5.10), β is unique for z, so that we put $\beta = \beta(z)$. Then we can define a homomorphism φ of R into $C^{\times}$ by $\varphi(z) = (\beta(z)/f(z))_{\infty 1}$, and further a Grössen-character ψ of $k_A^{\times}$ by $\psi = \varphi \circ N_{k/K'}$, since $N_{k/K'}(k_A^{\times}) \subset R$. If the map α of Theorem 2 is defined for the present (A, θ) over k, then $\alpha(x) = \beta(N_{k/K'}(x))$, so that ψ *is exactly the Grössen-character determined by* (A, θ).

Let us now consider the one-dimensional case. As mentioned in *Remark* 1, we have $K' = K$, and can assume that Φ is the identity map, so that f is also the identity map. Given a lattice $\mathfrak{a}$ in K, let j be the invariant of an elliptic curve isomorphic to $C/\mathfrak{a}$, and let $L = K(j)$, $M = Q(j)$. Then

(5.13) $$P = K^{\times} \cdot N_{L/K}(L_A^{\times}) = K^{\times} \cdot K_\infty^{\times} \cdot U(\mathfrak{a}) .$$

For a Grössen-character φ of $K_A^{\times}$, we consider the following two conditions:

(5.14) $\varphi(y) = 1/y$ *for every* $y \in K_\infty^{\times}$;

(5.15) $\varphi(z) \in Z(\mathfrak{a})$ *if* $z \in U(\mathfrak{a})$.

PROPOSITION 8. *The character φ satisfies (5.14) and (5.15) if and only if $\varphi \circ N_{L/K}$ satisfies (2.2) and (2.3) with $k = L$. Moreover, for any h as in (5.2'), φ_h satisfies (5.14) and (5.15). Conversely, the restriction to P of any φ satisfying (5.14) and (5.15) can be obtained as φ_h from a map h as in (5.2'). Such an h is unique for φ, and actually is the restriction of φ to $U(\mathfrak{a})$.*

PROPOSITION 9. *Let $\psi = \varphi_h \circ N_{L/K}$ with any h as in (5.2'). Let γ be the generator of $\mathrm{Gal}(L/M)$. Then the following three conditions are equivalent:*

(i) $\psi(x^\gamma) = \psi(x)^\rho$ *for all* $x \in L_A^\times$;

(ii) $\varphi_h(x^\rho) = \varphi_h(x)^\rho$ *for all* $x \in P$;

(iii) $h(x^\rho) = h(x)^\rho$ *for all* $x \in U(\mathfrak{a})$.

These two propositions can be proved in a straightforward way, due to (5.13). Thus, if $n = 1$, the assertions of Theorem 11 can be obtained, by virtue of these propositions, from corollaries of Theorems 6 and 10.

Example 3. Let K be an imaginary quadratic field whose discriminant is >3 and divisible by a prime $q \equiv 3 \bmod (4)$. Let $\mathfrak{q}$ be the prime ideal in K such that $N(\mathfrak{q}) = q$. If ν denotes the class number of K, there exist exactly ν Grössen-characters φ of $K_A^\times$ with conductor $\mathfrak{q}$ satisfying (5.14) and such that $\varphi(\mathfrak{o}_q^\times) = \{\pm 1\}$, where $\mathfrak{o}_q$ denotes the ring of q-adic integers. These φ satisfy (5.15) with any $\mathfrak{a}$, since $Z(\mathfrak{a}) = \{\pm 1\}$. Moreover $\varphi(x^\rho) = \varphi(x)^\rho$ for every $x \in U(\mathfrak{a})$. Therefore, by virtue of Theorem 11, we find an elliptic curve E which is isomorphic to $C/\mathfrak{a}$, and defined over its field of moduli, and such that its points of finite order are rational over K_{ab}.

In general, a map h satisfying (5.2') and (5.5') may not necessarily exist. For example, let $K = Q(\sqrt{-l})$ with a prime $l \equiv 1 \bmod (8)$, and let $\mathfrak{a}$ be a fractional ideal in K. Suppose that there is a map h of $U(\mathfrak{a})$ into $Z(\mathfrak{a}) = \{\pm 1\}$ as in (5.2') satisfying (5.5'). For every prime ideal $\mathfrak{p}$ in K, let $\eta_\mathfrak{p}$ denote the element of $K_A^\times$ whose $\mathfrak{p}$-component is -1, and other components are 1. Then $\prod_\mathfrak{p} h(\eta_\mathfrak{p}) = -1$. Since $h(\eta_\mathfrak{p}) = \pm 1$, the number of $\mathfrak{p}$ for which $h(\eta_\mathfrak{p}) = -1$ must be odd. If $\mathfrak{p}^\rho = \mathfrak{q}$, we have $h(\eta_\mathfrak{p}) = h(\eta_\mathfrak{q})$, so that there exists at least one $\mathfrak{p}$ such that $\mathfrak{p}^\rho = \mathfrak{p}$ and $h(\eta_\mathfrak{p}) = -1$. Let p be the rational prime divisible by $\mathfrak{p}$, and $K_\mathfrak{p}$ the $\mathfrak{p}$-completion of K. If $N(\mathfrak{p}) = p^2$, then p is odd, and $p^2 \equiv 1 \bmod (4)$, so that there is an element x of $K_\mathfrak{p}$ such that $x^2 = -1$. Then $h(\eta_\mathfrak{p}) = h(x)^2 = (\pm 1)^2 = 1$. We have $\mathfrak{p}^\rho = \mathfrak{p}$ and $N(\mathfrak{p}) = p$ only for $p = 2$ or l. Since $l \equiv 1 \bmod (4)$, we have again an element x of Q_l such that $x^2 = -1$, hence $h(\eta_\mathfrak{p}) = 1$. Let $p = 2$. Since $-l \equiv -1 \bmod (8)$, we have $K_\mathfrak{p} = Q_2(\sqrt{-l}) = Q_2(\sqrt{-1})$, so that $h(\eta_\mathfrak{p}) = 1$ for the same reason. Thus $h(\eta_\mathfrak{p}) = 1$ for all $\mathfrak{p}$ such that $\mathfrak{p}^\rho = \mathfrak{p}$, a contradiction. Therefore if $K = Q(\sqrt{-l})$ with a prime $l \equiv 1 \bmod (8)$, there is no elliptic curve E which is isomorphic to $C/\mathfrak{a}$ and defined over its field of moduli, and whose points of finite order are rational over K_{ab}.

6. The zeta-function of A over a smaller field of rationality, relative to a subfield of K

We shall now generalize the results in the preceding sections by replacing F by a more general type of subfield of K. For a given $(A, \mathcal{C}, \theta)$, define the symbols S, T, θ_X, M_X, and M as in §3. As is observed there, for $\sigma \in \mathrm{Aut}(C/M)$, we obtain an automorphism $\pi(\sigma)$ of S by

$$\theta(a)^\sigma \circ \eta = \eta \circ \theta(a^{\pi(\sigma)}) \qquad (a \in S)$$

with any isomorphism η of $(A, \mathcal{C})$ onto $(A, \mathcal{C})^\sigma$. Then $\sigma \mapsto \pi(\sigma)$ defines an isomorphism of $\mathrm{Gal}(M_S/M)$ into $\mathrm{Aut}(S)$. Let us now fix any subgroup G of $\mathrm{Gal}(M_S/M)$, and put

$$V = \{x \in S \mid x^{\pi(\sigma)} = x \text{ for all } \sigma \in G\},$$
$$D' = \{x \in K' \mid x^\sigma = x \text{ for all } \sigma \in G\}.$$

By (3.2), K' is normal over $M \cap K'$, and $\mathrm{Gal}(K'/M \cap K')$ is isomorphic to $\mathrm{Gal}(M_S/M)$. We see easily that $M_V = D'M$. Now we assume the following condition:

(6.1) *K has a subfield D such that $K = DS$ and $V = D \cap S$.*

For example, this is satisfied if $V = T$ and $D = F$. If A is simple, we have $K = S$, so that (6.1) is satisfied with $D = V$.

Fix D as in (6.1). Then $\mathrm{Gal}(K/D)$ is isomorphic to $\pi(G) = \mathrm{Gal}(S/V)$. By the same reasoning as in the proofs of Propositions 2 and 3, we can verify

(6.2) $M_K = K'M_D$;

(6.3) *If $(A, \mathcal{C}, \theta_D)$ is rational over a field k_0, then $K'k_0$ is the smallest field of definition, containing k_0, for $(A, \mathcal{C}, \theta)$.*

Now let us assume that $(A, \mathcal{C}, \theta_D)$ is defined over an algebraic number field k_0 of finite degree such that

(6.4) $k_0 \cap K' = D'$.

Then $\mathrm{Gal}(K'k_0/k_0)$, $\mathrm{Gal}(M_S/M_V)$, and $\mathrm{Gal}(K'/D')$ are all isomorphic. Hereafter we use the same symbol G to denote these groups, and $\pi(G)$ for $\mathrm{Gal}(K/D)$ and $\mathrm{Gal}(S/V)$. Then π is an isomorphism of $\mathrm{Gal}(K'k_0/k_0)$ onto $\mathrm{Gal}(K/D)$, and

(6.5) $$\theta(a)^\sigma = \theta(a^{\pi(\sigma)}) \qquad (a \in K, \sigma \in G).$$

Therefore, if $\mathfrak{o} = \theta^{-1}[\mathrm{End}(A) \cap \theta(K)]$, then

(6.6) $\mathfrak{o}^{\pi(\sigma)} = \mathfrak{o}$ *for every* $\sigma \in G$.

Let $\mathfrak{o}_D$ denote the ring of all algebraic integers in D. We assume

(6.7) $\theta(\mathfrak{o}_D) \subset \mathrm{End}(A)$.

Then we can define the zeta-function of A over k_0 relative to D as in [5, § 7.6], which is denoted by $\zeta(s; A/k_0, D)$. The condition (6.7) is not so essential, since a given (A, C, θ_D) rational over k_0 is always isogenous, over k_0, to a structure satisfying (6.7) (see [6, § 7, Prop. 7]).

THEOREM 12. *Let (A, C, θ) be of type $(K, \Phi; \mathfrak{a}, \varepsilon)$. Suppose that $\theta(\mathfrak{o}_D) \subset \mathrm{End}(A)$, and (A, C, θ_D) is defined over a field k_0 satisfying (6.4). Let ψ be a Grössen-character of $(K'k_0)_A^\times$ determined by (A, C, θ). Then $\zeta(s; A/k_0, D)$ coincides with $L(s, \psi)$ up to finitely many possible bad Euler factors. More precisely, let $\mathfrak{q}$ be a prime ideal in k_0. Then A has good reduction modulo $\mathfrak{q}$ if and only if $\mathfrak{q}$ is unramified in $K'k_0$, and ψ is unramified at the prime factors of $\mathfrak{q}$ in $K'k_0$. For such a $\mathfrak{q}$, the Euler $\mathfrak{q}$-factor of $\zeta(s; A/k_0, D)$ coincides with the product of the Euler $\mathfrak{p}$-factors of $L(s, \psi)$ for all prime factors $\mathfrak{p}$ of $\mathfrak{q}$ in $K'k_0$.*

A further discussion of "bad" Euler factors will be given in Proposition 10.

Proof. Put $k = K'k_0$. Denote by the same symbol π the composed map of π with the natural map of $\mathrm{Gal}(k_{ab}/k_0)$ onto $\mathrm{Gal}(k/k_0)$. Define $\omega\colon K_{\mathbf{R}}/\mathfrak{a} \to A$ as in § 1. Let $\sigma \in \mathrm{Gal}(k_{ab}/k_0)$. Then

$$u \mapsto \omega^{-1}\bigl(\omega(u^{\pi(\sigma)})^{\sigma^{-1}}\bigr) \qquad (u \in K/\mathfrak{a}^{\pi(\sigma)^{-1}})$$

defines an $\mathfrak{o}$-linear isomorphism of $K/\mathfrak{a}^{\pi(\sigma)^{-1}}$ into $K/\mathfrak{a}$. Therefore we find an element z of $K_A^\times$ such that $z\mathfrak{a}^{\pi(\sigma)^{-1}} = \mathfrak{a}$, and $\omega^{-1}\bigl(\omega(u^{\pi(\sigma)})^{\sigma^{-1}}\bigr) = zu$, i.e.,

$$(6.8) \qquad \omega(v) = \omega(zv^{\pi(\sigma)^{-1}})^\sigma \qquad (v \in K/\mathfrak{a}) .$$

Define $\alpha(x)$ for $x \in k_A^\times$ as in Theorem 2. Then $\sigma[x^\sigma, k] = [x, k]\sigma$, and $N_{k/K'}(x^\sigma) = N_{k/K'}(x)^\sigma$. By Proposition 1, we have

$$(6.9) \qquad g(x^\sigma) = g(x)^{\pi(\sigma)} \qquad (x \in k_A^\times)$$

for the map g of (1.4). Further

$$(6.10) \qquad \alpha(x)^{\pi(\sigma)} = \alpha(x^\sigma) \qquad (x \in k_A^\times, \ \sigma \in \mathrm{Gal}(k/k_0)) .$$

Let $\mathfrak{p}$ be a prime ideal in k unramified over k_0, and $\mathfrak{q}$ the restriction of $\mathfrak{p}$ to k_0. Suppose that A has good reduction modulo $\mathfrak{q}$, so that ψ is unramified at $\mathfrak{p}^\sigma$ for all $\sigma \in \mathrm{Gal}(k/k_0)$. Denote by $(\tilde{A}, \tilde{\theta})$ the couple obtained from (A, θ) by reduction modulo $\mathfrak{p}$. Let $f_{\mathfrak{p}}$ and $f_{\mathfrak{q}}$ denote the Frobenius endomorphisms of $\tilde{A}$ of degree $N(\mathfrak{p})$ and $N(\mathfrak{q})$, respectively. Let $c_{\mathfrak{p}}$ be a prime element in the $\mathfrak{p}$-completion of k, considered as an element of $k_A^\times$, and let $\beta = \alpha(c_{\mathfrak{p}}) = \psi(c_{\mathfrak{p}})$. As is seen in the proof of [5, Th. 7.42], $f_{\mathfrak{p}} = \tilde{\theta}(\beta)$. Let γ denote the Frobenius element of $\mathrm{Gal}(k/k_0)$ for $\mathfrak{p}$. From the equality $\theta(a)^\gamma = \theta(a^{\pi(\gamma)})$, we obtain

$$f_{\mathfrak{q}} \circ \tilde{\theta}(a) = \tilde{\theta}(a^{\pi(\gamma)}) \circ f_{\mathfrak{q}} \qquad (a \in \mathfrak{o}) .$$

By (6.10), we have

$$(6.11) \qquad \psi(c_\mathfrak{p}^\sigma) = \alpha(c_\mathfrak{p}^\sigma) = \beta^{\pi(\sigma)} \qquad (\sigma \in \mathrm{Gal}(k/k_0)) ,$$

so that $\beta^{\pi(\gamma)} = \beta$. (Note that $\psi(c_\mathfrak{p})$ does not depend on the choice of $c_\mathfrak{p}$, since ψ is unramified at $\mathfrak{p}$.) Let $\mathfrak{S}$ be the subalgebra of $\mathrm{End}_Q(\tilde{A})$ generated by $\tilde{\theta}(K)$ and $f_\mathfrak{q}$, and

$$Y = \{\sigma \in \mathrm{Gal}(k/k_0) \,|\, \mathfrak{p}^\sigma = \mathfrak{p}\} .$$

Then γ is a generator of Y, and $\mathfrak{S}$ is a central simple algebra of degree ν^2 over the fixed subfield of K by $\pi(Y)$, where ν is the order of Y, i.e., the degree of $\mathfrak{p}$ relative to $\mathfrak{q}$. Now, for a prime ideal $\mathfrak{l}$ in D prime to $N(\mathfrak{p})$, define an $\mathfrak{l}$-adic representation $R_\mathfrak{l}$ of $\mathrm{End}_Q(\tilde{A})$ (see [5, § 7.6]). By [5, Prop. 7.22], the restriction of $R_\mathfrak{l}$ to $\mathfrak{S}$ is equivalent to a reduced representation of $\mathfrak{S}$ over $\tilde{\theta}(D)$, since the degree of $R_\mathfrak{l}$ is $[K:D]$, and $\tilde{\theta}(K)$ is a maximal subfield of $\mathfrak{S}$. Therefore, since $f_\mathfrak{q}^\nu = f_\mathfrak{p} = \tilde{\theta}(\beta)$, we have

$$\det[1 - R_\mathfrak{l}(f_\mathfrak{q})X] = \prod_\tau (1 - \beta^\tau X^\nu)$$

where τ runs over a set of representatives for $\pi(G)/\pi(Y)$, and X is an indeterminate. By (6.11), we obtain

$$\det[1 - R_\mathfrak{l}(f_\mathfrak{q})N(\mathfrak{q})^{-s}] = \prod_\sigma [1 - \psi(c_\mathfrak{p}^\sigma)N(\mathfrak{p}^\sigma)^{-s}] ,$$

where σ runs over a set of representatives for G/Y. Thus the Euler $\mathfrak{q}$-factor of $\zeta(s; A/k_0, D)$ coincides with the product of the Euler $\mathfrak{p}$-factors of $L(s, \psi)$ for all the prime ideals $\mathfrak{p}$ in k dividing $\mathfrak{q}$. Now our assertion concerning good reduction of A can be proved exactly in the same manner as [5, Prop. 7.47], by means of the result of Serre and Tate [2].

PROPOSITION 10. *Let $\mathfrak{p}$ be a prime ideal in k ramified over k_0, I the inertia group of $\mathfrak{p}$, and W the fixed subfield of K by $\pi(I)$. If W is not totally imaginary, then any A' isomorphic to A over k has bad reduction modulo $\mathfrak{p}$.*

Proof. Assume that such an A' has good reduction modulo $\mathfrak{p}$. Let μ be an isomorphism of A to A' rational over k. Put $\lambda_\sigma = \mu^\sigma \circ \mu^{-1}$ for every $\sigma \in \mathrm{Gal}(k/k_0)$. Then λ_σ is an isomorphism of A' onto A'^σ rational over k, and $\lambda_\sigma^\tau \circ \lambda_\tau = \lambda_{\sigma\tau}$. Define an isomorphism θ' of K into $\mathrm{End}_Q(A')$ by $\theta'(a) = \mu \circ \theta(a) \circ \mu^{-1}$. Then we obtain the same $(\tilde{A}', \tilde{\theta}')$ from $(A', \theta')^\sigma$ for all $\sigma \in I$, by reduction modulo $\mathfrak{p}$. Moreover, for every $\sigma \in I$, $\tilde{\lambda}_\sigma$ is an automorphism of $\tilde{A}'$, and $\tilde{\lambda}_\sigma \tilde{\lambda}_\tau = \tilde{\lambda}_{\sigma\tau}$ for $\sigma, \tau \in I$. Now we have $\theta'(a)^\sigma \circ \lambda_\sigma = \lambda_\sigma \circ \theta'(a^{\pi(\sigma)})$, so that

$$\tilde{\theta}'(a) \cdot \tilde{\lambda}_\sigma = \tilde{\lambda}_\sigma \cdot \tilde{\theta}'(a^{\pi(\sigma)}) \qquad (\sigma \in I, a \in K) .$$

Therefore, we see that $\tilde{\theta}'(K)$ and the $\tilde{\lambda}_\sigma$ for all $\sigma \in I$ generate a subalgebra of $\mathrm{End}_Q(A')$ isomorphic to the total matrix algebra of degree d over W, where $d = [K:W] = [I:1]$. Then $\tilde{A}'$ must be isomorphic to a product of d copies of an abelian variety B such that $\mathrm{End}_Q(B)$ contains a subalgebra isomorphic to W. Since $\dim(B) = n/d = [W:Q]/2$, this contradicts the following lemma, if W is not totally imaginary, q.e.d.

LEMMA 5. *Let P be an algebraic number field of degree $2m$ isomorphic to a subfield of $\mathrm{End}_Q(Y)$ with an abelian variety Y of dimension m defined over a field of any characteristic. Then P is totally imaginary.*

Proof. If η is an isomorphism of P into $\mathrm{End}_Q(Y)$, then η maps the identity element of P to the identity element of $\mathrm{End}_Q(Y)$, and for every $x \in P$, $N_{P/Q}(x)$ coincides with the degree of $\eta(x)$ (see [6, § 5.1, Prop. 1, Prop. 2]). Therefore $N_{P/Q}(x) \geqq 0$ for every $x \in P$. Such an algebraic number field P is obviously totally imaginary.

Let us now consider the special case in which $D = F$. Then G is of order 2, so that $I = G$, $W = F$. Since F is totally real, A' has bad reduction modulo $\mathfrak{p}$. Therefore we obtain Theorem 7. In general, if D is not totally imaginary, and $[K:D]$ is a prime, then, for the same reason, one has $\zeta(s; A/k_1, D) = L(s, \psi)$.

Remark 7. It can easily be seen that every subfield of a CM-field is either totally real or a CM-field. Therefore, in Proposition 10, the condition that W is not totally imaginary simply means that W is totally real. We shall later show that the assertion of Proposition 10 is not necessarily true without that condition.

Example 4. Let A be the jacobian of $y^2 = 1 - x^p$ as considered in Remark 3. In this case, we can put $K = K' = S$. Now take Q as D. Then $\zeta(s; A/Q, Q)$ is exactly the zeta-function of A over Q. The inertia group of a (unique) ramified prime in K is the whole Galois group $\mathrm{Gal}(K/Q)$, so that $W = Q$. Therefore the zeta-function of A over Q is exactly $L(s, \psi)$ with the Grössen-character ψ of $K_A^\times$ determined by A over K. The zeta-functions of the curves of the form $ax^m + by^n = c$ were determined by Weil [9], without the discussion of bad primes. Here we have considered the simplest case, but our method is applicable to a more general case.

Let us now discuss the generalizations of Theorems 8, 9, 11, without going into details of proofs, since it is obvious how to modify the original argument.

First, Theorem 8 is true, with D in place of F, and with k_0 satisfying (6.4) in place of k_1. To generalize Theorem 9, let Σ_D be the set of all $(A, \mathcal{C}, \theta_D)$ rational over k_0 such that $(A, \mathcal{C}, \theta)$ is of type $(K, \Phi; \mathfrak{a}, \varepsilon)$ and determines ψ over $K'k_0$. If Σ_D is not empty, the number of isomorphism-classes over k_0 in

532   GORO SHIMURA

Σ_D is the order of the (Galois) cohomology group $H^1(\pi(G), Z(\mathfrak{a}))$.

As for Theorem 11, we assume that $M_D \cap K' = D'$, $\mathfrak{a}$ is a fractional ideal in K, and the following condition is satisfied for the map h of (5.2'):

$$(6.12) \qquad\qquad h(x^\tau) = h(x)^\tau \qquad\qquad (x \in U(\mathfrak{a}),\ \tau \in \pi(G)).$$

Then Ker(h) is stable under $\pi(G)$. Under this condition, modifying the proof of Theorem 11 in an obvious way, we obtain $(A, \mathcal{C}, \theta_D)$ rational over M_D.

Example 5. Let $F = Q(\zeta + \zeta^{-1})$ with $\zeta = e^{2\pi i/7}$, $K = F(\sqrt{-11})$, $D = Q$. Then $[K:Q] = 6$. We can find Φ so that (K, Φ) is primitive. Then $K' = K$. Consider $(K, \Phi; \mathfrak{o}_K, \varepsilon)$ with the maximal order $\mathfrak{o}_K$ in K, and an element ε such that: (i) $\varepsilon^\rho = -\varepsilon$; (ii) $\mathrm{Im}(\varepsilon^\sigma) > 0$ for every $\sigma \in \mathrm{Gal}(K/Q)$ belonging to Φ; (iii) $\varepsilon^{-1}\mathfrak{o}_K$ is the different of K relative to Q. Then there exists $(A, \mathcal{C}, \theta)$ of type $(K, \Phi; \mathfrak{o}_K, \varepsilon)$. The condition (iii) for ε implies that the polarization $\mathcal{C}$ is of principal type, i.e., $(A, \mathcal{C})$ is self-dual. Moreover, K is of class number one, and ε is unique up to multiplication by the square of a unit of $\mathfrak{o}_K$. From these facts one can easily derive that the field of moduli of $(A, \mathcal{C})$ is Q. Now there is a unique prime ideal $\mathfrak{p}$ in K such that $N(\mathfrak{p}) = 11^3$. Then we can find a unique continuous homomorphism $h: U(\mathfrak{o}_K) \to Z(\mathfrak{o}_K) = \{\pm 1\}$ with $\mathfrak{p}$ as its "conductor". Since $h(x^\tau) = h(x)$ for all $\tau \in \mathrm{Gal}(K/Q)$, we obtain a structure $(A, \mathcal{C})$ rational over Q, by taking Q to be D in the above discussion. Now the Grössen-character φ_h is ramified exactly at $\mathfrak{p}$. Take a prime ideal $\mathfrak{q}$ in K dividing 7. Then $\mathfrak{q}$ is ramified over Q, but φ_h is unramified at $\mathfrak{q}$. We observe that the fixed subfield of the inertia group of $\mathfrak{q}$ is $Q(\sqrt{-11})$, which is totally imaginary. Thus the assertion of Proposition 10 is false if W is totally imaginary.

Princeton University

REFERENCES

[1] M. Deuring, *Die Zetafunktion einer algebraischen Kurve vom Geschlechte Eins*, I, II, III, IV, Nachr. Akad. Wiss. Göttingen (1953), 85–94, (1955), 13–42, (1956), 37–76, (1957), 55–80.

[2] J.-P. Serre and J. Tate, *Good reduction of abelian varieties*, Ann. of Math. **88** (1968), 492–517.

[3] G. Shimura, *On the field of definition for a field of automorphic functions*, I, II, Ann. of Math. **80** (1964), 160–189, **81** (1965), 124–165.

[4] ———, *On canonical models of arithmetic quotients of bounded symmetric domains*, Ann. of Math. **91** (1970), 144–222.

[5] ———, *Introduction to the arithmetic theory of automorphic functions*, to appear.

[6] G. Shimura and Y. Taniyama, Complex multiplication of abelian varieties and its applications to number theory, Publ. Math. Soc. Japan No. 6, 1961.

[7] Y. Taniyama, *L-functions of number fields and zeta functions of abelian varieties*, J. Math. Soc. Japan **9** (1957), 330–366.

[8] A. WEIL, Basic number theory, Grundl. Math. Wiss. 144, Springer, 1967.

[9] ————, *Jacobi sums as "Grossencharaktere"*, Trans. Amer. Math. Soc. **73** (1952), 487–495.

[10] ————, Variétés abéliennes et courbes algébriques, Paris, 1948.

(Received December 25, 1970)

Class fields over real quadratic fields in the theory of modular functions

In *Several Complex Variables* II, Maryland 1970,
Lecture notes in mathematics 185 (1971), 169-188*

1. Introduction

It is well-known that elliptic modular functions or forms are closely connected with imaginary quadratic fields in various ways. For example, as Hecke showed, the L-function of such a field with a Grössen-character can be associated, through Mellin transformation, with a few cusp forms belonging to a certain congruence subgroup. Further, the class-fields over an imaginary quadratic field can be generated by special values of elliptic or elliptic modular functions. Compared with this, the connection of real quadratic fields with modular functions is not so close, although we know at least the following two facts: (i) the L-function of a real quadratic field with a character of finite order corresponds to a holomorphic modular form of weight 1 under Mellin transformation (Hecke [3]); (ii) the L-function of a real quadratic field with a Grössen-character corresponds to non-holomorphic automorphic forms under Mellin transformation (Maass [5]). However, no theory analogous to complex multiplication has been known for real quadratic fields (see Note 1 at the end of the paper).

The purpose of this lecture is to show that the arithmetic of a real quadratic field k is closely connected with certain *holomorphic* modular forms of weight $w \geq 2$, and also that some class fields over k can be generated by points of finite order on certain abelian varieties obtained from such modular forms. These do not quite parallel the known results for imaginary fields mentioned above, but still there is some similarity.

To be more explicit, we fix a positive integer $N > 1$, and a non-trivial quadratic character χ of $(\mathbf{Z}/N\mathbf{Z})^{\times}$, and put

$$\Gamma_0(N) = \left\{ \begin{bmatrix} a & b \\ c & d \end{bmatrix} \in SL_2(\mathbf{Z}) \,\middle|\, c \in N\mathbf{Z} \right\},$$

$$\Gamma_\chi = \left\{ \begin{bmatrix} a & b \\ c & d \end{bmatrix} \in \Gamma_0(N) \,\middle|\, \chi(a) = 1 \right\}.$$

Here and throughout the paper, for an associative ring R with identity element we denote by $R^\times$ the group of all invertible elements of R. For a positive integer $w > 1$, let $S_{w,\chi}$ denote the vector space of all cusp forms g satisfying

$$g\left(\frac{az+b}{cz+d}\right)(cz+d)^{-w} = \chi(d)g(z) \quad \text{for all} \quad \begin{bmatrix} a & b \\ c & d \end{bmatrix} \in \Gamma_0(N).$$

For an obvious reason, we assume

$$(1.1) \qquad \chi(-1) = (-1)^w.$$

1

After Hecke, we define a **C**-linear endomorphism $T(n)$ of $S_{w,\chi}$, for a positive integer n, by

$$(1.2) \qquad (g|T(n))(z) = n^{w-1} \sum_{a>0,\,ad=n} \sum_{b=0}^{d-1} \chi(a) g\!\left(\frac{az+b}{d}\right) d^{-w} \qquad (g \in S_{w,\chi}),$$

and a **C**-linear automorphism H of $S_{w,\chi}$ by

$$(1.3) \qquad (g|H)(z) = g(-1/Nz)\big(N^{1/2}z\big)^{-w}.$$

Hecke proved that the operators $T(n)$, for all n, form a commutative algebra $\mathcal{R}_{\mathbf{C}}$ over **C**, whose rank is exactly the dimension of $S_{w,\chi}$ over **C**. Further we have

$$(1.4) \qquad T(n)H = \chi(n)HT(n) \quad \text{if } n \text{ is prime to } N.$$

Hereafter we make the following assumption on N and χ, though this is not absolutely necessary:

(1.5) $\mathcal{R}_{\mathbf{C}}$ *is generated by the* $T(n)$ *for all* n *prime to* N.

This is satisfied, for instance, if χ is a primitive character modulo N (Hecke [4, Satz 24]). Under (1.5), $\mathcal{R}_{\mathbf{C}}$ is semi-simple (but the converse if false). Taking **Q** instead of **C**, we obtain

Proposition 1. *Let* $\mathcal{R}$ *denote the algebra generated by the* $T(n)$ *over* **Q** *for all* n*. Then* $\mathcal{R}_{\mathbf{C}} = \mathcal{R} \otimes_{\mathbf{Q}} \mathbf{C}$*. Moreover, under (1.5),* $\mathcal{R}$ *is a direct sum of fields, each of which is a totally imaginary quadratic extension of a totally real algebraic number field.*

Let $\mathcal{R} = \mathcal{K}_1 \oplus \cdots \oplus \mathcal{K}_r$ be the direct sum decomposition of $\mathcal{R}$ with fields $\mathcal{K}_1, \ldots, \mathcal{K}_r$. Let K_ν be a subfield of **C** isomorphic to $\mathcal{K}_\nu$, and e_ν the identity element of $\mathcal{K}_\nu$. Then we can find an element

$$f_\nu(z) = \sum_{n=1}^{\infty} a_{\nu n} e^{2\pi i n z}$$

of $e_\nu \cdot S_{w,\chi}$ with the following properties:

(1.6) K_ν *is generated by the numbers* $a_{\nu n}$ *over* **Q***;*

(1.7) *The functions* $f_{\nu\sigma}(z) = \sum_{n=1}^{\infty} a_{\nu n}^{\sigma} e^{2\pi i n z}$*, for all the isomorphisms* σ *of* K_ν *into* **C***, form a basis of* $e_\nu \cdot S_{w,\chi}$ *over* **C***;*

(1.8) $f_{\nu\sigma}|T(n) = a_{\nu n}^{\sigma} f_{\nu\sigma}$ *for every* n*.*

The fields K_ν are unique up to conjugacy over **Q**.

By the Hecke theory, the Dirichlet series $L(s, f_{\nu\sigma}) = \sum_{n=1}^{\infty} a_{\nu n}^{\sigma} n^{-s}$ associated with $f_{\nu\sigma}$ has the following Euler Product

$$(1.9) \qquad L(s, f_{\nu\sigma}) = \prod_p \big[1 - a_{\nu p}^{\sigma} p^{-s} + \chi(p) p^{w-1-2s}\big]^{-1}.$$

In (1.2) and (1.9) we put $\chi(n) = 0$ if n is not prime to N. If ρ denotes the restriction of complex conjugation to K_ν, then

$$(1.10) \qquad a_{\nu n}^{\sigma} = \chi(n) a_{\nu n}^{\sigma \rho} \quad \text{for every } n \text{ prime to } N.$$

Let K'_ν be the maximal (totally) real subfield of K_ν, which is meaningful in view of Proposition 1. Then $a_{\nu n} \in K'_\nu$ if $\chi(n) = 1$.

Let us now assume that w is even, so that $\chi(-1) = 1$ by (1.1). (We shall make a brief discussion about the case of odd w in §4.) Then the character χ corresponds to a unique real quadratic field k. For example, $k = \mathbf{Q}(\sqrt{N})$ if $\chi(a) = \left(\dfrac{a}{N}\right)$, N is square-free, and $N - 1 \in 4\mathbf{Z}$. Now our main interest lies in the following relation between the arithmetic of k and the Fourier coefficients $a_{\nu p}$ for rational primes p not dividing N.

(I) *There is a prime ideal $\mathfrak{b}_\nu$ in K_ν, ramified over K'_ν, with the following properties:*

(Ia) $a_{\nu p} \equiv 0 \pmod{\mathfrak{b}_\nu}$ *if* $\chi(p) = -1$.

(Ib) *If* $\chi(p) = 1$ *and* $p = N_{k/\mathbf{Q}}(\alpha)$ *with a totally positive algebraic integer α of k, then* $\mathrm{Tr}_{k/\mathbf{Q}}(\alpha^{w-1}) \equiv a_{\nu p} \pmod{\mathfrak{b}_\nu \cap K'_\nu}$.

(Ic) *If u is the fundamental unit of k, then* $\mathrm{Tr}_{k/\mathbf{Q}}(u^{w-1}) \in N(\mathfrak{b}_\nu)\mathbf{Z}$.

This is neither a theorem nor a conjecture, but merely a somewhat vague statement of what we find empirically from the numerical values of $a_{\nu p}$, and what we can prove for some smaller N and w. Because of lack of data, it is safe to restrict the validity of (I) to the case where k has class number one. Probably it is necessary to modify the satement in the case of larger class numbers.

Actually the existence of $\mathfrak{b}_\nu$ satisfying (Ia) can easily be shown if we assume that $a_{\nu p}$, for at least one p, generates a prime ideal $\mathfrak{b}_\nu$ ramified over K'_ν. Indeed, (Ia) then follows from (1.10). It is properties (Ib) and (Ic) which are really striking, and by which relatively trivial (Ia) can acquire a new significance.

As a simple example, let us consider the case $N = 29$, $w = 2$. Then $S_{w,\chi}$, with $\chi(a) = \left(\dfrac{a}{29}\right)$, is of dimension 2. As Hecke showed (Werke, pp. 903-905), there are two cusp forms $\sum_n a_n e^{2\pi i n z}$ and $\sum_n a_n^\rho e^{2\pi i n z}$ with a_n in $K = \mathbf{Q}(\sqrt{-5})$; further $a_p \equiv 0 \pmod{(\sqrt{-5})}$ if $\chi(p) = -1$. Now the fundamental unit u of $k = \mathbf{Q}(\sqrt{29})$ is given by $u = (5 + \sqrt{29})/2$, so that $\mathrm{Tr}_{k/\mathbf{Q}}(u) = 5$. Thus (Ia) and (Ic) are true with $\mathfrak{b}_\nu = (\sqrt{-5})$. (In this case $\mathcal{R}$ is isomorphic to K.) We can actually prove (Ib) in this case. More examples will be discussed later.

(II) *If $w = 2$, then the maximal ray class field over k modulo $N(\mathfrak{b}_\nu)P_\infty$ can be generated over k by the coordinates of certain points of finite order on the Jacobian variety of the modular function field for Γ_χ, where P_∞ denotes the product of the two archimedean primes of k. Moreover, the above relations (Ia, Ib) describe the reciprocity law in this class field.*

Again this is neither a theorem nor a conjecture. We can prove the statement under certain conditions on N and $N(\mathfrak{b}_\nu)$. It is at least true for the above special case $N = 29$, $w = 2$.

2. Preliminary considerations about class fields over a real quadratic field

A real quadratic field k has "less" class fields than imaginary quadratic fields, since it has a unit of infinite order. To explain this in a quantitative way, let $\mathfrak{o}_k$ denote the ring of algebraic integers in k, $\mathfrak{a}$ an integral ideal in k, h the class number of k, and u the fundamental unit of k. Further let P_∞ be as in (II), and $F_\mathfrak{a}$ the maximal ray class field over k modulo $\mathfrak{a}P_\infty$. Then we have

$$[F_\mathfrak{a} : k] = 2h \cdot \varphi(\mathfrak{a})/\mu,$$

where $\varphi(\mathfrak{a})$ denotes the order of the group $(\mathfrak{o}_k/\mathfrak{a})^\times$, and μ the smallest positive integer such that u^μ is totally positive and $\equiv 1 \pmod{\mathfrak{a}}$. Therefore, if μ is small, we obtain a relatively large class field. For a fixed μ, we have a very restrictive choice for $\mathfrak{a}$, since it must divide $u^\mu - 1$. This is why we have "less" class fields over k. It often happens even that $F_\mathfrak{a}$ is the composite of a cyclotomic field with k. A more precise statement can be proved, for example, in the following form:

Proposition 2. *Let q be a rational prime unramified in k, m a positive integer, and μ the smallest positive integer such that u^μ is totally positive and $u^\mu - 1 \in q^m\mathfrak{o}_k$. Suppose that $q^m > 2$ and $u^\mu - 1 \notin q^{m+1}\mathfrak{o}_k$. Then the maximal abelian extension of k in which only the factors of $q \cdot P_\infty$ are ramified can be generated by the q^n-th roots of unity for all $n \in \mathbf{Z}, > 0$, over $F_\mathfrak{a}$, where $\mathfrak{a} = q^m\mathfrak{o}_k$.*

The class field we shall obtain in our later discussion is exactly the extension $F_\mathfrak{a}$ of this type. We note that if $N_{k/\mathbf{Q}}(u) = -1$, then $u^2 - 1 = u \cdot \mathrm{Tr}_{k/\mathbf{Q}}(u)$, so that relation (Ic) with $w = 2$ implies that $u^2 \equiv 1 \pmod{N(\mathfrak{b}_\nu)}$. Therefore, if $N(\mathfrak{b}_\nu)$ is an odd prime and $u^2 - 1$ is not divisible by $N(\mathfrak{b}_\nu)^2$, then the maximal ray class field over k modulo $N(\mathfrak{b}_\nu)P_\infty$ has a special meaning as described in the above proposition. For example, in the case $N = 29$ and $w = 2$, we obtain the maximal ray class field F over $\mathbf{Q}(\sqrt{29})$ of conductor $5 \cdot P_\infty$, which generates, together with the 5^n-th rooots of unity for all n, the maximal abelian extension of k in which only the prime factors of $5 \cdot P_\infty$ are ramified.

3. Detailed discussion in the case $w = 2$

In this section we assume throughout $w = 2$. Let $\mathfrak{H}$ denote the upper half complex plane $\{ z \in \mathbf{C} \mid \mathrm{Im}(z) > 0 \}$. Then the quotient $\Gamma_\chi\backslash\mathfrak{H}$, suitably compactified, is isomorphic to a projective non-singular curve V defined over $\mathbf{Q}$. (There is a "standard" way to find a model for $\Gamma_\chi\backslash\mathfrak{H}$ defined over $\mathbf{Q}$. For the detailed proofs of this and other facts stated in this section, the reader is referred to [10, Chap. 7].) Let J denote the Jacobian variety of V, also defined over $\mathbf{Q}$. The generator of $\Gamma_0(N)/\Gamma_\chi$ defines an automorphism γ of J defined over $\mathbf{Q}$. Put $J_\chi = (1-\gamma)J$. Then J_χ is an abelian subvariety of J, defined over $\mathbf{Q}$, whose space of holomorphic 1-forms can be identified with $S_{2,\chi}$. Moreover, J is isogenous to the product of J_χ and the Jacobian variety of $\Gamma_0(N)\backslash\mathfrak{H}$.

The operator $T(n)$ on $S_{2,\chi}$ defines an endomorphism of J_χ in a natural way, which we denote by $\tau(n)$. The subspace $e_\nu \cdot S_{2,\chi}$ of $S_{2,\chi}$ corresponds to an abelian subvariety A_ν of J_χ. Further we can define an isomorphism θ_ν of K_ν

into $\mathrm{End}(A_\nu) \otimes \mathbf{Q}$ so that $\theta_\nu(a_{\nu n})$ is the restriction of $\tau(n)$ to A_ν for all n. We note that

$$[K_\nu : \mathbf{Q}] = \dim(A_\nu), \qquad J = (1+\gamma)J + J_\chi, \qquad J_\chi = A_1 + \cdots + A_r.$$

It can be shown that A_ν and the elements of $\mathrm{End}(A_\nu) \cap \theta_\nu(K_\nu)$ are all defined over $\mathbf{Q}$.

Now the operator H of (1.3) defines an automorphism η of J_χ of order 2, defined over k, satisfying

$$(3.1) \qquad \eta \cdot \tau(n) = \chi(n)\tau(n)\eta \quad \text{if } n \text{ is prime to } N,$$

$$(3.2) \qquad \eta^\varepsilon = -\eta,$$

where ε is the generator of $\mathrm{Gal}(k/\mathbf{Q})$. Put

$$(3.3) \qquad A'_\nu = (1+\eta)A_\nu.$$

Then A'_ν is an abelian subvariety of A_ν rational over k, and

$$(3.4) \qquad A_\nu = A'_\nu + (A'_\nu)^\varepsilon, \qquad (A'_\nu)^\varepsilon = (1-\eta)A_\nu.$$

Let $a \in K_\nu$ and $\theta_\nu(a) \in \mathrm{End}(A_\nu)$. We see from (3.1) that

$$(3.5) \quad \theta_\nu(a) \text{ sends } A'_\nu \text{ into } \frac{A'_\nu}{(A'_\nu)^\varepsilon}, \text{ and } (A'_\nu)^\varepsilon \text{ into } \frac{(A'_\nu)^\varepsilon}{A'_\nu}, \text{ if } \begin{array}{l} a^\rho = a \\ a^\rho = -a. \end{array}$$

(As defined in §1, ρ is complex conjugation and $K'_\nu = \{\, x \in K_\nu \,|\, x^\rho = x \,\}$.) Therefore we can define an isomorphism θ'_ν of K'_ν into $\mathrm{End}(A'_\nu) \otimes \mathbf{Q}$ so that $\theta'_\nu(a)$ is the restriction of $\theta_\nu(a)$ if $a \in K'_\nu$.

Now the congruence relations for $\tau(p)$ enable us to determine the zeta functions of A_ν and A'_ν:

Proposition 3. *The zeta function of A_ν over $\mathbf{Q}$ and the zeta function of A'_ν over k coincide, up to finitely many Euler factors, with the product $\prod_\sigma L(s, f_{\nu\sigma})$ taken for all the isomorphisms σ of K into $\mathbf{C}$.*

Although this is implicit in the previous works by Eichler and the author, the present formulation (in a more general case) is due to Miyake [6] and the author [10, Chap. 7]. A result of Igusa, combined with a result about good reduction of factors of an abelian variety, implies that the bad Euler factors may exist only for the prime factors of N. In view of this, the above proposition is essentially equivallent with the following statement:

Let p be a rational prime not dividing N, and Φ a Frobenius automorphism of the algebraic closure $\overline{\mathbf{Q}}$ of $\mathbf{Q}$ for any prime divisor of $\overline{\mathbf{Q}}$ extending p. Then

$$(3.6) \qquad t^\Phi + \chi(p)p t^{\Phi^{-1}} = \theta_\nu(a_{\nu p})t$$

for every point t of A_ν whose order is finite and prime to p.

From now on, we fix our attention on one ν, and write $K_\nu, K'_\nu, A_\nu, A'_\nu, \theta_\nu, \theta'_\nu$, and $f_\nu = \sum a_{\nu n} e^{2\pi i n z}$ simply as $K, K', A, A', \theta, \theta'$, and $f = \sum a_n e^{2\pi i n z}$, by dropping the subscript ν. Let $\mathfrak{o}$ (resp. $\mathfrak{o}'$) be the ring of algebraic integers in

6

K (resp. K'). Changing A and A' for some abelian varieties isogenous to them, if necessary, we may assume that $\theta(\mathfrak{o}) \subset \mathrm{End}(A)$ and $\theta'(\mathfrak{o}') \subset \mathrm{End}(A')$. It can be shown that this change can be made without interfering with (3.2–3.6).

Let $\mathfrak{b}$ denote the ideal of $\mathfrak{o}$ generated by the elements a of $\mathfrak{o}$ such that $a^\rho = -a$. We assume

(3.7) $\mathfrak{b} \neq \mathfrak{o}$ and $\mathfrak{b}$ is prime to 2.

Then $\mathfrak{b}$ is a square-free product of several prime ideals of $\mathfrak{o}$ ramified over K'. Therefore we find a unique ideal $\mathfrak{c}$ of $\mathfrak{o}'$ such that $\mathfrak{co} = \mathfrak{b}^2$ and $\mathfrak{c} = \mathfrak{b} \cap \mathfrak{o}'$; then $\mathfrak{c}$ is also square-free. We can prove

Proposition 4. $-N(\mathfrak{b}) \in N_{k/\mathbf{Q}}(k^\times)$.

For a prime N, Hecke proved that $a_N a_N^\rho = N$, so that $N \in N_{K/K'}(K^\times)$. The above result is "reciprocal" to this result.

Now our first main result can be stated as follows:

Theorem 1. Let $A[\mathfrak{b}] = \{\, t \in A \mid \theta(\mathfrak{b})t = 0 \,\}$ and let F be the extension of k generated by the coordinates of the points of $A[\mathfrak{b}]$. Then the following assertions hold:

(1) F is an abelian extension of k, which is unramified at every prime of k not dividing $N(\mathfrak{b})N$.

(2) If m is a positive rational integer prime to $N(\mathfrak{b})N$ and $\sigma = \left(\dfrac{F/k}{(m)} \right)$, then $x^\sigma = mx$ for every $x \in A[\mathfrak{b}]$.

(3) If $\mathfrak{c} \cap \mathbf{Z} = q\mathbf{Z}$ with a positive integer q and ζ is a primitive q-th root of unity, then $\zeta \in F$ and $F \neq k(\zeta)$.

The fact that F is abelian can be shown as follows. Let $Y = A[\mathfrak{b}] \cap A'$ and $Y_\varepsilon = A[\mathfrak{b}] \cap (A')^\varepsilon$. Then assumption (3.7) and some elementary observations imply that $A[\mathfrak{b}] = Y + Y_\varepsilon$, and both Y and Y_ε are isomorphic to $\mathfrak{o}'/\mathfrak{c}$ as $\mathfrak{o}'$-modules. (Note that $\mathfrak{o}'/\mathfrak{c}$ is $\mathfrak{o}'$-isomorphic to $\mathfrak{o}/\mathfrak{b}$.) Every element of $\mathrm{Gal}(F/k)$ induces an automorphism of each of Y and Y_ε. Since the group of all automorphisms of the $\mathfrak{o}'$-module $\mathfrak{o}'/\mathfrak{c}$ is isomorphic to $(\mathfrak{o}'/\mathfrak{c})^\times$, we thus obtain a (canonically defined) homomorphism

(3.8) $$\mathrm{Gal}(F/k) \to (\mathfrak{o}'/\mathfrak{c})^\times \times (\mathfrak{o}'/\mathfrak{c})^\times,$$

which is obviously injective. Therefore F is abelian over k. The proof of the remaining assertions is somewhat more involved, but follows, in essence, from (3.6).

We can obtain a more definite result about the conductor of F over k, under the following set of assumptions:

(3.9) (i) $N(\mathfrak{c})$ is a prime (so that $N(\mathfrak{c}) = q$); (ii) N is a square-free product of odd primes which do not divide $q(q-1)$; (iii) $\chi(a) = \left(\dfrac{a}{N} \right)$.

Then $k = \mathbf{Q}(\sqrt{N})$.

Theorem 2. *Under the above assumptions, the conductor of F over k is exactly $q \cdot P_\infty$.*

In view of Proposition 4, we see that the prime q becomes the product of two primes $\mathfrak{q}$ and $\mathfrak{q}^\varepsilon$ in k. If $\mathfrak{o}_k$ denotes the ring of algebraic integers in k, then both $\mathfrak{o}_k/\mathfrak{q}$ and $\mathfrak{o}_k/\mathfrak{q}^\varepsilon$ are isomorphic to $\mathbf{Z}/q\mathbf{Z}$. Now, to each totally positive element α of $\mathfrak{o}_k$, we associate an element $\left(\dfrac{F/k}{\alpha\mathfrak{o}_k}\right)$ of $\mathrm{Gal}(F/k)$. Since every element of $\mathfrak{o}_k/q\mathfrak{o}_k = \mathfrak{o}_k/\mathfrak{q} \oplus \mathfrak{o}_k/\mathfrak{q}^\varepsilon$ can be represented by such an α, we thus obtain a sequence of homomorphisms

$$(3.10) \qquad (\mathbf{Z}/q\mathbf{Z})^{\times 2} \to (\mathfrak{o}_k/q\mathfrak{o}_k)^\times \to \mathrm{Gal}(F/k) \to (\mathfrak{o}'/\mathfrak{c})^{\times 2} \to (\mathbf{Z}/q\mathbf{Z})^{\times 2}.$$

The first and last arrows are isomorphisms, unique up to the exchange of the factors. Therefore we obtain a homomorphism

$$(3.11) \qquad (\mathbf{Z}/q\mathbf{Z})^{\times 2} \ni (x,\, y) \mapsto \big(g(x,\, y),\, h(x,\, y)\big) \in (\mathbf{Z}/q\mathbf{Z})^{\times 2},$$

which is canonically defined up to the exchange of the factors. Now we can naturally ask the following question:

(Q) *Is the map (3.11) the identity map, up to the exchange of x and y ?*

It is very likely that this is so if k has class number one and N is a prime. Probably we will have to modify the formulation in a more general case. We can at least prove:

$$g(x,\, x) = h(x,\, x) = x, \qquad g(x,\, y)h(x,\, y) = xy, \qquad g(x,\, y) = h(y,\, x).$$

Further we obtain

Proposition 5. *Let u be the fundamental unit of k. If the map of (3.11) is an isomorphism, $N_{k/\mathbf{Q}}(u) = -1$, and the class number of k is one, then F is the maximal ray class field over k of conductor $q \cdot P_\infty$ and $u^2 - 1$ is divisible by q.*

Let p be a rational prime such that $\chi(p) = 1$, and α a totally positive element of $\mathfrak{o}_k$ such that $N_{k/\mathbf{Q}}(\alpha) = p$. If $\alpha \equiv x \pmod{\mathfrak{q}}$ and $\alpha^\varepsilon \equiv y \pmod{\mathfrak{q}}$ with rational integers x and y, then relation (3.6) implies

$$(3.12) \qquad g(x,\, y) + h(x,\, y) \equiv a_p \pmod{\mathfrak{c}}.$$

From this fact we easily obtain

Theorem 3. *The answer to (Q) is affirmative if and only if*

$$(3.13) \qquad \mathrm{Tr}_{k/\mathbf{Q}}(\alpha) \equiv a_p \pmod{\mathfrak{c}}$$

for every rational prime p not dividing qN and for every totally positive element α of $\mathfrak{o}_k$ such that $\chi(p) = 1$ and $N_{k/\mathbf{Q}}(\alpha) = p$. Moreover, (3.13) is satisfied by all such p and α, if it is satisfied by at least one α such that $\alpha/\alpha^\varepsilon$ is of order $q - 1$ in $(\mathfrak{o}_k/\mathfrak{q})^\times$.

It should be noted that (3.13) is a special case of (Ib).

We now give a table of K, q, and u for several small primes N.

Table I: $w = 2$; $N \equiv 1 \pmod 4$. (See Note 2 at the end.)

N	dim $S_{2,\chi}$	K	q	u
29	2	$\mathbf{Q}(\sqrt{-5})$	5	$(5 + \sqrt{29})/2$
37	2	$\mathbf{Q}(\sqrt{-1})$	?	$6 + \sqrt{37}$
41	2	$\mathbf{Q}(\sqrt{-2})$	?	$32 + 5\sqrt{41}$
53	4	$\mathbf{Q}(\sqrt{-3 + \sqrt{2}})$	7	$(7 + \sqrt{53})/2$
61	4	$\mathbf{Q}(\sqrt{-4 + \sqrt{3}})$	13	$39 + 5\sqrt{61})/2$
73	4	$\mathbf{Q}(\sqrt{(-19 + \sqrt{5})/2})$	89	$1068 + 125\sqrt{73}$
89	6	$t^6 + 17t^4 + 83t^2 + 125$	5 (125?)	$500 + 53\sqrt{89}$
97	6	$t^6 + 27t^4 + 204t^2 + 467$	467	$5604 + 569\sqrt{97}$
101	8	$t^8 + 13t^6 + 51t^4 + 67t^2 + 20$	5	$10 + \sqrt{101}$
109	8	$\mathbf{Q}(\sqrt{-3})$ / $t^6 + 12t^4 + 39t^2 + 29$	3 / 29	$(261 + 25\sqrt{109})/2$

Each polynomial has a root that generates K over $\mathbf{Q}$.

At the beginning of this section we decomposed J_χ into the sum of abelian subvarieties $A_1, \ldots, A_r$. For each prime $N \le 101$ we have $J_\chi = A = A' + (A')^\varepsilon$, so that there is only one K. However, for $N = 109$ we have $J_\chi = A_1 + A_2$, $A_1 = A_1' + (A_1')^\varepsilon$, $A_2 = A_2' + (A_2')^\varepsilon$, $\dim(A_1') = 1$, $\dim(A_2') = 3$; the above table indicates that $\mathrm{End}(A_1) \otimes \mathbf{Q}$ contains $\mathbf{Q}(\sqrt{-3})$, and $\mathrm{End}(A_2) \otimes \mathbf{Q}$ contains a field of degree 6.

Now applying the last assertion of Theorem 3 to these cases, we obtain

Theorem 4. *The following assertions hold (at least) for $N = 29, 53, 61, 73, 89$, and 97.*
   (1) *The map of (3.11) is the identity map, up to the exchange of x and y.*
   (2) *F is the maximal class field over k of conductor $q \cdot P_\infty$.*
   (3) *There is no abelian variety defined over $\mathbf{Q}$ which is isogenous to A' over* **C**.
   (4) *A' is simple and $\mathrm{End}(A') \otimes \mathbf{Q} = \theta'(K')$.*

The class number of k is one for these six values of N. At present, the existence of α in the last assertion of Theorem 3 can be verified only by the numerical values of a_p. This is why we have to restrict the above theorem only to those N. We can naturally conjecture that the assertions of Theorem 4 hold for many more, probably infinitely many, N.

Recently W. Casselman [1] has proved that the abelian variety A' for each N of Theorem 4 has good reduction for all prime ideals of k. As a consequence of this result, we can show that if $N = 29$, the zeta-function of A' over k is exactly (i.e., without exceptional Euler factors) $L(s, f)L(s, f_\rho) = \sum_{n=1}^{\infty} a_n n^{-s} \sum_{n=1}^{\infty} a_n^\rho n^{-s}$.

4. Generalizations and comments

It is possible to make the same type of discussion for cusp forms of weight $w > 2$ at least if w is even. In fact we can consider the ℓ-adic representations obtained by P. Deligne [2] in place of the ℓ-adic representations on A. His result, suitably generalized and modified, seems to indicate the following (see Note 3 at the end).

For a fixed N and for each rational prime ℓ, there exists a free $\mathbf{Z}_\ell$-module W_ℓ, with the action of Hecke operators $T(n)$ and $\mathrm{Gal}(\overline{\mathbf{Q}}/\mathbf{Q})$, satisfying the following conditions, where $\mathbf{Z}_\ell$ denotes the ring of ℓ-adic integers and $\overline{\mathbf{Q}}$ the algebraic closure of $\mathbf{Q}$:

(4.1) $[W_\ell : \mathbf{Z}_\ell]$ is twice the dimension of the vector space $S_w(\Gamma_\chi)$ of all cusp forms of weight w with respect to Γ_χ.

(4.2) The action of $T(n)$ commutes with the action of every element of $\mathrm{Gal}(\overline{\mathbf{Q}}/\mathbf{Q})$.

(4.3) If p is a prime not dividing ℓN and Φ a Frobenius element of $\mathrm{Gal}(\overline{\mathbf{Q}}/\mathbf{Q})$ for any prime divisor of $\overline{\mathbf{Q}}$ dividing p, then there exists an endomorphism Φ^* of W_ℓ such that

$$\Phi + \Phi^* R_p = T(p) \quad \text{and} \quad \Phi\Phi^* = p^{w-1},$$

where R_p is the action of an element γ of $\Gamma_0(N)$ such that $p\gamma \equiv \begin{bmatrix} 1 & 0 \\ 0 & p^2 \end{bmatrix}$ (mod N).

(4.4) The action H of $\begin{bmatrix} 0 & -1 \\ N & 0 \end{bmatrix}$ on W_ℓ can be defined, and

$$H^2 = (-N)^{w-2}, \quad \Phi^* R_p = H\Phi^* H^{-1}.$$

(4.5) If δ is an element of $\mathrm{Gal}(\overline{\mathbf{Q}}/\mathbf{Q})$, then $\delta H = H\delta$ or $\delta H = -H\delta$ according as δ is trivial on k or not.

(4.6) $\det(x - \Phi)$ coincides with the determinant of $x^2 - T(p)x + R_p p^{w-1}$ on $S_w(\Gamma_\chi)$.

Now assuming the existence of such a $\mathbf{Z}_\ell$-module W_ℓ, we can develop, for every even w, a theory parallel to the case $w = 2$. In fact, we fix an eigenform $f(z) = \sum_{n=1}^{\infty} a_n e^{2\pi i n z}$ in $S_{w,\chi}$ and the field K generated by the a_n over $\mathbf{Q}$. Then we define $\mathfrak{b}$ and $\mathfrak{c}$ in the same way. Assuming $N(\mathfrak{c})$ to be an odd prime q, we consider the space W_ℓ with $\ell = q$. Put

$$X = \left\{ t \in e \cdot W_q \,\middle|\, R_p t = \chi(p)t, \; \mathfrak{c}t = 0 \right\},$$

where e is the identity element of the simple component of $\mathcal{R}$ corresponding to K. Further put

$$Y = (1 + H)X, \qquad Y_\varepsilon = (1 - H)X.$$

Then Y and Y_ε are stable under $\mathrm{Gal}(\overline{\mathbf{Q}}/k)$ by virtue of (4.5). Let M be the subgroup of $\mathrm{Gal}(\overline{\mathbf{Q}}/k)$ consisting of the elements which induce the identity map on $Y + Y_\varepsilon$, and let F be the subfield of $\overline{\mathbf{Q}}$ corresponding to M. Then F is abelian over k; also, there is an injective homomorphism

$$\mathrm{Gal}(F/k) \longrightarrow (\mathfrak{o}'/\mathfrak{c})^{\times 2},$$

from which we obtain, by a procedure similar to (3.10), a homomorphism $(x, y) \mapsto (g_w(x, y), h_w(x, y))$ of $(\mathbf{Z}/q\mathbf{Z})^{\times 2}$ into itself. Then question (Q) can be generalized to the following form (when k is of class number one):

(Q_w) *Do we have $g_w(x, y) = x^{w-1}$ and $h_w(x, y) = y^{w-1}$ up to the exchange of x and y ?*

It is then easy to prove a generalization of Theorem 3, i.e., that (Ib) is equivalent to the affirmative answer to (Q_w).

Let us now present some numerical evidence.

Table II: w even, ≥ 4; $N \equiv 1 \pmod{4}$.

N	w	$\dim S_{2,\chi}$	K	q	u^{w-1}
5	6	2	$\mathbf{Q}(\sqrt{-11})$	11	$(11 + 5\sqrt{5})/2$
5	8	2	$\mathbf{Q}(\sqrt{-29})$	29	$(29 + 13\sqrt{5})/2$
5	10	4	$\mathbf{Q}(\sqrt{-854 + 30\sqrt{809}})$	19	$38 + 17\sqrt{5}$
13	4	2	$\mathbf{Q}(\sqrt{-1})$	3?	$18 + 5\sqrt{13}$
17	4	4	$\mathbf{Q}(\sqrt{-37 + 3\sqrt{33}})$	67	$268 + 65\sqrt{17}$

We notice that

(Ic) $$\mathrm{Tr}_{k/\mathbf{Q}}(u^{w-1}) \equiv 0 \pmod{q}$$

in all these cases. By means of the same principle as in Theorem 3, we can prove, assuming (4.1 – 4.6), that

$$\mathrm{Tr}_{k/\mathbf{Q}}(\alpha^{w-1}) \equiv a_p \pmod{q}$$

for every totally positive algebraic integer α such that $N_{k/\mathbf{Q}}(\alpha) = p$ and $\chi(p) = 1$, at least in the following two cases: $N = 5$, $w = 6$; $N = 17$, $w = 4$. Then the field F is the maximal class field over k of conductor $q \cdot P_\infty$.

Table III: w odd, ≥ 3; $N \equiv 3 \pmod{4}$; u is the fundamental unit of $\mathbf{Q}(\sqrt{N})$.

N	w	$\dim S_{2,\chi}$	K	$u^{(w-1)/2}$
3	9	2	$\mathbf{Q}(\sqrt{-14})$	$97 + 56\sqrt{3}$
3	11	2	$\mathbf{Q}(\sqrt{-5})$	$362 + 209\sqrt{3}$
3	13	3	$\mathbf{Q}(\sqrt{-26})$	$1351 + 780\sqrt{3}$
7	7	3	$\mathbf{Q}(\sqrt{-510})$	$2024 + 765\sqrt{7}$
11	5	3	$\mathbf{Q}(\sqrt{-30})$	$199 + 60\sqrt{11}$
19	3	3	$\mathbf{Q}(\sqrt{-13})$	$170 + 39\sqrt{19}$

If w is odd and N is a prime $\equiv 3 \pmod{4}$, then $S_{w,\chi}$ may contain cusp forms corresponding to a Grössen-character of the *imaginary* field $\mathbf{Q}(\sqrt{-N})$.

Table III excludes the simple components of $\mathcal{R}$ corresponding to them. For such an N, $\mathbf{Q}(\sqrt{-N})$ is a reasonable field to take as the basic field. However, as the above table shows, there is still some relation between the field K and the power $u^{(w-1)/2}$ of the fundamental unit u of the *real* field $Q(\sqrt{N})$. Namely, if $u^{(w-1)/2} = a + b\sqrt{N}$ with rational integers a and b, then b and the discriminant of K over $\mathbf{Q}$ have a non-trivial common divisor, except in the case $N = 3$, $w = 11$.

We mention that there is room for improvement on the formulation. In the above discussion we have started with an ideal $\mathfrak{b}$ generated by the integers x in K such that $x^\rho = -x$. It may be better to take, instead of $\mathfrak{b}$, the ideal $\mathfrak{b}^*$ generated by the elements a_p for all primes p such that $\chi(p) = -1$. Clearly $\mathfrak{b}^* \subset \mathfrak{b}$. For example, in the case $N = 37$, $w = 2$ (resp. $N = 13$, $w = 4$), it seems that a_p is a multiple of $2i$ (resp. $3i$) for all such p. Further, in the case $N = 89$, $w = 2$, there is a possibility that one should consider 5^3 instead of $q = 5$. Also, it is desirable to loosen assumptions (3.7) and (3.9), which are too restrictive.

It should be noted that relations (Ia, Ib) are analogous to the known relations for the coefficients of the zeta-function with a Grössen-character ψ of an *imaginary* quadratic field M. Suppose that M has class number one, and let $g(z) = \sum_{n=1}^{\infty} b_n e^{2\pi i n z}$ be the Mellin inverse transform of the L-function $L(s) = \sum_{\mathfrak{a}} \psi(\mathfrak{a}) N(\mathfrak{a})^{-s+(w-1)/2}$. If $\mathfrak{f}$ is the conductor of ψ, we have $\psi(\mu \mathfrak{o}_M) = (\mu/|\mu|)^{w-1}$ for every $\mu \in \mathfrak{o}_M$ such that $\mu \equiv 1 \pmod{\mathfrak{f}}$, where $\mathfrak{o}_M$ denotes the ring of algebraic integers in M. Therefore we see that

$$b_p = 0 \quad \textit{if } p \textit{ remains prime in } M,$$

$$b_p = \mathrm{Tr}_{M/\mathbf{Q}}(\mu^{w-1}) \quad \textit{if } p = \mu\bar{\mu} \textit{ with } \mu \in \mathfrak{o}_M, \ \mu \equiv 1 \pmod{\mathfrak{f}}.$$

Thus, in this case, we have the exact equalities instead of the congruences in (Ia, Ib).

Also, we may notice that Proposition 4 resembles the relation between a CM-type and its reflex in the theory of complex multiplication of abelian varieties.

Instead of quadratic characters, we can consider characters of higher order, although things are rather obscure in this case. To be more specific, for a character χ of $(\mathbf{Z}/N\mathbf{Z})^\times$ of an arbitray order > 1, we define Γ_χ, $S_{w,\chi}$, $T(n)$, and $\mathcal{R}_\mathbf{C}$ as in Section 1. Take again a common eigen-function $f(z) = \sum_{n=1}^{\infty} a_n e^{2\pi i n z}$ in $S_{w,\chi}$ for all $T(n)$, and let K denote the field generated by the coefficients a_n over $\mathbf{Q}$. If condition (1.5) is satisfies, we can prove that K is a totally imaginary quadratic extension of a totally real algebraic number field K'. The field K contains the values $\chi(n)$ for all n, and $a_n = \chi(n)a_n^\rho$ for all n. Therefore, assuming that χ is of order > 2, we easily see that

$$a_n/[1 + \chi(n)] \in K' \quad \text{if} \ \ \chi(n) \neq -1,$$

$$a_n/[\omega - \omega^{-1}] \in K' \quad \text{if} \ \ \chi(n) = -1,$$

where ω is any root of unity, other than ± 1, contained in K. Therefore, if we let b_n denote the ideal in K generated by the algebraic integers x in K satisfying $x = \chi(n)x^\rho$, then b_n is a divisor of either $1 + \chi(n)$ or $\omega - \omega^{-1}$ according as $\chi(n) \neq -1$ or $\chi(n) = -1$. This phenomenon makes the case of characters of higher order quite different from that of quadratic characters. However, we can at least prove the following: If $N = 13$, $w = 2$, and χ is of order 6, then $\dim(S_{2,\chi}) = 1$, so that there is a unique eigenform $f(z) = \sum_{n=1}^{\infty} a_n e^{2\pi i n z}$ in $S_{2,\chi}$. Then $K = \mathbf{Q}(\zeta)$ with $\zeta = e^{2\pi i/3}$, and we have

$a_p \equiv 0 \pmod{(1 - \zeta)}$ *if p remains prime in $k = \mathbf{Q}(\sqrt{13}\,)$,*

$a_p \equiv \chi(c)\mathrm{Tr}_{k/\mathbf{Q}}(\alpha) \pmod{(3)}$ *if $p = N_{k/\mathbf{Q}}(\alpha) \equiv c^2 \pmod{13}$ with a totally positive integer α of k and a rational integer c.*

The fundamental unit of $\mathbf{Q}(\sqrt{13}\,)$ is $(3 + \sqrt{13}\,)/2$. Further we obtain the maximal ray class field over k of conductor $3 \cdot P_\infty$ from certain points of finite order on the Jacobian variety of $\Gamma_\chi \backslash \mathfrak{H}$. It is an open question whether or not this is an exceptional case.

Finally a word about further generalizations. One can also discuss automorphic forms with respect to unit groups in a quaternion algebra, whose relation with zeta-functions curves was given by the author [8] and T. Miyake [6]. Actually Miyake has obtained the decomposition of J and a generalization of Proposition 3 in the most general case with such automorphic forms and with characters of arbitrary order. In this case, k is replaced by an algebraic number field of higher degree, so that one has to deal with groups of units with more than one generator.

Note 1. Except for [7], [8], and [9]. In these papers, it was shown that class fields over a totally real algebraic number field occur as fields of definition, or fields of moduli, of arithmetic quotients of bounded symmetric domains. However, this is completely different from the subject of the present article.

Note 2. Tables I, II, and III are based on the numerical values of the eigenvalues of Hecke operators for some prime degrees, which have been obtained by K. Doi and H. Naganuma (by hand), and H. Trotter (by computer). It is a pleasure for me to express my thanks to them.

Note 3. At least one has to verify relations (4.4) and (4.5) which are not given in [2], although these do not seem difficult. there is also a different method of constructing ℓ-adic representations, on which I gave a leclture at the conference *Automorphic functions for arithmetically defined groups*, Oberwolfach, Germany, July28–August 3, 1968. (See [**68c**] in these volumes.) In this theory the existence of W_ℓ satisfying (4.1) through (4.5) can be shown, but there is some difficulty in proving (4.6).

REFERENCES

1. W. Casselman, Some new abelian vareities with good reduction, to appear.

2. P. Deligne, Formes modulaires et représentations ℓ-adiques, Séminaire Bourbaki, février 1969, exposé 355.

3. E. Hecke, Zur Theorie der elliptischen Funktionen, Math. Ann. **97** (1926), 210–242 (= Werke, 428–460).

4. ______, Analytische Arithmetik der positiven quadratischen Formen, Danske Vid. Selsk. Mat.-Fys. Medd. XVII, 12 (= Werke 789–918).

5. H. Maass, Über eine neue Art von nichtanalytischen automorphen Funktionen und die Bestimmung Dirichletscher Reihen durch Funktionalgleichungen, Math. Ann. **121** (1949), 141–183.

6. T. Miyake, Decomposition of Jacobian varieties and Dirichlet series of Hecke type, to appear.

7. G. Shimura, Class-fields and automorphic functions, Ann. of Math. **80** (1964), 444–463.

8. ______, Construction of class fields and zeta functions of algebraic curves, Ann. of Math. **85** (1967), 58–159.

9. ______, On canonical models of arithmetic quotients of bounded symmetric domains, Ann. of Math. **91** (1970), 144–222.

10. ______, Introduction to the arithmetic theory of automorphic functions, to appear.

Princeton University, Princeton, New Jersey 08540

On elliptic curves with complex multiplication as factors of the Jacobians of modular function fields

Nagoya Mathematical Journal, 43 (1971), 199-208

1. As Hecke showed, every L-function of an imaginary quadratic field K with a Grössen-character λ is the Mellin transform of a cusp form $f(z)$ belonging to a certain congruence subgroup Γ of $SL_2(\mathbf{Z})$. We can normalize λ so that

$$\lambda((\alpha)) = \alpha^\nu \quad \text{for} \quad \alpha \in K, \ \alpha \equiv 1 \ \mathrm{mod}^\times \mathfrak{c}$$

with a positive integer ν, where $\mathfrak{c}$ is the conductor of λ, and $\mathrm{mod}^\times \mathfrak{c}$ means the multiplicative congruence modulo $\mathfrak{c}$. Then $f(z)$ is of weight $\nu+1$, i.e.,

$$f((az + b)/(cz + d)) = f(z)(cz + d)^{\nu+1} \quad \text{for} \quad \begin{bmatrix} a & b \\ c & d \end{bmatrix} \in \Gamma,$$

and Γ is given by

$$\Gamma = \left\{ \begin{bmatrix} a & b \\ c & d \end{bmatrix} \in SL_2(\mathbf{Z}) \,\middle|\, a \equiv d \equiv 1, \ c \equiv 0 \ \mathrm{mod} \ (D \cdot N(\mathfrak{c})) \right\},$$

where $-D$ is the discriminant of K. If $\nu = 1$, $f(z)dz$ is a differential form of the first kind on the compactification $(H/\Gamma)^*$ of the quotient H/Γ, where H denotes the upper half complex plane. Denote by $\mathrm{Jac}\,(H/\Gamma)$ the jacobian variety of $(H/\Gamma)^*$, and identify the tangent space of $\mathrm{Jac}\,(H/\Gamma)$ at the origin with the space of all differential forms of the first kind on $(H/\Gamma)^*$. Let A be the smallest abelian subvariety of $\mathrm{Jac}\,(H/\Gamma)$ that has $f(z)dz$ as a tangent at the origin. Then the first main result of this paper can be stated as follows:

The abelian variety A is a product of copies of an elliptic curve whose endomorphism algebra is isomorphic to K.

Hecke [3] proved this fact in the case where $K = Q(\sqrt{-q})$ with a prime $q > 3$, $\equiv 3 \ \mathrm{mod} \ (4)$ and $\mathfrak{c} = (\sqrt{-q})$. In the general case, he showed only that

---

Received January 16, 1971.

199

the periods of $f(z)dz$ belong to a certain class field over K. His proof requires rather deep arithmetic results of complex multiplication. Ours is simpler, and based on the following

LEMMA 1. *Let X be an abelian variety of dimension n defined over C, and h an injective homomorphism of K into $\mathrm{End}_Q(X)$. Suppose that the representation of K, through h, on the tangent space of X at the origin is equivalent to n copies of the identity injection of K into C. Then X is isogenous to a product of n copies of an elliptic curve E such that $\mathrm{End}_Q(E)$ is isomorphic to K.*

Here and henceforth we denote by $\mathrm{End}(X)$ the ring of all endomorphisms of X over C, and put $\mathrm{End}_Q(X) = \mathrm{End}(X) \otimes Q$.

Our next purpose is to show that every elliptic curve E defined over Q with complex multiplication is isogenous over Q to a factor of $\mathrm{Jac}\,(H/\Gamma')$ for some Γ' in the following way. By virtue of Deuring's result [1], if K is isomorphic to $\mathrm{End}_Q(E)$, the zeta-function of E over Q is exactly the L-function of a certain Grössen-character λ of K. Then we obtain an abelian variety A by the procedure described above, i.e.,

elliptic curve $E \to$ zeta-function with a Grössen-character λ
$\to$ cusp form $f(z) \to$ abelian subvariety A of $\mathrm{Jac}\,(H/\Gamma')$.

In this situation, we shall prove:

A is an elliptic curve isogenous to E over Q.

This is an easy consequence of the results in the previous articles [7], [8]. If $-D$ is the discriminant of K, and $\mathfrak{c}$ is the conductor of λ, the group Γ' is of the form

$$\Gamma' = \left\{ \begin{bmatrix} a & b \\ c & d \end{bmatrix} \in SL_2(Z) \,\middle|\, c \equiv 0 \ \mathrm{mod}\ (D \cdot N(\mathfrak{c})) \right\}.$$

2.   Let us first prove the above lemma. Although it is a special case of [6, Prop. 14], we give here a direct proof for the reader's convenience.

Identify X with a complex torus C^n/L with a lattice L. Let $Q \cdot L$ denote the Q-linear span of L. Then K acts, through h, on $Q \cdot L$, so that there exists a K-linear isomorphism p of K^n onto $Q \cdot L$, where K^n is the submodule of C^n consisting of the vectors whose components belong to K. Since $C^n = K^n \otimes_Q R = (Q \cdot L) \otimes_Q R$, we can extend p to an R-linear automorphism of C^n, which we denote again by p. By our assumption, we may assume that the action of an element α of K on X is represented by the complex linear transformation $u \longrightarrow \alpha u$ $(u \in C^n)$ of C^n. We can find a real number r and an element α of K so that $r \cdot \alpha = \sqrt{-1}$. Now p is K-linear and R-linear, hence p commutes with the map $u \to \sqrt{-1} \cdot u$, i.e., p is C-linear. Take any free Z-submodule $\mathfrak{a}$ of rank 2 in K. Then p gives an isogeny of $C^n/\mathfrak{a}^n = (C/\mathfrak{a})^n$ onto C^n/L. This proves the lemma, since $C/\mathfrak{a}$ is an elliptic curve with K as its endomorphism algebra.

3. For a function $f(z)$ on H and $\xi = \begin{bmatrix} a & b \\ c & d \end{bmatrix} \in GL_2(R)$ with $\det(\xi) > 0$, we define a function $f|[\xi]_k$ on H by

$$(f|[\xi]_k)(z) = \det(\xi)^{k/2} \cdot (cz + d)^{-k} \cdot f((az + b)/(cz + d)).$$

For an arbitrary positive integer N, put

$$\Gamma_0(N) = \left\{ \begin{bmatrix} a & b \\ c & d \end{bmatrix} \in SL_2(Z) \,\middle|\, c \equiv 0 \mod (N) \right\},$$

$$\Gamma_1(N) = \left\{ \begin{bmatrix} a & b \\ c & d \end{bmatrix} \in \Gamma_0(N) \,\middle|\, a \equiv 1 \mod (N) \right\}.$$

Further, for a complex-valued character ε of $(Z/NZ)^{\times}$,[1] we denote by $S_k(N, \varepsilon)$ the vector space of all the cusp forms $f(z)$ satisfying

$$f|[\gamma]_k = \varepsilon(d) \cdot f$$

for every $\gamma = \begin{bmatrix} a & b \\ c & d \end{bmatrix} \in \Gamma_0(N)$.

LEMMA 2. *Let* $f(z) = \sum_{n=1}^{\infty} a_n e^{2\pi i n z}$ *be an element of* $S_k(N, \varepsilon)$, r *a positive integer,* M *a common multiple of* Nr *and* r^2, *and let*

$$g(z) = \sum_{(n,\, r)=1} a_n e^{2\pi i n z}.$$

Then $g \in S_k(M, \varepsilon')$, *where* ε' *is the restriction of* ε *to* $(Z/MZ)^{\times}$.

Proof. Put $\zeta = e^{2\pi i/r}$, $\eta_u = \begin{bmatrix} r & u \\ 0 & r \end{bmatrix}$ for $u \in Z$, and $\Gamma = \Gamma_1(N)$. We see easily that $\Gamma \eta_u = \Gamma \eta_v$ if and only if $u \equiv v \mod (r)$. We can find numbers x_u of $Q(\zeta)$ for $u \in Z$ such that

$$x_u = x_v \quad \text{if} \quad u \equiv v \mod (r),$$

$$\sum_{u=0}^{r-1} x_u \zeta^{un} = \begin{cases} 1 & \text{if} \quad (n, r) = 1, \\ 0 & \text{otherwise.} \end{cases}$$

We see easily that $g(z) = \sum_{u=0}^{r-1} x_u \cdot f|[\eta_u]_k$.   Further, it can be seen that

(1) $x_u = x_{au}$ if $(a, r) = 1$,

and x_u is invariant under $\mathrm{Gal}\,(Q(\zeta)/Q)$, hence $x_u \in Q$.   Now $g(z)$ is a cusp form of level Nr^2 (see for example [7, Prop. 2.4, Lemma 3.9]).   Therefore, to prove our assertion, it is sufficient to check the behavior of g under an element $\gamma = \begin{bmatrix} a & b \\ Mc & d \end{bmatrix}$ of $\Gamma_0(M)$.   We have

$$\begin{bmatrix} r & u \\ 0 & r \end{bmatrix}\begin{bmatrix} a & b \\ Mc & d \end{bmatrix} = \begin{bmatrix} a' & b' \\ Mc & d' \end{bmatrix}\begin{bmatrix} r & d^2u \\ 0 & r \end{bmatrix}$$

with $a' = a + cuM/r$, $b' = b + du(1 - a'd)/r$, $d' = d - cd^2u\,M/r$.   Note that $a' \equiv a$, $d' \equiv d$ mod $(N)\cap(r)$, and $a'd \equiv ad \equiv 1$ mod (r).   Therefore, putting $v = d^2u$, we have $f|[\eta_u\gamma]_k = \varepsilon(d)\cdot f|[\eta_v]_k$.   In view of (1), we obtain $g|[\gamma]_k = \varepsilon(d)\cdot g$, q.e.d.

4.   For our purpose, it is necessary to consider Grössen-characters which are not necessarily "primitive".   To define them, let $\mathfrak{m}$ be an integral ideal in K, and $I_\mathfrak{m}$ the group of all fractional ideals in K prime to $\mathfrak{m}$. Let $W_\mathfrak{m}$ denote the group of all elements α of $K^\times$ such that $\alpha \equiv 1$ mod$^\times$ $\mathfrak{m}$, i.e., $\alpha - 1$ is $\mathfrak{p}$-integral and divisible by $\mathfrak{m}_\mathfrak{p}$ for all prime factors $\mathfrak{p}$ of $\mathfrak{m}$, where $\mathfrak{m}_\mathfrak{p}$ is the $\mathfrak{p}$-closure of $\mathfrak{m}$.   Further let $P_\mathfrak{m}$ denote the subgroup of $I_\mathfrak{m}$ consisting of all principal ideals (α) with $\alpha \in W_\mathfrak{m}$.   For a positive integer ν, let $\Lambda_\mathfrak{m}^\nu$ denote the set of all homomorphisms λ of $I_\mathfrak{m}$ into $C^\times$ such that $\lambda((\alpha)) = \alpha^\nu$ for every $\alpha \in W_\mathfrak{m}$.   Such a λ is called a Grössen-character of K defined modulo $\mathfrak{m}$.   Obviously, $\Lambda_\mathfrak{m}^\nu$ is not empty if and only if the following condition is satisfied:

(2)   *If ζ is a root of unity in K and $\zeta \equiv 1$ mod $\mathfrak{m}$, then $\zeta^\nu = 1$.*

For each $\lambda \in \Lambda_\mathfrak{m}^\nu$, there is a unique divisor $\mathfrak{c}$ of $\mathfrak{m}$ such that: (i) λ is the restriction of an element of $\Lambda_\mathfrak{c}^\nu$; (ii) no proper divisor of $\mathfrak{c}$ has the property (i).   Then $\mathfrak{c}$ is called the *conductor* of λ.   We call λ *primitive* if $\mathfrak{m}$ is the conductor of λ.

We can associate with every $\lambda \in \Lambda_\mathfrak{m}^\nu$ an L-function $L(s, \lambda)$ and a function $f_\lambda(z)$ on H by

$$L(s, \lambda) = \sum_\mathfrak{x} \lambda(\mathfrak{x}) N(\mathfrak{x})^{-s} \qquad\qquad (s \in C),$$

$$f_\lambda(z) = \sum_\mathfrak{x} \lambda(\mathfrak{x}) e^{2\pi i N(\mathfrak{x})z} \qquad\qquad (z \in H),$$

where each sum is taken over all integral ideals $\mathfrak{x}$ in $I_\mathfrak{m}$. Under the assumption (2), let $V_\mathfrak{m}^\nu$ be the vector space spanned by the f_λ over C for all $\lambda \in \Lambda_\mathfrak{m}^\nu$. For λ, $\mu \in \Lambda_\mathfrak{m}^\nu$, we see easily that $f_\lambda = f_\mu$ if and only if $\lambda = \mu$. Moreover, we shall see later that the f_λ for $\lambda \in \Lambda_\mathfrak{m}^\nu$ are linearly independent over C. Therefore $V_\mathfrak{m}^\nu$ is of dimension $[I_\mathfrak{m} : P_\mathfrak{m}]$.

Fix any set S of representatives for $I_\mathfrak{m}$ modulo $P_\mathfrak{m}$, whose members are prime to $\mathfrak{m}$, and put, for each $\mathfrak{a} \in S$,

$$(3) \qquad g_\mathfrak{a}(z) = \sum_{(\alpha)} \alpha^\nu \cdot e^{2\pi i N(\alpha)z/N(\mathfrak{a})},$$

where the sum is taken over all ideals (α) such that $\alpha \in W_\mathfrak{m} \cap \mathfrak{a}$. We have then

$$f_\lambda = \sum_{\mathfrak{a} \in S} \lambda(\mathfrak{a})^{-1} \cdot g_\mathfrak{a},$$

so that the functions $g_\mathfrak{a}$, for $\mathfrak{a} \in S$, form a basis of $V_\mathfrak{m}^\nu$ over C. Hecke [2] proved that $g_\mathfrak{a}$ is a cusp form belonging to a certain congruence subgroup. We can state this fact in the following form.

LEMMA 3. *Let* $-D$ *be the discriminant of* K, *and let* $\lambda \in \Lambda_\mathfrak{m}^\nu$, $M = D \cdot N(\mathfrak{m})$. *Then* f_λ *is an element of* $S_{\nu+1}(M, \varepsilon)$, *where* ε *is the character of* $(\mathbf{Z}/M\mathbf{Z})^\times$ *defined by*

$$\varepsilon(a) = \left(\frac{-D}{a}\right) \cdot \frac{\lambda((a))}{a^\nu} \qquad (a \in \mathbf{Z}, \ (a, M) = 1).$$

Proof. If λ is primitive, our assertion can be proved by examining the functional equations of $L(s, \lambda)$ and

$$L(s, \lambda, \chi) = \sum_{\mathfrak{x}} \lambda(\mathfrak{x}) \chi(N(\mathfrak{x})) N(\mathfrak{x})^{-s}$$

with primitive characters χ of $(\mathbf{Z}/p\mathbf{Z})^\times$ for all rational primes p not dividing M, and applying the principle of Weil [9]. Although [9, Satz 2] is concerned with $S_k(M, \varepsilon)$ for real characters ε, the result can easily be extended to the case of an arbitrary character ε. Let us now prove the general case by induction on $N(\mathfrak{c}^{-1}\mathfrak{m})$, where $\mathfrak{c}$ is the conductor of λ. Suppose that $\mathfrak{c}^{-1}\mathfrak{m}$ has a prime factor $\mathfrak{p}$, and put $\mathfrak{n} = \mathfrak{p}^{-1}\mathfrak{m}$. Let μ be the element of $\Lambda_\mathfrak{n}^\nu$ whose restriction to $\Lambda_\mathfrak{m}^\nu$ is λ. By the induction assumption, f_μ belongs to $S_{\nu+1}(D \cdot N(\mathfrak{n}), \varepsilon)$. Put $q = N(\mathfrak{p})$. Then

$$f_\mu(qz) = \sum_{(\mathfrak{x}, \mathfrak{n})=1} \mu(\mathfrak{x}) e^{2\pi i N(\mathfrak{p}\mathfrak{x})z},$$

hence

(4) $$f_\mu(z) - \mu(\mathfrak{p})f_\mu(qz) = \sum_{(\mathfrak{k},\,\mathfrak{m})=1}\mu(\mathfrak{k})e^{2\pi i N(\mathfrak{k})z} = f_\lambda(z),$$

where we understand that $\mu(\mathfrak{p}) = 0$ if $\mathfrak{p}$ divides $\mathfrak{n}$. Since we have

$$\begin{bmatrix} q & 0 \\ 0 & 1 \end{bmatrix}\begin{bmatrix} a & b \\ qc & d \end{bmatrix} = \begin{bmatrix} a & qb \\ c & d \end{bmatrix}\begin{bmatrix} q & 0 \\ 0 & 1 \end{bmatrix},$$

it can easily be verified that $f_\mu(qz)\in S_{\nu+1}(q\cdot D\cdot N(\mathfrak{n}),\varepsilon)$. Therefore the equality (4) implies that $f_\lambda\in S_{\nu+1}(q\cdot D\cdot N(\mathfrak{n}),\varepsilon)$, q.e.d.

The symbols λ, M, and ε being as above, put $f_\lambda(z) = \sum_n a_n e^{2\pi i n z}$. Then the L-function $L(s,\lambda)$ has an Euler product:

$$L(s,\lambda) = \Pi_p(1 - a_p p^{-s} + \varepsilon(p)p^{\nu-2s})^{-1},$$

where the product is taken over all rational primes p ; $\varepsilon(p) = 0$ for every prime factor p of M. Therefore, by Hecke [4, II, Satz 42] (see also [7, Th. 3.43]), f_λ must be a common eigen-function of all Hecke operators. Thus the functions f_λ, for $\lambda\in\Lambda^\nu_\mathfrak{m}$, are distinct eigen-functions whose first Fourier coefficients are 1. Therefore they are linearly independent over C.

5. Let us now consider a projective non-singular curve C_M biregularly isomorphic to the compactification of the quotient $H/\Gamma_1(M)$ for a positive integer M. There is a "standard" way to define C_M rational over Q, up to biregular isomorphisms over Q. (One can define, for instance, the function field of C_M to be the field of all $\Gamma_1(M)$-invariant modular functions whose Fourier expansions with respect to $e^{2\pi i z}$ have rational coefficients. See also [5], [7, §6.7, §6.3].) Then the jacobian variety Jac(C_M) of C_M can naturally be defined over Q. We denote by τ_n the endomorphism of Jac(C_M) corresponding to the Hecke operator of degree n.

Let $\lambda\in\Lambda^1_\mathfrak{m}$, $M = D\cdot N(\mathfrak{m})$, and $f_\lambda(z) = \sum_n a_n e^{2\pi i n z}$. Further let k_λ denote the field generated over Q by the numbers a_n for all n. Since f_λ is a common eigen-function of all Hecke operators, we obtain, by virtue of [7, Th. 7.14], a couple $(A_\lambda,\theta_\lambda)$ satisfying the following three conditions:

(i) A_λ *is an abelian subvariety of* Jac(C_M) *of dimension* $[k_\lambda : Q]$.

(ii) θ_λ *is an isomorphism of* k_λ *into* $End_Q(A_\lambda)$ *such that* $\theta_\lambda(a_n)$ *is the restriction of* τ_n *to* A_λ *for all* n.

(iii) A_λ *is rational over* Q.

Moreover, $(A_\lambda,\theta_\lambda)$ is unique for f_λ under the conditions (i) and (ii).

For an automorphism σ of the algebraic closure of $\mathbf{Q}$, we define an element λ_σ of $\Lambda^1_{\mathfrak{m}^\rho}$ by $\lambda_\sigma(\xi) = \lambda(\xi^\sigma)^\sigma$. If $f_\lambda(z) = \sum_n a_n e^{2\pi i n z}$, we see that $f_{\lambda_\sigma}(z) = \sum_n a_n^\sigma e^{2\pi i n z}$. Now identify the tangent space of $\mathrm{Jac}\,(C_M)$ at the origin with the space of all cusp forms of weight 2 with respect to $\Gamma_1(M)$. Then the proof of [7, Th. 7.14] shows that the tangent space of A_λ at the origin can be identified with the vector space spanned by all distinct f_{λ_σ}. Therefore our result mentioned at the beginning of this paper follows from the following

THEOREM 1. *The abelian variety A_λ is isogenous to a product of copies of an elliptic curve whose endomorphism algebra is isomorphic to K.*

Proof. (I) First let us assume that $\mathfrak{m}$ is divisible by $\sqrt{-D}$, and $\mathfrak{m} = \mathfrak{m}^\rho$, where ρ denotes the complex conjugation. Put

$$\Gamma = \Gamma_1(M), \quad \delta = \begin{bmatrix} 1 & 1/d \\ 0 & 1 \end{bmatrix}.$$

We can let $\Gamma\delta\Gamma$ act on the vector space of cusp forms with respect to Γ (see [7, §3.4]). Denote the action by $[\Gamma\delta\Gamma]_2$. Take a disjoint coset decomposition $\Gamma\delta\Gamma = \bigcup_{i=1}^{e}\Gamma\delta\gamma_i$ with $\gamma_i \in \Gamma$. Let $g_\mathfrak{a}$ be as in (3). Then, by definition,

$$g_\mathfrak{a}|[\Gamma\delta\Gamma]_2 = \bigcup_{i=1}^{e} g_\mathfrak{a}|[\delta\gamma_i]_2.$$

If $\alpha, \beta \in W_\mathfrak{m} \cap \mathfrak{a}$, we have

$$N(\alpha)/N(\mathfrak{a}) \equiv N(\beta)/N(\mathfrak{a}) \bmod (D),$$

so that, if $\zeta_D = e^{2\pi i/D}$,

$$g_\mathfrak{a}|[\delta]_2 = \zeta_D^{N(\alpha)/N(\mathfrak{a})} \cdot g_\mathfrak{a}$$

with any fixed α contained in $W_\mathfrak{m} \cap \mathfrak{a}$. Therefore

$$(5) \qquad g_\mathfrak{a}|[\Gamma\delta\Gamma]_2 = \kappa \cdot \zeta_D^{N(\alpha)/N(\mathfrak{a})} \cdot g_\mathfrak{a}.$$

Thus $[\Gamma\delta\Gamma]_2$ maps $V^1_\mathfrak{m}$ onto itself. Let A' be the abelian subvariety of $\mathrm{Jac}\,(C_M)$ generated by the A_λ for all $\lambda \in \Lambda^1_\mathfrak{m}$. Since $\mathfrak{m} = \mathfrak{m}^\rho$, $V^1_\mathfrak{m}$ can be identified with the tangent space of A' at the origin. Let ω denote the endomorphism of A' obtained from $[\Gamma\delta\Gamma]_2$. The relation (5) shows that the representation of ω on the tangent space has characteristic roots $\kappa \cdot \zeta_D^{N(\alpha)/N(\mathfrak{a})}$, where α must be fixed for each $\mathfrak{a} \in S$. Put $\chi(r) = \left(\dfrac{-D}{r}\right)$. Then we see that

$N(\alpha)/N(\mathfrak{a})$ is prime to D, and $\chi(N(\alpha)/N(\mathfrak{a})) = 1$. We can define an embedding h of $\boldsymbol{Q}(\zeta_D)$ into $\mathrm{End}_{\boldsymbol{Q}}(A')$ by $h(\zeta_D) = \kappa^{-1}\omega$. If σ is an automorphism of $\boldsymbol{Q}(\zeta_D)$ such that $\zeta_D^\sigma = \zeta_D^r$ with $\chi(r) = 1$, then the restriction of σ to K is the identity map. Therefore applying Lemma 1 to A', we see that A' is isogenous to a product of copies of an elliptic curve with K as its endomorphism algebra.

(II) Next assume that λ is primitive, and put $\mathfrak{m}' = \mathfrak{m}\mathfrak{m}^\rho\cdot(\sqrt{-D})$, $M' = N(\mathfrak{m}')\cdot D$, $\eta_u = \begin{bmatrix} M & u \\ 0 & M \end{bmatrix}$ for $u \in \boldsymbol{Z}$. Then $M' = M^2$ and $\mathfrak{m}' = \mathfrak{m}'^\rho$. Define, as in the proof of Lemma 2, rational numbers x_u so that

$$\textstyle\sum_{u=0}^{M-1} x_u \zeta_M^{un} = \begin{cases} 1 & \text{if } (n, M) = 1, \\ 0 & \text{otherwise,} \end{cases}$$

where $\zeta_M = e^{2\pi i/M}$. Take a positive integer t so that tx_u is an integer for every u. Put $\xi = \sum_{u=0}^{M-1} tx_u \cdot [\eta_u]_2$. For every

$$f(z) = \textstyle\sum_n a_n e^{2\pi i n z} \in S_2(M, \varepsilon),$$

we have, by Lemma 2 and its proof,

$$f|\xi = t \cdot \textstyle\sum_{(n, M)=1} a_n e^{2\pi i n z} \in S_2(M', \varepsilon).$$

Especially $f_\lambda|\xi = t \cdot f_\mu$ if μ is the restriction of λ to $I_{\mathfrak{m}'}$. Let V_λ be the subspace of $V_{\mathfrak{m}}^1 + V_{\mathfrak{m}'}^1$ spanned by all distinct f_{λ_σ} with automorphisms σ of the algebraic closure of $\boldsymbol{Q}$. Since λ is primitive, we see that ξ maps V_λ *injectively* into $V_{\mathfrak{m}'}^1$. (This is not necessarily true if λ is not primitive.) Since $\eta_u \cdot \Gamma_1(M')\eta_u^{-1} \subset \Gamma_1(M)$, the action $[\eta_u]_2$ defines a homomorphism of $\mathrm{Jac}(C_M)$ into $\mathrm{Jac}(C_{M'})$, hence ξ defines a homomorphism ξ^* of $\mathrm{Jac}(C_M)$ into $\mathrm{Jac}(C_{M'})$. Then the restriction of ξ^* to A_λ is an isogeny onto an abelian subvariety of A'', where A'' is the sum of A_μ for all $\mu \in \Lambda_{\mathfrak{m}'}^1$. By the result in the case (I), A'' is isogenous to a product of copies of an elliptic curve with K as its endomorphism algebra. Therefore A_λ has the same property.

(III) Finally let us consider the general case with no assumption on $\mathfrak{m}$. Let $\mathfrak{c}$ be the conductor of λ. To prove our assertion by induction on $N(\mathfrak{c}^{-1}\mathfrak{m})$, suppose that $\mathfrak{c}^{-1}\mathfrak{m}$ has a prime factor $\mathfrak{p}$, and put $\mathfrak{n} = \mathfrak{p}^{-1}\mathfrak{m}$, $q = N(\mathfrak{p})$, $L = q^{-1}M$, $\beta = \begin{bmatrix} q & 0 \\ 0 & 1 \end{bmatrix}$. Since $\beta\Gamma_1(M)\beta^{-1} \subset \Gamma_1(L)$, $[\beta]_2$ defines an endomorphism ψ of $\mathrm{Jac}(C_L)$ into $\mathrm{Jac}(C_M)$. Let φ be the natural map of $\mathrm{Jac}(C_L)$ into $\mathrm{Jac}(C_M)$ corresponding to $[1]_2$. If μ is the element of $\Lambda_{\mathfrak{n}}^1$ whose restriction to $I_{\mathfrak{m}}$ is λ, we have $f_{\lambda_\sigma} = f_{\mu_\sigma} - s \cdot f_{\mu_\sigma}|[\beta]_2$ with a constant s, by virtue of (4), for every automorphism σ of the algebraic closure of Q. This shows that $A_\lambda \subset \varphi(A_\mu) + \psi(A_\mu)$. Therefore our assertion about A_λ follows from that about A_μ, which is ensured by induction.

Remark. We have thus shown that the center $\mathfrak{Z}$ of $\mathrm{End}_Q(A_\lambda)$ is isomorphic to K. It should be noted here that $\mathfrak{Z}$ *is not contained in* $\theta_\lambda(k_\lambda)$. This follows from either of the following two facts:

(i) The elements of $\theta_\lambda(k_\lambda) \cap \mathrm{End}(A_\lambda)$ are rational over Q (see [7, pp. 182-183]), while K is the smallest field of definition for any generator of $\mathfrak{Z}$ contained in $\mathrm{End}(A_\lambda)$.

(ii) The representation of k_λ, through θ_λ, on the tangent space of A_λ at the origin is equivalent to a regular representation over Q.

6. Let E be an elliptic curve defined over Q such that $\mathrm{End}_Q(E)$ is isomorphic to K. (This can happen if and only if the class number of K is one.) By the result of Deuring [1], the zeta-function of E over Q coincides exactly with $L(s, \lambda)$ with some primitive Grössen-character λ of K. Let $\mathfrak{c}$ be the conductor of λ, and $M = D \cdot N(\mathfrak{c})$. Then we obtain an element f_λ of $S_2(M, \varepsilon)$ as before. If $f_\lambda(z) = \sum_n a_n e^{2\pi i n z}$, we have

$$(6) \qquad L(s, \lambda) = \Pi_p(1 - a_p p^{-s} + \varepsilon(p) p^{1-2s})^{-1}.$$

Since E is defined over Q, we see that $a_n \in Q$, and ε is the trivial character, so that f_λ is a cusp form invariant under $\Gamma_0(M)$. Therefore we can take $\mathrm{Jac}\,(H/\Gamma_0(M))$ (of course defined over Q) instead of $\mathrm{Jac}\,(H/\Gamma_1(M))$ in the above discussion, and define A_λ as an abelian subvariety of $\mathrm{Jac}\,(H/\Gamma_0(M))$. Since $k_\lambda = Q$, A_λ is an elliptic curve defined over Q.

THEOREM 2. *The elliptic curve A_λ is isogenous to E over Q.*

Proof. By [7, Th. 7.15], the zeta-function of A_λ over Q coincides, up to finitely many Euler factors, with (6). On the other hand, by Theorem 1, $\mathrm{End}_Q(A_\lambda)$ is isomorphic to K, so that the zeta-function of A_λ over Q is $L(s, \mu)$ with a primitive Grössen-character μ of K. Thus $L(s, \lambda)$ coincides with $L(s, \mu)$ up to finitely many Euler factors. It follows that $\lambda(\mathfrak{p}) = \mu(\mathfrak{p})$ or $\lambda(\mathfrak{p}) = \mu(\mathfrak{p}^\rho)$ for almost all prime ideals $\mathfrak{p}$ in K. If $\mathfrak{m}$ is a common multiple of the conductors of λ and μ, we have $\lambda((\alpha)) = \alpha = \mu((\alpha))$ for $\alpha \in K$, $\alpha \equiv 1$ $\mathrm{mod}^\times \mathfrak{m}$. Therefore we must have $\lambda(\mathfrak{p}) = \mu(\mathfrak{p})$, so that $\lambda = \mu$. Thus E and

A_1 determine the same Grössen-character of K. By [8, Th. 8], they must be isogenous over $\mathbf{Q}$.

It should be noted that E has good reduction modulo a rational prime p if and only if p does not divide $D \cdot N(\mathfrak{c})$. This is due to Deuring [1, IV] (see also [8] for a simpler proof).

REFERENCES

[1] M. Deuring, Die Zetafunktion einer algebraischen Kurve vom Geschlecht Eins, I, II, III, IV, Nachr. Akad. Wiss. Göttingen, (1953) 85–94, (1955) 13–42, (1956) 37–76, (1957) 55–80.

[2] E. Hecke, Zur Theorie der elliptischen Modulfunktionen, Math. Ann., 97 (1926), 210–242 (=Math. Werke, 428–460).

[3] E. Hecke, Bestimmung der Perioden gewisser Integrale durch die Theorie der Klassenkörper, Math. Zeitschr., 28 (1928), 708–727 (=Math. Werke, 505–524).

[4] E. Hecke, Über Modulfunktionen und die Dirichletschen Reihen mit Eulerscher Produktentwicklung I, II, Math. Ann., 114 (1937), 1–28, 316–351 (=Math. Werke, 644–707).

[5] G. Shimura, Correspondances modulaires et les fonctions ζ de courbes algébriques, J. Math. Soc. Japan, 10 (1958), 1–28.

[6] G. Shimura, On analytic families of polarized abelian varieties and automorphic functions, Ann. of Math., 78 (1963), 149–192.

[7] G. Shimura, Introduction to the arithmetic theory of automorphic functions, Publ. Math. Soc. Japan, No. 11, 1971.

[8] G. Shimura, On the zeta-function of an abelian variety with complex multiplication, to appear.

[9] A. Weil, Über die Bestimmung Dirichletscher Reihen durch Funktionalgleichungen, Math. Ann., 168 (1967), 149–156.

Princeton University

On the field of rationality for an abelian variety

Nagoya Mathematical Journal, 45 (1972), 167-178

The purpose of this paper is to prove the following two facts:

(I) *Every generic polarized abelian variety of odd dimension has a model rational over its field of moduli.*

(II) *No generic principally polarized abelian variety of even dimension has a model rational over its field of moduli.*

In both statements and throughout the paper, we assume that the universal domain is of characteristic 0. We call a polarized abelian variety *generic* if its field of moduli has the maximum transcendence degree (i.e., $n(n+1)/2$ if the variety is of dimension n) over the rational number field.

It is well-known that an elliptic curve has a model rational over its field of moduli. However, no general result, not even a counter-example, seems to have been obtained in the higher-dimensional case. In a previous paper [5], we have shown that a polarized abelian variety with sufficiently many complex multiplications, under a certain condition, has a model rational over its field of moduli. We discuss here the other extreme case in which varieties are generic. A negative answer similar to (II) will be given also for abelian varieties with a certain type of polarization which is not necessarily principal, and for hyperelliptic curves of even genera.

It is still an open question to obtain a criterion under which an arbitrarily given polarized abelian variety has a model rational over its field of moduli. The above two statements combined together seem to indicate a rather complicated nature of the problem, which almost defies conjecture. A new viewpoint is certainly necessary to understand the whole situation.

Since the proofs are not so long, we do not try to explain the main ideas at this point, except the following two general remarks.

Received June 17, 1971.

(A)   The parity of dimensionality intervenes in the problem in the following way. If ω is a holomorphic n-form on an abelian variety of dimension n, then the automorphism -1 sends ω to $(-1)^n \cdot \omega$. Our proof of (I) relies heavily on this fact. It is not clear, however, whether or not this connection of differential forms with the problem is essential.

(B)   The existence of a model is closely connected with the problem of extending a given Galois extension to a larger Galois extension with a preassigned Galois group. But our proof is rather implicit in this respect. Actually we could make it more explicit at the cost of lengthening the paper. It may be interesting to investigate this point in full generality.

1. Preliminaries

Since our treatment is restricted to the case of characteristic 0, we may assume, without losing generality, that the universal domain is the complex number field C. Let $P = (A, W)$ be a polarized abelian variety, i.e., a structure formed by an abelian variety A and a polarization W of A. Consider a structure

$$Q = (A, W, t_1, \cdots, t_m)$$

with points $t_1, \cdots, t_m$ of A of finite order. Then we can define, in a natural manner, the notion of an isomorphism of Q onto another structure of the same type, and also the transform Q^σ of Q by an isomorphism σ of a field of rationality for Q into C. For details, see [1], [2, II, 1.1]. By *the field of moduli of Q*, we mean the subfield k of C which is characterized by the following property:

For every automorphism σ of C, Q^σ is isomorphic to Q if and only if σ is the identity map on k.

Such a field k always exists and is unique for Q (see [1, § 2], [2, II, 1.4]). Obviously the field of moduli depends only on the isomorphism-class of Q, and is contained in any field of rationality for Q. We identify (A, W) with $(A, W, 0)$, and define *the field of moduli of* (A, W) to be that of $(A, W, 0)$. We call any structure isomorphic to Q a *model* of Q. Our construction of a model over a desired field is based on the following proposition which follows immediately from the result of Weil [8].

PROPOSITION 1.   *Let Q be rational over a finite Galois extension K of a field*

F. Suppose there is, for each $\sigma \in \mathrm{Gal}\,(K/F)$, *an isomorphism* λ_σ *of* Q *onto* Q^σ, *satisfying* $\lambda_{\sigma\tau} = \lambda_\sigma^\tau \circ \lambda_\tau$ *for all* σ, $\tau \in \mathrm{Gal}\,(K/F)$. *Then there is a structure* Q' *rational over* F *and an isomorphism* μ *of* Q' *onto* Q *rational over* K *such that* $\mu^\sigma = \lambda_\sigma \circ \mu$ *for all* $\sigma \in \mathrm{Gal}\,(K/F)$.

PROPOSITION 2. *Let* Q *be as above, and* λ *an isomorphism of* Q *onto another structure* Q'. *Suppose that* Q *has no automorphisms other than the identity map. Then* λ *is rational over any field of rationality for* Q *and* Q'.

Proof. Let σ be an automorphism of C over a field of rationality for Q and Q'. Then $\lambda^{-1} \circ \lambda^\sigma$ is an automorphism of Q, hence $\lambda^{-1} \circ \lambda^\sigma = 1$, so that $\lambda^\sigma = \lambda$, q.e.d.

PROPOSITION 3. *If the identity map is the only automorphism of* Q, *then* Q *has a model rational over its field of moduli.*

For the proof, see [2, II, 1.5].

Let us now fix $P = (A, W)$, and put, for a positive integer r,

$$(1) \qquad\qquad T_r = \{t \in A \mid rt = 0\}.$$

Let $\{t_1, \cdots, t_m\}$ be a set of generators of the module T_r, and let K_r denote the field of moduli of $Q_r = (A, W, t_1, \cdots, t_m)$. Then K_r depends only on P and r; it is independent of the choice of $t_1, \cdots, t_m$. Moreover, K_r is normal over K_1. Of course K_1 is the field of moduli of P.

PROPOSITION 4. *Suppose that* P *has no automorphisms other than* ± 1. *Then, for every integer* $r > 1$, Q_r *has a model rational over* K_r.

Proof. If $r > 2$, this follows immediately from Prop. 3. To prove the case $r = 2$, we start with a structure $Q_4 = (A, W, t_1, \cdots, t_m)$ rational over K_4. Let G denote the group of all automorphisms α of the module T_4 such that $\alpha(t) = t$ if $2t = 0$. We see easily that $\alpha^2 = 1$ for every $\alpha \in G$. Therefore G is a product of several cyclic groups of order 2. Therefore $G = \{\pm 1\} \cdot H$ with a suitable subgroup H not containing -1. (Of course -1 means the element of G that maps t onto $-t$.) Let $\sigma \in \mathrm{Gal}\,(K_4/K_2)$. Then there is an isomorphism λ of Q_2 onto Q_2^σ. By Prop. 2, λ is rational over K_4. We observe that $\lambda^{-1}(t^\sigma) = \alpha(t)$ for all $t \in T_4$ with an element α of G. Changing λ for $-\lambda$ if necessary, we may assume that $\alpha \in H$. Under the condition $\alpha \in H$, λ is unique for σ. Putting $\lambda = \lambda_\sigma$, we can easily verify $\lambda_{\sigma\tau} = \lambda_\sigma^\tau \circ \lambda_\tau$ for all σ, $\tau \in \mathrm{Gal}\,(K_4/K_2)$. By Prop. 1. Q_2 has a model rational over K_2.

PROPOSITION 5. *Let P be rational over a field F. Suppose that (i) P has no automorphisms other than ± 1; (ii) $[F : K_r \cap F]$, with an integer $r > 2$, is odd. Then P has a model rational over $K_r \cap F$.*

Proof. Put $M = K_r \cap F$. Since $K_1 \subset M$, K_r is normal over M, so that K_r and F are linearly disjoint over M. For simplicity, let us hereafter write K for K_r. Consider $Q = (A, W, t_1, \cdots, t_m)$ with (A, W) rational over F and with a set of generators $\{t_1, \cdots, t_m\}$ of T_r. Let S be the smallest field of rationality for Q containing F. Then $K \subset S$. By Prop. 4, Q has a model $Q' = (A', W', t_1', \cdots, t_m')$ rational over K. Let λ be an isomorphism of Q onto Q'. By Prop. 2, λ is rational over S. Obviously S is normal over F. Put $L = F \cdot K$. If $\sigma \in \mathrm{Gal}\,(S/L)$, there is an isomorphism ε of Q onto Q^σ. Since $P^\sigma = P$, we have $\varepsilon = \pm 1$. It follows that $[S : L] \leq 2$. Now we divide our discussion into two cases according to $[S : L]$.

(I) $S = L$. In this case, $\mathrm{Gal}\,(S/F)$ can be identified with $\mathrm{Gal}\,(K/M)$. Put $P' = (A', W')$ and $\mu_\sigma = \lambda^\sigma \circ \lambda^{-1}$ for every $\sigma \in \mathrm{Gal}\,(S/F)$. Then μ_σ is an isomorphism of P' onto P'^σ. By Prop. 2, μ_σ is rational over K. Since $\mu_{\sigma\tau} = \mu_\sigma^\tau \circ \mu_\tau$ for $\sigma, \tau \in \mathrm{Gal}\,(K/M)$, we obtain, by Prop. 1, a model rational over M.

(II) $[S : L] = 2$. Let $G = \mathrm{Gal}\,(S/F)$, and let π be the generator of $\mathrm{Gal}\,(S/L)$. Then $G/\{1, \pi\}$ can be identified with $\mathrm{Gal}\,(L/F)$ and $\mathrm{Gal}\,(K/M)$. Take an element v of S so that $S = L(v)$, $v^2 \in L$. Put $x_\sigma = v^\sigma/v$ for every $\sigma \in G$. Then $x_\pi = -1$, $x_\sigma \in L$, and $x_{\sigma\tau} = x_\sigma^\tau x_\tau$. Put $y_\sigma = N_{L/K}(x_\sigma)$, and $z = N_{L/K}(v^2)$. Then $y_\sigma^2 = z^\sigma/z$ and $y_{\sigma\tau} = y_\sigma^\tau y_\tau$. Since $[L : K]$ is odd, $y_\pi = -1$. Let w be a square root of z. We discuss two cases $w \in K$ and $w \notin K$ separately.

(II$_a$) Suppose $w \in K$. Put $f(\sigma) = y_\sigma w/w^\sigma$. Since $y_\sigma^2 = (w^\sigma/w)^2$, we have $f(\sigma) = \pm 1$. Therefore f is a cocycle with values ± 1, so that it must be a character of G. Furthermore f is non-trivial, since $f(\pi) = -1$. Let H be the kernel of f. Then $G = H \cdot \{1, \pi\}$, and H can be identified with $\mathrm{Gal}\,(K/M)$. For every $\sigma \in H$, put $\mu_\sigma = \lambda^\sigma \circ \lambda^{-1}$. We see again that μ_σ is an isomorphism of P' onto P'^σ rational over K, and $\mu_{\sigma\tau} = \mu_\sigma^\tau \circ \mu_\tau$. Therefore P' has a model rational over M.

(II$_b$) Suppose $w \notin K$. Extend an element of $\mathrm{Gal}\,(K/M)$ to an automorphism α of C. If an element ξ of G coincides with α on K, then $(w^\alpha)^2 = z^\xi = y_\xi^2 w^2$, so that $w^\alpha = \pm y_\xi w \in K(w)$. Therefore $K(w)$ is normal over M. Now,

462

for every $\sigma \in G$, we can define an element $[\sigma]$ of $\mathrm{Gal}\,(K(w)/M)$ by $w^{[\sigma]} = y_\sigma w$, and $[\sigma] = \sigma$ on K. Then we see easily that $\sigma \longmapsto [\sigma]$ is an isomorphism of G onto $\mathrm{Gal}\,(K(w)/M)$. Put $\mu_{[\sigma]} = \lambda^\sigma \circ \lambda^{-1}$ for every $\sigma \in G$. Then $\mu_{[\sigma]}$ is an isomorphism of P' onto P'^σ, rational over K, and

$$\mu_{[\sigma][\tau]} = \lambda^{\sigma\tau} \circ \lambda^{-1} = (\lambda^\sigma \circ \lambda^{-1})^\tau \circ (\lambda^\tau \circ \lambda^{-1}) = \mu_{[\sigma]}^{[\tau]} \circ \mu_{[\tau]}.$$

Therefore, by Prop. 1, we obtain a model rational over M.

2. The odd dimensional case

Let $P = (A, W)$ and $P' = (A', W')$ be two polarized abelian varieties of dimension n, rational over a field K, and λ an isomorphism of P onto P', not necessarily rational over K. Take non-zero holomorphic differential n-forms ω on A and ω' on A', rational over K. Denote by $\omega' \circ \lambda$ the transform of ω' by λ. Then $\omega' \circ \lambda = c \cdot \omega$ with a constant c. If τ is an automorphism of C over K, we have $\omega' \circ \lambda^\tau = c^\tau \cdot \omega$, hence $\omega' \circ \lambda^\tau \circ \lambda^{-1} = c^\tau c^{-1} \omega'$. Now we impose the following two conditions on P:

(2) *P has no automorphisms other than ± 1.*

(3) *n is odd.*

Then $\lambda^\tau \circ \lambda^{-1} = \pm 1$, and $\omega' \circ (-1) = (-1)^n \omega' = -\omega'$. Therefore $c^\tau = \pm c$, and $\lambda^\tau = (c^\tau/c) \cdot \lambda$. It follows that $K(c)$ is the smallest field of rationality for λ containing K, $c^2 \in K$, and $[K(c) : K] \leqq 2$. After this preliminary consideration, we now prove

PROPOSITION 6. *Suppose that $P = (A, W)$ satisfies* (2) *and* (3). *Let K be a field of rationality for P, and σ an automorphism of K of order 2 such that P^σ is isomorphic to P. Then P has a model rational over the fixed subfield of K by σ.*

Proof. Let $F = \{x \in K \,|\, x^\sigma = x\}$, and let λ be an isomorphism of P onto P^σ. Take a non-zero holomorphic n-form ω on A rational over K, and put $\omega^\sigma \circ \lambda = c \cdot \omega$ with a constant c. Extend σ to an automorphism of C, and denote it again by σ. Then $\omega \circ \lambda^\sigma = c^\sigma \cdot \omega^\sigma$, and $\omega \circ \lambda^\sigma \circ \lambda = c^\sigma c \cdot \omega$. Since $\lambda^\sigma \circ \lambda$ is an automorphism of P, we have $\lambda^\sigma \circ \lambda = \pm 1$, and $c^\sigma c = \pm 1$. By the above consideration, λ is rational over $K(c)$, and $c^2 \in K$. Now $N_{K/F}(c^2) = 1$, so that $c^2 = b/b^\sigma$ with an element b of K. Let d be a square root of b. Suppose $d \in K$. Then $c^2 = (d/d^\sigma)^2$, so that $c = \pm d/d^\sigma \in K$, hence $c^\sigma c = 1$. Therefore λ is rational over K, and $\lambda^\sigma \circ \lambda = 1$. By Prop. 1, P has a model rational over F. Next suppose $d \notin K$. Then we can find a polarized abelian variety P'

$= (A', W')$ rational over K and an isomorphism μ of P to P' rational over $K(d)$ such that $\mu^\alpha = -\mu$ if α is the generator of $\mathrm{Gal}\,(K(d)/K)$. (This is another application of Prop. 1.) Let ω' be a non-zero holomorphic n-form on A' rational over K. Then $\omega' \circ \mu = e \cdot \omega$ with a constant e such that $K(e) = K(d)$ and $e^\alpha = -e$. Then $d/e \in K$. Changing ω' for $(d/e) \cdot \omega'$, we may assume that $\omega' \circ \mu = d \cdot \omega$. Put $\lambda' = \mu^\sigma \circ \lambda \circ \mu^{-1}$. Then λ' is an isomorphism of P' onto P'^σ, and $\omega'^\sigma \circ \lambda' = \omega'^\sigma \circ \mu^\sigma \circ \lambda \circ \mu^{-1} = (d^\sigma c/d) \cdot \omega'$. Now $(d^\sigma c/d)^2 = c^2 b^\sigma / b = 1$, hence $d^\sigma c/d = \pm 1$. Therefore λ' is rational over K, and $\omega' \circ (\lambda'^\sigma \circ \lambda') = \omega'$, hence $\lambda'^\sigma \circ \lambda' = 1$. By Prop. 1, P' has a model rational over F. This completes the proof.

PROPOSITION 7. *Suppose that P satisfies* (2) *and* (3). *Let K_r be defined for P as in* §1. *Let r be a positive integer > 2, and F a finite normal extension of K_1 over which P has a model. Then P has a model rational over $K_r \cap F$.*

Proof. Let H be a 2-Sylow subgroup of $\mathrm{Gal}\,(F/K_1)$, and M the subfield of F corresponding to H. The field F can be obtained from M by successive quadratic extensions. Therefore, by Prop. 6, P has a model rational over M. Now $K_1 \subset M \cap K_r \subset M$, and $[M : K_1]$ is odd. By Prop. 5, P has a model rational over $K_r \cap M$, hence over $K_r \cap F$, q.e.d.

As an immediate consequence, we obtain

PROPOSITION 8. *Suppose that P satisfies* (2) *and* (3). *Let r and s be positive integers such that $r > 2$ and $s > 1$. Then P has a model rational over $K_r \cap K_s$.*

Let us now consider a generic $P = (A, W)$ of even or odd dimension. Define K_r for P as above. The structure of $\mathrm{Gal}\,(K_r/K_1)$ has been determined in [2] and [4]. To explain the result, take a basic polar divisor X belonging to W, let $T_r = \{t \in A | rt = 0\}$, and define the symbol $e_{X,r}(s, t)$ for $(s, t) \in T_r \times T_r$ as in Weil [7, §XI]. Let G denote the group of all automorphisms α of T_r such that

$$e_{X,r}(\alpha(s),\ \alpha(t)) = e_{X,r}(s,\ t)^{c(\alpha)}$$

with an integer $c(\alpha)$ prime to r. Further let L denote the algebraic closure of K_1. For every $\sigma \in \mathrm{Gal}\,(L/K_1)$, take an isomorphism λ of P onto P^σ. Then $t \longmapsto \lambda^{-1}(t^\sigma)$ is an automorphism of T_r, which can be shown to belong to G. Writing it α_σ, we have

$$(4) \qquad\qquad t^\sigma = \lambda(\alpha_\sigma(t)) \qquad (t \in T_r).$$

The element α_σ of G is uniquely determined for σ by (4), up to the factor -1. Then the map $\sigma \longmapsto \pm \alpha_\sigma$ gives an injection of $\mathrm{Gal}\,(K_r/K_1)$ into $G/\{\pm 1\}$. This much is true for any P, not necessarily generic, whose automorphisms are ± 1. Suppose that r is prime to the degree of the polarization. (By the *degree* of the polarization W, we understand the degree of the isogeny of A onto its Picard variety associated with X.) Then G is isomorphic to a group $\mathfrak{G}_r$ defined by

$$\mathfrak{G}_r = \{S \in GL_{2n}(\mathbf{Z}/r\mathbf{Z}) \mid {}^t S J S = c(S) J \text{ with } c(S) \in \mathbf{Z}/r\mathbf{Z}\},$$

$$J = \begin{bmatrix} 0 & -1_n \\ 1_n & 0 \end{bmatrix}, \quad n = \dim\,(A).$$

If P is generic, the map of $\mathrm{Gal}\,(K_r/K_1)$ into $\mathfrak{G}_r/\{\pm 1\}$ is surjective. (For the proof, see [2, II, § 7], where a more general result is given with no restriction on r, nor on the type of polarization. See also [4, 9.5, 9.8].)

PROPOSITION 9. *Suppose P is generic. If r and s are relatively prime, and rs is prime to the degree of the polarization, then $K_r \cap K_s = K_1$.*

Proof. It can easily be seen that $\mathfrak{G}_{rs} = \mathfrak{G}_r \times \mathfrak{G}_s$, and K_r, as a subfield of K_{rs}, corresponds to the subgroup $\{\pm 1\} \times \mathfrak{G}_s$ of $\mathfrak{G}_{rs}$, and K_s to $\mathfrak{G}_r \times \{\pm 1\}$, hence $K_r \cap K_s = K_1$.

(Note that $[K_{rs} : K_r K_s] = 2$ if both r and s are > 2.)
Combining Prop. 9 with Prop. 8, we obtain

THEOREM 1. *Every generic polarized abelian variety of odd dimension has a model rational over its field of moduli.*

3. The even dimensional case

Let $\mathfrak{H}_n$ denote the Siegel upper half space of degree n, i.e., the set of all complex symmetric matrices with positive definite imaginary parts. Let δ be a diagonal matrix of degree n whose diagonal elements are non-zero rational integers with no common divisors other than ± 1. For each $z \in \mathfrak{H}_n$, we consider an $(n \times 2n)$-matrix

$$\Omega(z) = (z \ \delta)$$

and a lattice $L(z)$ in C^n generated by the column vectors of $\Omega(z)$. Then the complex torus $C^n/L(z)$ has a structure of an abelian variety. Moreover, we obtain a Riemann form Φ_z on it by

$$\Phi_z(\Omega(z)x,\ \Omega(z)y) = {}^t x \Delta y \qquad (x,\ y \in \boldsymbol{R}^{2n}),$$

$$\Delta = \begin{bmatrix} 0 & -\delta \\ \delta & 0 \end{bmatrix}.$$

Here we consider the elements of $\boldsymbol{R}^{2n}$ as column vectors. Each point z of $\mathfrak{H}_n$ determines in this way an isomorphism class of polarized abelian varieties over C. We take any model in the class, and call it $P_z = (A_z,\ W_z)$. The degree of W_z coincides with $\det(\Delta)$.

It is well-known that an arbitrary polarized abelian variety P is isomorphic to P_z for some δ and z; the invariant factors of δ are completely determined by P; z is unique for P modulo a certain discontinuous group commensurable with $\mathrm{Sp}\,(2n,\ \boldsymbol{Z})$.

Now we assume that n is even, and put $n = 2m$. Furthermore we consider the following condition on the type of polarization.

(5) *Each invariant factor of δ occurs with an even multiplicity. In other words, there exist two elements α and β of $GL_n(\boldsymbol{Z})$ such that $\alpha\delta\beta = \begin{bmatrix} d & 0 \\ 0 & d \end{bmatrix}$ with a diagonal matrix d of size m.*

If this is satisfied, we may assume, by transforming Δ by $\begin{bmatrix} {}^t\alpha & 0 \\ 0 & \beta \end{bmatrix}$, that δ itself is of the form $\begin{bmatrix} d & 0 \\ 0 & d \end{bmatrix}$.

PROPOSITION 10. *Suppose that $\delta = \begin{bmatrix} d & 0 \\ 0 & d \end{bmatrix}$ with a diagonal matrix d of size m. Let $j = \begin{bmatrix} 0 & -1_m \\ 1_m & 0 \end{bmatrix}$, and let $\mathfrak{Y}$ be the set of all $z \in \mathfrak{H}_n$ such that $jz = -z^\rho j$. Then, for every $z \in \mathfrak{Y}$, there is an isomorphism λ of P_z onto P_z^ρ such that $\lambda^\rho \circ \lambda = -1$.*

Here and henceforth, ρ denotes the complex conjugation.

Proof. Let f be a holomorphic map of C^n onto A_z which induces an isomorphism of $C^n/L(z)$ onto A_z. Then we can define a holomorphic map f' of C^n onto A_z^ρ with kernel $L(z)^\rho$ by $f'(u) = f(u^\rho)^\rho$ for $u \in C^n$. Suppose $z \in \mathfrak{Y}$. Then

$$j \cdot \Omega(z) = \Omega(-z^\rho)J', \qquad J' = \begin{bmatrix} j & 0 \\ 0 & j \end{bmatrix}.$$

Since $L(z)^\rho = L(-z^\rho)$, the automorphism of C^n obtained from j gives an isomorphism of $C^n/L(z)$ onto $C^n/L(z)^\rho$. Therefore we can define an isomorphism λ of A_z onto A_z^ρ by $\lambda(f(u)) = f'(ju)$ for $u \in C^n$. Then

$$\lambda^\rho(f'(u)) = \lambda^\rho(f(u^\rho)^\rho) = f'(ju^\rho)^\rho = f(ju),$$

so that $\lambda^\rho(\lambda(f(u))) = \lambda^\rho(f'(ju)) = f(j^2 u) = -f(u)$, hence $\lambda^\rho \circ \lambda = -1$. Let X be a polar divisor of A_z corresponding to Φ_z with respect to f. Define the symbol $e_{X,N}(s, t)$ as in Weil [7, § XI] for every positive integer N. By [6, p. 25, (7)], we have

$$e_{X,N}(f(u), f(v)) = \exp[-2\pi i \cdot N \cdot \Phi_z(u, v)] \qquad (u, v \in N^{-1}L(z)).$$

Applying ρ to this relation, we obtain

$$e_{X^\rho,N}(f'(u), f'(v)) = \exp[2\pi i \cdot N \cdot \Phi_z(u^\rho, v^\rho)] \qquad (u, v \in N^{-1}L(-z^\rho)).$$

Put $S = \begin{bmatrix} -1_n & 0 \\ 0 & 1_n \end{bmatrix}$, $u = \Omega(-z^\rho)x$, $v = \Omega(-z^\rho)y$ with $x, y \in R^{2n}$. Then $u^\rho = \Omega(z)Sx$, $v^\rho = \Omega(z) Sy$, so that

$$\Phi_z(u^\rho, v^\rho) = {}^t x \cdot {}^t S \Delta S y = -{}^t x \Delta y = -\Phi_w(u, v), \quad w = -z^\rho.$$

Therefore

$$e_{X^\rho,N}(f'(u), f'(v)) = \exp[-2\pi i \cdot N \cdot \Phi_w(u, v)].$$

This implies that X^ρ is a divisor of A_z^ρ corresponding to Φ_w with respect to f'. Since $j \cdot \Omega(z) = \Omega(-z^\rho)J'$ and ${}^t J' \Delta J' = \Delta$, we see that λ sends W_z onto W_z^ρ, q.e.d.

PROPOSITION 11. *The set $\mathfrak{Y}$ of Prop. 10 is non-empty. Moreover, let g be a holomorphic function defined on a connected domain D contained in $\mathfrak{H}_n$. If $g = 0$ on a non-empty open subset of $D \cap \mathfrak{Y}$, then g is identically 0 on D.*

Proof. It can easily be verified that every point z of $\mathfrak{Y}$ can be written in the form

$$z = \begin{bmatrix} a & b \\ b^\rho & -a^\rho \end{bmatrix}$$

with complex matrices a and b of size m such that ${}^t a = a$, ${}^t b^\rho = b$. Conversely, such a z, whose imaginary part is positive definite, belongs to $\mathfrak{Y}$. If we put $a = x + iy$ and $b = r + is$ with real matrices x, y, r, s, then the conditions become as follows:

$${}^t x = x, \quad {}^t y = y, \quad {}^t r = r, \quad {}^t s = -s;$$

$$\begin{bmatrix} y & s \\ {}^t s & y \end{bmatrix} \text{ is positive definite.}$$

Therefore $\mathfrak{Y}$ is non-empty. Our second assertion follows from the following well-known facts:

(i) A holomorphic function $h(z)$ in one complex variable z is identically 0, if $h(x) = 0$ for all real x.

(ii) A holomorphic function $h(z, z')$ in two complex variables z and z' is identically 0, if $h(z, z^\rho) = 0$ for all complex z.

Of course the function in each case must be defined on a connected domain for which the condition $h(x) = 0$ or $h(z, z^\rho) = 0$ is meaningful.

PROPOSITION 12. *If P_z, with $z \in \mathfrak{Y}$, has no automorphisms other than ± 1, then P_z has no model rational over its field of moduli.*

Proof. Since P_z is isomorphic to P_z^ρ, the field of moduli of P_z is contained in $\boldsymbol{R}$. Assume that P_z has a model P rational over $\boldsymbol{R}$, and let μ be an isomorphism of P onto P_z. Then $\lambda^{-1} \circ \mu^\rho \circ \mu^{-1}$, with λ as in Prop. 10, is an automorphism of P_z, so that $\lambda^{-1} \circ \mu^\rho \circ \mu^{-1} = \pm 1$, hence $\lambda = \pm \mu^\rho \circ \mu^{-1}$. But this contradicts the equality $\lambda^\rho \circ \lambda = -1$.

THEOREM 2. *Let P be a generic polarized abelian variety of even dimension. If the polarization satisfies the condition (5), then P has no model rational over its field of moduli.*

Proof. As mentioned above, there is a discrete subgroup Γ of $\mathrm{Sp}(2n, \boldsymbol{R})$ such that P_z is isomorphic to P_w if and only if $\gamma(z) = w$ for some $\gamma \in \Gamma$. Moreover $\mathfrak{H}_n / \Gamma$ is isomorphic to a Zariski open subset V of a projective variety. Let φ be a holomorphic map of $\mathfrak{H}_n$ onto V which induces an isomorphism of $\mathfrak{H}_n / \Gamma$ onto V. By [3], we can take V and φ so that the following conditions are satisfied:

(i)   *V is defined over the rational number field $\boldsymbol{Q}$;*

(ii) *$\boldsymbol{Q}(\varphi(z))$ is the field of moduli of P_z for every $z \in \mathfrak{H}_n$;*

(iii) *For an automorphism σ of C, P_z is isomorphic to P_w^σ if and only if $\varphi(z) = \varphi(w)^\sigma$.*

Thus P_z is generic if and only if $\varphi(z)$ is generic on V over $\boldsymbol{Q}$. Therefore if

P_z and P_w are generic, there is an automorphism σ of C such that P_z is isomorphic to P_w^σ. For this reason and by virtue of Prop. 12, to prove our theorem, it is sufficient to find a generic P_z with z in $\mathfrak{Y}$. This can be done as follows. Let S be the set of all homogeneous polynomials with rational coefficients viewed as functions in the projective space in which V is situated. Let S' be the subset of S consisting of all the $f \in S$ such that $f \circ \varphi$ is not identically 0. Put, for each $f \in S'$,

$$X_f = \{z \in \mathfrak{Y} \mid f(\varphi(z)) = 0\}.$$

Then X_f is a closed subset of $\mathfrak{Y}$, which contains no non-empty open subset of $\mathfrak{Y}$ by Prop. 11. Since S' is a countable set, $\mathfrak{Y}$ cannot be covered by the X_f. Therefore $\mathfrak{Y}$ has a point z for which $f(\varphi(z)) \neq 0$ for all $f \in S'$. Then $\varphi(z)$ is generic on V over $\mathbf{Q}$, q.e.d.

A counter-example of the same nature can be obtained also for hyperelliptic curves of even genera. In fact consider a hyperelliptic curve U of genus $m - 1$ defined by

$$y^2 = a_0 x^m + \sum_{r=1}^{m} (a_r x^{m+r} + (-1)^r a_r^\rho x^{m-r}), \quad a_m = 1,$$

where a_0 is a real number and $a_1, \cdots, a_{m-1}$ are complex numbers. Suppose m is odd. Then we can define a birational map μ of U onto U^ρ by

$$\mu(x,\ y) = (-x^{-1},\ i \cdot x^{-m} y).$$

We see that $\mu^\rho \circ \mu$ maps $(x,\ y)$ onto $(x,\ -y)$. Take the a_r so that $a_0,\ a_1,\ \cdots,\ a_{m-1},\ a_1^\rho,\ \cdots, a_{m-1}^\rho$ are algebraically independent over $\mathbf{Q}$. Then U has no automorphisms other than the obvious two. Therefore, for the same reason as in the proof of Prop. 12, U cannot have a model rational over $\mathbf{R}$. Thus we obtain

THEOREM 3. *No generic hyperelliptic curve of even genus has a model rational over its field of moduli.*

REFERENCES

[1] G. Shimura, On the theory of automorphic functions, Ann. of Math., **70** (1959), 101–144.

[2] G. Shimura, On the field of definition for a field of automorphic functions: I, II, Ann. of Math., **80** (1964), 160–189, **81** (1965), 124–165.

[3] G. Shimura, Moduli and fibre systems of abelian varieties, Ann. of Math., **83** (1966), 294–338.

[4] G. Shimura, Algebraic number fields and symplectic discontinuous groups, Ann. of Math. **86** (1967), 503–592.

[5] G. Shimura, On the zeta-function of an abelian variety with complex multiplication, to appear.

[6] G. Shimura and Y. Taniyama, Complex multiplication of abelian varieties and its applications to number theory, Publ. Math. Soc. Japan, No. **6**, 1961.

[7] A. Weil, Variétés abéliennes et courbes algébriques, Hermann, Paris, 1948.

[8] A. Weil, The field of definition of a variety, Amer. J. Math., **78** (1956), 509–524.

Princeton University

72b

Class fields over real quadratic fields and Hecke operators

Annals of Mathematics, 95 (1972), 130-190

Introduction

This is a continuation of my previous investigation [9, § 7.7], [10] concerning the relation between the arithmetic of real quadratic fields and the cusp forms of "Neben"-type in Hecke's sense. It was shown in these articles that the eigen-values of Hecke operators for such forms are closely connected with the reciprocity law in certain abelian extensions of a real quadratic field k, and moreover, such extensions can be generated by the coordinates of certain points of finite order on an abelian variety associated with the cusp forms. However, the discussion was restricted, for the most part, to the case in which k is of class number one, and the level is a prime $\equiv 1 \bmod (4)$. The purpose of the present paper is: (i) to give a more comprehensive treatment with a somewhat modified formulation of the problem; (ii) to discuss more examples, especially those for which k is of class number > 1, and those of composite level; and (iii) to present a method of constructing an abelian variety of the same nature independent of modular forms.

Our object of study is an abelian variety B, rational over k, whose transform B^{ε} by the non-trivial automorphism ε of k is isogenous to B. Such a B can be obtained from a certain cusp form which is a common eigen-function of Hecke operators. We consider the intersection $\mathfrak{h}$ of the kernels of certain isogenies of B onto B^{ε}, and the field $k(\mathfrak{h})$ generated over k by the coordinates of the points of $\mathfrak{h}$. It can be shown that $k(\mathfrak{h})$ is abelian over k. From the action of $\mathrm{Gal}(k(\mathfrak{h})/k)$ on $\mathfrak{h}$, we obtain several characters of certain ideal groups of k with values in finite fields. The first section of the paper is devoted to a general discussion of cusp forms of "Neben"-type, and elementary properties of abelian varieties associated with them. We shall prove in § 2 some fundamental theorems about those characters. Theorems 2.2 and 2.3 are re-formulations of the results of [9, § 7.7]; Theorem 2.8 is completely new, and seems important. Various examples will be discussed in § 3 through § 8. Especially, the example of § 4 concerns an eight-dimensional B rational over $k = Q(\sqrt{229})$,

* During the preparation of this paper, the author was a John Simon Guggenheim Fellow, and partially supported by NSF Grant GP-23855.

for which $k(\mathfrak{y})$ is shown to contain the maximal unramified abelian extension of k. (Note that the class number of $Q(\sqrt{229})$ is 3.) In the last two sections, we shall consider the variety B, or rather an abelian variety isogenous to $B \times B^t$, in a more axiomatic way, independent of modular forms. It will be shown that many of the results of §2 are applicable to such a general case, and further that one can construct infinitely many one-dimensional B satisfying the "axiom", in a rather explicit manner.

In the present paper, as well as in [9, §7.7] and [10], we are interested only in the finite extensions of the basic real quadratic field generated by some special points of finite order on a certain abelian variety, as we explained above. For the proper understanding of the whole structure, one should undoubtedly consider the infinite extension generated by all points of finite order on the variety. If so, we may hope that the present investigation will form an interesting, possibly important, part of the theory, which is yet to be established.

In the discussion of examples, we use some numerical eigen-values of Hecke operators, which have been computed, by means of Eichler's trace formula [1], [2], by K. Doi, N. Iwasaki, H. Naganuma, M. Yamauchi, and H. Wada. Also H. Trotter helped the author to determine the irreducible factors of the characteristic polynomials of Hecke operators. To all of them, the author wishes to express his deep gratitude.

CONTENTS

Notation and terminology

We denote by Z, Q, R, and C, respectively, the ring of rational integers, the rational number field, the real number field, and the complex number field. The algebraic closure of Q in C is denoted by $\bar{Q}$. If x is a complex number or a complex valued function etc., then x^ρ denotes its complex conjugate.

For an associative ring P with identity element, $P^\times$ denotes the group of all invertible elements, and $M_n(P)$ the ring of all matrices of size n with coefficients in P.

Let K be a field, and F a subfield of K. Then $\mathrm{Aut}(K/F)$ denotes the group of all automorphisms of K over F. If K is a Galois extension of F, we write $\mathrm{Gal}(K/F)$ for $\mathrm{Aut}(K/F)$. For $x \in K$ and $\sigma \in \mathrm{Aut}(K/F)$, the image of x by σ is denoted by x^σ under the rule $(x^\sigma)^\tau = x^{\sigma\tau}$.

Let k be an algebraic number field of finite degree. For a fractional ideal $\mathfrak{x}$ in k, $N(\mathfrak{x})$ denotes the absolute norm of $\mathfrak{x}$, which is a positive rational number. For a formal product $\mathfrak{f}$ of an integral ideal $\mathfrak{f}_0$ in k and archimedean primes of k, we denote by $I(\mathfrak{f})$ the group of all fractional ideals in k prime to $\mathfrak{f}_0$, and by $P(\mathfrak{f})$ the subgroup of $I(\mathfrak{f})$ consisting of all the principal ideals generated by the elements α of k such that α is positive at all archimedean primes involved in $\mathfrak{f}$, and $\alpha \equiv 1 \bmod \mathfrak{f}_\mathfrak{p}$ for all finite prime factors $\mathfrak{p}$ of $\mathfrak{f}$, where $\mathfrak{f}_\mathfrak{p}$ is the $\mathfrak{p}$-closure of $\mathfrak{f}_0$ in the $\mathfrak{p}$-completion of k.

We shall be discussing homomorphisms χ of the ideal group $I(\mathfrak{f})$ into a finite group $\mathfrak{g}$ whose kernel contains $P(\mathfrak{f})$. The *conductor* of χ is defined to be the divisor $\mathfrak{f}'$ of $\mathfrak{f}$ characterized by: (i) χ is trivial on $I(\mathfrak{f}) \cap P(\mathfrak{f}')$; (ii) no proper divisor of $\mathfrak{f}'$ has the property (i). Then χ can be extended uniquely to a homomorphism χ' of $I(\mathfrak{f}')$ into $\mathfrak{g}$ whose kernel contains $P(\mathfrak{f}')$. We often denote χ' by the same symbol χ.

For a prime ideal $\mathfrak{p}$ in k prime to 2 and an element x in k prime to $\mathfrak{p}$, we define the quadratic residue symbol $\left(\dfrac{x}{\mathfrak{p}}\right)$ in a natural way, namely $\left(\dfrac{x}{\mathfrak{p}}\right) = 1$ or -1 according as $x \bmod p$ is a square in the residue field modulo $\mathfrak{p}$ or not. Similarly, if $\mathfrak{p}_\infty$ is a real archimedean prime of k, $\left(\dfrac{x}{\mathfrak{p}_\infty}\right) = 1$ or -1 according as x is positive or negative at $\mathfrak{p}_\infty$.

An algebraic variety (especially an algebraic curve or an abelian variety) is always a subvariety of a projective space, often identified with the set of all points rational over a fixed universal domain. Let V and W be algebraic varieties, and f a rational map of V into W. If k is a field of rationality for V, W, and f, and σ an isomorphism of k onto a field k', we denote by V^σ, W^σ, and f^σ the transforms of V, W, and f by σ. (If any of these objects V, W, and f is defined by a set of polynomial equations $\sum a_{(i)h} X_0^{i_0} \cdots X_n^{i_n} = 0$ ($h = 1, 2, \cdots$) with $a_{(i)h} \in k$, then its transform by σ is defined by the set of equations $\sum a_{(i)h}^\sigma X_0^{i_0} \cdots X_n^{i_n} = 0$.) Thus V^σ, W^σ, and f^σ are rational over k', and f^σ is a rational map of V^σ into W^σ. The transform of a differential form on V by σ is defined in a similar way.

For an abelian variety A, we denote by $\mathrm{End}(A)$ the ring of all endomorphisms of A (rational over the universal domain), and put $\mathrm{End}_Q(A) = \mathrm{End}(A) \otimes Q$.

1. Cusp forms and abelian varieties associated with them

Throughout the paper, we shall denote by $\mathfrak{H}$ the complex upper half plane:

$$\mathfrak{H} = \{z \in C \,|\, \mathrm{Im}(z) > 0\} .$$

For a positive integer κ, a holomorphic function $g(z)$ on $\mathfrak{H}$, and an element $\alpha = \begin{bmatrix} a & b \\ c & d \end{bmatrix}$ of $GL_2(R)$ with $\det(\alpha) > 0$, we define a function $g\,|[\alpha]_\kappa$ by

$$(1.1) \qquad g\,|[\alpha]_\kappa = g\big((az + b)/(cz + d)\big) \cdot \det(\alpha)^{\kappa/2} \cdot (cz + d)^{-\kappa} .$$

Let N be a positive integer, and ψ an arbitrary $C^\times$-valued character of $(Z/NZ)^\times$ such that

$$(1.2) \qquad \psi(-1) = (-1)^\kappa .$$

Define subgroups $\Gamma_0(N)$ and $\Gamma_1(N)$ of $SL_2(Z)$ by

$$\Gamma_0(N) = \left\{ \begin{bmatrix} a & b \\ c & d \end{bmatrix} \in SL_2(Z) \,\Big|\, c \equiv 0 \bmod NZ \right\} ,$$

$$\Gamma_1(N) = \left\{ \begin{bmatrix} a & b \\ c & d \end{bmatrix} \in \Gamma_0(N) \,\Big|\, a \equiv d \equiv 1 \bmod NZ \right\} .$$

For any congruence subgroup Γ of $SL_2(Z)$, we denote by $S_\kappa(\Gamma)$ the vector space of all cusp forms of weight κ with respect to Γ, and by $S_\kappa(N, \psi)$ the subspace of $S_\kappa(\Gamma_1(N))$ consisting of the functions g such that $g\,|[\gamma]_\kappa = \psi(d)g$ for all $\gamma = \begin{bmatrix} a & b \\ c & d \end{bmatrix} \in \Gamma_0(N)$. Then we can define, as in Hecke [3], the Hecke operator $T(n)_N$ on $S_\kappa(\Gamma_1(N))$ for each positive integer n. We denote by $T(n)_{N,\psi}$ the restriction of $T(n)_N$ to $S_\kappa(N, \psi)$. To be more explicit, we have

$$g\,|\,T(n)_{N,\psi} = n^{\kappa-1} \cdot \sum_{\substack{a>0 \\ ad=n}} \sum_{b=0}^{d-1} \psi(a) \cdot g\big((az + b)/d\big) \cdot d^{-\kappa} .$$

Further we can define an automorphism H of $S_\kappa(\Gamma_1(N))$ which maps $S_\kappa(N, \psi)$ onto $S_\kappa(N, \bar{\psi})$ by

$$H = \left[\begin{bmatrix} 0 & -1 \\ N & 0 \end{bmatrix}\right]_\kappa , \qquad \text{i.e.,} \qquad g\,|\,H = (N^{1/2}z)^{-\kappa} \cdot g(-1/Nz) .$$

Then

$$T(n)_{N,\psi} \cdot H = \psi(n) \cdot H \cdot T(n)_{N,\bar{\psi}} \qquad\qquad \text{if } n \text{ is prime to } N .$$

Now we can define the "essential part" $S_\kappa^0(N, \psi)$ of $S_\kappa(N, \psi)$ as follows. First let $S_\kappa^1(N, \psi)$ denote the subspace of $S_\kappa(N, \psi)$ spanned by all the functions of the form $g(mz)$, where $g \in S_\kappa(M, \psi)$ for a divisor $M\,(<N)$ of N,

modulo which ψ can be defined, and m is a positive divisor of N/M. Then we let $S_\kappa^0(N, \psi)$ denote the orthogonal complement of $S_\kappa^1(N, \psi)$ in $S_\kappa(N, \psi)$ with respect to the Petersson metric.

LEMMA 1.1 (Miyake [5]). (i) *The operator H maps $S_\kappa^0(N, \psi)$ onto $S_\kappa^0(N, \bar\psi)$.*

(ii) *Let $T(n)_{N,\psi}^0$ denote the restriction of $T(n)_{N,\psi}$ to $S_\kappa^0(N, \psi)$, and $\mathfrak{A}$ the algebra generated over C by the $T(n)_{N,\psi}^0$ for all n. Then $\mathfrak{A}$ is commutative and semi-simple. Moreover, $\mathfrak{A}$ is generated by the $T(p)_{N,\psi}^0$ for almost all primes p.*

These assertions are special cases of the results of [5].

Now let us fix a non-zero element $f(z) = \sum_{n=1}^{\infty} a_n e^{2\pi i n z}$ of $S_\kappa^0(N, \psi)$, which is a common eigen-function of $T(n)_{N,\psi}$ for all n. Replacing f by its suitable constant multiple, we can assume that $a_1 = 1$. Then

$$f \mid T(n)_{N,\psi} = a_n \cdot f \qquad \text{for all } n .$$

Then we define a zeta-function with an Euler product:

$$(1.3) \qquad L(s, f) = \sum_{n=1}^{\infty} a_n n^{-s} = \prod_p \left(1 - a_p p^{-s} + \psi(p)p^{1-2s}\right)^{-1} ,$$

where we understand that $\psi(p) = 0$ if p divides N.

Let K denote, for a fixed f, the subfield of C generated over Q by the coefficients a_n for all n. (Of course K depends on f.) Suppose $\kappa \geq 2$. By [9, Th. 3.48], K is an algebraic number field of finite degree. Moreover, from the Euler product, we see that K contains the roots of unity $\psi(n)$ for all n.

PROPOSITION 1.2. *Let $\mathfrak{I}$ denote the set of all isomorphisms of K into C. Suppose that $\kappa \geq 2$. Then, for every $\sigma \in \mathfrak{I}$, there exists an element f_σ of $S_\kappa^0(N, \psi^\sigma)$ such that $f_\sigma \mid T(n)_{N,\psi^\sigma} = a_n^\sigma \cdot f_\sigma$ for all n, and $f_\sigma(z) = \sum_{n=1}^{\infty} a_n^\sigma e^{2\pi i n z}$, where ψ^σ is defined by $\psi^\sigma(n) = \psi(n)^\sigma$.*

The proof will be given in Appendix I. We call f_σ a *companion* of f. Presumably the assertion is true even if $\kappa = 1$.

Let ρ denote the complex conjugation. By [9, Prop. 3.57], we have

$$(f_\sigma \mid H) \mid T(n)_{N,\psi^{\sigma\rho}} = a_n^{\sigma\rho} \cdot (f_\sigma \mid H) = f_{\sigma\rho} \mid T(n)_{N,\psi^{\sigma\rho}} \qquad \text{if } (n, N) = 1 .$$

By (ii) of Lemma 1.1, this shows that

(1.4) $f_\sigma \mid H$ *is a constant multiple of* $f_{\sigma\rho}$.

Therefore, if W is the vector subspace of $S_\kappa(\Gamma_1(N))$ spanned by the companions of f, then H maps W onto itself.

PROPOSITION 1.3. *If ψ is a trivial character, K is totally real. If ψ is not trivial, K is a CM-field.*

Here and henceforth, we understand by a *CM-field* a totally imaginary quadratic extension of a totally real algebraic number field.

Proof. By [9, Prop. 3.56], we have $a_n^\sigma = \psi(n)^\sigma a_n^{\sigma\rho}$ for all n prime to N. Therefore we see that $K^\rho = K$, and $a_n^{\sigma\rho} = a_n^{\rho\sigma}$ for every $\sigma \in \mathfrak{F}$, hence we obtain our assertion.

Now let us restrict ourselves to the case $\kappa = 2$. The vector space $S_2(\Gamma_1(N))$ can be identified with the vector space of all holomorphic differential forms on the compactification $(\Gamma_1(N)\backslash\mathfrak{H})^*$ of the quotient $\Gamma_1(N)\backslash\mathfrak{H}$. Consider a projective non-singular curve V isomorphic to $(\Gamma_1(N)\backslash\mathfrak{H})^*$. We can take as V a "standard model" defined over $\mathbf{Q}$, which is unique up to biregular isomorphisms over $\mathbf{Q}$, see [9, §7.3]. Let J be the jacobian variety of V naturally defined over $\mathbf{Q}$. We can identify $S_2(\Gamma_1(N))$ with the tangent space of J at the origin. The operator $T(n)_N$ naturally defines an endomorphism ξ_n of J. In [9, §§ 7.3–7.5], we have shown that ξ_n is rational over $\mathbf{Q}$. For a fixed common eigen-function f of $T(n)_N$ in $S_2^0(N, \psi)$ as above, there is a couple (A, θ) with the following properties:

(1.5) *A is an abelian subvariety of J of dimension $[K:\mathbf{Q}]$.*

(1.6) *θ is an isomorphism of K into $\mathrm{End}_0(A)$ such that $\theta(a_n)$ is the restriction of ξ_n to A for all n.*

(1.7) *A and the elements of $\theta(K) \cap \mathrm{End}(A)$ are rational over $\mathbf{Q}$.*

(1.8) *The tangent space of A at the origin can be identified with the subspace of $S_2(\Gamma_1(N))$ spanned by all the companions of f.*

The couple is unique for f under the conditions (1.5) and (1.6). The existence as well as the unicity of (A, θ) has been proved in [9, Th. 7.14]. From (1.4), we obtain

(1.9) *H induces an automorphism of A.*

Let η denote the automorphism of A given by H. From (1.4) we obtain

$$(1.10) \qquad\qquad \eta \circ \theta(a) = \theta(a^\rho) \circ \eta \qquad\qquad \text{for all } a \in K \,.$$

THEOREM 1.4. *Let A be the abelian variety associated with a common eigen-function f of Hecke operators as above. Then the zeta-function of A over $\mathbf{Q}$ coincides, up to finitely many Euler factors, with $\prod_{\sigma \in \mathfrak{F}} L(s, f_\sigma)$. Moreover, $\zeta(s; A/\mathbf{Q}, K)$ coincides, up to finitely many Euler factors, with $L(s, f)$.*

For the definition of $\zeta(s; A/\mathbf{Q}, K)$, see [9, §7.6].

Proof. For ψ not trivial, the first assertion was proved in [9, Th. 7.16] under a certain condition [9, (7.5.8)] on $T(n)_{N,\psi}$, which was necessary to ensure the property (1.9) of A. In the present setting, we have replaced the condition by a more reasonable assumption that f belongs to $S_2^0(N, \psi)$, from which we obtain (1.9), hence our first assertion by the same reasoning as in [9, Th. 7.11, Th. 7.16].

To prove the second assertion, take a rational prime p modulo which A has good reduction. Fix a prime factor $\mathfrak{p}$ of p in $\mathbf{Q}(e^{2\pi i/N})$, and consider reduction modulo $\mathfrak{p}$. Indicate reduced objects by putting tildes. Let π_p be the Frobenius endomorphism of $\mathrm{End}_Q(\tilde{A})$ of degree p, and π_p^* its adjoint. From [9, (7.5.1), (7.5.2)], we obtain

$$(1.11) \qquad \begin{aligned} \pi_p + \pi_p^* \cdot \tilde{\theta}\big(\psi(p)\big) &= \tilde{\theta}(a_p) \,, \\ \pi_p^* \cdot \tilde{\theta}\big(\psi(p)\big) &= \tilde{\eta}^{-1} \pi_p^* \tilde{\eta} \,. \end{aligned}$$

Let $\mathfrak{l}$ be a prime ideal in K not dividing p, and consider the $\mathfrak{l}$-adic representation $R_\mathfrak{l}$ on $\tilde{A}$. Since $\dim(\tilde{A}) = [K:\mathbf{Q}]$, $R_\mathfrak{l}$ maps the commutor of $\tilde{\theta}(K)$ in $\mathrm{End}_Q(\tilde{A})$ into the matrix algebra of degree 2 over the $\mathfrak{l}$-completion of K. Now our assertion follows from

$$(1.12) \qquad \det[X - R_\mathfrak{l}(\pi_p)] = X^2 - a_p X + \psi(p)p \,,$$

where X is an indeterminate. To show this, first assume that $\pi_p = \tilde{\theta}(c)$ with an element c of K. Then the characteristic polynomial of $R_\mathfrak{l}(\pi_p)$ must be $(X - c)^2$. On the other hand, since ρ induces a unique positive involution on any subfield of K, we have $\pi_p^* = \tilde{\theta}(c^\rho)$, so that $\pi_p^* \cdot \tilde{\theta}\big(\psi(p)\big) = \tilde{\eta}^{-1} \cdot \tilde{\theta}(c^\rho) \cdot \tilde{\eta} = \tilde{\theta}(c)$, hence $\tilde{\theta}(a_p) = \tilde{\theta}(2c)$. It follows that

$$X^2 - a_p X + \psi(p)p = (X - c)^2 \,,$$

which proves (1.12). Next assume that $\pi_p \notin \tilde{\theta}(K)$. Identify K with $\tilde{\theta}(K)$. Observe that $K[\pi_p]$ is a commutative semi-simple algebra, since it is a homomorphic image of $K \otimes_Q \mathbf{Q}[\pi_p]$. Moreover, $R_\mathfrak{l}(\pi_p)$ satisfies a quadratic equation with coefficients in K, so that $[K[\pi_p]:K] = 2$. Therefore a quadratic equation for π_p over K must be unique. Since π_p is a root of the polynomial

$$X^2 - a_p X + \psi(p)p = (X - \pi_p)\big(X - \pi_p^* \cdot \psi(p)\big) \,,$$

we obtain (1.12). This completes the proof.

We shall now show that if A has an abelian subvariety of CM-type, then A is isogenous to a product of elliptic curves with complex multiplication.[1] Here we say that an abelian variety is of CM-type, if it is isogenous to a product $A_1 \times \cdots \times A_s$ with abelian varieties A_i such that $\mathrm{End}_Q(A_i)$ is iso-

---

[1] W. Casselman has informed the author of this fact with a sketch of proof. In the following proposition, we prove his result in a somewhat different formulation.

morphic to a CM-field of degree $2 \cdot \dim(A_i)$ for each i. This is so if and only if $\mathrm{End}_Q(A)$ has a commutative semi-simple algebra of rank $2 \cdot \dim(A_i)$ over Q (see [13, § 5.1]).

PROPOSITION 1.5. *Let A be an abelian variety of dimension n, K an algebraic number field of degree n over Q, and θ an isomorphism of K into $\mathrm{End}_Q(A)$ which maps the identity element of K to the identity map of A. If A has an abelian subvariety of CM-type, then A itself is of CM-type. If further A and the elements of $\theta(K) \cap \mathrm{End}(A)$ are rational over Q, then A is isogenous to a product of several copies of an elliptic curve with complex multiplication.*

Proof. Let S be the center of $\mathrm{End}_Q(A)$. Then S is a direct sum of fields $S_1, \cdots, S_r$. Let e_i be the identity element of S_i. Take a positive integer h so that $h \cdot e_i \in \mathrm{End}(A)$ for every i, and put $A_i = he_i(A)$. By our assumption, at least one of the A_i, say A_1, is of CM-type. Observe that if L is a field of definition for A and the elements of $\theta(K) \cap \mathrm{End}(A)$, every element of $\mathrm{Aut}(C/L)$ permutes $e_1, \cdots, e_r$. Changing the ordering if necessary, assume that $\mathrm{Aut}(C/L)$ maps e_1 onto $e_1, \cdots, e_t$. Put $A' = A_1 + \cdots + A_t$, $A'' = A_{t+1} + \cdots + A_r$. Then A' and A'' are rational over L, and $\theta(K) \cap \mathrm{End}(A)$ acts L-rationally on A' and A''. By [13, § 5.1, Prop. 2], both $\dim(A')$ and $\dim(A'')$ are multiples of $n/2$. Therefore we have either $A = A'$ or $\dim(A') = \dim(A'') = n/2$. In any case A is of CM-type. By [13, § 8.5, Prop. 30], the latter case can happen only when L contains a CM-field. Now suppose that L contains no CM-fields, and identify K with $\theta(K)$. Further assume that $S \subset K$. Then S is a field, so that A is isogenous to a product of several copies of a simple abelian variety B such that $\mathrm{End}_Q(B)$ is isomorphic to S. Let Φ (resp. Ψ) be the representation of S on the space of linear holomorphic differential forms on A (resp. B). Then (S, Ψ) is a CM-type, and Φ is equivalent to several copies of Ψ. Since $\theta(K) \cap \mathrm{End}(A)$ is rational over L, $\mathrm{tr}\,\Phi(x) \in L$ for every $x \in S$. This is a contradiction, since the numbers $\mathrm{tr}\,\Psi(x)$, for all $x \in S$, generate a CM-field. Therefore $S \not\subset K$. Let T be the subalgebra of $\mathrm{End}_Q(A)$ generated by S and K. Then T is commutative, and being a homomorphic image of $K \otimes_Q S$, must be semi-simple. By [13, § 5.1, Prop. 1], we have $[T:Q] = 2n$. Now the action of $\mathrm{Aut}(C/L)$ induces an automorphism of T over K, which must be of order 1 or 2, since T is either a field, or the direct sum of two copies of K. Therefore every element of S is defined over a quadratic extension M of L. Now A_i is isogenous to $B_i^{m_i}$ with a simple abelian variety B_i such that $\mathrm{End}_Q(B_i)$ is isomorphic to S_i, and S_i is a CM-field. Let Φ_i (resp. Ψ_i) be the representation of S_i on the space of linear holomorphic differential forms on A (resp. B_i). Then Φ_i is equivalent to the sum of m_i copies of Ψ_i and possibly a zero representation. Let (S_i', Ψ_i') be the reflex of (S_i, Ψ_i). Then S_i' is generated by $\operatorname{tr} \Psi_i(x)$ for $x \in S_i$, so that $S_i' \subset M$. Therefore $[S_i'L : L] \leq 2$. Especially, if $L = \mathbf{Q}$, we have $[S_i' : \mathbf{Q}] \leq 2$, so that $[S_i : \mathbf{Q}] \leq 2$. Thus each B_i is an elliptic curve with complex multiplication. Since A_i is the image of A_1 by an element of $\operatorname{Aut}(C)$, we see that S_i is isomorphic to S_1. This completes the proof.

Remark. The above proof gives some information about the structure of A or $\operatorname{End}_Q(A)$ for a specifically given field L which is not necessarily $\mathbf{Q}$. But we shall not discuss this problem in the present paper.

PROPOSITION 1.6. *Let f be an element of $S_2^0(N, \psi)$, and (A, θ) the couple associated with f. Suppose that A has an abelian subvariety of CM-type. Then A is a product of several copies of an elliptic curve E with complex multiplication. Moreover, if S is the imaginary quadratic field isomorphic to $\operatorname{End}_Q(E)$, then the Mellin transform of f is the L-function with a primitive Grössen-character of S.*

Proof. The first assertion is an immediate consequence of Proposition 1.5. Put $SK = S \otimes_Q K$. Extend θ to an injection θ' of SK into $\operatorname{End}_Q(A)$ so that $\theta'(S)$ is the center of $\operatorname{End}_Q(A)$. Then by an argument similar to [11, Prop. 3], we see that (A, θ') is rational over S. Now we have the following result:

(1.13) $\zeta(s; A/\mathbf{Q}, K)$ *coincides, up to finitely many Euler factors, with an L-function with a Grössen-character χ of S.*

This will be proved afterwards. We can take χ to be primitive. Put

$$g(z) = \sum_{\mathfrak{a}} \chi(\mathfrak{a}) e^{2\pi i N(\mathfrak{a})z} \, ,$$

where the sum is taken over all the integral ideals $\mathfrak{a}$ in S for which χ is defined. By [12, Lemma 3], g belongs to $S_2(N', \psi')$ for some N' and ψ'. By virtue of the results of Miyake [5], it can be seen that $g \in S_2^0(N', \psi')$, if N' and ψ' are determined for χ as in [12, Lemma 3]. On the other hand, by Theorem 1.4, $\zeta(s; A/\mathbf{Q}, K)$ is, up to finitely many Euler factors, given by $L(s, f)$. Therefore $L(s, f)$ coincides with $L(s, g)$ up to finitely many Euler factors. Again by Miyake [5], this implies $f = g$, $N = N'$, $\psi = \psi'$.

Proof of (1.13). This follows from [11, Th. 12] if SK is a field. Therefore suppose that SK is not a field. Then $S \subset K$, and SK can be identified with $K \oplus K$ by the map $h \colon SK \to K \oplus K$ such that

$$h(a \otimes b) = (ab, ab^\rho) \qquad\qquad (a \in S, \, b \in K) \, .$$

Put $(SK)_R = (SK) \otimes_Q R$. The variety A, as a real torus with the action of SK, is isomorphic to $(SK)_R/\mathfrak{a}$ with a lattice $\mathfrak{a}$ in SK. Fix an isomorphism $\omega: (SK)_R/\mathfrak{a} \to A$. Let $S_A^\times$ denote the idele group of S. Then we can define a map $\alpha: S_A^\times \to (SK)^\times$ such that

$$(1.14) \qquad \omega(v)^{[x,S]} = \omega(\alpha(x)x^{-1}v) \qquad (x \in S_A^\times,\ v \in SK/\mathfrak{a}),$$

where $[x, S]$ denotes the automorphism of the maximal abelian extension of S corresponding to x. The existence of such an α can be proved by the same argument as in [9, Prop. 7.40], on account of [8, I, 4.3]. Let π be the nontrivial automorphism of SK over K. Then $\theta'(a)^\rho = \theta'(a^\pi)$ for $a \in SK$. Therefore, by the same reasoning as in the proofs of [9, Th. 7.46] and [11, Prop. 1], we obtain

$$(1.15) \qquad \alpha(x)^\pi = \alpha(x^\rho).$$

Let $\alpha_1(x)$ and $\alpha_2(x)$ be the projections of $\alpha(x)$ to the first and second direct summands of SK, i.e., let $h(\alpha(x)) = (\alpha_1(x), \alpha_2(x))$. Then (1.15) implies

$$(1.16) \qquad \alpha_1(x^\rho) = \alpha_2(x)^\rho.$$

Let $\mathfrak{p}$ be a prime ideal in S for which A has good reduction, and p the rational prime divisible by $\mathfrak{p}$. Consider reduction modulo $\mathfrak{p}$, and indicate the reduced objects by putting tildes. Let φ_p (resp. $\varphi_\mathfrak{p}$) the Frobenius endomorphism of $\tilde{A}$ of degree p (resp. $N(\mathfrak{p})$), and $c_\mathfrak{p}$ a prime element of the $\mathfrak{p}$-completion of S. Now we can repeat the last half of the proof of [9, Th. 7.46], by taking Q, K, KS, and S in place of the symbols k_0, F, K, and K^* there. In fact, let $\mathfrak{l}$ be a prime ideal in K not dividing p, and $R_\mathfrak{l}$ the $\mathfrak{l}$-adic representation of $\mathrm{End}_Q(\tilde{A})$. Taking $c_\mathfrak{p}$ to be x in (1.14), we find that $\varphi_\mathfrak{p} = \tilde{\theta}'(\alpha(c_\mathfrak{p}))$. If $N(\mathfrak{p}) = p^2$, we have $\alpha(c_\mathfrak{p}) \in K$, and

$$(1.17) \qquad \det[1 - R_\mathfrak{l}(\varphi_p)X] = 1 - \alpha(c_\mathfrak{p})X^2 = 1 - \alpha_1(c_\mathfrak{p})X^2,$$

exactly in the same manner as in [9, p. 219]. If $\mathfrak{p} \neq \mathfrak{p}^\rho$, we have

$$\det[1 - R_\mathfrak{l}(\varphi_p)X] = (1 - \alpha(c_\mathfrak{p})X)(1 - \alpha(c_\mathfrak{p})^\pi X).$$

In view of (1.15, 16), it can easily be seen that the right hand side, as a polynomial with coefficients in K, equals

$$(1.18) \qquad (1 - \alpha_1(c_\mathfrak{p})X)(1 - \alpha_1(c_\mathfrak{p}^\rho)X).$$

Now we can define a Grössen-character χ of S so that $\chi(\mathfrak{p}) = \alpha_1(c_\mathfrak{p})$ for all prime ideals $\mathfrak{p}$ which are unramified over Q, and for which A has good reduction. (It corresponds to the character χ_0 of $S_A^\times$ defined by $\chi_0(x) = \alpha_1(x) \cdot x_\infty^{-1}$, where x_∞ denotes the archimedean component of x.) Taking the product of (1.17, 18) with $X = \mathfrak{p}^{-s}$ for all "good $\mathfrak{p}$'s", we obtain the desired conclusion.

Remark 1.7. In [12], we have proved that, if f is the Mellin transform of an L-function with a Grössen-character χ of an imaginary quadratic field S, then A is a product of copies of an elliptic curve whose endomorphism algebra is isomorphic to S. The above proposition gives the converse of this fact. The level N and the character ψ can be determined by [12, Lemma 3]. This gives a necessary condition for A to have an abelian subvariety of CM-type. For example, if N is a prime, A has no abelian subvariety of CM-type. (See also § 8 below.)

2. Formulation of the main problems

Let us now recall the theory developed in [9, § 7.7] with a few changes of notation and formulation. We fix an element

$$f(z) = \sum_{n=1}^{\infty} a_n e^{2\pi i n z}$$

of $S_2^0(N, \psi)$ that is a common eigen-function of $T(n)_{N,\psi}$ for all n, and consider the field K and the couple (A, θ) associated with f as above. Hereafter till the end of the paper, we assume that ψ is a non-trivial real character, so that $\psi^2 = 1$. The condition (1.2) becomes

(2.1) $\psi(-1) = 1$.

Let k denote the quadratic extension of $\mathbf{Q}$ corresponding to the character ψ. By (2.1), k must be real. We denote by ε the generator of $\mathrm{Gal}(k/\mathbf{Q})$. By (1.9), H induces an automorphism η of A. Then η has the following properties:

(2.2) η *is rational over* k, *and* $\eta^{\varepsilon} = -\eta$;

(2.3) $\eta^2 = 1$;

(2.4) $\eta \cdot \theta(a) = \theta(a^{\varepsilon}) \cdot \eta$ *for every* $a \in K$.

Define an abelian subvariety B of A by

(2.5) $$B = (1 + \eta)A .$$

Then B is rational over k, $A = B + B^{\varepsilon}$, $B^{\varepsilon} = (1 - \eta)A$, and $B \cap B^{\varepsilon}$ is a finite group annihilated by 2. By Proposition 1.3, K must be a CM-field. Let F denote the maximal real subfield of K. Then we can define an injection θ_F of F into $\mathrm{End}_{\mathbf{Q}}(B)$ so that $\theta_F(a)$ is the restriction of $\theta(a)$ to B for every $a \in F$.

Let $\mathfrak{o}_K$ and $\mathfrak{o}_F$ denote the rings of all algebraic integers in K and in F, respectively. Changing (A, θ) by an isogeny over $\mathbf{Q}$ if necessary, we may assume

(2.6) $\theta(\mathfrak{o}_K) \subset \mathrm{End}(A)$, $\theta_F(\mathfrak{o}_F) \subset \mathrm{End}(B)$

(see [9, pp. 198–199]).

Let $\mathfrak{b}_0$ denote the ideal of $\mathfrak{o}_K$ generated by all x in $\mathfrak{o}_K$ such that $x^{\varepsilon} = -x$. Then we can define the "odd part" $\mathfrak{b}$ of $\mathfrak{b}_0$ by the properties: (i) $\mathfrak{b}$ *is a divisor of* $\mathfrak{b}_0$ *prime to* 2; (ii) $N(\mathfrak{b}^{-1}\mathfrak{b}_0)$ *is a power of* 2. Then $\mathfrak{b}_0$ divides the different of K relative to F, $\mathfrak{b}$ is square-free, $\mathfrak{b}^\rho = \mathfrak{b}$, and $\mathfrak{b}^2 = \mathfrak{c}\mathfrak{o}_K$ with a square-free integral ideal $\mathfrak{c}$ in F. Obviously $\mathfrak{o}_K/\mathfrak{b}$ is $\mathfrak{o}_F$-isomorphic to $\mathfrak{o}_F/\mathfrak{c}$.

We are interested in the points of A annihilated by $\mathfrak{b}$. Namely, put

$$(2.7) \qquad \mathfrak{x} = \{t \in A \,|\, \theta(\mathfrak{b})t = 0\} \,,$$

$$(2.8) \qquad \mathfrak{y} = B \cap \mathfrak{x} \,, \qquad \mathfrak{z} = B' \cap \mathfrak{x} \,.$$

Then $\mathfrak{x} = \mathfrak{y} \oplus \mathfrak{z}$, and both $\mathfrak{y}$ and $\mathfrak{z}$ are isomorphic to $\mathfrak{o}_F/\mathfrak{c}$ as $\mathfrak{o}_F$-modules. Now let $k(\mathfrak{x})$ denote the field generated over k by the coordinates of the points of $\mathfrak{x}$. It is our main objective to study the structure of $k(\mathfrak{x})$ in connection with the cusp form f. If $\mathfrak{b} = \mathfrak{o}_K$, we have $\mathfrak{x} = \{0\}$, so that $k(\mathfrak{x}) = k$. To exclude this trivial case, we hereafter assume that $\mathfrak{b} \neq \mathfrak{o}_K$. The discussion of [9, § 7.7] has been made under the assumption that $\mathfrak{b}_0 = \mathfrak{b}$. (The notation $\mathfrak{b}$ was used there instead of $\mathfrak{b}_0$.) It can easily be seen that the whole argument and results of [9, § 7.7] are valid under the present definition of $\mathfrak{x}$ and $k(\mathfrak{x})$.

PROPOSITION 2.1. *If $N(\mathfrak{c})$ is square-free as a rational integer, then every rational prime factor of $N(\mathfrak{c})$ unramified in k decomposes into two prime factors in k.*

Proof. We have shown in [9, Prop. 7.28] that $-N(\mathfrak{b}_0) \in N_{k/Q}(k^\times)$. Since $N(\mathfrak{c})$ is the "odd part" of $N(\mathfrak{b}_0)$, we obtain our assertion.

Now $k(\mathfrak{x})$ is an abelian extension of k (see [9, Th. 7.30]). Moreover, letting $\mathrm{Gal}(k(\mathfrak{x})/k)$ act on $\mathfrak{y}$ and $\mathfrak{z}$, we obtain an injective homomorphism

$$(2.9) \qquad \mathrm{Gal}(k(\mathfrak{x})/k) \longrightarrow (\mathfrak{o}_F/\mathfrak{c})^\times \times (\mathfrak{o}_F/\mathfrak{c})^\times \,.$$

Let $\mathfrak{f}[k(\mathfrak{x})/k]$, or simply $\mathfrak{f}$, denote the conductor of $k(\mathfrak{x})$ as a class-field over k. For any integral ideal $\mathfrak{m}$ in k, let $I(\mathfrak{m})$ denote the group of all fractional ideals in k prime to $\mathfrak{m}$. It has been shown that every finite prime factor of $\mathfrak{f}$ divides $N(\mathfrak{c}) \cdot N$. For $\mathfrak{a} \in I(\mathfrak{f})$, let $(r(\mathfrak{a}), s(\mathfrak{a}))$ denote the image of $\left(\dfrac{k(\mathfrak{x})/k}{\mathfrak{a}}\right)$ by the map (2.9). Then r and s can be considered "characters" of the ideal group $I(\mathfrak{f})$ with values in $(\mathfrak{o}_F/\mathfrak{c})^\times$. Let $\mathfrak{o}_k$, or simply $\mathfrak{o}$, denote the ring of all algebraic integers in k. Then we have

THEOREM 2.2. *The characters r and s have the following properties:*
(i) $r(\mathfrak{a}) = s(\mathfrak{a}^\iota)$ *for every* $\mathfrak{a} \in I(\mathfrak{f})$.
(ii) $r(m\mathfrak{o}) = s(m\mathfrak{o}) = \left(\dfrac{m}{p_\infty}\right) \cdot (m \bmod \mathfrak{c})$ *for every* $m \in Z$, $(m, \mathfrak{f}) = 1$, *where*

p_∞ *is the archimedean prime of* Q, *and* $\left(\dfrac{}{p_\infty}\right)$ *is as in Notation.*
(iii) $r(\mathfrak{a})s(\mathfrak{a}) = (N(\mathfrak{a}) \bmod \mathfrak{c})$ *for every* $\mathfrak{a} \in I(\mathfrak{f})$.
(iv) *If p is a rational prime in k prime to $N(\mathfrak{c}) \cdot N$, and $p\mathfrak{o} = \mathfrak{p}\mathfrak{p}^\iota$ with*

two prime ideals $\mathfrak{p}$ and $\mathfrak{p}^{\epsilon}$ in k, then

$$r(\mathfrak{p}) + s(\mathfrak{p}) = (a_p \bmod \mathfrak{c}) ,$$

where a_p is the p-th Fourier coefficient of the cusp form $f(z)$.

This is a restatement of what was proved in [9, Th. 7.30] and its proof. We shall prove a stronger Theorem 2.3, of which Theorem 2.2 is an obvious consequence.

For various reasons, it is necessary and natural to localize r and s at the prime factors of $\mathfrak{c}$. To do this, for every prime factor $\mathfrak{l}$ of $\mathfrak{c}$ in F, put

(2.10) $\mathfrak{y}_{\mathfrak{l}} = \{u \in \mathfrak{y} \mid \theta(\mathfrak{l})u = 0\} , \qquad \mathfrak{z}_{\mathfrak{l}} = \{u \in \mathfrak{z} \mid \theta(\mathfrak{l})u = 0\} .$

Then $\mathfrak{y}$ and $\mathfrak{z}$ are respectively the direct sums of the $\mathfrak{y}_{\mathfrak{l}}$ and the $\mathfrak{z}_{\mathfrak{l}}$ for all prime factors $\mathfrak{l}$ of $\mathfrak{c}$. Both $\mathfrak{y}_{\mathfrak{l}}$ and $\mathfrak{z}_{\mathfrak{l}}$ are isomorphic to $\mathfrak{o}_F/\mathfrak{l}$ as $\mathfrak{o}_F$-modules. Taking $\mathfrak{y}_{\mathfrak{l}}$ instead of $\mathfrak{y}$, we can define a homomorphism $r_{\mathfrak{l}}$ of $I(\mathfrak{f})$ into $(\mathfrak{o}_F/\mathfrak{l})^{\times}$. In other words, $r_{\mathfrak{l}}$ is the composed map of r with the natural map of $\mathfrak{o}_F/\mathfrak{c}$ onto $\mathfrak{o}_F/\mathfrak{l}$. Thus $r_{\mathfrak{l}}$ may be called *the $\mathfrak{l}$-component of r.* We can define the conductor $\mathfrak{f}[r_{\mathfrak{l}}]$ of $r_{\mathfrak{l}}$ in a natural way (see Notation), and extend $r_{\mathfrak{l}}$ to a homomorphism of $I(\mathfrak{f}[r_{\mathfrak{l}}])$ into $(\mathfrak{o}_F/\mathfrak{l})^{\times}$, which we denote again by $r_{\mathfrak{l}}$. Obviously $\mathfrak{f}[r_{\mathfrak{l}}]$ is the conductor of the abelian extension of k generated by the coordinates of the points in $\mathfrak{y}_{\mathfrak{l}}$. Therefore every finite prime factor of $\mathfrak{f}[r_{\mathfrak{l}}]$ divides $N(\mathfrak{l}) \cdot N$. We can define $s_{\mathfrak{l}}$ and $\mathfrak{f}[s_{\mathfrak{l}}]$ in a similar way with $\mathfrak{z}_{\mathfrak{l}}$ in place of $\mathfrak{y}_{\mathfrak{l}}$. Then we see that $\mathfrak{f}[k(\mathfrak{x})/k]$ is the least common multiple of $\mathfrak{f}[r_{\mathfrak{l}}]$ and $\mathfrak{f}[s_{\mathfrak{l}}]$ for all prime factors $\mathfrak{l}$ of $\mathfrak{c}$.

THEOREM 2.3. *The characters $r_{\mathfrak{l}}$ and $s_{\mathfrak{l}}$ have the following properties:*

(i) $\mathfrak{f}[s_{\mathfrak{l}}] = \mathfrak{f}[r_{\mathfrak{l}}]^{\epsilon}.$

(ii) $r_{\mathfrak{l}}(\mathfrak{a}) = s_{\mathfrak{l}}(\mathfrak{a}^{\epsilon})$ *for every* $\mathfrak{a} \in I(\mathfrak{f}[r_{\mathfrak{l}}]).$

(iii) $r_{\mathfrak{l}}(m\mathfrak{o}) = s_{\mathfrak{l}}(m\mathfrak{o}) = \left(\dfrac{m}{p_{\infty}}\right) \cdot (m \bmod \mathfrak{l})$ *for every* $m \in \mathbf{Z}$ *prime to* $\mathfrak{f}[r_{\mathfrak{l}}]$,

where p_{∞} is the archimedean prime of $\mathbf{Q}$.

(iv) $r_{\mathfrak{l}}(\mathfrak{a})s_{\mathfrak{l}}(\mathfrak{a}) = (N(\mathfrak{a}) \bmod \mathfrak{l})$ *for all* $\mathfrak{a} \in I(\mathfrak{f}[r_{\mathfrak{l}}]) \cap I(\mathfrak{f}[s_{\mathfrak{l}}]).$

(v) *If p is a rational prime that is prime to $N(\mathfrak{l}) \cdot N$, and that decomposes into two distinct prime ideals $\mathfrak{p}$ and $\mathfrak{p}^{\epsilon}$ in k, then*

$$r_{\mathfrak{l}}(\mathfrak{p}) + s_{\mathfrak{l}}(\mathfrak{p}) = (a_p \bmod \mathfrak{l}) .$$

Proof. Let $\mathfrak{b}_{\mathfrak{l}}$ be the prime ideal in K such that $\mathfrak{b}_{\mathfrak{l}}^2 = \mathfrak{l}\mathfrak{o}_K$. Then we see easily that

(2.11) $\mathfrak{y}_{\mathfrak{l}} + \mathfrak{z}_{\mathfrak{l}} = \{u \in A \mid \theta(\mathfrak{b}_{\mathfrak{l}})u = 0\} .$

Let p be a rational prime not dividing $N(\mathfrak{l}) \cdot N$, $\mathfrak{p}$ a prime ideal in k dividing p, and $\mathfrak{P}$ a prime divisor of $\bar{\mathbf{Q}}$ which extends $\mathfrak{p}$. Consider reduction modulo $\mathfrak{P}$ and indicate reduced objects by putting tildes. Let π_p denote the Frobenius endomorphism of $\tilde{A}$ of degree p, π_p^* its adjoint, and σ a Frobenius element of $\mathrm{Gal}(\bar{Q}/k)$ for $\mathfrak{P}$. Then $\sigma = \left(\dfrac{k(\mathfrak{x})/k}{\mathfrak{p}}\right)$ on $k(\mathfrak{x})$.

Suppose $\psi(p) = -1$. From (1.11) we obtain

$$\pi_p^2 - p = \pi_p(\pi_p - \pi_p^*) = \pi_p \cdot \tilde{\theta}(a_p) .$$

Since $a_p^\rho = -a_p$, we have $a_p \in \mathfrak{b}_0 \subset \mathfrak{b}$, hence $\pi_p^2 \tilde{x} - p\tilde{x} = 0$ for every $x \in \mathfrak{y}_\iota + \mathfrak{z}_\iota$. Since reduction modulo $\mathfrak{P}$ defines an isomorphism of $\mathfrak{y}_\iota + \mathfrak{z}_\iota$ onto $\tilde{\mathfrak{y}}_\iota + \tilde{\mathfrak{z}}_\iota$, we obtain $x^\sigma = px$ for such x, so that

$$(2.12) \qquad\qquad r_\iota(p\mathfrak{o}) = s_\iota(p\mathfrak{o}) = (p \bmod \mathfrak{l}).$$

Next suppose $\psi(p) = 1$. Let R denote the $\mathfrak{b}_\iota$-adic representation of $\mathrm{End}_Q(\tilde{A})$. By (1.12), we have

$$\det[X - R(\pi_p)] = X^2 - a_p X + \psi(p)p .$$

In view of (2.11), we see that the action of σ on $\mathfrak{y}_\iota + \mathfrak{z}_\iota$, reduced modulo $\mathfrak{P}$, is represented by $R(\pi_p) \bmod \mathfrak{b}_\iota$. Therefore we obtain

$$R(\pi_p) \bmod \mathfrak{b}_\iota = \begin{bmatrix} r_\iota(\mathfrak{p}) & 0 \\ 0 & s_\iota(\mathfrak{p}) \end{bmatrix},$$

and hence

$$(2.13) \qquad\qquad (X^2 - a_p X + p) \bmod \mathfrak{l} = \big(X - r_\iota(\mathfrak{p})\big)\big(X - s_\iota(\mathfrak{p})\big) .$$

This proves (v). Let δ be the element of $\mathrm{Gal}(\bar{Q}/Q)$ which extends ε. Then $\mathfrak{y}_\iota^\delta = \mathfrak{z}_\iota$ and $\delta^{-1}\sigma\delta = \left(\dfrac{k(\mathfrak{x})/k}{\mathfrak{p}^\varepsilon}\right)$. Therefore, for every $x \in \mathfrak{y}_\iota$, we have $x^\delta \in \mathfrak{z}_\iota$, and $s_\iota(\mathfrak{p})x^\delta = x^{\delta\sigma} = (x^{\delta\sigma\delta^{-1}})^\delta = r_\iota(\mathfrak{p}^\varepsilon)x^\delta$, so that $s_\iota(\mathfrak{p}) = r_\iota(\mathfrak{p}^\varepsilon)$. This together with (2.12), proves (i) and (ii). Further we obtain, from (2.13),

$$r_\iota(p\mathfrak{o}) = r_\iota(\mathfrak{p})r_\iota(\mathfrak{p}^\varepsilon) = r_\iota(\mathfrak{p})s_\iota(\mathfrak{p}) = (p \bmod \mathfrak{l}) ,$$

which together with (2.12) proves (iii). To be more precise, we first obtain the relations (i–iii) for the ideals prime to $N(\mathfrak{l}) \cdot N$. Then it is obvious that they hold for the ideals prime to the conductor. The relation (iv) follows immediately from (ii) and (iii).

The structure of $k(\mathfrak{x})$ and its reciprocity law over k in connection with the cusp form f can be completely determined by r and s. Thus our main problem is reduced to the study of the characters r and s, or their components r_ι and s_ι.

PROPOSITION 2.4. *Let $\mathfrak{p}$ be a prime ideal in k ramified over Q. If $\mathfrak{p}$ is prime to $N(\mathfrak{l}) \cdot (N(\mathfrak{l}) - 1)$, then $\mathfrak{p}$ does not divide $\mathfrak{f}[r_\iota]$.*

This can be proved by the same argument as in the proof of [9, Th. 7.33].

PROPOSITION 2.5. *The conductor $\mathfrak{f}[r_\iota]$ is divisible by exactly one of the two archimedean primes of k.*

Proof. Let $\mathfrak{p}_\infty$ and $\mathfrak{p}'_\infty$ be the archimedean primes of k, and $\mathfrak{f}_0$ the finite part of $\mathfrak{f}[r_\iota]$. Then we can write

$$r_\iota(\alpha\mathfrak{o}) = \left(\frac{\alpha}{\mathfrak{p}_\infty}\right)^a \cdot \left(\frac{\alpha}{\mathfrak{p}'_\infty}\right)^b \cdot \lambda(\alpha \bmod \mathfrak{f}_0)$$

for every α in k prime to $\mathfrak{f}_0$, with a homomorphism λ of $(\mathfrak{o}/\mathfrak{f}_0)^\times$ into $(\mathfrak{o}_F/\mathfrak{l})^\times$, and a, b are 0 or 1. Take α to be a rational integer m so that $m \equiv 1 \bmod (N(\mathfrak{l}) \cdot N(\mathfrak{f}_0))$. By (iii) of Theorem 2.3, we obtain $\left(\frac{m}{\mathfrak{p}_\infty}\right)^{a+b} = \left(\frac{m}{\mathfrak{p}_\infty}\right)$. Therefore exactly one of the exponents a and b is 1, q.e.d.

Remark 2.6. The proof of Proposition 2.5 is not applicable to r in place of r_ι, if $\mathfrak{c}$ is not a prime ideal. In fact, if $\mathfrak{o}_F/\mathfrak{c}$ is not a field, an element of $(\mathfrak{o}_F/\mathfrak{c})^\times$ of order 2 is not unique. Therefore it is an open question whether or not $\mathfrak{f}[r]$ is divisible by only one archimedean prime of k.

Now we define a formal Dirichlet series $\mathfrak{L}(s, r_\iota)$ with coefficients in $\mathfrak{o}_F/\mathfrak{l}$ by

$$(2.14) \qquad \mathfrak{L}(s, r_\iota) = \sum_\mathfrak{a} r_\iota(\mathfrak{a}) \cdot N(\mathfrak{a})^{-s} = \prod_\mathfrak{p} [1 - r_\iota(\mathfrak{p}) \cdot N(\mathfrak{p})^{-s}]^{-1} ,$$

where the sum is extended over all integral ideals $\mathfrak{a}$ in $I(\mathfrak{f}[r_\iota])$, and the product over all prime ideals $\mathfrak{p}$ prime to $\mathfrak{f}[r_\iota]$.

PROPOSITION 2.7. *There is a formal congruence between certain partial sums of $L(s, f)$ and $\mathfrak{L}(s, r_\iota)$, namely*

$$\sum_{(n, N(\mathfrak{l})N)=1} a_n n^{-s} \equiv \sum_{(\mathfrak{a}, N(\mathfrak{l})N)=1} r_\iota(\mathfrak{a}) \cdot N(\mathfrak{a})^{-s} \quad \bmod \mathfrak{b}_\iota ,$$

where $\mathfrak{b}_\iota$ is the prime factor of $\mathfrak{l}$ in K such that $\mathfrak{b}_\iota^2 = \mathfrak{l}\mathfrak{o}_K$. (In other words, $\mathfrak{b}_\iota$ is the "$\mathfrak{l}$-part" of $\mathfrak{b}$.)

Proof. Let p be a rational prime not dividing $N(\mathfrak{l}) \cdot N$. If $p\mathfrak{o} = \mathfrak{p}\mathfrak{p}^\iota$ in k, we have $\psi(p) = 1$, and

$$(1 - r_\iota(\mathfrak{p})X)(1 - r_\iota(\mathfrak{p}^\iota)X) \equiv 1 - a_p X + \psi(p)pX^2 \quad \bmod \mathfrak{b}_\iota$$

by (ii, iv, v) of Theorem 2.3. If p remains prime in k, we have $\psi(p) = -1$, so that $a_p^2 = -a_p$, hence $a_p \in \mathfrak{b}_0 \subset \mathfrak{b} \subset \mathfrak{b}_\iota$. Therefore, by (iii) of Theorem 2.3,

$$1 - r_\iota(p\mathfrak{o})X^2 \equiv 1 - pX^2 \equiv 1 - a_p X + \psi(p)pX^2 \quad \bmod \mathfrak{b}_\iota .$$

Taking the product of all these congruences with $X = p^{-s}$, we obtain our assertion.

We can actually conjecture that the congruence holds not only between partial sums, but also between the entire sums:

$$(2.15) \qquad\qquad\qquad L(s, f) \equiv \mathfrak{L}(s, r_\iota) \quad \bmod \mathfrak{b}_\iota ,$$

i.e., the congruence holds also between the Euler p-factors for the primes dividing $N(\mathfrak{l}) \cdot N$. If p divides N, the question is related to whether or not

B has good reduction modulo the prime factors of p in k. For this question, see Remark 8.9 below, [9, Remark 7.27], and the discussion of [9, p. 210].

As for the prime factors of $N(\mathfrak{l})$, we can prove the following theorem, which not only gives a partial answer to the question about the congruence, but also is very useful for the determination of the character $r_{\mathfrak{l}}$, at least in the case of prime level (see the next section.)

THEOREM 2.8. *Let q be a rational prime which divides $N(\mathfrak{f}[r_{\mathfrak{l}}])$ but not N. Suppose that $\psi(q) = 1$, and a_q is prime to $\mathfrak{l}$. Then $\mathfrak{f}[r_{\mathfrak{l}}]$ is divisible by only one of the two prime factors of q in k. Moreover, if $\mathfrak{q}$ denotes that factor of q, then*

$$r_{\mathfrak{l}}(\mathfrak{q}^{\mathfrak{c}}) \equiv (a_q \bmod \mathfrak{l}) \ .$$

Since $\mathfrak{f}[r_{\mathfrak{l}}]$ consists of prime factors of $N(\mathfrak{l}) \cdot N$, we see that $\mathfrak{l}$ divides q, hence the last relation implies a formal congruence

$$1 - r_{\mathfrak{l}}(\mathfrak{q}^{\mathfrak{c}})q^{-\mathfrak{s}} \equiv 1 - a_q q^{-\mathfrak{s}} + q \cdot q^{-2\mathfrak{s}} \quad \bmod \mathfrak{l} \ ,$$

which supports the conjecture (2.15).

Proof. Let $\mathfrak{q}$ be a prime ideal in k which divides both q and $\mathfrak{f}[r_{\mathfrak{l}}]$. Take a prime factor $\mathfrak{Q}$ of $\mathfrak{q}^{\mathfrak{c}}$ in $k(\mathfrak{x})$, and consider reduction modulo $\mathfrak{Q}$. We indicate reduced objects by putting tildes. Let π denote the Frobenius endomorphism of $\tilde{A}$ of degree q, and π^* its adjoint. By (1.11), we have $\pi + \pi^* = \tilde{\theta}(a_q)$, $\pi\pi^* = q$. If π has a totally real characteristic root λ, then $\lambda = \pm q^{1/2}$, so that $a_q^2 = 4q$, which contradicts the assumption that a_q is prime to $\mathfrak{l}$. Thus no characteristic roots of π can be totally real. Identify K and F with subfields of $\mathrm{End}_Q(\tilde{A})$ through $\tilde{\theta}$. Since A and $\theta(K)$ are rational over Q, π commutes with every element of $\tilde{\theta}(K)$. Therefore $F(\pi)$ is a totally imaginary quadratic extension of F. Since

$$X^2 - a_q X + q \equiv X(X - a_q) \ , \qquad a_q \not\equiv 0 \quad \bmod \mathfrak{l} \ ,$$

we see that $\mathfrak{l}$ decomposes into two primes $\mathfrak{l}_1$ and $\mathfrak{l}_2$ in $F(\pi)$ such that $\pi \in \mathfrak{l}_1$, $\pi^* \in \mathfrak{l}_2$, $\pi \notin \mathfrak{l}_2$. On the other hand, $\mathfrak{l}$ is ramified in K, hence $K(\pi)$ is a quadratic extension of K. Then we see that $\mathfrak{l}_1 = \mathfrak{L}_1^2$ and $\mathfrak{l}_2 = \mathfrak{L}_2^2$ with two prime ideals $\mathfrak{L}_1$ and $\mathfrak{L}_2$ in $K(\pi)$. Now let $\mathfrak{O} = \mathrm{End}(\tilde{A}) \cap K(\pi)$. Then $\mathfrak{O}$ may not be a maximal order in $K(\pi)$, but we have at least

(2.16) *The conductor of $\mathfrak{O}$ is prime to $\mathfrak{l}$,*

which will be proved afterwards. Put $\mathfrak{L}_i' = \mathfrak{L}_i \cap \mathfrak{O}$ for $i = 1, 2$. Then $\mathfrak{b}_{\mathfrak{l}}\mathfrak{O} = \mathfrak{L}_1'\mathfrak{L}_2'$. Let $\tilde{A}[\mathfrak{a}]$ denote the module consisting of the points of $\tilde{A}$ annihilated by $\tilde{\theta}(\mathfrak{a})$. Then we have

$$\tilde{\mathfrak{y}}_{\mathfrak{l}} + \tilde{\mathfrak{z}}_{\mathfrak{l}} = \tilde{A}[\mathfrak{b}_{\mathfrak{l}}] = \tilde{A}[\mathfrak{L}_1'] + \tilde{A}[\mathfrak{L}_2'] \ .$$

Now $\pi \in \mathcal{S}_1'$ and $\pi \notin \mathcal{S}_2'$. Therefore $\tilde{A}[\mathcal{S}_1'] = \{0\}$ and

(2.17)　$\tilde{A}[\mathcal{S}_2'] \neq \{0\}$,

which will also be proved afterwards. Now $\tilde{\mathfrak{y}}_{\mathfrak{l}}$, $\tilde{\mathfrak{z}}_{\mathfrak{l}}$, and $\tilde{A}[L_2']$ are all at most one-dimensional over the finite field $\mathfrak{o}_F/\mathfrak{l}$, and

$$\tilde{\mathfrak{y}}_{\mathfrak{l}} \cap \tilde{\mathfrak{z}}_{\mathfrak{l}} \subset \tilde{\mathfrak{y}} \cap \tilde{\mathfrak{z}} \subset (1 + \eta)\tilde{A} \cap (1 - \eta)\tilde{A} \subset \{x \in \tilde{A} \,|\, 2x = 0\} \,,$$

hence $\tilde{\mathfrak{y}}_{\mathfrak{l}} \cap \tilde{\mathfrak{z}}_{\mathfrak{l}} = \{0\}$, since $\mathfrak{l}$ is prime to 2. Therefore either $\tilde{\mathfrak{y}}_{\mathfrak{l}}$ or $\tilde{\mathfrak{z}}_{\mathfrak{l}}$ must be $\{0\}$. One has actually

(2.18)　$\tilde{\mathfrak{z}}_{\mathfrak{l}} = \{0\}$, and $\tilde{\mathfrak{y}}_{\mathfrak{l}} = \tilde{A}[L_2'] \neq \{0\}$.

To show this, assume, on the contrary, $\tilde{\mathfrak{z}}_{\mathfrak{l}} \neq \{0\}$. Take u so that $\mathfrak{z}_{\mathfrak{l}} = \theta_F(\mathfrak{o}_F)u$. Since $\mathfrak{q}$ divides $\mathfrak{f}[r_{\mathfrak{l}}]$, the conductor of $k(u)$ is divisible by $\mathfrak{q}^\mathfrak{e}$, so that $\mathfrak{q}^\mathfrak{e}$ is ramified in $k(u)$. Hence there exists a non-trivial automorphism τ of $k(u)$ over k under which $\mathfrak{O} \cap k(u)$ is invariant, and

$$a^\tau \equiv a \quad \bmod \mathfrak{O} \cap k(u)$$

for all integers a in $k(u)$. Then $\tilde{u}^\tau = \tilde{u}$. Since $\tilde{\mathfrak{z}}_{\mathfrak{l}} \neq \{0\}$, reduction modulo $\mathfrak{O}$ defines an isomorphism of $\mathfrak{z}_{\mathfrak{l}}$ onto $\tilde{\mathfrak{z}}_{\mathfrak{l}}$, hence $u^\tau = u$. It follows that τ is the identity map, a contradiction. This proves (2.18). Thus we have $\tilde{\mathfrak{y}}_{\mathfrak{l}} = \tilde{A}[\mathcal{S}_2'] \neq \{0\}$. Let t be a point of $\mathfrak{y}_{\mathfrak{l}}$ such that $\mathfrak{y}_{\mathfrak{l}} = \theta_F(\mathfrak{o}_F)t$. Assume that $\mathfrak{f}[r_{\mathfrak{l}}]$ is divisible by $\mathfrak{q}^\mathfrak{e}$. Then $\mathfrak{q}^\mathfrak{e}$ is ramified in $k(t)$. Applying to $\mathfrak{y}_{\mathfrak{l}}$ the above reasoning used in the proof of (2.18), we obtain a contradiction. Therefore $\mathfrak{f}[r_{\mathfrak{l}}]$ is not divisible by $\mathfrak{q}^\mathfrak{e}$. Let $\sigma = \left(\dfrac{k(t)/k}{\mathfrak{q}^\mathfrak{e}}\right)$. Then

$$\tilde{t}^\sigma = \pi\tilde{t} = (\tilde{\theta}(a_q) - \pi^*)\tilde{t} = \tilde{\theta}(a_q)\tilde{t} \,,$$

since $\pi^* \in \mathcal{S}_2'$ and $\tilde{t} \in \tilde{A}[\mathcal{S}_2']$. Now reduction modulo $\mathfrak{O}$ defines an isomorphism of $\mathfrak{y}_{\mathfrak{l}}$ onto $\tilde{\mathfrak{y}}_{\mathfrak{l}}$. Therefore we obtain $t^\sigma = \theta(a_q)t$, so that $r_{\mathfrak{l}}(\mathfrak{q}^\mathfrak{e}) = (a_q \bmod \mathfrak{l})$. This completes the proof.

Proof of (2.16). Let $\mathfrak{O}_1$ be the maximal order in $K(\pi)$. Since $\mathfrak{o}_K \subset \mathfrak{O}$, we have $\mathfrak{O} = \mathfrak{o}_K + \mathfrak{t}\mathfrak{O}_1$ with an integral ideal $\mathfrak{t}$ in K. (This well-known fact holds for every order in a quadratic extension of a number field containing the maximal order of the basic field.) Then $\mathfrak{t}\mathfrak{O}_1$ is the conductor of $\mathfrak{O}$. Suppose that $\mathfrak{t}$ is divisible by $\mathfrak{b}_{\mathfrak{l}}$. Since $\pi \in \mathfrak{O}$, we have $\pi - s \in \mathfrak{b}_{\mathfrak{l}}\mathfrak{O}_1$ with an element s of $\mathfrak{o}_K$. Applying the non-trivial automorphism of $K(\pi)$ over K to this relation, we obtain $\pi^* - s \in \mathfrak{b}_{\mathfrak{l}}\mathfrak{O}_1$, so that $q \equiv s^2 \bmod \mathfrak{b}_{\mathfrak{l}}$, hence $s \in \mathfrak{b}_{\mathfrak{l}}$. But then $a_q = \pi + \pi^* \equiv 2s \bmod \mathfrak{b}_{\mathfrak{l}}$, so that $a_q \in \mathfrak{l}$, a contradiction. Therefore $\mathfrak{t}$ must be prime to $\mathfrak{l}$.

Proof of (2.17). Let $\kappa = \mathbf{Z}/p\mathbf{Z}$, and let w be a generic point of $\tilde{A}$ over κ. Since $\mathfrak{t}$ is prime to $\mathcal{S}_2$, we can find a positive integer m so that $\mathcal{S}_2^m = \gamma\mathfrak{O}_1$ with an integer γ of $K(\pi)$ such that $\gamma \equiv 1 \bmod \mathfrak{t}\mathfrak{O}_1$. Then $\gamma \in \mathfrak{O}$ and $\mathfrak{L}_2'^m = \gamma\mathfrak{O}$. Assume that $\tilde{A}[\mathfrak{L}_2'] = \{0\}$. Then $\tilde{A}[\gamma] = \{0\}$. It follows that $\kappa(w)$ is purely inseparable over $\kappa(\gamma w)$, hence $\kappa(\gamma w) \supset \kappa(\pi^n w)$ for some positive integer n. Then we can define an endomorphism δ of $\tilde{A}$ by $\delta\gamma w = \pi^n w$, i.e., $\delta\gamma = \pi^n$. We see that $\delta \in \mathrm{End}\,(\tilde{A}) \cap K(\pi) = \mathfrak{O}$, hence $\pi^n \in \gamma\mathfrak{O}_1 \subset \mathfrak{L}_2^m$, so that $\pi \in \mathfrak{L}_2$, a contradiction.

Remark 2.9. Suppose that $N(\mathfrak{f}[r_1])$ is divisible by a rational prime p which remains prime in k, and which does not divide N. Then $\psi(p) = -1$, and $a_p \in \mathfrak{b}_1$. Therefore we have a formal congruence

$$1 - a_p p^{-s} + \psi(p)p \cdot p^{-2s} \equiv 1 \bmod \mathfrak{b}_1 .$$

Again this supports the conjecture (2.15).

Remark 2.10. We have defined $\mathfrak{b}_0$ to be the ideal generated by the integers x of $\mathfrak{o}_K$ such that $x^\rho = -x$, and taken the "odd part" $\mathfrak{b}$ of $\mathfrak{b}_0$. It seems more reasonable to consider, instead of $\mathfrak{b}_0$ or $\mathfrak{b}$, the ideal $\mathfrak{b}^*$ generated by the numbers a_n for all n such that $\psi(n) = -1$. Obviously $\mathfrak{b}^* \subset \mathfrak{b}_0 \subset \mathfrak{b}$. In the above treatment we formulated everything with respect to $\mathfrak{b}$ simply because our discussion becomes more difficult with $\mathfrak{b}^*$. For instance, $\mathfrak{b}$ is always square-free, but $\mathfrak{b}^*$ may not. Further, the factors of 2 are always troublesome to treat for obvious reasons. In the next section, we shall give some numerical evidence which suggests that $\mathfrak{b}_0 \neq \mathfrak{b}^*$ in certain cases.

3. The case of prime level

Throughout this section, we assume that N is a prime. Accordingly $\psi(a) = \left(\dfrac{a}{N}\right)$, and the condition (2.1) implies that $N \equiv 1 \bmod (4)$, so that $k = Q(\sqrt{N})$. Let u_0 be the fundamental unit of k. Then $N_{k/Q}(u_0) = -1$, hence $u_0^2 - 1 = u_0 \cdot Tr_{k/Q}(u_0)$. As was mentioned in [9, § 7.7] and [10], two rational integers $N(\mathfrak{c})$ and $Tr_{k/Q}(u_0)$ have non-trivial common factors, and this fact has a class field theoretical meaning. Some numerical data for $29 \leq N \leq 233$ are given in Tables I_a, I_b. (See also [9, p. 207] for the values of a_p for $29 \leq N \leq 97$.) Note that $S_2(N, \psi) = S_2^0(N, \psi)$ if N is a prime, and $S_2(N, \psi) = \{0\}$ for all primes $N < 29$. Except in the cases $N = 109, 157, 229$ (within the limit of the tables), the space $S_2(N, \psi)$ is spanned by *one* cusp form f and its companions f_σ. In each of the three cases, there are exactly two non-companionate cusp forms, from which we obtain two different systems of $\{(A, \theta), K/F\}$.

Now we can make the following empirical observations, although we do not know to what extent these are true in general.

(I)[2] $N(\mathfrak{c})$ *and* $Tr_{k/Q}(u_0)$ *consist of the same prime factors, if we disregard* 2 *and* 3.

(II) $N(\mathfrak{c})$ *is prime to* N.

(III) $\psi(q) = 1$ *for every prime factor* q *of* $N(\mathfrak{c})$.

(IV) *All the prime factors of* $\mathfrak{c}$ *(in* F*) are of degree* 1 *if the class number of* k *is one.*

Within the limit of the tables, k is of class number > 1 only in the case $N = 229$, of which a detailed discussion will be made in the next section.

Obviously $Tr_{k/Q}(u_0)$ is prime to N, so that (I) implies (II). Further, by Proposition A.II.1 of Appendix II, we have $\psi(q) = 1$ for every odd prime factor q of $Tr_{k/Q}(u_0)$, hence (III) is true if $N(\mathfrak{c})$ divides a power of $Tr_{k/Q}(u_0)$. If $N(\mathfrak{c})$ is square-free, (III) follows also from (II) by Proposition 2.1.

Now consider (A, θ), K, F, $\mathfrak{c}$, etc. for a fixed $f(z) = \sum_{n=1}^{\infty} a_n e^{2\pi i n z}$, and let q be a common prime factor of $N(\mathfrak{c})$ and $Tr_{k/Q}(u_0)$. By Proposition A.II.1, q is prime to N, and decomposes into two prime ideals in k. Let $\mathfrak{l}$ be a common prime factor of $\mathfrak{c}$ and q in F. Consider the character $r_{\mathfrak{l}}$ and its conductor $\mathfrak{f}[r_{\mathfrak{l}}]$ defined in §2. Then we can make the following conjecture:

(3.1) *The finite part of* $\mathfrak{f}[r_{\mathfrak{l}}]$ *is exactly a prime ideal dividing* q.

PROPOSITION 3.1. *The notation being as above, suppose that* a_q *is prime to* q, *and* N *is prime to* $N(\mathfrak{l}) - 1$. *Then* (3.1) *is true.*

Proof. In general, the finite part of $\mathfrak{f}[r_{\mathfrak{l}}]$ consists of the prime factors of qN. By Proposition 2.4, $\mathfrak{f}[r_{\mathfrak{l}}]$ is prime to N. Therefore, by Theorem 2.8, the finite part of $\mathfrak{f}[r_{\mathfrak{l}}]$ is of the form $\mathfrak{q}^m$ with a prime ideal $\mathfrak{q}$ dividing q. By [9, Lemma 7.32], $m \leq 1$. One can not have $m = 0$ on account of (iii) of Theorem 2.3.

PROPOSITION 3.2. *Suppose* (3.1) *is true. Then* $\mathfrak{f}[r_{\mathfrak{l}}] = \mathfrak{q} \cdot \mathfrak{p}_\infty$ *with a prime factor* $\mathfrak{q}$ *of* q *in* k *and an archimedean prime* $\mathfrak{p}_\infty$ *of* k. *The prime* $\mathfrak{p}_\infty$ *is uniquely determined for* $\mathfrak{q}$ *by the condition that* $v > 0$ *at* $\mathfrak{p}_\infty$ *for every* $v \in \mathfrak{o}^\times$ *such that* $N_{k/Q}(v) = -1$ *and* $v \equiv 1 \bmod \mathfrak{q}$. *Moreover, if* μ *denotes a unique embedding of* $\mathfrak{o}/\mathfrak{q}$ *into* $\mathfrak{o}_F/\mathfrak{l}$, *then*

(3.2) $$r_{\mathfrak{l}}(\alpha\mathfrak{o}) = \left(\frac{\alpha}{\mathfrak{p}_\infty}\right) \cdot \mu(\alpha \bmod \mathfrak{q})$$

for every α *in* k *prime to* $\mathfrak{q}$.

---

[2] It is also plausible that $N(\mathfrak{c})$ is divisible by 3 if and only if $Tr_{k/Q}(u_0)$ is divisible by 9. A similar statement will probably hold for the divisibility of $N(\mathfrak{b}^*)$ by a power of 2. See also Propositions A.II.2, A.II.3 of Appendix II.

TABLE I$_a$

N: prime, $\equiv 1 \bmod (4)$.

N	dim $S_2(N, \psi)$	$[F: \mathbf{Q}]$	$N(c)$	$Tr_{k/\mathbf{Q}}(u_0)$
29	2	1	5	5
37	2	1	1	$2^2 \cdot 3$
41	2	1	1	2^6
53	4	2	7	7
61	4	2	13	$3 \cdot 13$
73	4	2	89	$2^3 \cdot 3 \cdot 89$
89	6	3	5	$2^3 \cdot 5^3$
97	6	3	467	$2^3 \cdot 3 \cdot 467$
101	8	4	5	$2^2 \cdot 5$
109	8	1	3	$3^2 \cdot 29$
		3	29	
113	8	4	97	$2^4 \cdot 97$
137	10	5	109	$2^5 \cdot 109$
149	12	6	61	61
157	12	1	1	$3 \cdot 71$
		5	71	
173	14	7	13	13
181	14	7	$3 \cdot 5 \cdot 29$	$3^2 \cdot 5 \cdot 29$
193	14	7	147011?	$2^3 \cdot 3 \cdot 147011$
197	16	8	7	$2^2 \cdot 7$
229	18	1	5	$3 \cdot 5$
		8	5^2	
233	18	9	$7 \cdot 827$?	$2^3 \cdot 7 \cdot 827$

TABLE I_b

N	p	$\left(\dfrac{p}{N}\right)$	characteristic polynomial of $T(p)_{N,\psi}$
101	2	$-$	$x^8 + 13x^6 + 51x^4 + 67x^2 + 20$
	5	$+$	$(x^4 - 11x^2 - 9x + 9)^2$
109	2	$-$	$(x^2 + 3)(x^6 + 12x^4 + 39x^2 + 29)$
	3	$+$	$(x - 2)^2(x^3 + 2x^2 - 3x - 3)^2$
	5	$+$	$(x + 3)^2(x^3 - 4x^2 + x + 3)^2$
113	2	$+$	$(x^4 + x^3 - 5x^2 - 4x + 3)^2$
	3	$-$	$x^8 + 19x^6 + 122x^4 + 297x^2 + 194$
137	2	$+$	$(x^5 + 2x^4 - 5x^3 - 8x^2 + 5x + 3)^2$
	3	$-$	$x^{10} + 23x^8 + 188x^6 + 670x^4 + 989x^2 + 436$
149	2	$-$	$x^{12} + 20x^{10} + 148x^8 + 499x^6 + 766x^4 + 465x^2 + 61$
	5	$+$	$(x^6 - 2x^5 - 17x^4 + 20x^3 + 78x^2 - 18x - 27)^2$
157	2	$-$	$(x^2 + 1)(x^{10} + 17x^8 + 103x^6 + 266x^4 + 273x^2 + 71)$
	3	$+$	$(x + 2)^2(x^5 + x^4 - 9x^3 - x^2 + 12x - 5)^2$
173	2	$-$	$x^{14} + 20x^{12} + 151x^{10} + 542x^8 + 972x^6 + 833x^4 + 276x^2 + 13$
181	2	$-$	$x^{14} + 23x^{12} + 210x^{10} + 974x^8 + 2441x^6 + 3234x^4 + 2030x^2 + 435$
	3	$+$	$(x^7 + x^6 - 13x^5 - 10x^4 + 41x^3 + 25x^2 - 26x - 4)^2$
193	2	$+$	$(x^7 - x^6 - 8x^5 + 8x^4 + 15x^3 - 13x^2 - 4x + 3)^2$
	3	$+$	$(x^7 + x^6 - 10x^5 - 10x^4 + 25x^3 + 16x^2 - 20x + 1)^2$
197	2	$-$	$x^{16} + 24x^{14} + 228x^{12} + 1095x^{10} + 2834x^8 + 3942x^6 + 2795x^4 + 925x^2 + 112$
229	2	$-$	$(x^2 + 5)(x^{16} + 24x^{14} + 229x^{12} + 1122x^{10} + 3063x^8 + 4723x^6 + 3942x^4 + 1575x^2 + 225)$
	3	$+$	$(x - 1)^2(x^8 - 17x^6 - 5x^5 + 84x^4 + 44x^3 - 99x^2 - 60x + 4)^2$
	5	$+$	$(x - 3)^2(x^8 + 6x^7 - 50x^5 - 52x^4 + 93x^3 + 129x^2 + 24x - 9)^2$
233	2	$+$	$(x^9 + 2x^8 - 11x^7 - 20x^6 + 38x^5 + 60x^4 - 43x^3 - 53x^2 + 15x + 9)^2$

Proof. The first assertion follows immediately from Proposition 2.5. Now for α in k prime to $\mathfrak{q}$, we can find an integer m such that $m\alpha > 0$ at $\mathfrak{p}_\infty$ and $m \equiv \alpha \bmod \mathfrak{q}$. Then

$$r_{\mathfrak{l}}(\alpha\mathfrak{o}) = r_{\mathfrak{l}}(m\mathfrak{o}) = \left(\frac{m}{\mathfrak{p}_\infty}\right) \cdot \mu(m \bmod \mathfrak{q}) = \left(\frac{\alpha}{\mathfrak{p}_\infty}\right) \cdot \mu(\alpha \bmod \mathfrak{q}) \,,$$

which proves the last assertion. If $v \in \mathfrak{o}^\times$ and $N_{k/\mathbb{Q}}(v) = -1$, then $v = \pm u_0^n$ with an odd integer n, so that $v^2 - 1$ is divisible by $\mathfrak{q}\mathfrak{q}^\iota$. By Proposition A.II.1 of Appendix II, $\mathfrak{q}$ divides exactly one of $v - 1$ or $v + 1$. If $v \equiv 1 \bmod \mathfrak{q}$, we have

$$1 = r_{\mathfrak{l}}(v\mathfrak{o}) = \left(\frac{v}{\mathfrak{p}_\infty}\right) \cdot \mu(v \bmod \mathfrak{q}) \,,$$

so that $v > 0$ at $\mathfrak{p}_\infty$, which proves the second assertion.

As an immediate consequence, we obtain

COROLLARY 3.3. *The notation and the assumption being as in Proposition 3.2, suppose that the class number of k is one. Then the field generated over k by the coordinates of the points of $\mathfrak{h}_{\mathfrak{l}}$ (resp. $\mathfrak{h}_{\mathfrak{l}} + \mathfrak{z}_{\mathfrak{l}}$) is exactly the maximal ray class field over k of conductor $\mathfrak{q} \cdot \mathfrak{p}_\infty$ (resp. $\mathfrak{q}\mathfrak{q}^\iota \cdot \mathfrak{p}_\infty\mathfrak{p}_\infty^\iota$).*

In [9, § 7.7], we discussed (under a certain set of conditions [9, (7.7.15)]) the structure of $k(\mathfrak{x})$, r, s in connection with the congruence

$$(3.3) \qquad\qquad Tr_{k/\mathbb{Q}}(\gamma) \equiv a_p \bmod \mathfrak{l}$$

for a totally positive algebraic integer γ such that $N_{k/\mathbb{Q}}(\gamma)$ is a rational prime p not dividing qN. The relation (3.2) is essentially equivalent to (3.3). In [9, Prop. 7.38], we showed that (3.3) holds for all p prime to qN if and only if it holds for some γ such that γ^ι/γ generates $(\mathfrak{o}/\mathfrak{q})^\times$. The above Proposition 3.1 gives another criterion for the validity of (3.2) and (3.3). Thus, by examining either the existence of such a γ, or whether a_q is prime to q, we can verify that (3.1) is true (and hence (3.2) and (3.3) hold). That is actually so at least in all computable cases.

Examples. (1) $N = 181$. As Table I_a shows, $N(\mathfrak{c}) = 3 \cdot 5 \cdot 29$. From the equations for a_2 and a_3 given in Table I_b, we see that $\mathfrak{c} = \mathfrak{l}_3\mathfrak{l}_5\mathfrak{l}_{29}$ with prime ideals $\mathfrak{l}_q$ in F with norm q for $q = 3, 5, 29$. Moreover, a_3 is prime to 3, $a_3 \equiv 1$ mod $\mathfrak{l}_5$, and $a_3 \equiv 6$ mod $\mathfrak{l}_{29}$. Therefore we can apply Propositions 3.1 and 3.2 to this case with $q = 3$. As for $q = 5$ and 29, we lack the information about a_5 and a_{29}, but we can apply [9, Prop. 7.38] (or rather, its modification) to the present case, and conclude that (3.1) is true also for $q = 5, 29$. Thus $k(\mathfrak{x})$ is the maximal class field over k of conductor $3 \cdot 5 \cdot 29 \cdot \mathfrak{p}_\infty \cdot \mathfrak{p}_\infty^\iota$.

(2) $N = 193$. In this case, $Tr_{k/Q}(u_0) = 2^3 \cdot 3 \cdot 147011$. Table I_b tells us only the equation for a_2 and a_3. Since $\psi(2) = \psi(3) = 1$, $N(\mathfrak{c})$ cannot be determined from these data. However, if $\Phi_p(x) = 0$, for $p = 2, 3$, is the equation for a_p, then we can verify that

$$\Phi_p(Tr_{k/Q}(\alpha_p)) \equiv 0 \bmod (147011)$$

for a totally positive element α_p of $\mathfrak{o}$ such that $N_{k/Q}(\alpha_p) = p$. (Actually $Tr_{k/Q}(\alpha_2) \equiv 56445$, $Tr_{k/Q}(\alpha_3) \equiv 28 \bmod (147011)$.) As for the prime factor 3 of $Tr_{k/Q}(u_0)$, we can easily check that the divisibility of $N(\mathfrak{c})$ by 3 contradicts the assertion of Theorem 2.8. Therefore it is almost certain that $N(\mathfrak{c}) = 147011$.

(3) $N = 89$. One has $N(\mathfrak{c}) = 5$. On the other hand $u_0 = 500 + 53\sqrt{89}$, so that $Tr_{k/Q}(u_0)$ is divisible by 5^3. From the numerical data about a_p for $p = 2, 3, 5, 7, 11, 17, 47, 53$ (not given here), we may conjecture that, for a rational prime p,

$$a_p \equiv \begin{cases} 0 \\ Tr_{k/Q}(\alpha) \end{cases} \bmod \mathfrak{c}^3 \qquad \text{if } \psi(p) = \begin{cases} -1 \\ 1, \end{cases}$$

where α is a totally positive integer in $\mathfrak{o}$ such that $N_{k/Q}(\alpha) = p$. If we define $\mathfrak{b}^*$ as in Remark 2.10, then probably $\mathfrak{b}^{*2} = \mathfrak{c}^3 \mathfrak{o}_K$.

4. The case $N = 229$

The character $r_\mathfrak{l}$ is completely determined by (3.2) if the class number of k is one, but if not, (3.2) gives only incomplete information about $r_\mathfrak{l}$. Therefore it is very interesting to see what happens in the case of class number > 1. We shall now study the case $N = 229$, which is the smallest prime $N \equiv 1 \bmod (4)$ such that $Q(\sqrt{N})$ is of class number > 1.

First let us make a simple observation. Take, as in §3, a common prime factor q of $N(\mathfrak{c})$ and $Tr_{k/Q}(u_0)$, and a common prime factor $\mathfrak{l}$ of $\mathfrak{c}$ and q in F. Let h be the class number of k, and $\mathfrak{a}$ an element of $I(\mathfrak{f}[r_\mathfrak{l}])$. Then $\mathfrak{a}^h = \beta\mathfrak{o}$ with a totally positive element β of k. If (3.2) is true, we have

$$r_\mathfrak{l}(\mathfrak{a})^h = \mu(\beta \bmod \mathfrak{q}) .$$

If the order of $(\mathfrak{o}_F/\mathfrak{l})^\times$ is prime to h, one has a unique h-th root of every element x of $(\mathfrak{o}_F/\mathfrak{l})^\times$. Denoting it by $x^{1/h}$, one has

$$r_\mathfrak{l}(\mathfrak{a}) = \mu(\beta \bmod \mathfrak{q})^{1/h} ,$$

which completely determines $r_\mathfrak{l}$. However, if the order of $(\mathfrak{o}_F/\mathfrak{l})^\times$ is not prime to h, a certain new phenomenon takes place, as we are going to see in the following discussion.

Let us now consider the special case $N = 229$. The class number of $k = Q(\sqrt{229})$ is 3, and the fundamental unit is given by $u_0 = (15 + \sqrt{229})/2$. As Table I_b shows, the characteristic polynomials Ψ_p of $T(p)_{N,\psi}$ for $p = 2, 3, 5$ are given by

$$\Psi_2(x) = (x^2 + 5)\cdot\Phi_2(x^2) ,$$
$$\Psi_3(x) = (x - 1)^2\cdot\Phi_3(x)^2 ,$$
$$\Psi_5(x) = (x - 3)^2\cdot\Phi_5(x)^2 ,$$

with irreducible polynomials Φ_p of degree 8. Therefore we obtain two non-companionate eigen-functions

$$f_\nu = \sum_{n=1}^{\infty} a_{\nu n} e^{2\pi i n z} \qquad\qquad (\nu = 1, 2) ,$$

each of which defines a system $\{(A_\nu, \theta_\nu), K_\nu/F_\nu\}$. The above decomposition of Ψ_p shows that

$$[K_1 : Q] = \dim (A_1) = 2 ,$$
$$[K_2 : Q] = \dim (A_2) = 16 .$$

(I) We first discuss A_1 which is simpler. Let us fix our attention to $A_1, K_1, f_1 = \sum_{n=1}^{\infty} a_{1n} e^{2\pi i n z}$, etc., and denote them, for simplicity, by $A, K, f = \sum a_n e^{2\pi i n z}$, etc., by dropping the subscript 1. Define $\mathfrak{b}$ and $\mathfrak{c}$ in the present case. Then we see that $K = Q(\sqrt{-5})$, $F = Q$, $\mathfrak{b} = (\sqrt{-5})$, $\mathfrak{c} = 5Z$, and $a_5 = 3$. Also we notice that $Tr_{k/Q}(u_0)$ is divisible by 5. Therefore we can apply Proposition 3.1 and Proposition 3.2 to the present case with $\mathfrak{l} = \mathfrak{c} = 5Z$ and a prime factor $\mathfrak{q}$ of 5 in k. Moreover, the class number of k is prime to the order of $(\mathfrak{o}_F/\mathfrak{l})^\times$, hence, as we remarked above, the character $r_{\mathfrak{l}}$ is completely determined by

$$r_{\mathfrak{l}}(\mathfrak{a}) = \left(\frac{\alpha}{\mathfrak{p}_\infty}\right)\cdot\mu(\alpha \bmod \mathfrak{q})^{1/3} .$$

Here $\mathfrak{a} \in I(\mathfrak{q})$, $\mathfrak{a}^3 = \alpha\mathfrak{o}$ with an element α of k, and $\mathfrak{p}_\infty$ is determined as in Proposition 3.2. It follows that $k(\chi)$ is the class field over k corresponding to the ideal group

$$\{\mathfrak{a} \in I(5\mathfrak{o}) \mid \mathfrak{a}^3 \in P(5\cdot\mathfrak{p}_\infty\mathfrak{p}'_\infty)\} ,$$

where $P(\ )$ is as in Notation.

(II) Next let us consider A_2, K_2, f_2, etc. Again we denote them simply by A, K, f, etc., dropping the subscript 2. Define $\mathfrak{b}_0, \mathfrak{b}$, and $\mathfrak{c}$. We shall now state various facts of which the proofs will be given afterwards.

(4.1) $\mathfrak{b}_0 = \mathfrak{b}$, $N(\mathfrak{c}) = 5^2$; $\mathfrak{c}$ *is a unique common prime factor of* a_2^2 *and* 5 *in* F.

Now $Tr_{k/Q}(u_0)$ is divisible by 5. Since the constant term of Φ_5 is -9, we see that a_5 is prime to 5. Therefore, by Proposition 3.1 and Proposition 3.2 with $I = c$ and $q = 5$, we can conclude that the conductor of r $(=r_t)$ is $q \cdot \mathfrak{p}_\infty$, where q is a prime factor of 5 in $\boldsymbol{Q}(\sqrt{229})$, and (3.2) holds. Since $(\mathfrak{o}_F/c)^\times$ is a cyclic group of order 24, there are exactly two injective homomorphisms λ of the ideal class group of k into $(\mathfrak{o}_F/c)^\times$. We consider such a λ as a homomorphism of the ideal group of k into $(\mathfrak{o}_F/c)^\times$ in an obvious way.

PROPOSITION 4.1. *The notation being as above, with a suitable choice of λ, one has*

$$r(\mathfrak{a}) = \left(\frac{\alpha}{\mathfrak{p}_\infty}\right) \cdot \lambda(\mathfrak{a}) \cdot \mu((\alpha \bmod q)^{1/3})$$

for every $\mathfrak{a} \in I(q)$, where α is an element of k such that $\mathfrak{a}^3 = \alpha \mathfrak{o}$, and μ is a unique injection of $\mathfrak{o}/q$ into $\mathfrak{o}_F/c$.

As an immediate consequence, we obtain

PROPOSITION 4.2. *The field $k(\mathfrak{x})$ is the maximal class field over k of conductor $5 \cdot \mathfrak{p}_\infty \mathfrak{p}_\infty^\iota$.*

By the same reasoning as in [9, Proposition 7.34] (cf. also [10, Proposition 2]), we can easily show that $k(\mathfrak{x})$ has the following property.

PROPOSITION 4.3. *Let M be the maximal abelian extension of k in which only the prime factors of $5 \cdot \mathfrak{p}_\infty \mathfrak{p}_\infty^\iota$ are ramified. Further let M' denote the field generated over k by the 5^n-th roots of unity for all positive integers n. Then $M = M' \cdot k(\mathfrak{x})$, and $M' \cap k(\mathfrak{x}) = k(e^{2\pi i/5})$.*

Let us now prove (4.1) and Proposition 4.1. Put $a_2^2 = \xi$. The explicit form of Φ_2 given in Table I_b shows that $N_{F/Q}(\xi) = 3^2 \cdot 5^2$. Hence there is a valuation v of F such that $v(5) = 1$ and $v(a_2) > 0$. Write the equation $\Phi_2(\xi) = 0$ in the form

$$(4.2) \qquad -3^2 \cdot 5^2 - 3^2 \cdot 5^2 \cdot 7 \cdot \xi = 3942 \cdot \xi^2 + 4723 \cdot \xi^3 + \cdots$$

and take the value of v. Then we find $2 = v(\xi^2)$, so that $v(\xi) = 1$. Dividing (4.2) by 5^2, we find that

$$v(3^2 + 3942 \cdot (\xi/5)^2) > 0 .$$

Since -3942 is not a quadratic residue modulo 5, the residue field of v is at least of degree 2 over $\boldsymbol{Z}/5\boldsymbol{Z}$. On the other hand, $N_{F/Q}(\xi) = 3^2 \cdot 5^2$, hence the degree must be exactly 2. Thus there is a unique prime ideal c which divides both ξ and 5. Since $a_2^2 = \xi$ is divisible by c but not by c^2, c must be ramified in K.

Similarly we see that there is a valuation w of F such that $w(3) = 1$ and $w(a_2) > 0$. Write the equation $\Phi_2(\xi) = 0$ in the form

$$(4.3) \qquad -3^2 \cdot 5^2 - 3^2 \cdot 5^2 \cdot 7 \cdot \xi - 2 \cdot 3^3 \cdot 73 \cdot \xi^2 = 4723 \cdot \xi^3 + 3063 \cdot \xi^4 + \cdots$$

and take the value of w. Then we find $3 \cdot w(\xi) = 2$. If $\mathfrak{b}$ denotes the prime ideal of $\mathfrak{o}_F$ associated with w, we see that the ramification index of $\mathfrak{b}$ over Q is at least 3. Since $N(\xi) = 3^2 \cdot 5^2$ and ξ is divisible by $\mathfrak{b}^2$, we can conclude that $N(\mathfrak{b}) = 3$, and the ramification index is exactly 3. Thus we have $\xi\mathfrak{o}_F = \mathfrak{c}\mathfrak{b}^2$, so that $\mathfrak{b}$ is unramified in K. Now $\mathfrak{b}_0$ is a common divisor of a_2 and the relative different of K over F. If $y \in \mathfrak{b}_0$, we have $y = a_2 z$ with $z \in F$, so that $y^2 = \xi z^2$. Since z^2 is divisible by an even power of $\mathfrak{c}$, this shows that $\mathfrak{b}_0 \neq \mathfrak{o}_K$. Therefore we have $\mathfrak{b}_0^2 = \mathfrak{b}^2 = \mathfrak{c}\mathfrak{o}_F$ which completes the proof of (4.1).

Now we have $3^3 = N_{k/Q}(16 + \sqrt{229})$, $5^3 = N_{k/Q}((27 + \sqrt{229})/2)$. Therefore $\mathfrak{q}^3 = \alpha\mathfrak{o}$, where α is either $(27 + \sqrt{229})/2$ or $(27 - \sqrt{229})/2$. In any case, $\alpha^\iota \equiv 2 \bmod \mathfrak{q}$. On the other hand, $\Phi_5 \bmod (5)$ has the following decomposition into irreducible factors:

$$\Phi_5(x) = (x^2 - 2x - 1)(x^6 - 2x^5 + 2x^4 + 2x^3 - x^2 - 2x - 1) \bmod (5) .$$

Thus 5 has exactly two prime factors in F: one is $\mathfrak{c}$, and the other is of degree 6. Further, $a_5 \bmod \mathfrak{c}$ satisfies the equation $x^2 - 2x - 1 = 0$ considered in $\mathfrak{o}_F/\mathfrak{c}$. Therefore we have

$$a_5 \bmod \mathfrak{c} = 3\omega$$

with a cubic root of unity ω in $\mathfrak{o}_F/\mathfrak{c}$. By Theorem 2.8, we have

$$\begin{aligned} r(\mathfrak{q}^\iota) &= (a_5 \bmod \mathfrak{c}) = 3\omega \\ &= \mu(\alpha^\iota \bmod \mathfrak{q})^{1/3} \cdot \omega . \end{aligned}$$

Now we see easily that $\mathfrak{q}^\iota$ generates $I(\mathfrak{q})/P(\mathfrak{q} \cdot \mathfrak{p}_\infty)$. Therefore defining λ so that $\lambda(\mathfrak{q}^\iota) = \omega$, we obtain Proposition 4.1.

Remark. Instead of applying Theorem 2.8 to a_5, we can use the equation for a_3 and (iv) of Theorem 2.2 to prove Proposition 4.1.

5. The case of composite level

If N has more than one prime factor, $\Gamma_0(N)$ is normalized by some nontrivial elements other than $\begin{bmatrix} 0 & -1 \\ N & 0 \end{bmatrix}$, which often induce the decomposition of the abelian variety B. This phenomenon makes the case of composite level somewhat different from the case of prime level. In this section we shall discuss some general properties of those elements normalizing $\Gamma_0(N)$ and their effect on A, B, K, F, and a few further examples in the next two sections.

Let $N = L \cdot M$ with two positive integers L and M which are relatively prime. (We are primarily interested in the case where both L and M are

> 1, but the following discussion holds even in the case $L = 1$. If $M = 1$, everything becomes trivial.) Choose and fix two integers x and y so that $Mx + Ly = 1$, and put

$$(5.1) \qquad \delta = \begin{bmatrix} M & -1 \\ Ny & Mx \end{bmatrix} .$$

Then we see that $\det (\delta) = M$, and

$$(5.2) \qquad M^{-1}\delta^2 = \begin{bmatrix} M - Ly & -1 - x \\ Ny(1 + x) & Mx^2 - Ly \end{bmatrix} \in \Gamma_0(N) .$$

If $\begin{bmatrix} a & b \\ c & d \end{bmatrix} \in M_2(\mathbf{Z})$ and $c \equiv 0 \bmod (N)$, we have

$$\begin{bmatrix} a' & b' \\ c' & d' \end{bmatrix} = \delta \begin{bmatrix} a & b \\ c & d \end{bmatrix} \delta^{-1} \in M_2(\mathbf{Z})$$

with

$$(5.3) \qquad \begin{aligned} & c' \equiv 0 \bmod (N) , \\ & a' \equiv a, \; d' \equiv d \bmod (L) , \\ & a' \equiv d, \; d' \equiv a \bmod (M) . \end{aligned}$$

Therefore $\delta \Gamma_0(N) \delta^{-1} = \Gamma_0(N)$, $\delta \Gamma_1(N) \delta^{-1} = \Gamma_1(N)$. Moreover $\Gamma_1(N)\delta$ does not depend on the choice of x and y.

Let $\psi = \psi_L \cdot \psi_M$ with a character ψ_L of $(\mathbf{Z}/L\mathbf{Z})^\times$ and a character ψ_M of $(\mathbf{Z}/M\mathbf{Z})^\times$. Define a character ψ' of $(\mathbf{Z}/N\mathbf{Z})^\times$ by

$$(5.4) \qquad \psi' = \psi^{-1}\psi_L^2 = \psi_L \cdot \psi_M^{-1} .$$

Obviously $\psi'_L = \psi_L$, $\psi'_M = \psi_M^{-1}$, so that $(\psi')' = \psi$. The action of $[\delta]_\kappa$ defined by (1.1) maps $S_\kappa(\Gamma_1(N))$ onto itself, and $S_\kappa(N, \psi)$ onto $S_\kappa(N, \psi')$. Further it can easily be verified that it maps $S_\kappa^0(N, \psi)$ onto $S_\kappa^0(N, \psi')$. From (5.2), we obtain

$$(5.5) \qquad f \,|\, [\delta]_\kappa^2 = \psi_M(-1) \cdot \psi_L(M)^{-1} \cdot f \qquad\qquad (f \in S_\kappa(N, \psi)) .$$

Let α be an element of $M_2(\mathbf{Z})$ such that

$$\det (\alpha) = n > 0, \quad (n, N) = 1, \quad \alpha \equiv \begin{bmatrix} 1 & 0 \\ 0 & n \end{bmatrix} \bmod (N) .$$

Put $\beta = \delta \alpha \delta^{-1}$. Then $\beta \in M_2(\mathbf{Z})$, and

$$\beta \equiv \begin{bmatrix} a' & * \\ 0 & d' \end{bmatrix} \bmod (N) ,$$

$$a' \equiv 1, \; d' \equiv n \bmod (L) ,$$

$$a' \equiv n, \; d' \equiv 1 \bmod (M) .$$

Let γ be an element of $\Gamma_0(N)$ such that

$$\gamma \equiv \begin{bmatrix} a' & 0 \\ 0 & a'^{-1} \end{bmatrix} \bmod (N) \ .$$

Then $\beta \equiv \gamma\alpha \bmod (N)$. Observe that β and α have the same set of elementary divisors. Therefore, putting $\Gamma = \Gamma_1(N)$, we have $\Gamma\beta\Gamma = \Gamma\gamma\alpha\Gamma$ by [9, Lemma 3.29, (2)], so that, by [9, Proposition 3.7],

$$(\Gamma\delta\Gamma)\cdot(\Gamma\alpha\Gamma) = (\Gamma\beta\Gamma)\cdot(\Gamma\delta\Gamma) = (\Gamma\gamma\Gamma)\cdot(\Gamma\alpha\Gamma)\cdot(\Gamma\delta\Gamma) \ .$$

Now the action of $\Gamma\alpha\Gamma$ on $S_\kappa(\Gamma_1(N))$ is $T(n)_N$. Therefore

(5.6) $$[\delta]_\kappa \cdot T(n)_N = \psi_M(n)^{-1} \cdot T(n)_N \cdot [\delta]_\kappa \qquad\qquad \text{on } S_\kappa(N, \psi) \ .$$

Therefore, if $f \mid T(n)_N = a_n f$ with an eigen-value a_n, we have

(5.7) $$(f \mid [\delta]_\kappa) T(n)_N = \psi_M(n)^{-1} \cdot a_n \cdot (f \mid [\delta]_\kappa) \ .$$

Further, if $\tau = \begin{bmatrix} 0 & -1 \\ N & 0 \end{bmatrix}$, then

$$\delta\tau\delta^{-1}\tau^{-1} = \begin{bmatrix} L + M & y - x \\ Ny - Nx & Mx^2 + Ly^2 \end{bmatrix} \in \Gamma_0(N) \ ,$$

hence

(5.8) $$[\delta]_\kappa \cdot H = \psi_M(L)^{-1} \cdot \psi_L(M)^{-1} \cdot H \cdot [\delta]_\kappa \qquad\qquad \text{on } S_\kappa(N, \psi) \ .$$

Let us now study the automorphisms of $(\Gamma\backslash\mathfrak{H})^*$ and A obtained from δ. For that purpose, we have to recall how the model V of $(\Gamma\backslash\mathfrak{H})^*$ has been defined in [9, § 7.3]. Let G_A denote the adelization of $GL_2(Q)$, and let $U_p = GL_2(Z_p)$ for each rational prime p. Define subgroups U, U', and S of G_A by

$$U = G_{\infty+} \times \prod_p U_p \ ,$$

$$G_{\infty+} = \{x \in GL_2(R) \mid \det(x) > 0\} \ ,$$

$$U' = \left\{ \begin{bmatrix} a & b \\ c & d \end{bmatrix} \in U \,\middle|\, d \equiv 1,\ c \equiv 0 \bmod (N) \right\} \ ,$$

$$S = Q^\times \cdot U' \ .$$

Then V can be defined as the curve V_S of [9, §§ 6.7, 7.3]. We see easily that

$$\delta^{-1}U'\delta = \left\{ \begin{bmatrix} a & b \\ c & d \end{bmatrix} \in U \,\middle|\, d \equiv 1 \bmod (L),\ a \equiv 1 \bmod (M),\ c \equiv 0 \bmod (N) \right\} \ .$$

Therefore the subgroup $Q^\times \cdot \det(S \cap \delta^{-1}S\delta)$ of $Q_A^\times$ corresponds to $Q(e^{2\pi i/M})$.

PROPOSITION 5.1. *Let $X_{ss}(\delta)$ be the automorphism of the curve V_S associated with δ as in [9, § 7.3]. Let γ_n, for a positive integer n prime to N, be an element of $SL_2(Z)$ such that*

$$\gamma_n \equiv \begin{bmatrix} n^{-1} & 0 \\ 0 & n \end{bmatrix} \bmod (N) \, ,$$

and $J_{SS}(\gamma_n)$ *the automorphism of* V_S *defined in* [9, § 6.7]. *Then* $X_{SS}(\delta)$ *is rational over* $Q(\zeta_M)$ *with* $\zeta_M = e^{2\pi i/M}$, *and*

$$X_{SS}(\delta) = J_{SS}(\gamma_m) \circ X_{SS}(\delta)^{\sigma_n} \circ J_{SS}(\gamma_n) \, ,$$

where m *is an integer such that* $m \equiv 1 \bmod (M)$, $m \equiv n \bmod (L)$, *and* σ_n *is the automorphism of* $Q(\zeta_M)$ *such that* $\zeta_M^{\sigma_n} = \zeta_M^n$.

Proof. The first assertion follows from [9, Proposition 7.2, (1)]. Let y be the element of U whose p-component y_p is given by

$$y_p = \begin{cases} \begin{bmatrix} 1 & 0 \\ 0 & n \end{bmatrix} & \text{if } p \mid N \, , \\ 1 & \text{if } p \nmid N \text{ or } p = \infty \, . \end{cases}$$

Put $z = \delta y \delta^{-1}$. Then

$$z_p \begin{cases} \equiv \begin{bmatrix} n & * \\ 0 & 1 \end{bmatrix} \bmod M \cdot M_2(Z_p) & \text{if } p \mid M \, , \\ \equiv \begin{bmatrix} 1 & * \\ 0 & n \end{bmatrix} \bmod L \cdot M_2(Z_p) & \text{if } p \mid L \, , \\ = 1 & \text{if } p \nmid N \text{ or } p = \infty \, . \end{cases}$$

By [9, Proposition 7.2, (7), (8)], we have

$$J_{SS}(z) \circ X_{SS}(\delta) = X_{SS}(z\delta) = X_{SS}(\delta y) = X_{SS}(\delta)^{\sigma(y)} \circ J_{SS}(y) \, .$$

Since $\gamma_n^{-1} y \in U'$ and $\gamma_m^{-1} z \in U'$, we have $J_{SS}(y) = J_{SS}(\gamma_n)$ and $J_{SS}(z) = J_{SS}(\gamma_m)$. Further $\sigma(y) = \sigma_n$ on $Q(\zeta_M)$, hence the formula of our proposition.

Especially if $n \equiv 1 \bmod (L)$, we have $\gamma_m \in \Gamma_1(N)$, so that we have

$$(5.9) \qquad X_{SS}(\delta) = X_{SS}(\delta)^{\sigma_n} \circ J_{SS}(\gamma_n) \qquad \text{if } n \equiv 1 \bmod (L) \, .$$

Fix again an eigen-function $f(z) = \sum a_n e^{2\pi i n z}$ belonging to $S_2^0(N, \psi)$, and define (A, θ), K, F as in § 1. Suppose $\psi^2 = 1$, $\psi_L \neq 1$, $\psi_M \neq 1$. Then $\psi = \psi'$, so that $f \mid [\delta]_2$ is an eigen-function belonging to $S_2^0(N, \psi)$ with eigen-values $\psi_M(n)a_n$. Now we make the following assumption:

(5.10) *There is an isomorphism* σ_M *of* K *into* C, *different from the identity map, such that* $a_n = \psi_M(n)a_n^{\sigma_M}$ *for all* n *prime to* N, *i.e.,* $f \mid [\delta]_2$ *is a constant multiple of* f_{σ_M}.

We shall later give a few examples for which this does or does not hold. If (5.10) is satisfied, we see that σ_M gives an automorphism of K of order 2. For every isomorphism τ of K into C, we have $a_n^\tau = \psi_M(n)a_n^{\sigma_M \tau}$, so that by

(5.7), $[\delta]_2$ sends $f_{\sigma_M \tau}$ to a constant multiple of f_τ. It follows that $[\delta]_2$ induces an automorphism of A, which we write η_M. We see that $\eta_M = \eta$ if $M = N$.

PROPOSITION 5.2. *Let k_M be the subfield of $\mathbf{Q}(\zeta_M)$ corresponding to the character ψ_M, and ε_M the generator of $\mathrm{Gal}(k_M/\mathbf{Q})$. Then η_M is rational over k_M, and $\eta_M^{\varepsilon_M} = -\eta_M$.*

Proof. For every $\sigma \in \mathrm{Gal}(\mathbf{Q}(\zeta_M)/\mathbf{Q})$, we can find a positive integer n such that $\zeta_M^\sigma = \zeta_M^n$ and $n \equiv 1 \bmod (L)$. By (5.9), we have $\eta_M = \psi(n)\eta_M^\sigma = \psi_M(n)\eta_M^\sigma$, which proves our proposition.

Now let us make, besides (5.10), another assumption:

$$(5.11) \qquad \psi_M(-1)\psi_L(M) = 1.$$

On account of (5.5), this implies that $\eta_M^2 = 1$. Therefore, if we put

$$(5.12) \qquad B_M = (1 + \eta_M)A ,$$

then B_M is an abelian subvariety of A of dimension $\dim(A)/2$, rational over k_M, and

$$(5.13) \qquad A = B_M + B_M^{\varepsilon_M} , \qquad B_M^{\varepsilon_M} = (1 - \eta_M)A .$$

Let F_M denote the subfield of K defined by

$$(5.14) \qquad F_M = \{x \in K \mid x^{\sigma_M} = x\} .$$

Then we can define an injection

$$\theta_M \colon F_M \longrightarrow \mathrm{End}_Q(B_M)$$

so that $\theta_M(a)$ is the restriction of $\theta(a)$ for $a \in F_M$. From (5.7) we obtain

$$(5.15) \qquad \eta_M \cdot \theta(a) = \theta(a^{\sigma_M}) \cdot \eta_M \qquad\qquad (a \in K) .$$

Now we can develop a theory similar to that of §2 by taking k_M, η_M, B_M, F_M, etc., in place of k, η, B, F, etc.. (If $M = N$, the new symbols coincide with the old ones. Note that k_M may or may not be real; F_M is a CM-field unless $M = N$.) First we have

THEOREM 5.3. *The notation being as above, $\zeta(s; B_M/k_M, F_M)$ coincides with $L(s, f)L(s, f_{\sigma_M})$, up to a finite number of Euler factors occurring only for the primes dividing N.*

Proof. Let p be a rational prime unramified in k_M, $\mathfrak{p}$ a prime factor of p in k_M, and $\mathfrak{m}$ a prime ideal in F_M not dividing p. Consider reduction modulo $\mathfrak{p}$, and indicate reduced objects by putting tildes. Let R (resp. R') denote the $\mathfrak{m}$-adic representation of $\mathrm{End}_Q(\tilde{A})$ (resp. $\mathrm{End}_Q(\tilde{B}_M)$), and $\varphi_\mathfrak{p}$ (resp. $\varphi'_\mathfrak{p}$) the Frobenius endomorphism of $\tilde{A}$ (resp. $\tilde{B}_M$) of degree $N(\mathfrak{p})$. Since $\tilde{A}$ is isogenous to $\tilde{B}_M \times \tilde{B}_M$ over the residue field modulo $\mathfrak{p}$, we have

$$\det\left[X - R'(\varphi'_\mathfrak{p})\right]^2 = \det\left[X - R(\varphi_\mathfrak{p})\right],$$

where X is an indeterminate. Let π_p denote the Frobenius endomorphism of $\tilde{A}$ of degree p. Since K is a quadratic extension of F_M, we see from (1.12) that, if $N(\mathfrak{p}) = p$, then $a_p \in F_M$, and

$$\det\left[X - R(\varphi_\mathfrak{p})\right] = \det\left[X - R(\pi_p)\right] = N_{K/F_M}\left(X^2 - a_p X + \psi(p)p\right)$$
$$= \left(X^2 - a_p X + \psi(p)p\right)^2.$$

Similarly, if $N(\mathfrak{p}) = p^2$, we have $a_p^{\sigma_M} = -a_p$, and

$$\det\left[X^2 - R(\varphi_\mathfrak{p})\right] = \det\left[X^2 - R(\pi_p^2)\right] = \det\left[X - R(\pi_p)\right]\cdot\det\left[X + R(\pi_p)\right]$$
$$= N_{K/F_M}\left[\left(X^2 - a_p X + \psi(p)p\right)\left(X^2 + a_p X + \psi(p)p\right)\right]$$
$$= \left(X^2 - a_p X + \psi(p)p\right)^2\left(X^2 - a_p^{\sigma_M} X + \psi(p)p\right)^2.$$

Therefore

$$\det\left[X - R'(\varphi'_\mathfrak{p})\right] = X^2 - a_p X + \psi(p)p \qquad\qquad \text{if } N(\mathfrak{p}) = p,$$
$$\det\left[X^2 - R'(\varphi'_\mathfrak{p})\right] = \left(X^2 - a_p X + \psi(p)p\right)\left(X^2 - a_p^{\sigma_M} X + \psi(p)p\right)$$
$$\text{if } N(\mathfrak{p}) = p^2.$$

Our assertion follows from these equalities.

Now we consider the ideal $\mathfrak{b}_{M0}$ generated by the elements x of $\mathfrak{o}_K$ such that $x^{\sigma_M} = -x$, and its "odd part" $\mathfrak{b}_M$. Put $\mathfrak{o}_M = \mathfrak{o} \cap F_M$. Then $\mathfrak{b}_M^2 = \mathfrak{c}_M \mathfrak{o}_K$ with a square-free integral ideal $\mathfrak{c}_M$ in F_M, and $a_p \in \mathfrak{b}_M$ if $\psi_M(p) = -1$. Define submodules $\mathfrak{x}_M, \mathfrak{y}_M, \mathfrak{z}_M$ of A by

$$\mathfrak{x}_M = \left\{t \in A \,|\, \theta(\mathfrak{b}_M)t = 0\right\},$$
$$\mathfrak{y}_M = B_M \cap \mathfrak{x}_M,\ \mathfrak{z}_M = (B_M)^{\sigma_M} \cap \mathfrak{x}_M.$$

Let $k_M(\mathfrak{x}_M)$ denote the extension of k_M generated by the coordinates of the points of $\mathfrak{x}_M$. Then $\mathrm{Gal}\left(k_M(\mathfrak{x}_M)/k_M\right)$ is isomorphic to a subgroup of $(\mathfrak{o}_M/\mathfrak{c}_M)^{\times 2}$, so that $k_M(\mathfrak{x}_M)$ is abelian over k_M. Let $\mathfrak{f}_M$ be the conductor of $k_M(\mathfrak{x}_M)$ over k_M. By the same procedure as in §2, we can define, for every prime factor $\mathfrak{l}$ of $\mathfrak{c}_M$ (in F_M), two characters $r_\mathfrak{l}^M$ and $s_\mathfrak{l}^M$ of $I(\mathfrak{f}_M)$ with values in $(\mathfrak{o}_M/\mathfrak{l})^\times$, and their conductors $\mathfrak{f}[r_\mathfrak{l}^M]$ and $\mathfrak{f}[s_\mathfrak{l}^M]$. Then we obtain a generalization of Theorem 2.3 as follows:

THEOREM 5.4. *The characters $r_\mathfrak{l}^M$ and $s_\mathfrak{l}^M$ have the following properties:*

(i) $\mathfrak{f}[s_\mathfrak{l}^M] = \mathfrak{f}[r_\mathfrak{l}^M]^{\sigma_M}.$

(ii) $r_\mathfrak{l}^M(\mathfrak{a}) = s_\mathfrak{l}^M(\mathfrak{a}^{\sigma_M})$ *for every* $\mathfrak{a} \in I(\mathfrak{f}[r_\mathfrak{l}^M]).$

(iii) $r_\mathfrak{l}^M((m)) = s_\mathfrak{l}^M((m)) = \left(\dfrac{m}{p_\infty}\right) \cdot \psi_L(m) \cdot (m \bmod \mathfrak{l})$ *for every* $m \in \mathbf{Z}$ *prime to* $\mathfrak{f}[r_\mathfrak{l}^M].$

(iv) $r_\mathfrak{l}^M(\mathfrak{a})s_\mathfrak{l}^M(\mathfrak{a}) = \psi_L(N(\mathfrak{a})) \cdot (N(\mathfrak{a}) \bmod \mathfrak{l})$ *for all* $\mathfrak{a} \in I(\mathfrak{f}[r_\mathfrak{l}^M]) \cap I(\mathfrak{f}[s_\mathfrak{l}^M]).$

(v) *If p is a rational prime that is prime to $N(\mathfrak{l}) \cdot N$, and that decomposes into two distinct prime ideals $\mathfrak{p}$ and $\mathfrak{p}^{\epsilon_M}$ in k_M, then*

$$r_\mathfrak{l}^M(\mathfrak{p}) + s_\mathfrak{l}^M(\mathfrak{p}) = (a_p \bmod \mathfrak{l}) .$$

The proof can be given in the same manner as in Theorem 2.3.

We can also generalize Propositions 2.4 and 2.5 to the present setting: we only have to replace k, $\mathfrak{o}_F$, $r_\mathfrak{l}$, etc., by k_M, $\mathfrak{o}_M$, $r_\mathfrak{l}^M$, etc.

Remark 5.5. From (5.8), we obtain

$$\eta_M \cdot \eta = \psi_M(L) \cdot \psi_L(M) \cdot \eta \cdot \eta_M .$$

Therefore, if $\psi_M(L)\psi_L(M) = 1$, we see that η induces an automorphism of B_M, $\neq \pm 1$, if $\psi_L \neq 1$. (In fact, if $\eta = 1$ or -1 on B_M, we have $B_M = B$ or B', so that B is defined over $Q = k_M \cap k$. This is a contradiction, since $A = B + B'$.) Thus $B_M = (1 + \eta)B_M + (1 - \eta)B_M$, hence B_M is not simple if $\psi_L \neq 1$ and $\psi_M(L)\psi_L(M) = 1$.

Next suppose $\psi_M(L)\psi_L(M) = -1$ and $\psi_L \neq 1$. Then η_M maps B' onto B. Let c be a non-zero element of $\mathfrak{o}_K$ such that $c^\rho = -c$. (One can take, for instance, c to be a_p with $\psi(p) = -1$.) Let ξ be the restriction of $\eta_M \circ \theta(c)$ to B. Then $\xi \in \mathrm{End}(B)$, $\xi^2 = \theta_F(c^{\sigma_M}c) \in \theta_F(F \cap F_M)$, and $\xi \circ \theta_F(a) = \theta_F(a^{\sigma_M}) \circ \xi$ for $a \in F$. Identify F with $\theta_F(F)$. Then F and ξ generate a subalgebra of $\mathrm{End}_Q(B)$, which is a quaternion algebra over $F \cap F_M$. Since F is totally real, this quaternion algebra is totally indefinite. We shall later discuss an example for which the quaternion algebra is actually a division algebra and forms the whole $\mathrm{End}_Q(B)$.

6. The case $N = 5 \cdot 29$

Let us now discuss a special case $N = 5 \cdot 29$ with $\psi(a) = \left(\dfrac{a}{5 \cdot 29}\right)$. We

	K	a_2	a_3	a_7	a_{17}	a_{37}	a_{59}
$f^{(1)}$	$Q(\sqrt{5}, \sqrt{-1})$	$\sqrt{5}$	0	$2\sqrt{-1}$	$2\sqrt{5}$	$-2\sqrt{5}$	4
$f^{(2)}$	$Q(\sqrt{17}, \sqrt{-3})$	1	$(-1+\sqrt{17})/2$	$2\sqrt{-3}$	-2	$3-\sqrt{17}$	$3+\sqrt{17}$
$f^{(3)}$	$Q(\sqrt{17}, \sqrt{-3})$	-1	$(1-\sqrt{17})/2$	$2\sqrt{-3}$	2	$-3+\sqrt{17}$	$3+\sqrt{17}$

p	2	3	7	17	37	59
$\left(\dfrac{p}{5}\right)$	$-$	$-$	$-$	$-$	$-$	$+$
$\left(\dfrac{p}{29}\right)$	$-$	$-$	$+$	$-$	$-$	$+$
$\left(\dfrac{p}{5 \cdot 29}\right)$	$+$	$+$	$-$	$+$	$+$	$+$

have $\dim S_2(N, \psi) = 12$ in this case. There are exactly three non-companionate eigen-functions $f^{(1)}$, $f^{(2)}$, $f^{(3)}$, which, together with their companions, form a basis of $S_2(N, \psi)$. The field K and the eigen-values a_n for each $f^{(i)}$ are given in the table of the preceding page.

Now we fix our attention to $f^{(1)}$ and consider the objects δ and η_M of §5. Then we obtain the following table for all possible choices of L and M (excluding the trivial case $M = 1$).

L	M	F_M	c_M	k_M
5	29	$Q(\sqrt{-1})$	(5)	$Q(\sqrt{29})$
29	5	$Q(\sqrt{-5})$	(1)	$Q(\sqrt{5})$
1	5·29	$Q(\sqrt{5})$	(1)	$Q(\sqrt{5 \cdot 29})$

Observe that (5.10) is satisfied in each case. Now in the case $L = 5$, $M = 29$, we have $c_M = 5 o_M = I \cdot I^\rho$ with $I = (2 + \sqrt{-1})$. Define r_I^M as in §5 in the present case. Then the finite part of $\mathfrak{f}[r_I^M]$ is divisible only by the prime factors of 5 in k_M, on account of the above-mentioned generalization of Proposition 2.4.

PROPOSITION 6.1. *The notation being as above, suppose* $I = (2 - \sqrt{-1})$. *Then, with a suitable choice of a prime factor* q *of 5 in* $k_M = Q(\sqrt{29})$ *and an archimedean prime* $\mathfrak{p}_\infty$ *of* k_M, *one has*

$$(1) \qquad r_I^M((\alpha)) = \left(\frac{\alpha}{\mathfrak{p}_\infty}\right) \cdot \left(\frac{\alpha}{q^e}\right) \cdot \mu(\alpha \bmod q) \,,$$

where μ *is the isomorphism of the residue field modulo* q *onto* o_M/I. *Similarly, if* $I = (2 + \sqrt{-1})$, *with a prime factor* q *of 5 in* k_M *and an archimedean prime* $\mathfrak{p}_\infty$ *of* k_M *(which may or may not be the same as the above* $\mathfrak{p}_\infty$ *and* q), *one has*

$$(2) \qquad r_I^M((\alpha)) = \left(\frac{\alpha}{\mathfrak{p}_\infty}\right) \cdot \left(\frac{\alpha}{q}\right) \cdot \mu(\alpha \bmod q) \,,$$

where μ *is the same as above. In* (1) *and* (2), q *is determined by* $\mathfrak{p}_\infty$ *by the condition that* $v > 0$ *at* $\mathfrak{p}_\infty$ *for every unit* $v \equiv 1 \bmod q$.

Proof. Let q be a prime factor of 5 in k_M. Since $(o_M/I)^\times$ is of order 4, $\mathfrak{f}[r_I^M]$ may be divisible by q, but not by q^2. Now $7 = N_{k_M/Q}(\beta)$ with $\beta = 6 + \sqrt{29}$. By Theorem 5.4, $r_I^M((\beta))$ satisfies

$$(*) \qquad\qquad X^2 - a_7 X - 7 \equiv 0 \bmod I \,.$$

Suppose $I = (2 - \sqrt{-1})$. Since $a_7 = 2\sqrt{-1}$, the roots of $(*)$ are 1 and 3. On the other hand, q is generated by $(7 + \sqrt{29})/2$ or $(7 - \sqrt{29})/2$, so that the second formula of (6.1) holds for $\alpha = \beta$. By (iii) of Theorem 5.4, the relation (1) holds for $\alpha = m$ with any rational integer m prime to 5. Now $I(5)/P(5\mathfrak{p}_\infty\mathfrak{p}_\infty')$ can be generated by (β) and (m) for all positive integers m prime to 5. Therefore (1) is true for any α prime to 5. Let v be a unit of k_M such that $N(v) = -1$. Since $\mathfrak{q}$ divides exactly one of the two numbers $v - 1$ and $v + 1$, $\mathfrak{q}$ is determined by $\mathfrak{p}_\infty$ as stated above. Formula (2) can be proved by the same type of consideration.

Remark 6.2. In view of Remark 5.5, we see that the abelian variety A associated with $f^{(1)}$ is isogenous to the product of four copies of an elliptic curve defined over $Q(\sqrt{5}, \sqrt{29})$.

Let us now consider $f^{(2)}$. This case is similar to the case $N = 229$. If we define δ of § 5 with $L = 5$ and $M = 29$, then $[\delta]_2$ sends $f^{(2)}$ to $f^{(3)}$. Define (A, θ), K, F, etc. with $f^{(2)}$ as f. Then we see that $\mathfrak{c} = 3\mathfrak{o}_F$, and $\mathfrak{c}$ is a prime ideal in $F = Q(\sqrt{17})$, hence $(\mathfrak{o}_F/\mathfrak{c})^\times$ is of order 8. By Proposition 2.4, the characters r and s in the present case are ramified only at the factors of 3 and the archimedean primes of $k = Q(\sqrt{145})$. The prime 3 decomposes into two primes in k; the fundamental unit u_0 of k is given by $u_0 = 12 + \sqrt{145}$; the ideal class group of k is a cyclic group of order 4. We note that $N_{k/Q}(u_0) = -1$, and $Tr_{k/Q}(u_0)$ is divisible by 3.

PROPOSITION 6.3. *The notation being as above, one has* $\mathfrak{f}[r] = \mathfrak{q}\cdot\mathfrak{p}_\infty$ *with a prime factor* $\mathfrak{q}$ *of 3 in k and an archimedean prime* $\mathfrak{p}_\infty$ *of k;* $\mathfrak{p}_\infty$ *is determined by the property that $v > 0$ at $\mathfrak{p}_\infty$ for every $v \in \mathfrak{o}^\times$ such that $N_{k/Q}(v) = -1$ and $v \equiv 1 \bmod \mathfrak{q}$. The character r gives an isomorphism of $I(\mathfrak{q})/P(\mathfrak{q}\mathfrak{p}_\infty)$ onto $(\mathfrak{o}_F/\mathfrak{c})^\times$. The field $k(\mathfrak{x})$ is the maximal class field over k of conductor $3\cdot\mathfrak{p}_\infty\mathfrak{p}_\infty'$.*

Proof. Observe that a_3 is prime to 3. Therefore, by Theorem 2.8, the finite part of $\mathfrak{f}[r]$ is of the form $\mathfrak{q}^m$ with a prime factor $\mathfrak{q}$ of 3 in k. By [9, Lemma 7.32] and (iii) of Theorem 2.3, we have $m = 1$. Then our first assertion can be proved by the same reasoning as in the proof of Proposition 3.2. Now we have $3^4 = N_{k/Q}(49 + 4\sqrt{145})$. Therefore $\mathfrak{q}^4 = \alpha\mathfrak{o}$ with $\alpha = 49 + 4\sqrt{145}$ or $49 - 4\sqrt{145}$. On the other hand, $a_3 = (-1 + \sqrt{17})/2$, so that

$$a_3^2 + a_3 - 1 \equiv 0 \bmod 3\mathfrak{o}_F ,$$

hence $a_3^4 \equiv -1 \bmod 3\mathfrak{o}_F$. By Theorem 2.8,

$$r(\mathfrak{q}^4) = (a_3 \bmod 3\mathfrak{o}_F) .$$

Since $\mathfrak{q}^4$ generates $I(\mathfrak{q})/P(\mathfrak{q}\cdot\mathfrak{p}_\infty)$, we obtain the second assertion, from which the last assertion follows immediately.

7. The case of level qq', $q \equiv q' \equiv 3 \bmod (4)$

Let us now consider the case where $N = qq'$ with two primes q and q' which are $\equiv 3 \bmod (4)$, and $\psi(a) = \left(\dfrac{a}{qq'}\right)$. We normalize q and q' so that $\left(\dfrac{q'}{q}\right) = 1$, by exchanging q and q' if necessary. The space $S_2(N, \psi)$ may contain cusp forms corresponding to Grössen-characters of $\mathbf{Q}(\sqrt{-q})$, which we shall discuss in the next section. The purpose of the present section is to investigate the eigen-functions *not* corresponding to such Grössen-characters.

By our choice of ψ, we have $k = \mathbf{Q}(\sqrt{qq'})$. Therefore, if u_0 is the fundamental unit of k, we have $N_{k/\mathbf{Q}}(u_0) = 1$. Moreover, by Proposition A.II.4 of Appendix II, $N_{k/\mathbf{Q}}(u_0 - 1)$ is divisible by q. Table II gives some numerical data for a few small $N = qq'$. In all cases of the table, the class number of k is one. Moreover, in each case, $S_2(N, \psi)$ is spanned by the companions of *one* eigen-function not corresponding to Grössen-characters, and possibly some cusp forms corresponding to Grössen-characters. Therefore we obtain only one K in each case. Furthermore, if we define δ_q as in § 5 with q as M, then (5.10) is satisfied in all cases. Since $\left(\dfrac{q'}{q}\right) = 1$, we see that (5.11) is satisfied. Therefore the objects B_q, k_q, F_q, c_q, etc., can be defined as in § 5 with q and q' as M and L. Note that $k_q = \mathbf{Q}(\sqrt{-q})$. Within the limit of the table, we observe that $N(c)$ and $N_{k/\mathbf{Q}}(u_0 - 1)$ consist of the same prime factors, if we exclude the primes 2, 3, and q.

TABLE II

$N = qq'$: $q \equiv q' \equiv 3 \bmod (4)$

q	q'	$[K: \mathbf{Q}]$	$-N_{k/\mathbf{Q}}(u_0 - 1)$	$N(c)$	$N(c_q)$
3	19	4	$2^2 \cdot 3 \cdot 5^2$	5^2	1
3	31	8	3^3	?	?
3	43	12	$2^2 \cdot 3 \cdot 53^2$	53^2	1
47	3	4	$2^2 \cdot 47$	1	$3^2 \cdot 5^2$
7	11	4	7	1	5
7	23	12	$2^2 \cdot 7 \cdot 29^2$	29^2	?
31	7	12	$2^2 \cdot 3^2 \cdot 31 \cdot 83^2$	83^2	97^2

We now discuss two examples which seem to be fairly typical. The first example is the case $N = 57$, $q = 3$, $q' = 19$. Then $\dim S_2(N, \psi) = 4$. There are no cusp forms corresponding to Grössen-characters. The eigen-functions are companions of a cusp form $f(z) = \sum a_n e^{2\pi i n z}$ whose first few eigen-values are as follows.

p	a_p	$\left(\dfrac{p}{3}\right)$	$\left(\dfrac{p}{19}\right)$	$\left(\dfrac{p}{57}\right)$
2	$\sqrt{2}$	$-$	$-$	$+$
5	$\sqrt{-5}$	$-$	$+$	$-$
7	1	$+$	$+$	$+$
11	$-\sqrt{-5}$	$-$	$+$	$-$
13	$\sqrt{-10}$	$+$	$-$	$-$
43	-5	$+$	$+$	$+$

Therefore we see that $K = Q(\sqrt{2},\ \sqrt{-5})$, $F = Q(\sqrt{2})$, $\mathfrak{c} = (5)$. Observe that 5 remains prime in both F and $k = Q(\sqrt{57})$. Define the $(\mathfrak{o}_F/\mathfrak{c})^\times$-valued character r as in § 2.

PROPOSITION 7.1. *The conductor of r is $5\cdot\mathfrak{p}_\infty$ with an archimedean prime $\mathfrak{p}_\infty$ of k. Moreover, there is an isomorphism μ of $\mathfrak{o}/5\mathfrak{o}$ onto $\mathfrak{o}_F/5\mathfrak{o}_F$ such that*

$$r(\alpha\mathfrak{o}) = \left(\frac{\alpha}{\mathfrak{p}_\infty}\right)\cdot\mu(\alpha \bmod 5\mathfrak{o})$$

for every α in k prime to 5.

Proof. The finite part of $\mathfrak{f}[r]$ is divisible by the prime factors of $3\cdot5\cdot19$. Now $5\cdot19$ is prime to the order of $(\mathfrak{o}_F/5\mathfrak{o}_F)^\times$, so that by virtue of [9, Lemma 7.32] and Proposition 2.4, the finite part of $\mathfrak{f}[r]$ is of the form $5\cdot\mathfrak{b}^\nu$, where $\mathfrak{b}$ is the prime ideal of k such that $\mathfrak{b}^2 = 3\mathfrak{o}$. Therefore, on account of Proposition 2.5, one has

$$(*)\qquad r(\alpha\mathfrak{o}) = \left(\frac{\alpha}{\mathfrak{p}_\infty}\right)\cdot\lambda(\alpha \bmod \mathfrak{b}^\nu)\cdot\mu(\alpha \bmod 5\mathfrak{o})$$

with characters λ of $(\mathfrak{o}/\mathfrak{b}^\nu)^\times$ and μ of $(\mathfrak{o}/5\mathfrak{o})^\times$ (both with values in $(\mathfrak{o}_F/5\mathfrak{o}_F)^\times$). Let U be the group of all $\mathfrak{b}$-units in the $\mathfrak{b}$-completion of k, and let $U_n = \{u \in U \mid u \equiv 1 \bmod \mathfrak{b}^n\}$ for every integer $n \geq 0$. Define an $(\mathfrak{o}_F/5\mathfrak{o}_F)$-valued character π of U by $\pi(\alpha) = \lambda(\alpha \bmod \mathfrak{b}^\nu)$. Then ν is the smallest integer n such that $U_n \subset \mathrm{Ker}(\pi)$. Since $(\mathfrak{o}_F/5\mathfrak{o}_F)^\times$ is of order 24, we have $\pi^{24} = 1$. Now it can easily be verified that

$$U_4 \subset \{u^{24} \mid u \in U\}\ .$$

Therefore $\nu \leq 4$. Observe that U_1/U_4 is the product of cyclic groups of order 3 and 9. Therefore U_1/U_4 contains an element z of order 9 such that z^3 generates U_3/U_4. Since $\pi^{24} = 1$, we have $\pi(z)^3 = 1$, so that $U_3 \subset \mathrm{Ker}(\pi)$. Now U_2/U_3 is generated by a positive integer $m \equiv 1 \bmod (3)$. Take m so that $m \equiv 1 \bmod (5)$. By (ii) of Theorem 2.2, we have $1 = r(m\mathfrak{o}) = \pi(m)$. This proves that $U_2 \subset \mathrm{Ker}(\pi)$. Observe that $43 = N_{k/Q}(\alpha)$ with $\alpha = 10 + \sqrt{57}$. By Theorem 2.2, $r(\alpha\mathfrak{o})$ and $s(\alpha\mathfrak{o})$ satisfy the congruence

$$x^2 - a_{43}x + 43 \equiv 0 \bmod 5\mathfrak{o}_F .$$

Since $a_{43} = -5$, we see easily that $r(\alpha\mathfrak{o})^8 = s(\alpha\mathfrak{o})^8 = 1$. On the other hand, we have $\alpha^8 \equiv 1 \bmod 5\mathfrak{o}$, and $\alpha^8 \equiv 1 - \sqrt{57} \bmod 3\mathfrak{o}$. Therefore $1 = r(\alpha^8\mathfrak{o}) = \pi(\alpha^8)$, and α^8 generates U_1/U_2, so that $U_1 \subset \mathrm{Ker}(\pi)$. Now U/U_1 is generated by a positive integer $m \equiv 2 \bmod (3)$. Take m so that $m \equiv 1 \bmod (5)$. By (ii) of Theorem 2.2, we have $1 = r(m\mathfrak{o}) = \pi(m)$, which shows that π is trivial, i.e., $\mathfrak{b}$ does not divide $\mathfrak{f}[r]$. To complete the proof, take $\beta = 8 + \sqrt{57}$ with $N_{k/Q}(\beta) = 7$. Since $a_7 = 1$, $r(\beta\mathfrak{o})$ satisfies, by Theorem 2.2,

$$x^2 - x + 7 \equiv 0 \bmod 5\mathfrak{o}_F ,$$

so that $r(\beta\mathfrak{o})$ is $3 \pm \sqrt{2} \bmod 5\mathfrak{o}_F$. We see easily that $\beta\mathfrak{o}$ and $m\mathfrak{o}$ with $m \in \mathbf{Z}$ generate $I(5)/P(5\mathfrak{p}_\infty)$, and two isomorphisms of $\mathfrak{o}/5\mathfrak{o}$ onto $\mathfrak{o}_F/5\mathfrak{o}_F$ map $8 + \sqrt{57}$ to $3 \pm \sqrt{2}$. Therefore we obtain our proposition.

PROPOSITION 7.2. *The abelian variety B in the present case $N = 57$ is simple. Moreover $\mathrm{End}_Q(B)$ is an indefinite quaternion algebra over $\mathbf{Q}$ of discriminant $2 \cdot 5$.*

Proof. Consider the automorphism η_M of A defined in §5 with $M = 3$. Let ξ be the restriction of $\eta_M \circ \theta(\sqrt{-5})$ to B. As is seen in Remark 5.5, $\xi^2 = \theta_F(5)$ and $\xi \circ \theta_F(a) = \theta_F(a^\sigma) \circ \xi$ for $a \in \mathbf{Q}(\sqrt{2})$, where σ is the generator of $\mathrm{Gal}(\mathbf{Q}(\sqrt{2})/\mathbf{Q})$. It can easily be verified that ξ and $\theta_F(\mathbf{Q}(\sqrt{2}))$ generate a quaternion algebra over $\mathbf{Q}$ of discriminant $2 \cdot 5$. Suppose that $\mathrm{End}_Q(B)$ is larger than the quaternion algebra. Then B is isogenous to the product of two copies of an elliptic curve with complex multiplication. By Proposition 1.6, the Mellin transform of our cusp form f must be the L-function with a primitive Grössen-character χ of an imaginary quadratic field S. But this is impossible for the following reason. Let $-D$ be the discriminant of S and $\mathfrak{m}$ the conductor of χ. By [12, Lemma 3] (see also the next section), one has $3 \cdot 19 = D \cdot N(\mathfrak{m})$, and

$$\psi(a) = \left(\frac{-D}{a}\right) \cdot \chi((a))/a \qquad\qquad (a \in \mathbf{Z}, (a, 57) = 1) .$$

Then we must have $D = 3$, $N(\mathfrak{m}) = 19$, and $\chi((a))/a = \left(\frac{a}{19}\right)$. But it can easily be seen that $\mathbf{Q}(\sqrt{-3})$ has no such Grössen-character χ. Therefore $\mathrm{End}_Q(B)$ is exactly the quaternion algebra, so that B is simple.

As the second example, consider the case $N = 7 \cdot 11$, $q = 7$, $q' = 11$. Then $\dim S_2(N, \psi) = 6$. There are two eigen-functions corresponding to

Grössen-characters of $Q(\sqrt{-7})$. The remaining 4 eigen-functions are companions of a cusp form $f(z) = \sum a_n e^{2\pi i n z}$, whose first few eigenvalues are given in the following table.

p	a_p	$\left(\dfrac{p}{7}\right)$	$\left(\dfrac{p}{11}\right)$	$\left(\dfrac{p}{77}\right)$
2	$\sqrt{-2}$	$+$	$-$	$-$
3	$\sqrt{-5}$	$-$	$+$	$-$
5	$-\sqrt{-5}$	$-$	$+$	$-$
13	$-2\sqrt{10}$	$-$	$-$	$+$
17	0	$-$	$-$	$+$
19	$\sqrt{10}$	$-$	$-$	$+$
23	-3	$+$	$+$	$+$

We see that $K = Q(\sqrt{-2}, \sqrt{-5})$, $F = Q(\sqrt{10})$, $\mathfrak{c} = (1)$. Take q and q' as M and L of § 5. Observe that (5.10) and (5.11) are satisfied. Therefore we can define F_M, B_M, $\mathfrak{c}_M$, etc. as in § 5. Then $k_M = Q(\sqrt{-7})$, $F_M = Q(\sqrt{-2})$, $\mathfrak{c}_M = (5)$. Note that the prime 5 remains prime in both k_M and F_M. Define r_I^M with $I = (5)$. Let $\mathfrak{o}'$ denote the ring of algebraic integers in k_M.

PROPOSITION 7.3. *The conductor of r_I^M is $5 \cdot q'$ with a prime factor q' of 11 in $Q(\sqrt{-7})$. Moreover, for every α in $Q(\sqrt{-7})$ prime to $5 \cdot q'$, one has*

$$r_I^M(\alpha\mathfrak{o}') = \left(\frac{\alpha}{q'}\right) \cdot \mu(\alpha \bmod 5\mathfrak{o}') \,,$$

where μ is one of the two isomorphisms of $\mathfrak{o}'/5\mathfrak{o}'$ onto $\mathfrak{o}_M/5\mathfrak{o}_M$.

PROPOSITION 7.4. *The abelian variety B_M is simple. Moreover $\mathrm{End}_Q(B_M)$ is an indefinite quaternion algebra over Q of discriminant $2 \cdot 5$.*

We omit the proofs of these two propositions, since they can be proved by the same type of reasoning as in the proofs of Propositions 7.1 and 7.2.

8. The abelian varieties associated with Grössen-characters of an imaginary quadratic field

Throughout this section, we denote by S an imaginary quadratic field. As Hecke showed, the Mellin transform of an L-function of S with a Grössen-character λ is a cusp form $f(z)$ belonging to $S_\kappa(N, \psi)$ for some κ, N, ψ. The abelian variety A associated with this type of f, in the case $\kappa = 2$, is actually isogenous to the product of several copies of an elliptic curve E such that $\mathrm{End}_Q(E)$ is isomorphic to S (see [12]). Therefore all points of finite order on A are rational over the maximal abelian extension of

S (see below). In this sense, one may say that the study of A belongs to the theory of complex multiplication. However, for a suitably chosen λ, it can happen that ψ is a real character of order 2 associated with a real quadratic field k. In such a case, we are still able to obtain from A a non-trivial class field over k and its reciprocity-law, a detailed discussion of which is the purpose of this section.

Let us begin by recalling some basic definitions and properties of Grössen-characters. For our present need, it is more convenient to regard them as characters of the *ideal* group rather than as characters of the *idele* group. Let m be an integral ideal in S, and $I(m)$ the group of all fractional ideals in S prime to m. Fix a positive integer ν, and let Λ_m^ν denote the set of all homomorphisms λ of $I(m)$ into C such that

$$(8.1) \qquad\qquad \lambda(\alpha o_S) = \alpha^\nu \qquad\qquad \text{for } \alpha \in S^\times,\ \alpha \equiv 1 \bmod^\times m ,$$

where o_S denote the ring of algebraic integers in S, and $\bmod^\times m$ means the usual multiplicative congruence modulo m. We call such a λ a Grössen-character of S defined modulo m. For each $\lambda \in \Lambda_m^\nu$, there is a unique divisor c of m such that: (i) λ is the restriction of an element of Λ_c^ν; (ii) no proper divisor of c has the property (i). Then c is called the *conductor* of λ. We call λ *primitive* if m is the conductor of λ. With every $\lambda \in \Lambda_m^\nu$, we associate a function f_λ on $\mathfrak{H}$ by

$$f_\lambda(z) = \sum_x \lambda(x) e^{2\pi i N(x) z} \qquad\qquad (z \in \mathfrak{H}) ,$$

where x runs over all the integral ideals prime to m. Let $-D$ be the discriminant of S. In [12], we have shown that f_λ belongs to $S_{\nu+1}(N, \psi)$ with

$$N = D \cdot N(m) ,$$
$$\psi(a) = \left(\frac{-D}{a}\right) \cdot \lambda(a o_S)/a^\nu \qquad\qquad (a \in Z,\ (a, N) = 1) .$$

One can also show that $f_\lambda \in S_{\nu+1}^0(N, \psi)$ *if and only if* λ *is primitive.* This follows from the result of Miyake [5], on account of the functional equation of the corresponding L-function

$$L(s, \lambda) = L(s, f_\lambda) = \sum_x \lambda(x) N(x)^{-s} .$$

Let us hereafter assume that $\nu = 1$ and λ is primitive, so that $f_\lambda \in S_2^0(N, \psi)$. Let K and (A, θ) be defined for this f_λ as in §1. As is shown in [12], A is isogenous to a product of copies of an elliptic curve E such that $\mathrm{End}_Q(E)$ is isomorphic to S. Therefore we can define an isomorphism θ_S of S onto the center of $\mathrm{End}_Q(A)$ so that the representation of S through θ_S on the tangent space of A at the origin is equivalent to the sum of copies of the identity injection of S into C. Now the elements of $\theta(K) \cap \mathrm{End}(A)$ are defined over Q, while any generator of $\theta_S(S)$ belonging to $\mathrm{End}(A)$ is defined over S, but not over Q. Therefore we can extend θ and θ_S to an isomorphism of $K \otimes_Q S$ into $\mathrm{End}_Q(A)$, which we denote by θ'. Note that $K \otimes_Q S$ may or may not be a field. At any rate, $K \otimes_Q S$ is a commutative semi-simple algebra of rank $2 \cdot \dim (A)$ over Q, and every element of $\mathrm{End}(A) \cap \theta'(K \otimes_Q S)$ is rational over S. It follows easily that *every point of finite order on A is rational over the maximal abelian extension of S.*

Now let us consider the case where ψ is a real character of order 2. Although it seems possible to discuss the most general case without great difficulty, we treat here only the following two simplest cases:

(I) $S = Q(\sqrt{-q})$, $N(m) = q'$, $q \equiv q' \equiv 3 \bmod (4)$, $q > 3$.

(II) $S = Q(\sqrt{-qq'})$, $N(m) = q'$, $q \equiv 1$, $q' \equiv 3 \bmod (4)$.

In both cases, q and q' are rational primes. First let us prove a lemma, which is useful for the determination of K.

LEMMA 8.1. *Let p be a prime ideal in S, t the order of p in the ideal class group of S, and α an element of S such that $p^t = (\alpha)$. Then the polynomial $X^t - \alpha$ is irreducible over S. Moreover, if β is a t-th root of α and $p^\rho \neq p$, then $S(\beta)$ is a CM-field, $\beta\beta^\rho = p$, $S(\beta) = Q(\beta)$, $[Q(\beta): Q] = 2t$, and*

$$[Q(\beta - \beta^\rho): Q] = \begin{cases} t & \text{if } t \text{ is even}, \\ 2t & \text{if } t \text{ is odd}. \end{cases}$$

Furthermore, $S \not\subset Q(\beta - \beta^\rho)$ if t is even.

Proof. Suppose that

$$X^t - \alpha = \left(\sum_{i=0}^{u} b_i X^i\right)\left(\sum_{j=0}^{t-u} c_j X^j\right)$$

is a non-trivial decomposition of $X^t - \alpha$ with b_i and c_j in o_S. Then $(b_0 c_0) = (\alpha) = p^t$, so that we may assume $(b_0) = p^t$, and $(c_0) = o_S$. Let μ be the smallest subscript such that b_μ is not divisible by p. Then $0 < \mu \leqq u < t$, and

$$b_\mu c_0 + b_{\mu-1} c_1 + \cdots + b_0 c_\mu = 0 ,$$

where we understand that $c_j = 0$ for $j > t - u$. Since $b_0, \cdots, b_{\mu-1}$ are divisible by p, we see that b_μ is divisible by p, a contradiction. Therefore $X^t - \alpha$ is irreducible, and $[S(\beta): S] = t$. Suppose $p^\rho \neq p$. Then $N(p)$ is a rational prime p, and, for every isomorphism σ of $S(\beta)$ into C, one has $(\beta^\sigma \beta^{\sigma\rho})^t = \alpha^\sigma \alpha^{\sigma\rho} = p^t$, so that $\beta^\sigma \beta^{\sigma\rho} = p$. It follows that $S(\beta)$ is stable under ρ, and $\sigma\rho = \rho\sigma$, hence $S(\beta)$ is a CM-field. A conjugate of β over Q is either of the form $\beta\zeta$ or $\beta^\rho\zeta$ with a t-th root of unity ζ. Conversely every element of the form $\beta\zeta$ or $\beta^\rho\zeta$ is a conjugate of β over $\mathbf{Q}$. Since $p \neq p^\rho$, these $2t$ elements are all different, so that $[\mathbf{Q}(\beta):\mathbf{Q}] = 2t$, $\mathbf{Q}(\beta) = S(\beta)$. Let σ be an isomorphism of $S(\beta)$ into C such that $\beta - \beta^\rho$ is invariant under σ. We are going to show that $\beta^\sigma = \beta$ or $\beta^\sigma = -\beta^\rho$; the latter case can happen only if t is even. Suppose $\beta^\sigma = \beta\zeta$ with $\zeta \neq 1$ or $\beta^\sigma = \beta^\rho\zeta$ with $\zeta \neq -1$. Then we have

$$\beta - \beta^\rho = (\beta - \beta^\rho)^\sigma = \begin{cases} \beta\zeta - \beta^\rho\zeta^\rho \,, \text{ or} \\ \beta^\rho\zeta - \beta\zeta^\rho \,, \end{cases}$$

so that β/β^ρ equals $(1 - \zeta^\rho)/(1 - \zeta)$ or $(1 + \zeta)/(1 + \zeta^\rho)$, respectively. Since the last two quotients are units, $\alpha/\alpha^\rho = (\beta/\beta^\rho)^t$ is also a unit, which contradicts the assumption $p^\rho \neq p$. Therefore, if t is odd, $\beta - \beta^\rho$ is invariant under σ only when $\beta^\sigma = \beta$, hence $[\mathbf{Q}(\beta - \beta^\rho):\mathbf{Q}] = 2t$. If t is even, $\beta - \beta^\sigma$ is invariant under σ such that $\beta^\sigma = -\beta^\rho$, hence $[\mathbf{Q}(\beta - \beta^\rho):\mathbf{Q}] = t$. For this σ, we have $\alpha^\sigma = \alpha^\rho \neq \alpha$, hence S is not contained in $\mathbf{Q}(\beta - \beta^\rho)$. This completes the proof.

Now let us start the discussion of the case (I). Let q and q' be two different rational primes such that $q > 3$, $q \equiv q' \equiv 3 \bmod (4)$, and $\left(\dfrac{q'}{q}\right) = 1$. Put $S = \mathbf{Q}(\sqrt{-q})$, and let h denote the class number of S. It is well-known that h is odd. Take a prime ideal $\mathbf{q}'$ in S such that $N(\mathbf{q}') = q'$. Then there are exactly h elements λ of $\Lambda^1_{\mathbf{q}'}$ such that

$$\lambda(\alpha o_s) = \left(\frac{\alpha}{\mathbf{q}'}\right)\cdot\alpha$$

for every α in S prime to $\mathbf{q}'$. All such λ are primitive. Define f_λ as above, and put $f_\lambda(z) = \sum_{n=1}^\infty a_n e^{2\pi i n z}$. Then $f_\lambda \in S_2^0(qq', \psi)$ with $\psi(a) = \left(\dfrac{a}{qq'}\right)$. Define the symbols K, $\mathfrak{b}$, $\mathfrak{c}$, and (A, θ) as in § 2. To make our treatment easier, we impose the following condition on S:

(8.2) *The ideal class group of S is cyclic.*

LEMMA 8.2. *There exist a rational prime p and a prime ideal $\mathbf{p}$ in S such that $N(\mathbf{p}) = p$, $\left(\dfrac{p}{q}\right) = 1$, $\left(\dfrac{p}{q'}\right) = -1$, and $\mathbf{p}$ generates the ideal class group of S.*

Proof. Let T be the maximal unramified abelian extension of S. Since T does not contain $\sqrt{-q'}$, $\mathrm{Gal}(T(\sqrt{-q'})/S)$ is a cyclic group of order $2h$. Let σ be the generator of this cyclic group. Observe that $T(\sqrt{-q'})$ is normal over $\mathbf{Q}$. We can find a prime ideal P in $T(\sqrt{-q'})$, for which the Frobenius element in $\mathrm{Gal}(T(\sqrt{-q'})/\mathbf{Q})$ is σ. Let $\mathbf{p} = P \cap S$ and $p = N(\mathbf{p})$. Then $\mathbf{p}$ and p have the required properties.

Take $\mathbf{p}$ and p as in the above lemma, and take $\alpha \in S$ so that $\mathbf{p}^h = (\alpha)$

and $(\alpha/q') = 1$. Put $\beta = \lambda(p)$. Then $\beta^h = \alpha$, $\beta\beta^\rho = p$, and $\lambda(pp^\rho) = \lambda(po_S) = -p$, so that $\lambda(p^\rho) = -\beta^\rho$. Therefore the Euler p-factor of $L(s, \lambda)$ is

$$[1 - \lambda(p)p^{-s}]^{-1}[1 - \lambda(p^\rho)p^{-s}]^{-1} = [1 - (\beta - \beta^\rho)p^{-s} - p^{1-2s}]^{-1} .$$

In other words, $a_p = \beta - \beta^\rho$. By Lemma 8.1, we have $\boldsymbol{Q}(a_p) = S(\beta)$. If $a \in I(q')$, we have $a = \gamma p^m$ for some $\gamma \in S$ and an integer m, so that $\lambda(a) = \pm\gamma\beta^m \in S(\beta)$. This implies that

$$K = \boldsymbol{Q}(a_p) = S(\beta), \quad \dim(A) = 2h .$$

Our next question is about the ideals $\mathfrak{b}$ and $\mathfrak{c}$. Let F be as before the maximal real subfield of K. Then $K = FS$, so that the different of K relative to F is non-trivial, and consists of prime factors of q. On the other hand, p, α, and β being as above, we have

$$N_{F/Q}(a_p^2) = N_{K/S}(a_p^2) = \prod_{n=0}^{h-1}(\beta\zeta^n - \beta^\rho\zeta^{-n})^2 = (\alpha - \alpha^\rho)^2 \qquad (\zeta = e^{2\pi i/h}) .$$

If $\alpha = (x + y\sqrt{-q})/2$, we obtain $N_{F/Q}(a_p^2) = -y^2q$. Let us now make another assumption on S:

(8.3) *There is a prime ideal p satisfying the conditions of Lemma 8.2 such that $p^h = ((x + y\sqrt{-q})/2)$ with $x, y \in \boldsymbol{Z}$, $(y, q) = 1$.*

It is unknown whether this is always the case, although no counter-example has been found. Under this assumption, we have $N(\mathfrak{c}) = q$, since $N(\mathfrak{c})$ must divide y^2q. Define the $(o_F/\mathfrak{c})$-valued characters r and s in the present case. Note that $o_F/\mathfrak{c}$ is isomorphic to $\boldsymbol{Z}/q\boldsymbol{Z}$.

PROPOSITION 8.3. *Let* $\mathfrak{q}$ *be the prime ideal in* $k(= \boldsymbol{Q}(\sqrt{qq'}))$ *such that* $\mathfrak{q}^2 = qo_k$. *Then* $\mathfrak{f}[r] = \mathfrak{q}\cdot p_\infty$ *with an archimedean prime* p_∞ *of* k. *Moreover, let* j *be the class number of* k, *and* μ *the isomorphism of* $o_k/\mathfrak{q}$ *onto* $o_F/\mathfrak{c}$. *Then for every* $a \in I(\mathfrak{q})$, $a^j = \alpha o_k$ *with* $\alpha \in k$, *one has*

$$r(a)^j = \left(\frac{\alpha}{p_\infty}\right)\cdot\mu(\alpha \bmod \mathfrak{q}) ,$$

$$r(a)^2 = \left(\frac{N(a)}{\mathfrak{q}}\right)\cdot\mu(N(a) \bmod \mathfrak{q}) .$$

It is well-known that j is odd. Therefore r is completely determined by these two relations.

Proof. The finite part of $\mathfrak{f}[r]$ consists of prime factors of qq'. Since $\left(\dfrac{q}{q'}\right) = -1$, q' does not divide $q - 1$. Therefore, by Proposition 2.4, $\mathfrak{f}[r]$ is prime to q'. By [9, Lemma 7.32], the finite part of $\mathfrak{f}[r]$ is exactly $\mathfrak{q}$. Therefore our first assertion follows from Proposition 2.5. Let $a \in I(\mathfrak{q})$, $a^j = \alpha o$

with $\alpha \in k$. Take $m \in Z$ so that $m \equiv \alpha$ mod q and $\left(\dfrac{m\alpha}{\mathfrak{p}_\infty}\right) = 1$. By (iii) of Theorem 2.3,

$$r(\mathfrak{a})^j = r(\alpha \mathfrak{v}) = r(m\mathfrak{v}) = \left(\dfrac{m}{\mathfrak{p}_\infty}\right) \cdot (m \bmod \mathfrak{c}) = \left(\dfrac{\alpha}{\mathfrak{p}_\infty}\right) \cdot \mu(\alpha \bmod \mathrm{q}) \, .$$

Next take a prime ideal $\mathfrak{p}$ of degree one so that $\mathfrak{p}^{-1}\mathfrak{a} \in P(\mathrm{q}\mathfrak{p}_\infty)$. Then $N(\mathfrak{p}) \equiv N(\mathfrak{a})$ mod (q). By Theorem 2.2,

$$[X - r(\mathfrak{p})][X - s(\mathfrak{p})] \equiv X^2 - a_p X + p \bmod \mathfrak{b}$$

$$\equiv \begin{cases} [X - \lambda(\boldsymbol{p})][X - \lambda(\boldsymbol{p}^\rho)] \bmod \mathfrak{b} \text{ if } \left(\dfrac{p}{q}\right) = 1 \text{ and } p\mathfrak{o}_S = \boldsymbol{p}\boldsymbol{p}^\rho \text{ in } S \, , \\ X^2 - \lambda(p\mathfrak{o}_S) \bmod \mathfrak{b} \text{ if } \left(\dfrac{p}{q}\right) = -1 \, . \end{cases}$$

If $\left(\dfrac{p}{q}\right) = 1$, we have $\left(\dfrac{p}{q'}\right) = 1$ and $\lambda(\boldsymbol{p})\lambda(\boldsymbol{p}^\rho) = \lambda(p\mathfrak{o}_S) = p$. Since $|\lambda(\boldsymbol{p})|^2 = p$, we have $\lambda(\boldsymbol{p}^\rho) = \lambda(\boldsymbol{p})^\rho$, so that $\lambda(\boldsymbol{p}) \equiv \lambda(\boldsymbol{p}^\rho)$ mod $\mathfrak{b}$, hence $r(\mathfrak{p}) = s(\mathfrak{p})$. If $\left(\dfrac{p}{q}\right) = -1$, we have $r(\mathfrak{p}) + s(\mathfrak{p}) = 0$. Therefore $r(\mathfrak{p}) = \left(\dfrac{p}{q}\right) \cdot s(\mathfrak{p})$ in either case, so that

$$r(\mathfrak{a})^2 = r(\mathfrak{p})^2 = \left(\dfrac{p}{q}\right) \cdot r(\mathfrak{p})s(\mathfrak{p}) = \left(\dfrac{N(\mathfrak{p})}{q}\right) \cdot (N(\mathfrak{p}) \bmod \mathfrak{c})$$

by (iv) of Theorem 2.3. This completes the proof, since $N(\mathfrak{p}) \equiv N(\mathfrak{a})$ mod (q).

Remark. Let u_0 be the fundamental unit of k. Then $N_{k/Q}(u_0) = 1$. Therefore we can take u_0 so as to be totally positive. By Proposition A.II.4 of Appendix II, we have $u_0 \equiv 1$ mod q. This agrees with the formulas of the above proposition.

The case (II) is concerned with $S = Q(\sqrt{-qq'})$ with two primes $q \equiv 1$, $q' \equiv 3$ mod (4), which is somewhat more interesting than the above case (I). We take the prime ideal $\boldsymbol{q}'$ in S such that $N(\boldsymbol{q}') = q'$, and consider the Grössen-characters λ of $\Lambda^1_{q'}$ such that

$$(8.4) \qquad\qquad \lambda(\alpha \boldsymbol{o}_S) = \left(\dfrac{\alpha}{\boldsymbol{q}'}\right) \cdot \alpha$$

for every α prime to q'. Then $f_\lambda \in S_2^0(qq'^2, \psi)$, where $\psi(a) = \left(\dfrac{a}{q}\right)$, so that $k = Q(\sqrt{q})$ in this case. Let h denote the class number of S. Since the discriminant of S is qq', h is even.

LEMMA 8.4. *The notation being as above, one has* $\left(\dfrac{q}{q'}\right) = (-1)^{h/2}$.

This is a special case of the result of Rédei and Reichardt [6].

Again, for simplicity, we make the assumption that the ideal class group of S is cyclic. Take a prime ideal $\mathfrak{p}_0$ of degree one in S that generates the ideal class group, and put $p_0 = N(\mathfrak{p}_0)$. Then

$$(8.5) \qquad \left(\frac{p_0}{q}\right) = \left(\frac{p_0}{q'}\right) = -1 \;.$$

In fact, $S(\sqrt{q})$ is unramified over S, so that $\mathfrak{p}_0$ remains prime in $S(\sqrt{q})$. Therefore p_0 remains prime in $Q(\sqrt{q})$, hence $\left(\frac{p_0}{q}\right) = -1$. Since $\left(\frac{p_0}{qq'}\right) = 1$, we obtain (8.5).

Put $f_\lambda(z) = \sum_{n=1}^{\infty} a_n e^{2\pi i n z}$, and $\lambda(\mathfrak{p}_0) = \beta_0$. Then $\mathfrak{p}_0^h = (\beta_0^h)$, $\beta_0 \beta_0^\rho = p_0$, $\lambda(\mathfrak{p}_0)\lambda(\mathfrak{p}_0^\rho) = -p_0$, hence $\lambda(\mathfrak{p}_0^\rho) = -\beta_0^\rho$, $a_{p_0} = \beta_0 - \beta_0^\rho$. By Lemma 8.1, $[Q(a_{p_0}): Q] = h$, so that $[K: Q] \geq h$. On the other hand, it can easily be seen that every companion of f_λ is obtained from an element of Λ_q^1, satisfying (8.4), so that $[K: Q] \leq h$. Therefore $[K: Q] = h$. Moreover $S(\beta_0) = SK$ and $S \not\subset K$ by Lemma 8.1.

Let p be a rational prime such that $\left(\frac{p}{q}\right) = \left(\frac{p}{q'}\right) = -1$, and $\mathfrak{p}$ a prime factor of p in S. Put $\beta = \lambda(\mathfrak{p})$. Then $a_p = \beta - \beta^\rho$. We have $\mathfrak{p} = \gamma \mathfrak{p}_0^\nu$ with $\gamma \in S$ and an integer ν, so that $\beta = \gamma \beta_0^\nu$, changing γ for $-\gamma$ if necessary. Put $\zeta = e^{2\pi i/h}$, $\alpha = \beta^h$, $m = h/2$. Since the conjugates of β_0 over S are exactly $\beta_0 \zeta^a$ for $a = 0, 1, \cdots, h-1$, we have, if ν is prime to h,

$$\begin{aligned} N_{K/Q}(a_p) = N_{SK/S}(\beta - \beta^\rho) &= \prod_{a=0}^{h-1} (\beta \zeta^a - \beta^\rho \zeta^{-a}) \\ &= -(\beta^m - \beta^{\rho m})^2 = 2p^m - \alpha - \alpha^\rho \;. \end{aligned}$$

Therefore $N_{K/Q}(\mathfrak{b})$ is a common divisor of the numbers $2p^m - \alpha - \alpha^\rho$ for all such primes p. By checking this common divisor for a few small qq', we find the following empirical result, which is probably true in general:

(8.6) $N_{K/Q}(\mathfrak{b}) = q'$ or q according as $\left(\frac{q}{q'}\right) = 1$ or -1.

(One can at least show, by virtue of Lemma 8.4, that $N_{K/Q}(a_p)/q'$ or $N_{K/Q}(a_p)/q$ is a square of a rational integer according as $\left(\frac{q}{q'}\right) = 1$ or -1.)

Assuming (8.6) to be true, define $(\mathfrak{o}_F/\mathfrak{c})$-valued characters r and s as before.

PROPOSITION 8.5. *The notation being as above, one has*

$$r(\mathfrak{a})^2 = \left(\frac{N(\mathfrak{a})}{q'}\right) \cdot \left(N(\mathfrak{a}) \bmod \mathfrak{c}\right)$$

for every ideal $\mathfrak{a}$ prime to $\mathfrak{f}[r]$.

Proof. It is sufficient to prove the formula in the case where $\mathfrak{a}$ is a prime ideal $\mathfrak{p}$ in k prime to qq'. Let p be the rational prime divisible by $\mathfrak{p}$.

If $N(\mathfrak{p}) = p^2$, the formula follows from (ii) of Theorem 2.2. Suppose $N(\mathfrak{p}) = p$, i.e., $\left(\dfrac{p}{q}\right) = 1$. If p remains prime in S, then $a_p = 0$, so that $r(\mathfrak{p}) = -s(\mathfrak{p})$, hence $r(\mathfrak{p})^2 = -r(\mathfrak{p})s(\mathfrak{p}) = \left(\dfrac{p}{q'}\right) \cdot (p \bmod \mathfrak{c})$ by (iii) of Theorem 2.2. Next suppose that p decomposes into two primes $\boldsymbol{p}$ and $\boldsymbol{p}^\rho$ in S. Then $p = \gamma \boldsymbol{p}_0^\nu$ with $\gamma \in S$ and $\nu \in \boldsymbol{Z}$. Changing γ for $-\gamma$ if necessary, we have $\lambda(\boldsymbol{p}) = \gamma \beta_0^\nu$. Since $\lambda(\boldsymbol{p}\boldsymbol{p}^\rho) = p$, we have $\lambda(\boldsymbol{p}^\rho) = \lambda(\boldsymbol{p})^\rho$, so that

$$(*) \qquad X^2 - a_p X + p = (X - \gamma \beta_0^\nu)(X - \gamma^\rho \beta_0^{\rho\nu}) \, .$$

Let w be a valuation of KS whose valuation ideal contains $\mathfrak{b}$. Then $w(qq') > 0$, $w(\gamma) = 0$, $w(\gamma - \gamma^\rho) > 0$, $w(\beta_0 - \beta_0^\rho) > 0$, hence

$$w(\gamma \beta_0^\nu - \gamma^\rho \beta_0^{\rho\nu}) > 0 \, .$$

This shows that the polynomial $(*)$ has multiple roots modulo $\mathfrak{b}$, hence $r(\mathfrak{p}) = s(\mathfrak{p})$, so that $r(\mathfrak{p})^2 = (N(\mathfrak{p}) \bmod \mathfrak{c})$ by (iii) of Theorem 2.2. This completes the proof.

PROPOSITION 8.6. *Suppose $\left(\dfrac{q}{q'}\right) = 1$. Then $\mathfrak{f}[r] = q' \cdot \mathfrak{p}_\infty$ with an archimedean prime $\mathfrak{p}_\infty$ of $k = \boldsymbol{Q}(\sqrt{q})$. Moreover, with a suitable choice of a prime factor $\mathfrak{q}'$ of q' in k, one has*

$$r(\alpha\mathfrak{o}) = \left(\dfrac{\alpha}{\mathfrak{p}_\infty}\right) \cdot \mu(\alpha \bmod \mathfrak{q}')^{(q'+1)/4} \cdot \mu(\alpha^\iota \bmod \mathfrak{q}')^{(3-q')/4}$$

for every α in k prime to $\mathfrak{q}'$, where μ is the isomorphism of $\mathfrak{o}/\mathfrak{q}'$ onto $\mathfrak{o}_F/\mathfrak{c}$.

The relations of Propositions 8.5 and 8.6 combined together completely determine r. In fact, let j be the class number of k. It is well-known that j is odd. Since $\mathfrak{a}^j$ is a principal ideal, $r(\mathfrak{a})^j$ is given by Proposition 8.6, while $r(\mathfrak{a})^2$ is given by Proposition 8.5, so that $r(\mathfrak{a})$ can be determined.

Proof. We can find a positive integer m such that the finite part of $\mathfrak{f}[r]$ divides $(qq')^m$. Let δ be a totally positive element in $\mathfrak{o}$ prime to qq' such that $\delta \equiv 1 \bmod (q')$. We are going to show that $r(\delta\mathfrak{o}) = 1$. Since q' is odd, there exists an element δ_1 of $\mathfrak{o}$ prime to q such that $\delta_1 \equiv 1 \bmod q'\mathfrak{o}$, $\delta_1^2 \equiv \delta \bmod q'^m\mathfrak{o}$. Put $T = \boldsymbol{Q}(\sqrt{-qq'}, \sqrt{q})$. Let $\mathfrak{Q}_1$ and $\mathfrak{Q}_2$ be the prime factors of q in T. By virtue of the generalized Dirichlet theorem, we can find a prime ideal of degree one in T generated by an element ξ such that

$$\xi \equiv \delta_1 \bmod (q')^m \, ,$$
$$\xi \equiv \delta \bmod \mathfrak{Q}_1^{2m} \, ,$$
$$\xi \equiv 1 \bmod \mathfrak{Q}_2^{2m} \, .$$

Put $\beta = N_{T/k}(\xi)$, $\gamma = N_{T/S}(\xi)$, $p = N_{T/Q}(\xi)$. Then β and γ generate prime ideals of degree one in k and in S respectively; β is totally positive; $\beta \equiv \delta$ mod $(qq')^m$, $\gamma \equiv 1$ mod (q'), $p \equiv 1$ mod (q'). Further $\lambda(\gamma o_S) = \gamma$, so that $a_p = \gamma + \gamma^\rho \equiv 2$ mod (q'), hence

$$X^2 - a_p X + p \equiv (X - 1)^2 \text{ mod } (q') \ .$$

Therefore, $r(\beta o)$, being a root of this polynomial modulo c, must be equal to 1. Since $r(\delta o) = r(\beta o)$, this shows that the finite part of $\mathfrak{f}[r]$ divides q'. Thus we can write

$$r(\alpha o) = \left(\frac{\alpha}{\mathfrak{p}_\infty}\right) \cdot \mu_1(\alpha \text{ mod } \mathfrak{q}') \cdot \mu_2(\alpha \text{ mod } \mathfrak{q}'')$$

for every α prime to q', with $(o_F/c)^\times$-valued characters μ_1 of $(o/\mathfrak{q}')^\times$ and μ_2 of $(o/\mathfrak{q}'')^\times$. Identify $o/\mathfrak{q}'$, $o/\mathfrak{q}''$, and o_F/c with $Z/q'Z$. Then μ_1 and μ_2 are endomorphisms of $(Z/q'Z)^\times$, so that $\mu_1(x) = x^a$, $\mu_2(y) = y^b$ with integers a and b. Proposition 8.5 implies $x^{2a} y^{2b} = (xy)^{(q'-1)/2} \cdot xy$, so that

$$2a \equiv 2b \equiv 1 + (q' - 1)/2 \text{ mod } (q' - 1) \ .$$

Also, by (iii) of Theorem 2.3, $x = x^{a+b}$, hence $a + b \equiv 1$ mod $(q' - 1)$. Therefore, changing q' for q'' if necessary, we have

$$a \equiv (q' + 1)/4, \ b \equiv (3 - q')/4 \text{ mod } (q' - 1) \ ,$$

which completes the proof.

Remark 8.7. Let u_0 be the fundamental unit of k. Then $u_0 u_0' = -1$. Substituting u_0 for α in the first formula of Proposition 8.5, we have

$$1 = \left(\frac{u_0}{\mathfrak{p}_\infty}\right) \cdot (-1)^{(3-q')/4} \cdot \mu(u_0^{(q'-1)/2} \text{ mod } \mathfrak{q}') \ .$$

In view of Proposition A.II.1 of Appendix II, we see that the primes q' and $\mathfrak{p}_\infty$ are determined by each other by this relation.

PROPOSITION 8.8. *Suppose* $\left(\dfrac{q}{q'}\right) = -1$. *Let* $\mathfrak{q}$ *be the prime ideal in* k *generated by* $\sqrt{q}$. *Then* $\mathfrak{f}[r] = \mathfrak{p}_\infty \cdot q' \cdot \mathfrak{q}$ *with an archimedean prime* $\mathfrak{p}_\infty$. *Moreover one has*

$$r(\alpha o) = \left(\frac{\alpha}{\mathfrak{p}_\infty}\right) \cdot \mu(\alpha \text{ mod } \mathfrak{q}) \cdot \mu'(\alpha \text{ mod } q'o)$$

for every α *in* k *prime to* qq', *where* μ *is the isomorphism of* $o/\mathfrak{q}$ *onto* o_F/c, *and* μ' *is a homomorphism of* $(o/q'o)^\times$ *into* $(o_F/c)^\times$ *such that*

$$\mu'(\alpha \text{ mod } q'o)^2 = \left(\frac{\alpha \alpha'}{q'}\right) \ .$$

The last relation implies $\mu'^4 = 1$. The character r can be completely determined by these formulas and Proposition 8.5 for the same reason as in Proposition 8.6.

Proof. We have $N(\mathfrak{c}) = q$, so that $[k(\chi): k]$ divides $(q - 1)^2$. Since $\left(\dfrac{q}{q'}\right) = -1$, q' does not divide $q - 1$. Therefore, by [9, Lemma 7.32], the finite part of $\mathfrak{f}[r]$ is square-free, so that it is a divisor of $q'\mathfrak{q}$. By Proposition 2.5 and Proposition 8.5, we obtain $\mathfrak{f}[r] = q'\mathfrak{q}\mathfrak{p}_\infty$ with a suitable choice of $\mathfrak{p}_\infty$. Therefore we can write

$$r(\alpha\mathfrak{o}) = \left(\frac{\alpha}{p_\infty}\right) \cdot \mu(\alpha \bmod \mathfrak{q}) \cdot \mu'(\alpha \bmod q'\mathfrak{o})$$

for every α prime to qq', with $(\mathfrak{o}_F/\mathfrak{c})^\times$-valued characters μ of $(\mathfrak{o}/\mathfrak{q})^\times$ and μ' of $(\mathfrak{o}/q'\mathfrak{o})^\times$. Then we obtain the formula for μ'^2 from Proposition 8.5. Now every element of $(\mathfrak{o}/\mathfrak{q})^\times$ can be represented by a positive integer m such that $m \equiv 1$ mod (q'). Substituting m for α, we obtain, by (iii) of Theorem 2.3,

$$\dot{\mu}(m \bmod \mathfrak{q}) = m \bmod \mathfrak{c} \, ,$$

which completes the proof.

Remark 8.9. As is mentioned above, A is a product of copies of an elliptic curve whose endomorphism algebra is isomorphic to S, and θ can be extended to an injection θ' of SK into $\mathrm{End}_Q(A)$. By (1.13), $\zeta(s; A/Q, K)$ is, up to finitely many Euler factors, an L-function $L(s, \chi)$ with a Grössen-character χ of S. As is shown in the proof of Proposition 1.6, χ must coincide with the character λ with which we have defined f_λ and A. Now λ is ramified at q', hence A has bad reduction modulo q'. Since q' is unramified in the extension $S(\sqrt{q}\,)$, A has bad reduction modulo the prime factors of q' in $S(\sqrt{q}\,)$. It follows that B has bad reduction modulo the prime factors of q' in k. This corresponds to the fact that the q'-Euler factor of $L(s, f_\lambda)$ is trivial, which supports the conjecture (2.15).

9. An axiomatic method

Our investigation so far is concerned with the couple (A, θ) obtained from a cusp form. In this section, we shall show that a part of the theory can be developed in a more abstract setting, independent of Hecke operators. Although it is quite possible that this may eventually turn out to be the study of the same objects from a different angle, the author believes that the abstraction can reveal some new aspects of the theory in a wider outlook.

Let F be a totally real algebraic number field of finite degree, and $\mathfrak{R}$ a quaternion algebra over F, which may or may not be a division algebra. We fix a maximal subfield K of $\mathfrak{R}$, an element h of $\mathfrak{R}$, and an element d of F, such that

$$\mathfrak{R} = K + Kh \, ,$$
$$ha = a^\rho h \qquad \text{for all } a \in K \, ,$$
$$h^2 = d \, ,$$

where ρ denotes the generator of $\mathrm{Gal}(K/F)$. Thus $\mathfrak{R}$ is the cyclic algebra determined by (K, ρ, d). We assume that K is a CM-field, and d is a totally positive algebraic integer, so that $\mathfrak{R}$ is totally indefinite. Let $\mathfrak{o}_K$ and $\mathfrak{o}_F$ denote respectively the maximal orders in K and in F. Then $\mathfrak{o}_K + \mathfrak{o}_K h$ is an order in $\mathfrak{R}$, which may or may not be maximal. For a fixed $\mathfrak{R}$ and (K, ρ, d), we consider a couple (A, θ) and a (real or an imaginary) quadratic extension k of $\boldsymbol{Q}$ satisfying the following conditions (9.1–4).

(9.1) *A is an abelian variety of dimension $[K: \boldsymbol{Q}]$.*

(9.2) *θ is an injection of $\mathfrak{R}$ into $\mathrm{End}_0(A)$ which maps the identity element of $\mathfrak{R}$ to the identity map of A.*

(9.3) *A and the elements of $\theta(K) \cap \mathrm{End}(A)$ are rational over $\boldsymbol{Q}$.*

(9.4) *$\theta(h)$ is contained in $\mathrm{End}(A)$ and rational over k, and $\theta(h)^\varepsilon = -\theta(h)$, where ε denotes the generator of $\mathrm{Gal}(k/\boldsymbol{Q})$.*

Changing A by an isogeny over $\boldsymbol{Q}$, we can assume

(9.5) $\theta(\mathfrak{o}_K) \subset \mathrm{End}(A)$.

This can be shown by the same argument as in [9, § 7.7, pp. 198–199]. Therefore hereafter we always assume (9.5). It is also natural to consider the existence of a polarization $\mathcal{C}$ of A satisfying

(9.6) *$\mathcal{C}$ is rational over $\boldsymbol{Q}$; if the map $\lambda \mapsto \lambda^*$ is the involution of $\mathrm{End}_0(A)$ determined by $\mathcal{C}$, then $\theta(a)^* = \theta(a^\rho)$ for all $a \in K$, and $\theta(h)^* = \theta(h)$.*

However, we shall not need this assumption in the following discussion, although our proofs may be simplified under it.

The couple (A, θ) obtained from a cusp form satisfies, after changed by an isogeny over $\boldsymbol{Q}$, the conditions (9.1–5) with $\mathfrak{R} = M_2(\boldsymbol{Q})$, $d = 1$, and $\theta(h) = \eta$. In this case k is real. Moreover, A has a polarization satisfying (9.6) (see [9, § 7.6, pp. 193–194]). In the next section, we shall explain another method of constructing (A, θ) without cusp forms.

Fix (A, θ) satisfying (9.1–5), let p be a rational prime unramified in k, modulo which A has good reduction. Consider reduction modulo a prime factor of p in k. Indicate reduced objects by putting tildes. Let π_p denote the Frobenius endomorphism of $\tilde{A}$ of degree p, and π_p^* its adjoint. Further, let D be the discriminant of k, and ψ the quadratic character of $(\boldsymbol{Z}/D\boldsymbol{Z})^\times$ defined by $\psi(m) = \left(\dfrac{D}{m}\right)$.

THEOREM 9.1. *The notation being as above, one has $\pi_p + \psi(p)\pi_p^* = \tilde{\theta}(a_p)$ with an element a_p of K such that $a_p^\rho = \psi(p)a_p$. Moreover, the Euler p-factor of $\zeta(s; A/\boldsymbol{Q}, K)$ is $1 - a_p p^{-s} + \psi(p)p^{1-2s}$.*

Proof. The first assertion in the case $\psi(p) = 1$ follows immediately from Proposition A.III.6 of Appendix III. Suppose $\psi(p) = -1$. Identify F, K, $\mathcal{R}$ with their images by $\tilde{\theta}$. If $\pi_p = \pi_p^*$, it is sufficient to put $a_p = 0$. Therefore assume $\pi_p \neq \pi_p^*$. Then $\pi_p \notin F$. From (9.4) we see that $\pi_p h = -h\pi_p$. Take an element c of $\mathfrak{o}_K$ such that $c^\rho = -c \neq 0$, and put $b = c \cdot (\pi_p - \pi_p^*)$. Then b commutes with h and the elements of K, so that b belongs to the commutor of $\mathcal{R}$. By Proposition A.III.5, the commutor of $\mathcal{R}$ in $\mathrm{End}_Q(\tilde{A})$ is either a totally imaginary quadratic extension of F or a totally definite quaternion algebra over F. Therefore $F[b]$ is either F or a totally imaginary quadratic extension of F. Now observe that $K[\pi_p]$ is a homomorphic image of $K \otimes_Q Q[\pi_p]$, so that $K[\pi_p]$ is a direct sum of CM-fields. It follows that $K[\pi_p]$ has a unique positive involution, which induces a unique positive involution to any subalgebra of $K[\pi_p]$. Especially it coincides with ρ on K, and maps π_p to π_p^*. Now b is stable under this involution. Since $F[b]$ is either F or a totally imaginary quadratic extension of F, we see that $b \in F$, hence $\pi_p - \pi_p^* \in c^{-1}F$, which completes the proof of the first assertion. Now, from (9.4), we obtain $\pi_p \cdot \tilde{\theta}(h) = \psi(p) \cdot \tilde{\theta}(h) \cdot \pi_p$, so that

$$(9.7) \qquad \psi(p) \cdot \pi_p^* = \tilde{\theta}(h)^{-1} \cdot \pi_p^* \cdot \tilde{\theta}(h) \ .$$

Our second assertion can be proved in the same manner as in the proof of (1.12), by means of (9.7) and the equality $\pi_p + \psi(p)\pi_p^* = \tilde{\theta}(a_p)$.

Define the ideals $\mathfrak{b}_0$, $\mathfrak{b}$, and $\mathfrak{c}$, exactly in the same manner as in §2, for the present F and K. Then we put

$$\mathfrak{x} = \{t \in A \,|\, \theta(\mathfrak{b})t = 0\} \ .$$

By [9, Prop. 7.20], $\mathfrak{x}$ is $\mathfrak{o}_K$-isomorphic to $(\mathfrak{o}_K/\mathfrak{b})^2$. Since $\mathfrak{b}^\rho = \mathfrak{b}$, $\theta(h)$ acts on $\mathfrak{x}$ as an endomorphism. Now let us assume

(9.8) $\quad d \equiv e^2 \bmod \mathfrak{c}$ *for some element e of $\mathfrak{o}_F$ prime to* $\mathfrak{c}$.

(This is of course a condition on F, K, and d.) This is satisfied, for example, if $d = 1$. With such an e, put

$$\mathfrak{y} = \{t \in \mathfrak{x} \,|\, \theta(h - e)t = 0\} \ ,$$
$$\mathfrak{z} = \{t \in \mathfrak{x} \,|\, \theta(h + e)t = 0\} \ .$$

Then the following proposition can easily be verified.

PROPOSITION 9.2. *The submodules $\mathfrak{y}$ and $\mathfrak{z}$ of $\mathfrak{x}$ are $\mathfrak{o}_F$-isomorphic to $\mathfrak{o}_F/\mathfrak{c}$, and $\mathfrak{x} = \mathfrak{y} \oplus \mathfrak{z}$.*

Now let $k(\mathfrak{x})$ denote the smallest extension of k over which the points of $\mathfrak{x}$ are rational. Then $k(\mathfrak{x})$ is a Galois extension of k, and $\mathrm{Gal}(k(\mathfrak{x})/k)$ acts on $\mathfrak{y}$ and on $\mathfrak{z}$. Therefore we can define two $(\mathfrak{o}_F/\mathfrak{c})^\times$-valued characters r_t and s_t for every prime factor $\mathfrak{l}$ of $\mathfrak{c}$, exactly in the same manner as in § 2.

Let p be a rational prime modulo which A has good reduction, and $\mathfrak{p}$ a prime factor of p in k. Consider reduction modulo $\mathfrak{p}$, and indicate reduced objects by putting tildes. Let $\mathfrak{b}_{\mathfrak{l}}$ be the unique prime factor of $\mathfrak{l}$ in K, and R the $\mathfrak{b}_{\mathfrak{l}}$-adic representation of $\mathrm{End}_Q(\tilde{A})$. Further let $\varphi_{\mathfrak{p}}$ denote the Frobenius endomorphism of A of degree $N(\mathfrak{p})$. Suppose p is prime to $\mathfrak{l}$. Then the last assertion of Theorem 9.1 implies that, if $N(\mathfrak{p}) = p$, one has

$$\det\left[X - R(\varphi_{\mathfrak{p}})\right] = X^2 - a_p X + p\,,$$

where X is an indeterminate. Observe that

$$\{t \in \mathfrak{x} \,|\, \theta(\mathfrak{l})t = 0\} = \{t \in A \,|\, \theta(\mathfrak{b}_{\mathfrak{l}})t = 0\}\,.$$

Now the action of $\left(\dfrac{F/k}{\mathfrak{p}}\right)$ on this module coincides, after reduced modulo $\mathfrak{p}$, with the action of $\varphi_{\mathfrak{p}}$. Therefore, identifying $\mathfrak{o}_F/\mathfrak{l}$ with $\mathfrak{o}_K/\mathfrak{b}_{\mathfrak{l}}$, we obtain

$$(9.9) \qquad [X - r_{\mathfrak{l}}(\mathfrak{p})][X - s_{\mathfrak{l}}(\mathfrak{p})] \equiv X^2 - a_p X + p \bmod \mathfrak{l}\,.$$

This is a generalization of (v) of Theorem 2.3. It is now an easy matter to see that all other statements (i-iv) of Theorem 2.3 are true in the present situation; in fact, the original proof is applicable with almost no change. One has to be careful about the property (iii) of Theorem 2.3, since we have started, in this section, with a quadratic field k which is either real or imaginary. Actually we have

PROPOSITION 9.3. *If* $\mathfrak{c} \neq \mathfrak{o}_F$, k *must be real.*

Proof. By means of the same reasoning as in the proof of Theorem 2.3, we first obtain

$(9.10) \quad r_{\mathfrak{l}}(m\mathfrak{o}) = s_{\mathfrak{l}}(m\mathfrak{o}) = (m \bmod \mathfrak{l})$ *for every positive* $m \in \mathbf{Z}$ *prime to* $\mathfrak{f}[r_{\mathfrak{l}}]$.

Suppose k is imaginary. Take a positive integer m so that

$$m \equiv -1 \bmod N(\mathfrak{l}) \cdot N(\mathfrak{f}[r_{\mathfrak{l}}])\,.$$

Then $1 = r_{\mathfrak{l}}(m\mathfrak{o}) = (m \bmod \mathfrak{l}) = -1$, which is a contradiction, since $\mathfrak{l}$ is prime to 2.

Now we see easily that Propositions 2.4 and 2.5 hold in the present case. Furthermore, Theorem 2.8 can be generalized with the following modification: the assumption that "q does not divide N" must be replaced by "A has good reduction modulo q".

10. Construction of (A, θ) from an elliptic curve

The abelian varieties obtained from an element of $S_2^0(N, \psi)$ are examples of (A, θ) satisfying the conditions (9.1–4). One can naturally ask a question: Is there any other way to construct (A, θ)? The purpose of this section is to answer this question affirmatively in the simplest case in which $\dim(A) = 2$, $d = 1$, and $\mathscr{R} = M_2(\boldsymbol{Q})$. In this case, if we put $B = (1 + \theta(h))A$, then A is isogenous to the product $B \times B$. Therefore one can try to obtain A from an elliptic curve with some special properties. To materialize the idea, we consider an elliptic curve E defined over a real quadratic field k satisfying the following condition:

(10.1) *If ε is the generator of $\mathrm{Gal}(k/\boldsymbol{Q})$, then there is an isogeny μ of E onto E^ε rational over k such that $\mu^\varepsilon \circ \mu = -c \cdot \mathrm{id}_E$ with a positive integer c, where id_E denotes the identity element of $\mathrm{End}(E)$.*

With such an E, let f be an isomorphism of $E \times E^\varepsilon$ onto $(E \times E^\varepsilon)^\varepsilon = E^\varepsilon \times E$ defined by $f(x, y) = (y, x)$ for $x \in E$, $y \in E^\varepsilon$. Then $f^\varepsilon \circ f$ is the identity map of $E \times E^\varepsilon$. By Weil [15], there exists an abelian variety A rational over $\boldsymbol{Q}$ and an isomorphism g of $E \times E^\varepsilon$ onto A rational over k such that $g = g^\varepsilon \circ f$. Consider the algebra $\mathscr{R} = K + Kh$ as in §9 with $K = \boldsymbol{Q}(\sqrt{-c})$ and $h^2 = 1$. Now we can define $\theta_0 \colon \mathscr{R} \to \mathrm{End}_Q(E \times E^\varepsilon)$ and $\theta \colon \mathscr{R} \to \mathrm{End}_Q(A)$ as follows:

$$\theta_0(h)(x, y) = (x, -y) \,,$$
$$\theta_0(\sqrt{-c})(x, y) = (\mu^\varepsilon y, \mu x) \qquad\qquad (x \in E, \, y \in E^\varepsilon) \,,$$
$$\theta(a) = g \circ \theta_0(a) \circ g^{-1} \qquad\qquad (a \in \mathscr{R}) \,.$$

It can easily be seen that θ is actually an injective ring-homomorphism, and the elements of $\theta(\mathscr{R}) \cap \mathrm{End}(A)$ are rational over k. Moreover, we see that $\theta_0(\sqrt{-c}) = f^\varepsilon \circ \theta_0(\sqrt{-c})^\varepsilon \circ f$, so that

$$\theta(\sqrt{-c})^\varepsilon = g^\varepsilon \circ \theta_0(\sqrt{-c})^\varepsilon \circ (g^\varepsilon)^{-1} = g \circ f^\varepsilon \circ \theta_0(\sqrt{-c})^\varepsilon \circ f \circ g^{-1} = \theta(\sqrt{-c}) \,.$$

Therefore the elements of $\theta(K) \cap \mathrm{End}(A)$ are rational over $\boldsymbol{Q}$. Further, $\theta_0(h) = -f^\varepsilon \circ \theta_0(h)^\varepsilon \circ f$, so that

$$\theta(h)^\varepsilon = g^\varepsilon \circ \theta_0(h)^\varepsilon \circ (g^\varepsilon)^{-1} = -\theta(h) \,.$$

Thus (A, θ) satisfies (9.1–4) with $\mathscr{R} = K + Kh = M_2(\boldsymbol{Q})$, $d = 1$. Changing (A, θ) by an isogeny over $\boldsymbol{Q}$ if necessary, we obtain a couple satisfying (9.1–5).

Thus our next question is to find an elliptic curve E satisfying (10.1). For that purpose, we first make a simple preliminary observation about the field of rationality for the isogenies of elliptic curves. Let E and E' be two elliptic curves rational over a subfield F of $\boldsymbol{C}$, and λ an isogeny of E onto E', not necessarily rational over F. Let ω and ω' be non-zero holomorphic differential forms on E and on E', respectively, rational over F. Let $\omega' \circ \lambda$ denote the transform of ω' by λ. Then $\omega' \circ \lambda = s\omega$ with a constant $s \in \boldsymbol{C}$. Suppose $\mathrm{End}(E)$ is isomorphic to $\boldsymbol{Z}$. Then

(10.2) *$F(s)$ is the smallest field of definition for λ containing F. Moreover, if $s \notin F$, one has $s^2 \in F$, and $s^\tau = -s$ for the generator τ of $\mathrm{Gal}\big(F(s)/F\big)$.*

In fact, let τ be an automorphism of C over F. Since $\mathrm{End}(E) = \mathbf{Z}$, we have $m\lambda^\tau = n\lambda$ with non-zero integers m and n. Then $m^2 \cdot \deg(\lambda^\tau) = n^2 \cdot \deg(\lambda)$, hence $m = \pm n$, so that $\lambda^\tau = \pm\lambda$. Therefore $\pm s\omega = \omega' {\circ} \lambda^\tau = (\omega' {\circ} \lambda)^\tau = s^\tau \omega$, hence $s^\tau = s$ or $-s$ according as $\lambda^\tau = \lambda$ or $-\lambda$, which proves (10.2).

PROPOSITION 10.1. *Let k be a real quadratic field, E an elliptic curve rational over k, and c a positive integer. Suppose that the following three conditions are satisfied:*

(1) *There exists an isogeny λ of E onto E^ι of degree c (not necessarily rational over k).*

(2) *E has no complex multiplication.*

(3) *$-c \in N_{k/Q}(k^\times)$.*

Then there exists an elliptic curve E' rational over k, isomorphic to E over C, and an isogeny μ of E' onto E'^ι rational over k such that $\mu^\iota {\circ} \mu = -c \cdot \mathrm{id}_{E'}$.

Proof. Let L be the smallest field of definition for λ containing k, and ω a non-zero holomorphic differential form on E rational over k. Put $\omega^\iota {\circ} \lambda = s\omega$ with a constant s. By (10.2), $L = k(s)$, and $s^2 \in k$. Extend ε to an automorphism σ of $\bar{Q}$. By the assumptions (1) and (2), we have $\lambda^\sigma {\circ} \lambda = \pm c \cdot \mathrm{id}_E$, so that $s^\sigma s = \pm c$. Then $N_{k/Q}(-s^2/c) = 1$, hence $-s^2/c = b/b^\iota$ for some $b \in k$. By the assumption (3), $-c = N_{k/Q}(z)$ for some $z \in k$. Let e be a square root of bz. By Weil [15], there exists an elliptic curve E' rational over k, and an isomorphism π of E onto E' rational over $k(e)$ such that $\pi^\tau = -\pi$ if τ is the non-trivial automorphism of $k(e)$ over k. (If $k(e) = k$, we take $E' = E$ and $\pi = \mathrm{id}_E$.) Let ω' be a non-zero holomorphic differential form on E' rational over k. By (10.2), $\omega' {\circ} \pi = fe\omega$ with $f \in k$. Changing ω' for $f^{-1}\omega'$, we may assume $\omega' {\circ} \pi = e\omega$. Put $\mu = \pi^\sigma {\circ} \lambda {\circ} \pi^{-1}$. Then $\omega'^\sigma {\circ} \mu = (se^\sigma/e)\omega'$. Now $(se^\sigma/e)^2 = s^2 b^\sigma z^\sigma / bz = -cz^\sigma/z = (z^\sigma)^2$, hence $se^\sigma/e = \pm z^\sigma \in k$. By (10.2), μ is rational over k. Since $\omega'^\sigma {\circ} \mu = \pm z^\sigma \omega'$, we see that $\omega' {\circ} (\mu^\sigma {\circ} \mu) = N_{k/Q}(z)\omega' = -c\omega'$, hence $\mu^\sigma {\circ} \mu = -c$, which completes the proof.

Remark 10.2. The condition (3) of the above proposition is *necessary* for the existence of E and μ satisfying (10.1). To see this, assume that E and μ satisfy (10.1), and take a holomorphic differential form ω on E rational over k. Put $\omega^\iota {\circ} \mu = t\omega$ with a constant t. Then $t \in k$, $\omega {\circ} \mu^\iota = t^\iota \omega^\iota$, so that $-c\omega = \omega {\circ} \mu^\iota {\circ} \mu = tt^\iota \omega$. Therefore $-c = tt^\iota \in N_{k/Q}(k^\times)$.

Our next task is to find an elliptic curve E satisfying the conditions (1–3) of Proposition 10.1. Consider the standard modular function $J(z)$ normalized by $J(\sqrt{-1}) = 12^3$, $J(e^{2\pi i/3}) = 0$, $J(\infty) = \infty$. Let V be a curve rational over $\boldsymbol{Q}$ which gives a model for the compactification of $\mathfrak{H}/\Gamma_0(c)$. We take the function field of V over $\boldsymbol{Q}$ to be $\boldsymbol{Q}(J(z), J(cz))$. Suppose $c > 1$. From $\begin{bmatrix} 0 & -1 \\ c & 0 \end{bmatrix}$, we obtain an automorphism α of V rational over $\boldsymbol{Q}$, of order 2. Let V' be the quotient of V by α. Obviously α interchanges $J(z)$ and $J(cz)$. Now take a point x on V' rational over $\boldsymbol{Q}$ and consider the points y and $\alpha(y)$ on V lying on x. In general, y is rational over a quadratic field k, and $\alpha(y) = y^\varepsilon$ with the generator ε of $\mathrm{Gal}(k/\boldsymbol{Q})$. If z_0 is a corresponding point on $\mathfrak{H}$, we see that both $J(z_0)$ and $J(cz_0)$ are rational over k, and $J(z_0)^\varepsilon = J(cz_0)$. Take an elliptic curve E rational over k with the invariant $J(z_0)$. Then E^ε has the invariant $J(cz_0)$, so that there is an isogeny of E onto E^ε of degree c. If the conditions (2) and (3) are satisfied, we obtain, by Proposition 10.1, E and μ satisfying (10.1). As to the condition (2), we note that there are only finitely many points z on $\mathfrak{H}$ modulo $\Gamma(1)$ for which the corresponding elliptic curve has complex multiplication, and $J(z)$ is quadratic over $\boldsymbol{Q}$. Therefore, if V' is of genus 0, we can assure the condition (2) by excluding finitely many points on V'. We shall later show another method of verifying (2).

The procedure of finding $J(z_0)$ is practicable at least when c is small. For example, let us consider the case $c = 5$. Put

$$\Delta(z) = e^{2\pi i z} \cdot \prod_{n=1}^{\infty} (1 - e^{2\pi i n z})^{24} ,$$
$$g(z) = 5^3 \cdot [\Delta(5z)/\Delta(z)]^{1/4}$$
$$= 5^3 \cdot e^{2\pi i z} \cdot \prod_{n=1}^{\infty} (1 - e^{2\pi i 5 n z})^6 (1 - e^{2\pi i n z})^{-6} .$$

It can easily be seen that $\boldsymbol{Q}(J(z), J(5z)) = \boldsymbol{Q}(g(z))$ and $g(z) \cdot g(-1/5z) = 5^3$. (One may write $g \circ \alpha$ for $g(-1/5z)$ with the above automorphism α.) Moreover, one has

$$(10.3) \qquad J = (g^2 + 10g + 5)^3/g, \quad J - 12^3 = (g^2 + 22g + 5^3)(g^2 + 4g - 1)^2/g$$

(see Klein-Fricke [4, p. 61, (11)]). Put $s(z) = g(z) + g(-1/5z)$. Then

$$(10.4) \qquad\qquad\qquad g^2 - sg + 5^3 = 0 ,$$

and s generates the function field of V' over $\boldsymbol{Q}$. Therefore, solving the quadratic equation (10.4) with any rational value of s, and substituting the value of g into (10.3), we obtain the value $J(z_0)$.

For instance, let M be a rational prime such that

$$(10.5) \qquad\qquad\qquad M \equiv 1 \bmod (4) , \quad \left(\frac{M}{5}\right) = 1 .$$

Then there are infinitely many rational solutions (u, v) of the Diophantine equation

$$(10.6) \qquad 4 \cdot 5^{h+2} = u^2 - v^2 M ,$$

where h is the class number of $Q(\sqrt{M})$. For any solution (u, v), put $g_0 = 5^{(1-h)/2}(u + v\sqrt{M})/2$, $s_0 = 5^{(1-h)/2}u$. (Note that h is always odd.) Then $(g, s) = (g_0, s_0)$ is a set of solutions of (10.4). From the second formula of (10.3), we obtain

$$(10.7) \qquad J(z_0) = (22 + s_0)(g_0^2 + 4g_0 + 1)^2 + 12^3 ,$$

so that $J(z_0)$ is an algebraic integer if $s_0 \in Z$. In this case, $k = Q(\sqrt{M})$.

Let us now show that an elliptic curve E with the invariant $J(z_0)$ has no complex multiplication. Assume, on the contrary, that $\mathrm{End}(E)$ is isomorphic to an order o in an imaginary quadratic field S. Then the class number of o is 2, and $S(\sqrt{M})$ is the "Ringklassenkörper" of o. If S is of class number one, the conductor L of o must be a multiple of M, and the class number of o is

$$(10.8) \qquad [o_S^\times : o^\times]^{-1} L \cdot \prod_{p|L}\left(1 - \left(\frac{S}{p}\right)p^{-1}\right) ,$$

where o_S is the maximal order of S, and $\left(\dfrac{S}{p}\right) = 0, 1,$ or -1 according as a prime p is ramified, decomposes, or remains prime in S. Since $M \geq 29$, the number (10.8) cannot be 2. Therefore the class number of S is 2, $S(\sqrt{M})$ is unramified over S, and the conductor of o is 1 or 2. Then $S = Q(\sqrt{-nM})$ with a square-free positive integer n prime to M. If n is divisible by 5, the discriminant of S has at least three distinct prime factors, so that the class number of S is divisible by 4, a contradiction. Therefore n is prime to 5. The existence of μ implies the existence of a non-principal integral o-ideal a in S such that $N(a) = 5$. Put $a^2 = wo$ with $w \in o$, and $w = (x + y\sqrt{-nM})/2$ with rational integers x and y. Since 5 is unramified in S, we see that $w \notin Q$, hence $y \neq 0$. Now we have $100 = x^2 + y^2 nM$. But it can easily be verified that there are no such x and y. This proves that E has no complex multiplication.

As for the condition (3) of Proposition 10.1, the assumption (10.5) implies that $-5 \in N_{k/Q}(k^\times)$. Thus, for each solution (u, v) of (10.6), we obtain an elliptic curve satisfying the conditions of Proposition 10.1, and hence E and μ satisfying (10.1) with $k = Q(\sqrt{M})$. Then we obtain a two-dimensional (A, θ) satisfying (9.1-5) with $\mathfrak{R} = M_2(Q)$, $d = 1$, and $K = Q(\sqrt{-5})$. In this case $\mathfrak{c} = 5Z$, so that one can define $(Z/5Z)$-valued characters r and s. Define a_p as in Theorem 9.1, and suppose that E has good reduction modulo a common prime factor $\mathfrak{p}$ of 5 and g_0 in k. Then, from (10.7), we obtain $J(z_0) \equiv s_0 \bmod \mathfrak{p}$. It can easily be seen that $a_5 = 0$ if and only if $s_0 \equiv 0 \bmod (5)$. Therefore, if s_0 is not divisible by 5, and E has good reduction modulo $\mathfrak{p}$, then we can apply

Theorem 2.8 to the present (A, θ), as is mentioned at the end of §9. Thus, if an explicit form of E is given, it is possible to determine the character r by means of Theorem 2.3 and Theorem 2.8. One can of course compute a_p for a few p, and use them for the determination of r. For example, one obtains elliptic curves for the values $g_0 = (\pm 23 + \sqrt{29})/2$ and $g_0 = (\pm 27 + \sqrt{229})/2$ with $M = 29$ and $M = 229$. Are these elliptic curves isogenous to those obtained in §3 in the cases $N = 29$ and $N = 229$? More generally, one can ask whether an elliptic curve satisfying (10.1) is always obtained from an element of $S_2^0(N, \psi)$ with suitable N and ψ. Anyhow a further investigation of elliptic curves of this type, or more generally, of abelian varieties of §8, will lead to results of considerable interest.

Remark 10.3. One can show that the function

$$G(z) = [\Delta(cz)/\Delta(z)]^{1/m}$$

is a modular function invariant under $\Gamma_0(c)$ belonging to $Q(J(z), J(cz))$, if $c \equiv 1 \bmod (m)$ and m is one of the integers 1, 2, 3, 4, 6, or 12. Moreover, $G(z)G(-1/cz) = c^{-12/m}$. Therefore, taking m to be 4, we see that, if $c \equiv 1 \bmod (4)$, then $c \in N_{k/Q}(k^\times)$ for any real quadratic field k obtained as above. In such a case, the condition (3) of Proposition 10.1 is satisfied if and only if $-1 \in N_{k/Q}(k^\times)$. It can easily be seen that $-1 \in N_{k/Q}(k^\times)$ if and only if k has an integral ideal whose square is a principal ideal generated by an element with negative norm.

Appendix I

PROOF OF PROPOSITION 1.2

Fix an integer $\kappa \geq 2$ and a character ψ of $(Z/NZ)^\times$, and denote by Ψ the set of all characters of the form ψ^σ with $\sigma \in \mathfrak{F}$. Let $S_{N,\Psi}$ (resp. $S_{N,\Psi}^0$, $S_{N,\Psi}^1$) denote the subspace of $S_\kappa(\Gamma_1(N))$ generated by $S_\kappa(N, \chi)$ (resp. $S_\kappa^0(N, \chi)$, $S_\kappa^1(N, \chi)$) for all $\chi \in \Psi$, and $S_{N,\Psi}^\perp$ the orthogonal complement of $S_{N,\Psi}$ in $S_\kappa(\Gamma_1(N))$ with respect to the Petersson inner product. Let $\mathfrak{A}$ (resp. $\mathfrak{A}_c$) be the algebra generated over Q (resp. C) by the $T(n)_N$ for all n. By [9, Theorem 3.51], we have $\mathfrak{A}_c = \mathfrak{A} \otimes_Q C$. We first prove

PROPOSITION A.I.1. *Let $\mathfrak{B}$ (resp. $\mathfrak{B}_c$) be the algebra generated over Q (resp. C) by the restriction of the elements of $\mathfrak{A}$ to $S_{N,\Psi}^0$. Then there is a $\mathfrak{B}$-stable submodule W of $S_{N,\Psi}^0$ such that $S_{N,\Psi}^0 = W \otimes_Q C$. Moreover $\mathfrak{B}_c = \mathfrak{B} \otimes_Q C$.*

Proof. Let Y be a set of representatives for $(Z/NZ)^\times$. We can find algebraic numbers c_n such that

$$\sum_{n \in Y} c_n \cdot \chi(n) = \begin{cases} 1 & \text{if } \chi \in \Psi, \\ 0 & \text{if } \chi \notin \Psi, \end{cases}$$

for characters χ of $(\mathbf{Z}/N\mathbf{Z})^{\times}$. Transforming this by an automorphism of the algebraic closure of $\mathbf{Q}$, we see easily that $c_n \in \mathbf{Q}$. For each $n \in Y$, let γ_n be an element of $SL_2(\mathbf{Z})$ such that

$$n \cdot \gamma_n \equiv \begin{bmatrix} 1 & 0 \\ 0 & n^2 \end{bmatrix} \mod (N) \ .$$

Then the action of $[\gamma_n]_\kappa$ defines an element of $\mathfrak{A}$. (The operator $[\gamma_n]_\kappa$ coincides with $T'(n, n)_{N,\chi}$ of [9, § 3.5] on $S_\kappa(N, \chi)$ up to a constant factor $n^{\kappa-2}$.) Since $f \,|\, [\gamma_n]_\kappa = \chi(n)f$ for $f \in S_\kappa(N, \chi)$, if we put

$$(1) \qquad\qquad P = \sum_{n \in Y} c_n \cdot [\gamma_n]_\kappa \ ,$$

then $P \in \mathfrak{A}$, and P is the projection map of $S_\kappa(\Gamma_1(N))$ to $S_{N,\Psi}$ with respect to the decomposition $S_\kappa(\Gamma_1(N)) = S_{N,\Psi} \oplus S_{N,\Psi}^{\perp}$. In [7] (cf. also [9, § 8.4]), we have constructed a lattice L_M in $S_\kappa(\Gamma_1(M))$ with the following two properties:

(2) *If $\alpha \in M_2(\mathbf{Z})$ and $\det(\alpha) > 0$, then $[\Gamma_1(M)\alpha\Gamma_1(N)]_\kappa$ maps L_M into L_N.* (For the notation $[\Gamma_1(M)\alpha\Gamma_1(N)]_\kappa$, see [9, § 3.4].)

(3) *If (f, g), for $f, g \in S_\kappa(\Gamma_1(M))$, denotes the Petersson inner product, then* $\mathrm{Re}\,((f, i^{\kappa-1}g))$ *is rational for $f, g \in L_M$.*

(This was shown in [7] for even κ, but the argument is valid even in the case of odd κ.) Now let U_M be the $\mathbf{Q}$-linear span of L_M. Then $S_\kappa(\Gamma_1(M)) = U_M \otimes_{\mathbf{Q}} \mathbf{C}$. For each divisor $M\ (<N)$ of N modulo which ψ can be defined, we consider the projection map P_M of $S_\kappa(\Gamma_1(M))$ to $S_{M,\Psi}$ defined by (1) with M in place of N. For every positive divisor m of N/M, put $\alpha_m = \begin{bmatrix} m & 0 \\ 0 & 1 \end{bmatrix}$, and consider the image of $P_M(U_M)$ by $[\Gamma_1(M)\alpha_m\Gamma_1(N)]_\kappa$. Let V denote the sum of these images for all possible M and m. Then V is a vector subspace of U_N, and $V \otimes_{\mathbf{Q}} \mathbf{C} = S_{N,\Psi}^{\perp}$. Let $W = U_N \cap S_{N,\Psi}^0$. In view of (3), we see that W is the orthogonal complement of V in U_N with respect to a non-degenerate $\mathbf{Q}$-bilinear form. Therefore $S_{N,\Psi}^0 = W \otimes_{\mathbf{Q}} \mathbf{C}$. Obviously W is $\mathfrak{B}$-stable, and $\mathfrak{B}_c = \mathfrak{B} \otimes_{\mathbf{Q}} \mathbf{C}$.

After this preparation, let us now prove Proposition 1.2. Since $\mathfrak{B}_c$ is commutative and semi-simple by Lemma 1.1, $\mathfrak{B}$ is also commutative and semi-simple. Let K and f be as in Proposition 1.2. Then we obtain a homomorphism h of $\mathfrak{B}$ onto K by $f \,|\, b = h(b) \cdot f$ for $b \in \mathfrak{B}$. Since $S_{N,\Psi}^0 = W \otimes_{\mathbf{Q}} \mathbf{C}$, and W is a $\mathfrak{B}$-module, there is an element g of $S_{N,\Psi}^0$, for every $\sigma \in \mathfrak{S}$, such that $g \,|\, b = h(b)^\sigma \cdot g$ for all $b \in \mathfrak{B}$. Then $g \,|\, T(n)_N = a_n^\sigma \cdot g$. By [9, Prop. 3.53], $g \in S_\kappa(N, \psi')$ for some ψ'. Obviously $\psi' = \psi^\sigma$. Replacing g by its suitable constant multiple, we obtain $g(z) = \sum_{n=1}^{\infty} a_n^\sigma e^{2\pi i n z}$. This completes the proof.

Appendix II

We prove here some elementary congruences for the units in real quadratic fields. Most of them are probably well-known, but we have been unable to find appropriate references. We denote by k a real quadratic field, by $\mathfrak{o}$ the ring of algebraic integers in k, by u the fundamental unit of k, and by ε the generator of $\mathrm{Gal}(k/\mathbf{Q})$. As usual we normalize u so that $u > 1$.

PROPOSITION A.II.1. *Let v be an element of $\mathfrak{o}^{\times}$ such that $N_{k/\mathbf{Q}}(v) = -1$, and p an odd rational prime which divides $N_{k/\mathbf{Q}}(v-1)$. Then p decomposes into two prime ideals in k. Moreover, $v - 1$ is divisible by only one of the two prime factors of p in k.*

Proof. Let $\mathfrak{p}$ be a prime ideal in k dividing p. Assume that $\mathfrak{p}^{\varepsilon} = \mathfrak{p}$. Then $\mathfrak{p}$ divides both $v - 1$ and $v^{\varepsilon} - 1$. Since $v + 1 = v(1 - v^{\varepsilon})$, we have $2 = v + 1 - (v - 1) \equiv 0 \bmod \mathfrak{p}$, a contradiction. Therefore $\mathfrak{p} \neq \mathfrak{p}^{\varepsilon}$. Assume that both $\mathfrak{p}$ and $\mathfrak{p}^{\varepsilon}$ divides $v - 1$. Again we see that $\mathfrak{p}$ divides both $v - 1$ and $v^{\varepsilon} - 1$, which leads to a contradiction.

PROPOSITION A.II.2. *Let $k = \mathbf{Q}(\sqrt{N})$ with a positive square-free integer $N \equiv 1 \bmod (4)$. Suppose $N_{k/\mathbf{Q}}(u) = -1$. Then $N \equiv 1 \bmod (8)$ if and only if $Tr_{k/\mathbf{Q}}(u) \equiv 0 \bmod (8)$. Moveover, if $N \equiv 1 \bmod (8)$, one has $u = x + y\sqrt{N}$ with rational integers x and y such that $x \equiv 0 \bmod (4)$, $y \equiv 1 \bmod (2)$. If $N \equiv 5 \bmod (8)$, $Tr_{k/\mathbf{Q}}(u)$ is either odd or divisible by 4.*

Proof. Put $u = (a + b\sqrt{N})/2$ with integers a and b. Then a and b are either both even or both odd. Suppose $N \equiv 1 \bmod (8)$. Then $-4 \equiv a^2 - b^2 \bmod (8)$. Therefore a and b must be both even, so that $u = x + y\sqrt{N}$ with integers x and y. Since $-1 \equiv x^2 - y^2 \bmod (8)$, we see that x is even and y is odd, hence $x^2 \equiv 0 \bmod (8)$. It follows that x is divisible by 4, hence $Tr_{k/\mathbf{Q}}(u) = 2x \equiv 0 \bmod (8)$. Conversely, if $Tr_{k/\mathbf{Q}}(u) \equiv 0 \bmod (8)$, we have $u = x + y\sqrt{N}$ with integers x and y such that $x \equiv 0 \bmod (4)$ and $y \equiv 1 \bmod (2)$. Then $-1 = x^2 - y^2 N \equiv -N \bmod (8)$, hence $N \equiv 1 \bmod (8)$. Next suppose $N \equiv 5 \bmod (8)$ and $Tr_{k/\mathbf{Q}}(u)$ is even. Then we have $u = x + y\sqrt{N}$ with integers x and y. Since $-1 \equiv x^2 + 3y^2 \bmod (8)$, x must be even. This completes the proof.

PROPOSITION A.II.3. *Let $k = Q(\sqrt{N})$ with a positive square-free integer $N \equiv 1 \bmod (4)$. Suppose $N_{k/\mathbf{Q}}(u) = -1$. Then $N \equiv 1 \bmod (3)$ if and only if $Tr_{k/\mathbf{Q}}(u) \equiv 0 \bmod (3)$.*

Proof. Put $u = (a + b\sqrt{N})/2$ with integers a and b. Then $-4 = a^2 - b^2 N$. Obviously $b \not\equiv 0 \bmod (3)$, so that $b^2 \equiv 1 \bmod (3)$, hence $-1 \equiv a^2 - N \bmod (3)$. Therefore $N \equiv 1 \bmod (3)$ if and only if $a \equiv 0 \bmod (3)$.

PROPOSITION A.II.4. *Let $k = \mathbf{Q}(\sqrt{pq})$ with two rational primes p and q such that $p \equiv q \equiv 3 \bmod (4)$, and let $\mathfrak{p}$ and $\mathfrak{q}$ be the prime ideals in k such that*

$\mathfrak{p}^2 = p\mathfrak{o}$, $\mathfrak{q}^2 = q\mathfrak{o}$. *Suppose* $\left(\dfrac{q}{p}\right) = 1$. *Then* $u \equiv 1 \bmod \mathfrak{p}$, $u \equiv -1 \bmod \mathfrak{q}$. *Moreover,* $\mathfrak{p} = \alpha\mathfrak{o}$, $\mathfrak{q} = \beta\mathfrak{o}$ *with elements* α, β *of* $\mathfrak{o}$ *such that* $N_{k/Q}(\alpha) < 0$, $N_{k/Q}(\beta) > 0$.

Proof. Put $K = k(\sqrt{-q})$. Then every prime ideal in k is unramified in K. Since $\left(\dfrac{K/k}{\mathfrak{p}}\right) = \left(\dfrac{-q}{p}\right) = -1$, $\mathfrak{p}$ cannot be generated by any totally positive element of $\mathfrak{o}$. Let $\mathfrak{a} = (u - 1)\mathfrak{o}$. Since $N_{k/Q}(u) = 1$, we have $u - 1 = -u(u - 1)^{\iota}$, so that $\mathfrak{a} = \mathfrak{a}^{\iota}$, hence $\mathfrak{a} = m \cdot \mathfrak{p}^{e}\mathfrak{q}^{f}$ with a rational integer m and exponents e, f which are 0 or 1. If $\mathfrak{a} = m\mathfrak{o}$ or $\mathfrak{a} = m\mathfrak{p}\mathfrak{q} = m\sqrt{pq} \cdot \mathfrak{o}$, then $u - 1$ is mv or $m\sqrt{pq} \cdot v$ with a unit v. Then $u = -(u - 1)/(u - 1)^{\iota} = \pm v/v^{\iota} = \pm v^2$, a contradiction. Theorefore $\mathfrak{a} = m\mathfrak{p}$ or $m\mathfrak{q}$. If $\mathfrak{a} = m\mathfrak{q}$, we have $\mathfrak{p} = (m\sqrt{pq}/(u - 1))\mathfrak{o}$. This is a contradiction, since $\sqrt{pq}/(u - 1)$ is totally positive. Therefore $\mathfrak{a} = m\mathfrak{p}$, hence $u \equiv 1 \bmod \mathfrak{p}$, $\mathfrak{p} = \alpha\mathfrak{o}$ with $\alpha = (u - 1)/m$. Then $N_{k/Q}(\alpha) < 0$. Taking $u + 1$ instead of $u - 1$, we find, by a similar argument, that $(u + 1)\mathfrak{o} = m'\mathfrak{q}$ with an integer m', which completes the proof.

PROPOSITION A.II.5. *Let* k *be either* $Q(\sqrt{p})$ *or* $Q(\sqrt{2p})$ *with a rational prime* $p \equiv 3 \bmod (4)$, *and let* $\mathfrak{p}$ *and* $\mathfrak{r}$ *be the prime ideals in* k *such that* $\mathfrak{p}^2 = p\mathfrak{o}$ *and* $\mathfrak{r}^2 = 2\mathfrak{o}$. *Then* $u \equiv \left(\dfrac{2}{p}\right) \bmod \mathfrak{p}$, *and* $\mathfrak{r} = \xi\mathfrak{o}$ *with an element* ξ *of* $\mathfrak{o}$ *such that* $\left(\dfrac{2}{p}\right) \cdot N_{k/Q}(\xi) > 0$.

The proof can be given by the same method as in the preceding proposition, because the class number of k is odd.

Appendix III

The purpose of this part is to prove a few propositions concerning the commutor algebra of a subalgebra in the endomorphism algebra of an abelian variety. We denote by p the characteristic of the universal domain, which may be 0 or positive. We fix a totally real algebraic number field F of degree g, and an algebra U belonging to either of the following two types:

(I) $U = F$;

(II) a totally indefinite quaternion algebra over F, i.e., an algebra U over F such that $U \otimes_Q R$ is isomorphic to the direct sum of g copies of $M_2(R)$. (U is not necessarily a division algebra.)

Our discussion will often be divided according to these two types. Put $[U : F] = d^2$. Then $d = 1$ or 2 in (Case I) or (Case II), respectively. Now our object of study is an abelian variety A of dimension dg such that $\mathrm{End}_Q(A)$ contains U as a subalgebra with the same identity element. We put, for simplicity, $E = \mathrm{End}_Q(A)$, and denote by W the commutor of U in E, i.e.,

$$W = \{x \in E \,|\, xy = yx \text{ for all } y \in U\} \,.$$

LEMMA A.III.1. *Let P be an algebraic number field of degree $2m$ isomorphic to a subfield of $\mathrm{End}_0(Y)$ with an abelian variety Y of dimension m. Then P is totally imaginary.*

The proof is given in [11, Lemma 5].

PROPOSITION A.III.2. *W is a division algebra, and E is a simple algebra.*

Proof. If $0 \neq \gamma \in W \cap \mathrm{End}(A)$, $\mathrm{End}_0(\gamma(A))$ contains an isomorphic image of U. By [13, p. 39, Prop. 2], $\dim(\gamma(A))$ is divisible by $dg/2$. If $\gamma(A) \neq A$, we have $\dim(\gamma(A)) = dg/2$. Since F is totally real, this is impossible if $d = 1$, in view of Lemma A.III.1. If $d = 2$, we can find a totally real subfield of U of degree $2g$ and apply the same argument to it. Therefore we have $\gamma(A) = A$, so that W is a division algebra. The center of E, being contained in W, must be a field. Hence E is simple.

Let UW denote the subalgebra of E generated by U and W. The above proposition shows that $UW = U \otimes_F W$. For our purpose, it is convenient to state a well-known theorem of ring-theory:

LEMMA A.III.3. *Let R be a central simple algebra over a field F, S a simple subalgebra of R containing F, and T the commutor of S in R. Then:*

(i) *T is simple, and $S \cap T$ is the common center of S and T.*

(ii) *S is the commutor of T in R.*

(iii) *$[R:F] = [S:F][T:F]$.*

(iv) *If S^{-1} denotes the algebra anti-isomorphic to S, then $R \otimes_F S^{-1}$ and T belong to the same Brauer algebra class over $S \cap T$.*

Let $\mathfrak{q}$ be a prime ideal in F which does not divide p if $p > 0$. We denote by $F_\mathfrak{q}$ the $\mathfrak{q}$-completion of F, and put $S_\mathfrak{q} = S \otimes_F F_\mathfrak{q}$ for any algebra S over F. Then we can define the $\mathfrak{q}$-adic representation

$$\mathfrak{R}_\mathfrak{q} : (UW)_\mathfrak{q} \longrightarrow M_{2d}(F_\mathfrak{q}) \,,$$

which is an $F_\mathfrak{q}$-linear injection equivalent to a multiple of a reduced representation of $(UW)_\mathfrak{q}$ over $F_\mathfrak{q}$ (see [9, §7.6, Prop. 7.22]).

PROPOSITION A.III.4. *The degree $[W:F]$ is either 1 or 2 or 4. According to the degree, the following assertions hold:*

(1) *If $W = F$, the commutor of F in E is U.*

(2) *Suppose $[W:F] = 2$. Then W is a totally imaginary field; the commutor of W in E is UW; if $d = 2$, $W_\mathfrak{q}$ is $E_\mathfrak{q}$-linearly embeddable in $U_\mathfrak{q}$ for every prime ideal $\mathfrak{q}$ in F, not dividing p when $p > 0$.*

(3) *Suppose $[W:F] = 4$. Then $p > 0$; W is a totally definite quaternion*

algebra over F; the commutor of W in E is U; W_q is F_q-isomorphic to $M_2(F_q)$ or U_q according as $d = 1$ or 2, for every prime ideal q in F, not dividing p when $p > 0$.

Proof. Let C be the center of E. If $W = F$, we have $C \subset F$, hence (1) follows from (ii) of the above lemma. Suppose $W \neq F$. Since $\mathfrak{R}_q$ maps $(UW)_q$ into $M_{2d}(F_q)$, W is either a quadratic extension of F, or a quaternion algebra. If $d = 2$, W_q is mapped into the commutor of the image of U_q in $M_4(F_q)$, which is isomorphic to U_q. Let F' be either F or a totally real maximal subfield of U, according as $d = 1$ or 2. Suppose $[W : F] = 2$. We can take F' so that $W \otimes_F F'$ is a field. By Lemma A.III.1, $W \otimes_F F'$ is totally imaginary, hence W is totally imaginary. Obviously $C \subset W$, and W is the commutor of UW in E. By (ii) of Lemma A.III.3, UW is the commutor of W in E. Next suppose $[W : F] = 4$. For the same reason as above, every maximal subfield of W is totally imaginary, hence W is totally definite. We have $C \subset W$, hence $C \subset F$. By (ii) of Lemma A.III.3, U is the commutor of W in E. If $p = 0$, E has a faithful rational representation of degree $2dg$, which is impossible if $[W : F] = 4$. In fact, if $[W : F] = 4$, $UW = M_d(H)$ with a division quaternion algebra H over F. This completes the proof.

PROPOSITION A.III.5. *Suppose A is defined over a finite field. Then $[W : F] > 1$, and the following statements hold:*

(1) *If $[W : F] = 2$, the center of E is a totally imaginary subfield of W.*

(2) *If $[W : F] = 4$, then A is isogenous to the product of dg copies of a super-singular elliptic curve B, and $\mathrm{End}_0(B) \otimes_Q U$ is isomorphic to $M_d(W)$. (Therefore W is uniquely determined, up to isomorphisms, by U and p.)*

Proof. Let C be the center of E. Then C is either totally real or totally imaginary. Take a positive integer m so that A and the elements of $\mathrm{End}(A)$ are rational over a finite field with p^m elements. Let π denote the Frobenius endomorphism of A of degree p^m. Then $\pi \in C$. Suppose C is totally real. Then $\pi^2 = p^m$. By Tate [14], A is isogenous to the product of dg copies of a super-singular elliptic curve B, hence E can be identified with $M_{dg}(S)$, where $S = \mathrm{End}_0(B)$. By (iii) of Lemma A.III.3, $4d^2g^2 = [M_{dg}(S) : Q] = [U : Q][W : Q] = d^2g^2[W : F]$, so that $[W : F] = 4$. Moreover, by (iv) of Lemma A.III.3, $S \otimes_Q U$ and W belong to the same Brauer class. If $[W : F] \neq 4$, C must be totally imaginary. Since $C \subset W$, this completes the proof.

PROPOSITION A.III.6. *Suppose that A and the elements of $U \cap \mathrm{End}(A)$ are rational over a finite field with p^m elements. Let π be the Frobenius endomorphism of A of degree p^m, and π^* the element of $\mathrm{End}(A)$ such that*

$\pi\pi^* = p^m$. *Then* $\pi \in W$, $\pi + \pi^* \in F$, *and* $\left(X^2 - (\pi + \pi^*)X + p^m\right)^d$ *is the characteristic polynomial of* $\mathfrak{R}_q(\pi)$.

Proof. The assertion $\pi \in W$ is obvious. Since F is totally real, if $\pi \in F$, we have $\pi = \pi^*$ and $\mathfrak{R}_q(\pi) = \pi \cdot 1_{2d}$, so that our assertions are obvious. Suppose $\pi \notin F$. From the above propositions, we see that $F(\pi)$ and $Q(\pi)$ are totally imaginary subfields of W. Moreover, $\pi \mapsto \pi^*$ defines a positive involution of $Q(\pi)$. Then $\pi + \pi^*$ must be totally real, so that $\pi + \pi^* \in F$, since F is the maximal real subfield of $F(\pi)$. Obviously $X^2 - (\pi + \pi^*)X + p^m$ is an irreducible polynomial over F with π as a root, hence it is the minimum polynomial of $\mathfrak{R}_q(\pi)$. This proves the last assertion.

PRINCETON UNIVERSITY, PRINCETON, N. J.

REFERENCES

[1] M. EICHLER, *Über die Darstellbarkeit von Modulformen durch Thetareihen*, J. reine angew. Math. **195** (1956), 156-171.

[2] ———, *Quadratische Formen und Modulfunktionen*, Acta Arithmetica **4** (1958), 217-239.

[3] E. HECKE, *Über Modulfunktionen und die Dirichletschen Reihen mit Eulerscher Produktentwicklung* I, II, Math. Ann. **114** (1937), 1-28, 316-351 (= Math. Werke, 644-707).

[4] F. KLEIN and R. FRICKE, Vorlesungen uber die Theorie der Elliptischen Modulfunctionen, II, Leipzig, 1892.

[5] T. MIYAKE, *On automorphic forms on GL_2 and Hecke operators*, Ann. of Math. **94** (1971), 174-189.

[6] L. RÉDEI and H. REICHARDT, *Die Anzahl der durch 4 teilbaren Invarianten der Klassengruppe eines beliebigen quadratischen Zahlkorpers*, J. reine angew. Math. **170** (1934), 69-74.

[7] G. SHIMURA, *Sur les intégrales attachées aux formes automorphes*, J. Math. Soc. Japan **11** (1959), 291-311.

[8] ———, *On canonical models of arithmetic quotients of bounded symmetric domains*, I, II, Ann. of Math. **91** (1970), 144-222, **92** (1970), 528-549.

[9] ———, Introduction to the arithmetic theory of automorphic functions, Publ. Math. Soc. Japan, No. 11, 1971.

[10] ———, *Class fields over real quadratic fields in the theory of modular functions*, in Several Complex Variables II, Lecture Notes in Math. **185** (1971), 169-188.

[11] ———, *On the zeta-function of an abelian variety with complex multiplication*, Ann. of Math. **94** (1971), 504-533.

[12] ———, *On elliptic curves with complex multiplication as factors of the jacobians of modular function fields*, Nagoya Math. J. **43** (1971).

[13] G. SHIMURA and Y. TANIYAMA, Complex multiplication of abelian varieties and its applications to number theory, publ. Math. Soc. Japan, No. 6, 1961.

[14] J. TATE, *Endomorphisms of abelian varieties over finite fields*, Inventiones Math. **2** (1966), 134-144.

[15] A. WEIL, *The field of definition of a variety*, Amer. J. Math. **78** (1956), 509-524.

(Received July, 2, 1971)

On modular forms of half integral weight

Annals of the Mathematics, 97 (1973), 440-481

Introduction

The recent development of the theory of modular forms and associated zeta functions, together with all its arithmetic significance, is quite pleasing, and our knowledge in this field is evergrowing, but the forms of half integral weight have attracted only casual attention, in spite of their importance and ancientness. Indeed, the connection of such forms with zeta functions was never clarified. When Hecke developed his theory of Euler product for the forms of integral weight, he pointed out the impossibility of a similar theory for the forms of half integral weight, and that only partial information could be obtained for the Fourier coefficients of such forms (Werke, p. 639). He explained this in more detail in his last paper [3], which might have given a rather negative and somewhat misleading impression that one would not be able to do much except in some special cases. A treatment of a more general type of modular form was given by Wohlfahrt [12]. In fact, he defined Hecke operators whose degree is the square of a prime, and showed a certain multiplicative relation, as predicted by Hecke, for the Fourier coefficients, but discussed neither Euler product, nor connection with zeta functions.

In the present paper, we try to reveal a more affirmative aspect of the subject. To be specific, put, for each positive integer N,

$$\Gamma_0(N) = \left\{ \begin{bmatrix} a & b \\ c & d \end{bmatrix} \in SL_2(\mathbf{Z}) \mid c \equiv 0 \ (\mathrm{mod}\ N) \right\},$$

and define an automorphic factor $j(\gamma, z)$ by

$$j(\gamma, z) = \theta\big(\gamma(z)\big)/\theta(z) \quad \text{for} \quad \gamma \in \Gamma_0(4),$$
$$\theta(z) = \sum_{n=-\infty}^{\infty} \exp\left(2\pi i n^2 z\right),$$

where z is the variable on the upper half plane. It is known that

$$j(\gamma, z)^2 = \left(\frac{-1}{d}\right)(cz + d) \quad \text{if} \quad \gamma = \begin{bmatrix} a & b \\ c & d \end{bmatrix}.$$

Let κ be an odd positive integer, N a positive integer divisible by 4, and χ

---

* During the preparation of this paper, the author was partially supported by NSF Grant GP-29045.

a character modulo N. Then we consider a cusp form $f(z)$ satisfying

$$f(\gamma(z)) = \chi(d)j(\gamma, z)^\kappa f(z) \qquad \text{for all } \gamma = \begin{bmatrix} a & b \\ c & d \end{bmatrix} \in \Gamma_0(N) .$$

Denote by $S_\kappa(N, \chi)$ the complex vector space of all such f. We can then define a certain linear operator $T^N_{\kappa,\chi}(p^2)$ on $S_\kappa(N, \chi)$ for each prime p. If $f(z) = \sum_{n=1}^\infty a(n)e^{2\pi i n z}$ is a common eigen-function of those operators for all primes p, and $f \mid T^N_{\kappa,\chi}(p^2) = \omega_p f$, then for every square-free positive integer t, we have

$$\sum_{n=1}^\infty a(tn^2)n^{-s} = a(t) \cdot \prod_p \left[1 - \chi(p)\left(\frac{-1}{p}\right)^\lambda \left(\frac{t}{p}\right) p^{\lambda-1-s}\right]$$
$$\times \prod_p [1 - \omega_p p^{-s} + \chi(p)^2 p^{\kappa-2-2s}]^{-1} ,$$

where $\lambda = (\kappa - 1)/2$, and p runs over all primes (Theorem 1.9). Now our main theorem (§ 3) asserts that *if we put*

$$F(z) = \sum_{n=1}^\infty A_n e^{2\pi i n z} ,$$
$$\sum_{n=1}^\infty A_n n^{-s} = \prod_p [1 - \omega_p p^{-s} + \chi(p)^2 p^{\kappa-2-2s}]^{-1} ,$$

and if $\kappa \geq 3$, then F is an integral modular form satisfying

$$F(\gamma(z)) = \chi(d)^2(cz + d)^{\kappa-1}F(z) \qquad \text{for } \gamma = \begin{bmatrix} a & b \\ c & d \end{bmatrix} \in \Gamma_0(N_0)$$

with a certain positive integer N_0; F is a cusp form if $\kappa \geq 5$. The integer N_0 depends only on N and χ; all prime factors of N_0 divide N; in most cases, $N_0 = N/2$. We shall actually prove a more general result of the same type for the functions which are eigen-functions of $T^N_{\kappa,\chi}(p^2)$ only for some prime factors p of N.

Thus the forms of $S_\kappa(N, \chi)$ are closely connected with the forms of integral (actually even) weight belonging to the level N_0 and the character χ^2. Although we believe that this connection sheds a new light on the nature of the modular forms of half integral weight, a large part of this area of investigation is still covered by mysterious clouds. Some open questions will be discussed in § 4.

Notation and terminology

1. We denote, as usual, by **Z**, **Q**, **R**, and **C** the ring of rational integers, the rational number field, the real number field, and the complex number field. Also we put $T = \{z \in C \mid |z| = 1\}$. For a complex valued function f defined on some set, $\bar{f}$ denotes its complex conjugate. For $z \in C$, we put $e(z) = \exp(2\pi i z)$ with $i = \sqrt{-1}$, and define $\sqrt{z} = z^{1/2}$ so that $-\pi/2 < \arg(z^{1/2}) \leq \pi/2$. Further we put $z^{\kappa/2} = (z^{1/2})^\kappa$ for every $\kappa \in \mathbf{Z}$.

2. Let N be a positive integer and m an integer $\neq 0$. We write $m \mid N^\infty$ if every prime factor of m divides N. For two integers a and b, (a, b) denotes as usual their greatest common divisor.

Let r be a positive integer. By a *character modulo* r, we mean a complex valued function ψ on $\mathbf{Z}$ such that

$$\psi(n) = \begin{cases} 0 & \text{if } (n, r) \neq 1, \\ \psi_0(n \bmod r) & \text{if } (n, r) = 1, \end{cases}$$

with a homomorphism ψ_0 of the multiplicative residue class group modulo r into $\mathbf{T}$. *The conductor* of ψ is the smallest positive integer c such that $\psi(n)$ depends only on $n \bmod c$ when $(n, r) = 1$. We call ψ *primitive* if r is its conductor. For a primitive character ψ modulo r, we denote by $\mathsf{g}(\psi)$ the Gauss sum:

$$\mathsf{g}(\psi) = \sum_{k=1}^{r} \psi(k) e(k/r) .$$

If $r = 1$ and ψ is the trivial character, we understand that both the conductor of ψ and $\mathsf{g}(\psi)$ are equal to 1.

3. For an integer a and an odd integer $b \neq 0$, we denote by $\left(\dfrac{a}{b}\right)$ the "quadratic residue symbol" which is characterized by the following properties.

(i) $\left(\dfrac{a}{b}\right) = 0$ if $(a, b) \neq 1$.

(ii) If b is an odd prime, $\left(\dfrac{a}{b}\right)$ coincides with the ordinary quadratic residue symbol, i.e., it is one less than the number of solutions of $x^2 \equiv a$ (mod b).

(iii) If $b > 0$, the map $a \mapsto \left(\dfrac{a}{b}\right)$ defines a character modulo b.

(iv) If $a \neq 0$, the map $b \mapsto \left(\dfrac{a}{b}\right)$ defines a character modulo a divisor of $4a$, whose conductor is the conductor of $\mathbf{Q}(\sqrt{a})$ over $\mathbf{Q}$.

(v) $\left(\dfrac{a}{-1}\right) = 1$ or -1 according as $a > 0$ or $a < 0$.

(vi) $\left(\dfrac{0}{\pm 1}\right) = 1$.

It should be noted that our notation does not agree with the traditional symbol with the property $\left(\dfrac{a}{b}\right) = \left(\dfrac{a}{|b|}\right)$. In fact we have

$$\left(\frac{a}{b}\right) = \eta \cdot \left(\frac{a}{|b|}\right) \text{ with } \eta = \begin{cases} -1 & \text{if } a < 0 \text{ and } b < 0, \\ 1 & \text{if } a > 0 \text{ or } b > 0, \end{cases}$$

especially

$$\left(\frac{-1}{b}\right) = (-1)^{(b-1)/2}$$

for all positive or negative odd integers b.

4. For an odd integer m, we put $\varepsilon_m = 1$ or $\sqrt{-1}$ according as $m \equiv 1$ or $3 \pmod{4}$. Thus, for an odd prime p, we have a well known relation $\sum_{k=1}^{p}\left(\frac{k}{p}\right)e(k/p) = \varepsilon_p p^{1/2}$, which will often be needed in our later discussion.

1. Hecke operators and Euler products

The purpose of this section is to discuss a few elementary properties of modular forms with an automorphic factor $(cz + d)^{\kappa/2}$ and Hecke operators acting on them. We shall then construct an Euler product from the Fourier coefficients of a common eigen-function of the operators.

Let $GL_2^+(R)$ denote the group of all real matrices of degree 2 with positive determinant, and $\mathfrak{H}$ the upper half complex plane, i.e.,

$$\mathfrak{H} = \{z \in C \mid \mathrm{Im}\,(z) > 0\} .$$

We let an element $\alpha = \begin{bmatrix} a & b \\ c & d \end{bmatrix}$ of $GL_2^+(R)$ act on $\mathfrak{H} \cup R \cup \{\infty\}$ (or even on $C \cup \{\infty\}$) by $\alpha(z) = (az + b)/(cz + d)$. Here and henceforth, we use z as the standard variable on $\mathfrak{H}$, and put $z = x + iy$ with real x, y as usual. Let $\mathfrak{G}$ denote the set of all couples $(\alpha, \varphi(z))$ formed by an element $\alpha = \begin{bmatrix} a & b \\ c & d \end{bmatrix}$ of $GL_2^+(R)$ and a holomorphic function $\varphi(z)$ on $\mathfrak{H}$ such that

$$\varphi(z)^2 = t\cdot\det\,(\alpha)^{-1/2}(cz + d)$$

with $t \in T$. Defining a law of multiplication in $\mathfrak{G}$ by

$$(\alpha, \varphi(z))(\beta, \psi(z)) = (\alpha\beta, \varphi(\beta(z))\psi(z)) ,$$

we can make $\mathfrak{G}$ a group.[1] Let

$$P: \mathfrak{G} \longrightarrow GL_2^+(R)$$

be the natural projection map. Then Ker (P) is isomorphic to T. The center of $\mathfrak{G}$ contains Ker (P) and is isomorphic to $R^\times \times T$, where $R^\times = R - \{0\}$. Let $\xi = (\alpha, \varphi) \in \mathfrak{G}$. We define the action of ξ on $C \cup \{\infty\}$ to be the same as that of α. Furthermore, for a complex valued function $f(z)$ on $\mathfrak{H}$ and an integer κ, we define a function $f \mid [\xi]_\kappa$ on $\mathfrak{H}$ by

$$(f \mid [\xi]_\kappa)(z) = f(\xi(z))\varphi(z)^{-\kappa} .$$

---

[1] By a similar idea, one can define the universal covering of $SL_2(R)$ as the set of all (α, φ) with $\alpha = \begin{bmatrix} a & b \\ c & d \end{bmatrix} \in SL_2(R)$ and $\varphi(z)$ such that $\exp\,(\varphi(z)) = cz + d$, with the law of composition $(\alpha, \varphi)(\beta, \psi) = (\alpha\beta, \varphi(\beta(z)) + \psi(z))$.

It can easily be seen that $f \,|\, [\xi\eta]_\kappa = (f \,|\, [\xi]_\kappa) \,|\, [\eta]_\kappa$. For convenience, we put $\det(\xi) = \det(\alpha)$, and define a subgroup $\mathfrak{G}_1$ of $\mathfrak{G}$ by

$$\mathfrak{G}_1 = \{\xi \in \mathfrak{G} \mid \det(\xi) = 1\} \, .$$

By a *Fuchsian subgroup of* $\mathfrak{G}_1$, we understand a subgroup Δ of $\mathfrak{G}_1$ satisfying the following conditions (1.1–3).

(1.1)　$P(\Delta)$ *is a discrete subgroup of* $SL_2(\boldsymbol{R})$, *and* $P(\Delta)\backslash\mathfrak{H}$ *is of finite measure with respect to the invariant measure* $y^{-2}dxdy$.

(1.2)　P *gives a one-to-one map of* Δ *onto* $P(\Delta)$, *i.e.*, Δ *has no elements, other than the identity element, of the form* $(1, t)$ *with* $t \in \boldsymbol{T}$.

(1.3)　*If* $-1 \in P(\Delta)$, *then the inverse image of* -1 *by* P *in* Δ *is* $(-1, 1)$.

Let κ be an *odd* integer. We call a meromorphic function $f(z)$ on $\mathfrak{H}$ an *automorphic form of weight* $\kappa/2$ *with respect to* Δ if the following conditions (1.4–5) are satisfied.

(1.4)　$f \,|\, [\xi]_\kappa = f$ *for all* $\xi \in \Delta$.

(1.5)　f *is meromorphic at each cusp of* $P(\Delta)$.

The precise meaning of the second condition is as follows. Let s be a cusp of $P(\Delta)$, and let

$$\Delta_s = \{\xi \in \Delta \mid \xi(s) = s\} \, .$$

By (1.2) and (1.3), Δ_s is either free cyclic, or the product of a free cyclic group and a cyclic group of order 2 generated by $(-1, 1)$. Let η be an element of Δ_s which generates the free cyclic part, and ρ an element of $\mathfrak{G}_1$ such that $\rho(\infty) = s$. Then

$$\rho^{-1}\eta\rho = \left(\varepsilon \cdot \begin{bmatrix} 1 & h \\ 0 & 1 \end{bmatrix}, t\right)$$

with $\varepsilon = \pm 1$, $t \in \boldsymbol{T}$. Changing η for η^{-1} if necessary, we may assume $h > 0$. Under this condition, t depends only on the $P(\Delta)$-equivalence class of s, and not on the choice of η and ρ. Now the condition (1.5) means that

$$f \,|\, [\rho]_\kappa = \sum_{n=-\infty}^{\infty} c_n e\big((n + r)z/h\big) \qquad \text{(see *Notation* 1)}$$

with finitely many non-zero c_n for $n < 0$, where r is a real number determined by $t^\kappa = e(r)$, $0 \leqq r < 1$.

We denote by $G_\kappa(\Delta)$ the vector space of all such f which are holomorphic on $\mathfrak{H}$ and for which $c_n = 0$ if $n < 0$, and further by $S_\kappa(\Delta)$ the subspace of $G_\kappa(\Delta)$ consisting of all f for which $c_0 = 0$ if $r = 0$ at every cusp of $P(\Delta)$. The elements of $G_\kappa(\Delta)$ and $S_\kappa(\Delta)$ are called respectively *integral forms* and *cusp forms* of weight $\kappa/2$ with respect to Δ.

Let Δ_1 and Δ_2 be Fuchsian subgroups of $\mathfrak{G}_1$, and ξ an element of $\mathfrak{G}$

such that Δ_1 and $\xi\Delta_2\xi^{-1}$ are commensurable. Then $\Delta_1\xi\Delta_2$ can be expressed as a finite disjoint union:

$$\Delta_1\xi\Delta_2 = \bigcup_\nu \Delta_1\xi_\nu .$$

For $f \in G_\kappa(\Delta_1)$, we define a function $f \mid [\Delta_1\xi\Delta_2]_\kappa$ on $\mathfrak{H}$ by

$$f \mid [\Delta_1\xi\Delta_2]_\kappa = \det (\xi)^{(\kappa/4)-1} \cdot \sum_\nu f \mid [\xi_\nu]_\kappa ,$$

which is independent of the choice of the representatives ξ_ν. Obviously $f \mid [\Delta_1\xi\Delta_2]_\kappa$ belongs to $G_\kappa(\Delta_2)$. It can easily be seen that $[\Delta_1\xi\Delta_2]_\kappa$ maps $S_\kappa(\Delta_1)$ into $S_\kappa(\Delta_2)$. For a formal finite sum $X = \sum_m c_m \cdot \Delta_1\xi_m\Delta_2$ with $c_m \in C$, we define a linear map $[X]_\kappa$ by

$$[X]_\kappa = \sum_m c_m \cdot [\Delta_1\xi_m\Delta_2]_\kappa .$$

Suppose $\Delta_2\eta\Delta_3$ is a similar double coset. Then we can define the "product"

$$(\Delta_1\xi\Delta_2)\cdot(\Delta_2\eta\Delta_3) = \sum a_\zeta \cdot \Delta_1\zeta\Delta_3$$

as in [8, §§ 3.1, 3.4], which is compatible with their action on the automorphic forms.

Let Δ be a Fuchsian subgroup of $\mathfrak{G}_1$, and α an element of $GL_2^+(R)$ such that $\alpha\Gamma\alpha^{-1}$ is commensurable with Γ, where $\Gamma = P(\Delta)$. Denote by $L\colon \Gamma \to \Delta$ the inverse map of $P\colon \Delta \to \Gamma$. Let ξ be an element of $\mathfrak{G}$ such that $P(\xi) = \alpha$. Then we have

$$(1.6) \qquad L(\alpha\gamma\alpha^{-1}) = \xi \cdot L(\gamma) \cdot \xi^{-1} \cdot \bigl(1, t(\gamma)\bigr) \qquad \text{for } \gamma \in \Gamma \cap \alpha^{-1}\Gamma\alpha$$

with a homomorphism

$$t\colon \Gamma \cap \alpha^{-1}\Gamma\alpha \longrightarrow T ,$$

which depends only on Γ and α, and not on the choice of ξ.

PROPOSITION 1.0. *The notation being as above, one has* $\Delta \cap \xi^{-1}\Delta\xi = L(\mathrm{Ker}\,(t))$. *If* $\mathrm{Ker}\,(t)$ *is of finite index in* Γ, Δ *is commensurable with* $\xi\Delta\xi^{-1}$. *Moreover, if* t^κ *is non-trivial, then* $f \mid [\Delta\xi\Delta]_\kappa = 0$ *for all* $f \in G_\kappa(\Delta)$.

Proof. The first equality can be verified in a straightforward manner. Since L is one-to-one, we have

$$(1.7) \qquad [\Delta\colon \Delta \cap \xi^{-1}\Delta\xi] = [\Gamma\colon \mathrm{Ker}\,(t)] .$$

Now P gives an isomorphism of $\xi^{-1}\Delta\xi$ onto $\alpha^{-1}\Gamma\alpha$, hence

$$[\xi^{-1}\Delta\xi\colon \Delta \cap \xi^{-1}\Delta\xi] = [\alpha^{-1}\Gamma\alpha\colon \mathrm{Ker}\,(t)] .$$

Therefore Δ is commensurable with $\xi^{-1}\Delta\xi$ (and hence with $\xi\Delta\xi^{-1}$) if $[\Gamma\colon \mathrm{Ker}\,(t)]$ is finite. To prove the last assertion, let $\Gamma = \bigcup_\nu (\Gamma \cap \alpha^{-1}\Gamma\alpha)\gamma_\nu$ be a disjoint union, and $\{\delta_\mu\}$ a set of representatives for $\Gamma \cap \alpha^{-1}\Gamma\alpha$ modulo $\mathrm{Ker}\,(t)$. Then

$$\Delta = \bigcup_\nu L(\Gamma \cap \alpha^{-1}\Gamma\alpha)L(\gamma_\nu) = \bigcup_\nu L(\bigcup_\mu \mathrm{Ker}\ (t)\delta_\mu)L(\gamma_\nu)$$
$$= \bigcup_{\mu,\nu}(\Delta \cap \xi^{-1}\Delta\xi)L(\delta_\mu\gamma_\nu) \qquad \text{(disjoint)},$$

hence $\Delta\xi\Delta = \bigcup_{\mu,\nu}\Delta\xi \cdot L(\delta_\mu\gamma_\nu)$ is a disjoint union. By (1.6) we have

$$\xi \cdot L(\delta_\mu) = L(\alpha\delta_\mu\alpha^{-1}) \cdot \xi \cdot (1,\ t(\delta_\mu)^{-1}),$$

hence $f \mid [\xi \cdot L(\delta_\mu)] = t(\delta_\mu)^\kappa f \mid [\xi]_\kappa$ for $f \in G_\kappa(\Delta)$. Therefore

$$f \mid [\Delta\xi\Delta]_\kappa = \det (\xi)^{(\kappa/4)-1}\left(\textstyle\sum_\mu t(\delta_\mu)^\kappa\right)\left(\textstyle\sum_\nu f \mid [\xi \cdot L(\gamma_\nu)]_\kappa\right) = 0$$

if t^κ is non-trivial.

The above proposition naturally leads us to consider the following three conditions on α and ξ:

(1.8$_a$) $L(\Gamma \cap \alpha^{-1}\Gamma\alpha) = \Delta \cap \xi^{-1}\Delta\xi$.

(1.8$_b$) $L(\alpha\gamma\alpha^{-1}) = \xi \cdot L(\gamma) \cdot \xi^{-1}$ for all $\gamma \in \Gamma \cap \alpha^{-1}\Gamma\alpha$, i.e., the above map t is trivial.

(1.8$_c$) P gives a one-to-one map of $\Delta\xi\Delta$ onto $\Gamma\alpha\Gamma$.

If one of these conditions is satisfied, then it can easily be seen that it is satisfied by α and ξ' with any ξ' such that $P(\xi') = \alpha$. Further, as (1.8$_c$) indicates, we can view these as conditions on $\Gamma\alpha\Gamma$, by virtue of the following

PROPOSITION 1.1. *The above three conditions are equivalent. Moreover, when these are satisfied, $\Delta\xi\Delta = \bigcup_\nu \Delta\xi_\nu$ is a disjoint union if and only if $\Gamma\alpha\Gamma = \bigcup_\nu \Gamma \cdot P(\xi_\nu)$ is a disjoint union.*

Proof. The equivalence of (1.8$_a$) with (1.8$_b$) follows from the equality $\Delta \cap \xi^{-1}\Delta\xi = L(\mathrm{Ker}\ (t))$. Now, with the notation of the above proof, we have disjoint unions $\Delta\xi\Delta = \bigcup_{\mu,\nu} \Delta\xi \cdot L(\delta_\mu\gamma_\nu)$, $\Gamma\alpha\Gamma = \bigcup_\nu \Gamma\alpha\gamma_\nu$. Observe that P gives a one-to-one map of $\Delta\xi \cdot L(\delta_\mu\gamma_\nu)$ onto $\Gamma\alpha\gamma_\nu$. Therefore P gives a one-to-one map of $\Delta\xi\Delta$ onto $\Gamma\alpha\Gamma$ if and only if t is trivial. Thus (1.8$_c$) is equivalent to (1.8$_b$). The second assertion concerning the decomposition of $\Delta\xi\Delta$ and $\Gamma\alpha\Gamma$ follows immediately from (1.8$_c$).

We shall now specialize our discussion to the case where Δ is obtained from a congruence subgroup of $SL_2(Z)$ by the so-called "θ-multiplier system". For a positive integer N, put

$$\Gamma_0(N) = \left\{\begin{bmatrix} a & b \\ c & d \end{bmatrix} \in SL_2(Z) \mid c \equiv 0 \pmod{N}\right\},$$

$$\Gamma_1(N) = \left\{\begin{bmatrix} a & b \\ c & d \end{bmatrix} \in \Gamma_0(N) \mid a \equiv d \equiv 1 \pmod{N}\right\},$$

$$\Gamma(N) = \left\{\begin{bmatrix} a & b \\ c & d \end{bmatrix} \in \Gamma_1(N) \mid b \equiv 0 \pmod{N}\right\}.$$

Further put

$$(1.9_a) \qquad \theta(z) = \sum_{n=-\infty}^{\infty} e(n^2 z) \; ,$$

$$(1.9_b) \qquad j(\gamma, z) = \theta(\gamma(z))/\theta(z) \qquad\qquad \text{for } \gamma \in \Gamma_0(4).$$

It can be shown that (see § 2 or Hecke [3, (Werke) pp. 939-940])

$$(1.10) \qquad j\left(\begin{bmatrix} a & b \\ c & d \end{bmatrix}, z\right) = \varepsilon_d^{-1} \cdot \left(\frac{c}{d}\right) \cdot (cz + d)^{1/2} \; ,$$

where $\varepsilon_d = 1$ or i according as $d \equiv 1$ or $3 \pmod 4$, and $(cz + d)^{1/2}$ is defined as in *Notation* 1. We see easily that

$$(1.11) \qquad j\left(\begin{bmatrix} a & b \\ c & d \end{bmatrix}, z\right)^2 = \left(\frac{-1}{d}\right) \cdot (cz + d) \; .$$

(For the definition of the quadratic residue symbol $\left(\dfrac{c}{d}\right)$ or $\left(\dfrac{-1}{d}\right)$, see *Notation* 3.) By virtue of (1.9_b), if we put

$$(1.12) \qquad \gamma^* = (\gamma, j(\gamma, z)) \qquad\qquad (\gamma \in \Gamma_0(4)) \; ,$$

then the map $\gamma \mapsto \gamma^*$ defines an isomorphism of $\Gamma_0(4)$ into $\mathfrak{G}_1$. We denote by $\Delta_0(N)$, $\Delta_1(N)$, $\Delta(N)$ respectively the images of $\Gamma_0(N)$, $\Gamma_1(N)$, $\Gamma(N)$ under this isomorphism, for every positive integer N divisible by 4. Obviously $\Delta_0(N)$, $\Delta_1(N)$, $\Delta(N)$ are Fuchsian subgroups of $\mathfrak{G}_1$, and $\theta \in G_1(\Delta_0(4))$. To simplify our notation, we shall often write $[\gamma]_\kappa$ for $[\gamma^*]_\kappa$; thus

$$f \mid [\gamma]_\kappa = f(\gamma(z)) \cdot j(\gamma, z)^{-\kappa} \qquad\qquad (\gamma \in \Gamma_0(4)).$$

Let N be a positive multiple of 4, and χ a character modulo N. We denote by $S_\kappa(N, \chi)$ (resp. $G_\kappa(N, \chi)$) the set of all elements f of $S_\kappa(\Delta_1(N))$ (resp. $G_\kappa(\Delta_1(N))$) such that

$$f \mid [\gamma]_\kappa = \chi(d) \cdot f \qquad\qquad \text{for all } \gamma = \begin{bmatrix} a & b \\ c & d \end{bmatrix} \in \Gamma_0(N) \; .$$

Since $[-1]_\kappa$ gives the identity map, we see that $G_\kappa(N, \chi)$ is $\{0\}$ unless $\chi(-1) = 1$. Therefore we shall assume $\chi(-1) = 1$ whenever we discuss elements of $G_\kappa(N, \chi)$.

If $f \in G_\kappa(\Delta_1(N))$, we see that $f(z + 1) = f(z)$, since $\left(\begin{bmatrix} 1 & 1 \\ 0 & 1 \end{bmatrix}, 1\right) \in \Delta_1(N)$, hence the Fourier expansion of f has the form $f(z) = \sum_{n=0}^{\infty} a_n e(nz)$. If $f \in S_\kappa(\Delta_1(N))$, it can easily be seen that $\mathrm{Im}\,(z)^{\kappa/4} \cdot |f(z)|$ is bounded. Then, by a well known argument (see for example [8, p. 90]), we can show that if $f(z) = \sum_{n=1}^{\infty} a_n e(nz) \in S_\kappa(\Delta_1(N))$, then

$$(1.13) \qquad a_n = O(n^{\kappa/4}) \; .$$

PROPOSITION 1.2. *Let m and n be positive integers, and let*

$$\alpha = \begin{bmatrix} m & 0 \\ 0 & n \end{bmatrix}, \quad \xi = \left(\alpha,\ t\cdot(n/m)^{1/4}\right)$$

with any $t \in \boldsymbol{T}$. *Then*

$$(1.14) \qquad \xi\gamma^*\xi^{-1} = (\alpha\gamma\alpha^{-1})^* \cdot \left(1, \left(\frac{mn}{d}\right)\right) \qquad for\ \gamma = \begin{bmatrix} * & * \\ * & d \end{bmatrix} \in \Gamma_0(4) \cap \alpha^{-1}\Gamma_0(4)\alpha.$$

Especially,

$(1.15) \qquad$ *If mn is a square,* $\xi\gamma^*\xi^{-1} = (\alpha\gamma\alpha^{-1})^*$ *for all* $\gamma \in \Gamma_0(4) \cap \alpha^{-1}\Gamma_0(4)\alpha$.

PROPOSITION 1.3. *Let* $\sigma = \left(\begin{bmatrix} m & 0 \\ 0 & 1 \end{bmatrix}, m^{-1/4}\right)$ *with a positive integer m. Then $[\sigma]_\kappa$ sends $S_\kappa(N, \chi)$ (resp. $G_\kappa(N, \chi)$) into $S_\kappa(mN, \chi')$ (resp. $G_\kappa(mN, \chi')$), where* $\chi'(d) = \chi(d)\cdot\left(\dfrac{m}{d}\right)$.

These two propositions can easily be verified by virtue of (1.10). Note that (1.14) is a special case of (1.6).

PROPOSITION 1.4. *Put* $\beta_N = \begin{bmatrix} 0 & -1 \\ N & 0 \end{bmatrix}$, $\tau_N = \left(\beta_N, N^{1/4}(-iz)^{1/2}\right)$. *Then $[\tau_N]_\kappa^2$ gives the identity map on $G_\kappa(\Delta_1(N))$, and $[\tau_N]_\kappa$ sends $S_\kappa(N, \chi)$ (resp. $G_\kappa(N, \chi)$) onto $S_\kappa(N, \chi^*)$ (resp. $G_\kappa(N, \chi^*)$), where* $\chi^*(d) = \bar{\chi}(d)\cdot\left(\dfrac{N}{d}\right)$.

Proof. The first assertion is obvious. To prove the second one, for $\gamma = \begin{bmatrix} a & b \\ c & d \end{bmatrix} \in \Gamma_0(N)$, we have $\beta_N\gamma\beta_N^{-1} = \begin{bmatrix} d & -c/N \\ -Nb & a \end{bmatrix}$. Therefore it is sufficient to show

$$(1.16) \qquad \tau_N\gamma^*\tau_N^{-1} = (\beta_N\gamma\beta_N^{-1})^* \cdot \left(1, \left(\frac{N}{d}\right)\right) \qquad for\ \gamma = \begin{bmatrix} a & b \\ c & d \end{bmatrix} \in \Gamma_0(N).$$

This can be verified by checking "automorphic factors", but a simpler proof can be given as follows. Put $N = 4M$, and $\sigma = \left(\begin{bmatrix} M & 0 \\ 0 & 1 \end{bmatrix}, M^{-1/4}\right)$. Then $\tau_N = \tau_4\sigma$. Observe that $\theta \mid [\tau_4]_1 = \theta$. (This is well known; a proof will be given in the next section.) Put $h = \theta \mid [\sigma]_1$. By Proposition 1.3, $h \in G_1(N, \varphi)$ with $\varphi(d) = \left(\dfrac{M}{d}\right)$. Therefore

$$\theta \mid [\tau_N\gamma^*\tau_N^{-1}]_1 = h \mid [\gamma^*\tau_N^{-1}]_1 = \varphi(d)h \mid [\tau_N^{-1}]_1$$
$$= \varphi(d)\theta = \varphi(d)\theta \mid [\beta_N\gamma\beta_N^{-1}]_1,$$

hence the equality (1.16).

PROPOSITION 1.5. *Let m be a positive integer such that the conductor of $\boldsymbol{Q}(m^{1/2})$ divides N, and $m \mid N^\infty$ (see Notation 2). Put $\Delta_1 = \Delta_1(N)$, $\alpha = \begin{bmatrix} 1 & 0 \\ 0 & m \end{bmatrix}$, $\xi = (\alpha, m^{1/4})$. Then $[\Delta_1\xi\Delta_1]_\kappa$ sends $G_\kappa(N, \chi)$ into $G_\kappa(N, \chi')$ where $\chi'(d) = \chi(d)\cdot\left(\dfrac{m}{d}\right)$. Moreover, if $f(z) = \sum_{n=0}^{\infty} a(n)e(nz) \in G_\kappa(N, \chi)$, then*

$$(f \mid [\Delta_1 \xi \Delta_1]_\kappa)(z) = \sum_{n=0}^{\infty} a(mn)e(nz) .$$

Proof. Put $g = f \mid [\Delta_1 \xi \Delta_1]_\kappa$ with $f \in G_\kappa(N, \chi)$. Then $g \in G_\kappa(\Delta_1)$. Let $\gamma = \begin{bmatrix} a & b \\ c & d \end{bmatrix} \in \Gamma_0(mN)$ with $b \equiv 0 \pmod{mN}$, and let $\delta = \gamma^*$, $\varepsilon = \xi \delta \xi^{-1}$. Then $\varepsilon = (\alpha\gamma\alpha^{-1})^* \left(1, \left(\dfrac{m}{d}\right)\right)$, $\alpha\gamma\alpha^{-1} = \begin{bmatrix} a & b/m \\ mc & d \end{bmatrix} \in \Gamma_0(N)$, and $\delta\Delta_1\delta^{-1} = \varepsilon\Delta_1\varepsilon^{-1} = \Delta_1$. Therefore, by [8, Prop. 3.7],

$$(\Delta_1\xi\Delta_1)(\Delta_1\delta\Delta_1) = \Delta_1\xi\delta\Delta_1 = \Delta_1\varepsilon\xi\Delta_1 = (\Delta_1\varepsilon\Delta_1)(\Delta_1\xi\Delta_1) ,$$

hence $g \mid [\delta]_\kappa = (f \mid [\varepsilon]_\kappa) \mid [\Delta_1\xi\Delta_1]_\kappa = \left(\dfrac{m}{d}\right)\chi(d)g$. Since $g \in G_\kappa(\Delta_1)$ and $\Delta_0(N)$ can be generated by Δ_1 and the elements of the form γ^*, this proves that $g \in G_\kappa(N, \chi')$. If we put $\Gamma_1 = \Gamma_1(N)$, we have

$$\Gamma_1\alpha\Gamma_1 = \bigcup_{\nu=1}^{m}\Gamma_1\alpha\varepsilon_\nu, \quad \varepsilon_\nu = \begin{bmatrix} 1 & \nu \\ 0 & 1 \end{bmatrix} .$$

By (1.14), we see that α satisfies (1.8_b) with $\Gamma = \Gamma_1$. Therefore, by Proposition 1.1, we obtain a disjoint decomposition

$$\Delta_1\xi\Delta_1 = \bigcup_{\nu=1}^{m} \Delta_1\xi\varepsilon_\nu^* .$$

If $f(z) = \sum_{n=0}^{\infty} a(n)e(nz)$, we have

$$\begin{aligned}
f \mid [\Delta_1\xi\Delta_1]_\kappa &= m^{(\kappa/4)-1}\sum_{\nu=1}^{m}f \mid [\xi\varepsilon_\nu^*] \\
&= m^{(\kappa/4)-1}\sum_{\nu=1}^{m}f\big((z + \nu)/m\big)m^{-\kappa/4} \\
&= m^{-1}\sum_{n=0}^{\infty} a(n)\big(\sum_{\nu=1}^{m} e(\nu n/m)\big)e(nz/m) \\
&= \sum_{n=0}^{\infty} a(mn)e(nz) ,
\end{aligned}$$

which completes the proof.

PROPOSITION 1.6. *Let m and n be squares of positive integers, and let*

$$\alpha = \begin{bmatrix} 1 & 0 \\ 0 & m \end{bmatrix}, \quad \beta = \begin{bmatrix} 1 & 0 \\ 0 & n \end{bmatrix},$$
$$\xi = (\alpha, m^{1/4}) , \quad \eta = (\beta, n^{1/4}) .$$

Suppose either $(m, n) = 1$, or $m \mid N^\infty$. Let Δ be any one of $\Delta_0(N)$, $\Delta_1(N)$, and $\Delta(N)$, Then

$$(\Delta\xi\Delta)(\Delta\eta\Delta) = \Delta\xi\eta\Delta = (\Delta\eta\Delta)(\Delta\xi\Delta) .$$

Proof. Put $\Gamma = P(\Delta)$. By the results of [8, § 3.3], we have

$$(*) \qquad (\Gamma\alpha\Gamma)(\Gamma\beta\Gamma) = \Gamma\alpha\beta\Gamma = (\Gamma\beta\Gamma)(\Gamma\alpha\Gamma) .$$

By (1.15) and Proposition 1.1, we have disjoint unions of the form

$$\begin{aligned}
\Delta\xi\Delta &= \bigcup_{i=1}^{a}\Delta\xi_i, & \Gamma\alpha\Gamma &= \bigcup_{i=1}^{a}\Gamma\alpha_i, & \alpha_i &= P(\xi_i) , \\
\Delta\eta\Delta &= \bigcup_{j=1}^{b}\Delta\eta_j, & \Gamma\beta\Gamma &= \bigcup_{j=1}^{b}\Gamma\beta_j, & \beta_j &= P(\eta_j) , \\
\Delta\xi\eta\Delta &= \bigcup_{k=1}^{c}\Delta\zeta_k, & \Gamma\alpha\beta\Gamma &= \bigcup_{k=1}^{c}\Gamma\varepsilon_k, & \varepsilon_k &= P(\zeta_k) .
\end{aligned}$$

The equality (*) implies that there is exactly one (i, j) such that $\Gamma\alpha_i\beta_j = \Gamma\alpha\beta$, and $c = ab$, hence there is exactly one (i, j) such that $\Delta\xi_i\eta_j = \Delta\xi\eta$, so that $(\Delta\xi\Delta)(\Delta\eta\Delta) = \Delta\xi\eta\Delta$. Since $\xi\eta = \eta\xi$, we obtain our assertion.

To study the relation between Hecke operators and Fourier coefficients, let us now fix a positive multiple N of 4, and put $\Delta_0 = \Delta_0(N)$, $\Delta_1 = \Delta_1(N)$, $\Gamma_0 = \Gamma_0(N)$, $\Gamma_1 = \Gamma_1(N)$. Let m be the square of a positive integer, and let

$$\alpha = \begin{bmatrix} 1 & 0 \\ 0 & m \end{bmatrix}, \quad \xi = (\alpha, m^{1/4}) .$$

Put $\Delta_0\xi\Delta_0 = \bigcup_{\nu=1}^{r} \Delta_0\xi_\nu$ (disjoint), $\Gamma_0\alpha\Gamma_0 = \bigcup_{\nu=1}^{r} \Gamma_0\alpha_\nu$ with $\alpha_\nu = P(\xi_\nu)$. Then we define a linear operator $T_{\kappa,\chi}^N(m)$ on $G_\kappa(N, \chi)$ by

$$f \mid T_{\kappa,\chi}^N(m) = m^{(\kappa/4)-1}\sum_{\nu=1}^{r} \chi(a_\nu)\cdot f \mid [\xi_\nu]_\kappa \qquad (f \in G_\kappa(N, \chi)) ,$$

where $\alpha_\nu = \begin{bmatrix} a_\nu & * \\ * & * \end{bmatrix}$. It can easily be seen that $T_{\kappa,\chi}^N(m)$ coincides with the restriction of $[\Delta_1\xi\Delta_1]_\kappa$ to $G_\kappa(N, \chi)$, and it maps $G_\kappa(N, \chi)$ and $S_\kappa(N, \chi)$ into themselves. (If m is not a square and $(m, N) = 1$, then $[\Delta_1\xi\Delta_1]_\kappa$ is a zero operator on $G_\kappa(\Delta_1)$ by virtue of (1.14) and Proposition 1.0. This is why we consider $T_{\kappa,\chi}^N(m)$ only for square m.)

THEOREM 1.7. *Let p be a prime number, and f an element of $G_\kappa(N, \chi)$. Put*

$$f(z) = \sum_{n=0}^{\infty} a(n)e(nz) ,$$
$$(f \mid T_{\kappa,\chi}^N(p^2))(z) = \sum_{n=0}^{\infty} b(n)e(nz) .$$

Then

$$b(n) = a(p^2 n) + \chi_1(p)\left(\frac{n}{p}\right)p^{\lambda-1}a(n) + \chi(p^2)p^{\kappa-2}a(n/p^2) ,$$

where $\lambda = (\kappa-1)/2$, χ_1 is a character modulo N defined by $\chi_1(m) = \chi(m)\left(\dfrac{-1}{m}\right)^\lambda$; we understand that $a(n/p^2) = 0$ if n is not divisible by p^2, and $\chi_1(p)\left(\dfrac{n}{p}\right) = 0$ if $p = 2$.

As mentioned in the *Introduction*, a result of this type was first obtained by Wohlfahrt [12] in a formulation somewhat different from ours.

Proof. If p divides N, this follows immediately from Proposition 1.5. Therefore suppose p does not divide N, and put

$$\alpha = \begin{bmatrix} 1 & 0 \\ 0 & p^2 \end{bmatrix}, \quad \xi = (\alpha, p^{1/2}) .$$

Then a set of representatives for $\Gamma_0\backslash\Gamma_0\alpha\Gamma_0$ can be formed by the following $p^2 + p$ elements:

$$\alpha_b = \begin{bmatrix} 1 & b \\ 0 & p^2 \end{bmatrix} = \begin{bmatrix} 1 & 0 \\ 0 & p^2 \end{bmatrix}\begin{bmatrix} 1 & b \\ 0 & 1 \end{bmatrix} \qquad (0 \le b < p^2) \,,$$

$$\beta_h = \begin{bmatrix} p & h \\ 0 & p \end{bmatrix} = \begin{bmatrix} 1 & 0 \\ pNs & 1 \end{bmatrix}\begin{bmatrix} 1 & 0 \\ 0 & p^2 \end{bmatrix}\begin{bmatrix} p & h \\ -Ns & r \end{bmatrix} \qquad (0 < h < p) \,,$$

$$\sigma = \begin{bmatrix} p^2 & 0 \\ 0 & 1 \end{bmatrix} = \begin{bmatrix} p^2 & -t \\ N & d \end{bmatrix}\begin{bmatrix} 1 & 0 \\ 0 & p^2 \end{bmatrix}\begin{bmatrix} p^2 d & t \\ -N & 1 \end{bmatrix} \,.$$

As for β_h, we choose and fix, for each h, two integers r and s so that $pr + Nsh = 1$; for σ, we choose two integers t and d so that $p^2 d + Nt = 1$. Now, for $\gamma \alpha \delta \in \Gamma_0 \alpha \Gamma_0$ with $\gamma, \delta \in \Gamma_0$, put $(\gamma \alpha \delta)^* = \gamma^* \xi \delta^*$. By Proposition 1.1 and (1.15), $\eta \mapsto \eta^*$ gives a one-to-one map of $\Gamma_0 \alpha \Gamma_0$ onto $\Delta_0 \xi \Delta_0$. Then we have

$$\alpha_b^* = (\alpha_b, p^{1/2}) \,,$$

$$\beta_h^* = \left(\beta_h, \varepsilon_p^{-1}\cdot\left(\frac{-h}{p}\right)\right) \,,$$

$$\sigma^* = (\sigma, p^{-1/2}) \,.$$

The verification can be made by means of (1.10) and the above decompositions of α_b, β_h, and σ. By Proposition 1.1, these $p^2 + p$ elements α_b^*, β_h^*, σ^* form a set of representatives for $\Delta_0 \backslash \Delta_0 \xi \Delta_0$. Therefore

$$f \mid T_{\kappa,\chi}^N(p^2) = p^{(\kappa/2)-2}\{\textstyle\sum_b f \mid [\alpha_b^*]_\kappa + \chi(p) \sum_h f \mid [\beta_h^*]_\kappa + \chi(p^2) f \mid [\sigma^*]_\kappa\} \,.$$

The first term of the right hand side equals

$$\begin{aligned} p^{(\kappa/2)-2}\textstyle\sum_b f \mid [\alpha_b^*]_\kappa &= p^{(\kappa/2)-2}\sum_{b=0}^{p^2-1} f((z+b)/p^2)p^{-\kappa/2} \\ &= p^{-2}\sum_{n=0}^{\infty} a(n)(\sum_b e(bn/p^2))e(nz/p^2) \\ &= \sum_{n=0}^{\infty} a(p^2 n)e(nz) \,. \end{aligned}$$

The second term can be computed as follows:

$$\begin{aligned} \chi(p)p^{(\kappa/2)-2}\textstyle\sum_{h=1}^{p-1} f \mid [\beta_h^*]_\kappa &= \chi(p)p^{(\kappa/2)-2}\sum_h \varepsilon_p^\kappa\left(\frac{-h}{p}\right)f(z+(h/p)) \\ &= \chi(p)p^{(\kappa/2)-2}\left(\frac{-1}{p}\right)\varepsilon_p^\kappa \sum_{n=0}^{\infty} a(n)e(nz)\sum_{h=1}^{p-1}\left(\frac{h}{p}\right)e(nh/p) \\ &= \chi(p)p^{\lambda-1}\left(\frac{-1}{p}\right)^\lambda \sum_{n=0}^{\infty}\left(\frac{n}{p}\right)a(n)e(nz) \,. \end{aligned}$$

Finally we have

$$\chi(p^2)p^{(\kappa/2)-2}f \mid [\sigma^*]_\kappa = p^{\kappa-2}\chi(p^2)\sum_{n=0}^{\infty} a(n)e(np^2 z) \,.$$

Summing up all these, we obtain the formula for $b(n)$ as stated in our theorem.

COROLLARY 1.8. *Let t be a positive integer, p a prime, and $f(z) =*$

$\sum_{n=0}^{\infty} a(n)e(nz)$ *be an element of* $G_\kappa(N, \chi)$, *that is an eigen-function of* $T_{\kappa,\chi}^N(p^2)$ *with eigen-value* ω_p:

$$f \mid T_{\kappa,\chi}^N(p^2) = \omega_p f \ .$$

Suppose that either p *divides* N, *or* p^2 *does not divide* t. *Then*

(i)
$$\omega_p a(t) = a(p^2 t) + \chi_1(p)\Big(\frac{t}{p}\Big)p^{\lambda-1}a(t) \ ,$$

(ii)
$$\omega_p a(p^{2m}t) = a(p^{2m+2}t) + \chi_1(p^2)p^{\kappa-2}a(p^{2m-2}t) \qquad (m > 0) \ ,$$

where λ *and* χ_1 *are as in Theorem* 1.7. *Moreover, the formal Dirichlet series* $\sum_{n=1}^{\infty} a(tn^2)n^{-s}$ *can be factored as*

(iii)
$$\sum_{n=1}^{\infty} a(tn^2)n^{-s} = \{\textstyle\sum_{(n,p)=1} a(tn^2)n^{-s}\}$$
$$\times \Big[1 - \chi_1(p)\Big(\frac{t}{p}\Big)p^{\lambda-1-s}\Big]\cdot[1 - \omega_p p^{-s} + \chi(p)^2 p^{\kappa-2-2s}]^{-1} \ .$$

Proof. By Theorem 1.7, if n is prime to p, we have

(*)
$$\omega_p a(tn^2) = a(tp^2 n^2) + \chi_1(p)\Big(\frac{t}{p}\Big)p^{\lambda-1}a(tn^2) \ ,$$

(**)
$$\omega_p a(tp^{2m}n^2) = a(tp^{2m+2}n^2) + \chi(p)^2 p^{\kappa-2}a(tp^{2m-2}n^2) \qquad (m > 0) \ ,$$

of which the above (i) and (ii) are special cases. With an indeterminate x, put $H_n(x) = \sum_{m=0}^{\infty} a(tp^{2m}n^2)x^m$. Adding x times (*) and x^{m+1} times (**) for all $m > 0$, we obtain

$$\omega_p x \cdot H_n(x) = H_n(x) - a(tn^2) + \chi_1(p)\Big(\frac{t}{p}\Big)p^{\lambda-1}a(tn^2)x + \chi(p)^2 p^{\kappa-2}x^2 H_n(x) \ ,$$

hence

$$H_n(x) = a(tn^2)\Big[1 - \chi_1(p)\Big(\frac{t}{p}\Big)p^{\lambda-1}x\Big]\cdot[1 - \omega_p x + \chi(p)^2 p^{\kappa-2}x^2]^{-1} \ .$$

Since $\sum_{n=1}^{\infty} a(tn^2)n^{-s} = \sum_{(n,p)=1} H_n(p^{-s})n^{-s}$, we obtain our assertion.

Now consider the algebra $\mathfrak{A}$ generated by the operators $T_{\kappa,\chi}^N(p^2)$ for all primes p. By Proposition 1.6, $\mathfrak{A}$ is commutative, hence $G_\kappa(N, \chi)$ has a basis with respect to which every element $T_{\kappa,\chi}^N(p^2)$ is represented by an upper triangular matrix $\Lambda(p)$. Let $\omega_1(p), \cdots, \omega_k(p)$ denote the diagonal elements of $\Lambda(p)$, where $k = \dim\big(G_\kappa(N, \chi)\big)$. Then we can find k elements $f_1, \cdots f_k$ of $G_\kappa(N, \chi)$ such that

$$f_j \mid T_{\kappa,\chi}^N(p^2) = \omega_j(p)f_j \qquad (j = 1, \cdots, k)$$

for all rational primes p. (The existence of such f_j can be shown by the same argument as in [8, p. 82].) Thus we can meaningfully speak of a common eigen-function of $T_{\kappa,\chi}^N(p^2)$ for all p. As an immediate consequence of Corollary 1.8, we obtain

THEOREM 1.9. *Let $f(z) = \sum_{n=0}^{\infty} a(n)e(nz)$ be an element of $G_\kappa(N, \chi)$, that is a common eigen-function of $T_{\kappa,\chi}^{N}(p^2)$ for all primes p, and let $f \mid T_{\kappa,\chi}^{N}(p^2) = \omega_p f$. Further let t be a positive integer which has no square factor, other than 1, prime to N. Then the formal Dirichlet series $\sum_{n=1}^{\infty} a(tn^2)n^{-s}$ has the following Euler product:*

$$\sum_{n=1}^{\infty} a(tn^2)n^{-s}$$
$$= a(t)\cdot\prod_{p}\left[1 - \chi_1(p)\left(\frac{t}{p}\right)p^{\lambda-1-s}\right]\cdot[1 - \omega_p p^{-s} + \chi(p)^2 p^{\kappa-2-2s}]^{-1},$$

where the product is taken over all primes p, and λ and χ_1 are defined as in Theorem 1.7.

It should be observed that the "denominator part"

(1.17) $$\prod_{p}[1 - \omega_p p^{-s} + \chi(p)^2 p^{\kappa-2-2s}]^{-1}$$

is independent of t. We shall prove in §3 that if $f \in S_\kappa(N, \chi)$ with $\kappa \geq 5$, then (1.17) is the Euler product associated with an ordinary cusp form of weight $\kappa - 1$ (i.e., with the automorphic factor $(cz + d)^{\kappa-1}$).

The above Euler product implies

(1.18) $$a(tm^2)a(tn^2) = a(t)a(tm^2n^2) \text{ if } (m, n) = 1 .$$

One can also treat the case $(m, n) \neq 1$, but we shall not discuss it here. At any rate, such a formula can easily be derived from Corollary 1.8 and Theorem 1.9.

2. Theta series

The purpose of this section is to prove explicit transformation formulas for the theta functions of the form

$$\sum_{n=-\infty}^{\infty} \psi(n)n^\nu e(n^2 z) \qquad (\nu = 0, 1; \psi(-1) = (-1)^\nu)$$

by some elements of $\Gamma_0(4)$, where ψ is any primitive character modulo a positive integer. The formulas will be needed for the proof of the functional equation in the next section. We actually discuss a more general type of theta functions with positive definite quadratic forms, since they provide ample examples of modular forms considered in §1. The case of quadratic forms with an even number of variables was treated by Schoeneberg [7]. Then Pfetzer [5] generalized this so as to include the case with an odd number of variables. However, the formulas we need cannot be obtained just by quoting these works. Therefore we give here a detailed treatment of this topic, following the methods of Hecke [2] and Schoeneberg [7], with a few minor changes of formulation and simplification.

Let n be a positive integer, either odd or even, A a positive definite real symmetric matrix of size n, and N a positive integer. Further let ν be a non-negative integer, and P a spherical function of order ν with respect to A, i.e., a C-valued function in R^n which can be written as

$$P(x) = \begin{cases} \text{constant} & \text{if } \nu = 0 , \\ \sum \beta_q \cdot ({}^tqAx)^\nu & \text{if } \nu > 0 \end{cases} \qquad (x \in R^n)$$

with finitely many vectors $q \in C^n$ such that ${}^tqAq = 0$ if $\nu > 1$. (The condition ${}^tqAq = 0$ is unnecessary if $\nu = 1$. Here and in the following, all vectors in R^n, C^n, or Z^n will be considered column vectors, and the upper left t means "transpose" as usual.) Then, with a fixed element h of Z^n, we consider a theta-function, with variable $z \in \mathfrak{H}$, defined by

$$(2.0) \qquad \theta(z; h, A, N, P) = \sum_{m \equiv h(N)} P(m) \cdot e(z \cdot {}^tmAm/2N^2) ,$$

the summation being taken over all $m \in Z^n$ such that $m \equiv h \pmod{NZ^n}$. We always impose the following set of conditions on A, N, h:

(2.1) *Both A and NA^{-1} have coefficients in Z; $Ah \in NZ^n$.*

We do *not* assume that the diagonal elements of A are even. Our assumption implies

(2.2) $\det (A)$ *is a divisor of* N^n.

If $n = 1$, the only meaningful $P(x)$ are constant multiples of 1 or x with $\nu = 0$ or 1, respectively. In the following we put

(2.3) $D = \det (A)$, $\kappa = n + 2\nu$,

and denote the theta function by $\theta(z; h, A, N)$ (dropping P), since P will always be the same.

Now we start with the following three elementary transformation formulas:

$$(2.4) \qquad \theta(-1/z; h, A, N)$$
$$= (-i)^\nu D^{-1/2}(-iz)^{\kappa/2} \sum_{\substack{k \bmod N \\ Ak \equiv 0(N)}} e({}^tkAh/N^2)\theta(z; k, A, N) ,$$

$$(2.5) \qquad \theta(z + 2; h, A, N) = e({}^thAh/N^2) \cdot \theta(z; h, A, N) ,$$

$$(2.6) \qquad \theta(z; h, A, N) = \sum_{\substack{g \equiv h(N) \\ g \bmod cN}} \theta(cz; g, cA, cN) \qquad \text{if } c \in Z, \, c > 0 .$$

The last two formulas are obvious. The first one can be obtained by applying the Poisson summation to the function

$$P(x + u)e\big(z \cdot {}^t(x + u)A(x + u)/2\big)$$

in R^n with $u = N^{-1}h$ (and its Fourier transform).

For $\gamma = \begin{bmatrix} a & b \\ c & d \end{bmatrix} \in SL_2(Z)$, we have $c \cdot \gamma(z) = a - (cz + d)^{-1}$. Therefore, if $a = 2\alpha$, $d = 2\delta$ with integers α, δ, and if $c > 0$, we obtain, by applying the formulas (2.6), (2.5), and (2.4) in order,

$$\theta(\gamma(z); h, A, N)$$
$$= (-i)^{\nu} D^{-1/2} c^{-n/2} \big(-i(cz + d)\big)^{\kappa/2} \sum_{\substack{k \bmod cN \\ Ak \equiv 0(N)}} \Phi(h, k) \cdot \theta(cz; k, cA, cN)$$

where

$$\Phi(h, k) = \sum_{\substack{g \bmod cN \\ g \equiv h(N)}} e\big([\alpha \cdot {}^t g A g + {}^t k A g + \delta \cdot {}^t k A k]/cN^2\big) .$$

This expression depends only on k modulo N, since

$$\Phi(h, k) = e\big(-b(\delta \cdot {}^t k A k + {}^t k A h)/N^2\big) \cdot \Phi(h + 2\delta k, 0) .$$

Therefore, by (2.6),

$$\theta(\gamma(z); h, A, N) \cdot i^{\nu} D^{1/2} c^{n/2} \big(-i(cz + d)\big)^{-\kappa/2}$$
$$= \sum_{\substack{k \bmod N \\ Ak \equiv 0(N)}} \Phi(h, k) \cdot \theta(z; k, A, N) .$$

Substituting $-1/z$ for z, and using (2.4), we obtain

(2.7)
$$\theta\big((bz - a)/(dz - c); h, A, N\big) \cdot D c^{n/2} \big(-\operatorname{sgn}(d)i\big)^{-n} (dz - c)^{-\kappa/2}$$
$$= \sum_{\substack{l \bmod N \\ Al \equiv 0(N)}} \Big\{ \sum_{\substack{k \bmod N \\ Ak \equiv 0(N)}} e({}^t l A k/N^2) \Phi(h, k) \Big\} \cdot \theta(z; l, A, N) ,$$

if $a = 2\alpha$, $d = 2\delta$, and $c > 0$ where we put $\operatorname{sgn}(d) = 1$ or -1 according as $d \geqq 0$ or $d < 0$.

Suppose that $d \equiv 0 \pmod{2N}$, i.e., $\delta \equiv 0 \pmod{N}$. Then $\Phi(h, k) = e(-b \cdot {}^t k A h/N^2)\Phi(h, 0)$, hence the right hand side of (2.7) becomes

$$\Phi(h, 0) \cdot \sum_{\substack{l \bmod N \\ Al \equiv 0(N)}} \sum_{\substack{k \bmod N \\ Ak \equiv 0(N)}} e\big({}^t(l - bh)A k/N^2\big) \cdot \theta(z; l, A, N) .$$

Observe that

$$\sum_{\substack{k \bmod N \\ Ak \equiv 0(N)}} e\big({}^t(l - bh)A k/N^2\big) = \begin{cases} 0 & \text{if } l \not\equiv bh \pmod{N} , \\ \det(A) & \text{if } l \equiv bh \pmod{N} . \end{cases}$$

Therefore, writing $\begin{bmatrix} a & b \\ c & d \end{bmatrix}$ for $\begin{bmatrix} b & -a \\ d & -c \end{bmatrix}$, we obtain

(2.8)
$$\theta\big((az + b)/(cz + d); h, A, N\big)$$
$$= \big(-\operatorname{sgn}(c)i\big)^n (cz + d)^{\kappa/2} W \cdot \theta(z; ah, A, N),$$
$$W = |d|^{-n/2} \cdot \sum_{\substack{g \bmod dN \\ g \equiv h(N)}} e(\alpha \cdot {}^t g A g/|d|N^2) ,$$

if $b = -2\alpha$ with $\alpha \in \mathbf{Z}$, and $c \equiv 0 \pmod{2N}$ and $d < 0$. Since $ad \equiv 1 \pmod{N}$, we can put $g = adh + Nu$ with a variable u modulo $|d|$. Then

$$W = e(ab \cdot {}^t h A h/2N^2) \cdot w(\alpha, |d|) ,$$

where

$$w(\alpha, |d|) = |d|^{-n/2} \cdot \sum_{u \bmod |d|} e(\alpha \cdot {}^t u A u/|d|) .$$

If $c = 0$, we have $d = -1$, and $w(\alpha, |d|) = 1$. Therefore assume $c \neq 0$.

Substituting $z + 2m$ with $m \in Z$ for z in (2.8), we find

$$(2.9) \qquad w(\alpha, |d|) = w(\alpha - am, |d + 2cm|) \qquad (m \in Z,\ mc < 0) .$$

Take m so that $-d - 2cm$ is a prime p. Put $\beta = \alpha - am$. Then

$$w(\alpha, |d|) = w(\beta, p) = p^{-n/2} \cdot \sum_{u \bmod p} e(\beta \cdot {}^t u A u / p) .$$

Since $\begin{bmatrix} a & -2\beta \\ c & -p \end{bmatrix} \in SL_2(Z)$, we have $2\beta c - ap = 1$, hence p is prime to $2\beta N$. By (2.2), p is prime to $\det(A)$, hence there is an element S of $M_n(Z)$, whose determinant is prime to p, such that ${}^t SAS$ is congruent modulo p to a diagonal matrix. If $q_1, \cdots, q_n$ are the diagonal elements, we see that

$$w(\beta, p) = p^{-n/2} \prod_{i=1}^n \left(\sum_{x=1}^p e(\beta q_i x^2 / p) \right) = \varepsilon_p^n \cdot \left(\frac{\beta^n q_1 \cdots q_n}{p} \right) = \varepsilon_p^n \cdot \left(\frac{\beta^n D}{p} \right) ,$$

where ε_m is 1 or $\sqrt{-1}$ according as $m \equiv 1$ or 3 (mod 4). Now $D = D_0 K^2$ with a divisor D_0 of N, and $p \equiv -d$ (mod $4N$), hence $\left(\dfrac{D}{p} \right) = \left(\dfrac{D}{d} \right)$. Further $2\beta c - ap = 1$, $p \equiv -d$ (mod $2c$) and $c \equiv 0$ (mod 2), hence $\left(\dfrac{\beta}{p} \right) = \left(\dfrac{2c}{p} \right) = \left(\dfrac{2c}{-d} \right)$. Thus we obtain, since $d < 0$,

$$w(\alpha, |d|) = \varepsilon_{-d}^n \left(\frac{2c}{-d} \right)^n \left(\frac{D}{d} \right) = \varepsilon_d^{-n} (\operatorname{sgn}(c) i)^n \left(\frac{2c}{d} \right)^n \left(\frac{D}{d} \right) .$$

Therefore, from (2.8), we obtain the following proposition when $d < 0$.

PROPOSITION 2.1. *Let $\theta(z; h, A, N, P)$ be defined by* (2.0) *under the assumption* (2.1), *and let* $\gamma = \begin{bmatrix} a & b \\ c & d \end{bmatrix} \in SL_2(Z)$ *with $b \equiv 0$ (mod 2) and $c \equiv 0$ (mod $2N$). Then*

$$\theta(\gamma(z); h, A, N, P)$$
$$= e(ab \cdot {}^t hAh/2N^2) \left(\frac{\det(A)}{d} \right) \left(\frac{2c}{d} \right)^n \varepsilon_d^{-n} (cz + d)^{\kappa/2} \theta(z; ah, A, N, P) ,$$

where $\kappa = n + 2\nu$, and $\varepsilon_d = 1$ or i according as $d \equiv 1$ or 3 (mod 4).

If $d > 0$, our formula can be obtained by replacing γ by $-\gamma$. The factor $\varepsilon_d^{-1} \cdot \left(\dfrac{2c}{d} \right)$ is traditionally expressed as $\left(\dfrac{c}{d} \right) \cdot e^{\pi i (d-1)/4}$, but this is inconvenient and somewhat misleading.

One can easily verify that: (i) *if the diagonal elements of A are even, then the condition $b \equiv 0$ (mod 2) is unnecessary;* (ii) *if the diagonal elements of both A and NA^{-1} are even, then the above transformation formula holds for every* $\gamma = \begin{bmatrix} a & b \\ c & d \end{bmatrix} \in \Gamma_0(N)$.

PROPOSITION 2.2. *Let ψ be a primitive character modulo r, and let $\nu = 0$ or 1 be such that $\psi(-1) = (-1)^{\nu}$. Put*

$$j(\gamma, z) = \left(\frac{c}{d}\right)\varepsilon_d^{-1}\cdot(cz + d)^{1/2} \qquad for \ \gamma = \begin{bmatrix} a & b \\ c & d \end{bmatrix} \in \Gamma_0(4) \ ,$$

$$h_\psi(z) = h(z; \psi) = 2^{-1}\sum_{m=-\infty}^{\infty} \psi(m)m^\nu e(m^2 z) \ .$$

Then

$$h\big(\gamma(z); \psi\big) = \psi(d)\Big(\frac{-1}{d}\Big)^\nu j(\gamma, z)^{2\nu+1}h(z; \psi) \ for \ all \ \gamma \in \Gamma_0(4r^2) \ ,$$

$$h(-1/4r^2z; \psi) = (-i)^\nu r^{-1/2}\mathrm{g}(\psi)(-2riz)^{(2\nu+1)/2}h(z; \bar{\psi}) \ ,$$

where $\mathrm{g}(\psi)$ is the Gauss sum defined in Notation 2.

Proof. For an integer k, put

$$\theta(z; k, r) = \sum_{m\equiv k(r)} m^\nu e(m^2 z/2r) \ .$$

This is a special case of (2.0) with $n = 1$, $A = N = r$, and $P(m) = m^\nu$. We have then $h(z; \psi) = 2^{-1}\sum_{k=1}^{r} \psi(k)\cdot\theta(2rz; k, r)$. Therefore we obtain the first formula from Proposition 2.1, and the second one from (2.4) in a straightforward manner.

Remark. If we put $\psi_1(d) = \psi(d)\Big(\frac{-1}{d}\Big)^\nu$, the above proposition implies that

$$h_\psi \in G_1(4r^2, \psi) \quad \text{if } \nu = 0 \ ,$$
$$h_\psi \in S_3(4r^2, \psi_1) \quad \text{if } \nu = 1 \ .$$

If $r = 1$ and ψ is trivial, $2h_\psi$ is the classical theta-function $\sum_{m=-\infty}^{\infty}e(m^2 z)$, which we denoted by θ in § 1. This is not a cusp form. If $\psi(m) = \left(\frac{3}{m}\right)$, we see that

$$h_\psi(z) = \sum_{m=1}^{\infty} \left(\frac{3}{m}\right)\cdot e(m^2 z) = \eta(24z) \ ,$$

where

$$\eta(z) = e(z/24)\cdot\prod_{n=1}^{\infty} \big(1 - e(nz)\big) \ .$$

This fact was observed by Hecke [3].

PROPOSITION 2.3. *The notation being as in Proposition 2.2, let T be a positive integer, and u an integer. Then*

$$h_\psi\big(Tz/(4ruz + 1)\big)\cdot(4ruz + 1)^{-(2\nu+1)/2}$$
$$= (-1)^\nu r^{-1}T^{-1-\nu}2^{-1}\cdot\sum_{m=-\infty}^{\infty} m^\nu\cdot\xi(m)\cdot e(m^2 z/T) \ ,$$

where

$$\xi(m) = \sum_{k=1}^{r}\sum_{g=1}^{Tr} \psi(k)\cdot e\big((gm + Tgk - ug^2)/Tr\big) \ .$$

Proof. In general we have

$$\theta(Tz; k, A, N, P) = T^{-\nu}\cdot\theta(z; Tk, TA, TN, P)\ .$$

Apply the formula (2.7) to the right hand side with $\begin{bmatrix} b & -a \\ d & -c \end{bmatrix} = \begin{bmatrix} -1 & 0 \\ -2u & -1 \end{bmatrix}$, $n = 1$, $N = A = r$, and $P(m) = m^\nu$. Taking the sum over $k = 1, \cdots, r$, and substituting $2rz$ for z, we obtain the desired formula.

3. Main Theorem

Let M and k be positive integers, and φ a character modulo M. We denote by $\mathfrak{G}_k(M, \varphi)$ the vector space of all integral modular forms $g(z)$ satisfying

$$g\big((az + b)/(cz + d)\big) = \varphi(d)g(z)(cz + d)^k \ \text{ for all } \begin{bmatrix} a & b \\ c & d \end{bmatrix} \in \Gamma_0(M) ,$$

and by $\mathfrak{S}_k(M, \varphi)$ the subspace consisting of all cusp forms of $\mathfrak{G}_k(M, \varphi)$. (If we extend formally the definition of $G_\kappa(N, \chi)$ and $S_\kappa(N, \chi)$ in §1 to the case of even κ, we have $\mathfrak{G}_k(M, \varphi) = G_{2k}(M, \varphi')$ and $\mathfrak{S}_k(M, \varphi) = S_{2k}(M, \varphi')$ with $\varphi'(d) = \varphi(d)\left(\dfrac{-1}{d}\right)^k$, when M is divisible by 4.) Now our main theorem can be stated as follows.

MAIN THEOREM. *Let κ be an odd integer, N a positive integer divisible by 4, χ a character modulo N, and let $f(z) = \sum_{n=1}^{\infty} a(n)e(nz)$ be an element of $S_\kappa(N, \chi)$. Suppose $\kappa \geqq 3$, and put $\lambda = (\kappa - 1)/2$. Further, let t be a square-free positive integer, and χ_t the character modulo tN defined by*

$$\chi_t(m) = \chi(m)\left(\frac{-1}{m}\right)^\lambda\left(\frac{t}{m}\right) .$$

Define a function $F_t(z)$ on $\mathfrak{H}$ by

$$F_t(z) = \sum_{n=1}^{\infty} A_t(n)e(nz) ,$$

$$\sum_{n=1}^{\infty} A_t(n)n^{-s} = \left(\sum_{m=1}^{\infty} \chi_t(m)m^{\lambda-1-s}\right)\left(\sum_{m=1}^{\infty} a(tm^2)m^{-s}\right) .$$

Suppose that f is a common eigen-function of $T_{\kappa,\chi}^N(p^2)$ for all prime factors p of N not dividing the conductor of χ_t. Then F_t belongs to $\mathfrak{G}_{\kappa-1}(N_t, \chi^2)$ with a certain positive integer N_t. Moreover, if $\kappa \geqq 5$, F_t is a cusp form.

An explicit description of the integer N_t will be given as Proposition 3.7 at the end of this section. For the moment, we content ourselves with noting only that $N_t \mid N^\infty$; further one has $N_t = N/2$ if t is odd and every prime factor of N divides the conductor of χ_t.

COROLLARY. *The notation being as above, suppose that f is a common eigen-function of the operators $T_{\kappa,\chi}^N(p^2)$ for all primes p, and let $f \mid T_{\kappa,\chi}^N(p^2) =$*

$\omega_p f$. *Define a function F on $\mathfrak{H}$ by*

$$F(z) = \sum_{n=1}^{\infty} A(n)e(nz) ,$$

$$\sum_{n=1}^{\infty} A(n)n^{-s} = \prod_p [1 - \omega_p p^{-s} + \chi(p)^2 p^{\kappa-2-2s}]^{-1} ,$$

where the product is taken over all primes p. Let N_0 be the greatest common divisor of the integers N_t for all square-free t such that $a(t) \neq 0$. Then F belongs to $\mathfrak{G}_{\kappa-1}(N_0, \chi^2)$ if $\kappa \geq 3$, and to $\mathfrak{S}_{\kappa-1}(N_0, \chi^2)$ if $\kappa \geq 5$.

This is an immediate consequence of the Main Theorem and Theorem 1.9. Note that $a(t) \neq 0$ for at least one square-free t, if $f \neq 0$.

We shall prove the above theorem by means of the characterization of modular forms due to Weil [11]. Before stating it, let us first recall a well known fact concerning the functional equation of the Mellin transform of a modular form. Let $U(z) = \sum_{n=0}^{\infty} a_n e(nz)$ be an element of $\mathfrak{G}_k(M, \varphi)$, and for every primitive character ψ modulo r, put

$$R(s, U, \psi) = (r^2 N)^{s/2}(2\pi)^{-s}\Gamma(s) \sum_{n=1}^{\infty} \psi(n)a_n n^{-s} .$$

The last Dirichlet series is absolutely convergent for $\mathrm{Re}\,(s) > k$. If r is prime to M, $R(s, U, \psi)$ can be continued to a meromorphic function on the whole s-plane, and satisfies the functional equation

$$R(k - s, U, \psi) = \psi(M)\varphi(r)\mathsf{g}(\psi)^2 r^{-1} R(s, V, \bar{\psi})$$

with $V(z) = U(-1/Nz)N^{-k/2}(-iz)^{-k}$. (Further, $R(s, U, \psi)$ is everywhere holomorphic if $r > 1$. If $r = 1$ and ψ is trivial, it is holomorphic except at $s = 0$ or k, and $\lim_{s\to 0} sR(s) = -a_0$, $\lim_{s\to k} (k - s)R(s) = -b_0$, where b_0 is the constant term of the Fourier expansion of V.)

Now, Weil's characterization is, roughly speaking, the converse of this fact. More precisely, we have:

LEMMA 3.1. *Let $F(z) = \sum_{n=1}^{\infty} c_n e(nz)$ and $D(s) = \sum_{n=1}^{\infty} c_n n^{-s}$ with complex coefficients c_n. Let $\mathfrak{P}$ be a set of prime numbers which has non-empty intersection with every arithmetical progression $\{a, a + b, a + 2b, \cdots\}$ for which $(a, b) = 1$, $b > 0$. Then F belongs to $\mathfrak{G}_k(M, \varphi)$ if the following conditions (i–iii) are satisfied:*

(i) *$D(s)$ is absolutely convergent in some half plane $\mathrm{Re}\,(s) > \sigma$.*

(ii) *For every primitive character ψ modulo r such that $r \in \{1\} \cup \mathfrak{P}$, $(r, M) = 1$, put*

$$R(s, \psi) = (2\pi)^{-s}\Gamma(s)\cdot\sum_{n=1}^{\infty} \psi(n)c_n n^{-s} .$$

Then $R(s, \psi)$ can be continued to a holomorphic function in the whole s-plane, and is bounded in every vertical strip $\sigma_1 < \mathrm{Re}\,(s) < \sigma_2$.

(iii) *There exists a Dirichlet series* $D'(s) = \sum_{n=1}^{\infty} c'_n n^{-s}$, *absolutely convergent in some half plane* Re $(s) > \sigma'$, *satisfying the following functional equation for all characters ψ considered in* (ii):

$$R(k - s, \psi) = C_\psi \cdot (r^2 M)^{s-(k/2)} (2\pi)^{-s} \Gamma(s) \cdot \sum_{n=1}^{\infty} \bar{\psi}(n) c'_n n^{-s} ,$$

$$C_\psi = \psi(M) \cdot \varphi(r) \cdot \mathrm{g}(\psi)^2 / r .$$

Furthermore, F belongs to $\mathfrak{S}_k(M, \varphi)$ if $D(s)$ is absolutely convergent for some s with Re $(s) < k$.

Weil's theorem in [11] is actually stated only in the case where φ is a *real* character and $c_n = c'_n$, but one can easily generalize it to the above form, making obvious modifications in the proof.

Thus our aim is to prove that $\sum_{n=1}^{\infty} A_t(n) n^{-s}$ has all the required properties of Lemma 3.1. We shall take $\mathfrak{P}$ to be the set of all primes p not dividing tN. As to the convergence, we have $a(n) = O(n^{\varepsilon/4})$ by (1.13), so that $\sum_m a(tm^2) m^{-s}$ is absolutely convergent for $\mathrm{Re}(s) > (\kappa + 2)/2$. Since the L-series

$$\sum_{m=1}^{\infty} \chi_t(m) m^{\lambda-1-s}$$

is absolutely convergent for Re $(s) > \lambda = (\kappa - 1)/2$, we see that $\sum_n A_t(n) n^{-s}$ is absolutely convergent for Re $(s) > (\kappa + 2)/2$. Thus the condition (i) is satisfied. If $\kappa \geq 5$, we have $(\kappa + 2)/2 < \kappa - 1$. Therefore, by the last statement of Lemma 3.1, F_t becomes a cusp form, once (ii, iii) are proved.

The proof of the properties (ii, iii) for $\sum_n A_t(n) n^{-s}$ is rather long. We first prove a few necessary lemmas.

LEMMA 3.2. *Let N be a positive integer, and N' a positive divisor of N. Put*

$$\Gamma = \left\{ \begin{bmatrix} a & b \\ c & d \end{bmatrix} \in SL_2(\mathbf{Z}) \mid b \equiv 0 \ (\mathrm{mod}\ N'),\ c \equiv 0 \ (\mathrm{mod}\ N) \right\} ,$$

$$\Gamma_\infty = \left\{ \pm \begin{bmatrix} 1 & mN' \\ 0 & 1 \end{bmatrix} \middle| m \in \mathbf{Z} \right\} .$$

Let W_N be the set of all couples $\{c, d\}$ of integers such that

$$(c, d) = 1, \quad c \equiv 0 \ (\mathrm{mod}\ N) ,$$

$$c > 0 \ \text{if} \ c \neq 0 ,$$

$$d = 1 \ \text{if} \ c = 0 .$$

Then every coset in $\Gamma_\infty \backslash \Gamma$ contains an element $\begin{bmatrix} a & b \\ c & d \end{bmatrix}$ with $\{c, d\}$ in W_N, and the map $\Gamma_\infty \begin{bmatrix} a & b \\ c & d \end{bmatrix} \mapsto \{c, d\}$ gives a one-to-one correspondence between $\Gamma_\infty \backslash \Gamma$ and W_N.

This is well known and can be verified in a straightforward manner.

LEMMA 3.3. *Let α be a non-negative integer, A a positive integer, and φ a primitive character modulo A, and let*

$$H_\alpha(s, z, \varphi) = \pi^{-s}\Gamma(s)y^s \sum_{m,n} \varphi(n)(mAz + n)^\alpha \,|\, mAz + n \,|^{-2s} \,,$$

where the summation is taken over all $(m, n) \in \mathbf{Z}^2$ other than $(0, 0)$, and $z = x + iy \in \mathfrak{H}$. Suppose that either $\alpha > 0$ or $A > 1$. Then the infinite series is absolutely convergent for $\mathrm{Re}\,(s) > (\alpha/2) + 1$, and moreover the function $H_\alpha(s, z, \varphi)$ can be continued to the whole s-plane as an entire function satisfying the functional equation

$$(*) \qquad H_\alpha(\alpha + 1 - s, z, \varphi) = (-1)^\alpha g(\varphi) A^{3s-\alpha-2} z^\alpha H_\alpha(s, -1/Az, \bar{\varphi}) \,.$$

Furthermore, let $f \in S_\kappa(\Delta(B))$, $g \in G_\mu(\Delta(B))$ with a multiple B of $4A$, and let D be a fundamental domain for $\Gamma(B)\backslash\mathfrak{H}$. Suppose $\alpha = (\kappa - \mu)/2 \geqq 0$. Then the integral

$$(**) \qquad \int_D f(z)\bar{g}(z)y^{\mu/2}H_\alpha(s, z, \varphi)y^{-2}dxdy$$

is absolutely convergent for any s, and, as a function in s, bounded in every vertical strip $\sigma_1 < \mathrm{Re}\,(s) < \sigma_2$.

Note that $H_\alpha(s, z, \varphi)$ is identically 0 unless $\varphi(-1) = (-1)^\alpha$. Therefore, in $(*)$, the factor $(-1)^\alpha$ can be replaced by $\varphi(-1)$.

Proof. Our function is a type of Eisenstein-Epstein function, whose behavior is rather "well known". The results stated as above can be proved by modifying the arguments of Rankin [6]. Also, a general treatment is given in Siegel [9, pp. 60–73]. However, we shall give a self-contained (if somewhat sketchy) proof, since the whole lemma (both the functional equation and the convergence of the integral) plays an essential role in our later discussion.

First we note the equality

$$\int_{-\infty}^{\infty}\int_{-\infty}^{\infty} (uz + v)^\alpha \exp\left(-\pi t \,|\, uz + v \,|^2/y\right)e(ur + vs)dudv$$
$$= (-1)^\alpha t^{-\alpha-1}(r - sz)^\alpha \exp\left(-\pi \,|\, r - sz \,|^2/yt\right) \,,$$

where t, u, v, r, s are real variables, and $t > 0$. If $\alpha = 0$, this follows easily from the self-reciprocity of $\exp\left(-\pi(u^2 + v^2)\right)$. The case with $\alpha > 0$ can be obtained by applying $z \cdot (\partial/\partial r) + \partial/\partial s$ successively. Substituting $(u + A^{-1}p, v + A^{-1}q)$ for (u, v), and using the Poisson summation formula, we get

$$(1) \quad \sum_{(m,n) \equiv (p,q) \bmod A} (mz + n)^\alpha \exp\left(-\pi t \,|\, mz + n \,|^2 / A^2 y\right)$$
$$= (-A)^\alpha t^{-\alpha-1} \sum_{(m,n) \in Z^2} e\big((qm - pn)/A\big)(mz + n)^\alpha \exp\left(-\pi \,|\, mz + n \,|^2 / yt\right) .$$

Call this function $\xi\big(t, z, (p, q)\big)$. Then (1) implies

$$(2) \quad \xi\big(t, z, (p, q)\big) = (-A)^\alpha t^{-\alpha-1} \sum_{(a,b) \bmod A} e\big((qa - pb)/A\big)\xi\big(A^2 t^{-1}, z, (a, b)\big) .$$

Put $\eta\big(t, z, (p, q)\big) = \sum_{k=1}^A \mathcal{P}(k)\xi\big(t, z, k(p, q)\big)$. We assume $(p, q) \not\equiv (0, 0) \ (\bmod A)$ if $A > 1$. From (2), we obtain

$$(3) \quad \begin{aligned} & \eta\big(t^{-1}, z, (p, q)\big) \\ & = (-A)^\alpha t^{\alpha+1}\mathfrak{g}(\mathcal{P}) \sum_{(a,b) \bmod A} \bar{\mathcal{P}}(qa - pb)\xi\big(A^2 t, z, (a, b)\big) . \end{aligned}$$

Further we see easily that

$$\eta\big(t, \gamma(z), (p, q)\big)(cz + d)^\alpha = \eta\big(t, z, (p, q)\gamma\big) \text{ for } \gamma = \begin{bmatrix} a & b \\ c & d \end{bmatrix} \in SL_2(Z) .$$

Since $\alpha > 0$ or $A > 1$, the term $m = n = 0$ does not appear in both sides of (3). Now we see, at least formally, that

$$(4) \quad \begin{aligned} & \int_0^\infty \eta\big(t, z, (p, q)\big)t^{s-1}dt \\ & = A^{2s}\pi^{-s}y^s\Gamma(s) \sum_{k=1}^A \mathcal{P}(k) \sum_{m,n} (mz + n)^\alpha \,|\, mz + n \,|^{-2s} , \end{aligned}$$

where (m, n) runs over Z^2 under the condition $(m, n) \equiv k(p, q) \bmod A$. By a well known convergence property of Eisenstein series, we see that the right hand side of (4) is absolutely convergent for $\mathrm{Re}\,(2s) > \alpha + 2$. Now, for a fixed (p, q) and z, we see from (1) that

$$(5) \quad |\,\eta\big(t, z, (p, q)\big)\,| \leq \begin{cases} Me^{-ct} & \text{if } t > 1 , \\ M't^{-\alpha-1}e^{-c'/t} & \text{if } t < 1 \end{cases}$$

with positive constants M, M', c, c' depending only on z and (p, q). (We shall actually prove stronger results (10), (11) afterwards.) Therefore, splitting the integral (4) into two integrals over $(0, 1)$ and $(1, \infty)$, we see that the integral (4) is convergent for all s, and defines an entire function in s. Further, by virtue of the convergence of the "Eisenstein-series" mentioned above, the equality (4) holds for $\mathrm{Re}\,(2s) > \alpha + 2$. Putting $(p, q) = (0, 1)$, we obtain

$$A^{2s}H_\alpha(s, z, \mathcal{P}) = \int_0^\infty \eta\big(t, z, (0, 1)\big)t^{s-1}dt .$$

Therefore, substituting $\alpha + 1 - s$ for s, we have

$$A^{2(\alpha+1-s)}H_\alpha(\alpha + 1 - s, z, \mathcal{P}) = \int_0^\infty \eta(t, z, (0, 1))t^{\alpha-s}dt$$

$$= \int_0^\infty \eta(t^{-1}, z, (0, 1))t^{s-\alpha-2}dt$$

$$= (-A)^\alpha g(\mathcal{P}) \sum_{(a,b)\bmod A} \bar{\mathcal{P}}(a) \int_0^\infty \xi(A^2t, z, (a, b))t^{s-1}dt$$

$$= (-A)^\alpha g(\mathcal{P})y^s\pi^{-s}\Gamma(s) \sum_{(m,n)\in Z^2} \bar{\mathcal{P}}(m)(mz + n)^\alpha \,|\, mz + n\,|^{-2s}$$

$$= (-1)^\alpha g(\mathcal{P})A^{\alpha+s}z^\alpha H_\alpha(s, -1/Az, \bar{\mathcal{P}}) \,,$$

which is the desired functional equation (*).

To prove the convergence of (**), we majorize the function

$$F(u, z) = \sum{}' \,|\, mz + n\,|^\alpha \exp\left(-u\,|\, mz + n\,|^2\right)$$

where u is real positive, and $\sum'$ is the sum over all $(m, n)\in Z^2$, $\neq (0, 0)$. (For this, cf. Hardy [1, p. 177].) Let β be the smallest integer $\geq \alpha/2$. Then the terms with $m = 0$ give

$$(6)\qquad 2\sum_{n=1}^\infty n^\alpha \exp\left(-n^2u\right) \leq 2\sum_{k=1}^\infty k^\beta e^{-ku} = e^{-u}P(e^{-u})/(1 - e^{-u})^{\beta+1}$$

with a polynomial P. Therefore (6) is $O(e^{-u})$ when $u \to \infty$, and $O(u^{-\beta-1})$ when $u \to 0$, since $(1 - e^{-u})/u \to 1$ when $u \to 0$. Therefore

$$(7)\qquad \sum_{n=1}^\infty n^\alpha \exp\left(-n^2u\right) \leq M_\alpha u^{-\beta-1}e^{-u/2}$$

with a constant M_α depending only on α. Next, with $m \neq 0$ and $z = x + iy$ fixed, we can find a positive integer k for each n so that $k-1 \leq |\, mx + n\,| < k$. Therefore the terms with $m \neq 0$ produce

$$(8)\quad \begin{aligned} &2\sum_{m=1}^\infty\sum_{n=-\infty}^\infty \,|\, mz + n\,|^\alpha \exp\left(-u\,|\, mz + n\,|^2\right)\\ &\leq 4\sum_{m=1}^\infty\sum_{k=1}^\infty (m^2y^2 + k^2)^\beta \exp\left(-u(m^2y^2 + (k - 1)^2)\right)\\ &\leq 4\sum_{m=1}^\infty\sum_{\nu=0}^\beta \binom{\beta}{\nu}(m^2y^2)^{\beta-\nu} \exp\left(-um^2y^2\right) \sum_{k=1}^\infty k^{2\nu} \exp\left(-u(k - 1)^2\right)\,. \end{aligned}$$

Applying (7) to the last sum, we find that (8) is

$$\leq M' \sum_{\mu+\nu=\beta} y^{2\mu}(uy^2)^{-\mu-1} \exp\left(-uy^2/2\right)(1 + 2^{2\nu}M_{2\nu}u^{-\nu-1}e^{-u/2})\,,$$

where M' is a constant depending only on α. Therefore, for any $c > 0$,

$$(9)\qquad F(\pi t/A^2y, z) \leq Ly^{\beta+1}e^{-ht/y} \qquad \text{if } t > 1 \text{ and } y = \operatorname{Im}(z) > c > 0\,,$$

where L and h are positive constants depending only on α, A, and c. It follows that

$$(10)\qquad |\,\eta(t, z, (p, q))\,| \leq F(\pi t/A^2y, z) \leq Ly^{\beta+1}e^{-ht/y} \qquad \text{if } t > 1,\ y > c\,.$$

Similarly, we obtain, from (3),

$$(11)\qquad |\,\eta(t^{-1}, z, (p, q))\,| \leq L't^{\alpha+1}y^{\beta+1}e^{-h't/y} \qquad \text{if } t > 1,\ y > c\,,$$

with positive constants L' and h' depending only on α, A, and c.

Now let f, g, and D be as stated in our lemma. Then we can find finitely many elements γ_k of $SL_2(Z)$ and a positive constant c so that

$$D \subset \bigcup_k \gamma_k(U) , \quad U = \{x + iy \mid y > c , \ 0 \leq x \leq B\} .$$

Put $f_k = f \mid [\gamma_k]_\kappa$, $g_k = g \mid [\gamma_k]_\mu$. Then

$$\int_D |fg y^{\mu/2} H_\alpha(s, z, \varphi)| \, y^{-2} dx dy$$

$$\leq \sum_k \int_U |f_k g_k| \, y^{\mu/2} \, |c_k z + d_k|^\alpha \, |H_\alpha(s, \gamma_k(z), \varphi)| \, y^{-2} dx dy$$

where (c_k, d_k) is the lower part of γ_k. Now with $\sigma = \mathrm{Re}\,(s)$, we have

$$A^{2\sigma} |c_k z + d_k|^\alpha \, |H_\alpha(s, \gamma_k(z), \varphi)|$$

$$\leq \int_0^\infty |\eta(t, \gamma_k(z), (0, 1))(c_k z + d_k)^\alpha| \, t^{\sigma-1} dt$$

$$= \int_0^\infty |\eta(t, z, (0, 1)\gamma_k)| \, t^{\sigma-1} dt$$

$$= \int_1^\infty |\eta(t, z, (0, 1)\gamma_k)| \, t^{\sigma-1} dt + \int_1^\infty |\eta(t^{-1}, z, (0, 1)\gamma_k)| \, t^{-\sigma-1} dt$$

$$\leq L \int_1^\infty y^{\beta+1} e^{-ht/\nu} t^{\sigma-1} dt + L' \int_1^\infty y^{\beta+1} e^{-h't/\nu} t^{\alpha-\sigma} dt$$

by (10) and (11). If $\sigma \geq 1$, we can replace $\int_1^\infty$ by $\int_0^\infty$, and $t^{\sigma-1}$ by 1 if $\sigma < 1$. Thus the first integral of the last sum is

$$\leq L \cdot h^{-\sigma} \Gamma(\sigma) y^{\beta+1+\sigma} \ \text{if} \ \sigma \geq 1, \ \text{and} \ \leq (L/h) \cdot y^{\beta+2} \ \text{if} \ \sigma < 1 .$$

Similarly the second integral is

$$\leq L' \cdot h'^{\sigma-\alpha-1} \cdot \Gamma(\alpha - \sigma + 1) y^{\beta+\alpha-\sigma+2} \ \text{or} \ (L'/h') \cdot y^{\beta+2}$$

according as $\sigma \leq \alpha$ or $\sigma > \alpha$. Therefore, in every vertical strip $\sigma_1 < \mathrm{Re}\,(s) < \sigma_2$, we have

$$\int_D |fg y^{\mu/2} H_\alpha(s, z, \varphi)| \, y^{-2} dx dy \leq L'' \sum_k \int_U |f_k g_k y^\tau| \, y^{-2} dx dy$$

with positive constants L'' and τ which depend on σ_1 and σ_2. Now $f_k g_k$ is a cusp form of level B, so that $|f_k(z) g_k(z)| = O(e^{-2\pi y/B})$ when $y \to \infty$. Therefore the last integrals are convergent, which completes the proof.

LEMMA 3.4. *Let* $g \in G_\kappa(\Delta_1(N))$, *and let* $N = MK$ *with positive integers* M *and* K. *For every integer* v, *put*

$$g_v = g \mid [\xi_v]_\kappa , \quad \xi_v = \left(\begin{bmatrix} 1 & 0 \\ Mv & 1 \end{bmatrix}, \ (Mvz + 1)^{1/2} \right) .$$

Suppose that either M or K is divisible by 4. Then $g_v(z + K) = g_v(z)$, so that g_v has a Fourier expansion of the form

$$g_v(z) = \sum_{n=0}^{\infty} \alpha_v(n)e(nz/K) .$$

Moreover, if $g \in G_\kappa(N, \chi)$, $\begin{bmatrix} a & b \\ Mc & d \end{bmatrix} \in \Gamma_0(M)$, $d > 0$, and $(d, N) = 1$, then

$$g\big((az + b)/(Mcz + d)\big)(Mcz + d)^{-\kappa/2} = \chi(d)\Big(\frac{Mc}{d}\Big)\varepsilon_d^{-\kappa} g_w(z) ,$$

where w is any integer such that $dw \equiv c \pmod{K}$.

Proof. Put $\delta = \begin{bmatrix} 1 - Nv & K \\ -MNv^2 & 1 + Nv \end{bmatrix}$. Then $\delta \in \Gamma_1(N)$, and

(*) $$\xi_v \cdot \begin{bmatrix} 1 & K \\ 0 & 1 \end{bmatrix}^* = \delta^* \xi_v .$$

Indeed, the projection of (*) to $SL_2(\mathbf{R})$ is certainly true. The automorphic factors of both sides differ only by $\varepsilon_k^{-1} \cdot \Big(\dfrac{-K}{k}\Big)$, where $k = 1 + Nv$. If 4 divides M, then $k \equiv 1 \pmod{4K}$, hence $\varepsilon_k^{-1} \cdot \Big(\dfrac{-K}{k}\Big) = 1$. Obviously, this holds also if $K \equiv 0 \pmod 4$, hence we obtain (*) in either case. Applying (*) to g, we see that $g_v(z + K) = g_v(z)$. To prove the second assertion, let $dw = c - Kc'$ and $a' = a - bMw$. Then $\begin{bmatrix} a & b \\ Mc & d \end{bmatrix} = \begin{bmatrix} a' & b \\ Nc' & d \end{bmatrix}\begin{bmatrix} 1 & 0 \\ Mw & 1 \end{bmatrix}$. Observe that $\Big(\dfrac{Nc'}{d}\Big) = \Big(\dfrac{MKc'}{d}\Big) = \Big(\dfrac{Mc}{d}\Big)$ since $d > 0$ and $Kc' \equiv c \pmod d$. Therefore we obtain the desired equality by verifying the automorphic factors.

Remark. The above lemma is not necessarily true without the divisibility of M or K by 4. In fact, if $M = K = 2$ and $v = 1$, one has $g_v(z + 2) = -g_v(z)$.

LEMMA 3.5. *Let m be an integer, r an odd prime, and ψ a primitive character modulo r. Then*

$$\sum_{u,v}' \sum_{k=1}^{r} \sum_{t=1}^{r} \Big(\frac{u}{r}\Big)\psi(t)e\big([m^2 v - km - kt + uk^2]/r\big) = \varepsilon_r r^{3/2}\psi(-2m) ,$$

where $\sum_{u,v}'$ is the sum taken over all u, $v \pmod r$ under the condition $4uv \equiv 1 \pmod r$.

Proof. Let x be an integer such that $m + t \equiv 2ux \pmod r$. Then

$$\sum_{k=1}^{r} e\big([uk^2 - km - kt]/r\big) = \sum_{k=1}^{r} e\big(u(k - x)^2/r\big) \cdot e(-ux^2/r)$$

$$= \Big(\frac{u}{r}\Big)r^{1/2}\varepsilon_r e(-ux^2/r) .$$

Now we have $ux^2 \equiv 4uv \cdot ux^2 \equiv v \cdot (2ux)^2 \equiv v \cdot (m + t)^2 \pmod r$, hence the sum in question equals

$$r^{1/2}\varepsilon_r \sum_{t=1}^{r}\sum_{v=1}^{r-1} \psi(t)e\big(v[m^2 - (m+t)^2]/r\big) .$$

Since $\sum_{t=1}^{r}\psi(t) = 0$, we can replace $\sum_{v=1}^{r-1}$ by $\sum_{v=1}^{r}$. Then, non-vanishing terms occur only for $m^2 \equiv (m+t)^2 \pmod r$, i.e., $t \equiv 0$ or $t \equiv -2m$, hence the desired result.

LEMMA 3.6. *Let* $f(z) = \sum_{n=0}^{\infty} a_n e(nz) \in G_\kappa(N, \chi)$, *and let* ψ *be a primitive character modulo* r. *Let* s *be the conductor of* χ, *and* M *the least common multiple of* N, r^2, *and* rs. *Put*

$$f^*(z) = \sum_{n=0}^{\infty} \psi(n)a_n e(nz) .$$

Then f^* *belongs to* $G_\kappa(M, \psi^2\chi)$.

Proof. Put $\xi(u) = \left(\begin{bmatrix} 1 & u/r \\ 0 & 1 \end{bmatrix}, 1\right)$ for an integer u. Then $g(\bar{\psi})f^* = \sum_{u=1}^{r}\bar{\psi}(u)f\,|\,[\xi(u)]_\kappa$. For $\gamma = \begin{bmatrix} a & b \\ Mc & d \end{bmatrix} \in \Gamma_0(M)$ we have

$$(1) \qquad \begin{bmatrix} 1 & u/r \\ 0 & 1 \end{bmatrix}\begin{bmatrix} a & b \\ Mc & d \end{bmatrix} = \begin{bmatrix} a' & b' \\ Mc & d' \end{bmatrix}\begin{bmatrix} 1 & d^2u/r \\ 0 & 1 \end{bmatrix}$$

with integers a', b', d'. Especially $d' = d - cd^2uM/r$. Put $\delta = \begin{bmatrix} a' & b' \\ Mc & d' \end{bmatrix}$. Then

$$(2) \qquad \xi(u)\gamma^* = \delta^*\xi(d^2u) .$$

In fact, both sides have the same projection to $SL_2(R)$ by (1), and their automorphic factors differ only by $\varepsilon_{d'}\varepsilon_d^{-1}\left(\dfrac{Mc}{dd'}\right)$. Suppose r is odd. Then $M = 4r^2t$ with an integer t, and $d \equiv d' \pmod{4tc}$, hence $\varepsilon_d = \varepsilon_{d'}$ and $\left(\dfrac{Mc}{d}\right) = \left(\dfrac{tc}{d}\right) = \left(\dfrac{tc}{d'}\right) = \left(\dfrac{Mc}{d'}\right)$. Next suppose r is even. Then $r = 2^k r'$ with $k > 1$ and an odd integer r', hence $M = 2^{2k}r'^2t$ with an integer t. Therefore $d \equiv d' \pmod{4tc}$, and we again obtain the same equalities. Thus the automorphic factors are always the same, hence (2). Then our assertion can be obtained by the same argument as in [8, Prop. 3.64].

Let us now proceed to the proof of our main theorem. First we consider the case $t = 1$. Let $f(z) = \sum_{n=1}^{\infty} a(n)e(nz) \in S_\kappa(N, \chi)$, and define a character χ_1 modulo N by $\chi_1(m) = \chi(m)\left(\dfrac{-1}{m}\right)^\lambda$, where $\lambda = (\kappa - 1)/2$. Let r be either 1 or an odd prime not dividing N, and ψ a primitive character modulo r. With the notation of the Main Theorem, we have

$$\sum_{n=1}^{\infty} \psi(n)A_1(n)n^{-s} = \left(\sum_{m=1}^{\infty} \psi(m)\chi_1(m)m^{\lambda-1-s}\right)\left(\sum_{m=1}^{\infty} \psi(m)a(m^2)m^{-s}\right) .$$

Our aim is to prove that this Dirichlet series has the properties (ii, iii) of

Lemma 3.1. The main idea is an adaptation of Rankin's method [6]. Fix ψ, and put

$$h(z) = h(z; \bar{\psi}) = 2^{-1} \sum_{m=-\infty}^{\infty} \bar{\psi}(m) m^\nu e(m^2 z) ,$$

where $\psi(-1) = (-1)^\nu$ with $\nu = 0$ or 1. By Proposition 2.2, $h \in G_{2\nu+1}(4r^2, \bar{\psi}_1)$, where $\psi_1(d) = \psi(d)\left(\frac{-1}{d}\right)^\nu$. Since

$$f(z)\bar{h}(z) = 2^{-1} \sum_{n,m} a(n)\psi(m)m^\nu e\big((n - m^2)x\big) \exp\big(-2\pi(n + m^2)y\big)$$

for $z = x + iy$, we have

$$\int_0^1 f(z)\bar{h}(z)dx = \sum_{m=1}^{\infty} \psi(m)a(m^2)m^\nu \exp\big(-4\pi m^2 y\big) ,$$

hence, at least formally,

$$(3.1) \qquad \int_0^\infty \left\{\int_0^1 f\bar{h}dx\right\}y^{s-1}dy = (4\pi)^{-s}\Gamma(s) \sum_{m=1}^{\infty} \psi(m)a(m^2)m^{\nu-2s} .$$

Since fh is a cusp form of weight $(\kappa + 2\nu + 1)/2$, we see that $|fh| = O(e^{-2\pi y})$ when y is large, and $|fh| = O(y^{-(\kappa+2\nu+1)/4})$ when y is small, so that (3.1) holds for $\mathrm{Re}\,(s) > 1 + (\kappa + 2\nu + 1)/4$. Put

$$\Gamma = \Gamma_0(Nr^2) , \quad \Gamma_\infty = \left\{\pm\begin{bmatrix}1 & n \\ 0 & 1\end{bmatrix} \,\Big|\, n \in \mathbf{Z}\right\} ,$$

$$Y = \{x + iy \mid 0 < x \leqq 1 , \ 0 < y\} ,$$

$$B(z) = B(z, s) = f(z)\bar{h}(z)y^{s+1} ,$$

$$d_0 z = y^{-2}dxdy .$$

Then Y is a fundamental domain for $\Gamma_\infty\backslash\mathfrak{H}$, so that the integral (3.1) equals

$$(3.2) \qquad \int_Y B(z)d_0 z = \int_\Phi \sum_{\gamma \in R} B(\gamma(z))d_0 z$$

with any set R of representatives for $\Gamma_\infty\backslash\Gamma$ and any fundamental domain Φ for $\Gamma\backslash\mathfrak{H}$. Since $f \in S_\kappa(N, \chi)$ and $h \in G_{2\nu+1}(4r^2, \bar{\psi}_1)$, we have

$$B(\gamma(z)) = \varphi(d)B(z)(cz + d)^{\lambda-\nu}|cz + d|^{2\nu-1-2s} \qquad \text{for } \gamma = \begin{bmatrix}a & b \\ c & d\end{bmatrix} \in \Gamma ,$$

where φ is a character modulo Nr^2 defined by

$$\varphi(d) = \chi(d)\psi(d)\left(\frac{-1}{d}\right)^\lambda .$$

By our agreement $\chi(-1) = 1$ (otherwise $f = 0$, see § 1), we have

$$(3.3) \qquad\qquad\qquad \varphi(-1) = (-1)^{\lambda-\nu} .$$

By Lemma 3.2, the integral (3.2) becomes

$$\int_{\Phi} B(z)\{\sum_{\{c,d\}} \varphi(d)(cz + d)^{\lambda-\nu} \mid cz + d \mid^{2\nu-1-2s}\}d_0 z$$

where the sum is taken over all $\{c, d\}$ such that $(c, d) = 1$, $c \equiv 0 \pmod{Nr^2}$, $c > 0$ if $c \neq 0$, $d = 1$ if $c = 0$. Put

$$D(s, \psi) = \sum_{n=1}^{\infty} \psi(n)A_1(n)n^{-s},$$

$$L(s, \varphi) = \sum_{n=1}^{\infty} \varphi(n)n^{-s},$$

$$C(z, s) = \sum_{m=-\infty}^{\infty} \sum_{n=-\infty}^{\infty} \varphi(n)(Nr^2 mz + n)^{\lambda-\nu} \mid Nr^2 mz + n \mid^{2\nu-1-2s}.$$

Then we obtain, in view of (3.3),

$$2 \cdot (4\pi)^{-s}\Gamma(s)D(2s - \nu, \psi)$$
$$= 2 \cdot L(2s - \lambda - \nu + 1, \varphi)(4\pi)^{-s}\Gamma(s) \sum_{m=1}^{\infty} \psi(m)a(m^2)m^{\nu-2s}$$
$$= \int_{\Phi} f\bar{h}y^{s+1}C(z, s)d_0 z.$$

To simplify our notation, put

$$\alpha = \lambda - \nu, \quad \beta = (2\nu + 1)/2, \quad 2t = 2s - 2\nu + 1.$$

Then $s + 1 = t + \beta$. Let φ_0 (resp. χ_0) denote the primitive character associated with φ (resp. χ_1), and M the conductor of χ_1. Then Mr is the conductor of φ_0. Put $N = MK$. Let μ denote the Moebius function, and let E be the product of all prime factors of N not dividing M. Then E is a divisor of K, and $\varphi(n) = \varphi_0(n)$ or 0 according as $(n, E) = 1$ or $\neq 1$. Therefore

$$C(z, s) = \sum_{m,n} \left(\sum_{k\mid(n,E)} \mu(k)\right)\varphi_0(n)(Nr^2 mz + n)^{\alpha} \mid Nr^2 mz + n \mid^{-2t}$$
$$= \sum_{k\mid E} \mu(k)\varphi_0(k)k^{\alpha-2t} \sum_{m,n} \varphi_0(n)(Mk'r^2 mz + n)^{\alpha} \mid Mk'r^2 mz + n \mid^{-2t}$$
$$(kk' = K),$$

where (m, n) runs over all elements of $\mathbf{Z}^2$ other than $(0, 0)$, and k over all positive divisors of E. Put

$$C^*(z, t) = \pi^{-t}\Gamma(t)y^t C(z, s),$$

$$H_{\alpha}(t, z, \varphi_0) = \pi^{-t}\Gamma(t)y^t \sum_{m,n} \varphi_0(n)(Mrmz + n)^{\alpha} \mid Mrmz + n \mid^{-2t}.$$

The latter function has been defined in Lemma 3.3. Then we have

$$C^*(z, t) = (Kr)^{-t} \sum_{k\mid E} \mu(k)\varphi_0(k)k^{\alpha-t} H_{\alpha}(t, Krz/k, \varphi_0),$$

$$(3.4) \qquad 2 \cdot (4\pi)^{-s}\Gamma(s)\pi^{-t}\Gamma(t)D(2s - \nu, \psi) = \int_{\Phi} f\bar{h}y^{\beta} C^*(z, t)d_0 z.$$

Since $\nu = 0$ or 1, we have

$$(3.5) \qquad \Gamma(s)\Gamma(t) = 2\pi^{1/2}2^{\nu-2s}\Gamma(2s - \nu),$$

hence the left hand side of (3.4) equals

$$(3.6) \qquad 2^{2-2s}(2\pi)^{\nu-2s}\Gamma(2s - \nu)D(2s - \nu, \psi).$$

Our assumption $\kappa \geq 3$ implies $\alpha \geq 0$, and $\alpha = 0$ occurs only when $\kappa = 3$ and $\nu = 1$. If $\kappa = 3$, we have $\chi_1(d) = \chi(d)\left(\dfrac{-1}{d}\right)$, so that $\chi_1(-1) = -1$, hence $M > 1$. Therefore, by Lemma 3.3, the function (3.6) can be continued to a holomorphic function in the whole s-plane, and bounded in every vertical strip $\sigma_1 < \mathrm{Re}\,(s) < \sigma_2$. Thus $D(s, \psi)$ has the property (ii) of Lemma 3.1.

Call $\Omega(s)$ the function of (3.4). If we substitute $\lambda + \nu - s$ for s, then t is substituted by $\alpha + 1 - t$, so that

$$\Omega(\lambda + \nu - s) = \int_{\Phi} f\bar{h}y^{\beta}C^{*}(z, \alpha + 1 - t)d_0 z ,$$

and by Lemma 3.3,

$$C^{*}(z, \alpha + 1 - t) = \mathcal{P}(-1)\mathrm{g}(\varphi_0)z^{\alpha}(Mr)^{3t-\alpha-2}(Kr)^{t-1}$$
$$\times \sum_{k \mid E} \mu(k)\varphi_0(k)k^{t-\alpha-1}H_{\alpha}(t, -k/Nr^2 z, \bar{\varphi}_0) .$$

Let us now make the transformation $z \mapsto -1/Nr^2 z$ in the above integral. Put

$$g(z) = f(-1/Nz)N^{-\kappa/4}(-iz)^{-\kappa/2} .$$

Then $g \in S_{\kappa}(N, \chi')$ with $\chi'(d) = \bar{\chi}(d)\left(\dfrac{N}{d}\right)$ by Proposition 1.4, and

$$f(-1/Nr^2 z) = g(r^2 z)N^{\kappa/4}(-ir^2 z)^{\kappa/2} .$$

By Proposition 2.2, putting $N = 4P$, we have

$$h(-1/Nr^2 z) = (-i)^{\nu}r^{-1/2}g(\bar{\psi})(-2rPiz)^{\beta}h_{\psi}(Pz) ,$$

hence

$$(f\bar{h}y^{\beta})(-1/Nr^2 z) = T \cdot z^{\alpha}g(r^2 z)\bar{h}_{\psi}(Pz)y^{\beta} ,$$

where

$$T = 2^{\alpha}r^{\lambda+\alpha}P^{\kappa/4}i^{\nu-\alpha}\overline{\mathrm{g}(\bar{\psi})} .$$

Since the transformation $z \mapsto -1/Nr^2 z$ maps a fundamental domain for $\Gamma \backslash \mathfrak{H}$ onto another fundamental domain, we obtain

$$(3.7) \qquad \Omega(\lambda + \nu - s) = S \cdot \sum_{k \mid E} \mu(k)\varphi_0(k)k^{t-\alpha-1}I_k(t) ,$$

where

$$I_k(t) = \int_{\Phi} g(r^2 z)\bar{h}_{\psi}(Pz)y^{\beta}H_{\alpha}(t, kz, \bar{\varphi}_0)d_0 z ,$$

$$S = T \cdot \mathrm{g}(\varphi_0)r^{4t-3\alpha-3}M^{3t-2\alpha-2}K^{t-\alpha-1} .$$

Now we want to transform the integral on Φ to that on Y by the principle described by (3.2). As it will turn out, this does not go so smoothly. Anyhow, put

$$B'(z) = g(r^2z)\bar{h}_\psi(Pz)y^{s+1} ,$$

$$J(\gamma, z) = \bar{\varphi}_0(d)(cz + d)^\alpha \mid cz + d \mid^{-2t} \qquad \text{for } \gamma = \begin{bmatrix} a & b \\ c & d \end{bmatrix} \in \Gamma_0(Mr) .$$

Then $B'(\gamma(z)) = B'(z)J(\gamma, z)$ for every $\gamma \in \Gamma_0(Nr^2)$. Before proceeding further, let us discuss the simplest case in which $r = 1$ and $N = M$. In such a case, we have $\nu = 0$, and

$$\Omega(\lambda - s) = S \cdot \int_\Phi g(z)\bar{h}_\psi(Pz)y^s H_\alpha(t, z, \bar{\varphi}_0)d_0z$$

$$= S \cdot \pi^{-t}\Gamma(t) \int_\Phi B'(z) \sum_{m,n} \bar{\varphi}_0(n)(Nmz + n)^\alpha \mid Nmz + n \mid^{-2t} d_0z$$

$$= 2S \cdot \pi^{-t}\Gamma(t)L(2t - \alpha, \bar{\varphi}_0) \int_\Phi B'(z)\{\sum_{\gamma \in R} J(\gamma, z)\}d_0z$$

with any set of representatives R for $\Gamma_\infty \backslash \Gamma_0(N)$. Therefore, if $g(z) = \sum_{n=1}^\infty b(n)e(nz)$, we have

$$\Omega(\lambda - s) = 2S \cdot \pi^{-t}\Gamma(t)L(2t - \alpha, \bar{\varphi}_0) \int_Y g(z)\bar{h}_\psi(Pz)y^{s-1}dxdy$$

$$= 2S \cdot \pi^{-t}\Gamma(t)L(2t - \alpha, \bar{\varphi}_0)(4P\pi)^{-s}\Gamma(s) \sum_{m=1}^\infty b(Pm^2)m^{-2s} ,$$

which is a special case of the desired functional equation. This almost straightforward computation does not work if either $r \neq 1$ or $N \neq M$. We overcome this difficulty by reducing the integral on Φ to that on a fundamental domain for $\Gamma_0(Mr)$, under the assumption

$$(3.8) \qquad\qquad N \mid M^\infty, \text{ i.e., } E = 1 .$$

(The general case will be reduced to this special case.)

With any set R' of representatives for $\Gamma_\infty \backslash \Gamma_0(Mr)$, put

$$\eta(z) = \sum_{\gamma \in R'} J(\gamma, z) .$$

Then, assuming (3.8), we have

$$(3.9) \qquad \Omega(\lambda + \nu - s) = 2S \cdot \pi^{-t}\Gamma(t)L(2t - \alpha, \bar{\varphi}_0) \int_\Phi B'(z)\eta(z)d_0z .$$

Since $N \mid M^\infty$, we have the coset decomposition

$$\Gamma_0(Mr) = \bigcup_{u=1}^{Kr} \Gamma_0(Nr^2)\gamma_u , \quad \gamma_u = \begin{bmatrix} 1 & 0 \\ Mru & 1 \end{bmatrix} .$$

Let Ψ be any fundamental domain for $\Gamma_0(Mr)$. Then we can take $\bigcup_u \gamma_u(\Psi)$ as Φ, so that

$$(3.10) \qquad \int_\Phi B'(z)\eta(z)d_0z = \int_\Psi \sum_{u=1}^{Kr} B'(\gamma_u(z))\eta(\gamma_u(z))d_0z$$

$$= \int_\Psi \mathfrak{A}(z)\eta(z)d_0z ,$$

where

$$\mathfrak{A}(z) = \sum_{u=1}^{Kr} B'\big(\gamma_u(z)\big)J(\gamma_u, z)^{-1} \,.$$

It can easily be seen that $\mathfrak{A}\big(\gamma(z)\big) = \mathfrak{A}(z)J(\gamma, z)$ for every $\gamma \in \Gamma_0(Mr)$, hence (3.10) equals

$$\sum_{\tau \in R'} \int_\Psi \mathfrak{A}\big(\gamma(z)\big)d_0z = \int_Y \mathfrak{A}(z)d_0z \,.$$

To evaluate the last integral, we decompose $\mathfrak{A}$ into two parts:

$$\mathfrak{A}(z) = \mathfrak{A}_1(z) + \mathfrak{A}_2(z) \,,$$

$$\mathfrak{A}_1(z) = \sum_{1 \le u \le Kr,\, (u,r)=1} B'\big(\gamma_u(z)\big)J(\gamma_u, z)^{-1} \,,$$

$$\mathfrak{A}_2(z) = \sum_{u=1}^{K} B'\big(\gamma_{ru}(z)\big)J(\gamma_{ru}, z)^{-1} \,.$$

($\mathfrak{A}_2$ is unnecessary if $r = 1$.) Our next task is to obtain "explicit" Fourier expansions of $\mathfrak{A}_1$ and $\mathfrak{A}_2$. For an integer v, put

$$\delta_v^* = \left(\begin{bmatrix} 1 & 0 \\ Mv & 1 \end{bmatrix}, (Mvz + 1)^{1/2}\right) \,,$$

$$g_v = g \mid [\delta_v^*]_\kappa \,.$$

Then g_v depends only on v modulo K. Since M is the conductor of χ_1, M is divisible by 4 by virtue of (3.8). Therefore, by Lemma 3.4, we can write

$$(3.11) \qquad\qquad g_v(z) = \sum_{n=1}^{\infty} c_v(n)e(nz/K)$$

with complex coefficients $c_v(n)$. For an integer u prime to r, choose integers p and q so that $rp + Nuq = 1$ and $p > 0$. Then

$$\begin{bmatrix} 1 & 0 \\ Mru & 1 \end{bmatrix} = \begin{bmatrix} r^{-1} & 0 \\ 0 & r \end{bmatrix}\begin{bmatrix} r & -Kq \\ Mu & p \end{bmatrix}\begin{bmatrix} 1 & Kq/r \\ 0 & 1 \end{bmatrix} \,.$$

Therefore, by Lemma 3.4, we have

$$g\Big(\frac{r^2z}{Mruz + 1}\Big) = g\Big(\frac{r(z + Kq/r) - Kq}{Mu(z + Kq/r) + p}\Big)$$

$$= \chi'(p)\Big(\frac{Mu}{p}\Big)\varepsilon_p^{-\kappa}(Muz + r^{-1})^{\kappa/2}g_v(z + Kq/r)$$

with an integer v such that $pv \equiv u \pmod{K}$. Since $pr \equiv 1 \pmod{N}$, we have $\chi'(p) = \bar{\chi}'(r)$, $\varepsilon_p = \varepsilon_r$. Further we have $pr \equiv 1 \pmod{MKu}$, and M is divisible by 4, so that $\Big(\frac{Mu}{p}\Big) = \Big(\frac{Mu}{r}\Big)$, and $v \equiv ru \pmod{K}$. Therefore, under the condition $(r, u) = 1$, we obtain

(3.12)
$$g\big(r^2z/(Mruz + 1)\big)$$
$$= \bar{\chi}'(r)\Big(\frac{Mu}{r}\Big)\varepsilon_r^{-\kappa}r^{-\kappa/2}(Mruz + 1)^{\kappa/2}g_{ru}(z + Kq/r) \ .$$

Note that q satisfies

(3.13)
$$Nuq \equiv 1 \pmod r \ .$$

If $u = rw$ with an integer w, we have

(3.14)
$$g\big(r^2z/(Mruz + 1)\big) = g\big(r^2z/(Mwr^2z + 1)\big)$$
$$= g_w(r^2z)(Mruz + 1)^{\kappa/2} \ .$$

Next, to obtain the transformation of $h_\psi(Pz)$ by γ_u, put $M = 4Q$. By Proposition 2.3, taking T and z to be K and Qz, we obtain

(3.15)
$$h_\psi\big(Pz/(Mruz + 1)\big) = h_\psi\big(K \cdot Qz/(4ru \cdot Qz + 1)\big)$$
$$= (-1)^\nu K^{-1-\nu}r^{-1}(Mruz + 1)^\beta 2^{-1}\sum_{m=-\infty}^{\infty} m^\nu\zeta(m)e(m^2Qz/K) \ ,$$

where

$$\zeta(m) = \sum_{j=1}^{r}\sum_{k=1}^{Kr}\psi(j)e\big([km + Kjk - uk^2]/Kr\big) \ .$$

Combining this with (3.12), we obtain

$$\mathfrak{A}_1(z) = V \cdot y^{s+1}\sum_{u=1}^{Kr}\sum_{n=1}^{\infty}2^{-1}\sum_{m=-\infty}^{\infty}\Big(\frac{u}{r}\Big)c_{ru}(n)e(nq/r)$$
$$\times\ e(nz/K)m^\nu\overline{\zeta(m)}e(-m^2Q\bar{z}/K) \ ,$$

where $V = \chi(r)\Big(\dfrac{K}{r}\Big)r^{-1-\kappa/2}(-1)^\nu K^{-1-\nu}\varepsilon_r^{-\kappa}$. Since $\mathfrak{A}(z + 1) = \mathfrak{A}(z)$, we have

$$\int_Y \mathfrak{A}(z)d_0z = \int_0^\infty\Big\{K^{-1}\int_0^K[\mathfrak{A}_1(z) + \mathfrak{A}_2(z)]dx\Big\}y^{-2}dy \ .$$

In the integral $\displaystyle\int_0^K\mathfrak{A}_1(z)dx$, non-vanishing terms can occur only for $n = m^2Q$. Thus

$$K^{-1}\int_0^K\mathfrak{A}_1(z)dx = Vy^{s+1}2^{-1}\sum_{m=-\infty}^{\infty}\omega(m)m^\nu\exp{(-\pi m^2My/K)} \ ,$$

where

$$\omega(m) = \sum_{u=1}^{Kr}\sum_{j=1}^{r}\sum_{k=1}^{Kr}c_{ru}(m^2Q)\Big(\frac{u}{r}\Big)\bar{\psi}(j)e(m^2qQ/r)e\big([uk^2 - km - Kjk]/Kr\big) \ .$$

Assuming $r > 1$, put $k = Ka + rb$ with $a \pmod r$ and $b \pmod K$, and let $Ku \equiv v \pmod r$, $u \equiv w \pmod K$. Then $k^2 \equiv K^2a^2 + r^2b^2 \pmod{Kr}$, hence

$$\omega(m) = \Big(\frac{K}{r}\Big)\sum_{w=1}^{K}c_{rw}(m^2Q)c_w^* \ ,$$

where

$$c_w^* = \sum_{v=1}^{r} \sum_{j=1}^{r} \sum_{a=1}^{r} \sum_{b=1}^{K} \left(\frac{v}{r}\right) \bar{\psi}(j) e([m^2 qQ - am - Kaj + va^2]/r)$$
$$\times e([rwb^2 - mb]/K) .$$

By (3.13), we have $Mqv \equiv 1 \pmod{r}$. Therefore if $4vv' \equiv 1 \pmod{r}$, we have $m^2 qQ \equiv m^2 v' \pmod{r}$, so that

$$c_w^* = \sum_{b=1}^{K} e([rwb^2 - mb]/K) \sum_{j=1}^{r} \sum_{a=1}^{r} \sum_{v,v'} \left(\frac{v}{r}\right) \bar{\psi}(j)$$
$$\times e([m^2 v' - am - Kaj + va^2]/r) ,$$

where the sum $\sum_{v,v'}$ is taken over all v, $v' \bmod r$ under the condition $4vv' \equiv 1 \pmod{r}$. Applying now Lemma 3.5 to the last sum, we obtain

$$(3.16) \qquad \omega(m) = \left(\frac{K}{r}\right) \psi(K) \varepsilon_r r^{3/2} \bar{\psi}(-2m) \omega'(m) ,$$

where

$$\omega'(m) = \sum_{b=1}^{K} \sum_{w=1}^{K} c_w(m^2 Q) e([wb^2 - mb]/K) .$$

Observe that $\omega'(-m) = \omega'(m)$ and $\omega'(0) = 0$. Therefore

$$K^{-1} \int_0^K \mathfrak{A}_1(z)dx = W \cdot y^{s+1} \sum_{m=1}^{\infty} \bar{\psi}(m) m^\nu \omega'(m) \exp(-\pi m^2 My/K) ,$$

where $W = (-1)^\nu \chi_1(r) \psi(K) \bar{\psi}(-2) r^{-j} K^{-\nu-1}$. Thus we obtain

$$(3.17) \qquad K^{-1} \int_0^\infty \int_0^K \mathfrak{A}_1(z) y^{-2} dx dy = W \cdot \pi^{-s} M^{-s} K^s \Gamma(s) \sum_{m=1}^{\infty} \bar{\psi}(m) \omega'(m) m^{\nu-2s} .$$

If $r = 1$ and ψ is trivial, our consideration becomes simpler. Indeed, we can put $q = 0$, and obtain (3.16) immediately, so that (3.17) is true also in the case $r = 1$.

Next let us consider $\mathfrak{A}_2$ when $r > 1$. By (3.14) and (3.15),

$$K^{-1} \int_0^K \mathfrak{A}_2(z)dx = C \cdot y^{s+1} 2^{-1} \sum_{m=-\infty}^{\infty} \omega^*(m) m^\nu \exp(-\pi r^2 m^2 My/K)$$

with a constant C independent of m, and

$$\omega^*(m) = \sum_{v=1}^{K} c_v(m^2 Q) \sum_{j=1}^{r} \sum_{k=1}^{Kr} \bar{\psi}(j) e([rvk^2 - krm - Kjk]/Kr) .$$

Let $k \equiv -a \pmod{r}$ and $k \equiv b \pmod{K}$. Then

$$\omega^*(m) = \sum_{v=1}^{K} c_v(m^2 Q) \sum_{j=1}^{r} \sum_{a=1}^{r} \sum_{b=1}^{K} \bar{\psi}(j) e([vb^2 - mb]/K) e(aj/r) = 0 ,$$

since $\sum_{a=1}^{r} \sum_{j=1}^{r} \bar{\psi}(j) e(aj/r) = 0$. Therefore (3.10) coincides with (3.17), hence we obtain

$$(3.18) \qquad \Omega(\lambda + \nu - s) = U \cdot L(2t - \alpha, \bar{\varphi}_0) \sum_{m=1}^{\infty} \bar{\psi}(m) \omega'(m) m^{\nu-2s}$$

with $U = 2SWM^{-s} K^s \pi^{-s-t} \Gamma(s) \Gamma(t)$. To compute U, note that

$$g(\mathcal{P}_0) = \psi(M)\chi_0(r)g(\chi_0)g(\psi) ,$$

and $\overline{g(\overline{\psi})} = \psi(-1)g(\psi)$. Therefore, by (3.5),

$$U = U_0\chi(r)^2\psi(N/2)g(\psi)^2 r^{-1}r^{4s-2\lambda-2\nu}N^{2s}(2\pi)^{\nu-2s}\Gamma(2s - \nu)$$

with a constant U_0 independent of r and ψ. Since $\omega'(m)$ is independent of r and ψ, and $\mathcal{P}_0 = \chi_0\psi$, (3.18) can be written as

$$\Omega(\lambda + \nu - s) = U\cdot\sum_{n=1}^{\infty} \overline{\psi}(n)A^*(n)n^{\nu-2s}$$

with coefficients $A^*(n)$ independent of r and ψ. Put

$$C_\psi = \psi(N/2)\chi(r)^2 g(\psi)^2/r ,$$

$$R(s, \psi) = (2\pi)^{-s}\Gamma(s) \sum_{n=1}^{\infty} \psi(n)A_1(n)n^{-s} .$$

Then $R(s, \psi) = 2^{s+\nu-2}\Omega((s + \nu)/2)$, hence

$$R(2\lambda - s, \psi) = 2^{2\lambda-s+\nu-2}\Omega(\lambda + \nu - (s + \nu)/2)$$
$$= V\cdot C_\psi(r^2N/2)^{s-\lambda}(2\pi)^{-s}\Gamma(s) \sum_{n=1}^{\infty} \overline{\psi}(n)A^*(n)n^{-s}$$

with a constant V independent of r and ψ. This shows that our Dirichlet series $\sum_{n=1}^{\infty} A_1(n)n^{-s}$ satisfies the condition (iii) of Lemma 3.2 with $N/2$ and χ^2 as M and $\mathcal{P}$. Thus we have completed the proof of the Main Theorem in the case $t = 1$ under the assumption (3.8).

To prove the most general case, define χ_t, $A_t(n)$, and F_t as in the Main Theorem, with a fixed square-free positive integer t. Put

$$f_t = f(tz) = \sum_{n=1}^{\infty} a(n)e(tnz) = \sum_{m=1}^{\infty} b(m)e(mz) ,$$
$$\sum_{n=1}^{\infty} A'(n)n^{-s} = \left(\sum_{m=1}^{\infty} \chi_t(m)m^{\lambda-1-s}\right)\left(\sum_{m=1}^{\infty} b(m^2)m^{-s}\right) ,$$
$$R'(s, \psi) = (2\pi)^{-s}\Gamma(s) \sum_{n=1}^{\infty} \psi(n)A'(n)n^{-s} ,$$
$$R_t(s, \psi) = (2\pi)^{-s}\Gamma(s) \sum_{n=1}^{\infty} \psi(n)A_t(n)n^{-s} ,$$

where ψ is a primitive character modulo r; r is either 1 or a prime not dividing tN. By Proposition 1.3, $f_t \in S_\kappa(tN, \eta)$ with $\eta(d) = \chi(d)\left(\dfrac{t}{d}\right)$. Therefore, applying our result obtained for f to f_t, we see that $R'(s, \psi)$ has the property (ii) of Lemma 3.1. (For this we do not need (3.8).) Now $b(m) = a(m/t)$ or 0 according as m is divisible by t or not. It follows that $\sum_{m=1}^{\infty} b(m^2)m^{-s} = \sum_{n=1}^{\infty} a(tn^2)(tn)^{-s}$, hence $R_t(s, \psi) = \psi(t)^{-1}t^s R'(s, \psi)$. Therefore $R_t(s, \psi)$ has the property (ii) of Lemma 3.1.

Thus our remaining task is to prove the functional equation for $R_t(s, \psi)$. Since f_t may not satisfy the condition (3.8), we have to introduce a function satisfying it. For this purpose, let M_t be the conductor of χ_t, and K the product of all prime factors of N not dividing M_t. Now we can find a primitive character ξ modulo an integer H such that:

(i) $H = \prod_{p|K} p^{\nu(p)}$, $\nu(p) = 1$ if $p \neq 2$, $\neq 3$; $\nu(2) = 4$, $\nu(3) = 2$ (if 2 or 3 divides K).

(ii) The conductor of ξ^2 is $H/(H, 2)$.

Decompose t into the product $t = t_1 t_2$ with positive integers t_1 and t_2 such that $(t_1, N) = 1$, $t_2 \mid N$. By Proposition 1.5,

$$\sum_{n=1}^{\infty} a(t_2 n) e(nz) \in S_\kappa(N', \chi') \ ,$$

where $\chi'(d) = \chi(d)\left(\dfrac{t_2}{d}\right)$, and

$$N' = \begin{cases} 2N & \text{if } 2 \mid t_2 \text{ and } 8 \nmid N \, , \\ N & \text{otherwise} \, . \end{cases}$$

Let M' be the conductor of χ', and N^* the least common multiple of N', H^2, $M'H$. By Lemma 3.6,

$$\sum_{n=1}^{\infty} \xi(n) a(t_2 n) e(nz) \in S_\kappa(N^*, \xi^2\chi') \ .$$

Now put

$$f^*(z) = \sum_{n=1}^{\infty} \xi(n) a(t_2 n) e(t_1 nz) = \sum_{n=1}^{\infty} a^*(n) e(nz) \ .$$

By Proposition 1.3, $f^* \in S_\kappa(N^* t_1, \chi^*)$, where

$$\chi^*(d) = \xi^2(d)\chi(d)\left(\frac{t}{d}\right) = \xi^2(d)\chi_t(d)\left(\frac{-1}{d}\right)^{\lambda}.$$

Put $H' = H/(H, 2)$. Then the conductor of $\xi^2\chi_t$ is $H'M_t$, and $N^* t_1 \mid (H'M_t)^\infty$. Therefore the condition (3.8) is satisfied for f^*, so that our result obtained above for f is applicable to f^*. Namely, put

$$\sum_{n=1}^{\infty} A^*(n) n^{-s} = \left(\sum_{m=1}^{\infty} \chi^*(m)\left(\frac{-1}{m}\right)^{\lambda} m^{\lambda-1-s}\right)\left(\sum_{m=1}^{\infty} a^*(m^2) m^{-s}\right) \ ,$$

$$F^*(z) = \sum_{n=1}^{\infty} A^*(n) e(nz) \ .$$

Then F^* belongs to $\mathfrak{S}_{\kappa-1}(N^* t_1/2, \chi^{*2})$. Now we see that

$$\sum_{m=1}^{\infty} a^*(m^2) m^{-s} = \xi(t_1) t_1^{-s} \sum_{n=1}^{\infty} \xi(n)^2 a(tn^2) n^{-s} \ ,$$

hence

$$A^*(m) = \begin{cases} 0 & \text{if } t_1 \nmid m \, , \\ \xi(m^2/t_1) A_t(m/t_1) & \text{if } t_1 \mid m \, . \end{cases}$$

Put $F_0(z) = \xi(t_1) \sum_{n=1}^{\infty} \bar{\xi}(n)^2 A^*(n) e(nz)$. Observe that if L denotes the conductor of χ^{*2}, then $N^*/2$ is a common multiple of H'^2 and $H'L$. Therefore, by [8, Prop. 3.64], $F_0 \in \mathfrak{S}_{\kappa-1}(N^* t_1/2, \bar{\xi}^4\chi^{*2})$. Now we see that $\bar{\xi}^4(d)\chi^{*2}(d) = \chi^2(d)$ for $(d, tN) = 1$, and

$$F_0(z) = \sum_{(n, H)=1} A_t(n) e(t_1 nz) \ .$$

As remarked at the beginning of this section, if we put

$$R_0(s, \psi) = (2\pi)^{-s}\Gamma(s) \sum_{(n,H)=1} \psi(t_1 n) A_t(n)(t_1 n)^{-s} ,$$

with a character ψ modulo r, it satisfies the functional equation

$$R_0(\kappa - 1 - s, \psi) = \psi(N^* t_1/2)\chi(r)^2 g(\psi)^2 r^{-1}$$
$$\times (r^2 N^* t_1/2)^{s-\lambda}(2\pi)^{-s}\Gamma(s) \sum_{n=1}^{\infty} \bar{\psi}(n) B_n n^{-s}$$

with a certain Dirichlet series $\sum_n B_n n^{-s}$, provided $(r, tN) = 1$. Now we need the assumption that f is an eigen-function of $T_{\kappa,\chi}^N(p^2)$ for every prime factor p of H. Put $f \mid T_{\kappa,\chi}^N(p^2) = \omega_p f$. By Corollary 1.8,

$$\sum_{n=1}^{\infty} a(tn^2)n^{-s} = \prod_{p|H} (1 - \omega_p p^{-s})^{-1} \cdot \sum_{(n,H)=1} a(tn^2)n^{-s} .$$

Therefore we have

$$R_t(s, \psi) = \bar{\psi}(t_1) t_1^s \cdot \prod_{p|H} (1 - \psi(p)\omega_p p^{-s})^{-1} R_0(s, \psi) .$$

Let K_0 be the product of all prime factors p of H for which $\omega_p \neq 0$. Then

$$R_t(\kappa - 1 - s, \psi) = C \cdot \psi(N^*/2K_0)\chi(r)^2 g(\psi)^2 r^{-1}(r^2 N^*/2K_0)^{s-\lambda}$$
$$\times (2\pi)^{-s}\Gamma(s) \prod_{p|K_0} (1 - \bar{\psi}(p)\omega_p^{-1} p^{\kappa-1-s})^{-1} \sum_{n=1}^{\infty} \bar{\psi}(n) B_n n^{-s}$$

with a constant C independent of ψ. Thus R_t satisfies the condition (iii) of Lemma 3.1. This completes the proof of the Main Theorem in the most general case.

The above proof shows that one can take $N^*/2K_0$ as the integer N_t of the Main Theorem. Now it can easily be verified that N^* is the least common multiple of N, H^2, and M'. Therefore we obtain

PROPOSITION 3.7. *The notation being as in the Main Theorem, let M_t be the conductor of χ_t, and M' the conductor of the character χ' defined by* $\chi'(d) = \chi(d)\left(\frac{t_2}{d}\right)$, *where t_2 is the greatest common divisor of t and N. Define a positive integer H by*

$$H = \prod_{p|N, p\nmid M_t} p^{\nu(p)} \qquad (p: prime)$$

with $\nu(p) = 1$ if $p > 3$, and $\nu(2) = 4$, $\nu(3) = 2$. Let K_0 be the product of all prime factors p of H for which $\omega_p \neq 0$, where $f \mid T_{\kappa,\chi}^N(p^2) = \omega_p f$. Further let N^* be the least common multiple of N, H^2, and M'. Then $N^*/2K_0$ can be taken as the integer N_t of the Main Theorem.

4. Open questions and examples

The theory developed in §1 and the Main Theorem lead us to a few natural questions, which we shall now discuss in order.

(A) Probably the integer $N^*/2K_0$ given in Proposition 3.7 is not best possible. For example, one can ask:

Can one always take $N/2$ as N_t, or as N_0 of Corollary to the Main Theorem?

Assuming this to be true, say, for some N and χ, one can then ask:

Can every element of $\mathfrak{S}_{\kappa-1}(N/2, \chi^2)$ be obtained from an element of $S_\kappa(N, \chi)$ as described in the Main Theorem? If so, to what extent is the correspondence one-to-one?

The answers to both questions seem to be on the affirmative side. One can at least compare the dimensionality of two vector spaces in question. For simplicity, let us consider here only the case of the trivial character χ, and write $S_\kappa(N)$ (resp. $\mathfrak{S}_k(N)$) for $S_\kappa(N, \chi)$ (resp. $\mathfrak{S}_k(N, \chi)$) with trivial χ. By means of a well-known argument using the Riemann-Roch theorem, we can easily compute $\dim\big(S_\kappa(N)\big)$ and $\dim\big(\mathfrak{S}_k(N)\big)$ in terms of the genus of $\Gamma_0(N)$, the number of cusps, and the orders of elliptic elements of $\Gamma_0(N)$. Then we find

$$(4.1) \qquad \dim\big(S_\kappa(N)\big) = \dim\big(\mathfrak{S}_{\kappa-1}(N/2)\big)$$

if $\kappa \geq 5$, and $N/4$ is either 1 or a prime. However, if $N/4$ has a square factor, there is a difference, which has the following rather interesting form:

$$(4.2) \qquad \dim\big(S_\kappa(4p^2)\big) - \dim\big(\mathfrak{S}_{\kappa-1}(2p^2)\big) = (-1)^{(\kappa-1)/2}(p \pm 1)/4 \,,$$

if p is an odd prime and $p \pm 1 \equiv 0 \pmod 4$. The difference is probably explained by the duplication of modular forms of the type $f(z)$, $f(pz)$, and $f(p^2z)$, or something similar to it.

It should also be noted that $\chi^2 = \chi'^2$ for two different characters χ and χ'. This point must be taken into account when one tries to solve questions of the above type.

To illustrate the correspondence between $\mathfrak{S}_{\kappa-1}(N/2)$ and $S_\kappa(N)$, let us consider the simplest case $N = 4$. Put

$$\theta(z) = \sum_{n=-\infty}^{\infty} e(n^2 z) \,, \quad \eta(z) = e(z/24) \prod_{n=1}^{\infty} \big(1 - e(nz)\big) \,,$$

$$f(z) = \theta(z)^{-3} \eta(2z)^{12} = \sum_{n=1}^{\infty} a(n) e(nz) \,.$$

Then f generates $S_9(4)$. Put

$$\sum_{n=1}^{\infty} A_n n^{-s} = \big(\sum_{n=1}^{\infty} (2n-1)^{3-s}\big)\big(\sum_{n=1}^{\infty} a(n^2) n^{-s}\big) \,.$$

By Theorem 1.9, we have

$$\sum_{n=1}^{\infty} A_n n^{-s} = (1 - A_2 2^{-s})^{-1} \prod_{p \cdot \mathrm{odd}} (1 - A_p p^{-s} + p^{7-2s})^{-1} \,,$$

$$(4.3) \qquad A_2 = a(2^2), \quad A_p = a(p^2) + p^3 \qquad\qquad (p \geq 3) \,.$$

Our main theorem or rather its corollary implies that $\sum_{n=1}^{\infty} A_n e(nz)$ belongs to $\mathfrak{S}_8(2)$. Actually $\mathfrak{S}_8(2)$ is generated by $(\eta(z)\eta(2z))^8$, hence

$$(\eta(z)\eta(2z))^8 = \sum_{n=1}^{\infty} A_n e(nz) \ .$$

Thus the theorem specialized to this case implies a set of equalities (4.3) between the coefficients of $\theta(z)^{-3}\eta(2z)^{12}$ and $(\eta(z)\eta(2z))^8$.

Similar examples for higher level can be obtained by considering products of $\theta(mz)$ and $\eta(kz)$ with various positive integers k and m. For instance, $\theta(3z)^{-1}\eta(2z)^3\eta(6z)^3$ generates $S_5(12)$, and corresponds to the generator $[\eta(z)\eta(2z)\eta(3z)\eta(6z)]^2$ of $\mathfrak{S}_4(6)$.

(B) Given an element $f(z) = \sum_{n=1}^{\infty} a(n)e(nz)$ of $S_\kappa(N, \chi)$, our theory tells us the mutual relation of the Fourier coefficients $a(tn^2)$ for $n = 1$, $2, \cdots$, with a fixed square-free t. Then one can ask:

What is the nature of the coefficients $a(t)$ as a whole for all square-free t?

If one starts from an element $F(z) = \sum_{n=1}^{\infty} A(n)e(nz)$ of $\mathfrak{S}_{\kappa-1}(N/2, \chi^2)$, a similar question can be asked with a somewhat different nuance. Namely, if F is obtained from f as described in the Main Theorem, then the coefficients $a(t)$, for square-free t, are a sort of "interpolation" for the original coefficients $A(n)$; for instance, one may symbolically put $a(p) = A(\sqrt{p})$ for a prime p. Then *what is the meaning of all these $a(t)$ for the Dirichlet series $\sum_{n=1}^{\infty} A(n)n^{-s}$?* In a certain special case, $\sum_{n=1}^{\infty} A(n)n^{-s}$ becomes an L-function with a Grössen-character of an imaginary quadratic field. *Is there any relation between the $a(t)$ and this character?*

(C) Our main theorem is stated under the restriction $\kappa \geqq 3$ or $\kappa \geqq 5$. When $\kappa = 3$, the situation will probably be as follows. Let U be the sub-space of $S_3(N, \chi)$ spanned by the functions of the form $h_\psi(kz)$, where k is a positive integer, and $h_\psi(z) = \sum_{m=1}^{\infty} \psi(m)me(m^2z)$ with a primitive character ψ modulo a positive integer r such that $\psi(-1) = -1$. $\Big($Of course $4kr^2$ must be a divisor of N, and $\chi(d) = \psi(d)\Big(\dfrac{-k}{d}\Big)\Big)$. Further let V be the orthogonal complement of U in $S_3(N, \chi)$ with respect to the Petersson inner product. Then one can conjecture that, *with the notation of the Main Theorem $F_t \in \mathfrak{S}_2(N_t, \chi^2)$ if and only if $f \in V$.* (Obviously F_t is not a cusp form if $f \in U$.)

As to the case $\kappa = 1$, one can ask *whether $G_1(N, \chi)$ can be spanned by the functions of the form $h_\psi(kz)$ with a positive integer k and a primitive character ψ such that $\psi(-1) = 1$.* More generally, *is the space $S_\kappa(N, \chi)$ for $\kappa \geqq 5$ spanned by the theta-series discussed in § 2?*

(D) The significance of coverings of SL_2, over local fields and adeles, in the theory of theta-series, was pointed out by Weil [11]. Actually, our group $\mathfrak{G}_1$ of §1 is the simplest case of the metaplectic group introduced in the paper. Also, the transformation formulas of theta series considered in §2 can be obtained by the methods of [11]. Now it is very likely that one can develop a theory which connects our results with the representation-theory of coverings of GL_2, similar to what is done in Jacquet-Langlands [4]. This might lead to answers to some of the above questions. One can also try to obtain a trace-formula for the Hecke operators on $S_\kappa(N, \chi)$. It is almost needless to say that all these should be studied in the framework with an arbitrary global field as the ground field.

5. More functional equations

If $f(z) = \sum_{n=1}^{\infty} a_n e(nz) \in S_\kappa(N, \chi)$, one can naturally consider a Dirichlet series

$$(5.1) \qquad \sum_{n=1}^{\infty} \psi(n) a_n n^{-s}$$

with any character ψ modulo some positive integer. As shown in Lemma 3.6, $\sum_{n=1}^{\infty} \psi(n) a_n e(nz) \in S_\kappa(N', \chi')$ for some N' and χ'. Therefore one can prove the analytic continuation and the functional equation of (5.1) by the standard method. Especially, if $f = h_\varphi$ with a character φ as in Proposition 2.2, the series (5.1) is the L-function with the character $\psi^2\varphi$, but for a more general type of f, the nature of (5.1) is obscure. However, one can at least prove the characterization of f by such functional equations, which is similar to Weil's criterion [11] (see Lemma 3.1 above). Although neither the significance of this characterization nor its relation with our main theorem is clear, it seems worth while discussing it, since there are some new features in this case. We start with

PROPOSITION 5.1. *Let* $f(z) = \sum_{n=0}^{\infty} a_n e(nz) \in G_\kappa(N, \chi)$, *and let* ψ *be a primitive character modulo* r. *Suppose that* r *is prime to* N, *and put* $\lambda = (\kappa - 1)/2$,

$$f^*(z) = \sum_{n=0}^{\infty} \psi(n) a_n e(nz) ,$$

$$g(z) = f(-1/Nz) N^{-\kappa/4} (-iz)^{-\kappa/2} = \sum_{n=0}^{\infty} b(n) e(nz) ,$$

$$g^*(z) = \sum_{(u,r)=1} \psi(u) \left(\frac{u}{r}\right) g(z + u/r) ,$$

where u *runs over all irreducible residue classes modulo* r. *Then*

$$f^*(-1/Nr^2 z)(Nr^2)^{-\kappa/4}(-iz)^{-\kappa/2}$$
$$= \left(\frac{-1}{r}\right)^\lambda \chi(r)\left(\frac{N}{r}\right) \psi(-N)\varepsilon_r^{-1} g(\bar{\psi})^{-1} g^*(z) .$$

Proof. For an integer u prime to r, take integers d and w so that $dr - Nuw = 1$. Then

$$\begin{bmatrix} 1 & u/r \\ 0 & 1 \end{bmatrix}\begin{bmatrix} 0 & -r^{-1} \\ rN & 0 \end{bmatrix} = \begin{bmatrix} 0 & -1 \\ N & 0 \end{bmatrix}\begin{bmatrix} r & -w \\ -Nu & d \end{bmatrix}\begin{bmatrix} 1 & w/r \\ 0 & 1 \end{bmatrix}.$$

Let $\xi(u) = \left(\begin{bmatrix} 1 & u/r \\ 0 & 1 \end{bmatrix}, 1\right)$, and $\tau = \left(\begin{bmatrix} 0 & -1 \\ Nr^2 & 0 \end{bmatrix}, (Nr^2)^{1/4}(-iz)^{1/2}\right)$. Checking automorphic factors carefully, we see that

$$f \mid [\xi(u)\tau]_\kappa = \chi(r)\left(\frac{N}{r}\right)\varepsilon_r^{-\kappa}\left(\frac{w}{r}\right)g(z + w/r) \,.$$

Since $\mathfrak{g}(\bar{\psi})f^* = \sum_u \bar{\psi}(u)f \mid [\xi(u)]_\kappa$, we obtain the desired result.

Now suppose that r is a prime, and put $\varphi(u) = \left(\dfrac{u}{r}\right)$. Then $\varphi\psi$ is either a primitive character modulo r, or the identity character modulo r. Therefore

$$g^*(z) = \begin{cases} r\cdot\sum_{n=0}^{\infty} b(rn)e(rnz) - g(z) & \text{if } \psi = \varphi \,, \\ \mathfrak{g}(\varphi\psi)\cdot\sum_{n=1}^{\infty} \bar{\psi}(n)\left(\dfrac{n}{r}\right)b(n)e(nz) & \text{if } \psi \neq \varphi \,. \end{cases}$$

Combining this with Proposition 5.1, we obtain

$$(5.2)\qquad f^* \mid [\tau]_\kappa = \begin{cases} C'_\psi\cdot\sum_{n=1}^{\infty} \bar{\psi}(n)\left(\dfrac{n}{r}\right)b(n)e(nz) & \text{if } \psi \neq \varphi \,, \\ C'_\psi\left\{r\cdot\sum_{n=0}^{\infty} b(rn)e(rnz) - \sum_{n=0}^{\infty} b(n)e(nz)\right\} & \text{if } \psi = \varphi \,, \end{cases}$$

where

$$C'_\psi = \begin{cases} \chi(r)\left(\dfrac{N}{r}\right)\left(\dfrac{-1}{r}\right)^{\lambda}\psi(-N)\varepsilon_r^{-1}\mathfrak{g}(\varphi\psi)\mathfrak{g}(\bar{\psi})^{-1} & \text{if } \psi \neq \varphi \,, \\ \chi(r)\left(\dfrac{-1}{r}\right)^{\lambda}r^{-1/2} & \text{if } \psi = \varphi \,. \end{cases}$$

Actually we can write

$$C'_\psi = \chi(r)\left(\frac{N}{r}\right)\left(\frac{-1}{r}\right)^{\lambda}\psi(N/4)\mathfrak{g}(\psi^2)r^{-1/2} \quad \text{if } \psi \neq \varphi.$$

This is due to the following

PROPOSITION 5.2. *Let ψ be a primitive character modulo an odd prime r, and let $\varphi(u) = \left(\dfrac{u}{r}\right)$. If $\psi \neq \varphi$, one has*

$$\mathfrak{g}(\psi\varphi)/\mathfrak{g}(\bar{\psi}) = \varphi(-1)\bar{\psi}(-4)\mathfrak{g}(\psi^2)/\mathfrak{g}(\varphi) \,.$$

Proof. It seems possible to prove this by means of Jacobi sums. Here we deduce it by taking the above f to be the function $2^{-1}\theta(z) = (1/2) + \sum_{n=1}^{\infty} e(n^2z)$ with $N = 4$. Then $g = 2^{-1}\theta$, $f^* = h_{\psi^2}$, $g^* = \mathfrak{g}(\varphi\psi)h_{\bar{\psi}^2}$

with the notation of Proposition 2.2. Therefore our assertion can be obtained by comparing the transformation formula of that proposition with (5.2).

Suppose that f is a cusp form, and put

$$R(s, f, \psi) = (2\pi)^{-s}\Gamma(s) \sum_{n=1}^{\infty} \psi(n)a_n n^{-s} .$$

Since $R(s, f, \psi) = \int_0^{\infty} f^*(iy)y^{s-1}dy$, we obtain, by making the variable change $y \mapsto 1/r^2 Ny$, the analytic continuation and the functional equation

$$R(\kappa/2 - s, f, \psi) = \begin{cases} C_{\psi}'(r^2 N)^{s-\kappa/4}R(s, g, \bar{\psi}) & \text{if } \psi \neq \varphi , \\ C_{\psi}'(r^2 N)^{s-\kappa/4}(2\pi)^{-s}\Gamma(s) \\ \qquad \times \{\sum_{m=1}^{\infty} b(rm)r^{1-s}m^{-s} - \sum_{n=1}^{\infty} b(n)n^{-s}\} & \text{if } \psi = \varphi . \end{cases}$$

Now these functional equations characterize the elements of $S_\kappa(N, \chi)$ exactly in the same manner as in Lemma 3.1, although we do not give here an explicit statement, which is rather obvious; the proof can also be given by the same ideas as in Weil [11].

PRINCETON UNIVERSITY

REFERENCES

[1] G. H. HARDY, *Ramanujan*, Cambridge, 1940.

[2] E. HECKE, Zur Theorie der elliptischen Modulfunktionen, Math. Ann. **97** (1926), 210-242 (=Werke, 428-460).

[3] ————, Herleitung des Euler-Produktes der Zetafunktion und einiger L-Reihen aus ihrer Funktionalgleichung, Math. Ann. **119** (1944), 266-287 (=Werke, 919-940).

[4] H. JACQUET and R. P. LANGLANDS, *Automorphic forms on GL(2)*, lecture notes in mathematics, Springer, 1970.

[5] W. PFETZER, Die Wirkung der Modulsubstitutionen auf mehrfache Thetareihen zu quadratischen Formen ungerader Variablenzahl, Archiv der Math. **4** (1953), 448-454.

[6] R. A. RANKIN, Contributions to the theory of Ramanujan's function $\tau(n)$ and similar arithmetical functions I, II, Proc. Cambridge Phil. Soc. **35** (1939), 351-372.

[7] B. SCHOENEBERG, Das Verhalten von mehrfachen Thetareihen bei Modulsubstitutionen, Math. Ann. **116** (1939), 511-523.

[8] G. SHIMURA, *Introduction to the arithmetic theory of automorphic functions*, Iwanami Shoten and Princeton Univ. Press, 1971.

[9] C. L. SIEGEL, *Lectures on advanced analytic number theory*, Tata Institute of Fundamental Research, Bombay, 1961.

[10] A. WEIL, Sur certains groupes d'opérateurs unitaires, Acta Math. **111** (1964), 143-211.

[11] ————, Über die Bestimmung Dirichletscher Reihen durch Funktionalgleichungen, Math. Ann. **168** (1967), 149-156.

[12] K. WOHLFAHRT, Über Operatoren Heckescher Art bei Modulformen reeller Dimension, Math. Nachr. **16** (1957), 233-256.

(Received May 18, 1972)

73d

On the factors of the jacobian variety of a modular function field

Journal of Mathematical Society of Japan, 25 (1973), 523-544

(Received Nov. 13, 1972)

§ 0. Introduction.

For a positive integer N, put

$$\Gamma_0(N) = \left\{ \begin{bmatrix} a & b \\ c & d \end{bmatrix} \in SL_2(\mathbf{Z}) \mid c \equiv 0 \pmod{N} \right\},$$

$$\Gamma_1(N) = \left\{ \begin{bmatrix} a & b \\ c & d \end{bmatrix} \in \Gamma_0(N) \mid a \equiv d \equiv 1 \pmod{N} \right\}.$$

We consider any group Γ such that $\Gamma_1(N) \subset \Gamma \subset \Gamma_0(N)$, and call it *a group of level N*. Let J denote the jacobian variety of the compact Riemann surface $\mathfrak{H}^*/\Gamma$, where $\mathfrak{H}^*$ means the union of the upper half plane

$$\mathfrak{H} = \{ z \in \mathbf{C} \mid \mathrm{Im}\,(z) > 0 \}$$

and the cusps of Γ. Further let $S_k(\Gamma)$ be the vector space of all holomorphic cusp forms of weight k with respect to Γ. Then, with each common eigenfunction $f(z)$ in $S_2(\Gamma)$ of the Hecke operators T_n for all n, one can associate an abelian variety A that is a "factor" of J. The purpose of this note is to consider a few arithmetical questions concerning the correspondence between f and A. Besides, as an application of our methods, we shall give a proof of Dirichlet's class number formula for an imaginary quadratic field, without using the residue technique.

We start our treatment by proving that A can naturally be obtained as a *quotient* of J by an abelian subvariety rational over $\mathbf{Q}$ (Theorem 1). Actually in [11, § 7.5], we gave a formulation with such a factor as a *subvariety* of J. The two formulations are essentially equivalent, but there is a subtle difference. At any rate, they are connected by the following fact: there is a canonical C-linear isomorphism of $S_2(\Gamma)$ onto the tangent space of J at the origin, which has a certain commutative property with the action of Hecke operators. Such an isomorphism was used in the proofs of [11, Th. 7.14, Prop. 7.19] and also in [12], but not explicitly given. This point will be clarified in § 2. It will be shown in § 3 that A can be obtained as a complex torus whose periods are those of f and some other cusp forms. We shall consider

574

in §4 how the map

$$(*) \qquad f(z) = \sum_{n=1}^{\infty} a_n e^{2\pi i n z} \longmapsto \sum_{n=1}^{\infty} \chi(n) a_n e^{2\pi i n z}$$

with a numerical character χ, can be reflected by geometric objects. If χ is a character of order 2, it corresponds to a homomorphism of A rational over the quadratic field $\mathfrak{k}$ associated with χ. Under certain conditions, A is similar to abelian varieties of the type discussed in [13, §9]. Therefore the coordinates of some specific points of finite order on such an A can generate an abelian extension of $\mathfrak{k}$. (The field $\mathfrak{k}$ can be either real or imaginary.) Recently K. Doi and M. Yamauchi have found some interesting arithmetical relations for the Fourier coefficients of certain eigen-functions in $S_2(\Gamma_0(p^3))$ with a prime p. It is expected that these will be understood in the framework of [13, §9], under the formulation of §4 of the present paper.

The next §5 is devoted to a proof of Dirichlet's formula for the class number $h(-q)$ of an imaginary quadratic field $Q(\sqrt{-q})$ of discriminant $-q$. In the previous papers [12], [13], we showed that A has complex multiplication if and only if the Mellin transform of f is an L-function with a Grössencharacter of an imaginary quadratic field. When the field is $Q(\sqrt{-q})$, the number of such characters, under certain conditions, is $h(-q)$ times a simple factor. Combining these facts with some other observations about eigenfunctions, we find a relation between $h(-q)$ and the trace of the map $(*)$ on $S_2(\Gamma)$ for a certain Γ (Proposition 10). Then Dirichlet's formula can be obtained by computing the trace by the Riemann-Roch theorem or by the Selberg-Eichler trace formula. The whole idea is under the influence of Hecke's works [4], [5], in which Hecke computed the multiplicities of the irreducible representations of $PSL_2(Z/qZ)$ in a certain space of cusp forms, when q is a prime. However, here we need no information about such representations. Although the proof is by no means simple nor elementary, yet the author thinks that this method is natural and opens some possibilities of generalization.

In the final §6, we shall briefly explain a method of determining the zeta-function of A over a certain type of non-abelian extension of Q.

§1. The factor A as a quotient of J.

Let us start with some definitions and notational convention. For an abelian variety B defined over C, we denote by D_B, or $D(B)$, the vector space over C of all holomorphic 1-forms on B, and by Y_B, or $Y(B)$, the tangent space of B at the origin. Then D_B is dual to Y_B, so that we have a C-bilinear pairing

$$(\, , \,)_B : \quad D_B \times Y_B \longrightarrow C .$$

If α is a homomorphism of B to another abelian variety B', one can naturally define two maps

$$d\alpha : \quad Y_B \longrightarrow Y_{B'} ,$$

$$\delta\alpha : \quad D_{B'} \longrightarrow D_B ,$$

which satisfy

$$(1.1) \qquad (\delta\alpha u, v)_B = (u, d\alpha v)_{B'} \qquad (u \in D_{B'}, \, v \in Y_B) .$$

By a *quotient of B by an abelian subvariety*, say C, *of* B, we understand a couple (A, ν) formed by an abelian variety A which canonically represents B/C and a natural map $\nu : B \to A$ with $C = \mathrm{Ker}\,(\nu)$ (cf. Chow [3]). We see easily that

$$(1.2) \qquad Y_C = \{ v \in Y_B \mid (u, v)_B = 0 \text{ for all } u \in \delta\nu(D_A) \} .$$

Let Γ be a group of level N, and J the jacobian variety of $\mathfrak{H}^*/\Gamma$. Let $\psi : \mathfrak{H}^* \to J$ be the map obtained by composing the natural map $\mathfrak{H}^* \to \mathfrak{H}^*/\Gamma$ with a canonical map of $\mathfrak{H}^*/\Gamma$ into J. Then there is an isomorphism $\mu : S_2(\Gamma) \to D_J$ defined by $\delta\psi(\mu(f)) = f(z)dz$ for $f \in S_2(\Gamma)$, where $\delta\psi$ is the " pull back " associated with ψ, and z is a standard variable on $\mathfrak{H}$.

Let Γ' be a group of level M, and let

$$\tilde{\Gamma} = \{ \alpha \in \boldsymbol{R} \cdot GL_2(\boldsymbol{Q}) \mid \det(\alpha) > 0 \} .$$

Define, for $\alpha \in \tilde{\Gamma}$, an algebraic correspondence $X(\Gamma'\alpha\Gamma)$ on $\mathfrak{H}^*/\Gamma \times \mathfrak{H}^*/\Gamma'$ as in [11, §7.2]. (Briefly, it is the locus of $z \times \alpha(z)$ modulo $\Gamma \times \Gamma'$.) Let J' and μ' denote the corresponding geometric objects defined with Γ' in place of Γ. One can attach to $X(\Gamma'\alpha\Gamma)$ a homomorphism of J into J' as in Weil [15, § VI], which we denote by $\{\Gamma'\alpha\Gamma\}$. Now the commutative diagrams (7.2.2) and (7.2.6) of [11] imply

$$(1.3) \qquad \mu \circ [\Gamma'\alpha\Gamma]_2 = \delta\{\Gamma'\alpha\Gamma\} \circ \mu' ,$$

where $[\Gamma'\alpha\Gamma]_k$ is the map $S_k(\Gamma') \to S_k(\Gamma)$ defined in [11, § 3.4]. Then, from (1.1) and (1.3), we obtain

$$(1.4) \qquad (\mu'(f), d\{\Gamma'\alpha\Gamma\}v)_{J'} = (\mu(f \mid [\Gamma'\alpha\Gamma]_2), v)_J \qquad (f \in S_2(\Gamma'), \, v \in Y_J) .$$

For a positive integer n, let q be the largest divisor of n prime to N, and put $n = mq$. Then we define T_n and ξ_n by $T_n = T_m T_q$, $\xi_n = \xi_m \xi_q$, $T_m = [\Gamma\alpha\Gamma]_2$, $\xi_m = \{\Gamma\alpha\Gamma\}$ with $\alpha = \begin{bmatrix} 1 & 0 \\ 0 & m \end{bmatrix}$, and $T_q = \sum_\beta [\Gamma\beta\Gamma]_2$, $\xi_q = \sum_\beta \{\Gamma\beta\Gamma\}$, where $\sum_\beta$ is extended over all distinct $\Gamma\beta\Gamma$ with $\det(\beta) = q$, $\beta \equiv \begin{bmatrix} 1 & 0 \\ 0 & q \end{bmatrix} \pmod{N}$. Then $\mu \circ T_n = \delta\xi_n \circ \mu$.

As shown in [11, §7.3], there is a "standard model" of $\mathfrak{H}^*/\Gamma$ defined over Q (see also §4 below), so that J can naturally be defined over Q. Let $f(z) = \sum\limits_{n=1}^{\infty} a_n e^{2\pi i n z}$, with $a_1 = 1$, be an element of $S_2(\Gamma)$, that is a common eigen-function of T_n for all n (i. e., $f | T_n = a_n f$). Fix f, and let K be the subfield of C generated over Q by the complex numbers a_n for all n. (Of course K depends on f.) Then we have

THEOREM 1. *There exists a triple (A, ν, θ) formed by the objects satisfying the following conditions:*

(i) *(A, ν) is a quotient of J by an abelian subvariety rational over Q.*

(ii) *θ is an isomorphism of K into $\mathrm{End}\,(A) \otimes Q$ such that $\nu \circ \xi_n = \theta(a_n) \circ \nu$ for all n. (This implies especially $\theta(1) = id_A$.)*

(iii) *$\dim(A) = [K : Q]$.*

Moreover, let I denote the set of all isomorphisms of K into C. Then, for every $\sigma \in I$, there exists an element f_σ of $S_2(\Gamma)$ such that $f_\sigma | T_n = a_n^\sigma f$ for all n, and $f_\sigma(z) = \sum\limits_{n=1}^{\infty} a_n^\sigma e^{2\pi i n z}$. With these f_σ, one has

(iv) *$\delta \nu(D_A) = \mu(\sum\limits_{\sigma \in I} C f_\sigma)$.*

Under the conditions (i, ii), the triple (A, ν, θ) is unique up to isomorphisms over Q. Furthermore, the one-dimensional part of the zeta function of A over Q coincides, up to finitely many Euler factors for bad primes, with $\prod\limits_{\sigma \in I} \left(\sum\limits_{n=1}^{\infty} a_n^\sigma n^{-s} \right)$, provided that $\Gamma = \Gamma_0(N)$, or $\sum\limits_{\sigma \in I} C f_\sigma$ is stable under the map $h(z) \mapsto h(-1/Nz)/z^2$.

PROOF. Let $\mathfrak{T}$ be the subalgebra of $\mathrm{End}\,(J) \otimes Q$ generated by the ξ_n for all n, and $\mathfrak{R}$ the radical of $\mathfrak{T}$. Then $\mathfrak{T} = \mathfrak{R} \oplus \mathfrak{S}$ with a commutative semi-simple algebra $\mathfrak{S}$, whose simple components we denote by $\mathfrak{R}_1, \cdots, \mathfrak{R}_r$. By (1.3), we see that $\delta \xi_n$ maps $\mu(f)$ onto $a_n \cdot \mu(f)$, hence we can define a homomorphism ρ of $\mathfrak{T}$ onto K by $\rho(\xi_n) = a_n$. We may assume, changing the order if necessary, that ρ gives an isomorphism of $\mathfrak{R}_1$ onto K. Let $\rho': K \to \mathfrak{R}_1$ denote its inverse map, and put $\mathfrak{U} = \mathfrak{R}_2 + \cdots + \mathfrak{R}_r + \mathfrak{R}$. Let C be the abelian subvariety of J generated by $\alpha(J)$ for all $\alpha \in \mathfrak{U} \cap \mathrm{End}\,(J)$. Since the elements of $\mathfrak{T} \cap \mathrm{End}\,(J)$ are rational over Q, we can construct a quotient (A, ν) of J by C rational over Q. Observe that $\mathfrak{U}$ is an ideal of $\mathfrak{T}$, hence C is stable under $\mathfrak{R}_1 \cap \mathrm{End}\,(J)$. Therefore we can define a homomorphism $\theta: K \to \mathrm{End}\,(A) \otimes Q$ such that $\nu \circ \rho'(a) = \theta(a) \circ \nu$ for all $a \in K$. Since $\xi_n - \rho'(a_n) \in \mathfrak{U}$, we obtain $\nu \circ \xi_n = \theta(a_n) \circ \nu$ for all n. To prove $J \neq C$, put $\mathfrak{T}_C = \mathfrak{T} \otimes_Q C$. By [11, Th. 3.51], $S_2(\Gamma)$ is iso-morphic to $\mathfrak{T}_C$ as a $\mathfrak{T}_C$-module, hence D_J is (via δ) $\mathfrak{T}_C$-isomorphic to $\mathfrak{T}_C$. If $J = C$, we have

$$\{0\} = \{x \in D_J \mid \delta\alpha(x) = 0 \text{ for all } \alpha \in \mathfrak{U}\} ,$$

so that the $\mathfrak{T}_C$-isomorphism between D_J and $\mathfrak{T}_C$ implies

(*) $$\{0\} = \{\xi \in \mathfrak{T}_C \mid \mathfrak{U}\xi = 0\} .$$

Take the largest non-negative integer q such that $\Re_1 \Re^q \neq \{0\}$. (We understand that $\Re^0 = \Im$.) Then the right hand side of (*) contains $\Re_1 \Re^q$, a contradiction. Thus $J \neq C$, so that $\dim(A) > 0$, and θ is injective. Now suppose that (A', ν', θ') satisfies (i, ii). Then the elements of $\theta'(K) \cap \mathrm{End}(A')$ are rational over $\boldsymbol{Q}$, hence the representation of K on $D_{A'}$ via θ' is a multiple of a regular representation of K over $\boldsymbol{Q}$. Therefore, if m is the multiplicity, $D_{A'}$ has a basis $\{w_{\sigma j} \mid \sigma \in I, \ 1 \leq j \leq m\}$ over C such that $\delta\theta'(a)(w_{\sigma j}) = a^\sigma w_{\sigma j}$ for all $a \in K$. Let $f_{\sigma j} = \mu^{-1}(\delta\nu'(w_{\sigma j}))$. Then $f_{\sigma j} \mid T_n = a_n^\sigma f_{\sigma j}$ for all n. It follows that $m = 1$, hence $\dim(A') = [K : \boldsymbol{Q}]$. This shows also the existence of f_σ as stated in our theorem, and $\delta\nu'(D_{A'}) = \mu(\sum_{\sigma \in I} Cf_\sigma)$. By (1.2), the tangent space of $\mathrm{Ker}(\nu')$ at the origin is the annihilator of $\mu(\sum_{\sigma \in I} Cf_\sigma)$. This proves the uniqueness of (A, ν, θ) and the property (iv). The assertion concerning the zeta-function of A can easily be proved by shifting the congruence relation [11, (7.5.1, 2)] to A.

As to the zeta-function, we have somewhat more generally

PROPOSITION 1. *Suppose* $\Gamma = \Gamma_0(N)$. *Let* (A, ν) *be a quotient of* J *by an abelian subvariety rational over* $\boldsymbol{Q}$. *Then* $\mu^{-1}(\delta\nu(D_A))$ *is stable under* T_n *for all n prime to N. Moreover, if* T_n' *denotes the restriction of* T_n *to* $\mu^{-1}(\delta\nu(D_A))$, *then the one-dimensional part of the zeta-function of* A *over* $\boldsymbol{Q}$ *coincides, up to finitely many Euler factors, with* $\det(\sum_{(n, N)=1} T_n' n^{-s})$.

PROOF. By [11, Prop. 7.19], we see that $\mathrm{Ker}(\nu)$ is stable under ξ_n for all n prime to N. Then (1.2) shows that $\delta\nu(D_A)$ is stable under $\delta\xi_n$ for all such n, hence the first assertion. Also, we can define an endomorphism ξ_n', for such an n, of A such that $\xi_n' \circ \nu = \nu \circ \xi_n$. Shifting the congruence relation [11, (7.5.1)] to A, we obtain the second assertion.

§2. The factor A_0 as a subvariety of J and a canonical isomorphism of $S_2(\Gamma)$ onto Y_J.

Let $\alpha = \begin{bmatrix} a & b \\ c & d \end{bmatrix} \in GL_2(\boldsymbol{R})$, $\det(\alpha) > 0$. For a complex valued function $f(z)$ on $\mathfrak{H}$ and a positive integer k, we define a function $f \mid [\alpha]_k$ on $\mathfrak{H}$ by

$$(f \mid [\alpha]_k)(z) = \det(\alpha)^{k/2}(cz+d)^{-k} f(\alpha(z)),$$

where $\alpha(z) = (az+b)/(cz+d)$ for $z \in \mathfrak{H}$. Put $\omega_N = N^{-1/2} \begin{bmatrix} 0 & 1 \\ N & 0 \end{bmatrix}$, and define a function $f \mid [\omega_N]_k$ on $\mathfrak{H}$ by

$$(f \mid [\omega_N]_k)(z) = N^{-k/2} z^{-k} \bar{f}(1/N\bar{z}),$$

where bars mean the complex conjugation. Then one can easily verify:

$$(2.1) \qquad (f \mid [\alpha]_k) \mid [\omega_N]_k = (f \mid [\omega_M]_k) \mid [\omega_M^{-1} \alpha \omega_N]_k,$$

$$(2.2) \qquad f \,|\, [\omega_N]_k^2 = f ,$$

$$(2.3) \qquad \omega_N^{-1} \Gamma \omega_N = \Gamma \ \text{for any group } \Gamma \text{ of level } N .$$

It follows that $[\omega_N]_k$ induces an anti-C-linear automorphism of $S_k(\Gamma)$. Moreover, let Γ' be a group of level M, and let $\alpha \in \tilde{\Gamma}$. Then

$$(2.4) \qquad (f \,|\, [\Gamma'\alpha\Gamma]_k) \,|\, [\omega_N]_k = (f \,|\, [\omega_M]_k) \,|\, [\Gamma'\omega_M^{-1}\alpha\omega_N\Gamma]_k \qquad (f \in S_k(\Gamma')) .$$

If $\langle \, , \, \rangle$ denotes the Petersson inner product on $S_k(\Gamma)$, then

$$(2.5) \qquad \langle f, g \,|\, [\omega_N]_k \rangle = \langle g, f \,|\, [\omega_N]_k \rangle .$$

Let $\mu : S_2(\Gamma) \to D_J$ be the map defined in §1. We can now define a C-linear isomorphism $\lambda : S_2(\Gamma) \to Y_J$ by

$$(2.6) \qquad (\mu(f), \lambda(g))_J = i \cdot \langle f, g \,|\, [\omega_N]_2 \rangle \qquad (f, g \in S_2(\Gamma)) .$$

The constant factor i is not absolutely necessary, but will make a later discussion smooth. Put $Y' = Y(J')$ with the jacobian variety J' of $\mathfrak{H}^*/\Gamma'$.

Define $\lambda' : S_2(\Gamma') \to Y'$ in the same manner. Combining (2.4) and (1.4) with [11, (3.4.5)], we obtain a relation

$$(2.7) \qquad \lambda(f \,|\, [\Gamma'\alpha\Gamma]_2) = d\{\Gamma\omega_N^{-1}\alpha^\iota\omega_M\Gamma'\}(\lambda'(f)) \qquad \text{for } f \in S_2(\Gamma') ,$$

where ι denotes the main involution, i.e., $\begin{bmatrix} a & b \\ c & d \end{bmatrix}^\iota = \begin{bmatrix} d & -b \\ -c & a \end{bmatrix}$.

Consider the special case in which $M = N$, $\Gamma' = \Gamma$, and $\alpha = \begin{bmatrix} 1 & 0 \\ 0 & n \end{bmatrix}$ with a positive integer n, or $\alpha \equiv \begin{bmatrix} 1 & 0 \\ 0 & q \end{bmatrix} \pmod{N}$, $\det(\alpha) = q$, $(q, N) = 1$. Then (2.7) implies

$$(2.8) \qquad \lambda(f \,|\, T_n) = d\xi_n(\lambda(f)) \qquad \text{for } f \in S_2(\Gamma) .$$

Further let q be an integer prime to N, σ_q an element of $SL_2(Z)$ such that $q \cdot \sigma_q \equiv \begin{bmatrix} 1 & 0 \\ 0 & q^2 \end{bmatrix} \pmod{N}$, and let $\tau_N = \begin{bmatrix} 0 & -1 \\ N & 0 \end{bmatrix}$. Then we obtain from (2.7) the following relations:

$$(2.9) \qquad \lambda(f \,|\, [\sigma_q]_2) = d\{\Gamma\sigma_q\Gamma\}(\lambda(f)) \qquad \text{for } f \in S_2(\Gamma) ,$$

$$(2.10) \qquad \lambda(f \,|\, [\tau_N]_2) = d\{\Gamma\tau_N\Gamma\}(\lambda(f)) \qquad \text{for } f \in S_2(\Gamma) .$$

The formula (2.7) and its specializations (2.8—10) are the commutative property of the isomorphism λ mentioned in the *Introduction*.

Now define K for a common eigen-function $f(z) = \sum_{n=1}^{\infty} a_n e^{2\pi i n z}$, with $a_1 = 1$, as in §1. Then [11, Th. 7.16] can be re-stated as follows.

THEOREM 2. *There exists an abelian subvariety A_0 of J and an isomorphism θ_0 of K into $\mathrm{End}(A_0) \otimes Q$ with the following properties:*

(i) $\dim(A_0) = [K : \mathbf{Q}]$;

(ii) $\theta_0(a_n)$ is the restriction of ξ_n to A_0 for all n ;

(iii) A_0 is rational over $\mathbf{Q}$.

(iv) *If f_σ for $\sigma \in I$ is defined as in Theorem 1, then $Y(A_0) = \lambda(\sum_{\sigma \in I} Cf_\sigma).$*
Moreover the couple (A_0, θ_0) is unique under (i) *and* (ii).

The last property (iv) of A_0 is not explicitly stated in [11, Th. 7.16], but follows immediately from its proof.

The formulation of the results of [12] and their proofs require the isomorphism λ: Let us now clarify a few points in [12], where the explanation may be somewhat vague. In the proof of [12, Th. 1], we should put $\delta = \begin{bmatrix} 1 & 1/D \\ 0 & 1 \end{bmatrix}$, which is erroneously printed as $\begin{bmatrix} 1 & 1/d \\ 0 & 1 \end{bmatrix}$. Then the endomorphism ω of A' mentioned there is $\{ \Gamma \varepsilon \Gamma \}$ with $\varepsilon = \begin{bmatrix} 1 & 0 \\ -M/D & 1 \end{bmatrix}$. Similarly the homomorphism ξ^*, ϕ, and φ are given by $\xi^* = \sum_u t x_u \cdot \{ \Gamma_1(M') \omega_{M'}^{-1} \eta_u^t \omega_M \Gamma_1(M) \}$, $\phi = \{ \Gamma_1(M) \omega_M^{-1} \beta^t \omega_L \Gamma_1(L) \}$, $\varphi = \{ \Gamma_1(M) \omega_M^{-1} \omega_L \Gamma_1(L) \}$. One can also formulate the results of [12] in terms of (A, ν, θ) of Theorem 1.

PROPOSITION 2. *Let (A, ν, θ) be as in Theorem 1, and (A_0, θ_0) be as in Theorem 2. Suppose that $\sum_{\sigma \in I} Cf_\sigma$ is stable under $\begin{bmatrix} 0 & -1 \\ N & 0 \end{bmatrix}_2$. Then the restriction of ν to A_0 is an isogeny of A_0 onto A, and $\nu \circ \theta_0(a) = \theta(a) \circ \nu$ for all $a \in K$.*

PROOF. Put $X = \sum_{\sigma \in I} Cf_\sigma$. For $h(z) = \sum_n b_n e^{2\pi i n z}$, put $h^*(z) = \sum_n \bar{b}_n e^{2\pi i n z}$. Then we see easily that X is stable under the map $h \mapsto h^*$, and $h \vert [\omega_N]_2 = (h \vert [\tau_N]_2)^*$. Therefore if $g = (h \vert [\tau_N]_2)^*$, $h \neq 0$, we obtain, from (2.5), $(\mu(h), \lambda(g))_J \neq 0$. This implies that the bilinear form $(,)_J$ is non-degenerate on $\mu(X) \times \lambda(X)$. It follows that $d\nu$ maps $Y(A_0)$ onto $Y(A)$, hence ν maps A_0 onto A. The relation $\nu \circ \theta_0(a) = \theta(a) \circ \nu$ is obvious from (ii) of Theorem 1 and (ii) of Theorem 2.

§3. Periods of cusp forms.

We first consider a general case in which Γ is not necessarily a congruence subgroup of $SL_2(\mathbf{Z})$. Let Γ be a discrete subgroup of $SL_2(\mathbf{R})$ such that $\mathfrak{H}/\Gamma$ is of finite measure, and $\mathfrak{H}^*$ the union of $\mathfrak{H}$ and the cusps of Γ. Further let J be the jacobian variety of the compact Riemann surface $\mathfrak{H}^*/\Gamma$. We can then define D_J, Y_J, $S_2(\Gamma)$, and $\mu : S_2(\Gamma) \to D_J$ in the same fashion as in §1. Put

$$[\gamma, g] = \int_z^{\gamma(z)} g(z) dz$$

for $g \in S_2(\Gamma)$ and $\gamma \in \Gamma$ with any $z \in \mathfrak{H}^*$. Note that this does not depend on the choice of z. Let L_J be the submodule of Y_J generated over $\mathbf{Z}$ by the linear maps

$$\eta \longmapsto [\gamma, \mu^{-1}(\eta)] \qquad (\eta \in D_J)$$

for all $\gamma \in \Gamma$. Then J can be identified with Y_J/L_J. Fix a point z_0 of $\mathfrak{H}^*$, and define, for $z \in \mathfrak{H}^*$, an element φ_z of Y_J by

$$\varphi_z(\mu(g)) = \int_{z_0}^{z} g(z)dz \qquad (g \in S_2(\Gamma)).$$

Then the map $z \mapsto \varphi_z$ induces a canonical map of $\mathfrak{H}^*/\Gamma$ into $J = Y_J/L_J$.

Let (A, ν) be a quotient of J by an abelian subvariety of J. Take any C-basis $\{f_1, \cdots, f_m\}$ of $\mu^{-1}(\delta\nu(D_A))$. Let $W = Y(\mathrm{Ker}\,(\nu))$. Then

$$W = \{v \in Y_J \mid (\delta\nu(D_A), v)_J = 0\}\,,$$

and A can be identified with $Y_J/(W + L_J)$. Now $\mu(f_1), \cdots, \mu(f_m)$ define a C-linear map $F: Y_J \to C^m$ whose kernel is W. In this situation, we have

PROPOSITION 3. *Let P be the submodule of C^m generated over Z by the vectors $([\gamma, f_1], \cdots, [\gamma, f_m])$ for all $\gamma \in \Gamma$. Then A is isomorphic, via F, to C^m/P.*

PROOF. From the above definition of L_J, we see easily that $F(L_J) = P$, hence our assertion.

REMARK. If we put

$$\Phi(z) = \left(\int_{z_0}^{z} f_1(z)dz, \cdots, \int_{z_0}^{z} f_m(z)dz \right) \pmod{P}\,,$$

then Φ defines a morphism of $\mathfrak{H}^*/\Gamma$ into C^m/P. Obviously $\Phi(z) = F(\varphi_z)$, i. e., Φ is the map composed by a chain of maps

$$\mathfrak{H}^*/\Gamma \xrightarrow{\varphi} J \xrightarrow{\nu} A \xrightarrow{F} C^m/P.$$

The above discussion naturally applies to a group Γ of level N and (A, ν, θ) defined as in Theorem 1. In this case, we can take $\{f_\sigma \mid \sigma \in I\}$ as $\{f_i\}$. Thus we obtain the abelian variety A from the periods of the cusp forms f_σ.

PROPOSITION 4. *Let λ be the map of §2 defined with respect to a group Γ of level N. Then $\lambda^{-1}(L_J)$ consists of all the elements g of $S_2(\Gamma)$ such that $\mathrm{Re}\,[\gamma, g] \in Z$ for all $\gamma \in \Gamma$, where Re means the real part.*

PROOF. First we note a relation

$$(3.1) \qquad -\overline{[\gamma, g \mid [\omega_N]_2]} = [\omega_N \gamma \omega_N^{-1}, g] \qquad (\gamma \in \Gamma,\ g \in S_2(\Gamma))$$

which can easily be verified. Now it is well known that

$$-i\langle g, h \rangle = \int_{\mathfrak{H}/\Gamma} g(z)dz \wedge \mathrm{Re}\,(h(z)dz)$$

$$= \sum_{j=1}^{p} \{[\alpha_j, g] \cdot \mathrm{Re}\,[\beta_j, h] - [\beta_j, g] \cdot \mathrm{Re}\,[\alpha_j, h]\}\,,$$

where p is the genus of $\mathfrak{H}^*/\Gamma$, and $\{\alpha_1, \cdots, \alpha_p, \beta_1, \cdots, \beta_p\}$ is a set of elements of Γ corresponding to a standard set of generators of the homology group of $\mathfrak{H}^*/\Gamma$. Moreover, every C-linear map of $S_2(\Gamma)$ into C can be given as

$$g \longmapsto \sum_{j=1}^{p}(a_j[\alpha_j, g]+b_j[\beta_j, g])$$

with real numbers a_j and b_j, which are unique for the map. Our proposition now follows immediately from these facts and (3.1).

Put $L' = \lambda^{-1}(L_J)$. Let A_0 be any abelian subvariety of J, and let $U = \lambda^{-1}(Y(A_0))$. Then A_0 is isomorphic, via λ, to $U/(U \cap L')$. Let $\{g_1, \cdots, g_m\}$ be a C-basis of U, and let P_0 be the submodule of C^m generated over Z by the vectors $([\gamma, g_1], \cdots, [\gamma, g_m])$ for all $\gamma \in \Gamma$. Then

PROPOSITION 5. A_0 *is dual to* C^m/P_0.

PROOF. Define a C-linear isomorphism $G: C^m \to U$ by $G(w_1, \cdots, w_m) = \sum_j w_j g_j$. By Proposition 4, $w = (w_1, \cdots, w_m) \in G^{-1}(U \cap L')$ if and only if $\operatorname{Re}(\sum_j w_j[\gamma, g_j]) \in Z$ for all $\gamma \in \Gamma$, i. e., $\operatorname{Re}(\sum_j w_j v_j) \in Z$ for all $(v_1, \cdots, v_m) \in P_0$. It follows that $C^m/G^{-1}(U \cap L')$ is dual to C^m/P_0, q. e. d.

PROPOSITION 6. *Define* (A, ν, θ) *and* (A_0, θ_0) *as in Theorems* 1 *and* 2 *with the same eigen-function* f. *Then* A *is dual to* A_0.

PROOF. In this case $\lambda^{-1}(Y(A_0)) = \sum_{\sigma \in I} C f_\sigma = \mu^{-1}(\delta\nu(D_A))$, so that we can take $P = P_0$. Then our assertion is immediate from Propositions 3 and 5.

§4. A twisting operator R and its geometric meaning.

Let χ be a primitive character modulo a positive integer r, and let $\alpha_u = \begin{bmatrix} 1 & u/r \\ 0 & 1 \end{bmatrix}$ for $u \in Z$. Define an operator R by

$$(4.1) \qquad f \mid R = \sum_{u=1}^{r} \bar{\chi}(u) f \mid [\alpha_u]_2 \qquad (f \in S_2(\Gamma_1(N))).$$

By [11, Prop. 3.64], R maps $S_2(\Gamma_1(N))$ into $S_2(\Gamma_1(r^2 N))$. If $f(z) = \sum_n a_n e^{2\pi i n z}$, then $f \mid R = g(\bar{\chi}) \sum_n \chi(n) a_n e^{2\pi i n z}$, where $g(\bar{\chi}) = \sum_{u=1}^{r} \bar{\chi}(u) e^{2\pi i u/r}$. The purpose of this section is to study the geometric meaning of R in connection with the quotients (A, ν) of J.

Let s be a positive divisor of N, and $\mathfrak{h}$ a subgroup of $(Z/sZ)^\times$. Define a group Γ of level N by

$$\Gamma = \left\{ \begin{bmatrix} a & b \\ c & d \end{bmatrix} \in \Gamma_0(N) \mid d \pmod{s} \in \mathfrak{h} \right\}.$$

Let M be a common multiple of N, r^2, and rs. Let Γ' be a group of level M such that

$$\Gamma' \subset \left\{ \begin{bmatrix} a & b \\ c & d \end{bmatrix} \in \Gamma_0(M) \mid d \pmod{s} \in \mathfrak{h}, \ a \equiv d \pmod{r} \right\}.$$

Then $\alpha_u \Gamma' \alpha_u^{-1} \subset \Gamma$, and R maps $S_2(\Gamma)$ into $S_2(\Gamma')$.

Define J, J', μ, μ', λ, λ' as before with respect to the present Γ and Γ'. Since $f|[\alpha_u]_2 = f|[\Gamma \alpha_u \Gamma']_2$ for $f \in S_2(\Gamma)$, we have

$$(4.2) \qquad \mu'(f|R) = \sum_{u=1}^{r} \bar{\chi}(u) \cdot \delta\{\Gamma \alpha_u \Gamma'\}(\mu(f))$$

$$(f \in S_2(\Gamma)),$$

$$(4.3) \qquad \lambda'(f|R) = \sum_{u=1}^{r} \bar{\chi}(u) \cdot d\{\Gamma' \beta_u \Gamma\}(\lambda(f))$$

where $\beta_u = \omega_M^{-1} \alpha_u^{\iota} \omega_N$.

PROPOSITION 7. *Let* $f(z) = \sum_n a_n e^{2\pi i n z}$, *with* $a_1 = 1$, *be a common eigenfunction of* T_n *in* $S_2(\Gamma)$ *for all* n. *Define* (A, ν, θ) *and* (A', ν', θ') *for* f *and* $h(z) = \sum_n \chi(n) a_n e^{2\pi i n z}$, *respectively, as in Theorem* 1. *Then* A' *is a homomorphic image of the product of* r *copies of* A.

PROOF. Let K'' be the subfield of C generated by the Fourier coefficients a_n and the values $\chi(n)$ for all n, and let I'' be the set of all isomorphisms of K'' into C. For $\sigma \in I''$, put $f_\sigma(z) = \sum_n a_n^\sigma e^{2\pi i n z}$, $h_\sigma(z) = \sum_n \chi(n)^\sigma a_n^\sigma e^{2\pi i n z}$. Then $\mu(\sum_{\sigma \in I''} C f_\sigma) = \delta \nu(D_A)$, and

$$(4.4) \qquad \delta \nu'(D_{A'}) = \mu'(\sum_{\sigma \in I''} C h_\sigma) \subset \sum_{u=1}^{r} \delta\{\Gamma \alpha_u \Gamma'\} \delta \nu(D_A).$$

Define a homomorphism $\xi: J' \to A \times \cdots \times A$ (r copies) by $\xi(x) = (\nu(\{\Gamma \alpha_1 \Gamma'\} x),$ $\cdots, \nu(\{\Gamma \alpha_r \Gamma'\} x))$ for $x \in J'$. Then (4.4) shows that the identity component of $\mathrm{Ker}\,(\xi)$ is contained in $\mathrm{Ker}\,(\nu')$. Therefore we can find a homomorphism $\beta: A \times \cdots \times A \to A'$ such that $\beta \circ \xi = m\nu'$ with a positive integer m. This proves our proposition.

Let us now assume that χ is a non-trivial real character. Put

$$(4.5) \qquad \eta_0 = \sum_{u=1}^{r} \chi(u)\{\Gamma \alpha_u \Gamma'\}.$$

Then $\eta_0 \in \mathrm{Hom}\,(J', J)$, and

$$(4.6) \qquad \mu'(g|R) = \delta \eta_0(\mu(g)) \qquad (g \in S_2(\Gamma)).$$

PROPOSITION 8. *Let* A *and* A' *be as in Proposition* 7. *Then there exists a homomorphism* η *of* A' *into* A *such that* $\eta \circ \nu' = \nu \circ \eta_0$. *The homomorphisms* η_0 *and* η *are defined over the quadratic extension* $\mathfrak{k}$ *of* Q *corresponding to* χ, *and* $\eta_0^\varepsilon = -\eta_0$, $\eta^\varepsilon = -\eta$ *if* ε *is the generator of* $\mathrm{Gal}\,(\mathfrak{k}/Q)$.

PROOF. We need the precise definition of "the standard models" for $\mathfrak{H}^*/\Gamma$ and $\mathfrak{H}^*/\Gamma'$. Using the notation of [11, § 7.3], let us define subgroups U_N and S of G_A by

$$U_N = \left\{ x \in \prod_p GL_2(Z_p) \times G_{\infty+} \,\middle|\, x_p \equiv \begin{bmatrix} * & * \\ 0 & 1 \end{bmatrix} \pmod{N} \right\},$$

$$S = Q^\times U_N \Gamma.$$

Similarly put $S' = Q^{\times} U_M \Gamma'$. Then $G_Q \cap S = Q^{\times}\Gamma$, $G_Q \cap S' = Q^{\times}\Gamma'$; we take V_S, $V_{S'}$ of [11, p. 155] as models of $\mathfrak{H}^*/\Gamma$, $\mathfrak{H}^*/\Gamma'$; and $X(\Gamma\alpha_u\Gamma') = X_{SS'}(\alpha_u)$. Now one can easily verify that

$$Q^{\times} \cdot \det(S' \cap \alpha_u^{-1} S \alpha_u) \supset Q^{\times} \cdot R_+^{\times} \cdot \{(v_p) \in \prod_p Z_p^{\times} \mid v_p \equiv 1 \pmod{r}\} .$$

By [11, Prop. 7.2], this shows that $X_{SS'}(\alpha_u)$ is rational over $Q(\zeta_r)$, where $\zeta_r = e^{2\pi i/r}$. For an integer q prime to M, let σ_q be an element of $SL_2(Z)$ such that $q \cdot \sigma_q \equiv \begin{bmatrix} 1 & 0 \\ 0 & q^2 \end{bmatrix} \pmod{M}$. If $g \in S_2(\Gamma)$ and $g|[\sigma_q]_2 = \varphi(q)g$ with a character φ modulo s, then, by [11, Prop. 3.64], one has $(g|R)|[\sigma_q]_2 = \varphi(q)g|R = (g|[\sigma_q]_2)|R$. This shows that

$$(4.7) \qquad \eta_0 \circ \{\Gamma'\sigma_q\Gamma'\} = \{\Gamma\sigma_q\Gamma\} \circ \eta_0 .$$

On the other hand, let $y = (y_p)$ be an element of G_0 such that $y_p = \begin{bmatrix} 1 & 0 \\ 0 & q \end{bmatrix}$ or 1 according as p divides M or not. Then $\sigma_q^{-1}y \in U_M$ so that $J_{SS}(\sigma_q) = J_{SS}(y)$, $J_{S'S'}(\sigma_q) = J_{S'S'}(y)$. Since $\alpha_u \begin{bmatrix} 1 & 0 \\ 0 & q \end{bmatrix} = \begin{bmatrix} 1 & 0 \\ 0 & q \end{bmatrix}\alpha_{qu}$, we obtain, by [11, Prop. 7.2],

$$\sum_{u=1}^{r} \chi(qu) X_{SS'}(\alpha_u)^{\sigma(y)} \circ J_{S'S'}(y) = J_{SS}(y) \circ \left(\sum_{u=1}^{r} \chi(qu) X_{SS'}(\alpha_{qu}) \right),$$

where $\sigma(y)$ is the element of $\mathrm{Gal}(Q_{ab}/Q)$ defined by [11, (6.4.1)], hence $\chi(q)\eta_0^{\sigma(y)} \circ \{\Gamma'\sigma_q\Gamma'\} = \{\Gamma\sigma_q\Gamma\} \circ \eta_0$. Combining this with (4.7), we obtain

$$\eta_0^{\sigma(y)} = \chi(q)\eta_0 .$$

Observe that $\sigma(y)$ sends $\zeta_M = e^{2\pi i/M}$ onto ζ_M^q. Therefore η_0 is rational over the quadratic field $\mathfrak{k}$ corresponding to χ, and $\eta_0^{\varepsilon} = -\eta_0$ for the generator ε of $\mathrm{Gal}(\mathfrak{k}/Q)$. Now R maps $\sum_{\sigma \in I'} Cf_\sigma$ into $\sum_{\sigma \in I''} Ch_\sigma$, so that $\delta\eta_0$ maps $\delta\nu(D_A)$ into $\delta\nu'(D_{A'})$. Therefore we can define a homomorphism η of A' into A, rational over $\mathfrak{k}$, so that $\eta \circ \nu' = \nu \circ \eta_0$. Then the relation $\eta^{\varepsilon} = -\eta$ is obvious.

Let us now assume the following set of conditions:

$(4.8) \qquad$ (i) *N is a common multiple of r^2 and rs;* (ii) $a \equiv d \pmod{r}$ *for every* $\begin{bmatrix} a & b \\ c & d \end{bmatrix} \in \Gamma$.

Then we can take $M = N$, $\Gamma' = \Gamma$, $J' = J$, so that η_0 is an endomorphism of J, and $\eta_0^2 = \chi(-1)r\eta_0$.

Fix an eigen-function $f(z)$ as in Proposition 6, and define K and (A, ν, θ) as before. Suppose that the following condition is satisfied:

$(4.9) \qquad$ *There is an automorphism ρ of K, other than the identity map, such that $\chi(n)a_n = a_n^{\rho}$ for all n.* (This implies especially that $a_n = 0$ if $(n, r) \neq 1$.)

Then $\rho^2 = 1$, and $\chi(n)a_n^\sigma = a_n^{\rho\sigma}$ for any $\sigma \in I$, so that

$$(4.10) \qquad\qquad f_\sigma | R = \mathfrak{g}(\chi)f_{\rho\sigma} .$$

PROPOSITION 9. *Under the assumptions (4.8) and (4.9), η is an endomorphism of A satisfying*

(i) $\eta^2 = \chi(-1)r \cdot id_A$,

(ii) $\eta \circ \theta(a) = \theta(a^\rho) \circ \eta$ *for every* $a \in K$.

PROOF. Observe that the diagram

$$
\begin{array}{ccccc}
S_2(\Gamma) & \xrightarrow{\ \mu\ } & D_J & \xleftarrow{\ \delta\nu\ } & D_A \\
\Big\downarrow{\scriptstyle R} & & \Big\downarrow{\scriptstyle \delta\eta_0} & & \Big\downarrow{\scriptstyle \delta\eta} \\
S_2(\Gamma) & \xrightarrow{\ \mu\ } & D_J & \xleftarrow{\ \delta\nu\ } & D_A
\end{array}
$$

is commutative, and another diagram with T_n, $\delta\xi_n$, $\delta\theta(a_n)$ in place of R, $\delta\eta_0$, $\delta\eta$ is commutative, too. Let w_σ be the element of D_A such that $\delta\nu(w_\sigma) = \mu(f_\sigma)$. Then $\{w_\sigma | \sigma \in I\}$ is a C-basis of D_A, and the commutativity of the diagram implies $\delta\eta(w_\sigma) = \mathfrak{g}(\chi)w_{\rho\sigma}$, and $\delta\theta(a)(w_\sigma) = a^\sigma w_\sigma$ for $a \in K$. Therefore $\delta\eta^2(w_\sigma) = \mathfrak{g}(\chi)^2 w_\sigma$, and $(\delta\eta \circ \delta\theta(a))(w_\sigma) = \mathfrak{g}(\chi)a^\sigma w_{\rho\sigma} = (\delta\theta(a^\rho) \circ \delta\eta)(w_\sigma)$, hence our proposition.

Thus, if an eigen-function $f(z)$ satisfying (4.9) exists, then the corresponding (A, θ) and η form a system similar to that of [13, §9]. The field K in the present case is not necessarily a CM-field as assumed there, but one can still develop, by the same ideas as in [13, §9], a theory of construction of class fields over the quadratic field $\mathfrak{k}$. As mentioned in the Introduction, K. Doi and M. Yamauchi have found a few examples of $f(z)$, satisfying (4.9), with $N = p^3$ and $r = p$ for a prime p.

§5. A proof of Dirichlet's class number formula.

Let $Q(\sqrt{-q})$ be an imaginary quadratic field with discriminant $-q$, and $h(-q)$ the class number of $Q(\sqrt{-q})$. Then the Dirichlet formula is, if $q > 4$,

$$(5.1) \qquad\qquad h(-q) = -\sum_{n=1}^{q}\left(\frac{-q}{n}\right)n/q .$$

The purpose of this section is to prove this as an application of the results of previous sections, without using the residue of the zeta function of $Q(\sqrt{-q})$. The proof is inspired by a result of Hecke's in [5]; we shall explain its connection with our proof afterwards.

For simplicity, we assume $q > 4$, although the cases $q = 3, 4$ can be included, without much trouble, in the following treatment. Put $\chi(n) = \left(\frac{-q}{n}\right)$, and

$$\Gamma = \left\{ \begin{bmatrix} a & b \\ c & d \end{bmatrix} \in \Gamma_0(q^2) \,\middle|\, a \equiv d \equiv 1 \ (\mathrm{mod}\ q) \right\}.$$

Define the operator R of §4 for this χ with $N = q^2$, $r = q$. Since the present Γ satisfies (4.8), R maps $S_2(\Gamma)$ into itself, and

$$\left(\sum_{n=1}^{\infty} a_n e^{2\pi i n z} \right)\Big| R = \sqrt{-q}\, \sum_{n=1}^{\infty} \chi(n) a_n e^{2\pi i n z}$$

Here we recall a well-known formula of Gauss

$$(5.2) \qquad \sum_{n=1}^{q} \chi(n) e^{2\pi i n k/q} = \sqrt{-q} \cdot \chi(k).$$

We understand that $\sqrt{-q}$ has positive imaginary part.

PROPOSITION 10. *One has* $\mathrm{tr}\,(R) = \sqrt{-q} \cdot h(-q) \cdot \varphi(q)/2$, *where φ is Euler's function.*

PROOF. By virtue of the results of Miyake [8] and Casselman [2] (which generalize Atkin-Lehner [1]), we can find a basis of $S_2(\Gamma)$ formed by the elements

$$g_1, \cdots, g_s, g_1(mz), \cdots, g_s(mz),$$

where g_k belongs to "the essential part" of $S_2(\Gamma_1(N_k))$ for a divisor N_k of q^2, and m runs over all non-trivial positive divisors of q^2/N_k. (In other words, g_k is a "new form".) We take g_k to be a common eigen-function of all Hecke operators of level N_k, with the first Fourier coefficient 1. If $g_k(z) = \sum_{n=1}^{\infty} a_n e^{2\pi i n z}$, we see that the $g_k(mz)$, for a fixed k and for all positive divisors of q^2/N_k, span, over C, the vector space

$$\{ f \in S_2(\Gamma) \mid f|T_n = a_n f \text{ for all } n \text{ prime to } q \}.$$

Observe that R maps $g_k(mz)$ onto 0 if $m > 1$, and g_k onto a common eigen-function of all Hecke operators of level q^2, since $\sum_n \chi(n) a_n n^{-s}$ has an Euler product. Such an eigen-function must be of the form $\sum_m c_m g_j(mz)$ for some j, where m divides q^2/N_j. The comparison of the first Fourier coefficients yields $c_1 = \sqrt{-q}$. Therefore $\mathrm{tr}\,(R)/\sqrt{-q}$ is the number of indices k such that

$$g_k|R = \sqrt{-q} \cdot g_k + \sum_m c_m g_k(mz) \qquad (m > 1,\ m|N_k^{-1}q^2).$$

Put $g_k^*(z) = \sum_{(n,q)=1} a_n e^{2\pi i n z}$. Since $(g_k|R)|T_n = a_n \cdot (g_k|R)$ for $(n, q) = 1$, and the n-th Fourier coefficient of $g_k|R$ is 0 for $(n, q) \neq 1$, we have $g_k|R = g_k^*|R = \sqrt{-q} \cdot g_k^*$. Define (A, ν), I, and f_σ as in Theorem 1 with g_k^* as f. By Proposition 8, there is an endomorphism η of A such that

$$\delta\eta(\delta\nu^{-1}(\mu(f_\sigma))) = \delta\nu^{-1}(\mu(f_\sigma|R))$$

536 G. Shimura

for all $\sigma \in I$. Since $f_\sigma | R = \sqrt{-q} \cdot f_\sigma$, we see that $\delta\eta$ is $\sqrt{-q}$ times the identity map on D_A. Therefore, by [12, Lemma 1], A is isogenous to the product of copies of an elliptic curve, say E, with complex multiplication, such that $\mathrm{End}(E) \otimes Q$ is isomorphic to $Q(\sqrt{-q})$. Now define an abelian variety B for the function g_k (at level N_k) by Theorem 1. By Proposition 8, B is isogenous to A, hence B is also isogenous to the product of copies of E. Then, by [13, Prop. 1.6], g_k must be the Mellin (inverse) transform of an L-function with a primitive Grössen-character λ of $Q(\sqrt{-q})$. This means that $g_k(z) = \sum_{\mathfrak{a}} \lambda(\mathfrak{a}) e^{2\pi i N(\mathfrak{a})z}$, where $\mathfrak{a}$ runs over all integral ideals in $Q(\sqrt{-q})$ prime to the conductor, say $\mathfrak{c}$, of λ, and $\lambda((\alpha)) = \alpha$ for $\alpha \equiv 1 \bmod \mathfrak{c}$. Moreover the functional equation of the L-function implies that g_k belongs to the essential part of $S_2(\Gamma_1(q \cdot N(\mathfrak{c})))$, i. e., it is a "new form". Therefore $N(\mathfrak{c})$ divides q, hence $\mathfrak{c}$ divides the ideal $(\sqrt{-q})$. It can easily be seen that there are exactly $h(-q) \cdot \varphi(q)/2$ primitive Grössen-characters ψ whose conductor $\mathfrak{c}$ divides $\sqrt{-q}$ and such that $\psi((\alpha)) = \alpha$ for $\alpha \equiv 1 \bmod \mathfrak{c}$. For any such ψ, put $f_\psi(z) = \sum_{\mathfrak{a}} \psi(\mathfrak{a}) e^{2\pi i N(\mathfrak{a})z}$, where $\mathfrak{a}$ runs over all integral ideals of $Q(\sqrt{-q})$ prime to the conductor of ψ. Then the f_ψ, for all such ψ, are linearly independent, and belong to $S_2(\Gamma)$, by [12, Lemma 3]. Further each f_ψ must coincide with one of the g_k, and obviously $f_\psi | R = \sqrt{-q} \cdot f_\psi^*$. Thus we have shown that $\mathrm{tr}(R)/\sqrt{-q}$ is the number of the characters ψ, which completes the proof.

To compute $\mathrm{tr}(R)$ in a different manner, put $\alpha = \begin{bmatrix} 1 & 0 \\ 0 & q \end{bmatrix}$, $\varepsilon = \begin{bmatrix} 1 & 1 \\ 0 & 1 \end{bmatrix}$, and

$$\Gamma(q) = \left\{ \begin{bmatrix} a & b \\ c & d \end{bmatrix} \in SL_2(Z) \,\middle|\, \begin{bmatrix} a & b \\ c & d \end{bmatrix} \equiv \begin{bmatrix} 1 & 0 \\ 0 & 1 \end{bmatrix} \ (\mathrm{mod}\ q) \right\}.$$

Then $\Gamma = \alpha \Gamma(q) \alpha^{-1}$, $\alpha \varepsilon \alpha^{-1} = \begin{bmatrix} 1 & 1/q \\ 0 & 1 \end{bmatrix}$. Therefore $\mathrm{tr}(R)$ is equal to the trace of the operator

$$f \longmapsto \sum_{u=1}^{q} \chi(u) f \,|\, [\varepsilon^u]_2$$

on the space $S_2(\Gamma(q))$ of cusp forms of weight 2 with respect to $\Gamma(q)$. Let ε_0 be the automorphism of $\mathfrak{H}^*/\Gamma(q)$ obtained from ε. The fixed points of ε_0 occur only at cusps. Indeed, suppose $\varepsilon(z) = \gamma(z)$ for some $z \in \mathfrak{H}$ and some $\gamma = \begin{bmatrix} a & b \\ c & d \end{bmatrix} \in \Gamma(q)$. Then $\varepsilon^{-1}\gamma = \begin{bmatrix} a-c & b-d \\ c & d \end{bmatrix}$ is elliptic, hence $|\mathrm{tr}(\varepsilon^{-1}\gamma)| < 2$. But $a - c + d \equiv 2 \ (\mathrm{mod}\ q)$, which is a contradiction, since $q > 4$.

Now let r/s, with integers r and s such that $(r, s) = 1$, be a cusp representing a fixed point of ε_0. By [11, Lemma 1.42], this is so if and only if $\begin{bmatrix} r \\ s \end{bmatrix} \equiv \pm \begin{bmatrix} r+s \\ s \end{bmatrix} \ (\mathrm{mod}\ q)$. Therefore ε_0 has exactly $\varphi(q)/2$ fixed points represented by r/q with $(r, q) = 1$, $0 < r < q/2$. For such an r, take an element β

of $SL_2(\mathbf{Z})$ of the form $\beta = \begin{bmatrix} u & v \\ -q & r \end{bmatrix}$. Then β maps r/q to ∞, and

$$\beta^{-1} \begin{bmatrix} 1 & m \\ 0 & 1 \end{bmatrix} \beta = \begin{bmatrix} 1-mqr & mr^2 \\ -mq^2 & 1+mqr \end{bmatrix}.$$

Choose an integer m so that $mr^2 \equiv 1 \pmod{q}$. Then $\beta^{-1} \begin{bmatrix} 1 & m \\ 0 & 1 \end{bmatrix} \beta \varepsilon^{-1} \in \Gamma(q)$, so that $\beta^{-1} \begin{bmatrix} 1 & m \\ 0 & 1 \end{bmatrix} \beta$ represents the automorphism ε_0, and moreover it has r/q as a fixed point. Therefore, if t is the local parameter on $\mathfrak{H}^*/\Gamma(q)$ around r/q defined by

$$t(z) = \exp\left[2\pi i \cdot \beta(z)/q\right],$$

one has $t \circ \varepsilon_0 = \zeta^m t$ with $\zeta = e^{2\pi i/q}$. Let $\mathrm{tr}\,(\varepsilon_0)$ denote the trace of $[\varepsilon]_2$ on $S_2(\Gamma(q))$. Applying the Selberg-Eichler trace formula to ε_0, we obtain

$$(5.3) \qquad \mathrm{tr}\,(\varepsilon_0^u) - 1 = 2^{-1} \sum_{(r,q)=1} \zeta^{r^2 u}/(1-\zeta^{r^2 u}),$$

for $(u, q) = 1$. Therefore

$$\mathrm{tr}\,(R) = \sum_{u=1}^{q-1} \chi(u)\,\mathrm{tr}\,(\varepsilon_0^u) = 2^{-1}\varphi(q) \sum_{m=1}^{q-1} \chi(m)\zeta^m/(1-\zeta^m).$$

To compute the last sum, take an indeterminate w, and observe that

$$\sum_{m=1}^{q-1} \chi(m)\zeta^m/(1-\zeta^m w) = \sum_{k=1}^{\infty} w^{k-1} \sum_{m=1}^{q-1} \chi(m)\zeta^{mk}$$

$$= \sqrt{-q} \sum_{k=1}^{\infty} \chi(k) w^{k-1}$$

$$= \sqrt{-q} \left(\sum_{k=1}^{q-1} \chi(k) w^{k-1} \right)/(1-w^q),$$

hence

$$\mathrm{tr}\,(R) = (\sqrt{-q} \cdot \varphi(q)/2) \cdot \lim_{w \to 1} \left(\sum_{k=1}^{q-1} \chi(k) w^k \right)/(1-w^q)$$

$$= -(\sqrt{-q} \cdot \varphi(q)/2q) \sum_{k=1}^{q-1} \chi(k) k,$$

by differentiating the numerator and the denominator. Combining this with Proposition 10, we obtain Dirichlet's formula (5.1).

Especially, suppose q is a prime, and denote by U the right hand side of (5.3). Then $U + \bar{U} = -\varphi(q)/2$, and

$$U - \bar{U} = \sum_{m=1}^{q-1} \chi(um)\zeta^m/(1-\zeta^m) = -\sqrt{-q} \cdot \chi(u) \sum_{k=1}^{q-1} \chi(k) k/q,$$

so that

$$(5.4) \qquad \mathrm{tr}\,(\varepsilon_0^u) - 1 = -\varphi(q)/4 - \sqrt{-q} \cdot \chi(u) \sum_{k=1}^{q-1} \chi(k) k/2q.$$

Let us now clarify the relation between our proof and Hecke's results in [4], [5]. Put $\mathfrak{M}_q = SL_2(\mathbf{Z}/q\mathbf{Z})/\{\pm 1\}$, and suppose that q is a prime such that $q \equiv 3 \pmod 4$ and $q > 3$. Then $\mathfrak{M}_q$ has an irreducible representation $\mathfrak{G}$ of degree $(q-1)/2$ such that

$$\operatorname{tr}(\mathfrak{G}(\varepsilon)) = (-1 + \sqrt{-q})/2 .$$

Let $\mathfrak{D}$ be the representation of $\mathfrak{M}_q$ on $S_2(\Gamma(q))$ via the map $\gamma \mapsto [\gamma]_2$ for $\gamma \in SL_2(\mathbf{Z})$. Let y_1 and y_2 be the multiplicities of $\mathfrak{G}$ and its complex conjugate $\bar{\mathfrak{G}}$ in $\mathfrak{D}$, respectively. In [5], Hecke proved

$$(5.5) \qquad\qquad y_1 - y_2 = h(-q)$$

by first showing

$$(5.6) \qquad\qquad y_1 - y_2 = -\sum_{n=1}^{q-1} \chi(n) n/q ,$$

and employing Dirichlet's formula (5.1). Actually Proposition 10, combined with a simple observation, leads to (5.5) with neither Eichler-Selberg's trace formula nor Dirichlet's formula, but with the knowledge of all irreducible representations of $\mathfrak{M}_q$. To see this, put $\omega = \begin{bmatrix} 1 & 1/q \\ 0 & 1 \end{bmatrix}$, and take a basis of $S_2(\Gamma)$ formed by eigen-functions of $[\omega]_2$, and let f be a member of the basis. Then we have $f|[\omega]_2 = \zeta^m f$ with $\zeta = e^{2\pi i/q}$ and $0 \leq m < q$, hence

$$(5.7) \qquad\qquad f|R = \sum_{k=1}^{q-1} \chi(k) \zeta^{mk} f = \chi(m) \sqrt{-q} \cdot f .$$

Now, for any irreducible representation $\mathfrak{G}^*$ of $\mathfrak{M}_q$ other than $\mathfrak{G}$ and $\bar{\mathfrak{G}}$, $\operatorname{tr}(\mathfrak{G}^*(\varepsilon))$ is a rational integer (see, e. g., Hecke [4, p. 529 (Werke)]). Therefore we have

$$(5.8) \qquad \operatorname{tr}([\omega]_2) = \operatorname{tr}(\mathfrak{D}(\varepsilon))$$

$$= y_1 \cdot \sum_{\chi(m)=1} \zeta^m + y_2 \cdot \sum_{\chi(m)=-1} \zeta^m + a\zeta^0 + b\sum_{m=1}^{q-1} \zeta^m$$

with non-negative integers a and b. Here each term ζ^m $(m \geq 0)$ corresponds to a member of our basis, and by (5.7), contributes $\chi(m)\sqrt{-q}$ to $\operatorname{tr}(R)$, hence

$$\operatorname{tr}(R) = (y_1 - y_2)\sqrt{-q}\,(q-1)/2 .$$

Comparing this with Proposition 10, we obtain (5.5).

We also notice that (5.6) can be derived from (5.4) and (5.8) by comparing the imaginary parts, since $\operatorname{tr}(\varepsilon_0) = \operatorname{tr}(\mathfrak{D}(\varepsilon))$. Actually Hecke obtained (5.6) by computing the dimension of the vector space

$$(5.9) \qquad \{f \in S_2(\Gamma(q)) \mid f|[\varepsilon]_2 = \zeta^v f\} \qquad (0 \leq v < q)$$

by means of the Riemann-Roch theorem. For the reader's convenience, let us now compute $\mathrm{tr}\,(R)$ by Hecke's technique, using the Riemann-Roch theorem instead of the Selberg-Eichler trace formula.

Let λ_v denote the dimension of (5.9). Then

$$(5.10) \qquad \mathrm{tr}\,(R) = \sum_{k=1}^{q-1} \chi(k)\,\mathrm{tr}\,(\varepsilon_0^k) = \sum_{k=1}^{q-1}\sum_{v=0}^{q-1} \chi(k)\zeta^{kv}\lambda_v = \sqrt{-q}\cdot\sum_{v=1}^{q-1}\chi(v)\lambda_v\,.$$

Put $W=\mathfrak{H}^*/\Gamma(q)$ and $W'=\mathfrak{H}^*/\Gamma_1(q)$. For every positive divisor n of q, let P_i^n, for $i=1,\cdots,c_n$, denote all the points of W' with ramification index q/n, and $Q_{i1}^n,\cdots,Q_{in}^n$ the points of W lying above P_i^n. Put $q_n=q/n$. Let $\mathfrak{g}$ and $\mathfrak{g}'$ be the genera of W and W', respectively. Then

$$2\mathfrak{g}-2 = (2\mathfrak{g}'-2)q + \sum_{n\mid q}(q_n-1)nc_n\,.$$

We can find a $\Gamma(q)$-invariant automorphic form f_0 of weight 2 such that $f_0|[\varepsilon]_2=\zeta^v f_0$ with any integer v. Let $\mathrm{div}\,(X)$ resp. $\mathrm{div}'\,(X)$ denote the divisor of a function or a differential form X on W resp. W'. Then

$$\mathrm{div}\,(f_0(z)dz) = D + \sum_{n\mid q}\sum_{i=1}^{c_n} k_{ni}(Q_{i1}^n + \cdots + Q_{in}^n)$$

with integers k_{ni} and a divisor D on W not involving Q_{ij}^n. Let t_{ni} be a local parameter on W at Q_{i1}^n. Then

$$t_{ni}\circ\varepsilon_0^n = \zeta^{nh}t_{ni}+(\text{higher terms})$$

with an integer h prime to q_n. Then we see that $nv \equiv nh(k_{ni}+1) \pmod{q}$. Take an integer s_{ni} so that $s_{ni}h\equiv 1 \pmod{q_n}$. Note that s_{ni} is independent of v. Then $k_{ni}\equiv s_{ni}v-1 \pmod{q_n}$. Especially, if $n=1$, the above study of the fixed points of ε_0 shows that $c_1=\varphi(q)/2$, and P_i^1 is represented by r/q under the conditions $(r,q)=1$, $0<r<q/2$. For this point, we can take $s_{11}=r^2$, as seen above. Now the map $f\mapsto F=f_0^{-1}f$ sends the vector space (5.9) isomorphically onto the vector space of all $\Gamma_1(q)$-invariant modular functions F such that

$$\mathrm{div}\,(F) \geqq -\mathrm{div}\,(f_0 dz)\,.$$

We see easily that this is so if and only if

$$(5.11) \qquad \mathrm{div}'\,(F) \geqq -q^{-1}D' - \sum_{n\mid q}\sum_{i=1}^{c_n}[k_{ni}/q_n]P_i^n\,,$$

where D' is the projection of D onto W', and $[x]$ is the largest integer $\leq x$. Put $\{x\}=x-[x]$. Let μ_v denote -1 times the degree of the right hand side of (5.11). Assume that $(v,q)=1$. Then $s_{ni}v\not\equiv 0 \pmod{q_n}$, hence $\{k_{ni}/q_n\}=\{s_{ni}v/q_n\}-1/q_n$. Therefore

$$\mu_v = q^{-1} \deg \left(\mathrm{div}\, (f_0 dz) \right) - \sum_{n,\iota} \{k_{n\iota}/q_n\}$$

$$= (2g-2)/q - \sum_{n,\iota} \{k_{n\iota}/q_n\}$$

$$= 2g' - 2 + \sum_{n,\iota} (1 - \{s_{n\iota}v/q_n\})$$

$$> 2g' - 2 .$$

Thus, by the Riemann-Roch theorem, one has

$$\lambda_v = g' - 1 + \sum_{n,i} (1 - \{s_{n\iota}v/q_n\}) .$$

Observe that $\sum_{v=1}^{q-1} \chi(v)\{sv/q_n\} = 0$ for any s if $n > 1$. Therefore by (5.10) we obtain

$$\mathrm{tr}\,(R) = -\sqrt{-q} \cdot \sum_{r} \sum_{v=1}^{q-1} \chi(v)\{vr^2/q\} = -\sqrt{-q} \cdot (\varphi(q)/2) \cdot \sum_{k=1}^{q-1} \chi(k)k/q ,$$

which is the desired result.

§ 6. The zeta-function of A over a non-abelian extension of Q.

Let Γ be a group of level N, and let A be defined for a common eigen-function f in $S_2(\Gamma)$ as in Theorem 1. Then one can naturally ask about (the one-dimensional part of) the zeta-function of A over an arbitrary finite algebraic extension F of Q. This problem can easily be settled (always up to finitely many bad Euler factors), if F is abelian over Q, as shown in [11, § 7.9] (see also the discussion below). We shall now show that one can determine the zeta-function for a certain class of non-abelian extensions.

Let us denote by $Z(s, A/F)$ the one-dimensional part of the zeta-function of A over F, and by $\zeta_F(s)$ the Dedekind zeta-function of F. We can write

$$\zeta_F(s) = \prod_{p} \det \left[1 - \Psi_p p^{-s} \right]^{-1} ,$$

$$Z(s, A/Q) = \prod_{p} \det \left[1 - \Phi_p p^{-s} \right]^{-1} \qquad \text{(the product for all good primes)}$$

with complex matrices Φ_p and Ψ_p given for each prime p; the size of Φ_p (resp. Ψ_p) is $2 \cdot \dim(A)$ (resp. $[F:Q]$); the matrices are invertible for almost all p. Then we see easily that

$$Z(s, A/F) = \prod_{p} \det \left[1 - (\Phi_p \otimes \Psi_p) p^{-s} \right]^{-1} .$$

This applies to any abelian variety A defined over Q.

Now take a common eigen-function $f(z) = \sum_{n} a_n e^{2\pi i n z}$, with $a_1 = 1$, of all Hecke operators in $S_k(\Gamma)$, with any positive integer k, and put

$$L(s, f) = \sum_{n=1}^{\infty} a_n n^{-s} = \prod_p (1 - a_p p^{-s} + \varepsilon(p) p^{k-1-2s})^{-1}$$

$$= \prod_p \det [1 - \Xi_p p^{-s}]^{-1}$$

with complex matrices Ξ_p of size 2. Here $\varepsilon(p)$ is a character of $(\mathbf{Z}/N\mathbf{Z})^{\times}$ such that

$$f \Big\vert \Big[\begin{bmatrix} a & b \\ c & d \end{bmatrix}\Big]_k = \varepsilon(d) f \quad \text{for all} \quad \begin{bmatrix} a & b \\ c & d \end{bmatrix} \in \varGamma_0(N).$$

Then we define an Euler product

$$L(s, f, F) = \prod_p \det [1 - (\Xi_p \otimes \varPsi_p) p^{-s}]^{-1},$$

which obviously converges in some half plane. The last part of Theorem 1 implies that, if $k=2$ and A corresponds to f, then, under the condition stated there, $Z(s, A/F)$ coincides with $\prod_{\sigma \in I} L(s, f_\sigma, F)$, up to finitely many Euler factors. It is natural to conjecture that if $k > 1$, $L(s, f, F)$ can always be continued to a holomorphic function on the whole s-plane satisfying a functional equation. (If $k=1$, one will have to allow the function to have finitely many poles.) Thus our purpose is to show that this is so at least for a certain class of extensions F of $\mathbf{Q}$. To define it, we consider the following condition on a finite group G:

(X) *Every irreducible representation of G by complex matrices is either one-dimensional or induced from a one-dimensional representation of a subgroup of index 2.*

It can easily be shown that the following types of groups satisfy the condition: abelian group; dihedral group; generalized quaternion group (i. e., a group of order $4m$ generated by two elements r and s with the relations $r^2 = s^m$, $s^{2m} = 1$, $rsr^{-1} = s^{-1}$); a homomorphic image of a group satisfying (X); the product of an abelian group and a group satisfying (X).

Let E be a Galois extension of $\mathbf{Q}$ such that $\mathrm{Gal}(E/\mathbf{Q})$ satisfies (X), and let F be any subfield of E. It is this type of extension F, for which we are going to show that $L(s, f, F)$ can be determined. Write simply G for $\mathrm{Gal}(E/\mathbf{Q})$, and let $L(s, E/\mathbf{Q}, \varphi)$ denote the Artin L-function with a character φ of G. By virtue of (X), $\zeta_F(s)$ can be written as a finite product

$$\zeta_F(s) = \prod_\mu L(s, E/\mathbf{Q}, \varphi_\mu) \times \prod_\nu L(s, E/\mathbf{Q}, \psi_\nu),$$

where φ_μ is a one-dimensional character of G, and ψ_ν is a character induced from an abelian character of a subgroup H_ν of G of index 2. Each φ_μ corresponds to a primitive character φ_μ^* of $(\mathbf{Z}/m_\mu \mathbf{Z})^{\times}$ with a positive integer m_μ, and

$$L(s, E/\mathbf{Q}, \varphi_\mu) = \prod_p (1 - \varphi_\mu^*(p) p^{-s})^{-1} .$$

As to ϕ_ν, take the subfield M_ν of E corresponding to H_ν. Then $[M_\nu : \mathbf{Q}] = 2$, and $L(s, E/\mathbf{Q}, \phi_\nu)$ coincides with an L-function

$$L(s, M_\nu, \chi_\nu) = \prod_{\mathfrak{p}} (1 - \chi_\nu(\mathfrak{p}) N(\mathfrak{p})^{-s})^{-1}$$

of M_ν with a primitive character χ_ν of finite order, where the product is extended over all prime ideals $\mathfrak{p}$ of M_ν prime to the conductor of χ_ν. Put

$$L(s, M_\nu, \chi_\nu) = \sum_n b_{\nu n} n^{-s} = \prod_p \det[1 - \Psi_{\nu p} p^{-s}]^{-1}$$

with complex numbers $b_{\nu n}$ and complex matrices $\Psi_{\nu p}$ of size 2. Then

$$L(s, f, F) = \prod_\mu L(s, f, \varphi_\mu^*) \times \prod_\nu L(s, f, M_\nu, \chi_\nu) .$$

where

$$L(s, f, \varphi_\mu^*) = \prod_p \det[1 - \varphi_\mu^*(p) \mathcal{Z}_p p^{-s}]^{-1} = \sum_n \varphi_\mu^*(n) a_n n^{-s} .$$

(6.1) $$L(s, f, M_\nu, \chi_\nu) = \prod_p \det[1 - (\mathcal{Z}_p \otimes \Psi_{\nu p}) p^{-s}]^{-1} .$$

By an easy formal computation, we see that

$$L(s, f, M_\nu, \chi_\nu) = \Big(\sum_{n=1}^\infty a_n b_{\nu n} n^{-s}\Big) \Big(\sum_{n=1}^\infty \Big(\frac{D_\nu}{n}\Big) \chi_\nu((n)) \varepsilon(n) n^{k-1-2s}\Big) ,$$

where D_ν is the discriminant of M_ν. Note that $\det(\Psi_{\nu p}) = \Big(\frac{D_\nu}{p}\Big) \chi_\nu((p))$, and $\det(\mathcal{Z}_p) = \varepsilon(p) p^{k-1}$. The Dirichlet series $L(s, f, \varphi_\mu^*)$ can be continued to a holomorphic function on the whole s-plane satisfying a functional equation, since $\sum_n \varphi_\mu^*(n) a_n e^{2\pi i n z}$ is an element of $S_k(\Gamma_1(N_\mu))$ for some N_μ (see [11, Prop. 3.64]).

To study the nature of $L(s, f, M_\nu, \chi_\nu)$, we first observe that $L(s, M_\nu, \chi_\nu)$ is the Mellin transform of an automorphic form on $GL_2(\mathbf{Q}_A)$. This is essentially due to Hecke and Maass, and included, as a special case, in the results of Jacquet and Langlands [6], and Weil [17]. Especially, if M_ν is imaginary, and if we put $g_\nu(z) = \sum_n b_{\nu n} e^{2\pi i n z}$, then g_ν is a modular form of weight 1 satisfying

(6.2) $$g_\nu \Big| \Big[\begin{matrix} a & b \\ c & d \end{matrix}\Big]\Big]_1 = \Big(\frac{D_\nu}{d}\Big) \chi_\nu((d)) g_\nu \qquad \text{for all } \Big[\begin{matrix} a & b \\ c & d \end{matrix}\Big] \in \Gamma_0(D_\nu \cdot N(\mathfrak{c}_\nu)) ,$$

where $\mathfrak{c}_\nu$ is the conductor of χ_ν. Therefore $L(s, f, M_\nu, \chi_\nu)$ is similar to the Dirichlet series considered by Rankin [10]. Recently Jacquet [7] has given a general treatment of Dirichlet series of type (6.1). Thus it is virtually possible to obtain the desired property, or at least something very close to it, of $L(s, f, M_\nu, \chi_\nu)$ or of $L(s, f, F)$, though one may have to work out the details to get a precise statement.

For example, consider the simplest non-abelian case in which E is a non-abelian extension of Q of degree 6, whose unique quadratic subfield is imaginary, and take F to be a cubic subfield of E. Then $\zeta_F(s) = \zeta_Q(s)L(s, M, \chi)$ with the L-function $L(s, M, \chi)$ of M for a character χ of order 3, corresponding to the extension E/M, hence

$$L(s, f, F) = L(s, f)L(s, f, M, \chi).$$

Put $L(s, M, \chi) = \sum_n b_n n^{-s}$, $g(z) = \sum_n b_n e^{2\pi i n z}$. Then g is a modular form of weight 1, satisfying a relation of the type (6.2). (In this special case, one has $\chi((d)) = 1$.) Applying the method of [14, §3] to f and g, we can show that $L(s, f, M, \chi)$ can be continued to an entire function in s, if $k > 1$. To obtain the exact form of the functional equation, one has to assume that $f\left|\left[\left[\begin{smallmatrix} 0 & -1 \\ N & 0 \end{smallmatrix}\right]\right]\right._k$ is a common eigen-function of Hecke operators, and make considerations similar to Ogg [9] and the author [14, §3]; but we shall not treat this problem further in this paper.

References

[1] A. O. L. Atkin and J. Lehner, Hecke operators on $\Gamma_0(m)$, Math. Ann., 185 (1970), 134–160.

[2] W. Casselman, On some results of Atkin and Lehner, to appear.

[3] W.-L. Chow, On the quotient variety of an abelian variety, Proc. Nat. Acad. Sci. U. S. A., 38 (1952), 1039–1044.

[4] E. Hecke, Über ein Fundamentalproblem aus der Theorie der elliptischen Modulfunktionen, Abh. Math. Sem. Hamburg, 6 (1928), 235–257 (=Math. Werke, 525–547).

[5] E. Hecke, Über das Verhalten der Integrale 1. Gattung bei Abbildungen, insbesondere in der Theorie der elliptischen Modulfunktionen, Abh. Math. Sem. Hamburg, 8 (1930), 271–281 (=Math. Werke, 548–558).

[6] H. Jacquet and R. P. Langlands, Automorphic forms on $GL(2)$, Lecture notes in mathematics, 114, Springer, 1970.

[7] H. Jacquet, Automorphic forms on $GL(2)$, Part II, Lecture notes in mathematics, 278, Springer, 1972.

[8] T. Miyake, On automorphic forms on GL_2 and Hecke operators, Ann. of Math., 94 (1971), 174–189.

[9] A. P. Ogg, On a convolution of L-series, Invent. math., 7 (1969), 297–312.

[10] R. A. Rankin, Contributions to the theory of Ramanujan's function $\tau(n)$ and similar arithmetical functions I, II, Proc. Cambridge Phil. Soc., 35 (1939), 351–372.

[11] G. Shimura, Introduction to the arithmetic theory of automorphic functions, Publ. Math. Soc. Japan, No. 11, Iwanami Shoten and Princeton University Press, 1971.

[12] G. Shimura, On elliptic curves with complex multiplication as factors of the jacobians of modular function fields, Nagoya Math. J., 43 (1971), 199–208.

[13] G. Shimura, Class fields over real quadratic fields and Hecke operators, Ann. of Math., **95** (1972), 130–190.

[14] G. Shimura, On modular forms of half integral weight, Ann. of Math, **97** (1973), 440–481.

[15] A. Weil, Variétés abéliennes et courbes algébriques, Hermann, Paris, 1948.

[16] A. Weil, Über die Bestimmung Dirichletscher Reihen durch Funktionalgleichungen, Math. Ann., **168** (1967), 149–156.

[17] A. Weil, Dirichlet series and automorphic forms, Lecture notes in mathematics, 189, Springer, 1971.

Goro SHIMURA

Department of Mathematics
Princeton University
Princeton, New Jersey
U. S. A.

On the trace formula for Hecke operators

Acta mathematica, 132 (1974), 245-281

The formula to be proved in this paper has roughly the following form:

$$\operatorname{tr}\,(\Gamma\alpha\Gamma\,|\,A_m) - \operatorname{tr}\,(\Gamma\alpha^{-1}\Gamma\,|\,B_{2-m}) = \sum_C J(C).$$

Here Γ is a discrete subgroup of $SL_2(\mathbf{R})$ such that $SL_2(\mathbf{R})/\Gamma$ is of finite measure, m an arbitrary rational number, A_m the space of cusp forms of weight m with respect to Γ on the upper half complex plane $\mathfrak{H}$, B_{2-m} the space of integral forms of weight $2-m$ with respect to Γ, α an element of $SL_2(\mathbf{R})$ such that Γ and $\alpha^{-1}\Gamma\alpha$ are commensurable, and $J(C)$ a complex number defined for each class C of elements of $\Gamma\alpha\Gamma$ under a certain equivalence. The double cosets $\Gamma\alpha\Gamma$ and $\Gamma\alpha^{-1}\Gamma$ act on A_m and B_{2-m} respectively, under some conditions. An *integral form* of weight m is a holomorphic function $f(z)$ on $\mathfrak{H}$ which satisfies $f(\gamma(z))/f(z) = t(\gamma)(d\gamma(z)/dz)^{-m/2}$ for every $\gamma \in \Gamma$ with a certain constant factor $t(\gamma)$, and which is holomorphic at every cusp; an integral form is called a *cusp form* if it vanishes at every cusp.

If m is an integer >2, then $B_{2-m} = \{0\}$. The formula in this case was obtained by Selberg [5] and Eichler [2]. If $m=2$, B_{2-m} consists of the constants, and therefore $\operatorname{tr}\,(\Gamma\alpha^{-1}\Gamma\,|\,B_{2-m})$ is simply the number of right or left cosets in $\Gamma\alpha^{-1}\Gamma$. This case is also included in [2]. It should also be mentioned that the generalized Riemann-Roch theorem of Weil [8] is closely related to the above formula when α belongs to the normalizer of Γ.

Although our formula is given for an arbitrary rational m, the cases of integral and half integral weight with respect to an arithmetic Γ seem most significant. If m is a half integer >2, we have again $B_{2-m} = \{0\}$, and the formula is of the same nature as in the case of integral $m>2$. However, if $m=3/2$, both A_m and B_{2-m} can be non-trivial. Especially if Γ is a congruence subgroup of $SL_2(\mathbf{Z})$, it is conjecturable that $B_{\frac{1}{2}}$ is spanned by theta series of the type

$$\Sigma_n \psi(n) \exp\,(2\pi i n^2 r z)$$

with a rational number r and a character ψ modulo a positive integer. This is at least true for the groups of low level. In such a case, $\mathrm{tr}\,(\Gamma\alpha\Gamma\,|\,A_{3/2})$ is effectively computable.

The non-triviality of both A_m and B_{2-m} occurs also when $m=1$. In this case, however, the formula does not seem to bring forth any new information about the modular forms of weight 1. We can only compute the trace on the space of Eisenstein series, and also rediscover the forms whose Mellin transforms are L-functions of an imaginary quadratic field. (See 5.8 for a more detailed discussion.) Thus our formula for $m=1$ is not so effective in this sense, but it tells at least of what the trace formula should be in the extreme case $m=1$.

To prove the formula, we adapt to our setting the methods of Kappus [4] and Eichler [3], in which the forms of even positive weight were treated. In § 1, we consider automorphic forms in an axiomatic way, and then construct, in § 2, an algebraic analogue of kernel function. We work on the product of two copies of an algebraic curve, while the authors of [3] and [4] considered the composite of two copies of an algebraic function field. Although the theory of §§ 1, 2 as well as a part of later sections seems developable for the curves defined over a field of positive characteristic, our discussion is restricted to the case of characteristic 0, mainly for the sake of simplicity. In § 3, we prove the first formulation of the trace formula, which is algebro-geometric in the sense that the right hand side is expressed in terms of the fixed points of the algebraic correspondence attached to $\Gamma\alpha\Gamma$. A more group-theoretical formulation will be given in § 4. In the final § 5, we make a few remarks and discuss some features peculiar to the cases $m=3/2$ and $m=1$.

One remark, though obvious, may be added: A formula of the same type will undoubtedly be proved for higher dimensional manifolds instead of a curve. For example, we note that if Γ has neither parabolic nor elliptic elements and $\alpha\Gamma\alpha^{-1}=\Gamma$, then the above formula follows directly from the fixed point formula of Atiyah and Bott. Although such a special case is not important from an arithmetical viewpoint, the Atiyah-Bott formula will suggest a plausible form in a more general case. It should also be mentioned that we do not put any emphasis on our choice of method. The framework of the present paper has a natural limitation, while it enables us to obtain a fairly general and practical formula in the one-dimensional case with a relatively small amount of complexity. (At least we have dispensed with any discussion of convergence or limit process.) Therefore any method, either analytic, geometric or group-theoretical, may be adopted on its own merit in proving the higher dimensional generalization.

1. Axioms of automorphic forms

1.1. We fix a "universal domain" Ω of characteristic 0 in the sense of Weil's Foundations [9], and consider algebro-geometric objects rational over subfields of Ω, denoted by k, k', k_0, etc. We take these fields so that Ω has an infinite transcendence degree over them.

Let V be a complete non-singular curve, which will be fixed throughout the first two sections. If V is defined over a field k, we denote by $k(V)$ the field of k-rational functions on V, by $\Phi(k)$ the module of k-rational differential 1-forms on V, and by $D(k)$ the module of all k-rational divisors of V. The unions of $k(V)$, $\Phi(k)$, $D(k)$ for all fields k of rationality for V will be denoted by $\Omega(V)$, Φ, and D, respectively. It is necessary for our purpose to consider divisors with fractional coefficients. Therefore we put $D_{\mathbf{Q}} = D \otimes_{\mathbf{Z}} \mathbf{Q}$, $D_{\mathbf{Q}}(k) = D(k) \otimes_{\mathbf{Z}} \mathbf{Q}$, and $\deg(\Sigma_i c_i x_i) = \Sigma_i c_i$ for $\Sigma_i c_i x_i \in D_{\mathbf{Q}}$ with $c_i \in \mathbf{Q}$, $x_i \in V$. An element of $D_{\mathbf{Q}}$ is called *k-rational* if it belongs to $D_{\mathbf{Q}}(k)$. For $0 \neq \alpha \in k(V)$ and $0 \neq \omega \in \Phi(k)$, we can define their divisors (which are of course elements of $D(k)$) as usual, and denote them by $\operatorname{div}(\alpha)$ and $\operatorname{div}(\omega)$. Let $P(k)$ denote the set of all k-rational prime divisors of V. For each $p \in P(k)$, we can define a discrete valuation ν_p of $k(V)$ in a natural manner so that $\operatorname{div}(\alpha) = \Sigma_p \nu_p(\alpha) p$. We use the symbol ν_p also for the map $D_{\mathbf{Q}}(k) \to \mathbf{Q}$ defined by $\mathfrak{a} = \Sigma_p \nu_p(\mathfrak{a}) p$ for $\mathfrak{a} \in D_{\mathbf{Q}}$, and put $\nu_p(\omega) = \nu_p(\operatorname{div}(\omega))$ for $0 \neq \omega \in \Phi(k)$. For $\mathfrak{a}$, $\mathfrak{b} \in D_{\mathbf{Q}}(k)$, we write $\mathfrak{a} \geqslant \mathfrak{b}$ if $\nu_p(\mathfrak{a} - \mathfrak{b}) \geqslant 0$ for all $p \in P(k)$.

1.2. To discuss automorphic forms in an axiomatic way, we consider a system $\mathfrak{F} = \{F, F', Z, \mathfrak{z}\}$ formed by the objects satisfying the following axioms (A_{1-4}).

(A_1) *F and F' are one-dimensional vector spaces over $\Omega(V)$.*

(A_2) *To each non-zero element f of F or F', one can assign an element of $D_{\mathbf{Q}}$ denoted by* $\operatorname{div}(f)$ *satisfying*

$$\operatorname{div}(hf) = \operatorname{div}(h) + \operatorname{div}(f) \quad \text{for } h \in \Omega(V), f \in F \text{ or } F'.$$

(A_3) *Z is a non-degenerate $\Omega(V)$-bilinear map $F \times F' \to \Phi$.*

(A_4) *$\mathfrak{z}$ is an element of $D_{\mathbf{Q}}$ such that*

$$\operatorname{div}(Z(f, g)) = \mathfrak{z} + \operatorname{div}(f) + \operatorname{div}(g) \quad (0 \neq f \in F, 0 \neq g \in F').$$

For $0 \neq f \in F$ and $u \in F$, we can define $h = f^{-1}u = uf^{-1} = u/f$ to be the element of $\Omega(V)$ such that $hf = u$. This applies also to the elements of F'.

To make our notation more suggestive, we use a symbol dz instead of Z, and write

$$fg\,dz = Z(f, g) \quad (f \in F, g \in F'),$$
$$\operatorname{div}(dz) = \mathfrak{z}.$$

Then we have
$$\operatorname{div}(fg\,dz) = \operatorname{div}(f) + \operatorname{div}(g) + \operatorname{div}(dz).$$

For the moment, dz is merely a symbol replacing Z; it has no meaning as the differential of z until § 3, where we take F and F' to be the modules of automorphic forms of weight m and $2-m$, respectively, and dz as the differential of the variable on the upper half plane.

To define "k-rational elements" of F and F', fix any non-zero element w of F. Let k_0 be a field of rationality for V, $\operatorname{div}(w)$, and $\mathfrak{z}$. Take a non-zero element η of $\Phi(k_0)$. There is a uniquely determined element v of F' such that $wv\,dz = \eta$. For any field k of rationality for V containing k_0, put
$$F(k) = k(V)w, \quad F'(k) = k(V)v.$$
Then we see that

(A_5) Z *maps* $F(k) \times F'(k)$ *onto* $\Phi(k)$; $\operatorname{div}(f)$ *is* k-*rational if* $f \in F(k)$ *or* $F'(k)$; $F(k_1) = F(k) \otimes_k k_1$, $F'(k_1) = F'(k) \otimes_k k_1$ *if* $k \subset k_1$.

Hereafter we fix k_0, and consider only the fields k containing k_0 as basic fields. Such a field k will be called a field of rationality for $\mathfrak{F}$. For each $p \in P(k)$ and $f \in F$ or F', we define a rational number $\nu_p(f)$ by $\operatorname{div}(f) = \Sigma_p \nu_p(f)p$.

A simple example of $\mathfrak{F}$ is obtained by taking $F = \Phi$, $F' = \Omega(X)$, $Z(f, g) = fg$, and $\mathfrak{z} = 0$.

Remark. The modules $F(k)$ and $F'(k)$ depend on the choice of w. Instead of taking w, one could start with (A_5) as an additional axiom.

1.3. Let us now introduce a module $R(k)$, which may be called the module of "$F(k)$-valued adeles" in a weak sense. To be precise, we consider a map $b: P(k) \to F(k)$ which assigns to each $p \in P(k)$ an element b_p of $F(k)$, such that $\nu_p(b_p) \geqslant 0$ for all except a finite number of p's. We denote by $R(k)$ the module of all such b, addition being defined by $(b+c)_p = b_p + c_p$. We write $b = (b_p)$, and call b_p *the p-component of b*. For $a \in k(V)$ and $b \in R(k)$, we can define an element ab of $R(k)$ by $(ab)_p = a \cdot b_p$. Each $c \in F(k)$ defines an element of $R(k)$ whose p-component equals c for every $p \in P(k)$. In this way $F(k)$ can be identified with a submodule of $R(k)$.

Now for $\mathfrak{a} \in D_Q(k)$, put
$$R(\mathfrak{a}, k) = \{b \in R(k) \mid \nu_p(b_p) \geqslant -\nu_p(\mathfrak{a}) \text{ for all } p \in P(k)\},$$
$$F(\mathfrak{a}, k) = \{f \in F(k) \mid \operatorname{div}(f) \geqslant -\mathfrak{a}\}$$
$$= R(\mathfrak{a}, k) \cap F(k),$$
$$F(\mathfrak{a}) = \{f \in F \mid \operatorname{div}(f) \geqslant -\mathfrak{a}\}.$$

Taking F' in place of F, we define $R'(k)$, $R'(\mathfrak{a}, k)$, $F'(\mathfrak{a}, k)$ and $F'(\mathfrak{a})$ in the same manner. Also we put

$$L(\mathfrak{a}, k) = \{f \in k(V) \,|\, \mathrm{div}\,(f) \geq -\mathfrak{a}\},$$

$$L(\mathfrak{a}) = \{f \in \Omega(V) \,|\, \mathrm{div}\,(f) \geq -\mathfrak{a}\},$$

$$l(\mathfrak{a}) = \dim L(\mathfrak{a}).$$

Here dim stands for the dimension of a vector space over Ω. If $0 \neq u \in F(k)$, then

$$(1.3.1) \qquad F(\mathfrak{a}, k)u^{-1} = L(\mathfrak{a} + \mathrm{div}\,(u), k), \quad F(\mathfrak{a})u^{-1} = L(\mathfrak{a} + \mathrm{div}\,(u)),$$

hence $F(\mathfrak{a}, k)$ is finite dimensional over k, and

$$\dim F(\mathfrak{a}) = l(\mathfrak{a} + \mathrm{div}\,(u)),$$

$$F(\mathfrak{a}) = F(\mathfrak{a}, k) \otimes_k \Omega.$$

1.4. For $p \in P(k)$ and $\omega \in \Phi(k)$, we define the residue of ω at p, denoted by $\mathrm{Res}_p(\omega)$ as follows. Put $p = p_1 + \ldots + p_s$ with points p_i on V. Let $\bar{k}$ be the algebraic closure of k, and t an element of $\bar{k}(V)$ such that $v_{p_i}(t) = 1$. Define $\mathrm{Res}_{p_i}(\omega)$ as usual to be the coefficient of t^{-1} in the power-series expansion of ω/dt in t with coefficients in $\bar{k}$. Then we put

$$\mathrm{Res}_p(\omega) = \sum_{i=1}^{s} \mathrm{Res}_{p_i}(\omega).$$

We see easily that $\mathrm{Res}_p(\omega) = \mathrm{Tr}_{k(p_1)/k}(\mathrm{Res}_{p_1}(\omega))$, and as is well known,

$$\sum_{p \in P(k)} \mathrm{Res}_p(\omega) = 0 \quad \text{for all } \omega \in \Phi(k).$$

1.5. PROPOSITION. *Let $\mathfrak{a}$ and $\mathfrak{b}$ be two elements of $D_{\mathbf{Q}}$ such that*

$$(1.5.1) \qquad \mathfrak{a} + \mathfrak{b} = \mathfrak{z}; \quad \mathfrak{a} + \mathrm{div}\,(f) \in D \text{ for every non-zero } f \in F.$$

Then $\mathfrak{b} + \mathrm{div}\,(g) \in D$ for every non-zero $g \in F'$. Moreover, if $\mathfrak{a}$ and $\mathfrak{b}$ are k-rational, the vector space $F(\mathfrak{a}, k)$ is dual to $R'(k)/[R'(\mathfrak{b}, k) + F'(k)]$ by the k-bilinear pairing

$$(f, v) \to \langle f, v \rangle = \sum_{p \in P(k)} \mathrm{Res}_p\,(fv_p dz)$$

for $f \in F(\mathfrak{a}, k)$, $v = (v_p) \in R'(k)$.

Proof. First note that $\mathfrak{a} + \mathrm{div}\,(f) \in D$ holds for all non-zero $f \in F$ if it holds for at least one f. Now let $0 \neq g \in F'$. Then $fg\,dz \in \Phi$, so that $\mathrm{div}\,(fg\,dz) \in D$. Subtracting $\mathfrak{a} + \mathrm{div}\,(f)$ from $\mathrm{div}\,(fg\,dz)$, one finds $\mathrm{div}\,(g) + \mathfrak{b} \in D$. Now the duality in the case $F = \Phi$, $F' = \Omega(V)$, $\mathfrak{z} = 0$ is well known. In fact, let $R_0(k)$ (resp. $R_0(\mathfrak{b}, k)$) denote the module $R'(k)$ (resp. $R'(\mathfrak{b}, k)$) defined with $F' = \Omega(V)$. In this special case, we see that $\mathfrak{b} \in D$, and

$$F(\mathfrak{a}, k) = \{\omega \in \Phi(k) \,|\, \mathrm{div}\,(\omega) \geq \mathfrak{b}\},$$

and this vector space is dual to $R_0(k)/[R_0(\mathfrak{b}, k) + k(V)]$. (See e.g. Chevalley [1], especially

p. 30, Th. 2. See also Weil [8], pp. 58–59, and Eichler [3], p. 177.) In the general case, take any non-zero $w \in F'(k)$. Then

$$F(\mathfrak{a}, k)w\,dz = \{\eta \in \Phi(k) \,|\, \mathrm{div}\,(\eta) \geqslant \mathfrak{b} + \mathrm{div}\,(w)\},$$

$$w^{-1}R'(\mathfrak{b}, k) = R_0(\mathfrak{b} + \mathrm{div}\,(w), k).$$

Further for $f \in F(k)$ and $v \in R'(k)$, we have $\mathrm{Res}_p (fv_p dz) = \mathrm{Res}_p (fww^{-1}v_p dz)$. Therefore our assertion for F and R' reduces to the above special case.

1.6. Let $\mathfrak{a}, \mathfrak{b}$ be as in Proposition 1.5 under the assumption (1.5.1). Let $0 \neq f \in F$, $0 \neq g \in F'$, $\omega = fg\,dz$. Then

$$\mathfrak{a} + \mathfrak{b} + \mathrm{div}\,(f) + \mathrm{div}\,(g) = \mathrm{div}\,(\omega),$$

$$\dim F(\mathfrak{a}) = l(\mathfrak{a} + \mathrm{div}\,(f)),$$

$$\dim F'(\mathfrak{b}) = l(\mathfrak{b} + \mathrm{div}\,(g)) = l(\mathrm{div}\,(\omega) - \mathfrak{a} - \mathrm{div}\,(f)),$$

hence by the Riemann-Roch theorem, we obtain

(1.6.1) $$\dim F(\mathfrak{a}) - \dim F'(\mathfrak{b}) = \deg\,(\mathfrak{a} + \mathrm{div}\,(f)) - \mathfrak{g} + 1,$$

where $\mathfrak{g}$ is the genus of V.

1.7. Let k_1 be an extension (either algebraic or transcendental) of k. Then we can embed $R'(k)$ into $R'(k_1)$ as follows. To each $b = (b_p) \in R'(k)$, we assign $b^* = (b_q^*) \in R'(k_1)$ by

$$b_q^* = \begin{cases} b_p & \text{if } q \leqslant p, \\ 0 & \text{if } q \leqslant p \text{ for no } p \in P(k). \end{cases}$$

This embedding maps $R'(\mathfrak{b}, k) + F'(k)$ into $R'(\mathfrak{b}, k_1) + F'(k_1)$. (Note that $F'(k)$ is not necessarily mapped into $F'(k_1)$.) Further it maps $R'(k)/[R'(\mathfrak{b}, k) + F'(k)]$ injectively into $R'(k_1)/[R'(\mathfrak{b}, k_1) + F'(k_1)]$, and the latter can be identified with the tensor product of the former with k_1 over k. This is also compatible with the duality with $F(\mathfrak{a}, k_1) = F(\mathfrak{a}, k) \otimes_k k_1$ explained in Proposition 1.5.

1.8. Let $\mathfrak{a}$ and $\mathfrak{b}$ be elements of D_Q satisfying (1.5.1). Take a field k of rationality for $\mathfrak{F}$ and $\mathfrak{a}$. Take also a non-zero element v of $F'(k)$ and a prime divisor $q \in P(k)$ of degree one (i.e., a k-rational point of V) that is disjoint with $\mathrm{div}\,(v)$, $\mathfrak{a}$, and $\mathfrak{z}$. Let t_q be an element of $k(V)$ such that $v_q(t_q) = 1$. (We call such a t_q a *k-rational local parameter at q*.) Then we can find a basis $\{w_1, \ldots, w_n\}$ of $F(\mathfrak{a}, k)$ such that

$$w_i v\,dz = (t_q^{\alpha_i} + c_{i1} t_q^{\alpha_i + 1} + \ldots)\,dt_q,$$

(1.8.1) $$0 \leqslant \alpha_1 < \ldots < \alpha_n$$

with $c_{ij} \in k$. Subtracting a suitable linear combination $\Sigma_{j>i} b_j w_j$ from w_i, we may assume that

(1.8.2) $\qquad\qquad$ *The coefficient of $t_q^{\alpha_j} dt_q$ in $w_i v dz$ is 0 if $j > i$.*

We call $\{w_1, ..., w_n\}$ a *q-basis of $F(\mathfrak{a}, k)$ relative to v and t_q*, if (1.8.1) and (1.8.2) are satisfied.

Now define an element u_i of $R'(k)$ so that

$$u_{ip} = 0 \quad \text{for} \quad p \neq q,$$
$$u_{iq} = t_q^{-\alpha_i - 1} v.$$

Then $\langle w_i, u_j \rangle = \mathrm{Res}_q\, (w_i u_{jq} v^{-1} v dz) = \delta_{ij}$, by virtue of (1.8.2). Therefore $u_1, ..., u_n$ form a basis of $R'(k)/[R'(\mathfrak{b}, k) + F'(k)]$, hence every element of $R'(k)$ is congruent to a linear combination of the form $\Sigma_i c_i u_i$ with c_i in k modulo $R'(\mathfrak{b}, k) + F'(k)$. We state this fact as

1.9. Proposition. *Let q, t_q, and v be as above. Then, for every $r \in R'(k)$, there exists an element s of $R'(k)$ such that*

$$r - s \in R'(\mathfrak{b}, k) + F'(k),$$
$$s_p = 0 \quad \text{for } p \neq q,$$
$$s_q = \left(\textstyle\sum_{i=1}^n c_i t_q^{-\alpha_i - 1}\right) v$$

with $c_i \in k$.

2. An algebraic kernel function

2.1. Let V and $\mathfrak{F}$ be the same as before, and k a field of rationality for $\mathfrak{F}$. The purpose of this section is to construct an "algebraic kernel function" which will play an essential role in the computation of the trace of a Hecke operator in the next section. In the construction we shall be considering "generic points" of V over k in the sense of [9]. If x is a generic point of V over k, then $k(x)$ is a subfield of Ω, isomorphic to $k(V)$ over k by the map $g \mapsto g(x)$ for $g \in k(V)$. Here $g(x)$ is the value of g at the point x. For our purpose, it is absolutely necessary to distinguish $k(V)$ from $k(x)$. (Note that $k(V)$ is linearly disjoint with Ω over k.) It is also necessary to consider the functions and the divisors on the product $V \times V$, which is a non-singular surface rational over k. There are three types of k-rational prime divisors of $V \times V$:

(i) $\quad p \times V$ with $p \in P(k)$,

(ii) $\quad V \times p$ with $p \in P(k)$,

(iii) a k-rational prime divisor of $V \times V$ which has a non-trivial intersection with any divisor of the above two types.

A prime divisor of the type $p \times V$ or $V \times p$ is called *left constant* or *right constant*, respectively. A divisor of the third type is called *non-constant*. Let $k(V \times V)$ denote the field of k-rational functions on the surface $V \times V$. For each prime divisor $\mathfrak{P}$ of the above three types, we can define a discrete valuation $v_{\mathfrak{P}}$ of $k(V \times V)$.

Now we identify $k(V) \otimes_k k(V)$ with a subring of $k(V \times V)$ in a natural manner. Namely, for $\alpha \in k(V)$, $\beta \in k(V)$, we view $\alpha \otimes \beta$ as the element of $k(V \times V)$ defined by $(\alpha \otimes \beta)(x, y) = \alpha(x)\beta(y)$ with a generic point (x, y) of $V \times V$ over k. Then we define a module $E(k)$ by

$$E(k) = k(V \times V) \otimes_K (F(k) \otimes_k F'(k)) \quad (K = k(V) \otimes_k k(V)).$$

To be more explicit, $E(k)$ is a one-dimensional vector space over $k(V \times V)$ formed by all the expressions of the form

$$W = H \otimes f \otimes g$$

with $H \in k(V \times V)$, $f \in F(k)$, $g \in F'(k)$, under the rule

$$H \otimes \alpha f \otimes \beta g = (\alpha \otimes \beta) H \otimes f \otimes g$$

for $\alpha \in k(V)$, $\beta \in k(V)$. For a k-rational prime divisor $\mathfrak{P}$ of $V \times V$, we define $v_{\mathfrak{P}}(W)$ as follows:

$$v_{p \times V}(W) = v_{p \times V}(H) + v_p(f),$$

$$v_{V \times p}(W) = v_{V \times p}(H) + v_p(g),$$

$$v_{\mathfrak{P}}(W) = v_{\mathfrak{P}}(H) \quad \text{if } \mathfrak{P} \text{ is non-constant.}$$

To express W, it is often convenient to use the notation

$$W(x, y) = H(x, y) f(x) g(y)$$

with a generic point (x, y) of $V \times V$ over k. For example, given $f \in F(k)$, $g \in F'(k)$ and an element η of $k(x, y)$, we shall be speaking of the element W of $E(k)$ defined by

$$W(x, y) = \eta f(x) g(y).$$

This means $W = H \otimes f \otimes g$ with the element H of $k(V \times V)$ defined by $H(x, y) = \eta$. Here the symbols $f(x)$, $g(y)$ are meaningless only by themselves; x and y are merely to indicate "the left and right variables".

2.2. Let (x, y) be a generic point of $V \times V$ over k, and let $k_1 = k(x)$. For every $H \in k(V \times V)$, define an element H_1 of $k_1(V)$ by $H_1(y) = H(x, y)$. (Note that y is generic on V over k_1.) Then $H \mapsto H_1$ gives an isomorphism of $k(V \times V)$ onto $k_1(V)$. Take a non-constant prime divisor $\mathfrak{P}$ of $k(V \times V)$. As a k-rational algebraic cycle, $\mathfrak{P}$ has a generic point of the form (x, y') over k, where $y' \in V$ and $k(x, y')$ is algebraic over $k(x) = k_1$. Let $\mathfrak{y}'$ be the k_1-rational prime divisor of V, that is the sum of all conjugates of y' over k_1. Then we see that the isomorphism $H \mapsto H_1$ sends the "place" $\mathfrak{P}$ of $k(V \times V)$ to the "place" $\mathfrak{y}'$ of $k_1'(V)$, since $H_1(y') = H(x, y')$. More symbolically, one has

$$H_1 \bmod \mathfrak{y}' = H \bmod \mathfrak{P}.$$

Especially we have

$$(2.2.1) \qquad \qquad \nu_{\mathfrak{P}}(H) = \nu_{\mathfrak{y}'}(H_1).$$

As a special case, take as $\mathfrak{P}$ the locus Δ of (x, x) on $V \times V$ over k. We call Δ *the diagonal* on $V \times V$. In this case, we consider the k_1-rational prime divisor consisting of the point x, which we denote by the same letter x. Then we have

$$(2.2.2) \qquad \qquad \nu_{\Delta}(H) = \nu_x(H_1).$$

2.3. Now we consider two elements $\mathfrak{a}$ and $\mathfrak{b}$ of $D_{\mathbf{Q}}$ under a set of conditions

$$(2.3.0) \qquad \mathfrak{a} + \mathfrak{b} = \mathfrak{z}; \quad \mathfrak{a} + \operatorname{div}(f) \in D \text{ for } 0 \neq f \in F; \quad \mathfrak{b} + \operatorname{div}(g) \in D \quad \text{for } 0 \neq g \in F'.$$

As seen in Proposition 1.5, the last two conditions are equivalent. These $\mathfrak{a}$, $\mathfrak{b}$ will be always the same throughout this section.

Take any field k of rationality for $\mathfrak{F}$, $\mathfrak{a}$, $\mathfrak{b}$. Let x be a generic point of V over k, and let $k_1 = k(x)$. Define the k_1-rational prime divisor x as above, and take a k_1-rational local parameter t_x at x of the form $t_x = \tau - \tau(x)$ with a non-constant element τ of $k(V)$. (This special form of t_x will simplify our later discussion.) Let $\{f_1, ..., f_n\}$ be a basis of $F(\mathfrak{a}, k)$ over k, and take non-zero elements u of $F(k)$ and v of $F'(k)$. These f_i, u, and v will be fixed throughout this section. Consider the power-series expansion of f_i/u:

$$f_i/u = \sum_{r=0}^{\infty} \varphi_{ir} t_x^r$$

with $\varphi_{ir} \in k_1$. Since f_i/u is k-rational, we see that f_i/u is finite at x, and $\varphi_{i0} = (f_i/u)(x) \neq 0$. Now take an x-basis $\{w_1, ..., w_n\}$ of $F(\mathfrak{a}, k_1)$ relative to t_x and v in the sense of 1.8. Then

$$w_i v dz = (t_x^{\beta_i} + ...) dt_x \quad (i = 1, ..., n; \ 0 \le \beta_1 < ... < \beta_n) \qquad (2.3.1)$$

with non-negative integers β_i. Since $fvdz$ is finite and $\neq 0$ at x for every non-zero $f \in F(k)$, we have $\beta_1 = 0$. Put $w_i = \sum_{j=1}^n c_{ij} f_j$, with $c_{ij} \in k_1$, and $\zeta = uvdz/dt_x$. Since $dt_x = d\tau$, we see that $\zeta \in k(V)$, and

$$w_i v dz = (\sum_{j=1}^n c_{ij} f_j/u) \zeta dt_x = (\sum_{j,r} c_{ij} \varphi_{jr} t_x^r) \zeta dt_x.$$

Therefore

$$(2.3.2) \qquad \sum_{j=1}^n c_{ij} \varphi_{jr} = \begin{cases} 0 & \text{if } r < \beta_i, \\ \zeta(x)^{-1} & \text{if } r = \beta_i. \end{cases}$$

Take elements $g_1, ..., g_n$ of $R'(k)$ so that they form a basis of $R'(k)/[R'(\mathfrak{b}, k) + F'(k)]$ dual to $\{f_1, ..., f_n\}$. Consider the embedding of $R'(k)$ into $R'(k_1)$ defined in 1.7, and denote by g_i^* the image of g_i by this embedding. Now define an element G_0 of $R'(k_1)$ by

$$(2.3.3) \qquad G_0 = \sum_{i=1}^n (f_i/u)(x) \, g_i^*.$$

By Proposition 1.9, there exists an element s of $R'(k_1)$ such that

$$(2.3.4) \qquad \begin{cases} s_p = 0 \quad \text{for all } p \in P(k_1) \quad \text{other than } x, \\ s_x = \left(\sum_{i=1}^n a_i t_x^{-\beta_i - 1}\right) v \end{cases}$$

with $a_i \in k_1$, and

$$(2.3.5) \qquad G_0 - s \in R'(\mathfrak{b}, k_1) + F'(k_1).$$

Then $\operatorname{Res}_x (f_i s_x dz) = \langle f_i, s \rangle = \langle f_i, \sum_j (f_j/u)(x) g_j^* \rangle = (f_i/u)(x)$. Therefore $\operatorname{Res}_x (f s_x dz) = (f/u)(x)$ for all $f \in F(\mathfrak{a}, k_1)$. Substituting $w_i = \sum_j c_{ij} f_j$ for f, we obtain

$$\operatorname{Res}_x (w_i s_x dz) = \sum_j c_{ij} (f_j/u)(x) = \sum_j c_{ij} \varphi_{j0} = \begin{cases} 0 & \text{if } i > 1, \\ \zeta(x)^{-1} & \text{if } i = 1, \end{cases}$$

by virtue of (2.3.2). On the other hand,

$$\begin{aligned} \operatorname{Res}_x (w_i s_x dz) &= \operatorname{Res}_x (w_i v \cdot v^{-1} s_x dz) \\ &= \operatorname{Res}_x \left[(t_x^{\beta_i} + \ldots)\left(\sum_j a_j t_x^{-\beta_j - 1}\right) dt_x \right] = a_i. \end{aligned}$$

Therefore
$$a_i = \begin{cases} 0 & (i > 1), \\ \zeta(x)^{-1} & (i = 1), \end{cases}$$

hence (2.3.4) becomes

$$(2.3.6) \qquad \begin{cases} s_p = 0 \quad \text{for } x \neq p \in P(k_1), \\ s_x = \zeta(x)^{-1} t_x^{-1} v. \end{cases}$$

2.4. By (2.3.5), we have

$$(2.4.1) \qquad G_0 - s = A_0 + B_0$$

with $A_0 \in R'(\mathfrak{b}, k_1)$ and $B_0 \in F'(k_1)$. Define an element B of $E(k)$ by

$$B(x, y) = (B_0/c)(y) u(x) c(y)$$

with any non-zero $c \in F'(k)$. Note that $B_0/c \in k_1(V)$, and $(B_0/c)(y)$ is an element of the field $k_1(y) = k(x, y)$. Similarly, for every $p \in P(k_1)$, we can define elements G_p, A_p, and S_p of $E(k)$ by

$$\begin{aligned} G_p(x, y) &= (G_{0p}/c)(y) u(x) c(y), \\ A_p(x, y) &= (A_{0p}/c)(y) u(x) c(y), \\ S_p(x, y) &= (s_p/c)(y) u(x) c(y). \end{aligned}$$

Obviously B, G_p, A_p, S_p do not depend on the choice of c. (As for u, it has been fixed at the beginning.) From our definition of G_0, we obtain

$$(2.4.2) \qquad G_p(x, y) = \sum_{i=1}^n f_i(x) g_{ip}^*(y) \quad (p \in P(k_1)).$$

With these elements of $E(k)$, the equality (2.4.1) becomes

$$(2.4.3) \qquad G_p - S_p = A_p + B \quad (p \in P(k_1)).$$

More precisely, one has

$$(2.4.4) \qquad \sum_{i=1}^{n} f_i(x) g_{ip}^*(y) = A_p(x, y) + B(x, y) \quad \text{for } x \neq p \in P(k_1),$$

$$(2.4.5) \qquad -S_x(x, y) = A_x(x, y) + B(x, y).$$

Since $A_0 \in R'(\mathfrak{b}, k_1)$, we have

$$(2.4.6) \qquad \nu_{V \times p}(A_p) \geqslant -\nu_p(\mathfrak{b}) \quad (p \in P(k)),$$

hence

$$(2.4.7) \qquad \nu_{V \times p}(B) \geqslant -\nu_p(\mathfrak{b}) \quad (p \in P(k)),$$

unless $\nu_p(g_{ip}) < -\nu_p(\mathfrak{b})$ for some i. Furthermore, by virtue of (2.2.2), we have

$$(2.4.8) \qquad \nu_\Delta(A_x) = \nu_x(A_{0x}) \geqslant 0.$$

Now let $\mathfrak{P}$ be a non-constant k-rational prime divisor of $V \times V$, with a generic point (x, y') over k. Let $\mathfrak{y}'$ be the k_1-rational prime divisor of V corresponding to the point y' as defined in 2.2. Since $-s_{\mathfrak{y}'} = A_{0\mathfrak{y}'} + B_0$, we have, by (2.2.1),

$$\nu_\mathfrak{P}(B) = \nu_{\mathfrak{y}'}(B_0/v) = \nu_{\mathfrak{y}'}((s_{\mathfrak{y}'}/v) + (A_{0\mathfrak{y}'}/v)).$$

Since $A_0 \in R'(\mathfrak{b}, k_1)$, we have $\nu_{\mathfrak{y}'}(A_{0\mathfrak{y}'}/v) \geqslant 0$. If $\mathfrak{P} \neq \Delta$, we have $s_{\mathfrak{y}'} = 0$, while if $\mathfrak{P} = \Delta$, we have $\mathfrak{y}' = x$ and $\nu_x(s_x/v) = \nu_x(s_x) = -1$. Thus

$$(2.4.9) \qquad \nu_\Delta(B) = -1; \quad \nu_\mathfrak{P}(B) \geqslant 0 \quad \text{\textit{for every non-constant} } \mathfrak{P} \neq \Delta.$$

We are going to normalize B so that it has a pole only at some pre-assigned constant divisors. To do this, we have to impose the following conditions on $\mathfrak{a}$:

$$(2.4.10) \qquad F(\mathfrak{a} - p) \neq F(\mathfrak{a}) \textit{ for every point } p \textit{ of } V;$$

$$(2.4.11) \qquad F'(\mathfrak{b} + p) = F'(\mathfrak{b}) \quad \textit{for every point } p \textit{ of } V.$$

Note that $\dim F(\mathfrak{a} - p) \geqslant \dim F(\mathfrak{a}) - 1$. By (1.6.1), we have

$$\dim F(\mathfrak{a} - p) - \dim F'(\mathfrak{b} + p) = \dim F(\mathfrak{a}) - \dim F'(\mathfrak{b}) - 1,$$

hence (2.4.10) is equivalent to (2.4.11).

Let us now prove a few lemmas which are necessary for our process of normalization.

2.5. Lemma. *Let $0 \neq \xi \in k(V \times V)$. Suppose that the pole of ξ consists of the diagonal Δ with multiplicity one, and possibly right or left constant divisors. Let p be a k-rational point of V such that $p \times V$ is not contained in the pole of ξ, and let η be the element of $k(V)$ defined*

by $\eta(y) = \xi(p, y)$ *with any generic point y of V over k. Then $v_q(\eta) \geq v_{V \times q}(\xi)$ for $p \neq q \in P(k)$, and*
$v_p(\eta) \geq v_{V \times p}(\xi) - 1$.

Proof. Take an element π of $k(V)$ so that $v_q(\pi) = v_{V \times q}(\xi)$. Define an element ξ' of $k(V \times V)$ by $\xi'(x, y) = \pi(y)^{-1} \xi(x, y)$ with a generic point (x, y) of $V \times V$ over k. Then $v_{V \times q}(\xi') = 0$. If $q \neq p$, we see that ξ' is finite at (p, q) and $\xi'(p, y) = \pi(y)^{-1} \eta(y)$. Therefore $v_q(\pi^{-1}\eta) \geq 0$, hence $v_q(\eta) \geq v_q(\pi) = v_{V \times q}(\xi)$. Next take an element γ of $k(V)$ so that $v_p(\gamma) = 1$. Define two elements α and β of $k(V \times V)$ by $\alpha(x, y) = \gamma(x)$, $\beta(x, y) = \gamma(y)$. Put $e = v_{V \times p}(\xi)$. Since $v_\Delta(\alpha - \beta) = 1$, we see that $\beta^{-e}(\beta - \alpha)\xi$ is finite at (p, p). (In fact, $v_Q(\beta^{-e}(\beta - \alpha)\xi) \geq 0$ for all k-rational prime divisors Q of $V \times V$ passing through (p, p).) Therefore, specializing $\beta^{-e}(\beta - \alpha)\xi$ to $p \times V$, we find $v_p(\gamma^{1-e}\eta) \geq 0$, so that $v_p(\eta) \geq e - 1$, q.e.d.

2.6. **LEMMA** *Put $r = \dim F'(\mathfrak{b})$. Let $q_1, \ldots, q_s$ be independent generic points of V over a field of rationality for $\mathfrak{F}$ and $\mathfrak{a}$. Then*

$$\dim F(\mathfrak{a} + \textstyle\sum_{i=1}^{s} q_i) = \begin{cases} \dim F(\mathfrak{a}) & \text{if } s \leq r, \\ \dim F(\mathfrak{a}) + s - r & \text{if } s > r. \end{cases}$$

Proof. Let $0 \neq f \in F(k)$, $0 \neq g \in F'(k)$ with a field k of rationality for $\mathfrak{F}$ and $\mathfrak{a}$. Put $\omega = fg\,dz$ and $\mathfrak{b} = \mathfrak{a} + \operatorname{div}(f)$. Then

$$\dim F(\mathfrak{a} + \textstyle\sum_{i=1}^{s} q_i) = l(\mathfrak{b} + \textstyle\sum_{i=1}^{s} q_i),$$

$$r = l(\operatorname{div}(\omega) - \mathfrak{b}).$$

Therefore our assertion can be written as

$$l(\mathfrak{b} + \textstyle\sum_{i=1}^{s} q_i) = \begin{cases} l(\mathfrak{b}) & \text{if } s \leq r, \\ l(\mathfrak{b}) + s - r & \text{if } s > r, \end{cases}$$

which is nothing else than Weil [10, p. 11, Prop. 8].

2.7. **LEMMA.** *Let k be a field of rationality for $\mathfrak{F}$; p a k-rational point of V; t_p a k-rational local parameter at p; w an element of $F(k)$ such that $0 \leq v_p(w) < 1$; $\mathfrak{b}$ an element of $D_Q(k)$ such that $\operatorname{div}(w) + \mathfrak{b} \in D$. Further let c be an integer such that*

$$(\ast) \qquad\qquad \deg(\operatorname{div}(w) + \mathfrak{b}) + c - 1 > 2\mathfrak{g} - 2,$$

where $\mathfrak{g}$ is the genus of V, and let $e = v_p(\mathfrak{b}) - v_p(w)$. Then there exists an element f of $F(k)$ such that $v_p(f - t_p^{e-c}w) \geq v_p(w) + e - c + 1$, and $v_q(f) \geq -v_q(\mathfrak{b})$ for $p \neq q \in P(k)$.

Proof. By (1.3.1), $a \mapsto aw$ gives a k-linear isomorphism of $L(\operatorname{div}(w) + c'p + \mathfrak{b}, k)$ onto $F(c'p + \mathfrak{b}, k)$ for any integer c'. By $(\ast)$, we have

$$l(\operatorname{div}(w) + cp + \mathfrak{b}) = \deg(\operatorname{div}(w) + \mathfrak{b}) + c - \mathfrak{g} + 1,$$

and this holds also when c is replaced by $c-1$. Therefore

$$F(cp+\mathfrak{b},\,k)/F((c-1)p+\mathfrak{b},\,k)$$

is one-dimensional, hence our assertion.

2.8. Let us now fix a field k_0 of rationality for $\mathfrak{F}$ and $\mathfrak{a}$, and take an extension k of k_0, which is algebraically closed, and which has an infinite transcendence degree over k_0. Hereafter we take this k as our basic field. Since k is algebraically closed, $P(k)$ can be identified with the set of all k-rational points of V.

Now, with this k, we fix u, v, f_i, g_i, x, t_x, and define G_p, S_p, A_p, and B as in 2.4. We shall now show that under the condition (2.4.10), A_p and B can be chosen so that

$$(2.8.1) \qquad v_{p\times V}(B) \geqslant -v_p(\mathfrak{a}) \quad \text{for every } p \in P(k) - \{q_1, \ldots, q_r\},$$

$$(2.8.2) \qquad v_{q_i\times V}(B) \geqslant -1 \quad \text{for } i=1, \ldots, r,$$

with r independent generic points $q_1, \ldots, q_r$ of V over k_0 rational over k, where $r = \dim F'(\mathfrak{b})$.

To show this, we start from any choice of A_p and B as in 2.4, and observe that $v_{q\times V}(B)+v_q(\mathfrak{a}) \in \mathbf{Z}$ for all $q \in P(k)$ by virtue of (2.3.0). Let us fix one $p \in P(k)$, and put $v_p(\mathfrak{a}) = -e-e'$ with $e \in \mathbf{Z}$ and $0 \leqslant e' < 1$, $v_{p\times V}(B)-e-e' = -c$. Then $c \in \mathbf{Z}$. Assume that (2.8.1) is not satisfied for this p, i.e., $c > 0$. (The number of such points p is of course finite.) Take a non-zero element w of $F(k)$ so that $v_p(w) = e'$, and take a k-rational local parameter t_p at p. Put, for each $q \in P(k)$,

$$A_q(x,\,y) = t_p(x)^{e-c}a_q(x,\,y)w(x)v(y),$$

$$B(x,\,y) = t_p(x)^{e-c}b(x,\,y)w(x)v(y)$$

with elements a_q and b of $k(x,\,y)$. Then

$$(2.8.3) \qquad t_p(x)^{c-e} \textstyle\sum_{i=1}^n (f_i/w)(x)(g_{iq}/v)(y) = a_q(x,\,y)+b(x,\,y),$$

$$(2.8.4) \qquad v_{p\times V}(b) = v_{p\times V}(B)-e-e'+c = 0.$$

Consider (2.8.3) as an equality about the elements of $k(V \times V)$, and take it modulo $p \times V$. Since $f_i \in F(\mathfrak{a})$, we have $v_p(f_i/w) \geqslant e$, hence the left hand side is 0 modulo $p \times V$. This together with (2.8.4) shows

$$(2.8.5) \qquad v_{p\times V}(a_q) = 0.$$

By (2.4.6), we have $\qquad v_{V\times q}(a_q) \geqslant -v_q(\mathfrak{b})-v_q(v).$

Let β be the element of $k(V)$ defined by $\beta(y) = b(p,\,y)$. Then β is exactly b mobulo $p \times V$, which is the same as $-a_q$ modulo $p \times V$. By (2.4.9), Δ is the only non-constant pole of b, so that (2.8.3) shows that a_q has the same property. By (2.8.4) and (2.8.5), we can apply Lemma 2.5 to b and a_q modulo $p \times V$. Then we find

$$\nu_q(\beta) \geqslant \begin{cases} \nu_{V \times q}(a_q) \geqslant -\nu_q(\mathfrak{b}) - \nu_q(v) & \text{if } q \neq p, \\ \nu_{V \times p}(a_p) - 1 \geqslant -\nu_p(\mathfrak{b}) - \nu_p(v) - 1 & \text{if } q = p. \end{cases}$$

By (2.4.11), we have $\operatorname{div}(\beta v) \geqslant -\mathfrak{b}$. By Lemma 2.7, we can find an element f of $F(k)$ such that

$$\nu_p(f - t_p^{e-c} w) \geqslant e + e' - c + 1,$$

$$\nu_q(f) \geqslant -\nu_q(\mathfrak{a}) \quad \text{for } q \neq p, p_1, \ldots, p_\lambda,$$

$$\nu_{p_i}(f) \geqslant -1 \quad \text{for } i = 1, \ldots, \lambda,$$

if we choose sufficiently many k-rational points $p_1, \ldots, p_\lambda$ of V not involved in $\mathfrak{a}$. (We take $\mathfrak{a} + p_1 + \ldots + p_\lambda$ to be the divisor $\mathfrak{b}$ of Lemma 2.7.) Then we define an element H_0 of $F'(k_1)$ and an element H of $E(k)$ by

$$H_0 = (f/u)(x)\beta v,$$

$$H(x, y) = f(x)\beta(y)v(y) = (H_0/v)(y)u(x)v(y).$$

Then $H_0 \in F'(\mathfrak{b}, k_1)$, and $\nu_{V \times q}(H) = \nu_q(\beta v) \geqslant -\nu_q(\mathfrak{b})$ for every $q \in P(k)$. Further we have

$$B(x, y) - H(x, y) = t_p(x)^{e-c}[b(x, y) - (t_p^{c-e} f/w)(x)\beta(y)]w(x)v(y),$$

hence
$$\nu_{p \times V}(B - H) \geqslant e + e' - c + 1.$$

We see also that

$$\nu_{q \times V}(B - H) \geqslant \operatorname{Min}(\nu_{q \times V}(B), -\nu_q(\mathfrak{a})) \quad \text{for } q \in P(k) - \{p, p_1, \ldots, p_\lambda\}.$$

The points p_i can be chosen so that $\nu_{p_i \times V}(B) \geqslant 0$. Then we have

$$\nu_{p_i \times V}(B - H) \geqslant -1.$$

Now replace B_0 by $B_0 - H_0$. Then B and $A_{\mathfrak{z}}$ are replaced by $B - H$ and $A_{\mathfrak{z}} + H$ for every $\mathfrak{z} \in P(k_1)$. Apply the same procedure to the new B with the same point p if still $\nu_{p \times V}(B) - e - e' < 0$, or with other p for which (2.8.1) does not hold. After a finite number of steps, we can now assume that B has the following properties:

$$\nu_{p \times V}(B) \geqslant -\nu_p(\mathfrak{a}) \quad \text{if } p \in P(k) - M,$$

$$\nu_{p \times V}(B) = -1 \quad \text{if } p \in M,$$

with a finite set M of independent generic points over k_0, all rational over k.

Take r independent generic points $q_1, \ldots, q_r$ of V over k_0, independent of the points of M. Apply again the procedure of taking (2.8.3) modulo $p \times V$ for each $p \in M$. Let w_p, b_p, and β_p denote the functions w, b, and β defined above for this p (with respect to the new B). Since $e = e' = 0$ and $c = 1$ in the present case, we have $B(x, y) = t_p(x)^{-1} b_p(x, y) w_p(x) v(y)$; β_p is b_p modulo $p \times V$, and div $(\beta_p v) \geqslant -\mathfrak{b}$. By Lemma 2.6,

$$F(\mathfrak{a} + p + q_1 + \ldots + q_r)/F(\mathfrak{a} + q_1 + \ldots + q_r)$$

is one-dimensional, hence there exists, for each $p \in M$, an element f_p of $F(k)$ such that

$$\nu_p(f_p - t_p^{-1} w_p) \geqslant 0$$

$$\nu_q(f_p) \geqslant -\nu_q(\mathfrak{a}) \quad \text{for } q \in P(k) - \{p, q_1, \ldots, q_r\},$$

$$\nu_{q_i}(f_p) \geqslant -1 \quad \text{for } i = 1, \ldots, r.$$

Define an element J of $E(k)$ by

$$J(x, y) = \Sigma_{p \in M} f_p(x) \beta_p(y) v(y).$$

Then $\nu_{V \times q}(J) \geqslant -\nu_q(\mathfrak{b})$ for all $q \in P(k)$, and

$$\nu_{q \times V}(B - J) \geqslant 0 \quad \text{for } q \in M,$$

$$\nu_{q_i \times V}(B - J) \geqslant -1 \quad \text{for } i = 1, \ldots, r,$$

$$\nu_{q \times V}(B - J) \geqslant -\nu_q(\mathfrak{a}) \quad \text{for } q \in P(k) - M \cup \{q_1, \ldots, q_r\}.$$

Therefore, replacing B and A_3 by $B - J$ and $A_3 + J$, we obtain the desired properties (2.8.1, 2), retaining the properties (2.4.2–9). It is this B which was to be constructed and which may be called an algebraic kernel function. We conclude this section by proving two propositions concerning the behavior of B at its pole.

2.9. PROPOSITION. *Let (x, y) be a generic point of $V \times V$ over k with the same x as in 2.4, and let $k_2 = k(y)$, $0 \neq w \in F'(k)$. Define an element B_w of $k_2(V)$ by $B(x, y) = B_w(x) u(x) w(y)$. Then $\nu_y(B_w uw\, dz) = -1$, and $\mathrm{Res}_y (B_w uw\, dz) = 1$.*

Proof. Let us first consider the case where w is the element v with which we constructed B. Take the local parameter $t_x = \tau - \tau(x)$ with $\tau \in k(V)$ as in 2.3. Let s_x and S_x be as in 2.3 and 2.4, and put $\zeta = uv\, dz/dt_x$. Then $\zeta \in k(V)$, $S_x(x, y) = (s_x/v)(y) u(x) v(y)$, and $s_x = \zeta(x)^{-1} t_x^{-1} v$ by (2.3.6). Define an element S^* of $k_2(V)$ by $S^*(x) = (s_x/v)(y)$. Define also a k_2-rational local parameter t_y at y by $t_y = \tau - \tau(y)$. Then

$$S^*(x) = \zeta(x)^{-1} t_x(y)^{-1} = \zeta(x)^{-1} [\tau(y) - \tau(x)]^{-1} = -\zeta(x)^{-1} t_y(x)^{-1},$$

hence $S^* = -1/\zeta t_y$, and $S^* uv\, dz = -t_y^{-1} dt_y$. Therefore $\nu_y(S^*) = -1$ and $\mathrm{Res}_y(S^* uv\, dz) = -1$. By (2.4.5), we have $-S_x = A_x + B$. Define an element A^* of $k_2(V)$ by $A_x(x, y) = A^*(x) u(x) v(y)$. Then $-B_v = A^* + S^*$. By (2.2.2) and (2.4.8), we have $\nu_y(A^*) = \nu_\Delta(A_x) \geqslant 0$. Therefore

$v_y(B_v) = v_y(S^*) = -1$, and $B_v uv\,dz$ has the same residue as $-S^* uv\,dz$ at y. This settles the case $w = v$. In the general case, put $w = \alpha v$ with $\alpha \in k(V)$. Then $B_w = \alpha(y)^{-1} B_v$, so that $B_w uw\,dz = \alpha(y)^{-1} \alpha \cdot B_v uv\,dz$, hence our assertion.

2.10. PROPOSITION. *Let the notation be the same as in Proposition 2.9. Suppose $r > 0$, $w \in F'(\mathfrak{b}, k)$, and $v_{q_i}(w) = 0$ for $i = 1, ..., r$. Then $v_{q_i}(B_w uw\,dz) = -1$ for $i = 1, ..., r$, and $B_w uw\,dz$ has a pole only at y, q_1, ..., q_r. Moreover, if c_i is the element of $k(V)$ defined by*

$$c_i(y) = \operatorname{Res}_{q_i}(B_w uw\,dz) \quad (i = 1, ..., r),$$

then $c_1 w$, ..., $c_r w$ form a basis of $F'(\mathfrak{b}, k)$ over k, and for every $g \in F'(\mathfrak{b})$, one has

(*) $$g = -\textstyle\sum_{i=1}^{r}(g/w)(q_i) \cdot c_i w.$$

Note that, since $r = \dim F'(\mathfrak{b})$, the assumption $r > 0$ implies the existence of a non-zero element of $F'(\mathfrak{b}, k)$. Moreover, since the q_i are generic over a field k_0 of rationality for $\mathfrak{b}$, we have $v_{q_i}(w) = 0$ for every non-zero $w \in F'(\mathfrak{b}, k_0)$.

Proof. Define an element B' of $k(V \times V)$ by $B(x, y) = B'(x, y) u(x) w(y)$. Then $v_{p \times V}(B') = v_{p \times V}(B) - v_p(u) \geqslant -v_p(\mathfrak{a}) - v_p(u)$ for $p \in P(k) - \{q_1, ..., q_r\}$, and $v_{q_i \times V}(B') \geqslant -1$. By Lemma 2.5, we have $v_{q_i}(B_w) \geqslant -1$, $v_y(B_w) \geqslant -1$, and $v_q(B_w) \geqslant v_{q \times V}(B')$ for all $q \in P(k_2) - \{y, q_1, ..., q_r\}$. Therefore, for every $g \in F'(\mathfrak{b})$, we have

$$\operatorname{div}(B_w ug\,dz) \geqslant -y - \textstyle\sum_{i=1}^{r} q_i + \operatorname{div}(u) + \operatorname{div}(g) + \mathfrak{z} - \mathfrak{a} - \operatorname{div}(u)$$
$$\geqslant -y - \textstyle\sum_{i=1}^{r} q_i.$$

This shows that the differential form $B_w ug\,dz$ has no pole except at y, q_1, ..., q_r. Especially this applies to the case $g = w$. By Proposition 2.9, we have

$$\operatorname{Res}_y(B_w ug\,dz) = \operatorname{Res}_y((g/w) \cdot B_w uw\,dz) = (g/w)(y).$$

By our definition of c_i,

$$\operatorname{Res}_{q_i}(B_w ug\,dz) = \operatorname{Res}_{q_i}((g/w) \cdot B_w uw\,dz) = (g/w)(q_i) \cdot c_i(y).$$

Since the sum of all residues is 0, we obtain

$$(g/w)(y) + \textstyle\sum_{i=1}^{r}(g/w)(q_i) c_i(y) = 0,$$

which proves the equality (*). This shows also that $F'(\mathfrak{b})$ is contained in the Ω-linear span of $c_1 w$, ..., $c_r w$. Since $r = \dim F'(\mathfrak{b})$, the $c_i w$ must form a basis of $F'(\mathfrak{b})$ over Ω. It follows that $c_i \neq 0$, hence $v_{q_i}(B_w uw\,dz) = -1$. This completes the proof.

3. The trace formula: first formulation

3.1. For $\alpha = \begin{bmatrix} a & b \\ c & d \end{bmatrix} \in GL_2(\mathbb{C})$ and $z \in \mathbb{C} \cup \{\infty\}$, we put, as usual,

$$\alpha(z) = (az+b)/(cz+d),$$

and denote by $\mathfrak{H}$ the complex upper half plane:

$$\mathfrak{H} = \{z \in \mathbb{C} \mid \operatorname{Im}(z) > 0\}.$$

We are going to discuss automorphic forms of an arbitrary rational weight. Fix a "weight" m which is a rational number, and consider the set $\mathfrak{G}_m$ consisting of all couples $(\alpha, h(z))$ formed by an element $\alpha = \begin{bmatrix} a & b \\ c & d \end{bmatrix}$ of $SL_2(\mathbf{R})$ and a holomorphic function $h(z)$ on $\mathfrak{H}$ of the form $h(z) = t \cdot (cz+d)^m$ with $t \in \mathbb{C}$, $|t| = 1$. We make $\mathfrak{G}_m$ a group by defining the law of multiplication by

$$(\alpha, h(z))(\beta, j(z)) = (\alpha\beta, h(\beta(z))j(z)).$$

Let $\tau = (\alpha, h(z)) \in \mathfrak{G}_m$. For a meromorphic function f on $\mathfrak{H}$, we define a function $f|\tau$ by

$$(f|\tau)(z) = f(\alpha(z))h(z)^{-1}.$$

Let Γ be a discrete subgroup of $SL_2(\mathbf{R})$ such that $\mathfrak{H}/\Gamma$ is of finite measure with respect to $y^{-2}dx\,dy$. (We denote the quotient space by $\mathfrak{H}/\Gamma$ although we let Γ act on the left of $\mathfrak{H}$.) By a *proper lifting of Γ of weight m*, we understand a map $L: \Gamma \to \mathfrak{G}_m$ satisfying the following conditions (3.1.1–3):

(3.1.1) *L is an injective homomorphism of Γ into $\mathfrak{G}_m$ such that $P \circ L$ is the identity map of Γ, where P is the natural projection map of $\mathfrak{G}_m$ onto $SL_2(\mathbf{R})$.*

(3.1.2) $L(-1) = (-1, 1)$ *if* $-1 \in \Gamma$.

(3.1.3) *If* $\gamma = \begin{bmatrix} a & b \\ c & d \end{bmatrix} \in \Gamma$, $L(\gamma) = (\gamma, t \cdot (cz+d)^m)$, *then* $t^n = 1$ *for some positive integer n.*

Since Γ is finitely generated, we can take n to be independent of γ.

Let us fix such a lifting $L: \Gamma \to \mathfrak{G}_m$, and put

$$L(\gamma) = (\gamma, j(\gamma, z)) \quad (\gamma \in \Gamma).$$

Then $j(\gamma\delta, z) = j(\gamma, \delta(z))j(\delta, z)$. Now we can define a proper lifting L' of Γ of weight $2 - m$ by

$$L'(\gamma) = (\gamma, j'(\gamma, z)),$$

$$j'(\gamma, z) = j(\gamma, z)^{-1}(cz+d)^2$$

$$= j(\gamma, z)^{-1}(dz/d\gamma(z)) \qquad \text{for } \gamma = \begin{bmatrix} a & b \\ c & d \end{bmatrix} \in \Gamma.$$

3.2. Let $\mathfrak{H}^*$ denote the union of $\mathfrak{H}$ and the cusps of Γ. Then $\mathfrak{H}^*/\Gamma$ has a natural structure of a compact Riemann surface. We take a projective non-singular curve V which is complex analytically isomorphic to $\mathfrak{H}^*/\Gamma$, and fix a Γ-invariant holomorphic map $\varphi\colon \mathfrak{H}^* \to V$ through which $\mathfrak{H}^*/\Gamma$ is isomorphic to V. (We call such (V, φ) a *model of* $\mathfrak{H}^*/\Gamma$.) Then $\mathbf{C}(V)\circ\varphi$ is the field of all Γ-automorphic functions: we identify $\mathbf{C}(V)\circ\varphi$ with $\mathbf{C}(V)$ if there is no fear of confusion.

For fixed L and L' as above, let F (resp. F') denote the module of all meromorphic functions f on $\mathfrak{H}$ which satisfy $f|L(\gamma)=f$ (resp. $f|L'(\gamma)=f$) for all $\gamma\in\Gamma$, and which are meromorphic at every cusp of Γ in the sense explained below. We see that F is either $\{0\}$ or one-dimensional over $\mathbf{C}(V)$. In the following treatment, let us simply assume that F is not $\{0\}$, without discussing the condition for the non-triviality of F. Then F' is also non-trivial, and we obtain a system $\mathfrak{F}=\{F, F', Z, \mathfrak{z}\}$ satisfying the axioms (A_{1-4}) of 1.2 as follows. For $(f, g)\in F\times F'$, define $Z(f, g)$ to be the differential form on V which can be identified with $fg\,dz$. The divisor $\mathrm{div}\,(f)=\Sigma_p\nu_p(f)p$ for $f\in F$ or $f\in F'$ can be defined in the following way. Let $p=\varphi(z_0)\in V$ with $z_0\in\mathfrak{H}^*$. If $z_0\in\mathfrak{H}$, consider the expansion of f at z_0:

$$(3.2.1) \qquad f(z) = c_k(z-z_0)^k + c_{k+1}(z-z_0)^{k+1} + \dots, \quad c_k \neq 0.$$

Then we put $\nu_p(f)=k/e_p$, where e_p is the order of the group

$$(3.2.2) \qquad \{\gamma\in\Gamma\,|\,\gamma(z_0)=z_0\}/(\Gamma\cap\{\pm 1\}).$$

If z_0 is a cusp of Γ, the last group is free cyclic. Take an element δ of Γ that generates this free cyclic group, and take an element $\varrho^*=(\varrho, \xi(z))$ of $\mathfrak{G}_m$ so that $\varrho(\infty)=z_0$. Then

$$(3.2.3) \qquad \varrho^{*-1}L(\delta)\varrho^* = \left(\varepsilon\begin{bmatrix}1 & h\\ 0 & 1\end{bmatrix},\ e^{2\pi i r}\right)$$

with $\varepsilon=\pm 1$, $h\in\mathbf{R}$, and $0\leqslant r<1$. By (3.1.3), r must be rational. Changing δ for δ^{-1} if necessary, we may assume $h>0$. Now we say that f is *meromorphic at* z_0 if

$$f|\varrho^* = \sum_{n\in\mathbf{Z}} c_n \exp\left[2\pi i(n+r)z/h\right]$$

with only finitely many non-zero c_n for $n<0$. Then we put $\nu_p(f)=n+r$ with the smallest n for which $c_n\neq 0$. Finally we put

$$\mathfrak{z} = \mathrm{div}\,(dz) = -\sum_{p\in R}(1-e_p^{-1})\,p,$$

where R is the set of all the points of V corresponding to the elliptic points and the cusps of Γ; e_p denotes the order of the group (3.2.2) for each point $p=\varphi(z_0)\in V$; especially $e_p=\infty$ if p corresponds to a cusp. It is now easy to verify that $\mathfrak{F}$ actually satisfies (A_{1-4}) of 1.2. (As for (A_4), see for example [6, § 2.4].)

3.3. For each $p \in V$, there is a unique rational number μ_p' such that

$$(3.3.1) \qquad 0 \leqslant \mu_p' < 1, \quad v_p(g) \equiv \mu_p' \bmod \mathbf{Z} \quad \text{for } 0 \neq g \in F'.$$

This is because F' is one-dimensional over $C(V)$. We see that $e_p \mu_p' \in \mathbf{Z}$ (if $e_p < \infty$), hence $\mu_p' \leqslant 1 - e_p^{-1}$. Put $\mu_p = 1 - e_p^{-1} - \mu_p'$ for each $p \in V$, and define two elements $\mathfrak{a}$ and $\mathfrak{b}$ of $D_{\mathbf{Q}}$ by

$$(3.3.2) \qquad \mathfrak{a} = -\Sigma_{p \in R} \mu_p p, \quad \mathfrak{b} = -\Sigma_{p \in R} \mu_p' p.$$

Then we see that $\mathfrak{a}$ and $\mathfrak{b}$ satisfy the condition (2.3.0), and we have

$$(3.3.3) \qquad 0 \leqslant \mu_p \leqslant 1, \quad v_p(f) \equiv \mu_p \bmod \mathbf{Z} \quad \text{for } 0 \neq f \in F.$$

Obviously $\mu_p = \mu_p' = 0$ if $p \notin R$. Moreover one has

$$(3.3.4) \qquad F'(\mathfrak{b}) = F'(0),$$

so that $F'(\mathfrak{b})$ consists of all the elements of F' which are holomorphic on $\mathfrak{H}$ and also holomorphic at every cusp. We have $0 \leqslant \mu_p \leqslant 1 - e_p^{-1}$ if p corresponds to an elliptic point. If p corresponds to a cusp, define r by (3.2.3) under the condition $0 \leqslant r < 1$. Then $\mu_p = r$ or 1 according as $r > 0$ or $= 0$. Therefore $F(\mathfrak{a})$ consists of all the elements of F which are holomorphic on $\mathfrak{H}$ and vanish at every cusp. Thus $F(\mathfrak{a})$ is the vector space of all cusp forms with respect to the automorphic factor $j(\gamma, z)$.

Let $\mathfrak{g}$ denote the genus of V. Define $v(\mathfrak{H}/\Gamma)$ by

$$v(\mathfrak{H}/\Gamma) = (2\pi)^{-1} \int_{\mathfrak{H}/\Gamma} y^{-2} dx\, dy.$$

It is well known that $v(\mathfrak{H}/\Gamma) = 2\mathfrak{g} - 2 + \Sigma_{p \in R}(1 - e_p^{-1})$. Moreover one has

$$(3.3.5) \qquad \deg\,(\operatorname{div}\,(f)) = (m/2) \cdot v(\mathfrak{H}/\Gamma) \quad \text{for } 0 \neq f \in F.$$

To see this, take a positive integer n so that $mn \in \mathbf{Z}$ and $j(\gamma, z)^{2n} = (d\gamma(z)/dz)^{-mn}$ for all $\gamma \in \Gamma$. Such an integer n always exists by virtue of (3.1.3). Then $\operatorname{div}\,(f) = (2n)^{-1} \operatorname{div}\,(f^{2n})$, hence (3.3.5) follows from [6, Prop. 2.16].

If $0 \neq f \in F$ and $0 \neq g \in F'$, we have

$$\deg\,(\operatorname{div}\,(f) + \mathfrak{a}) = (m-1)v(\mathfrak{H}/\Gamma)/2 + \Sigma_{p \in R}\{(1 - e_p^{-1})/2 - \mu_p\} + \mathfrak{g} - 1,$$

$$\deg\,(\operatorname{div}\,(g) + \mathfrak{b}) = (1 - m/2)v(\mathfrak{H}/\Gamma) - \Sigma_{p \in R} \mu_p'.$$

By (1.6.1), we obtain

$$(3.3.6) \qquad \dim F(\mathfrak{a}) - \dim F'(\mathfrak{b}) = (m-1)v(\mathfrak{H}/\Gamma)/2 + \Sigma_{p \in R}\{(1 - e_p^{-1})/2 - \mu_v\}.$$

Further it can easily be verified that

(3.3.7) *If* $(m/2-1)v(\mathfrak{H}/\Gamma)+\Sigma_{p\in R}\mu'_p>1$, *then* $F'(\mathfrak{b})=\{0\}$, *and the condition* (2.4.10) *is satisfied.*

3.4. LEMMA. *Let* $(\beta,h)\in\mathfrak{G}_m$ *and* $\beta(z_0)=z_0$ *with* $z_0\in\mathfrak{H}$. *Put* $\sigma=\begin{bmatrix}\bar{z}_0 & z_0\\ 1 & 1\end{bmatrix}$. *Then* $\sigma^{-1}\beta\sigma=\begin{bmatrix}\zeta & 0\\ 0 & \zeta\end{bmatrix}$, $h(z_0)=\eta$ *with complex numbers* ζ *and* η *such that* $|\zeta|=|\eta|=1$. *If* $\beta=\begin{bmatrix}a & b\\ c & d\end{bmatrix}$ *and* $h(z)=t\cdot(cz+d)^m$, *then* $\zeta=cz_0+d$, $\eta=t\zeta^m$. *Especially if* $\beta\in\Gamma$ *and* $(\beta,h)=L(\beta)$, *then* $\eta=j(\beta,z_0)=\zeta^{-2e\mu}$, *where* $e=e_p$ *and* $\mu=\mu_p$ *with* $p=\varphi(z_0)$.

Proof. Almost all assertions can be verified in a straightforward manner. The relation $\eta=\zeta^{-2e\mu}$ can be obtained by considering the expansion (3.2.1) for $0\neq f\in F$ and the equality $f|L(\beta)=f$. (This is a consequence of our assumption $F\neq\{0\}$.)

3.5. Hereafter till the end of this section, we fix an element $\tau=(\alpha,h)$ of $\mathfrak{G}_m$ satisfying the following two conditions:

(3.5.1) $\qquad\qquad\qquad\Gamma$ *and* $\alpha^{-1}\Gamma\alpha$ *are commensurable;*

(3.5.2) $\qquad\qquad L(\alpha\gamma\alpha^{-1})=\tau\cdot L(\gamma)\tau^{-1}$ *for all* $\gamma\in\Gamma\cap\alpha^{-1}\Gamma\alpha$.

Put $\Gamma^*=L(\Gamma)$. Then, by [7, Prop. 1.1], the projection map $\Gamma^*\tau\Gamma^*\to\Gamma\alpha\Gamma$ is one-to-one. Let β^* denote the element of $\Gamma^*\tau\Gamma^*$ corresponding to an element β of $\Gamma\alpha\Gamma$, and put

(3.5.3) $\qquad\qquad\qquad \beta^*=(\beta,h(\beta,z))\quad(\beta\in\Gamma\alpha\Gamma)$.

Especially $\tau=\alpha^*=(\alpha,h(\alpha,z))$. By [7, Prop. 1.0, Prop. 1.1], Γ^* is commensurable with $\tau\Gamma^*\tau^{-1}$. Moreover, if $\Gamma\alpha\Gamma=\bigcup_\nu\Gamma\alpha_\nu$ is a disjoint union, then $\Gamma^*\alpha^*\Gamma^*=\bigcup_\nu\Gamma^*\alpha_\nu^*$ is a disjoint union.

Now we define a linear transformation $[\Gamma\alpha\Gamma]^*$ on F by

$$f|[\Gamma\alpha\Gamma]^*=\Sigma_\nu f|\alpha_\nu^*=\Sigma_\nu f(\alpha_\nu(z))h(\alpha_\nu,z)^{-1}\quad(f\in F).$$

It can easily be verified that $[\Gamma\alpha\Gamma]^*$ maps $F(\mathfrak{a})$ into itself.

Furthermore, put $\Gamma_*=L'(\Gamma)$, and

(3.5.4) $\qquad\qquad\qquad \beta_*=(\beta,h'(\beta,z)),$

$$h'(\beta,z)=h(\beta^{-1},\beta(z))(d\beta(z)/dz)^{-1}\quad\text{for }\beta\in\Gamma\alpha^{-1}\Gamma.$$

Then it can easily be seen that α_*^{-1} satisfies (3.5.1, 2) with respect to L'. Therefore, with a disjoint union $\Gamma\alpha^{-1}\Gamma=\bigcup_\nu\Gamma\beta_\nu$, we define a linear transformation $[\Gamma\alpha^{-1}\Gamma]_*$ on F' by

$$f|[\Gamma\alpha^{-1}\Gamma]_*=\Sigma_\nu f|\beta_{\nu*}=\Sigma_\nu f(\beta_\nu(z))h'(\beta_\nu,z)^{-1}\quad(f\in F').$$

This maps $F'(\mathfrak{b})$ into itself. Our main purpose is to prove a trace-formula of the form

$$\text{tr} \left([\Gamma \alpha \Gamma]^* \,|\, F(\mathfrak{a}) \right) - \text{tr} \left([\Gamma \alpha^{-1} \Gamma]_* \,|\, F'(\mathfrak{b}) \right) = \Sigma_{\xi \in \Xi}\, I(\xi),$$

where $I(\xi)$ is a certain complex number defined for each "fixed point" ξ of $\Gamma \alpha \Gamma$ on V, whose precise description will be given in the next paragraph.

3.6. Define an algebraic curve $T = T(\Gamma \alpha \Gamma)$ on $V \times V$ by

$$T = \{\varphi(z) \times \varphi(\alpha(z)) \,|\, z \in \mathfrak{H}^*\},$$

where (V, φ) is the model of $\mathfrak{H}^*/\Gamma$ fixed in 3.2. Let us now assume the following condition on $\Gamma \alpha \Gamma$:

(3.6.1) *If π denotes the natural map of $SL_2(\mathbf{R})$ onto $SL_2(\mathbf{R})/\{\pm 1\}$, one has*

$$\pi(\alpha^{-1} \Gamma \alpha \cap \Gamma) = \pi(\alpha^{-1} \Gamma \alpha) \cap \pi(\Gamma).$$

This is satisfied whenever $-1 \in \Gamma$. Let $\Gamma \alpha \Gamma = \bigcup_{\nu=1}^{d} \Gamma \alpha_\nu$ be a disjoint union. Then we write $d = \deg (\Gamma \alpha \Gamma)$. Under the assumption (3.6.1), d is the degree of the covering

$$\mathfrak{H}^*/(\alpha^{-1} \Gamma \alpha \cap \Gamma) \to \mathfrak{H}^*/\Gamma,$$

and the algebraic correspondence T maps a point $\varphi(z)$ onto the points $\varphi(\alpha_\nu(z))$. Let us further assume that $\alpha \notin \{\pm 1\} \Gamma$. Then $\pm 1 \notin \Gamma \alpha \Gamma$, and T is different from the diagonal.

A point $\varphi(z)$ on V with $z \in \mathfrak{H}^*$ may be called a "fixed point" of T if (and only if) $z \in \Gamma \alpha \Gamma z$. However, we have to take account of "the branches of the correspondence" T passing through $\varphi(z)$. Therefore we consider all $z_0 \in \mathfrak{H}^*$ such that $z_0 \in \Gamma \alpha \Gamma z_0$, and fix a complete set of representatives Ξ_0 for such z_0 under Γ-equivalence. Then let $\Xi = \Xi(\Gamma \alpha \Gamma)$ denote the set of all couples $(z_0, \Gamma \beta)$ with $z_0 \in \Xi_0$ and $\Gamma \beta \subset \Gamma \alpha \Gamma$ such that $\Gamma \beta z_0 = \Gamma z_0$. We call Ξ a *representative set of fixed points of* $\Gamma \alpha \Gamma$. (The number of elements of Ξ is not necessarily equal to the intersection number of T with Δ.)

We are going to define a complex number $I(\xi)$ for each $\xi = (z_0, \Gamma \beta) \in \Xi$. Choose β so that $\beta(z_0) = z_0$, and call ξ *elliptic, hyperbolic,* or *parabolic,* according to the type of β (which depends only on ξ).

(I) *Elliptic case.* Put $\sigma = \begin{bmatrix} \bar{z}_0 & z_0 \\ 1 & 1 \end{bmatrix}$. By Lemma 3.4, we have $\sigma^{-1} \beta \sigma = \begin{bmatrix} \bar{\lambda} & 0 \\ 0 & \lambda \end{bmatrix}$, $h(\beta, z_0) = \eta$ with $|\lambda| = |\eta| = 1$. Then we put

$$I(\xi) = \eta^{-1} \lambda^{-2e\mu}/(1 - \lambda^{-2e}),$$

where $e = e_p$, $\mu = \mu_p$ with $p = \varphi(z_0)$. Note that $\lambda^{2e} \neq 1$ since $\beta \notin \{\pm 1\} \cdot \Gamma$. By virtue of Lemma 3.4, $I(\xi)$ depends only on ξ, and not on the choice of β.

18 – 742909 *Acta mathematica* 132. Imprimé le 19 Juin 1974

(II) *Hyperbolic case.* Let β be an arbitrary hyperbolic element of $SL_2(\mathbf{R})$, and z_0 a fixed point of β on $\mathbf{R} \cup \{\infty\}$. Take $\varrho \in SL_2(\mathbf{R})$ so that $\varrho(\infty) = z_0$. Then $\varrho^{-1}\beta\varrho = \begin{bmatrix} \lambda^{-1} & x \\ 0 & \lambda \end{bmatrix}$ with real numbers λ and x. We call z_0 *the upper fixed point* or *the lower fixed point of* β, according as $|\lambda| > 1$ or $|\lambda| < 1$. This does not depend on the choice of ϱ. If z_0 is the upper fixed point of β, then the other fixed point is the lower fixed point.

Now suppose that $\beta \in \Gamma\alpha\Gamma$, and z_0 is a cusp of Γ. Take an element ϱ^* of $\mathfrak{G}_m$ with ϱ as its projection to $SL_2(\mathbf{R})$. Then we have

$$\varrho^{*-1}\beta^*\varrho^* = \left(\begin{bmatrix} \lambda^{-1} & x \\ 0 & \lambda \end{bmatrix}, \eta \right)$$

with a complex number η such that $|\eta| = |\lambda|^m$. Now, for $\xi = (z_0, \Gamma\beta)$, we put

$$I(\xi) = \begin{cases} -\eta^{-1} & \text{if } \mu_p = 1 \text{ and } |\lambda| > 1, \\ 0 & \text{otherwise,} \end{cases}$$

where $p = \varphi(z_0)$. Since Γ and $\beta\Gamma\beta^{-1}$ are commensurable, λ^2 must be a rational number. (See also Lemma 4.2 below.)

(III) *Parabolic case.* Let z_0 be a cusp, and let δ be an element of Γ that generates $\{\gamma \in \Gamma \,|\, \gamma(z_0) = z_0\}/(\Gamma \cap \{\pm 1\})$. Take the above ϱ so that $\varrho^{-1}\delta\varrho = \varepsilon \begin{bmatrix} 1 & 1 \\ 0 & 1 \end{bmatrix}$ with $\varepsilon = \pm 1$, and take an element ϱ^* of $\mathfrak{G}_m$ whose projection to $SL_2(\mathbf{R})$ is ϱ. Then

$$\varrho^{*-1}L(\delta)\varrho^* = \left(\varepsilon \begin{bmatrix} 1 & 1 \\ 0 & 1 \end{bmatrix}, e^{2\pi i \mu} \right),$$

where $\mu = \mu_p$ with $p = \varphi(z_0)$. Now let $\xi = (z_0, \Gamma\beta)$ with a parabolic β such that $\beta(z_0) = z_0$. Then

$$\varrho^{*-1}\beta^*\varrho^* = \left(c \begin{bmatrix} 1 & x \\ 0 & 1 \end{bmatrix}, \eta \right)$$

with $c = \pm 1$, $x \in \mathbf{R}$, $|\eta| = 1$. We put then

$$I(\xi) = \eta^{-1}e^{2\pi i \mu x}/(1 - e^{2\pi i x}).$$

In each of the three cases, the number $I(\xi)$ is independent of the choice of β.

We are now ready to state our main result:

3.7. **THEOREM.** *Let $[\Gamma\alpha\Gamma]^*$ and $[\Gamma\alpha^{-1}\Gamma]_*$ be as in 3.5, under the assumption* (3.6.1). *Suppose that $\alpha \notin \{\pm 1\}\Gamma$, and the divisor $\mathfrak{a}$ defined in 3.3 satisfies* (2.4.10). *Then*

$$\mathrm{tr}\,([\Gamma\alpha\Gamma]^*\,|\,F(\mathfrak{a})) - \mathrm{tr}\,([\Gamma\alpha^{-1}\Gamma]_*\,|\,F'(\mathfrak{b})) = \Sigma_{\xi\in\Xi}\,I(\xi),$$

where Ξ and $I(\xi)$ are defined as in 3.6.

The proof will be completed in 3.11.

3.8. As a preliminary step, let us make a few observations about the field of rationality for automorphic forms, although these are actually dispensable. Let $\Gamma_1 = \Gamma \cap \alpha^{-1}\Gamma\alpha$, and let (V_1, φ_1) be a model of $\mathfrak{H}^*/\Gamma_1$. Then V_1 is birationally equivalent with the curve T. Obviously the restriction of L or L' to Γ_1 are also a proper lifting of Γ_1, hence we can define $\mathfrak{F}_1 = \{F_1, F_1', Z_1, \mathfrak{z}_1\}$ for V_1 and Γ_1. Now there are two projection maps π and π' of V_1 onto V defined by $\pi\circ\varphi_1 = \varphi$ and $\pi'\circ\varphi_1 = \varphi\circ\alpha$. Fix any non-zero $f_0 \in F$ and $f_1 \in F_1$. Then we see that both f_1/f_0 and $(f_0\,|\,\alpha^*)/f_0$ belong to $\mathrm{C}(V_1)\circ\varphi_1$. Fix any field of rationality k_0 for $\mathfrak{F}$, $\mathfrak{F}_1$, π, π', f_0, f_1/f_0, and $(f_0\,|\,\alpha^*)/f_0$. Then if $k_0 \subset k$, we see that

$$(3.8.1) \qquad\qquad F(k) = F \cap F_1(k), \quad F'(k) = F' \cap F_1'(k);$$

$$(3.8.2) \qquad\qquad f \in F(k) \Rightarrow f\,|\,\alpha^* \in F_1(k).$$

In fact, if $f \in F(k)$, then $f = (r\circ\varphi)f_0$ with $r \in k(V)$, so that $f = (r\circ\pi\circ\varphi_1)f_1 \cdot (f_1/f_0)^{-1} \in F_1(k)$ and $f\,|\,\alpha^* = (r\circ\varphi\circ\alpha)(f_0\,|\,\alpha^*)f_0^{-1}f_0 = (r\circ\pi'\circ\varphi_1)(f_0\,|\,\alpha^*)f_0^{-1}f_0 \in F_1(k)$, q.e.d.

Let $\Gamma\alpha\Gamma = \bigcup_\nu \Gamma\alpha_\nu$ be a disjoint union. Then we see easily that

$(3.8.3)$ *For $f \in F(k)$, let r be an element of $k(V_1)$ such that $r\circ\varphi_1 = (f\,|\,\alpha^*)/f$. Then $(\mathrm{Tr}_{k(V_1)/k(V)}(r)\circ\varphi)f = f\,|\,[\Gamma\alpha\Gamma]^*$, where Tr is defined with respect to the injection $k(V) \to k(V)\circ\pi \subset k(V_1)$.*

This shows especially that $[\Gamma\alpha\Gamma]^*$ maps $F(k)$ into itself.

3.9. For each field k of rationality for $\mathfrak{F}$, define $E(k)$ as in 2.1, and let E denote the union of $E(k)$ for all fields k of rationality for $\mathfrak{F}$. Then E is a one-dimensional vector space over $\mathrm{C}(V \times V)$. With each $X = A \otimes f \otimes g \in E$ with $A \in \mathrm{C}(V \times V)$, $f \in F$, and $g \in F'$, we associate a meromorphic function $X(z, w)$ on $\mathfrak{H} \times \mathfrak{H}$ by

$$X(z, w) = A(\varphi(z), \varphi(w))f(z)g(w) \quad ((z, w) \in \mathfrak{H} \times \mathfrak{H}).$$

This does not depend on the choice of A, f, g for a given X, and

$$X(\gamma(z), \delta(w)) = X(z, w)j(\gamma, z)j'(\delta, w) \quad \text{for } (\gamma, \delta) \in \Gamma \times \Gamma.$$

In this way E can be identified with the set of all meromorphic functions $X(z, w)$ on $\mathfrak{H} \times \mathfrak{H}$ such that $X(z, w)f(z)^{-1}g(w)^{-1}$ is an element of $\mathrm{C}(V \times V)$ for $0 \neq f \in F$, $0 \neq g \in F'$. Let $\Gamma\alpha\Gamma = \bigcup_\nu \Gamma\alpha_\nu$ be as before. We now define $X\,|\,T$ to be an element of E such that

$$(X\,|\,T)\,(z, w) = \Sigma_\nu X(\alpha_\nu(z), w)h(\alpha_\nu, z)^{-1}.$$

More algebraically, we have

$$X\,|\,T = \mathrm{Tr}_{\mathbf{C}(V_1 \times V)/\mathbf{C}(V \times V)}(A') \otimes f \otimes g,$$

where A' is an element of $\mathbf{C}(V_1 \times V)$ such that

$$A'(\varphi_1(z), \varphi(w)) = A(\varphi(\alpha(z)), \varphi(w))\,(f\,|\,\alpha^*)\,(z)/f(z).$$

In view of (3.8.3), this shows that

(3.9.1) $X\,|\,T \in E(k)$ *if* $X \in E(k)$ *and* k *contains the field* k_0 *of* 3.8.

Suppose that the diagonal of $V \times V$ is not contained in the pole of X. Then we see that $X(z, z)$ is a Γ-automorphic form of weight 2 in the ordinary sense. Therefore $X(z, z)dz$ can be viewed as a differential form on V, hence the residue $\mathrm{Res}_p(X(z, z)dz)$ at each $p \in V$ is meaningful. We write

$$X(z, z)dz = X_{z-w}dz.$$

It can easily be seen that

(3.9.2) $X_{z-w}dz$ *is* k-*rational if* $X \in E(k)$.

3.10. We take the field k_0 of 3.8 so that the points of R, $\mathfrak{a}$, $\mathfrak{b}$, and T are all rational over k_0, and take an extension k of k_0 which is algebraically closed and has an infinite transcendence degree over k_0. With this k as the basic field, we define objects f_i, g_j, u, v, G_p, S_p, A_p, B, and q_i as in § 2. Put

$$n = \dim F(\mathfrak{a}), \quad r = \dim F'(\mathfrak{b}).$$

In § 2, we chose an arbitrary $\{g_i\}$ dual to $\{f_i\}$. Here we fix a k_0-rational point q of $V - R$, which is neither a fixed point of T, nor contained in the image or the inverse image of R by T. Then we choose $\{g_i\}$ so that

(3.10.1) $$g_{jp} = 0 \quad \text{for } q \neq p \in P(k).$$

This is possible by virtue of Proposition 1.9. Note also that the set of points $\{q_i\}$ is disjoint with the image and the inverse image of $\{q\} \cup R$ by T and also with the fixed points of T, since the q_i are generic points of V over k_0. We can also choose u and v so that

(3.10.2) $$\nu_p(u) = \mu_p, \nu_p(v) = \mu_p' \quad \text{for all } p \in R \cup \varphi(\Xi_0).$$

For brevity, let us write $T^* = [\Gamma\alpha\Gamma]^*$. To compute $\mathrm{tr}\,(T^*\,|\,F(\mathfrak{a}))$, put $f_j\,|\,T^* = \Sigma_{i=1}^n a_{ij}f_i$ with $a_{ij} \in k$. Since $\{g_i\}$ is dual to $\{f_i\}$, we have

$$a_{ij} = \Sigma_{p \in P(k)}\,\mathrm{Res}_p\,((f_j\,|\,T^*)g_{ip}dz) = \mathrm{Res}_q\,((f_j\,|\,T^*)g_{iq}dz)$$

by (3.10.1). By (2.4.4), we have

(3.10.3) $$\Sigma_{i=1}^n f_i \otimes g_{iq} = A_q + B.$$

By (2.4.9), a non-constant divisor $\mathfrak{P}$ of $k(V \times V)$ is contained in the pole of B if and only if $\mathfrak{P}$ is the diagonal Δ. By (3.10.3), A_q has the same property. Since T is different from the diagonal, both $(B|T)_{z=w}dz$ and $(A_q|T)_{z=w}dz$ are meaningful, hence

$$(3.10.4) \qquad \mathrm{tr}\,(T^*|F(\mathfrak{a})) = \Sigma_{i=1}^n a_{ii} = \mathrm{Res}_q\,((A_q|T)_{z=w}dz) + \mathrm{Res}_q\,((B|T)_{z=w}dz).$$

Now by (2.4.7), (2.4.9), (2.8.1), (2.8.2), we have

$$(3.10.5_a) \qquad \qquad \nu_{V \times p}(B) \geqslant 0 \qquad (p \in V - R \cup \{q\}),$$

$$(3.10.5_b) \qquad \qquad \nu_{V \times p}(B) \geqslant \mu_p' \qquad (p \in R),$$

$$(3.10.5_c) \qquad \qquad \nu_{p \times V}(B) \geqslant 0 \qquad (p \in V - R \cup \{q_1, ..., q_r\}),$$

$$(3.10.5_d) \qquad \qquad \nu_{p \times V}(B) \geqslant \mu_p \qquad (p \in R),$$

$$(3.10.5_e) \qquad \qquad \nu_{q_i \times V}(B) \geqslant -1 \quad (i = 1, ..., r),$$

$$(3.10.5_f) \qquad \qquad \nu_\Delta(B) = -1 \quad (\Delta\text{: diagonal}),$$

$$(3.10.5_g) \qquad \qquad \nu_{\mathfrak{P}}(B) \geqslant 0 \quad (\mathfrak{P} \text{ non-contant}, \,\neq \Delta).$$

(In § 2, we considered only k-rational prime divisors. However, since B is k-rational, we see easily that the above inequalities hold for any points or divisors which are not necessarily k-rational.)

By (2.4.6), we have $\nu_{V \times q}(A_q) \geqslant 0$. Now let $q = \varphi(z_0)$ with a point z_0 of $\mathfrak{H}$. For any $\beta \in \Gamma \alpha \Gamma$, put $p = \varphi(\beta(z_0))$. By (3.10.3) and $(3.10.5_c)$, we have $\nu_{p \times V}(A_q) \geqslant 0$. It follows that $A_q(\beta(z), z)$ is finite at $z = z_0$ for every $\beta \in \Gamma \alpha \Gamma$. (One cannot have $p = q$ because of our choice of q.) Therefore $\mathrm{Res}_q\,((A_q|T)_{z=w}dz) = 0$. On the other hand, $(B|T)_{z=w}dz$ is a differential form on V, hence

$$\Sigma_{p \in V}\,\mathrm{Res}_p\,((B|T)_{z=w}dz) = 0.$$

Therefore (3.10.4) becomes

$$(3.10.6) \qquad \mathrm{tr}\,(T^*|F(\mathfrak{a})) = \mathrm{Res}_q\,((B|T)_{z=w}dz) = -\Sigma_{p \neq q}\,\mathrm{Res}_p\,((B|T)_{z=w}dz).$$

(By (3.8.1, 2), $(B|T)_{z=w}dz$ is k-rational, but we do not need this fact.)

3.11. Our task is thus to compute $\mathrm{Res}_p\,((B|T)_{z=w}dz)$ for each $p \neq q$. Let us first show that the residue can be non-trivial only when either p is a fixed point of T, or p belongs to the inverse image of $\{q_1, ..., q_r\}$ under T. Let $p = \varphi(z_0)$, $p' = \varphi(\beta(z_0))$ with any $z_0 \in \mathfrak{H}^*$ and any $\beta \in \Gamma \alpha \Gamma$. Suppose $p \neq q$ and $p' \notin \{p, q_1, ..., q_r\}$. Take $u_1 \in F$ and $v_1 \in F'$ so that $\nu_{p'}(u_1) = \mu_{p'}$ and $\nu_p(v_1) = \mu_p'$. Put $B(z, w) = B_1(z, w)u_1(z)v_1(w)$. Then

$$(3.11.0) \qquad B(\beta(z), z)h(\beta, z)^{-1}dz = B_1(\beta(z), z)(u_1|\beta^*)(z)v_1(z)dz.$$

By (3.10.5), B_1 is finite at $(\beta(z_0), z_0)$. If z_0 is not a cusp, we have $\nu_p(v_1 dz) = -\mu_p > -1$, and

$u_1|\beta^*$ is finite at z_0. If z_0 is a cusp, then $\nu_p(v_1\,dz) = -\mu_p \geq -1$, and $u_1|\beta^*$ vanishes at z_0, since $\beta(z_0)$ is also a cusp, and $\nu_{p'}(u_1) = \mu_{p'} > 0$. Therefore, in either case, the form (3.11.0) measured by a local parameter on V at p has order > -1, hence the desired conclusion.

To compute the residue at a fixed point p of T, take $z_0 \in \Xi_0$ so that $p = \varphi(z_0)$, and consider $\xi = (z_0, \Gamma\beta) \in \Xi$ such that $\beta(z_0) = z_0$.

(I) First suppose that β is elliptic, hence $z_0 \in \mathfrak{H}$. Let $\mathfrak{D}$ denote the unit disc, and put

$$\sigma = \begin{bmatrix} \bar{z}_0 & z_0 \\ 1 & 1 \end{bmatrix}, \ \sigma(s) = (\bar{z}_0 s + z_0)/(s+1) \quad \text{for } s \in \mathfrak{D},$$

and define a holomorphic function $\varkappa$ on $\mathfrak{D}$ by

$$\varkappa(0) = 1, \ \varkappa(s) = (s+1)^m \quad (s \in \mathfrak{D}).$$

Then σ maps $\mathfrak{D}$ onto $\mathfrak{H}$, and $\sigma(0) = z_0$. By Lemma 3.4, if $\beta \in \Gamma\alpha\Gamma$ and $\beta(z_0) = z_0$, we have

$$\sigma^{-1}\beta\sigma = \begin{bmatrix} \lambda & 0 \\ 0 & \bar{\lambda} \end{bmatrix}, h(\beta, z_0) = \eta$$

with $|\lambda| = |\eta| = 1$. Moreover we can easily verify that

$$(3.11.1) \qquad\qquad h(\beta, \sigma(s)) = \eta \cdot \varkappa(\lambda^{-2}s)/\varkappa(s).$$

Let us write e, μ, μ' for e_p, μ_p, μ_p' with $p = \varphi(z_0)$. By (3.10.2), we can put $u(\sigma(s)) = s^{e\mu}u_0(s)$, $v(\sigma(s)) = s^{e\mu'}v_0(s)$ with functions u_0 and v_0 which are holomorphic and $\neq 0$ at the origin. Put $B = B_0 \otimes u \otimes v$ with $B_0 \in \mathbb{C}(V \times V)$. Further put $\psi = \varphi \circ \sigma$ and

$$D(s, t) = (s^e - t^e)s^{-e\mu}t^{-e\mu'}B(\sigma(s), \sigma(t))$$
$$= (s^e - t^e)B_0(\psi(s), \psi(t))u_0(s)v_0(t) \quad ((s, t) \in \mathfrak{D} \times \mathfrak{D}).$$

Now s^e is a local parameter at z_0. Therefore, by (3.10.5), we see that $D(s, t)$ is holomorphic at $(0, 0)$. Consider the differential form

$$B_0(\psi(s), \psi(t))u(\sigma(s))v(\sigma(s))d\sigma(s) = (s^e - t^e)^{-1}D(s, t)s^{e\mu+e\mu'}v_0(s)v_0(t)^{-1}d\sigma(s).$$

By Proposition 2.9, viewing t as a constant, the residue of this form at $s^e = t^e$ is 1. Since $e\mu + e\mu' = e - 1$ we have

$$(3.11.2) \qquad\qquad s^{e\mu+e\mu'}d\sigma(s) = e^{-1}\sigma'(s)d(s^e),$$

hence $e^{-1}\sigma'(t)D(t, t) = 1$, especially

$$(3.11.3) \qquad\qquad e^{-1}\sigma'(0)D(0, 0) = 1.$$

Putting $z = \sigma(s)$, we have, by (3.11.1, 2),

$$B(\beta(z),\, z)h(\beta,\, z)^{-1}dz = B(\sigma(\lambda^{-2}s),\, \sigma(s))h(\beta,\, \sigma(s))^{-1}d\sigma(s)$$

$$= (\lambda^{-2e}s^e - s^e)^{-1}(\lambda^{-2}s)^{e\mu}s^{e\mu'}D(\lambda^{-2}s,\, s)\eta^{-1}\varkappa(\lambda^{-2}s)^{-1}\varkappa(s)d\sigma(s)$$

$$= \eta^{-1}\lambda^{-2e\mu}(\lambda^{-2e}-1)^{-1}s^{-e}e^{-1}\sigma'(s)\,D(\lambda^{-2}s,\, s)\varkappa(\lambda^{-2}s)^{-1}\varkappa(s)d(s^e).$$

By (3.11.3), the residue of the last form at $s^e = 0$ is $-I(\xi)$ with $\xi = (z_0,\, \Gamma\beta)$, hence

(3.11.4)
$$\operatorname{Res}_p\,[(B\,|\,T)_{z=w}dz] = -\Sigma_\xi\,I(\xi),$$

the sum being taken over all $\xi = (z_0,\, \Gamma\beta)$ with the fixed point z_0 in question.

(II) Suppose $p = \varphi(z_0)$ with a cusp z_0 of Γ, $\beta(z_0) = z_0$ with a hyperbolic element $\beta \in \Gamma\alpha\Gamma$. We may assume $z_0 = \infty$, and take $t(z) = e^{2\pi i z}$ as a local parameter. Again we write μ and μ' for μ_p and μ_p'. By virtue of (3.10.5), if we put

$$H(t(z),\, t(w)) = t(\mu z)^{-1}t(\mu'w)^{-1}(t(z) - t(w))\,B(z,\, w),$$

then H is holomorphic at $(0, 0)$. Define B_0 as in (I). Then

$$B_0(\varphi(z),\, \sigma(w))u(z)v(z)dz$$

$$= (t(z) - t(w))^{-1}H(t(z),\, t(w))v(z)t(\mu'z)^{-1}v(w)^{-1}t(\mu'w)(2\pi i)^{-1}dt(z).$$

Viewing $t(w)$ as a constant, this has residue 1 at $t(z) = t(w)$, by virtue of Proposition 2.9, hence $H(t(w),\, t(w)) = 2\pi i$, especially

(3.11.5)
$$H(0,\, 0) = 2\pi i.$$

Now we can put $\beta = \begin{bmatrix} \lambda^{-1} & 0 \\ 0 & \lambda \end{bmatrix}$ and $h(\beta,\, z) = \eta$ with $\lambda \in \mathbf{R}$ and $\eta \in \mathbf{C}$. We have seen that λ^2 is a rational number. Put $\varkappa = \lambda^{-2}$. Then $\beta(z) = \varkappa z$, and

(3.11.6) $\quad B(\beta(z),\, z)h(\beta,\, z)^{-1}dz$

$$= \eta^{-1}t(z)^{-1}t((\varkappa\mu + \mu')z)(t(\varkappa z) - t(z))^{-1}H(t(\varkappa z),\, t(z))\,(2\pi i)^{-1}dt(z).$$

If $\varkappa < 1$, we have

$$t((\varkappa\mu + \mu')z)/(t(\varkappa z) - t(z)) = t(\mu'(1-\varkappa)z)/(1 - t((1-\varkappa)z)),$$

hence the residue, or more precisely the coefficient of $t(z)^{-1}dt(z)$ of (3.11.6), is either 0 or η^{-1} according as $\mu' > 0$ or $\mu' = 0$, by virtue of (3.11.5). If $\varkappa > 1$,

$$t((\varkappa\mu + \mu')z)/(t(\varkappa z) - t(z)) = -t((\varkappa - 1)\mu z)/(1 - t((\varkappa - 1)z)),$$

hence the "residue" of (3.11.6) is 0. Thus (3.11.4) holds also for hyperbolic ξ.

(III) Still with $z_0 = \infty$, suppose β parabolic. We can put

$$\beta^* = \left(\varepsilon\begin{bmatrix} 1 & x \\ 0 & 1 \end{bmatrix},\ \eta \right)$$

with $\varepsilon = \pm 1$, $x \in \mathbf{R}$, $|\eta| = 1$. Put $\zeta_1 = e^{2\pi i x}$, $\zeta_2 = e^{2\pi i \mu x}$. With the same $t(z)$ and H as in (II), we have

$$B(\beta(z), z) h(\beta, z)^{-1} dz = \eta^{-1}(\zeta_1 t - t)^{-1} \zeta_2 H(\zeta_1 t, t)(2\pi i)^{-1} dt,$$

hence the residue at $t = 0$ is $\eta^{-1}\zeta_2/(\zeta_1 - 1) = -I(\xi)$.

(IV) Suppose $r(= \dim F'(\mathfrak{b})) > 0$, and p belongs to the inverse image of $\{q_1, ..., q_r\}$ under T. Let $q_j = \varphi(z_j)$ for $j = 1, ..., r$. Then the sum of $\mathrm{Res}_p [(B|T)_{z=w}dz]$ at all such p is equal to

$$\sum_{j=1}^r \sum_{\Gamma\delta w = \Gamma z_j} \mathrm{Res}_w[B(\delta(z), z) h(\delta, z)^{-1} dz],$$

where the second Σ is extended over all $\Gamma\delta \subset \Gamma\alpha\Gamma$ such that $\Gamma\delta w = \Gamma z_j$; w is any point satisfying $\Gamma\delta w = \Gamma z_j$. Take a set of representatives $\{\beta\}$ so that $\Gamma\alpha^{-1}\Gamma = \cup \Gamma\beta = \cup \beta\Gamma$. Then $\Gamma\alpha\Gamma = \cup\Gamma\beta^{-1}$, and the above sum becomes

$$(3.11.7) \qquad \sum_{j=1}^r \sum_\beta \mathrm{Res}_{\beta(z_j)}[B(\beta^{-1}(z), z) h(\beta^{-1}, z)^{-1} dz]$$
$$= \sum_{j=1}^r \sum_\beta \mathrm{Res}_{z_j}[B(z, \beta(z)) h'(\beta, z)^{-1} dz],$$

where h' is defined by (3.5.4). Fix an element g of $F'(\mathfrak{b})$ such that $v_{q_i}(g) = 0$ for $i = 1, ..., r$, and define an element B_g of $C(V \times V)$ by $B = B_g \otimes u \otimes g$, and further define $c_j \in C(V)$ as in Proposition 2.10 with g in place of w. Put

$$H_j(z, w) = (z - z_j) B_g(\varphi(z), \varphi(w)) u(z) g(w).$$

Since $\varphi(z_j) \neq \varphi(\beta(z_j))$, we see, by (3.10.5), that $H_j(z, w)$ is holomorphic at $(z_j, \beta(z_j))$. Therefore, by Proposition 2.10, viewing w as a constant, we obtain

$$c_j(\varphi(w)) = \mathrm{Res}_{z_j}[(z - z_j)^{-1} H_j(z, w) g(z) g(w)^{-1} dz],$$

hence $H_j(z_j, w) g(z_j)/g(w) = c_j(\varphi(w))$, especially, putting $a_j(z) = c_j(\varphi(z)) g(z)$, we have

$$(3.11.8) \qquad\qquad H_j(z_j, \beta(z_j)) = a_j(\beta(z_j))/g(z_j).$$

Therefore (3.11.7) equals

$$
\begin{aligned}
(3.11.9) \qquad & \sum_{j=1}^r \sum_\beta \mathrm{Res}_{z_j}[(z - z_j)^{-1} H_j(z, \beta(z)) h'(\beta, z)^{-1} dz] \\
&= \sum_{j=1}^r \sum_\beta H_j(z_j, \beta(z_j)) h'(\beta, z_j)^{-1} \\
&= \sum_{j=1}^r g(z_j)^{-1} \sum_\beta a_j(\beta(z_j)) h'(\beta, z_j)^{-1} \\
&= \sum_{j=1}^r g(z_j)^{-1} b_j(z_j),
\end{aligned}
$$

where we put $b_j = a_j|[\Gamma\alpha^{-1}\Gamma]_*$. By Proposition 2.10, $\{a_j\}$ is a basis of $F'(\mathfrak{b})$, and $b_i = -\sum_{j=1}^r (b_i/g)(z_j) \cdot a_j$, hence

$$\mathrm{tr}\,([\Gamma\alpha^{-1}\Gamma]_*|F'(\mathfrak{b})) = -\sum_{j=1}^r (b_j/g)(z_j),$$

which is exactly (-1) times (3.11.9).

Combining the results of (I, II, III, IV) together, we obtain Theorem 3.7.

3.12. *Remark.* In this section we have considered only a special type of divisors $\mathfrak{a}$ and $\mathfrak{b}$, while a more general case was discussed in § 2. Actually we could state our theorem in such a general case, provided that $[\Gamma\alpha\Gamma]^*$ (resp. $[\Gamma\alpha^{-1}\Gamma]_*$) maps $F(\mathfrak{a})$ (resp. $F'(\mathfrak{b})$) into itself. In general, however, it is not easy to obtain a criterion for this requirement. A discussion is given in Eichler [3] for a question of the same type in a somewhat different formulation.

4. The trace formula: second formulation

4.1. We shall now express the sum $\Sigma_{\xi\in\Xi} I(\xi)$ of Theorem 3.7 in a more group-theoretical fashion. We do this not only for its own sake, but also to weaken the condition (2.4.10) under which the formula was proved. We shall introduce certain equivalence classes C in $\Gamma\alpha\Gamma$, and define a complex number $J(C)$ for each $C\subset\Gamma\alpha\Gamma$. Then the sum $\Sigma_{\xi\in\Xi} I(\xi)$ will be expressed as $\Sigma_{C\subset\Gamma\alpha\Gamma} J(C)$. To define $J(C)$, first put, for each $\beta\in\Gamma\alpha\Gamma$,

$$Z_\Gamma(\beta) = \{\gamma\in\Gamma \,|\, \gamma\beta = \beta\gamma\}.$$

Let $\Phi(\Gamma\alpha\Gamma)$ denote the subset of $\Gamma\alpha\Gamma$ consisting of:

all scalar elements of $\Gamma\alpha\Gamma$,

all elliptic elements of $\Gamma\alpha\Gamma$,

all hyperbolic elements of $\Gamma\alpha\Gamma$ whose upper fixed points are cusps of Γ (see 3.6, (II)),

all parabolic elements of $\Gamma\alpha\Gamma$ whose fixed points are cusps of Γ.

We call two elements β and β' of $\Phi(\Gamma\alpha\Gamma)$ *equivalent* if:

$\beta = \beta'$ when β and β' are scalars,

$\gamma\beta\gamma^{-1} = \beta'$ for some $\gamma\in\Gamma$, when β and β' are elliptic or hyperbolic,

$\gamma\beta'\gamma^{-1}\in Z_\Gamma(\beta)\beta$ for some $\gamma\in\Gamma$, when β and β' are parabolic.

We denote by $\Phi(\Gamma\alpha\Gamma/\Gamma)$ the set of all equivalence classes in $\Phi(\Gamma\alpha\Gamma)$ in this sense. Let $\beta\in\Phi(\Gamma\alpha\Gamma)$. If β is elliptic or parabolic, then β has a unique fixed point z_0 in $\mathfrak{H}^*$. Then

$$Z_\Gamma(\beta) = \{\gamma\in\Gamma \,|\, \gamma(z_0) = z_0\}.$$

If β is hyperbolic, one has $Z_\Gamma(\beta) = \Gamma\cap\{\pm 1\}$.

Now we define, for each $C\in\Phi(\Gamma\alpha\Gamma/\Gamma)$, a complex number $J(C)$ as follows:

$$J(C) = \begin{cases} [\Gamma\cap\{\pm 1\}:1]^{-1}\eta^{-1}2^{-1}(m-1)\,v(\mathfrak{H}/\Gamma) & \text{if } \beta^* = (\pm 1, \eta), \\ [Z_\Gamma(\beta):1]^{-1}\eta^{-1}/(1-\lambda^{-2}) & \text{if } \beta \text{ is elliptic,} \\ -[\Gamma\cap\{\pm 1\}:1]^{-1}\eta^{-1}/(1-\lambda^{-2}) & \text{if } \beta \text{ is hyperbolic,} \\ \eta^{-1}e^{2\pi i\mu x}(2^{-1}-\mu) & \text{if } \beta \text{ is parabolic and } \beta\in\{\pm 1\}\cdot\Gamma, \\ \eta^{-1}e^{2\pi i\mu x}/(1-e^{2\pi i x}) & \text{if } \beta \text{ is parabolic and } \beta\notin\{\pm 1\}\cdot\Gamma. \end{cases}$$

In each case we pick any β from C, and define λ, η, μ, and x for β as in 3.6, (I, II, III). We have $|\lambda| > 1$ for hyperbolic β, since we consider only the upper fixed point of β. Obviously $J(C)$ does not depend on the choice of β. Note also that $J(C) \neq 0$ even if β is hyperbolic and $\mu < 1$.

4.2. LEMMA. *Let β be a hyperbolic element of $\Gamma \alpha \Gamma$ with a cusp z_0 as its fixed point. Let δ be an element of Γ that generates*

$$\{\gamma \in \Gamma \,|\, \gamma(z_0) = z_0\}/(\Gamma \cap \{\pm 1\}).$$

Let ϱ^* be an element of $\mathfrak{G}_m$ whose projection ϱ to $SL_2(\mathbf{R})$ is such that $\varrho^{-1}\delta\varrho = \varepsilon \begin{bmatrix} 1 & 1 \\ 0 & 1 \end{bmatrix}$ with $\varepsilon = \pm 1$, and put

$$\varrho^{*-1}\delta^*\varrho^* = \left(\varepsilon \begin{bmatrix} 1 & 1 \\ 0 & 1 \end{bmatrix}, e^{2\pi i \mu} \right),$$

$$\varrho^{*-1}\beta^*\varrho^* = \left(\begin{bmatrix} \lambda^{-1} & x \\ 0 & \lambda \end{bmatrix}, \eta \right).$$

Then λ^2 is a rational number. Moreover, put $\lambda^2 = s/t$ with positive integers s and t such that $(s, t) = 1$. Then

   (i) $\beta\delta^s = \varepsilon^{t-s}\delta^t\beta$;

   (ii) $s - t$ *is even, if* $-1 \notin \Gamma$ *and* $\varepsilon = -1$;

   (iii) $(s - t)\mu \in \mathbf{Z}$.

Proof. We have

$$\varrho^{-1}\beta^{-1}\varrho \begin{bmatrix} 1 & 1 \\ 0 & 1 \end{bmatrix} \varrho^{-1}\beta\varrho = \begin{bmatrix} 1 & \lambda^2 \\ 0 & 1 \end{bmatrix},$$

hence the rationality of λ^2 follows from the commensurability of Γ with $\beta^{-1}\Gamma\beta$. Then (i) is immediate. If $-1 \notin \Gamma$, $\varepsilon = -1$, and $s - t$ is odd, then $\delta^{-t}\beta\delta^s = -\beta$, which contradicts the assumption (3.6.1). To prove (iii), we may assume $\beta\delta^s = \delta^t\beta$. (If $\varepsilon = -1$ and $s - t$ is odd, then $-1 \in \Gamma$. Take $-\delta$ in place of δ.) Then $\varrho^{*-1}(\delta^{-t}\beta\delta^s)^*\varrho^* = \varrho^{*-1}\beta^*\varrho^*$, hence $e^{2\pi i \mu(s-t)} \cdot \eta = \eta$, which proves (iii). (Note that (iii) is a consequence of (3.5.2).)

4.3. LEMMA. *Let x be an indeterminate, and let ζ be a primitive k-th root of unity with a positive integer $k > 1$. Then, for $b = 0, 1, \ldots, k-1$, one has*

$$\sum_{a=1}^{k} \zeta^{-ab}/(1 - \zeta^a x) = kx^b/(1 - x^k),$$

$$\sum_{a=1}^{k-1} \zeta^{-ab}/(1 - \zeta^a) = (k-1)/2 - b.$$

The proof is easy, and therefore omitted.

4.4. For any subgroup Γ_1 of Γ of finite index, we can consider the restrictions of L and L' to Γ_1. Then we can define objects F_1, F_1', $\mathfrak{a}_1$, $\mathfrak{b}_1$ with respect to Γ_1 corresponding to F, F', $\mathfrak{a}$, $\mathfrak{b}$. If an element $\tau = (\alpha, h)$ of $\mathfrak{G}_m$ satisfies (3.5.1, 2), then it satisfies the same conditions with Γ_1 in place of Γ. Therefore $[\Gamma_1 \alpha \Gamma_1]^*$ and $[\Gamma_1 \alpha^{-1} \Gamma_1]_*$ are meaningful.

4.5. **Theorem.** *Let* $\tau = (\alpha, h)$ *be an element of* $\mathfrak{G}_m$ *satisfying* (3.5.1, 2) *and* (3.6.1). *Suppose that* Γ *has a normal subgroup* Γ_1 *of finite index with the following properties:*

(i) $\deg(\Gamma \alpha \Gamma) = \deg(\Gamma_1 \alpha \Gamma_1)$;

(ii) $\Gamma \alpha \Gamma_1 = \Gamma_1 \alpha \Gamma = \Gamma \alpha \Gamma$;

(iii) $F_1(\mathfrak{a}_1 - p) \neq F_1(\mathfrak{a}_1)$ *for every* $p \in V_1 = \mathfrak{H}^* / \Gamma_1$;

(iv) Γ_1 *and* α *satisfy* (3.6.1).

Then, without assuming (2.4.10) *for* $F(\mathfrak{a})$, *one has*

$$\operatorname{tr}([\Gamma \alpha \Gamma]^* | F(\mathfrak{a})) - \operatorname{tr}([\Gamma \alpha^{-1} \Gamma]_* | F'(\mathfrak{b})) = \Sigma_{C \in \Phi(\Gamma \alpha \Gamma / \Gamma)} J(C).$$

Proof. Let us first prove the case $\alpha = \pm 1$. Put $\alpha^* = (\alpha, t)$ with $|t| = 1$. Then $(\alpha^{-1})_* = (\alpha, t)$, hence $f | [\Gamma \alpha \Gamma]^* = t^{-1} f$, $g | [\Gamma \alpha^{-1} \Gamma]_* = t^{-1} g$. Therefore our formula follows from (3.3.6) and Lemmas 3.4, 4.3.

Next let us prove the case $\Gamma = \Gamma_1$, assuming $\alpha \notin \{\pm 1\} \cdot \Gamma$. In this case, our task is to transform the sum $\Sigma_{\xi \in \Xi} I(\xi)$ into $\Sigma_C J(C)$. Let $\xi = (z_0, \Gamma \beta)$ be as in 3.6, and suppose that ξ is elliptic and $\beta(z_0) = z_0$ with $z_0 \in \mathfrak{H}$. Let γ be a generator of $Z_\Gamma(\beta)$, and put $\sigma = \begin{bmatrix} \bar{z}_0 & z_0 \\ 1 & 1 \end{bmatrix}$. By Lemma 3.4, we can put

$$\sigma^{-1} \gamma \sigma = \begin{bmatrix} \bar{\zeta} & 0 \\ 0 & \zeta \end{bmatrix}, \; j(\gamma, z_0) = \zeta^{-2e\mu},$$

$$\sigma^{-1} \beta \sigma = \begin{bmatrix} \bar{\lambda} & 0 \\ 0 & \lambda \end{bmatrix}, \; j(\beta, z_0) = \eta,$$

where $e = e_p$, $\mu = \mu_p$ with $p = \varphi(z_0)$. Let C_a denote the class containing $\gamma^a \beta$ for $a = 1, ..., k$, where $k = [Z_\Gamma(\beta):1]$. Thus ξ corresponds exactly to these k classes C_a. Now $k = 2e$ or e according as k is even or odd, and in both cases one has

$$\Sigma_{a=1}^{k} J(C_a) = \Sigma_{a=1}^{k} k^{-1} \eta^{-1} \zeta^{2ae\mu} / (1 - \lambda^{-2} \zeta^{-2a})$$

$$= \eta^{-1} \lambda^{-2e\mu} / (1 - \lambda^{-2e}) = I(\xi)$$

by Lemma 4.3.

Next suppose that β is hyperbolic. Without losing generality, we may assume ∞ is the upper fixed point of β. Define δ, ε, λ, η, s, and t as in Lemma 4.2.

626

$\left(\text{We may assume } \varrho^* = 1, \text{ so that } \delta = \varepsilon \begin{bmatrix} 1 & 1 \\ 0 & 1 \end{bmatrix}.\right)$ Put $\Gamma_\infty = \{\gamma \in \Gamma \mid \gamma(\infty) = \infty\}$. We have $s > t$, since $|\lambda| > 1$. Let us first assume that $-1 \notin \Gamma$ and $\varepsilon = -1$. Then

$$\{\sigma \in \Gamma \beta \delta^n \mid \sigma(\infty) = \infty\} = \{\delta^m \beta \delta^k \mid m \in \mathbf{Z}\} = \Gamma_\infty \beta \delta^k.$$

Consider all $\xi \in \Xi$ of the form $\xi = (\infty, \Gamma \beta \delta^k)$. Now $\delta^m \beta \delta^k = \delta^{-k} \delta^{m+k} \beta \delta^k$, hence we obtain from such a ξ a class C containing elements of the form $\delta^m \beta$. Suppose $\gamma \delta^m \beta \gamma^{-1} = \delta^n \beta$ with $\gamma \in \Gamma$. Then $\gamma \in \Gamma_\infty$, and such a γ exists if and only if $m \equiv n \pmod{s-t}$, by virtue of Lemma 4.2, (i), (ii). Thus there are exactly $s - t$ classes C_n represented by $\delta^n \beta$ with $n = 1, \ldots, s - t$. On the other hand, if $\gamma \delta^m \beta \gamma^{-1}$ has ∞ as its upper fixed point, γ must be contained in Γ_∞, so that $\Gamma \gamma \delta^m \beta \gamma^{-1} = \Gamma \beta \delta^n$ with $n \in \mathbf{Z}$. Now $\Gamma \beta \delta^n = \Gamma \beta \delta^m$ if and only if $m \equiv n \pmod{s}$. Thus there are exactly s different $\xi_k = (\infty, \Gamma \beta \delta^k)$ for $k = 1, \ldots, s$ corresponding to the C_n. Since $\lambda^{-2} - 1 = (t-s)/s$, we have

$$\sum_{n=1}^{s-t} J(C_n) = \sum_{n=1}^{s-t} \eta^{-1} e^{-2\pi i \mu n}/(\lambda^{-2} - 1) = \begin{cases} -s\eta^{-1} & \text{if } \mu = 1, \\ 0 & \text{if } \mu < 1, \end{cases}$$

by virtue of Lemma 4.2, (iii). Thus $\Sigma_n J(C_n) = \Sigma_k I(\xi_k)$. The same conclusion holds also in the case $-1 \in \Gamma$ or $\varepsilon = 1$, by a similar and simpler argument.

Still with $\Gamma = \Gamma_1$, consider a parabolic $\xi = (z_0, \Gamma \beta)$. Then there is a unique C in $\Phi(\Gamma \alpha \Gamma/\Gamma)$ containing β, and conversely C determines ξ uniquely. According to our definition, we have $J(C) = I(\xi)$ trivially. This completes the proof in the case $\Gamma = \Gamma_1$.

Now let us consider the general case assuming $\alpha \notin \{\pm 1\} \cdot \Gamma$. Fix a normal subgroup Γ_1 of Γ satisfying the conditions (i–iv). Let S be a set of representatives for Γ/Γ_1. Define $P: F_1 \to F$ and $P': F_1' \to F'$ by

$$P = [\Gamma: \Gamma_1]^{-1} \Sigma_{\gamma \in S} L(\gamma),$$
$$P' = [\Gamma: \Gamma_1]^{-1} \Sigma_{\gamma \in S} L'(\gamma).$$

We see that, for any $\gamma \in \Gamma$, (3.5.1, 2) and (3.6.1) are satisfied by $\alpha^* \gamma^*$ and Γ_1, hence $[\Gamma_1 \alpha \gamma \Gamma_1]^*$ and $[\Gamma_1 \gamma^{-1} \alpha^{-1} \Gamma_1]_*$ are meaningful, and $[\Gamma_1 \alpha \gamma \Gamma_1]^* = [\Gamma_1 \alpha \Gamma_1]^* L(\gamma)$, $[\Gamma_1 \gamma^{-1} \alpha^{-1} \Gamma_1]_* = L'(\gamma^{-1})[\Gamma_1 \alpha^{-1} \Gamma_1]_*$. Since our formula is true for Γ_1, we have, for every $\gamma \in S$,

$$\text{tr}\,([\Gamma_1 \alpha \gamma \Gamma_1]^* \mid F_1(\mathfrak{a}_1)) - \text{tr}\,([\Gamma_1 \gamma^{-1} \alpha^{-1} \Gamma_1]_* \mid F_1'(\mathfrak{b}_1)) = \Sigma_{C_1} J(C_1),$$

where C_1 runs over all classes in $\Phi(\Gamma_1 \alpha \gamma \Gamma_1/\Gamma_1)$. By our assumptions (i, ii), $[\Gamma \alpha \Gamma]^*$ (resp. $[\Gamma \alpha^{-1} \Gamma]_*$) is the restriction of $[\Gamma_1 \alpha \Gamma_1]^*$ (resp. $[\Gamma_1 \alpha^{-1} \Gamma_1]_*$) to F (resp. F'). Furthermore, P (resp. P') defines a projection map of $F_1(\mathfrak{a}_1)$ onto $F(\mathfrak{a})$ (resp. $F_1'(\mathfrak{b}_1)$ onto $F'(\mathfrak{b})$). Therefore

$$\text{tr}\,([\Gamma \alpha \Gamma]^* \mid F(\mathfrak{a})) - \text{tr}\,([\Gamma \alpha^{-1} \Gamma]_* \mid F'(\mathfrak{b}))$$
$$= [\Gamma: \Gamma_1]^{-1} \Sigma_{\gamma \in S} \{\text{tr}\,([\Gamma_1 \alpha \gamma \Gamma_1]^* \mid F_1(\mathfrak{a}_1)) - \text{tr}\,([\Gamma_1 \gamma^{-1} \alpha^{-1} \Gamma_1]_* \mid F_1'(\mathfrak{b}_1))\}$$
$$= [\Gamma: \Gamma_1]^{-1} \Sigma_D J(D),$$

where D runs over all classes in $\bigcup_{\gamma \in S} \Phi(\Gamma_1 \alpha \gamma \Gamma_1 / \Gamma_1)$. Observe that $\Gamma \alpha \Gamma = \bigcup_{\gamma \in S} \Gamma_1 \alpha \gamma \Gamma_1$, and this is a disjoint union by (i, ii). Let $C \in \Phi(\Gamma \alpha \Gamma / \Gamma)$. If C is elliptic or hyperbolic, it can easily be seen that C contains exactly $[\Gamma : \Gamma_1 Z_\Gamma(\beta)]$ classes D of $\bigcup_{\gamma \in S} \Phi(\Gamma_1 \alpha \gamma \Gamma_1 / \Gamma_1)$, where $\beta \in C$. Now we have

$$[\Gamma : \Gamma_1] = [\Gamma : \Gamma_1 Z_\Gamma(\beta)][Z_\Gamma(\beta) : 1][Z_{\Gamma_1}(\beta) : 1]^{-1},$$

hence $J(C) = [\Gamma : \Gamma_1]^{-1} \Sigma_{D \subset C} J(D)$.

It remains to prove the last equality for parabolic C. Let $\beta \in C \in \Phi(\Gamma \alpha \Gamma / \Gamma)$ with a parabolic β. We may assume $\beta(\infty) = \infty$. Put

$$\Gamma_\infty = \{\gamma \in \Gamma \mid \gamma(\infty) = \infty\}, \quad \Gamma_{1\infty} = \Gamma_\infty \cap \Gamma_1.$$

Let us first consider the case $-1 \notin \Gamma$. Then we may assume that Γ_∞ is generated by an element δ of the form $\delta = \varepsilon \begin{bmatrix} 1 & 1 \\ 0 & 1 \end{bmatrix}$, with $\varepsilon = \pm 1$. Put $k = [\Gamma_\infty : \Gamma_{1\infty}]$, and $\beta^* = \left(c \begin{bmatrix} 1 & x \\ 0 & 1 \end{bmatrix}, \eta \right)$. Let D be an element of $\bigcup_{\gamma \in S} \Phi(\Gamma_1 \alpha \gamma \Gamma_1 / \Gamma_1)$ contained in C. Then D contains an element of the form $\gamma \delta^m \beta \gamma^{-1}$ with $\gamma \in \Gamma$. It can easily be seen that $\gamma \delta^m \beta \gamma^{-1}$ and $\gamma' \delta^n \beta \gamma'^{-1}$ belong to the same D if and only if $\gamma^{-1} \gamma' \in \Gamma_1 \Gamma_\infty$ and $m \equiv n \pmod{k}$. Let P be a set of representatives for $\Gamma / \Gamma_1 \Gamma_\infty$. Then $[\Gamma : \Gamma_1]$ elements

$$\gamma \delta^n \beta \gamma^{-1} \quad (\gamma \in P; n = 1, \ldots, k)$$

form a set of representatives for all the classes D contained in C. If $\gamma \delta^n \beta \gamma^{-1} \in D$, then

$$J(D) = \eta^{-1} e^{-2\pi i n \mu} e^{2\pi i \mu_1 (x+n)/k} / (1 - e^{2\pi i (x+n)/k}),$$

where μ_1 is defined with respect to Γ_1. We can put $k\mu - \mu_1 = b$ with an integer b such that $0 \leqslant b < k$. Then

$$\begin{aligned}
\Sigma_{D \subset C} J(D) &= [\Gamma : \Gamma_1 \Gamma_\infty] \eta^{-1} e^{2\pi i (\mu - b/k) x} \sum_{n=1}^{k} e^{-2\pi i n b/k} / (1 - e^{2\pi i (x+n)/k}) \\
&= k \cdot [\Gamma : \Gamma_1 \Gamma_\infty] \eta^{-1} e^{2\pi i \mu x} / (1 - e^{2\pi i x}) \\
&= [\Gamma : \Gamma_1] J(C)
\end{aligned}$$

by Lemma 4.3. The case $-1 \in \Gamma$ can be treated in a similar way, which concludes our proof.

5. Supplementary results and remarks

5.1. We observe that the couple $(F(\mathfrak{a}), F'(\mathfrak{b}))$ is almost symmetric. Therefore if $F'(\mathfrak{b})$ satisfies (2.4.10), i.e., if $F'(\mathfrak{b} - p) \neq F'(\mathfrak{b})$ for every $p \in V$, then we can repeat the whole discussion interchanging $F(\mathfrak{a})$ and $F'(\mathfrak{b})$, and obtain a formula of the type

$$\mathrm{tr}\,([\Gamma \alpha^{-1} \Gamma]_* \mid F'(\mathfrak{b})) - \mathrm{tr}\,([\Gamma \alpha \Gamma]^* \mid F(\mathfrak{a})) = \Sigma_{C'} J'(C'),$$

where the sum is taken over all $C' \in \Phi(\Gamma\alpha^{-1}\Gamma/\Gamma)$. Let us now show that this becomes exactly -1 times the previous formula.

First we prove a formula corresponding to 3.7, with the sum $\Sigma_{\xi'} I'(\xi')$ extended over all $\xi' \in \Xi(\Gamma\alpha^{-1}\Gamma)$. In this case, since $0 \leqslant \mu'_p < 1$, we have to define $I'(\xi')$ for $\xi' = (z_0, \Gamma\beta)$ with a parabolic β by

$$I'(\xi') = \begin{cases} \eta^{-1} & \text{if } \mu'_p = 0 \text{ and } |\lambda| < 1, \\ 0 & \text{otherwise,} \end{cases}$$

where $\varrho_*^{-1}\beta_*\varrho_* = \left(\begin{bmatrix} \lambda^{-1} & 0 \\ 0 & \lambda \end{bmatrix}, \eta \right)$ with a suitable $\varrho_* \in \mathfrak{G}_{2-m}$. Then we can repeat the discussion of § 4, and arrive at the desired conclusion. As a consequence, we obtain

5.2. THEOREM. *The formula of 4.5 holds also when the condition* (iii) *is replaced by the following*

(iii') $$F'_1(\mathfrak{b}_1 - p) \neq F'_1(\mathfrak{b}_1) \quad \text{for all } p \in V_1 = \mathfrak{H}^*/\Gamma_1.$$

5.3. As a simple example, take the case where $m = 2$, and L is defined by $L(\gamma) = (\gamma, (cz+d)^2)$ for $\gamma = \begin{bmatrix} a & b \\ c & d \end{bmatrix} \in \Gamma$. Then we see that $L'(\gamma) = (\gamma, 1)$, $\mathfrak{b} = 0$, $F'(\mathfrak{b}) = \mathbb{C}$, and $F'(\mathfrak{b} - p) = \{0\} \neq F'(\mathfrak{b})$ for every $p \in V$. Therefore (iii') is satisfied with $\Gamma_1 = \Gamma$, and the trace-formula is valid. In this case, $F(\mathfrak{a})$ is exactly the space of cusp forms of weight 2 in the ordinary sense. Therefore (2.4.10) is satisfied if and only if $F(\mathfrak{a}) \neq \{0\}$. Thus our discussion shows that the trace-formula holds even if $F(\mathfrak{a}) = \{0\}$.

5.4. There is still another symmetry between $F(\mathfrak{a})$ and $F'(\mathfrak{b})$. First, to indicate that $\mathfrak{a}$ and $\mathfrak{b}$ are defined with respect to L and L', put $\mathfrak{a} = \mathfrak{a}(L)$ and $\mathfrak{b} = \mathfrak{b}(L')$. Now, interchanging L and L', we can define $\mathfrak{a}(L')$ and $\mathfrak{b}(L)$. More explicitly,

$$\mathfrak{a}(L') = \mathfrak{b}(L') - \Sigma_{p \in S} p,$$
$$\mathfrak{b}(L) = \mathfrak{a}(L) + \Sigma_{p \in S} p,$$

where S is the set of all cusps $p \in R$ for which $\mu_p = 1$ (i.e., $\mu'_p = 0$). Then $F(\mathfrak{b}(L))$ is the space of all integral forms with respect to L, and $F'(\mathfrak{a}(L'))$ is the space of all cusp forms with respect to L'. Our formula applied to this case gives the difference

(5.4.1) $$\text{tr}\,([\Gamma\alpha\Gamma]^* \,|\, F(\mathfrak{b}(L))) - \text{tr}\,([\Gamma\alpha^{-1}\Gamma]_* \,|\, F'(\mathfrak{a}(L'))).$$

We have of course $F(\mathfrak{a}(L)) \subset F(\mathfrak{b}(L))$ and $F'(\mathfrak{a}(L')) \subset F'(\mathfrak{b}(L'))$; the complementary parts are spanned by Eisenstein series. Therefore (5.4.1) gives the sum of the value given in

Theorem 4.5 and the traces of $[\Gamma\alpha\Gamma]^*$ and $[\Gamma\alpha^{-1}\Gamma]_*$ on Eisenstein series with respect to L and L'. As a special case of this fact, we obtain, from (3.3.6),

$$(5.4.2) \qquad \dim F(\mathfrak{b}(L)) - \dim F(\mathfrak{a}(L)) + \dim F'(\mathfrak{b}(L')) - \dim F'(\mathfrak{a}(L'))$$

$$= \text{the number of cusps } p \text{ on } V \text{ for which } \mu'_p = 0.$$

5.5. Let us now consider the case of modular forms of half integral weight. For a positive integer N, put

$$\Gamma_0(N) = \left\{ \begin{bmatrix} a & b \\ c & d \end{bmatrix} \in SL_2(\mathbf{Z}) \,\Big|\, c \equiv 0 \pmod{N} \right\},$$

$$\Gamma_1(N) = \left\{ \begin{bmatrix} a & b \\ c & d \end{bmatrix} \in \Gamma_0(N) \,\Big|\, a \equiv d \equiv 1 \pmod{N} \right\},$$

and define functions $\theta(z)$ and $j(\gamma, z)$ for $\gamma \in \Gamma_0(4)$ by

$$\theta(z) = \sum_{n=-\infty}^{\infty} \exp(2\pi i n^2 z),$$

$$j(\gamma, z) = \theta(\gamma(z))/\theta(z) \qquad (\gamma \in \Gamma_0(4)).$$

Then $j(\gamma, z)^4 = (cz+d)^2$ for $\gamma = \begin{bmatrix} a & b \\ c & d \end{bmatrix} \in \Gamma_0(4)$, and hence the map $\gamma \mapsto (\gamma, j(\gamma, z)) \in \mathfrak{G}_{\frac{1}{2}}$ defines a proper lifting of $\Gamma_0(4)$ of weight $\frac{1}{2}$. (For this and other facts on modular forms of half integral weight, the reader is referred to [7].) Now fix an odd positive integer k, a positive multiple N of 4, and a character χ modulo N such that $\chi(-1) = 1$; put then

$$\begin{aligned} L(\gamma) &= (\gamma, \chi(d) j(\gamma, z)^k) \\ L'(\gamma) &= (\gamma, \chi(d)^{-1} j(\gamma, z)^{4-k}) \end{aligned} \quad \text{for } \gamma = \begin{bmatrix} a & b \\ c & d \end{bmatrix} \in \Gamma_0(N).$$

These are obviously proper liftings of $\Gamma_0(N)$ of weight $k/2$ and $(4-k)/2$, respectivey. The elements of F and F' defined with these L and L' are exactly the modular forms considered in [7]. In this case as well as in the case of ordinary modular forms of integral weight, $[\Gamma\alpha\Gamma]^*$ has a certain commutative property with the map $f(z) \mapsto \overline{f(-\bar{z})}$, from which we can deduce a somewhat simpler form for the trace-formula; but we shall not go into details of this topic.

Let us now fix our attention to the case $k=3$, which is of special interest because both $F(\mathfrak{a})$ and $F'(\mathfrak{b})$ can be non-trivial. To simplify our discussion, we consider only the case $N=4M$ with an odd prime M.

5.6. PROPOSITION. *If $k=3$ and $N=4M$ with an odd prime M, then the condition* (iii') *of 5.2 is satisfied by $\Gamma_1 = \Gamma_1(N)$ and L' defined as above.*

Proof. Let (V_0, φ_0), (V, φ), and (V_1, φ_1) be models of $\mathfrak{H}^*/\Gamma_0(4)$, $\mathfrak{H}^*/\Gamma_0(N)$, and $\mathfrak{H}^*/\Gamma_1(N)$, respectively. Note that $F_1'(\mathfrak{b}_1) = F_1'(0)$ and $F_1'(\mathfrak{b}_1 - p) = F_1'(-p)$ for every $p \in V_1 = \mathfrak{H}^*/\Gamma_1(N)$. Therefore it is sufficient to show that for every $p \in V_1$, there is an element h of $F_1'(\mathfrak{b}_1)$ such that $v_p(h) < 1$. Let div_0, div, div_1 denote the divisors measured on V_0, V, V_1, respectively. Now $\Gamma_0(4)$ has three inequivalent cusps 0, ∞, $\frac{1}{2}$, but no elliptic elements. By (3.3.5), we have $\deg(\mathrm{div}_0(\theta)) = \frac{1}{4}$, from which we can easily conclude that $\mathrm{div}_0(\theta) = (\frac{1}{4}) \cdot \varphi_0(\frac{1}{2})$, which is actually a well known classical fact. There are exactly two points $\varphi(\frac{1}{2})$ and $\varphi((2M)^{-1})$ on V lying above $\varphi_0(\frac{1}{2})$ with ramification index 1 and M, respectively. Further, above each one of them, there are exactly $(M-1)/2$ points on V_1 with ramification index 2. Therefore

$$\mathrm{div}_1(\theta) = \sum_{i=1}^{t}((1/2)\,p_i + (M/2)\,q_i) \qquad (t = (M-1)/2)$$

with these points p_i and q_i. Put $g(z) = \theta(-1/Nz)z^{-\frac{1}{2}}$. By [7, Prop. 1.4], $g \in F_1'(\mathfrak{b}_1)$, and

$$\mathrm{div}_1(g) = \sum_{i=1}^{t}((M/2)\,p_i + (1/2)\,q_i).$$

Therefore, for every $p \in V_1$, we have either $v_p(\theta) < 1$ or $v_p(g) < 1$, q.e.d.

5.7. Let n be a positive integer, and let

$$\alpha = \begin{bmatrix} n^{-1} & 0 \\ 0 & n \end{bmatrix}, \quad \alpha^* = \tau = (\alpha, 1) \in \mathfrak{G}_{3/2}.$$

Then we see that the conditions (3.5.1, 2) and (3.6.1) are satisfied by α, τ, and $\Gamma = \Gamma_0(N)$ with the above L. Moreover, (i, ii, iv) of 4.5 are satisfied by $\Gamma_1 = \Gamma_1(N)$. Therefore, by 5.2, the trace-formula holds for $[\Gamma\alpha\Gamma]^*$ and $[\Gamma\alpha^{-1}\Gamma]_*$ in the present case with $k = 3$. The operators $[\Gamma\alpha\Gamma]^*$ and $[\Gamma\alpha^{-1}\Gamma]_*$ differ from $T_{3,\chi}^N(n^2)$ and $T_{1,\bar{\chi}}^N(n^2)$ of [7] only by constant factors. In this case, it is plausible that $F'(\mathfrak{b})$ is one-dimensional and spanned by θ if χ is trivial and $N/4$ is a prime. In such a case, $\mathrm{tr}\,([\Gamma\alpha\Gamma]^* \,|\, F(\mathfrak{a}))$ is effectively computable.

5.8. We conclude our study by making some observations in the case $m = 1$. Consider a lifting of the type

$$L(\gamma) = (\gamma, \chi(d)(cz+d)) \quad \left(\gamma = \begin{bmatrix} a & b \\ c & d \end{bmatrix} \in \Gamma_0(N)\right)$$

with a character χ modulo N such that $\chi(-1) = -1$. By an argument similar to the proof of 5.6, we can show that our trace formula holds for the ordinary Hecke operators on the space of modular forms of weight 1 with respect to L. Unfortunately, however, it can be verified that the difference

$$\mathrm{tr}\,([\Gamma\alpha\Gamma]^* \,|\, F(\mathfrak{a})) - \mathrm{tr}\,([\Gamma\alpha^{-1}\Gamma]_* \,|\, F'(\mathfrak{b}))$$

with a natural choice of α, say $\alpha = n^{-\frac{1}{2}}\begin{bmatrix} 1 & 0 \\ 0 & n \end{bmatrix}$, produces nothing particularly significant: it shows either that something which must be 0 is actually 0, or that the trace on the space of Eisenstein series is computable, which we could do anyway without the trace-formula. (Note that this is so even if $\chi^2 \neq 1$.) If we take an element of the form $\alpha\beta$ instead of α with a suitable element β of the normalizer of $\Gamma_0(N)$, say $\beta = N^{-\frac{1}{2}}\begin{bmatrix} 0 & -1 \\ N & 0 \end{bmatrix}$, then the formula becomes somewhat more non-trivial. But still this gives only the trace on the space of cusp forms corresponding to L-functions of imaginary quadratic fields with abelian characters. In this way one can obtain at least, or at most, a certain characterization of such cusp forms.

References

[1]. CHEVALLEY, C. *Introduction to the theory of algebraic functions of one variable.* Amer. Math. Soc., Math. Surveys No. 6, 1951.

[2]. EICHLER, M., Eine Verallgemeinerung der Abelschen Integrale. *Math. Zeitschr.*, 67 (1957), 267–298.

[3]. —— *Einführung in die Theorie der algbraischen Zahlen und Funktionen.* Birkhäuser, 1963. (English edition, Academic Press, 1966.)

[4]. KAPPUS, H., Darstellungen von Korrespondenzen algebraischer Funktionenkörper und ihre Spuren. *J. Reine Angew. Math.*, 210 (1962), 123–140.

[5]. SELBERG, A., Harmonic analysis and discontinuous groups in weakly symmetric Riemannian spaces with applications to Dirichlet series, *J. Indian Math. Soc.*, 20 (1956), 47–87.

[6]. SHIMURA, G., *Introduction to the arithmetic theory of automorphic functions.* Iwanami Shoten and Princeton Univ. Press, 1971.

[7]. —— On modular forms of half integral weight. *Ann. of Math.*, 97 (1973), 440–481.

[8]. WEIL, A. Généralisation des fonctions abéliennes. *J. Math. Pure Appl.*, [IX] 17 (1938), 47–87.

[9]. —— *Foundations of Algebraic Geometry.* Amer. Math. Soc. Coll. Publ. No. 29, 2nd ed., 1962.

[10]. —— *Sur les courbes algébriques et les variétés qui s'en déduisent.* Hermann, 1948.

Received August 13, 1973

19 – 742909 *Acta mathematica* 132. Imprimé le 25 Juin 1974

On the holomorphy of certain Dirichlet series

Proceedings of the London Mathematical Society, 3rd ser. 31 (1975), 79-98

[Received 15 April 1974]

Introduction

Let $f(z)$ be a cusp form of weight w satisfying

$$f((az+b)/(cz+d)) = \chi(d)(cz+d)^w f(z) \quad \text{for all } \begin{pmatrix} a & b \\ c & d \end{pmatrix} \in \Gamma_0(M),$$

where w and M are positive integers, χ is a Dirichlet character modulo M, and

$$\Gamma_0(M) = \left\{ \begin{pmatrix} a & b \\ c & d \end{pmatrix} \in \mathrm{SL}_2(\mathbf{Z}) : c \equiv 0 \ (\mathrm{mod}\ M) \right\}.$$

We naturally assume $\chi(-1) = (-1)^w$. Further let $f(z) = \sum_{n=1}^{\infty} c(n)e^{2\pi i n z}$ be its Fourier expansion. Suppose that $c(1) = 1$, and $f(z)$ is a common eigenfunction of all Hecke operators of level M. Then we have an Euler product

$$\sum_{n=1}^{\infty} c(n)n^{-s} = \prod_p [1 - c(p)p^{-s} + \chi(p)p^{w-1-2s}]^{-1},$$

where the product is taken over all rational primes p; we put $\chi(n) = 0$ for $(n, M) = 1$. Now decompose each Euler factor in the form

$$1 - c(p)p^{-s} + \chi(p)p^{w-1-2s} = (1 - \alpha_p p^{-s})(1 - \beta_p p^{-s})$$

with two complex numbers α_p and β_p, and define, with an arbitrary primitive Dirichlet character ψ, a new Euler product $D(s)$ by

$$(0.1) \quad D(s) = D(s, f, \psi)$$

$$= \prod_p [(1 - \psi(p)\alpha_p^2 p^{-s})(1 - \psi(p)\alpha_p\beta_p p^{-s})(1 - \psi(p)\beta_p^2 p^{-s})]^{-1}.$$

If p divides M, either or both of α_p and β_p may be 0. It can easily be seen that

$$(0.2) \quad D(s) = L(2s - 2w + 2, \chi^2\psi^2) \sum_{n=1}^{\infty} \psi(n)c(n^2)n^{-s}$$

$$= \prod_p [1 - \psi(p)c(p^2)p^{-s} + \psi(p)^2\chi(p)c(p^2)p^{w-1-2s}$$

$$- \psi(p)^3\chi(p)^3 p^{3w-3-3s}]^{-1},$$

where

$$L(s, \chi^2\psi^2) = \sum_{n=1}^{\infty} \chi(n)^2\psi(n)^2 n^{-s}.$$

Obviously the Euler product for $D(s)$ and the Dirichlet series $\sum c(n^2)n^{-s}$ are absolutely convergent for sufficiently large $\mathrm{Re}(s)$. Now the main purpose of the present paper is to prove

THEOREM 1. *Put*

$$(0.3) \qquad R(s) = \pi^{-3s/2}\Gamma(s/2)\Gamma((s+1)/2)\Gamma((s-w+2-\lambda_0)/2)D(s),$$

where λ_0 is 0 or 1 according as $\chi(-1)\psi(-1) = 1$ or -1. Then $R(s)$ can be continued to a meromorphic function on the whole s-plane, which is holomorphic except for possible simple poles at $s = w$ and $s = w-1$.

We can actually show that $R(s)$ is often an entire function. For example, R has no poles if all prime factors of Mr divide the conductor of $\chi^2\psi^2$, where r is the conductor of ψ. This is obviously so if $M = r = 1$. More general conditions under which R is holomorphic at $s = w$ or $s = w-1$ will be given as Theorem 2 in §5.

Our results are closely related to the work of Rankin ([6]) and Jacquet ([2]). In fact, we have

$$(0.4) \quad \sum_{n=1}^{\infty} \chi(n)\psi(n)n^{w-1-s}D(s) = L(2s-2w+2, \chi^2\psi^2) \sum_{n=1}^{\infty} \psi(n)c(n)^2 n^{-s}.$$

Here notice the difference of $c(n)^2$ from $c(n^2)$. Especially, if $M = r = 1$, we have

$$\zeta(s-w+1)D(s) = \zeta(2s-2w+2) \sum_{n=1}^{\infty} c(n)^2 n^{-s},$$

where $\zeta(s)$ denotes the Riemann zeta function. In [6], Rankin showed that the function

$$R^*(s) = (2\pi)^{-2s}\Gamma(s)\Gamma(s-w+1)\zeta(2s-2w+2) \sum_{n=1}^{\infty} c(n)^2 n^{-s}$$

can be continued to a meromorphic function on the whole s-plane, which is holomorphic except for simple poles at $s = w$ and $s = w-1$; moreover, it satisfies a functional equation

$$R^*(2w-1-s) = R^*(s).$$

This was generalized by Jacquet ([2]) for all automorphic forms on $GL(2)$ over any global field. From this result and the functional equation of $\zeta(s)$, one can easily derive that the function R, in the case $M = r = 1$, is meromorphic on the whole plane, and satisfies

$$R(2w-1-s) = R(s),$$

since $R(s) = 2^{w+1}\pi^{(w+1)/2}R^*(s)/\xi(s-w+1)$, where $\xi(s) = \pi^{-s/2}\Gamma(s/2)\zeta(s)$. This division by $\xi(s-w+1)$, however, does not guarantee the holomorphy of R. Our theorems assert that R is actually an entire function.†

In the proofs of our theorems, Siegel's generalization of Eisenstein–Epstein series ([11], [12]) plays an essential role. We shall actually make an explicit computation of the Fourier coefficients (Proposition 1), and prove a uniform boundedness of the series (Proposition 3). A certain Dirichlet series formed by these coefficients has a simple expression (Proposition 2) similar to the Dirichlet series formed by the Fourier coefficients of a modular form of half integral weight as considered in our previous paper [8]. We should also mention papers by Maass ([3]) and Petersson ([5]) on some cognate topics, although their connection with our present subject is rather indirect.

Notation and terminology. For a complex number s, we denote its real part, imaginary part, and complex conjugate by $\mathrm{Re}(s)$, $\mathrm{Im}(s)$, and $\bar{s}$, respectively. For complex numbers α and $v \neq 0$, we define v^{α} by

$$v^{\alpha} = e^{\alpha \log(v)}, \quad -\pi < \mathrm{Im}(\log(v)) \leqslant \pi.$$

A *Dirichlet character*, or simply a *character*, modulo a positive integer N, means a complex valued function ω defined for all rational integers such that $\omega(ab) = \omega(a)\omega(b)$, $\omega(a) = \omega(b)$ if $a \equiv b \pmod{N}$, and $\omega(a) \neq 0$ if and only if $(a, N) = 1$. We can then define the conductor of ω, and speak of primitive and imprimitive characters as usual. For a character ω modulo N, we put

$$L(s, \omega) = \sum_{n=1}^{\infty} \omega(n)n^{-s}.$$

If ω' is the primitive character such that $\omega'(a) = \omega(a)$ for $(a, N) = 1$, we see that

$$L(s, \omega') = L(s, \omega)\prod_{p|N}[1 - \omega'(p)p^{-s}].$$

To emphasize the possible missing factors, we also write $L_N(s, \omega')$ or $L_N(s, \omega)$ for $L(s, \omega)$, thus

$$L_N(s, \omega') = L_N(s, \omega) = L(s, \omega) = \sum_{(n,N)=1} \omega'(n)n^{-s}.$$

We denote the upper-half complex plane by H, that is,

$$H = \{z \in \mathbf{C} : \mathrm{Im}(z) > 0\},$$

and we let every element $\gamma = \begin{pmatrix} a & b \\ c & d \end{pmatrix}$ of $\mathrm{SL}_2(\mathbf{R})$ act on H by the rule $\gamma(z) = (az+b)/(cz+d)$ for $z \in H$.

---

† This result was announced in [9] without details of proof.

1. The integral expression for $D(s)$

With the same notation as in the Introduction, put

$$h(z) = \frac{1}{2} \sum_{n=-\infty}^{\infty} \bar{\psi}(n)n^{\nu}\exp(2\pi i n^2 z),$$

where $\nu = 0$ or 1 according as $\psi(-1) = 1$ or -1. Then with $z = x+iy$, we have

$$\int_0^1 f(z)\bar{h}(z)\,dx = \sum_{n=1}^{\infty} \psi(n)c(n^2)n^{\nu}\exp(-4\pi n^2 y),$$

hence

$$(1.1) \quad \int_0^{\infty}\left\{\int_0^1 f(z)\bar{h}(z)\,dx\right\}y^{(s/2)-1}\,dy = (4\pi)^{-s/2}\Gamma(s/2)\sum_{n=1}^{\infty}\psi(n)c(n^2)n^{\nu-s}.$$

By [8], Proposition 2.2, we have

$$h(\gamma(z)) = \bar{\psi}(d)\left(\frac{-1}{d}\right)^{\nu}j(\gamma,z)^{2\nu+1}h(z) \quad \text{for all } \gamma = \begin{pmatrix} a & b \\ c & d \end{pmatrix} \in \Gamma_0(4r^2),$$

where r is the conductor of ψ, and

$$(1.2) \quad j\left(\begin{pmatrix} a & b \\ c & d \end{pmatrix}, z\right) = \left(\frac{c}{d}\right)\varepsilon_d^{-1}(cz+d)^{1/2},$$

$$\varepsilon_d = \begin{cases} 1 & \text{if } d \equiv 1 \pmod 4, \\ i & \text{if } d \equiv 3 \pmod 4. \end{cases}$$

For this and other facts on modular forms of half integral weight, the reader is referred to [8]. Put $A(z,s) = f(z)\bar{h}(z)y^{(s/2)+1}$. Then we have

$$A(\gamma(z),s) = A(z,s)\omega(d)j(\gamma,z)^{2w-2\nu-1}|j(\gamma,z)|^{4\nu-2-2s}$$

$$\text{for all } \gamma = \begin{pmatrix} a & b \\ c & d \end{pmatrix} \in \Gamma_0(N),$$

where N is the least common multiple of M and $4r^2$, and ω is the character modulo N defined by

$$(1.3) \quad \omega(d) = \left(\frac{-1}{d}\right)^{w+\nu}\chi(d)\psi(d).$$

Take a fundamental domain Φ for $\Gamma_0(N)\backslash H$, and also take a set W of representatives for $\Gamma_{\infty}\backslash\Gamma_0(N)$, where

$$\Gamma_{\infty} = \left\{\pm\begin{pmatrix} 1 & m \\ 0 & 1 \end{pmatrix}\,\middle|\, m \in \mathbf{Z}\right\}.$$

Then the integral (1.1) can be transformed to the form

$$\int_{\Phi} A(z,s)\left\{\sum_{\gamma \in W}\omega(d_{\gamma})j(\gamma,z)^{2w-2\nu-1}|j(\gamma,z)|^{4\nu-2-2s}\right\}y^{-2}\,dx\,dy,$$

where d_γ is the lower right entry of γ. Therefore, putting

$$(1.4) \qquad E(z,s) = E(z,s,k,\omega)$$

$$= y^{s/2} \sum_{\gamma \in W} \omega(d_\gamma) j(\gamma,z)^k \,|\, j(\gamma,z)|^{-2s} \quad (z \in H,\ s \in \mathbf{C}),$$

we obtain, with $k = 2w - 2v - 1$,

$$(1.5)\ \ (4\pi)^{-s/2}\Gamma(s/2)D(s-v)$$

$$= \int_\Phi f\bar{h} y^{v+(1/2)} L(2s - 2w - 2v + 2, \chi^2\psi^2) E(z, s - 2v + 1) y^{-2}\, dx\, dy.$$

Note that $E(z,s)$ is convergent for $\mathrm{Re}(s) > (k+4)/2$, and the last equality holds at least for $\mathrm{Re}(s) > w + v + 1$.

Now our aim is to show that the function

$$L(2s - 2w - 2v + 2, \chi^2\psi^2) E(z, s - 2v + 1)$$

multiplied by a suitable Γ-factor and a linear factor can be continued to a holomorphic function on the whole s-plane, and the integral (1.5), when the integrand is multiplied by the same factors, is absolutely convergent for all s.

2. Preliminary analytic considerations

To obtain the Fourier expansion of $E(z,s)$, let us introduce, after Siegel ([11]), a confluent hypergeometric function $\sigma(y,\alpha,\beta)$ by

$$(2.1) \qquad \sigma(y,\alpha,\beta) = \int_0^\infty (u+1)^{\alpha-1} u^{\beta-1} e^{-yu}\, du$$

$$= y^{-\beta} \int_0^\infty (1 + y^{-1}t)^{\alpha-1} t^{\beta-1} e^{-t}\, dt,$$

where y is real positive, and α, β are complex variables. This is convergent for $\mathrm{Re}(\beta) > 0$. It can easily be verified that

$$(2.2) \qquad (e^{2\pi i\beta} - 1)\sigma(y,\alpha,\beta) = y^{-\beta} \int_\infty^{(0+)} (1 + y^{-1}t)^{\alpha-1} t^{\beta-1} e^{-t}\, dt$$

with the usual notation for contour integrals. This shows that

$$(e^{2\pi i\beta} - 1)\sigma(y,\alpha,\beta)$$

can be analytically continued to a function holomorphic for all $(\alpha,\beta) \in \mathbf{C}^2$, and

$$(2.3) \qquad y^\beta \Gamma(\beta)^{-1}\sigma(y,\alpha,\beta) = \frac{e^{-\pi i\beta}}{2\pi i} \Gamma(1-\beta) \int_\infty^{(0+)} (1 + y^{-1}t)^{\alpha-1} t^{\beta-1} e^{-t}\, dt.$$

LEMMA 1. *Let α and β be complex numbers such that $\mathrm{Re}(\alpha) > 0$, $\mathrm{Re}(\beta) > 0$, and $\mathrm{Re}(\alpha+\beta) > 1$. Then, for $z = x + iy \in H$, we have*

$$\sum_{m=-\infty}^{\infty} (z+m)^{-\alpha}(\bar{z}+m)^{-\beta} = \sum_{n=-\infty}^{\infty} \tau_n(y,\alpha,\beta)e^{2\pi inx},$$

where $\tau_n(y,\alpha,\beta)$ is given by

$$i^{\alpha-\beta}(2\pi)^{-\alpha-\beta}\Gamma(\alpha)\Gamma(\beta)\tau_n(y,\alpha,\beta) = \begin{cases} n^{\alpha+\beta-1}e^{-2\pi ny}\sigma(4\pi ny,\alpha,\beta) & (n > 0), \\ |n|^{\alpha+\beta-1}e^{-2\pi|n|y}\sigma(4\pi|n|y,\beta,\alpha) & (n < 0), \\ \Gamma(\alpha+\beta-1)(4\pi y)^{1-\alpha-\beta} & (n = 0). \end{cases}$$

This and the following lemmas are stated in [11]. For the reader's convenience, we give a proof here.

According to our definition of v^α in the Notation, we have $v^{\alpha+\beta} = v^\alpha v^\beta$, $v^{m\alpha} = (v^\alpha)^m$ for $m \in \mathbf{Z}$, and $(uv)^\alpha = u^\alpha v^\alpha$ if $\arg(u)$, $\arg(v)$, and $\arg(u) + \arg(v)$ are all contained in the interval $(-\pi, \pi]$ for suitable choices of $\arg(u)$ and $\arg(v)$. These points must be kept in mind in the following computation. Now put $f(x) = z^{-\alpha}\bar{z}^{-\beta}$ for $z = x + iy$ with a fixed y. By the Poisson summation formula, we have, for $\mathrm{Re}(\alpha+\beta) > 1$,

$$\sum_{m=-\infty}^{\infty} (z+m)^{-\alpha}(\bar{z}+m)^{-\beta} = \sum_{m=-\infty}^{\infty} f(x+m) = \sum_{n=-\infty}^{\infty} f^*(n)e^{2\pi inx},$$

where f^* is the Fourier transform of f, that is,

$$f^*(t) = \int_{-\infty}^{\infty} z^{-\alpha}\bar{z}^{-\beta}e^{-2\pi itx}\,dx.$$

This integral is absolutely convergent for $\mathrm{Re}(\alpha+\beta) > 1$. Putting $v = i\bar{z} = y + ix$, we have

$$f^*(t) = i^{\beta-\alpha-1}e^{2\pi ty}\int_{y-i\infty}^{y+i\infty} v^{-\beta}(2y-v)^{-\alpha}e^{-2\pi tv}\,dv$$

$$= i^{\beta-\alpha-1}e^{2\pi ty}\Gamma(\alpha)^{-1}\int_{y-i\infty}^{y+i\infty} v^{-\beta}e^{-2\pi tv}\left\{\int_0^\infty e^{-\xi(2y-v)}\xi^{\alpha-1}\,d\xi\right\}dv$$

$$= i^{\beta-\alpha-1}e^{2\pi ty}\Gamma(\alpha)^{-1}\int_0^\infty \xi^{\alpha-1}e^{-2y\xi}\left\{\int_{y-i\infty}^{y+i\infty} v^{-\beta}e^{(\xi-2\pi t)v}\,dv\right\}d\xi$$

if $\mathrm{Re}(\alpha) > 0$. Now, if $\mathrm{Re}(\beta) > 0$, we have

$$\int_{y-i\infty}^{y+i\infty} v^{-\beta}e^{\lambda v}\,dv = \begin{cases} 2\pi i\Gamma(\beta)^{-1}\lambda^{\beta-1} & (\lambda > 0), \\ 0 & (\lambda \leqslant 0). \end{cases}$$

Hence, putting $\xi = 2\pi p$ and $u = \max(0,t)$, we obtain

$$f^*(t) = (2\pi)^{\alpha+\beta}i^{\beta-\alpha}\Gamma(\alpha)^{-1}\Gamma(\beta)^{-1}e^{2\pi ty}\int_u^\infty p^{\alpha-1}(p-t)^{\beta-1}e^{-4\pi py}\,dp.$$

This holds when $\mathrm{Re}(\alpha) > 0$, $\mathrm{Re}(\beta) > 0$, and $\mathrm{Re}(\alpha + \beta) > 1$. Putting $p - t = tq$ or $-tq$ according as $t > 0$ or $t < 0$, we obtain

$$\int_u^\infty p^{\alpha-1}(p-t)^{\beta-1}e^{-4\pi p v}\,dp = \begin{cases} t^{\alpha+\beta-1}e^{-4\pi t y}\sigma(4\pi t y, \alpha, \beta) & (t > 0), \\ |t|^{\alpha+\beta-1}\sigma(4\pi|t|y, \beta, \alpha) & (t < 0). \end{cases}$$

Finally, if $t = 0$, the integral becomes

$$\int_0^\infty p^{\alpha+\beta-2}e^{-4\pi p v}\,dp = \Gamma(\alpha+\beta-1)(4\pi y)^{1-\alpha-\beta}.$$

This completes the proof.

LEMMA 2. *The function* $y^\beta\Gamma(\beta)^{-1}\sigma(y,\alpha,\beta)$ *is invariant under the transformation* $\alpha \mapsto 1-\beta$, $\beta \mapsto 1-\alpha$.

Proof. Consider the Mellin transform of $\sigma(y,\alpha,\beta)$:

$$\int_0^\infty \sigma(y,\alpha,\beta)y^{s-1}\,dy = \Gamma(s)\int_0^\infty (u+1)^{\alpha-1}u^{\beta-s-1}\,du$$

$$= \Gamma(s)\Gamma(\beta-s)\Gamma(1-\alpha-\beta+s)/\Gamma(1-\alpha).$$

This holds if $\mathrm{Re}(s) > 0$, $\mathrm{Re}(\beta - s) > 0$, and $\mathrm{Re}(1-\alpha-\beta+s) > 0$. By the Mellin inversion formula, we have

$$\Gamma(1-\alpha)\sigma(y,\alpha,\beta) = \frac{1}{2\pi i}\int_{c-i\infty}^{c+i\infty}\Gamma(s)\Gamma(\beta-s)\Gamma(1-\alpha-\beta+s)y^{-s}\,ds$$

if $c > 0$ and $\mathrm{Re}(\beta) > c > \mathrm{Re}(\alpha+\beta-1)$. Putting $S = s - \beta$, we obtain

$$(2.4) \quad y^\beta\Gamma(\beta)^{-1}\sigma(y,\alpha,\beta) = \frac{1}{2\pi i}\int_{-p-i\infty}^{-p+i\infty}\frac{\Gamma(-S)\Gamma(S+\beta)\Gamma(S+1-\alpha)}{\Gamma(1-\alpha)\Gamma(\beta)}y^{-S}\,dS$$

if $0 < p < \min\{\mathrm{Re}(1-\alpha), \mathrm{Re}(\beta)\}$. Such a number p can be found whenever $\mathrm{Re}(\beta) > 0$ and $\mathrm{Re}(1-\alpha) > 0$. The last integrand is obviously invariant under the transformation $\alpha \mapsto 1-\beta$, $\beta \mapsto 1-\alpha$; hence our lemma follows.

Actually (2.4) can be improved as follows:

$$(2.5) \quad y^\beta\Gamma(\beta)^{-1}\sigma(y,\alpha,\beta) = \sum_{m=0}^{[q]}\frac{\Gamma(m+1-\alpha)\Gamma(m+\beta)}{\Gamma(m+1)\Gamma(1-\alpha)\Gamma(\beta)}(-y)^{-m}$$

$$+\frac{1}{2\pi i}\int_{q-i\infty}^{q+i\infty}\frac{\Gamma(-S)\Gamma(S+1-\alpha)\Gamma(S+\beta)}{\Gamma(1-\alpha)\Gamma(\beta)}y^{-S}\,dS.$$

This holds for $\mathrm{Re}(1-\alpha) > -q$, $\mathrm{Re}(\beta) > -q$, where q is a positive number that is not an integer and $[q]$ is the greatest integer less than q. In fact, if $\mathrm{Re}(1-\alpha) > 0$ and $\mathrm{Re}(\beta) > 0$, (2.5) can be obtained from (2.4) by shifting the line of integration from $\mathrm{Re}(S) = -p$ to $\mathrm{Re}(S) = q$. In this process, the integral acquires the residues at $S = 0, 1, \ldots, [q]$, which form the first sum of the right-hand side of (2.5). The asymptotic behavior of $\Gamma(S)$ for large $\mathrm{Im}(S)$ justifies this argument. By analytic continuation, we see that (2.5) holds whenever $\mathrm{Re}(\beta) > -q$ and $\mathrm{Re}(1-\alpha) > -q$.

3. The Fourier expansion of $E(z,s)$

In the above work, we started from a cusp form f and a character ψ. However, we can actually define $E(z,s)$ by (1.4) with an arbitrary character ω modulo N and an odd (positive or negative) integer k. Naturally we assume that N is a multiple of 4, and $\omega(-1) = 1$. In this and the next sections we shall discuss $E(z,s)$ without any reference to $f(z)$ and $D(s)$.

For the computation of the Fourier coefficients, it is more convenient to consider, instead of $E(z,s)$, another function $E'(z,s)$ defined by

$$(3.1) \qquad E'(z,s) = E'(z,s,k,\omega) = E(-1/Nz,s)(-iz\sqrt{N})^{k/2}.$$

By [8], Lemma 3.2, we can choose a set W of representatives for $\Gamma_\infty \backslash \Gamma_0(N)$ so that the mapping with domain W

$$\begin{pmatrix} a & b \\ c & d \end{pmatrix} \mapsto \{c,d\} \in W_0$$

is one-to-one, where W_0 is the set of all ordered couples of integers $\{c,d\}$ such that $(c,d) = 1, c \equiv 0 \pmod{N}, d > 0$. Therefore, by virtue of the explicit expression (1.2) for $j(\gamma,z)$, we obtain

$$(3.2) \quad N^{(2s-k)/4} i^{k/2} y^{-s/2} E'(z,s)$$

$$= \sum_{d=1}^{\infty} \sum_{b=-\infty}^{\infty} \omega(d)\left(\frac{-Nb}{d}\right)\varepsilon_d^{-k}(dz+b)^{k/2}|dz+b|^{-s}.$$

Let us denote the last sum simply by $E^* = E^*(z,s) = E^*(z,s,k,\omega)$, and consider its Fourier expansion. Put $b = dl+m$ with $l \in \mathbf{Z}$ and $1 \leqslant m \leqslant d$. Then, by Lemma 1,

$$(3.3) \quad E^* = \sum_{d=1}^{\infty} \omega(d)\varepsilon_d^{-k}d^{(k/2)-s} \sum_{m=1}^{d} \left(\frac{-Nm}{d}\right) \sum_{l=-\infty}^{\infty} \left(z+\frac{m}{d}+l\right)^{k/2}\left|z+\frac{m}{d}+l\right|^{-s}$$

$$= \sum_{n=-\infty}^{\infty} \alpha(n,s)e^{2\pi i n x} \tau_n(y,(s-k)/2,s/2),$$

where

$$(3.4) \qquad \alpha(n,s) = \sum_{d=1}^{\infty} \left(\frac{-N}{d}\right)\varepsilon_d^{-k}\omega(d)d^{(k/2)-s} \sum_{m=1}^{d} \left(\frac{m}{d}\right)e^{2\pi i n m/d}.$$

Put

$$(3.5) \quad G_n(d) = \sum_{m=1}^{d} \left(\frac{m}{d}\right)e^{2\pi i m n/d} \quad \text{and} \quad H_n(d) = \varepsilon_d^{-k}G_n(d)$$

$$(d \text{ odd and positive}).$$

Then we see easily that

$$(3.6) \qquad G_n(ab) = \left(\frac{a}{b}\right)\left(\frac{b}{a}\right) G_n(a)G_n(b) \quad \text{if } (a,b) = 1,$$

and $\left(\dfrac{a}{b}\right)\left(\dfrac{b}{a}\right) = \varepsilon_{ab}{}^k \varepsilon_a{}^{-k}\varepsilon_b{}^{-k}$; hence

$$(3.7) \qquad H_n(ab) = H_n(a)H_n(b) \quad \text{if } (a,b) = 1.$$

Therefore $\alpha(n,s)$ can be decomposed into the product of its 'p-parts' for all primes p (not dividing N). However, we can compute $\alpha(n,s)$ more directly by means of the following

LEMMA 3. *Let χ_0 be a primitive character modulo r, and χ a character modulo rs such that $\chi(n) = \chi_0(n)$ for $(n,s) = 1$. Then, for any integer q, we have*

$$\sum_{n=1}^{rs} \chi(n)e^{2\pi i n q/rs} = \left(\sum_{m=1}^{r} \chi_0(m)e^{2\pi i m/r}\right)\left(\sum_{c|(s,q)} c\mu(s/c)\chi_0(s/c)\bar\chi_0(q/c)\right),$$

where μ is the Moebius function.

Proof. First observe that, for any positive integer c, we have

$$(3.8) \qquad \sum_{n=1}^{rc} \chi_0(n)e^{2\pi i n q/rc} = \begin{cases} 0 & \text{if } c \nmid q, \\[2mm] c\bar\chi_0(q/c) \displaystyle\sum_{m=1}^{r} \chi_0(m)e^{2\pi i m/r} & \text{if } c \mid q. \end{cases}$$

Now we have

$$\sum_{n=1}^{rs} \chi(n)e^{2\pi i n q/rs} = \sum_{n=1}^{rs}\left(\sum_{d|(n,s)} \mu(d)\right)\chi_0(n)e^{2\pi i n q/rs}$$

$$= \sum_{d|s} \mu(d)\chi_0(d) \sum_{m=1}^{rc} \chi_0(m)e^{2\pi i m q/rc}$$

(putting $n = md$ and $s = cd$). Applying (3.8) to the last sum, we obtain the desired expression.

Now the explicit form of $\alpha(n,s)$ can be given as follows.

PROPOSITION 1. *Let t be a (positive or negative) square-free integer. Put $\lambda = (k+1)/2$ and define primitive characters ω_1 and ω_2 by*

$$\omega_1(a) = \left(\frac{-1}{a}\right)^{\lambda}\left(\frac{tN}{a}\right)\omega(a) \quad \text{for } (a,tN) = 1,$$

$$\omega_2(a) = \omega(a)^2 \qquad\qquad \text{for } (a,N) = 1.$$

Then, for $n = tm^2$ with a positive integer m, we have

$$L_N(2s - 2\lambda, \omega_2)\alpha(n,s) = L_N(s - \lambda, \omega_1)\beta(n,s),$$

$$\beta(n,s) = \sum \mu(a)\omega_1(a)\omega_2(b)a^{\lambda - s}b^{k+2-2s},$$

where the last sum is extended over all positive integers a, b prime to N such that ab divides m, and μ denotes the Moebius function.

Proof. Let r be a square-free odd positive integer and u an odd positive integer. By Lemma 3,

$$G_n(ru^2) = \varepsilon_r r^{1/2} \sum_c c\mu(u^2/c)\left(\frac{u^2/c}{r}\right)\left(\frac{n/c}{r}\right) \quad (c\,|\,(u^2, n)).$$

Put $u^2 = ac$. Since $\mu(a) = 0$ unless a is square-free, we can put $u = ab$ with a positive integer b. Therefore

$$G_n(ru^2) = \varepsilon_r r^{1/2} \sum_{a,b} ab^2\mu(a)\left(\frac{n/b^2}{r}\right) \quad (ab = u,\ ab^2\,|\,n);$$

hence

$$\alpha(n, s) = \sum_{r,u} \left(\frac{-N}{ru^2}\right)\omega(ru^2)\mu(r)^2(ru^2)^{(k/2)-s}\varepsilon_r^{-k}G_n(ru^2)$$

$$= \sum_{r,a,b} \left(\frac{-N}{r}\right)\varepsilon_r^{1-k}\mu(r)^2\omega(ra^2b^2)(ra^2b^2)^{(k/2)-s}r^{1/2}ab^2\mu(a)\left(\frac{n/b^2}{r}\right)$$

$$((rab, N) = 1, ab^2\,|\,n).$$

Let $n = tm^2$. Then $ab^2\,|\,n$ only if $b\,|\,m$. Put $m = bh$. Then $a\,|\,th$ if $\mu(a) \neq 0$. Put $\omega'(r) = \left(\frac{-1}{r}\right)^{\lambda}\left(\frac{tN}{r}\right)\omega(r)$. Then we see that

$$\varepsilon_r^{1-k}\left(\frac{-N}{r}\right)\left(\frac{n/b^2}{r}\right)\omega(r) = \begin{cases} 0 & \text{if } (th, r) \neq 1, \\ \omega'(r) & \text{if } (th, r) = 1. \end{cases}$$

Therefore

$$\alpha(n, s) = \sum_{b|m} \omega(b)^2 b^{k+2-2s} \sum_{a|th} \mu(a)\omega(a)^2 a^{2\lambda-2s} \sum_{(r,th)=1} \mu(r)^2\omega'(r)r^{\lambda-s}.$$

Now we have

$$\sum_{a|th} \mu(a)\omega(a)^2 a^{2\lambda-2s} = \prod_{p|th} (1 - \omega(p)^2 p^{2\lambda-2s}),$$

$$\sum_{(r,th)=1} \mu(r)^2\omega'(r)r^{\lambda-s} = \prod_{p \nmid thN} (1 + \omega_1(p)p^{\lambda-s})$$

$$= \frac{L_N(s-\lambda, \omega_1)}{L_N(2s-2\lambda, \omega_2)} \prod_{\substack{p|th \\ p \nmid N}} \frac{1 - \omega_1(p)p^{\lambda-s}}{1 - \omega_2(p)p^{2\lambda-2s}}.$$

Note that $\omega_1(p) = 0$ if $p\,|\,t$ and $p \nmid N$. Therefore we obtain

$$L_N(2s - 2\lambda, \omega_2)\alpha(n, s) = L_N(s-\lambda, \omega_1)\sum_{b|m} \omega(b)^2 b^{k+2-2s} \prod_{\substack{p|h \\ p \nmid N}} (1 - \omega_1(p)p^{\lambda-s});$$

hence the desired result follows.

The case $n = 0$ is much simpler. In fact, if we denote Euler's function by φ, we have

$$(3.9) \qquad \alpha(0, s) = \sum_u \omega(u^2) u^{k-2s} \varphi(u^2)$$

$$= L_N(2s - k - 2, \omega_2)/L_N(2s - 2\lambda, \omega_2).$$

The above proposition shows that the Fourier coefficients $\alpha(tm^2, s)$ $(m = 1, 2, \ldots)$ with the same t are mutually related. We can express this fact by forming a new Dirichlet series as follows.

PROPOSITION 2. *Let the notation be as in Proposition 1, and S a complex variable. Then, for every primitive character θ modulo a positive integer P, we have, formally,*

$$\frac{L_N(2s - 2\lambda, \omega_2)}{L_N(s - \lambda, \omega_1)} \sum_{m=1}^{\infty} \alpha(tm^2, s)\theta(m)m^{s-S} = \frac{L_{NP}(S + s - k - 2, \theta\omega_2)L(S - s, \theta)}{L_{NP}(S - \lambda, \theta\omega_1)}.$$

Proof. Putting $m = abc$, we obtain

$$\sum_{m=1}^{\infty} \beta(tm^2, s)\theta(m)m^{s-S}$$

$$= \sum_{(a,N)=1} \mu(a)\theta(a)\omega_1(a)a^{\lambda-S} \sum_{(b,N)=1} \theta(b)\omega_2(b)b^{k+2-s-S} \sum_{c=1}^{\infty} \theta(c)c^{s-S}$$

$$= L_{NP}(S - \lambda, \theta\omega_1)^{-1} L_{NP}(S + s - k - 2, \theta\omega_2)L(S - s, \theta).$$

Remark. The right-hand side of the above equality has a symmetry under the transformation $s \mapsto k + 2 - s$, especially if $\theta = \bar{\omega}$. It should also be noted that the equality is similar to that of [8], Theorem 1.9. This point becomes clearer when s is specialized to 0. In fact, for an odd integer $\kappa \geqslant 5$, we have

$$E^*(z, 0, -\kappa, \omega) = \sum_{d=1}^{\infty} \sum_{b=-\infty}^{\infty} \omega(d)\left(\frac{-Nb}{d}\right)\varepsilon_d^{\kappa}(dz + b)^{-\kappa/2}.$$

This is an Eisenstein series of level N of weight $\kappa/2$. The Fourier expansion (3.3) specialized to $s = 0$ can be written as

$$E^*(z, 0, -\kappa, \omega) = (-2\pi i)^{\kappa/2}\Gamma(\kappa/2)^{-1} \sum_{n=1}^{\infty} a(n)e^{2\pi inz},$$

where $a(n) = \alpha(n, 0)n^{(\kappa/2)-1}$. Therefore, for a square-free t, we have, by Proposition 2,

$$(3.10) \qquad \sum_{m=1}^{\infty} a(tm^2)\theta(m)m^{-S} = t^{(\kappa/2)-1} \sum_{m=1}^{\infty} \alpha(tm^2, 0)\theta(m)m^{\kappa-2-S}$$

$$= t^{(\kappa/2)-1}\frac{L_N(\delta, \omega_1)}{L_N(\kappa - 1, \omega_2)} \frac{L_{NP}(S, \theta\omega_2)L(S - \kappa + 2, \theta)}{L_{NP}(S + 1 - \delta, \theta\omega_1)},$$

where $\delta = (\kappa-1)/2$. This was given in [9], pp. 70–71, without detailed proof. The reader is referred to papers by Maass ([3]) and Petersson ([5]) for the discussion of Fourier coefficients of the same type.

4. Analytic continuation and boundedness of E'

We first prove two lemmas.

LEMMA 4. *For every compact subset K of $\mathbf{C}^2$, there exist two positive constants A and B depending only on K such that*

$$|y^\beta \Gamma(\beta)^{-1}\sigma(y,\alpha,\beta)| \leqslant A.\max(y^{-B},1) \quad \text{if } (\alpha,\beta) \in K.$$

Proof. This follows immediately from (2.5). One can also prove the estimate by using (2.1) or (2.3) according to the magnitude of $\mathrm{Re}(\alpha-1)$ and $\mathrm{Re}(\beta)$, although this requires a somewhat lengthy computation.

LEMMA 5. *Let χ be a non-trivial primitive character modulo r, and let $R(s,\chi) = (r/\pi)^{(s+\nu)/2}\Gamma((s+\nu)/2)L(s,\chi)$, where $\nu = 0$ or 1 according as $\chi(-1) = 1$ or -1. Then, for any compact subset J of R, there exists a constant C_J independent of r and χ such that $|R(s,\chi)| \leqslant C_J r^{(|\sigma|/2)+2}$ for $\mathrm{Re}(s) = \sigma \in J$.*

Proof. If $\mathrm{Re}(s) = \sigma > 1$, we have

$$|R(1-s,\bar\chi)| = |R(s,\chi)| \leqslant (r/\pi)^{(\sigma+\nu)/2}\Gamma((\sigma+\nu)/2)\zeta(\sigma).$$

Therefore it is sufficient to prove the desired estimate for $-1 < \sigma < 2$. For this we use a well-known expression

$$R(s,\chi) = P(s,\chi)+c(\chi)P(1-s,\bar\chi),$$

$$P(s,\chi) = \int_1^\infty g(y,\chi)y^{(\nu+s-2)/2}\,dy,$$

where $g(y,\chi) = \sum_{n=1}^\infty \chi(n)n^\nu \exp(-\pi n^2 y/r)$ and $c(\chi)$ is a constant of absolute value 1. Observe that

$$|g(y,\chi)| \leqslant \sum_{n=1}^\infty ne^{-\pi n y/r} = e^{-\pi y/r}(1-e^{-\pi y/r})^{-2}.$$

Substituting tr/π for y, we obtain

$$|P(s,\chi)| \leqslant (r/\pi)^{(\sigma+\nu)/2}\int_{\pi/r}^\infty e^{-t}(1-e^{-t})^{-2}t^{(\sigma+\nu-2)/2}\,dt.$$

To prove our estimate, we may assume $r > \pi$. Decompose the last integral into two parts over the intervals $(1,\infty)$ and $(\pi/r, 1)$. The first part is a continuous function in σ independent of r and χ. As for the second part, we have $e^{-t}(1-e^{-t})^{-2} \leqslant At^{-2}$ for $0 \leqslant t \leqslant 1$ with a constant A, whence

$$\int_{\pi/r}^1 e^{-t}(1-e^{-t})^{-2}t^{(\sigma+\nu-2)/2}\,dt \leqslant A\int_{\pi/r}^1 t^{((\sigma+\nu)/2)-3}\,dt \leqslant B+Cr^{2-(\sigma+\nu)/2}$$

if $-1 \leqslant \sigma \leqslant 2$, with constants B and C independent of r and χ. Therefore $|P(s,\chi)| \leqslant Dr^2$ for $-1 \leqslant \sigma \leqslant 2$, with a constant D independent of r and χ. This completes the proof.

PROPOSITION 3. *For $z \in H$, $s \in \mathbf{C}$, put*

$$F'(z,s) = F'(z,s,k,\omega)$$

$$= \begin{cases} \Gamma(s/2)\Gamma((s-\lambda-\lambda_0+1)/2)L_N(2s-2\lambda,\omega_2)E'(z,s) & (k \geqslant -1), \\ \Gamma((s-k)/2)\Gamma((s-\lambda+\lambda_0)/2)L_N(2s-2\lambda,\omega_2)E'(z,s) & (k \leqslant 1), \end{cases}$$

where $\lambda = (k+1)/2$, and $\lambda_0 = 0$ or 1 according as λ is even or odd. Then $(s-\lambda-1)F'(z,s)$ can be continued to a holomorphic function in the whole s-plane. Moreover, for any compact subset K of $\mathbf{C}$, there exist two positive constants u and v depending on K such that

$$|(s-\lambda-1)F'(z,s)| \leqslant u(y^v + y^{-v}) \quad (y = \mathrm{Im}(z))$$

for all $s \in K$ and all $z \in H$. The factor $s-\lambda-1$ is unnecessary either if $(|k|+1)/2$ is even or if ω^2 is non-trivial.

Remark. If $k = \pm 1$, the two expressions for $F'(z,s)$ are identical.

Proof. Let us first consider the case $k \geqslant 1$. From the Fourier expansion for E', we obtain, by Proposition 1,

$$(4.1) \quad (2\pi/\sqrt{N})^{(k/2)-s}y^{-s/2}F'(z,s) = \sum_{n=-\infty}^{\infty} e^{2\pi(inx-|n|y)}|n|^{s-(k/2)-1}A(n,y,s),$$

where

$$A(n,y,s) = L_N(s-\lambda,\omega_1)\beta(n,s)\Gamma((s-\lambda-\lambda_0+1)/2)\Gamma((s-k)/2)^{-1}$$

$$\times \begin{cases} \sigma(4\pi ny,(s-k)/2,s/2) & (n > 0), \\ \sigma(4\pi|n|y,s/2,(s-k)/2) & (n < 0), \end{cases}$$

$$A(0,y,s) = L_N(2s-k-2,\omega_2)\Gamma((s-\lambda-\lambda_0+1)/2)\Gamma((s-k)/2)^{-1}$$

$$\times \Gamma(s-(k/2)-1)(4\pi y)^{1+(k/2)-s}.$$

Now we have

$$(4.2) \quad \Gamma((s-\lambda-\lambda_0+1)/2)\Gamma((s-k)/2)^{-1} = \begin{cases} 2^{-c}\prod_{a=1}^{c}(s-\lambda-\lambda_0+1-2a) & (k \geqslant 3), \\ 1 & (k = 1), \end{cases}$$

where $c = (\lambda-\lambda_0)/2$, and

$$\Gamma(s-(k/2)-1)L_N(2s-k-2,\omega_2)$$

$$= \Gamma(s-(k/2)-1)L(2s-k-2,\omega_2)\prod_{p|N}[1-\omega_2(p)p^{k+2-2s}].$$

Therefore we see that $A(0,y,s)$ is meromorphic on the whole s-plane. The only possible pole may occur at $s = \lambda + 1/2$ or $\lambda + 1$ when ω_2 is trivial. The pole at $s = \lambda + 1/2$ is cancelled by $1 - 2^{k+2-2s}$; the pole at $s = \lambda + 1$ is cancelled by the factor $s-\lambda-1$ of (4.2) if λ is even. Thus $(s-\lambda-1)A(0,y,s)$ is entire and bounded by $g(y^h+y^{-h})$ with constants g and h depending only on K. The factor $s-\lambda-1$ is unnecessary if λ is even or if ω_2 is non-trivial.

To study $A(n,y,s)$ for $n\neq 0$, first note that $\beta(n,s)$ is entire, and $|\beta(n,s)|\leqslant\gamma|n|^{\delta|\mathrm{Re}(s)|+\varepsilon}$, with constants $\gamma,\delta,\varepsilon$ independent of n. Let $n=tm^2$ as in Proposition 1, and let η be 0 or 1 according as $\omega_1(-1)=1$ or -1. Then $\lambda-\eta$ is even or odd according as n is positive or negative. Suppose $n>0$. Then

$$(4.3)\quad A(n,y,s)=\Gamma((s-\lambda+\eta)/2)\beta(n,s)L(s-\lambda,\omega_1)\prod_{p|N}(1-\omega_1(p)p^{\lambda-s})$$

$$\times\Gamma(s/2)^{-1}\sigma(4\pi ny,(s-k)/2,s/2)$$

$$\times\Gamma(s/2)\Gamma((s-\lambda+\eta)/2)^{-1}\Gamma((s-\lambda-\lambda_0+1)/2)\Gamma((s-k)/2)^{-1}.$$

Since $\lambda-\eta$ is even, the last line of factors is equal to 1 or

$$2^{-c-d}\prod_{b=1}^{d}(s-2b)\prod_{a=1}^{c}(s-\lambda-\lambda_0+1-2a),$$

where $d=(\lambda-\eta)/2$, according as $k=1$ or $k>1$. By Lemma 4, the second line of factors, when $s\in K$, is bounded by

$$C(4\pi ny)^{-\mathrm{Re}(s/2)}\max[1,(4\pi ny)^B],$$

with two constants B and C which depend on K but not on n. The first line of factors is holomorphic on the whole s-plane, except when ω_1 is trivial and $s=\lambda$ or $s=\lambda+1$. (This can happen only if λ is even.) But either pole is cancelled by the factor $1-2^{\lambda-s}$ or $s-\lambda-1$. Therefore $A(n,y,s)$ is entire, and by Lemma 5,

$$|A(n,y,s)|\leqslant un^v(y^w+y^{-w})$$

for $s\in K$, with constants u,v,w depending only on K.

Next suppose $n<0$. Then $\lambda-\eta$ is odd, whence $\lambda_0+\eta=1$. Therefore

$$A(n,y,s)=\Gamma((s-\lambda+\eta)/2)L(s-\lambda,\omega_1)\prod_{p|N}(1-\omega_1(p)p^{\lambda-s})\beta(n,s)$$

$$\times\Gamma((s-k)/2)^{-1}\sigma(4\pi|n|y,s/2,(s-k)/2).$$

By the same reasoning as above, we see that $(s-\lambda-1)A(n,y,s)$ is entire, and

$$|(s-\lambda-1)A(n,y,s)|\leqslant u'|n|^{v'}(y^{w'}+y^{-w'})$$

for $s\in K$ with constants u',v',w' depending only on K. The factor $s-\lambda-1$ is necessary only when ω_1 is trivial, which can happen if λ is odd and $\omega(m)=\left(\dfrac{-1}{m}\right)^{\lambda}\left(\dfrac{tN}{m}\right)$. Taking the infinite sum of (4.1), we obtain the desired result for $k\geqslant 1$.

The case $k < 0$ can be treated in a similar fashion. However, we can easily verify that

$$E(z, s, k, \omega) = y^{k/2}\bar{E}(z, \bar{s}-k, -k, \bar{\omega}),$$

$$E'(z, s, k, \omega) = y^{k/2}\bar{E}'(z, \bar{s}-k, -k, \bar{\omega}),$$

$$F'(z, s, k, \omega) = y^{k/2}\bar{F}'(z, \bar{s}-k, -k, \bar{\omega}).$$

By means of these formulas, we can reduce the case of negative k to that of positive k.

PROPOSITION 4. *For each* $\gamma = \begin{pmatrix} a & b \\ c & d \end{pmatrix} \in \mathrm{SL}_2(\mathbf{R})$, *put*

$$F'_\gamma(z, s) = F'(\gamma(z), s)(cz+d)^{k/2}$$

with the function F' *of Proposition 3. Then, for every compact subset* K *of* $\mathbf{C}$ *and every positive integer* ξ, *there exist constants* u *and* v *such that*

$$|(s-\lambda-1)F'_\gamma(x+iy, s)| \leqslant uy^v$$

for $s \in K$, $0 \leqslant x \leqslant \xi$, $y \geqslant 1$.

Proof. This is obvious if $c = 0$. Therefore assume $c \neq 0$. Since $\mathrm{Im}(\gamma(z)) = y/|cz+d|^2$, we have, by Proposition 3,

$$|(s-\lambda-1)F'_\gamma(z, s)| \leqslant 2uy^{-v}|cz+d|^{2v+(k/2)}$$

if $s \in K$ and $\mathrm{Im}(\gamma(z)) \leqslant 1$. Observe that $|cz+d|/cy \to 1$ as $z \to \infty$ under the condition $0 \leqslant \mathrm{Re}(z) \leqslant \xi$. Then our assertion is obvious.

5. Proof of Theorem 1

Let us now come back to the situation of §1. Then the character ω is defined by (1.3), and $k = 2w - 2v - 1$; hence $\lambda = w - v$. Since $w \geqslant 1$, we have $k \geqslant -1$. The case $k = -1$ can happen only when $w = v = 1$. Put

$$F(z, s) = \Gamma(s/2)\Gamma((s-\lambda-\lambda_0+1)/2)L_N(2s-2\lambda, \omega_2)E(z, s).$$

Then, from (1.5), we obtain

$$(4\pi)^{(-v-s)/2}\Gamma(s/2)\Gamma((s+1)/2)\Gamma((s-w+2-\lambda_0)/2)D(s)$$

$$= \int_\Phi f\bar{h}y^{v-(3/2)}F(z, s-v+1)\,dx\,dy \times \begin{cases} 1 & \text{if } Mr \text{ is even,} \\ [1 - \omega_2(2)2^{2w-2-2s}]^{-1} & \text{if } Mr \text{ is odd.} \end{cases}$$

Now we can find a finite subset T of $\mathrm{SL}_2(\mathbf{Z})$ and a positive number p so that $\Phi \subset \bigcup_{\gamma \in T} \gamma(U)$, where

$$U = \{x+iy : 0 \leqslant x \leqslant N, y \geqslant p\},$$

and $\{\gamma(\infty) : \gamma \in T\}$ is a set of representatives for $\Gamma_0(N)$-equivalence classes of cusps. Then the convergence of the last integral can be reduced to that of the integrals

$$(5.1) \qquad \int_{\gamma(U)} f\bar{h}y^{\nu-(3/2)}F(z, s-\nu+1)\,dx\,dy \qquad (\gamma \in T).$$

For each $\alpha = \begin{pmatrix} a & b \\ c & d \end{pmatrix} \in \mathrm{SL}_2(\mathbf{R})$, put

$$f_\alpha(z) = f(\alpha(z))(cz+d)^{-w}, \quad h_\alpha(z) = h(\alpha(z))(cz+d)^{-(2\nu+1)/2},$$

$$F_\alpha(z, s) = F(\alpha(z), s)(cz+d)^{k/2}.$$

Then (5.1) is equal to

$$(5.2) \qquad \int_U f_\gamma \bar{h}_\gamma y^{\nu-(3/2)}F_\gamma(z, s-\nu+1)\,dx\,dy.$$

Define $F'_\gamma(z, s)$ as in Proposition 4. Then we see that $F_\gamma(z, s)$ coincides with a constant times $F'_{\tau\gamma}(z, s)$ with $\tau = N^{-1/2}\begin{pmatrix} 0 & -1 \\ N & 0 \end{pmatrix}$. Since f_γ is a cusp form, $f_\gamma(x+iy) = O(e^{-ay})$ with a positive constant a. Therefore, by Proposition 4, we see that the integral (5.2) multiplied by $s-w$ is absolutely convergent for all $s \in \mathbf{C}$. This proves that the function $R(s)$ of Theorem 1 is a meromorphic function on the whole plane whose only possible poles are:

(1) $s = w$; this may happen only when $(|k|+1)/2$ is odd and ω^2 is the identity character;

(2) those s for which $1 = \omega_2(2)2^{2w-2-2s}$; this may happen only when Mr is odd.

Actually the poles in the latter case cannot happen except for $s = w-1$. More precisely we have

THEOREM 2. *The notation being the same as in Theorem 1, let r be the conductor of ψ. Then $R(s)$ has a pole at $s = w$ if and only if the following two conditions are satisfied:*

(i) *$\chi\psi$ is a non-trivial character of order 2;*

(ii) *$\int_{\Phi'} f\bar{g}y^{w-2}\,dx\,dy \neq 0$, where $g(z) = \sum_{n=1}^\infty \bar{\psi}(n)\bar{c}(n)e^{2\pi inz}$, and Φ' is a fundamental domain for $\Gamma_0(Mr^2)\backslash H$.*

The pole of $R(s)$ at $s = w-1$ occurs only when the following two conditions are satisfied:

(i') *Mr is odd and has a prime factor not dividing the conductor of $\chi\psi$;*

(ii') *$\chi(2)\psi(2) = \pm 1$.*

Remarks. (1) If $\chi^2\psi^2$ is trivial, we see, in view of the following lemma, that the integral of (ii) is the Petersson inner product of f and g. Since f and g are eigenfunctions of Hecke operators with eigenvalues $c(n)$ and

$\bar{\psi}(n)\bar{c}(n)$, respectively, (ii) implies

$$(5.3) \qquad \bar{c}(n) = \psi(n)c(n) \quad \text{for all } n \text{ prime to } Mr.$$

(2) If R has a pole at $s = w$, $(|k|+1)/2$ must be odd, that is, $w - \nu$ is odd or $w = \nu = 1$. Therefore, conditions (i), (ii) of Theorem 2 imply that $w - \nu$ is odd or $w = \nu = 1$.

To prove the above theorem, we need two lemmas.

LEMMA 6. *The function $g(z)$ defined in* (ii) *of Theorem 2 is a cusp form satisfying*

$$g(\gamma(z)) = \bar{\psi}(d)^2\bar{\chi}(d)(cz+d)^w g(z) \quad \text{for all } \gamma = \begin{pmatrix} a & b \\ c & d \end{pmatrix} \in \Gamma_0(Mr^2).$$

Proof. Put $f^*(z) = \bar{f}(-\bar{z})$. Then $f^*(z) = \sum_{n=1}^{\infty} \bar{c}(n)e^{2\pi i n z}$, and

$$f^*(\gamma(z)) = \bar{\chi}(d)(cz+d)^w f^*(z)$$

for all $\gamma = \begin{pmatrix} a & b \\ c & d \end{pmatrix} \in \Gamma_0(M)$. Our assertion follows immediately from this and [7], Proposition 3.64.

LEMMA 7. *Let ξ be a (primitive or an imprimitive) character modulo a positive integer A, and let*

$$H(z, s; \xi, A) = (y/\pi)^s\Gamma(s)\sum_{m,n} \xi(n)|Amz+n|^{-2s} \quad (z \in H,\ s \in \mathbf{C}),$$

where (m, n) runs over $\mathbf{Z}^2 \setminus (0, 0)$. Then H, as a function of s, can be continued to a meromorphic function on the whole plane. Moreover,

(i) *H is entire if ξ is not trivial,*

(ii) *if $A = 1$, H is holomorphic on the whole plane except for simple poles at $s = 0$ and $s = 1$,*

(iii) *if $A > 1$ and ξ is trivial, then H is holomorphic on the whole plane except for a simple pole at $s = 1$.*

This is well known (Rankin, [6], and Siegel, [10], for example). The case of imprimitive ξ can be reduced to that of primitive ξ. In fact, if B is the conductor of ξ, $A = BC$, and ξ_0 denotes the primitive character associated with ξ, then

$$(5.4) \qquad H(z, s; \xi, A) = \sum_{t|C} \mu(t)\xi_0(t)(tC)^{-s}H(t^{-1}Cz, s; \xi_0, B).$$

To see (iii), suppose that $A > 1$ and ξ is trivial. Note that $H(z, s; \xi_0, B)$ in this case has residues 1 and -1 at $s = 1$ and $s = 0$, respectively. Now we have

$$\sum_{t|A} \mu(t)t^{-s} = \prod_{p|A}(1-p^{-s}),$$

which is 0 at $s = 0$ and differs from 0 at $s = 1$; hence (iii) follows.

Now with g as in our theorem, we have

$$\int_0^1 f\bar{g}\,dx = \sum_{n=1}^{\infty} \psi(n)c(n)^2 e^{-4\pi n y};$$

hence, by (0.4),

$$L(2s-2w+2,\chi^2\psi^2)\int_0^{\infty}\left(\int_0^1 f\bar{g}\,dx\right)y^{s-1}\,dy = (4\pi)^{-s}\Gamma(s)L(s-w+1,\chi\psi)D(s)$$

when $\mathrm{Re}(s)$ is sufficiently large. By Lemma 6, we have

$$(f\bar{g}y^{s+1})\circ\gamma = \chi(d)^2\psi(d)^2|cz+d|^{2w-2-2s} \quad \text{for all } \gamma = \begin{pmatrix} a & b \\ c & d \end{pmatrix} \in \Gamma_0(Mr^2).$$

Therefore, putting

$$G(z,s) = \sum_{\gamma\in W'}\chi(d)^2\psi(d)^2|cz+d|^{-2s} \quad \left(\gamma = \begin{pmatrix} * & * \\ c & d \end{pmatrix}\right)$$

with a set W' of representatives for $\Gamma_{\infty}\backslash\Gamma_0(Mr^2)$, we obtain

$$\int_0^{\infty}\left(\int_0^1 f\bar{g}y^{s-1}\,dx\right)dy = \int_{\Phi'} f\bar{g}y^{s-1}G(z,s-w+1)\,dx\,dy$$

with a fundamental domain Φ' for $\Gamma_0(Mr^2)\backslash H$. Observe that

$$2(y/\pi)^s\Gamma(s)L(2s,\chi^2\psi^2)G(z,s) = H(z,s;\chi^2\psi^2,Mr^2),$$

hence

$$(5.5) \qquad 2\pi^{w-1-s}(4\pi)^{-s}\Gamma(s)\Gamma(s-w+1)L(s-w+1,\chi\psi)D(s)$$
$$= \int_{\Phi'} f\bar{g}y^{w-2}H(z,s-w+1;\chi^2\psi^2,Mr^2)\,dx\,dy$$

for sufficiently large $\mathrm{Re}(s)$. Now the last integral is absolutely convergent for all $s\in\mathbf{C}$ if the integrand is multiplied by $(s-w)(s-w+1)$. This can be shown by establishing the estimate of the function H similar to Propositions 3 and 4, by the same technique as in the proofs of these propositions or by the methods of [8], Lemma 3.3.

Call $R_1(s)$ the function expressed by (5.5). We know that R_1 is holomorphic on the whole plane except for possible simple poles at $s = w$ and $s = w-1$, which can occur only when $\chi^2\psi^2$ is trivial as explained in Lemma 7. Moreover, from the integral expression of (5.5), we see that the residues are non-vanishing only when

$$\int_{\Phi'} f\bar{g}y^{w-2}\,dx\,dy \neq 0.$$

Put $\eta(s) = \pi^{-s/2}\Gamma((s+\lambda_0)/2)L(s,\chi\psi)$ with $\lambda_0 = 0$ or 1 as in Theorem 1, and observe that $R_1(s)$ is a constant times $\eta(s-w+1)R(s)$. Suppose $\chi\psi$ is not trivial. Then $\eta(1)$ is finite and non-zero. Therefore R has a pole at

$s = w$ if and only if R_1 has a pole at $s = w$. If $\chi\psi$ is trivial, η has a pole at $s = 1$. Since the pole of R_1 at $s = w$ is simple, R has no pole at $s = w$. This proves the assertion of Theorem 2 concerning the pole at $s = w$.

To prove the remaining part, we may assume that Mr is odd. Then there is a possibility that R has a pole at a point s_0 such that $1 = \omega_2(2)2^{2w-2-2s_0}$. Put $\omega_2(2) = e^{2\pi i u}$. Then

$$s_0 = w - 1 - i\pi(m - u)/\log 2, \quad m \in \mathbf{Z}.$$

Let θ denote the primitive character associated with $\chi\psi$, and K the product of all prime factors of Mr not dividing the conductor of θ. Then

$$L(s, \chi\psi) = L(s, \theta) \prod_{p|K} [1 - \theta(p)p^{-s}].$$

It is well known that $\Gamma((s + \lambda_0)/2)L(s, \theta) \neq 0$ for $\mathrm{Re}(s) = 0$. Therefore if $\eta(s_0 - w + 1) = 0$, we have $1 = \theta(p)p^{w-1-s_0}$ for some $p | K$. Then we see that $(m - u)\log(p)/\log(2)$ is rational. This can happen only if $m = u$ since p is odd. Therefore $\eta(s_0 - w + 1) \neq 0$ unless $s_0 = w - 1$. Now $\eta(s - w + 1)R(s)$ has no pole if $\mathrm{Im}(s) \neq 0$, hence $s_0 = w - 1$ is the only possibility. If $K = 1$, we have $\eta(0) \neq 0$; hence R has a pole at $s = w - 1$ only if R_1 has a pole at $s = w - 1$, which can happen only when $Mr^2 = 1$. However, if $Mr^2 = 1$, $\eta(s - w + 1)$ has a pole at $s = w - 1$, and the pole of R at $s = w - 1$ produces a pole of R_1 at $s = w - 1$ of order greater than 1, a contradiction. This completes the proof of Theorems 1 and 2.

Let us now give an example of $D(s)$ which has a pole at $s = w$. Let $K = Q(\sqrt{(-\Delta)})$ be an imaginary quadratic field with discriminant $-\Delta$, and λ a primitive Grössen-character of K of conductor c such that $\lambda((\alpha)) = \alpha^{w-1}$ for $\alpha \in K, \alpha \equiv 1 \pmod{c}$, with an integer $w > 1$. Put $f(z) = \sum_{\mathfrak{a}} \lambda(\mathfrak{a})e^{2\pi i N(\mathfrak{a})z}$, where $N(\mathfrak{a})$ denotes the norm of $\mathfrak{a}$, and $\mathfrak{a}$ runs over all integral ideals in K prime to c. We can take this f as our cusp form of weight w with $M = N(c)\Delta$ and a character χ such that

$$\chi(n) = \left(\frac{-\Delta}{n}\right)\lambda_0(n), \quad \lambda_0(n) = \lambda((n))^{1-w} \text{ for } (n, M) = 1.$$

Now it can easily be verified that

$$(5.6) \qquad D(s) = L_M(s - w + 1, \psi\lambda_0) \sum_{\mathfrak{a}} \psi(N(\mathfrak{a}))\lambda(\mathfrak{a})^2 N(\mathfrak{a})^{-s}.$$

Therefore if $\psi\lambda_0$ is a trivial character, R has a pole at $s = w$.

It should be mentioned that our theorems are not formulated in the best possible form if $Mr > 1$. For example, in the expression (5.6), the characters $\psi\lambda_0$ and $(\psi \circ N_{K/Q})\lambda^2$ may not be primitive. In general, even if f is not related to a Grössen-character, it is necessary to modify $D(s)$ by some factors, which guarantee the functional equation in a natural form. This was remarked by Ogg ([4]) in the case of square-free level. Obviously the best formulation will be obtained in the framework of representations of GL(2) as given by Jacquet and Langlands ([1]) and Jacquet ([2]).

REFERENCES

1. H. Jacquet and R. P. Langlands, *Automorphic forms on* GL(2). Lecture notes in mathematics 114 (Springer, Berlin, 1970).
2. —— *Automorphic forms on* GL(2), *Part* II, Lecture notes in mathematics 278 (Springer, Berlin, 1972).
3. H. Maass, 'Konstruktion ganzer Modulformen halbzahliger Dimension mit ϑ-Multiplikatoren in einer und zwei Variablen', *Abh. Math. Sem. Univ. Hamburg* 12 (1937) 133–62.
4. A. P. Ogg, 'On a convolution of L-series', *Invent. Math.* 7 (1969) 297–312.
5. H. Petersson, 'Über die systematische Bedeutung der Eisensteinschen Reihen', *Abh. Math. Sem. Univ. Hamburg* 16 (1949) 104–30.
6. R. A. Rankin, 'Contributions to the theory of Ramanujan's function $\tau(n)$ and similar arithmetical functions, I, II', *Proc. Cambridge Philos. Soc.* 35 (1939) 351–72.
7. G. Shimura, *Introduction to the arithmetic theory of automorphic functions*, Publ. Math. Soc. Japan, No. 11 (Iwanami Shoten and Princeton University Press, 1971).
8. —— 'On modular forms of half integral weight', *Ann. of Math.* 97 (1973) 440–81.
9. —— 'Modular forms of half integral weight', *Proc. Int. Summer School on modular functions* (Antwerp, 1972); Lecture notes in mathematics 320 (Springer, Berlin, 1973), pp. 57–74.
10. C. L. Siegel, *Lectures on advanced analytic number theory* (Tata Institute of Fundamental Research, Bombay, 1961).
11. —— 'Die Funktionalgleichungen einiger Dirichletscher Reihen', *Math. Z.* 63 (1956) 363–73 (= Abh. III, 228–38).
12. —— 'A generalization of the Epstein zeta function', *J. Indian Math. Soc.* 20 (1956) 1–10 (= Abh. III, 239–48).

Princeton University

75b

On the real points of an arithmetic quotient
of a bounded symmetric domains

Mathematische Annalen, 215 (1975), 135-164

Introduction

As shown in some of our previous papers and also by K. Miyake, the quotient of a bounded symmetric domain by an arithmetic discontinuous group has in many cases a model defined over an algebraic number field and characterized by certain number-theoretical properties. It is natural to ask whether the model has rational points over a given number field. This is obviously a difficult question, but it turns out, as will be shown in this paper, that localization at an archimedean prime brings forth surprisingly strong results in the situations where a more number-theoretical approach such as non-archimedean localization, which looks more natural, may fail. Indeed, some of those models defined over the rational number field have no real points.

More generally, given an algebraic variety V defined over C, say projective non-singular, one can ask a few natural questions:

(1) Does V have a model over R?

(2) When V is actually defined over R, does V have a real point?

(3) How many connected components are there in the set V_R of all real points of V?

(4) How many of the components of V_R are homologically independent?

(5) What is the structure of each component as a topological or a real analytic manifold?

(6) How do the answers to these questions vary with the deformation of V?

Further questions may be asked about the real singularities of V, if one allows V to have such.

The study of V_R was suggested, if in a somewhat different vein, by Hilbert as the sixteenth of his problems, but few investigations have been made on this subject. The purpose of this paper is to present some new ideas and results in this field with emphasis on the number-theoretical aspect. Although our methods are applicable to varieties of a more general type, we content ourselves with treating only abelian varieties and quotients by discontinuous groups of symplectic type.

Let us now describe more precisely the groups and indicate some of our results. Let B denote, throughout the paper, a quaternion algebra over a totally real algebraic number field F of finite degree. We include the case where B is the total matrix algebra $M_2(F)$ of degree 2 over F. Put $g = [F:Q]$. For a positive integer n, define an algebraic subgroup G of GL_{4ng} defined over Q so that

$$G_Q = \{\alpha \in GL_n(B) \mid \alpha \cdot {}^t\alpha^\iota = \nu(\alpha)1_n \text{ with } \nu(\alpha) \in F^\times\},$$

653

where ι denotes the main involution of B. Put

$$G^u = \{\alpha \in G \mid \nu(\alpha) = 1\}.$$

Then $G^u_{\mathbf{R}}$ modulo a maximal compact subgroup can be identified with $\mathfrak{H}^r_n$, the product of r copies of the Siegel upper half space $\mathfrak{H}_n$ of degree n, where r is the number of archimedean primes of F unramified in B. Naturally we assume $r > 0$. If Γ is a discrete subgroup of G^u defined by some congruence conditions, we understand by a *model* of $\Gamma \backslash \mathfrak{H}^r_n$ a couple (V, φ) formed by a Zariski open subset V of a projective variety and a Γ-invariant holomorphic map φ of $\mathfrak{H}^r_n$ onto V that gives an isomorphism of $\Gamma \backslash \mathfrak{H}^r_n$ onto V. The existence of such a model is a special case of the theorem of Baily-Borel [1]. It was shown in [7, 9, 10] that $\Gamma \backslash \mathfrak{H}^r_n$ has a "canonical model" (V, φ) characterized by certain arithmetic properties. The variety V is defined over an abelian extension of another totally real algebraic number field F' determined by B. Now one of our main purposes is to give a necessary and sufficient condition for V to have real points. Obviously this must be preceded by a criterion for the field of definition for V to be real. Since the statements of our results in the most general case are lengthy, we give here only a typical result in a special case. Let $\mathfrak{o}$ be a maximal order in B, and let $\Gamma_1 = G^u \cap GL_n(\mathfrak{o})$. Suppose B is totally indefinite, i.e., $r = g$. Then the canonical model V for $\Gamma_1 \backslash \mathfrak{H}^g_n$ is defined over $\mathbf{Q}$, and actually a variety of moduli for a certain family of abelian varieties. Now we have

Theorem 0. *The variety V has real points if and only if either n is even or $B = M_2(F)$.*

Let us now give a brief summary of contents. After preliminary considerations in the first three sections, we shall prove various results of the above type in §§ 4, 5, 6. The condition for the existence of real points in the case $r < g$ involves arithmetic properties of the units of F. Then in §§7, 8, we study the structure of the set $V_{\mathbf{R}}$. Under certain conditions, we shall show that each connected component of $V_{\mathbf{R}}$ is an arithmetic quotient space associated with a unitary group of a quadratic extension K of F, which is not totally imaginary. If $n = 1$, this quotient space is an r-dimensional real torus, and the number of connected components of $V_{\mathbf{R}}$ can be expressed in terms of the class numbers of such extensions K. We shall determine in § 9 all $\mathbf{R}$-isomorphism classes of polarized abelian varieties defined over $\mathbf{R}$ by specifying the corresponding points on $\mathfrak{H}_n$ together with some additional data. Then a formula for the number of components of $A_{\mathbf{R}}$ for an abelian variety A will be given. The last § 10 concerns real $\varphi(z)$ for an isolated fixed point z of elements of $G_{\mathbf{Q}}$. Such a point $\varphi(z)$ is a generalization of a real singular value of the classical modular function j.

While it may be said that we touch here only the surface of the number-theoretical analysis of $V_{\mathbf{R}}$, the topological aspect also offers many interesting problems, on which little is discussed in the present paper.

Notation

We use, for the most part, the same notation as in [10]. For example, if S is an associative ring with identity element, $S^{\times}$ denotes the group of all invertible elements of S, $M_n(S)$ the matrix algebra over S of degree n, and $GL_n(S) = M_n(S)^{\times}$.

The identity element of $M_n(S)$ will be denoted by 1_n, often simply by 1. If K is an algebraic number field, $K_A^\times$ denotes the idele group of K, K_{ab} the maximal abelian extension of K, and K_c the kernel of the canonical isomorphism of $K_A^\times$ onto $\mathrm{Gal}(K_{ab}/K)$. If σ is an isomorphism of a field, the image of an element x by σ is denoted by x^σ, under the rule $(x^\sigma)^\tau = x^{\sigma\tau}$. The image of an object y (complex number, matrix, variety, etc.) by the complex conjugation is denoted by $\bar{y}$, with an exception $\overline{\mathscr{G}}_+$, which is the closure of a group $\mathscr{G}_+$.

The symbols F, B, g, r, G, G^u, ι, v, $\mathfrak{H}_n$ of the Introduction will be retained in the text.

1. Clarification of the Main Ideas

Define subgroups $Gp(n, R)$ and $Sp(n, R)$ of $GL_{2n}(R)$ by

$$Gp(n, R) = \{U \in GL_{2n}(R) \mid {}^tU J_n U = v(U) J_n \text{ with } v(U) \in R\},$$

$$Sp(n, R) = \{U \in Gp(n, R) \mid v(U) = 1\},$$

where

$$J_n = \begin{bmatrix} 0 & 1_n \\ -1_n & 0 \end{bmatrix}.$$

Let $\mathfrak{H}_n$ (resp. $\overline{\mathfrak{H}}_n$) denote the space of all symmetric complex matrices of degree n with positive (negative) definite imaginary part. Then for $z \in \mathfrak{H}_n \cup \overline{\mathfrak{H}}_n$ and $U = \begin{bmatrix} a & b \\ c & d \end{bmatrix} \in Gp(n, R)$ with a, b, c, $d \in M_n(R)$, we can define $U(z)$ by $U(z) = (az + b)(cz + d)^{-1}$. Obviously U maps $\mathfrak{H}_n$ onto $\overline{\mathfrak{H}}_n$, and $\overline{\mathfrak{H}}_n$ onto $\mathfrak{H}_n$ if $v(U) < 0$.

Let Γ be a discrete subgroup of $Sp(n, R)$ with no elements of finite order other than ± 1, such that $\Gamma \backslash \mathfrak{H}_n$ has a model (W, ψ). Further let $\Gamma^* = \Gamma \cdot \{\pm 1\}$, and

$$\Xi = \{\beta \in Gp(n, R) \mid v(\beta) = -1, \beta^{-1} \Gamma^* \beta = \Gamma^*, \beta^2 \in \Gamma^*\}.$$

Now our first observation is that if W is defined over R, the complex conjugation map on W can be obtained from an anti-holomorphic automorphism of $\mathfrak{H}_n$. More precisely we have

Proposition 1.1. *For each $\beta \in \Xi$, there is a model (V, φ) of $\Gamma \backslash \mathfrak{H}_n$ such that V is defined over R and $\overline{\varphi(z)} = \varphi(\beta(\bar{z}))$ for all $z \in \mathfrak{H}_n$. Conversely, if (V, φ) is a model of $\Gamma \backslash \mathfrak{H}_n$ and V is defined over R, there is an element β of Ξ such that $\varphi(\beta(\bar{z})) = \overline{\varphi(z)}$ for all $z \in \mathfrak{H}_n$. Moreover $\beta \mapsto (V, \varphi)$ gives a one-to-one correspondence between $\Gamma^* \backslash \Xi$ and the R-isomorphism classes of (V, φ).*

Especially $\Gamma \backslash \mathfrak{H}_n$ has a model rational over R if and only if $\Xi \neq \emptyset$.

Proof. Changing Γ for Γ^* if necessary, we may assume that $-1 \in \Gamma$. Take any model (W, ψ) of $\Gamma \backslash \mathfrak{H}_n$ and an element α of $Gp(n, R)$ such that $v(\alpha) = -1$. Define $\psi' : \mathfrak{H}_n \to \overline{W}$ by $\psi'(z) = \overline{\psi(\alpha(\bar{z}))}$. Then $(\overline{W}, \psi')$ is a model of $\alpha^{-1} \Gamma \alpha \backslash \mathfrak{H}_n$. If $\overline{W} = W$, there is an element γ of $Sp(n, R)$ such that $\gamma \Gamma \gamma^{-1} = \alpha^{-1} \Gamma \alpha$ and $\psi = \psi' \circ \gamma$. Put $\beta = \alpha\gamma$. Then $\overline{\psi(z)} = \psi(\beta(\bar{z}))$. We have $\psi(z) = \psi(\beta^2(z))$, so that $\beta^2 \in \Gamma$. Conversely let $\beta \in \Xi$. Define $\psi^* : \mathfrak{H}_n \to \overline{W}$ by $\psi^*(z) = \overline{\psi(\beta(\bar{z}))}$. Then $(\overline{W}, \psi^*)$ is a model of $\Gamma \backslash \mathfrak{H}_n$. Therefore there is an isomorphism λ of W onto $\overline{W}$ such that $\psi^* = \lambda \circ \psi$. Since $\beta^2 \in \Gamma$, we see that $\overline{\lambda} \circ \lambda = \mathrm{id}_W$. By Weil's criterion [13], there is a variety V

defined over R and an isomorphism μ of W onto V defined over C such that $\mu = \bar{\mu} \circ \lambda$. Put $\varphi = \mu \circ \psi$. Then (V, φ) is a model of $\Gamma \backslash \mathfrak{H}_n$ and $\varphi(\beta(\bar{z})) = \overline{\varphi(z)}$. This completes the proof except the last assertion, which can be verified in a straight-forward manner.

Actually we shall not need this proposition in our later discussion. But it will explain the following basic principle: *When (V, φ) corresponds to β as above, a point $\varphi(z)$ is real if and only if $z = \gamma\beta(\bar{z})$ for some $\gamma \in \Gamma^*$.* Therefore if we put, for $\alpha \in \Xi$,

$$X(\alpha) = \{z \in \mathfrak{H}_n \,|\, \alpha(\bar{z}) = z\},$$

then $V_{\boldsymbol{R}} = \bigcup\limits_{\alpha \in \Gamma^* \beta} \varphi(X(\alpha))$. What we shall do is an elaboration of this principle in the case where Γ is a congruence subgroup of G. The same idea is of course applicable to a more general type of variety or domain, often under a milder condition. In fact we have used the technique in [12, §3] for abelian varieties. This topic will be taken up again in §9.

Let us now prove the "only if"-part of Theorem 0 in the case $n = 1$. This is again unnecessary in our later proof of the general case, but will clarify a basic idea which may be obscured by minor technical details in the general treatment.

If $n = 1$, we have $G_{\boldsymbol{Q}} = B^{\times}$, $v(\alpha) = \alpha\alpha^{\iota}$ for $\alpha \in B$, and the group Γ_1 of the Introduction becomes

$$\Gamma_1 = \{\gamma \in \mathfrak{o} \,|\, v(\gamma) = 1\} ;$$

$\mathfrak{H}_1$ is the usual upper half plane. With r and g as in the Introduction, we can identify $B_{\boldsymbol{R}} = B \otimes_{\boldsymbol{Q}} R$ with $M_2(R)^r \times H^{g-r}$, where H denotes the Hamilton qua-ternion algebra. Let $\tau_1, \ldots, \tau_g$ denote the isomorphisms of F into R. We arrange them so that $\tau_1, \ldots, \tau_r$ correspond to the factors $M_2(R)$, and $\tau_{r+1}, \ldots, \tau_g$ to H. For $\alpha \in B$, let α_i denote the projection of α to the i-th factor $(M_2(R)$ or $H)$ of $B_{\boldsymbol{R}}$. Then we define the action of α on $(\mathfrak{H}_1 \cup \bar{\mathfrak{H}}_1)^r$ by

$$\alpha(z_1, \ldots, z_r) = (\alpha_1(z_1), \ldots, \alpha_r(z_r)) \quad (z_i \in \mathfrak{H}_1 \cup \bar{\mathfrak{H}}_1).$$

Now the general theory [10] tells that $\Gamma_1 \backslash \mathfrak{H}_1^r$ has a canonical model (V, φ), and V is defined over Q if $r = g$.

Proposition 1.2. *Let α be an element of $B^{\times}$ such that α^2 has a fixed point on $\mathfrak{H}_1^r$, but α itself has no fixed point on $(\mathfrak{H}_1 \cup \bar{\mathfrak{H}}_1)^r$. Then $\alpha^2 \in F$ and $\alpha^{\iota} = -\alpha$.*

Proof. By [7, Proposition 2.6], $F[\alpha^2] = F$, or $F[\alpha^2]$ is totally imaginary. Since α has no fixed point, $F[\alpha] \neq F[\alpha^2]$. Then $F[\alpha^2] = F$, since $[F[\alpha] : F] \leq 2$. Therefore $\alpha^2 = c$ with $c \in F$. If B is a division algebra, $F[\alpha]$ is a quadratic extension of F, hence $\alpha^{\iota} = -\alpha$. If $B = M_2(F)$, we see that $X^2 - c$ is the characteristic poly-nomial of α, so that $\mathrm{tr}(\alpha) = 0$ and $\alpha^{\iota} = -\alpha$.

Proposition 1.3. *Suppose B is totally indefinite, i.e., $r = g$. Let β be an element of $\mathfrak{o}$ such that $v(\beta) = -1$. Then $\varphi(\beta(\bar{z})) = \overline{\varphi(z)}$ for all $z \in \mathfrak{H}_1^g$.*

This is a special case of a theorem which will be discussed in later sections. The existence of an element β of $\mathfrak{o}$ such that $v(\beta) = -1$ is guaranteed by Lemma 1.4 below. It should also be mentioned that Proposition 1.3 is on a far deeper level than Proposition 1.1, because it concerns a specific model (V, φ) and a specific element β.

Now the proof of the "only if"-part of Theorem 0 in the case $n = 1$ can be given as follows. Suppose B is a totally indefinite division algebra, and $\varphi(w)$ is real for some $w \in \mathfrak{H}_1^g$. Then $\gamma\beta(\bar{w}) = w$ for some $\gamma \in \Gamma_1$. Put $\alpha = \gamma\beta$. Then $v(\alpha) = -1$, $\alpha(w) = \bar{w}$, $\alpha^2(w) = w$. By Proposition 1.2, $\alpha^\iota = -\alpha$, so that $\alpha^2 = -\alpha\alpha^\iota = 1$. Since B is a division algebra, we have $\alpha = \pm 1$, a contradiction.

We insert here a special case of Eichler's theorem [3, Satz 5] as

Lemma 1.4. *Let c be an algebraic integer of F such that $c^{\tau_i} > 0$ for all $i > r$. Then every maximal order in B contains an element d such that $v(d) = c$.*

2. The Behavior of Arithmetic Automorphic Functions under Anti-Holomorphic Automorphisms

Now our task is to generalize the above consideration to the case $n > 1$ or $r < g$. This naturally involves the complex conjugation in an algebraic number field. To discuss it in a general situation, let k be an algebraic number field of finite degree, and ξ an isomorphism of k into $\boldsymbol{R}$. Extend ξ to an isomorphism ξ' of k_{ab} into $\boldsymbol{C}$, and put $\eta = \xi' \cdot (\text{complex conjugation}) \cdot \xi'^{-1}$. Then η is an automorphism of k_{ab} of order 2. Moreover we can easily verify:

(2.1) *η depends only on ξ, and is independent of the choice of the extension ξ' of ξ to k_{ab};*

(2.2) *If x is an element of $k_\infty^\times$ whose ξ-component is negative and other components are positive, then η is the image of x under the canonical map $k_A^\times \to \mathrm{Gal}(k_{ab}/k)$.*

We call an abelian extension k_1 of k *real with respect to ξ* if η is the identity map on k_1, and η *the complex conjugation with respect to ξ.*

Let us now recall some important symbols introduced in [10]. Let F' be the field generated over $\boldsymbol{Q}$ by $\sum_{i=1}^{r} x^{\tau_i}$ for all $x \in F$. Then we can define a map $\lambda : F_A'^\times \to F_A^\times$ as follows. Let K be the Galois closure of F over $\boldsymbol{Q}$, and let H and H' be the subgroups of $\mathrm{Gal}(K/\boldsymbol{Q})$ corresponding to F and F', respectively. Extend τ_i to an element of $\mathrm{Gal}(K/\boldsymbol{Q})$ and denote it again by τ_i. Then $\lambda(x) = \prod_{j=1}^{s} x^{\sigma_j}$ with elements σ_j of $\mathrm{Gal}(K/\boldsymbol{Q})$ such that

$$(2.3) \qquad \bigcup_{i=1}^{r} \tau_i^{-1} H = \bigcup_{j=1}^{s} H' \sigma_j \qquad (\text{disjoint union}).$$

For details, see [7, § 5], [9, § 4.10], [10, I, § 2.2].

Here we may consider F as a subfield of $\boldsymbol{R}$ for convenience, but this is not absolutely necessary. However, F' is defined as a subfield of $\boldsymbol{R}$, because it is generated by $\sum_{i=1}^{r} x^{\tau_i}$ with embeddings τ_i of F into $\boldsymbol{R}$. Therefore we can speak of the complex conjugation and real subfields of F_{ab}' with respect to this natural embedding $F' \to \boldsymbol{R}$.

Let G_A, G_0, G_∞, and $G_{\infty+}$ denote the adelization of G, the non-archimedean part of G_A, the archimedean part of G_A, and the identity component of G_∞, respectively. For every $x \in G_A$, we denote by x_0 and x_∞ the projection of x to G_0 and to G_∞, respectively. We define subgroups G_{A+}, $\mathscr{G}_+$, and $\bar{\mathscr{G}}_+$ by

$$G_{A+} = G_0 G_{\infty+},$$
$$\mathscr{G}_+ = \{x \in G_{A+} \mid v(x) \in \lambda(F_A'^\times) F^\times F_{\infty+}^\times\},$$
$$\bar{\mathscr{G}}_+ = \{x \in G_{A+} \mid v(x) \in \lambda(F_A'^\times) F_c\}$$
$$= \text{the closure of } \mathscr{G}_+ \text{ in } G_A.$$

Let $\mathfrak{f}^*$ be the subfield of F_{ab}' corresponding to the subgroup

$$\{x \in F_A'^\times \mid \lambda(x) \in F_c\}.$$

Then we can define a homomorphism

$$\varrho : \bar{\mathscr{G}}_+ \to \mathrm{Gal}(\mathfrak{f}^*/F')$$

in a natural manner [10, II, 3.10].

Let us now denote by G_Q^{neg} the set of all elements α of G_Q such that $v(\alpha)^{\tau_i} < 0$ for all $i \leq r$. [Note that $v(\alpha)^{\tau_i} > 0$ for $i > r$ and for every $\alpha \in G_Q$.]

Proposition 2.1 (K.-y. Shih). *Let $\alpha \in G_Q^{\mathrm{neg}}$. Then α_0 (the projection of α to G_0) belongs to $\mathscr{G}_+$, and $\varrho(\alpha_0)$ is the complex conjugation on $\mathfrak{f}^*$.*

Proof. Let v be a negative element of F' whose all other conjugates are positive. Put $w = \lambda(v)$. Let H, H' and σ_j be as in (2.3). Then $w^{\tau_i} = \prod_{j=1}^{s} v^{\sigma_j \tau_i}$. If $v^{\sigma_j \tau_i} < 0$, then $\sigma_j \tau_i \in H'$, hence $H' \sigma_j = H' \tau_i^{-1}$. This cannot happen if $i > r$. Therefore $w^{\tau_i} > 0$ for $i > r$. If $i \leq r$, we have $H' \sigma_j = H' \tau_i^{-1}$ for a unique $j \leq s$, and $v^{\sigma_j \tau_i} > 0$ for all other j, hence $w^{\tau_i} < 0$. In brief, $w^{\tau_i} < 0$ or > 0 according as $i \leq r$ or $i > r$. Let $\alpha \in G_Q^{\mathrm{neg}}$. Then

$$v(\alpha_0) = v(\alpha) v(\alpha_\infty^{-1}) \lambda(v_\infty)^{-1} \lambda(v_\infty) \in F^\times F_{\infty+}^\times \lambda(F_A'^\times).$$

This shows $\alpha_0 \in \mathscr{G}_+$. According to our definition of ϱ, $\varrho(\alpha_0)$ is the restriction to $\mathfrak{f}^*$ of the image of v_∞^{-1} under the map of $F_A'^\times$ to $\mathrm{Gal}(F_{ab}'/F')$. Therefore, by (2.2), $\varrho(\alpha_0)$ is the complex conjugation on $\mathfrak{f}^*$.

Let $\mathfrak{R}$ be the field of arithmetic automorphic functions on $\mathfrak{H}_n^r$ with respect to G in the sense of [10, II, 6.1], whose definition we shall recall in §4. Then we have an exact sequence

$$1 \to F_c G_{\infty+} \to \bar{\mathscr{G}}_+ \xrightarrow{\tau} \mathrm{Aut}(\mathfrak{R}/F').$$

The map τ is actually surjective if $r = 1$; in general $\tau(\bar{\mathscr{G}}_+)$ is an open subgroup of finite index of $\mathrm{Aut}(\mathfrak{R}/F')$ (see [10, II, §6] and Miyake [6]). The field $\mathfrak{f}^*$ is contained in $\mathfrak{R}$, and $\tau(x) = \varrho(x)$ on $\mathfrak{f}^*$ for every $x \in \bar{\mathscr{G}}_+$.

Let $\alpha \in G_Q$. As shown in [9, 4.4], G_∞ has $Gp(n, \mathbf{R})^r$ as a factor. Let $(\alpha_1, \ldots, \alpha_r)$, with $\alpha_i \in Gp(n, \mathbf{R})$, be the projection of α to this factor. Then we can define the action of α on $(\mathfrak{H}_n \cup \bar{\mathfrak{H}}_n)^r$ by

$$\alpha(z_1, \ldots, z_r) = (\alpha_1(z_1), \ldots, \alpha_r(z_r)) \quad (z_i \in \mathfrak{H}_n \cup \bar{\mathfrak{H}}_n),$$

where each action α_i on $\mathfrak{H}_n \cup \bar{\mathfrak{H}}_n$ is defined as in §1. If $\alpha \in G_Q^{\mathrm{neg}}$, this maps $\mathfrak{H}_n^r$ onto $\bar{\mathfrak{H}}_n^r$, and $\bar{\mathfrak{H}}_n^r$ onto $\mathfrak{H}_n^r$. Now we can make the following

Conjecture. $h^{\tau(\alpha_0)}(z) = \overline{h(\alpha(\bar{z}))}$ *for all* $h \in \mathfrak{R}$, *all* $\alpha \in G_Q^{\mathrm{neg}}$, *and all* $z \in \mathfrak{H}_n^r$.

This is a generalization of [11, Theorem 6.38] which concerns the special case where $B = M_2(Q)$ and $n = 1$. It has been verified by K.-y. Shih and the author that the conjecture is "almost true" in the sense described by

Theorem A. *There exists an element t of $F_A^\times$ such that $t^2 = 1$, $\varrho(t) = 1$, and*

$$(2.4) \qquad\qquad h^{\tau(t\alpha_0)}(z) = \overline{h(\alpha(\bar{z}))}$$

for all $\alpha \in G_Q^{\mathrm{neg}}$, *all* $h \in \mathfrak{R}$, *and all* $z \in \mathfrak{H}_n^r$. *Moreover if B is totally indefinite, t can be taken to be* 1.

As we shall see later, Proposition 1.3 is a special case of (a paraphrase of) this theorem. The detailed proof will appear in a forthcoming paper by Shih.

3. Propositions Concerning the Elements of G_Q^{neg}

Let us now generalize Proposition 1.2 to the case $n > 1$, and study the algebraic nature of the elements of G_Q^{neg}. Put $X = B^n$, and consider an element of X as a row vector with components in B. Then X becomes a left B-module and a right $M_n(B)$-module in a natural manner. Further put

$$h(x, y) = x \cdot {}^t y^\iota \quad \text{for} \quad (x, y) \in X \times X .$$

Then G_Q consists of all $\alpha \in GL_n(B)$ such that $h(x\alpha, y\alpha) = v(\alpha) h(x, y)$ with $v(\alpha) \in F$. If Y is a B-submodule of X on which h is non-degenerate, then there is an isomorphism p of Y onto B^m for some m such that

$$h(u, v) = p(u) \cdot {}^t p(v)^\iota \quad \text{for all} \quad (u, v) \in Y \times Y .$$

This follows easily from the Hasse principle, as remarked in [10, I, 2.1]. (This is so even if $B = M_2(F)$. See [9, 3.2].) We put $\alpha^\sigma = {}^t \alpha^\iota$ for every $\alpha \in M_n(B)$. Then

$$h(x\alpha^\sigma, y) = h(x, y\alpha) \quad \text{for all} \quad (x, y) \in X \times X .$$

Lemma 3.1. *Let Y be a subalgebra of $M_n(B)$ with the same identity element as $M_n(B)$, and let $S = \{\alpha \in Y \otimes_Q R \mid \alpha\alpha^\sigma = 1\}$. If σ induces a positive involution on Y, then the elements of S have a common fixed point on $\mathfrak{H}_n^r$.*

Proof. Put $Y_R = Y \otimes_Q R$, and let Tr denote the reduced trace from Y_R to R. If σ is positive on Y, $\mathrm{Tr}(xx^\sigma)$ for $x \in Y_R$ is a positive definite quadratic form. It follows that S is compact, and therefore contained in a maximal compact subgroup of G_R^u, hence our assertion.

Proposition 3.2. *Let c be an element of F such that $c^{\tau_i} > 0$ for $i > r$ and $c^{\tau_j} < 0$ for some $j \leq r$. Then G_Q has an element α such that $v(\alpha) = c$ and α^2 has a fixed point on $\mathfrak{H}_n^r$ if and only if one of the following conditions is satisfied:*
(i) *n is even.*
(ii) *B has an element u such that $u^\iota = -u$ and $u^2 = -c$.*

Remark. The second condition is satisfied if $B = M_2(F)$. In fact, we can put

$$u = \begin{bmatrix} 0 & -1 \\ c & 0 \end{bmatrix}.$$

Proof. Let us first prove the "if"-part. Since $c^{r_i} > 0$ for $i > r$, we find an element v of B such that $vv^\iota = c$. Suppose n is even, and put

$$\alpha = \begin{bmatrix} 0 & v1_m \\ v^\iota 1_m & 0 \end{bmatrix}, \quad m = n/2.$$

Then $\alpha^2 = \alpha \cdot {}^\iota\alpha^\iota = c1_n$, hence $\alpha \in G_Q$, $v(\alpha) = c$, and α^2 has a fixed point on $\mathfrak{H}_n^r$. Next take an element u as in (ii). Put $\alpha = u1_n$. Then $\alpha \in G_Q$, $v(\alpha) = c$, and $\alpha^2 = -c1_n$.

To prove the converse part, let $\alpha \in G_Q$ with $v(\alpha) = c$, and suppose α^2 has a fixed point on $\mathfrak{H}_n^r$. If n is even or $B = M_2(F)$, the converse is trivially true. Therefore let us assume that n is odd and B is a division algebra. Our task is to show the existence of an element u of B as in (ii). Put $\alpha^2 = \beta$. Then $v(\beta) = c^2$, and σ gives a positive involution of $F[\beta]$, by virtue of [9, Proposition 4.5]. Therefore both $F[\alpha]$ and $F[\beta]$ are semi-simple. Let $\varepsilon_1, \ldots, \varepsilon_s$ be the mutually orthogonal idempotents of $F[\alpha]$ such that $F[\alpha] = F[\alpha]\varepsilon_1 + \cdots + F[\alpha]\varepsilon_s$, and $F[\alpha]\varepsilon_i$ is a field for each i. Now σ gives a permutation of $\varepsilon_1, \ldots, \varepsilon_s$ of order 1 or 2. In general, for every $\gamma \in M_n(B)$, $X\gamma$ and $X\gamma^\sigma$ have the same rank over B. Therefore there exists an i such that $X\varepsilon_i$ has an odd rank over B and $\varepsilon_i^\sigma = \varepsilon_i$. We may assume that this is so for $i = 1$. Put $\alpha_1 = \alpha\varepsilon_1$. Note that h is non-degenerate on $X\varepsilon_1$. Changing the coordinate system of X by an element of G_Q^u, we may assume that $\alpha = \begin{bmatrix} \alpha_1 & 0 \\ 0 & \alpha' \end{bmatrix}$ with $\alpha_1 \in M_m(B)$ and $\alpha' \in M_{n-m}(B)$, where m is the rank of $X\varepsilon_1$ over B. (Note that σ commutes with an inner automorphism of $M_n(B)$ by an element of G_Q.) Since σ gives a positive involution on $F[\alpha_1^2]$, we see from Lemma 3.1 that α_1^2 has a fixed point on $\mathfrak{H}_m^r$. Considering α_1 instead of α, we can now assume that $F[\alpha]$ is a field. Suppose $F[\alpha] = F[\beta]$. Then σ is positive on $F[\alpha]$. This is a contradiction, since $\alpha\alpha^\sigma = c$, and c is not totally positive. Therefore $F[\alpha]$ is a quadratic extension of $F[\beta]$. Suppose $\beta^\sigma \neq \beta$. Then $[F[\alpha] : F]$ is divisible by 4. Consider X as a module over $B \otimes_F F[\alpha]$ by the action $(a \otimes b)x = axb$ for $a \in B$, $b \in F[\alpha]$, and $x \in X$. Every irreducible module over $B \otimes_F F[\alpha]$ is of dimension $4[F[\alpha] : F]$ or $2[F[\alpha] : F]$ over F. Therefore the dimension of X over F is divisible by 8, so that X is of an even rank over B, a contradiction. Therefore $\beta^\sigma = \beta$, and $\beta^2 = \beta\beta^\sigma = c^2$. It follows that $\alpha^2 = \pm c$, and so $F[\alpha]$ is a quadratic extension of F. Consider again X as a module over $B \otimes_F F[\alpha]$. Since X is of odd rank over B, $B \otimes_F F[\alpha]$ cannot be a division algebra. Therefore B contains an element u such that $u^\iota = -u$ and $u^2 = c$ or $-c$ according as $\alpha^2 = c$ or $-c$. Since $F[\alpha]$ and $F[u1_n]$ are isomorphic subfields of $M_n(B)$, there exists an element γ of $GL_n(B)$ such that $\gamma\alpha\gamma^{-1} = u1_n$. Let N denote the reduced norm of $M_n(B)$ to F. Then $N(\alpha) = N(u1_n) = (uu^\iota)^n$. If $\alpha^2 = c$, we have $uu^\iota = -u^2 = -c$, so that $N(\alpha) = -c^n$, since n is odd. But this contradicts the following lemma. Therefore $u^2 = -c$, and the proof is completed.

Lemma 3.3. *If N denotes the reduced norm of $M_n(B)$ to F, then $N(\alpha) = v(\alpha)^n$ for every $\alpha \in G_Q$.*

Proof. Let K be the algebraic closure of F. Then we can find an F-linear injection p of $M_n(B)$ into $M_{2n}(K)$ such that $p({}^\iota\xi^\iota) = J_n^{-1} \cdot {}^t p(\xi) J_n$. (This is easy if $n = 1$. The general case can be seen by rearranging matrices.) Let $\alpha \in G_Q$, $v(\alpha) = a^2$ with $a \in K$, and $\beta = a^{-1}p(\alpha)$. Then ${}^t\beta J_n\beta = J_n$. It is well known that $\det(\beta) = 1$. Therefore $N(\alpha) = \det(p(\alpha)) = a^{2n} = v(\alpha)^n$.

We can obtain a canonical form of an element α of the above type under certain conditions:

Proposition 3.4. *Let c be an element of F such that $c^{\tau_i} > 0$ for all $i > r$ and $c^{\tau_i} < 0$ for all $i \leq r$. Let α be an element of G_Q such that $v(\alpha) = c$ and $\alpha^2 \in F^\times$. Then $\alpha^\sigma = \pm\alpha$, and the following assertions hold.*

(I) If $\alpha^\sigma = \alpha$, then n is even. Moreover, for any element b such that $bb^\iota = c$, there exists an element γ of G_Q^u such that

$$\gamma\alpha\gamma^{-1} = \begin{bmatrix} 0 & b1_m \\ b^\iota 1_m & 0 \end{bmatrix} \quad (m = n/2).$$

(II) Suppose $\alpha^\sigma = -\alpha$ and n is odd. Then B has an element u such that $u^2 = -c$ and $u^\iota = -u$. Moreover, for any such element u, there exists an element δ of G_{Q+} such that $\delta\alpha\delta^{-1} = u1_n$.

Remark. Since $c^{\tau_i} > 0$ for $i > r$, B has an element b such that $bb^\iota = c$.

Proof. Since $\alpha^2 \in F$, we have $(\alpha^2)^2 = v(\alpha^2) = c^2$, hence $\alpha^2 = \pm c$ and $\alpha^\sigma = \pm\alpha$.

Case I. $\alpha^\sigma = \alpha$, $\alpha^2 = c$. In this case $F[\alpha]$ is a quadratic extension of F, since $c^{\tau_i} < 0$ for $i \leq r$. Put $K = F[\alpha]$ and $A = B \otimes_F K$. Denote the main involution of A also by ι. Consider X as an A-module by the action $(b \otimes k)x = bxk$ for $b \in B$, $k \in K$ and $x \in X$. Put $f(x, y) = \mathrm{Tr}_{B/F}(h(x, y))$. Since $\alpha^\sigma = \alpha$, we see that $f(ax, y) = f(x, a^\iota y)$ for every $a \in A$. Therefore, by [8, 1.6], there exists an A-valued ι-hermitian form $g(x, y)$ on X such that $2\mathrm{Tr}_{B/F}(h(x, y)) = \mathrm{Tr}_{A/F}(g(x, y))$. Here Tr denotes the reduced trace. It can easily be seen that g is non-degenerate. Moreover an archimedean prime ω of K is ramified in A if and only if ω is an extension of τ_i with $i > r$. Then $\mathrm{Tr}_{K/F}(g(x, x)k^2)^{\tau_i} = 2h(xk, xk)^{\tau_i} > 0$ for all $0 \neq x \in X$ and for all $k \in K^\times$. This shows that $g(x, x)^\omega > 0$. Therefore, by the Hasse principle [10, I, (2.1.4)], we can find an A-basis $\{x_1, \ldots, x_m\}$ of X such that $g(x_i, x_j) = \delta_{ij}$, where $m = n/2$. As remarked above, this is so even if $A = M_2(K)$. With an element b of B such that $bb^\iota = c$, put $x_{i+m} = b^{-1}x_i\alpha$ for $i = 1, \ldots, m$. Then $\{x_1, \ldots, x_n\}$ is a B-basis of X and $h(x_i, x_j) = \delta_{ij}$. With respect to this basis, α is represented by

$$\begin{bmatrix} 0 & b1_m \\ b^\iota 1_m & 0 \end{bmatrix}.$$

This proves the assertion (I).

Case II. $\alpha^\sigma = -\alpha$, $\alpha^2 = -c$. The last part of the proof of Proposition 3.2 shows the existence of u and an element γ of $GL_n(B)$ such that $\gamma^{-1}\alpha\gamma = u1_n$, if B is a division algebra. This holds also when $B = M_2(F)$ with $u = \begin{bmatrix} 0 & -1 \\ c & 0 \end{bmatrix}$, if $F[\alpha]$ is a field. Assuming this, put $K = F[u]$ and $\beta = u1_n$. Then $\alpha\gamma = \gamma\beta$ and $\gamma^\sigma\alpha = \beta\gamma^\sigma$, so that $\gamma^\sigma\gamma$ commutes with β, hence $\gamma^\sigma\gamma \in M_n(K)$. Here we consider $M_n(K)$ as a subalgebra of $M_n(B)$ through the natural injection $K \to B$. Put $\xi = \gamma^\sigma\gamma$. Then $\xi^\sigma = \xi$, so that ξ is a hermitian matrix in $M_n(K)$. We see that ξ is positive definite at τ_i for $i > r$. Put $v = \det(\xi)$. Then $v^{\tau_i} > 0$ for $i > r$. Since n is odd,

$$\det(v^{-1}\xi) = v^{1-n} \in N_{K/F}(K^\times).$$

Applying the Hasse principle to ξ and $v1_n$, we find an element η of $GL_n(K)$ such that $\eta^\sigma \xi \eta = v1_n$. Now we can find an element $w \in K$ such that $N_{K/F}(w)v$ is totally positive. Put $\delta = \gamma\eta \cdot w1_n$. Then $\delta \in G_{Q+}$ and $\delta^{-1}\alpha\delta = u1_n$ as desired.

It remains to consider the case in which $B = M_2(F)$ and $F[\alpha]$ is not a field. Then $c = -a^2$ with $a \in F$. Since $\alpha^\sigma = -\alpha$, we see that the characteristic roots a and $-a$ of α appear with the same multiplicity. Therefore we obtain again an element γ of $GL_n(B)$ such that $\gamma^{-1}\alpha\gamma = u1_n$ with $u = \begin{bmatrix} 0 & -1 \\ c & 0 \end{bmatrix}$ $\left(\text{or } u = \begin{bmatrix} a & 0 \\ 0 & -a \end{bmatrix} \right)$. Two "non-degenerate hermitian forms" over $K = F[u] = F \oplus F$ of n variables are always equivalent, and therefore we find $\eta \in GL_n(K)$ such that $\eta^\sigma \xi \eta = 1_n$ (even if n is even). Then $\delta = \eta\gamma$ has the required property.

Remark 3.5. As the above proof shows, the assertion (II) of Proposition 3.4 holds also when n is even, if $B = M_2(F)$ and $c = -a^2$ with $a \in F$.

4. The Real Points on Canonical Models in the General Case

Let $\mathscr{L}^*$ denote the set of all open subgroups S of $\bar{\mathscr{G}}_+$ containing $F_c G_{\infty+}$ such that $S/F_c G_{\infty+}$ is compact. For each $S \in \mathscr{L}^*$, put $\Gamma_S = G_Q \cap S$. Then as in [10, II], we obtain a "canonical model" (V_S, φ_S) of $\Gamma_S \backslash \mathfrak{H}_n^r$. Recall that V_S is defined over the subfield k_S of F'_{ab} corresponding to the subgroup

$$\{u \in F'^\times_A \mid \lambda(u) \in v(S)F_c\}$$

of $F'^\times_A$.

Let K_S denote the field of all functions on V_S, in the sense of algebraic geometry, rational over k_S, and let

$$\mathfrak{R}_S = \{f \circ \varphi_S \mid f \in K_S\}.$$

Then the union of the fields $\mathfrak{R}_S$ for all $S \in \mathscr{L}^*$ is the field $\mathfrak{R}$ mentioned in § 2. Recall that

$$S = \{x \in \bar{\mathscr{G}}_+ \mid \tau(x) = \text{id. on } \mathfrak{R}_S\},$$

and $\mathrm{Gal}(\mathfrak{R}/\mathfrak{R}_S)$ is isomorphic to $\tau(S)$. In view of our definition of $\tau: \bar{\mathscr{G}}_+ \to \mathrm{Aut}(\mathfrak{R}/F')$ in [10, II, 6.1], formula (2.4) is equivalent to

(4.1) *If $\alpha \in G_Q^{\mathrm{neg}}$, $S \in \mathscr{L}^*$, and $T = \alpha^{-1}S\alpha$, then $\overline{\varphi_S(\alpha(\bar{z}))} = J_{ST}(t\alpha_0)[\varphi_T(z)]$ for all $z \in \mathfrak{H}_n^r$.*

We put, for each $S \in \mathscr{L}^*$,

$$\Gamma_S^{\mathrm{neg}} = \{\alpha \in G_Q^{\mathrm{neg}} \mid \alpha_0 \in S\},$$

where α_0 means the projection of α to the non-archimedean part G_0 of G_A. Obviously $\Gamma_S^{\mathrm{neg}} = \alpha\Gamma_S$ with any $\alpha \in \Gamma_S^{\mathrm{neg}}$, if $\Gamma_S^{\mathrm{neg}} \neq \emptyset$.

Proposition 4.1. *The field k_S is real if and only if $\Gamma_S^{\mathrm{neg}} \neq \emptyset$.*

Proof. Since $\varrho(x) = \text{id. on } k_S$ for every $x \in S$, the "if"-part follows from Proposition 2.1. To prove the converse, let v be a negative element of F' whose all other conjugates are positive. Suppose k_S is real. Then, by (2.2), $\lambda(v_\infty) \in v(S)F_c$. By [10, II, Lemma 2.2], we have $\lambda(v_\infty) = v(s)abc$ with $s \in S$, $a \in \bar{E}_0$, $b \in F^\times$, $c \in F^\times_{\infty+}$, where $\bar{E}_0$

is defined in that lemma. We may assume that $s_\infty = 1$. As shown in the proof of Proposition 2.1, $\lambda(v)^{r_i} < 0$ or >0 according as $i \leq r$ or $i > r$. Then we see that $b^{r_i} > 0$ or <0 according as $i > r$ or $i \leq r$. Therefore we can find an element β of G_Q^{neg} such that $v(\beta) = b$. Then $v(s\beta)_0 a = 1$. This means that for any open compact subgroup W of the finite part of $F_A^\times$, we can find a totally positive unit e of F such that $(v(s\beta) e)_0 \in W$. Since $e \in v(G_{Q+})$, this means that for any such W, there exists an element α of G_Q^{neg} such that $v(s\alpha)_0 \in W$. Let T be an open compact subgroup of G_0 such that $T \cap \bar{\mathscr{G}}_+ \subset S$. Observe that $v(T) F_{\infty+}^\times$ is an open subgroup of $F_A^\times$. Therefore we can find, an element α of G_Q^{neg} such that $v(s\alpha)_0 \in v(T)$, i.e., $v(s\alpha)_0 = v(x)$ for some $x \in T$. By [10, I, Proposition 3.4], we have

$$G_A^u = G_Q^u (G_A^u \cap T G_{\infty+}),$$

so that $(\alpha s x^{-1})_0 = \gamma u$ with $\gamma \in G_Q^u$ and $u \in G_A^u \cap T G_{\infty+}$. Since $G_A^u \subset \bar{\mathscr{G}}_+$ and $(s\alpha)_0 \in \bar{\mathscr{G}}_+$, we see that $(ux)_0 \in \bar{\mathscr{G}}_+ \cap T$, and so $(\gamma^{-1}\alpha)_0 = (uxs^{-1})_0 \in S$. Therefore $\gamma^{-1}\alpha \in \Gamma_S^{\mathrm{neg}}$, which completes the proof.

Let t be the element of $F_A^\times$ as given in Theorem A. We call a member S of $\mathscr{L}^*$ *real* if it satisfies the following set of conditions:

$$(4.2) \qquad\qquad \text{(i)} \quad t \in S; \quad \text{(ii)} \quad \Gamma_S^{\mathrm{neg}} \neq \emptyset.$$

By the above proposition, k_S is real if S is real. Now Proposition 1.3 can be generalized as follows.

Proposition 4.2. *Let S be a real member of $\mathscr{L}^*$. Then $\overline{\varphi_S(z)} = \varphi_S(\alpha(\bar{z}))$ for all $z \in \mathfrak{H}_n^r$ and all $\alpha \in \Gamma_S^{\mathrm{neg}}$.*

This is an immediate consequence of (4.1). Thus Γ_S^{neg} represents the anti-holomorphic automorphisms of $\mathfrak{H}_n^r$ which give the complex conjugation map on V_S.

Let us denote by $V_S(R)$ the set of all real points of V_S. For each $\alpha \in G_Q^{\mathrm{neg}}$, put

$$(4.3) \qquad\qquad X(\alpha) = \{z \in \mathfrak{H}_n^r \mid \alpha(\bar{z}) = z\}.$$

A point of $X(\alpha)$ may be called a *false fixed point* of α. Further put

$$(4.4) \qquad\qquad \Delta_S = \{\alpha \in \Gamma_S^{\mathrm{neg}} \mid X(\alpha) \neq \emptyset\}.$$

It can happen that Δ_S is empty even if Γ_S^{neg} is not empty. Now the principle explained in § 1 together with Proposition 4.2 implies

Proposition 4.3. *Let S be a real member of $\mathscr{L}^*$. Then*

$$V_S(R) = \bigcup_{\alpha \in \Delta_S} \varphi_S(X(\alpha)).$$

Especially V_S has a real point if and only if Δ_S is not empty.

Proposition 4.4. *Let S be a real member of $\mathscr{L}^*$. Suppose that n is odd, and*

$$v(\Gamma_S^{\mathrm{neg}}) \cap \{-u^2 \mid u \in B, u^\iota = -u\} = \emptyset.$$

Then V_S has no real points.

Proof. If V_S has a real point, we have $\alpha(\bar{z}) = z$ for some $z \in \mathfrak{H}_n^r$ and some $\alpha \in \Gamma_S^{\mathrm{neg}}$. Then α^2 has z as a fixed point. Therefore our assertion follows immediately from Proposition 3.2.

5. The Case of Level 1

To obtain a more transparent condition, let us consider an S of a more specific type. Let $\mathfrak{o}$ be a maximal order in B, and let

$$L = \mathfrak{o}^n = \{(x_1, \ldots, x_n) \in B^n \mid x_i \in \mathfrak{o}\},$$

(5.0) $$U_L = \{y \in G_A \mid Ly = L\},$$

(5.1) $$S = F_c \cdot (U_L \cap \bar{\mathscr{G}}_+).$$

By [10, II, 3.6, 3.7], we have $S = F^\times \cdot (U_L \cap \bar{\mathscr{G}}_+)$, and $\Gamma_S = F^\times \cdot (G_{Q+} \cap GL_n(\mathfrak{o}))$. We can also easily verify

(5.2) $$\Gamma_S^{\text{neg}} = F^\times \cdot (G_Q^{\text{neg}} \cap GL_n(\mathfrak{o})).$$

Theorem 5.1. *Let E_θ denote the group of all units e of F such that $e^{\tau_i} > 0$ for $i \leq r$ and $e^{\tau_i} < 0$ for $i > r$. Then k_S is real if and only if E_θ is non-empty. Suppose k_S is real. Then the following assertions hold.*

(I) If n is even or $B = M_2(F)$, V_S has a real point.

(II$_a$) Suppose n is odd and B is a division algebra. If V_S has a real point, B splits over $F(e^{1/2})$ for some $e \in E_\theta$.

(II$_b$) Suppose n is odd and B is a division algebra. If $\mathfrak{o}$ has an element δ such that $\delta^\iota = -\delta$ and $\delta^2 \in E_\theta$, then V_S has a real point.

Proof. Let t be as in Theorem A. Since $t \in F_A^\times$ and $t^2 = 1$, we see that $t \in U_L \cap \bar{\mathscr{G}}_+$, hence $t \in S$. Let us now prove

(5.3) $$v(\Gamma_S^{\text{neg}}) = \{-c^2 e \mid c \in F^\times, e \in E_\theta\}.$$

The inclusion $\subset$ follows immediately from (5.2). To prove the opposite inclusion, let $e \in E_\theta$. By Lemma 1.4, $\mathfrak{o}$ has an element δ such that $\delta\delta^\iota = -e$. Put $\alpha = \delta 1_n$. Then, for every $c \in F^\times$, we have $c\alpha \in \Gamma_S^{\text{neg}}$ and $v(c\alpha) = -c^2 e$. This proves (5.3). Therefore, by Proposition 4.1, k_S is real if and only if $E_\theta \neq \emptyset$. Suppose n is odd, $E_\theta \neq \emptyset$, and V_S has a real point. By Proposition 4.4, we have $v(\alpha) = -u^2$ and $u^\iota = -u$ for some $\alpha \in \Gamma_S^{\text{neg}}$ and some $u \in B$. Then $u^2 = c^2 e$ with $c \in F^\times$ and $e \in E_\theta$. This implies that B is either $M_2(F)$ or a division algebra with a quadratic subfield isomorphic to $F(e^{1/2})$. This proves (II$_a$).

Conversely, suppose B is a division algebra, and $\mathfrak{o}$ has an element δ such that $\delta^2 = e \in E_\theta$ and $\delta^\iota = -\delta$. Put $\alpha = \delta 1_n$. Then $v(\alpha) = -e$, and $\alpha \in \Gamma_S^{\text{neg}}$. Let δ_h be the projection of δ to the h-th factor $M_2(\mathbf{R})$ of $B_{\mathbf{R}}$. We may, without losing generality, assume that $\delta_h = \begin{bmatrix} c_h & 0 \\ 0 & -c_h \end{bmatrix}$ where c_h is a real number such that $c_h^2 = e^{\tau_h}$. By [9, (4.4.5)], the projection of α to the h-th factor $G_p(n, \mathbf{R})$ is

$$\begin{bmatrix} c_h 1_n & 0 \\ 0 & -c_h 1_n \end{bmatrix}.$$

Therefore if $z = (y_1 i, \ldots, y_r i)$ with $i = \sqrt{-1}$ and positive definite real symmetric matrices y_h, then $\alpha(\bar{z}) = z$. By Proposition 4.3, $\varphi_S(z)$ is real for such a point z.

Next suppose $B = M_2(F)$. We may assume that

$$\mathfrak{o} = \left\{ \begin{bmatrix} a & b \\ c & d \end{bmatrix} \,\middle|\, a \in \mathfrak{r},\ d \in \mathfrak{r},\ b \in \mathfrak{a},\ c \in \mathfrak{a}^{-1} \right\},$$

where $\mathfrak{r}$ is the maximal order of F, and $\mathfrak{a}$ is a fractional ideal of F. Taking the element $\begin{bmatrix} 1 & 0 \\ 0 & -1 \end{bmatrix}$ of $\mathfrak{o}$ as δ, we can repeat the above reasoning to obtain a real point of V_S.

Finally suppose n is even. Let e be an element of E_θ. By Lemma 1.4, $\mathfrak{o}$ has an element ε such that $\varepsilon\varepsilon' = -e$. Put $m = n/2$, and

$$\alpha = \begin{bmatrix} 0 & \varepsilon 1_m \\ \varepsilon' 1_m & 0 \end{bmatrix}.$$

Then $\alpha \in \Gamma_S^{neg}$ and $v(\alpha) = -e$. Let $\varepsilon_h = \begin{bmatrix} a_h & b_h \\ c_h & d_h \end{bmatrix}$ be the projection of ε to the h-th factor $M_2(\mathbf{R})$ of $B_{\mathbf{R}}$. Then the projection of α to the h-th factor of $G_{\mathbf{R}}$ is

$$\begin{bmatrix} 0 & a_h 1_m & 0 & b_h 1_m \\ d_h 1_m & 0 & -b_h 1_m & 0 \\ 0 & c_h 1_m & 0 & d_h 1_m \\ -c_h 1_m & 0 & a_h 1_m & 0 \end{bmatrix}.$$

Therefore if $z = (z_1, \ldots, z_r)$, $z_h = \begin{bmatrix} u_h 1_m & 0 \\ 0 & v_h 1_m \end{bmatrix}$ with $u_h, v_h \in \mathfrak{H}_1$, then $\alpha(\bar{z}) = (w_1, \ldots, w_r)$ with

$$w_h = \begin{bmatrix} \varepsilon_h(\bar{v}_h)\, 1_m & 0 \\ 0 & \varepsilon_h^t(\bar{u}_h)\, 1_m \end{bmatrix}.$$

Consider the case $u_h = \varepsilon_h(\bar{v}_h)$. Then $\varepsilon_h^t(\bar{u}_h) = v_h$, and so $\alpha(\bar{z}) = z$, which produces a real point of V_S. This completes the proof.

In the case $n = 1$, we have more clearly

Theorem 5.2. *Suppose that $n = 1$, and the set E_θ of Theorem 5.1 is not empty. Put*

$$\Delta = \{\alpha \in \mathfrak{o} \mid \alpha' = -\alpha,\ \alpha^2 \in E_\theta\}.$$

For each $\alpha \in \Delta$, put $X(\alpha) = \{z \in \mathfrak{H}_1^r \mid \alpha(\bar{z}) = z\}$. Then $\Delta_S = F^\times \Delta$, and

$$V_S(\mathbf{R}) = \bigcup_{\alpha \in \Delta} \varphi_S(X(\alpha)).$$

Especially V_S has a real point if and only if Δ is not empty.

Proof. By (5.2) and Proposition 4.3, we have $\Gamma_S^{neg} = F^\times \cdot (\mathfrak{o}^\times \cap B^{neg})$, where B^{neg} is the set of all $\alpha \in B$ such that $(\alpha\alpha')^{r_i} < 0$ for $i \leq r$. If $\alpha(\bar{z}) = z$ and $\alpha \in \mathfrak{o}^\times \cap B^{neg}$, we have $\alpha^2(z) = z$, so that $\alpha' = -\alpha$ and $\alpha^2 \in F$ by Proposition 1.2. Then $\alpha^2 = -\alpha\alpha' \in E_\theta$. Conversely if $\alpha \in \mathfrak{o}$, $\alpha' = -\alpha$ and $\alpha^2 \in E_\theta$, then $\alpha\alpha' = -\alpha^2$, so that $\alpha \in \mathfrak{o}^\times \cap B^{neg}$. Now for such an α, we see that $X(\alpha) \neq \emptyset$ from the proof of Theorem 5.1. Actually $X(\alpha)$ is the transform of $\{(y_1 i, \ldots, y_r i) \mid y_v > 0\}$ by an element of $SL_2(\mathbf{R})^r$. Therefore we obtain our assertions.

The statement $(\mathrm{II_b})$ of Theorem 5.1 is not the exact converse of $(\mathrm{II_a})$. This point leads to the following natural question:

(5.4) *Given an element e of E_θ, assuming that B splits over $F(e^{1/2})$, can we find an element α of $\mathfrak{o}$ such that $\alpha^\iota = -\alpha$ and $\alpha^2 = e$?*

Without fixing $\mathfrak{o}$, this is trivial: there is always a maximal order containing an element α such that $\alpha^\iota = -\alpha$ and $\alpha^2 = e$. But with a fixed order $\mathfrak{o}$, this seems a rather difficult question. Anaway, if the answer is affirmative, say for a given F, then we can conclude that V_S has a real point if and only if n is even or B splits over $F(e^{1/2})$ for some $e \in E_\theta$. The following two lemmas will give several conditions under which the answer to (5.4) is affirmative.

Lemma 5.3. *Let $\mathfrak{o}$ be a maximal order in B, M a quadratic extension of F that splits B, and $\mathfrak{g}$ an order in M containing the maximal order of F. Suppose that the conductor of $\mathfrak{g}$ is prime to the discriminant of B over F. Then there exists an F-linear embedding f of M into B such that $f(\mathfrak{g}) = f(M) \cap \mathfrak{o}$ if any one of the following conditions is satisfied:*

(i) The type number of B is one.

(ii) M is ramified over F at a finite prime or an infinite prime corresponding to τ_i for some $i \leq r$.

(iii) There is a finite prime of F ramified in B.

Proof. The condition on the conductor of $\mathfrak{g}$ is necessary and sufficient for the existence of a maximal order $\mathfrak{o}'$ in B and an F-linear injection h of M into B such that $h(\mathfrak{g}) = h(M) \cap \mathfrak{o}'$ (see e.g. [4, Satz 6]). Therefore the sufficiency of (i) is obvious. The remaining part is essentially the same as the proof of [7, Proposition 2.8]. Take a right $\mathfrak{o}'$-ideal $\mathfrak{x}$ that is a left $\mathfrak{o}$-ideal. Now condition (ii) implies that for every ideal $\mathfrak{a}$ in F, we can find an ideal $\mathfrak{b}$ in M such that $c\mathfrak{a} = N_{M/F}(\mathfrak{b})$ with $c \in F$, $c^{\tau_i} > 0$ for $i > r$. We may assume that $\mathfrak{b}$ is integral and prime to the conductor of $\mathfrak{g}$. Put $\mathfrak{b}_1 = \mathfrak{b} \cap \mathfrak{g}$. Taking $\mathfrak{a}$ to be $N_{B/F}(\mathfrak{x})$, we have $c \cdot N_{B/F}(\mathfrak{x}) = N_{B/F}(h(\mathfrak{b}_1)\mathfrak{o}')$. By Eichler's theorem, we have $\alpha\mathfrak{x} = h(\mathfrak{b}_1)\mathfrak{o}'$ with $\alpha \in B$. Define $f : M \to B$ by $f(a) = \alpha^{-1}h(a)\alpha$. Then it can easily be verified that $f(\mathfrak{g}) = f(M) \cap \mathfrak{o}$. Next suppose that M is ramified over F only at infinite primes corresponding to $\tau_{r+1}, \ldots, \tau_g$. Suppose also that a prime ideal $\mathfrak{p}$ in F is ramified in B. Then $\mathfrak{p}$ must remain prime in M. Now, for every ideal $\mathfrak{a}$ in F, we have either $c\mathfrak{a} = N_{M/F}(\mathfrak{b})$ as above or $c\mathfrak{a}\mathfrak{p} = N_{M/F}(\mathfrak{b})$ with an integral ideal $\mathfrak{b}$ in M prime to the conductor of $\mathfrak{g}$ and $c \in F$, $c^{\tau_i} > 0$ for $i > r$. Take $\mathfrak{a}$ to be $N_{B/F}(\mathfrak{x})$. If $c\mathfrak{a} = N_{M/F}(\mathfrak{b})$, we can reason as before. Suppose $c \cdot N_{B/F}(\mathfrak{x})\mathfrak{p} = N_{M/F}(\mathfrak{b})$. Now there is a two-sided $\mathfrak{o}$-ideal $\mathfrak{q}$ such that $N_{B/F}(\mathfrak{q}) = \mathfrak{p}$. Then $c \cdot N_{B/F}(\mathfrak{q}\mathfrak{x}) = N_{B/F}(h(\mathfrak{b}_1)\mathfrak{o}')$ with $\mathfrak{b}_1 = \mathfrak{b} \cap \mathfrak{g}$. By Eichler's theorem, we have $\alpha\mathfrak{q}\mathfrak{x} = h(\mathfrak{b}_1)\mathfrak{o}'$ with $\alpha \in B$. Defining again f by $f(a) = \alpha^{-1}h(a)\alpha$, we obtain $f(\mathfrak{g}) = f(M) \cap \mathfrak{o}$.

Lemma 5.4. *Suppose $r < g$ and there exists a unit u of F such that $u^{\tau_{r+1}} < 0$ and $u^{\tau_i} > 0$ for all $i > r+1$ (after a suitable re-arrangement of $\tau_{r+1}, \ldots, \tau_g$). Then, for every $e \in E_\theta$, $F(e^{1/2})$ is ramified over F at a prime factor of 2 in F; moreover, every maximal order in B has an element α such that $\alpha^2 = e$ and $\alpha^\iota = -\alpha$, if B splits over $F(e^{1/2})$.*

Remark. If $r = g - 1$, one can take $u = -1$. Therefore the assumption is always satisfied if $r = g - 1$.

Proof. Let K be an arbitrary finite abelian extension of F unramified except at $\tau_{r+1}, \ldots, \tau_g$. The existence of u implies that K is also unramified at τ_{r+1}. (This can be seen, for example, by considering the corresponding characters of $F_A^\times$.) Suppose $F(e^{1/2})$, for some $e \in E_\theta$, is unramified at every prime factor of 2 in F. Then $F(e^{1/2})$ is unramified at τ_{r+1}, a contradiction, since $e^{\tau_{r+1}} < 0$. The last assertion follows from Lemma 5.3.

Example 5.5. Suppose $r = g$, and B is a division algebra. Then E_θ is the set of all totally positive units of F, and therefore non-empty. Actually $k_S = Q$ as will be seen in the next section. Moreover condition (iii) of Lemma 5.3 holds. Therefore V_S has a real point if and only if either n is even or B splits over $F(e^{1/2})$ for some $e \in E_\theta$. This case will be considered again in the next section.

Example 5.6. Suppose $[F : Q] = 2$ and $r = 1$. We can identify F' with F via τ_1. Let e_1 be a fundamental unit of F such that $e_1^{\tau_1} > 0$. Then E_θ is non-empty if and only if $N_{F/Q}(e_1) = -1$. If so, $E_\theta = \{e_1^{2m+1} | m \in Z\}$. We see that k_S is the Hilbert class field over F, and totally real. Again by the above lemmas, V_S has a real point if and only if either n is even, or B splits over $F(e_1^{1/2})$.

6. The Case of Totally Indefinite Algebras

Let us assume, throughout this section, that B is totally indefinite, i.e., $r = g$. In this case, it is more natural to consider a group

$$(6.1) \qquad \Gamma_1 = \{\gamma \in GL_n(\mathfrak{o}) | \gamma \cdot {}^t\gamma^\iota = 1\}$$

instead of the group Γ_S of § 5. Here $\mathfrak{o}$ is as before a maximal order in B. Since $r = g$, we have $F' = Q$, and λ is the natural injection of $Q_A^\times$ into $F_A^\times$. Further we have $\bar{\mathscr{G}}_+ = F_c G_{Q+} D_+$ with

$$D_+ = \{x \in G_{A+} | v(x) \in Q_A^\times\} \qquad \text{(see [10, II, Proposition 3.4])}.$$

Proposition 6.1. *Put* $T = F_c \cdot (U_L \cap D_+)$ *with* U_L *of* (5.0). *Then* $T \in \mathscr{Z}^*$, $\Gamma_T = F^\times \Gamma_1$, $k_T = Q$, *and* $v(\Gamma_T^{\mathrm{neg}}) = \{-a^2 | a \in F^\times\}$.

Proof. Put $P = (U_L \cap D_+) G_{\infty+}$. Then P belongs to the class $\mathscr{Z}$ defined in [10, I, 2.3], and $T = F_c P$. Therefore $T \in \mathscr{Z}^*$ by [10, II, 3.6]. Obviously $F^\times \Gamma_1 \subset \Gamma_T$. To prove the opposite inclusion, let $\gamma \in \Gamma_T$ and put $\gamma = ab$ with $a \in F_c$ and $b \in U_L \cap D_+$. Then $v(b) \in Q_A^\times \cap F_c = Q^\times Q_{\infty+}^\times$. Since $b \in U_L$, the non-archimedean component of $v(b)$ is 1. Now, by Chevalley [2], there exists a positive integer c such that

$$\{e \in \mathfrak{r}^\times | e \equiv 1 \,(\mathrm{mod}\, c\mathfrak{r})\} \subset \{e^2 | e \in \mathfrak{r}^\times\},$$

where $\mathfrak{r}$ is the maximal order of F. Take a positive integer m so that $u^m \equiv 1 \,(\mathrm{mod}\, c\mathfrak{r}_p)$ for every $u \in \mathfrak{r}_p^\times$, where $\mathfrak{r}_p = \mathfrak{r} \otimes Z_p$ for each rational prime p. By [10, II, 2.2], we have $a = x^m yz$ with $x \in \bar{E}_0$, $y \in F^\times$, $z \in F_{\infty+}^\times$. Then $v(y^{-1}\gamma) = x^{2m} z^2 v(b)$, so that $v(y^{-1}\gamma) \equiv 1 \,(\mathrm{mod}\, c\mathfrak{r})$. Then $v(y^{-1}\gamma) = e^2$ with $e \in \mathfrak{r}^\times$. Now $e^{-1}y^{-1}\gamma = e^{-1}x^m zb \in U_L$ and $v(e^{-1}y^{-1}\gamma) = 1$, hence $e^{-1}y^{-1}\gamma \in \Gamma_1$. This proves $\Gamma_T = F^\times \Gamma_1$. Now the p-part of $v(U_L \cap D_+)$ contains $Z_p^\times$, hence $Q_A^\times \subset v(T) F_c$. It follows that $k_T = Q$. By Lemma 1.4, $\mathfrak{o}$ contains an element ε such that $\varepsilon\varepsilon^\iota = -1$. Put $\alpha = \varepsilon 1_n$.

Then $\alpha \in \Gamma_T^{neg}$ and $v(\alpha) = -1$. Therefore $v(\Gamma_T^{neg}) = v(\alpha\Gamma_T) = v(\alpha F^\times \Gamma_1)$. This proves the last assertion.

The construction of models in [10, I] shows that V_T has a natural structure of a variety of moduli of the family of abelian varieties (PEL-structures) as described in [10, I, § 5]. Let us now restate Theorem 0 as

Theorem 6.2. *Suppose B is totally indefinite. Then the variety V_T with $T = F_c$ $\cdot (U_L \cap D_+)$ as above has a real point if and only if either n is even, or $B = M_2(F)$.*

Proof. If B is a division algebra, we see that

$$\{-a^2 \mid a \in F^\times\} \cap \{-u^2 \mid u \in B, u^t = -u\} = \emptyset.$$

Therefore the "only if"-part follows immediately from Proposition 4.4 and Proposition 6.1. The "if"-part can be proved in exactly the same manner as in the proof of Theorem 5.1. In fact, if $B = M_2(F)$, we consider an element $\alpha = \delta 1_n$ with $\delta = \begin{bmatrix} 1 & 0 \\ 0 & -1 \end{bmatrix}$. If n is even, we consider $\alpha = \begin{bmatrix} 0 & \varepsilon 1_m \\ \varepsilon^t 1_m & 0 \end{bmatrix}$ with an element ε of $\mathfrak{o}$ such that $\varepsilon\varepsilon^t = -1$, whose existence is guaranteed by Lemma 1.4. In either case, $\alpha \in \Gamma_S^{neg}$, and we find points z of $\mathfrak{H}_n^r$ such that $\alpha(\bar{z}) = z$.

Corollary 6.3. *The notation and assumptions being as above, suppose that n is odd and B is a division algebra. Let w be an arbitrary point of V_T. Then $Q(w)$ has no isomorphic image in R. Especially if $Q(w)$ is algebraic over Q, then $Q(w)$ is totally imaginary.*

This follows immediately from the above theorem, because for every injection μ of $Q(w)$ into C, w^μ is a point of V_T.

Let us now compare Theorem 6.2 with Theorem 5.1 (in the case $r = g$). Let E denote the group of all units in F, E_+ the subgroup of E consisting of all totally positive units, and let $E^2 = \{e^2 \mid e \in E\}$. Further let $\Gamma' = G_{Q_+} \cap GL_n(\mathfrak{o})$. We have $\Gamma_S = F^\times \Gamma'$ with the member S of $\mathscr{Z}^*$ defined in § 5. We see easily that $\Gamma'/E\Gamma_1$ is isomorphic to E_+/E^2. Thus V_T is a covering of V_S of degree $[E_+ : E^2]$, and the projection map $J_{ST}(1)$ is rational over Q. Therefore the result of Example 5.5 is included in Theorem 6.2 if $E_+ = E^2$.

It should be mentioned that V_T is only one of the many possible models of $\Gamma_1 \backslash \mathfrak{H}_n^g$, although it is a natural one. (The same remark applies to the model V_S of $\Gamma' \backslash \mathfrak{H}_n^r$.) To obtain a different model V_R with real points, suppose B has an element b such that $b^t = -b$, $b \mathfrak{o} b^{-1} = \mathfrak{o}$, and b^2 is a totally positive element of F. Then $\alpha = b 1_n$ belongs to G_Q^{neg}, $\alpha T \alpha^{-1} = T$, $\alpha^2 \in T$, and $X(\alpha) \neq \emptyset$. Put

$$R = R' \cup R'\alpha_0,$$

$$R' = F_c \cdot \{x \in U_L \cap D_+ \mid v(x) \in Q^\times N_{K/Q}(K_A^\times)\}$$

with any imaginary quadratic extension K of Q. It can easily be seen that $k_{R'} = K$, $k_R = Q$, $\Gamma_R = \Gamma_T = \Gamma_{R'}$, $\Gamma_R^{neg} = \Gamma_R \alpha$. Thus V_R has real points. By Lemma 5.3, such an element b exists, if there is a totally positive algebraic integer c of F such that B splits over $F(c^{1/2})$ and every prime factor of c is ramified in B. For example, if $F = Q$, we can take c to be the product of all primes ramified in B.

7. The Structure of $V_S(R)$ in the Case $n=1$

So far we have been considering only the existence or non-existence of real points on V_S. It is actually a very interesting problem to investigate the structure of the whole set $V_S(R)$ when it is non-empty. Let us first discuss the case $n=1$ in this section, and then the general case in the next section.

If $n=1$, we have $G_Q = B^\times$, and $v(\alpha) = \alpha\alpha'$ for $\alpha \in B$. Define B^{neg} and B_+ by

$$
(7.1) \qquad
\begin{aligned}
B^{\mathrm{neg}} &= \{\alpha \in B^\times \mid v(\alpha)^{\tau_i} < 0 \text{ for } i \leqq r\}, \\
B_+ &= \{\alpha \in B^\times \mid v(\alpha)^{\tau_i} > 0 \text{ for all } i\}.
\end{aligned}
$$

Then $G_Q^{\mathrm{neg}} = B^{\mathrm{neg}}$ and $G_{Q+} = B_+$. For a maximal order $\mathfrak{o}$ in B, put

$$
(7.2) \qquad \Gamma(\mathfrak{o}) = \mathfrak{o}^\times \cap B_+, \qquad \Gamma(\mathfrak{o})^{\mathrm{neg}} = \mathfrak{o}^\times \cap B^{\mathrm{neg}}.
$$

Define S as in §5 with $n=1$, i.e.,

$$
S = F_c \cdot \{y \in B_A^\times \mid y\mathfrak{o} = \mathfrak{o},\ v(y) \in \lambda(F_A'^\times) F_c,\ v(y_\infty) \in F_{\infty+}^\times\}.
$$

Then $\Gamma_S = F^\times \Gamma(\mathfrak{o})$, $\Gamma_S^{\mathrm{neg}} = F^\times \Gamma(\mathfrak{o})^{\mathrm{neg}}$ as seen in §5. If $r=1$, the model (V_S, φ_S) in the present case is the same as the model (V, φ) of [7, Main Theorem 1]. We consider again the group of units E of F, and the subgroup E_+ of all totally positive elements of E as in §6. We put also as in Theorems 5.1 and 5.2,

$$
(7.3) \qquad
\begin{aligned}
E_\theta &= \{e \in E \mid e^{\tau_i} > 0 \text{ or } < 0 \text{ according as } i \leq r \text{ or } > r\}, \\
E^2 &= \{e^2 \mid e \in E\}, \\
\Delta &= \Delta(\mathfrak{o}) = \{\alpha \in \mathfrak{o} \mid \alpha' = -\alpha,\ \alpha^2 \in E_\theta\}, \\
X(\alpha) &= \{z \in \mathfrak{H}_1^r \mid \alpha(\bar{z}) = z\} \qquad (\alpha \in \Delta).
\end{aligned}
$$

Now we take a finite set of representatives Z such that

$$
\{\alpha^2 \mid \alpha \in \Delta\} = \bigcup_{c \in Z} E^2 c.
$$

The cardinality of Z is $\leqq [E_+ : E^2]$. For each $c \in Z$, let Δ_c denote the set of elements α of Δ such that $\alpha^2 = c$. We have obviously $\gamma(X(\alpha)) = X(\gamma\alpha\gamma^{-1})$ for $\gamma \in \Gamma(\mathfrak{o})$, and therefore by Theorem 5.2,

$$
(7.4) \qquad V_S(R) = \bigcup_{c \in Z} \bigcup_{\alpha \in R_c} \varphi_S(X(\alpha)),
$$

where R_c is a set of representatives for Δ_c modulo inner automorphisms given by the elements of $\Gamma(\mathfrak{o})$. These sets $\varphi_S(X(\alpha))$ are not necessarily disjoint. To study the mutual relationship, suppose $\varphi_S(X(\alpha)) \cap \varphi_S(X(\alpha')) \neq \emptyset$ for some α and α' in Δ. Then $\alpha(\bar{z}) = z$, $\alpha'(\bar{z}') = z'$, $\varphi_S(z) = \varphi_S(z')$ for some z and z' of $\mathfrak{H}_1^r$. The last equality implies $z' = \gamma(z)$ for some $\gamma \in \Gamma(\mathfrak{o})$. Put $\beta = \alpha^{-1}\gamma^{-1}\alpha'\gamma$. Then $\beta \in \Gamma(\mathfrak{o})$ and $\beta(z) = z$. Therefore $\beta \in E$, or β is elliptic. Let $\alpha^2 = c$ and $\alpha'^2 = c'$. Then $(\alpha\beta)^2 = (\gamma^{-1}\alpha'\gamma)^2 = c'$, so that $\alpha\beta\alpha^{-1} = \alpha\beta\alpha\beta\beta^{-1}\alpha^{-2} = c^{-1}c'\beta^{-1}$, hence $\alpha\beta\alpha^{-1} \in F[\beta]$. Therefore $\alpha\beta\alpha^{-1} = \beta'$ if β is elliptic. Thus we have two possible cases:

(i) $\gamma^{-1}\alpha'\gamma = e\alpha$ *with* $e \in E$.

(ii) $\gamma^{-1}\alpha'\gamma = \beta\alpha$, $\alpha\beta\alpha^{-1} = \beta'$, *and* $F[\beta]$ *is totally imaginary.*

In the first case, we have $c' = e^2 c$. Therefore if we take c and c' from Z, we have $c = c'$ and $e = \pm 1$, thus $\gamma^{-1}\alpha'\gamma = \pm\alpha$.

In the second case, β is an elliptic element of $\Gamma(\mathfrak{o})$, so that $\beta^m \in E$ with a positive integer $m > 1$. [An example for such a β is given by a root of unity times an element of E. But this is not always the case. For example, $\beta^2 = -2 - \sqrt{3}$ with $F = Q(\sqrt{3})$.] Since $\alpha\beta\alpha^{-1} = \beta^\iota$, we see that our quaternion algebra B is expressed as a cyclic algebra $(F[\beta], c)$. This imposes a rather restrictive condition on the possibility for β.

Let us now count the number of connected components of $V_S(R)$. Although the treatment in the most general case seems feasible with no great difficulty, and may be a worthwhile problem, we consider here only a special case, making the following assumption:

(7.5) $\Gamma(\mathfrak{o})/E$ *has no elements of finite order other than the identity element.*

This implies that B is a division algebra, and V_S is a non-singular projective variety. Moreover at least one finite prime of F is ramified in B. (Otherwise, by Lemma 5.3, a root of unity of order 3 or 4 can be found in $\mathfrak{o}$.) If $\alpha \in R_c$, $\alpha' \in R_{c'}$ with $c, c' \in Z$ and $\varphi_S(X(\alpha)) \cap \varphi_S(X(\alpha')) \neq \emptyset$, then $c = c'$ and $\gamma^{-1}\alpha'\gamma = \pm\alpha$ with $\gamma \in \Gamma(\mathfrak{o})$ as above. Conversely, if $\gamma^{-1}\alpha'\gamma = \pm\alpha$ with $\gamma \in \Gamma(\mathfrak{o})$, then obviously $\varphi_S(X(\alpha)) = \varphi_S(X(\alpha'))$. Let us now show that $\gamma^{-1}\alpha\gamma = -\alpha$ is impossible. If $\gamma^{-1}\alpha\gamma = -\alpha$, we see that γ^2 commutes with both α and γ so that $\gamma^2 \in F \cap \Gamma(\mathfrak{o}) = E$. By (7.5), we have $\gamma \in E$, which contradicts $\gamma^{-1}\alpha\gamma = -\alpha$. These observations lead to the following conclusion: the number of connected components of $V_S(R)$ is half the sum of the numbers of elements of R_c for all $c \in Z$. This is preliminary to the following

Theorem 7.1. *Let E'_θ be the set of all units e of F such that B splits over $F(e^{1/2})$ and $e^{\tau_i} > 0$ or < 0 according as $i \leq r$ or $> r$. Let Y be a set of representatives for E'_θ modulo multiplication by the squares of units of F. Then, under (7.5), the number of connected components of $V_S(R)$ is*

$$\sum_{c \in Y} 2^{s(c)-1} \sum_{\mathfrak{g}} [E_1 : N_{K/F}(\mathfrak{g}^\times) E_+] \cdot h(\mathfrak{g})/h(B) \quad (K = F(c^{1/2})).$$

Here E_1 is the group of all units such that $e^{\tau_i} > 0$ for $i > r$; E_+ the group of all totally positive units of F; $s(c)$ the number of all prime ideals of F ramified in B but unramified in $F(c^{1/2})$; $\mathfrak{g}$ runs over all orders in $K = F(c^{1/2})$ containing $\mathfrak{r}[c^{1/2}]$ whose conductor is prime to the discriminant of B over F, where $\mathfrak{r}$ denotes the maximal order of F; $h(\mathfrak{g})$ denotes the class number of $\mathfrak{g}$; $h(B)$ the class number of B, i.e. the number of ideal classes in F modulo the product of all archimedean primes of F ramified in B.

Proof. Fix an element c of Y, and put $b = c^{1/2}$, $K = F(b)$. For any $\alpha \in \Delta(\mathfrak{o})$ such that $\alpha^2 = c$, we can define an F-linear embedding f of K into B by $f(b) = \alpha$. Put $\mathfrak{g} = f^{-1}[f(K) \cap \mathfrak{o}]$. Then $\mathfrak{g}$ is an order in K containing $\mathfrak{r}[b]$. We obtain the same $\mathfrak{g}$ from $\gamma^{-1}\alpha\gamma$ with $\gamma \in \mathfrak{o}^\times$. The conductor of $\mathfrak{g}$ must be prime to the discriminant of B. Conversely, given an order $\mathfrak{g}$ in K containing $\mathfrak{r}[b]$ with such a conductor, there exists, by Lemma 5.3, an F-linear embedding $f : K \to B$ such that $f(\mathfrak{g}) = f(K) \cap \mathfrak{o}$. (It is crucial that this holds for *every* maximal order $\mathfrak{o}$ in B.) Call two such embeddings f and f' *equivalent* (resp. *strongly equivalent*) if $f'(x) = \gamma^{-1} f(x) \gamma$ with some $\gamma \in \mathfrak{o}^\times$ [resp. $\gamma \in \Gamma(\mathfrak{o})$]. Let $n(\mathfrak{g})$ [resp. $n'(\mathfrak{g})$] the number of equivalence (resp. strong equivalence) classes of such f. Then we see that the number of connected components of $V_S(R)$ is $2^{-1} \sum_{c \in Y} \sum_g n'(\mathfrak{g})$, where $\mathfrak{g}$ runs over all orders in K as specified in our assertion. Now one can show, by the same method as in [7, 2.16, 2.17], using Eichler's lemma [4, Satz 7], that $n(\mathfrak{g}) = 2^{s(c)} h(\mathfrak{g})/h(B)$. Also it can easily be seen that

$$n'(\mathfrak{g}) = [E_1 : N_{K/F}(\mathfrak{g}^\times) E_+] \cdot n(\mathfrak{g}),$$

hence our theorem.

Proposition 7.2. *With $\mathfrak{o}$ and S as before, if $\Gamma(\mathfrak{o})/E$ has no element of finite order other than the identity element, then each connected component $\varphi_S(X(\alpha))$ of $V_S(R)$ is isomorphic, as a real analytic manifold, to a torus $(R/Z)^r$ for every $\alpha \in \Delta$.*

Proof. Suppose $\gamma(X(\alpha)) \cap X(\alpha) \neq \emptyset$ for $\gamma \in \Gamma(\mathfrak{o})$. Then $\gamma^{-1}\alpha\gamma = \alpha$ as seen above. Put $\alpha^2 = c$ and $K = F(b)$ with $b^2 = c$ as in the above proof, and define $f : K \to B$ by $f(b) = \alpha$. Put $\mathfrak{g} = f^{-1}[f(K) \cap \mathfrak{o}]$ and

$$(7.6) \qquad\qquad E_\mathfrak{g} = \{u \in \mathfrak{g}^\times \mid N_{K/F}(u) \in E_+\}.$$

Then $\gamma \in F[\alpha] \cap \Gamma(\mathfrak{o}) = f(E_\mathfrak{g})$. Therefore $\varphi_S(X(\alpha))$ can be identified with $f(E_\mathfrak{g}) \backslash X(\alpha)$. Extend τ_ν, for each $\nu \leq r$, to two injections of K into R, and denote them by τ_ν and τ'_ν. We may assume that the projection of α to the ν-th factor $M_2(R)$ of B_R is $\begin{bmatrix} b^{\tau_\nu} & 0 \\ 0 & -b^{\tau_\nu} \end{bmatrix}$. Then $X(\alpha) = \{(y_1 i, \ldots, y_r i) \mid y_\nu > 0\}$. If $u \in E_\mathfrak{g}$, we have

$$f(u)(y_1 i, \ldots, y_r i) = ((u^{\tau_1}/u^{\tau'_1}) y_1 i, \ldots, (u^{\tau_r}/u^{\tau'_r}) y_r i).$$

Now the map $E_\mathfrak{g} \ni u \to (\log(u^{\tau_1}/u^{\tau'_1}), \ldots, \log(u^{\tau_r}/u^{\tau'_r})) \in R^r$ gives an isomorphism of $E_\mathfrak{g}/E$ onto a lattice Λ in R^r. Therefore the map

$$X(\alpha) \ni (y_1 i, \ldots, y_r i) \to (\log y_1, \ldots, \log y_r) \in R^r$$

gives an isomorphism of $\varphi_S(X(\alpha)) = f(E_\mathfrak{g}) \backslash X(\alpha)$ onto R^r/Λ.

It is natural to consider the case of groups other than $\Gamma(\mathfrak{o})$, or other models. For instance, it can be shown that the set of real points of the standard model of $\Gamma_0(N) \backslash \mathfrak{H}_1$, for a prime N, is connected, where

$$\Gamma_0(N) = \left\{ \begin{bmatrix} a & b \\ c & d \end{bmatrix} \in SL_2(Z) \,\middle|\, c \equiv 0 (\bmod N) \right\}.$$

In general, the number of components is closely related to the class numbers of quadratic extensions of F, for the reason explained in Theorem 7.1 and its proof. This applies especially to the models of the type V_R discussed at the end of § 6.

8. The Structure of $V_S(R)$ in the General Case

To study the nature of the set $X(\alpha)$, let us now give a "canonical form" of an element of Δ_S in $Gp(n, R)$:

Proposition 8.1. *Let W be an element of $Gp(n, R)$ such that $\nu(W) = -1$ and $W(\bar{z}) = z$ for a point z of $\mathfrak{H}_n$. Then there exists an element T of $Sp(n, R)$ such that*

$z = T(i1_n)$ *and*

$$(8.1) \qquad T^{-1}WT = \begin{bmatrix} q & 0 \\ 0 & -q \end{bmatrix}$$

with a real orthogonal matrix q. *Conversely if* W *is given in the form* (8.1) *with* $T \in Sp(n, \mathbf{R})$ *and a real orthogonal* q, *then* $W(\bar{z}) = z$ *with* $z = T(i1_n)$.

Proof. Let $W(\bar{z}) = z$ and $v(W) = -1$. Take an element P of $Sp(n, \mathbf{R})$ so that $z = P(i1_n)$, and put

$$(8.2) \qquad Y = \begin{bmatrix} 1_n & 0 \\ 0 & -1_n \end{bmatrix} P^{-1}WP.$$

Then $Y \in Sp(n, \mathbf{R})$, $Y(i1_n) = i1_n$. Therefore $Y = \begin{bmatrix} a & b \\ -b & a \end{bmatrix}$ with $a, b \in M_n(\mathbf{R})$ such that $a + ib$ is a unitary matrix. Put $u = a + ib$ and $R = \begin{bmatrix} 1_n & -i1_n \\ 1_n & i1_n \end{bmatrix}$. Then $RP^{-1}WPR^{-1} = \begin{bmatrix} 0 & \bar{u} \\ u & 0 \end{bmatrix}$. By Lemma 8.2 below, we have $u = c^{-1}q\bar{c}$ with a unitary c and a real orthogonal q. Put $S = R^{-1} \begin{bmatrix} \bar{c} & 0 \\ 0 & c \end{bmatrix} R$. Then we obtain the first assertion with $T = PS^{-1}$. The converse part is obvious. In this proof we needed

Lemma 8.2. *Every unitary matrix can be written as* $\bar{P}QP^{-1}$ *with a unitary matrix* P *and a real orthogonal matrix* Q.

Proof. In the vector space $\mathbf{C}^n$, define a hermitian form (x, y) and a real symmetric form $[x, y]$ by

$$(x, y) = {}^t x\bar{y}, \qquad [x, y] = Re(x, y) \qquad (x, y \in \mathbf{C}^n),$$

where the elements of $\mathbf{C}^n$ are considered column vectors. Given a unitary matrix U, define an anti-$\mathbf{C}$-linear map $T : \mathbf{C}^n \to \mathbf{C}^n$ by $T(x) = U\bar{x}$. Then $(T(x), T(y)) = (y, x)$ and $[Tx, Ty] = [x, y]$. Therefore the following two cases can happen:

(I) $T(w) = \pm w$ for some $w \neq 0$.

(II) There exist two vectors w and z such that $[w, w] = [z, z] = 1$, $[w, z] = 0$, $T(w) = aw + bz$, $T(z) = -bw + az$ with an orthogonal $\begin{bmatrix} a & b \\ -b & a \end{bmatrix} \neq \pm 1$.

In Case II, we have $(z, w) = (T(w), T(z)) = (aw + bz, -bw + az) = a^2(w, z) - b^2(z, w)$, hence $(w, z) = 0$. Therefore w, z are linearly independent over $\mathbf{C}$. Put

$$Y = \begin{cases} \{x \in \mathbf{C}^n \mid (x, w) = 0\} & \text{(Case I)} \\ \{x \in \mathbf{C}^n \mid (x, w) = (x, z) = 0\} & \text{(Case II)}. \end{cases}$$

Then Y is stable under T. By induction we can thus find an orthonormal $\mathbf{C}$-basis $\{p_1, \ldots, p_n\}$ of $\mathbf{C}^n$ such that $T(p_j) = \sum_{i=1}^{n} q_{ij}p_i$ with real q_{ij}. Let P denote the $n \times n$ matrix whose columns are $\bar{p}_1, \ldots, \bar{p}_n$, and let $Q = (q_{ij})$. Then $UP = \bar{P}Q$ as desired.

Given an element W as in Proposition 8.1, the set $\{z \in \mathfrak{H}_n \,|\, W(\bar{z}) = z\}$ is the transform by T of the set $\{z \in \mathfrak{H}_n \,|\, -q\bar{z}q^{-1} = z\}$, which is a real analytic submanifold of $\mathfrak{H}_n$ of dimension $\leq n(n+1)/2$; here the strict inequality can happen (see below). If W belongs to a discrete subgroup of $GL_{2n}(R)$, then W is of finite order, and so is q.

Let S be a real member of $\mathscr{Z}^*$. We shall now examine the structure of $\varphi_S(X(\alpha))$ for each $\alpha \in \Delta_S$ under the following assumptions:

(8.3)   *$\Gamma_S/F^\times$ has no elements of finite order other than the identity element.*

(8.4)   *There is an element β of G_{Q+} and an element δ of B such that $\beta\alpha\beta^{-1} = \delta 1_n$ and $\delta' = -\delta$.*

(8.5)   *There is no $\gamma \in \Gamma_S$ such that $\gamma^{-1}\alpha\gamma = -\alpha$.*

The first condition implies the second one if n is odd. In fact, α^2 is an element of Γ_S with a fixed point, so that $\alpha^2 \in F$. Therefore we obtain β and δ as in (8.4) by virtue of Proposition 3.4. We shall later give sufficient conditions under which (8.5) is satisfied.

Put $v(\alpha) = c$. To discuss the structure of $\varphi_S(X(\alpha))$, we may, by (8.4), assume that $\alpha = \delta 1_n$ with an element δ of B such that $\delta^2 = -c$ and $\delta' = -\delta$. Put $K = F[\delta]$. Then K is either a quadratic extension of F or a direct sum of two copies of F. Embed $M_n(K)$ into $M_n(B)$ in a natural manner. For every $\xi \in B$ or $\in M_n(B)$, let ξ_v denote the projection of ξ to the v-th factor of B_R or of $M_n(B_R)$. We may assume that

$$\delta_v = \begin{bmatrix} a_v & 0 \\ 0 & -a_v \end{bmatrix}, \qquad a_v^2 = -c^{\tau_v} \qquad (v = 1, \ldots, r).$$

Extend τ_v to two homomorphisms τ_v and τ'_v of K into R by $\delta^{\tau_v} = a_v$ and $\delta^{\tau'_v} = -a_v$. Then, by [9, (4.4.5)], for every $A \in M_n(K)$, we have

$$A_v = \begin{bmatrix} A^{\tau_v} & 0 \\ 0 & A^{\tau'_v} \end{bmatrix} \qquad (v = 1, \ldots, r).$$

Now ι induces the non-trivial automorphism of K over F. Define algebraic subgroups P and P^u of G defined over Q by

$$P_Q = \{A \in GL_n(K) \,|\, {}^t A^\iota A = \mu 1_n \text{ with } \mu \in F^\times\},$$
$$P^u = P \cap G^u.$$

Then we have a commutative diagram:

$$\begin{array}{ccc} P_R^u & \longrightarrow & GL_n(R)^r \times U(n)^{g-r} \\ \downarrow & & \downarrow \psi \\ G_R^u & \longrightarrow & Sp(n, R)^r \times Sp(n)^{g-r} \end{array}$$

Here $U(n)$ and $Sp(n)$ denote the ordinary unitary group and the quaternion unitary group of degree n, respectively; the horizontal arrows are isomorphisms; ψ is defined so that

$$\psi(A_1, \ldots, A_r, \ldots) = \left(\begin{bmatrix} A_1 & 0 \\ 0 & {}^t A_1^{-1} \end{bmatrix}, \ldots, \begin{bmatrix} A_r & 0 \\ 0 & {}^t A_r^{-1} \end{bmatrix}, \ldots \right).$$

Now we see that $X(\alpha)$ consists of all the points of the form $(iy_1, \ldots, iy_r)$ with positive definite real symmetric matrices y_v. Put $z_0 = (i1_n, \ldots, i1_n)$. Then the map

$$P_{\mathbf{R}}^u \ni \xi \mapsto \xi(z_0) \in X(\alpha)$$

gives an isomorphism of $P_{\mathbf{R}}^u/[O(n)^r \times U(n)^{g-r}]$ onto $X(\alpha)$, where $O(n)$ is the orthogonal group of degree n.

Suppose $\gamma(X(\alpha)) \cap X(\alpha) \neq \emptyset$ for some $\gamma \in \Gamma_S$. Then $\gamma^{-1}\alpha\gamma = \pm\alpha$ for the same reason as explained in § 7. By (8.5), we have $\gamma^{-1}\alpha\gamma = \alpha$, so that $\gamma \in M_n(K) \cap G_Q = P_Q$. Now put $\Gamma' = \Gamma_S \cap P_Q$. Then we can conclude that $\varphi_S(X(\alpha))$ is isomorphic to $\Gamma' \backslash X(\alpha)$, or rather to

$$\Gamma' \backslash P_{\mathbf{R}}^u/[O(n)^r \times U(n)^{g-r}].$$

The fact that a quotient space of this type is embeddable in $\Gamma_S \backslash \mathfrak{H}_n^r$ may perhaps have been observed by many people. But that this occurs as a connected component of $V_S(\mathbf{R})$ does not seem to have been noticed before. If $n = 1$, this result is essentially the same as Proposition 7.2, except that the assumptions made here are somewhat different from those of § 7.

Let us now show that (8.5) follows from the following two conditions, if n is odd:

(8.6) $v(\Gamma_S) \subset N_{K/F}(K)$. (Obviously this is satisfied if $\Gamma_S \subset F^\times G_Q^u$.)

(8.7)   *B has no element ε such that $\varepsilon\delta\varepsilon^{-1} = -\varepsilon$ and $\varepsilon^2 = -1$, i.e. B is not isomorphic to the cyclic algebra $(F(\sqrt{-c}), -1)$.*

Suppose $\gamma^{-1}\alpha\gamma = -\alpha$. We can find an element π of B such that $\pi^{-1}\delta\pi = -\delta$. Then we see that $\gamma\pi^{-1}$ commutes with α, hence $\gamma = \beta\pi$ with $\beta \in M_n(K)$. By Lemma 3.3, $v(\beta)^n = N_{K/F}(\det(\beta))$. If n is odd, this implies $v(\beta) \in N_{K/F}(K)$. By (8.6), we have $v(\gamma) \in N_{K/F}(K)$, so that $v(\pi) \in N_{K/F}(K)$. Now $\pi^2 = -v(\pi)$, so that B is isomorphic to the cyclic algebra $(K, -v(\pi))$, hence to $(K, -1)$, which contradicts (8.7).

In a more general case where $\Gamma_S/F^\times$ has elements of finite order, the variety V_S may have singularities, and the structure of $V_S(\mathbf{R})$ seems considerably more complicated than the above case. Let us now briefly touch this problem by considering the Siegel modular case.

If $B = M_2(Q)$, it is natural to identify G_Q with the group

$$\{\alpha \in GL_{2n}(Q) \mid {}^t\alpha J_n \alpha = v(\alpha) 1_n \text{ with } v(\alpha) \in Q^\times\}.$$

Let d be a diagonal matrix of size n whose diagonal elements are positive rational integers. Put

(8.8)
$$\delta = \begin{bmatrix} 1_n & 0 \\ 0 & d \end{bmatrix}, \quad \varepsilon = \begin{bmatrix} -1_n & 0 \\ 0 & 1_n \end{bmatrix},$$

and define a member S of $\mathscr{X}^*$ by

(8.9)
$$S = Q^\times U_d$$
$$U_d = \{y \in G_{A+} \mid \delta y_p \delta^{-1} \in GL_{2n}(Z_p) \text{ for all } p\},$$

where Z_p denotes the ring of all p-adic integers with a rational prime p. Then $k_S = Q$, $\Gamma_S = Q^\times \Gamma_d$, $\Gamma_S^{\mathrm{neg}} = \Gamma_S \varepsilon$, where

$$(8.10) \qquad \Gamma_d = Sp(n, R) \cap \delta^{-1} GL_{2n}(Z)\, \delta .$$

Moreover V_S is the variety of moduli for the maximal family of abelian varieties of dimension n whose polarization type is $\begin{bmatrix} 0 & -d \\ d & 0 \end{bmatrix}$. The determination of the set Δ_S is a problem of an elementary nature, but seems rather complicated. For example, if $n = 2$, it can be verified that, with the notation of Proposition 8.1, we have the following possibilities for q: $q = 1_2, q^2 = -1_2, q^3 = 1_2, q^4 = -1_2, q^6 = -1_2$. Thus $V_S(R)$ for $n = 2$ would be determined if we could classify all elements α of $\varepsilon \Gamma_d$ modulo conjugacy by the elements of Γ_d, according to each type of q. This is not so difficult at least when $d = 1_2$, i.e. when $\Gamma_d = Sp(2, Z)$. One can show that there is only one class in each of the cases $q^2 = -1_2, q^4 = -1_2, q^6 = -1_2$; there are two classes in the case $q^3 = 1_2$. If $q = 1_2$, there are exactly 4 classes represented by $\alpha = \begin{bmatrix} -1 & \sigma \\ 0 & 1 \end{bmatrix}$, where $\sigma = 0, 1_2, \begin{bmatrix} 1 & 0 \\ 0 & 0 \end{bmatrix}, \begin{bmatrix} 0 & 1 \\ 1 & 0 \end{bmatrix}$, as will be shown in Proposition 9.2 of the next section. The set $X(\alpha)$ has dimension 3 in the cases $q = 1_2$ and $q^2 = -1_2$; otherwise the dimension is 1.

If n becomes large, we obtain an element α of a large order. Then the number of classes involves the class number of a cyclotomic field.

9. Abelian Varieties Defined over R and Their Real Points

With a fixed diagonal matrix d as above, define δ and ε by (8.8) and Γ_d by (8.10). Let $\omega_d(z)$ for $z \in \mathfrak{H}_n \cup \bar{\mathfrak{H}}_n$ denote the $n \times 2n$ matrix obtained by juxtaposing z and d, and $L_d(z)$ the lattice in C^n generated over Z by the columns of $\omega_d(z)$. Put

$$\Phi_z(\omega_d(z)x,\ \omega_d(z)\, y) = {}^t x D y, \qquad D = \begin{bmatrix} 0 & -d \\ d & 0 \end{bmatrix} \qquad (x, y \in R^{2n}).$$

Then Φ_z defines a Riemann form on $C^n/L_d(z)$. Take any projective model $A_d(z)$ of $C^n/L_d(z)$, and let $P_d(z)$ denote the polarized abelian variety $A_d(z)$ with the polarization determined by Φ_z. Now $\varphi_S(z)$ is the "modulus" of $P_d(z)$. If $P_d(z)$ has a model over R, then $\varphi_S(z)$ is real, but the converse is not necessarily true as remarked in [12]. To determine all $P_d(z)$ with models over R, put

$$(9.1) \qquad \Delta_d^* = \{\alpha \in \Gamma_d \varepsilon \mid \alpha^2 = 1\}.$$

Consider all couples (z, α) with $\alpha \in \Delta_d^*$ and $z \in X(\alpha)$, and call (z, α) and (z', α') *equivalent* if $z' = \gamma(z)$ and $\alpha' = \gamma \alpha \gamma^{-1}$ for some $\gamma \in \Gamma_d$.

Theorem 9.1. *For every equivalence class of (z, α) as above, there exists a polarized abelian variety $P = (A, C)$ isomorphic to $P_d(z)$ and defined over R, and a holomorphic map $f : C^n \to A$ that gives an isomorphism of $C^n/L_d(z)$ onto A, such that*

$\overline{f(u)} = f(\overline{\lambda u})$ *for all* $u \in \mathbf{C}^n$, *where* λ *is an element of* $GL_n(\mathbf{C})$ *determined by*

$$\lambda \omega_1(z) = \omega_1(\overline{z}) \cdot {}^t\alpha .$$

Moreover, $(z, \alpha) \mapsto P$ *gives a one-to-one correspondence between the equivalence classes of* (z, α) *and the* $\mathbf{R}$-*isomorphism classes of polarized abelian varieties defined over* $\mathbf{R}$ *of polarization type* d.

Especially, $P_d(z)$ has a model over $\mathbf{R}$ if and only if $\alpha(\overline{z}) = z$ with $\alpha \in \Gamma_d \varepsilon$, $\alpha^2 = 1$.

Proof. Consider $P = (A, C)$ isomorphic to $P_d(z)$ with an arbitrary $z \in \mathfrak{H}_n$. Let f be a holomorphic map of $\mathbf{C}^n$ onto A that gives an isomorphism of $\mathbf{C}^n/L_d(z)$ onto A. Put $f'(u) = \overline{f(\overline{u})}$ for $u \in \mathbf{C}^n$. Then f' is an isomorphism of $\mathbf{C}^n/L_d(-\overline{z})$ onto $\overline{A}$. Thus $A_d(-\overline{z})$ is isomorphic to $\overline{A_d(z)}$. Moreover it can be checked that $P_d(-\overline{z})$ is isomorphic to $\overline{P_d(z)}$. (For the proof, see [12, p. 175].) Now suppose that P is defined over $\mathbf{R}$. Then there exists an element λ of $GL_n(\mathbf{C})$ and an element β of $GL_{2n}(\mathbf{Z})$ such that $f(u) = f'(\lambda u)$, $\lambda \omega_d(z) = \omega_d(-\overline{z})\beta$, and ${}^t\beta D\beta = D$. Put $\alpha = \delta^{-1} \cdot {}^t\beta \varepsilon \delta$. Then $\alpha \in \Gamma_d \varepsilon$, $\lambda \omega_1(z) = \omega_1(\overline{z}){}^t\alpha$, and $z = \alpha(\overline{z})$. Moreover, from $f(u) = f'(\lambda u)$, we see that $\overline{\lambda}\lambda = 1$, and therefore $\alpha^2 = 1$. Conversely suppose $\alpha(\overline{z}) = z$ with $\alpha \in \Delta_d^*$. Put $\beta = {}^t(\delta \alpha \delta^{-1} \varepsilon)$. Then we find an element λ of $GL_n(\mathbf{C})$ such that $\lambda \omega_d(z) = \omega_d(-\overline{z})\beta$. We see that λ gives an isomorphism of $\mathbf{C}^n/L_d(z)$ onto $\mathbf{C}^n/L_d(-\overline{z})$ preserving the polarization. Then we obtain an isomorphism μ of P onto $\overline{P}$ such that $\mu(f(u)) = f'(\lambda u)$. Since $\alpha^2 = 1$, we have $\overline{\lambda}\lambda = 1$, so that $\overline{\mu} \circ \mu = 1$. Applying the criterion of Weil [13] to P and μ, we obtain a polarized abelian variety P' defined over $\mathbf{R}$ and an isomorphism ξ of P' to P such that $\overline{\xi} = \mu \circ \xi$. Put $g = \xi^{-1} \circ f$. Then g gives an isomorphism of $P_d(z)$ to P' and $g(\overline{\lambda u}) = \overline{g(u)}$. It remains to show that the $\mathbf{R}$-isomorphism class of P corresponds to the equivalence class of (z, α). But this can be done in a straightforward manner.

Thus the moduli of the polarized abelian varieties with models over $\mathbf{R}$ correspond to the set

$$(9.2) \qquad \{\varphi_S(z) \mid \alpha(\overline{z}) = z \text{ for some } \alpha \in \Gamma_d \varepsilon, \alpha^2 = 1\} .$$

As shown in [12], this set is smaller than $V_S(\mathbf{R})$ at least if n is even and each invariant factor of d occurs with an even multiplicity.

Proposition 9.2. *Let* $\Delta^* = \{\alpha \in Gp(n, \mathbf{R}) \cap GL_{2n}(\mathbf{Z}) \mid v(\alpha) = -1, \alpha^2 = 1\}$. *Then*

$$\Delta^* = \left\{ \gamma^{-1} \begin{bmatrix} -1_n & \sigma \\ 0 & 1_n \end{bmatrix} \gamma \;\middle|\; \gamma \in Sp(n, \mathbf{Z}), {}^t\sigma = \sigma \in M_n(\mathbf{Z}) \right\} .$$

Moreover, let σ *and* σ' *be symmetric matrices of* $M_n(\mathbf{Z})$. *Then* $\begin{bmatrix} -1_n & \sigma \\ 0 & 1_n \end{bmatrix}$ *and* $\begin{bmatrix} -1_n & \sigma' \\ 0 & 1_n \end{bmatrix}$ *are conjugate under an element of* $Sp(n, \mathbf{Z})$ *if and only if* ${}^tp\sigma p \equiv \sigma' \pmod{2}$ *with an element* p *of* $GL_n(\mathbf{Z})$.

Proof. Define an alternating form $J(x, y)$ on $\mathbf{Q}^{2n}$ by

$$J(x, y) = {}^txJ_ny \qquad (x, y \in \mathbf{Q}^{2n}),$$

where the elements of $\mathbf{Q}^{2n}$ are considered column vectors. Put $L = \mathbf{Z}^{2n}$ and $M = \{x \in L \mid \alpha x = -x\}$. Then L/M is torsion-free, so that $L = M \oplus N$ with a submodule N. By Remark 3.5, α is conjugate to ε in G_Q, so that M and N have rank n over Z. Take bases $\{x_i\}$ and $\{y_i\}$ of M and N, respectively. Since $v(\alpha) = -1$, we see that $J(x, x') = 0$ for $x, x' \in M$. Therefore J is represented with respect to $\{x_i, y_j\}$ by a matrix of the form

$$\begin{bmatrix} 0 & A \\ -{}^tA & B \end{bmatrix}$$

with $A, B \in M_n(Z)$. Since $\det(J_n) = 1$, we have $\det(A) = \pm 1$. Changing $\{x_i\}$ suitably, we may assume $A = 1_n$. Since $B = -{}^tB$, we can find an element C of $M_n(Z)$ so that $B = {}^tC - C$. Then

$$\begin{bmatrix} 1 & 0 \\ C & 1 \end{bmatrix}\begin{bmatrix} 0 & 1 \\ -1 & B \end{bmatrix}\begin{bmatrix} 1 & {}^tC \\ 0 & 1 \end{bmatrix} = \begin{bmatrix} 0 & 1 \\ -1 & 0 \end{bmatrix} = J_n .$$

This means that we can change N and $\{y_i\}$ suitably so that J is represented by J_n with respect to $\{x_i, y_j\}$ (without changing M). Since $\alpha x = -x$ for $x \in M$, this implies that there is an element γ of $Sp(n, Z)$ such that $\gamma^{-1}\alpha\gamma = \begin{bmatrix} -1 & \sigma \\ 0 & \tau \end{bmatrix}$ with $\sigma, \tau \in M_n(Z)$. Since $v(\alpha) = -1$, we see that $\tau = 1_n$ and ${}^t\sigma = \sigma$. This proves the nontrivial part of the first assertion. The converse part is obvious. To prove the second assertion, suppose

$$(9.3) \qquad \beta\begin{bmatrix} -1 & \sigma \\ 0 & 1 \end{bmatrix} = \begin{bmatrix} -1 & \sigma' \\ 0 & 1 \end{bmatrix}\beta$$

with $\beta \in Sp(n, Z)$. Then it can easily be seen that $\beta = \begin{bmatrix} p & q \\ 0 & {}^tp^{-1} \end{bmatrix}$ with $p \in GL_n(Z)$ and $q \in M_n(Z)$ such that $p^tq = q^tp$, $2q = \sigma'{}^tp^{-1} - p\sigma$. Therefore $p\sigma^tp \equiv \sigma'$ (mod 2). Conversely if $p\sigma^tp \equiv \sigma'$ (mod 2) with $p \in GL_n(Z)$, we can put $\sigma'{}^tp^{-1} - p\sigma = 2q$ with $q \in M_n(Z)$. Then $2q^tp = \sigma' - p\sigma^tp = 2p^tq$, so that if we put $\beta = \begin{bmatrix} p & q \\ 0 & {}^tp^{-1} \end{bmatrix}$, then $\beta \in Sp(n, Z)$, and (9.3) holds. This completes the proof.

Let Y_n denote the set of all positive definite real symmetric matrices of size n. If $\alpha = \begin{bmatrix} -1 & \sigma \\ 0 & 1 \end{bmatrix}$ as above, we see that

$$X(\alpha) = \{(1/2)\,\sigma + iy \,|\, y \in Y_n\} .$$

Therefore if $d = 1_n$, the set (9.2) can be expressed as

$$(9.4) \qquad \bigcup_\sigma \varphi_S(\{(1/2)\,\sigma + iy \,|\, y \in Y_n\}),$$

where σ runs over a set of representatives of all equivalence classes modulo 2 of symmetric matrices in $M_n(Z)$ as explained in the above proposition. The geometric meaning of σ will be clarified by Theorem 9.4 below.

Theorem 9.3. *Let A be an abelian variety of dimension n defined over R, A_R the set of all real points on A, and D the connected component of A_R containing 0. Then*

D is a real n-dimensional torus, and A_R/D is isomorphic to $(Z/2Z)^m$ with a non-negative integer $m \leq n$. Moreover $\{u \in A_R \mid 2u = 0\}$ is isomorphic to $(Z/2Z)^{m+n}$.

Proof. Let $P = (A, C)$ with a polarization C of A rational over R. Then P is isomorphic to $P_d(z)$ for some d and z. By Theorem 9.1, we have a map $f : C^n \to A$, an element α of Δ_d^*, and an element λ of $GL_n(C)$ such that $\lambda \omega_1(z) = \omega_1(\bar{z})^t \alpha$ and $\overline{f(u)} = f(\bar{\lambda} \bar{u})$. By Remark 3.5, there exists an element $\beta = \begin{bmatrix} p & q \\ r & s \end{bmatrix} \in G_{Q+}$ such that $\beta \alpha \beta^{-1} = \varepsilon$. Put $w = \beta(z)$ and $\kappa = {}^t(rz + s)$. Then $\lambda \kappa \omega_1(w) = \bar{\kappa} \omega_1(-\bar{w})$, so that $\lambda \kappa = \bar{\kappa}$ and $w = iy$ with $y \in Y_n$. Now $f(u)$ is real if and only if $\bar{\lambda} \bar{u} - u \in L_d(z)$. Putting $u = \kappa v$, we obtain

$$A_R = \{f(\kappa v) \mid v \in C^n, \bar{v} - v \in \kappa^{-1} L_d(z)\} .$$

Put $M = {}^t \beta^{-1} \delta Z^{2n}$. Then $\kappa^{-1} L_d(z) = \omega_1(w) M$. Decompose Q^{2n} into the direct sum $Q^n \oplus Q^n$ by taking the first and the last n components of a column vector, and let π (resp. π') be the projection map of M into the first (resp. second) summand Q^n. Since $w = iy$, we see that

$$A_R = \{f(\kappa v) \mid v \in C^n, \bar{v} - v \in iy \cdot \mathrm{Ker}(\pi')\} .$$

For each $t \in 2^{-1} \mathrm{Ker}(\pi')$, put

$$X_t = \{\kappa(x + iyt) \mid x \in R^n\} .$$

Then we can easily verify that

$$f(X_t) = f(X_s) \Leftrightarrow f(X_t) \cap f(X_s) \neq \emptyset \Leftrightarrow t - s \in \pi(M) .$$

Since $\delta \alpha \delta^{-1} \in GL_{2n}(Z)$ and $\varepsilon = \beta \alpha \beta^{-1}$, we see that $\varepsilon M = M$. Therefore if $\begin{pmatrix} a \\ b \end{pmatrix} \in M$, then $\begin{pmatrix} -a \\ b \end{pmatrix} \in M$, so that $\begin{pmatrix} 2a \\ 0 \end{pmatrix} \in M$. In other words, $2 \cdot \pi(M) \subset \mathrm{Ker}(\pi')$. Let T be a set of representatives for $2^{-1} \mathrm{Ker}(\pi')/\pi(M)$. Then A_R is a disjoint union of $f(X_t)$ for all $t \in T$, and $f(X_0)$ is isomorphic to $R^n/\mathrm{Ker}(\pi)$ via the map $x \mapsto f(\kappa x)$. Obviously $f(X_t)$ is a translation of $f(X_0)$ by $f(\kappa iyt)$, and $[2^{-1} \mathrm{Ker}(\pi') : \pi(M)] = 2^m$ with $0 \leq m \leq n$. Further $f(X_0)$, being a real n-dimensional torus, has 2^n points annihilated by 2. Combining these facts, we obtain our theorem.

A more definite information about the number of components can be obtained if $d = 1_n$:

Theorem 9.4. *Let $P = (A, C)$ be a principally polarized abelian variety of dimension n defined over R, corresponding to (z, α) in the sense of Theorem 9.1. Let σ be a symmetric matrix of $M_n(Z)$ such that $\gamma^{-1} \alpha \gamma = \begin{bmatrix} -1 & \sigma \\ 0 & 1 \end{bmatrix}$ with $\gamma \in Sp(n, Z)$ as in Proposition 9.2. Then A_R has $2^{n-\varrho}$ connected components, where $\varrho = \mathrm{rank}(\sigma \bmod 2)$. Moreover suppose $\mathrm{Aut}(P) = \{\pm 1\}$, and let $P' = (A', C')$ be defined over R and isomorphic to P over C. Then A_R' has the same number of connected components as A_R.*

Proof. In the proof of Theorem 9.3, take $d = 1_n$ and $\alpha = \begin{bmatrix} -1 & \sigma \\ 0 & 1 \end{bmatrix}$. Then we have $\beta\alpha\beta^{-1} = \varepsilon$ with $\beta = \begin{bmatrix} 1 & -\sigma/2 \\ 0 & 1 \end{bmatrix}$. Therefore

$$M = {}^t\beta^{-1}Z^{2n} = \left\{ \begin{pmatrix} a \\ b + \tfrac{1}{2}\sigma a \end{pmatrix} \Big| a \in Z^n, b \in Z^n \right\},$$

$$\mathrm{Ker}(\pi') = \left\{ \begin{pmatrix} a \\ 0 \end{pmatrix} \Big| a \in Z^n, \sigma a \in 2Z^n \right\}, \qquad \pi(M) = \left\{ \begin{pmatrix} a \\ 0 \end{pmatrix} \Big| a \in Z^n \right\},$$

$$\mathrm{Ker}(\pi) = \left\{ \begin{pmatrix} 0 \\ b \end{pmatrix} \Big| b \in Z^n \right\}, \qquad \pi'(M) = \left\{ \begin{pmatrix} 0 \\ b \end{pmatrix} \Big| b \in Z^n + 2^{-1}\sigma Z^n \right\}.$$

Let $\varrho = \mathrm{rank}(\sigma \bmod 2)$. Then both $2^{-1}\mathrm{Ker}(\pi')/\pi(M)$ and $2^{-1}\mathrm{Ker}(\pi)/\pi'(M)$ have $2^{n-\varrho}$ elements. Therefore A_R has $2^{n-\varrho}$ connected components. This proves the first assertion. To prove the second one, let

$$A_R^- = \{p \in A_C \,|\, \bar{p} = -p\}.$$

Then by the same type of reasoning as in the proof of Theorem 9.3, we can show that

$$A_R^- = \{f(\kappa v) \,|\, \bar{v} + v \in \mathrm{Ker}(\pi)\}.$$

For each $s \in 2^{-1}\mathrm{Ker}(\pi)$, put $W_s = \{\kappa(s + iyx) \,|\, x \in R^n\}$. We see that $2 \cdot \pi'(M) \subset \mathrm{Ker}(\pi)$, and if S is a set of representatives for $2^{-1}\mathrm{Ker}(\pi)/\pi'(M)$, A_R^- is a disjoint union of $f(W_s)$ for all $s \in S$. Therefore A_R^- has $2^{n-\varrho}$ components. Now suppose $\mathrm{Aut}(P) = \{\pm 1\}$, and let μ be an isomorphism of P to $P' = (A', C')$ over C, where P' is rational over R. Then $\bar{\mu} = \pm\mu$. If $\bar{\mu} = \mu$, A_R' is isomorphic to A_R. Suppose $\bar{\mu} = -\mu$. Then μ gives an isomorphism of A_R^- onto A_R'. This completes the proof.

In the simplest case $n = 1$, where A is an elliptic curve, $V_S(R)$ coincides with the set (9.2) and hence with (9.4), by virtue of Proposition 1.2. It corresponds to the imaginary axis and the boundary of the ordinary fundamental domain for $SL_2(Z)\backslash\mathfrak{H}_1$. The number of connected components of the set of real points of an elliptic curve is 2 on the imaginary axis, and 1 otherwise, except at $z = i$. If $z = i$, we obtain 2 components with $\alpha = \begin{bmatrix} -1 & 0 \\ 0 & 1 \end{bmatrix}$, but 1 component with $\alpha = \begin{bmatrix} 0 & 1 \\ 1 & 0 \end{bmatrix}$, since $\begin{bmatrix} 1 & 0 \\ 1 & 1 \end{bmatrix}\begin{bmatrix} 0 & 1 \\ 1 & 0 \end{bmatrix}\begin{bmatrix} 1 & 0 \\ 1 & 1 \end{bmatrix}^{-1} = \begin{bmatrix} -1 & 1 \\ 0 & 1 \end{bmatrix}$. This shows that the condition $\mathrm{Aut}(P) = \{\pm 1\}$ cannot be dropped from the last assertion of Theorem 9.4.

We conclude this section by proving a generalization of a theorem of [12].

Theorem 9.5. *Let P be a polarized abelian variety of dimension n defined over a field of characteristic 0, and k a field containing the field of moduli of P and $\det(\alpha)$ for all $\alpha \in \mathrm{Aut}(P)$, where $\det(\alpha)$ denotes the representation of α on the space of n-forms of the first kind on P. If the map*

$$(9.5) \qquad\qquad \mathrm{Aut}(P) \ni \alpha \mapsto \det(\alpha) \in k$$

is injective, P has a model over k.

Proof. We can find a finite Galois extension K of k such that P is defined over K and all isomorphisms of P to P^σ for every $\sigma \in \mathrm{Gal}(K/k)$ are defined over K. [Take for example K to be the composite of k and the field of moduli of $(P, t_1, ..., t_s)$ with all points t_i on P such that $Nt_i = 0$ with an integer $N > 2$.] Put $G = \mathrm{Gal}(K/k)$. For each $\sigma \in G$, choose an isomorphism λ_σ of P to P^σ. Then $\lambda_{\sigma\tau} = \lambda_\sigma^\tau \lambda_\tau \alpha_{\sigma,\tau}$ with $\alpha_{\sigma,\tau} \in \mathrm{Aut}(P)$. Take a non-zero n-form ξ on P of the first kind, rational over K. Then $\xi^\sigma \circ \lambda_\sigma = c_\sigma \xi$ with $c_\sigma \in K$, and $c_{\sigma\tau} = c_\sigma^\tau c_\tau \det(\alpha_{\sigma,\tau})$. Let h be the order of $\mathrm{Aut}(P)$. Then $c_{\sigma\tau}^h = (c_\sigma^h)^\tau c_\tau^h$, so that there is an element b of K such that $c_\sigma^h = b/b^\sigma$ for all $\sigma \in G$. Consider an extension $K(d)$ with an element d such that $d^h = b$. Then $K(d)$ is normal over k. Put $G^* = \mathrm{Gal}(K(d)/k)$, and let $r : G^* \to G$ be the natural homomorphism. For every $\sigma \in G^*$, we see that $c_{r(\sigma)} d^\sigma/d$ is an h-th root of unity, hence there is a unique $\alpha_\sigma \in \mathrm{Aut}(P)$ such that $c_{r(\sigma)} \det(\alpha_\sigma) = d/d^\sigma$. Put $\mu_\sigma = \lambda_{r(\sigma)} \alpha_\sigma$. Then $(d\xi)^\sigma \circ \mu_\sigma = d^\sigma \xi^\sigma \circ \lambda_{r(\sigma)} \circ \alpha_\sigma = d^\sigma c_{r(\sigma)} \det(\alpha_\sigma) \xi = d\xi$. It follows from this and the injectivity of (9.5) that $\mu_{\sigma\tau} = \mu_\sigma^\tau \mu_\tau$ for all $\sigma, \tau \in G^*$. Applying Weil's criterion [13] to P and μ_σ, we obtain a model of P rational over k.

As an immediate consequence, we obtain

Corollary 9.6. *Let P be a polarized abelian variety of odd dimension defined over a field of characteristic 0. If $\mathrm{Aut}(P) = \{\pm 1\}$, P has a model over its field of moduli.*

10. Real Singular Moduli

In this section we shall study the points $\varphi_S(z)$ with isolated fixed points z in the case $n = 1$. The reader is reminded that such points generate certain class fields. For a maximal order $\mathfrak{o}$ in B, define $\Gamma(\mathfrak{o})$ and $\Gamma(\mathfrak{o})^{\mathrm{neg}}$ by (7.2) and $\Delta(\mathfrak{o})$ by (7.3). Let us write $(V_\mathfrak{o}, \varphi_\mathfrak{o})$ for the model (V_S, φ_S) considered in § 7. Let K be a totally imaginary quadratic extension of F which splits B, and $\mathfrak{r}_K$ the maximal order of K. For every prime ideal $\mathfrak{p}$ in F, we denote by $F_\mathfrak{p}$ the $\mathfrak{p}$-completion of F, and put $B_\mathfrak{p} = B \otimes_F F_\mathfrak{p}$ and $K_\mathfrak{p} = K \otimes_F F_\mathfrak{p}$. Let $f : K \to B$ be an F-linear embedding. Then the elements of $f(K^\times)$ have a unique fixed point on $\mathfrak{H}_1^r$, which we denote by z_f (cf. [7, 2.6]). Let us now examine when $\varphi_\mathfrak{o}(z_f)$ is real.

Proposition 10.1. *Suppose the set E_Θ of Theorem 5.1 is non-empty. Then $\varphi_\mathfrak{o}(z_f)$ is real if and only if there exists an element β of $\Delta(\mathfrak{o})$ such that $\beta f(x) \beta^{-1} = f(\bar{x})$ for all $x \in K$.*

Proof. If $\varphi_\mathfrak{o}(z_f)$ is real, we have $\beta(\bar{z}_f) = z_f$ for some $\beta \in \Delta(\mathfrak{o})$. Now we have $f(K^\times) = \{\alpha \in B_+ \,|\, \alpha(z_f) = z_f\}$, hence $\beta f(x) \beta^{-1} = f(\bar{x})$. The converse part is straightforward.

Proposition 10.2. *Suppose the set E_Θ of Theorem 5.1 is non-empty, and let $f : K \to B$ be an F-linear embedding. Then there exists a maximal order $\mathfrak{o}$ such that $f(\mathfrak{r}_K) \subset \mathfrak{o}$ and $\varphi_\mathfrak{o}(z_f)$ is real if and only if the following condition is satisfied:*

(10.1)   *E_Θ has an element e such that $e \in N_{K/F}(K_\mathfrak{p})$ if and only if $B_\mathfrak{p}$ splits.*

It can easily be seen that this condition implies

(10.2)   *Every prime ideal of F ramified in B is also ramified in K.*

Proof. Suppose $f(\mathfrak{r}_K) \subset \mathfrak{o}$ and $\varphi_\mathfrak{o}(z_f)$ is real. By Proposition 10.1, $f(\bar{x}) = \beta^{-1} f(x) \beta$ for all $x \in K$ and some $\beta \in \Delta(\mathfrak{o})$. Then B can be given as the cyclic algebra (K, c), where $c = \beta^2$. We see that $c \in N_{K/F}(K_\mathfrak{p})$ if and only if $B_\mathfrak{p}$ splits. This proves the "only if"-part. Conversely, let e be as in (10.1). Then B is given as the cyclic algebra (K, e), i.e., B has an element α such that $\alpha^2 = e$ and $f(\bar{x}) = \alpha^{-1} f(x) \alpha$ for all $x \in K$. Then $f(\mathfrak{r}_K) + f(\mathfrak{r}_K) \alpha$ is an order in B. Take a maximal order $\mathfrak{o}$ containing this order. Then $\alpha \in \Delta(\mathfrak{o})$. By Proposition 10.1, $\varphi_\mathfrak{o}(z_f)$ is real.

Theorem 10.3. *Suppose $E_\Theta \neq \emptyset$, and (10.1) is satisfied. Let f and f' be F-linear embeddings of K into B such that $f(\mathfrak{r}_K) \subset \mathfrak{o}$ and $f'(\mathfrak{r}_K) \subset \mathfrak{o}'$ with maximal orders $\mathfrak{o}$ and $\mathfrak{o}'$ in B. Let α be an element of B such that $f'(x) = \alpha^{-1} f(x) \alpha$ for all $x \in K$, and let $\mathfrak{x}$ be a right $\mathfrak{o}$-ideal that is a left $\mathfrak{o}'$-ideal. Then there exists an ideal $\mathfrak{a}$ in K such that $\alpha\mathfrak{x} = f(\mathfrak{a}) \mathfrak{o}$. Moreover suppose $\varphi_\mathfrak{o}(z_f)$ is real. Then $\varphi_{\mathfrak{o}'}(z_{f'})$ is real if and only if $\bar{\mathfrak{a}}$ belongs to the same ideal class as $\mathfrak{a}$ in K.*

This is a generalization of the classical result about the singular j-values (see e.g. [11, (5.4.3)]). See also [7, 2.16, 2.17] for the significance of the ideal $\mathfrak{a}$ such that $\alpha\mathfrak{x} = f(\mathfrak{a}) \mathfrak{o}$.

Proof. The existence of $\mathfrak{a}$ is shown in [7, 2.12, 2.16]. If $\varphi_\mathfrak{o}(z_f)$ is real, we have, by Proposition 10.1, $f(\bar{x}) = \beta^{-1} f(x) \beta$ with $\beta \in \Delta(\mathfrak{o})$. Put $\gamma = \alpha^{-1} \beta\alpha$. Then $f'(\bar{x}) = \gamma^{-1} f'(x) \gamma$. Now $\gamma\mathfrak{x} = \alpha^{-1} \beta f(\mathfrak{a}) \mathfrak{o} = \alpha^{-1} f(\bar{\mathfrak{a}}) \beta\mathfrak{o} = \alpha^{-1} f(\bar{\mathfrak{a}}) \mathfrak{o}$. Suppose $\varphi_{\mathfrak{o}'}(z_{f'})$ is real. Then $f'(\bar{x}) = \delta^{-1} f'(x) \delta$ with $\delta \in \Delta(\mathfrak{o}')$. We see that $\gamma = f'(h) \delta$ with $h \in K$. Then $\gamma\mathfrak{x} = f'(h) \delta\mathfrak{x} = f'(h) \mathfrak{x} = f'(h) \alpha^{-1} f(\mathfrak{a}) \mathfrak{o} = \alpha^{-1} f(h\mathfrak{a}) \mathfrak{o}$. Thus $\alpha^{-1} f(\bar{\mathfrak{a}}) \mathfrak{o} = \alpha^{-1} \cdot f(h\mathfrak{a}) \mathfrak{o}$, and so $\bar{\mathfrak{a}} = h\mathfrak{a}$ by [7, (3.20.9)]. This proves the "only if"-part. Conversely suppose $\bar{\mathfrak{a}} = h\mathfrak{a}$ with $h \in K$. Then $\gamma\mathfrak{x} = \alpha^{-1} f(h\mathfrak{a}) \mathfrak{o} = \alpha^{-1} f(h) \alpha\mathfrak{x} = f'(h) \mathfrak{x}$. Put $\delta = f'(h)^{-1} \gamma$. Then $f'(\bar{x}) = \delta^{-1} f'(x) \delta$, $\delta\mathfrak{x} = \mathfrak{x}$, and $\delta^2 \in F$, hence $\delta \in \Delta(\mathfrak{o}')$. This completes the proof.

References

1. Baily, W. L., Borel, A.: Compactification of arithmetic quotients of bounded symmetric domains. Ann. of Math. **84**, 442—528 (1966)
2. Chevalley, C.: Deux théorèmes d'arithmétique. J. Math. Soc. Japan 3, 36—44 (1951)
3. Eichler, M.: Allgemeine Kongruenzklasseneinteilungen der Ideale einfacher Algebren über algebraischen Zahlkörpern und ihre L-Reihen. J. Reine Angew. Math. 179, 227—251 (1938)
4. Eichler, M.: Zur Zahlentheorie der Quaternionen-Algebren. J. Reine Angew. Math. 195, 127—151 (1956)
5. Miyake, K.: On models of certain automorphic function fields. Acta Math. **126**, 245—307 (1971)
6. Miyake, T.: On automorphism groups of the fields of automorphic functions. Ann. of Math. 95 243—252 (1972)
7. Shimura, G.: Construction of class fields and zeta functions of algebraic curves. Ann. of Math. **85**, 58—159 (1967)
8. Shimura, G.: Discontinuous groups and abelian varieties. Math. Ann. **168**, 171—199 (1967)
9. Shimura, G.: Algebraic number fields and symplectic discontinuous groups. Ann. of Math. **86**, 503—592 (1967)

10. Shimura, G.: On canonical models of arithmetic quotients of bounded symmetric domains. I, Ann. of Math. **91**, 144—222 (1970) II, **92**, 528—549 (1970)
11. Shimura, G.: Introduction to the arithmetic theory of automorphic functions. Iwanami Shoten and Princeton Univ. Press, 1971.
12. Shimura, G.: On the field of rationality for an abelian variety. Nagoya Math. J. **45**, 167—178 (1972)
13. Weil, A.: The field of definition of a variety. Amer. J. Math. **78**, 509—524 (1956)

G. Shimura
Department of Mathematics
Princeton University
Princeton, N.J. 08540, USA

(Received November 25, 1974)

On some arithmetic properties of modular forms of one and several variables

Annals of Mathematics, 102 (1975), 491-515

The purpose of this paper is twofold. We first present a principle that a certain non-holomorphic derivative of a Hilbert modular form with algebraic Fourier coefficients, divided by another modular form, takes an algebraic value at every point with 'complex multiplication'. Second, we investigate the Hilbert or Siegel modular forms with cyclotomic Fourier coefficients in the framework of canonical models as developed in our previous papers.

The first principle, even specialized to the one-dimensional case, seems new and is as follows. Let $f(z)$ and $g(z)$ be modular forms of weight r and $r + 2n$, respectively, with respect to a congruence subgroup Γ of $SL_2(\mathbf{Z})$, and put

$$h(z) = (2\pi i)^{-n}\left(\frac{r + 2n - 2}{2iy} + \frac{\partial}{\partial z}\right) \cdots \left(\frac{r + 2}{2iy} + \frac{\partial}{\partial z}\right)\left(\frac{r}{2iy} + \frac{\partial}{\partial z}\right)f,$$

where $z = x + iy$ and $\partial/\partial z = (\partial/\partial x - i\partial/\partial y)/2$; r and n are arbitrary positive integers. Further let w be an element of an imaginary quadratic field K such that $\mathrm{Im}(w) > 0$ and $g(w) \neq 0$. *Then $g(w)^{-1}h(w)$ generates an abelian extension of K provided that the Fourier coefficients of f and g at $i\infty$ are cyclotomic.* It may be noted that

$$h((az + b)/(cz + d)) = h(z)(cz + d)^{r+2n} \quad \text{for all } \begin{pmatrix} a & b \\ c & d \end{pmatrix} \in \Gamma .$$

Thus the function h behaves almost like a modular form of weight $r + 2n$ both arithmetically and analytically.

The generalization of this fact to the Hilbert modular case will be given as three main theorems. The first two of them are elementary and weaker versions of the last one, which tells the behavior of $g(w)^{-1}h(w)$ under Frobenius substitutions. As an application, we shall show that a few special values of a certain L-function $L(s, \chi)$ multiplied by a constant of the type $g(w)^{-1}$ and a power of π are algebraic. Here χ is an 'algebraic valued' Hecke character of infinite order of a totally imaginary quadratic extension of a totally real algebraic number field. For each χ, we shall give finitely many integers or half integers s for which $L(s, \chi)$ has that property. This generalizes the result of Damerell [2] which concerns the case of an imaginary quadratic field.

It is indispensable for the proofs of the main theorems to make a detailed study of the second theme. We shall show, among other things, that:

(i) a compactified canonical model for an arithmetic quotient can be obtained whenever a non-compactified canonical model exists;

(ii) a Hilbert or Siegel modular function of any level is the quotient of two modular forms with cyclotomic Fourier coefficients if and only if it is arithmetic in the sense of [11];

(iii) the vector space of Hilbert or Siegel modular forms with respect to a principal congruence subgroup is spanned by those with rational Fourier coefficients.

Results of the same nature as (iii) have been obtained by Baily for $\mathrm{Sp}(n, \mathbf{Z})$ and some of its subgroups.

Obviously there is much room for further generalization and elaboration of our results. In the present paper, however, we have confined ourselves within a rather modest range, and have tried to make our presentation as raw and elementary as possible whenever such is feasible, in order not to obscure the ideas by a more sophisticated formulation, which may eventually be needed but seems disadvantageous at the first stage. Exceptions to this rule are made in later sections, especially in Section 5, where the adele language is most effective.

1. The first two main theorems

To make our expression brief, for $r = (r_1, \cdots, r_m) \in \mathbf{Z}^m$ and any point with indexed coordinates $z = (z_1, \cdots, z_m) \in \mathbf{C}^m$, we write

$$(1.1) \qquad\qquad z^r = \prod_{\nu=1}^m z_\nu^{r_\nu} \,,$$

$$(1.2) \qquad\qquad \{r\} = \sum_{\nu=1}^m r_\nu \,.$$

The functions to be studied are those defined on $\mathfrak{H}^m$, the product of m copies of the upper half complex plane

$$\mathfrak{H} = \{z \in \mathbf{C} \mid \mathrm{Im}(z) > 0\} \,.$$

Let $\mathrm{GL}_2^+(\mathbf{R}) = \{\gamma \in \mathrm{GL}_2(\mathbf{R}) \mid \det(\gamma) > 0\}$. We let every element $\alpha = (\alpha_1, \cdots, \alpha_m)$ of $\mathrm{GL}_2^+(\mathbf{R})^m$ with $\alpha_\nu = \begin{pmatrix} a_\nu & b_\nu \\ c_\nu & d_\nu \end{pmatrix}$ act on $\mathfrak{H}^m$ by

$$\alpha(z_1, \cdots, z_m) = (\alpha_1(z_1), \cdots, \alpha_m(z_m)) \,, \quad \alpha_\nu(z_\nu) = (a_\nu z_\nu + b_\nu)/(c_\nu z_\nu + d_\nu) \,.$$

For a complex-valued function f on $\mathfrak{H}^m$ and r as above, we define a function $f|_r \alpha$ on $\mathfrak{H}^m$ by

$$(1.3) \qquad (f \mid_r \alpha)(z) = f(\alpha(z)) \prod_{\nu=1}^{m} j(\alpha_\nu, z_\nu)^{-r_\nu} ,$$

where $j(\alpha_\nu, z_\nu) = \det(\alpha_\nu)^{-1/2}(c_\nu z_\nu + d_\nu)$. If we use the notation (1.1), the right-hand side of (1.3) can be written as $f(\alpha(z))j(\alpha, z)^{-r}$, or as $f(\alpha(z))(cz + d)^{-r}$ if $\det(\alpha_\nu) = 1$ for all ν.

Now we define a differential operator $D_{\nu,t}$, for $\nu = 1, \cdots, m$ and $t \in \mathbf{C}$, acting on functions f on $\mathfrak{H}^m$ by

$$(1.4) \qquad D_{\nu,t}f = \frac{1}{2\pi i}\left(\frac{t}{2iy_\nu} + \frac{\partial}{\partial z_\nu}\right)f$$

whenever this is meaningful, where $z_\nu = x_\nu + iy_\nu$ and $\partial/\partial z_\nu = (\partial/\partial x_\nu - i\partial/\partial y_\nu)/2$ as usual. The functions f to which $D_{\nu,t}$ is applied will be polynomials in $y_1, \cdots, y_m, y_1^{-1}, \cdots, y_m^{-1}$ whose coefficients are meromorphic functions in z. Let us temporarily call such functions *reasonable*. We can easily verify the following properties of $D_{\nu,t}$:

$$(1.5) \qquad D_{\nu,t+s}(fg) = D_{\nu,t}(f)g + fD_{\nu,s}(g) ;$$

$$(1.6) \qquad D_{\nu,t+s}(y_\nu^{-t}f) = y_\nu^{-t}D_{\nu,s}(f) ;$$

$$(1.7) \qquad D_{\nu,t}D_{\mu,s} = D_{\mu,s}D_{\nu,t} \quad \text{if} \quad \mu \neq \nu ;$$

$$(1.8) \qquad D_{\nu,r_\nu}(f \mid_r \alpha) = (D_{\nu,r_\nu}f) \mid_{r'} \alpha$$

for every $\alpha \in \mathrm{GL}_2^+(\mathbf{R})^m$, where

$$r = (r_1, \cdots, r_m) \in \mathbf{Z}^m \quad \text{and} \quad r' = (r_1, \cdots, r_{\nu-1}, r_\nu + 2, r_{\nu+1}, \cdots, r_m) .$$

Let F denote, throughout the paper, a totally real algebraic number field of degree m, and let $a \mapsto a_\nu$ ($\nu = 1, \cdots, m$) be the embeddings of F into $\mathbf{R}$. Then we embed F into $\mathbf{R}^m$ by $a \mapsto (a_1, \cdots, a_m)$, and $\mathrm{GL}_2(F)$ into $\mathrm{GL}_2(\mathbf{R})^m$ in a natural manner; we then put $\mathrm{GL}_2^+(F) = \mathrm{GL}_2(F) \cap \mathrm{GL}_2^+(\mathbf{R})^m$ and $\mathrm{GL}_2^+(\mathfrak{o}) = \mathrm{GL}_2(\mathfrak{o}) \cap \mathrm{GL}_2^+(F)$, where $\mathfrak{o}$ denotes the maximal order of F. We call a subgroup Γ of $\mathrm{GL}_2^+(\mathfrak{o})$ a *congruence subgroup* if there is a positive integer N such that $\Gamma_N \subset \Gamma$, where

$$(1.9) \qquad \Gamma_N = \{\gamma \in \mathrm{SL}_2(\mathfrak{o}) \mid \gamma - 1 \in N \cdot M_2(\mathfrak{o})\} .$$

For such a Γ, let $B_r(\Gamma)$ denote the set of all reasonable functions f on $\mathfrak{H}^m$ satisfying $f \mid_r \gamma = f$ for all $\gamma \in \Gamma$. Then (1.8) shows that D_{ν,r_ν} maps $B_r(\Gamma)$ into $B_{r'}(\Gamma)$. Hereafter we let D_{ν,r_ν} act on the elements of $B_r(\Gamma)$ only when r_ν is actually the νth component of r. Therefore we simply write D_ν for D_{ν,r_ν}, and define a symbol D^e for every $e = (e_1, \cdots, e_m) \in \mathbf{Z}^m$ with $e_\nu \geq 0$ by $D^e = D_1^{e_1} \cdots D_m^{e_m}$. For example, if $e = (3, 2, 0, \cdots, 0)$ and $f \in B_r(\Gamma)$, then

$$D^e f = D_{1,r_1+4}D_{1,r_1+2}D_{1,r_1}D_{2,r_2+2}D_{2,r_2}f .$$

We see easily that $D^{e+e'} = D^e D^{e'}$, and D^e maps $B_r(\Gamma)$ into $B_{r+2e}(\Gamma)$.

A holomorphic function f on $\mathfrak{H}^m$ is called a *Hilbert modular form of weight r with respect to* Γ if $f|_r \gamma = f$ for all $\gamma \in \Gamma$ and f is finite at every cusp of Γ. If $m = 1$, this amounts to the definition of classical modular forms. If $m > 1$, as is well-known, the holomorphy and the Γ-invariance of f guarantee the Fourier expansion of f of the form

$$f(z) = \sum_\xi c(\xi) e^{2\pi i \operatorname{tr}(\xi z)} \; ,$$

with $c(\xi)$ in $\mathbf{C}$, where ξ runs over 0 and all totally positive elements of a lattice $\mathfrak{a}$ in F, and $\operatorname{tr}(\xi z) = \xi_1 z_1 + \cdots + \xi_m z_m$; the lattice $\mathfrak{a}$ depends on Γ. We denote by $\mathfrak{M}_r(\Gamma)$ the vector space of all Hilbert modular forms of weight r with respect to Γ, and by $\mathfrak{M}_r$ the union of $\mathfrak{M}_r(\Gamma)$ for all congruence subgroups Γ. For any subfield R of $\mathbf{C}$, we denote by $\mathfrak{M}_r(\Gamma, R)$ (resp. $\mathfrak{M}_r(R)$) the set of all $f \in \mathfrak{M}_r(\Gamma)$ (resp. $\mathfrak{M}_r$) whose Fourier coefficients $c(\xi)$ at $(i\infty, \cdots, i\infty)$ belong to R, and by $\mathcal{G}_r(R)$ the set of all meromorphic functions of the form f/g with $f \in \mathfrak{M}_{r+s}(R)$ and $g \in \mathfrak{M}_s(R)$ with any s in $\mathbf{Z}^m$; further we denote by $\mathcal{G}_r(\Gamma, R)$ the set of all $f \in \mathcal{G}_r(R)$ such that $f|_r \gamma = f$ for all $\gamma \in \Gamma$, and put

$$\mathcal{G}_r(\Gamma) = \mathcal{G}_r(\Gamma, \mathbf{C}) \; , \quad \mathcal{G}_r = \mathcal{G}_r(\mathbf{C}) \; .$$

We see that $\mathfrak{M}_r(\Gamma, R)$ is the set of all holomorphic elements of $\mathcal{G}_r(\Gamma, R)$, and $\mathcal{G}_0(\Gamma)$ is the field of all Hilbert modular functions with respect to Γ in the ordinary sense. Now we let the operator D^e act on $\mathcal{G}_r$. If R contains the Galois closure of F over $\mathbf{Q}$, then D_ν maps $\mathcal{G}_0(R)$ into $\mathcal{G}_s(R)$, where $s = (0, \cdots, 0, 2, 0, \cdots, 0)$ with 2 at the νth component. For some technical reasons, we restrict the weight r to the set

$$(1.10) \qquad Y = \{(r_1, \cdots, r_m) \in \mathbf{Z}^m \mid r_1 \equiv \cdots \equiv r_m \pmod{2}\} \; .$$

Taking R to be the algebraic closure $\bar{\mathbf{Q}}$ of $\mathbf{Q}$, we can state our first main theorem as follows.

MAIN THEOREM I. *Let $f \in \mathcal{G}_r(\bar{\mathbf{Q}})$ and $g \in \mathcal{G}_{r+2e}(\bar{\mathbf{Q}})$ with $r \in Y$ and $e = (e_1, \cdots, e_m) \in \mathbf{Z}^m$, $e_\nu \geq 0$ for all ν. Further let w be a point of $\mathfrak{H}^m$ fixed by a non-scalar element of $\operatorname{GL}_2^+(F)$. Suppose that both f and g are holomorphic at w and $g(w) \neq 0$. Then $(g^{-1} D^e f)(w) \in \bar{\mathbf{Q}}$.*

Proof. We first make a few statements of which the proofs will be given in later sections.

(1.11_a) $\mathcal{G}_r(\bar{\mathbf{Q}}) \neq \{0\}$ *if* $r \in Y$.

(1.11_b) *If* $f \in \mathcal{G}_r(\bar{\mathbf{Q}})$, $r \in Y$, *and* $\alpha \in \operatorname{GL}_2^+(F)$, *then* $f|_r \alpha \in \mathcal{G}_r(\bar{\mathbf{Q}})$. (We assume $r \in Y$ in our theorem merely because we do not know whether $(1.11_{a,b})$ are true without this condition.)

(1.12) *There is an element* $\alpha = \begin{pmatrix} a & b \\ c & d \end{pmatrix}$ *of* $\operatorname{SL}_2(F)$ *such that* $\alpha(w) = w$ *and*

$(cw + d)^{2e} \neq 1$.

(1.13) $f(w) \in \bar{\mathbf{Q}}$ *for every* $f \in \mathcal{G}_0(\bar{\mathbf{Q}})$ *holomorphic at* w.

Now we prove our theorem by induction on $\{e\} = e_1 + \cdots + e_m$. If $\{e\} = 0$, our assertion amounts to (1.13), since $g^{-1}f \in \mathcal{G}_0(\bar{\mathbf{Q}})$. To prove the case $\{e\} > 0$, we may assume that $f(w) \neq 0$. In fact, take any non-trivial element, say φ, of $\mathcal{G}_r(\bar{\mathbf{Q}})$. Since the images of w under the elements of $\mathrm{GL}_2^+(F)$ are dense in $\mathfrak{H}^m$, we have $\varphi(\beta(w)) \neq 0$ for some $\beta \in \mathrm{GL}_2^+(F)$. Put $\psi = \varphi |_r \beta$. Then $\psi \in \mathcal{G}_r(\bar{\mathbf{Q}})$ by (1.11$_b$). If $f(w) = 0$, we consider $f + \psi$ and ψ in place of f. Since $f = (f + \psi) - \psi$, our assertion can be reduced to the case $f(w) \neq 0$. Therefore assuming $f(w) \neq 0$, take $\alpha = \begin{pmatrix} a & b \\ c & d \end{pmatrix} \in \mathrm{SL}_2(F)$ so that $\alpha(w) = w$ and $\alpha \neq \pm 1$, and put $h = (f |_r \alpha)/f$. Then $h \in \mathcal{G}_0(\bar{\mathbf{Q}})$ by (1.11$_b$), and h is holomorphic at w. We see easily that $h(w) = \lambda^{-r}$, where $\lambda = (\lambda_1, \cdots, \lambda_m)$, $\lambda_\nu = c_\nu w_\nu + d_\nu$. Applying D_ν to the equality $f |_r \alpha = fh$, we obtain

$$(1.14) \qquad (D_\nu f) |_s \alpha = (D_\nu f)h + f \cdot (D_\nu h) \quad \left(s = (r_1, \cdots, r_\nu + 2, \cdots, r_m)\right)$$

by (1.5) and (1.8). Evaluate both sides at w. Then

$$(1.15) \qquad D_\nu f(w)(\lambda_\nu^{-2} - 1)\lambda^{-r} = f(w)(D_\nu h)(w) .$$

Observe that $f \cdot D_\nu h \in \mathcal{G}_s(\bar{\mathbf{Q}})$. Therefore if $g \in \mathcal{G}_s(\bar{\mathbf{Q}})$ and $g(w) \neq 0$, then $g^{-1}f \cdot D_\nu h \in \mathcal{G}_0(\bar{\mathbf{Q}})$, so that $(g^{-1}f \cdot D_\nu h)(w) \in \bar{\mathbf{Q}}$ by (1.13) and thus $(g^{-1}D_\nu f)(w) \in \bar{\mathbf{Q}}$ by (1.15). This proves the case $\{e\} = 1$. In the general case, apply D^e to $f |_r \alpha = fh$. Then we obtain, by (1.5) and (1.8),

$$(D^e f) |_{r+2e} \alpha = (D^e f)h + \sum_p C_p (D^{e-p} f)(D^p h)$$

with $C_p \in \mathbf{Z}$, where $p = (p_1, \cdots, p_m)$ runs over the elements of $\mathbf{Z}^m$ such that $0 \leqq p_\nu \leqq e_\nu$ and $\{p\} > 0$. Evaluate this at w. Then

$$D^e f(w)(\lambda^{-2e} - 1)\lambda^{-r} = \sum_p C_p D^{e-p} f(w) D^p h(w) .$$

Take $g_p \in \mathcal{G}_{2p}(\bar{\mathbf{Q}})$ so that $g_p(w) \neq 0$. (Such a g_p is guaranteed by (1.11$_{a,b}$) and the denseness of $\{\beta(w) \mid \beta \in \mathrm{GL}_2^+(F)\}$.) Then

$$(1.16) \qquad (g^{-1}D^e f)(w)(\lambda^{-2e} - 1)\lambda^{-r} = \sum_p C_p (g_p g^{-1} D^{e-p} f)(w)(g_p^{-1} D^p h)(w) .$$

Applying induction to $D^{e-p}f$, we know that $(g_p g^{-1} D^{e-p} f)(w) \in \bar{\mathbf{Q}}$. As for $g_p^{-1} D_p h$, we have $D^p h = D^q(D_\nu h)$ for suitable q and ν. Since $\{q\} = \{p\} - 1 < \{e\}$, our induction is applicable to $D^p h$ too. Obviously the induction process is not disturbed by our reduction to the case $f(w) \neq 0$. By (1.12), we can take α so that $\lambda^{2e} \neq 1$. Thus relation (1.16) completes our proof.

Remark. We could have used, instead of $D_{\nu,t}$, the operator

$$\delta_{\nu,t} = (2\pi i)^{-1}(t + 2iy_\nu \cdot \partial/\partial z_\nu) = 2iy_\nu D_{\nu,t} .$$

It can easily be verified, by virtue of (1.6), that

$$\delta_{\nu,t+n} \cdots \delta_{\nu,t+1}\delta_{\nu,t} = (2iy_\nu)^{n+1}D_{\nu,t+2n} \cdots D_{\nu,t+2}D_{\nu,t}$$

for every positive integer n. We preferred $D_{\nu,t}$ to $\delta_{\nu,t}$ simply in order to avoid the automorphic factors $c_\nu\bar{z}_\nu + d_\nu$. It should be noted that $2\pi i\delta_{\nu,t}$ was introduced by Maass in [9] as K_t. It is also an inhomogeneous version of $\bar{\omega}_1\partial/\partial\omega_1 + \bar{\omega}_2\partial/\partial\omega_2$, which Weil uses in [18] to give a proof of the result of Damerell [2]. In fact, if $\varphi(\omega_1, \omega_2) = \omega_2^{-t}f(\omega_1/\omega_2)$ in the case $m = 1$, we have

$$(\bar{\omega}_1\partial/\partial\omega_1 + \bar{\omega}_2\partial/\partial\omega_2)\varphi = -2\pi i\bar{\omega}_2\omega_2^{-t-1}(\delta_t f)(\omega_1/\omega_2) \ .$$

Let w be a fixed point of a non-scalar element of $\mathrm{GL}_2^+(F)$ as above. Then there is a totally imaginary quadratic extension K of F and an F-linear embedding θ of K into the matrix algebra $M_2(F)$ such that

$$\theta(K^\times) = \{\beta \in \mathrm{GL}_2^+(F) \mid \beta(w) = w\} \ .$$

Moreover, the embedding $a \mapsto a_\nu$ of F into $\mathbf{R}$ can be extended to an embedding τ_ν of K into $\mathbf{C}$ so that

$$(1.17) \qquad d/dz_\nu[\theta(\xi)_\nu(z_\nu)]_{z_\nu=w_\nu} = \xi^{\tau_\nu\rho}/\xi^{\tau_\nu} \qquad \text{for all } \xi \in K^\times ,$$

where ρ denotes the complex conjugation. Then (K, Φ) with $\Phi = \{\tau_1, \cdots, \tau_m\}$ is a CM-type. For details, the reader is referred to [10, 2.6, 2.7]. If $\theta(\xi) = \begin{pmatrix} a & b \\ c & d \end{pmatrix}$, we have $\xi^{\tau_\nu} = c_\nu w_\nu + d_\nu$. Note that there is a unique $w_0 \in K$ such that $w_0^{\tau_\nu} = w_\nu$ for all ν. Now take $\xi \in K^\times$ so that $\prod_{\nu=1}^m (\xi^{\tau_\nu\rho}/\xi^{\tau_\nu})^{2e_\nu} \neq 1$, and put $\alpha = \theta(\xi^\rho/\xi)$. Then α has the properties of (1.12). The existence of such a ξ is obvious, because $\xi \mapsto (\xi^{\tau_1}, \cdots, \xi^{\tau_m})$ maps K into a dense subset of $\mathbf{C}^m$.

Let K' be the field generated over $\mathbf{Q}$ by $\sum_{\nu=1}^m \xi^{\tau_\nu}$ for all $\xi \in K$. Then a stronger version of (1.13) holds in the following form:

(1.18) $f(w) \in K'_{\mathrm{ab}}$ *for every* $f \in \mathcal{C}_0(\mathbf{Q}_{\mathrm{ab}})$ *holomorphic at* w.

Here X_{ab} denotes the maximal abelian extension of X. With this result in mind, we can now state a stronger form of the above theorem as

MAIN THEOREM II. *The notation and the assumptions being as in Main Theorem I, suppose $f \in \mathcal{C}_r(\Lambda)$ and $g \in \mathcal{C}_{r+2e}(\Lambda)$ with a subfield Λ of $\bar{\mathbf{Q}}$. Then $(g^{-1}D^e f)(w) \in K'_{\mathrm{ab}}\Lambda$.*

We shall actually prove a still stronger version as Main Theorem III in Section 5. We can often take $\mathbf{Q}$ or $\mathbf{Q}_{\mathrm{ab}}$ as Λ if $r_1 = \cdots = r_m$ and $e_1 = \cdots = e_m$. Actually the condition $\mathcal{C}_r(\Lambda) \neq \{0\}$ with a small Λ imposes restrictions on r. In fact, if $\mathcal{C}_r(\Lambda) \neq \{0\}$ and there is an element σ of $\mathrm{Gal}\,(\bar{\mathbf{Q}}/\Lambda)$ such that $a_\nu^\sigma = a_\mu$ for all $a \in F$, then $r_\nu = r_\mu$. This fact will be stated and proved as Proposition 7, (iv) in Section 5. It should also be mentioned that our main theorems can easily be generalized to the case of modular forms of 'half-integral weight', and also to automorphic forms with respect to arithmetic discontinuous groups other than the Hilbert modular groups. We shall not, however, discuss such generalizations in the present paper.

2. Eisenstein series

To simplify our notation, we write u for the element of $\mathbf{C}^m$ whose components are all equal to 1; thus

$$(2.1) \qquad tu = (t, t, \cdots, t) \qquad (t \in \mathbf{C}) .$$

Let κ be a positive integer, and $r = (r_1, \cdots, r_m) \in \mathbf{Z}^m$ with $r_\nu \geq 0$. Further let $\mathfrak{x}$ be a fractional ideal in F, and $\mathfrak{n}$ an integral ideal in F. With a, b in $\mathfrak{x}$ and a complex variable s, we define an Eisenstein series

$$(2.2) \quad \begin{aligned} &E_{\kappa,r}(z, s; a, b; \mathfrak{x}, \mathfrak{n}) \\ &= (z - \bar{z})^{-r}(2\pi i)^{-\kappa m - |r|} \sum_{c,d} \left(\frac{c\bar{z} + d}{cz + d} \right)^r (cz + d)^{-\kappa u} |cz + d|^{-su} , \end{aligned}$$

where (c, d) runs over all *equivalence classes* of pairs of elements of $\mathfrak{x}$ such that $(c, d) \neq (0, 0)$, $c \equiv a$, $d \equiv b \pmod{\mathfrak{n}\mathfrak{x}}$; (c, d) and (c', d') are said to be *equivalent* if there is a totally positive unit ε of F such that $c' = c\varepsilon$, $d' = d\varepsilon$, and $\varepsilon \equiv 1 \pmod{\mathfrak{n}}$; the notation $(z - \bar{z})^{-r}$, $(cz + d)^{\kappa u}$, etc. should be understood in the sense of (1.1). The series is absolutely convergent for $\mathrm{Re}\,(s) + \kappa > 2$. Moreover, it has a Fourier expansion of the form

$$(2.3) \qquad E_{\kappa,r}(x + iy, s; a, b; \mathfrak{x}, \mathfrak{n}) = \sum_{\alpha \in \mathfrak{y}} e^{2\pi i \mathrm{tr}(\alpha x) - 2\pi \mathrm{tr}(|\alpha|y)} H_\alpha(y, s) ,$$

where $\mathfrak{y} = \mathfrak{b}^{-1}\mathfrak{n}^{-1}$ with the different $\mathfrak{b}$ of F over $\mathbf{Q}$, and $H_\alpha(y, s)$ can be explicitly expressed by confluent hypergeometric functions. If $r = 0$, as shown by Hecke [3] and Kloosterman [7], the series on the right-hand side defines a meromorphic function in s, holomorphic at $s = 0$. The same is true also for $r \neq 0$, and therefore we can define a function $E_{\kappa,r}(z; a, b; \mathfrak{x}, \mathfrak{n})$ to be the value of (2.3) at $s = 0$.

Let us fix $a, b, \mathfrak{x}$ and $\mathfrak{n}$ for the moment, and denote the function (2.3) and its value at $s = 0$ simply by $E_{\kappa,r}(z, s)$ and $E_{\kappa,r}(z)$. It can easily be verified that $E_{\kappa,r} |_{\kappa u + 2r} \gamma = E_{\kappa,r}$ for all γ in a suitable conguence subgroup. Moreover a direct computation shows that

$$(2.4) \qquad D_{\nu, p + (s/2)} E_{\kappa,r}(z, s) = (\kappa + r_\nu + (s/2)) E_{\kappa, t}(z, s) ,$$

where $p = \kappa + 2r_\nu$ and $t = (r_1, \cdots, r_{\nu-1}, r_\nu + 1, r_{\nu+1}, \cdots, r_m)$. Substituting 0 for s, we obtain, at least formally,

$$(2.5) \qquad D_{\nu, p} E_{\kappa,r}(z) = (\kappa + r_\nu) E_{\kappa, t}(z) .$$

This is obviously true if $\kappa \geq 3$. Even if κ is either 1 or 2, (2.5) is valid, because we have

$$[(\partial/\partial z_\nu)E_{\kappa,r}(z,\,s)]_{s=0} = \partial E_{\kappa,r}(z)/\partial z_\nu\,,$$

which can be seen by letting $\partial/\partial z_\nu$ act on the right-hand side of (2.3) term by term, and observing that the differentiated series are absolutely and uniformly convergent when $(z,\,s)$ stays in a compact set. (This is more or less elementary, but it may be worth noting that, with the notation of [13], one has

$$(\partial/\partial y)\Gamma(\beta)^{-1}\sigma(y,\,\alpha,\,\beta) = -\Gamma(\beta)^{-1}\sigma(y,\,\alpha,\,\beta+1)\,;$$

also [13, Lemma 4] can be used to majorize $H_\alpha(y,\,s)$ and its derivatives.)

Now, in the case $r = 0$, $E_{\kappa,0}(z)$ belongs to $\mathfrak{M}_{\kappa u}(\mathbf{Q}_{ab})$ except for the case $m = 1$ and $\kappa = 2$. The $\mathbf{Q}_{ab}$-rationality of the Fourier coefficients of $E_{\kappa,0}$ was shown by Klingen [5], [6][1] and Siegel [15], [16].

THEOREM 1. *Let w and K' be as in Section 1. Further let g be an element of $\mathcal{Q}_{\kappa u+2r}(\Lambda)$ with a subfield Λ of $\bar{\mathbf{Q}}$ such that $g(w)$ is finite and $\neq 0$. Then $g(w)^{-1}E_{\kappa,r}(w) \in K'_{ab}\Lambda$.*

Proof. From (2.5), we obtain easily

$$(2.6) \qquad\qquad D^r E_{\kappa,0}(z) = B \cdot E_{\kappa,r}(z)$$

with a positive integer B. Therefore, if $m > 1$ or $\kappa \neq 2$, our assertion follows immediately from Main Theorem II. Suppose $m = 1$ and $\kappa = 2$. Without losing generality, we may assume that $\mathfrak{x} = \mathbf{Z}$ and $\mathfrak{n} = N\mathbf{Z}$ with a positive integer N. By Hecke [3, Werke, p. 469], we have

$$E_{2,0}(z) = -[2\pi i N^2(z - \bar{z})]^{-1} + \sum_{n=0}^{\infty}\mu_n q^{n/N} \qquad\qquad (q = e^{2\pi i z})$$

with $\mu_n \in \mathbf{Q}_{ab}$. Denote by $E(z)$ this function in the case $N = 1$. Then

$$\begin{aligned}
E(z) &= (4\pi y)^{-1} - 12^{-1} + 2\sum_{n=1}^{\infty}\left(\sum_{c|n} c\right)q^n \\
&= -(24\pi i)^{-1}\left((12/2iy) + d\log\Delta(z)/dz\right) \\
&= -(12\Delta)^{-1}D\Delta\,,
\end{aligned}$$

where $\Delta(z) = q\prod_{n=1}^{\infty}(1 - q^n)^{24}$. We see that $N^2 E_{2,0}(z) - E(z)$ belongs to $\mathfrak{M}_2(\mathbf{Q}_{ab})$. Therefore, in view of Main Theorem II, it is sufficient to show that $(g^{-1}D^r E)(w) \in K_{ab}\Lambda$ if $g \in \mathcal{Q}_{2+2r}(\Lambda)$ and $g(w) \neq 0$. Now

$$-12D^r E = D^r(\Delta^{-1}D\Delta) = \sum_{b=0}^{r}\binom{r}{b}D^{r-b}(\Delta^{-1})D^{b+1}\Delta\,,$$

so that

$$-12(g^{-1}D^r E)(w) = \sum_{b=0}^{r}\binom{r}{b}(g^{-1}g_b D^{r-b}\Delta^{-1})(w)(g_b^{-1}D^{b+1}\Delta)(w)$$

---

[1] In [6], the weight κ is assumed to be ≥ 2. However, the case $\kappa = 1$ can be reduced to the case $m = 1$ and $\kappa \geq 2$, by putting $z_1 = \cdots = z_m$.

with $g_b \in G_{14+2b}(\mathbf{Q})$, $g_b(w) \neq 0$. (The existence of such g_b is obvious.) Therefore our assertion follows from Main Theorem II.

3. The special values of certain L-functions

We now apply the above result to certain Hecke L-functions with algebraic-valued (Grössen-) characters. Let K be a totally imaginary quadratic extension of F, and χ a primitive Hecke character of K (considered an ideal character) of conductor $\mathfrak{f}$ such that

$$\chi((a)) = \prod_\sigma (a^\sigma/|a^\sigma|)^{\mu_\sigma} \qquad \text{for } a \equiv 1 \pmod{\mathfrak{f}}$$

with $\mu_\sigma \in \mathbf{Z}$, where σ runs over all embeddings of K into $\mathbf{C}$. Then we can choose a suitable set $\Phi = \{\tau_1, \cdots, \tau_m\}$ of m embeddings of K into $\mathbf{C}$ so that

$$(3.1) \qquad \chi((a)) = \prod_{\nu=1}^m (a^{\tau_\nu\rho}/|a^{\tau_\nu}|)^{t_\nu} \qquad \text{for } a \equiv 1 \pmod{\mathfrak{f}}$$

with non-negative integers t_ν, and $\{\tau_1, \cdots, \tau_m, \tau_1\rho, \cdots, \tau_m\rho\}$ is the set of all embeddings of K into $\mathbf{C}$, where ρ denotes the complex conjugation. We define as usual the L-function with character χ by $L(s, \chi) = \sum_\mathfrak{a} \chi(\mathfrak{a})N(\mathfrak{a})^{-s}$, where $\mathfrak{a}$ runs over all integral ideals of K prime to $\mathfrak{f}$. Take a set of representatives B for the ideal classes modulo $\mathfrak{f}$, consisting of integral ideals. Then

$$L(s, \chi) = \sum_{\mathfrak{b} \in B} \chi(\mathfrak{b})^{-1}N(\mathfrak{b})^s \sideset{}{'}\sum_{a \in \mathfrak{b}} \chi((a))N(a)^{-s},$$

where a runs over all elements of $\mathfrak{b}$ such that $0 \neq a \equiv 1 \pmod{\mathfrak{f}}$, modulo multiplication by units ε of K such that $\varepsilon \equiv 1 \pmod{\mathfrak{f}}$. Let $U_\mathfrak{f}$ denote the group of all such units, and $U_\mathfrak{f}^0$ the group of all totally positive elements of $U_\mathfrak{f} \cap F$. Take an element w of K so that $\mathrm{Im}(w^{\tau_\nu}) > 0$ for all ν. We can find an ideal $\mathfrak{x}$ and an integral ideal $\mathfrak{n}$ in F so that

$$\mathfrak{n}\mathfrak{x}w + \mathfrak{n}\mathfrak{x} \subset \mathfrak{f}\mathfrak{b} \subset \mathfrak{b} \subset \mathfrak{x}w + \mathfrak{x} \qquad \text{for all } \mathfrak{b} \in B.$$

Then $\mathfrak{n} \subset \mathfrak{f}$ and

$$\sideset{}{'}\sum_{a \in \mathfrak{b}} \chi((a))N(a)^{-s} = [U_\mathfrak{f} : U_\mathfrak{n}^0]^{-1} \sum_{c \in R_\mathfrak{b}} \sum_{a \sim c} \chi((a))N(a)^{-s},$$

where we write $a \sim c$ if $a \equiv \varepsilon c \pmod{\mathfrak{n}\mathfrak{x}w + \mathfrak{n}\mathfrak{x}}$ for some $\varepsilon \in U_\mathfrak{n}^0$, and $R_\mathfrak{b}$ is a set of representatives for the elements a of $\mathfrak{b}$ such that $a \equiv 1 \pmod{\mathfrak{f}}$ modulo the equivalence $\sim$; a must be taken modulo $U_\mathfrak{n}^0$.

Now we arrange τ_ν so that $x^{\tau_\nu} = x_\nu$ for $x \in F$. Suppose $t_1 \equiv \cdots \equiv t_m$ (mod 2). Let κ be a positive integer such that $\kappa \equiv t_\nu$ (mod 2) and $\kappa \leq t_\nu$ for all ν. Put $r = (r_1, \cdots, r_m)$ with $r_\nu = (t_\nu - \kappa)/2$. Then, identifying every $a \in K$ with the point $(a^{\tau_1}, \cdots, a^{\tau_m})$ of $\mathbf{C}^m$, we have

$$\chi((a))N(a)^{-(\kappa+s)/2} = (\bar{a}/a)^r a^{-\kappa u}|a|^{-su}$$

if $a \equiv 1 \pmod{\mathfrak{f}}$. Therefore we obtain

$$(w - \bar{w})^{-r}(2\pi i)^{-\kappa m - \{r\}} L((\kappa + s)/2, \chi)$$
$$= [U_{\mathfrak{f}} : U_{\mathfrak{n}}^0]^{-1} \sum_{\mathfrak{b} \in B} \chi(\mathfrak{b})^{-1} N(\mathfrak{b})^{(\kappa+s)/2} \sum_{c \in R_{\mathfrak{b}}} E_{\kappa,r}(w, s; p_c, q_c; \mathfrak{x}, \mathfrak{n}),$$

where $c = p_c w + q_c$ with p_c, q_c in $\mathfrak{x}$. Both sides are meaningful at $s = 0$, hence

$$(w - \bar{w})^{-r}(2\pi i)^{-\kappa m - \{r\}} L(\kappa/2, \chi)$$
$$= [U_{\mathfrak{f}} : U_{\mathfrak{n}}^0]^{-1} \sum_{\mathfrak{b} \in B} \chi(\mathfrak{b})^{-1} N(\mathfrak{b})^{\kappa/2} \sum_{c \in R_{\mathfrak{b}}} E_{\kappa,r}(w; p_c, q_c; \mathfrak{x}, \mathfrak{n}).$$

Applying Theorem 1 to the right-hand side, we obtain

THEOREM 2. *Let χ be a Hecke character of K satisfying (3.1) with exponents t_ν such that $t_1 \equiv \cdots \equiv t_m \pmod 2$. Let κ be a positive integer such that $\kappa \equiv t_\nu \pmod 2$ and $\kappa \leq t_\nu$ for all ν. Then, for every $g \in \mathcal{A}_t(\bar{\mathbf{Q}})$ such that $g(w) \neq 0$, the value $\pi^{-\lambda} g(w)^{-1} L(\kappa/2, \chi)$ is algebraic, where $\lambda = (\kappa m + \sum_{\nu=1}^{m} t_\nu)/2$, and w is any element of K such that $\mathrm{Im}(w^{r_\nu}) > 0$ for all ν, identified with the point $(w^{r_1}, \cdots, w^{r_m})$ of $\mathfrak{H}^m$.*

As mentioned in the Introduction, Theorems 1 and 2 in the case $m = 1$ were first proved by Damerell [2]; recently Weil gave a simplified proof in [18].

A consequence of the above result, though easy, may be worth noting. Let k be an abelian extension of K, and ψ a Hecke character of k such that $\psi(\mathfrak{a}) = \chi(N_{k/K}(\mathfrak{a}))$ with a character χ of the above type. Then we have

$$L(s, \psi) = \prod_\varphi L(s, \varphi\chi),$$

where φ runs over all characters of finite order corresponding to the extension k/K. Therefore we can conclude that $(\pi^\lambda g(w))^{-[k:K]} L(\kappa/2, \psi)$ is algebraic. The zeta-function of an abelian variety with complex multiplication provides a typical example of such $L(s, \psi)$.

4. The Hilbert modular functions that are arithmetic

We now consider a canonical system with respect to $\mathrm{GL}_2(F)$ in the sense of [11]. More precisely, we take an algebraic subgroup G of GL_{2m} defined over $\mathbf{Q}$ so that $G_{\mathbf{Q}}$ can be naturally identified with $\mathrm{GL}_2(F)$. Then we consider its adelization G_A and define subgroups G_{A+} and $\bar{\mathcal{G}}_+$ by

$$G_{A+} = \{x \in G_A \mid x_\infty \in \mathrm{GL}_2^+(\mathbf{R})^m\},$$
$$\bar{\mathcal{G}}_+ = \{x \in G_{A+} \mid \det(x) \in \mathbf{Q}_A^\times \overline{F^\times F_{\infty+}^\times}\},$$

where x_∞ denotes the archimedean component of x. We can also speak of arithmetic automorphic functions on $\mathfrak{H}^m$ with respect to G as defined in [11, II, §6.1]. They form a field $\mathfrak{K}$ which contains $\mathbf{Q}_{ab}$ and is linearly disjoint with $\mathbf{C}$ over $\mathbf{Q}_{ab}$; the composite $\mathbf{C}\mathfrak{K}$ is the field of all Hilbert modular functions with respect to all congruence subgroups of $SL_2(\mathfrak{o})$. Moreover, $\bar{\mathcal{G}}_+$ acts on $\mathfrak{K}$ as a group of automorphisms. Especially, if $\alpha \in GL_2^+(F)$, then the action of α as an automorphism of $\mathfrak{K}$ is given by $f \mapsto f \circ \alpha$. If w, K, and K' are as in Section 1, then, by [11, II, (6.2.3)], we have

$$(4.1) \qquad f(w) \in K'_{ab} \text{ for every } f \in \mathfrak{K} \text{ holomorphic at } w .$$

Therefore (1.13) and (1.18) follow from

THEOREM 3. *Let Λ be a subfield of $\mathbf{C}$ containing $\mathbf{Q}_{ab}$. Then $\mathcal{C}_0(\Lambda) = \Lambda\mathfrak{K}$.*

The proof will be completed after a few propositions and theorems. We reduce part of the problem to the one-dimensional case. Let us denote the objects G, $\mathfrak{K}$, $\mathfrak{M}_r(\Lambda)$, $\mathcal{C}_r(\Lambda)$ in the case $F = \mathbf{Q}$ by G^1, $\mathfrak{K}^1$, $\mathfrak{M}_r^1(\Lambda)$, $\mathcal{C}_r^1(\Lambda)$. By [12, Ch. 6], we know that $\mathfrak{K}^1 = \mathcal{C}_0^1(\mathbf{Q}_{ab})$ (but not yet $\Lambda\mathfrak{K}^1 = \mathcal{C}_0^1(\Lambda)$). Now consider the diagonal embedding $P \colon \mathfrak{H} \to \mathfrak{H}^m$ defined by $P(z) = (z, \cdots, z)$. Obviously $P \circ \alpha = \alpha \circ P$ for every $\alpha \in GL_2^+(\mathbf{Q})$.

THEOREM 4. *If $f \in \mathfrak{K}$ and $f \circ P$ is defined, then $f \circ P \in \mathfrak{K}^1$, and $(f \circ P)^v = f^v \circ P$ for every $v \in G_{\Lambda+}^1$, where the upper right v means the action of v as an automorphism of $\mathfrak{K}$ and of $\mathfrak{K}^1$.*

This can easily be proved by the same technique as in [11, I, 8.8–8.12; II, 7.1], where a more difficult case is treated. Especially Theorem 8.9 and the reasoning in 8.12 of [11, I] are applicable to the present case with an obvious translation of the notation.

In the following three propositions, Λ always denotes a subfield of $\mathbf{C}$ containing $\mathbf{Q}_{ab}$.

PROPOSITION 1. *Let A be a subset of $GL_2^+(F)$ such that $\bigcup_{\alpha \in A} \alpha(\mathfrak{H})$ is dense in $\mathfrak{H}^m$. Then an element f of $\mathbf{C}\mathfrak{K}$ belongs to $\Lambda\mathfrak{K}$ if and only if $f \circ \alpha \circ P \in \Lambda\mathfrak{K}^1$ for every $\alpha \in A$ such that $f \circ \alpha \circ P$ is defined.*

Proof. The 'only if'-part is included in Theorem 4. To prove the converse, let $f \in \mathbf{C}\mathfrak{K}$ and express f as $f = (\sum a_n g_n)/(\sum a_n h_n)$ with finitely many elements a_n of $\mathbf{C}$ and g_n, h_n of $\Lambda\mathfrak{K}$. We may assume that the a_n are linearly independent over Λ. Let B be the subset of A consisting of all α such that $f \circ \alpha \circ P$, $g_n \circ \alpha \circ P$, $h_n \circ \alpha \circ P$ are all meaningful. Then $\bigcup_{\alpha \in B} \alpha(\mathfrak{H})$ is still dense in $\mathfrak{H}^m$. Now we have

$$\sum a_n[(f \circ \alpha \circ P)(h_n \circ \alpha \circ P) - (g_n \circ \alpha \circ P)] = 0 .$$

Suppose $f \circ \alpha \circ P \in \Lambda\mathfrak{K}^1$ for all $\alpha \in B$. Since $\Lambda\mathfrak{K}^1$ and $\mathbf{C}$ are linearly disjoint over Λ, we have $(f \circ \alpha \circ P)(h_n \circ \alpha \circ P) = g_n \circ \alpha \circ P$ for every n, i.e., $fh_n = g_n$ on $\bigcup_{\alpha \in B} \alpha(\mathfrak{H})$. Therefore $fh_n = g_n$ on the whole $\mathfrak{H}^m$, and hence $f \in \Lambda\mathfrak{K}$.

For example, we can take A to be the group

$$(4.2) \qquad \left\{ \begin{pmatrix} a & b \\ 0 & a^{-1} \end{pmatrix} \,\middle|\, a \in F^{\times},\, b \in F \right\}.$$

We now consider $\mathfrak{M}_r(\Lambda)$ with $r = (r_1, \cdots, r_m) \in \mathbf{Z}^m$. Observe that $f \circ P \in \mathfrak{M}^1_{\{r\}}(\Lambda)$ if $f \in \mathfrak{M}_r(\Lambda)$.

PROPOSITION 2. *If $f \in \mathfrak{M}_r(\Lambda)$, then $f \circ \alpha \circ P \in \mathfrak{M}^1_{\{r\}}(\Lambda)$ for every α of $\mathrm{GL}_2^+(F)$ of the form $\alpha = \begin{pmatrix} a & b \\ 0 & d \end{pmatrix}$. Conversely, an element f of $\mathfrak{M}_r$ belongs to $\mathfrak{M}_r(\Lambda)$ if $f \circ \beta \circ P \in \mathfrak{M}^1_{\{r\}}(\Lambda)$ for all β of the form $\beta = \begin{pmatrix} 1 & b \\ 0 & 1 \end{pmatrix}$ with $b \in F$.*

Proof. Let $f \in \mathfrak{M}_r$ and let $f(z) = \sum_{\xi \in \mathfrak{a}} \lambda(\xi) e^{2\pi i \mathrm{tr}(\xi z)}$ with a lattice $\mathfrak{a}$ in F. If $\alpha = \begin{pmatrix} a & b \\ 0 & d \end{pmatrix} \in \mathrm{GL}_2^+(F)$, then

$$(4.3) \qquad f(\alpha(z)) = \sum_{\xi \in \mathfrak{a}} \lambda(\xi) e^{2\pi i \mathrm{tr}(\xi d^{-1} b)} e^{2\pi i \mathrm{tr}(\xi d^{-1} a z)}.$$

Therefore the first assertion is obvious. Especially if $\beta = \begin{pmatrix} 1 & b \\ 0 & 1 \end{pmatrix}$, we have

$$f \circ \beta \circ P = \sum_{v \in \mathbf{Q}} \mu(v) e^{2\pi i v z}, \quad \mu(v) = \sum_{\xi \in \mathfrak{a},\, \mathrm{tr}(\xi) = v} \lambda(\xi) e^{2\pi i \mathrm{tr}(\xi b)}.$$

Therefore the converse follows immediately from

LEMMA. *For every non-empty finite subset X of F, there exists a subset Y of F with the same number of elements as X such that*

$$\det \left(e^{2\pi i \mathrm{tr}(\xi \eta)} \right)_{\xi \in X,\, \eta \in Y} \neq 0.$$

Proof. Put $\varphi_\xi(x) = e^{2\pi i \mathrm{tr}(\xi x)}$ for ξ, $x \in F$. Then the φ_ξ for $\xi \in X$ define different characters of a finite additive group $\mathfrak{b}/\mathfrak{c}$ for suitable lattices $\mathfrak{b}$ and $\mathfrak{c}$ in F. Therefore the existence of Y is obvious.

PROPOSITION 3. $\mathcal{A}_0(\Lambda) \subset \Lambda \mathfrak{R}$ *if* $\mathcal{A}_0^1(\Lambda) \subset \Lambda \mathfrak{R}^1$. *Especially* $\mathcal{A}_0(\mathbf{Q}_{ab}) \subset \mathfrak{R}$.

Proof. Let $f = g/h$ with g and h in $\mathfrak{M}_r(\Lambda)$. Take $\alpha = \begin{pmatrix} a & b \\ 0 & d \end{pmatrix} \in \mathrm{SL}_2(F)$ so that $h \circ \alpha \circ P \neq 0$. By Proposition 2, both $g \circ \alpha \circ P$ and $h \circ \alpha \circ P$ belong to $\mathfrak{M}^1_{\{r\}}(\Lambda)$, and hence $f \circ \alpha \circ P \in \mathcal{A}_0^1(\Lambda)$. Therefore our assertions follow from Proposition 1.

PROPOSITION 4. (i) *Let σ be a field-automorphism of $\mathbf{C}$, and $g(z) = \sum_\xi c(\xi) e^{2\pi i \mathrm{tr}(\xi z)}$ be an element of $\mathfrak{M}_{\kappa u}$ with a positive integer κ, where $\kappa u = (\kappa, \cdots, \kappa)$ as in (2.1). Then there exists an element g^σ of $\mathfrak{M}_{\kappa u}$ such that $g^\sigma(z) = \sum_\xi c(\xi)^\sigma e^{2\pi i \mathrm{tr}(\xi z)}$.*

(ii) *Let $f = g/h$ with g and h in $\mathfrak{M}_{\kappa u}(\mathbf{Q}_{ab})$, and let $v = \begin{pmatrix} 1 & 0 \\ 0 & t \end{pmatrix}^{-1}$ be an element of $\mathrm{GL}_2(\mathbf{A})\,(= G_{\mathbf{A}}^1)$ with $t \in \prod_p \mathbf{Z}_p^\times$. Then $f^v = g^t/h^t$, where we consider t as an element of $\mathrm{Gal}(\mathbf{Q}_{ab}/\mathbf{Q})$ in a natural manner.*

Proof. We postpone the proof of (i) and prove (ii) assuming (i). In the case $m = 1$, assertion (ii) follows immediately from [12, 6.2, 6.5] (though the fact is not explicitly stated). Put

$$g(z) = \sum_\xi c(\xi)e^{2\pi i \operatorname{tr}(\xi z)} \quad \text{and} \quad h(z) = \sum_\xi d(\xi)e^{2\pi i \operatorname{tr}(\xi z)} .$$

Suppose $h \circ \beta \circ P \neq 0$ for an element $\beta = \begin{pmatrix} a & b \\ 0 & a^{-1} \end{pmatrix} \in \mathrm{SL}_2(F)$. Now we can find a positive integer N so that g and h are Γ_N-invariant and f is invariant under every element of the subgroup

$$\{(x_p) \in \textstyle\prod_p \mathrm{SL}_2(\mathfrak{o}_p) \mid x_p \equiv 1 \ N \cdot M_2(\mathfrak{o}_p)\}$$

of $\bar{\mathfrak{S}}_+$, where $\mathfrak{o}_p = \mathfrak{o} \otimes_{\mathbf{Z}} \mathbf{Z}_p$ for each rational prime p. We can further find a positive integer s so that $ab(t_p - s) \in N\mathfrak{o}_p$ for all p, where t_p is the p-component of t. Then we have $v\beta = w\gamma v$ with

$$\gamma = \begin{pmatrix} a & sb \\ 0 & a^{-1} \end{pmatrix} \quad \text{and} \quad w = \begin{pmatrix} 1 & ab(t - s) \\ 0 & 1 \end{pmatrix} .$$

Therefore $f^{v\beta} = f^{\gamma v} = (f \circ \gamma)^v$, so that, by Theorem 4,

$$f^v \circ \beta \circ P = (f \circ \gamma)^v \circ P = (f \circ \gamma \circ P)^v = (g \circ \gamma \circ P)^t/(h \circ \gamma \circ P)^t ,$$

since our assertion is true in the one-dimensional case. Now

$$(g \circ \gamma \circ P)^t = \sum_\xi c(\xi)^t (e^{2\pi i \operatorname{tr}(\xi abs)})^t e^{2\pi i \operatorname{tr}(a^2 \xi)z} ,$$
$$g^t \circ \beta \circ P = \sum_\xi c(\xi)^t e^{2\pi i \operatorname{tr}(\xi ab)} e^{2\pi i \operatorname{tr}(a^2 \xi)z} .$$

Since ξ runs over the elements of a lattice in F, we see, with a choice of s sufficiently close to t, that $g^t \circ \beta \circ P = (g \circ \gamma \circ P)^t$. The same is true for h. Therefore we have

$$f^v \circ \beta \circ P = (g^t \circ \beta \circ P)/(h^t \circ \beta \circ P) = (g^t/h^t) \circ \beta \circ P$$

for all $\beta = \begin{pmatrix} a & b \\ 0 & a^{-1} \end{pmatrix}$ whenever $h \circ \beta \circ P \neq 0$ and $h^t \circ \beta \circ P \neq 0$. Hence $f^v = g^t/h^t$.

Remark. The above proof shows that, *without assuming* (i), *we have* $(g/h)^v = g^t/h^t$ *as long as* g^t *and* h^t *define elements of* $\mathfrak{M}_{\kappa u}$. This is so, for example, if $c(\xi)$ and $d(\xi)$ are invariant under t. Also, *without assuming* (i), *suppose* $f^v = f$ *and* $d(\xi)^t = d(\xi)$ *for all* ξ. *Then* $c(\xi)^t = c(\xi)$ *for all* ξ. Indeed, in the above proof, we have an equality of formal power series:

$$(g \circ \beta \circ P)/(h \circ \beta \circ P) = f \circ \beta \circ P = f^v \circ \beta \circ P = (f \circ \gamma \circ P)^v$$
$$= (g \circ \gamma \circ P)^t/(h \circ \gamma \circ P)^t = (g \circ \gamma \circ P)^t/(h \circ \beta \circ P) ,$$

hence $g \circ \beta \circ P = (g \circ \gamma \circ P)^t$. Take $a = 1$. Then the last (formal) equality implies, for every $\rho \in \mathbf{Q}$, that

$$\sum_{\operatorname{tr}(\xi) = \rho} c(\xi)e^{2\pi i \operatorname{tr}(b\xi)} = \sum_{\operatorname{tr}(\xi) = \rho} c(\xi)^t e^{2\pi i \operatorname{tr}(b\xi)} .$$

Therefore $c(\xi)^t = c(\xi)$ by the lemma. Obviously these facts are true even for $\mathfrak{M}_r$ with an arbitrary $r \in \mathbf{Z}^m$ instead of κu, although (i) in general is false (see Proposition 7 below).

Let $\{V_S,\ \varphi_S,\ J_{TS}(x)\ (S,\ T \in \mathfrak{Z}^*;\ x \in \bar{\mathfrak{G}}_+)\}$ be a canonical system for G in the sense of [11]. Here $\mathfrak{Z}^*$ is the set of all open subgroups S of $\bar{\mathfrak{G}}_+$ containing $F_c G_{\infty+}$ such that $S/F_c G_{\infty+}$ is compact, where F_c is the closure of $F^\times F_{\infty+}^\times$ in $F_\mathbf{A}^\times$ (see [11, I, Notation; II, 2.1, 3.5]). We put $\Gamma_S = S \cap G_\mathbf{Q}$. We recall that φ_S is a Γ_S-invariant holomorphic map of $\mathfrak{H}^m$ onto V_S which gives a biregular isomorphism of $\mathfrak{H}^m/\Gamma_S$ onto V_S; further V_S is defined over the subfield k_S of $\mathbf{Q}_{ab}$ corresponding to the subgroup $\det(S)F_c \cap \mathbf{Q}_\mathbf{A}^\times$ of $\mathbf{Q}_\mathbf{A}^\times$. Let $\mathbf{C}(V_S)$ denote the field of all 'functions' on V_S in the sense of algebraic geometry, and $k_S(V_S)$ its subfield consisting of all the k_S-rational elements. The map $p \mapsto p \circ \varphi_S$ for $p \in \mathbf{C}(V_S)$ gives a $\mathbf{C}$-linear isomorphism of $\mathbf{C}(V_S)$ onto $\mathcal{A}_0(\Gamma_S)$.[2] We shall often identify p with $p \circ \varphi_S$ in the following treatment. Then the union of $k_S(V_S)$ for all $S \in \mathfrak{Z}^*$ is exactly $\mathfrak{R}$.

Now $\mathfrak{H}^m/\Gamma_S$ has a natural compactification $(\mathfrak{H}^m/\Gamma_S)^*$ that is isomorphic, as a complex analytic space, to a normal projective variety W. Then V_S is isomorphic to the image of $\mathfrak{H}^m/\Gamma_S$ in W. The construction of V_S in [11] does not immediately imply that W is isomorphic to the Zariski closure of V_S. We can prove, however,

THEOREM 5. *The canonical models $(V_S,\ \varphi_S)$ for all $S \in \mathfrak{Z}^*$ can be chosen so that* (i) *the projective variety V_S^* of which V_S is a Zariski open subset is actually isomorphic to $(\mathfrak{H}^m/\Gamma_S)^*$ through φ_S;* (ii) *the morphism $J_{ST}(x)$ of V_T onto $V_S^{\rho(x)}$ can naturally be extended to a morphism of V_T^* onto $V_S^{*\rho(x)}$.*

Proof. This is obvious if $m = 1$; so assume $m > 1$. Take m algebraically independent elements $f_1,\ \cdots,\ f_m$ of $k_S(V_S)$, and put
$$g(z) = c \cdot \partial(f_1,\ \cdots,\ f_m)/\partial(z_1,\ \cdots,\ z_m)$$
with any non-zero constant c. Then $g \in \mathcal{A}_{2u}(\Gamma_S,\ \mathbf{C})$, and
$$g(z)dz_1 \wedge \cdots \wedge dz_m = c \cdot df_1 \wedge \cdots \wedge df_m.$$
Let div ( ) denote the divisor of a function or a differential form on V_S. Put $X = \mathrm{div}\,(df_1 \wedge \cdots \wedge df_m)$, and for each positive integer μ,
$$L(\mu X) = \{h \in \mathbf{C}(V_S) \mid \mathrm{div}\,(h) \geqq -\mu X\},$$
$$L(\mu X,\ k_S) = L(\mu X) \cap k_S(V_S).$$

By Weil [17, p. 265, Cor. 1], we have $L(\mu X) = L(\mu X,\ k_S) \otimes_{k_S} \mathbf{C}$. Observe that $L(\mu X)$ is $\mathbf{C}$-linearly isomorphic to $\mathfrak{M}_{2\mu u}(\Gamma_S)$ through the map $h \mapsto hg^\mu$.

---

[2] Strictly speaking, Γ_S is not a 'congruence subgroup'. But we define $\mathcal{A}_r(\Gamma_S)$, $\mathfrak{M}_r(\Gamma_S)$, etc. in the same fashion as in Section 1.

Take a basis $\{h_j\}$ of $L(\mu X, k_S)$ over k_S. Then $\{h_j g^\mu\}$ is a basis of $\mathfrak{M}_{2\mu u}(\Gamma_S)$ over $\mathbf{C}$. Now Theorem 10.11 of Baily-Borel [1] (combined with Theorem 10.14) shows that, for a suitably large μ, such a basis defines a biregular projective embedding of the compactification of $\mathfrak{H}^m/\Gamma_S$. At the same time, it gives the projective embedding of V_S by the linear system $L(\mu X, k_S)$, which is k_S-rational. Therefore we obtain the desired model by changing V_S for its image by the embedding. To prove (ii), it is sufficient to consider the case $\dot{x}Tx^{-1} = S$. Put $p_j = f_j^{\rho(x)} \circ J_{ST}(x)$ and $Y = \mathrm{div}\,(dp_1 \wedge \cdots \wedge dp_m)$. Then $J_{ST}(x)$ maps Y onto $X^{\rho(x)}$. Therefore the projective embedding of V_T by $L(\mu Y)$ is isomorphic to that of $V_S^{\rho(x)}$ by $L(\mu X^{\rho(x)})$. Therefore assertion (ii) is obvious.

We shall now show that $\mathfrak{M}_{2\mu u}(\Gamma_S)$ is spanned by $\mathfrak{M}_{2\mu u}(\Gamma_S, k_S)$ for a certain type of S. (This is false in general.) First we put

$$(4.4) \qquad \Delta_S = \left\{ \begin{pmatrix} 1 & 0 \\ 0 & t \end{pmatrix} \,\middle|\, t \in \prod_p \mathbf{Z}_p^\times,\ t = \mathrm{id.\ on}\ k_S \right\} \qquad (S \in \mathfrak{Z}^*)\,.$$

PROPOSITION 5. *If $\Delta_S \subset S$, we have $\mathfrak{a}_0(\Gamma_S, k_S) \subset k_S(V_S)$.*

Proof. Let $f \in \mathfrak{a}_0(\Gamma_S, k_S)$. Since $f \in \mathfrak{K}$ by Proposition 3, we can find $T \in \mathfrak{Z}^*$ so that $f \in k_T(V_T)$ and $T \subset S$. Let $w \in S$. Then we can find $v = \begin{pmatrix} 1 & 0 \\ 0 & t \end{pmatrix}$ with $t \in \prod_p \mathbf{Z}_p^\times$ so that v has the same action as w on k_T. By [11, II, (3.10.3)], we have $w = vy\alpha$ with $y \in T$ and $\alpha \in \mathrm{GL}_2^+(F)$. Now t acts as the identity on k_S. Therefore $f^v = f$ by (ii) of Proposition 4. (See Remark after Proposition 4.) Now $\alpha = y^{-1}v^{-1}w \in S \cap G_{\mathbf{Q}+} = \Gamma_S$, so that $f^\alpha = f$. Therefore $f^w = f$, q.e.d.

PROPOSITION 6. *For each positive integer N, put*

$$R_N = G_{\infty+} \cdot \left\{ x \in \prod_p \mathrm{GL}_2(\mathfrak{o}_p) \,\middle|\, x_p \equiv \begin{pmatrix} 1 & 0 \\ 0 & a_p \end{pmatrix} (\mathrm{mod}\ N\mathfrak{o}_p)\ with\ a_p \in \mathfrak{o}_p \right\}\,,$$
$$P_N = \{x \in R_N \mid \det(x) \in \mathbf{Q}_\mathbf{A}^\times\}\,,$$
$$T_N = F_c P_N\,,$$

where $\mathfrak{o}_p = \mathfrak{o} \otimes_{\mathbf{Z}} \mathbf{Z}_p$ for each rational prime p, and x_p is the p-component of x. Then $T_N \in \mathfrak{Z}^*$, $k_{T_N} = \mathbf{Q}$, and $\Gamma_{T_N} = F^\times \Gamma_N$.

Proof. Let us simply write T for T_N, fixing N. Observe that $G_{\infty+} \subset F_c P_N$. Since $P_N G_{\infty+}$ belongs to the class $\mathfrak{Z}$ of [11, I, 2.3], T belongs to $\mathfrak{Z}^*$ by [11, II, 3.6]. Now every element $\begin{pmatrix} 1 & 0 \\ 0 & t \end{pmatrix}$ with $t \in \prod_p \mathbf{Z}_p^\times$ belongs to T, hence $k_T = \mathbf{Q}$. Obviously $F^\times \Gamma_N \subset \Gamma_T$. To prove the opposite inclusion, let $\gamma \in \Gamma_T$ and $\gamma = ab$ with $a \in F_c$, $b \in P_N$. Then $\det(b) \in F_c \cap \mathbf{Q}_\mathbf{A}^\times = \mathbf{Q}^\times \mathbf{Q}_{\infty+}^\times$, so that the non-archimedean component of $\det(b)$ must be 1. By a well-known theorem of Chevalley, there exists a positive multiple M of N such that

$$\{e \in \mathfrak{o}^\times \mid e \equiv 1 \;(\mathrm{mod}\; M\mathfrak{o})\} \subset \{e^2 \mid e \in \mathfrak{o}^\times,\; e \equiv 1 \;(\mathrm{mod}\; N\mathfrak{o})\} \;.$$

Take a positive integer h so that $u^h \equiv 1 \;(\mathrm{mod}\; M\mathfrak{o}_p)$ for every $u \in \mathfrak{o}_p^\times$. By [11, II, 2.2], we have $a = x^h y z$ with $x \in \bar{E}_0$, $y \in F^\times$, $z \in F_{\infty+}^\times$ using the notation there. Then $\det(y^{-1}\gamma) = x^{2h} z^2 \det(b)$, so that $\det(y^{-1}\gamma) \equiv 1 \;(\mathrm{mod}\; M\mathfrak{o})$. Now $y^{-1}\gamma = x^h z b \in R_N$, so that $\det(y^{-1}\gamma) \in \mathfrak{o}^\times$. Therefore $\det(y^{-1}\gamma) = e^2$ with $e \in \mathfrak{o}^\times$, $e \equiv 1 \;(\mathrm{mod}\; N\mathfrak{o})$. Then $e^{-1}y^{-1}\gamma \in R_N$, hence $e^{-1}y^{-1}\gamma \equiv \begin{pmatrix} 1 & 0 \\ 0 & a_p \end{pmatrix} \;(\mathrm{mod}\; N\mathfrak{o}_p)$ with $a_p \in \mathfrak{o}_p$ for each p. Then $a_p \equiv \det(e^{-1}y^{-1}\gamma) = 1 \;(\mathrm{mod}\; N\mathfrak{o}_p)$, so that $e^{-1}y^{-1}\gamma \in \Gamma_N$. This proves $\Gamma_T \subset F^\times \Gamma_N$, and completes the proof.

THEOREM 6. *Suppose the set Δ_S defined by (4.4) is contained in S. Then we have*

(1) $\mathcal{Q}_0(\Gamma_S, k_S) = k_S(V_S)$;

(2) $\mathfrak{M}_{\kappa u}(\Gamma_S) = \mathfrak{M}_{\kappa u}(\Gamma_S, k_S) \otimes_{k_S} C$ *for every even positive integer κ;*

(3) $\mathfrak{M}_{\kappa u}(\Gamma) = \mathfrak{M}_{\kappa u}(\Gamma, k_S) \otimes_{k_S} C$ *for an odd positive integer κ, if Γ is a congruence subgroup such that $F^\times\Gamma = \Gamma_S$ and $\mathcal{Q}_{\kappa u}(\Gamma, k_S) \neq \{0\}$.*

Proof. We first take the group $T = T_N$ as S. For every positive integer κ, put $E_\kappa(z) = d_F^{1/2} E_{\kappa,0}(z; 0, 0; \mathfrak{o}, \mathfrak{o})$, where d_F is the discriminant of F. By Klingen [6], E_κ has rational Fourier coefficients. Moreover, by Maass [8] and Klingen [5], we can find m algebraically independent functions among $E_\kappa^{-\lambda}/E_{\kappa\lambda}$ with positive integers κ and λ.

Let $f_1, \cdots, f_m$ be such m functions. They are Γ_1-invariant and hence belong to $Q(V_T)$ by Proposition 5. Put

$$(4.5) \qquad g(z) = (2\pi i)^{-m} d_F^{1/2} \partial(f_1, \cdots, f_m)/\partial(z_1, \cdots, z_m) \;.$$

Then we see that $g = p/q$ with $p \in \mathfrak{M}_{\kappa u}(\Gamma_1, Q)$ and $q \in \mathfrak{M}_{(\kappa-2)u}(\Gamma_1, Q)$ for some κ, hence $g \in \mathcal{Q}_{2u}(\Gamma_1, Q)$. Let us now assume $m > 1$ and consider again $L(\mu X, Q)$ of the proof of Theorem 5. Let $h \in L(\mu X, Q)$. Then for $\beta = \begin{pmatrix} 1 & b \\ 0 & 1 \end{pmatrix}$, we see that $(hp^\mu) \circ \beta \circ P \in \mathfrak{M}_{\mu\kappa m}^1(Q_{ab})$, since both $h \circ \beta \circ P$ and $p^\mu \circ \beta \circ P$ have Fourier coefficients in Q_{ab}; hence $hp^\mu \in M_{\mu\kappa u}(Q_{ab})$ by Proposition 2. By Remark after Proposition 4, we see that $hp^\mu \in \mathfrak{M}_{\mu\kappa u}(\Gamma_N, Q)$; hence $hg^\mu = hp^\mu/q^\mu \in \mathcal{Q}_{2\mu u}(\Gamma_N, Q) \cap \mathfrak{M}_{2\mu u} = \mathfrak{M}_{2\mu u}(\Gamma_N, Q)$. This proves

$$(4.6) \qquad g^\mu L(\mu X, Q) \subset \mathfrak{M}_{2\mu u}(\Gamma_N, Q) \;.$$

Since $\mathfrak{M}_{2\mu n}(\Gamma_N) = g^\mu L(\mu X)$, this shows that $\mathfrak{M}_{2\mu u}(\Gamma_N, Q)$ spans $\mathfrak{M}_{2\mu u}(\Gamma_N)$ over C. Let $g_1, \cdots, g_n$ be linearly independent elements of $\mathfrak{M}_{2\mu u}(\Gamma_N, Q)$ over Q. By Proposition 5, $g_i/g_1 \in Q(V_T)$. Since $Q(V_T)$ and C are linearly disjoint over Q, the g_i/g_1 are linearly independent over C. Therefore $g_1, \cdots, g_n$ are linearly independent over C, which proves

$$(4.7) \qquad \mathfrak{M}_{2\mu u}(\Gamma_N) = \mathfrak{M}_{2\mu u}(\Gamma_N, Q) \otimes_Q C \;,$$

and also that (4.6) is actually an equality. Now every element of $\mathcal{C}_0(\Gamma_N)$ $(=C(V_T))$ is a quotient of two elements of $\mathfrak{M}_{2\mu\kappa}(\Gamma_N)$ for some μ. Therefore (4.7) shows that $\mathcal{C}_0(\Gamma_N)$ is generated by $\mathcal{C}_0(\Gamma_N, \mathbf{Q})$ over C. Recall again that $Q(V_T)$ is linearly disjoint with C over $\mathbf{Q}$. This combined with Proposition 5 proves that $\mathcal{C}_0(\Gamma_N, \mathbf{Q}) = \mathbf{Q}(V_T)$.

Suppose $\mathcal{C}_{\kappa\mu}(\Gamma_N, \mathbf{Q})$ with an odd κ contains an element $\psi \neq 0$. (This is so for sufficiently large N.) Observe that each component of div $(\psi^2 g^{-\kappa}) + \kappa X$ has an even coefficient (as an analytic divisor, and hence as an algebraic divisor). Therefore we have a divisor Y on V_T such that $2Y = \text{div } (\psi^2 g^{-\kappa}) + \kappa X$. Since $\psi^2 g^{-\kappa} \in \mathcal{C}_0(\Gamma_N, \mathbf{Q}) = \mathbf{Q}(V_T)$, Y is $\mathbf{Q}$-rational. Now we have $\mathfrak{M}_{\kappa\mu}(\Gamma_N) = \psi L(Y)$. By the same reasoning as above, we obtain (3) in the case $S = T_N$ and $\Gamma = \Gamma_N$.

If $m = 1$, we have to take cusps and elliptic points into account. But still we obtain a $\mathbf{Q}$-rational divisor Y_κ and an element g_κ of $\mathcal{C}_\kappa(\Gamma_N, \mathbf{Q})$ such that $g_\kappa L(Y_\kappa) = \mathfrak{M}_\kappa(\Gamma_N)$. Then the above reasoning is valid in this case with Y_κ in place of μX or Y.

Next we consider an arbitrary S such that $\Delta_S \subset S$. Take N so that $\Gamma_N \subset \Gamma_S$, and put $R = S \cap T$ with $T = T_N$. It can easily be seen that $k_R = k_S$, $k_S(V_S) \subset k_R(V_R)$, and $F^\times \Gamma_N \subset \Gamma_R \subset \Gamma_T = F^\times \Gamma_N$, hence $\Gamma_R = F^\times \Gamma_N$. Therefore

$$k_R(V_R) = k_R \mathbf{Q}(V_T) = k_R \mathcal{C}_0(\Gamma_N, \mathbf{Q}) \subset \mathcal{C}_0(k_R) ,$$

hence $k_S(V_S) \subset \mathcal{C}_0(k_S)$. This together with Proposition 5 proves (1). To prove (3), suppose $\Gamma_S = F^\times \Gamma$ and $\mathcal{C}_{\kappa\mu}(\Gamma, k_S) \ni \theta \neq 0$. We may assume that $\Gamma_N \subset \Gamma$. Take a set of representatives B for $\Gamma_N \backslash \Gamma$. Since $[S: R] = [k_R(V_R): k_S(V_S)] = [\Gamma_S: \Gamma_R]$, we have $S = R\Gamma$, so that B is a 'multiple' of a set of representatives for $R \backslash S$. Put $U(f) = \sum_{\beta \in B} f|_{\lambda\mu} \beta$ for $f \in \mathcal{C}_{\lambda\mu}(\Gamma_N)$, $\lambda \in \mathbf{Z}$. Then U maps $\mathfrak{M}_{\kappa\mu}(\Gamma_N)$ onto $\mathfrak{M}_{\kappa\mu}(\Gamma)$. Now let $f \in \mathfrak{M}_{\kappa\mu}(\Gamma_N, k_S)$. Then $U(f) = U(f/\theta)\theta$. Since $f/\theta \in \mathcal{C}_0(\Gamma_N, k_R) = k_R(V_R)$, we see that $U(f/\theta) \in k_S(V_S) = \mathcal{C}_0(\Gamma, k_S)$, hence $U(f) \in \mathfrak{M}_{\kappa\mu}(\Gamma, k_S)$. Since $\mathfrak{M}_{\kappa\mu}(\Gamma_N, \mathbf{Q})$ spans $\mathfrak{M}_{\kappa\mu}(\Gamma_N)$ over C, this proves (3). Assertion (2) can be proved in the same fashion, since if we define g by (4.5) with algebraically independent $f_1, \cdots, f_m$ in $k_S(V_S)$, then $0 \neq g^\mu \in \mathcal{C}_{2\mu\mu}(\Gamma_S, k_S)$. (We can also use the linear system of div (g^μ) on V_S if $m > 1$.)

The non-vanishing of $\mathcal{C}_{\kappa\mu}(\Gamma, k_S)$ can often be verified by constructing either an Eisenstein series or a theta series. We know in this way at least that $\mathcal{C}_{\kappa\mu}(\Gamma_N, \mathbf{Q}) \neq \{0\}$ for sufficiently large N. Therefore $\mathfrak{M}_{\kappa\mu} = \mathfrak{M}_{\kappa\mu}(\mathbf{Q}) \otimes_{\mathbf{Q}} C$ for every positive integer κ. We can derive, from this,

THEOREM 7. *Let Λ be an arbitrary subfield of* C. *Then* $\mathfrak{M}_{\kappa\mu}(\Lambda) = \mathfrak{M}_{\kappa\mu}(\mathbf{Q}) \otimes_{\mathbf{Q}} \Lambda$ *for every odd and every even positive integer κ.*

Assertion (i) of Proposition 4 is an immediate consequence of this theorem. We have proved in the above that $k_T(V_T) \subset \mathcal{C}_0(k_T)$. Therefore $\mathfrak{K} \subset \mathcal{C}_0(\mathbf{Q}_{ab})$, so that $\Lambda\mathfrak{K} \subset \mathcal{C}_0(\Lambda)$ for any subfield Λ of $\mathbf{C}$ containing $\mathbf{Q}_{ab}$. Now let $g, h \in \mathfrak{M}_\kappa^1(\Lambda)$. From Theorem 7 we see that $g/h \in \Lambda\mathcal{C}_0^1(\mathbf{Q}) \subset \Lambda\mathfrak{K}^1$. This proves $\mathcal{C}_0^1(\Lambda) \subset \Lambda\mathfrak{K}^1$, so that $\mathcal{C}_0(\Lambda) \subset \Lambda\mathfrak{K}$ by Proposition 3. This completes the proof of Theorem 3.

5. Main Theorem III

Let σ be an arbitrary field-automorphism of $\mathbf{C}$. If $a \in F$ and $a_1, \cdots, a_m$ are its images in $\mathbf{R}$ as before, σ gives a permutation of $a_1, \cdots, a_m$, or rather a permutation $\nu \mapsto \mu$ of the indices $1, \cdots, m$ defined by $a_\nu^\sigma = a_\mu$. Now we can let σ act on $\mathbf{Z}^m$ by this permutation; namely, we write $r^\sigma = s$ for $r = (r_1, \cdots, r_m) \in \mathbf{Z}^m$ if $r_\nu = s_\mu$.

We now consider the ring Ψ of all *formal* series of the form

$$(5.1) \qquad \sum_\xi c(\xi) e^{2\pi i \, \mathrm{tr}(\xi z)}$$

with $c(\xi) \in \mathbf{C}$, where ξ runs over 0 and all totally positive elements in a lattice in F; the lattice may depend on each series. It can easily be seen that Ψ is an integral domain. Therefore we can form its quotient field Ω. Now embed $\bigcup_{r \in \mathbf{Z}^m} \mathcal{C}_r$ into Ω in a natural way. For σ as above, we extend it to Ψ by defining the image of (5.1) to be $\sum_\xi c(\xi)^\sigma e^{2\pi i \, \mathrm{tr}(\xi z)}$, and then to a field-automorphism of Ω. The image of $f \in \Omega$ under σ will be denoted by f^σ. By Proposition 4, we have $\mathfrak{M}_{\kappa u}^\sigma = \mathfrak{M}_{\kappa u}$ for every positive integer κ. Now $\mathcal{C}_0$ is the set of quotients g/h with $g \in \mathfrak{M}_{\kappa u}$ and $h \in \mathfrak{M}_{\kappa u}$ for some κ, and $\mathcal{C}_{\lambda u} = \mathcal{C}_0 f^\varepsilon$ with $0 \neq f \in \mathfrak{M}_{|\lambda| u}$ and $\varepsilon = \lambda/|\lambda|$. Therefore $\mathcal{C}_{\lambda u}^\sigma = \mathcal{C}_{\lambda u}$ for every $\lambda \in \mathbf{Z}$; moreover, by Proposition 4, $f^v = f^\sigma$ for every $f \in \mathfrak{K}$, if $v = \begin{pmatrix} 1 & 0 \\ 0 & t \end{pmatrix}^{-1}$ with $t \in \prod_p \mathbf{Z}_p^\times$ and $\sigma = t$ on $\mathbf{Q}_{ab}$. To study $\mathcal{C}_r$ in a more general case, we restrict r to the set Y defined by (1.10).

PROPOSITION 7. (i) $\mathcal{C}_r^\sigma = \mathcal{C}_{r^\sigma}$ if $r \in Y$.

(ii) $(f |_r \alpha)^\sigma = f^\sigma |_{r^\sigma} \alpha$ for every $\alpha \in \mathrm{SL}_2(F)$, $f \in \mathcal{C}_r$, $r \in Y$, if $\sigma = \mathrm{id.}$ on $\mathbf{Q}_{ab}$.

(iii) $(D_\nu f)^\sigma = D_\mu(f^\sigma)$ for $f \in \mathcal{C}_0$ if $a_\nu^\sigma = a_\mu$ for all $a \in F$.

(iv) If $0 \neq f \in \mathcal{C}_r(\Theta)$, $r \in Y$, and $\sigma \in \mathrm{Aut}\,(\mathbf{C}/\Theta)$ with a subfield Θ of $\mathbf{C}$, then $f^\sigma = f$ and $r^\sigma = r$.

Proof. Take $g \in \mathfrak{M}_{pu}(\Gamma_1, \mathbf{Q})$ and $h \in \mathfrak{M}_{qu}(\Gamma_1, \mathbf{Q})$ with any positive integers p and q so that h^p/g^q is not a constant.[3] Put

$$B_\nu = (2\pi i)^{-1}\big(pg \cdot (\partial h/\partial z_\nu) - qh \cdot (\partial g/\partial z_\nu)\big) \qquad (\nu = 1, \cdots, m) \ .$$

Then $B_\nu \in \mathfrak{M}_{t_\nu}(\Gamma_1, F_\nu)$, where F_ν is the νth conjugate of F, and $t_\nu = (p + q)u +$

---

[3] It should be noted that if $f \in \mathcal{C}_0$ is not a constant, then $\partial f/\partial z_\nu \neq 0$ for *all* ν.

$(0, \cdots, 0, 2, 0, \cdots, 0)$ with 2 at the νth component. Observe that $B_\nu^a = B_\mu$ if $a_\nu^a = a_\mu$ for $a \in F$ as above. Put $B^e = \prod_{\nu=1}^m B_\nu^{e_\nu}$ for $e = (e_1, \cdots, e_m) \in \mathbf{Z}^m$. Given $r \in Y$, we can find $e \in \mathbf{Z}^m$ so that $r = 2e + nu$ with $n \in \mathbf{Z}$. Then $\mathcal{A}_r = B^e \mathcal{A}_{\kappa u}$ with $\kappa = n - \{e\}(p+q)$, from which we obtain (i). Assertions (iii) and (iv) now follow directly from (i) and our definition of σ on Ω. To prove (ii), suppose $f \in \mathcal{A}_r$, $r \in Y$, and $\sigma \in \mathrm{Aut}\,(\mathbf{C}/\mathbf{Q}_{\mathrm{ab}})$. In the relation $\mathcal{A}_r = B^e \mathcal{A}_{\kappa u}$, we can choose e so that $\kappa > 0$. Now we can find a non-zero $\varphi \in \mathfrak{M}_{\kappa u}(\mathbf{Q}_{\mathrm{ab}})$ such that $\varphi|_{\kappa u}\, \varepsilon$ belongs to $\mathfrak{M}_{\kappa u}(\mathbf{Q}_{\mathrm{ab}})$ for $\varepsilon = \begin{pmatrix} 0 & -1 \\ 1 & 0 \end{pmatrix}$. (Take, for example, an Eisenstein series or a theta series.) Then $f = B^e \varphi \psi$ with $\psi \in \mathcal{A}_0$. Observe that $\sigma = \mathrm{id.}$ on $\mathfrak{K}$. Since $\mathcal{A}_0 = \mathbf{C}\mathfrak{K}$, we see that $(\psi \circ \alpha)^\sigma = \psi^\sigma \circ \alpha$ for every $\alpha \in \mathrm{GL}_2^+(F)$. Now $\varphi^\sigma = \varphi$ and $B^e|\varepsilon = B^e$. Combining all these, we can easily verify that $(f|_r \varepsilon)^\sigma = f^\sigma|_{r^\sigma} \varepsilon$. Put $f = P/Q$ with $P \in \mathfrak{M}_{r+s}$ and $Q \in \mathfrak{M}_s$ for some s. Since $f = B^e \varphi \psi$, we may assume that $s \in Y$, $P^\sigma \in \mathfrak{M}_{r^\sigma+s^\sigma}$, $Q^\sigma \in M_{s^\sigma}$. Let $\alpha = \begin{pmatrix} a & b \\ 0 & d \end{pmatrix} \in \mathrm{SL}_2(F)$. Then, by (4.3), we see easily that $(Q|_s \alpha)^\sigma = Q^\sigma|_{s^\sigma} \alpha$, and the same is true for P. Thus $(f|_r \alpha)^\sigma = f^\sigma|_{r^\sigma} \alpha$. Now $\mathrm{SL}_2(F)$ is generated by such α and ε, hence we obtain (ii).

The above construction of B^e proves (1.11_a). Now (1.11_b) follows from

PROPOSITION 8. *Let Λ be a subfield of $\mathbf{C}$ containing both $\mathbf{Q}_{\mathrm{ab}}$ and the Galois closure of F over $\mathbf{Q}$. If $f \in \mathcal{A}_r(\Lambda)$ and $r \in Y$, then $\det(\alpha)^{-r/2} f|_r \alpha \in \mathcal{A}_r(\Lambda)$ for every $\alpha \in \mathrm{GL}_2^+(F)$. Especially if $r_1 = \cdots = r_m$, then*

$$\det(\alpha)^{-r/2} f|_r \alpha \in \mathcal{A}_r(\Lambda)$$

even if Λ does not contain the Galois closure of F. Moreover these assertions hold with $\mathfrak{M}_r(\Lambda)$ instead of $\mathcal{A}_r(\Lambda)$.

Proof. Given $f \in \mathcal{A}_r(\Lambda)$, let $f = B^e \varphi \psi$ as in the proof of Proposition 7. Then $\psi \in \mathcal{A}_0(\Lambda) = \Lambda \mathfrak{K}$. Now $\mathfrak{K}$ is stable under ε, hence $\psi \circ \varepsilon \in \Lambda \mathfrak{K}$. Therefore $f|_r \varepsilon = B^e(\varphi|_{\kappa u} \varepsilon)(\psi \circ \varepsilon) \in \mathcal{A}_r(\Lambda)$. From (4.3), we see, for every $g \in \mathfrak{M}_r(\Lambda)$ and $\alpha = \begin{pmatrix} a & b \\ 0 & d \end{pmatrix}$, that $\det(\alpha)^{-r/2} g|_r \alpha = d^{-r} g \circ \alpha \in \mathfrak{M}_r(\Lambda)$. Therefore the same is true for $\mathcal{A}_r(\Lambda)$. Now $\mathrm{GL}_2^+(F)$ is generated by such α and ε, hence we obtain our first assertion. If $r_1 = \cdots = r_m$, we have $B^e \in \mathcal{A}_{r-\kappa u}(\mathbf{Q})$ and $d^{-r} = N_{F/\mathbf{Q}}(d)^{-r_1} \in \mathbf{Q}$ for every $d \in F^\times$, which are the only points necessary for the second assertion. Since $\mathfrak{M}_r(\Lambda)$ is the set of holomorphic elements of $\mathcal{A}_r(\Lambda)$, we obtain the same results for $\mathfrak{M}_r(\Lambda)$.

To state Main Theorem III, we have to define an action of $\bar{\mathcal{G}}_+$ on $\bar{\mathcal{A}}_r(\mathbf{Q})$. First we consider the map $\rho: \bar{\mathcal{G}}_+ \to \mathrm{Gal}\,(\mathbf{Q}_{\mathrm{ab}}/\mathbf{Q})$ defined in [11, II, 3.10]. Recall that, for every $v \in \bar{\mathcal{G}}_+$, $\rho(v)$ is the restriction to $\mathbf{Q}_{\mathrm{ab}}$ of the action of v on $\mathfrak{K}$. Put

$$(5.2) \qquad \Delta = \left\{ \begin{pmatrix} 1 & 0 \\ 0 & t \end{pmatrix} \in G_{\mathbf{A}} \,\middle|\, t \in \prod_p \mathbf{Z}_p^{\times} \right\}.$$

Then $\rho(\Delta) = \mathrm{Gal}\,(\mathbf{Q}_{ab}/\mathbf{Q})$. We now define a subgroup $\mathfrak{G}$ of $\bar{\mathfrak{S}}_{+} \times \mathrm{Gal}\,(\bar{\mathbf{Q}}/\mathbf{Q})$ by

$$(5.3) \qquad \mathfrak{G} = \{(x, \sigma) \in \bar{\mathfrak{S}}_{+} \times \mathrm{Gal}\,(\bar{\mathbf{Q}}/\mathbf{Q}) \mid \rho(x) = \sigma \text{ on } \mathbf{Q}_{ab}\}.$$

THEOREM 8. *Put* $Y' = 2\mathbf{Z}^m$. *Then* $\mathfrak{G}$ *acts on* $\sum_{r \in Y'} \mathcal{C}_r(\bar{\mathbf{Q}})$ *as a group of 'automorphisms' with the following properties:*

(1) (x, σ) *maps* $\mathcal{C}_r(\bar{\mathbf{Q}})$ *onto* $\mathcal{C}_{r\sigma}(\bar{\mathbf{Q}})$;

(2) $(af + bg)^{(x,\sigma)} = a^{\sigma} f^{(x,\sigma)} + b^{\sigma} g^{(x,\sigma)}$ *for* $a, b \in \bar{\mathbf{Q}}$, $f, g \in \mathcal{C}_r(\bar{\mathbf{Q}})$;

(3) $(fg)^{(x,\sigma)} = f^{(x,\sigma)} g^{(x,\sigma)}$, $(f/g)^{(x,\sigma)} = f^{(x,\sigma)}/g^{(x,\sigma)}$;

(4) $(f^{(x,\sigma)})^{(x',\sigma')} = f^{(xx',\sigma\sigma')}$;

(5) $f^{(x,\sigma)} = f^{\sigma}$ *(as in Proposition 7) if* $x \in \Delta$;

(6) $f^{(\alpha,1)} = f|_r \alpha$ *if* $\alpha \in G_{\mathbf{Q}+}$ *and* $f \in \mathcal{C}_r(\bar{\mathbf{Q}})$;

(7) $(D_\nu f)^{(x,\sigma)} = D_\mu(f^{(x,\sigma)})$ *if* $f \in \mathcal{C}_0(\bar{\mathbf{Q}})$ *and* $a_\nu^{\sigma} = a_\mu$ *for* $a \in F$;

(8) $f^{(x,\sigma)} = f$ *if* $f \in \mathcal{C}_{\kappa u}(\Gamma_S, k_S)$, κ *even*, $\Delta_S \subset S$, *and* $x \in S$;

(9) $\mathfrak{M}_{\kappa u}(\mathbf{Q}_{ab})$ *is stable under every* (x, σ) *of* $\mathfrak{G}$.

Proof. Since $\mathfrak{K}$ is linearly disjoint with $\bar{\mathbf{Q}}$ over $\mathbf{Q}_{ab}$, we can define the action of (x, σ) on $\mathcal{C}_0(\bar{\mathbf{Q}}) = \bar{\mathbf{Q}}\mathfrak{K}$ to be the same as x on $\mathfrak{K}$ and σ on $\bar{\mathbf{Q}}$. Before proceeding further, we prove

PROPOSITION 9. *For every non-constant* f, g *in* $\mathcal{C}_0(\bar{\mathbf{Q}})$ *and* $(x, \sigma) \in \mathfrak{G}$, *one has* $(D_\nu f/D_\nu g)^{(x,\sigma)} = D_\mu(f^{(x,\sigma)})/D_\mu(g^{(x,\sigma)})$ *if* $a_\nu^{\sigma} = a_\mu$ *for* $a \in F$.

Proof. Put $h = D_\nu f/D_\nu g$. Take $S \in \mathfrak{Z}^*$ so that f, g, $h \in \bar{\mathbf{Q}}(V_S)$, and take $v \in \Delta$ so that $\rho(v) = \sigma$ on $\mathbf{Q}_{ab}$. Then $\rho(xv^{-1}) = 1$, so that, by [11, II, (3.10.3)], $xv^{-1} = s\alpha$ with $s \in S$ and $\alpha \in G_{\mathbf{Q}+}$. Then $\rho(s) = 1$, hence $(x, \sigma) = (s, 1)(\alpha, 1)(v, \sigma)$. By our definition, $(s, 1)$ acts trivially on $\bar{\mathbf{Q}}(V_S)$. Therefore, by Proposition 4,

$$h^{(x,\sigma)} = (h \circ \alpha)^{\sigma}, \quad f^{(x,\sigma)} = (f \circ \alpha)^{\sigma}, \quad g^{(x,\sigma)} = (g \circ \alpha)^{\sigma}.$$

By Proposition 7 (iii), we have

$$D_\mu(f^{(x,\sigma)}) = (D_\nu(f \circ \alpha))^{\sigma} = ((D_\nu f) \mid \alpha)^{\sigma},$$

so that

$$D_\mu(f^{(x,\sigma)})/D_\mu(g^{(x,\sigma)}) = ((D_\nu f) \mid \alpha)^{\sigma}/((D_\nu g) \mid \alpha)^{\sigma} = (h \circ \alpha)^{\sigma} = h^{(x,\sigma)},$$

which proves our assertion.

Now take a non-constant f of $\mathcal{C}_0(\bar{\mathbf{Q}})$. For every $r \in Y'$, put $C_r(f) = \prod_{\nu=1}^m (D_\nu f)^{r_\nu/2}$. Then $0 \neq C_r(f) \in \mathcal{C}_r(\bar{\mathbf{Q}})$, so that $\mathcal{C}_r(\bar{\mathbf{Q}}) = \mathcal{C}_0(\bar{\mathbf{Q}})C_r(f)$. Let $h = \varphi C_r(f) \in \mathcal{C}_r(\bar{\mathbf{Q}})$ with $\varphi \in \mathcal{C}_0(\bar{\mathbf{Q}})$. We then define $h^{(x,\sigma)}$ by $h^{(x,\sigma)} = \varphi^{(x,\sigma)} C_{r\sigma}(f^{(x,\sigma)})$. By virtue of Proposition 9, we see that this does not depend on the choice of

f. Then properties (1-7) can be easily verified. To prove (8), suppose $h \in \mathcal{C}_{\kappa u}(\Gamma_S, k_S)$, κ even, $\Delta_S \subset S$, and $x \in S$. Take a non-constant f in $k_S(V_S)$. Observe that $C_{\kappa u}(f) \in \mathcal{C}_{\kappa u}(\Gamma_S, k_S)$. Put $h = \varphi C_{\kappa u}(f)$. Then $\varphi \in \mathcal{C}_0(\Gamma_S, k_S) = k_S(V_S)$ by Theorem 6. Now (x, σ) acts trivially on $k_S(V_S)$, and hence on φ and f, which proves (8). Since $\mathfrak{M}_{\kappa u}(\mathbf{Q}_{ab}) = \bigcup_{N=1}^{\infty} \mathfrak{M}_{\kappa u}(\Gamma_N, \mathbf{Q}_{ab})$ and $\bar{\mathfrak{G}}_+ = T_N G_{\mathbf{Q}+}$ for every N, assertion (9) follows from Propositions 4 and 8.

We cannot extend the action of $\mathfrak{G}$ to $\mathcal{C}_r(\bar{\mathbf{Q}})$ for $r \notin Y'$ keeping properties (4), (5), (6). In fact, let $\alpha = \begin{pmatrix} a & 0 \\ 0 & 1 \end{pmatrix}$ with $a \in F$ and $x \in \Delta$. If (4), (5), (6) hold for $f \in \mathcal{C}_{\kappa u}(\bar{\mathbf{Q}})$ with odd κ and if $\sigma = \rho(x)$ on $\mathbf{Q}_{ab}$, then

$$\big(N(a)^{\kappa/2}\big)^\sigma f^\sigma(az) = f^{(\alpha,1)(x,\sigma)} = f^{(x,\sigma)(\alpha,1)} = N(a)^{\kappa/2} f^\sigma(az) ,$$

which is false unless $\big(N(a)^{\kappa/2}\big)^\sigma = N(a)^{\kappa/2}$. This difficulty is avoidable by changing (6) for $f^{(\alpha,1)} = \det(\alpha)^{-r/2} f|_r \alpha$. But the change will cause another difficulty of finding the action of an element (x, σ) of general type. There is actually a certain way of defining a reasonable associative action by restricting (x, σ) to the subgroup

$$\{(x, \sigma) \in \mathfrak{G} \mid \det(x) \in \mathbf{Q}_\mathbf{A}^\times\} ,$$

or considering only $\mathcal{C}_{\kappa u}(\bar{\mathbf{Q}})$ with $\kappa \in \mathbf{Z}$. But we define here the action of $\mathfrak{G}$ on the whole $\sum_{r \in Y} \mathcal{C}_r(\bar{\mathbf{Q}})$ by weakening the associativity (4).

Take a non-zero $\psi \in \mathfrak{M}_{\lambda u}(\Gamma_M, \mathbf{Q})$ with any M and any odd λ. Then every $f \in \mathcal{C}_r(\bar{\mathbf{Q}})$ with $r \in Y - Y'$ can be written as $f = g\psi$ with $g \in \mathcal{C}_{r-\lambda u}(\bar{\mathbf{Q}})$, $r - \lambda u \in Y'$. Now every (x, σ) of $\mathfrak{G}$ can be written as $(x, \sigma) = (s, \sigma)(\alpha, 1)$ with $s \in T_M$ and $\alpha \in G_{\mathbf{Q}+}$, since $\bar{\mathfrak{G}}_+ = T_M G_{\mathbf{Q}+}$ by [11, II, (3.10.3)]. Then we define $f^{(x,\sigma)} = g^{(x,\sigma)}(\psi |_{\lambda u} \alpha)$. This definition depends on the choices of ψ, s, α. If $(x, \sigma) = (\alpha, 1)$ with $\alpha \in G_{\mathbf{Q}+}$, we have to define $f^{(x,\sigma)} = g^{(\alpha,1)}(\psi |_{\lambda u} \alpha)$ to keep (3); similarly, if $x \in \Delta$, we put $f^{(x,\sigma)} = g^{(x,\sigma)}\psi = f^\sigma$. In this way we obtain, for each individual $(x, \sigma) \in \mathfrak{G}$, its action on $\sum_{r \in Y} \mathcal{C}_r(\bar{\mathbf{Q}})$ satisfying (1), (2), (3), (5), (6), (7), (9); (8) must be replaced by

$$(8') \qquad f^{(x,\sigma)} = \pm f \text{ if } f \in \mathcal{C}_{\kappa u}(\Gamma_N, k_S), \ \kappa \ odd, \ \Gamma_N \subset S \subset T_N, \ \Delta_S \subset S, \ x \in S .$$

Obviously we have at most two ways to define the action; if one action is defined, the other (possible) action is $f \mapsto -f^{(x,\sigma)}$. Therefore we have, instead of (4),

$$(4') \qquad (f^{\mathfrak{x}})^{\mathfrak{y}} = c(\mathfrak{x}, \mathfrak{y}) f^{\mathfrak{x}\mathfrak{y}} \quad for \ \mathfrak{x}, \mathfrak{y} \in \mathfrak{G}, \ f \in \mathcal{C}_r(\bar{\mathbf{Q}}), \ r \in Y - Y' ,$$

with a constant $c(\mathfrak{x}, \mathfrak{y}) = \pm 1$, *that is independent of f and r.*

This may look artificial, but is quite satisfactory for our need in the following theorem.

Now given a fixed point w as in Section 1, take a totally imaginary quadratic extension K of F and an F-linear embedding θ of K into $M_2(F)$ so that $\theta(K^\times) = \{\beta \in \mathrm{GL}_2^+(F) \mid \beta(w) = w\}$. Define $\{\tau_1, \cdots, \tau_m\}$ by (1.17), and let $(K', \sum_{\lambda=1}^n \sigma_\lambda)$ be the reflex (or dual) of $(K, \sum_{\nu=1}^m \tau_\nu)$ in the sense of [10], [11]. Define $\eta: K'^\times \to \mathrm{GL}_2(F)$ by $\eta(a) = \theta(\prod_{\lambda=1}^n a^{\sigma_\lambda})$. Then an explicit reciprocity-law at the point w is as follows (see [11, II, (6.2.3)]).

(5.4) *If $b \in K_A'^\times$ and $x = \eta(b)^{-1}$, then $f(w)^b = f^x(w)$ for every f of $\Re$ holomorphic at w.*

This can be generalized in the context of our operator D^e:

MAIN THEOREM III. *With f, g, and w as in Main Theorem I, define K' and η as above. Let $b \in K_A'^\times$ and $\sigma \in \mathrm{Gal}(\bar{\mathbf{Q}}/K')$ be such that σ coincides with the action of b on K_{ab}'. Put $\mathfrak{x} = (\eta(b)^{-1}, \sigma)$. Then*

$$(g^{-1}D^e f)(w)^\sigma = [(g^{\mathfrak{x}})^{-1} D^{e^\sigma}(f^{\mathfrak{x}})](w) .$$

We first prove

PROPOSITION 10. *If $\psi \in \mathcal{C}_r(\bar{\mathbf{Q}})$ is holomorphic at w, so is $\psi^{\mathfrak{x}}$.*

Proof. Put $x = \eta(b)^{-1}$ and suppose $r = 0$. Write $\psi = p \circ \varphi_S$ with $p \in \bar{\mathbf{Q}}(V_S)$ for a suitable $S \in \mathfrak{Z}^*$. By our definition of the action of x on $\Re$ in [11, II, 6.1] and that of $\mathfrak{x}$ on $\mathcal{C}_0(\bar{\mathbf{Q}})$, we have $\psi^{\mathfrak{x}} = p^\sigma \circ J_{ST}(x) \circ \varphi_T$, where $T = x^{-1}Sx$. If ψ is holomorphic at w, then p is finite at $\varphi_S(w)$, so that p^σ is finite at $\varphi_S(w)^\sigma$. By (5.4), or rather by [11, II, (5.2.4)], $\varphi_S(w)^\sigma = J_{ST}(x)[\varphi_T(w)]$. Therefore $\psi^{\mathfrak{x}}$ is holomorphic at w, and $\psi^{\mathfrak{x}}(w) = p^\sigma(\varphi_S(w)^\sigma) = \psi(w)^\sigma$. This proves the case $r = 0$. To prove the general case, take a non-constant $f_\nu \in \mathcal{C}_0(\bar{\mathbf{Q}})$ for each ν so that $D_\nu f_\nu$ is finite and $\neq 0$ at w. Put $\psi^2 = g \prod_{\nu=1}^m (D_\nu f_\nu)^{r_\nu}$ with $g \in \mathcal{C}_0(\bar{\mathbf{Q}})$. If ψ is finite at w, so is g. Since $(\psi^{\mathfrak{x}})^2 = g^{\mathfrak{x}} \prod_{\nu=1}^m (D_\mu(f_\nu^{\mathfrak{x}}))^{r_\nu}$ if $a_\nu^\sigma = a_\mu$ for $a \in F$, $\psi^{\mathfrak{x}}$ must be finite at w.

We now prove Main Theorem III by induction on $\{e\}$. If $\{e\} = 0$, then $g^{-1}f \in \mathcal{C}_0(\bar{\mathbf{Q}})$. Taking $g^{-1}f$ as ψ of the above proof, we obtain $g^{-1}f(w)^\sigma = \psi^{\mathfrak{x}}(w) = ((g^{\mathfrak{x}})^{-1}f^{\mathfrak{x}})(w)$ as desired. To prove the case $\{e\} > 0$, we may assume that both f and $f^{\mathfrak{x}}$ are finite and $\neq 0$ at w. In fact, take $\psi \in \mathcal{C}_r(\bar{\mathbf{Q}})$ so that it is finite and $\neq 0$ at w, and put $p = a\psi + b\psi^{\mathfrak{y}}$ and $\mathfrak{y} = \mathfrak{x}^{-1}$ and $a, b \in \mathbf{Q}$. By Proposition 10, p is finite at w. For a suitable choice of (a, b), all of p, $p^{\mathfrak{x}}$, $p + f$, $p^{\mathfrak{x}} + f^{\mathfrak{x}}$ take non-zero finite values at w. Then it is enough to prove our assertion for p and $p + f$ in place of f. Assuming thus $f(w) \neq 0$ and $f^{\mathfrak{x}}(w) \neq 0$, let α, h, and λ be as in the proof of Main Theorem I, and let $g \in \mathcal{C}_s(\bar{\mathbf{Q}})$ be finite and $\neq 0$ at w, where $s = (r_1, \cdots, r_\nu + 2, \cdots, r_m)$. Since $\sigma = $ id. on K', $\{\tau_1\sigma, \cdots, \tau_m\sigma\}$ coincides with $\{\tau_1, \cdots, \tau_m\}$ as a whole. If $\tau_\nu\sigma = \tau_\mu$, we have $\lambda_\nu^\sigma = \lambda_\mu$. Multiply (1.15) by $g(w)^{-1}$ and apply σ. Then we obtain

$$(*) \qquad \begin{aligned} (g^{-1}D_\nu f)(w)^\sigma &= (\lambda_\mu^{-2} - 1)^{-1}\lambda^t (g^{-1}f D_\nu h)(w)^\sigma \\ &= (\lambda_\mu^{-2} - 1)^{-1}\lambda^t [(g^\iota)^{-1} f^\iota D_\mu h^\iota](w) \,, \end{aligned}$$

where $t = r^\sigma$. Now $\alpha \in \theta(K^\times)$, so that $(\alpha, 1)$ commutes with $\mathfrak{x} = (\eta(b)^{-1}, \sigma)$. Therefore, by (4'), we have

$$h^\iota = (f^{(\alpha,1)}/f)^\iota = (f^{(\alpha,1)})^\iota/f^\iota = \zeta(f^\iota)^{(\alpha,1)}/f^\iota = \zeta(f^\iota |_\iota \alpha)/f^\iota$$

with $\zeta = \pm 1$. Evaluate both sides at w. Then $h^\iota(w) = \zeta\lambda^{-t}$. On the other hand, applying our result in the case $\{e\} = 0$ to h, we find $h^\iota(w) = h(w)^\sigma = \lambda^{-t}$, and hence $\zeta = 1$. Therefore, substituting f^ι and h^ι for f and h in (1.15), we see that $(*)$ is equal to $[(g^\iota)^{-1}D_\mu(f^\iota)](w)$. This proves the case $\{e\} = 1$. Applying σ to (1.16), we can prove the general case in a similar way by induction on $\{e\}$. We need only to observe that if we write $C(e, p)$ for the coefficient C_p of (1.16), then $C(e, p) = C(e^\sigma, p^\sigma)$.

Now with the notation of Main Theorem II, take any σ of $\mathrm{Gal}\,(\bar{\mathbf{Q}}/K'_{ab}\Lambda)$. By (iv) of Proposition 7, we have $f^\sigma = f$, $g^\sigma = g$, $r^\sigma = r$, and $(r + 2e)^\sigma = r + 2e$. Therefore, taking $b = 1$ in the above theorem, we see that $(g^{-1}D^e f)(w)$ is invariant under σ. This proves Main Theorem II.

6. The Siegel modular functions that are arithmetic

The methods of Section 4 are applicable to the arithmetic subgroups of algebraic groups of a more general type. Especially the principle of Theorem 5 is valid whenever cusps and elliptic loci are of codimension > 1. We also relied heavily on the Fourier expansions of automorphic forms, which in general are more complicated. But we can at least treat the symplectic groups over $\mathbf{Q}$ by the same technique. We shall now discuss this case briefly. With a positive integer $n > 1$, let G be an algebraic subgroup of GL_{2n} defined over $\mathbf{Q}$ such that

$$G_\mathbf{Q} = \{\alpha \in \mathrm{GL}_{2n}(\mathbf{Q}) \mid {}^t \alpha J\alpha = \nu(\alpha)J \text{ with } \nu(\alpha) \in \mathbf{Q}\} \,,$$

where $J = \begin{pmatrix} 0 & -1_n \\ 1_n & 0 \end{pmatrix}$. Then $\mathrm{Sp}(n, \mathbf{Q}) = \{\alpha \in G_\mathbf{Q} \mid \nu(\alpha) = 1\}$. Put

$$G_{\mathbf{Q}+} = \{\alpha \in G_\mathbf{Q} \mid \nu(\alpha) > 0\} \,,$$
$$\Gamma_N = \{\gamma \in \mathrm{Sp}(n, \mathbf{Z}) \mid \gamma \equiv 1_{2n} \pmod{N}\}$$

for every positive integer N. Let $\mathfrak{H}_n$ denote the space of all complex symmetric matrices with positive definite imaginary part. We let every element $\alpha = \begin{pmatrix} a & b \\ c & d \end{pmatrix} \in G_{\mathbf{Q}+}$ act on $\mathfrak{H}_n$ by $\alpha(z) = (az + b)(cz + d)^{-1}$. For a meromorphic function f on $\mathfrak{H}_n$ and $r \in \mathbf{Z}$, we put

$$(f |_r \alpha)(z) = \nu(\alpha)^{rn/2} f(\alpha(z)) \det (cz + d)^{-r} \,.$$

By a *congruence subgroup*, we mean a subgroup Γ of Γ_1 containing Γ_N for some N. Then we denote by $\mathfrak{M}_r(\Gamma)$ the set of all holomorphic functions $f(z)$ on $\mathfrak{H}_n$ such that $f|_r \gamma = f$ for all $\gamma \in \Gamma$. Such an f always has a Fourier expansion of the form

$$f(z) = \sum_\tau p(\tau)e^{2\pi i \mathrm{tr}(\tau z)/N}$$

with $p(\tau) \in C$, where τ runs over all semi-integral semi-definite symmetric matrices of degree n. Now for a subfield Λ of C, we denote by $\mathfrak{M}_r(\Gamma, \Lambda)$ the set of all $f \in \mathfrak{M}_r(\Gamma)$ with $p(\tau) \in \Lambda$. Then we define $\mathfrak{M}_r$, $\mathfrak{M}_r(\Lambda)$, $\mathcal{Q}_s$, $\mathcal{Q}_s(\Lambda)$, $\mathcal{Q}_s(\Gamma)$, $\mathcal{Q}_s(\Gamma, \Lambda)$ $(r, s \in Z; r \geq 0)$ in exactly the same fashion as in Section 1.

We now consider a canonical system and the field $\mathfrak{K}$ of all arithmetic automorphic functions with respect to G in the sense of [11]. Also we can define an embedding $P: \mathfrak{H}_1 \to \mathfrak{H}_n$ by $P(z) = z1_n$, which is compatible with a homomorphism $\begin{pmatrix} a & b \\ c & d \end{pmatrix} \mapsto \begin{pmatrix} a1_n & b1_n \\ c1_n & d1_n \end{pmatrix}$ of $GL_2(Q)$ into G_Q. Then all the Propositions and Theorems of Section 4, without exception, have natural analogues in the present case, which can be proved in almost the same manner. For example, we have $\mathcal{Q}_0(Q_{ab}) = \mathfrak{K}$, and $\mathfrak{M}_r(\Gamma_N) = \mathfrak{M}_r(\Gamma_N, Q) \otimes_Q C$ for all N and all even r, and also for odd r provided that $\mathcal{Q}_r(\Gamma_N, Q) \neq \{0\}$. The exact statements of other theorems are obvious and therefore may be left to the reader. We mention here only two points at which the treatment is somewhat different from the Hilbert modular case.

(1) If $f_1, \cdots, f_l$ $(l = n(n + 1)/2)$ are algebraically independent elements of $\mathcal{Q}_0(\Gamma, \Lambda)$ with a subfield Λ of C and if

$$(2\pi i)^l g(z) \wedge_{p \leq q} dz_{pq} = df_1 \wedge \cdots \wedge df_l \qquad (z = (z_{pq})) \,,$$

then $g \in \mathcal{Q}_{n+1}(\Gamma, \Lambda)$.

(2) Such f_j can be found in $\mathcal{Q}_0(\Gamma_1, Q)$ by taking the quotients of the Eisenstein series of Siegel [14]. The non-vanishing of $\mathcal{Q}_r(Q)$ for any (odd) $r \in Z$ can be shown by theta-functions.

Princeton University

REFERENCES

[1] W. L. BAILY and A. BOREL, Compactification of arithmetic quotients of bounded symmetric domains, Ann. of Math. **84** (1966), 442-528.

[2] R. M. DAMERELL, L-functions of elliptic curves with complex multiplication, I, II, Acta Arith. **17** (1970), 287-301; **19** (1971), 311-317.

[3] E. HECKE, Analytische Funktionen und Algebraische Zahlen, II, Abh. Math. Sem. Hamb. **3** (1924), 213-236 (=Werke, No. 20).

[4] ———, Theorie der Eisensteinschen Reihen höherer Stufe und ihre Anwendung auf Funktionentheorie und Arithmetik, Abh. Math. Sem. Hamb. **5** (1927), 199-224 (=Werke, No. 24).

[5] K. KLINGEN, Über die Werte der Dedekindschen Zetafunktion, Math. Ann. **145** (1962), 265-272.

[6] ———, Über den arithmetischen Charakter der Fourierkoeffizienten von Modulformen,

Math. Ann. **147** (1962), 176-188.

[7] H. D. KLOOSTERMAN, Theorie der Eisensteinschen Reihen von mehreren veränderlichen, Abh. Math. Sem. Hamb. **6** (1928), 163-188.

[8] H. MAASS, Zur Theorie der automorphen Funktionen von n Veränderlichen, Math. Ann. **117** (1940), 538-578.

[9] ————, Die Differentialgleichungen in der Theorie der Elliptischen Modulfunktionen, Math. Ann. **125** (1953), 235-263.

[10] G. SHIMURA, Construction of class fields and zeta functions of algebraic curves, Ann. of Math. **85** (1967), 58-159.

[11] ————, On canonical models of arithmetic quotients of bounded symmetric domains, I, II, Ann. of Math. **91** (1970), 144-222; **92** (1970), 528-549.

[12] ————, *Introduction to the arithmetic theory of automorphic functions*, Iwanami Shoten and Princeton Univ. Press, 1971.

[13] ————, On the holomorphy of certain Dirichlet series, Proc. London Math. Soc. (3) **31** (1975), 79-98.

[14] C. L. SIEGEL, Einführung in die Theorie der Modulfunktionen n-ten Grades, Math. Ann. **116** (1939), 617-657 (=Abh. II, No. 32).

[15] ————, Berechnung von Zetafunktionen an ganzzahligen Stellen, Göttingen Nachr. Akad. Wiss. 1969, 87-102.

[16] ————, Über die Fourierschen Koeffizienten von Modulformen, Göttingen Nachr. Akad. Wiss. 1970, 15-56.

[17] A. WEIL, *Foundations of Algebraic Geometry*, Amer. Math. Soc. Coll. Publ. 29, 1962 (2nd ed.).

[18] ————, *Elliptic functions according to Eisenstein and Kronecker*, Springer, to appear.

(Received April 23, 1975)

75d

On the Fourier coefficients of modular forms
of several variables

Nachrichten der Akademie der Wissenschaften in Göttingen,
Mathematisch-Physikalische Klasse, 1975, 261-268

Let $f(z)$ be a modular form of some weight on the Siegel upper half space $\mathfrak{H}_n$ of degree n with respect to

$$\Gamma_N = \{\gamma \in Sp(n, \mathbf{Z}) \mid \gamma \equiv 1_{2n} \ (\mathrm{mod}\ N)\}$$

with a positive integer N. Then f has a Fourier expansion

$$(1) \qquad f(z) = \sum_{\xi} a(\xi) e^{2\pi i \operatorname{tr}(\xi z)/N}$$

with $a(\xi) \in \mathbf{C}$, where ξ runs over all semi-integral semi-definite symmetric matrices of degree n. The main purpose of this paper is to show that the vector space of such f can be spanned by those with Fourier coefficients $a(\xi)$ in $\mathbf{Z}$. In the case $N = 1$, this has been proved by Baily [3]. Our proof, like that of [3], is based on the result of Siegel that the Fourier coefficients of the Eisenstein series

$$(2) \qquad \Psi_\mu(z) = \sum_{c,\,d} \det (cz + d)^{-\mu}$$

defined by himself are rational numbers with bounded denominators ([11, II, p. 132; III, p. 443]; cf. also [12] by the same author which gives some results of a similar nature). It may be noted that our methods are very simple and have the following advantages: (i) the information about the Fourier coefficients of Eisenstein series is needed only at level 1; (ii) no explicit set of generators for the field of modular functions (of any level) is necessary. Therefore one can apply the same technique to discontinuous groups of different types, whenever canonical models in the sense of [8] can be established, and the functions similar to Ψ_μ are obtained at lower level. This is so at least in the Hilbert modular case. In the latter half of the paper, we make some additional remarks about the Siegel modular forms with cyclotomic Fourier coefficients in connection with canonical models. This part may be regarded as a supplement to [10].

Notation. The letters $\mathbf{Z}$, $\mathbf{Q}$, $\mathbf{R}$, and $\mathbf{C}$ denote the ring of rational integers, the fields of rational numbers, real numbers, and complex numbers, respec-

[1]

708

tively. If $\Re$ is a ring with identity element, we denote by $\Re^{\times}$ the group of all invertible elements of $\Re$, by $M_n(\Re)$ the ring of all $n \times n$ matrices with entries in $\Re$, and put $GL_n(\Re) = M_n(\Re)^{\times}$. The identity element of $M_n(\Re)$ is denoted by 1_n, and the transpose of a matrix α by ${}^t\alpha$.

1. We first prove a lemma in algebraic geometry, using the terminology "generic point, specialization, etc." in the sense of Weil [14]. Let k be a field of an arbitrary characteristic. We call an indexed set of elements $(y_0, \ldots, y_m)$ of an extension of k *homogeneous over* k, if the ideal of polynomials $F(X_0, \ldots, X_m)$ in indeterminates $X_0, \ldots, X_m$ with coefficients in k such that $F(y_0, \ldots, y_m) = 0$ is generated by homogeneous polynomials. This is so if and only if each non-zero y_i is transcendental over the field $k(y_0/y_i, \ldots, y_m/y_i)$.

Lemma 1. *Let V be a subvariety defined over k of a projective space P of dimension m, and $(y_0, \ldots, y_m)$ a homogeneous set of elements over k that represents a generic point of V over k. Let H_i, for $1 \leq i \leq r$, denote a hyperplane of P defined by $\sum_{j=0}^{m} c_{ij} X_j = 0$ with $c_{ij} \in k$, where $(X_0, \ldots, X_m)$ is the coordinate system of P. Suppose $\bigcap_{i=1}^{r} H_i \cap V = \emptyset$. Then $y_0, \ldots, y_m$ are integral over $k[z_1, \ldots, z_r]$ where $z_i = \sum_{j=0}^{m} c_{ij} y_j$.*

Proof. This is a well-known principle due to Weil [13, p. 5]. For the reader's convenience, we reproduce here the proof given in [7] in a more general case. Take an arbitrary specialization of the form $(a_0, \ldots, a_m, 0, \ldots, 0)$ of $(y_0, \ldots, y_m, z_1, \ldots, z_r)$ over k, where a_i may be finite or ∞. Suppose $(a_0, \ldots, a_m) \neq (0, \ldots, 0)$. Then we can find an index j such that $y_j \neq 0$ and $(y_0/y_j, \ldots, y_m/y_j, z_1/y_j, \ldots, z_r/y_j)$ has a finite specialization $(b_0, \ldots, b_m, 0, \ldots, 0)$ compatible with the first specialization. Since y_j is transcendental over $k(y_0/y_j, \ldots, y_m/y_j)$, we have a specialization

$$(y_j, y_0/y_j, \ldots, y_m/y_j, z_1/y_j, \ldots, z_r/y_j) \to (1, b_0, \ldots, b_m, 0, \ldots, 0)$$

over k. Then we see that $(b_0, \ldots, b_m, 0, \ldots, 0)$ is a specialization of $(y_0, \ldots, y_m, z_1, \ldots, z_r)$ over k. The assumption $\bigcap_{i=1}^{r} H_i \cap V = \emptyset$ implies $(b_0, \ldots, b_m) = (0, \ldots, 0)$, which is a contradiction, since $b_j = 1$. Therefore $(a_0, \ldots, a_m) = (0, \ldots, 0)$. It follows that, for every i, $(1, 0, \ldots, 0)$ cannot be a specialization of $(y_i, z_1, \ldots, z_r)$ over k. Observe that $(y_i, z_1, \ldots, z_r)$ is homogeneous over k. Therefore we can find a homogeneous polynomial $F(X_0, \ldots, X_r)$ of degree, say d, such that $F(y_i, z_1, \ldots, z_r) = 0$ and $F(1, 0, \ldots, 0) \neq 0$. Put $F(X_0, X_1, \ldots, X_r) = \sum_{s=0}^{d} F_s(X_1, \ldots, X_r) X_0^s$. Then F_s is of degree $d - s$, and $F_d \neq 0$. This shows that y_i is integral over $k[z_1, \ldots, z_r]$.

[2]

2. Let G be an algebraic subgroup of GL_{2n} defined over $\mathbf{Q}$ such that

$$G_{\mathbf{Q}} = \{\alpha \in GL_{2n}(\mathbf{Q}) \mid {}^t\alpha\,\varepsilon\,\alpha = \nu(\alpha)\varepsilon \text{ with } \nu(\alpha) \in \mathbf{Q}\},$$

where $\varepsilon = \begin{pmatrix} 0 & -1_n \\ 1_n & 0 \end{pmatrix}$. Then $\mathrm{Sp}\,(n,\mathbf{Z}) = \{\alpha \in G_{\mathbf{Q}} \cap GL_{2n}(\mathbf{Z}) \mid \nu(\alpha) = 1\}$. Put $G_{\mathbf{Q}+} = \{\alpha \in G_{\mathbf{Q}} \mid \nu(\alpha) > 0\}$. We let every element $\alpha = \begin{pmatrix} a & b \\ c & d \end{pmatrix}$ of $G_{\mathbf{Q}+}$ (with a, b, c, d in $M_n(\mathbf{Q})$) act on $\mathfrak{H}_n$ by $\alpha(z) = (az + b)(cz + d)^{-1}$. For a meromorphic function f on $\mathfrak{H}_n$ and $r \in \mathbf{Z}$, we put

$$f \mid_r \alpha = \nu(\alpha)^{rn/2} \det\,(cz + d)^{-r} f(\alpha(z)).$$

Let Γ be a subgroup of $G_{\mathbf{Q}+}$, containing Γ_N for some N, such that $\Gamma/(\Gamma \cap \mathbf{Q}^{\times})$ is commensurable with Γ_1. We denote by $\mathfrak{M}_r(\Gamma)$ the set of all modular forms f on $\mathfrak{H}_n$ such that $f \mid_r \gamma = f$ for all $\gamma \in \Gamma$. Such a function f has a Fourier expansion of the form (1). Now for a subring $\mathfrak{R}$ of $\mathbf{C}$, we denote by $\mathfrak{M}_r(\Gamma, \mathfrak{R})$ the set of all $f \in \mathfrak{M}_r(\Gamma)$ with Fourier coefficients $a(\xi)$ in $\mathfrak{R}$, and put $\mathfrak{M}_r(\mathfrak{R}) = \overset{\infty}{\underset{N=1}{\mathbf{U}}} \mathfrak{M}_r(\Gamma_N, \mathfrak{R})$. Further we denote by $\mathfrak{A}_s(\mathfrak{R})$, for $s \in \mathbf{Z}$, the set of all meromorphic functions of the form f/g with $f \in \mathfrak{M}_{r+s}(\mathfrak{R})$, $0 \neq g \in \mathfrak{M}_r(\mathfrak{R})$ (with any r), and by $\mathfrak{A}_s(\Gamma, \mathfrak{R})$ the set of all $h \in \mathfrak{A}_s(\mathfrak{R})$ such that $h \mid_s \gamma = h$ for all $\gamma \in \Gamma$. Then $\mathfrak{A}_0(\Gamma, \mathbf{C})$ is the field of all modular functions on $\mathfrak{H}_n$ with respect to Γ. We denote by $(\mathfrak{H}_n/\Gamma)^*$ the Satake compactification of the quotient space $\mathfrak{H}_n/\Gamma$.

Theorem 1. *Let $\overline{\mathbf{Q}}$ denote the algebraic closure of $\mathbf{Q}$, and $\mathfrak{O}$ the ring of all algebraic integres in $\overline{\mathbf{Q}}$. Then, for every $g \in \mathfrak{M}_r(\Gamma, \overline{\mathbf{Q}})$, there exists a positive rational integer c such that $cg \in \mathfrak{M}_r(\Gamma, \mathfrak{O})$.*

Proof. It is sufficient to prove this for $\Gamma = \Gamma_N$. We consider the series Ψ_μ of (2) defined by Siegel [11, II, p. 125]. Now Baily showed the existence of finitely many Ψ_μ that have no common zeros on $(\mathfrak{H}_n/\Gamma_1)^*$. (This is an easy consequence of the fact that if $\sum_{m=1}^{\infty} a_m$ is an absolutely convergent series of complex numbers a_m, not all 0, then $\sum_{m=1}^{\infty} a_m^t \neq 0$ for some positive integer t. For Baily's result in the most general case, see [2].) Therefore, given $g \in \mathfrak{M}_r(\Gamma_N, \overline{\mathbf{Q}})$, we can find a multiple s of r so that: (i) (all monomials of Ψ_μ of weight s have no common zeros on $(\mathfrak{H}_n/\Gamma_N)^*$; (ii) a basis $B = \{g_0, \ldots, g_m\}$ of $\mathfrak{M}_s(\Gamma_N, \mathbf{C})$ defines a biregular projective embedding of $(\mathfrak{H}_n/\Gamma_N)^*$ onto a projective variety V. The latter is a special case of the main result of Baily and Borel [1]. Now $\mathfrak{M}_s(\Gamma_N, \overline{\mathbf{Q}})$ spans $\mathfrak{M}_s(\Gamma_N, \mathbf{C})$ as will be shown in Theorem 3 below. Therefore we may assume that $B \subset \mathfrak{M}_s(\Gamma_N, \overline{\mathbf{Q}})$, and V is defined over $\overline{\mathbf{Q}}$. We may also assume that the first several members $g_0, \ldots, g_\lambda$ of B are monomials of Ψ_μ with no common zeros on $(\mathfrak{H}_n/\Gamma_N)^*$. By [9, Lemma 6.5], there exists a point w of $\mathfrak{H}_n$ such that evaluation of functions at w defines an isomorphism of the field $\overline{\mathbf{Q}}(g_0, \ldots, g_m)$ onto $\overline{\mathbf{Q}}(g_0(w), \ldots, g_m(w))$. Applying Lemma 1 to V,

[3]

we see that $g_0(w), \ldots, g_m(w)$ are integral over $\overline{\mathbf{Q}}[g_0(w), \ldots, g_\lambda(w)]$; hence $g_0, \ldots, g_m$ are integral over $\overline{\mathbf{Q}}[g_0, \ldots, g_\lambda]$. Since $g^{s/r}$ is a $\overline{\mathbf{Q}}$-linear combination of the g_i, we thus obtain an equation

$$(3) \qquad g^d + \sum_{\nu=0}^{d-1} h_\nu g^\nu = 0$$

with positive d, where h_ν are polynomials in Ψ_μ with coefficients in $\overline{\mathbf{Q}}$. By virtue of the result of Siegel mentioned at the beginning, we can find a positive rational integer c such that $c h_\nu$ has all of its Fourier coefficients in $\mathfrak{O}$. Put $l = n(n+1)/2$. We can find l linearly independent positive definite rational symmetric matrices $\tau_1, \ldots, \tau_l$ so that $\mathrm{tr}\,(\xi\tau_\nu) \in \mathbf{Z}$ for all semi-integral symmetric matrices ξ. Then we can define a map ω of the graded ring $\Sigma_r \mathfrak{M}_r(\Gamma_N)$ into the formal power-series ring $\mathbf{C}[[X_1, \ldots, X_l]]$ in l indeterminates $X_1, \ldots, X_l$ by

$$\omega(f) = \Sigma_\xi a(\xi) X_\xi, \qquad X_\xi = \prod_{i=1}^{l} X_i^{\mathrm{tr}(\xi\tau_i)},$$

when f has the expansion (1). (Such a map was first considered by Siegel in [11, III, No. 65], and later used by Eichler in [4] and also by Baily in [3].) Obviously this is an injective ring-homomorphism. Then relation (3) shows that $\omega(cg)$ is integral over $\mathfrak{O}[[X_1, \ldots, X_l]]$. Since $\mathfrak{O}[[X_1, \ldots, X_l]]$ is integrally closed, $\omega(cg)$ must be contained in $\mathfrak{O}[[X_1, \ldots, X_l]]$, and therefore $cg \in \mathfrak{M}_r(\Gamma_N, \mathfrak{O})$ q. e. d. (See the footnote at the end of the paper.)

We note that an analogue of the above theorem can be proved for the Hilbert modular forms with an automorphic factor of the form $N(cz + d)^\varkappa$ with a positive integer $\varkappa$ (i. e., the elements of $\mathcal{M}_{\varkappa u}$ of [10]). Indeed, Klingen showed that the Fourier coefficients of the Eisenstein series in this case have bounded denominators [5, Satz 2]. It should be noted that plural cusps produce no difficulty. Therefore the above proof is valid for such modular forms with no essential change.

3. That $\mathfrak{M}_r(\Gamma_N, \mathbf{C})$ is spanned by $\mathfrak{M}_r(\Gamma_N, \mathbf{Q})$ is already mentioned in [10]. It seems, however, worth making precise statements of the results of this kind for congruence subgroups of a more general type. For this purpose we consider the canonical system with respect to G in the sense of [8]. Let $G_\mathbf{A}$ denote the adelization of G, G_0 the non-archimedean part of $G_\mathbf{A}$, G_∞ $(= G_\mathbf{R})$ the archimedean part of $G_\mathbf{A}$, and $G_{\infty+}$ the identity component of G_∞. We then put $G_{\mathbf{A}+} = G_0 G_{\infty+}$. The map $\nu : G_\mathbf{Q} \to \mathbf{Q}^\times$ can be extended to a continuous map of $G_\mathbf{A}$ into $\mathbf{Q}_\mathbf{A}^\times$, which we denote again by ν. Let $\mathscr{Z}$ be the set of all open subgroups S of $G_{\mathbf{A}+}$ containing $\mathbf{Q}^\times G_{\infty+}$ such that $S/\mathbf{Q}^\times G_{\infty+}$ is compact. For each $S \in \mathscr{Z}$, we put $\Gamma_S = G_\mathbf{Q} \cap S$. Let $\mathbf{Q}_{ab}$ be the maximal abelian extension of $\mathbf{Q}$. We denote by k_S the subfield of $\mathbf{Q}_{ab}$ corresponding to the subgroup $\mathbf{Q}^\times \cdot \nu(S)$ of $\mathbf{Q}_\mathbf{A}^\times$. By [8], for each $S \in \mathscr{Z}$, we have a canonical model (V_S, φ_S) of $\mathfrak{H}_n/\Gamma_S$. Here V_S is a variety defined over k_S, and φ_S is a Γ_S-invariant holomorphic

[4]

map of $\mathfrak{H}_n$ onto V_S that gives a biregular isomorphism of $\mathfrak{H}_n/\Gamma_S$ onto V_S. Moreover, as remarked in [10, § 6], we may assume that φ_S gives an isomorphism of $(\mathfrak{H}_n/\Gamma_S)^*$ onto the projective variety V_S^* of which V_S is a Zariski open subset. Let $\mathbf{C}(V_S)$ denote the field of all "functions" on V_S in the sense of algebraic geometry, and $k_S(V_S)$ its subfield consisting of all the k_S-rational elements of $\mathbf{C}(V_S)$. Define fields $\mathfrak{K}$, $\mathfrak{K}_S$ of meromorphic functions on $\mathfrak{H}_n$ by

$$(4) \qquad \mathfrak{K} = \bigcup_{S \in \mathscr{L}} \mathfrak{K}_S, \qquad \mathfrak{K}_S = \{ f \circ \varphi_S \mid f \in k_S(V_S) \}.$$

We recall that there is a commutative diagram

$$(5) \qquad \begin{array}{ccccccc} 1 & \longrightarrow & \mathbf{Q}^\times G_{\infty+} & \longrightarrow & G_{\mathbf{A}+} & \overset{\tau}{\longrightarrow} & \mathrm{Aut}\,(\mathfrak{K}) & \longrightarrow & 1 \\ & & \downarrow & & \downarrow{\scriptstyle r^{-1}} & & \downarrow{\scriptstyle \mathrm{res}} & & \\ 1 & \longrightarrow & \mathbf{Q}^\times \mathbf{Q}_{\infty+}^\times & \longrightarrow & \mathbf{Q}_{\mathbf{A}} & \overset{\sigma}{\longrightarrow} & \mathrm{Gal}\,(\mathbf{Q}_{ab}/\mathbf{Q}) & \longrightarrow & 1 \end{array} \quad \begin{array}{c} \text{(exact)} \\ \\ \text{(exact)} \end{array}$$

where $v^{-1}(x) = v(x)^{-1}$, res denotes restriction of the action of an automorphism of $\mathfrak{K}$ to its subfield $\mathbf{Q}_{ab}$, and $\mathrm{Aut}\,(\mathfrak{K})$ the group of all automorphisms of the field $\mathfrak{K}$ (see [8, 2.7, 2.8], T. Miyake [6]). Define subgroups Δ and Δ_S of $G_{\mathbf{A}+}$ by

$$(6) \qquad \Delta = \left\{ \begin{pmatrix} 1_n & 0 \\ 0 & t\,1_n \end{pmatrix} \Bigg| \, t \in \Pi_p \mathbf{Z}_p^\times \right\},$$

$$\Delta_S = \{ v \in \Delta \mid \tau(v) = \mathrm{id.\ on}\ k_S \},$$

where $\mathbf{Z}_p$ denotes the ring of p-adic integers for each rational prime p. Now we have

Theorem 2. (i) *Let ϱ be a field-automorphism of $\mathbf{C}$, and $g(z) = \Sigma_\xi a(\xi) e^{2\pi i \operatorname{tr}(\xi z)}$ an element of $\mathfrak{M}_r(\mathbf{C})$. Then there exists an element g^ϱ of $\mathfrak{M}_r(\mathbf{C})$ such that $g^\varrho(z) = \Sigma_\xi a(\xi)^\varrho e^{2\pi i \operatorname{tr}(\xi z)}$.*

(ii) *Let $f = g/h$ with g and h in $\mathfrak{M}_r(\mathbf{Q}_{ab})$, and let $v = \begin{pmatrix} 1_n & 0 \\ 0 & t^{-1} 1_n \end{pmatrix}$ be an element of Δ with $t \in \Pi_p \mathbf{Z}_p^\times$. Then $f^{\tau(v)} = g^{\sigma(t)}/h^{\sigma(t)}$, where σ and τ are the maps of the diagram (5).*

Theorem 3. *Let S be a member of $\mathscr{L}$ such that $\Delta_S \subset S$. Then we have*

(i) $\mathfrak{A}_0(\Gamma_S, k_S) = \mathfrak{K}_S$;

(ii) $\mathfrak{M}_r(\Gamma, \mathbf{C}) = \mathfrak{M}_r(\Gamma, k_S) \otimes_{k_S} \mathbf{C}$ *for a positive integer r, if Γ is a subgroup of Γ_S such that $\mathbf{Q}^\times \Gamma = \Gamma_S$ and $\mathfrak{A}_r(\Gamma, k_S) \neq \{0\}$.*

These theorems can be proved in exactly the same fashion as in Proposition 4 and Theorem 6 of [10], as remarked in § 6 of the same paper. It should be noted that $\mathfrak{A}_r(\Gamma_S, k_S) \neq \{0\}$ for every even r, if $\Delta_S \subset S$. Indeed, take any $f \in \mathfrak{K}_S = \mathfrak{A}_0(\Gamma_S, k_S)$, and put

$$(7) \qquad g(z) = (2\pi i)^{-n} \det(\eta_{\mu\nu} \partial f/\partial z_{\mu\nu}) \qquad (z = (z_{\mu\nu})),$$

$$(8) \qquad \eta_{\mu\nu} = \begin{cases} 1 & \text{if}\ \ \mu = \nu, \\ \tfrac{1}{2} & \text{if}\ \ \mu \neq \nu. \end{cases}$$

[5]

Then it can easily be verified that $g^r \in \mathfrak{A}_{2r}(\Gamma_S, k_S)$ for every $r \in \mathbf{Z}$. To ensure $g \neq 0$, take $f = \sum_{j=1}^{l} a_j f_j$ with $a_j \in \mathbf{Q}$ and l algebraically independent elements $f_1, \ldots, f_l$ of $\mathfrak{K}_S$, where $l = n(n+1)/2$. For a suitable choice of a_j, we obtain $g \neq 0$.

As a typical example of S, take any matrix q of $M_n(\mathbf{Z}) \cap GL_n(\mathbf{Q})$ whose entries have no common divisor > 1, and put, with $\beta = \begin{pmatrix} 1_n & 0 \\ 0 & q \end{pmatrix}$,

$$U_q = \{x \in G_{\mathbf{A}+} \mid \beta^{-1} x_p \beta \in GL_{2n}(\mathbf{Z}_p) \text{ for all } p\},$$

$$S_{N,q} = \mathbf{Q}^\times \cdot \left\{ x \in U_q \mid \beta^{-1} x_p \beta \equiv \begin{pmatrix} 1_n & 0 \\ 0 & a_p 1_n \end{pmatrix} (\text{mod } N \mathbf{Z}_p) \text{ with } a_p \in \mathbf{Z}_p \right\}$$

for every positive integer N, where x_p denotes the p-component of x. Then, for $S = S_{N,q}$, we have $\Delta \subset S$, $k_S = \mathbf{Q}$, and

$$\Gamma_S = \mathbf{Q}^\times \{\gamma \in \mathrm{Sp}\,(n, \mathbf{Q}) \mid \beta^{-1} \gamma \beta \in GL_{2n}(\mathbf{Z}), \ \beta^{-1} \gamma \beta \equiv 1_{2n} \ (\text{mod } N)\}.$$

Especially if $q = 1_n$, this is $\mathbf{Q}^\times \Gamma_N$. Combining Theorem 1 with Theorem 3 in this case, we obtain

$$(9) \qquad \mathfrak{M}_r(\Gamma_N, \mathbf{C}) = \mathfrak{M}_r(\Gamma_N, \mathbf{Z}) \otimes_{\mathbf{Z}} \mathbf{C}$$

for every even positive r, and also for an odd r provided that $\mathfrak{A}_r(\Gamma_N, \mathbf{Q}) \neq \{0\}$. We note also that, by the same reasoning as in [10, § 4], we obtain

$$(10) \qquad \mathfrak{A}_0(\Lambda) = \Lambda \mathfrak{K} \text{ for every subfield } \Lambda \text{ of } \mathbf{C} \text{ containing } \mathbf{Q}_{ab},$$

and especially $\mathfrak{A}_0(\mathbf{Q}_{ab}) = \mathfrak{K}$, which we already noted in [10, § 6].

4. Obviously $G_{\mathbf{Q}+}$ acts on $\mathfrak{A}_r(\mathbf{C})$ and $\mathfrak{M}_r(\mathbf{C})$. On the other hand, we can let every field-automorphism ϱ of $\mathbf{C}$ act on $\mathfrak{M}_r(\mathbf{C})$ as in (i) of Theorem 2. This can be extended to $\mathfrak{A}_s(\mathbf{C})$ by putting $(g/h)^\varrho = g^\varrho/h^\varrho$ for $g \in \mathfrak{M}_{r+s}(\mathbf{C})$ and $0 \neq h \in \mathfrak{M}_r(\mathbf{C})$. The mutual relationship of these two types of action can be best described by defining the action of $G_{\mathbf{A}+}$ on $\mathfrak{A}_s(\mathbf{Q}_{ab})$. We first prove an easy

Theorem 4. *If* $f \in \mathfrak{A}_r(\mathbf{Q}_{ab})$ *and* $\alpha \in G_{\mathbf{Q}+}$, *then* $f \mid_r \alpha \in \mathfrak{A}_r(\mathbf{Q}_{ab})$. *This assertion is also true with* $\mathfrak{M}_r(\mathbf{Q}_{ab})$ *in place of* $\mathfrak{A}_r(\mathbf{Q}_{ab})$.

Proof. Put $h = \theta^{2r}$ with $\theta(z) = \sum_m e^{\pi i \cdot {}^t m z m}$, where m runs over all (column) vectors in $\mathbf{Z}^n$. Then $h \in \mathfrak{A}_r(\mathbf{Q})$ and $h \mid_r \varepsilon = (-i)^{nr} h$, where $\varepsilon = \begin{pmatrix} 0 & -1_n \\ 1_n & 0 \end{pmatrix}$. Given $f \in \mathfrak{A}_r(\mathbf{Q}_{ab})$, put $g = f/h$. Then $g \in \mathfrak{A}_0(\mathbf{Q}_{ab}) = \mathfrak{K}$. Since $g \circ \alpha = g^{\tau(\alpha)} \in \mathfrak{K}$ for every $\alpha \in G_{\mathbf{Q}+}$ (see [8, 1, (2.7.2)]), we have $f \mid_r \varepsilon = (g \circ \varepsilon) h \mid_r \varepsilon \in \mathfrak{A}_r(\mathbf{Q}_{ab})$. Now for every $\Psi \in \mathfrak{M}_r(\mathbf{Q}_{ab})$ and $\alpha = \begin{pmatrix} a & b \\ 0 & d \end{pmatrix} \in G_{\mathbf{Q}+}$ with a, b, d in $M_n(\mathbf{Q})$, it can easily be seen that $\Psi \mid_r \alpha \in \mathfrak{M}_r(\mathbf{Q}_{ab})$, and therefore $f \mid_r \alpha \in \mathfrak{A}_r(\mathbf{Q}_{ab})$. Since

[6]

713

$G_{\mathbf{Q}+}$ is generated by such α and ε, we obtain our assertion for $\mathfrak{A}_r(\mathbf{Q}_{ab})$. The assertion for $\mathfrak{M}_r(\mathbf{Q}_{ab})$ is now obvious, since $\mathfrak{M}_r(\mathbf{Q}_{ab})$ is the set of holomorphic elements of $\mathfrak{A}_r(\mathbf{Q}_{ab})$.

Now we restrict the weight to even integers.

Theorem 5. *Each element x of $G_{\mathbf{A}+}$ acts on the graded ring $\sum_{r\in 2\mathbf{Z}} \mathfrak{A}_r(\mathbf{Q}_{ab})$ as an "automorphism" $f \mapsto f^x$ with the following properties:*

 (i) $\mathfrak{A}_r(\mathbf{Q}_{ab})$ *and* $\mathfrak{M}_r(\mathbf{Q}_{ab})$ *for each r are stable under the action;*

 (ii) $f^x = f^{\tau(x)}$ *if* $f \in \mathfrak{A}_0(\mathbf{Q}_{ab})$, *where τ is the map of (5);*

 (iii) $(f + g)^x = f^x + g^x$, $(fg)^x = f^x g^x$, $(f/g)^x = f^x/g^x$;

 (iv) $f^{xy} = (f^x)^y$;

 (v) $f^x = f^{\sigma(t)}$ *if* $x^{-1} = \begin{pmatrix} 1_n & 0 \\ 0 & t1_n \end{pmatrix}$ *with* $t \in \Pi_p \mathbf{Z}_p^\times$;

 (vi) $f^\alpha = f \mid_r \alpha$ *if* $\alpha \in G_{\mathbf{Q}+}$ *and* $f \in \mathfrak{A}_r(\mathbf{Q}_{ab})$;

 (vii) $(Df)^x = D(f^x)$ *for every* $f \in \mathfrak{A}_0(\mathbf{Q}_{ab})$, *where* $Df = (2\pi i)^{-n} \det(\eta_{\mu\nu}\partial f/\partial z_{\mu\nu})$ *with* $\eta_{\mu\nu}$ *as in* (8);

 (viii) $f^x = f$ *if* $f \in \mathfrak{A}_r(\Gamma_S, k_S)$, $x \in S$, *and* $\Delta_S \subset S$.

Proof. Let g and h be elements of $\mathfrak{A}_0(\mathbf{Q}_{ab})$ such that $Dh \neq 0$. Then we have $D(h^{\tau(x)}) \neq 0$ for every $x \in G_{\mathbf{A}+}$, and

$$(11) \qquad (Dg/Dh)^{\tau(x)} = D(g^{\tau(x)})/D(h^{\tau(x)}).$$

This can be proved in exactly the same manner as in [10, Prop. 9]. Now, given $f \in \mathfrak{A}_{2m}(\mathbf{Q}_{ab})$ with $m \in \mathbf{Z}$, define f^x by $f^x = (f/(Dh)^m)^{\tau(x)} D(h^{\tau(x)})^m$. By virtue of (11), this does not depend on the choice of h. Then we can easily verify all the properties (i—viii) (cf. the proof of [10, Theorem 8]).

References

[1] W. L. Baily and A. Borel, Compactification of arithmetic quotients of bounded symmetric domains, Ann. of Math. 84 (1966) 442—528.

[2] W. L. Baily, Eisenstein series on tube domains, Problems in Analysis, A symposium in honor of Salomon Bochner, Princeton Univ. Press, (1970) 139—156.

[3] W. L. Baily, Automorphic forms with integral Fourier coefficients, Several complex variables I, Maryland 1970, Lecture notes in math. 155, 1—8.

[4] M. Eichler, Zur Begründung der Theorie der automorphen Funktionen in mehreren Variablen, Aequationes Math. 8 (1969) 93—111.

[5] H. Klingen, Über den arithmetischen Charakter der Fourierkoeffizienten von Modulformen, Math. Ann. 147 (1962) 176—188.

[6] T. Miyake, On automorphism groups of the fields of automorphic functions, Ann. of Math. 95 (1972) 243—252.

[7] G. Shimura, A note on the normalization-theorem of an integral domain, Scientific Papers of the College of General Education, Univ. of Tokyo, 4 (1954) 1—8.

[7]

[8] G. Shimura, On canonical models of arithmetic quotients of bounded symmetric domains, I, II, Ann. of Math. **91** (1970) 144—222; **92** (1970) 528—549.

[9] G. Shimura, Introduction to the arithmetic theory of automorphic functions, Iwanami Shoten and Princeton Univ. Press, 1971.

[10] G. Shimura, On some arithmetic properties of modular forms of one and several variables, to appear in Ann. of Math.

[11] C. L. Siegel, Gesammelte Abhandlungen I—III, Springer, 1966.

[12] C. L. Siegel, Über Moduln Abelscher Funktionen, Nachr. Akad. Wiss. Göttingen, math.-phys. Kl. (1971) 79—96.

[13] A. Weil, Arithméthique et Géométrie sur les variétés algébriques, Actualités scientifiques et industrielles **206**, Hermann, Paris 1935.

[14] A. Weil, Foundations of Algebraic Geometry, Amer. Math. Soc. Coll. Publ. **29** (1962) (2nd ed.).

At the end of the proof of Theorem 1, it is tacitly assumed that ω (cg) belongs to the quotient ring of $\mathfrak{D}\,[[X_1, \ldots, X_l]]$. This can actually be proved in an elementary way, and will be explained in detail in a succeeding paper by the author.

[8]

Theta functions with complex multiplication

Duke Mathematical Journal, 43 (1976), 673-696

Introduction. The theory of complex multiplication of abelian varieties tells how the value of an abelian function at a point of finite order behaves under Frobenius substitutions. An abelian function can naturally be obtained as the quotient of two theta functions. In the present paper we shall show that a theta function with complex multiplication modified by a suitable exponential factor takes values of algebraic nature at the points commensurable with the periods. The images of the values under Frobenius substitutions can also be determined.

To be more explicit, let V be a vector space over $\mathbf{C}$ of dimension n, and L a lattice in V, i.e., a discrete subgroup of V isomorphic to $\mathbf{Z}^{2n}$. Then we consider a theta function f satisfying

$$(\mathrm{I}) \qquad f(u + l) = f(u)\psi(l)e\left(\frac{1}{2i}\,H\left(l, u + \frac{l}{2}\right)\right) \qquad (u \in V, l \in L)$$

with a hermitian form H and a certain map ψ of L into $\mathbf{C}$, where $e(x) = e^{2\pi i x}$ for $x \in \mathbf{C}$. Assume that $\psi(l)$ for each $l \in L$ is a root of unity, and put

$$(\mathrm{II}) \qquad f_*(u) = e\left(\frac{i}{4}\,H(u, u)\right)f(u).$$

Then it can easily be verified that

$$(\mathrm{III}) \qquad f_*(u + l) = \psi(l)e(\mathrm{Im}\ (H(l, u))/2)f_*(u) \qquad (l \in L).$$

This implies that f_* can be extended to a continuous function on the adelization of $\mathbf{Q}L$. Suppose that V/L has many complex multiplications. Then V/L determines a number field K' over which the values of abelian functions on $\mathbf{Q}L$ are of abelian nature. Now we call f *arithmetic* if $f_*(u)$ is abelian over K' for every $u \in \mathbf{Q}L$. We shall show that the space of theta functions satisfying (I) can be spanned by arithmetic ones. Moreover, we shall define a natural action of the idele group of K' on the arithmetic theta functions. The exact statement of the properties of the action will form our main theorem in §2.

If we regard a theta function as a function in two kinds of variables, one in the complex vector space and the other in the Siegel upper half space $\mathfrak{H}_n$, then $f_*(u)$ for $u \in \mathbf{Q}L$ is the value of a modular form at a point on $\mathfrak{H}_n$ corresponding to V/L. This fact is essential in the proof of the theorem. Indeed, the

Received March 8, 1976. Revision received June 7, 1976.

images of $f_*(u)$ under Frobenius substitutions can easily be established by means of the results of the previous papers [7], [11]. In this sense, the present paper may be regarded as an appendix to them, but here emphasis is laid on the explicit nature of the functions involved, and also on the fact that f_* is not a function on V/L but only has property (III).

It should be remarked that the value $f_*(u)$ in the one-dimensional case occurs in the Kronecker limit formula. We consider in §4 an arithmetic theta function which is implicitly defined without any coordinate system and which was studied by Kronecker in a special case. While this connection with the limit formula is one of the motives of our investigation, arithmetic theta functions appear naturally in the theory of automorphic forms with respect to certain unitary groups, which the author hopes to treat in detail in a subsequent paper.

NOTATION. We denote as usual by $\mathbf{Z}, \mathbf{Q}, \mathbf{R}$, and $\mathbf{C}$ the ring of rational integers, the fields of rational numbers, real numbers, and complex numbers, respectively; $\mathbf{Z}_p$ and $\mathbf{Q}_p$ denote, for a prime p, the p-adic completions of $\mathbf{Z}$ and $\mathbf{Q}$. For a positive integer n, $\mathbf{Z}^n, \mathbf{Q}^n$, etc. denote the module or vector space of n-dimensional column vectors with components in $\mathbf{Z}, \mathbf{Q}$, etc. We denote by $\mathfrak{H}_n$ the set of all complex symmetric matrices of degree n with positive definite imaginary part. For $z \in \mathbf{C}$, we denote by $\bar{z}$ the complex conjugate of z, by Im (z) the imaginary part of z, and put $e(z) = e^{2\pi i z}$. If Y is an associative ring with identity element, then $Y^\times$ denotes the group of all invertible elements of Y, and $M_n(Y)$ the ring of all matrices of size n with components in Y; the identity element of $M_n(Y)$ is denoted by 1_n or simply by 1; $GL_n(Y) = M_n(Y)^\times$. The transpose of a matrix x is denoted by $^t x$. If K is an algebraic number field, K_{ab} denotes the maximal abelian extension of K, and $K_A^\times$ the idele group of K. By class field theory, every element x of $K_A^\times$ acts on K_{ab} as an automorphism. The image of an element s of K_{ab} under x is denoted by s^x.

1. Reduced and classical theta functions.

Let V and L be as in the Introduction, with or without complex multiplication. By a *Riemann form* for V/L, we mean a positive semi-definite hermitian form $H(u, v)$ on $V \times V$, $\mathbf{C}$-linear in v, such that Im $(H(u, v)) \in \mathbf{Z}$ for all $(u, v) \in L \times L$. Given such an H, we put

$$(1) \qquad E(u, v) = \mathrm{Im}\,(H(u, v)),$$

and consider a *reduced theta function of type* (H, ψ) relative to L in the sense of Weil [12, Ch. VI, n°3], which is, by definition, a holomorphic function f on V satisfying

$$(2) \qquad f(u + l) = f(u)\psi(l)e\left(\frac{1}{2i} H\left(l, u + \frac{l}{2}\right)\right) \qquad (u \in V, d \in L).$$

Here ψ is a map of L into the group $\{z \in \mathbf{C}|\ |z| = 1\}$ such that

$$(3) \qquad \psi(l + m) = \psi(l)\psi(m)e(E(l, m)/2) \qquad (l, m \in L).$$

We denote by $T(H, \psi, L)$ the vector space of all such holomorphic functions on V. Then we can easily verify

PROPOSITION 1.1. *With any $f \in T(H, \psi, L)$ and any $w \in V$, put*

$$g(u) = e\left(\frac{-1}{2i} H\left(w, u + \frac{w}{2}\right)\right) f(u + w).$$

Then $g \in T(H, \psi', L)$, where $\psi'(l) = \psi(l)e(E(l, w))$.

Let us now identity V with $\mathbf{C}^n$, and suppose that H is non-degenerate. Then we can find a positive integer μ, a diagonal matrix ϵ with integral diagonal elements, and a complex $(n \times 2n)$-matrix Ω such that

$$(4) \qquad E(\Omega x, \Omega y) = \mu \cdot {}^t x J y \qquad ((x, y) \in \mathbf{R}^{2n} \times \mathbf{R}^{2n}),$$

$$(5) \qquad J = \begin{pmatrix} 0 & -1_n \\ 1_n & 0 \end{pmatrix},$$

$$(6) \qquad L = \left\{ \Omega \begin{pmatrix} a \\ b \end{pmatrix} \mid a \in \mathbf{Z}^n, b \in \epsilon\mathbf{Z}^n \right\},$$

$$(7) \qquad \epsilon = \begin{bmatrix} \epsilon_1 & & & \\ & \epsilon_2 & & \\ & & \ddots & \\ & & & \epsilon_n \end{bmatrix}, \qquad \epsilon_1 = 1, \qquad \epsilon_k \mid \epsilon_{k+1} \qquad (k = 1, \cdots, n - 1).$$

Write $\Omega = (\omega_1\ \omega_2)$ with square matrices ω_1 and ω_2, and put $z = \omega_2^{-1} \omega_1$. It is well known that $z \in \mathfrak{H}_n$; moreover, a simple calculation shows that

$$(8) \qquad \frac{1}{2i} H(u, v) = \mu \cdot {}^t\bar{u}(\omega_1 \cdot {}^t\bar{\omega}_2 - \omega_2 \cdot {}^t\bar{\omega}_1)^{-1} v$$

$$= \mu \cdot {}^t(\bar{\omega}_2^{-1}\bar{u})(z - \bar{z})^{-1}\omega_2^{-1} v \qquad (u, v \in \mathbf{C}^n),$$

and ψ must be of the form

$$(9) \qquad \psi(\omega_1 a + \omega_2 b) = e\left(\frac{\mu}{2} \cdot {}^t ab + {}^t rb + {}^t sa\right) \qquad (a \in \mathbf{Z}^n, b \in \epsilon\mathbf{Z}^n)$$

with elements r and s of $\mathbf{R}^n$. For any choice of r and s, we can define ψ by (9), with which $T(H, \psi, L)$ is non-trivial. We shall now study the relationship between the reduced theta functions of the above type and the classical theta functions. For $u \in \mathbf{C}^n$, $z \in \mathfrak{H}_n$, $r \in \mathbf{R}^n$, and $s \in \mathbf{R}^n$, put

$$(10) \qquad \theta(u, z; r, s) = \sum_{x \in \mathbf{Z}^n} e(\tfrac{1}{2}{}^t(x + r)z(x + r) + {}^t(x + r)(u + s)).$$

Then we can easily verify

$$(11) \qquad \theta(u + za + b, z; r, s)$$

$$= e\left(\frac{-1}{2} \cdot {}^t aza - {}^t a(u + b + s)\right)\theta(u, z; r + a, s + b) \qquad (a, b \in \mathbf{R}^n),$$

$$(12) \qquad \theta(u + za + b, z; r, s) = e\left(\frac{-1}{2}\cdot {}^t a z a - {}^t a u + {}^t r b - {}^t s a\right)\theta(u, z; r, s)$$

$$(a, b \in \mathbf{Z}^n),$$

$$(13) \qquad \theta(u, z; r + a, s + b) = e({}^t r b)\theta(u, z; r, s) \qquad (a, b \in \mathbf{Z}^n).$$

PROPOSITION 1.2. *For* $u \in \mathbf{C}^n$, $z \in \mathfrak{H}_n$, $r \in \mathbf{R}^n$, $s \in \mathbf{R}^n$, *and* $0 < \mu \in \mathbf{Z}$, *put*

$$\varphi(u, z; r, s) = e(\tfrac{1}{2}\cdot {}^t u(z - \bar{z})^{-1}u)\theta(u, z; r, s),$$

$$f(u) = f_\mu(u; r, s) = \varphi(\mu\omega_2^{-1}u, \mu\omega_2^{-1}\omega_1 ; r, s).$$

Then $f \in T(H, \psi, L)$, *where* H *is defined by* (8), *and* ψ *by*

$$\psi(\omega_1 a + \omega_2 b) \doteq e\left(\frac{\mu}{2}\,{}^t a b - {}^t s a + \mu\cdot {}^t r b\right) \quad . \quad (a \in \mathbf{Z}^n, b \in \epsilon\mathbf{Z}^n).$$

Moreover, let $\mathfrak{J}$ *be a complete set of representatives for* $\mu^{-1}\epsilon^{-1}\mathbf{Z}^n/\mathbf{Z}^n$. *Then the functions* $f_\mu(u; r + j, s)$ *with* $j \in \mathfrak{J}$ *and with fixed* r *and* s *form a basis of* $T(H, \psi, L)$ *over* $\mathbf{C}$.

The first assertion can easily be verified by means of (12). The second one follows from the classical fact that the functions $g_j(u) = \theta(\mu u, \mu z; r + j, s)$ with $j \in \mathfrak{J}$ form a basis for the vector space of holomorphic functions g satisfying

$$g(u + za + b) = g(u)e\left(\mu\left(\frac{-1}{2}\,{}^t a z a - {}^t a u + {}^t r b\right) - {}^t s a\right) \qquad (a \in \mathbf{Z}^n, b \in \epsilon\mathbf{Z}^n)$$

(cf. Weil [12, VI, $n^\circ 7$]).

It should be remarked that if $\mu \geq 3$, then a basis of $T(H, \psi, L)$ defines a projective embedding of V/L. This implies especially that, for any u, z, and $\mu \geq 3$, there exists an $r \in \mu^{-1}\mathbf{Z}^n$ such that $\theta(u, z; r, 0) \neq 0$.

Let G be the algebraic subgroup of GL_{2n} defined over $\mathbf{Q}$ such that

$$G_\mathbf{Q} = \{\alpha \in GL_{2n}(\mathbf{Q}) \mid {}^t\alpha J\alpha = \nu(\alpha)J \quad \text{with} \quad \nu(\alpha) \in \mathbf{Q}\},$$

where J is defined by (5). Then $Sp(n, \mathbf{Z}) = \{\alpha \in G_\mathbf{Q} \cap GL_{2n}(\mathbf{Z}) \mid \nu(\alpha) = 1\}$. Put $G_{\mathbf{Q}+} = \{\alpha \in G_\mathbf{Q} \mid \nu(\alpha) > 0\}$. We let every element $\alpha = \begin{pmatrix} a & b \\ c & d \end{pmatrix}$ of $G_{\mathbf{Q}+}$ act on $\mathfrak{H}_n$ by $\alpha(z) = (az + b)(cz + d)^{-1}$. For every positive integer N, we put

$$\Gamma(N) = \Gamma_N = \{\gamma \in Sp\,(n, \mathbf{Z}) \mid \gamma \equiv 1_{2n} \bmod N\cdot M_{2n}(\mathbf{Z})\}.$$

PROPOSITION 1.3. *For every*

$$\gamma = \begin{bmatrix} a & b \\ c & d \end{bmatrix} \in \Gamma_1 ,$$

one has

$$(14) \qquad \theta({}^t(cz + d)^{-1}u, \gamma(z); r, s)$$

$$= \zeta\cdot\det\,(cz + d)^{1/2}e(\tfrac{1}{2}\cdot {}^t u(cz + d)^{-1}cu)\theta(u, z; r'', s''),$$

with a constant ζ of absolute value 1 depending only on r, s, and γ (and also on the choice of the branch of $\det (cz + d)^{1/2}$*), and*

$$\binom{r''}{s''} = {}^t\gamma \binom{r}{s} + \frac{1}{2}\binom{\{{}^t ac\}}{\{{}^t bd\}},$$

where $\{S\}$ *denotes the column vector whose components are the diagonal elements of S. Further if* $\{{}^t ac\}$ *and* $\{{}^t bd\}$ *belong to* $2\mathbf{Z}^n$*, then*

$$(14') \qquad \theta({}^t(cz+d)^{-1}u, \gamma(z); r, s)$$
$$= \lambda_\gamma e(({}^t rs - {}^t r's')/2) \det (cz+d)^{1/2} e(\tfrac{1}{2} \cdot {}^t u(cz+d)^{-1}cu)\theta(u, z; r', s'),$$

where λ_γ is a constant of absolute value 1 depending only on γ, and

$$\binom{r'}{s'} = {}^t\gamma \binom{r}{s}.$$

These formulas are well known (see for example Krazer-Prym [2]). For the reader's convenience, we give here a short, if sketchy, proof. We first state another version of the transformation formula as

PROPOSITION 1.4. *Define $\varphi(u, z; r, s)$ as in Prop. 1.2. Suppose $mr \in \mathbf{Z}^n$ and $ms \in \mathbf{Z}^n$ with a positive integer m. Then, for every*

$$\gamma = \begin{pmatrix} a & b \\ c & d \end{pmatrix} \in \Gamma(2m^2),$$

one has

$$\varphi({}^t(cz+d)^{-1}u, \gamma(z); r, s) = \lambda_\gamma \det (cz+d)^{1/2}\varphi(u, z; r, s)$$

with the same λ_γ as in (14′), and $\lambda_\gamma^4 = 1$.

To prove the above formulas, we observe that $(cz+d)^{-1}c$ is symmetric, and (14) is equivalent to

$$(14_a) \qquad \varphi({}^t(cz+d)^{-1}u, \gamma(z); r, s) = \zeta \cdot \det (cz+d)^{1/2}\varphi(u, z; r'', s'').$$

This follows easily from a relation

$$(15) \qquad \nu(\gamma)(\gamma(z) - \gamma(\bar z))^{-1} = (c\bar z + d)(z - \bar z)^{-1} \cdot {}^t(cz+d) \quad \left({}^t\gamma = \begin{pmatrix} a & b \\ c & d \end{pmatrix} \in G_{\mathbf{Q}+}\right).$$

Also, from (11), we obtain

$$(16) \qquad e\left(-{}^t\bar v(z - \bar z)^{-1}\left(u + \frac{v}{2}\right)\right)\varphi(u + v, z; r, s)$$

$$= e\left(\frac{-1}{2} \cdot {}^t p(q + 2s)\right)\varphi(u, z; r + p, s + q) \qquad (v = zp + q; p, q \in \mathbf{R}^n).$$

Now we can easily verify (14_a) with $|\zeta| = 1$ if γ is of the form

$$\begin{pmatrix} 0 & -1 \\ 1 & 0 \end{pmatrix}, \begin{pmatrix} a & 0 \\ 0 & d \end{pmatrix}, \quad \text{or} \quad \begin{pmatrix} 1 & b \\ 0 & 1 \end{pmatrix}.$$

In particular, if

$$\gamma = \begin{pmatrix} 0 & -1 \\ 1 & 0 \end{pmatrix},$$

we obtain, by the Poisson summation formula,

$$(16')\qquad \theta(z^{-1}u, -z^{-1}; r, s) = e({}^{t}rs)\,\det\,(-iz)^{1/2}e({}^{t}uz^{-1}u/2)\theta(u, z; s, -r).$$

Therefore we obtain (14$_a$) for an arbitrary $\gamma \in \Gamma_1$ with some r'', s'', and a constant ζ of absolute value 1. Thus our task is to determine r'' and s''. Put $w = \gamma(z)$ and $f(u) = \varphi(u, w; r, s)$. By Prop. 1.2, f belongs to $T(H', \psi', L')$, where

$$L' = w\mathbf{Z}^n + \mathbf{Z}^n, \qquad \frac{1}{2i}H'(u, v) = {}^{t}\bar{u}(w - \bar{w})^{-1}v,$$

and

$$\psi'(wp + q) = e(\tfrac{1}{2}\cdot{}^{t}pq - {}^{t}sp + {}^{t}rq) \qquad (p, q \in \mathbf{Z}^n).$$

Put $P = {}^{t}(cz + d)^{-1}$ and $g(u) = f(Pu)$. Then $g \in T(H_1, \psi_1, L_1)$, where

$$L_1 = P^{-1}L' = z\mathbf{Z}^n + \mathbf{Z}^n, \psi_1(l) = \psi'(Pl), \quad \text{and} \quad H_1(u, v) = H'(Pu, Pv).$$

By (15), we have

$$\frac{1}{2i}H_1(u, v) = {}^{t}\bar{u}(z - \bar{z})^{-1}v.$$

Now a straightforward calculation shows that

$$\psi_1(zp + q) = e(\tfrac{1}{2}\cdot{}^{t}pq - {}^{t}s''p + {}^{t}r''q) \quad \text{for} \quad p, \quad q \in \mathbf{Z}^n,$$

with r'' and s'' as given in Prop. 1.3. This proves (14).

Next assume that both $\{{}^{t}ac\}$ and $\{{}^{t}bd\}$ belong to $2\mathbf{Z}^n$. Then we can put

$$(17)\qquad \varphi({}^{t}(cz + d)^{-1}u, \gamma(z); 0, 0) = \lambda_\gamma\,\det\,(cz + d)^{1/2}\varphi(u, z; 0, 0)$$

with a constant λ_γ of absolute value 1. By (16), we have

$$(17')\qquad \varphi(u, z; r, s) =$$

$$e({}^{t}rs/2)e\left(\frac{-1}{2}\cdot{}^{t}(\bar{z}r + s)(z - \bar{z})^{-1}\cdot(2u + zr + s)\right)\varphi(u + zr + s, z; 0, 0).$$

Combining this with (17), we obtain (14'). Now Prop. 1.4 follows immediately from (14') and

LEMMA 1. *Let λ_γ be as in Prop. 1.3, and let Γ_θ be the subgroup of $\mathrm{Sp}\,(n, \mathbf{Z})$ generated by all the elements of the forms:*

$$(*)\qquad \begin{pmatrix} 0 & -1_n \\ 1_n & 0 \end{pmatrix}, \begin{pmatrix} a & 0 \\ 0 & d \end{pmatrix}, \begin{pmatrix} 1 & 2b \\ 0 & 1 \end{pmatrix} (a = {}^{t}d^{-1} \in GL_n(\mathbf{Z}), {}^{t}b = b \in M_n(\mathbf{Z})).$$

Then $\Gamma_2 \subset \Gamma_\theta$, $\lambda_\gamma{}^{8} = 1$ for $\gamma \in \Gamma_\theta$, and $\lambda_\gamma{}^{4} = 1$ for $\gamma \in \Gamma_2$.

Proof. Let Γ_2' be the subgroup of Γ_1 consisting of all elements $\begin{pmatrix} a & b \\ c & d \end{pmatrix}$ such that $b \equiv c \equiv 0 \pmod 2$. Then Γ_2' is generated by the last two types of elements of (*) and the elements of the form $\begin{pmatrix} 1 & 0 \\ 2c & 1 \end{pmatrix}$ with ${}^t c = c \in M_n(\mathbf{Z})$. This can be seen by induction on n together with a standard reduction process. (One should note that if $0 \neq g \in \mathbf{Z}$ and $h \in \mathbf{Z}$, then $|h + 2kg| \leq |g|$ for some $k \in \mathbf{Z}$.) We have obviously $\Gamma_2 \subset \Gamma_2' \subset \Gamma_\theta$. Now we can easily verify $\lambda_\gamma{}^8 = 1$ if γ is one of the elements of (*), and hence $\lambda_\gamma{}^8 = 1$ for $\gamma \in \Gamma_\theta$. Put

$$(18) \qquad \theta(z) = \theta(0, z; 0, 0).$$

Making substitutions $z \to z - 2c$ and $z \to -z^{-1}$ in (16'), we obtain

$$\theta(z(2cz + 1)^{-1}) = \pm \det (2cz + 1)^{1/2}\theta(z) \qquad ({}^t c = c \in M_n(\mathbf{Z})).$$

This shows that

$$\lambda_\gamma{}^2 = 1 \quad \text{if} \quad \gamma = \begin{pmatrix} 1 & 0 \\ 2c & 1 \end{pmatrix}.$$

Since

$$\lambda_\gamma{}^4 = 1 \quad \text{if} \quad \gamma = \begin{pmatrix} a & 0 \\ 0 & d \end{pmatrix},$$

we obtain the last assertion.

For a positive integer k and a subring $\mathfrak{R}$ of $\mathbf{C}$, let $\mathfrak{M}_k(\Gamma_N, \mathfrak{R})$ denote the vector space of all modular forms f on $\mathfrak{H}_n$ such that

$$(19_a) \qquad f(\gamma(z)) = \det (cz + d)^k f(z) \quad \text{for all} \quad \gamma = \begin{pmatrix} a & b \\ c & d \end{pmatrix} \in \Gamma_N,$$

$$(19_b) \qquad f(z) = \sum_\xi A(\xi)e(\mathrm{tr}\,(\xi z)/N)$$

with $A(\xi) \in \mathfrak{R}$, where ξ runs over all semi-integral semi-definite symmetric matrices of degree n. We put $\mathfrak{M}_k(\mathfrak{R}) = \bigcup_{N=1}^\infty \mathfrak{M}_k(\Gamma_N, \mathfrak{R})$, and denote by $\mathfrak{a}_m(\mathfrak{R})$, for $m \in \mathbf{Z}$, the set of all meromorphic functions of the form g/h with $g \in \mathfrak{M}_{k+m}(\mathfrak{R})$, $0 \neq h \in \mathfrak{M}_k(\mathfrak{R})$ (with any k), and by $\mathfrak{a}_m(\Gamma_N, \mathfrak{R})$ the set of all $f \in \mathfrak{a}_m(\mathfrak{R})$ satisfying (19_a) with m in place of k. Then $\mathfrak{a}_0(\Gamma_N, \mathbf{C})$ is the field of all modular functions on $\mathfrak{H}_n$ with respect to Γ_N. If $n > 1$, we denote by $\mathfrak{M}_{k/2}(\mathfrak{R})$ (resp. $\mathfrak{a}_{k/2}(\mathfrak{R})$) the set of all holomorphic (resp. meromorphic) functions of the form $f\theta^k$ with $f \in \mathfrak{a}_0(\mathfrak{R})$, where θ is defined by (18); if $n = 1$, the same definition applies to $\mathfrak{a}_{k/2}(\mathfrak{R})$, but a customary condition at cusps must be imposed on the elements of $\mathfrak{M}_{k/2}(\mathfrak{R})$. Since $\theta^2 \in \mathfrak{M}_1(\mathbf{Q})$ as will be shown below, this definition is consistent with the above one when k is even.

Let $G_\mathbf{A}$ denote the adelization of our algebraic group G, G_0 the non-archimedean part of $G_\mathbf{A}$, $G_\infty(= G_\mathbf{R})$ the archimedean part of $G_\mathbf{A}$, and $G_{\infty+}$ the identity component of G_∞. We then put $G_{\mathbf{A}+} = G_0 G_{\infty+}$, and extend the multiplier map $\nu : G_\mathbf{Q} \to \mathbf{Q}^\times$ to a continuous map of $G_\mathbf{A}$ into $\mathbf{Q}_\mathbf{A}^\times$, which we denote again by ν.

Let $\Re$ be the field of all arithmetic automorphic functions with respect to G in the sense of [7, II, §6.1]. Then $\Re = \mathcal{A}_0(\mathbf{Q_{ab}})$ as shown in [10]. Moreover every element of $G_{\mathbf{A}+}$ acts on $\Re$ as an automorphism. If $x \in G_{\mathbf{A}+}$ and $f \in \Re$, we denote by f^x the image of f under x. We note especially that $f^\alpha = f \circ \alpha$ if $\alpha \in G_{\mathbf{Q}+}$. For details, the reader is referred to [7], [10], [11].

LEMMA 2. *The group* $G_{\mathbf{Q}+}$ *is generated by the elements of the form*

$$\begin{pmatrix} a & b \\ 0 & d \end{pmatrix} \quad and \quad \begin{pmatrix} 0 & -1 \\ 1 & 0 \end{pmatrix}.$$

Proof. Put

$$Y = \left\{ \begin{pmatrix} a & b \\ c & d \end{pmatrix} \in G_{\mathbf{Q}+} \mid \det(c) \neq 0 \right\}.$$

Then every element of $G_{\mathbf{Q}+}$ is a product of two elements of Y. This follows easily from [5, Lemma 6.1]. If $\det(c) \neq 0$, we have

$$\begin{pmatrix} 0 & -1 \\ 1 & 0 \end{pmatrix}\begin{pmatrix} 1 & -ac^{-1} \\ 0 & 1 \end{pmatrix}\begin{pmatrix} a & b \\ c & d \end{pmatrix} = \begin{pmatrix} -c & -d \\ 0 & b - ac^{-1}d \end{pmatrix},$$

which proves the lemma.

PROPOSITION 1.5. *For* $f \in \mathfrak{M}_{k/2}(\mathbf{Q_{ab}})$ *and*

$$\alpha = \begin{pmatrix} a & b \\ c & d \end{pmatrix} \in G_{\mathbf{Q}+},$$

put $g(z) = f(\alpha(z)) \det(cz + d)^{-k/2}$. *Then* $g \in \mathfrak{M}_{k/2}(\mathbf{Q_{ab}})$.

Proof. By Lemma 2, it is sufficient to prove this for

$$\alpha = \begin{pmatrix} 0 & -1 \\ 1 & 0 \end{pmatrix} \quad and \quad \alpha = \begin{pmatrix} a & b \\ 0 & d \end{pmatrix}.$$

Put $f = h\theta^k$ with $h \in \mathcal{A}_0(\mathbf{Q_{ab}})$. If $\alpha = \begin{pmatrix} 0 & -1 \\ 1 & 0 \end{pmatrix}$, then $g = \lambda \cdot (h \circ \alpha)\theta^k$ with an eighth root of unity λ, so that $g \in \mathfrak{M}_{k/2}(\mathbf{Q_{ab}})$, since $h \circ \alpha \in \mathcal{A}_0(\mathbf{Q_{ab}})$. Suppose $\alpha = \begin{pmatrix} a & b \\ 0 & d \end{pmatrix}$ and put $\varphi = (\theta \circ \alpha)/\theta$. Then $g(z) = \det(d)^{-k/2}(h \circ \alpha)\varphi^k\theta^k$. Therefore it is sufficient to show that $\varphi \in \mathcal{A}_0(\mathbf{Q_{ab}})$. Now, it can easily be seen from Lemma 1 that $\varphi \circ \gamma = \zeta(\gamma)\varphi$ for every $\gamma \in \Gamma_N$ with a fourth root of unity $\zeta(\gamma)$, where N is a positive integer such that $\Gamma_N \subset \Gamma_2 \cap \alpha^{-1}\Gamma_2\alpha$. Put

$$\Delta = \{\gamma \in \Gamma_N \mid \zeta(\gamma) = 1\}.$$

Then Δ is a subgroup of Sp $(n, \mathbf{Z})$ of finite index. If $n > 1$, Δ contains Γ_M for some M, by virtue of a result of Mennicke [4] and Bass, Milnor and Serre [1]. Thus $\varphi \in \mathcal{A}_0(\Gamma_M)$. As is well-known, this is so even if $n = 1$. Since $\alpha = \begin{pmatrix} a & b \\ 0 & d \end{pmatrix}$,

$\theta \circ \alpha$ has Fourier coefficients in $\mathbf{Q}_{ab}$. Put

$$\theta(\gamma(z))^2 \det (rz + s)^{-1} = \eta(\gamma)\theta(z)^2 \quad \text{for} \quad \gamma = \begin{pmatrix} p & q \\ r & s \end{pmatrix} \in \Gamma_2 .$$

Then η is a character of Γ_2 of order at most 2. Therefore, for the same reason as above, we see that $\theta^2 \in \mathfrak{M}_1(\mathbf{Q})$. Similarly $(\theta \circ \alpha)\theta \in \mathfrak{M}_1(\mathbf{Q}_{ab})$. Therefore $\varphi = (\theta \circ \alpha)\theta/\theta^2 \in \mathfrak{a}_0(\mathbf{Q}_{ab})$. This completes the proof.

PROPOSITION 1.6. *With p, q, r, s in $\mathbf{Q}^n$ and $0 < \xi \in \mathbf{Q}$, put*

$$f(z) = e\left(\frac{\xi}{2} \cdot {}^t pzp\right)\theta(\xi(zp + q), \xi z; r, s) \qquad (z \in \mathfrak{H}_n).$$

Then $f \in \mathfrak{M}_{1/2}(\mathbf{Q}_{ab})$.

Proof. By (11), we have

$$f(z) = e(-{}^t p(\xi q + s))\theta(0, \xi z; r + p, s + \xi q).$$

By Prop. 1.3, $\theta(0, z; r, s)$, as a function in z, belongs to $\mathfrak{M}_{1/2}(\mathbf{Q}_{ab})$. Therefore our assertion follows from Prop. 1.5.

PROPOSITION 1.7. *Let μ and m be positive integers, and put*

$$\Phi(z; r, s; r', s') = \theta(0, \mu z; r, s)/\theta(0, \mu z; r', s')$$

with r, s, r', s' in $m^{-1}\mathbf{Z}^n$. Then $\Phi(z; r, s; r', s')$, as a function in z, belongs to $\mathfrak{a}_0(\Gamma(2\mu m^2), \mathbf{Q}_{ab})$. Moreover, if x is an element of $G_{\mathbf{A}+}$ such that its p-component x_p belongs to $GL_{2n}(\mathbf{Z}_p)$ and

$$x_p \equiv \begin{pmatrix} 1_n & 0 \\ 0 & t1_n \end{pmatrix} \mod 2\mu m^2 M_{2n}(\mathbf{Z}_p)$$

for all rational primes p with a positive integer t, then

$$\Phi(z; r, s; r', s')^x = \Phi(z; r, ts; r', ts').$$

Proof. The first assertion follows easily from Propositions 1.4 and 1.6. Now, for every positive integer N, put $\zeta_N = e(1/N)$ and

$$(20) \quad R_N = \mathbf{Q}^{\times} \cdot \{a \in G_{\mathbf{A}+} \mid a_p \in GL_{2n}(\mathbf{Z}_p), a_p \equiv 1_{2n} \mod N \cdot M_{2n}(\mathbf{Z}_p) \quad \text{for all } p\}.$$

Then we see that $\mathfrak{a}_0(\Gamma_N , \mathbf{Q}(\zeta_N))$ is the subfield of $\mathfrak{K}$ consisting of all the R_N-invariant elements (see [11, Th. 3]). Therefore, if $N = 2\mu m^2$ and if we put $y = \begin{pmatrix} 1 & 0 \\ 0 & \nu(x) \end{pmatrix}$, then x and y have the same action on $\mathfrak{a}_0(\Gamma_N , \mathbf{Q}(\zeta_N))$. For two elements

$$f(z) = \sum_{\xi} b(\xi)e(\text{tr} (\xi z)) \quad \text{and} \quad g(z) = \sum_{\xi} c(\xi)e(\text{tr} (\xi z))$$

of $\mathfrak{M}_k(\Gamma_N , \mathbf{Q}(\zeta_N))$, we have, by [11], Th. 5, (v)],

$$(f/g)^\nu = (\sum_{\xi} b(\xi)^\nu e(\text{tr} (\xi z))/(\sum_{\xi} c(\xi)^\nu e(\text{tr} (\xi z)),$$

where σ is the automorphism of $\mathbf{Q}(\zeta_N)$ such that $\zeta_N{}^\sigma = \zeta_N{}^t$.   Now

$$\Phi(z; r, s; r', s') = \theta(0, \mu z; r, s)\theta(0, \mu z; r', s')^3/\theta(0, \mu z; r', s')^4;$$

and both the denominator and the numerator belong to $\mathfrak{M}_2(\Gamma_N, \mathbf{Q}(\zeta_N))$.   Therefore we obtain the second assertion.

PROPOSITION 1.8.   *With a fixed positive integer $\mu \geqq 3$ and a fixed $s \in \mathbf{Q}^n$, put*

$$\psi_{r,t,v,w}(z) = \theta(\mu(zv + w), \mu z; r, s)/\theta(\mu(zv + w), \mu z; t, s)$$

for $r \in \mu^{-1}\mathbf{Z}^n$, $t \in \mu^{-1}\mathbf{Z}^n$, $v \in \mathbf{Q}^n$, $w \in \mathbf{Q}^n$.   Then $\mathfrak{K}$ is generated by all such $\psi_{r,t,v,w}$ (with fixed μ and s) over $\mathbf{Q}$.

Proof.   Take any complete set of representatives $\mathfrak{J}$ for $\mu^{-1}\mathbf{Z}^n/\mathbf{Z}^n$, and put, for every $r \in \mathfrak{J}$,

$$\varphi_r(u, z) = \theta(\mu u, \mu z; r, s).$$

Denote by $\Theta(u, z)$ the point in a projective space with $\{\varphi_r(u, z)\}_{r \in \mathfrak{J}}$ as its homogeneous coordinates.   Then, for every $z \in \mathfrak{H}_n$, the map $u \to \Theta(u, z)$ gives a biregular embedding of $\mathbf{C}^n/L(z)$ onto an abelian variety $A(z)$, where $L(z)$ is the lattice generated by the columns of z and 1_n over $\mathbf{Z}$.   Now, by Prop. 1.6, we see that

$$\psi_{r,t,v,w}(z) = \varphi_r(zv + w, z)/\varphi_t(zv + w, z) \in \mathfrak{K}.$$

Let Ξ be a countable set of generators of $\mathfrak{K}$ over $\mathbf{Q}$.   Take a point z_0 of $\mathfrak{H}_n$ generic over $\mathbf{Q}$ for Ξ and $\psi_{r,t,v,w}$ in the sense of [8, §6.2, p. 137].   Let σ be an automorphism of $\mathbf{C}$ over the field generated over $\mathbf{Q}$ by $\psi_{r,t,v,w}(z_0)$ for all r, t, v, w as above.   Since the points $\Theta(z_0 v + w, z_0)$ with v and w in $\mathbf{Q}^n$ form the group of all points of finite order on $A(z_0)$, we have $A(z_0)^\sigma = A(z_0)$, and therefore σ is the identity map on the field of moduli of $A(z_0)$ with points of finite order of an arbitrary order.   This shows that σ is the identity map on $\mathbf{Q}(\xi(z_0) \mid \xi \in \Xi)$ so that Ξ is contained in $\mathbf{Q}(\psi_{r,t,v,w})$, which completes the proof.

Let us now clarify one point in our previous paper [11], which was claimed without detailed proof.   It is the assertion at the end of the proof of Theorem 1 of the paper which states that $\omega(cg)$ belongs to the quotient field of $\mathfrak{O}[[X_1, \cdots, X_l]]$.   Actually it is sufficient to show that $\omega(cg)$ belongs to the quotient field, say $\mathfrak{F}$, of $\mathfrak{O}[[X_1{}^{1/M}, \cdots, X_l{}^{1/M}]]$ with a sufficiently large positive integer M.   Since $\mathfrak{O}[[X_1{}^{1/M}, \cdots, X_l{}^{1/M}]]$ is integrally closed, $\omega(cg)$ must then be contained in $\mathfrak{O}[[X_1{}^{1/M}, \cdots, X_l{}^{1/M}]]$ and hence in $\mathfrak{O}[[X_1, \cdots, X_l]]$, which completes the proof of the theorem in question.

To show that $\omega(cg) \in \mathfrak{F}$, we first define $\omega(f)$ as an element of $\mathbf{C}[[X_1{}^{1/M}, \cdots, X_l{}^{1/M}]]$ for every holomorphic f on $\mathfrak{H}_n$ of the form (19_b) with $A(\xi) \in \mathbf{C}$ and ξ running over all semi-definite symmetric matrices in $M^{-1}M_n(\mathbf{Z})$.   Given $g \in M_r(\bar{\mathbf{Q}})$, we see from Prop. 1.8 that $g/\theta^{2r} = (\Sigma_\kappa a_\kappa \varphi_\kappa)/(\Sigma_\lambda a_\lambda{}'\varphi_\lambda{}')$ with a_κ, $a_\lambda{}' \in \mathbf{C}$ and monomials φ_κ, $\varphi_\lambda{}'$ in $e({}^t v \mu z v/2)\varphi_r(zv + w, z)$.   Taking a basis

$\{b_\nu\}$ of $\Sigma_\lambda \bar{Q} a_\kappa + \Sigma_\lambda \bar{Q} a_\lambda'$ over $\bar{Q}$, we obtain $\omega(g) = (\Sigma_\nu b_\nu \zeta_\nu)/(\Sigma_\nu b_\nu \zeta_\nu')$ with ζ_ν, ζ_ν' in $\mathfrak{O}[[X_1^{1/M}, \cdots, X_l^{1/M}]]$ for some M. Then $\Sigma_\nu b_\nu[\omega(g)\zeta_\nu' - \zeta_\nu] = 0$. Since the b_ν are linearly independent over $\bar{Q}$, we have $\omega(g)\zeta_\nu' = \zeta_\nu$ for every ν, hence $\omega(g) \in \mathfrak{F}$.

In [11], we also indicated that the Hilbert modular case could be treated by the same method. Indeed, the point corresponding to the fact $\omega(cg) \in \mathfrak{F}$ in the Hilbert modular case can be proved by considering an embedding of $\mathfrak{H}_1{}^n$ into $\mathfrak{H}_n$ compatible with the group action. To be more specific, let F be a totally real algebraic number field of degree n, and $\{p_1, \cdots, p_n\}$ a basis of F over Q. Further let $\tau_1, \cdots, \tau_n$ be the embeddings of F into $\mathbf{R}$, and P the element of $GL_n(\mathbf{R})$ with $p_\nu{}^{\tau_k}$ as (j, k)-component. Define a map $E : \mathfrak{H}_1{}^n \to \mathfrak{H}_n$ by

$$E(z_1, \cdots, z_n) = P \begin{pmatrix} z_1 & & \\ & \ddots & \\ & & z_n \end{pmatrix} {}^t P,$$

which is compatible with the injection of $SL_2(F)$ into $Sp\ (n, \mathbf{Q})$ given by

$$\begin{pmatrix} a & b \\ c & d \end{pmatrix} \mapsto \begin{pmatrix} P & 0 \\ 0 & {}^tP^{-1} \end{pmatrix} \begin{pmatrix} \Psi(a) & \Psi(b) \\ \Psi(c) & \Psi(d) \end{pmatrix} \begin{pmatrix} P^{-1} & 0 \\ 0 & {}^tP \end{pmatrix}, \qquad \Psi(a) = \begin{pmatrix} a^{\tau_1} & & \\ & \ddots & \\ & & a^{\tau_n} \end{pmatrix}.$$

If $f \in \mathfrak{M}_r(\mathfrak{O})$, then $f \circ E$ is a Hilbert modular form of weight r with Fourier coefficients in $\mathfrak{O}$. Since $\mathcal{Q}_0(\bar{Q})$ is generated by $\mathcal{Q}_0(\mathfrak{O})$ over $\bar{Q}$, we see that the same is true for the field of Hilbert modular functions. Therefore the proof of Theorem 1 of [11] is valid in the Hilbert modular case with no essential change.

2. Theta functions with complex multiplication.

We shall now consider elements of $T(H, \psi, L)$ assuming that $\psi(l)$ is a root of unity for every $l \in L$; H may or may not be degenerate; if ψ is defined by (9), then r and s must be contained in $\mathbf{Q}^n$. For each f in $T(H, \psi, L)$, define a (non-holomorphic) function f_* on V by

$$(21) \qquad f_*(u) = e\left(\frac{-1}{4i} H(u, u)\right) f(u).$$

Then we see that, for every $l \in L$,

$$(22) \qquad f_*(u + l) = \psi(l)e(E(l, u)/2)f_*(u).$$

Therefore, for every lattice M contained in QL, if we take a sufficiently small lattice M', then f_* defines a function on M/M'. Let $\mathfrak{W}$ denote the non-archimedean part of $QL \otimes_Q A$, i.e. the set of all elements (v_p) of $\prod_p QL \otimes_Q Q_p$ such that $v_p \in L_p$ for all except a finite number of p, where $L_p = L \otimes_Z Z_p$. For every $v = (v_p) \in \mathfrak{W}$, we can find a lattice M in QL such that $v_p \in M_p$ for all p. Take M' as above. Then there is an element w of M such that $w - v_p \in M_p'$

for all p. We define $f_*(v)$ to be the same as $f_*(w)$. Obviously this does not depend on the choice of M, M', and w. Thus the value of f_* on each point of $\mathfrak{W}$ is meaningful.

We now specialize this to the case where V/L has many complex multiplications. An abelian variety A defined over a subfield of $\mathbf{C}$ is said to have many complex multiplications if $\mathrm{End}\,(A) \otimes \mathbf{Q}$ contains a commutative semi-simple algebra over $\mathbf{Q}$ of rank $2 \cdot \dim\,(A)$. Such an A is isogenous to a product $A_1^{m_1} \times \cdots \times A_t^{m_t}$ with abelian varieties A_λ such that $\mathrm{End}\,(A_\lambda) \otimes \mathbf{Q}$ is isomorphic to a number field K_λ of degree $2 \cdot \dim\,(A_\lambda)$. Put

$$(23) \qquad Y = M_{m_1}(K_1) \oplus \cdots \oplus M_{m_t}(K_t),$$

and fix an anti-isomorphism ι of Y onto $\mathrm{End}\,(A) \otimes \mathbf{Q}$. Let Φ_λ be the representation of K_λ through ι on the Lie algebra of A_λ, and $(K_\lambda', \Phi_\lambda')$ the reflex of $(K_\lambda, \Phi_\lambda)$ in the sense of [7, I, §1.3]. Further let K' be the composite of $K_1', \cdots,$ K_t'. Take a complex torus V/L isomorphic to A. With H and ψ as above, we call an element f of $T(H, \psi, L)$ *arithmetic* if $f_*(u) \in K_{ab}'$ for every $u \in \mathbf{Q}L$, and denote by $T_a(H, \psi, L)$ the set of all such f.

PROPOSITION 2.1. *For $f \in T_a(H, \psi, L)$ and $w \in \mathbf{Q}L$, put*

$$g(u) = e\left(\frac{-1}{2i} H\left(w, u + \frac{w}{2}\right)\right) f(u + w).$$

Then $g \in T_a(H, \psi', L)$, where $\psi'(l) = \psi(l)e(E(l, w))$, and $g_*(u) = e(E(u, w)/2)f_*(u + w)$.

This follows easily from Prop. 1.1. and a simple calculation.

Identify V with $\mathbf{C}^n$ and fix a holomorphic map j of $\mathbf{C}^n$ onto A that gives an isomorphism of $\mathbf{C}^n/L$ onto A. Then Y acts on $\mathbf{C}^n$ and on $\mathbf{Q}L$; more explicitly, we obtain an anti-representation $\Phi : Y \to M_n(\mathbf{C})$ such that $\iota(a)j(u) = j(\Phi(a)u)$ for $a \in Y$, $u \in \mathbf{C}^n$. Put $U = K_1^{m_1} \oplus \cdots \oplus K_t^{m_t}$ and consider the elements of $K_\lambda^{m_\lambda}$ as row vectors. We let Y act on U by right matrix multiplication. Then we obtain an isomorphism $\mathfrak{y}$ of U onto $\mathbf{Q}L$ such that $\mathfrak{y}(wa) = \Phi(a)\mathfrak{y}(w)$ for $w \in U$, $a \in Y$. Now suppose that H is non-degenerate. Then we can define an involution δ of Y by

$$H(\Phi(a)u, v) = H(u, \Phi(a^\delta)v) \qquad (a \in Y).$$

Take $\Omega = (\omega_1\ \omega_2)$ so that (4), (5), (6), (7) hold. We can then define an injection ξ of Y into $M_{2n}(\mathbf{Q})$ by

$$(24) \qquad \Phi(a)\Omega = \Omega \cdot {}^t\xi(a) \qquad (a \in Y).$$

It can easily be verified that $\xi(a^\delta) = J^{-1} \cdot {}^t\xi(a)J$. Therefore if we put

$$S = \{\alpha \in Y^\times \mid \alpha\alpha^\delta \in \mathbf{Q}^\times\},$$

and $z_0 = \omega_2^{-1}\omega_1$, then $\xi(S) = \{\alpha \in G_{\mathbf{Q}+} \mid \alpha(z_0) = z_0\}$. Moreover, Φ_λ is exactly the representation of K_λ defined at the point z_0 as in [6, §4.9] for the reason explained in [6, §9.3]. We use z_0 to indicate that the point corresponds to the fixed abelian variety with complex multiplication, and retain z for the variable on $\mathfrak{H}_n$.

Let us now define a map $\eta : K_{\mathbf{A}}'^{\times} \to Y_{\mathbf{A}}^{\times}$ by

$$(25) \qquad \eta(x) = \sum_{\lambda=1}^{t} \det\,(\Phi_\lambda{}'(N_{K'/K_\lambda}(x)))e_\lambda \qquad (x \in K_{\mathbf{A}}'^{\times}),$$

where e_λ is the identity element of $M_{m_\lambda}(K_\lambda)$ in the direct sum (23), and consider the map

$$(26) \qquad\qquad\qquad \xi \circ \eta : K_{\mathbf{A}}'^{\times} \to G_{\mathbf{A}+} .$$

Note that $\nu(\xi(\eta(x))) = N_{K'/\mathbf{Q}}(x) \in \mathbf{Q}_{\mathbf{A}}^{\times}$ (see [6, §4.10], [7, I, §2.4]). We let $Y_{\mathbf{A}}^{\times}$ act on $\mathfrak{W}$ in a natural manner. Now we have

PROPOSITION 2.2. *Let z_0 be as above. Then $f(z_0) \in K_{\mathrm{ab}}'$ for every $f \in \mathfrak{K}$. Moreover, for every $x \in K_{\mathbf{A}}'^{\times}$, we have $f(z_0)^x = f^y(z_0)$, where $y = \xi(\eta(x))^{-1}$.*

PROPOSITION 2.3. *Let P be the polarization of A determined by H, σ an arbitrary automorphism of $\mathbf{C}$ over K', and x an element of $K_{\mathbf{A}}'^{\times}$ whose action on K'_{ab} coincides with σ. Then there is an isomorphism j' of $\mathbf{C}^n/\eta(x)^{-1}L$ onto A^σ such that $j(u)^\sigma = j'(\eta(x)^{-1}u)$ for every $u \in QL$, and P^σ corresponds to $N(\mathfrak{x})H$, where $\mathfrak{x}$ is the ideal in K' associated with x, and $N(\mathfrak{x})$ its absolute norm.*

These two propositions are restatements of [7, I, (2.7.3); II, (6.2.3)] and [7, I, Th. 4.3] (cf. also [9, Th. 1]), respectively. It should be noted that for every $s \in Y_{\mathbf{A}}^{\times}$, $s^{-1}L$ is meaningful as a lattice in QL, and $s^{-1}u$ is meaningful as an element of $QL/s^{-1}L$ for every $u \in QL$. (More precisely, $s^{-1}L$ and $s^{-1}u$ must be written as $\Phi(s^{-1})L$ and $\Phi(s^{-1})u$, but Φ is dropped for the sake of simplicity.)

PROPOSITION 2.4. *Define f as in Prop. 1.2 with r and s in $\mathbf{Q}^n$. Then $h(z_0)^{-1}f \in T_a(H, \psi, L)$ for every $h \in \mathcal{G}_{1/2}(\mathbf{Q}_{\mathrm{ab}})$ such that $h(z_0) \neq 0$, where ψ is as in Prop. 1.2.*

Proof. By Prop. 1.2, we have $f \in T(H, \psi, L)$. For $u = \omega_1 a + \omega_2 b$ with a and b in $\mathbf{Q}^n$, we have $\omega_2^{-1}u = z_0 a + b$, so that, by (8) and (11),

$$(27) \qquad f_*(u) = e\!\left(\frac{\mu}{2}\,{}^t a(z_0 a + b)\right)\theta(\mu(z_0 a + b),\, \mu z_0\,;\, r, s)$$

$$= e\!\left(-\frac{\mu}{2}\cdot{}^t ab - {}^t as\right)\theta(0,\, \mu z_0\,;\, r + a,\, s + \mu b).$$

This is the value at z_0 of a function belonging to $\mathfrak{M}_{1/2}(\mathbf{Q}_{\mathrm{ab}})$. Therefore our assertion follows from Prop. 2.2.

From this and Prop. 1.2, we obtain

PROPOSITION 2.5. *The notation being as in Propositions 1.2 and 2.4, $T_a(H, \psi, L)$ can be spanned by $h(z_0)^{-1}f_\mu(u, r + j, s)$ with $j \in \mathfrak{J}$ over K'_{ab}.*

Now we are ready to state and prove

MAIN THEOREM. *Every element x of $K'_\mathbf{A}{}^\times$ defines a $\mathbf{Q}$-linear map $f \to f^x$ of $T_a(H, \psi, L)$ onto $T_a(N(\mathfrak{x})H, \psi', \eta(x)^{-1}L)$ satisfying the following conditions (i –vi), where $\psi'(l) = \psi(\eta(x)l)^x$, and $N(\mathfrak{x})$ is the same as in Prop. 2.3:*

    (i) $(af + bg)^x = a^x f^x + b^x g^x$ *for* $a, b \in K'_{ab}$, *and* $f, g \in T_a(H, \psi, L)$;

    (ii) $(f^x)^y = f^{xy}$ *for* $x, y \in K'_\mathbf{A}{}^\times$;

    (iii) $f_*(u)^x = (f^x)_*(\eta(x)^{-1}u)$ *for all* $u \in \mathbf{Q}L$;

    (iv) $f^x(u) = f(\eta(x)u)$ *if* $x \in K'^\times$;

    (v) $f^x = f$ *if* $x \in K'_\infty{}^\times$;

    (vi) $\{x \in K'_\mathbf{A}{}^\times \mid f^x = f\}$ *is an open subgroup of* $K'_\mathbf{A}{}^\times$ *for each fixed* f.

Proof. We first prove that, for given f and x, there exists an element g of $T_a(N(\mathfrak{x})H, \psi', \eta(x)^{-1}L)$ such that $g_*(u) = f_*(\eta(x)u)^x$ for all $u \in \mathbf{Q}L$. Such a g is obviously unique. Therefore, defining f^x to be g, we can easily verify (i) and (ii). Also, it is sufficient to prove the existence of g for functions of the type of Prop. 2.4. More specifically, with fixed μ and h such that $h(z_0) \neq 0$, and with variable $\Omega = (\omega_1\ \omega_2)$, put $z = \omega_2^{-1}\omega_1$, $v = \omega_2^{-1}u$, and

$$(28) \qquad \varphi(u, \Omega; r, s) = h(z)^{-1}e\left(\frac{\mu}{2} \cdot {}^t v(z - \bar{z})^{-1}v\right)\theta(\mu v, \mu z; r, s).$$

As h, we take $h(z) = \theta(0, \mu z; r', 0)$ with a suitable r' for the reason explained just after Prop. 1.2. We see easily from Prop. 1.4 that

$$(29) \qquad \varphi(u, \Omega \cdot {}^t\gamma; r, s) = \varphi(u, \Omega; r, s)$$

for every $\gamma \in \Gamma(2\mu m^2)$, where m is a positive integer such that r, r', s belong to $m^{-1}\mathbf{Z}^n$. Now put

$$(30) \qquad F(z; r, s) = F_{r,s}(z) = \varphi(0, \Omega; r, s) = \theta(0, \mu z; r, s)/\theta(0, \mu z; r', 0).$$

Then $F_{r,s} \in \mathcal{Q}_0(\Gamma_N, \mathbf{Q}_{ab})$ if N is a multiple of $2\mu m^2$. We are going to prove the existence of g for $f(u) = \varphi(u, \Omega; r, s)$ with Ω specialized to the period matrix of A. From (27), we obtain

$$(31) \qquad f_*(\omega_1\kappa + \omega_2\lambda) = e\left(\frac{-1}{2} \cdot {}^t\kappa(\mu\lambda + 2s)\right)F(z_0\ ; r +_{.}\kappa, s + \mu\lambda) \qquad (\kappa, \lambda \in \mathbf{Q}^n).$$

Define R_N by (20), and put

$$\Delta = \left\{\begin{pmatrix} 1_n & 0 \\ 0 & t1_n \end{pmatrix} \mid t \in \prod_\nu \mathbf{Z}_\nu^\times\right\}.$$

By [7, I, 3.4; II, (3.10.3)], we have $G_{\mathbf{A}+} = R_N\Delta G_{\mathbf{Q}+}$ for every N. Take a positive integer m so that $r, r', s, \kappa, \lambda$ are all contained in $m^{-1}\mathbf{Z}^n$, and let N be a multiple of $2\mu m^2$. Put

$$y = \xi(\eta(x))^{-1} = k\begin{pmatrix} 1 & 0 \\ 0 & t \end{pmatrix}\alpha$$

with $k \in R_N$, $t \in \prod_p \mathbf{Z}_p^{\times}$, and $\alpha \in G_{\mathbf{Q}+}$; define g by

$$g(u) = \varphi(u, \Omega \cdot {}^t\alpha : r, \tau s),$$

where τ is a positive integer such that $\tau \equiv t_p \pmod{N\mathbf{Z}_p}$ for all p. By virtue of (13) and (29), g does not depend on the choice of α, t, τ. Now for

$$w = \Omega \cdot {}^t\alpha \begin{pmatrix} \kappa \\ \lambda \end{pmatrix} ,$$

we have, by (27) or (31),

$$(32) \qquad g_*(w) = e\!\left(\frac{-1}{2} \cdot {}^t\kappa(\mu\lambda + 2\tau s)\right) F(\alpha(z_0); r + \kappa, \tau s + \mu\lambda).$$

Now we can take N so that $f_*(u + u') = f_*(u)$ for all $u \in m^{-1}\Omega\mathbf{Z}^{2n}$ and all $u' \in Nm^{-1}\Omega\mathbf{Z}^{2n}$. Observe that

$$\eta(x)w = \Omega \cdot {}^t\xi(\eta(x)) \cdot {}^t\alpha \begin{pmatrix} \kappa \\ \lambda \end{pmatrix} = \Omega \cdot {}^tk^{-1}\begin{pmatrix} 1 & 0 \\ 0 & t \end{pmatrix}^{-1} \equiv \Omega\begin{pmatrix} \kappa \\ \tau'\lambda \end{pmatrix} \bmod Nm^{-1}\Omega\mathbf{Z}^{2n},$$

where τ' is a positive integer such that $\tau' \equiv t_p^{-1} \pmod{N\mathbf{Z}_p}$ for all p. Therefore, by (31),

$$f_*(\eta(x)w) = f_*(\omega_1\kappa + \omega_2\tau'\lambda)$$

$$= e\!\left(\frac{-1}{2} \cdot {}^t\kappa(\mu\tau'\lambda + 2s)\right) F(z_0 \; ; r + \kappa, s + \mu\tau'\lambda).$$

By Prop. 2.2, we have $F_{r,s}(z_0)^x = (F_{r,s})^y(z_0)$. Since

$$y = k\begin{pmatrix} 1 & 0 \\ 0 & t \end{pmatrix}\alpha,$$

we have, by Prop. 1.7, $(F_{r,s})^y(z) = F(\alpha(z); r, \tau s)$. If $\zeta = e(1/(2m^2))$, we have $\zeta^x = \zeta^r$, so that

$$f_*(\eta(x)w)^x = e\!\left(\frac{-1}{2} \cdot {}^t\kappa(\mu\lambda + 2\tau s)\right) F(\alpha(z_0); r + \kappa, \tau s + \mu\tau\tau'\lambda) = g_*(w)$$

by (32) and (13). This completes the proof of the existence of f^x satisfying (iii) of our theorem.

To show that $f^x \in T_a(N(\mathfrak{x})H, \psi', \eta(x)^{-1}L)$, let $\mathfrak{F}$ be the same as in Prop. 1.2, and take m so that $\mu\epsilon_n$ divides m. Then, for each $j \in \mathfrak{F}$, x maps $\varphi(u, \Omega; r + j, s)$ onto $\varphi(u, \Omega \cdot {}^t\alpha; r + j, \tau s)$. Also we see that

$$\eta(x)^{-1}L = \Omega \cdot {}^t\alpha \begin{pmatrix} 1 & 0 \\ 0 & \epsilon \end{pmatrix}\mathbf{Z}^{2n}.$$

Put $\Omega \cdot {}^t\alpha = (\omega'_1 \, \omega'_2)$, $z' = \omega'^{-1}_2 \, \omega'_1$, and define a Riemann form H' by (8) with ω'_1 , ω'_2 in place of ω_1 , ω_2 . Since $z' = \alpha(z_0)$, we have, by (15), $H' = \nu(\alpha)^{-1}H$. Observe that $\nu(\alpha)^{-1} = N(\mathfrak{x})$. Therefore f^x belongs to $T_a(N(\mathfrak{x})H, \psi', \eta(x)^{-1}L)$,

where

$$\psi'(\omega_1'a + \omega_2'b) = e\left(\frac{\mu}{2}\cdot{}^t ab - r\cdot{}^t sa + \mu\cdot{}^t rb\right) \qquad (a \in \mathbf{Z}^n,\ b \in \epsilon\mathbf{Z}^n).$$

Take the above κ and λ in $\mathbf{Z}^n$ and in $\epsilon\mathbf{Z}^n$, respectively. Then

$$\psi(\eta(x)w) = e\left(\frac{\mu}{2}\cdot{}^t\kappa\tau'\lambda - {}^t s\kappa + \mu\cdot{}^t r\tau'\lambda\right),$$

so that $\psi(\eta(x)w)^x = \psi'(w)$.

Now let $\beta \in Y^\times$, $c = \beta\beta^\delta$, and suppose $c \in \mathbf{Q}$. For $f \in T(H, \psi, L)$, put $f_1(u) = f(\Phi(\beta)u)$ and $\psi_1(m) = \psi(\Phi(\beta)m)$ for $m \in \Phi(\beta)^{-1}L$. Then $f_1 \in T(cH, \psi_1, \Phi(\beta)^{-1}L)$ and $f_{1*}(u) = f_*(\Phi(\beta)u)$. If $x \in K'^\times_\mathbf{A}$, we have $\eta(x) \in Y^\times$ and $\eta(x)\eta(x)^\delta = N_{K'/\mathbf{Q}}(x)$. Therefore we obtain (iv) from (iii); property (v) follows also immediately from (iii). Finally if x belongs to a sufficiently small open subgroup of $K'^\times_\mathbf{A}$, the above α and τ can be chosen so that $\alpha \in \Gamma_N$ and $\tau = 1$. Then $\varphi(u, \Omega\cdot{}^t\alpha; r, \tau s) = \varphi(u, \Omega; r, s) = f$ by virtue of (13) and (29). This proves (vi) and completes the proof.

Remark 1. One can also determine ψ' by applying x to relation (22) and comparing the result with the same relation (22) with f^x in place of f.

Remark 2. Our main theorem can be viewed as a refinement of Prop. 2.3. Indeed, fix $\mu \geqq 3$, and take $\mathfrak{J}$ as in Prop. 1.2. Denote by $\Theta(u, \Omega; r, s)$ the point in a projective space whose homogeneous coordinates are

$$\{\theta(\mu\omega_2^{-1}u, \mu z; r + j, s)\}_{j\in\mathfrak{J}}$$

with fixed r and s in $\mathbf{Q}^n$. Then $u \to \Theta(u, \Omega; r, s)$ gives a projective embedding of $\mathbf{C}^n/L$ onto an abelian variety $A(z)$, whose hyperplane sections correspond to H. Now, if $A(z)$ has many complex multiplications, then the above theorem shows that

$$(33) \qquad \Theta(u, \Omega; r, s)^x = \Theta(\eta(x)^{-1}u, \Omega'; r, \tau s) \qquad (x \in K'^\times_\mathbf{A},\ u \in \mathbf{Q}L),$$

with Ω' and τ determined as in the above proof. The factor

$$h(z)^{-1}e\left(\frac{\mu}{2}\cdot{}^t(v - \bar{v})(z - \bar{z})^{-1}v\right)$$

is ignored in (33), which concerns points in a projective space. For this reason, the above theorem contains more information than (33), while (33) is stronger than Prop. 2.3.

Remark 3. Suppose $Y = K_1$ (i.e. $t = m_1 = 1$), and let k be a finite algebraic extension of K'. The existence of a model for V/L over k whose zeta function is the L-function with a given Hecke character of $k^\times_\mathbf{A}$ was proved by Casselman (see [9, Th. 6]). By means of the above theorem, we can give another proof as follows. Take μ even, $\geqq 4$, and $\psi = 1$. Let T_b be a k_{ab}-linear span of $T_a(H,$

ψ, L). If a Hecke character of $k_A^{\times}$ satisfying [9, (1.12), (1.13)] is given, we can define an action of Gal (k_{ab}/k) on T_b. Let U be the set of elements of T_b invariant under the action. Then $T_b = U \otimes_k k_{ab}$. A basis of U gives a projective embedding of V/L, whose image is the desired model. This method works also in the case $n_1 > 1$.

Remark 4. Without assuming complex multiplication, we can still consider algebraic valued theta functions whenever the abelian variety has a model defined over an algebraic number field. The last condition is satisfied if and only if $\varphi(z_0) \in \bar{Q}$ for every $\varphi \in \Re$, where z_0 is a point of $\mathfrak{H}_n$ corresponding to the variety, and $\bar{Q}$ is the algebraic closure of Q. In such a case we denote by $T'(H, \psi, L)$ the set of all $f \in T(H, \psi, L)$ such that $f_*(u) \in \bar{Q}$ for all $u \in QL$. Then it can easily be seen that the assertions of Propositions 2.4, 2.5, and 4.2 below hold with T' and $\bar{Q}$ in place of T_a and K_{ab}'.

Remark 5. Put

$$\varphi_*(u, \Omega; r, s) = e\left(\frac{-1}{4i} H(u, u)\right)\varphi(u, \Omega; r, s).$$

From (17′), we easily obtain

$$(34) \qquad \varphi_*(u, \Omega; r, s) = e(E(u, u')/2)e('rs/2)\varphi_*(u + u', \Omega; 0, 0),$$

where $u' = \omega_1 r + \mu^{-1}\omega_2 s$. Therefore if we are interested only in the special values and not in the functions, then it is sufficient to consider the values of $\varphi_*(u, \Omega; 0, 0)$ times roots of unity. We also note

$$(35) \qquad \varphi_*(\omega_1 p + \omega_2 q, \Omega; 0, 0)$$

$$= h(z)^{-1}e\left(\frac{\mu}{2} {}'p(zp + q)\right)\theta(\mu(zp + q), \mu z; 0, 0) \qquad (p, q \in R^n).$$

Remark 6. In the above we interpreted $f_*(u)$ as the value of a Siegel modular function at z_0 as shown by (31). It is also practicable and even natural to regard it as the value of a Hilbert modular function, by restricting V/L to those whose endomorphism algebras contain a fixed totally real algebraic number field of degree n.

3. The degenerate case.

We shall now discuss the case of degenerate H. Take V, H, ψ, L as in §1 without assuming H non-degenerate. Put

$$V^0 = \{x \in V \mid H(x, V) = 0\}, \qquad L^0 = V^0 \cap L,$$

$$V_0 = V/V^0, \qquad L_0 = (L + V^0)/V^0.$$

Then L_0 is a lattice in V_0. Moreover we can define a non-degenerate Riemann form H_0 for V_0/L_0 by $H_0(p(u), p(v)) = H(u, v)$, where p denotes the natural map of V onto V_0. Let $f \in T(H, \psi, L)$. Then we can define f_0 on V_0 and ψ_0

on L_0 by $f_0 \circ p = f$ and $\psi_0 \circ p = \psi$, and thus $f_0 \in T(H_0, \psi_0, L_0)$. For details, the reader is referred to Weil [12, p. 113, Prop. 5, pp. 121–122].

Now suppose that V/L has many complex multiplications. Then so does V_0/L_0; and we see easily that $f_0{}_* \circ p = f_*$, and

$$T_a(H, \psi, L) = \{f \circ p \mid f \in T_a(H_0, \psi_0, L_0)\}.$$

Denote by K_0' and η_0 the objects K' and η defined with respect to V_0/L_0. Then we see easily that $K_0' \subset K'$ and $\eta_0(N_{K'/K_0'}(x)) \circ p = p \circ \eta(x)$ for every $x \in K'_{\mathbf{A}}{}^{\times}$. Define f^x for $f = f_0 \circ p \in T_a(H, \psi, L)$ and $x \in K'_{\mathbf{A}}{}^{\times}$ by $(f_0 \circ p)^x = f_0{}^{x_0} \circ p$, where $x_0 = N_{K'/K_0'}(x)$. This reduces the question in the degenerate case to the non-degenerate case. Then we can easily verify that the main theorem holds also for any degenerate H.

4. Intrinsic theta functions.

The theta functions $\theta(u, z; r, s)$ and $\varphi(u, \Omega; r, s)$ depend on the choice of the coordinate system in the vector space or the period matrix. We shall now define some theta functions in a more intrinsic way and discuss their relationship with the classical ones. Let V, L, H, E, and ψ be as in §1. Without identifying V with $\mathbf{C}^n$, suppose that H is non-degenerate, and put

$$(40) \qquad g_v(u) = g(u, v) = g(u, v; H, \psi, L)$$

$$= \sum_{l \in L} \psi(l)^{-1} e\left(\frac{-1}{2i} H\left(l, u + \frac{l}{2}\right) - \frac{1}{2i} H(v, u + l)\right) \qquad (u, v \in V).$$

Obviously the series is convergent, holomorphic in u, and anti-holomorphic in v. Put

$$\beta(l, u, v) = \psi(l) e\left(\frac{1}{2i} H\left(l, u + \frac{l}{2}\right) + \frac{1}{2i} H(v, u + l)\right).$$

Then we obtain, from (3),

$$\beta(l + m, u, v) = \beta(l, u + m, v)\psi(m) e\left(\frac{1}{2i} H\left(m, u + \frac{m}{2}\right)\right)$$

$$= \beta(l, u, v + m)\psi(m) e\left(\frac{1}{2i} H\left(v + \frac{m}{2}, m\right)\right).$$

Therefore we see that $g_v \in T(H, \psi, L)$ for each $v \in V$, and

$$(41) \qquad g(u, v + m) = g(u, v)\psi(m) e\left(\frac{1}{2i} H\left(v + \frac{m}{2}, m\right)\right) \qquad (m \in L).$$

By Prop. 1.2, g_v must be a linear combination of the functions $f(u; r, s)$, when V is identified with $\mathbf{C}^n$. This can be expressed explicitly as follows.

PROPOSITION 4.1. *If H and ψ are defined by (8) and (9) with $r \in \mathbf{R}^n$, $s \in \mathbf{R}^n$, and a positive integer μ, then*

$$\det(\epsilon)\det\left(-i(z-\bar{z})/\mu\right)^{-1/2}g(\omega_2 u,\omega_2 v) = e\left(\frac{\mu}{2}\left\{{}^t u(z-\bar{z})^{-1}u + {}^t\bar{v}(z-\bar{z})^{-1}\bar{v}\right\}\right)$$

$$\times \sum_{j\in\mathfrak{J}} \theta(\mu u,\mu z; j+\mu^{-1}r, -s)\theta(\mu\bar{v}, -\mu\bar{z}; j+\mu^{-1}r, s),$$

where $z = \omega_2^{-1}\omega_1$, and $\mathfrak{J}$ is a complete set of representatives for $\mu^{-1}\epsilon^{-1}\mathbf{Z}^n/\mathbf{Z}^n$.

Proof. Put $z = x + iy$. Then

$$g(\omega_2 u,\omega_2\bar{v}) = \sum_{a,b} e\left(\frac{-\mu}{2}\cdot{}^t a\epsilon b - {}^t r\epsilon b - {}^t sa\right)$$

$$\times e(-\mu\cdot{}^t(\bar{z}a+\epsilon b)(4iy)^{-1}(2u+za+\epsilon b) - \mu\cdot{}^t v(2iy)^{-1}(u+za+\epsilon b)),$$

where a and b run over $\mathbf{Z}^n$. To simplify our notation, put $A[u] = {}^t uAu$ for $u\in\mathbf{C}^n$ and $A = {}^t A\in M_n(\mathbf{C})$. Then

$$(*)\qquad e(-\mu\{(4iy)^{-1}[u] + (4iy)^{-1}[v]\})g(\omega_2 u,\omega_2\bar{v})$$

$$= \sum_a e\left(\frac{-\mu}{4i}\{y[a] + 2i\cdot{}^t a(v-u)\}\right)e(-{}^t sa)X_a ,$$

where

$$X_a = \sum_b e\left(\frac{-\mu}{2}\,{}^t a\epsilon b - {}^t r\epsilon b\right)e(-\mu(4iy)^{-1}[u+v+xa+\epsilon b]).$$

Now the Poisson summation formula shows that if A is real symmetric and positive definite, then

$$\sum_{b\in\mathbf{Z}^n} e\left(\frac{i}{2}A[p+b] - {}^t cb\right) = \det(A)^{-1/2}\sum_{m\in\mathbf{Z}^n} e\left({}^t p(m+c) + \frac{i}{2}A^{-1}[m+c]\right)$$

for p and c in $\mathbf{C}^n$. Putting

$$A = \mu\epsilon y^{-1}\epsilon/2,\qquad p = \epsilon^{-1}(xa+u+v),\qquad c = \epsilon\left(\frac{\mu}{2}a+r\right),$$

we find

$$X_a = \det(\epsilon)^{-1}\det(2y/\mu)^{1/2}$$

$$\times \sum_{m\in\mathbf{Z}^n} e\left({}^t(xa+u+v)\left(\epsilon^{-1}m+\frac{\mu}{2}a+r\right) + \frac{i}{\mu}y\left[\epsilon^{-1}m+\frac{\mu}{2}a+r\right]\right).$$

Therefore $\det(\epsilon)\det(2y/\mu)^{-1/2}$ times $(*)$ is equal to $\sum_{a,m} e(Y_{a,m})$, where

$$Y_{a,m} = -{}^t sa + \frac{1}{2\mu}z[\epsilon^{-1}m + r + \mu a] + {}^t u(\epsilon^{-1}m + r + \mu a)$$

$$- \frac{1}{2\mu}\bar{z}[\epsilon^{-1}m + r] + {}^t v(\epsilon^{-1}m + r).$$

Put $m = \mu\epsilon(k + j)$ with $k \in \mathbf{Z}^n$ and $j \in \mathfrak{J}$. Then

$$Y_{a,m} = \frac{\mu}{2} z[k + a + j + \mu^{-1}r] + {}'(k + a + j + \mu^{-1}r)(\mu u - s)$$

$$- \frac{\mu}{2} \bar{z}[k + j + \mu^{-1}r] + {}'(k + j + \mu^{-1}r)(\mu v + s).$$

Hence we obtain our assertion.

The above proposition is a generalization of a formula of Kronecker [3, IV, p. 355, (C_0); V, pp. 52–54], which can be written as

$$(42) \qquad e(\tau^2(z + w)/2)\theta_1(\sigma + \tau z, z)\theta_1(\sigma - \tau w, w) = (-iC)^{1/2} \sum_{(m,n) \in \mathbf{Z}^2} (-1)^{mn+m+n}$$

$$\cdot e(m\sigma + n\tau + (m^2 A + mn B + n^2 C)/2) \qquad (z, w \in \mathfrak{H}_1 \; ; \; \sigma, \tau \in \mathbf{C}),$$

where $A = zw/(z + w)$, $B = (z - w)/(z + w)$, $C = -1/(z + w)$, $\theta_1(u, z) = \theta(u, z; 1/2, -1/2)$ in the one-dimensional case. In fact, put $n = \mu = 1$, $\epsilon = 1$, $r = -s = 1/2$, $u = \sigma + \tau z$, $-\bar{v} = \sigma + \tau\bar{z}$ in Prop. 4.1. Then we obtain (42) for $w = -\bar{z}$. If a holomorphic function $\varphi(z, w)$ in two variables z and w vanishes for $w = -\bar{z}$, it must vanish everywhere. This proves (42).

Now we put

$$(43) \qquad g(H_2, L) = g(0, 0; H_2, 1, L) = \sum_{l \in L} e\left(-\frac{1}{4i} H_2(l, l)\right),$$

where H_2 denotes the form H with $\mu = 2$. Note that we can take ψ trivial if μ is even, and also that $g(H_2, L) > 0$. Further put

$$(44) \qquad \mathfrak{g}(u, v; H, \psi, L) = g(H_2, L)^{-1} e\left(\frac{-1}{4i} H(v, v)\right) g(u, v; H, \psi, L).$$

From (41), we obtain

$$(45) \qquad \mathfrak{g}(u, v + l; H, \psi, L) = g(u, v; H, \psi, L)\psi(l)e(E(v, l)/2) \qquad (l \in L).$$

Therefore $\mathfrak{g}$ behaves, as a function of v, like f_* of §2.

Proposition 4.2. *Suppose that V/L has many complex multiplications. Then $\mathfrak{g}(u, v; H, \psi, L)$ as a function of u belongs to $T_a(H, \psi, L)$ for each $v \in \mathbf{Q}L$.*

Proof. We identify V with $\mathbf{C}^n$ and consider $\Omega = (\omega_1 \; \omega_2)$ and $z = \omega_2^{-1}\omega_1$ as in §1. Taking a suitable coordinate system, we can even assume $\omega_2 = 1$. Put $\Omega^* = (-\bar{\omega}_1 \; 1)$ and denote by H^* the form (8) defined with Ω^* in place of Ω. Observe that $H^*(\bar{u}, \bar{v}) = H(v, u)$. Then Prop. 4.1 shows

$$(46) \qquad \det(\epsilon)\det(-i(z - \bar{z})/\mu)^{-1/2}h(z)^{-1}h(-\bar{z})^{-1}g(u, v)$$

$$= \sum_{j \in \mathfrak{J}} \varphi(u, \Omega; j + \mu^{-1}r, -s)\varphi(\bar{v}, \Omega^*; j + \mu^{-1}r, s),$$

where $h(z) = \theta(0, \mu z; r', 0)$ as in the proof of the main theorem, and φ is defined by (28). Especially, taking $\mu = 2$, we have

$$(47) \qquad \mathrm{dct}\ (\epsilon)\ \det\ (-i(z - \bar{z})/2)^{-1/2} h_2(z)^{-1} h_2(-\bar{z})^{-1} g(H_2\,, L)$$

$$= \sum_{,\in\mathfrak{J}_2} \varphi_2(0,\ \Omega;\ j,\ 0)\varphi_2(0,\ \Omega^*;\ j,\ 0)$$

$$= \sum_{,\in\mathfrak{J}_2} |\theta(0,\ 2z;\ j,\ 0)/\theta(0,\ 2z;\ r'',\ 0)|^2$$

$$= \sum_{j\in\mathfrak{J}_2} |F_{,.0}{}^{(2)}(z)|^2,$$

where $\mathfrak{J}_2$, φ_2, $F^{(2)}$ denote the symbols $\mathfrak{J}$, φ, F defined with $\mu = 2$; $h_2(z) = \theta(0,\ 2z;\ r'',\ 0)$ with a suitable r''; $F_{r,.}$ is defined by (30). Call the last sum $P(z)$, and put

$$\varphi_*(u,\ \Omega;\ r,\ s) = e\left(-\frac{\mu}{2}\,{}^t\bar{u}(z - \bar{z})^{-1}u\right)\varphi(u,\ \Omega;\ r,\ s).$$

Then we have

$$(48) \qquad \mathfrak{g}(u,\ v;\ H,\ \psi,\ L)$$

$$= P(z)^{-1}\ |h(z)/h_2(z)|^2 \sum_{,\in\mathfrak{J}} \varphi(u,\ \Omega;\ j + \mu^{-1}r,\ -s)\varphi_*(\bar{v},\ \Omega^*;\ j + \mu^{-1}r,\ s).$$

Since h/h_2 and $F_{j.0}{}^{(2)}$ belong to $\mathfrak{R}$, our assertion follows from Prop. 2.4.

As to the factor $\det\ (-i(z - \bar{z})/2)^{1/2} h(z)h(-\bar{z})$, we have

PROPOSITION 4.3. *Let $z_0 = \omega_2^{-1}\omega_1$ be defined as in §2 for V/L with many complex multiplications. Then the following assertions hold:*

(i) *z_0 has coefficients in K';*

(ii) *there exists an element α of $\mathrm{Sp}\ (n,\ \mathbf{Q})$ such that $\alpha(z_0) = -\bar{z}_0$;*

(iii) *$\overline{h(z_0)}/h(z_0)$ and $g(H_2\,, L)/h(z_0)^2$ belong to K'_{ab} for every $h \in \mathfrak{a}_{1/2}(\mathbf{Q}_{ab})$ such that $h(z_0)$ is finite and $\neq 0$.*

Proof. Let $\{\tau_{\lambda 1}\,,\ \cdots,\ \tau_{\lambda n}\}$ be the injections of K_λ into $\mathbf{C}$ which constitute Φ_λ, where $2n_\lambda = [K_\lambda : \mathbf{Q}]$. Take a basis $\{a_i\}$ of K_λ over $\mathbf{Q}$, and let Ω_λ be the $(n_\lambda \times 2n_\lambda)$-matrix whose $(i,\ j)$-component is $a_i{}^{\tau_{\lambda i}}$. Further let L_λ be the lattice generated by the columns of Ω_λ over $\mathbf{Z}$. Then $\mathbf{C}^{n_\lambda}/L_\lambda$ is of type $(K_\lambda\,,\ \Phi_\lambda)$. Therefore if we put

$$\Omega' = \begin{bmatrix} \Omega_1 & & & & & \\ & \ddots & & & & \\ & & \Omega_1 & & & \\ & & & \ddots & & \\ & & & & \Omega_t & \\ & & & & & \ddots \\ & & & & & & \Omega_t \end{bmatrix}$$

with m_λ copies of Ω_λ, then $\Omega = (\omega_1\ \omega_2)$ is "isogenous" to Ω'; namely $X\Omega = \Omega'Y$ with $X \in \mathrm{GL}_n(\mathbf{C})$ and $Y \in \mathrm{GL}_{2n}(\mathbf{Q})$. Now let σ be an automorphism of $\mathbf{C}$ over K'. Then, for each λ, $\{\tau_{\lambda 1}\sigma, \cdots, \tau_{\lambda n}\sigma\}$ coincides with $\{\tau_{\lambda 1}, \cdots, \tau_{\lambda n}\}$ as a whole, so that σ gives a permutation of the rows of Ω_λ. Therefore $\Omega'^\sigma = P_\sigma\Omega'$ with $P_\sigma \in \mathrm{GL}_n(\mathbf{Z})$; hence $\Omega^\sigma = U_\sigma\Omega$ with $U_\sigma \in \mathrm{GL}_n(\mathbf{C})$. It follows that $z_0^\sigma = z_0$, which proves (i). To prove (ii), consider the map $\mathfrak{y} : U \to \mathbf{C}^n$ of §2, where $U = K_1^{m_1} + \cdots + K_t^{m_t}$. Let ρ denote the complex conjugation. For $a = \sum_\lambda a_\lambda \in Y$ with $a_\lambda \in M_{m_\lambda}(K_\lambda)$, we put $a^\rho = \sum_\lambda a_\lambda{}^\rho$, and define similarly the action of ρ on U. Then it can easily be seen that

$$E(\mathfrak{y}(\sum_\lambda v_\lambda), \mathfrak{y}(\sum_\lambda w_\lambda)) = \sum_{\lambda=1}^{t} \mathrm{Tr}_{K_\lambda/\mathbf{Q}}(v_\lambda s_\lambda \cdot {}^t w_\lambda{}^\rho) \qquad (v_\lambda, w_\lambda \in K_\lambda^{m_\lambda})$$

with $s_\lambda = -{}^t s_\lambda{}^\rho \in \mathrm{GL}_{m_\lambda}(K_\lambda)$. We can find an element $x = \sum_\lambda x_\lambda$ of $Y^\times$ so that $x_\lambda s_\lambda \cdot {}^t x_\lambda{}^\rho$ is diagonal. Put $r_\lambda = x_\lambda s_\lambda \cdot {}^t x_\lambda{}^\rho$. Then

$$E(\Phi(x)\mathfrak{y}(\sum_\lambda v_\lambda), \Phi(x)\mathfrak{y}(\sum_\lambda w_\lambda)) = \sum_{\lambda=1}^{t} \mathrm{Tr}_{K_\lambda/\mathbf{Q}}(v_\lambda r_\lambda \cdot {}^t w_\lambda{}^\rho).$$

Put $\Psi(a) = B\Phi(x^{-1}ax)B^{-1}$ for $a \in Y$ with a suitable B of $\mathrm{GL}_n(\mathbf{C})$; the choice of B can be done so that $\Psi(a^\rho) = \Psi(a)^\rho$ for all $a \in Y$. Put $\mathfrak{y}_1(v) = B\Phi(x)\mathfrak{y}(v)$ and $\mathfrak{y}_2(v) = \mathfrak{y}_1(v^\rho)^\rho$. Then $\mathfrak{y}_i(va) = \Psi(a)\mathfrak{y}_i(v)$ for $i = 1, 2$. Therefore, by [7, I, (4.1.11)], there exists an element D of $\mathrm{GL}_n(\mathbf{C})$ such that $\mathfrak{y}_2(v) = D\mathfrak{y}_1(v)$. Put $\mathfrak{y}_0(v) = \Phi(x)\mathfrak{y}(v)$ and $P = B^{-\rho}DB$. Then $\mathfrak{y}_0(v^\rho)^\rho = P\mathfrak{y}_0(v)$. Put $\Omega^* = (-\omega_1{}^\rho\ \omega_2{}^\rho)$ and define E^* by $E^*(\Omega^*x, \Omega^*y) = \mu \cdot {}^t xJy$ for $x, y \in \mathbf{R}^{2n}$. Then it can easily be verified that $E^*(u^\rho, v^\rho) = -E(u, v)$. Therefore

$$E^*(P\mathfrak{y}_0(\sum_\lambda v_\lambda), P\mathfrak{y}_0(\sum_\lambda w_\lambda)) = -\sum_\lambda \mathrm{Tr}_{K_\lambda/\mathbf{Q}}(v_\lambda{}^\rho r_\lambda \cdot {}^t w_\lambda)$$

$$= \sum_\lambda \mathrm{Tr}_{K_\lambda/\mathbf{Q}}(v_\lambda r_\lambda \cdot {}^t w_\lambda)$$

$$= E(\mathfrak{y}_0(\sum_\lambda v_\lambda), \mathfrak{y}_0(\sum_\lambda w_\lambda)),$$

or more simply

$$(49) \qquad E^*(Pu, Pu') = E(u, u') \qquad (u, u' \in \mathbf{C}^n).$$

Put $\mathfrak{a} = \mathfrak{y}_0^{-1}(L)$. Then $\mathfrak{a}$ is a lattice in U, and $PL = \mathfrak{y}_0(\mathfrak{a}^\rho)^\rho$ and $L^\rho = \mathfrak{y}_0(\mathfrak{a})^\rho$. Therefore L^ρ is commensurable with PL; hence there exists an element α of $\mathrm{GL}_{2n}(\mathbf{Q})$ such that $\Omega^* = P\Omega \cdot {}^t\alpha$. Relation (49) shows that $\alpha \in \mathrm{Sp}(n, \mathbf{Q})$. Thus we obtain $\alpha(z_0) = -z_0{}^\rho$. Take $r \in \mathbf{Q}^n$ so that $\theta(0, z_0; r, 0) \neq 0$ and put

$$\alpha = \begin{pmatrix} a & b \\ c & d \end{pmatrix},$$

$\psi(z) = \theta(0, z; r, 0)$, $g(z) = \psi(\alpha(z)) \det(cz + d)^{-1/2}/\psi(z)$, and $f = h/\psi$ for a given $h \in \mathcal{A}_{1/2}(\mathbf{Q}_{ab})$. Since $\psi(z)^\rho = \psi(-z^\rho)$, we have

$$h(z_0)^\rho/h(z_0) = \det(cz_0 + d)^{1/2}f(z_0)^{-1}f(z_0)^\rho g(z_0).$$

By Propositions 1.4 and 1.5, f and g belong to $\mathcal{Q}_0(\mathbf{Q}_{ab})$. Therefore (iii) follows from (i), Prop. 2.1 and (47).

5. The action of $K'_A{}^\times$ on the field of abelian functions.

Let us fix a set Λ of mutually commensurable lattices L in a complex vector space V such that V/L has many complex multiplications; we assume that Λ is maximal in the sense that Λ consists of *all* lattices commensurable with one of its members. For each $L \in \Lambda$, denote by $\mathfrak{F}_L$ the set of all L-invariant meromorphic functions on V, and put $\mathfrak{F}_\Lambda = \cup_{L \in \Lambda} \mathfrak{F}_L$. Define K' as in §2, and let $\mathfrak{G}$ be the subgroup of $K'_A{}^\times \times \mathrm{Aut}\,(\mathbf{C}/K')$ consisting of all the elements (x, σ) with $x \in K'_A{}^\times$ and $\sigma \in \mathrm{Aut}\,(\mathbf{C}/K')$ such that σ coincides with the action of x on K'_{ab}. We put $W_0 = \mathbf{Q}L$ with any $L \in \Lambda$; of course W_0 is independent of the choice of L. We now state a weaker version of the main theorem with abelian functions instead of theta functions.

PROPOSITION 5.1. *The group $\mathfrak{G}$ acts on the field $\mathfrak{F}_\Lambda$ as a group of automorphisms with the following properties:*
  (i) (x, σ) *coincides with σ on* $\mathbf{C}$;
  (ii) $f(u)^\sigma = f^{(x,\sigma)}(\eta(x)^{-1}u)$ *for every $f \in \mathfrak{F}_\Lambda$ and $u \in W_0$* ;
  (iii) $f^{(x,1)}(u) = f(\eta(x)u)$ *if $x \in K'^\times$*;
  (iv) $f^{(x,1)} = f$ *if $x \in K'_\infty{}^\times$*;
  (v) $(\mathfrak{F}_L)^{(x,\sigma)} = \mathfrak{F}_M$, *where $M = \eta(x)^{-1}L$.*

Proof. This follows easily from the main theorem. We can also prove it by means of Prop. 2.3, without theta functions, as follows. Take any $(x, \sigma) \in \mathfrak{G}$, and let A, j, and j' be as in Prop. 2.3. Given $f \in \mathfrak{F}_L$, there is a function φ on A such that $f = \varphi \circ j$. Then φ^σ is meaningful as a function on A^σ and has the property $\varphi(a)^\sigma = \varphi^\sigma(a^\sigma)$ for every $a \in A$. Define $f^{(x,\sigma)}$ by $f^{(x,\sigma)} = \varphi^\sigma \circ j'$. Then Prop. 2.3 shows that $f^{(x,\sigma)}$ satisfies condition (ii), which determines $f^{(x,\sigma)}$ independently of the choice of L, j, A. The remaining properties are either obvious or can be verified in a straightforward way.

Remark. Let $\mathfrak{F}_{\Lambda,a}$ be the subfield of $\mathfrak{F}_\Lambda$ consisting of all the elements f such that $f(u) \in K'_{ab}$ for every $u \in W_0$. Then $\mathfrak{F}_\Lambda = \mathbf{C}\mathfrak{F}_{\Lambda,a}$; and the action of $K'_A{}^\times$ on $\mathfrak{F}_{\Lambda,a}$ can be defined in a natural manner. It should also be mentioned that the main theorem could have been stated in terms of the action of $\mathfrak{G}$ on $T(H, \psi, L)$ instead of the action of $K'_A{}^\times$ on $T_a(H, \psi, L)$. Although this would have had the drawback of deemphasizing the significance of $T_a(H, \psi, L)$, there are some problems which necessitate a wider class of functions than those of $T_a(H, \psi, L)$, as shown in Remark 3 of §2.

REFERENCES

1. H. BASS, J. MILNOR AND J-P. SERRE, *Solution of the congruence subgroup problem for SL_n $(n \geq 3)$ and Sp_{2n} $(n \geq 2)$*, Publ. Math. I.H.E.S. **33**(1967), 59–137.
2. A. KRAZER AND F. PRYM, *Neue Grundlagen einer Theorie der Allgemeinen Thetafunktionen*, Teubner, Leipzig, 1892.

3. L. Kronecker, *Mathematische Werke* I–V, Teubner, Leipzig, 1895–1931; Chelsea, New York, 1968.

4. J. Mennicke, *Zur Theorie der Siegelschen Gruppen*, Math. Ann. **159**(1965), 115–129.

5. G. Shimura, *Arithmetic of unitary groups*, Ann. of Math. **79**(1964), 369–409.

6. ———, *Algebraic number fields and symplectic discontinuous groups*, Ann. of Math. **86** (1967), 503–592.

7. ———, *On canonical models of arithmetic quotients of bounded symmetric domains*, I, II, Ann. of Math. **91**(1970), 144–222; **92**(1970), 528–549.

8. ———, *Introduction to the arithmetic theory of automorphic functions*, Iwanami Shoten and Princeton Univ. Press, 1971.

9. ———, *On the zeta-function of an abelian variety with complex multiplication*, Ann. of Math. **94**(1971), 504–533.

10. ———, *On some arithmetic properties of modular forms of one and several variables*, Ann. of Math. **102**(1975), 491–515.

11. ———, *On the Fourier coefficients of modular forms with several variables*, Göttingen, Nachr. Akad. Wiss. 1975, 261–268.

12. A. Weil, *Introduction à l'étude des variétés kählériennes*, Hermann, Paris, 1958.

Department of Mathematics, Princeton University, Princeton, New Jersey 08540

The special values of the zeta functions
associated with cusp forms

Communications on pure and applied Mathematics, 29 (1976), 783-804

1. Introduction

For a positive integer k and a Dirichlet character χ modulo a positive integer n such that $\chi(-1) = (-1)^k$, let $G_k(N, \chi)$ denote the vector space of all holomorphic modular forms $f(z)$ satisfying

$$f(\gamma(z)) = \chi(d)(cz + d)^k f(z) \quad \text{for} \quad \text{all} \quad \gamma = \begin{pmatrix} a & b \\ c & d \end{pmatrix} \in \Gamma_0(N),$$

where z is the variable on the upper half-plane, $\gamma(z) = (az + b)/(cz + d)$, and

$$\Gamma_0(N) = \left\{ \begin{pmatrix} a & b \\ c & d \end{pmatrix} \in SL_2(\mathbf{Z}) \mid c \equiv 0 \ (\text{mod } N) \right\}.$$

The subspace of $G_k(N, \chi)$ consisting of all cusp forms is denoted by $S_k(N, \chi)$. Further we put

$$G_k(N) = \sum_\chi G_k(N, \chi), \qquad S_k(N) = \sum_\chi S_k(N, \chi),$$

where χ runs over all characters modulo N. These are obviously the spaces of holomorphic modular forms and cusp forms of weight k with respect to the group

$$\Gamma_1(N) = \left\{ \begin{pmatrix} a & b \\ c & d \end{pmatrix} \in \Gamma_0(N) \mid a \equiv d \equiv 1 \ (\text{mod } N) \right\}.$$

Every element f of $G_k(N)$ has a Fourier expansion

$$f(z) = \sum_{n=0}^\infty a_n e(nz)$$

783

with complex coefficients a_n, where $e(z) = e^{2\pi i z}$. Put

$$D(s, f) = \sum_{n=1}^{\infty} a_n n^{-s}.$$

More generally, with an arbitrary Dirichlet character ψ, we put

$$D(s, f, \psi) = \sum_{n=1}^{\infty} \psi(n) a_n n^{-s}.$$

The main purpose of this paper is to investigate the values of $D(s, f, \psi)$ at the positive integers less than k, when f is a cusp form. To give a precise statement, we have to introduce a few more symbols. First, f and ψ being as above, we denote by K_f (respectively K_ψ) the field generated over the rational number field $\mathbf{Q}$ by the coefficients a_n (respectively the values $\psi(n)$) for all n. Next, we define the Gauss sum $g(\psi)$ by

$$g(\psi) = g(\psi_0) = \sum_{n=1}^{c} \psi_0(n) e(n/c),$$

where ψ_0 is the primitive character associated with ψ, and c its conductor. We put, for every positive integer $m < k$,

(1.1) $$A(m, f, \psi) = (2\pi i)^{-m} g(\psi)^{-1} D(m, f, \psi).$$

Further with another Dirichlet character ψ' and another positive integer $m' < k$, we set

$$B(m, m'; f; \psi, \psi') = A(m, f, \psi)/A(m', f, \psi'),$$

assuming $D(m', f, \psi') \neq 0$. We call f primitive if $a_1 = 1$ and it is an eigenfunction of Hecke operators, which cannot be obtained from the forms of lower level (see Section 2 for the precise definition). If $f(z) = \sum_{n=1}^{\infty} a_n e(nz)$ is primitive, then for every automorphism σ of the complex number field $\mathbf{C}$, we can define a primitive cusp form f^σ by

$$f^\sigma(z) = \sum_{n=1}^{\infty} a_n^\sigma e(nz).$$

Now our main result can be stated as follows.

THEOREM 1. *Let ψ and ψ' be primitive Dirichlet characters, and f a primitive cusp form belonging to $S_k(N, \chi)$. Let m and m' be two positive integers* *less than k such that $(\psi\psi')(-1) = (-1)^{m-m'}$ and $D(m', f, \psi') \neq 0$. If $k > 2$, $B(m, m'; f; \psi, \psi')$ belongs to $K_f K_\psi K_{\psi'}$; moreover, for every automorphism σ of $\mathbf{C}$, we have $D(m', f^\sigma, \psi'^\sigma) \neq 0$, and*

$$B(m, m'; f; \psi, \psi')^\sigma = B(m, m'; f^\sigma; \psi^\sigma, \psi'^\sigma).$$

Here ψ^σ is defined by $\psi^\sigma(n) = \psi(n)^\sigma$. If $k = 2$, the same assertions hold provided that f satisfies the following condition:

(1.2) *For a given integer t, there is a primitive character ξ such that $D(1, f, \xi) \neq 0$ and $\xi(-1) = (-1)^t$.*

It is highly probable that (1.2) is always satisfied. For a form of low level, we can at least verify it by a simple computation.

The author obtained in [15] a result of the above type for the classical cusp form Δ of weight 12, and indicated the practicability of the generalization by the same method. Fourteen years later, Manin [8] gave such a generalization in the case of level one. A few years before this work, the case where $D(s, f)$ is a Hecke L-function of an imaginary quadratic field was treated by Damerell [3] using a completely different idea. Recently the author generalized his result to the case of Hilbert modular forms [20].

The method of the present paper has little in common with those of all these previous works. We take an element $g(z) = \sum_{n=0}^{\infty} b_n e(nz)$ of $G_l(N)$ with $l < k$, and consider a Dirichlet series

$$D(s, f, g) = \sum_{n=1}^{\infty} a_n b_n n^{-s},$$

which can be expressed by an integral of Rankin-Selberg type. Evaluating the integral at $s = k - 1$, we can express $D(k-1, f, g)$ as a Petersson inner product. Actually this type of connection of the Petersson inner product with the special values of $D(s, f, g)$ or $D(s, f)$ when the level is 1 was already noticed by Rankin in [13], but its significance seems to have been hitherto overlooked. Now choosing a suitable Eisenstein series as g, we obtain $D(k-1, f, \psi)D(k-l, f, \psi')$ as a certain quantity whose behavior under an automorphism of $\mathbf{C}$ can be studied. Taking the quotient of two such quantities, we obtain the above theorem. In Section 2, we shall give three theorems concerning the relation between the inner product and the special values. Sections 3 and 4 are devoted to the proofs of the theorems. In the final section, we shall briefly discuss the case where $D(s, f)$ is the L-function of a Hecke character of an imaginary quadratic field.

Besides its straightforwardness, our method has the following two advantages: first, the way to our theorem, as we have just described, is explicit enough to compute the numerical value of $B(m, m'; \cdots)$ in a fairly effective fashion; second, more importantly, all our arguments can easily be generalized to the case of Hilbert modular forms, and perhaps to the automorphic forms of more general types. In this paper, however, we consider only the elliptic modular case.

2. The Relation Between the Special Values and the Petersson Inner Product

Throughout the paper, we retain the notation used in the introduction, and denote by $\mathfrak{H}$ the upper half-plane $\{z \in \mathbf{C} \mid \mathscr{I}m\,(z) > 0\}$. For two elements f and h of $G_k(N)$ such that fh is a cusp form, we define the Petersson inner product $\langle f, h \rangle$ by

$$(2.1) \qquad \langle f, h \rangle = m(\Phi)^{-1} \int_{\Phi} \overline{f(z)}\, h(z) y^{k-2}\, dx\, dy, \qquad z = x + iy,$$

where Φ is a fundamental domain for $\mathfrak{H}$ modulo $\Gamma_1(N)$, and $m(\Phi)$ is the measure of Φ with respect to $y^{-2}\, dx\, dy$; the bar denotes the complex conjugate. We have actually $m(\Phi) = (\frac{1}{3}\pi)[SL_2(\mathbf{Z}) : \Gamma_1(N)\{\pm 1\}]$. We also define $\langle f, h \rangle$ by (2.1) for continuous functions f and h on $\mathfrak{H}$ with the same automorphic property as the elements of $G_k(N)$ whenever the integral is convergent. The value $\langle f, h \rangle$ depends only on f and h, and is independent of the choice of N.

Let us now fix an element f of $S_k(N, \chi)$ and an element g of $G_l(N, \psi)$ with Fourier expansions

$$f(z) = \sum_{n=1}^{\infty} a_n e(nz), \qquad g(z) = \sum_{n=0}^{\infty} b_n e(nz).$$

Then we put

$$\mathfrak{D}_N(s, f, g) = L_N(2s + 2 - k - l, \chi\psi) D(s, f, g),$$

where $L_N(s, \omega)$ with a Dirichlet character ω modulo N is defined as usual by $L_N(s, \omega) = \sum_{n=1}^{\infty} \omega(n) n^{-s}$ with $\omega(n) = 0$ for $(n, N) \neq 1$. The subscript N is used in order to emphasize the last condition; it will be dropped if ω is primitive. To obtain an expression of $D(s, f, g)$ by an integral, put $f_\rho(z) = \overline{f(-\bar{z})} = \sum_{n=1}^{\infty} \bar{a}_n e(nz)$. Then $f_\rho \in S_k(N, \bar{\chi})$, and

$$\int_0^1 \overline{f_\rho(z)}\, g(z)\, dx = \sum_{n=1}^{\infty} a_n b_n e^{-4\pi ny}, \qquad z = x + iy,$$

and hence

$$(2.2) \qquad \int_0^\infty y^{s-1} \int_0^1 \bar{f}_\rho g \, dx \, dy = (4\pi)^{-s}\Gamma(s)D(s, f, g) \, .$$

This is valid at least for sufficiently large $\mathscr{R}e\,(s)$. Observe that

$$(\bar{f}_\rho g y^{s+1}) \circ \gamma = (\chi\psi)(d)(cz+d)^{l-k}\,|cz+d|^{2k-2-2s}\,\bar{f}_\rho g y^{s+1}$$

$$\text{for all} \quad \gamma = \begin{pmatrix} a & b \\ c & d \end{pmatrix} \in \Gamma_0(N) \, .$$

Let R be a complete set of representatives for $\Gamma_\infty \backslash \Gamma_0(N)$, where

$$\Gamma_\infty = \left\{ \pm \begin{pmatrix} 1 & m \\ 0 & 1 \end{pmatrix} \,\middle|\, m \in \mathbf{Z} \right\} \, .$$

For an integer $\lambda \geqq 0$ and a Dirichlet character ω modulo N such that $\omega(-1) = (-1)^\lambda$, put

$$E^*_{\lambda,N}(z, s, \omega) = \sum_{\gamma \in R} \omega(d)(cz+d)^{-\lambda}\,|cz+d|^{-2s}, \qquad \gamma = \begin{pmatrix} a & b \\ c & d \end{pmatrix} .$$

This is absolutely convergent for $\mathscr{R}e\,(2s) > 2 - \lambda$. Assuming $k \geqq l$, we can transform (2.2) to the form

$$(2.3) \quad (4\pi)^{-s}\Gamma(s)D(s, f, g) = \int_{\Phi_0} \bar{f}_\rho g \cdot E^*_{k-l,N}(z, s+1-k, \chi\psi)y^{s-1}\,dx\,dy \, ,$$

where Φ_0 denotes a fundamental domain for $\Gamma_0(N)\backslash\mathfrak{H}$. Now put

$$E_{\lambda,N}(z, s, \omega) = \sum_{(m,n)} \omega(n)(mNz+n)^{-\lambda}\,|mNz+n|^{-2s} \, ,$$

where the summation is taken over all $(m, n) \in \mathbf{Z}^2$, $\neq 0$. Then

$$E_{\lambda,N}(z, s, \omega) = 2L_N(2s+\lambda, \omega)E^*_{\lambda,N}(z, s, \omega) \, ,$$

and hence

$$(2.4) \quad 2(4\pi)^{-s}\Gamma(s)\mathfrak{D}_N(s, f, g) = \int_{\Phi_0} \bar{f}_\rho g E_{k-l,N}(z, s+1-k, \chi\psi)y^{s-1}\,dx\,dy \, .$$

The function $\mathfrak{E}(s) = \Gamma(s+\lambda)E_{\lambda,N}(z, s, \omega)$ can be continued to a meromorphic function on the whole s-plane. As to its poles, we know that:

(i) $\mathfrak{E}$ is entire if $\lambda \neq 0$ or ω is non-trivial;

(ii) if $N = 1$ and $\lambda = 0$, $\mathfrak{E}$ is holomorphic on the whole plane except for simple poles at $s = 0$ and $s = 1$;

(iii) if $N > 1$, $\lambda = 0$, and ω is trivial, then $\mathfrak{E}$ is holomorphic on the whole plane except for a simple pole at $s = 1$;

(iv) if $\lambda = 0$ and ω is trivial, $E_{0,N}^{*}(z, s, \omega)$ is holomorphic for $\mathcal{R}e\,(s) > 1$, and has a simple pole at $s = 1$ with residue

$$\frac{3}{N\pi y} \prod_{p \mid N} (1 + p^{-1})^{-1} .$$

Also the integral on the right-hand side of (2.4) is absolutely convergent when $E_{k-l,N}(z, s+1-k, \chi\psi)$ is holomorphic in s. These facts are well known. See, for example, Rankin [12], Selberg [14], Petersson [11], as well as [18], Lemma 3.3, and [19], p. 95. Anyway, $D(s, f, g)$ can be continued to a meromorphic function on the whole plane.

Now suppose $k = l$, and take the residue of (2.3) at $s = k$. Then we find

$$(2.5) \qquad \qquad \mathrm{Res}_{s=k}\, D(s, f, g) = (4\pi)^{k} \Gamma(k)^{-1} \langle f, g \rangle .$$

This is due to Petersson [11].

Next we consider the case $l < k$. Put, for $\lambda > 0$,

$$(2.6) \qquad \qquad E_{\lambda,N}^{*}(z, \omega) = E_{\lambda,N}^{*}(z, 0, \omega) ,$$

$$(2.7) \qquad \qquad E_{\lambda,N}(z, \omega) = E_{\lambda,N}(z, 0, \omega) .$$

As Hecke showed in [6], both $E_{\lambda,N}^{*}(z, \omega)$ and $E_{\lambda,N}(z, \omega)$ belong to $G_{\lambda}(N, \bar{\omega})$ except when $\lambda = 2$ and ω is trivial; if $\lambda = 2$ and ω is trivial, each function is a constant times y^{-1} plus a holomorphic function in z. Now we define differential operators δ_{λ} and $\delta_{\lambda}^{(r)}$ by

$$(2.8) \qquad \delta_{\lambda} = \frac{1}{2\pi i}\left(\frac{\lambda}{2iy} + \frac{\partial}{\partial z}\right), \qquad \qquad 0 < \lambda \in \mathbf{Z} ,$$

$$\delta_{\lambda}^{(r)} = \delta_{\lambda+2r-2} \cdots \delta_{\lambda+2}\, \delta_{\lambda} , \qquad \qquad 0 \leqq r \in \mathbf{Z} ,$$

where $\partial/\partial z = \frac{1}{2}(\partial/\partial x - i\, \partial/\partial y)$; we understand that $\delta_{\lambda}^{(0)}$ is the identity operator. It can easily be verified that if $h \in G_{\lambda}(N, \omega)$, then $\delta_{\lambda}^{(r)} h$ has the same automorphic property as the elements of $G_{\lambda+2r}(N, \omega)$. Moreover we have

$$(2.9) \qquad E^{*}_{\lambda+2r,N}(z, -r, \omega) = \frac{\Gamma(\lambda)}{\Gamma(\lambda+r)} (-4\pi y)^r \delta^{(r)}_{\lambda} E^{*}_{\lambda,N}(z, \omega)$$

(cf. [20], (2.4), (2.5)). Observing that (2.3) is holomorphic at $s = k-1-r$ if $l+2r < k$, we obtain

THEOREM 2. *Suppose* $f \in S_k(N, \chi)$, $g \in G_l(N, \psi)$, *and* $l+2r < k$ *with a non-negative integer* r. *Then*

$$D(k-1-r, f, g) = c\pi^k \langle f_\rho, g \cdot \delta^{(r)}_{\lambda} E^{*}_{\lambda,N}(z, \chi\psi) \rangle,$$

where $\lambda = k-l-2r$, *and*

$$c = \frac{\Gamma(k-l-2r)}{\Gamma(k-1-r)\Gamma(k-l-r)} \cdot \frac{(-1)^r 4^{k-1} N}{3} \cdot \prod_{p|N} (1+p^{-1}),$$

the product being taken over all prime factors p of N.

We are going to apply this to an eigenfunction of Hecke operators. First we define the notion of primitive form as follows. Let $S'_k(N)$ denote the subspace of $S_k(N)$ spanned by the functions $h(tz)$ with $h \in S_k(t^{-1}N)$ for all divisors t of N greater than 1. Let $S^0_k(N)$ denote the orthogonal complement of $S'_k(N)$ in $S_k(N)$. Then we call an element $f(z) = \sum_{n=1}^{\infty} a_n e(nz)$ of $S^0_k(N)$ a *primitive form* if $a_1 = 1$ and f is a common eigenfunction of all Hecke operators of level N. We call N the *conductor* of f. As to the basic properties of primitive forms, the reader is referred to Atkin-Lehner [1], Miyake [9], and Casselman [2].

Now $S_k(N)$ and $G_k(N)$ have bases whose members have rational coefficients. This follows easily from the fact that each space $S_k(N)$ or $G_k(N)$ is the set of all products $F\varphi$ with a fixed modular form φ with rational Fourier coefficients and functions F on $\Gamma_1(N)\backslash\mathfrak{H}$ such that $\mathrm{div}(F) \geqq -A$ with a divisor A rational over $\mathbf{Q}$ (cf. [20], Theorem 6, [21], Theorem 3.52). Therefore, for every field automorphism σ of $\mathbf{C}$, we can define a map $\varphi \mapsto \varphi^\sigma$ of $G_k(N)$ and $S_k(N)$ onto themselves by $\varphi^\sigma(z) = \sum_n c_n^\sigma e(nz)$ for $\varphi(z) = \sum_n c_n e(nz)$. By [17], Proposition 1.2, if φ is a primitive element of $S_k(N, \chi)$, then φ^σ is a primitive element of $S_k(N, \chi^\sigma)$, where $\chi^\sigma(n) = \chi(n)^\sigma$.

THEOREM 3. *Let f be a primitive element of $S_k(N)$, g an element of $G_l(N)$, and m a positive integer. (The conductor of f may be smaller than N.) Suppose*

$l < k$ and $\frac{1}{2}(k+l-2) < m < k$. *Then*

$$\pi^{-k} \langle f, f \rangle^{-1} D(m, f, g)$$

belongs to $K_f K_g$. *Moreover, for every automorphism* σ *of* **C**, *we have*

$$[\pi^{-k} \langle f, f \rangle^{-1} D(m, f, g)]^{\sigma} = \pi^{-k} \langle f^{\sigma}, f^{\sigma} \rangle^{-1} D(m, f^{\sigma}, g^{\sigma}).$$

THEOREM 4. *For a primitive cusp form* f *belonging to* $S_k(N, \chi)$, *two primitive Dirichlet characters* ψ *and* ψ', *and two integers* m *and* m', *put*

$$C(m, m'; f; \psi, \psi') = \frac{A(m, f, \psi) A(m', f, \psi')}{i^{1-k} \pi g(\chi) \langle f, f \rangle},$$

where $A(m, f, \psi)$ *is defined by* (1.1). *Suppose that* $0 < m < k$, $0 < m' < k$, *and* $(\psi \psi')(-1) = (-1)^{m-m'-1}$. *Then* $C(m, m'; f; \psi, \psi')$ *belongs to* $K_f K_\psi K_{\psi'}$. *Moreover, for every automorphism* σ *of* **C**, *we have*

$$C(m, m'; f; \psi, \psi')^{\sigma} = C(m, m'; f^{\sigma}; \psi^{\sigma}, \psi'^{\sigma}).$$

PROPOSITION 1. *Let* $f(z) = \sum_{n=1}^{\infty} a_n e(nz)$ *be a primitive element of* $S_k(N)$, *and* ψ *a Dirichlet character modulo a positive integer* r. *Define a cusp form* h *by* $h(z) = \sum_{n=1}^{\infty} \psi(n) a_n e(nz)$. *Then* $\langle h, h \rangle / \langle f, f \rangle$ *depends only on* f *and* r. *Put* $J(f, r) = \langle h, h \rangle / \langle f, f \rangle$. *Then* $J(f, r)$ *is contained in* K_f. *Moreover, for every automorphism* σ *of* **C**, *we have* $J(f, r)^{\sigma} = J(f^{\sigma}, r)$.

These theorems and proposition will be proved in the next two sections. We note that the last assertion of each theorem or proposition obviously implies the preceding one. Theorem 4 has a form complementary to Theorem 1. The rationality of $C(m, m'; \cdots)$ was first found by Rankin in [13] in the case where $N = 1$, $\psi = \psi' = 1$, $S_k(1)$ is one-dimensional, $m = k - 1$, and $\frac{1}{2}(k + 4) \leqq m' \leqq k - 4$.

3. Proofs of Theorem 3 and Proposition 1

We first prove a few lemmas, the first of which is purely formal.

LEMMA 1. *Suppose we have formally*

$$\sum_{n=1}^{\infty} a(n) n^{-s} = \prod_p [(1 - \alpha_p p^{-s})(1 - \alpha'_p p^{-s})]^{-1},$$

$$\sum_{n=1}^{\infty} b(n) n^{-s} = \prod_p [(1 - \beta_p p^{-s})(1 - \beta'_p p^{-s})]^{-1},$$

where p runs over all rational primes. For each p, put

$$X_p(s) = 1 - \alpha_p \alpha_p' \beta_p \beta_p' p^{-2s},$$

$$Y_p(s) = (1 - \alpha_p \beta_p p^{-s})(1 - \alpha_p \beta_p' p^{-s})(1 - \alpha_p' \beta_p p^{-s})(1 - \alpha_p' \beta_p' p^{-s}).$$

Then

$$\sum_{n=1}^{\infty} a(n)b(n)n^{-s} = \prod_p X_p(s) Y_p(s)^{-1}.$$

Moreover, for arbitrary positive integers v and w, we have

$$\sum_{n=1}^{\infty} a(n/v)b(n/w)n^{-s} = (vw/d)^{-s} \prod_p X_p^*(s) Y_p(s)^{-1},$$

where d is the greatest common divisor of v and w, and $X_p^*(s)$ is a polynomial in p^{-s} of degree at most 2, which coincides with $X_p(s)$ for all p not dividing vw; we understand that $a(t)$ or $b(t)$ is 0 if t is not an integer. The coefficients of X_p^* and Y_p belong to the field generated by $a(n)$ and $b(n)$ over $\mathbf{Q}$.

Proof: Put $v = dv'$ and $w = dw'$. Then

$$\sum_{n=1}^{\infty} a(n/v)b(n/w)n^{-s} = (dv'w')^{-s} \sum_{m=1}^{\infty} a(w'm)b(v'm)m^{-s}.$$

Obviously the last sum has an Euler product whose p-factor is of the form $\sum_{n=0}^{\infty} a(p^{n+t})b(p^n)p^{-ns}$ or $\sum_{n=0}^{\infty} a(p^n)b(p^{n+t})p^{-ns}$ with $t \geqq 0$. Now we have

$$a(p^n) = (\alpha_p^{n+1} - \alpha_p'^{n+1})/(\alpha_p - \alpha_p'), \quad b(p^n) = (\beta_p^{n+1} - \beta_p'^{n+1})/(\beta_p - \beta_p')$$

if $\alpha_p \neq \alpha_p'$ and $\beta_p \neq \beta_p'$. Therefore we obtain

$$Y_p(s) \sum_{n=0}^{\infty} a(p^{n+t})b(p^n)p^{-ns}$$

$$(3.1) \qquad = \begin{cases} a(p^t) - a(p^{t-1})b(p)\alpha_p\alpha_p' p^{-s} + a(p^{t-2})(\alpha_p\alpha_p')^2 \beta_p\beta_p' p^{-2s} & \text{if} \quad t \geqq 2, \\ a(p) - b(p)\alpha_p\alpha_p' p^{-s} & \text{if} \quad t = 1, \\ 1 - \alpha_p\alpha_p'\beta_p\beta_p' p^{-2s} & \text{if} \quad t = 0. \end{cases}$$

It can easily be seen that this is valid even if $\alpha_p = \alpha_p'$ or $\beta_p = \beta_p'$. This yields our lemma.

LEMMA 2. *Let f be a primitive cusp form. Then K_f is either totally real or a totally imaginary quadratic extension of a totally real algebraic number field. Especially, K_f is stable under the complex conjugation.*

For the proof, see [17], Proposition 1.3.

Remark. The part of [17], Proposition 1.3 asserting that K_f is totally imaginary if $f \in S_k(N, \chi)$ and χ is non-trivial is false. It is valid only when χ is *not real.* To obtain a counter-example when χ is real, let L be an imaginary quadratic field of class number h, and for every ideal $\mathfrak{a}$ of L, put $\lambda(\mathfrak{a}) = \alpha^w$ with an element α of L such that $\mathfrak{a}^h = (\alpha)$, where w is the number of roots of unity in L. Further put $f(z) = \sum_{\mathfrak{a}} \lambda(\mathfrak{a}) e(N(\mathfrak{a})z)$, where $\mathfrak{a}$ runs over all integral ideals in L. Then f is primitive and belongs to $S_{hw+1}(D, \chi)$, where $-D$ is the discriminant of L, and $\chi(n) = (-D/n)$ (see [16], Lemma 3). It is easy to see that $K_f = \mathbf{Q}$.

Proof of Proposition 1: From (2.5) and Lemma 1 we know that

$$\langle h, h \rangle / \langle f, f \rangle = [D(s, h_\rho, h)/D(s, f_\rho, f)]_{s=k} = \prod_{p|r} X_p(k)^{-1} Y_p(k) \,,$$

where X_p and Y_p are defined as in Lemma 1 with $a(n) = a_n$ and $b(n) = \bar{a}_n$. Since X_p and Y_p are polynomials in p^{-s} whose coefficients can be explicitly expressed in terms of a_n, we obtain Proposition 1.

LEMMA 3. *Let f be a primitive cusp form. Put $f^{(v)}(z) = f(vz)$ for a positive integer v. Then $\langle f^{(v)}, f^{(w)} \rangle / \langle f, f \rangle$ belongs to K_f for any two positive integers v and w.*

Proof: As in the above proof, we have, again by (2.5) and Lemma 1,

$$(3.2) \qquad \langle f^{(v)}, f^{(w)} \rangle = \langle f, f \rangle (vw/d)^{-k} \prod_{p|vw} X_p^*(k)/X_p(k) \,,$$

which proves our lemma.

LEMMA 4. *Let f be a primitive cusp form of weight k, and h an element of $G_k(M)$ with an arbitrary M. Then $\langle f, h \rangle / \langle f, f \rangle$ belongs to $K_f K_h$. Moreover, for every automorphism σ of $\mathbf{C}$, we have*

$$(\langle f, h \rangle / \langle f, f \rangle)^\sigma = \langle f^\sigma, h^\sigma \rangle / \langle f^\sigma, f^\sigma \rangle \,.$$

Proof: Take N so that both f and h belong to $G_k(N)$. Let Φ be the set of all primitive forms contained in $S_k(N)$. For each $\varphi \in \Phi$, let $c(\varphi)$ denote the conductor of φ. Put $\varphi^{(t)}(z) = \varphi(tz)$ for $0 < t \in \mathbf{Z}$. Then the set of all $\varphi^{(t)}$ with φ

in Φ and with all positive divisors t of $N/c(\varphi)$ forms a basis of $S_k(N)$ (see Miyake [9], Casselman [2]). Observe that $\langle \varphi^{(t)}, \psi^{(u)} \rangle = 0$ for any two different members φ and ψ of Φ. By Lemma 3 and the standard orthogonal projection, we can find a basis $\{\varphi_1, \varphi_2, \cdots\}$ of the vector space $\sum_t K_\varphi \varphi^{(t)}$ over K_φ such that $\varphi = \varphi_1$, $\langle \varphi_i, \varphi_j \rangle = 0$ for $i \neq j$, and $(\varphi_i)^\sigma = (\varphi^\sigma)_i$ for every automorphism σ of $\mathbf{C}$. The effect of σ can be verified by means of formulas (3.1) and (3.2). Thus we can construct an orthogonal (not necessarily orthonormal) basis of $S_k(N)$ of the form

$$\{\varphi_j \mid \varphi \in \Phi, j = 1, 2, \cdots\}$$

whose members are permuted under σ. Now, given f and h, we have $h = \sum_{\varphi, j} \alpha_{\varphi j} \varphi_j + p$ with an Eisenstein series p and $\alpha_{\varphi j} \in \mathbf{C}$. Since f is one of the φ_1, we have $\langle f, h \rangle = \alpha_{f1} \langle f, f \rangle$. Applying σ, we obtain $h^\sigma = \sum \alpha_{\varphi j}^\sigma \varphi_j^\sigma + p^\sigma$, so that $\langle f^\sigma, h^\sigma \rangle = \alpha_{f1}^\sigma \langle f^\sigma, f^\sigma \rangle$. This proves the second assertion, from which the first one follows immediately.

LEMMA 5. *Let ω be an arbitrary (primitive or imprimitive) Dirichlet character modulo N such that $\omega(-1) = (-1)^k$ with a positive integer k. Put $P_N(k, \omega) = (2\pi i)^{-k} g(\omega)^{-1} L_N(k, \omega)$. Suppose that either $k \neq 2$ or ω is non-trivial. Then $P_N(k, \omega)$ and the Fourier coefficients of $E_{k,N}^*(z, \omega)$ belong to K_ω. Moreover, we have $E_{k,N}^*(z, \omega)^\sigma = E_{k,N}^*(z, \omega^\sigma)$ and $P_N(k, \omega)^\sigma = P_N(k, \omega^\sigma)$ for every $\sigma \in \mathrm{Gal}\,(K_\omega/\mathbf{Q})$.*

Proof: Let ω_0 be the primitive character associated with ω, and C the conductor of ω. Put $N = CM$. Then we have

$$(3.3) \quad E_{k,N}(z, \omega) = 2L_N(k, \omega)E_{k,N}^*(z, \omega) = \sum_{t \mid M} \mu(t)\omega_0(t)t^{-k} E_{k,C}(t^{-1}Mz, \omega_0),$$

where μ denotes the Moebius function, and

$$(3.4) \quad E_{k,C}(z, \omega_0) = 2L(k, \omega_0) + \frac{2g(\omega_0)(-2\pi i)^k}{C^k(k-1)!} \sum_{n=1}^{\infty} \left\{ \sum_{d \mid n} \bar{\omega}_0(d)\, d^{k-1} \right\} e(nz)$$

(see Hecke [6]). Put $G(\omega_0) = (2\pi i)^k \, (\omega_0)$ and

$$G(\omega_0)^{-1} E_{k,C}(z, \omega_0) = \sum_{n=0}^{\infty} b_n(\omega_0)e(nz).$$

Note that

$$b_0(\omega_0) = 2P_C(k, \omega_0) = 2 \prod_{p \mid M} (1 - \omega_0(p)p^{-k})^{-1} P_N(k, \omega).$$

Now, for every automorphism σ of $\mathbf{C}$, we see from (3.4) that

$$G(\omega_0^\sigma)^{-1}E_{k,C}(z, \omega_0^\sigma) - [G(\omega_0)^{-1}E_{k,C}(z, \omega_0)]^\sigma = b_0(\omega_0^\sigma) - b_0(\omega_0)^\sigma .$$

Since this is a modular form, we have $b_0(\omega_0^\sigma) = b_0(\omega_0)^\sigma$ and

$$G(\omega_0^\sigma)^{-1}E_{k,C}(z, \omega_0^\sigma) = [G(\omega_0)^{-1}E_{k,C}(z, \omega_0)]^\sigma .$$

Therefore, $P_N(k, \omega)^\sigma = P_N(k, \omega^\sigma)$. From (3.3) we obtain

$$E_{k,N}^*(z, \omega) = [2P_N(k, \omega)]^{-1} \sum_{t|M} \mu(t)\omega_0(t)t^{-k}G(\omega_0)^{-1}E_{k,C}(t^{-1}Mz, \omega_0) ,$$

which proves the remaining part of our lemma.

LEMMA 6. *Suppose $f \in S_k(N, \chi)$, $g \in G_l(N, \bar\chi)$ and $k = l + 2r$ with a positive integer r. Then $\langle f_\rho, \delta_l^{(r)}g \rangle = 0$.*

Proof: In general we have

$$\delta_l^{(r)}g = (2\pi i)^{-r} \sum_{\nu=0}^{r} c_\nu (2iy)^{\nu-r}(d/dz)^\nu g$$

with $c_\nu \in \mathbf{Z}$. Therefore if $g(z) = \sum_{n=0}^\infty b_n e(nz)$, then

$$\delta_l^{(r)}g = \sum_{\nu=0}^{r} (-4\pi y)^{\nu-r}c_\nu \sum_{n=0}^\infty b_n n^\nu e(nz) ,$$

hence we obtain

$$(3.5) \qquad \int_0^\infty y^{s-1} \int_0^1 \bar f_\rho \, \delta_l^{(r)}g \, dx \, dy = (4\pi)^{-s}D(s-r, f, g) \sum_{\nu=0}^{r} (-1)^{\nu-r}c_\nu \Gamma(s+\nu-r) .$$

In the same fashion as in Section 2, we can transform the integral to

$$\int_{\Phi_0} \bar f_\rho \, \delta_l^{(r)}g \cdot E_{0,N}^*(z, s+1-k, \tau)y^{s-1} \, dx \, dy ,$$

where τ denotes the trivial character modulo N. This has $\langle f_\rho, \delta_l^{(r)}g \rangle$ times a non-zero constant as its residue at $s = k$. But the right-hand side of (3.5) is holomorphic at $s = k$; hence we obtain our lemma.

Now let us denote by A_r the set of all functions of the form $h(z) = \sum_{\nu=0}^{r} y^{-\nu}g_\nu(z)$ with holomorphic functions g_ν on $\mathfrak{H}$ which have Fourier expansions $g_\nu(z) = \sum_{n=0}^\infty b_{\nu n} e(nz)$. We note that g_ν is uniquely determined by h.

LEMMA 7. *Suppose that, with a positive integer $k > 2r$ and a Dirichlet character χ modulo N, an element h of A_r satisfies*

$$h(\gamma(z))(cz + d)^{-k} \in A_r \quad \text{for} \quad \text{all} \quad \gamma = \begin{pmatrix} a & b \\ c & d \end{pmatrix} \in SL_2(\mathbf{Z}),$$

$$h(\gamma(z))(cz + d)^{-k} = \chi(d)h(z) \quad \text{for} \quad \text{all} \quad \gamma = \begin{pmatrix} a & b \\ c & d \end{pmatrix} \in \Gamma_0(N).$$

Then $h(z) = \sum_{\nu=0}^r \delta_{k-2\nu}^{(\nu)} g_\nu$ with elements g_ν of $G_{k-2\nu}(N, \chi)$, which are uniquely determined by h.

Proof: Observe that, for every $\gamma = \begin{pmatrix} a & b \\ c & d \end{pmatrix} \in SL_2(\mathbf{Z})$, we have

(3.6) $$y^{-1} \circ \gamma = y^{-1} \cdot (cz + d)^2 - 2ci(cz + d).$$

Therefore, if $h(z) = \sum_{\nu=0}^r y^{-\nu} g_\nu$ and $h(\gamma(z))(cz + d)^{-k} = \sum_{\nu=0}^r y^{-\nu} f_\nu(z)$, we find, substituting (3.6) for $y^{-1} \circ \gamma$ and comparing the coefficients of y^{-r}, that $(cz + d)^{2r-k} g_r(\gamma(z)) = f_r(z)$. Especially if $\gamma \in \Gamma_0(N)$, we have $f_r = \chi(d) g_r$. This shows that $g_r \in G_{k-2r}(N, \chi)$. Now observe that $h - p \delta_{k-2r}^{(r)} g_r$ with a suitable constant p is an element of A_{r-1}. By induction on r, we obtain the desired expression for h. The uniqueness can be seen by comparing the coefficients of $y^{-\nu}$.

Proof of Theorem 3: It is sufficient to prove the theorem in the case where $g \in G_l(N, \psi)$ with a character ψ. Denoting by ρ the complex conjugation, we see that $f_\rho = f^\rho$, and $f^{\rho\sigma} = f^{\sigma\rho}$ for every automorphism σ of $\mathbf{C}$, by virtue of Lemma 2. Observe also that $\langle f^\rho, f^\rho \rangle = \langle f, f \rangle$, and σ sends $G_l(N, \psi)$ onto $G_l(N, \psi^\sigma)$. Our idea is to apply σ to the equality of Theorem 2, with $m = k - 1 - r$, divided by $\pi^k \langle f, f \rangle$. If $\chi\psi$ is non-trivial or $k - l \neq 2$, and $r = 0$, we obtain the desired result immediately from Lemmas 4 and 5. In the general case, we need Lemmas 6 and 7. Let τ denote the trivial character modulo N. Then, from Hecke's result in [6], we know that

$$E_{2,N}^*(z, \tau) = \frac{c}{4\pi y} + \sum_{n=0}^\infty c_n e(nz)$$

with rational coefficients c and c_n. Therefore we see that $g \cdot \delta_\lambda^{(r)} E_{\lambda,N}^*(z, \chi\psi)$ belongs to A_{r+1} if $\chi\psi = \tau$ and $\lambda = k - l - 2r = 2$; it belongs to A_r in the remaining cases. Therefore, by Lemma 7, we have

$$g \cdot \delta_\lambda^{(r)} E^*_{\lambda,N}(z, \chi\psi) = \sum_{\nu=0}^{t} \delta_{k-2\nu}^{(\nu)} h_\nu$$

with $h_\nu \in G_{k-2\nu}(N, \bar\chi)$, where t is r or $r+1$. Expressing both sides as polynomials in $1/(-4\pi y)$, we can easily verify that

$$(3.7) \qquad g^\sigma \delta_\lambda^{(r)} E^*_{\lambda,N}(z, \chi^\sigma\psi^\sigma) = \sum_{\nu=0}^{t} \delta_{k-2\nu}^{(\nu)} h_\nu^\sigma,$$

for every automorphism σ of $\mathbf{C}$. By Theorem 2 and Lemma 6, $\pi^{-k}\langle f, f\rangle^{-1} D(k-1-r, f, g)$ is a rational number times $\langle f_\rho, h_0\rangle/\langle f_\rho, f_\rho\rangle$. Applying σ to this relation, we obtain the desired result from Lemma 4 and (3.7).

4. Proofs of Theorems 1 and 4

PROPOSITION 2. *Let f be an element of $S_k(N, \chi)$ with leading coefficient 1 that is an eigenfunction of all Hecke operators of level N, and let $\lambda = \frac{1}{2}(k+1)$. Then $D(\lambda, f, \psi) \neq 0$ for every Dirichlet character ψ.*

Proof: If $k = 1$, this follows immediately from the result of Deligne-Serre [5], which asserts that $D(s, f)$ is an Artin L-function up to finitely many Euler factors. Therefore suppose $k > 1$. Put $f(z) = \sum_{n=1}^{\infty} a_n e(nz)$, $\mu = \frac{1}{2}(k-1)$, and

$$1 - a_p p^{-s} + \chi(p) p^{k-1-2s} = (1 - \xi_p p^{\mu-s})(1 - \eta_p p^{\mu-s}),$$

with complex numbers ξ_p and η_p for each prime p. By the result of Deligne [4], we have $|\xi_p| = |\eta_p| = 1$ for p not dividing N. If p divides N, we have at least $|a_p| \leq p^{k/2}$, since $|a_n| = O(n^{k/2})$. Therefore,

$$(4.1) \qquad D(s, f, \psi) \neq 0 \quad \text{for} \quad \mathscr{R}e\,(s) > \tfrac{1}{2}(k+1),$$

because of the absolute convergence of the Euler product. Also, for the discussion of non-vanishing of $D(\lambda, f, \psi)$, we can always disregard finitely many Euler factors. Especially we may assume ψ to be primitive. Put $g(z) = \sum_{n=1}^{\infty} \psi(n) a_n e(nz)$. Then $g \in S_k(r^2 N, \psi^2 \chi)$, where r is the conductor of ψ (cf. [21], Proposition 3.64). Moreover the Euler product for $D(s, f, \psi)$ implies that g is an eigenfunction of all Hecke operators of level $r^2 N$. Therefore, replacing f by g, we only need to show that $D(\lambda, f) \neq 0$. This was shown by Ogg [10] in the case where χ is trivial, by adapting a traditional proof of non-vanishing of Hecke L-functions to $D(s, f)$ (cf. also Rankin [12]).

The same method with a slight modification applies to the general case. For the reader's convenience, we give here the whole argument. Put

$$D(s) = \prod_{p \nmid N} (1 - \xi_p p^{-s})^{-1}(1 - \eta_p p^{-s})^{-1},$$

$$\bar{D}(s) = \prod_{p \nmid N} (1 - \bar{\xi}_p p^{-s})^{-1}(1 - \bar{\eta}_p p^{-s})^{-1},$$

$$C(s) = \prod_{p \nmid N} (1 - \xi_p \bar{\eta}_p p^{-s})^{-1}(1 - \bar{\xi}_p \eta_p p^{-s})^{-1}.$$

Then $\mathfrak{D}_N^*(s + 2\mu, f_\rho, f) = \zeta^*(s)^2 C(s)$, where the asterisk indicates the removal of the Euler p-factors for p dividing N. Put $A(s) = \zeta^*(s)^3 C(s) D(s) \bar{D}(s)$. Our aim is to show that $D(1) \neq 0$. We see from (2.4) that $\mathfrak{D}_N(s, f_\rho, f)$ is holomorphic on the whole plane except for a simple pole at $s = k$. Suppose $D(1) = 0$. Then $A(s)$ is entire. Now, for $\mathfrak{Re}(s) > 1$, we have

$$\log A(s) = \sum_{n=1}^{\infty} \sum_{p \nmid N} n^{-1} p^{-ns} |1 + \xi_p^n + \eta_p^n|^2.$$

The right-hand side is a Dirichlet series with non-negative coefficients. It is well-known that such a series is not holomorphic at the real point on the line of convergence. Suppose $A(s)$ has real zeros less than or equal to 1, and let τ be the largest one. Then the series for $\log A(s)$ is convergent at $s = \sigma > \tau$, and

$$\log A(\sigma) = \sum_{n=1}^{\infty} \sum_{p \nmid N} n^{-1} p^{-n\sigma} |1 + \xi_p^n + \eta_p^n|^2 > 0,$$

so that $A(\sigma) > 1$, which is a contradiction, since $A(\tau) = 0$. Thus $A(s)$ has no real zero less than or equal to 1. But, for every negative even integer m, we have $\zeta^*(m) = 0$ and hence $A(m) = 0$, a contradiction. This completes the proof.

LEMMA 8. *For Dirichlet characters* $\chi_1, \cdots, \chi_m$, *put*

$$c(\chi_1, \cdots, \chi_m) = g(\chi_1) \cdots g(\chi_m)/g(\chi_1 \cdots \chi_m).$$

Then $c(\chi_1, \cdots, \chi_m)$ *belongs to the composite* $K_{\chi_1} \cdots K_{\chi_m}$. *Moreover,* $c(\chi_1, \cdots, \chi_m)^\sigma = c(\chi_1^\sigma, \cdots, \chi_m^\sigma)$ *for every automorphism* σ *of* $\mathbf{C}$.

Proof: It is sufficient to consider the case $m = 2$. Write χ and ψ for χ_1 and χ_2. It is also sufficient to consider the case where χ and ψ are primitive characters whose conductors are powers of the same prime, say p. Let p^m, p^n, and p^r be the conductors of χ, ψ, and $\chi\psi$, respectively.

Case I. First assume $m > n$. Then $r = m$, and $\chi\psi$ is primitive. We have

$$g(\chi)g(\psi) = \sum_{x=1}^{p^m} \sum_{y=1}^{p^n} \chi(x)\psi(y)e(p^{-m}(x + p^{m-n}y))$$

$$= \sum_{z=1}^{p^m} \sum_{y=1}^{p^n} \chi(z - p^{m-n}y)\psi(y)e(p^{-m}z) .$$

The terms for z divisible by p vanish. Therefore, putting $y = zw$, we obtain

$$(4.2) \qquad g(\chi)g(\psi) = g(\chi\psi) \sum_{w=1}^{p^n} \chi(1 - p^{m-n}w)\psi(w) ;$$

hence the desired result follows.

Case II. Next assume $m = n = r > 0$. Then we obtain, in the same fashion as in Case I,

$$g(\chi)g(\psi) = g(\chi\psi) \sum_{w=1}^{p^m} \chi(1 - w)\psi(w) + \sum_{v=1}^{p^{m-1}} \sum_{y=1}^{p^m} \chi(pv - y)\psi(y)e(pv/p^m) .$$

The terms for y divisible by p vanish. Therefore putting $v = yu$, we see that the last sum equals

$$\sum_u \chi(pu - 1) \sum_y (\chi\psi)(y)e(puy/p^m) = 0 ;$$

thus (4.2) is again valid.

Case III. Finally assume $m = n > r$. Then

$$\chi(-1)p^m c(\chi, \psi)^{-1} = c(\chi^{-1}, \chi\psi) ,$$

and therefore the problem can be reduced to Case I.

We shall now prove Theorems 1 and 4. Let ψ and ψ' be primitive characters of conductor r and r', respectively. Given a positive integer $m < k$, we can find a primitive Dirichlet character ξ so that $(\xi\psi)(-1) = (-1)^{k-m}$. Let q be the conductor of ξ. Put $l = k - m$, and

$$L(s, \xi)L(s + 1 - l, \psi) = \sum_{n=1}^{\infty} b_n n^{-s} .$$

As shown by Hecke [7], Satz 44, we can define, with a suitable constant b_0, an element

$$g(z) = \sum_{n=0}^{\infty} b_n e(nz)$$

of $G_l(qr, \xi\psi)$, except when $l = 2$, $r = 1$ and ξ is the identity character (with $q \geqq 1$). For the same reason as in the proof of Lemma 5, all the coefficients b_n including b_0 belong to $K_\xi K_\psi$. Let f be a primitive cusp form of conductor N belonging to $S_k(N, \chi)$. By Lemma 1, we have

$$D(s, f, g) = D(s, f, \xi)D(s+1-l, f, \psi)/L_M(2s+2-k-l, \chi\xi\psi) ,$$

where $M = Nqr$. Evaluate this at $s = k - 1$. Then

$$(4.3) \qquad D(k-1, f, \xi)D(m, f, \psi) = L_M(m, \chi\xi\psi)D(k-1, f, g) .$$

Therefore, defining $P_M(m, \omega)$ as in Lemma 5, we obtain

$$(4.4) \quad \frac{A(k-1, f, \xi)A(m, f, \psi)}{i^{1-k}\pi g(\chi)\langle f, f\rangle} = \frac{g(\chi\xi\psi)P_M(m, \chi\xi\psi)D(k-1, f, g)}{g(\chi)g(\xi)g(\psi)2^{k-1}\pi^k\langle f, f\rangle} .$$

Suppose $(\psi\psi')(-1) = (-1)^{m-m'}$. Then $(\xi\psi')(-1) = (-1)^{k-m'}$. Put $M' = Nqr'$. Then we have (4.4) with M', m', g', and ψ' in place of M, m, g, and ψ, where g' is the modular form corresponding to

$$L(s, \xi)L(s+1-k+m', \psi') .$$

Assuming that

$$(4.5) \qquad D(k-1, f, \xi)D(m', f, \psi') \neq 0 ,$$

we have

$$B(m, m'; f; \psi, \psi') = \frac{g(\chi\xi\psi)g(\psi')P_M(m, \chi\xi\psi)\pi^{-k}\langle f, f\rangle^{-1}D(k-1, f, g)}{g(\chi\xi\psi')g(\psi)P_M(m', \chi\xi\psi')\pi^{-k}\langle f, f\rangle^{-1}D(k-1, f, g')} .$$

Therefore we obtain Theorem 1 from Theorem 3 and Lemmas 5, 8 under assumption (4.5). If $k - m = 2$ and $r = 1$, we have to take a non-trivial character as ξ.

Next suppose $(\psi\psi')(-1) = (-1)^{m-m'-1}$. Then, from (4.3), we obtain

$$C(m, m'; f; \psi, \psi') = B(m', k-1; f; \psi', \xi)\frac{g(\chi\xi\psi)P_M(m, \chi\xi\psi)D(k-1, f, g)}{g(\chi)g(\xi)g(\psi)2^{k-1}\pi^k\langle f, f\rangle} .$$

Suppose we can find a character ξ' such that $(\xi\xi')(-1) = -1$ and $D(k-1, f, \xi') \neq 0$. Then combining our result on $B(m', k-1; f; \psi', \xi)$ with Lemmas 5, 8, and Theorem 3, we obtain Theorem 4. This completes the proof in the case $k \geq 3$, since $D(k-1, f, \xi) \neq 0$ for every ξ, by virtue of Proposition 2 and (4.1). (If $k - m = 2$ and $r = 1$, we may take any non-trivial character as ξ.)

Now suppose $k = 2$. Then $m = m' = 1$ is the only choice for m and m'. Therefore Theorem 4 in the case $k = 2$ follows immediately from (4.3). As for Theorem 1, we need condition (1.2) to validate our proof.

It is conjecturable that condition (1.2) is always satisfied. Here we note only a weaker result as

PROPOSITION 3. *Let f be a non-zero cusp form of weight 2. Then, for every positive integer M, there is a prime number p and a primitive character ξ of conductor p such that $(p, M) = 1$ and $D(1, f, \xi) \neq 0$.*

Proof: Suppose $f(z) = \sum_{n=1}^{\infty} a_n e(nz) \in S_2(N)$, and put

$$U(t) = \int_t^{i\infty} f(z)\,dz = \int_0^{i\infty} f(z+t)\,dz, \qquad t \in \mathbf{R}.$$

Let p be an odd prime and ξ a primitive character modulo p. Then

$$\sum_{a=1}^{p-1} \xi(a)U(a/p) = g(\xi)\int_0^{i\infty} \sum_{n=1}^{\infty} a_n\bar{\xi}(n)e(nz)\,dz = i(2\pi)^{-1}g(\xi)D(1, f, \bar{\xi}).$$

Suppose $D(1, f, \bar{\xi}) = 0$ for all such ξ. Then, for $(b, p) = 1$, we have

$$0 = \sum_{\xi \neq 1} \bar{\xi}(b) \sum_{a=1}^{p-1} \xi(a)U(a/p) = (p-1)U(b/p) - \sum_{a=1}^{p-1} U(a/p).$$

Therefore, $U(a/p) = U(b/p)$ for any two integers a and b prime to p; hence

$$(4.6) \qquad \int_{a/p}^{b/p} f(z)\,dz = 0 \quad \text{if} \quad (a, p) = (b, p) = 1.$$

Now put $[\gamma, f] = \int_w^{\gamma(w)} f(z)\,dz$ for $\gamma \in \Gamma_1(N)$ with any point w on the upper half-plane including cusps. This is independent of the choice of w. Moreover, $[\gamma\delta, f] = [\gamma, f] + [\delta, f]$; $[\alpha, f] = 0$ if α is parabolic. We are going to show that $[\gamma, f] = 0$ for all $\gamma \in \Gamma_1(N)$ if (4.6) holds for all primes p not dividing a given integer M. Put $\alpha = \begin{pmatrix} 1 & 1 \\ 0 & 1 \end{pmatrix}$, $\beta = \begin{pmatrix} 1 & 0 \\ N & 1 \end{pmatrix}$. Let $\gamma \in \Gamma_1(N)$. We can find an integer m so that $\beta^m \gamma = \begin{pmatrix} a & b \\ Np & d \end{pmatrix}$ with a prime p not dividing M. Put $d = 1 - Ns$. If $s = pr$ with $r \in \mathbf{Z}$, then we see that $\alpha^{-ar-b}\beta^m \gamma \alpha^r = \beta^p$, and hence $[\gamma, f] = 0$. Therefore assume $(p, s) = 1$, and put $t = as + bp$. Then $\beta^m \gamma$ sends s/p to t/p, and $(p, t) = 1$. Consequently, (4.6) shows that $[\beta^m \gamma, f] = 0$, and hence $[\gamma, f] = 0$. Thus $[\gamma, f] = 0$ for all $\gamma \in \Gamma_1(N)$, which is a contradiction since $f \neq 0$.

EXAMPLE. Let us consider a primitive cusp form f of $S_2(15)$ defined by

$$f(z) = \eta(z)\eta(3z)\eta(5z)\eta(15z), \qquad \eta(z) = e(z/24) \prod_{n=1}^{\infty} (1 - e(nz)).$$

Also set

$$g(z) = \tfrac{1}{6} + \sum_{n=1}^{\infty} \sum_{d \mid n} \left(\frac{d}{3}\right) e(nz).$$

Put $\psi_0(d) = \left(\dfrac{d}{3}\right)$, and let ψ be the imprimitive character modulo 15 such that $\psi(d) = \left(\dfrac{d}{3}\right)$ for $(d, 5) = 1$. Then it can be verified that $E^*_{1,15}(z, \psi) = g(z) + 5g(5z)$ and

$$24g(z)E^*_{1,15}(z, \psi) = 15f + 3G_3(z) + 10G_5(z) + 25G_3(5z),$$

where G_q is the Eisenstein series defined by

$$G_q(z) = \frac{q-1}{24} + \sum_{n=1}^{\infty} \sigma_q(n)e(nz),$$

$\sigma_q(n)$ denoting the sum of all positive divisors of n prime to q. Therefore,

(4.7) $$\langle f, gE^*_{1,15}(z, \psi)\rangle / \langle f, f\rangle = \tfrac{5}{8}.$$

This shows that $D(1, f)D(1, f, \psi_0) \neq 0$; hence f satisfies condition (1.2). Combining (4.7) with Theorem 2 and (4.3) (or (4.4)), we obtain

(4.8) $$C(1, 1; f; 1, \psi_0) = -\tfrac{2}{3}.$$

Next put $\xi(d) = \left(\dfrac{d}{5}\right)$ and $h(z) = \sum_{m=1}^{\infty}\sum_{n=1}^{\infty} \left(\dfrac{m}{5}\right)\left(\dfrac{n}{3}\right) e(mnz)$. Then $f(z) = h(z)E^*_{1,15}(z, \xi\psi)$, so that $\langle f, hE^*_{1,15}(z, \xi\psi)\rangle = \langle f, f\rangle$. Therefore Theorem 2 together with (4.3) yields

$$C(1, 1; f; \xi, \psi) = -\tfrac{16}{15}.$$

Dividing this by (4.8), we obtain

$$B(1, 1; f; \xi, 1) = g(\xi)^{-1} D(1, f, \xi)/D(1, f) = \tfrac{8}{5}.$$

The idea of the above computation is of course applicable to the forms of higher weight.

5. *L*-Functions of an Imaginary Quadratic Field

Let F be an imaginary quadratic field of discriminant $-\Delta$, and λ a primitive or an imprimitive Hecke character of F defined modulo an integral ideal $\mathfrak{c}$ such that

$$(5.1) \qquad\qquad \lambda((a)) = a^{k-1} \quad \text{for} \quad a \equiv 1 \bmod \mathfrak{c},$$

with an integer $k > 1$. Put

$$f_\lambda(z) = \sum_\mathfrak{a} \lambda(\mathfrak{a}) e(N(\mathfrak{a})z),$$

$$L(s, \lambda) = \sum_\mathfrak{a} \lambda(\mathfrak{a}) N(\mathfrak{a})^{-s},$$

where $\mathfrak{a}$ runs over all integral ideals of F prime to $\mathfrak{c}$. It can be verified that f_λ belongs to $S_k(N(\mathfrak{c})\Delta, \chi)$, where

$$\chi(m) = m^{1-k} \lambda((m)) \left(\frac{-\Delta}{m}\right) \quad \text{for} \quad 0 \neq m \in \mathbf{Z}$$

(cf. [16], Lemma 3). Moreover, by virtue of a result of Miyake [9], we can show that if λ is primitive, then f_λ is primitive, and has conductor $N(\mathfrak{c})\Delta$. Obviously we have $D(s, f_\lambda) = L(s, \lambda)$, and more generally

$$(5.2) \qquad\qquad D(s, f_\lambda, \psi) = L(s, (\psi \circ N_{F/\mathbf{Q}})\lambda)$$

for every Dirichlet character ψ.

PROPOSITION 4. *The form f_λ satisfies, when $k = 2$, condition (1.2).*

Proof: By Proposition 3, we have $D(1, f_\lambda, \psi) \neq 0$ for a suitable primitive character ψ of conductor q with a prime q. Define a Dirichlet character ξ by $\xi(m) = \psi(m)\left(\frac{-\Delta}{m}\right)$. Then $D(s, f_\lambda, \xi)$ differs from $D(s, f_\lambda, \psi)$ only by the Euler p-factors for p dividing Δ. Therefore $D(1, f_\lambda, \xi) \neq 0$. Since $(\xi\psi)(-1) = -1$, this means that f_λ satisfies (1.2).

PROPOSITION 5. *Define f_λ as above with a primitive Hecke character λ satisfying (5.1) with an integer $k > 1$. Let h be a modular form of an arbitrary level and of weight $k - 1$ with algebraic Fourier coefficients at $i\infty$, and τ an element of F such that $\mathcal{I}m(\tau) > 0$ and $h(\tau) \neq 0$. Then*

    (i) *$\pi^{-m} h(\tau)^{-1} L(m, \lambda)$ is algebraic,*

    (ii) *$\pi h(\tau)^{-2} \langle f_\lambda, f_\lambda \rangle$ is algebraic.*

The first assertion is a result due to Damerell [3], for which Weil gave a simpler proof in [22]. The author extended it to the Hilbert modular case in [20]. Here we shall derive it from our theorems.

First suppose $k \neq 3$. Since $\bar{\alpha}^{k-1} N(\alpha)^{1-k} = \alpha^{1-k}$ for $0 \neq \alpha \in F$, we see easily that $L(k-1, \lambda)$ is a linear combination of Eisenstein series of weight $k-1$ evaluated at τ. (Cf. [20], p. 499. If $k = 3$, non-holomorphic functions can occur.) Therefore assertion (i) in the case $m = k - 1$ follows immediately from the fact that the value of a modular function at τ is algebraic if it has algebraic Fourier coefficients. Take Dirichlet characters ψ and ψ' so that $D(k-1, f_\lambda, \psi) D(k-1, f_\lambda, \psi') \neq 0$ and $(\psi\psi')(-1) = -1$. (We need Proposition 4 when $k = 2$.) By Theorem 4,

$$\pi^{1-2k} D(k-1, f_\lambda, \psi) D(k-1, f_\lambda, \psi') / \langle f_\lambda, f_\lambda \rangle$$

is algebraic. Therefore we obtain assertion (ii). Then assertion (i) for $m < k - 1$ can be derived from Theorem 1 or Theorem 4. Next consider the case $k = 3$. Let μ (respectively ν) be a primitive Hecke character such that $\mu((a)) = a^{l-1}$ (respectively $\nu((a)) = a^{m-1}$) for a congruent to 1 modulo its conductor, say $\mathfrak{d}$ (respectively $\mathfrak{e}$). Suppose $m < l$. We can define a Hecke character λ_1 defined modulo $\Delta \mathfrak{d}\bar{\mathfrak{e}}$ such that $\lambda_1(\mathfrak{a}) = \mu(\mathfrak{a})\nu(\bar{\mathfrak{a}})N(\mathfrak{a})^{1-m}$. Then $\lambda_1((a)) = a^{l-m}$ if $a \equiv 1 \bmod \Delta \mathfrak{d}\bar{\mathfrak{e}}$. By virtue of Lemma 1, we can easily verify that

$$D(s, f_\mu, f_\nu) L_M(2s + 2 - l - m, \xi) = L(s, \mu\nu) L(s + 1 - m, \lambda_1),$$

where $M = N(\mathfrak{d}\mathfrak{e})\Delta$, and $\xi(x) = x^{2-l-m}\mu((x))\nu((x))$. Putting $s = l - 1$, we have

$$(5.3) \qquad D(l-1, f_\mu, f_\nu) L_M(l - m, \xi) = L(l - 1, \mu\nu) L(l - m, \lambda_1).$$

Now, given a primitive character λ with $k = 3$, we can find μ and ν so that $l = 4$, $m = 2$, and λ is associated with λ_1. Therefore the desired result for $L(2, \lambda)$ follows from (5.3), Theorem 3, and what we have already proved in the case $k \neq 3$. The remaining assertions in the case $k = 3$ can be proved in exactly the same manner as above.

PROPOSITION 6.   *The notation being as in Proposition 5, let g be an arbitrary modular form of weight $l < k$ with algebraic Fourier coefficients, and m an integer such that $\frac{1}{2}(k + l - 2) < m < k$. Then $\pi^{1-k} h(\tau)^{-2} D(m, f_\lambda, g)$ is algebraic.*

This follows immediately from Theorem 3 and Proposition 5.

Bibliography

[1] Atkın, A. O. L., and Lehner, J., *Hecke operators on $\Gamma_0(m)$*, Math. Ann., 185, 1970, pp. 134–160.

[2] Casselman, W., *On some results of Atkin and Lehner*, Math. Ann., 201, 1973, pp. 301–314.

[3] Damerell, R. M., *L-functions of elliptic curves with complex multiplication*, I, II, Acta Arith., 17, 1970, pp. 287–301; 19, 1971, pp. 311–317.

[4] Deligne, P., *La conjecture de Weil, I*, Publ. Math. I.H.E.S. no. 43, 1974, pp. 273–307.

[5] Deligne, P., and Serre, J.-P., *Formes modulaires de poids 1*, Ann. Sci. Ec. Norm. Sup. (4^e) 7, 1974, pp. 507–530.

[6] Hecke, E., *Theorie der Eisensteinschen Riehen höherer Stufe und ihre Anwendung auf Funktionentheorie und Arithmetik*, Abh. Math. Sem. Hamburg, 5, 1927, pp. 199–224.

[7] Hecke, E., *Über Modulfunktionen und die Dirichletschen Reihen mit Eulerscher Produktentwicklung II*, Math. Ann., 114, 1937, pp. 316–351.

[8] Manin, Y. I., *Periods of parabolic forms and p-adic Hecke series*, Mat. Sbornik, 92, 134, 1973, pp. 376–401.

[9] Miyake, T., *On automorphic forms on GL_2 and Hecke operators*, Ann. of Math., 94, 1971, pp. 174–189.

[10] Ogg, A. P., *On a convolution of L-series*, Inv. Math. 7, 1969, pp. 297–312.

[11] Petersson, H., *Über die Berechnung der Skalarprodukte ganzer Modulformen*, Comm. Math. Helv., 22, 1949, pp. 168–199.

[12] Rankin, R. A., *Contributions to the theory of Ramanujan's function $\tau(n)$ and similar arithmetical functions*, 1, II, Proc. Cambridge Phil. Soc., 35, 1939, pp. 351–372.

[13] Rankin, R. A., *The scalar product of modular forms*, Proc. London Math. Soc., 3, 2, 1952, pp. 198–217.

[14] Selberg, A., *Bemerkungen über eine Dirichletsche Reihe die mit der Theorie der Modulformen nahe verbunden ist*, Arch. Math. Naturvid, 43, 1940, pp. 47–50.

[15] Shimura, G., *Sur les intégrales attachées aux formes automorphes*, J. Math. Soc. Japan, 11, 1959, pp. 291–311.

[16] Shimura, G., *On elliptic curves with complex multiplication as factors of the Jacobians of modular function fields*, Nagoya Math. J., 43, 1971, pp. 199–208.

[17] Shimura, G., *Class fields over real quadratic fields and Hecke operators*, Ann. of Math., 95, 1972, pp. 130–190.

[18] Shimura, G., *On modular forms of half integral weight*, Ann. of Math., 97, 1973, pp. 440–481.

[19] Shimura, G., *On the holomorphy of certain Dirichlet series*, Proc. London Math. Soc., 3, 31, 1975, pp. 79–98.

[20] Shimura, G., *On some arithmetic properties of modular forms of one and several variables*, Ann. of Math., 102, 1975, pp. 491–515.

[21] Shimura, G., *Introduction to the Arithmetic Theory of Automorphic Functions*, Publ. Math. Soc. Japan, No. 11, Iwanami Shoten and Princeton Univ. Press, 1971.

[22] Weil, A., *Elliptic Functions According to Eisenstein and Kronecker*, Ergeb. Math., 88, Springer, 1976.

Received June, 1976.

On abelian varieties with complex multiplication

Proceedings of the London Mathematical Society, 3rd ser. 34 (1977), 65-86

[Received 26 May 1975]

Let A be an abelian variety of dimension n defined over $\mathbf{C}$ whose endomorphism algebra contains an isomorphic image of a number field K of degree $2n$. The main theorem of complex multiplication gives the behaviour of A under an automorphism of $\mathbf{C}$ over a certain field K' which is determined by the representation Φ of K on the tangent space of A. The purpose of the present paper is to investigate the behaviour of A under an automorphism that is not the identity map on K'. To explain this problem in a different fashion, take any polarization C of A and let M_0 be the field of moduli of the polarized variety (A, C). Then, by means of the main theorem, one can determine $K'M_0$ as a class field over K', but one does not know how large or small M_0 is. Such information can be supplied by the behaviour of A under an automorphism of general type. If A is one-dimensional, then K is imaginary quadratic and $K' = K$. In this case, M_0 always has a real archimedean prime, and hence M_0 and K are linearly disjoint over $\mathbf{Q}$. This can easily be derived from the behaviour of A under the complex conjugation. In the higher-dimensional case, however, no such general answer seems possible. Indeed, we shall show that the behaviour of (A, C) and the size of M_0 depend, to a great extent, on the *isomorphism class* of (A, C), whereas $K'M_0$ is completely determined by (K, Φ) if $\mathrm{End}(A)$ is the maximal order of K. Also, the Galois-theoretical structure of (K, Φ) seems essential to the answers, if any, to our questions. We shall actually investigate in this paper, after some general preliminary considerations, the following three cases:

(1) the Galois closure of K over $\mathbf{Q}$ has a dihedral group of order $4n$ as its Galois group;

(2) K is cyclic over $\mathbf{Q}$;

(3) K is the composite of two totally imaginary quadratic extensions of a totally real algebraic number field.

Each case presents some interesting features that are all its own. We shall also examine the possibility of constructing a model of (A, C) over M_0 in Case (2). Since our results cannot easily be described, we mention here only that: (i) the odd-dimensional case is in general simpler than the even-dimensional case; (ii) Case (3) is closely related to the problem of determining the behaviour of arithmetic automorphic functions under anti-holomorphic automorphisms.

Notation. For an algebraic number field k, we let I_k denote the group of all fractional ideals in k, P_k denote the group of all principal ideals in k, h_k denote the class number of k, $\mathfrak{o}_k$ denote the ring of all algebraic integers in k, E_k denote the group of all units in k, E_k^+ denote the group of all the units that are positive at all real archimedean primes of k, and $\mathfrak{d}_k$ denote the different of k over $\mathbf{Q}$. If K is a finite extension of k, $\mathfrak{d}_{K/k}$ denotes the relative different of K over k. If K is abelian over k and $\mathfrak{x} \in I_k$, we denote by $[K/k, \mathfrak{x}]$ the Artin symbol which is the element of $\mathrm{Gal}(K/k)$ corresponding to $\mathfrak{x}$. For convenience, we consider all algebraic number fields as subfields of $\mathbf{C}$. We denote by ρ the complex conjugation in $\mathbf{C}$. If $x \in k$, $\alpha_i \in \mathrm{Aut}(k)$, and $m_i \in \mathbf{Z}$, we put $x^{\sum_i m_i \alpha_i} = \prod_i (x^{\alpha_i})^{m_i}$, and use similar symbols for ideals such as $\mathfrak{a}^{\rho-1} = \mathfrak{a}^\rho \mathfrak{a}^{-1}$ for $\mathfrak{a} \in I_k$ when $k^\rho = k$.

1. Throughout the paper, we denote by F a totally real algebraic number field, by n its degree over $\mathbf{Q}$, and by K a totally imaginary quadratic extension of F. Let (K, Φ) be a CM-type and (K', Φ') its reflex (or dual) in the sense of [**10**; **7**, I, 1.3]. Our object of study is a structure $P = (A, C, \theta)$ formed by an abelian variety A of dimension n defined over (a subfield of) $\mathbf{C}$, a polarization C of A, and an injection θ of K into $\mathrm{End}(A) \otimes \mathbf{Q}$. We say that (A, θ) is of type (K, Φ) if the representation of K, through θ, on the space of holomorphic 1-forms on A is equivalent to Φ. We always assume that $\theta(K)$ is stable under the involution of $\mathrm{End}(A) \otimes \mathbf{Q}$ determined by C. Let $\tau_1, \ldots, \tau_n$ be the n injections of K into $\mathbf{C}$ which form Φ. We then write $\Phi \sim \sum_{i=1}^{n} \tau_i$. Given $P = (A, C, \theta)$ of type (K, Φ), we can find a $\mathbf{Z}$-lattice $\mathfrak{a}$ in K such that A is isomorphic to $\mathbf{C}^n/\Phi(\mathfrak{a})u$ with a suitable vector u in $\mathbf{C}^n$. Moreover, if $R(x, y)$ is the Riemann form of a basic polar divisor in C, then there is an element t of K such that

$$(1.1) \qquad R(\Phi(x)u, \Phi(y)u) = \mathrm{Tr}_{K/\mathbf{Q}}(txy^\rho) \quad ((x, y) \in K \times K),$$

$$(1.2) \qquad t^\rho = -t, \quad \mathrm{Im}(t^{\tau_i}) > 0 \quad (i = 1, \ldots, n).$$

We then say that P *is of type* $(K, \Phi; t, \mathfrak{a})$, or simply *of type* $(t, \mathfrak{a})$ when (K, Φ) is fixed. We call $(t, \mathfrak{a})$ and $(t', \mathfrak{a}')$ *equivalent* if $t' = bb^\rho t$ and $\mathfrak{a}' = b^{-1}\mathfrak{a}$ with $b \in K$. Then the equivalence class of $(t, \mathfrak{a})$ is uniquely determined by the isomorphism class of P. Since we obtained t from a *basic* divisor, we have

$$(1.3) \qquad \mathrm{Tr}_{K/\mathbf{Q}}(t\mathfrak{a}\mathfrak{a}^\rho) = \mathbf{Z}.$$

Conversely, given $(t, \mathfrak{a})$ satisfying (1.2) and (1.3), we can always find a structure $P = (A, C, \theta)$ of type $(K, \Phi; t, \mathfrak{a})$, which is unique up to isomorphism. We have obviously

$$\mathrm{End}(A) \cap \theta(K) = \theta(\{c \in K \mid c\mathfrak{a} \subset \mathfrak{a}\}).$$

Hereafter we consider, except in § 7, only those P for which $\mathfrak{a}$ is a fractional ideal in K, that is, $\mathrm{End}(A) \cap \theta(K) = \theta(\mathfrak{o}_K)$.

With a basic polar divisor X in C, consider the isogeny φ_X of A onto its Picard variety which sends a point v of A to the class of the divisor $X_v - X$. Then it can be shown that

$$\mathrm{Ker}(\varphi_X) = \{y \in A \mid \theta(\mathfrak{f})y = 0\}$$

with an integral ideal $\mathfrak{f}$ in F such that $\mathfrak{f}\mathfrak{o}_K = t\mathfrak{d}_K \mathfrak{a}\mathfrak{a}^\rho$ (see [**10**, p. 118]). We write $\mathfrak{f} = \mathfrak{f}(P)$. Obviously condition (1.3) is equivalent to

$$(1.4) \qquad\qquad \mathrm{Tr}_{F/\mathbf{Q}}(\mathfrak{d}_F^{-1}\mathfrak{f}) = \mathbf{Z}.$$

Note that there are infinitely many integral ideals $\mathfrak{f}$ satisfying (1.4), if $F \neq \mathbf{Q}$. It may also be remarked that the $\mathfrak{o}_K$-module $\mathfrak{a}$ is obtained as the first homology group of A with the natural action of $\mathfrak{o}_K$, and t from the cohomology class of X. Moreover, the divisors of A modulo algebraic equivalence form a module isomorphic to $\mathfrak{f}$.

PROPOSITION 1. *Let* $\mathfrak{m} \in I_F$. *If* $\mathfrak{d}_{K/F} \neq \mathfrak{o}_K$, *there is a positive rational number s and a structure P of type* (K, Φ) *such that* $\mathfrak{f}(P) = s\mathfrak{m}$. *If* $\mathfrak{d}_{K/F} = \mathfrak{o}_K$, *such a P exists if and only if* $[K/F, \mathfrak{n}^{-1}\mathfrak{m}] = 1$, *where $\mathfrak{n}$ is an ideal in F such that* $\mathfrak{n}\mathfrak{o}_K = t\mathfrak{d}_K$ *with any t satisfying* (1.2). *In either case, s is uniquely determined by* $\mathrm{Tr}_{F/\mathbf{Q}}(s\mathfrak{d}_F^{-1}\mathfrak{m}) = \mathbf{Z}$.

Proof. Take any t satisfying (1.2). Then $t\mathfrak{d}_K = \mathfrak{n}\mathfrak{o}_K$ with $\mathfrak{n} \in I_F$. Our question is whether there exists an ideal $\mathfrak{a}$ in K and a totally positive element b of F such that $b\mathfrak{n}^{-1}\mathfrak{m} = N_{K/F}(\mathfrak{a})$. Therefore our assertions follow easily from class field theory.

REMARK. For example, we can always find a P such that $\mathfrak{f}(P) = \mathfrak{o}_F$, if $\mathfrak{d}_{K/F} \neq \mathfrak{o}_K$.

2. Let us now consider the group consisting of all pairs $\{b, \mathfrak{c}\}$ with a totally positive element b of F and $\mathfrak{c} \in I_K$, defining multiplication by $\{b, \mathfrak{c}\}\{b', \mathfrak{c}'\} = \{bb', \mathfrak{c}\mathfrak{c}'\}$. We take the factor group of this group by the subgroup consisting of all pairs $\{xx^\rho, x\mathfrak{o}_K\}$ with $x \in K^\times$, and denote by $(b, \mathfrak{c})$ the coset of $\{b, \mathfrak{c}\}$.

Let $P = (A, C, \theta)$ and $P' = (A', C', \theta')$ be of type $(t, \mathfrak{a})$ and $(t', \mathfrak{a}')$, respectively, with the same (K, Φ). Then there is an isogeny λ of (A, θ) to (A', θ'). Take a basic polar divisor X (respectively X') in C (respectively C'). Put $Y = \lambda^{-1}(X')$. Then $\varphi_Y = \varphi_X \theta(b)$ with a totally positive element

b of F. Moreover there is an ideal $\mathfrak{c}$ in K such that

$$\mathrm{Ker}(\lambda) = \{x \in A \mid \theta(\mathfrak{c})x = 0\}.$$

It can easily be shown that

(2.1) $$(b, \mathfrak{c}) = (t^{-1}t', \mathfrak{a}'^{-1}\mathfrak{a}),$$

(2.2) $$\mathfrak{f}(P)\mathfrak{f}(P')^{-1} = b^{-1}N_{K/F}(\mathfrak{c}).$$

We then write

(2.3) $$(P' : P) = (b, \mathfrak{c}).$$

Obviously $(P' : P) = 1$ if and only if P' is isomorphic to P. Note also that for given P and $(b, \mathfrak{c})$, there is always a P' such that $(P' : P) = (rb, \mathfrak{c})$ with a positive rational number r.

With $\{\tau_\nu\}$ as in §1, we put $x^\Phi = x^{\tau_1}\ldots x^{\tau_n}$ for $x \in K$. Then x^Φ belongs to K'. Also we define, for an ideal $\mathfrak{x} \in I_K$, an ideal $\mathfrak{x}^\Phi \in I_{K'}$ such that $\mathfrak{x}^\Phi \mathfrak{o}_L = \mathfrak{x}^{\tau_1}\ldots\mathfrak{x}^{\tau_n}\mathfrak{o}_L$ if L is the Galois closure of K over $\mathbf{Q}$. Similarly we can define, for an element or ideal $\mathfrak{y}$ in K', an element or ideal $\mathfrak{y}^{\Phi'}$ in K. Now let M be the field of moduli of P. Then M is an unramified abelian extension of K' corresponding to the ideal group $I_0(\Phi')$, which consists of all $\mathfrak{a} \in I_{K'}$ such that $\mathfrak{a}^{\Phi'} = b\mathfrak{o}_K$ and $N(\mathfrak{a}) = bb^\rho$ with $b \in K$. Moreover, for $\sigma \in \mathrm{Aut}(\mathbf{C}/K')$, one has

(2.4) $$(P^\sigma : P) = (N(\mathfrak{x}), \mathfrak{x}^{\Phi'})$$

if $\sigma = [M/K', \mathfrak{x}]$ on M with $\mathfrak{x} \in I_{K'}$. (See [**10**, p. 127, (1)]; or **8**, Theorem 1].) It should be noted that M depends only on (K, Φ), and is independent of $(t, \mathfrak{a})$.

Now it is one of our main problems to study P^σ for an automorphism σ of $\mathbf{C}$ which is not necessarily the identity map on K'. The analysis of P^σ for an *arbitrary* σ seems very difficult, and therefore we consider here only those σ for which there exists $\tau \in \mathrm{Aut}(K)$ such that $\mathrm{tr}\,\Phi(a)^\sigma = \mathrm{tr}\,\Phi(a^\tau)$ for all $a \in K$. Obviously such a σ induces an automorphism of K'. If (K, Φ) is primitive in the sense of [**8**, p. 508], τ is unique for the restriction of σ to K'. Let K_0' denote the largest subfield of K' on which all such σ act as the identity map. Take an intermediate field J' between K_0' and K' such that there is an injective homomorphism

(2.5) $$\mathrm{Gal}(K'/J') \ni \sigma \mapsto [\sigma] \in \mathrm{Aut}(K)$$

satisfying $\mathrm{tr}\,\Phi(a)^\sigma = \mathrm{tr}\,\Phi(a^{[\sigma]})$. If (K, Φ) is primitive, we can always take K_0' to be J'. With a fixed J' and such an isomorphism $\sigma \mapsto [\sigma]$, let J be the subfield of K corresponding to the image of $\mathrm{Gal}(K'/J')$. Then $\mathrm{Gal}(K/J)$ is isomorphic to $\mathrm{Gal}(K'/J')$. A typical example of this situation is given by the maximal real subfields F and F' of K and K', respectively.

Let us now fix J' and the map (2.5). For $\sigma \in \mathrm{Aut}(\mathbf{C}/J')$, we define $[\sigma]$ to be the same as the image $[\sigma']$ of the restriction σ' of σ to K'. For $P = (A, C, \theta)$ of type $(K, \Phi; t, \mathfrak{a})$ and $\sigma \in \mathrm{Aut}(\mathbf{C}/J')$, we define

$$P_\sigma = (A^\sigma, C^\sigma, \theta_\sigma)$$

by

(2.6) $$\theta_\sigma(a) = \theta(a^{[\sigma]^{-1}})^\sigma \quad (a \in K).$$

Then P_σ is of the same type (K, Φ), $\mathfrak{f}(P_\sigma) = \mathfrak{f}(P)^{[\sigma]}$, and $(P_\sigma)_\tau = P_{\sigma\tau}$. One can also easily verify the following proposition.

PROPOSITION 2. (i) *If* $(P' : P) = (b, \mathfrak{c})$, *then* $(P'_\sigma : P_\sigma) = (b^{[\sigma]}, \mathfrak{c}^{[\sigma]})$.
(ii) *If* $(P_\sigma : P) = (b, \mathfrak{c})$ *and* $(P_\tau : P) = (d, \mathfrak{e})$, *then* $(P_{\sigma\tau} : P) = (b^{[\tau]}d, \mathfrak{c}^{[\tau]}\mathfrak{e})$.
(iii) *If* $(P' : P) = (b, \mathfrak{c})$ *and* $(P_\sigma : P) = (x, \mathfrak{y})$, *then*

$$(P'_\sigma : P') = (b^{[\sigma]-1}x, \mathfrak{c}^{[\sigma]-1}\mathfrak{y}).$$

PROPOSITION 3. *With J and J' as above, let θ_J denote the restriction of θ to J, and let M_J denote the field of moduli of (A, C, θ_J). Then M is normal over J'. If (K, Φ) is primitive, then $J' \subset M_J$ and $M = M_J K'$; moreover an element σ of $\mathrm{Aut}(\mathbf{C}/J')$ is the identity map on M_J if and only if $(P_\sigma : P) = 1$.*

Proof. We see easily that the field of moduli of P_σ is M^σ for every $\sigma \in \mathrm{Aut}(\mathbf{C}/J')$. Now the field of moduli of P depends only on (K, Φ) and is independent of $(t, \mathfrak{a})$. Therefore $M^\sigma = M$, and hence the first assertion is true. The remaining part can be proved in the same manner as in [8, Proposition 2].

Since the isomorphism class of P depends only on the effect of σ on M, we use the notation P_σ also for $\sigma \in \mathrm{Gal}(M/J')$ when it is unnecessary to speak of a specific model of P. Thus $\mathrm{Gal}(M/J')$ acts on the isomorphism classes of P, which can be identified with the equivalence classes of $(t, \mathfrak{a})$. Now, assuming that (K, Φ) is primitive, one can naturally ask whether, for a fixed P, M_J is linearly disjoint with K' over J'. Suppose this is so, and let H be the subgroup of $\mathrm{Gal}(M/J')$ corresponding to M_J. Then $(P_\sigma : P) = 1$ for $\sigma \in H$, and hence

(2.7) $$\mathfrak{f}(P)^\gamma = \mathfrak{f}(P) \quad \text{for all } \gamma \in \mathrm{Gal}(K/J).$$

This is a necessary condition for the linear disjointness of M_J with K' over J'. For a fixed $\mathfrak{g} \in I_F$, the number of all isomorphism classes of P such that $\mathfrak{f}(P) = \mathfrak{g}$ is either 0 or $[R_K^+ : P_K][E_F^+ : N_{K/F}(E_K)]$, where R_K^+ is the group of all $\mathfrak{x} \in I_K$ such that $N_{K/F}(\mathfrak{x})$ is a principal ideal generated by a totally positive element of F. Suppose $\mathfrak{g}^\gamma = \mathfrak{g}$ for all $\gamma \in \mathrm{Gal}(K/J)$. Then $\mathrm{Gal}(M/J')$ acts on all such P. Suppose

(2.8) $$[M : K'] = [R_K^+ : P_K][E_F^+ : N_{K/F}(E_K)].$$

Then $\mathrm{Gal}(M/K')$ acts transitively on the isomorphism classes of such P. (This is trivially the case if $n = 1$.) Therefore, for any P in this family, if we put

$$H_P = \{\sigma \in \mathrm{Gal}(M/J')\,|\,(P_\sigma : P) = 1\},$$

then $[H_P : 1] = [K' : J']$, and M_J is linearly disjoint with K' over J'. In §4 and the Appendix, we shall give a few cases in which (2.8) holds.

3. Let us now consider the special case in which $J = F$ and $J' = F'$. We first note the following proposition.

PROPOSITION 4. *Let P be of type $(t, \mathfrak{a})$. Then P_ρ is of type $(t, \mathfrak{a}^\rho)$. If $(P_\rho : P) = 1$, P is of type $(s, \mathfrak{b})$ with an ideal $\mathfrak{b}$ such that $\mathfrak{b}^\rho = \mathfrak{b}$.*

Proof. The first assertion is a restatement of [8, Lemma 3]. If $(P_\rho : P) = 1$, one has $\mathfrak{a}^\rho = c\mathfrak{a}$ and $cc^\rho = 1$ with $c \in K$. Take $d \in K$ such that $c = d^{\rho-1}$. Then P is of type $(tdd^\rho, d^{-1}\mathfrak{a})$, and $(d^{-1}\mathfrak{a})^\rho = d^{-1}\mathfrak{a}$.

Let σ be an element of $\mathrm{Gal}(M/F')$ such that $\sigma = \rho$ on K'. Then $\sigma = \rho[M/K', \mathfrak{x}]$ with $\mathfrak{x} \in I_{K'}$. We have $\sigma^2 = [M/K', \mathfrak{x}\mathfrak{x}^\rho] = 1$, since $(\mathfrak{x}\mathfrak{x}^\rho)^{\Phi'} = (N(\mathfrak{x}))$. Moreover, by (2.4) and Proposition 4,

$$(P_\sigma : P) = (P_\rho : P)(N(\mathfrak{x}), \mathfrak{x}^{\Phi'}) = (N(\mathfrak{x}), \mathfrak{a}^{1-\rho}\mathfrak{x}^{\Phi'})$$

if P is of type $(t, \mathfrak{a})$. Therefore, from Proposition 3, we obtain the following proposition.

PROPOSITION 5. *Given $P = (A, C, \theta)$ of type $(K, \Phi; t, \mathfrak{a})$, suppose there exist an ideal $\mathfrak{z}$ of K' and an element w of K such that $\mathfrak{z}^{\Phi'} = w\mathfrak{a}^{\rho-1}$ and $N(\mathfrak{z}) = ww^\rho$. Then $(P_\sigma : P) = 1$ with $\sigma = \rho[M/K', \mathfrak{z}]$. Moreover suppose (K, Φ) is primitive and let M_1 be the field of moduli of (A, C, θ_F). Then $[M : M_1] = 2$, and*

$$M_1 = \{x \in M \,|\, x^\sigma = x\}.$$

This principle is obviously applicable to the case where $n = 1$. As a more interesting example, we take a Galois extension L of $\mathbf{Q}$ whose Galois group is a dihedral group of order $4n$ $(n > 1)$ generated by two elements α and β such that $\alpha^{2n} = \beta^2 = 1$ and $\beta\alpha\beta = \alpha^{-1}$. Let K and F be the subfields of L corresponding to the subgroups $\{1, \beta\}$ and $\{1, \beta, \alpha^n, \beta\alpha^n\}$. We naturally assume that F is totally real, K is totally imaginary, and $\rho = \alpha^n$ on L. Let us now consider (K, Φ) with $\Phi \sim \sum_{k=0}^{n-1} \alpha^k$. It can easily be verified that K' corresponds to $\{1, \beta\alpha^{n-1}\}$, $\Phi' \sim \sum_{k=0}^{n-1} \alpha^{-k}$, and (K, Φ) is primitive; K' is conjugate to K if and only if n is odd. Put $\psi = 1 + \alpha^{n-1}$ and $\psi' = 1 + \alpha^{n+1}$. Since $\psi\beta\alpha^{n-1} = \beta\psi$, we see that $x \mapsto x^\psi$ gives a homomorphism of $K^\times$ into $K'^\times$, and hence a homomorphism of I_K into $I_{K'}$.

Similarly ψ' maps $K'^{\times}$ into $K^{\times}$ and $I_{K'}$ into I_K. Moreover, we have

$$\psi \sum_{k=0}^{n-1} \alpha^{-k} = \sum_{k=0}^{n-1} \alpha^k \psi' = 1 - \rho + \sum_{k=1}^{2n} \alpha^k,$$

so that

(3.1) $\qquad (\mathfrak{a}^{\Phi})^{\psi'} = (\mathfrak{a}^{\psi})^{\Phi'} = N(\mathfrak{a})\mathfrak{a}^{1-\rho} \quad (\mathfrak{a} \in I_K),$

(3.2) $\qquad (\mathfrak{b}^{\psi'})^{\Phi} = (\mathfrak{b}^{\Phi'})^{\psi} = N(\mathfrak{b})\mathfrak{b}^{1-\rho} \quad (\mathfrak{b} \in I_{K'}).$

PROPOSITION 6. *Let (K, Φ) be defined as above, and let $P = (A, C, \theta)$ be of type $(K, \Phi; t, \mathfrak{a})$. Then $(P_\sigma : P) = 1$ with $\sigma = \rho[M/K', \mathfrak{a}^{\psi}]^{-1}$, and the field of moduli M_0 of (A, C) is given by $M_0 = \{x \in M \mid x^\sigma = x\}$.*

Proof. By (3.1), we see that $\mathfrak{z} = \mathfrak{a}^{-\psi}$ and $w = N(\mathfrak{a})^{-1}$ satisfy the requirement of Proposition 5. Now K' has only two automorphisms. Therefore, by Proposition 2 and (3.1) of [8], we see that $F' \subset M_0$, and M_0 coincides with the field of moduli of (A, C, θ_F). This combined with Proposition 5 proves the last assertion.

REMARKS. (I) Suppose n is odd, and put $n = 2m + 1$. Then we can consider (K, Φ) with $\Phi \sim \sum_{k=-m}^{m} \alpha^k$. It can easily be seen that (K, Φ) has itself as the reflex. In this case, we have (3.1) and (3.2) with $\psi = \alpha^m + \alpha^{-m}$. We note also that K is the composite of F and an imaginary quadratic field whenever n is odd.

(II) Let $n = 7$ and $\Phi \sim 1 + \alpha + \alpha^2 + \alpha^3 + \alpha^{11} + \alpha^5 + \alpha^6$. Then (K, Φ) is primitive and $K' = L$. This gives an example of K which is a proper subfield of K'.

(III) There are some interesting facts about the 'relative ideal class group' of K/F of the above type, which we shall discuss in the Appendix.

Coming back to the general case without any assumption on the Galois group, we always have $[K' : \mathbf{Q}] \leqslant 2^n$ as shown in [7, I, 1.9]. Now in the extreme case where $[K' : \mathbf{Q}] = 2^n$, one can prove that

(3.3) $\qquad (x^{\Phi})^{\Phi'} = (N_{K/\mathbf{Q}}(x)x^{1-\rho})^{2^{n-1}} \quad (x \in K^{\times}).$

However, no universal formula seems to exist for $(y^{\Phi'})^{\Phi}$ with $y \in K'$.

4. With a fixed $(K, \Phi), J, J'$, and the map (2.5), we now assume that

(4.1) $\quad \mathrm{Gal}(K/J)$ *and* $\mathrm{Gal}(K'/J')$ *are cyclic;*

(4.2) $\quad J \subset F$ *and* $J' \subset F'$.

Furthermore, we put $2m = [K : J]$, and impose either one of the following conditions:

(4.3$_\mathrm{a}$) $\quad \mathfrak{d}_{K/F}$ *has exactly one prime factor;*

(4.3$_\mathrm{b}$) $\quad \mathfrak{d}_{K/F} \neq (1)$ *and m is odd.*

Of course $m = n$ if $J = \mathbf{Q}$.

THEOREM 1. *Let τ be a generator of $\mathrm{Gal}(K/J)$, and let σ be an element of $\mathrm{Gal}(M/J')$ such that $[\sigma] = \tau$. Let $\mathfrak{g}$ be an ideal in F such that $\mathfrak{g}^{\tau} = \mathfrak{g}$ and $\mathrm{Tr}_{F/\mathbf{Q}}(\mathfrak{d}_F^{-1}\mathfrak{g}) = \mathbf{Z}$. Suppose $\sigma^m = \rho$ on M if (4.3_{a}) does not apply. Then, under (4.3_{a}) or (4.3_{b}), the following assertions hold.*

(i) *σ is of order $[K : J]$.*

(ii) *There is a $P = (A, C, \theta)$ of type (K, Φ) such that $(P_\sigma : P) = 1$ and $\mathfrak{f}(P) = \mathfrak{r}\mathfrak{g}$ with $\mathfrak{r} \in I_J$. In particular $\mathfrak{f}(P) = \mathfrak{g}$ if $J = \mathbf{Q}$.*

(iii) *If (K, Φ) is primitive, the field of moduli M_J of (A, C, θ_J) for such a P is given by $M_J = \{x \in M \mid x^\sigma = x\}$.*

Proof. By Proposition 1, there is a P' of type (K, Φ) such that $\mathfrak{g} = \mathfrak{f}(P')$. Put $(P'_\sigma : P') = (b, \mathfrak{c})$. By (2.2), we have $\mathfrak{c}\mathfrak{c}^\rho = (b)$. This implies that $b \in N_{K/F}(K_v^\times)$ for every finite prime v of F unramified in K. The same is true also for infinite primes, since b is totally positive. Therefore, under (4.3_{a}), we have $b = aa^\rho$ with $a \in K$. Then $a^{-1}\mathfrak{c} = \mathfrak{x}^{1-\rho}$ with $\mathfrak{x} \in I_K$. Put $\mathfrak{y} = \mathfrak{x}\mathfrak{x}^\tau \ldots \mathfrak{x}^{\tau^{m-1}}$. Then $\mathfrak{y}^{1-\tau} = \mathfrak{x}^{1-\rho}$. Now there is a P such that $(P : P') = (s, \mathfrak{y})$ with a rational number s. By (iii) of Proposition 2, we have $(P_\sigma : P) = 1$, and by (2.2), $\mathfrak{f}(P) = sN_{K/F}(\mathfrak{y})^{-1}\mathfrak{f}(P') = sN_{K/J}(\mathfrak{x})^{-1}\mathfrak{g}$. Next assume that m is odd and $\sigma^m = \rho$. By (ii) of Proposition 2,

$$(P'_\rho : P') = (bb^\tau \ldots b^{\tau^{m-1}}, \mathfrak{c}\mathfrak{c}^\tau \ldots \mathfrak{c}^{\tau^{m-1}}).$$

By Proposition 4, $(P'_\rho : P') = (1, \mathfrak{a}^{1-\rho})$ with $\mathfrak{a} \in I_K$; hence $N_{F/J}(b) = xx^\rho$ and $\mathfrak{c}\mathfrak{c}^\tau \ldots \mathfrak{c}^{\tau^{m-1}} = x\mathfrak{a}^{1-\rho}$ with $x \in K$. Then $N_{F/J}(b^{-m}xx^\rho) = 1$, so that $b^{-m}xx^\rho = y^{\tau-1}$ with $y \in F$. Put $d = b^{\frac{1}{2}(1-m)}x$. Then $b = dd^\rho y^{1-\tau}$ and $N_{K/J}(d^{-1}\mathfrak{c}) = (1)$. Therefore $d^{-1}\mathfrak{c} = \mathfrak{z}^{1-\tau}$ with $\mathfrak{z} \in I_K$. Thus

$$(b, \mathfrak{c}) = (y^{1-\tau}, \mathfrak{z}^{1-\tau}),$$

and

$$(y^{-1}\mathfrak{z}\mathfrak{z}^\rho)^{1-\tau} = b^{-1}\mathfrak{c}\mathfrak{c}^\rho = (1).$$

Observe that if $\mathfrak{q} \in I_F$ and $\mathfrak{q}^\tau = \mathfrak{q}$, then $\mathfrak{q}^m = \mathfrak{r}\mathfrak{o}_F$ with $\mathfrak{r} \in I_J$. Then $\mathfrak{q} = \mathfrak{r}(\mathfrak{q}^{\frac{1}{2}(1-m)})^2$. Therefore $y^{-1}\mathfrak{z}\mathfrak{z}^\rho = \mathfrak{r}\mathfrak{w}^2$ with $\mathfrak{r} \in I_J$ and $\mathfrak{w} \in I_F$, and $\mathfrak{w}^\tau = \mathfrak{w}$. Now take P such that $(P : P') = (sy, \mathfrak{w}^{-1}\mathfrak{z})$ with $s \in \mathbf{Q}$. By (iii) of Proposition 2, we have $(P_\sigma : P) = 1$, and by (2.2), $\mathfrak{f}(P) = s\mathfrak{r}^{-1}\mathfrak{g}$. Thus we obtain P as in (ii) under (4.3_{a}) or (4.3_{b}). If $J = \mathbf{Q}$, the condition $\mathrm{Tr}_{F/\mathbf{Q}}(\mathfrak{d}_F^{-1}\mathfrak{f}(P)) = \mathrm{Tr}_{F/\mathbf{Q}}(\mathfrak{d}_F^{-1}\mathfrak{f}(P')) = \mathbf{Z}$ implies that $\mathfrak{f}(P) = \mathfrak{g}$. Put $\alpha = \sigma^{2m}$. Then $P_\alpha = P^\alpha$ and $(P_\alpha : P) = 1$. Therefore $\alpha = \mathrm{id}_M$, hence (i) is true. Assertion (iii) follows immediately from Proposition 3.

PROPOSITION 7. *Let $\mathfrak{g}$, τ, and σ be as in Theorem 1. Then the number of isomorphism classes of P such that $\mathfrak{f}(P) = \mathfrak{g}$ and $(P_\sigma : P) = 1$ is either 0 or equal to the number of all classes $(d, \mathfrak{e})$ such that $d \in J$, $\mathfrak{e}^2 = (d)$, and $\mathfrak{e}^\tau = \mathfrak{e}$.*

Proof. Fixing one such P, let $(P' : P) = (b, \mathfrak{c})$. Then $\mathfrak{f}(P') = \mathfrak{g}$ and $(P'_\sigma : P') = 1$ if and only if $\mathfrak{c}\mathfrak{c}^\rho = (b)$ and $(b^{\tau-1}, \mathfrak{c}^{\tau-1}) = 1$. If this is so,

$\mathfrak{b}^{\tau-1} = xx^\rho$ and $\mathfrak{c}^{\tau-1} = (x)$ with $x \in K$. Then $N_{K/J}(x) = 1$, so that $x = y^{1-\tau}$ with $y \in K$. Put $\mathfrak{d} = \mathfrak{b}yy^\rho$ and $\mathfrak{e} = y\mathfrak{c}$. Then $(\mathfrak{b}, \mathfrak{c}) = (\mathfrak{d}, \mathfrak{e})$, $\mathfrak{d}^\tau = \mathfrak{d}$, and $\mathfrak{e}^\tau = \mathfrak{e}$. This proves our proposition.

PROPOSITION 8. *The notation and assumptions being as in Theorem* 1, *suppose that* $\sigma^m = \rho$, $\mathfrak{q}^{\tau-1} \in P_K$ *for every prime factor* $\mathfrak{q}$ *of* $\mathfrak{d}_{K/F}$, *and* F *satisfies the following conditions:*
(4.4)   *the class number of* F *is odd;*
(4.5) $E_F^+ = N_{K/F}(E_K)$ *(see Notation).*
Let P *be such that* $(P_\rho : P) = 1$ *and* $\mathfrak{f}(P)^\tau = \mathfrak{f}(P)$. *Then* $(P_\sigma : P) = 1$.

Proof. By Theorem 1, there is a P' such that $\mathfrak{f}(P) = \mathfrak{f}(P')\mathfrak{x}$ with $\mathfrak{x} \in I_J$ and $(P'_\sigma : P') = 1$. Put $(P : P') = (\mathfrak{b}, \mathfrak{c})$. By (iii) of Proposition 2, we have $1 = (P_\rho : P) = (1, \mathfrak{c}^{\rho-1})$. Therefore $\mathfrak{c}^{\rho-1} = (x)$, $xx^\rho = 1$ with $x \in K$. Take $y \in K$ so that $x = y^{1-\rho}$. Changing $(\mathfrak{b}, \mathfrak{c})$ for $(yy^\rho \mathfrak{b}, y\mathfrak{c})$, we may assume that $\mathfrak{c}^\rho = \mathfrak{c}$. Then $\mathfrak{c} = \mathfrak{m}\mathfrak{q}_1 \ldots \mathfrak{q}_r$ with $\mathfrak{m} \in I_F$ and prime factors $\mathfrak{q}_i$ of $\mathfrak{d}_{K/F}$. Again by (iii) of Proposition 2, we have $(P_\sigma : P) = (\mathfrak{b}^{\tau-1}, \mathfrak{c}^{\tau-1})$. By our assumption, we have $\mathfrak{c}^{\tau-1} = \mathfrak{d}\mathfrak{m}^{\tau-1}$ with $\mathfrak{d} \in K$. Since $\mathfrak{f}(P) = \mathfrak{f}(P')\mathfrak{x}$, we have $(\mathfrak{b}) = \mathfrak{x}\mathfrak{c}^2$, so that $\mathfrak{d}\mathfrak{d}^\rho(\mathfrak{m}^{\tau-1})^2 = (\mathfrak{c}^{\tau-1})^2 = (\mathfrak{b}^{\tau-1})$. By (4.4), we have $\mathfrak{m}^{\tau-1} = (a)$ with $a \in F$. Put $u = \mathfrak{b}^{\tau-1}(a^2\mathfrak{d}\mathfrak{d}^\rho)^{-1}$. Then $u \in E_F^+$ and

$$(P_\sigma : P) = (u, \mathfrak{o}_K) = 1$$

by virtue of (4.5).

Let us now restrict ourselves to the case where K is a cyclic extension of $\mathbf{Q}$, and (K, Φ) is primitive. Then $K' = K$. We take $\mathbf{Q}$ to be both J and J'. For $\alpha \in \mathrm{Aut}(\mathbf{C})$, $[\alpha]$ is the restriction of α to K, so that we can often write α for $[\alpha]$.

Let σ be an element of $\mathrm{Gal}(M/\mathbf{Q})$ whose restriction to K generates $\mathrm{Gal}(K/\mathbf{Q})$. We are interested in those P for which $(P_\sigma : P) = 1$. For such P, we have $\mathfrak{f}(P)^\sigma = \mathfrak{f}(P)$. Now observe that there are only finitely many ideals $\mathfrak{g}$ in F such that

(4.6) $\mathrm{Tr}_{F/\mathbf{Q}}(\mathfrak{d}_F^{-1}\mathfrak{g}) = \mathbf{Z}$ *and* $\mathfrak{g}^\gamma = \mathfrak{g}$ *for all* $\gamma \in \mathrm{Gal}(K/\mathbf{Q})$.

Fix such a $\mathfrak{g}$, and consider a P such that

(4.7) $\mathfrak{f}(P) = \mathfrak{g}$ *and* $(P_\rho : P) = 1$.

The existence of such a P is guaranteed by Theorem 1 if σ can be chosen so that $\sigma^n = \rho$, under (4.3$_{a,b}$). To obtain another type of existence theorem, let us assume that

(4.8) $E_F^+ = E_F{}^2 = \{u^2 \mid u \in E_F\}$,

which is stronger than (4.5).

PROPOSITION 9. *Let r be the number of prime ideals of F ramified in K. Under (4.4) and (4.8), there are exactly 2^{r-1} isomorphism classes of P satisfying (4.7) for each $\mathfrak{g} \in I_F$ such that $\mathrm{Tr}_{F/\mathbf{Q}}(\mathfrak{d}_F^{-1}\mathfrak{g}) = \mathbf{Z}$.*

In this proposition, assumptions (4.1)–(4.3) are unnecessary.

Proof. Fix an element t satisfying (1.2). We can put $t\mathfrak{d}_{K/F} = \mathfrak{m}\mathfrak{o}_K$ with $\mathfrak{m} \in I_F$. Under (4.4) and (4.8), we can find $\mathfrak{b} \in I_F$ and a totally positive $s \in F$ such that $\mathfrak{g} = s\mathfrak{m}\mathfrak{d}_F\mathfrak{b}^2$. Let P_0 be of type $(st, \mathfrak{b}\mathfrak{o}_K)$. Then P_0 satisfies (4.7). Now P satisfies (4.7) if and only if $(P : P_0) = (b, \mathfrak{c})$ with $\mathfrak{c}\mathfrak{c}^\rho = (b)$ and $\mathfrak{c} \in I_0(K/F)$, where

$$I_0(K/F) = P_K\{\mathfrak{a} \in I_K \mid \mathfrak{a}^\rho = \mathfrak{a}\}.$$

By (4.8), such a $(b, \mathfrak{c})$ is completely determined by the ideal class of $\mathfrak{c}$. Thus the number of isomorphism classes of such P is the number of ideal classes $c \in I_0(K/F)/P_K$ such that $c^2 = 1$. Since h_F is odd, this is equal to $[I_0(K/F) : I_F P_K]$. Therefore our assertion follows from (a.11) of the Appendix.

Now there is a question of finding an element σ of $\mathrm{Gal}(M/\mathbf{Q})$ such that $\sigma^n = \rho$ and $\mathrm{Gal}(K/\mathbf{Q})$ is generated by the restriction of σ to K. Before answering this, we first prove the following proposition.

PROPOSITION 10. *Suppose F has exactly one prime ideal $\mathfrak{p}$ ramified in K. Let S be the Hilbert class field over K, and let $T = \{x \in S \mid x^\rho = x\}$. Then $E_F^+ = E_F{}^2$ if and only if there exists a prime factor of $\mathfrak{p}$ in S ramified over T.*

Proof. Let R denote the maximal abelian extension of F in which only archimedean primes are ramified. Suppose $E_F^+ \neq E_F{}^2$. Then R is totally imaginary, so that $R \not\subset T$ and $S = RT$. Therefore no finite prime of T is ramified in S. To prove the converse, take any prime factor $\mathfrak{q}$ of $\mathfrak{p}$ in S. Then the inertia group of $\mathfrak{q}$ relative to F is generated by an element α of order 2. Since $\alpha = \rho$ on K, we have $\alpha = \rho[S/K, \mathfrak{a}]$ with $\mathfrak{a} \in I_K$. Suppose $E_F^+ = E_F{}^2$. Then R is totally real, so that ρ is the identity on R. Since $\mathfrak{p}$ is unramified in R, α is the identity on R, and so $[S/K, \mathfrak{a}]$ is the identity on R. Therefore $\mathfrak{a}\mathfrak{a}^\rho = (x)$ with a totally positive element x of F. For the same reason as in the proof of Theorem 1, we have $x = yy^\rho$ with an element y of K, and therefore $y^{-1}\mathfrak{a} = \mathfrak{b}^{\rho-1}$ with $\mathfrak{b} \in I_K$. Then $\alpha = \gamma\rho\gamma^{-1}$ with $\gamma = [S/K, \mathfrak{b}]$. This means that ρ belongs to the inertia group of $\mathfrak{q}^\gamma$, that is, $\mathfrak{q}^\gamma$ is ramified over T.

PROPOSITION 11. *Let $\mathfrak{p}$ be as in Proposition 10. Suppose that K is cyclic over $\mathbf{Q}$ and $\mathfrak{p}$ is completely ramified over $\mathbf{Q}$. Suppose further that either $[M : K]$ is odd or $E_F^+ = E_F{}^2$. Then there is a cyclic subgroup H of $\mathrm{Gal}(M/\mathbf{Q})$ of order $2n$ such that $\rho \in H$ and $\mathrm{Gal}(M/\mathbf{Q}) = H\,\mathrm{Gal}(M/K)$.*

Proof. Put $M_1 = \{x \in M \mid x^\rho = x\}$. By Proposition 10, our assumptions guarantee a prime factor q of p in M ramified over M_1. Then we can take the inertia group of q over **Q** as H.

PROPOSITION 12. *The subgroup H of Proposition 11 is unique if (it exists and) one of the following two conditions is satisfied:*

(i) $[M : K]$ *is odd;*

(ii) *conditions* (4.3$_a$), (4.4), *and* (4.8) *are satisfied.*

Proof. Suppose $[M : K]$ is odd. Let σ be a generator of H. We have $\sigma^n = \rho$ as $\rho \in H$. Let H' be another subgroup of order $2n$ such that $\rho \in H'$ and $\mathrm{Gal}(M/\mathbf{Q}) = H' \, \mathrm{Gal}(M/K)$. Then H' contains an element τ such that $\tau = \sigma$ on K. Put $\tau = \sigma[M/K, \mathfrak{a}]$ with $\mathfrak{a} \in I_K$, and

$$\mathfrak{b} = \mathfrak{a}\mathfrak{a}^\sigma \ldots \mathfrak{a}^{\sigma^{n-1}}.$$

Then $\rho = \tau^n = \sigma^n[M/K, \mathfrak{b}]$, so that $\mathfrak{b}^{\Phi'} = (c)$ and $N(\mathfrak{b}) = cc^\rho$ with $c \in K$. Applying $1 - \sigma$, we obtain $(\mathfrak{a}^{1-\rho})^{\Phi'} = (c^{1-\sigma})$ and $c^{(1-\sigma)(1+\rho)} = 1$, so that $[M/K, \mathfrak{a}^{1-\rho}] = 1$. Now $[M/K, \mathfrak{a}\mathfrak{a}^\rho] = 1$, and hence $[M/K, \mathfrak{a}]^2 = 1$. Since $[M : K]$ is odd, we have $[M/K, \mathfrak{a}] = 1$, so that $H = H'$. Next assume that condition (ii) holds. Take any g satisfying (4.6). By Proposition 9, there is a unique $P = (A, C, \theta)$ satisfying (4.7). By Proposition 8, we have $(P_\sigma : P) = 1$ for $\sigma \in H$. By Proposition 3, H is the subgroup of $\mathrm{Gal}(M/\mathbf{Q})$ corresponding to the field of moduli of (A, C). Since P is unique, H must be unique.

5. To explain the above results in a more explicit form, let us consider the special case where K is a cyclic extension of **Q** whose conductor is a power of a prime $l \geqslant 2$. We exclude the case where $K = \mathbf{Q}(\sqrt{-1})$ to make our treatment uniform. (We could have excluded the case $n = 1$ at the beginning.) Let I_F (respectively I_K) denote the prime factor of l in F (respectively K). It can easily be verified that an ideal g in F satisfies (4.6) if and only if $\mathfrak{g} = I_F^e$ with $0 \leqslant e < n$. Note also that $N_{K/F}(E_K) = E_F^2$ as will be shown in Proposition A.5 below. We can find an element z of K such that $z^\rho = -z$ and z generates I_K. In fact, if $l = 2$, we can put $K = \mathbf{Q}(\zeta - \zeta^{-1})$ with $\zeta = \exp(2^{-\nu}\pi i)$. Then $z = \zeta - \zeta^{-1}$ has the required property. If l is odd, let $K \subset \mathbf{Q}(\zeta)$ with $\zeta = \exp(2\pi i/l^\nu)$. Since K is imaginary, $[\mathbf{Q}(\zeta) : K]$ is odd. Then we can put $z = N_{\mathbf{Q}(\zeta)/K}(\zeta - \zeta^{-1})$. Now put $y = zz^\rho$ and $z\mathfrak{d}_K = I_K^{2m}$. For each e such that $0 \leqslant e < n$, let P be of type $(uzy^{e-m}, \mathfrak{o}_K)$ with $u \in E_F$. Then

$$(5.1) \qquad \mathfrak{f}(P) = I_F^e, \quad (P_\rho : P) = 1,$$

and P is of type (K, Φ) if $\mathrm{Im}((uz)^{\tau_i}) > 0$ for all i. Such a unit u can be found for every Φ if $E_F^+ = E_F^2$. By Proposition 9, under (4.4) and (4.8),

P is characterized by (5.1), and by Proposition 8, $(P_\sigma : P) = 1$ for all σ in the subgroup H as in Proposition 11. We note also that the number of isomorphism classes considered in Proposition 7 is exactly

$$[E_F^+ : N_{K/F}(E_K)].$$

Suppose (4.4) and (4.8) are satisfied. Then we can determine $(P'_\gamma : P')$ for an arbitrary P' and $\gamma \in \mathrm{Gal}(M/\mathbf{Q})$ as follows. Let P' be of type $(t, \mathfrak{a})$, and put $t = suz$ with $s \in F$. Take P as above. Then $(P' : P) = (sy^{m-e}, \mathfrak{a}^{-1})$, and therefore, by (iii) of Proposition 2, we have

$$(5.2) \qquad\qquad (P'_\sigma : P') = (s^{\sigma-1}, \mathfrak{a}^{1-\sigma}) \quad (\sigma \in H),$$

noting that $y^{\sigma-1} = z^{(\sigma-1)(1+\rho)} \in N_{K/F}(E_K)$. Thus P'_σ is of type $(s^\sigma uz, \mathfrak{a}^\sigma)$. We can state this fact more symbolically as

$$(5.3) \qquad\qquad (suz, \mathfrak{a})_\sigma = (s^\sigma uz, \mathfrak{a}^\sigma) \quad (\sigma \in H).$$

Combining this with (2.4), we obtain P'_γ for every $\gamma \in \mathrm{Gal}(M/\mathbf{Q})$.

We now discuss a few examples.

(I) First suppose only that K is cyclic over $\mathbf{Q}$ and has exactly one prime ideal ramified over F, discarding other conditions for the moment. Let $\Phi = \sum_{i=0}^{n-1} \tau^{-i}$ with a generator τ of $\mathrm{Gal}(K/\mathbf{Q})$. Then (K, Φ) is primitive. Let $P = (A, C, \theta)$, and let P and P' be of type (K, Φ). If $\mathfrak{f}(P) = \mathfrak{f}(P')$, then $(P' : P) = (1, \mathfrak{x}^{1-\rho})$ with $\mathfrak{x} \in I_K$ as shown in the proof of Theorem 1. Put $\mathfrak{y} = \mathfrak{x}^{1-\tau}$. Then $\mathfrak{y}^{\Phi'} = \mathfrak{x}^{1-\rho}$. By (2.4), we have $(P' : P^\alpha) = 1$ if $\alpha = [M/K, \mathfrak{y}]$. This means that condition (2.8) holds in the present case. Therefore, if M_0 denotes the field of moduli of (A, C), then $[M : M_0] = [K : \mathbf{Q}]$ whenever $\mathfrak{f}(P)^\gamma = \mathfrak{f}(P)$ for all $\gamma \in \mathrm{Gal}(K/\mathbf{Q})$. Moreover, the fields M_0 for all such P with the same $\mathfrak{f}(P)$ are transformed to each other by elements of $\mathrm{Gal}(M/\mathbf{Q})$. Therefore, under (4.4) and (4.8), M_0 has a real archimedean prime. However, there is a counter-example, if (4.8) is dropped. In fact, let $K = \mathbf{Q}(\zeta)$ with $\zeta = e^{2\pi i/29}$. Then h_F is odd (Kummer [3]), but it can be verified that $[E_F^+ : E_F^2] = 8$. Then we can show that M_0 has no real archimedean prime for any P (with Φ of the present type). Thus there is no subgroup H as described in Proposition 11 in this case.

(II) Next we consider the jacobian A_0 of a curve $y^2 = 1 - x^l$ with an odd prime l. In this case, $K = \mathbf{Q}(\zeta)$ with $\zeta = e^{2\pi i/l}$. Define

$$\theta_0 \colon K \to \mathrm{End}(A_0) \otimes \mathbf{Q}$$

such that $\theta_0(\zeta)$ corresponds to the automorphism $(x, y) \mapsto (\zeta x, y)$. Take the canonical polarization C_0 of A_0, and put $P_0 = (A_0, C_0, \theta_0)$. As shown in [10, §8.4], we have $\Phi = \sum_{\nu=1}^n \tau_\nu$ and $\Phi' = \sum_{\nu=1}^n \tau_\nu^{-1}$, where $\zeta^{\tau_\nu} = \zeta^\nu$ and $n = \frac{1}{2}(l-1)$; (K, Φ) is primitive. Since (A_0, C_0) is self-dual, we have

$\mathfrak{f}(P_0) = (1)$. Obviously (A_0, C_0) is rational over $\mathbf{Q}$, so that $M = K$, and $(N(\mathfrak{x}), \mathfrak{x}^{\Phi'}) = 1$ for every $\mathfrak{x} \in I_K$. By taking an explicit basis of the homology group of the curve, we can easily show that P_0 is of type

$$((\zeta - \zeta^{-1})/l, \mathfrak{o}_K).$$

(Under conditions (4.4) and (4.8), this follows from Proposition 9.) Now take an arbitrary $\mathfrak{a} \in I_K$. Then we can find a P such that

$$(P : P_0) = (1, \mathfrak{a}^{1-\rho}).$$

Then $\mathfrak{f}(P) = (1)$, and by (iii) of Proposition 2, we have

$$(P_\gamma : P) = (1, \mathfrak{a}^{(1-\rho)(\gamma-1)})$$

for every $\gamma \in \mathrm{Gal}(K/\mathbf{Q})$. We can actually find examples of $\mathfrak{a}$ such that $\mathfrak{a}^{(1-\rho)(\gamma-1)}$ is principal only when $\gamma = 1$ (see Appendix). For such a $P = (A, C, \theta)$, the field of moduli M_0 of (A, C) is K. Thus the behaviour of P under $\mathrm{Gal}(K/\mathbf{Q})$ is completely different from that of P_0, despite the fact that $\mathfrak{f}(P) = \mathfrak{f}(P_0) = (1)$.

One can also consider the jacobian of a curve $x^l = y^i(y - 1)$, or more generally the factors of the jacobian of a curve $ax^m + by^n + c = 0$ with various m and n, as considered by Weil [11]. It may be an interesting problem to determine to what $(t, \mathfrak{a})$ they belong.

6. Still assuming that K is cyclic over $\mathbf{Q}$ and (K, Φ) is primitive, we now consider the problem of finding a model of (A, C) defined over its field of moduli, for a given $P = (A, C, \theta)$ of type (K, Φ). We let M denote the field of moduli of P as before, $M_{\mathbf{A}}^\times$ denote the idele group of M, and put $g(x) = N_{M/K}(x)^{\Phi'}$ for $x \in M_{\mathbf{A}}^\times$.

PROPOSITION 13. *Suppose $(P_\sigma : P) = 1$ for an element σ of $\mathrm{Gal}(M/\mathbf{Q})$ whose restriction to K generates $\mathrm{Gal}(K/\mathbf{Q})$ and such that $\sigma^n = \rho$. Let ψ be a Hecke character of $M_{\mathbf{A}}^\times$ satisfying the following three conditions:*
(6.1) $\psi(x)^\sigma = \psi(x^\sigma)$ *if* $x \in M_{\mathbf{A}}^\times$ *and* $x_\infty = 1$;
(6.2) $\psi(x_\infty) = 1/g(x)_{\infty 1}$ *for all* $x \in M_{\mathbf{A}}^\times$;
(6.3)   *if* $x \in M_{\mathbf{A}}^\times$ *and* $x_\infty = 1$, *then* $\psi(x) \in K^\times$, $\psi(x)\psi(x)^\rho = N(x\mathfrak{o}_M)$, *and* $\psi(x)\mathfrak{o}_M = g(x)\mathfrak{o}_M$.
Here x_∞ denotes the archimedean part of x, and $y_{\infty 1}$, for $y \in K_{\mathbf{A}}^\times$, denotes the component of y at the archimedean prime corresponding to the natural injection $K \to \mathbf{C}$. Then there is a model of (A, C) rational over

$$M_0 = \{x \in M \mid x^\sigma = x\}$$

such that (A, C, θ) determines ψ in the sense of [8].

Proof. Let $M_1 = \{x \in M \mid x^\rho = x\}$. By [8, Theorem 10], we can choose a model of (A, C, θ_F) rational over M_1 such that $P = (A, C, \theta)$ determines ψ.

Then $P_\rho = P$. By [8, Proposition 1], P_σ determines the same ψ. Let λ be an isomorphism of P to P_σ. By [8, Theorem 5], λ is rational over M. Put $\mu = \lambda^{\sigma^{n-1}} \circ \ldots \circ \lambda^\sigma \circ \lambda$. Then μ is an automorphism of P, so that $\mu = \theta(\xi)$ with a root of unity ξ in K. Then $\lambda^{\sigma^{2n-1}} \circ \ldots \circ \lambda^\sigma \circ \lambda = \mu^\rho \circ \mu = \theta(\xi^\rho \xi) = 1$. Applying the Weil criterion [12] to (A, C) and λ, we obtain the desired conclusion.

Thus our next problem is to construct a character satisfying (6.1)–(6.3). We consider here only the special case where $K = \mathbf{Q}(\zeta)$, and $\zeta = e^{2\pi i/l}$ with an odd prime l, although our method is applicable to some other cases. Let $\mathfrak{l} = (1 - \zeta)$, and

$$W = \{x \in M_{\mathbf{A}}^\times \mid x_\mathfrak{l} \equiv 1 \ (\mathrm{mod}\ \mathfrak{l}^2)\},$$

where $x_\mathfrak{l}$ is the $\mathfrak{l}$-component of x (which is of course meaningful). Then $M_{\mathbf{A}}^\times = M^\times W$. Now we define a homomorphism $\alpha \colon M_{\mathbf{A}}^\times \to K$ as follows. First we put $\alpha(x) = g(x)$ for $x \in M^\times$. Next let $x \in W$. Since M is the class field over K corresponding to $I_0(\Phi')$ of §2, there is an element β of K such that $g(x)\mathfrak{o}_K = \beta \mathfrak{o}_K$ and $N(x\mathfrak{o}_M) = \beta\beta^\rho$. Now

$$\beta\beta^\rho = N(N_{M/K}(x)\mathfrak{o}_K) \equiv 1 \ (\mathrm{mod}\ l\mathfrak{o}_\mathfrak{l}),$$

so that $\beta \equiv \pm 1 \ (\mathrm{mod}\ l\mathfrak{o}_\mathfrak{l})$, where $\mathfrak{o}_\mathfrak{l}$ is the $\mathfrak{l}$-completion of $\mathfrak{o}_K$. Changing β for its suitable multiple by a root of unity $\pm \zeta^\nu$, we may assume that $\beta \equiv 1 \ (\mathrm{mod}\ \mathfrak{l}^2\mathfrak{o}_\mathfrak{l})$. Such a β is uniquely determined for x. Then we put $\beta = \alpha(x)$. One can easily verify the compatibility of α on $M^\times \cap W$ and the continuity of α on $M_{\mathbf{A}}^\times$. Now we define $\psi \colon M_{\mathbf{A}}^\times \to \mathbf{C}$ by $\psi(x) = (\alpha(x)/g(x))_{\infty 1}$. Then ψ is a Hecke character of $M_{\mathbf{A}}^\times$ satisfying (6.2) and (6.3) (cf. [8, §1]). Moreover, we have $g(x)^\sigma = g(x^\sigma)$ and $\alpha(x)^\sigma = \alpha(x^\sigma)$ for every $x \in M_{\mathbf{A}}^\times$, so that ψ satisfies (6.1).

We can actually apply the results of [8, §§5, 6] to the present situation, if $l \equiv 3 \ (\mathrm{mod}\ 4)$. To see this, let Z_K denote the group of all roots of unity in K. Then there is a unique homomorphism

$$s \colon (\mathfrak{o}_K/\mathfrak{l}^2)^\times \to Z_K$$

such that $s(\xi \bmod \mathfrak{l}^2) = \xi$ for all $\xi \in Z_K$. Let U be the maximal compact subgroup of the finite part of $K_{\mathbf{A}}^\times$, and define $h \colon U \to Z_K$ by

$$h(x) = s(x_\mathfrak{l} \bmod \mathfrak{l}^2).$$

Then h satisfies condition (5.2′) of [8], and $h(x)^\tau = h(x^\tau)$ for all $x \in U$ and all $\tau \in \mathrm{Gal}(K/\mathbf{Q})$. Therefore, as remarked in [8, p. 532], if

$$[M : M_0] = [K : \mathbf{Q}]$$

for the field of moduli M_0 of (A, C), then (A, C) has a model over M_0, even if M_0 has no real archimedean prime. Finally it may be mentioned that a model over the field of moduli is often guaranteed by [9, Theorem 9.5] when n is odd.

7. There is one more interesting class of CM-type, for which the behaviour under an automorphism can be determined in a natural way. We start from two arbitrary CM-types (K, Φ_1) and (L, Ψ_1) with the same maximal real subfield F. We assume that $K \neq L$, and let S denote the composite field KL. Let $\{\tau_1, \ldots, \tau_n\}$ and $\{\chi_1, \ldots, \chi_n\}$ be isomorphisms of K and L into $\mathbf{C}$ which form Φ_1 and Ψ_1, respectively. We arrange them so that $\tau_i = \chi_i$ on F. Now take an integer r such that $1 \leqslant r < n$, and define isomorphisms α_i, β_i of S into $\mathbf{C}$ by

$$\alpha_i = \begin{cases} \tau_i & \text{on } K, \\ \chi_i & \text{on } L, \end{cases} \qquad \beta_i = \begin{cases} \tau_i \rho & \text{on } K, \\ \chi_i & \text{on } L \end{cases} \qquad (1 \leqslant i \leqslant r),$$

$$\alpha_i = \begin{cases} \tau_i & \text{on } K, \\ \chi_i & \text{on } L, \end{cases} \qquad \beta_i = \begin{cases} \tau_i & \text{on } K, \\ \chi_i \rho & \text{on } L \end{cases} \qquad (r < i \leqslant n).$$

Let $\Phi_0 \sim \sum_{i=r+1}^n \tau_i$, $\Psi_0 \sim \sum_{i=1}^r \chi_i$, $\Xi \sim \sum_{i=1}^n (\alpha_i + \beta_i)$ with the notation of [7, I, 1.1]. Then (S, Ξ) is a CM-type. Let (K', Φ_0'), (L', Ψ_0'), and (S', Ξ') be the reflexes of (K, Φ_0), (L, Ψ_0), and (S, Ξ), respectively. Then $S' = K'L'$, and

$$(7.1) \qquad x^{\Xi'} = N_{S'/K'}(x)^{\Phi_0'} N_{S'/L}(x)^{\Psi_0'} \qquad (x \in S').$$

For the proof, see [5, 5.16].

Let us now consider a structure (A, C, θ) of type $(S, \Xi; t, \mathfrak{a})$ with $t \in S$ and $\mathfrak{a} \subset S$. Here we drop the assumption that $\mathfrak{a}$ is a fractional ideal, and take an arbitrary lattice $\mathfrak{a}$ in S. Moreover, take finitely many points $\{x_j\}$ on A of finite order, and consider a structure $Q = (A, C, \theta, \{x_j\})$. We say that Q is of type $(S, \Xi; t, \mathfrak{a}, \{q_j\})$ with $q_j \in S/\mathfrak{a}$ if the isomorphism $\mathbf{C}^n/\Xi(\mathfrak{a})u \to A$ considered in §1 maps $\Xi(q_j)u$ onto t_j. We see, from (7.1), that the field of moduli of Q is contained in $K'_{\mathrm{ab}}L'_{\mathrm{ab}}$, where X_{ab} denotes the maximal abelian extension of X. For every $\sigma \in \mathrm{Aut}(\mathbf{C})$, we put $Q^\sigma = (A^\sigma, C^\sigma, \theta^\sigma, \{x_j^\sigma\})$.

Theorem 2. *Let σ be an element of $\mathrm{Aut}(\mathbf{C}/L')$ such that $\sigma = \rho$ on K'. Take $v \in K_{\mathbf{A}}'^{\times}$ and $w \in L_{\mathbf{A}}'^{\times}$ such that $\sigma\rho = [v, K']$ on K'_{ab} and $\sigma = [w, L']$ on L'_{ab}. Put $y = v^{\Phi_0'} w^{\Psi_0'}$. Then Q^σ is of type*

$$(7.2) \qquad (S, \Xi^\sigma; \lambda t, y^{-1}\mathfrak{a}, \{y^{-1}q_j\}),$$

where λ is a totally real element of S whose description will be given below.

Here $[u, X]$ denotes the element of $\mathrm{Gal}(X_{\mathrm{ab}}/X)$ corresponding to $u \in X_{\mathbf{A}}^{\times}$.

To determine λ, we consider (F, Θ) with $\Theta \sim \sum_{i=1}^r \tau_{i0}$, where τ_{i0} denotes the restriction of τ_i to F. Let (F', Θ') be the reflex of (F, Θ). Then it can easily be seen (cf. [5, (5.14.9)]) that

$$(7.3) \qquad x^{\Psi_0'(1+\rho)} = N_{L'/F'}(x)^{\Theta'} \quad (x \in L'),$$

$$(7.4) \qquad x^{\Phi_0'(1+\rho)} = N_{K'/\mathbf{Q}}(x)/N_{K'/F'}(x)^{\Theta'} \quad (x \in K').$$

Let s denote the element of $F_\infty'^\times$ of which the component at the natural embedding $F' \to \mathbf{R}$ is -1 and all other components are 1. Then $[s, F'] = \rho$ on F_{ab}', so that $\sigma = [sN_{K'/F'}(v), F'] = [N_{L'/F'}(w), F']$ on F_{ab}'. Therefore $N_{L'/F'}(w) = sgN_{K'/F'}(v)$ with $g \in F_c'$, where $X_c = \{u \in X_{\mathbf{A}} \mid [u, X] = 1\}$. With y as in our theorem, we have

$$(7.5) \qquad yy^\rho = N_{K'/\mathbf{Q}}(v)(sg)^{\Theta'} = N_{L'/\mathbf{Q}}(w)N_{F'/\mathbf{Q}}(sg)^{-1}(sg)^{\Theta'},$$

so that $s^{\Theta'}yy^\rho \in \mathbf{Q}_{\mathbf{A}}^\times F_c$. Let T be the maximal real subfield of S. We now define a homomorphism

$$(7.6) \qquad \mu_{T,U} : \mathbf{Q}_{\mathbf{A}}^\times T_c \to T^\times/(E_T^+ \cap U)$$

for every open subgroup U of $T_{\mathbf{A}}^\times$ of the type $U = U_0 T_{\infty+}^\times$ with an open compact subgroup U_0 of the finite part of $T_{\mathbf{A}}^\times$. First, for $a \in \mathbf{Q}_{\mathbf{A}}^\times$, take $\alpha \in \mathbf{Q}^\times$ such that $a\mathbf{Z} = \alpha\mathbf{Z}$ and $(\alpha a)_\infty > 0$, and put $\mu_{T,U}(a) = \alpha$. Next, observe that $T_c \subset T^\times U$. For $b \in T_c$, put $b = \beta u$ with $\beta \in T$ and $u \in U$. Then we put $\mu_{T,U}(b) = \beta \bmod E_T^+ \cap U$. Now the constant λ of (7.2) can be determined by

$$(7.7) \qquad \lambda = -\mu_{T,U}(s^{\Theta'}yy^\rho)$$

with a sufficiently small U. The smallness of U is enough if

$$E_T^+ \cap U \subset \{N_{S/T}(\varepsilon) \mid \varepsilon \in E_S, \varepsilon\mathfrak{a} = \mathfrak{a}, (\varepsilon-1)q_j \in \mathfrak{a} \text{ for all } j\}.$$

(Such a U of course exists.) Since $s^{\Theta'}yy^\rho \in \mathbf{Q}_{\mathbf{A}}^\times F_c$, we can also put $\lambda = -\mu_{F,V}(s^{\Theta'}yy^\rho)$ with a sufficiently small open subgroup V of $F_{\mathbf{A}}^\times$.

We note also that $s^{\Theta'}$ is the element of $F_\infty^\times$ whose component at τ_{i0} is -1 or 1 according as $i \leqslant r$ or $i > r$.

The above theorem is closely connected with a theorem on the behaviour of arithmetic automorphic functions under an anti-holomorphic automorphism (Shih [4, Theorems 2, 2']), which was stated as a conjecture in [9]. Actually this theorem and the above theorem are interdependent, and the proofs for both theorems must be given at the same time. For details, the reader is referred to [4]. We mention here only a direct proof of a weaker result that *the statement of Theorem 2 holds at least with $y^{-1}z$ in place of y^{-1} in (7.2), where z is an element of $F_{\mathbf{A}}^\times$ such that $z^2 = 1$*. (This itself is closely connected with [4, Proposition 3] (which is the same as [9, Theorem A]).) To prove this, let θ_X denote the restriction of θ to a subfield X of S, and let $Q_X = (A, C, \theta_X, \{x_j\})$. Then Q_L (respectively Q_K, Q_T) belongs to an $(n-r)$- (respectively r-, n-) dimensional family of abelian varieties whose behaviour under $\mathrm{Gal}(L'_{\mathrm{ab}}/L')$ (respectively $\mathrm{Gal}(K'_{\mathrm{ab}}/K')$, $\mathrm{Gal}(\mathbf{Q}_{\mathrm{ab}}/\mathbf{Q})$) can be explicitly given. Moreover the behaviour of Q_K under ρ can easily be described. Combining all these, we can determine Q^σ up to the ambiguity of z. It is noteworthy that the analysis of the behaviour of infinitely many different isogeny classes of abelian varieties is necessary for the proof of Theorem 2, or even of the weaker result, which concerns a single isogeny class.

As remarked in [6] (especially on p. 172), an abelian variety of the above type is a special member of a family of which even a generic member behaves in a peculiar manner under the complex conjugation. It is desirable to have deeper understanding of this phenomenon, possibly in connection with the above theorem.

Appendix

We collect here some results concerning the ideal class group and related objects of an algebraic number field, especially of the type K as in the text. Most of the results unrelated to CM-types are perhaps well known (cf. Hasse [2], for example), but we selected and organized them so as to satisfy our need and to fit our formulation in the text.

PROPOSITION A.1. *Let K be a cyclic extension of an algebraic number field k, let τ be a generator of $\mathrm{Gal}(K/k)$, and let*

$$I(K/k) = \{\mathfrak{a} \in I_K \mid \mathfrak{a}^{1-\tau} \in P_K\},$$
$$I_0(K/k) = P_K\{\mathfrak{a} \in I_K \mid \mathfrak{a}^\tau = \mathfrak{a}\}.$$

Then

(a.1) $I(K/k)/I_0(K/k) \simeq [E_k \cap N_{K/k}(K^\times)]/N_{K/k}(E_K),$

(a.2) $[I_0(K/k) : P_K] = h_k[K : k]^{-1}[E_k : N_{K/k}(E_K)]^{-1}2^r \prod_{\mathfrak{p}} e(\mathfrak{p}),$

where r is the number of archimedean primes of k ramified in K, $\prod_{\mathfrak{p}}$ is the product taken over all finite primes $\mathfrak{p}$ of k ramified in K, and $e(\mathfrak{p})$ is the ramification index at $\mathfrak{p}$.

This is well known in the classical proof of class field theory. For the proof, see Chevalley [1, pp. 402–6].

Now with F and K as in the text, let P_F^+ denote the subgroup of P_F consisting of all principal ideals generated by totally positive elements. Then $[P_F : P_F^+] = [E_F^+ : E_F{}^2]$, where $E_F{}^2 = \{e^2 \mid e \in E_F\}$. Put

(a.3) $$2^s = [E_F^+ : E_F{}^2].$$

Now $(I_F \cap P_K)/P_F$ is the kernel of the natural map of I_F/P_F into I_K/P_K. If $\mathfrak{a} \in I_F \cap P_K$ and $\mathfrak{a}\mathfrak{o}_K = b\mathfrak{o}_K$ with $b \in K$, then $b^{1-\rho}$ is a root of unity, and the map $\mathfrak{a} \mapsto \mathfrak{b}^{1-\rho}$ gives an injection $(I_F \cap P_K)/P_F \to Z_K/E_K{}^{1-\rho}$, where Z_K denotes the group of all roots of unity in K. Put

(a.4)
$$2^\eta = [I_F \cap P_K : P_F].$$

Since $E_K{}^{1-\rho}$ contains all squares of roots of unity in K, we have $\eta \leqslant 1$, and obviously

(a.5)
$$[I_K : I_F P_K] = 2^\eta h_K/h_F.$$

We now consider the map $N_{K/F} : I_K \to I_F$, and put

(a.6)
$$2^\delta = [I_F : P_F^+ N_{K/F}(I_K)],$$

(a.7)
$$R_K = \{\mathfrak{a} \in I_K \mid N_{K/F}(\mathfrak{a}) \in P_F\},$$

(a.8)
$$R_K^+ = \{\mathfrak{a} \in I_K \mid N_{K/F}(\mathfrak{a}) \in P_F^+\}.$$

Then $\delta = 0$ or 1 according as $\mathfrak{d}_{K/F} \neq (1)$ or $\mathfrak{d}_{K/F} = (1)$. Since

$$I_F = P_F N_{K/F}(I_K),$$

we see that $[R_K : R_K^+] = 2^{\varepsilon - \delta}$, and $N_{K/F}$ gives an isomorphism of I_K/R_K onto I_F/P_F, which maps $I_F R_K$ onto all the square classes in I_F/P_F. Let 2^φ denote the number of elements of order 1 or 2 in I_F/P_F. Then $[I_K : I_F R_K] = 2^\varphi$. Further $2^{\varepsilon - \delta}[R_K^+ : P_K] = [R_K : P_K] = h_K/h_F$. From these and (a.5), we obtain the following proposition.

PROPOSITION A.2. *If h_K/h_F is odd, then $\varepsilon = \delta$ and $\varphi \leqslant \eta$. Especially, if further $\eta = 0$, then h_F is odd.*

This, combined with Proposition A.5 below, generalizes the classical result of Kummer (cf. Hass [2, Satz 45]). We have also obviously the following proposition.

PROPOSITION A.3. *If h_F is odd, then $\eta = 0$, $I_K = I_F R_K$, and*

$$P_K = I_F P_K \cap R_K.$$

Now put

(a.9)
$$2^\beta = [I(K/F) : I_0(K/F)], \quad 2^\gamma = [I_0(K/F) : I_F P_K].$$

From (a.1) and (a.2), we obtain

(a.10)
$$\beta \leqslant \varepsilon,$$

(a.11)
$$2^\gamma = 2^{\eta + t - 1}[N_{K/F}(E_K) : E_F{}^2],$$

where t is the number of prime ideals of F ramified in K.

PROPOSITION A.4. *If $2^{-\eta} h_F$ is odd, then $2^{\beta + \gamma}$ is the number of elements of order 1 or 2 in I_K/P_K.*

Proof. If $\mathfrak{a} \in I(K/F)$, then $\mathfrak{a}^2 = \mathfrak{a}^{1+\rho}\mathfrak{a}^{1-\rho} \in I_F P_K$. Conversely, suppose $\mathfrak{a}^2 \in I_F P_K$. Then $\mathfrak{a}^{1-\rho} = \mathfrak{a}^2 \mathfrak{a}^{-1-\rho} \in I_F P_K$, so that $\mathfrak{a}^{1-\rho} = bc$ with $b \in K$ and $c \in I_F$. Then $c^2 \in P_K$. If $[I_F P_K : P_K]$ is odd, we see that $c \in P_K$, and so $\mathfrak{a}^{1-\rho} \in P_K$. This shows that

$$I(K/F) = \{\mathfrak{a} \in I_K \mid \mathfrak{a}^2 \in I_F P_K\}.$$

Since $I_K/I_F P_K$ is a quotient of I_K/P_K by a subgroup of odd order, we obtain our assertion.

Let us now consider the map $1-\rho : I_K/P_K \to R_K^+/P_K$. The kernel is $I(K/F)/P_K$, so that

$$2^{\eta-\beta-\gamma} h_K/h_F = [I_K : I(K/F)] \leqslant [R_K^+ : P_K] = 2^{\delta-\varepsilon} h_K/h_F.$$

It follows that $[R_K^+ : I_K^{1-\rho} P_K] = 2^{\beta+\gamma+\delta-\varepsilon-\eta}$ and

(a.12) $$\varepsilon + \eta \leqslant \beta + \gamma + \delta.$$

PROPOSITION A.5. *Suppose K has exactly one prime ideal ramified over F. Then the map $1-\rho : I_K/P_K \to R_K^+/P_K$ is surjective, $\delta = 0$, $\varepsilon + \eta = \beta + \gamma$, $E_F^+ \subset N_{K/F}(K^\times)$, and $2^\beta = [E_F^+ : N_{K/F}(E_K)]$. Moreover, if the ramified prime is principal (that is, if $\gamma = 0$), then $\eta = 0$, $\beta = \varepsilon$, $N_{K/F}(E_K) = E_F^2$, and $E_K = Z_K E_F$.*

Proof. Obviously $\delta = 0$. By (a.1), (a.11), and (a.12), we have

$$[E_F^+ : E_F^2] = 2^\varepsilon \leqslant 2^{\beta+\gamma-\eta} \leqslant 2^\beta [N_{K/F}(E_K) : E_F^2] = [E_F^+ \cap N_{K/F}(K^\times) : E_F^2],$$

from which we obtain the first series of assertions. (One can prove them also by the same idea as in the proof of Theorem 1.) The remaining part follows from (a.11) in a straightforward fashion.

In particular, Proposition A.5 applies to any imaginary cyclotomic field whose conductor is a power of a prime. By Proposition A.4, assuming h_F is odd, one knows that h_K/h_F is even if and only if $E_F^+ \neq E_F^2$.†

Put $C_K = I_K/P_K$ and $C_F = I_F/P_F$. If $\eta = 0$, C_F can be embedded in C_K, so that $I_K/I_F P_K$ can be identified with C_K/C_F, and may be called *the relative ideal class group of K/F.*

We now consider a CM-type (K, Φ) and its reflex (K', Φ'). Put

$$I(\Phi') = \{\mathfrak{a} \in I_{K'} \mid \mathfrak{a}^{\Phi'} \in P_K\},$$

$$I_0(\Phi') = \{\mathfrak{a} \in I_{K'} \mid \mathfrak{a}^{\Phi'} = b\mathfrak{o}_K, N(\mathfrak{a}) = bb^\rho \text{ with } b \in K\}.$$

Then $I_{F'} P_{K'} \subset I_0(\Phi') \subset I(\Phi')$. If $\mathfrak{a} \in I(\Phi')$ and $\mathfrak{a}^{\Phi'} = (b)$ with $b \in K$, then $N(\mathfrak{a})^{-1} bb^\rho \in E_F^+$, and the map $\mathfrak{a} \mapsto N(\mathfrak{a})^{-1} bb^\rho$ gives an injective homomorphism

(a.13) $$I(\Phi')/I_0(\Phi') \to E_F^+/N_{K/F}(E_K).$$

† This fact was kindly pointed out to the author by K. Iwasawa.

We note also that Φ' maps $I_{K'}/I(\Phi')$ into R_K^+/P_K. As mentioned in §2, $I_0(\Phi')$ is the ideal group corresponding to the field of moduli M. Since $[R_K^+ : P_K] = 2^{\delta-s}h_K/h_F$, we obtain the following proposition.

PROPOSITION A.6. *Condition* (2.8) *is equivalent to*

(a.14) $\quad [I_0(\Phi') : I_{F'}P_{K'}] = 2^{\eta'-\delta}(h_{K'}/h_{F'})(h_K/h_F)^{-1}[N_{K/F}(E_K) : E_F^2],$

where η' is defined by $2^{\eta'} = [I_{F'} \cap P_{K'} : P_{F'}].$

As a simple example, let us consider the case where K is cyclic over **Q** and $\Phi \sim \sum_{i=0}^{n-1} \tau^{-i}$ with a generator τ of $\mathrm{Gal}(K/\mathbf{Q})$. Since $\mathfrak{a}^{\Phi'(1-\tau)} = \mathfrak{a}^{1-\rho}$, we see easily that $I(\Phi') \subset I(K/F)$ and $I_0(\Phi') \subset I_0(K/F)$. For example, if K has exactly one prime ideal ramified over F, and this is principal, then $I_0(\Phi') = I_F P_K$, and (a.14) holds by virtue of Proposition A.5. As another example, we obtain the following proposition.

PROPOSITION A.7. *Let (K,Φ) be defined as in §3 with K contained in a Galois extension L of **Q** such that $\mathrm{Gal}(L/\mathbf{Q})$ is a dihedral group of order $4n$. Let d_X denote the discriminant of a number field X. Then the following assertions hold.*

(i) $d_K/d_F = d_{K'}/d_{F'}$ *and* $[E_K : E_F]^{-1}h_K/h_F = [E_{K'} : E_{F'}]^{-1}h_{K'}/h_{F'}.$

(ii) $I_0(\Phi') \subset I_0(K'/F')$ *and* $I(\Phi') \subset I(K'/F')$; $I(\Phi') \subset I_0(K'/F')$ *if $n = 2$.*

(iii) *If n is odd, $n = 2$, or $\mathfrak{d}_{K/F}$ does not divide 2, then $\eta = \delta = 0$ and $E_K = Z_K E_F$.*

Proof. If n is odd, K/F is conjugate to K'/F' over **Q**, so that (i) is obvious. Suppose n is even, and let $L(s,\chi)$ be the L-function of F with the character χ of order 2 corresponding to the extension K. Now the character of $\mathrm{Gal}(L/\mathbf{Q})$ induced from χ does not depend on the choice of K. (This is false if n is odd.) Observe that

$$L(1,\chi) = 2^{n-1}\pi^n|d_F/d_K|^{\frac{1}{2}}[E_K : E_F]^{-1}h_K/h_F$$

and $|d_K/d_F|^{\frac{1}{2}s}$ occurs in the Γ-factor of $L(s,\chi)$. Then we obtain (i). To prove (ii), let $\mathfrak{a} \in I_{K'}$ and $\mathfrak{a}^{\Phi'} = (b)$ with $b \in K$. By (3.2), $\mathfrak{a}^{1-\rho} = (c)$ with $c = N(\mathfrak{a})^{-1}b^\psi$, hence $I(\Phi') \subset I(K'/F')$. If $n = 2$ or $bb^\rho = N(\mathfrak{a})$, we have $cc^\rho = 1$. Writing $c = d^{\rho-1}$ with $d \in K'$, we have $(d\mathfrak{a})^\rho = d\mathfrak{a}$. This proves (ii). To prove (iii), first note that the order of Z_K is 2, 4, or 6, and the latter two cases can occur only when n is odd. Suppose n is odd and Z_K is of order 4. Let $\mathfrak{a} \in I_F$ and $\mathfrak{a}\mathfrak{o}_K = b\mathfrak{o}_K$ with $b \in K$. Then $b^{1-\rho} \in Z_K$. Changing b for bi if necessary, we may assume that $b^{1-\rho} = 1$ or i. Suppose $b^{1-\rho} = i$. Then $b = c(1+i)$ with $c \in F$, so that $(c^{-1}\mathfrak{a})^2 = (2)$, and hence $N(c^{-1}\mathfrak{a})^2 = 2^n$, a contradiction. Therefore $b^{1-\rho} = 1$, which proves that $I_F \cap P_K = P_F$. Next let $u \in E_K$. Then $u^{1-\rho} \in Z_K$. For the same reason as above, $u^{1-\rho} = \pm i$ leads to a contradiction. Therefore $u^{1-\rho} = \pm 1$, which proves that u or $ui \in E_F$, and hence $E_K = Z_K E_F$. The remaining part of (iii) can be proved either in the same manner or in a straightforward way, noting especially that if $n = 2$, we have $K = F((-w)^{\frac{1}{2}})$ with a totally positive element w of F such that $N_{F/\mathbf{Q}}(w)$ is not a square (otherwise K is normal over $\mathbf{Q}$).

Still with the same (K, Φ), suppose that n is odd, $n = 2$, or $\mathfrak{d}_{K/F}$ does not divide 2. Then (a.14) (and hence (2.8)) is equivalent to

$$(a.15) \qquad\qquad I_0(\Phi') = I_{F'} P_{K'}.$$

By (a.11), we have $\gamma = t - 1$. Therefore (a.15) and (2.8) hold if K' has only one prime ideal ramified over F'. By examining the images of the factors of $\mathfrak{d}_{K'/F'}$ under Φ', one can obtain weaker conditions under which (a.15) holds, and also some examples of K' for which (a.15) does not hold. Now, as mentioned above, Φ' gives an injection of $I_{K'}/I(\Phi')$ into R_K^+/P_K. The latter group contains an isomorphic image of $I_K/I(K/F)$ as a subgroup of index $2^{\beta+\gamma-\varepsilon}$. Therefore Φ' gives an isomorphism of 'the odd part' of $C_{K'}/C_{F'}$ onto 'the odd part' of C_K/C_F.

We conclude the Appendix by finding a certain ideal in a cyclotomic field. Let $K = \mathbf{Q}(\zeta)$, $F = \mathbf{Q}(\zeta + \zeta^{-1})$, where $\zeta = e^{2\pi i/l}$ with an odd prime l. Let S be a subfield of K. Then $I_0(K/S) = P_K I_S$ and $E_S^+ \subset N_{K/S}(K^\times)$, so that (a.1) and (a.2) can be written as

$$(a.16) \qquad\qquad I(K/S)/P_K I_S \simeq E_S^+/N_{K/S}(E_K),$$

$$(a.17) \qquad [I(K/S) : P_K] = \begin{cases} h_S & \text{if } S \text{ is imaginary,} \\ h_S[E_S^+ : E_S{}^2] & \text{if } S \text{ is real.} \end{cases}$$

If we put $g_S = [P_K \cap I_S : P_S]$, then we obtain, from (a.16) and (a.17),

$$(a.18) \qquad g_S = \begin{cases} [E_S : N_{K/S}(E_K)] & \text{if } S \text{ is imaginary,} \\ [E_S{}^2 : N_{F/S}(E_F{}^2)] & \text{if } S \text{ is real,} \end{cases}$$

since $E_S{}^2 \supset N_{F/S}(E_F{}^2) = N_{F/S}(N_{K/F}(E_K)) = N_{K/S}(E_K)$ if $S \subset F$. In particular we have $g_{\mathbf{Q}} = 1$ and $I(K/\mathbf{Q}) = P_K$. Suppose S is imaginary and $[S : \mathbf{Q}] = 2$. Then $[K : S]$ is odd and $N_{K/S}(E_K) = E_S$, so that $g_S = 1$ and $I(K/S) = P_K I_S$.

Let us now give an example of an ideal $\mathfrak{a}$ in K such that $\mathfrak{a}^{1-\rho} \notin I(K/S)$ for any proper subfield S of K. For this we assume that h_K is odd, and $\frac{1}{2}(l-1)$ is an odd prime. Then $E_F^+ = E_F{}^2$; moreover, $I(K/S) = P_K I_S$ and $I(K/S)/P_K \simeq I_S/P_S$ for all real and imaginary subfields S. Suppose $\mathfrak{a}^{1-\rho} \in I(K/S)$ for a proper subfield S of K. If $S \subset F$, we have $\mathfrak{a}^{1-\rho} \in P_K$ by Proposition A.3. Since $R_K = I_K{}^{1-\rho} P_K$ and $[R_K : P_K] = h_K/h_F$, we obtain a desired $\mathfrak{a}$ if $h_K/h_F > h_S$ for the quadratic imaginary subfield S of K, which is the case, for example, for $l = 47, 59$ as can be seen from the table of Hasse [2].

REFERENCES

1. C. CHEVALLEY, 'Sur la théorie du corps de classes dans les corps finis et les corps locaux', *J. Fac. Sci. Univ. Tokyo* 2 (1933) 365–476.
2. H. HASSE, *Über die Klassenzahl Abelscher Zahlkörper* (Akademie-Verlag, Berlin, 1952).
3. E. KUMMER, 'Über eine Eigenschaft der Einheiten der aus den Wurzeln der Gleichung $\alpha^\lambda = 1$ gebildeten complex Zahlen und über den zweiten Faktor der Klassenzahl', *Monatsb. Deutsch. Akad. Wiss. Berlin* (1970) 855–80.
4. K.-Y. SHIH, 'Anti-holomorphic automorphisms of arithmetic automorphic function fields', *Ann. of Math.* 103 (1976) 81–102.
5. G. SHIMURA, 'Construction of class fields and zeta functions of algebraic curves', ibid. 85 (1967) 58–159.
6. —— 'Discontinuous groups and abelian varieties', *Math. Ann.* 168 (1967) 171–99.
7. —— 'On canonical models of arithmetic quotients of bounded symmetric domains, I', *Ann. of Math.* 91 (1970) 144–222; II, 92 (1970) 528–49.
8. —— 'On the zeta-function of an abelian variety with complex multiplication', ibid. 94 (1971) 504–33.
9. —— 'On the real points of an arithmetic quotient of a bounded symmetric domain', *Math. Ann.* 215 (1975) 135–64.
10. —— and Y. TANIYAMA, 'Complex multiplication of abelian varieties and its applications to number theory', *Publ. Math. Soc. Japan* 6 (1961).
11. A. WEIL, 'Jacobi sums as "Grössencharaktere"', *Trans. Amer. Math. Soc.* 73 (1952) 487–95.
12. —— 'The field of definition of a variety', *Amer. J. Math.* 78 (1956) 509–24.

Department of Mathematics
Fine Hall, Box 37
Princeton University
Princeton, N.J. 08540

Unitary groups and theta functions

Proceedings of the International Symposium on Algebraic Number Theory,
Kyoto, 1976 (1977), 195-200

Our first theme is the theta functions which occur as the Fourier coefficients of automorphic forms. This will eventually lead to the second one: the special values of various zeta functions, especially those associated with cusp forms. The point of contact of these two themes is a certain Eisenstein series, of which the Fourier coefficients are "arithmetic theta functions", and which, evaluated at some points, gives special values of a certain zeta function.

First we define, with a fixed imaginary quadratic field K, a unitary group U by

$$U = \{X \in SL_m(K) \mid {}^t\bar{X}RX = R\} ,$$

$$R = \begin{pmatrix} 0 & 0 & 1_q \\ 0 & S & 0 \\ -1_q & 0 & 0 \end{pmatrix}, \qquad -{}^t\bar{S} = S \in GL_n(K) ,$$

where $m = n + 2q$ with positive integers n and q. We assume that $-iS$ is positive definite. The group acts naturally on the space $\mathfrak{Z}$ consisting of all elements (z, w) of $M_q(C) \times M_{n,q}(C)$ such that $i({}^t\bar{z} - z + {}^t\bar{w}Sw)$ is positive definite, where $M_{n,q}(F)$, for any field F, denotes the set of all $n \times q$ matrices with coefficients in F, and $M_n(F) = M_{n,n}(F)$ as usual. If $(z, w) \in \mathfrak{Z}$ and

$$\alpha = \begin{pmatrix} a_1 & b_1 & c_1 \\ a_2 & b_2 & c_2 \\ a_3 & b_3 & c_3 \end{pmatrix} \in U ,$$

then the image of (z, w) under α is defined by

$$\alpha(z, w) = ((a_1z + b_1w + c_1)(a_3z + b_3w + c_3)^{-1} ,$$
$$(a_2z + b_2w + c_2)(a_3z + b_3w + c_3)^{-1}) .$$

Put $j(\alpha; z, w) = \det(a_3z + b_3w + c_3)$. Then the jacobian of α is $j(\alpha; z, w)^{-m}$.

For any congruence subgroup Γ of U and a positive integer k, we understand, by an *automorphic form on $\mathfrak{H}$ of weight k with respect to Γ*, a holomorphic function f on $\mathfrak{H}$ satisfying

$$(1) \qquad f(\gamma(z, w)) = f(z, w) j(\gamma ; z, w)^k \qquad \text{for all } \gamma \in \Gamma .$$

Such a function f has an expansion

$$(2) \qquad f(z, w) = \sum_{0 \leqslant r \in B} g_r(w) e(\mathrm{tr}(rz)) ,$$

with holomorphic functions g_r on $M_{n,q}(C)$, where $e(x) = e^{2\pi i x}$, and B is a lattice in the vector space

$$\{ x \in M_q(K) \, | \, {}^t \bar{x} = x \} ;$$

we write $0 \leqslant r$ to indicate that r is non-negative. Moreover, each g_r satisfies

$$(3) \qquad g_r(w + l) = e\left(\frac{1}{2i} H_r\left(l, w + \frac{1}{2} l \right) \right) g_r(w) \qquad \text{for all } l \in L ,$$

where L is a lattice in $M_{n,q}(K)$ depending only on Γ, and H_r is the hermitian form defined by

$$H_r(u, v) = -2i \cdot \mathrm{tr}\,(r \cdot {}^t \bar{u} S v) \qquad ((u, v) \in M_{n,q}(C) \times M_{n,q}(C)) .$$

Thus the "Fourier coefficients" of f are theta functions on $M_{n,q}(C)$ with respect to L. Now, in the theory of elliptic modular functions, the modular forms that are important from the number-theoretical viewpoint are those with cyclotomic (or more generally algebraic) Fourier coefficients. This leads us to the following natural question:

Can one define the notion of "arithmetic theta functions" in such a way that the forms f with arithmetic Fourier coefficients behave like elliptic modular forms with cyclotomic Fourier coefficients?

We shall actually show that the answer is affirmative. To define "arithmetic theta functions" in a more general setting, let V be a finite dimensional complex vector space, L a lattice in V, and $H(u, v)$ a non-negative hermitian form on V. We put $E(u, v) = \mathrm{Im}\,(H(u, v))$, and assume that $E(L, L) \subset Z$, i.e., H is a Riemann form. Further let ψ be a map of L into the group of all roots of unity such that

$$\psi(l + l') = \psi(l)\psi(l') e(E(l, l')/2) .$$

Then we consider a "reduced theta function" of type (H, ψ, L) in the sense of Weil [8], which is a holomorphic function $g(u)$ on V satisfying

$$g(u + l) = \psi(l)e\left(\frac{1}{2i}H\left(l, u + \frac{1}{2}l\right)\right)g(u) \qquad \text{for all } l \in L \,.$$

Let $T(H, \psi, L)$ denote the set of all such g. For each $g \in T(H, \psi, L)$, put

$$g_*(u) = e\left(\frac{i}{4}H(u, u)\right)g(u) \,.$$

Then g_* is not holomorphic unless $g = 0$, but it satisfies

$$g_*(u + l) = \psi(l)e(E(l, u)/2)g_*(u) \qquad \text{for all } l \in L \text{ and } u \in V \,.$$

Therefore the restriction of g_* to QL can be extended to a function on $QL \otimes_Q A$, where A denotes the ring of adeles. It should be noted that g_* is not a function on V/M for any lattice M commensurable with L.

Now suppose that V/L is an abelian variety and $\mathrm{End}\,(V/L) \otimes Q$ is isomorphic to $M_h(F)$ with a CM-field F, where $h = 2 \cdot \dim\,(V)/[F:Q]$. Then we obtain a CM-type (F, Φ) such that h times Φ is the representation of F on V. Let (F', Φ') be the reflex of (F, Φ), and let $T_a(H, \psi, L)$ be the set of all g in $T(H, \psi, L)$ such that $g_*(u) \in F'_{ab}$ for all $u \in QL$, where F'_{ab} denotes the maximal abelian extension of F'. (It is also meaningful and even advantageous to consider all g such that $g_*(u)$ is algebraic for all $u \in QL$.) We call the elements of $T_a(H, \psi, L)$ *arithmetic theta functions*. It can be shown that $T(H, \psi, L)$ can be spanned by $T_a(H, \psi, L)$ over C. Let $F'^\times_A$ and $F^\times_A$ denote the idele groups of F' and F, respectively. We can define a map η of $F'^\times_A$ into $F^\times_A$ by $\eta(x) = \det\,(\Phi'(x))$.

Theorem 1. *Every element x of $F'^\times_A$ defines a Q-linear map $g \mapsto g^x$ of $T_a(H, \psi, L)$ onto $T_a(N(x)H, \psi', \eta(x)^{-1}L)$ satisfying $g_*(u)^x = (g^x)_*(\eta(x)^{-1}u)$ for all $u \in QL$, where $\psi'(l) = \psi(\eta(x)l)^x$, and $N(x)$ is the norm of the ideal associated with x.*

Here c^x for $c \in F'_{ab}$ denotes the image of c under the element of $\mathrm{Gal}\,(F'_{ab}/F')$ corresponding to x; one should also note that every element of $F^\times_A$ acts on $QL \otimes_Q A$, and therefore $(g^x)_*(\eta(x)^{-1}u)$ and $\psi(\eta(x)l)$ are meaningful. We can actually generalize the theorem to the case of an abelian variety which is isogenous to the product of several varieties of the above type with different F's. The details will be given in [5].

Let us now come back to the function f of (2) and its Fourier coefficients g_r satisfying (3). Putting $V = M_{n,q}(C)$, we observe that $\mathrm{End}\,(V/L) \otimes Q$ is isomorphic to $M_{nq}(K)$, so that the above definitions and results are applicable.

In this case, we have $F = F' = K$ and η is the identity map. Let $\mathscr{M}_k(\Gamma)$ denote the vector space of all f satisfying (1). We call an element f of $\mathscr{M}_k(\Gamma)$ *arithmetic* if $g_r \in T_a(H_r, 1, L)$ for all r. Let $\mathscr{M}_k^a(\Gamma)$ denote the set of all such f, and $\mathscr{M}_k^a$ the union of $\mathscr{M}_k^a(\Gamma)$ for all congruence subgroups Γ. Further let $\mathfrak{R}$ denote the field of all arithmetic automorphic functions with respect to the algebraic group

$$G = \{X \in GL_m(K) \,|\, {}^t\overline{X}RX = \nu(X)R \quad \text{with } \nu(X) \in Q\}$$

in the framework of canonical models as considered in [1], [2]. Then we have

Theorem 2. *The field $K_{ab}\mathfrak{R}$ consists of all the quotients f/g with automorphic forms f and g of the same weight belonging to $\bigcup_{k=1}^{\infty} \mathscr{M}_k^a$.*

We can also prove some theorems concerning the explicit action of a certain subgroup of the adelization of G on $\mathfrak{R}$ and on $\mathscr{M}_k^a$, similar to those in [3], [4].

Let us now consider Eisenstein series in the present setting. Suppose $q = 1$; let Λ be a lattice in $M_{1,m}(K)$, and let μ be an element of $M_{1,m}(K)$. Put

$$X = X(\Lambda, \mu) = \{\lambda \in M_{1,m}(K) \,|\, \lambda R^{-1} \cdot {}^t\bar{\lambda} = 0 \,, \, 0 \neq \lambda \equiv \mu \bmod \Lambda\} \,.$$

Then we define an Eisenstein series $E_{X,k}$ by

$$E_{X,k}(z, w) = \sum_{(a,b,c) \in X} (az + bw + c)^{-k} \qquad ((z, w) \in \mathfrak{Z}) \,,$$

where $a, c \in K$, and $b \in M_{1,n}(K)$. If $k > 2n + 2$, this is convergent and defines an element of $\mathscr{M}_k(\Gamma)$, where

$$\Gamma = \{\gamma \in U \,|\, \Lambda\gamma = \Lambda \,, \, \mu\gamma \equiv \mu \bmod \Lambda\} \,.$$

Our main result about $E_{X,k}$ is

Theorem 3. *Let h be an elliptic modular form of weight k with rational Fourier coefficients at $i\infty$, and τ an element of K with positive imaginary part such that $h(\tau) \neq 0$. Then $\pi^{-k}h(\tau)^{-1}E_{X,k}$ is arithmetic.*

This fact implies some interesting results about the special values of a new type of zeta function of the form

$$\zeta(s; P, d, \mathfrak{a}, k) = \sum x^k |x|^{-2s} \,.$$

Here s is a complex variable, P is the composite of K with a totally real algebraic number field P_0 of degree m, and d is a negative element of P_0 whose all other conjugates are positive; the sum is extended over all x belonging to a lattice $\mathfrak{a}$ in P such that $\mathrm{Tr}_{P_0/Q}(dx\bar{x}) = 0$. The series is convergent for sufficiently large $\mathrm{Re}(s)$, and can be continued to a meromorphic function on the whole s-plane. One interesting feature of this function is that $|x|^2$ is not necessarily a rational integer. Similar zeta functions can be defined with a direct sum of fields in place of P.

We finally mention a result on the special values of the zeta function

$$D(s, f, \varphi) = \sum_{n=1}^{\infty} \varphi(n)a_n n^{-s} ,$$

where φ is a primitive Dirichlet character, and f is a cusp form which has an expansion $f(z) = \sum_{n=1}^{\infty} a_n e(nz)$ and satisfies

$$f(\beta(z)) = \chi(d)(cz + d)^k f(z)$$

for all $\beta = \begin{pmatrix} a & b \\ c & d \end{pmatrix} \in SL_2(\mathbf{Z})$ such that $c \equiv 0 \,(\mathrm{mod}\, N)$ with a positive integer N; χ is a Dirichlet character modulo N. Suppose that $a_1 = 1$, and f is an eigenfunction of all Hecke operators of level N; suppose further that f is "primitive" in the sense that it cannot be obtained from the forms of lower level. Define the Gauss sum $\mathfrak{g}(\omega)$ by

$$\mathfrak{g}(\varphi) = \sum_{n=1}^{c} \varphi(n)e(n/c) ,$$

where c is the conductor of φ. Then we obtain

Theorem 4. *Let φ and φ' be primitive Dirichlet characters, and f a primitive cusp form as above. Further let m and m' be two positive integers less than k such that $(\varphi\varphi')(-1) = (-1)^{m-m'}$ and $D(m', f, \varphi') \neq 0$. Then*

$$(2\pi i)^{m'-m} \cdot \frac{\mathfrak{g}(\varphi')D(m, f, \varphi)}{\mathfrak{g}(\varphi)D(m', f, \varphi')}$$

is an algebraic number belonging to the field generated over $\mathbf{Q}$ by $a_n, \varphi(n)$, and $\varphi'(n)$ for all n.

There is also a complementary result which concerns the case where $(\varphi\varphi')(-1) = (-1)^{m-m'-1}$. For details, the reader is referred to [6], [7].

The proof of Theorem 4 relies on a certain result concerning the special values of a zeta function

$$D(s, f, g) = \sum_{n=1}^{\infty} a_n b_n n^{-s} ,$$

which is defined with another modular form $g(z) = \sum_{n=0}^{\infty} b_n e(nz)$. The pullback of $E_{\chi,k}$ to the upper half plane evaluated at a point belonging to K is a constant times $D(k-1, f, g)$, where f is the Mellin transform of an L-function of K with a Grössen-character and g is a product of two theta series. Although this phenomenon is not so important, nor needed for the proof of the above theorem, it shows at least how these values are intricately connected with each other.

One final remark may be added: the whole theory of arithmetic automorphic forms with respect to a unitary group can be generalized to the case where the basic field K is a totally imaginary quadratic extension of a totally real algebraic number field; also Theorem 4 can be generalized to the zeta functions associated with Hilbert modular forms.

References

[1] Miyake, K., Models of certain automorphic function fields, Acta Math. **126** (1971), 245–307.

[2] Shimura, G., On canonical models of arithmetic quotients of bounded symmetric domains, I, II, Ann. of Math. **91** (1970), 144–222; **92** (1970), 528–549.

[3] ——, On some arithmetic properties of modular forms of one and several variables, Ann. of Math. **102** (1975), 491–515.

[4] ——, On the Fourier coefficients of modular forms of several variables, Göttingen Nachr. Akad. Wiss. 1975, 261–268.

[5] ——, Theta functions with complex multiplication, Duke Math. J. **43** (1976), 673–696.

[6] ——, The special values of the zeta functions associated with cusp forms, Comm. Pure and Appl. Math. **29** (1976), 783–804.

[7] ——, On the periods of modular forms, to appear in Math. Ann.

[8] Weil, A., Introduction à l'étude des variétés kählériennes, Hermann, Paris, 1958.

Department of Mathematics
Princeton University
Princeton, New Jersey 08540
U.S.A.

On the derivatives of theta functions and modular forms

Duke Mathematical Journal, 44 (1977), 365-387

Dedicated to Kenkichi Iwasawa on the occasion of his sixtieth birthday

The algebraicity of an analytically defined object is a fascinating subject both in number theory and in algebraic geometry, but has attracted unaccountably few researchers. For instance, it seems that there are more mathematicians who deal with the transcendency of the special values of analytic functions than those who prove the algebraicity. In the present work, we shall discuss several problems of algebraicity which involve differentiation. We divide the paper into four sections, the first of which is an investigation of differential forms and derivations on an abelian variety A parametrized by classical theta functions. We first observe that if a point w on the Siegel upper half space and a matrix δ of elementary divisors which determines the type of polarization are given, then the variety A is defined over the field $\Re_s[w]$ generated by the values $f(w)$ for all f in a certain field $\Re_s$ of Siegel modular functions. Here $\Re_s$ consists of the quotients of modular forms with rational Fourier coefficients with respect to a certain congruence subgroup Γ_δ of Sp $(n, \mathbf{Z})$, depending on δ. Next, we shall give an explicit basis of holomorphic 1-forms on A rational over $\Re_s[w]$ by means of the derivatives of theta functions. This will lead to an answer to the question concerning the periods of integrals rational over a field of definition, which Weil posed in his recent article [8].

The second section, for which the first one may be considered preliminaries, is a continuation of the previous paper [7], where we showed that a classical theta function with complex multiplication, modified by a certain exponential factor and a constant factor, takes algebraic values on the points commensurable with the periods. This result will be extended to certain non-holomorphic derivatives of such functions. The behavior of the special values under automorphisms will also be determined. These are closely connected with the results of another previous paper [5], in which certain non-holomorphic derivatives of Hilbert modular forms are discussed. The connection will be clarified in the third section, by viewing A as a member of the family of abelian varieties with a fixed totally real algebraic number field as their endomorphism algebras. We shall first prove the analogues of the theorems of §1 for such A. Then it will be shown that the heat equation satisfied by theta functions implies the coincidence of the two types of derivatives at the origin of A. An observation of the same kind will be made for Siegel modular forms in the last section, in

Received December 6, 1976. Supported by NSF Grant MCS 76-11376.

which we shall also prove a theorem similar to the main theorems I, II of [5]
for such forms.

It is obvious that the non-holomorphic derivatives of the same kind can be
defined on a much wider class of varieties. In this paper, however, we do not
intend to present any general treatment, but try to reveal some new aspects
of the topic in the same unsophisticated spirit as in [5].

1. Derivations and differential forms on an abelian variety

Throughout the paper, we use the same notation as in [7]. In particular,
G denotes an algebraic subgroup of GL_{2n} rational over $\mathbf{Q}$ such that

$$G_X = \{\alpha \in GL_{2n}(X) \mid {}^t\alpha J \alpha = \nu(\alpha)J \text{ with } \nu(\alpha) \in X\},$$

where

$$J = \begin{bmatrix} 0 & -1_n \\ 1_n & 0 \end{bmatrix},$$

and X is a "variable field (or ring)", which we specialize to $\mathbf{Q}$, $\mathbf{R}$, or the ring
of adeles $\mathbf{A}$. Then $G_{\mathbf{Q}+}$, $G_{\mathbf{A}+}$, etc. can be defined as in [4], [5], [6]. We denote
by $\mathfrak{H}_n$ the Siegel upper half space of degree n, and by Γ_N or $\Gamma(N)$ the principal
congruence subgroup of $\mathrm{Sp}\,(n, \mathbf{Z})$ of level N (see [6], [7]). For a subring $\mathfrak{R}$
of $\mathbf{C}$ and a congruence subgroup Γ of $G_{\mathbf{Q}+}$, we define $\mathfrak{M}_k(\Gamma, \mathfrak{R})$, $\mathfrak{A}_k(\Gamma, \mathfrak{R})$,
$\mathfrak{M}_k(\mathfrak{R})$, and $\mathfrak{A}_k(\mathfrak{R})$ as in [7], and put $\mathfrak{K} = \mathfrak{A}_0(\mathbf{Q}_{ab})$. As shown in [4] and [6],
every element of $G_{\mathbf{A}+}$ acts on the field $\mathfrak{K}$ as an automorphism. For every open
subgroup S of $G_{\mathbf{A}+}$, we put

$$\mathfrak{K}_S = \{f \in \mathfrak{K} \mid f^s = f \text{ for all } s \in S\} \qquad (\text{cf. } [4, \text{II}, \S6]).$$

Further, for each point w of $\mathfrak{H}_n$, we denote by $\mathfrak{K}[w]$ and $\mathfrak{K}_S[w]$ the fields consisting of the values $f(w)$ for all elements f, finite at w, of $\mathfrak{K}$ and $\mathfrak{K}_S$, respectively.
Obviously $\mathfrak{K}[w] = \mathfrak{K}[\alpha(w)]$ for all $\alpha \in G_{\mathbf{Q}+}$. We shall be dealing with two types
of theta functions θ and φ which are defined, for $u \in \mathbf{C}^n$, $z \in \mathfrak{H}_n$, $r \in \mathbf{R}^n$, and
$s \in \mathbf{R}^n$, by

$$(1.1) \qquad \theta(u, z; r, s) = \sum_{x - r \in \mathbf{Z}^n} e(\tfrac{1}{2} \cdot {}^t x z x + {}^t x(u + s)),$$

$$(1.2) \qquad \varphi(u, z; r, s) = e(\tfrac{1}{2} \cdot {}^t u(z - \bar{z})^{-1} u)\theta(u, z; r, s),$$

where $e(v) = e^{2\pi i v}$ for $v \in \mathbf{C}$.

Let L be a lattice in $\mathbf{C}^n$, and H a non-degenerate Riemann form for $\mathbf{C}^n/L$ as
considered in [7]. Then we find a positive integer μ, a diagonal matrix ϵ, a
"period matrix" $\Omega = (\omega_1\ \omega_2)$, and a point $z_0 = \omega_2^{-1}\omega_1$ of $\mathfrak{H}_n$, which satisfy
(4), (5), (6), (7) of [7]; we recall in particular that $L = (\omega_1\ \omega_2\epsilon)\mathbf{Z}^{2n}$ and

$$\mathrm{Im}\,(H(\Omega x, \Omega y)) = \mu \cdot {}^t x J y \qquad (x, y \in \mathbf{R}^{2n}).$$

To simplify our notation, let us put $z = \mu z_0$,

$$(1.3) \qquad \delta = \mu\epsilon, \qquad L(z, \delta) = (z\ \delta)\mathbf{Z}^{2n}.$$

Obviously the linear map $u \mapsto \mu\omega_2^{-1}u$ gives an isomorphism of $\mathbf{C}^n/L$ onto $\mathbf{C}^n/L(z, \delta)$; H corresponds to the Riemann form H_1 for $\mathbf{C}^n/L(z, \delta)$ defined by

$$(1.4) \qquad H_1(u, v) = {}^t\bar{u}(z - \bar{z})^{-1}v \qquad (u, v \in \mathbf{C}^n).$$

As noted in [7] and as is well known, if $\mu \geq 3$, and $\mathfrak{J}$ is a complete set of representatives for $\delta^{-1}\mathbf{Z}^n/\mathbf{Z}^n$, then the map

$$(1.5_a) \qquad u \mapsto (\theta(\mu\omega_2^{-1}u, z; j, 0))_{j\in\mathfrak{J}} \qquad (u \in \mathbf{C}^n)$$

defines a biregular projective embedding of $\mathbf{C}^n/L$, which is essentially the same as the embedding

$$(1.5_b) \qquad u \mapsto \Theta(u) = \Theta_z(u) = (\theta(u, z; j, 0))_{j\in\mathfrak{J}} \qquad (u \in \mathbf{C}^n)$$

of $\mathbf{C}^n/L(z, \delta)$. Let $A(z, \delta)$ denote the image variety. Our first task is to determine a field of rationality for $A(z, \delta)$. For this purpose, we first introduce a function

$$(1.6) \qquad f(z; p, q; r, s) = \varphi(zp + q, z; r, s) \qquad (z \in \mathfrak{H}_n \,; p, q, r, s \in \mathbf{R}^n),$$

which gives a coordinate of the point on $A(z, \delta)$ corresponding to $zp + q$, when $s = 0$ and $r \in \mathfrak{J}$. Now observe that, if

$$\gamma = \begin{pmatrix} a & b \\ c & d \end{pmatrix} \in \Gamma_1 \,,$$

then $\{{}^tac\} \equiv \{{}^tbd\} \equiv 0 \pmod{2\mathbf{Z}^n} \Leftrightarrow \{a \cdot {}^tb\} \equiv \{c \cdot {}^td\} \equiv 0 \pmod{2\mathbf{Z}^n}$, where $\{X\}$ means the column vector consisting of the diagonal elements of X. Therefore we can define a subgroup Γ_θ of Γ_1 by

$$(1.7) \qquad \Gamma_\theta = \left\{ \begin{pmatrix} a & b \\ c & d \end{pmatrix} \in \Gamma_1 \mid \{{}^tac\} \equiv \{{}^tbd\} \equiv 0 \pmod{2\mathbf{Z}^n} \right\}.$$

By [7, (14'), (14$_a$)], we have

$$(1.8) \qquad f(\gamma(z); p, q; r, s) = e(U)\lambda_\gamma \det(cz + d)^{1/2} f(z; p', q'; r', s')$$

$$\text{if} \quad \gamma = \begin{pmatrix} a & b \\ c & d \end{pmatrix} \in \Gamma_\theta \,,$$

where λ_γ is the constant explained in [7, Prop. 1.3], and

$$\begin{pmatrix} p' \\ q' \end{pmatrix} = {}^t\gamma \begin{pmatrix} p \\ q \end{pmatrix}, \qquad \begin{pmatrix} r' \\ s' \end{pmatrix} = {}^t\gamma \begin{pmatrix} r \\ s \end{pmatrix},$$

$$U = \frac{-1}{2}({}^tra \cdot {}^tbr + {}^tsc \cdot {}^tds) - {}^trb \cdot {}^tcs.$$

Define a subgroup Γ_δ of Γ_θ by

$$\Gamma_\delta = \left\{ \begin{pmatrix} a & b \\ c & d \end{pmatrix} \in \Gamma_\theta \cap \beta^{-1}\Gamma_\theta\beta \mid a - 1 \in \delta \cdot M_n(\mathbf{Z}) \right\},$$

$$(1.9)$$

$$\beta = \begin{pmatrix} \delta^{-1} & 0 \\ 0 & \delta \end{pmatrix}.$$

From [7, (13)] and (1.8), we obtain

$$(1.10) \qquad f(\gamma(z); p, q; r, 0) = \lambda_\gamma \det(cz + d)^{1/2} f(z; p', q'; r, 0)$$

$$\text{if} \quad r \in \delta^{-1} \mathbf{Z}^n \quad \text{and} \quad \gamma = \begin{pmatrix} a & b \\ c & d \end{pmatrix} \in \Gamma_\delta .$$

Let R_θ be the subgroup of $G_{\mathbf{A}+}$ consisting of the elements $x = \begin{pmatrix} a & b \\ c & d \end{pmatrix}$ such that $x_p \in GL_{2n}(\mathbf{Z}_p)$ for all primes p, and

$$\{{}^t a_2 c_2\} \equiv \{{}^t b_2 d_2\} \equiv 0 \pmod{2\mathbf{Z}_2{}^n},$$

where x_p, a_p, etc. denote the p-components of x, a, etc. Put

$$(1.11) \qquad R_\delta = \left\{ \begin{pmatrix} a & b \\ c & d \end{pmatrix} \in R_\theta \cap \beta^{-1} R_\theta \beta \mid a_p - 1 \in \delta \cdot M_n(\mathbf{Z}_p) \text{ for all } p \right\},$$

$$S = S_\delta = \mathbf{Q}^\times R_\delta .$$

Then $\mathfrak{R}_S$ is meaningful, and

$$(1.12) \qquad G_\mathbf{Q} \cap S = \mathbf{Q}^\times \Gamma_\delta .$$

Therefore, by [6, Th. 3], we have

$$(1.13) \qquad \mathfrak{R}_S = \mathfrak{a}_0(\Gamma_\delta, \mathbf{Q}).$$

Theorem 1.1. *Let $\delta = \mu\epsilon$ with an integer $\mu \geq 3$, and let $A(z, \delta)$ denote the image abelian variety obtained by the projective embedding (1.5_b) of $\mathbf{C}^n/L(z, \delta)$. Then, for each point w of $\mathfrak{H}_n$, $A(w, \delta)$ and its origin are rational over $\mathfrak{R}_S[w]$. Moreover, $\mathfrak{R}[w]$ is the smallest field of rationality for all the points of finite order on $A(w, \delta)$.*

Proof. The last assertion follows immediately from [7, Prop. 1.8] and its proof. This shows in particular that $A(w, \delta)$ is rational over $\mathfrak{R}[w]$. Now, as shown in [4, I, Prop. 7.6], $\mathfrak{R}[w]$ is an infinite Galois extension of $\mathfrak{R}_S[w]$; moreover, for every element σ of Gal $(\mathfrak{R}[w]/\mathfrak{R}_S[w])$, there exists an element x of R_δ such that

$$(1.14) \qquad h(w)^\sigma = h^x(w) \quad \text{for all} \quad h \in \mathfrak{R}.$$

This can be obtained by combining [4, I, Prop. 7.6] with the definition of x on $\mathfrak{R}$ given by [4, I, 2.7; II, 6.1]. For each positive integer N, put

$$T_N = \{ x \in R_\delta \mid x_p - 1 \in N \cdot M_{2n}(\mathbf{Z}_p) \text{ for all } p \}.$$

By [4, I, 3.4; II, (3.10.3)], we have

$$x = \begin{pmatrix} 1 & 0 \\ 0 & \nu(x) \end{pmatrix} y\alpha \quad \text{with} \quad y \in T_N , \quad \nu(y) = 1, \quad \text{and} \quad \alpha \in G_{\mathbf{Q}+} .$$

We apply (1.14) to a function h defined by

$$h(z) = \theta(zq_1 + q_2, z; r', 0)/\theta(zq_1 + q_2, z; r, 0)$$

$$= f(z; q_1, q_2; r', 0)/f(z; q_1, q_2; r, 0)$$

with r, r' in $\mathfrak{F}$ and q_1, q_2 in $m^{-1}\mathbf{Z}^n$ with a positive integer m. By [7, (11)], we have

$$h(z) = \theta(0, z; r' + q_1, q_2)/\theta(0, z; r + q_1, q_2).$$

Therefore, by [6, Th. 5, (v)] or by [7, Prop. 1.7], if

$$v = \begin{pmatrix} 1 & 0 \\ 0 & \nu(x) \end{pmatrix}$$

and t is a positive integer such that $t \equiv \nu(x)_p \pmod{\det(\delta)m^2 \mathbf{Z}_p}$ for all p, then

$$h^v(z) = \theta(0, z; r' + q_1, tq_2)/\theta(0, z; r + q_1, tq_2)$$

$$= f(z, q_1, tq_2; r', 0)/f(z; q_1, tq_2; r, 0).$$

If we take a sufficiently large multiple N of $m^2 \det(\delta)$, then h^v is invariant under y, and hence

$$h(w)^v = h^{vy\alpha}(w) = f(\alpha(w); q_1, tq_2; r', 0)/f(\alpha(w); q_1, tq_2; r, 0).$$

Since

$$\alpha = y^{-1}\begin{pmatrix} 1 & 0 \\ 0 & \nu(x) \end{pmatrix}^{-1} x \in R_s \cap G_{\mathbf{Q}} = \Gamma_s,$$

we have, by (1.10),

$$h(w)^v = f(w; q_1', q_2'; r', 0)/f(w; q_1', q_2'; r, 0),$$

where

$$\begin{pmatrix} q_1' \\ q_2' \end{pmatrix} = {}^t\alpha\begin{pmatrix} q_1 \\ tq_2 \end{pmatrix}.$$

We can express this result as

$$(1.15) \qquad \Theta(wq_1 + q_2)^v = \Theta(wq_1' + q_2'),$$

where q_1' and q_2' are elements of $\mathbf{Q}^n$ satisfying

$$\begin{pmatrix} q_1' \\ q_2' \end{pmatrix} \equiv {}^t x\begin{pmatrix} q_1 \\ q_2 \end{pmatrix} \mod \begin{pmatrix} \mathbf{Z}^n \\ \delta \mathbf{Z}^n \end{pmatrix}.$$

Relation (1.15) shows that the variety $A(w, \delta)$ as a whole and its origin $\Theta(0)$ are stable under every automorphism of $\mathfrak{K}[w]$ over $\mathfrak{K}_s[w]$. This proves the theorem.

Our next task is to find a basis of $\mathfrak{K}_s[w]$-rational differential forms of the first kind on $A(w, \delta)$. For this purpose we introduce a certain class of matrix-valued modular forms. Suppose $n > 1$, and let $\mathfrak{M}_k'(\Gamma)$ denote, for a congruence subgroup Γ of $G_{\mathbf{Q}+}$ and $0 \le k \in \mathbf{Z}$, the set of all holomorphic maps P of $\mathfrak{H}_n$ into $M_n(\mathbf{C})$ such that

$$(1.16) \quad P(\gamma(z)) = \nu(\gamma)^{-(1+nk)/2}(cz + d)P(z) \det(cz + d)^k \text{ for all } \gamma = \begin{pmatrix} a & b \\ c & d \end{pmatrix} \in \Gamma.$$

Each component of P has a Fourier expansion of the same type as those of modular forms. We denote by $\mathfrak{M}_k'(\Gamma, \mathfrak{R})$, for a subring $\mathfrak{R}$ of $\mathbf{C}$, the set of all P of $\mathfrak{M}_k'(\Gamma)$ with Fourier coefficients in $\mathfrak{R}$, and put $\mathfrak{M}_k'(\mathfrak{R}) = \bigcup_{N=1}^{\infty} \mathfrak{M}_k'(\Gamma_N, \mathfrak{R})$. Further we denote by $\mathfrak{a}_m'(\mathfrak{R})$, for $m \in \mathbf{Z}$, the set of all meromorphic maps of the form $g^{-1}P$ with $0 \ne g \in \mathfrak{M}_k(\mathfrak{R})$ and $P \in \mathfrak{M}_{k+m}'(\mathfrak{R})$ (with any k), and by $\mathfrak{a}_m'(\Gamma, \mathfrak{R})$ the set of all P in $\mathfrak{a}_m'(\mathfrak{R})$ satisfying (1.16) with m in place of k. If $n = 1$, we define $\mathfrak{M}_k'(\Gamma, \mathfrak{R})$, $\mathfrak{a}_m'(\mathfrak{R})$, etc. to be simply $\mathfrak{M}_{k+1}(\Gamma, \mathfrak{R})$, $\mathfrak{a}_{m+1}(\mathfrak{R})$, etc.

PROPOSITION 1.2. *For every point w of $\mathfrak{H}_n$, there exists an element P of $\mathfrak{a}_0'(\Gamma_i, \mathbf{Q})$ such that P is holomorphic at w and $\det(P(w)) \ne 0$.*

Proof. With φ as in (1.2), put

$$(1.17) \qquad \psi(u, z; r, s) = \frac{1}{2\pi i} \begin{bmatrix} \partial\varphi/\partial u_1 \\ \vdots \\ \partial\varphi/\partial u_n \end{bmatrix}.$$

From [7, (14′)], we obtain

$$(1.18) \quad \psi(u, \gamma(z); r, s) = \lambda_\gamma e(({}^t rs - {}^t r's'))(cz + d)\det(cz + d)^{1/2}$$

$$\cdot \psi({}^t(cz + d)u, z; r', s') \quad \text{if} \quad \begin{pmatrix} r' \\ s' \end{pmatrix} = {}^t\gamma\begin{pmatrix} r \\ s \end{pmatrix} \quad \text{and} \quad \gamma = \begin{pmatrix} a & b \\ c & d \end{pmatrix} \in \Gamma_\theta.$$

Denote the elements of $\mathfrak{F}$ by $r_1, \cdots, r_M$, and put

$$\begin{aligned} \varphi_m(u, z) &= \varphi(u, z; r_m, 0) \\ \psi_m(u, z) &= \psi(u, z; r_m, 0) \end{aligned} \qquad (m = 1, \cdots, M).$$

Since $u \mapsto (\varphi_m(u, z))$ defines a biregular embedding, we have

$$\operatorname{rank} \begin{bmatrix} \varphi_1(u, z) & \cdots & \varphi_M(u, z) \\ \psi_1(u, z) & \cdots & \psi_M(u, z) \end{bmatrix} = n + 1$$

for every $(u, z) \in \mathbf{C}^n \times \mathfrak{H}_n$. Therefore, changing the order of r_m, we may assume that

$$\det(\psi_1(0, w) \cdots \psi_n(0, w)) \ne 0.$$

Also we can find an index j such that $\varphi_j(0, w) \ne 0$. Define P by

$$(1.19) \qquad P(z) = \varphi_j(0, z)^{-1}(\psi_1(0, z) \cdots \psi_n(0, z)).$$

Then P has the required properties.

Remark 1.3. Since $\varphi = \theta$ and $\partial\varphi/\partial u_k = \partial\theta/\partial u_k$ at $u = 0$, we have

$$(1.20) \qquad P(z) = \frac{1}{2\pi i \theta_{,}(0, z)} \begin{bmatrix} \partial\theta_1(u, z)/\partial u_1 & \cdots & \partial\theta_n(u, z)/\partial u_1 \\ \cdots & \cdots & \cdots \\ \partial\theta_1(u, z)/\partial u_n & \cdots & \partial\theta_n(u, z)/\partial u_n \end{bmatrix}_{u=0},$$

where $\theta_m(u, z) = \theta(u, z; r_m, 0)$.

PROPOSITION 1.4. *Given* $T \in \mathcal{Q}_k'(\mathbf{Q}_{ab})$ *and*

$$\alpha = \begin{pmatrix} a & b \\ c & d \end{pmatrix} \in G_{\mathbf{Q}+} ,$$

put

$$U(z) = (cz + d)^{-1} T(\alpha(z)) \det (cz + d)^{-k}.$$

Then $U \in \mathcal{Q}_k'(\mathbf{Q}_{ab})$.

Proof. By [7, Lemma 2], it is sufficient to prove this for $\alpha = \begin{pmatrix} a & b \\ 0 & d \end{pmatrix}$ and $\alpha = \begin{pmatrix} 0 & -1 \\ 1 & 0 \end{pmatrix}$. The case of $\begin{pmatrix} a & b \\ 0 & d \end{pmatrix}$ can be verified in a straightforward way. As for $\alpha = \begin{pmatrix} 0 & -1 \\ 1 & 0 \end{pmatrix}$, define P by (1.19) or (1.20), and put $W = P^{-1}T$. Then the components of W belong to $\mathcal{Q}_k(\mathbf{Q}_{ab})$, and

$$z^{-1}T(-z^{-1}) \det (z)^{-k} = z^{-1}P(-z^{-1})W(-z^{-1}) \det (z)^{-k}.$$

By [6, Th. 4] (cf. also [7, Prop. 1.5]), the components of $W(-z^{-1}) \det (z)^{-k}$ belong to $\mathcal{Q}_k(\mathbf{Q}_{ab})$. Therefore it is sufficient to show that $z^{-1}P(-z^{-1}) \in \mathcal{Q}_0'(\mathbf{Q}_{ab})$. But this follows immediately from [7, (16')] and (1.18) with $\gamma = \begin{pmatrix} 0 & -1 \\ 1 & 0 \end{pmatrix}$.

Let $\mathfrak{F}$ denote the field of all $\mathfrak{R}_S[w]$-rational meromorphic functions on $A(w, \delta)$ viewed as functions on $\mathbf{C}^n$ through Θ. Then $\mathbf{C}\mathfrak{F}$ is the field of all meromorphic functions on $\mathbf{C}^n/L(z, \delta)$. With $c_{,}(u)$ in $\mathbf{C}\mathfrak{F}$, we can define a derivation

$$\mathfrak{d} = \sum_{r=1}^{n} c_{,}(u) \, \partial/\partial u_{,}$$

of $\mathbf{C}\mathfrak{F}$. We say that $\mathfrak{d}$ is $\mathfrak{R}_S[w]$-rational if $\mathfrak{d}$ maps $\mathfrak{F}$ into $\mathfrak{F}$. If $c_{,}(u)$ are constants, this is so if and only if $\mathfrak{d}(f)_{u=0} \in \mathfrak{R}_S[w]$ for all $f \in \mathfrak{F}$ finite at 0.

THEOREM 1.5. *Let* w *be a point of* $\mathfrak{H}_n$, *and let* P *be an element of* $\mathcal{Q}_0'(\Gamma_\delta , \mathbf{Q})$ *holomorphic at* w *such that* $\det (P(w)) \neq 0$ (*cf. Prop. 1.2*).

(i) *Let* $\xi_1 , \cdots , \xi_n$ *be 1-forms on* $\mathbf{C}^n/L(w, \delta)$ *defined by*

$$\begin{bmatrix} \xi_1 \\ \vdots \\ \xi_n \end{bmatrix} = 2\pi i \cdot {}^t P(w) \begin{bmatrix} du_1 \\ \vdots \\ du_n \end{bmatrix},$$

where $u_1 , \cdots , u_n$ *are the standard variables on* $\mathbf{C}^n$. *Then* $\xi_1 , \cdots , \xi_n$, *viewed as*

1-forms on $A(w, \delta)$ through Θ, form a basis of holomorphic 1-forms rational over $\Re_s[w]$.

(ii) *Let* $\mathfrak{d}_1, \cdots, \mathfrak{d}_n$ *be the derivations defined by*

$$\begin{bmatrix} \mathfrak{d}_1 \\ \cdot \\ \cdot \\ \cdot \\ \mathfrak{d}_n \end{bmatrix} = [2\pi i P(w)]^{-1} \begin{bmatrix} \partial/\partial u_1 \\ \cdot \\ \cdot \\ \cdot \\ \partial/\partial u_n \end{bmatrix}.$$

Then $\mathfrak{d}_1, \cdots, \mathfrak{d}_n$ *form a basis of* $\Re_s[w]$*-rational derivations of the field of mero-morphic functions on* $A(w, \delta)$.

Proof. Since (i) and (ii) are obviously equivalent, we prove only (i). Fixing an index l such that $\varphi_l(0, w) \neq 0$, put

$$(1.21) \qquad q_k(u) = \varphi_k(u, w)/\varphi_l(u, w) \qquad (k = 1, \cdots, M),$$

and $'\xi = (\xi_1 \cdots \xi_n)$ with ξ_r defined as above. By Th. 1.1, $\mathfrak{F}$ is generated over $\Re_s[w]$ by $q_1, \cdots, q_M$. Now we have

$$(1.22) \qquad dq_k = {}^t s_k \xi,$$

where $s_k(u) = \sigma_k(u, w)$ with a function σ_k defined by

$$(1.23) \quad \sigma_k(u, z) = \varphi_l(u, z)^{-2}[\varphi_l(u, z)P(z)^{-1}\psi_k(u, z) - \varphi_k(u, z)P(z)^{-1}\psi_l(u, z)].$$

Evaluate this at $u = 0$. Then we see that the components of $\sigma_k(0, z)$ belong to $\mathcal{Q}_0(\Gamma_\delta, \mathbf{Q}) = \Re_s$, and hence their values at w, $s_k(0)$, belong to $\Re_s[w]$. Since the forms dq_k generate $\Re_s[w]$-rational 1-forms, and since $\xi_1, \cdots, \xi_n$ are invariant under translation, (1.22) shows that $\xi_1, \cdots, \xi_n$ are rational over $\Re_s[w]$, Q.E.D.

COROLLARY 1.6. *Let* $\mathfrak{X}$ *be a subfield of* $\mathbf{C}$ *containing* $\Re_s[w]$, *and let* P *be as in the above theorem. Then:* (i) *a 1-form* $2\pi i \sum_{r=1}^n a_r \, du_r$ *with* $a_r \in \mathbf{C}$ *on* $A(w, \delta)$ *is* $\mathfrak{X}$*-rational if and only if* $(a_1, \cdots, a_n)^t P(w)^{-1}$ *has components in* $\mathfrak{X}$; (ii) *a deriva-tion* $(2\pi i)^{-1} \sum_{r=1}^n b_r \, \partial/\partial u_r$ *with* $b_r \in \mathbf{C}$ *is* $\mathfrak{X}$*-rational if and only if* $(b_1, \cdots, b_n)P(w)$ *has components in* $\mathfrak{X}$.

This is simply a paraphrase of the above theorem.

Remark 1.7. Instead of $(1.5_{a,b})$, one can also consider a map

$$u \mapsto (\theta(\lambda u, z; r, s))$$

with any fixed integer $\lambda \geq 2$, where (r, s) runs over a complete set of repre-sentatives for $(\lambda\epsilon)^{-1}\mathbf{Z}^n \times \lambda^{-1}\mathbf{Z}^n$ modulo $\mathbf{Z}^n \times \mathbf{Z}^n$. This defines a biregular projective embedding of $\mathbf{C}^n/L(z, \epsilon)$. In fact, the functions $\theta(\lambda u, z; r, s)$ with such r and s form a basis of holomorphic functions g on $\mathbf{C}^n$ satisfying

$$g(u + za + b) = g(u)e(\lambda^2(-\tfrac{1}{2} \, {}^t aza + {}^t au)) \qquad (a \in \mathbf{Z}^n, b \in \epsilon\mathbf{Z}^n).$$

Obviously the above theorems can be formulated in terms of this embedding. We have only to take $\mathcal{Q}_0(\Gamma, \mathbf{Q}(e(\lambda^{-2}\epsilon_n^{-1})))$ instead of $\mathcal{Q}_0(\Gamma_\delta, \mathbf{Q})$ with a suitable congruence subgroup Γ, where ϵ_n is the last diagonal element of ϵ.

Remark 1.8. Let $\{c_1, \cdots, c_{2n}\}$ be the basis of the first homology group of $A(w, \delta)$ corresponding to the standard basis of $\mathbf{Z}^{2n}$ through the map

$$\mathbf{R}^{2n} \ni x \mapsto \Theta((w\ \delta)x) \in A(w, \delta).$$

With $\xi_1, \cdots, \xi_n$ and P as above, let $p_{\kappa\lambda}$ denote the (κ, λ)-component of P. Then we have

$$\int_{c_\kappa} \xi_\lambda = 2\pi i \sum_{\nu=1}^{n} p_{\nu\lambda}(w) \int_{c_\kappa} du_\nu,$$

so that

$$(1.24) \qquad \begin{bmatrix} \int_{c_1} \xi_1 & \cdots & \int_{c_{2n}} \xi_1 \\ \cdots & \cdots & \cdots \\ \int_{c_1} \xi_n & \cdots & \int_{c_{2n}} \xi_n \end{bmatrix} = 2\pi i \cdot {}^t P(w)(w\ \delta).$$

If we take P to be the matrix defined by (1.20), this equality gives an expression of the periods of holomorphic 1-forms rational over the field of definition by means of "Thetanullwerte". (It should be noted that $\Re_S[w]$ is algebraic over the field of moduli of $A(w, \delta)$ polarized by hyperplane sections.) Thus (1.24) may be regarded as an answer to the question posed by Weil in [8].

We now study the representation of an isogeny with respect to the basis $\{\xi_\nu\}$ of Th. 1.5. Let α be an element of $G_{\mathbf{Q}+}$. Put $\alpha^{-1} = \begin{pmatrix} a & b \\ c & d \end{pmatrix}$ and $\beta = \begin{pmatrix} 1 & 0 \\ 0 & \delta \end{pmatrix} \alpha \begin{pmatrix} 1 & 0 \\ 0 & \delta \end{pmatrix}^{-1}$. Then

$$ {}^t(cw + d)^{-1}(w\ \delta) = (\alpha^{-1}(w)\ \delta) \cdot {}^t\beta.$$

Suppose $\beta \in M_{2n}(\mathbf{Z})$. Then the linear map $u \mapsto {}^t(cw + d)^{-1}u$ of $\mathbf{C}^n$ onto itself gives an isogeny λ of $A(w, \delta)$ onto $A(\alpha^{-1}(w), \delta)$. Take P as in Th. 1.5, and suppose that $\det (P(\alpha^{-1}(w)) \neq 0$. Define 1-forms ξ_ν on $A(w, \delta)$ as in Th. 1.5, and similarly ξ_ν' on $A(\alpha^{-1}(w), \delta)$ with $P(\alpha^{-1}(w))$ in place of $P(w)$. Since λ is represented by ${}^t(cw + d)^{-1}$, we obtain

$$(1.25) \qquad \begin{bmatrix} \xi_1' \circ \lambda \\ \vdots \\ \xi_n' \circ \lambda \end{bmatrix} = {}^t P(\alpha^{-1}(w)) \cdot {}^t(cw + d)^{-1} \cdot {}^t P(w)^{-1} \begin{bmatrix} \xi_1 \\ \vdots \\ \xi_n \end{bmatrix}.$$

Thus ${}^t[P(w)^{-1}(cw + d)^{-1}P(\alpha^{-1}(w))]$ represents λ with respect to the bases $\{\xi_\nu\}$ and $\{\xi_\nu'\}$.

2. Differential operators $\mathfrak{d}_H$

Throughout this section, we fix a complex n-dimensional vector space V. Let $\mathfrak{D}$ denote the vector space of all differential operators $\mathfrak{d}$ of the form

$$(2.1) \qquad \mathfrak{d} = \sum_{\nu=1}^{n} c_\nu \, \partial/\partial u_\nu.$$

with $c_v \in \mathbf{C}$, where $\{u_1 , \cdots , u_n\}$ is a complex coordinate system of V. The map $c = (c_1 , \cdots , c_n) \mapsto \mathfrak{d}$ gives a $\mathbf{C}$-linear isomorphism of V onto $\mathfrak{D}$, which is independent of the choice of coordinate system. Indeed, $\mathfrak{d}$ corresponds to $c \in V$ if $\mathfrak{d}p = p(c)$ for every $\mathbf{C}$-linear map p of V into $\mathbf{C}$. We can of course call $\mathfrak{D}$ the (complex) Lie algebra of V, or of V/L for any lattice L in V. Further, for every $\mathbf{C}$-linear endomorphism β of V, we can define an element $\beta\mathfrak{d}$ of $\mathfrak{D}$ by

$$(2.2) \qquad (\beta\mathfrak{d})(f) \circ \beta = \mathfrak{d}(f \circ \beta)$$

for functions f on V. If $\mathfrak{d}$ corresponds to c, then $\beta\mathfrak{d}$ corresponds to βc.

Now we fix a hermitian form $H(u, v)$ on V, $\mathbf{C}$-linear in v, and define various operators relative to H. First, for an element $\mathfrak{d}$ of $\mathfrak{D}$ corresponding to $c \in V$, we define $\mathfrak{d}_H$ by

$$(2.3) \qquad (\mathfrak{d}_H f)(u) = -\pi H(u, c)f(u) + (\mathfrak{d}f)(u) \qquad (u \in V).$$

The functions f to which we apply $\mathfrak{d}$ and $\mathfrak{d}_H$ will be polynomials in $\bar{u}_1 , \cdots , \bar{u}_n$ of the form

$$(2.4) \qquad f(u) = \sum_a \bar{u}_1^{a_1} \cdots \bar{u}_n^{a_n} g_a(u)$$

with meromorphic functions g_a on V. For such an f, we define functions f_H and $S_H^v f$ on V, for each $v \in V$, by

$$(2.5) \qquad f_H(u) = e\!\left(\frac{i}{4} H(u, u)\right)\!f(u),$$

$$(2.6) \qquad (S_H^v f)(u) = e\!\left(\frac{i}{2} H(v, u + \tfrac{1}{2}v)\right)\!f(u + v).$$

Then we can easily verify the following four formulas:

$$(2.7) \qquad \mathfrak{d}_H S_H^v = S_H^v \mathfrak{d}_H ,$$

$$(2.8) \qquad (\mathfrak{d}_1 \cdots \mathfrak{d}_m f)(0) = (\mathfrak{d}_{1H} \cdots \mathfrak{d}_{mH} f)(0), \qquad (\mathfrak{d}_1 , \cdots , \mathfrak{d}_m \in \mathfrak{D}),$$

$$(2.9) \qquad (\mathfrak{d}_1 \cdots \mathfrak{d}_m S_H^v f)(0) = (\mathfrak{d}_{1H} \cdots \mathfrak{d}_{mH} f)_H(v),$$

$$(2.10) \qquad (S_H^v f)_H(u) = e(E(u, v)/2)f_H(u + v),$$

where we put

$$(2.11) \qquad E(u, v) = \mathrm{Im}\,(H(u, v)).$$

Next, we fix a lattice L in V, and consider a map ψ of L into $\mathbf{C}$ such that

$$\psi(l + m) = \psi(l)\psi(m)e(E(l, m)/2) \qquad (l, m \in L).$$

Then we denote by $\mathfrak{T}(H, \psi, L)$ the set of all functions of the above type satisfying

$$f(u + l) = f(u)\psi(l)e\!\left(\frac{1}{2i} H(l, u + \tfrac{1}{2}l)\right) \qquad (u \in V, l \in L).$$

PROPOSITION 2.1. *If $f \in \mathfrak{T}(H, \psi, L)$, $\mathfrak{d} \in \mathfrak{D}$, and $v \in V$, we have*

$$(2.12) \qquad \mathfrak{d}_H f \in \mathfrak{T}(H, \psi, L),$$

$$(2.13) \qquad S_H{}^v f \in \mathfrak{T}(H, \psi', L), \quad where \quad \psi'(l) = \psi(l)e(E(l, v)),$$

$$(2.14) \qquad f_H(u + l) = \psi(l)e(E(l, u)/2)f_H(u) \qquad (u \in V, l \in L).$$

These formulas can be verified in a straightforward way.

We now specialize our discussion by assuming that V/L is an abelian variety with many complex multiplications in the sense of [7, §2], H is a Riemann form, and $\psi(l)$ is a root of unity for every $l \in L$. The form H is then non-negative; we do not assume that H is positive definite (cf. [7, §3]). As in [7, §2], we can define an algebra

$$(2.15) \qquad Y = M_{m_1}(K_1) + \cdots + M_{m_t}(K_t)$$

acting on V, which is identified with End $(V/L) \otimes \mathbf{Q}$. Further, a *CM*-field K' and a map η of $K_{\mathbf{A}}'^{\times} \to Y_{\mathbf{A}}^{\times}$ can be defined. We denote by $T(H, \psi, L)$ the vector space of all holomorphic elements of $\mathfrak{T}(H, \psi, L)$, and by $T_a(H, \psi, L)$ the set of all elements f of $T(H, \psi, L)$ such that $f_H(u) \in K_{ab}'$ for all $u \in \mathbf{Q}L$. (In [7], we denoted f_H by f_*.) If $f \in T_a(H, \psi, L)$, then $S_H{}^v f \in T_a(H, \psi', L)$ for every $v \in \mathbf{Q}L$ with ψ' of (2.13), as observed in [7, Prop. 2.1].

Instead of fixing a single lattice L, it is more convenient to consider the set Λ of all lattices commensurable with a given lattice. Then we put

$$(2.16) \qquad \begin{aligned} T(H) &= \bigcup_{L \in \Lambda} \bigcup_{\psi} T(H, \psi, L), \\ T_a(H) &= \bigcup_{L \in \Lambda} \bigcup_{\psi} T_a(H, \psi, L), \end{aligned}$$

where ψ runs over all possible ones whose values are roots of unity. For each $L \in \Lambda$, take an abelian variety A, say projective, isomorphic to V/L such that all points of A of finite order are rational over K_{ab}'. (For instance, we can take $A(w, \delta)$ of §1 defined for the present V/L.) Let $\mathfrak{F}_{L,a}$ be the field of all K_{ab}'-rational functions on A viewed as functions on V. This is the same as the field of all L-invariant meromorphic functions f on V such that $f(u) \in K_{ab}'$ for all $u \in \mathbf{Q}L$. Then we put

$$(2.17) \qquad \mathfrak{F}_{\Lambda.a} = \bigcup_{L \in \Lambda} \mathfrak{F}_{L.a} .$$

Now we call an element $\mathfrak{d}$ of $\mathfrak{D}$ *arithmetic*, or K_{ab}'-*rational*, if $\mathfrak{d}$ maps $\mathfrak{F}_{\Lambda.a}$ into itself. This is so if and only if $\mathfrak{d}\mathfrak{F}_{L.a} \subset \mathfrak{F}_{L.a}$ for at least one member L of Λ. Let $\mathfrak{D}_a$ denote the set of all such $\mathfrak{d}$. Obviously $\mathfrak{D}_a$ is an n-dimensional vector space over K_{ab}'. In view of (2.2), $\mathfrak{D}_a$ is stable under the action of elements of Y.

THEOREM 2.2. *Let $\mathfrak{d}_1, \cdots, \mathfrak{d}_m$ be elements of $\mathfrak{D}_a$, and let $f \in T_a(H)$. Then $(\mathfrak{d}_{1H} \cdots \mathfrak{d}_{mH}f)_H(v) \in K_{ab}'$ for every $v \in \mathbf{Q}L$.*

Proof. Take an element $\beta = b_1 1_{m_1} \oplus \cdots \oplus b_t 1_{m_t}$ of the center of Y with $b_\nu \in K$, such that $b_1 \bar{b}_1 = \cdots = b_t \bar{b}_t = 1$. Then we have $H(\beta u, \beta v) = H(u, v)$ (even if H is degenerate). By virtue of (2.9), it is sufficient to prove

$$(2.18) \qquad (\mathfrak{d}_1 \cdots \mathfrak{d}_m f)(0) \in K_{ab}'$$

for every $f \in T_a(H)$, $\neq 0$. We can also assume that $f(0) \neq 0$. Indeed, if $f(0) = 0$, we take $v \in QL$ so that $f(v) \neq 0$, and put $g = f + S_H{}^v f$. Then $g \in T_a(H)$ and $g(0) \neq 0$. Expressing f as $f = g - S_H{}^v f$ and proving (2.18) for g and $S_H{}^v f$, we obtain the desired result for f. Thus assuming that $f(0) \neq 0$, put $g(u) = f(\beta u)/f(u)$. Then $g \in \mathfrak{F}_{A,a}$ and $g(0) = 1$. Applying $\mathfrak{d}$ to the relation $f(\beta u) = f(u)g(u)$, we have

$$(2.19) \qquad ((\beta \mathfrak{d})f) \circ \beta = g \, \mathfrak{d}f + f \, \mathfrak{d}g.$$

Let $\{\mathfrak{d}^1, \cdots, \mathfrak{d}^n\}$ be a basis of $\mathfrak{D}_a$ over K_{ab}'. Then $\beta \mathfrak{d}^\mu = \sum_{\nu=1}^n c_{\mu\nu} \mathfrak{d}^\nu$ with $c_{\mu\nu} \in K_{ab}'$. Putting $C = (c_{\mu\nu})$, we obtain, from (2.19),

$$(2.20) \qquad (C - 1)\begin{bmatrix} \mathfrak{d}^1 f(0) \\ \vdots \\ \mathfrak{d}^n f(0) \end{bmatrix} = f(0)\begin{bmatrix} \mathfrak{d}^1 g(0) \\ \vdots \\ \mathfrak{d}^n g(0) \end{bmatrix}.$$

Observe that $f(0) \in K_{ab}'$ as $f \in T_a(H)$, and $\mathfrak{d}^\nu g(0) \in K_{ab}'$ as $\mathfrak{d}^\nu g \in \mathfrak{F}_{A,a}$. Now the eigenvalues of the matrix C are those of $\Phi_1(b_1), \cdots, \Phi_t(b_t)$, where Φ_λ is the representation of K_λ on V defined as in [7, §2]. Take all b_ν to be different from 1. Then (2.20) shows that $\mathfrak{d}^\nu f(0) \in K_{ab}'$ for all ν. This proves our theorem in the case $m = 1$. To prove the general case, we let $\mathfrak{d}^\nu$ act on (2.19) successively and apply induction on m. Then we find an equality of the form

$$(2.21) \qquad (C \otimes \cdots \otimes C - 1)X = K_{ab}'\text{-rational vector},$$

where X is the vector whose components are $(\mathfrak{d}^{\nu_1} \cdots \mathfrak{d}^{\nu_m} f)(0)$. Since $(K_\lambda, \Phi_\lambda)$ is a CM-type, we can choose β so that no product of m eigenvalues of C become 1, and thereby obtain the desired result.

In [7], we defined an action of $K_A'^\times$ on $T_a(H)$ and also an action of $K_A'^\times$ on $\mathfrak{F}_{A,a}$. For $x \in K_A'^\times$ and $f \in T_a(H)$ or $f \in \mathfrak{F}_{A,a}$, let f^x denote the image of f under x. For $\mathfrak{d} \in \mathfrak{D}_a$, we can define an element $\mathfrak{d}^x$ of $\mathfrak{D}_a$ by

$$(2.22) \qquad \mathfrak{d}^x(f^x) = (\mathfrak{d}f)^x \qquad (f \in \mathfrak{F}_{A,a}).$$

By means of [7, Prop. 5.1, (iii), (iv)], we can easily verify

$$(2.23) \qquad \mathfrak{d}^x = \eta(x)^{-1}\mathfrak{d} \quad \text{if} \quad x \in K'^\times,$$

$$(2.24) \qquad \mathfrak{d}^x = \mathfrak{d} \qquad \text{if} \quad x \in K_\infty'^\times.$$

For simplicity, we write $\mathfrak{d}_H{}^x$ for $(\mathfrak{d}^x)_H$.

THEOREM 2.3. *Let x be an element of $K_A'^\times$ and b the absolute norm of the ideal in K' associated with x. Further let η be the map of $K_A'^\times$ into $Y_A^\times$ defined* *by* [7, (25)]. *Then, for every* $f \in T_a(H)$, $v \in QL$, *and* $\mathfrak{d}_1, \cdots, \mathfrak{d}_m$ *in* $\mathfrak{D}_a$, *we have*

$$(\mathfrak{d}_{1H} \cdots \mathfrak{d}_{mH} f)_H(v)^x = (\mathfrak{d}_{1H'}{}^x \cdots \mathfrak{d}_{mH'}{}^x f^x)_{H'} \cdot (\eta(x)^{-1}v),$$

where $H' = bH$.

Proof. First we observe that if $f \in T(H, \psi, L)$ and $l \in L$, then

$$(2.25) \qquad S_H{}^{v+l} f = \psi(l)e(E(l, v)/2)S_H{}^v f.$$

Therefore, if $v \in QL$, $S_H{}^v f$ depends only on the class of v modulo a sufficiently small lattice. Next, if $y = N_{K'/Q}(x)$, we have $y_\infty > 0$ and $yZ = bZ$, so that $e(r)^x = e(by^{-1}r)$ for every $r \in Q$. Now, by the main theorem of [7], if $f \in T_a(H)$ and $u, v \in QL$, then $f^x \in T_a(H')$ and $f_H(u)^x = (f^x)_{H'} \cdot (\eta(x)^{-1}u)$, and hence, by (2.10), we have

$$(S_H{}^v f)_H{}^x(u) = (S_H{}^v f)_H(\eta(x)u)^x = [e(E(\eta(x)u, v)/2)f_H(\eta(x)u + v)]^x$$

$$= e(bE(u, \eta(x)^{-1}v)/2)f_{H'}{}^x(u + \eta(x)^{-1}v) = (S_{H'}{}^w f^x)_{H'}(u)$$

with a point w of QL sufficiently close to $\eta(x)^{-1}v$ in the sense explained in [7, §2]. By virtue of the remark at the beginning of the proof, w can be chosen independently of u. Thus $(S_H{}^v f)^x = S_{H'}{}^w f^x$. Combining this with (2.9), we see that our theorem follows from

$$(2.26) \qquad (\mathfrak{d}_1 \cdots \mathfrak{d}_m f)(0)^x = (\mathfrak{d}_1{}^x \cdots \mathfrak{d}_m{}^x f^x)(0) \qquad (f \in T_a(H)).$$

This can be obtained by applying x to (2.20) and (2.21), and using induction on m. The idea is similar to the proof of the main theorem III of [5].

We have defined $\mathfrak{b}^x$ in the above rather formally by (2.22). Let us now give a more explicit description of $\mathfrak{b}^x$ with the same ideas as in Th. 1.5. We first identify V/L with $\mathbf{C}^n/L(w, \delta)$ as in §1 for some w and δ. The above Riemann form H of V/L may or may not be the same as the form H_1 of $\mathbf{C}^n/L(w, \delta)$ defined by (1.4) (with of course w as z). Then we consider the embedding Θ_w of $\mathbf{C}^n/L(w, \delta)$ onto $A(w, \delta)$ defined by (1.5_b). For every algebro-geometric object B rational over K_{ab}' and $x \in K_A'{}^\times$, we let B^x denote the image of B under the automorphism of K_{ab}' corresponding to x. In particular, $A(w, \delta)^x$ is meaningful. If g is a K_{ab}'-rational function on $A(w, \delta)$, then g^x is defined as a function of $A(w, \delta)^x$. On the other hand, since $g \circ \Theta_w \in \mathfrak{F}_{A,a}$, $(g \circ \Theta_w)^x$ is also defined. The relationship between these will be given by (2.29) below. Now we fix, as in [7, §2], an anti-representation Φ of Y into $M_n(\mathbf{C})$ that gives the action of Y on $\mathbf{C}^n$, and define an injection q of Y into $M_{2n}(\mathbf{Q})$ by

$$\Phi(a)(w\ 1_n) = (w\ 1_n) \cdot {}^t q(a) \qquad (a \in Y).$$

THEOREM 2.4. *Let* $x \in K_A'{}^\times$. *Then* $q(\eta(x))^{-1} = k\alpha$ *with* $k \in R_s$ *and* $\alpha \in G_{Q+}$, *where* R_s *is defined by* (1.11). *With any choice of such an* α, *put* $\alpha = \begin{pmatrix} a & b \\ c & d \end{pmatrix}$ *and* $\lambda = {}^t(cw + d)^{-1}$. *Then we have*

$$(2.27) \qquad A(w, \delta)^x = A(\alpha(w), \delta),$$

(2.28) $\Theta_w(u)^x = \Theta_{\alpha(w)}(\lambda \cdot \Phi(\eta(x)^{-1})u)$ *for every* $u \in \mathbf{Q} \cdot L(w, \delta)$,

(2.29) $(g \circ \Theta_w)^x = g^x \circ \Theta_{\alpha(w)} \circ \lambda$ *for every* K_{ab}'-*rational function* g *on* $A(w, \delta)$.

Moreover, let P *be an element of* $\mathcal{C}_0'(\Gamma_\delta, \mathbf{Q})$ *holomorphic at* w *and such that* $\det(P(w)) \neq 0$, *and define 1-forms* ξ_r *on* $A(w, \delta)$ *and elements* $\mathfrak{d}_r$ *of* $\mathfrak{D}_a$ *by*

$$\begin{bmatrix} \xi_1 \\ \cdot \\ \cdot \\ \cdot \\ \xi_n \end{bmatrix} \circ \Theta_w = 2\pi i \cdot {}^t P(w) \begin{bmatrix} du_1 \\ \cdot \\ \cdot \\ \cdot \\ du_n \end{bmatrix}, \qquad \begin{bmatrix} \mathfrak{d}_1 \\ \cdot \\ \cdot \\ \mathfrak{d}_n \end{bmatrix} = [2\pi i P(w)]^{-1} \begin{bmatrix} \partial/\partial u_1 \\ \cdot \\ \cdot \\ \partial/\partial u_n \end{bmatrix}.$$

Then $\det(P(\alpha(w))) \neq 0$, *and*

$$\begin{bmatrix} \xi_1{}^x \\ \cdot \\ \cdot \\ \xi_n{}^x \end{bmatrix} \circ \Theta_{\alpha(w)} = 2\pi i \cdot {}^t P(\alpha(w)) \begin{bmatrix} du_1 \\ \cdot \\ \cdot \\ du_n \end{bmatrix}, \qquad \begin{bmatrix} \mathfrak{d}_1{}^x \\ \cdot \\ \cdot \\ \mathfrak{d}_n{}^x \end{bmatrix} = [2\pi i \cdot {}^t\lambda P(\alpha(w))]^{-1} \begin{bmatrix} \partial/\partial u_1 \\ \cdot \\ \cdot \\ \partial/\partial u_n \end{bmatrix}.$$

Here $\mathfrak{d}^x$ is defined by (2.22); it is *not* the transform under x of the derivation of the function field of $A(w, \delta)$ as an algebro-geometric object. The difference will be clarified in the following proof.

Proof.   The decomposition $q(\eta(x))^{-1} = k\alpha$ follows immediately from [4, I, 3.4; II, (3.10.3)]. To prove the main part of our theorem, let us first state a basic principle:

(2.30)   *If an element* g *of* $\mathcal{C}_0(\Gamma_\delta, \mathbf{Q})$ *is finite at* w, *then* $g \circ \alpha$ *is also finite at* w *and* $g(w)^x = g(\alpha(w))$.

This is an obvious consequence of [4, I, (2.7.3)] and (1.13). Put $\theta(z) = \theta(0, z; 0, 0)$. Take $h \in \theta \cdot \mathcal{C}_0(\Gamma_\delta, \mathbf{Q})$ so that $h(w) \neq 0$ (e.g. $h(z) = \theta(0, z; t, 0)$ with a suitable $t \in \delta^{-1}\mathbf{Z}^n$), and put

$$f_r(u, z) = \theta(u, z; r, 0)/h(z),$$

$$g_r(z) = f_r(0, z) = \theta(0, z; r, 0)/h(z) \qquad (r \in \delta^{-1}\mathbf{Z}^n).$$

Then $g_r \in \mathcal{C}_0(\Gamma_\delta, \mathbf{Q})$, and hence g_r is finite at $\alpha(w)$ by (2.30). Since

$$0 \neq \theta(0, \alpha(w); r, 0) = h(\alpha(w))g_r(\alpha(w))$$

for at least one $r \in \delta^{-1}\mathbf{Z}^n$, we see that $h(\alpha(w)) \neq 0$. Put $f_r(u) = f_r(u, w)$. Then $f_r \in T_a(H_1)$ by [7, Prop. 2.4]. Now we have $f_r{}^x(u) = f_r(\lambda u, \alpha(w))$. This is essentially included in the proof of the main theorem of [7]. Since the embedding Θ_w is defined by

$$u \mapsto (f_r(u, w))_{r \in \mathfrak{F}},$$

we obtain (2.27) and (2.28); (2.29) follows from (2.28) and the definition of the action of x on $\mathfrak{F}_{A,a}$ in [7, §5]. (Note also that (2.28) is a special case of [7, (33)].) Next define q_k, s_k, and σ_k as in the proof of Th. 1.5, and put

$$q_k{}'(u) \;=\; \varphi_k(u,\,\alpha(w))/\varphi_l(u,\,\alpha(w));$$

define a vector ξ' of 1-forms on $A(\alpha(w),\,\delta)$ by

$$\xi' \circ \Theta_{\alpha(w)} \;=\; 2\pi i \cdot {}^t\!P(\alpha(w)) \begin{bmatrix} du_1 \\ \vdots \\ du_n \end{bmatrix}.$$

Here we use the same l as in (1.21). In fact, we have $\varphi_l(0,\,\alpha(w)) \neq 0$ by virtue of Lemma 2.5 below. Also $\det\,(P(\alpha(w))) \neq 0$ for the same reason. Put, for simplicity, $\Theta = \Theta_w$ and $\Theta' = \Theta_{\alpha(w)}$. Now the proof of Th. 1.5 shows that $dq_k = {}^t s_k(\xi \circ \Theta)$ and $dq_k{}' = {}^t s_k{}'(\xi' \circ \Theta')$ with $s_k{}'(u) = \sigma_k(u,\,\alpha(w))$. The relation $f_r{}^z(u) = f_r(\lambda u,\,\alpha(w))$ implies that $q_k{}^z = q_k{}' \circ \lambda$. Let $\bar{s}_k$, $\bar{q}_k$, and $\bar{s}_k{}'$ denote the functions on $A(w,\,\delta)$ and $A(\alpha(w),\,\delta)$ such that $s_k = \bar{s}_k \circ \Theta$, $q_k = \bar{q}_k \circ \Theta$, and $s_k{}' = \bar{s}_k{}' \circ \Theta'$. By (2.29), we have $q_k{}' \circ \lambda = \bar{q}_k{}^z \circ \Theta' \circ \lambda$, so that $q_k{}' = \bar{q}_k{}^z \circ \Theta'$. Now $d\bar{q}_k = {}^t\bar{s}_k\xi$, and hence

$$(2.31) \qquad ({}^t s_k{}^z \circ \lambda^{-1})(\xi^z \circ \Theta') = ({}^t\bar{s}_k{}^z\xi^z) \circ \Theta' = (d\bar{q}_k)^z \circ \Theta' = dq_k{}' = {}^t s_k{}'(\xi' \circ \Theta').$$

Since the components of $\sigma_k(0,\,z)$ belong to $\mathcal{C}_0(\Gamma_\delta,\,\mathbf{Q})$, we have, by (2.30),

$$s_k{}^z(0) \;=\; s_k(0)^z \;=\; \sigma_k(0,\,w)^z \;=\; \sigma_k(0,\,\alpha(w)) \;=\; s_k{}'(0).$$

This together with (2.31) proves that $\xi^z = \xi'$. For each $\mathfrak{b} \in \mathfrak{D}_a$, let $\bar{\mathfrak{b}}$ denote the derivation of the function field of $A(w,\,\delta)$ defined by

$$(2.32) \qquad\qquad (\bar{\mathfrak{b}}g) \circ \Theta \;=\; \mathfrak{b}(g \circ \Theta).$$

Our result about ξ^z shows that

$$\left\{ \begin{bmatrix} \bar{\mathfrak{b}}_1{}^z \\ \vdots \\ \bar{\mathfrak{b}}_n{}^z \end{bmatrix} p \right\} \circ \Theta' \;=\; [2\pi i P(\alpha(w))]^{-1} \begin{bmatrix} \partial/\partial u_1 \\ \vdots \\ \partial/\partial u_n \end{bmatrix} (p \circ \Theta')$$

for functions p on $A(\alpha(w),\,\delta)$. On the other hand, we obtain, from (2.32),

$$(\bar{\mathfrak{b}}^z g^z) \circ \Theta' \circ \lambda \;=\; \mathfrak{b}^z(g^z \circ \Theta' \circ \lambda),$$

and hence the desired result concerning $\mathfrak{b}_r{}^z$ follows.

LEMMA 2.5. *If an element h of $\theta^k \mathcal{C}_0(\Gamma_\delta,\,\mathbf{Q})$, with $k \in \mathbf{Z}$, is finite at w and $h(w) \neq 0$, then h is finite at $\alpha(w)$ and $h(\alpha(w)) \neq 0$ for every α as in Theorem 2.4.*

Proof. This is similar to [5, Prop. 10], and can be proved by the same idea in a more general case. Here we give a different proof. If $k = 1$, our assertion is included in the above proof of Th. 2.4. To prove the case $k \neq 1$, take $g \in \theta \cdot \mathcal{C}_0(\Gamma_\delta,\,\mathbf{Q})$ so that $g(w) \neq 0, \neq \infty$, and put $f = g^{-k}h$. Then the desired result follows from (2.30) and the special case $k = 1$.

Remark 2.6. In the proof of the main theorem of [7], we used, without proof, the fact that $\varphi(u, \Omega \cdot {}^t\alpha, r, \tau s)$ does not become infinity. This is in fact guaranteed by the above lemma.

3. The Hilbert modular case

Let F be a totally real algebraic number field of degree n, and $\tau_1, \cdots, \tau_n$ the injections of F into $\mathbf{R}$. We denote by $GL_2{}^+(F)$ the subgroup of $GL_2(F)$ consisting of all α such that det (α) is totally positive. We embed F and $GL_2(F)$ into $\mathbf{R}^n$ and $GL_2(\mathbf{R})^n$ through the map

$$F \ni a \mapsto (a^{\tau_1}, \cdots, a^{\tau_n}) \in \mathbf{R}^n,$$

and define an action of $GL_2{}^+(F)$ on $\mathfrak{H}_1{}^n$. For a lattice $\mathfrak{b}$ in F,

$$v \in \mathbf{C}^n, \quad z = (z_1, \cdots, z_n) \in \mathfrak{H}_1{}^n, \quad \rho = (\rho_1, \cdots, \rho_n) \in \mathbf{R}^n,$$
$$\text{and} \quad \sigma = (\sigma_1, \cdots, \sigma_n) \in \mathbf{R}^n,$$

we define functions $\mathfrak{f}$ and t by

$$(3.1) \qquad \mathfrak{f}(v, z; \rho, \sigma; \mathfrak{b}) = e(\tfrac{1}{2} \operatorname{Tr} ((z - \bar{z})^{-1}v^2))t(v, z; \rho, \sigma; \mathfrak{b}),$$

$$(3.2) \qquad t(v, z; \rho, \sigma; \mathfrak{b}) = \sum_{x - \rho \in \mathfrak{b}} e(\operatorname{Tr} (\tfrac{1}{2}zx^2 + x(v + \sigma))),$$

where we understand that $\operatorname{Tr} (ab^m) = \sum_{\nu=1}^n a^{\tau_\nu}b_\nu{}^m$ for $a \in F$, $b = (b_1, \cdots, b_n) \in \mathbf{C}^n$, and $m \in \mathbf{Z}$. To see that $\mathfrak{f}$ is actually the pull-back of the function φ of (1.2), take a basis $\beta_1, \cdots, \beta_n$ of $\mathfrak{b}$ over $\mathbf{Z}$, and put

$$(3.3) \qquad B = \begin{bmatrix} \beta_1{}^{\tau_1} & \cdots & \beta_n{}^{\tau_1} \\ \cdots & \cdots & \cdots \\ \beta_1{}^{\tau_n} & \cdots & \beta_n{}^{\tau_n} \end{bmatrix},$$

$$(3.4) \qquad W_B(z) = {}^tB \begin{bmatrix} z_1 & & \\ & \ddots & \\ & & z_n \end{bmatrix} B \qquad (z = (z_1, \cdots, z_n) \in \mathfrak{H}_1{}^n).$$

Then W_B maps $\mathfrak{H}_1{}^n$ into $\mathfrak{H}_n$. As noted in [7], this is compatible with the map I of $SL_2(F)$ into $G_\mathbf{Q}$ defined by

$$(3.5) \qquad I\left[\begin{pmatrix} a & b \\ c & d \end{pmatrix}\right] = \begin{bmatrix} {}^tB & 0 \\ 0 & B^{-1} \end{bmatrix} \begin{bmatrix} \Phi(a) & \Phi(b) \\ \Phi(c) & \Phi(d) \end{bmatrix} \begin{bmatrix} {}^tB^{-1} & 0 \\ 0 & B \end{bmatrix}, \qquad \Phi(a) = \begin{bmatrix} a^{\tau_1} & & \\ & \ddots & \\ & & a^{\tau_n} \end{bmatrix},$$

in the sense that if $\gamma = \begin{pmatrix} a & b \\ c & d \end{pmatrix} \in SL_2(F)$ and $\alpha = I(\gamma) = \begin{pmatrix} p & q \\ r & s \end{pmatrix}$, then $\alpha(W_B(z)) = W_B(\gamma(z))$, and

$$(3.6) \qquad r \cdot W_B(z) + s = B^{-1} \begin{bmatrix} c^{\tau_1}z_1 + d^{\tau_1} & & \\ & \ddots & \\ & & c^{\tau_n}z_n + d^{\tau_n} \end{bmatrix} B.$$

We see easily that

$$(3.7) \qquad \mathfrak{f}(v, z; \rho, \sigma; \mathfrak{b}) = \varphi({}^t Bv, W_B(z); B^{-1}\rho, {}^t B\sigma).$$

Now, with δ as in (1.3), let $\beta_1{}^*, \cdots, \beta_n{}^*$ be the elements of F determined by

$$(3.8) \qquad \mathrm{Tr}\,(\beta_\kappa\beta_\lambda{}^*) = (\kappa, \lambda)\text{-component of } \delta,$$

and let $\mathfrak{b}^* = \sum_{\lambda=1}^n \mathbf{Z}\beta_\lambda{}^*$. Further, define a lattice $L(z; \mathfrak{b}, \mathfrak{b}^*)$ in $\mathbf{C}^n$ by

$$(3.9) \qquad L(z; \mathfrak{b}, \mathfrak{b}^*) = \left\{ \begin{bmatrix} b^{\tau_1}z_1 + c^{\tau_1} \\ \vdots \\ b^{\tau_n}z_n + c^{\tau_n} \end{bmatrix} \ \middle|\ b \in \mathfrak{b}, c \in \mathfrak{b}^* \right\} \qquad (z \in \mathfrak{H}_1{}^n).$$

Then we have

$$(3.10) \qquad {}^t B \cdot L(z; \mathfrak{b}, \mathfrak{b}^*) = L(W_B(z), \delta),$$

so that ${}^t B$ gives an isomorphism of $\mathbf{C}^n/L(z; \mathfrak{b}, \mathfrak{b}^*)$ onto $\mathbf{C}^n/L(W_B(z), \delta)$. Thus $A(W_B(z), \delta)$ is the projective embedding of $\mathbf{C}^n/L(z; \mathfrak{b}, \mathfrak{b}^*)$ by means of the functions $\mathfrak{f}$. Our present aim is to reformulate Th. 1.5 in terms of the coordinate functions on $\mathbf{C}^n/L(z; \mathfrak{b}, \mathfrak{b}^*)$. For this purpose, define $\mathfrak{M}_r(\mathfrak{X})$ and $\mathfrak{a}_r(\mathfrak{X})$ as in [5] for $r = (r_1, \cdots, r_n) \in \mathbf{Z}^n$ and a subfield $\mathfrak{X}$ of $\mathbf{C}$, and denote by $\{e_1, \cdots, e_n\}$ the standard basis of $\mathbf{Z}^n$; namely $e_1 = (1, 0, \cdots, 0), \cdots, e_n = (0, \cdots, 0, 1)$. Further, for each $z_0 \in \mathfrak{H}_1{}^n$, we denote by $\mathfrak{K}[z_0]$ the field generated by the values $\mathfrak{f}(z_0)$ for all $\mathfrak{f} \in \mathfrak{a}_0(\mathbf{Q}_{ab})$. We see easily that $\mathfrak{K}[z_0] = \mathfrak{R}[W_B(z_0)]$. As in [5, §5], we let every automorphism σ of $\mathbf{C}$ act on $\mathfrak{a}_r(\mathbf{C})$. In particular, if $f(z) = \sum_{x \in F} c_x\,e(\mathrm{Tr}\,(xz)) \in \mathfrak{M}_r(\mathbf{C})$, we have

$$(3.11) \qquad f^\sigma(z) = \sum_x c_x{}^\sigma e(\mathrm{Tr}\,(xz)).$$

PROPOSITION 3.1. *For each point z_0 of $\mathfrak{H}_1{}^n$, there exist n functions $h_1, \cdots, h_n$ with the following properties:*

   (i) $h_\nu \in \mathfrak{a}_{e_\nu}(F^{\tau_\nu})$;
   (ii) $h_\nu{}^\sigma = h_\mu$ *if σ is an isomorphism of F^{τ_ν} onto F^{τ_μ} such that $\tau_\nu\sigma = \tau_\mu$* ;
   (iii) h_ν *is holomorphic at z_0 and $h_\nu(z_0) \neq 0$ for every ν.*

Proof. Take P as in Th. 1.5 with $w = W_B(z_0)$, and put

$$\begin{bmatrix} h_1(z) \\ \vdots \\ h_n(z) \end{bmatrix} = B \cdot P(W_B(z))q \qquad (z \in \mathfrak{H}_1{}^n)$$

with any $q \in \mathbf{Q}^n$. We can easily verify that $h_1, \cdots, h_n$ satisfy (i) and (ii); a suitable choice of q guarantees (iii).

Let $v_1, \cdots, v_n$ be the standard coordinate functions of $\mathbf{C}^n$. We view $dv_1, \cdots, dv_n$ as 1-forms on $A(W_B(z), \delta)$ through the map of $\mathbf{C}^n/L(z; \mathfrak{b}, \mathfrak{b}^*)$ onto $A(W_B(z), \delta)$ mentioned above.

THEOREM 3.2. *Let z_0 be a point of $\mathfrak{H}_1^n$, and let $h_1, \cdots, h_n$ be functions satisfying the conditions of Proposition 3.1. Then $2\pi i \cdot h_\nu(z_0) \, dv_\nu$ and $[2\pi i \cdot h_\nu(z_0)]^{-1} \, \partial/\partial v_\nu$ are rational over $F^{\tau_\nu}\mathcal{K}[z_0]$. Moreover, if σ is an automorphism of $\mathbf{C}$ over $\mathcal{K}[z_0]$ such that $\tau_\nu\sigma = \tau_\mu$ on F, then*

$$(2\pi i \cdot h_\nu(z_0) \, dv_\nu)^\sigma = 2\pi i \cdot h_\mu(z_0) \, dv_\mu \,,$$

$$([2\pi i \cdot h_\nu(z_0)]^{-1} \, \partial/\partial v_\nu)^\sigma = [2\pi i \cdot h_\mu(z_0)]^{-1} \, \partial/\partial v_\mu \,.$$

Proof. With P as in the proof of Prop. 3.1, put

$$(3.12) \qquad Q(z) = \begin{bmatrix} h_1(z) & & \\ & \ddots & \\ & & h_n(z) \end{bmatrix}^{-1} BP(W_B(z)) \qquad (z \in \mathfrak{H}_1^n).$$

If $u = {}^t Bv$, we have

$$(3.13) \qquad 2\pi i \begin{bmatrix} h_1(z) \, dv_1 \\ \vdots \\ h_n(z) \, dv_n \end{bmatrix} = {}^t Q(z)^{-1} 2\pi i \cdot {}^t P(W_B(z)) \begin{bmatrix} du_1 \\ \vdots \\ du_n \end{bmatrix} = {}^t Q(z)^{-1} \begin{bmatrix} \xi_1 \\ \vdots \\ \xi_n \end{bmatrix},$$

where ξ_ν is defined for the point $W_B(z)$ as in Th. 1.5. Put ${}^t Q(z)^{-1} = (q_{\mu\nu}(z))$. Then

$$2\pi i \cdot h_\nu(z_0) \, dv_\nu = \sum_{\lambda=1}^n q_{\nu\lambda}(z_0)\xi_\lambda \,.$$

From (3.12), we see that $q_{\nu\lambda} \in \mathcal{Q}_0(F^{\tau_\nu})$ and $q_{\nu\lambda}{}^\sigma = q_{\mu\lambda}$ if σ is an automorphism of $\mathbf{C}$ such that $\tau_\nu\sigma = \tau_\mu$ on F. If, moreover, σ is the identity map on $\mathcal{K}[z_0]$, then $q_{\nu\lambda}(z_0)^\sigma = q_{\mu\lambda}(z_0)$, since $\mathcal{Q}_0(F^{\tau_\nu}) = F^{\tau_\nu}\mathcal{Q}_0(\mathbf{Q})$. This proves the part of our theorem concerning differential forms. The remaining part about derivations is merely the dual of this result.

One can of course replace $\mathcal{K}[z_0]$ by a much smaller field, and also discuss the field of rationality for the endomorphisms corresponding to the elements of F; we shall not, however, go into details here.

Remark 3.3. Let $\{c_1, \cdots, c_{2n}\}$ be the basis of the first homology group of $A(W_B(z_0), \delta)$ over $\mathbf{Z}$ corresponding to the basis $\{\beta_\kappa, \beta_\lambda{}^*\}$ of $\mathfrak{b} + \mathfrak{b}^*$ in an obvious sense. Then, the notation being as in Th. 3.2, we have

$$(3.14) \qquad \int_{c_\kappa} 2\pi i \cdot h_\nu(z_0) \, dv_\nu = 2\pi i \cdot h_\nu(z_0)\beta_\kappa{}^{\tau_\nu} z_{0\nu} \qquad (\kappa = 1, \cdots, n),$$

$$\int_{c_{n+\lambda}} 2\pi i \cdot h_\nu(z_0) \, dv_\nu = 2\pi i \cdot h_\nu(z_0)\beta_\lambda{}^{\tau_\nu} \qquad (\lambda = 1, \cdots, n),$$

where $z_0 = (z_{01}, \cdots, z_{0n})$. These give the analogue of (1.24) in the Hilbert modular case.

Remark 3.4. Let A be an abelian variety of dimension n defined over $\mathbf{C}$, such that there is an injection ι of F into End $(A) \otimes \mathbf{Q}$. Then we can find a basis $\{\omega_\nu\}$ of holomorphic 1-forms on A such that $\omega_\nu \circ \iota(a) = a^{\iota\nu}\omega_\nu$ for every $a \in F$. Each ω_ν is uniquely determined by this condition up to constant factors. If, especially, A is defined over $\bar{\mathbf{Q}}$, the algebraic closure of $\mathbf{Q}$, then we can take ω_ν to be rational over $\bar{\mathbf{Q}}$. Now there is an isogeny λ of A onto $A(W_B(z_0), \delta)$ for some z_0 and δ, which commutes with the action of F. Then ω_ν differs from $2\pi i h_\nu(z_0)\, dv_\nu \circ \lambda$ by an algebraic factor. Consequently, (3.14) shows that

$$\int_c \omega_\nu / [\pi \cdot h_\nu(z_0)] \in \bar{\mathbf{Q}} z_{0\nu} + \bar{\mathbf{Q}}$$

for every 1-cycle c on A. In particular, if A has many complex multiplications, then $z_{0\nu}$ is algebraic, and hence

$$\int_c \omega_\nu / [\pi \cdot h_\nu(z_0)]$$

is algebraic for every 1-cycle c on A. Deligne has informed the author that he proved this by a different method. The same result is essentially included in [8], in which Weil gives also an explicit expression for the periods by means of the special values of the gamma function, when A is the jacobian of a curve $y^l = x^a(1 - x)$. In the one-dimensional case, the above facts about periods can be easily seen, for example, from the Weierstrass equation for an elliptic curve. It should also be noted that $\pi \cdot h_\nu(z_0)$ is transcendental for at least one ν. This follows from a theorem of Lang [1], which generalizes earlier results of Siegel and Schneider.

Remark 3.5. By virtue of Prop. 3.1 and (an analogue of) Prop. 1.4, we can now remove the assumption

$$r_1 \equiv \cdots \equiv r_m \pmod 2$$

from the main theorems of [5].

Let us now show that the derivations of Th. 3.2 are intimately connected with the differential operators considered in [5], which are defined as follows. For $r \in \mathbf{R}$, $\nu = 1, \cdots, n$, and $0 \leq m \in \mathbf{Z}$, we put

$$(3.15) \qquad D_{\nu,r} = \frac{1}{2\pi i}\left(\frac{r}{z_\nu - \bar{z}_\nu} + \frac{\partial}{\partial z_\nu}\right),$$

$$(3.16) \qquad D_{\nu,r}{}^m = \begin{cases} \text{the identity operator} & \text{if } m = 0, \\ D_{\nu,r+2m-2} \cdots D_{\nu,r+2}D_{\nu,r} & \text{if } m > 0. \end{cases}$$

THEOREM 3.6. *Let $\mathfrak{f}(v, z)$ and $\mathfrak{t}(v, z)$ denote the functions defined by (3.1) and (3.2) with some fixed ρ, σ, and $\mathfrak{b}$. Then, for every integer $m \geq 0$, we have*

$$2^{-m}(2\pi i)^{-2m}\frac{\partial^{2m}\mathfrak{f}}{\partial v_\nu{}^{2m}}(0, z) = D_{\nu,1/2}{}^m[\mathfrak{t}(0, z)],$$

$$2^{-m}(2\pi i)^{-2m-1}\frac{\partial^{2m+1}\mathfrak{f}}{\partial v_\nu{}^{2m+1}}(0, z) = D_{\nu,3/2}{}^m\left[\frac{1}{2\pi i}\frac{\partial \mathfrak{t}}{\partial v_\nu}(0, z)\right].$$

Before proving these relations, we note that

$$(3.17) \qquad t(0, z) = \sum_{x - \rho \in \mathfrak{b}} e(\mathrm{Tr}\,(\tfrac{1}{2}zx^2 + x\sigma)),$$

$$(3.18) \qquad \frac{1}{2\pi i}\frac{\partial t}{\partial v_\nu}(0, z) = \sum_{x - \rho \in \mathfrak{b}} x^\nu e(\mathrm{Tr}\,(\tfrac{1}{2}zx^2 + x\sigma)).$$

If ρ and σ are contained in F, these are Hilbert modular forms of weight $(1/2, \cdots, 1/2)$ and $(1/2, \cdots, 1/2) + e_\nu$, respectively. Therefore our theorem together with (2.9) shows the close relation between the value $(\mathfrak{d}_{1H} \cdots \mathfrak{d}_{mH}f)_H(v)$ of Th. 2.2 and the value of $D^\prime f$ considered in the main theorems of [5]. (In [5], we dealt with the modular forms of integral weight. However, as remarked in the paper, the case of half-integral weight can easily be treated by the same methods.)

Proof. Observe that t satisfies the "heat equation"

$$\partial^2 t/\partial v_\nu^2 = 4\pi i\,\partial t/\partial z_\nu\,.$$

Since our assertions concern the variables v_ν and z_ν with a fixed ν, we can reformulate them as follows:

If a holomorphic function $g(v, z)$ in two complex variables v and z satisfies $\partial^2 g/\partial v^2 = 4\pi i\,\partial g/\partial z$, and if $f(v, z) = e(\tfrac{1}{2}(z - \bar z)^{-1}v^2)g(v, z)$, then

$$(3.19) \qquad 2^{-m}(2\pi i)^{-2m}(\partial^{2m}f/\partial v^{2m})(0, z) = D_{1/2}{}^m[g(0, z)],$$

$$(3.20) \qquad 2^{-m}(2\pi i)^{-2m}(\partial^{2m+1}f/\partial v^{2m+1})(0, z) = D_{3/2}{}^m[(\partial g/\partial v)(0, z)].$$

Here $D_\nu{}^m$ is the operator $D_{\nu,\nu}{}^m$ defined as above with z as z_ν. Once stated, this is an easy excercise of differential calculus; so we omit the detailed proof.

4. Differential operators on $\mathfrak{H}_n$

We can also find a relation, similar to Th. 3.6, between the value $(\mathfrak{d}_{1H} \cdots \mathfrak{d}_{mH}f)_H$ (v) and a certain derivative of a modular form on $\mathfrak{H}_n$. First we define an operator Δ, which assigns to each C^∞-function f on $\mathfrak{H}_n$ a matrix valued function Δf on $\mathfrak{H}_n$, by

$$(4.1) \qquad (\Delta f)(z) = \left(\frac{1 + \delta_{\mu\nu}}{2}\frac{\partial f}{\partial z_{\mu\nu}}\right) \qquad (z = (z_{\mu\nu}) \in \mathfrak{H}_n),$$

where $\delta_{\mu\nu}$ is Kronecker's symbol. It can easily be verified that

$$(4.2) \qquad (\Delta f) \circ \alpha = \nu(\alpha)^{-1}(cz + d)\,\Delta(f \circ \alpha) \cdot {}^t(cz + d) \quad \text{if} \quad \alpha = \begin{pmatrix} a & b \\ c & d \end{pmatrix} \in G_{\mathbf{R}+},$$

$$(4.3) \qquad \Delta(\det(cz + d)) = \det(cz + d) \cdot (cz + d)^{-1}c \quad \text{if} \quad \alpha = \begin{pmatrix} a & b \\ c & d \end{pmatrix} \in G_{\mathbf{R}+},$$

$$(4.4) \qquad \Delta(\det(z - \bar z)) = \det(z - \bar z) \cdot (z - \bar z)^{-1}.$$

Furthermore, we define, for each $\rho \in \mathbf{R}$, an operator D_ρ by

$$(4.5) \qquad (D_\rho f)(z) = (2\pi i)^{-1}((z - \bar{z})^{-1}\rho f(z) + \Delta f(z)).$$

Then we have

$$(4.6) \quad D_\rho(f|_\rho \alpha) = \nu(\alpha)(cz + d)^{-1}[(D_\rho f)|_\rho \alpha] \cdot {}^t(cz + d)^{-1} \quad \text{if} \quad \alpha = \begin{pmatrix} a & b \\ c & d \end{pmatrix} \in G_{\mathbf{R}+},$$

where $f|_\rho \alpha$ is defined by

$$(4.7) \qquad (f|_\rho \alpha)(z) = \nu(\alpha)^{n\rho/2} \det(cz + d)^{-\rho} f(\alpha(z)).$$

Operators of this kind were introduced by Maass in [2]. Now the well known differential equation

$$(4.8) \qquad (\partial^2/\partial u_\mu \, \partial u_\nu)\theta(u, z; r, s) = 2\pi i(1 + \delta_{\mu\nu})(\partial/\partial z_{\mu\nu})\theta(u, z; r, s)$$

holds for any r and s. Therefore we have

$$(4.9) \qquad (2\pi i)^{-2}\left(\sum_{\mu=1}^n p_\mu \, \partial/\partial u_\mu\right)\left(\sum_{\nu=1}^n q_\nu \, \partial/\partial u_\nu\right)\varphi(u, z; r, s)\big|_{u=0}$$

$$= 2 \cdot {}^t p D_{1/2}[\theta(0, z; r, s)]q$$

for every p and q in $\mathbf{C}^n$ with components p_ν and q_ν. Specialize z to a point w such that $A(w, \delta)$ has many complex multiplications. Take P as in Th. 5, and put $p = {}^t P(w)^{-1}a$, $q = {}^t P(w)^{-1}b$ with vectors a and b with components in K_{ab}'. Let h be an element of $\mathcal{G}_{1/2}(\mathbf{Q}_{ab})$ such that $h(w) \neq 0$. By [7, Prop. 2.4], $h(w)^{-1} \cdot \varphi(u, w; r, s)$ is arithmetic. Therefore, by Cor. 1.6 and Th. 2.2, the values of the components of (4.9) at w divided by $h(w)$ belongs to K_{ab}'. This is actually a special case of Th. 4.1 below. To state the theorem, we first consider, for a congruence subgroup Γ of $G_{\mathbf{Q}+}$, a Γ-invariant vector field

$$(4.10) \qquad \sum_{\mu \leq \nu} s_{\mu\nu}(z) \, \partial/\partial z_{\mu\nu} = \text{tr}\,(S(z)\,\Delta)$$

on $\mathfrak{H}_n$ with meromorphic functions $s_{\mu\nu}$, where $S = {}^t S = (s_{\mu\nu})$. The Γ-invariance implies

$$(4.11) \qquad S(z) = \nu(\gamma)^{-1} \cdot {}^t(cz + d)S(\gamma(z))(cz + d) \quad \text{for every} \quad \gamma = \begin{pmatrix} a & b \\ c & d \end{pmatrix} \in \Gamma.$$

Now, for $k \in \mathbf{Z}$, let $\mathfrak{M}_k''(\Gamma)$ denote the set of all holomorphic maps U of $\mathfrak{H}_n$ into $M_n(\mathbf{C})$ satisfying ${}^t U(z) = U(z)$ and

$$(4.12) \qquad U(\gamma(z)) = \nu(\gamma)^{1-(nk/2)} \cdot {}^t(cz + d)^{-1}U(z)(cz + d)^{-1} \det(cz + d)^k$$

$$\text{for every} \quad \gamma = \begin{pmatrix} a & b \\ c & d \end{pmatrix} \in \Gamma,$$

and finite at cusps. We can then define $\mathfrak{M}_k''(\mathfrak{R})$, $\mathcal{G}_k''(\mathfrak{R})$, etc. in the same fashion as for $\mathfrak{M}_k'(\mathfrak{R})$, $\mathcal{G}_k'(\mathfrak{R})$, etc. Given a derivation (4.10) with $S \in \mathcal{G}_0''(\mathbf{C})$ and $\rho \in \mathbf{R}$, we define an operator S_ρ by

$$(4.13) \qquad S_\rho = \text{tr}\,(SD_\rho).$$

This is analogous to $\mathfrak{d}_H$. Obviously $S_0 = (2\pi i)^{-1} \operatorname{tr} (S\Delta)$. For $\alpha = \begin{pmatrix} a & b \\ c & d \end{pmatrix} \in G_{\mathbf{R}+}$, we define $S \mid \alpha$ by

$$(4.14) \qquad (S \mid \alpha)(z) = \nu(\alpha)^{-1} \cdot {}^t(cz + d) S(\alpha(z))(cz + d).$$

We see easily that

$$(4.15) \qquad (S \mid \alpha)_\rho(f|_\rho \alpha) = (S_\rho f)|_\rho \alpha \qquad (\alpha \in G_{\mathbf{R}+}),$$

$$(4.16) \qquad S_{\rho+\sigma}(fg) = S_\rho(f)g + fS_\sigma(g) \qquad (\rho, \sigma \in \mathbf{R}).$$

Obviously ${}^tP^{-1}P^{-1} \in \mathcal{Q}_0''(\mathbf{Q})$ if $P \in \mathcal{Q}_0'(\mathbf{Q})$ and $\det (P) \neq 0$. Therefore the analogues of Propositions 1.2 and 1.4 hold for $\mathcal{Q}_k''$.

Next we recall the fact that $A(w, \delta)$ has many complex multiplications if and only if there exists an element of $G_{\mathbf{Q}+}$ which has w as its unique fixed point; moreover, given such a point w, we can define the field K' either by considering the variety $A(w, \delta)$, or by means of the representation of the isotropy subgroup of $G_{\mathbf{Q}+}$ at w on the tangent space of $\mathfrak{H}_n$ at w. For details, the reader is referred to [3, §§4, 6]. Such a point w is called an *isolated fixed point* of $G_{\mathbf{Q}+}$.

THEOREM 4.1. *Let w be an isolated fixed point of $G_{\mathbf{Q}+}$ on $\mathfrak{H}_n$. Further let $T^{(1)}, \cdots, T^{(m)}$ be elements of $\mathcal{Q}_0''(\mathbf{Q}_{ab})$, f an element of $\mathcal{Q}_\rho(\mathbf{Q}_{ab})$, and g an element of $\mathcal{Q}_{-\rho}(\mathbf{Q}_{ab})$, with $\rho \in (1/2)\mathbf{Z}$, all holomorphic at w. Then*

$$(4.17) \qquad (gT_\rho^{(1)} \cdots T_\rho^{(m)} f)(w)$$

belongs to K_{ab}', where K' is the CM-field defined at w as above.

Proof. If $n = 1$, this is included, in essence, in the main theorems I, II of [5] as a special case. The general case $n \geq 1$ can be proved in the same way, except that here we have to deal with the difference between $cz + d$ and $\det (cz + d)$ as well as the involvement of $T^{(1)}, \cdots, T^{(m)}$. As in the proof of the main theorem I of [5], we first reduce the problem to the case where $f(w) \neq 0$. Consider the variety $A(w, \delta)$ and define Y and K_λ as in (2.15) for $A(w, \delta)$. Let Φ denote the map $Y \to M_n(\mathbf{C})$ obtained from the action of Y on $\mathbf{C}^n$. Take $\beta = b_1 1_{m_1} + \cdots + b_t 1_{m_t}$ as in the proof of Th. 2.2, and define an element α of $GL_{2n}(\mathbf{Q})$ by

$$\Phi(\beta)(w\ 1_n) = (w\ 1_n) \cdot {}^t\alpha.$$

Then $\alpha \in G_{\mathbf{Q}+}$, $\nu(\alpha) = 1$, and $\alpha(w) = w$. Put $X = {}^t\Phi(\beta)$, $\alpha = \begin{pmatrix} a & b \\ c & d \end{pmatrix}$, and $h = (f|_\rho \alpha)/f$. Then $cw + d = X$, $h(w) = \det (X)^{-\rho}$, and $h \in \mathcal{Q}_0(\mathbf{Q}_{ab})$ by [7, Prop. 1.5]. Now $\mathcal{Q}_0''(\mathbf{Q}_{ab})$ is a vector space over $\mathfrak{K}$ of dimension M, where $M = n(n + 1)/2$. Moreover, we can find a basis $\{S^{(\nu)}\}$ of $\mathcal{Q}_0''(\mathbf{Q}_{ab})$ over $\mathfrak{K}$ such that $\sum_{\nu=1}^{M} p_\nu S^{(\nu)}$ with p_ν in $\mathfrak{K}$ is finite at w if and only if p_ν is finite at w for every ν. Observe that $\{S^{(\nu)} \mid \alpha^{-1}\}$ is a basis of $\mathcal{Q}_0''(\mathbf{Q}_{ab})$ over $\mathfrak{K}$, and $S \mid \alpha^{-1}$ is finite at w if and only if S is finite at w. Therefore

$$(4.18) \qquad (S^{(\mu)} \mid \alpha^{-1})(w) = \sum_{\nu=1}^{M} q_{\mu\nu} S^{(\nu)}(w) \qquad (\mu = 1, \cdots, M)$$

with $q_{\mu\nu} \in K_{ab}'$. Since $(S \mid \alpha^{-1})(w) = {}^t X^{-1} S(w) X^{-1}$, we see that every eigenvalue of the matrix $(q_{\mu\nu})$ is the product of two eigenvalues of X^{-1}. By (4.15) and (4.16), we have

$$(4.19) \qquad ((S \mid \alpha^{-1})_\rho f) \mid_\rho \alpha = S_\rho(f \mid_\rho \alpha) = h S_\rho f + f S_0 h,$$

so that

$$(4.20) \qquad \sum_{\nu=1}^{M} q_{\mu\nu}(g S_\rho^{(\nu)} f)(w) - (g S_\rho^{(\mu)} f)(w) = (h^{-1} g f S_0^{(\mu)} h)(w).$$

We can take β so that none of the eigenvalues of $(q_{\mu\nu})$ is 1. Then (4.20) shows that $(g S_\rho^{(\nu)} f)(w)$ is K_{ab}'-rational for every ν. This proves the case $m = 1$. To prove the general case, we let $S^{(\nu)}$ act successively on (4.19), and obtain a relation

$$[(S^{(\nu_1)} \mid \alpha^{-1})_\rho \cdots (S^{(\nu_m)} \mid \alpha^{-1})_\rho f] \mid_\rho \alpha = S_\rho^{(\nu_1)} \cdots S_\rho^{(\nu_m)}(f \mid_\rho \alpha)$$
$$= h S_\rho^{(\nu_1)} \cdots S_\rho^{(\nu_m)} f + \cdots,$$

similar to [5, (1.16)], which together with induction yields an equality similar to (2.21) in the present case. Then the choice of a "sufficiently general" β as in the proof of Th. 2.2 concludes our proof.

We can of course prove a theorem about the action of $K_A'^\times$ on the value (4.17) similar to Th. 2.3 and the main theorem III of [5], at least when ρ is an integer. Both its statement and proof may be left to the reader, since they are obvious modifications of those theorems.

REFERENCES

1. S. LANG, *Transcendental points on group varieties*, Toplogy, 1(1962), 313–318.
2. H. MAASS, *Die Differentialgleichungen in der Theorie der Siegelschen Modulfunktionen*, Math. Annalen, 126(1953), 44–68.
3. G. SHIMURA, *Algebraic number fields and symplectic discontinuous groups*, Ann. of Math. 86(1967), 503–592.
4. ———, *On canonical models of arithmetic quotients of bounded symmetric domains*, I, II, Ann. of Math. 91(1970), 144–222; 92(1970), 528–549.
5. ———, *On some arithmetic properties of modular forms of one and several variables*, Ann. of Math. 102(1975), 491–515.
6. ———, *On the Fourier coefficients of modular forms of several variables*, Göttingen Nachr. Akad. Wiss. 1975, 261–268.
7. ———, *Theta functions with complex multiplication*, Duke Math. J. 43 (1976), 673–696.
8. A. WEIL, *Sur les périodes des intégrales abéliennes*, Comm. Pure and Appl. Math. 29(1976), 813–819.

DEPARTMENT OF MATHEMATICS, PRINCETON UNIVERSITY, PRINCETON, N. J. 08540

On the periods of modular forms

Mathematische Annalen, 229 (1977), 211-221

On the Periods of Modular Forms

Goro Shimura

Department of Mathematics, Princeton University, Fine Hall – Box 37, Princeton, NJ 08540, USA

Our purpose is to supplement the previous paper [12] concerning the special values of two types of series

$$D(s, f, \psi) = \sum_{n=1}^{\infty} \psi(n) a_n n^{-s},$$

$$D(s, f, g) = \sum_{n=1}^{\infty} a_n b_n n^{-s}$$

defined with a Dirichlet character ψ and cusp forms

$$f(z) = \sum_{n=1}^{\infty} a_n e(nz), \quad g(z) = \sum_{n=1}^{\infty} b_n e(nz) \tag{1}$$

of weight k and l respectively, belonging to a congruence subgroup of $SL_2(\mathbf{Z})$, where $e(z) = e^{2\pi i z}$. We first give a reformulation of Theorems 1 and 4 of [12], and then show that, given such an $f \neq 0$ of weight 2, there are infinitely many characters ψ such that $D(1, f, \psi) \neq 0$ and $\psi(-1)$ has a given signature. Consequently, Theorem 1 of [12] holds unconditionally for every weight ≥ 2. Next we shall discuss the connection of the value $D(1, f, \psi)$ with the periods of $f(z)dz$ in the case where $k = 2$. When the coefficients a_n are all rational, this has been done by Birch [1], Manin [2], and Mazur and Swinnerton-Dyer [3]. In §3, we improve on Theorem 3 of [12] which concerns the special values of $D(s, f, g)$ when $l < k$. A few remarks will be added in the last section, where we shall also discuss some open questions.

We should mention that the special values of $D(s, f, \psi)$ are also treated in the recent papers by Razar [5] and Weil [15]. Especially, the former proves, among other things, Assertion (ii) of Theorem 1 below in the case where f belongs to $\Gamma_0(N)$ under a certain condition on ψ, by a different method.

1. Reformulation of Theorems 1 and 4 of [12]

We use the same notation as in [12]. In particular, the Gauss sum $g(\psi)$ of a Dirichlet character ψ is defined to be the same as the Gauss sum $g(\psi_0)$ of the primitive character ψ_0 associated with ψ; then we put

$$A(m, f, \psi) = (2\pi i)^{-m} g(\psi)^{-1} D(m, f, \psi) ;$$

K_ψ and K_f denote the fields generated over Q by the values $\psi(n)$ and the coefficients a_n, respectively.

Let f be a primitive cusp form of weight k with Fourier expansion (1), belonging to $S_k(N, \chi)$. Then K_f is generated by a_p for almost all primes p. This follows trivially, for example, from [4, Theorem B, (b)]. We denote by I_f the set of all injections of K_f into C. For every automorphism σ of C or for $\sigma \in I_f$, we can define a primitive cusp form f^σ by $f^\sigma(z) = \sum_{n=1}^\infty a_n^\sigma e(nz)$. Now we have

Theorem 1. *There exist non-zero complex numbers u_τ^+ and u_τ^- depending on f and $\tau \in I_f$ with the following properties:*

    (i) *If ϱ denotes the complex conjugation, $(u_\tau^\pm)^\varrho = \pm u_{\tau\varrho}^\pm$.*

    (ii) $A(m, f, \psi) \in \begin{cases} u_1^+ K_f K_\psi & \text{if} \quad \psi(-1) = (-1)^m, \\ u_1^- K_f K_\psi & \text{if} \quad \psi(-1) = (-1)^{m-1}, \end{cases}$

for every positive integer $m < k$, where subscript 1 means the identity map.

    (iii) *If σ is an automorphism of C and τ its restriction to K_f, then*

$$[A(m, f, \psi)/u_1^\pm]^\sigma = A(m, f^\tau, \psi^\sigma)/u_\tau^\pm \qquad (0 < m < k),$$

where the choice of $\pm$ is determined as in (ii).

    (iv) *Put $E(f) = i^{1-k} \pi g(\chi) \langle f, f \rangle$, where $\langle\,,\,\rangle$ is the Petersson inner product defined by [12, (2.1)]. Then $E(f)/(u_1^+ u_1^-) \in K_f$, and $[E(f)/(u_1^+ u_1^-)]^\tau = E(f^\tau)/(u_\tau^+ u_\tau^-)$ for every $\tau \in I_f$.*

If $k > 2$, we can define $u_\tau^\pm$ by

$$u_\tau^+ = A(k-1, f^\tau, \varphi), \tag{2}$$

$$u_\tau^- = A(k-1, f^\tau, \varphi') \tag{3}$$

with any fixed real characters φ and φ' such that $\varphi(-1) = (-1)^{k-1}$ and $\varphi'(-1) = (-1)^k$. By [12, Proposition 2], we have $u_\tau^\pm \neq 0$, and obtain the above properties from Theorems 1 and 4 of [12]. We observe that if χ is the identity character (and hence k is even and K_f is totally real), then u_τ^+ is real and u_τ^- pure imaginary.

In the case $k = 2$, to prove that condition (1.2) of [12, Theorem 1] is always satisfied, we first note

Lemma 1. *For a given integer $M > 0$, let G_M denote the set of all elements γ of $\Gamma_1(N)$ such that $\gamma(s/p) = r/p$ for at least one prime p not dividing M and for two integers r and s prime to p. Then, for any fixed M, $\Gamma_1(N)$ can be generated by $\begin{pmatrix} 1 & 1 \\ 0 & 1 \end{pmatrix}, \begin{pmatrix} 1 & 0 \\ N & 1 \end{pmatrix}$, and G_M.*

This fact is included in the proof of [12, Proposition 3].

Theorem 2. *Let f be a primitive cusp form belonging to $S_2(N)$, and M an arbitrary positive integer. Then there exist a prime number p not dividing M and a primitive character ψ whose conductor divides p such that $D(1, f, \psi) \neq 0$ and $\psi(-1)$ has a given signature.*

Notice that if $\psi(-1)=1$, ψ may be the identity character (modulo 1), or has conductor p.

Proof. Put, for every $t \in Q$ and every $F \in S_2(N)$,

$$U(t, F) = -\int_t^{i\infty} F(z)dz = -\int_0^{i\infty} F(z+t)dz, \tag{4}$$

$$[\gamma, F] = \int_w^{\gamma(w)} F(z)dz \quad (\gamma \in \Gamma_1(N)) \tag{5}$$

with any point w that is either a cusp or belongs to $\mathfrak{H}$; this does not depend on the choice of w. Further, we put $F^\varrho(z) = \overline{F(-\bar{z})}$. Then

$$\overline{U(t, F)} = -U(-t, F^\varrho). \tag{6}$$

If ξ is a primitive character modulo a positive integer c, we have

$$\sum_{x=1}^c \bar{\xi}(x)U(x/c, f) = (2\pi i)^{-1} g(\bar{\xi})D(1, f, \xi) = \xi(-1)c \cdot A(1, f, \xi). \tag{7}$$

If ξ is imprimitive, the formula is somewhat more complicated, but at least when c is an odd prime p, it takes the form

$$\sum_{x=1}^{p-1} \bar{\xi}(x)U(x/p, f) = \begin{cases} \xi(-1)p \cdot A(1, f, \xi) & \text{if} \quad \xi \neq 1, \\ [a_p - 1 - \chi(p)](2\pi i)^{-1}D(1, f) & \text{if} \quad \xi = 1, \end{cases} \tag{8}$$

where $D(s, f) = \sum_{n=1}^\infty a_n n^{-s}$, and χ is the character modulo N such that $f \in S_2(N, \chi)$. Let X_p and Y_p denote the sets of all characters ξ modulo p such that $\xi(-1)=1$ and $\xi(-1)=-1$, respectively. Write the number expressed by (8) simply as $V_p(f, \xi)$. Then we have

$$U(b/p, f) + U(-b/p, f) = (2/(p-1)) \sum_{\xi \in X_p} \xi(b)V_p(f, \xi), \tag{9}$$

$$U(b/p, f) - U(-b/p, f) = (2/(p-1)) \sum_{\xi \in Y_p} \xi(b)V_p(f, \xi) \tag{10}$$

for $(b, p)=1$. Put $g = f + f^\varrho$ and $h = i(f - f^\varrho)$. Now suppose $D(1, f, \xi) = 0$ for all $\xi \in Y_p$ and all p prime to a given M. Then, from (6) and (10), we obtain $\mathrm{Re}[U(b/p, g)] = 0$ for all b prime to p. It follows that

$$\mathrm{Re} \int_{b/p}^{c/p} g(z)dz = 0$$

for $(b, p) = (c, p) = 1$, $p \nmid M$. By Lemma 1, this implies that $\mathrm{Re}[\gamma, g] = 0$ for all $\gamma \in \Gamma_1(N)$, and hence $g = 0$. Similarly we obtain $h = 0$, so that $f = 0$, a contradiction. We can apply the same reasoning to X_p with the imaginary part instead of the real part, and therefore obtain the desired conclusion.

Remark. The above proof shows that for every non-zero $f \in S_2(N)$ which may not be primitive, there is a character ψ modulo p for some prime p not dividing M, such that $\psi(-1) = -1$ and $D(1, f, \psi) \neq 0$. Then, applying the same result to the cusp form

$$g(z) = \sum_{n=1}^\infty \psi(n)a_n e(nz)$$

which is obviously different from 0, we find another character ξ modulo a prime q not dividing pM such that $\xi(-1)=-1$ and

$$D(1, f, \xi\psi)=D(1, g, \xi)\neq 0.$$

Notice that $\xi\psi$ is a primitive character modulo pq such that $\xi\psi(-1)=1$.

To prove Theorem 1 in the case where $k=2$, put

$$u_\tau^+ = U(b/p, f^\tau) - U(-b/p, f^\tau) \qquad (\tau \in I_f) \tag{11}$$

for some b/p such that $u_1^+ \neq 0$. The existence of such b/p is guaranteed by the above proof. Then $A(1, f, \varphi) \neq 0$ for some $\varphi \in Y_p$, and

$$u_1^+/A(1, f, \varphi) = (2p/(1-p)) \sum_{\xi \in Y_p} \xi(b) A(1, f, \xi)/A(1, f, \varphi).$$

By [12, Theorem 1], for every automorphism σ of C, we have $A(1, f^\sigma, \varphi^\sigma) \neq 0$, and

$$[u_1^+/A(1, f, \varphi)]^\sigma$$

$$= (2p/(1-p)) \sum_{\xi \in Y_p} \xi^\sigma(b) A(1, f^\sigma, \xi^\sigma)/A(1, f^\sigma, \varphi^\sigma)$$

$$= u_\tau^+/A(1, f^\sigma, \varphi^\sigma)$$

if τ is the restriction of σ to K_f. Therefore $u_\tau^+ \neq 0$ for all $\tau \in I_f$. Putting either

$$u_\tau^- = i\pi g(\chi^\tau) < f^\tau, f^\tau > /u_\tau^+ \quad \text{or}$$

$$u_\tau^- = U(b'/p', f^\tau) + U(-b'/p', f^\tau)$$

with a suitable b'/p', we obtain the assertions of Theorem 1 from [12, Theorems 1 and 4].

2. The Periods of Cusp Forms of Weight 2

The numbers $u_\tau^\pm$ defined as above when $k=2$ are obviously connected with the periods $[\gamma, f]$ with $\gamma \in \Gamma_1(N)$. Let us now clarify the connection in terms of the abelian variety associated to f which is defined in [7, §7.5] and [10]. Let f be a primitive cusp form of weight 2 and of conductor N with Fourier expansion (1), and J the Jacobian variety of the compactification of $\mathfrak{H}/\Gamma_1(N)$. [Of course here and in the following treatment, $\Gamma_1(N)$ can be replaced by $\Gamma_0(N)$ if f is a form belonging to $\Gamma_0(N)$.] We fix f and write simply K and I for K_f and I_f. As shown in [10, Theorem 1], there is a triple (A, v, θ) formed by a quotient abelian variety A of J, a natural map v of J onto A, and an injection θ of K into $\text{End}(A)\otimes Q$ satisfying the following conditions:

$$\dim(A) = [K:Q] \,; \tag{12}$$

$$\theta(a_n)\circ v = v\circ \xi_n$$

for all n, where ξ_n is the endomorphism of J corresponding to the Hecke operator of degree n and of level N ;   (13)

    A, v, and $\theta(a)$ for $a\in K$ are all defined over Q ;   (14)

the space of holomorphic 1-forms on A can be identified with

$$\sum_{\tau \in I} C f^{\tau} \quad \text{via the map of } \mathfrak{H} \text{ into } J \text{ combined with } v. \tag{15}$$

Let $[\gamma]$ denote the element $([\gamma, f^{\tau}])_{\tau \in I}$ of C^I, and let P be the submodule of C^I formed by $[\gamma]$ for all $\gamma \in \Gamma_1(N)$. As shown in [10], P is a lattice of C^I, and there is an isomorphism λ of A onto C^I/P such that the map $\mathfrak{H} \to J$ combined with $\lambda \circ v$ is given by

$$z \mapsto \left(\int_{i\infty}^{z} f^{\tau}(z) dz \right)^{\tau \in I} \quad (z \in \mathfrak{H}).$$

We let K act on C^I via λ. Obviously the action of an element b of K on C^I is represented by the diagonal matrix $\Phi(b)$ whose diagonal elements are $b^{\tau}, \tau \in I$. Let P_Q denote the Q-linear span of P. Then P_Q is a vector space over K of dimension 2. Now let μ denote the linear transformation of C^I that is obtained from the permutation $\tau \mapsto \tau \varrho$ of I, where ϱ denotes the complex conjugation. If K is totally real, μ is the identity map; if K is imaginary, μ is represented by

$$\begin{pmatrix} 0 & 1_n \\ 1_n & 0 \end{pmatrix} \quad (2n = [K:Q])$$

with a suitable arrangement of the elements of I. Obviously $\Phi(a)\mu = \mu\Phi(a^{\varrho})$ for $a \in K$. Now we see easily from (6) that, if $\gamma = \begin{pmatrix} a & b \\ c & d \end{pmatrix} \in \Gamma_1(N)$ and $\delta = \begin{pmatrix} a & -b \\ -c & d \end{pmatrix}$, then $\mu\overline{[\gamma]}$ $= -[\delta]$. Therefore P is stable under the map $x \mapsto \mu\bar{x}$. Put

$$R^+ = \{x \in P_Q | \mu\bar{x} = x\},$$
$$R^- = \{x \in P_Q | \mu\bar{x} = -x\}.$$

Then both R^+ and R^- are stable under K, and thus

$$P_Q = R^+ + R^-, \quad R^+ = \Phi(K)v^+, \quad R^- = \Phi(K)v^-$$

with two elements $v^{\pm}$ of P_Q. Let $v_{\tau}^{\pm}$ denote the τ-component of $v^{\pm}$. None of these $v_{\tau}^{\pm}$ is 0, since P_Q is dense in C^I. We have obviously

$$(v_{\tau}^{\pm})^{\varrho} = \pm v_{\tau\varrho}^{\pm}. \tag{16}$$

Theorem 3. *The numbers $v_{\tau}^{\pm}$ can be chosen as $u_{\tau}^{\pm}$ of Theorem 1 in the case $k=2$.*

Proof. Let G_M be as in Lemma 1, and let $\gamma \in G_M$ and $\gamma(s/p) = r/p$ as stated there. Put $x = [\gamma]$. By virtue of (6), we see that

$$x + \mu\bar{x} = (U(r/p, f^{\tau}) - U(-r/p, f^{\tau}))_{\tau \in I}$$
$$- (U(s/p, f^{\tau}) - U(-s/p, f^{\tau}))_{\tau \in I},$$
$$x - \mu\bar{x} = (U(r/p, f^{\tau}) + U(-r/p, f^{\tau}))_{\tau \in I}$$
$$- (U(s/p, f^{\tau}) + U(-s/p, f^{\tau}))_{\tau \in I}.$$

From (9), (10), and Theorem 1, we obtain

$$x + \mu\bar{x} \in \Phi(K)u^+, \quad x - \mu\bar{x} \in \Phi(K)u^-,$$

where $u^{\pm}$ are the vectors with components $u_{\tau}^{\pm}$. Therefore $R^{\pm} = \Phi(K)u^{\pm}$, which proves our assertion.

3. The Special Values of $\mathfrak{D}_N(s, f, g)$

We shall now give a reformulation of Theorem 3 of [12] which concerns the special values of

$$D(s, f, g) = \sum_{n=1}^{\infty} a_n b_n n^{-s}$$

defined with a cusp form f of weight k as above and another modular form $g(z) = \sum_{n=0}^{\infty} b_n e(nz)$ of weight l. To obtain a transparent result, let us assume that f is, as before, a primitive element belonging to $S_k(N, \chi)$, and $g \in G_l(N, \psi)$, with characters χ and ψ defined modulo N. (The conductor of f may be smaller than N.) Then we put

$$\mathfrak{D}_N(s, f, g) = L_N(2s + 2 - k - l, \chi\psi)D(s, f, g), \tag{17}$$

where

$$L_N(s, \chi\psi) = \sum_{\substack{n=1 \\ (n,N)=1}}^{\infty} (\chi\psi)(n)n^{-s}.$$

Theorem 4. *The notation being as above, suppose that $l < k$, and put*

$$T(m, f, g) = (2\pi i)^{l-1-2m} g(\psi)^{-1} \mathfrak{D}_N(m, f, g)$$

for an integer m such that $l \leq m < k$. Let $u_\tau^{\pm}$ be the numbers with the properties of Theorem 1 for f. Then

$$T(m, f, g)/(u_1^+ u_1^-) \in K_f K_g.$$

Moreover, for every automorphism σ of C, we have

$$[T(m, f, g)/(u_1^+ u_1^-)]^\sigma = T(m, f^\tau, g^\sigma)/(u_\tau^+ u_\tau^-),$$

where τ is the restriction of σ to K_f.

Before proving this, we note that $\mathfrak{D}_N(s, f, g)$ has a simple Euler product, if g is a primitive cusp form. In fact, put

$$\sum_{n=1}^{\infty} a_n n^{-s} = \prod_p (1 - \alpha_p p^{-s})^{-1} (1 - \alpha'_p p^{-s})^{-1},$$

$$\sum_{n=1}^{\infty} b_n n^{-s} = \prod_p (1 - \beta_p p^{-s})^{-1} (1 - \beta'_p p^{-s})^{-1}$$

with algebraic numbers $\alpha_p, \alpha'_p, \beta_p, \beta'_p$, which may be 0 when p divides N. Take N to be the least common multiple of the conductors of f and g. Then, by [12, Lemma 1], we have

$$\mathfrak{D}_N(s, f, g)$$
$$= \prod_p [(1 - \alpha_p \beta_p p^{-s})(1 - \alpha_p \beta'_p p^{-s})(1 - \alpha'_p \beta_p p^{-s})(1 - \alpha'_p \beta'_p p^{-s})]^{-1}. \tag{18}$$

If $(k + l - 2)/2 < m < k$, the assertion of the theorem follows immediately from [12, Theorem 3, Lemma 8], (iv) of Theorem 1, and the well known property of the special values of Dirichlet L-functions.

To prove the remaining case, with an arbitrary Dirichlet character ω modulo a positive integer C and a non-negative integer α such that $\omega(-1)=(-1)^\alpha$, put

$$F_{\alpha,C}(z,s,\omega)=\pi^{-s}y^s\Gamma(s+\alpha)E_{\alpha,C}(z,s,\omega) \qquad (y=\mathrm{Im}(z)),$$

$$J_{\alpha,C}(z,s,\omega)=\sum_{0\neq(m,n)\in\mathbf{Z}^2}\omega(m)(mz+n)^{-\alpha}|mz+n|^{-2s},$$

where $E_{\alpha,C}$ is defined as in [12, §2]. If $\alpha>0$, $F_{\alpha,C}$ is an entire function in s; moreover, if ω is primitive, it satisfies a functional equation

$$F_{\alpha,C}(z,1-\alpha-s,\omega)$$
$$=g(\omega)C^{3s+\alpha-2}z^{-\alpha}F_{\alpha,C}(-1/Cz,s,\bar\omega)$$
$$=g(\omega)C^{2s+\alpha-2}\pi^{-s}y^s\Gamma(s+\alpha)J_{\alpha,C}(z,s,\bar\omega). \tag{19}$$

These follow, for example, from [8, Lemma 3.3]. Take ω to be the primitive character associated with $\chi\psi$. Let $M=N/C$ with the conductor C of ω. Then

$$E_{\alpha,N}(z,s,\chi\psi)$$
$$=\sum_{t|M}\mu(t)\omega(t)t^{-\alpha-2s}E_{\alpha,C}(t^{-1}Mz,s,\omega), \tag{20}$$

where μ denotes the Moebius function. Now let $\alpha=k-l$, and substitute $s+1-k$ for s in (19). Then we obtain, from (20),

$$F_{\alpha,N}(z,l-s,\chi\psi)$$
$$=g(\omega)\pi^{k-1-s}\Gamma(s+1-l)y^{s+1-k}N^{2s-k-l}M$$
$$\cdot\sum_{t|M}\mu(t)\omega(t)t^{-1}J_{\alpha,C}(t^{-1}Mz,s+1-k,\bar\omega). \tag{21}$$

After these preparations, we use the integral expression [12, (2.4)] for $\mathfrak{D}_N(s,f,g)$, which can be written as

$$(2\pi)^{-2s}\Gamma(s)\Gamma(s+1-l)\mathfrak{D}_N(s,f,g)$$
$$=2^{-1}\pi^{1-k}\int_{\Phi_0}\bar f_\varrho g F_{\alpha,N}(z,s+1-k,\chi\psi)y^{k-2}dxdy. \tag{22}$$

Denote by $R(s,f,g)$ the function expressed by (22). As remarked in [12, §2], R is entire, and by (21), we have

$$R(k+l-1-s,f,g)$$
$$=2^{-1}g(\omega)\pi^{-s}\Gamma(s+1-l)N^{2s-k-l}M$$
$$\cdot\sum_{t|M}\mu(t)\omega(t)t^{-1}\int_{\Phi_0}\bar f_\varrho g J_{\alpha,C}(t^{-1}Mz,s+1-k,\bar\omega)y^{s-1}dxdy.$$

With the differential operator $\delta_\lambda^{(r)}$ of [12, (2.8)], we have

$$J_{\lambda+2r,C}(z,-r,\omega)=\frac{\Gamma(\lambda)}{\Gamma(\lambda+r)}(-4\pi y)^r\delta_\lambda^{(r)}J_{\lambda,C}(z,0,\omega)$$

for $0<\lambda\in Z$ and $0\leq r\in Z$. This is similar to [12, (2.9)]. Now let m be an integer such that $l\leq m<(k+l)/2$, and let $n=k+l-1-m$. Then $(k+l-2)/2<n<k$. Putting $s=n$, $r=m-l$, and $\lambda=k+l-2m$, we obtain

$$\mathfrak{D}_N(m,f,g)$$

$$=g(\omega)\pi^{4m+2-k-2l}\sum_{t\mid M}\omega(t)A_t\langle f_\varrho, g\cdot\delta_\lambda^{(r)}J_{\lambda,C}(t^{-1}Mz,0,\bar{\omega})\rangle$$

with rational numbers A_t depending on t, k, l, m, N, C. If $\lambda\neq 2$ or ω is non-trivial, $(\pi i)^{-\lambda}J_{\lambda,C}(z,0,\omega)$ is an element of $G_\lambda(C,\omega)$ with Fourier coefficients in K_ω; moreover an automorphism σ of C sends it to $(\pi i)^{-\lambda}J_{\lambda,C}(z,0,\omega^\sigma)$. If $\lambda=2$ and ω is trivial, it is a well known Eisenstein series with a non-holomorphic term $-\pi/y$. Anyway, by means of the same reasoning as in the proof of [12, Theorem 3], we see that

$$\pi^{l-2-2m}i^{2m-k-l}g(\omega)^{-1}\mathfrak{D}_N(m,f,g)/\langle f,f\rangle\in K_f K_g,$$

and it is sent under σ to the same type of quotient with ω^σ, f^σ, g^σ instead of ω, f, g. This combined with (iv) of Theorem 1 and [12, Lemma 8] completes the proof.

4. Miscellaneous Remarks

I. The proof of [12, Lemma 8] could have been given more simply by observing that $g(\chi^\sigma)=\chi(s)^\sigma g(\chi)^\sigma$ for every automorphism σ of C if $e(1/c)^\sigma=e(s/c)$, where c is the conductor of χ.

II. Let Γ be a congruence subgroup of $SL_2(Z)$, p the genus of Γ, and $\{\alpha_1,...,\alpha_p,\beta_1,...,\beta_p\}$ a set of elements of Γ corresponding to a canonical set of generators of the first homology group of the compactification of $\mathfrak{H}/\Gamma$. Then, for every $f\in S_2(\Gamma)$, we have

$$(2\pi i/3)[SL_2(Z):\Gamma\{\pm 1\}]\langle f,f\rangle$$

$$=\int_{\mathfrak{H}/\Gamma}\overline{f(z)dz}\wedge f(z)dz=\sum_{j=1}^{p}(\overline{[\alpha_j,f]}[\beta_j,f]-\overline{[\beta_j,f]}[\alpha_j,f]).$$

This shows that (iv) of Theorem 1 in the case $\Gamma=\Gamma_0(N)$, $k=2$ follows from Theorem 3. The same type of relation holds for the forms of weight greater than 2, as shown in [6] and [7, §8.2].

III. If K_f is imaginary, the variety A of §2 is isogenous to a product of two copies of an abelian variety B which is defined by $B=(1+\beta)A$, where β is an automorphism of A obtained from $\begin{pmatrix} 0 & -1 \\ N & 0 \end{pmatrix}$. As shown in [7, §7.5], B is defined over $Q(\zeta+\zeta^{-1})$, where $\zeta=e(1/N)$, and $End(B)\otimes Q$ contains an isomorphic image of the maximal real subfield of K_f. It can be shown that $\begin{pmatrix} 0 & -1 \\ N & 0 \end{pmatrix}$ sends f^τ to $\varepsilon_\tau f^{\tau\varrho}$ with an algebraic number ε_τ of absolute value 1. In general, ε_τ is not necessarily a root of unity. Using these ε_τ, one can obtain a certain relation between the "periods" of B and the numbers $u_\tau^\pm$, though the result is not so transparent as in the case of real K_f.

IV. Let f be a primitive form of weight 2. Then $2\pi i f(z)dz$ is "rational over K_f". Thus $2\pi i[\gamma, f]$ for $\gamma \in \Gamma_1(N)$ is a period of a holomorphic 1-form rational over K_f. In particular, if $K_f = Q$ and hence A is an elliptic curve, $w = u_1^+/u_1^-$ is a point of $\mathfrak{H}$ that gives a modulus of A. (Take $-u_1^+$ instead of u_1^+ if necessary.) Therefore, if h is a modular form of weight one with rational Fourier coefficients (say $h = \eta^2$), then $u_1^-/h(w)$ is algebraic. The same type of assertion holds for a higher-dimensional A with real K_f with a Hilbert modular form instead of an elliptic modular form h (this follows from an unpublished result due to Deligne, Weil, and the author). One can naturally ask whether there is a similar relation in the case of weight greater than 2. This is so when the Mellin transform of f is an L-function of an imaginary quadratic field (see [12, Proposition 5]).

V. We have evaluated $D(s, f, \psi)$ for the positive integers smaller than the weight. One may naturally be interested in the values outside this range. For this question, we first observe that $(2\pi)^{-s}\Gamma(s)D(s, f, \psi)$ is an entire function and satisfies a functional equation for $s \mapsto k - s$. Therefore $D(m, f, \psi) = 0$ if m is an integer ≤ 0. In this sense, the values $D(m, f, \psi)$ for $m \geq k$ are similar to the values of the Riemann zeta function for odd positive integers. Thus perhaps we should be content with the values for $0 < m < k$ until some new information about the values $\zeta(m)$ for odd positive m can be found.

VI. We now comment on Theorem 4. Given two primitive cusp forms f and g of weight k and l respectively, and both of level 1, put

$$R(s, f, g) = (2\pi)^{-2s}\Gamma(s)\Gamma(s + 1 - l)\mathfrak{D}(s, f, g)$$

$$\mathfrak{D}(s, f, g) = \zeta(2s + 2 - k - l)D(s, f, g).$$

From the expression (22), we see that R is entire if $f \neq g$; if $f = g$, it is holomorphic on the whole plane except for simple poles at $s = k - 1$ and $s = k$. Moreover it satisfies a functional equation

$$R(k + l - 1 - s, , g) = R(s, f, g).$$

Suppose $l < k$ and let n be an integer $< l$. Then $\mathfrak{D}(n, f, g) = 0$, since $\Gamma(s + 1 - l)$ has a pole at $s = n$. Such an integer n is sent by the map $s \mapsto k + l - 1 - s$ to the integers $\geq k$. Now, by virtue of the inequality $|a_p| \leq 2p^{(k-1)/2}$ due to Deligne and the corresponding fact for g, we see that the infinite product (18) is convergent for $\mathrm{Re}(s) > (k + l)/2$, and hence the functional equation implies that $R(s, f, g) \neq 0$ for $\mathrm{Re}(s) < (k + l - 2)/2$. If $m \in Z$ and $l \leq m < (k + l - 2)/2$, then $\zeta(2s + 2 - k - l) = 0$ for $s = m$, and hence $D(s, f, g)$ has a simple pole at $s = m$.

Next suppose that $k = l$ and $f \neq g$. Then again $\mathfrak{D}(m, f, g) = 0$ for $m \in Z$ and $m < k$. These integers are sent to those $\geq k$ by the map $s \mapsto 2k - 1 - s$. Therefore it seems that no general result can be expected in this case, for the same reason as in V.

Finally if $f = g$, it is more natural to consider the functions

$$C(s) = \zeta(s - k + 1)^{-1}\mathfrak{D}(s, f, f)$$

$$= \prod_p [(1 - \alpha_p^2 p^{-s})(1 - \alpha_p\alpha_p' p^{-s})(1 - \alpha_p'^2 p^{-s})]^{-1},$$

$$Q(s) = \pi^{-3s/2}\Gamma(s/2)\Gamma((s + 1)/2)\Gamma((s - k + 2)/2)C(s).$$

As shown in [11], $Q(s)$ is entire, and $Q(2k-1-s)=Q(s)$. In this case, the gamma factor is not zero for $0<m<k$, $m\equiv 1\pmod 2$, so that no obvious vanishing of C occurs for such m. In fact, Sturm [14] has obtained a result for the special values of C for such m similar to Theorem 4.

VII. Let F be a cusp form of weight $\kappa/2$ for an odd integer $\kappa\geq 3$ with Fourier expansion

$$F(z)=\sum_{n=1}^{\infty} c(n)e(nz).$$

It was shown in [9] that if F is an eigenfunction of Hecke operators, then there is a modular form f of weight $\kappa-1$ such that

$$c(t)D(s,f)=L(s-(\kappa-3)/2,\omega_t)\sum_{n=1}^{\infty} c(tn^2)n^{-s}$$

for every square-free positive integer t, where $L(s,\omega_t)$ is a Dirichlet L-function with a character ω_t depending on F and t. Now it is highly probable that the Fourier coefficients $c(t)$ for square-free t are closely connected with the values $D(m,f)$ for $0<m<k$. (It should be remarked that such coefficients for an Eisenstein series of weight $\kappa/2$ can be expressed by means of the special values of Dirichlet L-functions (see, for example, [11, (3.10)] and [9, p. 71]).) In fact, Shintani [13] obtained an expression for such $c(t)$ in terms of certain integrals of $f(z)z^m dz$ for $0\leq m<\kappa-2$. It will be a most intriguing theme to investigate further in this direction.

Corrections to [12]

P. 783, Line 6: Read "N" for "n".
P. 788, Line 17, formula (2.5): Read "$\langle f_\varrho,g\rangle$" for "$\langle f,g\rangle$".
P. 789, Lines 12–14: $S_k'(N)$ should be the space spanned by all $h(tz)$ with $h\in S_k(M)$, $tM|N$, $M<N$.
P. 800, Line 16: Read "Q" for "R".
P. 803, Line 5: Insert "if $m\in Z$ and $0<m<k$;" at the end of the line.

References

1. Birch,B.J.: Elliptic curves over Q: a progress report. Number Thoery Institute 1969, Proc. Symp. pure math., Vol. XX. Amer. Math. Soc. 396—400 (1971)
2. Manin,Y.I.: Parabolic points and zeta functions of modular curves. Izv. Akad. Nauk, SSSR, Ser. Mat. **36**, 19—66 (1972)
3. Mazur,B., Swinnerton-Dyer,P.: Arithmetic of Weil curves. Inv. math. **25**, 1—61 (1974)
4. Miyake,T.: On automorphic forms on GL_2 and Hecke operators. Ann. of Math. **94**, 174—189 (1971)
5. Razar,M.J.: Dirichlet series and Eichler cohomology. Preprint, Univ. of Maryland 1976
6. Shimura,G.: Sur les intégrales attachées aux formes automorphes. J. Math. Soc. Japan **11**, 291—311 (1959)
7. Shimura,G.: Introduction to the arithmetic theory of automorphic functions. Iwanami Shoten (Tokyo) and Princeton University Press 1971
8. Shimura,G.: On modular forms of half integral weight. Ann. of Math. **97**, 440—481 (1973)

9. Shimura,G.: Modular forms of half integral weight. Proc. Int. Summer School on modular functions. Antwerpen 1972. Lecture Notes in Mathematics 320. Berlin, Heidelberg, New York: Springer 1973
10. Shimura,G.: On the factors of the jacobian variety of a modular function field. J. Math. Soc. Japan **25**, 523—544 (1973)
11. Shimura,G.: On the holomorphy of certain Dirichlet series. Proc. London Math. Soc. **31**, 79—98 (1975)
12 Shimura,G.: The special values of the zeta functions associated with cusp forms. Comm. Pure Appl. Math. **29**, 783—804 (1976)
13. Shintani,T.: On construction of holomorphic cusp forms of half integral weight. Nagoya Math. J. **58**, 83—126 (1975)
14. Sturm,J.: Thesis. Princeton University 1977
15. Weil,A.: Remarks on Hecke's lemma and its use. To appear in Proceedings of International Symposium on Algebraic Number Theory, Kyoto, 1976

Received February 7, 1977

Notes II

Numbers in brackets set in boldface such as [**79a**] mean items in the list of articles; those in roman such as [17] and [94b] are references originally cited in the article to which notes are given. Some of the corrections were already made in the original articles, but they are superseded by the corrections made in these notes.

67b. Construction of class fields and zeta functions of algebraic curves

p. 62, line 3: For "ralation" read "relation".

p. 66: In Proposition 2.6, before "Conversely" insert "Moreover, every element of B^+ leaving this point fixed belongs to $f(M)$." This fact can be proved as follows. Let α and z be as in the proof. It is well-known that for any $w \in \mathfrak{H}$ the group $\{\xi \in GL_2(\mathbf{R}) \,|\, \det(\xi) > 0, \xi(w) = w\}$ is commutative. Therefore, if $\beta \in B^+$ and $\beta(z) = z$, then $\beta^{(v)}$ commutes with $f(a)^{(v)}$, so that β belongs to the commutor of $f(a)$ in B, which is $f(M)$. Thus $\beta \in f(M)$. See also [**67d**, Theorem 4.7, Propositions 4.5 and 4.8], which include Proposition 2.6 and the proved fact as special cases.

p. 77, line 2: For "$U(\epsilon)$" read "$U_0(\epsilon)$".

p. 77, line 13: For "non-zere" read "non-zero".

p. 78, line 16: For "V_μ to V_λ" read "V_λ to V_μ".

p. 80, line 6: For "$N(\mathfrak{x}_{v\mu})$" read "$N_{B/F}(\mathfrak{x}_{v\mu})$".

p. 81, line 8 from the bottom: For "elements" read "subgroups".

p. 83, line 1: , For "2.17" read "2.17, 3.17,".

p. 92, line 11 from the bottom: The second "$\mathfrak{N}/\mathfrak{M}$" should read "$\mathfrak{M}$".

p. 98 line 3 from the bottom: For "$E \cdot T$" read "$E \cdot T'$".

p. 99, line 7: For "L" read "F".

p. 100, line 6: For "$\mathfrak{r}$" read "L".

p. 100, line 16: For "Γ" read "Γ_0".

p. 101, line 1: Insert " where H is the subgroup of G corresponding to K" at the end of the line.

p. 102, line 2: For "$K'' \subset K'''$" read "$K'' \subset K'$".

p. 108, line 12 from the bottom: For "A" read "A_ι".

p. 110, line 19: For "$\mu_i^{-1}(X_\iota)$" read "$\mu_\iota^{-1}(X_\iota^\sigma)$".

p. 110, lines 26, 27: For "Applying ... described in (5.24.2)" read "Take μ_1 so as to satisfy (5.24.2) with this $\mathfrak{p}$ and $i = 1$. Take any isogeny ζ_ι of (A_1, θ_1) to (A_ι, θ_ι). By [15, Chapter II, Propositions 11 and 23], there exists an isogeny μ_ι of A_ι to A_ι^σ such that $\mu_\iota \zeta_\iota = \zeta_\iota^\sigma \mu_1$. Since $\mathrm{End}_Q(A_\iota) = \theta_\iota(K)$, we easily see that each μ_ι satisfies (5.24.2), and moreover $\mu_\iota \alpha = \alpha^\sigma \mu_j$ for every isogeny α of A_j to A_ι,".

p. 112 line 3 from the bottom; p. 113, lines 13, 15; p. 116, lines 4, 6, 12, 26; p. 118, lines 6, 9: For "B" read "$\mathfrak{B}$".

p. 118, line 14: For "$\varphi_\nu(x)$" read "$\varphi_\nu\big(\iota(x)\big)$".

p. 118, line 16: For "$\mathfrak{H}$" read "$\mathcal{H}$".

pp. 124–127: In the definition of Ω of (8.3.2), one should take *four* elements $x_1, \ldots, x_4$ instead of two elements x_1, x_2 so that $\mathfrak{c}^{-1}\mathfrak{a}\mathfrak{M}_\lambda = \mathfrak{a}\mathfrak{M}_\lambda + \sum_{i=1}^4 \mathfrak{r}_K x_\iota$ in (8.3.3). Similarly we have to take $\{y_i\}_{i=1}^4$ instead of y_1, y_2 in the paragraph following (8.3.3). These changes have no effect on the validity of the discussion in §§8.3–8.9. See [**67d**, §7.7].

p. 125: One should distinguish two weak PEL-types even if they are equivalent. Therefore the phrase "up to equivalence" in line 9 should be modified auitably. See [**67d**, §7.7].

p. 126, line 12: Insert "on $C(M, \mathfrak{c})$" after "$\big[C(M, \mathfrak{c})/M, \mathfrak{g}\big]$".

p. 127, line 21: For "$\Omega_{\mu\iota}^\sigma = \Omega_{\lambda_j}$" read "$\Omega_{\mu\iota}^\sigma$ is equivalent to Ω_{λ_j}".

p. 130, line 12: For "we have $(\Omega_{\mu\iota})^\xi = \Omega_{\lambda_j}$" read "$(\Omega_{\mu i})^\xi$ is equivalent to Ω_{λ_j}".

p. 132, line 1: For "the" read "then".

p. 138, line 2 from the bottom: For "φ" read "φ_λ" (two places).

p. 143, line 4 from the bottom: For "$\sum_{\iota=1}^\infty$" read "$\sum_{n=1}^\infty$".

p. 149, line 24: For the definition of Ω_λ, one should take *four* elements in place of $y_{\lambda 1}, y_{\lambda 2}$ in the same sense as in the above correction to pp. 124-127.

p. 158, Reference 4: For "algebraic groups" read "arithmetic groups".

67c. Number fields and zeta functions associated with discontinuous groups and algebraic varieties

This is the text of my invited lecture at the International Congress of Mathematicians in Moscow, 1966. The editor eliminated many sentences, paragraphs and references, and made meaningless rhetorical changes. These were done after my proof-reading. The page numbers in the proofs had been 290–298, which were changed to 100–107. Also, there were many typographical errors in the printed version. It should be noted that the whole volume of proceedings consists of less than 200 pages. No reprints were sent to me. There was one more interesting fact: the letter of invitation was a carbon copy.

Here I presented the original version, restoring the suppressed sentences, paragraphs, and references. As to Remark 2, some more related facts were given in [**96b**].

67d. Algebraic number fields and symplectic discontinuous groups

p. 543, line 12: For "Sence" read "Since".
p. 549, lines 9, 11 from the bottom: For "K" read "F".
p. 552, line 1: For "prime ideal" read "prime factor of p".
p. 552, lines 10, 11: For "the $\mathfrak{d}(P/F)^{\chi_\nu}$ for $\nu = \ldots$ relatively prime" read "$D(P/F)^{\chi_\nu}$ is divisible by $\mathfrak{p}^{\chi_\nu}$, but prime to $\mathfrak{p}^{\chi_\mu}$ if $\mu \neq \nu$".
p. 552, line 3 from the bottom: For "L_n" read "L^n".
p. 570, line 5 from the bottom: For "**Q**" read "**C**".
p. 585, line 13 from the bottom: For "$\mathfrak{G}(\mathfrak{K}/k(y))$" read "$G(\mathfrak{K}/k(y))$".
p. 588, line 10 from the bottom: For "$M_\mathfrak{p}$" read "$\mathfrak{M}_\mathfrak{p}$".
p. 589, lines 11, 13: For "$t_{w\nu}^\sigma$" read "$(t_{w\nu}^m)^\sigma$".

In §7.6 we first choose K depending on Z, and then change $\mathfrak{c}$ so that (7.6.2) and (7.6.3) hold. However, later in §§8.2 and 8.5, we take Z depending on $\mathfrak{c}$. Therefore our arguments must be modified so that K, Z, and $\mathfrak{c}$ can be chosen at the same time. Though this should be feasible, we shall not try to do it here, since a better approach was taken and such interdependence was avoided in the paper [**70**], which includes the main theorems of [**67d**] in improved forms.

68c. An ℓ-adic method in the theory of automorphic forms

This is an edited version of the text of a lecture at the conference *Automorphic functions for arithmetically defined groups*, Oberwolfach, Germany, July 28–August 3, 1968. The original unedited version was circulated as a preprint, but was never published.

69. Local representations of Galois groups

p. 107, line 10 from the bottom: For "GL" read "GL_r".
p. 120, line 10 from the bottom: For "$\mathrm{End}_Q(A)$" read "$\mathrm{End}_Q(\widetilde{A})$".

70a. On canonical models of arithmetic quotients of bounded symmetric domains

p. 154, line 9: Insert "Θ'" after "class".
p. 160, (2.6.2): A better statement including this as a special case can be given as follows: *If $T \subset S \subset \mathcal{G}$ and $k_S = k_T$, then $S = \Gamma_S T$.* The proof given in §3.6 applies to this statement.
p. 165, line 6 from the bottom: For "$\mathfrak{H}_n^r$" read "$\mathfrak{H}_n^-$".
p. 173, line 12 from the bottom: For "$\sigma(\omega)$" read "$\sigma(w)$".
p. 176, line 4 from the bottom: For "(3.5.5)" read "(2.5.5)".

p. 185, line 9: Here θ is taken to be an isomorphism. The case of an anti-isomorphism should (and can) be included, with obvious modifications, since such is later needed in §§5.2 and 6.7 (θ on p. 190, line 14 and θ^* on p. 196, line 11 from the bottom).

p. 196, line 11 from the bottom: For "A_z" read "A_y".

p. 220, line 12: For "$G_{\mathbf{Q}+}$" read "$G_{\mathbf{Q}+}$".

70b. On canonical models of arithmetic quotients of bounded symmetric domains: II

p. 537, line 13 from the bottom: Insert "on F" after $\alpha\tau_\nu = \tau_\mu$.

71a. On arithmetic automorphic functions

p. 346, line 8 from the bottom: For "$h \in \mathcal{K}$" read "$h \in \mathfrak{K}$".

p. 347, line 1: For "$\mathcal{H}_n$" read "$\mathfrak{H}_n$".

71b. On the zeta-function of an abelian variety with complex multiplication

p. 509, line 7: For "3" read "4".

p. 511, line 15: For the second and third ω^* read ω.

p. 511, line 3 from the bottom: Move "Morover," to the beginning of the next line.

p. 513, line 2: Insert "with invariant j" after "curve".

p. 515, line 8 from the bottom: For "K" read "S".

p. 519, last line: Insert "with invariant j" before "that".

71c. Class fields over real quadratic fields in the theory of modular functions

This is the text of my lecture at the International Mathematical Conference, College Park, Maryland, April 6–17, 1970.

71e. On elliptic curves with complex multiplication as factors of the Jacobians of modular function fields

p. 205, line 13: For "$1/d$" read "$1/D$".

p. 205, line 18: For "$\bigcup$" read "$\sum$".

72a. On the field of rationality for an abelian variety

p. 177, line 10 from the bottom: This line should read "we obtain".

Theorem 1 on page 173 can be generalized and proved in a simpler way; see [**75b**, Theorem 9.5 and Corollary 9.6].

72b. Class fields over real quadratic fields and Hecke operators

p. 134, (1.3): For "p^{1-2s}" read "$p^{\kappa-1-2s}$".
p. 134, last line: Insert "totally real or" after "K is".
p. 146, line 3: For "$\mathfrak{h}_1, \mathfrak{z}_1$" read "$\mathfrak{h}_\mathfrak{l}, \mathfrak{z}_\mathfrak{l}$".
p. 146, line 5: For "$\widetilde{\mathfrak{z}}_1$" read "$\widetilde{\mathfrak{z}}_\mathfrak{l}$".
p. 146, lines 3 and 8: For "L_2'" read "$\mathfrak{L}_2'$".
p. 156, line 3 from the bottom: For "$\begin{bmatrix} a' & * \\ 0 & d \end{bmatrix}$" read "$\begin{bmatrix} a' & b' \\ 0 & d \end{bmatrix}$".
p. 157, line 2: For "$\begin{bmatrix} a' & 0 \\ 0 & a'^{-1} \end{bmatrix}$" read "$\begin{bmatrix} a' & n^{-1}b \\ 0 & a'^{-1} \end{bmatrix}$".
p. 188, line 3 from the bottom: For "E_q" read "F_q".

73a. On modular forms of half integral weight

p. 449, line 8 from the bottom: The comma after $\Delta(N)$ should be a period.
p. 479, lines 1, 4: For "[11]" read "[10]".
p. 481, line 7: For "$\overline{\psi}$" read "$\varphi\overline{\psi}$".

A modular form of weight $\kappa/2$ is defined in this article with respect to the factor of automorphy $\varphi(z)^\kappa$. It is more natural, however, to define it with respect to $\varphi(z)(cz + d)^{(\kappa-1)/2}$. The latter definition can easily be generalized to the Hilbert modular case and also to the case of $Sp(n, F)$ with an arbitrary algebraic number field F; see [**87b**], [**93b**], [**93c**], and [**95b**]; see also [**74**, §3].

Some of the questions posed in Section 4 were affirmatively answered even in the Hilbert modular case; see [**87b**, Theorems 6.1, 6.2, and 6.4] and [**93c**].

74. On the trace formula for Hecke operators

p. 253, line 10 from the bottom: Cross-reference (2.3.1) should be placed at the beginning of the line.

75a. On the holomorphy of certain Dirichlet series

p. 84, line 4 from the bottom: For "$\mathrm{Re}(\beta) > 0$" read "$\mathrm{Re}(\beta) > 1$".
p. 85, line 2: For "$-tq$" read "$p = -tq$".

p. 91, (4.1): The term for $n = 0$ on the right-hand side means $A(0, y, s)$.

p. 97, line 9 from the bottom: For "$\lambda((n))^{1-w}$" read "$\lambda((n))n^{1-w}$".

75c. On some arithmetic properties of modular forms of one and several variables

p. 495, line 4: For $\{e)$ read $\{e\}$.

p. 495, line 7 from the bottom: For "D_p" read "D^p".

p. 503, line 7: Insert "mod" before N.

p. 506, line 19: For $E_\kappa^{-\lambda}$ read E_κ^λ.

p. 506, line 10 from the bottom: For $M_{\mu\kappa u}$ read $\mathcal{M}_{\mu\kappa u}$.

p. 506, line 6 from the bottom: For $\mathcal{M}_{2\nu n}$ read $\mathcal{M}_{2\mu u}$.

p. 514, Reference 5: For "K. Klingen" read "H. Klingen".

An elementary treatment of the series of type (2.2) is given in [**00**, Section 18]. This gives better results than [3] and [7] for such series.

Theorem 2 is actually true for every $\kappa \in \mathbf{Z}$ such that $-t_\nu < \kappa \le t_\nu$ and $\kappa - t_\nu \in 2\mathbf{Z}$ for every ν. This follows from Theorem 2 combined with the functional equation for $L(s, \chi)$. Also, the result can be generalized to the case of a Dirichlet series of a somewhat more general type; for these see [**00**, Theorem 18.16].

75d. On the Fourier coefficients of modular forms of several variables

p. 264, line 18: In order to say that $\omega(cg) \in \mathfrak{O}[[X_1, \ldots, X_l]]$, we have to know that $\omega(cg)$ is contained in the quotient field of $\mathfrak{O}[[X_1, \ldots, X_l]]$, as noted at the end of the paper. This is proved in the two paragraphs of [**76a**] starting from p. 682, line 13 from the bottom. Also, the Hilbert modular case is proved in the next paragraph.

76a. Theta functions with complex multiplication

p. 674, formula (2): For "$d \in L$" read "$l \in L$".

p. 681, line 2 from the bottom: Delete "]" after "11".

p. 692, line 6: For "(C_0)" read "$(\mathfrak{C}_0)$".

p. 696, Reference 11: For "with" read "of".

In the proof of the main theorem, we define φ by (28), which is meaningful only when $h(z) \ne 0$. We are choosing h so that $h(z_0) \ne 0$, so that (28) is well-defined for a fixed $\Omega = (\omega_1 \ \omega_2)$ such that $\omega_2^{-1}\omega_1 = z_0$. Since we have $\Omega \cdot {}^t\alpha$ on p, 687, line 2, in order to make $g(u)$ on the same line meaningful, we have to show that $h(\alpha(z_0)) \ne 0$. This is in fact true as proven in [**77c**, Lemma 2.5]; see also [**77c**, Remark 2.6]. The lemma can be generalized; see [**78a**, Proposition 3.13]. Using that proposition, the main theorem of the present article can be derived from a result in [**78a**]; see the last paragraph of [**78a**, p. 57].

76b. The special values of the zeta functions associated with cusp forms

p. 783, line 3 of the text: For "n" read "N".

p. 788, (2.5): For "$\langle f, g \rangle$" read "$\langle f_\rho, g \rangle$".

p. 789, lines 13, 14: For "$h \in S_k(t^{-1}N)\ldots$ greater than 1" read "$h \in S_k(M)$ with M such that $tM|N$ and $M < N$".

p. 793, line 4 from the bottom: For "$(2\pi i)^k\ (\omega_0)$" read "$(2\pi i)^k g(\omega_0)$".

p. 797, Lemma 8: This can be proved in a simpler way by observing that if χ is a Dirichlet character modulo M, then $g(\chi^\sigma) = \chi(r)^\sigma g(\chi)^\sigma$ with an integer r prime to M such that $e(1/M)^\sigma = e(r/M)$.

p. 800, line 16: For "$t \in \mathbf{R}$" read "$t \in \mathbf{Q}$".

p. 803, line 5: Insert "if $m \in \mathbf{Z}$ and $0 < m < k$" at the end of the line.

77a. On abelian varieties with complex multiplication

p. 69, line 16 from the bottom: For "P" read "P_σ".

p. 78, line 12: For "K" read "$K^\times$".

77c. On the derivatives of theta functions and modular forms

p. 370, line 3: For "1.16)" read "(1.16)".

p. 370, (1.18): Insert "2^{-1}" before "$({}^t rs$".

p. 372, line 4 from the bottom: For "$+{}^t au$" read "$-{}^t au$".

p. 385, line 6 from the bottom: For "$v(\gamma)^{1-(nk/2)}\,{}^t(cz+d)^{-1}$" read "$v(\gamma)^{1-(nk/2)} \cdot {}^t(cz+d)^{-1}$".

Theorem 2.4 and Lemma 2.5 can be generalized; see [**78a**, Theorem 3.12 and Proposition 3.13].

77d. On the periods of modular forms

p. 215, line 7: For $\big)^{\tau \in I}$ read $\big)_{\tau \in I}$.

p. 219, line 15 from the bottom: The left-hand side should read $R(k+l-1-s, f, g)$.

It seems that the article [5] never appeared.

If you have any concerns about our products,
you can contact us on
ProductSafety@springernature.com

In case Publisher is established outside the EU,
the EU authorized representative is:
Springer Nature Customer Service Center GmbH
Europaplatz 3, 69115 Heidelberg, Germany

Printed by Libri Plureos GmbH
in Hamburg, Germany